# HANDBOOK OF VIRTUAL ENVIRONMENTS
## Design, Implementation, and Applications

## HUMAN FACTORS AND ERGONOMICS

*Gavriel Salvendy, Series Editor*

---

**Bullinger, H.-J., and Ziegler, J.** (Eds.): *Human–Computer Interaction: Ergonomics and User Interfaces*

**Bullinger, H.-J., and Ziegler, J.** (Eds.): *Human–Computer Interaction: Communication, Cooperation, and Application Design*

**Stephanidis, C.** (Ed.): *User Interfaces for All: Concepts, Methods, and Tools*

**Smith, M. J., Salvendy, G., Harris, D., and Koubeck, R. J.** (Eds.): *Usability Evaluation and Interface Design: Cognitive Engineering, Intelligent Agents and Virtual Reality*

**Smith, M. J., and Salvendy, G.** (Eds.): *Systems, Social and Internationalization Design Aspects of Human–Computer Interaction*

**Stephanidis, C.** (Ed.): *Universal Access in HCl: Towards an Information Society for All*

**Meister, D., and Enderwick, T.**: *Human Factors in System Design, Development, and Testing*

**Stanney, K.** (Ed.): *Handbook of Virtual Environments: Design, Implementation, and Applications*

---

For more information on LEA titles, please contact Lawrence Erlbaum Associates, Publishers, at www.erlbaum.com.

# HANDBOOK OF VIRTUAL ENVIRONMENTS
## Design, Implementation, and Applications

Edited by

Kay M. Stanney
*University of Central Florida*

LEA LAWRENCE ERLBAUM ASSOCIATES, PUBLISHERS
2002 Mahwah, New Jersey London

Senior Acquisitions Editor: Anne Duffy
Editorial Assistant: Karin Wittig Bates
Cover Art Design: Branka Wedell, Graphics Artist, Winter Park, FL
Cover Design: Kathryn Houghtaling Lacey
Textbook Production Manager: Paul Smolenski
Full-Service Compositor: TechBooks
Text and Cover Printer: Hamilton Printing Company

This book was typeset in 10/12 pt. Times, Italic, Bold, Bold Italic. The heads were typeset in Helvetica Bold, and Helvetica Bold Italic.



Lawrence Erlbaum Associates, Inc., Publishers
10 Industrial Avenue
Mahwah, New Jersey 07430



**Library of Congress Cataloging-in-Publication Data**

Handbook of virtual environments : design, implementation, and applications / edited by Kay M. Stanney ; cover art by Branka Wedell.
p. cm.
ISBN 0-8058-3270-X (alk. paper)
1. Computer simulation—Handbooks, manuals, etc. 2. Human-computer interaction—Handbooks, manuals, etc. 3. Virtual reality—Handbooks, manuals, etc. I. Stanney, Kay M.

QA76.9.C65 H349 2002
04′.01′9—dc21 2001040284

Books published by Lawrence Erlbaum Associates are printed on acid-free paper, and their bindings are chosen for strength and durability.

Printed in the United States of America
10 9 8 7 6 5 4 3 2 1

To John—my everything
and
Sean, Ryan, and Michael—my sunshine

# Contents

# Series Foreword

With the rapid evolution of highly sophisticated computers, communications, service, and manufacturing systems, a major shift has occurred in the way people use and work with technology. The objective of this series on human factors and ergonomics is to provide researchers and practitioners alike with a platform through which to address a succession of human factors disciplines associated with advancing technologies, by reviewing seminal works in the field, discussing the current status of major topics, and providing a starting point to focus future research in these ever evolving disciplines. The guiding vision behind this series is that human factors and ergonomics should play a preeminent role in ensuring that emerging technologies provide increased productivity, quality, satisfaction, safety, and health in the context of the "Information Society."

The present volume is published at a very opportune time. Now more than ever technology is becoming pervasive in every aspect of the Information Society, both in the workplace and in everyday life activities. The field of virtual environments (VEs) emerged some 40 years ago as a very exotic, extremely expensive technology whose use was difficult to justify. The discipline has matured, and the cost of VE technology has decreased by over 100-fold, while computer speed has increased by over 1,000 fold, which makes it a very effective and viable technology to use in a broad spectrum of applications, from personnel training to task design. With this viability and broad potential application come numerous issues and opportunities, and a responsibility on the part of researchers, practitioners, designers, and users of this powerful technology to ensure that it is deployed appropriately.

The *Handbook of Virtual Environments* was guided by a distinguished advisory board of scholars and practitioners, who assisted the editor in ensuring a balanced coverage of the entire spectrum of issues related to VE technology, from fundamental science and technology to VE applications. This was achieved in a thorough and stimulating presentation, covered in 56 chapters, authored by 121 individuals from academia, industry, and government laboratories from Europe, Asia and the United States on topics of system requirements (including hardware and software), design and evaluation methods, and an extensive discussion of applications. All this was presented, after careful peer reviews, to the publisher in 1,911 manuscript pages, including 3,012 references for further in-depth reading, 255 figures, and 76 tables to illustrate concepts, methods, and applications. Thus, this handbook provides a most comprehensive account of the state of the art in virtual environments, which will serve as an invaluable source of reference for practitioners, researchers, and students in this rapidly evolving discipline.

This could not have been achieved without the diligence and insightful work of the editor and cooperative efforts of the chapter authors who have made it all possible. For this, my sincere thanks and appreciation go to all of you.

—Gavriel Salvendy
*Series Editor*

# Foreword

An explosion has occurred in recent years in our understanding of virtual environments (VEs) and in the technologies required to produce them. Virtual environments, as a way for humans to interact with machines and with complex information sets, will become commonplace in our increasingly technological world. In order for this to be practical, multimodal system requirements must be developed and design approaches must be addressed. Potential health and safety risks associated with VE systems must be fully understood and taken into account. Finally, ergonomic and psychological concerns must be investigated so people will enjoy using VE technology, be comfortable using it, and seek out its application.

This book provides an up-do-date discussion of the current research on virtual environments. It describes the current VE state of the art and points out the many areas where there is still work to be done. The *Handbook of Virtual Environments* provides an invaluable comprehensive reference for experts in the field, as well as for students and VE researchers. Both the theoretical and the practical side of VE technologies are explored.

The National Aeronautics and Space Administration (NASA) has long been interested in virtual environments. This interest arises from the need for humans to efficiently interact with complex spacecraft systems and to work with very large data sets generated by satellites. In the future, when humans travel beyond low Earth orbit to explore the universe, the relationship between the space-faring crew and the technologies they bring with them must be extremely intimate. The need for a small number of people to be able to work with a huge number of different technologies will be unprecedented. Virtual environment trainers, for example, are particularly attractive as a means of conducting just-in-time training before a crew member conducts a maintenance procedure not practiced for many months. In terrestrial applications, the complete life cycle for the design of complex systems such as aerospace vehicles will be completed virtually, before a single piece of metal is cut.

NASA's needs are unique in some respects but share much in common with other endeavors in today's world. Virtual environment technologies will also find military, medical, and commercial (e.g., in manufacturing and in entertainment) applications. As the science and technology of VEs progress, full-immersion technologies will likely become a standard interface between humans and machines. This book will help to make that vision a "real" reality.

—Guy Fogleman
*Acting Director, Bioastronautics Research Division*
*National Aeronautics and Space Administration*
*Washington, DC*

# Perspective

We read that we are in a new economy and that the pace of technological change is accelerating. This seems true because there are so many revolutionary innovations popping up. However, from the vantage of a given field, the picture can seem very different. It is over 35 years since Ivan Sutherland gave his address, "The Ultimate Display," at the National Computer Conference. Part of the delay is explained by the observation that problems that are not worked on do not get solved. Only a few years ago, I could argue that progress in the field was being held back because the enabling devices being used—the Polhemus magnetic tracker, the DataGlove, and the screens from handheld consumer television sets—were invented decades earlier. No one ever disagreed with those assertions. Today, it is different. In the best "divide and conquer" tradition, researchers have deployed themselves around every possible research problem. All of the easy problems have been run over and very smart people are working on most of the hard problems. In this book, those same people are reporting where things stand.

In fact, at least one hard problem has been solved, at least in preliminary form. Thirty years ago, I considered head-mounted displays (HMDs), but reasoned that they would be unacceptable unless they were wireless and permitted natural ambulation around a large area. I rejected the approach—not because I thought it was too hard but because I felt that the encumbering paraphernalia would be too awkward. I did not then appreciate how fast and how accurate the tracking would have to be and understood little about multipath transmission problems and diversity receivers. After 20 years of effort, the University of North Carolina recently demonstrated a wide-area tracking system with the needed performance and reported that its impact on the user's experience was every bit as powerful as one would hope it would be. Finally, it is possible to show what an HMD can do that no other display can.

A host of applications have been attempted in visualization, training, and entertainment. Even once unlikely opportunities have been gaining traction. In 1970, I received what I suspect was the first job offer in virtual reality. Dr. Arnold Ludwig came into my interactive installation and was so impressed with its impact on people's behavior that he wanted me to join him in the Department of Psychiatry at the University of Kentucky and to focus my interactions on psychotherapy. Today, virtual therapy is a thriving field with nothing to fear but the running out of phobias.

However, many of these "applications" have been motivated by the desire to do research rather than the expectation that the results would be immediately practical. Some of these systems are being used to do real work within the organizations that created them. An even smaller number have been sold to early adopters. But virtual environment (VE) applications are not yet being sucked into the marketplace by irresistible demand. Thus, while VE technology may be able to do a job, it is not yet recognized as the best or most cost effective way to do any job.

On the other hand, virtual environments have been unusually successful at spinning off its technology into adjacent fields even before it has gotten under way itself. Techniques for tracking human motion developed for VEs are now standard procedure in the film industry. Haptic devices are working their way into desktop and automotive systems. Virtual surgery, once a wild speculation, then a Defense Advanced Research Projects Agency (DARPA) program managed by Richard Satava, and now a nascent industry with publicly traded companies, is selling robotic surgical systems and surgical simulations to hospitals. Finally, the HMD has been incorporated into eyeglasses and has to be considered a competitor for the mobile display of the future.

In the background, Moore's law has continued to operate, assuring the increase in computer power and the advance of computer graphics. However, it is important to note that while the film industry routinely spends an hour computing a single frame of graphic animation, a VE system has only $\frac{1}{30}$ or $\frac{1}{60}$ of a second to create its necessarily simpler image. This is a 100,000:1 difference in time, or 16 doublings, or a 24-year lag between the time a state-of-the-art image appears in film and when a similarly complex image can be used in virtual environments. Mere doublings do not guarantee subjective improvement; only orders of magnitude are discernible. Nevertheless, we can be confident that ultimately the needed processing and graphics capacity will be readily affordable.

At that point, all that remains to be developed is the VE technology itself: the tracking, the displays, and an answer to the question that I posed well over a decade ago: Would you use it if it was free? Whatever you are willing to pay for a technology, once you own it, it is free. How much you choose to use it at this point determines its future as much as the economics of its purchase or the efficacy of its performance. If it is a pleasure to use, you will try to use it for everything you can think of. If it is awkward and uncomfortable, you will not use it for any task where it is not absolutely required. In fact, you will be looking for an alternative way to perform the tasks for which it is suited.

The question is: How good does it have to be? Not so long ago, I attended a conference at which the keynote speaker declared computer graphics to be the key to VE technology and that more realistic images would assure its success. I asked him, "If we turned out the lights, would we still be here?" The point was that graphic realism does not appear to be necessary or sufficient. It would seem simple to depict a dark, cloudy night—or just a dark room—with current graphic technology, but instinct suggests that the experience would not be convincing. There is still a long distance between depiction and illusion.

In visualization applications, believability is not so important. Training can be useful even if the experience is not totally persuasive. A game or a story can be entertaining even if the participant never suspends disbelief, but in each case, there is a threshold at which a technology goes beyond serviceable and becomes compelling. When this threshold is crossed, the technology is poised to go from being a niche solution to becoming a way of life.

It is not clear where critical mass will be reached first. Will one of the immersion applications take hold? Will a VE entertainment system supplant traditional video games? Or will the desire for portable wearable devices lead to the routine wearing of HMDs unobtrusively integrated with eyeglasses? At the moment, I would bet on the latter because the standards for success are so low. Text is easy to display. The screens on cell phones and handheld computers are too small. And while speech may be ideal for answering questions, it is too slow for presenting options. Only an eyeglass-mounted display can provide the full screen of information that we take for granted at the desktop. If popular, such limited devices would inevitably be used for gaming, just as cell phones are today. Head tracking would make those games more interactive, if not more convincing, and augmented reality applications could be implemented in specific locations like grocery stores. More immersive displays could then evolve over a period of time, always assured of this installed base.

Starting from the other direction, real breakthroughs are needed in HMD design to provide immersive experiences that are guaranteed to work. A wide field of view and minimal weight are required before we can be confident that the HMD wearer will forget the apparatus and embrace the virtual world.

Whatever the path or pace of the technology and its deployment, virtual reality will maintain its proper role as the best metaphor for the world that is evolving around us. It will continue to be depicted in films and incorporated into everyday thought to the point that it is so familiar as a concept that by the time the real thing seeps into our daily lives, we may barely notice.

—Myron Krueger
*President*
*Artificial Reality Corporation*

# Preface

When computers first permeated the public domain, thoughts of the Turing Test arose yet were quickly extinguished, as users labored over perplexing interfaces, which often left them bewildered and thoroughly frustrated. There had to be a better way, and so began the field of human–computer interaction (HCI). HCI efforts have substantially improved computer interaction, yet barriers to user friendliness still exist due to the abstract concepts that must be conquered to successfully use a computer. A user must work through an interface (e.g., window, menu, icon, or some other mechanism) to achieve desired goals. They cannot access these goals directly, but only through their interface surrogates. Until now.

Virtual environments (VEs) allow users to be immersed into three-dimensional digital worlds, surrounding them with tangible objects to be manipulated and venues to be traversed, which they experience from an egocentric perspective. Through the concrete and familiar, users can enact known perceptual and cognitive skills to interact with a virtual world; there is no need to learn contrived conventions. Virtual environments also extend the realm of computer interaction, from the purely visual to multimodal communication that more closely parallels human–human exchanges. VE users not only see visual representations, they can also reach out and grab objects, "feel" their size, rotate them in any given axis, hear their movement, and even smell associated aromas. Such experiences do not have to be in solitude, as VE users can take along artificial autonomous agents or collaborate with other users who also have representations within the virtual world. Taken together, this multisensory experience affords natural and intuitive interaction.

The paragraph above describes an ideal, but not, unfortunately, the current state of the art. In today's virtual environments, users are immersed into an experience with suboptimal visual resolution, inadequate spatialization of sound, encumbering interactive devices, and misregistration of tracking information. These issues are among the scientific and technological challenges that must be resolved to realize the full potential of VE technology, which were well defined by Nathaniel Durlach and Anne Mavor in their seminal work, *Virtual Reality: Scientific and Technological Challenges*. Chapter 1 of this handbook furthers this definitional effort by reviewing the recommendations set forth by Durlach and Mavor and identifying the current status of those objectives, the sine qua non being that VE technology, both hardware and software, has realized substantial gains in the past decade and is posed to support the next generation of highly sophisticated VE systems. However, psychological considerations and VE usability evaluation require additional study to identify how best to design and use VE technology. In addition, Durlach and Mavor (1995, p. 2) mentioned in their work the "inadequate terminology being used" to describe and communicate progress in VE technology. Chapter 2 thus presents definitions for many of the key terms used throughout this handbook, thereby providing users with a common understanding of concepts discussed.

After providing an introduction to VE technology, the handbook organizes the body of knowledge into five main sections and three subsections. In the first of these sections, "System Requirements," multimodal system requirements are addressed, including physiological characteristics that affect VE system design. This section is subdivided into hardware, software, and application requirements. The hardware subsection addresses visual, auditory, haptic, vestibular, motion and eye tracking, gesture, and locomotion interface design, specifying the state of the art and how it compares to user and application requirements. The software requirements subsection discusses virtual world modeling, design of interaction techniques, multimodal design considerations, autonomous agents design, and Internet-based VE design. The application requirements subsection covers structured development approaches and transfer-of-training issues.

"Design Approaches and Implementation Strategies" is the next main section. While conventional HCI design approaches can be used to some extent to develop VE system designs, their multimodal nature mandates that new approaches be developed that augment existing techniques. This section addresses cognitive design strategies, identifies perceptual illusions that can be leveraged in VE design, discusses navigational issues such as becoming lost within a virtual world, and provides insights into structured approaches to content design. This section also reviews technology management techniques and user acceptance concerns, uncovers means of addressing potential products liability issues related to the risks associated with VE exposure, and provides sales and market modeling strategies.

"Health and Safety Issues" is the next section. While repetitive use of conventional computer systems has led to a number of associated afflictions (e.g., carpal tunnel syndrome), VE technology brings its own set of additional physiological and psychological risks. This section covers the direct physiological effects of VE exposure, such as biological effects, trauma, and injury. It tackles issues associated with motion sickness, including signs and symptoms, as well as neurophysiology and physiological correlates, and provides potential approaches to treatment. The advantages and risks of perceptual and perceptual–motor adaptation are also discussed, including the ability of humans to overcome the deleterious effects and aftereffects frequently associated with current VE devices. This section also addresses social concerns, which have received far too little attention in the field, discussing both the antisocial and prosocial impact of VE exposure.

Evaluation approaches are presented in the next section. While VE system design can leverage existing usability engineering techniques, the novel aspects of human–VE interaction require these techniques to be modified and extended to better address VE evaluation. This section provides the state of the art in VE usability engineering, addresses human performance measurement in VEs, identifies usage protocols that can be used to manage the use of VE technology, and provides means of measuring and managing visual, proprioceptive, and vestibular aftereffects associated with VE exposure. Approaches to measuring and engendering sense of presence, which is often considered to be essential to VE usability, are discussed. Ergonomic concerns of fit and comfort are also addressed in this section.

The final main section, "Selected Applications of Virtual Environments," provides a compendium of VE applications. Virtual environment applications pioneers have set the stage for an exciting future. From its military roots, VE technology has branched out into numerous domains. Developments in the area of military defense have led to substantial gains in hardware and networking. Academic efforts are leading to new approaches to experiential learning. Medical VE applications are pushing the envelope in diagnosis, therapy, and rehabilitation, resulting in truly revolutionary advances. System design efforts are allowing models to be created in a more intuitive and natural manner, with multidisciplinary design teams communicating their ideas via the VE medium. Advances in information visualization are enabling dynamic investigation of multidimensional, highly complex data domains. Manufacturing VE applications have led to advances in the design of manufacturing activities and manufacturing facilities, execution of planning, control, and monitoring activities, and execution of physical processing activities. Likely the most popular of all VE applications, the entertainment industry is leading the way to truly innovative uses of the technology. From interactive arcades to cyber cafes, the entertainment industry has leveraged the unique characteristics of this communications medium, providing dynamic experiences to those who come along for the ride.

This handbook closes with a brief review of the history of VE technology, as we must acknowledge the pioneers whose innovativeness and courage provided the keystones for contemporary successes. The final chapter also provides information on the VE profession, providing those interested with a number of sources to further their quest for the keys to developing the ultimate virtual world.

The main objective of this handbook is to provide practitioners with a reference source to guide their development efforts. We have endeavored to provide a resource that not only addresses technology concerns but also tackles the social and business implications with which those associated with the technology are likely to grapple. While each chapter has a strong theoretical foundation, practical implications are derived and illustrated via the many tables and figures presented.

Taken together, the chapters present systematic and extensive coverage of the primary areas of research and development within VE technology. The handbook brings together a comprehensive set of contributed articles that address the principles required to define system requirements and design, build, evaluate, implement, and manage the effective use of VE applications. The scope and detail of the handbook are extensive, and no one person could possibly do justice to the breadth of coverage provided. Thus, the handbook leveraged authoritative specialists that were able to provide critical insights and principles associated with their given area of expertise. It is through the collective effort of the many contributing authors that such a broad body of knowledge was assembled.

—Kay M. Stanney, Ph.D.
*Orlando, Florida*
*February 2001*

# Acknowledgments

*If men will not act for themselves, what will they do when the benefit of their effort is for all?*
—Elbert Hubbard, *A Message to Garcia* (p. 23)

In the case of the many contributors to this handbook, the answer to this question is that they will selflessly endeavor to provide the insights and assistance required to realize this tremendous effort. In many ways I feel the creation of this handbook is "our Message to Garcia," one developed through altruistic dedication, the only impetus being that such a source is direly needed in the field. Many individuals openly gave of their time, energy, and knowledge in order to develop this handbook, often when they were fully loaded with their own responsibilities. The efforts of the many contributing authors and to the advisory board, which helped formulate the content coverage, are most sincerely appreciated.

To Gavriel Salvendy, who has provided me with many opportunities, including the invitation to edit this handbook, which have shaped and molded my career I am forever grateful. I have also been blessed with the finest of mentors, Robert S. Kennedy, who gives tirelessly of himself—thank you. I am greatly appreciative of the support of the National Science Foundation, Office of Naval Research, and Naval Air Warfare Center Training Systems Division, in particular Gary W. Strong, Helen M. Gigley, and Robert Breaux. The National Science Foundation CAREER Award and ONR Young Investigator Award have provided me with the opportunity to develop technical depth in human-computer interaction and virtual environment technology and fostered interchange with experts in the field, many of whom contributed chapters to this handbook.

Each chapter in the handbook was peer reviewed. I would like to thank the many advisory board members and chapter contributors who assisted with this process, as well as the following individuals who kindly gave of their time to the review process: Andi Cowell, Chuck Daniels, Nathaniel Durlach, Jason Fox, Thomas Furness, Phillip Hash, Susan Lanham, Dennis McBride, Dean Owen, Randy Pausch, Leah Reeves, Mario Rodriguez, Randy Stiles, and Mark Wiederhold.

For the persistent efforts and encouragement of Anne Duffy, our Lawrence Erlbaum senior editor, who stuck with me even as I missed deadlines and acted out of frustration, I am deeply grateful.

Much appreciation goes to Branka Wedell, who took my amorphous ideas and, through her inspired creativity, designed the striking cover art for this handbook.

The efforts of David Bush are greatly appreciated, as he assisted with many of the activities associated with the handbook and always with a smile on his face. I am also indebted to Kelly Kingdon, who assisted me with many of my personal responsibilities so that I had more time to dedicate to this effort.

To those individuals who constitute the fabric of my life, my parents who instilled the work ethic that allowed me to persevere and see this effort through to completion, my sisters and brother, my very best friends, my brother-in-law who introduced me to *A Message to Garcia* at an opportune moment, and my three sons, Sean, Ryan, and Michael, who fill my world with sunshine, I have been blessed by your encouragement and confidence.

Above all, I am deeply indebted to my husband, who not only encouraged me as I fretted that this handbook would remain a *virtual* reality, but also rolled up his sleeves and assisted with editing and proofreading. His love is my pillar and his steadfast support of my career is my forte.

—Kay M. Stanney

# Advisory Board

Terry Allard
Chief, Human Factors Division
Information Sciences and Technology Directorate
NASA Ames Research Center
Moffett Field, CA

Durand R. Begault
Researcher
Spatial Auditory Display Laboratory
Human Factors Division
NASA Ames Research Center
Moffett Field, CA

Robert Breaux
Program Manager, Virtual Reality
Naval Air Warfare Center
Training Systems Division
Orlando, FL

Frederick Brooks
Kenan Professor of Computer Science
University of North Carolina
Chapel Hill, NC

Steve Bryson
Research Scientist
Numerical Aerospace Simulation Division
NASA Ames Research Center
Moffett Field, CA

Grigore C. Burdea
Associate Professor
Department of Electrical and Computer Engineering
School of Engineering, Rutgers University
Piscataway, NJ

Malcolm M. Cohen
Chief
Human Information Processing Research Branch
Human Factors Division
NASA Ames Research Center
Moffett Field, CA

Carolina Cruz-Neira
Litton Associate Professor
Department of Electrical and Computer Engineering
Associate Director, Virtual Reality Applications Center
Iowa State University
Ames, IA

Rudy Darken
Assistant Professor
Department of Computer Science
The Moves Institute
Naval Postgraduate School
Monterey, CA

# About the Editor

Kay Stanney is an associate professor with the University of Central Florida's Industrial Engineering and Management Systems Department, which she joined in 1992. She is an editor of the *International Journal of Human–Computer Interaction*. She is a cofounder of the Virtual Environments Technical Group of the Human Factors and Ergonomics Society. Dr. Stanney has more than 100 scientific publications and has given numerous invited lectures and presentations. Her research into the after-effects associated with virtual environment exposure, which is funded by the National Science Foundation, Office of Naval Research, and National Aeronautics and Space Administration, has appeared on MTV Network's health show *Mega-Dose*, *NBC Nightly News*, the Canadian Broadcasting Company's *Undercurrents*, and NBC's local Orlando news, as well as receiving front-page coverage in various newspapers. Dr. Stanney received a bachelor of science in industrial engineering from the State University of New York at Buffalo in 1986, after which time she spent three years working as a manufacturing/quality engineer for Intel Corporation in Santa Clara, California. She received her master's degree and Ph.D. in industrial engineering, with a focus on human factors engineering, from Purdue University in 1990 and 1992, respectively.

# Contributors

Jan M. Allbeck
Research Assistant
Center for Human Modeling and Simulation
University of Pennsylvania.
Philadelphia, PA

David R. Badcock
Professor
Department of Psychology
University of Western Australia
Nedlands, Western Australia

Eric Badiqué
Project Officer
European Commission
IST Programme
Brussels, Belgium

Norman I. Badler
Director
Center for Human Modeling and Simulation
Professor, Computer and Information Science Dept
University of Pennsylvania
Philadelphia, PA

Kathleen M. Bartlett
Research Assistant
RSK Assessments, Inc.
Orlando, FL

Cagatay Basdogan
Senior Member of Technical Staff
JPL–Virtual Environments Laboratory
California Institute of Technology
Pasadena, CA

S. James Biggs
Postdoctoral Associate
MIT Touch Lab
Research Laboratory of Electronics
Massachusetts Institute of Technology
Cambridge, MA

Richard Blade
Professor
Department of Physics and Energy Science
University of Colorado, Colorado Springs
Colorado Springs, CO

James P. Bliss
Associate Professor
Department of Psychology
Old Dominion University
Norfolk, VA 23529

Robert S. Bolia
Computer Scientist
Air Force Research Laboratory
Wright-Patterson Air Force Base, OH

Doug A. Bowman
Assistant Professor
Department of Computer Science
Virginia Tech
Blacksburg, VA

Robert Breaux
Program Manager, Virtual Reality
Naval Air Warfare Center
Training Systems Division
Orlando, FL

Steve Bryson
Research Scientist
Numerical Aerospace Simulation Division
NASA Ames Research Center
Moffett Field, CA

J. Galen Buckwalter
Senior Investigator
Department of Research and Evaluation
Southern California Kaiser Permanente
Medical Group
Pasadena, CA

Grigore C. Burdea
Associate Professor
Department of Electrical and Computer Engineering
School of Engineering, Rutgers University
Piscataway, NJ

David R. Bush
Doctoral Student
Industrial Engineering and Management Systems
University of Central Florida
Orlando, FL

Sandra L. Calvert
Professor
Department of Psychology
Georgetown University
Washington, DC

Janis A. Cannon-Bowers
Senior Research Psychologist
Naval Air Warfare Center
Training Systems Division
Orlando, FL

Marc Cavazza
Professor
School of Computing and Mathematics
University of Teesside
Middlesbrough, United Kingdom

Ryad Chellali
Technical Head
TridimAge, Inc.
Assistant Professor of Robotics
Ecole des Mines de Nantes
Nantes Cedex, France

Kyung H. Chung
Research Assistant
Department of Industrial and Systems Engineering
Virginia Tech
Blacksburg, VA

Deborah M. Clawson
Associate Professor
Department of Psychology
The Catholic University of America
Washington, DC

Sue V. G. Cobb
Research Manager
Virtual Reality Applications Research Team (VIRART)
University of Nottingham
Nottingham United Kingdom

Joseph R. Coble
Chair and Professor
Psychology Department
Clark Atlanta University
Atlanta, Georgia

Philippe Coiffet
Director of Research, CNRS
Member of the French Academy of Technologies
Laboratoire de Robotique de Paris
Vélizy, France

Daniel Compton
Research Assistant
RSK Assessments, Inc.
Orlando, FL

Joanna K. Crosier
Research Scientist
Education Applications, Virtual Reality Applications Research Team (VIRART)
University of Nottingham
Nottingham, United Kingdom

Mirabelle D'Cruz
Research Scientist
Industrial Applications, Virtual Reality Applications Research Team (VIRART)
Institute for Occupational Ergonomics
University of Nottingham
Nottingham, United Kingdom

Leonard Daly
President
Daly Realism.com
Van Nuys, CA

Rudy Darken
Assistant Professor
Department of Computer Science
The MOVES Institute
Naval Postgraduate School
Monterey, CA

Eleni M. Daskarolis
Research Assistant
Department of Psychology
University of Central Florida
Orlando, FL

Roy C. Davies
Research Associate
Department of Design Sciences
Division of Ergonomics and Aerosol Technology
Lund Institute of Technology
Lund University
Lund, Sweden

Paul DiZio
Associate Professor
Volen Center for Complex Sysytems and Ashton Graybiel Spatial Orientation Laboratory
Brandeis University
Waltham, MA

John V. Draper
Research Scientist
Robotics and Process Systems Division
Oak Ridge National Laboratory
Oak Ridge, TN

Mark H. Draper
Major, United States Air Force
Air Force Research Laboratory
Wright-Patterson Air Force Base, OH

Richard Eastgate
Technical Manager
Virtual Reality Applications Research Team (VIRART)
Institute for Occupational Ergonomics
University of Nottingham
Nottingham, United Kingdom

Clive Fencott
Senior Lecturer
School of Computing and Mathematics
University of Teesside
Middlesbrough, United Kingdom

Eric Foxlin
Chief Technology Officer
InterSense, Inc.
Burlington, MA

Joe Gabbard
Senior Research Associate
Systems Research Center
Virginia Tech
Blacksburg, VA

Stephen L. Goldberg
Chief, Simulator Systems Research Unit
US Army Research Institute
Orlando, FL

David A. Graeber
ONR/NAWC Fellow
Industrial Engineering and Management Systems
University of Central Florida
Orlando, FL

David Gross
Associate Technical Fellow of the Boeing Company
Boeing Company
Huntsville, AL

Deborah L. Harm
Research Scientist
NASA Johnson Space Center
Houston, TX

Michael Heim
Virtual Worlds Director
Art Center College of Design
Pasadena, CA

Lawrence J. Hettinger
Senior Human Factors Engineer
Northrop Grumman Information Technologies
Harvard, MA

Katherine A. Hickinbotham
Research Associate
Spatial Orientation Systems
Naval Aerospace Medical Research Laboratory
Pensacola, FL

Gerald A. Higgins
Vice President
Image Medical, Inc.
Silver Spring, MD

Deborah Hix
Research Computer Scientist
Department of Computer Science
Systems Research Center
Virginia Tech
Blacksburg, VA

Maureen K. Holden
Research Scientist
Department of Brain and Cognitive Sciences
Massachusetts Institute of Technology,
Cambridge, MA

John M. Hollerbach
Professor
University of Utah
School of Computing
Salt Lake City, UT

Charles E. Hughes
Professor
School of Electrical Engineering and Computer Science
University of Central Florida
Orlando, FL

George K. Hung
Professor
Department of Biomedical Engineering
Rutgers University
Piscataway, NJ

Jerry Isdale
Research Staff Scientist
HRL Laboratories, LLC
Malibu, CA

Shaun B. Jones
Associate Professor of Surgery
Uniformed Services University of Health Sciences
Program Manager, Pathogen Countermeasures
Defense Advanced Research Projects Agency
Washington, DC

David B. Kaber
Assistant Professor
Department of Industrial Engineering
North Carolina State University
Raleigh, NC

Manpreet Kaur
Usability Specialist
Idea Integration
Houston, TX

Kristyne E. Kennedy
Attorney at Law
RSK Assessments, Inc.,
Orlando, FL

Robert S. Kennedy
President
RSK Assessments, Inc.
Orlando, FL

G. Drew Kessler
Assistant Professor
Computer Science and Engineering Department
Lehigh University
Bethlehem, PA

Abderrahmane Kheddar
Assistant Professor
CEMIF–Complex Systems Laboratory
University of Evry Val-d'Essonne
Evry, France

Kelly Kingdon
Research Assistant
Industrial Engineering and Management Systems
University of Central Florida
Orlando, FL

Gudrun Klinker
Assistant Professor
Technische Universitaet Muenchen
Fachbereich Informatik
Munich, Germany

Bruce W. Knerr
Research Psychologist
US Army Research Institute
Orlando, FL

James Lackner
Professor
Volen Center for Complex Systems
Ashton Graybiel Spatial Orientation Laboratory
Brandeis University
Waltham, MA

Donald Ralph Lampton
Research Psychologist
U.S. Army Research Institute
Simulator Systems Research Unit
Orlando, FL

Corinna E. Lathan
President and CEO
AnthroTronix, Inc.
College Park, MD

Ben Lawson
Research Scientist
Spatial Orientation Systems
Naval Aerospace Medical Research Laboratory
Pensacola, FL

Gordon Mair
Director
Transparent Telepresence Research Group
University of Strathclyde
Glasgow, Scotland, United Kingdom

James G. May
Professor
Department of Psychology
University of New Orleans
New Orleans, LA

Pamela R. McCauley-Bell
Associate Professor
Department of Industrial Engineering and Management Systems
University of Central Florida
Orlando, FL

Andrew M. Mead
Research Scientist
Naval Air Warfare Center
Training Systems Division
Orlando, FL

Mark Mon-Williams
Lecturer
School of Psychology
University of St. Andrews
St. Andrews, Scotland, United Kingdom

Christina S. Morris
Research Administrator
Advanced Learning Technologies
Institute for Simulation and Training
Orlando, FL

J. Michael Moshell
Director
CREAT Digital Media Program and Professor of Computer Science
University of Central Florida
Orlando, FL

Allen Munro
Director
Behavioral Technology Laboratories
University of Southern California
Redondo Beach, CA

Eric Muth
Assistant Professor
Department of Psychology
Clemson University
Clemson, SC

Helen R. Neale
Research Scientist
Special Needs Applications, Virtual Reality Applications Research Team (VIRART)
University of Nottingham
Nottingham, United Kingdom

W. Todd Nelson
Senior Usability Engineer
divine/Whitman-Hart
Cincinnati, Ohio

Max M. North
Associate Professor
Computer Science and Information Systems
Kennesaw State University
Kennesaw, GA

Sarah M. North
Director
Human–Computer Interaction Group
Associate Professor
Computer and Information Sciences
Clark Atlanta University
Atlanta, Georgia

Randy L. Oser
Senior Research Psychologist
Naval Air Warfare Center
Training Systems Division
Orlando, FL

Mary Lou Padgett
President
Padgett Computer Innovations, Inc.
Auburn, AL

James Patrey
Research Scientist
Naval Air Warfare Center
Training Systems Division
Orlando, FL

Barry Peterson
Research Associate
Department of Computer Science
The MOVES Institute
Naval Postgraduate School
Monterey, CA

George V. Popescu
Research Staff Member
IBM Thomas J. Watson Research Center
Hawthorne, NY

Dean Reed
Systems Programmer
Visual Systems Laboratory
Institute for Simulation and Training
University of Central Florida
Orlando, FL

Albert A. Rizzo
Research Assistant Professor
Integrated Media Systems Center and School of Gerontology
University of Southern California
Los Angeles, CA

Wallace Sadowski
Advisory Human Factors Engineer
IBM Voice Systems
Boca Raton, FL

Eduardo Salas
Professor
Department of Psychology
Institute for Simulation & Training
University of Central Florida
Orlando, FL

Richard M. Satava
Professor of Surgery
Yale University School of Medicine
Program Manager, Advanced Biomedical Technologies
Defense Advanced Research Projects Agency
New Haven, CT

Marc M. Sebrechts
Professor and Chair
Department of Psychology
The Catholic University of America
Washington, DC

Elizabeth Sheldon
ONR/NAWC Fellow
Industrial Engineering and Management Systems
University of Central Florida
Orlando, FL

John P. Shewchuk
Associate Professor
Department of Industrial and Systems Engineering
Virginia Tech
Blacksburg, VA

Russell D. Shilling
Associate Professor
Operations Research
The MOVES Institute
Naval Postgraduate School
Monterey, CA

Barbara Shinn-Cunningham
Assistant Professor
Departments of Cognitive and Neural Systems and Biomedical Engineering
Hearing Research Center
Boston University
Boston, MA

Stephanie Sides
Research Associate
Spatial Orientation Systems
Naval Aerospace Medical Research Laboratory
Pensacola, FL

Mandayam A. Srinivasan
Director, Touch Lab
Department of Mechanical Engineering and Research Laboratory of Electronics
Massachusetts Institute of Technology
Cambridge, MA

Kay M. Stanney
Associate Professor
Department of Industrial Engineering and Management Systems
University of Central Florida
Orlando, FL

Thomas A. Stoffregen
Associate Professor
Director, Human Factors Research Laboratory
Division of Kinesiology
University of Minnesota
Minneapolis, MN

Robert J. Stone
Scientific Director
Virtual Presence
Manchester, United Kingdom

Russell L. Storms
Research Scientist
Army Research Laboratory
Georgia Institute of Technology
Atlanta, GA

G. M. Peter Swann
Professor of Economics
Manchester Business School
University of Manchester
Manchester, United Kingdom

Tony Sweeney
Deputy Director
National Museum of Photography, Film and Television
Bradford, United Kingdom

Daniel Thalmann
Professor
Computer Graphics Lab
Swiss Federal Institute of Technology
Lausanne, Switzerland

Nadia Magnenat-Thalmann
Professor
Miralab
University of Geneva
Geneva, Switzerland

Richard A. Thurman
Assistant Professor
Western Illinois University
Instructional Technology and Telecommunications
Macomb, IL

Emanuel Todorov
Senior Research Fellow
Gatsby Computational Neuroscience Unit
University College London
London, United Kingdom

Michael R. Tracey
Research Assistant
Department of Biomedical Engineering
The Catholic University of America
Washington, DC

Helmuth Trefftz
Graduate Assistant
Department of Electrical and Computer Engineering
School of Engineering
Rutgers University
Piscataway, NJ

Marilyn Mantei Tremaine
Research Professor
Center for Advanced Information Processing (CAIP)
Rutgers University
Piscataway, NJ

Matthew Turk
Associate Professor
Computer Science Department
University of California
Santa Barbara, CA

John M. Usher
Professor
Department of Industrial Engineering
Mississippi State University
Mississippi State, MS

Cheryl van der Zaag
NIA Fellow
School of Gerontology
University of Southern California
Los Angeles, CA

Erik Viirre
Senior Research Scientist
Human Interface Technology Laboratory
University of Washington
Seattle, WA

John Wann
Professor
Action Research Laboratory
Department of Psychology
University of Reading
Reading, United Kingdom

Robert B. Welch
Research Scientist
Human Factors Research and Technology Division
NASA Ames Research Center
Moffett Field, CA

Joseph Wilder
Research Professor
Center for Advanced Information Processing (CAIP)
Rutgers University
Piscataway, NJ

Robert C. Williges
Ralph H. Bogle Professor of Industrial and Systems Engineering
Virginia Tech
Blacksburg, VA

John R. Wilson
Professor of Occupational Ergonomics
Director of VIRART
Management and Human Factors Group
School of 4M
University of Nottingham
Nottingham, United Kingdom

Michael Zyda
Director
The MOVES Institute
Naval Postgraduate School
Monterey, CA

# PART I: Introduction

# 1

# Virtual Environments in the 21st Century

Kay M. Stanney[1] and Michael Zyda[2]
[1]*University of Central Florida*
*Industrial Engineering and Management Systems*
*4000 Central Florida Blvd.*
*Orlando, FL 32816-2450*
*stanney@mail.ucf.edu*
[2]*Naval Postgraduate School*
*Modeling, Virtual Environments*
*and Simulation (MOVES) Academic Group*
*Spanagel Hall 252, Code MOVES/mjz*
*Monterey, California 93943-5118*
*zyda@acm.org*

## 1. INTRODUCTION

*You see, then, that a doubt about the reality of sense is easily raised, since there may even be a doubt whether we are awake or in a dream. And as our time is equally divided between sleeping and waking, in either sphere of existence the soul contends that the thoughts which are present to our minds at the time are true; and during one half of our lives we affirm the truth of the one, and, during the other half, of the other; and are equally confident of both.*

—*Theaetetus*, Plato

As Plato so eloquently stated, that which is reality emanates from that which is present to our minds. In *Theaetetus*, Plato examines perception, knowledge, truth, and subjectivity. This work suggests that Forms (i.e., circularity, squareness, and triangularity) have greater reality than objects in the physical world. This reality is derived because Forms serve as models for our perceptions. So it is that a virtual environment (i.e., a modeled world) can represent a "truth" that can educate, train, entertain, and inspire. In their ultimate form, virtual environments (VEs) immerse users in a fantastic world, one that stimulates multiple senses and provides vibrant experiences that somehow transform those exposed (e.g., via training, educating, marketing, or entertaining).

Visions such as *The Matrix*, written and directed by Andy and Larry Wachowski, have elevated the status of VE to the level of pop iconography, and some of those associated with the technology have arguably risen to star status (e.g., Jaron Lanier). Yet, while one may speak of VE as contemporary, even in vogue, how far has the technology really come since the pioneering work of Ivan Sutherland, with his 1963 Sketchpad that provided the first interactive computer graphics, or Morton Heilig's 1956 engineering marvel Sensorama rambled through Brooklyn's streets and California's sand dunes (Rheingold, 1991; Sutherland, 1963)? Sensorama provided a multisensory experience of riding a motorcycle by combining three-dimensional (3-D) movies seen through a binocularlike viewer, stereo sound, wind, and enticing aromas (see chap. 56, this volume). Some aspects of the technology have improved substantially since Sketchpad and Sensorama, such as greater visual resolution (see chap. 3, this volume), spatialized audio (see chap. 4), and haptic interaction (e.g., the net force and torque feedback used in tool usage; see chaps. 5 and 6), while others have yet to make any significant strides. In particular, the small grills placed near the nose of Sensorama's passenger that emitted authentic aromas are arguably as sophisticated as today's olfactory technology (see chaps. 14, 21, and 40), although digiScents·com now promises to bring the sense of smell to our computer. In addition, the generation of tactile sensations (i.e., distribution of force fields on the skin during contact with objects) remains elusive (see chaps. 5 and 6).

Perhaps a more appropriate yardstick by which to judge the current state of the art in VE technology would be the agenda set by Durlach and Mavor (1995) a half decade ago in the seminal National Research Council (NRC) report *Virtual Reality: Scientific and Technological Challenges*. That report developed a set of recommendations that, if heeded, should assist in realizing the full potential of VE technology (see Table 1.1). While this work provided many suggestions, the importance of improved computer generation of multimodal images and advancements in hardware technologies that support interface devices were stressed, as was improvement in the general comfort associated with donning these devices. As the following sections will discuss, the former objectives have largely been met by astounding technological advances, yet the latter has yet to be fully realized, as VE users are still impeded by cumbersome devices and binding tethers (but that will soon change with technologies such as Bluetooth). This chapter focuses on a number of key recommendations put forth by Durlach and Mavor (1995), while many others are described in detail in other chapters in this handbook (see status notes in Table 1.1).

## 2. TECHNOLOGY

Virtual environments are driven by the technology that is used to design and build these systems. This technology consists of the human–machine interface devices that are used to present multimodal information and sense the virtual world, as well as the hardware and software used to generate the virtual environment. It also includes the techniques and electromechanical systems used in telerobotics, which can be transferred to the design of VE systems, as well as the communication networks that can be used to transform VE systems into shared virtual worlds.

### 2.1 Human–Machine Interface

Human–machine interfaces consist of the multimodal devices used to present information to VE users. For multimodal VE applications, advances in peripheral connections to the computer are the single largest issue. When an input device is connected, such as a body or limb tracker, a serial port is generally utilized, a port typically designed for character input and not high-speed data transfer. A solution to the input device connectivity issue that is available on commodity computing is the great unsolved problem. At some point, this input-port speed problem needs to be solved, and that resolution must be included on mass-marketed PCs or their descendents.

Visual displays, especially head-mounted displays (HMDs), have come down substantially in weight but are still hindered by cumbersome designs, obstructive tethers, suboptimal resolution, and insufficient field of view (see chap. 3). (Note: For an excellent comparative source on HMDs, see the "HMD/VR–Helmet Comparison Chart," Bungert, 2001.) Recent advances in wearable computer displays (e.g., Microvision, MicroOptical), which can incorporate miniature LCDs directly into conventional eyeglasses or helmets, should ease cumbersome design and further reduce weight (Lieberman, 1999). There are several low- to mid-cost HMDs (InterSense's InterTrax i-glasses, Olympus Eye-Trek FMD, Interactive Imaging Systems' VFX3D, Sony Cybermind, Sony Glasstron, and Kaiser ProViewXL) that are lightweight (approximately 39 g to 1,000 g) and provide a horizontal field of view (30 to 35 degrees per eye) and resolution (180 K to 2.4 M pixels/LCD) exceeding predecessor systems. While the resolution range looks impressive, most consumer-grade HMDs (those around 180 K pixels/LCD) use three pixels (red, green, and blue) to produce one colored pixel, providing a true resolution of only about 60 K pixels per LCD (Bungert, 2001).

Virtual Retinal Displays (VRDs) may bring truly revolutionary advances in display technology. VRD technology, which was invented in 1991 at the University of Washington's HIT (Human Interface Technology) Lab, holds the promise for greatly enhanced optics (Kleweno et al., 1998). With this technology, an image is scanned directly onto a viewer's retina using low-power red, green, and blue light sources, such as lasers or LEDs (Urey, Wine, & Lewis, online). The VRD system has superior color fidelity, brightness, resolution, and contrast compared to LCDs and CRTs, as it typically uses spectrally pure lasers as the light source.

With advances in wireless and laser technologies and miniaturization of LCDs, during the next decade visual display technology should realize the substantial gains necessary to provide high-fidelity virtual imagery in a lightweight noncumbersome manner.

In the area of virtual auditory displays there have also been tremendous gains (see chap. 4). For example, while early spatialized audio solutions (Blauert, 1997) were expensive to implement, it is currently feasible to include spatialized audio in most VE systems. (For an excellent source on spatialized audio, see "The Ultimate Spatial Audio Index," http://www.speakeasy.org/~draught/spataudio.html.) On the hardware side, systems are available that present multiple sound sources to multiple listeners using positional tracking. Technology for designing complex spatial audio scenarios, including numerous reflections, real-time convolution, and head tracking is currently under way. Software solutions are also under development, which provide low-level control of a variety of signal processing functions, including the number and position of reflections, as well as allowing for normal head-related transfer function (HRTF) processing (see chap. 4), manipulation of acoustic radiation patterns, spherical spreading loss, and atmospheric absorption. HRTFs have yet, however, to effectively include reverberation or echoes. Adding reverberation to a VE causes auditory sources to seem more realistic and provides robust information about relative source distance; thus, further research is needed in the development of tractable reverberation algorithms for real-time systems. Advances in HRTF individualization (i.e., to the physiological makeup of a listener's ear) are also of great importance for localizing sounds, especially for distinguishing front from back and up from down. In particular, means to tailor HRTFs to an individual listener without explicitly measuring HRTFs for that individual are needed. This may be possible, since the transfer functions of the external ear have been found to be similar across different individuals (Middlebrooks, Makous, & Green, 1989).

Current haptic technology provides net force and torque feedback (i.e., simulating tool usage) but has yet to develop effective tactile feedback (e.g., simulating skin contact or dynamic flexibility, such as the sensation of bumps, scratches, and deformations due to flexion of body

**TABLE 1.1**
Status of Durlach and Mavor's (1995) Recommendations for Advancing VE Technology

| *Area* | *Recommendation* | *Status*[1] |
|---|---|---|
| Technology: human–machine interface | • Address issues of information loss due to technology shortcomings (e.g., poor resolution, limited field of view, deficiencies in tracker technology) | S |
| | • Improvements in spatialization of sounds, especially sounds to the front of a listener and outside of the "sweet spot" surrounding a listener's head | M (see chap. 4) |
| | • Improvements in sound synthesis for environmental sounds | M |
| | • Improvements in real-time sound generation | M |
| | • Better understanding of scene analysis (e.g., temporal sequencing) in the auditory system | M |
| | • Improvements in tactile displays that convey information through the skin | L (see chaps. 5–6) |
| | • Better understanding of the mechanical properties of soft tissues that come in contact with haptic devices, limits on human kinesthetic sensing and control, and stimulus cues involved in the sensing of contact and object features | L |
| | • Improvements in locomotion devices beyond treadmills and exercise machines | M (see chap. 11) |
| | • Address fit issues associated with body-based linkage tracking devices; workspace limitations associated with ground-based linkage tracking devices; accuracy, range, latency, and interference issues associated with magnetic trackers; and sensor size and cost associated with inertial trackers | M (see chap. 8) |
| | • Improvements in sensory, actuator, and transmission technologies for sensing object proximity, object surface properties, and applying force | M (see chaps. 5–6) |
| | • Improvements in the vocabulary size, speaker independence, speech continuity, interference handling, and quality of speech production for speech communication interfaces | S |
| | • Improvements in olfactory stimulation devices | L |
| | • Improvements in physiological interfaces (e.g., direct stimulation and sensing of neural systems) | M (see chap. 7) |
| | • Address ergonomic issues associated with interaction devices (e.g., excessive weight, poor fit both mechanically and optically) | M (see chap. 41) |
| | • Better understanding of perceptual effects of misregistration of visual images in augmented reality | M (see chap. 37) |
| | • Better understanding of how multimodal displays influence human performance on diverse types of tasks | M (see chaps. 14, 21) |
| Technology: computer generation of virtual environments | • Improvements in techniques to minimize the load (i.e., polygon flow) on graphics processors | S (see chap. 12) |
| | • Improvements in data access speeds | S |
| | • Development of operating systems that ensure high-priority processes (e.g., user tracking) receive priority at regular intervals and provide time-critical computing and rendering with graceful degradation | L |

| | | |
|---|---|---|
| | • Improvements in rendering photorealistic time-varying visual scenes at high frame rates (i.e., resolving the trade-off between realistic images and realistic interactivity) | M |
| | • Development of navigation aids to prevent users from becoming lost | M (see chap. 24) |
| | • Improvements in ability to model psychological and physical models that "drive" autonomous agents | M (see chap. 15) |
| | • Improved means of mapping how user's control actions update the visual scene | M (see chaps. 12, 13) |
| | • Improvements in active mapping techniques (e.g., scanning-laser range finders, light stripes) | M (see chap. 8) |
| Technology: telerobotics | • Improvements in the ability to create and maintain accurate registration between the real and virtual worlds in augmented reality applications | M (see chap. 48) |
| | • Development of display and control systems that support distributed telerobotics | M (see chap. 48) |
| | • Improvements in supervisory control and predictive modeling for addressing transport delay issues | M (see chap. 48) |
| Technology: networks | • Development of network standards that support large-scale distributed VEs | M (see chap. 16) |
| | • Development of an open VE network | M (see chap. 16) |
| | • Improvements in ability to embed hypermedia nodes into VE systems | M (see chap. 16) |
| | • Development of wide-area and local-area networks with the capability (e.g., increased bandwidth, speed, and reliability, reduced cost) to support the high-performance demands of multimodal VE applications | L (see chap. 16) |
| | • Development of VE-specific applications-level network protocols | L (see chap. 16) |
| Psychological consideration | • Better understanding of sensorimotor resolution, perceptual illusions, human-information-processing transfer rates, and manual tracking ability | M (see chaps. 20, 22, 23) |
| | • Better understanding of the optimal form of multimodal information presentation for diverse types of tasks | M (see chap. 21) |
| | • Better understanding of the effect of fixed sensory transformations and distortions on human performance | M (see chap. 31) |
| | • Better understanding of how VE drives alterations and adaptation in sensorimotor loops and how these processes are affected by magnitude of exposure | M (see chaps. 31, 37–39) |
| | • Better understanding of the cognitive and social side effects of VE interaction | M (see chaps. 19, 20, 33) |
| Evaluation | • Establish set of VE testing and evaluation standards | M (see chap. 34) |
| | • Determine how VE hardware and software can be developed in cost-effective manner, taking into consideration engineering reliability and efficiency, as well as human perceptual and cognitive features | M (see chap. 28) |
| | • Identify capabilities and limitations of humans to undergo VE exposure | M (see chaps. 29–41) |
| | • Examine medical and psychological side effects of VE exposure, taking into consideration effects on human visual, auditory, and haptic systems, as well as motion sickness and physiological/psychological aftereffects | M (see chaps. 29–33, 37–39) |
| | • Determine if novel aspects of human–VE interaction require new evaluation tools | M (see chap. 34) |
| | • Conduct studies that can lead to generalizations concerning relationships between types of tasks, task presentation modes, and human performance | L (see chap. 35) |
| | • Determine areas in which VE applications can lead to significant gains in experience or performance | M (see chaps. 42–55) |

[1] L = limited to no advancement; M = modest advancement; S = substantial advancement.

segments, see chaps. 5 and 6). Srinvasan (see chap. 5) suggests that for the foreseeable future advances in haptic technology will be limited by the development of new actuator hardware. In addition, hardware required to simulate distributed forces on the skin may require substantial gains in miniature rotary and linear actuators or advances in alternative technologies, such as shape memory alloys, piezoelectrics, microfluidics, and other microelectromechanical systems.

Advances in tracking technology have been realized in terms of drift-corrected gyroscopic orientation trackers, outside-in optical tracking for motion capture, and laser scanners (see chap. 8). The future of tracking technology is likely hybrid tracking systems, with an acoustic-inertial hybrid on the market (see http://www.isense.com/products/) and several others in research labs (e.g., magnetic-inertial, optical-inertial, and optical-magnetic). In addition, ultrawideband radio technology holds promise for an improved method of omni-directional point-to-point ranging.

Led largely by the Information Society Directorate General of the European Union and the Information Science and Engineering Directorate of the National Science Foundation, the quality of speech recognition and synthesis systems have made substantial gains in the past half decade. Speaker-independent continuous speech recognition systems are currently commercially available (Germain, 1999; Huang, 1998); however, additional advances are needed in acoustic and language modeling algorithms to improve the accuracy, usability, and efficiency of spoken language understanding. Synthetic speech can now be produced that reasonably resembles the acoustic and prosodic characteristics of the original speaker; however, improvements are required in the areas of naturalness, flexibility, and intelligibility of synthesized speech (Institution of Electrical Engineers, 2000). Speech recognition and synthesis are not addressed in detail in this handbook, not due to any implied lack of importance but rather because there are many significant works whose sole focus is speech technology (see Gibbon, Mertins, Moore, 2000; Varile & Zampolli, 1998). (For an excellent information source on commercial speech recognition, see http://www.tiac.net/users/rwilcox/speech.html; see http://www.cs.bham.ac.uk/~jpi/museum.html for resources on speech synthesis systems; see http://research.microsoft.com/research/srg/ for the latest in Microsoft's speech technology efforts).

Taken together, these technological advancements, along with those poised for the near future, provide the infrastructure on which to build complex, immersive multimodal VE applications.

## 2.2 Computer Generation of Virtual Environments

Computer generation of VEs requires very large physical memories, high-speed processors, high-bandwidth mass storage capacity, and high-speed interface ports for input/output devices (Durlach & Mavor, 1995). Remarkable advances in hardware technologies have been realized in the past half decade that will better meet these VE demands. Moore's law (see chap. 28) is being surpassed, with processor speeds doubling in less than 12 months (Nicholls, 2000a). In addition, technological advances (e.g., tantalum oxide chip gates, extreme ultraviolet [EUV] lithography, new microarchitecture, and better insulation) may help microprocessor speeds reach 20 gigahertz within the next decade (Kanellos, 2000). The memory bandwidth problem is being assuaged by 100-(peak bandwidth 800 Mbyte/sec) and 133-MHz (peak bandwidth 1.1 Gbyte/sec) SDRAM, with promises of 400 MHz and higher-speed memory in the near term. The gigahertz barrier has been surpassed by both Advanced Micro Devices' Athlon's and Intel's Pentium III and Willamette, and more advances will soon be realized with Intel's Itanium and McKinley, Sun's Ultrasparc III, Hewlett-Packard's 8700, and IBM's Power4, the latter of which can build an eight-processor system in a hand-size module. Yet this is just the start. The future promises massive parallelism in computing as we approach the molecular-feature-size limits in integrated circuits (Appenzeller, 2000).

Software development of VE systems has progressed tremendously, from proprietary and arcane systems, to development kits that run on general-purpose operating systems, such as Windows in most of its flavors, while still allowing high-end development on Silicon Graphics workstations (Pountain, 1996). Virtual environment system components are becoming modular and distributed, thereby allowing VE databases (i.e., editors used to design, build, and maintain virtual worlds) to run independently of visualizer and other multimodal interfaces via network links. Standard application program interfaces (APIs; e.g., OpenGL, Direct-3D, Mesa) allow multimodal components to be hardware-independent. Virtual environment programming languages are advancing, with APIs, libraries, and particularly scripting languages allowing nonprogrammers to develop virtual worlds. Using these tools, commercial applications developers can build a range of VEs, from the most basic mazes to complex medical simulators, from low-end single-user PC platform applications to collaborative applications supported by client–server environments.

A number of 3-D modeling languages and tool kits are available which provide intuitive interfaces and run on multiple platforms and renderers (e.g., AC3D Modeler, Clayworks, MR Toolkit, MultiGen Creator and Vega, RealiMation, Renderware, VRML, WorldToolKit). Beyond these languages, which deal with display devices that paint pixels on the screen and define higher-level inputs via triangles and polygons, a new approach to the computer generation of VEs is to use a scene management engine (RealiMation, 2000). This approach allows programmers to work at a higher level, defining characteristics and behaviors for more holistic concepts (e.g., attacker, enemy), thereby enabling developers to concentrate on content design (see chap. 25) without being concerned about how that content is delivered to users.

Photorealistic rendering tools are evolving toward full-featured physics-based global illumination rendering systems (e.g., Raster3D—http://www.bmsc.washington.edu/raster3d/raster3d.html; RenderPark—http://www.cs.kuleuven.ac.be/cwis/research/graphics/RENDERPARK/; Heirich & Arvo, 1997; Merritt & Bacon, 1997). Such physically based rendering techniques allow quantitative prediction of the illumination in a virtual scene and generation of photorealistic computer images, in which illumination effects such as soft shadows and glossy reflections are reproduced with high fidelity (Suykens, 1999).

Computer generation of autonomous agents is a key component of many VE applications involving interaction with other entities, such as adversaries, instructors, or partners. There has been significant research and development in modeling embodied autonomous agents (see chap. 15). Notable in this area is a spin-off from the MIT Artificial Intelligence Laboratory, Boston Dynamics, Inc. (BDI, http://www.bdi.com/). BDI has adapted advances in robotics systems, such as motion caching, variable motion interpolation, and task-level control optimization techniques to display dozens of lifelike articulated agents at one time. BDI's products allow system developers to work directly in a 3-D database, interactively specifying agent behaviors, such as paths to traverse and sensor regions. The resulting agents move realistically, respond to simple commands, and travel about a VE as directed. Further integration of telerobotics techniques (see chap. 48) into autonomous agent design is certain to lead to even more impressive advances. While the aforementioned gains are noteworthy, there are still a number of unsolved problems in agent design and development (see Table 15.3 of chap. 15).

Research in VE navigation has led to the development of design guidelines and aids that enable wayfinding in virtual worlds. These aids include maps, landmarks, trails, and direction finding (see chap. 24). In addition, for closed VEs (e.g., buildings), tools that demonstrate the surrounding area (maps, exocentric 3-D views) are recommended if training or exposure time is short, while internal landmarks (i.e., along a route) are recommended for longer exposure durations (Stanney, Chen, & Wedell, 2000). For semiopen (e.g., urban areas) and open

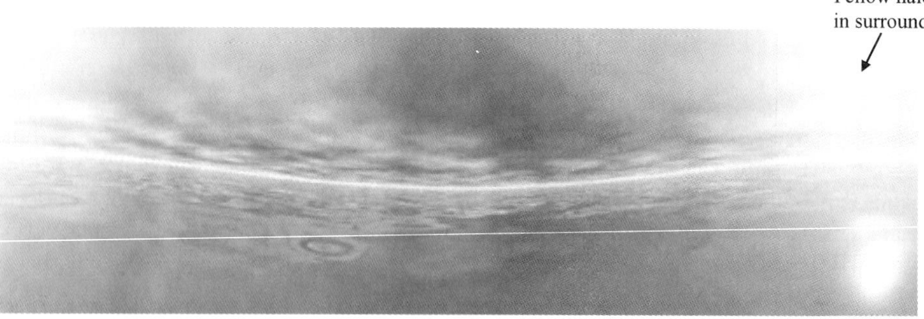


FIG. 1.1. Sea scene with window of normative color encircled by yellow halo indicating going off-course.

environments (e.g., sea, sky), demonstrating the surround is appropriate for short exposures, while use of external landmarks (i.e., outside a route) is recommended for long exposure times. Based on these and other guidelines, aids need to be developed to guide navigation in virtual environments. One such aid was designed by Stanney, Chen, and Wedell (2000), which provides wayfinders with a "window" of normative color shaded on the edges in a symbolic color (e.g., yellow or red), which appears when off-course (see Fig. 1.1). The scene appears in normative color when wayfinders are on-course, gradually changing to yellow and then to red as a wayfinder moves further off-course. While this is just an example, and one whose effectiveness has yet to be validated, it is shared here because it is an example of a collaborative effort between engineers and a graphic artist. Such multidisciplinary collaborations are likely to serve as the crux for truly innovative advances in VE design. More work is needed in the area of navigational aiding, as making one's way through a VE has been found to be one of the most significant usability issues influencing VE task performance (Ellis, 1993; Jul & Furnas, 1997).

The NRC report (Durlach & Mavor, 1995) indicated the need for a real-time operating system for virtual environments, but the expectation from that committee that such an effort would be funded was low. That proved to be an accurate assessment, as current operating systems (OSs) are perhaps less-supportive of VEs than six years ago, at the time of the NRC report's debut. Certainly there are a variety of Windows derivatives (Nicholls, 2000b), yet no one is convincingly arguing their appropriate use as OSs for VEs, with the exception that if one wants wide acceptance, a Windows variant allows for broad usage. Linux is available, which is a less-capable but open source derivative of Unix. At the same time, there is diminished use of Silicon Graphics (SGI) Irix. SGI Irix, to many of those on the bleeding edge of VE technology, was *the* operating system for developing VEs, and the many features of that system not found elsewhere are direly missed. So the right OS for developing VEs is still an open issue.

### 2.3 Telerobotics

Beyond the advantages to autonomous agent design discussed above, there are many areas (e.g., sensing, navigation, object manipulation) in which VE technology can prosper from the application of robotics techniques. Yet, if these techniques are to be adopted, issues of communication time delay (i.e., transport delay) and real-time control architecture design must be resolved. Chapter 48 discusses a number of techniques for addressing these issues. In that chapter, Kheddar, Chellali, and Coiffet note that

> a cleverly conceived yet "simple" VE intermediary representation contributes to solving the time delay problem, offers ingenious metaphors for both operator assistance and robot autonomy sharing problems, enhances operator sensory feedback through multiple sensory modalities admixtures,

> enhances operator safety, offers a huge possible combination of strategies for remote control and data feedback, shifts the well known antagonistic transparency/stability problem into an operator/VE transparency one without compromising the slave stability, offers the possibility to enhance—in terms of pure control theory—remote robot controllers, allows new human-centered teleoperation schemes, permits the production of advanced user-friendly teleoperation interfaces, makes possible the remote control of actual complex systems, such as mobile robots, nano and micro robots, surgery robots, etc.

To achieve these gains, however, advances in VE modeling techniques and means of addressing error detection and recovery inherent to VE–real environment discrepancies are needed (see discussion in chap. 48).

### 2.4 Networks

The NRC report (Durlach & Mavor, 1995) suggested that with improvements in communications networks, virtual environments would become shared experiences, in which individuals, objects, processes, and autonomous agents from diverse locations interactively collaborate. Advances in the Internet have been substantial in the time since that report, due particularly to the U.S. government's Next Generation Internet (NGI) effort and the University Corporation for Advanced Internet Development's (UCAID's) Internet2 (Langa, 2001). The NGI initiative (http://www.ngi.gov/) is connecting a number of universities and national labs at speeds 100 times faster than the 1996 Internet, and a smaller number of institutions at speeds 1,000 times faster in order to experiment with collaborative-networking technologies, such as high-quality video conferencing and audio and video streams. Of particular interest to VE developers, technologies have been developed to "mark" data streams as having specific characteristics (e.g., time-critical, lockstep) so that differentiated services can enable different types of data to be handled with different quality of service levels. Internet2 is using existing networks (e.g., the National Science Foundation's VBNS—Very-High-Speed Backbone Network Service) to determine the transport designs necessary to carry real-time multimedia data at high speed (http://apps.internet2.edu/). Networked VE applications, which require the ability to recognize and track the presence and movements of individuals as well as physical and virtual objects, while projecting them in realistic, multiple, geographically distributed immersive environments on stereo-immersive surfaces, are ideal for Internet2, as they leverage its special capabilities (i.e., high bandwidth, low latency, low jitter; Singhal & Zyda, 1999).

## 3. PSYCHOLOGICAL CONSIDERATION

There are a number of psychological considerations associated with the design and use of VE systems. Some of these focus on techniques and concerns that can be used to augment or enhance VE interaction and transfer-of-training (e.g., perceptual illusions, design based on human-information-processing transfer rates), while others focus on adverse effects due to VE exposure. In terms of the former, we know that perceptual illusions exist, such as auditory-visual cross-modal perception phenomena (see chap. 22), yet little is known about how to leverage these phenomena to reduce development costs while enhancing one's experience in a virtual environment. Perhaps the one exception is vection (i.e., the illusion of self-movement), which is known to be related to a number of display factors (see Table 23.1 of chap. 23). By manipulating these display factors, designers can provide VE users with a compelling illusion of self-motion throughout a virtual world, thereby enhancing their sense of presence (see chap. 40) often with the untoward effect of motion sickness, as well (see chap. 23). Other such illusions exist (e.g., visual dominance—see chap. 22) and could likewise be leveraged. While current knowledge of how such perceptual illusions occur is limited, it may be sufficient to know that they do occur in order to leverage them to enhance VE system design and reduce development costs. Substantially more research is needed in this area to identify perceptual and cognitive design principles that can be used to trigger and capitalize on these illusory phenomena.

Another psychological area in need of research is that of transfer of training. Stanney, Mourant and Kennedy (1998, p. 330) suggest that: "To justify the use of VE technology for a given task, when compared to alternative approaches, the use of a VE should improve task performance when transferred to the real-world task because the VE system capitalizes on a fundamental and distinctively human sensory, perceptual, information processing, or cognitive capability." Yet there is very limited understanding of the types of tasks or activities for which the unique characteristics of VEs (i.e., egocentric perspective, stereoscopic 3-D visualization, real-time interactivity, immersion, multisensory feedback) can be leveraged to provide significant gains in human performance, knowledge, or experience. One notable exception is the area of spatial knowledge, for which VE training has been found to be advantageous. For example, Darken and Banker (1998) conducted a study to evaluate whether or not a VE could serve as a useful tool in familiarizing individuals with unknown environments. In this experiment, three groups of participants (beginner, intermediate, and advanced orienteers) were tasked with learning a real-world environment by either: (1) studying an orienteering map; (2) exploring the real-world environment with an orienteering map; or (3) exploring a high-fidelity VE in addition to studying the orienteering map. Based on the results, Darken and Banker (1998) summarized that:

1. Experience level was a more important factor than training method in affecting navigation performance.
2. Intermediate orienteers benefited more from VE training than either advanced or beginner orienteers.
3. The VE allowed for time compression in training, permitting more area to be traversed in a shorter time.

Waller (1999) conducted related studies in a maze-shaped VE as well as in the real world. He examined a number of individual variables, including, spatial ability, real world presentation and navigational ability, gender, imaging strategy, computer experience and usage, the ability to acquire and transfer spatial knowledge from a VE, and proficiency with the navigational interface of the virtual environment. Variables that significantly contributed to spatial knowledge retrieval and transfer from the VE included: the proficiency with the navigational interface; spatial ability, especially spatial visualization and spatial orientation ability; and gender. Waller (1999) found that users' proficiency with the virtual interface (a joystick) was the most important factor in determining transferability of spatial knowledge. Those who experienced difficulties in operating the joystick usually retrieved less spatial knowledge from the virtual environment. Taken together, the Darken and Waller studies suggest that while VE training can enhance spatial knowledge of real-world environments, both individual and VE system factors may temper this transfer of knowledge. These findings provide impetus for further study of VE transfer of training. Without this knowledge, the unique benefits of VE technology may never be fully understood or realized.

In contrast to the limited knowledge concerning perceptual and cognitive design principles that augment or enhance VE interaction, more is known about identifying and controlling the adverse effects of VE exposure. Adverse effects are of particular concern because they can persist for some time after exposure, potentially predisposing those exposed to harm. These effects are both physiological (see chaps. 30, 31, 37–39) and psychological (see chap. 33), with considerable effort currently focused on the former and less on the latter. A survey of the current state of knowledge regarding physiological aftereffects concluded that the two most critical research issues in this area are: standardization and use of measurement approaches for VE aftereffects; and identification and prioritization of sensorimotor discordances that drive these aftereffects (Stanney, Salvendy et al., 1998). While significant gains have been made toward the former goal (see chaps. 30, 37–39), there is still insufficient knowledge concerning the latter, particularly in terms of prioritization (see chap. 31). Some are also concerned that exposure to VEs that portray violent content, as is often found in entertainment venues, could lead to aggressive, antisocial, or criminal behavior (see chap. 33). Thus, a proactive approach is needed that weighs the physiological and psychological risks and potential consequences associated with VE exposure against the benefits. Waiting for the onset of harmful consequences should not be tolerated.

Taken together, the research into psychological considerations of VE exposure indicates that more research is needed to derive perceptual and cognitive design strategies that enhance VE interaction, and that there are risks associated with VE exposure. However, usage protocols have been developed that, if successfully adopted, can assist in minimizing these risks (see chap. 36). Thus, VE technology is not something to be eschewed as it has many advantages for enticing and didactic experiences; it is rather something to leverage wholly yet vigilantly, taking care to address the associated risks.

## 4. EVALUATION

Most VE user interfaces are fundamentally different from traditional graphical user interfaces, with unique I/O devices, perspectives, and physiological interactions. Thus, when developers and usability practitioners attempt to apply traditional usability engineering methods to the evaluation of VE systems they find few if any that are particularly well suited to these environments (see chap. 34). Subsequently, very few principles for the design and evaluation of VE user interfaces exist, of which none are empirically derived or validated (Kaur, Maiden, & Sutcliffe, 1999).

While limited work on VE usability has been conducted to date, there are early works that have attempted to improve VEs from users' perspective by integrating a systematic approach to VE development and usability evaluation (Bowman, 1999; Gabbard & Hix, 1997; Kalawsky, 1999; Kaur, 1999; chap. 24). By addressing key characteristics unique to VEs (e.g., perceived presence and real-world fidelity, multidimensional interactivity, immersion), their work has identified limitations of existing usability methods for assessing VE systems (see Table 1.2).

**TABLE 1.2**
Limitations of Traditional Usability Methods
for Assessing Virtual Environments

- Traditional point-and-click interactions are not representative of the multidimensional object selection and manipulation characteristics of 3-D space.
- Quality of multimodal system output (e.g., visual, auditory, haptic) is not comprehensively addressed by traditional evaluation techniques.
- Means of assessing sense of presence and aftereffects have not been incorporated into traditional usability methods.
- Traditional performance measurements (i.e., time and accuracy) do not comprehensively characterize VE system interaction.
- Traditional single-user task-based assessment methods do not consider VE system characteristics in which two or more users interact in the same environment.

Based on these shortcomings, it is suggested that novel aspects of human–VE interaction require new evaluation tools. Toward this end, Gabbard and Hix (1997) have developed a taxonomy of VE usability characteristics that can serve as a foundation for identifying usability criteria that existing evaluation techniques fail to fully characterize. Stanney, Mollaghasemi, and Reeves (2000) used this taxonomy as the foundation on which to develop an automated system, *MAUVE* (Multi-Criteria Assessment of Usability for Virtual Environments), which organizes VE usability characteristics into two primary usability attributes (VE system usability and VE user considerations); four secondary attributes (interaction, multimodal system output, engagement, and side effects); and 11 tertiary attributes (navigation, user movement, object selection and manipulation, visual output, auditory output, haptic output, presence, immersion, comfort, sickness, and aftereffects). Similar to the manner in which traditional heuristic evaluations are conducted, *MAUVE* can be used at various stages in the usability engineering life cycle, from initial storyboard design to final evaluation and testing. It can also be used to compare system design alternatives. The results of a *MAUVE* evaluation not only identify a system's problematic usability components and techniques, but also indicate why they are problematic. Such results may be used to remedy critical usability problems, as well as to enhance the design for usability of subsequent system development efforts.

In terms of the cost-effectiveness of VE systems, models are currently available to evaluate the selling process of VE applications (see chap. 28). With these tools, developers can evaluate if VE technology offers financial advantages over current practices or technologies. This is an essential determination if VE technology is to thrive both commercially and in research domains.

Durlach and Mavor (1995) suggested that serious commercial applications would not be realized for 5 to 10 years (thus, sometime between 2000 and 2005), and those insights appear well founded. While viable applications exist today (see chaps. 42–55), most are still in the developmental stages. Further research is required to determine areas in which VE applications can lead to significant gains in experience or performance.

## 5. CONCLUSIONS

As seen from the above discussion and from the remaining chapters in this handbook, there has been considerable technological progress for the VE field since the advent of the NRC report (Durlach & Mavor, 1995). The NRC report was an inflection point in the VE field. Before the report, there were many technology futurists seeking headlines indicating that "VR [virtual reality] would destroy your television" and that VR was coming soon, perhaps even next week. The NRC report provided a more sobering assessment and indicated that, in some areas, there was as much as a decade or two of work yet to be done. Some of the prognosticators continue to this day with glowing reports and predictions, but we know that much careful work is required. The remainder of this text highlights some of the completed work and points out what is yet to be done. Those research directions will hopefully lead us to a platonic virtual reality that is sufficiently convincing to our minds.

## 6. ACKNOWLEDGMENTS

The authors would like to thank Nat Durlach and Tom Furness for their comments on an earlier version of this work. Their insights are greatly appreciated. This material is based in part on work supported by the Naval Air Warfare Center Training Systems Division (NAWC TSD) under contract No. N61339-99-C-0098. Any opinions, findings, conclusions, or recommendations expressed in this material are those of the authors and do not necessarily reflect the views or the endorsement of NAWC TSD.

## 7. REFERENCES


Appenzeller, T. (2000, May 1). The chemistry of computing: Computers made of molecule-size parts could build themselves. *U.S. News and World Report* [Online]. Available: http://www.usnews.com/usnews/issue/000501/chips.htm

Blauert, J. (1997). *Spatial hearing* (Rev. Ed.). Cambridge, MA: MIT Press.

Bowman, D. (1999). *Interaction techniques for common tasks in immersive virtual environments: Design, evaluation, and application*. Unpublished doctoral dissertation, Georgia Institute of Technology, Atlanta.

Bungert, C. (2001). *HMD/VR-helmet comparison chart* [Online]. Available: http://www.stereo3d.com/hmd.htm#resolution.

Darken, R. P., & Banker, W. P. (1998). Navigation in natural environments: Training transfer study. In *Proceedings of VRAIS '98*, IEEE Computer Society Press, 12–19.

Durlach, B. N. I., & Mavor, A. S. (1995). *Virtual reality: Scientific and technological challenges*. Washington, DC: National Academy Press.

Ellis, S. R. (1993). *Pictorial communication in virtual and real environments*. London: Taylor & Francis Press.

Gabbard, J. L., & Hix, D. (2000, July 31). *A taxonomy of usability characteristics in virtual environments* [Online]. Available: http://csgrad.cs.vt.edu/~jgabbard/ve/taxonomy/

Germain, A. H. (1999, June 21). How to win big with today's hottest emerging technologies. *VarBusiness* [Online]. Issue 1516. Available: http://www.techweb.com/se/directlink.cgi?VAR19990621S0027

Gibbon, D., Mertins, I., & Moore, R. (2000). *Handbook of multimodal and spoken dialogue systems: Resources, terminology and product evaluation*. Dordrecht, The Netherlands: Kluwer Academic Publishers.

Heirich, A., & Arvo, J. (1997). Scalable Monte Carlo image synthesis [Special issue]. *Parallel Computing, 23*(7), 845–859.

Huang, X. D. (1998). Spoken language technology research at Microsoft. *16th ICA and 135th ASA '98*, Seattle.

Institution of Electrical Engineers. (2000). *The State of the Art in Speech Synthesis* [Online]. Colloquium 058, held April 13, 2000, at Savoy Place, London. Available: http://www.iee.org.uk/Events/e13apr00.htm

Jul, S., & Furnas, G. W. (1997). Navigation in electronic worlds. CHI '97 workshop. *ACM SIGCHI Bulletin, 29*(4), 44–49.

Kalawsky, R. S. (1999). VRUSE—a computerized diagnostic tool: For usability evaluation of virtual/synthetic environments systems. *Applied Ergonomics, 30*, 11–25.

Kanellos, M. (2000, January 28). Chips embark on road to 20 gigahertz. *CnetNews.Com* [Online]. Available: http://news.cnet.com/news/0-1003-200-1534900.html

Kaur, K. (1999). *Designing virtual environments for usability*. Unpublished doctoral dissertation, City University, London.

Kaur, K., Maiden, N., & Sutcliffe, A. (1999). Interacting with virtual environments: An evaluation of a model of interaction. *Interacting with Computers, 11*(4), 403–426.

Kleweno, C., Seibel, E., Kloeckner, K., Burstein, B., Viirre, E., & Furness, T. (1998). *Evaluation of a scanned laser display as an alternative low vision computer interface* (Tech. Rep. No. R-98-39). Seattle, WA: University of Washington, Human Interface Technology Laboratory.

Langa, F. (2001, January 22). I want my Next Generation Internet. *Byte Magazine* [Online]. Available: http://byte.com/

Lieberman, D. (1999). Computer display clips onto eyeglasses. *Technology News* [Online]. Available: http://www.techweb.com/wire/story/TWB19990422S0003

Merritt, E. A., & Bacon, D. J. (1997). Raster3D: Photorealistic molecular graphics. *Methods in Enzymology, 277*, 505–524.

Middlebrooks, J. C., Makous, J. C., & Green, D. M. (1989). Directional sensitivity of sound-pressure levels in the human ear canal. *Journal of the Acoustical Society of America, 86*, 89–108.

Nicholls, B. (2000a, April 24). Interesting times, part 1: Hardware. *Byte Magazine* [Online]. Available: http://byte.com/column/BYT20000419S0005

Nicholls, B. (2000b, May 15). Interesting times, part 2: Operating systems. *Byte Magazine* [Online]. Available: http://www.byte.com/column/BYT20000510S0014

Pountain, D. (1996, July). VR meets reality: Virtual reality strengthens the link between people and computers in mainstream applications. *Byte Magazine* [Online]. Available: http://www.byte.com/art/9607/sec7/art5.htm

RealiMation. (2000). A scene data management approach to real-time 3D software applications [Tech. Rep. No. TP001 [Online]. Available: http://www.realimation.com/overview/technical/technical_papers.htm

Rheingold, H. (1991). *Virtual Reality*. London: Secker & Warburg.

Singhal, S., & Zyda, M. (1999). *Networked virtual environments—Design and implementation*. New York: ACM Press Books, SIGGRAPH Series.

Stanney, K. M., Chen, J., & Wedell, B. (2000). *Navigational metaphor design* (Final Rep. Contract No. N61339-99-C-0098). Orlando, FL: Naval Air Warfare Center Training Systems Division.

Stanney, K. M., Mollaghasemi, M., & Reeves, L. (2000). *Development of MAUVE, the multi-criteria assessment of usability for virtual environments system* (Final Rep., Contract No. N61339-99-C-0098). Orlando, FL: Naval Air Warfare Center Training Systems Division.

Stanney, K. M., Mourant, R., & Kennedy, R. S. (1998). Human factors issues in virtual environments: A review of the literature. *Presence: Teleoperators and Virtual Environments, 7*(4), 327–351.

Stanney, K. M., Salvendy, G., Deisigner, J., DiZio, P., Ellis, S., Ellison, E., Fogleman, G., Gallimore, J., Hettinger, L., Kennedy, R., Lackner, J., Lawson, B., Maida, J., Mead, A., Mon-Williams, M., Newman, D., Piantanida, T., Reeves, L., Riedel, O., Singer, M., Stoffregen, T., Wann, J., Welch, R., Wilson, J., & Witmer, B. (1998). Aftereffects and sense of presence in virtual environments: Formulation of a research and development agenda. *International Journal of Human–Computer Interaction, 10*(2), 135–187. (Report sponsored by the Life Sciences Division at NASA Headquarters)

Sutherland, I. E. (1963). Sketchpad—A man–machine graphical communication system. Tech. Rep. No. 296, MIT Lincoln Laboratory.

Suykens, F. (1999). *Bidirectional path tracing and improvements* [Online]. Available: http://www.cs.kuleuven.ac.be/~graphics/SEMINAR/. (Talk given at *Seminar on Physically Based Techniques and Applications*, Leuven, Belgium, December 15, 1999).

Urey, H., Wine, D. W., & Lewis, J. R. *Scanner design and resolution tradeoffs for miniature scanning displays* [Online]. Available: http://www.mvis.com/urey_phototonics_wp.htm

Varile, G., & Zampolli, A. (Eds.). (1998). *Survey of the state of the art in human language technology: Studies in natural language processing*. New York: Cambridge University Press.

Waller, D. A. (1999). *An assessment of individual difference in spatial knowledge of real and virtual environments*. Unpublished doctor dissertation, University of Washington.

# 2

# Virtual Environments Standards and Terminology

Richard A. Blade[1] and Mary Lou Padgett[2]

[1]*Department of Physics, University of Colorado at Colorado Springs*
*13631 E. Marina Dr., #302*
*Aurora, CO 80014*
*rblade@mail.uccs.edu*

[2]*Padgett Computer Innovations, Inc.*
*1165 Owens Road*
*Auburn, AL 36830*
*m.padgett@ieee.org*

## 1. INTRODUCTION

In just the past decade virtual environment (VE) applications have emerged in entertainment, training, education, and other areas (see Chapters 43–55, this volume). In that time extensive research in VE technology has also been conducted. However, the terminology used to characterize this technology is still evolving. In fact, Durlach and Mavor (1995, p. 2) indicate that "inadequate terminology [is] being used" to describe VE technology and its applications. It is thus important to describe the key terms that are used in this handbook. The objective is not to resolve differences between disparate uses (in fact often multiple, even conflicting definitions are presented) but rather to provide a coherent set of commonly used terms. While it is customary to present a glossary at the end of handbooks such as this one, this work starts out with a glossary so that readers may develop a common understanding of the terms used throughout the handbook. Paradoxically, the one term that remains particularly elusive is *virtual environment.* Many authors, especially those among the application chapters (see chaps. 43–55, this volume) have catered the definition of VE to fit the forms of the technology that best suit their needs. Perhaps this definitional multiplicity demonstrates the versatile nature of VE technology and its wide array of potential uses.

While the definitions in this chapter have been presented in a relatively informal manner, we continue to work toward a set of standard definitions through the VR Terminology Project. This involves a multiphase effort on the part of the IEEE Computer Society under the auspices of the IEEE Standards Association, of which the first phase is the establishment of a working group on virtual reality terminology (VI-1392).

## 2. IMPORTANCE OF STANDARDS

Standards are critical for systematic and robust development of any emerging technology. Specification standards provide for practical descriptions of product characteristics and limitations, critical to an end user. Interface standards allow for interchangability of components developed by different manufacturers, thus permitting specialization and robust competition in the marketplace. Safety standards ensure the health and safety of product users. Terminology standards ensure that technical terminology is used in a consistent and rigorous manner, thus preventing confusion and ambiguity in scientific and technical reports and specifications.

### 2.1 Process of Establishing Standards

The process of establishing standards is quite possibly as important as the standards ultimately produced, for that process establishes a forum for open and systematic dialog between researchers, developers, manufacturers, and end users on the current and future needs and directions of an industry. It often progresses at a "glacial pace," meaning it could take years to actually reach agreement on a standard, which ensures a systematic, objective, and rigorous examination of all aspects of an issue by all who have interest or involvement.

There are many standards setting organizations, each having its own rules and procedures. For electrical, electronic, and computer-related standards, the two primary international organizations are the International Electrotechnical Commission (IEC) and the International Organization for Standardization (ISO). Most standards related to VEs will originate in the IEEE Computer Society, which is a part of the Institute of Electrical and Electronic Engineers (IEEE). The IEEE Standards Association (IEEE-SA), in turn, belongs to the American National Standards Institute (ANSI). Standards organizations can and do work together, but there are no requirement that they do so. Rather, obtaining the broadest possible acceptance of standards is the motivation for cooperation and collaboration. For example, the ISO and IEC have formed the Joint Technical Committee Number One (JTC1) to deal with information technology standards. They also cooperate on issues involving safety, electromagnetic radiation, and the environment.

The first step to establishing one or more VE standards (usually a group of related standards) in the IEEE Computer Society is for an individual to prepare a project authorization request (PAR) for submission to the IEEE Standards Board. If approved, that PAR is given an identifying label (e.g., vi-1392 in the case of VE terminology), and a working group is set up under that label to study the problem and make recommendations. All interested parties are invited to participate, and every attempt is made to ensure a broad representation. After much informal discussion and deliberation, the working group prepares draft standards. A balloting group is then assembled to vote on the standards and/or modify them if necessary. In contrast to the working group, the process and membership of the balloting group become very formal to ensure a balanced and fair consideration of the issues and proposed standards. The last step in establishing a standard by the IEEE-SA is for the IEEE Standards Board to approve the submission of the balloting group. When the standards being proposed overlap between two or more existing groups, a standards coordinating committee (SCC) is involved to ensure proper coordination and collaboration. The work of the IEEE-SA is supported by its publication of the standards. Various procedures must be followed throughout the standards process to ensure that the publication rights to each standard produced belong to the IEEE.

## 3. IMPORTANCE OF OFFICIAL TERMINOLOGY

Any standard set by the IEEE or other standards development organizations contains definitions. When multiple uses of a term are found in the literature or are in common use, it should help to have a careful record of the variations, including recommendations and cautions. Consider, for example, the term "megabyte". It may mean $10^6 = 1{,}000{,}000$ bytes, or it may mean $2^{10} = 1{,}024 \times 1{,}024 = 1{,}048{,}576$ bytes, or even $1{,}000 \times 1{,}024 = 1{,}024{,}000$ bytes. A consumer needs to be aware of these disparities. One important task of the VR Terminology Project is to provide consumers, including government contractors, reliable definitions so they can properly evaluate product descriptions and proposals. If these definitions are readily available to users and producers, reliable commerce will be enhanced.

A short glossary of terms is given in the next section. Most are technical terms that were provided by the authors of other chapters in this handbook because of their occurrence in those chapters, and many appear in more than one chapter. Although all have definitions in the professional literature, almost none of those definitions have been universally adopted by any standards organization. In preparing this chapter much discussion took place with the authors, as well as between various authors of this handbook, to ensure a reasonable degree of clarity and consistency. However, in some cases we found it necessary to provide multiple, sometimes even contradictory, definitions reflecting current usage.

## 4. A SHORT GLOSSARY

**6DOF:** six degrees of freedom, typically specified by three spatial directions and three angles of orientation.

**Accommodation:** change in the focal length of the eye's lens to maintain focus on a moving close object.

**Aftereffect:** any effect of *VE* exposure that is observed after a participant has returned to the physical world.

**Archetypes:** prototypes that provide templates to guide learning, development, and the construction of the personality, or psyche.

**Attentional inertia:** attentional mechanism in which the person's visual system gets locked into engaging, interesting experiences.

**Attitudes:** description of team members' affective views, both individually and collectively, on their abilities and motivation to accomplish goals.

**Aubert effect or phenomenon:** the apparent displacement of an isolated vertical line in the direction opposite to which the observer is tilted. This happens when an observer has a large tilt, e.g., 90 degrees.

**Audification:** an acoustic stimulus involving direct playback of data samples. See also *Sonification*.

**Augmented reality:** the use of transparent glasses on which a computer displays data so the viewer can view the data superimposed on real-world scenes.

**Avatar:** character that represents a player in second-person virtual systems.

**Binocular:** displaying a different image to each eye for the purpose of stereographic viewing.

**Binocular parallax:** same as parallax. See also *Motion parallax*.

**Bi-ocular:** displaying the same image to each eye. Sometimes this is done to conserve computing resources when depth perception is not critical. See also *Stereopsis*.

**BOOM:** Fake Space Labs' Binocular Omni Orientation Monitor, which consists of a visual display supported on a pedestal; users hold the display head with their hands and adjust the view direction in a manner similar to that of a pair of binoculars.

**Bots:** robots or intelligent agents who roam MUDs and other virtual environments.

**CAVE:** Cave Automatic Virtual Environment—a SID where images are projected on the walls, floor, and ceiling of a room that surrounds a viewer. Oftentimes a principal viewer's head position is tracked to determine view direction and content, while other viewers "come along for the ride."

**Cognitive map:** mental representation of an environment (also referred to as a mental map).

**Communication channel:** when applied to HCI, it is a pathway between the user and the simulation that allows human–computer interaction.

**Computer assisted teleoperation (CAT):** bilateral control of teleoperation through computers, including computer assistance in both robot control and information feedback.

**Convergence:** occurs in stereoscopic viewing when the left and right eye images become fused into a single image.

**Convolve:** to filter and intertwine signals (e.g., sounds) and render them three-dimensional. Used in VE applications to recreate sounds with directional cues.

**Coriolis, or cross-coupling, effect:** effect resulting from certain kinds of simultaneous multi-axis rotations, especially making head movements while rotating. This illusion is characterized by a feeling of head or self-velocity in a curved path that is roughly orthogonal to the axes of both body and head rotation, which can lead to simulator sickness.

**Culling:** removing invisible pieces of geometry and only sending potentially visible geometry to the graphics subsystem. Simple culling rejects entire objects not in the view. More complex systems take into account occlusion of some objects by others, e.g., a building hiding trees behind it.

**Cutaneous senses:** skin senses, including light touch, deep pressure, vibration, pain, and temperature.

**Cybersickness:** sensations of nausea, oculomotor disturbances, disorientation, and other adverse effects associated with VE exposure.

**Cyberspace:** the universe of digital data.

**DataGlove:** VPL Research, Inc., device for sensing hand gestures, which uses fiber-optic flex sensors to track hand orientation and position, as well as finger flexure.

**Deformable object technology (DOT):** virtual objects, which bend and deform appropriately when touched.

**Depth cueing:** use of shading, texture mapping, color, interposition, or other visual characteristics to provide a cue for the distance of an object from the observer.

**Desktop virtual systems:** virtual experiences that are displayed on a two-dimensional desktop computer; the person can see through the eyes of the character on the screen, but the experience is not three-dimensional.

**Dolly shot:** display of a scene while moving forward or backward. See also *Pan shot* and *Track shot*.

**Dynamic accuracy:** system accuracy as a tracker's sensor is moved. See also *Static accuracy*.

**"E" effect:** same as the Muller effect.

**EBAT:** Event-Based Approach to Training—provides the strategies, methods, and tools that are essential for an effective learning environment in a structured and measurable format for training and testing specific KSAs.

**Egocenter:** sense of one's own location in a virtual environment.

**Embodiment:** being inside a body.

**Ergonomics:** study of human factors, i.e., the interaction between the human and his or her working environment.

**Exoskeletal device:** flexible interaction devices worn by users (e.g., gloves and suits) or rigid-link interaction systems (i.e., jointed linkages affixed to users).

**EyePhone:** A VPL Research, Inc., display device consisting of two tiny television monitors (one per eye), earphones, and a sensor for tracking head position and orientation.

**Fidelity:** degree to which a VE or SE duplicates the appearance and feel of operational equipment (i.e., physical fidelity) and sensory stimulation (i.e., functional fidelity) of the simulated context.

**Field of view (FOV):** angle in degrees of the visual field. Since a human's two eyes have overlapping 140-degree FOV, binocular or total FOV is roughly 180 degrees in most people. A FOV greater than roughly 60 to 90 degrees may give rise to a greater sense of immersion.

**Fish tank VE:** illusion of looking "through" a computer monitor to a virtual outside world using a stereoscopic display system. When looking "out" through the stereo "window," the observer imagines him/herself to be in something resembling a fish tank.

**Formative evaluation:** to assess, refine, and improve user interaction by iteratively placing representative users in task-based scenarios in order to identify usability problems, as well as to assess a design's ability to support user exploration, learning, and task performance.

**Force feedback:** output device that transmits pressure, force, or vibrations to provide a VE participant with the sense of resisting force, typically to weight or inertia. This is in contrast to tactile feedback, which simulates sensation to the skin.

**Formal features:** audiovisual production features that structure, mark, and represent media experiences.

**Fractal:** self-similar graphical pattern generated by using the same rules at various levels of detail, that is, a graphical pattern that repeats itself on a smaller and smaller scale.

**Frustum of vision:** three-dimensional field of view in which all modeled objects are visible.

**Functional fidelity:** see *Fidelity*.

**Graphical user interface (GUI):** user interface that presents information graphically, typically with draggable windows, menus, buttons, and icons, as opposed to a textual user interface, where information is presented on a text-based screen and commands are all typed.

**Gravitoinertial force:** resultant force combining gravity and virtual forces created by acceleration.

**GUI:** abbreviation for *graphical user interface*.

**Guidelines-based evaluation:** see *Heuristic evaluation*.

**Gustatory:** pertaining to the sense of taste.

**Haptic:** skin-related, or cutaneous, information. Sometimes refers to information obtained by active touch and manual exploration.

**Haptic interface:** interface involving the human hand and manual sensing and manipulation.

**HCI:** abbreviation for *human–computer interaction*.

**Heads-up display:** display device that allows users to see graphics superimposed on their view of the real world; a form of augmented reality.

**Heuristic evaluation:** cost-effective and popular method for evaluating a user interface design; the goal is to find usability problems in a design so that they can be attended to as part of an iterative design process; involves having a small set of evaluators examine the interface to determine its compliance with recognized usability principles (the "heuristics").

**Hidden surface:** surface of a graphics object that is occluded from view by intervening objects.

**HMD:** abbreviation for head-mounted display. Though not required, most HMDs include not only a visual display attached to the head, but position tracking to provide a computer with the location and orientation of the head.

**Human–computer interaction (HCI):** study of how people work with computers and how computers can be designed to help people effectively use them.

**Hypotension:** a condition in which the body has difficulty regulating the blood pressure, especially when one is upright. There seems to be a misconnection between the brain and the nerves that control the blood pressure and heart rate. The reason is unknown.

**Immersion:** experience of being physically immersed within a VE experience. The term is sometimes subcategorized into external and internal immersion, and sensory and perceptual immersion. See also *Presence*.

**Intelligent user interface:** user interface that is adaptive and has some degree of autonomy.

**Kinesthesia:** ability to perceive extent, direction, or weight of movement.

**Kinesthetic:** all muscle, joint, and tendon senses. Excludes skins senses, such as touch and vestibular, visual, etc. A subset of somatosensory, which is usually applied to limb position and movement but would also include non-vestibular sensation of head movement (e.g., via neck musculature).

**Kinesthetic dissonance:** mismatch between feedback or its absence from touch or motion during VE experiences.

**KSA:** team competencies of knowledge, skills, and attitudes.

**Latency:** lag between user motion and tracker system response. Delay between actual change in position and reflection by the program. Delayed response time.

**Level of detail (LOD):** model of a particular resolution among a series of models of the same object. Greater graphic performance can be obtained by using a lower LOD when the object occupies fewer pixels on the screen or is not in a region of significant interest.

**Locomotion:** means of travel restricted to self-propulsion. Motion interfaces can be subdivided into those for passive transport (inertial and noninertial displays) and those for active transport (locomotion interfaces).

**Magic wand:** three-dimensional input device used for pointing and interaction. A kind of three-dimensional mouse.

**Master:** force-reflecting device that can be used to feel and manipulate virtual objects.

**Master–slave system:** bilateral coupled teleoperator, where the master is the operator control handle and the slave is the remote executing device.

**Matching:** degree to which a VE mimics reality not only in form but also in terms of function and the behaviors the VE elicits from users.

**Meneire's disease:** physiological disorder due to abnormal pressure of the vestibular organ producing a wide range of symptoms, which can include turbulent eye movements, extreme vertigo, severe nausea, and distortions in hearing.

**Metaball:** surface defined about a point specified by a location, a radius, and an "intensity." When two metaballs come in contact, their shapes blend together.

**MOO:** MUD object-oriented. See *MUD*.

**Motion parallax:** means whereby the eyes can judge distance by noticing how closer objects appear to move more than distant ones when the observer moves. See also *Binocular parallax* and *Parallax*.

**Motion platform:** controlled physical system that provides real motion to simulate the displayed motion in a VE world.

**Move-and-wait strategy:** typical teleoperation control mode when time delays occur (with direct feedback).

**MUD:** a multiuser domain where users can jointly interact and play games, such as Dungeons and Dragons.

**Muller effect (or "E" effect):** tendency to perceive an objectively vertical line as slightly tilted in the same direction as the observer when the observer is tilted by only a moderate amount.

**Multimodal command:** command issued by the user to a computer simulation using several input communication channels.

**Multimodal system:** when applied to HCI, a system that allows communication with computers via several modalities, such as voice, gesture, gaze, touch, etc.

**Nanomanipulator:** a device that allows manipulation on a very microscopic scale.

**Navigation:** aggregate task of wayfinding and motion (i.e., motoric element of navigation).

**Neural interface:** ultimate human–computer interface, which connects directly to the human nervous system.

**Nystagamus:** reflexive eye movements that usually serve to keep the world steady on the retina during self-motion. These eye movements are driven by vestibulo-ocular, cervical-ocular, and optokinetic reflexes. A rapid sideways snap of the eye followed by a slow return to normal fixation or rapid oscillatory movements of the eye, as in following a moving target, in blindness, or after rotation of the body.

**Occipital cortex:** back of the brain receiving retinotopic projections of visual displays.

**Occlusion:** hiding an object or a portion of an object from sight by interposition of other objects.

**Oculogravic illusion:** illusion of visual target displacement that is actually caused by body acceleration. The visual component of the somatogravic illusion (e.g., when an aircraft accelerates and there is a backward rotation of the resultant force vector, in addition to feeling a "pitch up" sensation, the pilot may sense an apparent upward movement of objects in his or her visual field).

**Oculogyral illusion:** illusion of visual target displacement that is actually caused by body rotation (e.g., the illusory movement of a faint light in a darkened room following rotation of the body).

**Olfactory:** pertaining to the sense of smell.

**Orientation:** sense of up and down or north, south, east, and west. Allows an individual to determine: Where am I now? Where did I come from? Where do I want to go?

**Otoconia:** hair cell mechanoreceptors that are embedded in gelatinous membranes containing tiny crystals of calcium carbonate, through which the otolith organs detect changes in the magnitude or direction of linear acceleration vectors.

**Pan:** angular displacement of a view along any axis of direction in a three-dimensional world.

**Pan shot:** display of a scene while moving about any axis. See also *Dolly shot* and *Track shot*.

**Parallax:** difference in viewing angle created by having two eyes looking at the same scene from slightly different positions, thereby creating a sense of depth. Also referred to as binocular parallax. See also *Motion parallax*.

**Parietal cortex:** area of the brain adjacent and above the occipital cortex, which processes spatial location and direction information.

**Persona:** public display, or mask, of the self. Players create persona or characters for their personal use on Internet MUDs.

**Perspective:** rules that determine the relative size of objects on a flat viewing surface to give the perception of depth.

**Phase lag:** when output from computer generated images lags the actual position of tracking sensors.

**Phi phenomenon:** illusion of continuous movement through space induced by a sequence of discontinuous events (e.g., lights flashing in sequence, as on a marquee).

**Phong shading:** method for calculating the brightness of a surface pixel by linearly interpolating points on a polygon and using the cosine of the viewing angle. Produces realistic shading.

**Physical fidelity:** see *Fidelity*.

**Presence:** illusion of being part of a virtual environment. The more immersive a VE experience, often the greater the sense of being part of the experience. See also *Immersion*.

**Proprioceptive:** internal sense of body position and movement. Includes kinesthetic and vestibular senses, but excludes outwardly directed senses such as vision, hearing, etc. This term is not a subset of somatosensory, nor is it synonymous with kinesthesia.

**Prosocial behavior:** socially valued behaviors, such as helping, sharing, cooperating, and engaging in constructive imaginary activities.

**Pseudo:** false.

**Radiosity:** diffuse illumination calculation system for graphics based on energy balancing that takes into account multiple reflections off many walls.

**Ray tracing:** technique for displaying a three-dimensional object with shading and shadows by tracing light rays backward from the viewing position to the light source.

**Reality engine:** computer system for generating virtual objects and environments in response to user input, usually in real time.

**Refresh rate:** frequency with which an image is regenerated on a display surface.

**Registration:** correspondence between a user's actual position and orientation and that reported by position trackers.

**Render:** conversion of image data into pixels to be displayed on a screen.

**Retinal binocular disparity (RBD):** ratio of the convergence angle of an image to the convergence angle of an object.

**Scenes view:** virtual display viewed on a large screen or through a terminal window rather than with immersive devices.

**Scientific visualization:** graphical representation of complex physical phenomena in order to assist scientific investigation and make inferences that are not apparent in numerical form.

**SE:** common abbreviation for *synthetic experience*.

**Second-person virtual systems:** virtual environment experiences in which a person is represented by an on-screen avatar rather than being fully immersed in the environment.

**Semantic unification:** the process of integration and synchronization of information from several input modalities (gesture, speech, gaze, etc.).

**Semiocclusion:** occlusion to one eye only.

**Sensorial redundancy:** presenting same or related sensory information to a user using several communication channels.

**Sensorial substitution:** using a different sensory channel to present information normally presented in the replaced modality.

**Shared mental models:** overlapping knowledge or understanding a task, team, equipment, and situation between team members.

**Shutter glasses:** glasses that alternately block out the left and right eye views in synchrony with the computer display of left and right eye images to provide stereoscopic images on the computer screen.

**SID:** spatially immersive display—semisurrounding projected stereo displays.

**Side effects.:** See *Aftereffects*.

**Simulation overdose:** spending too much time in virtual environments.

**Simulator sickness:** various disturbances, ranging in degree from a feeling of unpleasantness, disorientation, and headaches to extreme nausea, caused by various aspects of a synthetic experience. Possible factors include sensory distortions such as abnormal movement of arms and heads because of the weight of equipment, long delays or lags in feedback, and missing visual cues from convergence and accommodation.

**Situational awareness:** perception of elements in an environment within a volume of time and space, comprehension of their meaning, and projection of their status in the near future; "up-to-the-minute" cognizance required to operate or maintain the state of a system.

**Somatogravic illusion:** illusions of false attitude occur mainly in pitch or roll. When lacking appropriate visual cues, the balance organs and brain are unable to sort out a force vector (such as delivered by motion of an aircraft), which differs in magnitude and/or direction from the gravitational vector.

**Somatosensory or somesthesia:** stimuli or senses arising from the cutaneous, muscle, and joint receptors. Touch and pressure cues. Includes all body senses of the skin, muscle, joint, and internal organs (including Mittelstaeidts kidney receptors) but excludes vestibular, visual, auditory, and chemical senses (taste and smell). Forms a superset of kinesthetic, tactile, and haptic stimuli or senses.

**Sonification:** data are used to control various parameters of a sound generator, thereby providing the listener with information about the controlling data. See also *Audification.*

**Sopite syndrome:** chronic fatigue, lethargy, drowsiness, nausea, etc.

**Static accuracy:** ability of a tracker to determine the coordinates of a position in space. See also *Dynamic accuracy.*

**Stereopsis:** binocular vision of images with different views by the two eyes to distinguish depth.

**Summative evaluation:** typically performed after a product or design is more or less complete. Its purpose is to statistically compare several different systems, for example, to determine which one is "better"—where better is defined in advance; involves placing representative users in task-based scenarios.

**Synthetic experience (SE):** experience created through a virtual environment. Some authors include passive virtual environments, such as a movie ride where there is no interaction, in the SE classification, while reserving the term *virtual environment* for active synthetic experiences, where the user interacts with the virtual world.

**Tactile:** sensory information arising from contact with an object, detected through nerves within the skin. Sometimes restricted to passive information, such as air currents over the skin, in contrast to haptic, which is then applied to active manual exploration. See also *Proprioceptive* and *Kinesthetic.*

**Tele-:** operating from a remote location.

**Teleoperation:** technology of robotic remote control.

**Teleoperation time delay:** communication delay necessary to transmit control data to a remote robot or to receive information feedback from the remote location to the operator.

**Teleoperator:** machine that operates on its environment and is controlled by a human at a distance. It is typically a combination of telesensor or telemanipulator.

**Telepresence:** perception that one is at a different location, created by sensory data transmitted from that location and possibly interaction with the environment at that location through telemanipulators.

**Texture mapping:** bitmap pattern added to an object to increase realism.

**Track shot:** rotating display of the same scene. See also *Dolly shot* and *Pan shot.*

**Transparency:** teleoperation performance measurement (i.e., fidelity of both information feedback and direct remote robot control).

**Update rate:** tracker's ability to output position and orientation data.

**Usability:** the effectiveness, intuitiveness, and satisfaction with which specified users can achieve specified goals in particular environments, particularly interactive systems. Effectiveness is the accuracy and completeness of goals achieved in relation to resources expended. Intuitiveness is the learnability and memorability of using a system. Satisfaction is the comfort and acceptability of using a system.

**Usability engineering:** methodical "engineering" approach to user interface design and evaluation involving practical, systematic approaches to developing requirements, analyzing a usability problem, developing proposed solutions, and testing those solutions.

**Usability evaluation:** any of a variety of techniques for measuring or comparing the ease of use of a computer system, including: usability inspection; user interface critiques; user testing of a wide variety of kinds; safety and stress testing; functional testing; and field testing.

**User-centered design:** design around the needs and goals of users and with users involved in the design process; design with usability as a primary focus.

**User-centered evaluation:** approach to evaluating (typically a computer's user interface) that employs representative or actual system users. The evaluation typically involves users performing representative task scenarios in order to reveal how well or poorly the interface supports a user's goals and actions.

**User task analysis:** process of identifying and decomposing a complete description of tasks, subtasks, and actions required to use a system, as well as other resources necessary for user(s) and systems to cooperatively perform tasks.

**VE:** common abbreviation for virtual environment.

**Vection:** illusion of self-motion, usually elicited by viewing a moving image, but also achievable through other sensory modalities (e.g., audition, somesthesia).

**Vestibular:** sensory structure of the labyrinth of the inner ear that reacts to head and gross bodily movement. The sense organs embedded in the temporal bone on each side of the head are known collectively as the labyrinthine organs or simply the labyrinths. They include the organ of hearing, or cochlea, the three semicircular canals, and the otolith organs, or utricle and saccule.

**Vestibular nucleus:** a brain stem structure responsible for processing real or apparent motion stimuli, and implicated in the maintenance of stable posture and reflexive gaze control. On each side of the brain stem there are four principal vestibular nuclei: the lateral, medial, superior, and inferior.

**Vestibulo-ocular reflex (VOR):** when head moves in any direction, the vestibular apparatus senses this movement and sends velocity information directly to the oculomotor system, which responds by driving the eyes (conjunctively) at an approximately equal rate but opposite direction to compensate for head movement and help keep the visual image stabilized on the retina.

**Vibratory myesthetic illusions:** illusory feeling of limb or body movement that may accompany certain vibrations of the skeletal muscle. For example, vibration of the biceps tendon with the subject's arm fixed and his eyes closed can elicit an illusion of arm extension.

**Virtual:** simulated or artificial.

**Virtual environment (VE):** three-dimensional data set describing an environment based on real-world or abstract objects and data. Usually *virtual environment* and *virtual reality* are used synonymously, but some authors reserve VE for an artificial environment that the user interacts with.

**Virtual heritage:** use of computer-based interactive technologies to record, preserve, or recreate artifacts and sites of historic, artistic, religious and cultural significance, and to deliver the results openly to a global audience in such a way as to provide a formative educational experience through electronic manipulations of time and space.

**Virtual mechanism:** passive artificial modeling structure used in a haptic controller; linked to both a master device (haptic feedback interface) and controlled robot (real or virtual) to allow intuitive and stable (since passive) control and feedback.

**Virtual prototyping:** product design that is based on VE technology; provides an alternative concept for the design-test-evaluation cycle by eliminating the fabrication of physical prototypes.

**Virtual reality:** See *Virtual environment.*

**Virtual team members (VTMs):** multifunctional autonomous agents or simulated images of humans within a VE that function in the role programmed to them.

**Visual suppression of the vestibulo-ocular reflex:** use of visual information to suppress the normal eye-beating reflex. An example would be reading text at the same time as undergoing a turning movement while riding on a bus.

**Voxel:** three-dimensional generalization of a pixel. An indivisible small cube that represents a quantum of volume.

**VR:** common abbreviation for *virtual reality*.

**Wayfinding:** the cognitive element of navigation.

## 5. WEB SITES FOR ADDITIONAL INFORMATION ON STANDARDS

American National Standards Institute (ANSI): http://web.ansi.org.
British Standards Institute (BSI): www.bsi-global.com.
Human Engineering Design Criteria for Military Systems, Equipment and Facilities (MIL-STD-1472D): http://tecnet0.jcte.jcs.mil:9000/htdocs/teinfo/directives/soft/humeng.html.
IEEE Computer Society: http://www.computer.org.
IEEE Standards Association (IEEE-SA): http://standards.ieee.org/. Details of the standards creation process are found at http://standards.ieee.org/resources/index.html.
IEEE Standards Process at a Glance: http://standards.ieee.org/resources/glance.html.
Institute of Electrical and Electronics Engineers (IEEE): http://www.ieee.org.
International Electrotechnical Commission (IEC): http://www.iec.ch/.
International Organization for Standardization (ISO): http://www.iso.ch/.
Italian National Standards Board (Ente Nazionale Italiano di Unificazione, or UNI): http://www.unicei.it.
Joint Technical Committee Number One (JTC1): http://www.iso.ch/meme/JTC1.html.
Man–Systems Integration Standards Handbook (NASA-STD-3000): http://iac.dtic.mil/hsiac/products/man_sys.html.

## 6. REFERENCES

Aukstakainis, S., & Blatner, D. (1992). *Silicon mirage*. Berkeley, CA; Carmel, IN: Peachpit Press.

Buie, E. (1999). HCI standards: A mixed blessing, *Interactions*, *VI2*, 36–42.

Delaney, B. (1994). Glossary and acronyms [Special Ed.]. *CyberEdge Journal* [Online]. Available: http://www.cyberedge.com/4a1.html

Durlach, N. I., & Mavor, A. S. (Eds.). (1995). *Virtual reality: Scientific and technological challenges*. Washington, DC: National Academy Press.

Hamit, F. (1993). *Virtual reality and the exploration of cyberspace*. Carmel, IN: Sams.

Hix, D., & Hartson, H. R. (1993). *Developing user interfaces: Ensuring usability through product and process*. New York: John Wiley & Sons.

Kalawsky, R. S. (1993). *The science of virtual reality and virtual environments*. Wokingham, England: Addison-Wesley.

Latham, R. (1991). *The dictionary of computer graphics and virtual reality*. New York: Springer-Verlag.

Manetta, C., & Blade, R. (1995). Glossary of virtual reality terminology. *IJVR 1–2* [Online]. Available: http://www.ijvr.com/ijvr/glossary/glossary.htm

Nielsen, J. (1994). Heuristic evaluation. In J. Nielsen & R. L. Mack (Eds.), *Usability inspection methods* (pp. 25–62). New York: John Wiley & Sons.

Psotka, J., & Davison, S. *Virtual reality terms* [Online]. Available: http://198.97.199.60/vrterms.html

Rada, R., & Ketchel, J. (2000). Standardizing the European information society. *Communications of the ACM*, *43*(3), 21–25.

Rheingold, H. (1991). *Virtual reality*. New York: Simon & Schuster.

Scriven, M. (1967). The methodology of evaluation. In R. E. Stake (Ed.), *Perspectives of curriculum evaluation* [American Educational Research Association Monograph]. Chicago: Rand McNally.

Sheridan, T. (1992). Defining our terms. *Presence*, *1*(2), 272–274.

Steuer, J. (1992). Defining virtual reality: Dimensions determining telepresence. *Journal of Communication*, *42*(4), 73–93. Also available: http://www.cyborganic.com/people/jonathan/Academia/Papers/Web/defining-vr1.html

Vince, J. (1995). *Virtual reality systems*. Wokingham, England: Addison Wesley. Also available: http://www.vrs.org.uk/VR/reference/glossary.html

# PART II: System Requirements

Hardware Requirements

# 3

# Vision and Virtual Environments

James G. May[1] and David R. Badcock[2]
[1]*University of New Orleans*
*Department of Psychology*
*Lakefront, New Orleans, LA 70148*
*jmay@uno.edu*
[2]*University of Western Australia*
*Department of Psychology*
*Nederlands, Western Australia 6009*

## 1. INTRODUCTION

This chapter is intended to provide the reader with knowledge of the pertinent aspects of human visual processing that are relevant to virtual simulation of various environments. Before considering how real-world vision is simulated, it is perhaps prudent to review the kinds of information that are usually extracted by the visual system. It would, of course, be fruitless to provide visual detail that is rarely or never available to the senses, and it may be fatal to the endeavor to omit detail that is crucial. Thus, an overview of normal visual capabilities and idiosyncrasies is provided in section 2. Section 3 reviews some of the ways that perceptual systems provide short cuts to simulating the visual world. The existence of these phenomena allows system developers to compensate for hardware shortcomings with user inferences. One of the most exciting advantages of virtual environment (VE) technology is that it allows a more elaborate and complex interaction between the VE and the observer. A later section reviews a number of the ways in which we interact with the world and how these mechanisms might augment and detract from virtual simulation. After the discussion of what vision entails, a discussion of techniques that use two-dimensional renditions of the visual world to simulate normal viewing of the three-dimensional (3-D) world is provided in Section 5. The emphasis here will be to address the design requirements of VE displays and to determine if existing displays are machine- or observer-limited. Considering what is optimally required by the user, a review of the adequacy of existing visual displays is also provided in section 6. Suggestions are made as to how existing limitations might be overcome and speculations are made concerning what new technology might allow in section 7.

## 2. WHAT ARE THE LIMITS OF NORMAL VISION? (USER REQUIREMENTS)

Research on human visual capacities is an active field of endeavor. There are still many aspects of visual performance that are not well understood, but knowledge has been accumulating at an accelerating pace throughout the last century. The inquiry continues, but it is important to set forth, from time to time, a compendium of what we think we know. The formal study of vision is perhaps one of the oldest disciplines, and the community of visual scientists is larger than for any other sense. Since the early psychophysicists initiated the systematic investigation of the senses, the notion has been to map the subjective sensations against the physical scales of interest and express the relationship as a "psychometric function" that defines the limits of sensitivity and the subjective scaling of the dimension. That approach will be employed here, but it is important to acknowledge that, in practice, we may utilize little information near the limits of visual abilities. It is conservative, in a sense, to employ this approach to ensure that display technologies meet these user capabilities, but we may learn to "cheat" on these limits if the task requirements allow.

A comprehensive attempt to list human performance capabilities was collated in 1986 and published as the *Engineering Data Compendium: Human Perception and Performance.* (Boff & Lincoln, 1988; see also Boff, Kaufman, & Thomas, 1986). In this chapter we will offer more in the way of explanation of the phenomena described, but the *Engineering Data Compendium* is still an excellent starting location for determining performance limits.

### 2.1 Luminance

The human visual system is sensitive to a broad range of ambient illumination, extending from absolute threshold ($\sim -6$ log cd/m$^2$) to levels where irreversible damage to the system can result ($\sim 8$ log cd/m$^2$). Many stimulus factors (e.g., size, wavelength, retinal location) have profound influences on the lower limits of this range (Hood & Finkelstein, 1979). The human visual system contains two types of photoreceptors (rods and cones) with significantly different sensitivities. The lowest absolute threshold is mediated by the more sensitive rods, and the cone threshold is achieved about 2.5 log units above that limit ($-3.5$ log cd/m$^2$). At about 1 log cd/m$^2$ the rod responses begin to saturate, and at higher luminance levels they provide no significant information because saturation is complete. The cones mediate vision above that point. Thus, the entire range can be broken up into a scotopic region (rods only), a mesopic region (rods and cones), and a photopic region (cones only). Many of the visual abilities reviewed below are different in rod- and cone-mediated vision. They also vary, with luminance level in the ranges subserved by each photoreceptor type. These adjustments take some time and are referred to as light (and dark) adaptation. For example, an observer moving from a sunny street into a dark room may take up to 30 min to reach maximum sensitivity for detecting the presence of a target.

### 2.2 Spatial Abilities

Much of human vision is concerned with discerning differences in luminance and color across space. The extent of the visual field of view, visual acuity, contrast sensitivity, and spatial position accuracy are reviewed in this section. The display factors that relate directly to providing adequate stimulation for such abilities are discussed later.

#### *2.2.1 Visual Fields*

The monocular visual field of view is determined by having a subject fixate a point in the center of a viewing area (180 deg H $\times$ 150 deg V) and presenting targets throughout the

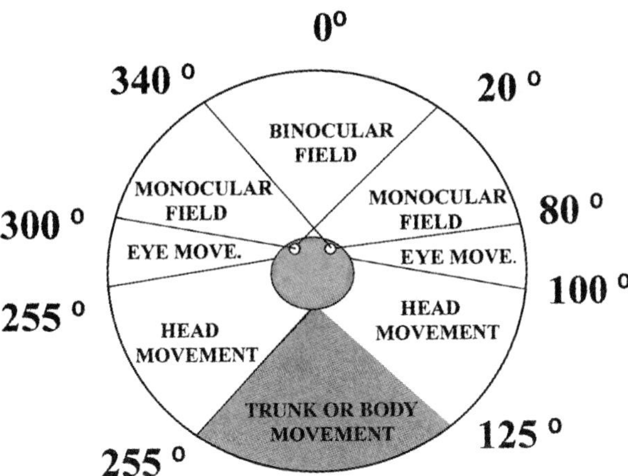


FIG. 3.1. A depiction of the horizontal monocular and binocular visual fields with straight ahead fixation, lateral eye movements, lateral head movement, and bodily rotation.

extent of the area. The limits of the field are defined as the most peripheral points at which target detection achieves some criterion (e.g., 75% correct). These limits vary somewhat with the choice of criterion and also as a result of stimulus factors such as the luminance of the background, the luminance of the target, the size of the target, and the wavelength of the target. In normal individuals with a target subtending approximately 20 arcmin, the horizontal extent is approximately 160 deg, and the vertical extent is approximately 120 deg (see Fig. 3.1). The shape is roughly circular, but is limited by the physiognomy of the nose, brows, and cheeks. Most fields show a restriction in the area of the lower nasal quadrant. A smaller target, subtending one arcmin, may only be detectable within a 20-deg-diameter area (Borish, 1970). The measured field size is also greatly reduced if the observer needs to name the hue of the light, rather than just detecting its presence. This effect also varies with the hue to be named. Arakawa (1953) found a 105-deg horizontal diameter when detecting blue, but only 75 deg for detecting red and 55 deg when detecting green. Similar changes occurred vertically. The consequence of this finding suggests that tasks requiring identification of hue should be centered within the visual field.

Careful probing of the area of the visual field about 13 to 18 deg temporal to the fixation point reveals the blind spot, a scotoma formed by the optic nerve head. It is roughly circular, with a diameter of about 5 deg. The areas of visibility that are common between the two monocular fields form what is termed the binocular visual field. It is roughly circular and is approximately 40 deg in diameter. Binocular viewing helps to compensate for the blind spot, because what is not visible in one eye is visible in the other. However, the disparity cues for depth (see section 2.3.4) are lost in this region.

### 2.2.2 Visual Acuity

The upper limit of the ability to resolve fine spatial detail is termed visual acuity. It has been measured with numerous techniques (Thomas, 1975), with the normal threshold for spatial resolution ranging between approximately 0.5 to 30 sec of arc, depending on the task. The two detection techniques that yield the lowest values are minimum visible and vernier acuity tasks. In the first task, the width of a single line is varied, and the minimally visible width (0.5 min of arc) is determined. In the second, the lateral offset between two vertically oriented line segments is determined (1–2 sec of arc). While these are both referred to as acuity tasks, they measure different abilities. The former measures contrast sensitivity (see section 2.2.3) because the change in line width has a greater impact on the retinal luminance-difference between the line and background than on the width of the retinal luminance-change distribution. This outcome is a result of the line-spread function of the eye. Even in a good optical system, a point of light is spread while passing through the optics. In the human system, this spread covers approximately 1 arcmin. Thus, reducing the width of a small bright line will change the total amount of light passing through the optics, but the spread of light on the retina will change very little. Consequently, a very high acuity for changes can be obtained if those changes create a threshold difference in luminance contrast. Levi, Klein, and Aitsebaomo (1985) used a target composed of five parallel lines in which one of the inner lines was displaced to the left or the right. The displacement varied the width of the gaps on either side of the central line. This produced a contrast difference between the gaps on the two sides of the line, which was detectable with only a 0.5-sec line shift—a remarkable spatial discrimination, but a normal contrast discrimination.

The vernier acuity task is more appropriately viewed as a position discrimination task. The misalignment of the bars may be detected by discriminating either a difference in the horizontal location of two vertical bars or a difference in the orientation of the overall figure (Watt, Morgan, & Ward, 1983). This task is discussed further later in the chapter.

Resolution tasks require the observer to determine the minimal separation in space of two or more luminance defined borders. Similar results are obtained with each of two types of commonly employed patterns: gratings and letters. The gratings are high-contrast black lines on a white background with 50% duty cycle, square-wave luminance profiles. The letters are created so that the width of the lines composing the letter is one-fifth of the overall letter size. The acuity measure is calculated from the size of the line width such that the visual angle equals atan ([line width]/[distance between observer and letter]).

Since different letters vary in the orientation and position of the lines, it is common to use Landolt rings instead. The rings have the same line width to overall size ratio, but also have a gap the same size as the stroke width of the ring. The observers must resolve the gap and identify its location on the ring. Grating tasks and the Landolt ring task can also be used as discrimination tasks by varying the orientation of the stimuli. These three tasks result in similar thresholds ($\sim$ 30 sec of arc in healthy young adult eyes). Numerous variables affect visual acuity performance. Visual acuity increases with luminance, from thresholds of about 1 min of arc at 1 cd/m$^2$ to 30 sec of arc at 100 cd/m$^2$ (Shlaer, 1937). The recommended standard for illumination for visual acuity measurement is 85 cd/m$^2$. Under photopic conditions, visual acuity is optimal for targets presented to the center of the visual field (the fovea) and falls off rapidly (see Fig. 3.2 for a demonstration). The minimum resolvable detail declines from 30 sec of arc at fixation to 20 min of arc at 60 deg eccentricity for Landolt ring acuity (Millodot, Johnson, Lamont, & Leibowitz, 1975). At scotopic levels, acuity is best about 4 deg from the fixation point. The estimates provided above assume that the targets are presented at the optimal distance. There is a limit to the ability to focus on nearby objects. Targets closer than this near point are blurred, and resolution is reduced. The near point recedes with age due to hardening of the lens (presbyopia). This distance may exceed 1 to 2 meters by age 60 and needs to be considered when providing artificial displays, because resolution cannot be good if the display is closer than the near point. Optical correction will frequently be needed for older observers.

### *2.2.3 Contrast Sensitivity*

While visual acuity has been one of the time-honored descriptors of human spatial ability, it only provides information about the extreme upper limit of the spatial dimension to which

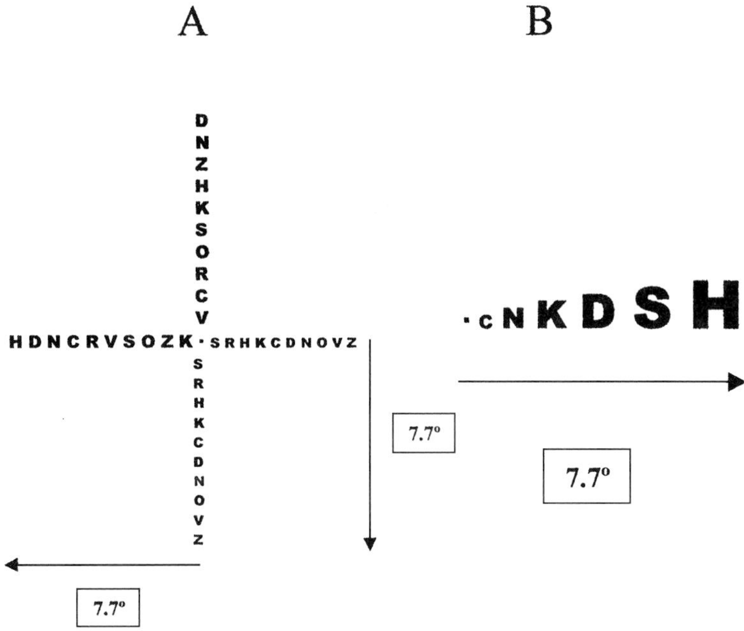


FIG. 3.2. The reader is invited to fixate the small dot in the center of the display at the left and, while maintaining fixation of the dot, take note of how the visibility of the letters declines as more peripheral letters are considered. The figure on the right attempts to compensate for the decline in visual acuity in the periphery by increasing the size of the letters proportionally.

humans are sensitive. Much of what humans see and use may not involve spatial detail near these limits. A popular approach to evaluating human sensitivity for objects of different sizes has been to specify the contrast necessary to detect them. In the simplest case, a light bar on a dark background, contrast (C) can be defined as $C = \Delta L/L$, where L is the background luminance and $\Delta L$ is luminance increment (or decrement) provided by the bar. Most visual stimuli are more complex, and thus a measure is needed that can provide a sensitivity estimate for these complex patterns. An alternate method takes into account the nature of the neural units early in the visual pathways. At the first stages of processing in the cortex, units are sensitive to short, oriented line segments. Most units give a strong response to a bar of the correct orientation that is surrounded by other bars of the opposite contrast polarity. The underlying weighting function for the cell's response to light can be approximated by a Gabor function that is produced by multiplying a cosinusoidal luminance variation by a Gaussian envelope (Field & Tolhurst, 1986). An image of a Gabor contains a set of bars that alternate between high and low luminance, with contrast that is high in the center of the patch and reducing to zero at the edge of the envelope. When the Gabor is used to represent a weighting function, the bright bars represent excitatory regions and the dark bars inhibitory regions. Individual neurons may vary in the spatial extent of the envelope, the orientation of the cosine, the cosine's spatial frequency, and the cosine's position in the envelope (spatial phase). Since these are the early detectors, one common strategy has been to measure the minimum amount of contrast required for these detectors to reach threshold. Typically, cosinusoidal patterns are employed as stimuli, and with these pattern's contrast is defined as $(L_{MAX} - L_{MIN})/(L_{MAX} + L_{MIN})$, where $L_{MAX}$ and $L_{MIN}$ are the maximum and minimum luminance of the image, respectively. With centrally viewed patches of grating of limited envelope size (e.g., 5 deg of arc) and a large surround (e.g., 30 deg) of the same average luminance, the function is band-limited from approximately 0.5 to 60 cycles per degree (c/deg). The peak sensitivity (the reciprocal of the contrast at thresholds of 0.0033 and 0.0025) is approximately 300 to 400 between 3 and 10 c/deg.

One of the important motivations for this approach to describing visual sensitivity was the power of the tools of Fourier analysis and now Wavelet analysis (Press, Teukolsky, Vetterling, & Flannery, 1992). Using these tools it is possible to describe any image as a collection of sinusoidal gratings varying in orientation, spatial frequency, phase, and contrast. The similarity between the basis functions employed for this analysis and the weighting functions of some visual neurons led to the suggestion that the visual system may be performing a similar analysis. While this may not be true, it is nevertheless valid to argue that the neurons will only respond to that restricted range of spatial frequencies and orientations within an image to which its receptive field is tuned. Thus, if one could determine the sensitivity of a unit and the amount of contrast in the passband of that unit within an image, it should be possible to predict whether that unit will respond to the particular image. In many cases such prediction is possible (Campbell & Robson, 1968; Graham, 1980). However, the utility of this approach is limited by the lack of generality of the foveal contrast sensitivity function. In the periphery of the visual field, high spatial frequencies are increasingly poorly resolved, and thus a different contrast sensitivity function is obtained at each eccentricity. Contrast sensitivity declines with luminance (Van Ness & Bouman, 1967), and varies with target motion (Kelly & Burbeck, 1980; Robson, 1966), orientation (Mitchell & Ware, 1974) and chromaticity (Metha, Vingrys, & Badcock, 1993; Mitchell & Ware, 1974; Mullen, 1985). While these limitations are significant, the contrast sensitivity function is a more comprehensive measure of visual sensitivity than spatial acuity alone, and is used extensively to characterize the performance of the visual system.

### *2.2.4 Spatial Position*

The ability to accurately localize image features is an important precursor to object recognition, shape discrimination, and interaction with a cluttered environment. The appreciation of a form presupposes the accurate relative localization of elements of the form, and the perception of peripherally viewed stimuli serves to guide eye movements and locomotion. Approaches to the characterization of such ability in the frontal plane involve: relative judgements of the spatial position of two or more objects (Badcock, Hess, & Dobbins, 1996; Westheimer & McKee, 1977); bisection of visual space (Levi et al., 1985); and localization of briefly presented target positions (Solman & May, 1990; Watt, 1987). In central vision, relative spatial position thresholds are on the order of 0.05 to 1.0 sec of arc (see vernier acuity previously mentioned) and once again, the threshold depends on the spatial scale of the pattern. If Gaussian luminance increments or Gabor patches are employed, the threshold is proportional to the standard deviation of the Gaussian envelope (Toet, van-Eekhout, Simons, & Koenderink, 1987). These thresholds can be 10 times higher when targets are presented eccentrically by just 10 deg (Westheimer, 1982), a decline that is substantially more rapid than that obtained for visual acuity estimates (Levi, Klein, & Aitsebaomo, 1985).

Partitioning studies also indicate that, at eccentricities of 10 deg, position thresholds vary with the size of the area to be partitioned, increasing from about 0.05 deg at 0.50 deg separation to 0.11 deg at 1.1 deg separation and remaining constant thereafter (Levi & Klein, 1990). Spatial location discrepancies for briefly presented, single targets increase with eccentricity from about 0.04 deg at 2 deg eccentricity to 3.5 deg at 12 deg eccentricity (Solman & May, 1990; Solman, Dain, Lim, & May, 1995). It is clear that the relative spatial sense measured with two or more objects, or partitioning, is more precise than is the localization of targets with more limited spatial landmarks (Matin, 1986). Apparent position is also influenced by the spatio-temporal proximity of nearby objects. Badcock and Westheimer (1985) showed that objects closer than 3-arcmin result in an apparent shift of the target toward the distractor's spatial position, whereas larger distances cause repulsion in apparent position (see also Fendick, 1983;

Rivest & Cavanagh, 1996). A similar phenomenon occurs over large distances, where spatial intervals appear smaller if a larger interval is present nearby, and vice versa (Burbeck & Hadden, 1993; Hess & Badcock, 1995).

## 2.3 Depth

Theoretically, detection of targets in depth is limited by the spatial resolution of the visual system. If the target is large enough (e.g., our moon) an individual can easily see it at a great distance (245,000 miles). At great distances, two objects (e.g., our sun and our moon) subtending the same visual angle (0.5 deg ) appear to be the same size and an individual cannot discern which is closer to them (except during an eclipse). At near distances, however, an individual has little difficulty determining the relative distances of two objects because of the presence of numerous depth cues. The cues are traditionally discussed in terms of those available with monocular and binocular viewing.

### 2.3.1 Monocular Cues

Monocular cues are those that would be available to an observer using just one eye.

*2.3.1.1 Pictorial Cues.* Pictorial cues are spatial arrangements that convey relative differences in depth. They are employed by artists adept at conveying 3-D impressions with 2-D depictions. They are:

1. Relative Size: In the absence of information about the absolute size of an object, its apparent size influences judgments of its distance. The zebras in Fig. 3.3 (A–F) appear to be at increasingly greater distances largely because of their progressively smaller sizes.
2. Height Relative to the Horizon: For objects below the horizon (the mice in Fig. 3.4), objects lower in the visual field appear closer than objects higher in the visual field. For objects above the horizon (the butterflies in Fig. 3.4), the reverse is true.
3. Interposition or Occlusion: When objects are opaque, those nearby occlude the view of parts of those objects that are further away and convey an immediate sense of depth. For example, the hill in Fig. 3.4 is interposed between the viewer and the bottom of the trees.
4. Shadows and Shading: A directed light source will strike the nearest parts of an object in its path and be prevented from striking other features on the same path. Thus,

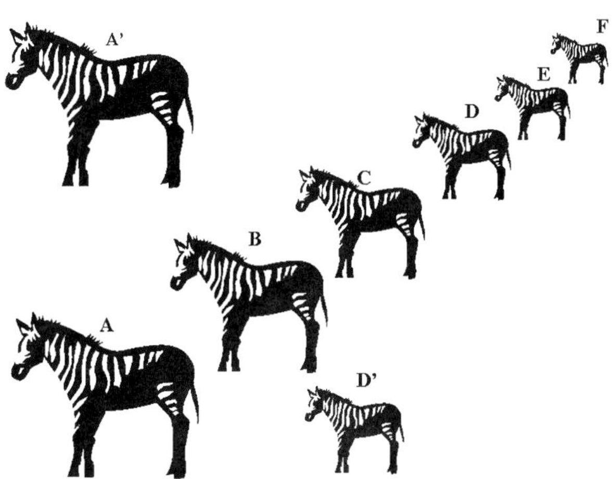


FIG. 3.3. These images convey differences in apparent depth from decreasing size. The smallest zebra appears farther away than the larger ones.

FIG. 3.4. This scene includes demonstrations of height relative to the horizon and interposition or occlusion cues in determining relative apparent depth. (See text for explanation.)

differences in the intensity of the light reflected from object parts can contribute to the appreciation of depth and 3-D shape (see Fig. 3.5a,b). In situations where the position of the light source is not detectable from the image, the visual system resolves ambiguous shading cues by assuming that the light source is above the object (see Fig. 3.5c). In addition, objects often cast shadows on surfaces near them, adding additional cues to figure ground configurations by occlusion or interpolation. The degree of occlusion of a shadow by the object casting it can serve as a cue to the distance between an object and the background. These cues are conveyed in 2-D objects by shading (see Fig. 3.5 d–f).

5. Aerial Perspective: The light reflected from objects is both scattered and absorbed by particles in the medium through which it is observed, causing near objects to be perceived as brighter, sharper, and more saturated in color than far objects. Thus, one cue to distance is the brightness of objects (compare objects A and B in Fig. 3.5).

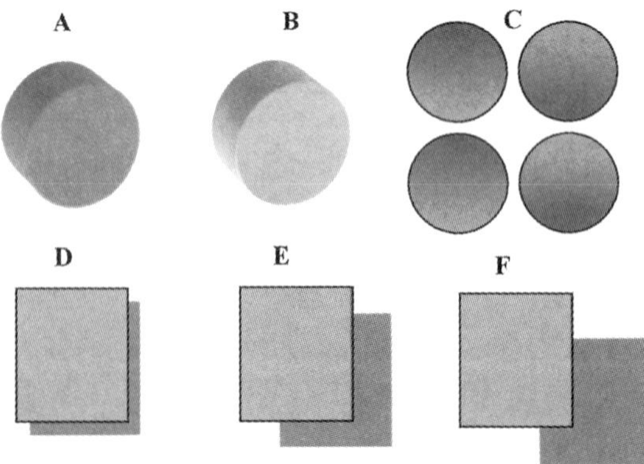


FIG. 3.5. Figures A and B differ in brightness, causing figure B to appear to be somewhat closer than A. The other figures offer other examples of how shadows and shading offer cues to depth. (See text for detail.)

FIG. 3.6. The sidewalk on the left is an example of how texture differences convey depth information. The image on the right is an inverted image of the left panel. In this case depth is still conveyed, but the image is more easily seen as a textured ceiling.

6. Linear Perspective: When parallel lines recede in distance from an observer (e.g., railroad tracks) their projection on the 2-D retina produces a convergence of the lines as distance increases. The contours do not have to be straight, but merely equally spaced as they become more distant (e.g., a winding roadway). This change in perspective with distance provides a compelling cue for depth and is a consequence of representing 3-D scenes on 2-D surfaces.
7. Texture Gradients: Most natural objects are visibly textured. When the image of those objects is projected onto a 2-D surface like the retina, the density of the texture of the surface increases with distance between the surface and the observer. These gradients of texture density are highly salient cues to relative depth (see Fig. 3.6).

### 2.3.2 Accommodation and Vergence

The human visual system contains an elastic lens that can change curvature and refractive index. If an individual looks at an object positioned within about 3 meters, the curvature of the lens is adjusted to focus the image sharply on the retina, and the muscular contraction necessary for that adjustment can serve as a cue for the distance of the object. Theoretically, differences in these muscle contractions and differences in sharpness for objects at different distances might also be used as depth cues. With binocular viewing, the two eyes converge when viewing near objects and diverge when viewing objects farther away. Information from the extrinsic eye muscle tension is thought to convey information about eye position to the brain, and it might also serve as a cue to depth. Accommodation and vergence are normally linked and covary with the depth plane of the target observed. Some doubt the utility of these potential physiological cues to depth, but these variables may be quite important for virtual simulations where depth may be produced by disparity cues between two screens that are near the eyes. Under these circumstances, the necessary accommodation and convergence may not be congruent with normal visual experience. The consequences of this are discussed in chapter 37 (this volume) and later in this chapter.

### 2.3.3 Motion Parallax

All of the cues so far discussed have concerned static scenes and a stationary observer, but additional depth information is conveyed when the scene is moving relative to the observer, or the observer is moving relative to a static scene. The typical example is an observer on a moving train viewing the landscape as it passes. If the observer fixates a point midway between the train and the horizon, two movement-related phenomena can be observed. First, objects beyond the plane of fixation appear to move in the direction that the observer is moving, while objects closer than the fixation plane move in the direction opposite of the observer's movement. Second, the retinal speed of objects is proportional to the distance of the object from the fixation point. Not only do these phenomena provide important information about the depth of objects in the field, but they also tell us about our movement relative to the environment (Rogers & Graham, 1982). A special case of motion parallax, known as the kinetic depth effect, occurs when an individual views a moving 3-D object. Imagine a sphere constructed of a meshwork of fine wire. When it is stationary it may be difficult to appreciate it's shape, but when it is rotated the contours near the observer move in one direction, whereas the contours farther away move in the opposite direction. This shape information can even be recovered from a shadow cast by the sphere onto a 2-D screen.

### 2.3.4 Binocular Cues

Although most people view the world through two eyes, they usually see a single unified view of the world. By viewing a scene alternatively with one eye and then the other, an individual can appreciate the differences in the two views. Ignoring the differences in field of view provided, the individual can also notice slight differences in the relative position of objects within the overlapping regions of the two monocular views. These differences are fused to form a singular view when the individual uses two eyes. The horopter (see Fig. 3.7) is a hypothetical surface in space determined by the point in depth for which the eyes are converged (and accommodated). For a given depth of focus, all the points on the horopter are associated with homologous pairs of points on the two retinae. The theoretical horopter defines the curved

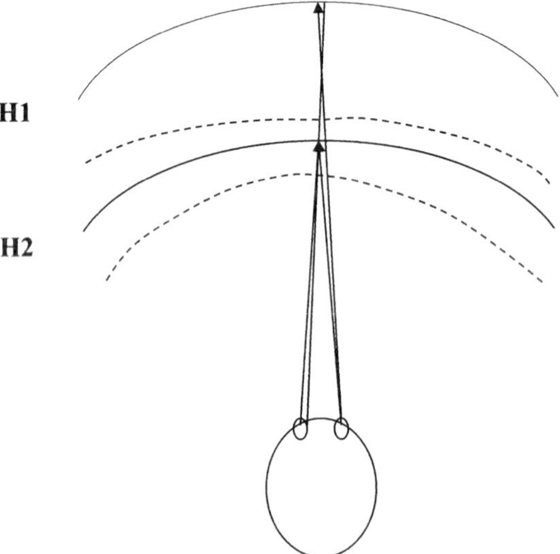


FIG. 3.7. A depiction of two horopters (H1 and H2) associated with fixation in two different depth planes. The existence of Panum's fusional area (dotted lines) is depicted only for H2.

region in which there is no disparity between the two monocular views. The area in front of the horopter contains disparities that are said to be "crossed" (because it would be necessary to cross the eyes to bring the points onto homologous retinal areas), and the area beyond the horopter contains disparities, which are said to be "uncrossed."

The region of binocular fusion is a zone approximately centered on the horopter. The reader can appreciate this fact by viewing two objects (e.g., fingers) in different depth planes. If the near object is fixated, attending to the far object reveals double images of that object (diplopia), but a single image of the fixated object. If the far object is fixated, attending to the near object reveals double images of that object (diplopia), but a single image of the fixated object. When fixating 3-D objects, the binocular image contains some areas that fall in front of or behind the horopter and must therefore contain retinal disparities, but these double images are fused and seen as one object. The region in depth over which disparities are fused and singular vision is perceived is termed Panum's fusional area (dotted lines in Fig. 3.7). This area encompasses the horopter and includes areas behind and in front of it.

Studies that have involved dichoptic viewing (presenting separate images to the two eyes) have revealed that subjects will strive to overcome diplopia with vergence eye movements that seek to reduce retinal disparity. If two images can be brought into register with only slight disparity between aspects of the two views, then fusion is achieved and singular vision is experienced. If, however, the disparities are too great, diplopia ensues and binocular rivalry between the two views is often experienced (seeing one view but not the other, or seeing part of one view and part of the other).

Fusion of disparate images brings with it not only singularity of view, but also more importantly, a vivid sense of depth. For example, imagine viewing dichoptically, two separate vertical, square-wave gratings. When fused they will appear as a single grating in the frontal plane without any depth differences. If the spatial frequency of one of the gratings is shifted slightly lower, the image remains fused but appears to rotate in depth. Retinal disparity is ubiquitously present in our binocular view of the 3-D world, and it may be simulated in 2-D to produce depth information (Howard & Rogers, 1995). The smallest disparity that provides depth information is termed the threshold for stereopsis. This threshold is smallest when the target is viewed foveally and on the horopter (Badcock & Schor, 1985). In the fovea, this threshold may be only a few seconds of arc, but the thresholds increase with the eccentricity of the target to about 300 sec at 8 deg.

### 2.4 Color Vision

In addition to our ability to discern differences in luminance across space and time, humans may also discriminate between the wavelength of light across these dimensions. The history of research on color vision is, perhaps, the most extensive consideration of any human ability and our understanding of the mechanisms by which humans appreciate color differences reflects that impressive effort. The concern here, however, is to define the limits of that human ability and not to explain the mechanisms underlying them. Humans are sensitive to electromagnetic radiation in the range of 370 to 730 nm (Wandell, 1995). Various light sources (e.g., the sun, tungsten bulbs, fluorescent lights) provide a broad band of radiation across and beyond this visible spectrum. Traditional methods of studying color vision have involved various methods of restricting the spectrum to narrow bandwidths. Transparent devices (e.g., filters, diffraction gratings, prisms) are characterized in terms of spectral transmittance, while opaque media (e.g., paper, paint) are described in terms of their spectral reflectance.

The first question addressed concerning sensitivity to chromaticity involved determining the threshold for detecting the presence of a light composed of only a narrow range of wavelengths. Measures at scotopic levels (rod-mediated vision) revealed that the sensitivity varied with a

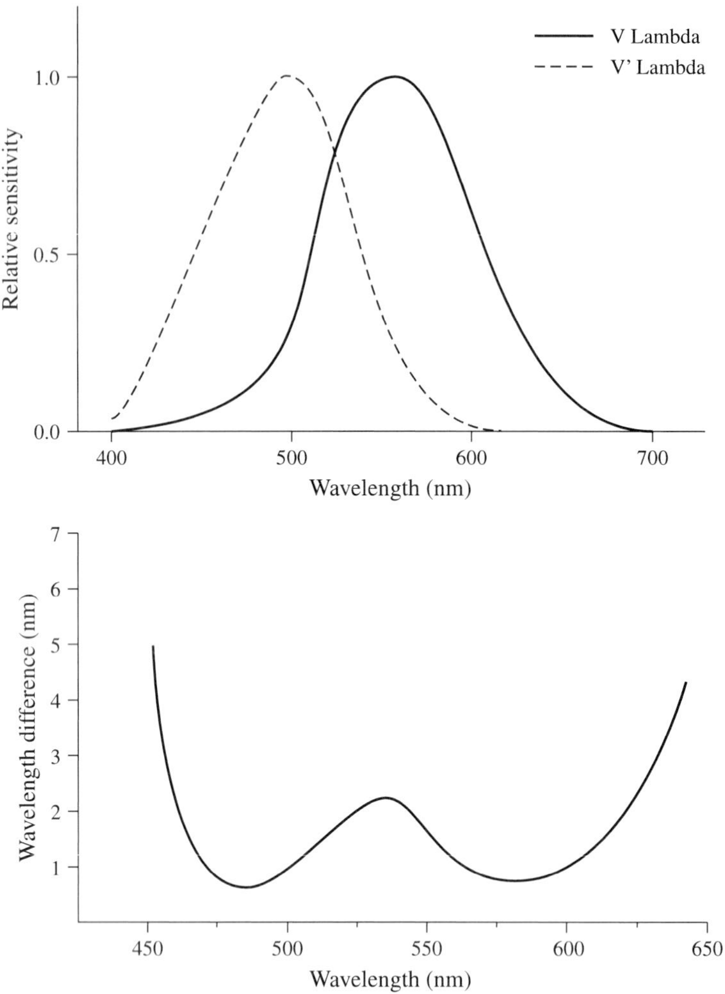


FIG. 3.8. The upper panel contains the function depicting the spectral sensitivity of the human eye under scotopic (A) and photopic (B) conditions, CIE V' and CIE V respectively. The two functions have been normalized to remove the overall increased sensitivity of the scotopic system. The lower panel contains the function that describes wavelength discrimination across the spectrum (from Hurvich, L. M. [1981]. Color Vision, Sinauer and Assoc., Inc., Sunderlund).

peak at about 500 nm (the Commission Internationale de l'Eclairage, CIE, V' function; see upper panel of Fig. 3.8). At photopic levels, a similar function was observed, with a peak at 555 nm (the CIE V function; see upper panel of Fig. 3.8 ). At mesopic levels the measured functions are a combination of the other two. It must be noted that these curves only reflect the sensitivity to light of different wavelengths. They do not address the ability to identify or discriminate between colors. Color vision is defined as the ability to discriminate between stimuli of equal brightness on the basis of wavelenth alone. Under scotopic conditions this is not possible, but under photopic conditions the full extent of color vision is measurable.

Wavelength discrimination tasks involve asking the observer to vary the wavelength of a comparison stimulus until it can be differentiated from a standard that is of some given wavelength. The increment in wavelength necessary to discern a difference when the stimuli are of equal brightness is plotted against wavelength (see lower panel of Fig. 3.8). The wavelength varies between about 1 and 13 nm, and the function contains two minima in the ranges of 450 to 500 and 550 to 600 nm. This kind of discrimination varies with a number of parameters. Chromatic discrimination ability decreases with luminance, especially at lower wavelength (Brown, 1951; Clarke, 1967; Farnsworth, 1955; Siegal & Siegal, 1972), but remains fairly constant above 100 td. (Trolands are a measure of retinal luminance that incorporate the effects of changes in pupil diameter: $I[\text{Trolands}] = R \cdot [\text{pupil area in mm}^2] \cdot [\text{luminance in cd} \cdot \text{m}^{-2}]$, where $R = 1 - 0.0106d^2 + 0.0000416d^2$ and $d =$ pupil diameter in mm). Discrimination is reduced with field size below 10 deg (Brown, 1951; Wyszecki & Stiles, 1967) but is relatively unaffected by separation between comparison fields (Boynton, Hayhoe, & MacLeod, 1977). If comparison fields are presented successively, discrimination is unaffected until a stimulus onset asynchrony (SOA) of 60 ms, but declines at higher intervals (Uchikawa & Ikeda, 1981). Discrimination for some wavelengths (400 and 580 nm) asymptotes at 100 msec, whereas other wavelengths (480 and 527 nm) require 200 msec for best performance (Regan & Tyler, 1971). Discrimination is poor at eccentricities beyond about 8 deg.

In addition to discrimination of hue, human observers can discern differences in the purity of white light. The threshold for colormetric purity is defined as the amount of chromatic light that is added to white light to produce a just noticeable difference (JND). Additional JNDs can be observed with increasing steps in the chromatic additive. The number of JNDs observed varies with the wavelength of the additive and is least at about 570 nm. The term *saturation* is used to describe the subjective correlate of colormetric purity. A highly saturated light appears to have little white light "contamination".

If the JND is taken as the step size in the dimensions of hue, brightness, and saturation, Gouras (1991) has suggested that there are about 200 JNDs for hue, 500 JNDs for brightness, and 20 JNDs for saturation. This suggests that human color vision capability involves the discrimination of about two million chromatic stimuli ($200 \times 500 \times 20 = 2{,}000{,}000$).

The visual subsystem that processes chromatic information differs in a number of significant ways from the subsystem that encodes luminance variation. Critically, for artificial systems where bandwidth limitations are significant, the chromatic system has substantially poorer spatial (Mullen, 1985) and temporal (Cropper & Badcock, 1994) resolution, and thus chromatic properties can be rendered more coarsely in both space and time. Mullen (1985) estimates that the spatial resolution for red-green gratings is three times worse than for achromatic gratings (10 to 12 c/deg instead of 60 c/deg), whereas Cropper and Badcock (1994) reported a twofold reduction in temporal resolution.

## 2.5 Motion

Humans can see objects moving with respect to themselves, whether they are stationary or moving, and they can detect their own movements through space in a static environment or with objects moving around them. Information about one object's movement relative to another (exocentric motion) is obtained from the image of the objects moving across the retina. An observer may attribute motion to the object that is actually moving or may see movement of a stationary object if the frame around it is displaced (induced movement). Information about object movement relative to an observer's own position or movement (egocentric motion) comes from translations in the retinal image in relation to nonvisual senses (e.g., vestibular and kinesthetic senses) that are involved in body, head, and eye movements. Mere translation of the object on the retina cannot provide information about egocentric information. Humans do not perceive object motion relative to themselves if they move their head or eyes while looking at or away from a stationary object (position constancy).

There are several ways in which one might measure a minimum motion threshold. First, one could measure the minimum distance a feature has to be displaced in order for the direction of motion to be detected. When a visual reference is present, displacements as small as 10 arcsec are sufficient for this judgment (Westheimer, 1978). Second, one could measure the minimum temporal frequency required to detect movement in a temporally extended motion sequence. This threshold depends on both the contrast and the spatial structure of the moving pattern (Derrington & Badcock, 1985). The minimum temporal frequency decreases as contrast increases for low-spatial-frequency periodic sinusoidal grating patterns, but is constant for high-spatial-frequency patterns. Third, one could measure the minimum number of dots needed to move in a common direction in a field of randomly moving dots for that direction to be discernible. Edwards and Badcock (1993) have found that as few as 5% to 10% of dots moving in a common direction is sufficient for either frontoparallel motion or for expanding/contracting patterns designed to mimic motion in depth.

Whereas these are impressive abilities, it is also the case that the perception of motion is influenced by a number of stimulus properties. The perceived speed of an object slows if the luminance contrast of the object is reduced (Cavanagh, & Favreau, 1985; Thompson, 1982), and observers are unable to see very high spatial frequency repetitive patterns move at all (Badcock & Derrington, 1985). Objects appear to move more slowly through smaller apertures and at greater distances (Rock, 1975).

Finally, even relatively brief periods of exposure to continuous motion (e.g., 30 sec to 1 minute) produces substantial motion aftereffects, which causes stationary patterns to appear to drift in the direction opposite to the adapted direction (Mather, Verstraten, & Anstis, 1998). In computer environments, these aftereffects commonly arise due to smooth scrolling of stimulus displays, and they can produce compelling impressions of vection (see later discussion and chap. 23, this volume). A common example is observed when trains pull into a station. The continuous visual adaptation to forward motion creates the egocentric impression of rolling backward when the train is stationary.

## 2.6 Motion in Depth

Regan and Beverly have developed the notion that motion in depth is mediated by at least two mechanisms: changing size, (Regan & Beverly, 1978a, 1978b) and changing retinal disparity (Regan & Beverly, 1973). Changing size is a monocular mechanism and is characterized by thresholds of about 1 minute of arc. The velocity of the size change is proportional to the perceived motion in depth. Changing retinal disparity is a stereoscopic cue with approximately the same thresholds for detection as changing size. Figure 3.9 summarizes the movement on the retina caused by monocular and binocular viewing conditions. In A, motion of an object toward the viewer results in the edges of the object moving in opposite directions (expansion) on the retina (and vice versa for movement away from the observer, not illustrated). In B, motion toward the observer results in both object expansion and movement of the edges of the object in opposite directions (translation) on the two retinas. In C, motion of an object toward the left eye will result in only expansion in that eye, but translation and expansion in the right eye. In addition to the differences in direction of movement in the binocular case, differences in velocity occur on the retinas when the object is not moving on a trajectory aimed midway between the eyes. Comparisons of monocular and binocular sensitivities to motion in depth reveal the binocular sense to be slightly more sensitive than the monocular case, as in spatial vision (McKee & Levi, 1987). Both mechanisms appear to provide input to a single motion in depth stage, however, because motion perceived in terms of changing size can be "nulled" with conflicting disparity manipulations. This information concerning motion in depth appears to be uniquely derived because this ability cannot be accounted for by static stereoacuity performance and the visual fields for stereo-motion in depth are constricted relative to those observed for static stereopsis. However, receptive fields for motion in depth

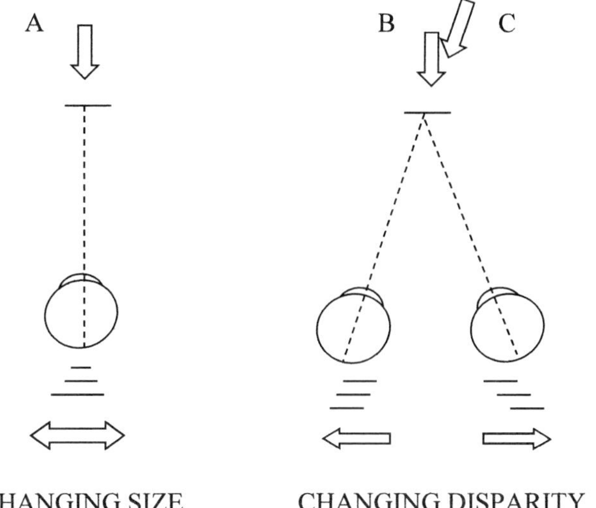


FIG. 3.9. The diagram on the left indicates how the edges of an object on a single retina move in opposite directions under expansion and contraction conditions produced by an object moving toward or away from the observer. The diagram on the right indicates how opposite object motion occurs on the two retinas when it is stereoscopically viewed. (See text for elaboration.)

signaled by optic flow without stereo cues are very large compared to those for stereopsis. Burr, Morrone, and Vaina (1998) have shown signal integration with stimuli up to 70 deg in diameter in psychophysical experiments, and Duffy and Wurtz (1991) have reported single cells with fields up to 100 deg in diameter in cortical area MT. These mechanisms seem ideal for detecting the optic flow produced by locomotion and appear to be centrally involved in its control. Interestingly though, Edwards and Badcock (1993) have shown that observers are more sensitive to contraction than expansion, which is the kind of flow produced by walking backward.

## 3. WHAT DO WE INFER FROM WHAT WE SEE? (SHORT-CUTS)

Previously, the various visual capabilities with regard to detection and discrimination along a number of stimulus dimensions were enumerated. That approach to visual perception assumes a relatively passive observer and defines the abilities in terms of the physical limitations of the visual system. But the human observer is rarely passive and usually interacts with the environment with expectations, goals, and purposes. These psychological aspects of perception have been the focus of numerous lines of research and have provided insight into various phenomena that serve to facilitate and disambiguate information-processing tasks. In this section, we consider these propensities and the implications for perception within the simulated environment.

### 3.1 Figure/Ground Organization and Segmentation

Glance at Fig. 3.10a and ask yourself what you saw first. You might say lots of Ss, or Hs made of Ss, or an E made of Hs made of Ss. Whatever it was that you perceived first, it is easily possible to see the other possibilities with further examination. While humans rarely think about it, similar perceptual processes are employed in the analysis of all images. An observer

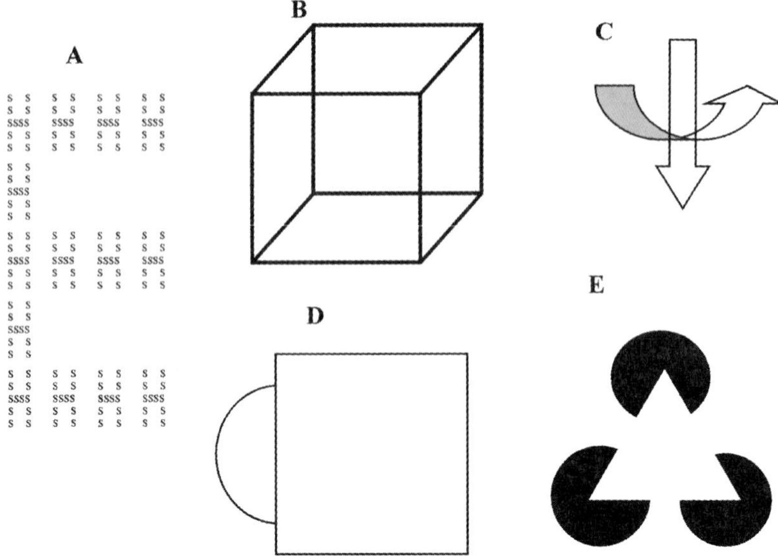


FIG. 3.10. Stimuli that provide examples of perceptual organization and the Gestalt laws that have been proposed to facilitate it. (See text for details.)

might stare at a forest and see its global outline against the sky, or look at a particular tree as an element in the forest, or attend to a leaf on one of it's branches. In each case, the observer can describe the object of their attention as an object perceived against a background (or ground). Since nothing need change in the scene in order for the observer to experience these different perceptual figure–ground organizations, they are forced to conclude that the experiences are not driven by stimulus change but are determined either physiologically (through adaptation) or psychologically, through directed attention and/or perceptual set. Consider Fig. 3.10b, the Necker cube. Although the figure is static, an observer can perceive two possible perceptual organizations involving different depth planes, one with the box protruding to the upper right, the other with the box protruding to the lower left. Much of the visual work accomplished is cognitively driven wherein the observer searches for objects or serially processes objects (e.g., reading) against a background. Early Gestalt psychologists (e.g., Koffka, 1922) proposed a set of "laws" which suggest what stimulus configurations facilitate perceptual organization. Recent work has been attempting to determine the algorithmic and physiological bases for some of these phenomena. They are still poorly understood but have proven to be useful for scene segmentation in computational models. The factors are:

1. Proximity: Given an array of elements, those closest together are more likely to be grouped into a global structure. Elements that are close to one another tend to be seen as an object (e.g., the Ss are grouped to form the *H* in Fig. 3.10a).
2. Similarity: Similar objects tend to be grouped together to form a global structure (e.g., the *H* in Fig. 3.10a).
3. Good Continuation: Lines are perceived as continuing with the minimum angular deviation necessary. In Fig. 3.10c the vertical lines of the vertical arrow could deviate sharply along the horizontal arrow but the default grouping avoids this solution. Field, Hayes, and Hess (1993) have described an association field that quantifies the likelihood of binding elements on a smooth path as a function of angular deviation of the path.

4. Common Fate: Elements of an image that undergo a common transformation over time are frequently grouped as one object. The most extensively studied transformation is lateral motion, which produces rapid segmentation of complex scenes. The precise mechanisms are yet to be revealed, but single cells in the medial temporal cortical area (MT) have stimulus preferences for dot patterns moving in different directions to their surroundings and thus may form a neural substrate for this segmentation (Allman, Miezin, & McGuinness, 1985).
5. Closure: When a form is drawn with broken or incomplete contours, the perceptual system perceives that form that would result if the contours were complete (Rock, 1975). This is not to say that the breaks in the contours are not detectable, but rather that those breaks do not prevent grouping for object recognition (see Fig. 3.10d).

These principles may be applied to images containing a broad array of stimulus attributes. Interestingly, once grouping has occurred quite distinct boundaries may be formed that are not physically present in the luminance profile of the image. For example, Fig. 3.10e appears to be interpreted by the visual system as a white triangle overlapping the black circles. An illusory brightness difference is usually perceived between the inferred center of the triangle and the surrounding area. The border between these two zones can appear to be very sharp. Some of these borders may be seen because they stimulate nonlinear texture detection processes within the visual system (Wilson, Ferrera, & Yo, 1992), but no adequate model exists for the full array of illusory contours at this stage.

## 3.2 Perceptual Constancies

Although the 2-D representation of objects on the retina may undergo considerable translations in lightness, color, orientation, size and position, humans maintain an appreciation for these basic properties of the object. Thus, a white piece of paper (on a gray background) viewed in direct light and in a cast shadow appears to be the same brightness despite the fact that one reflects more light than the other. Likewise, a piece of colored paper (on a gray background) viewed under different colored lights will appear a very similar color despite the differences in hue of the illuminant. This page is viewed as rectangular whether viewed from directly above or from a 45-deg angle and it is perceived to be the same size whether viewed from 18 inches or 18 feet. The page is seen to be in the same position in space whether an observer looks directly at it or gazes to the side. While the mechanisms whereby these constancies are achieved are still the subject of active research, most explanations emphasize a ratio principle. For lightness and color constancy, it is proposed that the ratio of the lightness or the hue of both the object and the background is extracted and if the ratio is the same, constancy is achieved. For shape and size constancy, the critical ratio refers to the size and orientation of elements in the object relative to size and orientation in the background under different viewing conditions. See Palmer (1999) for a recent review of these mechanisms. Position constancy will be elaborated upon later.

## 3.3 Exocentric Motion

Consider a small, stationary point of light in a totally dark room. That object appears to move in a quasi-random fashion and subjects find it quite difficult to specify its position (by pointing or touching its location). This illusion of exocentric motion is termed the autokinetic effect (Sandström, 1951), and it indicates how difficult it is to determine the motion of an object without a visual frame of reference. The effect is actually caused by implicit eye, head, and body movement in the absence of the stabilizing effect of visual information and is really an egocentric illusion. In a lighted room with referential stimuli, the object is seen as stationary (Mack, 1986). The next question concerns how much that object must be displaced for the observer to detect movement. With a frame of reference, the displacement threshold has been estimated to be 1 to 2 min, but that value increased tenfold when the reference was removed (Aubert, 1886; Bourdon, 1902; Grim, 1911). With real movement, the moving object is always visible. It moves a given distance (or is visible for a given time), and its movement generally has a fixed velocity (in the simple case). The upper threshold for the detection of movement depends on many factors (luminance, field of view, spatial pattern), but long before an object moves too fast to be seen, the form of the object is obscured. One estimate of this threshold point is termed dynamic visual acuity, and it indicates that form vision as measured by Landolt C acuity is compromised above about 60 to 100 deg/sec (Ludvigh & Miller, 1958). However, very large objects may be detectable up to 3000 deg/sec provided the temporal frequency of the stimuli falls inside the critical flicker fusion limit (Burr & Ross, 1982). There is a belief that, at moderate velocities, displacements are perceived as movement, but at lower velocities motion is based on change in position (Anstis, 1980; Exner, 1875). Some suggest that with brief durations ($< 0.25$ sec) only velocity is sensed and only at long durations would velocity thresholds be improved by a framework (Gregory, 1964; Liebowitz, 1955). Velocity thresholds increase in the periphery (Aubert, 1886). There is also a common misconception that the peripheral retina is better at motion detection than the central retina. On the contrary, Tynan and Sekuler (1982) reported minimum movement thresholds to be on the order of 0.03 deg/sec in the fovea, increasing to 0.45 deg/sec at 30 deg eccentricity.

When two objects are moved in the visual field, the difference threshold for velocity has been estimated to be in the range of 30 sec/sec (Graham, Baker, Hecht, & Lloyd, 1948) to 2 min/sec (Aubert, 1886). But with two objects, induced movement and motion contrast effects come into play. Induced movement is an illusion of motion in which the displacement of some elements results in the attribution of movement in other elements. For example, moving a frame around a stationary object often results in an apparent movement of the object in the direction opposite to frame movement. Motion contrast is a type of induced movement that occurs at the borders of adjacent fields of moving elements. Adjacent elements moving in different directions and/or at different velocities can induce illusory directional and velocity attributes (Levi & Schor, 1984).

## 3.4 Egocentric Motion

Motion in the visual field due to one's own movement is easily discriminated from motion in the visual field due to object motion. The mechanisms that underlie this ability involve a comparison between the movement sensed on the retina and nonvisual signals associated with one's own movements. When movement of an object on the retina is negatively correlated with eye, head, or body movement (later discussed), position constancy holds and objects are seen as stationary. Any mismatch between the visual and nonvisual information signals movement of the object relative to the observer. The limits of position constancy define the precision of these mechanisms. For head movements, position constancy breaks down with displacement ratios (retinal movement/head movement) of 1.5% (Wallach & Kravitz, 1965). While humans are quite accurate at detecting displacements during fixation (2–3 min), displacements as large as 6 min go undetected if they occur during a saccadic eye movement (Stark, Kong, Scwartz, Hendry, & Bridgeman, 1976). Target displacements of less than 1/3 saccadic displacements are rarely seen (Bridgeman, Hendry, & Stark, 1975). Thus, the resolution of the position constancy mechanism during saccades is degraded relative to fixation. When a moving target is tracked with smooth-pursuit eye movements, its perceived speed has been reported to be 63% of that reported when it is moved through the field when the eyes are fixated (Dichgans, Koener, & Voight, 1969; Turano & Heidenreich, 1999). In addition, stationary objects in the field appear to move (the Filehne Illusion) when the eyes track another moving object (Filehne, 1922). Wertheim (1994) argues that reference signals are generated through visual, oculomotor, and vestibular interactions, and it is the comparison of the retinal and reference signal that mediates the perception of self and object motion. He suggests that while oculomotor-generated signals encode how the eyes move in orbits, vestibular signals must be added to form these reference signals to encode how the eyes move in space.

## 3.5 Visually Induced Self-motion

Various senses contribute to human's perception of motion through the environment (see chaps. 7 and 23, this volume), but one of the most important inputs is from motion in the visual array. When motion information from nonvisual inputs is in correspondence to visual information, humans perceive themselves to be moving in a more or less veridical fashion. There are, however, instances when nonvisual inputs might be expected to signal no self-motion, but the visual array contains scene translations that would signal that the entire environment is moving relative to the observer (a situation that rarely occurs in nature without bodily movement). According to the notions about egocentric motion discussed above, one might expect this situation to result in the perception that one is stationary and the world is moving, but in many cases one perceives an illusory sense of self-motion. Parametric studies which have sought to outline the stimulus conditions that result in such illusions offer important knowledge about how visual inputs provide information about self-motion as opposed to scene movement, and how self-motion might be simulated with purely visual displays.

Mach (1875) was the first to investigate illusory self-motion. He noted that if an observer, seated on a stationary platform, is placed inside a rotating drum, the internal surface of which is covered with vertical stripes, they often report a vivid perceptual illusion of self-rotation in the direction opposite to that of the drum rotation. He referred to this illusion as circular vection (CV). Dichgans and Brandt (e.g., 1973) have provided the most extensive series of studies on the topic. They noted that such stimulation also causes optokinetic nystagmus (OKN—see section 4.1.2 below), regular eye-movement patterns in which smooth pursuits in one direction are interleaved with rapid return flicks. Stimulus parameters that effect the preponderance of CV are: field of view (FOV); fixation of a stationary target; cylinder rotational velocity; and the spatial frequency of the stimulus contours. The latency to onset of CV seems to map onto the differences in stimulation from the vestibular and visual system during real bodily rotation. In that situation, the two senses agree at the onset of motion, but as rotation ensues at a constant angular velocity, the vestibular system ceases to respond and only visual evidence of rotation continues. Studies of illusory CV indicate that at high levels of drum rotation (> 5 deg /sec) CV latency increases, but below that level the sensation is almost immediate. That finding is commensurate with the fact that vestibular stimuli are not effective at low levels of acceleration and would not be expected to provide a stabilizing source of stimulation. The velocity of CV is proportional to the velocity of drum rotation up until about 90 deg /sec, when the duration of CV declines.

Depth cues also affect CV (Brandt, Wist, & Dichgans, 1975). Displays farther away from the subject are more effective at producing illusory self-motion. Some argue that CV is dependent on peripheral-visual-field stimulation, whereas OKN requires only central field stimulation (Dichgans & Brandt, 1973), but Post (1988) reported that CV can be obtained with a restricted FOV. Young, Dichgans, Murphy, and Brandt (1973) showed that CV could be augmented or suppressed with bodily accelerations in the same or different directions, respectively, and that the threshold for detection of bodily movement was raised when the visual and vestibular stimuli were in conflict. Stern, Hu, Anderson, and Leibowitz (1990) reported that subjects who viewed rotating stripes with fixation reported somewhat less CV, and a restricted FOV (15 deg) resulted in greatly reduced CV. Brandt, Dichgans, and Koenig, (1973) found that CV was still present in a situation where stripes in the whole visual field moved in one direction, whereas OKN was elicited by stripes moving in the opposite direction in a small central field.

In addition to illusions of continuous self-rotation, displays that lead to CV also can elicit illusions of bodily tilt in the direction opposite to drum rotation (Held, Dichgans, & Bauer, 1975). The individual does not experience continuous "barrel rotation," but rather experiences a static illusion of tilt that increases with the velocity of the drum up to a value of about 15 deg. This illusion is more influenced by peripheral stimulation and has a latency of about 18 sec. These sensations are exacerbated by inclination of the head during stimulation (Young, Oman, & Dichgans, 1975). Illusions of "pitch" fore and aft are created with displays that rotate around the horizontal axis of the body. In these situations, observers experience a constant feeling of rotation in the direction opposite to drum rotation, with a concomitant static illusion of body inclination in the direction of illusory self-motion (Young et al., 1975).

Stimuli that involve movement in depth often give rise to illusory sensations of linear vection (LV). These stimuli also elicit vergence eye movements (Collewijn, Erkelens, & Steinman, 1995; Erkelens & Regan, 1986). Most stimulus conditions that produce the illusion of LV involve wide FOV; however, Anderson and Braunstein (1985) were able to produce strong sensations of LV with a small FOV. These authors and others (Telford & Frost, 1993) emphasize the strong influence of depth cues (kinetic occlusion, optical velocity, and size change). They suggest that the central visual field conveys a different kind of movement information (detailed depth information) than the periphery, and that if both mechanism are stimulated, synergistic or antagonistic influences on self-motion may be obtained. Many individuals become motion sick with prolonged exposure to vection-inducing stimulation (see chaps. 30, 32, and 36, this volume).

## 4. HOW DO WE LOOK AT AND MANIPULATE OBJECTS? (MOVEMENT OF THE OBSERVER)

Interaction with the real world is rarely passive, but instead involves purposive movement of the eyes, head, arms, legs and body. The way in which an individual accomplishes this involves complex sensory-motor systems that are characterized by reflexive and volitional components. These processes must be considered when simulating the environment, especially when attempting to incorporate simulation of the results of bodily movement within a virtual environment. Some of the systems involved in visual inspection of the world and their implications for man-machine interface are now considered.

### 4.1 Eye Movements

There are a variety of eye movements, and they may be described generally as being conjunctive or disjunctive. Conjunctive eye movements involve the eyes moving in the same direction (OKN, smooth pursuit and saccades), whereas disjunctive eye movements involve the eyes moving in opposite directions (convergence and divergence). A major consideration in ocular control concerns the ability to control the eyes when examining objects. While this is often done while the head and object are stationary (fixation), it is also done while the object moves (smooth pursuit) or when the head moves (see the vestibulo-ocular reflex—VOR—chap. 7, this volume). Visual capabilities are quite different during these oculomotor behaviors.

#### *4.1.1 Fixation*

When observers look at objects in the environment they position their eyes so that the object of interest falls on the area of the retina that has the best visual acuity (the central 5 deg around the optic axis of the eye termed the macula). When this is achieved the observer is said to have fixated the object. With a stationary target, fixation stability may be defined as the degree to which the eye is stable with reference to some fixation point on the object. Fixation stability is surprisingly poor. The eye may drift as much as a degree without the observer being aware of the drift in fixation. Even with well-controlled fixation, the eye is in constant motion, with microsaccades (movements) of many arcsec (Riggs, Armington, & Ratliff, 1954). These ocular tremors are necessary for continuous viewing of an object. Artificial conditions that render the stimulus stabilized on the retina result in disappearance of the stimulus (Kelly, & Burbeck, 1980). Attempts at maintaining fixation during image movement (see smooth pursuit, section 4.1.3) or head movement involve interactive inputs to and from the vestibular system.

#### *4.1.2 Optokinetic Nystagmus*

If a contoured visual field is rotated before a stationary observer, OKN is elicited. The eyes drift in the direction, of field rotation and then snap back in the opposite direction, and this cyclic pattern of eye movements is repeated in bursts, interrupted by short periods of relative gaze stability. The initial tracking response (slow phase) is seen as a period of smooth pursuit and the compensatory snap back (fast phase) as a corrective saccade (see section 4.1.4). The existence of this reflex is seen as evidence that there are visual mechanisms that take into account the effect of movement of stationary aspects of the visual field on the retina as the head or the eyes move, and that these mechanisms can provide, together with the VOR, additional complementary information about how the eye should move during head movements to provide gaze stability. Unlike the VOR, this response does not abate with continuous stimulation. While the VOR is thought to adapt because the vestibular end organs cease to respond at constant velocities, OKN is seen as a compensatory response for this failing. Another difference between the two responses is that the VOR has a short latency and a slow decay, while OKN has a long latency and slow buildup. While OKN is seen as a reflex, it also can be suppressed with fixation of a stationary target.

#### *4.1.3 Smooth Pursuit*

Observers are quite capable of tracking moving objects with the head held stationary. This is generally considered a voluntary response engaging the smooth pursuit system, which must calculate the speed of moving objects to maintain fixation. This system requires a moving object. Smooth-pursuit eye movements can not be made in total darkness or in a visual field devoid of movement. It is assumed that this system requires volition, because it is also possible to fixate a point when other moving objects are present. The upper limit of this response is about 100 deg/sec.

#### *4.1.4 Saccades*

Another response, which has generally been considered voluntary, is the saccade. This is a ballistic movement between one fixation and another. Its latency ranges from about 150 to 200 milliseconds, and its velocity is proportional (within limits) to the distance the eye must be moved. Saccades can occur with speeds up to 900 deg/sec. Whereas saccades usually are made from one target to another in space, saccades can be executed with high deg of precision to spatial positions defined cognitively (signals in other modalities, verbal commands, memories of spatial locations). Whereas this suggests voluntary control, aspects of saccades (direction, velocity or amplitude) can not be changed after they are initiated. However, saccadic latency can be shortened significantly if a presaccadic fixation target is removed shortly before the saccadic target is presented, indicating that disengagement from one target is necessary prior to a saccade (Abrams, Oonk, & Pratt, 1998; Forbes & Klein, 1996). Another characteristic of vision during saccades is that visibility of achromatic, but not chromatic, stimuli are suppressed throughout the eye movement (Burr, Morrone, & Ross, 1994). Chromatic stimuli are usually not perceived, however, because the retinal velocities exceed the resolution limits of the chromatic pathways.

#### *4.1.5 Vergence/Accommodation*

All of the eye movements so far considered are conjugate, but vergence movements are also necessary to focus on objects in depth. Accommodation of the crystalline lens is linked to vergence in normal observers, and changes in the refractive power of the eyes are correlated such that objects are maintained in sharp focus and registered with retinal disparities on the order of min of arc.

### 4.2 Head and Body Movements

One of the most important aspects of contemporary virtual simulations is the ability to provide shifts in the scene contingent on head, hand, or body movement. Position tracking technology provides a major contribution to sense of presence (see chap. 40, this volume) and is described elsewhere in this volume (see chap. 8, this volume). The aim of the present section is to review the implications of the visual forcing functions on head and body movement that might be expected to influence the ability to provide realistic scene translations.

Postural stability is maintained through the vestibular reflexes acting on the neck and limbs. These reflexes are under the control of three classes of sensory inputs: muscle proprioceptors, vestibular receptors, and visual inputs. As previously discussed, a comparison of the vestibular and visual inputs is necessary to determine if the observer or the environment is moving. Visual inputs that give rise to vection have been shown to result in bodily sway and falling (Lee, 1980). It is reasonable to expect that such stimulation may result in reflexive neck muscle activity that can lead to head movements. When all or a large part of the visual field is filled with moving contours, standing observers lean in the direction of scene motion (Lee & Lishman, 1975; Lestienne, Soechting, & Berthoz, 1977). There is a compensatory leaning in the opposite direction upon the cessation of stimulation that may last for many seconds (Reason, Wagner, & Dewhurst, 1981). If the virtual scene is driven by head movement, such unintentional reflex head and body movements (and the aftereffects thereof, see chaps. 37–39, this volume) may serve to impede or interfere with scene shifts intended to simulate voluntary attempts at navigation through the virtual environment. In some cases, reflex movements can be suppressed, but this may imply that some learning must occur to achieve efficient and unnatural adjustments that are necessary to facilitate optimal performance.

## 5. HOW DO WE DEPICT THE 3-D WORLD IN 2-D DISPLAYS? (HARDWARE REQUIREMENTS)

Modern VE technology is an extension of all past endeavors to depict renditions of the visual world. There has been an orderly progression in the mastery of monocular cues for depth in 2-D static displays to the movement and disparity cues for depth that are so compelling in dynamic displays. The advent of computer-controlled displays has provided the possibility of greater user interaction with these virtual depictions, and with it comes an illusory sense of presence (see chap. 40, this volume) and some adaptational problems for the user (see chap. 31, this volume).

## 5.1 Static 2-D Displays

A first approximation to simulating the 3-D world is the traditional 2-D display ubiquitously embodied in drawings, paintings, and photographs. What can be provided relative to the capabilities of the human visual system? Two-dimensional displays can provide fine spatial detail, variations in contrast, differences in hue, the spatial position of objects in the frontal plane, and indications of 3-D using the pictorial cues for depth (relative size, height in the field, occlusion, shading, texture, and linear and aerial perspective). Using a single image on paper, canvas or film, one cannot provide the stimulus conditions to provoke motion parallax, kinetic depth, motion, or retinal disparity. The field of view is usually restricted and viewed within a 3-D framework provided by the rest of the real world. Some have argued that because of human sensitivity to the actual flatness of the 2-D plane on which 3-D simulations are viewed, the pictorial cues are in conflict with other cues provided simultaneously in the 3-D world and are rendered less effective (Ames, 1925; Nakayama, Shimojo, & Silverman, 1989).

As previously noted, in real 3-D views the ability to focus and converge on different depth planes may provide reafferant information for depth judgments, but these are missing in a depiction of 3-D in only one depth plane. Can these other important visual capabilities be recruited with other manipulations of static 2-D displays? This is possible only if stereoscopic viewing conditions are provided—presenting two versions of the same scene with fusible retinal disparities. This is done in various ways. The first stereo displays were achieved with stereoscopes that allowed viewing of two disparate pictures superimposed by means of prismatic lenses. The two pictures were often obtained by simultaneous photography of the same scene with two cameras in slightly different positions. The anaglyphic approach is to render the two members of the stereo pair in different wavelengths (e.g., red and green) and let the observer view the two disparate scenes with glasses that segregate the visual images with different chromatic filters in front of the eyes. The two images can then be fused to provide rather striking depth information and limited motion parallax. A similar technique involves the use of polarizing filters. Two disparate scenes are presented through orthogonally polarized filters. Each eye of the observer views through a polarizer matched to only one of the scenes. The visual system combines the two views and is able to detect the disparity cues. The stereoscopic images so popular in the comics section of Sunday papers use an anaglyphic approach, which does not require chromatic glasses, but requires the observer to converge or diverge the eyes to achieve fusion. The motivation to attempt such viewing is to find the hidden figure not visible in the monocular or nonfused binocular view. While the gain in level of quality of 3-D simulation provided by stereo viewing is impressive, the static 2-D display lacks the vividness that scene motion conveys.

## 5.2 Dynamic 2-D Displays

In the early part of this century, considerable amusement was provided to those fortunate enough to own a stereoscope. The excitement generated by such devices paled in comparison to the first moving pictures, however. Movement in nature is an extremely compelling sensation, providing immediate impressions of one's position relative to other objects in the world. The simulation of movement in static displays was first accomplished with a deck of cards, wherein an ordered sequence of pictures denoting differences in scene position was presented in rapid succession by slight of hand. The deck was bent and individual cards were allowed to slip off the thumb to reveal a simulation of motion. The motion picture projector was developed using the same rudimentary principle, rapidly sequencing a series of still pictures. These pictures were captured on filmstrips with cameras capable of rapid photography. The current standard for film projection is 24 frames/sec, and each frame is transilluminated for about 30 msec.

A rotary shutter blade is interposed while the film is advanced (taking about 10 msec), yielding a frame rate of about 24 frames/sec. This simple arrangement provides a series of still pictures, but also produces considerable detectable flicker. By increasing the rate of flicker (to 50–60 Hz), the achromatic critical flicker-fusion limit for the human observer is surpassed and an acceptable sensation of smooth movement is attained (Cropper & Badcock, 1994).

While these developments added realistic movement to simulations of the real world, stereoscopic and motion parallax cues were still not available. In the 1950s, the first 3-D movies were screened, and this technology combined the anaglyphic technique for stereo vision with the cinematic technique for motion, providing quite compelling perception of motion in depth and some limited motion parallax information. Modern cinematography has become a ubiquitous source of entertainment and education in this century, but it does not approach a true simulation of our normal experiences with the 3-D world. In most cases, the observer is quite passive and relatively immobile. That is also the case with standard video simulations, which although technically more advanced electronically, suffer from the same inadequacies. Video, however, affords the possibility to interface with sophisticated computer systems that offer the possibility to incorporate user movements into the simulated environment.

## 5.3 Electronic Displays and the User Interface

The cathode ray tube (CRT) began as a method of rendering simple electronic activity immediately visible. For example, oscilloscopes are often used to display voltage changes over time. The early versions involved an electron "gun" aimed at a phosphorescent surface. Deflection plates or coils controlled the direction of the gun, and the intensity of fluorescence was varied via accelerating plates that determined the rate of electron flow. Originally, control of the gun was accomplished with "vector scanning" that addressed only the points (pixels) on the screen necessary to "draw" the image in question. This proved adequate for simple geometric shapes, but more complex scenes required the more elaborate "raster scanning" approach wherein every pixel was addressed. In raster scanning, the stream of electrons is swept across the screen horizontally from left to right and then snapped back and down one line in order to paint the next row. This procedure is repeated until the entire screen has been painted and the gun is turned off (blanked) during retrace (snap back). One complete painting of the screen is termed a frame, and frame rates above the human flicker fusion point (50–60 Hz) result in a flicker-free sequence of pictures. This technology provided the basis for black-and-white television. This relatively simple system has been expanded to provide color television by including more guns and colored (red, green and blue) phosphors. One problem with CRT displays is that the tube length increases as the tube width and height increase, resulting in large displays that are quite bulky. More recent advances in display technology involve attempts to reduce the size of the apparatus. Liquid crystal, plasma, and electroluminescent displays are steps in this direction, but have disadvantages associated with resolution, color or expense; these shortcomings are rapidly being corrected. The discussion of relevant parameters for meeting user specifications will be undertaken with regard to CRT displays, but the principles will apply to all display technologies. The parameters of importance for meeting the user requirements are provided in TABLE 3.1.

### *5.3.1 Spatial Parameters*

The field of view for modern displays varies widely, from the tiny screens produced for miniature television to the panoramic screens employed in IMAX and movie theaters. The small screens have been incorporated into head-mounted displays (HMDs), with optics designed to

**TABLE 3.1**
Relevant Display Factors Relating to Human Visual Abilities

| | | | | |
|---|---|---|---|---|
| **Spatial vision** | Pixel size and spacing | Pixel number | Screen size | Intensity resolution (bit depth) |
| **Color vision** | Phosphor types | Phosphor number | Gun independence and stability | Intensity resolution (bit depth) |
| **Image motion** | Phosphor decay | Raster rate | Pixel size and spacing | Intensity resolution (bit depth) |
| **Stereopsis** | Interlaced frames | Frame disparity | Monocular monitors | Intensity resolution (bit depth) |
| **Observer motion** | Scene update rate | Field of view | Frame rate | Intensity resolution (bit depth) |

allow sharp focus of images within the near point of vision and to provide a large field of view. Video projection systems can fill the human field of view but often present problems of spatial resolution, image distortion, and are limited to relatively low luminance levels (although this is rapidly improving). CRT-based HMDs currently have a limited field of view (< 80 deg ), and LCD displays offer somewhat larger fields of view (105 deg ). Although the peripheral retina has poor spatial resolution, a wide field of view provides valuable information for visually guided behavior and for producing the illusion of vection (see chap. 23, this volume).

The spatial resolution of most displays does not approach the limits of human visual acuity or spatial positioning ability unless viewing distance is extended, which, of course reduces field of view. This trade-off between field of view and spatial resolution is an important limitation for visual simulation, but solutions may soon be forthcoming. For projection systems and large screen devices, resolution must be increased, and for HMDs both spatial resolution and display size must be increased. Few visual tasks push the limits of visual acuity, and the major cost of less than optimal resolution is a sacrifice of texture and the sudden appearance of objects at simulated distance. This latter characteristic is potentially very serious. Castiello, Badcock, and Bennett (1999) have shown that when observers are reaching to grasp an object, a suddenly illuminated distractor object changes the motor aspects of reaching behavior. This change only occurs if the illumination is sudden; gradual onset has no effect, and if a spotlight suddenly illuminates the peripheral scene but there is no object the motor movement isn't affected. Thus, in VE setups, suddenly occurring peripheral objects would be expected to be the most disruptive of all possible object appearance modes if an observer is trying to make motor movements in the setup. Higher resolution renditions are needed to overcome this problem.

To match or exceed the contrast sensitivity of the human visual system, the luminance steps (gray scale) must be small enough to provide differences in luminance to which humans are insensitive. For many natural scenes, 24-bit (8 bits on each of the red, green, and blue guns) graphics systems are adequate. However, if fine contrast discrimination is required most observers will require either higher contrast resolution or that the full range of steps be compressed into the near threshold range of contrasts being presented. The latter has the consequence of restricting the maximum contrast range available at any instant but does allow small contrast steps to be presented. Computers that boast 16-bit digital-to-analog converters (DACs) for each gun are preferable, but expensive and uncommon. Many commercially available monitors provide adequate luminance (up to 100 cd/m$^2$) and support photopic viewing levels. Thus, the current limitation in intensity resolution is usually provided by the DACs employed.

### 5.3.2 Color

Many modern monitors are capable of providing high spatial resolution (1,600 × 1,200 pixels) and moderate temporal resolution (60–150-Hz frame rates) while also providing a colored image. The monitors contain three electron guns, which are each aimed at a different phosphor type so that in near spatial proximity on the screen, red, green, and blue signals can all be produced. There are a number of important monitor characteristics that are required for high-fidelity color rendition.

The range of achievable colors depends on the chromaticity coordinates of the phosphors used. However, current monitors are only capable of producing a small part of the full range of chromaticities detectable by the human visual system. Figure 3.11 depicts the full color gamut specified in the CIE 1931 xy chromaticity space for normal human vision, with the gamut available on a typical RGB monitor plotted as a dark triangle.

Within the available gamut, accurate color rendition requires high gun stability over time and gun independence, but both of these factors need to be rated against the intended use of the display. Gun stability is often a problem during the first hour a monitor is turned on. During this period, the luminance and chromaticity of a display will vary, even in very good monitors. Metha, Vingrys, and Badcock (1993) provide some representative data and a method for evaluating the visual salience of these changes.

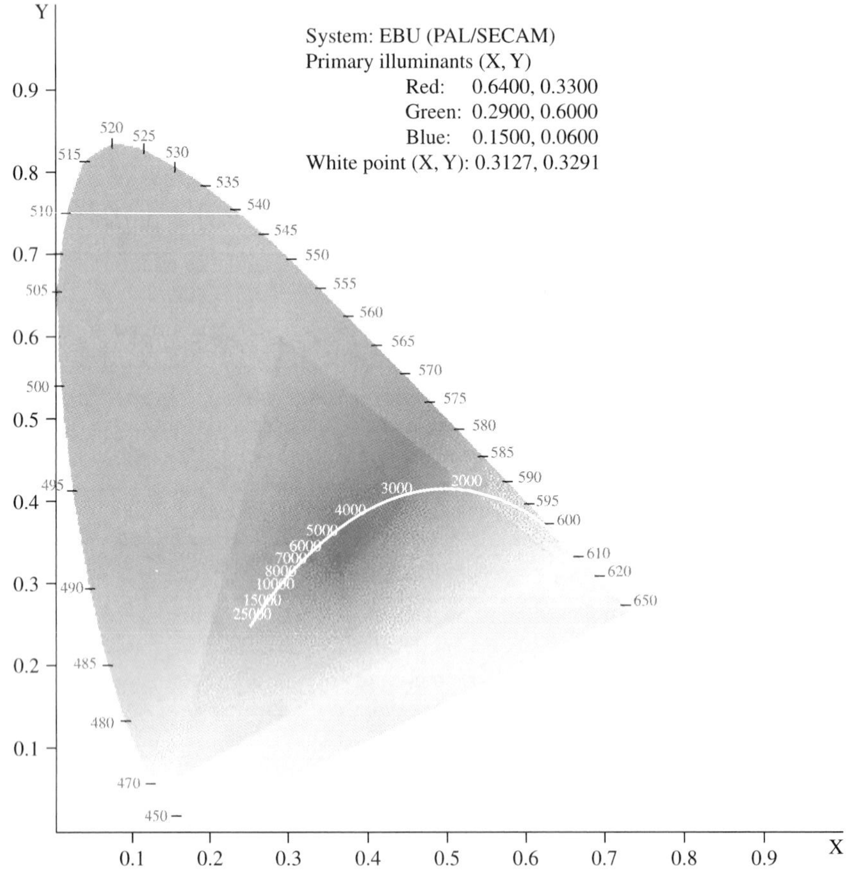


FIG. 3.11. The CIE (1931) xy chromaticity space for normal human vision is depicted with the gamut of color available, with the typical RGB monitor depicted as a dark triangle.

Gun interactions are more problematic. On less expensive monitors the intensity of the output from one gun will vary as the output of the other guns increases. The consequence is that there is not a fixed relationship between a given intensity level specified in the driving software and the output of a particular gun. The output will vary with different images. Observers are quite tolerant of these changes for noncritical applications (e.g., word processing), but the interactions are likely to be destructive when accurate color discriminations are required. Expensive monitors are more likely to avoid this problem, but the only way to be sure is to measure the light output for each gun independently and compare the output for several levels of activation of the other guns (see Metha et al., 1993). This check has become more critical recently, as many manufacturers of monitors are deliberately building in a gun interaction that affects the upper half of the intensity range to minimize the amount of flashing when users move pointers from one window to another in graphical user interfaces.

The final issue of significance for rendering color is the variation in light output across the monitor screen. Some variation is due to nonuniformity of the phosphor deposition during manufacture, but most of it is due to the design characteristics of the shadow mask technology. All such monitors are brightest in the center of the screen, and luminance drops considerably toward the periphery. In part, this is due to the smaller effective aperture in the shadow mask at shallower angles of incidence, but the loss is also due to the polarity of the emitted light. The output is much more intense in the direction continuing the path of the electron beam, a direction that points to one side of an observer aligned with the center of the screen. Metha et al. (1993) found this loss to affect luminance more than chromaticity but ideally stimuli requiring accurate luminance and chrominance discriminations should be small and placed centrally on the screen to minimize spatial variations due to the monitor.

### 5.3.3 Image Motion

Two factors that have a strong influence on simulated motion are phosphor persistence and frame rate. Phosphor persistence refers to how quickly the phosphorescence fades after it has been energized. Long persistence phosphors reduce flicker but will also result in image smear when image movement is simulated. Short persistence phosphors require higher update rates for pixels that convey the presence of static parts of the image and, if the rate is too low, flicker may be a problem. In practice, it is quite easy to determine whether the quantization of motion will be detectable. If the temporal frequency components introduced by the quantization fall within the range detectable by an observer, then the observer will be able to discriminate between smooth and sampled motion sequences (Cropper & Badcock, 1994; Watson, Ahumada, & Farrell, 1986). Other factors may be critical if either speed difference judgments or very slow speeds need to be produced. In both these cases, the achievable steps in speed will depend on a combination of the frame rate of the monitor and the spatial resolution. The quantization in both time and space limits the minimum speed and minimum difference in speed achievable.

### 5.3.4 Stereopsis

Stereoscopic vision is currently supported by either providing separate monitors for each eye (e.g., HMDs and boom-mounted devices), shutter glasses, anaglyphic glasses, or projectors and autostereoscopic displays. Dichoptic presentation using two monitors presents two views of the same scene, one to each eye, and depends on binocular fusion to yield stereopsis. Electronic shutter glasses present alternating views of the display synchronized to the frame rate such that one interleaved frame in each pair is presented to each eye. The display provided in each set of odd or even frames contains a disparity between the two eyes and binocular fusion provides stereopsis. Anaglyphic glasses filter the images provided to the two eyes with chromatic (usually red and green), lenses and the display contains disparate images rendered in the two colors. Autostereoscopic systems involve lenses imposed between the viewer and display or use a haploscopic arrangement in which dichoptic stimulation is achieved using mirrors to align a different display with each eye. While most would agree that stereoscopic vision is extremely important for the feeling of immersion and presence (see chap. 40, this volume) that the VE seeks to convey, all current techniques involve unwanted attributes. HMDs are bulky, have insufficient field of view, and can cause eyestrain if the optics create competition between the accommodative and vergence systems (see chap. 37, this volume). Electronic shutter glasses are less cumbersome, but are also restricted in field of view and require higher frame rates to avoid flicker. With flicker fusion thresholds near 60 Hz at current display luminances, a frame rate of 120 Hz or greater is required for flicker-free vision. The anaglyphic techniques are compromised because color is used to separate the signals for the two eyes and cannot simultaneously render a true-color image. They are also limited by the restrictions of field of view imposed by glasses. Most autostereoscopic systems offer only small display sizes. Stereoscopic capability is desirable, however, because it provides binocular depth cues through motion parallax and motion in depth.

## 6. WHAT ARE THE SUCCESSES AND FAILURES OF OUR ATTEMPTS TO SIMULATE THE "REAL WORLD"? (THE STATE OF THE ART AND BEYOND)

### 6.1 Successes

As observed above, the spatial aspects of visual simulation seem to be adequate for representing small regions of real-world scenes. Emerging technology promises higher resolution and more encompassing field of view. HMDs should benefit from less mass and projection systems should involve less peripheral distortion. As previously discussed, the depiction of color approximates or exceeds the limits of most human abilities. The temporal characteristics of most displays are adequate and might be expected to improve with increased raster rates and reduced persistence. These improvements will support more realistic depiction of object motion and smooth scene translation. As simulation of motion profiles improves, more pronounced vection and a more realistic sense of presence can be expected. Improving the sense of presence, however, might produce more rather than fewer problems for users.

### 6.2 Failures

One of the most problematic areas in VE system design concerns the simulation of environment motion relative to the observer as the observer's head or body moves in the simulator. The relevant parameters appear to be related to display update rates that are limited by the speed with which graphic depictions can be generated in accordance with movement signals. This limit depends in part on the graphics detail in the scenes depicted. The best contemporary systems boast 40 msec update rates, but these technologies achieve these rates at a sacrifice of resolution. In many systems with inadequate update rates, disorientation and simulator sickness abound (see chaps. 30–32, this volume). These are produced by the mismatch between visual and vestibular cues to motion. The lag in updating the scene means that a vestibular cue to head movement is followed by, rather than being coincident with, retinal-image motion. This particular problem can potentially be overcome by increased computational power that may eventually allow the time lag to become imperceptibly small. A much more difficult problem is the need to provide vestibular input more generally. When observers are free to navigate around a scene, there is often a mismatch between visually indicated changes and cues from the vestibular system. In the extreme, vision may indicate locomotion while the vestibular system indicates that the body is stationary. This mismatch frequently produces feelings of nausea. The relationship between visual and vestibular inputs is also under constant recalibration. Thus, periods of mismatch can lead to recalibration, which will maintain feelings of nausea in the natural environment until the normal relationship has been relearned (see chap. 31, this volume). Improving visual display technology will not reduce this particular problem. Providing appropriate vestibular inputs might reduce it (see chap. 7, this volume).

While current projection displays do afford FOVs that approach the limits of human observers, HMDs are limited by the size of small monitors relative to the interocular distance and the optics necessary to allow clear vision within the near point. Many optical compensatory arrangements result in eye strain and visual aftereffects. Research on the aftereffects generated with the use of HMDs has indicated another conflict situation in apparent motion situations. Stereoscopic displays that require near viewing often give rise to induced binocular stress. Mon-Williams, Wann, and Rushton (1993) reported loss of visual acuity and eye muscle imbalance after only a 10-min exposure to such displays, with some of the symptoms associated with motion sickness. They suggest that these aftereffects stem from adaptation of the accommodation–vergence system due to the disparity between the stereoscopic depth cues provided and the image focal depth. They later reported that displays that preserve the concordance between vergence and accommodation do not produce these aftereffects with viewing times of 30 min (Rushton, Mon-Williams, & Wann, 1994).

## 7. NEW RESEARCH DIRECTIONS

The development of VE systems has proceeded without an abiding concern for the intricacies of human sensory systems. As these factors are considered in simulator design, the dearth of knowledge of many aspects of human perception is highlighted. In many respects, efforts to create realistic VE serves to drive and motivate basic research efforts, but it also points up the limitations of classical research settings. One major problem in the approach to the study of human abilities centers on the way in which efforts have been compartmentalized (Kelly & Burbeck, 1984). The use of VE technology has forced, and will continue to force, a broadening in the scope of interests and considerations in the study of sensory motor interactions. As researchers begin to delve more deeply into these problems, the apparatus necessary to address these issues will require higher resolution on both spatial and temporal dimensions, and become more expensive. This may dictate that research goals must be narrowed. It may be laudable, but not practical, to attempt to generate realistically natural displays that are valid in every VE application.

One way that research tasks may be limited is to address perceptual problems of the particular device in question. While one can predict, to some extent, how well a display might meet the requirements of the human user, a more empirical approach might be to address the adequacy of the instrument with psychophysical experiments in the simulator at the outset. This approach may also reveal idiosyncratic problems that emerge through the use of the device and belie concerns based on theoretical predictions.

### 7.1 Limitations of Classical Research

Much of the research reviewed above was generated in laboratories developed for the study of a particular visual capability (e.g., color, motion) without regard to how that ability might be compromised when other sensory and motor systems are in operation. In many experimental situations, the aim was to study the ability in isolation to avoid sources of psychological interference (e.g., arousal, attention, expectation). The aim in VE, however, is often quite different. The user is often actively engaged in tasks where many systems are employed.

Perhaps the greatest research impetus from the use of VE technology has been to emphasize sensory–sensory and sensory–motor interactions. There is a pressing need for increased information with regard to visuo-vestibular interactions. In part, this is to alleviate some of the negative effects of using VE technology (e.g., nausea) but also because vestibular inputs can change basic properties of visual neurons, such as their preferred orientation of line stimuli (Kennedy, Magnin & Jeannerod, 1976). It will be important to determine to what extent vision is identical when vestibular inputs are decorrelated.

#### 7.1.1 Sensory and Sensory Motor Interactions

The synthesis of classical theories of color vision into the current view of multistage processing is perhaps a model for how researchers will proceed in elaborating knowledge of visual processing relative to nonvisual sensory and motor processes. While considerable psychophysical and physiological evidence supported both the trichromatic and opponent process views, the controversy cooled when we learned how these mechanisms were staged and integrated. In similar fashion, our understanding of vision during various eye movements (previously discussed) is a step toward understanding how vision is compromised during such activity, but we have yet to fully appreciate how visual scene movement might elicit reflexive eye, head, and body reactions. Becker (1989) reported that head movements are a regular feature of gaze shifts greater than 20 deg. Some of the studies that have addressed coincident eye, head, and hand movements, have done so while the subjects performed a volitional task (e.g., Pelz, Ballard, & Hayhoe, 1993), and they reveal the sequential ordering of gaze, head, and hand movements. With the use of head trackers and other pointing devices, it may be the case that these systems are requiring learned modifications to the natural interaction of the sensory and motor systems.

#### 7.1.2 Esoteric Devices and Generalization

If it is the case that users are forced to adapt to the idiosyncrasies of a particular VE, then the traditional laboratory work from which researchers hope to predict visual performance may not provide the basis for legitimate generalization. Studies of basic human visual function may not tell about the human capability to cope with unique sensory–motor rearrangements. Investigations designed to study these abilities (e.g., chap. 31, this volume) cannot anticipate all the kinds of interactive adaptations that might be required by new technology. For this reason, systematic research on basic visual function and sensory–motor interactions should be incorporated into the design and testing of new VE systems at the outset.

### 7.2 Basic Research in Simulators

Many of the more sophisticated VE systems in use today were developed to address practical problems in training and performance. Various vehicular simulators, for example, are used to train operators initially, and to help them maintain their skills. Since this represents the primary mission of the apparatus, little or no time is allotted for investigation of many of the human-interface problems that might exist unless they severely compromise that mission. It is important for the successful development of future systems to document such problems in existing systems and to identify causal variables and usage protocols to prevent them. The work of Kennedy and others (see chap. 36, this volume) serves as an exemplary first step in this direction.

#### 7.2.1 Machine- or Observer-Limited

The basic question asked of vision scientists concerning VE systems is: Will this system meet or exceed the visual requirements of the human user? This chapter has been concerned with the answer to that question, and has attempted to review visual capabilities and relate them to technical specifications. While such an approach has considerable heuristic value, many assumptions are made and conclusions based on such analyses are, in a sense, still quite theoretical. A more direct and empirical approach is to measure visual function within the VE system in question. Some things are readily predictable from extant data. For example, one might ask if the contrast scaling of a display is fine enough to exceed the contrast discrimination capacity of the user. However, the relationship between perceived distance traveled using visual information and the simulated distance traveled is less predictable (Harris, 1999). The most straightforward way to answer that question would be to measure the function of interest inside the VE system and compare the results to measures obtained with more standard methods. If similar findings are obtained with the two methods, the system could be described as observer-limited and adequate in that particular dimension. If performance on that measure was poorer in the VE system than by conventional testing, performance could be termed machine-limited, and this finding would indicate that the system falls short of the user's capability. It is an open question at the moment, as to whether VE systems need always be observer-limited to be adequate for a given application. Many visual tasks are not performed with stimuli at the limits of visual capability, but this is sometimes not clear until the tasks are performed with observer-limited and machine-limited systems.

### 7.2.2 Adaptation and Emerging Problems

Many early and continuing studies of human perception indicate that considerable adaptation occurs under conditions of perceptual rearrangement. Such adaptation can result in a rescaling of the perceptual systems and to corresponding changes in perceptual-motor responses. With brief exposures (< 10 min) to such stimuli, adaptation occurs but recovery is relatively fast. With prolonged exposures (> 10 min to several days), such adaptation might be more permanent and require readaptation over an extended period to restore normal perception and performance (Kohler, 1972; see also chap. 31, this volume). One notion is that as more realistic visual simulations are developed, less adaptation might occur and this may alleviate the problems associated with VE aftereffects (Stanney, et al., 1998). While this seems a reasonable hypothesis for visual adaptation, it may not hold true for adaptation to mismatches between the senses or the perceptual-motor linkages.

In traditional psychophysical experiments, it is often the case that the variance of the measurements is initially quite high and reduces with continued data collection. It is recognized that subjects often need experience in the paradigm to adopt the best perceptual set and response strategy. Thus, some forms of adaptation to the VE might be expected to occur through perceptual-motor learning and some through changing cognitive problem-solving strategies. These modifications in perceptual set and response strategy would not be expected to result in debilitating aftereffects, and might best be considered VE-specific effects. Little research has been carried out on these important but indirect effects on visual processing in VE systems.

## 8. REFERENCES

Abrams, R. A., Oonk, H. M., & Pratt, J. (1998). Fixation point offsets facilitate endogenous saccades. *Perception and Psychophysics, 60*, 201–208.

Allman, J., Miezin, F., & McGuinness, E. (1985). Direction- and velocity-specific responses from beyond the classical receptive field in the middle temporal visual area (MT). *Perception, 14*, 105–126.

Ames, A., Jr. (1925). Illusion of depth from single pictures. *Journal of the Optical Society of America, 10*, 137–148.

Anderson, G. J., & Braunstein, M. L. (1985). Induced self-motion in central vision. *Journal of Experimental Psychology: Human Perception and Performance, 11*(2), 122–132.

Anstis, S. M. (1980). The perception of apparent movement. In H. C. Longuet-Higgins & N. S. Sutherland (Eds.), *The Psychology of Vision*. London: Royal Society.

Arakawa, Y. (1953). Quantitative measurements of visual fields for colors with a direct-current method. *American Journal of Ophthalmology, 36*, 1594–1601.

Aubert, H. (1886). Die Bewegungsempfindung. *Archiv fur die Gesamte Physiologie, 39*, 347–370.

Badcock, D. R., & Derrington, A. M. (1985). Detecting the displacement of periodic patterns. *Vision Research, 25*, 1253–1258.

Badcock, D. R., Hess, R. F., & Dobbins, K. (1996). Localization of element clusters: multiple cues. *Vision Research, 36*, 1467–1472.

Badcock, D. R., & Schor, C. (1985). The depth increment detection function within spatial channels. *Journal of the Optical Society of America, A2*, 1211–1216.

Badcock, D. R., & Westheimer, G. (1985). Spatial location and hyperacuity: The centre/surround localization contribution function has two substrates. *Vision Research, 25*, 1259–1267.

Becker, W. (1989). Metrics. In M. E. Goldburg & R. H. Wurtz (Eds.), *The Neurobiology of Saccadic Eye Movements*. Elsevier Science Publishers.

Boff, K. R., Kaufman, L., & Thomas, J. P. (Eds.). (1986). *Handbook of Human Perception* (2 Vols.). New York: John Wiley & Sons.

Boff, K. R., & Lincoln, J. E. (1988). *Engineering Data compendium: Human Perception and Performance*. , OH: Armstrong Aerospace Medical Research Laboratory.

Borish, I. (1970). *Clinical refraction* (3rd ed). Chicago: Professional Press.

Bourdon, B. (1902). *La perception visuelle de l'espace*. Paris: Schleicher Freres.

Boynton, R. M., Hayhoe, M. M., & MacLeod, D. I. A. (1977). The gap effect: Chromatic and achromatic visual discrimination as affected by field separation. *Optica Acta, 24*, 159–177.

Brandt, T., Dichgans, J., & Koenig, E. (1973). Differential effects of central versus peripheral vision on egocentric and exocentric motion perception. *Experimental Brain Research, 16*, 476–491.

Brandt, T., Wist, E. R., & Dichgans, J. (1975). Foreground and background in dynamic spatial orientation. *Perception and Psychophysics, 17*, 497–503.

Bridgeman, B., Hendry, D., & Stark, L. (1975). Failure to detect displacement of the visual world during saccadic eye movements. *Vision Research, 15*, 719–722.

Brown, W. R. J. (1951). The influence of luminance level on visual sensitivity to color differences. *Journal of the Optical Society, 41*, 684–688.

Burbeck, C. A., & Hadden, S. (1993). Scaled position integration areas: accounting for Weber's law for separation. *Journal of the Optical Society of America, A,10*, 5–15.

Burr, D. C., Morrone, M. C., & Ross, J. (1994). Selective suppression of the magnocellular visual pathway during saccadic eye movements. *Nature, 371*, 511–513.

Burr, D. C., Morrone, M. C., & Vaina, L. M. (1998). Large receptive fields for optic flow detection in humans. *Vision Research, 38*, 1731–1743.

Burr, D. C., & Ross, J. (1982). Contrast sensitivity at high velocities. *Vision Research, 22*, 479–484.

Campbell, F. W., & Robson, J. G. (1968). Application of Fourier analysis to the visibility of gratings. *Journal of Physiology* [London], *197,* 551–566.

Castiello, U., Badcock, D. R., & Bennett, K. (1999). Sudden and gradual presentation of distractor objects: Differential interference effects. *Experimental Brain Research, 128*(4), 550–556.

Cavanagh, P., & Favreau, O. E. (1985). Color and luminance share a common motion pathway. *Vision Research, 25*, 1595–1601.

Clarke, F. J. J. (1967, April). The effect of field-element size on chromaticity discrimination. *Proceedings of the Symposium on Colour Measurement in Industry, 1*, The Colour Group, April.

Collewijn, H., Erkelens, C. J., & Steinman, R. M. (1995). Voluntary binocular gaze-shifts in the plane of regard: Dynamics of version and vergence. *Vision Research, 35*, 3335–3358.

Cropper, S. J., & Badcock, D. R. (1994). Discriminating smooth from sampled motion: Chromatic and luminance stimuli. *Journal of the Optical Society of America, 11*, 515–530.

Derrington, A. M., & Badcock, D. R. (1985). Separate detectors for simple and complex grating patterns? *Vision Researchl, 25*, 1869–1878.

Dichgans, J., & Brandt, T. (1973). Optokinetic motion sickness and pseudo-Coriolis effects induced by moving visual stimuli. *Acta Otolaryngologica, 76*, 339–348.

Dichgans, J., Koener, F., & Voigt, K. (1969). Verleichende sdalierung des afferenten und efferenten bewegungssehen beim menschen: Lineare funktionen mit verschiedener ansteigssteilheit. *Psychologische Forschung, 32*, 277–295.

Duffy, C. J., & Wurtz, R. H. (1991). Sensitivity of MST neurons to optic flow stimuli. I. A continuum of response selectivity to large-field stimuli. *Journal of Neurophysiology, 65*, 1329–1345.

Edwards, M., & Badcock, D. R. (1993). Asymmetries in the sensitivity to motion in depth: A centripetal bias. *Perception, 22*, 1013–1023.

Erkelens, C. J., & Regan, D. (1986). Human ocular vergence movements induced by changing size and disparity. *Journal of Physiology* [London], *379*, 145–169.

Exner, S. (1875). Uber das sehen von bewegungen und die theorie des zusammengesetzen auges. *Sitzungsberichts Akademie Wissen Schaft Wein, 72*, 156–190.

Farnsworth, D. (1955). Tritanomalous vision as a threshold function. *Die Farbe, 4*, 185–197.

Fendick, M. G. (1983). Parameters of the retinal light distribution of a bright line which correspond to its attributed spatial location. *Investigative Ophthalmology and Visual Science, 24*, 92.

Field, D. J., Hayes, A., & Hess, R. F. (1993). Contour integration by the human visual system: Evidence for a local "association field." *Vision Research, 33*, 173–193.

Field, D. J., and Tolhurst, D. J. (1986). The structure and symmetry of simple-cell receptive-field profiles in the cat's visual cortex. *Proc R Soc Lond B Biol Sci., 228*, 379–400.

Filehne, W. (1922). Ueber das optische wahenehmen von bewegungen. *Zeitschrift fur Sinnesphysiologie, 53*, 134–145.

Forbes, B., & Klein, R. M. (1996). The magnitude of the fixation offset effect with endogenously and exogenously controlled saccades. *Journal of Cognitive Neuroscience, 8*, 344–352.

Graham, C. H., Baker, K. E., Hecht, M., & Lloyd, V. V. (1948). Factors influencing thresholds for monocular movement parallax. *Journal of Experimental Psychology, 38*, 205–223.

Graham, N. (1980). Spatial frequency channels in human vision: Detecting edges without edge detectors. In C. Harris (Ed.), *Visual coding and adaptability* (pp. 215–262). Hillsdale, NJ: Lawrence Erlbaum Associations.

Gregory, R. L. (1964). Human perception. *British Medical Bulletin, 20*, 21–26.

Grim, K. (1911). Uber diie genauigkeit der wahrnehmung und ausfuhrung von augenbeweg unger. *Zeitschrift fur Sinnesphysiologie, 45*, 9–26.

Gouras, P. (1991). The perception of color. *Vision and visual dysfunction* (Vol. 6), London: Macmillan.

Harris, L. R. (1999). Conference symposium on head-mounted displays. Association for Research in Vision and Ophthalmology, Fort Lauderdale, FL.

Held, R., Dichgans, J., & Bauer, J. (1975). Characteristics of moving visual areas influencing spatial orientation. *Vision Research, 15*, 357–365.

Hess, R. F., & Badcock, D. R. (1995). Metric for separation discrimination by the human visual system. *Journal of the Optical Society of America, A12*, 3–16.

Hood, D. C., & Finkelstein, M. A. (1979). Comparison of changes in sensitivity and sensation: Implications for the response-intensity function of the human photopic system. *Journal of Experimental Psychology: Human Perception and Performance, 5*, 391–405.

Howard, I. P., & Rogers, B. J. (1995). Binocular vision and stereopsis. Oxford, England: Oxford University Press.

Kelly, D. H., & Burbeck, C. A. (1980). Motion and vision. III. Stabilized pattern adaptation. *Journal of the Optical Society of America, 70*, 1283–1289.

Kelly, D. H., & Burbeck, C. A. (1984). Critical problems in spatial vision. *Critical Reviews in Biomedical Engineering, 10*, 125–177.

Kennedy, H., Magnin, M., & Jeannerod, M. (1976). Receptive field response of LGB neurons during vestibular stimulation in awake cats. *Vision Research, 16*, 119–120.

Koffka, K. (1922). Perception: An introduction to Gestalt theorie. *Psychological Bulletin, 19*, 531–585.

Kohler, I. (1972). Experiments with goggles. In R. Held & W. Richards (Eds.), *Perception: Mechanisms and models* (p. 390). San Francisco: W. H. Freeman and Co.

Lee, D. N. (1980). The optic flow field: The foundation of vision. *Philosophical Transactions of the Royal Society of London, 290*, 169–179.

Lee, D. N., & Lishman, J. R. (1975). Visual proprioceptive control of stance. *Journal of Human Movement Studies, 1*, 89–95.

Leibowitz, H. W. (1955). The relation between the rate threshold for the perception of movement and luminance for various durations of exposure. *Journal of Experimental Psychology, 49*, 209–214.

Levi, D. M., & Klein, S. A. (1990). The role of separation and eccentricity in encoding position. *Vision Research, 30*, 557–585.

Levi, D. M., Klein, S. A., & Aitsebaomo, A. P. (1985). Vernier acuity, crowding and cortical magnification. *Vision Research, 25*, 963–977.

Levi, D. M., & Schor, C. M. (1984). Spatial and velocity tuning of processes underlying induced motion. *Vision Research, 24*, 1189–1195.

Lestienne, F., Soechting, J. & Berthoz, A. (1977) Postural readjustments induced by linear motion of visual scenes. *Experimental Brain Research, 28*, 363–384.

Ludvigh, E., & Miller, J. W. (1958). Study of visual acuity during the ocular pursuit of moving test objects I. Introduction. *Journal of the Optical Society of America, 48*, 799–802.

Mach, E. (1875). *Grundlinien der Lehre von der Bewegungsempfindungen*. Leipzig, Germany: Engelmann.

Mack, A. (1986). Perceptual aspects of motion in the frontal plane. In K. R. Boff, L. Kaufman, & J. P. Thomas (Eds.), *Handbook of perception and human performance: sensory processes and perception* (Vol. 1). New York: John Wiley & Sons.

Mather, G., Verstraten, F., & Anstis, S. (Eds.). (1998). *The motion aftereffect: a modern perspective*. London: MIT Press.

Matin, L. (1986). Visual localization and eye movements. In K. R. Boff, L. Kaufman, & J. P. Thomas (Eds.), *Handbook of Human Perception and Performance* (Vol. 1, pp. 1–45). New York: John Wiley & Sons.

Metha, A. B., Vingrys, A. J., & Badcock, D. R. (1993). Calibration of a colour monitor for visual psychophysics. *Behavior Research Methods, Instruments, and Computers, 25*(3), 371–383.

McKee, S. P., & Levi, D. M. (1987). Dichoptic hyperacuity: the precision of nonius alignment. *Journal of the Optical Society of America, A4*, 1104–1108.

Millodot, M., Johnson, C. A., Lamont, A., & Leibowitz, H. A. (1975). Effect of dioptrics on peripheral visual acuity. *Vision Research, 15*, 1357–1362.

Mitchell, D. E., & Ware, C. (1974). Interocular transfer of a visual after-effect in normal and stereoblind humans. *Journal of Physiology* [London], *236*, 707–721.

Mon-Williams, M., Wann, J. P., & Rushton, S. (1993). Binocular vision in a Virtual world: Visual deficits following the wearing of a head-mounted display. *Ophthalmic and Physiological Optics 13,* 387–391.

Mullen, K. T. (1985). The contrast sensitivity of human colour vision to red–green and blue–yellow chromatic gratings. *Journal of Physiology, 359*, 381–400.

Nakayama, K., Shimojo, S., & Silverman, G. H. (1989). Stereoscopic depth: Its relation to image segmentation, grouping and the recognition of occluded objects. *Perception, 18*, 55–68.

Palmer, S. (1999). *Vision Science: Photons to Phenomonology*. Cambridge, MA: MIT Press.

Pelz, J. B., Ballard, D. H., & Hayhoe, M. M. (1993). Memory use during the performance of natural visuo-motor tasks. *Investigative Ophthalmology and Visual Science* [Suppl.], *34* (4).

Pitts, D. G. (1982). The effects of aging on selected visual functions: Dark adaptation, visual acuity, stereopsis, and brightness contrast. In R. Sekuler, D. W. Kline, & K. Dismulkes (Eds.), *Aging in human visual functions* (pp. 131–160). New York: A. R. Liss.

Post, R. B. (1988). Circular vection is independent of stimulus eccentricity. *Perception, 17*, 737–744.

Press, W. H., Teukolsky, S. A., Vetterling, W. T., & Flannery, B. P. (1992). Numerical recipes in C: The art of scientific computing. Cambridge, England: Cambridge University Press.

Reason, J., Wagner, H. & Dewhurst, D. (1981). A visually driven postural aftereffect. *Acta Psychologica, 48*, 241–251.

Regan, D., & Beverley, K. I. (1973). The dissociation of sideways movements from movements in depth: Psychophysics. *Vision Research, 13*, 2403–2415.

Regan, D., & Beverley, K. I. (1978a). Illusory motion in depth: Aftereffect of adaptation to changing size. *Vision Research, 18*, 209–212.

Regan, D., & Beverley, K. I. (1978b). Looming detectors in the human visual pathway. *Vision Research, 18*, 415–421.

Regan, D., & Tyler, C. W. (1971). Temporal summation and its limit for wavelength changes: An analog of Bloch's law for color vision. *Journal of the Optical Society of America, 61*, 1414–1421.

Riggs, L. A., Armington, J. C., & Ratliff, F. (1954). Motions of the retinal image during fixation. *Journal of the Optical Society of America, 44*, 315–321.

Rivest, J., & Cavanagh, P. (1996). Localizing contours defined by more than one attribute. *Vision Research, 36*, 53–66.

Robson, J. G. (1966). Spatial and temporal contrast-sensitivity functions of the visual system. *Journal of the Optical Society of America, 56*, 1141–1142.

Rogers, B., & Graham, M. (1982). Similarities between motion parallax and stereopsis in human depth perception. *Vision Research, 22*, 261–270.

Rock, I. (1975). *An introduction to perception*. London: Macmillan.

Rushton S., Mon-Williams, M., & Wann, J. (1994). Binocular vision in a bi-ocular world: New generation head-mounted displays avoid causing visual deficit. *Displays, 15*, 255–260

Sandström, C. I. (1951). *Orientation in the Present Space*. Stockholm: Almgvist & Wicksell.

Shlaer, S. (1937). The relation between visual acuity and illumination. *Journal of General Physiology, 21*, 165–188.

Siegal, M. H., & Siegal, A. B. (1972). Hue discrimination as a function of luminance. *Perception and Psychophysics, 12*, 295–299.

Solman, R. T., Dain, S. J., Lim, H. S., & May, J. G. (1995). Reading-related wavelength and spatial frequency effects in visual spatial location. *Ophthalmic and Physiological Optics, 15*, 125–132.

Solman, R. T., & May, J. G. (1990). Spatial localization discrepancies: A visual deficiency in poor readers. *American Journal of Psychology, 103*(2), 243–263.

Stark, L., Kong, R., Schwartz, S., Hendry, D., & Bridgeman, B. (1976). Saccadic suppression of image displacement. *Vision Research, 16*, 1185–1187.

Stanney, K. M., Salvendy, G., Deisigner, J., DiZio, P., Ellis, S., Ellison, E., Fogleman, G., Gallimore, J., Hettinger, L., Kennedy, R., Lackner, J., Lawson, B., Maida, J., Mead, A., Mon-Williams, M., Newman, D., Piantanida, T., Reeves, L., Riedel, O., Singer, M., Stoffregen, T., Wann, J., Welch, R., Wilson, J., Witmer, B. (1998). Aftereffects and sense of presence in virtual environments: Formulation of a research and development agenda. Report sponsored by the Life Sciences Division at NASA Headquarters. *International Journal of Human–Computer Interaction, 10*(2), 135–187.

Stern, R. M., Hu, S., Anderson, R. B., & Leibowitz, H. W. (1990). The effects of fixation and restricted visual field on vection-induced motion sickness. *Aviation, Space, and Environmental Medicine, 61*, 712–715.

Telford, L., & Frost, B. J. (1993). Factors affecting the onset and magnitude of linear vection. *Perception and Psychophysics, 53*, 682–692.

Thomas, J. P. (1975). Spatial resolution and spatial interaction. In E. C. Carterette & M. P. Friedman (Eds.), *Handbook of perception* (Vol. 5). New York: Academic Press.

Thompson, P. (1982). Perceived rate of movement depends on contrast. *Vision Research, 22*, 377–380.

Toet, A., van-Eekhout, M. P., Simons, H. L., & Koenderink, J. J. (1987). Scale invariant features of differential spatial displacement discrimination. *Vision Research, 27*, 441–451;

Turano, K. A., & Heidenreich, S. M. (1999). Eye movements affect the perceived speed of visual motion. *Vision Research. 39*, 1177–1188.

Tynan, P., & Sekuler, R. (1982). Motion processing in peripheral vision: Reaction time and perceived velocity. *Vision Research, 22*, 61–68.

Uchikawa, K., & Ikeda, M. (1981). Temporal deterioration of wavelength discrimination with successive comparison method. *Vision Research, 21*, 591–595.

Van Ness, F. L. and Bouman, M. A. (1967). Variation of contrast sensitivity with luminance. *Journal of the Optical Society of America, 57*, 401–406.

Wallach, H., & Kravitz, J.-H. (1965). Rapid adaption in the constancy of visual direction with active and passive rotation. *Psychonomic-Science, 3*(4), 165–166.

Wandell, B. A. (1995). *Foundations of vision*. Sunderland, MA: Sinauer Associates.

Watt, R. J. (1987). Scanning from coarse to fine spatial scales in the human visual system after the onset of a stimulus. *Journal of the Optical Society of America, A4*, 2006–2021.

Watt, R. J., Morgan, M. J., & Ward, R. M. (1983). The use of different cues in vernier acuity. *Vision Research, 23*, 991–995.

Watson, A. B., Ahumada, A. J., & Farrell, J. E. (1986). Window of visibility: a psychophysical theory of fidelity in time-sampled visual motion displays. *Journal of the Optical Society of America A-Optics and Image Science, 3*, 300–307.

Wertheim, A. H. (1994) Motion perception during self-motion: The direct versus inferential controversy revisited. *Behavioral and Brain Sciences, 17*, 293–355.

Westheimer, G. (1978). Spatial phase sensitivity for sinusoidal grating targets. *Vision Research, 18*, 1073–1074.

Westheimer, G. (1982). The spatial grain of the perifoveal visual field. *Vision Research, 22*, 157–162.

Westheimer, G., & McKee, S. P. (1977). Spatial configurations for visual hyperacuity. *Vision Research, 17*, 941–947.

Wilson, H. R., Ferrera, V. P., & Yo, C. (1992). A psychophysically motivated model for two-dimensional motion perception. *Visual Neuroscience, 9*, 79–97.

Wyszecki, G., & Stiles, W. S. (1967). *Color science: Concepts and methods, quantitative data and formulae*. New York: John Wiley & Sons.

Young, L. R., Dichgans, J., Murphy, R., & Brandt, T. (1973). Interaction of optokinetic and vestibular stimuli in motion perception. *Acta Otolaryngologica, 76*, 24–31.

Young, L. R., Oman, C. M., & Dichgans, J. M. (1975). Influence of head orientation on visually induced pitch and roll sensation. *Aviation, Space, and Environmental Medicine, 46*, 264–269.

# 4

# Virtual Auditory Displays

Russell D. Shilling[1] and Barbara Shinn-Cunningham[2]
[1]*Naval Postgraduate School*
*Operations Research/Modeling, Virtual Environments and Simulation (MOVES)*
*Monterey, CA 93943*
*shilling@cs.nps.navy.mil*
[2]*Boston University*
*Depts. of Cognitive and Neural Systems and Biomedical Engineering*
*Hearing Research Center*
*677 Beacon St.*
*Boston, MA 02215*
*shinn@cns.bu.edu*

## 1. INTRODUCTION

Auditory processing is often given minimal attention when designing virtual environments (VEs) or simulations. This lack of attention is unfortunate, because auditory cues play a crucial role in everyday life. Auditory cues increase awareness of surroundings, cue visual attention, and convey a variety of complex information without taxing the visual system. The entertainment industry has long recognized the importance of sound to create ambience and emotion, aspects that are often lacking in virtual environments. In short, placing someone in a virtual world with an improperly designed auditory interface is equivalent to creating a "virtual" hearing impairment for the user.

Auditory perception, especially localization, is a complex phenomenon affected by physiology, expectation, and even the visual interface. This chapter will consider different methods for creating auditory interfaces. As will be discussed, spatialized audio using headphones is the only audio technique that is truly "virtual" because it reproduces azimuth, elevation, and distance, and offers the sound engineer the greatest amount of control over the auditory experience of the listener. For many applications, especially using projection screens, speaker systems may be simpler to implement and provide benefits not available to headphone systems. Properly designed speaker systems, especially using subwoofers, may contribute to emotional context. The positives and negatives associated with each option will be discussed.

It is impossible to include everything that needs to be known about designing auditory interfaces in a single chapter. The current aim is to provide a starting point and essential theory behind implementing sound in a VE without overwhelming the novice designer. Instead of trying to review all perceptual and technical issues related to creating virtual auditory displays, this chapter focuses on the essential aspects of spatial auditory perception and the generation of spatial auditory cues as they relate to virtual environments. In addition, unlike the visual channel, very little effort has been put into formulating theories concerning the creation of synthetic sound sources in virtual environments; consequently, the question of how to generate realistic sounds (rather than using sources from some precomputed, stored library of source sounds) is not discussed in this chapter.

Physical properties of sound and basic psychoacoustics are discussed with an emphasis on spatial hearing. Finally, general techniques for producing auditory stimuli (with and without spatial cues) will be discussed. A complete lexicon for understanding and developing auditory displays can be found in Letowski et al. (2000). The technology involved in producing spatialized audio is rapidly changing, and new products are continually introduced to the market while others are removed. Any specific recommendations would quickly be dated. However, a brief overview of current technology and solutions is discussed at the conclusion of the chapter.

## 1.1 Why Are Virtual Auditory Interfaces Important?

### *1.1.1 Environmental Realism and Ambience*

If it does nothing else, an auditory interface should convey basic information about the VE to the user. For instance, in the real world, dismounted soldiers are aware of the sound of their own footsteps, the sounds of other soldiers and the natural environment, and mechanical sounds from various types of equipment such as jeeps, artillery, and rifles. In "control room" situations such as nuclear power plants, air traffic control centers, or the bridge of a ship, sounds such as alarms, switches being toggled, and verbal communications with other people in the room (including sounds of stress or uncertainty) provide vital information for participants. The location of these voices, switches, and alarms also provides information concerning their function and importance. In the absence of these basic auditory cues, situational awareness is severely degraded. The same is true in virtual environments.

The entertainment industry has recognized that sound is a vital aspect of creating ambience and emotion for films. George Lucas, best recognized by the public for stunning visual effects in his movies, has stated that sound is 50% of the movie experience (THX, 2000). In virtual environments, the argument is often erroneously made that sound is secondary. Again, a compelling visual representation of soldiers slogging through the mud, traversing a jungle, or going house-to-house can be created. However, unless appropriate background sounds are provided (machinery, artillery, animals, footsteps, etc), the participant will probably feel detached from the action. The sound of footsteps is different depending on whether you are in grass, on pavement, or in a hallway. Likewise, the action of an M-16 sounds different depending on whether you are inside a room or outside. These are the types of things that create ambience and emotion in film; the same should hold true in virtual environments.

### *1.1.2 Presence/Immersion*

Presence (see chap. 40, this volume) can be defined as the "sense of being immersed in a simulation or virtual environment." Such a nebulous concept is difficult to quantify. Although definitive evidence is lacking, it is generally believed that the sense of presence is dependent on auditory, visual, and tactile fidelity (Sheridan, 1996). Referring back to the previous section, it can be inferred that as environmental realism increases, the sense of presence increases. However, although realism probably contributes to the sense of presence, the inverse is not true. It has been demonstrated that, although virtual or spatial audio does not necessarily increase the perceived realism of a virtual environment, it does increase the sense of presence

(Hendrix & Barfield, 1996). Thus, if implemented properly, appropriately designed audio increases the overall sense of presence in a VE or simulation.

#### *1.1.3 Collision/Tactile Cueing*

Auditory cues in a system using head-mounted displays (HMDs) are particularly important and should be given close attention during the design phase. In limited field-of-view (FOV) HMDs, sounds associated with collision detection may play a major role in the user being able to successfully move through the simulation. These auditory collision cues can be used to substitute for tactile collision cues when appropriate.

#### *1.1.4 Cross-modal Enhancement*

Response times to visual targets associated with localized auditory cues have been shown to decrease (Perrott, Saberi, Brown, & Strybel, 1990; Perrott, Sadralodabai, Saberi, & Strybel, 1991). It has also been shown that the latency of saccadic eye movements toward a target is reduced when the visual target is aligned with an auditory cue (Frens, Opstal, & Willigen, 1995). In this manner, a properly designed auditory interface may be used to enhance the FOV of an HMD by cueing the user to locations outside the limited FOV of the HMD. Appropriate care must be taken to properly design the auditory interface, as auditory location has been shown to affect perceived visual target location when the visual target is presented on a nontextured background (Radeau & Bertelson, 1976).

#### *1.1.5 Cocktail Party Effect*

In a multisource sound environment, it is easier to discriminate and comprehend sounds if they are separated in space. This enhancement in comprehension with spatially disparate sound sources is often called the *cocktail party effect* (Yost, 1994). The ability to understand multiple speakers simultaneously can be useful when applied to multiuser environments such as teleconferencing (Begault, 1999) or multichannel radio communications (Begault, 1993; Begault & Wenzel, 1992; Haas, Gainer, Wightman, Couch, & Shilling, 1997). Even when the perception of location is not optimal, communication is improved in multichannel situations when spatialized auditory displays are employed (Drullman & Bronkhorst, 2000).

#### *1.1.6 Sonification of Data*

Auditory cues can be used to represent information that is not normally available in the real world. For instance, if a user makes a response error, or is slow in responding, auditory icons can be created to indicate the deficiency to the user (Gaver, 1986). On the other hand, using a technique called *sonification,* complex information presented visually can be simplified by supplementing it with an auditory representation correlated with some dimension or dimensions of the data set (Rabenhorst, Farrell, & Jameson, 1990). For example, in simulations designed to teach operators how to interpret complex controls and displays (radar, sonar, etc), multidimensional data can be "sonified" to make it easier to interpret. Likewise, sonification could be used as a learning aid to illustrate principles using multidimensional data in physics and chemistry.

#### *1.1.7 Supernormal Auditory Localization*

When designing auditory interfaces, it is possible to exaggerate normal auditory cues so that the listener is able to localize sound in the virtual world with better resolution than in the real world. One way of achieving supernormal localization is to exaggerate the normal head and ear size cues available to the listener (e.g., see Shinn-Cunningham, Durlach, & Held, 1998a, 1998b). Supernormal auditory localization may be of best use in teleoperator applications, and VEs where there is a need to compensate for a limited-FOV HMD. For example, spatialized auditory displays have been designed that exaggerate normal auditory cues and zoom the auditory display as the magnification on the visual display is increased (Shilling, Wightman, Couch, Beutler, & Letowski, 1998; Shilling & Letowski, 2000). Of course, experience with such supernormal systems affects both the accuracy and resolution of perception, and both effects must be evaluated in order to determine the overall cost and benefit of such techniques for a given application (e.g., see Shinn-Cunningham, 2000a).

#### *1.1.8 Virtual Auditory Displays (VADs)*

While graphical displays are an obvious choice for displaying spatial information to a human operator (particularly after considering the spatial acuity of the visual channel), the visual channel is often overloaded, with operators monitoring a myriad of dials, gauges, and graphic displays. In these cases, spatial auditory cues can provide invaluable information to an operator, particularly when the visual channel is saturated (Begault, 1993; Bronkhorst, Veltman, & van Breda, 1996; Shilling & Letowski, 2000). Spatial auditory displays are also being developed for use in applications for which visual information provides no benefit, for instance, in limited FOV applications or when presenting information to the blind. In command/control applications, the primary goal is to convey unambiguous information to the human operator. Realism, per se, is not useful, except to the extent that it makes the operator's task easier (i.e., reduces the workload). Conversely, spatial resolution is critical. In these applications, signal-processing schemes that could enhance the amount of information transferred to the human operator may be useful, even if the result is "unnatural," as long as the user is able to extract this information (e.g., see Durlach, Shinn-Cunningham, & Held, 1993). It should be noted that when designing spatialized auditory displays for noisy environments such as cockpits, electronic noise cancellation technology should be employed and user's hearing loss taken into account to make certain the displays are localizable (Begault, 1996). Also, for high-g environments, more work needs to be conducted to discover the contribution of g-forces to displacements in sound localization, the so-called *audiogyral illusion* (Clark & Graybiel, 1949; DiZio, Held, Lackner, Shinn-Cunningham, & Durlach, 2000).

#### *1.1.9 Enhancement of Perceived Quality of a Simulation*

The importance of multimodal interactions involving the auditory system cannot be ignored (see chaps. 14, 21, and 22, this volume). It has been shown that using medium- and high-quality auditory displays can enhance the perception of quality in visual displays. Inversely, using low-quality auditory displays reduces the perceived quality of visual displays (Storms, 1998; see also chap. 22, this volume).

## 2. PHYSICAL ACOUSTICS

### 2.1 Properties of Sound

Simply put, sound is a pressure wave produced when an object vibrates rapidly back and forth. The diaphragm of a speaker produces sound by pushing against molecules of air, thus creating an area of high pressure (*condensation*). As the speaker's diaphragm returns to its resting position, it creates an area of low pressure (*rarefaction*). This localized disturbance travels through the air as a wave of alternating low pressure and high pressure at approximately 344 m/sec or 1128 ft/sec (at 70 °F) depending on temperature and humidity.

### 2.1.1 Frequency

If the musical note "A" is played as a pure sinusoid, there will be 440 condensations and rarefactions per second. The distance between two adjacent condensations or rarefactions equals the wavelength of the sound wave and is typically represented by the symbol $\lambda$. The velocity at which the sound wave is traveling is denoted as $c$. The time one full oscillatory cycle (condensation through rarefaction) takes is called the frequency ($f$) and is expressed in Hertz or cycles per second. The relationship between frequency, velocity, and wavelength is given by $f = c/\lambda$.

From a modeling standpoint, this relationship is important when considering Doppler Shift. As a sound source is moving toward a listener, the perceived frequency increases because the wavelength is compressed as a function of the velocity ($v$) of the moving source. This compression can be explained by the equation $\lambda = (c - v) / f$. For negative velocities (i.e., for sources moving away), this expression describes a relative increase in the wavelength (and a concomitant decrease in frequency).

### 2.1.2 Intensity

The amplitude of the waveform determines the intensity of a sound stimulus. Intensity is measured in decibels (dB). Decibels give the level of sound (on a logarithmic scale) relative to some reference level. One common reference level is $2 \times 10^{-5}$ N/m$^2$. Decibels referenced to this value are commonly used to describe sound intensity expressed in units of dB sound pressure level (SPL). The sound level in dB SPL can be computed by the following equation:

$$\text{dB } SPL = 20 \log_{10} \left( \frac{\text{RMS sound pressure}}{20 * 10^{-5}\text{N/m}^2} \right)$$

The threshold of hearing is in the range of 0 to 10 dB SPL for most sounds, although the actual threshold depends on the spectral content of the sound. When measuring sound intensity in the "real world," intensity is measured using a sound pressure meter. Most sound pressure meters allow one to collect sound level information using different scales which weight energy in different frequencies differently in order to approximate the sensitivity of the human auditory system to sound at low-, moderate-, or high-intensity levels. These scales are known as A, B, and C weighted scales, respectively. The B scale is rarely used; however, the C scale (dBC) is useful for evaluating noise levels in high intensity environments such as traffic noise and ambient cockpit noise. The frequency response of the dBC measurement is closer to a flat-response than dBA. In fact, when conducting "sound surveys" in a complex noise environment, it is prudent to measure sound level in both dBA and flat-response (or dBC) to make an accurate assessment of the audio environment.

Frequency, intensity, and complexity are physical properties of an acoustic waveform. The perceptual analogues for frequency, intensity, and complexity are pitch, loudness, and timbre, respectively. This distinction between physical and perceptual measures of sound properties is an important one. Thus, it is critical to consider both physical and perceptual descriptions when designing auditory displays.

## 3. PSYCHOPHYSICS

The basic sensitivity of the auditory system is reviewed in detail in a number of textbooks (e.g., see Gelfand, 1998; Moore, 1997; Yost, 1994). This section provides a brief overview of some aspects of human auditory sensitivity that are important to consider when designing auditory virtual environments.

### 3.1 Frequency Analysis in the Auditory System

In the cochlea, acoustic signals are broken down into constituent frequency components by a mechanical Fourier-like analysis. Along the length of the cochlea, the frequency to which that section of the cochlea responds varies systematically from high to low frequencies. The strength of neural signals carried by the auditory nerve fibers arrayed along the length of the cochlea varies with the mechanical displacement of the corresponding section of the cochlea. As a result, each nerve fiber can be thought of as a frequency channel that conveys information about the energy and timing of the input signal within a restricted frequency region. At all stages of the auditory system, these multiple frequency channels are in evidence.

Although the bandwidth changes with the level of the input signal and with input frequency, to a crude first order approximation, one can think of the frequency selectivity of the auditory system as constant on a log-frequency basis (approximately one/third–octave wide). Thus, a particular auditory nerve responds to acoustic energy at and near a particular frequency.

Humans are sensitive to acoustic energy at frequencies between about 20 Hz and 22,000 Hz. Absolute sensitivity varies with frequency. Humans are most sensitive to energy at frequencies around 2000 Hz, and are less sensitive for frequencies below and above this range.

The fact that input waveforms are deconstructed into constituent frequencies affects all aspects of auditory perception. Many behavioral results are best understood by considering the activity of the auditory nerve fibers, each of which responds to energy within about a third of an octave of its particular "best frequency." For instance, the ability to detect a sinusoidal signal in a noise background degrades dramatically when the noise spectrum is within a third octave of the sinusoid frequency. When a noise is spectrally remote from a sinusoidal target, it causes much less interference with the detection of the sinusoid. These factors are important when one considers the spectral content of different sounds that are to be used in an auditory virtual environment. For instance, if one must monitor multiple kinds of alerting sounds, choosing signals that are spectrally remote from one another will improve a listener's ability to detect and respond to different signals.

### 3.2 Intensity Perception

Listeners are sensitive to sound intensity on a logarithmic scale. For instance, doubling the level of a sound source causes roughly the same perceived change in the loudness independent of the reference level. This logarithmic sensitivity to sound intensity gives the auditory system a large dynamic range. For instance, the range between just detectable sound levels and sounds that are so loud that they cause pain is roughly 110 to 120 dB (i.e., an increase in sound pressure by a factor of a million). The majority of the sounds encountered in everyday experience span a dynamic intensity range of 80 to 90 dB. Typical sound reproduction systems use 16 bits to represent the pressure of the acoustic signal (providing a useful dynamic range of about 90 dB), which is sufficient for most simulations.

While sound intensity (a physical measure) affects the loudness of a sound (a perceptual measure), loudness does not grow linearly with intensity. In addition, the same decibel increase in sound intensity can result in different increments in loudness, depending on the frequency content of the sound. Thus, intensity and loudness, while closely related, are not equivalent descriptions of sound.

### 3.3 Masking Effects

As mentioned above, when multiple sources are presented to a listener simultaneously or in rapid succession, the sources interfere with one another in various ways. For instance, a tone that is audible when played in isolation may be inaudible when a loud noise is presented simultaneously. Such effects (known as "masking" effects) arise from a variety of mechanisms, from physical interactions of the separate acoustic waves impinging on the ear to high-level, cognitive factors. For a more complete description of these effects than is given below, see Yost (1994, pp. 153–167) or Moore (1997, pp. 111–120).

*Simultaneous masking* occurs when two sources are played concurrently. However, signals do not have to be played simultaneously for them to interfere with one another perceptually. For instance, both "forward" masking (in which a leading sound interferes with perception of a trailing sound) and "backward" masking (in which a lagging sound interferes with perception of a leading sound) occur. Generally speaking, many simultaneous and forward masking effects are thought to arise from peripheral interactions that occur at or before the level of the auditory nerve. For instance, the mechanical vibrations of the basilar membrane are nonlinear, so that the response of the membrane to two separate sounds may be less than the sum of the response to the individual sounds. These nonlinear interactions can suppress the response to what would (in isolation) be an audible event. In other words, because of nonlinearities in the transduction of acoustic signals, the response of the auditory nerve to a given signal may be less robust when a second signal is present.

Other, more central factors influence masking as well. For instance, backward masking may reflect higher order processing that limits the amount of information extracted from an initial sound in the presence of a second sound. The term *informational masking* refers to all masking that cannot be explained by peripheral interactions in the transduction of sound by the auditory periphery. There is ample evidence that informational masking is an important factor in auditory perception. For instance, perceptual sensitivity in discrimination and detection tasks is often degraded when there is uncertainty about the characteristics of a target source (e.g., see Yost, 1994, pp. 219–220).

### 3.4 Pitch and Timbre

Just as sound intensity is the physical correlate of the percept of loudness, source frequency is most closely related to the percept of pitch. For sound waves that are periodic (including pure sinusoids, for instance), the perceived "pitch" of a sound is directly related to the inverse of the period of the sound signal. Thus, sounds with low pitch have relative long periods and sounds with high pitch generally have short periods. Many real-world sounds are not strictly periodic in that they have a temporal pattern that repeats over time, but has fluctuations from one cycle to the next. Examples of such pseudo-periodic signals include the sound produced by a flute or a vowel sound spoken aloud. The perceived pitch of such sounds is well predicted by the average period of the cyclical variations in the stimulus.

The percept of pitch is not uniformly strong for all sound sources. In fact, nonperiodic sources such as noise do not have a salient pitch associated with them. For relatively narrow sources that are aperiodic, a weak percept of pitch can arise that depends on the center frequency of the spectral energy of the signal. In fact, perceived pitch is affected by a wide variety of stimulus attributes, including temporal structure, frequency content, harmonicity, and even loudness. Although the pitch of a pure sinusoid is directly related to its frequency, there is no single physical parameter that can predict perceived pitch for more complex sounds. Nonetheless, for many sounds, pitch is a very salient and robust perceptual feature that can be used to convey information to a listener. For instance, in music pitch conveys melody. In speech, pitch conveys a variety of information (ranging from the gender of a speaker to paralinguistic, emotional content of a speech).

The percept of timbre is the sound property that enables a listener to distinguish an oboe from a trumpet. Like pitch, the percept of timbre depends on a number of physical parameters of sound, including spectral content and temporal envelope (such as the abruptness of the onset and offset of sound).

## 3.5 Temporal Resolution

The auditory channel is much more sensitive to temporal fluctuations in sensory inputs than the visual (see chap. 3, this volume) or proprioceptive (see chaps. 5 and 6, this volume) channels. For instance, the system can detect amplitude fluctuations in input signals up to 50 Hz (i.e., a duty cycle of 20 ms) very easily (e.g., see Yost, 1994, pp. 146–149). Sensitivity degrades slowly with increasing modulation rate, so that some sensitivity remains even as the rate approaches 1000 Hz (i.e., temporal fluctuations at a rate of 1 per ms). The system is also sensitive to small fluctuations in the spectral content of an input signal for roughly the same modulation speeds. Listeners not only can detect rapid fluctuations in an input stimulus, they can react quickly to auditory stimuli. For instance, reaction times to auditory stimuli are faster than for vision by 30 to 40 ms (an improvement of roughly 20%, e.g., see Welch & Warren, 1986).

## 3.6 Spatial Hearing

Spatial acuity of the auditory system is far worse than that of the visual (see chap. 3, this volume) or proprioceptive (see chaps. 5 and 6, this volume) systems. For a listener to detect an angular displacement of a source from the median plane, the source must be displaced laterally by about a degree. For a source directly to the side, the listener does not always detect a lateral displacement of 10 degrees. Auditory spatial acuity is even worse in other spatial dimensions. A source in the median plane must be displaced by as much as 15 degrees for the listener to perceive the directional change accurately. While listeners can judge relative changes in source distance, absolute distance judgments are often surprisingly inaccurate even under the best of conditions.

Functionally, spatial auditory perception is distinctly different from that of the other "spatial" senses of vision and proprioception. For the other spatial senses, position is neurally encoded at the most peripheral part of the sensory system. For instance, the photoreceptors of the retina are organized topographically so that a source at a particular position (relative to the direction of gaze) excites a distinct set of receptors (see chap. 3, this volume). In contrast, spatial information in the auditory signals reaching the left and right ears of a listener must be computed from the peripheral neural representations. The way in which spatial information is carried by the acoustic signals reaching the eardrums of a listener has been the subject of much research. This section provides a brief review of how acoustic attributes convey spatial information to a listener and how the perceived position of a sound source is computed in the brain (for more complete reviews, see Blauert, 1997; Middlebrooks & Green, 1991; Mills, 1972; Wightman & Kistler, 1993).

### *3.6.1 Binaural Cues*

The most robust cues for source position depend on differences between the signals reaching the left and right ears. Such *interaural* or *binaural* cues are robust specifically because they can be computed by comparing the signals reaching each ear. As a result, binaural cues allow a listener to factor out those acoustic attributes that arise from source content from those attributes that arise from source position.

Depending on the angle between the interaural axis and a sound source, one ear may receive a sound earlier than the other. The resulting interaural time differences (ITDs) are the main cue indicating the laterality (left/right location) of the direct sound. The ITD grows with the angle of the source from the median plane; for instance, a source directly to the right of a listener results in an ITD of 600 to 800 microseconds ($\mu$s) favoring the right ear. ITDs are most salient for sound frequencies below about 2 kHz, but occur at all frequencies in a sound. At higher frequencies, listeners use ITDs in signal *envelopes* to help determine source laterality, but are insensitive to differences in the interaural phase of the signal.

Listeners can reliably detect ITDs of 10 to 50 microseconds (depending on the individual listener), which grossly correspond to ITDs that would result from a source positioned 1 to 5 degrees from the median plane. Sensitivity to changes in the ITD deteriorates as the reference ITD gets larger. For instance, the smallest detectable change in ITD around a reference source with an ITD of 600 to 800 microseconds (corresponding to the ITD of a source far to the side of the head) can be more than a factor of 2 larger than for a reference with zero ITD.

At the high end of the audible frequency range, the head of the listener reflects and diffracts signals so that less acoustic energy reaches the far side of the head (causing an *acoustic head shadow*). Due to the acoustic head shadow, the relative intensity of a sound at the two ears varies with the lateral location of the source. The resulting interaural intensity differences (IIDs) generally increase with source frequency and angle between the source and median plane. IIDs are perceptually important for determining source laterality for frequencies above about 2 kHz.

When a sound source is within reach of a listener, extra large IIDs (at all frequencies) arise due to differences in the relative distances from source to left and right ears (e.g., see Brungart & Rabinowitz, 1999; Duda & Martens, 1998). These additional IIDs are due to differences in the relative distances from source to left and right ears and help to convey information about the relative distance and direction of the source from the listener (Shinn-Cunningham, Santarelli, & Kopco, 2000). Other low-frequency IIDs that may arise from the torso appear to help determine the elevation of a source (Algazi, Avendano, & Duda, 2001). Most listeners are able to detect IIDs of 0.5 to 1 dB, independent of source frequency.

The perceived location of a sound source usually is consistent with the ITD and IID information available. However, there are multiple source locations that cause roughly the same ITD and IID cues. For sources more than a meter from the head, the locus of such points is approximately a hyperbolic surface of rotation symmetric about the interaural axis that is known as the *cone of confusion* (see left side of Fig. 4.1). When a sound is within reach of the listener, extra large IIDs provide additional robust, binaural information about the source location. For a simple spherical head model, low-frequency IIDs are constant on spheres centered on the interaural axis (see right side of Fig. 4.1). The rate at which the extra-large IID changes with spatial location decreases as sources move far from the head or near the median plane. In fact, once a source is more than a meter or so from the head, the contribution of this "near-field" IID is perceptually insignificant (Shinn-Cunninhgam, Santarelli, & Kopco, 2000). In general, positions that give rise to the same binaural cues (i.e., the intersection of constant ITD and IID contours) form a circle centered on the interaural axis (Shinn-Cunningham et al., 2000). Since ITD and IID sensitivity is imperfect, the locus of positions that cannot be resolved from binaural cues may be more accurately described as a *torus of confusion* centered on the interaural axis. Such tori of confusion degenerate to the more familiar cones of confusion for sources more than about a meter from a listener.

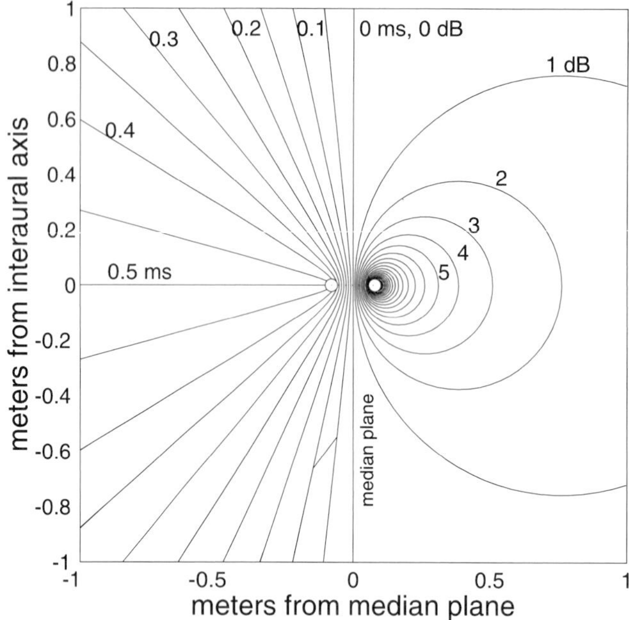


FIG. 4.1. Iso-ITD (left side of figure) and iso-IID (right side of figure) contours ofr sources near the head. On the left, sources at each location along a contour give rise to nearly the same ITD. On the right, sources at each location along a contour give rise to nearly the same unique near-field ILD component.

### 3.6.2 *Spectral Cues*

The main cue to resolve source location on the torus of confusion is the spectral content of signals reaching the ears. These spectral cues arise due to interactions of the outer ear (pinna) with the impinging sound wave that depend on the relative position of sound source and listener's head (Batteau, 1967). Spectral cues only occur at relatively high frequencies, generally above 6 kHz. Unlike interaural cues for source location, spectral cues can be confused with changes in the spectrum of the source itself. Perhaps because of this ambiguity, listeners are more likely to make localization errors in which responses fall near the correct torus of confusion, but are not in the right direction. Individual differences in spectral filtering of the pinnae are large and are important when judging source direction (e.g., see Wenzel, Arruda, Kistler, & Wightman, 1993).

### 3.6.3 *Anechoic Distance Cues*

In general, the intensity of the direct sound reaching a listener (i.e., sound that does not come off of reflective surfaces like walls) decreases with the distance of the source. In addition, the atmosphere absorbs energy in high, audible frequencies as a sound propagates, causing small changes in the spectrum of the received signal with changes in source distance. If a source is unfamiliar, the intensity and spectrum of the direct sound are not robust cues for distance because they can be confounded with changes in the intensity or spectral content (respectively) of the signal emitted from the source. However, even for unfamiliar sources, overall level and spectral content provide relative distance information (Mershon, 1997).

### 3.6.4 Reverberation

Reverberation (acoustic energy reaching a listener from indirect paths, via walls, floors, etc.) generally has little affect on or degrades the perception of source direction (e.g., see Begault, 1993; Hartmann, 1983; Shinn-Cunningham, Zurek, & Durlach, 1993; Shinn-Cunningham, 2000b). However, it actually aids in the perception of source distance (e.g., see Mershon, Ballenger, Little, McMurtry, & Buchanan, 1989; Shinn-Cunningham, Kopco, & Santarelli, 1999). At least grossly, the intensity of reflected energy received at the ears is independent of the position of the source relative to the listener (although it can vary dramatically from one room to another). As a result, the ratio of direct to reverberant energy provides an absolute measure of source distance for a given listening environment.

Reverberation not only provides a robust cue for source distance, it provides information about the size and configuration of a listening environment. For instance, information about the size and "spaciousness" of a room can be extracted from the pattern of reverberation in the signals reaching the ears. While many psychophysical studies of sound localization are performed in anechoic (or simulated anechoic) environments, reverberation is present (in varying degrees) in virtually all normal listening conditions. Anechoic environments (such as those used in many simulations and experiments) seem subjectively unnatural and strange to naive listeners. Conversely, adding reverberation to a simulation causes all sources to seem more realistic and provides robust information about relative source distance. While reverberation may improve distance perception and improve the realism of a display, it can decrease accuracy in directional perception, albeit slightly and may interfere with the ability to extract information in a source signal (e.g., degrade speech) and to attend to multiple sources (e.g., see section 3.6.8).

### 3.6.5 Dynamic Cues

In addition to "static" acoustic cues like ITD and IID, changes in spatial cues with source or listener movement also influence perception of source position and help to resolve torus-of-confusion ambiguities (e.g., see Wallach, 1940; Wightman & Kistler, 1989). For instance, either a source directly in front or directly behind a listener would cause near zero ITDs and IIDs; however, a leftward rotation of the head results in either ITDs and IIDs favoring the right ear (for a source in front) or the left ear (for a source behind).

While the auditory system generally has good temporal resolution, the temporal resolution of the binaural hearing system is much poorer. For instance, investigations into the perception of moving sound sources implies that binaural information averaged over a time window lasting 100 to 200 ms results in what has been termed *binaural sluggishness* (e.g., see Grantham, 1997; Kollmeier & Gilkey, 1990).

### 3.6.6 Effects of Stimulus Characteristics on Spatial Perception

Characteristics of a source itself affect auditory spatial perception in a number of ways. For instance, the bandwidth of a stimulus can have a large impact on the perceived location of a source. As a result, one must consider how nonspatial attributes of a source in a VE will impact spatial perception of a signal. In cases where one can design the acoustic signal (i.e., if the signal is a warning signal or some other arbitrary waveform), these factors should be taken into consideration when one selects the source signal.

For instance, the spectral filtering of the pinnae cannot be determined if the sound source does not have sufficient bandwidth. This makes it difficult to unambiguously determine the location of a source on the torus of confusion for a narrowband signal. Similarly, if a source signal does not have energy above about 5 kHz, spectral cues will not be represented in the signals reaching the ears and errors along the torus of confusion are more common (e.g., Gilkey & Anderson, 1995).

Ambiguity in narrowband source locations arises in other situations as well. For instance, narrowband, low-frequency signals in which ITD is the main cue can have ambiguity in their heard location because the auditory system is only sensitive to interaural phase. Thus, a low-frequency sinusoid with an ITD of half cycle favoring the right ear may also be heard far to the left side of the head. However, binaural information is integrated across frequency so that ambiguity in lateral location is resolved when interaural information is available across a range of frequencies (Brainard, Knudsen, & Esterly, 1992; Stern & Trahiotis, 1997; Trahiotis & Stern, 1989). When narrowband sources are presented, the heard location is strongly influenced by the center frequency of the source (Middlebrooks, 1997).

While spectral bandwidth is important, temporal structure of a source signal is also important. In particular, onsets and offsets in a signal make source localization more accurate, particularly when reverberation and echoes are present. A gated or modulated broadband noise will generally be more accurately localized in a reverberant room (or simulation) than a slowly gated broadband noise (e.g., see Rakerd & Hartmann, 1985; Rakerd & Hartmann, 1986).

### *3.6.7 Top-down Processes in Spatial Perception*

Experience with or knowledge of the acoustics of a particular environment also affects auditory localization (e.g., see Clifton, Freyman, & Litovsky, 1993; Shinn-Cunningham, 2000b). Results show that "top-down" processing of auditory information due to implicit learning and experience affects performance. In other words, spatial auditory perception is not wholly determined by stimulus parameters, but also by the state of the listener. Although such effects are not due to conscious decision, they can measurably alter auditory localization and spatial perception.

### *3.6.8 Benefits of Binaural Hearing*

Listeners benefit from receiving different signals at the two ears in a number of ways. As discussed above, ITD and IID cues allow listeners to determine the location of sound sources. However, in addition to allowing listeners to locate sound sources in the environment, binaural cues allow the listener to selectively to attend to sources coming from a particular direction. This ability is extremely important when there are multiple competing sources in the environment (e.g., see Bronkhorst, 2000).

Imagine a situation in which there is both a speaker (whom the listener is trying to attend) and a competing source (that is interfering with the speaker). If the speaker and competitor are both directly in front of the listener, the competitor degrades speech reception much more than if the competitor is off to one side, spatially separated from the speaker. This "binaural advantage" arises in part because when the competitor is to one side of the head, the energy from the competitor is attenuated at the far ear. As a result, the signal-to-noise ratio at the far ear is larger than when the competitor is in front. In other words, the listener has access to a cleaner signal in which the speaker is more prominent when the speaker and noise are spatially separated. However, the advantage of the spatial separation is even larger than can be predicted on the basis of energy: Spatial information can be used to "squelch" signals from directions other than the direction of interest.

A homologous benefit can be seen under headphones. In particular, if one varies the level of a signal until it is just detectable in the presence of a masker, the necessary signal level is much lower when the ITD of the signal and masker are different than when they are the same. The difference between these thresholds, referred to as the *masking level difference* (MLD), can be as large as 10 to 15 dB for some signals (e.g., see Durlach & Colburn, 1978; Zurek, 1993).

The binaural advantage affects both signal detection (e.g., see Gilkey & Good, 1995) and speech reception (e.g., see Bronkhorst & Plomp, 1988). It is one of the main factors contributing to the ability of listeners to monitor and attend multiple sources in complex listening environments (i.e., the "cocktail party effect"; see, for example, Yost, 1997). Thus, the binaural advantage is important for almost any auditory signal of interest. In order to get these benefits of binaural hearing, signals reaching a listener must have appropriate ITDs and/or IIDs.

#### 3.6.9 Adaptation to Distorted Spatial Cues

While a naive listener responds to ITD, IID, and spectral cues based on their everyday experience, listeners can learn to interpret cues that are not exactly like those that occur naturally. For instance, listeners can learn to adapt to unnatural spectral cues when given sufficient long-term exposure (Hofman, Van Riswick, & Van Opstal, 1998). Short-term training allows listeners to learn how to map responses to spatial cues to different spatial locations than normal (Shinn-Cunningham et al., 1998a). These studies imply that for applications in which listeners can be trained, "perfect" simulations of spatial cues may not be necessary. However, there are limits to the kinds of distortions of spatial cues to which a listener can adapt (Shinn-Cunningham et al., 1998b; Shinn-Cunningham 2000a).

#### 3.6.10 Intersensory Integration of Spatial Information

Acoustic spatial information is integrated with spatial information from other sensory channels (particularly vision) to form spatial percepts (e.g., see Welch & Warren, 1986). In particular, auditory spatial information is combined with visual (and/or proprioceptive) spatial information to form the percept of a single, multisensory event, especially when the inputs to the different modalities are correlated in time (e.g., see Warren, Welch, & McCarthy, 1981, see chap. 21, this volume). When this occurs, visual spatial information is much more potent than that of auditory information, so the perceived location of the event is dominated by the visual spatial information (although auditory information does affect the percept to a lesser degree, e.g., see Pick, Warren, & Hay, 1969; Welch & Warren, 1980; chap. 22, this volume. *Visual capture* refers to the perceptual dominance of visual spatial information, describing how the perceived location of an auditory source is captured by visual cues.

Summarizing these results, it appears that the spatial auditory system computes source location by combining all available acoustic spatial information. Perhaps even more importantly, a priori knowledge and information from other sensory channels can have a pronounced effect on spatial perception of auditory and multisensory events.

### 3.7 Auditory Scene Analysis

Listeners in real-world environments are faced with the difficult problem of listening to many competing sound sources that overlap in both time and/or frequency. The process of separating out the contributions of different sources to the total acoustic signals reaching the ears is known as *auditory scene analysis* (e.g., see Bregman, 1990).

In general, the problem of grouping sound energy across time and frequency to reconstruct each sound source is governed by a number of basic (often intuitive) principles. For instance, naturally occurring sources are often broadband, but changes in the amplitude or frequency of the various components of a single source are generally correlated over time. Thus, comodulation of sound energy in different frequency bands tends to group these signal elements together and cause them to fuse into a single perceived source. Similarly, temporal and spectral proximity both tend to promote grouping so that signals close in time or frequency are grouped into a single perceptual source (sometimes referred to as a stream). Spatial location also can influence auditory scene analysis such that signals from the same or similar locations are grouped into a single stream. Other factors affecting streaming include (but are not limited to) harmonicity, timbre, and frequency or amplitude modulation.

For development of auditory displays, these grouping and streaming phenomena are very important, because they can directly impact the ability to detect, process, and react to a sound. For instance, if a masker sound (comprised of a number of constituent modulated sinusoids) is played simultaneously with a target sinusoid that is also modulated, the ability to detect the target improves if the masker sinusoids are modulated with the same envelope (different from the target). This process cannot be explained by peripheral mechanisms, since the peripheral masking produced is independent of whether or not the masker components are temporally correlated. Thus, perceptual segregation of two perceptual streams reduces their interaction and interference, improving performance on signal detection, speech intelligibility, temporal discrimination, and other tasks.

### 3.8 Speech Perception

Arguably the most important acoustic signal is that of speech. The amount of information transmitted via speech is larger than for any other acoustic signal. For many applications, accurate transmission of speech information is the most critical component of an auditory display.

Speech perception is affected by many of the low-level perceptual issues discussed in previous sections. For instance, speech can be masked by other signals, reducing a listener's ability to determine the content of the speech signal. Speech reception in noisy environments improves if the speaker is located at a different position than the noise source(s), particularly if the speaker and masker are at locations giving rise to different interaural level differences. Speech reception is also affected by factors that affect the formation of auditory streams, such as comodulation, harmonic structure, and related features. However, speech perception is governed by many high-level, cognitive factors that do not apply to other acoustic signals. For instance, the ability to perceive a spoken word improves dramatically if it is heard within a meaningful sentence rather than in isolation. Speech information is primarily conveyed by sound energy between 200 and 5000 Hz. For systems in which speech communication is critical, it is important to reduce the amount of interference in this range of frequencies or it will impede speech reception.

## 4. SPATIAL SIMULATION

Spatial auditory cues can be simulated using headphone displays or loudspeakers. Headphone displays generally allow more precise control of the spatial cues presented to a listener, both because the signals reaching the two ears can be controlled independently and because there is no indirect sound reaching the listener (i.e., no echoes or reverberation). However, headphone displays are generally more expensive than loudspeaker displays and may be impractical for applications in which the listener does not want to wear a device on the head. While it is more difficult to control the spatial information reaching a listener in a loudspeaker simulation, loudspeaker-based simulations are relatively simple and inexpensive to implement and do not physically interfere with the listener.

Simulations using either headphones or speakers can vary in complexity from providing no spatial information to providing nearly all naturally occurring spatial cues. This section reviews both headphone and speaker approaches to creating spatial auditory cues.

## 4.1 Headphone Simulation

### 4.1.1 Diotic Displays

The simplest headphone displays present identical signals to both ears (*diotic* signals). With a diotic display, all sources are perceived as inside the head (not "externalized"), at midline. This internal sense of location is known as *lateralization* not *localization* (Plenge, 1974). While a diotic display requires no spatial auditory processing, it also provides no spatial information to a listener. Such displays may be useful if the location of an auditory object is not known or if spatial auditory information is unimportant. However, diotic displays are the least realistic headphone display. In addition, as discussed in section 3.6.8, benefits of spatial hearing can be extremely useful for detection and recognition of auditory information. For instance, when listeners are required to monitor multiple sounds sources, spatialized auditory displays are clearly superior to diotic displays (Hass, Gainer, Wightman, Couch, & Shilling, 1997).

### 4.1.2 Dichotic Displays

While normal interaural cues vary with frequency in complex ways, simple frequency-independent ITDs and IIDs affect the perceived lateral position of a sound source (e.g., see Durlach & Colburn, 1978). Stereo signals that only contain a frequency in dependent ITD and/or IID are herein referred to as *dichotic* signals (although the term is sometimes used to refer to any stereo signal in which left and right ears are different).

Generation of a constant ITD or IID is very simple over headphones since it only requires that the source signal be delayed or scaled (respectively) at one ear. Just as with diotic signals, dichotic signals result in sources that appear to be located on an imaginary line inside the head, connecting the two ears. Varying the ITD or IID causes the lateral position of the perceived source to move toward the ear receiving the louder and/or earlier-arriving signal. For this reason, such sources are usually referred to as *lateralized* rather than *localized.*

Dichotic headphone displays are simple to implement, but are only useful for indicating whether a sound source is located to the left or right of a listener. On the other hand, when multiple sources are lateralized at different locations (using different ITD and/or IID values), some binaural unmasking can be obtained (see section 3.6.8).

The left and right signals of commercial stereo recordings generally contain simple ITD and IID cues in direct sounds, but also contain reverberation and echoes. When these signals are played over headphones, sources are usually lateralized, but actually may seem more realistic than other signals due to reverberation.

### 4.1.3 Spatialized Audio

Using signal-processing techniques, it is possible to generate stereo signals that contain most of the normal spatial cues available in the real world. In fact, if properly rendered, spatialized audio can be practically indistinguishable from free-field presentation (Langendijk & Bronkhorst, 2000). When coupled with a head-tracking device, spatialized audio provides a true virtual auditory interface. Using a spatialized auditory display, a variety of sound sources can be presented simultaneously at different directions and distances. One of the early criticisms of spatialized audio was that it was expensive to implement; however, as hardware and software solutions have proliferated, it has become feasible to include spatialized audio in most systems. Spatialized audio solutions can be fit into any budget, depending on the desired resolution and number of sound sources required. Most HMDs are currently outfitted with headphones of sufficient quality to reproduce spatialized audio, making it relatively easy to incorporate spatialized audio in an immersive VE system.

*4.1.3.1 Head-Related Transfer Functions (HRTFs).* In order to simulate any source somewhere in space over headphones, one must simply play a stereo headphone signal that recreates at the eardrums the exact acoustic waveforms that would actually arise from a source at the desired location. This is generally accomplished by empirically measuring the transfer functions that describe how an acoustic signal at a particular location in space is transformed as it travels to and impinges on the head and ears of a listener. Then, in order to simulate an arbitrary sound source at a particular location, the appropriate transfer functions are used to filter the desired (known) source signal. The resulting stereo signal is then corrected to compensate for transfer characteristics of the display system (for instance, to remove any spectral shaping of the headphones) and presented to the listener.

The pairs of spatial filters that describe how sound is transformed as it impinges on a listener are known as *head-related transfer functions* (HRTFs). HRTFs describe how to simulate the direct sound reaching the listener from a particular position, but do not generally include any reverberant energy. Empirically measured HRTFs vary mainly with the direction from head to source, but also vary with source distance (particularly for sources within reach of the listener). For sources beyond about a meter away, the main effect of distance is to change the overall gain of the HRTFs. In the time domain, the HRTF pair for a particular source location provides the pressure waveforms that would arise at the ears if a perfect impulse were presented from the spatial location in question. Often, HRTFs are represented in the frequency domain by taking the Fourier Transform of time domain impulse responses.

HRTFs contain most of the spatial information present in real-world listening situations. In particular, ITD and IID are embodied in the relative phase and magnitude (respectively) of the linear filters for the left and right ears. Spectral cues and source intensity are present in the absolute frequency-dependent magnitudes of the two filters.

Figure 4.2 shows two HRTF pairs from a human subject in the time domain (left side of figure) and in the frequency domain (magnitude only, right side of figure). All panels are for a source at azimuth 90 and elevation 0. The top two panels show the HRTF for a source very close to the head (15 cm from the center of the head). The bottom two panels show the HRTF for a source 1 meter from the head. In the time domain, it is easy to see the interaural differences in time and intensity, while the frequency domain representation shows the spectral notches that occur in HRTFs, as well as the frequency-dependent nature of the interaural intensity difference. The IIDs are larger in all frequencies for the nearer source (top panels), as expected. In the time domain, the 1 m source must traverse a greater distance to reach the ears than the near source, resulting in additional time delay before the energy reaches the ears (note time onset differences in the impulse responses in the left top and left bottom panels).

*4.1.3.2 Room Modeling.* HRTFs generally do not include reverberation or echoes, although it is possible to measure transfer functions (know as *room transfer functions*) that incorporate the acoustic effects of a room. While possible, such approaches are generally not practical because such filters vary with listener and source position in the room as well as the relative position of listener and source to produce a combinatorially large number of transfer functions. In addition, such filters can be an order of magnitude longer than traditional HRTFs, increasing both computational and storage requirements of the system.

There has been substantial effort devoted to developing computational models for room reverberation (e.g., see discussion in Shinn-Cunningham, Lehnert, Kramer, Wenzel, & Durlach, 1997). The required computations are quite intensive; in order to simulate each individual echo, one must calculate the distance the sound wave has traveled, how the waveform was transformed by every surface on which it impinged, and the direction from which it is arriving at the head. The resulting waveform must then be filtered by the appropriate anechoic HRTF based on the direction of incidence with the head.

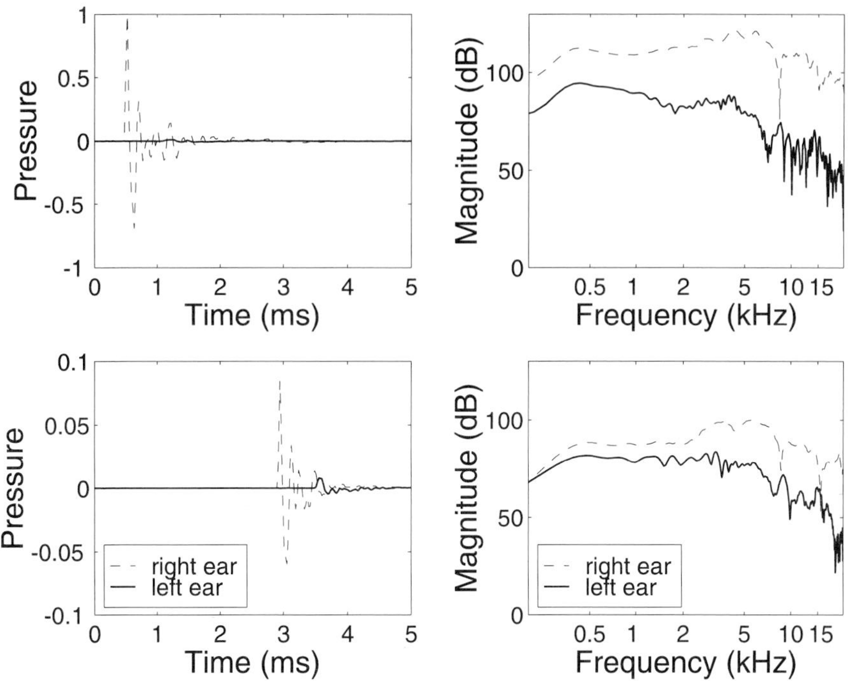


FIG. 4.2. Time domain (left panels) and magnitude spectra (right panels) representations of anechoic HRTFs for a human subject. All panels show a source at 90 azimuth, 0 elevation. Top panels are for a source at 15 cm, bottom panels for a source at 1 m.

If one looks at the resulting echoes as a function of time from the initial sound, the number of echoes in any given time slice increases exponentially with time (since the number of echoes grows exponentially with time as each echo impinges on multiple surfaces to effectively create new sources). At the same time the level of each individual echo decreases rapidly, both due to energy absorption at each reflecting surface and increased path length from source to ear. Many simulations only "spatialize" a small number of the loudest, earliest-arriving echoes, and then add random noise that dies off exponentially in time (uncorrelated at the two ears) to simulate later arriving echoes that are dense in time and arriving from essentially random directions. Even with such simplifications, the computations necessary to generate "realistic" reverberation (particularly in a system that tries to account for movement of a listener) can be overwhelming (e.g., see Shinn-Cunningham et al., 1997).

Figure 4.3 shows the room impulse response at the right ear for a source located at 45 azimuth, 0 elevation, and distance of 1 meter. This impulse response was measured in a moderate-size classroom in which significant reverberant energy persists for as long as 450 milliseconds. The initial few ms of the response are shown in the inset. In the inset, the initial response is that caused by the sound wave that travels directly from the source to the ear. The first reflection is also evident at the end of the inset, at a much reduced amplitude. In the main figure, the decay of the reverberant energy can be seen with time.

Development of tractable reverberation algorithms for real-time systems is an ongoing area of research. The extent to which listeners are sensitive to interaural and spectral details in reverberant energy is also not well understood and requires additional research. Nonetheless, there is clear evidence that inclusion of reverberation can have a dramatic impact on the subjective realism of a virtual auditory display and can aid in perception of source distance.

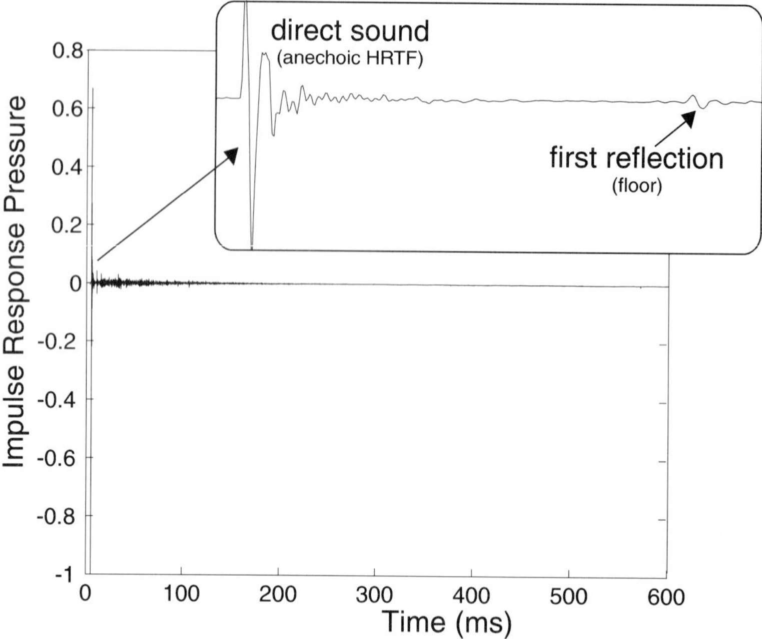


FIG. 4.3. Impulse response at the right ear for a source at 45 azimuth, 0 elevation, and distance 1 m in a standard classroom. Inset shows first 10 ms of total impulse response.

*4.1.3.3 Practical Limitations On Spatialized Audio.* While in theory, HRTF simulation should yield stimuli that are perceptually indistinguishable from natural experience, a number of practical considerations limit the realism of stimuli simulated using HRTFs. Measurement of HRTFs is a difficult, time-consuming process. In addition, storage requirements for HRTFs can be prohibitive. As a result, HRTFs are typically measured only at a single distance, relatively far from the listener, and at a relatively sparse spatial sampling. Changes in source distance are simulated simply by scaling the overall signal intensity. Because the HRTFs are only measured for a finite number of source directions at this single source distance, HRTFs are interpolated to simulate locations for which HRTFs are not measured. While this approach is probably adequate for sources relatively far from the listener and when some inaccuracy can be tolerated, the resulting simulation cannot perfectly recreate spatial cues for all possible source locations (Wenzel & Foster, 1993). Individual differences in HRTFs are very important for some aspects of sound source localization (particularly for distinguishing front/back and up/down). However, most systems employ a standard set of HRTFs that are not matched to the individual listener. Unfortunately, using these "nonindividualized" HRTFs reduces the accuracy and externalization of auditory images, but still results in useful performance increases (Begault & Wenzel, 1993). A great deal of ongoing research is focused on developing ways to tailor HRTFs to an individual listener without explicitly measuring HRTFs for that individual. For instance, researchers are exploring a variety of HRTF compression schemes in which individual differences are encoded in a small number of parameters that can be quickly or automatically fit to an individual (e.g., see Kistler & Wightman, 1991; Middlebrooks & Green, 1992). Nonetheless, many "typical" systems cannot simulate source position along a torus of confusion because they do not use individualized HRTFs.

The most sophisticated spatialized audio systems use trackers to measure movement of a listener and update the HRTFs in real time to produce appropriate dynamic spatial cues. The use of head tracking dramatically increases the accuracy of azimuthal localization (Sorkin,

Kistler, & Elvers, 1989). However, time lag in such systems (from measuring listener movement, choosing the new HRTF, and filtering the ongoing source signal) can be greater than 100 ms. While the binaural system is sluggish, the resulting delay can be perceptible. Real-time systems are also too complex and costly for some applications. Instead, systems may compute signals off-line and either ignore or limit the movement of the listener; however, observers may hear sources at locations inside or tethered to the head (i.e., moving with the head) with such systems.

Many simulations do not include any echoes or reverberation in the generated signals. Although reverberation has little impact (or degrades) perception of source direction, it is important for distance perception. In addition, anechoic simulations sound subjectively artificial and less realistic than do simulations with reverberation.

## 4.2 Simulation Using Speakers

The total acoustic signal reaching each ear is simply the sum of the signals reaching that ear from each source in an environment. Using this property, it is possible to vary spatial auditory cues (e.g., ITD, IID, and spectrum) by controlling the signals played from multiple speakers arrayed around a listener. In contrast with headphone simulations, the signals at the two ears cannot be independently manipulated; i.e., changing the signal from any of the speakers changes the signals reaching both ears. As a result, it is difficult to precisely control the interaural differences and spectral cues of the binaural signal reaching the listener to mimic the signals that would occur for a real-world source. However, various methods for specifying the signals played from each loudspeaker exist to simulate spatial auditory cues using loudspeakers.

To reduce the variability of audio signals reaching the ears, careful attention should be given to speaker placement and room acoustics. If speaker systems are not properly placed and installed in a room, even the best sound systems will sound inferior. Improperly placed speakers can reduce speech comprehension, destroy the sense of immersion, and dramatically reduce bass response (Holman, 2000). This is especially true when dealing with small rooms. If the system is installed properly, there will be a uniform (flat) frequency response at the listening area.

One example of a four speaker system (two front and two surround) is described in an International Telecommunications Union (ITU) recommendation and places speakers at ±30 degrees in front of the listener and at ±110 degrees behind the listener (ITU-R BS.775-1). It is further recommended that the signals emanating from the two surround channels be decorrelated to increase the sense of spaciousness. Correlated mono signals may give a sense of lateralization rather than localization. If a subwoofer is used, it is usually placed in front of the room. Placing the subwoofer too close to a corner may increase bass response, but may result in a muddier sound. The subwoofer should be moved to achieve the best response in the listening area. Unfortunately, speaker placement will vary depending on the dimensions and shape of the room, as well as the number of speakers employed. If the system is mobile, the sound system will have to be readjusted for every new location, unless the simulation incorporates its own enclosure. If the simulation will be housed in different sized rooms, the audio system (amplifiers and speakers) must have enough headroom (power) to accommodate both large enclosures as well as small. When possible, acoustical tile and diffusers should be employed where appropriate to reduce reverberation and echoes.

### *4.2.1 Nonspatial Display*

Many systems use free-field speakers in which each speaker presents an identical signal. Such systems are analogous to diotic headphone systems; although simple to develop, these displays (like diotic headphone displays) provide no spatial information to the listener. Such systems can be used when spatial auditory information is unimportant and when segregation of simultaneous auditory signals is not critical. For instance, if the only objects of interest are within the visual field and interference between objects is not a concern, this kind of simplistic display may be adequate.

### 4.2.2 Stereo Display

The analog of the dichotic headphone display presents signals from two speakers simultaneously in order to control the perceived laterality of a "phantom" source. For instance, simply varying the level of otherwise identical signals played from a pair of speakers can alter the perceived laterality of a phantom source. Most commercial stereo recordings are based on variations of this approach.

Imagine a listener sitting equidistant from two loudspeakers positioned symmetrically in front of the listener. When the left speaker is played alone, the listener hears a source in the direction of the left speaker (and ITD and IID cues are consistent with a source in that leftward direction). When the right speaker is played alone, the listener hears a source in the direction of the right speaker. When identical signals at identical levels are played from both speakers, each ear receives two direct signals, one from each of the symmetrically placed speakers. To the extent that the listener's head is left–right symmetric, the total direct sound in each ear is identical, and the resulting percept will be of a single source at a location that gives rise to zero ITD and zero IID (e.g., in the listener's median plane). Varying the relative intensity of otherwise identical signals played from the two speakers causes the gross ITD and IID cues to vary systematically, producing a phantom source whose location between the two speakers varies systematically with the relative speaker levels (e.g., see Bauer, 1961; Zurek & Shinn-Cunningham, under review 1997).

This simple "panning" technique produces a robust perception of a source at different lateral locations; however, it is nearly impossible to precisely control the exact location of the phantom image. In particular, the way in which the perceived direction changes with relative speaker level depends upon the location of the listener with respect to the two loudspeakers. As the listener moves outside a restricted area (the "sweet spot"), the simulation degrades rather dramatically. In addition, reverberation can distort the interaural cues, causing biases in the resulting simulation. Nonetheless, such systems provide some information about source laterality.

### 4.2.3 Lessons from the Entertainment Industry

The ability to generate an accurate spatial simulation using loudspeakers increases dramatically as the number of speakers used in the display increases. With an infinite number of speakers around the listener, one would simply play the desired signal from the speaker at the desired location of the source to achieve a "perfect" reproduction. Panning between multiple pairs of speakers (for instance, speakers arrayed in front of and behind the listener) is often used to improve spatial simulations using loudspeakers.

These Surround Sound technologies are primarily seen in the entertainment industry and are implemented via a variety of formats. Surround Sound systems find their genesis in a three-channel system created for the movie *Fantasia* in 1939. Fantasound speakers were located in front of the listener at left, middle, and right. The "surround" speakers consisted of approximately 54 speakers surrounding the audience and carried a mixture of sound from the left and right front speakers (i.e., it was not true stereo; Garity & Hawkins, 1941).

Currently, the most common Surround Sound format is the 5.1 speaker system, in which speakers are located at the left, middle, and right in front of the listener, and left and right behind the listener. The so-called 0.1 speaker is a subwoofer. The middle speaker, located in front of the listener, reproduces most of the speech information to the listener. Typical 5.1 Surround formats are Dolby Digital Surround and Digital Theater Systems (DTS).

Newer Surround Sound formats include Dolby Digital Surround EX, which is a 6.1 speaker system (adding a center speaker behind the listener). A 10.2 system is now on the horizon (Holman, 2000). In the future, even greater numbers of speakers and more complex processing are likely to become standard. However, it is important to note that adding additional speakers may be detrimental to producing a sense of immersion, especially in a small room. As the number of speakers increases, explicit care must be taken to assure that the sound field in the room is diffuse enough that the speakers themselves are not obtrusive. If the user notices the speakers in the room, the illusion of reality will be destroyed. To negate this possibility, extreme care must be taken to account for room acoustics, speaker design and placement, and the location of the listener in the room (Holman, 2000).

### *4.2.4 Cross-talk Cancellation and Transaural Simulations*

More-complex signal-processing schemes can also be used to control spatial cues using loudspeakers. In such approaches, the total signal reaching each ear is computed as the sum of the signals reaching that ear from each of the speakers employed. By considering the timing and content of each of these signals, one can try to reproduce the exact signal desired at each ear.

The earliest such approach attempted to recreate the sound field that a listener would have received in a particular setting from stereo recordings taken from spatially separated microphones. In the playback system, two speakers were positioned at the same relative locations as the original microphones. The goal of the playback system was to play signals from the two speakers such that the total signal at each ear was equal to the recorded signal from the nearer microphone. To the extent that the signal from the far speaker was acoustically cancelled, the reproduction would be accurate. Relatively simple schemes involving approximations of the acoustic alterations of the signals as they impinged on the head were used to try to accomplish this "cross-talk cancellation."

As signal-processing approaches have been refined and knowledge of the acoustic properties of HRTFs improves, more sophisticated algorithms have been developed. In particular, it is possible to calculate the contribution of each speaker to the total signal at each ear by considering the HRTF corresponding to the location of the speaker. The total signal at each ear is then the sum of the HRTF-filtered signals coming from each speaker. If one also knows the location and source of the signal that is to be simulated, one can write equations for the desired signals reaching each ear as HRTF-filtered versions of the desired source. Combining these equations yields two frequency-dependent, complex-valued equations that relate the signals played from each speaker to the desired signals at the ears. To the extent that one can find and implement solutions to these equations, it is possible (at least in theory) to recreate the desired binaural signal by appropriate choice of the signals played from each speaker. The problem with such approaches is that the simulation depends critically on the relative location of the speakers and the listener. In particular, if the listener moves his head outside of the sweet spot, the simulation degrades rapidly. It can be difficult to compute the required loudspeaker signals and the computations are not particularly stable, numerically. To the extent that the HRTFs used in the equations are not matched to the listener and reverberation distorts the signals reaching the ears, the derived solutions are even less robust.

Head trackers can be used in conjunction with multispeaker simulations in order to improve the simulation. However, this requires that computations be performed in real time and significantly increases the cost and complexity of the resulting system.

In general, it is possible to generate relatively realistic phantom sources using multiple loudspeakers whose lateral position changes with changes in the speaker waveforms. More complex simulations in which spectral cues are simulated are often too difficult to include.

## 5. DESIGN CONSIDERATIONS

### 5.1 Hardware

Despite the importance of audio cues in human existence, there are relatively few platforms available for easily implementing spatial audio into virtual environments. The most common platforms used today for implementing spatial audio in research and VE are the AuSim GoldServer, Lake Huron, and Tucker Davis Technologies PD-1 Power SDAC and expander.

The AuSim GoldServer is based on the original acoustetron II design pioneered by Crystal River Engineering (CRE; Wenzel, Wightman, & Foster, 1988). Although the original acoustetron II is no longer commercially available, AuSIM acquired the rights to the device and is now producing boxes and replacement parts for the original CRE devices. AuSIM also claims that their new product line will not only fill the niche for VE applications, but will also provide a more flexible research tool than its predecessor (William L. Chapin, personal communication, April, 2000). A typical system will cost between $12,000 and $18,000 and claims to be relatively easy to integrate into a variety of VE applications AuSim also produces a system for collecting customized HRTPs.

Hailing from Australia, Lake Technology offers a variety of hardware and software systems using either headphones or speakers. On the hardware side, the Huron System is a multiple DSP system that will present multiple sound sources to multiple listeners using positional tracking. The Huron can be configured using multiple speakers or headphones. Combined with an extensive set of software development tools, the Huron System is another good option for integrating sound in virtual environments. For instance, the software package Multiscape is advertised to be able to create a multiple user interactive auditory VE complete with geometric rooms with interconnecting doors. Finally, in cooperation with Dolby, Lake has recently helped develop, and now supports, headphone systems that simulate Dolby 5.1 Surround Sound.

The Tucker-Davis Technologies (TDT) PD-1 Power SDAC system is primarily a research tool. However, with some creative programming, it too can be used as a sound server for simulations (Shilling et al., 2000). The PD-1 combines up to 27 DSPs, allowing researchers to create complex spatial audio scenarios including numerous reflections and incorporating head tracking. It also provides real-time convolution, allowing the researcher to convolve "live" signals.

A software solution currently under development for research in spatial sound is Sound Lab (Wenzel, Miller, & Abel, 2000). Sound Lab (SLAB) is designed to run on a regular personal computer (PC) platform and provides low-level control of a variety of signal processing functions, including the number of reflections, number of filter taps, and positions of reflections. No special purpose hardware is required for the software to run. In addition to allowing normal HRTF processing and inclusion of reflections, the software allows for manipulation of acoustic radiation patterns, spherical spreading loss, and atmospheric absorption. Preliminary analysis of the dynamic performance of SLAB shows it achieves a very low 24 ms latency on a 450 MHz Pentium II, although the latency would be larger for a more complex acoustic model. At this time, SLAB continues to be developed.

### 5.2 Defining the Auditory Environment

When creating the auditory portion of a virtual environment, careful attention should be placed on what is absolutely essential for the task. The adage of motion picture sound designers, "see it, hear it" (Holman 1997; Yewdall, 1999), is also valid for designing audio for virtual environments. The sense of immersion experienced in a movie theater is a carefully orchestrated combination of expertly designed sound effects and skillfully applied auditory ambiences. It is also interesting to note that "realistically" rendered sound is often perceived as emotionally flat in motion pictures. Sound effects are often designed as exaggerated versions of reality to convey emotion or to satisfy the viewers' expectations of reality (Holman, 1997). Sound design in VE needs to balance the need for accurate reproduction with the need to make the user emotionally involved in a synthetic environment. When dealing with issues of emotionality in VE, sound should be considered as synonymous with emotion.

On the hardware side, a simple sound card solution may be adequate if spatialized audio and fidelity are not paramount. However, if there are multiple sound sources and/or a multiuser interface, an audio server such as the AuSIM Acoustetron or the Lake Huron System may be necessary. Just as photos and films of the visual environment are taken during the development process, it is a good idea to make audio recordings, including sound level measurements, when developing an auditory interface.

One of the current efforts in the Virtual Environments in Training Technology (VETT) program sponsored by the Office of Naval Research (see chap. 43, this volume) is to develop a systematic approach for obtaining baseline data concerning the content of an auditory environment. In addition to cataloguing the different sounds in a real environment, it is also important to systematically measure the intensity of sounds being experienced by the listener. In this manner, the VE developer has a detailed reference with which to compare the real world auditory environment with the virtual auditory environment. Two systems are currently being evaluated. The first system uses a portable Sony TCD-D8 DAT recorder coupled with Sennheisser microphone capsules (see Fig. 4.4, left). To produce a crude spatialized recording, the microphone capsules are inserted into an observer's auditory meatus (ear canal). In this manner, a complete spatialized recording can be made of the auditory environment, completely externalized with azimuth and elevation cues. The second system (see Fig. 4.4, right) is more robust, using a larger set of microphones produced by Core Sound, which can clip to a set of eyeglasses to produce a binaural recording. However, since the microphones are not inside the ear canals, this system will not include spectral cues. Although pinnae cues cannot be utilized, the latter system has more robust construction and will be more tolerant of extreme conditions, especially if recordings are made outdoors. Both systems can be clipped to a belt and will be used in conjunction with a real-time logging and event analyzer (CEL 593). The complete data set, including sound recordings and sound measurements, will be stored on CDROM for ease of use. Digital recordings also allow for spectral analyses to be conducted on specific auditory stimuli contained on tape so that synthesized versions of those stimuli can be constructed. If

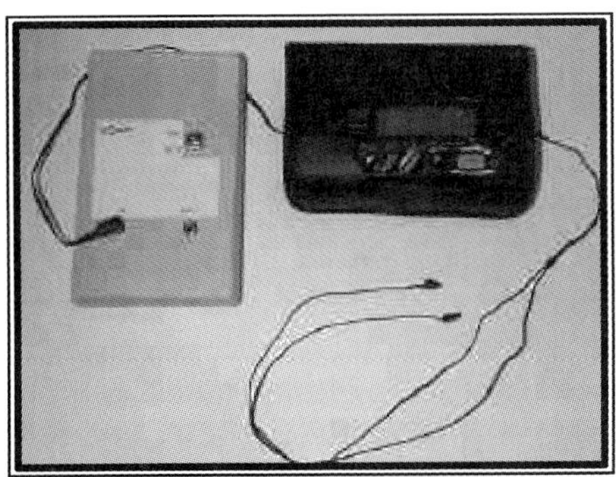

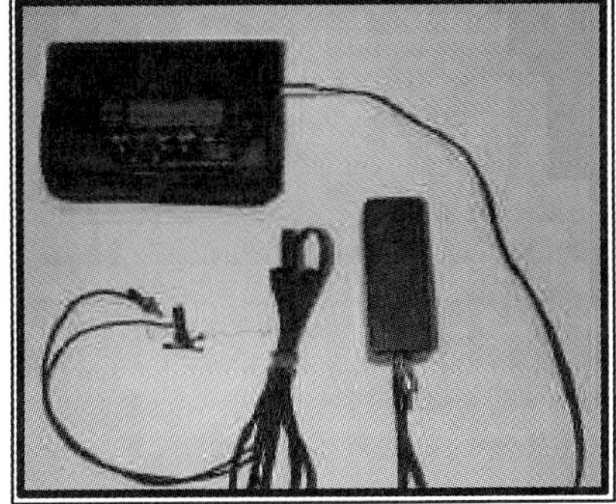

FIG. 4.4. Two portable systems for making binaural recordings. On the left is a system using microphone capsules placed in the auditory meatus using Sennheisser microphone capsules (Tucker Davis Technologies). On the right is a binaural recording system from CORE Sound.

concussive or low-frequency sounds need to be catalogued, microphones with wider frequency ranges should be employed.

Given the wide dynamic range involved with recording sounds ranging from concussive events to footsteps and the necessity that recordings be absolutely clean and accurate, the best solution may be to rely on professionals for making appropriate recordings. Different combinations of microphones and recording equipment produce vastly differing sound quality. Choosing the appropriate combination for a particular application is still more of an art than a science. In addition, there are many high-quality commercially available sound libraries for obtaining a wide variety of sound effects and ambiences (Holman, 1997; Yewdall, 1999).

## 5.3 How Much Realism Is Necessary?

Much of the research devoted to developing and verifying virtual display technology emphasizes the subjective realism of the display (see chap. 40, this volume); however, this is not the most important consideration for all applications. In some cases, signal processing that improves realism actually interferes with the amount of information a listener can extract. For instance, inclusion of echoes and reverberation can significantly increase the perceived realism of a display and improve distance perception. However, echoes can degrade perception of source direction. For applications in which information about the direction of a source is more important than the realism of the display or perception of source distance, including echoes and reverberation may be ill advised.

In headphone-based systems, realism is enhanced with the use of individualized HRTFs, particularly in the perception of up/down and front/back position. Ideally, HRTFs should also be sampled in both distance and direction at a spatial density dictated by human sensitivity. Thus, while the most "realistic" system would use individualized HRTFs that are sampled densely in both direction and distance, most systems use generic HRTFs sampled coarsely in direction and at only one distance.

If a particular application requires a listener to extract three-dimensional spatial information from the auditory display, HRTFs may have to be tailored to the listener to preserve directional information and reverberation may have to be included to encode source distance. On the other hand, if a particular application only makes use of one spatial dimension (for instance, to indicate the direction that a blind user must turn), coarse simulation of ITD and IID cues (even without detailed HRTF simulation) is probably adequate.

If information transfer is of primary importance, it may be useful to present acoustic spatial cues that are intentionally distorted so that they are perceptually more salient than more "realistic" cues. For instance, it may be useful to exaggerate spatial auditory cues to improve auditory resolution; however, such an approach requires that listeners are appropriately trained with the distorted cues (Shinn-Cunningham et al., 1998a).

The processing power needed to simulate the most realistic virtual auditory environment possible is not always cost-effective. For instance, the amount of computation needed to create realistic reverberation in a VE may not be justifiable when source distance perception and subjective realism are not important. Other acoustic effects are often ignored in order to reduce computational complexity of an acoustic simulation, including the nonuniform radiation pattern of a realistic sound source, spectral changes in a sound due to atmospheric effects, and Doppler shift of the received spectrum of moving sources. The perceptual significance of many of these effects is not well understood; further work must be done to examine how these factors affect the realism of a display, as well as what perceptual information such cues may convey.

In command-and-control applications (see chap. 43, this volume), the goal is to maximize information transfer to the human operator; subjective impression (i.e., realism) is unimportant. In these applications, both technological and perceptual issues must be considered to achieve this goal. If nonverbal warnings or alerts are created, the stimuli must be wideband enough to be localizable. In addition, stimuli should be significantly intense to be at least 15 dB above background noise level. Stimulus onset should be fairly gradual so as not to be excessively startling to the user. In many instances, one may want the stimuli to be aesthetically pleasing to the user. As can be imagined, creating acceptable spatialized auditory displays is no trivial chore and should involve formal evaluations to ensure perceptual accuracy and system usability. For applications in which speech is the main signal of interest, basic interaural cues are important for preserving speech intelligibility, particularly in noisy, multisource environments. On the other hand, there is probably little benefit gained from including the detailed frequency-dependence of normal HRTFs. In entertainment applications (see chap. 55, this volume), cost is the most important factor; the precision of the display is unimportant as long as the simulation is subjectively satisfactory. For scientific research (see chap. 53, this volume), high-end systems are necessary in order to allow careful examination of normal spatial auditory cues. In clinical applications (see chaps. 49 and 50, this volume), the auditory display must only be able to deliver stimuli that can distinguish listeners with normal spatial hearing from those with impaired spatial hearing. Such systems must be inexpensive and easy to use, but there is no need for a perfect simulation.

## 6. REFERENCES


Algazi, V. R., Avendano, C., & Duda, R. O. (2001). "Elevation localization and head-related transfer function analysis at low frequencies," *Journal of the Acoustical Society of America, 109* (3), 1110–1122.

Batteau, D. W. (1967). The role of the pinna in human localization. *Proceedings of the Royal Society Of London, B168*, 158–180.

Bauer, B. B. (1961). Phasor analysis of some stereophonic phenomena. *Journal of the Acoustical Society of America, 33* (11), 1536–1539.

Begault, D. R. (1993a). Head-up auditory displays for traffic collision avoidance system advisories: A preliminary investigation. *Human Factors, 35* (4), 707–717.

Begault, D. R. (1993b). *Call sign intelligibility improvement using a spatial auditory display*. NASA Technical Memorandum No. 104014. Moffet Fields, CA.

Begault, D. R. (1996, November). *Virtual acoustics, aeronautics, and communications*. Presented at the 101st convention of the Audio Engineering Society, Los Angeles.

Begault, D. R. (1999). Virtual acoustic displays for teleconferencing: Intelligibility advantage for "telephone-grade" audio. *Journal of the Audio Engineering Society, 47* (10), 824–828.

Begault, D. R., & Wenzel, E. M. (1992). Techniques and applications for binaural sound manipulation in human–machine interfaces. *The International Journal of Aviation Psychology, 2* (1), 1–22.

Begault, D. R., & Wenzel, E. M. (1993). Headphone localization of speech. *Human Factors, 35* (2), 361–376.

Blauert, J. (1997). *Spatial hearing* (2nd Ed.), Cambridge, MA: MIT Press.

Brainard, M. S., Knudsen, E. I., & Esterly, S. D. (1992). Neural derivation of sound source location: Resolution of ambiguities in binaural cues. *Journal of the Acoustical Society of America, 91*, 1015–1027.

Bregman, A. S. (1990). *Auditory scene analysis: The perceptual organization of sound*. Cambridge, MA: MIT Press.

Bronkhorst, A. W. (2000). The cocktail party phenomenon: A review of research on speech intelligibility in multiple-talker conditions. *Acustica, 86*, 117–128.

Bronkhorst, A. W., & Plomp, R. (1988). The effect of head-induced interaural time and level differences on speech intelligibility in noise. *Journal of the Acoustical Society of America, 83*, 1508–1516.

Bronkhorst, A. W., Veltman, J. A., & van Breda, L., (1996). Application of a three-dimensional auditory display in a flight task. *Human Factors, 38* (1), 23–33.

Brungart, D. S., & Rabinowitz, W. M. (in press). Auditory localization of nearby sources. I: Near-field head-related transfer functions. *Journal of the Acoustical Society of America*.

Clark, B., & Graybiel, A. (1949). The effect of angular acceleration on sound localization: The audiogyral illusion. *The Journal of Psychology, 28*, 235–244.

Clifton, R. K., Freyman, R. L., & Litovsky., R. L. (1993). Listener expectations about echoes can raise or lower echo threshold. *Journal of the Acoustical Society of America, 95*, 1525–1533.

DiZio, P., Held, R., Lackner, J. R., Shinn-Cunningham, B. G., & Durlach, N. I. (2001). "Gravitoinertial force magnitude and direction influence head-centric auditory localization," *Journal of Neurophysics, 85*, 2455–2560.

Drullman, R., & Bronkhorst, A. W. (2000). Multichannel speech intelligibility and talker recognition using monaural, binaural, and three-dimensional auditory presentation. *Journal of the Acoustical Society of America, 107* (4), 2224–2235.

Duda, R. O., & Martens, W. L. (1998). Range dependence of the response of a spherical head model. *Journal of the Acoustical Society of America, 104* (5), 3048–3058.

Durlach, N. I., & Colburn, H. S. (1978). Binaural phenomena. In E. C. Carterette and M. P. Friedman (Eds.), *Handbook of Perception* (Vol. 4, pp. 365–466). New York: Academic Press.

Durlach, N. I., Shinn-Cunningham, B. G., & Held, R. M. (1993). Supernormal auditory localization. I. General background. *Presence, 2* (2), 89–103.

Frens, M. A., van Opstal, A. J., & van der Willigen, R. F. (1995). Spatial and temporal factors determine auditory-visual interactions in human saccadic eye movements. *Perception and Psychophysics, 57*, 802–816.

Garity, W. E., & Hawkins, J. A. (1941). Fantasound. *Journal of the Society of Motion Picture Engineers,* August.

Gaver, W. W. (1986). Auditory icons: Using sound in computer interfaces. *Human-Computer Interaction, 2* (2), 167–177.

Gelfand, S. A. (1998). *Hearing: An introduction to psychological and physiological acoustics*. New York: Marcel Dekker, Inc.

Gilkey, R. H., & Anderson, T. R. (1995). The accuracy of absolute localization judgments for speech stimuli. *Journal of Vestibular Research, 5* (6), 487–497.

Gilkey, R. H., & Good, M. D. (1995). Effects of frequency on free-field masking. *Human Factors, 37* (4), 835–843.

Grantham, D. W. (1997). Auditory motion perception: Snapshots revisited. In R. Gilkey & T. Anderson, (Eds.), *Binaural and spatial hearing in real and virtual environments* (pp. 295–314). New York: Lawrence Erlbaum Associates.

Haas, E. C., Gainer, C., Wightman, D., Couch M., & Shilling, R. D. (1997). Enhancing system safety with 3-D audio displays. *Proceedings of the Human Factors and Ergonomics Society 41st Annual Meeting* (pp. 868–872), Albuquerque, NM.

Hartmann, W. M. (1983). Localization of sound in rooms. *Journal of the Acoustical Society of America, 74*, 1380–1391.

Hendrix, C., & Barfield, W. (1996). The sense of presence in auditory virtual environments. *Presence, 5* (3), 290–301.

Holman, T. (1997). *Sound for Film and Television*. Boston, MA: Focal Press.

Holman, T. (2000). *5.1 Surround Sound up and running*. Boston, MA: Focal Press.

Hofman, P. M., Van Riswick, J. G. A. & Van Opstal, A. J. (1998). Relearning sound localization with new ears. *Nature Neuroscience, 1* (5), 417–421.

International Telecommunications Union. (1994). *Multichannel stereophonic sound system with and without accompanying picture*. Recommendation No. ITU-R BS.775–1.

Kistler, D. J., & Wightman, F. L. (1991). A model of head-related transfer functions based on principal components analysis and minimum-phase reconstruction. *Journal of the Acoustical Society of America, 91*, 1637–1647.

Koehnke, J., Besing, J., Goulet, C., Allard, M., & Zurek, P. M. (1992). Speech intelligibility, localization, and binaural detection with monaural and binaural amplification. *Journal of the Acoustical Society of America, 92*, 24–34.

Koehnke, J., & Besing, J. M. (1996). A procedure for testing speech intelligibility in a virtual listening environment. *Ear and Hearing, 17* (3), 211–217.

Kollmeier, B., & Gilkey, R. H. (1990). Binaural forward and backward masking: Evidence for sluggishness in binaural detection. *Journal of the Acoustical Society of America, 87* (4), 1709–1719.

Langendijk, E. H., & Bronkhorst, A. W. (2000). Fidelity of three-dimensional-sound reproduction using a virtual auditory display. *Journal of the Acoustical Society of America, 107* (1), 528–537.

Letowski, T., Vause, N., Shilling, R., Ballas, J., Brungert, D., & McKinley, R. (in print). Human factors military lexicon: Auditory displays. (ARL Tech. Rep. APG [MD].

Litovsky, R. Y. (1998). Physiological studies of the precedence effect in the inferior colliculus of the kitten. *Journal of the Acoustical Society of America, 103* (6), 3139–3152.

Mershon, D. H. (1997). Phenomenal geometry and the measurement of perceived auditory distance. In R. Gilkey & T. Anderson (Eds.), *Binaural and spatial hearing in real and virtual environments* (pp. 257–274). New York: Lawrence Erlbaum Associates.

Mershon, D. H., Ballenger, W. L., Little, A. D., McMurty, P. L., & Buchanan, J. L. (1989). Effects of room reflectance and background noise on perceived auditory distance. *Perception, 18*, 403–416.

Middlebrooks, J. C. (1994). A panoramic code for sound location by cortical neurons. *Science, 264*, 842–844.

Middlebrooks, J. C. (1997). Spectral shape cues for sound localization. In R. Gilkey and T. Anderson (Eds.), *Binaural and spatial hearing in real and virtual environments* (pp. 77–98). New York: Lawrence Erlbaum Associates.

Middlebrooks, J. C., & Green, D. M. (1991). Sound localization by human listeners. *Annual Review of Psychology, 42*, 135–159.

Middlebrooks, J. C., & Green, D. M. (1992). Observations on a principal components analysis of head-related transfer functions. *Journal of the Acoustical Society of America, 92*, 597–599.

Miller, J. D., Abel, J. S., & Wenzel, E. M. (1999). Implementation issues in the development of a real-time, Windows-based system to study spatial hearing. *Journal of the Acoustical Society of America, 105*, 1193.

Mills, A. W. (1972). Auditory localization. In J. V. Tobias (Ed.), *Foundations of modern auditory theory* (pp. 303–348). New York, Academic Press.

Moore, B. C. J. (1997). *An introduction to the psychology of hearing* (4th ed.). San Diego, CA: Academic Press.

Perrott, D. R., Constantino, B., & Cisneros, J. (1993). Auditory and visual localization performance in a sequential discrimination task. *Journal of the Acoustical Society of America, 93* (4), 2134–2138.

Perrott, D. R., Saberi, K., Brown, K., & Strybel, T. (1990). Auditory psychomotor coordination and visual search behavior. *Perception and Psychophysics, 48*, 214–226.

Perrott, D. R., Sadralodabai, T., Saberi, K., & Strybel, T. (1991). Aurally aided visual search in the central visual field: Effects of visual load and visual enhancement of the target. *Human Factors, 33*, 389–400.

Pick, H. L., Warren, D. H., & Hay, J. C. (1969). Sensory conflict in judgements of spatial direction. *Perception and Psychophysics, 6*, 203–205.

Plenge, G. (1974). On the differences between localization and lateralization. *Journal of the Acoustical Society of America, 56* (3), 944–951.

Rabenhorst, D. A., Farrell, E. J., & Jameson, D. (1990). *Complementary visualization and sonfication of Multidimensional data* (IBM Tech. Rep. RC 15467 No. 68449).

Radeau, M., & Bertelson, P. (1976). The effect of a textured visual field on modality dominance in a ventriloquism situation. *Perception and Psychophysics 20* (4), 227–235.

Rakerd, B., & Hartmann, W. M. (1985). Localization of sound in rooms. II. The effects of a single reflecting surface. *Journal of the Acoustical Society of America, 78*, 524–533.

Rakerd, B., & Hartmann, W. M. (1986). Localization of sound in rooms. III. Onset and duration effects. *Journal of the Acoustical Society of America, 80*, 1695–1706.

Sheridan, T. B. (1996). Further musings on the psychophysics of presence. *Presence, 5* (2), 241–246.

Shilling, R. D., & Letowski, T. (2000). *Using spatial audio displays to enhance virtual environments and cockpit performance.* NATO Research and Technology Agency Workshop entitled What is essential for Virtual Reality to Meet Military Human Performance Goals, The Hague, The Netherlands.

Shilling, R. D., Wightman, D., Couch, M., Beutler, R., & Letowski, T. (1998). *The use of spatialized auditory displays in an aviation simulation.* Proceedings of the 16th Applied Behavioral Sciences Symposium, Colorado Springs, CO.

Shinn-Cunningham, B. G. (2000a). Adapting to remapped auditory localization cues: A decision-theory model. *Perception and Psychophysics, 62* (1), 33–47.

Shinn-Cunningham, B. G. (2000b). *Learning reverberation: Implications for spatial auditory displays.* Proceeding of the International Conference on Auditory Displays, Atlanta, GA., pp. 126–134.

Shinn-Cunningham, B. G., Durlach, N. I., & Held, R. (1998a). Adapting to supernormal auditory localization cues. 1: Bias and resolution. *Journal of the Acoustical Society of America, 103* (6), 3656–3666.

Shinn-Cunningham, B. G., Durlach, N. I., & Held, R. (1998b). Adapting to supernormal auditory localization cues. 2: Changes in mean response. *Journal of the Acoustical Society of America, 103* (6), 3667–3676.

Shinn-Cunningham, B. G., Kopco, N., & Santarelli, S. G. (1999). *Computation of acoustic source position in near-field listening.* Paper presented at the third International Conference on Cognitive and Neural Systems, Boston, MA.

Shinn-Cunningham, B. G., Lehnert, H., Kramer, G., Wenzel, E. M., & Durlach N. I. (1997). Auditory displays. In R. Gilkey and T. Anderson (Eds.), *Binaural and Spatial Hearing in Real and Virtual Environments* (pp. 611–663). New York: Lawrence Erlbaum Associates.

Shinn-Cunningham, B. G., Santarelli, S., & Kopco, N. (2000). Tori of confusion: Binaural localization cues for sources within reach of a listener. *Journal of the Acoustical Society of America, 107* (3), 1627–1636.

Shinn-Cunningham, B. G., Zurek, P. M., & Durlach, N. I. (1993). Adjustment and discrimination measurements of the precedence effect. *Journal of the Acoustical Society of America, 93*, 2923–2932.

Sorkin, R. D., Kistler, D. S., & Elvers, G. C. (1989). An exploratory study of the use of movement-correlated cues in an auditory head-up display. *Human Factors, 31* (2), 161–166.

Stern, R. M., & Trahiotis, C. (1997). *Binaural mechanisms that emphasize consistent interaural timing information over frequency.* Procedings of the eleventh International Symposium on Hearing: Auditory Physiology and Perception, Grantham, England.

Storms, Russell L. (1998). *Auditory-visual cross-modal perception phenomena.* Doctoral Dissertation. Naval Postgraduate School, Monterey, California.

THX Certified Training Program. (2000, June). Presentation Materials. San Rafael, CA.

Trahiotis, C., & Stern, R. M. (1989). Lateralization of bands of noise: Effects of bandwidth and differences of interaural time and phase. *Journal of the Acoustical Society of America, 86* (4), 1285–1293.

Wallach, H. (1940). The role of head movements and vestibular and visual cues in sound localization. *Journal of Experimental Psychology, 27*, 339–368.

Warren, D. H., Welch, R. B., & McCarthy, T. J. (1981). The role of visual-auditory "compellingness" in the ventriloquism effect: Implications for transitivity among the spatial senses. *Perception and Psychophysics, 30*, 557–564.

Welch, R. B., & Warren, D. H. (1980). Immediate perceptual response to intersensory discrepancy. *Psychological Bulletin, 88*, 638–667.

Welch, R., & Warren, D. H. (1986). Intersensory interactions. In K. R. Boff, L. Kaufman, & J. P. Thomas (Eds.), *Handbook of Perception and Human Performance* (Vol. 2, 25.1–25.36). New York: John Wiley & Sons. 25.1–25.36.

Wenzel, E. M., Arruda, M., Kistler, D. J., & Wightman, F. L. (1993). Localization using nonindividualized head-related transfer functions. *Journal of the Acoustical Society of America, 94*, 111–123.

Wenzel, E. M., & Foster S. H. (1993, October). *Perceptual consequences of interpolating head-related transfer functions during spatial synthesis*. Proceedings of the 1993 Workshop on the Applications of Signal Processing to Audio and Acoustics, New York.

Wenzel, E. M., Miller, J. D., & Abel, J. S. (2000, February). *Sound Lab: A real-time, software-based system for the study of spatial hearing*. Proceedings of the 108th Convention of the Audio Engineering Society, Paris.

Wenzel, E. M., Wightman, F. L., & Foster, S. H. (1988). Development of a three-dimensional auditory display system. *SIGCHI Bulletin, 20*, 52–57.

Wightman, F. L., & Kistler, D. J. (1989). Headphone simulation of free-field listening. II. Psychophysical validation. *Journal of the Acoustical Society of America, 85*, 868–878.

Wightman, F. L., & Kistler, D. J. (1993). Sound localization. In W. A. Yost, A. N. Popper, & R. R. Fay (Eds.), *Human psychophysics*. New York: Springer Verlag.

Wightman, F. L., & Kistler, D. J. (1999). Resolution of front–back ambiguity in spatial hearing by listener and source movement. *Journal of the Acoustical Society of America, 105* (5), 2841–2853.

Yewdall, D. L. (1999). *Practical Art of Motion Picture Sound*. Boston, MA: Focal Press.

Yost, W. A. (1994). *Fundamentals of hearing: An introduction* (3rd Ed.). San Diego, CA: Academic Press.

Yost, W. A. (1997). The cocktail party problem: Forty years later. In R. Gilkey & T. Anderson (Eds.), *Binaural and spatial hearing in real and virtual environments* (pp. 329–346). New York: Lawrence Erlbaum Associates.

Zurek, P. M. (1993). Binaural advantages and directional effects in speech intelligibility. In G. Studebaker & I. Hochberg (Eds.), *Acoustical factors affecting hearing aid performance*. Boston, MA: College-Hill Press.

Zurek, P. M., & Shinn-Cunningham, B. G. (under review). Localization cues in intensity-difference stereophony. *Journal of the Acoustical Society of America* Under Revised.

# 5

# Haptic Interfaces

S. James Biggs and Mandayam A. Srinivasan
*Laboratory for Human and Machine Haptics (The Touch Lab)*
*Massachusetts Institute of Technology*
*Cambridge, MA 02139*
*jbiggs@mit.edu, srini@mit.edu*

## 1. INTRODUCTION

Haptics is concerned with information acquisition and object manipulation through touch. Haptics is used as an umbrella term covering all aspects of manual exploration and manipulation by humans and machines, as well as interactions between the two, performed in real, virtual, or teleoperated environments. Haptic interfaces allow users to touch, feel, and manipulate objects simulated by virtual environments (VEs) and teleoperator systems (Salisbury & Srinivasan, 1992). The keyboard, mouse, and trackball are familiar, passive, haptic interfaces that sense a user's hand movements. Although they apply forces on the user's hand upon contact and consequently provide tactual sensation, the forces are not under program control. Active haptic interfaces, such as desktop robots and exoskeletal gloves with force feedback, are more sophisticated devices that have both sensors and actuators. In addition to transducing position and motion commands from the user, these devices can present controlled forces to the user, allowing him or her to feel virtual objects as well as control them. This chapter focuses on such devices.

This is an exciting time for the field of haptics. Within approximately 10 years of significant research activity, commercial efforts have brought simple, active haptic interfaces into mass production. Research efforts on a range of more sophisticated devices have intensified. The success of these endeavors depends on finding application tasks where haptics adds significant value and, from a design viewpoint, on achieving an optimal balance between the human haptic ability to sense object properties, fidelity of the interface device in delivering the appropriate mechanical signals, and computational complexity in rendering the signals in real time. Accordingly, this chapter discusses the usefulness of haptic displays in virtual environments (section 2), the human haptic system (section 3), and current interface hardware (section 4). Algorithms for estimating and rendering force feedback are covered in chapter 6 of this volume. VE-assisted teleoperation is treated specifically in chapter 48 of this volume.

Locomotion interfaces, which may actively display forces to the user, are covered in chapter 11 of this volume. Previous overviews of this field can be found in Burdea (1996), Srinivasan (1995), Youngblut, Johnson, Nash, Wienclaw, and Will (1996), and Srinivasan and Basdogan (1997).

## 2. ADVANTAGES OF ACTIVE HAPTIC INTERFACES

The gap between performance in the world and in a simulation is familiar to most computer users. In the real world, the placement, orientation, and scaling of a rectangle can be indicated in one quick gesture using the thumb and index fingers of both hands, and a rubber band to mark the perimeter. In a simulation (e.g., MacDraw) the same process must be performed in three steps using a mouse, requiring about 10 times longer (Fitzmaurice, Balakrishnan, Kurtenbach, & Buxton, 1999). A similar gap exists in VEs, both in terms of physical realism and task performance. For example, the phrase *virtual reality* typically conjures an image of a user with one passive VR glove that senses joint angles of a few fingers and position/orientation of the hand, through which the user can convey his or her intentions to the computer. However, in tasks such as surgical simulation or virtual sculpting, the glove would be inadequate. Two-way communication between the user and the computer enabled by force feedback would be absolutely necessary in order to simulate the "feel" of the organs or the clay as they are manipulated. In contrast to vision and hearing, haptics is the only modality that permits this bidirectional information transfer between the user and virtual environment. Current excitement in developing haptic interfaces arises from applications like these, and others like those listed below (from Srinivasan & Basdogan, 1997):

- *Medicine:* surgical simulators for medical training; manipulating micro and macro robots for minimally invasive surgery; remote diagnosis for telemedicine; aids for the disabled such as haptic interfaces for blind users (see chaps. 47–51, this volume).
- *Entertainment:* video games and simulators that enable the user to feel and manipulate virtual solids, fluids, tools, and avatars (see chap. 55, this volume).
- *Education:* giving students the feel of phenomena at nano, macro, or astronomical scales; "what if" scenarios for nonterrestrial physics; experiencing complex data sets (see chaps. 45–46, this volume).
- *Industry:* integration of haptics into CAD systems such that a designer can freely manipulate the mechanical components of an assembly in an immersive environment (see chaps. 52–54, this volume).
- *Graphic Arts:* virtual art exhibits, concert rooms, and museums in which the user can login remotely to play the musical instruments, and to touch and feel the haptic attributes of the displays; individual or cooperative virtual sculpting across the Internet.

The need for active haptic interfaces clearly depends on the task at hand, and can be classified as follows:

1. Active haptic interfaces are absolutely required for some tasks: Many medical procedures (for example, administering epidural anesthesia, palpating for cancerous lumps) are intrinsically haptic tasks. Haptic displays are required to simulate such tasks for training, because sensing of forces arising from tool-tissue interaction is critical for success. Another intrinsically haptic VE task is testing the ease of manual assembly of complex mechanisms before they are manufactured (Nahvi, Nelson, Hollerbach, & Johnson, 1998). In addition, active haptic

interfaces make VEs accessible to visually impaired users. Current VEs are almost entirely visual, therefore inaccessible to the roughly 0.75 million visually impaired users in the United States. The United States Congress has called for "every-citizen interfaces to the country's information infrastructure" (National Research Council, 1997). As VEs become more common in education and industry, it will be interesting to see whether the Americans with Disabilities Act is extended from the real environment to include virtual worlds.

2. Active haptic interfaces can improve a user's sense of presence: Haptic interfaces with 2 or fewer actuated degrees of freedom are now mass-produced for playing PC video games, making them relatively cheap (about $100 at the time of this writing), reliable, and easy to program. Although the complexity of the cues they can display is limited, they are surprisingly effective communicators. For example, if the joystick is vibrated when a player crosses a bridge (to simulate driving over planks) it can provide a landmark for navigation, and signal the vehicle's speed (vibration frequency) and weight (vibration amplitude). Haptic cues have also been developed to augment graphical user interfaces to windows operating systems, both Microsoft Windows (Immersion, 2000) and Linux/Unix (Miller & Zeleznik, 1999). Free source code for haptic effects for desktops and games are available. Most manufacturers of general-purpose, active interfaces also sell haptic authoring software. A few examples are the Ghost Toolkit for the Phantom (SensAble Technologies, 2000), Immersion Studio for the FeelIt mouse (Immersion, 2000), and the VirtualHand Studio for the CyberGrasp force feedback glove (Virtual Technologies, Inc., 2000). Just one haptic interaction (for example, handling a real plate in a VE) can significantly improve a user's expectations about the solidity and weight of all objects in a VE (Hoffman, 1998).

3. Active haptic interfaces can improve performance by providing natural constraints: In VEs, selecting and repositioning objects without haptic cues can be surprisingly difficult. Without force feedback, a user trying to set a simulated coffee mug down on a simulated table top is likely to merely push the mug through it. To overcome these difficulties, force-free interaction metaphors have been developed. For a review see Mine, Brooks, and Sequin (1997). The names of these metaphors generally suggest how they work (e.g., "extender grab," "spring widget," "virtual chopsticks"). Although these methods appear adequate for many tasks, they usually demand more visual attention than the same action would in the real world. A haptic interface is a more straightforward solution that may reduce the visual attention required of the user. Force feedback also can improve accuracy and rate of spatial input. For example, during a virtual a pick-and-place task force feedback cut positioning errors in half while speeding performance by about 20% (Noma, Miyasato, & Kishino, 1996).

4. Active haptic interfaces can reduce "information clutter": Unlike speakers and video monitors, haptic displays don't generally clutter a user's environment with unnecessary information. A good example of this property is a pager set to vibrate rather than beep. This haptic display provides only the right message ("You have a page"), to the right person (the owner), at the right time. This specificity is likely to become more important as embedded processors make more real-world objects intelligent and active. The same considerations suggest that haptic displays may reduce information clutter in VEs of increasing complexity.

## 3. HUMAN HAPTIC SYSTEM

In the real world, whenever an individual touches an object, forces are imposed on the skin. The net forces as well as the posture and motion of various limb segments are conveyed to the brain as *kinesthetic* information (the term *proprioceptive* is approximately equivalent; see endnote 2 in chap. 7, this volume), conveyed by multiple sources such as receptors in the joints, tendons, and muscles. This is the means by which the coarse properties of objects, such as large shapes and springlike compliances that require hand or arm motion in probing them, are sensed. In addition, the spatial and temporal variations of the force distributions within the contact region on the skin are conveyed as *tactile* information by several types of receptors embedded in the skin. Fine texture, small shapes, softness, and slipping of surfaces are felt through the tactile sensors. The temperature of the skin, which, in turn, is related to the temperature and thermal properties of the object, is also sensed through specialized tactile sensors.

In addition to the tactile and kinesthetic sensory subsystems, the human haptic system consists of the motor system that enables active exploration or manipulation of the environment and a cognitive system that can link sensations to perception and action. In general, a tactual image is composed of both tactile and kinesthetic sensory information, and is controlled by motor commands based on the user's intention. Because of the large number of degrees of freedom, multiplicity of the subsystems, spatially distributed heterogeneous sensory receptors, and the sensorimotor nature of haptic tasks, the human haptic abilities and limitations that prescribe the design specifications of haptic interfaces are difficult to quantify.

Haptic interfaces in VE or teleoperation systems receive the intended motor-action commands from the human and display tactual images to the human. A successful haptic interface represents a good match between the human haptic system and hardware for sensing and display. The primary input–output variables of the interfaces are displacements and forces, including their spatial and temporal distributions. Haptic interfaces can therefore be viewed as generators of mechanical impedances that represent a relationship between forces and displacements (and their derivatives) over different locations and orientations on the skin surface at each instant of time. In contact tasks involving finite impedances, either displacement or force can be viewed as the control variable, and the other is a display variable, depending on the control algorithms employed. However, consistency among free-hand motions and contact tasks is best achieved by viewing the position and motion of the hand as the control variable, and the resulting net force vector and its distribution within the contact regions as the display variables.

Because the human user is sensing and controlling the position and force variables of the haptic interface, the performance specifications of the interface are directly dependent on human abilities. In a substantial number of simple tasks involving active touch, one of the tactile and kinesthetic information classes is fundamental for discrimination or identification, whereas the other is supplementary. For example, in the discrimination of length of rigid objects held in a pinch grasp between the thumb and the forefinger (Durlach, Delhorne, Wong, Rabinowitz, & Holherbach, 1989), kinesthetic information is fundamental, whereas tactile information is supplementary. In such tasks, sensing and control of variables such as fingertip displacements are crucial. In contrast, for the detection of surface texture or slip, tactile information is fundamental, whereas kinesthetic information is supplementary (Srinivasan, Whitehouse, & LaMotte, 1990). Here, the sensing of spatiotemporal force distribution within the contact region provides the basis for inferences concerning contact conditions and object properties. Both classes of information are clearly necessary and equally important in more complex haptic tasks.

Detailed reviews of the human haptic system are available, focusing on position sense (Clark & Horch, 1986), skin sensitivity (Sherrick & Cholewiak, 1986), and perception (Loomis & Lederman, 1986). In this section a brief overview focused on issues relevant to haptic interfaces is provided. Sections 3.1 and 3.2 are excerpted from Srinivasan (1995) and summarize briefly the psychophysical results available on human haptic abilities in real environments at two levels: (1) sensing and control of interface variables and (2) perception of contact conditions and object properties. These results are also gathered in tables in the Appendix of this chapter. Although humans can feel heat, itch, pain, and so forth, through sensory nerve endings in the skin, these sensations are not discussed here because the availability of practical interface devices employing them is unlikely in the near future. This section concludes by emphasizing some features of the human haptic system that can provide guidance for haptic interface design (see section 3.3).

## 3.1 Sensing and Control of Interface Variables

### *3.1.1 Limb Position and Motion*

A large variety of psychophysical experiments have been conducted concerning the perception of limb position and motion (Clark & Horch, 1986; Jones & Hunter, 1992). It has been found that humans can detect joint rotations of a fraction of a degree performed over a time interval of the order of a second. The bandwidth of the kinesthetic sensing system has been estimated to be 20 to 30 Hz (Brooks, 1990). It is generally accepted that human sensitivity to rotations of proximal joints is higher than that of more distal joints. The just noticeable difference (JND) is about 2.5 degrees for the finger joints, 2 degrees for the wrist and elbow, and about 0.8 degrees for the shoulder (Tan, Srinivasan, Eberman, & Cheng, 1994). In locating a target position by pointing a finger, the speed, direction, and magnitude of movement, as well as the locus of the target, can all affect accuracy. In the discrimination of length of objects by the finger-span method (Durlach et al., 1989; Tan, Pang, & Durlach, 1992), the JND is about 1 mm for a reference length of 10 mm, and increases to 2 to 4 mm for a reference length of 80 mm, thus violating Weber's law (i.e., JND is not proportional to the reference length). In the kinesthetic space, psychophysical phenomena such as anisotropies in the perception of distance and orientation, apparent curvature of straight lines, non-Euclidean distance measures between two points, and others have been reported. For reviews, see Fasse, Kay, and Hogan (1990), Hogan, Kay, Fasse, and Mussa-Ivaldi (1990), and Loomis and Lederman (1986).

Investigations of the human ability in controlling limb motions have typically measured human tracking performance with manipulanda having various mass, spring, and damping characteristics (Brooks, 1990; Jones & Hunter, 1992; Poulton, 1974; Sheridan, 1992). The differential thresholds for position and movement have been measured to be about 8% (Jones & Hunter, 1992). Human bandwidth for limb motions is found to be a function of the mode of operation: 1 to 2 Hz for unexpected signals; 2 to 5 Hz for periodic signals; up to 5 Hz for internally generated or learned trajectories; and about 10 Hz for reflex actions. For a review see Brooks (1990).

### *3.1.2 Net Forces of Contact*

When a person contacts or presses objects through active motion of the hand, the contact forces are sensed by both the tactile and kinesthetic sensory systems. Overall contact force is probably the single most important variable that determines both the neural signals in the sensory system, as well as the control of contact conditions through motor action. It appears that the JND for contact force is 5% to 15% of the reference force value over a wide range of conditions involving substantial variation in force magnitude, muscle system, and experimental method, provided that the kinesthetic sense is involved in the discrimination task (Jones, 1989; Pang, Tan, & Durlach, 1991; Tan et al., 1992). In closely related experiments exploring human ability to distinguish among objects of different weights, a slightly higher JND of about 10% has been observed. For reviews see Clark and Horch (1986) and Jones (1986). An interesting illusion first observed in the late 19th century by Weber and reviewed more recently (Sherrick & Cholewiak, 1986) is that cold objects feel heavier than warm ones of equal weight. In experiments involving grasping and lifting of objects using a two-finger pinch grasp (Johansson & Westling, 1984) it has been shown that individuals have exquisite control over maintaining the proper ratio between grasping and lifting forces (i.e., the orientation of the contact force vector) so that the objects do not slip. However, when tactile information was blocked using local anesthesia, this ability deteriorated significantly because individuals could not sense contact conditions such as the occurrence of slip and hence did not apply appropriate compensating grasp forces. Thus, good performance in tasks involving contact requires the sensing of appropriate forces, as well as using them to control contact conditions. The maximum controllable force that can be exerted by a finger pad is about 100 N, and the resolution in visually tracking constant forces is about 0.04 N or 1%, whichever is higher (Srinivasan & Chen, 1993; Tan, Srinivasan, Eberman, & Cheng, 1994).

### 3.2 Perception of Contact Conditions and Object Properties

Although humans experience a large variety of tactile sensations when touching objects, these sensations are really combinations of a few building blocks or primitives. For simplicity, normal indentation, lateral skin stretch, relative tangential motion, and vibration are the primitives for conditions of contact with an object. Surface microtexture, shape (mm size), and compliance can be thought of as the primitives for the majority of object properties perceived by touch. The human perception of many of these primitives is through tactile information conveyed by mechanoreceptors in the skin.

Considerable research effort has been invested on psychophysics of vibration perception and electrocutaneous stimulation using single or multiple probes. For a review see Sherrick and Cholewiak (1986). These studies are mostly directed at issues concerned with tactile communication aids for individuals who are blind, deaf, or deaf and blind, areas that are beyond the scope of this chapter. A comprehensive list of references describing such tactile displays can be found in Kaczmarek and Bach-y-Rita (1993) and Reed, Durlach, and Braida (1982). In designing these devices, human perceptual abilities in both temporal and spatial domains are of interest. The human threshold for the detection of vibration of a single probe is about 28 dB (relative to 1 $\mu$m peak) for 0.4 to 3 Hz. It decreases at the rate of $-5$ dB/octave for 3 to 30 Hz, and decreases further at the rate of $-12$ dB/octave for 30 to about 250 Hz, after which the threshold increases for higher frequencies (Bolanowski, Gescheider, Verrillo, & Checkosky, 1988; Rabinowitz, Houtsma, Durlach, & Delhorne, 1987). Spatial resolution on the finger pad, as measured by the localization threshold of a point stimulus, is about 0.15 mm (Loomis, 1979), whereas the two-point limen is about 1 mm (Johnson & Phillips, 1981).

To answer questions concerning perception and neural coding of roughness or spatial resolution, precisely shaped rigid surfaces consisting of mm-sized bar gratings (Lederman & Taylor, 1972; Morley, Goodwin, & Darian-Smith, 1983; Phillips & Johnson, 1981a, 1981b; Sathian, Goodwin, John, & Darian-Smith, 1989), embossed letters (Phillips, Johnson, & Browne, 1983; Phillips, Johnson, & Hsiao, 1988), or Braille dots (Darian-Smith, Davidson, & Johnson, 1980; Lamb, 1983a, 1983b) have been used in psychophysical and neurophysiological experiments. For a review see Johnson and Hsiao (1992). The perception of surface roughness of gratings is found to be solely due to the tactile sense and is dependent on groove width, contact force, and temperature but not scanning velocity (Loomis & Lederman, 1986). Some of the salient results on the perception of slip, microtexture, shape, compliance, and viscosity are given below. Humans can detect the presence of a 2 $\mu$m high single dot on a smooth glass plate stroked on the skin, based on the responses of Meissner-type rapidly adapting fibers (RAs; LaMotte & Whitehouse, 1986; Srinivasan, Whitehouse, & LaMotte, 1990). Moreover, humans can detect a 0.075 $\mu$m high grating on the plate, owing to the response of Pacinian corpuscle fibers (LaMotte & Srinivasan, 1991). Among all the possible representations of the shapes of objects, the surface curvature distribution seems to be the most relevant for tactile sensing (LaMotte & Srinivasan, 1993; Srinivasan & LaMotte, 1991). Human discriminability of compliance of objects depends on whether the object has a deformable or rigid surface (Srinivasan & LaMotte, 1995). When the surface is deformable, the spatial pressure distribution within the contact region is dependent on object compliance, and hence information from cutaneous mechanoreceptors is sufficient for discrimination of subtle differences in compliance. When the surface is rigid, kinesthetic information is necessary for discrimination, and the discriminability is much poorer than that for objects with deformable surfaces. For deformable objects with rigid surfaces held in a pinch grasp, the JND for compliance is about 5% to 15% when the displacement range is fixed, increases to 22% when it is roved (varied randomly), and can be as high as 99% when cues arising out of mechanical work done are eliminated (Tan et al., 1992; Tan, Durlach, Shao, & Wei, 1993). Using a contralateral-limb matching procedure involving the forearm, it has been found (Jones & Hunter, 1992) that the differential thresholds for stiffness and viscosity are 23% and 34%, respectively. It has been found that a stiffness of at least 25 N/mm is needed for an object to be perceived as rigid by human observers (Tan et al., 1994). See the tables in the Appendix for a summary of these results.

## 3.3 Aspects of the Human Haptic System That Have Special Relevance to VE Hardware

A few aspects of the human haptic system that pertain to the design of haptic interface hardware deserve special attention. The following points are addressed:

1. A haptic precision gradient suggests that interface hardware deployed at distal body segments (e.g., fingertips) provides more benefit than interface hardware deployed proximally (e.g., shoulder).
2. A perceptual emphasis on transient stimuli suggests that users may tolerate considerable drift errors in haptic display hardware.
3. The human tendency to move the preferred hand with respect to the nonpreferred hand (rather than the world) suggests that two-handed interfaces offer significant advantages over one-handed.
4. The wide range of information transfer rates for different methods of manual communication suggests the importance of developing efficient "haptic languages" for interaction with virtual environments.

### *3.3.1 Distal to Proximal Gradient in Precision*

Given the importance of the fingertips for manipulation, it is reasonable to use them as a reference for describing more proximal body segments. Viewed this way, one finds a consistent gradient in performance, such that the skin and segments closer to the fingertip can be sensed and controlled more precisely than those closer to the trunk. This trend holds for detecting indentation of the skin and fingertip displacement, resolving position targets and rate of information transfer.

Gradient in tactile resolution: On the distal half of the fingertip humans can sense slow indentations of about 20 microns, sense vibrations ($\sim$250 Hz) of about 0.1 microns, distinguish separate points until they are within about 1 mm, and sense translations as small as 0.15 mm. At more proximal points on the limbs and trunk, sensitivity in all of these categories is poorer. For example, on the upper arm, vibration amplitude must be about 10 times larger to be detected, and slowly applied forces must be about 2 times larger. This distal to proximal gradient generally holds, but the tongue and lips are a notable exception. See Sherrick and Cholewiak (1986) for a review.

Gradient in detecting movement: A similar precision gradient is observed in detecting fingertip displacement. See Clark and Horch (1986) for a review. If the fingertip is moved passively at the speed humans normally use for pointing, then a displacement of about 0.3 mm can be detected if movement is constrained to the distal finger joint. If the movement is constrained to the elbow joint, detection requires about twice as much fingertip displacement (0.6 mm), and at the shoulder, twice as much again (1.2 mm; Hall & McCloskey, 1983). Thus, in terms of linear displacement of the fingertip, the distal segments again show higher sensitivity.

Gradient in resolving position targets: If participants are asked to discriminate or reproduce fingertip locations, they can typically distinguish about three targets when movement is constrained to 70 degrees of flexion of the middle joint of the index finger (Clark, Larwood, Davis, & Deffenbacher, 1995). This makes each target about 25 mm wide (one third of the corresponding range of fingertip displacement). Although participants can distinguish more targets over the angular range of motion of the more proximal joints, this precision does not compensate for the increased displacement of the fingertip these joints permit. Thus, flexion of the most proximal index finger joint allows participants to resolve fingertip targets about 40 mm wide, the wrist about 70 mm, the elbow about 100 mm, and the shoulder about 80 mm (Clark et al., 1995). Again, more distal segments generally show higher precision.

Gradient in pointing speed: If participants actively point with the limb, they can specify about 4.2 to 4.5 bit/sec using the finger joints with a tool held in a pen grasp (Balakrishnan & MacKenzie, 1997). If movement is constrained to a more proximal joint (wrist abduction), the rate decreases by about 30%, to about 4 bit/sec. Other studies report the same trend with far steeper gradients. An extreme example is Langolf (1976), in which they report 38 bit/sec with fingers only, 23 bit/sec with wrist and fingers, and 10 bit/sec with shoulder and elbow. The neck is estimated to provide only about 4 bit/sec (Card, Mackinlay, & Robertson, 1991). One suspects that this gradient is due to the increasing rotary inertia of limbs as the axis of rotation is shifted proximally, and also due to the fact that the human body seems to have more precise sensing and control hardware deployed at the distal body segments. In this case, more distal segments show higher rates of information transfer.

Implications for haptic interfaces: Since the bandwidth of human motor performance typically limits bandwidth of haptic input devices (Card et al., 1991), these results suggest that interfaces that sense finger movement may allow users to perform more quickly than they can with interfaces that only sense movement of the palm or arm. Human-computer interaction (HCI) developers have had mixed success with attempts to speed spatial input by including the fingers (Balakrishnan & MacKenzie, 1997; Zhai, Milgram, & Buxton, 1996).

A second consequence of the haptic precision gradient is that the quality of simulations is not degraded unduly when forces that ought to be grounded in the world are grounded on more proximal segments of the body. (See section 3.3.2 for a more detailed explanation of force grounding). CyberGrasp illustrates this body-grounded approach. This device renders forces on the fingertips that mimic what a user would experience handling objects in the real world. However, forces are grounded on the back of the hand, so the device fails to render appropriate torques about the wrist, elbow, or shoulder joints. Still, it provides a satisfactory haptic simulation, probably because it addresses the distal body segments (fingers), where the haptic system has greatest sensitivity. Experiments (Richard & Cutkosky, 1997) show that just rendering fingertip forces (i.e., grounding force on the middle phalanx) provides cues that lead participants to stop at virtual walls with penetration depths comparable to those achieved by world-grounded haptic interfaces that display appropriate torques around all joints.

### 3.3.2 Perceptual Importance of Change

Broadly speaking, haptic sensors in skin and muscle perform like systems of second order or higher. That is, the response of the sensor depends on both rate and magnitude of stimulus. Perhaps because haptic input emphasizes higher time derivatives, the precision of human haptic output also depends strongly on time. Over short intervals (< 1 sec), fine displacements can be sensed and controlled. However, over longer intervals (1 sec to 1 min) the haptic system alone does not generally notice or correct for substantial position errors (> 1 cm of fingertip displacement; Clark & Horch, 1986).

This emphasis on higher time derivatives is also apparent in the difference between the small, rapid displacements of a passive limb that can just be detected, and the 10- to 100-fold larger errors in actively reproducing pose of the same limb. The shoulder joint illustrates the trend. Here a fingertip displacement of about 1.2 mm can be detected (Hall & McCloskey, 1983), but for participants to reproduce target positions reliably with the shoulder, the targets at the fingertip must be about 70 times wider (80 mm; Clark, Larwood, Davis, & Deffenbacher, 1995).

Implications for haptic interfaces: This laxity in absolute position sensing may offer the haptic interface designer some leverage. For example, many active interfaces can probably be recentered at 1 to 10 mm/sec (up to a few centimeters) without attracting a user's attention or degrading the quality of the interaction. This could help compensate for the limited workspace of many devices.

### 3.3.3 Bimanual Frame of Reference

Humans naturally perform many manual tasks by setting a frame of reference with the nonpreferred hand (e.g., positioning a piece of paper) and then operating the preferred hand in this frame (e.g., writing on the paper; Guiard, 1987). This preference also affects the precision of a user's performance with haptic interfaces. For positioning tasks, errors of the preferred hand are typically about twofold smaller relative to the nonpreferred hand (~ 50 mm) than they are relative to the world (~ 100 mm; Mine, Brooks, & Sequin, 1997). It is not yet clear how much of this improvement is simply due to subtracting out sway of the torso and how much is due to the participant's experience with this sort of manipulation in the real world. Regardless of the source of improvement, the effect is robust. Participants can perform spatial input in about half as much time (Hinckley, Pausch, & Proffitt, 1997) when the preferred hand operates in the frame of the nonpreferred hand. These results suggest that two-handed interfaces offer about a two-fold improvement in speed and accuracy.

### 3.3.4 Factors That Determine Rates of Information Transfer

Coordination of movement: Since inertia of the limbs limits the rate of motor production to less than 5 Hz, and the precision of motor output typically limits information transfer to a few bits per joint, rapid information transfer depends on coordination of multiple joints. Work thus far (Zhai & Milgram, 1998) has shown that humans performing a one-handed 6 degree-of-freedom (DOF) docking task tend to set position of a grasped object first (simultaneously adjusting 3 DOF), and then specify orientation separately (simultaneously setting 3 more DOF). In these experiments, the haptic interface paradigm (position versus force input to the computer) did not have much effect on this basic division of labor. However, these efforts have provided some useful tools for haptic interface development—generic techniques for measuring how well humans can specify different kinematic parameters simultaneously. Advances in HCI require finding these kinematic parameters, and developing hardware that can transduce them.

Codes for information transfer: When a motor activity is more complicated than just pointing, the code used to transmit information has a large effect. The same fingers that are used to send Morse code at about 3 bit/sec can be used to type at about 14 bit/sec, transmitting information about 2,000-fold faster. See Appendix for information transfer rates for different modes of manual communication. Coding is comparably important for manual reception of information. Advances in HCI require finding rich codes for haptic input and output that users can learn quickly and easily.

## 4. CURRENT HARDWARE

In performing tasks with a haptic interface, the human user conveys desired motor actions by physically manipulating the interface, which, in turn, displays tactual sensory information to the user by appropriately stimulating his or her tactile and kinesthetic sensory systems. Thus, in general, haptic interfaces can be viewed as having two basic functions: (1) to measure the positions and contact forces (and time derivatives) of the user's hand (or other body parts) and (2) to display contact forces and positions (or their spatial and temporal distributions) to the user. Among these position (or kinematic) and contact force variables, the choice of which ones are the motor action variables (i.e., inputs to the computer) and which are the sensory display variables (i.e., inputs to the human) depends on the hardware and software design as well as the tasks for which the interface is employed. Typically, the user's hand position is sensed by the interface and contact forces computed by rendering algorithms are displayed to the user.

A primary classification of haptic interactions with real environments or VEs that affects interface design can be summarized as follows: (1) free motion, in which no physical contact is made with objects in the environment; (2) contact involving unbalanced resultant forces, such as pressing an object with a finger pad; (3) contact involving self-equilibrating forces, such as squeezing an object in a pinch grasp. Depending on the tasks for which a haptic interface is designed, some or all of these elements will have to be adequately simulated by the interface. For example, grasping and moving an object from one location to another involves all three elements. The design constraints of a haptic interface are strongly dependent on which of these elements it needs to simulate. Consequently, the interfaces can be classified according to whether they are force-reflecting or not, as well as by what types of motions (e.g., how many degrees of freedom) and contact forces they are capable of simulating.

An alternative but important distinction in haptic interactions with real environments or VEs is whether an object is touched, felt, and manipulated directly or with a tool. Which of these two types of interactions is supposed to be simulated seriously affects the complexity in the design of a haptic interface. Note that an ideal interface, designed to provide realistic simulation of direct haptic exploration and manipulation of objects, would be able to simulate handling with a tool as well. Such an interface would measure the position and posture of the user's hand, display forces to the hand, and make use of a single hardware configuration (e.g., an exoskeleton with force and tactile feedback) that could be adapted to different tasks by changes in software alone. For example, displaying forces on the fingers and palm when they were in proper position for grasping a hammer would simulate wielding this tool. However, the large number of degrees of freedom of the hand, extreme sensitivities of cutaneous receptors, together with the presence of mass, friction, and limitations of sensors and actuators in the interface make such an ideal impossible to achieve with current technology. In contrast, an interface in the form of a tool handle, for which reconfigurability within a limited task domain is achieved through both hardware and software changes, is quite feasible. Thus, one of the basic distinctions among haptic interfaces is whether they attempt to approximate the ideal exoskeleton or employ the tool-handle approach.

Another distinction concerning haptic interfaces has to do with whether the device "grounds" forces on the body or on the world. A "body grounded" device, such as a hand exoskeleton, is capable of simulating some forces (e.g., the resistance of a tennis ball to squeezing with the thumb and forefinger), but cannot simulate all forces (e.g., the torque about the user's shoulder due to the weight of the ball). In principle, a "world grounded" device, such as a desktop robot could simulate both types of forces, but is generally not portable. User performance with the two types of force grounding have been compared for some tasks (Richard & Cutkosky, 1997) and found roughly equivalent.

## 4.1 Current Technology

Compared to audio (see chap. 4, this volume) and video (see chap. 3, this volume) hardware, haptic interface hardware for VEs is in an early stage of development. Many of the devices available today have been motivated by needs predating those of VE technology. Simple position/motion-measuring systems have long been employed to provide control inputs to the computer. These have taken many forms, such as those that involve contact with the user without controlled force display (e.g., keyboards, computer mice, trackballs, joysticks, passive exoskeletal devices) and those that measure position/motion without contact (e.g., optical and electromagnetic tracking devices). Applications motivating development of these devices have ranged from the control of equipment (e.g., instruments, vehicles) to biomechanical study of human motion (e.g., gait analysis, time and motion studies).

The early developments in force-displaying haptic interfaces were driven by the needs of the nuclear energy industry and others for remote manipulation of materials (Sheridan, 1992). The force-reflecting teleoperator master arms in these applications were designed to communicate to the operator information about physically real tasks. The recognition of the need for good-quality force displays by early researchers (Goertz, 1964; Hill, 1979) continues to be relevant to today's VE applications. However, the dual challenges of making free motion feel unimpeded but making virtual surfaces feel stiff requires hardware with low friction, low apparent inertia, and very high bandwidth. Although Sutherland's (1965) pioneering description of VEs included force-reflecting interfaces, development of practical devices has proven difficult.

A wide variety of devices are under development in companies and universities worldwide. A rough breakdown of major types of haptic interfaces that are currently available or being developed is as follows:

1. Ground-based devices
   - Joysticks, mice, steering wheels, flight yokes
   - Tool-based (pen or instrument)
2. Body-based devices
   - Flexible (gloves and suits worn by user)
   - Exoskeletal (jointed linkages affixed to user)
3. Tactile displays

### 4.1.1 *Ground-Based Devices*

Joysticks are probably the oldest of these technologies and were originally conceived to control aircraft. Even the earliest of control sticks, connected by mechanical wires to the flight surfaces of aircraft, unwittingly presented force information about loads on flight surfaces to pilots. In general, these devices may be passive (i.e., not force reflecting), as in joysticks used for cursor positioning, or active (i.e., force reflecting), as in many of today's modern flight-control sticks. Many ground-based devices are now commercially available:

1. Force-reflecting joysticks are now commercially available in a wide range of prices and capabilities. Low-cost devices ($100–$1,000) with 2 actuated DOF are targeted primarily toward video games (Microsoft Sidewinder, Immersion Impulse Stick, I-Force). Devices with more DOF are produced in smaller quantities, generally have higher precision, and cost more ($1,000–$10,000). Joysticks with 3 actuated DOF include the Immersion Impulse Engine 3000 and Cybernet PER-Force 3DOF. More dof are available in the Cybernet PER-Force Handcontroller (6 DOF). Force-reflecting mice with 2 actuated DOF are also commercially available at low cost (e.g., Immersion FeelIt mouse, about $100). The video-game industry has also led to mass production of steering wheels with 1 actuated DOF, and flight yokes with 2 actuated DOF.
2. Pen-based force-reflecting interfaces are now mass-produced for general-purpose work. The Phantom (from SensAble Technologies) is a popular commercial desktop interface that comes in a variety of sizes, with either 3 or 6 actuated DOF. At the time of this writing, units are in the price range of about ($13,500–$61,000).
3. Force-reflecting surgery simulators are also in mass production. The Immersion Laparoscopic Impulse Engine drives the tips of surgical tools to simulate laparoscopic procedures. It offers 3 actuated DOF (5 sensed) for about $9,000. The Freedom-7 (with 7 actuated DOF) appears to be near market. Dedicated telesurgery systems (e.g., the Intuitive Surgical daVinci system) incorporate multiple hand-masters with 4 to 7 DOF per hand, but full force-feedback has not yet been implemented. The current price range is about three quarters to a million dollars.

These devices represent the commercial fruition of decades of research on teleoperation hand masters. For reviews of this field see Jacobus, Riggs, Jacobus, and Weinstein (1992), Meyer, H. L. Applewhite, and Biocca (1992), Brooks (1990), McAffee and Fiorini (1991), Honeywell, Inc. (1989), and Okamura, Smaby, and Cutkosky (2000). For reviews of the ergonometrics of hand controllers (shape, switch placement, motion and force characteristics, etc.) see Brooks and Bejczy (1985). For a review of actuator technologies see Hollerbach, Hunter, and Ballantyne (1992) and chapter 11 (this volume).

Notable applications of force-reflecting hand controllers to VEs include project GROPE at the University of North Carolina (Brooks, Ouh-Young, & Batter, 1990). In this simulator, the Argonne Mechanical Arm (ARM), and more recently the Phantom, were used successfully for force reflection during interactions with simulations of molecule docking. Haptic interactions with data from a scanning tunneling microscope have also been simulated (Taylor, 1994). The MIT Sandpaper is a 3-DOF joystick that is capable of displaying virtual textures (Minsky, Ouh-Young, Steele, Brooks, & Behensky, 1990). In Japan, notable desktop master manipulators have been built at Tsukuba University (Iwata, 1990; Noma & Iwata, 1993), ATR Laboratories in Kyoto (Noma et al., 1996), and Tokyo Institute of Technology (Walairacht, Koike, & Sato, 2000). At the University of British Columbia, high-performance hand controllers have been developed by taking advantage of magnetic levitation technology (Salcudean, Wong, & Hollis, 1992). At McGill University, the 2-DOF Pantograph has been developed for desktop applications, and the Freedom 7 has been developed for surgical simulation (Hayward, Gregorio, Astley, Greenish, & Doyon, 1997; Ramstein & Hayward, 1994). In conjunction with MPB technologies, the Freedom 6S hand controller has also been developed (MPB, 2000). PER-Force hand controllers were developed in conjunction with NASA

(Cybernet, 2000). The Phantom (Massie & Salisbury, 1994) was developed at MIT. Hand controllers that provide dynamically reprogrammable passive constraints have been developed at Northwestern University (Colgate, Peshkin, & Wannasuphoprasit, 1996) and Grenoble University (Troccaz-J & Delnondedieu-Y, 1996). A hand controller with very high peak stiffness has been developed at the University of Washington (Adams, Klowden, & Hannaford, 2000).

One of the most complex force-reflecting devices built to date is the Dexterous Teleoperation System Master designed by Sarcos, Inc., in conjunction with the University of Utah's Center for Engineering Design and the Naval Ocean Systems Center (NOSC). Although it is primarily ground-based, by having attachment points at the forearm and upper arm of the user it has the advantages of an exoskeleton, such as a large workspace comparable to that of the human arm. This device utilizes high-performance hydraulic actuators to provide a wide dynamic range of force exertion at relatively high bandwidth on a joint-by-joint basis for 7 DOF. Another high-performance force-reflecting master is a ground-based system for teleoperated eye surgery built by Hunter, Lafontaine, Nielsen, Hunter, & Hollerbach (1990). At Harvard, a planar manipulator has been developed to study precision teleoperation with a pinch grasp between the thumb and the index finger (Howe, 1992).

### *4.1.2 Body-based Devices*

Body-based devices fit over and move with the limbs or fingers of the user. Because they are kinematically similar to the arm and hands that they monitor and stimulate, they have the advantage of the widest range of unrestricted user motion. As position-measuring systems, body-based devices (gloves, suits, etc.) are relatively inexpensive and comfortable to use. A few of the more common commercially available gloves include the VTI CyberGlove, and the iReality 5$^{th}$ glove. Depending on the manufacturer and model, these typically resolve 5 to 23 DOF to approximately 1 degree. The FakeSpace Pinch Glove takes a different approach, detecting contact between the tips of two or more digits.

Body-based devices with rigid exoskeletons afford force display and slightly more accurate pose sensing, typically at the expense of greater bulk. For a review of the design issues in exoskeletal devices see Shimoga (1992). At the time of this writing, one force reflecting exoskeleton is in mass production (the VTI CyberGrasp). Research on body-based force reflecting hand exoskeletons has been ongoing for decades. See Okamura et al. (2000) for a review. Some current designs are notable. The Rutgers Portable Dextrous Master (Burdea, Zhuang, Roskos, Silver, & Langrana, 1992; Fabiani, Burdea, Langrana, & Gomez, 1996), is a light, simple exoskeleton that grounds fingertip forces on the palm rather than the back of the hand. A light, cable-based, 2-DOF, body-grounded hand controller (HapticGEAR) for immersive VEs has been developed at Tokyo University (Hirose, Ogi, Yano, & Kakehi, 1999). A force-reflecting hand master that senses and controls fingertip position (but not joint angles) is under development at Johnson Space Center (Sinah, Endsley, Riggs, & Millspaugh, 1999). Regardless of the exact mechanical design, providing force feedback with body-based hand controllers remains a difficult problem, placing great demands on minimizing actuator size to make the control bandwidth of the device commensurate with human haptic capabilities.

### *4.1.3 Tactile Displays*

While the display of net forces is appropriate for coarse object interaction, investigators have also recognized the need for more detailed displays within the regions of contact. In particular, the display of tactile information (e.g., force distributions for conveying information on texture and slip), though technically difficult, has long been considered desirable for remote manipulation (Bliss & Hill, 1971).

Very crude tactile displays for VEs are now in mass production. The Aurora Interactor uses a voice coil to display vibration through a vest or seat cushion. CyberTouch, by VTI, is another vibrotactile transducer made to mount on the backs of the fingers. A 1.25-inch diameter vibrotactile transducer has been developed by Engineering Acoustics, Inc., and applied to the torso for a situational awareness display (Raj, McGrath, Rochlis, Newman, & Rupert, 1998). The Displaced Temperature Sensing System from CM Research displays temperature changes to the fingertip. Although they are not designed expressly for VE applications, electromechanical Braille cells are also commercially available (American Foundation for the Blind, 2000) and are being actively developed (Petersen, 2000).

Research on tactile display systems in the last two decades has been motivated in part by efforts to convey visual and auditory information to deaf and blind individuals (Bach-y-Rita, 1982; Reed et al., 1982), and more recently to simulate tissue palpation during minimally invasive surgery (Howe, Peine, Kontarinis, & Son, 1995; Moy, Singh, Tan, & Fearing, 2000a).

Many of these displays have the general appearance of pin or balloon arrays, often with an antialiasing membrane between the pins and fingertip skin. A variety of actuators have been employed, including DC solenoids (Frisken-Gibson, Bach-Y-Rita, Tompkins, & Webster, 1987), shape memory alloys (Howe et al., 1995), compressed air (Moy, Wagner, & Fearing, 2000b), piezoelectric vibrators (Chanter & Summers, 2000), and electrorheological fluids (Monkman, 1992). Also under development are micromachined tactile displays that stimulate skin with tangential tactors (Ghodssi, Beebe, White, & Denton, 1996), electrical current (Beebe, Hymel, Kazcmarek, & Tyler, 1995), or tangential force due to electrostatic attraction (Tang & Beebe, 1998). A review of principles and technical issues in vibrotactile and electrotactile displays can be found in Kaczmarek and Bach-y-Rita (1993) and Shimoga (1992).

Some interesting alternatives to the rectangular array are under development. Researchers have developed an array of concentric rings that simulates compliant surfaces by controlling the rate at which the skin contact area spreads as normal force is applied to the display (Ambrosi, Bicchi, De Rossi, & Scilingo, 1999). A "grasp blob" is also under development (Aldridge, Carr, England, Meech, & Solomonides, 1996). This device consists of a grasped, fist-sized bag that can be depressurized rapidly, compacting the particles within it into a solid mass.

### 4.2 Hardware Summary

Computer keyboards, mice, and trackballs are the simplest haptic interfaces and are being widely used to interact with computers. Position-sensing gloves and exoskeletons without force reflection are also available on the market but are used mainly for research purposes. Among the force-reflecting desktop devices, joysticks, mice, small robots, and surgical simulators are commercially available. Force-reflecting exoskeletons are harder to design for adequate performance, and only a few are commercially available. Tactile displays offer particularly difficult design challenges because of the high density of receptors in the skin to which they must apply the stimulus. Basic (1-DOF) tactile stimulators are now available for VEs, whereas tactile arrays for VEs are still in development.

## 5. FUTURE WORK

Design specifications for haptic interfaces depend on the biomechanical, sensorimotor, and cognitive abilities of humans. Therefore, multidisciplinary studies involving biomechanical and psychophysical experiments together with computational models for both are needed in order to have a solid scientific basis for device design. Perhaps to a lesser extent, neurophysiological studies concerning peripheral and central neural representations and the processing of information in the human haptic system will also aid in design decisions concerning the kinds of information that need to be generated and how these should be displayed. A major barrier to progress from the perspectives of biomechanics, psychophysics, and neuroscience has been the lack of robotic stimulators capable of delivering a large variety of stimuli under sufficiently precise motion and force control.

## 5.1 Hardware Development

For the foreseeable future, it appears progress in haptics will be limited by the development of new actuator hardware. It is now clear that active interfaces with two and perhaps three DOF are commercially sustainable. It remains to be seen whether devices with 4 or more active DOF eventually become general purpose computer peripherals or remain limited to specialized applications like surgical simulation.

Hardware for displaying distributed forces on the skin remains an especially challenging problem. Exploration of novel technologies is needed for quantum improvements in miniature rotary and linear actuators. Shape memory alloys (SMAs), piezoelectrics, microfluidics, and other microelectromechanical systems (MEMS) for tactile display all warrant further investigation. Mechanical flexibility is another major challenge in the development of general-purpose wearable tactile displays. The human skin is a dynamic environment subject to bumps, scratches, and deformations due to flexion of the body segments. To put an array of actuators on it in a package that does not break or encumber an active user may require methods for manufacturing or embedding actuator arrays in flexible substrates, a challenge MEMS technology is only beginning to address.

## 5.2 Methods of Stimulation

The right balance of complexity and performance in system capabilities is generally task dependent. In particular, the fidelity with which tactual images have to be displayed, and motor actions have to be sensed by the interface depends on the task, stimulation of other sensory modalities, and interaction between the modalities. Experimenting with available haptic interfaces, in conjunction with visual and auditory interfaces, is necessary to identify needed design improvements. Design compromises and tricks for achieving the required task performance capabilities or telepresence (immersion) need to be investigated (see chaps. 21, 22, and 40, this volume). One of the tricks might be the use of illusions (such as visual dominance) to fool the human user into believing a less than perfect multimodal display (DiFranco, Beauregard, & Srinivasan, 1997; Srinivasan, Beauregard, & Brock, 1996). Techniques such as filtering the user's normal tremor or the use of sensory substitution within a modality (e.g., the use of tactile display to convey kinesthetic information) or among different modalities (e.g., visual display of a force) need to be developed to overcome limitations of devices and limitations of the human user, perhaps to achieve supernormal performance. To tackle the ever-present time delays, efficient and reliable techniques for running model-based and real-time controls concurrently are needed.

## 5.3 Evaluation of Haptic Interfaces

Evaluation of haptic interfaces is crucial to judge their effectiveness and to isolate aspects that need improvement. However, such evaluations performed in the context of teleoperation have been so task-specific that it has been impossible to derive useful generalizations and to form effective theoretical models based on these generalizations. There is a strong need to specify a set of elementary manual tasks (basis tasks) that can be used to evaluate and compare manual capabilities of a given system (human, robotic, VE) efficiently. Ideally, this set of basis tasks should be such that (1) knowledge of performance on these tasks enables one to predict performance on all tasks of interest and (2) it is the minimal set of tasks (in terms of time consumed to measure performance on all tasks in the set) that has this predictive power (Durlach, personal communication, 1990).

In this void, some tasks have become de facto standards, including point-to-point movements (i.e., Fitts' task, Balakrishnan & MacKenzie, 1997), target selection (Card et al., 1991), and docking (Zhai, Milgram, & Drascic, 1993), measured in terms of mean completion time, RMS error, and information transfer rates. For basic input devices like joysticks and mice, a task-based ISO9000 standard has been proposed (Douglas, Kirkpatrick, & MacKenzie, 1999). However, it is not clear how well these measurements extrapolate to devices intended for tasks more complicated than pointing.

Some progress has been made in developing a standard set of physical measurements for devices (e.g., workspace, degrees of freedom, force range, precision, etc.). A standard set of physical measurements is proposed by Hayward and Astley (1996). Although these physical measurements are clearly necessary, at this point in the development of haptic interfaces they are probably not sufficient to guide device development. For example, a device might perform brilliantly at displaying shoulder torque (a physical measurement), but if this parameter is not very effective at helping the user complete many tasks or feel immersed the device may not really be "good." Measures of the relative perceptual importance of force and position cues at different body sites during different tasks are also required in order to ascertain what stimuli are worth displaying to the user.

**TABLE 5.1**
Mechanical Properties of Human Upper Limb

| *Degrees of Freedom* | |
|---|---|
| Shoulder | 4 (shrug, flexion, abduction, rotation) |
| Elbow | 2 (flexion, pronation) |
| Wrist | 2 (flexion, pronation) |
| Each digit | 4 (abduction + 3 flexions) |
| Bandwidth of motor system[1] | |
| Unexpected signals | 1–2 Hz |
| Periodic signals | 2–5 Hz |
| Reflex actions | 10 Hz |
| Minimum hand closure time | 0.09 seconds |
| Fingertip forces[2] | |
| Typical pinch grip | 1–10 N |
| Controllable | Up to 100 N |
| Control resolution | 0.05 to 0.5 N |
| Grasp force range | 50–500 N |

[1](Brooks, 1990).
[2](Tan et al., 1994).

**TABLE 5.2**
Mechanical Impedance of Human Upper Limb

| | |
|---|---|
| *Passive elbow*[1] | |
| Up to 2 Hz | ~ 0.02 (N · m)/rad) |
| 10 Hz | ~ 0.1 (N · m)/rad) |
| 100 Hz | ~ 5 (N · m)/rad) |
| 1000 Hz | ~ 100 (N · m)/rad) |
| *Lumped response of index finger pad and segments* | |
| "Relax"[2] | 0.14–0.4 N/mm |
| "Resist force as much as possible"[2] | 0.30–0.86 N/mm |
| Straight[3] | 0.8–2.2 N/mm |
| Flexed[3] | 0.4–2.0 N/mm |
| *Lumped response of finger pad and digits during pen grasp*[2] | |
| "Relax" | 0.34–1.25 N/mm |
| "Resist force as much as possible" | 0.79–2.41 N/mm |
| *Lumped response of finger pad to pseudo-static plate indentor*[4] | |
| Physiologic range of finger pad indentation | ~ 3 mm |
| Stiffness at initial contact (0–1 mm) | ~ 0.1 N/mm |
| Upon further indentation (1–2 mm) | ~ 0.4 N/mm |
| Upon further indentation (2–3 mm) | ~ 1.0 N/mm |
| Indentations > 3 mm | 10–100 N/mm |

[1](Jones, Hunter, Lafontaine, Hollerbach, & Kearny, 1991).

[2](Buttolo, 1996) Measured 30 ms after start of force ramp at 0.1–0.5 N/s. Stiffness of pen grasp is six- to sevenfold greater than index finger alone. Precision of position control improves two- to threefold.

[3](Milner & Franklin, 1995).

[4](Gulati & Srinivasan, 1996).

**TABLE 5.3**
Tactile Sensation[1]

| | |
|---|---|
| *Indentation threshold* | |
| Static | 20 $\mu$m, 0.3 mN/mm$^2$ |
| 10 Hz | 10 $\mu$m |
| 250 Hz | 0.1 $\mu$m |
| *Feature detection* | |
| Texture | 0.075 $\mu$m (improvement due to lat scanning) |
| Single dot | 2 $\mu$m |
| Normal force when detecting features | 0.4–1.1 N (typical), 0.3–4.5 N (observed range) |
| *Temporal resolution* | |
| Time between successive taps | 10 ms |
| Bandwidth sensed | 0–1000 Hz |
| Reaction time | 70 ms to 500 ms (reflexive slip to threshold detection) |
| *Spatial resolution* | |
| Lateral localization | 0.15 mm |
| Lateral 2-point limen | 1 mm at fingertip |
| *Hot/cold* | |
| Detection threshold for temperature change | 0.01°C/s |
| Reaction time | 300–900 ms |
| Persistent hot | > 40°, (pain > 48°) |
| Persistent cold | < 20°, (pain < 15°) |

[1]For source details see Sherrick and Cholewiak (1986), Srinivasan (1995).

**TABLE 5.4**
Active Touch Including Tactile, Kinesthetic, and Motor Systems[1]

| *Parameter* | *(JND)* |
|---|---|
| Length | 10% or less |
| Velocity | 10% |
| Acceleration | 20% |
| Force | 7% |
| Compliance | 8% (rigid surface, piano key) |
| Compliance | 3% (deformable surface, rubber) |
| Viscosity | 14% |
| Mass | 21% |
| Rigidity perception | 25 N/mm or greater |

[1]For source details see Beauregard, Srinivasan, and Durlach (1995) and Chen and Srinivasan (1998)

**TABLE 5.5**
Information Transfer Rates Through the Hands[1]

| *Method for manual production* | Rate (bits/sec) |
|---|---|
| Morse code | 1.8–3.6 |
| Handwriting | 3.5–7 (estimated) |
| Finger spelling | 7–11 |
| Typing | 7.2–14.4 |
| Court stenography | 13.5–27 |
| Signing with ASL | 15–30 |
| *Method for manual reception* | |
| Kinesthetic Morse code | 0.9–1.4 |
| Tactual reception of finger spelling | 8.1–? |
| Tactual reception of spoken English (Tadoma) | 11.2–22.5 |
| Tactual reception of ASL | 11.7–23.4 |

[1]Gathered in Reed and Durlach (1998).

## 6. REFERENCES

Adams, R. J., Klowden, D., & Hannaford, B. (2000). Stable haptic interaction using the Excalibur force display. In *Proceedings of the IEEE International Conference on Robotics and Automation* (ICRA 2000). (Vol. 1, pp. 770–775). San Francisco: IEEE.

Aldridge, R. J., Carr, K., England, R., Meech, J. F., & Solomonides, T. (1996). Getting a grasp on virtual reality. In *Proceedings of the CHI 96* (pp. 229–230). Vancouver, DC Canada: ACM.

Ambrosi, G., Bicchi, A., De Rossi, D., & Scilingo, E. P. (1999). The role of contact area spread rate in haptic discrimination of softness. In *Proceedings of the 1999 IEEE International Conference on Robotics and Automation* (Vol. 1, pp. 305–310). Detroit, MI: IEEE.

American Foundation for the Blind. (2000, July). *Braille technology* [Online]. Available: http://www.afb.org/technology/fs_brailtech.html

Bach-y-Rita, P. (1982). Sensory substitution in rehabilitation. In L. Illis, M. Sedgwick, & H. Granville (Eds.), *Rehabilitation of the Neurological Patient* (pp. 361–383). Oxford, England: Blackwell Scientific.

Balakrishnan, R., & MacKenzie, I. S. (1997). Performance differences in the fingers, wrist and forearm in computer input control. In *Proceedings of the CHI 97* (pp. 303–310). Atlanta, GA: ACM.

Beauregard, G. L., Srinivasan, M. A., & Durlach, N. I. (1995). Manual resolution of viscosity and mass. In *Proceedings of the Proceedings of the ASME Dynamic Systems and Control Division* (DSC-Vol. 57-2, pp. 657–662). Chicago: ASME.

Beebe, D. J., Hymel, C., Kazcmarek, K., & Tyler, M. (1995). A polyimide-based electrostatic tactile display. In *Proceedings of the 17th Annual International Conference of the IEEE Engineering in Medicine and Biology Society*. Montreal, Canada: IEEE.

Bliss, J. C., & Hill, J. W. (1971). *Tactile perception studies related to teleoperator systems* (Final Rep., NASA-CR-114346). Menlo Park, CA: Stanford Research Institute.

Bolanowski, S. J., Gescheider, G. A., Verrillo, R. T., & Checkosky, C. M. (1988). Four channels mediate the mechanical aspects of touch. *Journal of Acoustical Society of America, 84*, 1680–1694.

Brooks, F. P., Ouh-Young, M., & Batter, J. (1990). Project GROPE: Haptic displays for scientific visualization. *Computer Graphics, 24*(4), 177–185.

Brooks, T. L. (1990). Telerobotic response requirements. In *Proceedings of the IEEE International Conference on Systems, Man and Cybernetics* (pp. 113–120). Los Angeles: IEEE.

Brooks, T. L., & Bejczy, A. K. (1985). *Hand controllers for teleoperation* (Tech. Rep. JPL Publication 85-11). Pasadena, CA: Jet Propulsion Laboratory.

Burdea, G. C. (1996). *Force and touch feedback for virtual reality*. New York: John Wiley & Sons.

Burdea, G. C., Zhuang, J., Roskos, E., Silver, D., & Langrana, N. (1992). A portable dexterous master with force feedback. *Presence, 1*(1), 18–28.

Card, S. K., Mackinlay, J. D., & Robertson, G. G. (1991). A morphological analysis of the design space of input devices. *ACM Transactions on Information Systems, 9*(2), 99–122.

Chanter, C., & Summers, I. (2000, July). *The Exeter fingertip stimulator array for virtual touch* [Online]. Available: http://newton.ex.ac.uk.medphys/pages/array1.html

Chen, J. S., & Srinivasan, M. A. (1998). *Human haptic interaction with soft objects: Discriminability, force control, and contact visualization* (Touch Lab Rep. 7, RLE TR-619). Cambridge, MA: Massachusetts Institute of Technology.

Clark, F. J., & Horch, K. W. (1986). Kinesthesia. In K. R. Boff, L. Kaufman, & J. P. Thomas (Eds.), *Handbook of Perception and Human Performance* (Vol. 1, pp. 13-1–13-62). New York: John Wiley & Sons.

Clark, F. J., Larwood, K. J., Davis, M. E., & Deffenbacher, K. A. (1995). A metric for assessing acuity in positioning joints and limbs. *Experimental Brain Research, 107*, 73–79.

Colgate, J. E., Peshkin, M. A., & Wannasuphoprasit, W. (1996). Nonholonomic haptic display. In *Proceedings of the 1996 IEEE International Conference on Robotics and Automation* (pp. 539–544). Minneapolis, MN: IEEE.

Cybernet. (2000, July). R & D Services [Online]. Available: http://www.cybernet.com

Darian-Smith, I., Davidson, I., & Johnson, K. O. (1980). Peripheral neural representation of spatial dimensions of a textured surface moving across the monkey's finger pad. *Journal of Physiology, 309*, 135–146.

DiFranco, D. E., Beauregard, G. L., & Srinivasan, M. A. (1997). The effect of auditory cues on the haptic perception of stiffness in virtual environments. In *Proceedings of the ASME Dynamic Systems and Control Division* (DSC-Vol. 61, pp. 17–22). Dallas, TX: ASME.

Douglas, S. A., Kirkpatrick, A. E., & MacKenzie, I. S. (1999). Testing pointing device performance and user assessment with the ISO 9241, Part 9 Standard. In *Proceedings of the CHI 99* (pp. 215–222). Pittsburgh, PA: ACM.

Durlach, N. I., Delhorne, L. A., Wong, A., Ko, W. Y., Rabinowitz, W. M., & Hollerbach, J. (1989). Manual discrimination and identification of length by the finger-span method. *Perception and Psychophysics, 46*(1), 29–38.

Fabiani, L., Burdea, G., Langrana, N., & Gomez, D. (1996). Human interface using the Rutgers master II force feedback interface. In *Proceedings of the VRAIS 1996* (pp. 54–59). Santa Clara, CA: ACM.

Fasse, E. D., Kay, B. A., & Hogan, N. (1990). Human haptic illusions in virtual object manipulation. In *Proceedings of the 14th Annual Conference of IEEE Engineering in Medicine and Biology Society* (pp. 1917–1918). Paris: IEEE.

Fitzmaurice, G. W., Balakrishnan, R., Kurtenbach, G., & Buxton, B. (1999). An exploration into supporting artwork orientation in the user interface. In *Proceedings of the CHI 99* (pp. 167–199). Pittsburgh, PA: ACM.

Frisken-Gibson, S. F., Bach-Y-Rita, P., Tompkins, W. J., & Webster, J. G. (1987). A 64-Solenoid, four-level fingertip search display for the blind. *IEEE Transactions on Biomedical Engineering BME, 34*(12), 963–965.

Ghodssi, R., Beebe, D. J., White, V., & Denton, D. D. (1996). Development of a tangential tactor using a LIGA/MEMS linear microactuator technology. In *Proceedings of the 7th Symposium on Micro-Mechanical Systems at the ASME Winter Annual Meeting* Atlanta, GA: ASME.

Goertz, R. (1964). Some work on manipulator systems at ANL: Past, present and a look at the future, *Seminars on Remotely Operated Special Equipment* (Vol. 1). Argonne National Laboratory: ANL.

Guiard, Y. (1987). Asymmetric division of labor in human skilled bimanual action: The kinematic chain as a model. *Journal of Motor Behavior, 19*(4), 486–517.

Gulati, R. J., & Srinivasan, M. A. (1996). *Determination of Mechanical Properties of the Human Fingerpad.* In *Vivo, Using a Tactile Stimulator* (Touch Lab Report 3, RLE TR-605). Cambridge, USA: MIT.

Hall, L. A., & McCloskey, D. I. (1983). Detections of movements imposed on finger, elbow, and shoulder joints. *Journal of Physiology, 335*, 519–533.

Hayward, V., & Astley, O. R. (1996). Performance measures for haptic interfaces. In *Proceedings of the Robotics Research: The 7th International Symposium* (pp. 195–207). Herrsching, Germany: Springer Verlag.

Hayward, V., Gregorio, P., Astley, O., Greenish, S., & Doyon, M. (1997). Freedom 7: a high fidelity seven axis haptic device with the application to surgical training. In *Proceedings of the ISER 97.* Barcelona, Spain:

Hill, J. W. (1979). *Study of modeling and evaluation of remote manipulation tasks with force feedback* (Final Rep., JPL Contract 95-5170). JPL.

Hinckley, K., Pausch, R., & Proffitt, D. (1997). Attention and visual feedback: The bimanual frame of reference. In *Proceedings of the 1997 Symposium on Interactive 3D Graphics* (pp. 121–126). Providence, RI: ACM.

Hirose, M., Ogi, T., Yano, H., & Kakehi, N. (1999). Development of wearable force display (HapticGEAR) for immersive projection displays. In *Proceedings of the VRAIS 1999* (pp. 79). Houston, TX: ACM.

Hoffman, H. G. (1998). Physically touching virtual objects using tactile augmentation enhances the realism of virtual objects. In *Proceedings of the IEEE Virtual Reality Annual International Symposium '98* (pp. 59–63). Atlanta GA: IEEE.

Hogan, N., Kay, B. A., Fasse, E. D., & Mussa-Ivaldi, F. A. (1990). Haptic illusions: Experiments on human manipulation and perception of "virtual objects". *Cold Spring Harbor Symposia on Quantitative Biology, 55*, 925–931.

Hollerbach, J. M., Hunter, I. W., & Ballantyne, J. (1992). A comparative analysis of actuator technologies for robotics. *The Robotics Review, 2*, 299–342.

Honeywell, Inc. (1989). *Hand controller commonality study* (Tech. Rep.). Clearwater, FL: Honeywell Inc., Avionics Division, for McDonnell Douglas Space Systems, Co.

Howe, R. D. (1992). A force-reflecting teleoperated hand system for the study of tactile sensing in precision manipulation. In *Proceedings of the 1992 IEEE International Conference on Robotics and Automation* Nice, France: IEEE.

Howe, R. D., Peine, W. J., Kontarinis, D. A., & Son, J. S. (1995). Remote palpation technology for surgical applications. *IEEE Engineering in Medicine and Biology Magazine, 14*(3), 318–323.

Hunter, I. W., Lafontaine, S., Nielsen, P. M. F., Hunter, P. J., & Hollerbach, J. M. (1990). Manipulation and dynamic mechanical testing of microscopic objects using a tele-micro-robot system. *IEEE Control Systems Magazine, 10*(2), 3–9.

Immersion. (2000, July). *Touch sense technology* [Online]. Available: http://www.immersion.com

Iwata, H. (1990). Artificial reality with force-feedback: Development of desktop virtual space with compact master manipulator. *Computer Graphics, 24*(4), 165–170.

Jacobus, H. N., Riggs, A. J., Jacobus, C. J., & Weinstein, Y. (1992). Implementation issues for telerobotic handcontrollers: Human-robot ergonomics. In M. Rahimi & W. Karwowski (Eds.), *Human-robot interaction* (pp. 284–314). New York: Taylor & Francis.

Johansson, R. S., & Westling, G. (1984). Roles of glabrous skin receptors and sensorimotor memory in automatic control of precision grip when lifting rougher or more slippery objects. *Experimental Brain Research, 56*, 550–564.

Johnson, K. O., & Hsiao, S. S. (1992). Neural mechanisms of tactual form and texture perception. *Annual Review of Neuroscience, 15*, 227–250.

Johnson, K. O., & Phillips, J. R. (1981). Tactile spatial resolution — I. Two point discrimination, gap detection, grating resolution and letter recognition. *Journal of Neurophysiology, 46*(6), 1177–1191.

Jones, L. A. (1986). Perception of force and weight: Theory and research. *Psychological Bulletin, 100*(1), 29–42.

Jones, L. A. (1989). Matching forces: Constant errors and differential thresholds. *Perception, 18*(5), 681–687.

Jones, L. A., & Hunter, I. W. (1992). Human operator perception of mechanical variables and their effects on tracking performance. In *Proceedings of the 1992 Advances in Robotics, ASME Winter Annual Meeting* (DSC Vol. 42, pp. 49–53). Anaheim, CA: ASME.

Jones, L. A., Hunter, I. W., Lafontaine, S., Hollerbach, J., & Kearny, R. (1991). Wide bandwidth measurements of human elbow joint mechanics. In *Proceedings of the 17th Canadian Medical and Biological Engineering Conference* (pp. 135–136). Banff, Canada: IEEE.

Kaczmarek, K. A., & Bach-y-Rita, P. (1993). Tactile displays. In W. Barfield & T. Furness, III (Eds.), *Virtual environments and advanced interface design.* Oxford, England: Oxford University Press.

Lamb, G. D. (1983a). Tactile discrimination of textured surfaces: Peripheral coding in the monkey. *Journal of Physiology, 338*, 567–587(a).

Lamb, G. D. (1983b). Tactile discrimination of textured surfaces: Psychophysical performance measurements in humans. *Journal of Physiology, 338*, 551–565.

LaMotte, R. H., & Srinivasan, M. A. (Eds.). (1991). *Surface microgeometry: Tactile perception and neural encoding.* London: Macmillan Press.

LaMotte, R. H., & Srinivasan, M. A. (1993). Responses of cutaneous mechanoreceptors to the shape of objects applied to the primate fingerpad. *Acta Psychologica, 84*, 41–51.

LaMotte, R. H., & Whitehouse, J. (1986). Tactile detection of a dot on a smooth surface. *Journal of Neurophysiology, 56*, 1109–1128.

Langolf, G. D. (1976). An investigation of Fitts' Law using a wide range of movement amplitudes. *Journal of Motor Behavior, 8*(2), 113–128.

Lederman, S. J., & Taylor, M. M. (1972). Fingertip force, surface geometry, and the perception of roughness by active touch. *Perception and Psychophysics, 12*, 401–408.

Loomis, J. M. (1979). An investigation of tactile hyperacuity. *Sensory Processes, 3*, 289–302.

Loomis, J. M., & Lederman, S. M. (1986). Tactual Perception. In K. R. Boff, L. Kaufman, & J. P. Thomas (Eds.), *Handbook of perception and human performance* (Vol. 1, pp. 31-1–31-41). New York: John Wiley & Sons.

Massie, T. H., & Salisbury, J. K. (1994). Probing virtual objects with the PHANToM haptic Interface. In *Proceedings of the ASME WAM*. Chicago: ASME.

McAffee, D. A., & Fiorini, P. (1991). Hand controller design requirements and performance issues in telerobotics. In *Proceedings of the International Conference on Advanced Robotics (ICAR). IEEE.*

Meyer, K., H. L. Applewhite, H. L., & Biocca, F. A. (1992). A survey of position trackers. *Presence, 1*(2), 173–200.

Milner, T. E., & Franklin, D. W. (1995). Two-dimensional endpoint stiffness of human fingers for flexor and extensor loads. In *Proceedings of the ASME Winter Annual Meeting, Dynamic Systems and Control Division* (DSC-Vol. 57-2, pp. 649–656). San Francisco: ASME.

Miller, T., & Zeleznik, R. (1999). The design of 3D haptic widgets. In *Proceedings of the 1999 Symposium on Interactive 3D Graphics* Atlanta, GA: ACM.

Mine, M. R., Brooks, F. P., & Sequin, C. H. (1997). Moving objects in space: Exploiting proprioception in virtual-environment interaction. In *Proceedings of the SIGGRAPH '97* (pp. 19–26). Los Angeles: ACM.

Minsky, M., M., Ouh-Young, M., Steele, O., Brooks, F. P. J., & Behensky, M. (1990). Feeling and seeing: Issues in force display. *Computer Graphics, 24*(2), 235–243.

Monkman, G. J. (1992). An electrorheological tactile display. *Presence, 1*(2), 219–228.

Morley, J. W., Goodwin, A. W., & Darian-Smith, I. (1983). Tactile discrimination of gratings. *Experimental Brain Research, 49*(291–299).

Moy, G., Singh, U., Tan, E., & Fearing, R. S. (2000a). Human psychophysics for teletaction system design. *Haptics-E The Electronic Journal of Haptics Research, 1*(3), 1–20.

Moy, G., Wagner, C., & Fearing, R. S. (2000b). A compliant tactile display for teletaction. In *Proceedings of the IEEE International Conference on Robotics and Automation* (ICRA 2000). San Francisco: IEEE.

MPB Technologies. (2000, November). *Freedom 6S force feedback hand controller* [Online]. Available: www.mpb-technologies.ca

Nahvi, A., Nelson, D. D., Hollerbach, J. M., & Johnson, D. E. (1998). Haptic manipulation of virtual mechanisms from mechanical CAD designs. In *Proceedings of the IEEE International Conference on Robotics and Automation* (pp. 375–380). Leuven, Belgium: IEEE.

National Research Council. (1997). More than screen deep: toward every-citizen interfaces to the nation's information infrastructure. In *Toward every-citizen interfaces to the nation's information infrastructure* (pp. 94–99). Washington DC: National Academy Press.

Noma, H., & Iwata, H. (1993). Presentation of multiple dimensional data by 6.D.O.F force display. In *Proceedings of the IEEE/RSJ International Conference on Intelligent Robots and Systems* (pp. 1495–1500). Yokohama, Japan: IEEE.

Noma, H., Miyasato, T., & Kishino, F. (1996). A palmtop display for dexterous manipulation with haptic sensation. In *Proceedings of the CHI '96* (pp. 126–137). Vancouver, Canada: ACM.

Okamura, A. M., Smaby, N., & Cutkosky, M. R. (2000). An overview of dexterous manipulation. In *Proceedings of the IEEE International Conference on Robotics and Automation* (Vol. 1, pp. 255–262). San Francisco: IEEE.

Pang, X. D., Tan, H. Z., & Durlach, N. I. (1991). Manual discrimination of force using active finger motion. *Perception and Psychophysics, 49*(6), 531–540.

Petersen, B. (2000, July). *Petersen braille display unit* [Online]. Available: http://members.aol.com/petersenrc/petersen.htm

Phillips, J. R., & Johnson, K. O. (1981a). Tactile spatial resolution—II. Neural representation of bars, edges and gratings in monkey primary afferents. *Journal of Neurophysiology, 46*(6), 1192–1203.

Phillips, J. R., & Johnson, K. O. (1981b). Tactile spatial resolution—III. A continuum mechanics model of skin predicting mechanoreceptor responses to bars, edges, and gratings. *Journal of Neurophysiology, 46*(6), 1204–1225.

Phillips, J. R., Johnson, K. O., & Browne, H. M. (1983). A comparison of visual and two modes of tactual letter recognition. *Perception and Psychophysics, 34*, 243–249.

Phillips, J. R., Johnson, K. O., & Hsiao, S. S. (1988). Spatial pattern representation and transformation in monkey somatosensory cortex. *Proceedings of the National Academy of Sciences* (USA), *85*(6), 1317–1321.

Poulton, E. C. (1974). *Tracking skill and manual control.* New York: Academic Press.

Rabinowitz, W. M., Houtsma, A. J. M., Durlach, N. I., & Delhorne, L. A. (1987). Multidimensional tactile displays: Identification of vibratory intensity, frequency, and contactor area. *Journal of the Acoustical Society of America, 82*(4), 1243–1252.

Raj, A. K., McGrath, B. J., Rochlis, J., Newman, D. J., & Rupert, A. H. (1998). The application of tactile cues to enhance situation displays. In *Proceedings of the 3rd Annual Symposium Exhibition on Situational Awareness in the Tactical Air Environment* (pp. 77–84). Patuxent River, MD:

Ramstein, C., & Hayward, V. (1994). The pantograph: a large workspace haptic device for a multi-modal human–computer interaction. In *Proceedings of the CHI 94* Boston, MA: ACM.

Reed, C. M., & Durlach, N. I. (1998). Note on information transfer rates in human communication. *Presence, 7*(5), 509–518.

Reed, C. M., Durlach, N. I., & Braida, L. D. (1982). *Research on tactile communication: A review* (Vol. 20). Rockville, MD: American Speech-Language-Hearing Association.

Richard, C., & Cutkosky, M. R. (1997). Contact force perception with an ungrounded haptic interface. In *Proceedings of the 1997 ASME IMECE 6th Annual Symposium on Haptic Interfaces* (pp. 181–187). Dallas, TX:

Salcudean, S. E., Wong, N. M., & Hollis, R. L. (1992). A force-reflecting teleoperation system with magnetically levitated master and wrist. In *Proceedings of the IEEE International Conference on Robotics and Automation* (pp. 1420–1426). Nice, France: IEEE.

Salisbury, J. K., & Srinivasan, M. A. (1992). *Sections on haptics, in virtual environment technology for training* (BBN Rep. No. 7661). Cambridge, MA: The Virtual Environment and Teleoperator Research Consortium (VETREC), affiliated with MIT.

Sathian, K., Goodwin, A. W., John, K. T., & Darian-Smith, I. (1989). Perceived roughness of a grating: Correlation with responses of mechanoreceptive afferents innervating the monkey's fingerpad. *Journal of Neuroscience, 9*, 1273–1279.

SensAble Technologies. (2000, July). *Introducing the freeform modeling system* [Online]. Available: http://www.sensable.com

Sheridan, T. B. (1992). *Telerobotics, automation, and supervisory control.* Cambridge, MA: MIT Press.

Sherrick, C. E., & Cholewiak, R. W. (1986). Cutaneous Sensitivity. In K. R. Boff, L. Kaufman, & J. P. Thomas (Eds.), *Handbook of human perception and human performance* (Vol. 1, pp. 12-1–12-57). New York: John Wiley & Sons.

Shimoga, K. B. (1992). Finger force and touch feedback issues in dexterous telemanipulation. In *Proceedings of the NASA-CIRSSE International Conference on Intelligent Robotic Systems for Space Exploration*. Troy, NY:

Sinah, V. K., Endsley, E. W., Riggs, A. J., & Millspaugh, B. K. (1999). Force-reflecting, finger-position-sensing mechanism. *NASA Tech Briefs, November*, 55–56.

Srinivasan, M. A. (1995). Haptic interfaces. In N. I. Durlach & A. S. Mavor (Eds.), *Virtual reality: Scientific and technical challenges*: (Report of the Committee on Virtual Reality Research and Development, National Research Council, National Academy Press).

Srinivasan, M. A., & Basdogan, C. (1997). Haptics in virtual environments: taxonomy, research status, and challenges. *Computers and Graphics, 21*(4).

Srinivasan, M. A., Beauregard, G. L., & Brock, D. L. (1996). The impact of visual information on haptic perception of stiffness in virtual environments. In *Proceedings of the ASME Dynamic Systems and Control Division* (DSC-Vol. 58, pp. 555–559). Fairfield, CA: ASME.

Srinivasan, M. A., & Chen, J. S. (1993). Human performance in controlling normal forces of contact with rigid objects. In *Proceedings of the 1993 Advances in Robotics, Mechatronics, and Haptic Interfaces* (DSC-Vol. 49). New Orleans, LA: ASME.

Srinivasan, M. A., & LaMotte, R. H. (Eds.). (1991). *Encoding of shape in the responses of cutaneous mechanoreceptors*. London: Macmillan Press.

Srinivasan, M. A., & LaMotte, R. H. (1995). Tactile discrimination of softness. *Journal of Neurophysiology, 73*(1), 88–101.

Srinivasan, M. A., Whitehouse, J. M., & LaMotte, R. H. (1990). Tactile detection of slip: Surface microgeometry and peripheral neural codes. *Journal of Neurophysiology, 63*(6), 1323–1332.

Sutherland, I. E. (1965). The ultimate display. In *Proceedings of the IFIP Congress* 2 (pp. 506–508). New York: International Federation for Information Processing.

Tan, H. Z., Durlach, N. I., Shao, Y., & Wei, M. (1993). Manual resolution of compliance when work and force cues are minimized. In *Proceedings of the 1993 Advances in Robotics, Mechatronics and Haptic Interfaces, ASME Winter Annual Meeting* (DSC-Vol. 49, pp. 99–104). New Orleans, LA: ASME.

Tan, H. Z., Pang, X. D., & Durlach, N. I. (1992). Manual resolution of length, force and compliance. In *Proceedings of the 1992 ASME Winter Annual Meeting* (DSC-Vol. 42, pp. 13–18). Anaheim, CA: ASME.

Tan, H. Z., Srinivasan, M. A., Eberman, B., & Cheng, B. (1994). Human factors for the design of force-reflecting haptic interfaces. *American Society of Mechanical Engineers* (DSC-Vol.) 55-1.

Tang, H., & Beebe, D. J. (1998). A microfabricated electrostatic haptic display for the visually impaired. *IEEE Transactions on Rehabilitation Engineering, 6*(3), 241–248.

Taylor, R. (1994). The nanomanipulator: A virtual-reality interface to a scanning tunneling microscope. Unpublished doctoral dissertation, University of North Carolina, Chapel Hill.

Troccaz-J, & Delnondedieu-Y. (1996). Semi-active guiding systems in surgery, A two-DOF prototype of the passive arm with dynamic constraints (PADyC). *Mechatronics, 6*(4), 399–421.

Virtual Technologies, Inc. (2000, July). *Simulation, M/CAD, digital prototype evaluation, & virtual reality R&D* [Online]. Available: www.virtex.com

Walairacht, S., Koike, Y., & Sato, M. (2000). String-based haptic interface device for multi-fingers. In *Proceedings of the VRAIS 2000* (pp. 293). New Brunswick, NJ: IEEE.

Youngblut, C., Johnson, R. E., Nash, S. H., Wienclaw, R. A., & Will, C. A. (1996). *Review of virtual environment interface technology*. Alexandria, VA: Institute for Defense Analyses.

Zhai, S., & Milgram, P. (1998). Quantifying coordination in multiple DOF movement and its application to evaluating 6 DOF input devices. In *Proceedings of the CHI '98* (pp. 320–327). Los Angeles: ACM.

Zhai, S., Milgram, P., & Buxton, W. (1996). The influence of muscle groups on performance of multiple degree-of-freedom input. In *Proceedings of the CHI '96* (pp. 308–315). Vancouver, Canada: ACM.

Zhai, S., Milgram, P., & Drascic, D. (1993). An evaluation of four 6 degree-of-freedom input techniques. In *Proceedings of the INTERACT '93 and CHI '93 conference companion on human factors in computing systems* (pp. 123–125). Amsterdam, the Netherlands: ACM.

# 6

# Haptic Rendering in Virtual Environments

Cagatay Basdogan[1] and Mandayam A. Srinivasan[2]
[1]*Jet Propulsion Laboratory*
*California Institute of Technology*
*Pasadena, CA 91109*
*Cagatay.Basdogan@jpl.nasa.gov*
[2]*Laboratory for Human and Machine Haptics (The Touch Lab)*
*Massachusetts Institute of Technology*
*Cambridge, MA 02139 USA*
*srini@mit.edu*

## 1. INTRODUCTION

The goal of *haptic rendering* is to enable a user to touch, feel, and manipulate virtual objects through a haptic interface. With the introduction of high-fidelity haptic devices (see chap. 5, this volume), it is now possible to simulate the feel of even fine surface textures on rigid complex shapes under dynamic conditions. Starting from the early 1990s, significant progress has occurred in our ability to model and simulate haptic interactions with three-dimensional (3-D) virtual objects in real time (Salisbury & Srinivasan, 1997; Srinivasan & Basdogan, 1997). The rapid increase in the number of workshops, conference sessions, community web pages, and electronic journals on haptic displays and rendering techniques* indicates growing interest in this exciting new area of research called *computer haptics*. Just as computer graphics is concerned with synthesizing and rendering visual images, computer haptics is the art and science of developing software algorithms that synthesize computer generated forces to be displayed to the user for perception and manipulation of virtual objects through touch. Various applications of computer haptics have been developed in the areas of medicine (e.g., surgical simulation, telemedicine, haptic user interfaces for blind people, rehabilitation of patients with neurological disorders; see chaps. 47–51, this volume), entertainment

*See the proceedings of Phantom Users Group Workshops starting from 1996 (http://www.ai.mit.edu/conferences/), ASME Dynamics Systems and Control (DSC) Division starting from 1993, IEEE International Conference on Robotics and Automation, and IEEE Virtual Reality Conference. Also visit the haptic community pages at http://www.sensable.com/community/index.htm and http://haptic.mech.nwu.edu/, and the hapticse-journal at http://www.haptics-e.org/.

(e.g., 3-D painting, character animation, morphing and sculpting; see chap. 55, this volume), mechanical design (e.g., path planning and assembly sequencing; see chap. 54, this volume), and scientific visualization (e.g., geophysical data analysis, molecular manipulation; see chap. 53, this volume). More applications are anticipated as devices and rendering techniques improve and computational power increases. This chapter will primarily focus on fundamental concepts of haptic rendering, with some discussion of implementation details. Although, it is impossible to cite all relevant work within the constraints of this chapter, an attempt has been made to cover the major references. Given that current technology is mature for net force and torque feedback (as in tool usage in the real world) and not tactile feedback (as in actual distribution of force fields on the skin during contact with real objects), the discussion is restricted to techniques concerning the former. In general, the concepts discussed in the chapter include: (1) *haptic interaction paradigms*, which define the nature of the "haptic cursor" and its interaction with virtual objects; and (2) *object property display algorithms*, which enable rendering of surface and material properties of objects and their behavior through repeated use of the haptic interaction paradigm.

## 2. PRINCIPLES OF HAPTIC RENDERING: OBJECT SHAPE

Typically, a haptic rendering algorithm is made of two parts: (a) collision detection; and (b) collision response (see Fig. 6.1). As a user manipulates the probe of a haptic device, the new position and orientation of the haptic probe are acquired and collisions with virtual objects are detected (i.e., *collision detection*). If a collision is detected, interaction forces are computed using preprogrammed rules for *collision response* and conveyed to the user through the haptic device to provide him or her with the tactual representation of 3-D objects and their surface details. Hence, a haptic loop, which updates forces around 1 kHz (otherwise, virtual surfaces feel softer, or, at worst, instead of a surface it feels as if the haptic device is vibrating), includes at least the following function calls:

- get_position (vector and position); // position and/or orientation of the end effector
- calculate_force (vector and force); // user-defined function to calculate forces
- send_force (vector force); // calculate joint torques and reflect forces back to the user

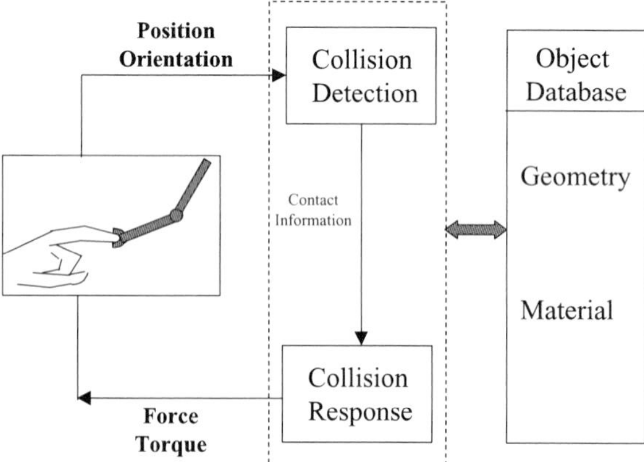


FIG. 6.1. A haptic interaction algorithm. Typically made of two parts: (a) collision detection; and (b) collision response. The haptic loop seen in the figure requires an update rate of around 1 kHz for stable force interactions. Computationally fast collision detection and response techniques are required to accommodate this requirement.

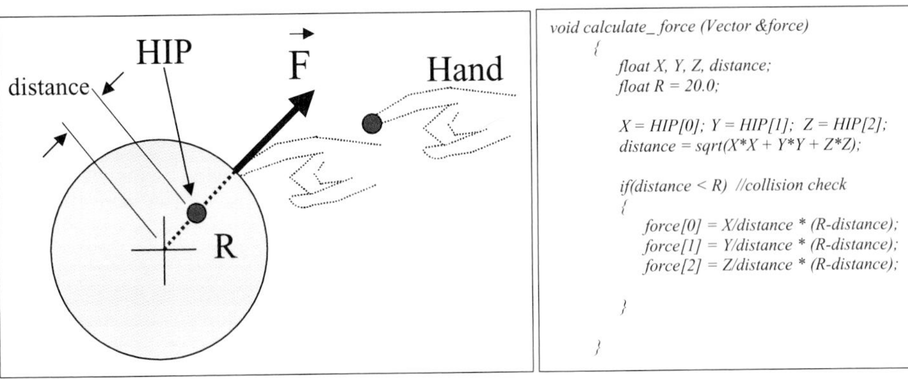
HIP
F
Hand
distance
R
void calculate_force (Vector &force)
{
float X, Y, Z, distance;
float R = 20.0;
X = HIP[0]; Y = HIP[1]; Z = HIP[2];
distance = sqrt(X*X + Y*Y + Z*Z);
if(distance < R) //collision check
{
force[0] = X/distance * (R-distance);
force[1] = Y/distance * (R-distance);
force[2] = Z/distance * (R-distance);
}
}

FIG. 6.2. Haptic rendering of a 3-D sphere in virtual environments. (The software code presented on the right-hand side calculates the direction and the magnitude of the reaction force for the sphere discussed in the example. The sphere has a radius of 20 units and is located at the origin of a 3-D virtual space.)

To describe the basic concepts of haptic rendering, consider a simple example, haptic rendering of a 3-D frictionless sphere, located at the origin of a 3-D virtual space (see Fig. 6.2). Assume that the user can only interact with the virtual sphere through a single point, which is the end point of the haptic probe, also known as the *haptic interaction point* (HIP). In the real world, this is analogous to feeling the sphere with the tip of a stick. As the 3-D space is freely explored with the haptic probe, the haptic device will be *passive* and will not reflect any force to the user until a contact occurs. Since the virtual sphere has a finite stiffness, HIP will penetrate into the sphere at the contact point. Once the penetration into the virtual sphere is detected and appropriate forces to be reflected back to the user are computed, the device will become *active* and reflect opposing forces to the user's hand to resist further penetration. The magnitude of the reaction force can easily be computed by assuming that it is proportional to the depth of penetration. Assuming no friction, the direction of this force will be along the surface normal as shown in Fig. 6.2.

As it can be seen from the example given above, a rigid virtual surface can be modeled as an elastic element. Then, the opposing force acting on the user during the interaction will be:

$$\vec{F} = k\Delta\vec{x}$$

where $k$ is the stiffness coefficient and $|\Delta\vec{x}|$ is the depth of penetration. Whereas keeping the stiffness coefficient low would make the surface perceived soft, setting a high value would make interactions unstable by causing undesirable vibrations. Figure 6.3 depicts changes in force profile with respect to position for real and virtual walls. Since the position of the probe tip is sampled digitally with a certain frequency during the simulation of a virtual wall, a "staircase" effect is observed. This staircase effect leads to energy generation (see the discussions and suggested solutions in Colgate & Brown, 1994, and Ellis, Sarkar, & Jenkins, 1996).

Although the basic recipe for haptic rendering of virtual objects seems easy to follow, rendering of complex 3-D surfaces and volumetric objects requires more sophisticated algorithms than the one presented for the sphere. The stringent requirement of updating forces around 1 kHz leaves very little CPU time for computing collisions and reflecting forces back to the user in real time when interacting with complex shaped objects. In addition, the algorithm given above for rendering of a sphere considered only "point-based" interactions (as if interacting with objects through the tip of a stick in the real world), which is far from what our hands are capable of in the real world. However, several haptic rendering techniques have been developed recently to simulate complex touch interactions in virtual environments (reviewed

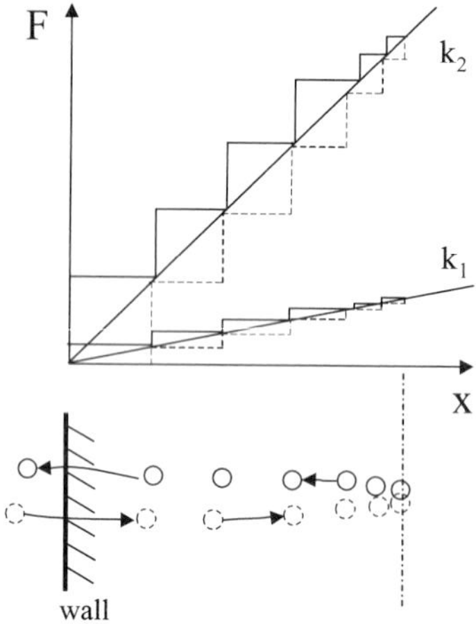


FIG. 6.3. Force-displacement curves for touch interactions with real and virtual walls. In the case of real walls, the force-displacement curve is continuous. However, a "staircase" effect is seen when simulating touch interactions with a virtual wall. This is due to the fact that haptic devices can only sample position information with a finite frequency. The difference in the areas enclosed by the curves that correspond to penetrating into and out of the virtual wall is a manifestation of energy gain. This energy gain leads to instabilities as the stiffness coefficient is increased (compare the energy gains for stiffness coefficients $k_2$ and $k_1$). On the other hand, a low value of the stiffness coefficient generates a soft virtual wall, which is not desirable either.

in Hollerbach & Johnson, 2001; Salisbury & Srinivasan, 1997; Srinivasan & Basdogan, 1997). The existing techniques for haptic rendering with force display can be distinguished based on the way the probing object is modeled: (1) a point (Adachi, Kumano, & Ogino, 1995; Avila & Sobierajski, 1996; Ruspini, Kalarov, & Khatib, 1997; Ho, Basdogan, & Srinivasan, 1999; Zilles & Salisbury, 1995); (2) a line segment (Basdogan, et al., 1997; Ho et al., 2000); or (3) a 3-D object made of a group of points, line segments, and polygons (McNeely et al., 1999). The type of interaction method used in simulations depends on the application.

In point-based haptic interactions, only the end point of the haptic device (i.e., the HIP) interacts with virtual objects (Fig. 6.4b). Each time a user moves the generic probe of the haptic device, the collision detection algorithm checks to see if the end point is inside the virtual object. If so, the depth of indentation is calculated as the distance between the current HIP and the corresponding surface point, also known as the *ideal haptic interface point* (IHIP; also called god-object, proxy point, or surface contact point; see Fig. 6.5). For exploring the shape and surface properties of objects in VEs, point-based methods are probably sufficient and could provide users with similar force feedback as what they would feel when exploring objects in

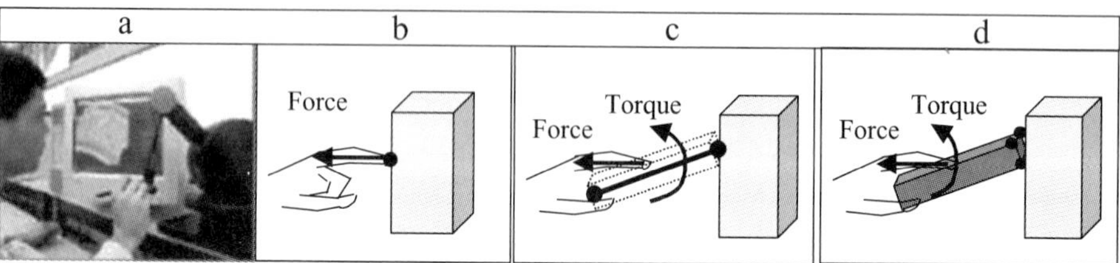


FIG. 6.4. Existing interaction methods for haptic rendering with force display can be distinguished based on the way the probing object is modeled: (b) a point; (c) a line segment; and (d) a 3-D object.

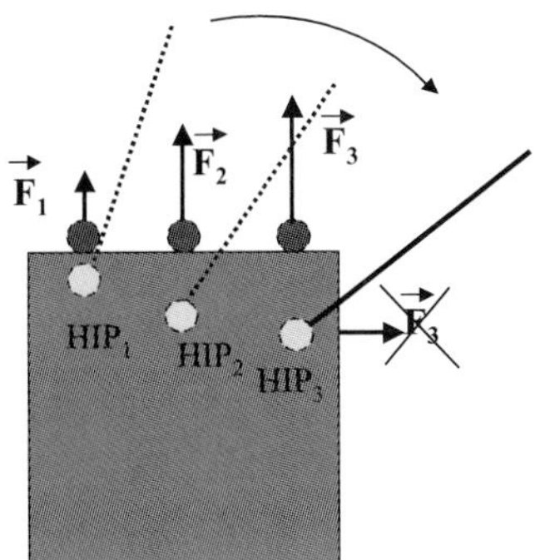


FIG. 6.5. A haptic rendering algorithm that is based on point interactions must calculate the IHIP for a given HIP at each rendering cycle. (IHIP can be considered as the surface point of the object with which the tool or finger would be in contact. The computation of this point relies on contact history. In the figure, for example, ignoring contact history and always choosing the closest point on the object surface as the new IHIP for a given HIP would make the user feel as if he is pushed out of the object as HIP moves from point 2 to point 3.

real environments with the tip of a stick. Using a point-based rendering technique, polyhedrons (Ho et al., 1999; Ruspini et al., 1997; Zilles & Salisbury, 1995). NURBS (Stewart, Chen, & Buttolo, 1997; Thompson, Johnson, & Cohen, 1997), implicit surfaces (Salisbury & Tarr, 1997), and volumetric objects (Avila & Sobierajski, 1996) have been successfully rendered. Point-based methods, however, are not capable of simulating more general *tool–object* interactions that involve single or multiple objects in contact with the tool at arbitrary locations of the tool. In such a context, both forces and torques displayed to the user need to be independently computed.

In ray-based interactions, the generic probe of the haptic device is modeled as a line segment whose orientation is taken into account, and collisions are checked between the finite line and objects. This approach enables the user to touch multiple objects simultaneously. In addition to forces, torque interactions can be simulated, which is not possible with point-based approaches (see Fig. 6.4c). Ray-based rendering can be considered as an approximation to long tools and as an intermediate stage in progress towards the full interaction between a 3-D cursor and 3-D objects. Also, if the geometric model of the probing object can be simplified to a set of connected line segments, then the ray-based rendering technique can be used and will be faster than simulation of full 3-D object interactions. For example, haptic interactions between a mechanical shaft and an engine block have been successfully simulated (Ho et al., 2000), as well as the interactions between laparoscopic surgical instruments and deformable objects (Basdogan, et al. 1998; Basdogan, Ho, & Srinivasan, 2001) using multiple line segments and the ray-based rendering technique. However, if the probing object has a complex geometry and cannot be easily modeled using a set of line segments, simulation of 6 degrees of freedom (DOF) object–object interactions has to be considered.

Simulation of haptic interactions between two 3-D polyhedra is desirable for many applications, but this is computationally more expensive than point-based and ray-based interactions (see Fig. 6.4d). Although a single point is not sufficient for simulating force and torque interactions between two 3-D virtual objects, a group of points, distributed over the surface of a probing object, has been shown to be a feasible solution. For example, McNeely et al. (1999) simulated touch interactions between the 3-D model of a teapot and a mechanical assembly.

Irrespective of the interaction method used, a haptic rendering algorithm must include both collision detection and collision response computations (see Fig. 6.1). Although collision detection has been extensively studied in computer graphics (Cohen, Lin, Manocha, & Ponamgi,

1995; Hubbard, 1995; Gottschalk, Lin, & Manocha, 1996; Lin, 1993), existing algorithms are not designed to work directly with haptic devices to simulate touch interactions in virtual environments. Moreover, merely detecting collisions between 3-D objects is not enough for simulating haptic interactions. How a collision occurs and how it evolves over time (i.e., *contact history*) are crucial factors (see the discussion in Fig. 6.5 on contact history) in haptic rendering to accurately compute the interaction forces that will be reflected to the user through a haptic device (Ho et al., 1999; Ho, Basdogan, & Srinivasan, 2000).

Although collision detection algorithms developed in computer graphics cannot be used in haptics directly, several concepts developed for fast collision detection can be ported to haptics. For example, haptic rendering algorithms can easily take advantage of: (1) space partitioning techniques (i.e., partitioning the space that encloses the object into smaller subspaces in advance for faster detection of first contact); (2) local search approaches (i.e., searching only the neighboring primitives for possible new contacts); and (3) hierarchical data structures (i.e., constructing hierarchical links between the primitives that make up the object for faster access to the contacted primitive). In addition to these improvements, a client–server model has to be considered to synchronize visual and haptic displays for achieving faster update rates. Using multithreading techniques, for example, one can calculate the forces at 1 kHz in one thread while updating the visual images at 30 Hz in another thread (Ho et al., 1999). If the forces cannot be computed at 1 kHz, a numerical scheme can be developed to extrapolate the new forces based on the previously computed ones to maintain the constant update rate of 1 kHz for stable interaction (Basdogan, 2000; Ellis, Sarkar, & Jenkins, 1996).

## 3. RENDERING OF SURFACE DETAILS: SMOOTH SHAPES, FRICTION AND TEXTURE

Quite frequently, it is desirable to display object surfaces as smooth and continuous, even when an underlying polyhedral representation is employed. In computer graphics, for example, illumination models are used to compute the surface color at a given point of a 3-D object (Foley, van Dam, Feiner, & Hughes, 1995). Then, shading algorithms are applied to shade each polygon by interpolating the light intensities (Gourand shading) or surface normals (Phong shading). Shading algorithms make the shared edges of the adjacent polygons invisible and provide the user with the display of visually smooth surfaces (Watt & Watt, 1992). One can integrate a similar approach into the study of haptics to convey the feel of haptically smooth object surfaces (Basdogan et al., 1997; Fukui, 1996; Morgenbesser & Srinivasan, 1996; Ruspini et al., 1997). By using the *force-shading* technique suggested by Morgenbesser and Srinivasan (1996), force discontinuities can be reduced and the edges of polyhedral object can be made to feel smoother. To implement force shading with polygonal surfaces (Basdogan et al., 1997), the surface normal at each vertex is precomputed by averaging the surface normals of neighboring polygons, weighted by their neighboring angles. During real-time computations, the collision point that divides the contacted triangle into three subtriangles is first detected. Then the surface normal ($\vec{N}_s$) can be calculated at the collision point by averaging the normals of the vertices ($\vec{N}_i$) of the contacted polygon, weighted by the areas $A_i$ of the three subtriangles:

$$\vec{N}_s = \frac{\sum_i^3 A_i \vec{N}_i}{\sum_i^3 A_i}$$

Haptic simulation of surface details such as friction and texture significantly improves the realism of virtual worlds. For example, friction is almost impossible to avoid in real-life

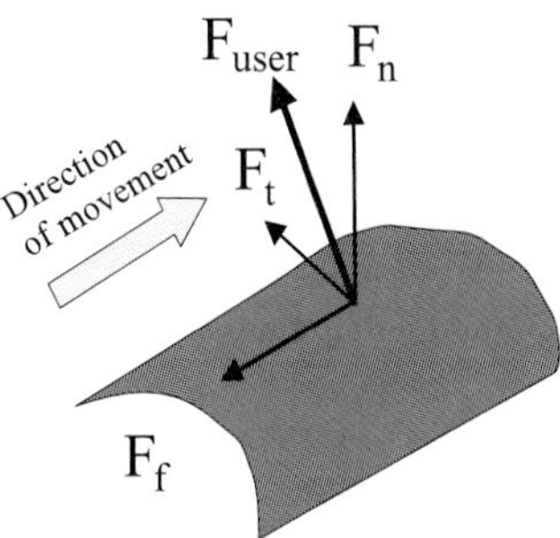


FIG. 6.6. Forces acting on the user ($F_{user} = F_n + F_t + F_f$) during the haptic simulation of friction and textures. (The normal force can be computed using a simple physics-based model such as Hooke's law ($F_n = k\,\Delta x$, where $|\Delta x|$ is the depth of penetration of the haptic probe into the virtual surface). To simulate coulomb friction, a force is created ($F_f = \mu F_n$, where $\mu$ is the coefficient of friction) that is opposite to the direction of the movement. To simulate textures, the magnitude and direction of the normal force vector are changed using the gradient vector of the texture field at the contact point.

and virtual surfaces without friction feel "icy-smooth" when they are explored with a haptic device. Similarly, most surfaces in nature are covered with some type of texture that is sensed and distinguished quite well by our tactile system (Srinivasan, Whitehouse, & LaMotte, 1990). Haptic texture is a combination of small-scale variations in surface geometry and its adhesive and frictional characteristics. However, displaying detailed geometry of textures would be computationally too expensive. Instead, both friction and texture can be simulated by appropriate perturbations of the reaction force vector computed using nominal object geometry and material properties. The major difference between friction and texture simulation via a haptic device is that the friction model creates only forces tangential to the nominal surface in a direction opposite to the probe motion, while the texture model can generate both tangential and normal forces in any direction (see Fig. 6.6).

Salcudean and Vlaar (1994) and Salisbury, Brock, Massie, Swarup, & Zilles, (1995) simulated static and dynamic friction such that the user feels the stick-slip phenomenon when the stylus of a haptic device is stroked over the surface of an object. By changing the mean value of the friction coefficient and its variation, more sophisticated frictional surfaces, such as periodic ones (Ho et al., 1999), and various grades of sandpaper (Green & Salisbury, 1997) can be simulated as well.

Haptic perception and display of textures in VEs require a thorough investigation, primarily because the textures in nature come in various forms. Luckily, graphics texturing has been studied extensively, and researchers can draw from that experience to simulate haptic textures in virtual environments. For example, bump mapping is a well-known technique in graphics for generating the appearance of a nonsmooth surface by perturbing surface normals. Blinn (1978) initially proposed this technique as a method of graphically rendering 3-D bumps for modeling clouds. His formulation depends on parameterization of the surface. Although the surface parameterization techniques have some advantages, they may generate uneven bump textures on the surface of complex shapes. Max and Becker (1994) suggested a direct method of mapping that does not require a transformation from global to parametric space. They developed a formulation that is purely based on the original surface normal and the local gradient of the height field that generates bumps on the surface of an object. Max and Becker utilized this technique to generate bumps on cloud contour surfaces. We used the same approach and perturbed the direction and magnitude of the surface normal ($\vec{N}_s$) to generate bumpy or textured surfaces that can be sensed tactually by users in virtual environments (Basdogan et al., 1997). The perturbed surface normals ($\vec{M}$) can be computed using the following formulation:

$$\vec{M} = \vec{N}_s - \nabla h + (\nabla h \cdot \vec{N}_s)\vec{N}_s$$
$$\nabla h = \frac{\partial h}{\partial x}\hat{i} + \frac{\partial h}{\partial y}\hat{j} + \frac{\partial h}{\partial z}\hat{k}$$

where, h(x,y,z) represents the texture field function and $\nabla h$ is the local gradient vector. One of the earliest methods of texture rendering that uses the force perturbation concept with a 2-DOF haptic device was developed by Minsky, Ming, Steele, Brook, & Behensky (1990). The formulation given above extends this concept to 3-D surfaces. In general, haptic texturing techniques can be classified into two parts: image-based and procedural.

### 3.1 Image-based Haptic Texturing

This class of haptic texturing deals with constructing a texture field from two-dimensional (2-D) image data. In computer graphics, digital images are wrapped around 3-D objects to make them look more realistic. While a graphical texture consists of 2-D texels with only color or gray scale intensities, a haptic texture should consist of texels with a height value. To display image-based haptic textures, the two-stage texture mapping technique of computer graphics (Bier & Sloan, 1986; Watt & Watt, 1992) can be followed. The first stage in this technique is to map the 2-D image to a simple intermediate surface such as a plane, cube, cylinder, or sphere. The second-stage maps the texels from the intermediate surface to the object surface. Following this stage, one can easily access the height value at any point on the object surface. In addition, the gradient vector at any point can be estimated using a finite difference technique and interpolation scheme. Then, the force perturbation concept described above can be used to display image-based haptic textures (see Ho et al., 1999, for implementation details).

### 3.2 Procedural Haptic Texturing

The goal of procedural haptic texturing is to generate synthetic textures using a mathematical function. The function usually takes the coordinate (x,y,z) as the input and returns the height value and its gradient as the output. For example, several investigators have implemented the well-known noise texture (Ebert et al., 1994; Perlin, 1985) to generate stochastic haptic textures (Basdogan et al., 1997; Fritz & Barner, 1996; Siira & Pai, 1996). Fractals are also appropriate for modeling natural textures since many objects seem to exhibit self-similarity (Mandelbrot, 1982). The fractal concept has been used in combination with other texturing functions, such as Fourier series and pink noise in various frequency and amplitude scales, to generate more sophisticated surface details (Ho et al., 1999). Similarly, several types of other graphical textures can be converted to haptic textures. For example, Ho et al. (1999) implemented haptic versions of reaction–diffusion textures (Turk, 1991; Witkin & Kass, 1991), spot noise (Wijk, 1991), and cellular texture (Worley, 1996).

## 4. RENDERING OF DEFORMABLE OBJECTS

Graphical display of deformable objects has been extensively studied in computer graphics. With the addition of haptic feedback, deformable objects have gained a new characteristic. Now, deformable models need to estimate not only the direction and amount of deformation of each node of an object but also the magnitude and direction of interaction forces that will be reflected to the user via a haptic device.

One way to categorize deformation techniques is according to the approach followed by researchers to deform surfaces: *geometry-* or *physics-based* deformations. In geometry-based deformations, the object or surrounding space is deformed, based purely on geometric manipulations. In general, the user manipulates vertices or control points that surround the 3-D object to modify the shape of the object. In contrast, physics-based deformation techniques aim to model the physics involved in the motion and dynamics of interactions. These models simulate physical behavior of objects under the effect of external and internal forces. Geometry-based deformation techniques are faster and are relatively easier to implement, but they do not necessarily simulate the underlying mechanics of deformation. Hence, the emphasis is on visual display and the goal is to make deformations more easily controllable and appear smooth to the user. Physics-based approaches, although necessary for simulating realistic behavior of deformable objects, are computationally expensive and not always suitable for real-time applications due to current limitations in computational power. In general, hybrid approaches that take advantage of both worlds seem to work well for many real-time applications.

Geometry- and physics-based deformation techniques used to render force-reflecting deformable objects can be further subgrouped as follows (see Basdogan, 1999, 2000, for more details):

1. Geometry-based Deformation Models
   - *Vertex-based:* The vertices of the object are manipulated to display the visual deformations. The reaction force can be modeled using Hooke's law, where depth of penetration can be computed based on the current and home positions of the vertex that is closest to the contact point. For example, in displaying soft tissue deformations graphically, polynomial functions that fit experimental data on finger pad deformation (Srinivasan, 1989) were used to remap the vertices of other organs (Basdogan, Ho, Srinivasan, Small, & Dawson, 1998).
   - *Spline-based:* Instead of directly transforming the vertices of the object, control points are assigned to a group of vertices and are manipulated to achieve smoother deformations. The concept of free-form deformation was originally suggested by Sederberg and Parry (1986) and extended by Hsu, Hughes, & Kaufman (1992) to direct free-form manipulation. The extension of this technique to haptic display of deformable objects with applications in medical simulation (Basdogan et al., 1998), computer-aided design (CAD), and haptic sculpting (Dachille, Qin, Kaufman, & El-sana, 1999; Edwards & Luecke, 1996) can be found in the literature. Here, the key advantages of using force feedback are to increase intuition, control deformations, and support implementation of various constraints.
2. *Physics-based* Deformation Models
   In comparison to geometry-based models, interaction forces are always part of the computation in physics-based models, and the developer does not need to consider a separate model for computing forces.
   - *Particle-based:* Particle systems consist of a set of point masses, connected to each other through a network of springs and dampers, moving under the influence of internal and external forces (see Witkin, Barraff, & Kass, 1998, for implementation details). In this model, each particle is represented by its own mass, position, velocity, and acceleration. This technique has been used extensively by computer graphics researchers in simulation of soft tissue and cloth behavior (Cover et al., 1993; Lee, Terzopoulos, & Waters, 1995; Ng & Grimsdale, 1996). Swarup (1995) demonstrated the implementation of this technique to haptic simulation of deformable objects.
   - *Finite element–based:* The volume occupied by a 3-D object is divided into finite elements, properties of each element are formulated, and the elements are assembled together to study deformation states for the given loads (Bathe, 1996). Due to the limited computational power that is available today, modeling simplifications have to be made to implement real-time finite element method with haptic displays

(see Cotin, Delingette, & Ayache, 1999; James & Pai, 1999, for static computations; see Basdogan et al., 2001, for dynamic computations). For example, De and Srinivasan (1999) have proposed a modeling approach in which organs are viewed as thin walled structures enclosing a fluid, thereby converting the 3-D problem into an effectively 2-D one (see also Balaniuk & Costa, 2000). In Basdogan et al. (2001), to compute dynamical deformations and interaction forces, a modal analysis approach was implemented such that only the most significant vibration modes of the object were selected. In this study, interactions between two deformable objects were also incorporated, one finite element based and the other particle based. Basdogan (2001) has also recently proposed the spectral Lanczos decomposition method to obtain explicit solutions of the finite element equations that govern the dynamics of deformations. Both methods (modal analysis and spectral Lanczos decomposition) rely on modeling approximations but generate dynamical solutions that are computationally much faster than the ones obtained through direct numerical integration techniques.

- *Meshless methods:* Because of the complexities of mesh generation and consequent constraints imposed on the computations, a meshless numerical technique called the method of finite spheres has recently been applied to physics-based soft-tissue simulation (De, Kim, & Srinivasan, 2001).

In earlier studies, a loose coupling between force and displacement computations has been proposed to take advantage of human perceptual limitations (Basdogan et al., 1998). In this approach, vertex-based or spline-based approaches have been used to generate smooth visual deformations, whereas interaction forces were computed and reflected to the user based on a simple spring and a damper model between the new and home positions of the contact point. Because human perception of position and motion is dominantly influenced by visual cues in interacting with objects (Srinivasan, Beauregard, & Brock, 1996), the loose coupling between force and displacement computations was not readily sensed by users during simulations (see Model A in Fig. 6.7). However, this architecture is limited to simulation of geometric-based deformations (e.g., sculpting and animation). Since force and displacement computations are

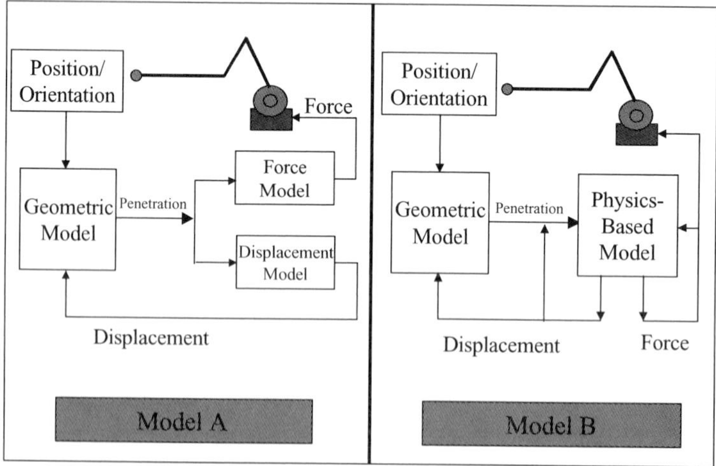


FIG. 6.7. Computational architectures for simulating force-reflecting deformable objects in virtual environments. In model A, force and displacement computations are loosely coupled. In model B, the physics-based model generates forces and displacements simultaneously, which are then fed back to the model to generate new values. Although this architecture is computationally expensive to implement, it is more comprehensive than architecture A (Basdogan, 1999).

usually tightly coupled in more physically based systems, a closed-loop architecture has to be developed to solve the equations that govern the physics of deformable behavior (see Model B in Fig. 6.7). In this architecture, forces and displacements computed in the previous cycle are continuously supplied back to the physics-based model to generate new values in the next cycle.

## 5. RENDERING OF DYNAMIC OBJECTS

Simulating haptic interactions with 3-D objects that translate and rotate over time is another challenge, which becomes computationally quite intensive if objects collide with each other, as well as with the haptic probe in real time. Dynamic equations have to be solved and the state of each object has to be updated at each iteration (see Fig. 6.8). Techniques to calculate forces and torques based on principles of rigid body dynamics (Baraff, 1994) and impulse mechanics (Mirtich & Canny, 1995; Mirtich, 1996) have been developed for graphical simulation of floating multiple objects. Latimer (1997) discusses the extension of these techniques to the haptics domain, whereas Colgate and Brown (1994) and Adams and Hannaford (1998) address the stability issues.

To describe the basic concepts of haptic interaction with dynamic objects, consider an example (see Fig. 6.8a). Assume that interactions are point-based again such that the user controls the dynamics of a floating object via only the tip point of the haptic probe. The new position and orientation of the object can be calculated using equations that govern the dynamics of rigid body motion. These equations can be solved in real time using a numerical

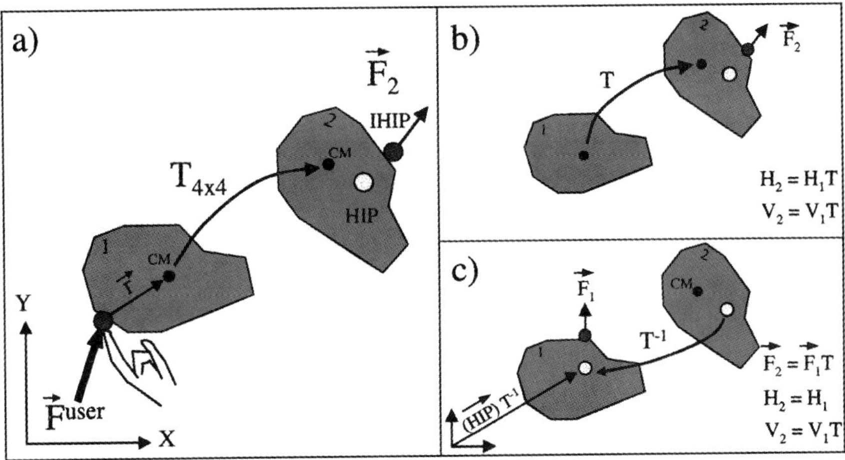


FIG. 6.8. Modeling point-based haptic interactions with a floating rigid object in virtual environments. (In the figure, T is the transformation matrix, $\vec{F}_2$ represents the force acting on the user through a haptic display, H and V represent the haptic and visual coordinates of the object respectively. Here, two different methods for computing the interaction forces when the object is transformed from State 1 to State 2 are discussed. Figure (b) describes a method in which both haptic and visual coordinates are updated and collisions are detected with the new haptic coordinates of the object to compute the new interaction force ($\vec{F}_2$) at State 2. In Figure (c), only visual coordinates are updated. First, HIP is multiplied with the inverse of the transformation matrix and collisions are detected with the original haptic coordinates. Then, the reaction force ($\vec{F}_1$) is computed relative to the original coordinates and then multiplied with the T matrix ($\vec{F}_2 = \vec{F}_1 \cdot \mathrm{T}$) to include the effects of transformation. Although both methods (b and c) are valid and return the same interaction force ($\vec{F}_2$) at the end, the method described in Figure (c) is computationally less expensive than the one described in Figure (b), because the haptic coordinates of the object are not required to be updated at each iteration when the object is transformed to a new state.

integration method, such as Euler integration, to update the state of the object at each cycle of the haptic loop:

| Constants: | Initial Conditions: | Integration over interval $\Delta t$: | Update Auxillary Quantities: |
|---|---|---|---|
| $I_{body}^{-1}$, $M$ | $\vec{v}_{t=0.0}$<br>$\vec{p}_{t=0.0}$<br>$R_{t=0.0}$<br>$\vec{L}_{t=0.0}$<br>$\vec{\omega}_{t=0.0} = I_{body}^{-1} \vec{L}_{t=0.0}$ | $\vec{v}_{t+\Delta t} = \vec{v}_t + \Delta t (M^{-1} \vec{F}_t^{user})$<br>$\vec{p}_{t+\Delta t} = \vec{p}_t + \Delta t \, \vec{v}_{t+\Delta t}$<br>$R_{t+\Delta t} = R_t + \Delta t \omega^* R_t$<br>$\vec{L}_{t+\Delta t} = \vec{L}_t + \Delta t \vec{T}_t^{user}$<br>where:<br>$\omega^* = \begin{bmatrix} 0 & -\omega_3 & \omega_2 \\ \omega_3 & 0 & -\omega_1 \\ -\omega_2 & \omega_1 & 0 \end{bmatrix}$ | $I_{t+\Delta t}^{-1} = R_t I_{body}^{-1} R_t^T$<br>$\vec{\omega}_{t+\Delta t} = I_{t+\Delta t}^{-1} \vec{L}_{t+\Delta t}$ |

where, $\vec{F}^{user}$ and $\vec{T}^{user}$ represent the total force and torque acting on the object, $\vec{v}$ and $\vec{p}$ represent the linear velocity and position vectors and $\vec{\omega}$ and $\vec{L}$ are the angular velocity and momentum vectors. In addition, $R$, $M$, and $I$ are the rotation, mass and inertia matrices of the object respectively and $\Delta t$ is the sampling time (e.g. $\Delta t = 0.001$ seconds if the haptic loop is updated at 1 kHz). The total force acting on the object ($\vec{F}^{user}$) is calculated based on the depth of penetration of the haptic probe into the object. The torque acting on the object will be the cross product of $\vec{r}$ and $\vec{F}^{user}$ vectors (see Fig. 6.8). The details of the formulation and the computation of mass and inertia matrices can be found in Baraff (1997).

Since the position and orientation of the object change at each time step, the updated coordinates of the object should be used to detect collisions with the new coordinates of the HIP. However, it would be computationally too expensive to update the object database at each time step to detect collisions with the new HIP. A better strategy is to compute everything relative to the original coordinates and then apply the effects of transformation later (see the description in the legend for Fig. 6.8).

## 6. RECORDING AND PLAYING-BACK HAPTIC STIMULI

The concept of recording and playing back haptic stimuli (MacLean, 1996) to the user has some interesting applications. For example, the user equipped with a desktop haptic device may touch and feel the prerecorded textures of a car seat that is advertised by a car company over the Internet (this concept is similar to downloading a movie file and playing it on screen). In another application, preprogrammed haptic devices can be used in neuro-rehabilitation (see chaps. 49 and 50, this volume) to improve the motor-control skills of patients with neurological disorders by displaying prerecorded haptic trajectories (Krebs et al., 1998). In both of these examples, the user will be a passive participant of the system, though he or she may have some control over the device. As the user holds the end-effector of the device gently, the device guides his or her hand along prerecorded shapes or paths.

To describe the concept of haptic "playback," consider a simple example. Assume that the goal is to program a haptic device such that it will display a square shaped frame to a passive user. Here, programming of a haptic device as a typical closed-loop control problem is considered. Although this is a well-studied problem in robotics, the following practical issues appear when it is implemented with haptic devices:

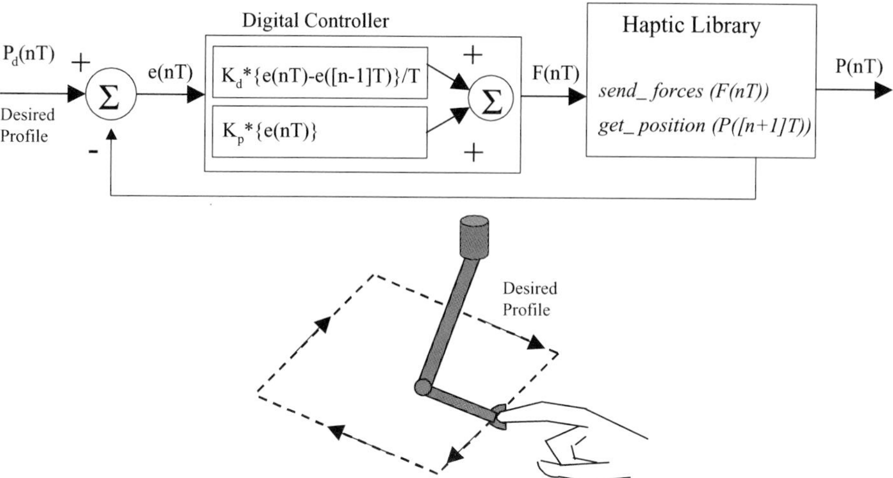


FIG. 6.9. Digital servo loop for commanding the haptic device to play prerecorded haptic stimuli (a frame in this example). P, e, and F represent position, error, and force vectors respectively. $K_d$ and $K_p$ are derivative and proportional control gains.

- Haptic devices are typically "open-loop" systems. The user manipulates the end-effector of the robot to control the end position, but the dynamics of interactions are not governed by the principles of closed-loop control systems. In other words, you need to design your own controller to command your haptic device to play back prerecorded haptic stimuli.
- Unless you build your own haptic device and do not purchase from a manufacturer, you most likely do not know the kinematics (e.g., mass and length of each link) and dynamic properties (e.g., friction, performance of motors) of your device exactly. Haptic devices, sold by a manufacturer, usually come with a library that includes function calls for acquiring position and orientation of the end-effector and for sending force commands to motors (see section 2). Hence, the goal is to use these function calls to control the movements of the haptic device and navigate from one point to another in 3-D space while compensating for positional errors. One can design a controller to achieve this goal (see Fig. 6.9; we use a PD controller to track the desired profile).

It is possible to play back prerecorded forces to the user even when the user is actively interacting with the environment by using, visual cues. Dang, Annaswamy, & Srinivasan (2001) have developed an epidural injection simulator in which a training feature is the capability to display an expert's path and forces to a novice in two ways. The expert's path is displayed visually either as a "tunnel" through which the novice could traverse or as a cursor moving along the prerecorded path that the trainee needs to follow closely to feel the prerecorded forces continuously.

## 7. SHARED HAPTICS: HAPTIC INTERACTION BETWEEN NETWORKED USERS

Another potentially promising area is haptic interaction among a group of users sharing the same VE over a network (see chaps. 16 and 44, this volume). Since the haptic loop requires a high update rate of around 1 kHz for stable interactions, and changes in the visual scene will require frequent update of the haptic database, developing a network protocol that can provide sufficient bandwidth with minimum latency to a group of distributed users is a quite challenging problem. For shared environments to be effective in performing collaborative tasks that involve haptic feedback, several network architectures have been proposed (Buttolo, Oboe, & Hannaford 1997; Wilson, Kline-Schoder, Kenton, & Hogan, 1999), and physics-based models that simulate haptic interactions among users have begun to be developed (Basdogan et al., 2000). Because of network time delays, the difficult nature of some collaborative tasks, and lack of knowledge on user abilities and behaviors, the problem of developing a universal model for shared haptics could be too complex. For example, it is obvious that time delays during the transfer and processing of data may easily result in unstable forces and can be harmful to the user (see chap. 41, this volume). Similarly, participants could follow several different strategies to manipulate virtual objects during the execution of a task if an interaction model is established. For example, one could talk about *sequential* versus *simultaneous* types of haptic manipulations. It is even possible to make one user "stronger" than the other (Basdogan et al., 2000). Since limited knowledge in this area makes it almost impossible to integrate all these features into a single interaction model, the type of model selected to simulate haptic interactions between participants in shared environments depend, at this stage, on the collaborative task itself. For example, a simple physics-based model has been developed to simulate haptic interactions between participants (Basdogan et al., 2000). In this model, each participant manipulates his or her own cursor through a stylus attached to the force feedback device placed next to his or her seat. When the participant manipulates the stylus of the haptic device with his or her dominant hand, the cursor moves in 3-D space so that the manipulated object translates or rotates depending on the task. In our experiments, a spring-damper model ($F = k\Delta p + b\Delta\dot{p}$, where $F$ is the force exerted by the user on the ring that is felt by his or her remote partner, $k$ and $b$ are the spring and the damping coefficients, and $\Delta p$ is the displacement of the participant's cursor) was used to control the impedance of interactions between participants and between the participant's cursor and the ring in the scene (see Fig. 6.10). This model simply

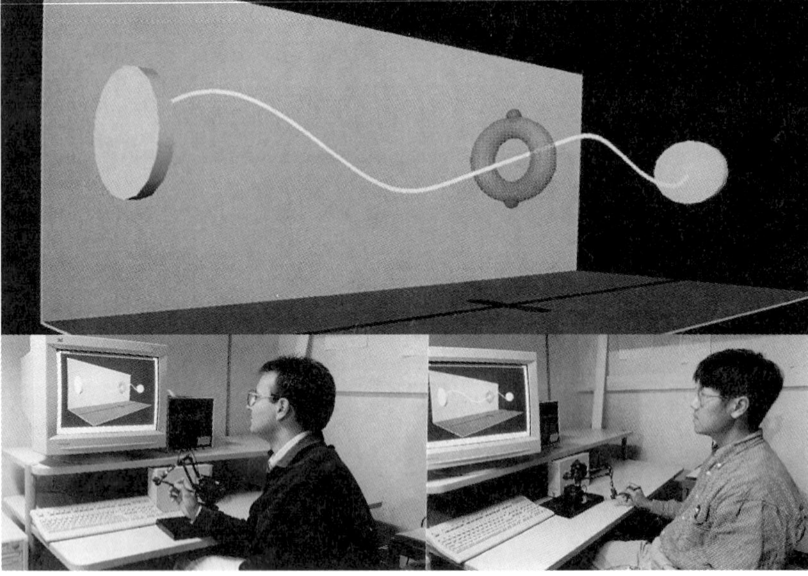

FIG. 6.10. We developed a shared VE setup that enables two people, at remote locations, to interact with each other through visual and haptic displays. An experiment was designed to investigate the role of haptics in collaborative manipulation. In our experiment, participants were asked to hold and move a ring on a wire in a collaborative manner, as depicted in this figure. To eliminate time delays due to network connections, we bifurcated the signals from a single host and displayed it on two separate monitors and haptic interfaces.

simulates the translational movements of the ring on a wire and pulling and pushing forces between the participants. Hence, if a participant pulls or pushes his own cursor the remote partner feels the forces. Visually, however, the ring remains rigid (i.e., no deformation of the ring is displayed graphically). The rotation of the ring due to unbalanced forces applied by the participants was prevented to make the task easier. Moreover, only "simultaneous" haptic interactions were supported such that the ring did not move if both participants did not apply sufficient forces to the ring at the same time.

## 8. SUMMARY AND FUTURE

Computer haptics is concerned with the development of software algorithms that enable a user to touch, feel, and manipulate objects in VEs through a haptic interface. It primarily deals with the computation of forces to be displayed to the user in response to user's actions. The demands placed on it depend on the capabilities and limitations of the interface hardware, computational engine, and user, together with the needs of the task to be accomplished. Current haptic interface technology is mature for net force and torque feedback, as in tool usage in the real world. Consequently, current haptic rendering algorithms have been developed mainly for simulating net forces and torques of interaction and give users the feeling of touching and manipulating objects through a stick or a rigid thimble. Point-based rendering, which views the haptic cursor as a point in computing point-object interaction forces, was developed first and is widely used. This was followed by ray-based rendering, which computes line–object interaction forces and torques. More recently, limited success has been achieved in simulating 3-D object–object interactions in which one of the objects is viewed as a collection of points. Using these haptic interaction paradigms, real-time haptic display of shapes, textures, and friction of rigid and deformable objects has been achieved. Haptic rendering of dynamics of rigid objects, and to a lesser extent, linear dynamics of deformable objects, has also been accomplished. Methods for recording and playing back haptic stimuli, as well as algorithms for haptic interactions between multiple users in shared VEs, are beginning to emerge.

In the future, capabilities of haptic interface devices are expected to improve primarily in two ways: (1) improvements in both desktop and wearable interface devices in terms of factors such as inertia, friction, workspace volume, resolution, force range, and bandwidth; (2) development of tactile displays to simulate direct contact with objects, including temperature patterns. These are expected to result in multifinger, multihand, and even whole body displays, with heterogeneous devices connected across networks. Even with current rapid expansion of the capabilities of affordable computers, the needs of haptic rendering with more complex interface devices will continue to stretch computational resources. Currently, even with point-based rendering, the computational complexity of simulating the nonlinear dynamics of physical contact between an organ and a surgical tool, as well as surrounding tissues is very high. Thus there will be continued demand for efficient algorithms, especially when the haptic display needs to be synchronized with the display of visual, auditory, and other modalities (see chap. 21, this volume). Similar to graphics accelerator cards used today, it is quite likely that repetitive computations will need to be done through specialized electronic hardware, perhaps through parallel processing. Given all the complexity and need for efficiency, in any given application the central question will be how good the simulation needs to be to achieve a desired goal.

## 9. REFERENCES

Adachi, Y., Kumano, T., & Ogino, K. (1995). Intermediate representation for stiff virtual objects. In *Proceedings of the IEEE Virtual Reality Annual International Symposium* (pp. 203–210). Research Triangle Park, NC: IEEE.

Adams, R., & Hannaford, B. (1998). A two-port framework for the design of unconditionally stable haptic interfaces. In *IEEE/RSJ International Conference on Intelligent Robots and Systems.* Victoria, Canada: IEEE.

Avila, R. S., & Sobierajski, L. M. (1996). A haptic interaction method for volume visualization. In *IEEE Proceedings of Visualization* (pp. 197–204).

Balaniuk, R., & Costa, I. F. (2000). LEM—An approach for physically based soft tissue simulation suitable for haptic interaction. In *Proceedings of the Fifth Phantom Users Group Workshop.*

Baraff, D. (1994). Fast contact force computation for nonpenetrating rigid bodies. *ACM (Proceedings of SIGGRAPH), 28,* 23–34.

Baraff, D. (1997). An introduction to physically based modeling: Rigid body simulation 1—Unconstrained rigid body dynamics. *SIGGRAPH'97 Tutorial Notes.*

Basdogan C. (1999, August). Force reflecting deformable objects for virtual environments. In *Haptics: From basic principles to advanced applications, SIGGRAPH '99 Tutorial Notes, 26th International Conference on Computer Graphics and Interactive Techniques, Course No. 38.*

Basdogan C. (2000). Course name: Simulating minimally invasive surgical procedures in virtual environments: From tissue mechanics to simulation and training. Medicine Meets Virtual Reality (MMVR '2000), Jan. 27–30, Irvine, CA, Available: http://www.amainc.com/MMVR/MMVR2000pro.html, http://eis.jpl.nasa.gov/~ basdogan

Basdogan, C. (2001). Real-time simulation of dynamically deformable finite element models using modal analysis and spectral lanczos decomposition methods. In *Medicine Meets Virtual Reality*, pp. 16–52.

Basdogan, C., Ho, C., & Srinivasan, M. A. (1997). A ray-based haptic rendering technique for displaying shape and texture of 3D objects in virtual environments. *ASME Winter Annual Meeting* (pp. 61, 77–84). Dallas, TX: ASME.

Basdogan, C., Ho, C., & Srinivasan, M. A. (2001). Virtual environments for medical training: Graphical and haptic simulation of common bile duct exploration. *IEEE/ASME Transactions on Mechatronics, 6*(3), 267–285.

Basdogan, C., Ho, C., Srinivasan, M.A., & Slater, M. (2000). An experimental study on the role of touch in shared virtual environments, In *ACM Human Computer Interactions,* Vol. 7, No. 4, pp. 440–463.

Basdogan C., Ho, C., Srinivasan, M. A., Small, S., & Dawson, S. (1998). Force interactions in laparoscopic simulations: Haptic rendering of soft tissues. *Proceedings of the Medicine Meets Virtual Reality VI Conference* (pp. 385–391).

Bathe, K. (1996). *Finite Element Procedures.* NJ: Prentice Hall.

Bier, E. A., & Sloan, K. R. (1986). Two-part texture mapping. *IEEE Computer Graphics and Applications*, 40–53.

Blinn, J. F. (1978). Simulation of wrinkled surfaces. *ACM (Proceedings of SIGGRAPH), 12*(3), 286–292.

Buttolo, P. (1996). *Characterization of human pen grasp with haptic displays.* Unpublished Ph.D. Dissertation, University of Washington, Seattle, WA.

Buttolo, P., Oboe, R., & Hannaford, B. (1997). Architectures for shared haptic virtual environments. *Computers and Graphics, 21,* 421–429.

Cohen, J., Lin, M., Manocha, D., & Ponamgi, K. (1995). I-COLLIDE: An interactive and exact collision detection system for large-scaled environments. *Proceedings of ACM Interactive 3D Graphics Conference* (189–196).

Colgate, J. E., & Brown, J. M. (1994). Factors affecting the z-width of a haptic display. *Proceedings of IEEE International Conference on Robotics and Automation* (pp. 3205–3210).

Cotin, S., Delingette, H., & Ayache, N. (1999). Real-time elastic deformations of soft tissues for surgery simulation. *IEEE Transactions on Visualization and Computer Graphics, 5*(1), 62–73.

Cover S. A., Ezquerra N. F., O'Brien J., Rowe R., Gadacz T., & Palm E. (1993). Interactively deformable models for surgery simulation. *IEEE Computer Graphic and Applications,* November, 65–78.

Dachille, F., Qin, H., Kaufman, A., & El-sana, J. (1999). Haptic sculpting of dynamic surfaces. In *ACM Symposium on Interactive 3D Graphics.*

Dang, T., Annaswamy, T. M., & Srinivasan, M. A. (in press). Development and evaluation of an epidural injection simulator with force feedback for medical training. In *Medicine Meets Virtual Reality.*

De, S., Kim, J., & Srinivasan, M. A. (2001). A meshless numerical technique for physically based real-time medical simulations. In *Medicine Meets Virtual Reality,* pp. 113–118.

De, S., & Srinivasan, M. A. (1999). Thin-walled models for haptic and graphical rendering of soft tissues in surgical simulations. In *Medicine Meets Virtual Reality* (Eds: Westwood, et al.), IOS Press.

Ebert, D. S., Musgrave F. K., Peachey D., Perlin K., & Worley S. (1994). *Texturing and Modeling.* Cambridge, MA: AP Professional.

Edwards, J., & Luecke, G. (1996). Physically based models for use in a force feedback virtual environment. *Japan/USA Symposium on Flexible Automation* (pp. 221–228). ASME.

Ellis, R. E., Sarkar, N., & Jenkins, M. A. (1996). Numerical methods for the haptic presentation of contact: Theory, simulations, and experiments. *Proceedings of the ASME Dynamic Systems and Control Division* (DSC-Vol. 58, pp. 413–420).

Foley, J. D., van Dam, A., Feiner, S. K., & Hughes, J. F. (1995). *Computer graphics: Principles and practice.* Addison-Wesley.

Fritz & Barner (1996). Haptic scientific visualization. In J. K. Salisbury & M. A. Srinivasan (Eds.), *Proceedings of the First PHANToM Users Group Workshop.* MIT-AI TR-1596 and RLE TR-612.

Fukui, Y. (1996). Bump mapping for a force display. *Proceedings of the First PHANToM User Group Workshop* [online]. Available: http://www.ai.mit.edu/conferences/pug/pug-proceedings.html

Gottschalk, S., Lin, M., & Manocha, D. (1996). OBB-Tree: A hierarchical Structure for Rapid Interference Detection. *ACM (Proceedings of SIGGRAPH).*

Green, D. F., & Salisbury, J. K. (1997). Texture sensing and simulation using the PHANToM: towards remote sensing of soil properties. In *Proceedings of the Second PHANToM Users Group Workshop.*

Ho, C., Basdogan, C., & Srinivasan, M. A. (1999). An efficient haptic rendering technique for displaying 3D polyhedral objects and their surface details in virtual environments. *Presence: Teleoperators and Virtual Environments, 8*(5), 477–491.

Ho, C., Basdogan, C., & Srinivasan, M. A. (2000). Modeling of force and torque interactions between a line segment and triangular surfaces for haptic display of 3D convex objects in virtual and teleoperated environments [Special issue] *International Journal of Robotics, 19*(7), 668–684.

Hollerbach J., & Johnson, D. (in press). In *Human and Machine Haptics* (Eds: Srinivasan, Cutkosky, Howe, and Salisbury).

Hubbard, P. (1995). Collision detection for interactive graphics applications. *IEEE Transactions on Visualization and Computer Graphics, 1*(3), 219–230.

Hsu, W. M., Hughes, J. F., & Kaufman, H. (1992). Direct manipulation of free-form deformations. *Computer Graphics (Proceedings of the SIGGRAPH), 26*(2), 177–184.

James, D., & Pai, D. (1999). ARTDEFO: Accurate real time deformable objects, In *Computer Graphics (SIGGRAPH '99 Conference Proceedings).*

Krebs, H. I., Hogan, N., Aisen, M. L., & Volpe, B. T. (1998). Robot-aided neurorehabilitation. *IEEE Transactions on Rehabilitation Engineering, 6*(1), 75–86.

Latimer, C. (1997). *Haptic interaction with rigid objects using real-time dynamic simulation.* Unpublished master's thesis, Massachusetts Institute of Technology.

Lee, Y., Terzopoulos, D., & Waters, K. (1995). Realistic modeling for facial animation. *Proceedings of the SIGGRAPH,* 55–62.

Lin, M. (1993). *Efficient collision detection for animation and robotics.* Unpublished doctoral dissertation, University of California, Berkeley.

MacLean, K. (1996). The "haptic camera": A technique for characterizing and playing back haptic properties of real environments. *Proceedings of ASME Dynamic Systems and Control Division* (DSC-Vol. 58, pp. 459–467).

Mandelbrot, B. (1982). *The Fractal Geometry of Nature.* San Francisco: W. H. Freeman.

Mark, W., Randolph S., Finch, M., Van Verth, J., & Taylor, R.M. (1996). Adding force feedback to graphics systems: issues and solutions. *Computer Graphics: Proceedings of SIGGRAPH '96,* 447–452.

Max, N. L., & Becker, B. G. (1994). Bump shading for volume textures. *IEEE Computer Graphics and Applications, 4,* 18–20.

McNeely, W. A., Puterbaugh K. D., and Troy, J. J. (1999). Six degree-of-freedom haptic rendering using voxel sampling. *Proceedings of SIGGRAPH,* 401–408.

Minsky, M., Ming, O., Steele, F., Brook, F. P., & Behensky, M. (1990). Feeling and seeing: issues in force display. *Proceedings of the Symposium on 3D Real-Time Interactive Graphics, 24,* 235–243.

Mirtich, B. (1996). *Impulse-based dynamic simulation of rigid body systems.* Unpublished doctoral dissertation, University of California, Berkeley.

Mirtich, B., & Canny, J. (1995). Impulse-based simulation of rigid bodies. *Proceedings of Symposium on Interactive 3D Graphics,* April.

Morgenbesser, H. B., & Srinivasan, M. A. (1996). Force shading for haptic shape perception. *Proceedings of the ASME Dynamic Systems and Control Division, 58,* 407–412.

Ng, H., & Grimsdale, R. (1996). Computer Graphics Techniques for Modeling Cloth. *IEEE Computer Graphic and Applications,* 28–41.

Oboe, R., & Fiorini, P. (1998). A design and control environment for internet-based telerobotics. *The International Journal of Robotics Research, 17*(4), 433–449.

Perlin, K. (1985). An image synthesizer. *ACM SIGGRAPH, 19*(3), 287–296.

Ruspini, D. C., Kolarov, K., & Khatib, O. (1996). Robust haptic display of graphical environments. In J. K. Salisbury & M. A. Srinivasan (Eds.), *Proceedings of the First PHANToM Users Group Workshop,* MIT-AI TR-1596 and RLE TR-612.

Ruspini, D. C., Kolarov, K., & Khatib, O. (1997). The haptic display of complex graphical environments. *ACM (Proceedings of SIGGRAPH),* 345–352.

Salcudean, S. E., & Vlaar, T. D. (1994). On the emulation of stiff walls and static friction with a magnetically levitated input/output device. *Proceedings of the ASME* DSC, *55*(1), 303–309.

Salisbury, J. K., Brock, D., Massie, T., Swarup, N., & Zilles, C. (1995). Haptic rendering: Programming touch interaction with virtual objects. In *Proceedings of the ACM Symposium on Interactive 3D Graphics.* Monterey, CA.

Salisbury, J. K., & Srinivasan, M. A. (1997). Phantom-based haptic interaction with virtual objects. *IEEE Computer Graphics and Applications, 17*(5).

Salisbury, J. K., & Tarr, C. (1997). Haptic rendering of surfaces defined by implicit functions. *Proceedings of the ASME* (DSC-Vol. 61, 61–67).

Sederberg, T. W., & Parry S. R. (1986). Free-form deformation of solid geometric models. *SIGGRAPH Proceedings on Computer Graphics, 20*(4), 151–160.

Siira, J., & Pai, D. K. (1996). Haptic texturing—A stochastic approach. *Proceedings of the IEEE International Conference on Robotics and Automation* (pp. 557–562). Minneapolis, MN: IEEE.

Srinivasan M. A. (1989). Surface deflection of primate fingertip under line load. *Journal of Biomechanics,* 22(4), 343–349.

Srinivasan, M. A. & Basdogan, C. (1997). Haptics in virtual environments: Taxonomy, research status, and challenges. *Computers and Graphics, 21*(4), 393–404.

Srinivasan M. A., Beauregard G. L., & Brock, D. L. (1996). The impact of visual information on haptic perception of stiffness in virtual environments. *Proceedings of the ASME Dynamic Systems and Control Division,* (DSC-Vol. 58, pp. 555–559).

Srinivasan M. A., Whitehouse J. M., & LaMotte, R. H. (1990). Tactile detection of slip: Surface microgeometry and peripheral neural codes. *Journal of Neurophysiology, 63*(6), 1323–1332.

Stewart, P., Chen, Y., & Buttolo, P. (1997). Direct integration of haptic user interface in CAD systems. *Proceedings of ASME* (DSC-Vol. 61, pp. 93–99).

Swarup, N. (1995). *Haptic interaction with deformable objects using real-time dynamic simulation.* Unpublished master's thesis, Massachusetts Institute of Technology.

Thompson, T. V., Johnson, D. E., & Cohen, E. (1997). Direct haptic rendering of sculptured models. *Proceedings of the Symposium on Interactive 3D Graphics* (pp. 1–10). Providence, RI:

Turk, G. (1991). Generating textures on arbitrary surfaces using reaction-diffusion. *ACM (Proceedings of SIGGRAPH), 25*(4), 289–298.

Watt, A., & Watt, M. (1992). *Advanced Animation and Rendering Techniques.* New York: Addison-Wesley.

Wilson J., Kline-Schoder, Kenton, M. A., & Hogan, N. (1999). Algorithms for network-based force feedback. In J. K. Salisbury & M. A. Srinivasan (Eds.), *Proceedings of the Fourth PHANToM Users Group Workshop,* (AI Lab Tech. Rep. No. 1675 and RLE Tech. Rep. No. 633). Cambridge, MA: Massachusetts Institute of Technology.

Wijk, J. J. V. (1991). Spot noise. *ACM (Proceedings of SIGGRAPH), 25*(4), 309–318.

Witkin A., Barraff D., & Kass, M. (1998). Tutorial notes on *"An Introduction to Physically-Based Modeling"*, SIGGRAPH '98.

Witkin, A., & Kass, M. (1991). Reaction-diffusion textures. *ACM (Proceedings of SIGGRAPH), 25*(4), 299–308.

Worley, S. (1996). A cellular texture basis function. *ACM (Proceedings of SIGGRAPH),* 291–294.

Zilles, C. B., & Salisbury, J. K. (1995). A constraint-based god–object method for haptic display. *IEEE International Conference on Intelligent Robots and System, Human Robot Interaction, and Co-operative Robots, 3,* 146–151.

# 7

# User Requirements for Perceiving Body Acceleration

Ben D. Lawson, Stephanie A. Sides, and K. Amy Hickinbotham
*Naval Aerospace Medical Research Laboratory*
*Spatial Orientation Systems*
*51 Hovey Road*
*Pensacola, FL 32508-1046*
*blawson@namrl.navy.mil*
*stephsides@hotmail.com*
*amyuwf@hotmail.com*

## 1. INTRODUCTION

This chapter describes various methods for eliciting perceptions of body acceleration. The focus of discussion is on the "user's requirements" for perceiving acceleration in a virtual environment (VE). The most direct means for eliciting an acceleration perception in a VE is by using a real acceleration stimulus that is meant to mimic critical aspects of the "virtual acceleration" perception being conveyed. A rudimentary understanding of human vestibular function is necessary in order to employ physical acceleration effectively as part of a virtual environment. This chapter provides an introduction to the "vestibular channel" of the VE stimulus and details different methods for exploiting knowledge about the vestibular modality to elicit perceptions of self-motion, self-tilt, and self-entry into unusual force environments. The authors compare physical acceleration of the user to other methods, such as visually induced illusions of self-motion or illusions of self-motion induced by locomotion without displacement (as occurs on treadmills). The authors conclude that the most compelling perceptions of body acceleration in VEs will be achieved through mutually enhancing combinations of vestibular, visual, and somesthetic stimuli. A particularly effective approach will be to use acceleration stimuli that duplicate the real acceleration profiles being simulated, but to do so in a different environment from the (virtual) environment where the user perceives himself to be located.

# 2. "USER REQUIREMENTS" FOR PERCEIVING BODY ACCELERATION

## 2.1 Purpose and Scope of This Chapter

The purpose of this chapter is to introduce various methods for eliciting acceleration perceptions. The acceleration perceptions one can elicit in a VE are not limited to feelings of velocity or displacement through space. Since gravity is one type of acceleration (according to Einstein's stated equivalence of inertial and gravitational force; see de Broglie, 1951), the perception of unusual G-forces on one's body qualifies as an acceleration perception. One might wish to elicit a perception of altered G-force in order to simulate travel in a high performance vehicle or into a nonterrestrial gravity environment.

Acceleration simulations that are designed correctly could be used to create experiences outside the user's immediate, veridical experience (e.g., training in a simulator building), but appropriate to the VE (e.g., flying in the cockpit of a high-performance airplane). Acceleration simulations need not be produced solely by VEs that elicit illusory perceptions of self-motion (such as occur when viewing a moving visual surround); they can be produced by actual body accelerations (such as occur with a "moving-base" simulator) or actual locomotions (such as on a treadmill—see Appendix A). The purpose of an acceleration simulation is to elicit a perception of acceleration appropriate to the VE being created by whatever combination of illusory or real motion is most effective.[1] It would be wrong to think that an acceleration display is only "virtual" when it does not employ an acceleration stimulus. Certainly no one would argue that a visual display is only virtual when it does not stimulate the retina with light. Similarly, the feeling of "virtual acceleration" refers not to the means by which the acceleration perception is generated, but rather to the resulting perceptual event created.

During normal daily activities, the integration of multimodal sensory information is critical to a human's appreciation of his motion, his position, and the forces acting on his body. The perception of body acceleration involves most of the senses, including vision, audition, vestibular sensation, and somesthesia (which refers to cutaneous, muscle, and joint receptors).[2] Similarly, during simulated activities, the aforementioned sensory modalities can be exploited to elicit the perception of acceleration. For example, the motion of large visual fields around the user is known to act upon the vestibular nucleus (and associated structures) and thereby to induce a compelling illusion of self-motion (Dichgans & Brandt, 1978). A visually induced illusion of self-motion is commonly referred to as "vection."[3] Vection can be considered an acceleration perception that indirectly implicates vestibular involvement. The use of vection as a means for simulating self motion in a VE is treated in detail in chapter 23 (this volume).

Just as visual stimuli can provide acceleration cues, so can somesthetic stimuli. For example, walking in place (as on a treadmill) can be exploited to enhance the perception of locomotion through a virtual display and can induce compensatory eye movements not unlike those elicited by real body motion (chap. 38, this volume; also see Bles, 1979; Brandt, Büchele, & Arnold, 1977; Lackner & DiZio, 1988). Like visual vection, walking in place can be considered an acceleration simulation. Some locomotion simulation devices are listed in Appendix A. The use of locomotion devices in VE is treated in detail in chapter 11 of this book.

The primary focus of the present chapter will be to describe the user requirements that must be met to elicit acceleration perceptions via information presented in the vestibular channel of a virtual environment. The most direct means for creating an acceleration perception in a VE is by using a real acceleration stimulus that is meant to mimic critical aspects of the acceleration scenario being simulated. This can be accomplished by allowing the user to move through a real space that serves as an ambient context for the virtual environment (see section 3.2 below) or by the use of moving simulators, centrifuges, and other acceleration devices (examples are provided in Appendix A). All of these "real-motion" methods tend to stimulate the vestibular organs, which are specifically evolved to transduce head acceleration. The vestibular modality should be exploited by VE designers to induce feelings of acceleration when they are desirable and to avoid feelings of acceleration when they are undesirable, yet are required because of the logistic constraints of the simulator and the space within which it operates. For example, "moving-base" simulators often employ a subthreshold motion stimulus during relatively quiescent periods in the simulation. This is done to slowly reposition the user and ready the motion platform to achieve the desired acceleration profile the next time a strong acceleration perception is required as part of the simulation. In discussing the vestibular channel in VEs, the essential qualities of the stimulus (acceleration) and the receptor (the vestibular end organs) will be briefly introduced. The authors will then introduce ways in which the vestibular system can be exploited to create, prevent, or modify acceleration perceptions.

### 2.2 Necessary Prerequisites of an Effective Simulation

To understand the simulation of a perception, one must start with a basic knowledge of how that perception is formed in the natural world. To understand how perceptions are formed, one must begin with the stimulus, i.e., the qualities in the physical world that are detected by the sensory system of interest. For example, the amplitude and frequency of vibration of air molecules is sensed by the auditory modality and then perceived by the listener as loudness and pitch. Similarly, if one wishes to meet the user's "visual requirements" with a display, one manipulates the intensity and wavelength of the light reaching the user's eye so as to match the stimuli that would be received if the viewer were physically present inside the virtual world.

Manipulating the quality of the stimulus is only part of a simulation. It is equally important to understand the functioning and limitations of the sensory modalities one is stimulating to evoke a synthetic experience effectively. Sometimes it may not be possible or desirable to recreate the stimuli the user would receive if he were actually inside the virtual world. In cases where it is desirable but does not appear possible, a solution might be finessed by exploiting known principles of sensory functioning.

Information about the functioning of the vestibular system is presented in this chapter. Such information can be exploited to enhance feelings of acceleration. In any VE, certain system requirements must be met to generate an appropriate stimulus, while critical user requirements dictated by the functional properties of the user's perceptual apparatus must be met for the VE to be compelling and effective. The acceleration stimulus and the functioning of the vestibular system will be the focus of the discussion below.

### 2.3 Introduction to the Acceleration Stimulus and Vestibular Receptors

The human vestibular modality is evolved to detect acceleration stimuli; it can do so in the absence of visual or somesthetic cues, but these cues are present most of the time. The sensory qualities to which the vestibular modality is sensitive include linear and angular accelerations due to self-motion and gravity. Linear and angular accelerations are detected by the otolith organs and the semicircular canals, respectively. These organs will be described below. For a detailed description, see Howard (1986, chap. 11).

The otolith organs of the inner ear detect linear accelerations of the head via the utricles and the saccules, and they also sense angular motion that produces changes in the angle of the earth's gravity vector relative to the perceiver's body. The otolith organs detect changes in the magnitude or direction of linear gravitoinertial force vectors via hair cell mechanoreceptors, which are embedded in gelatinous membranes containing tiny crystals of calcium carbonate called otoconia. The otoconia are more dense than the surrounding medium and hence lag behind motions of the head very slightly due to their inertia, causing a corresponding deformation of the hair cells.

Three semicircular canals located in each inner ear code angular motion in three roughly orthogonal axes. The semicircular canals are fluid filled rings with a gelatinous structure called the cupula obstructing each ring. The gelatinous cupula is embedded with hair cells that are deformed when the cupula is itself deformed by very small changes in fluid pressure caused by angular acceleration of the semicircular canals.

Collectively, the semicircular canals and otolith organs make up the equilibrium organs of the nonauditory portion of the human labyrinth. The semicircular canals and otolith organs interact with one another (and with other sensory modalities) during the processing of complex stimuli involving simultaneous angular and linear acceleration. The mutual interaction of otoliths and canals explains why an individual's perception of self-tilt during "off-center" rotation (i.e., rotation around a central axis that is located at some radial distance from one's own body) facing tangentially forward initially lags behind the stimulus given by the rate of change in the direction of the resultant gravito-inertial force vector (Graybiel & Brown, 1950; Lawson, Mead, Chapman, & Guedry, 1995). This lag effect is likely attributable to "suppression" of the otolith system by the angular acceleration signal being processed and transmitted by the semicircular canals. Numerous other intralabyrinthine interactions can be identified (Guedry & Benson, 1978; Reason & Brand, 1975), as well as important interactions among the labyrinthine and nonlabyrinthine sensory modalities, including vision (Dichgans & Brandt, 1978), audition (Lackner, 1977), and somesthesia (Bles, 1979; Lackner & Graybiel, 1978).

The vestibular system is involved in several aspects of human functioning besides the perception of acceleration, including the reflexive control of gaze stabilization, head righting, postural equilibrium, coordinated locomotion, and certain reaching behaviors. Functioning vestibular organs are probably critical to the elicitation of unpleasant symptoms such as motion sickness (so named by Irwin in 1881) and "cybersickness" (see chap. 30, this volume; Stanney et al., 1998). A comprehensive explanation of vestibular function, neurovestibular pathways, or the psychophysics of vestibular perception is not necessary for the reader to gain a rudimentary grasp of the practical options available for eliciting acceleration perceptions in virtual environments. (Appendix B provides a list of recommended readings for those who wish to learn more about vestibular function and psychophysics.) The most practical options available for eliciting acceleration perceptions are evaluated below.

## 2.4 Current Options Available for Creating an Acceleration Perception

Most of the methods for creating the perception of displacement, velocity, acceleration, or abnormal G-force upon one's body are enumerated in this section. In section 3, the focus of discussion is narrowed to those few methods that will be most helpful to the creation of controlled acceleration perceptions within a virtual environment.[4]

### *2.4.1 Methods Involving Acceleration of the User*

*2.4.1.1 Real Physical Motion.* The most obvious way to elicit the perception of dynamic body movement through a VE is by real physical motion of the user that is correctly perceived as such. Physical motion can be achieved "passively"[5] with moving devices such as simulators, centrifuges, and real vehicles (see sections 3.2.1 and 3.2.3). Physical motion can also be achieved by allowing the user to walk through a real environment that serves as an ambient context for the VE (see section 3.2.2). Real motion is a very effective and controllable way to create acceleration perceptions in a VE, provided the VE customer can afford it and the simulations are designed by individuals who understand the psychophysics of human acceleration sensations and the factors which give rise to motion induced discomfort.

*2.4.1.2 Addition or Subtraction of Felt Body Weight.* Physical acceleration of the user is not always perceived as such. Acceleration can lead to the perception of altered body weight instead. The external addition or subtraction of felt body weight during movement through space can be used to enhance the simulation of high-performance vehicles or nonterrestrial settings (Guedry, Richard, Hixson, & Niven, 1973; Guedry, 1974; Lackner & Graybiel, 1980).[6]

*2.4.1.3 Illusions of Self-motion Associated With Acceleration Stimuli.* Illusions of self-motion can accompany the cessation (or change in velocity) of certain real-motion stimuli, such as prolonged constant velocity self-motion (Guedry, 1974). This is a useful adjunct to method 2.4.1.1, above.

*2.4.1.4 Constant Velocity Rotation Followed By Head Movement.* Self-velocity perceptions can be produced by prolonged constant velocity rotation in one axis followed by a head movement in another, nonparallel axis (Guedry & Benson, 1978). This is a useful approach when the stimulus is relatively mild. Pronounced discomfort can arise among susceptible individuals making repeated head movements after prolonged body rotation, especially when body rotation occurs at high velocity. However, earth-referenced cues can lessen the unpleasant effects (Lawson, Guedry, Rupert, & Anderson, 1993; Lawson, Rupert, Guedry, Grissett, & Mead, 1997).

*2.4.1.5 Static Body Tilt.* Alterations in perceived orientation can be induced via static body tilt and the aftereffects thereof. This is a useful adjunct to other methods in this section, but simple body tilt by itself elicits effects that are restricted mostly to the visual modality (see section 4).

### *2.4.2 Methods Not involving Acceleration of the User*

*2.4.2.1 Visual Field Motion.* Acceleration perceptions can be elicited via tilting visual frames of reference, continuous whole-field visual motions, and the aftereffects thereof (Dichgans & Brandt, 1978). This is a promising method if applied with a full knowledge of the psychophysics of human perception and the factors that give rise to visually induced discomfort (see chap. 30, this volume).[7]

*2.4.2.2 Auditory Surround Motion.* Acceleration perceptions can be elicited via motion of auditory surrounds and the aftereffects thereof (Lackner, 1977). This is a promising method, especially when used in conjunction with confirmatory body motion (see section 2.4.1) or visual motion (see section 2.4.2.1).

*2.4.2.3 Tactile Surround Motion.* Acceleration perceptions can be elicited via moving touch and pressure cues and the aftereffects thereof (Chapter 38, this volume; see also Brandt, Büchele, & Arnold, 1977; Lackner & Graybiel, 1977; Lackner & DiZio, 1984). This is a promising method, especially when used in conjunction with confirmatory information from other sensory modalities (see sections 2.4.1, 2.4.2.1, 2.4.2.2).

*2.4.2.4 Voluntary Tactile/Kinesthetic Behaviors.* Acceleration perceptions can be elicited by voluntary tactile/kinesthetic behaviors such as locomotion without displacement, stationary pedaling, or simulated "hand walking" (as well as the aftereffects of these behaviors), as discussed in chapters 11 and 38 (this volume). (See also Brandt et al., 1977; Lackner & DiZio, 1988; Lackner & DiZio, 1993.) This is a most promising method, albeit limited to acceleration simulations involving limb movement.

*2.4.2.5 Vibration of Localized Body Regions.* Acceleration perceptions can be elicited by exploiting illusions of motion that accompany vibration of localized body regions. Vibration of skeletal muscles can elicit a variety of self-motion perceptions (Lackner & Levine, 1979; Levine & Lackner, 1979), as can vibration of the cranium (Lackner & Graybiel, 1974).[8] These are potentially useful methods that may augment other methods.

*2.4.2.6 Acoustic Vibrations.* Specific acoustic vibrations appear to act upon the vestibular organs (Minor, Solomon, Zinreich, & Zee, 1998; Van Campen, Dennis, King, Hanlin, & Velderman, 1999). However, this is currently not a promising method for eliciting acceleration perceptions, due to unpleasant symptoms induced and due to the unknown long-term consequences to auditory and vestibular organs.[9]

*2.4.2.7 Caloric Stimulation of Ear Canal.* Caloric stimulation of the ear canal (National Research Council, 1992) can elicit a sensation of movement. This is an effective method that can be applied unilaterally, but it is somewhat invasive and unpleasant, and the specific motion perception is difficult to alter from moment to moment.

*2.4.2.8 Direct Electrical Stimulation of Vestibular Nerve Fibers.* Direct electrical stimulation of the vestibular nerve fibers is a very effective way to elicit movement sensations, but when done invasively (as in the animal model), it is not a promising method for general (nonprosthetic) use. However, low-intensity noninvasive stimulation of the mastoids may be a method for achieving controlled-velocity perceptions (Quarck, Etard, Normand, Pottier, & Denise, 1998).

*2.4.2.9 Drug Treatments.* Dysfunctions in a given vestibular labyrinth or imbalances in the normal interaction of the respective labyrinths (such as may result from inner ear infections, Menier's syndrome, or ototoxic drug treatments) can create feelings of movement. It would be unethical to intentionally produce such invasive effects by similar means because of the long-term adverse effects. The vestibular effects of certain drugs such as alcohol (Fetter, Haslwanter, Bork, & Dichgans, 1999) are reversible, but not pleasant or easily controllable. Moreover, such drugs produce unpleasant side effects and raise ethical and safety concerns.

*2.4.2.10 Other Techniques.* Eliciting vertigo via looking over a visual cliff (Brandt, 1999), via postural hypotension (Baloh & Honrubia, 1979), or via atmospheric pressure changes (Benson, 1988) is effective for producing nonspecific feelings of instability, dizziness, or falling, but is not conducive to controllable motion perceptions and would give rise to safety concerns for certain individuals.

### 2.4.3 Combining Different Methods for Eliciting Acceleration Perceptions

A controlled feeling of acceleration within a VE can be achieved via a judicious combination of the more promising stimuli listed above. It is well-known that when motions of the visual field (see section 2.4.2.1) correspond with real motions of the user (see section 2.4.1.), feelings of acceleration are enhanced (Young, Dichgans, Murphy, & Brandt, 1973, 1988).

Multimodal inputs act in concert to create acceleration perceptions in the natural world. Similarly, compelling perceptions of acceleration in VE should be elicited by the introduction of mutually enhancing combinations of multimodal stimuli. The most promising methods will predictably combine real passive or active vestibular acceleration stimuli (see section 2.4.1) with visual field motion (see section 2.4.2.1.), auditory surround motion (see section 2.4.2.2.), or tactile surround motion (see section 2.4.2.3.). Vestibular motion stimuli and perceptions are described below.

## 3. FOCUS ON THE VESTIBULAR CHANNEL FOR CREATING ACCELERATION PERCEPTIONS

The vestibular end organs are stimulated directly by many of the methods of eliciting acceleration perceptions covered in section 2.4 above, including physical tilt or motion of the user, the aftereffects of prolonged rotation, caloric stimulation, low-frequency auditory vibrations, vibrations to the cranium, vestibular diseases, and drug or alcohol effects impacting the end organs. Many other acceleration perceptions arise by stimulating the vestibular nucleus (and associated structures in the brainstem) without directly stimulating the vestibular end organs. Illusions of self-motion can arise indirectly as a result of viewing a moving visual field or hearing a moving auditory surround stimulus. Several of the methods for inducing acceleration perceptions without using a direct stimulus to the vestibular end organ are promising and should be explored. Even if many of them do not prove to be viable ways of eliciting consistent and controllable acceleration perceptions, they must be understood if VE designers are to create effective acceleration simulations that are not inadvertently corrupted by interference from competing stimuli.

The best acceleration simulations will coordinate a variety of vestibular and nonvestibular stimuli so that different cues confirm one another and mutually enhance the simulation. Three of the most promising methods for eliciting acceleration perceptions are: (1) physical motion of the user (either passively or actively produced); (2) whole-field visual displays that induce illusions of self-motion (chap. 23, this volume); and (3) locomotion devices that induce illusions of self-motion without physical displacement of the user through space (chap. 11, this volume). This review focuses upon the use of "real-motion" stimuli (category A). However, it is instructive to compare the acceleration perceptions arising from real motion to those arising from visual-motion stimuli (2) or walking in place (3). This comparison is especially useful when one considers the likelihood that these three methods will be employed in combination, because it allows the reader to see where each method excels and thus to imagine how the three methods can complement one another. If virtual displays are to meet the promise of their initial "virtual reality" hype and become fully immersive and effective, a careful combination of the advantages of each of the three methods listed in Table 7.1 will be necessary.

### 3.1 Acceleration Perceptions Involving the Vestibular Channel

An important step toward developing a good acceleration simulation involving the vestibular modality is to gain some familiarity with the various vestibular acceleration sensations that are possible and how each one is generated. A wide variety of acceleration perceptions can be elicited via stimulation of the vestibular system, some of which are illusory perceptions. Below some acceleration perceptions involving the vestibular modality are listed, with a brief description of the stimuli that can cause them.[10] Most of the perceptions described below

**TABLE 7.1**
A Comparison of Three Methods for Eliciting Acceleration Perceptions

A. Passive or Active Body Acceleration (see sections 2.4.1 and 3.2, also Appendix A of this chapter)
B. Visually Induced Illusory Self Motion (see section 2.4.2.1, also chapter 23, this volume)
C. Illusory self Motion Induced by Locomotion without Displacement (see sections 2.4.2.4, Appendix A; chapters 11 & 38, this volume)

| *Points of Comparison* | *(A) Real-Body Acceleration (Passive or Active)* | *(B) Visually Induced Illusory Self-Motion* | *(C) Illusory Self-Motion Induced by Locomotion Without Displacement* |
|---|---|---|---|
| *Situation Flexibility* | | | |
| Range of perceived situations that can be simulated. | A vast number of passive transportation stimuli in 6 degrees-of-freedom can be simulated, as well as active locomotion, e.g., walking through a real space that serves as a context for a virtual space (see section 3.2.2). The range of simulations is very wide, with cost and space the main limitations. | A moderate number of virtual accelerations of predictable and uniform trajectory can be simulated if the motion is of sufficient duration and velocity is not too high. This method does not work well for all degrees of freedom and it will not work to simulate movement in darkness. | A wide number of active or semi-passive locomotion behaviors can be simulated without moving through space. These include active crawling, walking, running, climbing, peddling, paddling, hand cranking, and hand-over-hand locomotion as well as semi-passive stepping in time with a moving surface. |
| *Isomorphism* | | | |
| When simulation is the goal of the VE, isomorphism implies: (a) Presenting similar multi-modal stimuli in the VE as would have been present during the real event. (b) Demonstrating the absence of VE side effects or aftereffects (unless they are present in the real stimulus). | The acceleration effect is nearly isomorphic when the simulation requires limited displacement. When the desired movement exceeds the feasible distance possible with the moving device, simulation quality suffers and head or arm movements are no longer isomorphic. Illusions can be exploited at times to enhance the simulation, or real vehicles can be employed instead (see section 3.2). | The visual aspect of self motion through the world can be reproduced faithfully for unidirectional motion at constant velocity, but the quality of the acceleration perception will not be isomorphic during changes in simulation velocity or direction, and during the user's own actions (e.g., head movements, reaching behaviors). | The characteristics of human stride and energy expenditure during treadmill locomotion are not identical to real locomotion. Nevertheless, this remains the most isomorphic way to simulate active locomotion within a small space. |
| *compellingness* | | | |
| Indicated by short latency to onset of self motion perception, by degree of "saturation" of vection (perceived self-movement | Highly compelling, ranks first for vehicle motion simulation. Space required limits amplitude and duration of movement, however. When real vehicles are used for the simulation | Very compelling under the right circumstances. Does not work as well for rapidly changing or high velocity virtual motions. Rotational vection works best for | Very compelling for the simulation of active locomotion in a limited space (especially when coupled with confirming visual motion information—see column B of this Table). |

(Continued)

**TABLE 7.1**
(Continued)

| *Points of Comparison* | *(A) Real-Body Acceleration (Passive or Active)* | *(B) Visually Induced Illusory Self-Motion* | *(C) Illusory Self-Motion Induced by Locomotion Without Displacement* |
|---|---|---|---|
| without perceived surround motion), by high ratings of "presence," and by the inability to discriminate real from virtual stimuli. | (see section 3.2), the effect will be very compelling. When space is available, body acceleration via real locomotion (see section 3.2.2) will be superior to treadmill locomotion. | constant velocity visual surround rotation < 90 deg/sec or when the axis of rotation of the visual display is aligned with the earth vertical. (See Boff & Lincoln, 1988, for further details.) | |
| *Range of acceleration profiles possible* | | | |
| Range of perceived virtual velocities and accelerations possible. | Wide range of acceleration effects possible on moving devices, but effect duration is limited by the space within which the device operates. Simulation using real vehicles or active locomotion also requires a large space, but allows for additional acceleration simulations. When space is limited, perceived motion can be enhanced by vestibular illusions. | Works best at a moderate, constant velocity (e.g., below 90 deg/sec for rotational vection). No theoretical limit on duration of stimulus, however. | When used alone, will be limited by the speed, flexibility, and fitness of the human user, but this is not a problem since most of the simulations desired will be within normal human abilities. |
| *Temporal fidelity* | | | |
| How well the perception tracks changes in stimulus velocity or direction. | Often immediate onset or recovery of acceleration perception with changes in simulus. Perception stays in phase with the stimulus within normal frequencies of voluntary human motion. Vestibular illusions arise during passive motion outside the normal range. | Often delayed onset and will not readily follow sudden changes in the velocity of the visual stimulus. | Often delayed onset, but velocity of self motion perception probably stays roughly in phase with stepping. (Stepping direction can be de-coupled from speed or direction of visual motion, however). |
| *Usefulness for simulating altered force environments* | | | |
| Ability to simulate altered gravitoinertial force environments, such as occur on other planets or during high performance flight. | The only method that can generate a non-terrestrial force environment to simulate extraplanetary travel or the G-forces felt by operators of high performance vehicles. | Can enhance the feeling of falling, but is not ideal for simulating non-terrestrial gravity or motion induced G-forces. | Alone, no simulation of altered G is possible, but in conjunction with limb loading, body suspension, or certain cutaneous inputs, some subtle illusions may arise which are worth exploring. |

(Continued)

**TABLE 7.1**
(Continued)

| *Points of Comparison* | *(A) Real-Body Acceleration (Passive or Active)* | *(B) Visually Induced Illusory Self-Motion* | *(C) Illusory Self-Motion Induced by Locomotion Without Displacement* |
|---|---|---|---|
| *Behavioral flexibility* Degree to which the user's activities must be artificially restricted for the virtual acceleration perception to be elicited and maintained. | User should not be freely locomoting inside of a moving device at more than 2 g without precautions. Head, eye, and limb motions can alter the acceleration percept. Head movements can produce adverse symptoms. Restrictions during simulations exploiting real vehicles or real locomotion are the same as when no simulation is being created. | Works best when the user's head and eyes are relatively still. Probably won't work as well when simulating real situations where the field of view is limited and the sight of one's own body is limited (e.g., a virtual simulation of a view through a tank periscope). | Works best when the locomotion is at constant velocity and in a predictable direction. |
| *Discomfort* How disturbing or "sickening" the method can be (see chap. 30, this volume, for details) | Certain stimuli can be very disturbing, but these effects can be partially diminished by helpful visual or somatosensory inputs (Lawson et al., 1993, 1994, 1997). Drowsiness is one of the most common symptoms, while nausea contributes most to degraded feelings of overall well-being (see chap. 30, this volume, for details on symptomatology). Fewer adverse effects expected when using real vehicles or real locomotion. | Certain stimuli can be fairly disturbing and effects can persist after cessation of stimulus. However, the user need only close his eyes to avoid the stimulus. Even with eyes open, sudden cessation of prolonged visual flow velocity is not likely to be as sickening as sudden cessation of prolonged self rotation. "Head symptoms" may be somewhat more common during visually induced motion discomfort (see chap. 30, this volume). | Few symptoms arise when walking at a constant speed on a treadmill during conditions of constant velocity visual flow, but symptoms may arise when treadmills walkers are exposed to variations in visual flow velocity (Durlach & Mavor, 1995, chap. 6, 1995). |
| *Other undesirable effects* Extent and nature of other undesirable side and aftereffects besides motion discomfort. | May include postural dysequilibrium, altered eye–hand coordination, altered vestibulo-ocular and cervical reflexes. Fewer problems expected when a real vehicle or real locomotion is used. | Similar symptoms as above and at left, but often of milder severity than for real motion stimuli. Additional oculomotor effects may arise. | Altered limb control and altered gate on exiting treadmill simulator, possibly other aftereffects as well. |

(Continued)

**TABLE 7.1**
(Continued)

| *Points of Comparison* | *(A) Real-Body Acceleration (Passive or Active)* | *(B) Visually Induced Illusory Self-Motion* | *(C) Illusory Self-Motion Induced by Locomotion Without Displacement* |
|---|---|---|---|
| *Expense* of Building and Maintaining the Display. | The average "virtual accelerator" will be relatively expensive. Some moving devices, real vehicles, or real-locomotion simulations will be able to take advantage of existing infrastructure. | The average "visual-only" display will tend to be relatively inexpensive. Existing computer display infrastructure can be leveraged readily. Programming expense is the main financial burden. | The average cost will be moderate. The range of costs is likely to be very wide, ranging from stepping in place or walking on a treadmill (inexpensive) to sophisticated locomotion devices mounted to each limb separately (expensive). |
| *Technical difficulty* How difficult the virtual displays are to design, build, and program. | A difficult approach if taken without prior knowledge of acceleration psychophysics. Will be most feasible when "piggy-backing" virtual displays onto existing motion-capable devices or real vehicles. | Although still an unsolved challenge, the visual motion approach will be relatively easy to design, build, test, and modify. | Not very difficult when using simple treadmills to simulate constant velocity walking in one direction, but complicated when attempting to simulate more sophisticated behaviors. |
| *Size* Space occupied by the display. | Large "footprint" required for most moving based simulators; very large footprint for centrifuge based devices. Extremely large space needed when using real vehicles or active locomotion through a real space used to mimic the virtual space. | Very small space needed if head mounted, relatively small space needed if visual stimulus surrounds entire body. | Small space needed for simple treadmills and foot mounted devices; moderate space needed for omnidirectional and tiltable treadmill surfaces or when incorporating suspension/weighting systems to attempt altered G simulation. |
| Initial "*Market Niche*." | Costly government, sporting, and entertainment applications Especially useful for simulations of high-performance land, air, and sea vehicles. Will also be marketed for high-budget situations involving active locomotion through a real space used to mimic a virtual space (see section 3.2.2). In entertainment applications, start-up costs will be defrayed by user fees. | Lower-budget personal and home applications. Note, however, that high-budget applications will not be complete without visual displays of high quality, and hence this method will always have a key role to play in the market. | Mid-range budget applications requiring prolonged active locomotion perceptions within a small real space. Higher budget applications will include all limbs and all motions (not just walking, running, or pedaling). Promising approach for marketing portable units to augment training of athletic skills and training for dangerous activities such as close quarters combat. |

are especially salient when veridical information from nonvestibular modalities (e.g., helpful visual cues from the outside world) is absent.

### 3.1.1 Perception of Tilt

The simplest way of achieving the feeling of body tilt (i.e., misalignment with the earth vertical) in a VE is to physically tilt the user in the real world at the same speed and by the same amount. When this is not possible or the effect needs to be magnified, one can produce or enhance the feeling of tilt in various ways. An alternate way to produce a feeling of tilt is via the presence of a linear (including centripetal) acceleration vector not aligned with the gravitational acceleration vector. In a laboratory setting, stimuli for eliciting tilt perceptions include off-center rotation about the vertical axis (leading to a feeling of tilting away from the central axis of rotation) and horizontal linear oscillation at low frequency (leading to a feeling of tilting away from the gravity vector during each deceleration that must accompany a reversal in direction). In aerospace operations, strong accelerations of the aircraft (e.g., forward thrust during jet-assisted take-off) can induce a false perception of tilt known as the "somatogravic illusion" (Gillingham & Krutz, 1974). Similarly, an overall increase of the net +Gz[11] force (e.g., during a banking turn such as occurs in an aircraft) can cause a complex tilt sensation during head movement (known as the "G-excess effect"; see Chelette, Martin, & Albery, 1992; Guedry et al., 1973; Guedry & Rupert, 1991). Rolling back to the straight and level after a prolonged banking turn can cause a pilot to feel a false perception of tilt known as "the leans" (Gillingham & Krutz, 1974).[12] The feeling of self-inversion is a special case of perceived tilt. A perception of inversion can be induced in a VE by turning the user upside down, or it can be done by less direct means. Turning a user upside down in a VE could augment the "ground-based" simulation of the inversion illusion that occurs in aviation operations during – Gz (usually combined with +Gx) accelerations (e.g., rapidly leveling off following a fast, steep climb), which is a special case of the somatogravic illusion. The same technique could be used to simulate the inversion illusions that can occur during the sudden removal of +Gz that occurs upon transition to microgravity during parabolic flight or space flight (Nicogossian, Huntoon, & Pool, 1989).

### 3.1.2 Perception of Rotation

The simplest way to achieve a feeling of rotation (or other curvilinear motion) in a VE is to move the user in a rotating (or other curved) path that exactly duplicates the intended simulation. When an illusion of rotation is caused by rapid deceleration following prolonged unidirectional rotation in the darkness, it is called *after-rotation*. The after-rotation sensation involves a feeling of turning opposite to the direction of prior body rotation when the prior body rotation was passively generated, but when the body rotation is actively generated (by stepping around in a circle), the after-rotation illusion is diminished and can be felt in the same direction as prior rotation (Guedry et al., 1978).

When an illusion of rotation is caused by a sudden motion of the head in an axis not parallel to the central rotation axis of the body during prolonged rotation in the darkness, it is called a "Coriolis, cross-coupling" effect and is characterized by a feeling of head or self-velocity in a curved path that is predominantly[13] orthogonal to the axes of both body and head rotation. The aforementioned (see section 3.1.1) feeling of tilt vis *à* vis gravity that arises from a purely linear acceleration (usually via low-frequency oscillation) in a horizontal plane is also a perception that has an angular component, because the resultant of acceleration due to gravity and due to horizontal movement creates a rotating gravitoinertial acceleration vector that is perceived as an angular change in the direction of "up" when earth-referenced visual cues are absent or fixed relative to the observer (Howard, 1986).

### 3.1.3 Perception of Translation

When users' actions within a VE require them to move in a straight line, one can move them through the physical world in exactly the same manner. However, given the limited size of most buildings housing VE displays, often it is desirable to create the illusion of moving in a straight path over a longer time and distance than is actually occurring. This magnified movement perception can be accomplished via the use of a centrifuge, albeit with the introduction of some angular acceleration. The centrifuge technique is especially useful when the device can minimize the undesirable consequences of angular acceleration by having the longest possible rotation radius and by allowing for strategic combinations of radial linear translation (toward or away from the center of rotation) and capsule-centered counterrotation (on-axis capsule rotation opposite to the direction of rotation of the central axis of the centrifuge) during the off-center rotation stimulus. (For an explanation of such complex stimuli, see Hixson, Niven, & Correia, 1966.)

### 3.1.4 Illusory Absence of Tilt

In this case, one has the illusion of remaining in one's former orientation vis à vis gravity when one has actually tilted. This can occur when the tilt stimulus is below the threshold of detection, which is about 2.1 degrees (Mann, Dauterive, & Henry, 1949, 1974). Tilt can also fail to be perceived when the subject remains aligned with the resultant gravitoinertial force vector during off-center rotation, such as occurs during a banking turn in an airplane (Gillingham & Krutz, 1974). Finally, a failure to detect tilting of the resultant gravitoinertial force vector can occur when the individual remains seated upright while rotating off-center at a very low velocity (wherein the threshold for stimulation given by the oculogravic illusion would suggest that a tilt of the resultant that is less than 1.1 degrees would be undetectable, according to Graybiel & Patterson, 1955).

### 3.1.5 Illusory Absence of Rotation

In this situation, one is moving in a curved path but has the illusion of moving in a straight path or of not moving at all. This effect can occur during off center rotation below the threshold of the semicircular canals. The minimum threshold for rotation sensation is around 0.44 dg/s$^2$ using verbal reporting of self rotation, but can be as low as 0.11 dg/s$^2$ when detection of the oculogyral illusion is used as the threshold measure (Clark & Stewart, 1968). Even when the rotation stimulus is above threshold, it is possible to fail to perceive a part of the rotation stimulus during off-center angular acceleration with the nose (x-axis) aligned with the resultant of the tangential and centripetal acceleration vectors (Guedry, 1992). A failure to perceive self-rotation can also occur after prolonged constant velocity rotation, because the elastic properties of the cupula of the semicircular canal will cause it to return to resting position during prolonged constant velocity rotation. The user will feel stationary in this situation if there are no veridical visual (or other nonvestibular) cues to inform him that he is still turning. Finally, failure to perceive rotation can occur when rotating with the longitudinal z-axis of the body orthogonal to gravity (e.g., rotating as if on a horizontal barbecue spit). One variation of this stimulus results in a feeling of orbital motion of the body about a central point, with little or no change in the direction one's face seems to point, and hence no felt rotation about one's own longitudinal z-axis (Lackner & Graybiel, 1977).

### 3.1.6 Illusory Absence of Translation

In this instance, one has the illusion of not moving in a straight line when one is doing so. This effect can be produced by linearly translating an individual (without tilting him out of his former alignment vis-à-vis gravity) at a speed lower than the threshold of the otolith organs, which is about 6 cm/s$^2$ (Melville, Jones, & Young 1978, 1988), but may be slightly lower for subjects oriented sideways to the movement (i.e., translating along the y-axis, Melville et al., 1978; Travis & Dodge, 1928, 1988).

#### *3.1.7 Perception of Increased Weight or G-forces*

When driving a real vehicle such as a car or an airplane, one is subjected to G-forces during acceleration in various directions. However, in most current VE, these forces are absent during simulated driving or flying. This absence should be no surprise when one considers the simulation of a high-performance jet airplane. For example, one would need a simulator and a maneuvering space several hundred feet long to exactly reproduce the forces on a pilot's body during a catapult launch from an aircraft carrier. However, one can readily reproduce most of the perceptual aspects of a catapult launch from an aircraft carrier in a much smaller space by rotating the pilot off-center inside a centrifuge capsule while keeping his nose pointing into (i.e., his x-axis aligned with) the resultant of the tangential and centripetal acceleration vectors (Guedry, 1992). The predominant perception will be of rapid forward and upward acceleration while pitching back. Another way to produce the feeling of G-force during centrifugation is by exploiting the linear Coriolis acceleration that occurs when a body is moved radially in a rotating environment. Due to the inertia of the user's body, he or she will feel a sideways (tangentially directed) force act on the body during (actual) displacement either toward or away from the center of rotation.

### 3.2 Eliciting Acceleration Perceptions Via Isomorphic Motion Stimuli

Even the most careful orchestration of visual displays, locomotion devices, and man-moving machines will not provide an acceleration experience that can pass the Virtual Reality Turing Test (Barberi, 1992; Turing, 1950) for most of the synthetic experiences in a given VE's repertoire. However, there is a way of simulating real events that has the potential to generate isomorphic motion stimuli that elicit acceleration perceptions that are indistinguishable from reality. This method involves identically duplicating the real acceleration profiles being simulated, but doing so in an alternate location from the user's virtual location. This method is similar to the concept of augmented reality, wherein users see and move through a real environment while being able to view virtual information as well. However, the purpose of the type of augmented reality presently under consideration is not to provide additional information to a user who is otherwise engaged with the physical world. Rather, the goal is to provide a physical mock-up that serves as the ambient context for a synthetic event that would be difficult or impossible to experience in the physical world. What follows are several examples of potential future applications of isomorphic motion stimuli for enhancing flight training, law enforcement training, and sports training.

#### *3.2.1 Examples of Real Acceleration to Enhance Aviation Mission Rehearsal*

During advanced mission rehearsal, military aviators flying real jet aircraft would execute all the same maneuvers and actions as they would execute during a real combat mission, except they would fly in a designated friendly airspace. Thus, the flying would be real, but the targets, enemies, and ordnance would be simulated. Aviators would view simulated targets and simulated ground or air threats through virtual displays and neutralize them either through virtual representations of weapons systems or via live fire on veridical practice targets whose position in space coincides with the virtual image and virtual targeting information represented to the pilot. Opponents could also be real, but remote (e.g., aviators from allied countries coordinating in the exercise from their own friendly air space, located thousands of miles away). Safety of the crew and allied ground troops would be ensured by simulating (or deactivating) ordnance in all situations involving close support of ground troops, and by close monitoring from the ground and from an onboard safety observer who is not part of the virtual environment. This approach would allow for more realistic, varied, and challenging practice than is possible with current simulators or small unit training exercises. It would also provide many of the training benefits of full-scale war games on a more frequent basis, and without the same amount of preparation time, cost, danger, environmental impact, or political visibility (useful in cases where demonstration of force is not a mission goal).

### *3.2.2 Examples of Real Locomotion Through Simulated Environments as Part of Police Training*

Law enforcement officers serving on special weapons and tactics (SWAT) teams could go armed into a physical mock-up of a house and proceed to "clear" the armed and hostile criminals resisting within (e.g., as part of hostage crisis training). This simulation would be achieved via voluntary movement through the typical "fun house" or "Hogan's Alley" (a custom-built practice area that looks like a regular building but has bullet resistant backstops all around to enhance the safety of live-fire practice), coupled with head-mounted "look through" visual displays wherein virtual criminals and innocent people (either computer simulations or real-but-remote users sharing the virtual environment) appear suddenly and either respond to certain key verbal commands or attempt to attack the officers with various simulated weapons. Simulation of "return fire" from moving virtual opponents could be dictated probabilistically as a miss or a hit to various parts of the officer's body, based upon the officer's speed and skill in moving through the simulation (e.g., his skill with "quartering techniques" when rounding corners or entering rooms, his use of cover and concealment) and his speed in neutralizing threats with adequate force (e.g., his coordination with other officers, his skill with nonlethal force alternatives, the accuracy and speed of his shooting), and the likely degree of threat posed by the sort of criminals encountered in the scenario being practiced (e.g., first time mugger vs. professional terrorist; sober vs. drug influenced suspect).

Simulation of return fire could also be based on tracking technology on the bodies and weapons of two opposing teams of officers moving through one virtual space but physically located in different practice areas. The simplest example would be the case where one officer is locomoting through the Hogan's Alley and directing live fire against the virtual representation of another officer playing the role of a violent criminal in a different Hogan's Alley. The approach described is merely an extension of the training simulations already in use to allow live fire against projected images of actors portraying criminals (http://teams.drc.com/fast/index.html) and simulated fire against live opponents (www.simunition.ca/html/welcome_simunition.html). As the most useful features of these existing techniques are enhanced by the new virtual display features mentioned above, more sophisticated and realistic simulations will emerge. This outcome will yield obvious advantages for the training of law enforcement personnel and for certain national defense applications involving infantry soldiers (chap. 43, this volume).

In the better training devices available currently, self-motion is suggested solely by the visual stimulus, and "movement" is scripted in advance (http://teams.drc.com/fast/index.html). Even in some current devices where movement is not scripted, the user's navigation and visual reconnaissance may be accomplished via nonintuitive methods such as foot switches that initiate visual scene motion, the direction of which is then determined by head pointing (www.bdi.com/How_People_are_Using.html). One of the key features distinguishing future individual (or small unit) combat simulations will be that body movement through the VE will be real and naturally accomplished.[14]

#### *3.2.3 Examples of Real Acceleration as a Feature of Sports Simulations*

In sports as disparate as automobile racing, rock climbing, gymnastics, kayaking, and golf, VE that allow physical movement through nearly identical real spaces will provide a valuable training tool for Olympic and professional athletes. For example, race car drivers routinely practice alone on tracks to familiarize themselves before a race. Using virtual displays, it will be possible for drivers to do their practice runs against simulated competitors represented on an otherwise transparent ("look through") display while driving alone on a real track at a speed reduced just enough to keep the margins of exposure to monetary and personnel risk appropriate to the practice session. If the driver makes a driving error while maneuvering against a simulated (or real but remote) competitor, he will receive error feedback but will not suffer a dangerous and expensive car-to-car collision. This type of practice would form a bridge between lone practice on the track and live competition with other drivers.

### 3.3 Advantages and Limitations of Using Isomorphic Acceleration Stimuli

Real acceleration provides the most direct means for eliciting acceleration perceptions in VE and allows incredible flexibility and control concerning the types of accelerations perceptions that can be elicited at any given moment. Real acceleration that is isomorphic to the psychophysics of the VE should minimize the amount of interference a VE trainer creates regarding the transfer of psychomotor skills to a real situation. For example, if the airplane-based VE previously mentioned (see section 3.2.1) is designed and executed properly, it will be able to simulate a hazardous military combat mission over friendly air space, eliciting only those adaptive effects and aftereffects that occur during a comparable real flight.

The primary disadvantage of using real acceleration in VE is that this approach will require large and often costly practice spaces. In the case of military flight training, this concern is not insurmountable because the infrastructure for real flying (of simulated missions) is already in place. For police training, the infrastructure of "fun houses" is also in place among many of the larger agencies, although numerous software challenges remain in order to enhance realism and simulate the stress of actual operations. The cost associated with introducing real acceleration into VE training for the most popular professional or Olympic sports will probably be less of an impediment than it is for the military and law enforcement. However, the technical ease of implementation and the revenue available for implementation of "virtual acceleration" will be very different for different sports. For example, implementation of virtual acceleration training could be achieved readily for either automobile racing or whitewater kayaking, but would probably be implemented for automobile racing first (despite the fact that a larger practice area is necessary), because automobile racing has greater spectator and commercial support (and thus commands greater revenue). Also, virtual acceleration training during automobile racing can be accomplished on existing automobile tracks without substantive modifications to the track itself, whereas the simulation of whitewater kayaking requires the construction of a special recycling water course with programmable high-speed currents. Some sports will be very difficult to train for via real movement through virtual spaces. A good example would be equitation, which requires the mutual coordination of perception and action on the part of a horse and rider, each of whom has profoundly different user requirements.

## 4. VISUAL CONSEQUENCES OF ACCELERATION

This chapter is concerned with the perception of body acceleration. However, it should be noted that many acceleration stimuli can influence the perception of a visual display relative to the user. For example, simple body tilt in darkness can make it hard for an observer to set a line of light to align with the true earth vertical. The visual effects of acceleration stimuli include the following: the Müller (or "E") effect, the Aubert (or "A") effect (e.g., Bauermeister, 1964; Clark & Graybiel, 1963; Müller, 1916; Passey & Ray, 1950; Wade & Day, 1968), the oculogyral illusion (Graybiel & Hupp, 1946), the oculogravic illusion (Graybiel, 1952), the elevator illusion (Cohen, 1973; Whiteside, 1961), and the degradation of visual acuity that occurs when viewing a head fixed display while one's body is moving (Guedry, Lentz, Jell, & Norman, 1981).[15] Thus, even when a user is being exposed to a simple tilt, the VE designer must take into account the interactions between visual inputs and vestibular and somatosensory inputs for any task requiring visual estimates of the orientation of objects within the virtual scene or any tasks requiring eye-to-hand coordination. Fortunately, many of these effects will be minimized by the presence of a whole-field visual stimulus (rather than a simple visual target presented in the darkness, as in most of the laboratory studies cited). This will be true especially when the visual stimulus provides veridical information about self-orientation (Lawson et al., 1997).

## 5. SUMMARY

The perception of acceleration encompasses feelings of self-motion through space and self-tilt relative to the upright. The feeling that one is being subjected to unusual forces (such as occur during high-performance flight or in nonterrestrial settings) also qualifies as an acceleration perception. The techniques for eliciting acceleration perceptions in future VEs will employ either real motion or illusory motion. Real motion methods will accelerate the VE user aboard a moving device such as a centrifuge or will allow his to drive or actively locomote (as appropriate) within a real space that serves as an ambient context for the virtual environment. Illusory motion methods will induce an illusion of body motion by moving a visual, auditory, or somesthetic surround stimulus relative to the user or by having the user locomote (without displacement) on a treadmill or like device. Each of these methods has advantages and disadvantages. The methods for inducing acceleration perceptions via real motion will tend to generate a wider range of simulations and be more effective. The methods involving illusory motion will require less money and space to implement. Methods employing real or illusory motion are not mutually exclusive. When a person moves through space while receiving visual, auditory, or somesthetic information that confirms the movement, the resulting movement perception will tend to be stronger than when only one self-movement cue is present in isolation. Thus, combining different cues to elicit acceleration perceptions will be advisable whenever it is feasible. Such multisensory cueing is exactly what takes place whenever the reader goes for a stroll through the real world.

## 6. REFERENCES

Advisory Group for Aerospace Research and Development. (1988). Motion cues in flight simulation and simulator-induced sickness. In *Advisory Group for Aerospace Research and Development Conference Proceedings No. 433*. Neuilly-sur-Seine, France, AGARD.

Arenberg, I. K. (1993). *Dizziness and balance disorders*. New York: Kugler Publications.

Baloh, R. W., & Honrubia, V. (1979). *Clinical neurophysiology of the vestibular system.* Philadelphia: F. A. Davis.

Barberi, D. (1992). *The ultimate turing test* [Online]. Available: http://metalab.unc.edu/dbarberi/vr/ultimate-turing/, info@2meta.com

Bauermeister, M. (1964). Effect of body tilt on apparent verticality, apparent body position, and their relation. *Journal of Experimental Psychology, 67*(2), 142–147.

Benson, A. J. (1988). Motion sickness. In J. Ernsting & P. King (Eds.), *Aviation medicine* (pp. 318–493). London: Buttersworth.

Berthoz, A., & Melville Jones, G. (1985). *Adaptive mechanisms in gaze control: Facts and theories.* Amsterdam: Elsevier.

Bles, W. (1979). *Sensory interactions and human posture: An experimental study.* Amsterdam: Academische Pers.

Boff, K. R., & Lincoln, J. E. (Eds.). (1988). *Engineering data compendium: human perception and performance.* Wright-Patterson Air Force Base, OH: AAMRL.

Brandt, T. (1999). *Vertigo: Its multisensory syndromes.* New York: Springer-Verlag.

Brandt, Tü., Büchele, W., & Arnold, F. (1977). Athrokinetic nystagmus and ego-motion sensation. *Experimental Brain Research, 30*, 331–338.

Casali, J. G. (1985). *Vehicular simulation-induced sickness, Volume 1: An overview.* (IEOR Tech. Rep. No. 8501, NTSC-TR86-010). Orlando FL: Naval Training Systems Center.

Chelette, T. L., Martin, E. J., & Albery, W. B. (1992). *The nature of the g-excess illusion and its effect on spatial orientation* (Tech. Rep. No. AL-TR-1992-0182). Wright-Patterson Air Force Base, OH: Armstrong Laboratory.

Chien, Y. T., & Jenkins, J. (1994). *Virtual reality assessment.* Alexandria, VA: Institute for Defense Analyses. (A Report to the Task Group on Virtual Reality to the High Performance Computing and Communications and Information Technology subcommittee of the Information and Communications Research and Development Committee of the National Science and Technology Council.)

Clark, B., & Graybiel, A. (1963). Perception of the postural vertical in normals and subjects with labyrinthine defects. *Journal of Experimental Psychology: General, 65*, 490–494.

Clark, B., & Stewart, J. D. (1968). Comparison of three methods to determine thresholds for perception of angular acceleration. *American Journal of Psychology, 81*, 207–216.

Cohen, M. M. (1973). Elevator illusion: Influences of otolith organ activity and neck proprioception. *Perception and Psychophysics, 14*(3), 401–406.

Cohen, B., & Henn, V. (1988). Representation of three-dimensional space in the vestibular, oculomotor, and visual systems. *Annals of the New York Academy of Sciences, 545*, 239–247.

Cohen, H., & Keshner, E. A. (1988). Current concepts of the vestibular system reviewed: 2. Visual/vestibular interaction and spatial orientation. *The American Journal of Occupational Therapy, 43*(5), 331–338.

Cohen, B., Tomko, D. L., & Guedry, F. (Eds.). (1992). *Sensing and controlling motion: Vestibular and sensorimotor function.* New York: New York Academy of Sciences.

Correia, M. J., & Guedry, F. E. (1978). The vestibular system: Basic biophysical and physiological mechanisms. In R. B. Masterson (Ed.), *Handbook of Behavioral Neurobiology* (pp. 311–351). New York: Plenum Publishing Co.

Cronin, C. (Ed.). (1998). *Military psychology: An introduction.* Needham, MA: Simon & Schuster.

De Broglie, L. (1951). A general survey of the scientific work of Albert Einstein. In P. A. Schilpp (Ed.), *Albert Einstein: Philosopher-scientist* (pp. 107–127). New York: Tudor.

DeHart, R. L. (Ed.). (1996). *Fundamentals of aerospace medicine* (2nd Ed.). Baltimore, MD: Williams & Wilkins.

Dichgans, J., & Brandt, T. (1978). Visual–vestibular interaction: Effects on self-motion perception and postural control. In R. Held, H. W. Leibowitz, & H. L. Teuber (Eds.), *Handbook of sensory physiology* (Vol. 8, pp. 755–804). New York: Springer-Verlag.

Durlach, N. I., & Mavor, A. S. (Eds.). (1995). *Virtual reality—Scientific and technological challenges.* Washington, DC: National Academy Press.

Fetter, M., Haslwanter, T., Bork, M., & Dichgans, J. (1999). New insights into positional alcohol nystagmus using three-dimensional eye-movement analysis. *Annals of Neurology, 45*(2), 216–223.

Gillingham, K. K., & Krutz, R. W. (1974). *Aeromedical review—Effects of the abnormal acceleratory environment of flight* (Tech. Rep. Review 10–74). Brooks Air Force Base, TX: U.S. Air Force School of Aerospace Medicine, Aerospace Medical Division.

Gillingham, K. K., & Previc, F. H. (1996). Spatial orientation in flight. In R. L. DeHart (Ed.), *Aerospace Medicine* (2nd Ed., pp. 309–397). Baltimore, MD: Williams & Wilkins.

Graham, M. D., & Klemink, J. L. (1987). *The Vestibular System.* New York: Raven Press.

Graybiel, A. (1952). Oculogravic illusion. *Archives of Ophthalmology, 48*, 605–615.

Graybiel, A., & Brown, R. H. (1950). The delay in visual reorientation following exposure to a change in direction of resultant force on a human centrifuge. *Journal of General Psychology, 45*, 143–150.

Graybiel, A., & Hupp, D. I. (1946). The oculo-gyral illusion. A form of apparent motion which can be observed following stimulation of the semicircular canals. *Journal of Aviation Medicine, 17*, 2–27.

Graybiel, A., & Patterson, J. L., Jr. (1955). Thresholds of stimulation of the otolith organs as indicated by oculogravic illusion. *Journal of Applied Physiology, 7*, 666–670.

Guedry, F. E. (1968). The nonauditory labyrinth in aerospace medicine. In Naval Aerospace Medical Institute (Eds.), *U.S. Naval Flight Surgeon's Manual* (pp. 240–262). Washington, DC: U.S. Government Printing Office.

Guedry, F. E. (1974). Psychophysics of vestibular sensation. In H. H. Kornhuber (Ed.), *Handbook of sensory physiology* (pp. 1–154). New York/Heidelberg/Berlin: Springer-Verlag.

Guedry, F. E. (1992). Perception of motion and position relative to Earth: An overview. In B. Cohen, D. L. Tomko, & F. E. Guedry (Eds.), *Sensing and controlling motion: Vestibular and sensorimotor function* (pp. 315–328). New York: New York Academy of Sciences.

Guedry, F. E., Jr., & Benson, A. J. (1978). Coriolis cross-coupling effects: Disorienting and nauseogenic or not? *Aviation, Space, and Environmental Medicine, 49*, 29–35.

Guedry, F. E., & Correia, M. J. (1978). Vestibular function in normal and in exceptional conditions. In R. B. Masterson (Ed.), *Handbook of behavioral neurobiology* (pp. 317–325). New York: Plenum Publishing Corporation.

Guedry, F. E., Lentz, J. M., Jell, R. M., & Norman, J. W. (1981). Visual–vestibular interactions: The directional component of visual background movement. *Aviation, Space, and Environmental Medicine, 52*(5), 304–309.

Guedry, F. E., Mortenson, C. E., Nelson, J. B., & Correia, M. J. (1978). A comparison of nystagmus and turning sensations generated by active and passive turning. In J. D. Hood (Ed.), *Vestibular mechanisms in health and disease*. New York: Academic Press.

Guedry, F. E., Richard, D. G., Hixson, W. C., & Niven, J. I. (1973). Observation on perceived changes in aircraft attitude attending head movements made in a 2-G bank and turn. *Aerospace Medicine, 44*, 477–483.

Guedry, F. E., & Rupert, A. H. (1991). Steady state and transient g-excess effects, technical note. *Aviation, Space, and Environmental Medicine, 62*, 252–253.

Henn, V., Cohen, B., & Young, L. R. (1980). Visual-vestibular interaction in motion perception and the generation of nystagmus. *Neurosciences Research Program Bulletin, 18*(4), 457–651.

Hixson, W. C., Niven, J. I., & Correia, M. J. (1966). *Kinematics nomenclature for physiological accelerations: With special reference to vestibular applications* [Monograph No. 14]. Pensacola, FL: Naval Aerospace Medical Institute.

Howard, I. P. (1986). The vestibular system. In K. R. Boff, L. Kaufman, & J. P. Thomas (Eds.), *Handbook of perception and human performance* (pp. 11-1–11-30). New York: John Wiley & Sons.

Irwin, J. A. (1881). The pathology of seasickness. *Lancet, ii*, 907–909.

Kennedy, R. S., & Frank, L. H. (1985). *A review of motion sickness with special reference to simulator sickness* (Tech. Rep. NAVTRAEQUIPCEN 81-C-0105-16). Orlando, FL: Naval Training Equipment Center.

Keshner, E. A., & Cohen, H. (1989). Current concepts of the vestibular system reviewed: 1. The role of the vestibulospinal system in postural control. *American Journal of Occupational Therapy, 43*(5), 320–330.

Kolasinski, E. M. (1995). *Simulator sickness in virtual environments* (Tech. Rep. No. 1027). Alexandria, VA: U.S. Army Research Institute for the Behavioral and Social Sciences.

Lackner, J. R. (1977). Induction of illusory self-rotation and nystagmus by a rotating sound-field. *Aviation, Space, and Environmental Medicine, 48*, 129–131.

Lackner, J. R., & DiZio, P. (1984). Some efferent and somatosensory influences on body orientation and oculomotor control. In L. Spillman & B. R. Wooten (Eds.), *Sensory experience, adaptation, and perception* (pp. 281–301). Clifton, NJ: Lawrence Erlbaum Associates.

Lackner, J. R., & DiZio, P. (1988). Visual stimulation affects the perception of voluntary leg movements during walking. *Perception, 17*, 71–80.

Lackner, J. R., & DiZio, P. (1993). Spatial stability, voluntary action and causal attribution during self locomotion. *Journal of Vestibular Research, 3*, 15–23.

Lackner, J. R., & Graybiel, A. (1974). Elicitation of vestibular side effects by regional vibration of the head. *Aerospace Medicine, 45*, 1267–1272.

Lackner, J. R., & Graybiel, A. (1977). Somatosensory motion after-effect following Earth-horizontal rotation about the z-axis: A new illusion. *Aviation, Space, and Environmental Medicine, 48*, 501–502.

Lackner, J. R., & Graybiel, A. (1978). Postural illusions experienced during z-axis recumbent rotation and their dependence on somatosensory stimulation of the body surface. *Aviation, Space, and Environmental Medicine, 49*, 484–488.

Lackner, J. R., & Graybiel, A. (1980). Visual and postural motion aftereffects following parabolic flight. *Aviation, Space, and Environmental Medicine, 51*, 230–233.

Lackner, J. R., & Levine, M. S. (1979). Changes in apparent body orientation and sensory localization induced by vibration of postural muscles: Vibratory myesthetic illusions. *Aviation, Space, and Environmental Medicine, 50*, 346–354.

Lawson, B. D. (1995). Characterizing the altered perception of selfmotion induced by a Coriolis, cross-coupling stimulus. In *Proceedings of the Third International Symposium on the Head/Neck System.* Vail, CO: Brandeis University Waltham, MA.

Lawson, B. D., Guedry, F. E., Rupert, A. H., & Anderson, A. M. (1993). Attenuating the disorienting effects of head movement during whole-body rotation using a visual reference: further tests of a predictive hypothesis. In *Advisory Group for Aerospace Research and Development Conference Proceedings No. 541: Virtual Interfaces: Research and Applications.* Neuilly-Sur Seine, France AGARD.

Lawson, B. D., Guedry, F. E., Rupert, A. H., Anderson, A. M., & Tielemans, W. C. M. (1994). Multimodal influences on perception of spatial orientation during unusual vestibular stimulation. In *Proceedings of the Cognitive Neuroscience Society Inaugural Meeting.* San Francisco: Cognitive Neuroscience Society, Davis, CA.

Lawson, B. D., Mead, A. M., Chapman, J. E., & Guedry, F. E. (1996). Perception of orientation and motion during centrifuge rotation. In *Proceedings of the 66th Annual Meeting of the Aerospace Medical Association. Atlanta GA: Aviation, space, and Environmental Medicine, 67, p. 702.*

Lawson, B. D., Rupert, A. H., Guedry, F. E., Grissett, J. D., & Mead, A. M. (1997). The human–machine interface challenge of using virtual environment (VE) displays aboard centrifuge devices. In M. J. Smith, G. Salvendy, & R. J. Koubek (Eds.), *Design of computing systems: Social and ergonomic considerations* (pp. 945–948). Amsterdam: Elsevier.

Levine, M. S., & Lackner, J. R. (1979). Some sensory and motor factors influencing the control and appreciation of eye and limb position. *Experimental Brain Research, 36,* 275–283.

Ludel, J. (1978). *Introduction to sensory processes.* San Francisco: W. H. Freeman.

Mann, C. W., Dauterive, N. H., & Henry, J. (1949). The perception of the vertical: I. Visual and non-labyrinthine cues. *Journal of Experimental Psychology, 39,* 538–547.

Marcus, J. T. (1992). *Vestibulo-ocular responses in man to gravito-inertial forces.* Unpublished doctoral dissertation, University of Utrecht, The Netherlands.

Matlin, M. W., & Foley, H. J. (1992). *Sensation and Perception* (3rd Ed.). Needham, MA: Simon & Schuster.

McCauley, M. E. (Ed.). (1984). *Research issues in simulator sickness: Proceedings of a workshop.* Washington, DC: National Academy Press.

Melville-Jones, G., & Young, L. R. (1988). Subjective detection of vertical acceleration: A velocity-dependent response. In K. R., Boff, & J. E. Lincoln (Eds.). *Engineering data compendium: Human perception and performance.* Wright-Patterson Air Force Base, OH: AAMRL. (Original work published 1978)

Minor, L. B., Solomon, D., Zinreich, J. S., & Zee, D. S. (1998). Sound- and/or pressure-induced vertigo due to bone dehiscence of the superior semicircular canal. *Archive Otolaryngologica Head and Neck Surgery, 124*(3), 249–258.

Müller, G. E. (1916). Über das Aubertsche Phänomen. *Z. Sinnensphysiol., 49,* 109–244.

National Research Council. (1992). Evaluation of tests for vestibular function N[2, Suppl.]. *Aviation, Space, and Environmental Medicine, 63,* A1–A34. (Report of the Working Group on Evaluation of Test for Vestibular Function; Committee on Hearing, Bioacoustics, and Biomechanics, and the Commission on Behavioral and Social Sciences and Education).

National Research Council Committee on Vision and Working Group on Wraparound Visual Displays. (1990). *Motion sickness, visual displays, and armored vehicle design* (Tech. Rep. BRL-CR-629). Aberdeen Proving Ground, MD: Ballistic Research Laboratory.

Naunton, R. F. (1975). *The vestibular system.* New York: Academic Press.

Nicogossian, A. E., Huntoon, C. L., & Pool, S. L. (Eds.). (1989). *Space physiology and medicine.* Philadelphia: Lea & Febiger.

Passey, G. E., & Ray, J. T. (1950). *The Perception of the vertical: 10. Adaptation effects in the adjustment of the visual vertical.* The Tulane University of Louisiana under Contract N7onr-434. U.S. Naval school of Aviation Medicine, U.S. Navy Publication NM001 063.01.17.

Pausch, R., Crea, T., & Conway, M. (1992). A literature survey for virtual environments: Military flight simulator visual systems and simulator sickness. *Presence, 1*(3), 344–363.

Quarck, G., Etard, O., Normand, H., Pottier, M., & Denise, P. (1998). Low-intensity galvanic vestibulo-ocular reflex in normal subjects. *Neurophysiol. Clin., 28*(5), 413–422.

Reason, J. T., & Brand, J. J. (1975). *Motion sickness.* London: Academic Press.

Rupert, A. H., & Gadolin, R. E. (Eds.) (1993). *Motion and spatial disorientation systems: Special research capabilities.* (Tech. Rep. No. 93-1). Pensacola, FL: Naval Aerospace Medical Research Laboratory.

Stanney, K. M., Mourant, R. R., & Kennedy, R. S. (1998). Human factors issues in virtual environments: A review of the literature. *Presence, 7*(4), 327–351.

Stanney, K. M., Salvendy, G., Deisinger, J., DiZio, P., Ellis., S., Ellison, J., Fogelman, G., Gallimore, J., Hettinger, L., Kennedy, R., Lackner, J., Lawson, B., Maida, J., Mead, A., Mon-Williams, M., Newman, D., Piantanida, T., Reeves, L., Reidel, O., Singer, M., Stoffregen, T., Wann, J., Welch, R., Wilson, J., & Witmer, B. (1998). Aftereffects and sense of presence in virtual environments: Formulation of a research and development agenda. *International Journal of Human–Computer Interaction, 10*(2), 135–187.

Stark, L., & Bridgeman, B. (1983). Role of corollary discharge in space constancy. *Perception and Psychophysics, 34*, 371–380.

Teuber, H. L. (1960). Perception. In J. Field, H. W. Magoun, & V. E. Hall (Eds.), *Handbook of physiology* (Sec. 1, Vol. 3). Washington, DC: American Physiological Society.

Travis, R. C., & Dodge, R. (1988). Experimental analysis of the sensorimotor consequences of passive oscillation, rotary rectilinear. In K. R., Boff, & J. E. Lincoln (Eds.). *Engineering data compendium: Human perception and performance.* Wright-Patterson Air Force Base, OH: AAMRL. (Original work published 1928)

Tschermak, A. (1998). Opticher raumsinn [optical sense of space]. In A. Bethe, G. Bergmann, G. Emden, & A. Ellinger (Eds.), *Handbuch der Normalen und Opathologischen Physiologie*. In S. A. Jones, *Effects of restraint on vection and simulator sickness.* (Original work 1931)

Turing, A. (1950). Computing machinery and intelligence. *Mind, 59*, 433–60.

Van Campen, L. E., Dennis, J. M., King, S. B., Hanlin, R. C., & Velderman, A. M. (1999). One-year vestibular outcomes of Oklahoma City bombing survivors. *Journal of American Academy of Audiology,* 10, 467–483.

Wade, N. J., & Day, R. H. (1988). Apparent head position as a basis for a visual aftereffect of prolonged head tilt. In K. R. Boff and J. E. Lincoln (Eds.), *Engineering data compendium: Human perception and performance.* Wright-Patterson Air Force Base, OH: AAMRL. (Original work published 1968)

White, W. J. (1964). *A history of the centrifuge in aerospace medicine.* Santa Monica, CA: Douglas Aircraft Company, Inc.

Whiteside, T. C. D. (1961). Hand–eye coordination in weightlessness. *Aerospace Medicine, 32*, 719–725.

Wilson, V. J., & Melville-Jones, G. (1979). *Mammalian vestibular physiology.* New York: Plenum Press.

Young, L. R., Dichgans, J., Murphy, R., & Brandt, T. Interaction of optokinetic and vestibular stimuli on motion perception. In K. R., Boff, & J. E. Lincoln (Eds.). *Engineering data compendium: Human perception and performance.* Wright-Patterson Air Force Base, OH: AAMRL. (Original work published 1973)

## APPENDIX A: ACCELERATION AND LOCOMOTION DEVICES: TECHNOLOGIES FOR ELICITING ACCELERATION PERCEPTIONS

This appendix lists some of the current acceleration and locomotion devices available nationally, including simulators, centrifuges, and treadmills. Man-rated acceleration devices represent a fairly mature technology that is costly to implement. Hence, when sophisticated acceleration simulations are needed, new VE display and navigation systems may seek to interface with older human-rated acceleration devices already in operation. A partial list of organizations in the United States housing large acceleration devices is provided below, in Section 1 of this Appendix. (Additional information can be found in Cronin, 1998; Rupert & Gadolin, 1993; & White, 1964).

Information about locomotion devices suitable for VE applications is provided in section 2 below. The technology for building locomotion devices is not yet mature. Such devices are not as costly to build as simulators and centrifuges, but they are not cheap either.

### 1 Human-rated Acceleration Devices Suitable for VE Applications

#### 1.1 Commercial Agencies

Environmental Tectonics Corporation (ETC) Aeromedical Lab, Southampton, PA
Acceleration devices for flight training applications
www.etcusa.com

Wyle Laboratories, El Segundo, CA
Acceleration devices for flight training applications
www.wylelabs.com

Otis Elevator, Bristol, CT
A research elevator allowing large amplitude vertical oscillations
www.otis.nl

VEDA Corporation, Alexandria, VA
Acceleration devices for flight simulation
www.veda.com/moldel.htm

August Design, Inc., Ardmore, PA
A moving platform for ship simulation
www.august-design.com; e-mail: engineering@august-design.com

### 1.2 Government Agencies

National Air and Space Administration
Acceleration devices for flight simulation, space flight training, etc.

Johnson Space Center, Houston, TX
http://tommy.jsc.nasa.gov

NASA Ames, Moffet Field, CA
www.arc.nasa.gov

NASA Langley, Hampton, VA
www.larc.nasa.gov and www.sdb.larc.nasa.gov/Simulations/simulations.html

#### 1.2.1 Department of Defense

Naval Aerospace Medical Research Laboratory, Pensacola, FL
Acceleration devices for research on spatial orientation and motion adaptation syndromes
www.namrl.navy.mil

Naval Operation Medical Institute, Pensacola, FL
Acceleration devices for ground-based flight training and airsickness desensitization
www.nomi.navy.mil

Naval Air Warfare Center, Patuxent River, MD
Acceleration devices for human research, especially concerning effects of high-g
www.nawcad.navy.mil

Naval Air Weapons Center, Warminster, PA
"Johnsonville Centrifuge," with a 50-foot radius
www.crompton.com/wa3dsp/k3nal/astronauts/index.html

Naval Research Laboratory
Ship-motion simulator for research on human performance during motion
www.nrl.navy.mil

Wright-Patterson Air Force Base, Dayton, OH
Acceleration devices for human research, especially concerning aviation human factors
www.wpafb.af.mil

U.S. Army Tank—Automotive and Armaments Command, Warren, MI
Moving-base simulators, for reproducing ride dynamics of Army land-based vehicles
www.ihreport.com

U.S. Army Aeromedical Research Laboratory, Dothan, AL
Moving-base simulators, for human research and helicopter flight simulation
www.usaarl.army.mil

### 1.3 University Laboratories

Ashton Graybiel Spatial Orientation Laboratory, Brandeis University, Waltham, MA
Acceleration devices for research on spatial orientation, sensorimotor functioning, and adaptation to unusual force environments
http://graybiel.cc.brandeis.edu/userinfo.htm

Man-Vehicle Laboratory, Massachusetts Institute of Technology, Cambridge, MA
Acceleration devices for human research concerning spatial orientation and neurovestibular adaptation
http://web.mit.edu/afs/athena.mit.edu/dept/aeroastro/www/labs/MVL/NEW/MVL/home.html

National Biodynamics Laboratory, University of New Orleans, New Orleans, LA
Acceleration devices for research concerning effects of mechanical forces and motion discomfort
www.nbdl.org

## 2 Locomotion Devices Suitable for VE Applications

Kistler Corporation, Amherst, NY
www.kistler.com/biomech/biomechanics.htm

NeuroCom International, Inc., Clackamas, OR
www.onbalance.com

Quinton Instrument Company, Bothell, WA
www.quinton.com/fitness/fitness.html

NPSNET Research Group, Monterey, CA
www.npsnet.nps.navy.mil/Locomotion.html

Sarcos Engineering Corporation and the University of Utah, Salt Lake City, UT
www.sarcos.com (see especially www.cs.utah.edu/~jmh/Locomotion.html)

Naval Postgraduate School, Monterey, CA
www.npsnet.org ; e-mail: darken@cs.nps.navy.mil

Human Interface Technology Laboratory, University of Washington, Seattle, WA
www.hitl.washington.edu/publications/r-96-4/

# APPENDIX B: RECOMMENDED READINGS ON VESTIBULAR FUNCTION

## 1 Basic Introduction

For a general introduction to the perception of body position and movement via kinesthesia and vestibular sensation, the authors recommend the textbook by Ludel (1978).

## 2 General Scientific Review

Excellent scientific reviews on vestibular sensation were written by Howard (1986, chap. 11, from Boff et al.,) and by Guedry (1974). The chapter by Howard provides a lucid review of vestibular structure, dynamics, neural projections, and psychophysics, ranging afield of vestibular modality as well. The comprehensive paper by Guedry focuses on psychophysics of vestibular sensation during a wide variety of stimulus situations. Other reviews of interest have been contributed Correia and Guedry (1978), Guedry and Correia (1978), Cohen and Keshner (1988), Keshner and Cohen (1989), and others.

## 3 Introduction to Spatial Orientation in Flight

Two of the more readable introductory papers written on spatial disorientation vis-à-vis aerospace operations were contributed by Benson (1988) and by Gillingham and Krutz (1974). An updated version of the spatial disorientation section in Gillingham and Krutz (1974) is available in Gillingham and Previc (1996). These papers introduce the common illusions associated with spatial disorientation in flight. The original paper by Gillingham and Krutz covers acceleration nomenclature and human cardiovascular and vestibular reactions to unusual accelerations, especially high-G. Gillingham and Krutz (1974) illustrate their points vividly with 47 figures and four tables. Another excellent resource is the U.S. Naval Flight Surgeon's Manual, which has been published in many editions. Chapter 8 of the 1968 manual (Guedry, 1968) is a thorough effort with a clinical slant. (Some general information about acceleration and vibration can be found in the most recent online version of the manual at www.vnh.org//FSManual/02/SectionTop.html.)

## 4 Detailed Research Compendia Including Coverage of Special Topic Areas

Many of the works in this section are special edition compilations of papers by various scholars, published as the proceedings of a conference. Works of note include *Sensing and Controlling Motion,* edited by Cohen, Tomko, and Guedry (1992), *Visual-Vestibular Interaction in Motion Perception and the Generation of Nystagmus*, by Henn, Cohen, and Young (1980), *The Vestibular System: Neurophysiologic and Clinical Research*, edited by Graham and Klemink (1987), *Mammalian Vestibular Physiology*, by Wilson and Melville-Jones (1979), and *Adaptive Mechanisms of Gaze Control*, by Berthoz and Mellville-Jones (Vol. 1, 1985). The book by Cohen, Tomko, and Guedry is the most recent compendium mentioned above and covers a variety of topics on vestibular and sensorimotor function. The book by Henn, Cohen, and Young provides a succinct and well organized treatment of the neural substrate of human orientation and includes an appendix describing the major mathematical models of the era. The book edited by Graham and Klemink summarizes clinical and research findings concerning the structure, function, and disorders of the vestibular system, with special attention paid to testing, diagnosis, and research in weightlessness. Wilson and Melville-Jones discuss the peripheral and neuronal physiology of vestibular function, giving a lucid explanation of the biophysics of the peripheral end organs. Berthoz and Melville-Jones review oculomotor research relevant to the topic of adaptation of gaze control during sensory rearrangement. Other notable books include Cohen and Henn (1988), Arenberg (1993), Naunton (1975), and Marcus (1992). Of these, the book by Arenberg is the longest and the most recent, giving a thorough treatment of dizziness and balance disorders that would be a worthy alternative to Graham and Klemink (1987).

## 5 Works on Side Effects of Virtual Environments

Chapters 29 to 32 and 37 to 39 (this volume) discuss the signs and symptoms of exposure to VE and other synthetic experiences. Earlier sources for information on unpleasant vestibular side effects of VE include McCauley (1984), Casali (1985), Kennedy and Frank (1985), the

Advisory Group for Research and Development (1988), National Research Council technical report BRL-CR-629 (1990), Chien and Jenkins (1994), Durlach and Mavor (1995), Kolasinski (1995), Stanney, Mourant, and Kennedy (1998), and Stanney et al. (1998). An early issue of the journal *Presence* was dedicated to simulator sickness and cybersickness (Vol. 1, No. 3, 1992). In this issue, Pausch, Crea, and Conway (1992) contributed a literature review and several other authors contributed important papers that have shaped current thinking about cyber sickness. A comprehensive book on virtual environments was contributed by Durlach and Mavor (1995), which includes a discussion of acceleration perceptions and side effects in chapter 6.

### 6 A General Human-Factors Reference Work of Relevance to Spatial Orientation

The *Engineering Data Compendium: Human Perception and Performance*, is a multivolume set edited by Boff and Lincoln (1988). This compendium of human factors findings is not limited to spatial orientation, vestibular function, or simulation, but it has succinct sections devoted to each of these topics. It is a fine "one-stop" source of information on human perceptual function in all sensory modalities. The authors recommend it highly for anyone who anticipates doing research in human factors or allied fields.

## ENDNOTES

[1] To be deemed effective, the VE should produce no adverse side effects that are not also caused by the real event being simulated. Most importantly, the VE should provide good transfer of skills learned in VE training to performance in the situation being simulated. Provided side effects are appropriate and transfer of training is satisfactory, it would be advantageous for the VE to elicit consistent and controllable acceleration perceptions that are difficult to distinguish from the real event being simulated. (The stated criteria for an effective VE apply to VE that attempt to simulate real situations; criteria for effectiveness must be modified for VE that do not attempt to simulate real situations.)

[2] The authors prefer terms such as *somesthesia* or *somatosensation* to refer to cutaneous, muscle, and joint receptors collectively, instead of the commonly used terms *proprioception* or *haptic*. The word *proprioception* includes the vestibular modality and hence should not be used synonymously with somesthesia. The word *haptic* is applied too ambiguously at present to be meaningful. *Haptic* usually refers to "active touch", or cutaneous exploration of the world via manual exploration. This usage helps distinguish "active" from "passive" touch and presumably recognizes that manual exploration is informed by muscle and joint receptors as well as cutaneous receptors. However, the word *haptic* has come to be applied by the VE community to encompass information derived by manual exploration, locomotion, and even "passive" cutaneous sensations applied to the torso. For these reasons, it is difficult to understand what a particular researcher means when using the term *haptic*, unless the researcher provides his own definition.

[3] The term *vection* is attributed to Tschermak, 1931, 1998.

[4] The reviews recommended in Appendix B can provide the reader with additional ideas about how to simulate a perception of acceleration that is appropriate to the desired goal of a VE (see also Boff & Lincoln, 1988.).

[5] Such real motion is "passive" in the sense that it is not generated primarily by the individual's muscular efforts. However, the distinction between "passive" and "active" motion is more a continuum than a dichotomy, since driving a vehicle is less "passive" than riding as a passenger,

even though driving a vehicle could be considered an example of "passive acceleration." Similarly, coasting downhill on a bicycle is fairly "passive," while peddling a bicycle would be considered "active" acceleration. Both types of real acceleration ("active" and "passive") are lumped together in section 2.4.1.1.

[6] It is possible that whole-body acceleration perceptions can also be elicited by the aftereffects of carrying a load (or being partially suspended) for long periods of time within the normal terrestrial environment (i.e., at 1 g). However, the authors predict that the effects would be restricted mostly to changes in felt sense of force and a slight loss of position constancy of the visual environment.

[7] In addition to continuous visual motion, classical visual object motion aftereffects and illusory object motion sensations are elicited by the phi phenomenon (see Matlin & Foley, 1992); this phenomenon could be exploited to enhance feelings of self-motion, though it does not appear to be studied for this application at present.

[8] Note also that gently jiggling the eyeball (by pushing on the eyelid) elicits the perception that the visual world is jiggling (see Teuber, 1960; Stark & Bridgeman, 1983). Whereas this type of vibration does not elicit an illusion of self movement, it might be exploited to enhance synthetic perceptions of high-frequency body oscillation or ground shaking created by other means.

[9] However, if reversible effects can be demonstrated, acoustic energy weapons are likely to be exploited for inducing temporary vertigo, disorientation, and nausea in conflict situations requiring less-than-lethal force.

[10] A different method of categorizing acceleration stimuli and perceptions is presented in Boff and Lincoln (1988). For a detailed explanation of acceleration stimuli and acceleration perceptions, see Howard (1986, chap. 11); Benson (1988); Nicogossian, Huntoon, and Pool (1989); Guedry (1992); and Gillingham and Krutz (1974). An updated version of the spatial disorientation section in Gillingham and Krutz (1974) is available in Gillingham and Previc (1996). Many of the other works cited in this endnote are discussed in Appendix B (this chapter).

[11] Positive Gz refers to a downward force (caused by upward acceleration) in the longitudinal axis of the head and body. Negative Gz refers to an upward force (caused by downward acceleration) in the longitudinal axis. The three cardinal head axes used in vestibular research were established by Hixson et al. (1966) as follows: (1) the line corresponding with the z-axis is a dorsal–ventral (or cephalo–caudal line) traveling along the intersection of the midcoronal and midsagital planes and is commonly called *yaw* (for angular acceleration about this axis) or *heave* (for linear acceleration within this axis); (2) the line corresponding with the x-axis is an anterior–posterior (or naso-occipital) line traveling along the intersection of the midsagital and midhorizontal planes and commonly called *roll* (for angular acceleration) or *surge* (for linear acceleration). The y-axis is an interaural line traveling along the intersection of the midcoronal and midhorizontal planes and commonly called *pitch* (for angular acceleration) or *sway/side-slip* (for linear acceleration). This explanation suffices to define the cardinal axes used in this chapter; a diagram and full explanation of acceleration versus force nomenclature and of positive and negative signs given to linear and angular vectors is given in Hixson et al. (1966). The reader should note that the criteria therein differ slightly from aviation nomenclature regarding the positive direction assigned to the linear y-vector (+y projects from left ear in Hixson et al. [1996] and from right ear in aviation).

[12] Related phenomena such as the "graveyard spin," "graveyard spiral," and "giant hand phenomenon" are mentioned in the same review.

[13] Some preliminary evidence (Lawson, 1995) suggests that the perception of velocity in this situation is not as simple as described in the text. The actual percept may contain an additional velocity perception in the same axis as the head movement. Thus, an x-axis head movement

toward the right shoulder following prolonged counterclockwise rotation in the vertical z-axis (seated upright) seems to produce an illusory sensation of forward pitch in the y-axis and also may produce an illusory sensation of rolling in the same direction as the head movement (x-axis).

[14] The type of simulation described in this section could be made quite sophisticated and multisensory in ways not usually considered within the context of acceleration perceptions. For example, accelerations delivered to localized body parts could be employed to enhance the somesthetic aspects of the simulation. The somesthetic perception of weapons firing could be augmented via force feedback against the firearm to simulate recoil. This could be accomplished via a small off-center mass driven in an arc by an electric motor that is mounted on the weapon frame (and set in motion by trigger pull). Similarly, an acceleration to the chest of the officer could simulate a fraction of the force generated by a pistol shot impacting the officer's ballistic vest. This could be accomplished by small mechanical or pneumatic pistons under the officer's vest.

[15] The aforementioned illusions are not described herein because they fall outside the theme of this chapter. For a brief description, see chapter 3, this volume. For a full explanation, see the cited literature in this section and the recommended readings in Appendix B.

# 8

# Motion Tracking Requirements and Technologies

Eric Foxlin
*InterSense Inc.*
*73 Second Ave.*
*Burlington, MA 01803*
*ericf@isense.com*

## 1. INTRODUCTION

The science of motion tracking is fascinating because of its highly interdisciplinary nature and wide range of applications. This chapter will attempt to capture the interdisciplinary approach by organizing the subject differently from several excellent review articles already available (Bhatnagar, 1993; Ferrin, 1991; Meyer, Applewhite, & Biocca,1992; National Research Council, 1995). These reviews tend to break trackers into several technology categories and evaluate the merits of each technology by inferring from commonalties amongst the existing trackers in that category. This survey will instead focus on what capabilities are required for various applications and what are all the technologies and combinations that could potentially be used to realize these capabilities.

Separate examinations of tracking methods are given from the point of view of the physicist, who seeks to design new and better sensors, and the mathematician, who seeks to take whatever measurements are available from the physicist's sensors and calculate the best possible estimate (or prediction) of the object's motion. This dual taxonomy is necessary in order to support a new emphasis on hybrid tracking techniques. To simply append hybrid tracking as a new technology in addition to the usual four or five technologies would be a disservice; there are a combinatorial number of different hybrids possible, and each may behave quite differently. By categorizing physical sensing principles according to the type of observation they yield and understanding how multiple observations can be mathematically blended together while accounting for the quality of each individual observation, a rationale is provided for the design and evaluation of these many potential hybrid systems. The founding philosophy for InterSense was that in most difficult tracking applications one can obtain better results at lower cost and weight by fusing measurements from a larger or more varied set of midquality sensors, rather than a smaller or more homogeneous set of high-precision sensors. This bias will come through clearly in the emphasis and organization of this chapter.

The structure of this chapter basically parallels the development cycle of a tracking system. The rest of section 1 and section 2 discuss several categories of motion tracking applications that will be considered, and the required tracking fidelity and ergonomics associated with each of these categories. This corresponds to the identification of a market need and development of a specification for a tracker that will meet this need. Sections 3 and 4 discuss the physics and math that are used to design a tracking system to meet this specification. In section 5, consideration is given to the engineering trade-offs that are involved in the implementation of the design during the product development phase. Section 6 discusses real-time systems integration issues that must be handled correctly when interfacing a motion tracker into a larger system. If this integration is not done well the application will perform poorly no matter how good the tracking system is, so a large effort must typically be devoted to this back-end "applications engineering" support to make sure the end user can fully benefit from the tracking system.

## 1.1 Applications for Motion Tracking

Motion tracking has a much wider range of applications than what will be discussed herein. For example, it includes onboard navigation systems or external tracking systems for aircraft, missiles, space vehicles, ships, submarines, UAVs, mobile robots, land vehicles, smart bullets, hockey pucks, baseballs, fish, birds, mobile phones, stolen cars, soldiers in the field, or doctors in a hospital. Attention is focused here on tracking human head, limbs, or hand-held objects for purposes of interacting with three-dimensional (3-D) computer-generated displays or teleoperators. Within these computer-generated environments, motion-tracking devices have four main functions:

1. view control
2. locomotion/navigation
3. object selection/manipulation
4. avatar animation

Even with this seemingly narrow definition, there is a vast array of motion tracking applications, and it is useful to categorize them into five broad groups to facilitate the discussion of requirements in section 2:

1. Visual VEs using opaque head-mounted displays (HMDs)
2. Visual VEs using fixed-screen displays (FSDs)
3. Augmented reality using see-through HMDs
4. 3-D audio environments using headphones
5. Avatar animation

There are many types of visual display devices used in virtual environment (VE) systems, with new ones being invented every year (see chap. 3, this volume). Category 1 and 3 applications use head-mounted display (HMD)-type displays. The term HMD-type display shall be used in a broad sense to include both true HMDs and handheld relatives (e.g., virtual binoculars, palmtop or camcorder-style displays, telescope sights, etc.) or boom-mounted displays. If the display movement is tracked and the view is rendered without separately tracking the head, then it is an HMD-type display. The standard opaque HMD or handheld display is occluding, while the nonoccluding HMD-like displays include either optical see-through (OST-HMDs) or video see-through designs (VST-HMDs), as well as head-mounted projectors.

Category 2 applications use fixed-surface displays (FSDs). The FSD shall denote any stationary display surface from a desktop monitor to a virtual model display (VMD) such as the ImmersaDesk, Virtual Workbench or PowerWall on up to a full spatially immersive display (SID) like the CAVE or VisionDome. The applications we are considering use head-tracked FSDs, in which the virtual camera viewing parameters are controlled in real time by the user's tracked head position to achieve a first-person perspective, usually rendered in stereoscopic 3-D using shutter glasses, polarized glasses, autostereoscopic display techniques, or any other method. For an FSD, the display does not move, but the user's eye points must be tracked relative to the display surface in order to calculate a skewed through-the-window perspective projection. The FSD examples listed in the definition above are all of a direct-view type. Because the user's hands are in front of the display surface and visible, these are considered nonoccluding displays, which can sometimes have registration requirements akin to see-through HMDs. There are also reflected-view FSDs, in which an angled mirror is placed between the user's head and hands to occlude the hands and prevent them from blocking the display, especially if force-feedback devices are required which would block even more of the display area.

In application categories 4 and 5, tracking is not used for view control of a visual display. Category 4 is a pure auditory VE using tracked headphones to produce spatialized sounds that augment the user's perception of the real world. Category 5 concerns avatar–animation applications (see chap. 15, this volume), in which tracking is used to animate a virtual surrogate body (or part of one), usually for others to view. Although slightly outside the above-defined scope of 3-D interactive motion tracking, this application will be at least briefly discussed since third-person avatars sometimes play a role in multiplayer interactive environments.

## 2. HUMAN FACTORS OF MOTION TRACKING

In the previous section, the realm of VE-related motion tracking applications was divided into five major categories according to display type. The display type dictates the performance required from the head tracker just to view objects without any manual interaction. If manual interactions are used, this adds requirements for handheld tracking devices, and perhaps additional requirements on the head-tracking device. In deciding the quality of head or hand tracking required for an application, there are several possible metrics for "good enough":

1. User feels "presence" in the virtual world (see chap. 40, this volume).
2. Perceptual stability: Fixed virtual objects appear stationary, even during head motion.
3. No simulator sickness occurs (see chap. 30, this volume).
4. Task performance is unaffected by any tracking artifacts (see chap. 35, this volume).
5. Tracking artifacts below detection threshold of a user who is looking for them.

Clearly, all of these metrics are interrelated, and without additional clarification of experimental conditions, they are not even well-defined binary threshold tests. At the current level of understanding in the field, it is not even obvious how to rank these tests from least to most stringent. For example, keeping all tracking artifacts below detection threshold would mean that if the rest of the VE system were also perfect, the user would not be able to tell whether the scene was real or virtual. This would seem to be the most stringent metric possible for "good enough" tracking, but how can one be certain the system would not cause simulator sickness due to some subtle artifacts that the user cannot consciously discern? In addition, some of these metrics, especially presence, are highly dependent on factors other than tracking performance.

**TABLE 8.1**
Tracking Performance Specifications

I. Static

- Spatial distortion: Repeatable errors at different poses in the working volume, including effects of all sensor scale factors, misalignments, nonlinearity calibration residuals, and repeatable environmental distortions.
- Jitter: Noise in the tracker output that causes the perception of image shaking when the tracker is actually still.
- Stability or creep: Variations in the tracker output when still that are too slow to be seen as motion but which may be observed to cause a change in the mean output position of a stationary tracker over time. This may be caused by temperature drift or random processes affecting the sensors or by repeatability errors if the tracker is power-cycled or moved and returned to the same pose. There is not a clearly defined distinction between jitter and creep, as they represent the high and low frequency portions of a continuous noise spectrum. A reasonable cutoff might be to consider any motion slower than a minute hand (0.1 deg/sec) in orientation and slower than 1 mm/sec in translation to be creep, with everything else called jitter. Creep itself might be broken down into repeatability and short-term and long-term in-run stability.

II. Dynamic

- Latency: The mean time delay after a motion until the corresponding data is transmitted. It is possible to specify the latency of the tracker and other subsystems separately, but they do not simply add up (see section 6).
- Latency jitter: Any cycle-to-cycle variations in the latency. When moving, this will cause stepping, twitching, multiple image formation, or spatial jitter along the direction the image is moving. Again, this is a system specification and is discussed more fully in section 6.
- Dynamic error (other than latency): Any inaccuracies that occur during tracker motion that cannot be accounted for by latency or static inaccuracy (creep and spatial distortion). This might include overshoots generated by prediction algorithms or any additional sensor-error sources that are excited by motion.

There are several types of tracking errors that may contribute in varying degrees to destroying sense of presence or perceptual stability, causing sickness, or degrading task performance. Different authors or manufacturers have focused on different specifications or defined them differently, and every type of tracker has its own complicated idiosyncrasies that would require a thick document to characterize in complete detail. However, Table 8.1 presents six specifications that can capture the essential aspects of tracking performance that affect human perception of a VE system while a tracked object is still (static) or moving (dynamic).

These differ slightly from the traditional specifications of resolution, static accuracy and dynamic accuracy, but they provide a description of motion trackers that maps more readily to the psychophysics of viewing virtual environments. In tracking systems whose resolution is limited by noise rather than quantization, the classical resolution would be equivalent to jitter plus perhaps some of the short-term creep, depending on the time period of averaging that was used to make resolution discriminations. Static accuracy would include fixed spatial distortion as well as long-term creep, and dynamic accuracy would include everything on the list. Qualitative specifications, such as environmental robustness, range, line-of-sight requirements, and multiple-object tracking capability are important as well and must be considered in addition to the six error performance specifications listed above for determining the suitability of a tracker for an application.

Unfortunately, very little research has been completed so far about the effects of tracker errors on the five VE quality metrics listed above. More has been written about the effects of display parameters, image realism, and update rates on presence, and indeed it is impossible to quantify the effects of tracker parameters independent of these. Until a perfect display driven by a very fast image generator is developed, experiments to evaluate required tracking performance will not be simple. The sections below summarize the few results that have been published and offer some additional speculation, which remains to be evaluated by experiments.

## 2.1 Tracking Requirements for VE Applications Using Opaque Head-Mounted Displays (HMDs; Category 1)

### *2.1.1 Head-Tracking Requirements*

Opaque HMD-like displays range from highly immersive wide-FOV HMDs to less immersive hand-held devices such as binoculars or arms-length flat-panel displays. The highly immersive ones by definition strive for a magnification or zoom ratio of 1. Narrow-FOV binocular-style devices, on the other hand, generally have magnification much greater than 1, and "magic lens"–type flat-panel displays vary. The bulk of this discussion will focus on immersive HMDs, assumed to have a FOV and resolution matching the human visual system, with some commentary at the end about the implications of narrower FOV or higher magnification.

Tracker parameters that have been studied the most are latency and frame rate. So and Griffin (1995) studied how lag in tracked HMDs affected performance on a target tracking task. The task was to keep an HMD-fixed cross-hair reticle on a moving target that moves around randomly with a target root mean square (r.m.s.) velocity of 2, 3.5, or 5 deg/sec. The mean radial error in tracking performance increased linearly with increasing latency when additional 40-, 80-, 120- and 160-ms delays were added on top of the 40-ms base latency. The mean radial error increased more (about 50% increase) for faster moving targets than for slower targets (about 30%). The results demonstrated that delays as small as 80 ms can degrade task performance, although the effect of smaller delays could not be measured because of the inescapable base latency of 40 ms. Recently, a system with base latency of 27 ms has been built and used to test participants' ability to notice additional latency in increments of 16.7 ms (Ellis, Young, Adelstein, & Ehlrich, 1999). The psychometric functions for discrimination in a two-alternative forced-choice experiment were found to be independent of base latency; sensitivity to latency does not appear to follow a logarithmic Weber law, as do most other stimuli. Participants were just as able to discriminate a 16.7-ms increase in latency on top of a base latency of 27, 97, or 196 ms, from which one might extrapolate that they would also be able to discriminate a 16.7-ms latency from no latency. This is possibly bad news for those wishing to develop a VE that users cannot tell apart from reality, but it does not necessarily imply that there is also decreased task performance at such low levels of latency. Ellis, Adelstein, et al. (1999) studied the relative importance of latency, update rate, and spatial distortion to a variety of metrics related to tracking task performance, perceptual stability, realism, and simulator sickness. The task involved manually tracking a target in a virtual world presented via an immersive HMD, so it is not certain that the results are entirely due to head-tracking performance. It was found that frame rate and latency had a significant impact on task performance and perceptual stability, with latency having a much more dramatic effect. However, this study only went down to a minimum latency of 80 ms, which was already known to degrade tracking performance, so it does not answer the question of whether extremely small latency differences found to be subjectively noticeable are also detrimental to task performance.

Ellis, Adelstein, et al. (1999) also evaluated the impact of spatial distortion on task performance and other metrics. The spatial distortion studied in these experiments consisted of an upward curling of about 15 to 30 cm in the corners of a 1.2 × 1.8 m working volume. Since the distortion affected both head and hand trackers, participants would only perceive the difference between two relatively nearby points, and this difference was changing relatively slowly. Not surprisingly, this gradual change in head-to-hand transformation did not have a significant correlation with simulator sickness symptoms or major impact on performance, but it did create a small increase in the Cooper-Harper workload scale, which is highly correlated with perceptual stability and normalized r.m.s. tracking error. It can be conjectured that distortions with higher spatial frequencies would produce greater impacts on all metrics, because they would require more adaptation for a user moving around in space.

No research has been reported yet on the effects of jitter on VE users, although it seems obvious that jitter which is visible will reduce the illusion of presence and may even contribute to simulator sickness. The threshold of detectable jitter is not known. From experience, it seems that jitter of 0.05 degrees r.m.s. in orientation and 1-mm r.m.s. in position is generally unnoticeable in an HMD with magnification of 1, but becomes fairly visible in a virtual binoculars simulator with 7× zoom. Note that when viewing distant virtual objects, tracker position jitter becomes irrelevant, and orientation jitter multiplied by zoom factor is all that matters. For viewing close objects, translational jitter becomes dominant. It is an interesting question whether humans are sensitive to perceived jitter in world space or screen space (pixels) or some combination. If the former, than humans might be psychologically more forgiving of an object jittering 2 pixels at 1 meter apparent depth (4 mm in world space) than an object jittering 1 pixel at 10 meters apparent depth (2 cm in world space). Other factors that affect perception of jitter are display resolution and whether or not graphics are antialiased.

Latency jitter has also been largely neglected in works examining the effect of latency on presence, simulator sickness, and task performance. One study examined the effect of artificially imposed sinusoidally oscillating frame rate variations on task performance, using a task of grabbing a moving-target object and placing it on a pedestal in a head and hand-tracked HMD virtual environment (Watson, Spaulding, Walker, & Ribarsky, 1997). This study found that at "higher" frame rates of 20 fps, superimposing oscillating frame time variations of 10 to 20 ms amplitude on a 50-ms average frame time did not interfere significantly with task performance. However, at lower frame rates of 10 fps, varying the mean 100 ms frame time by +/− 60 ms did cause performance degradation if the oscillation was slow enough. This is probably because at slow oscillation frequencies, there would be quite a few frames in a row where the frame rate was only 6 to 8 fps. It should be noted that although the experiment introduced a high percentage of frame rate variation relative to the mean, the percentage of latency variation was much less because the system latency without artificially introduced delays was already 213 ms.

Although latency variations for a fast frame rate system were not found to degrade task performance much, our subjective experience is that they cause a variety of very distressing artifacts such as image and object stepping, multiple image formation, twitching, and jittering during head rotations. For a fixed latency system, the world appears to shift at the start and finish of a head rotation, but during the constant speed portion of the rotation the world appears stable, although displaced from its normal orientation by an angle proportional to the head speed times the latency. If there is latency jitter, then during the constant speed rotation the world will jitter by an amount proportional to the latency jitter times the speed of rotation. Interaction with the refresh rate of the display device will cause additional effects such as multiple imaging that are explained in more detail in section 6. More quantitative data is needed, but our experience with systems indicates that for very smooth apparent motion it is important to keep latency jitter below 1 ms.

### *2.1.2 Hand-Tracking Requirements*

The most typical interactive VE uses a head-tracker on the HMD for view control and a tracking device held in the hand for object selection and manipulation. The hand or device it holds (called manipulandum from here on) cannot be seen directly, but a graphical representation is provided that is used to select and manipulate virtual objects. If the graphical manipulandum is displaced from the real one, the absolute accuracy of tracking is relatively unimportant. It will suffice that translating or rotating the real manipulandum will cause the virtual one to follow in a smooth and predictable manner. With practice the user's eye–hand sensorimotor loop adapts to the displacement (see chap. 31, this volume). Good control and task performance can occur after adaptation is substantially complete, but the presence of significant latency interferes with this adaptation (Held, Efstathiou, & Green, 1966). Also, the user will be relatively inefficient or even disoriented during the period of adaptation and may spend a lot of time trying to locate the visual icon of the manipulandum. After adaptation, there may be negative aftereffects when returning to normal reality (see chap. 38 and 39, this volume). To avoid these problems, one may attempt to match the position of the visual representation to the actual position of the manipulandum so that the user can exploit natural proprioception without adaptation. This will produce a very natural and easy-to-learn interface if the absolute accuracy of the tracking system is better than the accuracy with which proprioception can sense hand position with the eyes closed. Due to visual capture (see chap. 22, this volume), the seen virtual manipulandum may be perceived as consistent with the felt real one even if tracking errors are somewhat large, but experiments need to be done to determine at what level of unpredictable tracking error this leads to noticeable sensory conflict and side effects. Whether the virtual manipulandum is colocated with or displaced from the real one, latency in tracking the manipulandum probably has equal consequences. Held et al. (1966), Ellis, Young, Adelstein and Ehrlich (1999), and Ware and Balakrishnan (1994) discuss the effects of this latency.

## 2.2 Tracking Requirements for VE Applications Using Fixed-Screen Displays (FSDs; Category 2)

### *2.2.1 Head-Tracking Requirements*

There is no experimental data yet, but theoretically, head-tracking requirements for FSDs are far less demanding than for HMDs (Cruz-Neira, Sandin, & DeFanti, 1993). The primary reason is that changing head orientation to first order does not change the displayed scene. With HMDs, the most noticeable tracking artifact is orientation latency, because every time the head turns by any amount the displays must be immediately updated with images corresponding to the new look direction. Even a very small orientation tracking latency causes the whole virtual world to first rotate with the head, then settle back to its expected position, resulting in a disturbing loss of perceptual stability. With an FSD, the appropriate images for all allowable look directions are already on the walls and can be seen immediately as soon as the head is turned toward them, without having to redraw anything. Head translation does require redrawing the screens to update the off-axis perspective projections, but these changes are slight for small head translations, so moderate translational tracking latency is not too noticeable unless one makes unusually quick translations. Sensitivity to translational tracking errors is less problematic in an FSD than an HMD when viewing close objects, but worse for distant objects (Cruz-Neira et al., 1993).

Virtual environment FSD displays are almost always rendered in stereo, so the real tracking requirement is to track the 3-DOF position of each eyeball in order to render the left and right eye perspective views. Since it is easier to attach a sensor to the head than to each eyeball, this is usually accomplished by tracking the position and orientation of the head and using this orientation to calculate the positions of the two eye points as displacements from the head position sensor. There is therefore some sensitivity of the displayed images to head orientation tracking errors, but these errors only cause changes in the separation between the virtual eye points used to create stereo. How much these temporary contractions in the virtual interocular distance affect stereoscopic fusion or depth perception for a given quality of orientation tracking still needs to be determined.

### 2.2.2 Hand-Tracking Requirements

While head-tracking requirements in FSDs are comparatively lenient if one only wishes to use the display for visualization, most FSDs are used for interactive design activities in which tracking one or more manipulanda is essential, and here the requirements are essentially the same as in category 1 applications. With reflected-view FSDs the real manipulandum is hidden behind the mirror and only a virtual representation can be seen, so tracking requirements should be equivalent to opaque HMD applications where the same conditions hold. In the direct-view arrangements such as a CAVE or Virtual Workbench, the virtual representation of the manipulandum is sometimes displaced from the real one so that it does not need to be accurately registered. In this case, the user is looking directly at the virtual manipulandum and the same requirements for latency, accuracy, and jitter should apply as in the previous cases. However, the real manipulandum may be visible in peripheral vision, and this may make the user more able to discern latency of the virtual relative to the real. Much more critical however are applications where an attempt is made to keep the virtual manipulandum precisely overlaid on the real manipulandum. Any errors in tracking are easily detected as visual misregistration, so the high tracking-accuracy requirements of augmented reality (AR) apply. It should be noted that such attempts are also frustrated by the impossibility of focusing the eyes simultaneously on both the real manipulandum held at arm's length and virtual one displayed on the screen way behind it. Therefore most practitioners display the virtual manipulandum at least slightly displaced from the real one, which also eases tracking requirements somewhat.

## 2.3 Tracking Requirements for AR Applications Using See-Through HMDs (Category 3)

### 2.3.1 Head-Tracking Requirements

Augmented reality systems, which overlay virtual annotations precisely on top of real objects in order to guide the user who is performing manual operations on these objects, are only useful to the extent that annotations are accurately registered with objects, and thus absolute accuracy of the tracking system takes primary importance. This is in sharp contrast with the situation for opaque HMDs, where jitter, latency, and latency jitter are crucial, but spatial distortion and creep are rarely noticed. For typical AR applications, the most important tracker specifications are latency, spatial distortion, and creep, because these cause visible misregistration. Jitter and latency jitter are also undesirable because they cause the annotations to shake or vibrate, but at least the entire world does not jitter as it does in immersive HMD applications, so the risk of simulator sickness is lower. The causes of registration error have been discussed and analyzed in the AR literature (e.g., Drascic & Milgram, 1996; Holloway, 1997). The consensus is that latency is usually the worst offender, but once that is addressed, optical distortion caused by the HMD optics must be tackled before millimeter-accurate tracking can be appreciated.

Not all AR systems require every virtual object to be precisely registered on a real object. Displaying virtual objects that appear to be floating in midair is useful for AR gaming, in which virtual beasts might jump through windows and attack the player, or for shared AR visualization, in which a 3-D model or data set might hover above a table while multiple participants view it from different angles. Nothing has been written about tracking performance requirements for this mode, but one would guess they will be slightly less demanding than for direct overlay applications because precise registration to the millimeter or even centimeter level may not be required. Thus a slight spatial distortion such as a systematic offset or nonlinearity may be less noticeable, but sensitivity to latency is probably nearly the same. The threshold for noticing latency in both types of see-through HMD applications is thought to be lower than for immersive HMDs because there are real objects having zero latency visible for comparison. On the other hand, unconscious effects of latency such as decreased presence or simulator sickness are probably worse in opaque HMDs because the whole world loses its perceived stability.

#### 2.3.2 Hand-Tracking Requirements

Tracked manipulanda are generally less discussed in the AR area, because the most common AR applications involve guiding a user to perform operations using a real tool on a real object. Since object–tool interaction provides its own visual and haptic feedback, virtual graphics are usually used just to annotate the object that needs to be acted upon. In this case the primary purpose of tracking the tool would be to provide feedback to the AR application software about whether an action has been completed. For example, if the AR program prompts the user to turn a certain bolt 90 degrees, it could determine, by tracking the wrench, when the bolt has been turned the correct amount, then automatically prompt the next action. To do this, the tool would need to be tracked with the same translational accuracy as the HMD in order to determine unambiguously when the actual tool has engaged the actual object. The orientation tracking requirement for the tool may be less stringent than for the HMD (which requires an orientation accuracy of about 0.1 degrees to register an annotation to within 1 mm at arm's length).

### 2.4 Tracking Requirements for Auditory VE Applications Using Headphones (Category 4)

The best 3-D-spatialized sound is generated using headphones to deliver to each ear specifically processed audio streams that vary as a function of the spatial relationship of the head to the sound source (see chap. 4, this volume). When headphones are used together with an HMD, the tracking requirements imposed by the visual display mode generally take precedence because visual acuity is higher than auditory localization acuity. However, in an FSD-presented VE or a real environment with no visual display, the need to track head orientation is driven by the need to make the sounds appear to come from a certain stable direction or object in space even as the head rotates. It is thus reasonable to ask what quality of head-tracking is required to fool the user into believing the sounds are coming from a fixed source location while the user moves through the space.

The directional resolution of binaural localization is best, about 1 degree in azimuth and 15 degrees in elevation, directly in front of the head (see chap. 4, this volume). This implies that orientation tracking jitter and short-term stability of better than 1 degree is all that is required to make sure a sound source does not appear to wobble around while the user's head is still. Accuracy of 1 degree is certainly sufficient to make a sound appear to come from the direction of a certain object, and may be needed if the user is blindfolded or cannot look at the source. However, if the listener can see the alleged sound source, then a phenomenon called visual capture or "the ventriloquism effect" makes the sound seem to emanate from the visual object, even if the auditory cues indicate a somewhat different direction (see chap. 22, this volume). Auditory depth perception is even weaker, so it follows that modest resolution and accuracy for both orientation and position are sufficient for pure auditory displays. Less is known about the temporal response of the binaural localization system. Logically, if there is a system latency of $\Delta t$ and the head pans at a constant angular rate $\omega$ the apparent location of all sound sources would shift by angle $\omega\Delta t$ in the direction of the head rotation, just as visual objects do in an HMD. Therefore, the orientation tracking latency requirement for perfect realism in an audio-only display may be just as critical as for an HMD. However, the ventriloquism effect may be able to compensate for this apparent auditory shift if it is only several degrees, so the detection threshold for latency in a spatialized audio system may actually be significantly higher than for an HMD. It is also unknown whether this auditory sensory conflict can contribute to simulator sickness.

### 2.5 Tracking Requirements for Avatar Generation Applications (Category 5)

Commercial full-body motion tracking systems are sold primarily for three application areas: biomechanics and gait analysis, motion capture, and performance animation. Avatar generation in VE applications (see chap. 15, this volume) is most similar to performance animation, which differs from the motion capture done for film special effects or medical diagnostics primarily in that it must be done in real time, and that motion that looks lifelike is usually sufficient even if not completely accurate. In VE applications, body avatars may be presented to the person being tracked, or to third-person viewers who may be other participants in a multiuser VE or nonparticipating trainers.

First-person avatars are unnecessary in category 2 and 3 applications because users can see their own real bodies. In category 1 displays, users who look down and see no legs may feel disembodied and lose some of their sense of presence in the virtual world. If a VE involves manual interaction, then at least avatars for the manipulanda (hands or handheld tools) must be presented. Motion tracking requirements for these manipulanda have already been discussed with respect to the manual interaction modes of category 1, 2, and 3 displays. In most systems, avatars of these manipulanda are presented as isolated floating objects, leaving the user the feeling of having invisible or nonexistent arms or even hands. This may not be important in a lot of applications, since visual feedback from the manipulanda is enough to accomplish interaction tasks. If it is psychologically desirable to have the hands and forearms that hold the tool visible, they can be drawn approximately most of the time without additional sensors by using inverse kinematics (IK, section 3.1.3). Nothing is really known about the required accuracy of representing these, but since the user's gaze and attention will usually be focused on the tip of the manipulandum, the forearm will only appear in the peripheral view and may not need to be drawn very accurately. On the other hand, if the user looks at an arm that is only tracked by IK and moves his elbow, he will see that the motion of the avatar does not follow correctly, which might defeat the purpose of rendering it. With the limited field of view available in current HMDs, the user will have great difficulty looking at his own torso, so it is rarely important to provide full-body tracking in these applications. The legs may be rendered if desired by a crude algorithm such as placing the feet directly under the head when standing, or in front of the chair when seated. Occasionally an application will warrant actually tracking the legs.

Avatar display to third-person viewers is more likely to provide a sufficient reason for full-body tracking. In a multiparticipant simulation, a given player knows basically what he is doing himself—standing, squatting, or crawling, for example—without visual feedback of his own body. Other players, however, have no such information, and it may be important. Is the other player aiming a gun at me or holding up both hands in surrender? Unlike the precise head and hand tracking that is required for VE navigation and interaction, this information requirement is often qualitative in nature. Duplicating the precision tracking sensors that are used for the head and hands on all other body segments may be more costly and cumbersome than necessary. Furthermore, line-of-sight problems, which are endemic to all high-accuracy tracking technologies, are multiplied when trying to put sensors all over the body. Sections 3.1.2, 3.1.3, and 3.2.4 present approaches, which could potentially provide lifelike motion with greater freedom and simplicity than today's prevalent optical and magnetic body-trackers.

## 3. PHYSICS OF MOTION TRACKING

The previous sections have focused on the first two stages of a motion-tracking-system life cycle: Identification of the type of application for which a tracking system is required, and analysis of the performance requirements for that application. This section focuses on the next phase: choosing appropriate basic physical sensing technologies to measure motion in the intended environment. Designing a sensor to make a specific measurement, for example, a distance between two points, is a matter of selecting an appropriate physical principle to exploit, so this section will summarize various physical laws and forces that are available and how they may be used.

Classical physics views the world as a collection of particles that interact with each other through four fundamental forces: strong, weak, electromagnetic, and gravitational. The primordial branch of physics, mechanics, studies the motion of material bodies (particles or systems of particles such as rigid bodies) in response to these forces. The study of mechanics includes kinematics, which aims to describe geometrically possible motions of objects without regard to the forces that cause those motions, and dynamics, which reveals the motions that will actually occur, given both kinematic constraints and existing forces and mass distributions.

For classical mechanics, which is adequate for describing the motion of human heads, limbs, and handheld objects, all motion follows from Isaac Newton's laws of motion. This discussion therefore starts by looking at the kinematics (section 3.1) and dynamics (section 3.2) of human motion from a simple Newtonian perspective to see what tracking methods they may reveal. The analysis of the mechanics of large numbers of colliding molecules in air leads to acoustics, which is examined in section 3.3, as another rich source of tracking technologies.

Of the four fundamental forces, strong and weak forces only operate over extremely short distances inside the nuclei of atoms, and therefore have no obvious application in motion tracking sensors. A third fundamental force, gravitation, is indeed usable for motion tracking systems. Unlike the other three forces, gravity affects every particle with mass and therefore plays a role in the dynamics of every system on the earth. Therefore it is impossible to discuss the dynamics of motion in section 3.2 without introducing the role of gravity, so the possible use of gravity in tracking systems is discussed there instead of creating a separate subsection. The final fundamental force, electromagnetism, has the richest variety of manifestations that are useful for remote sensing of object position and orientation. Section 3.4 discusses the proposed use of electric fields for tracking, and section 3.5 surveys the use of magnetic fields, which have been the most common means of tracking until recently. Sections 3.6 and 3.7 present techniques for exploiting electromagnetic waves in the lower and upper halves of the electromagnetic spectrum respectively.

### 3.1 Kinematics: Mechanical Tracking

In classical mechanics, a particle is represented as a point in space, with a constant mass, $m$, and a time-varying position, $r(t)$, specified by three Cartesian coordinates $x(t)$, $y(t)$, and $z(t)$. Tracking this particle would consist of reporting, whenever asked, the then-current 3 DOF of $r(t)$, and is therefore called a 3-DOF (position-only) tracking problem. In addition to its position $r(t)$, the particle has a velocity $v(t)$, which is the derivative of the position vector,

$v(t) = r'(t) = (x'(t), y'(t), z'(t))$. According to Newton's second law, this velocity evolves according to $v'(t) = a(t) = F(t)/m$, where $F(t)$ represents the sum of all the forces acting on the particle. In the absence of external forces, the velocity remains constant (Newton's first law), and one could track the particle with no further measurements once one had obtained the position and velocity at any point in time. This is a complete statement of the kinematics of an unconstrained point particle and captures the essence of the world's first motion tracking technique, *dead reckoning*. Sailor's slang for "deduced reckoning," dead reckoning was used by early mariners to calculate vessel position by knowing the starting point, velocity, and elapsed time.

Your head is hopefully not a point particle. It is better approximated as a *rigid body*, which is modeled as a system of many particles constrained to maintain constant distances between one another. A rigid body requires 6 DOF to specify its pose (position and orientation), from which one can calculate the 3-DOF positions of all point particles that make up the body. The kinematics of a rigid body are somewhat more complex than a point particle, including an additional set of three nonlinear equations to relate the time derivatives of the three euler angles that describe the orientation to the angular velocity components. Kinematics really gets interesting in the study of multiple rigid bodies interacting through constraint equations, such as a robotic arm or a human body. Kinematics of such linkage systems can be applied to motion tracking in three ways, outlined in the following three sections.

### *3.1.1 Fixed-Reference Forward Kinematics: Mechanical Linkages and Pull Strings*

Perhaps the most straightforward approach to head tracking is to make some kind of direct physical connection to the object being tracked, the displacement of which can be easily measured using potentiometers, optical encoders, rotary or linear variable differential transformers, or cable-extension transducers. The first HMD was tracked with such a mechanical contraption (Sutherland, 1968). Simplicity of mechanical design has dictated the two most common linkages for 6-DOF measurement: a single segment that extends telescopically, or two rigid segments jointed together at an "elbow." In either case, the arm is attached at one end with a 2-DOF "shoulder" joint to some fixed reference base, and at the other end is attached with a 3-DOF "wrist" to the object being tracked. The main causes of lag or inaccuracy are flexion of the linkages and transducer quality. The linkages can be made rigid if they are short, and mechanical encoders are available with extremely good precision and fast response.

The biggest problem with mechanical arm trackers is range. Making the arm segments longer lowers the mechanical resonance frequency, which may lead to unacceptable lag or oscillation. It also increases the inertia felt by the user. Even if one is willing to make such sacrifices to achieve larger range, the range of a two-segment arm is ultimately limited by ceiling clearance for the elbow as it folds upward. As Meyer et al. (1992) point out, they are not very "sociable," because multiple arms cannot be used easily in a shared space. Mechanical-arm tracking remains an attractive alternative where the infrastructure needs to be present anyway—boom supported displays (see chap. 3, this volume) and force feedback devices (see chaps. 5 and 6, this volume) are prime examples.

Another mechanical approach to head tracking uses pull-string encoders mounted on the wall or ceiling, with the free ends of the pull strings attached to the object being tracked. High-precision instrumented pull strings, called cable-extension transducers, are available at low cost for industrial distance measurement applications. There are two ways to measure head position using strings. The first method uses three pull strings with the free ends of all three strings attached to a common point. The Cartesian position of the point in space is found by trilateration. A second method would be to use a single pull string to measure distance and also direction. The direction the string departs the reel box could be measured by running the string through a hole in a very light, low-friction joystick on its way out from the box. It should be noted that to get faster response requires a tighter string-retraction spring, which might result in an annoying tugging if the forces from the various pull strings do not cancel each other out well. Like mechanical arms, pull-string arrangements have been used to provide force feedback (Ishii & Sato, 1994) and perhaps could also serve to help levitate a display.

### 3.1.2 Moving-Reference Forward Kinematics: Biokinematic Dead Reckoning

The mechanical trackers in the last section are grounded to a fixed reference base, thus limiting the range and introducing a cumbersome mechanical linkage system. When tracking creatures with rigid skeletons, there is a mechanical linkage system already present that could be used for tracking purposes. Most casual users will not allow you to instrument their skeleton directly with joint angle encoders, but you can approximate the joint angle measurement using strapped-on goniometers. Goniometers (joint angle measurement devices) may consist of two rigid plastic parts that strap on, for example to the thigh and shin, with an instrumented hinge between them that measures the knee angle. Alternatively, they may dispense with the rigid attachments and use a flexible bend sensing material based on fiber optics or conductive foam that changes resistance when it bends. In fact, virtual goniometers with no physical connection can be implemented by attaching sourceless orientation sensors to each side of the joint, or a magnetic source to one side and sensor to the other, and calculating the difference angle between them. Molet, Boulic, and Thalmann (1999) attached magnetic sensors to each body segment, then drove an animation of a human using only the angular differences between the sensors. They cite several advantages over the usual technique of using magnetically determined 6-DOF poses directly. In all cases, the tendency of strapped on devices to slip and the inevitable movement of flesh relative to the bones limit accuracy. To achieve higher accuracy, some vendors have connected from one goniometer to the next using rigid rods, producing in effect an exoskeleton for tracking. This is more cumbersome than isolated goniometers at each joint, but allows freer movement than a grounded mechanical tracker.

Whether measured by goniometers, an exoskeleton, electromyography (Suryanarayanan & Reddy, 1997), or a body-suit, the measured joint angles can be used to calculate the pose of each segment of the body, relative to a root node at the pelvis, using simple forward kinematics calculations. Joint angle data completely determines the *shape* of the whole body, but the position and orientation of the body as a whole must be determined by other means, for instance by tracking the 6-DOF pose of the root node. Unfortunately, this root-node tracking system turns an otherwise self-contained tracker into a system that must make reference to the external environment, introducing usual limitations such as range or line of sight.

It may be possible to track the position and orientation of the whole body just using self-contained goniometry if certain assumptions are made. For example, if it could be assumed that at least one foot is flat on the floor at all times, then that foot could be treated as the root node while it is in contact with the floor (perhaps determined by sensors on the soles). As long as the contact is maintained, the position and orientation of that foot are known to remain constant. During this time, the pelvis can be tracked relative to the grounded foot and the moving foot relative to the pelvis, all using just forward kinematics. When the moving foot lands it becomes the root node, and its current pose is maintained constant as the reference for the whole body until it loses that status. Errors will gradually accumulate as a percentage of the distance traveled, but it should work much better than an ordinary pedometer, in that:

- It can keep track of height changes as the user climbs stairs.
- It keeps track of direction changes and therefore position, not just total distance traveled.
- Each stride is individually measured, not just assumed to be the average stride for the user.

If long-term position accuracy must be maintained, then some method of correcting the accumulated position errors must be devised, perhaps based on map-correlation techniques or some simple light beams that are intercepted by photoelectric sensors on the ankles as the user walks through doorways or near landmarks.

This concept remains to be tried in practice. It may need to be augmented with some accurate foot-roll sensors to compensate for the heel-to-toe roll during normal walking, and the accuracy would depend critically on the quality of the ankle goniometers, because any error there would cause the whole body to be tilted incorrectly. It would fail if the user runs or jumps or lifts both feet while sitting. An enhanced version called biodynamic navigation is introduced in section 3.2, which in theory may be able to handle these cases better.

### *3.1.3 Inverse Kinematics*

A final application of kinematic calculations, again to the problem of whole-body tracking, is the use of inverse kinematics (IK) to deduce the pose of the whole body when only the poses of a few extremities, usually the head and hands, have been measured. Inverse kinematics problems are intrinsically more complex than forward kinematics, but a great deal of effort has been invested in developing algorithms to solve them, especially in the robotics community. In robotics, the desired pose for an end effector (i.e., robotic hand) to accomplish a task is specified, and the robot must figure how to control all the individual joint angles to get to the end effector into this pose. In a manipulator with redundant degrees of freedom, there may be a large number of different combinations of joint angles that result in the same end-effector pose, and the IK algorithm must solve for the combination that is optimal in terms of maximum speed, minimal work, or some other criterion. In the human-motion capture problem, the human brain has already solved this problem and controlled the joint angles to get the hand to a desired position. A sensor measures the pose of the hand and uses an IK algorithm to try to guess what is the most likely combination of poses for the rest of the body segments resulting in the measured end-effector pose. Discomfort factors may become part of this calculation. The amount of "guesswork" can be reduced by using more sensors (for example, by adding another sensor on the shoulder or the back), but the whole purpose of IK is to capture full-body motion using a minimal number of sensors. Badler, Hollick, and Graneri (1993) used a fairly minimal set of four sensors, which happens to be the number provided by many commercial 6-DOF tracking systems, to control an animated figure called Jack by inverse kinematics (see chap. 15, this volume). The sensors were placed on the head, hands, and waist. First, the spine was modeled as a single flexible rod with two axes of flexion and one of torsion, controlled by the waist and head sensors. Next, the arm positions were estimated by IK using the measured hand positions and shoulder positions consistent with the spine torsion. Finally, Jack's center of gravity was calculated from the updated upper body configuration, and the legs were moved along animated stepping sequences to positions that could support the body mass.

## 3.2 Dynamics: Gravimetric and Inertial Tracking

### *3.2.1 Accelerometers and Gravimetric Tilt Sensing*

An accelerometer is basically equivalent to a proof mass suspended in a housing by a spring and constrained to slide only along the sensitive axis. The displacement of the proof mass from the rest position provides a reading of the nongravitational acceleration acting on the accelerometer, traditionally called specific force, **f**. Because gravity affects both the proof mass and housing equally, the accelerometer does not directly sense the acceleration due to gravity (an accelerometer in free fall reads zero), but when it is not accelerating, such as when resting on a table, it reports an acceleration of 1 g upward, which is due to the upward push of the table to keep it from falling. Thus, a stationary accelerometer in Earth's gravity field produces an output proportional to the sine of its tilt angle with respect to the horizontal, and can be used as an inclinometer. A high-quality servo-accelerometer can have an accuracy of 10 $\mu$g, which is sufficient to measure a tilt angle of 0.0006 degrees, an order of magnitude smaller than the vertical deflections due to gravity anomalies. Even extremely low-cost silicon accelerometers based on micro-electro-mechanical-systems (MEMS) technology can measure pitch and roll of a static headset to within a degree or so. The so-called sourceless head-trackers that were included with consumer HMDs of the early 1990s were based on this principle, using an inclinometer to measure pitch and roll, and a compass for yaw. The problem is that any inclinometer, whether it is made from two horizontal accelerometers, a mechanical pendulum, or a fluid-filled bubble level, must be acceleration-sensitive in order to sense gravity. Any real horizontal accelerations will add to the apparent vertical component that cancels gravity, producing a resultant that is no longer vertical. Even moderate head motions produce "slosh" errors of 20 to 30 degrees. This is a physical limitation that cannot be overcome through improved sensor design. Judicious adjustment of damping, pendulosity, and rotational inertia can decrease the frequency response to horizontal accelerations, but the low-frequency motions characteristic of head translation will always come through if the device has fast enough dynamics to be useful. A sourceless orientation-tracking alternative without slosh is presented in the next section.

### *3.2.2 Gyroscopic Orientation Tracking*

Despite their tremendous potential performance advantages, the use of gyroscopes came relatively late to human-motion tracking (Foxlin, 1993). Before this, gyros were built with spinning wheels, or lasers and were too large and expensive for human motion tracking, and mechanical ones also produced distracting inertial reaction torques and noise. Motivated by the automotive market, the 1990s saw the commercial introduction of a new class of much smaller and cheaper gyros, now called coriolis vibratory gyroscopes (CVGs). A CVG is a kind of mechanical gyro (as opposed to optical gyros such as the ring laser gyroscope and the fiber-optic gyroscope), but it requires no spinning mass. Instead, a proof mass is made to oscillate at a fairly high frequency, usually in the tens of kHz, and a pickoff is provided to measure the secondary vibration mode caused by the Coriolis force $\mathbf{F} = \boldsymbol{\omega} \times \boldsymbol{\nu}$, which pushes the mass to vibrate in a direction perpendicular to the primary driven vibration. CVGs have been implemented with a wide variety of different geometries—vibrating rings, hemispherical shells, tuning forks, vibrating wheels, cylinders, and triangular or rectangular prisms. They have been made from quartz, ceramic, metal, and silicon elements, and the vibrations are caused and detected using piezoelectric, magnetic, electrostatic, or optical effects. An overview of the physics of CVGs is provided in Lynch (1998). In particular, the micromachined silicon CVGs have the potential in the long term to become extremely small and inexpensive, and possibly even combine three-axis sensing with electronics on a single chip. Such developments will make MEMS gyroscopes even more attractive for widespread deployment in human-motion tracking applications.

Some gyros measure orientation directly, as with vertical and directional spin-position gyros, which maintain spatial direction of the spin axis despite case motion by using isolation gimbals. Most modern gyros do not have gimbals, and they measure the angular rate of rotation around a sensitive axis. If three such rate gyros are mounted with orthogonal sensitive axes, the three angular rate signals can be integrated together to keep track of the current orientation of the sensor assembly relative to where it started.

Regardless of the sensor technology or orientation tracking mechanization, gyroscopic tracking provides a number of performance advantages that are crucially important to interactive graphics applications. Because they are self-contained, inertial sensors can track with undiminished performance over an unlimited range without any line-of-sight or interference concerns. This is a significant advantage over all externally referenced tracking technologies, which cannot increase the working volume without degrading the signal-to-noise ratio and interference susceptibility. A second crucial advantage is extremely low jitter. Gyroscopes measure angular rate with very low noise, approximately 0.001 deg/sec for midquality gyroscopes. This produces no visible angular jitter, even at very high zoom ratios. Even with very low-cost gyros, which currently have noise levels closer to 0.5 deg/sec, the angular rate signal needs to be integrated, and in the process the noise is attenuated by the low-pass filtering effect of an integrator, so that the resulting angular jitter of about 0.02 degrees is invisible in VE applications. A third important advantage of gyroscopic rotation tracking is speed. The outputs can be sampled and the orientation updated as often as desired. This can even be done right before a display refresh for final image shifting (see section 6.2). Because of the low jitter of gyroscopically measured orientation, there is never a need to perform low-pass filtering for noise reduction, and the latency can be a fraction of a millisecond. What's more, the inertial angular rates measured by gyros can be used to perform prediction with several times greater accuracy than without them (see section 4.4). Jitter and latency in orientation are the critical performance parameters for HMD tracking. Therefore, orientation tracking based on gyroscopic angular rate sensors should be used for any high-quality VE using an HMD, unless the tracking range is so small that an external tracking source can provide low enough jitter and latency in the intended environment.

The problem with tracking orientation using only gyros is drift. There are several causes of drift in a system that obtains orientation by integrating the outputs of angular rate gyros:

- Gyro bias, $\delta\omega$, when integrated causes a steadily growing angular error $\phi(t) = \delta\omega \bullet t$
- Gyro white noise, $\eta(t)$, when integrated leads to an angle random walk (Brownian Motion) process $\phi(t) = \int_0^t \eta(\tau)d\tau$, which has expected value zero, but a mean-squared error growing linearly in time
- Misalignment, nonlinearity, and scale factor calibration errors of the gyros lead to accumulation of additional drift proportional to the rate and duration of the motions
- Gyro bias instability, which means that even if initial gyro bias is known or can be measured and removed, the bias will gradually change, producing a residual bias that gets integrated to create a second-order random walk in angle. Bias stability is often the critical parameter for orientation drift performance, since constant gyro bias and deterministic scale factor errors can usually be calibrated and compensated effectively.

Gyro drift may be solved by using higher accuracy sensors and algorithms to keep the drift rate low, and requiring the user to return to a known position and restart after a certain period of

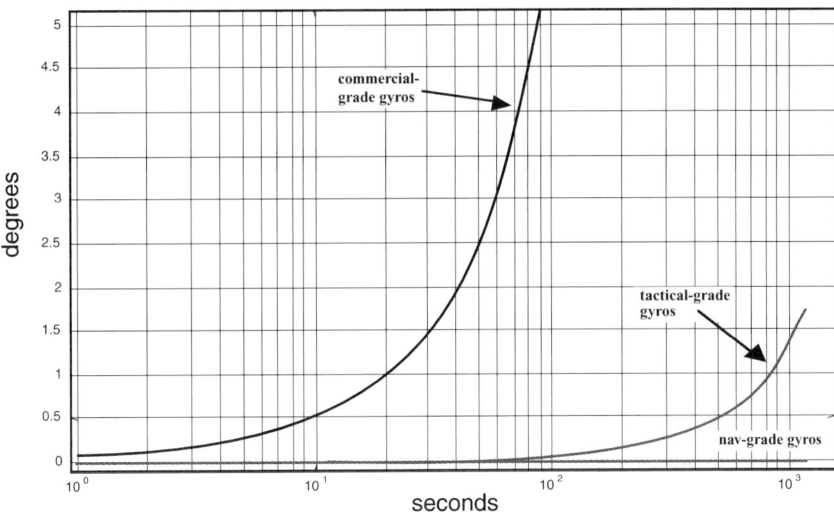


FIG. 8.1. Comparison of 1-$\sigma$ random orientation drift performance of various gyros over a 20-minute simulation.

time. Figure 8.1 illustrates the achievable drift performance for typical gyros of different grades. This simulation examines the effects of various gyro error sources over 20 minutes for an object that is not moving—scale-factor calibration errors are thus not included and would lead to additional error accumulation as a function of the particular motion trajectory of the object. The simulation assumes gyro bias stabilities of $1500°/hr/\sqrt{hr}$, $15°/hr/\sqrt{hr}$, and $0.015°/hr/\sqrt{hr}$ for commercial, tactical, and navigation grade gyros respectively, with initial gyro biases known to 150 deg/hr, 15 deg/hr, and 0.0015 deg/hr, respectively. Angle random walk caused by the gyro noise component has negligible effects over the simulation period, but it prevents the initial biases from being measured perfectly prior to the run. Initial bias uncertainties are the dominant cause of drift for the first 36 seconds, after which bias instability takes over.

An examination of Fig. 8.1 leads to the conclusion that today's commercial grade gyros (devices used in automobiles, camcorders, model helicopters, and low-cost head-trackers) can only be used for a minute or so before drift becomes distracting and the user needs to reset the orientation tracker. Head-orientation trackers may significantly extend this by deadbanding the gyro outputs to zero when they are below a certain threshold. This eliminates the random drift when the head is still, but introduces "slippage" error accumulation whenever the head rotates below the angular-rate threshold. On the other hand, tactical-grade gyros used in short-range missile guidance are good enough for uninterrupted head-tracking for more than 20 minutes, which may be quite acceptable for many uses, and navigation-grade gyros are complete overkill if only orientation is needed. Unfortunately, the price ratio between tactical and commercial gyros roughly follows the 100-fold performance ratio, and they are also too large and heavy for head-mounted use. However, there is optimism that MEMS gyroscopes over the coming decade will be gradually closing in on the performance of tactical grade gyros, and it may eventually become feasible to make a light and affordable head-orientation tracker out of gyros alone, requiring only occasional resets by the user.

Another solution is to correct the drift of gyros using occasional measurements from another source. If only orientation needs to be tracked, it is possible to do this while still maintaining the advantages of a sourceless tracker (Foxlin, 1993, 1997). Gravimetric tilt sensing, discussed in the previous section, corrects drift in pitch and roll, and a magnetic compass can be used to correct drift in yaw if necessary. The problems with slosh in the gravimetric tilt sensors can be overcome because the gyros do not need to be corrected very often. Clever algorithms determine moments when the horizontal acceleration is likely to be negligible, and measurements at these times are introduced with higher weighting factors, for example, using an adaptive Kalman filtering approach (Foxlin, 1996). Because there may be sustained periods of motion when no tilt measurements can be used, some orientation error may build up, and once calmness is restored the Kalman filter may try to correct this rapidly enough to create some perceptible motion. To mask this, a perceptual enhancement algorithm holds back the correction until the user again begins to move and applies it gradually proportional to the speed of head rotation.

Geomagnetic compassing provides a cheap and effectively sourceless way to correct drift in yaw; at least this source is provided by the earth and is present over most of its habitable surface. However, the accuracy of magnetic compasses in many environments is poor (see section 3.5.1). Temporary magnetic disturbances can be detected and prevented from entering using something similar to the antislosh techniques used to screen gravimetric tilt measurements. A degree of accuracy can be achieved that is suitable for many applications that just require the forward direction of a virtual world to remain consistent for a seated user. If the user is free to reorient toward the forward direction, or if the duration of use is limited, than the use of a compass may not even be necessary. However, in applications requiring registration accuracy, such as an outdoor AR application, magnetic compassing is not a suitable choice, and physics provides yet another option, still sourceless, called *gyrocompassing*. A mechanical gyrocompass makes use of the spin of the earth to cause the spin axis of a gyroscope to align itself toward true north through a damped gyroscopic precession. The same concept can be used with a stationary cluster of angular-rate sensors to detect the direction of the earth's angular velocity vector, then project it onto a horizontal plane to find north. This requires gyros with sensitivity that is a small fraction of Earth's 15 deg/hr rotation rate. To detect yaw with an accuracy of 0.2 degrees would require gyros good to 0.05 deg/hr. Unfortunately, gyros of this caliber won't be small enough for use on an HMD for a great many years, if ever.

### 3.2.3 Inertial Position Tracking

The previous section discussed the use of inertial angular-rate sensors (gyros) for orientation tracking, which offers great advantages due to the self-contained, fast, and noiseless measurement. In many applications it is also desirable to track position, and the aforementioned advantages would theoretically apply as well to a 6-DOF tracking system built with gyros to determine orientation and accelerometers to measure changes in position. In fact, this combination of sensors has been used successfully for inertial navigation systems (INS) in ships, airplanes and spacecraft since the 1950s. In this section we review the basic operating principles of inertial navigation, discuss the differences between human-scale inertial tracking and geographic-scale inertial navigation, and analyze the present and future options for use of pure or aided inertial tracking in human–machine interaction.

Originally navigation systems were built with a gimbaled platform (Fig. 8.2a) that was stabilized to a particular navigation reference frame by using gyros on the platform to drive the gimbal motors in a feedback loop. The platform-mounted accelerometers could then be individually double integrated to obtain position updating in each direction. Most recent systems are of a different type, called *strapdown INS* (Fig. 8.2b), which eliminates mechanical gimbals and measures the orientation of a craft by integrating three orthogonal angular-rate gyros strapped down to the frame of the craft. To get position, three linear accelerometers, also affixed to orthogonal axes of the moving body, measure the nongravitational specific force of the body relative to inertial space. This specific force vector, $\mathbf{f}^b$, measured in body

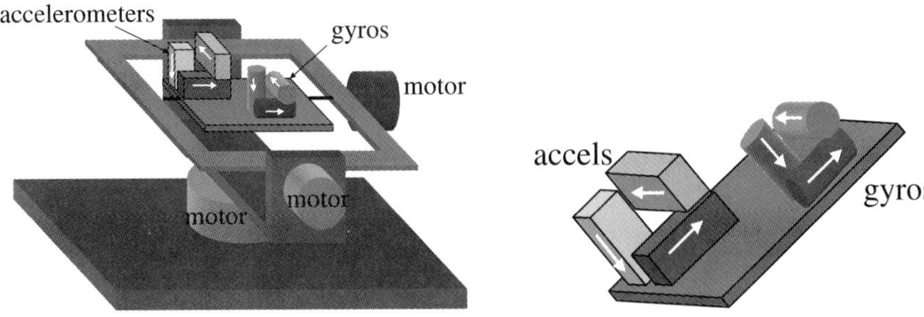


FIG. 8.2a Stable-platform INS. FIG. 8.2b Strapdown INS.

coordinates, is resolved into geodetic navigation coordinates using the current orientation of the body determined by the gyros. Position is then obtained by adding the local gravity to $\mathbf{f}^n$to get the total kinematic acceleration of the body $\mathbf{a}^n = \mathbf{f}^n + \mathbf{g}_l^n$, and then performing double integration starting from a known initial position. Figure 8.3 illustrates this flow of information.

Drift in linear position determined by an INS arises from several sources. First, there are accelerometer instrument errors corresponding to each of the four gyro-error types listed above. Since position is obtained by double integrating acceleration, a fixed accelerometer bias error results in a position drift error that grows quadratically in time. It is therefore critical to accurately estimate and eliminate any bias errors. An even more critical cause of error in position measurement is error in the orientation determined by the gyros. An error of $\delta\theta$ in tilt angle will result in an error of $1g \cdot \sin(\delta\theta)$ in the horizontal components of the acceleration calculated by the navigation computer. Thus, to take proper advantage of $\mu$g-accurate accelerometers, pitch-and-roll accuracy must be better than $1\mu$ rad $= 0.000057°$ for the duration of the flight, which puts a far more difficult task on gyros than on accelerometers. In practice, it is gyroscopes, not accelerometers, which limit the positional navigation accuracy of most INS systems, because the effects of gyroscopic tilt error will soon overtake any small accelerometer biases.

The scale of the human-motion-tracking problem is vastly different from that of global navigation. Tracking is only required over a small area, but it requires precision on the order of a centimeter or less, whereas with navigation a kilometer is often sufficient. The size and cost of sensors must also be scaled down tremendously for human body–mounted use. Thus, inertial human-motion tracking would need to achieve far higher accuracy using tiny sensors than navigation systems are able to achieve using instruments far larger and more costly. A reasonable question is whether or not purely inertial 6-DOF motion tracking will ever be viable.

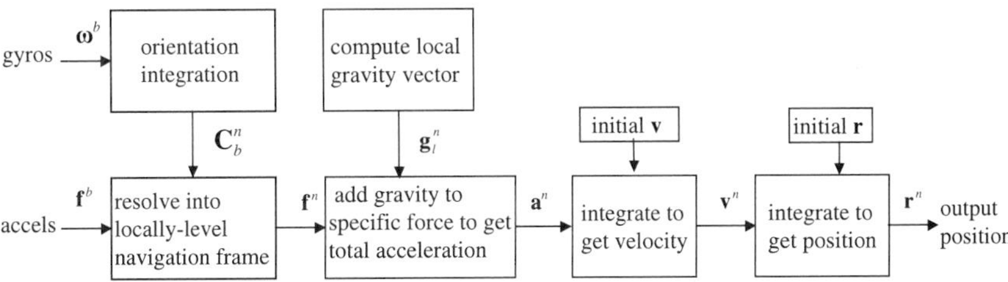


FIG. 8.3. Basic strapdown inertial navigation algorithm.

| | Commercial-Grade | Tactical-grade | Navigation-grade | Strategic-grade | Geophysical limit |
|---|---|---|---|---|---|
| Gyro bias stability | 1500°/hr/√hr | 15°/hr/√hr | 0.015°/hr/√hr | 0.000015°/hr/√hr | 0°/ hr/√hr |
| Gyro bias initial uncertainty | 150°/hr | 1.5°/hr | 0.0015°/hr | 0.0000015°/hr | 0°/hr |
| Accel bias stability | 1 mg/√hr | 100 μg/√hr | 10 μg/√hr | 0.5 μg/√hr | 0 μg/√hr |
| Accel bias initial uncertainty | 0.25 mg | 10 μg | 1 μg | 0.1 μg | 0.1 μg |
| Initial orientation alignment | 1 arcsecond | 1 arcsecond | 1 arcsecond | 0.1 μrad | 0.01 μrad |

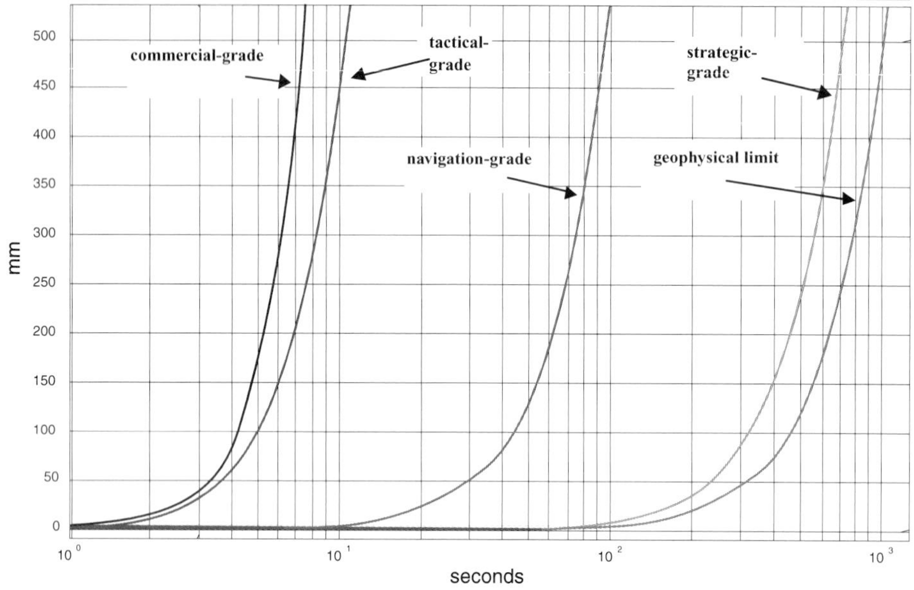

FIG. 8.4. Comparison of 1-$\sigma$ random-position drift performance of commercial, tactical, navigation, strategic-grade, and "perfect" inertial navigation systems over a 20-minute covariance simulation.

Figure 8.4 shows the results of a simulation designed to answer this question. It depicts positional drift rates of navigation systems using sensors of various grades. As with the orientation drift plotted in Fig. 8.1, the simulation only considers the effects of random error sources, predominantly those of the gyros, on drift accumulation for a stationary object over a 20-minute interval. If the object is moving, there will be additional drift due to scale factor calibration errors. The gyro and accel bias stability numbers listed in the figure represent typical values for the four grades of inertial measurement unit generally recognized in the industry. In addition, a drift curve is plotted for a theoretically perfect inertial measurement unit to show the physical limits of inertial sensing for small-scale position measurement. There are gravity anomalies, which create deflections of the local gravity field from the vertical by up to 30 arcseconds, corresponding to horizontal acceleration errors on the order of 150 $\mu$g. These have been mapped out in great detail by gravity surveys, and sophisticated inertial navigation systems are able to compute a local gravity vector with residual errors below 1 $\mu$g. It is impossible to drive these residual errors all the way to zero, though, because local gravity conditions are constantly changing due to earth tides, seismic activity, and cultural features such as buildings and trucks. All these effects produce gravitational variations on the order of 0.1 $\mu$g—approximately equivalent to the pull of a 1-meter-radius stone ball close to its surface. Because it would not be practical to map out all these extremely local spatiotemporal variations in gravity, the simulation assumes the geophysical limit on accuracy of inertial navigation to be 0.1 $\mu$g. Remarkably, strategic-grade systems come very close to reaching this limit for short-term navigation. However, MEMS gyros are currently below the tactical performance benchmark and may make it to navigation-grade performance levels over the next several decades, but they are unlikely to go beyond. Therefore, human-motion tracking systems that can maintain position to a few centimeters for more than a minute without external correction are not on the horizon. Nonetheless, 6-DOF inertial sensors are of prime importance in human-motion tracking as part of hybrid tracking systems in which they impart robustness, smoothness, and low latency, and reduce the requirements, and therefore cost, of other technologies with which they are combined (e.g., Foxlin et al., 1998). Techniques have been developed which allow inertial trackers to operate on motion-base simulators or moving vehicles without being disrupted by platform motions (Foxlin, 2000).

### *3.2.4 Body-dynamics-model-based Tracking*

The preceding discussion indicates that pure, unaided inertial navigation in its general form will never be an option for prolonged motion tracking on the human scale. The general form described above can track the motion of an arbitrary object moving along an arbitrary trajectory. All 6 DOFs may evolve independently according to any continuous functions of time, within the limits on angular velocities and linear accelerations imposed by the maximum ranges of the sensors. In other words, the only kinematic constraint assumed in the development of the strapdown INS algorithms is that the body being tracked is rigid. Nothing needs to be known about its mass distribution, the forces acting on it, or kinematic constraints of connection or contact with other rigid bodies. If significant kinematic constraints or a priori dynamics models can be exploited, there is a possibility of tracking for longer time periods without the need for external position information. To illustrate this idea, two hypothetical inertial tracking concepts are outlined below: zero-velocity-updating based on foot contact constraints, and body-dynamics-model-based tracking.

Researchers on the BodyLAN project at BBN have proposed a boot-mounted personal inertial navigation system (PINS) to help a soldier keep track of and report his own position during global positioning system (GPS) outages (R. LaRowe, personal communications, June, 1997). It would operate just like an ordinary INS, except every time it determined that the foot was in contact with the ground it would reset velocity to zero. Each minitrip would be just as accurate as the ones before it, and total positional error would only accumulate linearly with the number of steps taken. This may provide reasonable short-term accuracy using only one sensor on the shoe.

If full-body motion capture is the goal, a dynamical model for the body may go a long way toward reducing the amount of sensing hardware needed. Suppose the body is modeled (crudely) using 17 rigid segments and two 3-DOF "goosenecks" to represent the lower and upper spine. If each segment were independently tracked with a 6-DOF sensor there would be 102 degrees of freedom. However, kinematic constraints reduce this to about 39 independently controllable joint angles. Although there are a large number of muscles in the body, they work together in groupings to produce a net effect that can be approximately described by 39 torque values as a function of time. The complete state of the body can be described by 6-DOF for the root node plus 39 joint angles. These 45 generalized coordinates evolve according to a dynamical equation driven by the 39 muscle torques and constraint forces on any parts of the body that come in contact with the floor or other fixed objects. If the masses and moments of inertia of all the segments were approximately known (say by body scanning or just guesswork), it would be possible to produce lifelike animation by controlling the joint torques in a coordinated time sequence and enforcing constraints with techniques from physically based modeling. Conversely, estimating the motion of an actual human participant could potentially be reduced to estimating these 39 torques. By stacking generalized coordinates, their derivatives, and all causative forces and torques in the state vector of a dynamic system, it might be possible to develop a large centralized Kalman filter that can estimate the evolution of the joint angles from a set of indirect measurements such as angular velocities and/or linear accelerations. By modeling unknown muscle torques with an appropriate stochastic process, a fairly small number of sensors may suffice to achieve observability of the motion over time. Using exclusively inertial and gravimetric sensors would make this solution sourceless, so the range of motion would be unlimited. Unlike the biokinematic reckoning approach, ballistic motions of the body are captured in the dynamics model, so the participant can run and jump without losing tracking. The potential for optimal and self-consistent motion estimation using any desired, even seemingly incomplete, combination of sensors warrants further investigation.

## 3.3 Acoustic Waves

One of the earliest position tracker technologies, used by Ivan Sutherland in his pioneering HMD work (Sutherland, 1968) and widely available today in many commercial products, is ultrasonic time-of-flight (TOF) ranging in air. Acoustic trackers can be very inexpensive (witness the Mattel "PowerGlove" that was sold in toy stores in the early 1990s and included a 6-DOF ultrasonic tracker as a subsystem). Alternatively, they can have fairly large tracking range or high accuracy. Typical drawbacks are latency, update rate, and sensitivity to ultrasonic noises in the environment, but these can be overcome in hybrid configurations with inertial sensors.

All known commercial acoustic ranging systems operate by timing the flight duration of a brief ultrasonic pulse. In contrast, the system used by Sutherland employed a continuous-wave source, and determined range by measuring the phase shift between the transmitted signal and the signal detected at a microphone. Meyer et al. (1992) point out that this enables continuous measurement without latency, but can only measure relative distance changes within a cycle. To measure absolute distance, one needs to know the starting distance and then keep track of number of accumulated cycles. Another problem, which may be the reason no successful implementation of the "phase coherent" approach has been developed, is the effect of multipath reflections. *Multipath*, a term also associated with radio transmission, refers to the fact that the signal received is often the sum of the direct path signal and one or more reflected signals of longer path lengths. Since walls and objects in a room are extremely reflective of acoustic signals, the amplitude and phase of the signal received from a continuous wave acoustic emitter in a room will vary drastically and unpredictably with changes in position of the receiver.

An outstanding feature of pulsed TOF acoustic systems is that it is possible to overcome most of the multipath reflection problems by simply timing until detecting the first pulse that arrives, which is guaranteed to have arrived via the direct path unless it is blocked. The reason this simple method works for acoustic systems but not for RF and optical systems is the relatively slow speed of sound, allowing a significant time difference between the arrival of the direct path pulse and first reflection.

Point-to-point ranging for unconstrained 3-D tracking applications requires the use of transducers with radiation and sensitivity patterns that are as omnidirectional as possible, so that the signal can be detected no matter how the emitter is positioned or oriented in the tracking volume. The beamwidth of the sound emitted from a circular piston transducer is inversely proportional to $D/\lambda$, where $\lambda$ is the wavelength and $D$ is the diameter of the piston. To achieve an approximately hemispherical omnidirectional pattern ($+/-$ $60^\circ$ 3 dB points) requires $D/\lambda = 0.6$ (Baranek, 1954). For a typical ultrasonic tracker frequency of 40 kHz, $\lambda = 9$ mm, and one is required to use very tiny speakers and microphones with active surfaces of about 5.4-mm diameter. This is convenient for integration into human-motion tracking devices and helps reduce off-axis ranging errors, but the efficiency of an acoustic transducer is proportional to the active surface area, so these small devices cannot offer as much range as larger ones. To improve the range, most systems use highly resonant transducers and drive them with a train of about six or more electrical cycles right at the resonant frequency to achieve high amplitude. This results in a received waveform that "rings up" gradually for about 10 cycles to a peak amplitude then gradually rings down. For a typical pulse-detection circuit that stops the timer at the peak of the envelope, this means the point of detection is delayed about 10 cycles, or about 90 mm, from the beginning of the waveform.

It should now be clear that as long as the reflected path is longer than the direct path by more than 90 mm, the detection circuit will have already registered the peak of the first pulse and stopped the timer before the reflected signal begins to arrive. Formalizing this analysis, let us call the time delay from the beginning of the pulse waveform to the peak or whatever feature is detected by the receiver circuitry $t_d$. Thus the displacement of this detection point from the beginning of the waveform is $R_d = c_s t_d$, where $c_s$ is the speed of sound in air. Let $R_1$ be the distance from the emitter to a reflecting object, and $R_2$ the distance from the reflecting object to the receiver. Any reflection path length $R_1 + R_2$ that is longer than the direct path length $R_0$ by more than $R_d$ will not corrupt detection. Thus objects outside the ellipsoid defined by $R_1 + R_2 = (R_0 + R_d)$ with the emitter and receiver transducers as its foci cannot possibly cause multipath ranging error because signals reflected off of them will arrive after detection is complete. Conversely, objects inside this "ellipsoid of interference" risk causing reflections that interfere with the direct path signal to slightly distort range measurement. The major axis of this ellipse is of length $2a = (R_0 + R_d)$, and the distance between the foci is $2c = R_0$, so the minor axis is of length:

$$2b = 2\sqrt{a^2 - c^2} = \sqrt{(R_0 + R_d)^2 - R_0^2} = \sqrt{2R_0 R_d + R_d^2}$$

For a transmitter-receiver separation $R_0$ of 2 m, and the typical $R_d$ of 90 mm, the width of the ellipsoid in the middle would be 0.6 m. This allows significant opportunity for extraneous objects such as hands to come near enough to the line of sight between transmitter and receiver to produce reflections that corrupt the range measurement. Much of our own development work has therefore focused on the design of circuitry to reliably detect a sound burst much earlier than its peak. By detecting on the second or third cycle instead of the tenth, the volume of the "ellipsoid of interference" can be reduced by a factor of four or more, resulting in far fewer multipath errors in typical real-world situations. This phenomenon is little discussed in the literature, but in our experience it is one of the most important issues for accurate ultrasonic tracking outside of controlled laboratory settings.

There are of course many other design trade-offs and considerations dictated by the physics of ultrasonic waves in air and transducer design. Ultrasonic noise sources such as banging metal fall off rapidly with increasing frequency, so operating at a higher frequency is very beneficial for avoiding interference and also offers higher resolution due to shorter wavelengths. However, selecting a higher frequency also means less range due to problems with transducer size and frequency-dependent attenuation of sound in air. Attenuation of sound in air due to molecular absorption is basically negligible at 1 kHz, starts to play a significant role (compared to spherical spreading losses) by 40 kHz, and becomes the dominant factor in limiting range by 80 kHz. It also depends very significantly on relative humidity, with the humidity level that causes greatest attenuation shifting as a function of frequency (Baranek, 1954).

Since increasing the frequency much beyond 50 kHz is usually not an option due to range problems, other techniques may be considered to improve resolution and immunity to ambient noise sources. For resolution enhancement, a common trick is phase locking. Using some envelope-based technique to determine the rough TOF, the final TOF is determined by finding the zero-cross of the carrier wave nearest to the rough TOF detection point. Since the slope of the carrier wave is very steep as it crosses zero, the location of the zero-cross point is not much affected by additive noise, and resolutions of a small fraction of a millimeter are easily obtained. However, no phase-locking technique has been devised yet which can consistently pick out the same individual wave of the carrier every time no matter how the transducers move, so the technique occasionally produces 9-mm jumps in the output. To improve rejection of ambient noise, any communications engineer would probably suggest driving the emitter with a complex unique waveform and detecting it with a matched filter at the receiver. However, this is not easy to accomplish because: (1) the entire signature waveform has to be very short in order to avoid multipath by getting an early detection; and (2) the piezoelectric transducers have narrow bandwidth and cannot be made to transmit very nonsinusoidal waveforms.

The main factors limiting accuracy of ultrasonic ranging systems are wind (in outdoor environments) and uncertainty in the speed of sound. The update rate of acoustic systems is limited by reverberation. Depending on room acoustics, it may be necessary for the system to wait anywhere from 5 to 100 ms to allow echoes from the previous measurement to die out before initiating a new one, resulting in update rates as slow as 10 Hz. The latency to complete a given acoustic position measurement is the time for the sound to travel from the emitter to the receivers, or about 1 ms per foot of range. This is unaffected by room reverberation and is usually well under 15 ms worst case. However, in a pure acoustic system with a slow update rate, system latency is also affected by the need to wait for the next measurement. Hybrid tracking can improve both update rate and latency dramatically.

### 3.4 Electric Field Sensing

Before the remarkable successes of 19th-century physics, electricity, magnetism, radio, and light were all considered separate phenomena. In the wake of Einstein's special theory of relativity, one can see that there is a single fundamental force required to explain all these effects, i.e., electromagnetic force. Nonetheless, these four distinct manifestations of electromagnetic force behave quite differently, and different motion tracking possibilities are treated separately in this and the next three sections.

The electric field is the only one of the four that has not been routinely used for motion tracking, however, electric fields are theoretically detectable, and certain fish use them to sense object shape and distance. The use of static fields from charged objects is unlikely to produce practical tracking systems due to the difficulty of keeping the charge from leaking off, or other objects from getting charged, but oscillating electric fields bypass these problems. Zimmerman, Smith, Paradiso, Allport, & Gershenfeld (1995) have implemented a system for tracking hand motion or body location using capacitive sensing of electric fields. A radiating electrode and a ground electrode are set up in a workspace, and any conductive object that comes between or near them affects the capacitance between them. A human hand inserted between the plates acts as a conductive object connected to a large conductive mass that acts as a charge reservoir ground if it is outside of the capacitor gap. As the hand is inserted further into the capacitor gap, it shunts away more of the electric field lines that otherwise would have reached the ground electrode and therefore reduces capacitance. In another arrangement, an emitter electrode is placed close to or in contact with a person's body so that the excitation field is coupled into the person and their entire body becomes an electric field radiator. Then, the closer the hand approaches the ground electrode, the greater the capacitance.

Such a system is capable of low latency, high resolution, and can be built with lightweight low-power electronics that could be integrated into a palmtop or even wristwatch computer. Sensing is unaffected by nonconductive objects and requires no contact with the user's body. However, the electric field geometry in the dipole near-field regime is too complex for accurate analytical modeling, and some form of training or calibration procedure is necessary to convert the capacitance measurements from multiple electrodes into a position estimate. The potential for precision tracking with electric fields is not good. The system will track the whole human body as a "blob," or that part of the body that is inserted in the electric field region. It can therefore be used with large-scale electrodes to tell where people are in a room or with smaller scale electrode arrangements to track where a hand is when it is inserted into the region between electrodes. In this latter example, it may be useful for qualitative gesture tracking but not for quantitative precision pointing, because the indicated position will be affected by the shape of the hand, arm, and stance of the body if it is too close to the field region.

## 3.5 Magnetic Field Sensing

Unlike electric fields, magnetic fields are unaffected by the presence or absence of human bodies and other nonmetallic objects in the environment. This offers a tremendous opportunity, because it enables magnetic trackers to overcome the line-of-sight requirement that plagues acoustic, optical, and externally connected mechanical trackers. Magnetic tracking technologies have a long history and to date have been more widely deployed in human–machine interface tracking applications than any other technology.

### 3.5.1 Geomagnetic Sensing

Loadstone (magnetite) was known to the ancients, who eventually discovered that if suspended properly it tends to align itself toward the north, and the magnetic compass was invented. Thus, the world's first motion tracking system was a yaw direction indicator based on the earth's magnetic field. Modern navigators learned how to build electronic compasses with digital readout, and these became the head tracker of choice for consumer HMDs of the early 1990s, because their ultralow cost offset poor performance and lack of position tracking capability.

Earth's magnetic field from the surface outward approximates a magnetic dipole field (i.e., the magnetic field pattern produced by a small circulating current loop). The radial and tangential components of a dipole magnetic field are given by the equations:

$$B_r = \frac{2m\cos\theta}{r^3}$$

$$B_t = \frac{m\sin\theta}{r^3}$$

where $m$ is the magnetic dipole moment, $\theta$ is the angle away from the north pole, and $r$ is the radius from the center of the dipole source (derived from Purcell, 1965, p. 365).

Magnetic north deviates from true geographic north by an amount called the magnetic declination, $D$, which varies about $+/-\ 20^\circ$ across the United States because of the slant of the magnetic dipole axis, with anomalies as large as 60 degrees in certain regions due to ore deposits. To use a compass effectively for navigation, it is essential to correct the compass reading with the local value of $D$. Even with access to a good map of local declinations, the accuracy of a compass can be limited by changing magnetic disturbances caused by solar winds, which can cause hourly deviations of $+/-\ 0.3^\circ$ on magnetically turbulent days.

An electronic compass must find the direction of the horizontal field while disregarding the effect of the vertical field. Modern compasses, such as the ones used in HMDs, sense the full magnetic-field vector, then use pitch-and-roll information obtained by gravimetric and/or gyroscopic means to calculate the components of this vector in the horizontal and vertical directions. They then calculate heading using only the horizontal components.

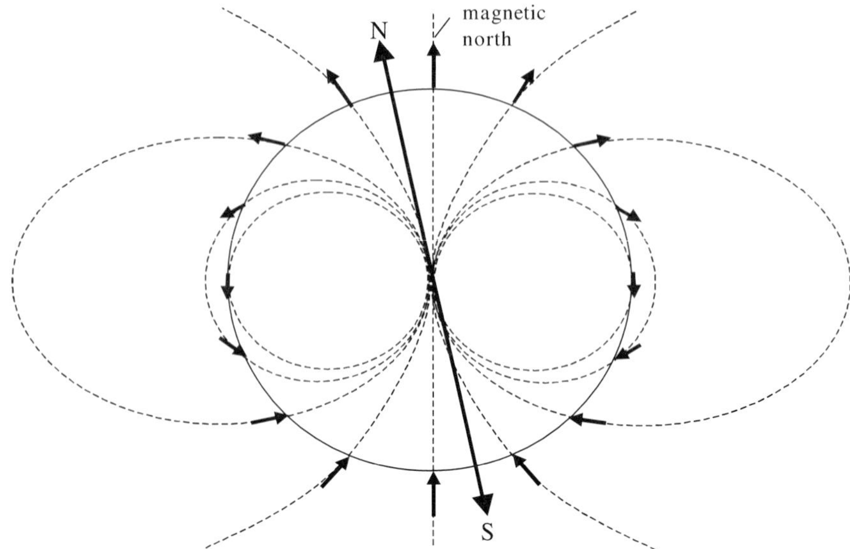


FIG. 8.5. Earth's magnetic field at the surface. The field is 0.3 gauss horizontal at the magnetic equator and 0.6 gauss vertical at the magnetic poles, which are inclined 11 degrees with respect to the spin axis.

Referring to Fig. 8.5, it can be seen that the dip-angle or magnetic inclination, *I*, varies from 0 degrees at the equator to 90 degrees at the poles. The error in compassing calculation due to a tilt measurement error $\varepsilon$ is:

$$\text{compass error} = \tan^{-1}[\sin(\varepsilon)\tan(I)]$$

so in Burlington, Massachusetts, where $I = 69.27^\circ$, every 1 degree of tilt error will cause a compass yaw error of 2.64 degrees. In northern Europe or Canada, the magnetic inclination *I* approaches 80 degrees, with tilt error amplification factors of 4 to 5. This means that even a state-of-the-art magnetometer that has been calibrated to 0.1 degrees may experience yaw errors of several degrees when used with a good gyro-stabilized inclinometer of 0.5 degrees peak tilt error. When used with a plain inclinometer that experiences many degrees of slosh error whenever a person moves, the yaw reading will twist several times as much as the pitch and the roll slosh. This is the effect that made the simple inclinometer/compass orientation trackers in early consumer HMDs virtually unusable. As a final warning, even if one could build a perfect gyro-stabilized inclinometer/compass, it would still be subject to yaw errors of typically 10 to 20 degrees in buildings due to furniture and building materials, and 3 to 5 degrees in urban outdoor areas due to nearby buildings, cars, and signposts. Clearly, geomagnetic sensing cannot be relied on as a primary yaw determination means in applications such as AR, which demand absolute accuracy better than a couple of degrees.

### *3.5.2 AC and DC Active-Source Systems*

A few years after Sutherland's early HMD-tracking experiments, Jack Kuipers of Polhemus Navigation Sciences invented a technique for tracking the position and orientation of a fighter pilot's helmet, using AC magnetic field coupling from a 3-axis source to a 3-axis sensor (Kuipers, 1975; Raab, Blood, Steiner, & Jones, 1979). More recently, similar systems using quasi-DC fields have been developed (Blood, 1989), and both technologies are now widely used in a broad range of human–machine interface applications.

In both systems, magnetic fields are generated by a source consisting of three orthogonal coils of wire activated in sequence by the electronics control unit to generate three orthogonal magnetic dipole fields similar in shape to Earth's dipole field, illustrated in Fig. 8.5 but on a much smaller scale. For AC systems, the source is activated with oscillating currents of 7 to 14 kHz, and the sensor consists of a similar triaxial coil assembly that is able to measure components of these oscillating magnetic fields by inductive pickup.

For the DC systems, the triaxial sensor assembly uses devices sensitive to small DC magnetic fields, namely any of the sensor types that are used in electronic compasses. Traditionally, the most popular such sensor was the fluxgate magnetometer, but recently magneto-resistive, magneto-inductive, and hall-effect sensors have been replacing fluxgates. After each source coil is activated with a DC current, the system waits for eddy currents induced in any nearby metal objects to die out before measuring the resulting field. This way, the DC tracker is unaffected by any nonmagnetic conductors, which cause trouble for AC trackers due to eddy currents.

The magnetic field magnitude, $b \equiv \sqrt{B_r^2 + B_t^2}$, falls off with the inverse cube of distance $r$ from a dipole source. Naively, one might assume that this means that magnetic tracker resolution and accuracy will degrade according to the cube of the transmitter-receiver separation $d_{tr}$. Unfortunately, Nixon et al. (1998) have shown analytically that at least position resolution must degrade as the fourth power of separation distance, and have confirmed experimentally that both position and orientation follow this trend for AC and DC magnetic trackers. The reason for the fourth power is that position resolution is not dependent on the magnitude of the magnetic field, but rather on the gradient of this magnitude with distance:

$$\Delta r = \frac{dr}{db} \Delta b \propto r^4 \Delta b.$$

This means a small disturbing field, $\Delta b$, will produce a position error component along the radial axis proportional to $d_{tr}^4$. It is not obvious from this analysis why the orientation errors should also grow proportional to $d_{tr}^4$, but the experimental data show that they do.

In addition to the dramatic effect of range on performance, which affects both AC and DC trackers the same way, there is a significant difference between the two in terms of sensitivity to external interference sources. The most common sources of interference are mains power wiring, appliances generating interference at 50 or 60 Hz, and computer monitors. Because the sensors in AC trackers only detect signals in a frequency band centered around typically 8, 10, 12, or 14 kHz, they are virtually immune to low-frequency-mains interference. AC trackers are able to operate at 0.2- to 0.25-mm resolution at a range of 600 mm in an ordinary room environment without special synchronization and filtering. DC sensors are sensitive at low frequencies and produce over 30 mm of position noise at the same range. To attenuate this down to a much more tolerable level, a DC tracker's sampling frequency should always be synchronized to twice the mains frequency, and then a filter employed to average two adjacent samples, thus canceling the interference.

Nixon, McCallum, Fright, & Price (1998) also analyzed the effects of metals on AC and DC trackers. Different types of metals were tried, and it was found that the DC tracker was completely unaffected by brass, aluminum, and stainless steel, but committed larger errors than the AC tracker in the presence of copper, ferrite, and mild steel. Presumably, the AC tracker performed better with ferromagnetic steel because the magnetic permeability decreases with frequency. They concluded that the best countermeasure is to keep the transmitter close to the receiver and metals far away from both of them.

The latency of magnetic tracking is limited only by the rate a system can cycle through three excitation states (plus a zero-excitation state in DC trackers to remove the effect of earth's magnetic field) and the need for noise reduction filtering. Adelstein et al. (1996) measured the latencies of modern DC and AC magnetic trackers using a carefully designed mechanical testbed and data analysis procedure to isolate the internal latencies of the trackers. With all filtering disabled and tracking a single receiver, they report latencies of 7.5 and 8.5 ms, respectively, for position, and less for orientation. In an environment with 60-Hz power-line noise, the DC tracker will normally need to be used with a two-tap averaging filter at 120 Hz, adding about 4 ms of additional latency. The main trade-off to consider is range versus resolution versus latency. If resolution is just acceptable at a range $r$, then at $2r$ with no filtering there will be 16 times as much noise. To filter that noise back to the original level using a simple rectangular moving average filter would require 256 taps, thus adding a latency of 128 times the sampling period. More sophisticated filters can, of course, be designed to accomplish the same noise reduction with fewer taps, but the example illustrates the general idea of the trade-offs involved.

## 3.6 Radio Waves, Microwaves, and Ultrawideband (UWB) Technology

Radio waves and microwaves have not been exploited much in tracking human motion, but they are widely used in navigation systems and various airport landing aids and radar systems. They have begun to find application in local positioning systems that find RF asset tags in warehouses or hospitals (Lanzl & Werb, 1998) and are likely to be used for human-motion tracking systems in the future as the precision improves and the technology becomes smaller and cheaper. Electromagnetic wave-based tracking techniques are capable of providing vastly greater range than quasi-static electromagnetic fields because radiated energy dissipates as $1/r^2$, whereas the dipole field strength gradient drops off as $1/r^4$. Furthermore, radio waves suffer negligible absorption losses in air and are virtually unaffected by wind and air temperature, so they are uncompromised outdoors or in large open spaces, where acoustic systems have difficulty. Unfortunately, radio waves are rapidly attenuated in water, so the human body is opaque to all radio frequencies useful for precision ranging.

### *3.6.1 Radio, Microwave, and Millimeter-wave (mmw) Ranging*

Most radionavigation systems operate on the principle of time-of-flight (TOF) rangefinding, much as described for acoustic ranging in section 3.3. The waves travel about a million times faster (roughly 1 ft/ns as opposed to 1 ft/ms for sound), making the task of measuring TOF with sufficient precision much more difficult. For example, ranging with 1 mm resolution would require a timer that can count at 300 GHz, implying expensive and power-consuming electronics based on gallium arsenide or indium phosphide semiconductors.

One technique to eliminate the requirement for such high-speed timers is interferometry using a continuous wave source. The transmitted and received waves are combined, and the phase difference between them can be measured to a small fraction of a wavelength. As with phase-based acoustic ranging, this approach suffers from an integer ambiguity problem: The distance between the transmitter and receiver can change by any integer multiple of the wavelength and the phase difference will be the same. It is therefore necessary to know initial position and keep track of the number of phase rollovers during tracking, then add the integer number of wavelengths to the fractional wavelength determined by phase interferometry to compute total distance. If the object moves too fast, gets temporarily blocked, or receives interference, there is a possibility of cycle slips leading to gross ranging errors.

Another approach to avoid the need for extremely high-speed digital counters is to use a combination analog/digital timer. A digital counter is run at a reasonable rate, say 1 GHz, and on each cycle it initiates an analog ramp signal. Reception of the pulse stops the ramp generator and digital timer, and the stored voltage on the ramping capacitor is sampled and used to interpolate between the counter stop value and the next count.

Another approach is to take advantage of the high speed of light to make ranging measurements at a very high repetition rate, and then average thousands of separate ranging measurements taken over a brief interval to produce a range measurement of higher resolution. Using conventional narrowband signals, only a limited number of separate bursts can be transmitted per second because each necessarily requires many cycles of the carrier to ring up and down again, but the next section describes technology for producing temporally shorter duration pulses.

A more sophisticated approach is the delay-locked loop (DLL) employed in global positioning system (GPS) receivers. A digital pseudo-random noise (PRN) code is modulated on the microwave carrier before transmission from the satellite. A replica of the code is played back in the receiver with a time delay $\tau$, which can be adjusted using a numerically controlled oscillator and multiplied by the incoming PRN code from the received signal. The resulting product is averaged over a period of time, yielding a signal representative of the correlation between the incoming PRN code and the local copy. The DLL uses a feedback loop to keep adjusting the delay $\tau$ to maximize this correlation signal. If the clock in the receiver were exactly synchronized with the clock in the satellite, the resulting time delay $\tau$ obtained by the DLL would provide a direct measure of the TOF from the satellite to the receiver. Since the receiver does not have an atomic clock, it is likely to have some clock bias $\Delta t_c$, and this must be estimated and added to $\tau$ to get the true range. In GPS, this is accomplished by measuring the "pseudoranges" from four satellites to solve for the four variables x, y, z, and clock bias $\Delta t_c$.

The autocorrelation of the PRN code is essentially a unit impulse so that multipath reflected copies of the code delayed by more than one chip have nearly zero correlation with the direct path signal and therefore don't disrupt the DLL's tracking loop. The chip length of the P-code signal, which is used for precise tracking after initial acquisition and lock are achieved, is 154 cycles (one period of the 10.23-MHz master oscillator), or about 29 meters long. Therefore, any multipath signals that travel paths longer than the direct path by more than 29 meters can usually be rejected. This provides a reasonable measure of multipath rejection for an airplane, because signals that bounce off the ground will be delayed by well over 29 meters before they get to the plane. However, for ranging between a "pseudolite" and receiver that are both in a building or near the ground, this provides almost no protection. Designing a custom spread-spectrum solution for VEs using higher frequency microwaves and fewer carrier cycles per chip might ultimately reduce the chip length to a few meters, but the analysis in section 3.3 for acoustic systems suggests that in human-scale tracking applications we need to reject all signals that are delayed more than a couple centimeters.

At the high-frequency extreme of the microwave spectrum, with wavelengths from 1 cm down to 1 mm, lie the millimeter waves, which exhibit behavior halfway between radio waves and light. They have recently become an active area of research, and practical applications in law enforcement are beginning to emerge (Williams, 1999). Most of the recent interest stems from the lightlike capabilities of the waves. Because of the very short wavelengths it is actually possible to build an imaging sensor in a man-portable size by packing an array of tiny antennas into the focal plane behind a plastic lens. Passive MMW imagers can visualize warm people through walls or detect cooler concealed weapons, even nonmetallic ones, beneath a person's clothing from a distance. Active MMW radar is under development to detect the distance of objects in a scene as well as their horizontal and vertical locations. Neither the passive nor active MMW imaging sensors offer enough cross-range resolution for precision HMD or tool tracking in virtual environments, but millimeter waves could also potentially be used for simple point-to-point TOF ranging between omni-directional transmit and receive antennas. Compared to lower microwave frequencies used in GPS and other RF tracking systems, millimeter wave electronics can operate with much wider bandwidth, which could potentially be used to achieve higher resolution and tighter rejection of multipath interference. In fact, since the wavelengths are of the same size as those used in acoustic ranging, the simple early-detection-of-first-arriving-wavefront strategy could possibly be implemented with minor modifications in a MMW system. The electronics are currently much more expensive and complex than audio frequency electronics, and the Federal Communications Commission (FCC) has not yet allocated any spectrum above 60 GHz, so the commercial deployment of such a solution is a long-term prospect.

### 3.6.2 Ultrawideband (UWB) Ranging

UWB ranging makes use of nonsinusoidal electromagnetic signals such as impulses. Because there is no carrier frequency these are sometimes called time-domain, carrierless, or baseband signals. These signals have been studied since the 1970s and applied to radar (Taylor, 1995) as well as communications (Win & Scholtz, 1998). Interest has increased lately due to the development of simple low-power electronic circuits for generating and timing short impulses, including the famous micro-impulse radar (MIR) from Lawrence Livermore Laboratory (McEwan, 1993), used in commercial applications such as studfinders and automobile warning radars.

Most UWB schemes use short pulses approximating impulse functions or doublets, such as a half-sine pulse or a Gaussian monocycle signal (derivative of a Gaussian function having a doubletlike shape with a positive excursion immediately followed by a negative one). By transmitting a sequence of such impulses with a random nonperiodic distribution in time, the frequency spectrum of the signal is kept flat like white noise, and no appreciable interference is caused in narrowband radio receivers. Likewise, the UWB receiver tunes in a specific UWB transmission by knowing in advance the PRN code for the expected distribution of pulses in time, and is therefore relatively immune to interference from narrowband transmitters because it is only receptive during occasional, narrow time windows. UWB cannot be allocated specific regions of the spectrum, as with conventional RF systems, because it necessarily emits energy across the whole spectrum, from DC to several GHz. However, its emissions look like very-low-level background noise and are therefore potentially interoperable with conventional systems. The FCC is currently considering whether to allow commercial UWB deployment and how to regulate it. Narrowband systems accommodate multiple users in a given area by assigning each transmitter a different frequency band (frequency-division multiple access, or FDMA), and spread-spectrum systems further allow multiple transmissions on the same frequency band by using different spreading codes (code-division multiple access, or CDMA). UWB instead accommodates different transmitters because each is transmitting pulses following different pseudo-random time-hopping patterns, with a low probability of two pulses colliding because they are so short. By spreading each bit of information or ranging operation over many pulses, even occasional collisions are tolerable.

The outstanding advantage of the UWB paradigm is the improved ability to reject multipath signals. With pulses as short as 200 ps, all reflection paths delayed by 6 cm or more can be easily disregarded. For this simple reason, it is this author's opinion that if precise and robust electromagnetic ranging in indoor environments ever happens, it will be based on UWB impulses. Logically, this ought to be achievable with much simpler electronics than those required to demodulate a complicated spread-spectrum signal. If this turns out to be true in practice, and if the FCC develops a policy that allows UWB transmissions without too many restrictions, then this may eventually become a preferable method of ranging in VE motion tracking systems.

## 3.7 Optical Tracking

There are a particularly large number of different designs that use visible or near-infrared light in some way to track motion. There is no apparent reason to try more exotic techniques using ultraviolet or far-infrared, and X rays with high enough energy to penetrate flesh and solve the line-of-sight problem are not safe, so the rest of this section will assume the use of visible or near-IR light. Most optical trackers use some form of bearing sensors (e.g., cameras, lateral-effect photodiodes, or quad cells) to track pointlike targets or beacons, as discussed in section 3.7.1. Other optical techniques are discussed in sections 3.7.2 and 3.7.3. Since most computer vision approaches involve identifying certain fiducials or landmark points, they are included in the beacon-tracking discussion.

### *3.7.1 Beacon Tracking*

Beacon trackers can be classified into *outside-in* and *inside-out* systems. Outside-in beacon-tracking is the simplest and most common arrangement. Two or more cameras are mounted on the walls or ceiling looking in on a workspace. The sensors detect the direction to the targets or beacons attached to the object being tracked, and a computer then triangulates the 3-D positions of the beacons using the bearing angles from the two nearest cameras. The biggest problem with outside-in systems is a trade-off between resolution and working volume. If sensors employ narrow FOV lenses the resolution is good, but the volume of intersection of the FOV is small. With wide-angle lenses, working volume can be increased at the expense of resolution. For example, to cover a $16 \times 16 \times 8$ ft working volume inside a $20 \times 20 \times 10$ ft room using four cameras mounted in the corners of the ceiling would require cameras with 78 degree horizontal and 76 degree vertical FOV. Assuming $1{,}000 \times 1{,}000$ pixel cameras and 0.1-pixel resolution for locating beacons, this would yield a resolution of about 0.7 mm. This is quite adequate positional resolution for most applications, but the orientation must be computed from the positions of three beacons mounted on a rigid triangle. A triangle of 15 cm per side would provide orientation resolution of about 0.4 degrees, which is too much jitter for some applications.

An alternative arrangement called inside-out optical tracking places bearing sensors on the object being tracked and the beacons at fixed locations on the ceiling or walls (Wang et al., 1990). This approach yields orientation resolution equivalent to the angular resolution of the bearing sensors, which is easily better than required even using modest-resolution sensors. However, to achieve position resolution comparable to an outside-in system requires multiple sensors looking out in different directions, which can be too heavy for some applications. Conceivably, one could use outside-in tracking to achieve good position resolution combined with a single outward looking camera to provide good orientation resolution.

Beacon trackers may be further classified according to whether they use imaging or non-imaging sensors for detecting the bearing angles to targets. Imaging sensors such as CCD or CMOS cameras require some digital computation to find locations of targets in the image. They have the advantages that they can find the locations of multiple targets in a single image and that the locations can be accurate even if there is background clutter as long as the image processing is smart enough to distinguish the actual targets from clutter. Nonimaging sensors such as quad cells (e.g., Kim et al., 1997) and lateral effect photo-diodes (LEPDs, e.g., Wang et al., 1990) are pure analog sensors that determine the centroid of all light in the FOV. They require no digital image processing, but care must be taken to ensure that the only light seen by the sensor at any given time is a single bright target. These sensors are therefore always used with active light-source targets that can be switched on one-at-a-time. In most cases, the targets are infrared LEDs and the sensor is equipped with an IR filter to block all visible light clutter. A background subtraction between the result with the LED on and the result with it off can be used to further reduce error caused by any IR sources other than the intended target. There is one type of error that even background subtraction cannot help: reflected light from the LED when it is turned on. For example, if the FOV of the sensor includes both an LED target on the ceiling and a portion of the wall, then some of the light from the LED will diffusely illuminate the wall and shift the centroid toward the reflection patch on the wall. To minimize this, the University of North Carolina inside-out optical tracker uses a cluster of outward looking LEPDs with only 6 degrees FOV each (Welch et al., 1999), so that when a particular sensor sights a target on the ceiling it is unlikely to also pick up a reflection on a wall. However, the system must use a very dense array of LEDs so that there will always be beacons available within the FOV of several of the sensors. Because outside-in systems require wide-FOV sensors to achieve good overlap volume, nonimaging sensors are not suitable for these systems.

Another type of nonimaging sensor is the one-dimensional (1-D) CCD array. These are often used with a cylindrical lens to measure a 1-D bearing angle with extremely high resolution. Arrays with 5,000 pixels or more are readily available, and extraction of the center of the target distribution in one dimension requires only minimal digital signal processing. Because there are discrete pixels and no analog centroid processing, they are not subject to the same problems with clutter and reflections, just described for quad cells and LEPDs. There are many commercial systems of quite similar form that have three 1-D bearing sensors mounted in a preassembled rigid bar of about a meter or more in length. In a typical case, the bar is mounted horizontally on the wall, and the LED targets are flashed one at a time. The two sensors at the ends of the bar are arranged horizontally and used for a simple 2-D triangulation to determine horizontal coordinates of the LED in the plane of the bar. The sensor in the center of the bar is arranged vertically and measures the elevation angle from the plane of the bar to the target, which is translated into the height of the target using the known horizontal location. Such systems provide high resolution and accuracy within a certain wedge-shaped volume in front of the tracking bar.

In contradistinction, imaging sensors can be used with active, retroreflective, or even passive targets. Many commercial videometric motion capture systems use cameras with a ring of LEDs around the lens to track balls coated with retroreflective film containing thousands of tiny corner-cube reflectors, which return light back along the direction it came. Because the light source is so close to the camera lens, the balls reflect all the light toward the camera and appear very bright in the video image relative to normal objects, which return only a small percentage of the illumination back toward the camera. The illumination scheme makes the targets so much brighter than the background that the only image processing required is to threshold the image and then find the centers of all the white circles. This technique can be used in both outside-in and inside-out tracking systems, but only works well indoors where ambient illumination is not too high. Passive targets such as printed fiducial marks or natural scene features require considerably more image processing computation to track. Since they are no brighter or darker than other white or black objects in the scene, they must be identified on the basis of size, shape, and/or location using computer vision algorithms. The relentless pace of microprocessor development is making this method increasingly viable for cost-effective real-time tracking. Soon there will be CMOS cameras with enough onboard image-processing functionality to perform certain target extraction algorithms. The potential advantages of a vision-based tracking method using passive landmarks are compelling, especially for inside-out systems. Advantages over an electro-optical system using active targets include:

- No need to wire up the ceiling with an array of active LEDs.
- Range can therefore be expanded at much lower cost.
- Large numbers of users can share the same set of landmarks with no scheduling conflicts.

- Wearable system is untethered, without requiring radio telemetry.
- They can use wide-FOV cameras without errors due to reflection, and therefore far fewer landmarks.

Advantages over a videometric system using retroreflective targets are:

- No need to carry an illumination source (and power for it) on the person being tracked.
- Targets are flat instead of spherical.
- Targets can be uniquely coded, and image processing can identify the location and identity of each.
- They can work indoors or outdoors.
- With increasing computer vision sophistication, there is the potential to track natural scene features as targets, and thus enable tracking in an arbitrary unprepared environment of unlimited range.

In light of these advantages, most recent research on tracking for AR has focused on vision-based tracking (e.g., Hoff, Nguyen, & Lyon, 1996; Koller et al., 1997; Mellor, 1995; Neumann & Cho, 1996). Since AR requires a self-contained wearable tracker that can operate over large areas with minimal preparation, inside-out vision-based tracking is a natural fit. So far, outward-looking vision alone has not yielded sufficient robustness, but hybrid techniques that combine inertial or magnetic tracking with vision are likely to succeed. For many VE applications the region of tracking is fairly defined, and data is needed off-body to drive a graphics workstation. In these conditions, an outside-in approach may be more natural, if optical tracking is needed at all.

### 3.7.2 Optical TOF Ranging

There are a variety of optical tracking techniques that do not entail finding the bearing angles from a sensor to certain target points. One such class of techniques involves optical ranging, in which the time of propagation of a light beam is used to measure the distance from a source to a detector much like the previously described acoustic and RF ranging methods. Both phase interferometry (of the carrier or of a slower modulation signal) and straightforward TOF counting for pulses exist. The most widely used optical ranging technique is lidar, in which distance along a laser beam to a reflecting target is measured based on round-trip TOF. The laser beam must be specifically pointed at the target. This is quite convenient for manual surveying applications with stationary targets. Automatic tracking systems for moving targets have been built by mounting the lidar on a servo-controlled pan-tilt mechanism programmed to continuously follow a given target once it has been locked on. The 3-DOF position of the target can then be directly read out in spherical polar coordinates using the current azimuth and elevation angles of the pan-tilt servo and the radius measured by the lidar. Such a system is accurate but very expensive and can only track one target at a time. If the line of sight gets temporarily blocked the system will lose lock, and reacquiring the target may require a time-consuming scan.

Ducharme, Baum, Wyntjes, Shepard, and Markos (1998) have prototyped an omnidirectional point-to-point optical ranging system analogous to the acoustic and RF approaches discussed above. The light from a laser diode is fanned out by a special lens to approximate a hemispherical point-source radiator. The laser diode is amplitude modulated by a 1-GHz sine wave, and a photodiode receiver within the illumination cone of the emitter produces a copy of this sine wave phase shifted by an amount proportional to the distance from the source. A digital phase processor circuit measures the phase difference between the transmitted and received signals and keeps track of phase wraparounds, which occur if the distance changes by more than 30 cm. Because of the omnidirectional emitter and receiver, the system will probably suffer from the same multipath issues discussed for similar acoustic and RF techniques, and the continuous-wave narrowband modulation scheme prevents the use of the multipath mitigation strategies described for those systems. However, under controlled laboratory conditions, the prototype exhibited peak ranging errors of +/− 0.2 mm over a range from 1- to 1.5-m separation. There are other point-to-point ranging techniques based on focus or intensity, but these are obscure and not very appropriate for VE motion tracking.

### *3.7.3 Structured Light*

The techniques discussed so far used either no light sources or approximations of point sources, possibly time modulated. There are many technologies that generate spatially-modulated light fields such as lines, grids, or even more complex patterns, often scanned or otherwise time varying. Most of these aim to recover the 3-D geometry of a scene, so this section will focus only on a few simple examples, which are concerned primarily with tracking.

The most common such technology is the laser scanner (Sorensen, Donath, Yang, and Starr, 1989), which is now commercially available in a variety of different configurations. In all of them, a laser beam is fanned out into a plane and then swept through the workspace by a spinning mirror. The time difference between the moment when the light plane crosses a reference detector mounted in the scanning mechanism and the moment it crosses a tracking detector in the workspace provides a measure of the bearing angle from the scanner head to the tracking detector. In a simplified configuration for illustrative purposes, there would be two scanners placed, say, in the front left and front right corners of a room. The two scanners would be synchronized so that a sequence of three nonoverlapping scans would occur for each revolution of the motors: a horizontal sweep from the left scanner followed by a horizontal sweep from the right scanner followed by a vertical sweep from the right scanner. (Two sweeps in different directions from a single scanning motor can be accomplished with a multifaceted mirror and/or multiple lasers.) The two horizontal bearing angles and the known baseline separation between the scanners would be used to triangulate the horizontal coordinates of the detector, then the vertical bearing angle would be used to calculate the height, just as with linear CCD-based trackers. In fact, the system can be construed as a form of beacon tracker in that it measures the bearing angle from one device, the scanner, to a point target, the detector.

State-of-the-art laser scanners can achieve bearing angle measurement resolution on the order of 0.1 milliradians—quite comparable to state-of-the-art camera/beacon systems. The major difference is that the sensor is located at the target rather than at the origin of the bearing angle. Mathematically, the configuration illustrated above is an outside-in system, but physically it is inside-out, with the data being measured and made available at a detector on the person. If the position data is needed onboard the moving object, as in robotic navigation, it can be computed autonomously on the robot using just the timing of the detection pulses without the need for any RF telemetry. Unlimited numbers of users can share the structured light fields without mutual interference. On the other hand, if the goal is to remotely track a moving user in a workspace, it may be more natural to use a camera-based outside-in tracker to avoid the need for data telemetry and computer circuitry on the person being tracked.

A variation on the scanner theme has been recently proposed that does not use any lasers, but rather an ordinary lightbulb inside a rotating cylindrical shadow mask to produce a structured light field (Palovuori, Vanhala & Kivikoski, 2000). The clear cylindrical drum is printed with a series of vertical black stripes whose varying widths and spacing form a pseudo-random bar code. A receiver contains matched-filter correlators for each of the shadow masks that are simultaneously rotating in the workspace, and thereby measures the code delay (proportional to bearing angle) from all the sources simultaneously and continuously through code-division multiple access. The potential advantage is continuous measurement, but it remains to be seen if the resolution will be sufficient.

Another class of structured-light devices uses projectors to paint patterns on the scene, and a camera viewing the patterns solves for its own pose if it knows the geometry of the surfaces on which the light is projected or is smart enough to recover the geometry from the distortions of the pattern. Livingston (1998) has proposed a very clever twist on this idea, in which the scene geometry does not need to be known or recovered in order to track the camera. The concept is based on algorithms from computer vision (Longuet-Higgins, 1981), which can solve for the pose of a second camera relative to a first camera given a number of corresponding points in the two images, even though these points are at arbitrary and unknown locations in 3-D. He replaces the first camera with a projector that flashes points one at a time into the scene so that correspondence with the points observed by the camera is easy. Since the projector uses the same projective geometry as a camera, the computer vision algorithms may be directly applied to compute the pose of the camera relative to the projector. The primary application for this technique is video-see-through AR, because the tracking can be done using the forward-looking cameras that are already part of a video-see-through HMD without having to mount 3-D fiducials inside of the object being looked at. If that object is a human body, mounting fixed fiducials inside is impossible.

We know of no example in the literature, but a conceivable structured light technique is the use of polarized light. For a simple example, consider a person wearing a lightbulb atop his helmet with a sheet of polarizing material over it. Above him on the ceiling is a photo detector with a sheet of polarizer spinning in front of it. The received intensity will oscillate with two peaks and two troughs per revolution of the motor, and the phase shift of this signal is a measure of the yaw direction of the person's head. A more elaborate scheme could be developed to measure multiple degrees of freedom.

## 4. MATHEMATICS OF MOTION TRACKING

### 4.1 Observables and Pose Recovery Algorithms

None of the sensing technologies described in the previous section directly measures position and orientation. Instead, each sensor measures certain *observables* which are functions of the desired position and/or orientation. Acoustic, RF, and certain optical methods measure *range* from one transducer to another. Most optical technologies, and some radar and sonar methods, measure *bearing angles* from a sensor to a target, while goniometers measure *joint angles* between connected rigid bodies. Other examples include homogeneous or dipole *field components*, angular or linear *velocities* or *accelerations*, and GPS *pseudo-ranges*.

Since a single measurement of any of these types does not in itself reveal the position and orientation (collectively called pose) of the object, calculations called *pose recovery algorithms* are used to solve for the pose from several measurements. Some classic pose recovery algorithms are *trilateration* to solve for position using three range measurements, and *triangulation* using bearing angles from two cameras. Although trilateration is just a matter of finding the intersection of spheres of known radius, it is surprisingly difficult to come up with an efficient, exact, and general algebraic solution in three dimensions, and this remains an active area of research (e.g., Manolakis, 1996).

Classical triangulation, where the bearing angles are measured from multiple known observation points to an unknown target, is a simple matter of intersecting rays or planes. However, there are some pose estimation problems in computer vision, which are closely related yet far more difficult. The most important of these is perhaps the perspective-n-point-problem (PnP), in which a camera observes n 3-D points and finds their corresponding 2-D projections on the image plane (basically bearing angles from the camera to the points). If the 3-D points are known and the goal is to determine camera pose, the problem is historically called the *exterior orientation problem* or *space resection problem*, but more simply should be called n-point camera pose estimation. The problem was first solved with three points in 1851 and has been solved many different ways since, which are reviewed and compared in Harelick et al. (1994). While direct three point algorithms exist, they produce four solutions, and a unique solution requires a minimum of four points. Until recently such multipoint overdetermined solutions were iterative or nonlinear (e.g., Longuet-Higgins, 1981). However, Quan and Lan (1999) just introduced a direct linear solution using only five points, and a two-step linear solution for four points, which also have fewer restrictions on the geometry of the points (e.g., the points may all be coplanar). Clearly, closed-form algebraic pose recovery algorithms are very complex, and the state of the art is still advancing.

### 4.2 Kalman Filtering and Multisensor Fusion

All of the pose recovery algorithms discussed in the previous section are designed to solve for the pose of an object at time t given a set of measurements that are functions of that pose at time t. In other words, they assume either that the object is not moving or that a complete set of measurements sufficient to determine the pose can be made simultaneously at time t. Even in the latter case, the traditional pose recovery algorithms are not optimal, because they reestimate the entire pose from scratch at every frame, throwing away any information in its past history.

In a seminal paper, Kalman (1960) combined a recursive formulation of least squares estimation with a state-space system dynamics model to develop a practical algorithm for computers to estimate the state of a dynamical system (e.g., pose of a moving object) by optimally combining past history, new measurements, and a priori models and information. The Kalman filter (KF) has become the foundation of modern multisensor data fusion and estimation because it provides the unique best estimate of the state of a linear dynamic system and can also be applied to nonlinear problems using the extended Kalman filter (EKF), often with near-optimal, robust, and computationally efficient performance. We will not derive the Kalman filter here; the reader is referred to the many excellent textbooks on the subject (e.g., Bar-Shalom & Li, 1993; Brown & Huang, 1992; Gelb, 1974). In a nutshell, the Kalman filtering procedure consists of two steps: a prediction step that extrapolates where the object will be at time $t_{k+1}$ based on its position at time $t_k$, and a correction step that is performed whenever there is a measurement available from a sensor that provides even partial or indirect information about the position.

An outstanding benefit of the Kalman filter is that it is very flexible about timing and the order in which it receives and processes measurements. There is no requirement for periodic measurements. This enables flexible integration of data from disparate sensors that are not even synchronized, making the design of hybrid trackers using multiple types of sensor technologies much easier. For example, an aided inertial navigation system may have signals from various navigation aids such as GPS, radio landing beacons, altimeter, radar, and so forth, arriving at different times. Each fix may have a different measurement model (1-D range measurement, 2-D bearing angle, 3-D GPS position fix, etc.) and is processed when and if it is available. A partial correction to the state is made immediately. This approach fits particularly well with inertial navigation systems, which by their very nature consist of a high-rate inertial integration process, which drifts and gets updated whenever a star fix or landmark sighting is made. However, the asynchronous updating capability is valuable in other applications too. For example, in ground-based aircraft tracking, bearings from multiple geographically distributed radars with unsynchronized scan rates are combined (Bar-Shalom & Li, 1995). Relying even more heavily on the incremental partial updating ability of the Kalman filter is a problem called bearings-only tracking, in which infrared or sonar bearing-angle measurements taken at different points along the trajectory of a moving vehicle are fused over time to yield the location of the target (e.g., Nardone, 1980). In active vision systems, a robot-mounted camera looks around the room and whenever it sees a feature it recognizes, it updates its estimate of its own location and that of the feature (e.g., Chenavier, 1992; Harris, 1992). Welch and Bishop (1997) coined the term SCAAT (single-constraint-at-a-time tracking) to refer to this type of updating and used it to improve the update rate and accuracy of the University of North Carolina optical ceiling tracker, which had previously used a batch pose-recovery algorithm. They point out quite rightly that this feature is particularly valuable in virtual environment systems, where high update rates and low latency are essential and could beneficially be applied to magnetic and acoustic trackers as well. In fact, the InterSense IS-900 takes advantage of sequential Kalman processing of individual range measurements to reverse the usual direction of flight and therefore eliminate the propagation latency of acoustic tracking (Foxlin, Harrington, & Pfeiffer, 1998).

Another major benefit of Kalman filtering theory is to provide a rigorous basis for the design of hybrid motion-tracking systems. The Kalman filter produces not only an estimate of the trajectory of a moving object, but also a covariance matrix that represents the statistical uncertainty of that estimate over time. In a simulation technique called *covariance analysis* (Gelb, 1974), one can evaluate the RMS accuracy that will be achieved by fusing together measurements from any particular combination of different sensors. Many of the different sensing principles discussed in section 3 have complementary properties. Covariance analysis can help determine which technologies have the greatest synergy in an application and can help optimize the trade-offs involved in the specification of hardware subsystems' performance and number and layout of sensors or targets. Covariance analysis results comparing many different combinations of inertial, ultrasonic, inside-out and outside-in optical sensors are used by Intersense to design customized hybrid configurations to meet customer requirements at lowest hardware complexity and cost.

### 4.3 Auto-mapping

In trackers built for short-range tracking, all the necessary sensors are usually built into one preassembled, precalibrated reference unit, which serves as the coordinate frame reference for the tracking data. For larger area tracking, a single sensor reference unit cannot provide enough coverage area, and multiple sensor units or landmarks must be spread throughout the work area. In this case a method is needed to auto-map the positions of the entire array of landmarks or sensors in a global reference frame by moving a calibration unit through the environment in advance. If the intended tracking environment is essentially unbounded, as in an outdoor wearable AR application, even this is impossible and it is necessary to discover and auto-map the landmarks on-the-fly during tracking. This is what we shall mean by auto-mapping from here on.

This problem has received a great deal of attention in the computer vision and mobile robotics navigation communities. In computer vision, the problem is referred to as *structure from motion* (SfM), and the primary goal is to reconstruct the geometry of a scene from moving-camera imagery. The camera motion trajectory must be determined in the process, but that is not the main goal. There are several approaches to the problem, some involving deducing the shapes of surfaces from optical flow and others involving auto-mapping the 3-D positions of a set of feature points and then connecting the dots. The modern EKF approach was introduced to the latter school by Broida et al (1990), who used 2-D bearing measurements over a sequence of frames to recursively estimate the camera pose and the 3-D locations of the observed feature points. Azarbayejani and Pentland (1995) refined the approach by simultaneously estimating focal length and changing the representation of the point positions into a 1-D format, which allows stable performance even if the initial feature positions are unknown.

The mobile robotic navigation community calls the problem *simultaneous localization and mapping* (SLAM) and considers a wider variety of sensors including sonar, radar, and lidar as well as vision. Unlike SfM, the primary goal of SLAM is often to know the location of the robot. Developing a map of the environment is a necessary means to that end. The robotics community has been working on the problem even longer, and an EKF approach was introduced by Smith, Self, and Cheeseman (1987) with similar structure. Like the early SfM papers, the early SLAM implementations employed a standard full-covariance Kalman filter. The position states of all N landmarks were appended to the vehicle position states to produce a giant augmented state vector. The covariance matrix of this augmented state vector maintains information about the cross-correlations between all the landmark and vehicle position states. Unfortunately, this rapidly becomes impractical for large numbers of landmarks, since the computational complexity increases with $N^2$. A naïve solution is to drop the cross-correlations between the landmark position estimates, decoupling the problem into separate small KFs for each landmark. This reduces the problem to order N, but in practice the estimates will eventually diverge because each new measurement, say to beacon n, is treated as independent information even though the position error of beacon n is actually highly correlated with that of a previously measured beacon $n-1$. Soon the vehicle will think it has been given a lot of independent information and knows precisely where it is, but in reality the positions of both beacons and the vehicle are all correlated and likely off in the same direction. Much current research focuses on trying to find a work-around to this problem. Some approaches use local submaps, each of which maintains cross-correlations, and then attempt to stitch them together to form a global map (Leonard, 2000) or build a relative map storing only interlandmark distances so that repeated measurements are uncorrelated (Csorba, Uhlmann, & Durrant-Whyte, 1997). Unfortunately, the latter approach does not even estimate the pose of the vehicle. Another very interesting approach is a new method of data fusion called *covariance intersection* (CI) which replaces the Kalman measurement update process with something that is suboptimal but does not assume each measurement to be uncorrelated with previous ones (Julier & Uhlmann, 1997). This is one of the few techniques available so far for handling very-large-scale auto-mapping problems in a very general mathematical framework.

### 4.4 Prediction

A good motion tracker should have the ability not just to accurately follow the current pose of the user, but also to predict motion enough to compensate for the end-to-end delay of a VE system caused by tracker latency, communications, rendering, and image scan-out on the display. Depending on the number of pipelined rendering stages and frame rate, system delay typically ranges from 25 ms to 150 ms or more. Obviously, longer predictions are less accurate, so prediction is not a panacea for slow virtual environment generators. However, when the total latency is less than about 60 to 80 ms, it can help dramatically.

All prediction is based on the Taylor Series expansion:

$$x(t+T_p) = x(t) + \dot{x}(t) \cdot T_p + \frac{1}{2}\ddot{x}(t) \cdot T_p^2 + \frac{1}{6}\dddot{x}(t) \cdot T_p^3 + \cdots$$

The more derivatives at time t are accurately known, the further into the future this prediction holds. In model-free prediction (most of the work to date) this formula is applied separately to each of the 6 degrees of freedom, as if they each could evolve independently. This is the most general approach but doesn't take advantage of any kinematic constraints that might be known to exist, such as attachment of a head to a neck.

Normally one only has samples of the position and orientation variables at discrete points in time and must obtain the derivatives by numerical differentiation, which is noisy. Early efforts to use prediction with magnetic trackers used a Kalman filter to estimate the derivatives (Friedmann, Starner, & Pentland, 1992; Liang, Shaw, & Green, 1991) with much less noise because it performs optimal smoothing based on an a priori model of head motion. Nonetheless, it is much better to actually measure the velocities or accelerations using inertial sensors. Azuma and Bishop (1994) showed two to five times higher accuracy with inertial sensors measuring the derivatives compared to a prediction method where these quantities had to be estimated from the optical position tracker data using a Kalman filter.

Whether using estimation or inertial sensing to obtain the derivatives, a model-based approach would allow longer prediction with the same accuracy from the same data. To see this, consider a hand which is constrained to remain attached to the end of a forearm, which is rotating about the elbow joint. Essentially, a model-free prediction would predict the future location along a straight-line projection of the current velocity of the hand, whereas a model-based method would predict the future location along the circular arc originating from the elbow. If predicting far enough into the future that the elbow would significantly rotate, the latter would be far more accurate. Akatsuka and Bekey (1998) proposed an extremely simplified model of head motion (a lollipop on a stick of length L, rotating about a fixed point at the base of the neck), and showed improved prediction compared to a model-free approach, although the paper doesn't explain how the model is used in the predictor. For human-motion prediction using a fairly simple model of 17 rigid objects, a model-free prediction would have $17 \times 6 = 102$ separate variables, whereas a model based approach would reduce this to predicting approximately 46 joint angles. By respecting these kinematic constraints much accuracy would be gained, but even more might be attainable by considering a dynamics model of the body including mass distributions and muscle forces. Hans Weber at the University of North Carolina, has been working primarily on the former, and Dov Adelstein at NASA Ames has an interest in the latter, but neither has yet published results (personal communications, June 5, 2000).

In addition to improving the kinematics models underlying prediction, there may be some benefits achievable with adaptive multiple-model filtering. In a Kalman filter, the state of the system is always modeled as a linear dynamic process driven by white noise. The white noise is used to model the unknown inputs, namely muscle forces (when no inertial sensors are present) and derivatives of these forces when inertial sensors have directly measured the effects of the forces. This isn't a very good model of human behavior, as it assumes random motion going on all the time with the same intensity. Real human-motion is much more episodic, having periods of stillness interspersed with bursts of motion activity. The onset of the motion is unpredictable, but once the motion starts it is likely to follow a certain course. Filters for tracking aircraft from radar observations make use of multiple-model techniques in which they use a second order kinematics model driven by white noise acceleration during straight and level flight and switch to a third-order model driven by white noise jerk during maneuvers that involve acceleration.

There are a variety of statistical techniques for detecting the onset of maneuvers and switching models or adjusting the blend of multiple concurrently running models (Bar-Shalom & Li, 1993). In addition, it would seem that model selection could be viewed as a classification problem and traditional classification algorithms such as neural nets or fuzzy logic could be usefully combined with the KF estimation paradigm, although this has not been discussed in the literature. Like aircraft motion, human-motion may also be classifiable into several distinct modes, but it will require model-switching on much shorter time frames, and the maneuvers are more complex and may require many more than two different modes. This is a fertile ground for research on prediction that has not yet been adequately explored. Short of this variable-state-dimension multiple-model approach, there are adaptive techniques for tuning the process noise to increase it during maneuvers. The best prediction methods of the future will probably combine all three techniques: inertial sensing, specific kinematic modeling, and adaptive stochastic models that adjust to the presence or absence of "maneuvers" such as gestures or visual pursuit patterns. Better yet, virtual environment designers will select rendering platforms that can render 60 frames per second without pipelining, and even simple prediction algorithms will be sufficient for the 20 to 30 ms of prediction that are required in such a system.

## 5. ENGINEERING OF MOTION TRACKING

To develop motion tracking systems into successful commercial products requires disciplines that are often quite foreign to the types of researchers who work with the physics and mathematics concepts outlined in the previous two sections. For such researchers, the end goal is often a proof of concept demonstration running on a UNIX or Windows computer, with sensor data read in through serial ports or data acquisition boards using vendor-supplied driver software over which the researcher does not have adequate control. Synchronization and timing may be dependent on a system clock with inadequate resolution, and the operating system may randomly interrupt tracking software to perform other tasks. Most obvious of all, the demonstration tracking system may require a lab cart to haul around and have so many boxes and cables that it is irreproducible and treacherously unreliable.

To go from this type of demo to a reliable, producible, and cost-effective device requires methodical software and hardware engineering practices and great attention to detail. Code must be written by professional real-time embedded programmers using a real-time operating system, or perhaps no operating system but a lot of careful benchmarking, timing diagram design, and conformance testing. Bug and feature tracking, and release-level testing, which were at best informal in the research laboratory take on highest priority and consume many times over the number of months that were required to build the original working prototype.

Besides robustness and performance, the obvious engineering goals are miniaturization, power reduction, cost reduction, and increased ease of use. A particular need that has not yet been addressed satisfactorily for most trackers is wireless multiuser tracking. For HMD-based systems, the limiting factor has been the difficulty of making the HMDs themselves wireless, because video imagery requires much higher bandwidth than does tracking data. However, FSD-based VEs are increasing in popularity, and stereoscopic viewing glasses are wireless. There is therefore a real need for wireless tracking in these environments. Certain tracking technologies have no electronics on the object being tracked and are therefore intrinsically wireless. These include computer vision techniques to directly track the image of the hands or their silhouettes (Leibe et al., 2000) or passive markers on handheld tools (Dorfmuller, 1999). For other tracking technologies, which require electronic sensors on the moving object, power consumption must be reduced to allow battery operation and infrared or RF telemetry provided to bring the data back to the host. Designing such telemetry links has previously required a great deal of RF engineering, but there is rapid progress recently in the development of embeddable RF modules or even single-chip radio solutions, driven by developments in the mobile computing and telecommunications sectors. Of particular interest is an emerging standard called Bluetooth, which is designed to facilitate hassle-free data exchange between cell phones, PDAs, notebooks, and other portable and office-based electronics (www.bluetooth.com). The

Bluetooth consortium members will soon begin to introduce a variety of low-cost single-chip radios that support data rates over 700 Kbit/sec, quite enough for several trackers in a VE.

## 6. SYSTEMS INTEGRATION OF MOTION TRACKING

After all of the best engineering practices have been followed, a wonderful motion tracker with low latency and state-of-the-art prediction algorithms gets shipped to a user who integrates it with a VE system and frequently gets miserable results. This happens because end-to-end system latency and image stepping are properties of the whole system and cannot be cured by the tracker without modifications to the rest of the system. The comments in the previous section about real-time programming apply equally to the development of VE system software, and this must be done in accordance with a synchronization policy that the motion tracking system supports.

Before discussing the solutions, consider some of the problems of poorly synchronized systems:

- Latency jitter, multiple images, image stepping
- Longer average latency
- Image shearing

In a typical asynchronous VE system, the tracker loop, graphics rendering, and display scan out all operate independently. The display refreshes at a constant rate, say 60 Hz, following a raster pattern from top to bottom. There are exceptions, such as calligraphic CRT displays and frameless rendering (Bishop, Fuchs, McMillan, & Scher-Zagier, 1994), but these are not in common use. The rendering process will have a nonconstant update rate that may vary from over 70 Hz down to less than 30 Hz, depending on the complexity of the part of the scene in view. The tracker runs internally at its own constant rate, say 130 Hz, and either spits data at the host computer continuously or provides the latest data record whenever it is polled. Figure 8.6 below illustrates a section of this asynchronous operation, with continuous mode tracker reporting, as the renderer frame rate drops from about 70 Hz to just below 30 Hz.

The light-colored bars in the top row indicate frames of the video that contain freshly rendered images; the dark bars on frames 5, 8, and 9 indicate "dropped frames" where there was no new rendering cycle computed in time so the frame buffer scanned out the same image as the last frame. Dropped frames cause the perception of multiple images (Moore, 1996). To see this, consider a 60-Hz display device driven by a graphics engine that renders at 30 Hz so that every other frame is dropped. As the eye tracks a moving object in the series of new frames and blends the sequence of discrete frames into an apparent continuous motion, the repeated

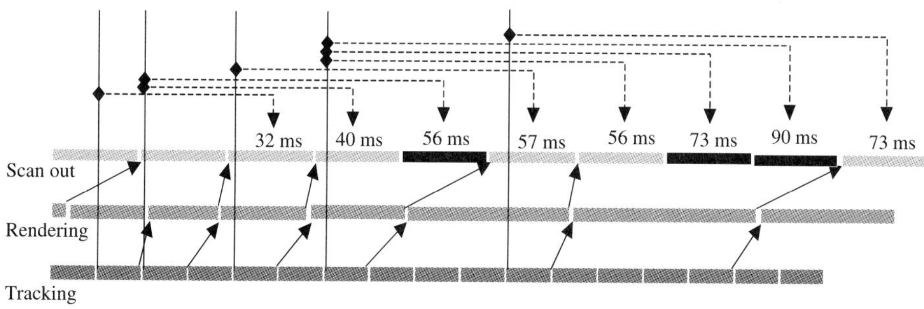


FIG. 8.6. Illustration of typical nonsynchronized VE system.

frames do not fit into this interpolated motion trajectory, but rather create a ghost image of the object lagging behind it by a distance proportional to the speed of motion. Likewise, a 20-Hz graphics update rate would produce a triple image. A similar triple image can be seen in field sequential color displays in which the same image is redrawn three times, and the image therefore separates into nonaligned red, green, and blue images even at field rates of 180 Hz. It is apparent from the figure that at rendering rates of 60 Hz or more there will be no dropped frames, from 30 to 60 Hz there will be single dropped frames, and as soon as the renderer dips below 30 Hz there is the possibility of occasionally dropping two frames in a row, as shown. If frames are dropped only sporadically they may appear as image twitching rather than a steady multiple image, but this can be equally annoying.

Even when no frames are being dropped and the image generator is rendering at a steady 60 Hz, there is a variable latency between the sampling of the head motion sensors and the display of the image. This latency ranges from 33.3 ms for best-case synchronization to 58.5 ms for worst case, for objects halfway down the raster display. When there is no synchronization policy, the latency will be varying across this range due to the beat frequencies between the display refresh, rendering cycle, and tracking loop. At a head rotation rate of 200 deg/sec, this latency variation of +/− 12.5 ms will cause +/−2.5 degrees peak spatial oscillation. Thus, in an unsychronized system, the effects of latency jitter are probably even more detrimental than the effects of the average latency itself, and average latency is also worse in this system than it needs to be. Finally, the effect of image shearing is caused by the 16-ms difference in latency between the top of the image and the bottom. This causes objects at the bottom of the image to lag behind those at the top and vertical lines to take on a slant proportional to the speed of panning. The remainder of this section discusses approaches to solving these problems.

## 6.1 Hard Genlock

Figure 8.7 illustrates a traditional and dependable synchronization policy that can be used to integrate a VE system with tracking. Used universally in professional video production and in many high-performance VE simulators, this system is based on a master "genlock" synchronization pulse at the video field rate, to which all other processes in the system are slaved. This imposes a requirement for the image generator to strictly maintain a 60-Hz frame rate. To accomplish this, the scene must be simplified until no portion of the scene causes "overloading," or else the rendering software must be provided with automatic level-of-detail control that checks every frame and merges polygons if necessary to keep the total load below the point of overloading.

‘ The case illustrated shows the lowest latency configuration with no pipelining in the renderer. This is called *guaranteed single-frame latency*, and is becoming the norm on modern personal computer graphics cards. Genlocking still works with pipelined graphics architectures:

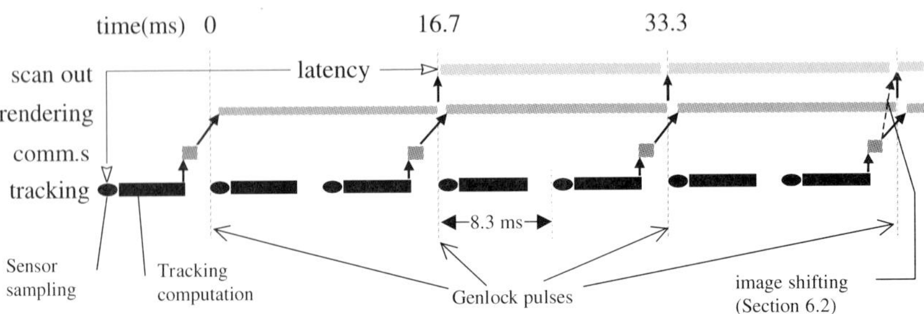


FIG. 8.7. Baseline synchronization policy using master "genlock" signal.

The latency will be larger but still constant. The tracker loop is shown running at twice the genlock rate. The first sensor sampling is initiated directly by the genlock pulse, and another one is internally initiated exactly half a frame period later. With many trackers it would be sufficient to sample at 60 Hz, but for inertial trackers a quicker update rate helps integration accuracy, and for DC magnetic trackers the 120-Hz rate allows the use of a two-tap filter for canceling 60-Hz interference, as discussed in section 3.5. With the inertial tracker, the latency from sensor sampling to the beginning of display screen scan out is 25 ms, as shown in the figure. This latency is so small that it can be compensated very effectively with prediction. In a pipelined rendering system, overall latency may be increased by one or more frames so that prediction algorithms may begin to demonstrate some noticeable overshoot. To overcome this, it is recommended that the motion tracker be sampled again right before the final stage of the rendering pipeline, with a newer and more accurate prediction used to drive final rendering operations. Because this view vector may be off by a degree or two from earlier predictions, it would be necessary to perform the earlier culling with a slightly enlarged viewing frustum, then narrow it down to final size on the last stage. Note from the drawing that the tight loop-timing shown is only possible if the communications of tracker data to the host can be completed before the beginning of the next frame. With an rs-232 serial port running at 115,200 bits/sec, a tracker datum containing six 4-byte floats can be transmitted in 2.3 ms. For longer data packets (e.g., containing multiple sensor data), less efficient encoding, or slower baud rates, either a faster communications link must be found or a longer latency endured.

## 6.2 Image Shifting

The effect of small rotation angles of a virtual camera in azimuth and elevation is nearly equivalent to a simple linear translation of a 2-D image. This fact makes it possible to perform a final adjustment of the scene using the very latest tracking data after the final stage of rendering, just before or during scan out. The last cycle in Fig. 8.7 has a dotted arrow showing how data from the tracker is used not only to start the next rendering cycle, but also to compute an image shift amount for the frame just rendered, right before scan out. The new tracker datum is based on sensor data sampled only 8.3 ms before scan out, and thus can be predicted forward with almost no error. Yaw and pitch values are compared to predictions that had been used for the rendering process, and the differences are used to determine the necessary horizontal and vertical shift values.

Image deflection was conceived by Rediffusion in the 1970s and implemented in the early 1980s by the Naval Air Warfare Center Training Systems Division (NAWCTSD) for a laser projector display (Breglia, 1981) and by the Institute of Sound and Vibration Research for a CRT in an HMD (So & Griffin, 1992; Wells & Griffin, 1984). In a CRT, it can be achieved with simple analog electronics, which add an additional offset current to the horizontal and vertical deflection yokes of the CRT. For LCD displays, they cannot be deflected this way, but the shifting can still be accomplished without bothering the main CPU by playing with the frame-buffer-addressing hardware. In both cases, the image has to be slightly overcomputed to allow shifting. The amount of overcomputed area can be greatly reduced by using the best possible prediction technique before rendering.

Even with a system that is tightly genlocked and designed to maintain a steady 60-Hz image rendering rate, there is always a concern that it may at some point become overloaded and drop some frames. Image shifting can also be used to help compensate for lost frames by predicting and shifting the previous frame until a newly rendered frame is available (Moore, Pope, & Foxlin, 1998). In a system where 60-Hz rendering is not possible due to cost limitations but the display must be run at 60 Hz to prevent flicker, image shifting can help reduce the multiple images effect, although it does have some side effects for moving objects in the scene (Moore, 1996).

A more sophisticated technique, which is basically an extrapolation of the image shifting concept, is the "address recalculation pipeline" architecture proposed by Regan and Pose (1994). Instead of just overcomputing the frame a little on each side, they compute a complete surrounding world, rendered on six faces of a cube around the user's current head position. Once this is rendered, the user with an HMD can look all around, and the orientation tracker is used to read out an appropriate portion of the six-sided frame buffer, automatically undistorted by hardware in real time at 60 Hz. Latency with respect to head-orientation changes is therefore eliminated. However, whenever the user translates in position, all six faces of the cube have to be rerendered. Since translational movement has relatively small effects on distant objects, closer objects get rerendered first, thus minimizing the translational latency penalty. By keeping sets of objects at different distances in different frame buffers, which are composited together during scan out, they can be rerendered at different rates.

All the prediction, synchronization, and shifting techniques discussed so far are designed to achieve consistent and minimal latency from the head motion to the beginning of image scan out. However, the scan out itself takes 16 ms, and so objects toward the bottom of the display screen will suffer more latency than objects at the top. The most obvious manifestation of this is during horizontal scanning, when everything in the frame appears to slant. A solution called just-in-time pixels (Olano et al., 1995) has been proposed, in which ideally each pixel is rendered separately with tracker data concurrent to its display time. Olano et al. (1995) suggest an approximation by using a separate viewing transformation for each line, or even just two transformations for the first and last pixel, with all others calculated by linear interpolation. In practice, the effect will be very nearly linear over the 16-ms scan period, so all that is needed is to measure the head rotation rate about the vertical axis with a gyro, or estimate it with a Kalman filter and then shift each scan line by an increasing amount proportional to this rate. Reichlen (1993) implemented a frame buffer with this shifting feature built in. It is of course also possible to predistort each polygon in software if it is known how fast it will be moving across the screen.

## 7. SUMMARY OF RECENT PROGRESS AND FUTURE POTENTIALS

Since Meyer et al.'s 1992 Survey of Position Trackers, the field has evolved and produced some new tracking options as well as some new demands. Drift-corrected gyroscopic orientation trackers have appeared, and where only orientation is needed they now provide an affordable use-anywhere solution with sufficient resolution, robustness, responsiveness, and sociability for HMD fly-through applications. Outside-in optical tracking has increasingly made inroads in motion capture and is beginning to reach a level of real-time performance and price suitable for performance animation. Laser scanners have come out of the laboratory and are now commercially available from several sources. The realization of the need for hybrid tracking has sprung up everywhere, with an acoustic-inertial hybrid on the market and research papers on magnetic-inertial, optical-inertial, and optical-magnetic combinations signaling a diverse future. GPS has proliferated throughout the world, and, driven by huge volumes, vendors have succeeded in shrinking complex receivers down into incredibly tiny packages. The field of machine vision, bolstered by industrial parts inspection applications and the ascent of PC computing power, has made great strides in bringing formerly expensive algorithms down to the level of routine use, including a new breed of compact vision systems and smart cameras suitable for embedding. Ultrawideband radio technology has appeared in the form of microimpulse radar products, and at least theoretically holds out the promise for an improved method of omnidirectional point-to-point ranging.

Two paradigm shifts in the VE field are happening which are creating new types of requirements for motion tracking systems. The first is a trend away from HMDs and towards FSDs, such as the CAVE, the virtual workbench, and personal minidome projectors. In simulation and training where a sense of presence is an important aspect of the application HMDs are still the norm, but for systems design (see chap. 52, this volume) and information visualization (see chap. 53, this volume), easier group dynamics tend to favor the headgear-free FSD paradigm. This reduces the need for high-resolution low-latency head-orientation tracking, but increases the emphasis on high-quality 6-DOF hand tracking. Furthermore, the user is no longer forced to don a heavy headset with a cable, so the requirement for a lightweight wireless tracking device is increasing. A second trend is the increasing interest in AR (see chap. 48, this volume). This creates requirements for tracking that is far more accurate and, at the same time, small enough to wear and able to operate over a large range in a cluttered environment. The consensus is that a hybrid of inside-out computer vision and inertial technology is likely to be the best fit for this problem, and that is unlikely to change since that is the solution that biological evolution has developed for the same problem. Initial offerings will require the use of artificial fiducial marks to simplify the computer vision requirements, but the long-term goal is to make use of natural features found in both indoor and outdoor scenery to allow unrestricted tracking in arbitrary unprepared environments.

This survey has attempted to overview physical principles that can be exploited for VE-style motion tracking applications. It is a testament to the diligence of the engineers in the field that almost all of them are already being used. However, a few ideas have been discussed that have not yet been developed as far as they could, or in some cases have not even been discussed in the literature. These represent the "low-hanging fruit" that may be able to yield new and useful tracking technologies for certain applications in the next few years. For example, the biokinematic reckoning approach of section 3.1.2 or the biodynamic model-based tracking of section 3.2.4 may yield new approaches to full-body avatar animation with significant mobility and cost advantages over the magnetic and optical systems that are prevalent today. The potentially reduced multipath incidence with ultrashort electromagnetic impulses mentioned in section 3.6.2 suggests that this could eventually become a better alternative than acoustic ranging in aided inertial trackers, especially outdoors. The polarized light technique in section 3.7.3 might make an alternative to the compass for correcting inertial yaw drift in confined metallic environments such as inside a vehicle. The vision-aided inertial approach, with GPS priming in outdoor applications, will eventually find a wide audience as portable and wearable systems become common. The research groundwork in these areas is being laid today, and it is possible to imagine in the foreseeable future ubiquitous computing devices offering new human–machine interface capabilities based on position and orientation tracking.

## 8. REFERENCES


Adelstein, B. D., Johnston, E. R., & Ellis, S. R. (1996). Dynamic response of electromagnetic spatial displacement trackers. *Presence: Teleoperators and Virtual Environments, 5*(3), 302–318

Akatsuka, Y., & Bekey, G. (1998). Compensation for end-to-end delays in a VR system. In *Proceedings of VRAIS 98 Conference* (pp. 156–159). Atlanta, GA: IEEE Computer Society Press.

Azarbayejani, A., & Pentland, A. P. (1995). Recursive estimation of motion, structure, and focal length. *IEEE Transactions on Pattern Analysis and Machine Intelligence, 17*(6), 562–575

Azuma, R., & Bishop, G. (1994). Improving static and dynamic registration in an optical see-through HMD. In *SIGGRAPH 94 Conference Proceedings*. Orlando, FL: ACM Press.

Badler, N., Hollick, M. J., & Graneri, J. P. (1993). Real-time control of a virtual human using minimal sensors. *Presence: Teleoperators and Virtual Environments, 2*(1), 82–86.

Bar-Shalom, Y., & Li, X. R. (1993). *Estimation and tracking principles, techniques, and software.* Boston: Artech House.

Bar-Shalom, Y., & Li, X. R. (1995). *Multitarget–multisensor tracking: Principles and techniques.* ISBN 0-9648312-0-1.

Baranek, L. L. (1954). *Acoustics.* New York: McGraw-Hill.

Bhatnagar, D. K. (1993). Position trackers for head mounted display systems: A survey. Chapel Hill, NC: University of North Carolina.

Bishop, G. Fuchs, H., McMillan, L., & Scher-Zagier, E. J. (1994). Frameless rendering: Double buffering considered harmful. In *SIGGRAPH 94 Conference Proceedings.* Orlando, FL: ACM Press.

Blood, E. B. (1989). Device for quantitatively measuring the relative position and orientation of two bodies in the presence of metals utilizing direct current magnetic fields. U.S. Patent No. 4,849,692.

Breglia, D. (1981). Helmet-mounted laser projector. In *Proceedings of the Third I/ITSEC Conference* (pp. 8–18). Orlando.

Broida, T. J., Chandrashekhar, S., & Chellappa, R. (1990). Recursive estimation of 3D motion from a monocular image sequence. *IEEE Transactions on Aerospace and Electronics Systems, 26*(4), 639–656

Brown, R. G., & Hwang, P. Y. C. (1992). *Introduction to random signals and applied Kalman filtering.* New York: John Wiley & Sons.

Chenavier, F., & Crowley, J. L. (1992). Position estimation for a mobile robot using vision and odometry. In IEEE International Conference on Robotics and Automation (pp. 2588–2593). Nice, France: IEEE.

Cruz-Neira, C., Sardin, P., & DeFante, T. (1993). Surround-screen projection-based virtual reality: The design and implematation of the CAVE. In *SIGGRAPH 93 conference proceedings,* Onahium, ACM Press.

Csorba, M., Uhlmann, J. K., & Durrant-Whyte, H. F. (1997). A suboptimal algorithm for automatic map building. In *Proceedings of the American Control Conference* (pp. 537–541). Albuquerque, NM: Omini Press.

Dorfmuller, K. (1999). An optical tracking system for VR/AR-applications. In *Proceedings of the Eurographics Virtual Environments '99 Workshop* (pp. 33–42), Vienna: Springer-Verlag.

Drascic, D., & Milgram, P. (1996). Perceptual issues in augmented reality. In *Proceedings of SPIE: Vol. 2653. Stereoscopic Displays and Virtual Reality Systems III* (pp. 123–134). SPIE

Ducharme, A. D., Baum, P. N., Wyntjes, G., Shepard, O., & Markos, C. T. (1998). Phase-based optical metrology system for helmet tracking. In *Proceeding of SPIE: Vol. 3362. Helmet and Head-Mounted Displays III, AeroSense 98.* Orlando, FL: SPIE

Ellis, S. R., Adelstein, B. D., Baumeler, S., Jense, G. J., & Jacoby, R. H. (1999). Sensor spatial distortion, visual latency and update rate effects on 3D tracking in virtual environments. In *Proceedings of VRAIS '99 Conference* (pp. 218–221). Houston, TX: IEEE Computer Society Press.

Ellis, S. R., Young, M. J., Adelstein, B. D., & Ehrlich, S. M. (1999). Discrimination of changes in latency during voluntary hand movement of virtual objects. In *Proceedings of the Human Factors and Ergonomics Society.* Houston, TX:

Ferrin, F. J. (1991). Survey of helmet tracking technologies. In *Proceedings of SPIE. Vol. 1456* (pp. 86–94).

Foxlin, E. (1993). *Inertial head-tracking.* Unpublished master's thesis, Massachusetts institute of technology.

Foxlin, E. (1996). A complementary separate-bias Kalman filter for inertial head-tracking. In *Proceedings of VRAIS 96.* Santa Clara, CA: IEEE Computer Society Press.

Foxlin, E. (1997). Inertial orientation tracker apparatus having automatic drift compensation for tracking human head and other similarly sized body. U.S. Patent No. 5,645,077. Filed June 16, 1994.

Foxlin, E. (2000). Head-tracking relative to a moving vehicle or simulator platform using differential inertial sensors. In *Proceedings of SPIE: Vol. 4021. Helmet and Head-Mounted Displays V, AeroSense Symposium.* Orlando, FL: SPIE.

Foxlin, E., Harrington, M., & Pfeiffer, G. (1998). Constellation: A wide-range wireless motion tracking system for augmented reality and virtual set applications. In *SIGGRAPH 98 Conference Proceedings,* Orlando, FL: ACM Press.

Friedmann, M., Starner, T., & Pentland, A. (1992). Device synchronization using an optimal filter. In *Proceedings of the 1992 Symposium on Interactive 3D Graphics.* Cambridge, MA: ACM Press.

Gelb, A., (Ed.). (1974). *Applied optimal estimation.* Cambridge, MA: MIT Press.

Harelick, R. M., Lee, C. N., Ottenberg, K., & Nolle, M. (1994). Review and analysis of solutions of the three point perspective pose estimation problem. *International Journal of Computer Vision, 13*(3), pp. 331–356

Harris, C. (1992). Geometry from visual motion. In A. Blake & A. Yuille (Eds.), *Active Vision* (pp. 263–284) Cambridge, MA: MIT Press.

Held, R., Efstathiou, A., & Greene, M. (1966). Adaptation to displaced and delayed visual feedback from the hand. *Journal of Experimental Psychology, 72,* 887–891.

Hoff, W., Nguyen, K., & Lyon, T. (1996). Computer vision-based registration techniques for augmented reality. In *Proceedings of SPIE: Vol. 2904. Intelligent Robots and Computer Vision XV* (pp. 538–548). Boston, MA: SPIE.

Holloway, R. L. (1997). Registration error analysis for augmented reality. *Presence: Teleoperators and Virtual Environments, 6*(4), 413–432.

Ishii, M., & Sato, M. (1994). A 3D space interface device using tensed strings. *Presence: Teleoperators and Virtual Environments, 3*(1), 81–86.

Julier, S. J. & Uhlmann, J. K. (1997). A non-divergent estimation algorithm in the presence of unknown correlations. In *Proceedings of the 1997 American Control Conference.* Albuquerque, NM: Omni Press.

Kalman, R. E. (1960). A new approach to linear filtering and prediction problems. *ASME Transactions Journal of Basic Engineering, 82*(1), 35–45.

Kim, D., Richards, S. W., & Caudell, T. P. (1997). An optical tracker for augmented reality and wearable computers. In *Proceedings of IEEE Virtual Reality Annual International Symposium* (pp. 146–150). Albuquerque, NM: IEEE Computer Society Press.

Koller, D., Klinker, G., Rose, E., Breen, D., Whitaker, R., & Tuceryan, M. (1997). Real-time vision-based camera tracking for augmented reality applications. In *Proceedings of ACM VRST 97 Conference.* Lausanne, Switzarland: ACM.

Kuipers, J. (1975). Object tracking and orientation determination means, system and process. U.S. Patent No. 3,868,565.

Lanzl, C., & Werb, J. (1998). Position location finds applications. *Wireless Systems Design.*

Leibe, B., Starner, T., Ribarsky, W., Wartell, Z., Krum, D., Singletary, B., & Hodges, L. (2000). The perceptive workbench: Toward spontaneous and natural interaction in semi-immersive virtual environments. In *Proceedings of the Virtual Reality 2000 Conference* (pp. 13–20). New Brunswick, NJ: IEEE Computer Society Press.

Leonard, J. J., & Feder, H. J. S. (2000). A computationally efficient method for large-scale concurrent mapping and localization. In J. Hollerbach & D. Koditschek (Eds.), *Ninth International Symposium, Robotics Research:* Salt Lake City: Springer-Verlag.

Liang, J. D., Shaw, C., & Green, M. (1991). On temporal-spatial realism in the virtual reality environment. In *Proceedings of the Fourth Annual ACM Symposium on User Interface Software and Technology.* Hilton Head, SC: ACM.

Livingston, M. A. (1998). *Vision-based tracking with dynamic structured light for video see-through augmented reality.* Unpublished doctoral dissertation, University of North Carolina, Chapel Hill.

Longuet-Higgins, H. C. (1981). A computer program for reconstructing a scene from two projections. *Nature,* 293.

Lynch, D. (1998). Coriolis vibratory gyros. Annex B in IEEE Working Draft P1431/D16. *Standard Specification Format Guide and Test Procedure for Coriolis Vibratory Gyros.* IEEE Standards Department.

Manolakis, D. (1996). Efficient solution and performance analysis of 3-D position estimation by trilateration. *IEEE Trans. on Aerospace and Electronic Systems, 32*(4).

McEwan, T. (1993). Ultra-short pulse generator. U.S. Patent No. 5,274,271. Filed July 12, 1991.

Mellor, J. P. (1995). Real-time camera calibration for enhanced reality visualization. In *Proceedings of Computer Vision, Virtual Reality and Robotics in Medicine* (pp. 471–475). Nice, France: IEEE.

Meyer, K., Applewhite, H. L., & Biocca, F. A. (1992). A survey of position trackers. *Presence: Teleoperators and Virtual Environments, 1*(2), 173–200.

Molet, T., Boulic, R., & Thalmann, D. (1999). human-motion capture driven by orientation measurements. *Presence: Teleoperators and Virtual Environments, 8*(2), 187–203.

Moore, R. G. (1996). Multiple image suppression. In *Proceedings of the 18$^{th}$ I/ITSEC Conference.* Orlando, FL.

Moore, R. G., Pope, C. N., & Foxlin, E. (1998). Toward minimal latency simulation systems. In *Proceedings of American Institute of Aeronautics and Astronautics Conference* (Vol. 4176). Boston, MA: AIAA.

Nardone, S. C., & Aidala, V. J. (1980). *Necessary and sufficient observability conditions for bearings-only target Motion Analysis* (Techn. Rep.). Newport, RI: Naval Underwater Systems Center.

Neumann, U., & Cho, Y. (1996). A self-tracking augmented reality system. In *Proceedings of ACM VRST '96.* Hongkong: ACM.

Nixon, M. A., McCallum, B. C., Fright, W. R., & Price, N. B. (1998). The effects of metals and interfering fields on electromagnetic trackers. *Presence: Teleoperators and Virtual Environments, 7*(2), 204–218.

Olano, M., Coher, J., Mine, M., & Bishop, G. (1995). Canbatting rendering latency. In *Proceedings of Symposium on Interactive 3D graphics,* Monterey, CA: ACM Press.

Palovuori, K. T., Vanhala, J. J., & Kivikoski, M. A. (2000). Shadowtrack: A novel tracking system based on spread-spectrum spatio-temporal illumination. *Presence: Teleoperators and Virtual Environments, 9*(6).

Purcell, E. M. (1965). *Electricity and Magnetism.* New York: McGraw-Hill.

Quan, L., & Lan, Z. (1999). Linear n-point camera pose determination. *IEEE Transactions on Pattern Analysis and Machine Intelligence, 21*(8), 774–780.

Raab, F. H., Blood, E. B., Steiner, T. O., & Jones, H. R. (1979). Magnetic position and orientation tracking system. *IEEE Transactions on Aerospace and Electronic Systems, 15*(5), 709–718.

Regan, M., & Pose, R. (1994). Priority rendering with a virtual reality address recalculation pipeline. In *SIGGRAPH 94 Conference Proceedings,* Orlando, FL: ACM Press.

Reichlen, B. (1993). Sparcchair: A 100-million-pixel display. In *Proceedings of VRAIS,* Seattle, WA: IEEE.
So, R. H. Y., & Griffin, M. J. (1992). Compensating lags in head-coupled displays using head position prediction and image deflection. *Journal of Aircraft, 29*(6), 1064–1068.
So, R. H. Y., & Griffin, M. J. (1995). Effects of lags on human-operator transfer functions with head-coupled systems. *Aviation, Space, and Environmental Medicine, 66*(6), 550–556.
Sorensen, B., Donath, M., Yang, G. B., & Starr, R. (1989). The Minnesota scanner: A prototype sensor for three-dimensional tracking of human body segments. *IEEE Transactions on Robotics and Automation, 5*(4), 499–509.
Smith, R., Self, M., & Cheeseman, P. (1987). A stochastic map for uncertain spatial relationships. In *Fourth International Symposium on Robotics Research.* Cambridge, MA: MIT Press.
Suryanarayanan, S., & Reddy, N. (1997). EMG-based interface for position tracking and control in VR environments and teleoperation. *Presence: Teleoperators and Virtual Environments, 6*(3), 282–291.
Sutherland, I. E. (1968). A head-mounted three-dimensional display. *1968 Fall Joint Computer Conference, AFIPS Conference Proceedings, 33*, 757–764.
Taylor, J., (Ed.). (1995). *Introduction to ultra-wideband radar systems.* Boka Raton, Boston, London, NY, Washington DC: CRC Press.
Wang, J. F., Azuma R., Bishop, G., Chi, V., Eyles, J., & Fuchs, H. (1990). Tracking a head-mounted display in a room-sized environment with head-mounted cameras. In *Proceeding of SPIE: Vol. 1290. Helmet-Mounted Displays II,* Orlando, FL: SPIE.
Ware, C., & Balakrishnan, R. (1994). Reaching for objects in VR displays: Lag and frame rate. *ACM Transactions on Computer–Human Interaction, 1*(4), 331–356.
Watson, B., Spaulding, V., Walker, N., & Ribarsky, W. (1997). Evaluation of the effects of frame time variation on VR task performance. In *Proceedings of VRAIS '97 Conference.* Albnquerque, NM: IEEE Computer Society Press.
Welch, G., & Bishop, G. (1997). Single-constraint-at-a-time tracking. In *SIGGRAPH '97 Conference Proceedings.* Los Angeles. ACM Press.
Welch, G., Bishop, G., Vicci, L., Brumback, S., Keller, K., & Colluci, D. (1999). The HiBall tracker: High-performance wide-area tracking for virtual and augmented environments. In *Proceedings of VRST '99.* London: ACM.
Wells, M. J., & Griffin, M. J., (1984). Benefits of helmet-mounted display image stabilization under whole-body vibration. *Aviation, Space, and Environmental Medicine, 55*(1), 3–18.
Williams, T. (1999). Millimeter waves and the EHF bands. In *Proceedings of the 25th Eastern VHF/UHF Conference.* Vernon, CT: ARRL.
Win, M. Z., & Scholtz, R. A. (1998). Impulse radio: How it works. *IEEE Communications Letters, 2*(2), 36–38.
Zimmerman, T. G., Smith, J. R., Paradiso, J. A., Allport, D., & Gershenfeld, N. (1995). Applying electric field sensing to human-computer interfaces. In *Proceeding of the Computer–Human Interface Symposium '95.* Penves ACM Press.

# 9

# Eye Tracking in Virtual Environments

Joseph Wilder,[1] George K. Hung,[2]
Marilyn Mantei Tremaine,[1] and Manpreet Kaur[1]
*Rutgers University*
*[1]Center for Advanced Information Processing (CAIP)*
*[2]Department of Biomedical Engineering*
*96 Frelinghuysen Rd.*
*Piscataway, NJ 08854-8088*
*wilder@caip.rutgers.edu, shoane@rci.rutgers.edu,*
*tremaine@acm.org, manpreet.kaur@t1.com*

## 1. INTRODUCTION

Development of instruments for tracking eye movements has been under way for over 25 years (Cornsweet & Crane, 1973; Young & Sheena, 1975). Early eye trackers were cumbersome devices in which the user wore uncomfortable gear, such as bulky helmets or other equipment affixed to the head. Except under the most controlled conditions, these devices were not capable of maintaining their calibration because it was difficult to maintain strict positioning of the devices on the head. More recently, eye trackers have been developed that attempt to overcome these shortcomings. Ultralight headband-mounted trackers have been developed that reduce slippage. Also, newer devices, based on a tracking system mounted on the display screen rather than the head, permit the user to roam relatively freely over a comfortable region of space.

The primary goal of early eye trackers was to support research in human visual data acquisition. But as instrumentation technology continued to evolve, it eventually led to applications in a variety of settings where understanding of human perception, attention, search, tracking, and decision making are of great importance. This includes industrial inspection (Megaw & Richardson, 1979), medical image analysis (Reuter & Shenck, 1985), visual response to advertising (Lohse & Johnson, 1996; Russo & LeClerc, 1994), and analysis of the performance of airplane pilots (Sanders, Simmons, & Hoffman, 1979). More recently, eye-tracking measures have been used to understand the causes of cybersickness in flight simulators and virtual environment (VE) devices (Kaiser, 1999) and to pinpoint theories of language processing (Eberhard, Spivey-Knowlton, Sedivy, & Tanenhaus, 1995). At the same time, there have been efforts to develop eye trackers as visual communication and control devices (Freedman, 1984; Jacob, 1991; Velichkovsky & Hansen, 1996), e.g., to aid disabled people (Yamada & Fukada, 1987). In particular, eye movement–based human–computer interaction (HCI) techniques have been studied to assess the ability of gaze, along with speech and tactile interactions, to provide more natural interactions with computers than the traditional mouse and keyboard.

This chapter discusses the use of gaze-based communication and control systems to support multimodal interactions, in real and virtual environments. Specifically, it first describes how these new trackers work (section 2), then, how they are used as input devices (section 3), how they are being integrated into VE systems (section 4), the human factors analyses that are underway to assess the effectiveness of eye tracking in comparison and in combination with other modalities (section 5), and concludes with an assessment of future directions in eye tracking for virtual environments (section 6).

## 2. HOW EYE TRACKERS FOR VE WORK

Eye trackers have been developed for measuring many properties of visual behavior, for example, saccade and smooth pursuit, as well as accommodation, vergence, and pupillary response. The key measurements for visual communication and control that have been used are saccades and pursuit, and the fixations between these movements.

Normal eye movements (e.g., in reading) consist of saccades, or jumps, from one fixation (stationary position) to another. Typically saccades range in amplitude from 1 to 20 degrees, corresponding to a duration of 30 to 70 msec, and peak velocities of 70 to 600 deg/sec, respectively (Bahill & Stark, 1979). When following slowly moving targets in the range of 1 to 30 deg/sec, the eye can track these movements with a smooth-pursuit behavior that appears to partially stabilize the image of the target on the retina (Young & Sheena, 1975).

Several basic techniques have been used for tracking eye position. They can be divided into contact and noncontact methods. The contact method uses magnetic induction of two sets of orthogonal induction coils, driven in quadrature, on a scleral ring worn on the eye of the participant. Rotation of the eye results in changes in the amounts of phase-locked horizontal and vertical induced currents picked up by the scleral ring, thereby providing a signal proportional to eye position. The search coil technique provides accurate measurement of two-dimensional (2-D) eye position, but requires local anesthesia that limits the experimental session to about 20 minutes (Remmel, 1984). Among the noncontract methods, the limbus eye tracker is the simplest and least expensive. Two infrared (IR) photo-emitters pulsed at 1 kHz are aimed at the iris (dark)–scleral (white) boundary on either side of the eye. More, or less, light will be reflected depending on the position of the eye relative to the emitter. A pair of infrared detectors picks up the reflected light from the emitters. The differential signal from the emitters is demodulated and filtered to provide a signal proportional to horizontal eye position. This technique provides a relatively easy-to-use recording method, but it is limited to horizontal eye movements. Another technique is the Purkinje eye tracker, which measures the relative displacement of the images formed by the reflection of a light source at the anterior corneal surface and the posterior lens surface, which are known as the first and fourth Purkinje images, respectively. Rotation of the eye results in a greater displacement of the first relative to the fourth Purkinje image, thereby providing a signal proportional to eye position. However, this device requires precise alignment and is not suitable for experiments that permit relatively free movement by the participant (Young & Sheena, 1975).

The video-based eye tracker is the most suitable for 2-D recording of eye movements of a participant who is relatively free to move about in a region of space. It is currently used for communication and control in virtual environments. The tracker captures a video image of the eye illuminated by a distant, low-power, infrared light source and creates an image that is seen as a highlight spot on the surface of the cornea. The image is analyzed by a computer, which calculates the centroid of the corneal reflection as well as the centroid of the pupil. The corneal reflection from the front spherical surface of the eye is insensitive to eye rotations, but it is sensitive to translational movements of the eye or head. On the other hand, the pupil center is sensitive to both rotation and translation. Thus, the difference between the pupil center and the corneal reflex provides a signal proportional to eye rotation, and thereby the direction of gaze, which is relatively free of translational artifacts.

In head-mounted eye trackers, the IR source, the camera that views the eye, and a second camera that views the observed scene are all mounted on the head. Additionally, as the user roams about in a prescribed space, a magnetic sensor is affixed to the head to record the position of the user in the space. The system computes the point of gaze in the image obtained by the scene camera. On the other hand, in more recently developed computer interface-type trackers, the IR source and one camera are gimbal-mounted on a stationary platform containing the computer display. The user is not required to wear any devices. After a simple calibration procedure, the tracker can be used to record eye gaze position in the monitor display. Moreover, the user would be able to use the eye as a pointer in the same fashion one would use a mouse. Both systems allow the user to operate in a hands-free mode.

Two methods of illuminating the eye are in common use, dark-pupil and bright pupil. In the dark-pupil technique, the IR source is off-axis, causing light to be trapped in the interior of the eye. The pupil appears very dark in the eye image. On the other hand, if the IR illumination is provided coaxial to the eye, the eye appears to glow brightly. The dark pupil provides somewhat less contrast than the bright pupil for detecting iris boundaries. However, the bright pupil may interfere with the corneal reflex for relatively larger eye movements. Moreover, the bright pupil technique requires precise alignment of the incident and reflected light (which is consistent with our experience at the Vision Laboratory of the Rutgers University Center for Advanced Information Processing [CAIP]). Therefore, for measurements that allow for relatively free movement of the participant, the dark pupil technique is preferred.

Video-based eye trackers capture images of the eye using a standard RS170, 30 frames/sec video camera. Each frame is composed of two $512 \times 240$ interlaced fields that result in a $512 \times 480$ pixel image. However, since the tracker analyzes each field independently, images can be captured at a field rate of 60 times/sec. This relatively rapid recording of data allows for separation of fixations from saccades. The eye movement results can be plotted as a scan path, that is, locations of the sequence of fixations in the scene and paths of the saccades connecting them. Scan path plots are useful as indicators of the human observer's regions of interest during a visual task. The fixation locations derived from the data are also useful for eye–mouse-type pointing and menu selection operations described in section 4.

To make these systems work reliably, a number of technical issues need to be addressed. For example, gimbal-mounted eye trackers attached to computer displays must:

1. Lock on to and track the eye as the user's head moves about in the space in front of the display (usually a rather confined volume of space).
2. Maintain the image of the eye in sharp focus for reliable computation of the centroids using some form of autofocus.
3. Measure and keep track of the distance of the eye from the screen in order to compute the visual angle subtended by the screen (required to maintain pointing accuracy).
4. Be easily calibrated.

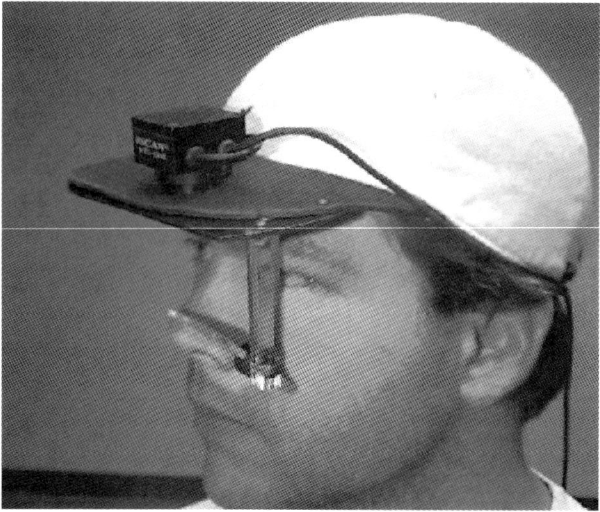

FIG. 9.1a. Head-mounted eye tracker.

Head-mounted trackers must:

1. Strictly maintain their position on the head or be able to recalibrate themselves with small shifts in position.
2. Provide for untethered operation (in most applications), either by RF links for the tracker and the magnetic head position sensor or by integration into wearable computers.

The value of this technology as an input mechanism will be closely linked with the degree to which problems associated with these issues are solved. Examples of head-mounted and computer-display-mounted eye trackers are shown in Fig. 9.1 (courtesy of ISCAN, Inc.).

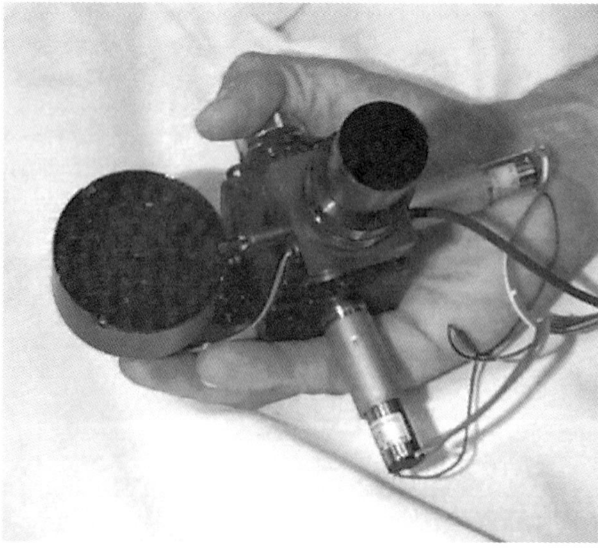

FIG. 9.1b. Gimbal-mounted eye tracker to be affixed to a computer display.

## 3. THE USE OF EYE TRACKERS AS A COMPUTER INPUT DEVICE

It is a common and often valid assumption that what a user is looking at on a computer screen is also the object the user wishes to select. Thus, people have attempted to use the eye tracker as an input device primarily for menu selection. Because the eye tends to rapidly saccade from place to place, object selection is defined by a dwell time of the eye resting on an object, usually 250 milliseconds. However, holding the eye in one position for 250 milliseconds is difficult for users. Dropping the dwell time to something more comfortable, e.g., 100 milliseconds leads to what is known as the "Midas" touch, that is, an overselection of screen objects. Notable successes have occurred with this type of input device for severely handicapped individuals, but everyday computer users find this form of input difficult and tiring (Jacob, 1991).

A key problem with using the eye as an input device for a computer system is that it already is being used as a visual input device for the human. As such, there are likely to be conflicts between the desired computer input behavior and the automatic human-eye behavior. Rather than requiring conscious control of the eye for selection, researchers have more recently looked at modeling the human-eye behavior in order to build intelligent heuristics for determining the computer input the user wants to make. A simple heuristic is that used by Zhai, Morimoto, and Ihde (1999). In their studies they assume that where the eye is looking is where the mouse cursor is best placed. No selection is made. Selection is still done by depressing the mouse button, but the user no longer has to move the mouse cursor a long distance to reach the desired selection target.

In another study by Tanriverdi and Jacob (2000), a threshold count of "landings" is used to deduce the desired selection object. It is assumed that if the eye saccades to a specific area of the screen a specified number of times, then the selection object in that area is the user's input. This heuristic is more complicated than simply counting the saccades over a selection object because a saccade may not even land on the desired object, only close enough to view it. Both of these mechanisms make inferences that may be undesired by the user so that, like the Midas touch of short eye dwell times, users may be faced with the extra effort of "undoing" the computer's decision making.

An alternate modality study, with speech recognition systems correcting speech recognition errors with additional speech, proved to be incredibly frustrating and time consuming for users (Karat, Halverson, Horn, & Karat, 1999). This is why both input designs (Tanniverdi & Jacob, 2000; Zhai, Morimoto, & Ihde, 1999) required a manual confirmation of a selection either with a mouse or Polhemus "click." A user could also gracefully exit from a computer inferred selection simply by looking elsewhere.

In addition to being a useful input mechanism when other methods are impractical or impossible (e.g., speech in a noisy environment), there is strong potential for combining input information from multiple modalities to further enhance the recognition of the information provided by any individual input device. For example, if gaze were combined with speech, it is possible that when a user says, "Put that there," the eye saccades to the location implied by "there" when the word is said. The timing interrelationship of the speech and the eye movement would then tell exactly where the "there" implied by the speech is located despite having a very noisy set of eye movements. The location of the eyes can also help untangle speech recognition. If a speaker said, "the red triangle," this could be understood as different from "the red quadrangle" simply by knowing that the speaker's eyes were saccading about a region of the screen containing the triangle not the quadrangle. Similar advantages might also be gleaned from looking at where a user is pointing or gesturing.

At this time, no natural, simple solution exists for using gaze alone for input selection. However, the above designs suggest that the use of gaze as an input mechanism is becoming more viable. Our own studies, and the two studies referenced above that make inferences from a user's natural eye behavior indicate that the design of a viable interaction mechanism is more complicated than simply using gross eye motor movements and dwell times. Such a mechanism may have to be trained to individual eye movement behaviors of the user as speech recognizers are now trained or be based on other basic eye movement behavior.

Head-mounted displays (HMDs) currently have limitations for gaze input because maintaining calibration is difficult due to slippage of the head-mounting device. Calibration is also a problem with a desk-mounted eye tracker, but the difficulties are not as severe. A gradual loss of calibration occurs over time, especially if the user is gazing at objects at the outer limits of the virtual scene. What is needed are mechanisms built into the VE that recalibrate the user periodically. Another calibration problem occurs if the user is moving to and fro during interaction. Face-tracking algorithms are being used (Yang, Stiefelhagen, Meier, & Waibel, 1998) to capture this motion and reset screen distance for the eye tracker. Overall, either head mounted or desk-mounted eye tracking devices are still difficult to use and take considerable fiddling, adjustment, and programming to interface with virtual environments. So, although promising as input devices, their English sports car tuning needs keep them from becoming over-the-counter input devices.

## 4. INTEGRATION OF EYE TRACKERS INTO MULTIMODAL VE SYSTEMS

Current human–machine communication systems rely, predominantly, on keyboard and mouse inputs that inadequately represent human capabilities for communication. More natural communication interfaces based on speech, sight, and touch can free computer users from the constraints of keyboard and mouse. Although these alternative modality interfaces are not currently sufficiently advanced to be used individually for robust human–machine interaction, they are sufficiently advanced to provide simultaneous multisensory exchange of information. Towards that end, the Rutgers University CAIP Center, under the National Science Foundation STIMULATE program (Flanagan et al., 1999), is conducting research to establish, quantify, and evaluate techniques for designing synergistic combinations of human–machine communication modalities. The modalities under study are sight, sound, and touch interacting in collaborative multiuser environments (see Fig. 9.2). The CAIP Center's goal is to design a multimodal HCI system with the following characteristics and components:

- Force-feedback tactile input and gesture recognition
- Automatic speech recognition, sound capture by microphone arrays and text to speech conversion
- Gaze and face tracking along with face verification for secure access
- language understanding and fusion of multimodal information (part of multimodal input management)
- Image understanding with region of interest detection and visual feature extraction
- Applications for collaborative work and design problems

As presently implemented, the current multimodal system incorporates the following components and characteristics:

- Force-feedback tactile input and gesture recognition
- Gaze tracking by a desk-mounted eye tracker, sound capture by a microphone array, automatic speech recognition, and a speech synthesizer
- Language understanding and fusion of multimodal information using a multimodal input manager

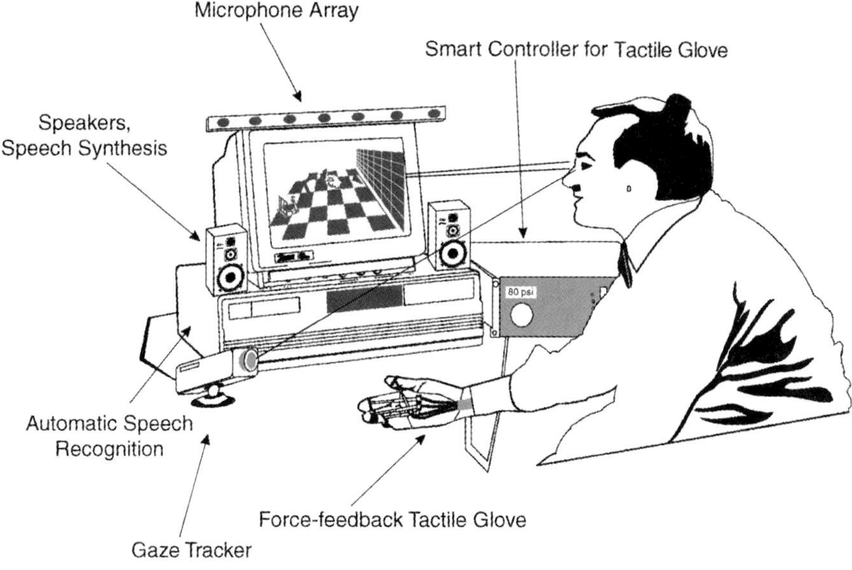


FIG. 9.2. CAIP Center multimodal system.

- Collaborative desktop used in an application requiring manipulation of 2-D and three-dimensional (3-D) graphical objects and icons on topographical maps (crisis management/disaster relief)
- Application for collaborative diagnosis of medical images

The integration of different input devices into one multimodal system also involves the integration of different platforms and languages. The contribution of gaze input to the multimodal system is to provide gaze directed hands-free visual communication. In the current implementation, gaze serves as a mouse-type pointer used in conjunction with the speech recognizer for issuing and carrying out commands like "Select it" and "Drop it." The gaze tracker used is a desk-mounted, nonintrusive ISCAN RK-726. It consists of a gimbal-mounted camera, an IR light source, and an ultrasonic depth sensor. The camera and IR source are used to compute the centroids of the pupil and the corneal reflection. The ultrasound sensor measures the distance of the eye from the screen to focus the camera lens automatically and to compute the visual angle subtended by the display screen.

The electronics for processing image data, carrying out tracking operations, and storing and transmitting data are mounted on a PCI bus circuit board installed in a Pentium personal computer (PC). This PC provides not only the coordinates of where the user is looking, but also the pupil diameter. It is possible to store gaze patterns and analyze them after the session. The gaze tracker PC is connected to a SUN Ultra workstation via a serial port connection. The main computer for the multimodal system is a Pentium Pro PC, which receives all multimodal inputs from the gaze tracker, tactile glove, and speech recognizer.

The main language used for the implementation of the multimodal system is Java because of its platform independence. An operating system (OS) level driver, Gaze Server, has been developed in the *C* programming language because of limited support for low-level programming in Java. The gaze server, which runs on the SUN Ultra workstation, reads the serial port from the eye tracker computer and writes gaze coordinates out to a socket. Since sockets are well supported in Java, a Java driver—Gaze Client—forms a TCP/IP (transmission control protocol/Internet protocol) connection with the gaze server to read the gaze coordinates. The coordinates of the eye position are then sent to the multimodal system after some preprocessing, which includes smoothing and translation of the coordinates to the appropriate screen location. The Gaze Handler in the modality manager and fusion agent uses these coordinates. The gaze handler, to give the user feedback on the accuracy of eye tracking, generates a gaze cursor on the screen. Similar OS level drivers, Java drivers, and modality handlers exist for all the independent modalities in the system.

One of the challenges faced in the use of the eye tracker as a mouselike pointing device is that the human eye does not move in the same calm and controlled fashion as the hand-controlled mouse. The eyes jump around from spot to spot. Special filters and averaging techniques have been developed to make the movements appear smooth and natural (Andre, 1998). These smoothing functions have the potential of interacting with the human's eye control, causing users to adjust their movements based on what they see is happening with the cursor on the screen.

Tracking the eye, determining gaze location on the monitor, and drawing the cursor on the monitor is carried out in real time. Although sockets are used for the connection, no noticeable delay is seen in the gaze cursor implementation for the multimodal application, though sometimes, due to high network traffic, system response time increases causing loss of eye position data and jumpiness of the gaze cursor. For this reason, a slightly different system integration approach has been used for human factors studies of the eye tracker as an input device. The human-factors testing program runs on a SUN workstation and reads directly from a serial port connection from the eye-tracker computer, thus avoiding network delays.

Calibration accuracy is another critical aspect of gaze tracking. Being 1 or 2 degrees off track is very distracting to the user and can result in mistaken commands. With desk-mounted eye trackers, the user is allowed to move the head but the degree of freedom is limited to about 1 cubic foot. The requirements for freedom of head movement and accuracy of tracking are obviously interdependent. Gaze tracking accuracy has been observed to be very sensitive to head movement, and it is very easy to lose calibration. One simple solution to this problem is to recalibrate. During experiments, it has been seen that calibration only takes about 10 seconds for an experienced user. This is a reasonably short period, so it is not a burden on the user or the application. However, for recalibration to be used effectively as a solution to the accuracy drift, the application has to provide specific commands to automate the process of calibration. In addition, data capturing is subject to human-perception inaccuracies. Although electronic performance characteristics of the eye tracker show that it is accurate to a fraction of a pixel, in practice, when humans are used to calibrate the system the accuracy is off by nearly as much as 5 pixels or 0.45 degrees.

Another problem faced in the implementation of the gaze tracker as a pointing device is that the presence of the gaze cursor causes what can be termed *Cursor drift*. This happens when the user is distracted by the presence of the gaze cursor on the screen, and starts tracking this cursor, thereby drifting further away from the target. If the eye tracker is always accurate, this problem is solved because the need for a gaze cursor disappears, improving the naturalness of the system.

For current eye-tracking systems, stable performance requires that all eye-tracker parameters (e.g., maximum and minimum pupil size, maximum and minimum corneal reflection size) have to be accurately adjusted for each user.

## 5. AN EVALUATION OF THE EFFECTIVENESS OF EYE TRACKING AS AN INPUT MECHANISM

Implied in the efforts to use gaze as an input medium is a belief that using the eye's positioning gives a user faster performance than other input devices. Certainly, eye saccades occur much quicker than hand motions or speech utterances used by more traditional input devices, but producing an effectively engineered gaze input device that actually takes advantage of the eye's speed may be difficult in practice. Early studies on the use of an eye tracker (Ware & Mikaelian, 1987) indicate that performance time is nearly one-half that of a mouse but with curious speed/accuracy trade-offs based on differences in the design of the target acceptance methods (pressing a button, dwelling on the target, etc.). Performance with any input device is dependent on differences in the design of the device not just the choice of input modality. For example, target acquisition times vary in a pointing task using a mouse depending on the shape and number of buttons on the mouse.

One can expect to uncover similar and often large differences when using eye-tracking hardware that varies in minor ways in the design of the user interface. In addition, the merging of gaze with a multimodal system with many types of input devices can also be expected to affect the overall performance of gaze input. We have therefore undertaken a series of experiments to capture the performance times of gaze in our multimodal setup for a simple target selection task. The next few paragraphs provide an overview of studies that were run and the results. The reader can find a more detailed presentation of these studies in Lin, Kaur, Tremaine, Hung, and Wilder (1999).

The experiment required users to position their eyes in a start circle until a target circle appeared at random positions on the screen. Users were then to saccade as rapidly as possible to the target circle and hold their eyes within the target until the software (200 ms) had recognized target acquisition. Users were first given a set of practice trials followed by 64 experiment trials. The direction of the target, target size, and distance to travel to the target were varied. A regression analysis was performed on the data to fit it to a log linear model based on target distance and target size. Such a model was developed for each of the four target directions used. Typical motor studies show that a user's performance is directly proportional to the distance traveled and inversely proportional to the target size, i.e.:

$$MT = a + b * \log_2 (2D/W) \text{ where}$$

$MT$ = movement time

$a$ and $b$ are constants dependent on the properties of the input device

$D$ = distance traveled

$W$ = width of the target

This model is known as Fitts' law (Fitts, 1966) and is the standard mechanism used to examine and compare performance characteristics of a multitude of computer input devices (Mackenzie, 1992). Because this model is based on an iterative succession of motor control signals that the brain sends out to guide motor function, it is not surprising that eye movement has been shown to follow the same model as other motor types of target acquisition tasks (Bahill & Stark, 1979). Fitting a target acquisition task to Fitts' law tells the input device designer many details about the user's performance and how the device design is affecting the performance. For example, it may be that a log linear model does not fit the data well and that a linear model does. This implies that the motion that the user is making is not ballistic (i.e., does not have an acceleration and deceleration component) and that the user is required to exercise more guided motor control throughout the task. This type of performance is seen with pull-down menus, where users need to keep the cursor within the confines of the menu. It may also be that the constant $a$ is very large. This implies that the device carries a cognitive load with it (e.g., the device may require the coordination of multiple muscle groups to start target acquisition, such as in using a glove to point to a target). Finally, the value $b$ may be high, indicating that the performance of the input device deteriorates quickly as either the distance traveled increases or the target size decreases. Such performance occurs, for example, when a user is required to do multiple motor tasks simultaneously (e.g., holding down a mouse button in order to drag an object to a target).

The data resulting from our experiment were examined to evaluate the ISCAN eye-tracker input device used. When the experimental data were fit to a log linear model, an extremely poor correlation of 30% was achieved. This correlation was so low compared to the 90% Fitts found in his motor tasks that the data were examined at a low level, looking at each point captured by the eye-tracking device instead of the start circle leaving and target circle arriving time. The data showed that fixations as well as saccades were being included in the measures, with the user making small fixations just outside the start circle and/or just before they reached the target circle. A velocity threshold of 13 cm/sec was used to distinguish saccades from fixations and the fixations were discarded as error data. Overall, eight types of errors that occurred in the experiment were categorized. These included:

- Overshoots: Participant leaves the target circle after entering it.
- Lost gaze position: Eye tracker lost the gaze position due to a hardware failure or because the participant looked out of the boundary of the screen.
- Tick larger than 100 milliseconds due to workstation slow response.
- No saccade at all (i.e., velocity never exceeds the threshold).
- Velocity stays above saccade threshold in the target circle.
- First fixation point outside start circle.
- Second fixation point outside end circle.
- Out of envelope.

"Out of envelope" referred to a "wandering eye" that moved to the target circle in a curved path, moving far outside an envelope of acceptable efficient movements. Removing these variations in the data improved the Fitts' law correlation to 65%. There was, however, wide variation among participants. The Fitts' law model does not fit gaze as well as other pointing devices because the data show that participants are not making single clean gaze movements as they do with other devices. The resulting model predicts that roughly, for distances of 10 cm, gaze is 1.7 times faster than recent mouse designs. For distances of 20 cm, gaze selection performance is 2.1 times faster. The performance advantage can be found in the speed of the saccades, not in the cognitive time it takes to begin the task. The values of $a$ for different mouse designs and gaze were comparable, as were the slopes of each performance model.

In summary, using gaze as an input mechanism is potentially quite promising. It is definitely faster than other input mechanisms, but its downside is that the eye, because of its dual usage, exhibits somewhat erratic performance. More detailed focus on eye behavior, such as using velocity thresholds to distinguish between saccades and fixations, is likely to improve the design of gaze input devices.

## 6. FUTURE DIRECTIONS

Approximately every 5 years a researcher investigates the use of eye tracking as an input device. Their conclusions are mixed about its viability because of the problems that are encountered in accurately interpreting data. Eye-tracking devices also improve significantly every 5 years so that both their cost and invasiveness are dropping. Our studies of gaze as an input mechanism, like prior studies, show large performance advantages for gaze. They also suggest that more considered design of input algorithms for processing eye-tracker data may circumvent problems facing the use of this mechanism as an input device. The combination of an eye tracker with other input modalities also indicates useful advantages that can be exploited in the interpretation of input from any one of the modalities.

## 7. CURRENT EYE-TRACKING RESEARCH IN VIRTUAL ENVIRONMENTS

Research in VEs is a growing and fluid field, so we hesitate to generate a list of research institutes or individuals that are conducting eye-tracking research in virtual environments. Also, given that this is a chapter in a book, no claims are made for such a list's validity over time. Below a set of key individuals and their research institutes currently involved in this work is provided. No claim is made either for the exhaustiveness or continued validity of this list. Individuals who are interested in obtaining further information on this area may want to look at the proceedings of the eye-tracking symposium that is jointly sponsored by ACM SIGCHI and ACM SIGGRAPH. The symposium takes place in late fall and the proceedings can be found in the Association for Computing Machinery's (ACM) digital library: http://www.acm.org/dl.

| *Researchers* | *Institute* |
|---|---|
| Colin Ware | Department of Computer Science<br>University of New Hampshire<br>United States |
| Robert Jacob | Department of Computer Science<br>Tufts University<br>United States |
| William Krebs | Operations Research Department<br>Naval Postgraduate School<br>United States |
| Robert S. Kennedy | RSK Assessments, Inc.<br>United States |
| Boris M. Velichovsky | Unit of Applied Cognitive Research<br>Dresden University of Technology<br>Germany |
| John Paulin Hanson | Risoe National Laboratory<br>Denmark |
| Roel Vertegaal | Department of Computer Science<br>Queens University<br>Canada |
| Jie Yang, Alex Waibel | College of Computing<br>Carnegie-Mellon University<br>United States |
| Andrew T. Duchowski | Department of Computer Science<br>Clemson University<br>United States |

## 8. REFERENCES

Andre, M. (1998). Multimodal human–computer interaction—The design and implementation of a prototype. Unpublished master's Thesis, Delft University of Technology, the Netherlands.

Bahill, T., & Stark, L. (1979). The trajectories of saccadic eye movements. *Scientific American, 240*(1), 108–117.

Cornsweet, T. N., & Crane, H. D. (1973). Accurate two-dimensional eye tracker using first and fourth Purkinje images. *Journal of the Optical Society of America, 63,* 921–928.

Eberhard, K. M., Spivey-Knowlton, M. J., Sedivy, J. C., & Tanenhaus, M. K. (1995). Eye movements as a window into real-time spoken language comprehension in natural contexts. *Journal of Psycholinguistic Research, 24*(6), 409–436.

Fitts, P. M. (1966). Cognitive aspects of information processing III: Set for speed versus accuracy. *Journal of Experimental Psychology, 71,* 849–857.

Flanagan, J., Kulikowski, C., Marsic, I., Burdea, G., Wilder, J., & Meer, P. (1999). Synergistic multimodal communication in collaborative multiuser environments. (Tech. Rep. NSF-STIMULATE). Rutgers University, CAIP Center.

Freedman, M. (1984). Eye-movement communication systems for computer control. In *Proceedings of the International Congress on Technology and Technology Exchange: Technology and the World Around Us* (pp. 34–35). Pittsburgh, PA: IEEE Press.

Jacob, R. J. K. (1991). The use of eye movements in human–computer interaction techniques: What you look at is what you get. *ACM Transactions on Information Systems, 9*(3), 152–169.

Kaiser, J. P. (1999). Sensory adaptation effects following exposure to a virtual environment. Unpublished master's thesis). Monterey, CA: Naval Postgraduate School.

Karat, C.-M., Halverson, C., Horn, D., & Karat, J. (1999). Patterns of entry and correction in large vocabulary continuous speech recognition system. In *Proceedings of the CHI '99 Conference on Human Factors in Computing Systems* (pp. 568–575). New York: Addison-Wesley/ACM Press.

Lin, W., Kaur, M., Tremaine, M., Hung, G., & Wilder, J. (1999). Performance analysis of an eye-tracker. In *Proceedings of the SPIE Conference on Machine Vision Systems for Inspection and Metrology VIII* (pp. 106–113).

Lohse, G. L., & Johnson, E. J. A.(1996). Comparison of two process-tracing methods for choice tasks. *Organizational Behavior and Human Decision Processes, 68*(1), 28–43.

Mackenzie, I. S. (1992). Fitt's law as a research and design tool in human–computer interaction. *Human–Computer Interaction, 7,* 91–139.

Megaw, E. D., & Richardson, J. (1979). Eye movements and industrial inspection. *Applied Ergonomics, 10,* 145–154.

Remmel, R. S. (1984). An inexpensive eye-movement monitor using the scleral search coil technique. *IEEE Trans. Biomed. Engin., 31,* 388–390.

Reuter, B. W., & Schenck, U. (1985). Data acquisition in visual screening of microscopical samples. In *Proceedings of Image Science '85 Acta Polytechnica Scandinavica, Applied Physics Series* (pp. 206–209).

Russo, J. E., & LeClerc, F. (1994). An eye-fixation analysis of choice processes for consumer nondurables. *Journal of Consumer Research, 21,* 274–290.

Sanders, M. G., Simmons, R. R., & Hoffman, M. A. (1979). Visual workload of the copilot/navigator during terrain flight. *Human Factors, 21*(3), 369–383.

Tanriverdi, V., & Jacob, R. J. K. (in press). Interacting with eye movements in virtual environments. In *Proceedings of the CHI 2000 Conference on Human Factors in Computing Systems,* New York: Addison-Wesley/ACM Press.

Velichkovsky, B. M., & Hansen, J. P. (1996). New technological windows into mind: There is more in eyes and brains for human–computer interaction. In *Proceedings of CHI '96: Human Factors in Computing Systems* (pp. 496–503). New York: ACM Press.

Ware, C., & Mikaelian, H. T. (1987). An evaluation of an eye tracker as a device for computer input. In *Proceedings of the ACM CHI+GI '87 Conference on Human Factors in Computing Systems* (pp. 183–188). New York: ACM Press.

Yamada, M., & Fukada, T. (1987). Eye word processor (EWP) and peripheral controller for ALS patient. In *Proceedings of IEEE, part A: Physical science, measurement and instrumentation, management and education, Reviews* (pp. 328–330).

Yang, J., Stiefelhagen, R., Meier, U., & Waibel, A. (1998). Visual tracking for multimodal human–computer interaction. In *Proceedings of CHI '98* (pp. 140–147).

Young, L. R., & Sheena, D. (1975). Methods and designs—Survey of eye-movement recording methods. *Behavior Research Methods and Instrumentation, 7*(5), 397–429.

Zhai, S., Morimoto, C., & Ihde, S. (1999). Manual and gaze input cascaded (MAGIC) pointing. In *Proceedings of the CHI '99 Conference on Human Factors in Computing Systems* (pp. 246–253). New York: Addison-Wesley/ACM Press.

# 10

# Gesture Recognition

Matthew Turk
*University of California*
*Computer Science Department*
*Santa Barbara, CA 93106*
*mturk@cs.ucsb.edu*

## 1. INTRODUCTION

A primary goal of virtual environments (VEs) is to provide natural, efficient, powerful, and flexible interaction. Gesture as an input modality can help meet these requirements. Human gestures are certainly natural and flexible, and may often be efficient and powerful, especially as compared with alternative interaction modes. This chapter will cover automatic gesture recognition, particularly computer vision based techniques that do not require the user to wear extra sensors, clothing, or equipment.

The traditional two-dimensional (2-D), keyboard- and mouse-oriented graphical user interface (GUI) is not well suited for virtual environments. Synthetic environments provide the opportunity to utilize several different sensing modalities and technologies and to integrate them into the user experience. Devices that sense body position and orientation, direction of gaze, speech and sound, facial expression, galvanic skin response, and other aspects of human behavior or state can be used to mediate communication between the human and the environment. Combinations of communication modalities and sensing devices can produce a wide range of unimodal and multimodal interface techniques. The potential for these techniques to support natural and powerful interfaces for communication in VEs appears promising.

If interaction technologies are overly obtrusive, awkward, or constraining, the user's experience with the synthetic environment is severely degraded. If the interaction itself draws attention to the technology rather than the task at hand, or imposes a high cognitive load on the user, it becomes a burden and an obstacle to a successful VE experience. Therefore, there is focused interest in technologies that are unobtrusive and passive.

To support gesture recognition, human position and movement must be tracked and interpreted in order to recognize semantically meaningful gestures. While tracking of a user's head position or hand configuration may be quite useful for directly controlling objects or inputting parameters, people naturally express communicative acts through higher level

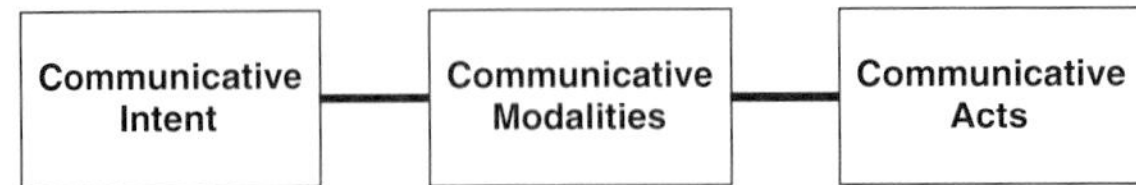


FIG. 10.1. Observable communication acts (such as gestures) are the result of expressing intent via communication modalities.

constructs, as shown schematically in Fig. 10.1. The output of position (and other) sensing must be interpreted to allow users to communicate more naturally and effortlessly through gesture.

Gesture is used for control and navigation in CAVEs (Cave Automatic Virtual Environments; Pavlovic et al., 1996; see chap. 11, this volume) and in other VEs, such as smart rooms, virtual work environments, and performance spaces. In addition, gesture may be perceived by the environment in order to be transmitted elsewhere (e.g., as a compression technique to be reconstructed at the receiver). Gesture recognition may also influence—intentionally or unintentionally—a system's model of the user's state. For example, a look of frustration may cause a system to slow down its presentation of information, or the urgency of a gesture may cause the system to speed up. Gesture may also be used as a communication *backchannel* (i.e., visual or verbal behaviors such as nodding or saying, "uh-huh" to indicate "I'm with you, continue," or raising a finger to indicate the desire to interrupt) to indicate agreement, participation, attention, conversation turn taking, and so forth.

Given that the human body can express a huge variety of gestures, what is appropriate to sense? Clearly the position and orientation of each body part—the parameters of an articulated body model—would be useful, as well as features that are derived from those measurements, such as velocity and acceleration. Facial expressions are very expressive. More subtle cues such as hand tension, overall muscle tension, locations of self-contact, and even pupil dilation may be of use.

Chapter 8 (this volume) covers technologies to track the head, hands, and body. These include instrumented gloves, body suits, and marker-based optical tracking. Most of the gesture recognition work applied to VEs has used these tracking technologies as input. Chapter 9 (this volume) covers eye-tracking devices and discusses their limitations in tracking gaze direction. This chapter covers interpretation of tracking data from such devices in order to recognize gestures. Additional attention is focused on passive sensing from cameras using computer vision techniques. The chapter concludes with suggestions for gesture-recognition system design.

## 2. THE NATURE OF GESTURE

Gestures are expressive, meaningful body motions—i.e., physical movements of the fingers, hands, arms, head, face, or body with the intent to convey information or interact with the environment. Cadoz (1994) described three functional roles of human gesture:

- Semiotic—to communicate meaningful information.
- Ergotic—to manipulate the environment.
- Epistemic—to discover the environment through tactile experience.

Gesture recognition is the process by which gestures made by the user are made known to the system. One could argue that in GUI-based systems, standard mouse and keyboard actions used for selecting items and issuing commands are gestures; here the interest is in less trivial cases. While static position (also referred to as posture, configuration, or pose) is not technically considered gesture, it is included for the purposes of this chapter.

In VEs, users need to communicate in a variety of ways, to the system itself and also to other users or remote environments. Communication tasks include specifying commands and/or parameters for:

- Navigating through a space
- Specifying items of interest
- Manipulating objects in the environment
- Changing object values
- Controlling virtual objects
- Issuing task-specific commands

In addition to user-initiated communication, a VE system may benefit from observing a user's behavior for purposes such as:

- Analysis of usability
- Analysis of user tasks
- Monitoring of changes in a user's state
- Better understanding a user's intent or emphasis
- Communicating user behavior to other users or environments

Messages can be expressed through gesture in many ways. For example, an emotion such as sadness can be communicated through facial expression, a lowered head position, relaxed muscles, and lethargic movement. Similarly, a gesture to indicate "Stop!" can be simply a raised hand with the palm facing forward or an exaggerated waving of both hands above the head. In general, there exists a many-to-one mapping from concept to gesture (i.e., gestures are ambiguous); there is also a many-to-one mapping from gesture to concept (i.e., gestures are not completely specified). And, like speech and handwriting, gestures vary among individuals, from instance to instance for a given individual, and are subject to the effects of coarticulation.

An interesting real-world example of the use of gestures in visual communications is a U.S. Army field manual (Anonymous, 1987) that serves as a reference and guide to commonly used visual signals, including hand and arm gestures for a variety of situations. The manual describes visual signals used to transmit standardized messages rapidly over short distances.

Despite the richness and complexity of gestural communication, researchers have made progress in beginning to understand and describe the nature of gesture. Kendon (1972) described a "gesture continuum," depicted in Fig. 10.2, defining five different kinds of gestures:

- *Gesticulation*—spontaneous movements of the hands and arms that accompany speech.
- *Languagelike gestures*—gesticulation that is integrated into a spoken utterance, replacing a particular spoken word or phrase.
- *Pantomimes*—gestures that depict objects or actions, with or without accompanying speech.
- *Emblems*—familiar gestures such as "V for victory," "thumbs up," and assorted rude gestures (these are often culturally specific).
- *Sign languages*—linguistic systems, such as American Sign Language, which are well defined.

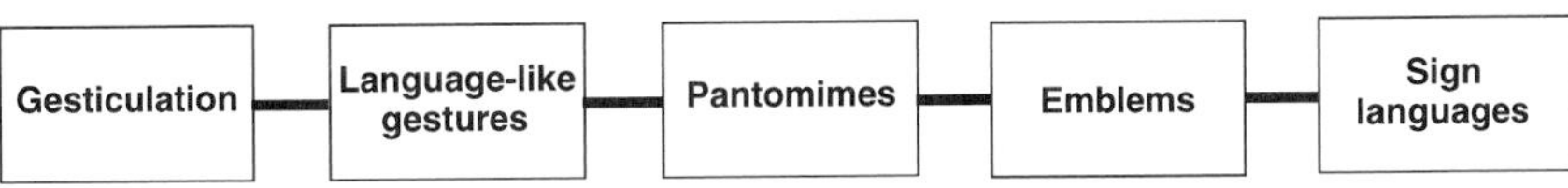


FIG. 10.2. Kendon's gesture continuum.

As the list progresses (moving from left to right in Fig. 10.2), the association with speech declines, language properties increase, spontaneity decreases, and social regulation increases.

Within the first category—spontaneous, speech-associated gesture—McNeill (1992) defined four gesture types:

- *Iconic*-representational gestures depicting some feature of the object, action, or event being described.
- *Metaphoric*—gestures that represent a common metaphor, rather than the object or event directly.
- *Beat*—small, formless gestures, often associated with word emphasis.
- *Deictic*—pointing gestures that refer to people, objects, or events in space or time.

These types of gesture modify the content of accompanying speech and may often help to disambiguate speech, similar to the role of spoken intonation. Cassell et al. (1994) describe a system that models the relationship between speech and gesture and generates interactive dialogs between three-dimensional (3-D) animated characters that gesture as they speak.

These spontaneous gestures (*gesticulation* in Kendon's continuum) make up some 90% of human gestures. People even gesture when they are on the telephone, and blind people regularly gesture when speaking to one another. Across cultures, speech-associated gesture is natural and common. For human–computer interaction (HCI) to be truly natural, technology to understand both speech and gesture together must be developed.

Despite the importance of this type of gesture in normal human-to-human interaction, most research to date in HCI, and most VE technology, focuses on the right side of Fig. 10.2, where gestures tend to be less ambiguous, less spontaneous and natural, more learned, and more culture-specific. Emblematic gestures and gestural languages, although perhaps less spontaneous and natural, carry more clear semantic meaning and may be more appropriate for the kinds of command-and-control interaction that VEs tend to support. The main exception to this is work in recognizing and integrating deictic (mainly pointing) gestures, beginning with the well-known "*Put That There*" system by Bolt (1980). The remainder of this chapter will focus on *symbolic gestures* (which includes emblematic gestures and predefined gesture languages) and *deictic gestures*.

## 3. REPRESENTATIONS OF GESTURE

The concept of gesture is loosely defined and depends on the context of the interaction. Recognition of natural, continuous gestures requires temporally segmenting gestures. Automatically segmenting gestures is difficult and is often finessed or ignored in current systems by requiring a starting position in time and/or space. Similar to this is the problem of distinguishing intentional gestures from other "random" movements. There is no standard way to do gesture recognition—a variety of representations and classification schemes are used. However, most gesture recognition systems share some common structure.

Gestures can be static, where the user assumes a certain pose or configuration, or dynamic, defined by movement. McNeill (1992) defines three phases of a dynamic gesture: prestroke, stroke, and poststroke. Some gestures have both static and dynamic elements, where the pose is important in one or more of the gesture phases; this is particularly relevant in sign languages. When gestures are produced continuously, each gesture is affected by the gesture that preceded it, and possibly by the gesture that follows it. These *coarticulations* may be taken into account as a system is trained.

There are several aspects of a gesture that may be relevant and therefore may need to be represented explicitly. Hummels and Stappers (1998) describe four aspects of a gesture which may be important to its meaning:

- Spatial information—where it occurs, locations a gesture refers to.
- Pathic information—the path that a gesture takes.
- Symbolic information—the sign that a gesture makes.
- Affective information—the emotional quality of a gesture.

In order to infer these aspects of gesture, human position, configuration, and movement must be sensed. This can be done directly with sensing devices such as magnetic field trackers, instrumented gloves, and datasuits, which are attached to the user, or indirectly using cameras and computer vision techniques. Each sensing technology differs along several dimensions, including accuracy, resolution, latency, range of motion, user comfort, and cost. The integration of multiple sensors in gesture recognition is a complex task, since each sensing technology varies along these dimensions. Although the output from these sensors can be used to directly control parameters such as navigation speed and direction or movement of a virtual object, here the interest is primarily in the interpretation of sensor data to recognize gestural information.

The output of initial sensor processing is a time-varying sequence of parameters describing positions, velocities, and angles of relevant body parts and features. These should (but often do not) include a representation of uncertainty that indicates limitations of the sensor and processing algorithms. Recognizing gestures from these parameters is a pattern recognition task that typically involves transforming input into the appropriate representation (feature space) and then classifying it from a database of predefined gesture representations, as shown in Fig. 10.3. The parameters produced by the sensors may be transformed into a global coordinate space, processed to produce sensor-independent features, or used directly in the classification step.

Because gestures are highly variable, from one person to another and from one example to another within a single person, it is essential to capture the essence of a gesture—its invariant properties—and use this to represent the gesture. Besides the choice of representation itself, a significant issue in building gesture recognition systems is how to create and update the database of known gestures. Hand-coding gestures to be recognized only works for trivial systems; in general, a system needs to be trained through some kind of learning. As with speech recognition systems, there is often a trade-off between accuracy and generality—the more accuracy desired, the more user-specific training is required. In addition, systems may be fully trained when in use, or they may adapt over time to the current user.

Static gesture, or pose, recognition can be accomplished by a straightforward implementation of Fig. 10.3, using template matching, geometric feature classification, neural networks,

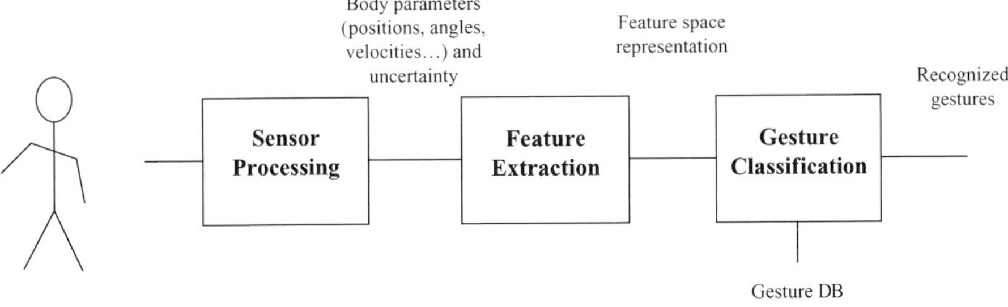


FIG. 10.3. Pattern recognition systems.

or other standard pattern recognition techniques to classify pose. Dynamic gesture recognition, however, requires consideration of temporal events. This is typically accomplished through the use of techniques such as time-compressing templates, dynamic time warping, hidden Markov models (HMMs), and Bayesian networks. Some examples will be presented in the following sections.

## 3.1 Pen-based Gesture Recognition

Recognizing gestures from 2-D input devices such as a pen or mouse has been considered for some time. The early Sketchpad system in 1963 (Johnson, 1963) used light-pen gestures, for example. Some commercial systems have used pen gestures since the 1970s. There are examples of gesture recognition for document editing, for air traffic control, and for design tasks such as editing splines. More recently, systems such as the OGI QuickSet system (Cohen et al. 1997) have demonstrated the utility of pen-based gesture recognition in concert with speech recognition to control a virtual environment. QuickSet recognizes 68 pen gestures, including map symbols, editing gestures, route indicators, area indicators, and taps. Oviatt (1996) has demonstrated significant benefits of using speech and pen gestures together in certain tasks. Zeleznik, Herndon, and Hughes (1996) and Landay and Myers (1995) developed interfaces that recognize gestures from pen-based sketching.

A significant benefit of pen-based gestural systems is that sensing and interpretation is relatively straightforward as compared with vision-based techniques. There have been commercially available personal digital assistants (PDAs) for several years, starting with the Apple Newton, and more recently the 3Com PalmPilot and various Windows CE devices. These PDAs perform handwriting recognition and allow users to invoke operations by various, albeit quite limited, pen gestures. Long, Landay, and Rowe (1998) survey problems and benefits of these gestural interfaces and provide insight for interface designers.

Although pen-based gesture recognition is promising for many HCI environments, it presumes the availability of, and proximity to, a flat surface or screen. In VEs this is often too constraining; techniques that allow the user to move around and interact in more natural ways are more compelling. The next two sections cover two primary technologies for gesture recognition in virtual environments: instrumented gloves and vision-based interfaces.

## 3.2 Tracker-based Gesture Recognition

There are a number of commercially available tracking systems (covered in chaps. 8 and 9), which can be used as input to gesture recognition, primarily for tracking eye gaze, hand configuration, and overall body position. Each sensor type has its strengths and weaknesses in the context of VE interaction. While eye gaze can be useful in a gestural interface, the focus here is on gestures based on input from tracking the hands and body.

### *3.2.1 Instrumented Gloves*

People naturally use their hands for a wide variety of manipulation and communication tasks. Besides being quite convenient, hands are extremely dexterous and expressive, with approximately 29 degrees of freedom (including the wrist). In his comprehensive thesis on whole-hand input, Sturman (1992) showed that the hand can be used as a sophisticated input and control device in a wide variety of application domains, providing real-time control of complex tasks with many degrees of freedom. He analyzed task characteristics and requirements, hand action capabilities, and device capabilities, and discussed important issues in developing whole-hand input techniques. Sturman suggested a taxonomy of whole-hand input that categorizes input techniques along two dimensions:

- Classes of hand actions: continuous or discrete.
- Interpretation of hand actions: direct, mapped, or symbolic.

The resulting six categories describe the styles of whole-hand input. A given interaction task can be evaluated as to which style best suits the task. Mulder (1996) presented an overview of hand gestures in human–computer interaction, discussing the classification of hand movement, standard hand gestures, and hand-gesture interface design.

For several years, commercial devices have been available which measure, to various degrees of precision, accuracy, and completeness, the position and configuration of the hand. These include "data gloves" and exoskeleton devices mounted on the hand and fingers (the term *instrumented glove* is used to include both types). Some advantages of instrumented gloves include:

- Direct measurement of hand and finger parameters (joint angles, 3-D spatial information, wrist rotation)
- Provide data at a high sampling frequency
- Easy to use
- No line-of-sight occlusion problems
- Relatively low cost versions available
- Data that is translation-independent (within the range of motion)

Disadvantages of instrumented gloves include:

- Calibration can be difficult.
- Tethered gloves reduce range of motion and comfort.
- Data from inexpensive systems can be very noisy.
- Accurate systems are expensive.
- The user is forced to wear a somewhat cumbersome device.

Many projects have used hand input from instrumented gloves for "point, reach, and grab" operations or more sophisticated gestural interfaces. Latoschik and Wachsmuth (1997) present a multiagent architecture for detecting pointing gestures in a multimedia application. Väänänen and Böhm (1992) developed a neural network system that recognized static gestures and allows the user to interactively teach new gestures to the system. Böhm et al. (1994) extend that work to dynamic gestures using a Kohohen Feature Map (KFM) for data reduction.

Baudel and Beaudouin-Lafon (1993) developed a system to provide gestural input to a computer while giving a presentation. This work included a gesture notation and set of guidelines for designing gestural command sets. Fels and Hinton (1995) used an adaptive neural network interface to translate hand gestures to speech. Kadous (1996) used glove input to recognize Australian sign language as did Takahashi and Kishino (1991) for the Japanese Kana manual alphabet. The system of Lee and Xu (1996) could learn and recognize new gestures online.

Despite the fact that many, if not most, gestures involve two hands, most of the research efforts in glove-based gesture recognition use only one glove for input. The features that are used for recognition, and the degree to which dynamic gestures are considered vary quite a bit.

The HIT Lab at the University of Washington developed GloveGRASP, a C/C++ class library that allows software developers to add gesture recognition capabilities to SGI systems, including user-dependent training and one- or two-handed gesture recognition. A commercial version of this system is available from General Reality.

### 3.2.2 Body Suits

It is well-known that by viewing only a small number of strategically placed dots on the human body, people can easily perceive complex movement patterns such as the activities, gestures, identities, and other aspects of bodies in motion. One way to approach the recognition of human movements and postures is to optically measure the 3-D position of several such markers attached to the body and then recover the time-varying articulated structure of the body. The articulated structure may also be measured more directly by sensing joint angles and positions using electromechanical body sensors. Although some of the optical systems only require dots or small balls to be placed on top of a user's clothing, all of these body motion capture systems are referred to herein generically as "body suits."

Body suits have advantages and disadvantages that are similar to those of instrumented gloves: they can provide reliable data at a high sampling rate (at least for electromagnetic devices), but they are expensive and very cumbersome. Calibration is typically nontrivial. The optical systems typically use several cameras and process their data offline, their major advantage is the lack of wires and a tether.

Body suits have been used, often along with instrumented gloves, in several gesture recognition systems. Wexelblat (1994) implemented a continuous gesture analysis system using a data suit, "data gloves," and an eye tracker. In this system, data from the sensors is segmented in time (between movement and inaction), key features are extracted, motion is analyzed, and a set of special-purpose gesture recognizers look for significant changes. Marrin and Picard (1998) have developed an instrumented jacket for an orchestral conductor that includes physiological monitoring to study the correlation between affect, gesture, and musical expression.

Although current optical and electromechanical tracking technologies are cumbersome and therefore contrary to the desire for more natural interfaces, it is likely that advances in sensor technology will enable a new generation of devices (including stationary field-sensing devices, gloves, watches, and rings) that are just as useful as current trackers but much less obtrusive. Similarly, instrumented body suits, which are currently exceedingly cumbersome, may be displaced by sensing technologies embedded in belts, shoes, eyeglasses, and even shirts and pants. While sensing technology has a long way to go to reach these ideals, passive sensing using computer vision techniques is beginning to make headway as a user-friendly interface technology.

Note that although some of the body tracking methods in this section use cameras and computer vision techniques to track joint or limb positions, they require the user to wear special markers. In the next section only passive techniques that do not require the user to wear any special markers or equipment are considered.

## 3.3 Passive Vision-based Gesture Recognition

The most significant disadvantage of the tracker-based systems in section 3.2 is that they are cumbersome. This detracts from the immersive nature of a VE by requiring the user to don an unnatural device that cannot easily be ignored and which often requires significant effort to put on and calibrate. Even optical systems with markers applied to the body suffer from these shortcomings, albeit not as severely. What many have wished for is a technology that provides real-time data useful for analyzing and recognizing human motion that is passive and nonobtrusive. Computer vision techniques have the potential to meet these requirements.

Vision-based interfaces use one or more cameras to capture images, at a frame rate of 30 Hz or more, and interpret those images to produce visual features that can be used to interpret human activity and recognize gestures. Typically the camera locations are fixed in the environment, although they may also be mounted on moving platforms or on other people. For the past decade, there has been a significant amount of research in the computer vision community on detecting and recognizing faces, analyzing facial expression, extracting lip and facial motion to aid speech recognition, interpreting human activity, and recognizing particular gestures.

Unlike sensors worn on the body, vision approaches to body tracking have to contend with occlusions. From the point of view of a given camera, there are always parts of the user's body that are occluded and therefore not visible, for erxample, the backside of the user is not visible when the camera is in front. More significantly, self-occlusion often prevents a full view of the fingers, hands, arms, and body from a single view. Multiple cameras can be used, but this adds correspondence and integration problems.

The occlusion problem makes full-body tracking difficult, if not impossible, without a strong model of body kinematics and perhaps dynamics. However, recovering all the parameters of body motion may not be a prerequisite for gesture recognition. The fact that people can recognize gestures leads to three possible conclusions: (1) the parameters that cannot be directly observed are inferred; (2) these parameters are not needed to accomplish the task; or (3) some are inferred and others are ignored.

It is a mistake to consider vision and tracking devices (such as instrumented gloves and body suits) as alternative paths to the same end. Although there is overlap in what they can provide, these technologies in general produce qualitatively and quantitatively different output which enable different analysis and interpretation. For example, tracking devices can in principle detect fast and subtle movements of the fingers while a user is waving his hands, whereas human vision in that case will at best get a general sense of the type of finger motion. Similarly, vision can use properties like texture and color in its analysis of gesture, whereas tracking devices do not. From a research perspective, these observations imply that it may not be an optimal strategy to merely substitute vision at a later date into a system that was developed to use an instrumented glove or a body suit—or vice versa.

Unlike special devices that measure human position and motion, vision uses a multipurpose sensor; the same device used to recognize gestures can be used to recognize other objects in the environment and also to transmit video for teleconferencing, surveillance, and other purposes. There is a growing interest in CMOS-based cameras, which promise miniaturized, low-cost, low-power cameras integrated with processing circuitry on a single chip. With its integrated processing, such a sensor could conceivably output motion or gesture parameters to the virtual environment.

Currently, most computer vision systems for recognition look something like Fig. 10.3. Analog cameras feed their signal into a digitizer board, or framegrabber, which may do a DMA transfer directly to host memory. Digital cameras bypass the analog-to-digital conversion and go straight to memory. There may be a preprocessing step, where images are normalized, enhanced, or transformed in some manner, and then a feature extraction step. The features—which may be any of a variety of 2-D or 3-D features, statistical properties, or estimated body parameters—are analyzed and classified as a particular gesture if appropriate.

Vision-based systems for gesture recognition vary along a number of dimensions, most notably:

- Number of cameras. How many cameras are used? If more than one, are they combined early (stereo) or late (multiview)?
- Speed and latency. Is the system real-time (i.e., fast enough, with low enough latency, to support interaction)?
- Structured environment. Are there restrictions on the background, lighting, speed of movement, and so forth?

- User requirements. Must the user wear anything special (e.g., markers, gloves, long sleeves)? Is anything disallowed (e.g., glasses, beard, rings)?
- Primary features. What low-level features are computed (edges, regions, silhouettes, moments, histograms, etc.)?
- Two- or three-dimensional representation. Does the system construct a 3-D model of the body part(s), or is classification done on some other (view-based) representation?
- Representation of time. How is the temporal aspect of gesture represented and used in recognition (e.g., via a state machine, dynamic time warping, HMMs, time-compressed template)?

### 3.3.1 Head and Face Gestures

When people interact with one another, they use an assortment of cues from the head and face to convey information. These gestures may be intentional or unintentional, they may be the primary communication mode or backchannels, and they can span the range from extremely subtle to highly exaggerated. Some examples of head and face gestures include:

- Nodding or shaking the head
- Direction of eye gaze
- Raising the eyebrows
- Opening the mouth to speak
- Winking
- Flaring the nostrils
- Looks of surprise, happiness, disgust, anger, sadness, etc.

People display a wide range of facial expressions. Ekman and Friesen (1978) developed a system called FACS for measuring facial movement and coding expression; this description forms the core representation for many facial expression analysis systems.

A real-time system to recognize actions of the head and facial features was developed by Zelinsky and Heinzmann (1996), who used feature template tracking in a Kalman filter framework to recognize thirteen head and face gestures. Moses, Reynard, and Blake (1995) used fast-contour tracking to determine facial expression from a mouth contour. Essa and Pentland (1997) used optical flow information with a physical muscle model of the face to produce accurate estimates of facial motion. This system was also used to generate spatiotemporal motion-energy templates of the whole face for each different expression. These templates were then used for expression recognition. Oliver, Pentland, and Bérard (1997) describe a real-time system for tracking the face and mouth that recognized facial expressions and head movements. Otsuka and Ohya (1998) model coarticulation in facial expressions and use an HMM for recognition.

Black and Yacoob (1995) used local parametric motion models to track and recognize both rigid and nonrigid facial motions. Demonstrations of this system show facial expressions being detected from television talk show guests and news anchors (in non–real time). La Cascia, Isidoro, and Sclaroff (1998) extended this approach using texture mapped surface models and nonplanar parameterized motion models to better capture the facial motion.

### 3.3.2 Hand and Arm Gestures

Hand and arm gestures receive the most attention among those who study gesture. In fact, many (if not most) references to gesture recognition only consider hand and arm gestures. The vast majority of automatic recognition systems are for deictic gestures (pointing), emblematic gestures (isolated signs), and sign languages (with a limited vocabulary and syntax). Some are components of bimodal systems, integrated with speech recognition. Some produce precise hand and arm configuration, whereas others only coarse motion.

Stark and Kohler (1995) developed the ZYKLOP system for recognizing hand poses and gestures in real time. After segmenting the hand from the background and extracting features such as shape moments and fingertip positions, the hand posture is classified. Temporal gesture recognition is then performed on the sequence of hand poses and their motion trajectory. A small number of hand poses comprises the gesture catalog, whereas a sequence of these makes a gesture. Similarly, Maggioni and Kämmerer (1998) described the GestureComputer, which recognized both hand gestures and head movements. Other systems that recognize hand postures amidst complex visual backgrounds are reported by Weng and Cui (1998) and Triesch and von der Malsburg (1996).

There has been a lot of interest in creating devices to automatically interpret various sign languages to aid the deaf community. One of the first to use computer vision without requiring the user to wear anything special was built by Starner and Pentland (1995), who used HMMs to recognize a limited vocabulary of ASL sentences. A more recent effort, which uses HMMs to recognize Sign Language of the Netherlands, is described by Assan and Grobel (1997).

The recognition of hand and arm gestures has been applied to entertainment applications. Freeman, Tanaka, Ohta, and Kyuma (1996) developed a real-time system to recognize hand poses using image moments and orientation histograms, and applied it to interactive video games. Cutler and Turk (1998) described a system for children to play virtual instruments and interact with lifelike characters by classifying measurements based on optical flow. A nice overview of work up to 1995 in hand gesture modeling, analysis, and synthesis is presented by Huang and Pavlovic (1995).

### *3.3.3 Body Gestures*

This section includes tracking full-body motion, recognizing body gestures, and recognizing human activity. Activity may be defined over a much longer period of time than what is normally considered a gesture; for example, two people meeting in an open area, stopping to talk, and then continuing on their way may be considered a recognizable activity. Bobick (1997) proposed a taxonomy of motion understanding in terms of:

- Movement—the atomic elements of motion.
- Activity—a sequence of movements or static configurations.
- Action—high-level description of what is happening in context.

Most research to date has focused on the first two levels.

The Pfinder system (Wren, Azarbayejani, Darrell, & Pentland, 1996) developed at the MIT Media Lab has been used by a number of groups to do body tracking and gesture recognition. It forms a 2-D representation of the body, using statistical models of color and shape. The body model provides an effective interface for applications such as video games, interpretive dance, navigation, and interaction with virtual characters. Lucente, Zwart, and George (1998) combined Pfinder with speech recognition in an interactive environment called Visualization Space, allowing a user to manipulate virtual objects and navigate through virtual worlds. Paradiso and Sparacino (1997) used Pfinder to create an interactive performance space where a dancer can generate music and graphics through their body movements, for example, hand and body gestures can trigger rhythmic and melodic changes in the music.

Systems that analyze human motion in VEs may be quite useful in medical rehabilitation (see chap. 49, this volume) and athletic and military training (see chap. 43, this volume). For example, a system like the one developed by Boyd and Little (1998) to recognize human gaits could potentially be used to evaluate rehabilitation progress. Yamamoto, Kondo, and Yamagiwa (1998) describe a system that used computer vision to analyze body motion in order to evaluate the performance of skiers.

Davis and Bobick (1997) used a view-based approach by representing and recognizing human action based on "temporal templates," where a single image template captures the recent history of motion. This technique was used in the KidsRoom system, an interactive, immersive, narrative environment for children. A nice online description of this project can be found at http://vismod.www.media.mit.edu/vismod/demos/kidsroom/.

Video surveillance and monitoring of human activity has received significant attention in recent years. For example, the $W^4$ system developed at the University of Maryland (Haritaoglu, Harwood, & Davis, 1998) tracks people and detects patterns of activity.

## 4. SUGGESTIONS FOR SYSTEMS DESIGN

There has been little work in evaluating the utility and usability of gesture recognition systems. However, those developing gestural systems have learned a number of lessons along the way. Here a few guidelines are presented in the form of "dos and don'ts" for gestural interface designers:

- **Do inform the user.** As discussed in section 2, people use different kinds of gestures for many purposes, from spontaneous gesticulation associated with speech to structured sign languages. Similarly, gesture may play a number of different roles in a virtual environment. To make compelling use of gesture, the types of gestures allowed and what they affect must be clear to the user.
- **Do give the user feedback.** Feedback is essential to let the user know when a gesture has been recognized. This could be inferred from the action taken by the system, when that action is obvious, or by more subtle visual or audible confirmation methods.
- **Do take advantage of the uniqueness of gesture.** Gesture is not just a substitute for a mouse or keyboard. It may not be as useful for 2-D pointing or text entry but great for more expressive input.
- **Do understand the benefits and limits of the particular technology.** For example, precise finger positions are better suited to instrumented gloves than vision-based techniques. Tethers from gloves or body suits may constrain the user's movement.
- **Do usability testing on the system.** Don't just rely on the designer's intuition (see chap. 34, this volume).
- **Do avoid temporal segmentation if feasible.** At least with the current state of the art, segmentation of gestures can be quite difficult.
- **Don't tire the user.** Gesture is seldom the primary mode of communication. When a user is forced to make frequent, awkward, or precise gestures, the user can become fatigued quickly. For example, holding one's arm in the air to make repeated hand gestures becomes tiring very quickly.
- **Don't make the gestures to be recognized too similar.** For ease of classification and to help the user.
- **Don't use gesture as a gimmick.** If something is better done with a mouse, keyboard, speech, or some other device or mode, use it—extraneous use of gesture should be avoided.
- **Don't increase the user's cognitive load.** Having to remember the whats, wheres, and hows of a gestural interface can make it oppressive to the user. The system's gestures should be as intuitive and simple as possible. The learning curve for a gestural interface

is more difficult than for a mouse and menu interface because it requires recall rather than just recognition among a list.

- **Don't require precise motion.** Especially when motioning in space with no tactile feedback, it is difficult to make highly accurate or repeatable gestures.
- **Don't create new, unnatural gestural languages.** If it is necessary to devise a new gesture language, make it as intuitive as possible.

## 5. CONCLUSIONS

Although several research efforts have been referenced in this chapter, these are just a sampling; many more have been omitted for the sake of brevity. Good sources for much of the work in gesture recognition can be found in the proceedings of the Gesture Workshops and the International Conference on Automatic Face and Gesture Recognition.

There is still much to be done before gestural interfaces, which track and recognize human activities, and can become pervasive and cost-effective for the masses. However, much progress has been made in the past decade and with the continuing march toward computers and sensors that are faster, smaller, and more ubiquitous, there is cause for optimism. As PDAs and pen-based computing continue to proliferate, pen-based 2-D gestures should become more common, and some of the technology will transfer to 3-D hand, head, and body gestural interfaces. Similarly, technology developed in surveillance and security areas will also find uses in gesture recognition for virtual environments.

There are many open questions in this area. There has been little activity in evaluating usability (see chap. 34, this volume) and understanding performance requirements and limitations of gestural interaction. Error rates are reported from 1% to 50%, depending on the difficulty and generality of the scenario. There are currently no common databases or metrics with which to compare research results. Can gesture recognition systems adapt to variations among individuals, or will extensive individual training be required? What about individual variation due to fatigue and other factors? How good do gesture recognition systems need to be to become truly useful in mass applications?

Each technology discussed in this chapter has its benefits and limitations. Devices that are worn or held—pens, gloves, body suits—are currently more advanced, as evidenced by the fact that there are many commercial products available. However, passive sensing (using cameras or other sensors) promises to be more powerful, more general, and less obtrusive than other technologies. It is likely that both camps will continue to improve and coexist, often be used together in systems, and that new sensing technologies will arise to give even more choice to VE developers.

## 6. REFERENCES

Anonymous (1987). Visual Signals. *U.S. Army Field Manual FM-2160* [online]. Available: http://155.217.58.58/atdls.html

Assan, M., & Grobel, K. (1997). Video-based sign language recognition using hidden Markov models. In I. Wachsmuth & M. Fröhlich (Eds.), *Gesture and sign language in human–computer interaction* (Proceedings of the International Gesture Workshop). Bielefeld, Germany: Springer-Verlag, Berlin.

Baudel, T., & Beaudouin-Lafon, M. (1993). CHARADE: remote control of objects using free-hand gestures. *Communications of the ACM, 36*(7), 28–35.

Black, M., & Yacoob, Y. (1995). Tracking and recognizing rigid and non-rigid facial motions using local parametric models of image motion. In *Proceedings of the International Conference on Computer vision* (pp. 374–381). Cambridge, MA:

Bobick, A. (1997). Movement, activity, and action: The role of knowledge in the perception of motion. *Royal Society Workshop on Knowledge-based Vision in Man and Machine*. London, England.

Böhm, K., Broll, W., & Solokewicz, M. (1994). Dynamic gesture recognition using neural networks: A fundament for advanced interaction construction. In S. Fisher, J. Merrit, & M. Bolan (Eds.), *Stereoscopic displays and virtual reality systems*. (SPIE Conference on Electronic Imaging Science and Technology, Vol. 2177). San Jose, CA:

Bolt, R. A. (1980). Put-That-There: Voice and gesture at the graphics interface. *Computer Graphics, 14*(3), 262–270.

Boyd, J., & Little, J. (1998). Shape of motion and the perception of human gaits. *IEEE Workshop on Empirical Evaluation Methods in Computer Vision*. Santa Barbara, CA, IEEE Computer Society Press.

Cadoz, C. (1994). *Les réalités virtuelles*. Dominos, Flammarion,

Cassell, J., Steedman, M., Badler, N., Pelachaud, C., Stone, M., Douville, B., Prevost, S., & Achorn, B. (1994). Modeling the interaction between speech and gesture. In *Proceedings of the Sixteenth Conference of the Cognitive Science Society*.

Cohen, P. R., Johnston, M., McGee, D., Oviatt, S., Pittman, J., Smith, I., Chen, L., & Clow, J. (1997). QuickSet: Multimodal interaction for distributed applications. In *Proceedings of the Fifth Annual International Multimodal Conference* (pp. 31–40). Seattle, WA: ACM Press.

Cutler, R., & Turk, M. (1998). View-based interpretation of real-time optical flow for gesture recognition. In *Proceedings of the Third International Conference on Automatic Face and Gesture Recognition*. Nara, Japan: IEEE Computer Society Press.

Davis, J., & Bobick, A. (1997). The representation and recognition of human movement using temporal trajectories. In *Proceedings of the IEEE Conference on Computer Vision and Pattern Recognition*. Puerto Rico, IEEE Computer Society Press.

Ekman, P., & Friesen, W. V. (1978). *Facial action coding system: A technique for the measurement of facial movement*. Palo Alto, CA: Consulting Psychologists Press.

Essa, I., & Pentland, A. (1997). Coding, analysis, interpretation and recognition of facial expressions. *IEEE Transactions on Pattern Analysis and Machine Intelligence* (Vol. 19, No. 7). IEEE Computer Society Press.

Fels, S., & Hinton, G. (1995). Glove-Talk II: An adaptive gesture-to-formant interface. *CHI '95*. Denver, CO.

Freeman, W., Tanaka, K., Ohta, J., & Kyuma, K. (1996). Computer vision for computer games. In *Proceedings of the Second International Conference on Automatic Face and Gesture Recognition*. Killington, VT: IEEE Computer Society Press.

Haritaoglu, I., Harwood, D., & Davis, L. (1998). W4: Who? When? Where? What? A real-time system for detecting and tracking people. In *Proceedings of the Third International Conference on Automatic Face and Gesture Recognition*. Nara, Japan: IEEE Computer Society Press.

Huang, T., & Pavlovic, V. (1995). Hand-gesture modeling, analysis, and synthesis. In *Proceedings of the International Workshop on Automatic Face- and Gesture-Recognition*. Zurich, Switzerland:

Hummels, C., & Stappers, P. (1998). Meaningful gestures for human–computer interaction: Beyond hand gestures. In *Proceedings of the Third International Conference on Automatic Face and Gesture Recognition*. Nara, Japan: IEEE Computer Society Press.

Johnson, T. (1963). Sketchpad III: Three-dimensional graphical communication with a digital computer. *AFIPS Spring Joint Computer Conference, 23*, 347–353.

Kadous, W. (1996). Computer recognition of Auslan signs with PowerGloves. In *Proceedings of the Workshop on the Integration of Gesture in Language and Speech,* Wilmington, DE:

Kendon, A. (1972). Some relationships between body motion and speech. In A. W. Siegman and B. Pope (Eds.), *Studies in dyadic communication*. New York: Pergamon Press.

La Cascia, M., Isidoro, J., & Sclaroff, S. (1998). Head tracking via robust registration in texture map images. In *Proceedings of the IEEE Conference on Computer Vision and Pattern Recognition*. Santa Barbara, CA: IEEE Computer Society Press.

Landay, J. A., & Myers, B. A. (1995). Interactive sketching for the early stages of user interface design. In *Proceedings of CHI '95* (pp. 43–50).

Latoschik, M., & Wachsmuth, I. (1997). Exploiting distant pointing gestures for object selection in a virtual environment. In I. Wachsmuth & M. Fröhlich (Eds.), *Gesture and sign language in human–computer interaction*. Bielefeld, Germany: International Gesture Workshop.

Lee, C., & Xu, Y. (1996). Online, interactive learning of gestures for human/robot interfaces. In *1996 IEEE International Conference on Robotics and Automation* (Vol. 4, pp. 2982–2987). Minneapolis, MN: IEEE Computer Society Press.

Long, A., Landay, J., & Rowe, L. (1998).*PDA and gesture uses in practice: Insights for designers of pen-based user interfaces* (Report CSD-97-976). Berkeley, CA: University of California, Berkeley, CS Division, EECS Department.

Lucente, M., Zwart, G., & George, A. (1998). *Visualization space: A testbed for deviceless multimodal user interface*. Intelligent Environments Symposium. Stanford, CA: AAAI Spring Symposium Series.

Maggioni, C., & Kämmerer, B. (1998). GestureComputer—history, design and applications. In R. Cipolla & A. Pentland (Eds.), *Computer vision for human–machine interaction.* Cambridge University Press, Cambridge, U.K.

Marrin, T., & Picard, R. (1998). The conductor's jacket: A testbed for research on gestural and affective expression. *Twelth Colloquium for Musical Informatics,* Gorizia, Italy.

McNeill. D. (1992). *Hand and mind: What gestures reveal about thought.* Chicago: University of Chicago Press.

Moses, Y., Reynard, D., & Blake, A. (1995). Determining facial expressions in real time. In *Proceedings of the Fifth International Conference on Computer Vision.* Cambridge, MA: IEEE Computer Society Press.

Mulder, A. (1996). Hand gestures for HCI (Tech. Rep. No. 96-1). Simon Fraser University, School of Kinesiology.

Oliver, N., Pentland, A., & Bérard, F. (1997). LAFTER: Lips and face real-time tracker. In *Proceedings of the IEEE Conference on Computer Vision and Pattern Recognition.* Puerto Rico: IEEE Computer Society Press.

Otsuka, T., & Ohya, J. (1998). Recognizing abruptly changing facial expressions from time-sequential face images. In *Proceedings of the IEEE Conference on Computer Vision and Pattern Recognition.* Santa Barbara, CA: IEEE Computer Society Press.

Oviatt, S. L. (1996). Multimodal interfaces for dynamic interactive maps. In *Proceedings of the CHI '96 Human Factors in Computing Systems* (pp. 95–102). New York: ACM Press.

Paradiso, J., & Sparacino, F. (1997). Optical tracking for music and dance Performance. *Fourth Conference on Optical 3-D Measurement Techniques.* Zurich, Switzerland.

Pavlovic. V, Sharma, R., & Huang, T. (1996). Gestural interface to a visual computing environment for molecular biologists. In *Proceedings of the Second International Conference on Automatic Face and Gesture Recognition.* Killington, VT: IEEE Computer Society Press.

Stark, M., & Kohler, M. (1995). Video-based gesture recognition for human–computer interaction. In W. D. Fellner (Ed.), *Modeling—Virtual Worlds—Distributed Graphics.*

Starner, T., & Pentland, A. (1995). Visual recognition of American Sign Language using hidden Markov models. In *Proceedings of the International Workshop on Automatic Face- and Gesture-Recognition.* Zurich, Switzerland:

Sturman, J. (1992). *Whole-hand Input.* Unpublished doctoral dissertation, MIT Media Labortory.

Takahashi, T., & Kishino, F. (1991). Gesture coding based in experiments with a hand-gesture interface device. *SIGCHI Bulletin, 23*(2), 67–73.

Triesch, J., & von der Malsburg, C. (1996). Robust classification of hand postures against complex backgrounds. In *Proceedings of the Second International Conference on Automatic Face and Gesture Recognition.* Killington, VT: IEEE Computer Society Press.

Väänänen, K., & Böhm, K. (1992). Gesture-driven interaction as a human factor in virtual environments—An approach with neural networks. In *Proceedings of the Virtual Reality Systems.* British Computer Society, Academic Press.

Weng, J., & Cui, Y. (1998). Recognition of hand signs from complex backgrounds. In R. Cipolla and A. Pentland (Eds.), *Computer vision for human–machine interaction.* Cambridge University Press.

Wexelblat, A. (1994). A feature-based approach to continuous-gesture analysis. *Unpublished master's thesis, MIT Media Labortory.*

Wren, C., Azarbayejani, A., Darrell, T., & Pentland, A. (1996). Pfinder: Real-time tracking of the human body. In *Proceedings of the Second International Conference on Automatic Face and Gesture Recognition.* Killington, VT: IEEE Computer Society Press.

Yamamoto, J., Kondo, T., Yamagiwa, T., & Yamanaka, K. (1998). Skill recognition. In *Proceedings of the Third International Conference on Automatic Face and Gesture Recognition.* Nara, Japan: IEEE Computer Society Press.

Zeleznik, R. C., Herndon, K. P., & Hughes J. F. (1996). Sketch: An interface for sketching 3D scenes. Proceedings of SIGGRAPH 96, Computer Graphics Proceedings, Annual Conference Series, pp. 163–170 (August 1996, New Orleans, Louisiana). Addison Wesley. Edited by Holly Rushmeier.

Zelinsky, A., & Heinzmann, J. (1996). Real-time visual recognition of facial gestures for human–computer interaction. In *Proceedings of the Second International Conference on Automatic Face and Gesture Recognition.* Killington, VT: IEEE Computer Society Press.

# 11

# Locomotion Interfaces

John M. Hollerbach
*University of Utah*
*School of Computing*
*50 S. Central Campus Drive, Rm. 3190*
*Salt Lake City, Utah 84112-9205*
*jmh@cs.utah.edu*

## 1. INTRODUCTION

A *motion interface* is the means by which a user travels through a virtual environment (VE). Motion interfaces are categorized as active or passive (Durlach & Mavor, 1995). Active motion interfaces are defined as *locomotion interfaces*, which require self-propulsion by a user. Passive motion interfaces transport a user through the VE without significant user exertion. They are further subdivided as inertial motion interfaces, where the body is moved as in a flight simulator on a Stewart platform, and noninertial motion interfaces, where the body is stationary as in the use of a joystick.

### 1.1 Features of Locomotion Interfaces

Ordinarily the use of our legs in walking is thought of as the meaning of locomotion, but a human-powered vehicle like a bicycle may be involved. Similarly, treadmills and stationary bicycles have served as locomotion interfaces. A locomotion interface approach that does not involve a physical device is walking in place. For greater generality, a more encompassing view of locomotion interface is taken to include any leg and arm combination for user self-propulsion, for example, rowing machines and Nordic skiing machines.

The key feature of locomotion interfaces that distinguishes them from passive motion interfaces is repetitive limb motion. Repetitive limb motion, or *gait*, results in self-propulsion. With such interfaces, tracking of a user's legs or body motion controls motion through a virtual environment. Due to repetitive limb motion, energy expenditure in gait is a key feature of locomotion interfaces. Some issues involved with these interfaces include:

- The effort in cycling the limbs is a basic physiological load.
- Body weight may have to be supported if the user is standing.

- The motion platform itself may require effort tied to a user's speed of motion. For example, a stationary bicycle may have a flywheel to simulate inertia and vanes to simulate viscous drag.

The energy extraction from the user by the locomotion interface can cause fatigue and affect decision-making processes, just as in the real world, where humans have to balance movement effort against movement goal.

Another feature of locomotion interfaces is the integration of proprioception with vision. Vision operates better for relative depth perception than for absolute distance judgments (see chap. 3, this volume). It has been shown that locomotion calibrates visual distance judgments (Reiser, Pick, Jr., Ashmead, & Garing, 1995). The implication is that appreciation of VE geometry and distance is enhanced by the ability to locomote. The features of energy extraction and sensorimotor integration are hypothesized to yield an increased sense of presence in the virtual environment (see chap. 40, this volume).

### 1.2 Workspace and Position Versus Rate Control

In passive-motion interfaces, a user is standing or seated, and manipulates some control or makes some motion that changes the user's position in the virtual environment. The amount of user motion typically controls the rate of moving through the virtual environment. An example is a joystick; the deflection angle from center is the rate command. Another example is a driving simulator; the amount by which the pedal is depressed controls the velocity of the vehicle. Rate control arises because the workspace of the control device is much smaller than the virtual space through which the user wishes to move. Because of the use of rate control rather than position control, repetitive motions are not required by the user to move through a virtual environment. Consequently, the user expends very little energy.

If such a device were used in position control mode, then the user would have to reindex to the center of the workspace before making the next excursion to move forward. This repeated cycling would according to our definition result in a locomotion interface. For example, repeated cycling of a joystick under position control would be like turning a crank. For haptic interfaces, the device's workspace directly matches that of the VE, and so position control may be used. Repeated cycling is not required because the workspaces match (scaling may be involved). More sophisticated controllers may use position control in central regions of the device's workspace for precise positioning and rate control near the workspace boundaries for slewing across larger distances.

In summary, the control mode and relation of the device workspace to the VE workspace leads to an approximate taxonomy for mechanical interfaces to VE systems, including:

- Rate control is the key feature of passive motion interfaces.
- Position control, where the device and VE workspaces match is the key feature of haptic interfaces.
- Position control with repeated cycling of the device (the *gait*) to cover the VE workspace is the key feature of locomotion interfaces.

### 1.3 Locomotion rendering

In the following section, different types of locomotion interfaces are surveyed that have been built to date. Notwithstanding the taxonomy and attempt at generality above, most of these locomotion interfaces involve walking while a few involve cycling.

Locomotion rendering may be defined similarly to haptic rendering, as the presentation of mechanical stimuli to simulate normal locomotion. Aspects of locomotion that might be rendered include the following:

1. *Forward motion.* Ideally, a user should be able to walk or run forward or backward at any speed. Some devices limit forward motion to a slow or moderate walk and constrain the step size. Other locomotion interfaces based on walking in place don't actually involve physical forward motion. When accelerating to a run, a user does not typically experience an inertial force because the locomotion interface keeps the body stationary with respect to the ground. One such device is capable of providing an artificial inertial force display.
2. *Turning.* Ideally, a user should be able to change the direction of walking arbitrarily. Some locomotion interfaces limit the range of turning, while other devices, which are essentially linear, use some control action to initiate turning that is somewhat artificial.
3. *Slope.* Arbitrary aspects of slope would be ideally displayed, such as walking on smooth slopes, including sideways traversal and uneven terrain such as stairs. Locomotion interfaces differ greatly in whether they address slope or to what extent.
4. *Obstacles.* Humans can't walk through real walls nor should they be able to walk through virtual walls. Wall constraints could be enforced merely by braking the locomotion interface, but certain devices display hitting a wall more directly to the body.
5. *Body postures.* Especially for walking simulators, there is an issue of the wide variety of postures and stepping patterns that one might adopt, such as crawling or sidling. What postures a locomotion interface can accommodate differ among the designs.

These aspects of locomotion rendering will serve to structure the discussion of locomotion interfaces below.

## 2. LOCOMOTION INTERFACES

A number of locomotion interfaces are derived from exercise machines, such as stationary bicycles, stair steppers, and treadmills. Others have been specifically designed toward a particular locomotion interface approach. The following is a taxonomy of locomotion interfaces along with a number of examples.

### 2.1 Pedaling Devices

Stationary bicycles represent a straightforward way to fashion a locomotion interface. Pedaling resistance can be achieved passively, by means of a friction brake, or more realistically by a flywheel and vanes to simulate inertia and viscosity. Simulating slope requires an electric motor to modulate the pedaling effort. Turning control is naturally achieved with handlebars. Position sensors on the pedals and on the handlebars measure the linear motion and direction of travel.

An elaboration is to place the bicycle on a motion platform. Brogan, Metoyer, and Hodgins (1998) employed a tilting platform with an up–down range of +/− 12 deg to represent hills for a racing bicycle simulator (Fig. 11.1A). The Sarcos Uniport is a unicycle on a turntable (Fig. 11.1B). The unicycle turns left or right based on user exertion against the seat, measured by load sensors. The user is therefore obtaining the appropriate vestibular cues regarding direction.

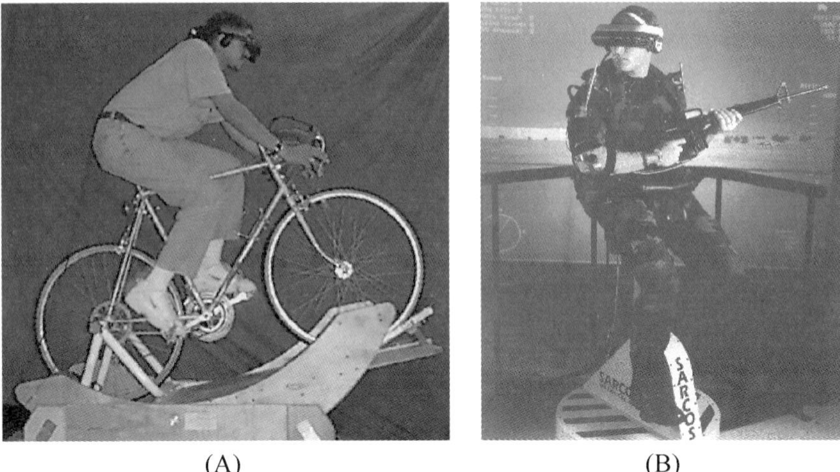


(A) (B)

FIG. 11.1. (A) Bicycle simulator on tilt platform. Photo courtesy of D. Brogan, R. Metoyer, & J. Hodgins, College of Computing, Georgia Institute of Technology. (B) The Sarcos Uniport.

## 2.2 Walking in Place

A motion platform may be entirely avoided by measuring walking in place. A different set of muscles is employed versus walking forward, but an advantage is a potentially lower cost system.

The Gaiter system (Templeman, Denbrook, & Sibert, 1999) employs magnetic trackers attached to the thighs just above the knees and force sensors in the foot pads (Fig. 11.2). Forward motion and turning are controlled by mapping knee height, rate, and direction. The foot pad sensors aid in segmenting the steps. Magnetic trackers are also located at the waist, head, and a handgrip. The waist sensor controls the position and orientation of the body; the head sensor controls the gaze direction in conjunction with a head-mounted daire (HMD) and the handgrip sensor is for auxiliary purposes.

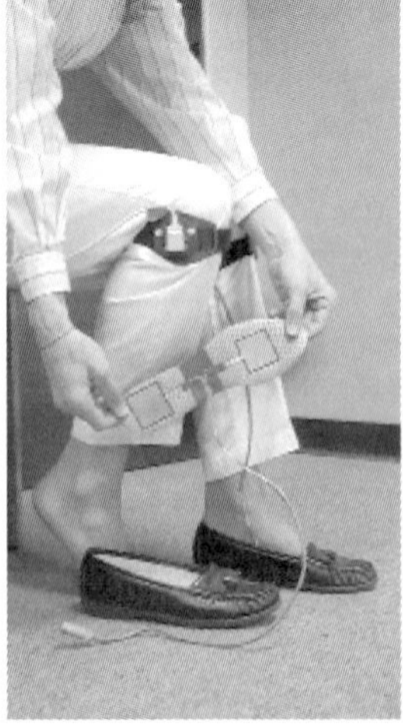

FIG. 11.2. The Gaiter system. Photo courtesy of J. Templeman.

Gestural knee actions, which are defined as excess motions that do not participate in physical displacement of the body, control the rate and direction of motion. Rocking the knee back then forward indicates walking backward. In addition to physically turning, it is possible to indicate turning while walking straight ahead and rocking the knees to the side.

In an alternative approach to a walking-in-place locomotion interface, Parsons et al. (1998) employ mechanical trackers on the ankles to trigger a step when the foot is raised enough. Slater, Steed, and Usoh (1995) inferred the footsteps by employing a neural network classifier on head bobbing as measured by a magnetic tracker; error rates of 10% were reported on correctly recognizing a step. They employ hand gestures to indicate going up or down a ladder.

With the foot simply being raised up and down, there is of course no sense of forward propulsion when walking in place. To give a sense of backward foot motion as if moving forward, one may use sliding between the feet and the ground. A series of different locomotion interface approaches by Iwata are summarized in Iwata (2000). One approach employed roller skates. The user was rigidly attached to a frame by a belt around the waist to absorb the forward reaction forces and to provide stability. Foot forces for climbing or descending were created by a single string belt for each foot, routed through pulleys to the bottom of each foot, and actuated by a DC motor. Subsequently, Iwata employed shoes with low-friction films on the soles. A brake pad on the toe allowed completion of the swing phase. The user was loosely constrained by a hoop frame. Grant and Magee (1997) utilized the amount of horizontal foot motion in a sliding-surface locomotion interface to drive the extent of forward motion.

## 2.3 Foot Platforms

A derivative of a stair-stepper exercise machine is individually programmable foot platforms, where each platform can be positioned in three dimensions. The Sarcos Biport (Fig. 10.3A) employs hydraulically actuated 3-degree-of-freedom (DOF) serial-link arms on which the user stands. The user's feet are attached to the platforms with releasable bindings. Force sensors are located near the attachment points and are employed in force control strategies and steering control. When the user lifts a foot, the attached arm must follow with zero force to avoid dragging the foot. When the user steps to contact a surface, the arm must be servoed to present a rigid surface.

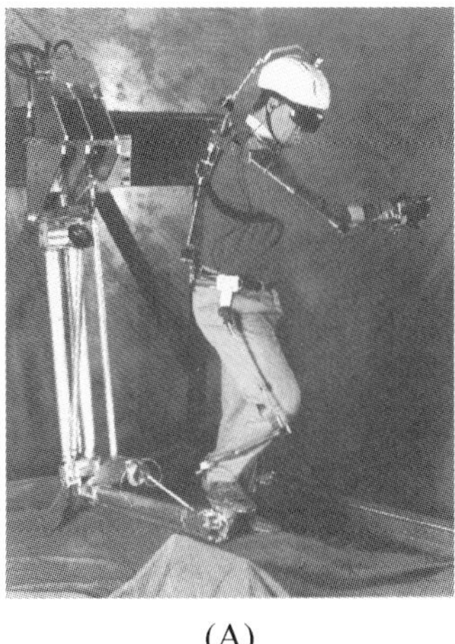

(A)

(B)

FIG. 11.3. (A) The Sarcos Biport. (B) The GaitMaster. Photo courtesy of H. Iwata.

A reverse centering motion is superimposed on the forward motion to attract the user back toward the center of the device. A foot mount permits the user to swivel left or right, and force sensors in the mounts signal the direction of motion. The platforms are servoed to follow the user's steps in a particular direction of motion. Moderate walking speeds can be supported. Speed limitations occur due to the structural stiffness of the present design. In addition to the presentation of uneven terrain, another advantage of such a device is the potential for simulating soft surfaces.

The Sarcos Biport's hydraulic arms are necessarily powerful to support the forces of walking. Safety is a major concern. A variety of safety devices have been implemented, including ceiling restraints, releasable bindings, and user-activated kill switches.

Iwata's GaitMaster (Fig. 10.3B) comprises two 3-DOF parallel drive platforms (Iwata & Yoshida, 1999). Each platform contains a passive spring-loaded yaw joint to allow some turning. The payload of each platform is 150 kg. A three-joint goniometer attaches a user's foot to the platform to measure foot motion.

The platforms are mounted on a turntable to accommodate turning. The platforms primarily accommodate forward or backward motion. The intent of this design is to avoid platform interference when stepping to walk in a sideways direction. This type of interference is a potential problem with the Sarcos Biport. Conversely, some modest oblique walking angles are possible with the Biport, which lead to more natural walking.

## 2.4 Treadmills

Initial research employed passive treadmills, where the belt motion is entirely caused by users pushing with their legs. The forward motion is directly indicated by the belt motion. The user has to brace against side bars or handle bars when pushing against the belt; handlebars allow turning control to be easily implemented. Passive treadmills are a low-cost and inherently safe platform, although the naturalness of locomotion is compromised. More recently, active treadmills have been employed where the belt speed is controlled by measurements of user position, either by optical or mechanical tracking.

### *2.4.1 The Sarcos Treadport*

The original Sarcos Treadport is comprised of a large 4-by-8 foot treadmill and a 6-axis mechanical tether that acts as a position tracker (Christensen, Hollerbach, Xu, & Meek, 2000). The tether is actuated along its linear axis to be able to push and pull on the user. A second-generation Sarcos Treadport has been constructed (Hollerbach, Xu, Christensen, & Jacobsen, 2000) featuring a larger running surface ($6 \times 10$ feet) and a fast tilt mechanism. Figure 10.4 shows the new Treadport in conjunction with a 3-wall CAVE (Cave Automatic Virtual Environment).

The Treadport's tether attaches to a user at the back of a body harness. The belt velocity is proportional to how far forward from center a user moves. Walking backward is also possible. An integral control term is added that gradually recenters the user to avoid the user getting too close to the front edge (Christensen, Hollerbach, Xu, & Meek, 2000). Natural forward motion speeds are supported: accelerations of 1 g and peak velocities of 12 mph. Due to the belt size, a variety of body postures can be supported including crouching and crawling.

Because the Treadport uses a linear treadmill, the issue of how to control turning arises. Body pose measurements from the mechanical tether are employed to control the rate of turning. Two control regimes are used. For stationary users, the amount of twist about the vertical axis controls the rate of turning. For rapidly walking or running users, the amount by which the user is displaced sideways from the treadmill center controls the rate of turning. To accommodate intermediate locomotion speeds, the two control regimes are blended. The use of rate control

FIG. 11.4. The second-generation Sarcos Treadport.

requires reindexing. The user has to move back to a center position to stop turning, then moves the opposite direction from center to turn the other way.

Treadmills can represent slope naturally by means of a platform tilt mechanism. Due to the platform mass and the slow actuation in commercial treadmills, the tilt cannot represent sudden slope changes. In the first-generation Sarcos Treadport, for example, the tilt rate is approximately 1 deg/sec. While this may suffice for an exercise function, the slow tilt rate is a limitation for virtual environments. The second-generation Treadport is specified for $+/-$ 20 deg in one second, although the tilt is not currently operational.

An alternative is provided by the active mechanical tether on the Sarcos Treadport. The tether can push or pull along its linear axis at the small of the back in a roughly horizontal direction. When walking uphill, the component $f$ of gravity force parallel to the slope opposing the user's motion is $f = m\ g\ sin\ \theta$, where $m$ is the user's mass, $g$ is gravity, and $\theta$ is the slope angle. This extra effort in going uphill can be simulated by the tether pulling on the user with a force $f = m\ g\ sin\ \theta$, requiring a greater user effort. For a user going downhill, the tether would push with that force, thereby requiring a lesser effort. User studies show the effectiveness of the mechanical tether force for slope display, but the empirical relation between walking on a slope and user-perceived equivalent pulling force was $f = 0.65\ m\ g\ sin\ \theta$, or two thirds of the expected force (Tristano, Hollerbach, & Christensen, 2000). It was hypothesized that the concentrated force application to just one point on the body rather than a distributed gravity load over the body was responsible for the fractional force perception.

When accelerating on treadmills, it has been found that runners exert 35% less energy than on the ground. The difference has been ascribed to an inertial force difference on the treadmill, essentially because the treadmill belt is servoed to keep the user's body stationary with respect to the ground. One approach is to vary the belt speed to decrease the difference from ground locomotion (Moghaddam & Buehler, 1993), but this difference can never be completely eliminated with this approach. Supposing a highly simplified dynamic model of the human as a cart with mass $m$, it would be predicted that the missing inertial force should be $f = ma$, where $a$ is the acceleration. Christensen et al. (2000) utilized the active tether of the Sarcos Treadport to provide this simulated inertial force. User studies showed a preference for such an inertial force display over conditions of no tether force, or of a springlike tether force. A fractional perception of roughly $f = 0.8\ ma$ was found, possibly due again to the lack of distributed force application over the body.

In the real world one cannot walk through walls, nor should one be able to walk through virtual walls. One alternative is to simply brake the locomotion interface, although the user might then stumble forward. The active tether of the Sarcos Treadport provides an alternative by applying a springlike penalty force to simulate hitting a wall. The treadmill is stopped simultaneously, but due to the active tether force the user does not stumble forward. This penalty force is similar to viscoelastic opposing forces applied by haptic interfaces when a user attempts to push into a hard surface.

Finally, it is desired to keep users away from workspace boundaries. When a user accelerates on a treadmill, it is possible to move off center because of limited bandwidth of the belt. Of course, it is not desirable to have users running off the end of a platform. In the Sarcos Treadport, an integral control term is added to draw the user back toward the center of the platform (Christensen et al., 2000). A software linear spring is also simulated by the active mechanical tether to provide a kinesthetic cue to the user about the amount of forward deviation from center. There are hard limit stops on the tether, which prevent movement beyond the front edge.

Another concern is sideways motion on the treadmill platform and the user potentially falling off the side. Hardware springs are provided on a base rotary joint and an attachment-end rotary joint on the mechanical tether of the Sarcos Treadport to provide kinesthetic cues about the amount of sideways deviation. The second-generation Sarcos Treadport has a large platform ($6 \times 10$ feet) primarily to provide extra safety margins for sideways or forward motion without falling off. In addition, crawling is facilitated. Another reason for a larger belt is the use of a dead zone around a center position on the belt. If the user is stationary, small motions should not cause the belt to move; otherwise it would be impossible to stand still. While a dead zone prevents this, the dead zone does remove some useable area of the belt.

### *2.4.2 ATR Locomotion Interfaces*

Two treadmill-style locomotion interfaces have been constructed at ATR by Noma, Sugihara, and Miyasato (2000). The ATLAS system (Fig. 10.5A) comprises a linear treadmill on a spherical joint that can pitch, yaw, and roll the platform. The pitching motion is the normal tilt motion found in many commercial treadmills. The yaw motion acts like a turntable to move the treadmill left and right, and is used for turning control. The roll motion allows the simulation of walking sideways along a slope. The belt area is 145 mm long and 55 mm wide, and the maximum belt speed is 4.0 m/sec. The yaw motion has a maximum rate of 1 rad/sec.

Motion control is achieved by optical foot tracking, employing a commercial video tracking system that tracks bright markers on the front of the subjects shoes at a 60-Hz rate. It was found that stance time was the best predictor of walking speed, and a curvilinear relation was established to generalize across all users. Turning is achieved by swiveling the treadmill in the direction in which the user is stepping. The lateral amount of foot motion is measured to estimate how much to rotate the platform. The platform cannot rotate continuously, and so must be reindexed to center beyond a certain angle. Noma et al. (2000) refer to this swiveling as turning cancellation, and to the function of a locomotion interface as motion cancellation in general. An HMD is employed for the visual display. A magnetic tracker is employed to detect head orientation.

The Ground Surface Simulator (GSS) is a system meant to simulate uneven or steplike terrain (Fig. 10.5B). The GSS employs a linear treadmill with a flexible belt that is deformed underneath by six vertical stages. The belt is 1.5 m long and 0.6 m wide. Each stage is 0.25 m

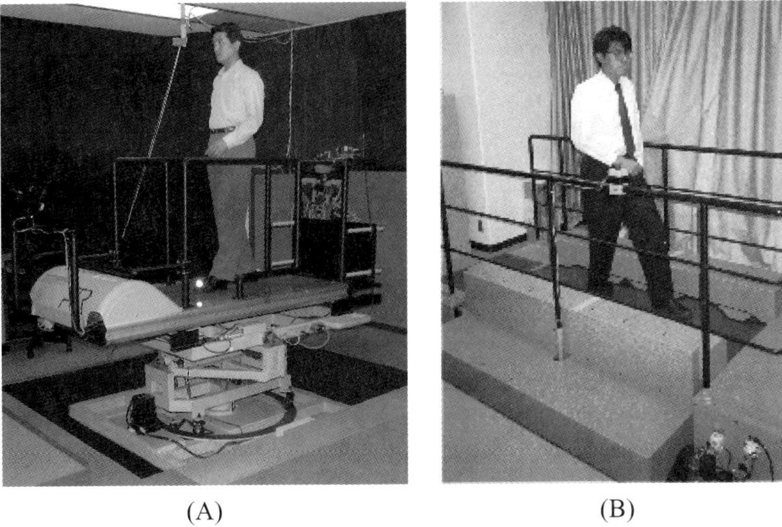

(A) (B)

FIG. 11.5. (A) The ATLAS system. (B) The Ground Surface Simulator. Photos courtesy of H. Noma.

long and has a stroke of 6 cm at a speed of 6 cm/sec. There are rollers on the support surface of each stage. Because the geometry of the belt changes when deformed by the stages, an active belt tensioning system is employed. A slope of 5 deg can be presented by the GSS. Future plans call for mounting the GSS on a motion stage as for the ATLAS system.

### 2.4.3 The Omni-Directional Treadmill

The Omni-Directional Treadmill (ODT), designed by Virtual Space Devices, Inc., provides a two-dimensional treadmill surface designed to facilitate turning (Fig. 10.6A). A two-orthogonal belt arrangement creates the two-dimensional surface. A top belt is comprised of rollers whose axes are parallel to the direction of rotation of that belt. These rollers are rotated underneath

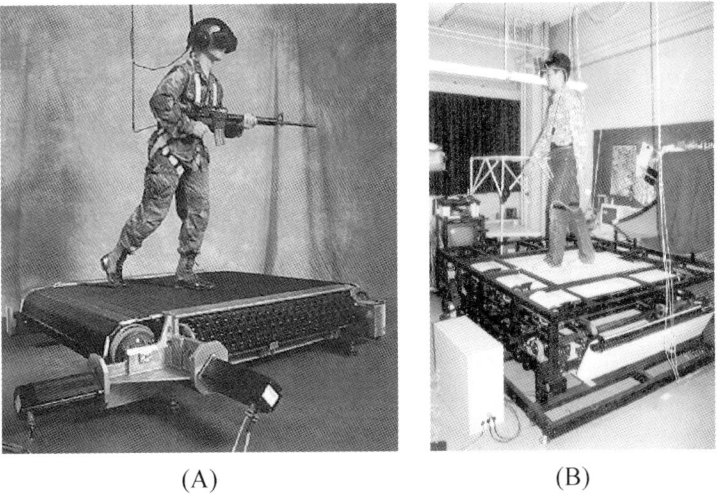

(A) (B)

FIG. 11.6. (A) The Omni-Directional Treadmill. From http://www.vsdevices.com/. (B) The Torus treadmill. Photo courtesy of H. Iwata.

by another belt orthogonal to the first. The active surface is 1.3 m by 1.3 m, and the peak speed is 3 m/sec.

A mechanical position tracker on an overhead boom attaches to a harness worn by the user in order to control the treadmill. The boom is active and may exert a force up to 89 N. Both an HMD and a CAVE-like display have been employed for the visuals.

Some unsteadiness in walking on the roller surfaces has been reported (Darken & Cockayne, 1997). A mismatch between a user's walking direction and the centering motion of the belt could occur that causes the user to stumble. The mismatch presumably arises due to system lags and bandwidth limitations that permit the user to move off center. This type of mismatch would seem to be a potential problem for any two-dimensional motion stage.

#### *2.4.4 The Torus Treadmill*

Another two-dimensional treadmill design is Iwata's Torus Treadmill (Fig. 10.6[B]), which employs 12 small treadmills connected side-by-side to form a large belt to allow arbitrary planar motion (Iwata & Yoshida, 1999). The walkable area is 1 m by 1 m, and the maximum treadmill speed is 1.2 m/sec. Motion control is achieved by foot tracking employing magnetic trackers. An HMD is provided where a magnetic tracker senses head orientation.

In the initial implementation of the Torus Treadmill concept, the speed and area limitations limit walking to a slow speed and relatively short steps. A larger design in the future would presumably overcome such limitations.

## 3. DISCUSSION

Locomotion interfaces represent a relatively new field in which there are currently few systems and researchers. The thrust of current research may be described as exploring the design space of possible devices and approaches. Many of these designs represent initial explorations and have not been optimized in any sense, so that judging performance limitations is premature. Utility, functionality, cost, and safety are all issues that will need to be considered in choosing among the options.

### 3.1 Forward Motion Display

Both leg-based tracking and body-based tracking have been employed to measure how much forward or backward motion a user has produced. Leg-based tracking is readily accomplished where a mechanically tracked device is coupled to a user's feet, as for bicycles, foot platforms, and passive treadmill belts. Direct leg measurement can also be achieved by magnetic sensors (Templeman et al., 1999) or by optical tracking (Noma et al., 2000). The use of the stance time as an indicator of speed appears to be a viable method (Noma et al., 2000). Foot force sensors can aid the segmentation of gait (Templeman et al., 1999).

For body-based tracking, a mechanical tracker attached to the user's body is employed in the Sarcos Treadport (Hollerbach, Xu, Christensen, & Jacobsen, 2000) and in the Omni-Directional Treadmill (Darken & Cockayne, 1997). As mentioned earlier for a walking-in-place locomotion interface (Slater, Steed, & Usoh, 1995), steps can also be approximately inferred from measurements of head bobbing. The inaccuracy reported for this technique limits its utility, however.

Optical tracking is attractive as opposed to mechanical tracking because the user is relatively unencumbered. Magnetic tracking is known to suffer from sensitivity to environmental metal, which may not make it attractive around the heavy metal structures of many locomotion interfaces. If one is to apply forces to the body, as with the Sarcos Treadport, then a mechanical connection is required.

Relative freedom of movement and unencumberance tend to be strengths of treadmill-style locomotion interfaces. For example, on the Sarcos Treadport a user can stride naturally and run at will. This kind of freedom has to be counterbalanced against movement restrictions due to optimizing some other aspect of locomotion rendering such as turning or uneven terrain. Foot platforms in particular impose some restrictions on forward motion speed.

Accelerating on a locomotion interface can never be the same as accelerating on the ground because the user's body is stationary, unless the missing inertial forces are applied to the body by some mechanical linkage such as the active mechanical tether of the Sarcos Treadport. In that sense, an active tether is demonstrably necessary to make ground and locomotion interface running the same. Psychological studies clearly show the preference of users for inertial force display (Christensen et al., 2000), but aside from preference it may be that displaying inertial force is not necessarily important for making the locomotion interface useful.

Walking-in-place simulators are a low-cost approach toward constructing locomotion interfaces. Comparisons with treadmill or foot platform systems should be made to determine the trade-offs in terms of loss of motion fidelity versus utility.

### 3.2 Turning Display

Turning intent can be measured through body tracking, leg tracking, or manipulation of a steering device. Steering bars are readily employed for turning control for stationary bicycles, and have also been employed in passive treadmills. Either rate control or position control in steering may be used.

Leg-based tracking is achieved optically by the ATLAS system (Noma et al., 2000). The amount of lateral motion in a step guides the swiveling of the treadmill by a turntable. A magnetic sensor on the knee is used in the Gaiter system (Templeman et al., 1999) to indicate turning by a rocking motion to the side; this represents a gestural control rather than a natural turning motion. Mechanical attachments of the legs to the foot platforms for the Sarcos Biport and for the GaitMaster (Iwata, 2000) guide turning. For the Sarcos Biport, force sensors on the attachment points are used to infer the direction of walking.

Body-based tracking can directly indicate the direction of tracking. For example, the magnetic trackers in walking-in-place locomotion interfaces (Slater et al., 1995; Templeman et al., 1999) measure rotation about the vertical axis. The Sarcos Uniport employs force sensors in the seat, and any sensed sideways twisting controls the rate of turning. The ODT and the Sarcos Treadport use mechanical tethers attached to body harnesses for direct sensing of motion direction.

The two-dimensional treadmill belt designs are attractive in terms of the naturalness of turning. Some control issues apparently must be addressed in the ODT because of reported unstable walking (Darken & Cockoyne, 1997). In particular, a mismatch between the direction of a centering motion and the direction of a user's walking can result in a sideways force that causes the user to stumble. The initial implementation of the Torus Treadmill (Iwata & Yoshida, 1999) has led to slow gait and short steps, but this could presumably be remedied in a redesign. There is an issue of the mechanical design complexity of the Torus Treadmill concept and the speed capabilities of the main belt, which itself is the juxtaposition of many small treadmills. Speed constraints, complexity, and cost are particularly pertinent factors to be investigated when the utility of such designs is judged.

Another approach is using a turntable as in the Sarcos Uniport, the ATLAS system (Noma et al., 2000), and the GaitMaster (Iwata, 2000). By swiveling the treadmill in the direction of walking, the ATLAS system achieves a natural gait pattern in turning. Due to lags in rotation, it is possible that footfall occurs at a slant relative to the desired walking direction, and so on the next step some correction by the user could be necessary. This might pose a problem when an HMD is employed, since users cannot see their orientation on the treadmill. In the GaitMaster, the user does not step into the turn, so the gait is not as natural as for the ATLAS.

The Sarcos Treadport is a linear treadmill without a turntable; hence turning is not as natural. Because of its width, some sideways motion is permissible, although a user would eventually have to recenter. At an obvious cost, the treadmill could be made even wider to accommodate sideways motion. Although rate control is currently used for turning, current research is addressing the use of proportional control within a certain angle. In proportional control mode, a user actually steps in the direction of desired motion, but again, a reindexing is eventually necessary.

The different turning approaches surveyed above may have implications for wayfinding (see chap. 24, this volume). Most natural locomotion occurs, of course, when visual, vestibular, and proprioceptive cues are consistent with normal turning, as in two-dimensional treadmills and the ATLAS system. The GaitMaster provides consistent vestibular and visual cues, although the proprioceptive cues are not veridical. The Sarcos Treadport provides only realistic visual cues. Interestingly, there have been no reported instances of simulator sickness (see chap. 30, 31 and 36, this volume) with the Treadport, even though the vestibular and visual cues are in conflict. This could be because the user is receiving proprioceptive cues for walking that although not realistic for turning, do indicate to users' sensory processes that they are moving.

For the Gaiter walking-in-place locomotion interface, learning to gesture with the knees to indicate turning is evidently an effective strategy (Templeman et al., 1999). Similarly, rate control for the Sarcos Treadport can be viewed as a form of gesturing. From observation of hundreds of people interacting with this system, ranging from elementary school children to adults of all ages, users have been found to adapt quite readily to this turning control mode. Often users discover how to control turning rapidly without instruction. Future research and experience will be required to decide how detrimental or effective different turning control modes are, particularly in the context of other tradeoffs. For example, the Sarcos Treadport excels at forward motion display at the cost of turning, whereas two-dimensional treadmills excel at turning at the cost of forward motion display.

Another issue is whether an active mechanical tether is used in conjunction with a CAVE display, as in the Sarcos Treadport. It has been argued that a tether is necessary for displaying inertial forces and unilateral constraints and is useful for displaying slope. The Treadport tether protrudes from the back of the treadmill and thus interferes with display screens if the treadmill is placed on a turntable. Two-dimensional treadmills also have problems providing the ability to apply forces in any direction. The overhead boom of the ODT would seem a better solution than the horizontal boom of the Treadport, but it might be difficult to get the same force output from such an arrangement.

### 3.3 Slope Display

Treadmills often are placed on tilting motion stages, so that walking up and down a smooth slope can be realistically displayed. The Sarcos Treadport is an example, but also the bicycle simulator of Brogan et al. (1998) has been placed on a tilting stage. The ATLAS system carries this concept one step further with a spherical joint mounting for the treadmill, which not only acts as a turntable and tilts but also rotates the treadmill sideways. This allows traversing sideways along a slope to be displayed.

The Sarcos Treadport can also simulate slope walking by pushing and pulling with a mechanical tether. Although the Treadport has a tilt capability, due to the large platform mass the tilt speed is limited and sudden slope transients cannot be displayed. The fast-acting mechanical tether has been reported to simulate slope reasonably well (Tristano et al., 2000), and so may provide a lower-cost alternative to having a tilting stage. One difficulty with a tilting stage is that when a CAVE display is used, larger portions of the screens have to be projected. Also, the belt of the Sarcos Treadport is white to allow for the possibility of floor projection. A tilted platform may complicate correct imagery for the floor.

Displaying uneven terrain such as steps is a particular strength of foot platforms. Soft ground could even be displayed in principle. Unless the foot platform stages tilt, the biomechanics of walking on a smooth slope will not be veridical. The Ground Surface Simulator (Noma et al., 2000) displays uneven terrain through the use of vertical motion stages that deform a treadmill belt. The same comment about differences from smooth slope walking will apply unless the belt is very stiff and does not deflect much in its unsupported regions. For the Sarcos Treadport, if the mechanical tether is effective at displaying rapidly changing slopes, then its force profile might be fashioned to give the illusion of uneven terrain. This is a topic of current research.

### 3.4 Unilateral Constraint Display

Along with inertial force display, a mechanical tether is demonstrably necessary to display unilateral constraints. Simply braking the locomotion interface is only a vague simulation of hitting a wall. With the current harness attachment of the Sarcos Treadport, the sensation is more akin to being pulled from the back rather than hitting something with the front of the body. An attachment to a plate in front of the body might simulate hitting a wall more realistically. There have been light-hearted suggestions on swinging sandbags down from the ceiling. It seems, nevertheless, that hitting a wall is not something we want to simulate with complete fidelity!

### 3.5 Body Posture

Relatively unencumbered walking and diverse gaits are strengths of treadmills. Most locomotion interfaces surveyed here only support upright stance. If prone postures such as crawling are important, then a large treadmill surface such as the Sarcos Treadport is the most realistic option.

### 3.6 Visual Display

CAVE-like displays and HMDs have both been employed with locomotion interfaces. The use of HMDs is particularly appropriate when there are obstacles to the projection screens of CAVEs, such as motion platforms that rotate or tilt treadmills or mechanical tethers. There are more safety concerns with HMDs because users cannot see their positions on the locomotion interface, which could cause them to fall off a treadmill or to lose balance (see chap. 41, this volume). The inconvenience of wearing an HMD is sometimes cited. The usual criticism of poor visual quality of HMDs also applies. A further problem with HMD technology is the lack of economic drivers to improve quality and reduce cost. Simulator sickness might be more of a problem with HMDs than with CAVEs.

One advantage of HMDs over CAVEs is that objects can be projected between a user's hand and line of sight. In locomotion displays one is typically dealing with far-field vision, that is, objects and terrain that are beyond the arm's reach. Therefore, displaying the arm behind an object, which is a strength of HMDs and a weakness of CAVEs, is not as important. Safety is facilitated by CAVEs, because users can see their positions, for example, on a treadmill belt and make appropriate adjustments. In a CAVE, users see their bodies, which might lead to a greater sense of immersion (see chap. 40, this volume), as opposed to HMDs where a user's body is occluded. The importance of self-vision is not yet clear. Economic drivers are very strong for improved CAVEs in the future because of the projection market, and so higher brightness, resolution, and image quality are to be expected.

There is not as much of an economic driver for stereoscopic projectors, which would be highly desirable for depth perception. Ground projection has been reported as being particularly important for CAVE displays. For the new Sarcos Treadport, the belt material was specified white to allow for the possibility of ground projection. Ideally, the ground projection should be in stereo for depth effects.

CAVEs can complicate the design of locomotion interfaces, because platform motion might interfere with the screens. As aforementioned, the mechanical tether of the Sarcos Treadport protrudes from the back, and so placement of a treadmill on a turntable is problematical. Such a tether on a 2-D treadmill is also problematic, and, as mentioned earlier, an overhead boom might be more appropriate but more difficult to design. If the platform tilts, then projection screens are covered or uncovered, and larger projected images might be required. There can be gaps between the screens and locomotion interfaces. The belt of a treadmill itself is a visual gap, unless there is floor projection. Issues of shadows cast by the user from a projector above and of visual deformation due to platform tilting would need to be addressed.

A final concern with visual displays is their cost, which can be on a par with mechanical displays. Quality HMDs are quite expensive. CAVEs require multiple high-quality projectors, special screens, and special mirrors. At the moment, special-purpose graphics engines are required to produce realistic displays, but the expense of graphics engines is expected to decline dramatically in the future. An indirect cost for CAVEs is the need for large rooms for placement of screens, mirrors, and projectors, including overhead space.

Besides vision, it would be desirable to include auditory displays for ambient sounds as well as footstep sounds. Stereophonic ambient sounds, for example, from walking in a forest, could aid the sense of immersion (see chap. 40, this volume) and spatial orientation (see chap. 24, this volume). In the haptics literature, it has been shown that the sound of contact can bias the perception of surface stiffness and material. Similarly, it is possible that different footstep sounds can bias the perception of the kind of ground one is walking on. Noise-cancelling headphones might be useful to filter out any sounds coming from a locomotion interface.

But why stop with audio? Olfactory displays are progressing, and it would be nice to smell the pine scent of forests we're walking through. Fans could be placed to simulate wind.

### 3.7 Safety

Because body weight and the forces of locomotion have to be supported, locomotion interfaces have to be powerful. Safety is potentially much more of a concern than with haptic interfaces. Programmable foot platforms are essentially powerful robot arms attached to the feet, and any malfunction could be painful (e.g., potential wishbone effect).

Dangers with treadmills are falling off the sides or stumbling. Ceiling restraints can help, as well as proprioceptive centering cues and sidebars. As mentioned above, the use of CAVEs is probably safer than HMDs. The active tether of the Sarcos Treadport is also a potential safety concern, although the tether forces are well within normal loads to the spine (Hollerbach et al., 2000).

Safety in locomotion interfaces will be enhanced through a judicious mixture of kill switches, watchdog timers, software constraints on motion, and hardware limit switches.

### 3.8 Locomotion and Haptics

Adding a haptic interface for manipulation to a locomotion interface will be quite a challenge because of the movement possible on a locomotion interface. A haptic interface would have to have a large workspace but be kept out of the way. Some sort of ceiling mount might be required.

One usually associates the forces experienced in manipulating via arms and hands with the term haptics. These days the term haptics seems to be generalized to include any force application to any part of the body. Thus Iwata (2000) has coined the term *foot haptics* for programmable foot platforms. Similarly, body haptics has been used to describe forces applied to the body by the active tether of the Sarcos Treadport. It is probably useful, though, to consider that locomotion interfaces are a distinct class of devices from traditional haptic interfaces.

### 3.9 Applications

Whereas concrete application areas for haptic interfaces are starting to emerge, such as for mechanical CAD interfaces and surgery training, concrete applications for locomotion interfaces are not well defined. Certainly, training is a strong candidate, such as in military rehearsals (Grant et al., 1997; Parsons et al. 1998) or fire fighter training (Iwata et al., 1999). Telecommunication and control of a remote mobile robot to serve as a physical avatar to allow one to experience walking in a remote place is an intriguing application suggested for the ATLAS (Noma et al., 2000). With terrain data now becoming available for moons and other planets of the solar system, one could simulate walking on Mars or Venus, which could serve an education function. Entertainment, exercise, recreation, and health rehabilitation are other suggested applications.

Finally, VE displays have a strong role to play in psychophysical research, because physical and visual stimuli are so conveniently synthesized. Rieser et al. (1995) towed a treadmill behind a car to decouple the optical flow from the kinesthetic sense of walking. Such decoupling could be much more conveniently achieved with a locomotion interface.

## 4. ACKNOWLEDGMENT

This work was supported by National Science Foundation grants IIS-9908675 and CDA-96-23614.

## 5. REFERENCES

Brogan, D. C., Metoyer, R. A., & Hodgins, J. K. (1998). Dynamically simulated characters in virtual environments. *IEEE Computer Graphics and Applications, 15*, 58–69.

Christensen, R., Hollerbach, J. M., Xu, Y., & Meek, S. (2000). Inertial force feedback for the Treadport locomotion interface. *Presence: Teleoperators and Virtual Environments, 9*, 1–14.

Darken, R. P., & Cockayne, W. R. (1997). The Omni-Directional Treadmill: A locomotion device for virtual worlds. In *Proc. UIST* (pp. 213–221).

Durlach, N. I., & Mavor, A. S. (Ed.). (1994). *Virtual reality: scientific and technological challenges.* Washington, DC: National Academy Press.

Grant, S. C., & Magee, L. E. (1997). Navigation in a virtual environment using a walking interface. In *NATO RSG-18 Workshop on Capability of Virtual Reality to Meet Military Requirements.* Orlando, FL:

Hollerbach, J. M., Xu, Y., Christensen, R., & Jacobsen, S. C. (2000). Design specifications for the second generation Sarcos Treadport locomotion interface. In *Proceeding of the ASME Dynamic Systems and Control Division.* Orlando, FL: ASME, pp. 1239–1298.

Iwata, H. (2000). Locomotion interface for virtual environments. In J. Hollerbach & D. Koditschek (Ed.), *Robotics Research: The Ninth International Symposium* (pp. 275–282). London: Springer-Verlag.

Iwata, H., & Yoshida, Y. (1999). Path reproduction tests using a Torus Treadmill. *Presence: Teleoperators and Virtual Environments, 8*, 587–597.

Moghaddam, M., & Buehler, M. (1993). Control of virtual motion systems. In *Proceedings of the IEEE/RSJ International. Conference on Intelligent Robots and Systems* (pp. 63–67). Yokohama, Japan: IEEE.

Noma, H., Sugihara, T., & Miyasato, T. (2000). Development of ground surface simulator for tel-E-merge system. In *Proceeding IEEE Virtual Reality* (pp. 217–224). New Brunswick, NJ: IEEE.

Parsons, J., Lampton, D. R., Parsons, K. A., Knerr, B. W., Russell, D., Martin, G., Daly, J., Kline, B., & Weaver, M. (1998). Fully immersive team training: A networked testbed for ground-based training missions. In *Proceedings of the 1998 Interservice/Industry Training, Simulation and Education Conference.* Orlando, FL:

Rieser, J. J., Pick, H. L., Jr., Ashmead, D. H., & Garing, A. E. (1995). The calibration of human locomotion and models of perceptual-motor organization. *Journal of Experimental Psychology: Human Perception and Performance, 21*, 480–497.

Slater, M., Steed, A., & Usoh, M. (1995). The virtual treadmill: A naturalistic metaphor for navigation in immersive virtual environments. In M. Gobel (Ed.), *Virtual Environments '95* (pp. 135–148). New York: Springer-Verlag.

Templeman, J. N., Denbrook, P. S., & Sibert, L. E. (1999). Maintaining spatial orientation during travel in an immersive virtual environment. *Presence: Teleoperators and Virtual Environments, 8*, 598–617.

Tristano, D., Hollerbach, J. M., & Christensen, R. (2000). Slope display on a locomotion interface. In P. Corke and J. Trevelyan (Ed.), *Experimental Robotics VI* (pp. 193–201). London: Springer-Verlag.

# 12

# Virtual Environment Models

G. Drew Kessler
*Lehigh University*
*Computer Science and Engineering Dept.*
*19 Memorial Drive West*
*Bethlehem, PA 18015*
*dkessler@eecs.lehigh.edu*

## 1. INTRODUCTION

Many of the design decisions of virtual environment (VE) applications are driven by the capabilities and limitations of users. These considerations affect both the design of hardware software. To be interactive, a VE software application must constantly present the current "view" of a computer-generated world and have that world quickly react to the user's actions. To be convincing, the presentation must provide enough detail to make objects easily recognizable and enough objects to give the user the sense of being in the world. To be useful, the environment must respond to the user. The user's location in the world should change when a navigation action is performed. Objects that the user can grab or nudge should move as expected. Manipulation of three-dimensional (3-D) interface elements, such as floating buttons, tabs, and sliders, should have the desired effect on the environment, perhaps by changing the appearance of an object.

An environment can be described as a set of geometry, the spatial relationship of the geometry to the user or users, and the change in the geometry in response to user actions and as time progresses. The set of geometry can be decomposed into geometric objects that may have a visual appearance, make a sound, or have an odor, taste, or texture to be felt. A geometric object can also be simply a point in space. This geometric environment is presented to each user utilizing output devices. Virtual environment systems usually provide a visual display, and many provide audio, although output for all of the far senses (i.e., sight, sound, and odor) has been shown to be effective (Dinh, Walker, Hodges, Song, & Kobayashi, 1999). Tactile and force feedback devices can also present an aspect of the world to the user and are currently in development. It is much more difficult, however, to mimic the touch sensation of arbitrary geometry than it is to provide a visual representation.

Given that a 3-D environment may occupy a large "space," providing a convincing illusion that the user is in a realistic world requires many geometric objects that constantly react to other objects and the users. However, current technology is not capable of handling the amount of information and processing needed to present such an environment. In addition, creating all of the pieces of the world at such a level of detail would be a significant effort. Luckily, many applications can be created with VE technology that do not require a full illusion, and a somewhat realistic world can be quite convincing. It is still a challenge, though, to create and process the description of an environment, which is the environment's model.

An environment model has at least three representations. One representation describes the model in a stored file. Another representation describes the model to a running application, which can be accessed and modified to produce complex behaviors that could not be described in the model itself. A final representation stores the model internally. This representation is optimized for the task of constantly supplying information to the output devices to render a view of the model, providing 3-D sound, and so on. At some point, the model is translated from one representation to another. For example, a file format that provides a stored model is the Virtual Reality Modeling Language (VRML; Ames, Nadeau, & Moreland, 1997). The Java3D package (Sowizral, Rushforth, & Deering, 2000) provides a run-time model for Java programs. Parts of the Java3D run-time model can be explicitly "compiled" to produce an optimized internal model at the cost of making modification of the environment model more difficult.

Some file representations are designed to translate easily to the internal representation, but cannot be easily translated to a model that can be modified at run time. Other file representations may translate quickly to a model that is easy to modify at run time, but require more processing to obtain an internal model that is optimized for interactive use. Over long execution times, a translation step from a file to an internal representation is cheap, as it occurs only once. On the other hand, if the run-time model is modified often, the translation from the run-time model to the internal model can become a critical performance bottleneck.

In this chapter, a discussion of how an environment model is described in terms of geometric primitives and behaviors is provided. This model is often used as the run-time representation. Next, this representation is related to common file formats used to describe a virtual environment, and to an internal representation that provides optimizations for efficient presentation of the model. Finally, a description of how the user can be incorporated into the model to provide an interface between the user and environment is provided.

## 2. GEOMETRY

For the most part, VEs are spatial in nature. A VE consists of properties associated with locations in three-space, such as sounds and appearance. A VE application, therefore, maintains a representation of spatial information and attempts to organize that information so that arbitrary views can be produced quickly and repetitively. Since visual perception is the primary channel used by sighted people to obtain information about the world around them, VE systems have focused on representing geometric models of visual objects and their placement in the environment.

A VE is usually described by low-level primitives such as polygons, lines, and text, and by higher level primitives such as spheres, boxes, cones, polygonal meshes, polyhedra, curves, and curved surfaces. Current graphics hardware renders low-level primitives, as well as polygonal meshes, very efficiently. However, many of the higher-level primitives, such as spheres, curves, and curved surfaces, cannot be precisely rendered quickly enough. Instead, they are converted to a polygonal representation when they are added to the environment. Advances in hardware, however, will likely allow higher level primitives to be rendered directly at sufficient speeds. A system that handles high-level primitives, by translating them to low-level primitives when storing the model in the internal representation, will be best positioned to take advantage of this type of technological advance.

Although the visual appearance of an environment can be completely described, albeit approximately in some cases, by a set of polygons, lines, and text images, it is more useful to organize the geometry into separate groups that are related to each other. This network of geometries and relationships is known as a *scene graph*. The scene graph contains nodes that define geometric primitives and properties, relationships between geometry, and perhaps behaviors of the geometry. In this section, the geometric primitives used in VE applications, how these primitives are organized into a scene graph, what relationships the scene graph provides between groups of primitives, and what properties may be stored in the scene graph that determine the appearance of primitives and the environment, such as materials and light sources, are described.

## 2.1 Geometric Primitives

Current computer graphics systems are optimized to display convex polygons and lines, and are capable of rendering millions of polygons per second to the screen. Therefore, geometry is decomposed into, or approximated by, a set of polygons or lines. Text is often treated as a special primitive, although it can be crudely represented by a set of lines, a set of polygons, or more nicely represented by an image with "see-through" areas.

Geometric primitives are described by point coordinates in a 3-D coordinate system. By convention, the coordinate system is right-handed (when curling the fingers of the right hand from one positive axis to another, the outstretched thumb points in the direction of the third positive axis). Whenever a group of primitives describe an object that has a particular orientation (for example, a tree generally grows "upward"), convention assigns the *y* coordinate axis to represent the "up" direction, the *x* coordinate axis to represent the "right" direction, and the z-axis to represent the "back" direction. This coordinate system is a remnant of models developed for display on a computer monitor, where the x-axis is the horizontal axis of the monitor, and the y-axis is the vertical axis of the monitor. (If only monitors had been placed face up on or in a desk—then the coordinate system convention might match the z-up convention of mathematicians and physicists!)

Polygons, also known as faces, are described by a list of coplanar points in three-space. Each pair of points on the list (and the pair containing the first and last points) defines an edge of the polygon. Graphics systems, however, may only be able to render convex polygons, and may just render triangles, as they are simple to process. Concave polygons or polygons with four or more edges are usually decomposed into triangles before they are rendered. Often, a graphics system needs to be instructed to spend the time to convert concave polygons correctly. Polygons are shown filled with some color or pattern, but are considered infinitely thin, meaning that, if they are viewed from the side, they will be invisible. Lines, however, are of a given width regardless of the viewing angle.

In addition to allowing for polygons and lines, modeling systems will generally allow users to create higher level geometries, such as boxes, spheres, cones, and cylinders. These shapes may be converted to polygons by the modeling system or may be stored as their shapes, and must be converted to a polygonal representation by the rendering system. Unfortunately, a polygonal representation is often a rough approximation of the shape. This is especially true of shapes with curved lines or surfaces. Since a computer image is an approximation itself, the error in the shape representation may be hidden by the error introduced in using discrete pixels to define an image. If the rendering system has the shape description, it can translate the shape to a representation that is as accurate as possible given its visibility, at the cost of extra processing time. In general, a more accurate representation requires more polygons or

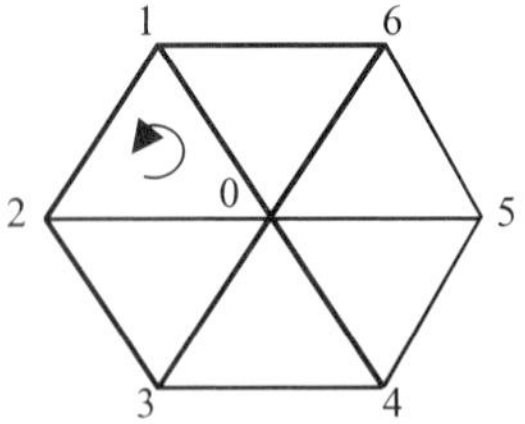


FIG. 12.1. A polygonal surface.

lines. Current modeling packages can produce polygonal representations of their models, but note that the primary application for modeling packages is to produce high-quality animation frames or still images. Thus, they often produce many more polygons than can be handled by an interactive rendering system. What's worse, some of these packages do not allow the user to control the polygonalization algorithm to put a greater weight on fewer polygons.

An environment generally consists of a set of geometric objects, where each object is described by a set of solids or lines, and each solid is described by a set of surfaces, which, in turn, are described by a set of edge-connected polygons (see Fig. 12.1). The viewpoint is assumed to always be outside of a solid. With this assumption, the rendering system can limit the polygons they draw to those on surfaces that face the viewer, since surfaces that don't face the viewer are on the other side of a solid and are obscured by the front surface of the solid. This technique is known as *back-face culling*. However, when the assumption fails (the viewpoint moves inside of a solid), then the effect of back-face culling is that the surfaces disappear altogether (all surfaces are facing away from the viewpoint). At the cost of deciding if the viewpoint is inside an object, a rendering system may display an "inside" color instead. However, this technique is expensive for arbitrary solids and large numbers of objects.

In order for the back-face culling technique to work, polygons of a surface must have a "front" and a "back." By convention, the order of the points that describe the polygon determines the front and back sides of the polygon: if the viewer is on the front side, then the points are given in counterclockwise order. If the points are given in clockwise order from the point of view of the viewer, the viewer is on the back side of the polygon (as shown in Fig. 12.1).

Surfaces can be approximated as a set of polygons, but surface representation methods that require less information can be loaded into the application faster and transmitted faster from the application to the graphics rendering system. One such method represents a surface that is defined by a set of height values along a regular grid. This method can represent terrain that contains no caves or outcroppings (i.e., all bumps go up, all holes go down). This method allows a surface to be described by the coordinates of the corners, the distance between grid points, and a height value for each grid point, roughly one third of the information needed if only points were used. In addition, the representation can be shrunk further using algorithms that can quickly approximate these surfaces with fewer polygons when the viewpoint is far away (Lindstrom et al., 1996).

A polygon mesh also describes a surface using less information than the same surface given as a set of distinct polygons. Graphics systems generally support meshes of triangles or quadrangles. A mesh is described by a list of polygons, where each successive polygon shares one edge completely with the previous polygon, and therefore shares two points. For example, a triangle mesh can be described by three points for the first triangle, then one more point for the next triangle, another point for the third triangle, and so on. A surface described in this manner uses roughly one third fewer points to describe a surface and does not have the restrictions of the elevation map. In fact, an elevation map will likely be translated to a triangle mesh before it is rendered.

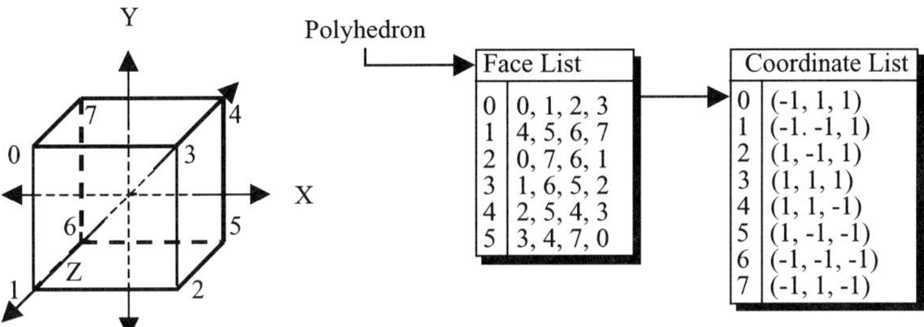


FIG. 12.2. Face and coordinate lists for a cube.

When a polyhedron is specified by a set of face polygons, each being described by a set of coordinates in three-space, many of the same coordinates are given more than once. This is a result of the surface polygons' common edges. To avoid storing redundant information, the geometric model is very often given as a set of labeled coordinates, and a set of faces that list the labels of the coordinates that describe the face's edges. This storage scheme is shown in Fig. 12.2 for a $2 \times 2 \times 2$ cube centered at the origin. Note that the order of the coordinates for each face are given in counterclockwise order from the point of view of being outside of the cube. A similar strategy can be used to store coordinates of a list of line segments.

Graphics hardware vendors tend to describe the performance of their display system in terms of how many polygons per second can be drawn. However, this measurement does not tell the full story. Depending on the hardware architecture, a graphics system may be polygon-bound or fill-bound. If a system is fill-bound, then the number of polygons that can be displayed to the screen is heavily dependent on how much screen space each polygon takes when rendered. In other words, the time taken to fill each polygon dominates the rendering time. If a system is polygon-bound, then the changes to the number of polygons have a more significant effect to rendering time than the size of the polygons. This difference in architecture has a profound effect on how models should be designed and built for VE applications. If a VE application will use a polygon-bound system, then the model should contain as few polygons as possible, no matter how large they are individually. Mostly flat surfaces could be replaced by single polygons, for example. If the system used is fill-bound, however, this change in the model will not have much of an effect on the rendering time. In fact, a system may be fill-bound because it is designed to handle many polygons in parallel, and works best when polygons take about the same amount of time to fill because the parallel processing is balanced. If this is the case, it would be a good idea to break up a large polygon (or, more precisely, one that will appear prominently on the display) into a set of smaller, coplanar polygons.

## 2.2 The Scene Graph

One could represent the geometric model of a VE with just a set of polygons, lines, and text. However, when a particular object, such as a chair, table, or vase, moved in the environment, then all of the coordinates of the polygons that represent the moving object would need to be changed, or transformed, in the same way. In addition, if the vase was conceptually "on top of" the table, the coordinates of the vase should move when the table moves. A more efficient representation of the model would group geometric primitives together into geometric objects, and would define "attachment" relationships between the objects.

An environment model, therefore, can be represented by a coordinate system graph, which is a directed acyclic graph. Each node of a coordinate system graph contains a set of geometric primitives, which usually represent a conceptual object or object part. The geometry of a node is described in terms of a local coordinate system. The edges of the graph define a transformation between the coordinate system of one node to the coordinate system of another node. Conceptually, the coordinate system of one node (the child) is contained in the coordinate system of another (the parent). If the parent's coordinate system is transformed, the child's coordinate system is transformed as well (unbeknownst to the child), but if the child's coordinate system is transformed, the parent's coordinate system is unaffected, and the child's transformation is visible in the parent's coordinate system.

A coordinate system graph does not contain cycles because it is unclear how a change in the relationship between objects in the cycle should change the relationships between the other objects in the cycle. Should all of the transformations be changed equally, or should just one other edge change? If just one edge changes, which one should it be? For example, if five marbles represented in a coordinate system graph were placed next to each other, and the graph contained edges between adjacent marbles to keep them spaced evenly, and an edge between the outer marbles to maintain a certain length for the marble line, how should the marbles move if the distance between the outer marbles changes? Or if the distance between two adjacent marbles changes? Should the marbles adjust to space out evenly again, or should just one other distance change?

Coordinate system transformations are described by a $4 \times 4$ matrix that operates on homogeneous coordinates, where a location in space is given by the 4-tuple (i.e., x, y, z, 1). A set of locations is transformed from one coordinate system to another by multiplying each location by the matrix transformation between the coordinate systems. Matrix transformations can be combined by multiplying them together, but matrix multiplications are not commutative, and a change in the order in which the transformations are given will change the total transformation (unless the transformations are all the *identity matrix*, which does nothing).

A matrix transformation in a coordinate system graph is often described as a combination of scale changes along the major axes, x, y, and z, rotations about the major axes, and translations along the major axes. However, the transformation may include shears, mirrors, or other affine transformations. Standard 3-D matrix transformations and the use of these transformations in a coordinate system graph are described in more detail by Foley, van Dam, Feiner, and Hughes (1996). Some systems or file formats (such as VRML) may describe the rotation component as a certain rotation around a given vector. This representation, known as a *quaternion*, describes a rotation more efficiently, and is better for interpolating between two rotations (smoothly changing from one rotation to another without experiencing gimbal lock). However, an extra step is required to incorporate a rotation in this form into a $4 \times 4$ matrix transformation (Watt & Watt, 1992).

A coordinate system graph is usually represented as a tree, where the root node contains geometry in a "world" coordinate system, and the child nodes have local coordinate systems that related to the world coordinate system. Therefore, points in the geometry of a child node can be transformed to the world coordinate system. Points further down a subtree of the root node can be transformed to their parent's coordinate system, and then to its grandparent's coordinate system, and so on until it is in the world coordinate system. Similarly, a point in the world coordinate system can be transformed to the local coordinate system of a descendant node by applying the inverse of each transformation on the path to that node. When rendering a scene for the display, geometry must be transformed to the world coordinate system, then transformed to a descendant node representing the coordinate system of the viewer, where the origin is at the eye point.

The coordinate system graph is a type of *scene graph*, as it describes the scene for an environment. Other types of scene graphs have nodes that do not represent geometry and do not associate a transformation for every edge in the graph. For example, the VRML file format (Ames et al., 1997) and the Cosmo3D run-time library (Eckel, 1998) provide a scene graph with many types of nodes, one of which is the "transform" node that defines a spatial relationship between geometry attached to it and the geometry it is attached to. Edges are still used to group geometry by defining, directly or indirectly, rigid attachments. Other nodes in the graph define light sources, appearance properties, and different types of grouping.

## 2.3 Material Properties

Surfaces in the real world have many properties that affect how they appear beyond their position and orientation. They may be rough or smooth. They may be dyed or painted with different colors, or they may contain grains. They may have fine detail that is too expensive to model geometrically, such as the texture of skin or hair. At a great distance, even features like brick grooves, windowsills, or door handles might be considered fine detail. These features can be described by their contribution to the color of a pixel on the display, if the feature is visible at all. In addition to the various properties of the face, this contribution will be affected by the light sources and global lighting model that is in effect for the environment.

The appearance of a face under a given lighting configuration is generally described by properties of the face as a whole and properties of the points on the face. Properties given to the entire face include the color of the face under no light, the color response to a diffuse light, the color response to a specular (mirror) reflection, and the transparency of the face. Properties of individual points on a face can be given by an image or texture map, which contains color and transparency information for points on a two-dimensional (2-D) grid. The color information of the image map may be combined with the face properties or may override them (the fountain scene in Fig. 12.3 uses texture maps for the floors and ceilings). As a whole, the material properties define a contribution of certain values of red, green, and blue to a pixel on the display, if the face is visible at that point on the display.

## 2.4 Lighting and Shading

The appearance of a geometric primitive depends not only on its material properties, but also on how light sources in the environment are positioned. An environment is assumed to have a

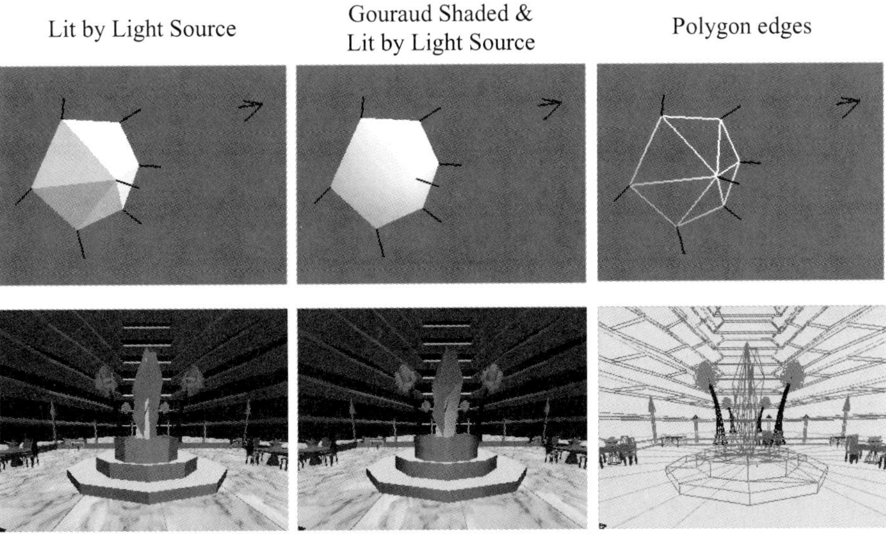


FIG. 12.3. The effects of a lights sources and Gouraud shading.

certain amount of low-level ambient light, and the material definition given to a face describes the color to use for the interior of the face in the absence of any light sources. For each light source that does exists, a proportional amount of diffuse color, from the material definition, is added to particular pixel colors for parts of the face that are visible. The amount of diffuse color is dependent on the incident angle of the ray from the light source onto the face. Light sources can be given at infinity, where all light rays are parallel and travel in a certain direction, or can be given a position from which the rays emanate. Systems that render the display at interactive rates generally do not consider obstructions to light rays, and therefore no shadows are shown. The left two screen-shots in Fig. 12.3 demonstrate the effect of light sources. The location of the light source for the six triangles in the top left image is shown in the top right corner of the image.

Light sources can be included as a node or with geometry in the scene graph, but the rendering system must only use the location and direction of the light in the coordinates in which objects are rendered. The coordinate system transform that is usually used to define a node's position in the world or relative to another node may include nonrigid transformations, such as scales and shears, which do not preserve angles or distances.

In the images on the left of Fig. 12.3, each polygon was considered individually. If a group of polygons approximates a single, curved surface, the interior of each polygon can be shaded so that the surface appears curved rather than faceted. When the color of a pixel is calculated, the normal used in the lighting calculations is the normal to the *surface* at the point which is projected to that pixel, rather than the normal to the polygon. Calculating a new normal for each pixel drawn when filling a polygon on the image can take a significant amount of time, however. The most common technique for "smooth shading," or Gouraud shading, uses the normal to the surface at each vertex of the polygons that make up the surface. Normal vectors for the interior pixels of the polygon are interpolated from the normal vectors at the corners (the value is a weighted average of the vertex normal vectors, based on the relative distance from each vertex). The middle images of Fig. 12.3 use the Gouraud shading technique. The images on the right show the edges of the polygons that are being drawn in the other images. The black lines in the top images show the normal vectors for each vertex. Foley et al. (1996) describe the Gouraud and other shading algorithms in more detail.

## 3. BEHAVIOR

A dynamic environment can be represented by a set of objects that have particular behaviors. These behaviors may be given as part of the stored model of the object, may be defined by the execution of the application code, or may be some combination of the two. An object may have a set of behaviors that it follows given its current context. These behaviors can be roughly categorized as *environment-independent behaviors*, which do not consider the current state of the other objects in the environment, and *environment-dependent behaviors*, which do consider other objects. Environment-independent behaviors include changes to the object that are solely based on the passing of time, or *time-based behaviors*, and changes that occur in a particular sequence, or *scripted behavior*. Environment-dependent behaviors include *event driven behaviors*, which respond to events initiated by users or other objects, and *constraint maintenance behaviors*, which react to changes of other objects to maintain defined constraints, such as relative placement, gravity, or penetration limits (see Fig. 12.4). The following subsections describe each type of behavior in more detail, concluding with a subsection on interactive behaviors encapsulated into *interactor* objects.

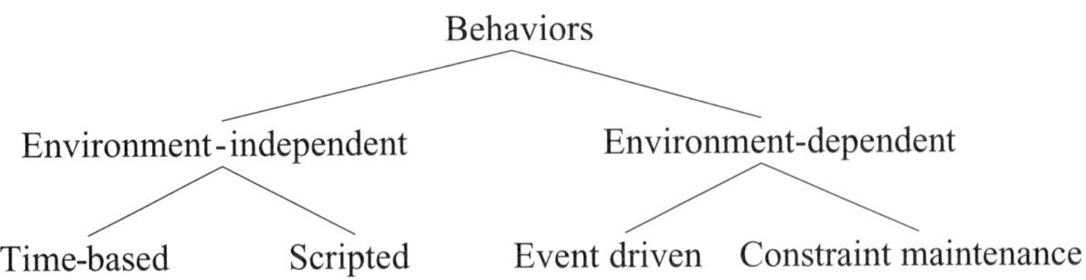


FIG. 12.4. Categories of behavior specification in virtual environments.

## 3.1 Time-based

An object's behavior may describe a change in one of the object's properties over time or at a certain point in time. Object properties that may be changed include the object's position, orientation, color, and visibility. A circling airplane may follow a particular path as is flies around a town, where its position and orientation are determined by the amount time that has passed from when it began flying. A distant mountain may change from a detailed model displayed during daylight hours, to a simpler model for dawn and dusk, to being invisible at night.

Time-based behaviors may be defined to use wall-clock time, also known as absolute time, which is the time frame of the user; or a time frame that is transformed from wall-clock time. Behaviors usually have a start and end time. A behavior may change a property continually over the time between start and end times. For convenience, the time values used to define such a change usually are zero at the start time and one at the end time. If a behavior should last, say, 5 minutes, then the behavior can be scaled appropriately.

Time frames can be used like one-dimensional coordinate system frames. A time frame can define a translation (time 0 to a start time) and a scale (time 1 to an end time). Time frames can be related to the wall-clock time, or they can be organized hierarchically, like a coordinate system graph. For example, in the Mirage system (Tarlton & Tarlton, 1992), an *activity* consists of the change of an object's property from a $t = 0$ to a $t = 1$ value. A single activity can be set to begin and end at certain times, or be combined with other activities by becoming siblings of a parent activity. The relationship between a parent activity and its children may specify that the child activities occur simultaneously, or one after the other. The parent activity will have a relationship with wall-clock time that will have it occur at a certain time and pace. With this structure, an activity could flow at various speeds, and could even be paused or reversed.

For example, Fig. 12.5 shows an activity hierarchy where a "Rolling Wheel" activity consists of a "Rotating Wheel" activity and a "Moving Wheel" activity. These child activities may be combined for a coordinated motion by making no transformation from the parent time frame to

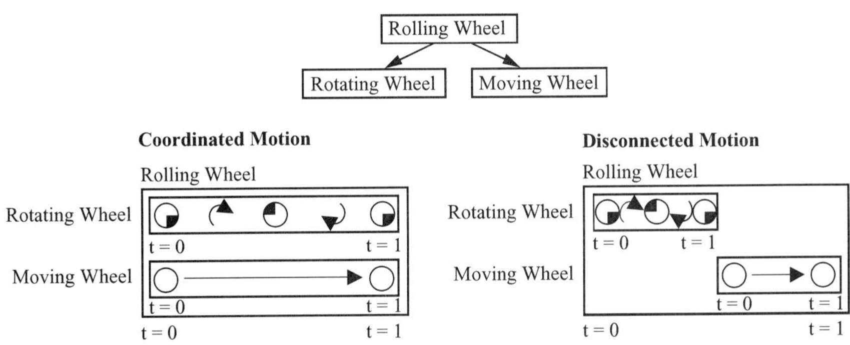


FIG. 12.5. Activity combinations (Adapted from Tarlton & Tarlton, 1992, with permission).

the time frame of each child, as is shown on the diagram on the left. Alternatively, the motions can be performed one after the other by transforming the 0 to 0.5 time values of the Rolling Wheel activity to the 0 to 1 time values of the Rotating Wheel activity, and transforming the 0.5 time values of the Rolling Wheel activity to the 0 to 1 time values of the Moving Wheel activity. This is shown in the diagram on the right of Fig. 12.5. The left structure of the figure produces a rolling wheel, and the right structure produces a wheel that rotates and then slides.

Individual activities may specify an instantaneous change in an object's property at the activity's start time, or may specify a continuous change of a property based on the time value. The continuous change may be given as a function of the time value, or may simply be specified as an *interpolation* between a starting and ending property value. Different interpolation methods may be used depending on the property type. For example, a linear interpolation is often used for position changes, while a color change may be a linear interpolation of colors in HSV (hue, saturation, value) space, having been been transformed from RGB (red, green, blue) space. Other interpolation methods can be used to produce a different transition between values. For example, an "ease-in, ease-out" method slowly increases the rate a value changes until the halfway point, then slowly decreases the rate back to zero, at which point the new value should be obtained. A common animation technique is to define particular property values, such as position and orientation, for objects at certain moments in time. A set of these values are known as a *key frame*, which define the start and ending values of interpolation activities that follow one another. (See Watt and Watt [1992] for a detailed description of how interpolation is used in animating object movements.)

### 3.2 Scripted

A scripted behavior defines a set of steps to be taken in order. The run-time system may make a new step every time it renders an image to the display, or one step after a specified number of image frames. The speed of the behavior, therefore, is directly tied to the frame rate of the application. Scripted behaviors may not appear smooth, because they are not tied to a constant passage of time. However, a scripted behavior can provide a transition or animation that shows every step to the user while completing as fast as possible. If the behavior is time-based, there is a chance that a step will not be seen when it occurs between the times an image frame is rendered.

### 3.3 Event-driven

The behavior of an object may be specified as a reaction to an event that occurs in the environment. An event may represent a user's action, such as pressing a button, or it may represent an incidence in the environment, such as the collision of two objects. Alternatively, it may represent a change to the property of another object, such as the other object's appearance or mass.

Events can be stored in a global *event queue* (which may actually be many queues), and dispatched to objects that have expressed an interest in them, or can be stored locally in an object's *out field*, and *routed* to the *in field* of other objects when such a route has been set up. Generally, systems that are designed to support event response in application code use the event queue model, whereas systems that use only behaviors defined in a file description, such as VRML (Ames et al., 1997), will use event routes. However, the Java3D package (Sowizral et al., 2000) uses an event route network, which eases the incorporation of VRML models into its run-time system. The event route model is not as flexible as the event queue model, as a route must have a single source, whereas an event can come from anywhere. It is not difficult for an object to listen for events from an event queue that come from any object, but it is difficult for an object to set up routes from many objects.

### 3.4 Constraint Maintenance

It is sometimes easier to describe a behavior declaratively, stating a set of relationships that should be maintained, rather than defining a procedure to enforce the relationships. This is akin to specifying what result you want, but not how to get the result. The positional relationship between a parent and child in a coordinate system graph is an example of a declarative constraint: The child should have a certain position in relation to the parent. Of course, the simplicity of describing a behavior with constraints comes with a cost. The run-time system must ensure that the constraints are maintained, and must deal with situations where the behavior is overconstrained or underconstrained. In addition, it is difficult to design a system of constraints of any complexity, as one change can have far-reaching, and sometimes puzzling, effects.

Constraints can be specified for any set of properties of the objects in the environment. For example, the TBAG tool kit (Elliot, Schechter, Yeung, & Abi-Ezzi, 1994) is a graphics library and run-time system that provides "constrainable" program objects that describe relationships between values that should be maintained. The TBAG tool kit uses the SkyBlue constraint satisfaction algorithm (Sannela, 1992) to generate the values that satisfy given constraints as best as possible. However, two types of constraints are found more often than others in VE run-time support systems: geometric constraint maintenance and physics-based motion.

A geometric constraint uses properties such as attachment and spatial relationships with other objects. Using the coordinate system graph as part of a scene graph, a run-time system automatically maintains a spatial relationship between child and parent objects in the graph. The run-time system may also maintain a direct spatial relationship between two objects that are not a child or parent of the other in the scene graph. For example, objects that represent 3-D interface components (buttons, menus, etc.) may be automatically arranged adjacent to each other so that they do not overlap even if components are moved or scaled. In another example, the WALKEDIT system (Bukowski & Sequin, 1995) allows users to quickly position objects to construct rooms with furniture, books, and other items. This task is facilitated by the enforcement of spatial constraints that come from abstract relationships or associations. Picture frames are constrained to vertical surfaces; cups sit on horizontal surfaces such as desks or floors, and so on.

A run-time system can reasonably maintain a list of one-way constraints, where a change in one property results in a change in another property, but not vice versa. For example, a change in the position of a parent node in the scene graph affects the position of the child node in the environment, but a change in the child node's position does not affect the parent node's position. It is much more difficult to support two-way constraints, or constraint dependencies that contain cycles, as a solution must be found by solving a system of simultaneous equations, and it may be over- or underconstrained (a solution does not exist, or many possible solutions exist). A run-time system that needs to enforce two-way constraints can implement constraint satisfaction algorithms such as SkyBlue.

A virtual environment can look more "realistic" if the objects in the environment follow the physical laws of the real world. Real-world objects have mass, may be rigid or flexible, solid or fluid, will fall at a constant acceleration when not supported, will move and twist when pushed by some force, and so on. However, simulating a world with many objects that obey such laws, and producing an accurate view of the objects at interactive rates, is a job beyond the ability of most computers in existence today. Therefore, current run-time systems generally do not try to enforce physical laws, but a few provide assistance to applications for a small subset of the objects in the environment. Objects may be given properties such as mass, velocity, angular velocity, acceleration, and angular momentum. In addition, some run-time systems can report when collisions occur between the geometries of two objects, or a separate support library can be used, such as I-COLLIDE (Cohen, Lin, Manocha, & Ponamgi, 1995). It is usually up to the application, however, to determine how objects might change form or trajectory due to a collision.

## 3.5 Interactors

As is true of 2-D graphical applications, a 3-D environment may contain geometric objects that belong to a class that interact with the user and other objects in a well-defined way. In 2-D interfaces, these objects are called widgets, components, or interactors, and include buttons, menus, scrollbars, and pointers. Each class of objects encapsulates an interactive behavior, parameterized to allow for multiple uses. Buttons, for example, have the same behavior when "pressed" by the mouse pointer, but have different labels and reactions to user actions. Since the number of ways a user interacts with a VE can be large and the interaction between 3-D objects complex, the implementation of such interactor classes is important to facilitate the development of interactive VE applications.

A 3-D interactor is generally a geometric object with a defined behavior when certain events occur and certain properties of the environment and objects in it change. For example, a 3-D button might react to the collision of a ray or associated hand object. Figure 12.6 shows a VE application with several 3-D interactors. An interactor can be defined as a specialized node in the scene graph, as is done in the VRML file format using PROTO nodes (Ames et al., 1997). These specialized nodes can be given fields to store state, and *in* and *out* connection points for event routes. A system, such as the Virtuality Builder II (VB2) (Gobbetti & Balaguer, 1993), which uses a constraint satisfaction algorithm, can enforce constraints between fields internal to an interactor node and between fields of different interactors. In addition, one-way constraints can constrain field values to user input, such as hand position.

An interactor framework that uses event routes to connect interactors together can be difficult to wire in, especially if the interactor reacts to a property for a set of objects (the hand position of any of a group of users, for example). In the Simple Virtual Interactor Framework and Toolkit (SVIFT), an extension to the SVE library, interactor classes can themselves identify the set of events and property changes that they are interested in, based on the parameters given at instantiation and the behavior desired (Kessler, 1999). For example, a "narrator" interactor may look for any "description" property that is stored with any object added to the scene graph and observe the gaze direction of the user. If the user looks in the direction of an object with a description for a set amount of time, the narrator object would display or "read" the

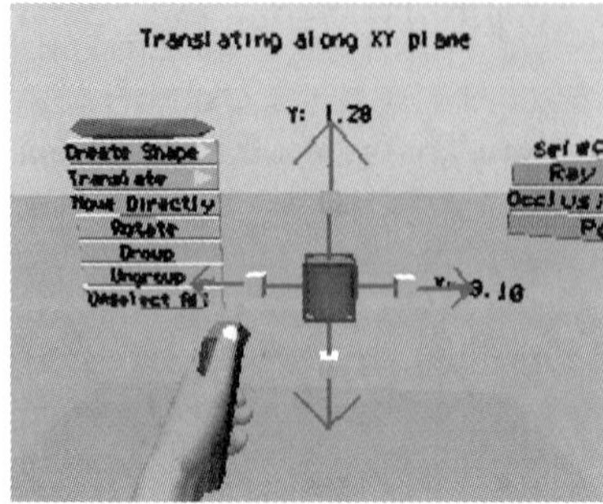

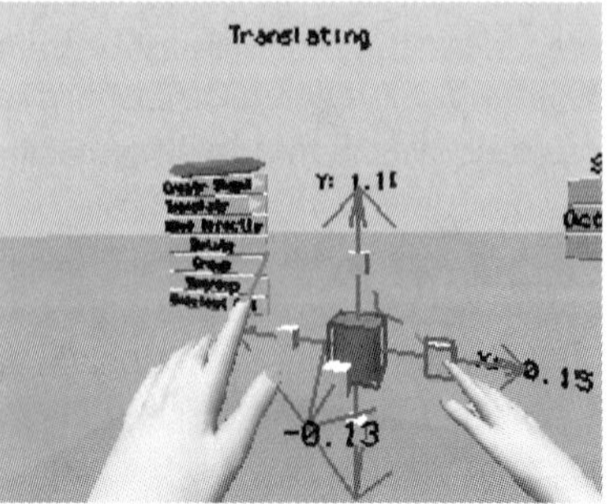


FIG. 12.6. Interactors in a shape manipulation task.

description. Figure 12.6 contains a set of SVIFT interactors, including buttons, menus, tabs that can be grabbed and moved (or bestow the grab and move ability to another object), constraint maintainers (to keep the manipulated object on a particular plane, for example), and selectors. The figure on the left is using a ray selector interactor, and the figure on the right is using a "poking" selector. Since the other interactors in the environment are defined to respond to any "selection" event that has them as the selected object, they can work with either selection mechanism.

## 4. FILE FORMATS

In the previous sections, the vocabulary with which VEs are described was defined. The geometry and behavior of a VE application may be given procedurally, using this vocabulary as a basis. A VE application may contain code that constructs geometry using a low-level graphics library, such as OpenGL, or a high-level library such as Java3D. The application may also contain a step-by-step procedure for how geometry changes over time or as a reaction to the user's actions.

A geometric model can generally be described more efficiently by its geometric primitives and properties, such as vertices, colors, and so on, than by giving a procedure for building the geometry. A file format can be designed to describe geometry and spatial relationships between geometries, and therefore allow for clear and concise descriptions of a geometric model. It becomes the task of a file loader and the run-time system to translate the model from the stored representation to the run-time model, and ultimately to the internal model for efficient rendering.

A file format might also allow simple, object-based behaviors to be defined. However, complex behaviors and behaviors involving many objects are usually best left to a programming language to describe the behavior procedurally. For example, the Alice system (Pausch et al., 1995) uses geometries that are described by model files, but object and environment behaviors are described using Python scripts.

Files that describe geometric models may be intended to be read and edited directly by developers, such as those defined by VRML (Ames et al., 1997), or be intended to be created and modified only by modeling applications, such as those defined by the Multigen OpenFlight format (MultiGen, Inc. 1998). A format that will be edited by a developer must define an organization of geometric primitives and components that match the developer's conceptual model of the environment's construction. In addition, it should use descriptive keywords and allow components to be named. A model format that is not intended to be edited by a person can be more compact, even though it can still be read. For example, the Wavefront obj format defines a single geometric object using vertices, vertex normals, texture coordinates, faces, and face materials, each of which are defined using a short keyword ("v" for a vertex, "vn" for a vertex normal, etc.). Numbers are represented as a series of numeric characters, rather than being stored as a binary value, allowing the file to be read (and edited if necessary). A Wavefront material file supplements an obj file to provide the definition of face materials.

## 5. INTERNAL MODEL

The internal model representation of the environment is designed to provide the rendering performance required for VE applications. The run-time representation of a geometric model could be passed directly to the low-level graphics library and audio libraries to present to the user, but the run-time representation is not likely to be the most efficient representation to use, especially for visual components. An internal representation, therefore, can be created that provides a more efficient transfer to the graphics hardware, and therefore allows for faster performance of the system.

The goal of almost all rendering optimizations is to reduce the amount of information passed from a software application to the rendering hardware. These optimizations can be automatic or application-assisted. One type of automatic optimization generates additional representations of a geometry from a run-time or stored model that uses less information to define the geometry. Another computes extra information that allows for run-time selection of parts of the geometry that can be ignored. Automatic optimizations include primitive sorting for reduced state changes, display list generation, and culling faces for the render list based on the user's viewpoint. Application-assisted optimizations include switching between different levels of detail and using "billboarded" geometries, described later. This section describes these optimizations in more detail. Many of these are implemented, for example, by the IRIS Performer tool kit, described by Rohlf and Helman (1994).

## 5.1 Primitive Sorting

A graphics rendering system uses a current *graphics state* to determine how primitives will appear when they are rendered. The current graphics state includes the coordinate system, material properties, current texture image, and global properties such as fog and light sources. When describing each primitive of a model to be rendered, an application could specify all properties of the graphics state required to render the primitive correctly. However, more often than not, a primitive is part of a greater whole and shares most of its properties with other primitives. For example, a set of triangles that describes the surface of a hood for a car model will share the same material and lighting properties, and will share the same lighting properties with the triangles that describe the hubcap. Repeating the current state for every polygon on a particular surface would be quite inefficient.

Systems that generate an internal model for rendering efficiency, such as the Performer tool kit, provide a framework that allows primitives to be sorted by their graphics state. Primitives that have the same material and global properties will appear together, while those with similar properties will be close to each other, and so on. This modified list will be used to describe the model to the renderer, and state changes will only be given when necessary. The result of this optimization is that much of the redundancy of the model is removed.

The most optimal list of primitives would be a sorted list of all of the primitives. However, this list would need to be updated whenever one of an object's properties changes. For example, since all of the primitives would need to have a common coordinate system, all of the primitives of an object would need to be transformed to a common, world coordinate system, taking into account the coordinate system transformations between the object and its parent and ancestors. If the object were to move, however, through the change of one of these transformations, all of the primitives in the list that are part of the object would need to updated with a new position. This would be a bookkeeping nightmare. Instead, lists of primitives that are sorted are usually limited to individual objects, as described in the next section. However, Durbin, Gossweiler, and Pausch (1995) describe a system that provides a list of primitives for each object which is either just the primitives of the object, or one that includes the primitives of the object and its children, if the children have not changed their state in a prescribed time. In this system, when a child changes its state and its primitives are part of a parent, the parent returns to its individual list until its children do not change for some specified time. Therefore, object trees that do not change often are rendered more efficiently (perhaps dramatically so, if the objects share many properties).

### 5.2 Display Lists

Another method to reduce the transfer of geometry and property changes through the graphics system is to, instead, create *display lists* representing the graphics commands required to create the geometry and to make property changes. A display list can be created and transferred once through the graphics system, then referred to by a unique ID each time the commands should be executed by the graphics system. The graphics system may perform the sorting optimization. Display lists can contain the commands to create a geometric object, or may simply consist of the commands to set up a particular material with a texture image to be used for a variety of primitives.

Like a sorted primitive list, a display list is only useful as long as the properties of the primitives do not change. When one property of a single primitive changes, the list needs to be recreated. Since this can cause delays, the Performer library does not use display lists for most primitives. Systems that do use display lists and sort primitives must catch changes to properties and perform a regeneration of the list, or simply prevent the change.

### 5.3 Face and Object Culling

Two common methods to reduce the amount of information transferred to the rendering system involve omitting faces that cannot be seen by the user. One method, *back-face culling*, makes the assumption that face primitives define the surface of a solid object and that the face will only be viewed from outside of the object. Therefore, the face is only rendered when the user's eye point is on the front side of the primitive, which is generally defined to be the side of the face for which the list of vertices is given in a counterclockwise manner. If the viewer is on the back side of the face, it is not rendered. If the object is one that the viewer can be inside of, such as a room, the faces must be made to be two-sided, usually by giving the same list of face vertices twice, in opposite orders. The back-face removal technique is usually a feature of the low-level graphics library.

The other method to reduce the amount of faces rendered involves the higher level representation of the objects in the environment. Objects are checked for an intersection or containment in the user's view volume frustum, which defines the volume of the environment that the user would see if there were not obstructing objects. This volume is defined by the position of the user's viewpoint and viewing window in the environment, and the near and far clipping planes. If an object is found to be outside of the viewing frustum, it is simply not rendered, saving the rendering system the effort of determining that each face of the object falls outside of the rendered display.

This method depends on having a simple geometric representation of each object that is a good approximation of the volume the object occupies. Common representations include spheres and boxes. It is easy to check for intersections between spheres and other spheres or plane-based volumes, but sphere boundaries often contain much more empty space than space occupied by the object. In addition, to make intersection checking fast, spheres must be defined in a common coordinate system, as nonuniform scale and shear transformations will change the spheres into nonaligned ellipsoids that are much more difficult to work with. A box volume can be aligned to a single world axis for easier intersection testing or aligned with the local coordinate system of the object it represents for a more accurate approximation (objects are often defined so that their widest length is parallel to a major axis).

Since a scene graph defines a hierarchy of objects, this hierarchy can be exploited for more effective view volume culling. The parent of a group of child objects might have a bounding volume that contains the geometry of the parent and the geometries of the group. Therefore, if the parent object's bounding volume is outside of the viewing volume, so are all of its children, and that whole subtree can be omitted from the rendering of the current frame. If the environment is sparsely populated by objects, then a more effective bounding volume hierarchy may be constructed separately from the scene graph based on grouping the closest objects, and then grouping the closest groups together, until one group contains all objects.

### 5.4 Level-of-Detail Switching

The amount of detail necessary for an object's description depends on how much detail the user can possibly notice. If the object is far away or in a foggy or low-light environment, the user will not notice much detail, while if the object is near by on a clear day, most detail will be perceived. A method of reducing the amount of information transferred to the rendering system involves selecting the object geometry from a list of progressively less detailed representations that has the least detail but will appear very similar to the object at its full detail. That object is used, potentially saving the wasted effort of rendering detail geometry that would not be noticed. This method of omitting detail can work for all presentation mediums involving the distance senses. Sights, sounds, and smells diminish in intensity and detail based on distance or other factors. In general, the application must supply different levels of detail for an object and give some measure of how it compares to a full detail model. A common approach, as seen in the *level of detail node* (LOD) of the VRML file format, is to associate a representation with a range of distances from the user's eye point for which that representation should be used. A system, such as Performer, can also automatically determine when an object's geometry is so small that it does not occupy enough of the screen to be noticed, and therefore can be omitted completely.

### 5.5 Billboarded Geometry

A common technique that is used to reduce the amount of detail needed for a geometric model is to use an image containing a picture of the detail. This technique works well for distant objects, as the absence of depth to parts of the model is not noticed. However, this technique alone will not work for objects that can be viewed from all directions, such as trees or clouds. Another technique can be used, in addition to an image, to give the appearance of depth to a model if the model can be assumed to appear the same from all directions. This technique is known as *billboarding*. The image is placed on a flat polygon, but the polygon automatically always faces the viewer, or always orients to the viewer while keeping one axis aligned (such as the "up" axis for standing objects). For example, a tree image will remain upright, but will appear to have a trunk and crown from any point of view (except from directly above or below). The tree will appear to be three dimensional, if a bit too symmetrical. This technique can be combined with changing the image based on the rotated angle to give an approximation of an object that is not symmetric around one or more axes.

## 6. INCLUDING THE USER

When creating 2-D images for interface and content, computer applications use a standard location for the viewer: some distance in front of the computer screen. In fact, the viewer is mathematically assumed to be at a great distance from the screen, as raised buttons appear the same whether they are on the left, right, top, or bottom. The geometric model of these applications is two dimensional, or an ordered composite of 2-D image layers, all of which are in front of the viewer. However, when the geometric model becomes truly three dimensional, the viewer can be placed in the model itself, giving the user a presence in the 3-D environment.

This allows users to view a 3-D model from any perspective, to get closer and even "fly" through the model. Including a representation of the user in the 3-D model itself clearly specifies the point of view for the image rendered for the user, and it allows the behavior of the model to affect the view: When the user is "attached" to a moving airplane object, the viewpoint moves with it. In addition, geometry can be "attached" to the user so that it moves when the view changes (through mouse manipulations or tracking device movements), as in a "head's up display." Finally, the user model can describe the type of display being used and specify the window in the 3D world that matches the display image edges. The image may be like glasses into the virtual world, moving as if attached to the user's head (a head-mounted display [HMD]), or a window into the world that moves independently of the user's head (a "fish tank," workbench, or handheld display). Figure 12.7 shows three different representations of the user (Kessler, Bowman, & Hodges, 2000). These three representations will be described in more detail below.

Placing a representation of the user in the 3-D model of a VE not only provides a convenient way to specify the user's view of the world, but it also allows for a unified method for handling the user's visible representation in the world and interaction with the world. Geometric objects can be given for important parts of the user's body, such as the arms and hands, which may be controlled by any positional input device (e.g. mouse, joystick, or 6-degree-of-freedom tracker) or scripted animations. The direct interaction of the user with the world can be handled just as the interaction between any two or more objects is handled. The following sections describe how the user can be represented in the environment model, how tracking input devices can control the user model, and finally, how the user model can generate the viewing parameters for the various display devices that the user may be utilizing.

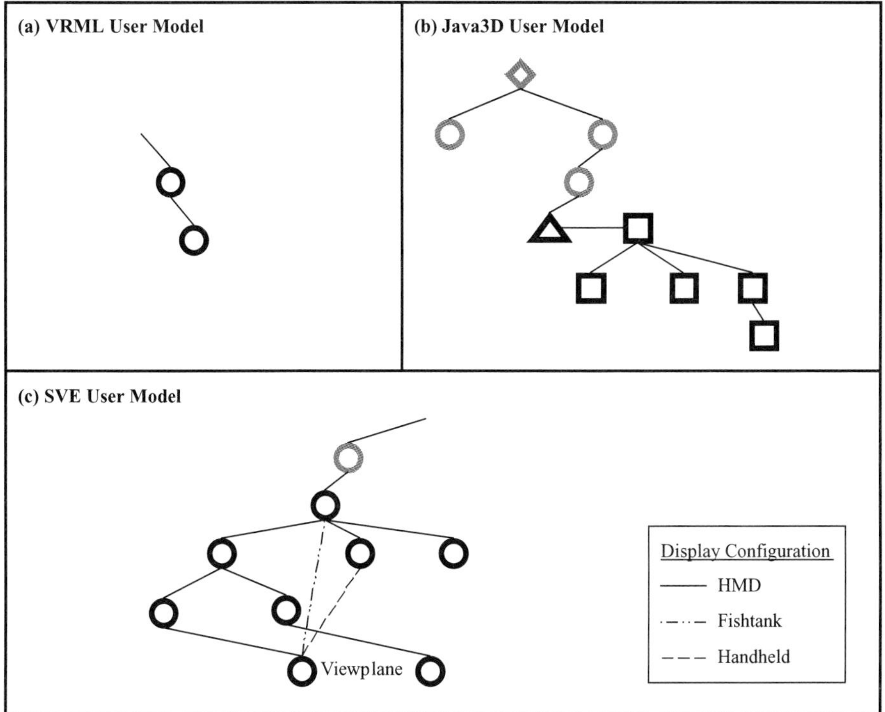


FIG. 12.7. User model in the scene graph for VRML and the Java3D and SVE libraries.

## 6.1 User Model

The simplest representation of a user in a 3-D environment includes a location, gaze direction, and "up" direction for the view. However, this "viewpoint" method conceptually places the user outside of the environment. A better method, common to most VE systems, is to represent the user by one entity or a group of entities in the environment model, which can be "attached" to other entities of the model and be manipulated like other entities. This method allows the user's viewpoint to follow an object that she is "in," such as an airplane or car. The VRML file format provides an example, shown in Fig. 12.7a, where the user is represented by a *Viewpoint* node (Ames et al., 1997). The Viewpoint node can be attached to a *Transform* or other *Group* node, which means that the user's viewpoint will change as the positions of the Transform nodes further up the tree change. The *NavigationInfo* node describes how the user's viewpoint can change within the coordinate system of the Viewpoint node, including the speed, and constraints on movement.

The viewpoint representation of the user also fails to include other important information, such as the position of the two eyes for stereo viewing; the position of hands, feet, torso, and other body parts; and the movement of digits. Information about the user is used to help determine the view presented, to provide a visible representation of the user in the environment, to allow direction interaction with the geometric objects in the environment, and to provide input to gesture recognition algorithms. Given that this information is spatial in nature and that it is all about one subject (the user, of which there may be more than one in the environment), it is natural to include the user data in the environment model.

Figure 12.7 shows the user representation in three different 3-D environment models, the VRML file format (Ames et al., 1997), the Java3D scene graph (Sowizral et al., 2000), and the SVE coordinate system graph (Kessler et al., 2000). In addition to the viewpoint being associated with a Viewpoint node in the VRML file format, an attached NavigationInfo node describes the size of the user's avatar in the environment. The VRML avatar is described as having a width, given as a radius from the user's position, a height, and a maximum step size. The width is used to determine collisions between the user and geometry in the world, the height is used to describe the distance the user should be above the ground if following the terrain, and the step size is used to determine if a height difference in the terrain is too large to allow the user to pass over it.

The representation of the user in the Java3D scene graph is similar to the VRML format. The location of the user in the 3-D environment is defined by the ViewPlatform node, which can be placed in the scene graph so that the user automatically moves with objects. However, Java3D has two additional program objects, the *PhysicalBody* and *PhysicalEnvironment* objects, which define the positions of the user's head, eye(s), and hands in the environment. These objects encapsulate the transformation from the ViewPlatform coordinate system to a "coexistence" coordinate system and describe the locations of user body parts in that coexistence coordinate system. The application programmer can use the calculated positions of the various body parts in *Behavior* nodes to define interactions between the user and the environment, and to have separate geometric objects follow the user's movements and provide a visible representation of the user in the environment.

In contrast to Java3D, the SVE coordinate system graph associates a separate geometric object in the coordinate system graph for each important body part, as is shown in Fig. 12.7c. A *User* object represents the user as a whole and can be attached to the environment model at any place, like the Viewpoint and ViewPlatform nodes. Attached to the User are a *Head* object and two *Hand* objects. Attached to the Head object are two *Eye* objects. Other user objects, such as feet, palms, finger joints, and so forth can be added to the representation as needed. The next section describes how the relationship between the physical locations of the user's body parts, if tracked, and location of a geometric object in the environment is maintained. These user objects may not have a geometry defined for them, simply representing a location in the world, or may have a geometry or child objects with geometry to give a visible representation of the user in the environment. An advantage to this user model is that the interaction of the user with the world can be handled in the same way as interactions between any set of objects in the environment are handled.

## 6.2 Incorporating Tracking Information

Tracking devices generally report the position and orientation of a "tracker" in relationship to a particular reference frame. If the tracker is at the end of a set of mechanical linkages, then the reference frame is the base of the linkages. If the tracker is an electromagnetic device, then the reference frame is the "transmitter" part of the device (often placed on the desk or hung above the user).

Information from the tracking device can be used in raw form or adjusted by a given offset to relate the tracking reference frame to the world coordinate system of the environment. This information can be used to determine the position of the viewer or be used to determine how the user interacts with the scene. However, if the tracking information is to be used in a coordinate system other than the world coordinate system, then the reference frame offset must be carefully calculated and updated when the scene graph changes. In fact, two offsets might need to be maintained: a transformation between the tracker reference frame and the environment, and a transformation between the tracking device and the reference point of the associated user's body part. For example, the transmitter may be mounted above the user's head, and a head tracker may be mounted on an HMD such that it is a few centimeters above and behind the top of the head and tilted.

The user model of Java3D defines a set of coordinate systems between the ViewPlatform node in the scene graph representation of the user's position (or, more precisely, the position of the user's feet, screen, or eye), and the user's head, eyes and hands. The ViewPlatform node defines where its origin is in "coexistence" coordinates, based on whether it represents the user's feet, screen, or eye. The PhysicalEnvironment object defines a transformation from the coexistence coordinates to the tracker reference frame, the "tracker base" (the first offset discussed above). Finally, the PhysicalBody object defines a transformation from the head position to the head tracker position (the second offset discussed above). As discussed in the next section, this model supports a variety of display configurations. The particular display configuration used can be defined separately from the Java3D scene graph, since the necessary parameters are in the View object and associated objects (shown as squares in Fig. 12.7b).

In the SVE library, tracking information is introduced into the user model to control a body part (or any other object in the environment) by adding two new geometric objects. One object is used to represent the reference frame itself, and it is set to be the child of the controlled object's parent. The other new object represents the tracker and is placed as a child of the reference frame object. The controlled object is then set to be the child of the tracker object. The tracker object's coordinate system transformation is set to the raw tracking information each frame. By initializing the coordinate system transform of the reference frame object correctly, the controlled object will move around its previous parent correctly. An added benefit to having a new reference frame object for each tracker is that different tracking devices can be used simultaneously, even if their tracker bases are at different locations and the tracker reference frames do not match. If a correction needs to be made for how the tracker is attached to the body part it is tracking (e.g., if it is 90 degrees off), then a correction can be placed in the transformation from the tracker to the controlled object. The SVE model provides the application with spatial information such as tracker reference frame and tracker position in the same way it provides spatial information about any other object, simplifying the implementation of user interaction and environment behaviors.

### 6.3 Viewing Model

One of the main tasks of the user model is to define the parameters necessary to generate the user's current view so that it can be rendered to the user's display. The VRML Viewpoint node provides a common representation for the user's view: that of a camera in the environment. The Viewpoint node provides a location, orientation, and a field of view for a camera model, which defines how geometry is projected onto the display toward an eye point. Two viewpoints can be used with a camera model when generating the images for a head-mounted display. In this case, however, precision requires that the optics of the display be accounted for, as described by Robinett and Holloway (1995). For example, the standard assumptions that each eye is centered on the image and that the display plane is perpendicular to the gaze direction are generally not quite correct.

The main assumption of the camera model is that the eye point is related to the display image by a constant offset. When the eye moves relative to the display, then the camera model no longer applies. This is the case for "fish tank" display configurations (Deering, 1992), which include desktop monitors and projected images on horizontal workbenches or on walls. For these configurations, the user model provides the position of the user's eye (or eyes, if a stereo image is displayed), and the location in the environment of the window that defines the view seen from each eye point. For example, the Java3D scene graph supports both HMD and fish tank display configurations by storing transformations in the PhysicalEnvironment, ViewPlatform, and Scene3D objects that relate the tracker reference frame (the "tracker base") to the displayed image. For HMD configurations, this relationship changes constantly. For fish tank configurations, it is constant. Regardless, the view is determined by the relationship between the screen and the eye point (which is controlled by the tracker that moves in the user model's tracker reference frame).

Following the philosophy that spatial objects in the user's environment should be represented by geometric objects in the coordinate system graph, the SVE user model includes a *Viewplane* object that represents the outline of the display. In SVE, a view is defined by one or two Eye objects, a Viewplane object, and the edges of the display, given in the coordinate system of the Viewplane object. Since the Eye objects and the Viewplane object are in the same coordinate system graph, they can be related spatially and determine the viewing parameters. Obviously, if the Viewplane and Eye are placed in the graph so that they can move independently, then this model can support fish tank display configurations as well as HMD configurations. In fact, since the Viewplane and the Eye can be moved independently of the tracker reference frame, the SVE user model supports a third display configuration: handheld displays (Rekimoto, 1997). In this configuration, the window to the virtual world is rendered to a handheld display that moves with the user's hand motions.

The Viewplane object is the representation of the projection plane for the default view of the environment and is placed in the coordinate system graph differently depending on the display configuration, as shown in Fig. 12.7c. For HMD configurations, the Viewplane object is positioned to correspond to the optical projection plane of the HMD (with window extent values that provide the correct vertical and horizontal field of view angles for the HMD) and is attached to the Head object. For fish-tank displays, the view plane represents the actual monitor screen in the real-world configuration. Therefore, the view plane does not move with the user's head but stays stationary in the user's reference frame. Thus, the Viewplane object is attached to the User or Workspace object. For fish tank displays, the view plane must be positioned carefully so that the position difference, or disparity, between left and right views of an object is correct (objects at the view plane should have no disparity, whereas objects further in front of or behind the view plane should have larger disparities).

## 7. SUMMARY

Virtual environment applications present 3-D environments to a user that is located in the environment. In this chapter, a description of how the environment is defined in terms of three representations was provided. One representation defines the model in a file or set of files. The format of these files is designed to provide a concise, geometrically organized description of geometric primitives and simple behaviors. Another representation is used by the application at run time to modify the environment, effect behavior of the objects in the environment, and to make the environment react to user interactions. The final environment representation is an internal representation that, when possible, distills the description of the environment to the information that is strictly necessary for processing, and provides that information to a rendering system to produce the image for the display. Many of the techniques described in this chapter are given in terms of a visual result, but they could also be adapted to the presentation for other senses, such as sound and smells.

## 8. REFERENCES

Ames, A. L., Nadeau, D. R, & Moreland, J. L. (1997). *VRML 2.0 sourcebook* (2nd ed.). New York: John Wiley & Sons.

Bukowski, R. W., & Sequin, C. H. (1995). Object associations: a simple and practical approach to virtual 3D manipulation. In *Proceedings of the 1995 Symposium on Interactive 3D Graphics* (pp. 131–138). Monterey, CA: ACM Press.

Cohen, J. D., Lin, M. C., Manocha, D., & Ponamgi, M. K. (1995). I-COLLIDE: An interactive and exact collision detection system for large-scale environments. In *Proceedings of the 1995 Symposium on Interactive 3D Graphics* (pp. 189–196). Monterey, CA: ACM Press.

Deering, M. (1992). High–resolution virtual reality. In *Proceedings of ACM SIGGRAPH '92* (pp. 195–202). Chicago: ACM Press.

Dinh, H. Q., Walker, N., Hodges, L. F., Song, C., & Kobayashi, A. (1999). Evaluating the importance of multi-sensory input on memory and the sense of presence in virtual environments. In *Proceedings of IEEE Virtual Reality '99* (pp. 222–228). Houston, TX: IEEE Press.

Durbin, J., Gossweiler, R, & Pausch, R. (1995). Amortizing 3D graphics optimization across multiple frames. In *Proceedings of the ACM Symposium of User Interface Software and Technology* (UIST '95). Pittsburg, PA: ACM Press.

Eckel, G. (1998). *Cosmo 3D programmer's guide* (Doc. No. 007-3445-002). Mountain View, CA: Silicon Graphics.

Elliot, C., Schechter, G., Yeung R., & Abi-Ezzi, S. (1994). TBAG: A high-level framework for interactive, animated, 3D graphics applications. In *Proceedings of ACM SIGGRAPH '94* (pp. 421–434), Orlando, FL: ACM Press.

Foley, J., van Dam, A., Feiner, S., & Hughes, J. (1996). Computer graphics: Principles and practice (2nd ed., pp. 222–226). Reading, MA: Addison-Wesley.

Gobbetti, E., & Balaguer, J. (1993). VB2: An architecture for interaction in synthetic worlds. In *Proceedings of the ACM Symposium on User Interface Software and Technology* (UIST '93, pp. 167–178), Atlanta, GA: ACM Press.

Kessler, G. D. (1999). A framework for interactors in immersive virtual environments. In *Proceedings of IEEE Virtual Reality '99* (pp. 190–197), Houston, TX: IEEE Press.

Kessler, G. D., Bowman, D. A., & Hodges, L. F. (2000). The simple virtual environment library: An extensible framework for building VE applications. *Presence, 9*(2), 187–208.

Lindstrom, P., Koller, D., Ribarsky, W., Hodges, L. F., Faust, N., & Turner, G. A. (1996). Real-time, continuous level of detail rendering of height fields. In *Proceedings of ACM SIGGRAPH '96* (pp. 109–118). New Orleans, LA: ACM SIGGRAPH.

MultiGen, Inc. (1998). OpenFlight Specification; v15.6.0 (7/31/98) San Jose, CA.

Pausch, R., Burnette, T., Capehart, A. C., Conway, M., Cosgrove, D., DeLine, R., Durbin, J., Gossweiler, R., Koga, S., & White. J. (1995). Alice: Rapid prototyping for virtual reality. *IEEE Computer Graphics and Applications* (pp. 8–11). IEEE Press.

Rekimoto, J. (1997). NaviCam: A magnifying glass approach to augmented reality. *Presence, 6*(4), 399–412.

Robinett, W., & Holloway, R. (1995). The visual display transformation for virtual reality. *Presence*, *4*(1), pp. 1–23.

Rohlf, J., & Helman, J. (1994). IRIS performer: A high-performance muliprocessing toolkit for real-time 3D graphics. In *Proceedings of ACM SIGGRAPH '94* (pp. 381–394). Orlando, FL: ACM Press.

Sannela, M. (1992). The skyblue constraint solver. (Tech. Rep. No. TR-92-07-02). Seattle, WA: University of Washington, Department of Computer Science.

Sowizral, H., Rushforth, K., & Deering, M. (2000). *The Java 3D API specification, version 1.2.* Sun Microsystems. Palo Alto, CA.

Tarlton, M. A., & Tarlton, P. N. (1992). A framework for dynamic visual applications. In *Proceedings of the 1992 Symposium on Interactive 3D Graphics* (pp. 161–164). Cambridge, MA: ACM Press.

Watt, A. H., & Watt, M. (1992). *Advanced animation and rendering techniques: Theory and practice*. Reading, MA: Addison-Wesley.

# 13

# Principles for the Design of Performance-oriented Interaction Techniques

Doug A. Bowman
*Virginia Polytechnic Institute and State University*
*Department of Computer Science (0106)*
*660 McBryde Hall, Blacksburg, VA 24061*
*bowman@vt.edu*

## 1. INTRODUCTION

Applications of virtual environments (VEs) are becoming increasingly interactive, allowing the user to not only look around a three-dimensional (3-D) world, but also to navigate the space, manipulate virtual objects, and give commands to the system. Thus, it is crucial that researchers and developers understand the issues related to 3-D interfaces and interaction techniques. In this chapter, the space of possible interaction techniques for several common tasks is explored and guidelines for their use in VE applications are offered. These guidelines are drawn largely from empirical research results.

### 1.1 Motivation

Interaction (communication between users and systems) in a 3-D VE can be extremely complex. Users must often control 6 degrees of freedom (DOFs) simultaneously, move in three dimensions, and give a wide array of commands to the system. To make matters worse, the standard and familiar input devices such as mice and keyboards are usually not present, especially in immersive VEs.

Meanwhile, VE applications are themselves becoming increasingly complicated. Once a technology only for interactively simple systems (those in which interaction is infrequent or lacks complexity) such as architectural walkthrough (Brooks et al., 1992) or phobia treatment (Hodges et al., 1995), VEs are now proposed for use in domains such as manufacturing, design, medicine, and education. These domains will require a much more active user, and therefore a more complex user interface.

One of the main concerns with the advent of these complex applications is interaction performance, defined broadly. Two aspects of performance will be considered in this chapter: task performance and technique performance with human effects. Task performance refers to the quality of task completion, such as time for completion or accuracy. It is usually measured quantitatively. Technique performance refers to the qualitative experience of the user during interaction, including ease of use, ease of learning, and user comfort. These factors are all related to the concept of usability. The focus is on design of interaction techniques (ITs) that maximize performance and the use of such techniques in interactively complex VE applications. This is an extremely important topic. Since VEs support human tasks, it is essential that VE developers show concern for human performance issues when selecting interaction techniques and metaphors for their systems. Until recently, however, most VE interaction design was done in an ad hoc fashion, because little was known about the performance characteristics of VE interaction techniques. Here, a wide range of techniques for the most common VE tasks (travel, selection, manipulation, and system control) is presented. Perhaps more importantly, a large number of design guidelines are provided. These guidelines are taken, where possible, from published empirical evaluation of VE interaction techniques (and in many other cases from personal experience and observation of user interaction within VE applications), and are meant to give VE developers practical and specific ways to increase the interaction performance of their applications. The guidelines are summarized in Table 13.1.

## 1.2 Methodology

Many of the results in this chapter stem from the use of a particular methodology (Bowman & Hodges, 1999) for the design, evaluation, and application of interaction techniques, with the goal of optimizing performance. In order to understand the context in which the guidelines presented here were developed, the parts of this methodology relating to design are briefly discussed.

Principled, systematic design and evaluation frameworks (e.g., Price, Baecker, & Small, 1993) give formalism and structure to research on interaction, rather than relying solely on experience and intuition. Formal frameworks provide not only a greater understanding of the advantages and disadvantages of current techniques, but also better opportunities to create robust and well-performing new techniques, based on the knowledge gained through evaluation. Therefore, several important design and evaluation concepts are elucidated in the following sections.

### *1.2.1 Initial Evaluation*

The first step toward formalizing the design of interaction techniques is to gain an intuitive understanding of the tasks involved with VE interaction and current techniques available for these tasks. This is accomplished through experience using ITs and through observation and evaluation of groups of users. Often in this phase, informal user studies or usability tests are performed in which users are asked what they think of a particular technique or observed trying to complete a given task with the technique under study. These initial evaluation experiences are drawn upon heavily for the process of creating taxonomies and categorizations of interaction techniques (see section 1.2.2). It is helpful to gain as much experience of this type as possible so that informed decisions can be made in the next phase of formalization.

### *1.2.2 Taxonomy and Categorization*

The next step in creating a formal framework for design and evaluation is to establish a *taxonomy* of interaction techniques for each interaction task (i.e., travel, selection, manipulation, and system control; see Fig. 13.1). Such taxonomies partition the tasks into separable subtasks, each of which represents a decision that must be made by the designer of a technique. Some of these subtasks are related directly to the task itself, whereas others may only be

**TABLE 13.1**
Summary of Guidelines for the Design of VE Interaction Techniques

| *Generic Guidelines (sec. 3)* | *Travel Guidelines (sec. 4.1.2)* | *Selection Guidelines (sec. 4.2.2)* | *Manipulation Guidelines (sec. 4.3.2)* | *System Control Guidelines (sec. 4.4.5)* |
|---|---|---|---|---|
| Practice user-centered design and follow well-known general principles from HCI research. | Make travel tasks simple by using target-based techniques. | Use the natural virtual-hand technique if all selection is within arm's reach. | Reduce the number of degrees of freedom to be manipulated if the application allows it. | Reduce the necessary number of commands in the application. |
| Use HMDs or SIDs when immersion within a space is a performance requirement. Use workbench displays when viewing a single object or set of objects from a third-person point of view. | If steering techniques are used, train users in strategies to acquire survey knowledge. Use target-based or route-planning techniques if spatial orientation is required but training is not possible. | Use ray-casting techniques if speed of remote selection is a requirement. | Provide general or application-specific constraints or manipulation aids. | When using virtual menus, avoid submenus and make selection at most a 2-D operation. |
| In SIDs, design the system to minimize the amount of indirect rotation needed. | Avoid the use of teleportation; instead, provide smooth transitional motion between locations. | Ensure that the chosen selection technique integrates well with the manipulation technique to be used. | Allow direct manipulation with the virtual hand instead of using a tool. | Indirect menu selection may be more efficient over prolonged periods of use. |
| Use an input device with the appropriate number of degrees of freedom for the task. | Use physical head motion for viewpoint orientation if possible. | Consider multimodal input for combined selection and command tasks. | Avoid repeated, frequent scaling of the user or environment. | Voice and gesture-based commands should include some method of reminding the user of the proper utterance or gesture. |
| Allow two-handed interaction for more precise input relative to a frame of reference. | Use non-head-coupled techniques for efficiency in relative motion tasks. If relative motion is not important, use gaze-directed steering to reduce cognitive load. | If possible, design the environment to maximize the perceived size of objects. | Use indirect depth manipulation for increased efficiency and accuracy. | Integrate system control with other interaction tasks. |
| Use absolute devices for positioning tasks and relative devices for tasks to control the rate of movement. | Consider integrated travel and manipulation techniques if the main goal of viewpoint motion is to maneuver for object manipulation. | | | |
| Take advantage of the user's proprioceptive sense for precise and natural 3-D interaction. | Provide wayfinding and prediction aids to help the user decide where to move, and integrate those aids with the travel technique. | | | |
| Use well-known 2-D interaction metaphors if the interaction task is inherently one- or two-dimensional. | | | | |
| Use physical props to constrain and disambiguate complex spatial tasks. | | | | |
| Provide redundant interaction techniques for a single task. | | | | |

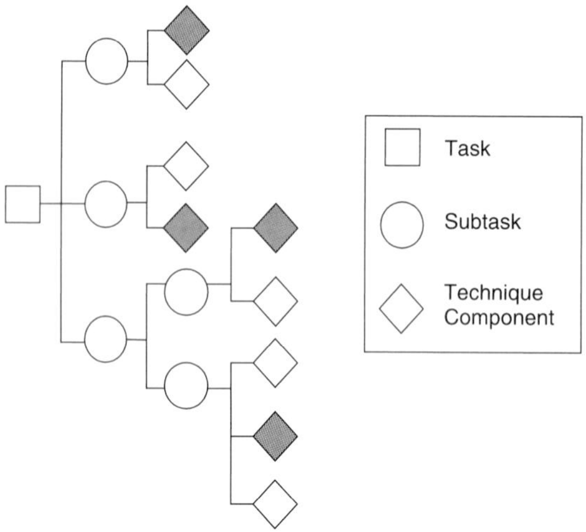


FIG. 13.1. General taxonomy format.

important as extensions of the metaphor on which the technique is based. In this sense, a taxonomy is the product of a careful task analysis. Once a task has been broken down to a sufficiently fine-grained level, listing possible methods (technique components) for accomplishing each of the lowest level subtasks completes the taxonomy. An interaction technique is made up of one technique component from each of the lowest level subtasks, such as the shaded components in Fig. 13.1.

Ideally, the taxonomies established for universal tasks need to be correct, complete, and general. Any IT that can be conceived for the task should fit within the taxonomy. Thus, the subtasks will necessarily be abstract. The taxonomy will also list several possible technique components for each of the subtasks but will not claim to list each conceivable component. For example, in an object-coloring task, a taxonomy might list touching the virtual object, giving a voice command, or choosing an item in a menu as choices for the color application subtask. However, this does not preclude a technique that applies the color by some other means, such as pointing at the object.

One way to verify the generality of taxonomies is through the process of *categorization*—defining existing ITs within the framework of the taxonomy. If existing techniques for the task fit well into a taxonomy, one can be surer of its correctness and completeness. Categorization also serves as an aid to evaluation of techniques. Fitting techniques into a taxonomy makes explicit their fundamental differences, thus the effect of choices can be determined in a more fine-grained manner.

### *1.2.3 Guided Design*

Taxonomies and categorization are good ways to understand the low-level makeup of ITs and to formalize the differences between them, but once they are in place, they can also be used in the design process. A taxonomy can be thought of not only as a classification but also as a design space. In other words, a taxonomy informs or guides the design of new ITs for the task, rather than relying on a sudden burst of insight.

Since a taxonomy breaks a task down into separable subtasks, a wide range of designs can be considered quite quickly, simply by trying different combinations of technique components for each of the subtasks. There is no guarantee that a given combination will make sense as a complete interaction technique, but the systematic nature of the taxonomy makes it easy to generate designs and to reject inappropriate combinations.

Categorization may also lead to new design ideas. Placing existing techniques into a design space allows designers to see "holes" that are left behind—combinations of components that have not yet been attempted. One or more of the holes may contain a novel, useful technique for the task at hand. This process can be extremely useful when the number of subtasks is small enough and the choices for each of the subtasks are clear enough to allow a graphical representation of the design space, as this makes the untried designs quite obvious (Card, Mackinlay, & Robertson, 1990).

### 1.3 Universal Interaction Tasks

What user tasks need to be supported in immersive VEs? At first glance, there appear to be an extremely large number of possible user tasks—too many, in fact, to think about scientific design and evaluation for all of them. However, as Foley (1979) has argued for two-dimensional (2-D) interaction, there is also a set of "universal" tasks (simple tasks that are present in most applications, which can be combined to form more complex tasks) for 3-D interfaces. These universal tasks include *navigation*, *selection*, and *manipulation*.

Navigation refers to the task of moving the viewpoint within a 3-D space, and includes both a cognitive component (wayfinding), and a motor component (travel, also called viewpoint motion control; see chap. 24, this volume). Selection refers to the specification of one or more objects from a set. Finally, manipulation refers to the modification of various object attributes (including position and orientation, and possibly scale, shape, color, texture, or other properties). Selection may be used on its own to specify an object to which a command will be applied (e.g., "delete the selected object"), or it might denote the beginning of a manipulation task.

These simple tasks are the building blocks from which more complex interactions arise. For example, the user of a surgery simulator might have the task of making an incision. This task might involve approaching the operating table (navigation), picking up a virtual scalpel (selection), and moving the scalpel slowly along the desired incision line (manipulation). One class of complex tasks, *system control*, involves the user giving commands to the system. For example, this might be accomplished by bringing a virtual menu into view (manipulation) and then choosing a menu item (selection). However, system control is so ubiquitous in VE applications that the design of system control techniques can be considered separately.

This chapter is targeted at developers and researchers who produce complete VE applications. It provides background information, a large set of potential techniques for universal interaction tasks, and guidelines to help in the choice of an existing technique or the design of a new technique for a particular system. Use of these guidelines should lead to more usable, useful, efficient, and effective virtual environments.

### 1.4 Performance Requirements

In order to determine whether or not a VE interaction technique exhibits good performance, metrics must be defined that capture performance. Metrics allow the performance of a technique to be quantified, the performance of competing techniques compared, and the interaction requirements of an application specified. Listed below are some (but certainly not a complete set) of the most common performance metrics for VE interaction, including metrics for task performance and technique performance. For each individual interaction task, the metrics may have slightly different meanings.

1. *Speed.* This refers to the classic quantifier of performance: task completion time. This efficiency metric will undoubtedly be important for many tasks, but should not be the only measurement considered.

2. *Accuracy*. Accuracy is a measurement of the exactness with which a task is performed. For travel or manipulation tasks, this will likely be measured by the distance of the user or object from the desired position or path. For selection, one might measure the number of errors that were made. Often, required accuracy is held constant in experiments while speed is measured, but the speed/accuracy trade-off should be fully explored, if possible.

3. *Spatial awareness*. A user's spatial awareness is related to his knowledge of the layout of a space and his own position and orientation within it. This may be an important performance requirement in large, highly occluded, or complex virtual environments. Most often, movement within the space (travel) affects spatial awareness, but other interaction tasks may also affect this metric.

4. *Ease of learning*. This is commonly discussed in the human–computer interaction (HCI) community, and refers to the ease with which a novice user can comprehend and begin to use a technique. It may be measured by subjective ratings, or the time for a novice to reach some level of performance, or by characterizing the performance gains by a novice as exposure time to the technique increases.

5. *Ease of use*. This is another HCI concept that may be difficult to quantify. It refers to the simplicity of a technique from the user's point of view. In psychological terms, this may relate to the amount of cognitive load induced upon the user of the technique. This metric is usually obtained through subjective self-reports, but measures of cognitive load may also indicate ease of use.

6. *Information gathering*. One of the goals of many immersive VEs is for the user to obtain information from or about the environment while in it. The choice of interaction techniques may affect the user's ability to gather information, and so measurement of this ability can be seen as an aspect of technique performance.

7. *Presence*. Another goal of VEs is to induce a feeling of presence ("being there," or immersion within the space) in users (see chap. 40, this volume). This quality is purported to lend more realism to a VE system, which may be desirable in systems for entertainment, education, or simulation. Presence may be affected by the interaction techniques in a system. It is usually measured by subjective reports and questionnaires (Slater, Usoh, & Steed, 1994; Witmer & Singer, 1998).

8. *User comfort*. Most of the interaction techniques discussed herein require activity on the part of the user (e.g., moving the arm, turning the head). It is important in systems that require a moderate to long exposure time that these motions do not cause discomfort in the user. Discomfort can range from classic simulator sickness to eye strain to hand fatigue and so on (see chaps. 29–32, this volume). Although VEs in general may induce some level of discomfort, this may be increased or decreased depending on the interaction techniques and input devices chosen. Comfort measurements are usually user self-reports (Kennedy et al., 1993).

9. *Expressiveness*. Virtual environment interaction techniques may be general in nature or quite domain-specific. The choice depends in part on the system's need for expressiveness. Expressiveness refers to the generality and flexibility of use of a given technique. For example, a travel technique that allows 3-D flying is more expressive than one that restricts the user to 2-D movement on a ground plane. Increased expressiveness is not always desirable because constraints can help to guide a user's actions. It is important for each application to carefully specify the level of expressiveness needed for a given interaction task.

10. *Unobtrusiveness*. An interaction technique is obtrusive if it interferes with the user's ability to focus on the task at hand. This metric will be most important for applications that

have repeated and frequent use of the same interaction technique. Such a technique will be required to be unobtrusive so users do not become quickly frustrated.

11. *Affordance*. Finally, a technique's performance can be described by the affordances that it presents for a task. An affordance (Norman, 1990) is simply a characteristic of a technique or tool that helps the user understand what the technique is to be used for and how it is to be used. For example, voice commands in general have little affordance because the user must know what the commands are. Listing the available commands on screen is an affordance that aids the user. Like expressiveness and unobtrusiveness, affordance is an innate characteristic of a technique that is not easily measured. Nonetheless, it must be taken into consideration when choosing techniques for a VE application.

## 2. NATURALISM VERSUS MAGIC

A common misconception about VEs is that, in order to be effective, they should work exactly the way the real world works, or at least as close as is practically possible (interaction with a VE application should be "natural"). The very term *virtual reality* promotes such a view—that virtual reality should be the same as "real reality." In fact, this is not always the case. It may be very useful to create VEs that operate quite differently from the physical world.

In chapter 11 (this volume), locomotion devices for VEs are discussed. These devices usually strive to reproduce a realistic or natural mode of travel. There are several advantages to such techniques. Natural mappings are present, so users can easily perform tasks based on principles they are familiar with from daily life. Also, this simulation of the physical world may create a greater sense of immersion or presence (see chap. 40, this volume) in the virtual world. Finally, realism may enhance the user experience.

However, there is an alternative to the naturalistic approach, which we'll call "magic" (Smith, 1987). In this approach, the user is given new abilities, and non-natural methods for performing tasks are used. Examples include allowing the user to change his scale (grow or shrink), providing the user with an extremely long arm to manipulate faraway objects, or letting the user fly like a bird. Magic techniques are less natural, and thus may require more explanation or instruction, but they can also be more flexible and efficient if designed for specific interaction tasks.

Clearly, there are some applications that need naturalism. The most common example is training, in which users are trained in a VE for tasks that will be carried out in a physical environment. Such applications have the *requirement* of natural interaction. On the other hand, applications such as immersive architectural design do not require complete naturalism—the user only has the goal of completing certain tasks, and the performance requirements of the system do not include naturalism.

This discussion focuses on techniques involving some magic or nonrealism, in the interest of optimizing performance. Such techniques may enhance the user's physical, perceptual, or cognitive abilities, and take advantage of the fact that the VE can operate in any definable fashion. No possible techniques are excluded from consideration as long as they exhibit desirable performance characteristics (task and technique performance).

## 3. GENERIC VE INTERACTION GUIDELINES

When attempting to develop guidelines that will produce high-performance VE interaction, both generic guidelines that inform interaction design at a high level, and specific guidelines for common tasks (i.e., travel, selection, manipulation, and system control) should be considered. The next two sections will cover these areas. The guidelines presented in this section are not intended to be exhaustive, but are limited to those that are especially relevant to enhancing performance and those that have been verified through formal evaluation. A large number of VE-specific usability guidelines can be found in Gabbard and Hix (1998).

## 3.1 Existing HCI Guidelines

The first thing to remember when developing interaction for VEs is that interaction is not new. The field of HCI has its roots in many areas, including perceptual and cognitive psychology, graphic design, and computer science, and has a long history of design and evaluation of 2-D computer interfaces. Through this process, a large number of general-purpose guidelines have been developed that have wide applicability to interactive systems, and not just the standard desktop computer applications with which everyone is familiar. This existing knowledge and experience can be leveraged in interaction design of virtual environments. If VE design does not meet these most basic requirements, then VE systems are sure to be unusable. Furthermore, the application of HCI principles to VEs may lead to VE-specific guidelines as well. These guidelines are well-known, if not always widely practiced, so they are not reviewed in detail here.

*Practice user-centered design and follow well-known general principles from HCI research.*

Two important sources for such general guidelines are Donald Norman's *The Design of Everyday Things* (Norman, 1990) and Jakob Nielsen's usability heuristics (Nielsen & Molich, 1992). These guidelines focus on high-level and abstract concepts such as making information visible (how to use the system, what the state of the system is, etc.), providing affordances and constraints, using precise and unambiguous language in labeling, designing for both novice and expert users, and designing for prevention of and recovery from errors.

Following such guidelines should lead to a more understandable, efficient, and usable system. However, because of their abstract nature, applying these principles is not always straightforward. Nevertheless, they must be considered as the first step toward a usable system.

## 3.2 Choice of Devices

A basic question one must ask when designing a VE system regards the choice of input and output devices to be used. Currently, little empirical data exists about relative interaction performance, especially for VE display devices. There are, however, a few general guidelines that can be posited here.

### *3.2.1 Display Devices*

Three common VE display devices, as described in chapter 3 (this volume), are head-mounted displays (HMDs), spatially immersive displays (SIDs, semisurrounding projected stereo displays, such as the CAVE™), and desktop stereo displays, such as the Responsive Workbench. These display types have very different characteristics, and interaction with these displays is likely to be extremely different as well.

*Use HMDs or SIDs when immersion within a space is a performance requirement.*

*Use workbench displays when viewing a single object or set of objects from a third-person point of view.*

These two guidelines are based on the essential difference between display types. HMDs and SIDs encourage an egocentric, inside-out point of view, and are therefore appropriate for first-person tasks such as walk-throughs or first-person gaming. Workbench displays support an outside-in point of view and therefore work well for third-person tasks such as manipulating a single object or arranging military forces on a 3-D terrain near the surface of the workbench. If objects are spatially located near to the user in a SID or workbench setup, however, there may be problems involving the user's hand occluding objects that should actually be in front of the hand.

*In SIDs, design the system to minimize the amount of indirect rotation needed.*

Most projected VEs do not completely surround the user. For example, a standard CAVE configuration places graphics on four surfaces of a six-sided cube (floor and three walls). Thus, the ceiling and back wall are not displayed. This means that for users to view parts of the VE directly behind or above them, they must rotate the environment indirectly, using some input device (e.g., pressing a button to rotate the scene 90 degrees rather than simply turning her head). In an application in which immersion is important, this indirect rotation will likely break the illusion of presence within the space to some degree. Recent research (Bakker, Werkhoven, & Passenier, 1998; Chance, Gaunet, Beall, & Loomis, 1998) has shown that physical turning produces better estimates of turn magnitude and direction to objects (indicating superior spatial orientation) than does indirect turning. One way to alleviate this problem is to adopt a vehicle metaphor for navigation so that the user is always facing the front wall and using the side walls for peripheral vision. With a steering wheel for choosing vehicle direction, indirect rotation seems much more natural. Note that fully immersive SIDs, such as a six-sided cube or spherical dome, do not suffer from this problem.

### 3.2.2 Input Devices

Input devices and their differences have been studied more extensively than display differences. Common VE input devices include 6-DOF trackers, continuous posture-recognition gloves, discrete event gloves, penlike devices, simple button devices, and special-purpose devices such as the Spaceball or force-feedback joysticks.

*Use an input device with the appropriate number of degrees of freedom for the task.*

Many inherently simple tasks become more complex if an improper choice of input device is made. For example, toggling a switch is inherently a 1-degree-of-freedom task (the switch is on or off). Using an interaction technique which requires the user to place a tracked hand within a virtual button (a 3-DOF task) makes it overly complex. A simple discrete event device, such as a pinch glove, makes the task simpler. Of course, one must trade off the reduced degrees of freedom with the arbitrary nature of the various actions the user must learn when using a pinch glove or other such device. In general, designers should strive to reduce unnecessary DOFs when it is practical (Hinckley, Pausch, Goble, & Kassell, 1994). If only a single input device is available, software constraints can be introduced to reduce the number of DOFs the user must control (see section 4.3.2).

*Use physical props to constrain and disambiguate complex spatial tasks.*

This guideline is related to the previous discussion about degrees of freedom. Physical props can help to reduce the number of DOFs that the user must control. For example, the pen-and-tablet interaction metaphor (Bowman, Wineman, Hodges, & Allison, 1998) uses a physical tablet (a 2-D surface) and a tracked pen. A 2-D interface is virtually displayed on the surface of the tablet, for tasks such as button presses, menu selection, and 2-D drag and drop (see Fig. 13.2). The physical props allow the user to do these tasks precisely, because the tablet surface guides and constrains the interaction to two dimensions.

Physical props can also make complex spatial visualization easier. For example, in the Netra system for neurosurgical planning (Goble, Hinckley, Pausch, Snell, & Kassell, 1995), it was found that surgeons had difficulty rotating the displayed brain data to the correct orientation when a simple tracker was used to control rotation. However, when the tracker was embedded

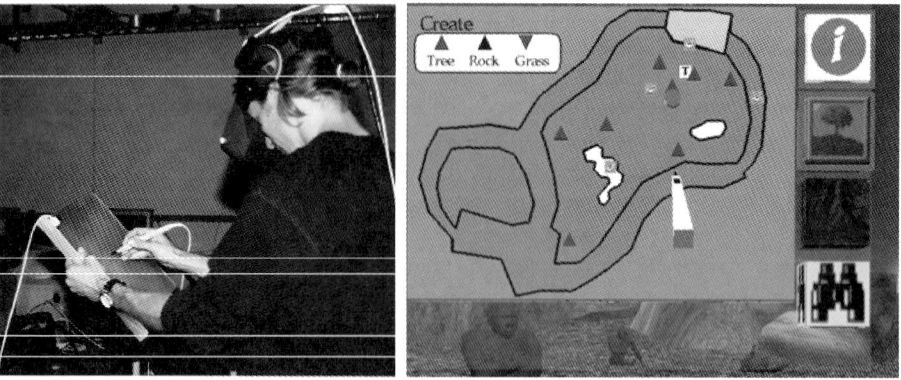


FIG. 13.2. Physical (*left*) and virtual (*right*) views of a pen-and-tablet system.

within a doll's head, the task became much easier because the prop gave orientation cues to the user.

*Use absolute devices for positioning tasks and relative devices for tasks to control the rate of movement.*

This guideline is well-known in desktop computing, but not always followed in virtual environments. Absolute positioning devices such as trackers will work best when their position is mapped to the position of a virtual object. Relative devices (devices whose positional output is relative to a center position that can be changed), such as joysticks, excel when their movement from the center point is mapped to the rate of change (velocity) of an object, usually the viewpoint. Interaction techniques which use absolute devices for velocity control or relative devices for position control will perform less efficiently and easily. Zhai (1993) has extended this idea by comparing isometric and isotonic devices in a 3-D-manipulation task.

## 3.3 Interacting in Three-dimensional Space

By its very nature, 3-D (also called spatial) interaction is qualitatively and quantitatively different than standard 2-D interaction. As discussed in the previous section, the choice of 3-D input devices can be quite important, but there are also other general principles related to the way 3-D interaction is implemented in software.

*Take advantage of the user's proprioceptive sense for precise and natural 3-D interaction.*

Proprioception is a person's sense of the location of the parts of his body, no matter how the body is positioned. For example, a driver can easily change gears without looking, because of his knowledge of his body and hand position relative to the gearshift. Mine, Brooks, and Sequin (1997) discuss how to take advantage of this sense in VEs by providing body-centered interactions. One possibility is to give users a virtual tool belt on which various tools (e.g., pointer, cutting plane, spray paint, etc.) can be hung. Because users know where the various tools are located on their body, they can interact and choose tools much more efficiently and easily without looking away from their work.

*Use well-known 2-D interaction metaphors if the interaction task is inherently one- or two-dimensional.*

It seems to be an unspoken rule among VE application developers that interaction techniques should be new and unique to virtual environments. This is as much a myth as the concept discussed earlier that all interaction should mimic the real world. In fact, there are many 2-D interaction metaphors that can be used directly in or adapted for use in virtual environments. Pull-down or pop-up menus, 2-D buttons, and 2-D drag-and-drop manipulation have all been implemented in VEs with success (e.g., Bowman, Hodges, & Bolter, 1998). With these interaction techniques, issues related to reducing the number of DOFs the user must control often arise. When 2-D interaction metaphors are used, the provision of a 2-D surface for interaction (such as the pen-and-tablet metaphor discussed above) can increase precision and efficiency.

*Allow two-handed interaction for more precise input relative to a frame of reference.*

Most VE interfaces "tie one hand behind the user's back," allowing only input from a single hand. This severely limits the flexibility and expressiveness of input. By using two hands in a natural manner, the user can specify arbitrary spatial relationships, not just absolute positions in space. However, it should not be assumed that both hands will be used in parallel to increase efficiency. Rather, the most effective two-handed interfaces are those in which the nondominant hand provides a frame of reference in which the dominant hand can do precise work (Hinckley, Pausch, Profitt, Patten, & Kassell, 1997).

*Provide redundant interaction techniques for a single task.*

One of the biggest problems facing evaluators of VE interaction is that the individual differences in user performance seem to be quite large relative to 2-D interfaces. Some users seem to comprehend complex techniques easily and intuitively, whereas others may never become fully comfortable. Work on discovering the human characteristics that cause these differences is ongoing, but one way to mitigate this problem is to provide multiple interaction techniques for the same task. For example, one user may think of navigation as specifying a location within a space and therefore would benefit from the use of a technique where the new location is indicated by pointing to that location on a map. Another user may think of navigation as executing a continuous path through the environment and would benefit from a continuous steering technique. In general, "optimal" interaction techniques may not exist even if the user population is well-known, so it may be appropriate to provide two or more techniques, each of which have unique benefits. Of course, the addition of techniques also increases the complexity of the system, and so this must be done with care and only when there is a clear benefit.

## 4. TECHNIQUES AND GUIDELINES FOR COMMON VE TASKS

### 4.1 Travel

Travel, also called viewpoint motion control, is the most ubiquitous and common VE interaction task—simply the movement of the user within the environment. Travel and wayfinding (the cognitive process of determining one's location within a space and how to move to a desired location—see chap. 24, this volume) make up the task of navigation. Chapter 11 (this volume) presents locomotion interface devices, which are physical devices supporting travel tasks, so in this section the focus is on passive movement, in which the user remains physically stationary while moving through the virtual space.

There are three primary tasks for which travel is used within a virtual environment. *Exploration* is travel which has no specific target, but which is used to build knowledge of the environment or browse the space. *Search tasks* have a specific target, whose location may be completely unknown (naïve search) or previously seen (primed search). Finally, *maneuvering tasks* refer to short, precise movements with the goal of positioning the viewpoint for another type of task, such as object manipulation. Each of these three types of tasks may require different travel techniques to be most effective, depending on the application.

#### *4.1.1 Technique Classifications*

Because travel is so universal, a multitude of techniques have been proposed (see Mine [1995] for a survey of early techniques). Many techniques have similar characteristics, so it

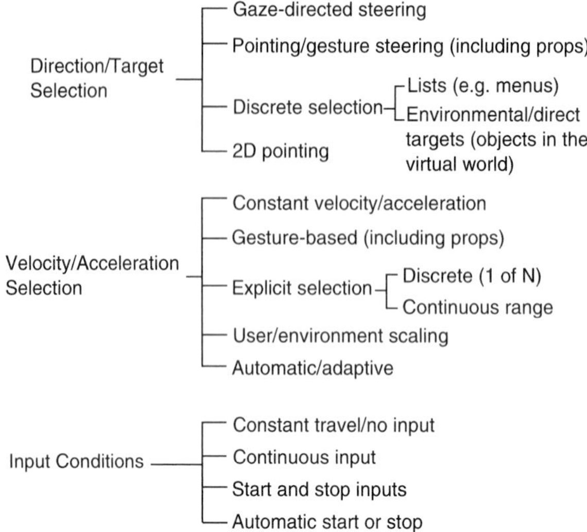


FIG. 13.3. Taxonomy of passive, first-person travel techniques.

will be useful to present classifications of techniques rather than discussing each technique separately.

A simple taxonomy of passive movement techniques was described in Bowman, Koller, and Hodges (1997) and is reproduced in Fig. 13.3. This taxonomy partitions a travel task into three subtasks: direction or target selection, velocity and/or acceleration selection, and conditions of input (specifying the beginning and end of movement).

Most techniques differ only in the direction or target specification subtask, and several common technique components are listed in the taxonomy. Gaze-directed steering uses the orientation of the head for steering, whereas pointing gets this information from the user's hand. The orientation of other body parts, such as the torso or foot, could also be used. Physical devices, such as a steering wheel, provide another way to specify direction. Other techniques specify only a target of motion, by choosing from a list, entering coordinates, pointing to a position on a map, or pointing at the target object in the environment.

The velocity/acceleration selection subtask has been studied much less than direction or target selection, but several techniques have been proposed. Many systems simply default to a reasonable constant velocity. Gesture-based techniques use hand or body motions to indicate velocity or acceleration (e.g., speed depends on the distance of the hand from the body). Again, physical props such as accelerator and brake pedals can be used. Velocity or acceleration could be chosen discretely from a menu. Finally, velocity and acceleration may be automatically controlled by the system in a context-sensitive fashion (e.g., depending on the distance from the target or the amount of time the user has been moving; Mackinlay, Card, & Robertson, 1990).

The conditions of input may seem trivial, but this simple subtask can have an effect on performance. Generally, the user simply gives a single input, such as a button press, to begin moving, and another to stop moving. There may also be situations when it is appropriate for the system to automatically begin and/or end the motion, or when the user should be moving continuously.

Another way to classify travel techniques relates to the amount of control that the user has over viewpoint motion. *Steering techniques* give the user full control of at least the direction of motion. These include continuous steering, such as gaze-directed or pointing techniques, and discrete steering, such as a ship steering technique in which verbal commands are interpreted to give the user control of the ship's rudders and engines (William Walker, personal communication, 1999). On the other end of the control spectrum, *target-based techniques* only allow the user to specify the goal of the motion, whereas the system determines how the viewpoint will move from the current location to the target. This requires the use of a selection technique (see section 4.2). *Route-planning techniques* represent a middle ground. Here the user specifies a path between the current location and the goal, and then the system executes that path. This might be implemented by drawing a path on a map of the environment or placing markers using some manipulation technique and having the system interpolate a spline between these control points.

Finally, there is a class of techniques that do not fit well into the above classifications that use manual manipulation to specify viewpoint motion. Ware and Osborne (1990) identified the "camera in hand" metaphor, where the user's hand motion above a map or model of the space specifies the viewpoint from which the scene will be rendered, and the "scene in hand" metaphor, in which the environment itself is attached to the user's hand position. Both of these techniques are exocentric in nature, but manual viewpoint manipulation can also be done from a first-person perspective. Any direct object manipulation technique (see section 4.3) can be modified so that the user's hand movements affect the viewpoint instead of the selected object. The selected object remains fixed in the environment, and the user moves around that object using hand motions. Such a technique might be extremely useful for maneuvering tasks where the user is constantly switching between travel and manipulation.

### 4.1.2 Guidelines for Designing Travel Techniques

*Make simple travel tasks simple by using target-based techniques.*

If the goal of travel is simply to move to a known location, such as moving to the location of another task, target-based techniques provide the simplest metaphor for the user to accomplish this task. In many cases, the exact path of travel itself is not important; only the goal is important. In such situations, target-based techniques make intuitive sense and leave the user's cognitive and motor resources free to perform other tasks. The use of target-based techniques assumes that the desired goal locations are known in advance or will always coincide with a selectable position in the environment. If this is not true (e.g., the user wishes to obtain a bird's-eye view of a building model), target-based techniques will not be appropriate. Users of these techniques should also pay attention to the guideline regarding teleportation below.

*Use physical head motion for viewpoint orientation if possible.*

Almost all immersive VE systems use head tracking to render the scene from the user's point of view. Since viewpoint motion control involves both viewpoint position and orientation, it makes sense to use the head tracking information for setting viewpoint orientation. However, in certain applications, especially those in which the user is seated, it might be tempting to specify viewpoint orientation indirectly, for example, by using a joystick. It has been shown (Chance, Gaunet, Beall, & Loomis, 1998) that such indirect orientation control, which does not take advantage of proprioception, has a damaging effect on the spatial orientation of the user. Therefore, when it is important for the user to understand the spatial structure of the environment, physical head motion should always be used.

*Avoid the use of teleportation; instead, provide smooth transitional motion between locations.*

Teleportation, or "jumping," refers to a target-based travel technique in which velocity is infinite, that is, the user is moved immediately from the starting position to the target. Such a technique seems very attractive from the perspective of efficiency. However, empirical results (Bowman et al., 1997) have shown that disorientation results from teleportation techniques.

Interestingly, all techniques that used continuous smooth motion between the starting position and the target caused little disorientation in this experiment, even when the velocity was relatively high. This effect may be lessened if a common reference, such as a map or World-in-Miniature showing the current position is used, but even with these techniques smooth transitional motion is recommended, if practical.

*If steering techniques are used, train users in strategies to acquire survey knowledge. Use target-based or route-planning techniques if spatial orientation is required but training is not possible.*

Spatial orientation (the users' spatial knowledge of the environment and their position and orientation within it) is critical in many large-scale VEs, such as those designed to train users about a real-world location. The choice of interaction techniques can affect spatial orientation. In particular, evaluation (Bowman, Davis, Hodges, & Badre, 1999) has shown that good spatial orientation performance can be obtained with the use of steering techniques, such as pointing in the desired direction of motion, where the user has the highest degree of control. Steering techniques only produce high spatial orientation, however, if sophisticated strategies are used (e.g., flying above the environment to obtain a survey view, moving in structured patterns). If such strategies are not used, steering techniques may actually perform worse, because users are concentrating on controlling motion rather than viewing the environment. Techniques where the user has less control over motion, such as target-based and route-planning techniques, provide moderate levels of spatial orientation due to the low cognitive load they place on the user during travel—the user can take note of spatial features during travel because the system is controlling motion.

*Consider integrated travel and manipulation techniques if the main goal of viewpoint motion is to maneuver for object manipulation.*

Manual viewpoint manipulation techniques use object manipulation metaphors (see section 4.3) to specify viewpoint position. Such techniques have been shown experimentally (Bowman, Johnson, & Hodges, 1999) to perform poorly on general travel tasks such as exploration and search. However, such techniques may prove quite useful if the main goal of travel is to maneuver the viewpoint during object manipulation. Manual viewpoint manipulation allows the use of the same technique for both travel and object manipulation tasks, which may be intermixed quite frequently in applications requiring complex manipulation.

*Use non-head-coupled techniques for efficiency in relative motion tasks. If relative motion is not important, use gaze-directed steering to reduce cognitive load.*

Relative motion is a common VE task in which the user wishes to position the viewpoint at a location in space relative to some object. For example, an architect wishes to view a structure from the proposed location of the entrance gate, which is a certain distance from the front door—movement must be relative to the door and not to any specific object. A comparison of steering techniques (Bowman et al., 1997) showed that a pointing technique performed much more efficiently on this task than gaze-directed steering. This is because pointing allows the user to look at the object of interest while moving, whereas gaze-directed steering forces the user to look in the direction of motion. Gaze-directed steering performs especially badly when motion needs to be away from the object of interest. Thus, techniques that are not coupled to head motion support relative-motion tasks. Non-head-coupled techniques include not only steering techniques such as pointing, but also manipulation-based techniques. For example, a technique in which the user grabs the object of interest then uses hand motion to move the viewpoint about it was shown to produce very efficient performance on a relative motion task in a small unpublished user study performed by the author. On the other hand, gaze-directed steering is slightly less cognitively complex than either pointing or manipulation techniques, so it may still be useful if relative motion is not an important task in an application.

*Provide wayfinding and prediction aids to help the user decide where to move, and integrate those aids with the travel technique.*

The design of interaction techniques for travel assumes that the user knows where to go and how to get there, but this is not always the case. Wayfinding aids (e.g., Darken, 1996) may be needed, especially in large-scale VEs where the user is expected to build survey knowledge of the space. Such aids include maps, signposts, compass markings, and paths. During travel, the user may need to know whether the current path will take them to the desired location. Predictor displays (Wickens, Haskell, & Harte, 1989) have long been used in aircraft simulation to show the pilot the result of the current heading, pitch, and speed. Such displays might also be useful in other VEs that involve high-speed 3-D motion.

## 4.2 Selection

Selection is simply the task of specifying an object or set of objects for some action. Most often, selection precedes object manipulation (see section 3.3), or specifies the object of some command (see section 3.4), such as "delete the selected object." In interactively complex VEs, selection tasks occur quite often, and therefore efficiency and ease of use are important performance requirements for this task.

### *4.2.1 Technique Classifications*

The most obvious VE selection technique is again the one that mimics real-world interaction—simply touching the desired object with a virtual hand. Within this general metaphor, there are several possible implementations, however. The virtual hand could simply be a rigid object controlled by a single 6-DOF tracker, or it could be controlled by a glove that recognizes a multitude of different hand postures for more precision. Another issue relates to the precision of selection. Can the user only select at the object level, or can specific points on an object's surface be selected? Finally, there is the issue of selection feedback. Most systems present simple graphical (e.g., highlight the object) or audio feedback to indicate touching, but haptic feedback (see chaps. 5 and 6) more closely simulates real-world touching and allows the user to select objects without looking at them.

No matter how implemented, this simple virtual hand metaphor suffers from a serious problem: The user can only select objects in the VE that are actually within arms reach. In many large-scale VEs, especially in the design or prototyping application domains, the user will wish to select remote objects—those outside of the local area surrounding the user. Therefore, a number of "magic" techniques have been developed for object selection. These fall into two main categories: arm extension and ray casting.

Arm-extension techniques still use a virtual hand to select objects via touching, but the virtual hand has a much greater range than the user's physical hand. The simplest example is a technique that linearly maps the physical hand movements onto the virtual hand movements, so that for each unit the physical hand moves away from the body, the virtual hand moves away N units. The Go-Go technique (Poupyrev, Billinghurst, Weghorst, & Ichikawa, 1996) takes a more thoughtful approach. It defines a radius around the user within which the physical hand is mapped directly to the virtual hand. Outside that radius, a nonlinear mapping is applied to allow the virtual hand to reach quite far into the environment, although still only a finite distance (see Fig. 13.4). Other techniques allow infinite arm extension, such as an indirect technique that uses two buttons to extend and retract the virtual hand. Such techniques are generally less natural and induce more cognitive load.

Ray-casting techniques move away from the object touching metaphor and instead adopt a pointing metaphor. A ray emanates from the user, with the user controlling its orientation

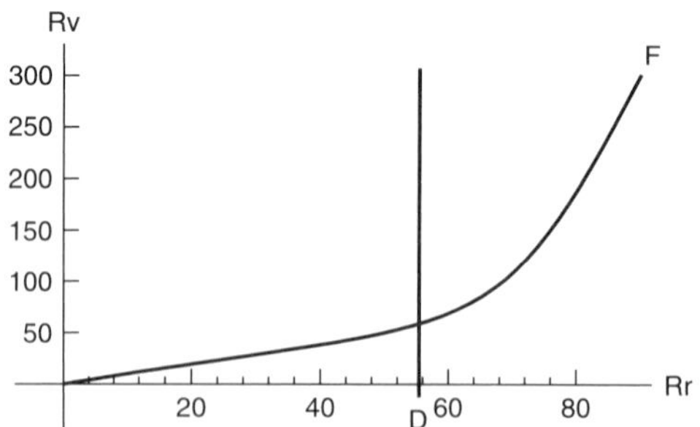


FIG. 13.4. Mapping function for Go-Go technique.

only, and the first object the ray intersects may be selected. The ray may be linear, or it may be cone-shaped so that small objects are more easily selected at a distance. The most common implementation of ray casting is to attach the ray to the user's virtual hand so that simple wrist movements allow pointing in any direction (Mine, 1995). Another class of techniques uses gaze direction for ray casting so that an object can be selected by placing it in the center of one's field of view. Finally, some techniques use a combination of eye position and hand position for selection (Pierce et al., 1997), with the ray emanating from the eye point and passing through the virtual-hand position (see Fig. 13.5). This is often called occlusion, or framing, selection.

Bowman and Hodges (1999) defined a taxonomy of selection techniques, which is presented at the top of Figure 13.6. Note that besides the subtask discussed thus far (indication of object), there are also feedback and indication to select subtasks. The latter refers to the event used to signal selection to the system, such as a button press, gesture, or voice command.

Finally, it should be noted that all of the techniques presented herein are designed for single-object selection only. Selection of multiple objects simultaneously has not been the subject of much research, but techniques from 2-D interfaces may work reasonably well in three dimensions. These include sequential selection with a modifier button pressed and rubberbanding, or "lassoing." Techniques such as rubberbanding, which must be extended to specify a 3-D volume, will present interesting usability challenges.

### 4.2.2 Guidelines for Designing Selection Techniques

*Use the natural virtual hand technique if all selection is within arms reach.*

The simple virtual-hand metaphor works well in systems where all interaction with objects is local. This usually includes VE applications implemented on a workbench display, where most objects lie on or above the surface of the table.

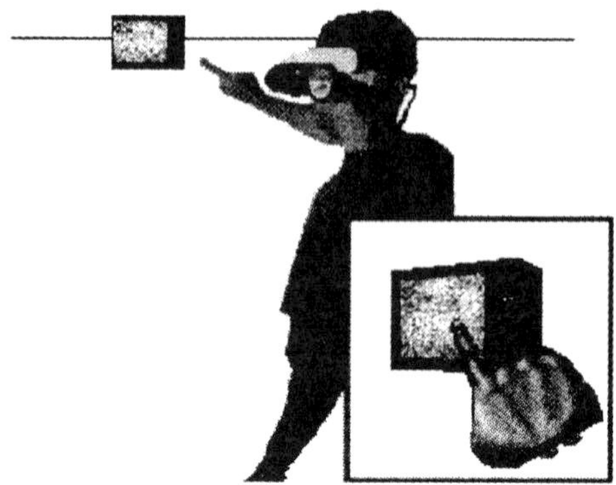

FIG. 13.5. Occlusion selection.

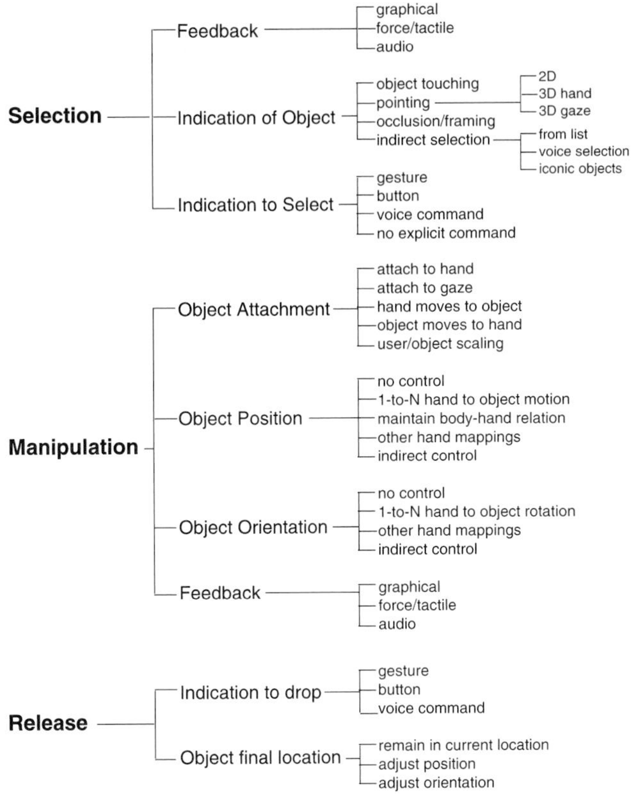


FIG. 13.6. Taxonomy of single-object selection and manipulation techniques.

*Use ray-casting techniques if speed of remote selection is a requirement.*

Evaluation (Bowman & Hodges, 1999) has shown that ray-casting techniques perform more efficiently than arm-extension techniques over a wide range of possible object distances, sizes, and densities. This is due to the fact that ray-casting selection is essentially 2-D (in the most common implementation, the user simply changes the pitch and yaw of the wrist).

*Ensure that the chosen selection technique integrates well with the manipulation technique to be used.*

Selection is most often used to begin object manipulation, and so there must be a seamless transition between the selection and manipulation techniques to be used in an application. Arm-extension techniques generally provide this transition, because the selected object is also manipulated directly with the virtual arm, and so the same technique is used throughout the interaction. As discussed below, however, it is possible to integrate ray-casting techniques with efficient manipulation techniques.

*Consider multimodal input for combined selection and command tasks.*

When selection is used in combination with system control tasks, it may be more efficient and natural to use multimodal interaction (Bolt, 1980; chap. 21, this volume). For example, one may point at an object and then give the voice command "Delete."

*If possible, design the environment to maximize the perceived size of objects.*

Selection errors are affected by both the size and distance of objects, using either ray-casting or arm-extension techniques (Bowman & Hodges, 1999). These two characteristics can be combined in the single attribute of visual angle or the perceived size of an object in an image. Unless the application requires precise replication of a real-world environment, manipulating the perceived size of objects will allow more efficient selection (Poupyrev, Weghorst, Billinghurst, & Ichikawa, 1997).

## 4.3 Manipulation

Manipulation goes hand in hand with selection. Manipulation refers broadly to modification of the attributes of the selected object. Attributes may include position, orientation, scale, shape, color, or texture. For the most part, research has only considered the manipulation of the position and orientation of rigid objects, although some special-purpose applications include object deformation or scaling. Object manipulation tasks have importance in such applications as design, prototyping, simulation, and entertainment, all which may require environments that can be modified by the user.

### *4.3.1 Technique Classifications*

The most common object manipulation technique is a natural one, in which the selected object is rigidly attached to the virtual hand and moves along with it until some signal is given to release the object. This technique is simple and intuitive, but certain object orientations may require the user to twist the arm or wrist to uncomfortable positions, and it does not use the inherent dexterity of the user's fingers. Recent research, then, has focused on more precise and dextrous object manipulation using fingertip control (Kitamura, Yee, & Kishino, 1998). This can be simulated to a degree using a rigid virtual hand if a clutching mechanism is provided. Researchers have also proposed two-handed object manipulation techniques, which often use the nondominant hand as a reference (e.g., a pivot point for rotation) and the dominant hand for fine, precise manipulation.

As with selection, natural manipulation techniques suffer from limitations of reach. Also, manipulating large objects within arms reach may occlude the user's view. Therefore, techniques for remote manipulation are also important, and several categories of such techniques have been proposed.

Arm-extension and ray-casting selection techniques can also be used for object manipulation. Arm-extension metaphors simply attach the object to the virtual hand and allow the user to control it using physical-to-virtual hand mapping (Poupyrev et al., 1996). Ray-casting techniques may attach the object to the ray itself, which allows intuitive, but limited and imprecise, manipulation (Bowman & Hodges, 1997). Because ray casting is so efficient as a selection mechanism, several researchers have attempted to increase the utility of ray casting for manipulation. One idea is to select using ray casting and then move the virtual hand to the selected object for direct manipulation (Bowman & Hodges, 1997). Another set of techniques scales the user or the environment so that the virtual hand, which was originally far from the selected object, is actually touching this object so that it can be manipulated directly (Pierce et al., 1997).

Another metaphor, called the world-in-miniature (WIM, Stoakley, Conway, & Pausch, 1995), solves the remote object manipulation problem by giving the user a small handheld copy of the environment. Direct manipulation of the WIM objects causes the larger environment objects to move as well. This technique is usually implemented using a small 3-D model of the space, but a 2-D interactive map can also be considered a type of WIM.

All of the techniques in the preceding section can be implemented in a general way—that is, allow objects to be positioned and oriented anywhere in the space—but most applications will benefit from the addition of special manipulation aids and constraints. One way to do this is through the use of snapping or gridlines (Mine, 1995) such that objects can only end in a discrete number of positions or orientations, or line up with other objects. Physical simulation can also be used, including gravity and impenetrable objects, so that manipulation results in a realistic configuration of objects. Physical simulation may be compute-intensive, however. Finally, objects themselves can be given intelligence (Bukowski & Sequin, 1995) so that coffee cups rest on their bottom surface and paintings hang on walls, for example.

The taxonomy presented by Bowman and Hodges (1999) also addresses object manipulation and release of objects after manipulation (see Fig. 13.6). The size of the manipulation taxonomy indicates that the design space is quite large. New techniques can be created using the process of guided design by combining components for each of the lowest level subtasks.

#### 4.3.2 Guidelines for Designing Manipulation Techniques

*Reduce the number of degrees of freedom to be manipulated if the application allows it.*

*Provide general or application-specific constraints or manipulation aids.*

These two guidelines address the same issue: reducing the complexity of interaction from the user's point of view. This can be done by considering the characteristics of the application (e.g., in an interior design task, the furniture should remain on the floor) by off-loading complexity to the computer (using constraints or physical simulation), or by providing widgets to allow the manipulation of one or several related DOFs (Conner et al., 1992; Mine, 1997). This also relates to the guideline concerning the DOFs of the input device to be used.

*Allow direct manipulation with the virtual hand instead of using a tool.*

Tools, such as a virtual light ray, may allow a user to select objects from great distances. However, the use of these same tools for object manipulation is not recommended due to the fact that positioning and orienting of the object is not direct—the user must map desired object manipulations to the corresponding tool manipulations. Manipulation techniques that allow the direct positioning and orienting of virtual objects with the user's hand have been shown empirically to perform more efficiently and to provide greater user satisfaction than techniques using a tool (Bowman & Hodges, 1999). For efficient selection *and* manipulation, then, a 2-D selection metaphor such as ray casting must be combined with a hand-centered, direct manipulation technique. This is the basis of techniques such as HOMER (Bowman & Hodges, 1997) and Sticky Finger (Pierce et al., 1997).

*Avoid repeated, frequent scaling of the user or environment.*

Techniques that scale the user or the world to allow direct manipulation have some desirable characteristics. The user's perception of the scene does not change at the moment of selection, and small physical movements can allow large virtual movements (Pierce et al., 1997). However, experimental data shows a correlation between the frequent use of such techniques and discomfort (dizziness and nausea) in users (Bowman & Hodges, 1999). Techniques that scale the user or environment infrequently and predictably should not suffer from these effects.

*Use indirect depth manipulation for increased efficiency and accuracy.*

Indirect control of object depth (e.g., using joystick buttons to move an object nearer to or farther away from the user) is completely unnatural and requires some training to be used well. However, once this technique is learned it provides more accurate object placement, especially if the target is far from the user (Bowman & Hodges, 1999). This increased accuracy leads to more efficient performance as well. Moreover, these techniques do not exhibit the arm strain that can result from the use of more natural arm-extension techniques.

### 4.4 System Control

Many of the other interactions found in VE applications fall under the heading of system control. This category includes commands, mode changes, and other modifications of system state. Often, system control tasks are composites of the other universal tasks. For example,

FIG. 13.7. Virtual pull-down menu (*left*), pen-and-tablet menu (*center*), and rotary menu (*right*).

choosing a menu item is a selection task, whereas dragging an object to a trash can for deletion is a manipulation task.

There has been little empirical evaluation of VE system control techniques, and no formal classification of which the author is aware. Therefore, in this section several categories of techniques are focused on including menus, voice commands, tools, and gestures.

#### *4.4.1 Virtual Menus*

Menus are the most common forms of system control found in VEs, and many of the virtual menu systems that have been developed are simple adaptations of menus from 2-D desktop systems.

The most simple menu system is a series of labeled buttons that appears in the virtual environment. These may be at a specific location in the environment, or they may be attached to the user for greater availability from any location. Slightly more complex are pull-down menus, which appear only as a label and whose items are revealed when the label is selected (Jacoby & Ellis, 1992). Pop-up menus have also been implemented, so that the menu appears at the location of the user's hand for easy access. Other implementations use menus on a virtual surface, such as in the pen-and-tablet metaphor or on the surface of a workbench display. Mine (1997) developed a rotary menu system in which items are chosen by rotating the wrist. This takes advantage of the fact that menu selection is essentially a one-dimensional task, so menu selection is done by changing only 1 DOF. TULIP menus (Bowman & Wingrave, 2001) map menu items to fingers, which are selected using Pinch Gloves. Figure 13.7 shows three example virtual menu systems.

Many virtual menu systems have faced a set of common problems. One is that the resolution of text, especially in HMDs, is low, so menus and labels must contain few items and take up considerable display space due to large font sizes. Also, input using trackers is imprecise, so menu items must be large and few submenus can be used. For a command-intensive application such as immersive design, these problems force designers to think of creative ways to issue commands.

#### *4.4.2 Voice Commands*

The use of voice as a command input is very popular in virtual environments. Voice has many advantages, including its simple input device (a microphone), freedom to use the hands for other operations, and flexibility of voice input to specify complex commands. Voice also has disadvantages, including limited recognition capability, forcing the user to remember arbitrary command utterances, inappropriateness for specifying continuous quantities, and the distraction to other users in the same room.

Voice has most often been used to implement simple, discrete commands such as "Save," "Delete," or "Quit," but it has also been used in more complex menu hierarchies. Darken (1994) combined voice input with visual menus so that recall of command names was eliminated. Currently, much voice research is focused on multimodal interaction (Bolt, 1980), where voice and gesture or other input modalities are combined to form a richer and more precise interaction.

#### 4.4.3 Gesture Commands

Many of the earliest VE systems used gloves as input, and glove gestures to indicate commands. Gesture recognition is covered in detail in chapter 10 (this volume). Advantages of using gestures include the flexibility and number of degrees of freedom of the human hand, the lack of a need for traditional input devices, and the ability to use two hands. However, gestures suffer from many of the same problems as voice, including forced user recall and poor recognition rates. For these reasons, most production VE systems use menu or button systems, or only primitive- and limited-gesture commands.

#### 4.4.4 Virtual Tools

Since having a large command space proves problematic for VE applications, developers have looked for ways to reduce the number of commands needed. One way to do this is through the use of virtual tools. Tools are common in many desktop applications as a more direct way to indicate an action to the system. For example, instead of selecting an area of the screen and choosing an "erase" command, the user can simply select an eraser tool and directly erase the desired areas.

There have been similar efforts in virtual environments. One example is Mine's ISAAC system (Mine, 1997), which makes use of a wide variety of tools to perform interactive modifications to a geometric scene. The Virtual Tricorder (Wloka & Greenfield, 1995) is a multipurpose tool that changes appearance and function depending on system mode and state.

#### 4.4.5 Guidelines for Designing System Control Techniques

Unlike the other three universal interaction tasks, there has been very little evaluation of VE system control techniques. Therefore, most of the guidelines below are drawn from experience and intuition. Much more research is needed in this area.

*Reduce the necessary number of commands in the application.*

In many cases, system control in VEs is awkward and distracts from the actual task at hand. If users are struggling with the menu interface, for example, their domain-specific goals will take much longer to realize. This seems to affect immersive VEs more than traditional desktop applications because of the narrow field of view (see chap. 3, this volume), fatigue from spatial interaction (see chap. 41, this volume), and adverse effects of extended VE use (see chaps. 29–32, this volume). Developers need to look closely at what commands are absolutely necessary and what actions can be done using direct manipulation or virtual tools instead.

*When using virtual menus, avoid submenus and make selection at most a 2-D operation.*

Input to VEs (usually based on 6-DOF trackers) can be quite imprecise. Requiring users to navigate more than one level of menus can lead to frustration and inefficiency. The same holds for menu systems that require 3-D input. The use of a physical surface such as the pen-and-tablet metaphor can alleviate some of these problems, as can a 1-D solution such as rotary menus.

*Indirect menu selection may be more efficient over prolonged periods of use.*

In a usability study of two handheld virtual menu systems (Bowman, Hodges, & Bolter, 1998), it was found that menu item selection for novice users was more efficient when the items were touched directly with a stylus. However, as users became more expert with the menu system, a technique using joystick buttons to navigate the menu performed better, because users could "click ahead" of the system and no large hand movements were needed.

*Voice and gesture-based commands should include some method of reminding the user of the proper utterance or gesture.*

Lack of affordances is a major problem for most gesture and voice command systems, unless the number of commands is very small. Therefore, small visual reminders of the allowed words or gestures (Darken, 1994) may greatly enhance performance.

*Integrate system control with other interaction tasks.*

If the user thinks of the command subsystem as a completely separate component, the level of presence (see chap. 40, this volume) in the VE may decrease and the level of distraction may increase with the use of the system control technique. As much as possible, system control should seem a part of the environment in which the user is working. For example, pull-down menus attached to the user's view seem unnatural and detached from the environment, whereas a handheld menu seems part of the environment. Also, if the devices and widgets used for system control can also be used for travel, selection, or manipulation, the use of these techniques will not be as distracting to the user.

## 5. EVALUATION AND APPLICATION

This chapter has focused on the design of VE interaction techniques with high performance, but design is not the end of the story. Evaluation and application are also necessary components to the entire process of determining the appropriate interaction techniques for a VE application.

Following the guidelines and principles in this chapter should lead to interaction techniques that are well suited for the performance requirements of VE applications. However, because of the youth of this research area, one cannot guarantee performance levels no matter how many guidelines are followed. For this reason, evaluation and assessment of VE interaction is essential. Chapter 34 (this volume) covers the topic of usability evaluation in detail. Application designers need to test their systems with members of the intended user population to ensure that required performance levels are met.

For those who are designing new VE interaction techniques and performing basic research, the aspect of application must be kept in mind. Future research should focus on developing techniques that meet the performance requirements of current and proposed real-world VE applications. For examples of such applications, see chaps. 42–55 (this volume).

## 6. CONCLUSIONS

This chapter has been intended as a practical guide both for researchers and developers of interactive VE applications. It is essential to remember that most of the guidelines and principles presented here are the results of formal evaluation and that further guidelines should also be the product of careful testing.

This chapter emphasized the concept of performance, with a broad definition. In this area, user comfort, satisfaction, ease of use, and other subjective parameters are in some cases more important than more traditional performance variables such as speed and accuracy. Application developers must carefully consider the performance requirements of their systems before choosing interaction techniques.

Another consideration is the required naturalism of the interface. Certain applications need interaction that closely matches interaction in the physical environment. However, other applications with different requirements may benefit from the use of magic techniques that extend the user's real-world capabilities.

Finally, the issue of standards should be addressed. There has been a great deal of discussion about the possibility of a standard interface or set of interaction techniques for VEs, much like the desktop metaphor for personal computers. Time may prove the author wrong, but the current situation appears to suggest that it will not be fruitful to pursue a standardized interface for VE interaction. Most of the types of VE applications that are currently in use or that have been proposed are highly specialized and domain-specific, and they exhibit a wide range of interaction performance requirements. There is currently no killer application for VEs that promises to put an HMD in every home. Furthermore, evaluation has shown time and time again that the optimal interaction technique is nonabsolute, but is instead application- and task-dependent. For these reasons, the author advocates the development of VE interaction techniques that are optimized for particular tasks and applied to particular systems.

## 7. ACKNOWLEDGMENTS

The author would like to thank the following people for their help, support, and discussion regarding the issues in this chapter: Larry Hodges, Donald Allison, Drew Kessler, Rob Kooper, Donald Johnson, Albert Badre, Elizabeth Davis, Ivan Poupyrev, Joseph LaViola, Ernst Kruijff, Mark Mine, Matthew Conway, Jeffrey Pierce, Andrew Forsberg, Ken Hinckley, the other members of the 3DUI mailing list, and Dawn Bowman.

## 8. REFERENCES

Bakker, N., Werkhoven, P., & Passenier, P. (1998). Aiding orientation performance in virtual environments with proprioceptive feedback. In *Proceedings of the Virtual Reality Annual International Symposium* (pp. 28–33). Atlanta: IEEE Computer Society Press.

Bolt, R. (1980). "Put-that-there": Voice and gesture at the graphics interface. In *Proceedings of SIGGRAPH* (pp. 262–270). ACM Press.

Bowman, D., Davis, E., Hodges, L., & Badre, A. (1999). Maintaining spatial orientation during travel in an immersive virtual environment. *Presence: Teleoperators and Virtual Environments, 8*(6), 618–631.

Bowman, D., & Hodges, L. (1997). An evaluation of techniques for grabbing and manipulating remote objects in immersive virtual environments. In *Proceedings of the ACM Symposium on Interactive 3D Graphics* (pp. 35–38). Providence, RI: ACM Press.

Bowman, D., & Hodges, L. (1999). Formalizing the design, evaluation, and application of interaction techniques for immersive virtual environments. *Journal of Visual Languages and Computing, 10*(1), 37–53.

Bowman, D., Hodges, L., & Bolter, J. (1998). The virtual venue: User-computer interaction in information-rich virtual environments. *Presence: Teleoperators and Virtual Environments, 7*(5), 478–493.

Bowman, D., Koller, D., & Hodges, L. (1997). Travel in immersive virtual environments: An evaluation of viewpoint motion control techniques. In *Proceedings of the Virtual Reality Annual International Symposium* (pp. 45–52). Albuquerque, NM: IEEE Computer Society Press.

Bowman, D., Johnson, D., & Hodges, L. (1999). Testbed evaluation of VE interaction techniques. In *Proceedings of the ACM Symposium on Virtual Reality Software and Technology* (pp. 26–33). London, ACM Press.

Bowman, D., Wineman, J., Hodges, L., & Allison, D. (1998). Designing animal habitats within an immersive VE. *IEEE Computer Graphics and Applications, 18*(5), 9–13.

Bowman, D., & Wingrave, C. (2001). Design and evaluation of menu systems for immersive virtual environments. In *Proceedings of IEEE virtual reality* (pp. 149–156). Yokohana, Japan: IEEE Computer Society Press.

Brooks, F. et al. (1992). *Final technical report: Walkthrough project.* National Science Foundation.

Bukowski, R., & Sequin, C. (1995). Object Associations: A simple and practical approach to virtual 3D manipulation. In *Proceedings of the ACM Symposium on Interactive 3D Graphics* (pp. 131–138). ACM Press.

Card, S., Mackinlay, J., & Robertson, G. (1990). The design space of input devices. In *Proceedings of CHI* (pp. 117–124). ACM Press.

Chance, S., Gaunet, F., Beall, A., & Loomis, J. (1998). Locomotion mode affects the updating of objects encountered during travel. *Presence: Teleoperators and Virtual Environments, 7*(2), 168–178.

Conner, B., Snibbe, S., Herndon, K., Robbins, D., Zeleznik, R., & van Dam, A. (1992). Three-dimensional widgets. In *Proceedings of the ACM Symposium on Interactive 3D Graphics* (pp. 183–188). ACM Press.

Darken, R. (1994). Hands-off interaction with menus in virtual spaces. In *Proceedings of SPIE, Stereoscopic Displays and Virtual Reality Systems, 2177* (pp. 365–371). SPIE.

Darken, R. (1996). Wayfinding behaviors and strategies in large virtual worlds. In *Proceedings of CHI* (pp. 142–149). ACM Press.

Foley, J. (1979). A standard computer graphics subroutine package. *Computers and Structures, 10*, 141–147.

Gabbard, J., & Hix, D. (1998). Usability engineering for virtual environments through a taxonomy of usability characteristics. Unpublished masters thesis, Virginia Tech.

Goble, J., Hinckley, K., Pausch, R., Snell, J., & Kassell, N. (1995, July). Two-handed spatial interface tools for neurosurgical planning. *IEEE Computer*, 20–26.

Hinckley, K., Pausch, R., Goble, J., & Kassell, N. (1994). Design hints for spatial input. In *Proceedings of the ACM Symposium on User Interface Software and Technology* (pp. 213–222). ACM Press.

Hinckley, K., Pausch, R., Profitt, D., Patten, J., & Kassell, N. (1997). Cooperative bimanual action. In *Proceedings of CHI* (pp. 27–34). ACM Press.

Hodges, L., Rothbaum, B., Kooper, R., Opdyke, D., Meyer, T., North, M., de Graff, J., & Williford, J. (1995). Virtual environments for treating the fear of heights. *IEEE Computer, 28*(7), 27–34.

Jacoby, R., & Ellis, S. (1992). Using virtual menus in a virtual environment. In *Proceedings of SPIE, Visual Data Interpretation, 1668* (pp. 39–48). SPIE.

Kitamura, Y., Yee, A., & Kishino, F. (1998). A sophisticated manipulation aid in a virtual environment using dynamic constraints among object faces. *Presence: Teleoperators and Virtual Environments, 7*(5), 460–477.

Kennedy, R., Lane, N., Berbaum, K., & Lilienthal, M. (1993). A simulator sickness questionnaire (SSQ): A new model for quantifying simulator sickness. *International Journal of Aviation Psychology, 3*(3), 203–220.

Mackinlay, J., Card, S., & Robertson, G. (1990). Rapid controlled movement through a virtual 3D workspace (Proceedings of SIGGRAPH). *Computer Graphics, 24*(4), 171–176.

Mine, M. (1995). *Virtual environment interaction techniques* (Tech. Rep. No. TR95-018). Chapel Hill: University of North Carolina.

Mine, M. (1997). ISAAC: A meta-CAD system for virtual environments. *Computer-Aided Design, 29*(8), 547–553.

Mine, M., Brooks, F., & Sequin, C. (1997). Moving objects in space: Exploiting proprioception in virtual-environment interaction (Proceedings of SIGGRAPH). ACM Press, 19–26.

Nielsen, J., & Molich, R. (1992). Heuristic evaluation of user interfaces. In *Proceedings of CHI* (pp. 249–256). ACM Press.

Norman, D. (1990). *The design of everyday things*. New York: Doubleday.

Pierce, J., Forsberg, A., Conway, M., Hong, S., Zeleznik, R., & Mine, M. (1997). Image plane interaction techniques in 3D immersive environments. In *Proceedings of the ACM Symposium on Interactive 3D Graphics* (pp. 39–44). Providence, RI: ACM Press.

Poupyrev, I., Billinghurst, M., Weghorst, S., & Ichikawa, T. (1996). The go-go interaction technique: Non-linear mapping for direct manipulation in VR. In *Proceedings of the ACM Symposium on User Interface Software and Technology* (pp. 79–80). ACM Press.

Poupyrev, I., Weghorst, S., Billinghurst, M., & Ichikawa, T. (1997). A framework and testbed for studying manipulation techniques for immersive VR. In *Proceedings of the ACM Symposium on Virtual Reality Software and Technology* (pp. 21–28). ACM Press.

Price, B., Baecker, R., & Small. I. (1993). A principled taxonomy of software visualization. *Journal of Visual Languages and Computing*. ACM Press.

Slater, M., Usoh, M., & Steed, A. (1994). Depth of presence in virtual environments. *Presence: Teleoperators and Virtual Environments, 3*(2), 130–144.

Smith, R. (1987). Experiences with the alternate reality kit: An example of the tension between literalism and magic. In *Proceedings of ACM CHI+GI* (pp. 61–67). ACM Press.

Stoakley, R., Conway, M., & Pausch, R. (1995). Virtual reality on a WIM: Interactive worlds in miniature. In *Proceedings of CHI* (pp. 265–272). ACM Press.

Ware, C., & Osborne, S. (1990). Exploration and virtual camera control in virtual three-dimensional environments (Proceedings of the ACM Symposium on Interactive 3D Graphics). *Computer Graphics*, *24*(2), 175–183.

Wickens, C., Haskell, I., & Harte, K. (1989). Ergonomic perspective of flight path displays. *IEEE Control Systems Magazine*,

Witmer, B. & Singer, M. (1998). Measuring presence in virtual environments: A presence questionare. Presence: *Teleoperators and virtual environments, 7*(3), 225–240.

Wloka, M. & Greenfield, E. (1995). The virtual tricorder: A unified interface for virtual reality. In *Proceedings of the ACM Symposium on User Interface Software and Technology* (pp. 39–40).

Zhai, S. (1993). Investigation of feel for 6-DOF inputs: Isometric and elastic rate control for manipulation in 3D environments. In *Proceedings of the Human Factors and Ergonomics Society 37th Annual Meeting*.

# 14

# Technological Considerations in the Design of Multisensory Virtual Environments: The Virtual Field of Dreams Will Have to Wait

W. Todd Nelson[1] and Robert S. Bolia[2]

*[1]divine, Inc.*
*5151 Pfeiffer Road, Suite 200*
*Cincinnati, OH 45242*
*todd.nelson@divine.com*
*[2]Air Force Research Laboratory—AFRL/HECP*
*2255 H Street*
*Wright-Patterson Air Force Base, Ohio 45433-7022*
*robert.bolia@wpafb.af.mil*

*It was the game, the parks, the smells, the sounds. Have you ever held a bat or a baseball to your face? The varnish, the leather. And it was the crowd, the excitement of them rising as one when the ball was hit deep. The sound was like a chorus . . .*
*. . . the chance to squint my eyes when the sky is so blue it hurts to look at it, and to feel the tingle that runs up your arms when you connect dead-on. The chance to run the bases, stretch a double to a triple, and flop face-first into third base, wrapping my arms around the bag.*

—W. P. Kinsella, *Shoeless Joe*

*The real activity was done with the radio—not the all-seeing, all falsifying television—and was the playing of the game in the only place it will last, the enclosed green field of the mind.*

—A. B. Giamatti, *A Great and Glorious Game: Baseball Writings of A. Bartlett Giamatti*

## 1. INTRODUCTION

The purpose of this chapter is to examine the technological challenges inherent in the design of multisensory virtual environments (VEs). In this context, multisensory VEs are viewed as closed-loop systems comprised of humans, computers, and the interfaces through which continuous streams of information flow. More specifically, VEs are distinguished from other simulator systems by their capacity to portray three dimensional (3-D) spatial information in a variety of modalities, their ability to exploit users' natural input behaviors for human-computer interaction, and their potential to "immerse" the user in the virtual world. Accordingly, the primary objective of multisensory VEs is to provide users with salient and meaningful information that affords veridical perception and permits adaptive, goal-directed behavior in the VE, ultimately enhancing operator efficiency in a wide variety of application domains, including training, information visualization, product development and testing, entertainment, medicine and health care, and teleoperation. However, despite significant advances in human–computer interface (HCI) technology over the past decade (e.g., head-mounted displays [HMDs], spatial audio displays, and haptic interfaces), compelling and useful multisensory VEs are sparse, and the purported advantages of the technology have yet to be realized.

The goal of this chapter is to address the technological challenges involved in designing and implementing multisensory virtual environments. Whereas other authors (Barfield & Furness, 1995; Durlach & Mavor, 1995; Melzer & Moffitt, 1997) have addressed these issues in detail, their approach has typically involved the decomposition of the VE into modality-specific technologies linked to various perceptual systems—e.g., chapters such as "The Design of Multidimensional Sound Interfaces" (Cohen & Wenzel, 1995) and "The Auditory Channel" (Durlach & Mavor, 1995). While the unisensory treatment is pedagogically convenient, it fails to provide an obvious framework within which multisensory VEs can be conceived and constructed. The approach used herein will be guided by two major themes: appropriate problem specification and selection; and critical evaluations of contemporary VE technologies as they relate to human perceptual and perceptual-motor capabilities.

There are several factors entwined in the specification and selection of a suitable problem. First, it should be well defined, i.e., it should be bounded by a finite set of rules, which constrain the types of events and interactions that may occur. Second, it should be inherently multisensory, meaning that it should mimic a real or fictitious environment that naturally demands a multimodal interface for achieving veridical perception, coordinated behaviors, and a sense of presence (see Held & Durlach, 1991, 1992; Sheridan, 1992; Steuer, 1992; and chap. 40, this volume, for reviews of the concept of presence). These requirements connote a highly interactive and immersive environment. Third, the problem should be nontrivial to the extent that it represents both an interesting and utile means of discussing the challenges associated with the design of multisensory virtual environments. It is clear that these three requirements do not uniquely specify a virtual environment. Indeed, there are numerous candidates, for example, surgical simulation, tactical air combat, planetary exploration, and atomic and molecular visualization. However, many of these require domain-specific expertise for both the design and use of the virtual environment. In order to make this discussion more generally accessible, a domain has been selected that satisfies the requirements set forth above and is both broadly understood and commercially attractive: the game of baseball.

Before providing an examination of some of the technological challenges involved in designing a multisensory virtual baseball environment, it will be instructive to consider several key parameters affecting the fidelity of VE interface technologies: (1) spatial resolution; (2) temporal resolution; (3) computational complexity; and (4) synchronization. The intent is not to enumerate exhaustively the manner in which these factors impact all possible display and control interfaces, but rather to introduce the concepts and terminology necessary to facilitate an analysis of the appropriateness of current technologies and their integration to support several of the key aspects of the game.

*Spatial resolution* refers to the fineness of spatial detail that can be discriminated in an image, without reference to a particular modality. The spatial resolutions of the most common perceptual systems (e.g., sight, hearing, and touch) have been well documented in texts on human perception (see Coren, Ward, & Enns, 1999; Sekuler & Blake, 1990) and elsewhere in this book (see chaps. 3–7, this volume), whereas Barfield and his colleagues (1995) have provided a direct comparison between human perceptual capabilities and current VE technology. The analysis provided by Barfield, Hendrix, Bjorneseth, Koczmarek, & Lotens (1995) raises a number of issues: (1) there exist numerous inter- and intra-modality discongruities in spatial resolution with respect to both perceptual systems and VE technologies; (2) capabilities of prevailing VE technologies and human perceptual systems are often mismatched; and (3) the effectiveness of multisensory VE for supporting veridical perception will depend to some extent on the number and magnitude of these disparities.

*Temporal resolution* is the time-domain analogue of spatial resolution, that is, the smallest temporal interval that can be discriminated by a perceptual system. This parameter is important for effective interaction of humans and VE technologies, as discrepancies may generate errors in perception that are manifested in modality-specific ways. For example, temporal mismatches between the human auditory system and spatial audio displays may result in inadequate spatial resolution and mislocalization of virtual sounds; comparable discrepancies in haptic displays may result in failure to accurately depict object properties such as texture and rigidity; and in the case of vision, insufficient temporal resolution can produce flicker (Durlach & Mavor, 1995). Additionally, when excessive time delays exist between a user's control inputs and concomitant changes in the displays (see Wloka, 1995), perceptual-motor disruptions may result.

*Computational complexity* occurs as the result of the spatial and temporal resolution requirements imposed by the perceptual capabilities of the human observer and the mathematical models required to specify the entities in the VE and their interactions. The virtual concert hall serves as a convenient example of the complexities involved in the implementation of an acoustic model. In any hall, there are perceptible higher order acoustic reflections off of the walls and other large objects within the space. However, given the state of current audio display technology, even first-order reflections are computationally expensive, meaning that only auditory models of the most trivial virtual worlds can be achieved in real time. Similar examples are easily found in other modalities. Two of the anticipated consequences of inadequate or underspecified models are the diminution of presence (see chap. 40, this volume) and, in the case of training simulators (see chap. 43, this volume), the acquisition of inappropriate behaviors. On the other hand, the primary cost associated with the implementation of a more fully specified model is the introduction of excessive time delay, which may result in perceptual adaptations and negative aftereffects (Held & Durlach, 1991; Held, Efstathioy, & Greene, 1966; Welch, 1986; chaps. 31, 37–39, this volume), degraded perceptual-motor skills (Poulton, 1974; Ricard, 1994, 1995; chap. 38, this volume), and even sickness (Hettinger & Riccio, 1992; Kennedy, Lane, Lilienthal, Berbaum, & Hettinger, 1992; chap. 30, this volume).

*Synchronization* is one of the key issues in the design of multisensory virtual environments. First, user motion must be synchronized with its portrayal in the VE or severe degradations in system usability may ensue (Friedmann, Starner, & Pentland, 1992; see chap. 34, this volume). Furthermore, the spatial and temporal attributes of objects, events, and their interactions in the VE must often be exquisitely coupled due to the relatively large variations in spatial and temporal resolutions across sensory modalities (see chap. 21, this volume). In the real world, spatial coordination of the senses is achieved by sensory integration and intersensory and sensorimotor coordination, which develop over the life span of the human (Gregory, 1987). Intersensory mismatches induced by lack of spatial or temporal synchronization in the VE may occasion unintended perceptual illusions (see chapter 22, this volume), or ineffective perceptual-motor interactions with objects in the VE (see chap. 13, this volume). These may result in undesirable behavioral adaptations and skill acquisitions, which may be inappropriately transferred to behaviors in the real environment (Kozak, Hancock, Arthur, & Chrysler, 1993; see chaps. 29–33, 37–39, this volume). They may also lead to simulator sickness (Kennedy Lane, Lilienthal, Berbaum, & Hettinger, 1992; see chap. 30, this volume).

## 2. TECHNOLOGICAL CHALLENGES INVOLVED IN DESIGNING A MULTISENSORY VIRTUAL ENVIRONMENT

One means of facilitating discussion of designing an immersive multimodal baseball environment would be to organize the following sections according to the key perspectives of the game: hitter, pitcher, fielder, umpire, and spectator. Each of these roles is unique in its requirements for providing users with a compelling and authentic simulation. However, due to the structural limitations imposed hereupon (this being a book *chapter*, and not a *book*), this discussion will concentrate primarily on the act of hitting, delving into other roles only as required, for example, to discuss displays for sensory modalities not typically required of the hitter.

### 2.1 The Hitter

The act of hitting consists of establishing a stance, gripping the bat, reading the pitch, swinging the bat, and running the bases. Indeed, it is a complex perceptual-motor task that demands input from the visual, tactile and kinesthetic, and auditory modalities. Perhaps one of the most complex aspects of simulating the act of hitting involves the haptic and tactile representation of the bat, and the muscular stimulation that results from the holding and swinging thereof. Clearly, the apparatus required to provide this information does not currently exist. Consider, for example, the trivial case in which the batter does not swing at all (i.e., a ball or a called strike). To create the illusion of holding the bat would require a means by which a multidimensional array of forces could be presented to the batter, including the force on the bat due to gravity, which affects the perception of the bat's weight and the batter's postural control, as well as the forces involved in grasping. While a crude representation of the latter may be possible with current tactile display technology (cf. Kaczmarek & Bach-Y-Rita, 1995; Kontarinis & Howe, 1995; see chaps. 5–6, this volume), it is difficult at this time to conceive of a stable or elegant kinesthetic apparatus to affect the former. Along these lines, Hannaford and Venema (1995) have identified several formidable challenges for the realization of high-fidelity kinesthetic simulation, including specification of contact with rigid surfaces and the complexities involved in the real-time modeling of physically energetic processes.

Given the problems associated with generating a realistic simulation of grasping, holding, and swinging a virtual bat, a more practical solution may involve the combination of a real baseball bat (i.e., a "prop") with a glove-based tactile interface. In this case, the forces experienced by grasping and swinging the bat would be real, whereas those generated by hitting a virtual baseball would be displayed through force-reflecting actuators in the gloves. A variety of techniques may be employed to portray contact information (Durlach & Mavor, 1995), including shape-changing, vibrotactile, and electrotactile actuators. Moreover, the haptic perception of contact with a virtual baseball may be enhanced by providing the hitter with a realistic auditory contact cue (i.e., the so-called crack of the bat), which may also be varied to convey the type of contact made with the bat (e.g., hit squarely on the barrel, up tight on the hands, etc.). The overall effectiveness of this contact display, however, would depend on accurate tracking of bat position in relation to the position of the virtual baseball, precise collision detection models, and synchronization between resultant visual, haptic, and auditory contact information.

As with other VE applications (e.g., medical, military, teleoperation), accurate tracking of object position and orientation is essential to the overall effectiveness of the system. In the case of the hitter, position and orientation trackers would be required to track the location of the bat and various parts of the hitter's body, including hands, arms, feet, hips, shoulders, eyes, and head, and data from the trackers would be used to update the multimodal display of objects in the environment. Specifically, data from the position and orientation trackers would be used collectively as inputs to determine whether the pitch was hit or missed, and, if hit, its trajectory and distance.

While significant advances in tracking technologies have occurred over the last several decades, (see Durlach & Mavor, 1995, for review; see chaps. 8–9, this volume), the spatial and temporal resolution required of tracking a hitter's swing are extreme and constrain the types of tracking technologies that may be employed. For example, Anderson (1993) has estimated that sampling rates of approximately 1000 Hz are required to track a baseball bat in a virtual batting cage in which the hitter swings at a virtual pitch displayed in an HMD. Accordingly, trackers of the electromagnetic variety, which range from 60 to 120 Hz, would be ineffective, and if employed would ultimately lead to an unusable virtual environment. One potential solution, noted by Durlach and Mavor (1995), would be to use optical tracking systems and/or inertial trackers that are capable of 1000 Hz sampling rates; however, the former are often cost prohibitive, whereas the accuracy of the latter has been noted to suffer from "drift," thus requiring continuous calibration. A similar state of affairs exists for accurate eye tracking (see chap. 9, this volume), which may also require bandwidths in the 1000 Hz range (Inchingolo & Spanio, 1985). Unfortunately, current noninvasive eye-tracking technologies tend to have much lower sampling rates and often suffer from suboptimal accuracy (see Borah, 1998, for review).

Given the sensitive perceptual-motor couplings required for the act of hitting, substantial errors in tracking, whether caused by inadequate sampling rates, interference, or suboptimal resolution, will disrupt the display of information to the hitter, subsequently rendering the task unachievable. This line of reasoning follows from the recognition that the act of hitting can be viewed as a high gain, closed-loop tracking task, which is particularly susceptible to the deleterious effects of time delay and perturbations (see Jagacinski, 1977; Pew & Rupp, 1971; Poulton, 1974; Ricard, 1994). For example, in the case of continuous tracking tasks, increases in time delay must be accompanied by decreases in gain in order to maintain tracking stability. If not, unstable tracking will occur, which in the case of hitting may be exhibited by swinging too high, low, early, or late. Conversely, if the hitter's gain is too low, tracking of the pitch will be sluggish, resulting in increased tracking error and poor hitting performance.

Position and orientation tracking of other aspects of the hitter's body—head, shoulders, and hips—may not be as severely constrained by bandwidth requirements; however, issues such as encumbrance, susceptibility to obfuscation and interference, and accuracy, may limit their effectiveness for supporting the hitting scenarios. Despite these caveats, potential candidates for body-position tracking may include commercially popular electromagnetic trackers (Ascension Flock of Birds and Fastrak), optical systems, and mechanical trackers, each of which vary in degree of accuracy, resolution, range, sampling rate, interference, and latency (Durlach & Mavor; 1995; Meyer, Applewhite, & Biocca, 1992; Nixon, McCallum, Fright & Price, 1998; Piantanida, Bowman, & Gille, 1993; Pimentel & Teixeira, 1993; Sutherland, 1970). For example, the position of the key body joints (e.g., ankles, knees, hips, hands, shoulders, head) could be tracked with either active or passive markers in conjunction with the appropriate optical systems (see Durlach & Mavor, 1995) or by wearing a suit that contains goniometers—mechanical trackers—at the body joints of interest. Electromagnetic trackers would also be appropriate for tracking body positions that do not require high sampling rates, but limitations such as interference and range should be recognized (Nixon et al., 1998). To summarize, the technological challenge imposed by tracking position and orientation information for the act of hitting are substantial. In all likelihood, the tracking system would comprise a collection of mechanical, inertial, optical, and electromagnetic trackers. Given that hitting is a tightly coupled, closed-loop system, any degradation of the temporal and spatial accuracy of the tracking information would severely impact the user's ability to hit the virtual pitch.

Of course, since the batter is not going to hit what cannot be seen, several important considerations exist for the selection of an appropriate visual display for the act of hitting. In addition to reading the pitch, detecting the pitcher's release, and predicting the ball's trajectory, the visual display must permit the interpretation of hitting signs from the coach, survey of the defensive configuration, and determination of ball's location and trajectory after being struck. Ideally, the visual display should endeavor to match the capabilities of the human visual system, that is, a 210-degree horizontal × 125-degree vertical field of view and resolution on the order of 0.5—1 minute of arc, allowing for depth perception resolution of 10 seconds of arc, and temporal resolution that exceeds critical flicker-fusion thresholds. In addition, to enhance the sense of presence and realism, the visual display should be capable of realistic color (approximately 300 chromaticities/luminance level) and photopic daylight luminance levels ($0.314 - 3.14 \times 10^6$ cd/m$^2$). Unfortunately, current VE visual display technologies are incapable of matching the capabilities of the human visual system (Barfield et al., 1995; Durlach & Mavor, 1995; see chap. 3, this volume).

Another important consideration is the actual display medium, that is, whether visual display should be head mounted, boom mounted, surface projected, or even retinally displayed. While HMDs have been used extensively in VEs, they suffer from narrow fields of view, insufficient resolution, small exit pupils, and problems resulting from size, weight, and fit. The latter severely limits this technology's usability, especially for long or extended sessions. In addition, weight and associated inertial liabilities have been noted as a factor contributing to simulator sickness and disorientation (see chap. 30, this volume). Moreover, since the ability of an HMD to support goal-oriented behavior is intimately tied to the performance of the tracking system, any deficiencies in the tracking data will only serve to exacerbate problems resulting from the HMD's limitations. Finally, sliding into base while wearing an HMD may prove unsafe for both user and apparatus.

An alternative to HMDs is the employment of an off-head or projected display, which is capable of producing relatively high-resolution stereoscopic visual images over large fields of view. Stereoscopic images are often achieved by employing infrared LCD shutter glasses to enable field synchronization (e.g., CAVE display system, Cruz-Neira, Sandin, DeFanti, Kenyon, & Hart, 1992). While the projection display alternative alleviates problems such as excessive helmet weight, issues of comfort, and narrow fields of view, it is important to recognize that the user will not be totally immersed in the VE, that is, aspects of the apparatus, such as cables, mechanical supports, housings, and other apparatus may be part of the user's visual field of regard. Another alternative is retinal display technology (Pryor, Furness, & Viirre, 1998), which may provide a compelling and effective visual medium. However, it is currently an immature technology with numerous technological limitations (e.g., small exit pupils, not robust movement, etc.) that make it inadequate for the hitting scenario.

## 2.2 Running the Bases

Running the bases is another essential element for simulating the act of hitting, and it too introduces some formidable technological challenges. Ideally, following a hit the player should be able to quickly accelerate down the first base line, deciding to either turn the corner and head toward second base, or simply to run out the single. As described by Durlach and Mavor (1995), the ground surface interface should allow the user to experience the active sensation of locomotion, permitting natural limb movement and providing feedback that matches the spatiotemporal characteristics of the simulated surface. It must also permit the collection of position and orientation data so that appropriate updates can be made to it and the other display interfaces. There are several key challenges for providing an effective ground interface display for running the bases, including, but not limited to, issues involving balance, change in direction, modeling surface characteristics (e.g., friction, rigidity, etc.), speed, acceleration and deceleration, noise, and capturing, modeling, and translating real leg movements into virtual leg motions.

One approach to providing ground surface interfaces for locomotion is the use of omnidirectional treadmills (see Darken, Cockayne, & Carmein, 1997; Iwata & Yoshida, 1999; and Templeman, Denbrook, & Sibert, 1999, for reviews; also see chap. 11, this volume). For example, the Torus Treadmill, described by Iwata and Yoshida (1999), comprises a set of 12 individual treadmills arranged in a toroidal configuration, affording motion in all directions. However, there are significant limitations in terms of both the speeds and accelerations required for activities such as base running. For example, a fast baserunner can achieve speeds in excess of 9.75 m/sec, whereas the Torus Treadmill has a maximum speed of 1.2 m/sec (Adair, 1994; Iwata & Yoshida, 1999). Clearly, this will not accommodate realistic baserunning, not to mention sliding into home, an activity that is constrained not only by the surface texture but also by the physical size of the device. Furthermore, Iwata and Yoshida (1999) point out that the veridical portrayal of the acceleration cues accompanying changes in direction require an additional layer of apparatus.

Another option for the simulation of locomotion, and hence baserunning, is a gesture-recognition-based input system such as that described by Templeman and his colleagues (1999), which uses leg movements to infer the direction, extent, and timing of the user's traversal of a virtual environment. Because this sensor-based system is not constrained by the physical apparatus associated with a motion platform, it appears that limitations on velocity and acceleration will not be an issue. Furthermore, since the system allows for actual steps as well as gestural steps, realistic baserunning should be attainable given an arbitrary region of sufficient extent. However, it must be noted that most of the technologies described up to this point demand a tethered user and thus do not support motions over large areas. Thus, one has recourse to the gesture-based system, which has its own problems, including under-specification of accelerative forces, necessitating the introduction of a virtual vestibular cue (an admittedly immature technology with its own set of problems; cf. Cress et al., 1996; see chap. 7, this volume), imperfect gesture classification algorithms, and problems generic to all tracking systems (see chaps. 8–9, this volume).

## 3. OTHER ROLES, OTHER MODALITIES

While the realization of the act of hitting a virtual baseball requires precise and accurate visual, haptic, and tactile displays, it may be that veridical portrayal to other modalities may be neglected without sacrificing fidelity, usability, or presence. This is certainly not true for all of the roles that must be considered in constructing the virtual baseball game. The auditory modality is particularly important for the fielding team, as well as for umpires calling the game (the first base umpire may hear, not see, when the ball reaches the first baseman's glove), while the role of the spectator may require audition (Gilkey and Weisenberger (1995) have suggested that audio-free VEs suffer a diminution of presence), as well as olfaction and potentially gustation (heeding the suggestion that a baseball game without a hot dog is not a baseball game).

Spatial audio displays (see chap. 4, this volume) are one of the more mature technologies required for the generation of multisensory VEs, though even they must be employed with some caution. The generation of veridical spatial audio cues is accomplished by measuring the acoustic waveform at or near the eardrum of a listener, and interpreting the difference between the spectra of the source and the resultant wave as a transfer function characterizing, for a given ear and a given source position, the frequency-dependent modifications of the incident waveform by the head, torso, and pinna. If this is done for both ears, then it is possible to encode all of the cues necessary for accurate auditory localization using two channels, one for each ear. The transfer functions for each ear, commonly referred to as head-related transfer functions, or HRTFs, can then be represented as digital filters and convolved with an appropriate audio signal to produce the sensation of an externalized sound source whose perceived direction is the same as that of the measured HRTF (Wightman & Kistler, 1989).

While this process may appear trivial, to the extent that it can be completely specified in one or two sentences, there are a number of computational issues that come into play when actually generating a multisource auditory virtual environment. For example, each source requires its own pair of HRTFs, each of which requires a certain amount of memory and digital signal processing. If each spectator in a stadium is equated with a single source, the problem becomes clear. There is also the issue of reflections, which are important when one is not simulating an anechoic environment. Specifically, each reflection requires its own HRTF, and, in a room with four walls, a ceiling, and a floor, each source will have six first-order reflections. In such an extensive environment as the typical baseball stadium, effects such as wind, atmospheric absorption, and Doppler effects might also need to be considered.

Another issue arising in the implementation of spatial audio displays is the fact, recently indicated by Brungart and his colleagues (Brungart, Durlach, & Rabinowitz, 1999; Brungart & Rabinowitz, 1999), that the physical cues subserving auditory localization and distance perception in the near field (i.e., distances less than 1 meter from the listener) are different from those used in the far field, and that listeners are much better at judging distances within this "peripersonal" region than they are at distance judgments in "extrapersonal" space. Thus, in order to faithfully represent auditory events occurring at distances within 1 meter of the listener (such as arguing with the man who just spilled beer on you), multiple sets of HRTFs are required.

One modality that has received little note from the designers of multisensory VEs is that of olfaction. This is surprising, given the recognition that compelling olfactory cues have the potential to enhance the sense of presence in a virtual environment (Barfield & Danas, 1996), invoke emotion (Corbin, 1982), and provide salient spatial cues (von Békésy, 1964). To date, the most comprehensive treatment of the olfactory system in the context of VEs has been provided by Barfield and Danas (1996), who review the physiological and psychological aspects of olfaction, discuss mechanisms for the presentation of odorants, and enumerate a number of olfactory analogs to visual parameters, such as field of smell (field of view), the great variety of smells (color), and spatial resolution. While other researchers have speculated on the utility of olfactory displays for entertainment applications (Cater, 1992, 1994), only recently have commercially available olfactory displays been offered (e.g., SENX Machines, iSmell Digital Scent Technology, Aromajet, etc.). While these and other olfactory displays might serve to enhance the presence of the spectator, Barfield and Danas (1996) suggest a number of caveats. Of particular concern are such stimulus-related issues as the precise control of intensity and duration, both functions of the volatility of the odorant, as well as perceptual issues such as habituation, individual differences, and pathologies in odor perception (as Barfield and Danas point out, approximately two million American adults suffer some form of olfactory dysfunction). Collectively, these challenges continue to limit the use of virtual olfactory displays.

## 4. CONCLUSIONS

The ambition of this chapter was to discuss technological challenges in the design of multisensory virtual environments. Toward this end, a general classification scheme was introduced that facilitates meaningful comparisons between current VE display and control technologies and users' perceptual and perceptual-motor requirements. The relatively simple taxonomy, i.e., spatial and temporal resolution, computational complexity, and synchronization, captures the necessary elements for a critical analysis of most, if not all, multisensory virtual environments.

As an example of the complexities and challenges involved in designing a usable multisensory VE, some of the technological issues in providing a multisensory virtual baseball game were considered, focusing primarily on the hitter. Indeed, even the most cursory analysis of this baseball application reveals numerous inadequacies of current multisensory VE technology. Specifically, significant advances in position and orientation tracking technologies (see chaps. 8–9, this volume), tactile and force feedback displays (see chaps. 5–6, this volume), and omnidirectional locomotion displays (see chap. 11, this volume) are needed to afford a compelling simulation of the act of hitting and running the bases. Even in those cases where VE technologies are sufficiently mature (e.g., various visual and auditory displays), the overall effectiveness of the system will be determined by the extent to which spatiotemporal disruptions of critical perceptual information are minimized. This conclusion is not unique to the baseball application, but rather, to any application domain that comprise tasks involving tightly coupled perceptual-motor skills.

It is also interesting to note that the effects of inadequate VE technology are highly task-dependent. For example, while excessive time delay for the hitter and baserunner would likely result in uncoordinated behavior, time delays in the spectator's displays may result in disturbing perceptual illusions (Azuma & Bishop, 1994; see chap. 22, this volume), spatial disorientation (see chap. 24, this volume), and possibly sickness (see chap. 30, this volume). In the case of the spectator, where the goal is to create presence (see chap. 40, this volume) by immersing the user in the sight, sounds, and smells of the virtual ball park, it appears that current multisensory technologies are not sufficiently mature. Consequently, perhaps the most effective means for immersing oneself in the game—other than going to a real ballpark—is to sit down with a radio and spectate within the "enclosed green field of the mind" (Giamatti, 1998).

The current discussion of technological challenges in designing usable multisensory VEs is limited in several ways. First, though not addressed here, it is important to point out that researchers and engineers have developed a wide variety of schemes designed to compensate for some of the technological limitations described herein, including: (1) offsetting time delays in the system via algorithmic prediction of input data; (2) reducing the computational complexity for generating visual scenes (e.g., texture mapping, restricting the resolution outside of the user's field of view, etc.); and (3) exploiting user's ability to adapt to spatial–temporal rearrangements in the virtual environment (see chap. 31, this volume).

As one might expect, each of these strategies is accompanied by a variety of advantages and disadvantages, hence the decision to incorporate these solutions is nontrivial and highly task-dependent. Second, the adequacy of VE technology as a training apparatus for hitting was not addressed. The key question here is whether or not VE technology facilitates positive transfer of training to the real world (i.e., would training in a VE batting cage transfer to batting in a real game?). Of particular concern is the issue of whether constraints imposed by inadequate VE technology, for example, time delay, compel users to develop and adopt control strategies that are specific to the VE, and, if so, whether or not these VE-specific skills negatively impact performance of tasks in the real world. Considerable research has been devoted to the transfer of training question in the area of flight simulation, and it will likely continue to be a relevant question in any application domain in which VE technology is considered for real-world skill acquisition (see chap. 19, this volume). Finally, although the current discussion was largely limited to the act of hitting, it should be realized that there are numerous other roles in baseball that warrant a similar examination.

## 5. ACKNOWLEDGMENTS

The authors would like to acknowledge Dr. Mark Draper of the Air Force Research Laboratory for his careful review of an earlier draft of this manuscript.

## 6. REFERENCES

Adair, R. K. (1994). *The physics of baseball*. New York: HarperCollins.

Andersen, R. L. (1993). A real experiment in virtual environments: A virtual batting cage. *Presence, 2*, 16–33.

Azuma, R., & Bishop, G. (1994). Improving static and dynamic registration in an optical see-through HMD. In *Proceedings of SIGGRAPH '94* (pp. 197–203). NEW YORK, NY: ACM. ACM SIGGRAPH.

Barfield, W., & Danas, E. (1996). Comments on the use of olfactory displays for virtual environments. *Presence*, *5*(1), 109–121.

Barfield, W., & Furness, T. A. (1995). *Virtual environments and advanced interface design*. New York: Oxford University Press.

Barfield, W., Hendrix, C., Bjorneseth, O., Kaczmarek, K. A., & Lotens, W. (1995). Comparison of human sensory capabilities with technical specifications of virtual environment equipment. *Presence, 1*, 329–356.

Békésy, G. von (1964). Olfactory analog to directional hearing. *Journal of Applied Physiology, 19*, 369–373.

Borah, J. (1998). Technology and application of gaze-based control. In *Proceedings of RTO Lecture Series 215: Alternative Control Technologies: Human Factors Issues* (pp. 3:1–3:10). Neuilly-sur-Seine Cedex, France: RTO/NATO.

Brungart, D. S., Durlach, N. I., & Rabinowitz, W. M. (1999). Auditory localization of nearby sources. II. Localization of a broadband source. *Journal of the Acoustical Society of America, 106*, 1956–1968.

Brungart, D. S., & Rabinowitz, W. M. (1999). Auditory localization of nearby sources. Head-related transfer functions. *Journal of the Acoustical Society of America, 106*, 1465–1479.

Cater, J. P. (1992). The nose have it! *Presence, 1*, 493–494.

Cater, J. P. (1994). Smell/taste: Odors in reality. In *1994 IEEE International Conference on Systems, Man, and Cybernetics* (p. 1871). San Antonio, TX: IEEE.

Cohen, M., & Wenzel, E. M. (1995). The design of multidimensional sound interfaces. In W. Barfield & T. A. Furness (Eds.), *Virtual Environments and Advanced Interface Design* (pp. 291–346). New York: Oxford University Press.

Corbin, A. (1982). *Le miasme et la jonquille: L'odorat et l'imaginaire social, XVIIIe–XIXe siècles*. Paris: Librairie Chapitre.

Coren, S., Ward, L. M., & Enns, J. T. (1999). *Sensation and perception*. New York: Harcourt College Publishers.

Cress, J. D., Hettinger, L. J., Cunningham, J. A., Riccio, G. E., McMillan, G. R., & Haas, M. W. (1996). An initial evaluation of a direct vestibular display in a virtual environment. In *Proceedings of the Human Factors and Ergonomics Society 40$^{th}$ Annual Meeting* (pp. 1131–1135).

Cruz-Neira, C., Sandin, D. J., DeFanti, T. A., Kenyon, R., & Hart, J. C. (1992, June). The CAVE audio visual experience automatic Virtual environment. *Communications of the ACM*, 64–72

Darken, R. P., Allard, T., & Achille, L. B. (1999). Spatial orientation and wayfinding in large-scale virtual spaces. II: Guest editors' introduction. *Presence*, *8*, iii–vi.

Darken, R. P., Cockayne, W. R., & Carmein, D. (1997). The omni-directional treadmill: A locomotion device for virtual worlds. In *Proceedings of the ACM USIT '97* (pp. 213–222). New York: ACM Press.

Durlach, N. I., & Mavor, A. S. (Eds.). (1995). *Virtual reality: Scientific and technological challenges*. Washington, DC: National Academy Press.

Friedmann, M., & Starner, T., & Pentland, A. (1992). Synchronization in virtual realities. *Presence, 1*, 139–144.

Giamatti, A. B. (1998). *A great and glorious game: Baseball writings of A. Bartlett Giamatti*. Chapel Hill, NC: Algonquin Books.

Gilkey, R. H., & Weisenberger, J. M. (1995). The sense of presence for the suddenly deafened adult: Implications for virtual environments. *Presence, 4*, 357–363.

Gregory, R. L. (1987). *The Oxford companion to the mind*. Oxford, England: Oxford University Press.

Hannaford, B., & Venema, S. (1995). Kinesthetic displays for remote virtual environments. In W. Barfield & T. A. Furness (Eds.), *Virtual Environments and Advanced Interface Design* (pp. 415–436). New York: Oxford University Press.

Held, R., & Durlach, N. (1991). Telepresence, time delay, and adaptation. In S. R. Ellis, M. K. Kaiser, & A. J. Grunwald (Eds.), *Pictorial communication in virtual and real environments* (pp. 232–246). London: Taylor & Francis.

Held, R. M., & Durlach, N. I. (1992). Telepresence. *Presence, 1*, 109–112.

Held, R., Efstathioy, A., & Greene, M. (1966). Adaptation to displaced and delayed visual feedback from the hand. *Journal of Experimental Psychology, 72*, 887–891.

Hettinger, L. J., & Riccio, G. E. (1992). Visually induced motion sickness in virtual environments. *Presence, 1*, 306–310.

Inchingolo, P., & Spanio, M. (1985). On the identification and analysis of saccadic eye movements—A quantitative study of the processing procedures. *IEEE Transactions on Biomedical Engineering, 32*, 683–695.

Jagacinski, R. J. (1977). A qualitative look at feedback control theory as a style of describing behavior. *Human Factors, 19*, 331–347.

Iwata, H., & Yoshida, Y. (1999). Path reproduction tests using a torus treadmill. *Presence, 8*, 587–597.

Kaczmarek, K. A., & Bach-Y-Rita, P. (1995). Tactile displays. In W. Barfield & T. A. Furness (Eds.), *Virtual Environments and Advanced Interface Design* (pp. 349–414). New York: Oxford University Press.

Kennedy, R. S., Lane, N. E., Lilienthal, M. G., Berbaum, K. S., & Hettinger, L. J. (1992). Profile analysis of simulator sickness symptoms: Application to virtual environment systems. *Presence, 1*, 295–301.

Kinsella, W. P. (1982). *Shoeless Joe*. New York: Ballantine.

Kontarinis, D. A., & Howe, R. D. (1995). Tactile display of vibratory information in teleoperation and virtual environments. *Presence, 4*, 387–402.

Kozak, J. J., Hancock, P. A., Arthur, E. J., & Chrysler, S. T. (1993). Transfer of training from virtual reality. *Ergonomics, 36*, 777–784.

Melzer, J. E., & Moffitt, K. (1996). *Head-mounted displays: Designing for the user*. New York: McGraw Hill.

Meyer, K., Applewhite, H. L., & Biocca, F. A. (1992). A survey of position trackers. *Presence, 1*, 173–200.

Nixon, M. A., McCallum, B. C., Fright, W. R., & Price, N. B. (1998). The effects of metals and interfering fields on electromagnetic trackers. *Presence, 7*, 204–218.

Pew, R. W., & Rupp, G. L. (1971). Two quantitative measures of skill development. *Journal of Experimental Psychology, 90*, 1–7.

Piantanida, T., Bowman, D. K., & Gille, J. (1993). Human perceptual issues and virtual reality. *Virtual Reality System, 1*, 43–52.

Pimentel, K., & Teixeira, K. (1993). *Virtual reality: Through the new looking glass*. New York: Intel/Windcres/McGraw-Hill.

Poulton, E. C. (1974). *Tracking skills and manual control*. New York: Academic Press.

Pryor, H. L., Furness, T. A., & Viirre, E. (1998). The virtual retinal display: A new display technology using scanned laser light. *Proceedings of Human Factors and Ergonomics Society 42nd Annual Meeting* (pp. 1570–1574). Santa Monica, CA: Human Factors and Ergonomics Society.

Ricard, G. L. (1994). Manual control with delays: A bibliography. *Computer Graphics, 28*, 149–154.

Ricard, G. L. (1995). Acquisition of control skill with delayed and compensated displays. *Human Factors, 37*, 652–658.

Richard, P., Birebent, G., Coiffet, P., Burdea, G., Gomez, D., & Langrana, N. (1996). Effect of frame rate and force feedback on virtual object manipulation. *Presence, 5*, 95–108.

Sekuler, R., & Blake, R. (1990). *Perception*. New York: McGraw-Hill.

Sheridan, T. B. (1992). Musings on telepresence and virtual presence. *Presence, 1*, 120–126.

Steuer, J. (1992). Defining virtual reality: Dimensions determining telepresence. *Journal of Communication, 42*, 73–89.

Sutherland, I. E. (1970). Computer displays. *Scientific American, 222*, 56–81.

Templeman, J. N., Denbrook, P. S., & Sibert, L. E. (1999). Virtual locomotion: Walking in place through virtual environments. *Presence, 8*, 598–617.

Welch, R. B. (1986). Adaptation of space perception. In K. R. Boff, L. Kaufman, & J. P. Thomas (Eds.), *Handbook of perception and human performance: Vol. I. Sensory processes and perception* (pp. 24-1–24-37). New York: John Wiley & Sons.

Wightman, F. L., & Kistler, D. J. (1989). Headphone simulation of free-field listening. I: Stimulus synthesis. *Journal of the Acoustical Society of America, 85*, 858–867.

Wloka, M. M. (1995). Lag in multiprocessor virtual reality. *Presence, 4*, 50–63.

# 15

# Embodied Autonomous Agents

Jan M. Allbeck and Norman I. Badler
*Center for Human Modeling and Simulation*
*University of Pennsylvania*
*200 S. 33rd St.*
*Philadelphia, PA 19104-6389*
*allbeck@graphics.cis.upenn.edu*
*badler@central.cis.upenn.edu*

## 1. INTRODUCTION

Creating embodied autonomous agents for a virtual environment combines research from computer graphics, animation, and artificial intelligence, as well as ideas from psychology, sociology, and cognitive sciences. In this chapter, some concepts that should be considered when designing and creating embodied agents for virtual environments (VEs) are discussed. Some of these concepts have been researched for many years, and some are just now being considered. Likewise, some of the associated research problems have more complete solutions than others. The chapter begins by defining embodied autonomous agents for virtual environments. Then agent appearances are discussed. Next, research on animating the actions of agents is reviewed. This is followed by a discussion of agent communication and mental processes. Finally, an outline of research in action selection is provided, concluding with a discussion of future research directions.

### 1.1 Definition

The term *agent* has received a lot of use in the past few years, but the definition of an agent is still unclear. In computer graphics an agent might mean any autonomous entity in the virtual environment. For example, any human or even gravity might be viewed as an agent. Agents might be defined in contrast to avatars, which are graphical representations of a user in the environment that are under his or her direct control: An agent would be any autonomous human actor.

Artificial intelligence has struggled even more with their definition of an agent. Although there is still no set definition of what an agent is, Wooldridge and Jennings (1995) have proposed a weak definition as a computer system with the following properties:

- *Autonomy*: Agents operate without the direct intervention of humans or others, and have some kind of control over their actions and internal state.
- *Social ability*: Agents interact with other agents (and possibly humans) via some kind of agent-communication language.
- *Reactivity*: Agents perceive their environment, and respond in a timely fashion to changes that occur in it.
- *Proactiveness*: Agents do not simply act in response to their environment; they are capable of goal-directed behavior by taking the initiative.

A stronger definition of an agent adds to these properties the idea that the system is described using concepts that are usually applied to humans, such as beliefs, knowledge, and intention.

In this chapter the term *agent* will be used to describe a synthetic or virtual human agent in a VE with some degree of autonomy, social ability, reactivity, and proactiveness. An agent might be a semiautonomous agent, which acts autonomously but can be corrected or redirected by a user. The user may input high-level goals for the agent to achieve, but how the agent achieves these goals is left to the agent. Adding the word embodied further focuses this chapter. All of the agents herein discussed will have a physical (geometric) representation in the environment.

## 2. APPEARANCE

Autonomous agents can take the form of human beings, faces, animals, or more exotic forms. Their appearance in the virtual world can range from cartoonish to a physiologically accurate model. The proper level of detail and style of the character depends on the application. This discussion concentrates on realistic looking (embodied) human, autonomous agents. A human's appearance can be characterized by its age, aesthetics, gender, body type, ethnicity, hair, skin, bone, and muscle modeling.

Generally, the appearance of a virtual human is designed offline and then displayed during live interactions. Body, face, and muscle models have been created by many groups, including Gourret, Magnenat-Thalmann and Thalmann (1989); Badler, Phillips and Webber (1993); Wilhelms and Van Gelder (1997); Capin, Pandzic, Noser, Thalmann, and Magnenat-Thalmann (1997); Scheepers, Parent, Carlson, and May (1997); and Ting (1998). In general, low-polygon-count models with simple shading are used for interactive applications such as virtual environments. For an overview of modeling realistic virtual humans, see Kalra et al. (1998).

Modeling of hair and skin adds to the realism of an agent's appearance. The complex task of hair modeling has been studied by Anjyo, Usami, and Kurihara (1992); Watanabe and Suenaga (1992); and Sourin, Pasko and Savchenko (1996). In recent years, much attention has also been given to the modeling of skin including wrinkles and textures. There are three basic models used to simulate skin deformations: geometric models; physically based models; and biomechanical models. Geometric models can be further broken down into parametric models (Parke, 1982) and muscle-based models (Magnenat-Thalmann, Primeau, & Thalmann, 1988; Waters, 1987). Ishii, Yasuda, Yokoi, Yoriwaki, and Toriwaki (1993) propose a geometric model of micro wrinkles. For expressing folds and ridge features, their model uses a curved surface on a base polygon. Viaud and Yahia (1992) proposed a hybrid model for expressive wrinkles. The bulges of the wrinkles are modeled as spline segments.

Physically based models of the skin include three layered deformable lattices (Lee, Waters, & Terzopoulos, 1995) and finite element methods (Koch et al., 1996; Larrabee, 1986; Pieper,

1992). Wu, Kalra, and Magnenat-Thalmann (1997) have created a biomechanical model in which a simulated muscle layer causes skin deformations. Several layers of color, bump, and displacement mapping are used to increase the realism of the wrinkles.

An agent's clothing also influences one's perception of an agent. Dressing a virtual human is not a simple task. Not only must one design the clothing shapes, one must also design the simulation of the cloth, handle collisions between the cloth and the body, and simulate folding, wrinkling, and crumpling. Most cloth models are physically based, using elastic deformations and collision response. Early models were based on spring and mass models (Haumann & Parent, 1988; Weil, 1986) and simulated flags or curtains. Terzopoulos, Platt, and Barr (1987) and Terzopoulos and Fleischer (1988) created more general elastic models, which modeled a range of deformable objects. Particle-system-based models, which use fabric deformation data to simulate simple cloth objects, are fast and flexible (Breen, House, & Getto, 1992; Eberhardt, Weber, & Strasser, 1996). Finite element models (Collier, Collier, O'Toole, & Sargand, 1991; Eischen, Deng, & Clapp, 1996; Kang & Yu, 1995) are accurate but slower and more complex.

Volino and Magnenat-Thalmann (1997) propose a cloth model that is robust and suited for interactive design and simulation. Their aim is to create a tool for interactively designing garments and dressing virtual humans in a natural way. They describe a mechanical model that simulates the elastic behavior of cloth, in which the cloth surface is broken into irregular triangle meshes. They build on previous work in an attempt to create a fast, flexible, and accurate cloth modeling system. Baraff and Witkin (1998) have also created a fast-cloth model using physically based constraints and implicit integration.

Although the appearance of agents has been researched for years, and Babski and Thalmann (1999) have made strides in fast skin modeling, there is still a need to make both skin and cloth deformations real time and also a need to show how the characteristics of appearance impact other parts of an agent model. For example, an agent's age may impact its locomotion, and its appearance or ethnicity could have an impact on its behaviors. An agent's appearance will influence how he or she is perceived because it is often an external indicator of status and role.

## 3. ANIMATING ACTIONS

In order to move agents in VEs one must animate actions such as posture changes, balance adjustments, reaching, pointing, grasping, giving–taking, other arm/hand gestures, locomotion, eye gaze, head gestures, facial expressions, and speech. These animations involve controlling parts at the graphical level via joint transformations (e.g., for limbs) or surface deformations (e.g., for face). There are two tool kits available to aid in the construction of actions for virtual humans: Motion Factory's Motivate and Engineering Animation, Inc.'s, Jack. Standards are also being designed, such as Humanoid Animation Group and MPEG-4. Many other approaches are described in Wilcox (1998). Methods for animating agents for VEs include frame-by-frame animation, key-frame animation, motion capture, kinematics, and dynamics.

### 3.1 Locomotion

Most methods for generating locomotion can be classified into kinematics (Boulic, Magnenat-Thalmann, & Thalmann, 1990; Bruderlin & Calvert, 1996;), dynamics, which are slower and more complex (Bruderlin & Calvert, 1989; Hodgins, Wooten, Brogan, & O'Brien, 1995; Laszlo, van de Panne, & Fiume, 1996; McKenna & Zelter, 1990), motion editing (Gleicher, 1997; Lee & Shin, 1999; Popovic & Witkin, 1999), and combinations of these methods (Anderson & Pandy, 1999; Girard & Maciejewski, 1985; Ko & Badler, 1996). Sun, Goswami,

Metaxas, and Bruckner (1999) have created a system that combines motion-captured data with a procedural model. Their system can scale the motion across different size figures, alter the motion stylistically, and allow the agent to perform curved-path locomotion on uneven terrain in near real time.

The inclusion of locomotion for agents in VEs brings about issues of path planning or navigation, and collision detection and avoidance (Becket, 1996; Kuffner & Latombe, 1999; Noser, Renault, Thalmann, & Magnenat-Thalmann, 1995; Pandzic, Capin, & Magnenat-Thalmann, 1997; Reich, 1997). For a good survey of research on locomotion see Multon, France, Cani-Gascuel, and Debunne (1999).

## 3.2 Body Actions

Body motion is crucial to agent behavior and is accordingly well studied. Research has been done on many areas of body actions, including gesture generation (Cassell et al., 1994), grasping (Gourret et al., 1989), reaching, lifting, posture changes, and more (Badler, Phillips, & Webber, 1993). Similar to locomotion, techniques for generating body actions can be classified as kinematic (Calvert, Chapman, & Patla, 1982; Tolani & Badler, 1996) or dynamic (Hodgins, Wooten, Brogan, & O'Brien, 1995; Kokkevis, Metaxas, & Badler, 1996). There is also work being done on editing motions (Wilhelms, 1986), and motion retargeting (Bindiganavale & Badler, 1998; Hodgins & Pollard, 1997; Gleicher, 1998).

Transitioning from one action to another is also very important to natural-looking behaviors (Boulic, Becheiraz, Emering, & Thalmann, 1997; Granieri, Crabtree, & Badler, 1995; Maiochhi & Pernici, 1990; Rose, Guenter, Bodenheimer, & Cohen, 1996; Wiley & Hahn, 1997). Sengers (1998) talks about action expression, which is the linking together of actions with transition actions to better describe the agent's cognitive state. These types of transition actions may allow human observers of agents' actions to better understand the agents' goals and activities.

In recent years, adding expressivity to human animation has been studied (Koga, Kondo, Kuffner, & Latombe, 1994; Perlin & Goldberg, 1996). In order to have believable agents, their movements should be individualized and based on their current emotional state. EMOTE (Chi, Costa, Zhao, & Badler, 2000) is a real-time, three-dimensional (3-D) character animation that modifies independently defined arm and torso movements through the specification of Effort and Shape parameters (Laban & Lawrence 1974, see Fig. 15.1). The underlying arm and torso movements are defined through key time and pose information, which may be created using key framing or generated synthetically by inverse kinematics, motion capture, or other prestored movement patterns. EMOTE provides an easy way to add emotion and drama to gestures, giving them a more natural appearance.

## 3.3 Facial Expressions

The face has a great deal of detail in the muscle, bone, and tissue. These structures influence deformations in facial features. There is no facial animation model that represents and simulates the complete, detailed anatomy. Approaches differ in their characterization of surface representation, sources of facial surface data, ability to generate varied geometric models, and animation capability. In addition to the physical structure of the face, there are also attributes of surface colors and textures. There has been significant effort in the development of computational models of the human face.

Parke (1972) proposed the first computer-generated facial model. It uses linear interpolation of a few expressive poses of a face, similar to key frame interpolation. The first physics-based face model was developed by Platt and Badler (1981). It included a simplified model of skin, muscle, and bone structure. The skin was represented by an interconnected elastic mesh, which

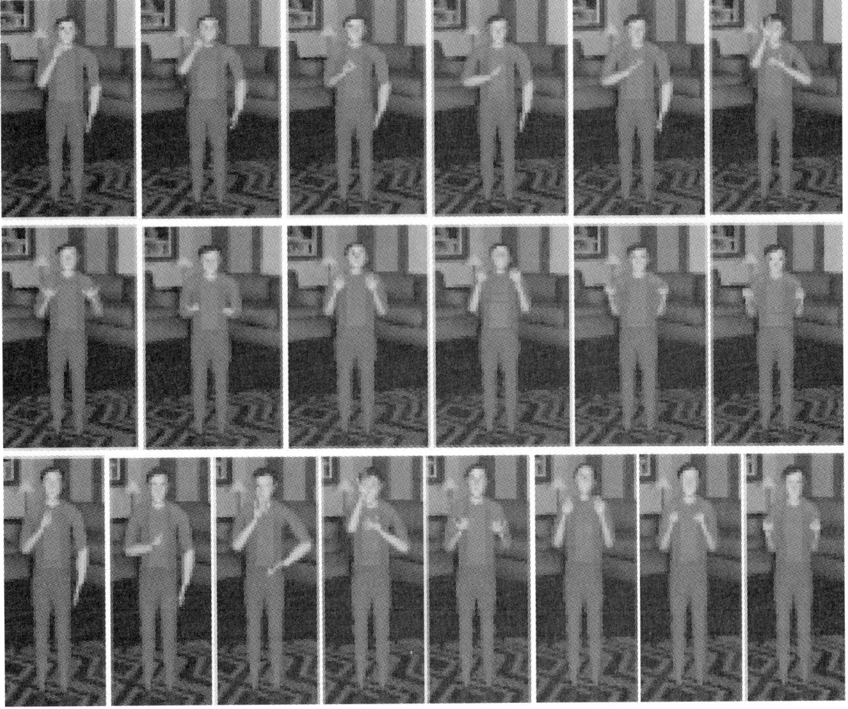

FIG. 15.1. Signing with the EMOTE system.

connected to inflexible bone by muscle arcs. Facial expressions were manipulated by applying forces to the skin mesh through the muscle arcs. The Facial Action Coding System (FACS), proposed by Ekman and Friesen (1977), was used as the notation scheme for the muscle actions. FACS represents facial expressions in terms of 66 action units, which involve one or more muscles and associated activation levels.

Waters (1987) simulated two types of facial muscles, linear muscles that pull and sphincter muscles that squeeze, by using a mass-and-spring model. In these models, the ability to manipulate facial expressions is based on simulating facial muscles.

Lee, Waters, and Terzopoulos (1995) use a physics-based approach for creating a head from laser scanner data. Their biomechanical model is more accurate than previous models, for it allows for the automatic placement of facial muscles and enables the use of a simple control process across different facial models. The physically based point-to-point control technique produces realistic results by offering subtle facial movements. Unfortunately, the equations used tend to be complex and therefore computationally intensive.

Kalra, Mangili, Magnenat-Thalmann, and Thalmann (1992) use a Rational Free Form Deformation (RFFD)–based approach to deform the facial skin surface. An expression editor allows the user to specify expressions at a high level without detailed knowledge of the anatomy of the face (see Fig. 15.2). Regrettably, the digitization process to construct the geometric model of the face is a labor-intensive job, and the approach does not produce realistic facial expressions and subtle facial motions. The interpolation-based animation approach used by this method requires manual creation of all facial expressions.

DeCarlo, Metaxas, and Stone (1998) present an anthropometry-based approach. They use anthropometric statistics to automatically produce varied geometric models of human faces. Compared to manual modeling techniques, this method provides sufficient generality to

FIG. 15.2. An angry facial expression.

describe a large sampling of face geometries, but because a single base shape was used, all of the sample models generated by their system have a similar shape. Blantz and Vetter (1999) describe a system for generating new morphable faces from linear combinations of prototypes. For an overview of facial animation, see Parke and Waters (1996) and Massaro (1998).

## 4. COMMUNICATION

Even though natural language in the form of spoken utterances is a natural and easy form of communication for humans, it is quite difficult for virtual humans. It involves solving many hard problems that are still being researched, including: dialogue planning, natural language processing, lip-synching, turn taking, and speech synthesis. Interactions between virtual autonomous agents and the human user in a VE go in both directions. Not only will the autonomous agent convey information through speech or body language, the agent will also receive information from the user by observing his or her avatar's facial expressions, body language, and attention.

It is often said that body language reveals as much if not more than verbal communication does. Cassell et al. (1994) and Cassell, Bickmore, Campbell, Vilhjalmsson, and Yan (2000) present a system that automatically generates and animates conversations between multiple agents. A dialogue planner creates the conversation and generates and synchronizes appropriate intonation, facial expressions, eye gaze, head motion, and arm gestures.

Gestural communication and body language have been studied by several groups (e.g., Kurlander, Skelly, & Salesin, 1996; Morawetz & Calvert, 1990). Becheiraz and Thalmann (1996) present a model of nonverbal communication where agents react to one another in a VE based on their postures. The agent's character influences how it reacts to the other agent's postures, and relationships between the agents evolve based on the perceptions of postures. Johnson and Rickel (1997) present an animated pedagogical agent that uses both gestures and attention to aid in the instruction of manual tasks. Andre, Rist, and Muller (1998) describe a presentation agent for use on the World Wide Web. Their interface agents present multimedia material to the user by following directives in a script. Every agent has its own persona and presents the material through moving and pointing. Chi, Costa, Zhao, and Badler's (2000) EMOTE animation system conveys emotional information through gesture.

Facial expressions are also a very important source of communicative information. Facial expressions can indicate what the agent is thinking and feeling. Facial expressions can be explicit or affective. Explicit expressions lack emotional content and include "fluttering of the eyelashes" and "pursing the lips." Affective expressions include "raising eyebrows expectantly" and "indulgent smiles."

Both Brand (1999) and Poggi and Pelachaud (2000) have researched ways of generating facial expressions for speech. Brand generates facial animation from information in an audio track. The facial control model is learned from real facial behavior. His voice puppet automatically incorporates facial and vocal dynamics, including coarticulation. Poggi and Pelachaud concentrate on the visual display of intentions through facial animation based on semantic data. They model performatives, which are the type of action a sentence performs, such as requesting or informing. They also discuss how degree of certainty, power relationship, type of social encounter, and affective state effect facial animation.

What the agent or user pays attention to is another form of communication. Like facial expressions, attention can be termed explicit or affective. Explicit attentions include "looking at weapons," "gazing," "staring," and "listening." Affective attentions include "eyeing him warily," "glaring," and "giving dirty looks." Vilhjalmsson and Cassell (1998) created an interface for chat room avatars that allows the user to give conversational cues through attention control. If a user sees an agent that he is talking to begin to look away from his avatar more and more, then this is probably an indication that the agent no longer wants to participate in the conversation. Thorisson (1998) describes an architecture for multimodal, face-to-face communication between a human and an agent.

Agents and users can also interact or communicate through actions and changes to the environment. For example, a user may direct his or her avatar to move a box to a tabletop. An agent may move the same box to the floor. The user may move it back to the table and the agent may move it back to the floor. This exchange of actions indicates each participant's desired location for the box. Even though this is not a direct interaction, it is an important interaction nonetheless. For more information on virtual agent communication see Cassell, Sullivan, Prevost, and Churchill (2000).

## 5. MENTAL PROCESSES

Mental processes are an autonomous agent's decision-making processes and other processes that contribute to, influence, and interfere with the decision-making process.

### 5.1 Personality

A believable autonomous agent should have a personality. Personality is a pattern of behavioral, temperamental, emotional, and mental traits for an individual, and it describes the long-term tendencies of the agent. Personalities can affect the way an agent perceives, acts, and reacts, and add to the believability of the character (Loyall & Bates, 1997; Perlin, 1995; Rousseau & Hayes-Roth, 1996). Most of the work in personality for virtual humans centers around five basic personality types: dominance, friendliness, conscientiousness, emotional stability, and openness (Tupes & Christal, 1961). For more information on research done on personality and agents, see Trappl and Petta (1997).

### 5.2 Emotion

Emotions are closely related to personality but have a shorter duration. The creation of emotion for autonomous agents has been studied by several research groups (Amaya, Bruderlin, & Calvert, 1996; Bates, Loyall, & Reilly, 1992; Bates, 1994; King, 1994; Perlin & Goldberg, 1996; Picard, 1997). Most simulations of emotions in autonomous agents are based on the work of Damasio (1994), Frijda (1987), or Ortony, Clore, and Collins (1988). In the Ortony, Clore, and Collins (OCC) model, emotions are generated based on three top-level constructs, namely the consequences of events, the actions of agents, and aspects of objects. There are a limited number of other influences considered under each of these areas including consequences for others and self, and prospects relevant and irrelevant. The model eventually is grounded in

22 different emotions: love, hate, pride, admiration, shame, reproach, joy, distress, happy-for, gloating, resentment, pity, satisfaction, relief, fears-confirmed, disappointment, gratification, gratitude, remorse, hope, fear, and anger. The first and most comprehensive implementation of the OCC model is the Affective Reasoner (Elliot, 1992). Emotions affect not only which actions the agent chooses to perform, but also the manner in which they are performed. For more information on emotions and autonomous agents, see Cassell, Sullivan, Prevost, and Churchill (2000).

### 5.3 Ethnicity and Culture

Cultural associations condition the way an agent perceives, acts, and reacts. For example, in Japan bowing is the accepted form of greeting, but in Western culture a handshake is preferred. These same norms should hold for autonomous, virtual agents. Culture helps in determining the importance and immediacy of the activities of life. Culture, along with personality and role, determines how emotions are expressed. Furthermore, when two people converse, culture influences more than just the language in which they speak. It is said that cultural information is a minimum prerequisite for human interaction, in the absence of such information communication becomes a trial and error process (Knapp & Hall, 1992). Culture influences how messages are sent, received, and interpreted. Cultural information pertains to a person's native language, myths, societal norms, and prevailing ideology. Different cultures have different distances for interacting; they also have different touching behaviors, gestures, and eye gaze patterns. However, it is interesting to note that there are some things that are similar across cultures (see Fig. 15.3). Studies have shown that the six basic facial expressions (happiness, anger, disgust, sadness, surprise, and fear) can be identified across cultures (Ekman & Friesen, 1971). Also, some behaviors have cross-cultural similarities. Among them are coyness, flirting, embarrassment, open-handed greetings, and a lowered posture for showing submission.

Skin color has been studied by Kalra, Mangili, Magnenat-Thalmann, and Thalmann (1991), but the behavior differences that accrue from ethnicity are just now receiving attention from various research groups.

### 5.4 Status

People in the theatrical arts have long known that status plays an important role in a scene. If one character has power over another, it will be reflected in what actions he or she performs and the manner in which they are performed. The same should be true of virtual actors. Hayes-Roth, van Gent, and Huber (1996) have explored the use of status with virtual humans in the form of a master–slave relationship. They illustrate how the postures and actions of the characters change as the servant becomes the dominant character in the environment and then returns to

FIG. 15.3. A sad agent.

his submissive role. Poggi and Pelachaud (2000) model status through facial expressions called performatives, which are facial expressions that accompany and add interpersonal relationship information to speech.

### 5.5 Role

Every character in a VE should have a role that it is playing, whether it is a professor of astrophysics, a tour guide, or just a man walking down the street. Roles involve expectations, both from the individual playing the role and from those interacting with the individual playing the role. One expects certain appearances, personalities, status, capabilities, resources, ages, actions, and goals. These expectations affect the way people interact. In order for a character in a VE to be believable, it must meet the expectations of the role it is playing (unless its personality or emotional state leads to a deviation from the expectation for dramatic effect).

Roles are learned, generalized guidelines for behavior. People play different roles in social contexts. The participants' expectations and perceptions define a role more than what is actually said or done. In fact, there are three types of roles: expected roles, perceived roles, and enacted roles. *Expected roles* are sets of behaviors that members of a group anticipate from an individual in a certain position. *Perceived roles* are sets of behaviors a person in a position believes he or she should exhibit. *Enacted roles* are the sets of behaviors actually performed. Among other things, a role can stem from an individual's occupation, kinship, age, gender, prestige, wealth, or associational grouping. In a situation, one participant normally establishes his or her role and the other participant(s) must either go along or counter with a different role definition. There must be an agreement on the roles to effectively interact, otherwise communication will break down.

Generally, people have multiple roles. They must be flexible and able to switch between their roles with ease. This, however, is not always the case and conflicts can arise. For example, workaholics often have a conflict between their role in professional life and their role in family life. Competent role performance involves experimentation, improvisation, and adjustment. Adopting a role over a long period of time may affect one's personality and identity.

The complex characteristics of roles and their interactions with other mental processes make them difficult to simulate well, but they may simplify the decision-making process of the agent by narrowing the number of action choices. Isbister and Hayes-Roth (1998) have explored roles in relation to intelligent interface agents. They found that making the role of an interface agent clear helps to constrain the actions users will take in their corresponding roles.

### 5.6 Situation Awareness

People all perceive situations differently and form different mental representations of the environment, people, and actions of a situation. This implies that their actions are predicated on their knowledge and understanding of the situation. To be able to create true individualized autonomous agents, situation awareness needs to be included.

An agent's representation of a situation depends on that agent's level of situation awareness. Situation awareness is a task-related understanding of the dynamic entities in an environment and their possible impact on agent actions. Situation awareness involves extracting useful information from the world while ignoring the overwhelming amounts of useless information. According to Endsley (1998), there are three cognitive levels involved in a person's situational awareness: observation, integration, and projection. A person perceives elements of the situation; then, he or she forms an understanding or mental image of the situation. Finally, he or she projects the status of the elements in the situation into the near future. Baxter and Bass (1998) indicate that there are three types of situations: the real world, the perceived situation, and the desired or expected situation. A key idea is that the situation representation can vary across agents, especially agents of differing cultures. It is possible that the represented information may even vary by role, gender, personality, and emotions.

Before one can even begin modeling situation awareness, the agents' perception of the environment must be modeled. Currently, many VE systems and computer games fake agent perception by allowing the agent code access to the data structures that hold the world model, making the agents omniscient. Doyle and Hayes-Roth (1998) describe an object annotated Multi-User Domain (MUD)–type world. Kallmann and Thalmann (1999) describe smart objects for virtual environments that provide agents with information about interaction. Both of these models limit an agent's perception of the world to what is annotated into the objects. Noser, Renault, Thalmann, and Magnenat-Thalmann (1995) have studied synthetic vision, which is a truer version of perception. Chopra-Khullar and Badler (1999) have studied attending behaviors. They indicate that attention is directed by volitional, goal-directed aims known as endogenous factors that correspond to the current tasks being performed. Exogenous factors are involuntary attentional capture, such as stimulated by peripheral motion or local feature contrast. Here, attentional behavior is generated automatically based not only on the motor actions the agent is performing, but also involuntary visual functions, such as visual pursuit and spontaneous looking.

Naturally, an individual's perception of the environment is a key element in its behavior. The following are examples of perceptual bases and corresponding reactions. The perception of formality can lead to less relaxed, more hesitant, stylized, and superficial actions. The perception of psychological warmth tends to make individuals be relaxed, comfortable, and linger longer. The perception of privacy also greatly impacts an individual's actions depending on whether the individual is alone or with another person, what his or her perceptions of the other person are, and the individual's other perceptions of the environment. If the environment is unfamiliar, a person will tend to be cautious, deliberate, and conventional. The perception of constraint is important; whether and how easily a person feels he or she can leave will influence his or her decisions and actions. Another important environmental perception is distance. When the setting forces people into close quarters with individuals they do not know, they try to psychologically increase the distance and decrease the feelings of intimacy by lessening eye contact, increasing body tenseness and immobility, becoming silent, laughing nervously, joking about the intimacy, and creating public conversations directed to all present.

People also perceive and act in terms of the label applied to a situation. For example, if a foreman told his workers that they were going to be working around empty steel drums they would react differently than if he told them they would be working around drums filled with radioactive waste.

Although situation awareness is an important feature for agents in virtual environments, it has not been heavily researched by the community. It requires attention, synthetic vision, a representation of the situation, and a way to determine what is important in the situation based on the agent's current goals, emotions, culture, role, personality, expectations, and more. Once the environment has been perceived and the situation represented, the representation can be used in projection or planning and decision making.

## 6. ACTION SELECTION

### 6.1 Planning, Decision Making, and Volition

Agents, however, should not merely react to the world they find themselves in. They should actively attempt to satisfy goals, execute instructions, interact with other agents, and otherwise change the environment. Since agents are to act in the world they need a way to create, retrieve, manage, and execute sets of actions. An autonomous agent must have some goal or goals and some mechanism for choosing actions that will lead to those goals. In choosing which actions to perform, the agent model should consider various properties of the actions. Are the actions fallible or infallible? What is the utility of the actions? What do the actions cost in terms of energy, time, money, and the agent's value system? Also, all of the characteristics outlined in previous sections should have an impact on the decision-making process (as should the agent's capabilities and resources). In addition, since most VEs are dynamic, continuous environments, in order for the agents to be realistic and believable, decisions and reactions must take place in real time.

How an agent behaves can be dictated through scripting, by using rules in a rule-based system, by specifying goals that are input to a planning or reactive system, or combinations of these. Scripting the actions of agents results in agents that can only be used for the scenario that they were scripted for. They lack robustness. Using rule-based systems or planning systems can add to the robustness, autonomy, reactivity, and proactiveness of the agent.

Other than scripting and artificial life methods, there are currently three strategies for building autonomous agents. Agent architectures can be deliberative, reactive, or a hybrid of the two. These architectures can include blackboards, rule-based systems, cased-based reasoning, finite-state machines, neural networks, genetic algorithms, and virtually any other tool developed for artificial intelligence applications.

### 6.1.1 Deliberative Architectures

The deliberative model contains an explicit symbolic representation of the world and the agent's actions. In this model, decisions are made through planning and logical reasoning. While these techniques work well on toy domains, it has been found that they do not scale well to time critical, real-world scenarios (Chapman, 1987).

References to deliberative systems include: GPS system (Newell & Simon, 1961); STRIPS planning (Fikes & Nilsson, 1971); hierarchical planning (Sacerdoti, 1974); nonlinear planning (Sacerdoti, 1975); AI planning (Allen, Hendler, & Tate, 1990); Algernon (Crawford & Kuipers, 1991); O-Plan (Currie & Tate, 1991); interleaving planning (Wood, 1993); regression planning (Kurlander & Ling, 1995); hierarchical decomposition (Russell & Norvig, 1995); and abstract plans (Pfleger & Hayes-Roth, 1998).

### 6.1.2 Reactive Architectures

In order to avoid the limitations of the deliberative model, some researchers turned to a reactive model. The idea here is to create agents that respond to their environment in time for those responses to be useful. Case-based reasoning is where all the expected situations and responses are characterized in advance. During execution, if an existing case is determined to match the current situation, then that case determines the next actions to be performed.

Rule-based reasoning is more robust than case-based reasoning. Rule-based reasoning uses mappings of complex relationships reduced to if–then parameterized production rules. Actions are the result of firing matched rules. Using rules for control may enable certain types of learning (e.g., from examples and possibly by parameter generalization). This model also has drawbacks. Because agents do not have a representation of their environment, there must be enough local information available from the environment for them to determine their next action. Also, the overall behavior of these agents emerges from its reaction to the environment. While this is a selling point for this model it is also a drawback, because it is difficult to determine the behavior of an agent ahead of time. In other words, it is difficult to engineer these agents to fulfill specific tasks. In addition, it remains difficult to construct agents with numerous and complex behaviors. Reactive systems are usually constructed with finite state machines or situation–action rules that map percepts to actions.

References to reactive systems include: the PENGI system (Agre & Chapman, 1987); subsumption architecture (Brooks, 1991); competence modules (Maes, 1994); Improv (Perlin & Goldberg, 1996); concurrent reactive agents (Costa & Feijo, 1996); JackMOO (Shi, Smith, Granieri, & Badler, 1999); and the L-system (Noser & Thalmann, 1999).

### *6.1.3 Hybrid Architectures*

In order to lessen the problems associated with both of these models, hybrid systems have been created. In these systems the agent is both data-driven and goal-driven. The agents are able to react to their environment while pursuing their own or externally supplied goals. Such systems have a tendency to be specialized to a certain environment. These architectures are normally layered, hierarchical systems. Each layer in the hierarchy deals with data from the environment at a different level of abstraction. Normally the lowest layer is reactive, the middle level layer is knowledge (Newell, 1982) and makes use of symbolic representations, and the highest layer is social knowledge (Jennings & Campos, 1997). Each layer may produce suggestions about the next action to perform. These suggestions are then mediated to ensure coherent behavior.

References to hybrid systems include: touring machines (Ferguson, 1992); RAP (Firby, 1994); 3T (Bonasso, Kortenkamp, Miller, & Slack, 1996); INTERRAP (Muller, 1997); STEVE (Johnson & Rickel, 1997); YMIR (Thorisson, 1998); CML (Funge, Tu, & Terzopoulos, 1999); and PAR (Bindiganavale et al., 2000).

A subclass of hybrid models is practical reasoning agents, which are also known as belief, desire, intention (BDI) agents. The construction of this type of agent is based on the pragmatic reasoning done by real humans. Beliefs relate to what the agents think about their environment. Desires roughly correspond to goals that the agent could pursue. Intentions are the desires to which the agent has committed resources to accomplishing.

References to practical reasoning agents include: the Procedural Reasoning System (Georgeff & Lansky, 1987); practical reasoning (Bratman, Israel & Pollack, 1988); agent-oriented programming (Shoham, 1993); and BDI (Kumar & Shapiro, 1994).

One way to improve an agent's decision-making performance, regardless of the architecture, is to create a layered architecture where each layer differs in level of detail of the action represented. For example, a planner or rule-based system may just decide that an agent needs to walk from point A to point B. The next layer in the architecture might decide the exact path the agent should take. Yet another layer might take care of obstacle avoidance and decide the placement of each footstep. In this way, each layer can be specialized and independently optimized. For more information on decision making, action selection, and agent architectures, see Russell and Norvig (1995) and Jennings, Sycara, and Wooldridge (1998).

## 6.2 Reflexes and Background Behaviors

There should also be a notion of reflexes, which bypass any planning or decision-making components and directly map situations to actions. These actions should take place immediately after a stimuli is sensed and include both body actions, such as removing one's hand from a hot stove, and facial expressions, such as a look of shock when seeing a pig fly. There may also be a notion of subconscious behaviors, which are actions that are not planned and the agents may not even be aware that they are doing. Many of these actions can be thought of as background behaviors, such as scratching the head, stretching, yawning, tapping nervously, and twiddling of hair. These sorts of actions do not overtly achieve a goal, but they greatly add to the realism of the character.

### 6.3 Goal Generation

Goals must also be generated in some way. Logan (1998) discusses some issues associated with agents and their goals, including: the autonomous generation of goals; achievement versus maintenance goals; single versus multiple goals; commitment to goals; utilities of goals; meta-goals (managing the time and resources used to achieve a goal); consistent beliefs; and certainty of beliefs.

Goals normally are created by the user or the designer of the VE in a programming environment. The Parameterized Action Representation (PAR) system (Bindiganavale et al., 2000) allows users to dynamically input immediate or persistent instructions during the simulation, thereby giving new goals to the agents or constraining the way previously established goals are achieved. The user inputs these instructions as natural-language statements. This allows users to dynamically refine the agents' behaviors in reaction to simulated stimuli without having to undertake a lengthy offline programming session. The input instructions can range from specific instantaneous instructions like "Stand up" to very general standing orders, like "Always drive an abandoned vehicle to the parking lot." The natural-language instructions are translated into PARs and then sent to the appropriate agent for execution. The details of how the actions are performed are left up to the agent, which allows a degree of individuality.

Table 15.1 summarizes many of the characteristics of agents and notes increasing levels of complexity for each characteristic. Table 15.2 compares how the complexity of each of these characteristics varies across current agent applications.

## 7. CONCLUSION

Creating agents for VEs can be a complex task, but when creating agents one should keep in mind what role each agent is playing and therefore where design efforts should be concentrated. For example, if one is creating an agent that will essentially be an "extra" that never comes close to the camera position, time and effort should not be spent creating a detailed appearance for the agent. Likewise, if the role of an agent is to sit on a bench and read a newspaper, there is

**TABLE 15.1**
Increasingly Complex Characteristics of Agents

| | |
|---|---|
| Appearance | 2-D drawings > 3-D wireframe > 3-D polyhedra > curved surfaces > free-form deformations > accurate surfaces > muscles and fat > biomechanics > clothing and equipment > physiological effects (perspiration, irritation, injury) |
| Function | Cartoon > jointed skeleton > joint limits > strength limits > fatigue > hazards > injury > skills > effects of loads and stressors > psychological models > cognitive models > roles > teaming |
| Time | Offline animation > interactive manipulation > real-time motion playback > parameterized motion synthesis > multiple agents > crowds > coordinated teams (time to create movement at the next frame) |
| Autonomy | Drawing > scripting > interacting > reacting > decisions making > communicating > intending > initiative taking > leading |
| Individuality | Generic character > hand-crafted character > cultural distinctions > gender and age > personality > psychological–physiological profiles > specific individual |

**TABLE 15.2**
Comparative Graphical Agents

| *Application* | *Appearance* | *Function* | *Time* | *Autonomy* | *Individuality* |
|---|---|---|---|---|---|
| Cartoons | high | low | long | low | high |
| Special effects | high | low | long | low | medium |
| Medical | high | high | medium | med | medium |
| Ergonomics | medium | high | medium | med | low |
| Games | high | low | short | med./high | medium |
| Military | medium | medium | short | med./high | low |
| Education | medium | low | short | med./high | medium |
| Training | medium | low | short | high | medium |

no need to create a complex goal or perceptual model for the agent. There are times when the simplest solution is the best solution. Similarly, different applications require different levels of concentration in different dimensions of agent design (see Table 15.2). The success of the agent depends on its purpose in the VE, whether it be to entertain, do job training, participate in team coordination, test manufacturing and maintenance procedures, educate, test emergency drills, or just be an "extra." When the purpose of the agent is to entertain it is often not necessary for the agent to have extremely complex, consistent behaviors. Human viewers have an imagination and will come up with explanations for the agent's persona, provided that the agent has at least a minimal amount of believability and coherence in its behavior.

However, when complex models are needed one should try to create autonomous agents with reuse in mind. One approach is to create the autonomous agent with as much modularity and parameterization as possible. This will allow designers to more easily adapt agents for other scenarios. For example, if an autonomous agent is created in which emotions, personality, role, status, culture, and other components are parameterized structures that are each modified by a corresponding process (which may reference the parameters of other components during its processing), the decision-making process can then also reference the parameters of the components when it is choosing the agent's next actions. This structure allows new autonomous agents to be created simply by changing the setting of the parameters and upgrading or modifying the complexity of an agent component by component. Different decision-making processes could be built with varying complexities and tools (i.e., planners, rule-based systems, neural networks, etc.) and then interchanged to create agents with even more individuality. Creating agents with individualized identities makes the VE more dynamic and interesting for the humans that interact with it.

With some architectures and representations it is also possible to incorporate learning (explanation, abstraction, caching, reinforcement, etc.) into agents. This can add to the robustness and naturalness of agents while providing another form of individualization (different learning rates). In order to include learning, some form of memory must be included. Memories can be declarative (I know something), procedural (I know how to do something), or episodic (I remember the event). For more information on learning and memory, see Russell and Norvig (1995).

As Table 15.3 shows, even though there has been significant research on embodied autonomous agents, there are still a number of unsolved problems in agent design and development.

**TABLE 15.3**
Some Unsolved Problems in Agent Design and Development

| | |
|---|---|
| Appearance | Real-time, realistic deformations of muscle and skin |
| | Real-time modeling of secondary movement of hair and clothing |
| Animating actions | Blending of actions and action transitions |
| | Actions that will work in any context (especially grasping) |
| | Actions that express both the cognitive and affective state of the agent |
| | Real-time lip-synching |
| Communication | Real-time robust dialogue planning |
| | Expressive, personalized gesture |
| | Proper synchronization of the different channels of communication |
| | Natural sounding and personalized speech synthesis |
| | Comprehensive model of perception and attention |
| Mental processes | Creation of individuals |
| | More comprehensive and integrated model of the mental processes |
| | Modeling of ethnicity and culture |
| | Modeling of situation awareness and its links to perception, attention, and decision making |
| Action selection | Systems that are more robust and always real time |
| | Systems that are both reactive and deliberative |
| | Systems that are easily portable to new domains (the agents have the necessary functionality) |
| | Systems that take into account comprehensive mental processes |
| | Systems that take into account physiological effects (e.g., fatigue, hunger) |
| | Systems that enable agents to be comprehensively autonomous (in reacting, decision making, communicating, intending, and leading) |

## 8. REFERENCES


Agre, P. E., & Chapman, D. (1987). PENGI: An implementation of a theory of activity. In *Proceedings of AAAI—'87* (pp. 268–272). Menlo Park, CA: AAAI Press.

Allen, J. F., Hendler, J., & Tate, A. (Eds.). (1990). *Readings in planning*. San Mateo, CA: Morgan Kaufmann Publishers.

Amaya, K., Bruderlin, A., & Calvert, T. (1996). Emotion from motion. In *Proceedings of Graphics Interface '96* (pp. 222–229). Toronto: Canadian Information Processing Society.

Anderson, F., & Pandy, M. (1999). A dynamic optimization solution for one complete cycle of human gait. In *Proceedings of the International Society of Biomechanics 17th Congress* (p. 381). Calgary, Canada.

Andre, E., Rist, T., & Muller, J. (1998). Integrating reactive and scripted behaviors in a life-like presentation agent. In *Proceedings of the Second International Conference on Autonomous Agents* (pp. 261–268). Minneapolis/St. Paul, MN: ACM Press.

Anjyo, K., Usami, Y., & Kurihara, T. (1992). A simple method for extracting the natural beauty of hair. In *Proceedings of SIGGRAPH '92* (pp. 111–120). Reading, MA: Addison-Wesley.

Babski, C., & Thalmann, D. (1999). A seamless shape for HANIM compliant bodies. In *Proceedings of VRML '99* (pp. 21–28). Paderborn, Germany: ACM Press.

Badler, N., Phillips, C., & Webber, B. (1993). *Simulating humans: Computer graphics, animation, and control.* New York: Oxford University Press.

Baraff, D., & Witkin, A. (1998). Large steps in cloth simulation. In *Proceedings of SIGGRAPH '98* (pp. 43–54). Reading, MA: Addison-Wesley.

Bates, J. (1994). The role of emotion in believable agents. *Communications of the ACM, 37*(7), 122–124.

Bates, J., Loyall, A., & Reilly, W. (1992). Integrating reactivity, goals, and emotion in a broad agent. In *Proceedings of the 14th Annual Conference of the Cognitive Science Society* (pp. 696–701). Hillsdale, NJ: Lawrence Erlbaum Associates.

Baxter, G. D., & Bass, E. J. (1998). Human error revisited: Some lessons for situation awareness. In *Proceedings of the Fourth Symposium on Human Interaction with Complex Systems* (HICS '98; pp. 81–87). Dayton, OH: Computer Society Publications.

Becheiraz, P., & Thalmann, D. (1996). A model of nonverbal communication and interpersonal relationship between virtual actors. In *Proceedings of Computer Animation* (pp. 58–67). Geneva, Switzerland: IEEE Press.

Becket, W. (1996). *Reinforcement learning of reactive navigation of computer animation of simulated agents.* Unpublished doctoral dissertation, University of Pennsylvania.

Bindiganavale, R., & Badler, N. (1998). Motion abstraction and mapping with spatial constraints. In *Workshop on Modeling and Motion Capture Techniques for Virtual Environments.* Geneva, Switzerland.

Bindiganavale, R., Schuler, W., Allbeck, J. M., Badler, N. I., Joshi, A. K., & Palmer, M. (2000). Dynamically altering agent behaviors using natural language instructions. In *Proceedings of Autonomous Agents 2000.* New York: ACM Press.

Blantz, V., & Vetter, T. (1999). A morphable model for the synthesis of 3D faces. In *Proceedings of SIGGRAPH '99* (pp. 187–194).

Bonasso, R. P., Kortenkamp, D., Miller, D. P., & Slack, M. (1996). Experiences with an architecture for intelligent, reactive agents. *LNAI, 1037* (pp. 187–202). Berlin, Germany: Springer–Verlag.

Boulic, R., Becheiraz, P., Emering, L., & Thalmann, D. (1997). Integration of motion control techniques for virtual human and avatar real-time animation. In *ACM Symposium on Virtual Reality Software and Technology.* New York: ACM Press.

Boulic, R., Magnenat-Thalmann, N., & Thalmann, D. (1990). A global human walking model with real-time kinematic personification. *The Visual Computer, 6,* 344–358.

Brand, M. (1999). Voice puppetry. In *Proceedings of SIGGRAPH '99.* (pp. 21–28). New York: ACM Press.

Bratman, M. E., Israel, D. J., & Pollack, M. E. (1988). Plans and resource-bounded practical reasoning. *Computational Intelligence, 4,* 349–355.

Breen, D. E., House, D. H., & Getto, P. H. (1992). A physical-based particle model of woven cloth. *The Visual Computer, 8*(5–6), 264–277.

Brooks, R. A. (1991). Intelligence without representation. *Artificial Intelligence, 47,* 139–159.

Bruderlin, A., & Calvert, T. W. (1989). Goal-directed, dynamic animation of human walking. In *Proceedings of SIGGRAPH '89* (pp. 233–242). Reading, MA: Addison-Wesley.

Bruderlin, A., & Calvert, T. (1996). Knowledge-driven, interactive animation of human running. In *Proceedings of Graphics Interface '96* (pp. 213–221). Toronto Canadian Information Processing Society.

Calvert, T. W., Chapman, J., & Patla, A. (1982). Aspects of the kinematic simulation of human movement. *IEEE Computer Graphics and Applications, 2,* 41–48.

Capin, T. K., Pandzic, I. S., Noser, H., Magnenat-Thalmann, N., & Thalmann, D. (1997). Virtual human representation and communication in VLNET networked virtual environments [Spectal Issue]. *IEEE Computer Graphics and Applications, 17*(2), 42–53.

Cassell, J., Bickmore, T., Campbell, L., Vilhjalmsson, H., & Yan, H. (2000). Human conversation as a system framework: Designing embodied conversational agents. In Cassell et al. (Eds.), *Embodied Conversational Agents* (pp. 29–63). Cambridge, MA: MIT Press.

Cassell, J., Pelachaud, C., Badler, N., Steedman, M., Achorn, B., Becket, T., Douville, B., Prevost, S., & Stone, M. (1994). Animated conversation: Rule-based generation of facial expression, gesture, and spoken intonation for multiple conversational agents. In *Proceedings of SIGGRAPH '94* (pp. 413–420). New York: ACM Press.

Cassell, J., Sullivan, J., Prevost, S., & Churchill, E. (Eds.). (2000). *Embodied conversational agents.* Cambridge, MA: MIT Press.

Chapman, D. (1987). Planning for conjunctive goals. *Artificial Intelligence, 32,* 333–378.

Chi, D., Costa, M., Zhao, L., & Badler, N. (2000). The EMOTE model for effort and shape. In *Proceedings of SIGGRAPH '00.* New Orleans, LA: ACM Press.

Chopra-Khullar, S. & Badler, N. (1999). Where to look? Automating attending behaviors of virtual human characters. In *Proceedings of the Third Annual Conference on Autonomous Agents* (pp. 16–23). New York: ACM Press.

Collier, J. R., Collier, B. J., O'Toole, G., & Sargand, S. M. (1991). Drape prediction by means of finite-element analysis. *Journal of the Textile Institute, 82*(1), 96–107.

Costa, M., & Feijo, B. (1996). An architecture for concurrent reactive agents in real-time animation. In *Proceedings of SIBGRAPHI IX* (pp. 281–288). Cexambu, MG.

Crawford, J. M., & Kuipers, B. J. (1991). Algernon—A tractable system for knowledge-representation. *SIGART Bulletin, 2*(3), 35–44.

Currie, K., & Tate, A. (1991). O-plan: The open planning architecture. *Artificial Intelligence, 52,* 49–86.

Damasio, A. (1994). *Descartes' error: Emotion, reason, and the human brain.* New York: G. P. Putnam.

DeCarlo, D., Metaxas, D., & Stone, M. (1998). An anthropometric face model using variational techniques. In *Proceedings SIGGRAPH '98* (pp. 67–74). New York: ACM Press.

Doyle, P., & Hayes-Roth, B. (1998). Agents in annotated worlds. In *Proceedings Autonomous Agents* (pp. 173–180). Minneapolis/St. Paul, MN: ACM Press.

Eberhardt, B., Weber, A., & Strasser, W. (1996). A fast, flexible, particle-system model for cloth draping. *IEEE Computer Graphics and Applications, 16*(5), 52–59.

Eischen, J. W., Deng, S., & Clapp, T. G. (1996). Finite-element modeling and control of flexible fabric parts. *IEEE Computer Graphics and Applications, 16*(5), 71–80.

Ekman, P., & Friesen, W. V. (1971). Constraints across cultures in the face and emotion. *Journal of Personality and Social Psychology, 17*, 124–129.

Ekman, P., & Friesen, W. V. (1977). *Manual for the facial action coding system*. Palo Alto, CA: Consulting Psychologists Press.

Elliot, C. (1992). *The Affective Reasoner: A process model of emotions in a multi-agent system*. Unpublished doctoral dissertation, Institute for the Learning Sciences, Northwestern University.

Endsley, M. (1998). Situation awareness, automation, and decision support: Designing for the future. *CSERIAC Gateway, 9* (1), 11–13.

Engineering Animation, Inc. http://www.transom.com/

Ferguson, I. A. (1992). *Touring machines: An architecture for dynamic, rational, mobile agents*. Unpublished doctoral dissertation, University of Cambridge.

Firby, R. J. (1994). Task networks for controlling continuous processes. In *Proceedings of the Second International Conference on AI Planning Systems*. Chicago, IL: AAAI Press.

Fikes, R. E., & Nilsson, N. (1971). STRIPS: A new approach to the application of theorem proving to problem solving. *Artificial Intelligence, 5*(2), 189–208.

Frijda, N. (1987). *Emotions*. Cambridge, England: Cambridge University Press.

Funge, J., Tu, X., & Terzopoulos, D. (1999). Cognitive modeling: knowledge, reasoning and planning for intelligent characters. In *Proceedings of SIGGRAPH '99* (pp. 29–38). Reading, MA: Addison-Wesley.

Georgeff, M. P., & Lansky, A. L. (1987). Reactive reasoning and planning. In *Proceedings of the Sixth National Conference on Artificial Intelligence* (AAAI–87; pp. 677–682). Menlo Park, CA: AAAI Press.

Girard, M., & Maciejewski, A. (1985). Computational modeling for the computer animation of legged figures. *ACM Computer Graphics, 19*(3), 263–270.

Gleicher, M. (1997). Motion editing with space-time constraints. In *Proceedings of SIGGRAPH '97* (pp. 139–148). Reading, MA: Addison-Wesley.

Gleicher, M. (1998). Retargeting motion to new characters. In *Proceedings of SIGGRAPH '98* (pp. 33–42). Orlando, FL: Addison-Wesley.

Gourret, J. P., Magnenat-Thalmann, N., & Thalmann, D. (1989). Simulation of object and human skin deformations in a grasping task. *ACM Computer Graphics, 23*(3), 21–30.

Granieri, J., Crabtree, J., & Badler, N. (1995). Off-line production and real-time playback of human figure motion for 3D virtual environments. In *Proceedings of VRAIS '95*. Los Alamitos, CA IEEE Press.

Haumann, D. R., & Parent, R. E. (1988). The behavioral test-bed: Obtaining complex behavior from simple rules, *The Visual Computer, 4*(6), 332–347.

Hayes-Roth, B., van Gent, R., & Huber, D. (1996). *Acting in character* (Tech. Rep. No. KSL-96-13). Stanford, CA: Knowledge Systems Laboratory, Stanford University.

Hodgins, J. K., & Pollard, N. S. (1997). Adapting simulated behaviors for new characters. In *Proceedings of SIGGRAPH '97* (pp. 153–162). Reading, MA: Addison-Wesley.

Hodgins, J., Wooten, W., Brogan, D., & O'Brien, J. (1995). Animating human athletics. In *Proceedings of SIGGRAPH '95* (pp. 71–78). Reading, MA: Addison-Wesley.

Humanoid Animation Group. http://www.h-anim.org

Isbister, K., & Hayes-Roth, B. (1998). *Social implications of using synthetic characters: An examination of a role-specific intelligent agent* (Tech. Rep. No. KSL-98-01). Stanford, CA: Knowledge Systems Laboratory, Stanford University.

Ishii, T., Yasuda, T., Yokoi, S., & Toriwaki, J. (1993). A generation model for human skin texture. In *Proceedings of CGI '93* (pp. 139–150). Berlin, Germany: Springer-Verlag.

Jennings, N. R., & Campos, J. R. (1997). Towards a social-level characterisation of socially responsible agents. *IEEE Transactions on Software Engineering, 144*(1), 11–15.

Jennings, N. R., Sycara, K., & Wooldridge, M. (1998). A roadmap of agent research and development. *Autonomous Agents and Multi-Agent Systems, 1*, 275–306.

Johnson, W. L., & Rickel, J. (1997). Steve: An animated pedagogical agent for procedural training in virtual environments. *ACM SIGART Bulletin, 8*(1–4), 18–21.

Kallmann, M., & Thalmann, D. (1999). A behvioral interface to simulate agent-object interactions in real time. In *Proceedings of Computer Animation* (pp. 138–146). Los Alamitos, CA: IEEE Press.

Kang, T. J., & Yu, W. R. (1995). Drape simulation of woven fabric by using the finite-element method. *Journal of the Textile Institute, 86*(4), 635–648.

Kalra, P., Magnenat-Thalmann, N., Moccozet, L., Sannier, G., Aubel, A., & Thalmann, D. (1998). Real-time Animation of Realistic Virtual Humans. *IEEE Computer Graphics and Applications, 18*(5), 42–55.

Kalra, P., Mangili, A., Magnenat-Thalmann, N., & Thalmann, D. (1991). Smile: A multilayered facial animation system. In *Proceedings of Modeling in Computer Graphics* (pp. 189–198). Berlin: Germany Springer-Verlag.

Kalra, P., Mangili, A., Magnenat-Thalmann, N., & Thalmann, D. (1992). Simulation of muscle actions using rational free form deformations. *Proceedings Eurographics '92, Computer Graphics Forum, 2*(3), 59–69.

King, W. J. (1994). Defining phenomena for an emotion state model in the human interface. In *Proceedings of the third IEEE International Workshop on Robot and Human Communication* (pp. 162–166). Piscataway, NJ: IEEE Press.

Knapp, M. L., & Hall, J. A. (1992). *Nonverbal communication in human interaction* ($3^{rd}$ *ed.*). Fort Worth, TX: Harcourt Brace Jovanovich.

Ko, H., & Badler, N. (1996). Animating human locomotion in real-time using inverse dynamics, balance, and comfort control. *IEEE Computer Graphics and Applications, 16* (2), 50–59.

Koch, R. M., Gross, M. H, Carls, F. R., von Buren, D. F., Fankhauser, G., & Parish, Y. I. H. (1996). Simulating facial surgery using finite element models. In *Proceedings of SIGGRAPH '96* (pp. 421–428). Reading, MA: Addison-Wesley.

Koga, Y., Kondo, K., Kuffner, J., & Latombe, J. C. (1994). Planning motions with intentions. In *Proceedings of SIGGRAPH '94* (pp. 395–408). Reading, MA: Addison-Wesley.

Kokkevis, E., Metaxas, D., & Badler, N. (1996). User-controlled physics-based animation for articulated figures. In *Proceedings of Computer Animation.*

Kuffner, J. J., & Latombe, J. C. (1999). Fast synthetic vision, memory, and learning models for virtual humans. In *Proceedings of Computer Animation '99* (pp. 118–127).

Kumar, D., & Shapiro, S. C. (1994). The OK BDI architecture. *International Journal of Artificial Intelligence, 3*(3), 349–366.

Kurlander, D., & Ling, D. T. (1995). Planning-based control of interface animation. In *Proceedings of CHI '95* (pp. 472–479).

Kurlander, D., Skelly, T., & Salesin, D. (1996). Comic chat. In *Proceedings of SIGGRAPH '96* (pp. 225–236). Reading, MA: Addison-Wesley.

Laban, R., & Lawrence, F. C. (1974). Effort: Economy in body movement. Boston: Plays, Inc.

Larrabee, W. (1986). A finite element model of skin deformation I: Biomechanics of skin and soft tissue: A review. *Laryngoscope, 96*, 399–419.

Laszlo, J. F., van de Panne, M., & Fiume, E. (1996). Limit cycle control and its application to the animation of balancing and walking. In *Proceedings of SIGGRAPH '96* (pp. 155–162). Reading, MA: Addison-Wesley.

Lee, J., & Shin, S. Y. (1999). A hierarchical approach to interactive motion editing for human-like figures. In *Proceedings of SIGGRAPH '99* (pp. 39–47).

Lee, Y., Waters, K., & Terzopoulos, D. (1995). Realistic modeling for facial animation. In *Proceedings SIGGRAPH '95* (pp. 55–62) Reading, MA: Addison-Wesley.

Logan, B. (1998). *Classifying agent systems*. In *Proceedings of AAAI Workshop on Software Tools for Developing Agents* (pp. 11–21). Menlo Park, CA: AAAI Press.

Loyall, A. B., & Bates, J. (1997). Personality-rich believable agents that use language. In *Proceedings of Autonomous Agents '97* (pp. 106–113). New York: ACM Press.

Maes, P. (1994). Agents that reduce work and information overload. *Communications of the ACM, 37*(7), 31–40.

Magnenat-Thalmann, N., Primeau, E., & Thalmann, D. (1988). Abstract muscle action procedures for human face animation. *The Visual Computer, 3*(5), 290–297.

Maiochhi, R., & Pernici, B. (1990). Directing an animated scene with autonomous actors. *The Visual Computer, 6*(6), 359–371.

Massaro, D. W. (1998). *Perceiving talking faces: From speech perception to a behavioral principle*. Cambridge, MA: MIT Press.

McKenna, M., & Zelter, D. (1990). Dynamic simulation of autonomous legged locomotion. In *Proceedings of SIGGRAPH '90* (pp. 29–38).

Morawetz, C. L., & Calvert, T. W. (1990). Goal-directed Human Animation of Multiple Movements. In *Proceedings of Graphics Interface '90* (pp. 60–67). San Francisco, CA: Morgan Kaufmann Publishers.

Motion Factory. http://www.motion-factory.com/

MPEG-4. http://drogo.cselt.stet.it/mpeg/standards/mpeg-4/mpeg-4.htm

Muller, J. P. (1997). The design of intelligent agents, *LNAI, 1117*, Berlin, Germany: Springer-Verlag.

Multon, F., France, L., Cani-Gascuel, M. P., & Debunne, G. (1999). Computer animation of human walking: A survey. *Journal of Visualization and Computer Animation, 10*, 39–54.

Newell, A. (1982). The knowledge level. *Artificial Intelligence, 18*(1), 82–127.

Newell, A., & Simon, H. A. (1961). GPS: A program that simulates human thought. In *Lernende Automaten*. R. Oldenbourg, KG. E. A. Feigenbaum and J. Feldman (Eds.)

Noser, H., Renault, O., Thalmann, D., & Magnenat-Thalmann, N. (1995). Navigation for digital actors based on synthetic vision, memory and learning. *Computers and Graphics, 19*(1), 7–19.

Noser, H., & Thalmann, D. (1999). A rule-based interactive behavioral animation system for humanoids. *IEEE Transactions on Visualization and Computer Graphics, 5*(4), 218–307.

Ortony, A., Clore, G. L., & Collins, A. (1988). *The cognitive structure of emotions*. Cambridge, England: Cambridge University Press.

Pandzic, I., Capin, T., & Magnenat-Thalmann, N. (1997). A versatile navigation interface for virtual humans in collaborative virtual environments. *ACM Symposium on Virtual Reality Software and Technology*.: ACM Press.

Parke, F. I. (1972). Computer-generated animation of faces. In *Proceedings of ACM National Conference* (pp. 451–457): ACM Press.

Parke, F. I. (1982). Parameterized models for facial animation. *IEEE Computer Graphics and Applications, 2*(9), 61–68.

Parke, F. I. & Waters, K. (1996). *Computer facial animation*. Wellesley, MA: A. K. Peters.

Perlin, K. (1995). Real-time responsive animation with personality. *IEEE Transactions on Visualization and Computer Graphics, 1*, 5–15.

Perlin, K., & Goldberg, A. (1996). IMPROV: A system for scripting interactive actors in virtual worlds. In *Proceedings of SIGGRAPH '96* (pp. 205–216). Reading, MA: Addison-Wesley.

Pfleger, K., & Hayes-Roth, B. (1998). *Using abstract plans to guide behavior* (Tech. Rep. No. KSL-98-02). Stanford, CA: Knowledge Systems Laboratory, Stanford University.

Picard, R. W. (1997). *Affective computing*. Cambridge, MA: MIT Press.

Pieper, S. D. (1992). *CAPS: Computer-aided plastic surgery*. Unpublished doctoral dissertation, Massachusetts Institute of Technology.

Platt, S. M. (1981). Animating facial expressions. *Computer Graphics 15*(3), 245–252.

Platt, S. M., & Badler, N. I. (1981). Animating facial expressions. *Computer Graphics, 15*(3), 245–252.

Poggi, I., & Pelachaud, C. (2000). Performative facial expressions in animated faces. In J. Cassell, J. Sullivan, S. Prevost, E. Cherchill (Eds.), *Embodied conversational agents* (pp. 155–188). Cambridge, MA: MIT Press.

Popovic, Z., & Witkin, A. (1999). Physically based motion transformation. In *Proceedings of SIGGRAPH '99* (pp. 11–20). Reading, MA: Addison-Wesley.

Reich, B. (1997). *An Architecture for Behavioral Locomotion*. Unpublished doctoral dissertation, University of Pennsylvania.

Rose, C., Guenter, B., Bodenheimer, B., & Cohen, M. (1996). Efficient generation of motion transitions using space–time constraints. In *Proceedings of SIGGRAPH 1996* (pp. 147–154). Reading, MA: Addison-Wesley.

Rousseau, D., & Hayes-Roth, B. (1996). *Personality in synthetic agents* (Tech. Rep. No. KSL-96-21). Stanford, CA: Knowledge Systems Laboratory, Stanford University.

Russell, S., & Norvig, P. (1995). *Artificial intelligence: A modern approach*. Englewood Cliffs, NJ: Prentice-Hall.

Sacerdoti, E. (1974). Planning in a hierarchy of abstraction spaces. *Artificial Intelligence, 5*, 115–135.

Sacerdoti, E. (1975). The non-linear nature of plans. In *Proceedings of IJCAI-75* (pp. 206–214), Stanford, CA.

Sengers, P. (1998). Do the thing right: An architecture for action-expression. In *Proceedings of the Second International Conference on Autonomous Agents* (pp. 24–31). Minneapolis/St. Paul, MN: ACM Press.

Scheepers, F., Parent, R. E., Carlson, W. E., & May, S. F. (1997). Anatomy-based modeling of the human musculature. In *Proceedings of SIGGRAPH '97* (pp. 163–172). Los Angeles, CA: Addison-Wesley.

Shi, J., Smith, T. J., Granieri, J. P., & Badler, N. I. (1999). Smart avatars in JackMOO. In *Proceedings of IEEE Virtual Reality Conference '99*. Houston, TX: IEEE Press.

Shoham, Y. (1993). Agent-oriented programming. *Artificial Intelligence, 60*(1), 51–92.

Sourin A., Pasko, A, & Savchenko, V. (1996). Using real functions with application to hair modeling, *Computers and Graphics, 20*(1), 11–19.

Sun, H., Goswami, A., Metaxas, D., & Bruckner, J. (1999). Cyclogram planarity is preserved in upward slope walking. In *Proceedings of the International Society of Biomechanics 17th Congress* (p. 514). Calgary, Canada.

Terzopoulos, D., & Fleischer, K. (1988). Modeling inelastic deformation: Viscoelasticity, plasticity, fracture. *Proceedings of SIGGRAPH '88, 22*, 269–278.

Terzopoulos, D., Platt, J. C., & Barr, H. (1987). Elastically deformable models. *Proceedings of SIGGRAPH '87, 21*, 205–214.

Thorisson, K. R. (1998). Real-time decision making in multimodal face-to-face communication. In *Proceedings of the Second International Conference on Autonomous Agents* (pp. 16–23). Minneapolis/St. Paul, MN: ACM Press.

Ting, B. J. (1998). *Real-time human model design*. Unpublished doctoral dissertation, CIS, University of Pennsylvania.

Tolani, D., & Badler, N. (1996). Real-time human arm inverse kinematics. *Presence, 5*(4), 393–401.

Trappl, R., & Petta, P. (Eds.). (1997). Creating personalities for synthetic actors: Towards autonomous personality agents. Berlin, Germany: Springer-Verlag.

Tupes, E., & Christal, R. (1961). Recurrent personality factors based on trait rating (Tech. Rep. No. ASD-TR-61-97). Lasckland Air Force Base, TX: U.S. Air Force.

Viaud, M., & Yahia, H. (1992). Facial animation with wrinkles. *Third Workshop on Animation, Eurographics '92*. Cambridge, England.

Vilhjalmsson, H. H., & Cassell, J. (1998). BodyChat: Autonomous communicative behaviors in avatars. In *Proceedings of the Second International Conference on Autonomous Agents* (pp. 269–276). Minneapolis/St. Paul, MN: ACM Press.

Volino, P., & Magnenat-Thalmann, N. (1997). Developing simulation techniques for an interactive clothing system. In *Proceedings of Virtual Systems and Multimedia* (pp. 109–118). Los Alamitos, CA: IEEE Press.

Watanabe, Y., & Suenaga, Y. (1992). A trigonal prism-based method for hair image generation. *IEEE Computer Graphics and Applications, 12*(1), 47–53.

Waters, K. (1987). A muscle model for animating three-dimensional facial expressions. *Proceedings of SIGGRAPH '87, 21*(4), 123–128.

Weil, J. (1986). The synthesis of cloth objects. *Proceedings of SIGGRAPH '86, 24*, 243–252.

Wilcox, S. K. (1998). *Guide to 3D avatars*. New York: Wiley Computer Publishing.

Wiley, D., & Hahn, J. (1997). Interpolation synthesis of articulated figure motion. *IEEE Computer Graphics and Applications, 17*(6), 39–45.

Wilhelms, J. (1986). VIRYA—A motion control editor for kinematic and dynamic animation. In *Proceedings of Graphics Interface '86* (pp. 141–146). San Francisco, CA: Morgan Kaufmann Publishers.

Wilhelms, J., & Van Gelder, A. (1997). Anatomically based modeling. In *Proceedings of SIGGRAPH '97* (pp. 173–180). Reading, MA: Addison-Wesley.

Wood, S. (1993). *Planning and decision making in dynamic domains*. Chichester, England: Ellis Horwood.

Wooldridge, M., & Jennings, N. R. (1995). Intelligent agents: Theory and practice. *The Knowledge Engineering Review, 10*(2), 115–152.

Wu, Y., Kalra, P., & Magnenat-Thalmann, N. (1997). Physically based wrinkle simulation and skin rendering. *Eurographics Workshop on Computer Animation and Simulation '97* (pp. 69–79). Berlin, Germany: Springer.

# 16

# Internet-based Virtual Environments

Charles E. Hughes,[1] J. Michael Moshell,[1] and Dean Reed[2]
*University of Central Florida*
*[1]School of Electrical Engineering and Computer Science*
*[2]Institute for Simulation and Training*
*4000 Central Florida Blvd.*
*Orlando, FL 32816-2362*
*ceh@cs.ucf.edu*
*moshell@cs.ucf.edu*
*dreed@ist.ucf.edu*

## 1. INTRODUCTION

Hosting virtual worlds on the Internet introduces many problems not inherent in private networks. These include security, reliability, speed, and differing agendas of participants. The purpose of this chapter is to review the history and present status of Internet-based virtual environments (VEs) and to discuss a number of efforts designed to meet perceived future needs.

## 2. INTERNET PROTOCOLS

### 2.1 The Basics

This section sets the context for the rest of the chapter. It is by no means a thorough journey into the world of Internet protocols. Those wishing such a presentation are referred to Stevens (1994).

Several conceptually simple properties make the Internet work. Its richness comes from building new protocols and paradigms on top of a well-conceived foundation. First, for a computer (host) to participate as one of the many nodes in the worldwide network known as the Internet, it must be assigned a unique Internet protocol (IP) address. An IP address (old form) is a 32-bit number, usually presented as a "dotted-decimal notation," which includes the decimal equivalents of the four bytes comprising the address, separated by periods. The great popularity of the World Wide Web has resulted in the need for a richer set of addresses. This is being met by a new standard (IPng) that extends addresses to 128 bits, represented by 8 sixteen-bit segments, separated by colons rather than periods.

The IP addresses are assigned statically to hosts that constantly reside on the Internet, or dynamically, from a collection of locally available addresses, whenever a transient host joins the Internet through some server.

Communication on the Internet occurs between two hosts, using their respective IP addresses. Long messages are broken up into shorter packets, source and destination addresses are appended, and the data is effectively and efficiently routed from source to destination.

Basic IP communication is then extended with a very thin layer to create UDP (the User Datagram Protocol). The only thing that UDP adds is the notion of ports and some minimal error checking. Ports are the logical channels through which peer applications communicate. A port is represented by a 16-bit address, with low-numbered (well-known) ports being reserved for standard kinds of interactions, e.g., file transfer (21), time-of-day (13), or Web services (80). Packets associated with UDP communication include the port numbers of the sender and receiver, in addition to the IP addresses of these participants as required in any IP packet.

UDP is employed for very simple kinds of interactions. For example to inquire about the time of day, a user need only send a message (any data will do) to port 13 of a designated host. The sender must establish a local port through which this communication occurs. The recipient will then send a response back to the sender using the port through which the sender started this interchange. The response is, of course, the time of day at the recipient's site.

UDP is inherently unreliable. There is no acknowledgement of the receipt of a message unless the protocol of the service requires a return message, e.g., as in a time service. Furthermore, messages, even when received, may not be in the order in which they were sent. This latter property is a direct consequence of the fact that IP routing can be dynamic, allowing the system to effectively manage load balancing and adjust for lost links in the network. The low overhead of UDP makes it desirable for cases where speed is paramount and some loss or reordering of data is acceptable. Examples of this are streaming audio or video, and a global clock. In fact, streaming audio or video would be greatly compromised by an obsession with reliable receipt; can you imagine the player producing a great silence while it waits for one missed note?

In support of shared virtual worlds, VEs often use UDP when messages need to be communicated quickly, and the effects of lost data are minimal. Examples of protocols built in this way are often called *self healing*. A standard example of this is in the communication of small movements of a character in a virtual world. Sending the new position of an object every time it changes keeps every one's view up-to-date. Loss of a packet of positional information is easily compensated when later position packets arrive in a timely manner.

The Transmission Control Protocol (TCP), or TCP/IP, as it is usually known, builds on top of UDP. This protocol adds, among other things, a sequence number to each message. The receiver uses the sequence number to determine if packets are missed or received out of order. Its obligation is to acknowledge receipt of messages. The sender must retransmit missed packets. The receiver must reorder out of sequence messages.

TCP also extends UDP by being a connection-based protocol. This means that TCP establishes contact and keeps the connection open until some two-way communication task is completed. In contrast, UDP is used for quick exchanges. Writers often equate UDP to letter sending; you may or may not get a response, but you at least let the receiver know your name (port) and address (IP address). TCP equates to phone conversations; you establish a connection, participate in a dialogue, and then close the connection when one of the parties deems it is time to do so.

TCP is preferred over UDP in virtual worlds when losses of messages can create critical inconsistencies in the views of participants. For instance, if I pick up an object and the message does not get to you, you may get frustrated trying to pick up this same object when it is no longer available.

In summary, the Internet moves information in packets of manageable size using the following tools: IP addresses, message routing, ports, and optional reliability. These features serve as the building blocks to support all communication between participants in Internet-based virtual worlds (VWs).

## 2.2 HTTP Protocol

The HyperText Transfer Protocol (HTTP) is built on top of TCP. This service, running on port 80 of a host, just listens to see if anyone wants to acquire hypertext files from this particular site. When a request comes in, the message within that request specifies what file is to be sent back. The initiating computer is then responsible to interpret the returned file, for example, by displaying text and/or images in a browser's window and by requesting additional data, as required in the page's hypertext description. This simple protocol is at the heart of the World Wide Web.

Note that HTTP is built on top of TCP, not UDP. This is because reliability is critical here; a document with phrases out of order or missing makes no sense. Many Internet-based VEs use HTTP to help participants find and join currently available virtual worlds. Dynamically created hypertext can provide up-to-date information on existing worlds and include the links needed to connect to these worlds. In addition, the advent of Java applets even makes it possible to dynamically provide participants with the most current programs they need to interact with these worlds. Users need never worry about having outdated VE software cluttering their machines because these virtual world engines will be delivered on demand and destroyed when no longer in use.

## 2.3 RMI, Tuplespace, and Jini

UDP, TCP, and even HTTP are often too low level for the needs of developers of virtual environments. It's not that one cannot build what is needed with these protocols; all Internet protocols are built on top of UDP. It's that developers are often forced to work at the wrong level of abstraction, reinventing interaction protocols.

Many protocols have been developed over the years to support the kind of distributed collaboration required by shared virtual worlds. This discussion will focus on just a few, all of which are currently Java-based, but not Java-dependent. CORBA, the mother of all such protocols, is not addressed, not because it's irrelevant, but only because it is not used in any of the systems herein discussed.

Remote Method Invocation (RMI) is a modern version of Remote Procedure Calls (RPC). RMI is built on TCP, extending it to allow communication between remote objects (software objects distributed across a network) using the standard object-oriented paradigm of messages being sent to objects that respond in ways appropriate to their class identity. In effect, this carries the object-oriented message-passing paradigm into the world of distributed computing, even supporting the transmission of software objects as raw data, followed by the subsequent reconstitution of these objects at the receiver.

RMI is a natural tool for VE developers. With its availability, VE systems can send messages by reference or by value directly to remote objects without having their designers develop arcane, hard-to-understand message passing schemes that turn out to be homegrown versions of RMI. In other words, VE developers are left to concentrate on issues central to their domains of interest, knowing that the existing infrastructure supports object-oriented message passing, even in the context of distributed objects.

Tuplespace is based on the conceptual framework of a globally shared memory (shared among a very large group of computers, potentially including all computers in the world) conceived in 1985 and refined in Gelernter (1991). His simple idea is to allow cooperating processes to write messages onto a global message board. These messages can then be read or consumed (read and then destroyed) by other processes, based on some pattern that matches the form of the original message. *Tuple* is a mathematical term for messages that are comprised of ordered lists of components. One can think of a tuple as a row in a relational database, except that no preordained schemes need to be used. Matching is then a form of "selection" from among those tuples that would "naturally" be organized into the same relational table. The absence of a matching tuple can block the requestor, thus providing coordination as well as communication.

Tuplespace was a theoretically interesting, but not heavily used concept up until very recently. With the advent of Java and its potential to make ubiquitous computing a reality, Sun and IBM, the two big Java players, were each seeking a vision for how data should reside in the massive network of the Internet. Both came to the conclusion that tuplespace is a very appealing option. IBM has developed an API called TSpaces (Wyckoff, McLaughry, Lehman, & Ford, 1998), whereas Sun has developed JavaSpaces (Freeman, Hupfer, & Arnold, 1999). At present, both of these are layered on top of RMI.

Tuplespace, like RMI, offers many benefits for VE developers. It raises the level of abstraction to allow one to think about communication, not with peers, but rather with a shared blackboard. This shared space can serve to communicate state information and coordinate critical activities, all without participating nodes needing to have any explicit information about the locale or even presence of other participants. Multiple tuplespaces can be used to categorize data, provide redundancy, or even maintain a history of all interactions that ever occurred in a virtual community. Such a history can be an archeological treasure, especially if data is maintained in XML, the extensible markup language in which data is marked with domain-specific tags.

Sun has further enhanced the desirability of JavaSpaces by having made it one of the core technologies in its Jini vision of the network as the computer (Edwards & Joy, 1999). The VE community's interest in Jini can easily be seen from Sun's own statement (Roberts & Byous, 1999) that "Jini technology provides simple mechanisms which enable devices to plug together to form an impromptu community—a community put together without any planning, installation, or human intervention. Each device provides services that other devices in the community may use. These devices provide their own interfaces, which ensures reliability and compatibility." If you replace the word device by citizen in a shared virtual community, the relevance to VEs is immediately apparent.

The goal of Jini is to make network computing painless—the service provider and service requestors take care of their own connectivity without human intervention. The protocols by which services are registered and found are key to the Jini vision. The registering of a service is initiated when a provider broadcasts a message seeking to discover a service broker. Jini objects that act as service brokers recognize these messages and report back their willingness to act as intermediaries between the service provider and its potential customers. After a bit of negotiation, the service provider sends information to one or more brokers describing the kind of service and its special attributes. So, for example, an instantiation of a virtual world might announce its availability, indicating that it is a virtual world with attributes of being "social" and focused on "Celtic music." Those wishing to participate in a virtual world can broadcast a message intended to find a broker who is acting as the intermediary for some virtual world service. This participant can specify required attributes, like "social." Any broker who "hears" the request will respond with an indication that it can negotiate on the user's behalf. A successful negotiation results in access to the service, perhaps employing RMI (the service provides a handle to a remote object) and/or the local instantiation of software objects on the client's machine.

Clearly, Jini and other systems that support directory services provide one part of the solution to helping users find virtual worlds that meet their needs. The very existence of such brokers highlights the need of VE designers to pay close attention to the kind of audience they intend to serve. Some of those considerations are discussed in more detail in the next section.

## 3. CHARACTERISTICS OF INTERNET VEs

### 3.1 Lack of a Dedicated Network

The fact that Internet VEs must operate in the wide-open spaces of the Internet has profound effects on their design, and often on their inherent limitations. Three of the primary factors that the Internet influences are speed, reliability, and security. In fact, these three factors often interact with each other in very negative ways.

Consider speed first. Traffic on the Internet does not just slow down message passing; it introduces unpredictability. Experiments in human factors have shown that variations in delays can often be more disconcerting to users than predictably slow performance. This issue of unpredictability becomes more profound as the number and geographical diversity of participants increases, unless the VE is designed from the outset with scalability in mind. Two systems reviewed here, NetEffects (Das, Singh, Mitchell, Kumar, & McGee, 1997) and Bang (Bjarnason, 1999), address this problem in effective, but radically different ways.

Reliability often directly conflicts with speed. For example, using TCP over UDP improves reliability at a cost in speed. One of the VEs reviewed here, DeepMatrix (Reitmayr, Carrol, Reitemeyer, & Wagner, 1999), originally used UDP where unreliability is tolerable and TCP where reliability is critical.

Virtual environment designers often ignore security, although it is critical to the acceptance of some systems (e.g., when they are employed in schools or business settings). Again, as with reliability, security often runs counter to speed. One particularly easy way to provide security is through the encapsulation mechanism of objects. This implies the use of RMI, or some other remote object passing system, adding another layer with its attendant delays to the network protocol. The use of this mechanism can be seen in VRMINet (Reed, Hughes, & Moshell, 1999).

### 3.2 Device Heterogeneity

The Internet playground was designed to support diversity among participating computers. This platform independence has many sides to it. Some of the most significant variables are computation power, display capabilities, rendering speed, operating system and media support available in the machine being used.

Independence from specific machine architectures makes the Internet more accessible, but it can create a burden on VE designers. For example, to reduce network traffic designers often place computational burdens on all participating nodes, whether they are high-end workstations or low-end PCs. A simple example of this is in the use of "dead reckoning" within ExploreNet (Hughes & Moshell, 1997) and NetEffects. The idea behind dead reckoning, a concept borrowed from the military shared VE Simnet (Calvin et al., 1993), is that detailed movement of objects does not need to be communicated. Approximate positions can be computed locally, so long as each participating node knows the destination or direction of movement of an object, and the speed and perhaps acceleration at which it is moving. The upside of this is a reduction in network traffic. The downside is that all nodes interested in this object's movement must now consume computational power, including that needed to avoid unintended object interactions (e.g., a character going through, rather than around, a wall).

In general, there is no easy answer to the dilemma of supporting a wide range of machine architectures, but Java platform independence does provide a seductive middle ground, one that is being adopted by many of the newer systems appearing today. This comes with the usual performance penalties one faces when choosing complete machine independence, although that is less of a consideration with "just-in-time" (JIT) compilers.

Supporting a wide variety of display capabilities can lead designers to target their products for either the lowest common denominator or toward only high-end users. Some systems, like ExploreNet, defer the decision to world builders. In this approach, the size of scene background images determines the amount of resolution needed on the displays of participants. Again, there is no "silver bullet" that solves this problem, although the device independence of Java is certainly a great help.

Device independence, however, does not resolve the issue of the diversity of rendering speeds. In many experiments we have run, the rendering component has turned out to be the bottleneck, more so than computational or network speed. The solution to this is generally out of the hands of VE designers, and more appropriately the concern of device and rendering engine designers. However, VE developers often feel forced into making trade-offs between procedural graphical scene description APIs like Java3D (Sowizral, Rushforth, & Deering, 1997), and declarative description languages like Virtual Reality Modeling Language (VRML; Web3D Consortium, 2000). A compromise is to allow VRML to be used for the external description and Java3D for the internal representation. Bang employs such a strategy.

Operating system issues have led some VE designers to abandon diversity and focus on specific platforms (e.g., Microsoft VWorlds group). Again, Java is viewed by many as the savior, especially with its Java Media Framework, which supports a common protocol for existing and projected media objects, and the performance improvements one gets through its HotSpot technology.

### 3.3 Community Size and Diversity

The Internet community is both large and diverse. Its size factors are ones of geographic separation as well as sheer numbers of potential participants. NetEffects addresses this problem through the use a three-tiered network system, in which participants are connected to a community server, which is in turn connected to a master server. Load balancing and relevance clustering are then used to keep the system responsive over time.

Diversity is an extremely challenging issue. On one level, this means that VEs should be designed with an international audience in mind. An orthogonal concern is accessibility (e.g., for those without sight or hearing). Apple, IBM, Microsoft and Sun, among others, provide APIs and guidelines to address these issues, but there is little actual employment of these standards within VE systems. Accessibility represents one of the most important research issues in the design of human–computer interfaces (Stephanidis, 1997).

## 4. STANDALONE SHARED VEs

Virtual environment systems developed before 1997 tended to be all-or-nothing, stand-alone packages. Here, the designers determined what features users would get and how they would be provided. Two representatives of this class of systems are described in detail, ExploreNet and NetEffect.

## 4.1 ExploreNet

ExploreNet was developed by two of the authors (Hughes and Moshell) at the University of Central Florida in 1994 based on the Virtual Academy model. Major revisions were made in 1996 and 1997 as part of an experiment in the Department of Defense Dependents Schools. The premier world hosted in ExploreNet is called AutoLand. Eighth-grade students at Maitland Middle School in Maitland, Florida, authored this, based on the requirements of third-grade teachers in Würzburg, Germany. The world was then used to help third graders understand social science concepts related to the acquisition, transportation and refinement of mineral resources. The software is available from the project Web site (Hughes & Moshell, 2000) as unsupported freeware. Unlike the other environments discussed here, ExploreNet hosts two-dimensional (2-D) worlds, designed specifically to support the constructionist model of learning. Chapter 45 (this volume) provides more information on how ExploreNet addresses educational issues.

A world in ExploreNet consists of a number of scenes. These scenes are connected through portals, although some characters can have behaviors that allow them to teleport. An avatar is always located in some scene. Props start out in a "home" scene but can be conveyed from place to place as possessions of, or attachments to, other objects. Props can also propagate, producing fully functional clones. Figure 16.1 shows a scene from AutoLand. Here, two avatars, Sandy and John (he's the truck driver) are in the same scene, Pittsburgh. There are three well-labeled portals to other scenes. A truck is attached to John, and a pickaxe is, in turn, attached to the truck. Sandy possesses a sifter. These props move from scene to scene with the corresponding avatars. Communication is through cartoonlike speech lines. Normal communication is visible to all users whose avatars are in the same scenes. Additionally, an avatar can page a single recipient or yell so all can hear.

The ExploreNet environment was developed in Smalltalk. The choice of Smalltalk led to a clean object-oriented design. Unfortunately, Smalltalk did not provide appropriate network and media interfaces, and so ExploreNet's developers had to build their own. Consequently, ExploreNet does not provide a rich set of mechanisms to address issues of security, nor does

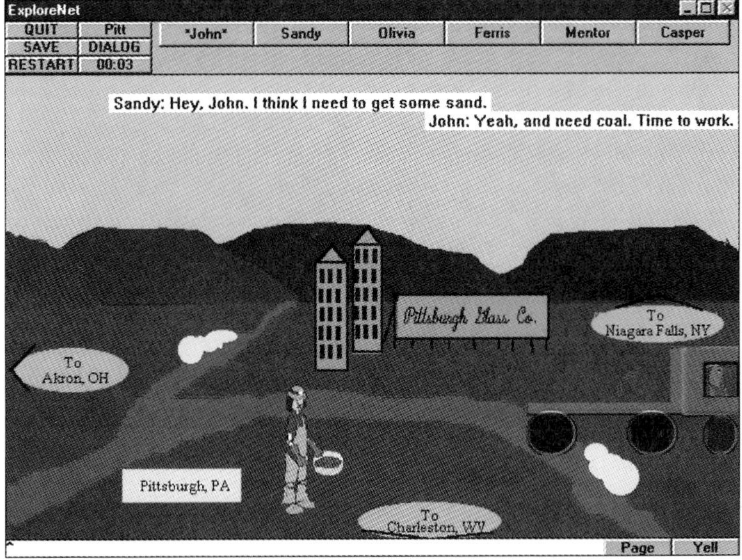


FIG. 16.1. AutoLand: a scene from an ExploreNet world.

it support communication beyond simple textual conversations among participants. Later versions of Smalltalk, like IBM's VisualAge Smalltalk, have addressed these issues, but ExploreNet's future development has been migrated to Java.

Explore Net employs a client–server architecture, but with a twist. The ExploreNet software supports three modes of operation: stand-alone, server, and client. If a user selects stand-alone, no sharing occurs—this case is ignored here. If the user selects server operation, he must then choose a world to offer to potential clients. This user also is a client; in essence, the server spawns a local client who is automatically connected to this world. When a user chooses the client mode, ExploreNet searches the net (using well-known or user selected sites) to find all available world servers. UDP is used here, since the servers need only report back the worlds they are offering. The client then filters the list of offered worlds to include only those for which descriptions are locally available (preinstalled on the client's machine). The worlds that meet these filtering constraints are then offered to the user. When a user selects a world, a TCP/IP connection is made to the associated server. This connection remains open until the user drops out or the server closes the world. The primary features of ExploreNet's protocols discussed below are:

- Affordances
- Ground truth
- Global consistency
- Client dropout
- Historical record

ExploreNet worlds consist of objects and their potential *affordances*. The central objects of a world are its scenes and avatars. Scenes, in turn, possess connecting portals and props. All objects have associated image sets, behavior scripts, and attributes, and can acquire additional images, behaviors, and attributes from their *prototypes*. The prototyping system allows, for instance, a character named Fred to acquire part or all of its behaviors from a prototype called Mentor and another called Human. Fred can, in turn, be a prototype for other characters.

Attributes are specified as names (e.g., hasChildren) and associated weights (e.g., 2). Scripts are written in a simple, homegrown, message-based language, with many primitive behaviors from which to build. Each behavior and portal has conditions that must be met in order for it to be active. These conditions can be based on the attributes of any object in the world, and/or the identities of the object, the initiator of the action, or even the relatives (prototypes) of any of these characters. The affordances of an object are its presently active behaviors. In addition to selectable behaviors, a world and any of its objects can have automatic behaviors that are triggered any time their conditions are met. These simple mechanisms of conditional and scheduled behavior activation, and conditional portals, along with means to trigger events randomly, make ExploreNet worlds appear to exhibit behaviors that are unpredictable and often interesting.

The evolutionary nature of ExploreNet worlds makes them interesting and useful for problem solving, as well as for quests and social interactions. Unfortunately, the system's primitive support for media, its nonstandard scripting language, and its lack of integration with Web browsers limits ExploreNet's general appeal and usefulness.

Over and above its use of a client–server model, ExploreNet employs several protocols worth discussing. In particular, each client maintains *ground truth* on the avatar(s) it controls. This means that critical decisions about an avatar's state are deferred to the client controlling that avatar, not to the server. However, the server always maintains ground truth on props. This approach is a compromise between the server-centric design seen in typical multiuser domains (MUDs) and the purely distributed approach taken in military distributed simulations.

Messages sent by clients are always to the server who provides *global consistency* to the ordering of events. In fact, clients are unaware of the identity of each other. This approach allows the server to impose an order on the receipt of messages by all clients (recall that TCP/IP is used here). The potential downside of making the server work too hard is alleviated by ExploreNet's protocol, which communicates a complex behavior as a simple message, not as a series of individual primitive actions. It is further alleviated by the fact that navigation is communicated as a destination, with the receiving clients responsible for path planning and actual articulated movement.

A world server expects regular communication from all its clients, and can use the lack of such communication to detect *client dropout*. The clients are correspondingly expected to send brief "I'm alive" messages even when they are otherwise passive. If a server observes that a client has been quiet for too long, the client is sent a "wake-up" message. Whenever a client fails to respond to three consecutive wake-up messages, the client's connection is closed and its avatars are made available to new clients. Similarly, a client expects regular communication from its server. If this fails to happen, the client can recast itself as a server, offering the world, in its current state, to other clients. Of course, that can lead to a world's replication if a server goes down and more than one client chooses to continue the world.

As a world evolves, each client, including the server's attached client, keeps an *historical record* of all messages that have been received. Clients are also free to take snapshots of this history at any time. The complete history, or any snapshot, can be used for retrospective analysis or to restart the world at any point in its evolution. In fact, retrospective analysis can be done from the perspective of any avatar, even with the ability to change the viewpoint from one avatar to another in mid story. Histories also allow late arrivers to a world to see how the world evolved to its current state.

### 4.2 NetEffects

NetEffects was developed by researchers at the National University of Singapore. The premier world hosted on NetEffects is called HistoryCity (Kent Ridge Digital Labs [KRDL], 1999). This world provides a shared community for 7- to 11-year-old children in Singapore to learn about the history and culture of their city-state. HistoryCity is a virtual Singapore of 1870, consisting of 24 communities complete with historical buildings, costumes, and objects. The software is presently available for download at the Kent Ridge Digital Labs at KRDL, 1999.

A world in NetEffect consists of a collection of communities. These communities have a hierarchy of venues, typically consisting of an outdoor area (e.g., a town square) and a number of buildings. The buildings, in turn, contain rooms. Avatars can move within a venue and from venue to venue based on geographical relationships. Visibility among objects is limited to those in the same venue. Movement can also occur between communities, but there are some limitations on this movement, as discussed below. Text communication is within a venue. Speech communication is a private conversation between any two avatars, whether or not they reside in the same venue or even the same community. Agents are objects in a venue that can provide content, for instance, in the form of stories, jokes, or sound effects, or services like acting as brokers for the exchange of possessions. Figure 16.2 is a snapshot of the Commercial Square venue in HistoryCity, inhabited by citizens and containing many buildings into which these citizens can move.

According to its authors, NetEffect is implemented largely in Java, communicating via TCP/IP. NetEffect employs a client/peer-server/master-server architecture. When a user enters a NetEffect world, she does so through a master server. Based on prior history (e.g., where this user was when she last left the world), the master server selects a peer server that is handling communication for this community, passing its address back to the user's client machine. The

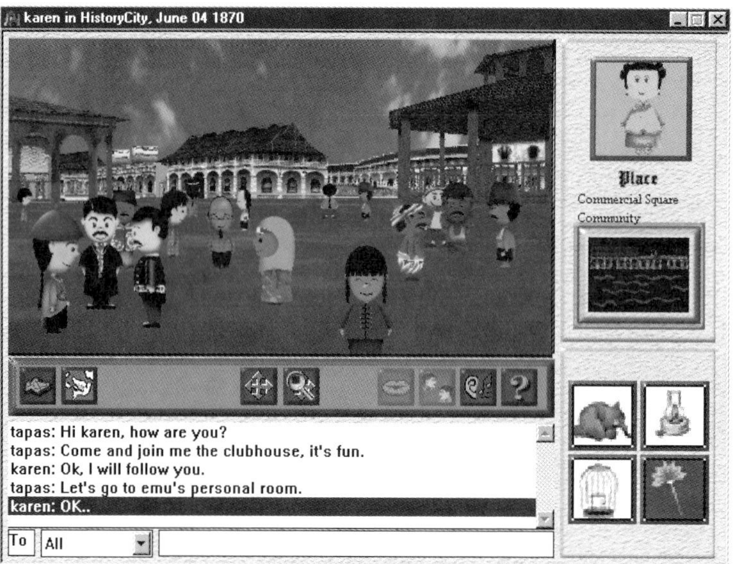


FIG. 16.2. HistoryCity: a scene from a NetEffect world.

client then connects to the community through the assigned peer server. The peer server provides the client with a community description, from which it constructs the three-dimensional (3-D) world. Users can navigate and interact within this community, or move into other communities through visible portals. The primary features of NetEffect's protocols discussed below are:

- Three-tiered networking
- Need-to-know updating
- Group dead-reckoning
- Dynamic load balancing

As noted above, NetEffect employs a three-tiered networking strategy. The center of the network is a single main server. As part of its start-up, the main server starts a number of peer server nodes, usually one per community. It then passes each of these peer servers a community allocation table so that each knows its responsibility. Having all servers aware of the location of every community also enables intercommunity communication.

The master server is also responsible for maintaining the personal databases of all the world's citizens. Since users always start at the master server, this provides a simple mechanism by which these users can resume wherever they left off in the world.

The periphery of the network consists of the client nodes on which users reside. Generally, a client node only talks to the peer node that supports the user's current community. When a user moves from one community to another, the client stops communicating with one peer server and starts communicating with another, unless of course both communities reside on the same peer server. The mechanism's operation is somewhat analogous to the way cellular telephone systems work.

The master server accepts TCP/IP connections on two ports, one for peer servers and the other for clients. A peer server keeps its connections open to the master service for as long as it is online, whereas a client drops its connection once it learns the identity (IP address, port number) of its peer server. Each peer server also accepts connections on two ports, one for other peer servers and the other for clients. Connections made by peer servers are persistent so long as the other server is online. Connections made by clients are broken when the client leaves or migrates to a community served by another peer. To migrate to another peer server, the client gets the IP address and port number of the new peer server from its current server. This is possible since every peer server knows about all the others. In fact, migration is only possible if the new peer reports that it can handle the additional load.

While this three-tiered network approach is an effective means of supporting NetEffect's goal to scale up to many users, congestion in a single community, and thus on a single peer server, is still possible. One might expect a situation in which all updates must be broadcast to all residents of a community. To alleviate this problem, NetEffect employs *need-to-know updating*. For instance, a user inside a room does not need to hear conversations or see movements anywhere but within this room. When the user leaves the room, information can quickly be passed that reflects the state of objects in the corridor into which the user just moved.

NetEffect also employs a technique called *group dead reckoning*. The idea here is to prevent a continuous flow of positional messages as a user moves throughout a community. ExploreNet uses a destination-based scheme to avoid this, whereas NetEffect uses an interval-based scheme. Users in NetEffect send their positions to their server every $C$ seconds, but only if they have moved. The peer server then accumulates this positional data from each user and broadcasts a composite packet to all group members every $P$ seconds. $C$ and $P$ are parameters with which an administrator can experiment. In the case of HistoryCity, they are set to 1 and 1.5 seconds, respectively. The notion of a group is also flexible. The group could be all users in a community or all users in a venue (e.g., room). HistoryCity uses the former, but NetEffect supports both protocols.

The final technique used to support NetEffect's scalability goal is *dynamic load balancing*. The scheme is rather simple, but effective. The master server periodically checks to see if the world load is balanced across the peer servers. A number of strategies can be employed here, from detailed traffic analysis to just counting the number of attached clients per server. When the master server recognizes an imbalance, it devises a new community allocation across peer servers and initiates the process of reassignment. During this time, all interactions by clients are stopped, so one hopes that this event is infrequent.

## 5. SHARED VRML WORLDS

Starting in 1997, VE developers switched from creating stand-alone monolithic systems to ones based on plug-ins and Java-enabled browsers. Much of the motivation for this change came from the emergence of the 3-D interactive Web standard VRML97 and its External Authoring Interface (EAI) specifications, which allow 3-D objects to be manipulated by external applications that are not written in the VRML language (Web3D Consortium, 2000).

VRML is based on the mature and powerful scene graph architecture defined in the Silicon Graphics Performer and Inventor technologies. VRML's shortcomings for large virtual worlds are its relatively low performance and the lack of a large body of ongoing development in substrate tools such as rendering engines, browsers, optimizers, and utility libraries. Low performance manifests itself in numerous ways, such as simplistic lighting models, unreliable texture handling, and rapid performance degradation as the world's polygon count increases. Performance degradation occurs because there is no effective continuous level-of-detail or load module control mechanism.

Despite VRML's drawbacks, its provision of an Open API rendering layer and its easy access to Java's rich networking capabilities afforded independent developers the tools needed to implement distributed virtual world systems. Further development of the Living Worlds specification (Web3D Consortium, 2000) provided guidance to anyone who wished to create

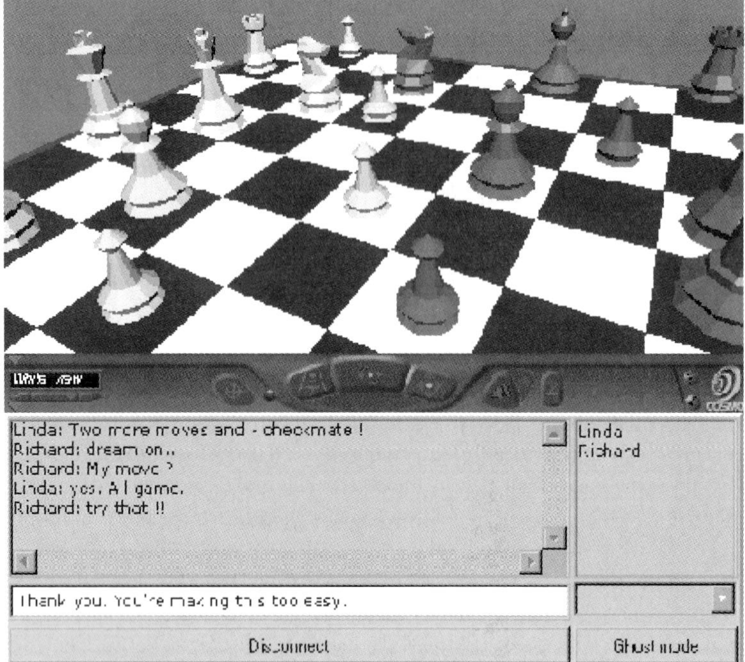


FIG. 16.3. Chess Board: a scene from a DeepMatrix world.

shared VRML worlds. Detailed descriptions are presented of two such systems, DeepMatrix and VRMINet.

## 5.1 DeepMatrix

DeepMatrix was developed in 1998 by a collaborative team from Vienna University of Technology, Geometrek in Nyon, Switzerland, and Arizona State University (Reitmayr et al., 1999). This system focuses on scalability and is optimized for low-bandwidth environments. It has no particular premier world, so Fig. 16.3 presents an image of a representative shared world in which participants can compete in chess.

DeepMatrix is written in Java, runs as an applet within a Java-enabled browser, employs a client–server model, and follows but does not completely implement the Living Worlds specification. All code is "semiopen" (Geometrek, 2000), with much of it having been previously implemented in VNet, the first publicly available implementation of the Living Worlds specification (White, 1997). The VRML plug-in provides basic navigation and display of the 3-D world. The chat window, list of participants, and the "Ghost mode" and "Disconnect" buttons complete the user interface. Each of these widgets is a standard visual component available in the Java Abstract Windowing Toolkit (AWT). With its simple textual chatting area, this system provides support for games like chess and many other forms of social interaction. The primary features of DeepMatrix's protocols discussed below are:

- Loading, rendering, navigation
- Room objects
- Real-time communication

DeepMatrix does not handle the loading, rendering, or navigation of a world's visual components. Rather, VRML code that defines the visual aspects of DeepMatrix worlds is transported using the HTTP protocol. DeepMatrix just passes information about the location of VRML code for its rooms and client objects by transmitting URLs pointing to VRML files. The VRML plug-in uses HTTP to acquire the files, creates a scene graph from the VRML description, and manages all rendering and navigation. DeepMatrix only requires that some simple changes be made to the VRML file so it can learn about gates (portals to other worlds), objects, and their behaviors, and so there is a "center" point in the world relative to which it can sense and communicate avatar movement to other clients. This requirement to sense movement is a consequence of the fact that DeepMatrix does not provide its own navigator. DeepMatrix also supports two optional fields, inMotionBehaviour and notInMotionBehaviour, in avatar prototypes. These specify behaviors that are triggered automatically when an avatar is moving or standing still. Their use facilitates articulated motion when the avatar is moving and fidgeting behaviors when it is at rest.

Scalability is addressed on a DeepMatrix system's server by defining *room objects*. A "room" defines a partition of the entire world, and clients receive messages only from the room that they are in. This partitioning of a single, large VE or several interconnected VEs dramatically reduces the overall messages that a client machine will receive. Partitioning in this manner can also reduce server load since separate machines can be responsible for different rooms.

The transport layer of DeepMatrix was originally a combination of the UDP and TCP/IP protocols. In these earlier versions, the streamlined UDP protocol was used for real-time communications (e.g., state updates of the avatars). Since reporting avatar state changes is the largest single source of network traffic, DeepMatrix's original use of UDP increased response time and used less bandwidth than a pure TCP/IP session. A TCP stream connection was implemented for control messages (e.g., logging in and logging out), managing of Avatar objects, and passing of chat messages.

DeepMatrix 1.1 incorporates a major change to the transport layer in that it no longer uses UDP, relying instead on a single TCP stream. Gerhardt Reitmayr, one of the authors of DeepMatrix, states that this change was necessitated by differences in Java implementations within different browsers (personal communication, May 12, 2000). Message filtering now removes "stale" messages (ones whose values are superceded by newer updates), in contrast to the UDP approach, which can sometimes flood a network with very small incremental updates.

### 5.2 VRMINet

VRMINet is a system developed in 1999 at the University of Central Florida (Reed et al., 1999). The name "VRMINet" is based on VRML + RMI (Remote Method Invocation) and also reminds one of the occasional use of "vrmin" as a humorous and somewhat pejorative term for geeks who have access to no professionally supported graphics systems. However, the choice of VRML in VRMINet was not driven by lack of alternatives, but by a desire to adhere to a component-based technology, wherein the descriptive language of VRML can be retained while the rendering component might later be replaced.

The premier world hosted in VRMINet is "Caracol Time Travel." This world was authored by an interdisciplinary group of 18 undergraduate students. The team spent two semesters designing and constructing an interactive drama-based educational role-playing game. The game is fashioned after an actual Central American archaeology project being carried out in Belize under the supervision of Diane and Arlen Chase (Caracol, 1999). The game was designed to teach basic concepts of archaeology and facts about the life of the ancient Maya, to a target audience of middle school students.

A world in VRMINet is a shareable VRML world with Java implemented behaviors. Figure 16.4 shows a scene in the Caracol world with three avatars (Ana, Pedro, and Dr. Smith). Participants are in a cave found at the site, studying a 3-D map of the ancient city. In addition

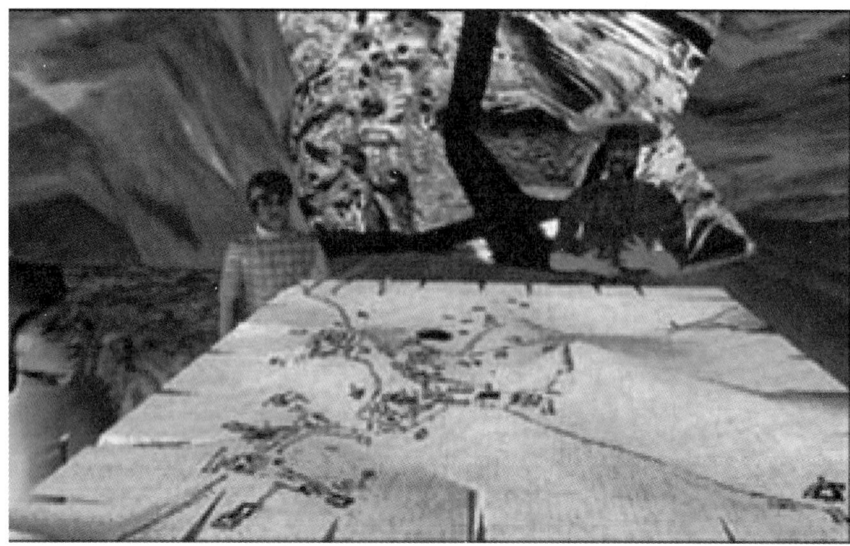

FIG. 16.4. Caana in the modern-day city of Caracol: a scene from a VRMINet world.

to the avatars, this scene is populated with many artifacts that have autonomic behaviors triggered by actions and/or states of the participants. One of the most complex behaviors in the Caracol world involves a flight by a Toucan, with all the avatars being dragged along for the ride from modern times at the archaeological dig into classic Maya times.

VRMINet is written entirely in Java. Software is available for educational use at Reed et al. (1999). The server is a Java application, whereas each participant merely references an html page that contains the client applet. Thus, the standard HTTP Web protocol is used to find a world. A user logs in, selecting a role to play in the world. When log-in is successful, the VRML scene and a Java applet dashboard are displayed. This dashboard allows many forms of interaction, but is not needed for basic navigation and thus can be minimized at any time.

Initially, the user is controlling the position of the avatar that was selected and seeing the world through its eyes. If the "out of body" button is clicked, this stations' viewpoint is detached from its avatar and can be pulled back to get an over-the-shoulder view, or maneuvered in such a way as to frame any desired view of the scene. The same button toggles the viewpoint back into the avatar.

An "online users" window provides the user with a list of the other humans in the world. The available sound menus contain, respectively, sound effects (or utterances such as "Hey!"), and short canned speeches (if desired). An "available gestures" menu selects behaviors a participant's avatar knows how to carry out. A "free bodies" window lets participants take into their possession any available objects in the scene. The primary features of VRMINet's protocols discussed below are:

- Behavior authoring
- Client–server architecture
- Distributed leases
- Remote method invocation
- Performance

Selecting a free object (which is not an avatar) in the scene either triggers an automatic default behavior or opens a menu, listing the optional behaviors one can specify for this object. This control is very similar to that provided by ExploreNet. However, *behavior authoring* in

VRMINet is very flexible, allowing any combination of VRML routes, VRML Script nodes, and Java methods to define behaviors. The component nature of this system even allows other scripting languages to be employed. In fact, JPython (a Java friendly variant of Python) has been successfully (actually effortlessly) used in several experimental worlds.

VRMINet employs strict client–server architecture. While this is necessitated by Java's security model, it also provides the message-ordering benefits discussed while describing ExploreNet. Thus, all clients receive messages in the order that the server handles them, and all clients have the opportunity to process these messages in the same order.

Ground truth in VRMINet is implemented through the concept of *distributed leases*. Java 2 (JDK 1.2) introduced the idea of leased resources in a Java network environment. The notion is similar to a landlord and tenant lease agreement. The landlord (server) grants the tenant (client) a lease on his property (behavior) if it is available. At some point in the future, the lease expires and the tenant and landlord must renegotiate the lease or the property (behavior) is no longer in the hands of the client. The client must vacate the property (stop executing the behavior) and regain the lease before taking possession of the property again. There are no guarantees that the lease will be renewed once it has expired, so the tenant should be ready to give up ownership of the property at any time.

Most IP-based shared virtual world systems are optimized for scalability and network performance only. These systems implement TCP/IP and/or UDP network communication with socket bases and a rigid protocol. The result is a highly scalable but inflexible system that requires protocol extension on client and server components anytime new features are added.

The Caracol project (the driving force behind VRMINet) required extensibility and flexibility, was operating in an environment supporting high network speed, and was willing to sacrifice scalability for the sake of the educational experiment. RMI was chosen as the transport layer due to its acceptable response time and flexibility. RMI's primary attractiveness came from the fact that it distributes objects among connected computers without the need for protocol extensions or explicit serialization. Serialization is the flattening of an object that makes it easy to transport across a network and then reconstruct at the receiving node.

Much concern over the network performance of VRMINet was raised during the initial trials with the Caracol project. Player avatars seemed to suffer from network delay when a large load was placed on the system. Observation later showed that delay in the system is largely related to rendering speed on the machine hosting a given client, and to the ways the networking architecture and browser generate motion events that have to be rendered.

Allocating more CPU time to the rendering thread and running the VRML browser under optimal display speed conditions gained some minor performance boost. Further improvement was realized by decreasing the granularity of message passing for movements. On fast machines, the VRML browser produces a large number of movement events for trivial moves. It was observed that for a move of 1 meter, some fast machines could produce over 50 messages that in turn must be sent across the network, executed by the receiving object and then rendered on the video display card. By reducing the granularity of motion, VRMINet decreases the total number of messages passed on the network and thus reduces the number of rendering calls to the receivers' browsers.

Major degradations in performance were noticed in worlds that were built using extremely complex structures or overabundant simpler structures. The VRML browser uses a great deal of resources when complex models are displayed, which leaves very few resources for the Java Virtual Machine (JVM) to execute. Continuous garbage collection can cause undesirable performance from the JVM, and hence the VRMINet system. Fortunately, Sun's HotSpot continuous garbage collection technology has made this a nonissue. Nonetheless, best results were obtained from worlds that were optimized by polygon count, model count, and texture complexity. Optimizing on these criteria resulted in performance boosts for the overall system.

## 6. THE FUTURE–COMPONENT-BASED VEs

Modern VEs have broken away from the monolithic, all-or-nothing closed systems described earlier in this chapter and have moved toward a component-based architecture. VRML, and its ability to plug-in to Web browsers, started this evolution. Strangely, this model (which provides no economic incentive for the construction of efficient plug-ins) may lead to VRML's own demise, or at least its relegation to a role as a scene description language, but not a renderer or navigator.

Properly designed VEs can and should allow different rendering systems to be "swapped" for the default system. For example, a system could support external scene description in pure VRML, but internal representation in Java3D. This design requires that the underlying state objects, which represent the 3-D on-screen objects, be loosely coupled with the rendering layer. Another requirement is that the 3-D content be specified in file formats that all available rendering layers understand.

Just as with the rendering layer, the network transport layer can be treated as a pluggable component. This means that raw UDP and TCP/IP might be used in some circumstances, and remote method invocation or even tuplespace in others. The choice will depend upon the requirements of the world and the characteristics of the network environment.

Two systems will be briefly discussed that are representative of component-based VEs, VRMINet (a second visit) and Bang.

### 6.1 VRMINet Revisited

VRMINet is an example of a VE that supports shared VRML worlds. That categorization, while correct, understates the component nature of this system. VRMINet really has no dependence on standard VRML rendering—its rendering component could easily be replaced. However, from another perspective, the renderer doesn't care about the network transport layer; it just wants to know about transformations and changes of viewpoint that occur. In fact, the system doesn't even depend on its current authoring tools; it was previously pointed out the ease with which JPython was incorporated as a scripting language.

### 6.2 Bang

Bang (Bjarnason, 1999) is a Java-implemented VE that can read VRML descriptions but maintains complete independence from VRML plug-ins. In fact, its internal scene graph representation is in Java3D, allowing content to be provided in either VRML or serialized Java3D. Bang is also one of the first VEs to use Sun's Jini to support the discovering and sharing of worlds. The persistent state of each world is maintained in a tuplespace, using Sun's JavaSpaces implementation. Basic communication is still done using open source from the DeepMatrix and VNet Living Worlds specification implementations.

Bang's use of JavaSpaces provides both persistency (world objects stay around after their creating processes leave) and scalability. The essence of Bang's approach to scalability is to divide the world into 100m × 100m × 100m-sized sectors that are used for loading and unloading objects and geometry. As a user moves from sector to sector, his client registers its presence in the new sector and unregisters its presence from the old one. Notifications of changes in geometry are only sent to those clients registered in the sector in which the change occurred. This approach is in sharp and obvious contrast to other VEs that manage large spaces through logical scenes or communities and portals, rather than physical regions. Bang also allows users great flexibility in their choice of sound and speech engines. It merely requires that chosen components implement the appropriate services defined in Java's Media Framework.

## 7. A FEW MORE SYSTEMS OF NOTE

Any page-limited review of IP-based VEs is destined to leave out many systems worthy of detailed study. We are clearly guilty of this crime of omission. The following are just a few of these systems. Readers are encouraged to pursue further study (and recreation) by following the leads provided below.

### 7.1 Noncommercial Systems

All but one (NetEffects) of the VEs previously discussed are free, noncommercial systems. Such software typically comes from universities or small groups of diehard developers. Free software, of course, comes with its risks. Over the last decade, most of the giveaways have died away, leaving their users feeling deserted. The recent trend of open software, in which users have access to the source code, as well as executables, is breathing new life into these systems. For instance, VNet, a student project at the University of Waterloo, has not been updated for several years, yet its code is still alive within DeepMatrix, VRMINet, and Bang. Among the many other systems that now provide open source, two are highlighted below, Bamboo and Virtual Worlds Platform.

#### *7.1.1 Bamboo*

Bamboo was developed at the Naval Postgraduate School, based on its many years of experience in the development and deployment of VEs (Singhal & Zyda, 1999). The goal of this system is to serve as an extensible environment using a generalization of the plug-in metaphor, whereby each plug-in can declare its interdependencies, triggering the loading of required cohorts. Bamboo is in effect a host for new VE technology and a VE system, in that it provides default components built on top of a very lightweight kernel. The ADAPTIVE Communication Environment (ACE), an open-source framework that implements design patterns for concurrent communication software, provides the networking infrastructure for Bamboo. The communication software tasks provided by ACE include event handling, signal handling, interprocess communication, shared memory management, message routing, dynamic (re)configuration of distributed services, concurrent execution, and synchronization. Using ACE, Bamboo can employ unicast, multicast, and broadcast message passing to build complex protocols, including HLA, the military "high-level architecture" for distributed interactive simulation. In this regard, Bamboo is the only one that addresses the needs of virtual worlds that must be physically correct.

#### *7.1.2 Microsoft VWorlds*

As in other areas of Internet computing, Microsoft has a major presence with both VE products and active research. The heart of this research is Microsoft's Virtual Worlds Group, which has developed a number of engaging virtual environments, ranging from the playful ComicChat to the experimental VWorlds system (Vellon, Marple, Mitchell, & Drucker, 1999). The latter VE integrates seamlessly with Internet Explorer and includes a suite of wizards for avatar and world creation. It also employs a novel event handler that automatically propagates event notifications to an object's contents (e.g., the passengers in a car), container (e.g., a scene), and peers (e.g., other objects in the scene). Source code, executable binaries, and sample worlds are available at the organization's Web site (Virtual Worlds Group, 2000).

## 7.2 Commercial Systems

The commercial world of VEs is a brutal one, where a typical system has a very short life span, occasionally followed by a reincarnation as a very specialized product. Some go on forever in a state of limbo, in the sense that they have a Web page but no content. Two of the survivors in this commercial arena are Active Worlds and Blaxxun.

### *7.2.1 Active Worlds*

Active Worlds provides a free browser, with additional privileges afforded to those who register for a nominal fee. Its server technology requires payment based on intended usage (personal or commercial) and intended size in inhabitants and virtual land. The main Active Worlds server hosts over 600 well-traveled virtual worlds, including a shopping mall. The granddaddy of these is AlphaWorld, with over 700,000 users and almost 50 million objects (sizes mentioned here are current as of February 2000). The AlphaWorld territory is slightly larger than the area of the state of California, and growing (Activeworlds, 2000).

### *7.2.2 Blaxxun*

Blaxxun also provides a free viewer, which is in fact a shared VRML browser that can be used as your primary VRML plug-in. Server and software development technology can be purchased. The primary world hosted by Blaxxun is called Cybertown. As with AlphaWorlds, there is a large virtual space for users to colonize. The metaphor to a real community is strong, in that citizens can develop unique identities, own homes and other possessions, organize activities, and, in general, go about the business of virtual life (Blaxxun, 2000).

# 8. CONCLUSIONS

This chapter has reviewed a number of shared VEs designed to interoperate using standard Internet protocols. We have not attempted to be complete, but rather to cover a number of contrasting systems, highlighting features that make them interesting objects of study.

Those wishing more detail are encouraged to visit the Web sites referred to herein. Of course, as with all things ephemeral, we cannot guarantee that these sites will still be alive and up-to-date if you try to access them years (or even months) from now. So, enjoy yourself, but come with a sense of humor and patience.

# REFERENCES

Activeworlds. (2000). *ActiveWorlds* [Online]. Available: http://www.activeworlds.com/ [2000, May 15].

Bjarnason, R. V. (1999). *Bang* [Online]. Available: http://this.is/bang/ [2000, May 15].

Blaxxun. (2000). *Blaxxun contact* [Online]. Available: http://www.blaxxun.com/ [2000, May 15].

Calvin, J., Dickens, A., Gaines, B., Metzger, P., Miller, M., & Owen, D. (1993). The SIMNET virtual world architecture. In *Proceedings of the 1993 IEEE Virtual Reality Annual International Symposium* (pp. 450–455). Seattle, Washington, September 18–22, 1993.

Caracol. (2000). *Caracol archaeological project* [Online]. Available: http://www.caracol.org/ [2000, May 15].

Das, T. K., Singh, G., Mitchell, A., Kumar, P. S., & McGee, K. (1997). NetEffect: A network architecture for large-scale multi-user virtual worlds. In *Proceedings of the Symposium on Virtual Reality Software and Technology* (pp. 157–163). Lausanne, Switzerland, September 15–17, 1997.

Edwards, W., & Joy, B. (1999). *Core JINI.* Upper Saddle River, NJ: Prentice-Hall.

Freeman, E., Hupfer, S., & Arnold, K. (1999). *JavaSpaces principles, patterns and practice.* Reading, MA: Addison-Wesley.

Gelernter, D. (1991). *Mirror worlds: Or the day software puts the universe in a shoebox ... how it will happen and what it will mean.* London: Oxford University Press.

Geometrek. (2000). *DeepMatrix* [Online]. Available: http://www.geometrek.com/ [2000, May 15].

Hughes, C. E., & Moshell, J. M. (2000). *ExploreNet* [Online]. Available: http://www.creat.cas.ucf.edu/projects/proj-explore00.html [2001, July 12].

Hughes, C. E., & Moshell, J. M. (1997). Shared virtual worlds for education: The ExploreNet experiment. *Multimedia Systems 5*(2), 145–154. Springer Berlin Heidelberg.

Kent Ridge Digital Labs. (2000). *HistoryCity singapore project* [Online]. Available: http://www.historycity.org.sg/ [2000, May 15].

Reed, D., Hughes, C. E., & Moshell, J. M. (1999). *VRMINet* [Online]. Available: http://www.creat.cas.ucf.edu/projectsVRMINet [2000, May 15].

Reitmayr, G., Carrol, S., Reitemeyer, A., & Wagner, M. G. (1999). DeepMatrix—An open technology based virtual environment system. *Visual Computer 15*(7/8), 395–412. Springer, Berlin, Heidelberg.

Roberts, S., & Byous, J. (1999). *Distributed events in JINI technology* [Online]. Available: http://developer.java.sun.com/developer/technicalArticles/jini/JiniEvents/ [2000, November 1].

Singhal, S., & Zyda, M. (1999). *Networked virtual environments: Design and implementation.* Reading, MA: Addison-Wesley.

Sowizral, H., Rushforth, K., & Deering, M. (1997). *The Java 3D API specification.* Reading, MA: Addison-Wesley.

Stephanidis, C. (1997). User interfaces for all: developing interfaces for diverse user groups. In Anthony Jameson, Cécile Paris, and Carlo Tasso (Eds.) *Proceedings of the Sixth international Conference on User Modeling* (pp. 443–444). Springer, Vienna, New York, 1997.

Stevens, W. R. (1994). *TCP/IP Illustrated* (Vol. 1). Reading, MA: Addison-Wesley.

Vellon, M., Marple, K., Mitchell, D., & Drucker, S. (1999). *The architecture of a distributed virtual worlds system* [Virtual Worlds Group, Microsoft Research, Online]. Available: http://www.research.microsoft.com/vwg/papers/oousenix.htm [2000, May 15].

Virtual Worlds Group. (2000). *Virtual worlds platform* [Online]. Available: http://www.vworlds.org/ [2000, May 15].

Web3D Consortium. (2000). [Online]. Available: http://www.web3d.org/ [2000, May 15].

White, S. F. (1997). *VNet* [Software developed by S. F. White and J. Sonstein at the University of Waterloo, Canada, Online]. Available: http://www.csclub.uwaterloo.ca/~sfwhite/vnet/ [2000, May 15].

Wyckoff, P., McLaughry, S. W., Lehman, T. J., & Ford, D. A. (1998). TSpaces. *IBM Systems Journal 37*(3). Armonk, New York.

# 17

# Structured Development of Virtual Environments

John R. Wilson, Richard M. Eastgate, and Mirabelle D'Cruz
*University of Nottingham*
*Virtual Reality Applications Research Team (VIRART)*
*Nottingham NG7 2RD UK*
*john.wilson@nottingham.ac.uk*
*richard.eastgate@nottingham.ac.uk*
*Mirabelle.d'cruz@nottingham.ac.uk*

## 1. INTRODUCTION

### 1.1 Background

Little systematic attention has been paid to the way in which virtual environments (VEs) should be designed. This may be because most early VEs were "proof of concept" demonstrations (with fairly crude object representations), entertainment games, or usability test prototypes. The first category, demonstrations, were produced at great speed, often by trial and error, to sell the idea of VEs to potential user organizations. Little time or thought was given to defining the development process. Often the VEs and development processes used were commercially or militarily confidential, and therefore publication was not an option. Second, VE games were often based on other computer games, and in any case their developers had no scientific reason to publish anything on their design and development process and had good commercial reasons not to. No published accounts therefore exist (although see the note on games development in the next paragraph). The third category of early VEs, simple usability test prototypes, allow ergonomists and psychologists to experiment with various aspects of the virtual experience and with factors of VE usability. Again, there would be little need for, or thought given to, the VE design process.

There is now some research under way that should lead to useful guidance on the whole VE development process; accounts exist which cover some of the development cycle (e.g., Grinstein & Southard, 1995; Green & Sun, 1995; Loftin, Bartasis, & Johnston, in press). Guidance on design for VE usability has been published, for instance from groups at City University in London (e.g., Kaur, 1998; Kaur Deol, Sutcliffe, & Maiden, 1999), at Virginia Polytechnic (e.g., Gabbard, Hix, & Swan, 1999; Hix et al., 1999; chap. 34, this volume) and chap. 46, this volume, at University of Nottingham (e.g., Neale, 2001; Tromp, 2001; Tromp & Fraser, 1998; Tromp et al., 1998a). Notwithstanding the somewhat anarchic nature of the PC games software industry, and especially the unstructured nature of their development alluded to above, there are moves to better understand best practice in games development. *Game Developer Magazine*, and especially its online archive at www.gamasutra.com, is a useful source of contributions. These are mostly at the level of general principles but do have content of relevance to VE development, to do with overall style, characters, levels of environment, menus and other interface elements, use of sound, animation, and the artwork.

Even recognizing the contributions identified above, much work of VE specification, building, and testing is characterized still by a lack of guidance across the whole process. Systematic guidance for VE development would reap dividends in terms of a more efficient design process and more functional and usable virtual environments. The allocation of limited processing resources, the nonintuitive nature of (most of) the physical interfaces, and the limits that must be applied in deciding what aspects of the real world to model are all issues that must be addressed within a structured VE development framework.

To some extent, VEs could be developed following design process guidance for any human–machine system or human–computer interaction (HCI), of which there are vast numbers (e.g., Newman & Lamming, 1995; Sanders & McCormick, 1993; Sutcliffe, 1995; Wickens, Gordan, & Liu, 1998). However, the effort required to model objects and actions of the real world and the nature of the cognitive interface (the participant being "within" the database) mean that VE development also requires guidance specific to its own special characteristics and needs. In any case, guidelines and models within the human factors and "conventional" HCI communities are notorious for not always being useful or usable by designers, engineers, or even ergonomists. Acceptable and useful guidelines to design for usability are not easy to produce in ergonomics generally. Designers might, of course, draw upon any similar development guidance in simulation design, but VE design has sufficiently different attributes and purpose for its own structured development framework and associated guidelines to be needed.

This chapter starts by describing such a framework—VEDS (the Virtual Environment Development Structure)—and subsequently expands in some detail upon its major stages of analysis, specification and storyboarding, building VEs, enhancing user performance, presence and interactivity, usability, and evaluation. Within the chapter the focus is on the third type of design as defined by Bolas (1994), namely design *of* VEs, in contrast to design *with* VEs (e.g., solving design problems) or design *for* VEs (VE system improvements); "design of virtual environments is perhaps the most exciting and challenging VR design task" (Bolas, 1994, p. 53).

The chapter is largely focused on VEs that model and give experience of interacting with a device, or that allow exploration of a space such as a factory or dwelling. Therefore the sorts of VE used for stress analysis or as a virtual wind tunnel, for instance, are not covered in terms of the engineering models on which they build. Nor are very large geographical representations addressed, for instance, as used in landscaping projects, harbor navigation, or city planning. Furthermore, support of groups of participants in collaborative virtual environments (CVEs) is not explicitly addressed. Many of these issues are taken up elsewhere throughout this volume (see chaps. 12, 14, 25, 34, and 41).

## 1.2 Barriers to VE Applications

In a United Kingdom (U.K.) research council–funded program of work from 1994 to 1995, examining potential industrial application for VEs, we collaborated directly with about a dozen U.K. companies and surveyed, in varying degrees of depth, over 250 more. Despite considerable enthusiasm for the potential of the technology (in planning, design, training, and layouts), internal "marketing" demonstrations, and exemplar VEs built by outside research or consultant teams, were the only industrial VE applications found (Wilson et al., 1996). This picture is confirmed by the authors in repeat exercises carried out from 1997 to 2000, and, despite some optimism from the German motor industry, in other sources (e.g. DTI, 1996; Financial Times, 1994; SUSECA, 1998). Across the world, there are few if any working applications of VEs built that have replaced another technology or are doing something new.

From work with potential users, four barriers have been identified from the state of VE technology, which meant that most companies were not then (1994–95) expecting take-up of the technology for at least another three to five years. The barriers were defined as: "technical" (particularly integration with other technologies and the implicit technical trade-offs in data processing); "usability" (the nonintuitive nature of the interfaces, both hardware and software, see chap. 34, this volume, and concerns about possible side and aftereffects, particularly with head-mounted displays [HMDs], see chaps. 30–32, this volume); "applications" (the paucity of serious applications working for real, and available for inspection); and "evaluation" (the poor evidence of added value from VE application; for some notable exceptions see chaps. 42–55, this volume).

Although the barriers were, and still are, real ones, they are not insurmountable, given likely VE technical developments over the next few years and assuming careful selection of appropriate applications. Potential industrial users raise the difficulty of effective data input and output and systems integration more frequently than any other, so this may be the key to mass industrial use. It is currently the focus of urgent development. The paucity of salient working demonstrations of VEs and poor evidence of added value from good quality evaluations was summed up by one manufacturing engineer attending one of a number of hands-on demonstration workshops we have run for industrialists (see Wilson, Cobb, D'Cruz, & Eastgate, 1996, p. 119): "I can't even begin to discuss use of VR sensibly with them [company engineers and decision makers] unless they sit with it, try it, use some worlds they can relate to." The needs for evaluated demonstrations of applications are related to the need for a structured methodology to guide application identification, building, and evaluation against operational criteria. Some means of assessing the usability of VE technology is required, embracing utility, usability, likeability, and cost, and addressing system integratability (with other technologies and company functions) or organizational usability. The last criterion, organizational usability, relates to the integration of VE with other technologies and to how VE changes existing company strategies and groupings.

To the four original limitations or barriers one can add a fifth, resulting from and leading to some of the others: the lack of a structured methodology by which organizations might establish a potential need for a VE solution, and then specify, develop, employ, and evaluate VEs in real applications. All five barriers are being addressed in the current large European Union Funded research project, "Virtual and Interactive Environments for Workplaces of the Future" (VIEW-project no IST-2000-26089).

## 2. VEDS—VIRTUAL ENVIRONMENT DEVELOPMENT STRUCTURE

Development of any product or system, including setting appropriate requirements and specifications, will be aided by guidance from a clear, flexible framework that defines what issues are relevant, what data and information are needed, what decisions must be made at what stages, and how these stages interconnect, including feedback loops and iterations. The human factors/ergonomics community is clear on the need for a process or framework to be established. Only in this way can early design decisions be taken to improve the usability, likeability, and general acceptability of a product or system in enough detail in order to influence the final outcomes. Virtual environments are no different than any other product, system, or human–computer interface, in that their development needs clear goal setting and constraints,

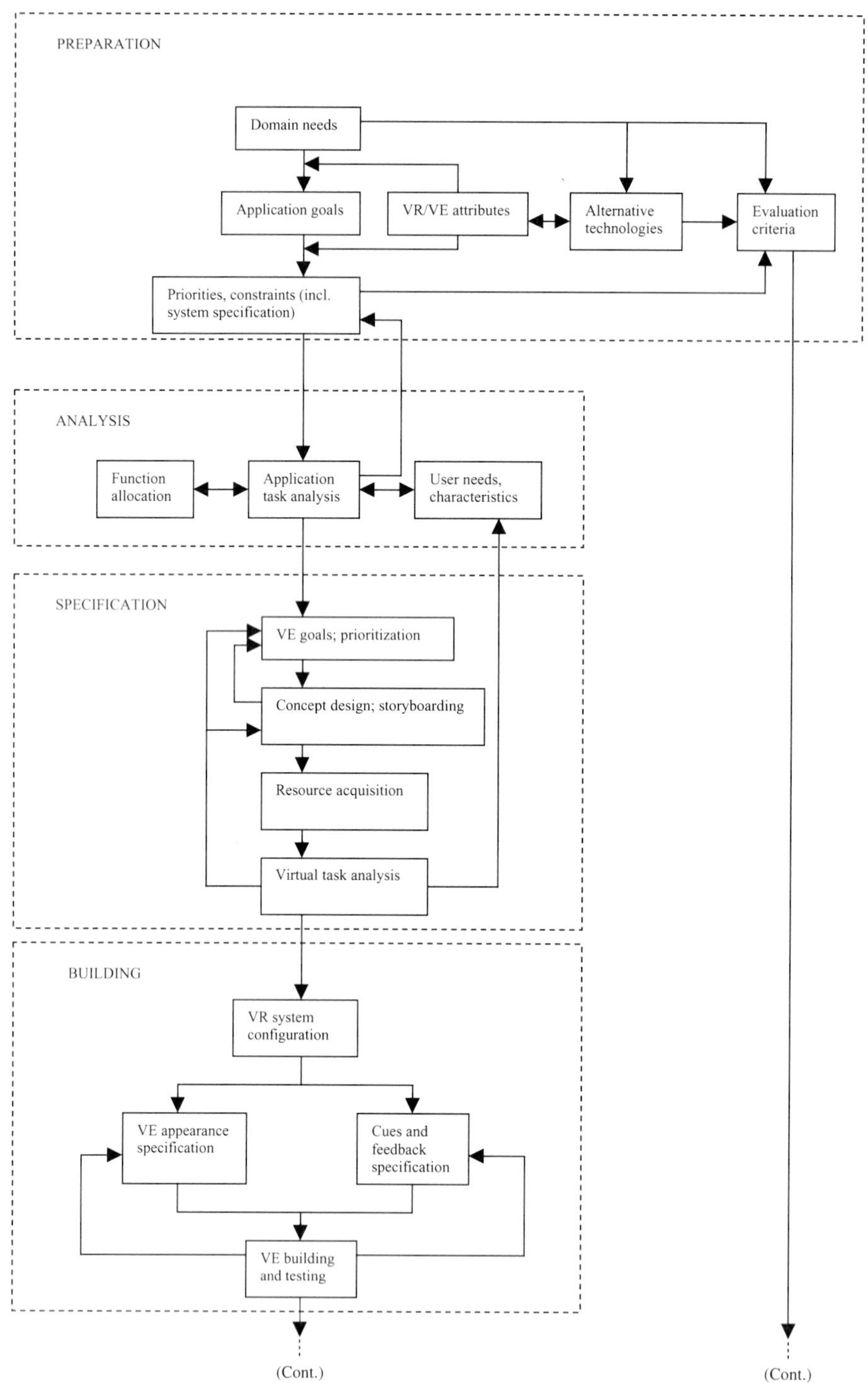


FIG. 17.1. VEDS: Virtual Environments Development Structure (adapted from Wilson et al., 1996, and Wilson, 1997). Note: Only major feedback loops are shown, for clarity; in fact, the whole process is an iterative one.

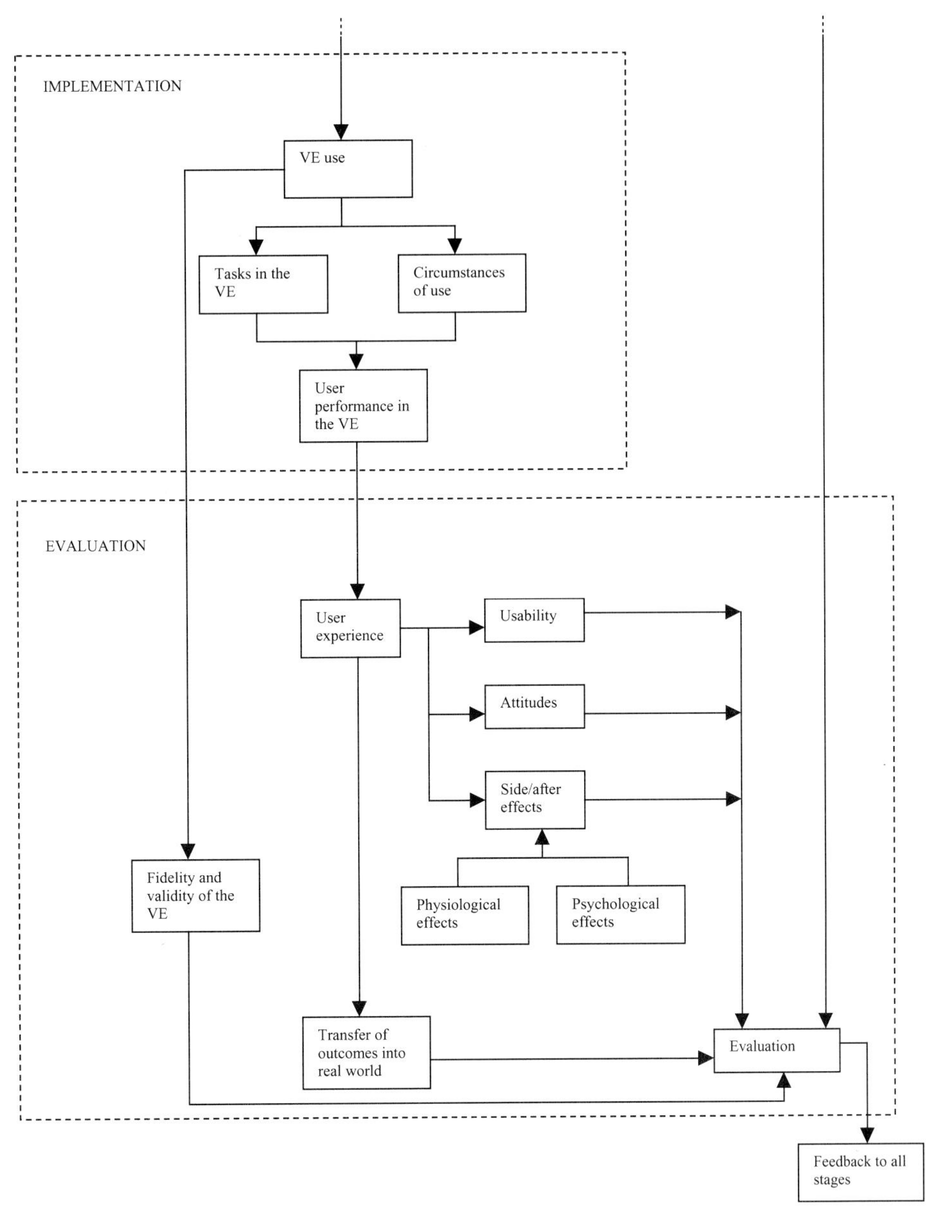


FIG. 17.1. (Continued)

requirements analysis, task and user analyses or models, appropriate interface guidelines, predictions of task performance, an iterative design/test cycle, and a clear evaluation process.

Over a number of years we have worked with potential users of virtual environments in industry, medicine, and education to produce a structured framework to guide VE development—the Virtual Environment Development Structure (VEDS). Various generations of this framework can be found in Wilson et al. (1996), Wilson (1997), and Wilson (1999). The current version is shown in Fig. 17.1. Before discussing VEDS, two points should be made. First, as drawn it appears as if only a top-down approach is suggested. Although a structured development process is perhaps best suited to top-down working, it is recognized that certain aspects of development will need to take place bottom-up. Bryson (1995) terms these as *design for the task and interface metaphor* (top-down) and *design for performance* (bottom-up). Second, only the main feedback loops are shown, but this is for clarity in the diagram. The whole process is iterative, embracing formative and summative evaluation and a design morphology as the VE (and test methods) becomes more sophisticated through design cycles within the process.

The key stages of VEDS are described in the rest of this chapter, but the whole process is summarized here. At the outset, a clear view of the attributes of VR/VE (what they are ready for and what not) and knowledge of the domain are needed to obtain agreement on goals for the application and to identify likely priorities and constraints. Considerable effort in task and user analysis and in function allocation (for instance, when it is preferable for the VE or intelligent agents within it to control a participant's options or actions) will pay off later in improved concept, prototype, and final design. Technical and systems limitations can be reduced in their effects on the application by appropriate choice of technology in the light of a task analysis, and by goal-oriented- and user-need-based world building. Concept design must be rational, realistic, and parsimonious, rejecting any elements not important to the core application goals and to support user performance. World building will involve choices of whether to provide the participant with extra functionality or higher levels of complexity and appearance of the environment. The number of objects modeled, and their level of detail and degree of associated intelligence, must also be considered carefully before including them within the virtual environment. As with any product or system, the participants' performance within the VE will be determined by the environment itself, personal characteristics, tasks to be performed in or with the VE, and the circumstances under which those tasks are carried out (for instance, without training, under time pressures, or with incentives). A virtual task analysis, which may build on the real-world task analysis carried out earlier, can support decisions within VE building. In addition to participants' performance within the VE and behavior with the VE system interface, subsequent outcomes might be manifest in the transfer of the participants' effects upon the real world (for instance, measured as design process effectiveness, better product designs, or training program efficiency), their reported and observable experiences, and any health-and-safety-related side effects. At the outset, when defining the application and activities, alternative solutions that do not employ VEs should be identified, and evaluation methods, measures, and criteria set. Although the evaluation may be somewhat crude at times, it must be carried out to assess the success of the VE in meeting its goals, examining both individual and organizational effects against the evaluation criteria.

In the rest of this chapter the different stages of VEDS are discussed, with particular emphasis on structured analysis, specification and building.

## 3. TASK AND USER ANALYSIS

The early process of VE development to meet the needs of users should differ little from that normally followed in ergonomics, namely to: (1) assess what tasks must be completed in the VE and in the real world subsequently; (2) determine what the user characteristics and needs are; and (3) produce a task analysis (or often synthesis, because these frequently will be new endeavors). Such a process should allow the client and the developer of the virtual world, together with the ergonomist or psychologist, to specify the goals to be achieved by the virtual environment. Preferably these goals should be prioritized, because only rarely can all be attained.

The classic starting point for application of ergonomics in systems design, following specification of goals and constraints, is the task and user analysis (Kirwan & Ainsworth, 1992). In developing VEs, if the intention really is to enable cognitive (if not always physical) replication of interaction with the real world, then the task analysis will be carried out twice, once for the actual tasks being modeled—the *application task analysis*—and secondly for the tasks to be carried out within the VE—the *virtual task analysis*. The first of these is axiomatic for human–machine systems development to understand what tasks must be carried out within the domain where the system will be used and the requirements or constraints this will impose upon the people involved. Such early task and user analysis will then enable decisions to be made about:

- Whether VE technology even has a role for the particular application and what alternative technologies there are
- Which functions within the application will be supported by using VE technology (in many applications a mixed setup, comprising virtual and hard elements, may be appropriate)
- How VE technology will be integrated with other technologies if there is a mixed environment
- How eventual use of the VE will be evaluated in terms of direct and indirect outcomes and effects
- Which VE system (hardware and software) is most appropriate for the tasks, task environment, and user group
- What elements of the real world and tasks should be included in early specification of the VE design

Some of the development decisions that might emanate from the task and user analysis are shown in Fig. 17.2 (see also Gabbard & Hix, 1997; chap. 34, this volume). Some of the decisions implied in Fig. 17.2 may not be possible at this early stage and their resolution may depend upon the virtual task analysis that is carried out later.

There are two distinct fundamental approaches to account for the user in VE development that relate to the philosophy of VE itself. These are "*VE as an alternative real world*" and

Degree of 'reality' desirable: real world replication, highly abstract

Type of presence: egocentric, egocentric

Participants: one, few, many

Temporal collaboration: none, synchronous, asynchronous

View of scale: close-up, arm's length, room sized, far horizon

Vision: monocular, bi-ocular, stereoscopic

Manipulation: none, one-hand, either hand, both hands, whole body

Virtual tool provision: none, separate palette, within the VE

Resolution: coarse, fine

Mobility: stationary, ground anchored, flying

Control: passive, active presentation of scene

World state: transient, persistent

Sensory cues: visual, auditory, haptic, olfactory

FIG. 17.2. Development decisions made on the basis of application task analysis and virtual task analysis.

"*VE as computer interface*." Virtual environments can be treated as alternative representations of "real worlds," and existing knowledge, models, criteria, and methods can be applied as appropriate. There are a number of problems with this approach. First, there is little agreement about the meaning of such concepts as spatial awareness and perception in the real world (e.g., Flach & Holden, 1998), let alone in a virtual one. Second, even if the software could provide a strong representation of equivalent perceptions in the real world, the peripherals that people need to experience the VE currently interfere with this parallel. Third, there is still considerable confusion about the underlying psychological mechanisms and processes involved in the use of virtual environments, and whether and how these might be similar to those in the real world. For instance, what is a sense of presence (see chap. 40, this volume), and is it equivalent to or related to attention (Zahorik & Jenison, 1998)?

The other approach is to treat the VE as simply another, if potentially richer, computer interface and apply existing HCI knowledge, models, criteria and methods. Again, there are a number of problems with this approach. The state of knowledge and coherence of the HCI field itself is open to criticism, and it is arguable that HCI as a discipline has already struggled to redefine itself, first with the almost universal move from command line to graphical user interfaces and then with the growth of various multimedia interfaces. Is HCI knowledge ready to be applied to virtual environments, especially those where a participant is navigating across the interface and within the VE itself? There are a large number of technical variables which are specific to VEs and not relevant to work elsewhere in HCI, including the impact of temporal lags or problems with the optics and field of view of HMDs (see chap. 3, this volume) and which leave developers with trade-offs between potential attributes of complexity, pictorial realism, and speed of update. Therefore, although VE developers and researchers might build on the 30 or more years research in HCI, any such models, criteria, or knowledge must be used with care. Their applicability needs testing by careful research.

The task analysis phase of development provides information, ideas, and a set of requirements on which to base conceptual design and VE specification and is covered in the next section.

## 4. SPECIFICATION: CONCEPT DESIGN, STORYBOARDING AND VIRTUAL TASK ANALYSIS

### 4.1 Basic Concept Design

The analysis phase will leave the development team in a position to prioritize the goals for the VE implied by the application goals. Some basic choices as to the structure and functionality of the VE might be made, in the light of the computer processing limitations for candidate VE systems and on the basis of the application task analysis and the virtual task analysis. (Note: Virtual task analysis is defined in Fig. 17.1 as occurring at the end of the VE specification phase. In fact, it can occur at any time during specification or even during the building phase, much as application task analysis can take place at any time in the early phase of development.) For instance, VE builders and their clients might choose to look for added value from interaction via "manipulation" of objects or via "exploration" around the virtual environment. Development of a concept design and then VE building will need to be robust enough to deal with both VEs built to provide the participant with the ability to manipulate (e.g., switch on, switch off, open, turn, etc.) elements or objects in the VE, and also those to walk through or around a large, rich, complex, but essentially unchanging virtual environment (see Fig. 17.3).

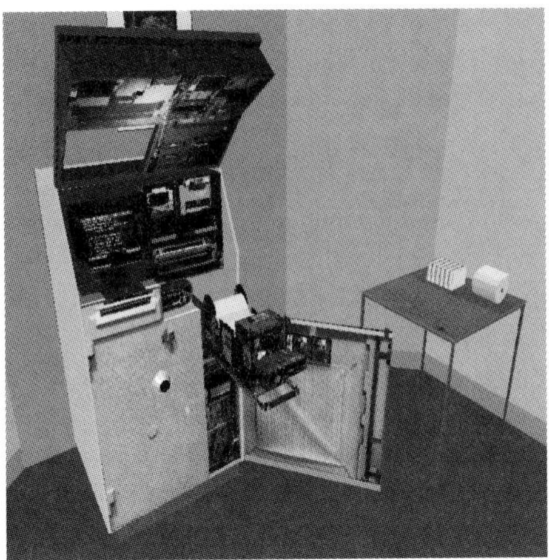

FIG. 17.3. An example of a VE that gives the participant the ability to manipulate elements within the VE as well as navigate through them. This shows a virtual ATM, or cash point, that can be "racked out" to practice day-to-day replenishment procedures by selecting buttons, catches and switches.

Out of the analyses, the target VE will be specified in terms of its goals, the expected user tasks, complexity, and the balance between interactivity and exploration afforded. This specification must then be agreed by the VE development team and client, the most common means being storyboard development.

## 4.2 Storyboarding

A storyboard is a set of frames, each using pictures and words to describe a stage in the experience offered by a virtual environment. It can differ from a storyboard for a film in that a VE may not be a straightforward narrative sequence, so the links and relationships between frames may be more complex. On the other hand, it may not be necessary to show the links between the frames if, for example, the sequence of events does not affect the outcome. In that case the frames will contain descriptions of interactive elements within that part of the virtual environment.

Virtual environments are often designed by more than one person, frequently by a group whose members work for different organizations and come from different backgrounds. Some of these people will have a good understanding of the possibilities offered by a virtual environment. Others will be experts in their own field and have a good idea of what they want the VE to do, but little idea of what is feasible. A storyboard is a method by which the different parties involved can describe, design, outline, and agree on the form the VE should take. The developer can use the storyboarding process to work out how to match the requirements of the client to the limiting factors of the VE system. The storyboard can then be used by the developer to build the VE, ensuring during the process that the design closely matches the requirements of the client.

It is important to get all stakeholders involved at the storyboarding stage. Experience has shown that if this is not done, a VE may be developed which does not meet the requirements of the one person who was not consulted, resulting in expensive and time-consuming re-development. Generally, the stakeholders in the storyboarding process are: the project manager (to oversee the process and keep it within budget); the VE developer(s) (to establish exactly what the client wants and to explain what is technically feasible and how long it will take); the client (the person or people who specified the VE to make sure the VE will fulfill the specified requirements); representatives of the user population or participants (to make sure the VE will be suited to their characteristics and needs); and a usability or HCI expert—who is possibly one of the developers (to guide discussion from a user perspective, instantiate this into the design, and incorporate best practice in interface design). It is often the case that one person may fill more than one of these roles, in which case the size of the storyboarding group will be reduced. Typically between two and five people can develop a storyboard. A storyboard can be graphical or textual (see Figs. 17.4 and 17.5), and its content will depend on the type of VE and size of project. A set of potential contents is shown in Fig. 17.6.

Once developers start building the VE they will be confronted with many unexpected and previously unforeseen problems at every stage. It may not be possible for developers to make some decisions, and they will have to be referred back to the relevant person in the storyboarding group, which can delay the development process. It is therefore important to establish as much detail as possible at the group storyboarding stage and to use the conceptualization that storyboarding allows to obtain consensus. Developers have to drive this process by asking questions about the design of the virtual environments.

## 4.3 Acquisition of Resources

Information must be acquired about objects to be modeled and their setting. As well as taking notes and making sketches, many developments will require object attributes that can be included from resources in the form of photographs, video, and sound recordings. At an early stage it may not be possible to determine exactly the way in which any particular object is going to be modeled, for instance, if it will be texture- or geometry-based. It is therefore important to have some guidelines to follow that will ensure that, regardless of the method employed, the data will be available. Table 17.1 shows how the nature of the geometry, color, and sounds of an object can indicate what data should be collected in order to model that object. Shortcuts can be taken in resource acquisition. One obvious source is computer-aided design (CAD) and other solid model (three-dimensional [3-D]) data. Translators may be required here and some precision in geometry may be lost, but this can allow much more rapid development. Other sources may be image and mapping agencies or the Internet for models or animations.

**TABLE 17.1**
Types of Information to Be Collected

| | *Simple* | *Complex* | *Dynamic* |
|---|---|---|---|
| Geometry | Measure dimensions (e.g., cube) | Measure dimensions and take photos (e.g., tree) | Measure dimensions, photograph and video (e.g., bicycle) |
| Color | Note color (e.g., painted wall) | Photograph orthogonally* (e.g., brick walls) | Video or photograph sequence (e.g., TV screen) |
| Sound | Record sound | Record sounds | Record sound and video or make notes on relationship between sound and object behavior (e.g., car accelerating) |

*If orthogonal photos cannot be obtained (e.g., due to obstructions or limited access), take the photos as square on to the surface as possible and then use a graphics package to "straighten" the image.

Screen layout
    Resolution 1024 x 768

VE window

- How big? 800 x 600
- Where? Top (center?)

Icons

- Company logos? VIRART, XXXX, Uni. of Nott
- Any buttons? Reset View (not world), Automated walk through
- Where? Lower screen

Help boxes
    Temporary over VE window (possibly only visible whilst right mouse button is held down?)

Instructions
    Automated walkthrough (in the form of an AVI?) shows users what to do

Initial scenario

Room layout
    Doors
- Door to XXX room

    Furniture
- Shelf
- Mat for HG testing
- Panel for WB testing
- Pictures on walls (XXXX, no smoking, ESD signs etc)

Clothing / equipment
- Wrist band (WB)

    On shelf
- Heel grounder (HG)

    On shelf
- Overalls

Actions

- No entry to XXX room without tested HGs and WBs and wearing overalls
- If attempt is made without them message tells user how to get them
- Click on HGs and WBs to select
- Feedback? Icon changes from bare wrist to wrist with WB

XXX room scenario
On entering the room a message appears;
There are ten ESD hazards in this room. Use the ESD meter to find them. Click on the ESD meter with the left mouse button to pick it up. Use the left mouse button to click on items that you think are an ESD hazard.

Room layout
    As built so far from electronic image provided by ****

    Door
- One entrance door

Furniture

- Benches
- Stuff on benches, e.g. soldering iron
- Shelves
- Swivel chairs
- Trolley

Hazards

- Plastic Cup
- Likely ESD voltage level (for ESD meter) ?
- Jacket on back of chair
- Likely ESD voltage level?

FIG. 17.4. Extract from a text-based storyboard.

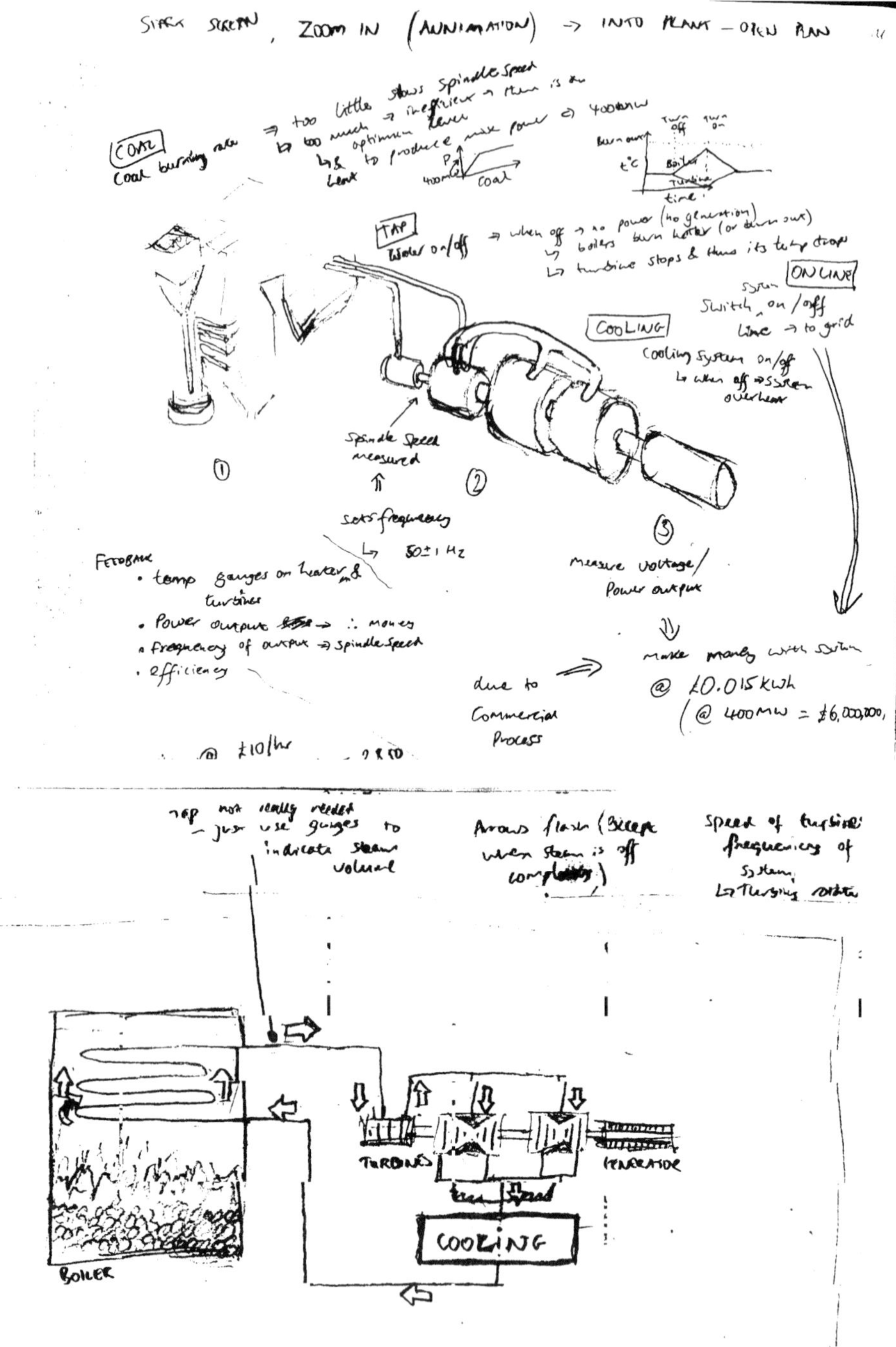


FIG. 17.5. An example of a picture-based storyboard (for an electricity generating plant demonstration).

## 4.4 Virtual Task Analysis

During or after specification of the VE there is the second application of task analysis: the analysis of the tasks to be performed by the participant within the virtual environment. This is a critical component of the VE development process for several reasons. First, there are trade-offs across the technical capabilities of the systems. For any particular level of system

| A wire-frame model or skeleton of the structure of the VE. |
|---|
| The initial scene presented when the participant first enters the environment. |
| The expected layout of the scene at subsequent points. |
| Any narration or instructions within the VE, whether aural or textual. |
| Any implicit or explicit signs or cues given to the participant. |
| Images to describe the activities in which the participant is expected to take part. |
| The broad class of control and input devices by which participants perform activities. |
| Images to show the consequences of various actions (both correct and incorrect). |
| The approximate sequence of events given an 'ideal' use (e.g. for training); where there is no sequence required (i.e. the user can perform the tasks in any order) this should be stated. |
| Any links with other software (e.g. multimedia clips). |
| A "formal end" to the virtual experience if this is needed (for instance when a VE lesson has been completed or a full escape route navigated). |

FIG. 17.6. Potential elements in a storyboard.

sophistication and cost, there must be prioritization among the variables of: VE complexity—and thus sensory richness; update rate—related to sense of presence and any disturbing effects; and interactivity—numbers of objects that can be "manipulated" in real time and how this is to be done. Analysis of the tasks that the participant must perform within the VE and any consequences and support needs can assist this prioritization. Second, a critical part of VE specification, assisted by the task analysis, is to define the minimum set of objects capable of interactivity and how cues to this interactivity might be provided (Eastgate et al., 1997, and later in this chapter). Third, by extension, there is the production (through prediction and observation) of task analyses for collaboration and task performance within collaborative virtual environments (Tromp et al., 1998a, p. 61). Finally, task analysis of what must be done within the VE to achieve set goals can be used to define behaviors which are expected (or not) to underpin successful use of the VE—for instance, in achieving learning objectives (e.g., Neale et al., 1999)—and so give a basis to evaluation of participation within a virtual environment.

Many of the decisions, which could have been made on the basis of the application task analysis (see Fig. 17.2) may, in fact, be made after the virtual task analysis or be revisited at this stage in the process. Some assistance for the process of virtual task analysis may be obtained by use of taxonomies (e.g. Gabbard & Hix, 1997) or else cognitive models. An example of the latter is the Resources Model reported by Smith, Duke, & Wrigth (1999). Kaur et al.'s (1999) model of interaction might also help with virtual task analysis.

## 5. APPROACHES TO BUILDING VIRTUAL ENVIRONMENTS

There are a number of factors that influence the way a VE is developed and, consequently, how it eventually performs. Some of these factors have a stronger influence than others, but all are interlinked in some way. These factors include: the purpose of the application; the target user population; the developer's ability and experience; the involvement (or not) of end users; the client–developer relationship; the VE system (hardware/software) available; the time available for development; and financing available. Also influential will be the extent of the developer's understanding of human factors and human performance in VEs and of how to enhance usability, which requires much better fundamental knowledge and practical guidance than is available at present. The understanding of performance in VEs will be addressed in the next main section. First, the act of building VEs will be examined in terms of strategies and techniques available to developers.

## 5.1 Basic Choices

After VE specification is complete, construction of the VE involves a series of hard choices over what can be done to meet the user and application needs. A fundamental choice facing the world developer is where to concentrate building effort (in time and creativity) and computer processing resources. Crudely, VEs may be designed to maximize the participant's ability to interact with them through exploration (i.e., visualization and walk-throughs), their interaction through manipulation (i.e., alterations in the state of objects), or to compromise by trying to meet needs for both. Technical limitations of VE technology, present in differing degrees in all systems from the simplest desktop to a Cave Automatic Virtual Environment (CAVE), usually result in the world builder having to make a number of trade-offs. The aim is to render recognizable 3-D representations in real time, updating synchronously with a participant's control inputs or movements in the virtual environment. The trade-offs will often involve choosing between providing the user with functionality, including a high degree and speed of interactivity, or producing a higher specification for the complexity and appearance of the environment itself. The total number of objects to be modeled and the level of detail and degree of associated intelligence for each must be considered carefully before including any within the virtual environment (see Eastgate & Wilson, 1994; Richard et al., 1996; Smets & Overbeeke, 1995). The application and virtual task analyses and prioritization exercises will be used for this. Some decisions over what to model, and how, can be made through modeling tools, for instance, extraction of geometry from photographs and drawings.

## 5.2 Programming Interactivity

One vital aspect of building VEs, and a major component of usability, is interactivity. Virtual environments have, of course, been typically defined in terms of possessing presence, autonomy, and interactivity (Zeltzer, 1992). Of these, interactivity is arguably the most critical attribute, as it clearly defines the difference between VE and other 3-D modeling systems. Here, the term *interactivity* is used broadly to mean any action on the part of the participant that results in a change in the virtual environment. An obvious case is selecting one object to initiate change in another object or in the scene, but interactivity can also include navigation where a user moves through a VE while it updates and changes according to one's movements and position. There are many computer applications that offer varying amounts of interactivity, but it is reasonable to suppose that VE technology can offer more representations of the types of interactivity commonly encountered in real life.

As with all facets of VE development, the technical limitations of any particular system and user interface mean that there will be limits on the amount and quality of interactivity available. In an ideal world the participant would feel that they are interacting directly with the virtual objects in the VE even while they are actually interacting with peripherals in the interface. However, due to the limitations of current VE technology, the user interface is more likely to form a barrier between the user and the VE, whatever the nature of the sensors and effectors and whichever of the user's senses are engaged. One of the jobs of the VE developer is to make this barrier seem minimal within the constraints of the application.

There are two extremes in VE interactivity. The first is to aim to accurately model as many of the characteristics of the real application as possible. The second approach is to deliberately create an approximation of reality, where the result is a crude representation of salient characteristics rather than an accurate simulation. The first approach may end up with a VE that is too complex for the processor to update as fast as required for exploration and manipulation interactivity, whereas the second approach may result in an impoverished or very simple VE, insufficient for the requirements of the application user. Therefore, a compromise between the two extremes of modeling interactivity will generally be used to achieve the best solution for any given application. The model will be built to mimic the behavior of the real world, behaving "correctly" in a variety of circumstances. Object-oriented design, maximizing object autonomy, will make the VE easier to adapt or expand in the future if the need arises. Where accurate models of the structure or dynamics of an object or process cannot be justified on the grounds of programming time or processing overhead, they can be replaced by, for example, a texture, an animation, or a model based on a simplified algorithm.

### 5.3 Choices Over Object "Intelligence"

One way in which VEs may not match the real world, at least in their underlying construction, is in the selection of which objects have the capability to be "intelligent," to change configuration or else determine the form of the total virtual environment. For example, materials or objects being worked on or manipulated during a manufacturing process are generally "dumb" in that they have no influence over the process. It is the manufacturing machines or systems that can influence the position, orientation, deformation, and joining of the materials or objects. Although veridical representation means we should attach code to give intelligence to the machines, it is sometimes simpler to attach intelligence to the objects being manipulated without affecting the visual representation of the process. In this way, the real world can be represented in the most efficient way possible, rather than exactly simulating it. For example, where a robot has to move components from a conveyor onto an assembly, in principle it is one "intelligent" object working with an unknown number of "dumb" objects of known type. It is often simpler to create one component which has a program associated with it, such that it can react correctly to the robot, and then to duplicate that component (and associated program) as necessary, than it is to program the robot to cope correctly with an unknown number of components. This effectively takes the "intelligence" from the robot and assigns it to the component.

This approach, however, has disadvantages. If an object has been programmed to sense when a robot is picking it up and to behave accordingly, it may be given code only to be able to recognize one robot or type of robot. Also the robot will be incapable of manipulating other objects for which the user would expect it to be suitable but which do not have the required program associated with them.

### 5.4 Implementation of Physical Properties

Virtual environment software usually gives developers limited ability to assign some physical properties to virtual objects, particularly dynamic characteristics approximating those expected in the real world. One factor limiting the "realism" from doing this is a lack of temporal consistency in some 3-D visualization systems. Any factor that decreases the rendering speed, such as visual complexity, can result in a slowing down of the dynamic properties of moving objects. To get around this, it is usually possible to link the dynamic properties to the processor's real-time clock. However, it is not possible to increase the rendering speed using this method; instead, the step size of the movement of the objects is altered, so if the rendering speed becomes very slow a fast-moving object may appear to jump across the screen. This can become particularly pronounced in collaborative virtual environments, where the collaboration is taking place over a network with its inherent potential for lag. Local activity will render smoothly, but remote activity will jump about as packets of information arrive over the network. Thus, to achieve smooth but temporally accurate dynamics, the developer has to be aware of the potential bottlenecks in the processing and rendering of the virtual world on particular VE systems.

Some VE development systems provide facilities to give objects physical properties, for example, gravity, coefficient of friction, coefficient of restitution, and velocity, both linear and angular. These are not generally designed to simulate the laws of physics, but rather to give VE developers shortcuts to making objects appear to obey the laws of physics. Coefficients of friction and restitution are determined by the properties of both of the materials which are coming into contact and cannot be allocated to an object in isolation. One physical property which is not generally available is center of gravity. Without this property is not possible to make an object topple realistically.

Dynamic characteristics not provided by a VE development system can sometimes be programmed by developers using a relevant programming or scripting language. By assigning a piece of program code, any virtual object can be programmed to behave in a particular way independent of other virtual objects in the same environment. Alternatively, objects can have their separate codes linked (e.g., via common variables) such that one object's behavior depends upon another's.

Any use of simplified physical constraints can make an interaction more realistic without sacrificing real-time updating and can provide a more intuitive interface (Papper & Gigante, 1993).

## 5.5 Overcoming Limiting Factors in VE Development

The combination of VE complexity and finite system processing power will limit the rendering speed of a VE presentation. It is important that rendering speeds are sufficient to avoid causing usability, likeability, and utility problems. When faced with restrictions, there are various possibilities available to the developer. More than one of these techniques would normally be used in combination. The facilities offered by the VE system software, and the developer's ability and preferences are most likely to influence the choice (see Grinstein & Southard, 1995).

### *5.5.1 Texturing*

A large amount of 3-D detail can be replaced by a bitmap texture pasted onto a single facet in a virtual environment. Depending on the subject matter textures can increase rendering speeds and will generally be quicker to create than 3-D geometry. Whereas textures are a good way of adding realism to a VE, they can only be used in certain circumstances. A new development, which offers a compromise between the increased speed from textures and the interactivity available with modeled objects, are algorithms that automatically replace textures with modeled objects according to the participant's apparent distance away from an object (e.g., Rosenblum, Burdea, & Tachi, 1998).

### *5.5.2 Level-of-Detail Management*

Level-of-detail (LOD) management, or distancing, is a widely used technique, whereby if a virtual object is more than a specified distance from the viewpoint of the participant it is replaced by a simpler model or else disappears altogether, thus reducing the amount of processing time required to render that object. Extensions of this technique can be found in layered depth images, and, for one specific problem, use of portal culling and portal textures (e.g., Rafferty, Aliaga, Popescu, & Lastra, 1998).

#### 5.5.3 Reducing the Screen Resolution

Reducing the resolution of a display is an indiscriminate solution that results in a reduction in the amount of detail visible to the participant. This will make some tasks in the VE more difficult (e.g., perceiving details) and will also reduce the photorealism of the VE, possibly making the experience less meaningful and enjoyable. It is possible to reduce the screen resolution selectively around the periphery of the screen while maintaining a high resolution in the center (Watson, Walker, & Hodges, 1995).

#### 5.5.4 Selectively Including Objects

Reducing the number of objects in an environment will increase rendering speed and decrease development time. The decisions as to which objects to include should be based on the task analyses and rational thought about which objects are really needed, but tend also to be made on the basis of the size of objects, ease of programming, and individual programmers' preferences. In the worst case such decisions are made arbitrarily, with little understanding of how the choice will affect participants.

A technique that can both decrease development time and increase rendering speeds is to selectively include objects. It is applicable across all hardware and software platforms and will be compatible with future VE systems. The resulting VE should not be significantly inferior (at achieving its specified purpose) to a hypothetical one in which all objects are modeled. Indeed, it may be found that omitting some objects may increase the effectiveness of a virtual environment. A parallel is the use of a stylized graphic drawing in a book of instructions rather than a photograph, to avoid confusion through unnecessary detail. What is required is a set of guidelines to help developers through the process of deciding which objects to include and which to omit. This guidance will have to be developed with reference to the specified purpose of a VE and user behavior within the virtual environment.

## 6. UNDERSTANDING HUMAN PERFORMANCE IN VEs

The goal of VE world developers is to create effective, safe, usable virtual environments that enable participants to achieve goals in a better, more motivational, and cost-effective fashion than by using alternative technologies or by using the real situation. Producing VEs should thus ideally be based on an understanding of how people will behave and carry out tasks within the virtual environments. Developers need general guidance on human performance within VEs and specific guidance on enhancing VE usability (see chap. 34, this volume). Clarity about what determines usability of VEs will be enhanced with better understanding of participant performance.

Task and user analyses will allow some identification and categorization of participants' requirements, and storyboarding will outline how the VE should be specified and designed to meet those requirements. A more detailed examination should be made of how participants will respond to different elements of the VE, utilize all its functionality, and be able to comprehend and use interface elements to meet application goals. As part of this, developers need to know how to encourage participants to explore the VE and enable them to understand which elements can be interacted with, identify how this interaction might be achieved, and minimize dysfunctional participant behavior and serious errors. In other words, an improved understanding of human behavior and performance within a VE is needed, but such understanding is in its infancy.

There are a number of candidate theories and approaches that might help. For instance, behavior of groups of workers in complex technical and organizational environments can be interpreted within a framework of distributed cognition (e.g., Hutchins, 1995). In such a view, the handling of information and decision making that characterize work in transport control, for instance, is distributed in time and space across teams of people and various computer and telecommunication systems. An ethnographic approach may then be taken to study and interpret work behavior and skills. This may be a framework within which to study performance in networked VEs or CVEs.

Another possibility is to build understanding of participation in VEs, and thus guidance for their design, around the notion of situated action (Suchman, 1987). In this view, work cannot be understood merely in terms of individual skills and institutional norms, but is the product of continually changing people, technology, information and space. Further, people do not approach tasks with a clear set of plans in mind, but build and modify their plans in the course of situated action.

A third possibility for understanding performance in VEs is to utilize the notion of situation awareness. People perceive cues from their environment and use them to make sense of the current state of the world and to project this understanding into the future (Endsley, 1995). Various measurement methods are available to assess situation awareness that might have value in understanding and measuring performance in virtual environments. Also, there is a strong connection between the role played by situation awareness and the participant's mental model. Guidance on design usability then, would concentrate on those VE elements expected to help build a strong and "correct" mental model for the participant or to match their existing ones, and to contribute to high levels of situation awareness.

Finally, developers could borrow from a large variety of models and frameworks used to understand, predict, and reduce human error, generally in safety critical systems. A particular possibility here is Rasmusson's (1986) well-known SRK model of skill-, rule-, and knowledge-based behavior. If the types of tasks in the VE can be defined as to whether they are skill-, rule-, or knowledge-based, one might be able to distinguish between the types of design guidance offered for each. More usefully, characterization of real-world tasks as skill-, rule-, or knowledge-based, and particular variants of these, could allow association of different ways to deliver their representations in the virtual environment. The SRK model is often used in conjunction with Reason's (1990) typology of human error—slips, mistakes, and violations—and the underlying psychological error mechanisms involved. Again, adaptation of these to the VE domain could allow a better understanding of the types of behavior exhibited by VE participants and, subsequently, lead to production of useful development guidance.

## 7. USABILITY OF VEs

The notion of usability of VEs and usability engineering is explored in depth elsewhere in this volume (see chap. 34). Included here are only the major issues of usability that must be addressed in the development process.

A NASA-sponsored workshop on VE research requirements started with a brief that the most important issues to discuss were aftereffects and sense of presence. Twenty-five specialists, from Germany, the United States, and United Kingdom, prioritized problems and research needs and established interconnections between issues. Many concerns were with systems design, for instance, display design, latency, and real-time interaction. Beyond these, the workshop participants agreed with the importance of minimizing aftereffects, the need for better understanding of participants' sense of presence, and on the need for a substantial human-factors research agenda to address VE usability concerns (Stanney et al., 1998).

Defining usability of any product is not easy, particularly if this definition must be translated into appropriate design parameters and evaluation metrics. Also, most accepted definitions of usability refer to it as being related to "specified users" with "defined goals" in a "specified context of use," rather than as a generic context-free attribute. Furthermore, in HCI for instance, attributes of usability—leading to measurement methods, measures, and criteria—might include effectiveness, efficiency, usefulness, learnability, long-term performance, memorizability, flexibility, feature usage, acceptability, likeability, and satisfaction, meaning that usability is a multidimensional concept. An understanding of how people operate in VEs, and of what concepts such as presence (see chap. 40, this volume) or navigation (see chap. 24, this volume) actually mean in a synthetic environment is not yet advanced enough to operationally define such usability attributes in order to generate metrics and explicit design guidelines. Some exceptions in part include contributions from Gabbard and Hix (1997), Nichols, Haldane, and Wilson (2000), Pausch, Proffitt, and Williams (1997), Singer and Witmer (1999), Vinson (1999), Watson, Walker, Ribarsley, and Spaulding (1998) and Welch (1999), from all of which some metrics and some guidelines might be extracted. Our knowledge at this time would also allow some usability heuristics and structures for cognitive walk-through assessments to be established and run, which in turn might be translated into a first set of guidelines.

One of the barriers to the proliferation of VE technology identified earlier was widespread concern over the usability of both the VE equipment and of the VE itself. The equipment concerns will limit the freedom of systems developers. They manifest themselves in terms of the design (see chap. 25, this volume), comfort and fit of HMDs and handheld devices (see chap. 41, this volume), temporal and spatial resolution limitations, field of view restrictions (see chap. 3, this volume), and generally the limitations implicit in communication using visual and auditory information only. Some of these systems concerns translate into potential problems of usability within the VE itself. Others might be minimized by careful design of the virtual environment.

There are a large number of usability issues related to successful implementation and use of VE technology (see Barfield & Furness, 1995; Draper, Kaber, & Usher, 1998; Wann & MonWilliams, 1996; Wilson, 1997, 1999). Few have been satisfactorily addressed as yet, although the work of Gabbard and Hix (see chap. 34, this volume), Kaur (1998), and Tromp (2000) is helping to redress this. For each usability issue one must ask: Is it a significant issue for VE development and for its safe and effective use? Are the usability issues unique? Can we utilize data and criteria from other domains and from knowledge of performance of related tasks? Broad usability issues to be considered during VE development include:

- Forms of representation of the participant within the VE: none, limbs, full block, full lifelike
- Modeling of avatars and mannequins: sizes, shapes, appearance, movements, facial expressions (see chap. 15, this volume)
- Supporting navigation and orientation within VEs: interface tools and other aids, shortcuts, familiarization routines, optimum world sizes (see chap. 24, this volume)
- Understanding and enhancing presence and involvement in VEs: balancing pictorial realism, size and complexity of the VE, update rules; use of sound and shadowing; enhancing interest (see chap. 40, this volume)
- Requirements for cues and feedback to assist the participant (see later discussion)
- Minimizing any side and aftereffects: sickness, performance decrement, and physiological change (see chaps. 29–32, this volume)
- Providing interface support and tools for interactivity: mixed-reality design, metaphors, interface elements, tool kits, and so forth

## 8. CUES AND FEEDBACK

One particular aspect of usability that is intimately linked to design choices made by developers, and to the notion of interactivity addressed earlier, is the provision of cues and feedback to the participant. When participants enter a VE it is important that they soon perform some activities that affect the VE to motivate them in further exploration. In order that they can successfully perform these activities, the VE should offer some cues that tell participants what types of activity are afforded. In this sense, VEs are no different to human–computer interfaces or to consumer products generally, in terms of the concept and importance of affordances (e.g., Norman, 1988). Cues to assist the participant could be in the form of navigational signs to aid their movement through the VE and to help them recognize their current location. Alternatively, cues could suggest that a certain type of control action would afford a particular response. The participant will benefit from knowing which objects afford interaction, whether they are available for interaction at that moment, what method of interaction is appropriate, and what effect interaction with that object might have on the virtual environment. Then, having interacted with a chosen object, they will want to know whether that interaction was successful. Tables 17.2 to 17.4 illustrate simple "look up table" guidance being produced by our group at VIRART to support world developers in improving the interactivity and feedback in their virtual environments. Many of the potential problems can exist also when interacting with objects in the real world, but some, especially those relating to explicit, formal (or "designed in") feedback may be exacerbated in a badly designed virtual environment.

The design choice outlined earlier—of whether to design to model realism or some abstraction of this, or even to model a deliberate distortion of reality—is as valid for interactivity as for visual appearance. The points of interaction, the cues for these and the feedback given might all interfere with, or detract from, any visual fidelity in the scene and objects. Points of interaction may be of a size, shape, and color very different from their real-world counterparts; feedback and cues may need to be explicit in the form of text or audio prompts and messages. Such elements may strongly influence any usability/presence trade-off in the virtual environment.

**TABLE 17.2**
Possible Combinations of Cue for, and Potential for, Interaction

<table>
<tr><th></th><th>No Interaction Possible</th><th>Expected Interaction Not Available at This Time</th><th>Affords Expected Interaction</th></tr>
<tr><td>Little or no cue</td><td>User assumes correctly that the object is noninteractive.</td><td rowspan="3">User may not attempt interaction, however, if interaction is attempted, and with appropriate feedback, the user will realize the potential for interaction.</td><td>User may not attempt interaction.</td></tr>
<tr><td>Inappropriate cue</td><td>User may try in vain to interact with the object.</td><td>Unsuccessful interaction is likely.</td></tr>
<tr><td>Appropriate cue</td><td>User will correctly not try to interact with the object.</td><td>Successful interaction is likely.</td></tr>
</table>

**TABLE 17.3**
Possible Combinations of Potential for Interaction and Resulting Feedback

<table>
<tr><th></th><th>No Interaction Possible</th><th>Expected Interaction Not Available at This Time</th><th>Affords Interaction</th></tr>
<tr><td>No immediate feedback</td><td>User assumes correctly that the object is noninteractive.</td><td rowspan="2">User may not realize that interaction with this object can, but currently does not, have the desired effect.</td><td>User assumes wrongly that the object is noninteractive.</td></tr>
<tr><td>Inappropriate feedback</td><td>Not applicable.</td><td>User does not know what they have done.</td></tr>
<tr><td>Appropriate feedback</td><td>Not applicable.</td><td>User realizes that this action can, but currently does not, have the desired effect.</td><td>User knows what they have done.</td></tr>
</table>

An example of the use of decision tables within a real VE development project, provided as part of VE development guidance for models of an automatic teller machine, is shown in Table 17.5.

## 9. EVALUATION

For all VE applications, evaluations should be made of both the environments themselves and also their use and usefulness. Such evaluations can be divided into examinations of validity, outcomes, user experience, and process (see also Tromp, Istance, Hand, Steed, & Kaur, 1998b).

Developers should seek to address the validity of VEs in every project, at the very least by "walking through" them with the client or other experts. The potential usefulness of VEs

**TABLE 17.4**
Possible Combinations of Cue for, and Feedback from, Interaction

<table>
<tr><th></th><th>No Immediate Feedback</th><th>Inappropriate Feedback</th><th>Appropriate Feedback</th></tr>
<tr><td>Little or no cue</td><td>User assumes that the object is noninteractive.</td><td>User may not attempt interaction.</td><td>User may not attempt interaction.</td></tr>
<tr><td>Inappropriate cue</td><td rowspan="2">User may try to interact with object, but will not know whether that interaction was successful.</td><td>Unsuccessful interaction is likely.</td><td>Initially unsuccessful interaction is likely.</td></tr>
<tr><td>Appropriate cue</td><td>Successful interaction may be aborted.</td><td>Successful interaction is likely.</td></tr>
</table>

**TABLE 17.5**
Decision Table for Interaction Points—The Case of a Virtual ATM

| *Point of Interaction* | *Cue to Possible Interaction* | *Method of Interaction* | *Effect of Interaction* | *Main Feedback* | *Availability of Interaction When Object Is Visible* | *Feedback If Interaction Is Currently Unavailable* |
|---|---|---|---|---|---|---|
| All room doors | Implicit | Collision | Door opens | Visual | Always | N/A |
| Lock | Hidden | Mouse | Console becomes unlocked | Audible click | Sometimes | None |
| Console | Obscure | Mouse | Console opens if closed, closes if open | Visual | Sometimes | Textual prompt |
| Mode panel | Color coded | Mouse | Slides out if in, in if out | Visual | Always | N/A |
| Mode switch | Implicit | Mouse | Toggles display screen between normal and supervisor mode | Visual | Always | N/A |
| Outer door catch | Color coded | Mouse | Outer door opens | Visual | Sometimes | None |
| Outer door | Implicit | Mouse | Outer door closes | Visual | Sometimes | Textual prompt |
| Receipt printer release catch | Color coded | Mouse | Allows receipt printer module to be racked out | Small visual | Sometimes | None |
| Receipt printer handle | Color coded | Mouse | Racks out or replaces receipt printer module | Visual | Sometimes | Textual prompt |
| Receipt paper feed switch | Implicit | Mouse | Feeds paper through | Visual | Sometimes | None |
| Receipt paper spool spindle | Color coded | Mouse | Multiple effects depend on current state of system | Visual | Sometimes | Textual prompt |
| Receipt paper | Obscure | Mouse | Multiple effects depend on current state of system | Visual | Sometimes | Textual prompt |
| Safe door Combination lock | Implicit | Mouse | Unlocks safe door | None | Sometimes | None |
| Safe door handle | Implicit | Mouse | Opens safe door | Visual | Sometimes | None |
| Safe door | Implicit | Mouse | Closes safe door | Visual | Sometimes | Textual prompt |
| Currency cassette release catch | Color coded | Mouse | Allows currency cassette to be removed | None | Sometimes | None |
| Currency cassette handle | Color coded | Mouse | Removes or replaces currency cassette | Visual | Sometimes | Textual prompt |
| Currency cassette lid release catch | Color coded | Mouse | Opens or close currency cassette | Visual | Sometimes | Textual prompt |
| Currency holder | Color coded | Mouse | Fills or empties currency cassette | Visual | Always | N/A |

has been somewhat harder to evaluate sensibly since few, if any, are sufficiently developed in a form whereby potential users can directly see how they might be applied and carry out test applications. Most structured evaluation work has been of the immediate outcomes of use of a virtual environment, and not surprisingly—given the better developed nature of measures in this area—work has concentrated on education or training applications (e.g., Crosier, Cobb, and Wilson 2000; D'Cruz, 1999; Darken & Banker, 1998; Hall, Stiles, & Horwitz, 1998). The user experience can be evaluated in terms of performance measures within the VE (reaction times, navigation or selection tasks such as in the VEPAB from Lampton et al., 1994; chap. 35, this volume), side and aftereffects (e.g., Cobb et al., 1999; Nichols, Cobb, & Wilson, 1997; Nichols et al., 2000, as well as chaps. 29–32 and 36–39, this volume), and participant attitudes and usability of the interface and virtual environment (see chap. 34, this volume). Finally, there is evaluation of the process of both building and using virtual environments. One example of this is evaluation of how potential users will help specify, build and evaluate VEs within a participatory design process (Neale, Brown, Cobb, & Wilson, 1999); another is the evaluation of the process whereby VEs are used in participatory work redesign in control rooms (Wilson, 1999).

## 10. CONCLUSIONS

This chapter has argued for VEs to be developed in a systematic and structured fashion. It describes one framework to support development in this way, VEDS (Virtual Environment Development Structure). This framework has been used to support development of VEs in industrial use (e.g., Wilson, Cobb, D'Cruz, & Eastgate, 1996), industrial training (D'Cruz, 1999) and education (see chaps. 45–46, this volume). By careful, but flexible, use of such a framework, the consequences of some of the barriers to VE application can be minimized. Technical limitations can be circumvented or their effects reduced by appropriate choice of technology and careful specification of the minimum VE size and least complexity required to meet user goals. Trade-offs, such as choices between visual fidelity and interactivity, can more easily be resolved. Usability will be enhanced through selection of appropriate hardware and development of cognitive interface tools and aids, all on the basis of the task and user analysis and storyboarding. Real working applications will only be achieved through selection of those domains to be modeled that make the most of the attributes of VE, using a thorough understanding of what constraints will apply and of how to overcome them in specifying and building the virtual environment. Finally, these applications will best show added value if a rational evaluation process is established at the outset on the basis of realistic and achievable targets for use of VEs and for their incorporation into industrial and commercial practice, education curricula, or training programs.

## 11. ACKNOWLEDGMENTS

The work reported in this chapter has been carried out over the years with a number of collaborators, most notably Sue Cobb, Helen Neale and Sarah Nichols. Jolanda Tromp, Jerry Isdale, and anonymous reviewers commented on an earlier version and we are grateful for their help. The research on which the chapter is based has been funded over time by the EPSRC, the ESRC and various industrial clients, and most recently under European Commission grant IST-2000-26089.

## 12. REFERENCES


Barfield, W., & Furness, III, T. (1995). *Virtual environments and advanced interface design*. New York: Oxford University Press.

Barfield, W., Zeltzer, D., Sheridan, T., & Slater, M. (1995). Presence and performance within virtual environments. In W. Barfield & T. Furness, III (Eds.), *Virtual environments and advanced interface design* (pp. 473–513). New York: Oxford University Press.

Bolas, M. (1994). Designing virtual environments. In C. Loeffler & T. Anderson (Eds.). *The Virtual Reality Casebook* (pp. 49–55). New York: Van Nostrand Rheingold.

Bryson, S. (1995). Approaches to the successful design and implementation of VR applications. In R. Earnshaw, J. Vince, & H. Jones, *Virtual Reality Applications* (pp. 3–15). London: Academic Press.

Cobb, S. V. G., Nichols, S. C., Ramsey, A. D., & Wilson, J. R. (1999). Virtual reality–induced symptoms and effects (VRISE). *Presence: Teleoperators and Virtual Environments*, *8*(2), 169–186.

Crosier, J. K., Cobb, S. V., & Wilson, J. R. (2000). Experimental comparison of virtual reality with traditional teaching methods for teaching radioactivity. *Education and Information Technologies*, *5*(4), 329–343.

Darken, R. P., & Banker, W. P. (1998). Navigating in natural environments: A virtual environment training transfer study. In *Proceedings of the IEEE Virtual Reality Annual International Symposium* (pp. 12–19).: IEEE Press.

D'Cruz, M. D. (1999). *Structured evaluation of training in virtual environments*. Unpublished doctoral dissertation, Virtual Reality Applications Research Team, University of Nottingham.

Draper, J. V., Kaber, D. B., & Usher, J. M. (1998). Telepresence. *Human Factors*, *40*, 354–375.

DTI (1996). A Study of the virtual reality market. Information society instructive report. Department of Trade and Industry. London: HMSO.

Eastgate, R. M., Nichols, S., & D'Cruz, M. (1997). Application of human performance theory to virtual environment development. In D. Harris (Ed.), *Engineering psychology and cognitive ergonomics: Vol. 2. Job design and product design* (pp. 467–475). Ashgate: Aldershot, UK.

Eastgate, R. M., & Wilson, J. R. (1994). Virtual worlds. *EXE*: *The Software Developers' Magazine*, *9*(7), 14–16.

Endsley, M. (1995). Toward a theory of situation awareness in dynamic systems. *Human Factors*, *37*, 32–64.

Financial Times (1994). Virtual reality: Video game or business tool. Financial Times. Business Information. London. Sept.

Flach, J. M., & Holden, J. G. (1998). The reality of experience: Gibson's way. *Presence*: *Teleoperators and Virtual Environments*, *7*, 90–95.

Gabbard, J. L., & Hix, D. (1997). A taxonomy of usability characteristics in virtual environments [Tech. Rep., Online]. Available :http://csgrad.cs.vt.edu/~jgabbard/ve/taxonomy

Gabbard, J. L., & Hix, D., & Swan, J. E. (1999). User-centred design and evaluation of virtual environments. *IEEE Computer Graphics and Applications*, *19*, 51–59.

Green, M., & Sun, H. (1995). Computer graphics modelling for virtual environments. In W. Barfield & T. Furness, III (Eds.), *Virtual environments and advanced interface design*. New York: Oxford University Press.

Grinstein, G. G., & Southard, D. A. (1995). Rapid modelling and design in virtual environments. *Presence: Teleoperations and Virtual Environments, 5*, 146–158.

Hall, C. R., Stiles, R. J., & Horwitz, C. D. (1998). Virtual reality for training = evaluating knowledge retention. In *Proceedings of the IEEE Virtual Reality Annual International Symposium* (pp. 184–189).: IEEE Press.

Hendrix, C., & Barfield, W. (1996a). Presence within virtual environments as a function of visual display parameters. *Presence: Teleoperators and Virtual Environments, 5*, 274–289.

Hix, D., Swan, J. E., Gabbard, J. L., McGee, M., Durbin, J., & King, T. (1999). User-centered design and evaluation of a real-time battlefield visualisation virtual environment. In *Proceedings of VR '99 Conference* (pp. 96–103).: IEEE Press.

Hutchins, E. (1995). *Cognition in the wild*. Cambridge, MA: MIT Press.

Kaur, K. (1998). Designing virtual environments for usability. Unpublised doctoral dissertation, Centre for HCI Design, City University, London.

Kaur, K., Maiden, N., & Sutcliffe, A. (1999). Interacting with virtual environments: An evaluation of a model of interaction. *Interacting with Computers, 11*, 403–426.

Kaur Deol, K., Sutcliffe, A., & Maiden, N. (1999). A design advice tool presenting usability guidance for virtual environments. In *User-Centred Design and Implementation of Virtual Environments* (Proceedings of Workshop at York, England, pp. 15–21). London: British HCI Group.

Kirwan, B., & Ainsworth, L. K. (Eds.). (1992). *A guide to task analysis*. London: Taylor & Francis.

Lampton, D. R., Knerr, B. W., Goldberg, S. L., Bliss, J. P., Moshell, J. M., & Blau, B. S. (1994). The virtual environment performance assessment battery (VEPAB): Development and evaluation. *Presence: Teleoperators and Virtual Environments*, *3*, 145–157.

Loftin, R. B., Bartasis, J. A., & Johnston, R. B. E. (1999). The design and implementation of immersive environments for training and education. In L. Hettinger & M. Haas (Eds), *Psychological issues in the design and use of adaptive virtual interfaces*. Mahwah, NJ: Lawrence Erlbaum Associates.

Neale, H., Brown, D. J., Cobb, S. V. G., & Wilson, J. R. (1999). Structured evaluation of virtual environments for special needs education. *Presence: Teleoperators and Virtual Environments*, *8*(3), 264–282.

Neale, H. (2001). Virtual environments in special education: Considering users in design. Unpublished doctoral dissertation, VIRART, University of Nottingham.

Newman, D. A. (1988). *The psychology of everyday things*. New York: Basic Books.

Newman, W. M., & Lamming, M. G. (1995). *Interactive Systems Design*. Reading, MA: Addison-Wesley.

Nichols, S. C., Cobb, S. V. G., & Wilson, J. R., (1997). Health and safety implications of virtual environments: Measurement issues. *Presence: Teleoperators and Virtual Environments, 6*(6), 667–675.

Nichols, S. C., Haldane, C., & Wilson, J. R. (2000). Measurement of presence and its consequences in virtual environments. *International Journal of Human–Computer Studies, 52*(3), 471–491.

Niclohs, S. C., Ramsey, A. D., Cobb, S. V., Ncale, H, D'Croz, M. D., & Wilson, J. R. (2000). Incidence of virtual reality induced symptoms and effects (VRISE) in desktop and projection screen systems. HSE Research Publications. Contract Research SYSECA (1998). VR for Europe: Industrial Application of VR. Report 1786-Doc-97-D-2700-C001765 for the European commission, DC-III, April.

Norman, D. A. (1988). The psychology of everyday things. New York.: Basic Books.

Papper, M. J., & Gigante, M. A. (1993). Using physical constraints in a virtual environment. In R. Earnshaw, M. Gigante, & H. Jones (Eds.), *Virtual reality systems* (pp. 107–118). London: Academic Press.

Pausch, R., Proffitt, D., & Williams, G. (1997). Quantifying immersion in virtual reality. In *ACM Computer Graphics Annual Conference* (pp. 13–18).: ACM Press.

Rafferty, M. M., Aliaga, D. G., Popescu, V., & Lastra, A. A. (1998). Images for accelerating architectural walkthroughs. *IEEE Computer Graphics and Applications, 18*(6), 21–23.

Rasmussen, J. (1986). *Information processing and human–machine interaction*. Amsterdam, the Netherlands: North-Holland.

Reason, J. (1990). *Human error*. Cambridge: Cambridge University Press.

Richard, P., Birebent, G., Coiffet, P., Burden, G., Gomez, D., & Langrama, N. (1996). Effect of frame rate and force feedback on virtual object manipulation. *Presence: Teleoperators and Virtual Environments, 5*, 95–108.

Rosenblum, L., Burdea, G., & Tachi, S. (1998). VR reborn. *IEEE Computer Graphics and Applications, 18*(6), 21–23.

Sanders, M. S., & McCormick, E. J. (1993). *Human factors in engineering and design* ($7^{th}$ ed.). New York: McGraw-Hill.

Singer, M. J., & Witmer, R. G. (1999). On selecting the right yardstick. *Presence, 8*, 566–573.

Slater, M., Usoh, M., & Steed (1994). Depth of presence in virtual environments. *Presence, 3*, 130–144.

Smets, G. J. F., & Overbeeke, K. J. (1995, September). Trade-off between resolution and interactivity in spatial task performance. *IEEE Computer Graphics and Applications*, 46–51.

Smith, S., Duke, D., & Wright, P. (1999). Using the resources model in virtual environment design. In *User-centred design and implementation of virtual environments* (Proceedings of Workshop at York, England, pp. 57–72). London: British HCI Group.

Stanney, K. M., Salvendy, G., Deisigner, J., DiZio, P., Ellis, S., Ellison, E., Fogleman, G., Gallimore, J., Hettinger, L., Kennedy, R., Lackner, J., Lawson, B., Maida, J., Mead, A., Mon-Williams, M., Newman, D., Piantanida, T., Reeves, L., Riedel, O., Singer, M., Stoffregen, T., Wann, J., Welch, R., Wilson, J. R., Witmer, B. (1998). Aftereffects and sense of presence in virtual environments: Formulation of a research and development agenda. *International Journal of Human–Computer Interaction, 10*(2), 135–187.

Suchman, L. A. (1987). *Plans and situated actions*. Cambridge: Cambridge University Press.

Sutcliffe, A. G. (1995). *Human–computer interface design* ($2^{nd}$ ed.). London: McMillan.

Tromp, J., Bullock, A., Steed, A., Sadagic, A., Slater, M., & Frcon, E. (1998). Small-group behaviour experiments in the COVEN project. *IEEE Computer Graphics and Applications, 18*(6), 53–63.

Tromp, J. G., & Fraser, M. C. (1998). Designing flow of interaction for virtual environments. In J. Tromp, H. Istance, C. Hand, A. Steed, & K. Kaur (Eds.), *Proceedings of the First International Workshop on Usability Evaluation for Virtual Environments* (pp. 162–170). Leicester, England.

Tromp, J., Istance, H., Hand, C., Steed, A., & Kaur, K. (1998). In *Proceedings of the First International Workshop on Usability Evaluation for Virtual Environments*. Leicester, England: IEEE Press.

Tromp, J. G. (2001). Systematic design for usability in collaborative virtual environments. Unpublished doctoral dissertation, Communications Research Group, University of Nottingham.

Vinson, N. G. (1999). Design guidelines for landmarks to support navigation in virtual environments. In *Proceedings of CHI '99 Conference on Human Factors in Computing Systems* (pp. 278–285). : ACM Press.

Wann, J., & Mon-Williams, M. (1996). What does virtual reality really NEED? Human factors issues in the design of three-dimensional computer environments. *International Journal of Human–Computer Studies, 44*, 829–847.

Watson, B., Walker, N., & Hodges, L. F. (1995). A user study evaluating level of detail degradation in the periphery of head-mounted displays. In M. Slater (Ed.), *Proceedings Conference of the FIVE Working Group*. London: Queen Mary and Westfield College, University of London.

Watson, B., Walker, N., Ribarsley, W., & Spaulding, V. (1998). Effects of variation system responsiveness on user performance in virtual environments. *Human Factors, 40*, 403–414.

Welch, R. B. (1999). How can we determine if the sense of presence affects task performance? *Presence, 8*, 574–577.

Wickens, C. D., Gordan, S. E., & Liu, Y. (1998). *An introduction to human factors engineering*. New York: Longman.

Wilson, J. R. (1996). Effects of participating in virtual environments: A review of current knowledge. *Safety Science, 23*(1), 39–51.

Wilson, J. R. (1997). Virtual environments and ergonomics: Needs and opportunities. *Ergonomics, 40*(10), 1057–1077. London: Taylor & Francis.

Wilson, J. R. (1999). Virtual environments applications and applied ergonomics. *Applied Ergonomics, 30*(1), 3–9.

Wilson, J. R., Cobb, S. V. G., D'Cruz, M. D., & Eastgate, R. M. (1996). *Virtual reality for industrial application: opportunities and limitations*. Nottingham, England: Nottingham University Press.

Zahorik, P., & Jenison, R. L. (1998). Presence as being-in-the-world. Presence: *Teleoperators and Virtual Environments, 7*(1), 78–89.

Zeltzer, D. (1992). Autonomy, interaction and presence. *Presence: Teleoperators and Virtual Environments, 1*, 127–132.

# 18

# Influence of Individual Differences on Application Design for Individual and Collaborative Immersive Virtual Environments

David B. Kaber,[1] John V. Draper,[2] and John M. Usher[3]

[1]*North Carolina State University*
*Department of Industrial Engineering*
*Raleigh, NC*
*dbkaber@eos.ncsu.edu*
[2]*Oak Ridge National Laboratory*
*Robotics and Process Systems Division*
*Oak Ridge, TN*
*draperjv@ornl.gov*
[3]*Mississippi State University*
*Department of Industrial Engineering*
*Mississippi State, MS*
*usher@engr.msstate.edu*

## 1. INTRODUCTION

In this chapter a discussion of how individual differences and features of virtual environment (VE) technology affect VE application design is provided. This discussion is motivated by the fact that differences among individual users often account for more variability in performance than system design factors. Individual differences overshadowing system design features in terms of their influence on human–machine system performance is particularly troubling with respect to achieving the primary objective of human factors research, that is, to determine the aggregate effect of a particular system variable on human performance and behavior for the purposes of formulating future system design guidelines to optimize safety, performance, reliability, comfort, ease-of-use, and so forth. Individual differences can severely inhibit a designer's ability to design for a "typical" effect of a particular design factor. With this in mind, addressing user characteristics through VE design may produce substantial improvements in, for example, virtual task performance and the effectiveness of VE trainers. Specifically, user issues that have been hypothesized to be important to collaborative virtual environment (CVE) design are considered, including the number of virtual participants (i.e., whether the environment is to be shared or structured for individual use), and the location of users. The impact of both these issues on virtual task performance is evaluated. With respect to the latter, the implications of remote versus proximate access to a VE and the application design are examined.

Educated speculation on the role of user characteristics that have been found to be critical in human-computer interaction (HCI) is offered, including experience, age, and disabilities in the design of VE applications. Projections are made on user performance given different experience levels, and other criteria. Because few, if any, experiments have specifically investigated the effect of individual differences on VE system configuration and VE design, this chapter develops VE application design guidelines to address individual differences based on results of recent VE research amalgamated with inferences based on seminal HCI research into psychological and somatic factors. It is important to consider with caution any inferences on research and to be careful in construing opinions on design as definitive policies.

Given the maturity and flexibility of VE technology and the substantial impact that individual differences may have on system performance, methods for identifying the causes of individual differences in VE use need to be formulated, and knowledge of common user issues should be reflected in VE system design. Addressing these needs will make it possible to develop virtual task performance standards and to ensure acceptable levels of functioning in the use of VEs by broad populations. To this end, this chapter provides general VE design recommendations (at the close of various subsections), on the basis of literature review, to account for individual differences.

Very preliminary research has been conducted to identify the underlying somatic and psychological factors that may affect human behavior in VE applications. This research has primarily focused on VE system development for individual use. It has concentrated on properties of immersion, which have been hypothesized to be linked to performance in VE applications. These properties include the degree of matching between user behavior in the real world and their behavior in a VE, the form of self-representation provided to the user in experiencing a VE, and the extent to which VEs appear to be "worlds" in, and of, themselves (i.e., independent of anything real). Using one set of terminology or another, these elements of immersion have been discussed at length and extant speculation has been offered as to how they may be influential in the design of individual immersive VEs (see Durlach & Slater (1998) for coining of the term *individual* in this context). Convincing anecdotal evidence of their apparent importance in individual VEs has also been offered.

However important these factors may actually be, as in reality, each individual may experience a VE differently because of their background, interests, and other factors. Therefore, individual differences may serve to moderate the effects of the properties of immersion and other VE system design factors on user performance and experiences in virtual environments. Specific psychological or psychophysical factors have also been considered as to their potential impact on user performance in stand-alone VE applications. These factors include user susceptibility to presence experiences, as dictated by imagination, concentration and attention, and self-control (Psotka & Davison, 1993), and user familiarity with the real environments or systems modeled in VEs (Held & Durlach, 1992). Psychophysical factors identified as being potentially influential in performance in VEs have included user age and disabilities (Stanney et al., 1998).

It seems reasonable to assume that many of the individual differences that have been found to be important in HCI may be relevant to human–virtual environment interaction (HVEI). It also seems possible that the properties of immersion that have been established as being important in the design of individual or stand-alone VE systems for enhancing user performance may be applicable to CVE systems. For example, VE application design factors identified as being relevant to user experiences of presence in individual VEs, such as display field-of-view, simulation update rate, and others may be relevant to CVE design for encouraging shared presence among users. Such contentions are assessed analytically in this chapter. Additionally, hypotheses are formulated on the effects of user experience, age, and abilities in CVE performance. Finally, theoretical guidelines for collaborative immersive VE design are developed considering both user differences and properties of immersion. Some examples are offered to support guideline formulation for effective VE application design.

As a point of clarification of terminology, throughout the chapter, "individual differences" is used to refer to individual variation within user characteristics. For example, human users predominately have two eyes (a user characteristic), but the interpupillary distance may be greater for some users than others (an individual difference). Addressing user characteristic effects on VE application performance seems to have little utility for design purposes. For example, because humans have two eyes, a design guideline that it is necessary to present images to both eyes in display design seems obvious. This is not to say that identifying user characteristics that play a critical role in designing VEs is not an important aspect of the design process. However, in a handbook chapter this work is ultimately encompassed in addressing individual difference issues. This is because in a design context identifying user characteristics is, in fact, a description of the subset of individual differences one expects to affect the application design process. Or, the effect a particular user characteristic has on application performance may ultimately be attributed to the individual differences within the characteristic that produce the effect. Finally, a quick perusal of the HCI literature (e.g., Egan, 1988) on individual difference effects in conventional computer application design reveals the terminology *individual differences* and *user characteristics* to be used interchangeably. Certainly, this does not justify the use of this terminology in the same manner in the VE literature, in general; however, HCI research is supportive of the approach used herein.

As an additional note prior to discussing the effects of specific individual factors on VE application effectiveness, in general, the impact of individual differences on virtual task performance and measures of human cognition in many experimental studies has appeared to be similar to individual difference effects on real-world task performance. Specifically, in studies of human performance in complex tasks, such as human control of teleoperators, simulated air traffic management, and simulated aircraft piloting, whether participants have direct control over a system through real interfaces, interact with virtual models of a real system, or use a VE as an interface to a real system, significant effects of individual differences within experimental groups have been observed. Similar individual difference effects on task performance across reality and synthetic environments have been observed even when dealing with participant samples from similar populations (e.g., undergraduate students, graduate students, highly trained remote robotic system operators at Oak Ridge National Laboratory). These observations seem sensible in that the experience of reality is different for everyone, and if VEs are truly unique environments functioning (for the most part) independently of reality, then why should VE experiences not vary significantly from one person to the next? Unfortunately, these observations do not make the jobs of VE researchers and practitioners easier.

## 2. INDIVIDUAL DIFFERENCES INFLUENCING VE PERFORMANCE

Individual differences may affect the experience and outcomes of using a virtual environment. To the degree that these effects are important to VE users, user characteristics must be identified so VE system developers can effectively address them. As examples, the sense of presence is affected by user characteristics (Barfield & Weghorst, 1993), as is cybersickness (Parker

& Harm, 1992). Several individual differences that may affect experiences or outcomes are reviewed in this section.

## 2.1 User Experience

Experience with VE systems can enhance user knowledge of task- and system-relevant information and user skills and abilities that may affect performance. In HCI, it can be "the most powerful predictor of human performance" (Egan, 1988, p. 553). Experience has also been shown to affect performance during teleoperation (Draper, Omori, Harville, Wrisberg, & Handel, 1987a). It seems likely that experience with VE systems will influence exploration strategies (see chap. 24, this volume) and interaction skills (see chap. 13, this volume), perhaps, for example, through development of mental models of how one generally interacts with a virtual environment. In addition, experience within a specific VE system will likely influence task-specific information acquisition and decision making, for example, through development of a mental model of the specific virtual environment.

Virtual environment research has explored, to a very limited extent, the impact of user mental models of virtual tasks on performance in VE applications and the effects of user experience and aptitudes on virtual task performance. What research has been conducted suggests that experience may be an important individual difference in users' perception that a VE is realistic and, consequently, the sensation that they are present within the virtual environment as well as the effectiveness of the VE application. Egan (1988) summarized several HCI studies that describe experience effects on simple computer task performance. Most importantly, these studies demonstrate that small individual differences in practice or experience produce large differences in human–computer performance. It is possible that this may be the case for VE task performance as well. Further, experience in computer use has been found to play a significant role in user development of mental representations and understanding of systems and tasks (Egan, 1988). Experience in VEs may play a critical role in VE user perceptions of cognitive task load and development of situation awareness (SA) of virtual tasks and processes. Finally, user experience in HCI has been demonstrated to significantly interact with other user characteristics, including age and aptitudes, to affect performance (particularly in the case of novices; Egan, 1988). In VE use, the effect of age and aptitude on user ability to perform virtual tasks may be modulated by VE experience.

The level of experience users achieve in functioning in real environments that serve as subject matter for VE application development may have a significant effect on user performance in the virtual environment. More simply, it may have an effect on user acceptance of VE applications as surrogates to real tasks. This is particularly critical to VE training applications in which a sense of matching to the real-world target task may ensure skill transfer. According to Norman (1986), users form their mental models of tasks or systems based on the task images and information presented to them. In VEs for training, user mental conceptualizations of a task are predictably formed based on their familiarization with the real task. Individual differences in such conceptualizations may affect performance in virtual tasks to the extent that it is influenced by user expectations of consistency between the real and virtual forms of the task (e.g., the degree of matching expected by an expert versus a novice). Consistency is dependent on many factors, including level of VE detail, interactivity, user self-representations, and virtual task loads (i.e., comparison of virtual and real experiences). The latter factors are particularly important in shared VEs in which complex human interactions may need to be modeled.

Here, matching is discussed, including the physical fidelity of a VE, the functional fidelity of a user's activities, and the fidelity of virtual task loads placed on users, based on their mental representations of virtual tasks developed through real experiences. Matching can be formally defined as the degree to which a VE mimics reality not only in form, but also in terms of function and the behaviors the VE elicits from users. Physical fidelity refers to the extent to which the structure of a VE is matched to reality. In a reciprocal sense, the influence of VE structure and design on naive user development of concepts of real tasks and environments based on exposure to virtual surrogates are also discussed.

Virtual environment literature (Romano, Brna, & Self, 1998a) has identified individual and collaborative VEs to have the capability to reproduce high-risk, dynamic environments involving team communication, specifically for training purposes. Virtual tasks can also be designed to mimic the degree of cognitive load experienced by users. The realism of the VE and the fidelity of user interactions dictate this form of matching. Matching VEs to reality in terms of cognitive task loading can be critical for two reasons. First, the degree of virtual task difficulty may predict user involvement in, and concentration on, the VE. In fact, virtual task difficulty has been identified as a predictor of the level of presence experienced in a VE (Draper, Kaber, & Usher, 1998). For example, if virtual tasks are boring and mundane, it is likely that users may not attend to them and be readily distracted from performance by activities in the real world. Conversely, if virtual forms of real tasks are too difficult (e.g., using VE for above real-time training [ARTT]) users may give up in attempting to achieve virtual goals. The second important aspect of matching the cognitive load imposed on users in virtual tasks to their real-world counterparts is that loadings must be accurate to ensure skill transfer. Research on the use of VEs for ARTT (Miller, Stanney, Guckenberger, & Guckenberger, 1997) has found that pilots actually perceive flight simulations running at 1.5 to 1.75 times faster than real time to be more realistic than simulations running in real time. Furthermore, the authors state that ARTT improves transfer of training to real domains. Interestingly, through this research one can see that matching of VEs to reality is critically dependent upon users' perceptions and experience and may not always be the best design solution. In the above example, pilots found faster than real-time simulation speeds to be more representative of real flight than real-time simulation. This is apparently a case in which strict matching is undesirable. In designing VE applications it appears to be important not only to match cognitive loading to reality but also to consider user perceptions and experience as well.

Aside from predicting user perceptions of task loadings across virtual and real environments, the physical fidelity of VEs and user interactions in CVEs may also drive a user's ability to develop SA (i.e., to perceive information in the VE needed for virtual task performance and relate it to their virtual task goals). For example, if user information requirements for virtual task performance are not met as they are in the same task in reality (that users may be familiar with) negative transfer of real task performance behavior and skills may result. Furthermore, if specific task-relevant information is not available in VE surrogates to real scenarios, experienced users most likely will not be able to develop the same level of SA in virtual reality (VR) as in reality. Such a lack of matching between VEs and reality could be absolutely critical to the effectiveness of VEs for training applications, such as fighter pilot training in which developing and maintaining SA skills is fundamental to mission success. With this in mind, matching of SA requirements in VEs to reality may also be necessary in VE application design, specifically in VE trainer development for experienced users in order to ensure virtual skill transfer to reality.

The benefit of VEs appears dependent on individual user perceptions and behavioral issues. That is, some users may not perceive CVEs to be realistic because of interface design features, VE rendering capabilities of VE systems (e.g., low simulation update rates, low display pixel density) and their abilities to imagine being in a real environment. For example, Kaber, Riley, Zhou, & Draper (2000) conducted a study of the effects of the design of a VE interface for control of a real telerobot and lag between user actions at the interfaces and the response of the VE or real robot on system performance and user perceptions of presence. They compared a 3-D graphical model of the robot in a simulated task environment with an interface presenting

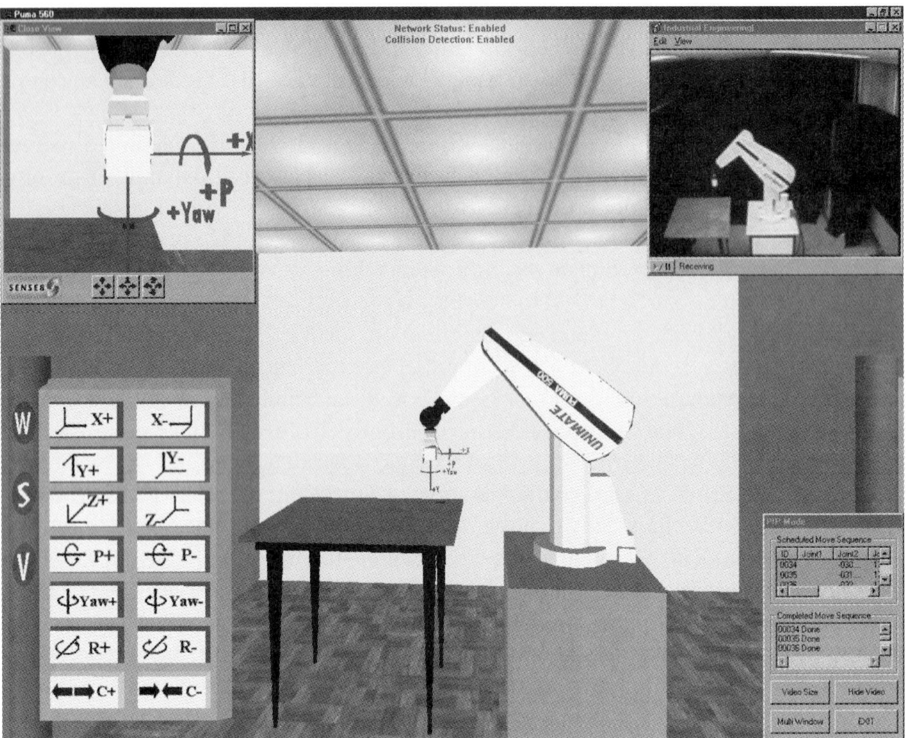


FIG. 18.1. A VE-based graphical preview control interface for a telerobot.

the same models along with live video feedback on the real robot being controlled through the VE (see Fig. 18.1). Control lags in using either interface ranged from 0 to 4 seconds. Both task performance and user presence were significantly influenced by individual differences among a sample of 30 participants; however, perceived realism of the task was found to be significantly greater with the interface integrating VE and video and system performance was best under the minimal lag condition. Along this line of thought, individual differences in perception and behavior may also be underlying predictors of cognitive load experience in VEs and SA acquisition. Consequently, individual differences may limit the capability of individual and CVEs to, for example, accurately replicate for all users realistic cognitive loadings in complex task training or the development of accurate SA in training simulations.

Related to this, much human factors research has demonstrated significant individual differences in perceptions of cognitive workload and the ability to develop good SA in complex task performance (Wickens, 1992; Endsley, 1995). More specifically, significant individual difference effects have been observed in subjective ratings of workload in a virtual task. Riley and Kaber (1999b) observed individual differences in cognitive load perceptions within subject groups in VE experiments using a traditional subjective workload measure (NASA—Task Load Index; Hart & Staveland, 1988) to be so significant that differences among interface technologies in terms of their effect on virtual task workload were not discernable. With respect to SA, Endsley (1995) has observed that often participant ratings of confidence in their own SA differs significantly from the level of SA they have actually developed in complex task performance (e.g., simulated fighter aircraft piloting, automobile navigation with intelligent vehicle highway systems).

It appears that individual differences in experiencing cognitive workload and developing SA may dictate the effectiveness of both individual and collaborative VE applications for training. That is, inaccurate perceptions of cognitive load in VEs may limit transferability of skills developed in virtual tasks to real-world tasks. Beyond this, Ellis (1993, see Romano et al., 1998a, for reference to Ellis' work) offered that internal (cognitive) processes (such as SA) might actually dictate human perception of sensory information in a VE and, consequently, one's sense of it being a physical reality. Therefore, differences in higher levels of individual cognition, which develop with task experience, may drive perceptual activity in VEs and, consequently, the effectiveness of an application.

#### *2.1.1 Relevance of Experience to CVE Design*

To this point the focus has largely been on the role of user experience and cognition in VE application design to individual virtual environments. Of course there are aspects of user experience that are particularly relevant to performance in CVEs and teaming in virtual tasks.

*2.1.1.1 Team Experience.* On the basis of research in team decision making (Orasanu & Salas, 1993), VE researchers (Romano, Brna, & Self, 1998b) have inferred that experience in team performance may be highly relevant to the success of CVE applications. Specifically, individuals who are accustomed to working in teams and, more specifically, working with a particular group of individuals may better collaborate in virtual groups as a result of being able to easily predict the actions of fellow team members. Novices to real-world teamwork and decision making may, for example, not engage in virtual group behavior in CVEs as readily as experienced team players and, consequently, virtual team interaction may be fragmented and interrupted.

*2.1.1.2 Virtual Environment Experience.* The impact of experience in VEs on user performance needs to be considered. For example, how effective can a person be in a virtual task if they have experience in using VE systems in general or if they have experience in the specific VE in which a task is to be presented? Observation of people using VE systems (or teleoperators) reveals differences in users' abilities to interact with such systems. Some people find interaction with a VE a simple task, whereas others find it a difficult one. There is a parallel phenomenon in teleoperation: although there are no systematic data, one of us with long experience with teleoperation (Draper) has observed that some people never become capable of performing teleoperation well, whereas some are good at it from their first experience. It is difficult to predict whether these differences will be ameliorated or compounded by greater immersion (i.e., introduction of a fuller range of sensory modalities and enhanced motor interaction in VE systems). It is therefore important to further develop an understanding of how a broad range of aptitudes, including motor and verbal abilities affect human–VE interaction.

Romano et al. (1998a) have argued that individual differences in user adjustment and adaptation to VEs is linked to their experiences of presence and virtual task performance in those environments. Kalawsky's (1993) and Lombard and Ditton's (1997) works are also supportive of this argument. Lombard and Ditton (1997) speculated that the intensity of experiences of presence in VEs is associated with duration and frequency of exposure to virtual environments. Therefore, individual differences in experience level with VEs may dictate user presence.

With this research in mind, CVE design for distributed user teamwork in virtual tasks should consider user experiences in real-world teamwork for forming effective virtual teams. For example, system administrators may need to ensure that users immersed in a CVE participating in a virtual group have a similar level of teamwork experience to promote collaborative virtual task performance. Similarly, CVE application designers should also consider the level of adaptation of users to specific VEs for the same reason.

## 2.2 Technical Aptitudes

Human–computer interaction research has demonstrated that different technical aptitudes (e.g., spatial, reasoning, verbal ability) are predictive of performance in certain types of computer-based tasks (see Egan, 1988). These studies seem to suggest that persons with low spatial aptitudes may struggle with system navigation issues, like becoming disoriented within hierarchical menu systems (Sellen & Nicol, 1990). This may be particularly relevant for VE systems that require navigation through a world. In teleoperation, McGovern (1993) found that remote vehicle operators can become lost within the remote world, even when provided with maps and landmarks, and remote manipulator operators can become confused in terms of display/control orientation if they are not provided with a view of the entire remote manipulator (a "worldview"). The former result might be expected given that knowledge acquisition from maps is associated with visual/spatial ability (Throndyke & Statz, 1980). In general, spatial memory appears to dictate user ability to find objects on a display. It seems reasonable to assume that specific aptitudes such as spatial memory, visualization, and orientation may be predictive of human performance in VEs as well, particularly given that an additional spatial dimension describes virtual task spaces in comparison to most conventional computer applications.

Stanney and Salvendy (1994) found that even though persons with poor spatial abilities were unable to internalize the structure of complex systems, they could recognize the structure of systems when they were well organized and when structure acquisition was made salient. This suggests an opportunity for VE system designers; the system can be tailored to accommodate differences among users, but even more powerfully to compensate for the weaknesses of a particular user. It also suggests an approach to training for VE use; persons with low spatial abilities could have their initial experiences constructed to emphasize spatial relationships within the virtual environment. These persons could be restrained from proceeding to tasks until they demonstrated some criterion knowledge of VE structure.

### *2.2.1 Spatial Aptitude*

Virtual environment research (e.g., Greenhalgh & Benford, 1995) has relied on the notion that a user's position and orientation in a VE, as in a real environment, will largely affect their perceptions of the VE, other users in a CVE, and their performance in virtual tasks and collaboration. Interestingly, individual interests and projections about events based on environmental information dictate a user's spatial position in any environment. On a deeper level, the background and experience differences among individual users may dictate their spatial abilities to satisfy perceptual interests in a VE and, consequently, their performance.

Obvious differences have been observed between novice and experienced VE users in moving about virtual worlds. Novices appear to have greater difficulty, or encounter problems more frequently, in navigating close virtual spaces (e.g., passing through doors) to see objects of interest (Greenhalgh & Benford, 1995). Therefore, novices may not always be able to perceive VEs as designers intended. Experienced user perceptual interests may be better satisfied and their perceptions of a VE may be more accurate.

As users develop more experience in using sophisticated immersive displays as part of VE systems for perceiving visual stimuli from a VE and they become more accustomed to using hand-controller interfaces to move in VEs that do not use natural human motion control mechanisms as metaphors, their abilities to accurately perceive VEs and to interact with other users in CVEs improves (cf. Hix et al., 1999). With this in mind, VE application designers must look to create more natural control interfaces to promote novice user spatial positioning performance or to take into account user experience in assigning virtual tasks or provide training in virtual motion (Kaber & Riley, in press).

#### 2.2.2 Reasoning Aptitude

Like spatial abilities, reasoning (based on mathematics and science knowledge) has been found to be an important aptitude for computer task performance. Results of HCI research have shown that users who are able to develop their own strategies for computer tasks through experience-based reasoning skills are faster than nontechnical users in error-free performance (Egan, 1988). It is likely that reasoning aptitudes (deductive, inductive, etc.) are important to HVEI as well, particularly in user understanding of the plot, or dynamic of a VE, and how to respond to independent behavior of virtual objects. In a VE that a user has not been exposed to before, the concept of a plot has been identified as being potentially important to providing a sense of involvement in the virtual environment (Allen, Bolton, Ehrlich, & McDonald, 1999). Consequently, reasoning aptitude may affect user perceptions of virtual task involvement.

Unfortunately, there is paucity of VE research relating specific technical aptitudes and user knowledge to experiences of presence in individual and collaborative VEs, or to user performance in virtual tasks. However, on the basis of the HCI literature, it can be hypothesized that different types of reasoning skills may be critical to user development of strategies for virtual task performance. The technical aptitude of computer users has been found to be critical to computer task performance. For example, scientists and engineers have been found to perform better at computer-based information search tasks than persons of a social science background (see Egan, 1998). User technical aptitude and experience may also be important in HVEI, specifically science and engineering learning may be more suited to logic-oriented virtual tasks, whereas other learning may be more appropriate for creative tasks, such as virtual art. With this in mind, it may be critical for VE application designers to consider user aptitudes in virtual task design or to consider the aptitudes of multiple users involved in collaborative virtual tasks to ensure specific aptitudes and experience are represented for error-free performance. It is worth noting that the designers of VEs are generally people with high technical aptitudes, whereas users may not be (depending on the specific application). This mismatch may result in the development of a VE that is not optimal for the target user population. This is a classic human factors problem in system design.

### 2.3 Age-related Considerations

Age has been identified as a predictor of the difficulty of learning to use computers (Egan, 1988). However, the impact of age seems to operate only across widely varying ages and when experience-related factors are controlled. Furthermore, the studies reviewed by Egan (1988) examined learning to use complex computer systems. In VE systems, although the systems may be more complex *per se* as computerized systems, learning to interact with them may often be simpler. That is, walking through a virtual forest may require a more complex computer hardware and software system than text editing, but the VE task (walking through a forest) may be more similar to mundane experiences than text editing. Therefore it is not clear how well HCI results apply to interaction with VE systems in regard to age-related considerations. The ways in which age affects human interaction with computers and VE systems is not yet fully understood and may involve a cultural bias. At the time of this writing, very few persons have as much as two decades of experience with computers, no matter what their age. The population that matriculated after about 1983 will have much greater experience with computers than the population prior to that time. It remains to be seen how (or if) greater lifelong exposure to HCI will moderate the relationship between age and learning for computers and VE systems.

However, there are physical changes that may occur with age, especially in terms of reductions in perceptual and cognitive capabilities that may influence VE system interaction (Birren

& Livingston, 1985; Fisk & Rogers, 1991; Hertzog, 1989; Salthouse, 1992). These changes may result in difficulty in sensing, perceiving, and understanding VE scenes commensurate with difficulties experienced in real-world scenes. Age-related degradations in visual acuity, hearing, memory, and range of motion are well documented and well-known colloquially. Helping users to cope with these phenomena is one of the potentially important capabilities of VE systems and at the same time a challenge for developing human–VE interfaces.

Many unique applications of VEs for elderly and disabled person use have been identified. These applications include using VE for continuing education and skill training of older persons remote from real educational facilities (Odegard & Oygard, 1997), providing disabled persons with access to virtual replicas of real environments, like clean-laboratory facilities, that would not otherwise be accessible to them (Nemire, Burke, & Jacoby, 1994), and VE training of the elderly and disabled in navigating replicas of foreign environments in advance of actually experiencing their real counterparts (cf. Cripps, 1995). These types of virtual experiences can be extremely valuable to the employability of these populations and the sustenance of a nation as the majority of its population ages.

Beyond these applications, networked VEs and CVEs can be used to provide elderly and disabled populations with access to commercial goods and services in a manner far exceeding that achievable through the Internet and World Wide Web, and that might not be possible in reality because of a person's physical limitations. For example, simple applications such as a virtual grocery store could significantly enrich a physically disabled person's life by allowing them to buy food for themselves and to have a multisensory experience of an everyday activity like an average person (e.g., enjoying the sights and smells of a bakery or delicatessen). This essentially means that networking capabilities for supporting CVEs must provide for near real-time application responsiveness and system synchronization.

Such practical applications of VE motivate the formulation of application design considerations of user age and physical abilities. Unfortunately, as with other user characteristics and individual differences identified in this chapter as being critical to VE application design, little structured, formal empirical research has been conducted into accommodating age and disabilities in VE application design.

User age may be important in VE application design based on learning abilities and difficulty experienced in understanding VEs (Egan, 1988; Stanney et al., 1998). Middle-aged persons and youngsters have been found to be similar in the stages of learning they exhibit in CVE experiences, including developing knowledge on motion and orientation in an environment, and then how to collaborate on interaction with virtual objects and tasks (Odegard & Oygard, 1997). However, developing knowledge on complex VEs may be more difficult for aged persons because of degenerative effects on physiology, particularly the senses (e.g., vision and hearing). This is potentially critical to VE design involving multimodal stimuli, including 3-D imagery and spatialized audio, or tactile feedback through force-reflecting control devices.

Virtual environment application design efforts need to be focused on age-related individual differences in vision and hearing. They may, however, be structured in an identical manner for all age groups according to the stages of user learning. For example, in acquainting both children and older adults with CVE applications, VE research (Odegard & Oygard, 1997) suggests presenting displays and interfaces that allow them to develop an understanding of spatial positioning and navigation in VEs in advance of presenting them with virtual task objects and showing them the representations of other users in the CVE.

The impact of age-related effects is, of course, not limited to the perceptual system, but higher level cognitive processes may also be affected. Human–computer interaction research has demonstrated that with age, user ability to allocate attentional resources to stimuli and divide attention across multiple sources of information degrades (Egan, 1988). User ability to generate complex computer commands and effectively search for information also becomes significantly worse with age. Differences have been informally observed in how youngsters and middle-aged persons approach virtual tasks. According to Odegard and Oygard (1997), older persons are much more analytical in their approach to the tasks they need to perform, yet they are less clever in formulating strategies to accomplish the tasks in comparison to young children experiencing the same virtual worlds. Youngsters appear to respond spontaneously to stimuli in VEs and are explorative in their approach to solving problems.

These effects of age on human information processing need to be considered in designing VE applications for user performance. Beyond considering attention allocation capabilities, creativity and long-term memory storage in application design, geriatric effects on spatial orientation and memory, as well as technical aptitudes for reasoning should be considered. These user characteristics have been discussed previously, and it has been suggested that they may interact with experience and affect user sensations of presence in VEs and virtual task performance.

### 2.4 Accommodating Disabilities

Although much speculation could be offered on how user disabilities might affect performance in VE, there is a paucity of research evidence to support specific VE application design guidelines. With this in mind, without attempting to reflect different user deficits in VE design, such applications may not realize their full potential. Like the degenerative effects of aging, deficits in physical and information processing capabilities should be considered to accommodate users in VE design for presence and performance. How specific deficits should be handled needs to be empirically researched.

Virtual environments provide unique capabilities for addressing physical disabilities of users, including hand-control navigation techniques and voice driven input. They also offer methods for dealing with perceptual-cognitive limitations, including multimodal representations of stimuli (e.g., visualizing sound for the hearing impaired). With these capabilities, VE system designers should be able to structure, for example, visual, auditory and tactile interfaces to allow all persons to interact effectively with virtual environments. In addition, displays should be structured to ensure user perception of stimuli from VEs for virtual task performance.

### 2.5 Personality Traits

Other user characteristics, such as personality, may affect performance in VEs (Stanney et al., 1998). Unfortunately, little empirical data has been developed that supports personality as being critical to HVEI. With this in mind, inferences on the potential effects of personality traits in VE task performance have been made on the basis of recent VE literature and results of HCI research. Specific traits related to computer task performance, in general, which may play a role in VE user performance, include self-concept and self-representation. Self-representation, such as via an avatar, could be critical in CVEs where interpersonal communication must be facilitated among participants. It may also be important to ensure consistent or recognizable representation of users in shared virtual environments.

Other influential personality traits may include user dedication to virtual tasks, the sense of being a part of a group of virtual collaborators and the willingness to collaborate. Human-computer interaction research has demonstrated that user passion for playing computer games is dependent on interface design and gender (Egan, 1988). System design factors and individual differences may also dictate user concentration on VE tasks. With respect to oneness with a group in CVEs, Slater and Wilbur (1997) offered that performance in shared VEs, unlike individual VEs, not only depends on whether the user feels part of the VE, but whether they also feel a part of a group of virtual actors. If users are compelled to a group experience, they may be more willing to collaborate, which has been found to be critical to performance in shared virtual environments. Romano et al. (1998b) stated that noncollaboration in CVEs degrades performance due to other users simply being distracters to individual task performance.

As a note, when we undertook the development of this chapter we speculated that the majority of theoretical research into individual difference effects on VE application effectiveness would be focused on the study of characteristics other than personality traits (or social skills) and their relevance to VE design. However, subsequent to reviewing the literature on CVEs, we found that many contemporary theoretical writings exist on this topic and some informal empirical investigations have been conducted with intriguing results.

### *2.5.1 Self-representation in CVEs and Matching VE Interfaces to Human Interfaces to Reality*

Virtual environment experiments (Fraser & Glover, 1998) have revealed that differences in interfaces to CVEs, in terms of system displays and hand controllers, as well as how users are virtually represented in a VE, significantly affect individual sensations of realism about the VE and the continuity of user interactions. For example, desktop VR has been compared with immersive VEs, facilitated through projection displays or Computer Automatic Virtual Environments, demonstrating that the transparency of the interface (i.e., the degree to which the user perceives virtual stimuli to be natural or unfiltered by visual display interfaces?) dictates whether task interruptions may occur and how fragmented interaction is among CVE participants (Fraser & Glover, 1998). Low transparency, common with desktop displays with which users feel as if they are looking through a window into the virtual world (Cruz-Neira, 1998), tends to promote interaction breakdown and poor task performance.

With respect to self-representation, recent research (Crampton & Hilton, 1999) has been conducted to develop human body scanning systems for developing virtual embodiments of users participating in CVEs. This work suggests that accurate representation of a user's physical form in VE may be important to participants in CVEs. Further, some VE research (Tromp, Bullock, Steed, Sadagic, Slater, & Frecon, 1998) has offered that CVE users may become embarrassed through their self-representation as a result of specific user–user interactions in CVEs.

Breakdowns in user interactions in CVEs also appear to result from disparities in the functional fidelity of VE from reality causing users to make incorrect assumptions on collaborative activities. Fraser and Glover (1998) found that user perceptions of the behavior of their self-representations in VEs must match, for example, the control interfaces they use to move their virtual embodiment. That is, if a user is provided with a realistic human representation in VE, the interfaces he or she uses to cause behavior of the virtual representation must resemble natural human interfaces to reality (i.e., hands, arms, etc.).

These findings are not particularly surprising, as physical and functional fidelity of VEs have been examined frequently in the literature as to their role in user presence in VE applications and their performance. However, the findings are important to this chapter in that individual differences may largely moderate user perceptions of interface transparency and consistency of virtual representation in a CVE with real interfaces to the virtual environment. For example, different users may perceive desktop VE systems to be more or less transparent depending on their immersive tendencies, including the ability to imagine oneself in another world and to concentrate on the presentation of that world to the extent that they ignore stimuli from reality (Witmer & Singer, 1994). Further, crude virtual block-humanoid self-representations may be acceptable to some participants in CVEs, whereas others may require high-resolution body scans be used to represent them, and that their virtual embodiments be clothed and in a manner representative of their professional or social position. Also, for some CVE users, a wand controller, for example, may be acceptable for facilitating virtual locomotion; however, others may require tracking of their actual body motions to direct the movements of their virtual embodiments. Therefore, appropriate user self-representation in VE applications may be in the eye of the beholder, as are beauty and fluidity of motion in realty. Unfortunately, to our knowledge there is no empirical evidence available at this point in time that maps individual sensitivities to interface quality and the form and function of self-representations in CVEs to the success of the collaboration measured in any terms (based on Fraser and Glover [1998]).

## 3. RELATIONSHIP OF PRESENCE TO VIRTUAL COLLABORATION AND TASK PERFORMANCE AND THE RELEVANCE OF USER COGNITION TO SHARED PRESENCE

In the VE literature, individual differences in personality traits seem to be discussed primarily because of a suspected influence on presence experiences in virtual environments. Many facets of human personality and aspects of cognition have been theoretically implicated in user presence experiences in individual VEs, and to a greater extent in CVEs because of the role of personality in human interpersonal communication. Durlach and Slater (1998) have suggested that user experiences of shared presence in CVEs are based mainly on two factors pervasive across users, including a sense of physical presence in a common virtual place and the ability to communicate with other participants. Therefore, the experience of personal presence may be a prerequisite to shared presence, and to achieve shared presence in CVEs user–user interaction must be possible. As a first thought, these seem to be relatively simplistic conditions for shared presence in CVEs as compared to the conditions speculated about in the literature (cf. Slater & Wilbur, 1997) for presence in stand-alone virtual environments. However, Durlach and Slater (1998) suggest that their achievement is critically dependent upon individual differences including a user's background, interests, viewpoints, and sensitivities, which play a role in virtual collaboration and individual presence as well as VE application effectiveness. Differences in perception and cognition may critically influence the effectiveness of CVE applications, or the use of common VEs, by limiting users' experiences of presence and their abilities to concentrate on virtual tasks. Of course, this issue is also relevant to individual VE effectiveness. Also, factors influencing acceptance of individual VEs for work, and presence experiences in stand-alone VEs, may be relevant to CVEs. In fact, recent research (Tromp et al., 1998) has demonstrated that individual presence and sensations of shared presence in CVEs are positively correlated. The potential factors underlying shared presence, and the conditions of its occurrence, seem equally complex as those associated with presence, if not more so.

One might ask at this point, why should we care about presence experiences in stand-alone VEs and shared presence in CVEs, and, furthermore, the role of user psychology in presence experiences in VEs? Romano et al. (1998b) offered that virtual task performance in individual or collaborative VEs is dependent on user ability to develop a sense of presence or shared presence, including the perception of being in an environment with others and engaging in group behavior. Shared presence in CVEs has also been linked to user collaboration. Specifically, it has been speculated that individual ability to achieve presence and sensations of presence in VEs may dictate whether collaboration occurs among users in the same VE (Romano et al., 1998a). Vice versa, empirical study (Romano et al., 1998b) of CVE gaming has shown that when users collaborated effectively, they shared a strong sense of presence in the CVE and performance was superior. However, whether presence produces collaboration and performance or whether collaboration promotes performance and presence still remains a circular argument.

Consequently, the success of a VE application may be indirectly dependent on individual differences in psychology and cognitive abilities including, for example, immersive tendencies, sociability, concentration, and imagination. Furthermore, Romano et al. (1998b) linked the enjoyment of virtual experiences to presence experiences; therefore, deriving pleasure from VEs, such as in using an entertainment application may also be dependent on individual differences. Finally, since CVEs can be used for unique applications involving human–human interaction and shared presence has been linked to the success of collaboration in VEs, VE researchers have speculated that it may be important to look at human factors in interpersonal communication that are apparent in reality, such as personality, as well as their role in presence experiences for ensuring application effectiveness.

### 3.1 Immersive Tendencies

Like individual subjective presence, the factors that influence a user's sense of shared presence with other participants in a common virtual place may be affected by individual immersive tendencies. Empirical studies of presence have demonstrated a significant correlation among subjective ratings of presence in VEs and user predisposition or susceptibility to immersion (Kaber et al., 2000; Riley & Kaber, 1999a; Witmer & Singer, 1994, 1998). These tendencies are thought to be dependent on aspects of human cognition and behavior, including concentration, imagination, and self-control (Psotka & Davison, 1993). Individual differences in immersive tendencies, particularly the capabilities of one's "imaginoscope," may dictate the degree to which a CVE seems real to a user and, consequently, the degree of association users perceive with a VE and the shared behavioral presence they exhibit (i.e., acting as if they were present in a real environment). For example, Witmer's and Singer's (1994) work suggests that daydreaming and becoming lost in science fiction novels, among other activities, may predict immersive tendencies and presence experiences in virtual environments. They have also considered abilities, such as remaining focused on television programs and performing mundane tasks, to be predictive of immersive tendencies. These seem to reflect psychological factors that may be indirectly related to VE application effectiveness.

### 3.2 Attention and Concentration

Recent research into the role of user cognition in performance in CVE tasks has also revealed that user abilities to allocate attention and to concentrate on stimuli from a VE may dictate their degree of presence (Draper et al., 1998), participation in virtual groups and, to some extent, the success of an application. Different degrees of presence among users of CVEs have been informally observed, including an extreme lack of presence. Greenhalgh and Benford (1995) noticed that some users of a collaborative virtual teleconferencing system became easily distracted with activities external to the CVE to the extent that the virtual representations of these users in the CVE did not respond to, or ignored, direct communication from other participants. This is an example of individual differences in attention allocation and concentration abilities playing a role in the effectiveness of a VE for collaborative activity.

The possible psychological factors in presence discussed here most likely represent a fraction of user characteristics that potentially moderate presence experiences in VEs and the effectiveness of applications, particularly VE trainers. However, the role of attention as a factor underlying the impact of a cluster of user characteristics on presence is a research topic that is, at the present time, both insufficiently pursued and potentially very important.

## 4. PERSONALITY TRAITS AND COLLABORATION IN VEs

The properties of immersion in VEs and individual psychology may not only indirectly affect CVE effectiveness through user presence experiences in virtual worlds, but they may affect it through virtual collaboration, and group behavior or teaming as well. This effect may occur directly through the roles users assume or do not assume as participants in CVEs on the basis of personality and social skills. Certainly, noncollaboration in CVEs is an option for users, and some may be inclined to function independently based on social disposition (e.g., antisocial) or personality type (e.g., dominant, reclusive). Others may be apt to collaborate in VEs through different roles.

Roles people adopt in VEs have been empirically observed (see Romano et al., 1998a) including "teacher," "pupil," "team player," and "dominator." As in reality, individual differences in personality traits may drive the adoption of these roles. People who are accustomed to doing most of the talking and guiding others may assume the role of "teacher." Persons who normally derive more confidence in their activities through reliance on others may tend to act as "pupils" in CVEs. Collaborative virtual environment participants that share resources and are comfortable with the process of reaching decisions by consensus usually play "team player" roles. "Dominators" are essentially nonteam players that tend to ignore the suggestions of others and attempt to keep physical and mental distance between themselves and other people. "Dominators" are, for the most part, noncollaborators.

Subjective evidence of these roles has been provided by Romano et al. (1998a), who showed that some participants in a CVE game felt that they could do better without a partner and that others felt their performance was enhanced when they could perceive a partner. Unfortunately, specific personality traits of the participants were not recorded and linked to the roles they assumed in virtual gaming.

With respect to the properties of immersion impacting VE effectiveness through virtual group behavior, VE research has shown that the roles persons adopt in CVEs may be dependent on the VE system interfaces they use in addition to user personality traits. Tromp et al. (1998) conducted experiments on small group behavior in CVEs revealing that the computational resources available to a user and the degree of user immersion in the CVE predicted the group leader in collaborative virtual tasks. Interestingly, persons who emerged as leaders in the VE applications did not adopt the same role in reality. This suggests that interfaces to VEs and the structures of VEs in comparison to real environments may be directly associated with user roles. It also suggests that interfaces and structures of CVEs interact with user characteristics to determine emergent roles in a way not predicted by observation of individual behavior in real-world groups.

Other factors beyond individual differences in personality and the capabilities of VE interfaces have been shown to be influential in the roles users adopt in CVEs and their social behavior. Tromp et al. (1998), in their CVE investigations, demonstrated that male users tended to adopt leadership roles in collaborative virtual tasks; however, at a similar frequency, male participants appeared to impose on themselves traditional social barriers observed in reality and did not engage in group behavior. Whether such findings are useful for VE application design purposes has yet to be seen. However, evidence of individual differences in role playing in VEs based on user psychology and gender, for example, can be considered as a testament to the unique capabilities of VEs to replicate real situations, or at least convey to some human users the sense of being in a real environment with others.

While user personality traits could affect VE application effectiveness through user role assumption, it seems important to identify exactly how this may occur. In team decision making, the decision-making power that team members have may be equal, or power may be limited to only a few participants or to the team leader. This is critical because the roles people adopt in groups have been related to the overall effectiveness of the group. Therefore, depending upon how decision-making power is allocated and the roles assumed by users in a CVE, user collaboration may or may not be successful. For example, if decision-making power is allocated to a virtual team leader and a person who is accustomed to a real leadership role adopts the same role in the CVE, the capability of that individual to exhibit autocratic behavior will largely predict the team's effectiveness.

## 5. ACCOMMODATING INDIVIDUAL DIFFERENCES IN DESIGNING VE APPLICATIONS FOR PRESENCE

Given the plethora of VE properties that may dictate presence and that may play a role in individual and collaborative VE presence, as well as the significant psychological individual differences possibly moderating presence experiences, it is important to identify VE application design methods to optimize presence. Unfortunately, at this time empirical VE research is in its infancy in terms of mapping relationships between VE system characteristics, human factors, and virtual task dimensions to user presence experiences in individual virtual environments. Furthermore, there is a paucity of research into shared presence experiences in CVEs and VE system and user factors underlying shared presence. Of course, VE application design guidelines for encouraging presence experiences are only important if presence plays a role in, for example, VE training effectiveness and performance with VE interfaces to real systems for real task performance. Also, a relationship between presence and performance in VEs has yet to be empirically established through objective measurement of presence and modeling of virtual task performance.

Even in light of all this, VEs have unique capabilities that could be exploited to resolve some of the effects of individual differences in perceptions of virtual worlds including, for example, the ability to provide a single user with multiple viewpoints of an environment, such as the viewpoint of a fellow collaborator. With this in mind, the capabilities of VEs may also be exploited to ensure shared user perspectives on virtual tasks and to encourage a sense of understanding and belonging among users, and shared presence in an application (Greenhalgh & Benford, 1995). These effects may be moderated based on user access, either remote or proximate.

## 6. REMOTE VERSUS PROXIMATE ACCESS EFFECTS

A VE allows human sensory, motor, and cognitive abilities to span distances and to enter where people may not physically intrude. This requires application of three critical subsystems: the remote portion, including sensors and actuators; the local portion, primarily a human–machine interface; and the communication subsystem, which transmits information back and forth between the user and the computer-mediated area. This section will explore some effects of the latter subsystem on performance within VEs and on how they interact with user characteristics.

For VEs, remote may mean a physical separation, as when a user in Raleigh, North Carolina, operates a VE in Oak Ridge, Tennessee; or it may mean a metaphorical separation, as when a user explores a VE that resides on files stored on her computer. However, physical separation *per se* is not the fundamental issue. Physical separation exercises its impact through the effects of distance on communication bandwidth, latency, and noise. Bandwidth tends to be reduced and latency and noise increased by distance. These critical features of a VE communication system are important whatever the physical separation between user and computer-mediated world. Remote versus proximate is best understood as the metaphorical separation between a user and a VE, a separation of sensorimotor capabilities within the VE from normal human capabilities. From this perspective, the degree of remoteness is the degree of perturbation of information flowing to and from the virtual environment.

### 6.1 Behavioral Cybernetics

The behavioral cybernetics (or psycho-cybernetic) model described by Smith and colleagues (Smith, 1972; Smith, Henning, & Smith, 1995a; Smith, Keran, Koehler, & Mathison, 1995b; Smith & Smith, 1962; Smith & Smith, 1987; Smith & Smith, 1988; and Smith & Smith,

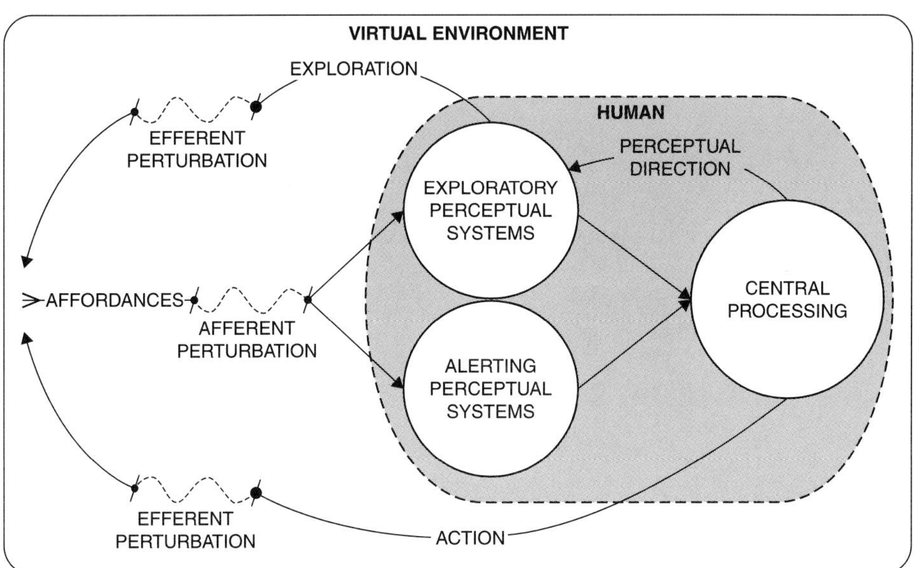


FIG. 18.2. Simple model illustrating perturbations of information flow in the human–VE interface.

1989) provides a perspective for examining the impact of proximate versus distal access. At the heart of this model is the notion that humans respond to information present in an environment (feedback) by manipulating that environment. This changes the feedback and elicits another controlled response. Therefore, humans manage the system with the aim of controlling feedback in a continuous or continuously sampled process that persists until reaching some feedback goal. Perturbations occur when the expected relationship between the feedback and the controlled response is altered as, for example, when their timing is different than expected.

During use of a synthetic environment, perturbations occur in three ways. First, temporal perturbations affect the timing of the feedback and control response. Time delay in signal transmission is an example. Second, spatial perturbations affect the spatial relationships between feedback and control, as when the viewpoint of a representation of one's hands is offset to one side or another in the synthetic environment. Finally, perturbation by filtering occurs when the amount of information in the feedback signal is in some way different than expected. An example is when users view a scene by television: because a television presents images as mosaics with a finite number of elements, the amount of information afforded is less than during direct viewing. In effect, the television is a low-pass filter on visual information. Figure 18.2 presents a simple model of perturbations in information flow in HVEI.

From the perspective of the behavioral cybernetic model, the goal of human factors in synthetic environment development is to identify perturbations, quantify their effects, and analyze the trade-offs of potential solutions. The aim of the process is to optimize the machine for its mission by reducing perturbations as far as is economically appropriate and providing the means for overcoming the effects of perturbations when it is not practical to remove them.

### 6.2 Common Sources of Perturbations

The issues critical to effective VE application design that depend on user location are predominately technological in nature. The central issue is communication bandwidth or capacity, which brings with it the subissues of application resolution, update rate or latency, and level of environment detail, among others. These factors dictate whether, for example, a 3-D user interface can be provided or spatialized sound can be used.

### 6.2.1 Communication Bandwidth

Bandwidth serves as a low-pass filter on the information from the remote world presented to the user. Certain displays have inherently limited bandwidth: Television, for example, reduces visual information presentation to a 525-line, raster-scanned mosaic image. In many cases, the tools used to transmit information sensed in a remote world to local displays also filter information, typically as low-pass filters.

### 6.2.2 Latency

Latency is the temporal perturbation that occurs when events in the remote world occur later than expected. There are two common sources: Latency often occurs when the computer system cannot update a virtual world quickly enough, and communication delays can be present between the remote world and the local controls and displays (the latter is commonly called lag).

Latency causes serious difficulties with VE use. For example, low update rates (below 12 Hz) can induce illusory perception of motion and simulator sickness, but images presented at higher rates seem smoother and support better depth perception (Durlach & Mavor, 1995, p. 117; Piantanida, Boman & Gille, 1993; chaps. 29–32, this volume). Time delays during teleoperation trigger adoption of a move-and-wait strategy, in effect shifting users out of real-time control, whether produced by communication delay (Ferrell, 1965, 1966) or poor timing between controller inputs and remote manipulator actions (Draper et al., 1987b).

### 6.2.3 Noise

Noise may be present in the remote world or produced during signal transmission. For example, suspended particles in seawater and marine growth may be considered noise during subsea teleoperation, and the additional bandwidth provided by stereoscopic television can help users see through the noise to view tasks (Smith, Cole, Merritt, & Pepper, 1979). Noise may be introduced into signals by interference, as when radio signals transmitted from earth surface to low earth orbit are affected by atmospheric conditions.

## 6.3 Remote Versus Proximate Access Effects on VE Application Design

From the preceding, it is more fruitful to examine remote versus proximate access in a metaphorical sense than in a literal sense. It makes no difference to the user if the "remote" world exists within the computer on her desktop or on the surface of the moon, as long as the literal distance does not affect bandwidth, latency, or noise. The literal distance is important only in terms of its affect on metaphorical distance, that is, if it creates perturbations.

But what is the impact of perturbations on requirements for VE design? Telerobotics is a cousin of VEs, differing mainly in that the remote world is a physical world for telerobotics, whereas it is a computer-generated one for VEs (Durlach & Mavor, 1995). Experience in telerobotics provides illustrative examples.

First, function allocation is affected. The recent success of the NASA Mars rover demonstrates how time delay affects function allocation. Without time delay, it is possible to control telerobotic vehicles in real time. However, with time delay, it is necessary to allocate functions differently, with the robot responsible for controlling its actions and the human adopting a more supervisory role. In VEs, one might expect a similar effect. Systems with time lags should probably be designed for interaction that is more symbolic and oriented toward strategies.

Second, the nature of displays and controls is affected. Relatively noisy, low-resolution displays may be enhanced by stereoscopic presentations, as was the case for subsea teleoperation (Smith, Cole, Merritt, & Pepper 1979). High-resolution visual images providing crisp edges and texture rendering may not be much enhanced by stereoscopic presentation (depending on the nature of the task and user characteristics). Avatars that have good responsiveness, where responsiveness is how well the VE "reproduces user . . . trajectories and impedance in time and space" (Draper, 1995, p. 71), are probably most appropriately controlled using position–position tracking schemes. Avatars with poor responsiveness may be controlled better using rate control joysticks.

## 7. SUMMARY AND CONCLUSIONS

The study of individual differences runs somewhat against the common grain of science, which more routinely observes individual cases and induces, from the aggregate, the general. The corollary of this in design is identification of shared characteristics, or a shared range on some characteristic, like stature, and designing toward the common. In science, the unusual is sometimes referred to as an outlier and removed from a data set; in systems operation the unusual must accommodate themselves to the design or create their own modifications. However, individual differences are more than just annoying departures from the norm. They are expressions of human variability that can be accommodated by a system if the technology is mature enough and if the designers are clever enough. In fact, systems designed to the aggregate do not allow the best performance on average; they only allow the best performance for a single instantiation of a design, on average. Systems designed to take advantage of individual differences may, in fact, allow optimal performance.

This chapter has identified some individual differences which designers should be aware of when developing virtual environments. It has paid particular attention to collaboration in VEs, which appears to be an important emerging application of that technology.

Physical individual differences are of course important in the ergonomics, strictly speaking, of any system, but the focus of this chapter has been on psychological differences. For example, individuals vary in their social and teamwork skills, and one can expect differences on these dimensions that are observed in the real world to carry over into the virtual. It appears that this is so, but there may be a moderating impact of other user characteristics, specifically, for example, technical aptitudes. Differences in attention allocation skills affect not only a user's own experience of a VE, but also the type of interactions that occur within a CVE and group performance.

It seems, therefore, that there exist two classes of individual differences that are important for VE development. First, there are individual differences related to how a person interacts with a virtual environment. These differences are, most importantly, cognitive and affect how the person functions within a virtual environment. Specific variables have been identified and include attention allocation, predisposition toward presence experiences, and technological aptitudes, among others. This class is generally important; that is, it applies to all VE applications. Second, there are individual differences related to how a person interacts with other persons within a CVE. These differences are, most importantly, social skills. These include interaction style, interpersonal skills, and tendencies toward particular roles within a group dynamic, among others. This class is important in the case of CVEs.

**TABLE 18.1**
Virtual Environment Application Design Recommendations

| *Design Features* | *Recommendation* |
|---|---|
| Realism | • Ensure VEs provide for rich multisensory experiences like reality (and VE system interfaces are suited to innate abilities of users).<br>• Ensure VE and its objects make up realistic or believable surroundings.<br>• Preserve features of real spaces to encourage user creativity in virtual task performance (Acker, 1995).<br>• Make CVE applications and user interactions as realistic as possible to ensure cognitive loadings in VE tasks are similar to loads experienced in real-world tasks<br>• Match cognitive loading to reality, in addition to considering user perceptions and experience.<br>• Minimize disparities in functional fidelity of VE from reality to prevent breakdowns in user interactions in CVEs and prevent users from making incorrect assumptions on collaborative activities. |
| Interface | • Create more natural control interfaces to promote novice user spatial positioning performance or to take into account user experience in assigning virtual tasks or provide training in virtual motion.<br>• Structure displays to ensure user perception of stimuli in VEs for task performance.<br>• Structure, for example, visual, auditory and tactile interfaces to allow all persons to interact effectively with virtual environments.<br>• Provide capability to address physical disabilities of users including hand-control navigation techniques and voice driven input.<br>• Carefully consider design of interfaces to CVEs (system displays, hand controllers, and user representation) as they affect individual sensations of realism about VEs and continuity of user interactions.<br>• Exploit unique capabilities of VEs to resolve some of effects of individual differences in perceptions of virtual worlds including: ability to provide single user with multiple viewpoints of environment, such as viewpoint of fellow collaborator |
| Individual differences | • Focus design efforts on age-related individual differences in vision and hearing.<br>• Consider user ability to develop knowledge on complex virtual environments.<br>• Consider degenerative effects on physiology, particularly the senses (e.g., vision and hearing) in VE application design.<br>• Present displays and interfaces to allow aged persons to develop understanding of spatial positioning and navigation in VEs in advance of presenting them with virtual task objects and showing them representations of other users in CVE.<br>• Consider user attention allocation capabilities, and creativity and long-term memory storage in application design.<br>• Consider geriatric effects on spatial orientation and memory, as well as technical aptitudes for reasoning in VE design.<br>• Consider interaction effect of user characteristics including age, information processing abilities, and experience on user sensations of presence in VEs and virtual task performance.<br>• Offer methods for dealing with perceptual-cognitive limitations including multimodal representations of stimuli (e.g., visualizing sound for the hearing impaired). |
| Matching | • Consider individual differences in culture to enhance matching.<br>• Match the structure of virtual interactions among CVE users to real human–human communication through consideration of: user self-representations, the interfaces to user virtual embodiments, the sensory modalities available for communication, and the clarity of communication provided.<br>• Match SA requirements in VE to reality, specifically in VE trainer development for experienced users toward ensuring virtual skill transfer to reality. |

(Continued)

**TABLE 18.1**
(Continued)

| *Design Features* | *Recommendation* |
|---|---|
| | • Match user perceptions of behavior of self-representations in VEs to, for example, the control interfaces they use to move their virtual embodiment. |
| VE technology/general | • Eliminate or reduce effects of (temporal, spatial, and filtering) perturbations. Factors include: application resolution, update rate, level of environment detail, and others.<br>• Design networking capabilities for supporting CVEs to provide near real-time application responsiveness and system synchronization, or at least meet requirements of most sensitive user during any given deployment. |
| Teaming/VE sociology | • Consider group dynamics in CVE application design.<br>• Configure virtual groups to prevent social issues from degrading organization.<br>• Exploit user social skills toward organizing virtual groups.<br>• Devise methods to ensure users of CVEs are aware of professional status of other participants if individual perceptions of relevance of professional hierarchies in virtual spaces are critical to application success.<br>• Design virtual tasks with user aptitudes in mind.<br>• Design CVE teams to include users with specific aptitudes and experience to ensure error-free performance.<br>• CVE design for distributed user teamwork in virtual tasks should consider user experiences in real world for forming effective virtual teams.<br>• Ensure that users immersed in CVE participating in virtual group have similar level of team experience to promote collaborative virtual task performance.<br>• Consider level of adaptation of users to VEs for promoting virtual task performance.<br>• Allocate decision-making power and virtual task roles to be assumed by users in a CVE to optimize user collaboration. |
| Presence | • Consider individual differences in designing for shared presence including: a user's background, interests, viewpoints, and sensitivities.<br>• For shared presence in CVEs, provide a means for users to experience sense of physical presence in common VE, as well as user–user interaction. |

Although it seems clear that individual differences are important to how people function within VEs and CVEs, it is not as clear how to design VEs to accommodate, or even take advantage of, those differences. It seems that VE affords designers with the flexibility to shape each user's experience to his or her individual needs, but it is not obvious exactly how to accomplish this objective. Does a person with limited attention span, for example, benefit from the appearance of a halo around the head of a collaborator in a CVE when that person speaks? Development of ways and means to classify individuals on the basis of their differences, and to modify a VE to best match those differences, is a research topic of great complexity but with concomitant potential benefit to VE users.

Table 18.1 summarizes the VE design features discussed in this chapter. The table offers general recommendations toward accounting for individual differences in VE application design through these features. Due to the relative immaturity of VE research into individual difference effects on application design, "hard-and-fast" rules for ensuring each recommendation is upheld cannot be offered, but rather rules must be formulated based on future empirical research. Virtual environment application parameters should be explored to learn specific acceptable ranges of, for example, virtual control latency, for different types of virtual tasks. These parameter ranges will serve to describe overall design features and systems capabilities.

## 8. REFERENCES


Acker, S. R. (1995). Space, collaboration, and the credible city: academic work in the virtual university. *Journal of Computer-Mediated Communication, 1*(1), 1–14.

Allen, R. C., Bolton, A. E., Ehrlich, J., & McDonald, D. P. (1999). The virtual ride questionnaire (VRRQ): A virtual reality entertainment evaluation tool. In *Proceedings of the 43rd Annual Meeting of the Human Factors and Ergonomics Society* (p. 1409). Santa Monica, CA: Human Factors and Ergonomics Society.

Barfield, W., & Weghorst, S. (1993). The sense of presence within virtual environments: a conceptual model. In G. Salvendy and M. Smith (Eds.), *Human–computer interaction: Software and hardware interfaces* (pp. 699–704). Amsterdam, The Netherlands: Elsevier Science Publishers.

Birren, J. E., & Livingston, J. (1985). *Cognition, stress, and aging*. Englewood Cliffs, NJ: Prentice-Hall.

Crampton, S., & Hilton, A. (1999). Populating the Web: Pioneering a paradigm for photorealistic avatars [Online]. Available: http://www.avatarme.com/concept/concept.htm [1999, September 1].

Cripps, D. (1995). Wheelchair virtual reality training for path following and autonomous obstacle avoidance [Online]. Available: http://www.engineering.usu.edu/Departments/ece/projects/csois/vrtctrl.html [1999, September 1].

Cruz-Neira, C. (1998). Applied virtual reality—Course 14 notes: Overview of virtual reality. In *Proceedings of the 25th International Conference on Computer Graphics and Interactive Techniques: SIGGRAPH 1998* (pp. 2-1–2-28). Orlando, FL: Association for Computer Machinery.

Draper, J. V. (1995). Teleoperators for advanced manufacturing: Applications and human factors challenges. *International Journal of Human Factors in Manufacturing, 5*, 53–85.

Draper, J. V., Handel, S., Sundstrom, E., Herndon, J. N., Fujita, Y., & Maeda, M. (1987b). *Final report: Manipulator comparative testing program* (Tech. Rep. No. ORNL/TM-10109). Oak Ridge, TN: Oak Ridge National Laboratory.

Draper, J. V., Kaber, D. B., & Usher, J. M. (1998). Telepresence. *Human Factors, 40*(3), 354–375.

Draper, J. V., Omori, E., Harville, D. L., Wrisberg, C. A., & Handel, S. (1987a). Test results: The manipulator operator skill test (Tech. Rep. No. ORNL/TM-10524). Oak Ridge, TN: Oak Ridge National Laboratory.

Durlach, N. I., & Mavor, A. S. (Eds.). (1995). *Virtual reality: Scientific and technological challenges*. Washington, DC: National Research Council, National Academy Press.

Durlach, N., & Slater, M. (1998). *Presence in shared virtual environments and virtual togetherness* [Online]. Paper presented at the Presence in Shared Virtual Environments Workshop, Ipswich, England. Available: http://www.cs.ucl.ac.uk/staff/m.slater/BTWorkshop/durlach.html [1999, September 1].

Egan, D. E. (1988). Individual differences in human-computer interaction. In M. Helander (Ed.), *Handbook of human-computer interaction* (pp. 543–568). Amsterdam, The Netherlands, North-Holland: Elsevier Science Publishers.

Ellis, S. R. (1993). Virtual presence. In R. S. Kalawsky (Ed.), *The science of virtual reality and virtual environments* (pp. 80–85). New York: Addison-Wesley.

Endsley, M. R. (1995). Toward a theory of situation awareness in dynamic systems. *Human Factors, 37*, 32–64.

Ferrell, W. R. (1965). Remote manipulation with transmission delay. *IEEE Transactions on Human Factors in Electronics* HFE-6(1), 24–32.

Ferrell, W. R. (1966). Delayed force feedback. *Human Factors, 8*(5), 449–455.

Fisk, A. D., & Rogers, W. A. (1991). Toward an understanding of age-related memory and visual search effects. *Journal of Experimental Psychology, 120*, 131–149.

Fraser, M., & Glover, T. (1998). Representation and control in collaborative virtual environments. [Online]. In *Proceedings of the 1998 United Kingdom Virtual Reality Special Interest Group Conference*, Exeter, England. Available: http://www.dcs.ex.ac.uk/ukvrsig98/pap3_03.htm [1999, September 1].

Greenhalgh, C., & Benford, S. (1995). MASSIVE: A collaborative virtual environment for teleconferencing. *ACM Transactions on Computer-Human Interaction, 2*(3), 239–261.

Hart, S. G., & Staveland, L. E. (1988). Development of NASA-TLX (Task Load Index): Results of empirical and theoretical research. In P. A. Hancock & N. Meshkati (Eds.), *Human mental workload* (pp. 139–183). Amsterdam, The Netherlands, North-Holland: Elsevier Science Publishers.

Held, R. M., & Durlach, N. I. (1992). Telepresence. *Presence, 1*(1), 109–112.

Hertzog, C. (1989). Influences of cognitive slowing in age differences in intelligence. *Developmental Psychology, 25*, 636–651.

Hix, D., Swan, J. E., Gabbard, J. L., McGee, M., Durbin, J., & King, T. (1999). User-centered design and evaluation of a real-time battlefield visualization virtual environment. In *Proceedings IEEE Virtual Reality '99* (pp. 96–103). Houston, TX: IEEE Press.

Kaber, D. B., & Riley, J. R. (in press). Virtual reality for scientific data visualization. In D. Harris (Ed.), *Proceedings of the third International Conference on Engineering Psychology and Cognitive Ergonomics*. Aldershot, England: Ashgate.

Kaber, D. B., Riley, J. M., Zhou, R., & Draper, J. V. (2000). Effects of visual interface design, control interface type, and control latency on performance, telepresence, and workload in a teleoperation task. In *Proceedings of the 14th*

*Triennial Congress of the International Ergonomics Association / 44$^{th}$ Annual Meeting of the Human Factors and Ergonomics Society* (pp. 503–506). Santa Monica, CA: Human Factors and Ergonomics Society.

Kalawsky, R. S. (1993). *The science of virtual reality and virtual environments*. New York: Addison-Wesley.

Lombard, D. M., & Ditton, T. (1997). At the heart of it all: The concept of telepresence [Online]. *Journal of Computer Mediated Communication, 3*(2). Available: http://www.ascusc.org/jcmc/vol3/issue2/ [1999, September 1].

McGovern, D. E. (1993). Experience and results in teleoperation of land vehicles. In S. Ellis (Ed.), *Pictorial communication in virtual and real environments* (pp. 182–195). London: Taylor & Francis.

Miller, L., Stanney, K., Guckenberger, D., & Guckenberger, E. (1997). Above real-time training. *Ergonomics in Design*, *5*(3), 21–24.

Nemire, K., Burke., A., & Jacoby, R. (1994). Human factors engineering of a virtual laboratory for students with disabilities. *Presence: Teleoperators and Virtual Environments*, *3*, 216–226.

Norman, D. A. (1986). Cognitive engineering. In D. A. Norman & Draper (Eds.), *User-centered system design* (pp. 30–61). Hillsdale, NJ Lawrence: Erlbaum Associates.

Odegard, O., & Oygard, K. A. (1997). Learning in collaborative virtual environments—Impressions from a trial using the Dovre framework. Presented at ESPRIT EMMSEC '97 Conference [Online]. Available: http://televr.fou.telenor.no/~olao/CyberCity.html [1999, September 1].

Orasanu, J., & Salas, E. (1993). Team decision-making in complex environments. In G. Klein, J. Orasanu, R. Calderwood, & C. E. Zsambok (Eds.). *Decision making in action: Models and methods* (pp. 327–345). Norwood, NJ: Ablex Publishing.

Parker, D. E., & Harm, D. L. (1992). Mental rotation: A key to mitigation of motion sickness in the virtual environment? *Presence*: *Teleoperators and Virtual Environments*, *1*(3), 329–333.

Piantanida, T., Boman, D. K., & Gille, J. (1993). Human perceptual issues and virtual reality. *Virtual Reality Systems*, *1*(1), 43–52.

Psotka, J., & Davison, S. (1993). *Cognitive factors associated with immersion in virtual environments*. Alexandria, VA: Army Research Institute.

Riley, J. M., & Kaber, D. B. (1999a). The effects of visual display type and navigational aid on presence, performance, and workload in virtual reality training of telerover navigation. In *Proceedings of the 43$^{rd}$ Annual Meeting of the Human Factors and Ergonomics Society* (pp. 1251–1255). Santa Monica, CA: Human Factors and Ergonomics Society.

Riley, J. M., & Kaber, D. B. (1999b). The telepresence and performance effects of visual display type and navigational aid in simulated rover navigation training through a virtual environment [CD-ROM]. In *Proceedings of the ANS Eighth Topical Meeting on Robotics and Remote Systems*. LaGrange Park, IL: American Nuclear Society.

Romano, D. M., Brna, P., & Self, J. A. (1998a). *Collaborative decision-making and presence in shared dynamic virtual environments* [Online]. Paper presented at the Presence and Shared Virtual Environments Workshop, Ispwich, England. Available: http://www.cbl.leeds.ac.uk/~daniela/Collaborative_DM_and_Presence.html [1999, September 1].

Romano, D. M., Brna, P., & Self, J. A. (1998b). *Influence of collaboration and presence on task performance in shared virtual environments* [Online]. Paper presented at the 1998 United Kingdom Virtual Reality Special Interest Group (UKVRSIG) Conference, Exeter, England. Available: http://www.dcs.ex.ac.uk/ukvrsig98/pap4_03.htm [1999, September 1].

Salthouse, T. A. (1992). Why do adult age differences increase with task complexity? *Developmental Psychology*, *28*, 905–918.

Sellen, A., & Nicol, A. (1990). Building user-centered on-line help. In B. Laurel (Ed.), *The art of human–computer interface design* (pp. 143–153). Reading, MA: Addison-Wesley.

Slater, M., & Wilbur, S. (1997). A framework for immersive virtual environments (FIVE): Speculations on the role of presence in virtual environments. *Presence*, *6*(6), 603–617.

Smith, D. C., Cole, R. E., Merritt, J. O., & Pepper, R. L. (1979). *Remote operator performance comparing mono and stereo TV displays: The effects of visibility, learning and task factors* (Tech. Rep. No. NOSC TR 380). San Diego, CA: Naval Ocean Systems Center.

Smith, K. U. (1972). Cybernetic psychology. In R. N. Singer (Ed.), *The psychomotor domain* (pp. 285–348). New York: Lea and Febiger.

Smith, K. U., & Smith, W. M. (1962). *Perception and motion: An analysis of space-structured behavior*. Philadelphia: Saunders.

Smith, T. J., Henning, R. A., & Smith, K. U. (1995a). Performance of hybrid automated systems—A social cybernetic analysis. *International Journal of Human Factors in Manufacturing*, *5*, 29–51.

Smith, T. J., Keran, C. M., Koehler, E. J., & Mathison, P. K. (1995b). Teleoperation of mobile equipment—A behavioral cybernetic analysis. In *Proceedings of the 25th International Conference on Environmental Systems* (pp. 1–20). San Diego, CA: SAE International.

Smith, T. J., & Smith, K. U. (1987). Feedback-control mechanisms of human behavior. In G. Salvendy (Ed.), *Handbook of human factors* (pp. 251–293). New York: John Wiley & Sons.

Smith, T. J., & Smith, K. U. (1988). The social cybernetics of human interaction with automated systems. In W. Karwowski, H. R. Parsaei, & M. R. Wilhelm (Eds.), *Ergonomics of Hybrid Automated Systems I* (pp. 691–711). Amsterdam, The Netherlands: Elsevier Science Publishers.

Smith, T. J., & Smith, K. U. (1989). The human factors of workstation telepresence. In S. Griffin (Ed.), *Third annual workshop on space operations automation and robotics* (Tech. Rep. No. 3059, pp. 235–250). Houston, TX: National Aeronautics and Space Administration, Johnson Space Flight Center.

Stanney, K. M., Mourant, R. R., & Kennedy, R. S. (1998). Human factors issues in virtual environments: A review of literature. *Presence: Teleoperators and Virtual Environments*, *7*(4), 327–352.

Stanney, K. M., & Salvendy, G. (1994). Effects of diversity in cognitive restructuring skills on human-computer performance. *Ergonomics*, *37*(4), 595–609.

Thorndyke, P. W., & Statz, C. (1980). Individual differences in procedures for knowledge acquisition from maps. *Cognitive Psychology*, *12*, 137–175.

Tromp, J., Bullock, A., Steed, A., Sadagic, A., Slater, M., & Frecon, E. (1998, November/December). Small group behavior experiments in the Coven project. *IEEE Computer Graphics and Applications*, 53–63.

Wickens, C. D. (1992). *Engineering psychology and human performance*. New York: HarperCollins.

Witmer, B. G., & Singer, M. J. (1994). *Measuring immersion in virtual environments* (Tech. Rep. No. 1014). Alexandria, VA: U.S. Army Research Institute for the Behavioral and Social Sciences.

Witmer, B. G., & Singer, M. J. (1998). Measuring presence in virtual environments: A presence questionnaire. *Presence*, *7*(3), 225–240.

# 19

# Using Virtual Environments as Training Simulators: Measuring Transfer

Corinna E. Lathan,[1] Michael R. Tracey,[2]
Marc M. Sebrechts,[3] Deborah M. Clawson,[3]
and Gerald A. Higgins[4]

*[1]AnthroTronix, Inc.*
*387 Technology Drive, College Park, MD 20742*
*lathan@alum.mit.edu*
*[2]Department of Biomedical Engineering*
*[3]Department of Psychology*
*The Catholic University of America, Washington, DC 20064*
*mtracey13@email.msn.com*
*sebrechts@cua.edu*
*clawson@cua.edu*
*[4]8003 Boulder Lane, Silver Spring, MD 20910*
*Telemed@erols.com*

## 1. INTRODUCTION

Skills are generally acquired by learning a set of facts about a task and developing appropriate procedures. For example, to learn to fly a plane, you first might learn the function of the flight controls and their locations and then practice manipulating the controls in appropriate sequences for a flight procedure. The practice part of training often cannot be done in the actual situation; a plane may not be available or the cost of the plane and an instructor on a regular basis may be prohibitive. In these situations, training is provided in alternative environments, or simulators, that mimic the target task. Designing simulators that will produce effective training has been a long-standing goal for those involved in skill acquisition, and the success of such simulations is generally measured by the effectiveness of subsequent performance on the actual task—the degree of "transfer" from the simulated to the real environment.

Among the most recent additions to training simulators are virtual environment (VE) systems that enable a person to interact with a computer-generated environment. Although there is a broad range of definitions describing this new technology, it is generally agreed that VE technology provides a new level of interaction and expands the application options that can be explored. As a computer-modeling technology that generates an immersive and interactive synthetic environment, VE is typically less dependent on physical properties of context and materials than most other simulations.

The purpose of this chapter is to discuss several issues related to the use of VE simulators for transfer of training. First, the meaning of transfer in an abbreviated historical context is examined. Second, one of the key differences between real-world and VE transfer is explored, the match to the actual task, known as the fidelity. Third, measures of transfer are discussed, concluding with an examination of several applications in which VE simulators play a role.

## 2. HISTORICAL CONTEXT OF TRAINING TRANSFER

A standard technique for determining the effectiveness of any training is the success in the application of that training in the actual performance environment. Does a particular flight simulator result in better in-flight performance? Does training in a simulated building improve firefighting skills? It is generally acknowledged that the answer to these questions will depend on properties of the training environment. In some instances, classroom training is assumed to provide an adequate learning context, combining descriptions and images with general principles. In the case of simulations it is often assumed that the rate of skill acquisition depends on what has previously been learned in "similar" contexts. It is generally believed, for example, that a pilot who has flown one aircraft is more likely to learn to fly a new plane more quickly than a novice. Nonetheless, there is substantial debate about what constitutes a "similar" context and how useful is such prior training. Virtual environments can be designed to test an operator's ability to perceive and act upon simulated information. But, how will the same operator perform when faced with similar data in a real environment? This is known as the problem of *transfer of training*.

One common belief about transfer is that mental performance can be improved by general "conditioning" of the mind, analogous to the way in which an athlete might condition his or her body. This mental-muscle view was prevalent among a number of psychologists in the early part of the 20th century and was known as the Doctrine of Formal Discipline (Angell, 1908). The mind was thought to have general faculties that when exercised appropriately would lead to fairly broad transfer. Many current educational practices are still based on this conception that training in basic skills will improve performance in a wide variety of domains. This was, and to some extent still is, the basis for emphasizing foreign language training as improving general education. It has proven difficult, however, to provide any evidence supporting this very general view of learning.

An opposing view of transfer, emphasizing the specificity of learning, was the *theory of identical elements* (Thorndike, 1906). This theory suggested that transfer occurred only when two tasks shared identical elements. The presence of specific common elements, it was suggested, led to the illusion of transfer based on general reasoning. A number of studies have supported the finding of specificity of transfer. Chase and Ericsson (1982), for example, trained a person to increase short-term digit recall from the typical digit span of about 7 or 8 items to 81 digits following some 200 hours of practice. However, they found that this memory span increase did not transfer from digits to letters.

In an extensive review of the transfer literature, Singley and Anderson (1989) suggest that the identical elements view is too restrictive. Although transfer does exhibit some specificity, rather than requiring a specific match of surface situation-response elements, transfer occurs when two skills exhibit the same logical structure. Thus, the ability to navigate through a town in one type of vehicle may improve the ability to navigate the same town in another vehicle.

## 3. TRAINING ENVIRONMENT FIDELITY

Transfer of training may in some ways depend on all aspects of a given context. The optimal way to learn a specific procedure, in this view, is to learn a procedure precisely as it will be tested, including matching the context in which learning occurred with that in which performance is measured (see Tulving, 1983, on encoding specificity). This is in part the rationale behind on-the-job training. At the same time, this ideal match rarely occurs, and the exact conditions of performance are usually not known in advance or stable across time. Firefighters cannot test their navigation in the actual building as it is burning, and surgeons cannot practice on actual patients. So the goal is to find conditions in which training provides an adequate match.

To the extent that skill acquisition reflects substantial specificity, training transfer should be best when training mimics performance as closely as possible. Virtual environments provide a means to fulfill the objective of training in the target environment without really being there. The extent of this match is the training environment's "fidelity". With perfect fidelity, a training environment (virtual or real) would be indistinguishable from the actual task environment. Thus, one set of questions about VEs is the extent to which they can be used to replace real environments for training and transfer. A closely related question concerns which aspects of a VE are essential and which are secondary for training a specified set of tasks.

Another set of questions of equal or greater import concerns how VEs can be transformed from the normal appearance and behavior of the real world in ways that might benefit learning and transfer. Because of their wide-ranging flexibility, VEs provide a powerful framework for assessing a range of possible applications. By providing new situations in which structures can vary from "reality," VE technology evokes a new set of design questions about developing transfer that is not strictly dependent on the degree of training-performance match.

### 3.1 VE as a Real-World Replacement

Few current VEs provide a sustained experience that is perceptually and experientially indistinguishable from reality. Typically, there is some trade-off among different aspects of fidelity. For example, high-quality visual images usually result in a longer lag in tracking (see chap. 21, this volume). How much fidelity is required for any given performance objective remains an empirical question. There is still much to be learned about how specific changes in properties of VEs influence training or transfer performance. In some ways, the key is to map the demands of any given VE to human-information processing constraints (Card, Moran, & Newell, 1983). In the limit case, for example, a lag of a few milliseconds should make little difference to the visual processor while walking through a virtual building, whereas a lag of half a second might result in a number of difficulties, including walking into walls (see chap. 3, this volume).

The requirements for training environment fidelity will also depend to a large extent on the attentional demands of the specific task being performed. For example, in navigating a simple building, wall texture might have little impact on training (see chap. 24, this volume). However, in a complex structure, in which hallways are differentiated primarily by their appearance, wall texture might have a substantial impact on learning. In our own studies, although the issue has not been studied systematically, a variety of textured surfaces have been used in evaluating route learning. Although greater fidelity is recognized and appreciated by participants, no convincing evidence has been found that it influences learning in the building environments examined. Results in other domains also suggest that high fidelity may not be essential to certain kinds of procedural or part-task training. Of course, the quality of the environment may also have other consequences because performance measures cannot be strictly separated from other measures of motivation or utilization (see chap. 35, this volume). If fidelity during training is poor, learners may refuse to use training materials unless required to do so by the job.

### 3.2 Going Beyond the Real World

Virtual environments can be transformed from the normal appearance and behavior of the real world in ways that might benefit learning and transfer. For example, flying through a building might provide additional information to that acquired from walking through it. Transparency of walls might permit the development of a better comprehension of spatial structures and relations. This is discussed further in section 5 in the context of learning to navigate through an environment.

A virtual simulation may also present some *parts* of a task that normally cannot be separately controlled in the physical world. For example, a flight simulator may allow an operator to land an airplane repeatedly or a surgical simulator may be set to just test suturing skills and not a whole medical procedure. However, it is important to keep the part-task learning in the context of the whole task, as mastering either task alone clearly would not satisfy the training needs of a pilot or a surgeon respectively.

## 4. MEASURES OF TRANSFER

The effectiveness of a simulator is determined by how much an operator's performance is improved on a real-world task due to training in a simulator. Despite the wide acceptance of simulators as valid training tools, few studies exist that actually measure this transfer. Measures of simulator effectiveness require collection of operator performance data. Often, it is necessary to collect performance data before simulator based training begins or measure performance in control groups that do not train in simulators. Table 19.1 presents methodologies for measuring data necessary to assess simulator effectiveness. The data contained in this table are taken largely from a study undertaken by the NATO Advisory Group for Aerospace Research and Development (1980).

From an experimental design perspective, the approach that offers the greatest rigor in assessing training transfer is the transfer of training (ToT) method. When using this method it is important to derive measures of how long it takes to learn the task, how effectively the task is subsequently performed in the real world, and how much training that is normally conducted in the real world is saved with the simulator. The transfer effectiveness ratio (Taylor, Lintern, Hulin, Talleur, Emmanuel, and Phillips, 1997) estimates simulator effectiveness as the difference between presimulator training performance and postsimulator training performance, divided by time spent in the simulator.

$$\text{Transfer effectiveness ratio (TER)} = \frac{\text{Presimulator performance (e.g. time for task completion)} - \text{Postsimulator performance}}{\text{Time spent in simulator}}$$

The transfer effectiveness ratio provides a measure of savings gained by using the simulator (e.g. a TER of 0.50 means that every hour in the simulator saves a half hour of real-world training). Another measure of effectiveness reports the percent of transfer by comparing control-operator performance to operators trained in a simulator. Both of these proportions require control and experimental data.

It is possible to have a decrease in performance following simulator training. This is indicative of a so-called negative transfer of training. In this case the experimental group exhibits performance that is significantly worse than the control group that did not train in the simulator. In the case of negative transfer, the introduction of a simulator, or modification to an existing simulator, is a detriment to existing training methods.

## 5. TRAINING APPLICATIONS: SPATIAL NAVIGATION

Since VEs make possible "presence" in physically distant locations (see chap. 40, this volume), they provide a natural context for learning about movement through unfamiliar spaces, ranging

**TABLE 19.1**
Methods for Measuring Training Transfer from VE-based Simulators to Real-World Performance

| *Method* | *Description* | *Remarks* |
|---|---|---|
| Operator opinion method | Operators, instructors, training specialists and students are asked to give their opinions on the perceived training value of a simulator, features of the simulators, or probable impact of simulator based training on subsequent real world performance. | Useful when operational training or performance testing is not feasible. However, assumes the operator, instructor, or trainee is able to assess objectively how much is learned from the simulator. May fail to recognize that such opinion is based on previous knowledge and experience. |
| Assessment of fidelity | Describes the physical similarity between the simulator and the real-world environment, equipment, interface, or facility. | This method assumes that higher fidelity will yield higher transfer. It is commonly used by the commercial aircraft industry. Regulatory authorities often impose high physical and dynamic fidelity requirements so that instrument rating and proficiency checks can be carried out in the simulator rather than in a real aircraft. In such a role simulators are being used as assessment tools, not training tools. Training may be possible with far less sophisticated devices. In addition, it appears that high fidelity generates user acceptance, but this does not of itself mean that a device is more effective at training operators. |
| Transfer of training (ToT) method | Involves two groups of trainees, an experimental group, which receives simulator training prior to further training for performance testing and a control group, which receives all of its training in the real world. Alternatively, the experimental group could be participants using a newly developed simulator, and the control group could use an existing simulator program. Groups must be equated in terms of relevant prior training and experience. | This model is generally most appropriate to determine whether simulator training has improved subsequent operational performance. |
| Self-control transfer method | This method uses an experimental group to serve as their own controls. Operational performance is assessed, simulator based training is introduced, then subsequent operational performance is assessed and compared to the first performance assessment. | When using this method, the time interval between performance assessments must be taken into account. One problem with this model is that it assumes the trainee's subsequent performance on the operational task has improved as a result of simulator training. |
| Preexisting control transfer of training method | Determines transfer of training using the new simulator, as performance data already exists for the control group. | Often a simulator is introduced after an established training protocol is in place. This model is then similar to the transfer of training model. |

(Continued)

**TABLE 19.1**
(Continued)

| *Method* | *Description* | *Remarks* |
|---|---|---|
| Uncontrolled transfer method | Simply determines if naive subjects can perform a particular task in an operational setting following simulator training. | A crude way to determine training transfer. This is useful to quickly evaluate features or improvements to a simulator but does not provide measures to accurately quantify transfer. |
| Inverse transfer of training method (or backward transfer method) | Experts at the operational task perform the same tasks, without practice, in a simulator. A positive result assumes that a suitable training program exists for the simulator. | The experienced operator is already proficient at the task and may have highly generalized skill. The simulator may be suitably designed for the evocation of a particular set of behaviors from a skilled operator. |
| Simulator-to-simulator transfer method | A lower fidelity simulator is used in part task training and is followed by whole-task testing (or part-task) on a higher fidelity simulator. | Valuable in reducing the use of more complex simulators. A related method, the simulator fidelity method, assumes that a higher fidelity simulator will yield higher transfer and a lower fidelity simulator will yield less or negative transfer. |

from a new office complex to a building from which hostages need to be rescued. In learning to navigate a space there are several different kinds of knowledge that are used, and these are typically described as successive components in learning (Siegel & White, 1975; see chap. 24, this volume). Landmarks are used to establish reference points without any specific connection. A path through a space is described as a route; such "route knowledge" is often fairly specific and may reflect a rigid procedural sequence. A more global grasp of a space provides greater flexibility and less dependence on a specific sequence. This "survey knowledge" is typically accompanied by the ability to identify the relative locations of objects separate from a specific path.

Two uses of VEs for transfer of training in such navigational contexts are explored in the following sections. First, can reasonable fidelity provide transfer of training from virtual to real environments? Second, how can transformations of the VE be used to modify the utility of training and transfer of navigational skill? The ToT experimental method of measuring transfer was used to address these questions. Control groups were trained and tested in a real-world environment while other groups were trained in VE or through other methods such as map reading.

## 5.1 Medium Fidelity Environments

Most research on navigation has focused on characterizing route and survey knowledge. We examined transfer of training based on these two aspects of navigational learning. The first issue is whether or not people can learn a specific route based on training in a virtual environment. The answer to this question depends both on the VE used and the task; in this case, the typical task is following a specific real-world route that matches one learned in the virtual environment. Successful transfer of such route training has been demonstrated using very-high-fidelity visual displays, combined with maps (Witmer, Bailey, Knerr, & Parsons, 1996). In our studies (Clawson, Miller, & Sebrechts, 1998) we have emphasized medium fidelity visual displays; in addition, in order to assess VE utility, providing supplementary map displays has generally been avoided. Consistent with other findings (e.g., Waller, Hunt, & Knapp, 1998), our results, based on measures of correct turns, hesitations, and distance traveled, have supported transfer of training from virtual to real environments comparable to that using maps or real world training. However, these results indicated that VE route training showed substantial specificity (see chap. 24, this volume). Participants hesitated significantly more when moving through a building in a direction opposite to that of VE training than when moving through the building in a direction aligned with VE learning. This specificity was not present in matched map or real-world learning conditions.

With respect to survey learning, research also shows reasonable transfer as well as specificity. When participants followed the route learned during VE training, their ability to identify the relative locations of real-world objects through pointing was comparable to that of participants who learned in the actual building. However, when testing was in the direction opposite to that of training, VE-trained participants' performance was substantially worse than that of real-building-trained participants. A similar specificity was also found for distance estimates; relative direction of testing had more of an effect for the VE-training group than for either map or real-world trained groups.

These results leave open some issues about fidelity. Since real-world route training does not produce the same specificity, there must be VE characteristics that produce differential performance. One possibility is that the way in which route following is conducted changes the way in which the space is perceived. Route-following in VE may induce constraints on the way in which the space is viewed. This was addressed in part by another set of our experiments that used exploratory learning instead of route following (Sebrechts, Mullin, Clawson, & Knott, 1999). In these studies, participants explored a building, locating fifteen different regions of the building using either a VE or a map. They then were required to move between specific locations in the real building using the most efficient path possible. In these studies, the VE-trained participants found significantly more efficient paths than the map-trained participants. A potential explanation for these results is that the exploratory learning task induces more scanning of the environment, thus reducing the transfer specificity previously found for VE relative to map or real-world training.

## 5.2 Transformed Environments

The majority of research on navigational training and transfer from VEs has emphasized varying degrees of fidelity, helping to define the conditions needed for transfer. However, one of the central benefits of a VE is that it facilitates deviation from the real or normal context. Some of these transformations are relatively minor. For example, in the exploratory studies described above, research indicated that people failed to fully explore a virtual building in the absence of additional cues. When virtual spheres were placed in the building, which spoke their location names and changed color after being encountered, completeness of exploration improved dramatically. Other transformations can be substantial, such as allowing a person to "fly" through a virtual building.

A number of our recent studies have focused on one particularly promising transformation—transparency (Sebrechts & Knott, 1998). Comparisons were made between participants walking through a virtual building with traditional "opaque" walls, and an identical building in which the walls had been made "transparent" such that objects could be seen through the walls, although their locations were still evident. After learning the location of 21 objects in the building to a specified criterion, the participants' mental models of the building were assessed using a task that required participants to identify the relative location of objects they had seen in the building. The mental models of participants who learned from an opaque VE reflected distances along a path through the building, described by a city-block or route-based metric. In contrast, mental models of participants who learned from a transparent VE reflected a straight-line or Euclidean distance in addition to a route-based distance.

Differences between training environments were also reflected in drawings of the building. Both groups easily identified the objects in the building; however, the transparent group was substantially more accurate in identifying the structural layout of the building including the relative size and positioning of rooms. Overall, these results suggest that a transparent VE may help to facilitate the acquisition of survey knowledge, a task that frequently requires substantial experience in the real world.

## 6. TRAINING APPLICATIONS: COMPLEX MOTOR LEARNING

Complex motor learning, such as learning to pilot an airplane or to perform surgery, poses another interesting set of challenges in using VE systems for training. Similar to the navigation task described above, the fidelity of the VE simulator is of importance. However, because of the complexity of the task, the interface is usually much more than a visual display and may include mechanical components as well. For example, a flight simulator may have a high-fidelity physical control panel, medium-fidelity visual graphics, and a low-fidelity mechanical simulation of aircraft movement.

The need to manipulate objects in the environment is another distinguishing aspect of learning a motor task versus the navigation tasks described above. This poses additional questions concerning the modality of the interface. Haptic (force) feedback may become increasingly important (see chaps. 5–6, this volume). Experiments examining transfer of training of manipulation tasks from VE using low-fidelity graphics and no force feedback present conflicting results (Kenyon & Afenya, 1995; Kozak, Hancock, Arthur, & Chrysler, 1993). Using a task that required picking up and placing a can in a target location, Kozak et al. (1993) found no significant transfer-of training benefits from VE training as compared to real-world and no training conditions. However, using a similar task, Kenyon and Afenya (1995), found that VE trained subjects outperformed untrained subjects on the first trial of the real-world task. Apparently the difference in outcome was a result of the improved training environment and its higher fidelity. Kozak's task used a virtual hand that had to be synchronized with the real hand, resulting in an average completion time of 63 seconds. Kenyon's task used an augmented reality setup in which participants could see their real hand moving virtual objects, resulting in an average completion time of 19.2 seconds. In both studies, task completion in the real world required roughly 6 seconds. Kenyon et al. make a compelling argument that limitations in fidelity of the VE produced these substantial time differences, and that the improved, but limited transfer, stems from a better match between the VE and real world tasks. Limitations in the VE, including time lags, may result in different task performance. As a consequence, many such environments do not yet show the expected benefits of VE training that more closely matches normal human performance (see Stanney, Mourant, & Kennedy, 1998).

A key to developing a simulator for training an operator on a complex motor task is defining the relevant performance metrics for evaluating success. For example, the medical community is looking at VE as a tool for continued medical education and even as an accreditation tool for medical students (see chap. 47, this volume). However, there is still much work to be done in understanding what performance metrics should be used in evaluating surgical effectiveness in general. In other words, short of a simulator that is indistinguishable from the real world for the learner, it is still not clear what exactly should be simulated.

### 6.1 Flight Skills and the Simulation Training Environment

Despite the several hundreds of articles since the 1950s on pilot training, a recent review (Carretta & Dunlap, 1998) found that there are fewer than 40 studies that addressed transfer of training from a VE-based flight simulator to the real world. However, flight simulators have gained enough acceptance that the Federal Aviation Administration (FAA) allows the use of certified simulators in many contexts. For example, simulator training can substitute for some hours of in-flight instrument certification.

Because of the existence of "certified" VE trainers, the simulator-to-simulator method of measuring transfer may be used to test the efficacy of a new or low-fidelity simulator. The inverse transfer of training method is also used quite often in high fidelity simulators to test their validity. Experienced pilots perform a task in the simulator and are expected to do well because they are top performers in the real world and in certified simulators. Degradation in performance is attributed to a lack of simulator fidelity in the trainer under evaluation.

Training landing skills is the most common objective in using a VE flight simulator (for review see Carretta & Dunlap, 1998). In landing-skills acquisition the results indicate that neither scene detail nor field of view influence effectiveness of transfer from the simulator to the aircraft. However, pilots receiving more simulator trials exhibit better landing skills than those who receive fewer simulator trials. Another study (Lintern, Roscoe, Koonce, & Segal, 1990) used a transfer of training methodology to test the landing capabilities of a group of flight students, first allowing them to practice landing using a simulator, whereas a follow-on control group did not train in a simulator. The experimental group required 1.5 fewer presolo flying hours than the control group. Generally, results indicated that performance improves after using a simulator and that successful transfer may not require high-fidelity simulators for whole task training.

Efforts have been made to validate low-cost desktop VE flight simulators because "certified" trainers are usually high-fidelity, high-cost, multisensory environments. Because of the frequent use of simulators in flight training, potential measures of transfer from VE to the real world have already been established. Using a desktop personal trainer, Taylor et al. (1997) showed TERs of 0.15 to 0.50. Practically speaking, this meant that 10 hours of simulator training would save approximately 1.5–5 hours of cockpit training time depending on the task.

### 6.2 Surgical Skills and the Simulation Training Environment

The medical community has much to gain from the development and use of VE simulators for training, and many prototype VE systems to train surgeons in surgical procedures have already been developed (Lathan, Cleary, & Traynor, 2000; see *Medicine Meets Virtual Reality Conference Series*, Westwood, Hoffman, Stredney, & Weghorst, 1994–2000; see chap. 47, this volume). Current training methods essentially consist of the "see one, do one, teach one" philosophy, supplemented with limited access to animal models and physical simulators such as mannequins. Assessment of skill level is primarily through structured observation. A digital or virtual patient simulation would allow the surgeon to train on a variety of scenarios using simulated patients differing substantially in physical characteristics.

However, many surgeons feel that successful surgical training must be based on highly realistic simulation trainers even though low-fidelity simulators have been effective for training in many other domains. This is problematic given the current state of computer graphics and haptics in VE technologies. In addition, concerns about a lack of face validity of the current generation of surgical simulators may slow adoption of this technology for medical training.

One approach to the development of surgical simulators has relied on methods that have been used for simulator development in air combat training (Hettinger, Nelson, & Haas, 1994). In this approach, identification of critical tasks, related skills, and candidate metrics for assessing skills are accomplished primarily by users and human factors experts. User-centered design teams composed of surgeons, simulation engineers, and experimental psychologists use task analysis to decompose prioritized procedures into elemental skill modules that can be developed into part-task trainers. Candidate metrics are empirically validated using the ToT method and simulator-to-simulator validation. Ultimately, correlation of performance on the part-task trainer and prediction of performance in the "real world" determines the acceptability of a metric.

One issue that has emerged from task analysis of surgical skill acquisition and performance is the quest for a common task for assessment of surgical performance. Rosser, Rosser, and Savalgi (1998) have argued that intracorporeal knot tying is a good indicator of surgical skill (for both laparoscopic and open surgery) and should be adopted as a reference standard for surgical performance. They have used a variety of VE systems for assessing laparoscopic skill, including: suturing, thread manipulation, spatial and localization tasks, and manipulation of blocks. Knot tying was found to be a sensitive indicator of surgical performance, and thus this task could potentially serve as a standard for assessment.

Cuschieri (2000) has suggested that performance tracking using simulation technology should focus on component abilities. Thus, simulation tasks such as suturing should emphasize abilities such as eye–hand coordination, spatial abilities, and motor coordination. In surgical training there are no well-established simulators as there are in flight training, so the development of VE simulations and of standards to evaluate transfer from simulators to the real world are taking place at the same time. The work on specifying key components of surgical skill will establish the measures that must be used to evaluate such transfer.

## 7. CONCLUSIONS

Several themes have run through this chapter: fidelity of the simulator, simulator task, and methods of determining transfer. Each of these issues needs to be considered when designing and testing a simulator system.

*Simulator Fidelity*: In general, it is assumed that higher fidelity means better transfer. Technology is not to the point of creating a simulation that is indistinguishable from the real world. However, advances in technology may change that, which will open up a host of opportunities in simulation. For example, there has been a growing interest in the use of haptic or force feedback to improve task performance. Haptic feedback has been shown to improve performance in teleoperation tasks, particularly when used in conjunction with visual feedback (Burdea, Richard Coiffet, 1996; see chap. 21, this volume). Effective integration of multimodal sensory inputs and understanding their interaction will be key to increasing simulator fidelity and human performance training. Increased fidelity will also afford new opportunities to train under heretofore unknown or unexpected conditions and situations. For complex tasks and tasks performed in extreme environments such as outer space or underwater, there will be a need for increasingly complex and realistic simulators. In addition, VEs that differ from the real world and that are designed not to mimic the real world in important ways, such as violating laws of physics or exploiting unreal transparency, may be equally important as high-fidelity virtual environments.

*Simulator Task*: Deciding what part-tasks should be executed through VE simulation and what tasks should be practiced in whole is one of the issues that will be addressed by VE simulation development. Is transfer better when an entire task is simulated or when the simulator is broken down into part tasks? Which part task would be most appropriate for VE simulation? Virtual environments provide the flexibility to discover some of the answers to these questions to develop an optimal training environment. In fields in which simulators have not been used routinely, VEs are being used as a tool for establishing which parts of a task can be effectively trained separately.

*Measuring Transfer*: Measurement of training transfer is not unique to VE simulation. Rigorous methods of testing transfer must be employed independent of training method. However, the very advantages of using VE, such as being able to simulate extreme environments, may limit the methods that can be used realistically to measure transfer. Every simulator does not need to be tested using the ToT method of experimental and control groups. Logistics such as cost or availability of the real-world environment may prohibit that level of examination. Regardless of method chosen, some operational outcomes should be documented, such as the following: 1. How long is the task performed in the simulator? 2. How effectively is the task subsequently performed in an operational setting (whether that setting is the target environment or one correlated with performance in the target environment.) and 3. What is gained by using the simulator?

Virtual environments provide enhanced simulation that promises to improve transfer of simulator training to operational performance. At the same time, by relaxing the inherent constraints of physical simulators they raise a number of new issues about design. This chapter begins to address these emerging research issues that make VE simulation fertile ground for future research.

## 8. REFERENCES

Advisory Group for Aerospace Research and Development, NATO. (1980). *Fidelity of simulation for pilot training. NATO.*

Angell, J. R. (1908). The doctrine of formal discipline in light of the principles of general psychology. *Journal of Experimental Psychology, 36*, 1–14.

Burdea, G., Richard, P., Coiffet, P. (1996). Multimodality virtual reality: Input–output devices, systems integration, and human factors [Special Issue]. *International Journal of Human–Computer Interaction, 8*(1), 5–24.

Card, S. K., Moran, T. P., & Newell, A. (1983). *The psychology of human–computer interaction*. Hillsdale, NJ: Lawrence Erlbaum Associates.

Carretta, T. R., & Dunlap, R. D. (1998). *Transfer of effectiveness in flight simulation: 1986 to 1997*. U.S.: Air Force Research Laboratory, NTIS.

Chase, W. G., & Ericsson, K. A. (1982). Skill and working memory. In G. H. Bower (Ed.), *The psychology of learning and motivation* (Vol. 16, pp. 1–58). New York: Academic Press.

Clawson, D. M., Miller, M. S., & Sebrechts, M. M. (1998). *Specificity to route orientation: Reality, virtual reality, and maps*. Poster presented at the American Psychological Society 10th Annual Convention, Washington, DC.

Cuschieri, A. (2000). *Human reliability assessment in surgery—a new approach for improving surgical performance and clinical outcome*. Ann R Coll Surg Engl. *82*(2), 83–87.

Hettinger, L. J., Nelson, W. T., & Haas, W. M. (1994). *Applying virtual environment technology to the design of fighter aircraft cockpits: Pilot performance and situation awareness in a simulated air combat task*. 1994 *Paper presented at 1998 Human Factors and Ergonomics Society Meeting,* Santa Monica, CA, Human Factors and Ergonomics Society.

Kenyon, R., & Afenya, M. (1995). Training in virtual and real environments. *Annals of Biomedical Engineering, 23*, 445–455.

Kozak, J. J., Hancock, P. A., Arthur, E. J., & Chrysler, S. T. (1993). Transfer of training from virtual reality. *Ergonomics, 36*(7), 777–784.

Lathan, C. E., Cleary, K., & Traynor, L. (2000). Human-centered design of a spine biopsy simulator and the effects of visual and force feedback on performance. *Presence: Teleoperators and Virtual Environments, 9*(4), 337–349.

Lintern, G., Roscoe, S. N., Koonce, J. M., & Segal, L. D. (1990). Transfer of landing skills in beginning flight training. *Human Factors, 32*, 319–327.

Rosser, J. C. J., Rosser L. E., & Savalgi, R. S. (1998). Objective evaluation of a laparoscopic surgical skill program for residents and senior surgeons. *Arch Surg Jun, 133*(6), 657–661.

Sebrechts, M. M., & Knott, B. A. (1998). *Learning spatial relations in virtual environments: Route and Euclidean metrics*. Poster presented at the American Psychological Society 10th Annual Convention, Washington, DC.

Sebrechts, M. M., Mullin, L. N., Clawson, D. M., & Knott, B. A. (1999). Virtual exploration effects on spatial navigation and recall of location. Poster presented at the 40th Annual Meeting of the Psychonomic Society, Los Angeles, CA.

Siegel, A. W., & White, S. H. (1975). The development of spatial representations of large-scale environments. In H. W. Reeve (Ed.), *Advances in child development and behavior* (Vol. 10, pp. 9–55). New York: Academic Press.

Singley, M. K., & Anderson, J. R. (1989). *The transfer of cognitive skill.* Cambridge, MA: Harvard University Press.

Stanney, K. M., Mourant, R., & Kennedy, R. S. (1998). Human factors issues in virtual environments: A review of the literature. *Presence: Teleoperators and Virtual Environments, 7*(4), 327–351.

Taylor, H. L., Lintern, G., Hulin, C. L., Talleur, D., Emanuel, T., & Philips, S. (1997). *Transfer of training effectiveness of personal computer–based aviation training devices.* Washington, DC: U.S. Department of Transportation.

Thorndike, E. L. (1906). *Principles of learning.* New York: A. G. Seiler.

Tulving, E. (1983). *Elements of episodic memory.* London: Oxford University Press.

Waller, D., Hunt, E., & Knapp, D. (1998). The transfer of spatial knowledge in virtual environment training. *Presence, 7,* 129–143.

Westwood, J. D., Hoffman, H. M., Stredney, D., & Weghorst, S. J. (1995). (Ed.). *Medicine meets virtual reality.* Amsterdam: IOS Press.

Witmer, B. G., Bailey, J. H., Knerr, B. W., & Parsons, K. C. (1996). Virtual spaces and real world places: Transfer of route knowledge. *International Journal of Human–Computer Studies, 45,* 413–428.

# PART III: Design Approaches and Implementations Strategies

# 20

# Cognitive Aspects of Virtual Environments Design*

Allen Munro,[1] Robert Breaux,[2] Jim Patrey,[2] and Beth Sheldon[2]

*[1] University of Southern California*
*250 No. Harbor Drive, Suite 309*
*Redondo Beach, CA 90277*
*Munro@usc.edu*

*[2] Naval Air Warfare Center, Training Systems Division*
*12350 Research Parkway*
*Orlando, FL 32826-3275*
*BreauxRB@navair.navy.mil*

## 1. INTRODUCTION—COGNITIVE ISSUES FOR VIRTUAL ENVIRONMENTS

The uses of virtual environments (VEs) raise a variety of cognitive issues. Virtual environments provide information to users, and, in many cases, they provide users with the ability to interact with the environment in response to the information that the environment conveys. Many different cognitive factors play roles in the use of VE applications. These include issues related to perception, attention, learning and memory, problem solving and decision making, and motor cognition. Research and experimental applications of virtual environments have clarified some cognitive aspects of VE usage, but many questions have been raised for further research.

### 1.1 Perception

Cognitive research has revealed that perception is an active process, not merely an automatic bottom-up conveyance of sensory data such as visual images to higher order cognitive centers. Expectations and experience exert a top-down contribution to perception, which results from the active interpretation of sensations in the context of these expectations and experience. A number of characteristics of presently available VEs raise issues for perception. Among these are the very poor resolution of displays (in terms of number of pixels presented per degree of visual angle, for example), the frequent problems with alignment and convergence, and

*The views of the authors do not necessarily represent the policy of the U.S. government.

the often primitive three-dimensional (3-D) models and texture maps presented on displays. These characteristics place substantive constraints on the role of bottom-up data processing in perception in virtual environments, and they therefore increase the importance of the top-down contributions of expectations and experience. A question that has not yet been resolved is the extent to which perceptual *experience* with VE representations is required to ensure accurate perceptions. People living in a culture with artifacts that include visual representations learn how to interpret those representations. In modern high-technology culture, people are able to interpret the graphical representations on television and on computer screens. To what extent must people learn to interpret—learn to *see*—what is presented in a virtual environment?

### 1.2 Attention

There are several aspects of VEs that have potentially important consequences for the role of attention processes during VE experiencesss. A VE system has access to clues about the focus of a user's attention that are not available in an ordinary screen-presented graphics system. Both the field of view and resolution provided by current VE systems are so limited that a user must often carry out navigation and orientation actions (see chap. 24, this volume) so that the matter of interest is brought into view. These actions constitute an announcement—a sort of nonverbal protocol—about what the user is currently attending. Additional precision could be obtained by including eye movement observation systems in VE, but, until higher resolutions are commonly available in virtual environments, for many VE applications it will be enough to know what is in the user's field of view.

### 1.3 Learning and Memory

A number of research projects have studied the application of VE technology to learning environments, including advanced technical training and intelligent tutoring systems. The results of these projects provide useful guidance for the developers of VE-based learning environments, but a number of important questions remain to be answered. Some of these questions relate closely to issues of perception and attention. How can student attention most effectively be directed in virtual environment? How can a tutoring system know whether a student has perceived a feature of an environment? How much fidelity is required to produce desired levels of learning transference to a real-world context, such as on-the-job task performance? To what extent do the special cognitive demands of the VE experience interfere with learning? Does VE-based learning offer special advantages for conveying some types of knowledge but not for others?

### 1.4 Problem Solving and Decision Making

The same concerns about the cognitive demands of VE experiences that are of concern for learning are also very relevant to problem solving and decision making in the context of virtual environments. A VE makes it possible for a user to quite naturally find new perspectives for a graphically represented problem or set of data; this offers the potential for VE-based aiding systems for analysts and decision makers. On the other hand, the additional cognitive processing required for perception and for the VE navigation actions could tie up cognitive processing resources, reducing the effective deployment of higher level cognitive processing functions in some contexts.

### 1.5 Motor Cognition

In many real-world tasks, people must move about and change their physical orientation in order to make observations and to carry out actions. Infants and children spend many years acquiring the motor skills that make it possible, in most contexts, to carry out these navigation and orientation actions in a largely or even completely unconscious manner. In many virtual environments, the full set of motor skills used to navigate and orient in the real world is not wholly transferable to the virtual environment. To what extent does the nonautomatic nature of navigation and orientation in some VEs interfere with other cognitive processes?

Much of what is known about the cognitive aspects of VE design is a result of work done on using VE for learning environments and for training. The next section of this chapter discusses cognitive issues in the context of VE training. Virtual environment technology has implications for human information processing, mental models, and metaphors and analogies for the design of training systems that make use of immersion. Cognitive issues in training are discussed, and recent VE systems used by the navy to teach cognitive tasks are described. The overall intent of this chapter is to give the reader some insight into the design of VE that lends itself to teaching cognitive skills. The work is in progress. There are many unanswered questions in this area. However, great progress has been made with the systems that have been tested so far. A series of applications will be discussed that have been sponsored by the Office of Naval Research, Code 342. These have examined increasingly more complex cognitive skills. Early work examined maintenance training in a laboratory setting. Harbor navigation was explored next, and a test bed was developed to evaluate VE in a training environment. Then, complex-maneuvering skills in the open ocean were examined. This led to development of objective performance measures. Virtual environment modalities included visual, audio and haptic. The chapter concludes with a set of guidelines for system design and conclusions about future work.

## 2. COGNITIVE ISSUES IN VIRTUAL ENVIRONMENT TRAINING

### 2.1 Knowledge Type and the Acquisition of Knowledge in VEs

The cognitive factors listed above must be considered with respect to the types of information that are being conveyed in a virtual environment. Consider these major types of information that may be presented in a VE: location knowledge, structural knowledge, behavioral knowledge, and procedural knowledge.

For each type of information that can be presented to or learned by a user of a virtual environment, there are potential issues to be addressed in determining how to best support the cognitive processes—perceptual, memory, decision making, and motor—that are brought into play in interacting with or learning the information. This can be viewed as a matrix of cognitive processing issues by types of information.

#### *2.1.1 Location Knowledge*

Several types of knowledge about location may be particularly amenable to conveying with VE rather than through more conventional means. Many VEs give their participants the facility to change their location and their orientation while observing objects in the environment. This type of experience can give the participant a richer set of information about the relative locations of simulated objects, about how to get to a location, bring objects into view, and access and manipulate them.

*2.1.1.1 Relative Position Knowledge.* To teach the relative location of objects in a two-dimensional (2-D) plane, it is enough to present a map or a picture that shows the layout of the objects in two dimensions. Drill and practice or mnemonic techniques can be used to fix the information in the student's memory, but his or her basic understanding of the position relationships in 2-D space is an immediate apperception when the map or image is presented and the relevant relationships are pointed out. When objects have more complex relative locations in a 3-D layout, a single visual map will not suffice ordinarily. Students often require a variety of visual and/or positional experiences in 3-D in order to understand relative positional relationships. Two-dimensional presentations are particularly insufficient when the parts of two complex objects overlap because they have complex relationships to each other in the third dimension. However, in a VE a student can move about while looking at such "intertwined" objects, and can, thereby, quickly develop an accurate understanding of their relative positions.

*2.1.1.2 Navigation Knowledge.* Regian and Yadrick (1994) have shown, under controlled experimental conditions, that VEs promote more rapid acquisition of personal navigation knowledge in a complex environment of corridors and stairways than can be accomplished with an otherwise similar 2-D graphical trainer. Results suggest that accurate textural detail in such environments may enhance personal navigation learning.

*2.1.1.3 "How to View" Knowledge.* Many real-world tasks require that the worker know where to position himself and how to orient in order to make a necessary observation. Viewing a conventional 2-D representation of the environment may not provide a learner with enough information about what can actually be seen from where. A VE provides students an opportunity to find available lines of sight for viewing indicators, etc.

*2.1.1.4 'How to Use' Knowledge.* Knowledge about how to use objects in an environment includes two characteristics. First, a student must know how the object can be accessed. Is it necessary to open a door? Must one reach through an opening or between two other objects? Second, the student must know how the object should be manipulated. Should it be pushed away, grasped and twisted, pulled toward one, pushed down or pulled up? Is some more complex manipulation required? Virtual environments offer good potential for teaching how to access objects that are to be manipulated. Given the limitations of current technologies for manipulation in VE, the potential for teaching the details of manipulation are somewhat limited. Many VE systems have only one or two types of possible manipulation action (such as "pinch" and "release"). In these virtual environments, the simulation software typically interprets the action that is taken as being the one appropriate to the device that is being touched. Of course, there is every reason to expect that finer motor performance will be possible in future VE systems.

#### 2.1.2 Structural Knowledge

Virtual environments have the potential to convey a rich variety of structural information, although only a subset of those possibilities have thus far been exploited.

*2.1.2.1 Part-Whole Knowledge.* Participants in VE rely on their real-world knowledge of the represented world to perceive part–whole relationships. If an environment includes a display panel with many lights and labels, an observer is likely to conclude that the labels and lights are part of the panel. When less-familiar complex objects are presented in an environment, it may be necessary to give the participant means for exploring part–whole relationships. One such means is to give the participant the ability to move objects. A set of objects that move together may reasonably be construed as the parts of a whole.

*2.1.2.2 Support-Depend Knowledge.* If object A is above and flush with object B, then a participant can draw on ordinary pragmatic knowledge to conclude that object B supports object A, at least in VEs that simulate conventional gravity conditions. If the environment allows the participant to move object B out from beneath object A, then the participant can test the hypothesis that B supports A.

*2.1.2.3 Containment Knowledge.* Those VEs that permit a participant to open one object and manipulate or move another object within the first one provide a means for learning containment relationships.

### *2.1.3 Behavioral Knowledge*

There are several types of behavioral knowledge that can be effectively conveyed in a virtual environment. By *behavioral knowledge*, we mean knowledge about how objects in the VE interact with each other and with participants.

*2.1.3.1 Cause-And-Effect Knowledge.* Cause-and-effect knowledge lets a participant predict that if one state change takes effect, then another state change will also occur. For example, "If the power switch is depressed when the system is functioning normally, then the 'Power On' light will be lit." Cause-and-effect knowledge can be acquired by observing a system as it changes. Such knowledge is more likely to be acquired, however, if the participant has the ability to effect state changes directly. This makes it possible for participants to observe cascades of state changes and to develop cause-and-effect knowledge as a result.

This does not mean that exploratory environments are enough, in themselves, to assure learning of cause-and-effect knowledge. For most students, it is necessary to guide exploration and to point out explicitly the causal relationships.

*2.1.3.2 Function Knowledge.* Function knowledge is related to cause-and-effect knowledge. It has to do with the use of an object, that is, with what the object is *for*. Given the actions that can be taken on an object, the sets of intended effects of those actions constitute the function of the object. Again, a VE with interactive behavior can offer participants the opportunity to learn by experimenting, but guidance and tutorial exposition will typically be required to ensure effective learning for many students.

*2.1.3.3 Systemic Behavior Knowledge.* Some VEs are designed to help students to come to appropriate generalizations about behavior. Simulations of Newtonian mechanics systems, for example, can be constructed so as to work with tutors that help students to understand Newton's model for the behavior of masses subjected to accelerations. In these VEs, students are expected not to learn simply the behaviors of particular objects, but rather the principles that explain the behaviors of whole classes of objects.

### *2.1.4 Procedural Knowledge*

Virtual environments may also be appropriate for conveying knowledge about how to carry out procedures. Behavioral knowledge alone is not enough to determine how procedures should be carried out. In many cases, there are many action sequences that will result in a desired goal, but some will be more desirable than others for reasons of cost-effectiveness, safety, or speed.

*2.1.4.1 Task Prerequisite Knowledge.* Before a procedure can be undertaken, an initial state must often be achieved. This state has characteristics that are the prerequisites for successfully carrying out the task. A VE is a particularly apt context for teaching task prerequisites based on the positions and orientations of objects or of the participants who are to carry out the task.

*2.1.4.2 Goal Hierarchy Knowledge.* Task knowledge can be viewed as knowledge of a goal hierarchy and sequences of actions that can achieve transitions to subgoal states. Again, VEs are particularly useful for teaching procedures in which there are subgoals that prescribe positions and orientations, or states of relative movement.

*2.1.4.3 Action Sequence Knowledge.* The simplest procedures can be represented simply by a sequence of actions and a final goal. The subgoals of more complex procedures also have associated action sequences that are capable of bringing about the subgoal states. When these action sequences include movements and/or orientation changes, whether of the participant or of objects under the participant's control, a VE may be an appropriate learning context.

Hall, Stiles, & Horwitz (1998) conducted a study that can serve as a standard for evaluating the effectiveness of competing designs for learning how to move about in a complex environment in order to carry out a procedure. Two sets of students were trained to carry out a moderately complex task that involved moving among electrohydraulic machines onboard a ship to make observations and to set controls. Students had to learn both the sequence of actions (observations and manipulations) that were to be carried out and where to go to carry out the actions. One group learned the task in the virtual environment technology (VET) facility (Johnson, Rickle, Stiles, & Munro, 1998; Stiles et al., 1996), while the other learned the task in a 2-D graphical simulation environment. In both cases, the simulation was under the control of the same simulation, which was developed with the Virtual Interactive Intelligent Tutoring System (ITS) Development Shell (VIVIDS) simulation development-and-delivery system (Munro & Pizzini, 1998). Students from both groups were evaluated by asking them to perform the task in a real-life mockup environment modeled after the actual shipboard systems. There were no significant performance differences between the two groups, suggesting that, at least for this procedure, simulation training in a 2-D graphical environment may be as effective as virtual environment–based training.

## 2.2 Tutor Architecture

Figure 20.1 below presents an architecture for VE tutoring systems (Munro, Surmon, Pizzini, & Johnson, 1997). On the left side are those components responsible for the user interface and for the interactive behavior of the environment. On the right are the components that manage the course of instruction, maintain a model of the student, and conduct the moment-by-moment details of tutorial interactions. Some of the text in the right side of the figure is obscured because it is notional text.

The User View has subcomponents that are the responsibility of the VE software. An example of these components in use is shown in Fig. 20.2 below.

A user view is responsible for everything that can be seen (or heard or touched, etc.) by a student during simulation learning. Its model view component is responsible for rendering the simulated world prescribed by the behavior model and for detecting user events and passing them on to the behavior model for semantic processing. Presentation channels may include text, speech, audio, HTML, and video presenters, to name a few. Entries are user interface components that students can use to answer questions that cannot be answered by actions in

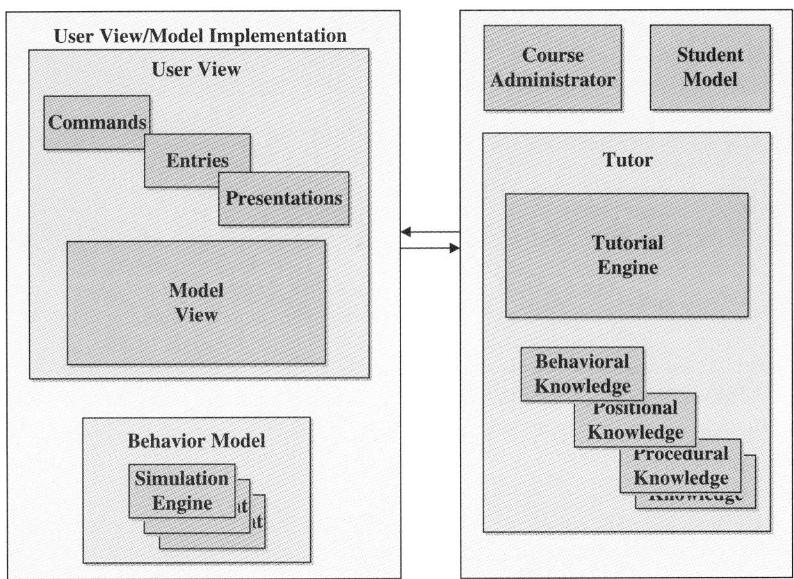


FIG. 20.1. A generic architecture for simulation-centered tutors.

the simulated model view. Commands are interfaces for higher level student interactions with the tutorial system, such as pacing instruction, asking for help, and so on.

A variant of this architecture, shown in Fig. 20.3 below, is capable of supporting implementations in which multiple students collaborate with each other in a simulation environment. Each student has a user view, including a model view that is under the control of a central behavior model. In this way, other students can view the cascading effects of actions taken by one student. The model views shown to different students need not depict the same scenes. Team members often work with different parts of a complex interconnected system and are able to see only one part of it at a time. The single behavior model determines the values of simulation attributes, and some of these value changes cause changes in what is displayed in the model views.

The instruction control components are responsible for the configuration of the complete learning environment, including the availability and positioning of the possible user interface elements. The tutorial engine component is a service requester for many of the other major components: It can ask a presentation channel to fetch and present a video. It can ask the

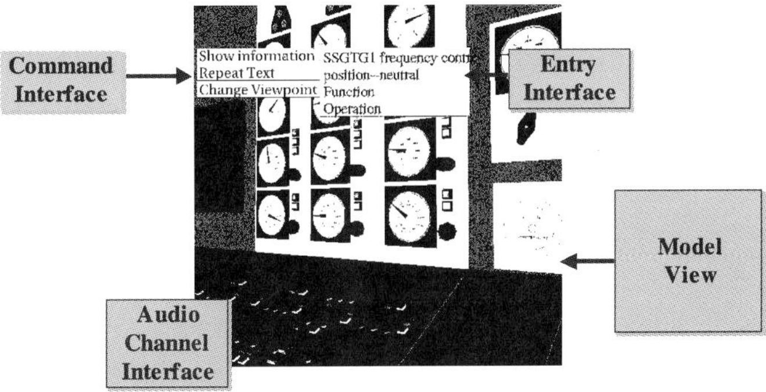


FIG. 20.2. A user view in a virtual environment.

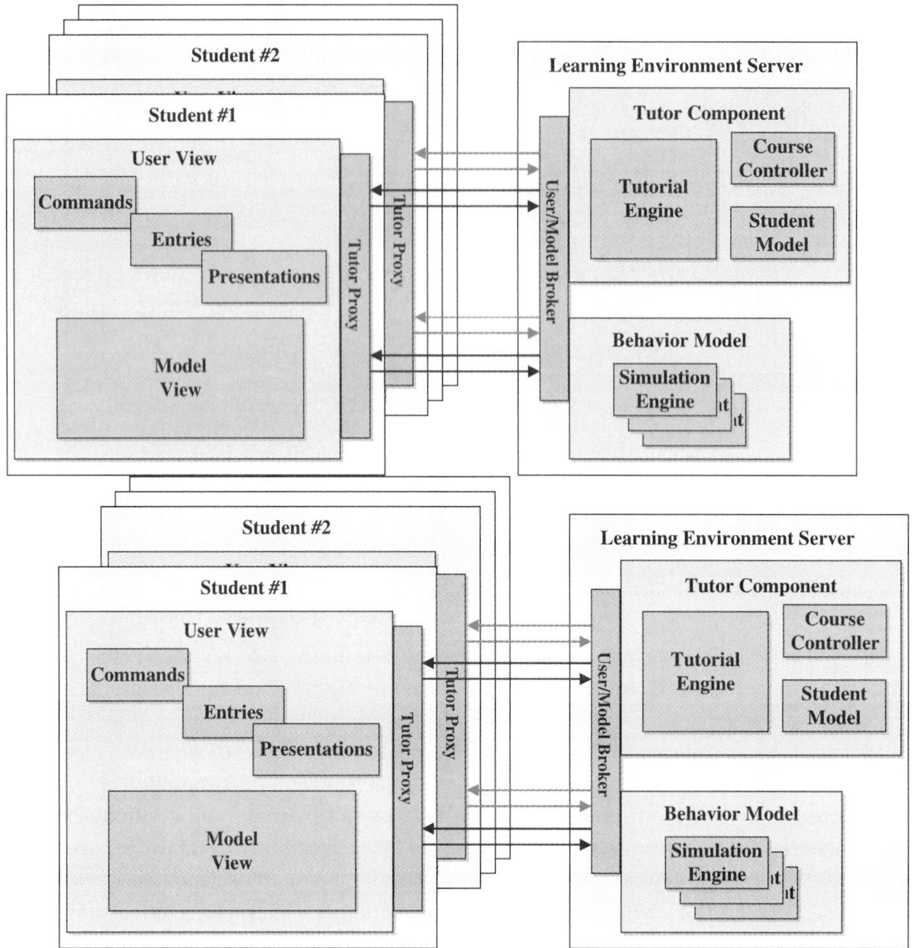


FIG. 20.3. One possible architecture for simulation-centered team training.

behavior model to put the simulation in a particular state. It can ask the model view to stop responding to user actions for a time so that the tutor can demonstrate a sequence of actions without student interference.

## 2.3 Task Analysis

Large-scale instructional systems developed for sophisticated clients must have a rigorous foundation in instructional design. A military or large business unit would no more contract for the development of a large training course without a detailed instructional design than a corporation would contract for a new headquarters building without an architectural plan. Instructional designs are typically constructed so as to realize objectives. In many cases, these objectives are best understood in terms of the desired quality of performance of a task. Students are to learn enough, as a result of the planned instruction, that they will be able to carry out some task at a specified performance level. This type of objective requires that the task that the trained student is to perform must be analyzed well enough so that the analysis can guide the design and development of the instruction. A number of approaches to task analysis for instructional design have been summarized in such works as those of Leshin, Pollock, and Reigeluth (1992), Jonassen, Hannum, and Tessmer (1989), and Jonassen (1999).

Several characteristics of VE training require special attention to certain task characteristics during the task analysis process. Virtual environments may be especially well suited to teaching students where to move and where to direct their gaze during a task. It is important that an analysis of a task should pay attention to elements of the task that require moving or orienting actions, so that those steps in the task will be effectively taught in the VE training system. If a task requires either that the student learn to move and/or orient himself or herself or that the student must learn to move and/or orient objects, then the VE training can be structured to explicitly teach the student to carry out these steps.

## 3. COGNITIVE TASKS APPROPRIATE FOR VEs

Although some progress has been made with applications, it is doubtful that all possible avenues of VE presentation have been explored. That is, if the logical assumption is true that increasing the bandwidth of information input will facilitate learning, then it follows that multimodal presentation will facilitate learning. If information can be received via vision, sound, and touch, then it seems to follow that the learner will experience a richer learning environment and will learn better, faster, with longer retention. Virtual environments are only beginning to be exploited as a multimodal training medium. Additional research is needed with olfactory, temperature, and large movement haptic displays.

### 3.1 Navigation and Locomotion in Complex Environments

Research conducted at the Air Force Research Laboratory (Regian, 1997) taught subjects to find a route through a complex building, using several modalities to carry out the instruction. One result of this work is that providing a richly decorated VE may have the effect of improving how well a student learns to navigate in the environment. If a student is to learn how to find his way in a building with many corridors and many turns, then the texture maps in the VE should include the pictures that can be found on the walls of the real building.

### 3.2 Learning Abstract Concepts with Spatial Characteristics

Several experiments on teaching scientific concepts in a VE context have been explored by Bowen Loftin and his colleagues. These include the Newtonian mechanics of NewtonWorld, the electromagnetism of MaxwellWorld, and the chemical bonding of PaulingWorld. One of the interesting findings of this work is that it may often be necessary in VE training to provide constraints on students' virtual locomotion (see chap. 11, this volume) and orientation changes (see chap. 24, this volume). In a virtual environment, students may miss a crucial demonstration because they face the wrong way or position themselves far away from what should be observed.

### 3.3 Complex Data Analysis

Some types of complex data can be presented in 3-D space in a manner that illuminates characteristics of the data that would not be evident in conventional 2-D presentations. Furthermore, users can be given the opportunity to manipulate the representation and to view it from any angle.

### 3.4 Manipulation of Complex Objects and Devices in 3-D Space

When there is a requirement to train someone on the physical manipulation of a complex object in 3-D space, training that makes use of the actual object under the conditions that prevail on the job offers the highest probability for good transfer to the actual task. In certain cases, however, the use of the actual object for training is inappropriate or impossible. If the training involves the handling of an extremely delicate and valuable object that can be damaged through inappropriate handling, it may be advantageous to train with a VE simulator. In other cases, danger to the operator or to others may argue for training in a simulation. For example, an inadequately trained crane operator on a construction site could carry out actions that would cause his crane to topple, possibly landing on other workers or bystanders, and pulling the operator's cab off its pedestal.

### 3.5 Decision Making

Where the domain of decision involves relationships that must be understood in three dimensions, and, especially, when spatial relationships are changing over time, VEs may prove extremely valuable. Ship-handling tasks require the conning officer to develop the "seaman's eye" (Crenshaw, 1975). Part of that skill appears to be recognition of relative motion of objects in the path of own ship. Another part of that skill appears to be maneuvering own ship to avoid obstacles. Now, wind and sea currents affect the direction the ship must point in order to achieve a desired course through the water. As they exert force upon the ship to push it in one direction and the ship's crew desires it to go in another, the heading of the ship must be adjusted slightly to compensate. It appears that one method to teach this compensation angle computation would be a mathematical formula. It also appears that one way of augmenting the concept might be to actually experience the forces of wind and current. There are a number of ways to accomplish this, which have been considered in our work. One would be to use VE to enlarge the user to the size of a giant so that they could actually hold the ship in their hand and feel the force of the wind and current acting on the force of the ship. This would be achieved both visually and with a haptic device for the hand. Other laws of force and motion might be taught by having one feel the forces with a haptic device. Current haptic devices do not appear to be robust enough for sophisticated forces acting in multiple dimensions. Thus, this area requires further work.

## 4. APPLICATIONS OF COGNITIVE TRAINING USING VEs

Martin and his colleagues (Breaux, Martin, & Jones, 1998) sought to lay a foundation for application to VE of a technique whereby the learner codes information both visually and verbally (see Pavio, 1991). This was intended to explain why VE immersion could be effective in developing a mental representation of an environment. Consider the VE that contains visual, haptic, and audio components to teach shiphandling skills. The intent of traditional training is to apply the concepts described by the teacher verbally and with mathematics in the classroom, and augmented with static depictions of course tracks. The real-world application is, instead, a dynamic, interactive environment of ships at sea maneuvering to avoid one another and to achieve a desired course to reach their destination. The intent of VE-based cognitive training is to solidify a relationship between the verbal concepts of the classroom with the dynamic visual environment of the simulation. Virtual environment training is intended as a method to coalesce the cognitive model from components of dynamic interactive objects of the visual scene with static linguistic concepts from the classroom. The model thereby developed by the learner is expected to be a richer, more robust model that generalizes to novel ship maneuvering situations. A more detailed explanation of specific tasks will follow.

Another challenge posed for VE by the work of Breaux et al. (1998) was the potential for VE to shape mental behavior. Let's say one seeks to teach a complex concept of adjusting the heading of a ship to correct its path through the water because wind and sea current forces acting to push the ship off course have disturbed it. If the student adjusts the course too much or too little, one may want to understand what it is about the student's mental concept that results in the errors. If a model of the student's behavior of adjusting too much or too little can be created, then one can claim to understand the misperception of the student. That model may be computationally intensive. However, if VE can project the path of the ship with the error introduced by the student, this may depict the current mental model being used by the student. The instructional intervention that VE allows is then to demonstrate visually and haptically how the forces are in fact acting upon the ship's path. Next, VE can dynamically transition the immersed student from the misperception to the correct perception by the use of visual and haptic displays to adjust what the student is experiencing visually and haptically. Consider the analogy of the teacher taking the hand of the student and moving it through the motions required in drawing, say, an image on paper. Now, extend that analogy to a VE adjusting the image and forces felt so that a student sees and feels the proper forces acting on a ship. This is what was intended by the concept espoused by Breaux et al. (1998) of morphing the student's misperception into the proper perception.

Let's move now to some specific applications of VE to training. Applications of submarine navigation, shiphandling, and electronic maintenance are considered.

### 4.1 VESUB Harbor Navigation

Hays, Vincenzi, Seamon, and Bradley (1998) provide a description of a VE training test bed, the Virtual Environment for SUBmarine (VESUB) training that evaluated the effectiveness of teaching harbor navigation using an immersive visual system with audio (see Fig. 20.4). The officer of the deck (OOD) stands in the open air while a submarine is on the surface and gives commands to the crew for speed and direction of the submarine. The surfaced submarine must navigate its way through the harbor, avoiding obstacles, and tie up at its pier. Current training is typically on the job, so this VE system was designed to provide concentrated training for a variety of situations, including various harbors, weather conditions, traffic, and time of day. Because of the expense of running aground or collision of the submarine with other objects, it is seldom that an inexperienced OOD has the opportunity to train in a real submarine. Therefore,

FIG. 20.4. VESUB harbor navigation training.

the system can provide much needed training for inexperienced officers as well as refresher training for the "old hands."

To make the system look like a real submarine, the visual system depicted a very detailed harbor and displayed various ship instruments on the bridge. A computer voice recognition system captured the commands from the student and translated them for use by the computer mathematical models driving the visual display. The computer models generated sounds of the water, foghorns, and other relevant audio cues as well. An instructor was present to create and observe a particular scenario for the student and to evaluate performance.

One of the features of this system is team communication. The computer expects the student officer to issue proper commands using proper terminology. Also, the instructor is available to provide feedback and manipulate the scenario to challenge the student's knowledge and skills. Unlike the real task where the commanding officer, acting as instructor, must be ever vigilant to the danger of an error that can be very expensive or even life threatening, the VE system can provide the instructor a less-intensive work pace with concentration on teaching the student.

Hays et al. (1998) reports that the evaluation of VESUB was quite successful. Two navy training facilities spent three weeks each evaluating the system, using 41 participants. Of the 15-shiphandling variables tested, improvement ranged from 13% to 57% on 11 of them. It was proposed that in order to perform the task of navigating the submarine through the harbor, one must develop a mental model of not only the task at hand, but also the relative motion of other water traffic and the effect of tides, current, and wind on the submarine. Some of the cognitive components of the task include: the relationship of what one sees in the environment to its representation on the harbor chart; the relative size and height of objects with the angle on the bow of the submarine; the relative motion of the objects; the prioritization of other water traffic that may conflict with the submarine's course; and how to maneuver the submarine out of danger.

### 4.2 COVE Underway Replenishment

Following the apparent success of the VESUB project, the Navy undertook application of VE to more complex cognitive skills. The Conning Officer Virtual Environment (COVE) project sought to combine VE with computer-generated performance assessment. In his thesis, Norris (1998) describes some of the cognitive components of the task of commanding a ship. The challenge for the COVE project was to ascertain and develop methods by which the VE could provide measures of the student's performance and then use those to deliver diagnosis, feedback, and prescriptive tasks to the student. COVE sought to divide the task into cognitive and perceptual components. The VE not only presented the environment to the student, but also allowed measures of the objects in the environment, indications of where the student looks, and computer speech recognition of the verbal commands that are given.

The task most studied in preparation for the COVE project was that of underway replenishment (UNREP). In this task, a supply ship is steaming a set course, and the ship to be replenished steams to an alongside position (see Fig. 20.5). Both ships then continue to maintain their course and speed while supplies are transferred across approximately 120 feet separating the two ships. The perceptual components include distance, speed, and relative position of the ships. The cognitive components include issuing verbal commands to guide the ship through four different phases of the task. Each of the four phases requires slightly different skills.
The following were proposed as phases of the UNREP task:

- Approach
- Slide-in
- Alongside
- Station
- Breakaway

FIG. 20.5. Under way replenishment.

The approach is defined as being from an arbitrary starting position behind the replenishment ship to a point where the officer must order an initial deceleration to slow to a speed matching that of the replenishment ship. Slide-in is from the initial deceleration to a point where the bow of the receiving ship crosses the stern of the replenishment ship. The alongside phase is from the bow–stern crossing to a point where the two ships are even with one another and at the same speed. The station phase is maintaining the position so that replenishment can occur. Breakaway phase is from separation of the two ships to the end of the scenario. The slide-in and alongside are transitional phases, the slide-in being the end of the approach and the alongside representing the beginning of station. It was proposed that the phases require different combinations of component skills, some of which include:

- Ship control
- Perception
- Decision making

Ship control refers to the ability to effectively control the speed and direction of the ship. Some implicit parameters include ship mass, hydrodynamics, and acceleration. Perceptual skill includes relative motion, the sensation of the visual scene flowing by the conning officer, and distance estimation (both range and lateral separation estimation). Decision making entails the processes involved in monitoring speed and position as well as the ability to stay ahead so as to prevent errors and to identify critical incidents. It was proposed that decision making, perception, and most particularly, ship dynamics best lend themselves to training in a VE.

These component skills were proposed to vary across the different phases. The approach and slide-in require more distal perceptual cues with larger, ballistic-type ship movements (and commands) and simpler, slower decisions. The alongside and station phases requiring more proximate perceptual cues with small, iterative ship movements (and commands) and quick decisions with foresight. Finally, the breakaway was proposed to involve transitioning from proximate cues and small, iterative movements to distal cues and large, ballistic movements. Decision making appeared to require analysis of errors and critical incidents, but unfortunately Norris (1998) did not provide operational definitions for errors and critical incidents.

At the time of this writing, the COVE project was in the formative stages. The benefits that were expected from a COVE system included:

- Training perceptual skills through multiple modalities in VE (visual and auditory)
- Tutor software to teach complex hydrodynamic concepts
- Techniques to teach spatial and temporal orientation of ship movements relative to one another

The preliminary work on planning the COVE project included development of measures of the cognitive components of the task. These will be discussed next.

### 4.2.1 Performance Measures

The underlying cognitive components of the COVE task were discussed above and include issuing verbal commands to guide the ship through four different phases of the task. In order to arrive at more specific components that would lend themselves to being useable for feedback, diagnosis, and remedial tasks, the preliminary COVE work sought to determine specific measures of cognitive ship-handling skills. Although Norris (1998) provided a task analysis, what were needed were specific measures that correlated with a "good UNREP."

Patrey, Breaux, Mead, and Sheldon (in press) proposed the application of a perceptual-action task model. This requires one to acquire and use spatial knowledge in order to be successful at ship handling. Actions that the officer of the receiving vessel can accomplish include giving heading and speed commands for the ship. This means that the officer must be aware of water traffic, some of which may be on a collision course, and a hypothetical position that would put the receiving ship in the proper position with the replenishment ship. Patrey et al. (in press) defined four characteristics of a ship's dynamics, position, heading, velocity and acceleration, as its "moments."

In their study, subjects spent about 1.5 hours to perform two UNREPs. They were evaluated by six expert shiphandlers and the average rating was correlated with 13 moments that were measurable in the virtual environment. Four of the moments proved significant: lateral separation of the two ships, bearing, speed, and acceleration. In addition, it appeared that 83% of the performance of the UNREP could be explained by these four moments at the point when the bow of the receiving ship passed the stern of the replenishment ship. This is the transition point between slide-in and alongside phases of the UNREP.

The conclusions drawn in this study contradicted current thinking about UNREP performance. Experts typically suggest that there is no single best way to UNREP. There are many variations that are just as good. However, Patrey et al. (in press) suggested that the fact that certain moments correlate so highly with performance rating indicates that it is possible to quantify what is traditionally thought of as a qualitative task. Further, typical training for UNREP done at sea is a trial and error proposition. This study suggested that there are in fact measures that may indicate the strength of the cognitive model used by students to perform the UNREP. Furthermore, it was suggested the VE training, by its ability to measure the entire environment, will allow more structured training of the cognitive model than will occur by trial and error training. One approach would be that the moments would be used to develop remedial exercises, or part-task exercises. For example, it may be that UNREP skills develop quicker if a student practices just the slide-in and approach phases until their performance is high and then moves on to approach, station, and breakaway. These hypotheses were under examination at this writing.

## 4.3 VETT Maintenance Training

Davidson (1996) describes a system designed to teach electronic maintenance skills using a commercial haptic display called the PHANToM, manufactured by Sensible Devices (see Fig. 20.6). The PHAMToM served as an electronics probe to test circuits. Traditional training

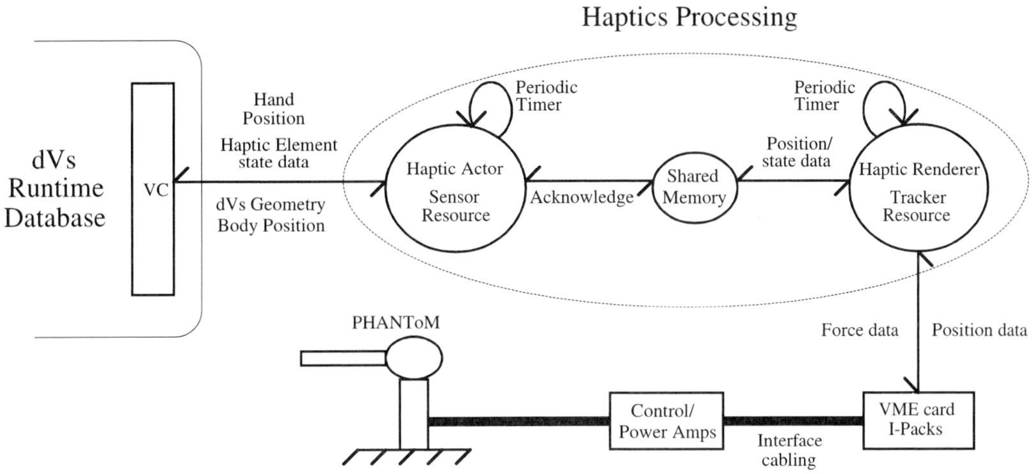

Haptics Processing
dVs
Runtime
Database
VC
Hand
Position
Haptic Element
state data
dVs Geometry
Body Position
Periodic
Timer
Haptic Actor
Sensor
Resource
Acknowledge
Shared
Memory
Position/
state data
Periodic
Timer
Haptic Renderer
Tracker
Resource
PHANToM
Force data
Position data
Control/
Power Amps
Interface
cabling
VME card
I-Packs

FIG. 20.6. Architecture of a VETT maintenance trainer.

uses the NIDA Model 130 Test Console, shown in Fig. 20.7 below. Various circuit cards can be inserted into the console, and the student must ascertain the fault that is contained on the card.

To replicate this task using VE, a visual system was devised as depicted below and interfaced with the PHANToM. This allowed a combination of seeing a simulated circuit card in a helmet display while at the same time feeling a simulated probe on the circuit card (see Fig. 20.8). In addition to simulated resistors, there were simulated toggle buttons and a simulated multimeter for test readings.

Unfortunately, this task was cancelled before a controlled experiment could be conducted. However, Davidson (1996) suggested that haptic interactions could provide a viable alternative to actual circuit cards for training troubleshooting concepts. Further, haptic displays can provide force feedback cues not available from other tracking systems.

## 4.4 Operations Training with VEs

In the Virtual Environments for Training Project (Johnson et al., 1998; Stiles et al., 1996), there was a concerted effort to generalize features of the training environment to exploit cognitive

*Test Console Controls*
*PC Card Interface*
*Fault Switches*

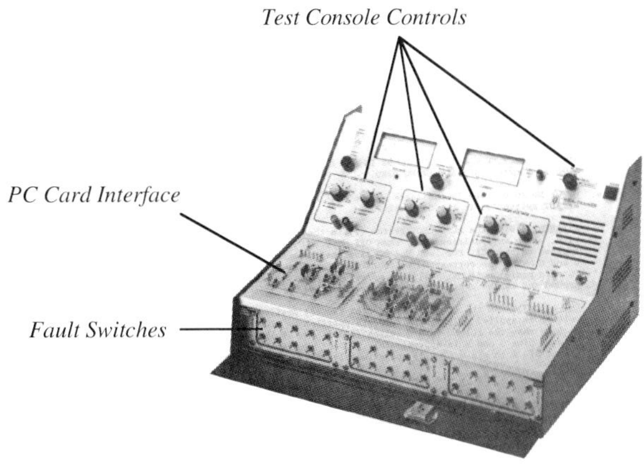

FIG. 20.7. NIDA Model 130 test console.

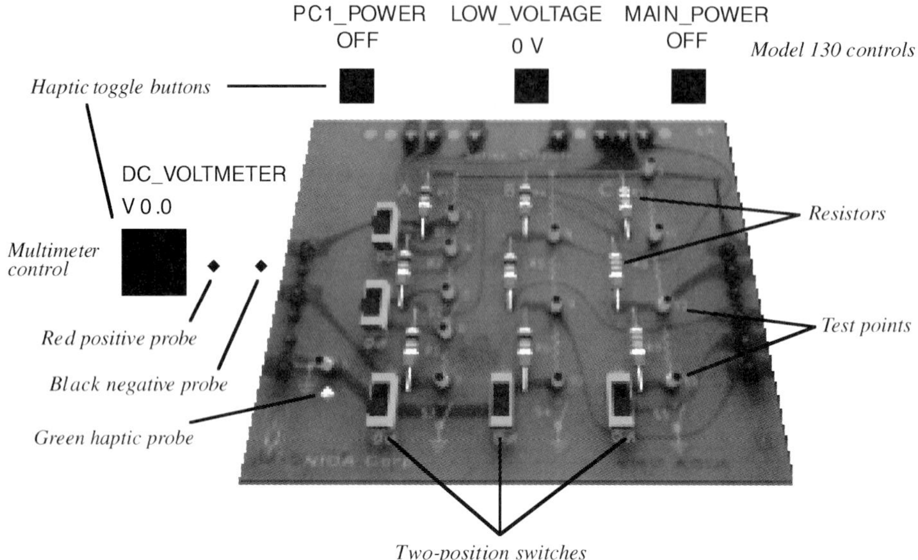


FIG. 20.8. A circuit card.

characteristics of operations training in a virtual environment. This project made use of Virtual Interactive ITS Development Shell (VIVIDS) (Munro et al., 1998; Munro & Pizzini, 1998), a tool for authoring interactive graphical simulations and tutorials that are delivered in the context of interactive graphical simulations. VIVIDS includes a deep representation of the notions of effectuating actions and observations. As a simulation author builds a simulation, the author designates those attributes that store the immediate results of effectuating actions and those attributes that store directly observable values in the simulation. Because the simulation is coded with these cognitively relevant concepts, it is possible to use the VIVIDS tool to rapidly author procedural lessons simply by carrying out the procedures that students are to learn.

VIVIDS provides a set of direct manipulation interfaces for building simulations and tutorials. A number of features support productive simulation authoring. Authors can select from libraries of behaving objects to add interactive components to a simulation. Alternatively, authors can draw new objects (or import object images) and then write rules that control the interactive behavior of the objects. A simulation engine that supports constraint-based effect propagation makes it possible to author complex simulation behaviors without attending to details of the flow of control of simulation effects. In the right portion of Fig. 20.9 below, a portion of the behavior-authoring interface is displayed. An object data view displays the attributes of a simulated shipboard throttle. Authors can add and delete attributes, change their values, and write constraint rules that prescribe the values of given attributes. Such rules specify values in terms of the values of other attributes and in terms of student actions. In this figure, the author has marked an attribute named "mode" as a *control attribute*, that is, an attribute whose value is directly determined by user actions. During instructional authoring and delivery, VIVIDS knows that it can make use of the values shown in control attributes to talk about student actions.

In the left portion of the figure is a snapshot of a portion of the VE that exhibits the authored behavior. The behavior of simulated devices such as the throttle shown here is determined by the behavior rules created in the VIVIDS authoring system.

VIVIDS supports productivity in tutorial development by exploiting the simulation author's designations of the cognitively relevant attributes for actions and observations. In the Fig. 20.10

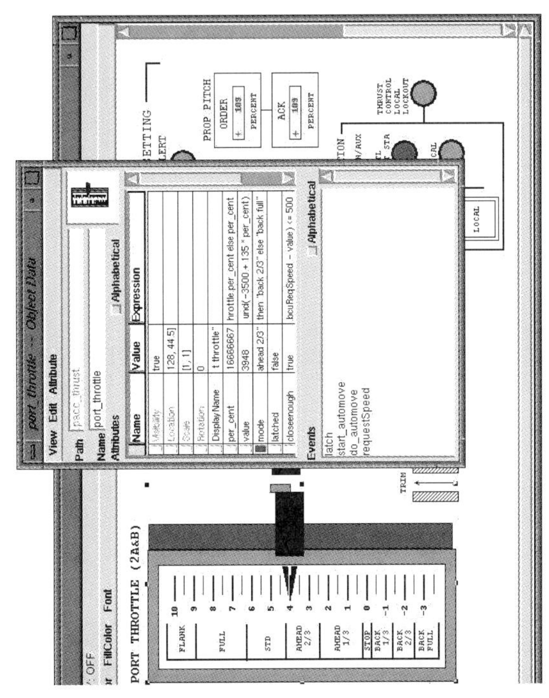

FIG. 20.9. VIVIDS' behavior-authoring interface.

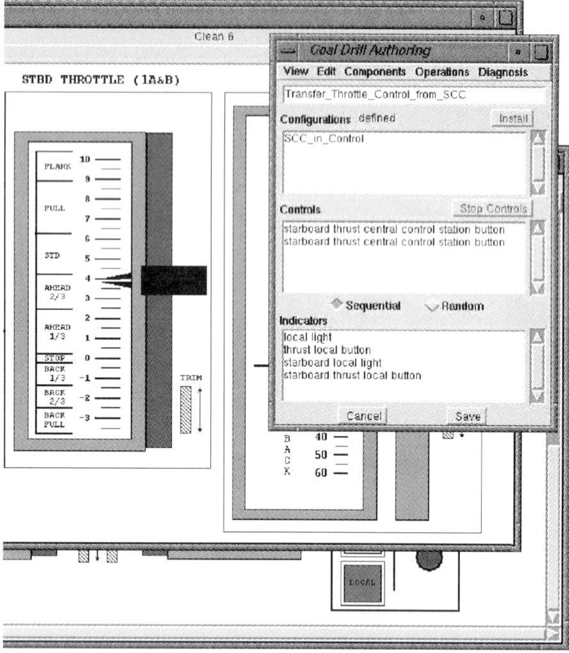


FIG. 20.10. Using VIVIDS to author a procedural tutorial.

below, for example, a simple procedural tutorial is being authored by carrying out a task and then pointing to the objects in the simulation that indicate that the task has been achieved. As the instructor carries out the steps in the procedure, the tutorial authoring system notes the actions and the associated changes in attribute values. It generates brief action descriptions, which appear in a list in the tutorial authoring view. When it is time to present the tutorial to the student, VIVIDS will be able to talk to the student in terms of these cognitively relevant action and indicator values.

The Virtual Environment Technology project supported more than one approach to tutoring procedures. An intelligent agent called *Steve*, developed by Lewis Johnson and Jeff Rickel atInformation Sciences Institute, could be used in conjunction with VIVIDS. At the cost of some expert knowledge engineering, Steve could be given a robust representation of the structure of as task in terms of plans and subgoals. During instruction, Steve could watch student actions and monitor changes in VIVIDS simulation attributes to determine where the student might be in the plan structure. Because Steve has a richer representation of the decision process, it is capable of generating richer tutorial dialogs during instruction than is the easily authored native VIVIDS procedure lesson. For example, Steve can describe the reasons for actions. In Fig. 20.11 below, Steve is shown demonstrating to a student how to carry out a step in a procedure.

## 5. CONCLUSIONS

Cognitive issues in the fields of perception, attention, learning and memory, motor cognition, and decision making are all relevant to virtual environments. Research on and prototype developments of VEs for learning have dealt with these issues in ways that may help to inform future VE projects of many types.

A number of research and practical issues remain to be addressed in future research. These include such questions as "How should a user's attention be drawn to objects or areas that are

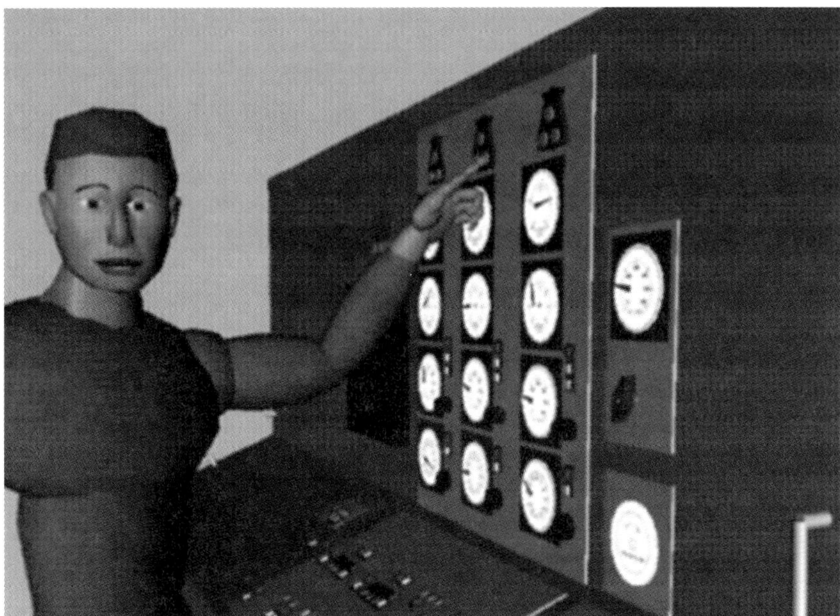

FIG. 20.11. Steve demonstrating a procedure.

currently outside of the student's field of view?" and "Under what conditions is it appropriate to assume that a user has perceived something that is in her field of view?"

Validation studies also remain to be conducted. Under what conditions does procedural expertise in a VE translate to competence in a corresponding real-world task?

As for training applications, where are VEs best suited? VE-based training may be appropriate for concepts that require mental images in conjunction with decisions about location, about relative location of things over time, or about the timing of location and direction changes. Virtual environments can help students to recognize, or visually imagine, patterns of time and location in tasks done before, so that the student can see a familiar pattern in a new situation and then do the job in the new situation.

Why should VE work well for training? The tutor can control the mental images, timing and location not only of objects, but also can control the location of the learner. This lets a learner explore relationships from different perceptual viewpoints so that the learner can later recognize the same kind of situation. Even if the situation looks like it is new, the learner can see that it is really the old situation presented at just a little different angle. The use of VEs encourages the presentation of many perceptual viewpoints to a learner so that an appropriate generalization of the concept can be rapidly acquired.

Another area of research to be explored is the further development of tools that encourage the production of VEs that take advantage of cognitive factors to ensure that such environments achieve their intended purpose in an effective and efficient manor. A key issue may be the development of familiar metaphors for VE that facilitate the learner's association of novel situations with more easily understood ones.

## 6. REFERENCES

Breaux, R., Martin, M., & Jones, S. (1998). Cognitive precursors to virtual reality applications. In *Proceedings of SimTecT '98* (pp. 21–26). Adelaide, Australia: .

Crenshaw, R. S., Jr. (1975). *Naval shiphandling*. Annapolis, MD: Naval Institute Press.

Davidson, S. (1996). *Software design of a virtual environment training technology testbed and virtual electronic systems trainer*. NAWCTSD TR-96-002.

Hall, C., Stiles, R., & Horwitz, C. D. (1998). *Virtual reality for training: Evaluating knowledge retention* VRAIS 1998.

Hays, R. T., Vincenzi, D. A., Seamon, A. G., & Bradley, S. K. (1998, May). *Training effectiveness evaluation of the VESUB technology demonstration system* [Naval Air Warfare Center Training Systems Division Tech. Rep. No. 98-003, Online] Available: http://www.ntsc.navy.mil/Programs/Tech/Virtual/Vesub/VESUB3b.ZIP.

Johnson, W. L., Rickle, J., Stiles, R., & Munro, A. (1998). Integrating pedagogical agents into virtual environments. *Presence: Teleoperators and Virtual Environments, 7*(6), 523–546.

Jonassen, D. H. (1999). *Task analysis methods for instructional design*. Mahwah, NJ: Lawrence Erlbaum Associates.

Jonassen, D. H., Hannum, W. H., & Tessmer, M. (1989). *Handbook of task analysis procedures*. New York: Praeger.

Leshin, C. B., Pollock, J., & Reigeluth, C. M. (1992). *Instructional design strategies and tactics*. Englewood Cliffs, NJ: Educational Technology Publications.

Munro, A., & Pizzini, Q. A. (1998). *VIVIDS reference manual*. Los Angeles: Behavioral Technology Laboratories, University of Southern California.

Munro, A., Johnson, M. C., Pizzini, Q. A., Surmon, D. S., Towne, D. M., & Wogulis, J. L. (1998). Authoring Simulation-Centered Tutors with RIDES. *International Journal of Artificial Intelligence in Education*.

Munro, A., Surmon, D. S., Pizzini, Q. A., & Johnson, M. C. (1997). Collaborative authored simulation-centered tutor components. In C. L. Redfield (Ed.), *Intelligent Tutoring System Authoring Tools* (Tech. Rep. FS-97-01). Menlo Park, CA: AAAI Press.

Norris, S. D. (1998). *A task analysis of underway replenishment for virtual environment ship-handling simulator scenario development*. Unpublished master's thesis, Naval Postgraduate School.

Patrey, J., Breaux, R., Mead, A. M., & Sheldon, E. M. (in press). Performance measurement in virtual reality In *Proceedings of RTA HFM-21* (What is Essential for Virtual Reality Systems to Meet Military Human Performance Goals?) The Hague, the Netherlands: North Atlantic Treaty Organization.

Regian, J. W., & Yadrick, R. (1994). Assessment of configurational knowledge of naturally and artificially acquired large-scale space. *Journal of Environmental Psychology, 14*, 211–223.

Regian, J. W. (1997). Virtual reality for training: Evaluating transfer. In Seidel and Chatelier (Eds.), *Virtual Reality, Training's Future?* (pp. 31–40). New York: Plenum Press.

Stiles, R., McCarthy, L., Munro, A., Pizzini, Q., Johnson, L., & Rickel, J. (1996). *Virtual environments for shipboard training*. Intelligent Ship Symposium. Pittsburgh, PA: American Society of Naval Engineers.

# 21

# Multimodal Interaction Modeling

George V. Popescu,[1] Grigore C. Burdea,[2]
and Helmuth Trefftz
*Rutgers University,*
*Department of Electrical and Computer Engineering,*
*96 Frelinghuysen Road,*
*Piscataway, NJ 08854.*
*popescu@us.ibm.com*
*burdea@vr.rutgers.edu*
*trefftz@caip.rutgers.edu*

## 1. INTRODUCTION

Virtual environments (VE) represent advanced, immerssive, human–computer interaction systems. Such interaction occurs through communication over several sensorial channels. Since the communication pathway between users and the simulation system groups several distinct channels, it is termed *multimodal*. The modalities used are primarily the visual, auditory, and haptic ones. The number, quality, and interaction between such modalities are key to the realism of the simulation and eventually to its usefulness. Thus the need for increased immersion and interaction motivates system designers to explore the integration of additional modalities in VE systems and to take advantage of cross-modal effects. The downside, of course, is increased system complexity, cost, and possible integration/synchronization problems. In addition, cross-modal interaction can lead to perceptual illusions that must be considered by system developers (see chap. 22, this volume).

Finding the best compromise between multimodal simulation realism and its cost and drawbacks requires a good understanding of each modality individually, and in combination. The first step is to have a model of the communication loop between the user and the computer(s) running the simulation, as illustrated in Fig. 21.1 (Schomaker et al., 1995). The Human Output Channels and Computer Input Modalities define the input flow into the simulation. The Computer Output Modalities and Human Input Channels define the feedback flow by which the user receives feedback from the virtual environment. The two communication channels form the human–VE interaction loop.

The processes involved at the human side of the interaction loop are perception, cognition, and control. *Perception* is the process of transforming sensorial information to higher

---

[1] Currently with IBM Co., T. J. Watson Research, Hawthorne NY.

[2] Author for correspondence.

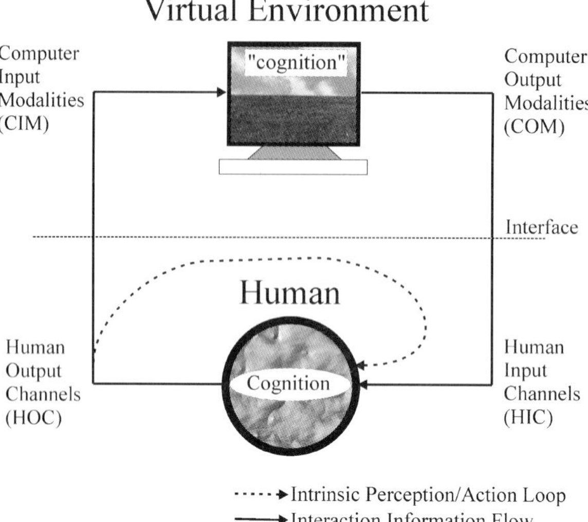


FIG. 21.1. Human–VE interaction model (adapted from Schomaker et al., 1995). Reprinted by permission.

level representations (cognitive and associative processes; Schomaker et al., 1995). In the model adopted here, perception involves the machine-to-human communication flow. The corresponding perception channels are visual, auditory, and somatic (haptic, olfactory, gustatory, vestibular). Interface devices that need to be matched to a user's input characteristics mediate the user's perception of the computer output. In order to guide the design of such interfaces, numerous studies have been dedicated to human sensory and motor physiology. A comprehensive comparison of human sensory capabilities with feedback device specifications can be found in Barfield, Hendrix, Bjorneseth, Kaczmarek, and Lotens (1995).

The *control* process, in the context of VE systems, represents the translation of human actions into task-related information for the computer. Thus control involves the human output channels and computer input modalities. The control modalities of gesture, speech, gaze, and touch are used in conjunction with hardware interfaces. Such devices convert a user's output to computer digital input data as accurately and as fast as possible. Control and perception are not totally independent processes. The dotted line in Fig. 21.1 accounts for I/O channel coupling of the human information processing system, which will be discussed later in this chapter.

The *cognition* process takes place both at the user's side and at the computer side of the system. Human cognition processes have a far greater complexity, however, and can only be approximated by simplistic computer "cognition" algorithms. As with the perception modalities, present limitations lay at the computer side of the interaction loop.

The capacity and bandwidth requirements for the VE system communication channels vary with human information processing capacity (visual, auditory, haptic, etc.) and with human motor performance (eye motion, body motion). Other communication parameters, such as time delays, interchannel lags, and channel interference, are also important for integrating all the above modalities. There are no models available, however, for describing multimodal integration due to the complexity of human information processing channels. The design guidelines for multimodal integration are mostly based on experimental studies (human factors). Typically these studies focus on either input or output modalities and do not evaluate integrated systems. Therefore the following sections will treat separately the two components of the VE loop (input and output). The discussion of input–output coupling is limited to a short final section.

## 2. MULTIMODAL INPUT

Development of multimodal input for VE systems is essential due to the increased user control and interactivity required by recent VE applications (e.g., medical VE, VRCAD). Performance-oriented interaction techniques (Bowman & Hodges, 1999; chap. 13, this volume) address this need by proposing gesture-based metaphors, which appeal to user imagination ("magic" interaction methods). Multimodal input (gesture plus speech and gaze) can also achieve a high level of interactivity while increasing the naturalness of the human VE interface.

Human input in VE is mostly achieved through gesture-controlled input devices. Even though many of these devices are intuitive and easy to use, they cannot provide a fluent dialogue between the user and virtual environment. Natural interfaces extract more information from human output (speech, gestures, gaze) and use "cognitive" models to respond intelligently to a user's input. These models integrate parallel streams of information from speech, gesture, and gaze, and provide real-time semantic interpretation. The present discussion on multimodal input starts with a presentation of performance and limitations of input modalities. This will highlight the need for integration of input modalities in order to improve the interface.

### 2.1 Input Modalities

The most prevalent input modality used in present VE systems is movement of the user's body. Such movement may be limited to the wrist (when a joystick or spaceball are used), the fingers (when sensing gloves are used), or even the whole body (when the user wears a sensing suit). Joysticks, mice, or trackballs are simpler and cheaper but constrain the user's freedom of motion, a very important ingredient for increased simulation realism (see chap. 40, this volume).

Sensing gloves and sensing suits preserve the user's freedom of motion (within the range of three-dimensional [3-D] trackers) and are more appropriate for modern large-volume virtual environments. Furthermore, sensing gloves allow dextrous interactions, overcoming the limited menu of actions of simpler devices (button clicks, etc.). They are, of course, more expensive and more complex to program.

Hand gesture has been extensively studied as an efficient input modality. Gesture languages with different levels of complexity (starting from grasp-release commands to American Sign Language) have been proposed for various applications (Billinghurst & Wang, 1999; Kjeldsen & Kender, 1997; Pavlovic, Sharma, & Huang, 1997; Starner & Pentland, 1995). Large vocabularies necessary for a fluent human–VE dialogue cannot be implemented through gesture only because this requires extensive user training and memorization.

The efficient use of gestures resides in their integration with speech input. The speech channel has great potential for control in virtual environments. Being the dominant channel of human–human communication, speech can convey levels of abstraction inaccessible to other input modalities. Giving input through speech relies on speech recognition capabilities of the computer(s) running the simulation. Current speech recognition algorithms use real-time Hidden Markov Models (HMM; Dragon Systems, 1999; HTK, 1999; IBM, 1999; Microsoft, 1999). These speech recognizers can achieve high recognition rates (over 95%) if the user employs a limited, application-dependent vocabulary. The limitations in current language understanding technology prevent widespread use of speech-based user control of virtual environments. Speech input thus should be combined with other modalities, such as gestures, to provide better spatiotemporal information (Oviatt, DeAngeli, & Kuhn, 1997). Gesture can also help disambiguate speech sentences.

Another approach to help speech input is to add gaze input. During control tasks, gaze automatically focuses on the VE region of interest (Jacob, 1995). This can be exploited for navigation or object selection tasks. More elaborate tasks, such as moving objects from one location in the VE to another, require user training in order to exercise eye movement control and to minimize eyestrain. Extracting the semantics of eye gestures is still a difficult problem. Eye movements represent a very fast way to communicate user intentions, but they also carry unintended (subconscious) content. These limitations make gaze input useful only in combination with speech and/or gesture.

The above discussion indicates why multimodal input is preferred to unimodal speech input. Furthermore, experiments done by Oviatt et al. (1997), Oviatt and Clow (1998), and Oviatt (1999) demonstrated that users have a strong preference for multimodal as compared to unimodal interactions, especially in the spatial domain. Experiments with a two-dimensional (2-D) map application showed that users were likely to express multimodal commands when describing spatial information (location, number, size, orientation, or shape of objects). User commands carrying spatial information were best sent to the computer through gestures. Simple speak-and-point interfaces were considered too rigid and of limited practical use.

The remainder of this section is organized around three aspects of multimodal input: namely software architecture; integration issues; and multimodal input applications. Interested readers should also consult Blattner and Glinert (1996) for a comprehensive review of multimodal input.

## 2.2 Multimodal Input Architecture

Virtual environments need to minimize time delays in system response, whether these delays are due to software processing, communication bottlenecks, or slow graphics. Real-time speech recognition, gesture recognition, and haptic rendering represent a significant computation load. This makes necessary a distributed computing platform, with modalities running on separate computers. The various input devices are therefore attached to their own computation platform running a driver and data processing software, and communicate through a network interface. The programming toolkits available to help developers integrate the different hardware interfaces are at the present time limited. Even in cases when the input device has an associated programming interface, integration is not easy. Most of the programming interfaces were not developed for multimodal input and have limitations when used as research tools.

Robust multimodal architectures integrate input modes synergistically in order to disambiguate input signals. This process takes advantage of modality-specific information content, redundancy, and complementary input modalities. Multimodal input software uses a layered architecture, with input devices at the lowest level and the application at the highest. Intermediate layers describe how input modalities are used and what tasks need to be accomplished. This modularity is important because it allows a better separation of application functionality from the user interface.

A layered model for modal input and interaction in VEs was presented in Schomaker et al. (1995) and is summarized in Table 21.1. The lower levels of the model contain the input devices and the associated events. The user VE interaction is realized through gesture, speech and gaze commands. When modalities are used in parallel, the interaction can be realized through multimodal commands. The multimodal input has uniform access to the pool of events received from input devices. The last two levels specify the application and the interaction tasks implemented by the command language. Tasks are modality independent, even though some modalities are better suited for certain tasks (e.g., gestures for object manipulation).

An example of a layered multimodal input platform is illustrated in Fig. 21.2 (Marsic, Medl & Flanagan, 2000; Sletterink, 1999). The first layer consists of modal software interfaces, which get data from input devices and detect input events (gestures, spoken units, mouse clicks,

**TABLE 21.1**
Layered Model for Interaction in VE (Adapted from Schomaker et al., 1995).
Reprinted by permission.

| | |
|---|---|
| Application | Virtual reality, CAD, visualization, architectural design |
| Interaction tasks | Navigation, manipulation (move, rotate, scale), identification, selection, VE editing |
| Interaction techniques | Gesture language (grabbing, releasing, pointing), 3-D menu, speech commands, Gaze commands, multimodal commands |
| Events | Hand and body gestures, 3-D Motion, button click, force, 2-D Motion, torque, spoken units, eye motion, gaze direction |
| Input devices | Sensing gloves, trackballs, 3-D mouse, 6-D tracker, eye tracker, joystick, microphone, tactile gloves |

etc.). These events are then transformed into text units and sent to the next layer. The second layer contains the central component of the software architecture, namely the *multimodal input manager* (MIM). The MIM receives streams of data from several modal interfaces and issues commands to the current application. The MIM in turn consists of four units, namely the *modality handler*, the *customizer*, the *multimodal integrator* and the *connection manager*. The modality handler allows adaptive use of available input modalities. It detects available modalities, decides which modalities are relevant for the current application and gets input streams from its corresponding modal interfaces. The connection manager handles automatic context by keeping track of a focused application (the selected window). Context switching increases the reliability of speech and gesture recognition modules by reloading the grammar and gesture sets through the customizer. Instead of implementing a large grammar that will reduce recognition performance, each application has its own grammar and a corresponding parser. The modal integrator implements the integration (fusion) and synchronization of input modalities, as described below.

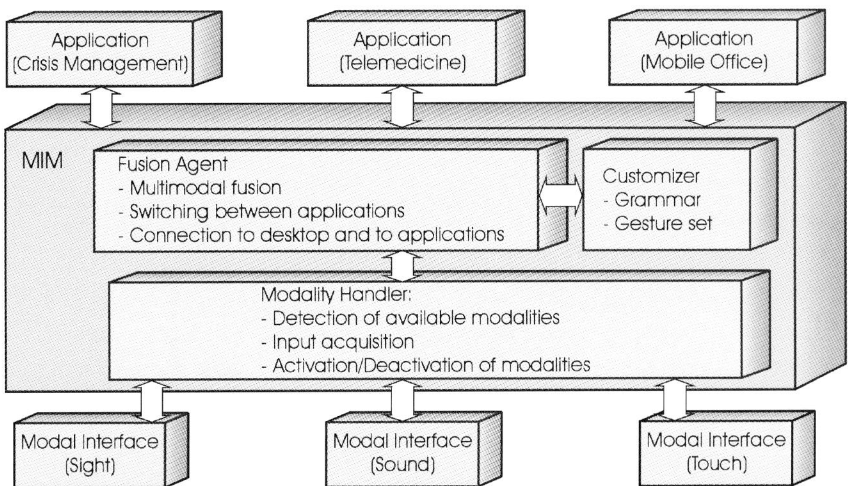


FIG. 21.2. Structure of the Multimodal Input Manager (Marsic, Medl & Flanagan, 2000). © IEEE. Reprinted by permission.

## 2.3 Multimodal Input Integration and Synchronization

Multimodal integration performs the mapping of modal inputs into a semantic interpretation. The main elements of multimodal integration are: unified representation synchronization and semantic analysis. *Unification of data representation* is the first step after capturing the data from human output. To obtain a unified representation, each input data stream is translated into a sequence of tokens contributing to the semantic. These tokens could be, for example, time-stamped text items.

The second step is the *synchronization of the input streams*. The input streams are first buffered and segmented in order to create a context for semantic analysis. Then the tokens are aligned on the time axis based on synchronization models. These models take into account the temporal precedence of modes and intermodal lags. The last step is semantic data fusion. The semantic analysis module receives the temporally aligned input tokens and extracts the control sentences. Natural language processing methods can be applied to extract the semantics of human–VE interactions when using a unified textual representation of modal input.

The steps described above are typically performed in order, but feedback loops from synchronization and semantic analysis can modify some input tokens during the integration process. Several mechanisms have been proposed for implementing multimodal integration, involving frames, neural nets, and agents. The frame–slot integration mechanism applies the well-known artificial intelligence method (slot-filler) for multimodal fusion. In this method, information necessary for command or language understanding is encoded in structures called "frames." For the multimodal input model shown in Fig. 21.2, a frame stores the information about multimodal input commands. The command frames are composed of slots, which are lexical units provided by the multimodal input. As an example, the "move" frame is composed of two slots: "object," to identify the object; and "where," to specify the final position.

A frame–slot-based method, with predefined and application-dependent frames was used in Medl et al. (1998). The multimodal architecture is illustrated in Fig. 21.3. The parser extracted the lexical units from different input modalities and filled the appropriate slots in the slot buffer. The concurrency of input modalities was handled on a first-come, first-served basis. The central component of the fusion agent was the slot buffer, which stored the information inputted by the user. This allowed back referencing of past lexical units (e.g., "it" could be used to reference the previously selected object), increasing the fluency and naturalness of the command language. Data fusion used lexical units drawn uniformly from input channels. The fusion agent continuously monitored the slot buffer checking for filled frames. Once a frame was filled (enough information to generate a command), the fusion agent sent it to be executed. The main advantage of this architecture is the uniform access of the input modes.

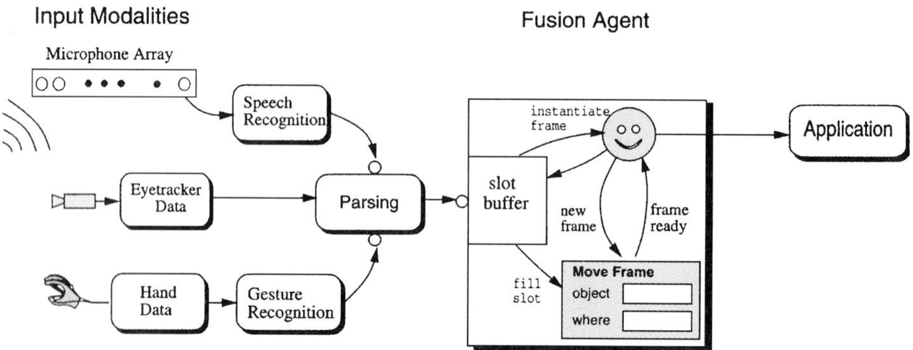


FIG. 21.3. Frame-based multimodal input integration (adapted from Medl et al. 1998). Reprinted by permission.

Neural networks are adaptive systems designed to model the way in which the brain performs a particular task. Their main quality is the capacity to learn complex relations between their inputs and outputs. This characteristic makes neural nets useful for multimodal input integration. Vo and Waibel (1993) were among the first to use neural networks for multimodal systems. Each input modality used time-delay neural nets (TDNN) for temporal alignment, which allowed simultaneous processing of data received over a period of time. The design exploited the temporal precedence of hand gesture (writing) over speech. Later designs used a connectionist network for multimodal semantic integration (Vo & Waibel, 1997). In this model the input modality channels were considered parallel streams that could be aligned in action frames and jointly segmented in parameter slots. Temporal alignment was based on time stamps associated with input tokens. The integration mechanism used a mutual information network architecture and a dynamic programming algorithm to generate an input segmentation. After obtaining the frames with the corresponding slot information, grammar-based algorithms were applied to capture syntactic elements and extract their meaning.

Another approach to multimodal input integration is the use of a set of hierarchically organized entities (agent-based algorithms). An agent implements a certain function (task) of the multimodal input manager. The agent has a specified set of sensors (inputs), a model to process the information, and a set of effectors (outputs). The agent-based multimodal input architecture contains three layers. The first layer consists of modal agents processing the data from input devices. The next layer contains language recognition agents. A language recognition agent gets input from a modal interface (speech, gesture, etc.) and generates a potential interpretation based on the specified language model. This layer is responsible also for unifying the modal data representation. The third and last layer contains a multimodal fusion agent, which receives language units from recognition agents and integrates them to produce the best possible interpretation of the user's input.

An agent-based fusion of speech and gesture modalities was presented in Johnston et al. (1997). Their multimodal architecture used six agents: interface client, speech recognition, natural language processing, gesture recognition, multimodal integration, and bridge (a command execution agent). The fusion agent used a probabilistic model. Probabilities were associated with each modal input, with the highest unified score interpretation being selected. This allowed grammar-based techniques employed in natural language processing to be used in the semantic integration process as well.

### 2.4 Examples of VE Applications Using Multimodal Input

Computer Aided Design (CAD) systems have a great need to replace the current windows-icons-menu-pointer (WIMP) paradigm with more-natural user interfaces. This is especially true in the early "concept" phase of the design process. The main limitations are related to the inability to create 3-D shapes intuitively and provide interactive 3-D visualization. Chu, Dani, and Gadh (1997) investigated the potential use of multimodal input for COVIRDS (conceptual virtual design)—a VE-based CAD system. Candidate input modalities were gaze, voice commands, and hand gestures. Multimodal interface requirements were analyzed for three stages of product design: part and assembly generation, part and assembly modification, and design review. Sets of interaction tasks (commands) were produced according to application requirements. Subjective evaluation tests showed that voice and hand gestures were the preferred modes of interaction, whereas gaze input was found useful for "select" commands. No comments referring to sensory input fusion or modalities concurrency were given in this study.

Speech and gesture inputs were subsequently used to develop an intuitive interface for concept shape creation (Chu, Dani, & Gadh, 1998). A series of tasks were implemented using different modalities (e.g., zoom-in, viewpoint translation and rotation, selection, resizing, translation). Evaluation of the interface was based on user questionnaires. Voice was intuitive to use in abstract commands like viewpoint zooming and object creation and deletion. Hand gestures were effective in spatial tasks (resizing, moving). Some tasks (resizing, zoom in a particular direction) were performed better when combining voice and hand input. The command language was very simple and the integration of modalities was implemented at syntax level. Therefore, in some cases users showed preference for a simple input device (a wand with five buttons) rather than for multimodal input.

Researchers at Beckman Institute (University of Illinois, Champaign-Urbana) developed MDscope—a VE testbed for visualization and interactive manipulation of complex molecular structures. Multimodal input (speech, hand gesture and gaze) was integrated in the VE in order to improve virtual object manipulation. The multimodal framework for object manipulation (Sharma, Huang, & Pavlovic, 1996) used a HMM speech recognizer. The hand gesture input module used a pair of cameras and HMM-based recognition software. Speech and gesture were integrated using a fixed syntax: <action>; <object>; and <modifier>. The user command language was rigid to allow easier synchronization of input modalities. The synchronization process assumed modality overlapping. The lag between speech and gesture input was considered to be at most one word. Gaze input was used to provide complementary information for gesture recognition. The direction of gaze, for example, was exploited in order to disambiguate object selection. Experimental work with the above testbed concluded that multimodal interfaces allow much better interactivity and user control compared to unimodal, joystick-based input.

The relationship between speech and gesture was also exploited by Cassell et al. (1994). Their application involved animated human figures capable of conversation in a virtual environment. The underlying assumption was that gesture and speech are "communicative representations of a single mental representation." The correlation between speech and gesture was analyzed and a set of rules to predict the location and type of gestures was given. The system provided a test bed where predictive gestures were computed during the dialogue. The implementation used the Jack virtual human toolkit, which was animated using PaT-Nets (Douville, Levison & Badler, 1996). Hand, wrist, beat (a specialized form of wrist motion), and arm motion were driven by computation modules that used information from the output of the speech synthesis module. Using this predictive approach, the VE was populated with autonomous animated conversational agents.

Another potential application area for multimodal input is military planning in which users perform very fast control tasks in a stressful and noisy environment. Such a system could possibly benefit from a multimodal input interface. Rosenblum and colleagues at the Navy Research Laboratory developed "Dragon 2," a VE research platform for battlefield planning and control (Rosenblum, Durbin, Doyle & Tate, 1997). The multimodal input was provided by QuickSet (Cohen et al., 1997, 1999), a pen-and-voice-based system that ran on handheld PCs and communicated via a wireless Local Area Network (LAN) with the host computer running the simulation. The multimodal input for Dragon 2 consisted of voice commands and a "flight stick," a commercial joystick modified to incorporate a Polhemus tracker. The flight stick was used for navigation, selection, and drawing on a 2-D or 3-D map. Users' speech and gesture input were recognized, parsed, and then fused via the QuickSet multimodal integration agent.

Another military planning application—3D Mission Control—was recently developed on top of the multimodal architecture shown in Fig. 21.3 (Medl et al., 1998). The system used a commercial speech recognition engine (Microsoft WHISPER), a haptic glove (the Rutgers Master II), and a gaze tracker (ISCAN). The data fusion mechanism used the slot-filler method described earlier in this section. The user saw a 3-D map of the area where the mission was being planned (shown in Fig. 21.4 Marsic, Medl & Flanagan, 2000). The scene was composed of 3-D terrain created from a digital terrain elevation data (DTED) map file, a virtual hand, and several 3-D objects representing "assets" (base camps, jeeps, trucks, helicopters, etc.). The user could add, select, move, and delete these objects in the 3-D environment. The manipulation

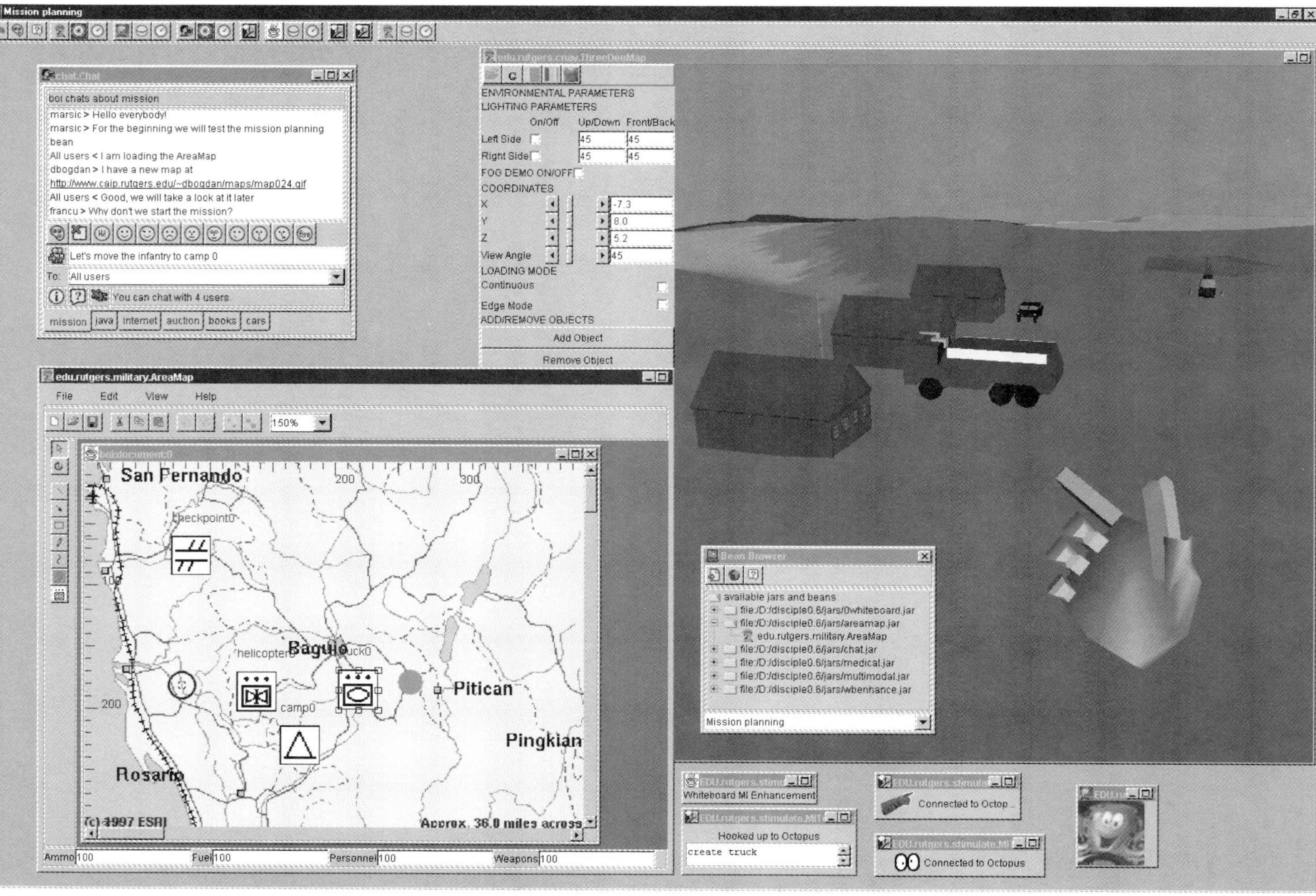


FIG. 21.4. "3-D Mission Control" (Marsic, Medl & Flanagan, 2000). © IEEE. Reprinted by permission.

of virtual objects relied on gesture input, whereas speech was most useful for object creation and deletion, and for viewpoint change. Feedback was provided to the user through visual, auditory, and haptic channels. Besides reinforcing the visual feedback, force feedback was used to map the physical properties of the objects in the virtual environment. For instance, the hardness of a transport vehicle indicated whether it was loaded or not. Synthesized voice feedback was used to confirm successful completion of a command, identify specific objects, and warn the user of errors.

### 2.5 Summary of Multimodal Input

Looking at the VE simulation loop from the user's perspective, multimodal input allows better control of the simulation through several communication channels. By using voice, gaze, and gesture, the simulation interactivity as well as the user's feeling of immersion (see chap. 40, this volume) into the VE may increase. Looking at the same simulation loop from the computer input perspective, multimodal input helps alleviate some very clear limitations of unimodal, voice-only input. Commands that otherwise may be difficult or impossible to understand can be disambiguated, objects can be placed in an environment more precisely, and navigation can be faster. Several approaches have been developed for the integration of the various input modalities, namely application-dependent frames, neural networks, and hierarchically organized agents. The limited data available from human factor evaluations of multimodal input suggests a clear subjective preference for multimodality versus unimodal input.

## 3. MULTIMODAL FEEDBACK

Multimodal feedback represents the computer response to the user's multimodal input to the simulation. Early VE systems had unimodal visual feedback through stereo or monoscopic graphics displayed on CRTs or head-mounted displays (HMDs). As research interest for other communication channels grew, bimodal VE systems (visual-auditory, visual-haptic feedback) have been developed and evaluated. Currently, the most frequently used feedback channels are visual, auditory, and haptic (force/tactile feedback). Additional channels such as olfactory feedback may significantly enhance user's immersion (see chaps. 14 and 40, this volume), but are less developed at the present time (note, however, that the technology for olfactory feedback is advancing, providing several commercially available systems including SENX Machines, iSmell Digital Scent Technology, and Aromajet).

The visual, auditory, and haptic feedback channels each use a different kind of interface. The information flow received through the visual sensorial channel is much larger than that corresponding to the haptic or auditory channels. Thus visual feedback takes the dominant role among the three primary VE feedback modalities. Stereoscopic vision allows 3-D perception of the left and right eye images rendered by the computer (see chap. 3, this volume). If the visual feedback modality is coupled with body posture (head motion) input, then the user controls the field of view to the simulation. This is thought to result in increased immersion and simulation realism (see chap. 40, this volume). Earlier HMDs were heavy and had low image resolution, problems that were compounded by low-accuracy tracking of the user's head position (Thorpe & Hodges, 1995). In recent years the problems with HMD weight and image resolution have been largely solved (Olympus, 2000). Furthermore, accurate and robust trackers are starting to immerge (InterSense, 2000).

Sounds are ubiquitous in the real world and therefore their presence in VEs should be beneficial to the simulation realism. Auditory displays can be classified as nonlocalized and localized (see chap. 4, this volume). Nonlocalized audio feedback is provided by multimedia-quality stereo sound, meant to increase user interaction. Localized (or spatialized) sound uses head tracking and head related transfer functions (HRTF) to display 3-D soundscapes. A localized sound source remains fixed in space regardless of the user's change in head position, similar to what happens in the real world (Barfield et al., 1995). Thus 3-D sound technology has significant potential for newer large-volume simulations. A current limitation is related to the inability to customize the HRTFs to the individual user's pine (external ear) characteristics, which results in less accurate mapping of sound spatial information in VEs than in the real world. Other limitations of earlier commercial products were the high computation load required and the high hardware cost. These problems have been largely addressed in recent years, with the introduction of significantly faster hardware, at lower costs. The customization of 3-D sound remains an area of active research.

Haptic feedback is considered an important modality for VE interactions involving active virtual object manipulation (Burdea, 1996; chaps. 5 and 6, this volume). Adding haptics increases user immersion and interactivity by allowing an object's physical characteristics (weight, compliance, inertia, surface smoothness, slippage, temperature, etc.) to be displayed. Haptic feedback is also needed when the visual channel is degraded or nonexistent. Current commercial haptic feedback interfaces suffer from large weight, small workspace, and large costs. Additionally, modeling physical interactions is complex and computationally intensive, which normally requires a multiprocessor host computer. The forces applied by the computer through the haptic interface are real, which makes user's safety a concern (see chap. 41, this volume).

### 3.1 Multimodal Feedback Integration

Each type of feedback interface mentioned above is the subject of other chapters in this handbook (see chaps. 3–6). By contrast, the focus here is on the integration of the separate feedback modalities, including such topics as sensorial redundancy and sensorial transposition. The user's perception of the VE spans the spatiotemporal and physical domains. The spatiotemporal domain refers to the perception of 3-D space and time. The physical domain describes such virtual object properties as weight, compliance, inertia, and surface temperature. The orthogonality of these domains simplifies the analysis of multimodal feedback.

The main aspects of feedback channel interaction are complementarity and modal redundancy. *Complementarity* refers to nonoverlapping information received through separate feedback channels. This allows the VE designer to increase simulation realism by simply adding more feedback modalities. *Modal redundancy* comes from overlapping the same information in the representation domain, fed back to the user through several sensorial channels. The VE designer can use such redundancy to reinforce perception of certain features of a VE as long as coherence constrains are imposed. Maintaining spatiotemporal and physical model coherence requires that feedback modalities be synchronized. Nonsynchronized events or objects affect drastically the VE illusion and reduce the user's participation. Thus it is important to analyze the complementarity and redundancy of the visual, auditory, and haptic feedback modalities.

#### *3.1.1 Visual-Auditory Feedback Interaction*

Sound is frequently used in combination with the visual channel because it provides additional simulation cues and increased user interactivity. The acoustic characteristics of virtual objects are their spatial location, temporal properties, timbre pitch intensity, and rhythm. Sounds can be mapped to physical objects or be independent of them. The visual and auditory channels are similar and complementary, as they can both represent spatiotemporal information. However, their spatiotemporal resolution differs. Sound spatial localization of objects is poor, whereas time accuracy associated with auditory feedback is superior to that of the visual channel. Therefore sound feedback will greatly enhance applications where the user's reaction time is critical.

The overlapping of visual and auditory information on the spatiotemporal domain results in sensorial redundancy as long as the two information flows are synchronized. This augments the perception of both modalities. For example, a graphics scene that has 3-D sound cues that are well mapped to graphical objects appears visually sharper to the user than the same scene without the associated sound. Sound can convey complementary spatial information when the object of interest is not in the field of view or when objects closer to the user occlude more distant ones. In the time domain, sound can provide very precise temporal stamps when needed. Sounds can be used to convey physical properties of objects as well, such as consistency (discriminate between an empty and a full container), weight (a heavy ball sounds different from a light ball bouncing off the floor), and elasticity (vibration of a string is correlated to its pitch). Sound can also act as a gaze direction guide by focusing the user's attention on a certain spatial event in the virtual environment. Three-dimensional sound cues can point the user to an object location (a telephone that rings) or help navigation tasks by suggesting a new direction to be explored. As a consequence of redundancy and complementarity, 3-D sound integrated with the visual display provides a more consistent spatiotemporal representation of a simulated environment.

### *3.1.2 Visuo-haptic Feedback Interaction*

The haptic channel can represent information related to the physical properties of virtual objects in addition to their spatiotemporal ones. Haptics is rarely used for spatial discrimination by itself (except in dark environments). In most cases haptic feedback is combined with information received on the visual channel as a redundant modality. The benefits of such redundancy were mentioned earlier, but a quantitative assessment can only be based on experimental data.

Human factors experiments were conducted by Richard et al. (1994) to assess the benefits of adding haptics to a partially immerssive environment in direct manipulation tasks. The experiments studied the influence of haptic feedback on dexterous manipulation of a plastically deformable ball. The performance measure was the amount of ball deformation (which had to be kept less than 10% of the ball radius). Experimental results showed increased spatial resolution (inversely proportional to ball deformation in excess of 10%) for VEs with haptic feedback. When the haptic modality was added to the visual feedback, the spatial resolution increased almost three times. Force resolution can therefore be used to increase spatial resolution when the relationship between space and force can be mentally evaluated by the user (e.g., a linear relationship—Hooke's law). This supports the idea that haptic feedback is an essential ingredient for direct manipulation tasks in virtual environments.

Coherence of spatiotemporal representation should be imposed for tactile and kinesthetic channels. For example, the "roughness" of a surface evaluated through visual inspection should be matched by the rugosity information provided by the tactile feedback interface. Large time lags between the graphics and haptic loops can confuse the user and may result in control instabilities. Information related to the physical properties of VE displayed on the haptic and visual channels needs to be coherent as well. Object surface deformation should be synchronized with force calculation to provide increased immersion in virtual environments. A "soft" ball (small forces applied to the user's finger when squeezing) should also be highly deformable. Virtual walls should resist with very high force when being pushed and should have no visual surface deformation. Plastically deformed objects should present a hysteresis behavior both in shape deformation (their surface shape remains deformed after the interaction) and in the associated force feedback profile produced by the haptic interface. When the deforming phase of such objects ends, the forces applied to the user's hand should drop immediately to zero.

Therefore, physical behavior of objects should be implemented both in graphics and haptics domains, and displayed synchronized to the user.

### 3.1.3 Haptic Channels Coupling

The haptic feedback channel constitutes a complex coupled system. There is a very tight coupling between its force and touch feedback components. Minsky, Ouh-young, Steele, Brooks, and Behensky (1990) demonstrated that high bandwidth (500–1000 Hz) force feedback displays could be used to render tactile information as well. Their method used the texture surface gradient and real-time physics (spring–damper model) to calculate the forces to be displayed to the user. The simulation integrated a high-bandwidth two-degree of freedom force feedback joystick that rendered various textured surfaces like sandpaper and elastic bumps. This research suggests that haptic displays can be classified as a function of bandwidth: low bandwidth corresponding to force feedback and high bandwidth to tactile feedback.

The benefits of superimposing vibratory feedback over force feedback for manipulation tasks were illustrated in Kotarinis and Howe (1995). This study identified the kind of tasks where high-frequency vibratory feedback is important (inspection, haptic exploration, direct manipulation). Additionally, it explored the use of tactile display for conveying task-related vibratory information. Their results showed that adding vibratory feedback to a force feedback system was very beneficial, resulting in increased performance in manipulation tasks, such as peg-in-hole insertion.

## 3.2 Sensorial Transposition

*Sensorial transposition* is the provision of feedback to the user through a different channel than the expected one. Sensorial transposition is typically used to substitute unavailable communication channels required by an application. For example, force feedback required in direct manipulation tasks could be substituted with visual, auditory, or tactile feedback. Sensorial transposition can be "simple," when one modality is replaced by another, or "complex," when one modality is substituted by multiple other types of feedback.

Sensorial transposition requires user adaptation. The level of user adaptation needed in the mappings involved in sensorial transpositions varies. Some mappings feel "natural," whereas others require more training. For easy user adaptation, the mapping should use the strongest representation domains (visual → spatial domain, auditory → temporal, tactile → temporal, etc.) of the transposed channel. Several examples of sensorial transposition and corresponding mapping domains are shown in Table 21.2.

**TABLE 21.2**
Examples of Sensorial Substitution Schemes.

| *Initial Channel* | *Input Domain* | *Transposed Channel* | *Mapping Domain* |
|---|---|---|---|
| Visual | Spatial | Tactile | Spatial |
| Auditory | Temporal (frequency) | Tactile | Sensorial intensity |
| Force feedback | Sensorial intensity | Auditory | Temporal (frequency) |
| Force feedback | Sensorial intensity | Auditory, tactile | Temporal (frequency), Sensorial intensity |
| Force feedback | Sensorial intensity | Auditory, tactile, visual | Temporal (frequency), Sensorial intensity, spatial |
| Force feedback | Sensorial intensity | Auditory, tactile | Temporal, spatial |

Sensorial transposition schemes can also provide *sensorial redundancy* in virtual environments. The same feedback information is mapped through a different channel in addition to the one normally used to communicate that type of feedback. This is done in order to reinforce the original message and can be used to increase user performance in complex tasks.

When using sensorial substitution to obtain redundancy, the substitution scheme should be chosen carefully in order to avoid sensorial contradictions or sensorial overload. Otherwise, instead of reinforcing the original signal, such a method may confuse the user and induce reaction delays as the user copes with unexpected sources of information. User sensorial overload resulting from too much feedback data may also decrease human performance in virtual environments.

Research on human psychology has produced several studies describing sensorial transposition effects (Kaczmarek & Bach-y-Rita, 1995). The VE literature focuses mostly on sensorial transposition and redundant feedback effects related to the haptics channel (Fukui & Shimojo, 1992; Massimo & Sheridan, 1993; Richard et al., 1994). This was motivated by the limitations in current haptic feedback devices, which are difficult and expensive to use. The tactile channel was used in many sensorial substitution schemes to display visual and auditory feedback information. For example, tactile-auditory substitution remaps of the frequency analysis performed by the ear to intensities of electrotactile stimulation of the skin have been used and are intended to be suitable for the hearing impaired. Such a system was used to discriminate phonemes by lip reading (Kaczmarek & Bach-y-Rita, 1995). The resolution of the electrotactile stimulus was much lower than the frequency discrimination capability of the ear. Therefore the approach was not beneficial in the discrimination of connected text, making it difficult to develop high-performance systems.

Kaczmarek and Bach-y-Rita (1995) also cite another sensorial transposition scheme, namely tactile substitution of visual data. This is based on mapping to the tactile display of spatiotemporal information normally received through the visual channel. The mapping represented pixel intensity as vibrotactile and electro-tactile pulse width and amplitude. Two-dimensional images (patterns of lines) were "projected" on the skin rather than on the retina. With practice, users were able to identify more complicated objects, such as faces. However, the limited spatial resolution and narrow dynamic range of the tactile display prevented the rendering of complex scenes.

Another tactile-visual sensorial transposition is described in Fritz and Barner (1996). Here proprioceptive information was combined with the tactile feedback produced by a PHANToM interface to substitute for visual information. Finger position information provided by the device was coupled with a haptic rendering to display textures in 3-D, as shown in Fig. 21.5. A stochastic modeling technique was used to generate random texture patterns in order to spatially display text data. High-resolution position detection, as well as the resolution of randomly generated haptic textures (256 for this application), were key in supporting a pattern discrimination task.

(a)

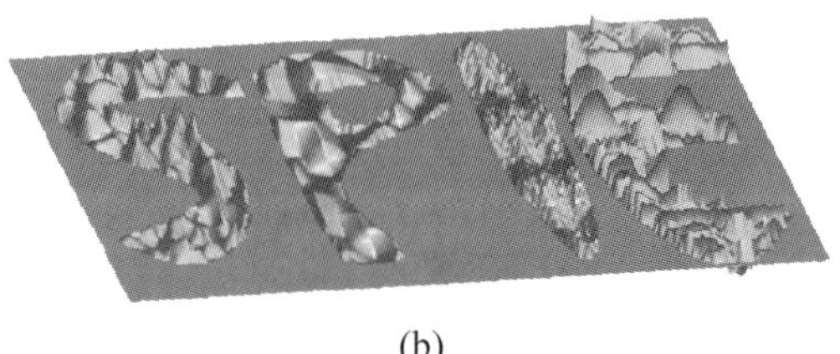

(b)

FIG. 21.5. (a) Original 2D image; (b) spatial textured pattern (Fritz & Barner, 1996). © SPIE. Reprinted by permission.

Redundant tactile feedback to reinforce the visual channel was described in Bouzit and Burdea (1998). The tactile feedback was used to increase pilot performance and reduce spatial disorientation when the visual feedback data was degraded. Airplane attitude orientation was mapped to pressure applied by pairs of air bellows on an elbow joint to resist flexion and extension movements. Tactile feedback was selected instead of force feedback due to its quicker response and unobtrusiveness of the interface. When asked to track a random attitude trajectory with degraded visual feedback, the user registered a 27.4-degree error without tactile feedback. When the tactile feedback was added to the simulation, performance increased threefold (error decreased to 8.9 degrees). This underscores the advantage of tapping into the underutilized haptic channel to convey spatial information when the visual channel is saturated. A similar example is the tactile mapping of aircraft instrumentation on pilot's torso using a tactile feedback vest under research at the Naval Aerospace Medical Research Laboratory (NAMRL, 1999).

Another area where sensorial substitution was shown to be beneficial is teleoperation with long time delays. Under such conditions the force feedback system that normally is beneficial becomes detrimental and may lead to system instabilities. Massimo and Sheridan (1993) studied the efficacy of using tactile and auditory substitution of the delayed teleoperation force feedback signal. Their experimental setup consisted of master and slave robotic manipulators, visual feedback of the remote site and additional tactile, and auditory feedback. In force-auditory feedback substitution, the intensity of the sound mapped the force magnitude. In force-vibrotactile feedback substitution, the vibration amplitude was proportional to the magnitude of force feedback. The experimental data for contact detection (taps) is illustrated in Fig. 21.6. Sensorial substitution schemes outperformed the visual display (worst) and the force feedback scheme. This experiment showed that for remote manipulation tasks the user could take advantage of the fast human response to auditory and tactile stimuli to reduce reaction time. In the second experiment, a peg-in-hole insertion task was executed with either zero or very large visual time delay. The auditory channel was used this time to convey spatial information in addition to force magnitude. Collisions with the left or right side of the hole were conveyed to the corresponding ear. Tactile mapping was similarly enhanced to display spatial information on the lower and upper part of the palm. Results showed that sensory substitution schemes allowed decreased manipulation time when no time delay was present. Auditory and vibratory substitution of forces could function independent of visual information and insure task completion for as much as 3 seconds time delay. Under such large time delay, sensorial substitution allowed teleoperation, which would otherwise have been impossible to perform.

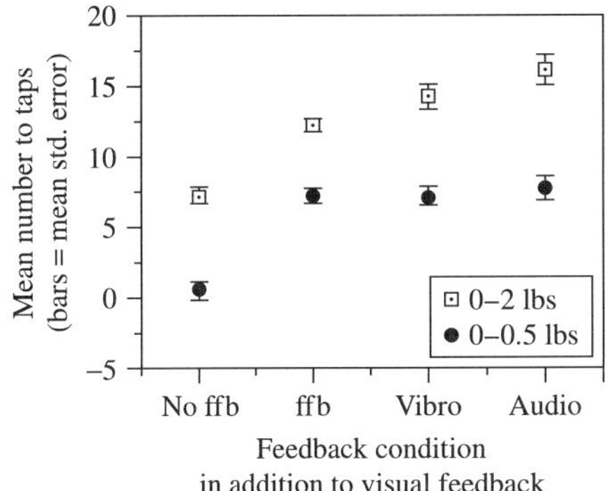


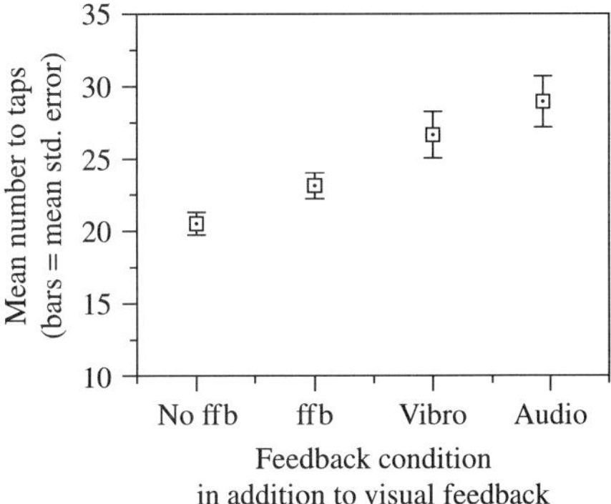


FIG. 21.6. Sensorial substitution of force feedback: (a) contact force detection; (b) magnitude of contact force (Massimo & Sheridan, 1993). © MIT Press Journals. Reprinted by permission.

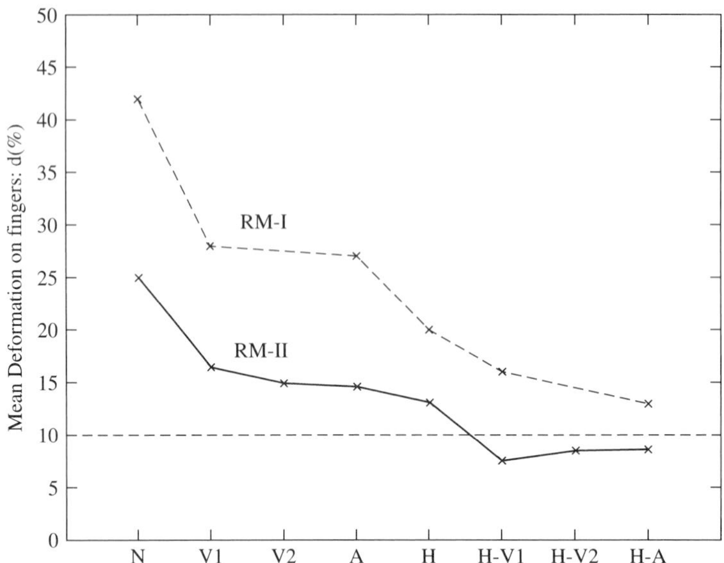


FIG. 21.7. Ball deformation for different force feedback modalities: N—graphics only; V—visual; A—auditory; H—haptic. The target was 10% deformation. Reprinted by permission from Fabiani et al. (1996). © 1996 IEEE.

Richard, Burdea, Gomez, and Coiffet (1994) and later Fabiani, Burdea, Langrana, and Gomez (1996) studied the influence of sensorial substitution and sensorial redundancy in a task involving dextrous virtual object manipulation. The experiment required the user to pick and place a plastically deformable virtual ball, as previously discussed. The substitution of force feedback through visual feedback used LED bargraphs (group V1) or bargraphs overimposed on the VE scene (V2). These were proportional with the contact forces that should have been applied to the user's fingertips by the Rutgers Master I glove. Ball deformation was also mapped on the auditory channel proportional to sound intensity (group A). Experimental results showed that redundant feedback (haptics H + another modality) provided through different mapping methods increased user performance (groups H-V and H-A; see Fig. 21.7). In the absence of sensorial redundancy, force feedback was better than methods using sensorial substitution approaches. Users were able to quickly adapt to the additional mappings available (LED and auditory) and use these successfully during the dexterous manipulation task. The overall relative performance structure remained the same in subsequent experiments with an improved haptic glove (RMII). However, absolute performance in each category improved due to the larger work envelope and dynamic range of RMII versus the older RMI.

### 3.3 I/O Channel Coupling

The above discussion treated multimodal input separate from multimodal feedback for ease of understanding. In reality, the human input and output systems are coupled. This was illustrated in Fig. 21.1 by the dashed line that closes the feedback loop "internally" as opposed to going through the simulator. The VE designer should not overlook this dependency.

Visuo-motor coupling is a well-known example: data from head-position trackers are used to compute the stereoscopic display images in immerssive virtual environments. Gaze tracking can be used to further improve the design of visual displays. The resolution of displayed images can be decreased as pixels get further away from gaze direction because the resolution of the retina decreases toward the periphery of field of view (the so-called foviating effect). Head orientation influences hearing, as described in the binaural model. Additional dynamic effects (acceleration, speed) should also be considered when the user moves in the virtual environment (Cohen & Wenzel, 1995).

The strongest form of integrated perception/control occurs in haptic perception. In direct manipulation tasks, tactile feedback is accompanied by muscle stimulation (force feedback). In order to increase immersion, tactile interfaces need to include force reflection devices. Otherwise the functionality of the tactual display is limited to a simple signaling mechanism (events in the VE can be associated with tactile feedback response). An example involving tactile, force feedback, and user control is grasping. An accurate model of object grasping in VE should include haptic feedback information in the hand-motion control loop. Otherwise the grasping is artificial (objects snap to the palm and force feedback is applied suddenly) and may result in control instabilities.

Hand–eye coordination is important in direct manipulation tasks. The mapping of the user's hand to a virtual hand depends on the position and orientation accuracy of the wrist tracking system. Misalignments between virtual and real hands (provided by the user's proprioceptive system) are resolved through users' adaptation mechanism (Groen & Werkhoven, 1998), but aftereffects can be observed (see chaps. 37–39, this volume). In addition to mapping, hand–eye coordination is influenced by the response time of the system. Early studies have shown that delay of hand movement in response to visual changes takes several hundred milliseconds. This limits the minimum refresh rate of the graphic scene (see chap. 3, this volume) required for preserving the same order of magnitude for the response time for hand–eye coordination in virtual environments.

Other forms of coordination are mentioned in (Burdea & Coiffet, 1994). Hand–hand coordination has three forms: coordination between hand sensing and hand force feedback; coordination between hand input and force feedback to other parts of the body; and two hand manipulation. The delay between sensing and feedback should be less than 100 msec in order to maintain coordination. Two-hand manipulation requires hand–hand coordination; one hand has a leading role while the other is playing an assisting role. Time delays between hand-relative movement, as well as hand misalignment, are important factors of two-hand coordination. The above discussion on hand–hand coordination applies to other parts of the body involving sensing and feedback.

Another important aspect of VE simulation is hand–ear coordination. This coordination depends on hand movement time delay in response to a sound event and on 3-D localization of the sound relative to hand position. Voice-ear and voice-eye are less important coordination modalities for a simulation. Voice synthesis is sometimes used to give feedback to users in response to their verbal commands. The lag between command and feedback is on the order of seconds. Changes in the visual feedback as result of a voice command have a response time of the same order of magnitude (seconds).

### 3.4 Summary of Multimodal Feedback

Looking again at Fig. 21.1 depicting the VE simulation loop, one realizes that multimodal feedback is key to simulation realism, and eventually to application usefulness. Multimodal feedback allows users immersed in a VE to not only see the graphics scene, but also to hear or feel virtual objects, in case such objects are manipulated. Multimodal feedback thus involves graphics, 3-D sound, and haptic (touch and force) feedback. These modalities can be used simultaneously to transmit modality-specific data independent of each other. Alternately, one feedback modality can be used to transmit data from another channel in a sensorial substitution arrangement. Such sensorial substitution may be needed when interfaces for another modality are expensive or difficult to use. Finally, several modalities can be used to convey a single event/data, such as contact between objects transmitted visually, through sound as well as hapticly. Such sensorial redundancy is beneficial in overcoming interface limitations (friction, limited dynamic range, etc.) but requires good synchronization of the redundant sensorial data. Otherwise sensorial feedback redundancy can become detrimental to the simulation. Human sensing and control are coupled such that the feedback loop has an internal component as well. This coupling involves hand–eye coordination, the proprioception component of the haptic loop, hand–hand and hand–ear coordination. Such coupling is another reason why synchronization of modalities and reduction in overall time delays are key to effective VE system design.

## 4. CONCLUSIONS AND FUTURE DIRECTIONS

The human–VE interface can be modeled as a bidirectional pathway containing perceptual as well as control processes. These processes are not independent, but interact with each other following complex models. These interactions manifest themselves through sensorial substitution and I/O coupling effects. Design of human–VE interfaces need not be limited to the use of a single input and/or output modality, but should exploit multimodal input and feedback.

Gesture and voice are the most popular inputs in VE, but they are typically not used together. Voice input is unreliable when using a large vocabulary. Multimodality allows erroneous speech recognition to be screened out and ambiguous gestures to be resolved, increasing application control and naturalness of human–machine dialogue. The increasing demand for interactivity and immersion (see chap. 40, this volume) will drive further development of multimodal input for human–VE interfaces. More experimental work is necessary to understand speech–gesture interaction. The studies should lead to guidelines for more efficient implementation of VE control.

Users' sensorial experience in a VE is a fusion of information received through the human perception channels (visual, auditory, somatic, etc.). While feedback in a VE is provided to users predominantly through the visual channel, additional modalities can increase the efficiency of VE simulations. Three-dimensional sound and haptic feedback increase the sense of presence in a VE with additional spatiotemporal and physical information. User immersion in a VE simulation results from rich sensorial interaction coupled with real-time simulation response. Multimodal feedback is not just the sum of visual, auditory, and somatic feedback, because there is redundancy and transposition in the human sensorial processes. Thus, to increase immersion and interactivity, VE designers can exploit sensorial redundancy and sensorial transposition effects.

Looking at the future, it is clear that the present state of multimodal interaction is rudimentary. Too few channels are tapped, and those that are are not fully utilized. Smell and taste are two feedback modalities that are currently missing. Large-volume simulations pose additional challenges to system designers, because sensing needs to be done at larger distances, and feedback interfaces need to be powerful, yet light and portable. New technologies are thus needed not only for user's multimodal input but also for multimodal feedback. The current human factors effort to quantify the usability of multimodal VEs (see chap. 34, this volume) needs to be accelerated and its scope broadened to cover the needs of large-volume virtual environments. On the theoretical front, more complex models of an intrinsically more complex system need to be developed. Such models will eventually lead to design guidelines and make system optimization efforts easier to undertake.

## 5. ACKNOWLEDGMENTS

The authors' research reported here was supported by grants from the National Science Foundation (NSF grant IRI-9618854—STIMULATE and BES-9708020) and from the CAIP Center at Rutgers University with funds provided by the New Jersey Commission on Science and Technology and by CAIP industrial members.

## 6. REFERENCES

Barfield, W., Hendrix, C., Bjorneseth, O., Kaczmarek, K., & Lotens, W. (1995). Comparison of human sensory capabilities with technical specifications of virtual environment equipment, *Presence, 4*(4), 329–356.

Billinghurst, M., & Wang, X. (1999). *GloveGRASP* [Online]. Available: http://www.hitl.washington.edu/people/grof/GestureGRASP.html [1999, September 20].

Blattner, M., & Glinert, E. (1996, Winter). Multimodal integration. *IEEE Multimedia*, 14–24.

Bowman, D., & Hodges, L. (1999). Formalizing the design, evaluation, and application of interaction techniques for immersive virtual environments. *The Journal of Visual Languages and Computing, 10*(1), 37–53.

Bouzit, M., & Burdea, G. (1998). Force feedback interface to reduce pilot's spatial disorientation, In *Third Annual Symposium and Exhibition on Situational Awareness in the Tactical Air Environment*, 69–76, Piney Point, MD:

Burdea, G. (1996). *Force and touch feedback for virtual reality*. New York: John Wiley & Sons.

Burdea G. (1999, June). Haptic interfaces for virtual reality. Keynote address in *Proceedings of International Workshop*, Laval, France.

Burdea, G., & Coiffet, P. (1994). *Virtual reality technology*. New York: John Wiley & Sons.

Cassell, J., Steedman, M., Badler, N., Pelachaud, C., Stone, M., Douville, B., Prevost, S., & Achron, B. (1994). Modeling interaction between speech and gesture. In *Proceedings of the 16th Annual Conference of the Cognitive Science Society*. Atlanta, GA: Georgia Institute of Technology.

Chu, C. P., Dani, T. H., & Gadh, R. (1997). Multimodal interface for a virtual reality–based computer-aided design system. In *Proceedings of the 1997 IEEE International Conference on Robotics and Automation*, 1329–1334, Albuquerque, NM: IEEE Press.

Chu, C. P., Dani, T. H., & Gadh, R. (1998). Evaluation of a virtual reality interface for product shape design. *IIE Transactions, 30*, 629–643.

Cohen, P. R., Johnston, M., McGee, D., Smith, I., Oviatt, S., Pittman, J., Chen, L., & Clow, J. (1997). QuickSet: Multimodal interaction for simulation set-up and control. In *Proceedings of the Fifth Applied Natural Language Processing Meeting*. Washington, DC: Association for Computational Linguistics.

Cohen, P., McGee, D., Oviatt, S., Wu, L., Clow, J., King, R., Julier, S., & Rosenblum L. (1999, July/August). Multimodal interaction for 2D and 3D environments. *IEEE Computer Graphics and Applications*, 10–13.

Cohen M., & Wenzel, E. (1995). The design of multidimensional sound interfaces. In W. Barfield, & Thomas A. Furness, III (Eds.), *Virtual environments and advance interface design*, 291–346. New York: Oxford University Press.

Dauville, B., Levinson, L., & Badler, N. (1996) Task-level object grasping for simeloted agents, *Presence, 5*(4), 416–430, Cambridge: MIT Press.

Dragon Systems. (1999). *Dragon Naturally Speaking Developer Suite* [Online]. Available: http://www.dragonsystems.com/products/developer/naturallyspeaking/index.html [1999, September 20].

Fabiani, L., Burdea, G., Langrana, N., & Gomez, D. (1996). Human performance using the Rutgers Master II force feedback interface. In *IEEE International Symposium on Virtual Reality and Applications* (VRAIS '96) 54–59. Santa Clara, CA: IEEE Press.

Flanagan, J., Kulikovski, C., Marsic, I., Burdea, G., Wilder, J., & Meer, P. (1999). *Synergistic multimodal communication in collaborative multiuser environments* (Ann. Rep.): National Science Foundation.

Fritz, J. P., & Barner, K. E. (1996). Stochastic models for haptic textures. In *Proceedings of Photonics East '96—the SPIE's International Symposium on Intelligent Systems and Advanced Manufacturing*, Boston, MA:.

Fukui, Y., & Shimojo, M. (1992). Differences in recognition of optical illusion using visual and tactual sense. *Journal of Robotics and Mechatronics*, 4910, 58–62.

Groen, J., & Werkhoven, P. (1998). Visuomotor adaptation to virtual hand position in interactive virtual environments. *Presence, 7*(5), 429–446.

HTK Limited. (1999). HTK speech platform [Online]. Available: http://www.htk.co.uk/ [1999, September 20].

IBM. (1999). ViaVoice SDK for Windows V1.5 [Online]. Available: http://www.software.ibm.com/speech/dev/ [1999, September 20].

InterSense. (2000). *InterSense IS-900 precision motion tracker* (Company Brochure). Burlington, MA: Intersense. Also at www.isense.com.

Jacob, L. (1995). Eye tracking in advance interface design. In W. Barfield & Thomas A. Furness, III (Eds.), *Virtual environments and advance interface design*, 258–290. New York: Oxford University Press.

Johnston, M., Cohen, P. R., McGee, D., Oviatt, S. L., Pittman, J. A., & Smith, I. (1997). Unification-based multimodal integration. In *Proceedings of the 35th Annual Meeting of the Association for Computational Linguistics*. Association for Computational Linguistics Press.

Kaczmarek, K. A., & Bach-y-Rita, P. (1995). Tactile displays. In W. Barfield & Thomas A. Furness, III (Eds.), *Virtual environments and advance interface design*, 349–414. New York: Oxford University Press.

Kjeldsen, R., & Kender, J. (1997). Interaction with on-screen objects using visual gesture recognition. In *Proceedings of the IEEE Computer Vision and Pattern Recognition*, 788–793, Son Juan, Puerto Rico: IEEE Press.

Kotarinis, D., & Howe, R. (1995). Tactile display of vibratory information in teleoperation and virtual environments. *Presence, 4*(4), 387–402.

Marsic, I., Medl, A., & Flanagan, J. (2000). Natural Communication with Information Systems. In *Proceeding of the IEEE, 88*(8), 1354–1366, IEEE Press.

Massimo, M., & Sheridan, T. (1993). Sensory substitution for force feedback in teleoperation. *Presence, 2*(4), 344–352.

Medl, A., Marsic, I., Andre, M., Liang, Y., Shaikh, A., Burdea, G., Wilder, J., Kulikowski, C., & Flanagan, J. (1998). *Multimodal man–machine interface for mission planning*. Paper presented at *Intelligent Environments–AAAI Spring Symposium*, Stanford, CA.

Microsoft. (1999). Windows highly intelligent speech recognizer: WHISPER [Online]. Available: http://research.microsoft.com/srg/

Minsky M., Ouh-young, M., Steele, O., Brooks F., Jr., & Behensky M., (1990). Feeling and seeing: Issues in force display. *Computer Graphics, 24*(2), 235–243.

Naval Aerospace Medical Research Laboratory. (1999). Accurate orientation information through a tactile sensory pathway in aerospace, land, and sea environments [Online]. Available: http://www.namrl.navy.mil/accel/TSAS/index.htm

Olympus Corporation of America. (2000). *Eye-trek specifications*. Available: http://www.eyetrek.com

Oviatt, S. L. (1999). Ten myths of multimodal interaction, *Communications of the ACM* [Online]. Available: http://church.cse.ogi.edu/CHCC/Personnel/oviatt.html [1999, September 20].

Oviatt, S. L., & Clow, J. (1998). An automated tool for analysis of multimodal system performance. In *Proceedings of the International Conference on Spoken Language Processing*.

Oviatt, S. L., DeAngeli, A., & Kuhn, K. (1997). Integration and synchronization of input modes during multimodal human–computer interaction. In *Proceedings of Conference on Human Factors in Computing Systems: CHI '97*. New York: ACM Press.

Pavlovic, V., Sharma, R., & Huang, T. S. (1997). Visual interpretation of hand gestures for human–computer interaction: A review. *IEEE Transactions on Pattern Analysis and Machine Intelligence, 19*(7), pp. 677–695.

Popescu, V., Burdea, G., & Bouzit, M. (1999). Virtual reality modeling for a haptic glove. In *Proceedings of Computer Animation '99*, 195–200. Geneva, Switzerland: IEEE Press.

Richard, P., Burdea, G., Gomez, D., & Coiffet, P. (1994). A comparison of haptic, visual and auditive force feedback for deformable virtual objects. In *Proceedings of ICAT 94 Conference*, 49–62. Tokyo, Japan:

Rosenblum, L., Durbin, J., Doyle, R., & Tate, D. (1997, July/August). Situational awareness using the VR responsive workbench. *IEEE Computer Graphics & Applications, 16*(4), 12–13.

Schomaker, L., Nijtmans, J., Camurri, A., Lavagetto, F., Morasso, P., Benoit, C., Guiard-Marigny, T., Le Goff, B., Robert-Ribes, J., Adjoudani, A., Defee, I., Munch, S., Hartung, K., & Blauert, J. (1995). *A taxonomy of multimodal interaction in the human information processing system* [Online]. Multimodal Integration for Advanced Multimedia Interfaces (MIAMI). ESPRIT III, Basic Research Project 8579. Available: http://hwr.nici.kun.nl/~miami/

Sharma, R., Huang, T. S. & Pavlovic, V. I. (1996). A multimodal framework for interacting with virtual environments. In C. A. Ntuen & E. H. Park (Eds.), *Human Interaction with complex systems: Conceptual Principles and Design Practice*, 53–71. Kluver Academic Publishers.

Sletterink, B. (1999). A managing agent for sharing multiple modalities. Unpublished master's thesis. Delft University of Technology.

Starner, T., & Pentland, A. (1995). Visual recognition of american sign language using hidden Markov models. In *Proceedings of International Workshop on Automatic Face and Gesture Recognition*. Zurich, Switzerland: 189–194.

Thorpe D., & Hodges, L. (1995). Human stereopsis, fusion, and stereoscopic virtual environments. In W. Barfield & Thomas A. Furness, III (Eds.), *Virtual Environments and Advance Interface Design*, 145–174. New York: Oxford University Press.

Vo, M., & Waibel, A. (1993). A multimodal human–computer interface: Combination of gesture and speech recognition. In *Adjunct Proceedings of InterCHI '93*.

Vo, M., & Waibel, A. (1997). Modeling and interpreting multimodal input: A semantic integration approach. (Tech. Rep. No. CMU-CS-97-192).

# 22

# Auditory–Visual Cross-Modality Interaction and Illusions

Russell L. Storms
*Georgia Institute of Technology*
*Army Research Laboratory*
*115 O'Keefe Building*
*Atlanta, GA 30332-0862*
*storms@airmics.gatech.edu*

## 1. INTRODUCTION

What are the fidelity requirements of a virtual environment (VE)? First and foremost (and sometimes neglected), the intended outcomes of the particular application ought to drive the fidelity requirements. System developers must not assume that the best possible fidelity is always needed. The particular application, overall sensory perception, and creative use of stimuli ought to drive fidelity requirements. For example, the visual fidelity of a VE intended to train surgeons in open-heart surgery probably needs to be greater than the visual fidelity of a VE intended to teach children how to read. Another consideration is that of the human sensory system: Fidelity requirements of VEs need not exceed that of the human perceptual system. Durlach and Mavor (1995) point out that any display resolution beyond the ability of human perception is wasteful. Barfield, Hendrix, Bjorneseth, Kaczmarek, and Lotens (1995) give a detailed description regarding humans' ability to detect and discriminate visual, auditory, tactile, and kinesthetic information along with corresponding technical specifications of VE equipment and emphasize the importance of using knowledge of human's sensory systems capabilities and limitations to direct the design of virtual worlds. But one must be careful not to evaluate a simulation or VE solely by the auditory, visual, and haptic technical rendering capabilities of the system. Virtual environment evaluation should be based on the degree to which auditory, visual, and haptic sensory modalities of the immersed participant, that is, an engaged human, are stimulated. Blattner and Glinert (1996) point out that we sometimes fail to appreciate the distinction between human and computer modalities. One factor in considering auditory and visual fidelity requirements is that of display resolution. In a VE, the auditory and visual resolutions ought to be properly matched. For example, Laurel (1993) mentions that designers would not want to couple simple visual displays such as cartoons with very realistic auditory displays such as environmental sounds, and likewise would not want to couple very-high-resolution visual animations with simple auditory beeps.

The National Research Council (NRC) Virtual Reality Research and Development Committee (Durlach & Mavor, 1995) recognizes that visual fidelity requirements are influenced by other modalities. They suggest that a greater understanding is needed in multimodal integration to determine how required visual display system parameters are affected by multimodal interaction, whether or not visual display system requirements can be relaxed during such interaction, and to determine the perceptual effects associated with merging multimodal display sources. Virtual environment system development must thus focus on multimodalities. This focus on the modalities need not only concentrate on the intrarelationships but also on the interrelationships. The NRC explains that study of both intrasensory and intersensory illusions is important because these illusions could help to develop more simplified and cost-effective virtual environments. Under the category of Psychological Considerations, the NRC recommends further study on visual dominance over audition and haptics (e.g. the ventriloquism effect) and how the auditory channel can be used to improve perception within visual and haptic domains. Since the technology to render very-high-quality auditory and visual displays is currently available, proper use of this technology must not neglect potential auditory and visual cross-modal perception phenomena. A goal for the VE community must be to verify and validate the existence and extent of auditory and visual cross-modal perception phenomena and its application to virtual environments. Chapter 21 (this volume) focuses on presentation of multimodal information, whereas this chapter focuses on auditory–visual cross-modal illusions.

## 2. INTERSENSORY RESEARCH

### 2.1 Theoretical Perspective

Ryan (1940) conducted a thorough literature survey on sensory interaction and concluded that certain perceptual qualities perceived by one sensory system can be influenced by the qualities of other sensory systems. Ryan also concludes that sensory systems must be considered as part of a whole unified system and not as individualized systems isolated from each other. Ryan ultimately concludes that the study of the interrelations among the senses is severely lacking.

Gilbert (1941) conducted an extensive literature review on intersensory facilitation and inhibition. He concluded that in terms of heteromodal (intersensory) stimulation, a stimulus in one modality can momentarily reduce sensitivity in another modality and after an optimal interval of about 1/2 second can then increase sensitivity in the other modality. Furthermore, a less-intense intersensory stimulation can momentarily increase sensitivity in another modality. Upon reviewing all intersensory research (through 1941), Gilbert, like Ryan, realized that the current view on the psychophysical aspect of intersensory interactions is lacking.

London (1954) presented his findings based on extensive intersensory research conducted in the Soviet Union. Upon review of numerous intersensory experiments, he concluded that the conditions that influence sensory interaction are best summarized as follows: strength, duration, and termination of accessory stimulation; excitatory state of sense organs; affectivity of stimulus; physiological state; diurnal variation; and summation, repetition, and cumulation of accessory effects. In reviewing London's research efforts, Stone and Pangborn (1968) suggested that a stimulus in one sense organ influences to a certain degree the sensitivity of other sense organs. Like Gilbert (1941) and Ryan (1940), Stone and Pangborn ultimately conclude that there is great need for intersensory studies.

Gibson (1966, 1979) suggests that within the natural environment, numerous senses respond to and interact with environmental stimulation. He believes that an organism initiates rather than reacts to certain events. As a result, intersensory perception is the rule of all perception and not just specialized higher-order complex reactions. In other words, it is the particular surrounding environment that determines how our senses respond and interact. Thus, sensory interaction must be based on the complexity of natural life events and not on simple isolated systems.

Marks (1978) provides a more modern view of sensory interaction. From a simple to a more complex perspective, Marks describes what he calls the Five Doctrines of sensory correspondence. Briefly, these five doctrines are outlined as follows:

1. Doctrine of Equivalent Information: Different senses can inform us about the same features of the external world.
2. Doctrine of Analogous Attributes and Qualities: Despite the salience of the phenomenal differences among qualities of various sense modalities, there are a few properties held in common.
3. Doctrine that Different Senses have Corresponding Psychophysical Properties: This theory proposes that at least some of the ways the senses behave and operate on impinging stimuli are general characteristics of sensory systems, similar from vision to hearing, from touch to olfaction.
4. Doctrine that Similar or Identical Neurophysiological Mechanisms Parallel Sensory Correspondence: This doctrine states that there is a neural analogue to each of the psychological doctrines (the first three doctrines).
5. Doctrine of the Unity of the Senses: This doctrine incorporates the first four, and suggests that several senses are interpreted as modalities of a general, perhaps more primitive sensitivity.

According to various intersensory research, Marks (1978) suggests that the dimension of quality appears to show the fewest similarities from modality to modality, but that intensity displays the strongest cross-modal similarity. Marks' findings indicate that meaningful perceptual interactions occur when concurrent information enters different sensory channels, but when stimuli presented to different senses bear no meaningful relation to each other, interaction often seems to be small or nonexistent. As such, Marks concludes that sensory interaction is highly stimuli dependent. An interesting point by Marks that deserves mentioning is that intersensory similarity must necessarily be one step removed from intrasensory similarity, because if there were truly continuity between modalities then there would be only one sense. Marks concedes that the entire area of sensory quality cross-modality comparisons has hardly been explored experimentally.

Based on her research with blind and normal children, Millar (1981) concludes that the sense modalities are neither separate nor unitary, but rather a combination of both. She believes that the sense modalities complement each other where information is flexibly passed among the different modalities, and that no global generalization can be made about the interrelationships of the sense modalities.

O'Connor and Hermelin (1981), having conducted experiments with children suffering from either specific perceptual or general cognitive handicaps, describe sensory integration through the concept of *sensory capture*. O'Connor and Hermelin found that when conflicting stimuli is presented to different sense modalities, an observer tends to resolve stimuli conflict by making the weaker sense impression conform to the dominant sense impression. As such, the dominant sense captures the weaker sense impression, suggesting that similar stimulus qualities tend to be perceived across various modalities.

Overall, sensory interaction has been studied for many years by researchers in numerous disciplines such as psychoacoustics, psychology, physiology, neurology, philosophy, musicology, ecology, and human–computer interaction, and by different organizations such as the Human

Factors and Ergonomics Society, Audio Engineering Society, Acoustical Society of America, Department of Defense, artistic community, and also the film and entertainment industry. Thus, there is a large amount of intersensory research, but this knowledge is often kept within the discipline from which it was derived. Consequently, there is little cross-disciplinary transfer of intersensory knowledge. Computer science in particular, having a significantly large number of VE developers, is severely lacking in its knowledge and use of intersensory phenomena.

## 2.2 Neurological Perspective

Stein and Meredith (1993) argue that limited research has focused on developing an understanding of the neural phenomena that make multisensory integration possible. Nevertheless, there has been a fair amount of research confirming locations within the brain where different input modalities converge, facilitating multisensory integration. One such place in the brain where visual, auditory, and somatosensory inputs converge is in the superior colliculus. In looking at the horizontal and vertical meridians of different sensory representations in the superior colliculus, one can see a similar common coordinate system. Stein and Meredith (1993) conclude that this common coordinate system suggests a representation of *multisensory space*. By examining the neurological responses of superior colliculus in various animals, primarily the cat, considerable evidence was found that supports the principles of multisensory convergence and interaction based on single-neuron-evoked potentials. Stein and Meredith (1993) suggest that neurological studies in other animals are very important and lead to a better understanding of human perception. Thus, based primarily on neurological studies of other animals, primarily cats, Stein and Meredith (1993, p. 172) outline rules in terms of space and time governing multisensory integration based on unimodal receptive field characteristics and conclude that "the spatial register among the receptive fields of multisensory neurons and their temporal response properties provide a neural substrate for enhancing responses to stimuli that covary in space and time and for degrading responses that are not spatially and temporally related."

Although they found considerable evidence supporting a neurological basis for sensory integration, they conclude that numerous challenges must be met before a better understanding of sensory integration is achieved.

One such challenge is gaining a better understanding of synesthesia. Cytowic (1989, p. 1) an expert in the study of synesthesia, defines synesthesia as "an involuntary joining in which the real information of one sense is accompanied by a perception in another sense. In addition to being involuntary, this additional perception is regarded by the synesthete as real, often outside the body, instead of imagined in the mind's eye."

The concept of synesthesia dates back over 200 years. (See Baron-Cohen and Harrison [1996] for an exhaustive survey of all classic and contemporary synesthesia literature dating back over the last 200 years.) It is estimated that synesthesia occurs in about one in 25,000 individuals (Cytowic, 1995). One of the most common forms of synesthesia is that of colored hearing. A synesthete experiences colored hearing when certain sounds (physical stimuli) evoke perceptions of various colors. For example, when listening to certain classical music, a synesthete might experience shades of blue and/or green. The validity of synesthesia, though, has suffered over the years for it is introspective in nature. However, Cytowic (1989, p. 176) has helped to validate synesthesia by examining the neural substrates of synesthesia and concludes that "The synesthetic experience may be a result of a fundamentally mammalian process in which the cortex briefly ceases to function in the modern manner, permitting the senses to fuse, or, rather, we should say, perceive fusion that may be there all along but that never arises to consciousness."

## 3. AUDITORY–VISUAL SENSORY INTERACTION

### 3.1 Auditory Scene Analysis

In terms of auditory–visual interaction, Bregman (1990) states that there are many similarities between visual and auditory perceptual groupings. He believes that the auditory and visual modalities seem to interact in order to specify the nature of certain events within a perceiver's environment. As opposed to the Gestalt point of view, which focuses on the similarities among modalities, Bregman presents an interesting ecological point of view focusing on differences of the modalities. One such difference between vision and audition is evidenced through the use of echoes. In audition, perceivers are mainly interested in the direct source of sound rather than its echoes, but they can also combine direct sound and indirect sound (echoes) to establish a mixed sound that still conveys information of the direct sound but with the additional properties (i.e. reverberation) of the indirect sound. However, with vision, perceivers are mainly concerned with the indirect image (echoes or reflections), and they are not able to combine direct and indirect images to establish a mixed visual image. Bregman (1990, p. 38) suggests that it is these ecological differences that might cause "apparent violations of the principle of exclusive allocation of sensory evidence."

### 3.2 The Entertainment Industry

For many years, the entertainment industry has realized the important relationship between visuals and sound. Even before sound was an integral part of film, silent movies were accompanied with specific music to enhance the mood of certain scenes. As Gary Rydstrom (1994, p. 5), of Skywalker Sound, explains, "When approached creatively, the combination of sound and image can bring something to vivid life, clarify the intent of the work, and make the whole experience more memorable."

Realizing this important relationship between visuals and sound in film, Lipscomb (1990) and Lipscomb and Kendall (1994) investigated the perceptual judgment of the relationship between musical and visual components in film. In their experiments, they took various motion picture sequences and manipulated their soundtracks. The motion picture sequence containing the original soundtrack along with the motion picture sequence containing various manipulated soundtracks were presented to participants. The task of the participants was to select the soundtrack that best fit the visuals of the film. Interestingly, the results indicated that the composer-intended musical score (the original score) was identified as the best fit by the majority of participants for all conditions. In a related experiment, they also found significant results strongly suggesting that a musical soundtrack can in fact change the perceived meaning of a film presentation.

### 3.3 Cross-modal Matching

Cross-modal matching is using information obtained through one sensory modality to make a judgment about an equivalent stimulus from another modality. Marks (1974, 1978, 1982, 1987, & 1989) and Marks, Szczesiul, and Ohlott (1986) have been studying auditory–visual cross-modal matching over the last 25 years. He has conducted several experiments, which suggest a strong auditory–visual cross-modal matching among brightness, pitch, and loudness. Marks (1974) had study participants match pure tones to the brightness of gray surfaces. Results indicated that most participants matched increasing auditory pitch to increasing visual brightness, and Marks concluded that the findings mimicked that of synesthesia. Marks (1982) conducted a series of four experiments in which subjects used scales of loudness, pitch, and brightness to evaluate the meanings of various auditory–visual synesthetic metaphors such as: sound of sunset, murmur of dawn, and bright whisper, to name a few. He found that loudness and pitch expressed themselves metaphorically as greater brightness, and likewise that brightness expressed itself metaphorically as greater loudness and as higher pitch. This series of experiments led Marks (1982, p. 177) to believe that: "The ways that people evaluate synesthetic metaphors emulate the characteristics of synesthetic perception, thereby suggesting that synesthesia in perception and synesthesia in language both may emulate from the same source—from a phenomenological similarity in the makeup of sensory experiences of different modalities."

Marks has also conducted experiments involving auditory–visual cross-modal perception of intensity (Marks et al., 1986), auditory–visual cross-modal similarities in *speeded discrimination* (Marks, 1987), and additional experiments concerning auditory–visual cross-modal similarities with pitch, loudness, and brightness (Marks, 1989). The results of these experiments are similar to his earlier ones and provide more evidence to support strong auditory–visual cross-modal matching among pitch, loudness, and brightness. In terms of cross-modal matching, one might conclude from Marks' findings that our senses are indeed influenced by one another.

## 3.4 Visual Dominance

### 3.4.1 Ventriloquism Effect

A well-known auditory–visual intersensory phenomenon is that of the *ventriloquism effect* (see Howard & Templeton, 1966). As the name implies, this phenomenon refers to the illusion created by a skilled ventriloquist when one think they hear the dummy talking when in fact they are actually hearing the altered voice of the ventriloquist. Not only does one perceive the dummy as talking, but they also actually think the sounds of the dummy are emanating from the dummy's mouth and not from the ventriloquist even though they know that the dummy cannot really talk. This effect demonstrates the strong spatial coupling that occurs between the auditory and visual senses, and as a result has been the topic of much research (see Bermant & Welch, 1976; Howard & Templeton, 1966; Pick, Warren, & Hay, 1969; Radeau & Bertelson, 1976; Ragot, Cave, & Fano, 1988; Stein & Meredith, 1993; Warren, Welch, & McCarthy, 1981). One reason why the ventriloquism effect occurs is that the visual sense is usually the dominant sense. Unless there are significant differences in the intensities of the information gathered via different modalities, visual stimuli has a greater influence on perception via other modalities as compared to the influence of other modalities on the visual sense (Stein & Meredith, 1993). Visual dominance is also explained by Wickens (1992, p. 108) who states that "if visual stimuli are appearing at the same frequency and providing information of the same general type or importance as auditory or proprioceptive stimuli, biases toward the visual source at the expense of the other two [auditory and proprioceptive] will be expected."

### 3.4.2 Experimental Results Supporting the Ventriloquism Effect

Radeau and Bertelson (1976) conducted an experiment on the effect of a textured visual field on modality dominance during the ventriloquism effect and concluded that visual texture affects the degree of auditory capture of vision, but not the degree of visual capture of audition. Bermant and Welch (1976) investigated the effect of degree of separation of an audiovisual stimulus and eye position on the spatial interaction of the ventriloquism effect and found that the ventriloquism effect is not dependent on the use of a visual source typically associated with the production of sounds. The role of auditory–visual *compellingness* in the ventriloquism effect was studied by Warren et al. (1981) where it was found that given a highly compelling stimulus situation, subjects showed a very high visual bias of audition, a significant auditory bias of vision, and a sum of bias effects indicating that subjects perceived a single perceptual event. Ragot et al. (1988) explored auditory and visual ventriloquism reciprocal effects and suggest that visual dominance appears when attention is divided between visual and auditory modalities, but seems to be absent when the subjects are asked to attend to one modality in the presence of another modality. Knudsen and Brainard (1995) present neurological evidence from studying the optic tectum (also referred to as the superior colliculus), indicating that bimodal tectal neurons are more sensitive to displacements of a visual stimulus than to displacements of an auditory stimulus. Thus, in regard to visual dominance over audition with respect to the ventriloquism effect, Knudsen and Brainard conclude that the location in the bimodal tectal map activated by visual and auditory stimuli may be more sensitive to the location of the visual stimulus than to the location of the auditory stimulus.

### 3.4.3 Divided Attention

During signal detection (temporal in nature and typically associated with sustained attention or vigilance), the auditory channel proves dominant over the visual channel, which is why warning signals are typically produced with auditory devices. However, in most other areas, human's visual sense dominates the hearing sense, as can be seen from the following experimental findings.

As part of a U.S. Air Force research investigation, Henneman and Long (1954, p. 13) released an extensive technical report comparing visual and auditory senses as channels for data presentation during cockpit crew coordination suggesting that "when a person is required to divide his attention or to shift back and forth between two tasks, one visually controlled, the other aurally controlled, either task can be made a "priority" task at the expense of the other. Sense channel as such does not determine this priority."

Henneman and Long (1954) indicate there has been very little experimental evidence comparing audition and vision as channels for data presentation and concludes that most auditory–visual intersensory studies have focused on sensory thresholds as opposed to suprathreshold levels that typify actual perceptual phenomena. Henneman et al. ultimately suggest that it might not be possible to make any reliable conclusions concerning perceptual phenomena from comparing auditory and visual perceptual judgments.

Colavita (1974) describes a series of experiments exploring *sensory dominance* in which participants responded to suprathreshold auditory and visual stimuli. The auditory stimuli consisted of tones, and the visual stimuli consisted of light flashes. The stimuli were randomly presented as auditory only, visual only, and combined auditory–visual. The participant's task was to identify which stimuli occurred. When participants were presented with the combined auditory–visual stimuli, the participants typically only responded that a visual light flash occurred and usually did not even notice that an auditory stimulus (tone) was present. Thus, in this task, the findings suggest visual dominance over the auditory sense.

In a study investigating the perceived duration of auditory and visual intervals, Behar and Bevan (1974) found that auditory intervals (white noise) were consistently judged to be about 20% longer than visual intervals (light from a neon glow-lamp) of the same duration. Behar and Bevan conclude that there are peripheral variables that must not be ignored when making psychophysical judgments.

Burrows and Solomon (1975) conducted an experiment investigating the ability to scan auditory and visual information in parallel. Participants were presented with pairs of letters, one being a visually presented letter and the other being an aurally presented letter. The pairs of letters were presented simultaneously or sequentially. The participants' efficiency of memory retrieval was measured in both conditions: (1) simultaneously presented letters or

(2) sequentially presented letters. The results indicated that parallel scanning is possible with a simultaneous presentation because of the continuous information from both auditory and visual stimuli, but it is not possible with sequential presentation because of the "dead time" needed to switch between modalities.

Egeth and Sager (1977) explored the locus of visual dominance over audition in which participants responded to suprathreshold stimuli consisting of an audio-only tone, a visual-only light flash, and a combined auditory-visual tone-light flash. Their findings suggest that perceptual judgment of the auditory tone is not affected by the light flash because visual dominance is nonsensory in locus and depends on the relevance of the associated visual stimulus.

Jones and Kabanoff (1975) conducted an experiment to determine if eye movements are a factor in auditory localization. They found that auditory localization accuracy is increased if participants move their eyes in the direction of the intended target. Their findings suggest that voluntary eye movement rather than a visual map is likely to provide an adequate framework for making spatial judgments (see chap. 24, this volume).

McGurk and MacDonald (1976) investigated the effect of seeing certain lip movements associated with hearing contradictory speech sounds. Participants were presented auditory-only speech sounds and mismatched auditory–visual (speech-lip movement) combinations. Their results were remarkable. During the combined auditory–visual mismatches, most participants were convinced they were hearing what they were seeing (lip movements), when in fact the lip movements were not the correct lip movements for the associated speech sound that they were hearing. Furthermore, even if one has prior knowledge of the auditory–visual mismatches, it does not preclude one from being convinced (incorrectly) they were hearing what they were seeing. The results of this experiment were so strong that it is commonly referred to as the *McGurk effect*. Interestingly, Stein and Meredith (1993) found that the visual cues of lip movement can actually modify auditory cortex activity to the extent of altering the signal-to-noise ratio of the auditory stimulus by 15 to 20 decibels.

Rosenblum and Fowler (1991) investigated if loudness judgments of speech are more closely related to the visual degree of exerted vocal effort than to the actual emitted acoustical properties of intensity. As in the McGurk effect, participants were presented conflicting audiovisual stimuli. Their findings suggest that when making loudness judgments of speech, the visual cues of vocal effort significantly outweigh the cues provided by the appropriate levels of acoustic intensity.

Hanson (1981) conducted an experiment to investigate if common processing of semantic, phonological, and physical systems were involved during reading and listening. Participants were simultaneously presented two words, one visually and one aurally, but were instructed to attend to only one modality and to make responses based on that attended modality. The results indicated that the unattended words had an influence on semantic and phonological decisions but had no influence on the physical task. (In the physical task, the visual words were presented in either small or capital letters, and the aural words were presented in either a male or female voice.) Hanson (1981, p. 99) concludes that the written and spoken words "share semantic and phonological processing but have separate modality-specific codes that operate on information prior to the convergence of information from visual and auditory inputs."

### *3.4.4 Haptic Perception of Stiffness*

A lesser-known area in which the visual sense can dominate is that of haptic perception. Because of technological constraints, it has not been feasible to adequately study haptic perception. However, recent technological advances have spurned significant new and exciting research into haptic perception (see chaps. 5 and 6, this volume). Srinivasan and Basdogan

(1997) outline recent advances and future challenges in haptic display technology. In terms of haptic perception of stiffness, Srinivasan, Beauregard, and Brock (1996) demonstrate the effect of visual dominance on haptic perception. In this experiment, participants had to discriminate the stiffness of two virtual springs. The two virtual springs were depicted visually on a computer monitor and force-feedback was rendered haptically via a Planar Grasper. The participants were presented randomized visual-haptic stiffness stimuli resulting in consistent (correct) and inconsistent (incorrect) visual–haptic stiffness stimuli parings. The results indicated that the perception of stiffness is greatly influenced by visual information. Specifically, when the visual stiffness stimuli conflicted with the haptic stiffness stimuli, participants relied on the visual stimuli to make judgments about the stiffness of the virtual spring. A counter example, that of haptic dominance, was provided by Proffitt and Kaiser (1995). They asked participants to estimate the incline of a hill while viewing it. The participants were asked to make their estimates verbally, visually (using a handheld modified compass to approximate the angle) or haptically (placing their hands on a board and moving the board until its position approximated the angle). The input of the "feel" of the angle in the haptic condition supplemented viewing to achieve the best estimates. These experiments are examples of how human senses involuntarily meld together forming an overall perception of the environment.

## 3.5 Threshold Perception Experiments

### 3.5.1 Overview

The body of evidence presented thus far clearly indicates that under certain conditions, auditory–visual perceptual phenomena do exist. However, most auditory–visual research has focused on threshold levels, absolute sensitivity, or just noticeable differences. As mentioned earlier, Gilbert (1941) and Ryan (1940) independently conducted exhaustive literature surveys covering these topics. Additional evidence supporting auditory–visual perceptual phenomena from threshold level stimuli can be found in the following references: Serrat and Karwoski (1935), Pratt (1936), London (1954), Thomas, Voss, and Brogden (1958); and Loveless, Brebner, and Hamilton (1970). For a better understanding of this type of threshold-level perception research, the findings of two experiments are presented showing auditory–visual perceptual phenomena from threshold-level stimuli.

### 3.5.2 Experimental Results

An example of the research reviewed by Gilbert and Ryan is that of Kravkov (1936), one of the early pioneers in the area of intersensory experimentation. Kravkov's experiment investigated the influence of sound on light and color sensitivity of the eye. In this experiment, three female participants were presented an auditory stimulus consisting of a 2100 Hz tone at 100 decibels for a duration of about 10 minutes. During these 10 minutes, measurements were made of color and light sensitivity. From the results of the experiment, Kravkov (1936, p. 351) concluded that: "The colour sensibility of the eye changes differently under the influence of sound, according to the wavelength of the stimulating light. . . . whereas the colour sensibility for green rises during the acoustic stimulation the colour sensibility for orange-red decreases."

Gregg and Brogden (1952) conducted an experiment on the effect of simultaneous visual stimulation on absolute auditory sensitivity. In their experiment, participants were presented an auditory tone along with an auxiliary light source. Their results indicate that when participants were asked to report the presence of a visual light source along with an auditory tone, the light stimulus decreased participant sensitivity to a 1000 Hz tone. However, when participants were only required to report the presence of an auditory tone, the light stimulus increased sensitivity to the auditory tone.

## 3.6 Suprathreshold Perception Experiments

### 3.6.1 Overview

When one talks about using both audio and visual displays for some kind of simulation, game, or VE, some people will say that the use of high quality sound positively influences their perception of the visual images, but that high-resolution visual images tend not to improve on the perception of sound quality (see Tierney, 1993). Stein and Meredith (1993) theorize that combinations of visual and auditory cues can indeed enhance one another and can also eliminate any ambiguity that may occur when cues are only available from one modality. Likewise, Murch (1973) suggests that under many conditions the encoding of visual-only material or auditory-only material involves the use of short-term storage of both systems. If auditory and visual displays can influence each other, then as Begault (1994) suggests, perhaps sound can be used for improving the perceived quality of visual displays. For example, Negroponte (1995) recounts a story of designing military tank simulators. The tank simulator designers were trying to make the visual display of the tank simulator very realistic by increasing the number of scan lines. As the simulator designers increased the scan lines the cost of the visual display increased. After increasing the number of scan lines resulted in a prohibitive price for the tank simulator, simulator designers decided to take an alternative approach. Instead of focusing on the visual information channel, they focused on the auditory and haptic channels of information. They introduced some tank motor and trend sounds and also added an inexpensive vibrating motion platform. As a result, Negroponte (1995) found that realism was enhanced to the point that designers were able to reduce the number of scan lines.

The empirical evidence supporting just how auditory and visual displays can influence the perception of each other is lacking. One reason for the lack of empirical evidence is that of specifying perceptually relevant dimensions for both auditory and visual modalities (Jones, 1981). Nevertheless, a few experiments have been conducted in which auditory displays influenced the perception of visual displays or visual displays influenced the perception of auditory displays.

### 3.6.2 Experimental Results

Neuman (1990) and Neuman, Crigler, and Bove (1991) conducted an experiment to measure the effect of changes in audio quality on visual perception on high-definition television (HDTV). The experimental design was to keep the quality of visual stimuli constant while only manipulating auditory stimuli. The auditory conditions were as follows: low fidelity (very low-quality speaker system) versus high fidelity (very-high-quality speaker system); monaural versus stereo; and three types of television programming: sports, situation comedy, and action-adventure. Participants were presented a short video clip along with one of the auditory conditions. The participants were then asked to rate: (1) their liking, (2) their level of interest, (3) their psychological involvement in the programming, (4) picture quality, (and 5) audio quality. Their results indicated that participants had difficulties distinguishing mono from stereo and low-fidelity from high-fidelity sound systems, but overall participants did prefer viewing the video clips when coupled with high-fidelity and stereo sound systems. Perhaps the most interesting finding was that a few participants perceived an increase in visual quality when coupled with better audio even though the visual quality remained constant throughout the experiment. This finding, however, was not statistically significant, and it only occurred in one of the three types of television programming presented.

Iwamiya (1992) investigated the effect of visual information on the impression of sound and the effect of auditory information on the impression of visual images when listening to music via audiovisual media. The factors used to evaluate the impression of both audio and visual images were: tightness, evaluation, brightness, uniqueness, and cleanness. Iwamiya suggests that these factors are considered to be the *intermodalities* between auditory and visual processing. Iwamiya found that the factors of brightness, tightness, and cleanness of the auditory images enhanced the perception of these factors of the visual images. Iwamiya concludes that better matched auditory and visual information will result in an overall higher evaluation of combined auditory and visual impression.

Hollier and Voelcker (1997) conducted an experiment investigating the influence of video quality on audio perception. Thirty-two participants watched video clips 10 seconds in duration with supporting audio (speech) commentaries. In total there were eight video-quality variations and four audio-quality variations. The results indicated that: (1) when no video was present, the perceived audio quality was always worse than if video was present, and (2) although only small differences were noted, a decrease in video quality corresponded to a decrease in perceived audio quality. They ultimately propose an algorithmic approach for the proper development of an auditory–visual cross-modal perceptual model.

Woszczyk, Bech, and Hansen (1995) and Bech, Hansen, and Woszczyk (1995) discuss the design and results of an experimental procedure examining the interaction between the auditory and visual modalities in the context of a home theater system. With the growing interest and development of VE systems, Woszczyk et al. identifies the need for testing the interaction of audio and visual displays in order to bring about substantial improvements in the audiovisual experience of viewers. They suggest that it is important to focus on the total perceptual experience and not on the individual auditory and visual modalities. In their experiments, participants assessed audio-visual reproductions using the subjective dimensions of action, space, mood, and motion while asking specific questions focusing on quality, magnitude, degree of involvement, and audiovisual balance. Quality was defined as: distinctness, clarity, and detail of the impression. One of their findings of particular interest is that both visual- and audio-perceived quality increased with increasing screen size. To further explore auditory–visual interaction, Bech (1997) conducted two more experiments to investigate the influence of stereophonic (audio) width on the perceived quality of an audio-visual presentation using multichannel Surround Sound systems. During the experiments, participants were asked to evaluate the quality (fidelity) of spatial information contained in audiovisual reproductions. The results indicate that the quality of perceived spatial reproduction increases linearly with an increase in stereophonic width.

Recent auditory–visual suprathreshold cross-modal research has been conducted by Storms (1998). In a series of experiments conducted at the Naval Postgraduate School (NPS) in Monterey, California, he found that by varying the quality (fidelity) of both auditory and visual displays it was possible to directly measure auditory–visual cross-modal perception phenomena. The basic idea of the experiments was to manipulate visual and auditory display parameters intramodally and intermodally, and to likewise measure visual and auditory display perception intramodally and intermodally. During the experiments, which each lasted approximately 30 minutes, a single participant wore headphones and sat in front of a 20-inch display monitor. The task of the participant was to rate the perceived quality of audio-only, visual-only, and combined audio-visual displays through Likert rating scales ranging from 1 to 7. The dependent variables were the perception of visual display quality and the perception of auditory display quality. It was hoped that by carefully varying the fidelity of both auditory and visual displays it would be possible to measure auditory–visual cross-modal perception phenomena. Specifically, the goal was to answer the following question: In an audiovisual display, what affect (if any) do various audio-quality levels have on the perception of visual quality and vice versa?

The overall conclusions from Storms' research suggest that (1) whether asked to specifically attend to both auditory and visual modalities or asked to attend to only one modality, (2) whether

**TABLE 22.1**
Characteristics of Auditory-Visual Suprathreshold Cross-modal Perceptual Evaluation

1. When attending only to the visual modality, a high-quality visual display coupled with either a medium- or high-quality auditory display causes an increase in the perception of visual quality relative to established baseline conditions derived from visual-only quality perception evaluations.
2. When attending only to the auditory modality, a low-quality auditory display coupled with either a medium- or high-quality visual display causes a decrease in the perception of auditory quality relative to established baseline conditions derived from auditory-only quality perception evaluations.
3. When attending to both auditory and visual modalities, a high-quality visual display coupled with a low-, medium-, or high-quality auditory display causes an increase in the perception of visual quality relative to established baseline conditions derived from visual-only quality perception evaluations.

manipulating visual display pixel resolution or Gaussian noise level, (3) whether manipulating auditory display sampling frequency or Gaussian noise level, or (4) whether an auditory–visual display is tightly or loosely coupled, cross-modal auditory–visual perception phenomena exist. From Storms' (1998) findings a number of characteristics of auditory–visual suprathreshold cross-modal perceptual evaluation have been identified (see Table 22.1).

Storms' results provide the empirical evidence to support what most people in the gaming business, multimedia industry, entertainment industry, and VE community have suspected all along: that audio can influence the quality perception of video and vice versa. The results also indicate that although humans can divide their attention between audition and vision, they are not consciously aware of any potentially significant intermodality effects.

Due to the multidisciplinary nature of Storms' research effort, the impact of the overall findings is far reaching, having both theoretical and commercial implications. In terms of the theoretical impact, because the overall findings indicate that auditory quality can influence visual quality perception and vice versa, some sort of sensory interaction must be taking place. These findings support the early intersensory research conclusions of both Ryan (1940) and Gilbert (1941). Also, O'Connor and Hermelin (1981) would likely argue that these findings support the concept of *sensory capture*. But how this sensory interaction occurs is still not known. Stein et al. (1993) might conclude that this interaction could be taking place at the neurological level. However, Gibson (1966, 1979) might argue that this sensory interaction is based on the complexity of natural life events. Furthermore, one of the overall findings of this research effort suggests that when attending only to the auditory modality, a low-quality auditory display coupled with either a medium- or high-quality visual display causes a decrease in the perception of auditory quality. The reason for degrading the perception of the auditory quality might be based on the concept of visual dominance. Perhaps at some higher cognitive level, the higher quality visual display is being compared with the lower quality auditory display. This unconscious comparison might cause one to perceive that the auditory quality is worse than it actually is because of the dominating nature of the visual modality.

In terms of the commercial impact, Storms' research suggests that when attending to both auditory and visual modalities, a high-quality visual display coupled with a low-, medium-, or high-quality auditory display causes an increase in the overall visual quality perception of an auditory–visual display. Thus, suppose a VE developer has been tasked to increase the realism (and perhaps presence) of a 3-D scene depicting a typical family living room. This virtual living room contains a TV and stereo system that is rendered using high-quality visual graphics. However, the living room scene does not have any associated sounds. Instead of increasing the pixel resolution of the living room scene, causing an unwanted increase in the visual rendering time of the scene, the VE developer adds high-quality music to the visual display of the stereo system and a motion picture expert group (MPEG) video sequence containing high-quality audio to the visual TV display. As a result, the perceptual visual quality of the scene ought to increase by simply adding the associated auditory displays without the need to manipulate any of the visual displays.

## 4. FUTURE NEEDS

Given that auditory–visual cross-modal perception phenomena exist, the next logical step is to incorporate these overall findings into some type of useful auditory–visual quantitative perceptual model similar to that proposed by Hollier and Voelcker (1997). This model can then be used to derive appropriate (quantitative) levels of auditory and visual fidelity for use by developers in the gaming business, multimedia industry, entertainment industry, VE community, and the Internet industry. For example, given a certain application, this auditory–visual quantitative perceptual model could help to derive the appropriate levels and specific amounts of visual display pixel resolution and auditory display sampling frequency as a function of visual-only, auditory-only, and/or combined auditory–visual media.

## 5. REFERENCES

Barfield, W., Hendrix, C., Bjorneseth, O., Kaczmarek, K. A., & Lotens, W. (1995). Comparison of human sensory capabilities with technical specifications of virtual environment equipment. *Presence: Teleoperators and Virtual Environments. 4*(4), pp. 329–356.

Baron-Cohen, S., & Harrison, J. E. (Eds.). (1996). *Synaesthesia: Classic and contemporary readings.* Oxford, England: Blackwell Publishers.

Bech, S. (1997, March). *The influence of stereophonic width on the perceived quality of an audio-visual presentation using a multichannel sound system.* Preprint No. 4432, presented at the 102nd Audio Engineering Society Convention, Munich, Germany.

Bech, S., Hansen, V., & Woszczyk, W. (1995, October). *Interaction between audio-visual factors in a home theater system: Experimental results.* Preprint No. 4096, presented at the 99th Audio Engineering Society Convention. New York.

Begault, D. R. (1994). 3-D sound for virtual reality and multimedia. Cambridge, MA: Academic Press.

Behar, I., & Bevan, W. (1961). "The perceived duration of auditory and visual intervals: Cross-modal comparison and interaction." *American Journal of Phychology, vol. 74*, pp. 17–26.

Bermant, R. I., & Welch, R. B. (1976). Effect of degree of separation of visual-auditory and eye position upon spatial interaction of vision and audition. *Perceptual and Motor Skills, 43*, 487–493.

Blattner, M. M., & Glinert, E. P. (1996, Winter). Multimodal integration. *IEEE Multimedia*, 14–24.

Bregman, A. S. (1990). *Auditory scene analysis.* Cambridge, MA: MIT Press.

Burrows, D., & Solomon, B. A. (1975). Parallel scanning of auditory and visual information. *Memory and Cognition, 3*(4), pp. 416–420.

Colavita, F. B. (1974). "Human sensory dominance." *Perception & Psychophysics, vol. 16*, pp. 409–412.

Cytowic, R. E. (1989). *Synesthesia: A union of the senses.* New York: Springer-Verlag.

Cytowic, R. E. (1995). Synesthesia: Phenomenology and neuropsychology. A review of current knowledge. *Psyche, 2*(10).

Durlach, N. I., & Mavor, A. S. (Eds.). (1995). *Virtual reality: Scientific and technological challenges.* Washington, DC: National Research Council, National Academy Press.

Egeth, H. E., & Sager, L. C. (1977). On the locus of visual dominance. *Perception and Psychophysics, 22*(1), 77–86.

Gibson, J. J. (1966). *The senses considered as perceptual systems.* Boston, MA: Houghton Mifflin.

Gibson, J. J. (1979). *The ecological approach to visual perception.* Boston, MA: Houghton Mifflin.

Gilbert, G. M. (1941). Inter-sensory facilitation and inhibition. *Journal of General Psychology, 24*, 381–407.

Gregg, L. W., & Brogden, W. J. (1952). The effect of simultaneous visual stimulation on absolute auditory sensitivity. *Journal of Experimental Psychology, 43*, 179–186.

Hanson, V. L. (1981). Processing of written and spoken words: Evidence for common coding. *Memory and Cognition, 9*(1), 93–100.

Henneman, R. H., & Long, E. R. (1954). *A comparison of the visual and auditory senses as channels for data presentation* (Tech. Rep. No. 54-363). Dayton, OH: U.S. Air Force, Wright Air Development Center.

Hollier, M. P., & Voelcker, R. (1997, September). Objective performance assessment: Video quality as an influence on audio perception. Preprint No. 4590, presented at the *103rd Audio Engineering Society Convention*. New York.

Howard, I. P., & Templeton, W. B. (1966). *Human spatial orientation.* New York: John Wiley & Sons.

Iwamiya, S. (1992, February). *Interaction between auditory and visual processing when listening to music via audio-visual media.* Paper presented at the *Second International Conference on Music Perception and Cognition*, Society for Music Perception and Cognition, University of California, Los Angeles.

Jones, B. (1981). The developmental significance of cross-modal matching. In Richard D. Walk & Herbert L. Pick (Eds.), *Intersensory perception and sensory integration* (pp. 109–136). New York: Plenum Press.

Jones, B., & Kabanoff, B. (1975). Eye movements in auditory space perception. *Perception and Psychophysics. 17*(3), 241–245.

Knudsen, E. I., & Brainard, M. S. (1995). Creating a unified representation of visual and auditory space in the brain. *Annual Review of Neuroscience, 18*, 19–43.

Kravkov, S. V. (1936). The influence of sound upon the light and color sensibility of the eye. *Acta Opthalmologica Scandinavica, 14*, 348–360.

Laurel, B. (1993). *Computers as theatre.* Reading, MA: Addison-Wesley.

Lipscomb, S. D. (1990). *Perceptual judgement of the symbiosis between musical and visual components in film.* Unpublished master's Thesis, University of California, Los Angeles.

Lipscomb, S. D., & Kendall, R. A. (1994). Perceptual judgment of the relationship between musical and visual components in film. *Psychomusicology, 13*, 60–98.

London, I. D. (1954). Research on sensory interaction in the Soviet Union. *Psychological Bulletin, 51*(6), 531–568.

Loveless, N. E., Brebner, J., & Hamilton, P. (1970). Bisensory presentation of information. *Psychological Bulletin, 73*(3), 161–199.

Marks, L. E. (1974). On associations of light and sound: The mediation of brightness, pitch, and loudness. *American Journal of Psychology, 87*, (1–2), 173–188.

Marks, L. E. (1978). *The unity of the senses: Interrelations among the modalities.* New York: Academic Press.

Marks, L. E. (1982). Bright sneezes and dark coughs, loud sunlight and soft moonlight. *Journal of Experimental Psychology: Human Perception and Performance, 8*(2), 177–193.

Marks, L. E. (1987). On cross-modal similarity: Auditory-visual interactions in speeded discrimination. *Journal of Experimental Psychology: Human Perception and Performance, 13*(3), 384–394.

Marks, L. E. (1989). On cross-modal similarity: The perceptual structure of pitch, loudness, and brightness. *Journal of Experimental Psychology: Human Perception and Performance, 15*(3), 586–602.

Marks, L. E., Szczesiul, R., & Ohlott, P. (1986). On the cross-modal perception of intensity. *Journal of Experimental Psychology: Human Perception and Performance, 12*(4), 517–534.

McGurk, H., & MacDonald, J. (1976). Hearing lips and seeing voices. *Nature, 264*, 746–748.

Millar, S. (1981). Crossmodal and intersensory perception and the blind. In Richard D. Walk & Herbert L. Pick, Jr. (Eds.), *Intersensory perception and sensory integration* (pp. 281–314). New York: Plenum Press.

Murch, G. M. (1973). *Visual and auditory perception.* Indianapolis, IN: Bobbs-Merrill.

Negroponte, N. (1995). *Being digital.* New York: Alfred A. Knopf.

Neuman, W., Crigler, A., & Bove, V. M. (1991). Television sound and viewer perceptions. In *Proceedings of the Audio Engineering Society Ninth International Conference,* 1/2, 101–104.

Neuman, W. R. (1990). *Beyond HDTV: Exploring subjective responses to very high definition television.* Cambridge, MA: MIT Media Library, Massachusetts Institute of Technology.

O'Connor, N., & Hermelin, B. (1981). Coding strategies of normal and handicapped children. In Richard D. Walk & Herbert L. Pick, Jr. (Eds.), *Intersensory perception and sensory integration* (pp. 315–343). New York: Plenum Press.

Pick, H. L., Jr., Warren, D. H., & Hay, J. C. (1969). Sensory conflict in judgements of spatial direction. *Perception and Psychophysics, 6*, 203–205.

Pratt, C. C. (1936). Interaction across modalities: I. Successive stimulation. *The Journal of Psychology, 2*, 287–294.

Proffitt, D. R., & Kaiser, M. K. (1995, March). Human factors in virtual reality development. Tutorial at the *Virtual Reality Annual International Symposium*, Research Triangle Park, NC.

Radeau, M., & Bertelson, P. (1976). The effect of a textured visual field on modality dominance in a ventriloquism situation. *Perception and Psychophysics, 20*(4), 227–235.

Ragot, R., Cave, C., & Fano, M. (1988). Reciprocal effects of visual and auditory stimuli in a spatial compatibility situation. *Bulletin of the Psychonomic Society, 26*(4), 350–352.

Rosenblum, L. D., & Fowler, C. A. (1991). Audiovisual investigation of the loudness-effort effect for speech and nonspeech events. *Journal of Experimental Psychology: Human Perception and Performance, 17*(4), 976–985.

Ryan, T. A. (1940). Interactions of the sensory systems in perception. *Psychological Bulletin, 37*, 659–698.
Rydstrom, G. (1994). *Film sound: How it's done in the real world.* Course number 12, Sound Synchronization and Synthesis for Computer Animation and VR, presented at SIGGRAPH '94, Orlando, FL.
Serrat, W. D., & Karwoski, T. (1936). "An Investigation of the effect of auditory stimulation on visual sensitivity." *Jonrnal of Experimental Psychology, vol. 19*, pp. 604–611.
Srinivasan, M. A., & Basdogan, C. (1997). Haptics in virtual environments: Taxonomy, research status, and challenges. *Computers and Graphics, 21*(4), 393–404.
Srinivasan, M. A., Beauregard, G. L., & Brock, D. L. (1996). The impact of visual information on the haptic perception of stiffness in virtual environments. *Proceedings of the ASME Dynamics Systems and Control Division, 58*, 555–559.
Stein, B. E., & Meredith, M. A. (1993). *The merging of the senses.* Cambridge, MA: MIT Press.
Stone, H., & Pangborn, R. M. (1968). *Intercorrelation of the senses. Basic Principles of Sensory Evaluation.* (Special Tech. Pub. No. 433, pp. 30–46). Philadelphia: American Society for Testing and Materials.
Storms, R. L. (1998). *Auditory-visual cross-modal perception phenomena.* Unpublished doctoral dissertation, Naval Postgraduate School.
Thompson, R. F., Voss, J. F., & Brogden, W. J. (1958). Effect of brightness of simultaneous visual stimulation on absolute auditory sensitivity. *Journal of Experimental Psychology, 55*(1), 45–50.
Tierney, J. (1993, September 16). Jung in motion, virtually and other computer Fuzz. *The New York Times*, C1, C9.
Warren, D. H., Welch, R. B., & McCarthy, T. J. (1981). The role of visual-auditory "compellingness" in the ventriloquism effect: Implications for transitivity among the spatial senses. *Perception and Psychophysics, 30*(6), 557–564.
Wickens, C. D. (1992). *Engineering psychology and human performance* (2nd ed.). New York: HarperCollins.
Woszczyk, W., Bech, S., & Hansen, V. (1995). *Interaction between audio-visual factors in a home theater system: Definition of subjective attributes.* Preprint No. 4133, presented at the 99th Audio Engineering Society Convention, New York.

# 23

# Illusory Self-motion in Virtual Environments

Lawrence J. Hettinger
*Logicon Technical Services, Inc.*
*57 Myrick Lane*
*Harvard, MA 01451*
*lhettinger@logicon.com*

## 1. INTRODUCTION

Current virtual environment (VE) systems can often induce compelling illusions of self-motion in users. In many instances, the effects of visually depicted motion in a VE can be strong enough to create an overwhelming sense of actual self-motion in physically stationary observers, along with pronounced postural adjustment and/or a strong sense of disequilibrium and, occasionally, motion sickness. Indeed, the ability to generate realistic sensations of self-motion within VEs might be considered an important element in the overall sense of "presence" (see chap. 40, this volume) imparted by a given system.

In all likelihood, future VE systems will be able to produce stronger and more durable illusions of self-motion. In some cases there may be compelling reasons for doing so, particularly if the attainment of larger human–VE system objectives is thereby aided. (Note: A *human–VE system objective* is defined as the intended, often measurable outcome of the VE system with respect to its human user. For instance, the objective of a VE-based training system is to promote the acquisition of a set of general or well-defined skills in its users. The experience of illusory self-motion, or any other component of presence, may or may not assist in the attainment of overall human–VE system objectives. Indeed, it may in some cases interfere with them [cf. Hettinger & Haas, in press].) In other cases, the benefits of these illusory experiences may be dramatically offset by the occurrence of negative side effects such as motion sickness (see chap. 30, this volume) and persistent perceptual–motor disruptions (see chaps. 37–39, this volume). In any event, rational design and usage decisions will need to be based on knowledge of the factors that promote (or inhibit) the occurrence of these illusions, their specific benefits with respect to the system user, and the potential risks that they entail.

The term *vection* is used to describe a broad class of illusory self-motion phenomena that are primarily induced by dynamic, large field-of-view optic flow patterns (Dichgans & Brandt, 1978; Howard, 1986a), but which have also been noted to occur with relatively small fields of view (Andersen & Braunstein, 1985; Andersen & Dyer, 1987). Although most commonly a visual phenomenon, vection illusions have also been reported as a result of exposure to specific classes of auditory stimulation ("audiokinetic vection"—e.g., Dodge, 1923; Lackner, 1977) and electrical vestibular stimulation (e.g., Cress et al., 1997; Dzendolet, 1963). Dichgans and Brandt (1978) describe a haptic form of vection ("haptokinetic vection") that produces a weak sensation of illusory self-motion induced through tactile motion stimulation of a large part of the body. Systematic examination of each of these nonvisual forms of vection has been very limited, and they will not be described in any further detail.

This chapter will examine the implications of the considerable experimental literature in this area for the design and use of VE systems. It provides a general summary of the experimental findings from the primary areas of research in this area (e.g., optical flow pattern characteristics of effective vection displays, neurophysiological correlates of vection, vection-induced motion sickness and disorientation) and describes their relevance to VE design and usage issues. One of the major concerns of the chapter is to heighten the reader's sensitivity to questions of whether the user's experience of vection in a given VE system in any way promotes its overall system objectives, under what circumstances and/or within which applications such benefits arise, and at what potential cost to the user's safety and well-being. With regard to the latter point, particular attention will be paid to the literature that has examined the negative side effects that have occasionally been associated with the vection illusion, particularly those involving motion sickness (see chap. 30, this volume) and disrupted perceptual–motor behavior or aftereffects (see chaps. 37–39, this volume).

The above questions are closely related to a major debate that has existed within the VE community almost since its inception, that centering around the potential benefits and costs of "presence" or "telepresence" in VE systems (cf. Held & Durlach, 1992; Slater & Steed, 2000; Zahorik & Jenison, 1998; Zeltzer, 1992). This debate echoes an essentially identical controversy that existed in the flight simulation community for many years, that of the role of "realism" in the design and use of these devices. Although difficult to define and even more so to measure (see chap. 40, this volume), presence and realism are too often assumed to be necessary in order for VE systems to be successful and/or accepted by users. The potential validity of this largely unexamined assumption may vary widely from one application to the next, but it is certain that it entails important cost and design implications for VE systems. In addition, in the case of vection, it also involves important issues of user safety and well-being.

While this chapter will not examine the notion of presence in depth, we do wish to emphasize that vection, as a key component of presence, may or may not meaningfully contribute to the system objectives of a human–VE system. It will, however, occasionally have undesired consequences for users. Therefore, it is important to understand the nature of the vection illusion and to approach its inclusion as a VE system design feature with something more than the simplistic assumption that "if it looks and feels real, then it must be good."

The strength and duration of the vection illusion, its role within and effect on the overall design goals of any given VE system, and the nature, strength, and duration of potential negative side effects are all dependent on a variety of factors to be discussed in this chapter. This chapter is intended to provide readers with a basic understanding of these phenomena, the factors that underlie their occurrence, and the degree to which they may help or hinder specific applications of VE technology.

This chapter will not provide an encyclopedic examination of all aspects of the vection illusion, nor will it be a complete and exhaustive review of the considerable experimental literature in the area. For excellent in-depth presentations of these aspects of vection, readers are referred to Dichgans & Brandt (1978) and Howard (1982, 1986a). This chapter will deal solely with summarizing the research to date on vection phenomena and describe its relevance for the effective design and use of VE systems.

## 2. GENERAL CHARACTERISTICS OF VECTION

Actual self-motion through the real world is visually specified by reliable and highly structured transformations in the pattern of optical information—the optic array—that directly correspond to the dynamics of specific forms of self-motion (e.g., running, crawling, driving in a car) through specific types of structured environments (Cutting, 1986; Gibson, 1958, 1961, 1979; Owen, 1990; Warren, 1976). Generally speaking, these patterns of optical transformations take place over a wide field of view (that corresponding to one's normal field of vision), and only occur when one is in fact moving. Therefore, when found in a situation in which large-field optical transformations of the sort that normally accompany and specify actual self-motion are made available, it is not surprising that one experiences a sense of being physically displaced even when in fact stationary.

Virtual environments, particularly those encompassing large visual fields, are particularly well suited to produce vection illusions. But how are these illusions produced? What controls their onset, compellingness, and duration? Why do these illusions appear to lead to motion sickness and perceptual–motor disturbances in some individuals? And, most importantly, of what potential benefit (if any) are vection illusions to the larger objectives of individual human–VE systems, and what are their potential costs?

Experimental psychologists and neurophysiologists have studied vection for well over a century. Mach (1875) was the first to systematically study vection in the laboratory. Tschermak (1931, cited in Dichgans & Brandt, 1978) appears to have been the first to refer to illusory self-motion as "vection." Although the empirical investigation of this phenomenon has primarily occurred within traditional laboratory settings using various types of specially designed devices such as vection cylinder "drums" and "moving rooms," the advent of high-fidelity flight simulation systems in the 1970s and 1980s extended its occurrence, as well as its study, into more "applied" domains (e.g., Frigon & Delorme, 1992; Hettinger, Berbaum, Kennedy, Dunlap, & Nolan, 1990; van der Steen, 2000; More recently, the increased development and use of VE systems for entertainment (see chap. 55, this volume), vehicular simulation (see chap. 42, this volume), and many other applications has produced an even greater incidence of vection illusions. Indeed, the tremendous popularity of "virtual reality rides" at many of the world's most popular amusement and entertainment parks is based in large part on the reliable production of vection in paying customers and has exposed very large segments of the population to its intended effects, and occasionally to its unintended negative side effects.

One of the earliest (and certainly most literary) accounts of vection is found in a description by Wood (1895) of the "Haunted Swing"—a popular attraction at the 1895 San Francisco Midwinter Fair and clearly a predecessor of today's VE-based entertainment devices. There is little doubt that the experience of illusory self-motion underlies the exhilarating and, apparently for some, slightly disturbing experience described by Wood. To experience this attraction, fairgoers entered a large room furnished in a more or less normal way, with the exception of a very large swing in the center of the room capable of holding up to 40 people and suspended from a large iron rod that passed through the center of the room. Visitors took their seats on the swing, and Wood describes what happened next as follows:

> We took our seats and the swing was set in motion, the arc gradually increasing in amplitude until each oscillation carried us apparently into the upper corners of the room. Each vibration of the swing caused those peculiar 'empty' sensations within which one feels in an elevator; and as we rushed backwards toward the top of the room there was a distinct feeling of "leaning forward," if I can so describe it—such as one always experiences in a backward swing, and an involuntary clutching at the seats to keep from being pitched out. We were then told to hold on tightly as the swing was going clear over, and sure enough, so it did … (p. 277).

The Haunted Swing did not, of course, traverse a 360-degree arc. In fact, it did not move at all (with the exception of "merely being joggled a trifle"). The entire effect was based on the presence of a vection illusion—the room itself (with its furnishings securely fastened to the floor and walls) moving back and forth around the astonished, yet physically stationary, observers.

Two further comments by Wood are particularly interesting in light of their relevance for the current discussion of vection: "The curious and interesting feature ... was that even though the action was fully understood, as it was in my case, it was impossible to quench the feeling of 'goneness within' with each apparent rush of the swing" (p. 277). As is common of contemporary vection illusions, even objective knowledge of the true state of affairs is insufficient to suppress the illusory sensation of motion. Finally, and somewhat more ominously: "Many persons were actually made sick by the illusion. I have met a number of gentlemen who said they could scarcely walk out of the building from dizziness and nausea" (pp. 277–278). Indeed, people are still being made dizzy and sick by similar phenomena.

Vection illusions occasionally occur in everyday life, outside of laboratories, simulators, and VE devices. One of the most frequently cited of these ordinary circumstances, perhaps more commonly experienced in Europe than in the United States, involves sitting in a stationary railroad car with another stationary train directly alongside. When the neighboring train begins to accelerate from a dead stop it is quite common for passengers in the stationary car to experience a sudden and somewhat arresting experience of self-motion, an experience that is particularly compelling when viewed in the periphery of the visual field. This experience manifests itself as a strong phenomenological sense of one's own motion relative to the adjacent train (which is experienced as stationary), accompanied by a distinct postural control motion as the observer "adjusts" to the experienced motion. The same phenomenon occurs in stationary cars, busses, airplanes, and other vehicles. Another common setting for self-motion illusions is the cinema, particularly the wide field-of-view setting of "cinemax" and I-Max theaters. Large, hemispheric dome displays, an attraction at several of the world's largest entertainment parks, on which compelling patterns of motion are projected, (e.g., fighter jets flying in formation) are designed to produce strong vection illusions. Wisely, railings are included for customers to hang onto and thereby avoid pitching over en masse in response to pronounced twists and turns of the motion pattern.

In most, if not all, VE applications there is a strong perceived need among designers to provide very-high-fidelity synthetic replications of the real (or, in some cases, imaginary) world. A corollary of this emphasis on design realism is the desire to provide high-fidelity representations of the multisensory consequences of users' motor behaviors and interactions with virtual entities. In other words, there is a perceived need in the VE design community to provide a highly realistic simulation of the "response" of the perceptual world to each significant activity undertaken by the user of the system.

For example, when a user initiates optical (and other sensory) transformations specifying self-motion in a VE (either by physically walking on a treadmill or by means of some sort of intermediary device such as a DataGlove or joystick) the pattern of visual (see chap. 3, this volume), auditory (see chap. 4, this volume), and in some cases haptic (see chaps. 5–6, this volume) and vestibular (see chap. 7, this volume) information made available to the user should correspond to the intended self-motion profile. In most cases, VE users will not in fact be physically displaced. However, the entire pattern of multisensory stimulation to which they are exposed will specify self-motion, and in many of these cases users will experience strong vection illusions. As seen from Woods' description above, this experience can be compelling and exhilarating, but also occasionally disturbing and even dangerous (depending on the activities the user engages in during exposure to the VE, and immediately upon leaving it). It is incumbent on designers of VE systems to understand the nature of this illusion in order to assure that the user's experience is safe and accomplishes its intended goals.

Before embarking on specific discussions of how an understanding of the vection illusion can substantially impact the design and use of VE systems, the major findings in vection research that are of most relevance to these applications will be reviewed. These findings are classified into the following general areas: (1) the nature of the effects experienced by observers and methods used to measure those effects; (2) the nature of the optical transformations that underlie the illusion; (3) the neurophysiological correlates of vection; and (4) motion sickness and perceptual–motor adaptation phenomena. Prior to discussing these areas, the various types of vection illusions that have received the most research attention will be reviewed.

## 2.1 Types of Vection

Vection can be induced across all six degrees of freedom of body motion (e.g., roll, pitch, yaw, and linear translations) and may be experienced as rotational motion, linear motion, or some combination of the two. In this section, the major forms of vection that have been identified and examined as a means of describing the conditions under which vection has traditionally been investigated and the general nature of the experimental findings that have been obtained will be discussed.

In describing these forms of vection, the following nomenclature will be used to describe the three principal physiocentric axes of real and illusory self-motion (see Fig. 23.1):

- Z-axis—refers to the vertical, spinal, or yaw axis of the body, or an imaginary line passing directly through the head, down through the body and out the soles of the feet.
- X-axis—refers to the roll axis of the body corresponding to the forward line of sight, or an imaginary line passing through the middle of the body from front to back.
- Y-axis—refers to the pitch axis of the body, or an imaginary line passing through the body from its right side to its left side.

### 2.1.1 Circular Vection

Dichgans and Brandt (1978) define circular vection as illusory rotation about one or more of the three body axes. Generally speaking, however, the term *circular vection* has primarily been

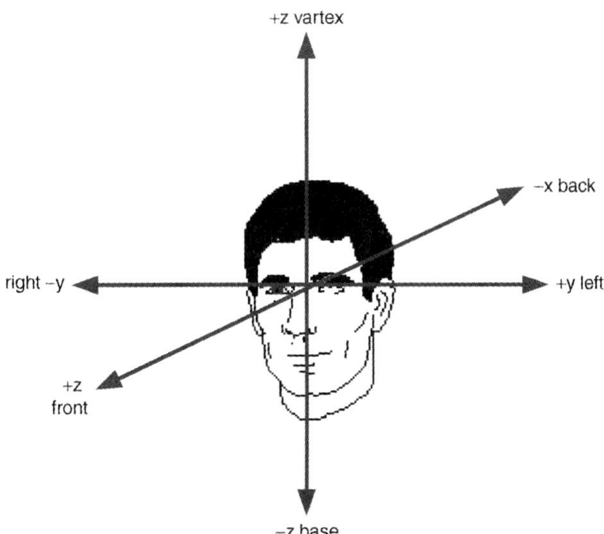


FIG. 23.1. Three body-centric axes used to classify forms of vection.

applied to illusory rotation about the upright body's z-axis, while rotation about the body's x and y axes are more commonly referred to as roll and pitch vection, respectively. Roll and pitch vection are described separately below.

Circular vection is perhaps the most thoroughly investigated of all illusory self-motion phenomena. Ernst Mach first studied it in the 19th century, using a rotating, striped cylinder that completely surrounded his subjects (Mach, 1875). He later used a large, effectively endless belt that moved across two rollers to induce the illusion (Mach, 1922, cited in Dichgans & Brandt, 1978). The former apparatus, frequently referred to as a "vection drum," has been the most common means of inducing and examining the illusion in laboratory settings over the years.

Though certain aspects of the experience of circular vection vary somewhat from one observer to the next (notably its latency of onset), the experience itself is quite uniform across individuals. Once the motion stimulus has been initiated (i.e., the vection drum is turned on and begins rotating about the observer) the common initial impression is of object or "surround" motion—that is, the observer perceives the drum as moving, and not himself. However, shortly thereafter a distinct transformation in the observer's phenomenological experience occurs as he begins to perceive himself as moving in the direction opposite to that of drum rotation, while the drum itself appears more or less stationary. *Saturated vection* refers to the condition in which the motion stimulus is perceived as completely stationary, with all motion experienced as self-motion.

As described in more detail below, the latency and strength of the circular vection illusion are dependent on a number key optical flow parameters. These include the velocity of the motion display, spatial and temporal frequency of the visual pattern, presence of foreground and background visual information, and size of the visual field of view. As Woods noted in reference to the "Haunted Swing" illusion, it is nearly impossible to suppress the illusion of illusory rotation given an optimal combination of these parameters. Objective knowledge of one's circumstances appears to have very little influence on any aspect of the user's experience of this, or any other form of vection. As noted by Brandt, Dichgans, and Koenig (1973), it is common for the experience of illusory circular self-motion to persist beyond termination of the optical motion, a phenomenon that may partially underlie problems with disrupted perceptual–motor behavior observed in some users of flight simulation and VE systems (Kennedy & Fowlkes, 1992).

### 2.1.2 Roll Vection

Roll vection is defined as illusory self-motion about the x or frontal axis of the body (Dichgans & Brandt, 1978). It has been studied using devices such as circular disks with a patterned surface that are positioned immediately in front of an observer or hollow spheres with a patterned inner surface that are set into motion around an observer (e.g., Dichgans, Held, Young, & Brandt, 1972; Held, Dichgans, & Bauer, 1975; Howard, Cheung, & Landolt, 1987). In each case the surface rotates (or in some cases oscillates) around the observer's x-axis. Recently, Allison, Howard, and Zacher (1999) examined roll vection and illusory self-tilt in a "tumbling room" very reminiscent of the "Haunted Swing," capable of a full 360-degree translation in the roll axis. When observers view such displays, they experience an illusory rotation and tilt in the direction opposite that of the motion of the display, the strength of the illusion is dependent on characteristics of the display described below.

The experience of vection in the roll axis is often quite paradoxical. As noted by Howard (1986a), it often seems as if the body is rotating continuously, but remaining at a more or less constant and fairly limited angle of tilt. As described below, this paradoxical phenomenon appears to be due to the influence of the vestibular otoliths (see chap. 7, this volume).

The same factors that influence the latency and strength of circular vection also appear to affect roll vection. For instance, Held et al. (1975) observed that the illusory self-tilt component of roll vection increased along with the size and retinal eccentricity of the visual display. Allison et al. (1999) observed that subjects were far more likely to report experiences of "tumbling sensations" (i.e., complete 360-degree apparent self-rotation) as a function of increased field-of-view size. This effect was also greater at the faster of the two rotational velocities tested (30 deg/sec versus 15 deg/sec). For smaller fields of view, observers reported sensations of constant tilt rather than full, 360-degree roll.

### *2.1.3 Pitch Vection*

Pitch vection occurs as a result of viewing motion patterns that rotate about the body's y-axis (Dichgans & Brandt, 1978; Young et al., 1975) and has been studied using devices similar to those employed in the study of roll vection, such as rotating disks or hollow spheres within which observers view patterns of moving visual stimuli (e.g., Howard et al., 1987).

The general characteristics of pitch vection are very similar in most respects to those of roll vection. That is, observers can be made to experience constant illusory self-motion, but only a limited degree of displacement or self-tilt (which tends to be less than that experienced in roll vection). Unlike roll vection, however, there appears to be a body-centric asymmetry in the degree of tilt experienced in pitch vection—specifically, smaller magnitude estimates of tilt are observed for apparent pitch-up conditions as compared to apparent pitch-down conditions (Young et al., 1975). Additionally, the stimulus parameters that underlie the experience of pitch vection are very similar to those of roll vection—in particular, larger visual fields of view tend to produce more compelling pitch illusions.

### *2.1.4 Linear Vection*

Linear vection can be produced along any of the three body axes described above (or as a vector sum of two or more of these axes), but has most often been examined as illusory self-motion along the body's x-axis. It has been studied using a broad array of devices, including moving rooms (e.g., Lee & Lishman, 1975), devices that incorporate projection of linear optical flow patterns onto the walls of a stationary room in which an observer is standing or seated (e.g., Berthoz, Lacour, Soechting, & Vidal, 1978; Lestienne, Soechting, & Berthoz, 1977), and frontal stereoscopic presentation of motion patterns (e.g., Ohmi & Howard, 1988). Flight simulators and numerous VE devices, particularly those with wide field-of-view display capabilities, are also capable of inducing compelling sensations of linear vection.

Postural sway responses to linear vection stimuli are often quite profound. Generally, observers will make a postural adjustment toward the direction of perceived self-motion in an apparent attempt to compensate for a perceived shift in the direction of gravito-inertial force vector. However, observers tend to show greater amplitude in backward as opposed to forward postural adjustments (Lestienne et al., 1977). It is not difficult to create conditions in a linear vection study that will cause observers to fall over, a finding of clear relevance for the safe design and operation of VE devices that incorporate such a capability. Indeed, a standing, unsupported observer in any compelling vection situation is at risk of falling. In addition, linear vection can induce sensations of disequilibrium that are at least as strong as those produced in other forms of the illusion (described in more detail below). For example, Lestienne et al. (1975) reported that of the 30 subjects in their study, 3 (10%) became so disoriented that they fainted while viewing a linear vection display.

## 2.2 Phenomenological and Behavioral Aspects of Vection

The most common phenomenological and behavioral manifestations of vection include: (1) a compelling illusory sense of self-motion in the absence of true inertial displacement; and (2) distinct adjustments in postural control activity that are closely coupled to the nature of the inducing stimulus. The majority of research described in this chapter employs metrics tied to these two classes of response, with particular emphasis placed on the former.

As aforementioned, when observers are first exposed to a vection-inducing display or apparatus, such as a rotating cylindrical drum in a circular vection setting, they typically first perceive the drum itself as being in motion. It is not unusual for a similar first impression to occur within a VE setting—that is, users perceive themselves as immersed in a computer-generated environment within which motion is being depicted, but the motion is not experienced as being their own. Shortly thereafter, however, the perception in most laboratory vection settings changes dramatically from one of object or surround motion to one of self-motion (assuming that conditions for inducing the experience of the illusion are appropriately configured)—the direction of self-motion being opposite in direction to that of the surround motion. Coincident with the onset of self-motion, surround motion may appear to effectively cease.

In the case of circular vection, the latency of this effect is on the order of 5 to 30 seconds (e.g., Dichgans & Brandt, 1973, 1978). This range is representative of all forms of vection, and specific values are influenced by a host of factors described in the following section. For example, onset of the vection illusion tends to occur sooner at lower optical accelerations (Howard, 1986a; Melcher & Henn, 1981). In addition, its latency tends to decrease as a direct function of increases in the size of the visual field of view (Dichgans & Brandt, 1978). In some cases the experience of vection may remain fairly constant once it has begun—that is, it will remain present until the motion stimulus is discontinued and may persist for a while afterward. However, in other cases the sensation of vection may come and go, and may also increase and decrease in intensity.

In measuring the phenomenological component of vection, two key questions that are typically addressed are: (1) Is vection present (i.e., is the subject in a vection experiment or the user of a VE system currently experiencing the vection illusion?); and (2) If present, how powerful is the phenomenological experience of the illusion? A number of approaches have been used over the years in experimental studies of vection, and they may be useful to VE system designers as they attempt to either enhance or diminish the presence and /or strength of this illusion.

As is often the case with phenomena that are largely subjective in nature (e.g., presence), the measurement of the phenomenological component of vection is of necessity largely an indirect process. A number of approaches derived from psychophysical measurement techniques have been used, including magnitude estimation of surround and/or self-motion, magnitude estimation of degree of perceived self-tilt, and subjective scaling of the strength of the illusion (cf. Diener, Wist, Dichgans & Brandt, 1976; Kennedy, Hettinger, Harm, Ordy, & Dunlap, 1996; Ohmi & Howard, 1988). These measures are widely accepted in the vection research community and have been shown to have a high degree of psychometric reliability (Kennedy et al., 1996).

Magnitude estimation techniques often require subjects to make estimates of the velocity of apparent self-motion in the case of continuous motion (e.g., in degrees/second), displacement (e.g., in degrees) in the case of discrete motion, or in estimating the illusory tilt so commonly perceived in roll and pitch vection. Subjects have occasionally been asked to provide magnitude estimates of their own apparent motion in addition to the perceived motion of the stimulus (e.g., Wong & Frost, 1978). In the latter case, the strength of the vection illusion can be inferred by greater magnitude estimates of self-motion relative to stimulus motion.

One of the more common methods of assessing the strength of the vection illusion involves the use of comparatively simple subjective scaling techniques. Typically these scales range from zero, corresponding to a rating of no illusory self-motion, to some upper limit meant to reflect a sense of "saturated" vection (cf. Hettinger et al., 1990; Howard et al., 1987).

As noted above, postural readjustment responses are also commonly observed in vection settings, and their characteristics can be used to assess the degree of vection experienced by an observer in a given situation. These types of metrics are derived from quantitative measures of the observer's reflexlike reactions to perceived motion in which body sway and/or other forms of postural control activity are the most common response (e.g., Bles & DeWit, 1976; Dichgans & Brandt, 1978; Stoffregen, Hettinger, Haas, Roe, & Smart, 2000). Under conditions of sufficient stimulus magnitude and/or duration it is not unusual to observe subjects losing functional postural control and eventually falling or collapsing (Lee & Aronson, 1974; Lestienne et al., 1977). As Kennedy and his colleagues (e.g., Kennedy, Drexler, Compton, Stanney, & Harm, in press) have noted, it is also not unusual for postural control disturbances to persist well beyond the cessation of the self-motion experience.

The precise, objective measurement of postural control responses during exposure to illusory self-motion stimulation is a far more complex and equipment-intensive procedure than is the use of psychophysical magnitude estimation or scaling techniques (see chap. 39, this volume). Typically, devices such as force platforms, body-mounted accelerometers, and calibrated photographic/video methods have been used to assess postural sway, and the data is then reduced to quantitative profiles of postural control activity. While this level of precision may require excessive instrumentation for most VE applications outside the laboratory, the simple observation of users' postural control behaviors in the presence of self-motion stimuli may be a useful converging measure for assessing the presence of vection if used in conjunction with the psychophysical techniques described above and may also provide information regarding early onset of the illusion. For example, Stoffregen (1985) found compensatory sway in a moving room to have a lower threshold than reports of motion.

One of the least understood aspects of vection involves the nature of differences between individuals in terms of their phenomenological and postural responses. In perhaps the only study of this aspect of the illusion, Kennedy and his colleagues (Kennedy et al., 1996) examined the psychometric reliability (within individuals) and variability (between individuals) of various measures of circular vection, such as latency and magnitude estimation of perceived self-motion. Results indicated that while the classes of measures used exhibited strong intraobserver reliability (i.e., observers were quite consistent in their responses from one session to the next), there were strong and consistent differences between individuals. This finding suggests that key aspects of the vection illusion may not be as uniformly experienced across individuals as previously believed and/or that observers may respond to aspects of vection-inducing situations differently. In addition, the authors hypothesize that those observers who exhibit greater sensitivity to vection may be among those most likely to experience occasional side effects and/or problems in readapting to normal perceptual-motor relationships once out of the vection environment. However, these hypotheses have yet to be tested.

## 2.3 Optical Determinants of Vection

In order to attempt to control (or perhaps, in some cases, prevent) the onset of vection illusions in VE devices, it is important to understand the nature of the optical information that influences its onset and strength. Most of these factors have been alluded to above, but are summarized in more detail in this section. It is interesting to note that many aspects of features that appear to enhance the vection illusion are the very factors that are widely sought in current VE systems,

**TABLE 23.1**
Visual Display Factors Affecting Vection

1. Size of the visual field of view.
2. Optical velocity (or temporal frequency) of the visual stimulus.
3. Spatial frequency of the stimulus pattern.
4. Presence of background and foreground information.

and which are generally thought to support the experience of presence. In other words, the visual display features that enhance vection, such as the size of the field of view, spatial and temporal frequency of the optical pattern (i.e., "scene detail"), and others described below are widely emphasized in the design of contemporary VE systems. Therefore, vection in VEs should become increasingly common as systems continue to more closely attain the engineering and display parameters that promote its occurrence.

Vection is strongly influenced by a number of visual display factors (see Table 23.1). The influence of most of these is summarized in Dichgans and Brandt (1978) and Howard (1986a). The vast majority of the vection literature is characterized by studies that focus on one form of vection at a time, and little work has been done to attempt to generalize findings across the various forms of the illusion. Recognizing that the critical elements underlying specific forms of vection may vary according to the form of the illusion under examination, it is nevertheless possible to make a number of general statements concerning visual display factors.

Despite several findings indicating that it is possible to induce sensations of vection with relatively small (less than 10 degrees of visual angle), centrally presented motion patterns (e.g., Andersen & Braunstein, 1986), there appears to be little doubt that peripheral visual stimulation is far more effective in inducing the illusion (and corresponding postural adjustments) than central visual stimulation. Simply put, wide fields of view are more effective for inducing vection than are comparatively narrow fields of view. This is almost certainly due to the greater connectivity between the vestibular system and peripheral retina in comparison to the central retina (see later discussion of neurophysiological correlates of vection).

Many studies have examined field of view effects on vection. The findings of Lestienne, Soechting, and Berthoz (1977) and Brandt, Dichgans, and Koenig (1973) are illustrative of this work in general. In their examination of linear vection, Lestienne and his colleagues observed that the amplitude of the postural sway response observed in their subjects depended primarily on the size of the motion pattern. Specifically, increases in postural sway amplitude were linearly related to field-of-view size—the larger the field of view, the more pronounced the postural sway response to the linear vection stimulus. This is consistent with the findings of Brandt et al. (1973), that the intensity (i.e., strength of the illusion) and velocity of illusory self-rotation in circular vection are strongly dependent on the size of the available field of view, with intensity being more adversely affected than apparent velocity at smaller field-of-view sizes. A centrally located field of view of 30 degrees (the lowest value tested) produced very low estimates of intensity compared to full-field stimulation. In a compelling illustration of the impact of the visual periphery on vection, Brandt et al. (1973) masked the central 120 degrees of the visual field so that subjects could only view the rotating motion stimulus in the far periphery. The intensity of the vection illusion was scarcely affected by this manipulation.

The relevance of these findings for VE system design is very clear. Larger fields of view will generally be more effective for eliciting vection illusions than will smaller fields of view. In some cases this may be advantageous (e.g., in helping to make a VE-based entertainment ride more exhilarating) but in other cases may result in excessive problems with motion sickness and/or postural disturbances. The addition of accurate representation of motion-in-depth information (e.g., radially expanding optical flow patterns characteristic of forward, linear self-motion), which Andersen and Braunstein (1985) assert to be the critical factor in supporting illusions of self-motion with small, centralized displays, should serve to enhance the field of view effects.

Optical velocity is another key visual display factor, influencing both the intensity and velocity of illusory self-motion. Howard (1986a) reports the general finding that the apparent velocity of illusory self-rotation in circular vection is directly proportional to optical velocity up to values of approximately 90 deg/sec, although this relationship is influenced by the spatial frequency (optical texture density) of the stimulus pattern. Similarly, Brandt et al. (1973) and Dichgans and Brandt (1974) observed that perceived velocity of illusory self-motion in circular vection is linearly related to stimulus velocity up to about 90 to 120 deg/sec, beyond which perceived illusory self-motion velocity lags behind stimulus velocity.

The spatial frequency of the visual stimulus, or what can generally be considered to be equivalent to the amount of "detail" presented in the display, also affects the perception of illusory self-motion. Diener, Wist, Dichgans, and Brandt (1976) observed that subjects' perceptions of the velocity of illusory self-rotation more closely matched the true velocity of the stimulus with increases in the spatial frequency of the visual display. These findings correspond to those obtained by Owen and his colleagues (e.g., Owen, Wolpert, & Warren, 1983; Warren, Owen, & Hettinger, 1982) who observed that "edge rate" information was more salient than global optical flow in affecting observers' judgments of self-motion.

One of the earliest and most common findings in the vection literature is that the perception of illusory self-motion is greatly enhanced when observers fixate a stationary target located in front of the visual motion pattern (e.g., Mach, 1875). By implication, this finding suggests that the inclusion of foreground visual information in VEs that requires sustained attention by the user will result in more pronounced sensations of vection, if a wide field-of-view pattern of motion is made available.

When there are multiple displays in view, each depicting patterns of motion and each at different distances from the observer, the most distant display has been observed to control the direction of illusory self-motion (Brandt, Wist, &Dichgans, 1975). Additionally, vection is suppressed by stationary objects seen beyond the moving display but not by stationary objects in the foreground (Howard & Howard, 1994).

## 2.4 Neurophysiological Correlates of Vection

Significant research attention has been devoted to the identification and description of the neurophysiological correlates of vection—that is, the nature of the activity that occurs in the central nervous system (CNS) in conjunction with exposure to vection-inducing displays. This work strongly suggests that the neurophysiological basis of the illusion lies in the activity of the vestibular nuclei (Daunton &Thomsen, 1978; Dichgans & Brandt, 1978; Waespe & Henn, 1977a), a CNS site whose activity is also strongly tied to actual self-motion. Such a convergence makes sense not only because vection and actual self-motion are very similar in a phenomenological sense, but also because they share many observable behavioral similarities (e.g., postural sway, motion sickness) and physiological similarities, such as the occurrence of optokinetic afternystagmus (the tendency of the eyes to exhibit a pattern of nystagmic motion following exposure to real or illusory self-motion), a phenomenon known to be directly affected by activity of the vestibular system (e.g., Waespe & Henn, 1977b, 1978).

Interactions between the visual and vestibular system are central to the vection illusion. Research in visual–vestibular interactions has its own long history (see Henn, Cohen, & Young,

1980; Howard, 1982, 1986b, 1993, for reviews of this literature), and much of the work either overlaps with research on vection or has direct relevance for it. Indeed, the desire to understand and explain the neurophysiology of the visual–vestibular system has always been a prime motivator of vection research (e.g., Dichgans & Brandt, 1978; Straube & Brandt, 1987).

One method used to examine the nature of visual–vestibular interactions in vection involved attempts to isolate components of this ordinarily highly interactive system. For instance, Howard et al. (1987) demonstrated that the paradoxical combination of continuous illusory self-rotation with limited, constant tilt in upright observers is a function of activity from the vestibular otoliths. Specifically, when observers lying on their back (effectively nulling inputs from the otoliths) view a roll vection stimulus, the experience is simply one of continuous self-rotation. This clearly suggests a strong role of the vestibular system in vection illusions.

A different approach for isolating visual and vestibular inputs from one another involves the examination of vection and other visual–vestibular phenomena in microgravity, where vestibular inputs are greatly reduced. For example, Young et al. (1986) studied roll vection in four crew members under microgravity conditions during a Space Shuttle mission, as well as during the microgravity phases of parabolic flight. Their findings indicated that perception of roll vection was enhanced in the early phases of weightlessness (i.e., early in the mission), possibly due to sharply reduced functionality of nonvisual cues for gravity and orientation. Results obtained during later phases of the mission were somewhat more variable, perhaps due to effective incorporation of nonvestibular cues for orientation.

As noted by Howard (1986a), it was originally thought that vection could be ascribed to the effects of optokinetic nystagmus (OKN)—the eye's reflexive tendency to alternative slow pursuit motions with rapid, saccadic return movements when viewing a scene or object (such as a train) moving in one direction (see chap. 3, this volume). However, Brandt et al. (1973) demonstrated that the phenomenological, vection-inducing effects of a large moving field are unaltered even upon adding a small central display of motion in the opposite direction. Whereas the latter changes the direction of OKN as long as the subject fixates on it, it has no effect on the vection experience. On the other hand, the duration of positive and negative aftereffects of illusory self-motion appear to be directly related to the time course of optokinetic afternystagmus, the tendency of the eye to show nystagmic patterns of motion following the cessation of a motion stimulus. (Note: *Positive aftereffects* are illusory motion sensations experienced as being in the same direction as the immediately preceding, stimulus-driven illusion. *Negative aftereffects* are illusory sensations experienced as being in the opposite direction.)

## 2.5 Motion Sickness and Adaptation Phenomena

One of the most widely studied areas in vection research, and one with tremendous relevance to VE design and usage issues, concerns the relation between vection, motion sickness, postural disturbances, and other negative perceptual–motor side effects. Circular vection has been shown to induce symptoms of motion sickness in approximately 60% of healthy human subjects (Stern et al., 1985; Stern, Koch, Stewart, & Lindblad, 1987), and, as previously noted, Lestienne et al. (1977) observed that 3 of the 30 subjects in their linear vection study were disoriented by the experience to the extent that they lost consciousness. Anyone who has been around flight simulators or vehicular simulators for very long is familiar with "simulator sickness" (Kennedy, Hettinger, & Lilienthal, 1990; see chap. 30, this volume), a phenomenon that appears to be spreading to the larger domain of VE systems in general.

Motion sickness resulting from vection has been discussed as a type of "visually induced motion sickness" (Hettinger & Riccio, 1992; McCauley & Sharkey, 1992). The underlying cause of sickness in these circumstances has traditionally been attributed to the existence of conflict between sources of sensory information specifying self-motion and orientation. The "sensory conflict theory" (Reason & Brand, 1975) is the most widely accepted theoretical explanation of motion sickness in real and virtual environments. However, over the years many concerns have been raised with the theory, including why only certain types of conflicts appear to be nauseogenic, why the same conflict might not reliably produce sickness across different individuals (or, indeed, even within the same individual across time), and how to attempt to quantify the amount of conflict present in a given situation and relate it to the frequency and severity of motion sickness. Stoffregen, Riccio, and their colleagues have proposed an alternative explanation, arguing that motion sickness, whether it occurs in virtual or real environments, results from prolonged disruptions in normal postural control activities (Riccio & Stoffregen, 1991; Stoffregen & Riccio; 1991; chap. 39, this volume). Results to date suggest that this is a promising theoretical approach (Smart, Pagulayan, & Stoffregen, 1998; Stoffregen & Smart, 1998; Stoffregen et al., 2000). However, the notion of sensory conflict still seems to be of value in explaining many of the situations that lead to prolonged postural disruptions and/or motion sickness. It may be that both views are to some extent correct, and that motion sickness, including that induced by vection stimuli, ultimately results from unusually provocative disruptions in the normal cycle of perceiving and acting.

One of the most vivid similarities between motion sickness observed in real and illusory motion conditions involves the presence of *Coriolis effects* in the former, and *pseudo–Coriolis effects* in the latter. Coriolis effects are powerful and well-known stimuli for motion sickness, and occur as a result of the cross-coupled angular accelerations that are produced within the vestibular system when the head is tilted out of the body's axis of rotation (cf. Benson, 1990). This is often done by placing subjects in a rotating "Barany chair"—a device that rotates the body about its z-axis like a barber chair might, and then asking the participant to tilt the head slightly out of the axis of that rotation. Intense nausea is quickly produced (cf. Melville-Jones, 1970) and vomiting will nearly always follow shortly afterward if the motion and/or cross-coupled stimulation are not ceased. Dichgans and Brandt (1973) demonstrated the existence of a pseudo–Coriolis effect through what one might today refer to as a virtual simulation of the Coriolis situation. By exposing subjects to a circular vection stimulus (a rotating cylindrical drum) and then asking them to tilt the head outside the axis of apparent rotation, they were able to produce the same sort of motion sickness symptomatology observed in actual Coriolis situations.

Stern, Koch, and their colleagues have pioneered the study of vection-induced motion sickness. In much of their research (e.g., Stern, Hu, Vasey, & Koch 1989; Stern et al., 1985) they have studied sickness by recording subjects' gastric myoelectric activity using an electrogastrogam (EGG). The EGG provides a very sensitive and reliable index of gastric motility associated with subjective symptoms of nausea and is often reflected by a shift from the normal gastric rhythm of 3 cpm to 4 to 9 cpm, referred to as *tachygastria* (Stern et al., 1985; see also chap. 32, this volume). The work of Stern, Koch, and their colleagues has demonstrated that motion sickness is a common side effect of the experience of vection, which appears to diminish in severity as a result of adaptation to repeated exposures (Hu, Grant, Stern, & Koch, 1991), concentrated focus on deep breathing (Jokerst, Gatto, Fazio, Stern, & Koch, 1999), and treatment with acupressure techniques (Hu, Stritzel, Chandler, & Stern, 1995).

There is nearly unanimous agreement among VE system designers, and especially users, that motion sickness is a distinctly unpleasant experience, the occurrence of which should be minimized to the greatest degree possible. (Note: An interesting exception is the NASA Preflight Adaptation Trainer, a VE-based system whose intent was to preadapt astronauts to conditions considered to be highly evocative of space motion sickness. This unique system's objective was, in fact, to induce feelings of queasiness and disorientation in its users, thereby inoculating them to such conditions in the microgravity environment of space [cf. Harm,

Zogofros, Skinner, & Parker, 1993; Parker, 1991]). However, the most urgent risk to users might not involve the feelings of discomfort that occasionally accompany use of the system. Rather the greatest risk might involve the disrupted perceptual-motor control that is often observed after use of the system (see chaps. 37–39, this volume). A point made repeatedly by Kennedy and his colleagues in their discussions of sickness in flight simulators and VE systems is that negative side effects experienced after exposure to a VE are of greater safety concern than those that occur during exposure (e.g., Kennedy & Fowlkes, 1992; Kennedy et al., 1990).

Aftereffects of exposure to vection-inducing events have been frequently noted in the experimental literature. For example, in his review of vection phenomena, Howard (1986a) notes that aftereffects can be readily demonstrated to occur in circular vection settings: "If the lights are put out after a moving scene has been observed for some time, the illusory self rotation continues as a *positive aftereffect* followed by a *negative aftereffect* in the opposite direction" (p. 28). Some form of adaptation to vection-inducing events may underlie these aftereffect phenomena, suggesting that on leaving a vection-inducing situation such as a VE system, flight simulator, or laboratory device, observers may need to undergo a process of readaptation and recalibration, during which time their normally effective perceptual–motor control might be disrupted.

## 3. RELEVANCE OF VECTION FOR VE SYSTEM DESIGN AND USE

An understanding of vection phenomena is important for the design of safe and effective VE technology for a number of reasons:

- Vection appears to be one of the key elements underlying the broader illusion of "presence" (see chap. 40, this volume) in a virtual environment. To the extent that presence is a desirable design feature of a particular VE device, then a proper understanding of the design elements that underlie vection is an important engineering issue.
- Vection may or may not contribute to the achievement of a VE device's "behavioral goals." To the extent that vection in fact facilitates the achievement of various desired outcomes with respect to the user of a device, it is important to understand the factors that promote its occurrence.
- Vection has frequently been identified as a correlate of, and perhaps a causal element in, the phenomenon of motion sickness in VEs—or *cybersickness*. Additionally, exposure to a compelling vection stimulus may be a crucial factor in a larger process of perceptual–motor adaptation (see chap. 31, this volume) to virtual environments. To the extent that adaptation does occur, a period of readaptation upon return to the normal spatiotemporal constraints of the actual world is necessarily required. These periods of adaptation and readaptation may expose users to a number of risks. To control these risks, it is necessary to understand the factors that create vection.

All of these concerns are directly related and cannot be considered in isolation from one another. There is little doubt that vection can play a key role in helping to produce a sense of presence in a VE system. An illusion of "being in" a computer-generated environment is clearly enhanced by a sense of illusory self-motion so compelling that it can hardly be distinguished from actual self-motion. Furthermore, the research literature provides relatively firm guidance, summarized in the preceding sections, concerning system design and usage guidelines for producing the illusion and enhancing its strength. However, much of this literature also describes the risks that system users may face as a result of exposure to self-motion illusions, most notably motion sickness and perceptual–motor disturbances.

The fundamental questions that VE system designers must address with respect to vection are as follows:

- In light of the potential risks to system users posed by vection, does the presence of illusory self-motion meaningfully contribute to the overall objectives of the human–VE system? In other words, are the overarching goals that the system is intended to accomplish being supported by producing compelling sensations of self-motion in users, are they being interfered with, or does the experience of vection have little or no impact on them?
- If a reasonable case can be made that vection is a key factor in promoting the overall objectives of the human–VE system, then the question becomes one of how to safeguard users from the risks associated with its negative side effects.

With regard to the first of these two points, there is little evidence to date that vection plays a critical role in the attainment of human–VE system objectives. However, there are situations when it is likely to play a key role, notably in the successful application of VE-based entertainment devices (see chap. 55, this volume) and potentially in certain VE-based training applications (see chaps. 43–44, this volume) in which skill in the perception and control of self-motion is critical. A major potential problem exists in the former application area, however, in that VE-based entertainment devices are intended to expose as many people as possible to the illusory perceptual features characteristic of this technology. This, of course, raises the risks of significant numbers of users being made ill and/or suffering postexposure side effects.

In many cases the occurrence of vection may actually interfere with the objectives of VE systems. "Self-motion artifacts" might arise in situations where no perception of self-motion is desired, but where the stimulus information is such that momentary (or longer) vection illusions arise. For instance, this could occasionally occur if motion of virtual objects in the periphery of the visual field gives rise to transient vection illusions, causing disruptions in both the phenomenological experience of the VE as well as psychomotor disturbances associated with postural readjustments.

Certainly the occurrence of vection-induced motion sickness and perceptual–motor disturbances can be expected to interfere dramatically with the objectives of most human–VE systems. Any benefit to be derived from the inclusion of vection-inducing stimuli in a VE system has to be carefully weighed against such potential risks. And if there is no benefit to be gained from the presence of self-motion (or, indeed, if there is significant risk associated with the occurrence of its side effects), then design and usage guidelines to minimize its occurrence should be adhered to. Familiarity with the stimulus factors that underlie the illusion, summarized above and described in detail in the literature cited in this chapter, can help to promote effective and safe system design. In addition, simulator and VE system usage guidelines developed by Kennedy and his colleagues (e.g., Kennedy et al., 1990) should be used to further protect users from the occasional negative side effects of exposure to simulated self-motion.

## 4. RESEARCH ISSUES

This section addresses research issues that may be of potential value in further illuminating our understanding of the relation between vection and virtual environments. These issues fall into two broad categories: research on vection and related phenomena that would be useful in enhancing the future safe and effective design and use of VE systems; and research on vection using VE systems as research tools.

As anyone who has ever experienced or worked with VEs knows, they are fascinating systems from many perspectives. As this chapter has discussed, they raise many important psychological, physiological, and behavioral issues as developers attempt to maximize their safe and effective use. However, they not only raise important research issues, they also provide new means for examining these issues. In other words, because of their compelling multisensory, interactive nature, and because they are programmable and permit the acquisition of very detailed information about user behavior (based, for instance, on data retrievable from body motion sensors—see chap. 8, this volume), they afford the means to examine important psychological and physiological issues in ways that have not been possible to date (cf. Durlach & Mavor, 1995). Even in restricting the present discussion to matters concerned with vection and self-motion perception, it is clear that VE technology offers many exciting possibilities for enhancing our knowledge.

For example, Oman, Howard, and their colleagues have been engaged in research on circular and linear vection in the microgravity environment of outer space (Oman et al., 2000) using a helmet-mounted display and computer-generated VE as a motion stimulus. As described previously, the microgravity environment is useful for the study of visual–vestibular interactions because it allows researchers to isolate the influence of the two systems. However, until the introduction of VE technology it was very difficult to examine vection phenomena in microgravity because of the severe weight and size limitations of the equipment involved. Oman's and Howard's work has provided important information on the relative roles of visual and proprioceptive sources of information for orientation in microgravity. Their results, indicating dramatically increased circular and linear vection after several days of exposure to microgravity, also provide a very useful source of converging information to those found by Young et al. (1986) for roll vection.

### 4.1 Effect of Complex Motion Patterns on Vection

Research conducted to date on vection has concentrated almost exclusively on relatively simple patterns of motion; specifically motion along or around a single axis. However, many current and future VE applications will require the depiction of far more complex patterns of self-motion, such as that involved in depicting the intricate and aggressive motion patterns of aircraft engaged in air combat (for training such skills) and any number of possible complex motion profiles for entertainment purposes.

Very little is known about illusory self-motion under such conditions, primarily because until very recently the technology for examining such issues has not existed. This is clearly an example of a research area that stands to benefit from the application of VE technology, and future applications of the technology will in turn benefit from the results of that research.

### 4.2 Multisensory Patterns of Information Specifying Self-motion

Virtual environments are fundamentally multisensory devices. Their intent is to take advantage of the information acquisition capabilities of as many of the sensory systems as possible to replicate the perceptual experience of the real world (or create compelling multisensory experiences of imagined or synthetic worlds). Research to date on the vection illusion has concentrated almost exclusively on the visual modality, with limited research attention having been devoted to audition (e.g., Lackner, 1977). In addition, several studies on electrical stimulation of the vestibular system have shown that it is possible to produce compelling vection illusions (primarily of roll) using such an approach (e.g., Cress et al., 1997; Dzendolet, 1963). However, there has been very little research to date examining the effects of multisensory information on vection. Cress and his colleagues (1997) contrasted the effects of visually specified roll vection, electrical stimulation of the vestibular (producing illusions of roll motion), and combined visual and electrical vestibular stimulation and found that the latter condition produced sensations of motion that were rated as significantly more realistic by observers than either of the two former conditions alone. However, no other formal work that we are aware of has been conducted in the area of multisensory effects on vection.

Research of this type would not only help to provide design and usage guidance for VE systems, but may even help to illuminate the role that sensory conflict plays (or does not play) in creating vection-induced motion sickness.

### 4.3 Adaptation, Readaptation, Transfer of Adaptation, and Virtual Environments

The problems that users occasionally experience when adapting to the rearranged perceptual–motor relationships of VEs and/or when readapting to normal actual environment spatiotemporal constraints after having spent time immersed in a VE have been described above and in chapter 31 (this volume). Not all of these problems are due to vection illusions of course, but motion sickness and adverse postexposure symptomatology have been repeatedly observed in vection studies and related settings with sufficient frequency to identify it as a probable key contributor to this problem.

As VE systems become more widely distributed and more commonly used in everyday work and entertainment settings, these problems can be expected to increase in frequency. In addition to being more widely available, the fidelity and resulting strength of vection illusions (indeed, the overall sense of presence) afforded by VE systems will almost certainly continue to increase for the foreseeable future. Each of these factors will combine to increase problems with undesired side effects unless research on their cause and amelioration is aggressively pursued.

One avenue of research that may help to alleviate the severity of these problems involves the examination of factors that promote functional adaptation across environments (actual and virtual) that vary in their spatiotemporal characteristics (cf. Hu, Stern, & Koch, 1991; Welch, in press). The identification of system design and usage features that promote the safe transition of users between the differing spatiotemporal arrangements of actual and virtual environments should remain a high priority in order to assure a safe and successful future for this technology.

### 4.4 Effect of Visual Frames of Reference on Vection and Orientation

Many sources of information interact to produce one's sense of static and dynamic orientation in the world, including the potentially powerful effect of "visual frames of reference." "Visual frames," as described by Howard and his colleagues, are sets of distinct horizontal lines and surfaces (e.g., elements of the optic array corresponding to walls, ceilings, pillars, etc.) whose orientation is highly linked to that of the gravito-inertial force vector (Allison et al., 1999; Howard & Childerson, 1994). Familiar objects (e.g., trees, furniture) within these frames of reference also exhibit an "intrinsic visual polarity" (Allison et al., 1999, p. 299). When the visual frame of reference and all objects contained within it are tilted with respect to gravity, the result is a compelling experience of self-tilt. Prothero and Parker (in press), in examining their "rest frame hypothesis," have identified strong interactive effects between observers' visual frames of reference, the perception of vection, and subsequent occurrences of motion sickness and disorientation.

As described above, Allison et al. (1999) studied the effect of field size, stimulus velocity, and visual fixation on roll vection and illusory self-tilt in a specially designed moving room reminiscent of the "Haunted Swing." While the results were clearly informative of the role of these visual factors on the strength and other characteristics of the illusion, there would seem to be distinct advantages in conducting such research in a virtual as opposed to an actual environment. The principal advantage may lie in the relative ease with which the key parameters of visual stimulation can be manipulated in the virtual environment. For instance, the relative effect of the visual polarity of familiar objects versus the effect of a more global frame of reference would seem to be much easier to manipulate. However, the problem at this stage in the development of the technology involves the relatively restricted field of view and limited resolution of most current VE systems compared to actual environments.

## 5. CONCLUSIONS

The purpose of this chapter has been to provide a summary of research performed to date on the vection illusion, and in so doing demonstrate its relevance, and the relevance of vection phenomena in general, to the design and use of effective VE systems. The illusion of self-motion is an increasingly ubiquitous aspect of modern VE systems, and it is likely to become even more characteristic of future systems. While there is little doubt that a compelling sense of vection can be a powerful contributor to an overall sense of presence in a VE system, it is by no means clear that it will always be of benefit in achieving the overall system objectives of a given device. Given the "cost" of the vection illusion—occasional, and occasionally severe, problems with motion sickness and perceptual-motor aftereffects—there is a clear need to understand the nature of this phenomenon and to incorporate knowledge of its stimulus characteristics and human performance implications into the design and use of VE systems.

## 6. ACKNOWLEDGMENTS

The very helpful comments of Robert S. Kennedy and Dean H. Owen on an earlier version of this chapter are gratefully acknowledged.

## 7. REFERENCES

Allison, R. S., Howard, I. P., & Zacher, J. E. (1999). Effect of field size, head motion, and rotational velocity on roll vection and illusory self-tilt in a tumbling room. *Perception, 28*, 299–306.

Andersen, G. J., & Braunstein, M. L. (1985). Induced self-motion in central vision. *Journal of Experimental Psychology: Human Perception and Performance, 11*, 122–132.

Andersen, G. J., & Dyer, B. P. (1987). *Induced roll vection from stimulation of the central visual field.* In *Proceedings of the Human Factors Conference 31st Annual Meeting* (pp. 263–265).

Benson, A. J. (1990). Sensory functions and limitations of the vestibular system. In R. Warren & A. H. Wertheim (Eds.), *Perception and control of self-motion.* Hillsdale, NJ: Lawrence Erlbaum Associates.

Berthoz, A., Lacour, M., Soechting, J. F., & Vidal, P. P. (1978). The role of vision in the control of posture during linear motion. In R. Granit & O. Pompeiano (Eds.), *Reflex control of posture and movement.* Amsterdam: Elsevier/North-Holland Biomedical Press.

Bles, W., & DeWit, G. (1976). Study of the effects of optic stimuli on standing. *Agressologie, 17*, 1–5.

Brandt, T., Dichgans, J., & Koenig, E. (1973). Differential effects of central versus peripheral vision on egocentric and exocentric motion perception. *Experimental Brain Research, 16*, 476–491.

Brandt, T., Wist, E. R., & Dichgans, J. (1975). Foreground and background in dynamic spatial orientation. *Perception and Psychophysics, 17*, 497–503.

Cheung, B. S. K., Howard, I. P., Nedzelski, J. M., & Landolt, J. P. (1989). Circularvection about earth-horizontal axes in bilateral labyrinthine-defective subjects. *Acta Otolaryngologica, 108*, 336–344.

Cress, J. D., Hettinger, L. J., Cunningham, J. A., Riccio, G. E., Haas, M. W., & McMillan, G. R. (1997). Integrating vestibular displays for VE and airborne applications. *IEEE Computer Graphics and Applications, 17* (6), 46–52.

Cutting, J. E. (1986). *Perception with an eye for motion.* Cambridge, MA: MIT Press.

Daunton, N., & Thomsen, D. (1979). Visual modulation of otolith-dependent units in cat vestibular nuclei. *Experimental Brain Research, 37*, 173–176.

Dichgans, J., & Brandt, T. (1973). Optokinetic motion sickness and pseudo-Coriolis effects induced by moving visual stimuli. *Acta Otolaryngologica, 76*, 339–348.

Dichgans, J., & Brandt, T. (1978). Visual–vestibular interaction: Effects on self-motion perception and postural control. In R. Held, H. Leibowitz, & H. L. Teuber (Eds.), *Handbook of sensory physiology: Vol. 8. Perception.* Berlin: Springer-Verlag.

Dichgans, J., Held, R., Young, L., & Brandt, T. (1972). Moving visual scenes influence the apparent direction of gravity. *Science, 178*, 1217–1219.

Diener, H. C., Wist, W. R., Dichgans, J., & Brandt, T. (1976). The spatial frequency effect on perceived velocity. *Vision Research, 16*, 169–176.

Dodge, R. (1923). Thresholds of rotation. *Journal of Experimental Psychology, 6*, 107–137.

Durlach, N. I., & Mavor, A. S. (1996). *Virtual reality: Scientific and technical challenges.* Washington, DC: National Academy Press.

Dzendolet, E. (1963). Sinusoidal electrical stimulation of the human vestibular system. *Perceptual and Motor Skills, 17*, 171–185.

Frigon, J. Y., & Delorme, A. (1992). Roll, pitch, longitudinal and yaw vection induced by optical flow in flight simulation conditions. *Perceptual and Motor Skills, 74*, 935–955.

Gibson, J. J. (1958). Visually controlled locomotion and visual orientation in animals. *British Journal of Psychology, 49*, 182–194.

Gibson, J. J. (1961). Ecological optics. *Vision Research, 1*, 253–262.

Gibson, J. J. (1979). *The ecological approach to visual perception.* Boston: Houghton-Mifflin.

Harm, D. L., Zogofros, L. M., Skinner, N. C., & Parker, D. E. (1993). Changes in compensatory eye movements associated with simulated conditions of space flight. *Aviation, Space, and Environmental Medicine, 64*, 820–826.

Held, R. M., Dichgans, J., & Bauer, J. (1975). Characteristics of moving visual scenes influencing spatial orientation. *Vision Research, 15*, 357–365.

Held, R. M., & Durlach, N. I. (1992). Telepresence. *Presence, 1*, 109–112.

Henn, V., Young, L. R., & Finley, C. (1974). Vestibular nucleus units in alert monkeys are also influenced by moving visual fields. Brain Research, *71*, 144–149.

Henn, V., Cohen, B., & Young, L. R. (1980). Visual–vestibular interaction in motion perception and the generation of nystagmus. *Neurosciences Research Progress Bulletin, 18*, 459–651.

Hettinger, L. J., Berbaum, K. S., Kennedy, R. S., Dunlap, W. P., & Nolan, M. D. (1990). Vection and simulator sickness. *Military Psychology, 2*, 171–181.

Hettinger, L. J., & Haas, M. W. (in press). Psychological issues in the design and use of virtual and adaptive environments. In L. J. Hettinger & M. W. Haas (Eds.), *Psychological issues in the design and use of virtual and adaptive environments.* Mahwah, NJ: Lawrence Erlbaum Associates.

Hettinger, L. J., & Riccio, G. E. (1992). Visually induced motion sickness in virtual environments. *Presence, 1*, 306–310.

Howard, I. P. (1982). *Human visual orientation.* London: John Wiley & Sons.

Howard, I. P. (1986a). The perception of posture, self motion, and the visual vertical. In K. R. Boff, L. Kaufman, & J. P. Thomas (Eds.), *Handbook of perception and human performance: Vol. I. Sensory processes and perception.* New York: John Wiley & Sons.

Howard, I. P. (1986b). The vestibular system. In K. R. Boff, L. Kaufman, & J. P. Thomas (Eds.), *Handbook of perception and human performance: Vol. I. Sensory processes and perception.* New York: John Wiley & Sons.

Howard, I. P. (1993). The optokinetic system. In J. A. Sharpe & H. O. Barber (Eds.), *The vestibulo-ocular reflex and vertigo.* New York: Raven Press.

Howard, I. P., Cheung, B., & Landolt, J. (1987). Influence of vection axis and body posture on visually induced self-rotation and tilt. In *Motion cues in flight simulation and simulator induced sickness* (AGARD Conference Proceedings No. 433). Neulliy-Sur-Seine, France: North Atlantic Treaty Organization, Advisory Group for Aerospace Research and Development.

Howard, I. P., & Childerson, L. (1994). The contribution of motion, the visual frame, and visual polarity to sensations of body tilt. *Perception, 23*, 753–762.

Howard, I. P., & Howard, A. (1994). Vection: The contributions of absolute and relative visual motion. *Perception, 23*, 745–751.

Hu, S., Grant, W. F., Stern, R. M., & Koch, K. L. (1991). Motion sickness severity and physiological correlates during exposures to a rotating optokinetic drum. *Aviation, Space, and Environmental Medicine, 62*, 308–314.

Hu, S., Stern, R. M., & Koch, K. L. (1991). Effects of pre-exposures to a rotating optokinetic drum on adaptation to motion sickness. *Aviation, Space, and Environmental Medicine, 62*, 53–56.

Hu, S., Stritzel, R., Chandler, A., & Stern, R. M. (1995). P6 acupressure reduces symptoms of vection-induced motion sickness. *Aviation, Space, and Environmental Medicine, 66*, 631–634.

Jokerst, M. D., Gatto, M., Fazio, R., Stern, R. M., & Koch, K. L. (1999). Slow deep breathing prevents the development of tachygastria and symptoms of motion sickness. *Aviation, Space, and Environmental Medicine, 70*, 1189–1192.

Kennedy, R. S., Drexler, J. M., Compton, D. E., Stanney, K. M., & Harm, D. L. (in press). Configural scoring of simulator sickness, cybersickness, and space adaptation syndrome: Similarities and differences. In L. J. Hettinger & M. W. Haas (Eds.) *Psychological issues in the design and use of virtual and adaptive environments.* Mahwah, NJ: Lawrence Erlbaum Associates.

Kennedy, R. S., & Fowlkes, J. E. (1992). Simulator sickness is polygenic and polysymptomatic: Implications for research. *International Journal of Aviation Psychology, 2*(1), 23–38.

Kennedy, R. S., Hettinger, L. J., Harm, D. L., Ordy, J. M., & Dunlap, W. P. (1996). Psychophysical scaling of circular vection (CV) produced by optokinetic (OKN) motion: Individual differences and effects of practice. *Journal of Vestibular Research, 6*, 331–341.

Kennedy, R. S., Hettinger, L. J., & Lilienthal, M. G. (1990). Simulator sickness. In G. H. Crampton (Ed.), *Motion and space sickness* (pp. 179–215). Boca Raton, FL: CRC Press.

Lackner, J. R. (1977). Induction of illusory self motion and nystagmus by a rotating sound field. *Aviation, Space, and Environmental Medicine, 44*, 129–131.

Lee, D. N., & Aronson, E. (1974). Visual proprioceptive control of standing in human infants. *Perception and Psychophysics, 15*, 529–532.

Lee, D. N., & Lishman, J. R. (1975). Visual proprioceptive control of stance. *Journal of Human Movement Studies, 1*, 87–95.

Lestienne, F., Soechting, J., & Berthoz, A. (1977). Postural readjustments induced by linear motion of visual scenes. *Experimental Brain Research, 28*, 363–384.

Mach, E. (1875). *Grundlinien der Lehre von der Bewegungsempfindungen.* Leipzig, Germany: Engelmann.

Mach, E. (1922). *Die Analyse der Empfindugnen.* Jena, Germany, G. Fischer.

McCauley, M. E., & Sharkey, T. J. (1992). Cybersickness: Perception of self-motion in virtual environments. *Presence, 1*, 311–318.

Melcher, G. A., & Henn, V. (1981). The latency of circular vection during different accelerations of the optokinetic stimulus. *Perception and Psychophysics, 30*, 552–556.

Melville-Jones, G. (1970). Origin, significance and amelioration of Coriolis illusions from semicircular canals: A non-mathematical appraisal. *Aerospace Medicine, 40*, 482–490.

Ohmi, M., & Howard, I. P. (1988). Effect of stationary objects on illusory forward self-motion induced by a looming display. *Perception, 17*, 5–12.

Oman, C. M., Howard, I. P., Carpenter-Smith, T., Beall, A. C., Natapoff, A., Zacher, J. E., & Jenkin, H. L. (2000). Neurolab experiments on the role of visual cues in microgravity spatial orientation. *Aviation, Space, and Environmental Medicine, 71*, 293.

Owen, D. H. (1990). Lexicon of terms for the perception and control of self-motion and orientation. In R. Warren and A. H. Wertheim (Eds.), *Perception and control of self-motion.* Hillsdale, NJ: Lawrence Erlbaum Associates.

Owen, D. H., Wolpert, L., & Warren, R. (1983). Effects of optical flow acceleration, edge acceleration, and viewing time on the perception of egospeed acceleration. In D. H. Owen (Ed.), *Optical flow and texture variables useful in detecting decelerating and accelerating self-motion* (AFHRL-TP-84-4). Williams AFB, AZ: Air Force Human Resources Laboratory. (NTIS No. AD-A148 718).

Parker, D. E. (1991). Human vestibular function and weightlessness. *The Journal of Clinical Pharmacology, 31*, 904–910.

Prothero, J. D., & Parker, D. E. (in press). A unified approach to presence and motion sickness. In L. J. Hettinger & M. W. Haas (Eds.), *Psychological issues in the design and use of virtual and adaptive environments.* Mahwah, NJ: Lawrence Erlbaum Associates.

Reason, J. T., & Brand, J. J. (1975). *Motion sickness.* London: Academic Press.

Riccio, G. E., & Stoffregen, T. A. (1991). An ecological theory of motion sickness and postural instability. *Ecological Psychology, 3*, 195–240.

Slater, M., & Steed, A. (2000). A virtual presence counter. *Presence, 9*, 413–434.

Smart, L. J., Pagulayan, R. J., & Stoffregen, T. A. (1998). Self-induced motion sickness in unperturbed stance. *Brain Research Bulletin, 47*, 449–457.

Stern, R. M., Hu, S., Vasey, M. W., & Koch, K. L. (1989). Adaptation to vection-induced symptoms of motion sickness. *Aviation, Space, and Environmental Medicine, 60*, 566–572.

Stern, R. M., Koch, K. L., Leibowitz, H. W., Lindblad, I. M., Shupert, C. L., & Stewart, W. R. (1985). Tachygastria and motion sickness. *Aviation, Space, and Environmental Medicine, 56*, 1074–1077.

Stern, R. M., Koch, K. L., Stewart, W. R., & Lindblad, I. M. (1987). Spectral analysis of tachygastria recorded during motion sickness. *Gastroenterology, 92*, 92–97.

Stoffregen, T. A. (1985). Flow structure versus retinal location in the optical control of stance. *Journal of Experimental Psychology: Human Perception and Performance, 11*, 554–565.

Stoffregen, T. A., Hettinger, L. J., Haas, M. W., Roe, M. M., & Smart, L. J. (2000). Postural instability and motion sickness in a fixed-base flight simulator. *Human Factors, 42*, 458–469.

Stoffregen, T. A., & Riccio, G. E. (1991). An ecological critique of the sensory conflict theory of motion sickness. *Ecological Psychology, 3*, 159–194.

Stoffregen, T. A., & Smart, L. J. (1998). Postural instability precedes motion sickness. *Brain Research Bulletin, 47*, 437–448.

Straube, A., & Brandt, T. (1987). Importance of the visual and vestibular cortex for self-motion perception in man (circularvection). *Human Neurobiology, 6*, 211–218.

Tschermak, A. (1931). Optischer Raumsinn. In A. Bethe, G. Bergmann, G. Emden, & A. Ellinger (Eds.), *Handbuch der normalen und pathologischen Physiologie.* Berlin: Springer.

Van der Steen, F. A. M. (1998). An earth-stationary perceived visual scene during roll and yaw motions in a flight simulator. *Journal of Vestibular Research, 8*, 411–425.

Van der Steen, F. A. M., & Brockhoff, P. T. M. (2000). Induction and impairment of saturated yaw and surge vection. *Perception and Psychophysics, 62*, 89–99.

Young, L. R., Oman, C. M., & Dichgans, J. M. (1975). Influence of head orientation on visually induced pitch and roll sensations. *Aviation, Space, and Environmental Medicine, 46*, 264–269.

Young, L. R., Shelhamer, M., & Modestino, O. (1986). MIT/Canadian vestibular experiments on the Spacelab-1 mission: 2. Visual vestibular interaction in weightlessness. *Experimental Brain Research, 64*, 299–307.

Waespe, W., & Henn, V. (1977a). Neuronal activity in the vestibular nuclei of the alert monkey during vestibular and optokinetic stimulation. *Experimental Brain Research, 27*, 523–538.

Waespe, W., & Henn, V. (1977b). Vestibular nuclei activity during optokinetic after-nystagmus (OKAN) in the alert monkey. *Experimental Brain Research, 30*, 323–330.

Waespe, W., & Henn, V. (1978). Reciprocal changes in primary and secondary optokinetic afternystagmus (OKAN) produced by repetitive optokinetic stimulation in the monkey. *Archiv fur Psychiatrie und Nervenkrankheiten, 225*, 23–30.

Warren, R. (1976). The perception of egomotion. *Journal of Experimental Psychology: Human Perception and Performance, 2*, 448–456.

Warren, R., Owen, D. H., & Hettinger, L. J. (1982). Separation of the contributions of optical flow rate and edge rate on the perception of egospeed acceleration. In D. H. Owen (Ed.), *Optical flow and texture variables useful in simulating self-motion (I)* (Interim Tech. Rep. for Grant No. AFOSH-81–0078). Columbus, OH: The Ohio State University, Department of Psychology, Aviation Psychology Laboratory. (NTIS No. AD-A117 016).

Welch, R. B. (in press). Adapting to telesystems. In L. J. Hettinger & M. W. Haas (Eds.), *Psychological issues in the design and use of virtual and adaptive environments.* Mahwah, NJ: Lawrence Erlbaum Associates.

Wong, S. C. P., & Frost, B. J. (1978). Subjective motion and acceleration induced by the movement of the observer's entire visual field. *Perception and Psychophysics, 24*, 115–120.

Wood, R. W. (1895). The "haunted swing" illusion. *The Psychological Review, 2*, 277–278.

Zahorik, P., & Jenison, R. L. (1998). Presence as being-in-the-world. *Presence, 7*, 78–89.

Zeltzer, D. (1992). Autonomy, interaction, and presence. *Presence, 1*, 127–132.

# 24

# Spatial Orientation, Wayfinding, and Representation

Rudolph P. Darken and Barry Peterson
*Naval Postgraduate School*
*Department of Computer Science*
*Monterey, California 93943-5118*
*darken@acm.org*
*peterson@cs.nps.navy.mil*

## 1. INTRODUCTION

Everyone has been disoriented at one time or another. It is an uncomfortable, unsettling feeling to be unfamiliar with your immediate surroundings and unable to determine how to correct the situation. Accordingly, one might think that the goal of navigation research in virtual environments (VEs) is to create a situation where everyone is oriented properly all the time and knows exactly where everything is and how to get there. This, however, may not be absolutely correct. Much is gained from the navigation process beyond just spatial knowledge. The path of discovery rarely lies on a known road. The experience of serendipitous discovery is an important part of human navigation and should be preserved. But how does one resolve the conflicts between this and the not-so-pleasant experience of "lostness"?

Navigation tasks are essential to any environment that demands movement over large spaces. However, navigation is rarely, if ever, the primary task. It just tends to get in the way of what you really want to do. The goal is to make the execution of navigation tasks as transparent and trivial as possible, but not to preclude the elements of exploration and discovery. Disoriented people are anxious, uncomfortable, and generally unhappy. If these conditions can be avoided, exploration and discovery can take place.

This chapter is about navigation in virtual environments—understanding how people navigate and how this affects the design of VE applications. The chapter begins with a clarification of terms and some theoretical background on navigation, primarily in the real world. A discussion of methods for navigation performance enhancement will follow. This is about how to improve performance in a virtual environment and differs from the next topic concerning the use of VEs as training tools, where the interest is in performance in the real world. The chapter concludes with a summary of principles for the design of navigable virtual environments.

### 1.1 Definition of Terms

One of the problems found in the literature is confusion over terms involving navigation. It is difficult to compare two studies that use different terms in different ways without unknowingly comparing apples to oranges. In this chapter specific terms with specific definitions are used and the research community is encouraged to adopt these.

*Wayfinding* is the cognitive element of navigation. It does not involve movement of any kind, but only the tactical and strategic parts that guide movement. As seen later in the chapter, wayfinding is not merely a planning stage that precedes motion. Wayfinding and motion are intimately tied together in a complex negotiation that is navigation. An essential part of wayfinding is the development and use of a *cognitive map,* also referred to as a *mental map.* Still poorly understood, a cognitive map is a mental representation of an environment. It has been called a "picture in the head," although there is significant evidence that it is not purely based on imagery but rather has a symbolic quality. The representation of spatial knowledge in human memory that constitutes a cognitive map will be an important part of this chapter.

*Motion* is the motoric element of navigation. A reasonable synonym for motion is *travel,* as used by Bowman et al. (1997; also see chap. 13, this volume). Durlach and Mavor (1995) subdivide this further into *passive transport,* such as a "point-and-fly" interface or other abstraction, and *active transport*, such as the Omni-Directional Treadmill (see chap. 11, this volume) or other literal motion interface (Darken, Cockayne, & Carmein, 1997). Active transport interfaces are often referred to as *locomotion interfaces. Maneuvering* is a subset of motion involving smaller movements that may not necessarily be a part of getting from "here" to "there," but rather adjusting the orientation of perspective, as in rotating the body, or sidestepping. This is an important distinction to make for the development of active transport interfaces for locomotion such as Gaiter (Templeman, Denbrook, & Sibert, 2000; also see chap. 11, this volume).

*Navigation* is the aggregate task of wayfinding and motion. It inherently must have both the cognitive element (wayfinding) and the motoric element (motion). Consequently, this term is used only when implying the aggregate task and not merely a part. The literature is replete with references to "navigation" that are only interested in novel motion techniques. This can be confusing and counterproductive to discussion and thus is avoided here.

It should also mention what is implied by *navigation performance*, as this is the metric used to determine the relative effectiveness of specific navigation tools and techniques. This is entirely dependent on the navigation task in question. Is the ability of a person to find an unknown object in a complex space being studied? If so, then search time might be an appropriate measure. Is the focus on the ability to find a known location in a complex space? Route following might then be appropriate. Or is the interest in determining a person's overall knowledge of the configuration of a space? A map drawing exercise might be appropriate. These issues will be discussed in more detail below.

### 1.2 Training Transfer or Performance Enhancement?

There are two primary classes of applications having to do with navigation in virtual environments. All VEs that simulate a large volume of space will have navigation problems of one sort or another. Typically, any space that cannot be viewed from a single vantage point will exhibit these problems as users move from one location to another. The need to maintain a concept of the space and the relative locations between objects and places is essential to navigation. This is called *spatial comprehension*, and like verbal comprehension, involves the ability to perceive, understand, remember, and recall for future use.

In applications where users tend to become disoriented or are unable to relocate previously visited points of interest, it is desirable to either redesign these applications so these problems do not appear or provide tools or mediators to help alleviate these problems. This class of application involves a need to enhance performance within the virtual environment. This distinction differentiates these applications from those where improved performance is required outside of the virtual environment, in the real world. These are training transfer applications which are discussed next.

The second class of applications that involves navigation has to do with the use of VEs as training aids for real-world navigation tasks. The fact that virtual representations of real environments can be constructed has led many to consider the use of VEs for much the same purposes that a conventional paper map might be used. While there are certainly similarities between these two classes of applications, the validation process is entirely different. If one wants to show that a visualization scheme in a virtual building walk-through, for example, can be used to lessen cases of disorientation, they need only show that users of the VE with this visualization scheme perform better on navigation tasks than users of an identical VE without the visualization scheme. However, if one wants to show that this same VE can be used as a training aid for navigation tasks in the actual building, navigation performance within the virtual environment, whereas interesting, does not prove this point. A training transfer study must be completed to show that users who trained in the VE navigate the building better than users who received some other form of training or possibly no training at all. This is an example of the use of a VE as a training aid for specific environments. Navigation performance is expected to improve in one specific real environment, and nowhere else.

Another interesting issue is the use of a VE as a training aid for general navigation tasks. If it could be shown that a VE could help people use paper maps more efficiently, or to select landmarks in an environment effectively, that performance increase would be expected to exist across physical spaces, possibly assuming some spatial similarities with that of the training environments (e.g., a VE that could be used to train a person to effectively navigate in a generic city would probably have little impact on that person's ability to navigate in a generic forest). This is beyond the scope of the current discussion but is an important use of VEs that has not yet been explored.

Whether or not a VE attempts to simulate the real world, one has to consider the fact that people are accustomed to navigation in physical spaces. Certainly, there are differences between regions of physical space that alter how navigation works, such as navigating in a forest versus navigating in a city, but there are assumptions that can be made based on past experience in real environments that are useful in any real space. While this is not a satisfactory reason to blindly copy the real world in every way possible, it is important to learn everything possible about how people relate to the physical world so designers can understand how to build better virtual environments.

## 1.3 Spatial Knowledge Acquisition

There are many ways to acquire spatial knowledge of any environment. The fundamental distinction between sources of spatial knowledge is whether the information comes directly from the environment (primary) or from some other source (secondary), such as a map. An issue specific to secondary sources has to do with whether or not the source is used inside or outside of the actual environment.

### *1.3.1 Direct Environmental Exposure*

When you navigate in an environment, you extract information both for use on whatever navigation task currently is being executed and for any subsequent navigation task. Exactly what information is useful for navigation? A person can't possibly attend to every stimulus and make use of it. Much of it is irrelevant or at least of lesser importance. If designers knew what information was the most important, this could be useful in designing virtual environments. Designers would know what to put where to help people find their way around. While there is no clear answer to this question, Kevin Lynch presents the most compelling, environment independent answer to date (Lynch, 1960).

In studying urban environments (Lynch was an urban planner), he found that there are certain similarities across cities. There are in fact "elements" of urban environments, or building blocks, that can be used to construct or decompose any city. He starts with landmarks, which are the most salient cues in any environment. They are also directional, meaning that a particular building might be a landmark from one side but not another. Then there are routes or paths that connect the landmarks. They don't necessarily connect them directly, but they move you through the city such that the spatial relationships between landmarks become known. A route isn't necessarily a road. It could also be a bike path or railroad. Cities tend to have complex road and rail structures. Interchanges or junctions between routes are called *nodes*. These are important because they are fundamental to the structure of the routes. This structure must be understood before proficient navigation can take place. Most cities have specific regions that are explicitly or implicitly separated from the rest of the city. These are *districts*. Landmarks and nodes typically are found within districts. Routes pass through districts and connect them. Finally, regions of the city, and in fact the city itself, are bounded by *edges*. Edges prevent or deter travel. A typical edge is a river or lake.

An interesting fact is that classifying some city object as one element or another does not preclude it being classified as something else in another context. To a pedestrian, a walking path is a route, whereas highways and railroads are edges. To a driver, the roads are routes and everything else is an edge. Classification depends on the mode of travel. Furthermore, the mode of travel also affects what gets encoded, not just how it gets encoded (Goldin & Thorndyke, 1982).

### *1.3.2 Map Usage*

There are a number of secondary sources that have been used for spatial knowledge acquisition. These include maps, photographs, videotape, verbal directions, and recently virtual environments. The most common of these is the map. For any map used in any environment, virtual or real, designers need to know when the map is used, or more appropriately, what tasks the map is to be used for. The critical issue is in whether or not the map is to be used preceding navigation or concurrent with navigation. This is important because maps that are used concurrently with navigation involve the placement of oneself on the map. The first part of any task of this nature is: Where am I? What direction am I facing? A transformation is required from the egocentric perspective to the geocentric perspective. If the map is used as a precursor to navigation, it is used only for planning and familiarization. No perspective transformation is required. Planning a trip is one example of such a geocentric navigation task. The planning is done outside of the environment so there is no perspective transformation needed. However, when such a transformation is required, as in using a map during navigation, the problem is more complicated.

When a perspective transformation must be performed, the rotation of the map can have a great effect on performance. Aretz and Wickens (1992) showed that maps used during navigation tasks (egocentric tasks) should be oriented "forward-up" (the top of the map shows the environment in front of the viewer), whereas maps used for planning or other geocentric tasks should be "north-up." Rossano and Warren (1989) reinforced this concept by demonstrating that judgments of direction are adversely affected by misaligned maps. Levine, Marchon, and Hanley (1984) go on to provide basic principles of map presentation, including aligning the map with the environment, always showing two concurrent points on the map so the viewer can triangulate position, and avoiding symmetry. Péruch, Pailhous, and Deutsch (1986) showed that redundancy of information is crucial to resolve conflicts between different frames of reference.

The key to map use for navigation is resolving the egocentric to geocentric perspective transformation. This certainly isn't the entire problem, but it is the biggest part of it. This involves the ability to perform a mental rotation. The easier this rotation is, the easier the task. Unfortunately, mental rotation is not a level playing field. Some of us are better at it than others, so much so that it affects the way individuals perform navigation tasks (McGee, 1979; Thorndyke & Stasz, 1980). Everyone knows individuals who maintain a high level of proficiency in navigating environments, including environments that they have never been in before. Alternatively, there are individuals who will get lost on the way home from work if they can't follow their usual route for some reason.

One way to simplify the transformation is to show the viewpoint position and orientation directly on a map. Until recent improvements in global positioning system (GPS) technologies, this was not possible to do in the real world. As of May 1, 2000, the United States no longer jams GPS signals for private use, so GPS is now 10 times more accurate than it was previously. While "dithered" GPS could only place a receiver in an area the size of a football field, it is now accurate to within the area of a tennis court. However, there are no commercially available databases available that make use of spatial data at this resolution. Even if there were, it is still not known what the best uses are for this data in the real world, so it will be some time before a payoff is realized from the declassification of GPS. Automobile navigation systems with moving maps are already commonplace. However, the interfaces to these systems are generally poor. One of the issues looked at later is the use of virtual maps and how they should be presented.

### 1.4 Representations of Spatial Knowledge

Arguably the most important part of this puzzle is the part understood the least. Once spatial knowledge is acquired, how is it organized in the mind for future use? Spatial knowledge must be organized in some way such that it can be used during navigation tasks. The term "cognitive map" was first used by Tolman (1948) to describe a mental representation of spatial information used for navigation. However, 50 years later, there are still no hard answers about the structure of spatial knowledge.

The representation of spatial knowledge is affected by the method used to acquire it. Knowledge acquired from direct navigation is different from knowledge acquired from maps. After studying a map of San Francisco before an initial visit there, information gleaned from the map is in a north-up orientation, just like the map. Consequently, when entering the city from the north, the mental representation of the city that has been acquired is upside down from what is being seen. This requires a 180-degree mental rotation before the cognitive map can be in line with the real environment. Spatial knowledge acquired from maps tends to be orientation specific (McNamara, Ratcliff & McKoon, 1984; Presson, DeLange & Hazelrigg, 1989; Presson & Hazelrigg, 1984). This implies that accessing information from a misaligned cognitive map is more difficult and error prone than if it were aligned (Boer, 1991; Rieser, 1989).

The longest standing model of spatial knowledge representation is the Landmark, Route, Survey (or LRS) model described by Siegel and White (1975) and Thorndyke and Goldin (1983). This model not only addresses spatial knowledge but also the development process. The theory states that first landmarks are extracted from an environment. These are salient cues but are static, orientation dependent, and disconnected from one another. Landmark knowledge is like viewing a series of photographs. Later, route knowledge develops as landmarks are connected by paths. These need not be optimal paths. At this point, an individual may know how to get from A to B and from B to C but probably won't know a direct route from A to C. Route knowledge can be thought of as a graph of nodes and edges that is constantly growing as more nodes and edges are added. Finally, survey knowledge (or configurational knowledge) develops as the graph becomes complete. At this point, even if an individual has not traversed every path through their environment, they can generate a path on the fly because they have the ability to estimate relative distances and directions between any two points. This model directly fits the elements of urban environments described by Lynch (1960).

The most important caveat to this development process has to do with the use of maps. Maps allow jumping over the route knowledge level and proceeding directly to survey knowledge because they afford a picture of the completed graph all at one time. However, survey knowledge attained from maps is inferior to survey knowledge developed from route knowledge and direct navigation because of the orientation specificity issue described earlier.

A second model of spatial knowledge is similar to the LRS model, but it is hierarchical (Colle & Reid, 1998; Stevens & Coupe, 1978). In some cases, direct exposure to an environment for extremely long durations never results in survey knowledge (Chase, 1983). In other cases, survey knowledge develops almost immediately. The model proposed by Colle and Reid suggests a dual-mode whereby survey knowledge can be acquired quickly for local regions and slowly for remote regions. The "room effect" comes from the ability to develop survey knowledge of a room rather quickly but survey knowledge of a series of rooms develops relatively slowly and with more errors.

### 1.5 Models of Navigation

Understanding how navigation tasks are constructed is useful in determining how best to improve performance. If it were possible to decompose navigation tasks in a general way, one might be able to determine where assistance is needed or where training can occur. Several attempts have been made at such a model (Chen & Stanney, 2000; Darken, 1996; Downs & Stea, 1977; Neisser, 1976; Passini, 1984; Spence, 1998), but most are either too specific to one type of environment or another or they do not capture the intricacies of the entire task.

The model proposed by Jul and Furnas (1997) is relatively complete in that it incorporates the motion component into the process in a way not attempted before (see Fig. 24.1). The model works as follows. I am at the shopping mall and decide I need a pair of shoes. I have just formulated a goal. Now, how should I go about finding shoes? I decide to try the department stores. Department stores are typically on the far points of the mall. I have just formulated a strategy. The next step is to gather information so I don't walk off in a random direction. I decide to seek out a map of the mall. I am acquiring information and scanning (perceiving) my environment. This is the tight wayfinding/motion loop referred too earlier. I view my surroundings, assess my progress toward my goal, and make judgments as to how to guide my movement. At any time in this loop, I may decide to stop looking for shoes and look for books instead. This is a change in goal. I could also decide to look for a small shoe store instead of a department store. This is a change of strategy. In any case, the task continues, shifting focus and process as necessary.

An important point is that navigation is a situated action (Suchman, 1987). Planning and task execution are not serial events but rather are intertwined in the context of the situation. It is neither possible nor practical to consider the task, the environment, and the navigator as separate from each other. Observable and measurable behavior is a product of these factors, yet the relationships between them are, as yet, poorly understood.

In the real world, this process is performed so often that it is typically automatic. When one knows where they want to go, they go there. When they don't know where to go, they ask someone or look for some other source of assistance. Virtual environments are entirely different.

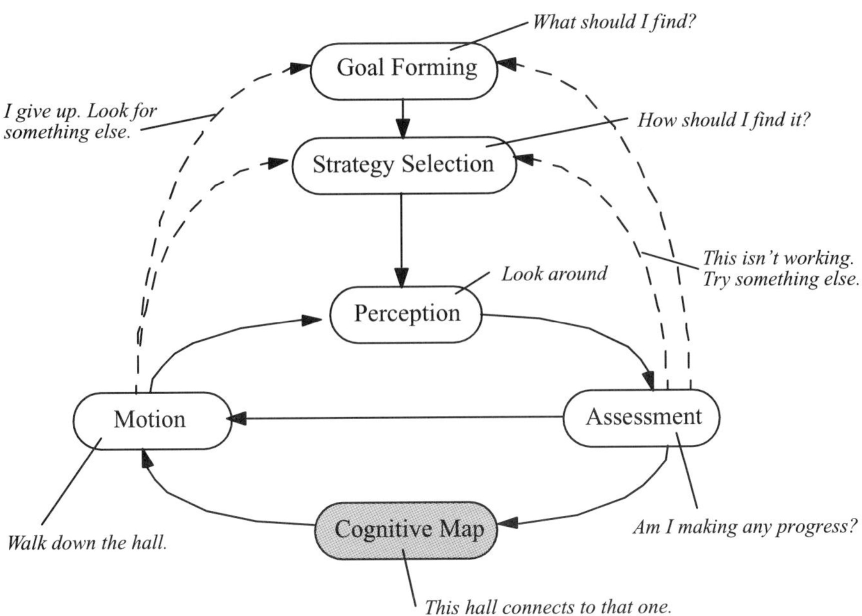


FIG. 24.1. A model of navigation adapted from Jul and Furnas (1997).

Movement is not typically so easy. It requires thought. Knowing where to go is problematic. Who can be asked for help? How can one get unlost? These are the issues discussed next.

## 2. NAVIGATION PERFORMANCE ENHANCEMENT

The ease with which one can navigation depends on the environment traversed. For example, many of the physical environments that the navy must operate in are inherently sparse and consequently difficult to navigate. Literal virtual representations of these environments are even harder to navigate considering that the number of cues available is diminished even further. Early prototypes provided valuable lessons in how difficult navigation can be. Whereas open-ocean naval applications might be on the more difficult end of the complexity scale, problems can arise in any application. This section will look at ways to help people navigate a virtual space without regard to whether or not their improved performance transfers outside of the virtual environment.

The most obvious way to address navigation problems in complex VEs is to provide some sort of tool or mediator that can be used directly on the task at hand. This, of course, is the history of the map, the compass, the sextant, and the chronometer, to name a few real-world tools. Used together, these tools are able to help their users determine their position in the environment, their direction of travel, and the relative position of other objects or places in the environment.

An alternative to mediators is the actual organization of the space itself. There are a few real-world vocations related to this including architectural design and urban planning, each of which has extensively studied the relationship between people and their environment. They are interested in much more than just navigation, of course, but there is much to learn from these disciplines about how to construct space in a meaningful way in which people can comprehend and operate effectively. In VEs where the contents of the environment cannot be changed, such as in a flight simulator, there is not much that can be done in the way of spatial organization.

However, in other applications, such as many scientific visualizations, it is possible to organize the data expressly for the purpose of navigability.

## 2.1 Navigation Tools and Mediators

This section discusses a number of navigation tools and mediators that have been investigated in our laboratory. Although they are similar to others in the literature, it should be noted that our intent is to investigate principles for the design of navigable virtual environments, not merely new techniques. It is unlikely that one of these tools as described here will perfectly match the needs of a real application, but by explaining how certain types of information affect behavior and its related performance, designers can mix and match the techniques described here and elsewhere to construct custom built tools specifically for the demands of their application.

### *2.1.1 Maps*

The navigation mediator people are most familiar with is the map. Maps are extremely powerful tools for navigation because of the wealth of information they can provide and the rate at which people can digest this information. However, their use in any VE application is not to be taken lightly. There are right ways and wrong ways to use maps. Maps come in a variety of forms, the differences usually being in terms of symbology or projection. However, VEs have certain qualities the real world does not have that make the use of maps in VEs different.

It is possible to navigate directly on a map in a virtual environment. Rather than use the map to determine where to go in the virtual environment, why not just point on the map to where you want to go? This has been attempted several times. The Worlds-in-Miniature (Stoakley, Conway, & Pausch, 1995) metaphor was one such implementation. In this case, a scaled-down version of the world, a virtual map, was held in the hand. Movement could be specified directly on the virtual map. This same behavior is possible in some video games. DOOM (Id Software, Inc.) allows movement to take place while viewing a map. The environment cannot be seen during this interaction. Similarly, the use of maps can be a moded or unmoded task in a virtual environment, meaning that its use can be in lieu of, or concurrent with, motion. In the real world, it is typically unmoded. Under certain conditions one might want to stop moving to read a map, but it is not required. Some games and VEs mode maps so that map use precludes motion.

For maps of very large virtual environments, there is a scaling problem. How do I view the map such that I can see the detail I need to navigate but still maintain a sense of the overall space? This is a classic problem of navigation in any problem domain (e.g., Donelson, 1978; Furnas, 1986). There are ways to zoom into a map or otherwise scale it to a usable level. One advantage of VEs is that often the problem only concerns maintaining one perspective—your own. In applications where I might need to not only know where I am but also where other people are as well, the problem gets considerably harder. Street Atlas USA (DeLorme, Inc.) is an excellent example of an application addressing this issue. As one zooms into the atlas to the street level, context is lost. Consequently, there is always an overview window available.

The last issue discussed about maps, and the primary focus of this section, has to do with the orientation of the map. As discussed earlier, the orientation of a map with respect to the viewer has a strong effect on the viewer's ability to perform the mental rotation required to use a map during navigation. However, if it is already known that forward-up maps are best for egocentric tasks and north-up maps are best for geocentric tasks, isn't this a moot point? The fact that most video games continue to use north-up maps is enough to warrant investigation.

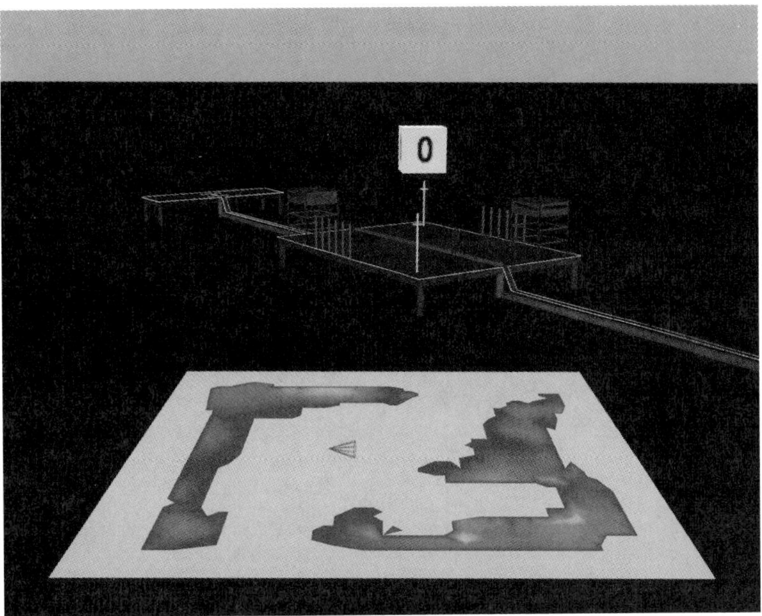


FIG. 24.2. The north-up map configuration has a directional cone for the you-are-here indicator. The top of the map always stays at the top. It is aligned with the viewer, not with the environment.

Is it possible to put enough redundant information on a north-up map to make it equivalent to a forward-up map?

To investigate this issue, we created two very large virtual environments. One was a sparse open-ocean environment and the other was an urban environment (Darken & Cevik, 1999). Both a head-mounted display and a Fakespace PUSH display were used with similar results. The results reported here are using the PUSH display. Participants were asked to locate objects in the spaces. Sometimes they were shown the targets on the map (a targeted search), sometimes they had to return to known targets that were not shown on the map (a primed search), and other times they had to locate a target not shown on the map and not seen before (a naive search). Performance was timed, and wrong turns where participants clearly moved away from a target rather than toward it were marked. At the conclusion of each trial, participants were asked to sketch a map of the environment from memory.

We created two virtual maps; one was in a north-up configuration (see Fig. 24.2), the other was in a forward-up configuration (see Fig. 24.3). Both maps had a "you-are-here" marker that dynamically moved across the map as the user moved through the environment. The only difference on the maps, besides their orientation, was that the you-are-here indicator on the north-up map was a cone, whereas the you-are-here indicator on the forward-up map was a small sphere. This was necessary because a north-up map does not indicate direction implicitly while the forward-up map does.

What we found was comparable to the results of Aretz and Wickens (1992) but to a lesser degree. The forward-up configuration seems to indeed be best for egocentric navigation tasks, whereas the north-up map is best for geocentric tasks. The results also indicated that individuals with high spatial abilities (as measured with the Guilford-Zimmerman standardized tests) are able to use either type of map better than participants with low spatial abilities on similar tasks (see also chap. 50, this volume). Furthermore, these principles were found to apply across different types of environments with vastly different spatial characteristics, but sparse

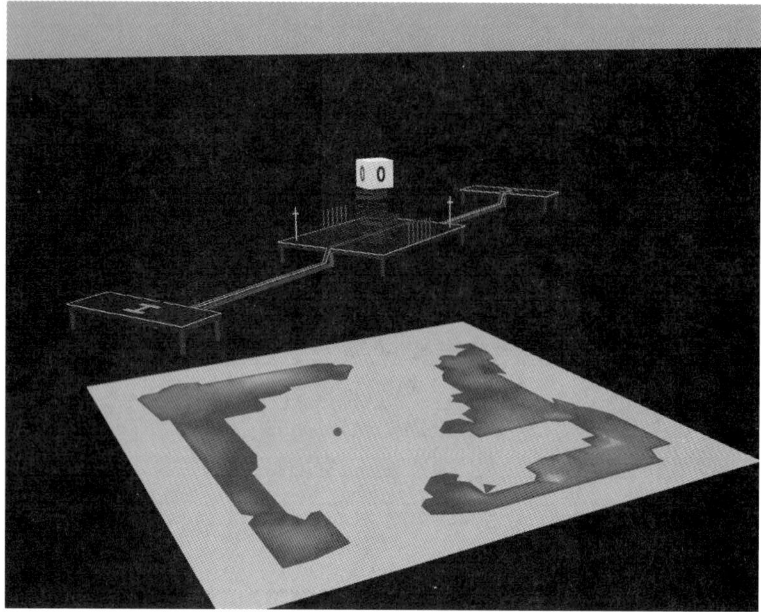

FIG. 24.3. The forward-up map configuration has a small sphere in place of the cone for a you-are-here indicator. This map rotates with respect to the viewer as the viewpoint is rotated in the environment. It is aligned to the environment, not the viewer.

environments seem to exhibit less of a performance difference than dense environments. Virtual environment designers should make virtual map decisions by carefully weighing the priorities of navigation tasks versus the spatial ability of their users.

Interestingly, in asking participant's subjective assessment, most preferred the north-up configuration even if their performance did not correlate. In postevaluation debriefing, most of these people reported maintaining a strong bias based on video game play. Regardless of how they perform, they like what they are comfortable with. This complicates the problem. This study seemed to indicate that while the forward-up map was better for egocentric tasks, it was not so much better that we would advise ignoring the preferences of users. It will be imperative in the future to examine the effects of exposure to video games on performance of navigation tasks in virtual environments.

### 2.1.2 Landmarks

Landmarks are extremely important to spatial knowledge acquisition and representation. As such, it might be useful to allow users to affect the placement of landmarks themselves. If landmarks are useful as "anchors" on which to relatively place other objects, what would happen if the locations of the objects remained constant but users were allowed to insert highly salient landmarks on which to anchor their locations?

Using the open-ocean environment as before, participants were given a set of ten different colored markers to place on the surface of the environment at any position. These markers were visible from a far greater distance than the objects themselves. Where might the markers be placed and how would this might affect performance?

Markers tended to be placed in-between objects such that as one marker would disappear from view, another would appear on the horizon. Moving between markers and objects, the participant could always have something to "hold on to" much like a "handrail" is used in

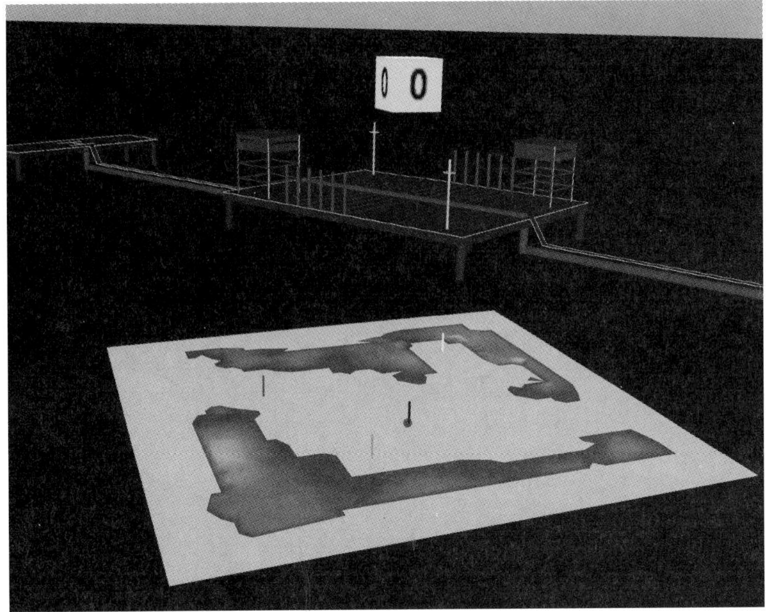


FIG. 24.4. This map shows a user returning from finding a number of objects. The markers are used like pushpins to show where targets were located. Before leaving the virtual world and beginning the map drawing task, the participant would simply remember the configuration of markers on the map.

sport orienteering. (Note: A *handrail* is a linear feature, such as a stream or road, that land navigators use to guide movement. They typically use it to constrain movement, keeping it to one side and traveling along it for some distance.) For most individuals, being in a void, even a partial void like sometimes occurred in this environment, is very uncomfortable. Most people need regular reassurance that they are not lost. Only the most advanced navigators do not seem to need this kind of assurance.

The same condition was studied in combination with a forward-up map (see Fig. 24.4). As participants would place a marker, it would appear both in the environment and on the map. Would the strategy remain the same or would the map override the utility of the markers?

This turned out to be unquestioningly the preferred condition of every approach explored to date. The reason is simple. Participants still were not given any help in locating target objects. As they found them, however, they no longer had to remember anything about them. They simply placed a marker at each object. The markers in the environment were not used at all, but the markers on the map were like colored pushpins. They simply had to remember what color coincided with the corresponding target, making the posttrial map drawing exercise trivial.

Certainly one might devise other techniques that may fare better than this condition, but it is our intent to study methods that can generalize to many navigation tasks in any type of environment rather than only to specific tasks and environments. The objective is to determine design principles so that application designers have a starting point from which to develop specific mediators for their specific tasks in their specific environments.

#### *2.1.3 Trails*

Since participants used the markers to create a sort of "trail" connecting objects with markers and other objects, it brought to mind an old idea. If a trail was left behind such that it specified

FIG. 24.5. In this example, the participant has moved backward and is looking toward the start object location. The footprints show the direction the user was looking at that point in time.

not only where the participant had been but what direction they were going at that time, this might be an even better tool than the markers (see Fig. 24.5). This was called the "Hansel and Gretel" technique (Darken & Sibert, 1993). A better analogy is that of footprints since footprints are directional and breadcrumbs are not.

This technique is useful for scanning space in an exhaustive search (e.g., a naive search). One of the problems in an exhaustive search is knowing if you have been in some place before. An optimal exhaustive search never revisits the same place twice, assuming the targets are not moving (ours weren't). However, it might be useful to retrace steps if disorientation occurred. This was not, however, found to be the case. As navigation proceeds, the environment becomes cluttered with footprints, so much so that the directional component is ignored. Also, when the user crosses paths, how does the user disambiguate which ingress route goes with which egress route?

As with the markers, this was tried with a forward-up map (see Fig. 24.6). Similarly, it was found that participants ignored the footprints in the environment, only using the trail left on the map to help direct the search.

### *2.1.4 Direction Finding*

So far, a number of tools and mediators have been discussed that deal with absolute position information. Maps show exactly where the user is at all times. Markers specify an exact location in an environment. Each footprint designates an exact location as well. What about orientation? How important is it to know absolute direction versus absolute position?

To study this, two simple tools were derived, both of which show nothing more than direction. Even though our environments did not have a "north," it was decided to make the initial view direction be north. It did not matter what was designated as north as long as it was done consistently. The compass in Fig. 24.7 always points to virtual north; it floats out in front of the viewpoint similarly to the maps discussed earlier.

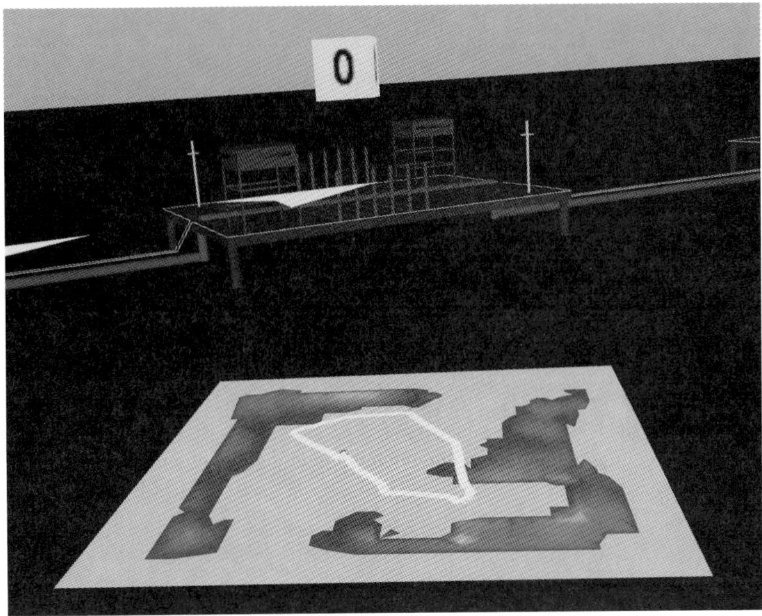


FIG. 24.6. This participant has made a cycle through the environment. The actual positions of the target objects are not shown on the map. The participant must remember where along the path they were located.

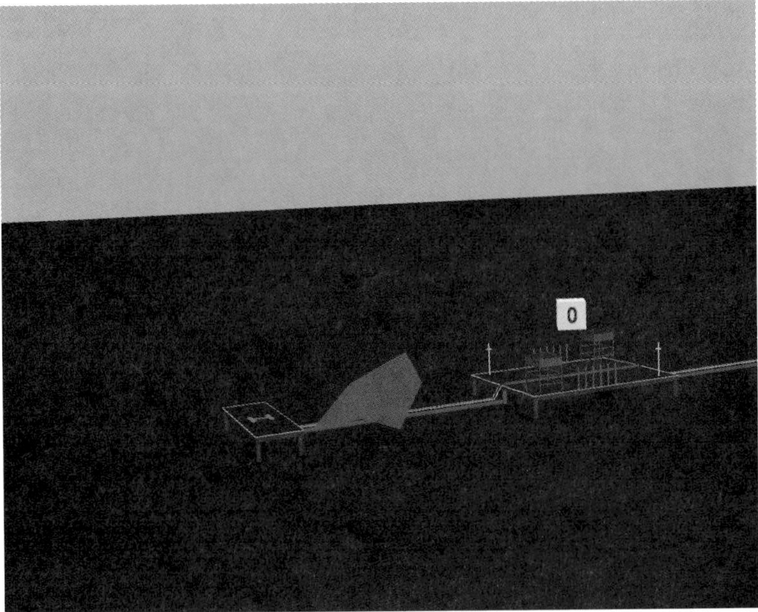


FIG. 24.7. The compass simply shows the direction that the participant was facing at the onset of the trial.

FIG. 24.8. The sun is positioned low on the horizon so directional information will be clear and unambiguous.

The second tool implemented was a virtual sun. This was placed on the horizon such that it would identify one direction only (see Fig. 24.8). It was up to the participant to call it east or west. This never seemed problematic.

The task was the same as it was for every other condition. The results indicated that the compass was preferable to the sun because it was in view all the time, versus the sun that required the participant to turn toward it. However, performance using both of these tools was very low as compared to the other conditions studied. In fact, performance on both of these conditions was not significantly better than the control condition where no assistance was given whatsoever. The reason for this is that while maintaining spatial orientation is essential to all navigation tasks, orientation without position information is not useful. Similarly, position information without orientation information is equally limited in utility as shown by the "coordinate feedback" tool described by Darken and Sibert (1993).

## 2.2 Organizational Remedies

It could be said that the tools and mediators described in the previous section should be the last recourse of the designer if all else fails. That is, if the constraints of your application are such that you simply cannot develop a navigable VE that does not need tools or mediators for usability, then you need to determine what tools are most appropriate to allow any user to navigate effectively. However, there are many applications where it is not appropriate or even possible to organize a space because the positions of objects in the space are constrained in some way. For example, if a VE driving simulation of metropolitan Los Angeles is being constructed, designers have no control over the positions and orientations of buildings and streets. If, alternatively, a VE visualization of the stock market is being constructed, designers would have control over how objects looked, where they were located, and how they were organized.

This section will discuss two primary methods for organizing spaces for navigability. The first relies heavily on Lynch's "elements" of urban landscape where the environment is implicitly organized (Lynch, 1960). The other involves a more explicit organizational technique where a specific pattern is forced on the space itself.

### *2.2.1 Implicit Sectioning*

Passini (1984) talks about the use of an "organizational principle" in architectural design. If a space has an understandable structure and that structure can be made known to the navigator, it will have a great influence on the strategies employed and resulting performance on navigation tasks. For example, knowing that Manhattan is generally a rectangular grid is of great benefit to any navigator. Given the knowledge of the grid and its orientation, words like "uptown" and "downtown" instantly have meaning. But while there is great power in using such an organizational principle, there is a danger that goes along with it. Violations to that principle will have a much greater negative effect than they might otherwise have. For example, Manhattan is generally a grid, but Broadway cuts through on an angle, thus violating the grid principle. A naive tourist thinking that the grid principle held throughout will probably be misled at some point. It is important to develop a clear organizational principle and stick to it throughout the environment. If it must be violated, it should be made clear where the violation occurs and that it is a violation; otherwise, the navigator may attempt to erroneously fit it into a cognitive representation.

Organizational principles can also add meaning to cues that might be seen during navigation. Again using the Manhattan example, if a tourist is looking for $57^{th}$ St., street signs for $44^{th}$ St., $45^{th}$ St., and so forth not only tell which is the correct direction but also give a rough estimate of distance as well.

There have been attempts to apply Lynch's elements of the urban landscape to virtual environments (Darken, 1996; Ingram & Benford, 1995). In principle, this will work, but implementations can seem contrived as if an inappropriate structure is being forced on abstract data. It may be that the very generic nature of VEs might demand a different set of elements similar to Lynch's but not necessarily identical.

Darken and Sibert (1996) commented that people inherently dislike a lack of structure. A person who is in an environment that is nearly void of useful cues and that does not suggest ways to move through it is generally uncomfortable. Users will grasp at anything they can view as structure whether or not the designer intended it to be that way. In our earlier experiments, participants were observed using coastlines and the edges of the world as paths even though this was not a particularly effective strategy toward completion of tasks. People adopted these cues because they were all they had to work with. If even a simple bit of structure is added such as a rectangular or radial grid (described in the next section), performance immediately improves. A path should suggest to the navigator that it leads to somewhere interesting and useful.

It is common to see urban metaphors applied to unstructured environments such as Web sites because the structure of a city can add meaning to information that may otherwise be viewed as amorphous. This works in many cases even when there is no obvious semantic connection between the information presented and the city metaphor. In other cases, this fails because the constraints implied by an urban landscape do not coincide with the constraints of the information presented. This should be used judiciously. Using a metaphor to simplify navigation may unknowingly inhibit some other piece of functionality. Therefore, it is important when studying user performance in these types of applications to study the whole task, not just the navigation component. Again, navigation is not an isolated action but is situated in some other higher order task.

### 2.2.2 Explicit Sectioning

In some cases, like our open-ocean environment, there are simply too few objects in the environment to organize into a navigable environment. In these cases, the space can be explicitly organized using explicit cues. We experimented with two sectioning schemes based on Lynch's ideas of "districts" and "edges" and also using a simple organizational principle as described by Passini.

The first scheme tried was a radial grid. A radial grid was placed over the open ocean environment. There were highly salient landmarks in each cardinal direction and a single landmark in the center. The second scheme used was basically the same idea but was a rectangular grid rather than radial. Again, highly salient landmarks were placed in the cardinal directions and the center.

The results indicated that search strategies for these two schemes were quite different but that performance between the two was not significantly different. The radial grid tended to elicit strategies having to do with "pie slices" or similar ideas (i.e., strategies within the radial grid condition tended to focus on the center landmark, moving through each "pie slice" sequentially until targets were found), whereas the rectangular grid produced more "back and forth" motions within bounded regions. The nature of the task or environment would probably dictate which was better in a particular situation. In our case, the difficulty of the task without any aids caused performance to increase for both grids. Observed strategies were clearly related to the type of grid presented. In a given situation, if a particular strategy is clearly preferable over alternatives, structure such as this can be used to guide the user toward the preferred method. As observed in our experiments, there will almost always be room for unexpected behavior. It is possible, given a well-defined task, to design in a desired tendency.

We then experimented with combining the grid overlays with the map tools previously described (see Fig. 24.9). This was found to be highly effective, particularly with the forward-up map because the map provided directional information and survey knowledge, whereas the

FIG. 24.9. The radial grid superimposed on a forward-up map. There is a wealth of information in this environment to help in navigation.

grid provided directional cues and landmark knowledge. Some participants tended to focus more on the map than the grid or vice versa but, in general, the two were used in combination as intended.

## 3. ENVIRONMENTAL FAMILIARIZATION

Imagine that you could visit places using a VE before you actually got there. If it were a vacation spot or resort, you might want to see if you like the views, beaches, or anything else that might catch your interest. If you were planning to drive into a city you have never been to before, you might plan a route using a conventional map and then drive your route in the VE to become familiar with distances and key landmarks. If you were a soldier about to enter a hostile environment, you might use a VE to rehearse a planned route and familiarize yourself with the area.

In each of these examples, the VE in question is not a mere abstraction but is a representation of a real environment. In each case, the interest in not in how well the user navigates the VE but rather in how they navigate the real world after exposure to the virtual environment. This section will examine several empirical studies involving the use of VEs for environmental familiarization. Studies of this kind are central to the topic of this chapter for two primary reasons:

1. Such investigations highlight the differences between navigation in the real world and in virtual environments, whereas studying VE navigation alone does not. Studying behavioral differences between virtual and real world navigation may provide insights to help us better understand and model human navigation, virtual or real. People know how to navigate in the real world. They bring this knowledge with them into virtual environments. While this does not mean that VE navigation must replicate real world navigation in every way, it makes sense that one needs to understand how humans navigate in the real world before they can optimize methods for navigation in the virtual environment.

2. Although the potential use of VEs to enhance training for real-world performance of many tasks has been touted from the inception of the technology, few application domains have clearly demonstrated a significant enhancement. Environmental familiarization (or mission rehearsal) is among these applications. In fact, a close inspection of the literature will lead to the conclusion that there is much confusion over whether or not VEs offer a significant enhancement over traditional methods of spatial knowledge acquisition. Furthermore, even if they were known to be useful, it is still not known exactly how they should be used to optimize the positive effects desired while minimizing negative effects (e.g., reverse training). Transfer of spatial information might be a near-term training domain that is within the reach of current technology. Environmental familiarization may represent a microcosm of the issues involved in virtual training of any knowledge domain.

A key aspect of the studies presented in this section is that performance measures are made in the real world to evaluate how much spatial information was acquired in the VE or how the VE may have affected behavior. This is not to say that measurements are never made in the virtual environment, only that without real-world measurements, one cannot know what affect, positive or negative, the VE tool may have had on participants.

### 3.1 Spatial Knowledge Transfer Studies

To ground this investigation, four actual studies are compared. There are many more studies of this type in the literature but a subset have been selected for presentation here. Each of these is a transfer study implying that there is a training phase involving a VE of some type and a testing phase involving transfer to the real environment. The focus here is only on studies about environmental familiarization rather than skill development so the VE must replicate a specific real environment rather than some generic real environment such as learning to navigate in a generic city.

The purpose of this comparison is to systematically point out the similarities and differences in the studies so that some statement can be made about what is currently known about the use of VEs for environmental familiarization. An issue that will become very clear is that there is little consistency in the literature about what to study or how to study it. Consequently, one sees a variety of experiments controlling a variety of parameters but in a way that it is difficult, if not impossible, to leverage what was done previously.

Four experiments will be briefly discussed, concluding with a discussion of environmental familiarization and how VEs might be used for these tasks. The experiments are:

- Witmer, Bailey, and Knerr (1996), *Virtual Spaces and Real World Places: Transfer of Route Knowledge*
- Darken and Banker (1998), *Navigating in Natural Environments: A Virtual Environment Training Transfer Study*
- Koh, von Wiegand, Garnett, Durlach, and Shinn-Cunningham (2000), *Use of Virtual Environments for Acquiring Configurational Knowledge About Specific Real-World Spaces: Preliminary Experiment*
- Waller, Hunt, and Knapp (1998), *The Transfer of Spatial Knowledge in Virtual Environment Training*

## 3.2 A Basis for Comparison

In an attempt to make a meaningful comparison between these experiments, each is looked at in terms of a structured set of parameters. The key elements for this comparison are: (1) the characteristics of the human participants; (2) the characteristics of the environment; (3) the characteristics of the tasks to be performed; (4) the characteristics of the human–computer interface; and (5) the characteristics of the experimental design to include dependent measures. Table 24.1 is a summary of the four experiments in terms of these elements.

Each experiment investigated human navigation performance, and the characteristics of the participant sample must be taken into account. Traditionally, important issues such as the quantity, age, and gender of the participants are considered. However, navigation is a specialized task, so two specific differences may be quite critical. First, some of the experiments situated the navigation task within a higher level task context. In those cases, participants' experience in the respective domain could be expected to influence both motivation to participate and task competence level. Second, individuals enter the experiment with a given level of innate spatial ability (see chap. 50, this volume). Although measures of individual spatial and navigation ability may provide ambiguous results, attempts to quantify and categorize participants based upon individual ability can help to explain differences in performance.

The real-world environment that is modeled in the VE influences navigation behavior. Just as individuals possess differing navigation ability, so do various environments afford different navigation experiences. Some real-world environments simply provide more navigation cues than others do. Furthermore, some environments lend themselves to a higher model fidelity level than others. How closely a VE matches its real counterpart is referred to as "environmental fidelity" (Waller et al., 1998). However, do not assume that higher environmental fidelity must correlate with higher performance. There are other issues that are equally important.

With so many varieties of VEs and their associated interfaces, it is necessary to be more descriptive in terms of the specific differences between them. The devices and interaction styles

**TABLE 24.1**
A Comparison of Training Transfer Studies

| *Experiment* | *Participants* | *Environment* | *Tasks* | *Interface* | *Measures* |
|---|---|---|---|---|---|
| Witmer, Bailey, & Knerr | No domain expertise | Architectural, structured | No maps used, 15-minute repeated exposure, route replication | Immersive display, buttons | Wrong turns, bearing/range estimation, time |
| Darken & Banker | Domain expertise | Natural, unstructured | Maps used, 60-minute exposure, route planning and execution | Keyboard, mouse, desktop display | Deviation from route |
| Koh, Durlach, & von Wiegand | No domain expertise | Architectural, structured | No maps used, 10-minute exposure, survey knowledge | Both immersive and desktop displays, joystick | Bearing/range estimation |
| Waller, Hunt, & Knapp | No domain expertise | Structured | No maps used, 2 or 5 minutes exposure, route replication, path integration | Both immersive and desktop displays, joystick | Time, bumps into walls |

used by the system provide differing levels of sensory stimulation to the user. How closely a VE interface matches the interface to the real world (e.g., walking, driving, etc.) is referred to as "interface fidelity" (Waller et al., 1998). Again, do not assume that higher interface fidelity equates to higher performance or training transfer. This has not yet been established, and it is unclear if this will eventually be found to be the case.

Desktop VEs channel visual output to a computer monitor that rests on the desktop (see chap. 12, this volume). Immersive VEs use a head-mounted unit or projection system to display the world to the user. Still, within both desktop and immersive categories, the mix of input and output devices requires a more detailed description. The power of the system to deliver high-fidelity multimodal output and monitor user input commands in real-time is a critical discriminator. The primary issue with interface fidelity in the studies concerned with here has to do with the motion technique, specifically, what interaction method is used to control speed and direction of travel (see chap. 11 and 13, this volume). Finally, some interfaces provide the user with special abilities and computer-generated tools that further differentiate one from another.

Experimental tasks, conditions, and standards differ across experiments. Since transfer studies consider both the training task and the testing task, both cases must be considered. The experimenter's instructions to the participant will constrain task performance. Two items of interest are the procedures and the dependent measures. Other items of consideration include the use of maps and exposure duration to the virtual environment.

Although the experiments investigate a wide range of issues related to knowledge transfer, the central issue is what is learned and how it is applied to the real environment. The level of spatial knowledge acquired is of particular interest. A study that develops route knowledge but then tests survey knowledge may mislead the reader to believe that some other factor was the cause of low performance. Even if a system had the right interface and the right level of fidelity for a given rehearsal task, it can be used incorrectly resulting in poor performance on the transfer task in the real world.

### 3.3 The Experiments

Witmer et al. (1995) were among the first to show that a VE could be useful for spatial knowledge acquisition. They compared a VE to the real world in an architectural walk-through application. They used an immersive display, and their population was a random sampling without any expertise on the task. Gaze-directed movement using buttons on the display controlled motion.

The experimental protocol divided the session into four stages: individual assessment, study, rehearsal, and testing. During the individual assessment stage, participants responded to numerous questionnaires, some of which probed their sense of direction and navigation experience level. Next, regardless of experimental treatment condition, every participant was given 15 minutes to study written step-by-step route directions and color photographs of landmarks. Half of the participants in each treatment condition were also provided a map of the building as a third study aid. Following the 15-minute study stage, each participant rehearsed the route three times, either in the VE or the real world depending on their group. All participants were required to identify six landmarks on the route, and researchers provided immediate correction if a participant made a wrong turn or misidentified a landmark. Finally, participants were tested in the real building, and were asked to replicate the route they had learned and to identify the six landmarks. The route replication measures included attempted wrong turns, route traversal time, route traversal distance, and misidentified landmarks. Configurational knowledge was tested by requiring participants to draw a line on a map from their known location to an unseen target.

This study effectively showed how landmark knowledge could become route knowledge, but survey knowledge was not given the opportunity to develop. Survey knowledge takes time to develop by primary exposure to an environment. Exposure times for this experiment were not long enough for this to occur. Nevertheless, this study effectively provided optimism that the technology could work for this purpose, but not how well or under what conditions.

Darken and Banker (1998) studied how a VE might be used as an augmentation to traditional familiarization methods. They compared performance of three groups: a map-only group; a VE group that also had the use of the map; and a real-world group that also had the use of the map. The interface to the VE was a desktop display controlled with a keyboard. The environment used was a natural region of central California with a few man-made structures but largely vegetation and rough paths (see Fig. 24.10). They used a participant population with specialized knowledge of the task, specifically sport orienteers and experienced military land navigators.

The experimental session was comprised of two phases, planning/rehearsal and testing. During the planning/rehearsal portion of the session, which lasted 60 minutes, participants studied and created a personal route from the starting point to nine successive control points. By the end of the planning phase, participants were required to draw their planned route on the map. Testing involved execution of the planned route in the real environment without the aid of the map or compass. As the participant navigated the real-world course, the researcher followed, videotaping participant behavior with a head-mounted camera. A differential GPS unit worn by the participant recorded their position every two seconds. This information was used to measure the quantity of unplanned deviations from the route and total distance traveled. In addition, the frequency of map-and-compass checks was recorded as a dependent measure.

This study is unique in many respects. It required the participants to plan their own routes rather than practice a given one. This serves to develop survey knowledge since alternative routes must be explored. They also attempted to introduce individual experience as a factor in addition to spatial ability. Since the task was specific to a particular domain, experience on these types of tasks should have an effect.

FIG. 24.10. The top image is a photograph from the testing area in the former Ft. Ord, California. The bottom image is a snapshot from virtual Ft. Ord at that exact location.

The results show that only intermediate participants seem to improve with the use of the virtual environment. Beginners have not yet developed enough skill at the task to be able to make use of the added information the VE offers, and advanced participants are so highly proficient at map usage that the VE simply does not add much information they cannot gain in other ways. In a complex environment such as the one studied here, even the hour of exposure provided did not seem to be enough. Given very short train-up times, maps still seem to be the best alternative for spatial knowledge acquisition.

Koh et al. (2000) expanded previous work by specifically looking toward the development of configurational knowledge in architectural environments. In their experiment, they compared a real-world group, an immersive VE group, a desktop VE group, and what they call a "virtual model" group that is similar to a noninteractive World-in-Miniature, which is held in the hand. They used a general population sample and varied the interface to the VE, as described by the group (immersive or desktop). In the two VE conditions, participants controlled their motion direction and speed with a joystick. The desktop group viewed a typical computer monitor. The immersive group members wore a head-mounted display. Their heads were tracked, and head rotation updated the visual scene, although gaze direction was not linked to the direction of travel.

The experimental sessions were split into three phases: administration, training, and testing. During the administrative phase, the participants were informed about bearing and distance estimation tasks, although the specific stations and targets were not disclosed. Participants in the three non-real-world groups then were provided a period of time to familiarize themselves with the interface. Training consisted of 10 minutes of free exploration under the assigned treatment conditions. The members of the real-world group explored the real building and members of the VE groups explored the synthetic building. Testing was specific to configurational knowledge.

Participants were asked to estimate distances and bearings to unseen targets in the real world while being transported from place to place while blindfolded in a wheelchair.

The purpose of this study was to specifically focus on the development of configurational knowledge. The experimenters were not interested in landmark or route knowledge, although they do reference the hierarchical landmark, route, and survey model. Furthermore, the researchers pronounced a bias that higher fidelity experiences may not necessarily lead to better transfer of knowledge, hence the use of three different fidelities in different configurations. Their results show that the VE conditions developed configurational knowledge at a comparable level to the real world.

Waller et al. (1998) used six different conditions based on the practice method. The conditions were real world, map, no-study, and three different virtual groups (desktop, short-duration immersive, and long-duration immersive). The environment was a maze constructed of full-length curtains with targets placed at selected locations. They also used a general population sample with no specific experience in these tasks.

All participants in VE conditions controlled their motion using a joystick, and the immersive groups both used a head-mounted display. The short-duration immersive group spent a total of 12 minutes in practice, while the long-duration immersive group spent a total of 30 minutes in practice over repeated trials.

The experimental protocol was comprised of four phases. The first was administrative and included proctoring of the Guilford-Zimmerman spatial abilities test. The second phase tested route knowledge by interleaving practice and testing six times. The participant would make a practice run followed immediately by a testing run. The third phase tested survey knowledge. The experimenter altered the configuration of the maze so that portions of the learned route were now blocked. The blindfolded participant then had to find a new route from one location to another. Dependent measures for both phases two and three included time to traverse the route and quantity of times the participant bumped into a wall. Finally, in the last phase, participants completed a pencil-and-paper test of their configurational knowledge.

This study showed that maps were just as effective as short durations of training but that if enough time was given, the VE did prove to be more effective, to the point of surpassing the real-world condition.

The graph in Fig. 24.11 attempts to explain these results. Note that these curves are hypothetical, because all the training transfer studies in the literature cannot be directly compared, and even if they could, there are far too few data points to establish the shape of the curves. Based on this research, however, it is suggested that given a short exposure duration, maps are better than any VE alternative simply because they do not overload the user with information that cannot be absorbed. However, the map is only useful to a point. Given enough

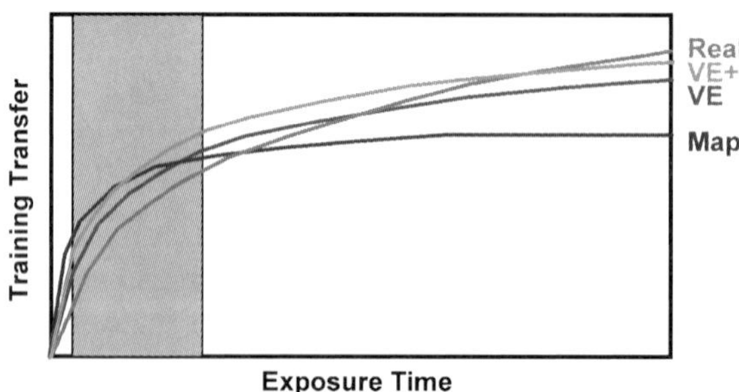


FIG. 24.11. This graph shows a hypothetical picture of how spatial knowledge might be acquired over time depending on the apparatus used. Maps are best for short term events, but the real world or virtual environments with training interventions (VE+) are best over time.

time, the added information a VE may provide may increase performance. The dark vertical bar represents the time slice that encompasses most of the studies in the literature. Very rarely are participants given enough time to develop survey knowledge in any meaningful way. The figure also differentiates a general virtual environment (VE) that is assumed to be only a virtual replication of the real space, from a specialized virtual environment (VE+) which might have added training features such as aerial views, transparent walls, or other features that may enhance the training effect. However, it is important to note that current research in this area is attempting to determine those features. In order to create VE systems for environmental familiarization that compare or even surpass the real world, an understanding of what tools to use under what conditions is eventually needed.

## 4. PRINCIPLES FOR THE DESIGN OF NAVIGABLE VIRTUAL ENVIRONMENTS

This final section will serve to summarize the chapter into a series of principles discussed in earlier sections. This is not to be interpreted as a design "cookbook" where designers can look to see how to select navigation aids for their application. Navigation is a highly aggregate task involving people, tasks, and environments. It is not possible to develop a design solution without addressing all three elements as a whole system and not as a set of parts. What is provided here is a set of guidelines based on the literature and on our experiments as a starting point for designers to address the important issues they may face. In some cases, one of these techniques may fit perfectly; in others, an adaptation may be needed. The key is in providing enough spatial information so that users can execute navigation tasks as demanded by the application without overconstraining the interface, thereby eliminating exploration and discovery.

### 4.1 Performance Enhancement

#### *4.1.1 Tools and Mediators*

##### *4.1.1.1 Map Usage.*

1. Maintain orientation but match map orientation to the task. Predominantly egocentric tasks like searching should use a forward-up map, whereas geocentric tasks such as exploration should use a north-up map.
2. Always show the user's position and view direction on the map and update dynamically.
3. The orientation problem will be more severe with a user population that includes individuals with low spatial abilities, such as mental rotation ability. Be aware of who your users are.
4. Video game play may have an effect on the selection of an appropriate map for a virtual environment. If the user population is largely a gaming community, their spatial abilities are likely to be high.
5. Use moded maps (e.g., where the use of the map precludes motion) only where appropriate. The default method should be unmoded.

##### *4.1.1.2 Landmarks.*

1. Allowing users to annotate the environment in some way to "personalize" the spatial cues they wish to use can be effective for complex spaces and is easily adaptable to a wide variety of navigation tasks.

2. Beware of giving the user the ability to clutter the space with excessive information. What they think will be helpful can become distracting noise.
3. Provide enough obvious landmarks (typically dependent on the context of the application and the task) so that the navigator has "reassurance" cues along a route to know that they are on the right path.
4. Make sure that landmarks are directional as well as salient. They should help provide orientation information to the navigator as well as position information.

#### *4.1.1.3 Trails or Footprints.*

1. Simply leaving a trail is marginally useful since it tends to clutter the space. Making the trail such that it disappears over time is better but can be confusing because it no longer tells the navigator that this place has been visited before, only that it has not been visited lately.
2. Trails can be particularly effective for exhaustive searches. It may be appropriate to use them in the context of a specific exhaustive search but then turn them off afterward.

#### *4.1.1.4 Directional Cues.*

1. Directional cues (e.g., sun, compass) alone will not be satisfactory as a navigation aid. They should be used with other techniques since they do not provide positional information.
2. Directional cues can be effective when moded.
3. Directional cues, when used with directional landmarks, are highly effective because they place landmarks in a global coordinate system.

### *4.1.2 Organizational Remedies*

#### *4.1.2.1 Implicit Sectioning.*

1. Use an organizational principle wherever possible and do not violate it.
2. If the organizational principle must be violated, make it obvious where and why the violation occurred so the navigator does not attempt to resolve the violation into the organization principle.
3. Match landmarks to the organizational principle whenever possible. They can be used to reinforce the shape of the space.
4. Do not blindly try to use the elements of urban design in any virtual environment. They might not be appropriate. Keep in mind the key concepts: provide useful paths, observable edges, and usable landmarks; and divide big, complex spaces into a number of smaller navigable spaces that are connected in some clear, understandable way.
5. Use an urban metaphor for abstract data judiciously. Make it clear where the metaphor fits and where it does not.

#### *4.1.2.2 Explicit Sectioning.*

1. Use explicit sectioning, particularly when implicit reorganization does not work for a particular environment.
2. Again, use an organizational principle and make it obvious.
3. Select a scheme for organizing your space based on which tasks people are likely to do there. If they are doing a lot of naive searches, for example, make sure there is a way for them too easily and systematically explore the entire space without repetition.

### 4.2 Environmental Familiarization

1. Do not assume that because someone can efficiently navigate a VE that they can navigate the real world as well. This is simply not the case.
2. Beware of creating performance "crutches" by adding features in the VE that enhance performance there but that are not available in the real world.
3. Given a short amount of time with which to familiarize someone with an environment, use a map and perhaps photos if they are available.
4. Beware of developing orientation-specific survey knowledge if only maps are used. Given enough time, a VE can be used to develop orientation-independent spatial knowledge, but it takes time.
5. Be careful when deciding how to use a VE for environmental familiarization because it is extremely difficult to compare studies in the literature. Make a decision based on the whole problem—the people you are training, the tasks they are doing, and the environment they are navigating in.

## 5. REFERENCES


Aretz, A. J., & Wickens, C. D. (1992). The mental rotation of map displays. *Human Performance, 5*(4), 303–328.

Boer, L. C. (1991). Mental rotation in perspective problems. *Acta Psychologica, 76*, 1–9.

Bowman, D., Koller, D., & Hodges, L. (1997). *Travel in immersive virtual environments: An evaluation of viewpoint motion control techniques.* Paper presented at the Virtual Reality Annual International Symposium (VRAIS), Albuquerque, NM.

Chase, W. G. (1983). Spatial representations of taxi drivers. In D. R. Rogers & J. A. Sloboda (Eds.), *Acquisition of symbolic skills.* New York: Plenum Press.

Chen, J. L., & Stanney, K. M. (2000). A theoretical model of wayfinding in virtual environments: Proposed strategies for navigational aiding. *Presence: Teleoperators and Virtual Environments, 8*(6), 671–685.

Colle, H. A., & Reid, G. B. (1998). The room effect: Metric spatial knowledge of local and separated regions. *Presence: Teleoperators and Virtual Environments, 7*(2), 116–128.

Darken, R. P. (1996). Wayfinding in large-scale virtual worlds. Unpublished doctoral dissertation, George Washington University.

Darken, R. P., & Banker, W. P. (1998). *Navigating in natural environments: A virtual environment training transfer study.* Paper presented at the IEEE Virtual Reality Annual International Symposium, Atlanta, GA.

Darken, R. P., & Cevik, H. (1999). *Map usage in virtual environments: Orientation issues.* Paper presented at the IEEE Virtual Reality '99, Houston, TX.

Darken, R. P., Cockayne, W. R., & Carmein, D. (1997). *The Omni-Directional Treadmill: A locomotion device for virtual worlds.* Paper presented at the ACM UIST '97, Banff, Canada.

Darken, R. P., & Sibert, J. L. (1993). *A toolset for navigation in virtual environments.* Paper presented at the ACM Symposium on User Interface Software and Technology, Atlanta, GA.

Darken, R. P., & Sibert, J. L. (1996). Wayfinding strategies and behaviors in large virtual worlds. Proceedings of *ACM SIGCHI '96*, 142–149.

Donelson, W. C. (1978). Spatial management of information. *Proceedings of ACM SIGGRAPH '78*, 203–209

Downs, R. M., & Stea, D. (1977). *Maps in minds: Reflections on cognitive mapping.* New York: Harper & Row.

Durlach, N., & Mavor, A. (Eds.). (1995).*Virtual reality: Scientific and technological challenges.* Washington, DC: National Academy Press.

Furnas, G. W. (1986). Generalized fisheye views. *Proceedings of ACM SIGCHI '86*, 16–23.

Goldin, S. E., & Thorndyke, P. W. (1982). Simulating navigation for spatial knowledge acquisition. *Human Factors, 24*(4), 457–471.

Ingram, R., & Benford, S. (1995). *Legibility enhancement for information visualisation.* Paper presented at Visualization 1995, Atlanta, GA.

Jul, S., & Furnas, G. W. (1997). Navigation in electronic worlds: A CHI '97 Workshop. *SIGCHI Bulletin, 29*(4), 44–49.

Koh, G., von Wiegand, T., Garnett, R., Durlach, N., & Shinn-Cunningham, B. (2000). Use of virtual environments for acquiring configurational knowledge about specific real-world spaces: Preliminary experiment. *Presence: Teleoperators and Virtual Environments 8*(6), 632–656.

Levine, M., Marchon, I., & Hanley, G. (1984). The placement and misplacement of you-are-here maps. *Environment and Behavior, 16*(2), 139–157.

Lynch, K. (1960). *The image of the city.* Cambridge, MA: MIT Press.

McGee, M. G. (1979). Human spatial abilities: Psychometric studies and environmental, genetic, hormonal, and neurological influences. *Psychological Bulletin, 86*(5), 889–918.

McNamara, T. P., Ratcliff, R., & McKoon, G. (1984). The mental representation of knowledge acquired from maps. *Journal of Experimental Psychology: Learning, Memory, and Cognition, 10*(4), 723–732.

Neisser, U. (1976). *Cognition and reality: Principles and implications of cognitive psychology.* New York: W. H. Freeman.

Passini, R. (1984). *Wayfinding in architecture.* New York: Van Nostrand Reinhold.

Péruch, P., Pailhous, J., & Deutsch, C. (1986). How do we locate ourselves on a map: A method for analyzing self-location processes. *Acta Psychologica, 61*, 71–88.

Presson, C. C., DeLange, N., & Hazelrigg, M. D. (1989). Orientation specificity in spatial memory: What makes a path different from a map of the path? *Journal of Experimental Psychology: Learning, Memory, and Cognition, 15*(5), 887–897.

Presson, C. K., & Hazelrigg, M. D. (1984). Building spatial representations through primary and secondary learning. *Journal of experimental psychology: Learning, memory, and cognition, 10*(4), 716–722.

Rieser, J. J. (1989). Access to knowledge of spatial structure at novel points of observation. *Journal of Experimental Psychology: Learning, Memory, and Cognition, 15*(6), 1157–1165.

Rossano, M. J., & Warren, D. H. (1989). Misaligned maps lead to predictable errors. *Perception, 18,* 215–229.

Siegel, A. W., & White, S. H. (1975). The development of spatial representations of large-scale environments. In H. Reese (Ed.), *Advances in child development and behavior* (Vol. 10). New York: Academic Press.

Spence, R. (1998). A framework for navigation (Tech. Rep. No. 98/2). London: Imperial College of Science, Technology, and Medicine.

Stevens, A., & Coupe, P. (1978). Distortions in judged spatial relations. *Cognitive Psychology, 10,* 422–437.

Stoakley, R., Conway, M. J., & Pausch, R. (1995). *Virtual reality on a WIM: Interactive worlds in miniature.* Paper presented at the Proceedings of ACM SIGCHI '95, Denver, CO.

Suchman, L. A. (1987). *Plans and situated actions: The problem of human–machine communication.* Cambridge, England: Cambridge University Press.

Templeman, J., Denbrook, P. S., & Sibert, L. E. (2000). Virtual locomotion: Walking-in-place through virtual environments. *Presence: Teleoperators and Virtual Environments, 8*(6), 598–617.

Thorndyke, P. W., & Goldin, S. E. (1983). Spatial learning and reasoning skill. In H. L. Pick & L. P. Acredolo (Eds.), *Spatial orientation: Theory, research, and application* (pp. 195–217). New York: Plenum Press.

Thorndyke, P. W., & Stasz, C. (1980). Individual differences in procedures for knowledge acquisition from maps. *Cognitive Psychology, 12*, 137–175.

Tolman, E. C. (1948). Cognitive maps in rats and men. *Psychological Review, 55*(4), 189–208.

Waller, D., Hunt, E., & Knapp, D. (1998). The transfer of spatial knowledge in virtual environment training. *Presence: Teleoperators and Virtual Environments, 7*(2) 129–143.

Witmer, B. G., Bailey, J. H., & Knerr, B. W. (1995a). *Training dismounted soldiers in virtual environments: Route learning and transfer* (Tech. Rep. No. 1022). Orlando, FL. U.S. Army Research Institute for the Behavioral and Social Sciences.

# 25

# Content Design for Virtual Environments

Jerry Isdale,[1] Clive Fencott,[2] Michael Heim,[3] and Leonard Daly[4]

*[1]HRL Laboratories LLC, 3011 Malibu Canyon Road Malibu, CA, 90265*
*isdale@acm.org*

*[2]University of Teesside, School of Computing and Mathematics Borough Road, Middlesbrough, Cleveland, England TS12 1PQ*
*p.c.fencott@tees.ac.uk*

*[3]Art Center College of Design, 2305 Ruhland Avenue Redondo Beach, CA 90278*
*mheim@artcenter.edu*

*[4]Daly Realism.com, 5843 Ranchito Ave. Van Nuys, CA 91401*
*daly@realism.com*

## 1. INTRODUCTION

Content development is the design and construction of the objects and environment that create a virtual experience. This includes the visual, auditory, interaction, and narrative aspects of the virtual environment (VE). While this new medium of expression has much in common with other media (e.g., computer animation and games), it has its own peculiar aspects and distinctive features. These include real-time network communications, interactive spaces, multiuser interaction, avatars, and multimodal interfaces.

Virtual environments have the revolutionary capability in human communication akin to that of the moving image in the last century. They may constitute the principal communications media of this century. We are just beginning to learn what it means to create a full sensory experience with control of view and narrative development. Exploring and expanding the limits of the technology requires a background in technical aspects of world creation. It also requires a background in aesthetic (perception of beauty) and metaphysical (nature of being) issues so as to best stimulate and hold a user's attention and effectively communicate ideas.

Discovery of proper aesthetics of a medium requires artistic experimentation: learning the technology; finding its limitations and unique aspects; and then turning those aspects into assets. Davies' (2001) Osmose and Ephemere and Laurel's Placeholder (Laurel, Strickland, & Tow, 1994) are early examples of artistic exploration of virtual environments. This chapter goes beyond the tools and techniques of content creation to outline aesthetics and design patterns that capture the distinguishing features of virtual environments.

## 2. TECHNICAL PRAGMATICS

Many of the pragmatic technical issues of content development for VEs are also found in computer animation and game development. Three-dimensional (3-D) modelers, textures, sound effects, and behavioral simulators are part of the common toolbox. Craftsmen skilled in their use can easily cross over between these disciplines. Technical and popular how-to handbooks for these tools are quite plentiful. Discussion of most of the pragmatics is left to those resources.

The development process for these media parallels standard engineering processes and benefits from the same techniques (e.g., planning, configuration management, usability testing). Several aspects of this development process are discussed in other chapters of this handbook (see chaps. 3–13, 17, 26, 28, and 34, this volume). Rollings and Morris (1999) argue convincingly for the use of these techniques in computer game development. Developers are however cautioned that a strictly engineering approach often shortchanges the content design and development aspects of a virtual environment. It is common for technically oriented developers to underestimate the importance or complexity of modeling and other content development tasks.

One pragmatic issue that does bear mentioning is the fidelity with which an environment is modeled. There are major tradeoffs in object and behavior complexity, development effort, rendering time and purpose (see chaps. 3 and 12, this volume). Photorealistic rendering, modeling of minute details, and exact physical simulation may be unnecessary for many purposes. Developers of the early military VE training system SIMNET (see chap. 43, this volume), faced with major budget restrictions, came up with the concept of "selective fidelity." They conducted a detailed analysis of the tasks of tank crews and carefully chose which parts of the physical interface and computer models would be recreated in high fidelity and which could be low fidelity or eliminated entirely. Many of the controls in these simulators are simply painted on, and multiple levels of detail are used throughout the computer models. Selective fidelity proved to be highly successful in reducing cost while preserving simulator effectiveness. This is an important lesson for VE designers.

## 3. LEARNING FROM OTHER MEDIA

Effective design of content for a medium comes from understanding something of its aesthetics. A new medium such as VE will have aesthetic properties that distinguish it from more established media. However, these aesthetic particulars are not readily knowable and designers often start with techniques from prior forms (Oren, 1990). The first film cameras were propped up immobile before a proscenium to record theater plays. Then filmmakers discovered the "camera eye" that could be moved around to pan, dolly, and so forth. There is now a large body of experience and accumulated theory for filmmaking. A number of authors (Clarke & Mitchell, 2000; Laurel, 1991; Lindley, 2000) have written on the applicability of film and theater theory to the design of virtual environments. However, a virtual world is not a broadcast medium like film, TV, or radio, where a story unfolds, told by one person to many. Virtual environments are at their heart an interactive medium, one in which the viewer has unprecedented control. The free-ranging participant, able to move and modify objects in the world, plays havoc with traditional narrative forms. Even the interactive techniques of multiform narrative (i.e., showing multiple points of view) and branching plotlines fail when the user can wander away from the scripted actions.

Interactivity and narration may be more or less important for a VE, depending on its purpose. Commerce and design review environments' need interaction for manipulation of viewpoint and the object under review. Narrative control is reduced to predefined viewpoints, collaborative baton passing, and perhaps modification of simulation parameters. A training and educational environment uses narrative control to lead the student to particular places of interest and induce them to undertake tasks. Community environments exist to foster the development and sustainability of virtual communities. They need to hold the attention and encourage participation of their residents. The architecture or composition of space within the VE plays a part in these more subtle narrative controls. This is an area where we may pick up some lessons from computer game design and architecture.

### 3.1 Computer Game Design

Computer games are a closer media relative of VE than film. Computer games are thus a fruitful area of study for those interested in understanding VE as a communications medium. Indeed, the line between a VE and a highly interactive computer game may be simply a difference in interface devices. Three-dimensional computer games are almost certainly the largest and most complex VEs in existence at the moment. Such games are designed to run using standard desktop environments, thus providing low levels of technological immersion (see chap. 40, this volume). This means games rely on content for their success and not technological sophistication.

The commercial nature of the computer games industry has led to a large body of knowledge being built up concerning what captures and retains player's attention. There is, however, little traditional experimental research on computer games. Johnson (1999) is one exception to this, though he is not a games developer. Most of the available material is in the form of accounts of the design process from specific companies and individuals. Saltzman (1999) devotes a chapter to accounts of the game design process from luminaries of the games world. There appear to be a number of common themes addressed by these and other writers on games design. Some of the most common are:

- A clear sense of purpose, mood and genre, to include setting emotional objectives for player's experience
- Verisimilitude or perceptual realism, for internal consistency of the game world is more important than adherence to real life
- Deep controls, an intuitive interface, easy to learn but with a steep learning curve before becoming an expert
- Multiple solution paths
- Flow, to include:
    - Challenges appropriate to player's skill levels
    - A balance of anxiety and reward
    - An almost unconscious flow of interaction

While not all application domains for VE will share all these themes, it is clear that computer game designers have a large and, probably, most critical audience, thus their insights should be of great benefit to the wider VE community.

A particularly rich resource on game design is the *Game Developer's Magazine* and its companion Web-site Gamasutra.com. The magazine has a regular design feature series with highly informative articles.

The computer games industry also has its own jargon, which clashes with that of the traditional VE community. For instance, immersion refers to what the VE industry calls presence (see chap. 40, this volume). Virtual reality in the games community refers to the technology of immersion (eg., headsets, gloves, and so on), which is what the VE community understands as immersion or the embodying interface (Biocca, 1997).

## 3.2 Architectural Design

Architects deal with freely moving visitors as they strive to create an experience in their design of spaces. Benedikt (1992) was an early exploration of the architectural aesthetics of virtual environments. Beckmann (1998) and Anders (1999) are more recent works relating architecture, aesthetics, and virtual environments. Anders presents VE as a space for perception and cognition, which spans physical presence (i.e., material artifacts, tools for symbolic work), perceivable representation (i.e., symbolic presence), and cognitive presence. Anders uses this scale of abstraction to characterize concrete and abstract attributes of artifacts, spaces, and motion. He suggests that on this scale, VEs are a cross between perceivable representations that provide symbolic presence and tools for symbolic work that provide physical presence. They present many kinds of perceived and cognitive movement and action, and their choreography should "be determined by an overarching unity—an art of aesthetics" (Anders, 1999, p. 101). Designers must thus construct VE experiences for consistency and grace, maintaining coherence between transitioning states and spaces. They may have to (Anders, 1999, p. 108) "think like the director of a film, unifying content, movement and transition through careful design."

Theme park designers are especially tasked with telling a story through the experience of a physical space. The colors, architecture, and sounds of a space can strongly affect an audience's emotions. One technique of designers is to infuse story into all elements of the environment, from the textures of walls to the scripts of theme park guides. The rules by which the imagined universe exists are very broad, and violating them will destroy a visitor's experience. Carson (2000a, 2000b) discussed the application of lessons from theme park design to computer games. Some of these lessons can be summarized as follows:

- Have a story, which provides the all encompassing idea of the world and/or rules that guide the design and project team to a common goal.
- Never break these rules.
- Provide localization—Where am I?
- Provide purpose—What is my relationship to this place?
- Use cause-and-effect vignettes to lead people to their own conclusions on previous events or suggest potential ahead.
- Use the familiar as anchors to draw people into the story.
- Less is more—don't clutter an environment with too much detail or too many choices.
- Use subintuitive design elements—underlying structure draws audience helplessly to a desired conclusion.
- Use "arrows" and "pathways" in visual design to guide the eye.
- Use physical archtypes to invoke desired mental and emotional responses.
- Use contrasting elements (light/dark, small/large).
- Give illusion of complexity—multiple paths to same goal, etc.
- Provide repeated gentle reminders of "The Story" (past, present, and future events/places).
- Use forced perspective.
- Frame it—guide audience with well framed glimpses of places to come.
- Observe "Past Art"—learn what worked and what did not work.

The convergence of theme park and VE design is evident in the work of the OZ Entertainment Company. They are developing both an international tourist destination and a Web-based virtual world based on "The Wonderful World of OZ" (http://www.worldofoz.com; E3D, 2000).

## 4. CHURCH–MURRAY AESTHETICS OF VE

The previous section discussed lessons from some other media. An emerging aesthetic particular to VEs is now discussed, which might be called the Church–Murray Aesthetics of virtual environments. It started as a combination of Murray's (1998) aesthetics for interactive media and Church's (1999) "Formal Abstract Design Tools" (FADTs) for games. It also draws from Slater's and Usoh's (1993) concept of copresence and Heim's (1999) "Transmogrification." The Church–Murray aesthetics of VE are characterized as agency, narrative potential, presence and copresence, and transformation.

### 4.1 Agency

The term *agency* is used by Murray (1998) to describe the pleasure of being, or appearing to be, in control through interacting with a medium as opposed to being a passive recipient. Church (1999) gives agency two components:

- Intention: Making an implementable plan in response to the current situation and an understanding of the available options.
- Perceivable Consequence: A clear reaction from the VE to the actions of the user.

Agency is a lot more than just firing a gun or opening a door. Agency is about users constructing links between cognitive responses to VE content and the way in which the consequences of their virtual actions match their mental model of what they thought would happen.

### 4.2 Narrative Potential

Traditional media designers invariably underestimate the way agency subverts traditional narrative forms. Agency gives a user the freedom to ignore some or all of the wonderful things the designer has so painstakingly built into the world. There are design tricks that entice users to follow the paths set for them and undertake the activities planned for them. If this is done well, users will come away from the VE able to tell appropriate stories that show the experiences they gained in the VE were more or less the ones designed in for them. This narrative potential is the glue that binds together the set of experiences offered through agency.

### 4.3 Presence and Copresence

Presence is the mental state of a VE user not noticing or choosing not to notice that the world, which they perceive, is artificially generated (see chap. 40, this volume). It is thus the sense of being mentally transported to the "place" of the virtual environment. Presence is thus an aesthetic pleasure of oral storytelling, painting, novels, television, and film, as well as virtual environments. Presence cannot be directly coded into a VE but arises from appropriately designed agency and narrative potential.

The related concept of copresence, being present with others, is also of interest (Slater & Usoh, 1993). By others, we mean animals and alien races, as well as humans, some or all of whom may be intelligent agents (see chap. 15, this volume). Copresence has been shown to be one of the primary aesthetic pleasures of digital media from e-mail to MUDs (multiuser domains), MOOs (MUD Object Oriented), and networked computer games.

### 4.4 Transformation

Heim's (1999) positions avatar design at the heart of VE design and relates this strongly to the pleasures of transmogrification (i.e., to change shape or form, particularly into one that is bizarre or fantastic). We use Murray's (1998) related term "transformation" as the change is often mundane. Transformation is a further by-product of agency, narrative potential, and the ability of the VE to allow players to feel themselves as part of the mediated environment. As a result of interacting with a VE, users may temporarily become someone else or even another species, either terrestrial or alien. But transformation is not just about taking on another character. It is about being able to do things one cannot normally do, having skills, power, and authority not normally possessed, and so on. This is very apparent in game and chat worlds. Participants normally do not want to look like themselves in these fantasy worlds—they choose nonrealistic avatars that project idealized self-images.

The Hubble VTE (Loftin & Kenney, 1994) is a good example of a less fantastical transformation experience. The flight team could become astronauts for a while even thought none of them, except the person that communicates directly with the flight crew, were likely to go into space themselves. Even grown up boys and girls might want to pretend to be astronauts.

## 5. PERCEPTUAL OPPORTUNITIES

This discussion of Church–Murray aesthetics has demonstrated that content for VEs is not just scene graphs nodes, objects, or audio assets. Content is about stimulating perceptions, which gives rise to meanings in the mind of the user and stimulates action, which in turn leads to further meaningful insights.

Perceptual opportunities (POs) offer a generic means of talking about VE content (Fencott, 1999b). They can also be used to look at the relationship between objects' various meanings and the way these affect users' behavior. Some objects seem to attract attention to the possibility of danger, reward, and so on. Identifying such objects offers opportunities to establish goals to further users' progress through a virtual environment. Other objects, or combinations of objects, are useful in helping users plan and achieve goals. Intense patterns of such objects form mini-missions and retain users' focus of activity in the cause of larger objectives. The keyword here is opportunity. Part of the art of VE design is providing users with carefully structured opportunities to allow them to explore, strategize, formulate and solve problems, and plan for and attain goals. This in turn allows users to feel some degree of control over what they are doing, allows them to creatively unfold the plot, feel present, and maybe become transformed in terms of skills and/or persona.

The PO model is a characterization of the roles objects are intended to play in establishing purposive experience. Figure 25.1 shows how POs may be broken down into three principal forms, each of which focuses on different kinds of meaning that objects may offer. Sureties deliver denotative meaning and collectively try to establish basic believability. Surprises seek to deliver connotative meaning and thus collectively seek to deliver purpose. Shocks are perceptual bugs that tend to negate the other two forms by breaking the illusion. Shocks will not be pursued here.

The relationships between POs can be documented using perceptual maps, which are a sort of grammatical structuring that seeks to ensure that users construct an appropriate temporal ordering over their attentions and activities within the virtual environment.

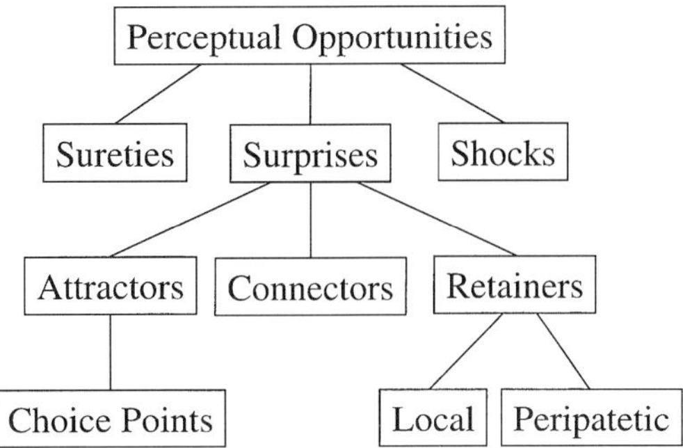


FIG. 25.1. Characterizing perceptual opportunities.

## 5.1 Sureties

Sureties provide certain kinds of basic information that support the main purpose of a virtual environment. They are mundane details that are somehow highly predictable—their attraction is their predictability. They should appear to arise quite naturally and are concerned with the logic of the environment unconsciously accepted. Sureties deliver denotative meaning and thus help users to accept the fundamental nature of the world or level. Recent research shows that much of what we know about the world we know unconsciously and that it is this knowledge that allows us to function from second to second.

Sureties should inform a user of such things as: How big am I? How fast am I moving? What do I look like? Have I been here before? Sureties also provide other reassuring information to do with such things as the physics of the world and the believability of other beings in terms of their avatars and behaviors. Furthermore, people are used to the real world being complex and cluttered, so it helps if the virtual world is as well. This is called redundant complexity or perceptual noise. A useful aphorism is that in interacting with the real world we are trying to make sense of too much information, whereas in a VE we are trying to make sense of too little.

Sureties succeed by not being noticed when they are there but being missed if they are not present. They are thus the basis on which a designer seeks to achieve the willful suspension of disbelief in the mind of users. If sureties are the basis of belief, then surprises are what really deliver the goods.

## 5.2 Surprises

Surprises are nonmundane details that are not always predictable but they do arise, however surprisingly, from the consciously accepted logic of a space. Surprises therefore are intended to deliver the memorable pleasures of the world by allowing users to accumulate conscious experience. Surprises are concerned with the connotative meaning of VE content. Surprises can be implausible but beneficial or totally plausible but unexpected, and there are three basic types: attractors, connectors, and retainers.

Perceptual opportunities can be both sureties and surprises, depending on the context in which they are offered—there is no mutual exclusivity between them. A fire escape can be both a surety—familiar objects that provide sureties for scale—and also a surprise—access to rooftops. Some things will be more or less surprising than others.

### *5.2.1 Attractors*

Attractors are POs that seek to draw users' attention directly to areas of interest or to situations that require action. Attractors are the means by which users are stimulated into setting goals for themselves. It is thus important that major attractors are associated with retainers, which reward users with things to do, remember, excite, and puzzle over, and which will allow them to feel they have attained the goal they set for themselves as a result of the attractor.

Attractors may be characterized according to the reasons they draw attention to themselves in several ways. First of all, there is the way they stimulate natural curiosity through mystery, movement, strangeness, and so on. An attractor may exhibit a combination of such characteristics. However, although attractors rely on peoples' natural curiosity they are also directly related to users' emotional involvement with the world. A second and very useful characterization concerns the connotative meaning users attach to attractors. Two of these that are important for computer games, for instance, are:

- Objects of desire: have some benign significance to users and to the task at hand.
- Objects of fear: have some malign significance to users and to the task at hand.

The purpose of attractors is to stimulate goal formation. Very often an attractor might have several possible goals associated with it and thus becomes a choice point—a source of great dramatic potential. When attractors are discussed, suggested goals with always be associated with them. Attractors are thus the means by which users are coaxed into following a particular course, choosing between possible courses, or changing course.

#### 5.2.2 Connectors

Connectors are POs that help users by supporting the planning used to achieve goals stimulated by attractors. Connectors are thus the means by which users make connections, both mental and "physical," between attractors and associated retainers, which allow users to achieve their goals and deliver objectives specific to the purpose of the world. The actual objective of a retainer might well be hidden or not clear from the point of view of its attractor(s), but lower level goal formation should lead users into situations where objectives can be realized.

#### 5.2.3 Retainers

Retainers are activities that seek to deliver specific objectives and rewards of the world and collectively, therefore, its purpose. The activity might be simply walking over some treasure to collect it, or it might be a whole mini-mission, such as a firefight. Retainers come in three forms: local, dynamic, and peripatetic. Local retainers seek to keep users in a particular place in a game. Users encounter dynamic retainers unexpectedly within the world. Peripatetic retainers are offered wherever users are in the world.

### 5.3 Perceptual Maps

Attractors, connectors, and retainers function together to deliver the aesthetics pleasures of a virtual experience. Surprises should work together in patterns to form possible temporal orders on retainers and thus the coherent set of purposive experiences that are intended to deliver the purpose of the world. These patterns are called *perceptual maps*, and they can be seen as a sort of invisible but comprehensible labyrinth that users will want to discover through their own agency (Murray, 1996). Perceptual maps have much in common with the way painters arrange the composition of a work so as to catch viewers' attention and lead them around the canvas in a particular way. This leads to the purposive accumulation of experience. A perceptual map is the framework within which narrative potential can be designed.

**TABLE 25.1**
Perceptual Map for a Hypothetical Game

| *Attractors* | *Connectors* | *Retainers* |
|---|---|---|
| Ricochets (Dynamic objects of fear) Goal is to find cover. | Plan is to make for cover. Uses doorways, walls, alleyway, etc. | Activity is to take cover. (Local) Reward is time to think, plan, etc. |
| Movement of opponent(s) (Dynamic object[s] of fear and desire—your opponent can fight back) Goal is to find cover. | Plan is to make for cover. Uses doorways, walls, alleyway, etc. | Activity is to take cover. (Local) Reward is time to think, plan, etc. |
| Movement of opponent(s) (Dynamic object[s] of fear and desire—your opponent can fight back) Goal is to frag opponent. | Plan is to take opponent by surprise. Uses guns and ammo and maybe cover. | Activity is to firefight (dynamic, peripatetic) Reward is fun and increase frag count. |

The simplest way of documenting a perceptual map is by way of a table with three columns, which relate attractor, connector, and retainer triples. Rows indicate suggested relationships left to right and cells give brief descriptions. For a hypothetical game, a partial table—just three of many entries—of surprises would look something like that shown in Table 25.1.

Of course, users might have a number of goals at any one time and will take notice of a host of attractors as they execute one or more of their current plans. Notice that the second two attractors are the same except for the goal. This is what Murray (1998) refers to as a choice point, which is identified as one of the most important dramatics structures in virtual environments. Comparative content analysis between various VE designs can be conducted by contrasting their perceptual maps (Fencott, 2000). This work is still the subject of research but involves comparing properties of these maps. For example:

- Complexity in terms of the number of rows in the table and resulting number of choice points, which gives an indication of the complexity of the decision making processes involved.
- The nature of the attractors, for example, their perceptual and emotional characteristics (e.g., shoot-em-ups will be largely composed of objects of desire and fear, whereas a virtual tourist site may be composed of objects of desire and curiosity).
- Goal stacking, which is the number of goals users work with at any one time; interestingly, shoot-em-ups seem to demand concurrent stacking of a large number of goals, whereas more contemplative VEs might only require one or two at a time.

Perceptual opportunities and maps have also been studied as part of a suggested VE development methodology (Fencott, 1999b), which looks, for instance, at the relationship between aesthetic design and engineering design.

Attractors should draw attention to sites of retainers and, if properly designed, lead users around a world in a meaningful way using connectors. Attractors may also themselves be retainers. Seen from a distance an animated object may act as an attractor but when experienced close up the object may be some sort of vehicle to ride in and control thus becoming a retainer.

Retainers are actually localized patterns of attractors and connectors. Early computer games, such as Viper and Breakout, can be viewed as retainers in this sense.

Some general rules govern the relationships between surprises:

- Users should be rewarded if they follow attractors.
- Retainers do not have to have attractors.
- Retainers can be their own attractors.
- Retainers can have multiple attractors.
- Connectors should lead to an attractor or directly to a retainer.

Thus connectors, like attractors, should be rewarded if followed.

Sureties and surprises should be designed to work together. If a perceptual map constitutes Murray's (1998) comprehensible labyrinth, then sureties are the means by which this is grounded, virtually, in a believable world.

In this section, the aesthetics of VE have been shown to be significantly different from the conventional aesthetics of media such as film and television. This stems largely from agency through which users and intelligent agents are empowered. The way VE content development is approached must take this into account. However, VE, perhaps more than any other media, offers the possibility to investigate the very nature of being.

## 6. DEEPER MOTIVATION OF DESIGN

Metaphysics is the systematic investigation of the nature of being or reality. It can inform and motivate a designer's aesthetic. Virtual environments expose the nature of reality to direct observation and experimentation. It is perhaps the first medium to do so and opens up a whole realm of exploration (Lauria, 1997, 2000).

### 6.1 The Nature of Being in a VE

Metaphysical questions do not immediately come to mind for someone looking at a medium for the first time. Yet such higher level questions allow us to experiment outside the box of previous media that rely heavily upon narrative story telling, one-way broadcasting, and the manipulation of attention (as opposed to interaction). Metaphysics takes us beyond conventional aesthetics, where one might hope to avoid the pitfalls of the early history of photography, film, and television. Those media remained stunted for decades by trying to reproduce the legacy content media of still painting, tripod cameras, and formally staged theater productions. "Content" is not something that can be simply dumped from one medium into another. The artistry in any medium lies in finding a harmony of content and form, something appropriate to the specific medium.

Plunging into the depths of metaphysics frees the imagination, but it does not take a big leap to get metaphysical about virtual environments. If metaphysics deals with the nature of being itself, then the virtual worlds designer faces that question head-on when first confronting virtual space. What does it take to make something "real" or to make it "be"? This is not an airy question when you design a virtual environment. It is a pressing decision. Designers have to decide what counts as a real thing or entity, and they need to also rank its level of reality (e.g., permanence, vivacity, porosity, intelligence, efficacy, agency) relative to other beings. Take "backgrounds," for instance. Is there to be a single, persistent texture that serves as the permanent backdrop for action, movement, and relations? Or should background textures change constantly and flow? If flow is chosen, then at what rate? How much permanence, if any, is needed for a particular world's background?

When William Bricken (Bricken & Coco, 1993) first conceived the Virtual Environment Operating System (VEOS), he drew on two main sources: the notion of Sunyata (emptiness) in Zen Buddhism, and the "Laws of Form," by George Spencer-Brown, who authored a book by that title. The Buddhist "void," or emptiness, offers a metaphysics of virtual space. This space can be understood as a fundamentally nondirectional, nonfixed emptiness, rife with fertile possibilities. From the void evoked by VEOS emerges any set of entities, their attributes, and the relationships among those entities. No substance precedes the void. Historically, Spencer-Brown's (1979) Laws of Form interpreted the *Principia Mathematica* written by Bertrand Russell and Albert North Whitehead (1910, 1912, 1913) at the beginning of the 20th century. The *Principia* did to logic theory what Newton's *Principia* did to physics. *Principia Mathematica* laid new foundations for mathematical logic, making possible the later information/noise theory of Claude Shannon (Shannon & Weaver, 1963). Spencer-Brown (1979) interpreted the *Principia* from a Zen point of view. The initial logical distinctions of the *Principia* became themselves grounded in void space. "Draw a distinction in empty space" became the first act of creativity. The drawing of a distinction between any *X* and non-*X* sprung from a timeless act, from emptiness or openness to all possibilities. By unifying these two intellectual foundations, Zen and the *Principia,* Bricken's VEOS applied general metaphysics to cyberspace, a synthesis which came to Bricken in the late 1960s. Today, virtual worlds designers still return to the void and the act of creating distinctions from the void as the primal principles of design.

The metaphysics of void space "grounds" cyberspace design because conceiving virtual space and creating virtual space are nearly identical. As a phenomenon of human perception, cyberspace requires no material substrate. As a software phenomenon, cyberspace of course requires algorithms, and its software requires a hardware base. For the designer, however, the thought of a distinction becomes the first act of creation. There is no material element to shape or take into consideration outside function and ergonomics, psychology, and the perceptions of the human user. While limitless on a conceptual level, there are the stubborn facts of software engineering and hardware input/output devices. These stubborn facts are less insistent than stone and are yielding to the social confluence between virtual world engineers and virtual world designers.

If a designer accepts the challenge of metaphysical design (i.e., designing from the void and Spencer-Brown's basic laws of form), then the designer must also take responsibility for communicating with software engineers. The basic ontology—the types and ranks of beings that can exist—must be designed at the most fundamental level of software design:

- How should a self be represented?
- Should avatars be anthropomorphic?
- How many degrees of freedom should movement have?
- What kinds of entities (models) should exist, and should they exist as independent entities or as arrays, ephemerals, or more substantive types?
- Should a world have multiple portals or a single ground zero?
- How much participation (editing) should be under the control of world participants (users)?

All these questions, and many more, will arise if the designer accepts the metaphysical responsibility of working with the software engineer to determine the world's ontology. In previous forms of technology—probably because of the substrate of expensive, resistant materials—the engineer maintained a certain distance from the art designer, and vice versa. When the conditions of cyberspace design are considered, however, this role separation becomes a distinction rather than a separation. The two will work together to mine the rich possibilities of the void.

### 6.2 Transmogrification

To "transmogrify" means to change into a different shape or form, especially one that is fantastic or bizarre. Transmogrification might prove to be a valuable style for designing VEs and avatar worlds. Virtual environments might succeed not as a replication of the physical world, nor as a direct representation of our personalities. Instead, avatar worlds might inject a touch of fantasy and a sense of fun into conventional activities like business meetings and instructional classes. The "disguise" quality of online virtual worlds suggests a playful attitude that adds a distinct quality to everyday activities.

*Avatar* in its Hindu origin means the incarnation of a deity, as in Vishnu taking the identity of Krishna in the Sanskrit poem the Bhagavad Gita. (The Sanskrit *avatârah* derives from *ava,* "down," and *tarati,* "he crosses," meaning "the crossing down.") The avatars of virtual worlds are placeholders for real-time human presence (see chaps. 14–15, 45 this volume). They are not—with the possible exception of "bots" or semi-intelligent surrogates—empty media receptors like answering machines or phone pagers. They are animistic spirit vessels in a vast system of digitally encoded events. As such, avatars maintain a remnant of fresh humanistic issues in an age of technical systems. If, as some smart observers suggest, these systems are "not really about people at all," then avatar presence can still be used to pervert the basic trajectory of systems whose teleology is immanent and whose tangential implications are antihumanistic, or at least hostile to humans insofar as human life becomes yet another artificial life form to assimilate. Avatars are more than subservient system input or maintenance attendants for the auto-poetic network.

Avatars tap into the age-old magic of transformation. Humans can, under the right conditions, take what lies immediately in front of them and transform it into something of cosmic significance. This transformative power is at the heart of ritual. The transformative power of the spirit animates avatars and confers on them presence at a distance (telepresence).

Likewise, when users put "on" their avatar, they also put "off" the habitual self. They accept a moment of transformation, shifting their shape in order to be who they are in different forms. They shed their form like a changeling. They lay aside the illusory fixity of being a hard ego encapsulated in a shell of flesh. Avatars allow users to engage a playful self, a self that does not let it be defined in narrow technical terms. This avatar self is a changeling, a joker-prankster who revives the human capacity to laugh and to laugh at oneself.

One day, the Chinese sage Chuang-Tzu had a dream, and he dreamt he was a butterfly. When he awoke, he was not so sure: Was he Chuang-Tzu dreaming he was a butterfly? Or was he a butterfly now dreaming he was Chuang-Tzu? The dreams people have show the expansive, tenuous quality of human's deep self-identity. A related Taoist practice is to go through an entire day nurturing a feeling of inner softness, blurring the outer events of life into a diaphanous, dreamlike pattern. In such a state, one's usually strongly invested attachment to the outcome of events recedes, and harsh reaction toward events fades. With the edgy ego momentarily disengaged, newly rewarding ways of responding to life events are often discovered. The diffuse ego flies free of being identified fully with either Chuang-Tzu or with the butterfly. Avatars can become the toys of Chuang-Tzu if designers use them rightly.

## 7. CONCLUSIONS

Art is often about diffusing the rigid ego so that we can move more freely through time and space. We should not let our pride in the new networks we have built override the inherent powers we have always enjoyed as natural teleoperatives. The challenges of world and avatar design remind us of the need for art to maintain our proper relationship to technology.

## 8. REFERENCES


Anders, P. (1999). *Envisioning cyberspace: Designing 3D electronic spaces.* New York: McGraw-Hill.

Beckmann, J. (Ed.). (1998). *The virtual dimension.* New York: Princeton Architectural Press.

Benedikt, M. (Ed.). (1992). *Cyberspace first steps.* Cambridge: MIT Press.

Biocca, F. (1997, September). Cyborg's dilemma: Progressive embodiment in virtual environments [Online]. In McLoughlin, M. & Rafaeli, S. (Ed.) *Journal of Computer Mediated Communication 3*(2). Available: http://jcmc.huji.ac.il/vol3/issue2

Bricken, W., & Coco, G. (1993). *The VEOS Project* [Human Interface Technology Lab Tech. Rep. No. TR-93-3, [Online]. University of Washington, Seattle. Available: http://www.hitl.washington.edu/publications/r-93-3/

Carson, D. (2000a). *Environmental storytelling part 1: Creating immersive 3D worlds using lessons learned from the theme park industry* [Online]. Available: http://www.gamasutra.com/features/20000301/carson_01.htm

Carson, D. (2000b). *Environmental storytelling part 2: Bringing theme park environment design techniques lessons to the virtual world* [Online]. Available: http://www.gamasutra.com/features/20000405/carson_01.htm

Church, D. (1999, August). Formal abstract design tools [Online]. *Games Developer Magazine,* 44–50. Available: http://www.gamasutra.com/features/19990716/design_tools_02.htm

Clarke, A., & Mitchell, G. (2000). Taking and making meaning: Semiotics and new media. In *Proceedings of workshop on computational semiotics for new media* [Online]. Available: http://www-scm.tees.ac.uk/users/p.c.fencott/newMedia/ clarkeMitchell.html

E3D. (2000, March). On the road to OZ: An interview with Dan Mapes [Online]. Available: http://www.e3dnews.com/e3d/Issues/200005-May/lead.html

Davies, C. (2001). Landscape, earth body, being, space and time in the immersive virtual environments osmose and ephemere [Online]. In J. Malloy (Ed.), *Women in new media* (forthcoming). Cambridge: MIT Press. Available: http://www.immersence.com

Fencott, C. (1999a). Content and creativity in virtual environment design. In *Proceedings of virtual systems and multimedia '99* (pp. 308–317). University of Abertay Dundee, Scotland. Available: http://www-scm.tees.ac.uk/users/p.c.fencott/vsmm99

Fencott, C. (1999b). Towards a design methodology for virtual environments. In *Proceedings of Workshop on User Centred Design and Implementation of Virtual Environments* (pp. 91–98). University of York, England.

Fencott, C. (2000). Comparative content analysis of virtual environments using perceptual opportunities. In *Proceedings of Digital Content Creation. National Museum of Photography, Film and Television.* Bradford, England.

Heim, M. (1999). *Transmogrification* [Online]. Available: http://www.mheim.com/html/transmog/transmog.htm

Johnson, C. (1999). *Taking fun seriously: Using cognitive models to reason about interaction with computer games.* Available: http://www.dcs.gla.ac.uk/~johnson/papers/um99/games.htm

Laurel, B. (1991). *Computers as theatre.* Addison-Wesley. Paperbound revised 1993. See additional material online: http://www.tauzero.com/Brenda_Laurel

Laurel, B., Strickland, R., & Tow, R. (1994). Placeholder: Landscape and Narrative in Virtual Environments. In *ACM Computer Graphics Quarterly*, Vol 28 No 2, May 1999 [Online]. Available: http://www.tauzero.com/Brenda_Laurel/severe/_Heads/CGQ_Placeholder.html

Lauria, R. (1997). Virtual reality: An empirical-metaphysical testbed [Online]. *Journal of Computer Mediated Communications.* Available: http://www.ascusc.org/jcmc/vol3/issue2/lauria.html

Lauria, R. (2000). *Virtuality*. Mahwah, NJ: Lawrence Erlbaum Associates.

Lindley, C. (2000). A computational semiotic framework for interactive cinematic virtual worlds. In *Proceedings of Workshop on Computational Semiotics for New Media* [Online]. Available: http://www-scm.tees.ac.uk/users/p.c.fencott/newMedia/Lindley.html

Loftin, R. B., & Kenney, P. J. (1994). *Virtual environments in training: NASA's Hubble space telescope mission* [Online]. Available http: //www.vetl.uh.edu/Hubble/shortpaper.html

Murray, J. H. (1998). *Hamlet on the Holodeck: The future of narrative in cyberspace.* MIT Press [outline online]. Avaliable: http://web.mit.edu/jhmurray/www/HOH.html

Oren, T. (1990). Designing a new medium. In B. Laurel (Ed.), *The art of human–computer interface design* (pp. 467–479). Reading, MA: Addison-Wesley.

Rollings, A., & Morris, D. (1999). *Game architecture and design.* Scottsdale, AZ: Coriolis Group.

Saltzman, M. (Ed.). (1999). *Games design: Secrets of the sages.* Macmillan. (Chapter 6 available online: http://www.-gamasutra.com/features/19990723/levdesign_chapter_01.htm)

Shannon, C., & Weaver, W. (1963). *The mathematical theory of communication.* University of Illinois Press. (First published in two parts in the July and October 1948 editions of the *Bell System Technical Journal.*)

Slater, M., & Usoh, M. (1993). Representations systems, perceptual position, and presence in immersive virtual environments. *Presence: Teleoperators and Virtual Environments, 2*(3), 221–233.

Spencer-Brown, G. (1979). *Laws of form.* New York: Dutton. (First ed., London, 1969.)

Whitehead, A. N., & Russell, B. (1910, 1912, 1913). *Principia mathematica* (3 Vols.). Cambridge, England: Cambridge University Press. Second edition, 1925 (Vol. 1), 1927 (Vols. 2, 3). Abridged as *Principia Mathematica,* Cambridge, England: Cambridge University Press, 1962.

# 26

# Technology Management and User Acceptance of Virtual Environment Technology

David Gross
*The Boeing Company*
*Mail Stop JR-80*
*499 Boeing Blvd.*
*Huntsville, AL 35824*
*David.C.Gross@Boeing.Com*

## 1. INTRODUCTION

Technology deployment, or transitioning innovation into use, is the natural objective of any technologist who develops an innovation. Despite the old saw that "build a better mousetrap, and the world will beat a path to your door," the particular power, or utility, of a technological innovation is just one of the criteria for successful technology deployment. Technology deployment takes places within a human society, which may (or may not) adopt the innovation. The economic circumstances of that society, as well as its mores (formal or informal moral attitudes), also play important roles in the success (or failure) of a technological deployment. Furthermore, the ongoing advance of technology, independent of the technology in question, as well as evolving demands from consumers in society, create forces that affect a society's adoption choice. Figure 26.1 illustrates the many forces affecting the success (or failure) of deploying a technological innovation.

Some of these forces tend to encourage technology deployment, but some discourage it because many of these forces create societal inertia. This resistance to change is a phenomenon so pervasive and widely recognized that it scarcely requires documentation (Mackie & Wylie, 1988). The forces resisting change can slow and stunt the adoption of new technologies, a classic example being the deployment of nuclear power technology.

Virtual environment (VE) technology, like other innovations before it, has vast potential for improving the way people work and live; therefore VE technologists seek to deploy it. However, fulfilling this potential will certainly require equally vast changes in the way these things are done now. Every significant technology deployment has changed the adopting society's way of life. Consider for example the deployment of automobile technology and the wide-ranging impact it has had on every aspect of modern life. Some of those changes were anticipated and appreciated; some were not. Whether desirable or not, change is a certainty with a new technology. The resistance to deploying VE technology can be reduced by

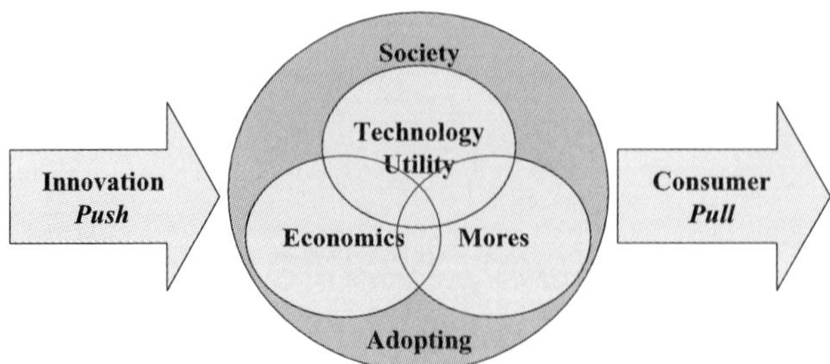


FIG. 26.1. Multiple effects on a technology's diffusion in society.

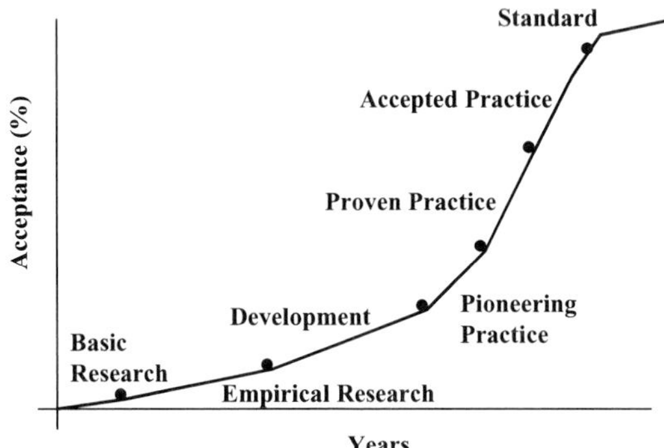


FIG. 26.2. Typical technology transition cycle.

understanding how organizations within a society come to understand and adopt technological innovations.

## 1.1 Technology Deployment and Adoption

Technologists can increase the rate of adoption if they understand the obstacles and concerns an adopting organization faces. Figure 26.2 suggests an expected timeframe for technology innovation adoption in software on the order of a decade or more (Utz & Water, 1992).

If technology deployment is the natural objective of a technology developer, then technology adoption is the choice of a technology user. Technology deployment can be very frustrating to the technology developer, who feels that the relative advantage of this technological innovation over present approaches is so overwhelming that only the naive would not immediately adopt it. However, potential adopters frequently see the technological innovation differently than developers. Adopters are concerned with issues such as an innovation's compatibility with the rest of an organization and with the complexity of using and maintaining the innovation. The reluctance of adopters to easily adopt innovation has led to what Rogers (1994) calls the *Innovativeness-Needs Parado,* in which individuals or other units in a system who most need the benefits of an innovation are generally the last to adopt it.

## 1.2 Toward an Understanding of Technology Adoption

One approach to understanding technology adoption is Innovation Diffusion Theory, especially as described by Rogers (1995). Technology adoption cannot occur unless the innovation is distributed or diffused through a society. The essence of the technology diffusion process is human interaction in which one person communicates a new idea to another person. Thus, at the most elemental level of conceptualization, the diffusion process consists of: (1) a new idea; (2) individual "A" who knows about the innovation; and (3) individual "B" who does not yet know about the innovation. The innovation diffusion model suggests a four-step process for technology diffusion: innovation, uncertainty, diffusion, and adoption/rejection. Although the technology itself plays an important role in this process, social relationships have a great deal to say about the conditions under which A will tell B about the innovation, and the results of this telling (Rogers, 1995). In this light, the factors that most affect a particular organization's decision to adopt a particular technology are internal to the organization, that is factors describing the social relationships in the organization. These factors can be grouped into three categories: (1) an organization's past experience with technology; (2) the organization's characteristics; and (3) the organization's pursued strategy (Lefebvre & Lefebvre, 1996).

While innovation diffusion theory has much explanatory power, the complaint has been made that it focuses too much on internal factors. Innovation diffusion theory models technology adoption based on a company's communication patterns in which prior adopters inform potential adopters and persuade them. One alternative is the theory of network externality, which argues that adopter behavior is strongly influenced by relevant parties outside the company (e.g., vendors, consultants, standards bodies). These external factors can be grouped into the following categories: (1) the industry level; (2) the macroeconomic environment; and (3) national policies (Lefebvre & Lefebvre, 1996).

If classical innovation diffusion theory focuses on internal factors, and network externality theory focuses on external factors, perhaps an approach that balances these factors is needed. Bhattacherjee and Gerlach (1998) suggest that while these approaches have explanatory power for relatively simply technology adoptions, they fail to explain more complicated pervasive technology adoptions such as software's object-oriented technology (OOT). The potential of VE technology certainly is as pervasive as the introduction of OOT was, so it is worth considering a more holistic approach. An emerging approach that balances the internal and external influences on adoption decisions is *Organizational Learning Theory*. Organizational Learning Theory suggests that organizations learn by a collective process very similar to how individuals learn; that is, organizational learning occurs when we learn collectively, when we act and reflect with others. Shibley (1998) divides an organization's learning into two parts, observation and reflection. Observation involves collection of the raw information of daily events and the discovery of patterns in those events. Reflection involves the attempt by organizations to place their observations in a useful context, either system structures, mental models, or corporate vision. Most often, observations are forced into preexisting structures, but valuable learning takes place when the raw data forces a change in these structures (Senge, 1990). An Organization Learning Theory approach to understanding technology adoption depicts an organization as attempting to integrate both the internal and external factors into a corporate view of a technological innovation, which dictates an adoption decision.

## 2. MANAGEMENT CHALLENGES

Managers of organizations considering adoption face real challenges in realizing the opportunities that the technological innovations afford them, while avoiding their pitfalls. This has led to difficulty in explaining experiences in technology adoption. For example, despite the widespread awareness of OOT's benefits for software development, and the commercial availability of reusable libraries, databases, methodologies, programming languages, and packaged applications, many businesses remain cautious about deploying OOT for major system development (Bhattacherjee & Gerlach, 1998). This experience may prove instructive for technologist concerned about deploying VE technologies in many of the same businesses that have been slow to adopt OOT.

## 2.1 Integrating Into Corporate Culture

Chief among management's concerns of about VE technological innovations is the difficulty of integrating new technologies into existing corporate culture. At companies like Boeing for example, introduction of personal computers capable of supporting technical documentation maintenance was slowed by concerns about what would happen to existing central word processing organizations. Of course, all adopting organizations are not the same; some will express innovation rapidly and easily, and some will not. Table 26.1 shows Roger's classification of different kinds of adopter organizations.

Even within these classifications, an organization may choose to have a more or less aggressive stance toward a particular technology, choosing, for example, to lead the market in one technology but simply defending its competitive position in another. Table 26.2 illustrates some objective organization characteristics, and whether or not they were significant factors affecting the adoption of a particular technology, namely Integrated Services Digital Network (ISDN). The data is an analysis of formal corporate surveys.

Understanding what kind of organization is considering adoption can help the technology developer plan for and accommodate the particular problems of an organization, thereby increasing the likelihood of success. For example, a VE technologist proposing a first VE project should realize from Tables 26.1 and 26.2 that organizations that commit significant resources toward a VE project and are at or near the forefront of their industrial domain are more likely to experience success. In contrast, the VE technologist should probably not be as concerned about the impact of the organization's degree of formality (rank, office, titles, etc.) or complexity (layers or styles of management) on success. In addition, a VE project proposal for a laggard organization will need plenty of support in the form of experience by other similar organizations.

**TABLE 26.1**
Classification of Innovation Adopters*

| *Adopter Categories* | *Share of Whole* | *Description* |
|---|---|---|
| Innovators | 2.5% | Venturesome, cosmopolite, networked with other innovators, financial resources, understand complex technical knowledge, cope with uncertainty |
| Early adopters | 13.5% | Respectable, more local than innovators, strong opinion leadership |
| Early majority | 34% | Interact frequently with peers, seldom hold positions of opinion leadership, interconnectedness to system's interpersonal networks, long period of deliberation before making an adoption decision |
| Late majority | 34% | Adoption might result from economic or social necessity due to the diffusion effect, skeptical and cautious, relatively scarce resources |
| Laggards | 16% | Most localized, point of reference is the past, suspicious of change agents and innovations, few resources |

* Adopted from Rogers (1994).

**TABLE 26.2**
Significance of Organization Factors in the Adoption of ISDN*

| *Organization Characteristic* | *Significant* | *Not Significant* |
|---|---|---|
| Adoption encouragement | X | |
| Open vs. closed organizations | X | |
| Slack (uncommitted) resources | X | |
| Large vs. small organization | X | |
| Norms encouraging change | | X |
| Organization centralization | | X |
| Organization formalization | | X |
| Organization complexity | | X |

* Adopted from Lai and Guynes (1997).

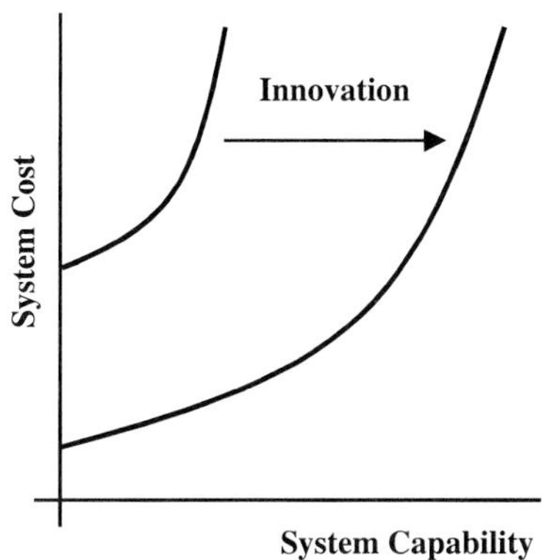


FIG. 26.3. Innovation increasing process productivity.

## 2.2 Need for Credible Productivity Gains

The second major challenge facing VE technology adopters' management is the need to address the "bottom line." Sometimes simplified as just showing a profit, the more accurate concern is to increase the process' productivity (return for resources invested). Figure 26.3 illustrates this desire in terms of moving an example process capability curve to a greater capability at a lower cost.

A legitimate claim that a VE project will improve productivity makes it easier for an organization to adopt the innovation. Can VE technology make this claim? The capability of business processes is a function of the knowledge, techniques, and tools that their employees apply. Human actions determine whether production is done efficiently and whether new and improved methods of production are introduced (Ostberg & Chapman, 1988). Virtual environment technology, integrating visualization and artificial intelligence, can legitimately claim to impact the productivity of business processes because it offers powerful opportunities for improvements in the way humans perceive and utilize knowledge (see Fig. 26.4).

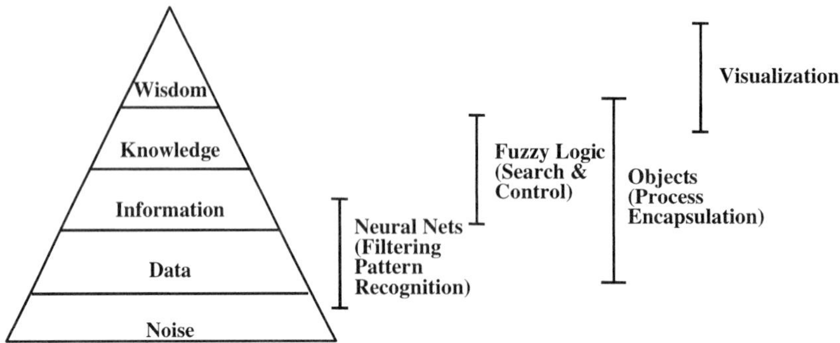


FIG. 26.4. Relevant merits of technologies to perceive and utilize knowledge (Hoffman & Wright, 1997).

### 2.3 Fear of the Operations and Sustainability Tail

The second major challenge facing VE technology adopters' management is the potential operations and sustainability tail. Major manufacturers such as Boeing have seen operations expense resulting from introduction of the personal computer far outstrip their acquisition cost as well as their productivity gain, particularly from earlier installations. Management has legitimate concerns about VE technology becoming another logistics nightmare, particularly considering the fragile nature of some of the equipment. When this is combined with the rapid evolution/revolution in VE devices and the rapidly changing nature of VE applications, this challenge can be a showstopper for deploying VE technology. Any VE technology deployment will have to address this challenge to survive scrutiny.

## 3. USER CHALLENGES

Management can be convinced, if the rank and file have powerful applications for technological innovation. The potential users of VE technology have challenges to overcome in their own right about discovering and explaining these applications.

### 3.1 The "Killer Application"

Worldwide, the market for business virtual reality services, software, hardware, and peripherals is seen as rising from 1995's level of 134.9 million to a little over $1 billion no later than 2001 (Boman, 1995). However, the marketplace is still hampered by the lack of a "killer application" (Stanney et al., 1998). The concept of a killer application dates from the development of the business spreadsheet, which motivated the purchase of millions of personal computers. Two years ago three-dimensional (3-D) and virtual technology on the Internet identified potential key applications: games, e-commerce, collaboration, education, training, and visualization. However, significant applications for VE in these fields are still just potential (Leavitt, 1999).

### 3.2 Physical Ergonomics

Cybersickness and related syndromes continue to be significant obstacles to widespread use of VE technology (see chaps. 29–32, and 37–39, this volume). At least for some users, cybersickness is a significant problem, and many organizations are reluctant to make significant commitment to technologies that clearly bypass significant portions of their population.

For example, wearers of head-mounted displays (HMDs) have been known to suffer stress because they are unable to utilize eye convergence and accommodation when viewing virtual objects at different distances (see chap. 37, this volume). In general, some people are very sensitive to unfamiliar motion stimuli, such as that generated in a VE, and these individuals may be discomforted by VE exposure (Vince, 1995).

Beyond cybersickness, much VE technology is uncomfortable or unpleasant to use (see chap. 41, this volume). Stuart (1999) notes some current VE technology limitations for users:

1. Position trackers with small working volumes, inadequate robustness, and problems of latency and poor registration.
2. HMDs with very poor resolution, somewhat limited field of view, and encumbering form factors (i.e., weight and tethers).
3. Virtual acoustic displays that require a great deal of computational resources in order to simulate a small number of sources.
4. Force and tactile displays, still in their infancies, with limited functionality.
5. Image generators that cannot provide low-latency rendering of head-tracked complex scenes, requiring severe trade-offs between performance and scene quality.

In addition to these limitations for users, Stuart notes developers of VEs experience limitations with the real-time capabilities of operating systems and inadequate software development tools.

### 3.3 Cognitive Ergonomics

As discussed, one of the primary benefits of VE systems is their ability to assist users in developing knowledge about subject matters portrayed in the virtual world. However, VEs developed to date deliver little of this promise. Stuart (1999) notes that in order for VEs to fulfill their promise they will have to deliver:

1. Sociability: the capability to permit multiple users to share an environment (see chap. 44, this volume).
2. Veridicality: the accuracy with which a VE reproduces its modeled environment (fidelity) (see chaps. 12 and 17, this volume).
3. Resolution: the level of precision and detail provided by a VE system (see chap. 3, this volume).
4. Immersion, presence: the extent to which users believe that they are "in" a virtual environment (see chap. 40, this volume).
5. Engagement: the extent to which users are deeply involved and occupied with what they are doing in the virtual environment (see chap. 40, this volume).
6. Reconfigurability: the capability to modify a VE model.
7. Responsiveness: the ability of a VE to respond to user inputs (lag, latency; see chap. 3, this volume).
8. Stability: lack of jitter or oscillation in the VE display (see chaps. 3, 11, 13, this volume).
9. Robustness: the ability to function correctly across all operational ranges.

## 4. DOMAIN-SPECIFIC CHALLENGES

### 4.1 Design and Testing

One of the first, and still most important application, of VEs has been for architectural design (see chap. 13, this volume). The success of this application has led to experimentation with VE in other kinds of product design and testing. The use of VE systems for designing and testing new products and systems is gaining acceptance, particularly as a complement to computer-aided design technology (Boman, 1995; chap. 52, this volume). The Boeing Company's 777 commercial airliner marked a significant milestone in the fielding of "paperless" design techniques, centered particularly on virtual representations of the aircraft. Less well-known is the application of these techniques to projects such as the international space station and the B-2 aircraft. Virtual representation of product design has been particularly helpful in the automobile industry in shortening product design cycle times.

### 4.2 Manufacturing

Virtual environments offer much potential for improving manufacturing productivity, particular in the emerging "mass-customization" marketplace (see chap. 54, this volume). Most applications to date have been with experiments in laying custom assembly instructions over a standardized work area. Breakthroughs in this domain await improvements in the physical ergonomics challenges discussed (see chap. 41, this volume).

### 4.3 Operations

Experiments have been successfully conducted in using VE technology to support operations, such as feeding firefighters real-time maps of buildings as they fight fires (Bliss, Tidwell, & Guest, 1997). Breakthrough applications in this domain await more robust (i.e., not fragile) VE equipment to support the harsh demands of varied operating environments.

### 4.4 Information Management

Successful VEs have been developed and fielded for information management problems, such as database maintenance (see chap. 53, this volume). This domain is limited by the cognitive ergonomics problem discussed earlier.

### 4.5 Entertainment

The broadest and most successful application of VE technologies has been for entertainment, both home-based and location-based systems (see chap. 55, this volume). The Disney Corporation has demonstrated its commitment to exploring the advantages of VE-based entertainment with its Aladdin VE ride at Epcot Center and Disney Quest arcade. However, the fact that 5% to 10% of Aladdin visitors experience dizziness or nausea is a reminder of the problems that may hinder mass appeal of VE systems (Boman, 1995).

### 4.6 Medicine

Extensive experimentation continues in the use of VEs for medicine, particular in the specialized field of medical training (see chap. 47, this volume). Beyond medical training, researchers are investigating the use of VE for aiding people with disabilities, cognitive rehabilitation, and telepresence for remote surgery (Boman, 1995). The National Institute's of Health's Visible Human Project is creating complete, anatomically detailed, 3-D representations of the male and female human body. The project's data sets are designed to serve as a common reference point for the study of human anatomy. The data sets are being applied to a wide range of educational, diagnostic, treatment planning, virtual reality, artistic, mathematical and industrial uses by over 1,000 licensees in 41 countries. But key issues remain in the development of methods to link such image data to text-based data. Standards do not currently exist for such linkages. Basic research is needed in the description and representation of image-based structures and to connect image-based structural-anatomical data to text-based functional–physiological data (Card, Mackinlay, & Shneiderman, 1999).

### 4.7 Education and Training

Many observers believe that education and training will be a major application domain for VE technologies (see chaps. 43–46, this volume). The few systems developed to date, however, have yet to demonstrate the kind of competitive advantage required for widespread adoption of this technology. Some specific applications include astronaut, military, and sports training (Boman, 1995).

## 5. CONCLUSIONS

Virtual environment technology offers exciting potential for revolutionizing the way work is performed in many domains. However, VE technology will fail to deliver on its potential if system developers do not begin to understand and grapple with the problems facing organizations desiring to adopt this technology. Chief among these challenges is the need for "killer" applications in varied domains, supported by robust and stable equipment. It is tempting to project the adoption of VE technology based on our experience with related technologies, such as the personal computer. However, any such projection would be conjecture as sufficient information about the degree to which VE is actually similar to such technologies, and ubiquitous applications do not yet exist.

## 6. REFERENCES

Bhattacherjee, A., & Gerlach, J. (1998). Understanding and managing OOT adoption. *IEEE Software, 15*(3), 91–96.

Bliss, J. P., Tidwell, P. D., & Guest, M. A. (1997). The effectiveness of virtual reality for administering spatial navigation training to firefighters. *Presence, 6*(1), 73–86.

Boman, D. (1995). *Virtual environment applications* (Tech. Rep. No. D95-1917). Menlo Park, CA: SRI Consulting.

Card, S. K., Mackinlay, J. D., & Shneiderman, B. (1999). *Readings in information visualization: Using vision to think.* San Francisco, CA: Morgan Kaufmann Publishers.

Hoffmann, M., & Wright, A. (1997). *Technology enablers for the intelligent enterprise.* (Tech. Rep. No. D97-2065). Menlo Park, CA: SRI Consulting.

Lai, V. S., & Guynes, J. L. (1997). An assessment of the influence of organizational characteristics information technology adoption decision: a discriminative approach. *IEEE Transactions on Engineering Management, 44*(2), 46–157.

Leavitt, N. (1999, July). Online 3D: Still waiting after all these years. *Computer*, pp. 4–7.

Lefebvre, É., & Lefebvre, L. A. (1996). *Information and telecommunication technologies: The impact of their adoption on small and medium-sized enterprises.* Ottawa, Canada: International Development Research Centre.

Mackie, R. R., & Wylie, C. D. (1988). Factors influencing acceptance of computer-based innovations. In M. Helander (Ed.), *Handbook of human–computer interaction* (pp. 1081–1106). Amsterdam: Elsevier Science Publishers.

Ostberg, O., & Chapman, L. (1988). Social aspects of computer use. In M. Helander (Ed.), *Handbook of human–computer interaction* (pp. 1033–1049). Amsterdam: Elsevier Science Publishers.

Rogers, E. M. (1995). *Diffusion of innovations* ($4^{th}$ ed.) New York: Free Press.

Senge, P. M. (1990). *The fifth discipline.* New York: Doubleday.

Shibley, J. J. (1998). *A primer on systems thinking and organizational learning* [Portland Learning Organization Group, Online]. Available: http://www.systemsprimer.com

Stuart, R. (1996). *The design of virtual environments.* New York: McGraw-Hill.

Stanney, K. M., Salvendy, G., Deisinger, J., DiZio, P., Ellis., S., Ellison, J., Fogelman, G., Gallimore, J., Hettinger, L., Kennedy, R., Lackner, J., Lawson, B., Maida, J., Mead, A., Mon-Williams, M., Newman, D., Piantanida,

T., Reeves, L., Reidel, O., Singer, M., Stoffregen, T., Wann, J., Welch, R., Wilson, J., & Witmer, B. (1998). Aftereffects and sense of presence in virtual environments: Formulation of a research and development agenda. *International Journal of Human–Computer Interaction, 10*(2), 135–187.

Utz, Jr., & W. J. (1992). *Software technology transitions: Making the transition to software engineering.* Englewood Cliffs, NJ: Prentice-Hall.

Vince, J. (1995). *Virtual reality systems.* Workingham, England: Addison-Wesley.

# 27

# Virtual Environments and Product Liability

Robert S. Kennedy, Kristyne E. Kennedy, and Kathleen M. Bartlett

*RSK Assessments, Inc.*
*1040 Woodcock Road, Suite 227*
*Orlando, FL 32803*
*6kennedy@bellsouth.net*

## 1. INTRODUCTION

Virtual environments (VE) employ dynamic images to produce the vicarious experience of immersion and presence in flight simulators and helmet-mounted displays (HMDs). However, these exposures often produce unwanted side effects such as motion sickness–like symptoms, balance disturbances, profound drowsiness, and coordination problems (see chaps. 29–32, this volume). Not everyone appears to be affected to the same extent, but such changes, when they occur, imply that the human nervous system has been temporarily recalibrated, or adapted, to the virtual world. This adaptation involves fundamental, significant changes in the basic function of an individual that may not be noticed when the individual exits the virtual environment (see chaps. 37–39, this volume). As individuals subsequently perform routine tasks, unaware of any lingering effects of VE exposure, negative consequences in the form of accidents may occur; thus, these aftereffects may have implications for safety and health. If a VE product occasions problems, the liability of VE developers could range from simple accountability (i.e., reporting what happened) to full legal liability (i.e., paying compensation for damages). In order to minimize their liability, manufacturers and corporate users of VE devices (i.e., companies) can proactively assess the potential risks associated with human factors by using a comprehensive systems safety approach. Human factors systems safety actions undertaken include the following: (1) systems should be properly designed; (2) aftereffects should be removed, guarded against, or warned against; or (3) adaptation methods should be developed; or (4) users should be certified to be at their preexposure levels; or (5) users should be monitored and debriefed.

## 2. AFTEREFFECTS FOLLOWING USAGE OF A PRODUCT

While most VE product manufacturers have given only slight, if any, consideration to issues of litigation involving aftereffects of VE exposure, the inevitable risks incurred by placing these products into wide circulation must be considered. It has been established, for example, that exposures of humans to simulated environments, for entertainment or education, produces motion sickness–like discomfort and aftereffects such as balance disturbances (Stanney, Mourant, & Kennedy, 1998; see chaps. 29–32, 37–39, this volume). The military flight simulators of the 1960s, which were the VEs of that day, first called our attention to this problem (Miller & Goodson, 1960). In those early days fielded devices had equipment limitations such as excessive transport delays, optical distortions, and flickering imagery; these equipment problems were considered to be the agents of the discomfort. Many of these equipment problems have been remedied in modern simulator and VE systems. At the same time, technological advances in simulation engineering have enabled newer systems to portray even more compellingly realistic dynamic scenarios. Yet varying degrees of motion sickness–like symptoms persist and have been reported in nearly every flight simulator fielded by the military services.

Although carefully controlled studies that employ comparable experimental design, kinematics (i.e., velocity and spatial frequency of a visual scene), equipment features, participants, durations of exposure, and other have not yet been conducted, it is known from survey work (Kennedy, 1998; Kennedy, Stanney, Drexler, Compton, & Jones, 1999) that the overall incidence of VE sickness (cybersickness, McCauley & Sharkey, 1992) appears to be greater than it ever was with flight simulators. In this chapter, some of these reports of motion sickness–like symptoms occasioned by VE exposures are described for purposes of illustration only. The reader is referred elsewhere in this volume for more complete treatments dealing with the cause and incidence of cybersickness (see chaps. 30–31, this volume). Here we merely wish to suggest that if sickness occurs, accidents could result, and if so there may be an issue of product liability. Specifically, it is known that VEs and simulators are products intended for use by adults and children and that they are designed to be manufactured and sold. Sickness, which may occur in connection with usage, may imply a hidden defect in the product or, alternatively, it may be an expected outcome of the product's usage. Also, in addition to the obvious symptoms of sickness, other adverse consequences of exposure may occur, and the symptoms of discomfort may only be a harbinger of other outcomes.

For the purposes of this discussion, the established definition (McCauley, 1984) that describes simulator sickness as the experience of symptomatology during and after the use of a VE that would not ordinarily be experienced if the same activity were carried out in the real world is followed. Obviously, when a real environment ordinarily occasions sickness, if a replicated VE also produces sickness then the simulator actually provides a "good" simulation.

Outcomes known to exist following exposure to VEs and simulators include nausea (up to 30% of exposures), eyestrain (up to 40% of exposures), drowsiness, salivation, sweating, headache, and dizziness/vertigo, as well as loss of postural stability (Kennedy, Fowlkes, & Lilienthal, 1993). Vomiting is infrequent (< 1%) and when it occurs, it is usually more than one hour after exposure. Long-term aftereffects, including visual flashbacks and disorientation, have been routinely reported to have occurred up to 12 hours after simulator exposure (Baltzley, Kennedy, Berbaum, Lilienthal, & Gower, 1989; Lilienthal, Kennedy, Berbaum, & Hooper, 1987), but such data are only now being collected experimentally. Sleepiness and fatigue are symptoms reported by large numbers, but the significance of this for subsequent performance is not well appreciated (see chap. 30, this volume). The incidence of all of these long-term aftereffects is generally low (< 10% of all exposures) and is correlated with the strength of aftereffects immediately after exposure and the duration of exposure. Although less than 10% with long-term aftereffects may seem like a low proportion, the huge number of persons who are exposed each day to such devices is staggering and may be expected to increase; 10% of a growing number of exposures represents a serious number of persons who may be expected to experience long-term aftereffects.

Visual and vestibular scientists have previously experimented with ancillary aftereffects when they were examining distortion due to tilted figures (Ebenholtz, 1988, 1992) and altered lenses (Wallach & Kravitz, 1965, 1968, 1969). These same distortions can be found in today's HMDs. The vestibulo-ocular reflex (Leigh, 1996; O'Leary, 1992) produces involuntary rotation of the eyes in the direction opposite to head movement in order to maintain fixation and reduce slippage of images across the retina (see chap. 30, this volume). The human internal calibration of this reflex seems to depend upon the recent experience of a relationship of head movement and the resulting image motion (Gonshor & Melville-Jones, 1976). If a new and consistent relationship between head movement and scene displacement is introduced, such as via an HMD, the scenery will at first appear to move because eye rotation will fail to stabilize the image (cf. Dizio & Lackner, 1997; see chap. 38, this volume). Apparent shifts in the scene with head movement may produce impressions of actual individual movement (vection), motion sickness, and eyestrain (Kennedy, Hettinger, Harm, Ordy, & Dunlap, 1996; see chaps. 30–32, this volume). Eventually, the individual will adapt so that the new relationship elicits eye rotation that yields a stable retinal image (see chaps. 37–39, this volume). However, the adapted individual will experience aftereffects; once adjusted to distortion, a head movement that produces the ordinary displacement of the visual scene will again appear to shift. Readaptation to the natural environment will be required. From the dynamics and kinematics of the helmet mounted VE displays, one can predict that such a recalibration is occurring in users, although this problem has only infrequently been studied (Prothero & Parker, 1999). Considerable research needs to be conducted in order to determine the size, time course, and duration of such aftereffects.

This peculiar kind of adaptation can be expected to occur with increasing frequency as VE systems are employed for longer exposure periods, the natural consequence of using them as training devices. This effect, when it occurs, can produce significant disability when entering the real world; indeed, initial exposure to VE would probably be a lot like getting used to a new pair of glasses and eventually could be no more troublesome than changing glasses. Nevertheless, this adaptation involves fundamental, significant changes in the function of the individual at a level at which he or she may be unaware. If developers take steps to address and prepare for the potential safety hazards involved with VE system interaction, they may be able to minimize harm. Companies that take such proactive measures and perform extensive preliminary safety analyses may circumvent product liability issues.

Other aftereffects are worth noting, particularly balance problems, which imply that the human nervous system has been temporarily recalibrated to the virtual world, and time may be required before the individual is ready to navigate safely in the "stable world." "Sea legs" or "le mal de debarquement" (Gordon, Spitzer, Shupak, & Doweck, 1992) is a related malady that also results in perturbed posture (see chap. 39, this volume), but such perceptual and motor aftereffects have long been known to occur. See studies of human perceptual distortion in connection with mirror and prism use (see chap. 31, this volume). They have also been reported following weightlessness. One could argue that if such effects outlast the period of time an individual is under the control of the VE device owner/manager, care for well-being should extend beyond the owner's property line. More is said about this later.

In one particularly interesting simulator study (Gower, Lilienthal, Kennedy, & Fowlkes, 1987), U.S. Army pilots who were exposed to the flight simulator daily for two weeks reported less sickness, as would be expected, with each subsequent exposure. But ironically, and perhaps with implications for safety, they also demonstrated progressively more balance problems over this same time frame and as adaptation ensued (Kennedy, 1993).

Similarly, in a remarkable case following VE exposure, posteffects were reported to have occurred in a middle-age man after he spent approximately 2 hours in a three-story VE entertainment facility (Kennedy, Stanney, & Fernandez, 1999). After exposure, the individual, who historically had not been judged suprasensitive to motion sickness, exhibited the characteristic signs and symptoms (viz., nausea, imbalance, dizziness, vertigo, sleep difficulties) of motion sickness and had to remain in bed for his first 48 hours at home. Subsequently, in medical examinations (neurologic and otolaryngologic) no organic cause for his symptoms could be found. The symptoms persisted for weeks, and then months, and he was still experiencing significant distress as much as six months after the initial exposure. Fortunately, by the ninth month he recovered, and now, several months later, he is considered to be symptom-free. If, during a bout of aftereffects, he were to become injured in an auto accident related to the persistent illusory perception of motion, it could be argued that the VE exposure caused the accident.

Note that the legal issue that surfaced here is not simply whether one can recover from cybersickness or any other form of motion sickness. The problem concerns the individual's ability to safely perform routine tasks. If his mental and physical functioning appears to have been perturbed because of his exposure-related distress, were this individual to fall from a roof, have an automobile accident, or participate in some other activity that he had not been warned about, then entities responsible for placing the VE device "in the stream of commerce" may be liable. For example, suppose that a driver of an automobile were to have a balance disturbance following a VE exposure, and suppose that his balance performance was equivalent to the disturbances seen with a .10% blood alcohol concentration level (Kennedy, Turnage, Rugotzke, & Dunlap, 1994). In this case, it could be argued that if postural equilibrium is disrupted, steering ability is also likely to be disrupted, and the cause of the degraded performance could be associated with a VE exposure. If that individual had an auto accident and had had no alcohol, he could argue that his imbalance might relate to the aftereffect experienced in the VE exposure.

Finally, profound drowsiness, sleepiness, and the experience of fatigue, frequently reported by-products of VE and simulator exposure, may occur during that period after a user is released from the VE device and is occupied with walking or driving home. These aftereffects can represent real hazards in the operation of motor vehicles and in certain types of work and recreational activities. Time may be required before the individual is ready to navigate safely in the "real world." Therefore, developers of VE systems should take steps to warn VE users about potential aftereffects and their implications for safety.

How long aftereffects may last after individuals leave the VE is an important question for which there are scant data. In one survey Baltzley et al. (1989) has shown that with 2- to 4-hour exposures in flight simulators, 1% to 10% of the operators had one or more symptoms that lasted a period of time after exposure and persisted after subjects left the simulator building. Careful analysis of problems associated with VE may help developers take proactive steps (such as the use of warnings, certification tests, and/or checklists) to assure that people who are exposed to VE may safely reenter the real world and to minimize developers' legal liability.

## 3. PRODUCT STANDARDS

With this potential for negative aftereffects from a product that may achieve wide usage in the private sector for education and training, as well as for entertainment, developers of VE systems should have concern over product liability issues. The law of product liability has its origins in tangible property, but has been extended beyond tangible goods to include intangibles such as electricity after it has been delivered to the consumer (Phillips, 1998). It has also been applied to other items such as natural products and real estate fixtures such as a house (Phillips, 1998). The rationale for imposing strict product liability rests on the following ideas: (1) that the manufacturer is in the best position to reduce the risk and insure against the risk; (2) that the manufacturer is the one responsible for the product being on the market; and (3) that the consumer lacks the means and skill to investigate the soundness of a product for herself. This section will deal with product liability issues and definitions of ordinary defects, unknowable defects, and product and consumer expectation, among others. The relevance of defective warnings and negligence on causation will be discussed.

If a commercial product is placed in the stream of commerce and injury occurs, then the product may be adjudged defective because it occasions injury, and the responsible person or entity who knew or should have known it was defective can be held liable (Phillips, 1998). This liability can be attached to any individual or business that profited from the defective technology. For legal liability to be found, a product "defect" that is "unreasonably dangerous to the user" must be proven (Phillips, 1998, p. 53). Liability may also be imposed for inadequate design and/or inadequate warnings for foreseeable risks of harm that could have been avoided by the adoption of a reasonable alternative design or by reasonable instructions or warnings (American Law Institute, Restatement 3rd of Torts, 1997). In the present case, the VE system must be considered "dangerous to the extent beyond that which would be contemplated by the ordinary consumer" (p. 11). If VE systems have this known propensity for aftereffects documented in published scientific literature, the ordinary consumer may not be aware of any of the effects, or of all of the effects. Furthermore, if they do not use reasonable care, the potentially adverse consequences of product liability may accrue, to persons involved in research and development of these systems for purposes of bringing them to consumer use.

Because of the postural unsteadiness, drowsiness, and the discomfort of vomiting and retching, it would appear likely that aftereffects could result in accidents after usage of a VE system (Kennedy & Stanney, 1996a). This is a likely scenario since, while many users of VE technology are aware of the possible side effects of cybersickness, and, when afflicted, these individuals may restrict their own movements, predictably few are aware of the potential for loss of balance and eye–hand incoordination effects (Kennedy, Stanney, Compton, Drexler, & Jones, 1999) as well as long-term aftereffects. As an extreme example, there are reports of sickness following exposure to immersive video games at commercial amusement parks where vomiting occurred less than 15 hours later; our own research has shown that vomiting can occur long after the experience of a 60-minute exposure to a VE display (Kennedy, Stanney, Dunlap, & Jones, 1996). Incidentally, a similar situation occurs with the posteffects following long durations in microgravity (Reschke et al., 1994), although the number of persons exposed to extended durations in space is far smaller than the anticipated number that will be exposed under training regimes planned for the military (see chap. 43, this volume) and other federal services and those exposed in educational (see chaps. 45–46, this volume) or recreational settings (see chap. 55, this volume). In our research, we advocate the use of a procedure to certify that an individual has recovered to at least preexposure levels. This process, if in force, can protect users, as well as vendors, manufacturers, and other purveyors of VE systems from potentially damaging lawsuits.

Manufacturers might wish to follow the trend in modern law to avoid costly, time-consuming litigation all together by anticipating and responding to any potentially compromising possibilities. Thus, in order to minimize legal liability, VE systems may need to be properly prepared and accompanied by warning labels and appropriate directions (i.e., debriefing protocols) to raise the level of consumer awareness of the potential for harm. In order to provide effective debriefing protocols, objective measures of adaptation resulting from VE exposure are needed. Specifically with respect to VE, rather than waiting for issues to arise, it might be prudent to address safety concerns before they result in crisis or harm. A proactive rather than reactive approach may allow researchers to identify and address potentially harmful side effects related to the use of VE technology. Standards or criteria need to be developed to guide VE developers' decisions, such as the principle of harm minimization. Under this principle, an activity should minimize harm to all exposed, which requires recognizing the concerns and interests of the various individuals involved (Kallman, 1993). More principles need to be established to guide and direct the safe development of virtual technology.

## 4. PRODUCT WARNINGS

Under a product liability theory, a manufacturer (or seller) of products may be liable for failure to provide adequate warnings as to dangers associated with its products. Generally, a manufacturer has a duty to warn against the latent dangers resulting from foreseeable uses of its products that it knew or should have known. An adequate warning is one that is reasonable under the circumstances (Lehto & Miller, 1986). To be adequate, a warning must describe the nature and extent of the danger involved (Cunitz, 1981). Often warning cases involve issues of providing adequate instructions for safe use of a product. A warning is distinguished from an instruction in that instructions are calculated primarily to secure the efficient use of a product, whereas warnings are designed to insure safe use (Phillips, 1998). A product distributed without adequate warnings or instructions is sometimes treated as one with a design defect. Therefore, a product may be faultlessly manufactured and designed, but may be defective when placed in the consumers' hands without an adequate warning concerning the manner in which to use the product safely. It is important to note that, ultimately, a manufacturer's liability for breach of duty to warn will depend on the plaintiff proving causation. This means the plaintiff must prove that he or she would not have suffered the harm in question if adequate warnings and instructions had been provided. Even though this element could be difficult to prove in many cases, it is of course in the manufacturer's best interest to protect from liability by giving adequate warnings. A warning must also be effectively communicated, and if it is, then failure to read the warning as given may bar recovery for a plaintiff (Lehto & Miller, 1986).

While there exists nothing in the case law addressing the situation of a consumer who is ultrasensitive to motion sickness, we do know that such persons exist in many environments (cf. Crampton, 1990). Furthermore, an analogy of hypersusceptibility can also be made to the allergic consumer where the duty normally owed to allergic users is one of warning, and then only when the plaintiff is a member of a substantial or appreciable number of persons subject to allergy, where the defendant should have known of this risk (Phillips, 1998). A seller may be required to warn that his product contains an ingredient that is a known allergen and may have to warn of the symptoms associated with an allergic reaction (Phillips, 1998). There is precedent for this action in that certain video games have been sold with labels that warn of the prospect of epileptiform seizures (see chap. 29, this volume) by susceptible individuals.

Furthermore, a "postsale" or "continuing" duty to warn generally exists where a manufacturer is responsible for warning of a defective product at any time after it has been manufactured and sold if the manufacturer becomes aware of the defect. This means that the manufacturer is under a duty to issue warnings or instructions as they later become known to the manufacturer, or even as they should become known, and that the manufacturer is expected to stay informed about advancements in the state of the art and/or of later accidents involving dangers in the product. For example, a drug manufacturer is treated as an expert and is under a continuing duty to keep abreast of scientific developments in regard to a product and to notify the medical profession of any additional side effects discovered from its use. Similarly, this idea could be applicable in the case of VE systems already in existence, imposing a continuing duty on the manufacturers to keep up with developments and to provide proper warnings (American Law Product Liability 3rd, 1988).

## 5. PRODUCT LIABILITY ISSUES

The increase in product liability claims and settlements witnessed in the last half of the 20th century may be linked to the development of more sophisticated products. The sheer complexity of modern products is a source of danger because consumers may not understand safe use; risks may not be readily apparent. Additionally, consumers may not possess the knowledge to judge the quality, efficacy, and safety of a product; thus, assurance of minimum standards of performance is demanded from manufacturers (Howells, 1993).

In recent years, product liability litigation has been extended into many areas, including recreational equipment such as pool tables, trampolines, amusement rides, and attractions. Cases have arisen involving injuries on amusement park rides, such as water slides and roller coasters, where the amusement park and/or manufacturer were sued under a product liability theory. In some cases, a higher standard of care is imposed on amusement park operators, as well as manufacturer-designers of amusement park rides, because of the risk created to thousands of passengers, many of them children, for monetary gain (this is similar to the higher standard of care imposed on a common carrier, who is required to exercise great care to protect passengers).

It is true that there is generally no duty to warn users of an open and obvious danger incident to a particular use of a recreational product that was not defectively designed or manufactured (for example, diving into a swimming pool). However, in the case of VE sickness, the duty to warn most likely will exist, because the danger is not open and obvious.

Product liability law involves four basic types of liability standards: contractual or warranty standards, which involve products that fail to meet the promised standard; negligence standards, which judge the conduct of the defendant in light of risks and benefits that result from his conduct; strict liability standards, which do not judge the conduct of the producer/supplier but rather assess the product; and absolute liability standards, which are based on proof of damage caused by the product (Howells, 1993). Although any of these standards might be the basis for litigation in a case involving VE sickness and manufacturers of VE systems, negligence standards, in particular, merit careful consideration.

Negligence has been defined as conduct that involves an unreasonably great risk of causing damage (Terry, 1915) and, alternatively, as conduct that "falls below the standard established by law for the protection of others against unreasonably great harm" (Restatement [Second] of Torts, 1965, sec. 282). To make out a cause of action in negligence, the plaintiff does not need to establish that the defendant either intended harm or acted recklessly. Negligence focuses on the conduct of the various participants who are responsible for the product's creation and entry into the marketplace. The judge balances the risks and benefits of the defendant's conduct; risks include those that the defendant knew of, or ought to have known of, at that time. Risks that become known at a later date are irrelevant unless they could have been discovered by the defendant via testing or research that he or she could be expected to have undertaken (Howells, 1993).

The reasonable manufacturer is expected to be aware of current industry-wide standards and technology. Although evidence that a manufacturer has met self-imposed standards, voluntary standards, or even statutory standards is significant, such evidence alone does not establish that the manufacturer is not negligent. An entire industry can be found negligent for continuing to apply unacceptable standards of manufacture (Hooper, 1932) or for failing to make currently available technological improvements. The flexibility of the negligence standard is achieved by permitting interplay among three factors. In every negligence action, the court must consider:

(1) the probability that harm would occur; (2) the gravity of the harm; and (3) the burden of taking precautions to protect against the harm—what the manufacturer might have done to minimize or eliminate the risk of harm.

Because all human activity involves an element of risk, a defendant's conduct is deemed negligent only when it falls below what a "reasonable" person would have done under similar circumstances (Brown, 1991). Risks to health justify more precautions than risks to property, and the more serious the risk, the greater the precautions required. Although this assessment is made at the time of the alleged negligent act, for example, the time of manufacture of a defective product, there may be a separate postmarketing duty to monitor the product. When risks are discovered it may be necessary to either order the recall of the product or to take other steps to avert any danger (Howells, 1993).

Even though such safety principles are developed, unintended outcomes may occur. Thus, after performing a safety analysis to minimize harm, the issue of who is accountable (i.e., the appropriate person to respond when something undesirable occurs), responsible, and/or strictly liable when harm occurs must be resolved (Johnson, 1993). Responsibility is broken into role responsibility (i.e., responsibility linked to an individual's duties or by virtue of occupying a certain role), causal responsibility (i.e., a person who does something or fails to do something and this causes something else to happen), blameworthy responsibility (i.e., a person who did something wrong that led to an event or circumstance), and legal liability responsibility (i.e., one is legally liable when it is claimed that the individual must pay damages or otherwise compensate those harmed). If a VE product malfunctions, the consequences for the VE developers could range from simple accountability (i.e., reporting what happened) to full legal liability (i.e., paying compensation for damages). If VE products are being employed for training purposes, whether or not the training is internal to the company or carried out under contract may also influence liability issues. In addition, a company's position on this liability continuum may be related to how proactive that company was in terms of product standards and how extensive a preliminary safety analysis was performed.

## 6. SYSTEMS SAFETY APPROACH

In order to minimize their liability, manufacturers and corporate users of VE devices (i.e., companies) can proactively assess the potential risks associated with human factors by using a comprehensive systems safety approach. The human factors system safety approach, as originally put forward by Christensen (1993), has five elements, including: (1) design, (2) remove, (3) guard, (4) warn, and (5) train, in general order of preference of application. We follow the Christensen model, and for purposes of the VE product safety rendition offered below, provide two new approaches which can usefully be added to the case for VE systems: (6) certification and (7) monitoring/debriefing (Kennedy & Stanney, 1996b).

1. *Design*. Design new products to be as free of hazard as possible. Obviously, the first method of choice would be to eliminate any known hazard, which requires knowledge of the exact causes of safety concerns. Unfortunately, the exact causes of simulator sickness associated with VE use are not currently well understood, which underscores the need for cautious introduction of new products.

2. *Remove*. Remove hazards from existing systems. The ability to fix existing systems is limited by the same lack of knowledge that hinders the ability to design safe new systems. Making changes in existing systems and observing the impact on sickness incidence represents the best methodology for gaining knowledge to deal with the problem.

3. *Guard*. Where hazards cannot be eliminated or removed, users should be prevented from coming into inadvertent contact with the hazard. Guards must be convenient, obviously

essential, and must not impair operations. If a guard is not effective in eliminating a problem or interferes with training it should be eliminated.

4. *Warn.* Warn users of remaining hazards. A simulator sickness field manual has been developed by the Department of Defense (NTSC, 1989) and is distributed at simulator sites. The scientific basis of this field manual is described in a "guideline" report (Kennedy et al., 1987). The "field manual" and "guideline" are used to teach instructors and trainees to recognize simulator sickness symptoms and have been distributed to the three military services and the U.S. Coast Guard; they are also available in NATO countries. The "field manual" warns trainees using military simulators about potential risks involved when resuming activities, including flying, immediately after a bout of simulator sickness. Similar manuals and guidelines might be provided to users of VE systems. It is important to note, however, that warning is a less preferable alternative than the first three approaches (i.e., effective design, removing hazards, and guarding, Christensen, 1993).

5. *Training.* Even the best products may require training for safe usage. Were humans not so adaptable to altered sensory environments, no one would be able to tolerate simulator training or VE devices. Much of the advice contained in the field manual described above has to do with how to facilitate sensory–motor adaptation to the simulator and readaptation once outside it.

Teaching people to adapt rapidly to the peculiarities of VEs will be the primary means of enhancing the benefits of these devices while reducing the discomfort and risk that they may produce. If the first four methods (i.e., design, remove, guard, warn) are not sufficiently effective, a program of training can be prepared so that users can be taught to avoid hazards. In the case of VE sickness, perceptual adaptation is likely to occur (cf. Kennedy, Smith, & Jones, 1991; Uliano, Kennedy, & Lambert, 1986; see chaps. 37–39, this volume) and can be employed to reduce symptoms if proper schedules are followed.

6. *Certification.* In connection with the five main human factors system safety approaches outlined by Christensen (1993), we believe two additional approaches should be added (Kennedy & Stanney, 1996b). The first concerns measurement of the effectiveness of the freedom from hazard. This is called *process certification.* Should human factors problems be suspected, VE systems should be measured to determine the level of hazard they are expected to represent. Until recently, self-report was the chief method for certifying that a simulator was safe. Currently, objective measures and tests of human balance are being developed that could be used for certification purposes (Kennedy & Lilenthal, 1994).

7. *Monitoring/Debriefing.* The second additional approach involves measuring individuals to determine if they have been adversely affected by exposure to VE systems (Kennedy & Stanney, 1996b). Even if a system "passes" its certification procedure, human variability is such that some individuals may still be adversely affected by VE system exposure. In order to limit one's liability, it may be prudent to establish monitoring procedures, such as a postexposure balance test. The same or similar tests of human balance used for certification could also be used for monitoring. For those individuals who "fail" or marginally "pass" the balance test, debriefing protocols could be developed that provide information to those affected so they can act accordingly to minimize harm.

## 7. CONCLUSIONS

Virtual environment systems, while holding much promise to educate, train, and entertain, also have the potential to harm. Individuals exposed to VE systems may experience malaise during exposure and adverse aftereffects that could linger days or even months after exposure. System developers can follow the systems safety certification approach discussed in this chapter to ensure they have exercised due care in protecting their patrons from harm.

## 8. REFERENCES


American Law Institute. (1965). *Restatement (second) of torts*, 291, 395, 402a: Author.

Baltzley, D. R., Kennedy, R. S., Berbaum, K. S., Lilienthal, M. G., & Gower, D. W. (1988). *Flashbacks and other delayed effects of simulator sickness: Incidence and implications.* Unpublished manuscript.

Baltzley, D. R., Kennedy, R. S., Berbaum, K. S., Lilienthal, M. G., & Gower, D. W. (1989). The time course of postflight simulator sickness symptoms. *Aviation, Space, and Environmental Medicine, 60*(11), 1043–1048.

Baltzley, F. A., Kennedy, R. S., Berbaum, K. S., Lilienthal, M. G., & Gower, D. W. (1989). The time course of postflight simulator sickness symptoms. *Aviation, Space, and Environmental Medicine, 60*, 1043–1048.

Brown, S. (Ed.). (1991). *The product liability handbook: Prevention, risk, consequence, and forensics of product failure.* New York: Van Nostrand Reinhold.

Christensen, J. M. (1993). Forensic human factors psychology: Part 2. A model for the development of safer products. *CSERIAC Gateway, 4*(3), 1–5. Dayton, OH: Crew System Ergonomics Information Analysis Center.

Crampton, G. (Ed.). (1990). *Motion and space sickness.* Boca Raton, FL: CRC Press.

Cunitz, R. J. (1981, May/June). Psychologically effective warnings. *Hazard Prevention*, 5–7.

DiZio, P., & Lackner, J. R. (1997). Circumventing side effects of immersive virtual environments. In M. J. Smith, G. Salvendy, & R. J. Koubek (Eds.), *Design of computing systems: Social and ergonomic considerations* (pp. 893–896). Amsterdam: Elsevier Science Publishers.

Ebenholtz, S. M. (1988). *Sources of asthenopia in navy flight simulators* (Accession No. AD-A212699). Alexandria, VA: Defense Logistics Agency, Defense Technical Information Center.

Ebenholtz, S. M. (1992). Motion sickness and oculomotor systems in virtual environments. *Presence, 1*(3), 302–305.

Gonshor, A., & Melville-Jones, G. (1976). Extreme vestibulo-ocular adaptation induced by prolonged optical reversal of vision. *J. Physiol., 256*, 381–414.

Gordon, C. R., Spitzer, O., Shupak, A., & Doweck, H. (1992). Survey of mal de débarquement. *BMJ, 304*, 544.

Gower, D. W., Lilienthal, M. G., Kennedy, R. S., & Fowlkes, J. E. (1987, September). Simulator sickness in U.S. Army and Navy fixed- and rotary-wing flight simulators. In *conference proceedings No. 433 of the AGARD Medical Panel Symposium on Motion Cues in Flight Simulation and Simulator-Induced Sickness* (pp. 8.1–8.20), Brussels, Belgium:

Hettinger, L. J., Berbaum, K. S., Kennedy, R. S., Dunlap, W. P., & Nolan, M. D. (1990). Vection and simulator sickness. *Military Psychology, 2*(3), 171–181.

Hooper, T. J. (1932). 60 Federal 2d 737 (Second Circuit Court of Appeals).

Howells, G. (1993). *Comparative product liability.* Aldershot, England: Dartmouth Publishing.

Johnson, D. (1993). *Computer ethics* (2nd ed.). Englewood Cliffs, NJ: Prentice-Hall.

Kallman, E. A. (1993). Ethical evaluation: A necessary element in virtual environment research. *Presence, 2*(2), 143–146.

Kennedy, R. S. (1993). *Device for measuring head position as a measure of postural stability* (Final Rep. No. 9260166). Washington, DC: National Science Foundation.

Kennedy, R. S. (1998, September). *Gaps in our knowledge about motion sickness and cybersickness.* Paper presented at the Motion Sickness, Simulator Sickness, Balance Disorders and Sopite Syndrome Conference, New Orleans, LA.

Kennedy, R. S., Berbaum, K. S., Lilienthal, M. G., Dunlap, W. P., Mulligan, B. E., & Funaro, J. F. (1987). *Guidelines for alleviation of simulator sickness symptomatology* (NAVTRASYSCEN TR-87-007). Orlando, FL: Naval Training Systems Center.

Kennedy, R. S., Fowlkes, J. E., & Lilienthal, M. G. (1993). Postural and performance changes in navy and marine Corps pilots following flight simulators. *Aviation, Space, and Environmental Medicine, 64*, 912–920.

Kennedy, R. S., Hettinger, L. J., Harm, D. L., Ordy, J. M., & Dunlap, W. P. (1996). Psychophysical scaling of circular vection (CV) produced by optokinetic (OKN) motion: Individual differences and effects of practice. *Journal of Vestibular Research, 6*(5), 331–341.

Kennedy, R. S., & Lilienthal, M. G. (1994). Measurement and control of motion sickness aftereffects. In *Virtual reality and medicine, The cutting edge* (pp. 111–119). New York: SIG-Advanced Applications.

Kennedy, R. S., Smith, M. G., & Jones, S. A. (1991). Variables affecting simulator sickness: Report of a semi-automatic scoring system. In *Proceedings of the Sixth International Symposium on Aviation Psychology* (pp. 851–856). Columbus, OH.

Kennedy, R. S., & Stanney, K. M. (1996a). Postural instability induced by virtual reality exposure: Development of a certification protocol. *International Journal of Human–Computer Interaction, 8*(1), 25–47.

Kennedy, R. S., & Stanney, K. M. (1996b). Virtual reality systems and products liability. *Journal of Medicine and Virtual Reality, 1*(2), 60–64.

Kennedy, R. S., Stanney, K. M., Compton, D. E., Drexler, J. M., & Jones, M. B. (1999). *Virtual environment adaptation assessment test battery* (Phase II Final Rep., Contract No. NAS9-97022). Houston, TX: NASA Lyndon B. Johnson Space Center.

Kennedy, R. S., Stanney, K. M., Drexler, J. M., Compton, D. E., & Jones, M. B. (1999). Computerized methods to evaluate virtual environment aftereffects. In *Proceedings of the Driving Simulation Conference "DSC '99"* (pp. 273–287). Paris: French Ministry of Equipment, Transport, and Housing.

Kennedy, R. S., Stanney, K. M., Dunlap, W. P., & Jones, M. B. (1996). *Virtual environment adaptation assessment test battery.* (Final Rep. No. NASA1-96-1, Contract No. NAS9-19453). Houston, TX: NASA Lyndon B. Johnson Space Center.

Kennedy, R. S., Stanney, K. M., & Fernandez, E. (1999). *Six months residual after effects from a virtual reality entertainment system.* Manuscript in preparation.

Kennedy, R. S., Turnage, J. J., Rugotzke, G. G., & Dunlap, W. P. (1994). Indexing cognitive tests to alcohol dosage and comparison to standardized field sobriety tests. *Journal of Studies on Alcohol, 55*(5), 615–628.

Lawyers Co-Operative Publishing Co. (1988, with 1999 supplement). *American Law of Products Liability 3rd*. Rochester, NY: Author.

Lehto, M. R., & Miller, J. M. (1986). *Warnings: Vol. 1. Fundamentals, design, and evaluation methodologies* (1st ed.). Ann Arbor, MI: Fuller Technical Publications.

Leigh, R. J. (1996). What is the vestibulo-ocular reflex and why do we need it? In R. W. Baloh & G. M. Halmagyi (Eds.), *Disorders of the vestibular system* (pp. 12–19). New York: Oxford University Press.

Lilienthal, M. G., Kennedy, R. S., & Hooper, J. (1987, November). *Vision-motion-induced sickness in navy flight simulators: Guidelines.* Paper presented at the ninth Interservice/Industry Training Systems Conference, Washington, DC.

McCauley, M. E. (Ed.). (1984). *Research issues in simulator sickness: proceedings of a workshop.* Washington, DC: National Academy Press.

McCauley, M. E., & Sharkey, T. J. (1992). Cybersickness: Perception of self-motion in virtual environments. *Presence, 1*(3), 311–318.

Miller, J. W., & Goodson, J. E. (1960). Motion sickness in a helicopter simulator. *Aerospace Medicine, 31*, 204–212.

NTSC. (1989, November). *Simulator Sickness Field Manual MOD 4.* Orlando, FL: Naval Training Systems Center, Human Factors Laboratory.

O'Leary, D. P. (1992). Physiological bases and a technique for testing the full range of vestibular function. *Revue Laryngol, 113*, 407–412.

Phillips, J. J. (1998). *Products liability in a nutshell.* St. Paul MN: West Group.

Prothero, J. D., & Parker, D. E. (1999). A unified approach to presence and motion sickness. In L. J. Hettinger & M. W. Haas (Eds.), *Virtual and adaptive environments: Psychological and human performance issues.* Mahwah, NJ: Lawrence Erlbaum Associates.

Reschke, M. F., Harm, D. L., Parker, D. E., Sandoz, G. R., Homick, J. L., & Vanderploeg, J. M. (1994). Neurophysiologic aspects: Space and motion sickness. In A. E. Nicogossian, C. L. Huntoon, & S. L. Pool (Eds.), *Space physiology and medicine* (3rd ed.), (pp. 228–260). Philadelphia: Lee & Febiger.

Stanney, K. M., Mourant, R. R., & Kennedy, R. S. (1998). Human factors issues in virtual environments: A review of the literature. *Presence, 7*(4), 327–351.

Terry, (1915). *Harvard Law Review, 40.*

Uliano, K. C., Kennedy, R. S., & Lambert, E. Y. (1986). Asynchronous visual delays and the development of simulator sickness. In *Proceedings of the Human Factors Society 30th Annual Meeting* (pp. 422–426). Dayton, OH: Human Factors Society.

Wallach, H., & Kravitz, J. H. (1965). Rapid adaptation in the constancy of visual direction with active and passive rotation. *Psychonomic Science, 3*, 165–166.

Wallach, H., & Kravitz, J. H. (1968). Adaptation in the constancy of visual direction tested by measuring the constancy of auditory direction. *Perception and Psychophysics, 4*(5), 299–303.

Wallach, H., & Kravitz, J. H. (1969). Adaptation to vertical displacement of the visual direction. *Perception and Psychophysics, 6*(2), 111–112.

# 28

# Virtually a Market? Selling Practice and the Diffusion of Virtual Reality

G. M. Peter Swann[1] and Robert J. Stone[2]
[1]*Manchester Business School*
*Booth Street West*
*Manchester M15 6PB UK*
*pswann@man.mbs.ac.uk*
[2]*Virtual Presence (UK)*
*Chester House, 79 Dane Road*
*Sale M33 7BP UK*
*robert.stone@musevp.com*

## 1. INTRODUCTION

It is now generally accepted that virtual environment (VE) technology has immense commercial and educational potential, but so far the development of truly focused markets for VE applications is at an early stage. This chapter describes some of the factors at this early stage that will influence the ultimate growth of the VE market. In particular, the chapter stresses that high-selling costs can constrain the growth of the market, to the near-term detriment of the predominantly small-to-medium-size enterprise vendor/developer base. These technology and service suppliers face high selling costs because they have to educate the user. Their efforts have to be spread across a wide range of sectors, and many users need to see sector-specific applications of VE to convince them that it is worthwhile to invest in the technology. In the face of these high selling costs, some suppliers may be tempted to "hype up" the product they offer, sometimes making claims beyond what can be delivered at present. While this may lead to additional sales in the short term, it can also lead to dissatisfied customers, and dissatisfied customers at an early stage can depress the growth of the market in the longer term. The chapter is based on a simulation model of the selling process in VE, which was originally featured in an unpublished work by Stone (1996a). The model contributed to a series of strategic recommendations on the future of British VE efforts as part of a market study conducted on behalf of the U.K. government (Stone, 1996b). The present paper builds upon this to explore some of the constraints on market growth and to draw out policy implications for a code of good selling practice for virtual environments.

Those in the VE business are understandably very enthusiastic about the future for their technology. Robert Voiers, the founder director of the EDS Detroit Virtual Reality Center, is quoted as saying, "Virtual Reality is beginning to change the way we engineer and manufacture our products. It's early days yet, but I believe we may be witnessing the start of a new industrial revolution".[1] However, these views are not only held within the VE industry. In their influential book, *Competing for the Future*, business school "gurus" Gary Hamel and C. K. Prahalad (1994, p. 103) described VEs, often known as virtual reality (VR) in commercial industries, as, "a technology with profound implications for almost every industry." Virtual reality, they concluded, "is a powerful perceptual tool."

Although the market for VE technology is still growing, a number of commentators have forecast a bright future (Bubley, 1994; Helsel & Jacobson, 1994; MAPS, 1995; Stone, 1996b; Thompson, 1993). Nine of the fifteen Technology Foresight sector panels set up by the British government highlighted VE as a technology with strong potential in their sectors: communications, construction, defense and aerospace; health and life sciences; information technology and electronics; leisure and learning; manufacturing production and business processes; transport; and financial services.

More and more companies are using VEs, at least in an exploratory way: at least 21 out of the top 30 from the U.K. Department of Trade and Industry's (DTI) World R&D Scoreboard. Recent market research commissioned by the U.K. VR Forum (Cydata, 1998, 1999) backed up to some extent by more international reviews (CyberEdge, 1997, 1999, 2000) found clear evidence that VE has now advanced far enough to enable a wide range of commercial applications, some of which are showing the prospect of substantial business benefits.

And yet the market for VE is still at an early stage. Given all this potential, why has it not developed further or at a more accelerated pace? Is it something to do with the particular character of the technology—the futuristic image of users sporting head-mounted displays (HMDs) and other forms of cyber gear? Or is it something to do with the early history of the market, with its high-tech bias toward leisure and entertainment? The aim of this chapter is to shed some light on these questions by examining what lessons business and economic thinking about the growth of markets generally has for the VE market in particular. In pursuing this aim, a model of the selling process in the VE business as developed by Stone (1996a) can be turned into a powerful simulation model to explore these questions.

It can be shown that selling strategies can have significant effects on the ultimate market potential. The simulation model described in this chapter is highly path dependent in the sense that the early history of selling strategies can have significant effects on the ultimate market potential. As Stone (1996a) emphasizes, selling costs can be very high at this formative stage of the market, essentially because customers know little about the technology and have to be taught—often at the expense of the pioneering producer. If companies try to reduce selling costs by cutting corners and perhaps by selling systems to customers who do not really stand to gain from the purchase, then this adverse experience can reduce the ultimate market potential for the technology. Good selling practice at an early stage may be costly for the pioneering producer, but by gradually building a community of highly satisfied users this can help to establish the long run market potential for the technology.

This chapter is organized as follows. Section 2 summarizes some of the key ideas in the economics of diffusion, that is, the rate at which new technologies are adopted by different types of end user. Then, section 3 sets out the background to the business process model of selling in VE, and section 4 sets out the basic structure of the model. Section 5 summarizes how the basic model is solved in each simulation, and what the structure of each simulation looks like, and section 6 summarizes some of the preliminary simulation results. Section 7 summarizes the conclusions for the future of the VE market.

[1] *VR News* 7(2), March 1998, p. 15

## 2. THE DEMAND FOR NEW PRODUCTS: LESSONS FROM ECONOMICS

Projecting the market for a new technology is probably one of the most difficult questions in economics. Thomas J. Watson's celebrated underestimate of the potential market for computers is one famous example. These difficulties emerge for at least three reasons:

- It is customary to assume that tomorrow's markets are just more developed versions of today's, but in fact new needs, applications, wants, and sources of competitive advantage may be evident tomorrow that are not present today.
- We usually fail to recognize how "path dependent" is the evolution of demand for a new technology—meaning that the minor historical details and events of today can have a decisive role in influencing the potential future market for the technology.
- Demand projections will depend on the evolution of this technology, but also of its rivals, and the latter is not always easy to foresee. Moreover, one of the regular features of competition in new technologies is that new rivals emerge from unexpected directions—especially when technologies are converging. This makes it very hard to know what will be the most important technological trends that influence demand for a particular technology.

In economics and other areas of business studies, writers use the term *diffusion* to describe the spread in use of a new technology. In this section some of the main economic thinking about the diffusion of new technologies is summarized, as this will be an essential background to what follows. Much of what follows is about the rate at which a new technology diffuses, but it also touches on the absolute level of diffusion that may be achieved in the long term.

### 2.1 "Epidemic" Models

Economists use the term *epidemic* to indicate that there is a certain similarity between the diffusion of a new technology amongst a population of users and the spread of an infectious disease amongst a population of people who do not have resistance. In its simplest possible format, the epidemic model assumes (as with the infectious disease) that the rate of new cases (of adoption) is proportional to the product of the number infected (who use the technology) and the number who are not infected, but could be (the potential future adopters). This simple form yields the familiar S-shaped diffusion curve, showing how the number of adopters increases over time. Simple mathematical considerations show that the rate of diffusion is fastest toward the middle of the diffusion process, and in the simplest version of the model the fastest rate of adoption per period is when 50% (exactly) of the target population are "infected."

Other, more subtle "epidemic" diffusion models generate slightly different patterns, but the basic story is the same. More complex models recognize that more complex patterns of interaction are required before one user will "catch" the technology from another. This leads to more detailed models of the buying and selling process.

The social process that spreads the use of a technology is taken to be the exchange of information amongst potential users, and the pioneer plays an important demonstrator and educational role in this. In more subtle models of this sort, it is not just the number of pioneers that count, but also their quality. Moreover, the effect that the pioneer has on subsequent adopters is also dependent on his experience with the technology: Good experiences promote diffusion, whereas bad experiences retard diffusion. This has important implications both for government policy and for the business strategy of VE companies.

In the very long term the level of diffusion approaches (asymptotically) a maximum or saturation level of diffusion. This technique is useful for projecting how fast the uptake will be, given a certain saturation level of diffusion, but it does not in itself help to estimate the saturation level. Moreover, speeds of diffusion will vary enormously from one technology to another—compare the diffusion of the car and a particular pop record, for example.

### 2.2 Diffusion Driven by Rising Income or Other Buyer Characteristics

Another force leading to diffusion is related to rising income, increasing turnover, or other characteristics of the buyer. As income per capita rises gradually over time, consumption of some products will rise. At an early stage, buyers may be limited to those on relatively high incomes, but as the average level of incomes rise, more and more buyers enter the market. The same principles can be applied to the diffusion of new technologies among firms. For some expensive new technologies, early buyers may be the larger firms (in a particular industrial sector), but as the smaller firms grow, they too would be potential buyers.

This approach to the analysis of diffusion can also be used with other consumer or firm characteristics. As customers become more educated they are more likely to buy certain products. Likewise, as firms become more concerned to seek new sources of competitive advantage in an ever more competitive market, they may be more likely to buy into some new technologies.

As far as VE is concerned, it appears that within a particular sector, and holding other factors constant, the larger companies are more likely to be early adopters. But the average size of companies in those industrial sectors that are pioneering the application of VE may not be as high as in some nonpioneering sectors.

In general, diffusion promoted by growth in incomes, revenues, or other buyer characteristics tends to be a slower process than diffusion attributable to other effects. The reason for this is simply that the rate of growth of incomes is typically quite a small percentage rate. Gross domestic product (GDP) per head rises at perhaps 3% per annum on average, and most consumers' incomes rise no faster than that. Large companies do well to grow any faster than that. Some smaller companies of course can grow a good deal faster, but they are outnumbered by companies that don't. And once again, this approach is useful as one of the factors determining the rate at which a technology diffuses, but does not in itself indicate the likely saturation level of diffusion.

### 2.3 Diffusion Driven by Falling Prices and Rising Quality and Performance

In many high-technology markets the main driving force behind the diffusion of a new technology is that as it becomes cheaper and/or better, ever more buyers are willing to adopt. Moore's law is a classic example of this. The law states that the rate of increase in the number of "gates" per semiconductor chip is doubling every 18 months or so, and also implies an exponential decline in the price per component. As the price per bit falls over time, so ever more users are drawn into the market. The speed of diffusion here depends on the speed of price decline and the variance of the willingness to pay of different buyers. As noted above, these rates of change provide a more rapid impetus to the diffusion of the technology than rising incomes or turnover.

A more subtle version of this story gives due account to the role of improvements in performance as well as reductions in price. The usefulness of this more subtle picture is that it can distinguish between the effects of improving quality (at given price) and falling price (at given quality). It can also show how the market is segmented between base-quality, superior-quality, and top-quality versions of the technology and how this segmentation would change over time. The rate of diffusion in this process, again, depends on the speed at which prices fall and quality improves.

### 2.4 Strategic Models of Diffusion

In the approaches described above, a company or consumer's decision to buy is assumed to be independent of what others decide. However, for both the individual customer and the industrial buyer, this need not be so. This observation opens up two important strands of diffusion analysis.

The first arises when the fact that others are already using a technology makes a new buyer more likely to adopt. This could arise when there are benefits from adopting a technology that has a large installed base of users. The existence of a large installed base is often a powerful signal to some buyers that this technology is no longer the risky prospect that it appeared at first. This feature makes for a very powerful and rapid diffusion process, where if one technology gets ahead of the field, the market will quite rapidly lock into that technology. This approach is used to model the emergence of *de facto standards* or *dominant designs* in markets.

This feature of rapid take off can also arise when firm B faces competitive pressure to adopt a new technology because their rival firm A is using it with good effect to take market share away from B. While firm B may not be especially interested in adopting the technology on its own merits, it may become interested if firm A exploits the technology for competitive advantage. This can lead to a "me too" attitude to adoption and a very rapid rate of diffusion.

This process of diffusion due to competitive pressure, operating in addition to the other processes described above, can further accelerate the process of diffusion. One major feature of technological change noted above is that firms tend to find new competitors emerging from new and unexpected directions. That means the technology adoption decisions of an ever greater number of firms can impinge on the decision faced by one particular firm. This proliferation of competitors can make diffusion due to competitive pressure a very pervasive as well as a very rapid process and will continue to raise the level at which the market reaches saturation.

So far, the case where the fact that others already use a technology makes the new buyer more likely to adopt has just been described. For completeness it should be mentioned that it is well recognized in consumer buying behavior that consumption may be as much influenced by the behavior of the *distinction group* from whom the consumer wishes to distinguish himself, as by the behavior of the *peer group* with which he wishes to associate. When this feature is important, it tends to suggest that there are likely to be clear upper limits to diffusion, covering far less than 100% of the potential population.

In principle these effects can also arise in the context of industrial buying, though it is not clear whether they are relevant in the context of virtual environments. They could be relevant if a group of pioneering VE buyers is (for whatever reason) seen as a distinction group by another large set of industrial buyers. This would happen if early applications in some sectors were generally thought to be unsuccessful and if other companies seek to distinguish themselves from the "tainted" early users. Or, as some VE vendors observe, the success of VE as a "games" technology may discourage some business users who see VE as a "toy" and not a serious business technology. This emphasizes again how important it can be that early applications of a technology are successful to maximize its ultimate diffusion.

## 3. BACKGROUND TO THE STONE MODEL

Stone (1996a) set out a detailed model of the buying and selling process that operates in virtual environments. In particular he charted a detailed description of some of the many stages between the first tentative meetings between the VE company and prospective buyers and the eventual adoption and use of the technology, with identified risks and nominal costs (labor rates and on-costs) to "VR users," "VR suppliers," and "VR developers." Consideration was given to such processes as evidence of relevant demonstrators, development of technology demonstrators, provision of relevant "educational" and informative material, recruitment, and training. It was shown that issues such as VE technology (hardware and software), education, standards and quality, future government roles, academic research, health and safety, plus other barriers to entry all affect the uptake of VE into a business and the further integration and development of the technology within that business (see also Appendix 3, which presents a checklist for potential VE users, published as part of a later "VR Awareness" CD by Virtual Presence [U.K.], under contract to the U.K.'s Department of Trade and Industry). Some of the flowcharts underpinning this model, together with a brief description of part of the initial ("Primary Education Process"), are included at the end of this chapter.

A prominent characteristic of the early markets for VE applications is that sellers face very high selling costs. For any new technology these selling costs can be high, but they do appear especially high in the case of virtual environments. As Stone (1996a) pointed out, sellers need to devote a high level of resources (especially the time of senior staff) toward educating prospective buyers.

These education costs can be expected to fall over time as customers become more familiar with the technology and as a larger base of established users is available to provide demonstrations of how the technology can be applied in a particular setting. While the purchase decision at an early stage may well involve a feasibility study, this may become less important at a later stage because the existence of established users able to provide relevant demonstrations may give the user sufficient confidence to proceed.

The model set out by Stone (1996a) could be placed in the "epidemic" tradition of diffusion models. As described above, these models have been so called because the diffusion of the new technology is a little like the spread of an infectious disease. Stone's model is a good deal more subtle than the basic epidemic model; in the latter, one meeting is sufficient to spread the disease, whereas in the case of a new technology many more stages are of course involved. Nevertheless, the basic structure of the model is the same, for it is as a result of a sequence of meetings and dialogues that the adoption or rejection decision is made.

It is also clear that in Stone's (1996a) model, that the quality of adoption—by which we mean the extent to which the adopted technology has been satisfactory or otherwise—can depend on the details of the selling process. To quote the terminology in Stone's model, if sellers who know little about the target business choose to "go in blind" rather than do some preliminary market research, then that can reduce the chances of a purchase. And if sellers who see that VE does not currently have a promising application in a user company decide to "come clean," then while this may reduce their sales revenue in the short term, it will also ensure that the percentage of satisfied adopters is kept up.

The preliminary model set out in Fig. 28.1 is a refinement of Table 1 in Stone's (1996a) paper. This corresponds to the Primary Education Process, from preliminary exploratory meetings up to the stage where the user company procures a VE system for in-house use. (Note: The model uses the more commercial term VR rather than VE) As such, it does not cover the subsequent stages when the company experiments with the in-house system before deciding whether to implement it long term. The preliminary model was cut off after this first stage of the process not because the subsequent stages are unimportant, nor because they are not amenable to analysis in this way, but simply because it was thought appropriate to test what is already a reasonably complex model before proceeding to build a more complex version.

As noted before, most of this model is in the *epidemic* tradition. But there also some important demonstration effects and *network externalities*, so that the probability of one user adopting is dependent on the number of others who have already adopted. For this reason, the

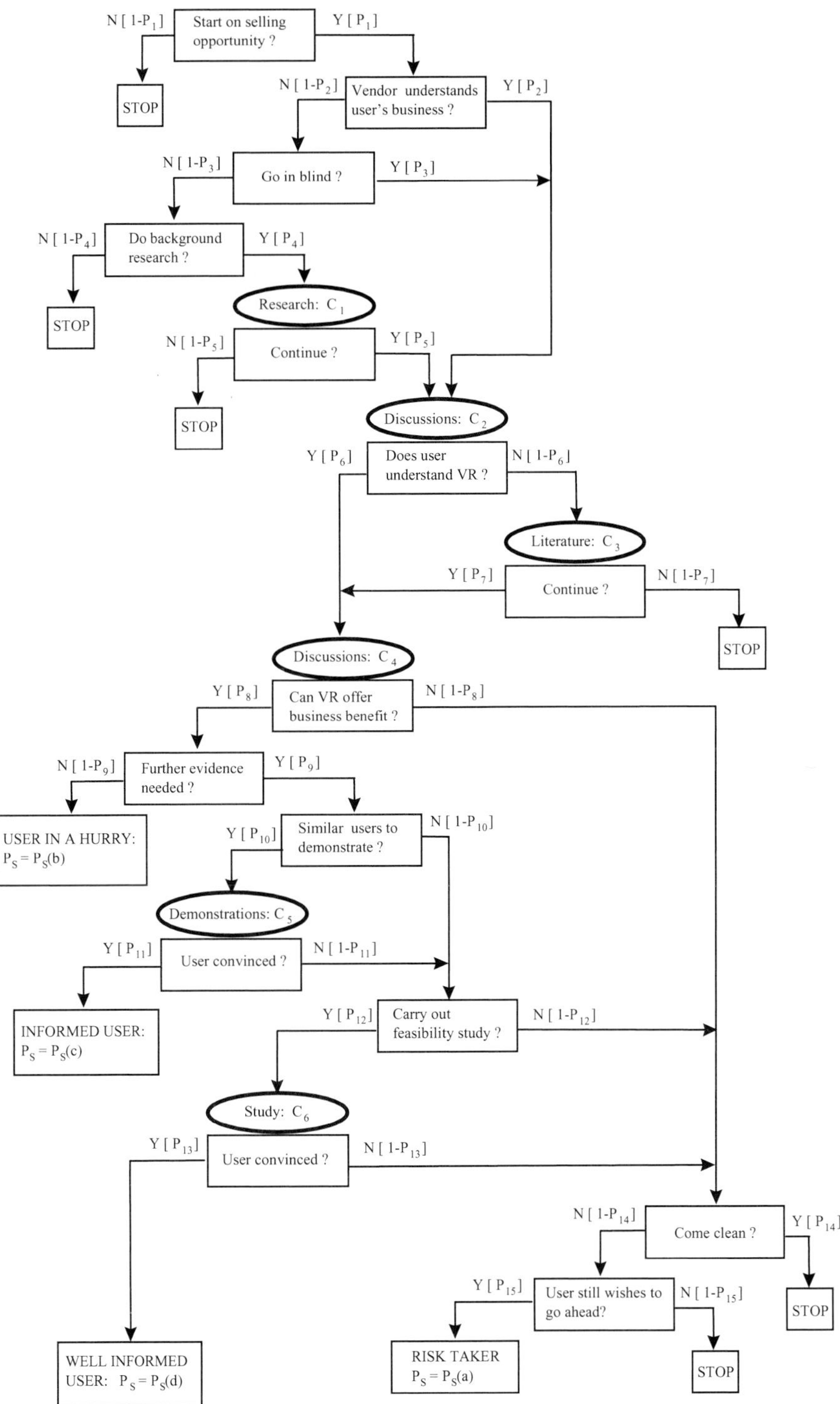


FIG. 28.1. Model of selling VR.

simulation model is able to make use of some of the features used in the simulation of market defined (or de facto) standards (Shurmer & Swann, 1995).

## 4. CREATING A SIMULATION MODEL

In this section the main relationship and parameters in the model are summarized. The full equations are set out in an appendix to this paper. Roughly speaking, each box in Fig.28.1 containing a question translates here into an equation yielding a probability. Thus for example, the box containing the question "Is the target business understood by the VR expert?" is turned into an equation stating the probability of a yes as a function of several variables and parameters. In particular, the probability of a yes is an increasing function of the sales "focus" of the producer, the accumulated knowledge of the producer, and the general level of diffusion of the technology.

The main components of the model are as follows. The first equation (Equation 1) simply reflects the fact that as the "focus" (or core competence) of producers narrows then the number of business opportunities they pursue will decline. On the other hand, as discussed below, an increased focus may improve the probability of success. If the producer does pursue the business opportunity, then the second equation (Equation 2) models the probability that the seller understands the user's business. Equation 3 models the decision of whether to "go in blind" or whether the vendor will consider doing background research on the user's business (Equation 4).

Equation 6 describes the probability that VR is actually understood by user's representatives. This is more likely as the general level of diffusion increases, but it also depends on the probability that the seller "goes in blind," knowing nothing about the user's business. When this happens, the chances that the user can understand the potential of VR will decline. If the user does not understand VR, then Equation 7 summarizes the chances that the seller's "literature" or other educational efforts will succeed in explaining to the potential user what VR can do. This probability is assumed to depend on the quality of the "literature" and also on the probability that the seller has adequately informed himself about the user's business before preparing the "literature."

If this educational process is successful, then the model proceeds to the next stage (Equation 8), which is to assess the probability that VR can offer a satisfactory business benefit to the user. This probability is assumed to follow an S-shaped function of the level of diffusion: the greatest competitive incentive for adoption accrues when other users have adopted. The analysis of strategic incentives for adoption recognizes that more subtle responses may be relevant here (previously mentioned). If the answer is yes that VR can offer a satisfactory payoff, then the next equation (Equation 9) asks whether it is likely that further evidence is needed before the user can make a decision about whether to procure a VR system for in-house use. This probability falls as the general level of diffusion increases, because the technology is perceived to be less risky. It also falls if sellers have chosen to inform themselves about the user's business (see above), for once again this means that the users sees the investment as a less risky one. If the user decides that no further evidence is needed to convince him that procuring a VR system is worthwhile, then he proceeds to the first "adoption" stage, and this is classified as a "user in a hurry."

If more evidence is needed, the first question (Equation 10) is whether there are similar users who have already adopted who could demonstrate the application of the technology to the prospective user. This probability increases as the general level of diffusion increases, and there is an important parameter in the equation describing the rate at which this probability increases. If the answer to this is yes, then the next question (Equation 11) is whether the user is convinced by such a demonstration. The probability of a convincing demonstration depends on the level of applications "focus" chosen by the seller (previously discussed): If this focus is high, then it is more likely that the demonstration will convince. This probability also depends on the proportion of satisfied adopters amongst the population of all adopters: The higher this is, the more convincing a demonstration will be. If the user is convinced, then he proceeds to the adoption stage and is classified an "informed user."

The next equation (Equation 12) in the model picks up the remaining unconvinced users. Do they ask for a feasibility study from the VR producer? This probability is proportional to the proportion of unsatisfied adopters amongst all adopters: The higher this is, the riskier is adoption, and so the greater the need for a feasibility study.

The next equation (Equation 13) models the probability that those who commission a feasibility study where there is a demonstrable commercial payoff to adoption will be persuaded by the feasibility study to proceed to procuring an in-house VR system. These are classified as well-informed users, but in the case of those who do not commission a feasibility study, and/or who are not convinced by the evidence, the next equation (Equation 14) models the probability that the seller chooses to "come clean." The vendor who "comes clean" would say that this investment does not look worthwhile, but those who don't may proceed with a sale, but one that has a smaller probability of a satisfactory application. This probability that the seller will "come clean" depends on the same parameter that influenced the seller's decision to "go in blind" (though inversely) and also increases as the general level of diffusion increases—because it is assumed that sellers have less incentive to make such sales when the market is larger. (In some cases we have made the rate of increase small or zero because the level of competition may also increase as the market size increases, and that could sustain the need for such sales.)

If the seller "comes clean," then it is assumed that the sale falls through; if he doesn't, then it is assumed that there is a 50% probability that the user will proceed and buy, and a 50% probability that he abandons the project. Those who buy are classified as "risk takers" because there is a reduced probability that adoption in such circumstances would lead to a satisfactory outcome.

In the model there are four routes to the decision to procure a VR system: the "well informed," the "informed," the "user in a hurry," and the "risk taker." It is assumed that the probability of success is highest in the first case. (Indeed, we set that probability at 1 and smallest in the last case. In the exploratory simulations it was set at 0.5, though in practice a lower figure may be relevant). There are also a number of "exit points," at which the user decides to abandon the project.

## 5. SOLUTION STRUCTURES

The model described above has been programmed into a simulation model. Each simulation involves the following steps:

1. For each possible sequence in the model, a probability is computed.
2. These probabilities are aggregated into five categories:
   - All paths leading to a "well-informed" decision to adopt
   - All paths leading to an "informed" decision to adopt
   - All paths leading to a "user in a hurry" decision to adopt
   - All paths leading to a "risk-taker's" decision to adopt
   - All paths leading to an abandonment of the VR project

**TABLE 28.1**
Parameter Values and Ranges

| *Parameter* | *Maximum* | *Minimum* | *Fixed Value* |
|---|---|---|---|
| $b_1$ ("focus") | 0.99 | 0 | |
| $b_2$ ("vendor knowledge") | 1 | 0 | |
| $b_3$ ("go in blind [1]") | 1 | 0 | |
| $b_4$ ("go in blind [2]") | 0.5 | 0.001 | |
| $b_5$ ("user understand") | 1 | 0 | |
| $b_6$ ("literature quality") | 1 | 0 | |
| $b_7$ ("business benefit") | 1 | 0.1 | |
| $b_8$ ("need for evidence") | 1 | 0 | |
| $b_9$ ("similar users") | 1 | 0 | |
| $b_{10}$ ("need for study") | 1 | 0 | |
| $b_{11}$ ("come clean") | 1 | 0 | |
| $P_s(a)$ ("risk-taker") | | | 0.5 |
| $P_s(b)$ ("user in a hurry") | | | 0.7 |
| $P_s(c)$ ("informed user") | | | 0.85 |
| $P_s(d)$ ("well-informed user") | | | 1.0 |
| $C_1$ ("research") | | | 20 |
| $C_2$ ("first discussions") | | | 10 |
| $C_3$ ("preparing literature") | | | 10 |
| $C_4$ ("second discussions") | | | 50 |
| $C_5$ ("demonstrations") | | | 50 |
| $C_6$ ("feasibility study") | | | 100 |
| $X(0)$ (total diffusion at $t = 0$) | | | 10 |
| $X_s(0)$ (satisfied diffusion at $t = 0$) | | | 5 |
| $X_u(0)$ (unsatisfied diffusion at $t = 0$) | | | 5 |

*For definition of model equations, see Appendix 2.

Simple probabilities are computed for each of these aggregates. In the case of the four paths that lead to adoption, the model assumes that the probabilities that adoption will be successful and that the user will be content depend on how well informed the user is at the time of adoption: the better informed, the higher the probability of success. This is captured in parameters $P_s(a)$ to $P_s(d)$, as shown in Fig.28.1 and defined in Table 28.1.

3. Each investment (or selling cost) on the part of the seller is given a weight or cost. These are the costs of: researching the users' business ($C_1$); discussions with the potential user about the applications of technology ($C_2$); preparing "literature" and other educational material for the inexperienced user ($C_3$); further discussions with the potential user about prospects to obtain a satisfactory payoff from VR investments ($C_4$); the costs of setting up demonstrations of other users' VR applications ($C_5$); those costs of setting up a feasibility study that fall on the seller ($C_6$).
4. The elementary probabilities for each path are weighted by the costs incurred along that path and summed to obtain an expected selling cost for each simulation.
5. Each simulation is run for 20 periods (where the length of the period is essentially indeterminate, though it is more likely to be a month than a year).

## 6. SOME SIMULATION RESULTS

Simulations have been run for a group of 5,000 parameter sets. Each simulation result consists of a set of solutions for each of the equations discussed above in section 4. Table 28.1 gives the range of values for those parameters that vary from simulation to simulation and also the (constant) values of parameters that do not vary across the simulations.

Before summarizing these simulation results in full, it is useful to explore some of the properties of the model by varying one parameter at a time and holding all other parameters constant. This is done by setting the values of the $b$ parameters equal to the midpoint of the range described in Table 28.1.

Figures 28.2 to 28.5 show how the model responds to variations in four of the parameters. Parameter $b_1$ describes the degree of focus in the vendor's selling strategy: the higher is $b_1$, the more the vendor focuses his selling on a particular set of end-user markets. The striking thing in Fig. 28.2 is that the "best" strategy—best, at least, in the sense that it minimizes selling costs per satisfied new customer—requires an intermediate degree of focus. If the seller is too focused, that is if $b_1 \rightarrow 100\%$, then sales are very low. If the seller is not focused, he does not know enough about the user's business to make a well-informed sale. He has either to incur research costs ($C_1$) or run the risk of making sales were the buyer is not well informed about VR—and such sales run the risk of building up a community of unhappy customers.

Fig. 28.3 summarizes what happens when parameter $b_3$ is varied, which could be described as the first "go in blind" parameter. It is the parameter in the numerator of Equation 3, and hence the higher it is, the more likely the vendor will "go in blind." Figure 28.3 shows a series of dots representing steps of 2.5% in the value of $b_3$. As this parameter increases from zero, total new diffusion rises steadily—even if little of this is to satisfied customers. Indeed, beyond about $b_3 = 0.5$, satisfied new diffusion actually falls even if total new diffusion continues to

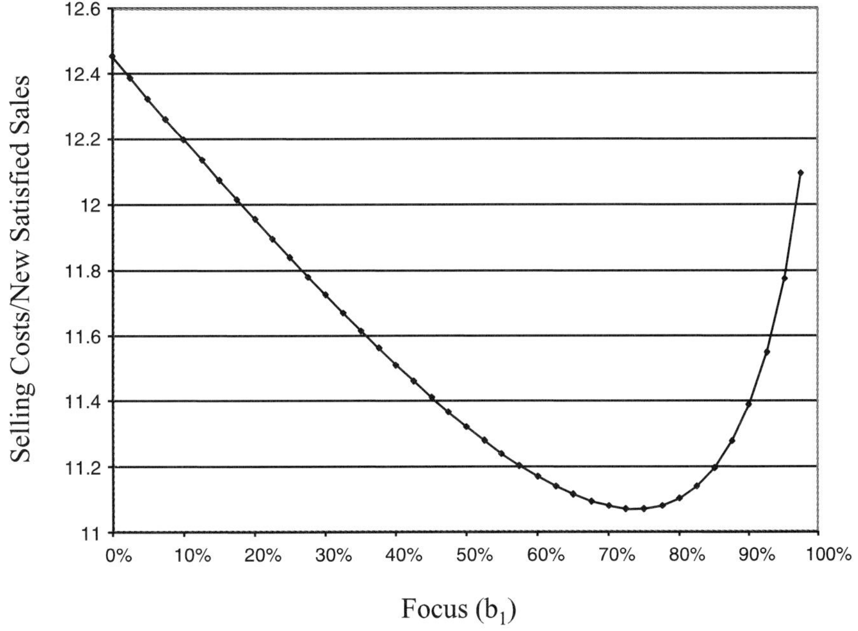


FIG. 28.2. Varying the "focus" parameter ($b_1$).

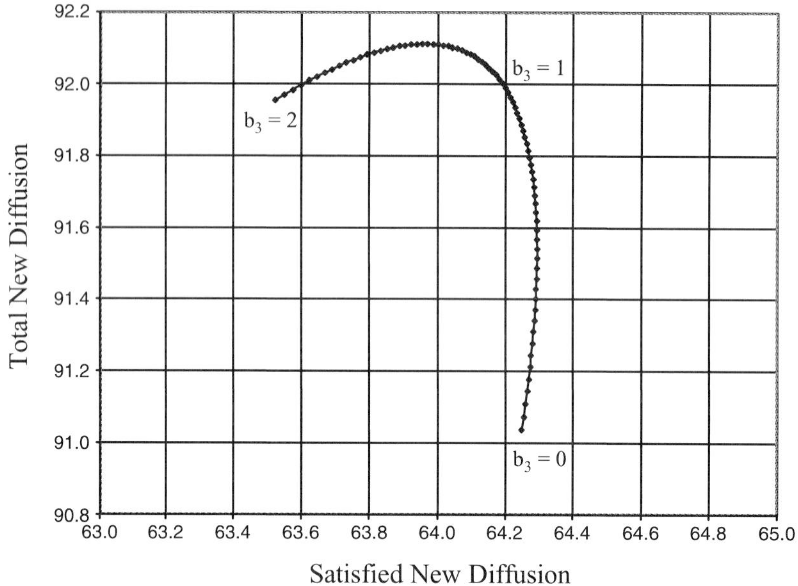


FIG. 28.3. Varying the "go in blind (1)" parameter ($b_3$).

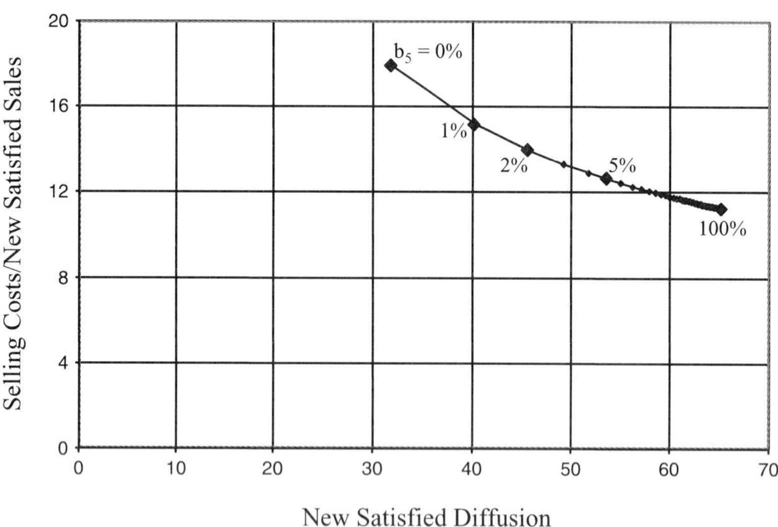


FIG. 28.4. Varying the "user understanding" parameter ($b_5$).

increase. Beyond about $b_3 = 1$, even the total new diffusion starts to tail off, and at very high values of $b_3$, total diffusion starts to fall. The implication of this is clear. A certain amount of "going in blind" may be essential to increase overall diffusion, but it does not do much for customer satisfaction. But too much "going in blind" will mean that the vendor does not collect any sales—not even unsatisfied customers.

Figure 28.4 shows a similar graph for different values of $b_5$, the parameter describing how the user's understanding of VR increases as total diffusion of VR increases. Starting from a

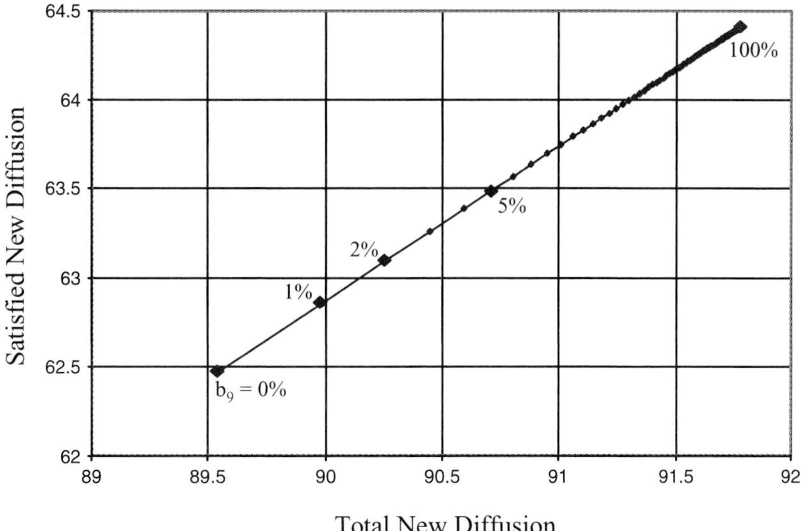


FIG. 28.5. Varying the "similar user" parameter ($b_9$).

base of $b_5 = 0$, it is clear that as this parameter increases the selling cost per satisfied customer falls and diffusion to satisfied customers increases. Indeed, for the particular set of values for all other parameters (see Table 28.1) it only takes a quite modest value of $b_5$ to generate these beneficial effects. The key point is that ideally the user should learn *something* from general diffusion of virtual reality.

Finally, Fig. 28.5 presents a similar picture for parameter $b_9$, which describes how the probability of finding a similar user to demonstrate an application to a prospective buyer increases as general diffusion increases. Again, as this parameter increases total diffusion and satisfied diffusion both increase. Indeed, it is the first few percentage points that are critical here: raising $b_9$ from 0% to 5% achieves about the same effect as increasing it further from 5% to 100%. The rate of increase in satisfied diffusion is not quite so high as total diffusion—suggesting a small amount of dissatisfied diffusion, but the former dominates.

Now the discussion turns to the full set of simulation results. As those familiar with simulation will know, it is easy to generate multidimensional complex simulation results and then lose sight of what the results indicate about the essential features of the model under evaluation. To avoid that, a procedure is adopted that has been used to good effect elsewhere (Shurmer & Swann, 1995), which is to summarize those simulations that are best at satisfying certain criteria. For the present study, the three criteria are as follows:

1. A high proportion of adopters are satisfied with the technology
2. High levels of total diffusion
3. Low selling costs per satisfied customer

Table 28.2 summarizes the simulation results as follows. Using each of the above criteria in turn, the simulation results are sorted into the "best 20%" and the "other 80%." Table 28.2 then shows the average value of each criterion and each parameter in the "best 20%" compared to that for the "other 80%."

The first two columns show the results when simulations are sorted by criterion (*a*). Here the "best 20%" contains the simulation results where the ratio of satisfied to unsatisfied customers

**TABLE 28.2**
Summary of Simulation Results

| | *Sort by Criterion (a)* | | *Sort by Criterion (b)* | | *Sort by Criterion (c)* | |
|---|---|---|---|---|---|---|
| | *Best 20%* | *Other 80%* | *Best 20%* | *Other 80%* | *Best 20%* | *Other 80%* |
| New satisfied/new unsatisfied | 2.70 | 2.33 | 2.32 | 2.42 | 2.35 | 2.41 |
| Total diffusion | 51.8 | 94.8 | 158.4 | 68.1 | 60.1 | 92.7 |
| Selling costs/satisfied customer | 14.6 | 12.3 | 12.6 | 12.8 | 10.9 | 13.2 |
| $b_1$ ("focus") | 68% | 45% | 12% | 59% | 67% | 45% |
| $b_2$ ("vendor knowledge") | 50% | 50% | 51% | 50% | 58% | 48% |
| $b_3$ ("go in blind [1]") | 38% | 52% | 50% | 50% | 55% | 48% |
| $b_4$ ("go in blind [2]") | 26% | 25% | 26% | 25% | 25% | 25% |
| $b_5$ ("user understand") | 49% | 51% | 54% | 49% | 61% | 48% |
| $b_6$ ("literature quality") | 49% | 50% | 53% | 49% | 55% | 48% |
| $b_7$ ("business benefit") | 62% | 54% | 57% | 55% | 70% | 52% |
| $b_8$ ("need for evidence") | 19% | 57% | 52% | 49% | 72% | 44% |
| $b_9$ ("similar users") | 53% | 49% | 51% | 50% | 49% | 50% |
| $b_{10}$ ("need for study") | 60% | 48% | 51% | 50% | 49% | 50% |
| $b_{11}$ ("come clean") | 53% | 50% | 50% | 51% | 49% | 51% |

is greatest. It is interesting to note that in these simulations, the typical total level of diffusion is considerably less than in the "other 80%." This reveals a typical (though not inevitable) trade-off in this simulation model between scale of diffusion and proportions of satisfied customers. And to achieve this degree of satisfaction requires slightly higher selling costs per satisfied customer than in the "other 80%" of simulations. The "best 20%" of simulations on this criterion have relatively high average values for $b$ parameters 1 ("focus"), 7 ("business benefit") and 10 ("need for feasibility study"), and relatively low values for $b$ parameters 3 ("go in blind [1]") and 8 ("need for evidence"). In short, the model generates outcomes with a high proportion of satisfied customers:

- When sellers pursue focused selling strategies where they are well informed about the prospective user's business and do not "go in blind"
- When sellers are prepared to help in the costs of feasibility studies when required
- When business benefits grow steadily as diffusion increases

Turning now to the second criterion, the second pair of columns shows that the "best 20%" of simulations achieve far higher levels of diffusion than the "other 80%." The proportion of satisfied customers is slightly lower in this "best 20%"—though not much—and selling costs per satisfied customer are indeed slightly lower in the "best 20%" than in the "other 80%". How is this? The main difference in parameter values between the "best 20%" and the "other 80%" here is that the simulations in the "best 20%" have far lower focus ($b_1$). Otherwise, there is little difference between the two sets of simulations, though the "best 20%" have higher values for the "user understanding" parameter. In short, an essential part of a strategy to achieve diffusion on a large scale is that selling should not be focused, but it also helps if users are quick to learn from the general diffusion of VR.

Finally, turning to the third criterion, the third pair of columns in Table 28.2 show that the "best 20%" can achieve substantially lower selling costs per satisfied customer, but only achieve this at substantially lower levels of diffusion. These lower selling costs also show up in a slightly lower proportion of satisfied customers. Here many parameters show a much higher average in the "best 20%" than in the "other 80%." These include *b* parameters 1 ("focus"), 2 ("vendor knowledge"), 3 ("go in blind (1)"), 5 ("user understand"), 6 ("literature quality"), 7 ("business benefit"), and 8 ("need for evidence"). In each case a higher value of the parameter helps to reduce the costs of selling per satisfied customer.

## 7. CONCLUSIONS AND POLICY DISCUSSION

This chapter has provided a brief summary of a simulation model of the VE selling process, based on the description of that process set out in Stone (1996a). Some preliminary simulations with the model have yielded the following preliminary and tentative conclusions.

Selling costs, which are high in the VE market at the stage of development, are seen to play an important role in the subsequent development of the market. To achieve a large proportion of satisfied customers in the diffusion process, it is necessary to incur high selling costs. It is not possible to achieve the highest proportions of satisfied customers with the lowest selling costs. And this satisfaction among early adopters can be important for the health of the subsequent diffusion process.

A focused selling strategy is effective if the desire is to keep selling costs low or if the desire is to achieve high proportions of satisfied customers. But to achieve the highest levels of diffusion a less focused selling strategy may be required. On the other hand, it generally pays if sellers as a group "do their homework" about the businesses of prospective users before trying to sell the technology, rather than "go in blind." In the same way, it is best if sellers as a group tend to "come clean" with users when prospects for a successful application of VE in their business look poor. The incentives that must exist for some sellers to break rank and behave "badly," but trusting "good" sellers to follow this public-spirited approach to selling have not, however, been explored here.

It is generally helpful for diffusion and user satisfaction if the pool of similar established users of virtual environments, who can demonstrate its potential, grows rapidly. This suggests that sector-specific case studies of the successful adoption of VE could have an important role in promoting diffusion. Moreover, for the model parameters selected for these simulations (and these may not be representative), high-quality educational "literature" seems to play a valuable role.

### 7.1 Policy Implications

Pandit et al. (1997) explored whether the especially high level of selling costs found in the formative stages of the VE market made a case for government activity to fund awareness activities. The essence of their argument was as follows. Traditionally, economists looking at microeconomic policies for microeconomic issues (as opposed to macroeconomic management) used the principle that policy intervention would be called for where there is market failure. The problems with this are:

1. Market failure is very common, but not always too serious.
2. Regulatory failure may be worse than market failure in some settings.
3. Some policies suffer from a lack of additionality. They cannot always distinguish between

> (1) subsidizing activities that would not have happened otherwise (additionality) and (2) subsidizing activities that would have happened anyway. Category (1) seems desirable while category (2) looks wasteful. Some skeptics fear that (2) may be quite large unless government agencies incur large monitoring costs.

In fact, even these observations do not tell the full story. Category (2) is an undesired transfer payment from the taxpayer to a company, but at least it does not distort incentives away from the "true" ones. Category (1) may induce some who should buy to do that, but may also persuade some to buy who probably should not. The implications of this are discussed later.

For these various reasons, the existence of market failure on its own is rarely a sufficient case for industrial policy. But the fact that pioneer producers in VE have to incur an educational cost in selling to inexperienced buyers is a potentially important case of market failure. The educational activities of pioneers convey benefits (externalities) on later entrants to the market. This is a part of the rationale for the DTI's current VR Awareness Programme.

Pandit, Swann, & Watts (1997) argued that a more compelling argument for industrial policy in this area is as follows. In those areas of industry where the market outcome is highly path dependent, a small and finite investment at the formative stages can shift the market onto a higher growth path. Indeed an increasing number of economists would now consider that most markets for new technologies show highly path-dependent behavior, meaning that the eventual market size is influenced by all kinds of second-order historical events throughout the life of the technology, and particularly at the formative stages. In particular, the analysis of standards and dominant designs above has stressed how small events at the formative stages of a market can influence which technology wins a standards race.

Some generic government policies have in the past been designed to try to promote diffusion by subsidizing pioneering users. The traditional measure of success here is that of *additionality*: A subsidy is only worthwhile if it promotes adoptions where adoption would not otherwise have happened. If the subsidy simply pays a company to do something that it would have done anyway, then it is at best irrelevant, and at worst an arbitrary and undesirable transfer payment.

The epidemic model described above and the Stone (1996a) model of the selling and buying process offer a somewhat different perspective. Additionality is not enough on its own; it must be additionality plus good experiences from adoption. If a subsidy creates an additional adopter who has good experiences from adopting the technology, then this is well and good. But if the subsidy creates an additional adopter who would not have adopted in the absence of a subsidy and who does not have good experiences from adoption then this is not just wasteful, but may also be damaging.

This suggests that any diffusion-oriented policy to promote pioneering users of VE should take care to ensure that these users' experiences will be as positive as possible. This suggests that such a policy should articulate a code of good practice in selling. Moreover, it may be best if resources are directed at increasing the information available to firms deciding whether or not to adopt virtual environments technology or applications. In that case, companies can make well-informed decisions taking into account local circumstances, rather than be swept along with a bandwagon that may not help them at all.

## 8. REFERENCES

Bubley, D. (1994). *Virtual reality: Video game or business tool?* (Management Rep.) London: Financial Times.

CyberEdge. (1997). *The market for visual simulation/virtual reality systems* (1st ed.).

CyberEdge. (1999). *The market for visual simulation/virtual reality systems* (2nd ed.).

CyberEdge. (2000). *The market for visual simulation/virtual reality systems* (3rd ed.).

Cydata. (1998). UK business potential for virtual reality [DTI Information Society Initiative, online]. Available: http://www.ukvrforum.org.uk/resources/index.htm

Cydata. (1999). *UK VR Market Awareness Survey* [UK VR Forum Publication, online]. Available: http://www.ukvrforum.org.uk/resources/index.htm

Hamel, G., & Prahalad, C. K. (1994). *Competing for the future.* Boston: Harvard Business School Press.

Helsel, S. K., & Jacobson, J. (1994). *Virtual reality market place.* Westport, Mecklermedia.

MAPS. (1995). *Strategic market report: Virtual reality 1995.* Market Assessment Publications.

Pandit, N., Swann, G. M. P., & Watts, T. (1997). High-technology small-firm marketing and customer education: A role for a technology awareness programme? *Prometheus, 15*(3), 293–308.

Pimentel, K., & Teixeira, K. (1995). *Virtual reality: Through the new looking glass* (2nd ed.). New York: Intel/McGraw-Hill.

Shurmer, M., & Swann, G. M. P. (1995). An analysis of the process generating de facto standards in the PC spreadsheet software market. *Journal of Evolutionary Economics, 5*(5), 119–132.

Stone, R. J. (1996a). *VR sales process model,* Unpublished Virtual Presence (U.K.) position paper, compiled in support of Stone (1996b).

Stone, R. J. (1996b). *A study of the virtual reality market.* Summary document prepared for the Department of Trade and Industry Communications and Information Industries Directorate, London, HMSO.

Swann, G. M. P. (1986). *Quality innovation: The economic analysis of rapid improvements in electronic components.* London: Pinter.

Swann, G. M. P., Pandit, N.R., & Watts, T. (1997). *The business payoff from VR applications: Pilot case studies of space planning and training.* Report for Department of Trade and Industry.

Thompson, J. (1993). *Virtual reality: An international directory of research projects.* Westport, Meckler.

U.K. VR Forum. (1998–2000). Available: http://www.ukvrforum.org.uk/

*VR News,* Several Issues: 1997–1998.

Watts, T. P., Swann, G. M. P., & Pandit, N. R. (1998). Virtual reality and innovation potential. *Business Strategy Review, 9* (3), 45–54.

## Appendix 1. Example Model Expansion: The Primary Education Process

At its highest level, the model developed by Stone (1996) is a six-stage process, with the following headings:

**Introduction:** whereby individuals or teams from a given industry sector first come into contact with VR.

**Primary Education:** the process whereby those individuals (technology "champions" by this stage—the power of ownership should not be underestimated) gain more information about the technology and make their initial commitments;

**Secondary Education:** typically quite a lengthy process whereby the technology champions, often with external support, organize demonstrators for exposure to their own local team(s) and to higher echelons within the organization. During this stage, much effort is expended in commercially justifying possible adoption of the technology:

**Adoption/Rejection:** Adoption can take two forms. One is relatively straightforward in that the organisation, having indulged in the secondary-education process, procures a turnkey VR system and begins to develop its own internal exploratory studies. The second involves continued VR development with external support (and frequent demonstrations), usually resulting in the delivery of a turnkey bespoke VR system. Rejection can, of course, occur if (a) little or no support is forthcoming from within the user organization, despite (sometimes many) demonstrations, or (b) if a VR system has been procured under false pretences (i.e., from the outset, VR stood little or no chance of solving the user's problem or delivering a quality application, but a system was, nonetheless, purchased by, or sold to him).

**Integration:** This is a process which has yet to be studied in detail, due mainly to the fact that few—if any—VR systems have been integrated within an organization's business practices with a suitable period of postintegration review. Also, no real methodologies or guidelines exist as yet to assist in the integration process.

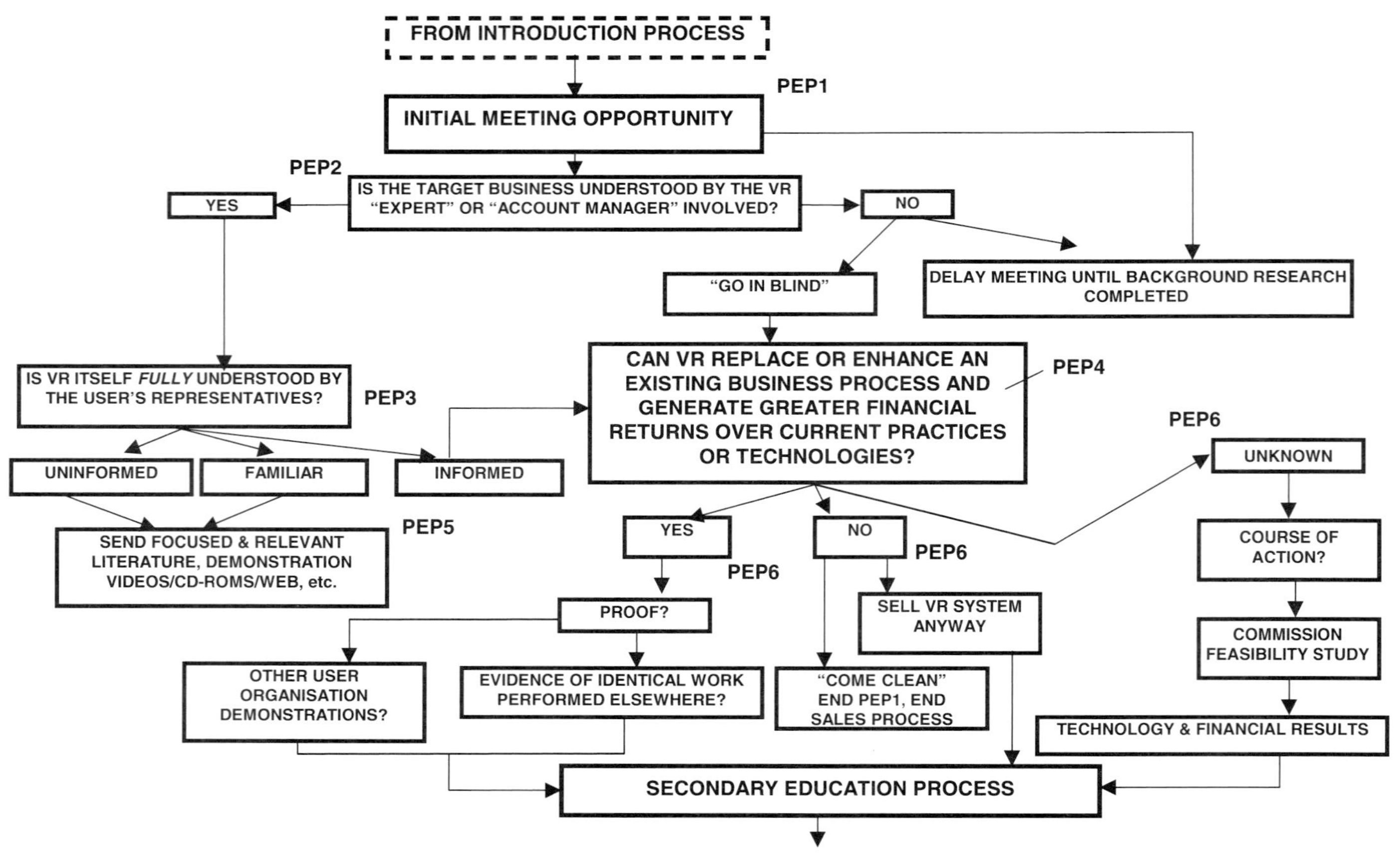


FIG. 28.6. "The Stone Sales Model": Primary Education Process (PEP).

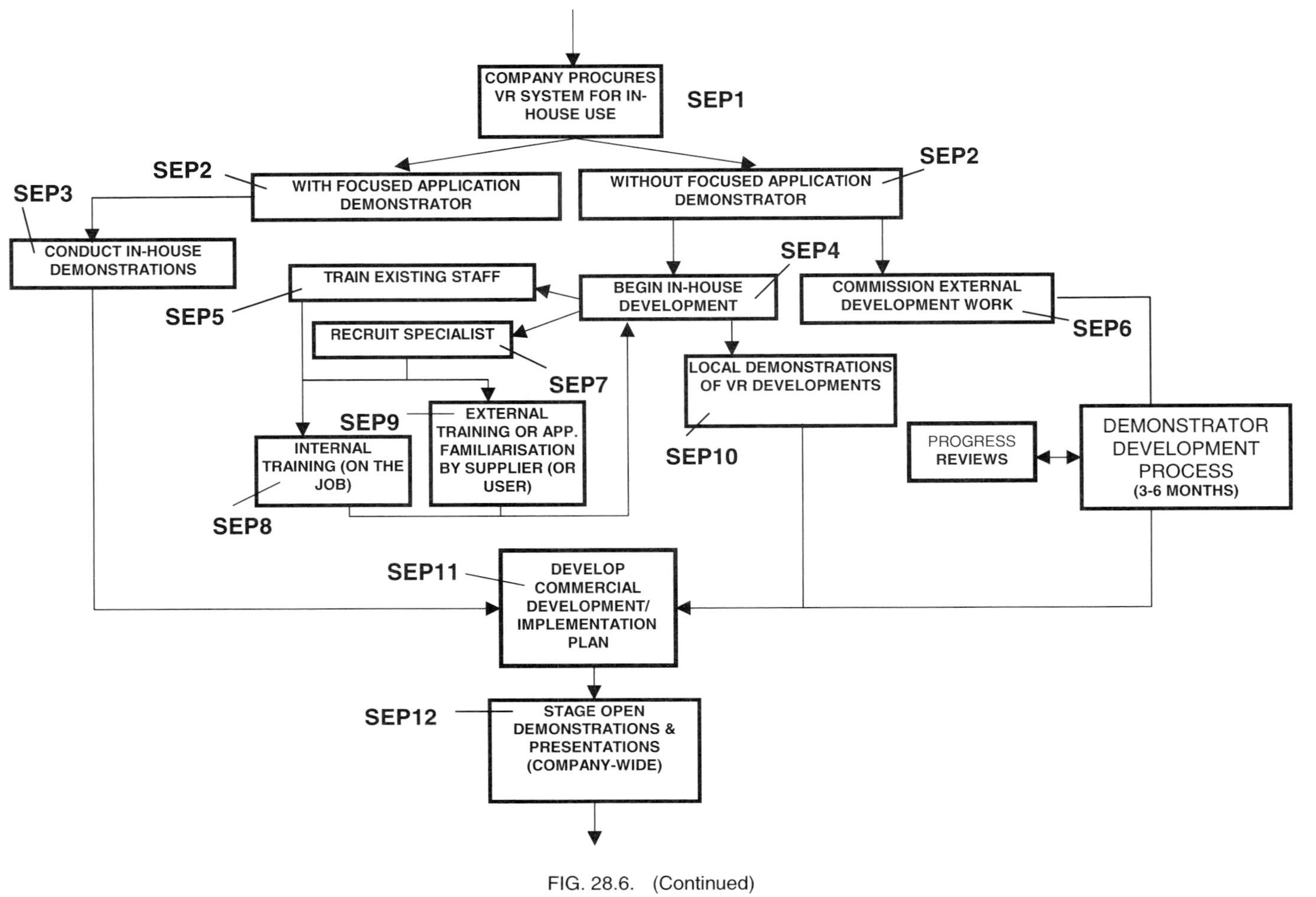


FIG. 28.6. (Continued)

**Development:** Development refers to how the organisation develops its business and growth strategies around the integration of the VR technology. Important issues include recruitment, training (and retraining), system enhancement (with, hopefully, original supplier and developer support), corporate and supplier communications and data sharing, quality, standardization, and so on.

Only the Primary Education Process is expanded here. See Fig. 28.6.

## PEP1: THE INITIAL MEETING OPPORTUNITY

Following the event at which the organization's VE champion first decides to take his initial lines of enquiry further (cards having been exchanged, inquiry forms filled out, telephone inquiries made, etc.), meetings are set up, either at the potential user's premises, at the offices and demonstration facilities of the supplier or developer, or both. Factors that influence the timing and location of the meetings are:

- Supplier/developer schedules and commitments
- The degree to which the supplier or developer perceives the user as a serious potential bringer of business
- The supplier's or developer's geographical location
- Availability of traveling salesmen ("account managers" in more recent terminology)

Note that the judgment made by the supplier or developer as to the perceived value of a user is not as clear-cut as one might expect (i.e., wealthy user = significant business). Even the most basic of contact with a user classified as having a "nonserious" application could bring short-term commercial value—in the form of future prestigious press releases, for instance—something VE companies are very responsive to.

## PEP2: IS THE TARGET BUSINESS UNDERSTOOD BY THE VE "EXPERT" OR "ACCOUNT MANAGER" INVOLVED?

This is a most significant question in terms of VE business development. Virtual environment technology is still very much a "technology push" rather than "market pull" domain. Yet, the salesman or account manager who makes no attempt to understand the business of the user he is about to visit (or be visited by), and entertains the user's representatives "blind," runs the risk of seriously damaging his company's chances of securing business, not to mention adversely affecting the still-fragile industrial and commercial status of the VE market. Virtual environment proponents or sales teams who have, for the sake of a quick sale, portrayed just the plus points of the technology (rather than "coming clean" about what the technology can and cannot do, vis-à-vis the user's business) have already caused problems, the effect of which has swept rapidly through both supplier or developer and user communities.

If the target user's business is understood, then an important follow-up question has to be asked:

## PEP3: IS VE ITSELF FULLY UNDERSTOOD BY THE USER'S REPRESENTATIVES?

Depending on how the user's representative first came into contact with VE technology, she will most likely have a preconceived idea as to what the technology is and what it might do for her organization. In the worst case, the user might still foster an impression that VE is quite a low quality technology, based on her exposure to advanced graphics from film and television advertising media (this suggests immediately that the user has not grasped the real-time significance of VE technology and the impact this driving feature has on the need to be "economical" with graphics quality). She might be further concerned about such issues as cost, quality, the "necessity" to wear HMDs or gloves, or the health and safety debate (see chap. 41, this volume). There may be a significant degree of skepticism (or even fear about loss of jobs) among her colleagues, particularly those involved in other electronic media, such as CAD, digital design and art, and so on.

As part of the supplier's/developer's role in the education process, it should be a mandatory requirement to provide the potential user with VE data sheets which are both informative and independent from any one product or solution. Should the user require further information, then this could be supplied in the form of video or CD-ROM applications examples and preagreed contact details with a range of other, more active users.

Whilst this list of "shoulds" is but an ideal at the moment, it is well within the VE community to produce something of relevance which could be held centrally for distribution by nominated "information brokers"—VE societies or respected magazines, such as *VR News* (see chap. 56, this volume). Until that happens, however, potential users will continue to be inundated with glossy brochures and press release statements from VE suppliers. In fact, these collections of often irrelevant data sheets actually do little to provide the user's representative with the information he needs, save some element of encouragement and motivation that "somebody else might doing something relevant out there." The documents typically fall into two categories. One category comprises technical data—polygon counts, rendering facts, chipset specifications, benchmarks, and so on. Often these may well confuse a practicing VE specialist, let alone a newcomer to the field. The other category comprises documents that are often major "cut-and-paste" affairs, featuring the same "pro-own-product" statements from the supplier's managing director and filtered "pro-their-product" statements from the "satisfied" user. It is amazing how many users actually express concerns other than those published when quizzed at a later date.

Whatever is forwarded to the user before a meeting or demonstration, it should be based on the supplier salesman's attempt at gaining some understanding of the business she is trying to penetrate, as highlighted in PEP2. Irrelevancy, as is often the case, can be as damaging as no information at all. An initial meeting should not normally be held until sufficient time has elapsed for the user's representative to digest all information.

Finally, the supplier or developer must, prior to the initial meeting, be thoroughly prepared for the following question:

## PEP4: CAN VE REPLACE/ENHANCE AN EXISTING BUSINESS PROCESS AND GENERATE GREATER FINANCIAL RETURNS OVER CURRENT PRACTICES OR TECHNOLOGIES?

Is there a significant amount of retraining necessary? Do specialists need to be recruited? What changes to the user's internal and external infrastructure need to be made? How much will it cost?

If the answer to the question posed under PEP4 is genuinely no, then two things might happen (PEP7). Either the salesman continues with the hard sell approach and convinces the user to purchase an entry level VE system, or the salesman can "come clean" and admit that he believes that VE is not yet ready to solve the commercial problems or enhance the organization's business process at this time. The effect of the former can be quite disastrous. Depending on the level at which the selling attempt is pitched, it has also been shown that overselling an inappropriate product can delay a company's adoption of VE for the better part of a year.

If there is still some uncertainty associated with the answer to PEP4, then one solution might be (and sometimes is, currently) to commission an off-site proof-of-concept demonstrator, which falls under the next main section, "Secondary Education." Another solution, which is becoming increasingly popular, is to commission a feasibility study (PEP8), addressing the specific problem in hand and proffering technical and fully costed solutions (typically with a range of options from low-level desktop VE implementations to full-blown immersive and projection systems), plus a semiformal proposal for the main phase of work, if successful. Depending on the customer, feasibility studies can cost anything from £UK3,500 (approximately \$5,000) to £UK25,000 (approximately \$36,250). Some 80% of feasibility studies result in paper report deliverables; the other 20% demand either a simple demonstration of the application on- or off-site. Some supplier companies, anxious to secure business, will carry out feasibility studies free of charge. Of course, the results of these studies tend to be completely biased towards the supplier's own product range. Also, any free VE modeling work carried out as part of the study tends to be put together extremely quickly (minimizing distractions to the supplier's core business), often resulting in a poor quality demonstration.

If the answer to PEP4 is a definite yes, then the responsibility falls on the supplier or developer to deliver "proof," typically in the form of a relevant case study conducted by or for another organization (typically delivered verbally, as few complete case studies yet exist), backed up by demonstrations (PEP6). Again, depending on the user company involved, this next step can sometimes be paid for (indeed at this stage, some companies are quite forthright in informing the user that the consultancy "clock" has been started). Alternatively, some of the larger supplier companies might, as recorded above, donate VE modeling and demonstration work for free. An interesting effect can occur at this point which has potentially damaging consequences for some user organizations. If the demonstrations are held in one of the (increasing number of) so-called reality or demo centers (in both commercial and academic circles), then there is evidence of an attitude among some user representatives that VE appears far too expensive to be even considered at this stage. The overemphasis on presentation quality—the demonstration environment, the method of display, even the decor—coupled with the rather glitzy entertainment flavor of the demonstrations presented, detracts from the serious message that VE can bring technical and commercial benefits. Worse still (and this is becoming increasingly evident in academia) is that the reality center or CAVE (Cave Automatic Virtual Environment, see chaps. 11 and 13, this volume) facility has no quality content (research or otherwise) associated with it, and the technology is being used to deliver rather poor "fly-throughs" or other developers' demonstrations.

## Appendix 2. Mathematical Model: Equations

(For interpretation of these equations, refer to section 4 and Fig. 28.1 in the chapter.)
Equation 1: Start on selling opportunity?

$$P_1 = 1 - b_1$$

Equation 2: Does VR seller understand user's business?

$$P_2 = b_1 \cdot \left(1 - \frac{1}{1 + b_2 X}\right)$$

Equation 3: Does VR seller elect to "go in blind?"

$$P_3 = \frac{b_3}{1 + b_4 X}$$

Equation 4: Does VR seller do background research on user's business?

$$P_4 = 1$$

Equation 5: Does VR seller continue with selling opportunity?

$$P_5 = 1$$

Equation 6: Does user representative fully understand VR?

$$P_6 = \left(1 - \frac{1}{1 + b_5 X}\right) \cdot [1 - (1 - P_2) \cdot P_3]$$

Equation 7: After reading literature, does user interest continue?

$$P_7 = b_6 \cdot [1 - (1 - P_2) \cdot P_3]$$

Equation 8: Can VR offer business benefits to the user?

$$P_8 = 1 - \frac{1}{1 + b_7 X}$$

Equation 9: Does the user require further evidence?

$$P_9 = \frac{1 - (1 - P_2) \cdot P_3}{1 + b_8 X}$$

Equation 10: Are there other similar users who can demonstrate?

$$P_{10} = 1 - \frac{1}{1 + b_9 X}$$

Equation 11: After seeing demonstrations, is user convinced?

$$P_{11} = b_1 \cdot \frac{X_s}{X}$$

Equation 12: Is it necessary to carry out a feasibility study?

$$P_{12} = b_{10} \cdot \frac{X_u}{X}$$

Equation 13: After seeing the feasibility study, is the user convinced?

$$P_{13} = 1$$

Equation 14: Does the vendor "come clean" that VR may be of limited use to this user?

$$P_{14} = (1 - b_3) \cdot \left(1 - \frac{1}{1 + b_{11} X}\right)$$

Equation 15: Does user still decide to go ahead?

$$P_{15} = 0.5$$

## Appendix 3. VR for Business Checklist

Does your organization fit into one or more of the categories tabled below? The row titles represent horizontal market sectors; the columns vertical market activities

| *Activities* → / ↓ *Sectors* | *Design (Physical/ Rapid Prototyping)* | *Design (Ergonomics)* | *Product Design-to-Market* | *Design (Content, inc. games)* | *Marketing & Sales* | *Web Design/CD Publication* | *Product or Service Supply* | *Factory/Plant/Site Planning* | *Health & Safety* | *Training* | *Data Visualization* | *Micro-engineering* | *Planning and Decision Making* | *Local/Global Communication (Networks)* |
|---|---|---|---|---|---|---|---|---|---|---|---|---|---|---|
| Entertainment | | | * | ** | ** | ** | * | | * | | | | | * |
| Tourism | | | * | ** | ** | ** | * | | | * | | | * | * |
| Museums/heritage | | | * | ** | ** | ** | ** | | | * | * | | * | ** |
| Education | * | * | | ** | | ** | ** | | * | ** | ** | * | * | ** |
| Aerospace | ** | ** | ** | ** | ** | * | ** | ** | ** | ** | ** | * | ** | ** |
| Marine | ** | ** | ** | ** | ** | * | ** | ** | ** | ** | ** | * | ** | ** |
| Defense | ** | ** | ** | ** | ** | * | ** | ** | ** | ** | ** | * | ** | ** |
| Nuclear | ** | ** | | ** | * | | * | ** | ** | ** | ** | * | ** | * |
| Petrochemical | ** | ** | ** | ** | * | | * | ** | ** | ** | ** | | ** | * |
| Automotive | ** | ** | ** | ** | ** | * | ** | ** | * | ** | * | * | * | ** |
| Medical | | * | * | ** | * | * | ** | * | ** | ** | ** | ** | * | * |
| Construction | ** | ** | ** | ** | ** | * | ** | ** | ** | ** | ** | | ** | * |
| Transportation | ** | ** | ** | ** | ** | * | ** | * | * | ** | * | | ** | * |
| Retail | ** | ** | ** | ** | ** | * | ** | ** | * | * | * | | ** | ** |
| Utilities | ** | * | * | * | ** | ** | * | ** | ** | ** | ** | | ** | * |
| Business/commerce | | | | ** | ** | ** | * | | | ** | ** | | ** | ** |
| Telecommunications | ** | ** | ** | ** | ** | * | * | * | | ** | ** | | ** | ** |
| TV and media | * | * | * | ** | ** | ** | * | * | | * | * | | * | * |
| Information Technology | | ** | ** | ** | ** | * | | | | ** | * | | * | ** |

*VR is relevant
**VR is highly relevant

## Appendix 3 (continued). Specific Business Questions

- Would you describe your organization as creative in the sense of how you design and create a visual impact for your product or service?
- Do you do your own marketing?
- Do you take part in a major exhibition, specific to your market, at least once a year?
- Do you supply to an organization already using or considering using VE technology
- Are your competitors evaluating VE (are you certain that your competitors aren't evaluating VE)?
- Do you already have an installed computer base?
- Are any of these computers Pentium II-33 or better?
- Are any of these computers equipped with special graphics boards (e.g., Permedia 2, Riva TNT, Voodoo, nVidia GeForce, or other Open GL-compliant cards)?
- Do you already use computer-based visual "techniques" for design, sales, marketing, training, or communication (e.g., MacroMedia, Alias, PhotoShop, 3DStudio, etc.)?
- Do you make the information technology (IT) investment decisions for your business? If not, how much influence do you have on those who do?
- Does your organization have an IT implementation or expansion plan for the next 2 to 5 years?
- Would you describe your organization's expenditure on IT research, investment, and regular upgrades as "healthy" (i.e., do you keep reasonably up-to-date with purchasing hardware, software or services of relevance to your market)?
- How receptive is your organization's board generally to advanced forms of IT? Are they aware of VE technology?
- Are board members likely to be more receptive to VE if you could show them something specific, relevant, and with evidence of commercial returns?
- Would your organization be prepared to sanction expenditure of, say:
- £UK7,000 (approximately $10,000) for an entry-level VE system (which, if unsuccessful, could be redeployed elsewhere in the organization)?
- £UK3,500 (approximately $5,000) for an independent review and presentation on VE and its relevance to your business?
- £UK5,000 (approximately $7,250) to allow one or more members of your organization to undergo basic VE training?
- Do you have access to the Web?
- Are you networked with local (office/building/national) parts of your business?
- Are you networked with remote (international) parts of your business?

# PART IV: Health and Safety Issues

# 29

# Direct Effects of Virtual Environments on Users

Erik Viirre[1,2] and David Bush[3]
[1] *University of Washington, Human Interface Technology Laboratory Seattle, Washington*
[2] *University of California, San Diego, Division of Otolaryngology School of Medicine, Perlman Ambulatory Care Center 9350 Campus Point Drive, 0970, La Jolla, CA 92037*
*erikv@hitl.washington.edu*
[3] *University of Central Florida Industrial Engineering and Management Systems Department 4000 Central Florida Blvd., Orlando, FL 32816-2450*
*roclore@hotmail.com*

## 1. INTRODUCTION

To deal with the effects that virtual environment (VE) technologies might have on users, the categorization of direct versus indirect effects is herein used. *Indirect effects* affect the user at a high functional level. Indirect effects include psychological effects, such as modification of phobias and enhancement or repression of emotions. Indirect effects also include neurological effects on the visual system (eyestrain, modification of stereoscopic vision, and visual acuity). Virtual environments can also affect the motion detection system and may result in imbalance, nausea, and motion sickness. Research and recommendations into indirect effects including eyestrain and motion sickness are reported in chapters 30 to 32 and 36 to 39 (this volume). *Direct effects* of VEs are less complex to study and control, but are potentially as great a hazard to users. The direct effects of a VE system are those that act at a direct tissue level, as opposed to the body systems level of the indirect effects. Direct effects are the influence of energy on body tissues from the technology and the risk of trauma because of encumbrances (e.g., weight of a helmet-mounted display [HMD]). Trauma is a particularly important issue, because the essential features of VEs are *interactivity* and *immersion*, where the user interacts in three dimensions with computer graphics. Irrational exuberance by users totally absorbed in the virtual experience might occur in some situations and result in injuries. Fortunately, dealing with direct effects of VEs is mostly a matter of common sense and awareness.

## 2. DIRECT TISSUE EFFECTS

### 2.1 Visual System

In the evolution of VE systems, considerable development efforts have gone into construction of visual displays. In the visible, ultraviolet, and infrared wavelengths of light, there are hazards to the eye. See Table 29.1 for a delineation of biological hazards to the eye from these light wavelengths.

Light energy can be concentrated enough to damage ocular structures directly; however, such light levels are unlikely to be used in VE systems, as they would be uncomfortable to look at and not serve any purpose in the display. There are a few possible exceptions. Note that eyestrain from visual displays will be discussed in chapter 37 (this volume). This chapter includes a discussion of seizures and migraine headaches induced by flashing lights and moving images.

#### *2.1.1 Visible Light*

A novel approach to development of visual displays is the use of scanned laser light directly onto the retina. In practice, as with the conventional displays described above, there should be little hazard. However, high-power light could potentially be introduced to the eye, so care must be taken (see Table 29.2). These light scanning systems use high-frequency vertical and horizontal scanners to scan tightly focused noncoherent light or coherent laser light directly onto the retina. The Virtual Retinal Display (VRD) (TM, Microvision Incorporated, Seattle, Washington) is one such display. The VRD was invented at the University of Washington in Seattle and uses laser light to scan images onto the retina. Practical use of the VRD in room light conditions has shown that images that appear very bright can be generated with very-low-power light levels, typically on the order of 100 to 500 nanowatts. The power limit for a Class 1 laser as defined by the American National Standards Institute (ANSI) is 400 nanowatts (ANSI Z136.1, 1993). Lasers with wavelengths only in the visible range that emit less than this power level can be viewed continuously without damage to the retina (ANSI Z136.1, 1993). Beyond this power level (Class 2 and above), damage can occur to the retina or the optics of the eye, depending on the wavelength, the duration of exposure, whether the exposure is pulsed or continuous, and the intensity. The ANSI standard explains how the risks can be determined.

The VRD has a potential risk when it is used as an augmented reality display, that is, where the computer-generated image is viewed superimposed on the ambient visual scene. If the

**TABLE 29.1**
Summary of Basic Biological Effects of Light on the Eye

| *Photobiological Spectral Domain* | *Eye Effects* |
|---|---|
| Ultraviolet C (0.200–0.280 $\mu$m) | Photokeratitis |
| Ultraviolet B (0.280–0.315 $\mu$m) | Photokeratitis |
| Ultraviolet A (0.315–0.400 $\mu$m) | Photochemical UV cataract |
| Visible (0.400–0.780 $\mu$m) | Photochemical and thermal retinal injury |
| Infrared A (0.780–1.400 $\mu$m) | Cataract, retinal burns |
| Infrared B (1.400–3.00 $\mu$m) | Corneal burn, aqueous flare, IR cataract |
| Infrared C (3.00–1000 $\mu$m) | Corneal burn only |

*Source:* OSHA technical manual.

**TABLE 29.2**
Laser Classifications—Summary of Hazards

| *Applies to Wavelength Ranges* | | | | | *Hazards* | | |
|---|---|---|---|---|---|---|---|
| *Class* | *UV* | *VIS* | *NIR* | *IR* | *Direct Ocular* | *Diffuse Ocular* | *Fire* |
| I | X | X | X | X | No | No | No |
| IA | — | X* | — | — | Only after 1000 sec | No | No |
| II | — | X | — | — | Only after 0.25 sec | No | No |
| IIIA | X | X** | X | X | Yes | No | No |
| IIIB | X | X | X | X | Yes | Only when laser output is near Class IIIB limit of 0.5 watt | No |
| IV | X | X | X | X | Yes | Yes | Yes |

*Key:* X indicates class applies in wavelength range.
* class IA applicable to lasers "not intended for viewing" only.
** CDRH Standard assigns Class IIIA to visible wavelengths only. ANSI Z 136.1 assigns Class IIIA to all wavelength ranges.

VRD were used in very-high-altitude aircraft, superimposition of the images on the ambient scene might require very-high light power levels to provide a perceived image of adequate brightness and contrast. These power levels could damage the eye. It should be noted, though, that such images would be uncomfortably bright and probably would not be usable without attenuation of the light from the external scene and from the VRD.

Safety of laser light can be analyzed through straightforward means and calculations using the ANSI standard. Interestingly, the same intensity limits might well be applied to noncoherent (nonlaser) light. Such analyses would be most relevant for displays using scanners.

### 2.1.2 Infrared and Ultraviolet Light

There are a few circumstances were infrared (IR) and ultraviolet (UV) light could be encountered in VE systems. The most likely reason would be incidental output of these wavelengths from the particular display technology. As with visible light, safety of IR and UV can be readily determined with simple measurements and calculations. These can also be found in the ANSI standard or in other sources. If there is no need for these wavelengths, inexpensive and effective optical coatings can be used to prevent their transmission into the eye.

Occasionally, UV or particularly IR can be useful. The technique of video-oculography (VOG) can measure the eye movements of a VE system user (see chap. 9, this volume). To obtain these measures, the user wears a head-mounted video camera, usually based on a micro CCD chip. Good illumination of the eye is necessary in order for the video image recognition technology to work. In order to avoid interfering with viewing, the illumination and video imaging are done in infrared. Cameras for capturing the images of the moving eye are mounted on the head. In a VE system, they can be mounted inside an HMD. The cameras are generally fitted with filters to block light above about 800 nanometers. The eye is illuminated with an infrared emitting LED, whose illumination is not visible to the eye. High-output infrared LEDs can be a hazard to the eye, as they can lead to cataracts. Furthermore, as the eye cannot see the illumination, the user is not aware of the hazard. Class 1 light emission device limits are generally a good place to start. The FDA regulation Part 1040—Performance Standards for Light Emitting Devices, describes the calculations. The most stringent standard is for continuous viewing: defined as longer that $10^4$ seconds. For example, for infrared light of 850 nm wavelength the permissible radiant power is $7.6 \times 10^{-5}$ W. The regulations and basis for calculations can be found at www.fda.gov/cdrh/radhlth/cfr/21cfr1040.10.pdf or through the Food and Drug Administration.

### 2.1.3 Photic Seizures

Statistics show that 1.1 in 10,000 people are prone to photic seizures (Quirk et al., 1995). A photic seizure is an epileptic event that is provoked by flashing lights. A bright light pulsing at the rate of 1 to 10 Hz can drive repetitive firing of neural cells in large groups. If enough of these cells begin firing in synchrony, the activity causes a chain reaction spreading throughout reasonably large portions of the brain. Individuals who experience such neural activity generally show a brief period of *absence*, where they are awake but do not respond. Rarely, photic seizures will become generalized convulsions. In standard neurological tests for epilepsy or the tendency for epilepsy, pulsed bright lights are used to see if seizure activity can be induced and recorded on an electroencephalograph. These types of events are at times triggered accidentally in a seizure prone driver passing a picket fence with the sun behind it, or with other sources of repetitive light flashes. Video images in the 50 to 60 Hz range showing striped patterns have been used to induce seizures (Kasteleijn-Nolst Trenite et al., 1999). Repeated seizures can lead to brain injury and a lower threshold for future episodes.

A well-publicized incident of large numbers of induced photic seizures occurred in Japan, where a cartoon television program featured a scene in which red and blue frames alternated at 12 Hz (Takada et al., 1999). Hundreds of children viewing this program had absence attacks. A similar event could occur, inadvertently or deliberately, if pulsing light was shown to a VE user. A plausible scenario involves transport delay: the processing time from the movement of a user's head to the movement of the scene in an HMD. If a complex scene is in the VE system, the computer processor may slow in the real-time reconstruction of the scene. If the slowing went to several frames per second, the potential for flashing in the range that provokes photic seizures might occur. As with most safety problems, this circumstance could be detected by an individual dedicated to safety monitoring from the company producing the images. Producers of VE display technologies and content would do well to have a reviewer monitor their products for prolonged flashing imagery. The National Society for Epilepsy Web site states photic seizures are triggered by flashing images or lights in the range of 5 to 30 Hz (5 to 30 flashes per second) (http://www.epilepsynse.org.uk/pages/info/leaflets/photo.html). It follows that designers and reviewers of VEs should avoid flashing images in the range of 5 to 30 Hz. The National Society for Epilepsy Web site also suggests if using a television, to utilize one that possesses a 100 Hz or greater rate of refresh. The Job Accommodation Network Web site recommends providing a high-resolution VGA monitor/flicker-free screen and glare guards or tinted computer glasses for photic seizure prone computer users (http://www.jan.wvu.edu).

### 2.1.4 Migraines

Migraines are a common phenomenon in the general population, occurring in 15% to 20% of women and approximately 10% of men (Stewart & Lipton, 1992). Migraine sufferers have a constellation of possible symptoms including the well-known headache, aurae (i.e., an ocular sensory phenomena immediately preceding the migraine attack), dizzy spells, and others. Migraine sufferers are prone to motion sensitivity, as well as light and sound sensitivity, particularly during an attack. Thus, the sensory experiences created in VE systems are the very phenomena that might make a migraine sufferer uncomfortable, if not ill. Fortunately, there are few serious consequences, but a bad experience could influence migraine sufferers to avoid

**TABLE 29.3**
Minimizing Migraine Stimulation in Virtual Environments

| *Sensory Stimulus* | *Control Measures* |
|---|---|
| Visual imagery | Minimize brightness, give brightness control to subjects |
| Visual motion | Minimize visual motion, especially point-of-view image motion (view from a head or body mounted camera viewpoint) |
| Auditory | Minimize loud pulsing sounds |
| Duration | Minimize duration, provide breaks, allow user control |

future VE exposures. Virtual environments should provide explanations of the experiences that they create. Migraine sufferers are often aware of their sensitivities, and so many would avoid experiencing something described as "the ultimate roller coaster" or "a sound and light explosion." As with the problems with photic seizures described above, a technology or content producer might want to have a reviewer consider the amount of visual motion, brightness and contrast, and loudness of sounds that their system produces. Particularly for systems intended for training or other general purposes where migraineurs might be required to work in a virtual environment, control of the amount and intensity of motion, sound, and light contours would be prudent. Control of the duration and repetition of intense sound and light experiences is also recommended. Some useful guidelines are provided in Table 29.3.

## 2.2 Auditory System

Sound is an essential part of VE experiences, not only for making more realistic sensations, but also for providing cues to actions in the environment (see chap. 4, this volume). Sounds for VE can be presented through speaker systems or through headsets worn over or in the ears. The realism of some VE experiences might depend on loud sounds (e.g., an aircraft) presented through a headset. Prolonged exposure to such sounds can result in noise levels that exceed the United States Federal Department of Labor, Occupational Safety and Health Administration's (OSHA's) permissible exposure limit (PEL; OSHA, Technical Manual). The OSHA PEL is 90 decibels, averaged over an 8–hour period on the A scale of a standard sound-level meter set on slow response (OSHA, Technical Manual). Noise levels exceeding OSHA's PEL will cause hearing damage (OSHA, Technical Manual). While no guidelines exist for VE systems, user guides for a typical Walkman-style audio player headset suggest avoiding continuous loud sounds, and that the headset should be removed every 30 minutes. Unfortunately, few users pay attention to such warnings. Manufacturers might thus consider limits on the intensity of sounds produced by their systems.

OSHA has established safety standards for sound exposures (OSHA regulations [Standards-29CFR], occupational noise exposure—1910.95). The standards establish sound level exposures that require monitoring or head protection for workers. The louder the average noise level, the lower the amount of time of exposure allowed. For example, an OSHA "action level" is 85 dBA averaged over an 8-hour workday. Exposure to noise of this level requires a hearing conservation program consisting mainly of regular hearing tests. If the sound level is over 90 dBA averaged over 8 hours, hearing protection is required. A level of 100 dBA exposure over 2 hours per day also requires hearing protection. These sound-level standards are for continuous

**TABLE 29.4**
Comparison Table of Duration per Day in Hours to Allowable Sound Level in Decibels, Measured on the Scale of a Standard Sound-Level Meter Set on Slow Response

| *Duration per Day, Hours* | *Sound Level, dBA, Slow Response* |
|---|---|
| 8 | 90 |
| 6 | 92 |
| 4 | 95 |
| 2 | 100 |
| 1 | 105 |
| 0.5 | 110 |
| 0.25 | 115 |

*Source:* 29 CFR 1910.95, Table G-16, OSHA technical manual.

sounds. It should be noted that repeated high-intensity pulsatile sounds can cause hearing loss as well. Industrial users of VE systems would do well to review the sound levels produced by their experiences and determine the amount and duration of exposures for users. These could then be compared to the guidelines set by OSHA to determine if the VE system should be used in moderation to prevent hearing damage (see Table 29.4 below).

Of note in VE systems is "spatialized" or three-dimensional (3-D) sound systems (see chap. 4, this volume). The same standards presented above apply to 3-D systems, but the variation in intensity over the virtual sound field should be considered. If loud sounds occur within a VE system, users might intentionally or unintentionally get close to the loud sound source. Controls to reduce prolonged exposure to such intense sounds may be needed. If a user is receiving a prolonged exposure to a loud sound, a safety circuit to automatically reduce the sound levels might be a desirable design feature.

## 2.3 Skin

The use of VE systems can injure the skin via body-worn technology. Sensible reviews of the technology should take place. For example, if the system produces X rays (e.g., some cathode ray tube), there is a risk of inducing carcinomatous changes in the skin (Cade, 1957). Similarly, systems producing high electromagnetic fields (EMF) at other frequencies (such as magnetic position trackers or infrared emitters) should be monitored for heat generation. Finally, there is a potential to transmit infective agents through objects worn on the skin of one user and transferred to the skin of another, such as bacteria, viruses, and fungi. In the absence of specific governmental guidelines in the workplace, contamination of shared devices will be reduced by the use of materials resistant to infective agents (e.g., metals and plastics, as opposed to porous materials) and materials that can be disinfected.

The skin is sensitive to prolonged physical pressure, which can result in skin irritation or breakdown. Furthermore, there is a risk of passing on skin contaminants such as bacteria between multiple users, as previously noted. Fortunately, there are numerous other types of devices with similar patterns of use as public VE systems already in practice (e.g., goggles for 3-D movies, public telephones), and there appears to be minimal risk associated with them. Developers of haptic feedback devices (see chaps. 5 and 6, this volume) and applications using them should be on the lookout for feedback or other unusual situations that could produce high pressure from their devices and cause injury. The devices for force generation should have limits on the maximal forces generated in order to minimize risk to hands and fingers.

## 3. TRAUMA

The encumbrances of many VE systems present a risk to users for physical injury. By its nature, VE has compelling visual imagery that obliterates or obscures the visual world. Furthermore, the scenes and sequences that occur are often intended to induce the user to move, rapidly at times. Sound systems can produce disorienting sensations and the myriad cables, weights, and tethers may cause a user to lose balance. For these reasons, system designers should watch ill effects of VE systems carefully. These are not "set and forget" kinds of systems. A "spotter" should watch users and be available to interact with them should ill effects arise. Finally, motion sickness does occur in VE systems (see chap. 30, this volume). Users should thus be monitored for imbalance or attendant symptoms of motion sickness, as its occurrence makes a fall more likely.

## 4. AVOIDANCE OF INJURY

Fortunately, the use of VE systems is not inherently unsafe. Most circumstances that may involve risk should be mitigated with a little forethought. An individual from VE system development teams should be dedicated to the task of monitoring risks. One important component of the monitoring should be reviews with developers of any hazards or sensations they notice. For example, if developers find they are experiencing motion sickness in the development stage, users should be expected to have the same sensation. Similarly, if developers repeatedly trip over a cable, a reconfiguration should occur to fix the problem. If there is dedicated responsibility for testing and monitoring risks, the possibility of injury will be reduced. The simple framework of this chapter provides an approach for risk review, summarized in Table 29.5.

The body has its own feedback for injury risk. If a light is uncomfortably bright, sound uncomfortably loud, gear uncomfortably tight, or system uncomfortably encumbering in the short term, the risk for long-term injury should be considered. Thus, a straightforward review by potential users should reveal a number of the risks involved with the use of VE technology. Some issues, however, such as those associated with repetitive use or those to which only a portion of the user population will be susceptible (e.g., seizures) should be monitored by safety engineers. In addition, users might need to be screened for their ability to wear and

**TABLE 29.5**
Minimizing Risk in Development of VE Systems

1. Assign responsibility to someone for risk review.
2. Monitor and review user and developer comments for discomfort.
3. Review visual stimuli for brightness and visual motion.
4. Review auditory levels.
5. Consider emissions from electronics (e.g., heat, ionizing radiation).
6. Review hazards from cables, tethers, and other devices.
7. Review the experience being provided.

appropriately interact with VE systems prior to allowing unlimited exposure. Warnings and restrictions might be needed in some situations (see chap. 27, this volume).

The type of experience being created for users should also be considered. There are potential indirect or psychological effects from being in virtual environments. The possibility of producing emotional and cognitive changes in users must be considered (see chap. 33, this volume). Indeed, the compelling nature of VE experiences is being used to affect the nervous systems and psyches of patients with a variety of disorders (see chaps. 50 and 51, this volume). Thus, the affects of battlefields or dramatic flight sequences on users should be considered. Indirect effects can combine with direct effects to produce injury. An anxious user might pull on cables or become uncomfortable with headgear. The best method of prevention is close monitoring of users by developers and manufacturers, especially naive users of an environment.

## 5. CONCLUSION

Virtual environment technology promises to provide a variety of experiences to users: bright visual displays, compelling sound environments, and dramatic interactivity. Further, the influences of VE interaction often result in changes in users' mind/brain via training or therapy. Virtual environment systems can potentially be used to improve the balance systems of patients with disturbed vestibular systems (see chap. 49, this volume), reduce the phobias of people afraid of heights, flying, or spiders (see chap. 51, this volume), and teach concepts of art or science to induce an emotional state or a level of education (see chaps. 43–46, this volume). Yet along with the promise of VE technology, one can postulate the potential negatives associated with VE exposure: bright lights, loud sounds, motion sickness, phobia induction, negative emotions, and destructive tendencies. While harnessing the potential of VE technology, developers cannot neglect to invoke common sense and research into new observations while determining the intent of technology creators to minimize the adverse effects of VE interaction.

## 6. REFERENCES

Cade, S. (1957). Radiation induced cancer in man. *Br. J. Radiol. 30*, 3939

Hoffman, H. M., Murray, M., Hettinger, L., & Viirre, E. (1998, in press). Assessing a VR-based learning environment for anatomy education. *Art, Science, Technology: Healthcare Evolution*, IOS Press.

Kasteleijn-Nolst Trenite, D. G., da Silva, A. M., Ricci, S., Binnie, C. D., Rubboli, G., Tassinari, C. A., & Segers, J. P. (1999). Video-game epilepsy. *Epilepsia, 40*(4 Suppl.), 70–74.

Institute ANS. (1993). *American national standard on the safe use of lasers* (ANSI Z136.1-1993). American National Standards Institute.

Quirk, J. A., Fish, D. R., Smith, S. J., Sander, J. W., Shorvon, S. D., & Allen, P. J. (1995). Incidence of photosensitive epilepsy: A prospective national study. *Electroencephalography Clinical Neurophysiology, 95*(4), 260–267.

Stewart, W. F., & Lipton, R. B. (1992). Prevalence of migraine headache in the United States: Relation to age, income range and other sociodemographic factors. *Journal of American Medical Association, 267*, 64–69.

Takada, H., Aso, K., Watanabe, K., Okumura, A., Negoro, T., & Ishikawa, T. (1999). Epileptic seizures, induced by animated cartoon, "Pocket Monster." *Epilepsia, 40*(7), 997–1002.

U.S. Department of Labor, Occupational Safety and Health Administration. (2000, November). *OSHA technical manual* [online]. Available: http://wwwosha-slc.gov/dts/osta.html

# 30

# Signs and Symptoms of Human Syndromes Associated with Synthetic Experiences*

Ben D. Lawson,[1] David A. Graeber,[2] Andrew M. Mead,[3] and Eric R. Muth[4]

[1] *Naval Aerospace Medical Research Laboratory Spatial Orientation Systems Department Pensacola, FL, 32508-1046 blawson@namrl.navy.mil*

[2] *University of Central Florida, Industrial Engineering and Management Systems Department, Orlando, FL, 32816-2450 dave_graeber@hotmail.com*

[3] *Naval Air Warfare Center, Training Systems Division Orlando, FL, 32826-3224 MeadAM@navair.navy.mil*

[4] *Clemson University, Psychology Department 410C Brackett Hall, Clemson, SC, 29634-1355, emuth@clemson.edu*

## 1. INTRODUCTION

This chapter describes the unpleasant side effects associated with human exposure to virtual environments (VEs) and other synthetic experiences (SEs). The authors discuss the incidence and nature of the discomfort associated with acceleration, unusual force environments, moving visual fields, flight simulators, and virtual environments. The most widely used scales for assessing the clinical signs and subjective symptoms of SE discomfort are explained, and the usual early warning signs are described. Other less obvious syndromes are mentioned that do not involve frank discomfort yet may pose a significant health and safety risk: These include the sopite syndrome, loss of visual acuity during head or body motion, and postural dysequilibrium. Situations conducive of SE-related syndromes are discussed, with a special emphasis on the aspects of visual field motion that are most disturbing. The evidence to date indicates that many SEs are capable of producing significant decrements in well-being. Careful assessment

*The views of the authors do not necessarily represent the policy of the U.S. government.

and correction of human factors problems with VEs will be necessary to minimize health and safety risks among users.

The purpose of this chapter is to describe the most common adverse signs and symptoms associated with exposure to synthetic experiences. The phrase *synthetic experience* encompasses technologies such as virtual environments, teleoperators, and augmented reality (Durlach & Mavor, 1995), while not excluding other technologically mediated experiences such as interactive flight simulation or noninteractive viewing of moving visual scenes such as wide-screen movies (Robinett, 1992). The term SE may also be applied to the wearing of optical prisms that alter the visual stimulus (Welch, 1978). Many of the principles of adaptation to prism goggles will be similar to those that apply to modern head mounted displays (HMDs; see chap. 31, this volume). Finally, the term *SE* could be applied to simulations that allow people to experience nonterrestrial forces or other feelings of body acceleration (see chaps. 7, 11, and 13, this volume).

The type of SEs known as a virtual environments (Durlach & Movor, 1995; Fisher, 1990) will be the focus of this chapter. The authors will emphasize the clinical symptomatology of the VE syndrome sometimes referred to as cybersickness (McCauley & Sharkey, 1992), focusing on classical symptoms such as nausea. Where cybersickness findings are scant, the authors will make conjectures based on similar syndromes such as simulator sickness,[1] motion sickness (Irwin, 1975), and visually induced motion discomfort (Crampton & Young, 1953). The present review will aid the reader in comparing human syndromes elicited across different studies by itemizing specific signs and symptoms rather than offering numerical summary scores of "overall sickness," because the computation and interpretation of summary scores varies from study to study (Lawson, 1993).

In the following pages, the authors describe the incidence of human syndromes associated with SEs, the nature of these syndromes (characteristic signs and symptoms), and the simplest, most reliable methods for measuring these syndromes. This chapter is intended as a practical guide; hence the causal factors of SE-related syndromes will be treated from the standpoint of known eliciting stimuli rather than theories of etiology (see Jones, 1998; Kennedy & Frank, 1985; Kolasinski, 1995; McCauley, 1984; Reason & Brand, 1975). The physiological pathogenesis of SE-related syndromes is discussed in chapter 32 (this volume) and by Crampton (1990) and Money, Lackner, and Cheung (1996). Sensorimotor side effects of VE (postural dysequilibrium and manual incoordination) are discussed in more detail in chapters 38 and 39 (this volume).[2] While the present chapter will focus on group responses to SE, individual differences in susceptibility and ability to adapt to SE are important topics reviewed by Kennedy, Dunlap, and Fowlkes (1990), Guedry (1991), and Kolasinski (1995).

## 2. TERMINOLOGY USED TO DENOTE THE DISCOMFORT ASSOCIATED WITH SYTHETIC EXPERIENCES

The term *cybersickness* was coined by McCauley and Sharkey (1992) to describe the motion-sickness-like symptoms associated with VE characterized by "far applications involving distant objects, that is, terrain, self-motion (travel) through the environment, and the illusion of self motion (vection)" (p. 313).[3] McCauley and Sharkey (1992) chose *cybersickness* as a more general term than *simulator sickness* (Havron & Butler, 1957) because they considered moving and fixed-base simulators to be a subset of virtual environments. The word *cybersickness* does not

[1] Researchers usually attribute "simulator sickness" to Havron and Butler (1957), who were the first to publish an observation of the phenomenon but did not employ the phrase *simulator sickness*.

[2] See also chapter 6 of Durlach and Mavor (1995), and Stanney, Kennedy, Drexler, and Harm (1999).

[3] The term *vection* is attributed to Tschermak (1931) in Jones (1998).

denote a diseased or pathological state, but rather a normal physiological response to an unusual stimulus. In fact, in the case of "classical motion sickness" (e.g., seasickness), failure to be susceptible to nauseogenic motion stimuli can indicate an abnormal state (e.g., Graybiel, 1967).[4]

Cybersickness and simulator sickness share much in common with motion sickness, but it is preferable to avoid such terms and instead describe these maladies as syndromes (Kennedy & Frank, 1985). For example, Benson (1988) noted that *motion maladaptation syndrome* was a more accurate phrase than *motion sickness*, a view that has been adopted by Probst and Schmidt (1998) and Finch and Howarth (1996). For convenience, Benson's apt phrase "motion maladaptation syndrome" could be shortened to "motion adaptation syndrome," roughly in accordance with the precedent of Nicogossian & Parker (1984), who recommended the phrase *space adaptation syndrome* in lieu of *space sickness*.[5]

## 3. INCIDENCE AND NATURE OF SYNDROMES ASSOCIATED WITH MOVING DEVICES AND SYNTHETIC EXPERIENCES

### 3.1 Acceleration and Unusual Force Environments

It is difficult to estimate the proportion of humans who are susceptible to nonterrestrial forces created by passive acceleration within a moving device or by travel into space. The incidence estimate will vary with the stimulus (e.g., frequency, duration), the individual exposed to the stimulus (e.g., susceptibility, experience), and the experimenter's measurement criteria. The authors estimate that about 10% of the population has not experienced significant nausea during transportation, whereas about 1% of the population will vomit or be made nauseated by vehicles as mild and ubiquitous as the automobile (Birren, 1949; Reason, 1975). The most provocative laboratory tests yield very high incidence estimates. Miller and Graybiel (1970b) found that 90% to 96% of participants will suffer from stomach symptoms by the time they reach the maximum number of head movements called for during a rotation protocol.[6] Over the years, laboratory tests and some very provocative field tests (e.g., Kennedy, Graybiel, McDonough, & Beckwith, 1968) have shown that persons with normal vestibular function can be made nauseated by motion, whereas participants without vestibular function cannot be made nauseated by motion (Money, 1970). Similarly, participants without vestibular function cannot be made nauseated by alternating periods of weightlessness and high-G (Graybiel, 1967), although at least 67% of shuttle astronauts feel some unpleasant symptoms during their first trip into space (Davis, Vanderploeg, Santy, Jennings, & Stewart 1988).

### 3.2 Moving Visual Fields

#### *3.2.1 Findings From the Literature*

Like motion stimuli, moving visual fields can cause significant discomfort in stationary observers who are healthy, but not in individuals without vestibular function (Chueng, Howard, &

---

[4]Similarly, if the natural experience being recreated in a simulator is not sickening, then the presence of "simulator sickness" implies that the *simulator* is sick, not the user (Kennedy & Frank, 1985).

[5]There are still some problems with phrases such as *space adaptation syndrome* or *motion adaptation syndrome*, but the phrases are both improvements on earlier phrases such as *space sickness* or *motion sickness*, and seem to have gained some acceptance among groups such as the Motion Sciences Consortium (MSC) whose website is under construction at http://ait.nrl.navy.mil/MSC. Please send questions or comments to web@ait.nrl.navy.mil and to blawson@namrl.navy.mil.

[6]In Miller and Graybiel, (1970b), 90% of 250 subjects tested to a prevomiting endpoint experienced at least minimal nausea, epigastric awareness, or epigastric discomfort, whereas another 96% of 25 subjects tested all the way to vomiting experienced one of the aforementioned symptoms of nausea syndrome. The maximum possible number of head movements in the protocol was a total of 204 carried out across ten steps of increasing chair rotation velocity.

Money, 1991). As early as 1895, Wood (1975) described the aftereffects of immersing nearly stationary observers into a swinging room: "I have met a number of gentlemen who said they could scarcely walk out of the building from dizziness and nausea." (Reason & Brand, 1975, p. 110). Using a visual surround that oscillated in the earth-vertical axis, Crampton and Young (1953) found that 46% of participants ($n = 26$) experienced dizziness without nausea, whereas 35% experienced nausea. Benfari (1964) found that all nine of his participants experienced dizziness while viewing a wide-screen film involving vehicle motion, two of whom experienced extreme dizziness and nausea.[7] Using an optokinetic drum, Lackner and Teixeira (1977) found that 40% of participants ($n = 10$) viewing a moving visual surround terminated the experimental trial because of aversive symptoms. Over all experimental conditions, the most common symptom during or after exposure was dizziness (mean 50%), followed by "epigastric disturbance" (mean 23%), "headache or eyestrain" (mean 20%), and drowsiness (mean 10%). In the head-fixed condition of Teixeira and Lackner (1979), dizziness was observed in 50% of participants ($n = 8$) and eyestrain in 13%. Using the same optokinetic drum as Teixeira and Lackner (1979) but a different velocity profile, Lawson (1993) found that 57% of participants ($n = 14$) experienced stomach symptoms (stomach awareness or discomfort or nausea) during viewing of a rotating visual surround. Using a smaller optokinetic drum (that surrounded the upper body) and summarizing across four previous studies, Stern, Hu, Anderson, Leibowitz, and Koch (1990) concluded that about 60% of people viewing a moving visual surround will report symptoms. The most common symptoms found in the experiment described in Stern et al. (1990) were dizziness, warmth, and nausea (during unrestricted viewing without fixation). Investigators have noticed that symptoms induced by moving visual fields seem to linger for some time after the stimulus has terminated (Crampton & Young, 1953; Havron & Butler, 1957; Lackner & Teixeria, 1977; Teixeira & Lackner, 1979). Similarly, lingering effects have been associated with visual "fixed-base" simulators (see section 3.3 below) and the wearing of scene-altering prism goggles (Dolezal, 1982; Welch, 1978). However, it should be noted that lingering symptoms are not limited to situations involving visual motion only, but can also occur at sea (Reason & Brand, 1975).[8]

### *3.2.2 Summary*

Generalizing across the different visual motion experiments mentioned above, it appears that while a few subjects are unaffected by moving visual fields, 50%–100% will experience dizziness and 20%–60% will experience stomach symptoms of some kind. The frequency of other symptoms cannot be estimated, but it seems that "oculomotor symptoms" such as eyestrain are a prominent feature of human responses to moving visual fields (Ebenholtz, 1992; McCauley, 1984; Teixiera & Lackner, 1979). It is possible that moving visual fields induce more "head" symptoms (Lackner, from NRC Report BRL-CR-629, 1990; Reason & Brand, 1975), such as eyestrain, headache, and dizziness. Such nongastric effects have been observed extensively in flight simulators, which have a strong visual component (e.g., Kennedy, Berbaum, & Lilienthal, 1997).

---

[7]Note: The experimental conditions also involved varying amounts of peripheral flicker.

[8]Up to 5% of those who put to sea fail to adapt to the motion of the sea throughout the duration of the voyage, and these "chronically motion-sick" individuals may show effects weeks after the end of a voyage (Schwab, 1943, from Reason and Brand, 1975; Tyler and Bard, 1949). Although the biochemical changes associated with repeated vomiting confound the interpretation of the lasting effects of ship motion, Gordon, Spitzer, Doweck, Melamed, and Shupak (1995) have found that *mal de débarquement* can strike persons of widely varying susceptibility after a sea voyage, resulting in feelings of postural instability and perceived instability of the visual field during self-movement (even among some individuals for whom fluid loss due to vomiting is not a concern).

## 3.3 Simulators

### 3.3.1 Findings From the Literature

Observing 36 trainees using a U.S. Navy helicopter visual simulator, Havron and Butler (1957) noticed problems in 78% of the participants. The most frequently reported effects were (in descending order) nausea, dizziness, vertigo, blurred vision, and headache. Some effects lingered long after participants left the simulator, particularly nausea, dizziness, and drowsiness. More than half of the participants reported sickness lasting an hour or longer, and 13.9% said the effects lasted overnight.[9]

Soon after the Havron and Butler study, researchers in Pensacola, Florida (Miller & Goodson, 1958, 1960), reported similar effects in 12% of student pilots and 60% of pilot instructors ($n = 36$), using the same model of simulator. Effects included disorientation, nausea, dizziness, vertigo, visual distortions, and headache. Miller and Goodson (1958) also noted that "even those individuals who did not become ill reported that they usually felt very tired after a run. This fatigued feeling lasted frequently throughout the day." (p. 208). Many symptoms lasted for hours after simulator exposure. Kellog, Castore, and Coward (1984) found that 88% of pilots reported adverse symptoms when exposed to a particularly disturbing simulator, experiencing visual flashbacks (sensations of climbing, turning, or visual inversion) as much as 8 to 10 hours afterexposure (thus confirming the observations of Miller & Goodson, 1958, 1960). In reviewing selected simulator findings between the years 1957 and 1982, McCauley (1984) noted an overall incidence ranging from 10% to 88% across a number of studies. Additional effects mentioned by McCauley during these years included the "leans" (McGuiness, Bouwman, & Forbes, 1981, 1984) and postural dysequilibrium (Crosby & Kennedy, 1984; for a more recent discussion see chap. 39, this volume).

Magee, Kantor, and Sweeney (1988, in AGARD CP-433) found that 95% of their participants ($n = 42$) reported at least one symptom following simulator exposure, with 83% of the symptoms classified as "simulator sickness" (according to the criteria of Kennedy, Dutton, Ricard, & Frank, 1984). The most common symptoms were eyestrain (27%) aftersensations of motion (25%), mental fatigue (22%), physical fatigue (21%), and drowsiness (17%). For 81% of the participants, symptoms lingered for a median of 2.5 hours postexposure, with the most common delayed symptoms being physical fatigue (19%), eyestrain (17%), and mental fatigue (16%). Chapelow (1988) also observed that physical and mental fatigue were the predominant symptoms of prolonged exposure to two different simulators.[10] Gower, Lilienthal, Kennedy, and Fowlkes (1988) estimated an overall "simulator sickness" of 44% across a number of simulator exposures ($n = 434$), with the most common symptoms being drowsiness/fatigue (43%), sweating (30%), eyestrain (29%), headache (20%), and difficulty concentrating (11%).

Kennedy, Lilienthal, Berbaum, Baltzey, and McCauley (1989) reviewed data from 1,186 U.S. Navy flight simulator training sessions involving both fixed and moving base simulators.[11] Drowsiness or fatigue occurred in 26% of the hops, eyestrain in 25%, headache in 18%, sweating in 16%, difficulty focusing in 11%, nausea in 10%, and difficulty concentrating in 10%. However, only 0.2% of the hops induced retching or vomiting. The raw frequency with which one or more symptoms characteristic of motion discomfort (i.e., vomiting, retching,

[9]Immediate or lingering symptoms are detailed on page 5 and page 8 of Appendix F from Havron and Butler (1957).

[10]The authors of this study pointed out that fatigue symptoms were far more prevalent in the simulator test that required long training exposures. This suggests that duration of stimulator exposure and common fatigue contributed to the symptoms observed.

[11]The Kennedy et al. (1989) database partly overlapped with that from the Gower et al. (1988) study mentioned in the preceding paragraph.

increased salivation, nausea, pallor, sweating, or drowsiness) was reported ranged from 10% to 60% across several (9) different simulators, with an average of 34.3%. In another analysis looking for lasting aftereffects, Baltzey, Kennedy, Berbaum, Lilienthal, and Gower (1989) found that of 700 pilots queried, 45% felt some symptoms after exposure. These effects lasted longer than 1 hour in 25% of the cases and longer than 6 hours in 8% of the cases.

Studies by Ungs (1988) and Kennedy, Massey, and Lilienthal (1995) indicate that the most frequently reported symptoms of simulator exposure are drowsiness and fatigue. In general, about 30% to 50% of people report fatigue or drowsiness (or both) upon exiting flight simulators (Kennedy et al., 1995). Ungs (1988) found that fatigue was not only the most common symptom after one's first simulator flight (occurring in 34% of participants), but also one of the most severe symptoms (along with sweating and nausea). Fatigue and sleeping problems are still present after several simulator flights, albeit reduced in frequency and severity. These effects seem to occur in both fixed-base and moving-base simulators (Kennedy et al., 1995), and may indicate the presence of the sopite syndrome (Graybiel & Knepton, 1976; Lawson & Mead, 1998), which is discussed in section 6.1.

#### *3.3.2 Summary*

The incidence and severity of the adverse effects associated with simulator exposure vary widely from one simulator study to the next. Adverse signs and symptoms are experienced by at least 10% and sometimes as many as 90% of simulator trainees; effects include (roughly in descending order of prevalence) eyestrain, fatigue, drowsiness, sweating, headache, and difficulty concentrating. The distribution of susceptibility to such effects is skewed, with approximately 40% of simulator pilots reporting no symptoms and 25% reporting only mild symptoms; however, about 5% of participants may be so severely disturbed that they should restrict subsequent activities until symptoms subside (Kennedy, 1996).

#### *3.3.3 Similarity Between Simulator Syndromes and Other Syndromes*

As with visually induced motion discomfort, simulator side effects can bother the sufferer long after simulator exposure ends, in which case the most common symptoms are fatigue and eyestrain. It is interesting to note that the proportion (about 10%) of individuals who do not report symptoms even in the most provocative simulator mentioned above is about the same as the proportion of people immune to real-motion stimuli. Similarly, the proportion of individuals who report profound and lasting symptoms (about 1%–5%) in simulators is roughly the same as the proportion of the population extremely susceptible to real motion (see section 3.1). Finally, the distributions for classical motion discomfort (during transportation) and for simulator discomfort appear to be skewed, with both syndromes showing more individuals with susceptibility below the middle of the range. Thus, the authors conclude that the incidence and nature of simulator syndromes is comparable to other real or apparent motion stimuli, even if different stimuli show different subfactor profiles as measured by the Simulator Sickness Questionnaire or SSQ (Kennedy & Stanney, 1997; Lane & Kennedy, 1998; Stanney & Kennedy, 1997; see sections 3.4.2 and 4.1).

### 3.4 Virtual Environments

#### *3.4.1 Preliminary Findings From the Literature*

The technological aspects of a VE display play an important role in the experience of the user. The fact that displays vary widely among VE devices (and within one manufacturer's device over time) may account for the range of findings reported in the research performed to date.

Wilson (1997) reviewed 12 previous experiments by his research group using 233 participants immersed in one of three HMDs for exposure times between 20 minutes and 2 hours. Wilson found that approximately 80% of the 233 participants experienced some symptoms. For most participants, the symptoms were mild and short lived, but for 5% they were so severe that they had to end their participation.

Specific signs and symptoms were detailed in one of the earliest published studies of the side effects of VEs, conducted by Regan and Price (1994). They reported that about 5% of their participants ($n = 146$) withdrew from the 20-minute experiment due to severe nausea or severe dizziness, whereas 61% reported adverse symptoms at some time during exposure, (25% of whom had nausea). When participants reported some symptoms but no nausea, the most frequently reported symptoms were dizziness, stomach awareness, headache, and eyestrain. Regan and Ramsey (1996) subsequently conducted a drug study during VE exposure using a similar protocol. They found that none of the 39 participants withdrew from this experiment due to severe discomfort, but 15 of the 20 participants in the placebo group reported symptoms at some point during or immediately after VE immersion, 5 of whom reported nausea. The most common symptoms other than nausea were dizziness, stomach awareness, disorientation, and headache.

Howarth and Finch (1999) asked participants to report symptoms every minute during VE use, up to 20 minutes. Fourteen participants completed both conditions of the experiment. It should be noted that 3 of the original 17 participants in the Howarth and Finch experiment chose to terminate their participation, one of whom continued to feel symptoms similar to car sickness for several hours after immersion. All 14 participants reported an increase in nausea, and 43% of the participants reported moderate nausea within the 20-minute immersion period. Howarth and Finch note that an unspecified number of participants who did not terminate their participation also experienced symptoms persisting for hours. The persisting symptoms mentioned included "general discomfort," "severe hangover," and "feeling vacant."

### 3.4.2 VE Syndromes Compared to Other Syndromes

Reviewing a number of VE studies, Stanney, Mourant, and Kennedy (1998) noted that 80% to 95% of participants exposed to a VE reported adverse symptoms, whereas 5% to 30% experienced symptoms severe enough to end participation. Stanney and colleagues observed "cybersickness" rates higher than had been observed in flight simulators, where they estimated 60% to 70% of users report side effects (Stanney, Kennedy, Drexler, & Harm, 1999). They cautioned that most simulator users are military aviators or aviator candidates, who are likely to be less susceptible to VE than the college students typically used in VE experiments. Of relevance to this assertion is a study by Regan and Price (1994), who exposed 150 participants to VE, 20 of whom were in the military. Eight participants quit prior to the end of the experiment, none of whom were military personnel. However, such a finding can be interpreted in a number of ways,[12] and it should be noted that malaise ratings were not significantly different between civilian and military participants. Further comparisons of different participant groups are needed to explain the apparent difference in prevalence of adverse effects in VE versus other devices (such as simulators). Further device comparisons are needed also, because the latest VE technology may yield a different estimate than was obtained with VE technology of the past decade.

[12]It is possible that people who are highly susceptible to motion are not as likely to join the military, where they may be exposed to a wide variety of challenging vehicle motions. It is also possible that military personnel acquire greater than average adaptation to challenging motions after joining. The observed result could also be explained by a higher level of commitment to finishing the experiment or simply by random chance (due to the relatively small sample of military subjects).

Differences exist in the Simulator Sicknes Questionnaire (SSQ) symptom clusters for simulator discomfort versus VE discomfort.[13] Kennedy and Stanney (1997; Stanney & Kennedy, 1997) have shown that immediate postexposure profiles usually indicate that VEs produce more of the Disorientation symptom cluster, whereas simulators produce more of the Oculomotor symptom cluster. Ranking the magnitude of each symptom cluster, they found that the SSQ profile for VEs is "Disorientation > Nausea > Oculomotor," whereas the profile for simulator discomfort is "Oculomotor > Nausea > Disorientation." For comparison, the profile of space adaptation syndrome is "Oculomotor > Disorientation > Nausea," whereas the profile aboard ships and planes is "Nausea > Disorientation > Oculomotor" (Kennedy, Lanham, Drexler, Massey, & Lilienthal, 1997; Kennedy & Stanney, 1997; Stanney & Kennedy, 1997). The distinct VE symptom profile (wherein the Disorientation clusterpredominates) has been replicated with five different VE display systems (Stanney, Mourant, & Kennedy, 1998).

#### *3.4.3 Summary*

At least 60% of VE users will report adverse symptoms during their first exposure. A conservative prediction that can be made from the limited VE research is that at least 5% of all users will not be able to tolerate prolonged use of current VE, whereas at least 5% of users will remain symptom free. The number of people highly susceptible to VE side effects during their first exposure appears to be about the same as the number of people highly susceptible to motion stimuli, moving visual fields, and simulators. The number of people immune to VE effects may be lower than for some of these aforementioned stimuli, but it is too early to be certain. Virtual environment stimuli seem to share in common with moving visual fields and simulators the tendency to elicit lasting symptoms following the cessation of the stimulus. Finally, both VE and optokinetic drums appear to produce discomfort during visual scene oscillation in pitch, roll, or yaw (Cheung et al., 1991; So & Lo, 1999). However, VE syndromes appear to be distinguishable from other syndromes via the ranking of symptom clusters from the SSQ (Kennedy & Stanney, 1997; Stanney & Kennedy, 1997), with the Disorientation symptom cluster most in evidence.

## 4. TECHNIQUES FOR ASSESSING SIGNS AND SYMPTOMS ASSOCIATED WITH MOTION AND SYNTHETIC EXPERIENCES

### 4.1 Development of the Most Widely Used Measures of Symptomatology

#### *4.1.1 Consensus Regarding Symptomatology and the Need for a Prevomiting End Point*

The clinical techniques for measuring dizziness, nausea, and similar effects associated with SEs derive from controlled laboratory studies. As pointed out by Durlach and Mavor (1995), motion discomfort is relatively easy to identify in the laboratory because fairly provocative stimulation is usually employed, multiple and systematic measurements of symptomatology are taken, and trained observers are present.

[13]The three symptom clusters of the SSQ include Nausea, Disorientation, and Oculomotor. The Nausea cluster is composed of "general discomfort," "increased salivation," "sweating," "nausea," "difficulty concentrating," "stomach awareness," and "burping." The Disorientation cluster is composed of "difficulty focusing," "nausea," "fullness of head," "blurred vision," "dizzy" (rated with eyes open and with eyes closed), and "vertigo." The Oculomotor cluster is composed of "general discomfort," "fatigue," "headache," "eye strain," "difficulty focussing," "difficulty concentrating," and "blurred vision." The four descriptors that appear across more than one symptom cluster are "nausea," "difficulty concentrating," "difficulty focussing," and "blurred vision."

Experts tended to agree about the chief characteristics of motion discomfort by the middle of the twentieth century. For example, Birren (1949), Tyler and Bard (1949), and Chinn and Smith (1955) all agreed that motion can elicit nausea, vomiting (or retching), (cold) sweating, and pallor. Tyler and Bard (1949) and Chinn and Smith (1955) additionally mentioned drowsiness, (increased) salivation or swallowing, epigastric awareness or discomfort (usually defined as a sensation just short of nausea), headache, and dizziness. In a comprehensive review, Money (1970) found that the four most frequently reported side effects of motion are pallor, cold sweating, nausea, and vomiting. (However, Money enumerated dozens of other characteristics that had been mentioned in the literature.)

Early researchers struggled with the identification of an end point for the termination of nauseogenic testing. Vomiting has the advantage of being a clearly observable end point that lends itself easily to ratio scaling. However, vomiting is not necessarily a good predictor of a participant's relative stress level, since some participants can be in great distress and still not vomit, although others vomit early on in the testing before nausea is well developed. When vomiting is used as an end point, systemic side effects (such as the accompanying fluid loss) can confound efforts at continued measurement of physiological responses or repeated testing of motion sickness susceptibility. Moreover, the unpleasantness of vomiting raises difficulties with participant recruitment, human-use approval, and the comfort and convenience of the participant and experimenter. Some of the early attempts to draw up a list of prevomiting signs and symptoms and scale them collectively were made by Hemingway (1943) and Alexander, Cotzin, Hill, Ricciuti, and Wendt (1945). These techniques obtained quantitative scores for an individual's overall discomfort based on a number of signs and symptoms besides vomiting, but they did not employ a checklist wherein data concerning each sign or symptom could be tabulated and analyzed separately before the final score was tabulated.

### *4.1.2 First Widely Used Multiple Symptom Checklists*

The first diagnostic categorization of multiple motion-related symptoms that saw widespread use was initiated by Ashton Graybiel and developed by a number of researchers associated with the Pensacola Naval Air Station,[14] mostly as part of studies of well-being and performance during adaptation to a rotating room. These efforts started with a rudimentary checklist that recorded the presence ("yes or no") of a number of distinct signs or symptoms, including general malaise, sweating, nausea, vomiting, dizziness, headache, apathy, and the number of hours slept during the day (Graybiel, Clark, & Zarriello, 1960; Kennedy & Graybiel, 1962). Soon after, the list of signs and symptoms was expanded and rated on a 0-to-3 scale (Graybiel & Johnson, 1963). Many related forms of this diagnostic categorization system were produced during these prolific years. For example, the Pensacola Motion Sickness Questionnaire, or MSQ (so named in Kennedy & Graybiel, 1965), consisted of more than 20 signs and symptoms initially presented in Hardacre and Kennedy (1963). The MSQ reached a relatively mature form in Kennedy, Tolhurst, and Graybiel (1965), which is one of the best of the early guides to the MSQ. Kennedy et al. (1965) categorized minor, major, and pathognomic diagnostic criteria, which were scored variously and tabulated as a single malaise score.

A shorter table was published around the same time by Wood, Graybiel, and McDonough (1966); it employed seven criteria, including the pathognomic sign of vomiting, the major sign of retching (later upgraded to a pathognomic sign in Miller & Graybiel, 1970b), and a number of other signs and symptoms whose classification as minor or major depended on

[14]The laboratory complexes housed at Pensacola Naval Air Station have been extant since 1939 and have gone by various names during that time, most recently the Naval Operational Medical Institute (www.nomi.navy.mil) and the Naval Aerospace Medical Research Laboratory (www.namrl.navy.mil).

their severity, including nausea, increased salivation, pallor, cold sweating, and drowsiness. These seven criteria contributed strongly to the participant's overall malaise score and were considered the cardinal characteristics of motion discomfort; the presence of one (or more) of these characteristics has also been used as the criterion for simulator discomfort (Kennedy et al., 1989).

### *4.1.3 Cardinal Signs and Symptoms*

Many investigators (e.g., Harm, 1990; Reason & Brand, 1975) restrict the list of seven "cardinal" signs and symptoms to those four mentioned most frequently in the motion discomfort literature (according to Money, 1970), namely, pallor, cold sweating, nausea, and vomiting. However, it should be remembered that two of these four cardinal characteristics (pallor and cold sweating) were not more heavily weighted in the Wood et al. (1966) scoring procedure than were the rest of the seven criteria. Moreover, Miller and Graybiel (1970b) found that flushing and feelings of warmth were more commonly reported than was increased salivation. Thus, decisions about the cardinal criteria of motion discomfort are not straightforward and will vary with the factors one considers most important. For example, cardinal criteria such as vomiting and retching are easily quantifiable and yield parametric data; furthermore, they may be the only obvious evidence of discomfort in some cases (Lackner & Graybiel, 1986). However, for stimuli that do not produce vomiting, the cardinal symptom of nausea is likely to account for most of the variance in estimates of overall well-being (Reason & Diaz, 1971). The picture changes again when the feeling of overall well-being is not the paramount criterion, but rather the likelihood of occurrence of a given symptom and the possible deleterious effects that symptom might have on performance of tasks. For example, drowsiness and fatigue are the most commonly reported symptoms of simulator exposure (Kennedy et al., 1995; Ungs, 1988, both from Lawson & Mead, 1998) and armored vehicle operations (Cowings, Toscano, DeRoshia, & Tauson, 1999), and such symptoms have been implicated in transportation accidents (Lawson & Mead, 1998).

Eventually, two mature forms of the diagnostic criteria for motion sickness emerged, as later described. One mature set of criteria was produced by Miller and Graybiel (1970b),[15] who presented a diagnostic categorization table along with an accompanying sheet for scoring signs and symptoms of motion discomfort. They also added a few new signs and symptoms of lesser importance to the list, including epigastric awareness, epigastric discomfort, flushing/subjective warmth, headache, and dizziness with eyes closed or open. Since the Miller and Graybiel table of diagnostic criteria can be somewhat difficult to apply correctly when presented in its original form, a simplified representation of the table of diagnostic criteria (Miller & Graybiel, 1970b) is presented in Table 30.1, rendered as an experimenter's checklist for the scoring of a single motion challenge (e.g., discomfort experienced during one sudden stop following rotation or one head movement while rotating).

The diagnostic criteria in Table 30.1 are derived from Miller and Graybiel (1970a, 1970b), with earlier contributions from Graybiel, Wood, Miller, and Cramer (1968), Graybiel (1969), and others. These authors devised a weighted scoring procedure that recognizes six distinct levels of severity for symptoms indicative of the nausea syndrome and two to four distinct levels of severity for nonnausea symptoms. Cumulative points scored for all symptoms are added to distinguish between different levels of general malaise. For example, a cumulative score of 8 could be reached by a subject reporting moderate or severe nausea, but an

[15]The diagnostic criteria which culminated in the Miller and Graybiel (1970b) paper have been referred to by various names, including the Pensacola Diagnostic Index, The Pensacola Diagnostic Criteria, and The Graybiel Scale, but the original publications regarding these diagnostic criteria did not usually employ these appellations.

**TABLE 30.1**
Diagnostic Criteria of Miller and Graybiel (1970b)

| | *None* | | *Minimal* I | *Minor* II | *Major* III | *Pathognomic* |
|---|---|---|---|---|---|---|
| *Nausea* | | *Epigastric Awareness* | *Epigastric Discomfort* | *Nausea 1 (Minimal)* | *Nausea 2–3 (Moderate to Severe)* | *Vomiting, Retching* |
| Point Value: | 0 | 1 | 2 | 4 | 8 | 16 |
| Other Symptoms | | | | | | |
| Flushing, warmth | 0 | (N.A.) | 0 | 1 | 1 | (N.A.) |
| Dizziness (eyes closed) | 0 | (N.A.) | 0 | 1 | 1 | (N.A.) |
| Dizziness (eyes open) | 0 | (N.A.) | 0 | 0 | 1 | (N.A.) |
| Headache | 0 | (N.A.) | 0 | 1 | 1 | (N.A.) |
| Drowsiness | 0 | (N.A.) | 2 | 4 | 8 | (N.A.) |
| Cold sweating | 0 | (N.A.) | 2 | 4 | 8 | (N.A.) |
| Pallor | 0 | (N.A.) | 2 | 4 | 8 | (N.A.) |
| Increased salivation | 0 | (N.A.) | 2 | 4 | 8 | (N.A.) |

8 would also be assigned to a subject whose only symptom was severe drowsiness. A total score of $\geq 16$ points = "frank sickness," 8–15 points = "severe malaise," 5–7 points = "moderate malaise A," 3–4 points = "moderate malaise B," and 1–2 points = "slight malaise."

Versions of the Miller and Graybiel scale have been used widely.[16] The scale has changed little over the years, although the way it is used and scored has varied (Lawson, 1993). By contrast, the Kennedy, Tolhurst, and Graybiel (1965) scale continued to undergo a series of further modifications over the years. A notable offshoot of the original Kennedy et al. (1965) scale was the Motion Sickness Severity scale of Wiker, Kennedy, McCauley, and Pepper (1979), which added several more signs and symptoms and offered one of the more convincing attempts to establish the reliability and validity of subjective motion discomfort reporting. The scales derived from Kennedy et al. (1965) eventually emerged in the form of the Simulator Sickness Questionnaire (SSQ), which has seen widespread use in navy simulator studies and has become the most popular measurement technique for the assessment of signs and symptoms associated with simulators and other SEs (e.g., Kennedy, Lane, Lilienthal, Berbaum, & Hettinger, 1992). One of the more recent versions of the 16-descriptor version of the SSQ is described in detail by Kennedy, Lane, Berbaum, and Lilienthal (1993).

## 4.2 Importance of Subjective Measures of SE Effects

A great deal can be learned by careful inquiry into the subjective aspects of motion discomfort. Researchers have yet to develop a complete description of the relation between the characteristics of different stimuli and the particular signs and symptoms elicited by those stimuli. Important steps in this direction have been taken by Kennedy and colleagues (e.g., Kennedy,

[16]The Miller and Graybiel (1970b) criteria have been used by researchers at the Pensacola Naval Air Station, Defense and Civil Institute of Environmental Medicine (Canada), National Aviation and Space Administration, Massachusetts Institute of Technology, Pennsylvania State University, Brandeis University, and other institutions.

Berbaum, & Lilienthal, 1992; Kennedy, Lanham, Drexler, Massey, & Lilienthal, 1997). To date, much inquiry has been restricted to a consideration of acute motion discomfort during short-term exposure to provocative stimuli. Little is known about such aspects of motion discomfort as the rate and retention of adaptation that occurs during prolonged or repeated exposures, or the distinct symptoms that can be elicited by long-term, low-grade exposure (e.g., the sopite syndrome described by Graybiel & Knepton, 1976). Graybiel and Lackner (1980) suggested that the aspect of motion discomfort that is most operationally relevant will depend upon the stimulus situation and the goals of the mission; for example, rate of adaptation to flying discomfort and length of retention of that adaptation may prove more important than an aviator's inherent susceptibility on initial exposure. A better understanding of these aspects of motion-related symptomatology may prove immensely useful in the future. Ultimately, the best approach to the quantification of signs and symptoms associated with SEs will not rely entirely on any one type of measure (subjective, physiological, performance), but will employ multiple cross-validated measures in order to get the best look at SE-related syndromes (see chaps. 29–32 and 37–39, this volume).

### 4.3 Modifications of the Original Motion Discomfort Scales

The authors have little reason to think that the motion discomfort criteria of Kennedy et al. (1965) or Miller and Graybiel (1970b) could be made much more reliable (Calkins, Reschke, Kennedy, & Dunlap, 1987; Kennedy, Dunlap, & Fowlkes, 1990; Miller & Graybiel 1970a, 1970b; Reason & Graybiel, 1970). However, there is some question as to the metric assumptions made in obtaining the numerical summary scores of overall motion discomfort yielded by these methods (Bock & Oman, 1982). Reason and Diaz (1971) were probably the first to publish a symptom checklist scoring procedure that used criteria nearly identical to those in Miller and Graybiel (1970b), but did not assume ratio scale data. During the development of the SSQ, Lane and Kennedy (1988) also dropped many of the metric assumptions in the original Kennedy et al. (1965) scale and opted for a simple 0-to-3 scale of severity for all signs and symptoms. Lawson (1993) similarly created a 0-to-3 version of the Miller and Graybiel (1970b) scale that minimized the unequal weighting of signs and symptoms of equal severity and eliminated doubling of points with increases in severity of certain symptoms (see Table 30.2). The Lane and Kennedy (1988) and Lawson (1993) scales showed good test–retest reliability and correlated well with the earlier scales that had made more assumptions. The Lane and Kennedy SSQ was specifically designed for the relatively nonprovocative (low vomiting incidence) situation of simulator exposures and is probably the most useful of the "Pensacola School" techniques, in that it allows for a three factor assessment of underlying symptom clusters. The Lawson (1993) adaptation of the Miller and Graybiel (1970b) criteria (see Table 30.2) is recommended for those investigators committed to using the original Miller and Graybiel criteria, but wishing to limit their metric assumptions to avoid making erroneous conclusions about the null hypothesis. The MSQ and SSQ (see Table 30.3) are recommended for investigators not committed to the Miller and Graybiel criteria. As administered in questionnaires, the MSQ and SSQ descriptors are usually preceded by a number of biographical questions. Table 30.3 shows only the Likert-type items from the SSQ and MSQ, suitable for administration immediately before and after a real or apparent motion stimulus. The first 16 descriptors listed in Table 30.3 are used in the SSQ to assess symptoms in situations that elicit low vomiting rates and tend to include a visual display, with or without a moving base. The entire list of 28 descriptors is used in the MSQ to assess situations where vomiting is more common and motion tends to be present, for example, in ships at sea. (Descriptors are not in the order shown in Table 30.3, but are mixed in the actual MSQ/SSQ.) Key publications

**TABLE 30.2**
Modified Diagnostic Criteria of Miller and Graybiel (1970b), Employing Fewer Metric Assumptions*

| *Symptoms* | *None* | *Minimal or Slight* | *Minor or Moderate* | *Major or Severe* |
|---|---|---|---|---|
| Nausea | 0 | 1 | 2 | 3 |
| Increased salivation | 0 | 1 | 2 | 3 |
| Cold sweating | 0 | 1 | 2 | 3 |
| Pallor | 0 | 1 | 2 | 3 |
| Drowsiness | 0 | 1 | 2 | 3 |
| Headache | 0 | 1 | 2 | 3 |
| Flushing/warmth | 0 | 1 | 2 | 3 |
| Dizziness (eyes: closed/open?) | 0 | 1 | 2 | 3 |
| Epigastric awareness or epigastric discomfort? (No/Yes) | No | Yes | | |
| Retching or vomiting? (No/Yes) | | | | |
| Number of such events: ________ | No | Yes | | |
| Number of minutes of stimulation tolerated: ________ | | | | |

*Adapted from Lawson (1993). Direct questions to blawson@namrl.navy.mil.

describing the development, administration, and scoring of the SSQ and MSQ were mentioned in section 4.1 above.

## 4.4 Other Motion Discomfort Scaling Techniques

Two of the most important motion discomfort scaling efforts initiated outside of Pensacola were the magnitude estimation procedure of Bock and Oman (1982) and the Nausea Profile of Muth, Stern, Thayer, and Koch (1996). The magnitude estimation procedure appears to yield data that are metrically superior to most other techniques, involving a prevomiting end point. The procedure can be explained to subjects easily and the estimates can be gathered from them rapidly once they have experienced the stimulus. The resulting data can be analyzed more easily than data from a symptom checklist. However, in its original form, the magnitude estimation procedure calls for previous exposure to the stimulus of interest (a requirement relaxed by Eagon, 1988) and does not generate multiple-factor solutions for exploring psychometric questions. It is also worth noting that the widely held assumption that magnitude estimation is superior to Likert-type rating has been questioned (Wills & Moore, 1994).

The Muth et al. (1996) Nausea Profile is the only procedure mentioned in this review that measures the construct of nausea per se, regardless of whether the nauseogenic stimulus derives from motion or from some other cause (e.g., illness). The Nausea Profile yields a multifactor solution that is likely to be applicable to any provocative stimulus strong enough to trigger the nausea syndrome. It has proven useful for demonstrating that the subjective sensation of nausea includes factors other than the experience of gastric discomfort. An especially interesting feature of the Nausea Profile is that the descriptors were generated by laymen

**TABLE 30.3**
The Simulator Sickness Questionnaire (items 1–16) and the Motion Sickness Questionnaire (items 1–28)[†]

| Circle how much each symptom below is affecting you right now. | |
|---|---|
| 1. General discomfort | None Slight Moderate Severe |
| 2. Fatigue | None Slight Moderate Severe |
| 3. Headache | None Slight Moderate Severe |
| 4. Eyestrain | None Slight Moderate Severe |
| 5. Difficulty focusing | None Slight Moderate Severe |
| 6. Increased salivation | None Slight Moderate Severe |
| 7. Sweating | None Slight Moderate Severe |
| 8. Nausea | None Slight Moderate Severe |
| 9. Difficulty concentrating | None Slight Moderate Severe |
| 10. Fullness of head | None Slight Moderate Severe |
| 11. Blurred vision | None Slight Moderate Severe |
| 12. Dizziness (eyes open) | None Slight Moderate Severe |
| 13. Dizziness (eyes closed) | None Slight Moderate Severe |
| 14. Vertigo* | None Slight Moderate Severe |
| 15. Stomach awareness** | None Slight Moderate Severe |
| 16. Burping | None Slight Moderate Severe |
| 17. Boredom | None Slight Moderate Severe |
| 18. Drowsiness | None Slight Moderate Severe |
| 19. Decreased salivation | None Slight Moderate Severe |
| 20. Depression | None Slight Moderate Severe |
| 21. Visual illusions*** | None Slight Moderate Severe |
| 22. Faintness | None Slight Moderate Severe |
| 23. Awareness of breathing | None Slight Moderate Severe |
| 24. Decreased appetite | None Slight Moderate Severe |
| 25. Increased appetite | None Slight Moderate Severe |
| 26. Desire to move bowels | None Slight Moderate Severe |
| 27. Confusion | None Slight Moderate Severe |
| 28. Vomiting | None Slight Moderate Severe |

*Vertigo is experienced as loss of orientation with respect to vertical upright.

**Stomach awareness is usually used to indicate a feeling of discomfort that is just short of nausea.

***Visual illusion of movement or false sensations of movement when not in the simulator, car, or aircraft.

†The entire questionnaire, with items in original order, can be obtained from Robert Kennedy, Ph.D. *Direct questions to 6kennedy@bellsouth.net.

rather than experts, thus enhancing the content validity of the scale and making it easier for nonexperts to administer.

## 4.5 Psychometric Topics for Future Emphasis

With the exception of the SSQ, most of the scaling techniques used to characterize motion discomfort have focused on situations of acute and highly provocative exposure. More emphasis should be given to the human response to mild or nonsickening motions and to chronic exposure situations. Understanding these situations may help elucidate the sopite syndrome (Graybiel & Knepton, 1976), an aspect of SE-related discomfort that is not explicitly measured in the three factor solution of the SSQ, and thus requires looking at greater than 3 factor solutions of the SSQ (Kennedy et al., 1995) or creating a new scale (Muth, Lawson, & Sausen, 1999; see section 4.3).

## 5. SYMPTOM PROGRESSION OF SYNDROMES ASSOCIATED WITH SYNTHETIC EXPERIENCES: PRELIMINARY INFERENCES FROM MOTION STUDIES

### 5.1 Limited Findings From the Literature

Most published accounts of simulator and VE exposure display quantitative estimates of total symptom severity before and after exposure without specifying the temporal progression of individual signs and symptoms. There is at least one published account of the time course of postflight simulator symptoms (Baltzey et al., 1989), but the present authors did not find a published account specifically focusing on the onset and time course of symptoms in virtual environments. Because the most thorough elucidation of temporal symptom progression in SEs has come from studies involving actual body rotation, these findings will be discussed below.

Reason and Graybiel (1975) exposed 41 participants to the Dial Test onboard a Slow Rotation Room. The Dial Test involves moving the head and arms in order to adjust dials in various positions around one's body during whole-body rotation in a moving room (with a participant-fixed visual surround). Participants rated their individual symptoms and also gave concurrent "overall well-being" ratings of 0 to 10 (0 = "I feel fine" to 10 = "I feel awful, just like I am about to vomit"). Looking at the overall well-being ratings of about 1 to 1.5 (the initial onset of effects), one finds that the earliest signs and symptoms during the Dial Test in the Slow Rotation Room are usually minimal dizziness and visible nystagmus. Next, at ratings of about 1.6 to 5.0, a number of symptoms appear, including (roughly in temporal order) bodily warmth, moderate dizziness, headache, minimal pallor, minimal cold sweating stomach awareness, and slightly increased salivation. The symptoms overlap quite a bit, but at ratings greater than 5.5, sweating and pallor tend to increase to moderate levels while stomach discomfort and unequivocal nausea builds.

Miller and Graybiel (1969) assessed symptom progression as well. They tracked the frequency of occurrence among 250 participants of specific symptoms associated with "Malaise I" or "mild malaise" during a now widely used rotation test. The most frequently reported symptoms by Malaise I were stomach awareness or discomfort (56%), minimal cold sweating (14%), and subjective warmth or flushing (9%). By the time participants experienced Malaise III ("severe malaise"), 70% of them felt subjective warmth/flushing and about a 60% felt stomach awareness/discomfort. Additionally, sweating, pallor, and nausea were often seen by the time Malaise III was reached.

### 5.2 Summary and Caveats

Great individual variability exists in the order of symptoms among participants. However, trained individuals are reliable within themselves regarding their responses to repeated provocative tests (DiZio & Lackner, 1991; Graybiel & Lackner, 1980). In the early stages of motion discomfort, the most likely symptoms to be reported by participants are subjective warmth/flushing, dizziness, and stomach awareness/discomfort. The final stages of motion discomfort during acute testing are usually marked by (at least) minimal nausea. Experimenters often terminate testing upon the appearance of moderate nausea, because an "avalanche effect" can occur where vomiting happens before the experiment can be stopped. In Reason and Graybiel (1975), participants who exhibited greater resistance to the stimulus also tended to proceed suddenly from well-being ratings of 2 to 4 all the way up to 10. The very act of stopping a rotating chair is a provocative stimulus; hence, adequate time must be allowed to bring the chair slowly to a halt before the participant feels worse.

Some caveats apply to the interpretation of Reason and Brand (1975) and Miller and Graybiel (1969). The methodology of Reason and Graybiel (1975) biased participants to focus on nausea at higher (worse) point levels of "well-being," to the exclusion of other symptoms, because the word "vomiting" was used to anchor the definition for a well-being rating of 10. Similarly, Miller and Graybiel (1969) used a weighted scoring system wherein some symptoms (especially nausea) contributed more than others to the Malaise rating (see Table 30.1). Finally, it should be remembered that although nausea and vomiting are very important aspects of motion discomfort, certain stimuli produce vomiting without marked nausea (Lackner & Graybiel, 1986).

### 5.3 Implications for Virtual Environments

Attempts should be made to evaluate temporal symptom progression in VE thoroughly to see how it compares to the motion responses summarized above. This would help identify similarities and differences between VE symptoms and symptoms associated with other SEs or with motion. It might also aid the development of health and safety warning labels for virtual environments (see chap. 27, this volume). Since individuals are likely to be fairly consistent in how they respond to a given VE, it is a realistic goal for persons whose jobs will require repeated VE training (e.g., future military pilots) to learn their own unique progression of signs and symptoms so they can pace themselves accordingly. Of course, persons exposed to VE routinely should also avail themselves of dual adaptation strategies that emerge from research (see chap. 31, this volume) as long as the strategies have been proven to enhance performance of the person's job duties.

## 6. OTHER SIDE EFFECTS OF MOTION AND SYNTHETIC EXPERIENCES

Nausea and vomiting are the obvious manifestations of motion discomfort. Nausea at sea or during space flight can be lessened using appropriate oral drugs, or in extreme cases, drug injections (Graybiel & Lackner, 1987; see chap. 32, this volume). While current interactive synthetic experiences (such as VE) may elicit the nausea syndrome, inexperienced users are not as likely to seek drug solutions as they would for a sea voyage because they may not anticipate that the VE stimulus is a sickening experience for which drug remedies are appropriate. Even experienced VE users may choose to avoid drug therapies because (unlike the occasional sea cruise) an experienced VE user is likely to subject himself to the stimulus frequently and for long durations, and because the magnitude of the stimulus is largely under the user's control (e.g., by restricting head movements, eye movements, or duration of apparent self-movement through the VE).

People who cannot adapt readily to the nausea created by VE are likely to limit themselves to brief exposures. However, the more subtle side-effects of VE syndromes will not deter such persons as strongly as nausea and so may contribute disproportionately to the future health and safety problems that arise among VE users. Some of these other side effects include sopite syndrome, degraded dynamic visual acuity, and postural dysequilibrium.

## 6.1 Sopite Syndrome

Pensacola researchers Graybiel and Knepton (1976) coined the phrase *sopite syndrome* to describe a sometimes sole manifestation of motion discomfort characterized by such symptoms as motion-induced drowsiness, difficulty concentrating, and apathy. Evidence that the sopite syndrome is somewhat distinct from the nausea syndrome and from fatigue due to nonmotion stimuli are discussed by Lawson and Mead (1998), who review what is known about the scope of the problem and its potentially insidious character.[17] A brief introduction to sopite syndrome will be provided in this section, emphasizing information not previously presented by Lawson and Mead (1998).

The term *sopite* derives from the same Latin root as the more common English words *sopor* and *soporific*, viz., the latin verb *sopire*, meaning "to lay to rest or put to sleep" (Neilsen et al., 1956). Since before recorded history, rocking back and forth has helped adults to relieve anxiety and tension (Watson, Wells, & Cox, 1998) and infants to fall asleep (Ter Vrugt & Peterson, 1973, from Leslie, Stickgold, DiZio, Lackner, & Hobson, 1997). Erasmus Darwin (1795) was probably the first person to publish the observation that motion induces drowsiness; he described a remarkable method that was related to him for procuring sleep by lying on a rotating mill wheel.[18] In 1894, De Zouche (1975, p. 53) described the effects of "chronic seasickness" as "great exhaustion . . . [and] . . . heavy sleepiness." In 1922, Quix (1976, p. 879) said the effects of motion discomfort include "slow ideation, lack of inclination to work, abulia [slow reaction], weakness, fatigue, feeling of uneasiness, and apathy that can lead to melancholy."

In one of many rotating room studies carried out at Pensacola Naval Air Station, participants lived aboard a 20-foot diameter rotating room for 12 days (Graybiel et al., 1965). The participants suffered few episodes of overt nausea or vomiting during the study; only one out of four participants vomited as late as day two and none vomited after that. However, the participants (and some onboard experimenters) yawned frequently and complained of strong fatigue and drowsiness throughout the experiment despite frequent naps and sometimes sleeping longer than usual. Participants showed little motivation for mental or physical work until the fifth day and had not fully recovered from the fatiguing effects of rotation by the end of the rotation period. A possible vestibular etiology for the sopite syndrome was inferred by Graybiel et al. (1965) because the participants restricted their head movements intentionally even after the cessation of nausea. Also, normal participants exposed to two days in the rotating room showed evidence of drowsiness and apathy at rotation rates as low as 1.71 to 3.82 rpm, while a control participant who had lost vestibular function was free of such symptoms under the same conditions (Graybiel et al., 1960).[19]

Although many researchers are aware of the earliest sopite syndrome paper written by Graybiel and Knepton (1976), fewer researchers know of the sopite-relevant studies that were carried out in the last decade. For example, Wright, Bose, and Stiles (1994) observed worse digit-span test performance among nauseated individuals ($n = 26$) following helicopter flight. However, the experimenters also observed worse digit-span test performance when participants had symptoms indicative of sopite syndrome and nausea was not prominent. This

---

[17]Evidence that the fatigue associated with motion is not entirely attributable to increased energy expenditure (e.g., due to the additional work performed by a person repeatedly traversing the deck of a moving ship) is to be found in Crossland and Lloyd (1993) and Lewis and Griffin (1998).

[18]Incidentally, such angular rotation would initially produce rhythmic eye movements known as *nystagmus*, a word derived from the Greek *nystagmos*, meaning "drowsiness" (Neufeldt & Guralnik, 1994).

[19]It is also interesting to note that at least one paper on the vestibular malady known as Menière's disease (Eklund, 1999) compares the aftereffects of an attack of rotational vertigo to sopite syndrome symptoms, which also can appear following motion (described as "late sopite" by Graybiel and Knepton, 1976).

degradation in performance would be attributable mostly to factors other than nausea, such as degraded attention and wakefulness. This notion is supported by studies that indicate low-frequency electroencephalographic (EEG) activity is associated with real or apparent motion. For example, preliminary studies by Miller (1995) and Yano et al. (1999) reported increased EEG alpha activity soon after entry into a driving situation, as well as longer gaze time and slower eye movements.[20] Chelen, Kabrisky, and Rogers (1993) and Hu et al. (1999) noted increased EEG delta activity during real or apparent motion.[21] Woodward, Taubar, Sprelmann, and Thorpy (1990) tested participants ($n = 8$) on two separate days using a parallel swing and found EEG evidence of altered sleep functioning on nights following the motion stimulus. Using a relatively mild stimulus, Leslie, Stickgold, DiZio, Lackner, and Hobson (1997) found that after 10 minutes of optokinetic stimulation, participants ($n = 14$) became significantly more drowsy than after a control condition (reading), but the participants ($n = 13$) from whom good EEG data could be obtained, did not show a significant decrease in sleep latency onset.[22]

Wood et al. (1990) observed a slowing of EEG alpha activity and emergence of some theta and delta waves in participants with symptoms of motion discomfort and identified some drugs that alleviate nausea but do not appear to alleviate symptoms of the sopite syndrome. This finding may have implications for the effort to develop drugs to alleviate motion discomfort, because the relief of motion discomfort clearly entails more than alleviating nausea. Earlier, Brandt, Dichgans, and Wagner (1974) had found that either dimenhydrinate or scopolamine were effective in reducing the incidence of vomiting during real or apparent motion stimuli, but that scopolamine subjects still exhibited fatigue, drowsiness, loss of concentration, and decreased performance (see also chap. 32, this volume). It is possible that some anti–motion-sickness drugs that are sedating are very effective in preventing emesis, but may mimic or exacerbate the symptoms associated with sopite syndrome.

An important nondrug therapy for alleviating motion discomfort is to adapt to the disturbing stimulus (see chap. 31, this volume). The U.S. Navy maintains a Self-Paced Airsickness Desensitization (SPAD) program wherein airsick flight students adapt to a rotation stimulus. This has proven to be effective in helping students get over their nausea and return to flight status. However, it appears that whether or not a student returns to flight status is unrelated to how much sopite syndrome he exhibits during desensitization training (Lawson, Flaherty, & Schmorrow, in preparation). Since sopite syndrome effects may linger (Graybiel & Knepton, 1976), it is possible that a few of the students who have successfully adapted to motion-induced nausea are returned to further flight training despite their tendency to exhibit greater drowsiness during and after motion stimuli (compared to their fellow students). This possibility should be investigated, because flight training itself is known to produce symptoms indicative of sopite syndrome. For example, Kay, Lawson, and Clarke (1998) found that two of the common effects that Naval Flight Officers in primary flight training experienced during flight days

[20]Miller (1995) observed EEG (and other) evidence of drowsiness in 80 truck drivers driving late at night. Of potential relevance to the sopite syndrome is Fig. 1 of Miller (1995), which shows the alteration of alpha and theta activity occurring within only about 30 to 40 minutes of transferring from a stationary rest break to a moving state. However, it is possible that the restorative effect of a rest break late at night is short-lived regardless of whether motion is present. Interestingly, the pilot data of Yano et al. (1997) indicated EEG alpha activity and eye movements associated with decreased arousal in three automobile drivers after only 30 minutes of daytime driving.

[21]Chelen, Kabrisky, and Rogers (1993) measured the EEG of 10 subjects during real motion, whereas Hu et al. (1999) measured the EEG of 52 subjects observing a moving optokinetic visual surround.

[22]Using an optokinetic drum (the same as in Teixeira & Lackner, 1979) and a brief (10-minute) period of exposure to a 60 deg/sec constant velocity optokinetic stimulus (as used in Lawson, 1993), Leslie et al. (1997) found that scores on the Median Sleep Latency Test dropped from 13.6 minutes to 12.7 mins; $F(1{-}12) = 3.53$, $p = 0.085$. (The aforementioned sample size of $n = 13$ is deduced from the degrees of freedom of the ANOVA.)

(vs. nonflight days) were "drowsiness despite adequate rest" and "persistent unexplained fatigue." Similar effects have been observed at sea. For example, Colwell (2000) recently completed an extensive Performance Assessment Questionnaire of individuals engaged in NATO North Atlantic Fleet Operations. Of the subjective variables surveyed, two of the most commonly reported items were "fatigue" and "difficulty sleeping" (the night before), which correlated with the amount of ship motion (as a function of sea state). (Also see Haward, Lewis, & Griffin, 2000; Wertheim, 1998).

Kennedy et al. (1995) have found ample reason to suspect that sopite syndrome is a feature of simulator exposure. The possibility for sopite syndrome as a result of VE exposure was first mentioned in an NRC report (BRL-CR-629, 1990) and has been recommended as a future VE research priority (Durlach & Mavor, 1995; Stanney, Salvendy et al., 1998), but to date, very little is known about sopite syndrome in virtual environments. For example, in some cases sopite syndrome may cause the sufferer to feel detached, distant, less communicative, and less willing to engage in group behavior (Graybiel & Knepton, 1976). Such an affective state would adversely affect the VE user's sense of presence (see chap. 40, this volume) and degree of performance in shared VEs (Durlach & Slater, 1995; see chapter 18, this volume). Such symptoms might also compromise crew coordination during some military operations, for example, teleoperation of unmanned aerial vehicles, highly mobile amphibious landings, or maneuvers aboard a crew transport vehicle (Cowings et al., 1999). Finally, Space Adaptation Syndrome has been suggested as a factor hindering communication between ground and space crews (Kelly & Kanas, 1993).

## 6.2 Loss of Visual Acuity During Head or Body Motion

Another likely side effect for VE users is the loss of visual acuity during head or body motion (see chaps. 37 and 39, this volume). Most whole-body motions will elicit a vestibulo-ocular reflex that helps gaze remain stable as the head or body moves. This natural coordination of visual and vestibular function is disrupted when one dons prism goggles that alter the amount of visual field movement corresponding to a given head movement (Gonshor & Mellville-Jones, 1976); it can also be altered when the user attempts to read a head-fixed display during prolonged whole body motion. In this latter situation, interpreting the display requires visual suppression of the ongoing vestibulo-ocular reflex. Anyone who has attempted to read fine print during turbulent flight aboard a commercial airplane will appreciate the difficulty and unpleasantness of the effort. Those who design VE will need to understand the conditions in which a viewer's ability to suppress his vestibulo-ocular reflex is enhanced without creating unpleasant symptoms (Lawson, Rupert, Gnedry, Grissett, & Mead, 1997).

## 6.3 Postural Disequilibrium

Postexposure disruptions of balance have been associated with motion and unusual force environments (see Baltzey et al., 1989; chap. 39, this volume). People can adapt to unusual gravitoinertial force environments such as micro-gravity (Nicogossian, Huntoon, & Pool, 1989), high G-force, ship motion, and rotating rooms (reviewed in Crampton, 1990). However, the acquired adaptation disrupts balance and coordination on returning to the stationary world. A familiar example is the way in which sailors must regain their "land legs" on return to shore (Gordon, Spitzer, Doweck, Melamed, & Shupak, 1995).

Postural disruption can also be caused by abnormally integrated visual-vestibular signals. Lee and colleagues (Lee & Aronson, 1974; Lee & Lishman, 1975) showed that a moving visual surround can "drive" postural sway in toddlers standing on a regular floor or in adults standing on compliant surfaces (which degrade ankle kinesthesia). A longitudinal study by Brandt and colleagues (1978) implied that the maintenance of posture within a moving visual field required the appreciation of the consequences of body movement. The causal and temporal relation among vection, postural sway, and feelings of discomfort is viewed differently by different researchers (see Jones, 1998, for a recent review). Regardless of the relations involved, it is clear from data gathered in navy ground-based simulators that significant ataxic aftereffects occur after short exposures to certain visual field motions (Kennedy, Lilienthal, Berbaum, Baltzley, & McCauley, 1989; Kennedy, Berbaum, & Lilienthal, 1997). Research by Kennedy and Stanney (1996) suggests that simple measures of postural instability can reveal the aftereffects of VE exposure. These various findings point to a potential health and safety risk associated with the prolonged use of "inertial" and "noninertial" virtual displays (Durlach & Mavor, 1995; Lawson, Rupert, Guedry, Grissett, & mead,1997; Stanney, Salvendy et al., 1998). Of particular concern is the fact that within another generation, the majority of the population will be routinely exposed to VEs and the segment of the population that is most vulnerable to falls, the senior females, will be larger than it is now (Fregly, 1974; Grabowski, 1996).

## 7. SITUATIONS MOST CONDUCIVE OF VE-RELATED SYNDROMES

### 7.1 Moving Visual Fields

#### *7.1.1 Relation Between Visually Induced Illusions of Self-motion and Feelings of Discomfort*

The role of visually induced illusions of self-motion (vection) in enhancing the experience of VE is treated in detail in chapter 23 (this volume) and mentioned in chapters 3 and 7 (this volume). The present section briefly summarizes the relation between vection and discomfort. McCauley and Sharkey (1992) and Hettinger, Berbaum, Kennedy, Dunlap, and Nolan (1990) hypothesized that simulator discomfort and VE discomfort should be correlated with the presence of a visually induced illusion of self motion, known as vection (Fischer & Kornmüller, 1930), and hence could be considered a type of visually induced motion discomfort. Hettinger et al. (1990) observed that 80% of participants ($n = 18$) who reported vection while viewing a moving visual field also experienced symptoms, whereas only 20% of those who did not report vection felt discomfort. Crampton and Young (1953) found that nine of their participants (i.e., 35% of $n = 26$) experienced nausea during visual field motion, five of whom perceived themselves to be moving. Studies by Dichgans and Brandt (1978) suggested that if strong vection was being experienced by the participant, head movements could elicit discomfort. However, studies by Lackner and Teixerira (1977) and Teixeira and Lackner (1979) indicated that if head movements were performed prior to the onset of vection they suppressed the vection illusion and prevented discomfort. Jones (1998) found a weak but significant correlation between magnitude of vection and "total sickness severity" score following a driving simulation task.[23]

The authors of this chapter conclude that moving visual fields often elicit vection, that behaviors (e.g., head movement) that affect vection are likely to affect discomfort, and that symptoms such as dizziness will be experienced in the majority of participants who report vection. However, it is important to note that participants can report vection without symptoms and symptoms without vection. Similarly, individuals without functional vestibular systems cannot be made sick by motion, yet can experience vection when viewing visual stimuli

[23]Pearson $r = 0.25$, $p < 0.03$, $n = 78$.

(Chueng et al., 1991). Finally, while feelings of vection should be positively correlated with feelings of presence in a SE, Witmer and Singer (1998) found a negative correlation between measures of presence and simulator discomfort such that greater presence resulted in less discomfort. Possibly, vection is correlated with visually induced symptoms, but it is not a necessary precursor of them (a view expressed by Hettinger & Riccio, 1992, and by Ebenholtz, 1992). A simple view of cybersickness as a subset of visually induced motion discomfort is unlikely to prove adequate in the long term because sensations of self-motion can be elicited by other than visual modalities (Bles, 1979; Lackner & DiZio, 1988), and future VE will not be limited to visual (see chap. 3, this volume) and auditory displays (see chap. 4, this volume), but will incorporate the cutaneous, kinesthetic (chaps. 5, 6, 11, and 38, this volume), and vestibular stimuli (see chap. 7, this volume) as well.

#### *7.1.2 Relation Between Field of View of a Moving Visual Display and Resulting Discomfort*

Throughout the literature on simulators, display field of view has been implicated in a trade-off between performance and side effects (e.g., Kennedy, Fowlkes, & Hettinger, 1989; Van Cott, 1990; Westra, Sheppard, Jones, & Hettinger, 1987). Although a larger field of view often enhances performance, it also increases the probability of experiencing side effects. DiZio and Lackner (1997) showed that by decreasing the field of view in a VE (and thus the lag in visual display update), they were able to decrease discomfort ratings.

Symptoms elicited by moving visual fields generated by rotating optokinetic drums are worse if one is viewing a large field-of-view display (Dichgans & Brandt, 1978); however, Andersen and Braunstein (1985) found that 17 of their participants (i.e., 31% of $n = 55$) reported adverse symptoms (9 felt dizzy, 2 had a headache, and 1 felt nauseated)[24] while viewing radially expanding patterns subtending only 7.5 to 21.2 dg of visual angle. One can conclude that although the Andersen and Braunstein Stimulus elicited a lower incidence of symptoms than previous experiments that had used a wider field of view (summarized above in section 3.2), it was a stimulus sufficient to induce 10% of the participants to terminate their participation.

#### *7.1.3 Other Aspects of Visual Motion That Cause Discomfort*

In simulators, the amount of movement of the visual scene accounts for 20% or more of the variance in simulator discomfort (Kennedy, Berbaum, Dunlap, & Hettinger, 1996). The velocity and spatial frequency of the visual display contributes to this effect (Dichgans & Brandt, 1978). The predominant frequency of oscillation of the visual display (Kennedy et al., 1996) is important to the resulting level of discomfort. It is well-known that real motion is most disturbing when it involves low frequencies of oscillation in the 0.2 Hz range of motion or 1 cycle every 5 seconds (e.g., McCauley & Kennedy, 1984); it is believed that the same generalization will apply to moving visual scenes (Hettinger, Berbaum, Kennedy, Dunlap, & Nolan, 1990; Stoffregen & Smart, 1997). Interestingly, in the case of very low frequency oscillation (0.02 Hz or 1 cycle every 50 seconds) of the visual scene, the visual stimulus can sometimes be more disturbing than real-body motion at the same frequency (Dichgans & Brandt, 1978). The amount of depth information represented by the moving visual display (Andersen & Braunstein, 1985) may also be a factor contributing to the perceived amount of visual motion and discomfort.

---

[24]Additional symptoms are mentioned for those five subjects who discontinued their participation in this experiment, but it is sufficient to note that four of them had nausea and one felt faint.

## 8. DISCOMFORT AS A FUNCTION OF INTERVAL BETWEEN EXPOSURES AND EXPOSURE DURATION

Kennedy, Lane, Berbaum, and Lilienthal (1993) found that repeated exposure in simulators was effective in promoting adaptation and reducing discomfort when the interval between exposures was 2 to 5 days. Watson (1998) obtained similar results in a driving simulator, noting that a 2-to-3 day interval between exposures was best.

It has been estimated that 20% to 50% of the variance in VE-related discomfort can be accounted for by the amount of time a person spends in a VE and the intersession interval (Kennedy, Stanney, & Dunlap, 2000). Kennedy, Jones, Stanney, Ritter, and Drexler (1996) found marked reductions in discomfort in the second of two 40-minute VE exposures. Cobb, Nichols, Ramsey, and Wilson (1999) looked at repeated 20-minute exposures to a passive virtual environment. They observed a reduction in self-reported discomfort over the three exposures spaced a week apart. (They noted that the greatest reduction occurred for the Disorientation symptom cluster of the SSQ). Cobb et al. (1999) reported a preliminary study of four participants interacting with a VE for up to 2 hours. They note that all four participants reported increased discomfort up to 1 hour, at which time two of the participants withdrew. However, the two participants who continued with the experiment reported decreasing levels of discomfort at 75 minutes. At the end of the 2-hour session, their reports of discomfort were back to preexposure levels.

## 9. DISCOMFORT DUE TO LACK OF CORRESPONDENCE BETWEEN VISUAL FIELD MOVEMENT AND HEAD MOVEMENT

The literature on simulators indicates that an asynchrony between inertial and visual motion will cause discomfort (Frank, Casali, & Wierwille, 1988; Uliano, Lambert, Kennedy, & Sheppard, 1986); this has been suggested as a problem for VE users also (Hettinger & Riccio, 1992). Recently, VE users were observed to have higher levels of disorientation, discomfort, and postural instability as lag increased in the update rate of the visual display (Cobb, Nichols, Ramsey, & Wilson, 1999; DiZio & Lackner, 1997). Problems stemming from a temporal discordance between head movement and visual-field movement are discussed in chapter 6 of Durlach and Mavor (1995). Improving the correspondence between head movement and visual field movement is an important area of VE development (Stanney, Salvendy et al., 1998).

## 10. DISCOMFORT AS A FUNCTION OF NAVIGATION CHARACTERISTICS

### 10.1 Extent of Control Over Navigation

Individuals who have little control over the movements of a simulator seem to experience greater discomfort than individuals who are controlling the simulator (Casali & Wierwille, 1998; Reason & Diaz, 1971). Similarly Stanney and Hash (1998) found that giving participants control over their actions will reduce their VE side effects. Stanney and Hash examined three conditions: (1) "active" (using a joystick to maneuver with six degrees of freedom), (2) "active–passive" (using a joystick to move forward and backward, side to side, up and down, and in specific circumstances yaw and pitch), and (3) "passive" (passively observing scripted movements). Stanney and Hash (1998) found that the average discomfort scores were highest for the "passive" condition. The "active" condition did not reduce the severity of the symptoms as much as the "active–passive" condition. Stanney and Hash suggest that in the "active" condition, the participants might not been able to adapt as quickly as in the "active–passive" condition, due to the abundant amount of sensory information gathered from unrestricted movements made in the "active" condition.

### 10.2 Discomfort Related to Navigation Using Head Versus Hand

Howarth and Finch (1999) found that head controlled navigation produced greater severity and sooner onset of nausea than hand controlled navigation, most likely due to greater system lag. Anecdotal evidence has suggested that as users gain experience with a VE, they tend to reduce the number and magnitude of their head movements to avoid discomfort (Cobb et al., 1999; Howarth & Finch, 1999). These findings suggest that exploring VEs via head movements is more provocative than using alternate input devices (see also Durlach & Mavor, 1995; chapter 13, this volume).

## 11. DISPLAY CHARACTERISTICS THAT CAN CAUSE DISCOMFORT WITHOUT REQUIRING WHOLE-FIELD VISUAL MOTION

Side effects of SEs involving real or apparent motion stimuli are the focus of this chapter. However, certain VE display characteristics can cause discomfort without requiring whole-field visual motion or other apparent motion stimuli. For example, some of the early head mounted VEs used low-resolution liquid crystal displays placed close to the eyes with insufficient contrast or illumination. Current stereoscopic VE displays require the user to accommodate to a particular depth plane while requiring vergence movements to a range of depths (see chaps. 3 and 37, this volume). As a result of such display factors, Wann and Mon-Williams (1997) observed changes in visual functioning that were correlated with reports of discomfort and have replicated these results with various head mounted displays (see chap. 37, this volume).

## 12. SUMMARY

Humans tend to experience a number of unwanted side effects during and after SEs involving real or apparent (usually visual) motion stimuli. Side effects of SEs are typically measured via symptom checklists, the most popular of which are described herein. The most common side effects of exposure to real motion stimuli, moving visual fields, or flight simulators include the following: dizziness, stomach symptoms (stomach awareness, stomach discomfort, or nausea), flushing/subjective warmth, eyestrain, fatigue, drowsiness, "cold" sweating, headache, or difficulty concentrating. Other classically defined cardinal signs or symptoms include facial pallor, increased salivation, retching, and vomiting. The side effects of VE exposure are not as well studied, but they seem to include many of the symptoms produced by other synthetic experiences. However, it is likely that different SEs produce somewhat different clusters of predominant symptoms.

The most conservative prediction that can be made from the limited VE research is that at least 5% of all users will not be able to tolerate prolonged use of current VEs; this is about the same as the proportion of people highly susceptible to motion stimuli, moving visual fields, and simulators. Virtual environment stimuli also seem to share in common with many other SEs the tendency to elicit symptoms that persist after exposure. Side effects such as stomach symptoms, dizziness, eyestrain, and headache among susceptible VE users will present an obstacle to widespread acceptance. Less obvious side effects may pose a significant risk to VE users who do not believe themselves to be susceptible; these include sopite syndrome, postural dysequilibrium, and loss of visual acuity during head or body motion.

Most of the techniques used to measure real or apparent motion discomfort have been developed and applied in situations of acute and provocative exposure. More emphasis should be given to stimuli that are mild or nonsickening and to chronic or repeated exposure situations. Careful evaluations of the temporal progression of VE side effects should be made in order to identify the early warning signs of an adverse reaction to a VE and to understand the time course of VE adaptation and postexposure recovery. Since the observable signs and subjective symptoms of real or apparent motion discomfort have proven to be valid and reliable measures of a person's state, they will continue to be important criteria for interpreting physiological and performance data regarding VE side effects.

## 13. REFERENCES


Advisory Group for Aerospace Research and Development (1988). Motion cues in flight simulation and simulator induced sickness. In *AGARD Conference Proceedings No. 433* (AD-A202 492). Brussels, Belgium: Author.

Alexander, S. J., Cotzin, M., Hill, C. J., Ricciuti, E. A., & Wendt, G. R. (1945). Wesleyan University studies of motion sickness: Vol. 4. The effects of waves containing two acceleration levels upon sickness. *Journal of Psychology, 20*, 9–18.

Andersen, G. J., & Braunstein, M. L. (1985). Induced self-motion in central vision. *Journal of Experimental Psychology: Human Perception and Performance, 11*, 122–132.

Baltzley, D. R., Kennedy, R. S., Lilienthal, M. G., & Gower, D. W. (1989). The time of postflight simulator sickness symptoms. *Aviation, Space, and Environmental Medicine, 60*(11), 1043–1048.

Benfari, R. C. (1964). Perceptual vertigo: A dimensional study. *Perceptual and Motor Skills, 18*, 633–639.

Benson, A. J. (1988). Motion sickness. In J. Ernsting & P. King (Eds.), *Aviation medicine* (pp. 318–493). London: Buttersworth.

Birren, J. E. (1949). Motion sickness: Its psychophysiological aspects. *A Survey Report on Human Factors in Undersea Warfare* (pp. 375–398). Washington, DC: Committee on Undersea Warfare.

Bles, W. (1979). *Sensory interactions and human posture: An experimental study*. Amsterdam: Academishe Pers.

Bock, O. L., & Oman, C. M. (1982). Dynamics of subjective discomfort in motion sickness as measured with a magnitude estimation method. *Aviation, Space, and Environmental Medicine, 53*, 773–777.

Brand, Th., Dichgans, J., & Wagner, W. (1974). Drug effectiveness on experimental optokinetic and vestibular motion sickness. *Aerospace Medicine, 45*(11), 1291–1297.

Calkins, D. S., Reschke, M. F., Kennedy, R. S., & Dunlap, W. P. (1987). Reliability of provocative tests of motion sickness susceptibility. *Aviation, Space, and Environmental Medicine, 58(9 Suppl.)*, A50–A54.

Casali, J. G., & Wierwille, W. (1998). Vehicular simulator-induced sickness (Tech. Rep. No. NTSC-TR86-012, AD-A173 266, Vol. 3, p. 155). Arlington, VA: Naval Training Systems Center. In K. M. Stanney, R. R. Mourant, & R. S. Kennedy. Human factors issues in virtual environments: A review of the literature. *Presence, 7*(4), 327–351.

Chapelow, J. W. (1988). Simulator sickness in the Royal Air Force: A survey. In *Advisory Group for Aerospace Research and Development, Motion cues in flight simulation and simulator-induced sickness* (AGARD Conference Proceedings No. 433, pp. 6-1–6-11).

Chelen, W. E., Kabrisky, M., & Rogers, S. K. (1993). Spectral analysis of the electroencephalographic response to motion sickness. *Aviation, Space, and Environmental Medicine 64*, 24–29.

Chinn, H. I., & Smith, P. K. (1955).[25] Motion sickness. *Pharmacol. Rev., 7*, 33–82.

Chueng, B. S., Howard, I. P., & Money, K. E. (1991). Visually induced sickness in normal and bilaterally labyrinthine-defective subjects. *Aviation, Space, and Environmental Medicine, 62*, 527–531.

Cobb, S., Nichols, S., Ramsey, A., & Wilson, J. R. (1999). Virtual reality–induced symptoms and effects. *Presence, 8*, 169–186.

Colwell, J. L. (2000, September). NATO questionnaire: Correlation between ship motions, fatigue, sea sickness, and


[25]The article "Motion Sickness," by Chinn and Smith, is dated 1953 in the reference section of Reason and Brand, (1975), but it is dated 1955 in Money (1970), and it is also dated 1955 in the first author's copy of the Chinn and Smith article, whose source was verified by the National Library of Medicine in 1961.

naval task performance. In *International Conference: Human Factors in Ship Design and Operation*, London: Royal Institution of Naval Architects.

Cowings, P. S., Toscano, W. B., DeRoshia, C., & Tauson, R. (1999). *Effects of command and control vehicle (C2V) operational environment on soldier health and performance* (Tech. Rep. No. NASA TM-1999-208786). Moffett Field, CA: National Aeronautics and Space Administration.

Crampton, G. H. (Ed.). (1990). *Motion and space sickness*. Boca Raton, FL: CRC Press.

Crampton, G. H., & Young, F. A. (1953). The differential effects of a rotary visual filed on susceptibles and nonsusceptibles to motion sickness. *Journal of Comparative and Physiological Psychology, 46*, 451–453.

Crosby, T. N., & Kennedy, R. S. (1982). *Proceedings*. Paper presented at the 53rd Annual Scientific Meeting of the Aerospace Medical Association. Cited in M. E. McCauley & T. J. Sharkey Cybersickness: Perception of self-motion in virtual environments. *Presence, 1*(3), 311–318.

Crossland, P., & Lloyd, A. R. J. M. (1993). *Experiments to quantify the effects of ship motion on crew task performance —Phase 1, motion-induced interruptions and motion-induced fatigue* (Tech. Rep. No. DRA/AWMH/TR/93025). Farnborough, England: Defence Research Agency.

Darwin, E. (1795). *Zoonomia*. London: J. Johnson. From chapter 1 of W. J. White (1964), *A history of the centrifuge in aerospace medicine*. Santa Monica, CA: Douglas Aircraft Company.

Davis, J. R., Vanderploeg, J. M., Santy, P. A., & Jennings, R. T., & Stewart, D. F. (1988). Space motion sickness during 24 flights of the Space Shuttle. *Aviation, Space, and Environmental Medicine, 59*, 1185–1189.

DeZouche, I. (1975). Quain's dictionary of medicine. In J. T. Reason & J. J. Brand, *Motion Sickness* (pp. 53). New York, NY: Academic Press Inc. (Original work published 1894.)

Dichgans, J., & Brandt, T. (1978). Visual-vestibular interaction: Effects on self-motion perception and postural control. In R. Held, H. W. Leibowitz, & H. L. Teuber (Eds.), *Handbook of sensory physiology* (Vol. 8, pp. 755–804). New York: Springer-Verlag.

DiZio, P., & Lackner, J. R. (1991). Motion sickness susceptibility in parabolic flight and velocity storage activity. *Aviation, Space, and Environmental Medicine, 62*, 300–307.

DiZio, P., & Lackner, J. R. (1997). Circumventing side effects of immersive virtual environments. In M. Smith, G. Salvendy, & R. Koubek (Eds.), *Design of computing systems: Social and ergonomic considerations* (pp. 893–896). Amsterdam: Elsevier Science Publishers.

Dolezal, H. (1982). *Living in a world transformed: Perceptual and performatory adaptation to visual distortion*. New York: Academic Press.

Durlach, N. I., & Mavor, A. S. (Eds.). (1995). *Virtual reality: Scientific and technological challenges*. Washington, DC: National Academy Press.

Durlach, N., & Slater, M. (1998). *Presence in shared virtual environments and virtual togetherness* [Online]. Paper presented at the Presence in Shared Virtual Environments Workshop, Ipswich, England. Available: http://www.cs.ucl.ac.uk/staff/m.slater/BTWorkshop/durlach.html [1999, September 1].

Eagon, J. C. (1988). *Quantitative frequency analysis of the electrogastrogram during prolonged motion sickness*. Unpublished master's thesis, Massachusetts Institute of Technology.

Ebenholtz, S. M. (1992). Motion sickness and oculomotor systems in virtual environments. *Presence, 1*(3), 302–305.

Eklund, S. (1999). *Gentamincin treatment and headache in Menière's disease*. Unpublished doctoral dissertation, Helsinki University Central Hospital.

Ernstein, J., & King, P. (Eds.). (1988). *Aviation medicine*. London: Buttersworth.

Finch, M., & Howarth, P. A. (1996). *A comparison between two methods of controlling movement within a virtual environment* (Tech. Rep. No. VISERG 9606). Leicestershire, England: Loughborough University, Department of Human Sciences.

Fischer, M. H., & Kornmüller, A. E. (1930). Perception of motion based on the optokinetic sense and optokinetic nystagmus. *Journal fuer Psychologie und Neurologie, 41*, 273–308.

Fisher, S. (1990). Virtual interface environments. In B. Laurel (Ed.), *The art of human–computer interface design*. Menlo Park, CA: Addison-Wesley.

Frank, L. H., Casali, J. G., & Wierwille, W. W. (1988). Effects of visual display and motion system delays on operator performance and uneasiness in a driving simulator. *Human Factors, 30*, 201–217.

Fregly, A. R. (1974). Vestibular ataxia and its measurement in man. In H. H. Kornhuber (Ed.), *Handbook of sensory physiology* (Vol. 6, no. 2). New York: Springer-Verlag.

Gordon, C. R., Spitzer, O., Doweck, I., Melamed, Y., & Shupak, A. (1995). Clinical features of mal de débarquement: Adaptation and habituation to sea conditions. *Journal of Vestibular Research, 5*, 363–369.

Goshnor, A., & Melville-Jones, J. G. (1976). Extreme vestibular-oculomotor adaptation induced by prolonged optical reversal of vision. (*Journal of Physiology, 256*), 381–414.

Gower, D. W., Lilienthal, M. G., Kennedy, R. S., & Fowlkes, J. E. (1988, September). Simulator sickness in U.S. Army and Navy fixed- and rotary-wing flight simulators. In *Advisory Group for Aerospace Research and Development, motion cues in flight simulation and simulator induced sickness* (AGARD Conference Proceedings No. 433, pp. 8.1–8.20).

Grabowski L. S. (1996). *Falls among the geriatric population* [Online]. Available: wwwsansumcom/highlite/1996/13_2_2html

Graybiel, A. (1967). Functional disturbances of vestibular origin of significance in space flight. In (Ed.), *Second International Symposium on Man in Space, Paris*. (pp. B8–B32). Wien, New York: Springer-Verlag.

Graybiel, A. (1969). Structural elements in the concept of motion sickness. Publisher: *Aerospace Medicine, 40*, 351–367.

Graybiel, A., Clark, B., & Zarriello, J. J. (1960). Observations on human subjects living in a "slow-rotation room" for periods of two days. *AMA Archives of Neurology, 3*, 55–73.

Graybiel, A., & Johnson, W. H. (1963). A comparison of the symptomatology experienced by healthy persons and subjects with loss of labyrinthine function when exposed to unusual patterns of centripetal force in a counter-rotating room. *Annls. Otol., Rhinol., and Laryngol., 72*, 1–17.

Graybiel, A., Kennedy, R. S., Knoblock, E. C., Guedry, F. E., Mertz, W., McLead, M. E., Colehour, J. K., Miller, E. F., & Fregly, A. R. (1965). Effects of exposure to a rotating environment (10 RPM) on four aviators for a period of twelve days. *Aerospace Medicine, 36*, 733–754.

Graybiel, A., & Knepton, J. (1976). Sopite syndrome: A sometimes sole manifestation of motion sickness. *Aviation, Space, and Environmental Medicine, 47*, 873–882.

Graybiel, A., & Lackner, J. R. (1980). A sudden-stop vestibulovisual test for rapid assessment of motion sickness manifestations. *Aviation, Space, and Environmental Medicine, 51*, 21–23.

Graybiel, A., & Lackner, J. R. (1987). Treatment of severe motion sickness with antimotion sickness drug injections. *Aviation, Space, and Environmental Medicine, 58*, 773–776.

Graybiel, A., Wood, C. D., Miller, E. F., & Cramer, D. B. (1968). Diagnostic criteria for grading the severity of acute motion sickness. *Aerospace Medicine, 39*, 4453–4455.

Guedry, F. E. (1991). Factors influencing susceptibility: Individual differences and human factors. In *Proceedings of motion sickness: Significance in aerospace operations and prophylaxis* (AGARD LS-175:5/1-5/18).

Hardacre, L. E., & Kennedy, R. S. (1963). Some issues in the development of a motion sickness questionnaire for flight students. *Aerospace Medicine, 34*, 401–402.

Harm, D. H. (1990). Physiology of motion sickness symptoms. In G. H. Crampton (Ed.), *Motion and space sickness* (pp. 153–177). Boca Raton FL: CRC Press.

Havron, M. D., & Butler, L. F. (1957). *Evaluation of training effectiveness of the 2-FH-2 helicopter flight training research tool* (Tech. Rep. No. NAVTRADEVCEN 20-OS-16, Contract 1915). Arlington, VA: U.S. Naval Training Device Center.

Haward, B. M., Lewis, C. H., & Griffin, M. J. (2000, September). Crew response to motions of an offshore oil production and storage vessel. In *International Conference: Human Factors in Ship Design and Operation*. London: Royal Institution of Naval Architects.

Hemmingway, A. (1943). *Adaptation to flying motion by airsick aviation students* (Research Rep. Project 170, Tech. Rep. No. 4). Randolph Field, TX: School of Aviation Medicine.

Hettinger, L. J., Berbaum, K. S., Kennedy, R. S., Dunlap, W. P., & Nolan, M. D. (1990). Vection and simulator sickness. *Military Psychology, 2*(3), 171–181.

Hettinger, L. J., & Riccio, G. E. (1992). Visually induced motion sickness in virtual environments. *Presence, 1*(3), 306–310.

Howarth, P. A., & Finch, M. (1999). The nauseogenicity of two methods of navigating within a virtual environment. *Applied Ergonomics, 30*, 39–45.

Hu, S., McChesney, K. A., Player, K. A., Bahl, A. A., Buchanan, J. B., & Scozzafava, J. E. (1999). Systematic investigation of physiological correlates of motion sickness induced by viewing an optokinetic rotating drum. *Aviation, Space, and Environmental Medicine, 70*, 759–765.

Hu, S., & Stern, R. M. (1998). Optokinetic nystagmus correlates with severity of vection-induced motion sickness and gastric tachyarrhythmia. *Aviation, Space, and Environmental Medicine, 69*, 1162–1165.

Irwin, J. A. (1975). The pathology of seasickness. In J. T. Reason & J. J. Brand, *Motion Sickness* (pp. 7). New York, NY: Academic Press Inc. (Original work published 1881.)

Jones, S. A. (1998). *Effects of restraint on vection and simulator sickness*. Unpublished doctoral dissertation, University of Central Florida.

Kay, D. L., Lawson, B. D., & Clarke, J. E. (1998). Airsickness and lowered arousal during flight training. In Proceedings of the 69th Annual Meeting of the Aerospace Medical Association. *Aviation, Space and Environmental Medicine, 69*, Number 3, p. 236. Seattle, WA.

Kellog, R. S., Castore, C. H., & Coward, R. E. (1984). Psychological effects of training in a full vision simulator. In M. E. McCauley (Ed.), *Research Issues in Simulator Sickness: Proceedings of a Workshop* (pp. 2,6). Washington, DC: National Academy Press. (Original work published in 1980)

Kelly, A. D., & Kanas, N. (1993). Communication between space crews and ground personnel: A survey of astronauts and cosmonauts. *Aviation, Space, and Environmental Medicine, 64*, 795–800.

Kennedy, R. S. (1996). Analysis of simulator sickness data (Tech. Rep. under Contract No. N61339-91-D-0004 with Enzian Technology, Inc.). Orlando, FL: Naval Air Warfare Center, Training Systems Division.

Kennedy, R. S., Berbaum, K. S., Dunlap, W. P., & Hettinger, L. J. (1996). Developing automated methods to quantify the visual stimulus for cybersickness. In *Proceedings of the Human Factors and Ergonomics Society 40th Annual Meeting* (pp. 1126–1130).

Kennedy, R. S., Berbaum, K. S., & Lilienthal, M. G. (1992). Human operator discomfort in virtual reality systems: Simulator sickness—causes and cures. In S. Kumar (Ed.), *Advances in industrial ergonomics and safety* (Vol. 4, pp. 1227–1234). Washington, DC: Taylor & Francis.

Kennedy, R. S., Berbaum, K. S., & Lilienthal, M. G. (1997). Disorientation and postural ataxia following flight simulation. *Aviation, Space, and Environmental Medicine, 68*(1), 13–17.

Kennedy, R. S., Dunlap, W. P., & Fowlkes, J. E. (1990). Prediction of motion sickness susceptibility. In G. H. Crampton (Ed.), *Motion and Space Sickness* (pages). City, State: CRC Press.

Kennedy, R. S., Dutton, B., Ricard, G. L., & Frank, L. H. (1984). *Simulator Sickness: A Survey of Flight Simulators for the Navy* (SAE Technical Paper Series No. 811597). Warrendale, PA.

Kennedy, R. S., Fowlkes, J. E., & Hettinger, L. J. (1989). Review of simulator sickness literature (Tech. Rep. No. NTSC TR89-024). Orlando, FL: Naval Training Systems Center.

Kennedy, R. S., & Frank, L. H. (1985). *A review of motion sickness with special reference to simulator sickness* (Tech. Rep. No. NAVTRAEQUIPCEN 81-C-0105-16). Orlando, FL: Naval Training Equipment Center.

Kennedy, R. S., & Graybiel, A. (1962). Symptomatology during prolonged exposure in a constantly rotating environment at a velocity of one revolution per minute. *Aerospace Medicine, 33*, 817–825.

Kennedy, R. S., & Graybiel, A. (1965). *The dial test: A standardized procedure for the experimental production of canal sickness symptomatology in a rotating environment* (Tech. Rep. No. NSAM-930). Pensacola, FL: Naval School of Aviation Medicine.

Kennedy, R. S., Graybiel, A., McDonough, R. G., & Beckwith, F. D. (1968). Symptomatology under storm conditions in the North Atlantic in control subjects and in persons with bilateral labyrinthine defects. *Acta Oto-Laryngologice, 66*, 533–540.

Kennedy, R. S., Hettinger, L. J., & Lilienthal, M. G. (1990). Simulator sickness. In Crampton G. H. (Ed.), *Motion and space sickness* (pp. 317–341). Boca Raton, FL: CRC Press.

Kennedy, R. S., Jones, M. B., Stanney, K. M., Ritter, A., & Drexler, J. M. (1996). *Human factors safety testing for virtual environment mission-operations training* (Tech. Rep. No. NASA1-96-2, Contract No. NAS9-19482). Houston, TX: NASA Lyndon B, Johnson Space Center.

Kennedy, R. S., Lane, N. E., Berbaum, K. S., & Lilienthal, M. G. (1993). Simulator sickness questionnaire: An enhanced method for quantifying simulator sickness. *International Journal of Aviation Psychology, 3*, 203–220.

Kennedy, R. S., Lane, N. E., Lilienthal, M. G., Berbaum, K. S., & Hettinger, L. J. (1993). Profile analysis of simulator sickness symptoms: Application to virtual environment systems. *Presence, 1*(3), 295–301.

Kennedy, R. S., Lanham, D. S., Drexler, J. M., Massey, C. J., & Lilienthal, M. G. (1997). A comparison of cybersickness incidences, symptom profiles, measurement techniques, and suggestions for further research. *Presence, 6*, 638–644.

Kennedy, R. S., Lilienthal, M. G., Berbaum, K. S., Baltzley, D. R., & McCauley, M. E. (1989). Simulator sickness in U.S. Navy flight simulators. *Aviation, Space, and Environmental Medicine, 60*, 10–16.

Kennedy, R. S., Massey, C. J., & Lilienthal, M. G. (1995, July). *Incidences of fatigue and drowsiness reports from three dozen simulators: Relevance for the sopite syndrome.* Paper presented at the First Workshop on Simulation and Interaction in Virtual Environments (SIVE '95), Iowa City, IA.

Kennedy, R. S., & Stanney, K. M. (1996). Postural instability induced by virtual reality exposure: Development of a certification protocol. *International Journal of Human–Computer Interaction 8*(1), 25–47.

Kennedy, R. S., & Stanney, K. M. (1997). Aftereffects of virtual environment exposure: Psychometric issues. In M. J. Smith, G. Salvendy, & R. J. Koubek (Eds.), *Design of computing systems: Social and ergonomic considerations* (pp. 897–900). Amsterdam: Elsevier Science Publishers.

Kennedy, R. S., Stanney, K. M., & Dunlap, W. P. (2000). Duration and exposure to virtual environments: Sickness curves during and across sessions. *Presence, 9*(5), 466–475.

Kennedy, R. S., Tolhurst, G. C., & Graybiel, A. (1965). *The effects of visual deprivation on adaptation to a rotating room.* Naval School of Aviation Medicine. Pensacola, FL: NSAM-918.

Kolasinski, E. M. (1995). *Simulator sickness in virtual environments* (Tech. Rep. No. 1027). Alexandria, VA: United States Army Research Institute for the Behavioral and Social Sciences.

Lackner, J. R., & DiZio, P. (1988). Visual stimulation affects the perception of voluntary leg movements during walking. *Perception, 17*, 71–80.

Lackner, J. R., & Graybiel, A. (1986). Sudden emesis following parabolic flight maneuvers: Implications for space motion sickness. *Aviation, Space, and Environmental Medicine, 57*, 343–347.

Lackner, J. R., & Teixeira, R. A. (1977). Optokinetic motion sickness: Continuous head movements attenuate the visual induction of apparent self-rotation and symptoms of motion sickness. *Aviation, Space, and Environmental Science, 48*(3), 248–253.

Lane, N., & Kennedy, R. S. (1988). *A new method for quantifying simulator sickness: Development and application of the Simulator Sickness Questionnaire* (SSQ), (EOTR 88-7). Orlando, FL: Essex Corporation.

Lawson, B. D. (1993). Human physiological and subjective responses during motion sickness induced by unusual visual and vestibular stimulation. *Dissertation Abstracts International, 54(4-B)*, 2249.

Lawson, B. D., Flaherty, D., & Schmorrow, D. (in preparation). *Sopite syndrome in operational flight training* (Tech. Rep.) Pensacola, FL: Naval Aerospace Medical Research Laboratory.

Lawson, B. D., & Mead, A. M. (1998). The sopite syndrome revisited: Drowsiness and mood changes during real of apparent motion. *Acta Astronautica, 43*(3–6), 181–192.

Lawson, B. D., Rupert, A. H., Guedry, F. E., Grissett, J. D., & Mead, A. M. (1997). The human–machine interface challenge of using virtual environment (VE) displays aboard centrifuge devices. In: M. J. Smith, G. Salvendy, & R. J. Koubek (Eds.), *Design of computing systems: Social and ergonomic considerations* (pp. 945–948). Amsterdam: Elsevier Science Publishers.

Lee, D. N., & Aronson, E. (1974). Visual proprioceptive control of standing in human infants. *Perception and Psychophysics, 15*, 529–532.

Lee, D. N., & Lishman, J. R. (1975). Visual proprioceptive control of stance. *Journal of Human Movement Studies, 1*, 87–95.

Leslie, K. R., Stickgold, R., DiZio, P., Lackner, J. R., & Hobson, J. A. (1997). The effect of optokinetic stimulation on daytime sleepiness. *Archives Italiennes de Biologie, 135*, 219–228.

Lewis, C. H., & Griffin, M. J. (1998). Modelling the effects of deck motion on energy expenditure and motion-induced fatigue. In *ISVR Contract Report No. 98/05*. Southampton, England: University of Southampton.

Magee, L. E., Kantor, L., & Sweeney, D. M. (1988). *Simulator-induced sickness among hercules aircrew*. In *Advisory Group for Aerospace Research and Development, Motion cues in flight simulation and simulator-induced sickness* (AGARD Conference Proceedings No 433, pp. 5-1–5-8): AGARD.

McCauley, M. E. (Ed.). (1984). *Research issues in simulator sickness: Proceedings of a workshop*. Washington DC: National Academy Press.

McCauley, M. E., & Kennedy, R. S. (1984). Recommended human exposure limits for very-low-frequency vibration (TP-76-36). Pacific Missile Test Center, Point Magu, CA. In M. E. McCauley (Ed.), *Research issues in simulator sickness: Proceedings of a workshop*. Washington DC: National Academy Press. (Original work published 1976)

McCauley, M. E., & Sharkey, T. J. (1992). Cybersickness: Perception of self-motion in virtual environments. *Presence, 1*(3), 311–318.

McGuiness, J., Bouwman, J. H., & Forbes, J. M. (1984). Simulator sickness occurance in the 2E6 air combat maneuvering simulator (ACMS). In In M. E. McCauley (Ed.), *Research Issues in Simulator Sickness: Proceedings of a Workshop* (pp. 3–7). Washington, DC: National Academy Press. (Original work published in 1981.)

Miller, E. F., & Graybiel, A. (1969). *A standardized laboratory means of determining susceptibility to Coriolis (motion) sickness* (Tech Rep. No. NAMI-1058). Pensacola, FL: Naval Aerospace Medical Institute.

Miller, E. F., & Graybiel, A. (1970a). *Comparison of five levels of motion sickness severity as the basis for grading susceptibility* (Bureau of Medicine and Surgery, NASA Order R-93). Pensacola, FL: Naval Aerospace Medical Institute.

Miller, E. F., & Graybiel, A. (1970b). A provocative test for grading susceptibility to motion sickness yielding a single numerical score. *Acta Otolarygologica* (274 Suppl).

Miller, J. C. (1995). Batch processing of 10,000 h of trucker driver EEG data. *Biological Psychology, 40*,(1–2), 209–222.

Miller, J. W., & Goodson, J. E. (1958). *A note concerning "motion sickness" in the 2-FH-2 Hover Trainer* Project No. 1 17 01 11, Subtask 3, Report No. 1. Pensacola, FL: Naval School of Aviation Medicine.

Miller, J. W., & Goodson, J. E. (1960). Motion sickness in a helicopter simulator. *Aerospace Medicine, 31*, 204–212.

Money, K. E. (1970). Motion sickness. *Physiolo. Rev., 50*, 1–39.

Money, K., Lackner, J., & Cheung R. (1996). Motion sickness and the autonomic nervous system. In A. Miller & W. Yates (Eds.), *Vestibular Autonomic Regulation* (pp. 147–173). City, State: CRC Press.

Muth, E. R., Lawson, B. D., & Sausen, K. (1999, May) The sopite profile: Preliminary characterization of the symptoms of the sopite syndrome. In *Proceedings of the Aerospace Medical Association* (Abstract 198). Detroit, MI:

Muth, E. R., Stern, R. M., Thayer, J. F., & Koch, K. L. (1996). Assessment of the multiple dimensions of nausea: The Nausea Profile. *Journal of Psychosomatic Research, 40*, 511–520.

National Research Council Committee on Vision and Working Group on Wraparound Visual Displays. (1990). *Motion sickness, visual displays, and armored vehicle design* (Tech. Rep. No. BRL-CR-629). Aberdeen Proving Ground, MD: Ballistic Research Laboratory.

Neufeldt, N., & Guralnik, D. B. (Eds). (1996). *Webster's new world dictionary* (3rd ed.) New York, NY: Macmillian USA.

Nicogossian, A. E., Huntoon, C. L., & Pool, S. L. (1989). *Space physiology and medicine*. Philadelphia: Lea & Febiger.

Nicogossian, A. E., & Parker, J. F. (Eds.). (1984). Space and physiology medicine. In M. E. McCauley (Ed.), *Research Issues in Simulator Sickness: Proceedings of a Workshop* (pp. 1). Washington, DC: National Academy Press. (Original work published in 1982.)

Neilson, W. A., Knott, T. A., & Carhart, P. W. (1956), *Webster's New International Dictionary of the English Language* (2nd ed.). Springfield, MA: G. & C. Merriam Company.

Probst, T. H., & Schmidt, U. (1998). The sensory conflict concept for the generation of nausea. *Journal of Psychophysiology, 12*, 34–49.

Quix, ??? (1976). In A. Graybiel & J. Knepton, Sopite Syndrome: A sometimes sole manifestation of motion stickness (pp. 879). *Aviation, Space, and Environmental Medicine, 47,* 873–882. (Original work published 1922.)

Reason, J. T., & Brand, J. J. (1975). *Motion sickness*. London: Academic Press.

Reason, J. T., & Diaz, E. (1971). *Simulator sickness in passive observers* (Tech. Rep. No. 1310). Flying Personnel Research Committee. London, Ministry of Defense.

Reason, J. T., & Graybiel, A. (1970). Changes in subjective estimates of well-being during the onset and remission of motion sickness symptomatology in the Slow Rotation Room. *Aerospace Medicine, 41*, 166–171.

Reason, J. T. (1975). An investigation of some factors contributing to individual variation to motion sickness susceptibility. In J. T. Reason & J. J. Brand., *Motion Sickness* (pp. 7). New York, NY: Academic Press Inc. (Original work published 1967)

Reason, J. T., & Graybiel, A. (1975). Changes in subjective estimates of well-being during the onset and remission of motion sickness symptomatology in the slow rotation room. In J. T. Reason & J. J. Brand, *Motion Sickness* (pp. ???). New York, NY: Academic Press Inc. (Original work published 1970)

Regan, E. C. & Price, K. R. (1994). The frequency of occurrence and severity of side-effects of immersion virtual reality. *Aviation, Space, and Environmental Medicine, 65*, 527–530.

Regan, E. C., & Ramsey, A. D. (1996). The efficacy of hyoscine hydrobromide in reducing side-effects induced during immersion in virtual reality. *Aviation, Space, and Environmental Medicine, 67*(3), 222–226.

Robinett, W. (1992). Synthetic experience: A proposed taxonomy. *Presence, 1*(2), 229–247.

Schwab, R. S. (1943). Chronic seasickness. *Amm. Int. Med., 19*, 28–35. In J. T. Reason & J. J. Brand (1975), *Motion sickness*. London: Academic Press.

So, R. H. Y., & Lo, W. T. (1999). Cybersickness: An experimental study to isolate the effects of rotational scene oscillations. In *Proceedings of the IEEE Virtual Reality Conference* (pp. 237–241). Los Alamitos, CA: IEEE Computer Society.

Stanney, K. M., & Hash, P. (1998). Locus of user-initiated control in virtual environments: Influences on cybersickness. *Presence, 7*, 447–459.

Stanney, K. M., & Kennedy, R. S. (1997). Cybersickness is not simulator sickness. In *Proceedings of the Human Factors and Ergonomics Society 41st Annual Meeting* (pp. 1138–1142). Santa Monica, CA: Human Factors and Ergonomics Society.

Stanney, K. M., Kennedy, R. S., Drexler, J. M., & Harm, D. H. (1999). Motion sickness and proprioceptive aftereffects following virtual environment exposure. *Applied Ergonomics, 30*, 27–38.

Stanney, K. M., Mourant, R. R., & Kennedy, R. S. (1998). Human factors issues in virtual environments: A review of the literature. *Presence, 7*(4), 327–351.

Stanney, K. M., Salvendy, G., Deisinger, J., DiZio, P., Ellis., S., Ellison, J., Fogelman, G., Gallimore, J., Hettinger, L., Kennedy, R., Lackner, J., Lawson, B., Maida, J., Mead, A., Mon-Williams, M., Newman, D., Piantanida, T., Reeves, L., Reidel, O., Singer, M., Stoffregen, T., Wann, J., Welch, R., Wilson, J., & Witmer, B. (1998). Aftereffects and sense of presence in virtual environments: Formulation of a research and development agenda. *International Journal of Human–Computer Interaction, 10*(2), 135–187.

Stern, R. M., Hu, S., Anderson, R. B., Leibowitz, H. W., & Koch, K. L. (1990). The effects of fixation and restricted visual field on vection-induced motion sickness. *Aviation, Space, and Environmental Medicine, 61*, 712–715.

Stoffregen, T. A., & Smart, L. J. (1998). Postural instability precedes motion sickness. *Brain Research Bulletin, 47*(5), 437–448.

Teixeira, R. A., & Lackner, J. R. (1979). Optokinetic motion sickness: Attentuation of visually induced apparent self-rotation by passive head movements. *Aviation, Space, and Environmental Medicine, 50*(3), 264–266.

Ter Vrugt, D., & Peterson, D. R. (1973). The effects of vertical rocking frequencies on the arousal level in two-month-old infants. *Child Development, 44*, 205–209.

Tyler, D. B., & Bard, P. (1949). Motion sickness. *Phycological Revice, 29*, 311–369.

Uliano, K. C., Lambert, E. Y., Kennedy, R. S., & Sheppard, D. J. (1986). *The effects of asynchronous visual delays on simulator flight performance and the development of simulator sickness symptomatology* (Tech. Rep. No. NAVTRASYSCEN 85-D-0026-1). Orlando, FL: Naval Training Systems Command.

Ungs, T. J. (1988). Simulator induced syndrome in Coast Guard aviators. *Aviation, Space, and Environmental Medicine, 59*, 267–272.

Van Cott, H. (1990). *Lessons from simulator sickness studies. Motion sickness, visual displays, and armored vehicle design* (pp. 76–84). Washington, DC: Ballistic Research Laboratory.

Wann, J. P., & Mon-Williams, M. (1997, May). Health issues with virtual reality displays: What we do know and what we don't. *Computer Graphics*, 53–57.

Watson, G. S. (1998). The effectiveness of a simulator screening session to facilitate simulator sickness adaptation for high-intensity driving scenarios. In *Proceedings of the 1998 IMAGE Conference*. Chandler, AZ: IMAGE Society.

Watson, N. W., Wells, T. J., & Cox, C. (1998). Rocking chair therapy for dementia patients: its effect on psychosocial well-being and balance, *American Journal of Alzheimer's Disease, 13*(6), 296–308.

Welch, R. B. (1978). Perceptual modification: Adapting to altered sensory environments. New York: Academic Press.

Wertheim, A. H. (1998). Working in a moving environment. *Ergonomics, 41*(12), 1845–1858.

Westra, D. P., Sheppard, D. J., Jones, S. A., & Hettinger, L. J. (1987). *Simulator design features for helicopter shipboard landings* (Tech. Rep. No. TR-87-041 AD-A203 992, p. 61). Orlando, FL: Naval Training Systems Center.

Wiker, S. F., Kennedy, R., McCauley, M. E., & Pepper, R. L. (1979). Susceptibility to seasickness: Influence of hull design and steaming direction. *Aviation, Space, and Environmental Medicine, 50*(10), 1046–1051.

Wills, C. E., & Moore, C. F. (1994). A controversy in scaling of subjective states: Magnitude estimation versus category rating methods. *Research in Nursing and Health, 17*, 231–237.

Wilson, J. R. (1997). Virtual environments and ergonomics: Needs and opportunities. *Ergonomics, 40*, 1057–1077.

Witmer, B. G., & Singer, M. J. (1998). Measuring Presence in Virtual Environments: A Presence Questionnaire. *Presence: Teleoperators and Virtual Environments*, 7(3), 225–240.

Wood, R. W. (1975). The "haunted swing" illusion. In J. T. Reason & J. J. Brand., *Motion Sickness* (pp. 110). New York, NY: Academic Press Inc. (Original work published 1895)

Wood, C. D., Graybiel, A., & McDonough, R. G. (1966). Clinical effectiveness of anti-motion-sickness drugs. *Journal of the American Medical Association,* February, 187–190.

Wood, C. D., Stewart, J. J., Wood, M. J., Manno, J. E., Manno, B. R., & Mims, M. E. (1990). Therapeutic effects of antimotion sickness medications on the secondary symptoms of motions sickness. *Aviation, Space, and Environmental Medicine, 61*, 157–161.

Woodward, S., Tauber, E. S., Spielmann, A. J., & Thorpy, M. J. (1990). Effects of otolithic vestibular stimulation on sleep. *Sleep, 13*(6), 533–537.

Wright, M. S., Bose, C. L., & Stiles, A. D. (1994). The incidence and effects of motion sickness among medical attendants during transport. *Journal of Emergency Medicine, 13*, 15–20.

Yates, B. J., & Miller, A. D. (Eds.). (1996). *Vestibular autonomic regulation*. New York: CRC Press.

Yano, F., Yazu, Y., Suziki, S., Kasamatsu, K., Idogawa, K., & Ninomija, S. P. (1997). Analysis of driver's eye directions at vehicle driving on the highway. In M. J. Smith and G. Salvendy *Proceedings from the 7th Annual Conference on Human–Computer Interaction* (p. 14). San Francisco: HCI International.

# 31

# Adapting to Virtual Environments

Robert B. Welch
*NASA—Ames Research Center*
*Moffett Field, CA 94035*
*rwelch@mail.arc.nasa.gov*

## 1. OVERVIEW

This chapter describes and illustrates a variety of procedures for helping virtual environment (VE) users overcome the deleterious sensory/perceptual, behavioral, and physical effects and aftereffects that result from current technological limitations of these devices. Many of the most important of these untoward consequences of interacting with VEs are due to the presence of *sensory rearrangements.* A sensory rearrangement (or distortion) exists when the sensory array and/or relationship between sensory systems differs from normal, as, for example, when viewing the hand through a wedge prism. For the observer, the initial exposure to a sensory rearrangement results in misperception, the surprise of violated expectations, and perceptual–motor disruption. Because laboratory experiments indicate that human beings are capable of adapting to a wide range of optical rearrangements, it is reasonable to assume that they are equally able to adapt to the sensory rearrangements found in VEs. It follows that the variables that have been found to influence adaptation to optical rearrangements, for example, active interaction, will apply as well to VE adaptation. Therefore, the VE-training procedures described in this chapter are based on the variables demonstrated by previous research to control or influence perceptual and perceptual–motor adaptation to rearranged sensory environments. It will be proposed that the problematical aftereffects that inevitably result from VE adaptation can be reduced or eliminated by means of systematic readaptation procedures and "dual adaptation" training (i.e., repeated alternation between adaptation and readaptation). Finally, the omnipresence of individual differences in adaptation to sensory rearrangements is acknowledged and its implications for the use of VE adaptation-training procedures considered.

## 2. THE COMPLAINTS OF VE USERS

Despite tremendous advances in the technology of virtual reality (e.g., Durlach & Mavor, 1995), one rarely mistakes it for reality. Thus, most VEs, including "augmented realities" (e.g., see-through displays), continue to be plagued by such deficiencies as poor lighting, unrealistic graphics, and a multitude of sensory rearrangements that render it a poor imitation of everyday experience. More importantly, these shortcomings are the source of a variety of sensory/perceptual, behavioral, and physical complaints, many of which afflict nearly all VE users to one degree or another (see Table 31.1). The last malady listed in Table 31.1, motion-sickness symptoms (variously referred to as simulator sickness, virtual environment sickness, and cybersickness) is dealt with in length in chapter 30 (this volume). Thus, although the occurrence of simulator sickness is typically considered indirect evidence of the presence of intersensory conflicts and its recovery an indicator that adaptation has taken place, these topics will, for the most part, be ignored here.

## 3. THE RESPONSE TO USERS' COMPLAINTS

Clearly, the adverse effects of VEs pose a serious obstacle to optimal task performance and training with these devices. Furthermore, as Biocca (1992) has noted, unpleasant experiences in VEs, especially if well publicized, are likely to have a chilling effect on the general diffusion of VE technology. In short, there is ample motivation to do whatever is necessary to overcome the problems that frequently disrupt the human–VE interface. There are two very different, albeit complementary, ways in which to respond to this imperative. The first is to modify (typically, improve) the VE to accommodate the user, and the second is to modify the user to accommodate the VE. Designers and implementers of VEs have typically chosen the first of these strategies. That is, they tend to seek engineering solutions (e.g., more powerful computers, faster tracking

**TABLE 31.1**
Sensory/Perceptual, Behavioral, and Physical Complaints Afflicting VE Users

Sensory/perceptual problems reported by VE users:

- Momentary reduction in binocular acuity (e.g., Mon-Williams, Rushton, & Wann, 1993)
- Misperception of depth (e.g., Roscoe, 1993)
- Changes in dark accommodative focus (Fowlkes, Kennedy, Hettinger, & Harm, 1993)
- Potentially dangerous "delayed flashbacks" (e.g., illusory experiences of climbing, turning, and inversion) that may not surface until several hours after user has left an airplane simulator or similar VE (e.g., Kennedy, Fowlkes, & Lilienthal, 1993)

Disruptive behavioral effects of VEs:

- Disturbed perceptual–motor (e.g., hand–eye) coordination (e.g., Biocca & Rolland, 1998)
- Locomotory and postural instability (e.g., DiZio & Lackner, 1997)
- Degraded task performance (e.g., Fowlkes et al., 1993)

Physical complaints reported by VE users:

- Eye strain, or "asthenopia" (e.g., Mon-Williams et al., 1993), which may be symptomatic of underlying distress of or conflict between oculomotor subsystems (e.g., Ebenholtz, 1992)
- Headaches (e.g., Mon-Williams et al., 1995)
- Cardiovascular, respiratory, or biochemical changes (e.g., Calvert & Tan, 1994)
- Motion-sickness symptoms (e.g., pallor, sweating, fatigue, and drowsiness, although rarely vomiting; e.g., Gower et al., 1987)

devices, and improved displays) for eliminating the deficiencies of their devices. According to the second and much less common tactic, the VE is left as it is, and users are provided with systematic training and/or deliberate strategies aimed at reducing, or perhaps merely circumventing, its shortcomings. It is generally agreed that the first of these two approaches will ultimately prevail as computer technology is perfected and becomes more generally affordable. However, the second strategy has the great advantage that it can be implemented now, without having to wait for those improvements to be realized. Some of the limitations of current VE technology, such as poor lighting, small field of view (FOV), and heavy headgear, can only be overcome by technological improvements. However, others, such as discrepancies between felt and seen limb position, delays of visual feedback, and distortions of perceived depth, represent the kinds of sensory rearrangements that have proven amenable to user-training procedures, in particular, those that promote adaptation. It is these procedures that represent the focus of this chapter.

Adaptation to sensory rearrangement is defined here as "a semi-permanent change of perception and/or perceptual-motor coordination that serves to reduce or eliminate a registered discrepancy between or within sensory modalities or the errors in behavior induced by this discrepancy" (Welch, 1978, p. 8). Adaptation is measured in two ways: (1) the reduction of observers' perceptual and/or perceptual–motor errors during exposure to the sensory rearrangement (the "reduction of effect"); and (2) postexposure "negative aftereffects" (errors in the direction opposite those initially elicited by the sensory rearrangement). Adaptation is automatic and "unthinking." For example, the negative aftereffects of hand–eye coordination occur even when the subject is aware that the sensory rearrangement has been removed. Perceptual–motor adaptation can occur even when perceptual adaptation has not, as illustrated by a study by Snyder and Pronko (1952) in which, for an entire month, a subject wore a pair of goggles that inverted his visual field. Despite the fact that by the end of this extended period the subject's perceptual–motor coordination was essentially error-free, the world still appeared to be upside-down to him. In general, it may be concluded that perceptual adaptation predicts behavioral adaptation but not vice versa. Before examining how research on adaptation can assist the design of procedures for training VE users to overcome the sensory rearrangements to which they may be exposed, it will be useful to specify what these sensory rearrangements are.

## 4. SENSORY REARRANGEMENTS (AND DISARRANGEMENTS) FOUND IN SOME VEs

There are five major deficiencies of VEs for which the processes of perceptual and perceptual-motor adaptation are relevant. These are (1) intersensory conflicts, (2) distortions of depth and distance, (3) distortions of form and size, (4) the loss of stability of the visual field due to visual feedback delays and asynchronies, and (5) sensory "disarrangement," that is, the presence of randomly varying distortions. This section will discuss each of these in turn.

### 4.1 Intersensory Conflicts

This category entails mismatches between paired spatial modalities—vision, audition, touch, proprioception, and the haptic and vestibular senses. Thus, the computer-generated surrogate of the observer's hand may be seen in one location while the actual hand is felt in another, an object may look rough but feel smooth, and a sound source may be seen in one place but heard elsewhere. Some intersensory discrepancies entail the *absence* of one member of a pair of normally correlated sensory impressions. For example, in fixed-base aircraft simulators, visual motion is often displayed without the vestibular and other inertial cues (e.g., tactual and somatosensory stimuli) that usually accompany it. Likewise, many VEs allow observers to manipulate a virtual object with their virtual hands, but provide no tactual or force feedback about these "contacts."

The type and severity of the effects of an intersensory conflict depend, in part, on which sensory modalities are involved. For example, conflicts between seen and felt position of the limb will produce reaching errors, particularly for very rapid or "ballistic" responses or when visual feedback is precluded (e.g., Welch, 1978, 1986). (Note: The virtual hand of the user will typically fail to look much like his or her own limb. It is not obvious, however, if this visual anomaly will cause any problems for observers except perhaps initial surprise and minor distraction.) On the other hand, the conflicts between visual and vestibular–inertial cues that characterize many flight simulators are likely to cause disorientation, postural instability (ataxia), and motion sickness–like symptoms (e.g., Oman, 1991).

Conflicts between spatial vision and vestibular sensations during movement of the observer's head or entire body are particularly problematic, both because of their unpleasant behavioral and gastrointestinal consequences and the fact that, at least at present, they are very difficult or impossible to avoid. Even very sophisticated VE technology may be incapable of preventing serious intersensory conflicts of this sort. For example, it has been argued (e.g., DiZio & Lackner, 1992) that the ability of VE technology to simulate inertial motion may never improve to the point where some adaptation will not be necessary. Thus, motion-based aircraft simulators may always be afflicted by significant discrepancies between the ranges and amplitudes of the applied inertial forces and the visual motions with which they are paired. As examples, the take-off of a fighter plane from an aircraft carrier or a violently maneuvering airplane will almost certainly produce substantial conflicts between the two senses because the appropriate gravitational forces exceed the capacity of the simulator. The fact that motion-based aircraft simulators tend to be more nauseogenic than their fixed-based counterparts (e.g., Gower, Lilienthal, Kennedy, & Fowlkes, 1987) may be due to the presence of such intersensory conflicts. Further limitations in attempting to simulate gravitational-inertial forces derive from the fact that sustained hypogravity (i.e., $G < 1.0$) cannot be simulated by earthbound VEs, whereas hypergravity, whether brief or prolonged, requires a human centrifuge (e.g., Cohen, Crosbie, & Blackburn, 1973), a device unavailable or unaffordable to most investigators.

A common instance of intersensory spatial conflict in a VE is when a normally correlated modality is simply not there. Here, too, the behavioral and physical effects are diverse. The absence of tactual or force feedback when attempting to grasp a virtual object, although unlikely to disturb gross reaching behavior, may interfere with fine manual control. On the other hand, when visual motion occurs without the normally attendant vestibular and other inertial cues, as in fixed-base aircraft simulators, the operator may experience a dramatic sense of virtual body motion known as vection. This illusion is nauseogenic for some individuals, as is the "freezing" of an aircraft simulator (or other VE) in an unusual orientation which can occur when the user has engaged in an extraordinary maneuver (e.g., McCauley & Sharkey, 1992).

## 4.2 Depth and Distance Distortions

One form of static depth distortion occurs in some VEs when visual objects that are being depicted as closer to the observer's viewpoint fail to occlude supposedly more distant objects. Another example is when visual objects whose edges are fuzzy due to poor resolution and which, on the basis of "aerial perspective," cues should thus appear far away, are instead presented with the saturated hue and optical sharpness of a much nearer object.

Moving the head or entire body while viewing a VE through a stereoscopic HMD frequently causes virtual objects to appear to change their relative positions (as well as their sizes and shapes) in the distance dimension. This form of dynamic depth distortion is caused because the VE system is unable to calculate correctly and instantly the images presented to the HMD display during the bodily movement. The same is true if the HMD wearer is stationary and the objects are moved around him or her. This movement-contingent depth distortion, referred to by perceptual psychologists as the "kinetic depth effect," is likely to disrupt the user's visual–motor interactions (e.g., pointing or reaching) with respect to these objects and may cause simulator sickness. Fortunately, this form of sensory rearrangement is subject to a certain amount of visual adaptation (e.g., Wallach & Karsh, 1963).

Besides such distortions of depth (i.e., relative distance) perception, the apparent absolute distance of objects may be under- or overestimated, as demonstrated, for example, by errors in visually "open-loop" pointing responses (i.e., hand–eye coordination in the absence of visual feedback). Furthermore, in situations that entail very few depth cues distance perception may be ambiguous, which is likely to result in substantial moment-to-moment variability in hand–eye coordination.

A common discrepancy observed with stereoscopic VE displays occurs between ocular vergence and the accommodation of the lens. Under everyday circumstances, fixating an object in the distance dimension entails convergence of the eyes to produce a single image, together with the appropriate shaping of the lens for fine focusing. However, in many stereoscopic VEs the convergence and accommodation appropriate for the visual display are placed in conflict. Thus, the stereoscopic stimuli may be set for a distance that differs from the one for which optimal focusing will occur. Besides leading to reaching errors, this discordance may cause diplopia, asthenopia, and even nausea (e.g., Ebenholtz, 1992). Unfortunately, it may be difficult or even impossible to overcome this problem by means of adaptation since, as Wann and Mon-Williams (1997) have argued, many current stereoscopic VE displays present the operator with a *range* of binocular targets to which they may verge. Clearly, such a situation fails to provide the consistent "rule" required for adaptation to take place.

Roscoe (1993) has demonstrated that pilots who are wearing see-through HUDs and viewing collimated virtual images will not focus their eyes on infinity, but instead toward their resting accommodation (approximately 1 meter away). The results of this misaccommodation are that: (1) the actual visual scene beyond the HUD appears minified; (2) a terrain or airport runway appears to be observed from a higher altitude than is actually the case; and (3) familiar objects look farther away than they really are. The behavioral consequences of these misperceptions are obvious and, according to Roscoe (1993), potentially *lethal*. Unfortunately, magnifying the images in an attempt to compensate for these illusions is not feasible because of the limitation this hardware modification places on display size (Roscoe, 1993). Fortunately, humans are capable of a certain amount of adaptation to such distance and size distortions. Interestingly, the evidence for this claim comes almost entirely from studies of adaptation to underwater distortions caused by the glass–water interface of the diver's mask (e.g., Ross, 1971).

### 4.3 Form and Size Distortions

An artifact of many VE systems is the illusory curvature of contours, most notably the so-called pincushion effect, in which the sides of a rectilinear shape (especially one that occupies most or all of the FOV) appear to be bowed inwardly. Magnification or minification of the visual scene or of objects within it can also occur with some VEs, especially those that disrupt perceived distance. Thus, in the example of the see-through HUD previously described (Roscoe, 1993), overestimation of the distance of familiar objects caused them to appear too large, presumably based on the mechanism of "misapplied size constancy scaling" often used to explain the "moon illusion" (e.g., Rock & Kaufman, 1962). The few published studies on optically induced size distortions have revealed only a modest amount of visual adaptation (Welch, 1978, chap. 8). This same resistance to adaptation applies to prismatically induced curvature, as seen most convincingly in an experiment by Hay & Pick (1966), in which subjects who wore prism goggles for an heroic 7 consecutive weeks underwent adaptation of only about 30% of the optical curvature to which they were exposed.

On the other hand, when head movements are made in the presence of a magnified or minified visual field, the resulting oculomotor disruption (and loss of visual position constancy) caused by the increased or reduced gain of the vestibular-ocular reflex (VOR), while likely to be nauseogenic, can apparently be completely overcome by adaptation (e.g., Collewijn, Martins, & Steinman, 1983). Even the drastic loss of visual position constancy due to head movements while wearing right–left reversing goggles can apparently be abolished (e.g., Stratton, 1897), although it is interesting to note that the resulting loss of VOR under these circumstances resists recovery (e.g., Gonshor & Mevill Jones, 1976).

## 4.4 Delays of Sensory Feedback

Perhaps the most serious and currently intractable flaw of many VEs is the presence of significant delays (lags) between the operator's movements and resulting visual, auditory, and/or somatosensory feedback. Such delays can occur because of insufficient refresh rates (which are exacerbated by the use of highly detailed graphics) and relatively slow position trackers. Studies of flight simulation have revealed that lags of as little as 50 msec can have a measurable effect on performance, whereas longer delays cause serious behavioral oscillations (e.g., Wickens, 1986).

An especially disruptive and unpleasant effect of visual delay is the interference of the VOR and concomitant illusory visual motion that occurs during head or entire body movements. Normally, a head movement in one direction causes an approximately equal eye movement in the opposite direction, effectively nulling motion (both real and perceived) of the visual field relative to the head. However, visual feedback delay causes the visual field to lag behind the head movement, initially making the VOR inappropriate and causing the visual world to appear to move in the same direction as the head. This loss of visual position constancy (or oscillopsia) is both behaviorally disruptive and nauseogenic. Furthermore, it has been suggested that these delays can cause simulator sickness and other physical problems even if they are below the observer's sensory threshold (Ebenholtz, 1992).

Visual feedback delays from motor movements that do not involve vestibular stimulation, such as reaching or pointing, rarely cause motion sickness symptoms. However, they can lead to other problems ranging from simple distraction to serious discoordination. Auditory feedback delays also cause difficulties, particularly with speech (e.g., Smith & Smith, 1962, chapter 13). Even more problematic perhaps is the presence of *variable* delays and asynchronies between the onsets of two (or more) sensory systems, as, for example, with some motion-based aircraft simulators involving mismatches between the onset times of visual and inertial stimuli. Kennedy (personal communication, May 15, 1994) believes that these problems may be an especially important factor in the etiology of simulator sickness.

## 4.5 Sensory "Disarrangement"

When the size and/or direction of a VE-induced sensory rearrangement vary from moment to moment (rather than remaining constant over an extended time period), a condition of sensory "disarrangement" is said to exist (e.g., Cohen & Held, 1960). Examples with respect to VEs include: (1) so-called "jitter" (jiggling of the visual image due to electronic noise in the position tracker and/or the image-generator system) and (2) moment-to-moment variability in the absolute and relative accuracy of position trackers in monitoring the operator's limbs and body transport. These limitations have been discussed by Meyer, Applewhite, and Biocca (1992). Such noise can cause the seen and felt positions of the hand to shift randomly relative to each other. Not surprisingly, unpredictably changing sensory conditions such as these resist adaptation since there is no constant compensatory "rule" on which such an adaptive process can be based . However, as discussed in a later section, although no change in *average* perceptual or perceptual–motor performance is likely to take place in the presence of a disarrangement, an increase in moment-to-moment behavioral variability (e.g., Cohen & Held, 1960) will occur, which represents a potentially serious degradation of the user's performance.

It is apparent that many of the sensory distortions found in current VEs are similar or identical to those that have been deliberately created by investigators by means of lenses, prisms, and mirrors (see Welch, 1978, 1986, for an extensive review of this literature). The sensory rearrangements examined include optically induced (1) lateral rotation (yaw) of the visual field (e.g., Held & Hein, 1958); (2) visual tilt (roll, e.g., Ebenholtz, 1969); (3) displacement in the distance dimension (e.g., Held & Schlank, 1959); (4) curvature (e.g., Hay & Pick, 1966); (5) right–left reversal (e.g., Kohler, 1964); (6) inversion (e.g., Stratton, 1897); (7) depth and distance distortion (e.g., Wallach, Moore, & Davidson, 1963); and (8) altered visual size (e.g., Rock, 1965). Acoustical rearrangements have entailed: (1) small (e.g., 10-deg) lateral rotations of the auditory field (e.g., Held, 1955); (2) right–left reversal (e.g., Young, 1928); and (3) functional increases in the length of the interaural axis (e.g., Durlach & Pang, 1986). Another atypical sensory environment to which humans have been exposed and adapted is altered gravitational-inertial force—hypergravity (e.g., Welch, Cohen, & DeRoshia, 1996), hypogravity (e.g., Lackner & Graybiel, 1983), and alternating hyper- and hypogravity (e.g., Cohen, 1992. Finally, underwater distortions caused by the air–glass interface of the diver's mask have been the subject of extensive investigation (e.g., Ross, 1971). On the basis of this vast literature on adaptation to sensory rearrangement, one conclusion is certain: Human beings (and a number of other mammalian species) are able to adjust their behavior and, to a lesser extent, their perceptions to any sensory rearrangement to which they are actively exposed, as long as this rearrangement remains essentially constant over the time interval in which systematic feedback is provided. It is important to note that adaptation is not a unitary process, but rather varies greatly in both acquisition rate and magnitude as a function of type of sensory rearrangement and the specific adaptive component (visual, proprioceptive, vestibular-ocular, etc.) under consideration. Furthermore, as mentioned previously, while human subjects are capable of adapting their *behavior* nearly completely to even the most dramatic visual rearrangements, such as 180-degree rotation of the visual field, adaptive changes in *visual perception* often fail to occur (e.g., Snyder & Pronko, 1952). On the other hand, a certain amount of visual adaptation does result from exposure to lesser distortions, such as prismatic displacement (e.g., Craske & Crawshaw, 1978) and curvature of the visual field (e.g., Hay & Pick, 1966).

Since the technological limitations of many VEs create many of the same sensory rearrangements imposed in laboratory experiments examining adaptation, it is reasonable to conclude that procedures that incorporate the variables that have been shown from these experiments to control or facilitate adaptation will be especially useful for the systematic training of VE users. Expediting this adaptive process should, in turn, facilitate task performance and transfer of training to the real-world task (if any) for which the VE is designed. Let us now look at the variables that have been demonstrated to influence or control adaptation to sensory rearrangement and consider how they might be integrated into VE adaptation/training procedures.

## 5. CONTROLLING AND FACILITATING VARIABLES FOR ADAPTING TO SENSORY REARRANGEMENTS AND THEIR POTENTIAL APPLICABILITY TO VE TRAINING PROCEDURES

### 5.1 A Stable Rearrangement

As indicated previously, exposure to a sensory rearrangement that is continuously changing in magnitude and/or direction will fail to produce adaptation. This condition is referred to as *disarrangement* and has been shown to have a degrading effect on hand–eye coordination, as seen by an increase in moment-to-moment variability of open-loop target pointing. This conclusion is illustrated by an experiment by Cohen and Held (1960) in which subjects viewed their actively moving hands through prisms whose strength varied continuously from 22 degrees leftward through no displacement to 22 degrees rightward and the reverse, at a rate of one cycle every 2 minutes. Although this experience had no effect on average target-pointing accuracy along the lateral dimension, (i.e., no adaptive shift), it greatly increased the variability of subjects' open-loop target-pointing behavior over repeated attempts. An analogous result has been reported for auditory disarrangement (e.g., Freedman & Pfaff, 1962) as measured by auditory localization errors in the right-left dimension. Another example of disarrangement is the situation in which delays of visual feedback from bodily movements vary from moment to moment, as may occur in some VEs. Visual delays are difficult enough to handle when they are constant over time, but represent a much more serious problem when they are inconsistent or variably asynchronous (e.g., Uliano, Kennedy, & Lambert, 1986).

In summary, sensory disarrangement represents a distortion whose effects cannot be ameliorated by adaptation-training procedures because the basic premise upon which adaptation is based (i.e., the presence of an unchanging sensory rearrangement) does not hold. It would appear, therefore, that VEs that suffer from this problem must await engineering solutions such as more stable position trackers.

### 5.2 Active Interaction

Given the presence of a stable sensory rearrangement, one can single out five major variables that control or facilitate adaptation. These are (1) active interaction, (2) error-corrective feedback, (3) immediate feedback, (4) incremental exposure to the rearrangement, and (5) the use of distributed practice.

It is generally agreed that the most powerful of all the controlling conditions for adaptation to sensory rearrangement is active interaction with the altered sensory environment, coupled with the visual consequences of these actions (*reafference*). In short, it has been demonstrated that passive exposure to a sensory rearrangement produces little or no adaptation (e.g., Held & Hein, 1958). Exactly why passive exposure is so ineffectual is, however, subject to serious debate (e.g., Welch, 1978, pp. 28–29). One possibility is that active interaction provides the observer with unmistakable information about the sensory rearrangement, which, in turn, initiates and/or catalyzes the adaptive process. For example, because felt limb position is more precise when bodily movement is self-initiated than when controlled by an external force (e.g., Paillard & Brouchon, 1968), the discrepancy between seen and felt limb position can be assumed to be particularly pronounced during active movement, thereby facilitating the adaptive process.

One caveat to the preceding endorsement of active visual–motor behavior as a facilitator of adaptation is that it is necessary that the motor intentions (*efference*) generating these bodily movements actually be in conflict with the resulting reafference. For example, when wearing prism goggles, subjects' active hand or head movements initially lead to visual consequences (e.g., errors in localizing a visual target) that conflict with those that usually occur. However, their eye movements will not be in error. Thus, although subjects who are wearing prism goggles and instructed to rapidly turn and face a visually perceived object will err in the direction of the prismatic displacement, they will have no difficulty turning their eyes to fixate the perceived locus of the object. The reason that the latter is true is that the prism goggles do not alter the relationship between retinal locus of stimulation and the appropriate fixating eye movements. On the other hand, if the prisms, instead of being attached to goggles, are affixed to contact lenses or are otherwise controlled by eye movements (e.g., White, Shuman, Krantz, Woods, & Kuntz, 1990), the situation is quite different. With this optical arrangement, not only will head movements be in error, but eye movements as well. The results of the few relevant studies using this unusual arrangement (e.g., Festinger, Burnham, Ono, & Bamber, 1967) suggest that observers who are so accoutered are not only able to acquire correct eye movements but also undergo much more visual adaptation to prismatically induced curvature of contours than is the case with traditional prism goggles. Perhaps adaptation to other discrepancies such as altered size and displacement would also be enhanced by such means. Therefore, for VEs that are designed to be controlled by the operator's eyes, it is possible that any visual discordance that may be present will undergo stronger adaptation than that found with VEs that are controlled by head movements (e.g., those involving an HMD).

It is clear from the preceding discussion that to produce substantial adaptation, VE users must be given the opportunity to actively interact with the sensory environment being displayed. This can be a two-edged sword, however, because before active movement can have this beneficial effect on adaptation it may first cause some users to get sick. For example, VEs that create a sense of gravitoinertial force by means of passive transport in a centrifuge or visual flow field are very likely to induce motion sickness symptoms as soon as the user engages in active bodily movements, particularly of the head (e.g., DiZio & Lackner, 1992). In such situations, those participants who cannot tolerate such unpleasant effects may exit the VE before much adaptation has had a chance to occur.

### 5.3 Error-corrective Feedback

The combination of making errors and then correcting for them when reaching for objects in a rearranged visual environment facilitates adaptation beyond the effect attributable to visual–motor activity alone. For example, Welch (1969) showed that when subjects were allowed to make error-corrective (visually closed-loop) target-pointing responses, they adapted significantly more than when they actively moved the visible hand in the same manner but without visual targets. Although the facilitating effect of error-corrective feedback on adaptation is now well established, it is not sufficient merely that targets are available in the environment. Rather to be most effective, target-pointing responses must be made either ballistically (i.e., so rapidly that their trajectories cannot be altered enroute to the target) or in such a manner that the outcome of a given attempt is not observable until it has reached its goal (often referred to as *terminal exposure*). The latter condition was used in the Welch (1969) experiment. If instead observers are allowed to make slow, visually guided reaching movements (*concurrent exposure*), they will almost certainly "zero in" on the target on each attempt and thus experience little or no error when the hand finally reaches its goal. The fact that concurrent exposure is thus less informative than terminal exposure is almost certainly the reason why adaptation is much greater in the latter condition (e.g., Welch, 1978, pp. 29–31). In sum, it can be concluded that VE exposure should, where possible, entail numerous targets with which to actively interact, together with unambiguous error-corrective feedback from these interactions, preferably at the termination of each response.

## 5.4 Immediate Sensory Feedback

Delays of motor-sensory feedback represent one of the most serious limitations of current virtual environments. Even with quite short delays, severe behavioral disruption occurs and is not amenable to adaptation as defined here (e.g., Smith & Smith, 1962). Such a state of affairs would seem to argue against the utility of adaptation-training procedures as a means of eliminating this problem. Fortunately, as will be seen below, there are some situations in which the effects of delayed feedback in VEs are not quite so dire.

The initial effects of visual feedback delays are quite different when the vestibular system and other sources of inertial information are involved than when they are not. An example of the first of these situations is when an HMD-wearing subject views the visual field while turning the head whose movements are being monitored by a relatively slow tracking device. Thus, in this situation the opposite sweep of the visual field begins slightly after head movement has begun and continues briefly after it is completed. An example of the second is rapidly moving the hand and seeing its virtual image lagging behind its true position because the refresh rate of the VE cannot keep pace. While the former condition is characterized by visual–spatial instability and possibly symptoms of motion sickness, the latter seriously interferes with hand–eye coordination but does not appear to be nauseogenic (e.g., Held, Efstathiou, & Greene, 1966).

A further complication involves the situation in which a sensory feedback delay is superimposed on a second intersensory rearrangement, such as a discrepancy between felt and seen limb position. In this case the ability to adapt to the latter condition may be severely impeded. For example, Held et al. (1966) showed that visual–motor adaptation to prismatic displacement is greatly reduced or even prevented by delays of visual feedback as short as 120 msec. Because many current VEs are subject to much greater delays than this, it might seem, as Held and Durlach (1993) have argued, that users will be unable to use adaptation to overcome any spatial (or other) discordance that may also be present. Fortunately, however, this pessimistic conclusion appears to be limited to the type of exposure condition typically preferred by Held and his colleagues in which subjects do not reach for specific targets, but merely view their hands as they move them from side to side before a visually homogeneous background. In contrast, it has been demonstrated that when subjects are provided with unambiguous error-corrective feedback from discrete target-pointing responses, substantial adaptation is possible despite the presence of significant visual feedback delays. For example, in an experiment by Rhoades (described by Welch, 1978, p. 105), subjects revealed significant hand–eye adaptation to prismatic displacement even when error-corrective feedback was delayed by as much as 8 sec. Furthermore, considerable adaptation has been reported in a number of published studies (e.g., Welch, 1972) in which for procedural reasons the investigators were forced to institute visual error-corrective feedback delays of up to approximately 1 second.

What, then, can we conclude about the role of sensory feedback delays for VE adaptation? First, visual feedback delays of a second or more seriously disrupt hand–eye coordination, a problem that VE users may, at best, only be able to circumvent by the use of conscious strategies such as deliberately moving very slowly (e.g., Smith & Smith, 1962). Second, if the feedback delay is superimposed on another discordance, such as a discrepancy between seen and felt limb position, and the observer's task does not entail target-pointing responses, not only will visual-motor performance be disturbed but adaptation to the second discordance is likely to be prevented, even for very short visual delays. Finally, if observers are allowed to engage in discrete perceptual–motor responses accompanied by error-corrective feedback, at least some adaptation to the discordance will occur even with feedback delays as long as several seconds.

### 5.5 Incremental Exposure

Another important variable for adapting to sensory rearrangement, and presumably to VEs, is the provision of incremental (rather than "all-at-once") exposure. It has been shown that if exposure to prismatic displacement (e.g., Lackner & Lobovits, 1978), optical tilt (e.g., Ebenholtz & Mayer, 1968), and slow-rotating rooms (e.g., Graybiel & Wood, 1969) entails gradual increases in the strength of the discordance, adaptation and/or the elimination of motion sickness symptoms is much greater than if the observer is forced to confront the full-blown sensory rearrangement from the start.

There are several ways in which this variable could be applied to VE adaptation. First, it has been advocated (e.g., Kennedy et al., 1987, p. 48) that aircraft simulator operators begin their training with short "hops" before attempting longer ones. Second, since intense, rapid actions are likely to cause greater delays of visual feedback than mild, slow ones, training should probably begin with the latter. Thus, perhaps by working up slowly to the longer delays, VE users will be better able to acquire the deliberate strategies (e.g., "wait and move") needed to make such delays more manageable. A third potentially useful form of incremental training would be to gradually increase the size of the FOV for HMD devices, rather than beginning with their maximum size. That is, because it is known that VEs with large FOVs produce better task performance than those with small FOVs but are also more nauseogenic (e.g., Pausch, Crea, & Conway, 1992), gradually increasing the size of the FOV may serve to reduce the simulator sickness symptoms users would otherwise experience and simultaneously improve their performance.

### 5.6 Distributed Practice

It has been demonstrated that providing periodic rest breaks facilitates adaptation to prismatic displacement (e.g., Cohen, 1974) and/or results in greater retention of prism adaptation (Dewar, 1970). Presumably, then, the same would be true for adaptation to VEs. Indeed, Kennedy, Lane, Berbaum, and Lilienthal (1993) determined that the ideal inter-"hop" interval for "inoculating" users against simulator sickness with respect to the aircraft simulators they examined was 2 to 5 days; shorter or longer intervals resulted in less adaptation. Thus, it appears that VE training should entail some sort of distributed practice regimen, although the ideal profile of "on" and "off" periods will almost certainly vary from one VE to another.

## 6. THE OPTIMAL ADAPTATION-TRAINING PROCEDURE AND HOW TO COUNTERACT THE DRAWBACKS OF MAXIMAL ADAPTATION

### 6.1 The Optimal Procedure

Awareness of the critical roles the preceding variables play in perceptual and perceptual–motor adaptation to sensory rearrangement provides a critical basis for the design and implementation of effective VE adaptation-training procedures. Thus, assuming a VE display whose discordance is constant rather than transient or variable, the ideal procedure would appear to entail: (1) providing operators with many opportunities to interact actively with the displayed environment; (2) providing error-corrective feedback and other salient sensory information about the discordance; (3) avoiding, when possible, significant delays of feedback, especially for conditions of continuous exposure that include no error-corrective feedback; (4) gradually incrementing the length of exposure and/or the magnitude of the discordance; and (5) using a practice regimen that entails frequent rest breaks distributed in an optimal manner.

## 6.2 Simulations of Naturally Rearranged Sensory Environments

Some VEs are designed (or could be designed) for the express purpose of simulating real-life sensory environments that are by their nature rearranged and to which individuals must adapt if they are to perform appropriately. In other words, for these VEs the presence of a particular intersensory discordance is deliberate rather than an unintended consequence of inadequate technology. An example is a VE designed to simulate the visual–inertial conflicts of hypogravity caused by the absence of the otolithic cues for gravity and used as a means of assisting astronaut trainees to adapt to this environment before actually entering it. This procedure, sometimes referred to as "preflight adaptation training" (PAT), has shown promising results (e.g., Harm & Parker, 1994 and see chap. 32, this volume). Presumably, PAT procedures could be used to generate adaptive immunity to other disruptive sensory environments such as rocking ocean vessels or the underwater visual world as viewed through a diving mask (e.g., Welch, 1978, chap. 12). Another example of a VE that has been deliberately designed to include a sensory rearrangement is that of Viirre, Draper, Gailey, Miller, and Furness (1998). These investigators proposed rehabilitating people who are suffering from chronically low VOR gains by using a VE to expose and adapt them to gradually incremented gain demands. Obviously, all of the facilitating procedures and conditions for adaptation that have been proposed in this chapter should prove useful in helping users adapt to this and any of the other deliberately imposed VE sensory rearrangements.

Alternatively, a VE might be designed to exaggerate an intersensory discordance on the belief that such a "super" discordance will serve a useful purpose. A good example is the magnification of the interaural axes as a potential means of enhancing the localizability of auditory objects (e.g., Durlach & Pang, 1986). It is important to note that with this arrangement one must hope that auditory adaptation does *not* occur because this would mean that observers' capacity to localize auditory objects had gone from supernormal back to normal, contrary to the aim of the device. On the other hand, it would probably be advantageous if the observers' *behaviors* (e.g., hand–ear coordination) did recover, presumably by means of adaptation. Otherwise, they might mis-reach for auditory objects, assuming that these responses were open-loop or ballistic. Whether it is possible in this situation to have things both ways remains to be seen.

## 6.3 Potential Drawbacks of Maximal Adaptation

### *6.3.1 Negative Aftereffects and Negative Transfer*

Implementation of the optimal training procedures prescribed by the principles of adaptation to sensory rearrangement should maximize adaptation to VEs and thus minimize the behavioral difficulties frequently experienced while using these devices. Naturally, in the case of VEs that serve as training devices (e.g., aircraft simulators) it is presumed, or at least hoped, that such adaptation training will also result in improved subsequent performance of the real-world task for which these devices have been designed. However, assuming that in real life this task does not include the intersensory discordances that are experienced in the VE, it is possible to imagine circumstances in which adaptation will actually make things *worse* for the user. Thus, although adaptation to a VE as evidenced by improved behavior and the reduction or elimination of simulator sickness indicates that users have become more comfortable with the device, it is still possible for this adaptation to cause them to be less capable of controlling the real-life devices they are being trained to use due to the presence of negative aftereffects and/or negative transfer. Obviously, only if the positive effects of training (i.e., generalization to the real-world task) are found to outweigh these potential deleterious consequences should one advocate and use the adaptation training regimens described here. These are, of course, empirical issues whose answers are likely to vary from one VE to another.

The unavoidable price for maximum adaptation is maximum aftereffects. (Note: It is important to distinguish between the aftereffects of exposure to a VE [or to any sensory rearrangement] and a mere perseveration of effects. Only if a particular effect [motion-sickness symptom, hand–eye miscoordination, etc.] has disappeared by the end of the VE exposure period (or perhaps never occurred in the first place), and then reappears [presumably in opposite form] at some time in the postexposure period, is it correct to say that an aftereffect has been observed. It is not accurate, for example, to refer to postexposure malaise as an aftereffect if the subject was experiencing such malaise just before leaving the VE. The reason why this distinction is important is the assumption, held here, that the presence of a true aftereffect is clear evidence that adaptation had occurred to the VE during the exposure period.) The potential danger of delayed "flashbacks" has already been noted. Furthermore, Biocca and Rolland (1998) raise the plausible example of a surgeon who, after using and adapting to a see-through HUD while operating on a human patient, removes the device to continue the operation and makes a serious mistake due to the negative aftereffect. Is there a way to retain the advantages of maximal VE adaptation while simultaneously avoiding its disruptive and potentially dangerous aftermath? Two solutions for ameliorating postexposure aftereffects may be proposed: (1) institute "readaptation" procedures immediately after the user has exited the VE; and (2) create "dual" (or "contingent") adaptation.

### *6.3.2 Eliminating the Aftereffects*

*6.3.2.1 "Readaptation" Procedures.* It is reasonable to assume that the unlearning of adaptation (i.e., readaptation to the normal sensory world) is controlled and facilitated by the same variables that operate during adaptation to the rearranged world. Given this premise, it follows that the optimal means of abolishing postexposure aftereffects is to require users to engage in the same activities after exiting the VE (i.e., in the absence of the sensory rearrangement) as they did when immersed in it. Studies of adaptation to the "traditional" forms of sensory rearrangement (e.g., prismatic displacement) have shown that postexposure aftereffects dissipate much more rapidly if subjects are allowed to interact actively with the nonrearranged visual world and receive visual feedback from this activity than if they simply sit immobile in the dark or even in a lighted environment (e.g., Welch, 1978, chap. 4). Consider a simulator designed to train people to pilot an oil tanker and which therefore deliberately duplicates the large delay between turning the wheel and the response of the simulated ship that exists for real oil tankers. Clearly, it would be ill-advised for well-trained users after leaving such a simulator to immediately operate an automobile or other vehicle. Rather, one would want to be certain that their presumed training aftereffects had been completely and permanently abolished before allowing them to leave the premises or, if not, to prohibit them from driving vehicles for a prescribed period of time. Reliance on the latter course of action appears to be based on the assumption (or hope) that, as a simple function of time away from the simulator, the aftereffects of training will disappear, either from "decay" or "unlearning/relearning" from random and unspecified perceptual–motor activities (or both). In contrast, the present proposal would have operators again pilot the virtual ship immediately after the simulator training session, but this time with the device arranged such that there is no delay between the user's actions and their visual consequences. Of course, for VEs with intersensory conflicts that cannot be eliminated, the preceding procedure is inappropriate. In the latter case, if subjects can at least be presented with a real situation that is similar to the one provided by the VE and then required to make the same visual–motor actions (e.g., hand–eye, head–eye) in which they engaged during VE exposure, readaptation should be accelerated.

It is unclear what causes delayed "flashbacks" from aircraft simulator training sessions. However, it is reasonable to assume that these triggers come about because the individual happens to encounter sensory cues in the environment (e.g., the feel of the automobile steering wheel, the lines on the highway) that are similar or identical to those with which certain perceptions were associated during the simulator training session. The great advantage of the unlearning/relearning procedures suggested here is that they are virtually guaranteed to decondition all such cues, whether or not we can identify them. It can be argued, therefore, that this proposed tactic represents a significant improvement over the current strategy for dealing with delayed "flashbacks" in which, pilots are grounded for 12 to 24 hours after an aircraft simulator training session.

*6.3.2.2 Creating "Dual" (or "Context-specific") Adaptation.* Adaptation, as evidenced either by its reduction of effects during exposure or its postexposure aftereffects, does not start from scratch each time one makes the transition between one sensory environment and another. Rather, both anecdotal and experimental observations indicate that frequent alternations between adapting to a rearranged sensory environment and readapting to the normal environment lead to decreased interference (e.g., aftereffects) at the point of changeover between the two situations and/or the more rapid reacquisition of the appropriate perceptions and behavior. This phenomenon of adapting separately to two (and perhaps more) mutually conflicting sensory environments has been referred to as *dual adaptation* (e.g., Welch, Bridgeman, Anand, & Browman, 1993) or *context-specific adaptation* (e.g., Shelhamer, Robinson, & Tan, 1992). An everyday example is adjusting to new prescription lenses in which, after repeatedly donning and doffing their spectacles, wearers report that the depth distortions, illusory visual motion, and behavioral difficulties they had experienced at the outset have now largely disappeared. Thus, as the result of this alternating experience, the presence or absence of the tactual sensations of the spectacles has become the discriminative cue for turning adaptation on or off.

Empirical evidence of dual or context-specific adaptation comes from a variety of sources. For example, it has been observed that those fortunate astronauts and cosmonauts who have gone into space and returned to Earth several times (thus repeatedly adapting to weightlessness and readapting to Earth's gravity) experience progressively less initial perceptual and perceptual interference when they first enter microgravity (or reenter 1 G) and/or are able to regain normal neurovestibular functioning more rapidly (e.g., Bloomberg, Peters, Smith, Heubner, & Reschke, 1997). Laboratory evidence of dual adaptation for both optically and gravitationally rearranged environments has been reported by, among others, Bingham, Muchisky, and Romack (1991), Cunningham and Welch (1994), Flook and McGonigle, 1977), McGonigle and Flook (1978), and Welch, Bridgeman, Williams, and Semmler (1998). It is important to understand that individuals who have acquired a dual adaptation do not actually remain adapted during the period between exposures to a given rearrangement, but rather maintain a *readiness* to adapt (or readapt). Thus, this proclivity (or "immunity") becomes manifest only in the presence of the sensory cues that signal entry into or departure from the sensory environment in question.

The importance of dual adaptation for adaptation to VEs is this: If users are systematically alternated between adapting to a VE and readapting to the normal environment (or to another VE), they should eventually be able to shift from one sensory environment to the other with little or no interference of perception, performance, or physical well-being. It is conceivable that, by means of such alternation training, operators can eventually acquire the ability to interact with a given VE without serious initial difficulty and then return to the normal world with little or no aftereffects. Presumably, the ideal dual adaptation-training regimen would use as its means of eliminating aftereffects the unlearning/relearning procedures advocated in the preceding section. Finally, the brain's "decision" about the form of adaptation to invoke in a given situation clearly requires the presence of one or more discriminative cues that reliably differentiate the two conditions. Therefore, it will be important for investigators to deliberately provide such cues for the observer and to make them both redundant and salient. Evidence that may outwardly appear to contradict the dual adaptation hypothesis has been reported by Gower et al., 1987, whose subjects revealed increasing, rather than the expected decreasing, postural aftereffects (ataxia) from repeated exposures to an aircraft simulator. Before accepting these data as evidence against the notion of dual adaptation, however, it should be determined if, despite the systematic increase in postexposure ataxia, the rate of readaptation to the normal environment changed. An increase in the latter from one simulator exposure to the next would support the dual adaptation hypothesis. Alternatively, the report by Kennedy and Stanney (1986) of progressively increasing aftereffects might signify that subjects' adaptation was accumulating from one adaptation session to the next, thereby causing the aftereffects to increase as well. According to the dual adaptation hypothesis, however, once adaptation has finally reached asymptote such that further exposures provide little or no increment, the size of the postexposure aftereffects should begin to decline on subsequent exposures, and the rate of readaptation should increase.

## 7. THE ROLE OF INDIVIDUAL DIFFERENCES

As with all measures of human perception and performance, the acquisition rate and magnitude of adaptation to sensory rearrangement vary widely from person to person (e.g., Welch, 1978, chap. 11). Therefore, it is reasonable to assume that VE users will differ reliably from one another with respect to such things as (1) the detectability of a given discordance; (2) the degree of interference the discordance causes them, and (3) how adaptable they are to it. Little is known about the causes or correlates of individual traits of adaptability.

Clearly, however, the existence of such individual differences in the response to current VEs means that some users will require more adaptation training than others to attain a given level of adaptation. Indeed, it is possible that those individuals who are particularly prone to the deleterious behavioral and perceptual effects of VEs will be the most benefited from adaptation training procedures. Furthermore, because of such individual differences it seems likely that, even after the major problems of current VEs have been overcome by engineering advances, at least some users will continue to register and be reactive to the small sensory and sensorimotor rearrangements that will almost certainly remain. Thus, adaptation training procedures of the sort advocated here will continue to have a place in VE use. However, even when all that is left are these idiosyncrasies, the likely response of the human-factors engineer would be to create a device that modified itself to conform to them, rather than employing the adaptive fine tuning suggested here. A potential problem with this solution, however, is that it entails confronting the user with a changing stimulus condition, which might cause problems of its own (M. Draper, personal communication, May 27, 1998).

## 8. SUMMARY AND CONCLUSIONS

Until the sensory and intersensory rearrangements that characterize many current VEs are corrected by means of improved design and technology, users will continue to experience disruptive perceptual, behavioral, and physiological effects during and after exposure to these devices. One important means of assisting users to deal with these problems is the application of training procedures based on well-established principles of perceptual and perceptual–motor adaptation to sensory rearrangement. The flipside of maximal adaptation is maximal aftereffects. However, the latter can be quickly, completely, and perhaps permanently abolished by means of the same procedures and conditions used to produce adaptation in the first place and by the induction of dual (or contingent) adaptation from repeated experience with the VE in question.

There is one serious caveat to the whole undertaking of training VE users to overcome the limitations of their devices. Namely, because it is possible for adaptation to inhibit transfer of training to the subsequent real-world task for which it is being used, it is important to balance carefully the benefits and potential disadvantages of adaptation for transfer of training. Even when engineering advances ultimately eliminate the major sensory and sensorimotor limitations of current VEs, it is likely that the adaptation training procedures proposed here will continue to be necessary as a means of accommodating the idiosyncrasies of the individual user. Finally, it should go with out saying that many of the arguments and training procedures proposed here are open to question and thus represent topics for empirical investigation with important implications for the VE community.

## 9. REFERENCES

Bingham, G. P., Muchisky, M., & Romack, J. L. (1991, November). *"Adaptation" to displacement prisms is skill acquisition.* Paper presented at the meetings of the Psychonomic Society, San Francisco, CA.

Biocca, F. (1992). Will simulator sickness slow down the diffusion of virtual environment technology? *Presence: Teleoperators and Virtual Environments, 1*, 334–343.

Biocca, F., & Rolland, J. (1998). Virtual eyes can rearrange your body: Adaptation to visual displacement in see-through, head-mounted displays. *Presence: Teleoperators and Virtual Environments, 7*, 262–277.

Bloomberg, J. J., Peters, B. T., Smith, S. L., Heubner, W. P., & Reschke, M. F. (1997). Locomotor head-trunk coordination strategies following space flight. *Journal of Vestibular Research, 7*, 161–177.

Calvert, S. L., & Tan, S. L. (1994). Impact of virtual reality on young adults' physiological arousal and aggressive thoughts: Interaction vs. observation. *Journal of Applied Developmental Psychology, 15*, 125–139.

Cohen, M. M. (1974). Visual feedback, distribution of practice, and intermanual transfer of prism aftereffects. *Perceptual and Motor Skills, 37*, 599–609.

Cohen, M. M. (1992). Perception and action in altered gravity. *Annals of the New York Academy of Sciences, 656*, 354–362.

Cohen, M. M., Crosbie, R. J., & Blackburn, L. H. (1973). Disorienting effects of aircraft catapult launchings. *Aerospace Medicine, 44*, 37–39.

Cohen, M. M., & Held, R. (1960, April). *Degrading visual-motor coordination by exposure to disordered re-afferent stimulation.* Paper presented at the Eastern Psychological Association, New York City.

Collewijn, H., Martins, A. J., & Steinman, R. M. (1983). Compensatory eye movements during active and passive head movements: Fast adaptation to changes in visual magnification. *Journal of Physiology, 340*, 259–286.

Craske, B., & Crawshaw, M. (1978). Spatial discordance is a sufficient condition for oculomotor adaptation to prisms: Eye muscle potentiation need not be a factor. *Perception and Psychophysics, 23*, 75–79.

Cunningham, H. A., & Welch, R. B. (1994). Multiple concurrent visual-motor mappings: Implications for models of adaptation. *Journal of Experimental Psychology: Human Perception and Performance, 20*, 987–999.

Dewar, R. (1970). Adaptation to displaced vision: The influence of distribution of practice on retention. *Perception and Psychophysics, 8*, 33–34.

DiZio, P., & Lackner, J. R. (1992). Spatial orientation, adaptation, and motion sickness in real and virtual environments. *Presence: Teleoperators and Virtual Environments, 3*, 319–328.

DiZio, P., & Lackner, J. R. (1997). Circumventing side effects of immersive virtual environments. In M. Smith, G. Salvendy, & R. Koubek (Eds.), *Design of computing systems: Social and ergonomic considerations* (pp. 893–896). Amsterdam: Elsevier Science Publishers.

Durlach, N. I., & Mavor, A. S. (1995). *Virtual reality: Scientific and technological challenges.* Washington, DC: National Academy Press.

Durlach, N. I., & Pang, X. D. (1986). Interaural magnification. *Journal of the Acoustical Society of America, 80*, 1849–1850.

Ebenholtz, S. M. (1969). Transfer and decay functions in adaptation to optical tilt. *Journal of Experimental Psychology, 81*, 170–173.

Ebenholtz, S. M. (1992). Motion sickness and oculomotor systems in virtual environments. *Presence: Teleoperators and Virtual Environments, 1*, 302–305.

Ebenholtz, S. M., & Mayer, D. (1968). Rate of adaptation under constant and varied optical tilt. *Perceptual and Motor Skills, 26*, 507–509.

Festinger, L., Burnham, C. A., Ono, H., & Bamber, D. (1967). Efference and the conscious experience of perception. *Journal of Experimental Psychology Monograph, 74*(4, Whole No. 637), 1-36

Flook, J. P., & McGonigle, B. O. (1977). Serial adaptation to conflicting prismatic rearrangement effects in monkey and man. *Perception, 6*, 15–29.

Fowlkes, J. E., Kennedy, R. S., Hettinger, L. J., & Harm, D. L. (1993). Changes in the dark focus of accommodation associated with simulator sickness. *Aviation, Space, and Environmental Medicine, 64*, 612–618.

Freedman, S. J., & Pfaff, D. W. (1962). The effect of dichotic noise on auditory localization. *Journal of Auditory Research, 2*, 305–310.

Gonshor, A., & Mevill Jones, G. (1976). Extreme vestibulo-ocular adaptation induced by prolonged optical reversal of vision. *Journal of Physiology* (London), *256*, 381–414.

Gower, D. W., Lilienthal, M. G., Kennedy, R. S., & Fowlkes, J. E. (1987, September). Simulator sickness in U.S. Army and Navy fixed- and rotary-wing flight simulators. In *Proceedings of the AGARD Medical Panel Symposium on Motion Cues in Flight Simulation and Simulator Induced Sickness*. Brussels, Belgium: AGARD.

Graybiel, A., & Wood, C. D. (1969). Rapid vestibular adaptation in a rotating environment by means of controlled head movements. *Aerospace Medicine, 40*, 638–643.

Harm, D. L., & Parker, D. E. (1994). Preflight adaptation training for spatial orientation and space motion sickness. *The Journal of Clinical Pharmacology, 34*, 618–627.

Hay, J. C., & Pick, H. L., Jr. (1966). Visual and proprioceptive adaptation to optical displacement of the visual stimulus. *Journal of Experimental Psychology, 71*, 150–158.

Held, R. (1955). Shifts in binaural localization after prolonged exposure to atypical combinations of stimuli. *American Journal of Psychology, 68*, 526–548.

Held, R., & Durlach, N. (1993). Telepresence, time delay and adaptation. In S. R. Ellis, M. K. Kaiser, and A. J. Grunwald (Eds.), *Pictorial communication in virtual and real environments* (2nd, ed. pp. 232–246). London: Taylor & Francis.

Held, R., Efstathiou, A., & Greene, M. (1966). Adaptation to displaced and delayed visual feedback from the hand. *Journal of Experimental Psychology, 72*, 887–891.

Held, R., & Hein, A. (1958). Adaptation of disarranged hand-eye coordination contingent upon re-afferent stimulation. *Perceptual and Motor Skills, 8*, 87–90.

Held, R., & Schlank, M. (1959). Adaptation to disarranged eye-hand coordination in the distance dimension. *American Journal of Psychology, 72*, 603–605.

Kennedy, R. S., Berbaum, K. S., Dunlap, M. P., Mulligan, B. E., Lilienthal, M. G., & Funaro, J. F. (1987). *Guidelines for alleviation of simulator sickness symptomatology* (NAVTRA SYSCEN TR-87-007). Orlando, FL: Navy Training Systems Center.

Kennedy, R. S., Fowlkes, J. E., & Lilienthal, M. G. (1993). Postural and performance changes following exposures to flight simulators. *Aviation, Space, and Environmental Medicine, 64*, 912–920.

Kennedy, R. S., Lane, N. E., Berbaum, K. S., & Lilienthal, M. G. (1993). Simulator Sickness Questionnaire: An enhanced method for quantifying simulator sickness. *The International Journal of Aviation Psychology, 3*, 203–220.

Kohler, I. (1964). The formation and transformation of the perceptual world. *Psychological Issues, 3*, 1–173.

Lackner, J. R., & Graybiel, A. (1983). Perceived orientation in free-fall depends on visual, postural, and architectural factors. *Aviation, Space, and Environmental Medicine, 54*, 47–51.

Lackner, J. R., & Lobovits, D. (1978). Incremental exposure facilitates adaptation to sensory rearrangement. *Aviation, Space and Environmental Medicine, 49*, 362–264.

McCauley, M. E., & Sharkey, T. J. (1992). Cybersickness: Perception of self-motion in virtual environments. *Presence: Teloperators and Virtual Environments, 1*(3), 311–318.

McGonigle, B. O., & Flook, J. P. (1978). Long-term retention of single and multistate prismatic adaptation by humans. *Nature, 272*, 364–366.

Meyer, K., Applewhite, H., & Biocca, F. (1992). A survey of position trackers. *Presence: Teleoperators and Virtual Environments, 1*, 173–201.

Mon-Williams, M., Rushton, S., & Wann, J. P. (1995). Binocular vision in stereoscopic virtual-reality systems. *Society for Information Display International Symposium Digest of Technical Papers, 25*, 361–363.

Mon-Williams, M., Wann, J. P., & Rushton, S. (1993). Binocular vision in a virtual world: Visual deficits following the wearing of a head-mounted display. *Ophthalmic and Physiological Optics, 13*, 387–391.

Oman, C. M. (1991). Sensory conflict in motion sickness: An Observer Theory approach. In S. R. Ellis (Ed.), *Pictorial communication in virtual and real environments* (pp. 362–376). New York: Taylor & Francis.

Paillard, J., & Brouchon, M. (1968). Active and passive movements in the calibration of position sense. In S. J. Freedman (Ed.), *The neuropsychology of spatially oriented behavior*. Homewood, IL: Dorsey Press.

Pausch, R., Crea, T., & Conway, M. (1992). A literature survey for virtual environments: Military flight simulator visual systems and simulator sickness. *Presence: Teleoperators and Virtual Environments, 1*(3), 344–363.

Rock, I. (1965). Adaptation to a minified image. *Psychonomic Science, 2*, 105–106.

Rock, I., & Kauffman, L. (1962). The moon illusion, II. *Science, 136*, 1023–1031.

Roscoe, S. N. (1993). The eyes prefer real images. In S. R. Ellis, M. K. Kaiser, & A. J. Grunwald (Eds.), *Pictorial communication in virtual and real environments* (2nd ed., pp. 577–585). London: Taylor & Francis.

Ross, H. E. (1971). Spatial perception underwater. In J. D. Woods & J. N. Lythgoe (Eds.), *Underwater science*. London: Oxford University Press.

Shelhamer, M., Robinson, D. A., & Tan, H. S. (1992). Context-specific adaptation of the gain of the vestibulo-ocular reflex in humans. *Journal of Vestibular Research, 2*, 89–96.

Smith, K. U., & Smith, W. K. (1962). *Perception and motion*. Philadelphia: Saunders.

Snyder, F. W., & Pronko, N. H. (1952). *Vision with spatial inversion*. Wichita, KS: University of Wichita Press.

Stratton, G. (1897). Vision without inversion of the retinal image. *Psychological Review, 4*, 341–460, 463–481.

Uliano, K. C., Kennedy, R. S., & Lambert, E. Y. (1986). Asynchronous visual delays and the development of simulator sickness. In *Proceedings of the Human Factors Society 30th Annual Meeting* (pp. 422–426). Dayton, OH: Human Factors Society.

Viirre, E. S., Draper, M. H., Gailey, C., Miller, D., & Furness, T. A. (1998). Adaptation of the VOR in patients with low VOR gains. *Journal of Vestibular Research*, 8, 331–334.

Wallach, H., & Karsh, E. B. (1963). Why the modification of stereoscopic depth-perception is so rapid. *American Journal of Psychology, 76*, 413–420.

Wallach, H., Moore, M. E., & Davidson, L. (1963). Modification of stereoscopic depth perception. *American Journal of Psychology, 76*, 191–204.

Wann, J. P., & Mon-Williams, M. (1997, May). Health issues with virtual reality displays: What we do know and what we don't. *Computer Graphics*, 53–57.

Welch, R. B. (1969). Adaptation to prism-displaced vision: The importance of target pointing. *Perception & Psychophysics, 5*, 305–309.

Welch, R. B. (1972). The effect of experienced limb identity upon adaptation to simulated displacement of the visual field. Perception & Psychophysics, 12, 453–456.

Welch, R. B. (1978). *Perceptual modification: Adapting to altered sensory environments*. New York: Academic Press.

Welch, R. B. (1986). Adaptation of space perception. In K. R. Boff, L. Kaufman, & J. P. Thomas (Eds.), *Handbook of perception and human performance*. New York: John Wiley & Sons.

Welch, R. B., Cohen, M. M., & DeRoshia, C. W. (1996). Reduction of the elevator illusion from continued hypergravity exposure and visual error-corrective feedback. *Perception and Psychophysics, 58*, 22–30.

Welch, R. B., Bridgeman, B., Anand, S., & Browman, K. E. (1993). Alternating prism exposure causes dual adaptation and generalization to a novel displacement. *Perception and Psychophysics, 54*, 195–204.

Welch, R. B., Bridgeman, B., Williams, J. A., & Semmler, R. (1998). Dual adaptation and adaptive generalization of the human vestibulo-ocular reflex *Perception and Psychophysics. 60*, 1415–1425.

White, K. D., Shuman, D., Krantz, J. H., Woods, C. B., & Kuntz, L. A. (1990, March). Destabilizing effects of visual environment motions simulating eye movements or head movements. In N. I. Durlach & S. R. Ellis (Eds.), *Human–machine interfaces for teloperators and virtual environments*, (NASA Conference Publication 10071) Santa Barbara, CA: NASA.

Wickens, C. D. (1986). The effects of control dynamics on performance. In K. R. Boff, L. Kaufman, & J. P. Thomas (Eds.), *Handbook of perception and human performance* (pp. 39–60). New York: John Wiley & Sons.

Young, P. T. (1928). Auditory localization with acoustical transposition of the ears. *Journal of Experimental Psychology, 11*, 399–429.

# 32

# Motion Sickness Neurophysiology, Physiological Correlates, and Treatment

Deborah L. Harm
*NASA Lyndon B. Johnson Space Center*
*2101 Nasa Rd. 1*
*Houston, TX 77058-3696*
*dharm@ems.jsc.nasa.gov*

## 1. INTRODUCTION

Cybersickness (a form of motion sickness) induced by exposure to virtual environments (VEs) is one of the most important health and safety issues that may influence the advancement of VE technology (Barfield & Weghorst, 1993; Gross, Yang, & Flynn, 1995; Kalawsky, 1993; Pausch, Crea, & Conway, 1992; Stanney et al., 1998). Research on space motion sickness (Oman, Lichtenberg, & Money, 1990; Reschke et al., 1994; Reschke, Kornilova, Harm, Bloomberg, & Paloski, 1996), and simulator sickness has shed light on the nature of cybersickness (Hettinger, Berbaum, Kennedy, Dunlap, & Nolan, 1990; Kennedy, Jones, Lilienthal, & Harm, 1993a; Pausch et al., 1992). As many as 80% to 95% of individuals exposed to VE systems report some sickness symptoms (Cobb, Nichols, Ramsey, & Wilson, 1999; Kennedy & Stanney, 1997), and 5% to 30% have symptoms severe enough to discontinue VE exposure (DiZio & Lackner, 1997; Wilson, Nichold, & Ramsey, 1995).

A brief history of motion sickness research with emphasis on theories developed to explain the symptomatology is presented in the next section. Section 3 focuses on what is known about the neural pathways and physiological mechanisms involved in the expression of motion sickness. It provides a general overview of the vestibular system, central nervous system (CNS), autonomic nervous system (ANS), and neuroendocrine system involvement in the motion sickness syndrome. Because the ANS plays a prominent role in the physiological expression of motion sickness, a brief review of its functions and characteristics is included as background for discussions in later sections.

Section 4 deals with signs, symptoms, and physiological correlates of motion sickness. The emphasis is on the physiology underlying the "cardinal" signs and symptoms and on measurement techniques. The most commonly measured physiological correlates are described within the context of the general "stress" or "alarm" response. Finally, section 5 provides a review of a variety of methods used in the prevention and treatment of motion sickness.

A number of excellent reviews are available in the literature that offer comprehensive descriptions of the signs, symptoms, and physiological correlates of motion sickness, and provide extensive bibliographies of supporting scientific research (Harm, 1990; Kohl, 1985; Money, Lackner, & Cheung, 1996; Money, 1970; Reason & Brand, 1975; Reschke et al., 1994; Reschke et al., 1996).

## 2. HISTORICAL BACKGROUND AND THEORIES OF ETIOLOGY

Nearly all definitions of motion sickness include a partial list of the signs and symptoms along with some description of the initiating stimulus, for example, "a condition characterized by pallor, cold sweating, nausea and vomiting that follows upon the perception of certain kinds of real or apparent motion" (Reason & Brand, 1975, p. 1).

> "Motion sickness is a clinical diagnostic term implying that certain criteria have been met to ensure validity. Thus, a close temporal order of exposure to "motion" generating stressful accelerations and the appearance either of vomiting or some combination of such cardinal symptoms as nausea, pallor, sweating, increased salivation and drowsiness constitutes, respectively, a pathognomonic or valid diagnosis" (Graybiel, 1969, p. 352).

As far back in time as the ancient Greeks, the father of medicine, Hippocrates, recognized the relationship between motion stimuli and physiological responses as evidenced in his assertion that "sailing on the sea proves that motion disorders the body" (Reason & Brand, 1975, p. 2).

Historically, the evolution of theories to explain the etiology of medical disorders begins with postulations targeting the organ system exhibiting the most salient signs and symptoms as the primary site affected by the causal factor. The development of theories concerning the etiology of motion sickness is no exception (Marti Usaney, 1954). Although only a brief description of the assertions of proponents of the major theories is possible here, there are a number of excellent reviews in the literature that offer more detailed historical accounts of motion sickness theories (Money, 1970; Reason & Brand, 1975; Tyler & Bard, 1949).

Theorists in the 19th century agreed that motion was the causal factor in the malady; however, there was considerable disagreement about which system was most disturbed by it. Based on the reasoning that the systems capable of independent movement should be the primary sites affected by external motion, the circulatory and gastrointestinal systems were the logical candidates. The general class of theories derived from this line of reasoning has been referred to as the "blood and guts" theories (Reason & Brand, 1975).

Two other classes of theories in this time period worth noting were the ANS theories and the vestibular theories. There was a strong tendency during this time to attribute the origin of the symptom of a variety of medical problems to the ANS (Marti Usaney, 1954). Hence, theories that suggested that motion resulted in a shock or concussion to the central ANS, paralyzed the sympathetic nervous system (SNS), or caused peripheral vagus nerve irritation enjoyed some acceptance in the medical and scientific community. The earliest theories providing the foundation for contemporary vestibular etiology theories of motion sickness were, however, largely ignored. Interestingly, the first theoretical contributions outlining the etiological role of the vestibular system in motion sickness (Irwin, 1881) were based on sound scientific evidence from neurophysiological studies, which established the function of the semicircular canals and otoliths. On the other hand, the more popular nonvestibular theories were based on little more than conjecture and speculation. More contemporary theories that have been proposed to explain the etiology of motion sickness include the *sensory conflict theory* (Oman,

1982; Reason & Brand, 1975), the *poison theory* (Money et al., 1996; Treisman, 1977), and the ecological theory (Riccio & Stoffregen, 1991).

Sensory conflict is currently the most widely accepted explanation for the initiation of motion sickness symptoms. The sensory conflict theory of motion sickness advanced by Reason and Brand (1975) best explains motion sickness, and is parsimonious. Briefly, the sensory conflict theory of motion sickness assumes that human orientation in three-dimensional space, under normal gravitational conditions, is based on at least four sensory inputs to the central nervous system. The otolith organs provide information about linear accelerations and tilt relative to the gravity vector; information on angular acceleration is furnished by the semicircular canals; the visual system provides information concerning body orientation with respect to the visual scene; and touch, pressure, and kinesthetic systems supply information about limb and body position. When the environment is altered in such a way that information from the sensory systems is not compatible and does not match previously stored neural patterns, motion sickness may result.

Total reliance on this theory, however, may limit our thinking. Sensory conflict explains everything in general, but little in the specific. Shortcomings of the sensory conflict theory include its lack of predictive power, inability to explain those situations where there is conflict but no sickness, inability to quantify conflict, failure to include sensory–motor conflict, inability to explain specific mechanisms by which conflict actually gives rise to vomiting, and failure to address the observation that adaptation is not possible without conflict.

Treisman (1977) suggested that the purpose of mechanisms underlying motion sickness, from an evolutionary perspective, was not to produce vomiting in response to motion but to remove poisons from the stomach. He believed that motion was simply an artificial stimulus that activated these mechanisms or, more specifically, that provocative motions act on mechanisms designed and developed to respond to minimal physiological disturbances produced by absorbed toxins. According to Treisman, neural activity to coordinate inputs from all the sensory systems in order to control limb and eye movements would be disrupted by the central effects of neurotoxins. Therefore, disruption of this activity by unnatural motions is interpreted as an early indication of the absorption of toxins, which then activates a mechanism to produce vomiting. Expanding on this theory, Money et al. (1996) argue that motion sickness is a poison response, initiated by the vestibular system, with two major phenomena: (1) a stomach emptying response (primarily parasympathetic control); and (2) a stress response (primarily sympathetic control). The vestibular system is involved in regulating the ANS, although the precise manner in which it does so is uncertain (Yates, 1992; Yates & Miller, 1996). One criticism of this theory is that the time required for a toxin to affect central vestibular mechanisms is too long for vomiting to be useful in removing toxins from the gastrointestinal tract, and thus would be one of the last lines of defense against poisoning (Yates, Miller, & Lucot, 1998a).

The ecological theorists argue that motion sickness is caused by prolonged postural instability (Riccio & Stoffregen, 1991) and adamantly reject the sensory conflict theory (Riccio & Stoffregen, 1991; Stoffregen & Riccio, 1991). This theory is based on an ecological approach to perception and action, where the interactions between the animal and environment are the critical unit of analysis. Riccio and Stoffregen (1991) acknowledge that there are changes in sensory stimulation in provocative situations, but argue that these changes are determined by changes in how the environment constrains the control of posture. Further, they argue that the length of time you are unstable and the magnitude of the instability are predictive of motion sickness and the intensity of symptoms. Some limitations of this theory include it's inability to fully explain why labyrinthine deficient individuals do not get motion sickness, the lack of potential neurologic pathways or mechanisms involved, and lack of a clear explanation of why postural instability causes motion sickness.

It is clear from the broad differences in these theories and the limitations of the theories to explain all aspects of motion sickness that this is a very complex phenomenon. Both the sensory conflict and the ecological theories emphasize the conditions under which motion sickness occurs, but not the reason for sickness, whereas the poison theory attempts to explain the evolutionary causes. Regardless of the theory to which one subscribes, the conditions that provoke motion sickness involve multimodal sensory information, where the information from all the sensory modalities is not in agreement. Exposure to altered or novel sensory conditions can lead to motion sickness as well as sensorimotor disturbances. The incidence, severity, and primary set of symptoms vary with the sensory conditions to which one is exposed (Kennedy et al., 1993a; McCauley & Sharkey, 1992; Reschke et al., 1994). Hence, different labels have been applied to indicate the provocative conditions, for example, terrestrial, space, simulator sickness, and cybersickness.

Some issues concerning susceptibility and prediction of motion sickness are addressed in chapter 30 (this volume), and countermeasures are presented in chapters 31 and 36 (this volume). These problems are covered in a more general sense in this chapter primarily from the perspective of research methodology and strategies, whereas the physiology of motion sickness is reviewed in somewhat greater detail.

## 3. NEUROPHYSIOLOGY

Despite the large body of literature containing numerous reports on signs, symptoms, and physiologic correlates of motion sickness, knowledge of the physiologic mechanisms and specific neural pathways involved in the manifestations of this syndrome in humans is still somewhat limited. The high degree of variability reported in the number, order, type, severity, and time-course of physiological responses across individuals and across motion sickness stimulus conditions attests to the fact that incredibly complex interactions among neurophysiological structures, pathways, and mechanisms are involved in the expression of motion sickness. As an indication of the complexity involved, Steele (1968, p. 90) offers the following summary from an earlier report (Rapoport, Horvath, Small, & Fox, 1967):

> The vestibular ganglia communicate with 14 specific neural structures having about 2 dozen mutual interconnections. These, in turn, communicate through approximately 120 identified channels to the next level consisting of 44 centers. Thus, going no farther than three steps from the sense organ, and without even considering the visual, auditory, proprioceptive, and other inputs, the system becomes quite unwieldy.

It is, however, widely accepted that the vestibular system is required, and both the CNS and ANS play important roles in the expression of motion sickness.

### 3.1 Autonomic Nervous System

The ANS is activated by brain stem, hypothalamic, and spinal cord centers. In addition, impulses transmitted from the cerebral cortex and, in particular, from the limbic system affect autonomic activities. More important, some information from various components within spatial orientation systems eventually passes through all or most of the central structures that can activate the ANS. Hence, neural inputs from spatial orientation systems can directly or indirectly influence autonomic activities.

Although it is not within the scope of this chapter to describe the ANS in detail, a basic description of its organization and its functions and characteristics is important for understanding motion sickness symptomatology.

### 3.1.1 General Organization and Description

The ANS consists of two components, the sympathetic nervous system (SNS) and the parasympathetic nervous system (PNS). Sympathetic pathways exit the CNS from the first thoracic to the third lumbar segments of the spinal cord, whereas parasympathetic pathways exit the CNS in the cranium (cranial nerves 3, 7, 9, and 10) and sacral spinal cord segments 2, 3, and 4. Sympathetic and parasympathetic preganglionic neurons exit the CNS and synapse onto postganglionic neurons that innervate visceral tissues. Sympathetic postganglionic neurons are located near the spinal column and project a distance to their targets, whereas parasympathetic postganglionic neurons lie in or near the tissue they innervate and have short projections. These efferent autonomic pathways innervate smooth muscle, cardiac muscle, and glands. Early researchers restricted their definition of the ANS to the peripheral autonomic nerves and their site of origin. Contemporary definitions of the ANS, however, often include visceral afferents and central neural circuitry (Sved & Ruggiero, 1996).

Sympathetic fibers at the effector site are primarily adrenergic, with norepinephrine (NE) being the major neurotransmitter. Parasympathetic fibers at the effector site are primarily cholingeric with acetylcholine (ACh) being the major neurotransmitter. However, a number of other neurotransmitters and peptides may be secreted by autonomic fibers at the effector. These include adenosine 5′-triphosphate, dopamine, 5-hydroxytryptamine, neuropeptide Y, enkephalin, somatostatin, and others (Money et al., 1996; Sved & Ruggiero, 1996).

### 3.1.2 Functions and Characteristics of the ANS

The principal function of the ANS is to maintain homeostatic conditions in the body, that is, conditions that provide an optimal internal environment for cellular functioning. Smooth muscle of the viscera and blood vessels, cardiac muscle, and glands are innervated by the ANS. Furthermore, most organs and glands receive fibers from both the sympathetic and parasympathetic divisions of the ANS. Examples of exceptions to this dual innervation feature are blood vessels and sweat glands, which receive fibers only from the SNS.

The SNS is responsible for evoking integrated response patterns that serve to mobilize body energy to prepare the organism and sustain it when there is a demand for physical activity; the parasympathetic nervous system evokes responses that promote digestion and conserve energy. In general, stimulation of the SNS tends to have diffuse physiological effects as compared to the more discrete effects of parasympathetic stimulation. Physiologic consequences of SNS stimulation also tend to be sustained for some time after the stimulus is removed. This is due to the fact that activation of the SNS causes the release of sympathetic neurotransmitters (epinephrine and norepinephrine) from the adrenal medulla, which serve to maintain sympathetic activity. The traditional view of ANS function proposed by Cannon (1939) suggested that the SNS functioned as a unit producing diffuse nonspecific responses, and the PNS produced discrete responses. However, depending on the stimulus and the CNS control center(s) called into play, sympathetic activation can produce changes in some parts of the system without affecting others. It is now known that very specific patterns of SNS activation can occur in response to certain stimuli, and that these patterned responses can include both increases and decreases in sympathetic outflow to specific target organs. Similarly, there are situations where PNS activity is more diffuse (Sved & Ruggiero, 1996). A good example of more discrete SNS effects is seen in thermoregulatory processes. When heat reduction is called for, sympathetic impulses to the cutaneous vasculature are inhibited to promote vasodilation and increased blood flow in the skin, causing heat loss by convection and radiation. Concomitantly, SNS outflow to sweat glands increases, promoting evaporative heat loss.

The SNS and PNS often function in a reciprocal fashion to each other. It should be pointed out, however, that one or the other division usually predominates in the control of a particular organ system or gland. Hence, stimulation, withdrawal, or inhibition of the efferent neural impulses from CNS control centers to the fibers of the predominant ANS branch can accomplish control over a wide range in the level of activity in a particular structure(s). An additional aspect of this feature is the differential stimulus intensity requirements of the two divisions to produce equal and opposite responses. The stimulus intensity from the nondominant division may have to be many times greater than that from the dominant division in order to produce a reciprocal change of a similar level of organ activity. For example, the PNS predominates in modulating the general activity level of the gastrointestinal tract; stimulation increases tone and motility. Intense SNS stimulation is required to inhibit normal tone and motility. Finally, stimulation of the SNS or PNS can produce either excitatory or inhibitory effects on the activities of the structure(s) they innervate. Whether the effector activity is increased or decreased by stimulation of one or the other branch of the ANS is dependent upon the neurotransmitter–receptor relationship. Most sympathetic postganglionic fibers release norepinephrine onto either alpha or beta adrenergic receptors, whereas parasympathetic postganglionic fibers release acetylcholine (ACh) onto muscarinic cholinergic receptors. An example of differential receptor responses to SNS activation is the regional vasomotor changes that occur in preparation for exercise. Blood vessels dilate in the skeletal muscles, and constrict in the skin and in organs not needed for the increase in physical activity.

The various features or characteristics of the ANS highlight the need to ask certain questions when attempting to interpret physiological responses to motion sickness provocation. Two questions come to mind immediately. The first derives from the characteristic predominance of one division over the other: Is an observed change in some parameter primarily due to modulation of activity within the predominant branch of the ANS or to strong stimulation in the nondominant branch? For example, inhibition of gastrointestinal (G-I) tone and motility could be due to either intense SNS stimulation or to inhibition of normal parasympathetic tone. The second question is related to the reciprocal nature of the two ANS divisions: Does the absence of a change in a physiological parameter indicate no change in autonomic activity, or is it the case, for example, that a reciprocal response in the parasympathetic branch is sufficient to override or resist an increase in the activity of the SNS? In other words, a stimulus activates some mechanism or control system that is effective in maintaining the parameter constant. The arguable point that recurs throughout this chapter is that these kinds of questions cannot be answered on the basis of observations of a single parameter or by examining responses independent of one another.

In summary, various ANS activities are controlled and/or modulated at all levels of the CNS. The specific type and intensity of the demand (stimulus) dictates the level or location of CNS control, which, in turn, dictates the mechanisms activated to accomplish the required physiological adjustments. Mechanisms called into play can range from relatively simple reflexes involving discrete effector sites, to a more complex set of reflex mechanisms involving multiple sites within a system, to incredibly complex patterns of responses involving virtually the entire organism. Given the many levels of neural control and the vast range of potential responses, it should not be surprising that physiological responses and response patterns during motion sickness provocation vary within and across individuals, as well as within and across different motion stimulus conditions.

### 3.2 Vestibular System and CNS Structures and Pathways

The sensory and CNS structures and pathways involved in motion sickness are described in detail in several recent publications (Kucharczyk, Stewart, & Miller, 1991; Tache & Wingate, 1991; Yates & Miller, 1996). Therefore, only a brief overview of structures and their neural connections involved in the production of motion sickness is presented here.

An intact vestibular system is required for motion sickness to occur. The vestibular end organs (semicircular canals and otoliths) project fibers, differentially, to four vestibular nuclei located in the brain stem approximately at the junction of the medulla and pons: lateral, medial, superior and inferior nuclei. Neural pathways from the vestibular nuclei project to areas in the medulla, the cerebellum (to oculomotor and spinal–motor control systems) and to the cerebral cortex. The vestibular nuclei and the vestibulo-cerebellum receive inputs from other sensory systems (visual, somatic, and proprioceptive) concerned with head–body orientation and movement.

In the 1950s it was believed that certain structures in the vestibulo-cerebellum, the uvula and nodulus, as well as the area postrema (chemoreceptor trigger zone), located in the dorsomedial medulla, were essential for motion induced vomiting. Although more recent work has shown that these areas are not essential (Borison & Borison, 1986; Miller, Nonaka, & Jakus, 1994; Miller & Wilson, 1983), structures in the vestibulo-cerebellum may still be involved in motion-induced vomiting. More recent work indicates that neurons involved in coordinating vomiting are distributed in an arc extending from the area postrema and nucleus tractus solitarius (NTS) through the lateral tegmental field of the reticular formation (RF) to the ventrolateral medulla (Miller & Ruggiero, 1994). Current research suggests that the NTS may serve as the final common pathway for all forms of emetic responses (Sved & Ruggiero, 1996; Yates et al., 1994). Although the NTS receives almost all types of visceral inputs and influences many output systems, the following discussion focuses on those inputs and outputs most relevant to vomiting and motion sickness.

Because most of the signs and symptoms of motion sickness are autonomically mediated, understanding the role of the vestibular system in autonomic regulation is essential to understanding motion sickness. The most direct pathway for vestibular modulation of autonomic responses involved in motion sickness is via projections from the medial and inferior vestibular nuclei to the NTS and the dorsal motor nucleus of the vagus (Sved & Ruggiero, 1996). In addition to vestibular inputs, the NTS receives inputs from other areas that can trigger vomiting, such as the area postrema and abdominal vagal afferents (Sved & Ruggiero, 1996; Yates et al., 1994). Gastrointestinal changes associated with vomiting are probably mediated by projections from the NTS to the dorsal motor nucleus of the vagus and the nucleus ambiguus (Miller & Grelot, 1996). Other motion sickness symptoms may be mediated, in part, through NTS projections to the parabrachial nucleus, which has extensive connections with other autonomic nuclei (Sved & Ruggiero, 1996).

The nucleus tractus solitarius plays an integral role in both gastric motility and emesis related to motion sickness (Ito & Honjo, 1990), and in cardiovascular regulation (Barron & Chokroverty, 1993; Onai, Takayama, & Miura, 1987). The NTS receives ascending carotid, aortic, and cardiopulmonary parasympathetic afferent fibers (Barron & Chokroverty, 1993), as well as fibers from the medial and inferior vestibular nuclei, the cerebellar fastigial nuclei, and certain reticular formation (RF) nuclei (Ito & Honjo, 1990; Onai et al., 1987). The NTS in turn influences (1) vagal efferent output to the stomach and heart (Barron & Chokroverty, 1993; Ito & Honjo, 1990) and (2) the output of certain brain stem RF nuclei that control the peripheral sympathetic nervous system via the intermediolateral cell column of the spinal cord (Barron & Chokroverty, 1993; Previc, 1993; Yates, 1992; Yates, Goto, & Bolton, 1993; Yates & Miller, 1996). Direct connections between the medial vestibular nuclei and brain stem cardiovascular output areas (such as the dorsolateral medullary RF and subretrofacial nuclei—a term mainly used to identify a part of the rostral ventrolateral medulla of the cat) also exist in experimental animals (Yates, 1992) as do otolith-subretrofacial connections (Previc, 1993; Yates et al., 1993). Otolithic inputs into the subretrofacial nucleus are seven times more plentiful than semicircular canal inputs (Previc, 1993; Yates et al., 1993).

Other pathways that may be involved include vestibular projections to the lateral tegmental field of the RF or to the caudal ventrolateral medulla (Yates, 1996). Finally, the cerebellum may be another route through which vestibular inputs may modulate autonomic activity (Balaban, 1996; Wood et al., 1994).

## 4. SIGNS, SYMPTOMS, AND PHYSIOLOGICAL CORRELATES

Individual reactions to real or apparent motion stimuli are as varied in number and complexity, as are the stimulus conditions that can produce them. With the exception of labyrinthine-defective individuals, anyone can be made sick given an appropriate stimulus of sufficient intensity and duration. Nonetheless, there are some signs and symptoms that occur with a high degree of regularity as part of the syndrome, which are commonly referred to as "cardinal" signs and symptoms, and are considered pathognomonic. These include pallor, cold sweating, nausea, and vomiting. Other reactions to motion stimuli, often referred to as associated or additional signs and symptoms, are much more variable in their occurrence and time course. A partial list includes such things as salivation, headache, drowsiness, dizziness, sensation of increased body temperature, general malaise, apathy and depression, and decreased motor coordination (Harm, 1990; Reason & Brand, 1975). Finally, changes in a wide range of physiological parameters in nearly every major system of the body have been observed during motion sickness provocation.

The following section offers some brief general comments and observations regarding each of the cardinal signs and symptoms, a review of the basic physiology underlying the responses, and techniques used to measure and record signs and symptoms. Physiological correlates are outlined as well. However, the number and variety of physiological parameters that have been examined in motion sickness research precludes reviewing the specific underlying physiology and measurement techniques for individual physiological variables.

### 4.1 Pallor and Cold Sweating

#### *4.1.1 General Comments and Observations*

Pallor and cold sweating are generally accepted as antecedents to nausea and often occur in close temporal relationship to early stomach symptoms of epigastric awareness/discomfort. In some motion sick subjects, a brief flushing of the face precedes pallor, and then pallor persists throughout the remainder of the period of motion sickness. Reports concerning the temporal relationships between the onset of cold sweating and subjective reports of stomach awareness or nausea are somewhat more equivocal than those concerning pallor. Some investigators have observed that cold sweating consistently preceded nausea, whereas others found the appearance of cold sweating to be quite variable in time (Reason & Brand, 1975), if it occurred at all. Differences in measurement techniques, ambient temperature and humidity conditions during testing, and the particular motion stimulus account, in part, for the equivocal nature of these findings.

#### *4.1.2 Physiology of Pallor*

Pallor is defined as a white or whitish-greenish hue to the skin. In motion sickness studies, it is most readily observed in the face; first it appears around the eyes, nose, and mouth (Reason & Brand, 1975) progressing in some individuals to the severe pallor associated with circulatory collapse. Skin color changes, such as pallor or flushing arise from altered vasomotor activity in the cutaneous circulation. Pallor is the result of vasoconstriction, whereas vasodilation produces flushing or reddening of the skin. Neural control of the skin vasculature is exclusively of sympathetic adrenergic origin; increased sympathetic activity causes vasoconstriction and inhibition or withdrawal of sympathetic activity results in vasodilation. CNS vasomotor control centers are located primarily in the ventral medulla and are in turn controlled by hypothalamic centers. In addition, the hypothalamic centers receive modulating inputs from limbic structures as well as from the cerebral cortex. Facial pallor may also be due, in part, to the increased levels of vasopressin (AVP) during motion sickness (Money et al., 1996). High levels of AVP produce vasoconstriction in the skin (Gilman, Rall, Nies, & Taylor, 1990). Finally, depending upon the type of stimulus, the vasoconstrictor response may be generalized, segmental, or regional (Brobeck, 1979).

#### *4.1.3 Pallor Measurement Techniques*

Visual observation techniques (i. e. color photographs; Crampton, 1955) or subjective impressions of discrete levels of pallor (Graybiel, Wood, Miller, & Cramer, 1968) are by far the most common methods of measuring changes in skin color during motion sickness. Three techniques that offer quantitative physiological indices of skin color have been employed in motion sickness provocation studies. Oman and Cook (1983) developed and employed an infrared reflectance plethysmographic technique that provided an index of blood volume in the skin, and Harm et al. (1987a) used transcutaneous oxygen ($P_{tc}O_2$), which provided an index of blood flow as indirect measures of skin color changes. In both studies, these quantitative measures of skin color showed a fairly close temporal relationship with increasing intensity of epigastric/stomach symptoms. However, the investigators observed differences in individual response patterns of skin color changes. Over repeated trials, some subjects consistently showed a transient blush/flush prior to pallor onset, whereas others responded exclusively with either blush or pallor (Oman & Cook, 1983). Harm and colleagues observed similar individual differences in $P_{tc}O_2$; however, the direction of the response depended, to a great extent, on the level of susceptibility in the subjects. Individuals categorized as susceptible exhibited increases in skin oxygen (flushing), whereas those categorized as insusceptible showed decreases in skin oxygen (pallor) throughout the motion test period. A third technique used to measure changes in skin circulation during motion sickness is with a laser Doppler flowmeter (Kolev, Moller, Nilsson, & Tibbling, 1997), which directly measures blood flow, as compared to the indirect measure ($P_{tc}O_2$) used by Harm et al. (1987a).

These techniques share several advantages over visual observation/detection techniques. First, they both offer means for continuous measurement that allows for greater precision in characterizing the time course of physiological changes associated with motion sickness. Second, skin color changes are likely to be detected earlier and more consistently during the development of the motion sickness syndrome than is possible with visual observation. Finally, they have greater potential for yielding physiologically relevant evidence of the underlying mechanisms responsible for motion sickness symptoms.

#### *4.1.4 Physiology of Sweating*

"Cold sweating is defined as that which occurs in the absence of an adequate thermal stimulus" (Reason & Brand, 1975, pp. 41–42). The sweat glands activated during cold sweating are termed *eccrine*. Eccrine sweat glands are located in large numbers on the palmar and plantar surfaces, as well as over most of the body. These glands are innervated exclusively by the sympathetic division of the ANS. They receive predominantly cholinergic innervation (the postganglionic synaptic neurotransmitter is ACh, but some adrenergic fibers are also in close proximity (Shields, MacDowell, Fairchild, & Campbell, 1987). Sweating is sometimes considered a parasympathetic response because it is mediated by cholingergic receptors, and because stimulation of CNS structures (premotor cortex and anterior regions of the hypothalamus) known to produce responses associated with generalized parasympathetic activation also elicit sweating.

Electrodermal or sweating responses measured on the palmar and plantar surfaces are primarily due to psychological stimuli or acute changes in sensory input, whereas responses from other sites (e.g., dorsal hand, forearm, and forehead) are generally due to thermal stimuli. The timing characteristics of responses, such as between palmar and dorsal surfaces, support this notion. Electrodermal responses from the palm usually have a short latency and rapid rise and recovery times indicative of reflexive orienting responses, compared to the longer latency and slow rise and recovery times observed from the dorsal surfaces. In fact, this is precisely what McClure Fregly, Molina, & Greybrel (1971) found in a study using cross-coupled angular accelerations to provoke motion sickness. The palmar response, that is, skin resistance, peaked on the first or second head movement and rapidly declined; the sweat response from the dorsal hand and forearm began later and increased more gradually. The observation that the responses were independent of one another and exhibited different timing characteristics implies activation of different control systems.

A variety of control systems are responsible for modulating electrodermal or sweat gland activity, including the premotor corticospinal system, the hypothalamic-limbic system, and the basal ganglia. Supportive evidence for the existence of multiple control systems having some shared and some independent neural pathways is comprehensively outlined elsewhere (Dawson, Schell, & Filion, 1990; Wang, 1964). More importantly, recognition of the various control systems and their independent neural pathways may provide some information concerning the neural pathways and mechanisms involved in cold sweating manifested as part of the motion sickness syndrome.

### *4.1.5 Sweat Measurement Techniques*

A wide variety of techniques, both subjective and objective, have been used to measure and record cold sweating. Qualitative scores of sweating are derived from either subjective reports offered by the subject or visual observations made by the experimenter (Graybiel et al., 1968). Quantitative assessments of sweat gland activity (or electrodermal activity) include measurements of skin resistance, skin conductance, and skin potential. The most strongly recommended of the three measures is skin conductance (Dawson et al., 1990). Dawson et al. (1990) have suggested that skin conductance level (SCL) and skin conductance responses (SCRs) are the two most sensitive and useful measures. SCL is defined as the tonic level of electrical conductivity of the skin (measurement unit is $\mu$mho or $\mu$S). SCRs are phasic waves superimposed on the tidal drifts in SCL. Skin conductance responses may occur spontaneously (usually 1–3 per minute in the resting subject), or in response to a specific stimulus. Such things as body movements and sighs also can elicit SCRs. These responses are typically evaluated in terms of their amplitude, frequency, and temporal characteristics (e.g., latency, rise and recovery time, and duration).

The technique used for recording skin conductance uses two electrodes placed on active sites (bipolar), and involves passing a small constant voltage (less than 0.05v) across the skin segment of interest. A resistor is placed in series with this circuit, and the voltage drop across the resistor is equal to the conductance in the segment of interest. The common measurement sites on the hand are between the first and second fingers at the distal or medial segments (volar surface) and between the thenar eminences of the palm.

Interpretation and characterization of electrodermal responses (i.e., latency, amplitude, rise and recovery time, duration, and polarity) depend upon a whole host of factors. These include the measurement site, type of electrode paste, stimulus characteristics, measurement technique selected, environmental conditions, and subject factors. Careful review of the methodological issues and measurement technique should be carried out before one undertakes recording of sweat responses in motion sickness research (see Dawson et al., 1990; Venables & Christie, 1980, for reviews).

## 4.2 Nausea and Vomiting

### *4.2.1 General Comments and Observations*

Nausea is undoubtedly the most frequently occurring, and probably the most profoundly unpleasant, symptom of motion sickness. Although the time course and progression of nausea and vomiting varies somewhat with the stimulus conditions, it usually begins with awareness of mild epigastric queasiness and steadily increases in intensity to severe nausea culminating in vomiting. Some individuals, however, will vomit without premonitory nausea, whereas others may experience severe nausea but are unable to vomit. Therefore, the person who vomits during motion provocation tests should not necessarily be considered "sicker" than one who becomes extremely nauseated without vomiting. In motion sickness provocation tests, nausea usually begins to subside immediately upon removal of the stimulus. Depending upon the stimulus intensity and duration, however, stomach symptoms may be completely absent within several minutes or may continue for several hours. A prolonged recovery is often characterized by waxing and waning in the intensity of nausea.

### *4.2.2 Physiology of Nausea and Vomiting*

Nausea is thought to be the awareness of unusual activity in the CNS circuitry involved in vomiting. At least two areas in the cerebral cortex are probably necessary for the sensation of nausea: the inferior frontal gyrus (Roberts, Miller, Rowley, & Kucharczyk, 1994) and the temporofrontal region (Chelen, Kabrisky, & Rogers, 1993). Autonomic visceral afferent inputs to the CNS may contribute to the sensation of nausea but are not required (Miller & Grelot, 1996). In fact, patients with total gastrectomy can still experience nausea (Cummins, 1958), and patients with bulimia can induce vomiting without experiencing nausea. Given that there are situations where nausea occurs without vomiting and vice versa, it seems likely that the two are mediated by different mechanisms. Finally, high levels of AVP also may contribute to nausea. During motion sickness, plasma levels of AVP can increase by as much as 20 to 30 times the baseline levels (Eversmann et al., 1978; Koch, Summy-Long, Bingaman, Sperry, & Stern, 1990; Kohl, Leach, Homick, & LaRochelle, 1983). Administration of AVP to humans is often accompanied by nausea and vomiting (Thomford & Sirinek, 1975), and administration of an AVP antagonist abolished motion-induced vomiting in squirrel monkeys (Cheung, Kohl, Money, & Kinter, 1994). It should be noted, however, that nausea and vomiting can occur in the absence of increased levels of AVP. For example, nausea and vomiting following ingestion of syrup of ipecac (Nussey, Hawthron, Page, Ang, & Jenking, 1988) and self-induced vomiting in patients with bulimia (Kaye, Gwirtsman, & George, 1989) are not accompanied by increases in AVP.

Vomiting is described as the forceful expulsion of stomach contents through the mouth, and is produced by complex, highly coordinated smooth and somatic motor responses. Visceral smooth muscle changes precede the somatic motor components and include decreased motility of the stomach and small intestine, a retrograde giant contraction (RGC) followed by phasic contractions of the stomach, and small intestine. The RGC moves the contents of the upper small intestine into the stomach, and the phasic contractions move the contents of the lower small intestine into the colon. Neural pathways from the NTS to the dorsal motor nucleus of the vagus and the nucleus ambiguus probably mediate the smooth muscle changes. Somatic motor components of vomiting begin with relaxation of the lower esophagus and movement of the upper stomach into the thorax, followed by coordinated contraction of the major respiratory muscles (abdominal, diaphragm, external and internal intercostal muscles). The respiratory motor components of vomiting are probably mediated, at least in part, via projections from the NTS to the ventrolateral medulla (see Miller & Grelot, 1996; Money et al., 1996, for reviews). More recent animal research provides strong evidence that the midline medulla plays an essential role in mediating the coordinated respiratory pattern involved in the vomiting process (Billig, Foris, Card, & Yates, 1999; Miller, Nonaka, Jakus, & Yates, 1996; Yates, Smail, Stocker, & Card, 1999).

### 4.2.3 G-I Response Measurement Techniques

Measurement of gastric symptoms is usually accomplished by having the subject report subjective estimates of stomach upset. One of the most commonly employed rating scales used in laboratory studies is the one incorporated in the Pensacola Diagnostic Criteria for motion sickness susceptibility (Graybiel et al., 1968), where each symptom level is assigned incremental point values. The first level of the scale is epigastric awareness, followed by epigastric discomfort, Nausea 1, 2, and finally Nausea 3 (vomiting). Subjective rating scales for simulator sickness have also been developed (Kennedy, Lane, Berbaum, & Lilenthal, 1993b) and have been used to measure cybersickness (Kennedy et al., 1993a). The simulator sickness questionnaire is described in more detail in chapter 30 (this volume). Other techniques employed in examining changes in gastrointestinal activity associated with motion sickness include radiographic and nuclear medicine techniques (Reid, Grundy, Khan, & Read, 1995; Stewart, Wood, Wood, & Mims, 1994), gastric balloon and intraluminal transducers (Crampton, 1955; Kolev & Altaparmakov, 1996), breath hydrogen to measure mouth to cecal transit time (MCTT) (Muth, Stern, & Koch, 1996), and electrogastrography (EGG); Andre, Muth, Stern, & Leibowitz, 1996; Harm, Sandoz, Stern, Koch, & Koslovskya, 1997; Harm, Stern, & Koch, 1987b; Muth et al., 1996; Stern, Koch, Stewart, & Lindbald, 1987; Stewart et al., 1994). The insertion of gastric balloons, intraluminal transducers, serosal electrodes, and radiographic techniques are not practical for routine use in motion sickness research, whereas EGG has the advantage of being a noninvasive technique, which greatly enhances its practical value. A brief description of EGG is provided below. For the history and a complete review of EGG, the interested reader is directed to the recent text by Chen and McCallum (1994) on the topic.

Electrogastrography is a technique for recording myoelectric activity of the stomach using surface electrodes and was first identified by Alvarez (1922). In humans, gastric smooth muscle pacesetter potentials oscillate around a frequency of 3 cpm (0.05 Hz), with an amplitude of 100 to 500 uV. Pacesetter potentials are propagated across the stomach from the antrum toward the pylorus. Gastric myoelectric activity is generally referred to as the basic electrical rhythm (BER) (Harm et al., 1987b; Rague & Oman, 1987) or electrical control activity (Harm et al., 1987b; Rague & Oman, 1987). Gastric myoelectric activity is responsible for coordinating stomach contractions and regulating gastric emptying. Gastric arrhythmias, including tachygastria (fast frequency waves), bradygastria (slow frequency waves), and dysrhythmias (no clear frequency peak) also can be detected accurately by EGG (Familoni, Abell, & Bowes, 1995). Gastric motility is mediated by both the sympathetic and parasympathetic branches of the ANS. Increased vagal (parasympathetic) activity augments gastric motility (Fujii & Mukai, 1985), whereas increased sympathetic activity or vagal withdrawal decreases gastric motility (Hall, El-Sharkawy, & Diamant, 1986).

The frequency and amplitude of the EGG have been studied in conjunction with measures of gastric motility. Chen et al. (1994) simultaneously recorded EGG and manometric activities of the distal stomach and upper small bowel in healthy subjects under fasted and fed conditions.

They observed that peak EGG power during the postprandial period was significantly greater than during motor quiescence in the fasted period, the dominant frequency was significantly higher during the postprandial period compared to periods of motor activity in the fasted state, and the instability coefficient of EGG power (ratio of the standard deviation to the mean value) was significantly higher during motor activity in the fasted state versus motor quiescence in the fasted state. Interestingly, a 1 cpm wave superimposed on the normal 3 cpm rhythm was noted in the power spectrum of the postprandial EGG. This low-frequency component was highly associated with intestinal contractions (Chen, McCallum, & Richards, 1993).

The EGG has been employed in studies of motion sickness induced by circular vection (Muth et al., 1996; Muth, Thayer, Stern, Friedman, & Drake, 1998; Stern et al., 1987), linear vection (Harm & Parker, 1994), cross-coupled Coriolis stimulation (Rague & Oman, 1987; Stewart, Wood, & Wood, 1989; Stewart et al., 1994), parabolic flight maneuvers (Harm et al., 1987b) and space flight (Harm et al., 1997). In these studies, gastric motion sickness symptoms were associated with the appearance of tachygastria and/or a decrease in the amplitude of the BER. In many instances of reduced amplitude, the reduction emerged before subjects reported awareness of any level of stomach symptoms. In a recent motion sickness adaptation study, tachygastria was associated with a subject's reports of stomach symptoms. This subject's gastric electrical rhythms gradually returned to the normal BER over several days of adaptation to the stimulus conditions (Harm & Parker, 1994).

## 4.3 Physiological Correlates

Physiological reactions to motion sickness provocation stimuli can, and often do, occur in nearly every system of the body. Most of the research in this area has been directed toward either identifying parameters that can be used to discriminate among susceptible and insusceptible individuals, or characterizing changes in physiological response variables. Although there are reports of statistically significant differences in a variety of physiological parameters as a function of susceptibility, investigators tend to interpret their findings with some caution because considerable variability is often observed within susceptibility categories. The reliability of physiologic correlates across repeated motion sickness tests was recently examined (Stout, Toscano, & Cowings, 1995). The results of this study suggested that autonomic responses measured during the last minute of the motion sickness test were moderately stable across repeated tests. A large number of physiological parameters in the cardiovascular, respiratory, gastrointestinal, and neuroendocrine systems have been examined during a variety of motion sickness provocation testing conditions. The most frequently measured parameters in each of these physiological systems, along with a few interesting but less common parameters, are outlined in Table 32.1.

At present there is no unifying hypothesis that explains all of the physiological manifestations of motion sickness. Most researchers would probably agree, however, that the motion sickness syndrome has many of the components characteristic of "stress" or "alarm" response patterns described in medical physiology textbooks. Most would also agree that this characterization of motion sickness is limited and should be viewed with some caution. It is applicable primarily to acute (short-duration) forms of laboratory-induced motion sickness or to the early stages of chronic (long-duration) forms before adaptive processes begin to exert their influence on physiological activities (see chap. 31, this volume). Moreover, it does not explain individual differences in specific physiological responses, nor does it explain some interesting paradoxical findings in the neuroendocrine literature. Before describing this research, some of the primary components of the "stress" response are presented.

The "stress" response consists of a generalized activation of the sympathetic nervous system that includes: (1) stimulation of the adrenal medulla to increase secretion of epinephrine and

**TABLE 32.1**
Physiological Correlates of Motion Sickness

| *Physiological System* | *Parameters* |
|---|---|
| Cardiovascular | Electrocardiogram, heart rate, cardiac interbeat interval, blood pressure, blood volume pulse, muscle blood flow, and skin blood flow, volume, and oxygen |
| Respiratory | Volume, rate and carbon dioxide |
| Gastrointestinal | Tonus, motility, and electrogastrogram, and mouth to cecal transit time |
| Neuroendocrine | *Adrenal factors*: steroid hormones and catecholamines. *Anterior pituitary factors*: GH, TSH, ACTH, FSH, LH, and prolactin. *Posterior pituitary factors*: ADH/AVP |
| Other | Skin conductance, dark focus, and pupil responses |

norepinephrine; (2) stimulation of the posterior and anterior pituitary gland (via hypothalamic input) to increase secretion of antidiuretic hormone or vasopressin (ADH or AVP, respectively), and adrenocorticotropin (ACTH), respectively (ACTH subsequently causes release of adrenocortical hormones such as cortisol); (3) redistribution of blood away from peripheral tissues and nonparticipating organs to skeletal muscle; and (4) stimulation of respiration to increase oxygen delivery and carbon dioxide removal. Admittedly, this is an oversimplified description of the "stress" response, but it should be adequate background for understanding the physiological stress responses already described (pallor, cold sweating, and neuroendocrine responses) and those presented next.

Real or apparent stressful motion activates the neuroendocrine system. Neurohumoral substances released by some CNS structures and subsequently by the various endocrine glands are responsible for regulating metabolic activity to meet the demands placed on the body. Observations of elevated levels of stress-related neuroendocrine hormones (Eversmann et al., 1978; Kohl, 1985) indicate the hypothalamus-pituitary-adrenal cortex (H-P-A) axis plays a role in the expression of motion sickness. The general neuroendocrine response to stressful motion and sensory-rearrangements includes elevated levels of ACTH, cortisol, prolactin, growth hormone, ADH or AVP, thyroid hormone, epinephrine, and norepinephrine (see Reschke et al., 1994; Reschke et al., 1996, for reviews). Stimulation of the hypothalamus initiates the release of pituitary hormones via its hormonal and neural connections with the pituitary gland. The anterior pituitary gland releases ACTH, prolactin, and growth hormone, as well as others. The posterior pituitary gland releases ADH. ACTH, in turn, stimulates the release of a variety of hormonal substances from the adrenal cortex (e.g., cortisol). More detailed descriptions of the H-P-A axis and endocrine responses to stress are available in standard medical physiology texts.

A number of investigators have found correlations between susceptibility to motion sickness and endogenous levels of neurohumoral substances. Differences between susceptibility subgroups and neuroendocrine responses to motion sickness stimulus conditions have also been observed. Kohl (1985) found those individuals resistant to motion sickness had higher endogenous levels of ACTH and exhibited higher levels of ACTH, epinephrine, and norepinephrine in response to motion sickness provocation. This finding led to the hypothesis that less susceptible individuals are better able to resolve sensory conflicts because the higher endogenous levels and higher responsivity of their endocrine systems make them more adaptable. In contrast, Eversmann et al. (1978) noted that subjects with additional symptoms (e.g., sweating, pallor) exhibited higher hormonal levels than those with vomiting/retching alone and proposed that increased hormonal secretion could be related to symptom severity. Grigoriev et al. (1988) found that hormonal levels in moderately susceptible subjects peaked immediately after rotation ceased, and those of more susceptible subjects peaked later. This group proposed that the stress of rotation stimulates an adequate hormonal response in moderately susceptible individuals (resulting in adaptation), whereas highly susceptible subjects experience a delayed adaptation response, CNS excitation, or both after the stimulus is terminated.

Generally, cardiovascular responses during motion sickness are consistent with a stress response. Decreased peripheral circulation is manifested in the cardinal sign of facial pallor, which was described in detail earlier. Increased skeletal muscle blood flow has been observed in cases of moderate to severe motion sickness. Skeletal muscle blood flow is most often measured in the forearm using a technique called venous occlusion plethysmography. In this technique, a pressure cuff blocks venous return, and the rate of swelling in the arm is the measure used to estimate blood flow. The rate of blood flow through the arm can increase two to three times in motion sick subjects and is generally correlated with symptom severity (Sunahara, Farewell, Mintz, & Johnson, 1987; Sunahara, Johnson, & Taylor, 1964).

The most commonly measured cardiovascular correlates of motion sickness are heart rate (HR), blood pressure (BP) and cardiac interbeat interval (R-R interval). Heart rate and blood pressure responses associated with motion sickness are, however, typically smaller than those seen in other conditions that produce a stress response, and they can increase, decrease, or stay the same (see Harm, 1990; Money et al., 1996, for reviews). The high levels of plasma AVP produced during motion sickness may partially counteract HR and BP increases normally exhibited as part of the stress response. Intravenous administration of high doses of AVP causes decreases in heart rate and blood pressure (Thomford & Sirinek, 1975). Also, the decrease in cutaneous blood flow and the increase in skeletal muscle blood flow that occurs with motion sickness may serve to stabilize or minimize BP changes during motion sickness (Money et al., 1996). The variability reported in the direction of changes in HR and BP can be due to a number of factors, including individual differences in responses to motion sickness, the specific stimulus conditions (duration, intensity, type of sensory conflict), and/or the severity of motion sickness at test termination.

Beat-by-beat variations in heart rate are mediated by both branches of the ANS. Spectral analyses of the frequency components of R-R interval data have identified three frequency bands that contribute to these variations. The higher of the three frequency bands is related to respiratory frequency. Respiratory frequency oscillations, often referred to as respiratory sinus arrhythmia (RSA), reflect parasympathetic vagal activity (Berger & Cohen, 1987; Chen, Berger, Saul, Stevenson, & Cohen, 1987). These changes in parasympathetic activity cause increases in the spectral power of the respiratory frequency band, which increases the overall R-R interval variability. The potential utility of cardiac variability as a measure of motion sickness severity was demonstrated in squirrel monkeys (Ishii, Igarashi, Patel, Himi, & Kulecz, 1987), where increased cardiac variability was related positively to symptom severity. In studies using a circular vection drum to induce motion sickness symptoms, Uijtdehaage, Stern, and Koch (1992) observed decreases in RSA during drum exposure. In addition, they found that higher levels of vagal tone (higher levels of RSA) prior to drum rotation predicted a low incidence of motion sickness symptoms and were associated positively with normal 3 cpm gastric BER. In a subsequent study, these investigators found that RSA level prior to drum rotation correctly predicted sickness level during the test in 76.7% of the subjects (significantly better than chance) with higher levels of vagal tone predicting fewer symptoms (Uijtdehaage,

Stern, & Koch, 1993). Other investigators also have found decreases in R-R interval responses during motion sickness (Doweck et al., 1997; Hu, Grant, Stern, & Koch, 1991), whereas another group found no change in heart rate variability during motion sickness (Mullen, Berger, Oman, & Cohen, 1998).

Directional differences in HR variability reported across the studies just described may be due to differences in: individual responses, stimulus conditions, severity of motion sickness, the specific measure of variability employed, and/or the methodology used for measuring and analyzing HR variability. Respiratory rate and volume contribute to HR variability. Studies that control respiratory rate and volume during HR measurement periods may show different results than those that do not control these respiratory factors. In addition, different measures of HR variability are used including the coefficient of variance (CV = standard deviation/mean), spectral power of the high-frequency band (RSA), total spectral power, or the ratio of low- to high-frequency power. Several reviews of R-R interval neurophysiology and measurement and analysis techniques are available (see Gottman, 1990; Kitney & Rompelman, 1980; Papillo & Shapiro, 1990; Porges & Bohrer, 1990).

Intuitively, one would expect a susceptible individual to exhibit the various signs, symptoms, and autonomic manifestations as well as higher levels of catecholamines and adrenocortical hormones than an insusceptible person. Although limited, the current findings indicate that while insusceptible individuals do not exhibit the usual symptom complex, their baseline values of neuroendocrine measures and the increase in these values during motion stimulation are higher than susceptible subjects. Kohl (1985, p. 1162) argues that "individuals possessing lower susceptibilities to stressful motion are more adaptive to environmental stressors or sensory conflicts by virtue of their higher baseline levels or activity of certain specific endocrine components, and particularly by the greater responsivity of these endocrine systems." Findings in studies concerned with physiological correlates, although limited, also indicate that insusceptibles show evidence of higher levels of sympathetic activation (elevated heart rate, cutaneous vasoconstriction), and earlier onset during motion sickness provocation than susceptibles (Cowings, Suter, Toscano, Kamiya, & Naifeh, 1986; Harm et al., 1987a). The neuropharmacological research concerned with motion sickness indicates that a combination of sympathomimetic and anticholinergic agents is more effective in preventing or reducing motion sickness symptomatology than either type of agent alone. These findings tend to support the notion that SNS activation provides some protection against motion sickness.

## 5. PREVENTION AND TREATMENT

By far, the most common method of preventing and treating motion sickness is the administration of medications. Other methods include biofeedback training, acupressure and acustimulation, and adaptation procedures. Adaptation and other techniques for managing cybersickness are dealt with in more detail in chapters 30, 31, and 36 (this volume).

### 5.1 Pharmacologic Agents

Many drugs have been tested for their effectiveness against motion sickness. Although some drugs have proven somewhat effective, no drug or drug combination has been identified that protects all individuals. In ranking drugs with respect to their effectiveness in preventing symptoms of motion sickness, Wood and Graybiel (1970) found that the drugs tended to be grouped according to their principal pharmacological action. The major classes of drugs that have been used in the treatment of motion sickness are anticholinergics (parasympatholytics), antihistamines, sympathomimetics, and sympatholytics. Medications from these classes have been evaluated for their effectiveness when given alone and when given together in various combinations. Antimotion sickness drug research has been reviewed extensively by Reschke et al. (1994), Wood (1979, 1990), and Yates, Miller, & Lucot (1998b).

Currently, the most effective antimotion sickness medications are scopolamine and promethazine, and both of these in combination with sympathomimetics (e.g., ephedrine and D-amphetamine). The most effective over-the-counter medication is Dramamine (Miller & Grelot, 1996). Numerous studies have found scopolamine, an anticholinergic (parasympatholytic) drug, to be effective in treating motion sickness (Graybiel & Lackner, 1987; Wood, 1990; Wood & Graybiel, 1968). Although most of the antihistamines tested for antimotion sickness properties have had some benefit, they generally provide less protection than scopolamine. Promethazine, the most effective of the antihistamines (that has the strongest central anticholinergic action of the antihistamines), approaches scopolamine in efficacy (Graybiel & Lackner, 1987), and is currently the drug of choice for treating space motion sickness (Davis, Jennings, Beck, & Bagian, 1993b; Miller & Grelot, 1996). The few sympatholytic drugs that are effective against motion sickness were found to have only marginal benefit and had less effect than the least effective antihistamine (Wood & Graybiel, 1970).

The ability of sympathomimetic drugs to prevent motion sickness was first discovered when amphetamine was combined with scopolamine to counteract the sedation caused by the latter. Experimental control subjects taking amphetamine alone also exhibited increased tolerance to motion (Wood & Graybiel, 1970). Wood and Graybiel (1970) found that combining a parasympatholytic drug (scopolamine) with a sympathomimetic produced either an additive effect (ephedrine) or a synergistic effect (amphetamine). These combinations were far more effective than any single drug. The combination of promethazine and ephedrine has also been found to be very effective in treating motion sickness (Wood, 1979).

Two newer classes of antiemetic drugs, 5HT3 and NK1 receptor antagonists, may be promising candidates for the treatment of motion sickness (Gardner et al., 1995; Koch, Xu, Bingaman, et al., 1994). The 5HT3 receptor antagonist drugs have been shown to be effective in treating nausea and vomiting associated with chemotherapy, anesthesia, and radiation (Fujii, Toyooka, & Tanaka, 1996; Kovac et al., 1996; Lehoczky, 1999). One drug in this class, ondansetron, was found ineffective in preventing motion sickness in humans in one study (Stott, Barnes, Wright, & Ruddock, 1989) but produced a small but significant improvement in resistance to motion sickness in another study (Koch et al., 1994). The drug was administered orally in the first study and intravenously in the second study. Additional research is needed to evaluate other medications in this class and to determine if higher dosages will be effective in treating motion sickness. The NK1 receptor antagonist drugs produce more broad-spectrum antiemetic activity than the 5HT3 antagonists do. Thus far, most of the research using this class of drugs has been done with animal models with one exception (Diemunsch et al., 1999). Although no studies have examined the effectiveness of NK1 receptor antagonists in preventing motion sickness in humans, Gardner et al. (1995) found an NK1 receptor antagonist was effective in preventing motion sickness in monkeys. A major advantage of these two classes of drugs is that they do not produce central nervous system depressant effects or degraded cognitive and psychomotor performance (Benline, French, & Poole, 1997).

### *5.1.1 Mechanisms of Action*

Given that the neural circuitry involved in motion sickness is not known, it is not surprising that the specific mechanism(s) of action of the effective antimotion sickness drugs is also unknown. However, a number of investigators have offered some general explanations, particularly for the effectiveness of the drug combinations. Motion sickness symptoms include both parasympathetic and sympathetic reactions (Money et al., 1996; Wood & Graybiel, 1972).

Wood and Graybiel (1972) postulated that exposure to stressful motion activates neurons in the central nervous system that respond to acetylcholine while simultaneously activating norepinephrine neurons. The enhanced effect of anticholinergic/sympathomimetic combinations would stem from paired cholinergic-blocking activity and norepinephrine activation. This explanation suggests a general nonspecific drug action on the ANS. Alternatively, the site of action of antimotion sickness medications may be in the vestibular nuclei (Yates et al., 1998b). The vestibular nuclei contain a number of the types of receptors acted on by antimotion sickness medications (Yates et al., 1998b). The site of action of broad-spectrum antiemetics is presumed to be somewhere in the final common pathway for vomiting. If this is the case, these drugs may not be effective against the nausea and autonomic symptoms of motion sickness.

In addition to the lack of knowledge of the neural circuitry involved in motion sickness, mechanisms of action of antimotion sickness medications are difficult to determine for several reasons. First, much of the research in this area is carried out on animals. There are considerable interspecies differences in responsiveness to different medications, and hence, results from this work may not generalize to humans. In addition, some drugs in a given class are effective, whereas others are not. Similarly, different drug combinations from the same two drug classes have varying degrees of effectiveness. For the interested reader, Yates et al. (1998b) recently published an excellent review of the pharmacology of motion sickness.

#### *5.1.2 Routes of Administration*

The vast majority of antimotion sickness drugs have been administered orally. Because the duration of action of these drugs is typically brief, frequent dosing is needed if the motion is expected to last for extended periods. An additional complication of oral medications is the reduction in gastric motility characteristic of acute motion sickness (Wood, Wood, Manno, Manno, & Redetzki, 1987); drugs must be given prophylactically to avoid decreases in drug absorption. For this reason, alternate routes of administration, such as transdermal application (Homick, Kohl, Reschke, Degioanni, & Cintron-Trevino, 1983; Levy & Rapaport, 1985; McCauley, Royal, Shaw, & Schmitt, 1979), suppositories (Davis et al., 1993a), and intramuscular injections (Davis et al., 1993b; Graybiel & Lackner, 1987) have been investigated. Another factor complicating the search for motion sickness remedies is the occurrence of side effects, which can preclude using the most effective dose or dosing schedule.

### 5.2 Biofeedback and Autogenic Feedback Training

In biofeedback training, instrumental information about selected autonomic activities is provided to the subject with a visual or auditory "reward" presented for producing the desired response (e.g., decreased heart rate). Autogenic training employs a collection of cognitive imagery techniques to produce the desired change in autonomic activity. Self-suggestion exercises are utilized to produce certain body sensations that correspond to the desired changes in physical parameters. Autogenic feedback training (AFT) combines the above techniques. Cognitive imagery is used to produce the desired changes with immediate sensory feedback on success via instrumental readouts (Cowings, 1990).

The basic rationale underlying the use of biofeedback or autogenic feedback training is that since the symptoms of motion sickness are autonomically mediated, self-regulation of ANS responses should lead to reduced symptoms during motion sickness–provoking situations. Cowings argues that nausea and vomiting are parasympathetic reactions to sympathetic activation. If so, then motion sickness symptoms should be prevented by training an individual to maintain his or her autonomic responses at baseline levels. Furthermore, since the perception of vestibular stimulation is unchanged by AFT, this type of training must interrupt the autonomic response after the sensory conflict has already occurred (Cowings, 1990). Raising the threshold for autonomic activation may inhibit development of motion sickness symptoms induced by sensory conflict (Cowings & Toscano, 1982).

More recent investigations, however, have found that autogenic feedback training is ineffective in reducing motion sickness symptoms (Dobie, May, Fischer, Elder, & Kubitz, 1987; Jozsvai & Pigeau, 1996). Jozsvai and Pigeau (1996) examined Coriolis-induced motion sickness severity and motion tolerance in three groups of subjects: (1) autogenic training plus true feedback; (2) autogenic training plus false feedback; and (3) a control group that received no treatment. These investigators found no differences between the two treatment groups in self-regulation of autonomic responses and no relationship between self-regulation ability and sickness severity or motion tolerance. Moreover, these authors suggest that treatment gains reported by others may result from placebo effect. The issue of biofeedback versus placebo effects is discussed in greater detail elsewhere (Furedy, 1985).

### 5.3 Electrostimulation and Acupressure

The use of electrical currents in the treatment of motion sickness has also been explored. Electroanalgesia or electrotranquilization uses two electrodes, one placed on the forehead and the other in the area of the mastoid process. Melnik et al. (1986) increased the current until the subject reported a sensation of warmth in the area of the electrodes with a session length of 30 to 60 minutes. Nekhayev, Vlasov, & Ivanov (1986) and Polyakov (1987) used a pulsed current during sessions lasting an hour. Electroanalgesia did not increase resistance to experimentally induced motion sickness when sessions were performed before stressful motion (Melnik et al., 1986; Polyakov, 1987). However, sessions conducted between two motion stressor tests reduced or eliminated the residual motion sickness symptoms from the first test (Melnik et al., 1986; Nekhayev et al., 1986) and increased tolerance to the motion sickness test performed following the electroanalgesia session (Melnik et al., 1986; Nekhayev et al., 1986; Polyakov, 1987). A second session of electroanalgesia following the second motion stressor test also improved recovery from symptoms induced by that test.

Ivanov and Snitko (1985) observed that motion sickness affected the conductivity along the standard acupuncture pathways regardless of symptom severity. Electroacupuncture was used successfully by this group to treat seasickness. Others have found electrical acustimulation and acupressure to be effective in reducing vection-induced nausea (Hu, Stern, & Koch, 1992; Hu, Stritzel, Chandler, & Stern, 1995). Acustimulation may work by enhancing the normal slow wave myoelectrical activity of the stomach (Lin et al., 1997). Subthreshold multichannel electrical stimulation of the antigravity group of cervical muscles has also been reported to achieve promising results as a countermeasure against motion sickness (Matveyev, 1987). Although electrical devices are reported to be an effective countermeasure to several forms of terrestrial motion sickness, they have not been tested as a countermeasure for cybersickness.

## 6. SUMMARY

Motion sickness induced by exposure to virtual environments may be a key factor in limiting widespread use of VE technology. As many as 30% of individuals exposed to VE systems have symptoms severe enough to discontinue use. Several theories have been proposed to explain the etiology of motion sickness; however, none can fully account for this complex malady. One reason for the lack of a more comprehensive theory is that the neural circuitry and mechanisms involved in motion sickness in humans is unknown. It is accepted that an intact vestibular system is required for motion sickness to occur and that there are both direct and indirect pathways from the vestibular nuclei to ANS centers that mediate many of the signs, symptoms, and physiologic correlates of motion sickness. Certain areas in the cerebral cortex are probably necessary for the sensation of nausea, and the NTS is believed to be the final common pathway for the vomiting response, regardless of its cause.

The most common method of preventing and treating motion sickness is the administration of medications. There are a number of drugs and drug combinations that are effective in treating motion sickness, although the specific mechanisms of action are unknown. Many of the more effective medications have a sedative effect, which raises concerns about safety and performance. Newer classes of drugs that hold promise in preventing and treating motion sickness without sedative effects still need to be evaluated more extensively in clinical trials. Other methods of preventing and treating motion sickness include biofeedback, acupressure and acustimulation, and adaptation techniques.

## 7. ACKNOWLEDGMENTS

I would like to thank all the reviewers for their thoughtful comments and suggestions, and express a special thanks to Dr. Bill Yates for his critical inputs to the neuroanatomy and neurophysiology sections.

## 8. REFERENCES

Alvarez, E. L. (1922). New methods of studying gastric peristalsis. *Journal of the American Medical Association, 79*, 1281.

Andre, J. T., Muth, E. R., Stern, R. M., & Leibowitz, H. W. (1996). The effect of tilted stripes in an optokinetic drum on gastric myoelectric activity and subjective reports of motion sickness. *Aviation, Space, and Environmental medicine, 67*(1), 30–33.

Balaban, C. (1996). The role of the cerebellum in vestibular autonomic regulation. In B. J. Yates & A. D. Miller (Eds.), *Vestibular Autonomic Regulation* (pp. 127–144). Boca Raton, FL: CRC Press.

Barfield, W., & Weghorst, S. (1993). The sense of presence within virtual environments: A conceptual model. In G. Salvendy & M. Smith (Eds.), *Human–computer interaction: software and hardware interfaces* (pp. 699–704). Amsterdam: Elsevier Science Publishers.

Barron, K. D., & Chokroverty, S. (1993). Anatomy of the autonomic nervous system: Brain and brainstem. In P. A. Low (Ed.), *Clinical autonomic disorders, evaluation and management* (pp. 3–15). Boston: Little, Brown.

Benline, T. A., French, J., & Poole, E. (1997). Anti-emetic drug effects on pilot performance: granisetron vs. ondansetron. *Aviation, Space, and Environmental Medicine, 68*(11), 998–1005.

Berger, R. D., & Cohen, R. J. (1987). Analysis of the response of the sinoatrial node to fluctuations in sympathetic and parasympathetic tone using broad band stimulation techniques. *Computers in Cardiology, 13*, 153–156.

Billig, I., Foris, J. M., Card, J. P., & Yates, B. J. (1999). Transneuronal tracing of neural pathways controlling an abdominal muscle, rectus abdominis, in the ferret. *Brain Research, 820,* 31–44.

Borison, H. L., & Borison, R. (1986). Motion sickness reflex arc bypasses the area postrema in cats. *Experimental Neurology, 92*, 723–737.

Brobeck, J. R. (1979). *Best and Taylor's Physiological Basis of Medical Practice* (10th ed.). Baltimore: Williams & Wilkins.

Cannon, W. B. (1939). *The Wisdom of the body*. New York: W.W. Norton.

Chelen, W. E., Kabrisky, M., & Rogers, S. K. (1993). Spectral analysis of the electroencephalographic response to motion sickness. *Aviation, Space, and Environmental Medicine, 64*, 24–29.

Chen, M. H., Berger, R. D., Saul, J. P., Stevenson, K., & Cohen, R. J. (1987). Transfer function analysis of the autonomic response to respiratory activity during random interval breathing. *Computers in Cardiology, 13*, 149–152.

Chen, J. Z., & McCallum, R. W. (1994). *Electrogastrography—Principles and applications*. New York: Raven Press.

Chen, J., McCallum, R. W., & Richards, R. D. (1993). Frequency components of the electrogastrogram and their correlations with gastrointestinal contractions in humans. *Medical Biological Engineering and Computing, 31*, 60–67.

Chen, J. D., Richards, R. D., & McCallum, R. W. (1994). Identification of gastric contractions from the cutaneous electrogastrogram. *American Journal of Gastroenterology, 89*, 79–85.

Cheung, B. S., Kohl, R. L., Money, K. E., & Kinter, L. B. (1994). Etiologic significance of arginine vasopressin in motion sickness. *The Journal of Clinical Pharmacology, 34*(6), 664–670.

Cobb, S. V., Nichols, S. C., Ramsey, A. D., & Wilson, J. R. (1999). Virtual reality induced symptoms and effects (VRISE). *Presence: Teleoperators and Virtual Environments, 8*, 169–186.

Cowings, P. S. (1990). Autogenic-feedback training: A treatment for motion and space sickness. In G. H. Crampton (Ed.), *Motion and space sickness* (pp. 353–372). Boca Raton, FL: CRC Press.

Cowings, P. S., Suter, S., Toscano, W. B., Kamiya, J., & Naifeh, K. (1986). General autonomic components of motion sickness. *Psychophysiology, 23*, 542–551.

Cowings, P. S., & Toscano, W. B. (1982). The relationship of motion sickness susceptibility to learned autonomic control for symptom suppression. *Aviation, Space, and Environmental Medicine, 53*, 570–575.

Crampton, G. H. (1955). Studies of motion sickness: Vol. 17. Physiological changes accompanying sickness in man. *Journal of Applied Physiology, 7*, 501–507.

Cummins, A. J. (1958). The physiology of symptoms: Vol. 3. Nausea and vomiting. *American Journal of Digestive Diseases, 3*, 710.

Davis, J. E., Jennings, R. T., & Beck, B. G. (1993a). Comparison of treatment strategies for space motion sickness. *Acta Astronautica, 29*, 587–591.

Davis, J. R., Jennings, R. T., Beck, B. G., & Bagian, J. P. (1993b). Treatment efficacy of intramuscular promethazine for space motion sickness. *Aviation, Space, and Environmental Medicine, 64*, 230–233.

Davis, J. R., Vanderploeg, J. M., Santy, P. A., Jennings, R. T., & Stewart, D. F. (1988). Space motion sickness during 24 flights of the space shuttle. *Aviation, Space, and Environmental Medicine, 59*, 1185–1189.

Dawson, M. E., Schell, A. M., & Filion, D. L. (1990). The electrodermal system. In J. T. Cacioppo & L. G. Tassinary (Eds.), *Principles of psychophysiology—Physical, social and inferential elements* (pp. 295–324). New York: Cambridge University Press.

Diemunsch, P., Schoeffler, P., Bryssine, B., Cheli-Muller, L. E., Lees, J., McQuade, B. A., & Spraggs, C. F. (1999). Antiemetic activity of the NK1 receptor antagonist GR205171 in the treatment of established postoperative nausea and vomiting after major gynaecological surgery. *British Journal of Anaesthesta, 82*(2), 274–276.

DiZio, P., & Lackner, J. R. (1997). Circumventing side effects of immersive virtual environments. In M. Smith, G. Salvendy, & R. Koubek (Eds.), *Design of computing systems: Social and ergonomic considerations* (pp. 893–896). Amsterdam: Elsevier Science Publishers.

Dobie, T. G., May, J. G., Fischer, W. D., Elder, S. T., & Kubitz, K. A. (1987). A comparison of two methods of training resistance to visually induced motion sickness. *Aviation, Space, and Environmental Medicine, 58*(9 Suppl.), A34–A41.

Doweck, I., Gordon, C. R., Shlitner, A., Spitzer, O., Gonen, A., Binah, O., Melamed, Y., & Shupak, A. (1997). Alterations in R-R variability associated with experimental motion sickness. *J Auton Nerv Syst, 67*(1–2), 31–37.

Eversmann, T., Gottsmann, M., Uhlich, E., Ulbrecht, G., von Werder, K., & Scriba, P. C. (1978). Increased secretion of growth hormone, prolactin, antidiuretic hormone, and cortisol induced by the stress of motion sickness. *Aviation, Space, and Environmental Medicine, 49*, 53–57.

Familoni, B. O., Abell, T. L., & Bowes, K. L. (1995). A model of gastric electrical activity in health and disease. *IEEE Transactions on Biomed Eng, 42*(7), 647–657.

Fujii, K., & Mukai, M. (1985). Neurohumoral mechanisms of excitation of gastric motility in the dog. In M. T. Y. Kasuya, F. Nagao, T. Matsuo (Eds.), *Gastrointestinal function, regulation and disturbances* (Vol. 3, pp. 15–27). Amsterdam: Excerpta Medica.

Fujii, Y., Toyooka, H., & Tanaka, H. (1996). Antiemetic effects of granisetron on postoperative nausea and vomiting in patients with and without motion sickness. *Canadian Journal of Anaesthesta, 43(2)*, 110–114.

Furedy, J. J. (1985). Specific vs. placebo effects in biofeedback: Science-based vs. snake-oil behavioral medicine. *Clinical Biofeedback and Health, 8*, 155–162.

Gardner, C. J., Twissell, D. J., Dale, T. J., Gale, J. D., Jordan, C. C., Kilpatrick, G. J., Bountra, C., & Ward, P. (1995). The broad-spectrum anti-emetic activity of the novel non-peptide tachykinin NK1 receptor antagonist GR203040. *British Journal of Pharmacology, 116*(8), 3158–3163.

Gilman, A. G., Rall, T. W., Nies, A. S., & Taylor, P. (1990). *Goodman and Gilman's the pharmacological basis of therapeutics* (8th ed.). New York: Pergamon Press.

Gottman, J. (1990). Time-series analysis applied to physiological data. In J. T. Cacioppo & L. G. Tassinary (Eds.), *Principles of psychophysiology—Physical, social, and inferential elements* (pp. 754–774). Cambridge, England: Cambridge University Press.

Graybiel, A. (1969). Structural elements in the concept of motion sickness. *Aerospace Medicine, 40*, 351–367.

Graybiel, A., & Lackner, J. R. (1987). Treatment of severe motion sickness with antimotion-sickness drug injections. *Aviation, Space, and Environmental Medicine, 58*, 773–776.

Graybiel, A., Wood, C. D., Miller, E. F., & Cramer, D. B. (1968). Diagnostic criteria for grading the severity of acute motion sickness. *Aerospace Medicine, 39*, 453–455.

Grigoriev, A. I., Nichiporuk, A. I., Yasnetsov, V. V., & Shashkov, V. S. (1988). Hormonal status and fluid electrolyte metabolism in motion sickness. *Aviation, Space, and Environmental Medicine, 59*, 301–305.

Gross, N., Yang, D. J., & Flynn, J. (1995, July 10). Seasick in cyberspace. *Business Week*, 110–113.

Hall, K. E., El-Sharkawy, T. Y., & Diamant, N. E. (1986). Vagal control of canine postprandial upper-gastrointestinal motility. *American Journal of Physiology, 250*, G501–G510.

Harm, D. L. (1990). Physiology of motion sickness symptoms. In G. H. Crampton (Ed.), *Motion and space sickness* (pp. 153–177). Boca Raton, FL: CRC Press.

Harm, D. L., Beatty, B. J., & Reschke, M. F. (1987a). Transcutaneous oxygen as a measure of pallor. *Aviation, Space, and Environmental Medicine, 58 (Abstract)*, 508.

Harm, D. L., & Parker, D. E. (1994). Preflight adaptation training for spatial orientation and space motion sickness. *The Journal of Clinical Pharmacology, 34*, 618–627.

Harm, D. L., Sandoz, G. R., Stern, R. M., Koch, K. L., & Koslovskya, I. (1997). *Gastric dysrhythmias associated with space motion sickness*. Paper presented at the International Workshop on Motion Sickness—Medical and Human Factors, Marbella, Spain.

Harm, D. L., Stern, R. M., & Koch, K. L. (1987b). *Tachygastria during parabolic flight*. Paper presented at the Space Life Science Symposium: Three Decades of Life Science Research in Space, Washington, DC.

Hemmingway, A. (1944). Cold sweating in motion sickness. *American Journal of Physiology, 141*, 172.

Hettinger, L. J., Berbaum, K. S., Kennedy, R. S., Dunlap, W. P., & Nolan, M. D. (1990). Vection and simulator sickness. *Military Psychology, 2*(3), 171–181.

Homick, J. L., Kohl, R. L., Reschke, M. F., Degioanni, J., & Cintron-Trevino, N. M. (1983). Transdermal scopolamine in the prevention of motion sickness: Evaluation of the time course of efficacy. *Aviation, Space, and Environmental Medicine, 54*, 994–1000.

Hu, S., Grant, W. F., Stern, R. M., & Koch, K. L. (1991). Motion sickness severity and physiological correlates during repeated exposures to a rotating optokinetic drum. *Aviation, Space, and Environmental Medicine, 62*(4), 308–314.

Hu, S., Stern, R. M., & Koch, K. L. (1992). Electrical acustimulation relieves vection-induced motion sickness. *Gastroenterology, 102*, 1854–1858.

Hu, S., Stritzel, R., Chandler, A., & Stern, R. M. (1995). P6 acupressure reduces symptoms of vection-induced motion sickness. *Aviation, Space, and Environmental Medicine, 66*(7), 631–634.

Irwin, J. A. (1881). The pathology of seasickness. *Lancet, 2*, 907.

Ishii, M., Igarashi, M., Patel, S., Himi, T., & Kulecz, W. B. (1987). Autonomic effects on R–R variations of the heart rate in squirrel monkey: An indicator of autonomic imbalance in conflict sickness. *American Journal of Otolaryngoloty, 3*, 144–148.

Ito, J., & Honjo, I. (1990). Central fiber connections of the vestibulo-autonomic reflex arc in cats. *Acta Oto-Laryngologica, 110*(5–6), 379–385.

Ivanov, A. L., & Snitko, V. M. (1985). Use of acupuncture in the prevention and cure of motion sickness. *Voenno-Meditisinskii Zhurnal*, 8, 56–57.

Jozsvai, E. E., & Pigeau, R. A. (1996). The effect of autogenic training and biofeedback on motion sickness tolerance. *Aviation, Space, and Environmental Medicine, 67*(10), 963–968.

Kalawsky, R. S. (1993). *The science of virtual reality and virtual environments*. Workingham, England: Addison-Wesley.

Kaye, W. H., Gwirtsman, H. E., & George, D. T. (1989). The effect of bingeing and vomiting on hormonal secretion. *Biological Psychology, 25*, 768–780.

Kennedy, R. S., Jones, M. B., Lilienthal, M. G., & Harm, D. L. (1993a). *Profile analysis of after-effects experienced during exposure to several virtual reality environments*. Paper presented at the AGARD 76th Aerospace Medical Panel/Symposium on Virtual Interfaces: Research and Application, Lisbon, Portugal.

Kennedy, R. S., Lane, N. E., Berbaum, K. S., & Lilenthal, M. G. (1993b). Simulator sickness questionnaire: An enhanced method for quantifying simulator sickness. *The International Journal of Aviation Psychology, 3*(3), 203–220.

Kennedy, R. S., & Stanney, K. M. (1997). Aftereffects of virtual environment exposure: Psychometric issues. In M. Smith, G. Salvendy, & R. Koubek (Eds.), *Design of computing systems: Social and ergonomic considerations* (pp. 897–900). Amsterdam: Elsevier Science Publishers.

Kitney, R. I., & Rompelman, O. (1980). *The Study of Heart Rate Variability*. Oxford, England: Clarendon Press.

Koch, K. L., Summy-Long, J., Bingaman, S., Sperry, N., & Stern, R. M. (1990). Vasopressin and oxytocin responses to illusory self-motion and nausea in man. *Journal of Clinical Endocrinology and Metabolism, 71*, 1269–1275.

Koch, K. L., Xu, L., Bingaman, S., et al. (1994). Effect of ondansetron on motion sickness, gastric dysrhythmias and plasma vasopressin. *Gastroenterology, 106*(4 Suppl.), A525.

Kohl, R. L. (1985). Endocrine correlates of susceptibility to motion sickness. *Aviation, Space, and Environmental Medicine, 56*, 1158–1165.

Kohl, R. L., Leach, C. S., Homick, J. L., & LaRochelle, F. T. (1983). Motion sickness susceptibility related to ACTH, ADH, and TSH. *Physiologist, 26*, S117–S118.

Kolev, O. I., & Altaparmakov, I. A. (1996). Changes in interdigestive migrating electric complex induced by motion sickness. *Functional Neurology, 11*(1), 29–33.

Kolev, O. I., Moller, C., Nilsson, G., & Tibbling, L. (1997). Responses in skin microcirculation to vestibular stimulation before and during motion sickness. *Can J Neurological Sciences, 24*(1), 53–57.

Kovac, A. L., Pearman, M. H., Khalil, S. N., Scuderi, P. E., Joslyn, A. F., Prillaman, B. A., & Cox, F. (1996). Ondansetron prevents postoperative emesis in male outpatients. S3A-379 study group. *Journal of Clinical Anesthesta, 8*(8), 644–651.

Kucharczyk, J., Stewart, D. J., & Miller, A. D. (1991). *Nausea and vomiting: Recent research and clinical Advances.* Boca Raton, FL: CRC Press.

Lehoczky, O. (1999). About the antiemetic effectivity of granisetron in chemotherapy-induced acute emesis: A comparison of results with intravenous and oral dosing (in process citation). *Neoplasma, 46*(2), 73–79.

Levy, G. D., & Rapaport, M. H. (1985). Transdermal scopolamine efficacy related to time of application prior to the onset of motion. *Aviation, Space, and Environmental Medicine, 56*, 591–593.

Lin, X., Liang, J., Ren, J., Mu, F., Zhang, M., & Chen, J. D. (1997). Electrical stimulation of acupuncture points enhances gastric myoelectrical activity in humans. *American Journal of Gastroenterology, 92*(9), 1527–1530.

Marti Usaney, F. (1954). Philosophical perspectives of motion sickness. *International Record of Medicine, 167*, 621.

Matveyev, A. D. (1987). Development of methods for the study of space motion sickness. *Kosmicheskaya Biologiya I Aviakosmicheskaya Meditsina, 21*(3), 83–88.

McCauley, M. E., Royal, J. W., Shaw, J. E., & Schmitt, L. G. (1979). Effect of transdermally administered scopolamine in preventing motion sickness. *Aviation, Space, and Environmental Medicine, 50*, 1108–1111.

McCauley, M. E., & Sharkey, T. J. (1992). Cybersickness: Perception of self-motion in virtual environments. *Presence: Teleoperators and Virtual Environments, 1*(3), 311–318.

McClure, J. A., Fregly, A. R., Molina, E., & Graybiel, A. (1971). *Response from arousal and thermal sweat areas during motion sickness* (NAMRL-1142). Pensacola, FL: Naval Aerospace Medical Research Laboratory.

Melnik, C. G., Shakula, A. V., & Ivanov, V. V. (1986). The use of the electrotranquilization method for increasing vestibular tolerance in humans. *Voenno-Meditisinskii Zhurnal, 8*, 42–45.

Miller, A. D., & Grelot, L. (1996). The neural basis of nausea and vomiting. In B. J. Yates & A. D. Miller (Eds.), *Vestibular Autonomic Regulation* (pp. 85–94). Boca Raton, FL: CRC Press.

Miller, A. D., Nonaka, S., & Jakus, J. (1994). Brain areas essential or non-essential for emesis. *Brain Research, 647*, 255–264.

Miller, A. D., Nonaka, S., Jakus, J., & Yates, B. J. (1996). Modulation of vomiting by the medullary midline. *Brain Research, 737*, 51–58.

Miller, A. D., & Ruggiero, D. A. (1994). Emetic reflex arc revealed by expression of the immediate-early gene C-Fos in the cat. *Journal of Neuroscience, 14*, 871–888.

Miller, A. D., & Wilson, V. J. (1983). Vestibular-induced vomiting after vestibulocerebellar lesions. *Brain Behavior Evolution, 23*, 26–31.

Money, K. E. (1970). Motion sickness. *Physiological Reviews, 50*, 1–39.

Money, K. E., Lackner, J., & Cheung, R. (1996). The autonomic nervous system and motion sickness. In B. J. Yates & A. D. Miller (Eds.), *Vestibular autonomic regulation* (pp. 147–173). Boca Raton, FL: CRC Press.

Mullen, T. J., Berger, R. D., Oman, C. M., & Cohen, R. J. (1998). Human heart rate variability relation is unchanged during motion sickness. *J Vestibulur Res, 8*(1), 95–105.

Muth, E. R., Stern, R. M., & Koch, K. L. (1996). Effects of vection-induced motion sickness on gastric myoelectric activity and oral-cecal transit time. *Digestive Diseases and Sciences, 41*(2), 330–334.

Muth, E. R., Thayer, J. F., Stern, R. M., Friedman, B. H., & Drake, C. (1998). The effect of autonomic nervous system activity on gastric myoelectrical activity: Does the spectral reserve hypothesis hold for the stomach? *Biological Psychology, 47*(3), 265–278.

Nekhayev, A. S., Vlasov, V. D., & Ivanov, V. V. (1986). Use of central electroanalgesia for functional recovery from motion sickness. *Kosmicheskaya Biologiya I Aviakosmicheskaya Meditsina, 20*(4), 42–44.

Nussey, S. S., Hawthron, J., Page, S. R., Ang, V. T. Y., & Jenking, J. S. (1988). Responses of plasma oxytocin and arginine vasopressin to nausea induced by apomorphine and ipecacuanha. *Clinical Endocrinology, 28*, 297–304.

Oman, C. M. (1982). *Space motion sickness and vestibular experiments in spacelab.* Paper presented at the Twelfth Intersociety Conference on Environmental Systems, San Diego, CA.

Oman, C. M., & Cook, W. J. C. (1983). *Dynamics of skin pallor in motion sickness as measured using an infrared reflectance technique.* Paper presented at the 54th Annual Aerospace Medical Association Meeting, Houston, TX.

Oman, C. M., Lichtenberg, B. K., & Money, K. E. (1990). Space motion sickness monitoring experiment: Spacelab-1. In G. W. Crampton (Ed.), *Motion and space sickness* (pp. 217–246). Boca Raton, FL: CRC Press.

Onai, T., Takayama, K., & Miura, M. (1987). Projections to areas of the nucleus tractus solitarii related to circulatory and respiratory responses in cats. *Journal of the Autonomic Nervous System, 18*(2), 163–175.

Papillo, J. F., & Shapiro, D. (1990). The cardiovascular system. In J. T. Cacioppo & L. G. Tassinary (Eds.), *Principles of psychophysiology—Physical, social, and inferential elements* (pp. 456–512). Cambridge, England: Cambridge University Press.

Pausch, R., Crea, T., & Conway, M. (1992). A literature survey for virtual environments: Military flight simulator visual systems and simulator sickness. *Presence, 1*(3), 344–363.

Polyakov, B. I. (1987). Discrete adaptation to sensory conflict. *Kosmicheskaya Biologiya I Aviakosmicheskaya Meditsina, 21*(5), 82–86.

Porges, S. W., & Bohrer, R. E. (1990). The analysis of periodic processes in psychophysiological research. In J. T. Cacioppo & L. G. Tassinary (Eds.), *Principles of psychophysiology—Physical, social, and inferential elements* (pp. 708–753). Cambridge, England: Cambridge University Press.

Previc, F. H. (1993). Do the organs of the labyrinth differentially influence the sympathetic and parasympathetic system? *Neuroscience and Biobehavioral Reviews, 17*, 397–404.

Rague, B. W., & Oman, C. M. (1987). *Detection of motion sickness onset using abdominal biopotentials.* Paper presented at the Space Life Sciences Symposium: Three Decades of Life Science Research in Space, Washington, DC.

Rapoport, A., Horvath, W. J., Small, R. B., & Fox, S. S. (1967). *The mammalian central nervous system as a network* (AMRL-TR-67-187). Dayton, OH: Wright Patterson AFB, Aerospace Medical Research Laboratories.

Reason, J. T., & Brand, J. J. (1975). *Motion sickness*. London: Academic Press.

Reid, K., Grundy, D., Khan, M. I., & Read, N. W. (1995). Gastric emptying and the symptoms of vection-induced nausea. *European Journal of Gastroenterology and Hepalotogy, 7*(2), 103–108.

Reschke, M. F., Harm, D. L., Parker, D. E., Sandoz, G. R., Homick, J. L., & Vanderploeg, J. M. (1994). Neurophysiologic aspects: Space motion sickness. In C. L. Huntoon . A. E. Nicogossian, & S. L. Pool (Eds.), *Space physiology and medicine* (3rd ed., pp. 228–260). Philadelphia: Lea & Febiger.

Reschke, M. F., Kornilova, L. M., Harm, D. L., Bloomberg, J. J., & Paloski, W. H. (1996). Neurosensory and sensory–motor function. In A. E. Nicogossian, S. R. Mohler, O. G. Gazenko, & A. I. Grigoriev (Eds.), *Space biology and medicine* (Vol. 3, pp. 135–193). Reston, VA: American Institute of Aeronautics and Astronautics.

Riccio, G. E., & Stoffregen, T. A. (1991). An ecological theory of motion sickness and postural instability. *Ecological Psychology, 3*(3), 195–240.

Roberts, T., Miller, A., Rowley, H., & Kucharczyk, J. (1994). Functional neural imaging of nausea in humans. *Society of Neuroscience Abstracts, 20*, 1417.

Shields, S. A., MacDowell, K. A., Fairchild, S. B., & Campbell, M. L. (1987). Is mediation of sweating cholinergic, adrenergic, or both? A comment on the literature. *Psychophysiology, 24*, 312–319.

Stanney, K. M., Salvendy, G., Deisigner, J., DiZio, P., Ellis, S., Ellison, E., Fogleman, G., Gallimore, J., Hettinger, L., Kennedy, R., Lackner, J., Lawson, B., Maida, J., Mead, A., Mon-Williams, M., Newman, D., Piantanida, T., Reeves, L., Riedel, O., Singer, M., Stoffregen, T., Wann, J., Welch, R., Wilson, J., Witmer, B. (1998). Aftereffects and sense of presence in virtual environments: Formulation of a research and development agenda (Report sponsored by the Life Sciences Division at NASA Headquarters). *International Journal of Human–Computer Interaction, 10*(2), 135–187.

Steele, J. E. (1968). *The symptomatology of motion sickness. In Fourth Symposium on the Role of the Vestibular Organs in Space Exploration* (NASA SP-187). Pensacola, FL: Naval Aerospace Medical Center.

Stern, R. M., Koch, K. L., Stewart, W. R., & Lindbald, I. M. (1987). Spectral analysis of tachygastria recording during motion sickness. *Gastroenterology, 92*, 92–97.

Stewart, J. J., Wood, M. J., & Wood, C. D. (1989). Electrogastrograms during motion sickness in fasted and fed subjects. *Aviation, Space, and Environmental Medicine, 60*, 214–217.

Stewart, J. J., Wood, M. J., Wood, C. D., & Mims, M. E. (1994). Effects of motion sickness and antimotion sickness drugs on gastric function. *Journal of Clinical Pharmacology, 34*(6), 635–643.

Stoffregen, T. A., & Riccio, G. E. (1991). An ecological critique of the sensory conflict theory of motion sickness. *Ecological Psychology, 3*, 151–194.

Stott, J. R., Barnes, G. R., Wright, R. J., & Ruddock, C. J. (1989). The effects on motion sickness and oculomotor function of GR 38032F, a 5-$HT_3$-receptor antagonist with anti-emetic properties. *British Journal of Clinical Pharmacology, 27*, 147–157.

Stout, C. S., Toscano, W. B., & Cowings, P. S. (1995). Reliability of psychophysiological responses across multiple motion sickness stimulation tests. *J Vestibular Res, 5*(1), 25–33.

Sunahara, F. A., Farewell, J., Mintz, L., & Johnson, W. H. (1987). Pharmacological interventions for motion sickness: Cardiovascular effects. *Aviation, Space, and Environmental Medicine, 58*, A270–A276.

Sunahara, F. A., Johnson, W. H., & Taylor, N. B. G. (1964). Vestibular stimulation and forearm blood flow. *Canadian Journal of Physiological Pharmacology, 42*, 199–207.

Sved, A., & Ruggiero, D. (1996). The autonomic nervous system structure and function. In B. J. Yates & A. D. Miller (Eds.), *Vestibular autonomic regulation* (pp. 25–51). Boca Raton, FL: CRC Press.

Tache, Y., & Wingate, D. (1991). *Brain-gut interactions*. Boca Raton, FL: CRC Press.

Thomford, N. R., & Sirinek, K. R. (1975). Intravenous vasopressin in patients with portal hypertension: advantages of continuous infusion. *Journal of Surgical Research, 18*, 113.

Treisman, M. (1977). Motion sickness: An evolutionary hypothesis. *Science, 197*, 493–495.
Tyler, D. B., & Bard, P. (1949). Motion sickness. *Physiological Reviews, 29*, 311.
Uijtdehaage, S. H. J., Stern, R. M., & Koch, K. L. (1992). Effects of eating on vection-induced motion sickness, cardiac vagal tone, and gastric myoelectric activity. *Psychophysiology, 29*, 193–201.
Uijtdehaage, S. H. J., Stern, R. M., & Koch, K. L. (1993). Effects of scopolamine on autonomic profiles underlying motion sickness susceptibility. *Aviation, Space, and Environmental Medicine, 64*, 1–8.
Venables, P. H., & Christie, M. J. (1980). Electrodermal activity. In I. Martin & P. H. Venables (Eds.), *Techniques in Psychophysiology*. New York: John Wiley & Sons.
Wang, G. H. (1964). *The neural control of sweating*. Madison: University of Wisconsin Press.
Wilson, J. R., Nichold, S. C., & Ramsey, A. D. (1995). Virtual reality health and safety: Facts, speculation and myths. *VR News, 4*, 20–24.
Wood, C. D. (1979). Antimotion sickness and antiemetic drugs. *Drugs, 17*, 471–479.
Wood, C. D. (1990). Pharmacological countermeasures against motion sickness. In G. H. Crampton (Ed.), *Motion and space sickness* (pp. 343–351). Boca Raton, FL: CRC Press.
Wood, C. D., & Graybiel, A. (1968). Evaluation of sixteen anti-motion sickness drugs under controlled laboratory conditions. *Aerospace Medicine, 39*, 1341–1344.
Wood, C. D., & Graybiel, A. (1970). Evaluation of antimotion sickness drugs: A new effective remedy revealed. *Aerospace Medicine, 41*, 932–933.
Wood, C. D., & Graybiel, A. (1972). Theory of antimotion sickness drug mechanisms. *Aerospace Medicine, 43*, 249–252.
Wood, C. D., Stewart, J. J., Wood, M. J., Struve, F. A., Straumanis, J. J., Mims, M. E., & Patrick, G. Y. (1994). Habituation and motion sickness. *Journal of Clinical Pharmacology, 34*(6), 628–634.
Wood, M. J., Wood, C. D., Manno, J. E., Manno, B. R., & Redetzki, H. M. (1987). Nuclear medicine evaluation of motion sickness and medications on gastric emptying time. *Aviation, Space, and Environmental Medicine, 58*, 1112–1114.
Yates, B. J. (1996). Vestibular influences on cardiovascular control. In B. J. Yates & A. D. Miller (Eds.), *Vestibular autonomic regulation* (pp. 97–111). Boca Raton, FL: CRC Press.
Yates, B. J. (1992). Vestibular influences on the sympathetic nervous system. *Brain Research Reviews, 17*, 51–59.
Yates, B. J., Goto, T., & Bolton, P. S. (1993). Responses of neurons in the rostral ventrolateral medulla of the cat to natural vestibular stimulation. *Brain Research, 601*(1–2), 255–264.
Yates, B. J., Grelot, L., Kerman, I. A., Balaban, C. D., Jakus, J., & Miller, A. D. (1994). Organization of vestibular inputs to nucleus tractus solitarius and adjacent structures in cat brain stem. *American Journal of Physiology, 267*(4, pt. 2), R974–R983.
Yates, B. J., & Miller, A. D. (1996). *Vestibular-autonomic regulation*. Boca Raton, FL: CRC Press.
Yates, B. J., Miller, A. D., & Lucot, J. B. (1998a). Physiological basis and pharmacology of motion sickness: An update. *Brain Research Bulletin, 47*(5), 395–406.
Yates, B. J., Miller, A. D., & Lucot, J. B. (1998b). Physiological basis and pharmacology of motion sickness: An update (in process citation). *Brain Research Bulletin, 47*(5), 395–406.
Yates, B. J., Smail, J. A., Stocker, S. D., & Card, J. P. (1999). Transneuronal tracing of neural pathways controlling activity of diaphragm motoneruons in the ferret. *Neuroscience, 90*(4), 1501–1513.

# 33

# The Social Impact of Virtual Environment Technology

Sandra L. Calvert
*Georgetown University*
*Department of Psychology*
*37th & O Streets, NW*
*Washington, DC 20057*
*calverts@georgetown.edu*

## 1. INTRODUCTION

*Alan Alda, an actor, throws a ball for Silas, the dog. Silas chases the ball and returns it to Alda who throws it for him again. Alda shifts his attention to the pet hamster, and Silas acts jealous. Alda scolds Silas and sends him to the other room for punishment. Silas then slinks away.*

—*(Calvert, 1999a, p. 236)*

Although Silas acts just like every other dog, he is not real. Silas is a virtual dog, programmed by Dr. Bruce Blumberg at the MIT Media Lab. Blumberg has been careful to program Silas to have the needs that other dogs have. Silas has about a dozen goals and motivations that he tries to satisfy (Blumberg, 1996). Autonomous agents like Silas T. Dog bring us one step closer in the quest for realistic, three-dimensional (3-D) virtual environments (VEs).

How will time spent in such VEs impact social skills? Will people better understand their real pets? Will something innately human be changed if people choose a virtual pet rather than a real one? Silas does not have to be walked (unless you feel like walking him), nor does he chew up your favorite shoes. He can provide company during the user's free time. But he is not a real dog either. Can Silas ever become man's best friend? Would watching Silas on television be any different from interacting with him in a virtual environment?

How will virtual experiences affect social relationships and social behaviors with people both inside virtual reality (VR) and outside in real life? Online spouses and friends are already part of the changing landscape of representational experiences (Turkle, 1995), and televised depictions of families have long been a staple of entertainment programming. Can virtual relationships substitute for, or supplement, face-to-face relationships, or will people feel lonely and isolated?

Who will people be in the information age? What will identities be like when people are no longer constrained by a physical body, when they are free to embody themselves in diverse characters and develop multiple facets of their identities? How will virtual actions impact real behavior?

Currently, there is limited empirical literature in the VE area to address these questions. Because media issues tend to recur (Calvert, 1999a), three primary sources of information will be examined in this chapter: (1) studies in virtual environments; (2) studies about the Internet; and (3) studies about television. Because VEs will be delivered across the Internet in the future (see chap. 16, this volume), current and future Internet social interactions are particularly germane to this review. Even so, answers to questions about the social impact of VE are preliminary and require additional empirical investigations.

## 2. THE FORM AND CONTENT OF INFORMATION TECHNOLOGIES

Knowledge is transmitted and must be decoded in all media experiences, including virtual environments. Like other technologies before it, VE simulations are comprised of both content and form. Content has to do with domains such as aggression, sexuality, prosocial behavior, and fantasy. Form has to do with the unique representational codes that are used to present content (Huston & Wright, 1998). More specifically, form refers to audiovisual production features, such as action, sound effects, and dialogue, that structure, mark, and represent various kinds of content (Calvert, 1999a). Virtual environments add interactive capabilities and a first-person perspective to early media experiences such as television.

### 2.1 The Form of Information Technologies

The world is a rich stream of changing visual, auditory, tactile, and olfactory sensations. The form of VE simulations reflects an ongoing evolution in the symbolic codes that are used to represent media experiences. More specifically, there has been a shift from the auditory spoken and written codes of the telephone, radio, and books to the audiovisual codes of television and the World Wide Web. With the Internet, particularly the World Wide Web, users can interact with visual and audio messages. Although embodied VE simulations will expand these representational experiences into olfactory and tactile modes of expression, VE currently relies on the visual system more than other sensory systems (Schroeder, 1996). Auditory dimensions are not that different from previous applications in the film, television, and video game areas. Virtual environments are unique in allowing users a first-person interaction with 3-D visual images in immersive environments.

In examining the impact of media on man, McLuhan (1964) argued that television was unique in its form, not its content. This statement is true of other media as well. Take, for instance, the incident where two Colorado adolescent boys killed their classmates, teachers, and themselves. Playing the video game Doom was implicated as one source of their antisocial acts. Apparently they learned a particular style of killing by playing this game. Perhaps they could have learned similar behaviors from reading books, newspapers, or watching television programs or films, but *how* the behavior was displayed as they pulled the triggers of their guns was similar to the form of how they fired the guns of their video game.

Embodiment, interactivity, perceptually salient production features, and immersion combine to make some VE simulations seem realistic. Embodiment involves the sensation that you are the character. In VE simulations, a player looks out of the eyes of a character, and the world looks similar to how it looks when a person looks out of their own body. Embodiment leads to a sensation known as *presence*, which occurs when a person suspends reality and feels like they are physically in a computer-generated environment (Barfield, Zeltzer, Sheridan,

& Slater, 1995; Stuart, 1996; see chap. 40, this volume). Virtual embodiment lends a first-person perspective to media experiences.

Presence is augmented by interactivity. In VE games, people interact with and control images much more so than when they are spectators viewing a prepackaged television program or even when they control a video game player but are not embodied in it (Calvert & Tan, 1994). When a person directly experiences a media form and is able to control it, there is a one-to-one correspondence between what is done and what happens. This increased sense of personal involvement and presence in the VE game may also increase player identification with characters, their perceptions of the vividness of events, and their perceptions of self-efficacy.

The realism of virtual images may be augmented further by the use of perceptually salient formal features. *Perceptual salience*, a term coined by Berlyne (1960), involves features that contain high levels of movement, contrast, incongruity, surprise, complexity, novelty, and the like. Perceptually salient television and computer features involve action (movement), sound effects (surprise, incongruity), and loud music (contrast, surprise). Immersion uses perceptual cues to trick the senses into feelings of presence in a computer-generated environment (Stuart, 1996). Because of primitive orienting responses that ensure our survival, salient features are likely to draw attention to information, thereby increasing the probability that the content will be processed (Calvert, 1999a).

When perceptually salient images are combined with interactivity, an embodied user feels immersed in these symbolic worlds, often making the virtual world very compelling. This personal experience of interacting with 3-D moving images is why some users want to run away when they are attacked in a VE game. The player personally feels attacked. As virtual interfaces are improved, these audio-visual representational interactions will provide a lifelike quality that is unparalleled in media experiences (Calvert, 1999a). This evolution of technology means that users can create and construct shared realities that are limited only by their imaginations and by the latest technological breakthroughs (Schroeder, 1994b).

Form and content are theoretically separate dimensions, but production practices link perceptually salient formal features with certain kinds of content. For instance, rapid action, sound effects, and loud music are often associated with action-adventure television programs containing high levels of violent content (Huston & Wright, 1998). These same perceptually salient features are used extensively in VE games (Calvert, 1999b), making these experiences attention getting and arousing to players. Even so, perceptually salient forms can be used to present prosocial content. When interaction, embodiment, and perceptual salience are used to present content, the potential for that content to be attention getting and acted on increases, be it antisocial or prosocial.

## 2.2 Content of Information Technologies

Formal features deliver content. Virtual environments offer the opportunity for users to experience any type of content that interests them. Content can range from realistic simulations where surgeons practice heart catherizations to fantasy simulations such as slink world. Slinks are yellow-faced characters with cylindrical bodies and distinct personalities who autonomously move, avoid collisions with other objects, and react to other slinks, the latter activity being based upon the slink's happy, angry, or uncaring moods (Green & Sun, 1995). Virtual environment games, models of buildings, artistic environments, visualization models, and training environments are the most commonly portrayed content in virtual worlds (Schroeder, 1996; see chaps. 42–55, this volume).

Obviously, the applications of VE are diverse, and the content of various applications varies to suit the particular goal of the simulation. Nonetheless, VE content reflects market trends, making games one of the most widespread applications for consumer use (Schroeder, 1996;

see chap. 55, this volume). The content of VE games partly reflects earlier media content, though its roots are in military simulation (Herz, 1997). Just as the entertainment industry relies on action and violence to attract viewers to television, film, and video game content, so too do many VE games (Calvert, 1999a). In fact, games are now spearheading VE development (Schroeder, 1996). Because the most common experience of people is currently VE games, games are a focal point here for understanding the social impact of VE technology.

Game content involves a player or players who at times assume various identities and perform certain activities within a virtual simulation. On the one hand, these games provide an avenue for aggressive interactions with other real or imaginary characters. On the other hand, these games provide an avenue for characters to create cognitive strategies and to work together collaboratively to win the game (Pearce, 1998). Therefore, both aggressive and prosocial interactions can occur within these imaginary scenarios.

The company Virtuality was one of the first producers of commercially available VE systems (Schroeder, 1996). They created adventure games such as Dactyl Nightmare and Legend Quest. Legend Quest uses archetypal fantasy images, such as the wizard, the warrior, and the elf. Narrative elements, comparable to that found in film, are central to this game (Schroeder, 1996). In Dactyl Nightmare, two cartoonlike characters gain points by shooting each other and a pterodactyl as they move through a maze of platforms suspended in space (Calvert & Tan, 1994). The goal is to get as many points as possible. Virtual Worlds Battle Tech Center, Magic Edge, and Fighter Town also provided early combat-based entertainment. Currently mainstream commercial virtual entertainment programs are offered by Disney Quest, which creates mostly nonviolent programs, and by Sony Metreon.

Immersive VE often connotes a space that arises when a person interacts with a computer with particular hardware, such as gloves and helmets, or in a CAVE (Cave Automatic Virtual Environment), where users move through a shared virtual space. MUDs (Multiuser domains, dungeons, or dimensions) are yet another type of virtual environment. MUDs, originally constructed around the idea of a real-life game called Dungeons and Dragons (Schroeder, 1997), are now found on the Internet. In time, MUDs will be delivered as VE simulations over the Internet (see chap. 16, this volume).

One current form of MUD is a text-based social virtual environment (Turkle, 1995). There are no particular goals or tasks that participants are expected to do in this MUD. Instead, MUD participants assume complex personae, such as a wizard or a warrior, and embark on various adventures in relatively unstructured situations. MOOs (multiuser object oriented environments) and MUSEs (multiuser simulated environments) are similar to MUDs except they are based on different kinds of software (Turkle, 1995).

MUDs have now evolved into visual, rather than solely textual, virtual spaces (Calvert, 1999c). These MUD applications, which are more sophisticated than text-based MUDs but less sophisticated than truly immersive VE simulations, include desktop VE and second person VE systems (Schroeder, 1997). In desktop VE systems, a person can look out of a character's perspective into a virtual world, but they are not immersed, as they would be when using head-mounted displays (HMDs) or in a CAVE. Second-person VE systems are even less immersive than desktop VE systems. In second person VE systems, the player is represented by an on-screen avatar (i.e., a figure or character). Users communicate through written text that appears on the screen. The player sees a computer-generated environment, but not through their own eyes (Schroeder, 1997). The experience is much like playing a video game, but it is often much less structured than a game.

In second-person VE systems, players represented through avatars roam visual spaces such as Alphaworld and Cybergate (Schroeder, 1997). Social interactions change substantially when visual images and personal characters are included in the simulation, rather than having to rely solely on text for communication. For example, verbal aggression expressed in words becomes physical aggression directed at avatars, sexual language becomes sexual interactions between avatars, and belonging to the group via text-based exchanges is expressed by physical proximity to other avatars (Schroeder, 1997). The additional visual information provided in these newer virtual spaces represents a fundamental leap in the experiences of players. Within the next decade, another leap will occur as computer networks allow truly immersive virtual interactions on the Internet (Adam, Awerbuch, Slonim, Wegner, & Yesha, 1997).

At present, however, social interactions with other users on the Internet are more common in these low-end VE applications than in immersive VE (Schroeder, 1997). Currently, MUDs include visual adventure games on the Internet where players slay beasts and triumph over evil (Sleek, 1998b) as well as social MUDs where players interact in relatively open spaces to construct whatever is of interest to them (Schroder, 1997; Turkle, 1995). The hundreds of thousands of current MUD users are primarily adolescent and adult males in their 20s, but there are also children playing Barbie and the Power Rangers online (Turkle, 1995). More than 250,000 people have visited Alphaworld alone (Schroeder, 1997). Through MUDs, players can recreate themselves, their identities, and be anyone that they want to be (Turkle, 1995).

Immersion influences the relative power of people who interact in virtual environments. For instance, small groups of young adults solved problems in a VE followed by a continuation of the task in the real world. Only one person was actually immersed. The immersed person was consistently the leader in the virtual environment, but this role did not generalize to the real world (Slater, Sadagic, Usoh, & Schroeder, 1999).

## 3. THEORETICAL PERSPECTIVES AND PREDICTIONS IN RELATION TO VIRTUAL ENVIRONMENT CONTENT

Virtual environments allow people to engage in various kinds of representational experiences in both realistic and imaginary simulations (Calvert, 1999b). Representational thought allows individuals to transcend the present reality, thereby extending the perceptual field in both time and space. Through representational experiences, memories of the past and fantasies about the future become part of current reality.

The impact of media on peoples' lives has been examined from various theories in the fields of psychology and communication. Various predictions are made about the impact of these experiences, depending on the particular theoretical perspective brought to bear on the information. These predictions are summarized in Table 33.1.

### 3.1 Social Cognitive Theory

Social cognitive theory (Bandura, 1997) focuses on the role that models play as influences on social behaviors. These models can be real people, or they can be symbolic media characters.

As a model displays behaviors, observers attend to and encode relevant information. If rewards are available for imitating the behavior, the observer will often do so. Imitation can occur for all kinds of behaviors, be they prosocial or antisocial (Calvert, 1999a). Models can also increase the probability that another person will exhibit antisocial actions through a process called disinhibition. People often inhibit behaviors that they would like to do because the behaviors are not socially acceptable. However, if a model performs antisocial behaviors and gets rewarded for them (or at least not punished), then the observer's internal controls are undermined or disinhibited (Bandura, 1997). Models can also increase the probability of socially desirable behaviors through a process called response facilitation. In response facilitation, infrequent yet desirable behaviors, such as buying breast cancer stamps at the post office, can be increased when a person observes a model perform them.

**TABLE 33.1**
Theoretical Predictions About the Impact of Information Technologies on Social Behavior

| *Theory* | *Predictions* |
|---|---|
| Social Cognitive Theory | After observing social models, observers will imitate the behaviors that they have observed others perform; they will be disinhibited from controlling antisocial and inappropriate behaviors; and response facilitation will increase the probability that they will enact socially desirable behaviors. *Self-efficacy*, the belief that you can control events, can be promoted by observing and participating in rewarding symbolic experiences. Media expose people to various social models. |
| Arousal Theory | Observing or interacting with exciting media gets a person pumped up. The behavior that follows is a function of what the environment cues. The personally felt experiences of VE increase a user's arousal levels. |
| Psychoanalytic Theory | |
| Freud | Sex and aggression are two major drives that people experience. If these drives cannot be satisfied directly, they can be harmlessly released via catharsis: when one uses fantasy to reduce a drive state. Sex and aggression are two content areas that are systematically used in media portrayals. |
| Jung | The personality, called the *psyche*, is constructed by developing various *archetypes* that are embedded in the *collective unconscious*. The self-archetype integrates information from the various archetypes, including the persona, the anima and animus, and the shadow. Archetypes, such as those of heroes and warriors, are regularly experienced in television, film, Internet, video game, and VE encounters, providing information that can be used in constructing the self-archetype. |
| Cultivation Theory | People construct a world view based upon the media experiences they encounter through processes of *mainstreaming* and *resonance*. In mainstreaming, people come to believe the dominant cultural messages they encounter. In resonance, effects are amplified because they parallel one's own unique experiences. Virtual environment players, particularly in MUDs, will create a shared reality that they construct together. |
| Information Processing Theory | People take in information through sensory systems and interpret it based on prior expectancies, or *schemas*. Virtual environments build on processing activities, such as perception, attention, and memory, in creating user-friendly interfaces. |
| Ethology | *Ethology* focuses on behaviors that have evolutionary significance, such as finding a mate and avoiding predators. Media allow people to experience these challenges of nature without experiencing real danger. |
| Perceptual Developmental Theory | Humans evolved in a perceptual world rich with meaning. As humans, one learns to perceive what is really there. Information technologies, such as virtual environments, can trick one's perceptual systems into believing some things are real when they are not. |
| Uses and Gratification Theory | Information technologies are used to fulfill various needs that people have, such as the need for entertainment, the need for companionship, or the need to escape from the demands of life. |

*Source*: Calvert (1999a). Adapted with permission of the McGraw-Hill Company.

Over time, Bandura (1997) increasingly focused on the concept of self-efficacy, the belief in personal control over life events. The subprocesses of attention, retention, production, and motivation occur as a person observes and later imitates a behavior. Self-efficacy regulates these subprocesses, particularly when a person fails and motivation to continue a task is weakened. If a person perceives that they have control over events, then they are more likely to continue to attend to, retain, and produce a behavior.

Virtual environments allow user control, a component of self-efficacy, via interaction. Experienced VE users take control of these environments more than inexperienced, novice users. For instance, novices observe more in simulations, whereas experts are more likely to act (Schroeder, 1997). Also, role playing, an important rehearsal mechanism for performing a behavior, is integrated into VE simulations through interaction. This means that the behaviors enacted (e.g., pulling the trigger of a gun) will be readily accessible to players after the VE interaction is over.

### 3.2 Arousal Theory

Autonomic arousal is often measured by heart rate, skin conductance, and blood pressure, whereas an electroencephalogram measures cortical arousal. Both kinds of arousal are related, but researchers in the media area typically study autonomic arousal (Calvert, 1999a). People who are exposed to aggressive, sexual, or scary content typically become more aroused and then direct that arousal in diverse ways, depending on what is cued by the environment. That is, arousal can be channeled into aggressive, sexual, or prosocial behaviors, regardless of the content, depending upon situational cues. After repeated exposure to aggressive or sexual content, the individual may become desensitized to that content and to the plights of others.

Certainly VE content is not always meant to be arousing. For instance, therapeutic applications (see chaps. 49–51, this volume) utilizing systematic desensitization procedures, such as a program to help a person overcome a fear of heights, focuses on how to get a person to relax in what is typically an arousing, stressful situation. By contrast, entertainment games (see chap. 55, this volume) focus on getting users to be excited, much like an amusement park ride. The arousing qualities of entertainment-driven VEs may increase attentiveness. Interaction and presence should increase users' heart rate and other indices of physiological arousal. The content that is being experienced in VE will be the immediate trigger for releasing arousal, but real-life events just after VE interactions can also trigger behavior. With repeated interaction, users should habituate to the content, as they have done in other exciting media experiences. That means that the content will have to be more vivid for arousal levels to return to prior levels (Calvert, 1999a).

### 3.3 Psychoanalytic Theory

Psychoanalytic theory focuses on human drives, particularly sex and aggression, as the reasons for our behaviors (Thomas, 1996). Most of the reasons for actions occur at an unconscious level, reflecting these basic drives. Although drives are a normal part of human experience, society often disapproves of, and therefore regulates, their expression. Consequently, drive reduction is accomplished in indirect ways. One example is catharsis. Instead of killing someone or having sex with someone, drives are released vicariously through a fantasy experience. Theoretically, VE games should yield the same cathartic releases that television, film, and video game experiences were expected to offer viewers or players.

In an extension of Freud's ideas, Jung (1959) proposed that humans share a collective unconscious, a repository of shared racial experiences that have been inherited. Archetypes, or primordial images, reside in the collective unconscious and are developed by individual experiences. These archetypes include mother, father, hero, trickster, persona, anima, animus, shadow, and self (Hall & Nordby, 1973). On the Internet, the persona, or mask, is a term that is used to describe a character's identity (Turkle, 1995). Moreover, men sometimes create and play a female persona online, thereby developing their anima or feminine side of the self, whereas women often create and play a male persona online, thereby developing their animus, or male side of the self. The shadow, or primitive side of human nature, protects the self but can overreact in very aggressive ways when threatened. The developmental task is to synthesize the various archetypes into ones self, the overarching archetype.

Most fantasies, including virtual ones, include many archetypal images. Playing out personae, or the archetypes, in VE games should lead to the development and differentiation of that archetype and the inclusion of it in the self-archetype. Put another way, virtual fantasy interactions may provide experiences that players can use to construct their identities.

## 3.4 Cultivation Theory

In cultivation theory, media present messages that form the basis for a shared, constructed reality (Calvert, 1999a). One cultivation effect, called mainstreaming, focuses on the dominant media images that people share, and therefore, come to believe to be true. For example, those who frequently view aggressive content on television come to believe that the world is a more aggressive place than it really is. These heavy viewers buy more guns, watchdogs, and locks than light viewers do (Gerbner, Gross, Morgan, & Signorielli, 1994). Another cultivation effect, called resonance, amplifies real-life experiences that overlap with media experiences.

Mainstreaming effects should become common for those who share the same virtual spaces day after day because they create a reality based on shared experiences. Resonance will occur when these experiences occur in real-life as well as in the virtual world. The implication is that experiences in the virtual world will influence people's views of reality, just as television viewing has.

## 3.5 Information Processing Theory

Information processing theory focuses on humans as information processing devices. Based on their experiences, people construct schemas, that is, cognitive structures that influence perception, attention, and memory. These schemas are then used to interpret future incoming information, thereby influencing what will be learned. In VE, input devices, such as helmets and goggles, alter the perceptual reality that is available for processing.

## 3.6 Ethology

Ethology focuses on behavior that has evolutionary significance for the species (Thomas, 1996). Activities such as finding food, avoiding predators, finding a mate, mating, protecting the young, and finding shelter are behaviors that ensure the survival of the species. Programs that build on basic human and animal needs, such as the program for Silas the dog, use ethology as a framework. These kinds of applications may be particularly useful in creating Abots, the robots who may eventually become indistinguishable from real users in simulated environments.

## 3.7 Perceptual Developmental Theory

In perceptual developmental theory, the focus is on how perceptual systems evolved to ensure survival (Thomas, 1996). People are active perceivers whose sensory systems evolved to pick out the salient and important environmental stimuli. The visual system is particularly important in evolution. Perceptually salient stimuli such as movement, contrast, incongruity, complexity, surprise, and novelty are features that are likely to elicit attention as well as improve the odds of survival (Berlyne, 1960).

In VE simulations, users are tricked into perceiving and seeing events that are not really there. Virtual environments accomplish this task, in part, by creating natural interactions where characters move and interact with each other without the use of artificial keyboards or textual commands (Bricken & Coco, 1995; see chaps. 11 and 13, this volume). Perceptual laws are obeyed in many of these simulations. For instance, in Slink World, slinks are programmed to avoid collisions with other moving objects. When rapidly approached, slinks pull back and act scared. But when slowly approached, slinks are likely to stay in their current position (Green & Sun, 1995).

### 3.8 Uses and Gratification Theory

Communication media are used by people to satisfy or gratify their needs (Rubin, 1994). Needs are diverse, ranging from sexual stimulation to relaxing to information seeking. In VE, users may play games to entertain themselves (see chap. 55, this volume), or they may interact with a clinical simulation to reduce phobias (see chaps. 50 and 51, this volume). Even so, many people use VE to satisfy entertainment needs, just like they used other media before it.

## 4. COGNITIVE AND BEHAVIORAL IMPACT OF VIRTUAL ENVIRONMENTS

According to uses and gratification theory, media fulfill certain needs that people have. Freud theorized that those needs involve sex and aggression, and the Nielsen ratings attest to the popularity of violent content. However, people have diverse needs for entertainment and for learning about others that extend well beyond violence and sex. Social interactions provide opportunities for constructive prosocial activity as well as for antisocial activity. Both antisocial and prosocial interactions are explored in this section.

### 4.1 Aggressive Interactions

Since 1953, violent crime has risen 600% in the United States, and homicide is the main cause of death for U.S. urban males who are 15 to 24 years of age (Coie & Dodge, 1998). American teenagers are at least four times more likely to be murdered than teenagers in 21 other industrialized nations; not surprisingly, Americans rank crime and violence as the most important problem being faced in this country. Media content, specifically violent television viewing, has been implicated as one reason for aggressiveness by American youth, in particular, and American society, in general (Coie & Dodge, 1998; Huston & Wright, 1998).

In an early examination of VE and aggression, Calvert and Tan (1994) examined the arousal levels and aggressive thoughts of college students in one of three conditions: (1) an immersive VE condition where they interacted in an aggressive game; (2) an observational condition where they viewed another person's aggressive VE game; and (3) a simulation experience where the players moved as if they were in a VE game, but in which no aggression was displayed (the control group). The game, Dactyl Nightmare, involved shooting an opponent and a pterodactyl that both periodically attack the player.

The results best supported arousal theory. Participants who played the game were more aroused, as measured by pulse rate, than those who observed another play the game or those who simulated game movements. Social cognitive theory received mixed support. Specifically, game players reported more aggressive thoughts than those who observed another person play the game or who simulated game movements; however, contrary to prediction, the observers reported no more aggressive thoughts than the game simulators. The psychoanalytic view that catharsis would release aggressive impulses safely via fantasy was not supported because players were more aroused and had more aggressive thoughts, not less. The results suggest that interactive VE experiences lead to stronger effects than simply observing another person perform an antisocial act.

According to Pearce (1998), the visceral kill or be killed formula created by game makers plays on primitive human instincts that are necessary for survival. Violence, as part of the human psyche, has been prevalent throughout time. War is expressed in simulations that were originally board games that became computer games as technology evolved. Thus, VE and video games, more generally, had their roots in military operations. In fact, the first VE games were designed to train military personnel. Through role playing, war strategies could be perfected. Virtual environment entertainment platforms adopted these earlier military game simulations.

A key question is the extent to which these aggressive encounters will generalize to real-life experiences. Those who view aggressive television content tend to be more aggressive in real life, and research in video games and initial VE games find similar results for aggressive behavior and aggressive thoughts, respectively (Calvert, 1999a). Estimates of 3% to 9% of unique variance are found in studies that link children's aggressive conduct to their exposure to television violence (Huston & Wright, 1998). Millions of dollars were invested in virtual and video game military simulations to impact real fighting skills, not just pretend ones (Herz, 1997).

There are also those who argue that violent media impact people in different ways. That is correct. Some people become more aggressive, but those who see themselves as victims become more afraid (Gerbner et al., 1994). Moreover, everyone who sees violent media learns about aggressive behavior, even if they don't act on it immediately, and they can act on it later if motivational incentives are provided (Bandura, 1965). If only a small percent of individuals spontaneously act more aggressively after exposure to media violence, the quality of life is impacted for all. The evidence suggests that a portion of the population does become more aggressive after viewing or interacting with violent media content (Huston & Wright, 1998).

## 4.2 Identity Construction

Constructing a personal identity is a life-long experience accomplished by day-to-day encounters, particularly those involving other people. Virtual environments will become a source of experiences for creating parts of identity. As individuals move between virtual worlds and real worlds, their sense of identity may become increasingly malleable. They can be a male, a female, or both, moving fluidly across varying MUDs. How will virtual construction impact the real-life construction of the self? Who will people be?

Historically, a debate involves whether identity is comprised of a single entity or multiple entities (Turkle, 1995). That is, does each individual have one and only one true self, or do individuals display multiple selves, depending upon the situation? Physical embodiment is a defining characteristic of the self (Harre, 1983), but people are free to use different bodies in virtual environments. What implication does the lack of real-life embodiment have for self-construction? Will people develop alternate and multiple selves that express hidden aspects of their character?

When children identify with other people, they selectively take in information about them and incorporate that person's characteristics and qualities into their own sense of self (Calvert, 1999c). Identification with televised and filmed models is one reason that viewers imitate certain characters: They are making that character's behaviors their own. The VE genre allows players to be, not just to observe, the characters in these narratives. Being an embodied character may allow players to incorporate personality characteristics and behaviors directly into their own personal repertoire (Calvert & Tan, 1994).

The opportunity to create different selves expands greatly in virtual environments. People are not constrained by their physical body in defining who they are. They can construct the specific body that they want, and they can be as many people as they want to be. They can also be animals, robots, or other nonhuman entities. Although they no longer have to use their own physical body in these virtual worlds, they do have a virtual body. Moreover, presence (see chap. 40, this volume) should increase their beliefs about the realism of these embodied entities. These personae may then become incorporated into their self-constructions.

For children, the role playing that takes place in imaginative play episodes is one way that they try out various perspectives, thereby coming to understand how and why others feel and act as they do (Calvert, 1999c). Similarly, VE immersions provide opportunities for both children and adults to take on different roles in a fantasy context, thereby mastering a role and understanding the perspective of that person in a safe environment.

In Jungian psychoanalytic theory, key archetypal images recur throughout history. Many of these images, embedded in the collective unconscious and displayed in fiction, involve the roles of hero, father, mother, child, wizard, sorceress, sorcerer, king, queen, emperor, princess, prince, wise old man, wise old woman (or crone), innocent, healer, trickster, witch, magician, persona, anima or animus, shadow, and self (Hall & Nordby, 1973). The self is a meeting place where various archetypes are integrated into one's personality (Turkle, 1995).

Archetypal images are passed from one generation to the next through the collective unconscious, a shared repository of ancestral images that are like negatives on a roll of undeveloped film (Hall & Nordby, 1973). Individual experiences then develop those latent images. The specific image that emerges is a function of a person's unique experiences. These experiences can be real, or they can be realistic or even imaginary, as is the case of VE immersions.

The persona is the public display, or the mask, of the self that is presented in public. The Internet is a place where personae are already being constructed in MUDs (Turkle, 1995; see chap. 16, this volume). Acting on information makes it your own (Calvert, 1999c). Thus, as people act on these archetypal images, they should be integrated within their own identities, thereby making their own sense of self much richer and more complex. The implication is that identity can potentially be altered as players enact various roles via these Internet personae, allowing the real person to integrate the archetypes that are being played out into his or her real personality. Personal views of a hero, or of any other archetypal image, should be influenced by role playing heroes in VE encounters. The role of hero, which has often been reserved for men, may become more available to women through VE games. One problem with these immersive identities is that an individual may lose track of who he or she really is. Fantasy can sometimes be compelling and even more appealing than reality. The person may choose to withdraw from real life to a more controlled imaginary persona and reality (Turkle, 1995). A person can be important in fantasy, a Walter Mitty so to speak, with all the embellishments made possible by being present in an electronically generated environment of that person's choosing. The virtual identity option may be particularly appealing to young adolescents who are struggling with an identity crisis. Negative identities can also be constructed. If a person chooses to play a mobster or other antisocial characters in virtual environments, antisocial behaviors are being practiced. Having such images and behaviors readily available can potentially increase antisocial behaviors and accentuate character flaws as seen in Littleton, Colorado.

In online symbolic encounters, people can also deceive others about who they really are (Calvert, 1999c). The use of temporary and fictitious identities, known as handles, are widespread in electronic communities (Mantovani, 1995). Is it ethical for a man to present himself as a woman, thereby gaining access to information that would not ordinarily be available to him? One man did, assuming an Internet persona of a woman who was helping other women with their emotional problems. Many women were upset that he deceived them (Turkle, 1995). Men also present themselves as women because they are more likely to be approached, and women present themselves as men to avoid being approached (Turkle, 1995). These gender switches are not uncommon online and will certainly be possible in VE as well.

Players can also confuse bots, which are online robots, with a real person. Or players can create a persona where they appear to be a bot, thereby gaining access to private conversations as cyberspace voyeurs (Turkle, 1995). One player created a passive character called Nibbet who was a rabbit. Others forgot he was there or assumed he was a bot, allowing him to listen to conversations where he would probably not be welcome.

In summary, virtual environments allow people to try out different facets of their identities and to rehearse behaviors in a safe place. Because players can be anyone they want to be, they can try out new ways of being in cyberspace. However, they can deceive others and themselves about their real selves. Identities can be enacted that are negative and that reinforce character flaws.

### 4.3 Sexual Interactions

Ongoing interest continues to be expressed in having virtual sex, particularly by men (Epley, 1993). Currently, these possibilities are limited because of tactile interface issues. However, once these telehaptic issues are resolved (see chaps. 5 and 6, this volume), there is clearly a paying audience for sexual virtual experiences. Computer software and Internet interfaces are already widely used for pornography, particularly by boys and men (Calvert, 1999a).

Opportunities for sexual enjoyment via VE offer positive as well as negative experiences for users. In text-based Internet MUD applications, many characters meet online and engage in virtual sex. Some even get married in virtual ceremonies (Turkle, 1995). These fantasy relationships provide an opportunity for safe sex because there is no danger of contracting or spreading a sexually transmitted disease. Users also are engaged in an experience with another person, allowing them to participate within the boundaries of a shared sexual fantasy rather than an individual one. By knowing how a partner feels and what a partner enjoys, a player may become better able to interact with real partners by understanding their needs. Players may even meet future real-life partners by exploring intimate experiences on the Internet. Sometimes people really marry a person that they were initially involved with in an Internet relationship, but other times real meetings are disappointing (Turkle, 1995).

On the other hand, sexual deviance and violations of another person's rights can also occur in virtual sexual experiences. For instance, some Internet sites are policed to prevent users from acting in sexually deviant ways on sites that are not meant to be sexual. MUD administrators for the Palace Internet site deal with issues such as pornography and profanity (Sleek, 1998b). Reports of cyber-rape have also occurred (Wallace, 1999). In one instance, a more sophisticated MUD user took control of another person's character, and he forced that character to have a violent sexual encounter (Turkle, 1995). As these sexual encounters become more realistic with the addition of immersive technology, how will the law deal with such verbal actions and symbolic violations? Words are deeds in virtual experiences (Turkle, 1995), making ethical issues a pressing concern. The anonymity afforded by cyberspace currently allows sexual deviants to act out with impunity (Wallace, 1999). Issues of imitation, disinhibition, and desensitization may become serious issues as sexual activity becomes an immersive, online option.

Ethical issues, such as marital fidelity, will also be experienced in virtual spaces. How will a person feel if their partner has virtual sex with an imaginary character, or with a character who is a real person in another location? Will betrayal and infidelity be experienced? *Mad About You*, a television program, played out the scenario of a fantasy partner a few years ago. Marital distress occurred when the husband had a virtual sexual encounter with the model Cindy Crawford, suggesting that these enacted fantasies are not harmless entertainment.

What will happen when a man pretends to be a woman or a woman pretends to be a man? Do partners have a right to know the real sex of that person, particularly if they are seeking an online sexual encounter? Is this kind of deception permissible, or is it unethical? Will the online partner care if they are being deceived? Bots, that is, cyberspace robots, are also mistaken for real people. In one instance, a man made repeated sexual advances toward Julia, an intelligent computer agent (Turkle, 1995). As these agents become more humanlike in their interactions with people, the line between the robot and the real person will be increasingly difficult for people to discern.

Finally, sexual predators may find a new home to prey on unsuspecting people, particularly children. Parents are particularly worried about these potential online dangers as Internet interactions can be used to create avenues for real-life meetings or for virtual online meetings (Turow, 1999).

## 4.4 Social Interaction or Social Isolation?

Humans are social beings who often seek the contact of others. Will computer simulations and online interactions promote friendships and close ties with others, or will simulations lead to social isolation and addiction? Although many parents are concerned that Internet interactions will lead to social isolation and potentially disrupt family interactions (Turow, 1999), the impact of online Internet interactions is mixed.

Kraut et al. (1998) examined the social and psychological impact of initial Internet use for 73 households. People used the Internet primarily to communicate. As Internet use increased, people reported that they interacted less with their families, had fewer people in their social circles, and felt more depressed and lonely. Similarly, Slater et al. (1999) found that young adults enjoyed working in groups, felt less isolated, felt more comfortable, and felt other group members were more cooperative after participating in a real, rather than a virtual environment, interaction; however, the design was flawed in that the real interaction always occurred after the virtual one. Sleek (1998a) argues that too much reliance on impersonal, weak, online interactions through e-mail correspondence and chat groups cannot replace the social support of real-life face-to-face interactions with family members and friends. As computer use increases through VE simulations, real face-to-face interactions may diminish even more.

The impersonal world of computers and business, coupled with the loss of social networks, has resulted in a loss of social skills and an increase in shyness for many users (Cooper, 1997). Part of this problem may also reflect the displacement of other activities. As one spends more time in virtual interactions, less time is available to participate in, and to practice, face-to-face social interactions. Thus, feelings of isolation and loneliness may increase as VE interactions become more commonplace. But if people do become too shy, they can then interact with a VE therapist in a systematic desensitization simulation. For instance, therapeutic interventions allow students with behavioral disorders to practice social skills (e.g., Muscott & Gifford, 1994).

As a participant observer, Schroeder (1997) explored two virtual worlds, Alphaworld and Cybegate, as a human and a cartoon avatar, respectively. Schroeder found that players created social stratification within these worlds. Insiders often developed a distinctive persona such as being a prankster, a bully, or a helpful guide, whereas outsiders acted more like a tourist. In fact, in Alphaworld, visitors or tourists literally wear a camera around their avatar's neck so that they are readily identifiable as outsiders. In these virtual worlds, insiders tend to explore and to create spaces with their friends in the far reaches of these worlds, whereas outsiders are less engaged with others, observe more, and stay in central locations where most of the avatars tend to be located. Overall, these two different groups had a different social status, marked by the distinctive roles and behaviors that separated them, and differential status emerged with varying levels of involvement and expertise in the virtual world.

Online VE interactions can eliminate the geographical constraints that disrupt human interactions by making friends and family members in far away places accessible in everyday social life. Players can meet and interact in virtual simulations, enacting fantasies when their real worlds impede actual physical contact. The loneliness associated with a move to a new place can be lessened by bringing close friends into one's symbolic space, even if it is not one's real space. While the experience is not the same as being face-to-face in the real world, having your characters face-to-face in an imaginary setting can also strengthen ties. Instant messaging allows ongoing online interactions with one's "buddies."

Online relationships can also lead to new friendships. For example, Parks and Roberts (1998) reported that 94% of a sample of 235 MOO participants, ranging in age from 13 to 74, had developed an ongoing personal relationship on a MOO. These included close friendships, friendships, and romantic relationships. Although offline relationships were better developed than MOO relationships, MOO relationships offered the possibility of creating new face-to-face friendships. One third of these MOO relationships had led to real-life meetings.

In a survey of 2,500 respondents, Katz and Aspden (1997) found that the Internet complements social connections. Friendships cultivated online often led to real-life meetings, family contact was enhanced, and a sense of community was fostered. There was no evidence that Internet users dropped out of real-life activities, such as religious, leisure, and community organizations. Eighty-eight percent of the respondents reported no change in the amount of time spent in face-to-face interactions or phone conversations with family and friends. Six percent of respondents spent more time interacting with or talking to family members, and 6% spent less time. Only 14% of the sample reported friendships created online, and 60% of those people had met at least one of their online friends. The authors estimate that about two million face-to-face meetings have taken place because of initial Internet interactions.

Addiction to computer interaction is one reason that people may have fewer face-to-face interactions (Turkle, 1995). For example, some players choose to interact with text-based VEs from the time that they wake up until they fall asleep (Turkle, 1995). Windows on computers allow them to participate in work and in fantasy text-based MUD worlds simultaneously, moving between various applications throughout the day.

In certain instances, however, online experiences may be engaging, not addictive. Consider a lesson from television. Addiction has often been used to describe those who cannot seem to quit looking at the television screen. Anderson and Lorch (1983), however, documented that such experiences actually involve an easy and engaged attentional state, known as attentional inertia, in which users get increasingly involved in the activity that they are doing. These activities can range from watching television to playing to reading a book. Moreover, attentional inertia has been documented from childhood through adulthood, making it a skill that all age groups use to sustain attention in an activity. Therefore, while some may describe rapt attention as addictive, it may just be a normal facet of attention: people become involved in interesting, responsive activities and environments. Virtual environments are certainly one kind of intrinsically interesting environment for many users.

The answer to the complex questions about social isolation and addiction ultimately depends on how a player treats online experiences (Turkle, 1995). The most extreme case involves those who spend most of their time in computer simulations. If those computer simulations become a permanent substitute for real life and the person pursues dysfunctional behaviors, then addiction to the computer world and isolation from real social relationships can become serious issues. By contrast, if computer simulations are a temporary working space to deal with adverse circumstances, the person may emerge healthier and more functional (Turkle, 1995).

In summary, while some studies find that users feel more lonely and isolated after engaging in Internet interactions, others find that the Internet does not substantially alter the amount of time spent in face-to-face interactions. In many instances, real face-to-face interactions occur because of initial Internet contact. For most people, virtual environment simulations will probably become a recreational activity, taking up part, but not all, of their leisure time experiences. In the latter situation, virtual environments have the potential to satisfy needs by supplementing the everyday social experiences of real life, just as other media experiences have done in the past. Nevertheless, heavy use of online computer technologies has the potential to disrupt social relationships, just as heavy television use can.

### 4.5 Prosocial Interactions

Just as media exposure can lead to antisocial activity, so too can exposure lead to constructive prosocial responses. Prosocial behavior involves socially constructive activities such as helping, sharing, cooperating, and creativity (Calvert, 1999a). There are numerous television studies demonstrating that prosocial television can enhance children's helpfulness, cooperation, empathic responsiveness, and delay of gratification (Huston & Wright, 1998). The same impact may occur for VE interactions, but studies in this area are rare at present.

Many VE games provide opportunities for players to interact with virtual actors, who can be either real or imaginary players. These games are often fantasies where the player assumes an archetypal role (e.g., hero or wizard) and then enters a VE simulation where other players assume the identities of other game characters. These games provide the opportunity for players to share an adventure where they work together and cooperate to overcome evil forces. Characters can talk to each other and plan group strategies to win the game. Prosocial and antisocial (e.g., aggressive) behaviors coexist in these VEs as they sometimes do in real life. Through role-play, characters enact the behaviors of others, thereby helping the player to control and to incorporate those qualities into their own identities (Calvert, 1999c). Self-efficacy, the belief in personal control, should increase as users role-play these ancient and powerful images.

Pearce (1998), who finds that women enjoy social interaction in computer games, created a nonviolent, story-based VE game called Virtual Adventures. Players compete against other teams and cooperate within teams. Prosocial actions, such as rescuing the Loch Ness Monster's eggs from bounty hunters, were incorporated in the game. The players became story characters that worked together to solve a common pressing goal. Women rated this software more favorably than men did. Her findings suggest that prosocial immersive programs can be entertaining to certain users.

A potential problem for VE players is that they may come to rely on others' fantasies rather than constructing their own. A constructed reality leaves little room for imagination (Mantovani, 1995). Heavy television viewing is associated with a decline in imaginative play and in creativity, but with increases in daydreaming (Valkenberg & van der Voort, 1994; van der Voort & Valkenberg, 1994). However, the kind of program being viewed is crucial in understanding how television impacts fantasy, imagination, and creativity. For instance, television programs that encourage imaginative activities lead to increases in creativity (Schmitt et al., 1997).

Interactive VE experiences may impact a player's imaginative world differently than observational media have, with daydreams taking on a new life as they are developed and enacted in online computer experiences. Unstructured situations also tend to enhance imaginative play more so than structured situations (Carpenter, Huston, & Holt, 1986). Therefore, there may be a difference in imaginative outcomes for those who enter relatively unstructured virtual spaces and create parts of that site when compared to those who just visit an unstructured site or for those who play structured VE games.

## 5. LIMITATIONS OF VIRTUAL ENVIRONMENTS

There are always limitations to any innovation, and VE is no exception. Recurring media concerns such as the displacement of other activities are already being documented. For example, some players spend 40 or more hours each week in text-based MUDs, an experience know as simulation overdose (Turkle, 1995). People who spend too much time in VE may head back to real life because they decide that the real world offers more than a simulated one. For instance, a real pet may ultimately be more fulfilling than a virtual one. Media research in other areas indicates that displacement usually occurs when a new technology is introduced, and is sustained only for activities that fill a similar need for people, such as entertaining them (Calvert, 1999a). Thus, it is likely that a novelty effect for VE experiences will eventually give way to a balanced use of the technology in daily life.

The distinction between what is real and what is pretend is another issue that will be faced. In the everyday world, children already have difficulty in telling the difference between a realistic, but fictional, event versus a real event portrayed on television (Huston & Wright, 1998). These issues are partly resolved though developmental changes in the way that children think, but the roles that television characters play can even confuse adults. For instance, people often expected actors like John Wayne to really be like the characters that he played. Perceptions of fantasy and reality should be explored, particularly as virtual simulations become more realistic in the future.

## 6. SOCIAL IMPLICATIONS FOR THE FUTURE OF VIRTUAL ENVIRONMENTS

The information age continues to hold new options and challenges for the people who participate. What will social life be like in this coming age of realistic VE immersions where virtually any behavior will be possible? The literature on this topic is very limited at present, and this issue requires additional study. However, when other media research from the Internet and television areas is included, the following observations can be made.

In virtual environments, the social behaviors that are played out in real life simply move to a new symbolic plane. The long history of media research can be used to predict that people will use VE to meet their needs, including their social ones, as they do with television and the Internet already. Virtual experiences will become integrated into the daily fabric of life, just as the technologies before it were. People will imitate the characters that they play and become desensitized to situations in which they overindulge themselves. But each new technology also adds its own unique imprint to how those representational experiences will affect users.

Virtual environments will allow people to participate in, and to exercise control over, online social interactions. In the future, friends will be virtual as well as real. People will assume alternate personae, both realistic and fantasy-driven. From the privacy of home, people will travel to and experience any place that they desire. They will experience intimacy. They will create and live out fantasies with real and imaginary people in virtual worlds. As is true of the technologies that came before it, people can use VE to enhance their real social lives, or they can use it to escape or to act out antisocial tendencies, augmenting the problems that they already have.

McLuhan (1964) told us that the medium was the message. Virtual environments represent a major step forward, bringing people one step closer to a social world where the lines between the symbolic and the real are merged.

## 7. REFERENCES


Adam, N., Awerbuch, B., Slonim, J., Wegner, P., & Yesha, Y. (1997). Globalizing business, education, culture, through the Internet. *Communication of the ACM, 40*, 115–121.

Anderson, D. R., & Lorch, E. P. (1983). Looking at television: Action or reaction? In J. Bryant & D. R. Anderson (Eds.), *Children's understanding of television: Research on attention and comprehension*. New York: Academic Press.

Bandura, A. (1965). Influence of models' reinforcement contingencies on the acquisition and performance of imitative responses. *Journal of Personality and Social Psychology, 1*, 589–595.

Bandura, A. (1997). *Self-efficacy: The exercise of control*. New York: W. H. Freeman.

Barfield, W., Zeltzer, D., Sheridan, T., & Slater, M. (1995). Presence and performance within virtual environments. In W. Barfield & T. A. Furness, III (Eds.), *Virtual environments and advanced interface design* (pp. 473–513). New York: Oxford University Press.

Berlyne, D. (1960). *Conflict, arousal, and curiosity*. New York: McGraw-Hill.

Bricken, W., & Coco, G. (1995). VEOS: The virtual environment operating shell. In W. Barfield & T. A. Furness, III (Eds.), *Virtual environments and advanced interface design* (pp. 102–142). New York: Oxford University Press.

Blumberg, B. (1996). *Old trick, new dogs: Ethology and interactive creatures*. Unpublished doctoral dissertation, Massachusetts Institute of Technology Media Lab.

Calvert, S. L. (1999a). *Children's journeys through the information age*. Boston: McGraw-Hill.

Calvert, S. L. (1999b). The form of thought. In I. Sigel (Ed.), *Theoretical perspectives in the concept of representation* (pp. 453–470). Hillsdale, NJ: Lawrence Erlbaum Associates.

Calvert, S. L. (1999c, May). *Identity on the Internet*. Paper presented at the Annenberg Public Policy Center Conference on the Family and the Internet, National Press Club, Washington, DC.

Calvert, S. L., & Tan, S. L. (1994). Impact of virtual reality on young adults' physiological arousal and aggressive thoughts: Interaction versus observation. *Journal of Applied Developmental Psychology, 15*, 125–139.

Carpenter, C. J., Huston, A. C., & Holt, W. (1986). Modification of preschool sex-typed behaviors by participation in adult-structured activities. *Sex Roles, 14*, 603–615.

Coie, J., & Dodge, K. (1998). Aggression and antisocial behavior. In W. Damon & N. Eisenberg (Eds.), *Handbook of child psychology: Vol. 3. Social, emotional, and personality development* (5th ed.). New York: John Wiley & Sons.

Cooper, G. (1997, July 16). Medium becomes message for the shy e-mailers. *The Independent* (London), News, p. 7.

Epley, S. (1993, May/June). A female's view: Sex in virtual reality. *CyberEdge, 15*.

Gerbner, G., Gross, L., Morgan, M., & Signorielli, N. (1994). Growing up with television: The cultivation perspective. In J. Bryant & D. Zillmann (Eds.), *Media effects: Advances in theory and research*. Hillsdale, NJ: Lawrence Erlbaum Associates.

Green, M., & Sun, H. (1995). Computer graphics modeling for virtual environments. In W. Barfield & T. A. Furness, III (Eds.), *Virtual environments and advanced design interface* (pp. 63–101). New York: Oxford University Press.

Hall, C., & Nordby, V. (1973). *A primer of Jungian psychology*. New York: Mentor Books.

Harre, R. (1983). *Personal being*. Oxford, England: Blackwell.

Herz, J. C. (1997). *Joystick nation*. Boston: Little, Brown.

Huston, A. C., & Wright, J. C. (1998). Mass media and children's development. In W. Damon, I. Sigel, & K. A. Renninger (Eds.), *Handbook of child psychology: Vol. 4. Child psychology in practice* (5th ed.). New York: John Wiley & Sons.

Jung, C. (1959). *The basic writings of C. G. Jung*. New York: Modern Library.

Katz, J., & Aspden, P. (1997). A nation of strangers? *Communications of the ACM, 40*, 81–87.

Kraut, R., Patterson, M., Lundmark, V., Kiesler, S., Mukophadhya, T., & Scherlis, W. (1998). Internet paradox: A social technology that reduces social involvement and psychological well-being. *American Psychologist, 53*, 1017–1031.

Mantovani, G. (1995). Virtual reality as a communication environment: Consensual hallucination, fiction, and possible selves. *Human Relations, 48*, 669–683.

McLuhan, H. M. (1964). *Understanding media: The extensions of man*. New York: McGraw-Hill.

Muscott, H., & Gifford, T. (1994). Virtual reality and social skills training for students with behavioral disorders: Applications, challenges, and promising practices. *Education and Treatment of Children, 17*, 417–434.

Parks, M., & Roberts, L. (1998). Making MOOsic: the development of personal relationships online and a comparison to their offline counterparts. *Journal of Social and Personal Relationships, 15*, 517–537.

Pearce, C. (1998). Beyond shoot your friends: A call to arms to battle against violence. In C. Dodsworth (Ed.), *Digital illusion: Entertaining the future*. Reading, Mass. Addison-Wesley.

Rubin, A. (1994). Media uses and effects: A uses-and-gratifications perspective. In J. Bryant & D. Zillmann (Eds.), Media effects: Advances in theory and research. Hillsdale, NJ: Lawrence Erlbaum Associates.

Schmitt, K., Linebarger, D., Collins, P., Wright, J., Anderson, D., Huston, A., & McElroy, E. (1997, April). Effects of preschool television viewing on adolescent creative thinking and behavior. Poster presented at the biennial meeting of the Society for Research in Child Development, Washington, DC.

Schroeder, R. (1994a). Cyberculture, cyborg post-modernism and the sociology of virtual reality technologies. *Futures, 26*, 519–528.

Schroeder, R. (1994b, June). Worlds in cyberspace: A typology of virtual realities and their social contexts. Paper presented at the BCS Displays Group Conference Applications of Virtual Reality, University of Leeds, Leeds, England.

Schroeder, R. (1996). *Possible worlds: The social dynamic of virtual reality technology*. Boulder, CO, & Oxford, England: Westview Press.

Schroeder, R. (1997). *Networked worlds: Social aspects of multi-user virtual reality technology [Online]. Sociological Research Online*, 2. Available: http://www.socresonline.org.uk/socresonline/2/4/5.html

Slater, M., Sadagic, M., Usoh, M., & Schroeder, R. (1999). *Small group behavior in a virtual and real environment: A comparative study [Online]*. Available: http://www.cs.ucl.ac.uk/staff/m.slater/BTWorkshop

Sleek, S. (1998a). Isolation increases with Internet use. *American Psychological Association Monitor, 29*, 30–31.

Sleek, S. (1998b). New cyber toast: Here's MUD in your psyche. *American Psychological Association Monitor, 29*, 30.

Stuart, R. (1996). *The design of virtual environments*. New York: McGraw-Hill.

Thomas, R. M. (1996). *Comparing theories of child development* (4th ed.). Pacific Grove, CA: Brooks Cole.

Turkle, S. (1995). *Life on the screen: Identity in the age of the Internet*. New York: Simon & Schuster.

Turow, J. (1999, May). The Internet and the family: The view from parents, the view from the press. Paper presented at the Annenberg Public Policy Center Conference on the Family and the Internet, National Press Club, Washington, D.C.

Valkenburg, P. M., & van der Voort, T. H. (1994). Influence of TV on daydreaming and creative imagination: A review of research. *Psychological Bulletin, 116*, 316–339.

Van der Voort, T. H. & Valkenburg, P. M. (1994). Television's impact on fantasy play: A review of research. *Developmental Review, 14*, 27–51.

Wallace, P. (1999). The *Psychology of the Internet*. Cambridge; New York: Cambridge University Press.

# PART V: Evaluation

# 34

# Usability Engineering of Virtual Environments

Deborah Hix and Joseph L. Gabbard
*Virginia Tech*
*Department of Computer Science/Systems Research Center*
*Blacksburg VA 24061*
*hix@vt.edu, jgabbard@vt.edu*

## 1. INTRODUCTION

*Usability engineering*, described rather simplistically, is the process by which usability is ensured for an interactive application at all phases in the development process. These phases include user task analysis, design of the user interaction, rapid prototyping, user-centered evaluation, and iterative redesign based on evaluation results. Usability engineering includes both design and evaluations with users; it is not just applicable at the evaluation phase. Usability engineering is not typically hypothesis-testing-based experimentation, but instead structured, iterative user-centered design and evaluation applied to all phases of the software development life cycle.

Until recently, the terms *usability engineering* and *virtual environments* were rarely, if ever, used in the same sentence or context. Virtual environment (VE) developers have focused largely on producing the next new "gee whiz" gadget and "way cool" interaction technique with too little attention given to how users will benefit (or not) from those gadgets and techniques. Admittedly, "gee whiz, way cool" exploration in a new realm such as VEs is necessary. However, VEs have been rather mature for nearly a decade now, with an ever expanding variety of possible applications.

Most extant usability engineering methods widely in current use were spawned by the development of traditional desktop graphical user interface (GUIs). Even when VE developers attempt to apply usability engineering methods, most VE user interfaces are so radically different that well-proven techniques that have produced usable GUIs may be neither particularly appropriate nor effective for virtual environments. Few principles for design of VE user interfaces exist, and almost none are empirically derived or validated. Use of usability engineering methods often results in VE designs that produce very unexpected reactions and performance of users, reaffirming the need for exactly such methods. Ultimately, researchers and developers of VEs should seek to improve VE applications *from a user's perspective*—ensuring their usability—by following a systematic approach to VE development such as offered from usability engineering methods.

To this end, we present several usability engineering methods, mostly adapted from GUI development, which have been successfully applied to VE development. These methods include *user task analysis*, *expert guidelines-based evaluation* (also sometimes called *heuristic evaluation* or *usability inspection*), and *formative usability evaluation*. Summative evaluation is also discussed because it is an important aspect of making comparative assessments of VEs from a user's perspective. Further, we postulate that, like GUI development, there is no single method for VE usability engineering. Thus, this chapter addresses how each of these methodologies supports focused, specialized design, measurement, management, and assessment techniques such as those presented in other chapters (see chaps. 11, 13, 16, 17, 21, 25, and 35, this volume). Our experiences with usability engineering of three different VEs are included in this discussion: a CAVE™ (Cave Automatic Virtual Environment)-based medical volume-imaging system; a Responsive Workbench–based military command and control application; and a new interaction technique for desktop virtual environments.

## 2. SETTING THE CONTEXT FOR VE USABILITY ENGINEERING

Before specific usability engineering methods are presented and examples of their application are given, it is important to set the context for usability engineering. Developers of interactive systems—systems with user interfaces—often confuse the boundaries between the software engineering process and the usability engineering process. This is due at least in part to a lack of understanding of techniques for usability engineering, as well as which of these techniques is appropriate for use at which stages in the development process. Software engineering has as a goal to improve software quality, but this goal, in and of itself, has little impact on the usability of the resulting interactive system, in this case, a virtual environment. For example, well-established "v&v" (validation and verification) techniques focus on software correctness, robustness, and so on, from a software developer's view, with little or no consideration of whether that software serves its users' needs. Thus, quality of the *user interface*—the usability—of an interactive system is largely independent of quality of the software for that system. Usability of the user interface is ensured by a user-centered focus on developing the *user interaction component*—the look, feel, and behavior as a user interacts with an application. The user interaction component includes all icons, text, graphics, audio, video, and devices through which a user communicates with an interactive system. Interaction designers and evaluators develop the user interaction component (see chaps. 11–13, this volume). Usability is unaffected by the *software component*, including that for both the user interface and the rest of the application (i.e., the non-user-interface software). Software engineers and systems engineers develop the software component of an interactive system.

Cooperation between usability engineers and software engineers is essential if VEs are to mature toward a truly user-centric work and entertainment experience. Thus, both the interaction component and the software component are necessary for producing any interactive system, including a VE, but the component that *ensures usability* is the user interaction component.

## 3. CURRENT USABILITY ENGINEERING METHODS

Current research at Virginia Tech aims to address the applicability of established usability engineering methods in conjunction with development of new usability engineering methods.

Simply put, our goal is to provide a methodology, or set of methodologies, to ensure usable and useful VE interfaces.

Usability engineering methods typically consist of one or more usability evaluation, or inspection, methods. Usability inspection is a process that aims to identify usability problems in user interaction design, and hence, end use (Nielsen & Mack, 1994).

As already mentioned, many of the techniques presented have been derived from usability engineering methods for GUIs. But from our own studies, as well as from collaboration with and experiences of other VE researchers and developers, GUI methods have been adapted, producing new methods for usability engineering of VE user interaction design. Adaptations and enhancements to existing methods have been made at two levels to evolve a usability engineering methodology applicable to virtual environments. Specific methods themselves had to be altered and extended to account for the complex interactions inherent in multimodal VEs (see chaps. 21–22, this volume), and various methods had to be applied in a meaningful sequence to both streamline the usability engineering process as well as provide sufficient coverage of the usability space.

To better understand the strengths and applicability of each individual GUI usability engineering method, we present a basic discussion of each method, to provide a brief overview of each. Following the discussion of individual methods, we discuss benefits and insights gained from sequential application of these adapted methods for VE user interaction development.

## 3.1 User Task Analysis

A *user task analysis* is the process of identifying a complete description of tasks, subtasks, and actions required to use a system as well as other resources necessary for user(s) and the system to cooperatively perform tasks (Hix & Hartson, 1993).

User task analyses follow a formal methodology, describing and assessing performance demands of user interaction and application objects. These demands are, in turn, compared with known human cognitive and physical capabilities and limitations, resulting in an understanding of the performance requirements of end users. User task analysis (Hackos & Redish, 1998) may be derived from several components of early systems analysis, and at the highest level, rely on an understanding of several physical and cognitive components. User task analyses are the culmination of insights gained through an understanding of user, organizational, and social work flow, needs analysis, and user modeling.

There are four generally accepted techniques for performing user task analysis: documentation review, questionnaire survey, interviewing, and observation (Eberts, 1999). Documentation review seeks to identify task characteristics as derived from technical specifications, existing components, or previous legacy systems. Questionnaire surveys are generally used to help evaluate interfaces that are already in use or have some operational component. In these cases, task-related information can be obtained by having domain experts such as existing users, trainers, or designers complete carefully designed surveys. Interviewing an existing or identified user base, along with domain experts and application "visionaries," provides very useful insight into what users need and expect from an application. Observation-based analysis, on the other hand, requires a user interaction prototype, resembling more the formative evaluation process than development of user task analysis, and as such, is used as a last resort. A combination of early analysis of application documentation and domain-expert and user interviewing typically provides the most useful and rich task analysis.

While user task analyses are typically performed early in the development process, it should be noted that, like all aspects of user interaction development, task analyses also need to be flexible and potentially iterative, allowing for modifications to performance and user interaction requirements during any stage of development. However, major changes to user task analysis during late stages of development can derail an otherwise effective development effort, and as such should only be considered under dire circumstances.

User task analysis generates critical information used throughout all stages of the application development life cycle. One such result is a top-down decomposition of detailed task descriptions. These descriptions serve, among other things, as an enumeration of desired functionality for designers and evaluators. Equally revealing results of task analysis include an understanding of required task sequences as well as sequence semantics. Thus, results of task analysis include not only identification and description of tasks, but ordering, relationships, and interdependencies among user tasks. These structured analytical results set the stage for other products of task analysis, including an understanding of information flow as users work through various task structures.

User task analysis also provides indications of where and how users contribute information to, and are required to make decisions that influence, user task sequencing. This information, in turn, can help designers identify which part(s) of the tasking process can be automated by computer (one of the original and still popular services a computer may provide), affording a more productive and useful work environment.

Without a clear understanding of user task requirements, both evaluators and developers are forced to "best guess" or interpret desired functionality, which inevitably leads to poor interaction design. Indeed, both user interaction developers and user interface software developers claim that poor, incomplete, or missing task analysis is one of the most common causes of both poor software and product design.

## 3.2 Expert Guidelines-based Evaluation

*Expert guidelines-based evaluation*, or *heuristic evaluation*, or *usability inspection* aims to identify potential usability problems by comparing a user interaction design, either existing or evolving, to established usability design guidelines. The identified problems are then used to derive recommendations for improving interaction design. This method is used by usability experts to identify critical usability problems early in the development cycle so that design issues can be addressed as part of the iterative design process (Nielsen, 1994b).

Expert guidelines-based evaluations rely on established usability guidelines to determine whether a user interaction design supports intuitive user-task performance (i.e., usability). Nielsen (1994a) recommends three to five evaluators for an expert guidelines-based evaluation because fewer evaluators generally cannot identify enough problems to warrant the expense, and more evaluators produce diminishing results at higher costs. It is not clear whether this recommendation is cost-effective for VEs, because more complex VE interaction designs may require more evaluators than GUIs. Each evaluator first inspects the design alone, independently of other evaluators' findings. Results are then combined, documented, and assessed as evaluators communicate and analyze both common and conflicting usability findings.

An expert guidelines-based evaluation session may last one to two hours or more, depending on the complexity of the design. Again, VE interaction designs may require more time for full exploration. Furthermore, expert guidelines-based evaluation can be done using a simple pencil and paper design (assuming that the design is mature enough to represent a reasonable amount of interaction components and interactions). This allows assessment to begin very early in the application development life cycle.

The output from an expert guidelines-based evaluation should not only identify problematic interaction components and interaction techniques, but should also indicate why a particular component or technique is problematic. Results from expert guidelines-based evaluation are subsequently used to remedy obvious and critical usability problems as well as to shape the design of subsequent formative evaluations (see section 3.3). Evaluation results further serve as both a working instructional document for user-interface software developers, and more importantly, as fundamentally sound, research-backed, design rationale.

Given that expert guidelines-based evaluations are based largely on a set of usability heuristics, it can be argued that these evaluations are only as effective and reliable as the guidelines themselves. Nielsen and Molich (1990) and Nielsen (1994a), respectively, present the original and revised set of usability heuristics for traditional GUIs. While these heuristics are considered to be the de facto standard for GUIs, they are too general, ambiguous, and high-level for effective and practical expert guidelines-based evaluation of virtual environments. Effectiveness is questioned on the simple fact that three-dimensional (3-D), immersive, VE interfaces are much more complex than traditional GUIs. The original heuristics (implicitly) assume traditional input–output devices such as keyboard, mouse, and monitor, and do not address the appropriateness of VE devices such as CAVEs, head-mounted displays (HMDs), haptic and tactile gloves, and various force feedback devices. Determining appropriate VE devices for a specific application and its user tasks is critical to designing usable virtual environments. Practicality is questioned due to the abstractness of the heuristics; VE evaluators need specific, concrete guidelines to apply in rapid and reliable evaluations.

It is well recognized that VE interfaces and VE user interaction are immature and currently emerging technologies for which standard sets of design, much less usability guidelines, do not yet exist. However, our recent research at Virginia Tech has produced a set of VE usability design guidelines, contained within a framework of usability characteristics (Gabbard, 1997). This framework document (see http://csgrad.sc.vt.edu/~jgabbard/ve/framework) provides a reasonable starting point for expert guidelines-based evaluation of virtual environments. The complete document contains several associated usability resources including specific usability guidelines, detailed context-driven discussion of the numerous guidelines, and citations of additional references.

The framework organizes VE user interaction design guidelines and the related context-driven discussion into four major areas: users and user tasks, input mechanisms, virtual model, and presentation mechanisms. The framework categorizes 195 guidelines covering many aspects of VEs that affect usability including navigation, object selection and manipulation, user goals, fidelity of imagery, input device modes and usage, interaction metaphors, and much more. The guidelines presented within the framework document are well suited for performing expert guidelines-based evaluation of VE user interaction because they provide both broad coverage of VE interaction/interfaces and are specific enough for practical application. For example, with respect to navigation within VEs, one guideline reads "provide information so that users can always answer the questions: Where am I now? What is my current attitude and orientation? Where do I want to go? How do I travel there?" Another guideline addresses methods to aid in usable object selection techniques stating "use transparency to avoid occlusion during selection."

These guidelines have been successfully used within context as a basis for expert guidelines-based evaluation of several VEs ranging from medical visualization to command and control situational awareness applications (see sections 4.1–4.3 for details).

### 3.3 Formative Usability Evaluation

The term *formative evaluation* was coined by Scriven (1967) to define a type of evaluation that is applied during evolving or formative stages of design. Scriven used this in the educational domain for instructional design. Williges (1984) and Hix and Hartson (1993) extended and refined the concept of formative evaluation for the human–computer interaction (HCI) domain.

The goal of formative evaluation is to assess, refine, and improve user interaction by iteratively placing representative users in task-based scenarios in order to identify usability problems, as well as to assess the design's ability to support user exploration, learning, and task performance (Hix & Hartson, 1993). Formative usability evaluation is an observational evaluation method that ensures usability of interactive systems by including users early and continually throughout user interface development. The method relies heavily on usage context (e.g., user task, user motivation) as well as a solid understanding of human–computer interaction (and in the case of VEs, human–VE interaction) and, as such, requires the use of usability experts (Hix & Hartson, 1993).

While the formative evaluation process was initially intended to support iterative development of instructional materials, it has proven itself to be a useful tool for evaluation of traditional GUI interfaces. Moreover, in the past few years, we have seen firsthand evidence indicating that the formative evaluation process is also an efficient and effective method of improving the usability of VE interfaces (Hix et al., 1999a).

The steps of a typical formative evaluation cycle begin with development of user task scenarios that are specifically designed to exploit and explore all identified tasks, information, and work flows. Representative users perform these tasks as evaluators collect both qualitative and quantitative data. These data are then analyzed to identify user interaction components or features that both support and detract from user task performance. These observations are in turn used to suggest user interaction design changes as well as formative evaluation scenarios and observation (re)design.

The formative evaluation process itself is iterative, allowing evaluators to continually refine user task scenarios in order to fine-tune both the user interaction and the evaluation process. Contrary to popular belief, the formative evolution process produces both qualitative and quantitative results collected from representative users during their performance of task scenarios (del Galdo, Williges, Williges, & Wixon, 1986). One type of qualitative data collected is termed *critical incidents* (del Galdo et al., 1986; Hix & Hartson, 1993). A critical incident is a user event that has a significant impact, either positive or negative, on users' task performance and/or satisfaction (e.g., a system crash or error, being unable to complete a task scenario, user confusion). Critical incidents that have a negative effect on users' work flow can drastically impede usability and may even have a dramatic effect on users' perceptions of application quality, usefulness, and reputation. As such, any obvious critical incidents are best discovered during formative evaluation phases as opposed to during consumer desktop use.

Equally important are the quantitative data collected during formative evaluation. These data include measures such as how long it takes a user to perform a given task, the number of errors encountered during task performance, and others. Collected quantitative data are then compared to appropriate baseline metrics, sometimes initially redefining or altering evaluators' perceptions of what should be considered baseline. Both qualitative and quantitative data are equally important because they each provide unique insight into an interaction design's strengths and weaknesses.

### 3.4 Summative Evaluation

*Summative evaluation*, in contrast to formative evaluation, is typically performed after a product or design is more or less complete. Its purpose is to statistically compare several different systems, for example, to determine which one is "better," where better is defined in advance. Another goal of summative evaluation is to measure and subsequently compare the productivity and cost benefits associated with various designs. In this fashion, evaluators are simply comparing the best of a few refined designs to determine which of the "finalists" is best-suited for delivery.

The term *summative evaluation* was also coined by Scriven (1967) for use in the instructional design field. As with the formative evaluation process, HCI researchers (e.g.,

Williges, 1984) have applied the theory and practice of summative evaluation to interaction design with surpassingly successful results.

In practice, summative evaluation can take on many forms. The most common are the comparative, field trial, and more recently, the expert review (Stevens, Frances, & Sharp, 1997). While both the field trial and expert review methods are well-suited for instructional content and design assessment, they typically involve assessment of single prototypes or field-delivered designs. In the context of VE design, we are mostly interested in assessing the quality of two or more user interaction designs, and as such, have focused on the comparative approach. Our experiences have found that this approach is very effective for analyzing the strengths and weaknesses of various well-formed, completed designs using representative user scenarios.

Stevens et al. (1997) present a short list of questions that summative evaluation should address. These questions have been modified to address summative, comparative evaluation of VE user interfaces. The questions include:

- What are the strengths and weaknesses associated with each user interaction design?
- To what extent does each user interaction design support overall user and system goal(s)?
- Did users perceive increased utility and benefit from each design? In what ways?
- What components of each design were most effective?
- Was the user interaction evaluation effort successful? That is, did it provide a cost-effective means of improving design and usability?
- Were the results worth the program's cost?

### 3.5 A Cost-effective Progression

As previously mentioned, one of our long-term research goals is to produce methodologies to improve the usability of VE user interaction designs. More specifically, the goal is to develop, modify, and fine-tune usability engineering techniques specifically for virtual environments. Techniques to aid in usability engineering of VEs will mature, and as such, the potential for conducting effective evaluations and delivering subsequent usable and useful VEs will increase.

Our current efforts are focusing on the combination of usability evaluation techniques described in sections 3.1 to 3.4. As depicted in Fig. 34.1, our applied research over the past several years has shown that, at a high level, progressing from user task analysis to expert guidelines-based evaluation to formative evaluation to summative evaluations is an efficient and cost-effective strategy for designing, assessing, and improving the user interaction—usability engineering—of a virtual environment (Gabbard, Hix, & Swan, 1999).

One of the strengths of this progression is the fact that it exploits a natural ordering or evolution of interaction design and prototyping, with each method generating a streamlined set of information that the next method utilizes. In this sense, each method is able to generate a much better starting point for subsequent methods than when applied in a stand-alone fashion. Moreover, simply applying more than one usability engineering method ensures more complete coverage of an interaction design's "usability space," each revealing its niche of particular usability problems and collectively shaping a more usable virtual environment. Finally, the progression of methods also produces a "paper trail" of persistent documentation that may serve as documented design rationale.

This progression is very cost-effective for assessing and improving virtual environments. For example, summative studies are often performed on VE interaction designs that have had little or no task analysis or expert guidelines-based or formative evaluation. This may result in a

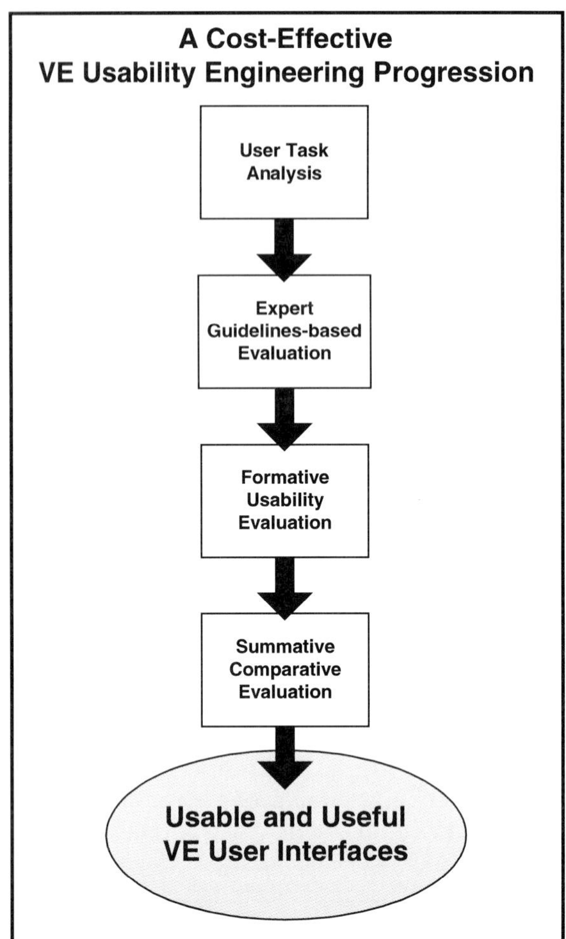


FIG. 34.1. A cost-effective progression of usability engineering methods.

situation where the expensive summative evaluation is essentially comparing "good apples" to "bad oranges" (Hix, Swan, Gabbard, McGee, Durbin, & King, 1999a). Specifically, a summative study of two different VEs may be comparing one design that is inherently better, in terms of usability, than the other one. When all designs in a summative study have been developed following the suggested progression of usability engineering, then the comparison is more valid. Experimenters will then know that the interaction designs are basically equivalent in terms of their usability, and any differences found among compared designs are, in fact, due to variations in the fundamental nature of the designs and not their usability.

## 4. APPLICATION OF USABILITY ENGINEERING METHODS

The following case studies present the application of usability engineering methods to interaction design of three different VE applications. The first, called Crumbs, is a medical imaging system developed to run in the CAVE. The next, called Dragon, is a military command and control application developed for the Responsive Workbench. The third, called pre-screen projection, is a new interaction technique developed for desktop virtual environment. The first two VE applications and their user interaction design are at different phases in their development and assessment. Specifically, Crumbs is in the formative evaluation phase and Dragon is in the summative evaluation phase. The discussion of pre-screen projection is the most detailed, because that work is essentially complete at the time of this writing. The important point is that for each of these VE applications, the usability engineering methods described above have been followed with great success in evolving the interaction design.

### 4.1 Crumbs

Crumbs is a VE application used as a tracking tool for biological and medical images, developed by the NCSA Biological Imaging Group at the University of Illinois (Brady et al., 1995). A goal of Crumbs is to facilitate identification of complex biological structures in images by allowing users to visualize dense volumetric data sets in an immersive VE, namely, the CAVE. Users can also mark (using a "bread crumb" metaphor) and measure individual structures in the data set they are viewing and manipulating. Magnetic resonance images and laser scanning confocal microscope images were predominantly used in Crumbs during the usability engineering activities herein described.

Crumbs was developed at NCSA, and Virginia Tech personnel were not involved from the beginning. The first collaboration between NCSA and Virginia Tech researchers was to conduct an expert guidelines-based evaluation of a fairly well-developed version of Crumbs. The framework of usability characteristics of VEs (Gabbard, 1997) discussed in section 3.2 was the basis for this evaluation. Proceeding methodically through the framework, a syntactic analysis of the Crumbs' user interaction design was performed. Evaluation was primarily focused on those features that could be determined by an expert inspection and comparison to the guidelines in the framework. The evaluators were not expert Crumbs users, so could not effectively perform an in-depth semantic analysis that would have determined more details about whether Crumbs' interaction design supported all appropriate user tasks that were needed. However, the syntactic analysis revealed numerous usability issues that were then addressed by evaluators collaboratively with the Crumbs developers.

Specifically, following the four main areas of the framework, VE users and user tasks as supported in Crumbs, the virtual model of Crumbs, the VE user interface input mechanisms, and the VE user interface presentation mechanisms were assessed. For example, in VE user interface presentation mechanisms, usability issues with objects including the color map, a slice plane, and a trashcan icon, particularly inconsistencies across these different objects and their manipulation were found. In VE user interface input mechanisms, pop-up menus were attached to the current location of the cursor and therefore often appeared very near the bottom of the front wall of the CAVE. This made it very difficult for users to make a selection. Further, numerous problems were uncovered with the wand device used to manipulate and navigate through the Crumbs application. The final report to Crumbs developers included more than a dozen suggestions for substantial changes to the Crumbs user interaction design (Swartz, 1998). This report included the framework usability design guideline that was violated, the potential usability problem therefore resulting, and one or more specific recommendations for fixing the usability problem. Most recommendations were incorporated before moving on to the current state of Crumbs development, which is the formative evaluation phase (Gabbard et al., 1999b; Swartz, Thakkar, Hix, & Brady, 2000).

Formative evaluations thus far have been performed in the NCSA CAVE; future ones will also be conducted in the Virginia Tech CAVE. In the first formative evaluation, users with a priori knowledge of Crumbs were chosen. This was, in part, to determine whether the changes made from the expert guidelines-based evaluation resulted in an improvement of the user interaction design. This assessment was largely anecdotal, because there were no metrics from use of the original Crumbs interaction design. A second reason for choosing participants already familiar with Crumbs was to avoid the fairly long learning curve needed to become familiar with a very complicated, very specialized application. Someone who knows nothing about magnetic resonance images or confocal microscope images would not be a representative user of Crumbs. The choice of effective participants was somewhat limited because of the complexity of the application domain and resulting scarcity of appropriate users. Although participants varied in their individual experience with Crumbs, their familiarity with it and related appropriate tasks provided adequate consistency across the participants chosen. Of the five participants in the first formative evaluation, two had largely used Crumbs to give demonstrations, whereas the other three had used Crumbs for "real work."

The first formative evaluation session was actually a pilot session to make sure the evaluation protocol was effective and reasonable. Some problems were found with the original protocol, which were amended in subsequent sessions. In particular, the plan was to have two evaluators present for each session, because of the large amount of data that are generated during a session. One evaluator was intended to be in the CAVE near the participant to serve as a facilitator and to keep the session moving along. The second evaluator was to be outside the CAVE, taking timings, counting errors, and collecting critical incidents that occurred and other qualitative data as a participant attempted to perform tasks with Crumbs. However, we quickly found that having an evaluator in the CAVE with the participant was very distracting for the participant. The evaluator often unintentionally got in the way as the participant moved around during task performance despite the evaluator's best efforts to stay out of the way. The evaluator also sometimes accidentally got tangled in wires across the floor of the CAVE, which further distracted the participant, who did not seem to get entangled. Perhaps the major problem was that the evaluator's presence was found to reduce the participant's sense of presence and immersion in the environment (see chap. 40, this volume), and therefore reduced the participant's productivity and focus on the Crumbs tasks. Modifications to the evaluation protocol were made so that all evaluators were located outside the CAVE, with the one serving as mediator nearest its open edge. For some sessions there were three evaluators, all of whom were kept busy collecting timing, error, and critical incident data in the complex Crumbs application.

Results from both guidelines-based and formative evaluations indicated numerous usability problems. For example, users encountered difficulties with use of buttons on the CAVE wand for various modes in Crumbs, especially a scale and mark mode. Several inconsistencies in the Crumbs interaction design also caused problems, including inconsistent use of directly manipulating various types of objects, and inconsistent ways of selecting objects. Users also had difficulty knowing where they were in Crumbs at any point in time, especially on task completion. A major usability issue was occlusion of 3-D objects by other objects, which contributed to multiple user errors and long task completion times. Further, some audio cues that were intended to help users actually annoyed them. For each usability issue, one or more redesign suggestions were made based on the framework of usability characteristics of virtual environments. Many of these redesign recommendations were incorporated into Crumbs by its developers at NCSA.

## 4.2 Dragon

Personnel at the Naval Research Laboratory's Virtual Reality Laboratory, again in collaboration with Virginia Tech researchers, have developed a VE for battlefield visualization, called Dragon (Hix et al., 1999a). Implemented on a Responsive Workbench, Dragon's metaphor for visualizing and interacting with 3-D computer-generated scenery uses a familiar tabletop environment. Applications in which several users collaborate around a work area, such as a table, are excellent candidates for the Workbench. This metaphor is especially familiar to Naval personnel and Marines, who have, for decades, accomplished traditional battlefield visualization on a tabletop. Paper maps of a battlespace are placed under sheets of acetate. As intelligence reports arrive from the field, technicians use grease pencils to mark new information on the acetate. Commanders then draw on the acetate to plan and direct various battlefield situations. Historically, before high-resolution paper maps, these operations were performed on a sandtable, literally a box filled with sand shaped to replicate battlespace terrain. Commanders moved around small physical replicas of battlefield objects to direct battlefield maneuvers. The fast-changing modern battlefield produces so much time-critical information that these cumbersome, time-consuming methods are inadequate for effectively visualizing and commanding a modern-day battlespace.

Dragon was developed on the Responsive Workbench to give a 3-D display for observing and managing battlefield information shared among technicians and commanders. Visualized information includes a high-resolution terrain map; entities representing friendly, enemy, unknown, and neutral units; and symbology representing other features such as obstructions or key map points. Users can navigate to observe the map and entities from any angle and orientation around the Workbench.

Dragon's early development was based on an admittedly cursory user task analysis, which drove early design. This was followed, however, by numerous cycles of expert guidelines-based evaluation as well as formative evaluations. Early in Dragon design, three general interaction methods for the Workbench were produced and assessed, any of which could have been used to interact with Dragon: hand gestures using a pinch glove (Obeysekare et al., 1996), speech recognition, and a handheld flightstick. Although it was an interesting possibility for VE interaction, formative evaluations found that speech recognition is still too immature for battlefield visualization. Further, the pinch glove was found to be too fragile and time-consuming to pass from user to user around the Workbench. It also worked best for right-handed users whose hands were approximately the same size. In contrast, formative evaluations revealed that the flightstick was robust, easily handed from user to user, and worked for both right- and left-handed users.

Based on these formative evaluations, a three-button game flightstick was modified by removing its base and placing a 6-degree-of-freedom position sensor inside. Initial designs used a laser pointer metaphor in which a laser beam appeared to come out the "head" of the flightstick as a user pointed it toward the VE map. When a beam intersected terrain or an object, a highlight marker appeared.

In early demonstrations of initial versions of Dragon in real military exercises for battlefield planning, users indicated they found Dragon's accurate and current visualization of the battlespace to be more effective and efficient than the traditional method of maps, acetate, and grease pencils. Following these successful demonstrations and positive feedback, we began intensive usability engineering of Dragon's user interaction design.

During early demonstrations and evaluations, user observation indicated that navigation (how users manipulate their viewpoint to move from place to place in a virtual world, in this case, the battlefield map), profoundly affects all other user tasks (see chap. 24, this volume). If a user cannot successfully navigate to move about in a virtual world, then other user tasks such as selecting an object or grouping objects cannot be performed. A user cannot query an object if the user cannot navigate through the virtual world to get to that object. Although a user task analysis was performed before guidelines-based and formative evaluations, these evaluations supported expectations of the importance of navigation.

Expert guidelines-based evaluation was done extensively for Dragon, prior to much formative evaluation. However, the framework of usability characteristics of VEs were being developed at the same the guidelines-based evaluation was performed, and so the guidelines used were much more ad hoc and less structured than those that eventually became the framework.

Nonetheless, even informal inspection of the evolving Dragon interaction design provided tremendous feedback in what worked and what did not, especially for use of the wand and other aspects of navigation in Workbench VE applications in general, and Dragon in particular. During these evaluations, VE user interaction design experts worked alone or collectively to assess the Dragon interaction design. In the earliest evaluations, the experts did not follow specific user task scenarios per se, but engaged simply in "free play" with Dragon using the wand. All experts knew enough about the purpose of Dragon as a battlefield visualization VE to explore the kinds of tasks that would be more important for Dragon users. During each session, one person was typically "the driver," holding the flightstick and generally deciding what and how to explore in the application. One and sometimes two other experts were observing and commenting as the "driver" worked, with much discussion during each session. Major design problems that were uncovered in the expert guidelines-based evaluation of Dragon included: poor mapping of navigation tasks (i.e., pan, zoom, pitch, heading) to flightstick buttons; missing functionality (e.g., exocentric rotate, terrain following); problems with damping of map movement in response to flightstick movement; and graphical and textual feedback to the user about the current navigation task. After these evaluations had revealed and remedied as many design flaws as possible, we moved on to formative evaluation.

The basic Dragon application was used to perform extensive formative evaluations, using anywhere from one to three users for each cycle of evaluation. A single evaluation session often uncovered design problems so serious that it was pointless to have a different user attempt to perform scenarios with the same design. The design was iterated, based on these observations, and then a new cycle of evaluation was commenced, undergoing four major cycles of iteration in all.

In designing scenarios for formative evaluation, coverage of specific usability issues related to navigation was carefully considered. For example, some of the tasks exploited an egocentric (user moves through the virtual world) navigation metaphor, whereas others exploited an exocentric (user moves the world) navigation metaphor. Some scenarios exercised various navigation tasks (e.g., degrees of freedom: pan, zoom, rotate, heading, pitch, roll) in the virtual map world. Other scenarios served as primed exploration or nontargeted searches for specific features or objects in the virtual world. Still others were designed to evaluate rate controlled interaction techniques versus position controlled interaction techniques (see chap. 13, this volume).

During each of six formative evaluation sessions, the participant was first asked to play with the flightstick to figure out which button activated which navigation task. Each user was timed as they attempted to determine this, and notes were taken on comments and any critical incidents that occurred. Once a user had successfully figured out how to use the flightstick, they formally performed the task scenarios. Only one user was unable to figure out the flightstick in less than 15 minutes; this user was told details they had not yet discovered and proceeded with the scenarios.

Time to perform the set of scenarios ranged from about 20 minutes to more than 1 hour. User performance of individual tasks and scenarios was timed and errors made during task performance were counted. A typical error was moving the flightstick in the wrong direction for the particular navigation metaphor (exocentric or egocentric) that was currently in use. Other errors involved simply not being able to maneuver the map (e.g., to rotate it) and persistent problems with mapping navigation tasks (degrees of freedom) to flightstick buttons, despite extensive prior evaluations to minimize this issue. During each formative evaluation session, at least two and often three evaluators were present. One served as the facilitator to interact with the participant and keep the session moving; the other one or two evaluators recorded times, counted errors, and collected critical incidents and other qualitative data. While these sessions (like those of Crumbs in the CAVE) seem personnel-intensive, with two or three evaluators involved, the quality and quantity of data collected by multiple evaluators was found to greatly outweigh the cost of those evaluators. A surprising amount of effort was spent on mapping flightstick buttons to navigation tasks (pan, zoom, rotate, heading, pitch, roll), but it was found to pay off with more effective, intuitive mappings.

As mentioned earlier, we went through four major iterations of the Dragon interaction design, based on our evaluations. The first iteration, the Virtual Sandtable, was an egocentric navigation metaphor based on the sandtable concept briefly described earlier. This was the version demonstrated in the military exercises also mentioned previously. A key finding of this iteration was that users wanted a terrain-following capability, allowing them to "fly" over the map, an egocentric design. Map-based navigation worked well when globally manipulating the environment and conducting operations on large-scale units. However, for small-scale operations, users wanted this "fly" capability to visually size up terrain features, entity placement, fields of fire, lines of sight, and other characteristics.

The second iteration, Point and Go, used the framework of usability characteristics of VEs to suggest various possibilities for an egocentric navigation metaphor design. This metaphor attempted to avoid having different modes (and flightstick buttons) for different navigation tasks because of known usability problems with moded interaction. Further, this decision was based on how a person often navigates to an object or location in the real world; namely, they point (or look) and then go (move) there. Our reasoning was that adopting this same idea to egocentric navigation would simplify the design and at least loosely mimic the real world. In this design, a user simply pointed the flightstick toward a location or object of interest and pressed the trigger to fly there. The single gesture to move about was not powerful enough to support the diverse, complicated variety of navigation tasks inherent in Dragon. Furthermore, a single gesture meant that all degrees of freedom were controlled by that single gesture. This resulted in, for example, unintentional rolling when a user only wanted to pan or zoom. Essentially, this presented a control versus convenience trade-off. Many navigation tasks (modes) were active simultaneously, which was convenient, but difficult to physically control for a user. With separate tasks (modes) there was less convenience, but physical control was easier because degrees of freedom were more limited in each mode. In addition to these serious problems, users wanted to rotate around an object, such as to move completely around a tank. This indicated that Dragon needed an exocentric rotation ability, which was added. This interesting finding showed that neither a pure egocentric nor a pure exocentric metaphor was desirable; each metaphor has aspects that are more or less useful depending on user goals. While this may seem obvious, it was confirmed by users through the formative evaluation approach. Further, the somewhat poor performance of what we thought was the natural "point and go" metaphor was rather counterintuitive, and further demonstrates that mimicking VE design after the real world is not always successful from the VE user's perspective.

The third iteration, Modal, went from the extreme of all navigation tasks coupled on a single button, as in the previous iteration, to a rather opposite design in which each navigation task was a separate mode. Specifically, as a user clicked the left or right flightstick button, Dragon cycled successively through the tasks of pan, zoom, pitch, heading, and exocentric rotate. A small textual indicator was displayed on the Workbench to show the current mode. Once a user had cycled to the desired task, the user simply moved the flightstick and that task was enabled with no need to push any flightstick button. As expected, it was very cumbersome for users to always have to cycle between modes, and it was obvious that a compromise between convenience and control still had not been achieved.

In the fourth iteration of the Dragon interaction design, Integrated Navigation, based on the framework of usability characteristics of VEs and user observations of what degrees of freedom could be logically coupled in the Dragon application, a hybrid design of the modeless/moded designs of prior iterations was produced. Specifically, pan and zoom were coupled onto the flightstick trigger, pitch, and heading onto the left flightstick button, and exocentric

**TABLE 34.1**
Design Parameters for Navigation, Organized by Framework of Usability Characteristics for VEs*

| *Design Parameters for Navigation Summative Evaluation* | | | |
|---|---|---|---|
| *User Tasks* | *Input Devices* | *Virtual Model* | *Presentation Devices* |
| User scenarios | Navigation metaphor | Mode switching | Visual presentation device |
| Navigation presets | Navigation degrees of freedom | Mode feedback | Stereopsis |
| | Gestures to trigger actions | Number of modes | |
| | Speech input | Visual navigation aids | |
| | Number of flightstick buttons | Data set characteristics | |
| | Input device type | Visual terrain representation | |
| | Movement dead space | Visual (battlefield) object representation | |
| | Movement damping | Visual input device representation | |
| | User gesture work | Size of (battlefield) objects | |
| | Volume | | |
| | Gesture mapping | Visual object relationship representation | |
| | Button mapping | Map constrained vs. floating | |
| | Head tracking | | |

* Adapted from Gabbard (1997).

rotate and zoom onto the right flightstick button. This fourth generation interaction design for Dragon finally achieved the desired convenience versus control compromise. Final formative evaluation studies revealed that this was a design for navigation that seemed to work well. The only usability problem observed was minor: Damping of map movement was too great and needed some adjustment, which was made.

As the formative evaluation cycles neared completion for Dragon, summative studies for the navigation design were planned. Design parameters that affect usability of the VE navigation metaphor to be studied in summative evaluations are those that, in general, could not be decided from the formative evaluations. Twenty-eight such design parameters were identified, as shown in Table 34.1, that potentially affect the usability of user navigation. The organization of these parameters is based on the framework of usability characteristics of virtual environments. The four main areas of this framework are the headings in Table 34.1, and design parameters for navigation are grouped based on these areas.

Based on observations during evaluations, extensive literature review, and our expertise in VE interaction design, the numerous variables have been narrowed to five that we feel are most critical for navigation, and therefore most important for the first cycle of summative evaluations. These five variables and their level of treatment in our summative study are:

- Navigation metaphor (ego- versus exo-centric)
- Gesture mapping (rate vs. position of hand movement)
- Visual presentation device (Workbench, desktop, CAVE™)
- Head tracking (present vs. absent)
- Stereopsis (present vs. absent)

This summative study has been performed on four presentation devices (i.e., Workbench, desktop, CAVE, one wall of CAVE) and we are analyzing results at the time of this printing. It is, we believe, one of the first formal studies to directly compare four different VE devices using the same application and the same user tasks.

### 4.3 Pre-screen Projection

*Pre-screen projection* is a new interaction technique (described below) developed at the Naval Research Laboratory's Human–Computer Interaction Laboratory (Hix, Templeman, & Jacob, 1995; Hix et al., 1999b), with strong collaboration by Virginia Tech researchers beginning very early in the development process and continuing throughout. An *interaction technique* is a way in which a human uses a physical input–output device to perform a task during human–computer interaction (Foley et al., 1990; see chap. 13, this volume). It abstracts a class of generic interactive tasks, for example, selecting an object on a screen by pointing and clicking with a mouse. A pop-up menu is an interaction technique. Interaction techniques are a useful research topic because they are specific enough to be studied, yet general enough to have practical applicability to a broad variety of interactive systems. But, like VEs, research specifically in interaction techniques often emphasizes technological creativity, while user-based evaluation of techniques is either cursory or nonexistent. Thus, over the years, a plethora of interaction techniques, both for GUIs and for VEs, have been developed, but to date too little is known about their impact on the usability of interactive systems. There is scant empirical evidence of whether they improve human performance and satisfaction.

Creating the new interaction technique of pre-screen projection involved many steps, beginning with its conceptualization, for example, the idea, the real-world metaphor, or whatever forms a technique's underlying motivation. User task analysis followed conceptualization to drive the design of the technique, going from general abstractions to specific details. Next came prototyping and extensive formative evaluation to determine what use of the technique "feels like." Eventually the technique was implemented in the context of user tasks and an application, on which summative evaluation was performed.

*Pre-screen projection* allows a user to pan and zoom integrally through a scene simply by moving the head relative to the screen. Its conceptualization is based on real-world visual perception (Gibson, 1986), namely, the fact that a person's view changes as the head moves. Pre-screen projection tracks a user's head in three dimensions and alters the display on the screen relative to head position, giving a natural perspective effect in response to a user's head movements. Further, projection of a virtual scene is calculated as if that scene were in front of the screen. As a result, the visible scene displayed on the physical screen expands (zooms) dramatically as a user moves closer. This is analogous to the real world, where the nearer an object is, the more rapidly it visually expands as a person moves toward it. Using pre-screen projection, a user wears a lightweight helmet or headband with a 3-D motion tracker mounted on the front. As the user moves from side to side, the display smoothly pans over the world view. As the user moves closer to or further from the screen, the display smoothly zooms in and out, respectively. The virtual scene is calculated to appear as if it were 20 inches *in front of* the physical screen. Thus, the scene is displayed on the physical screen, but its dynamic perspective from the user's viewpoint reacts to head movements as if the scene were in front of the screen. This causes the scene to enlarge more rapidly than the screen as a user moves toward it and therefore produces a dramatic zoom.

Having developed this concept to underlie pre-screen projection, we began the challenge of designing and instantiating the interaction technique, which included the usability engineering methods of user task analysis, detailed design, and formative evaluation followed by iterative refinement. The intent was to incorporate pre-screen projection into an application that would provide a realistic situation for its evaluation. Panning and zooming, which pre-screen projection supports, are inherently spatial, navigational information-seeking activities; that is, a user navigates through an information space. To design and evaluate pre-screen projection, we wanted to have user tasks that would place heavy demands on a user's ability to pan and zoom. Thus, a critical aspect of this research was the co-evolution of appropriate user tasks along with the interaction technique, to produce tasks for evaluating pre-screen projection, rather than for producing the best possible application. This approach reemphasizes the definition of an interaction technique as a means of performing tasks. Prior research at the Naval Research Laboratory's Human–Computer Interaction Laboratory (Jacob & Sibert, 1992) used rather simplistic, nonmilitary, low-fidelity domains and tasks. Because we wanted our work to be relevant to specific naval applications, we chose Naval command and control systems ($C^2$ systems) as a rich, realistic application domain for usability engineering of pre-screen projection. These systems, simply explained, support the planning, coordination, and execution of a military mission. Pre-screen projection was incorporated into a $C^2$-like testbed and task scenarios were created for evaluation.

$C^2$ is a highly diverse, extremely rich and demanding application area with a breadth and depth of tasks that its users perform. Based on interviews with appropriate military personnel, a detailed literature review, and other resources, a detailed user task analysis of such systems was performed to determine the design space from which to select possible tasks. Participants to whom access was available for evaluation were, in general, civilians with little or no military background. Thus, selecting and designing specific tasks to evaluate pre-screen projection that were rich enough, yet were unbiased and simple enough for users to learn quickly, was a crucial issue. The interaction technique and goals for its evaluation should drive task development, but in reality they are very closely coupled. Simple user tasks were often found to work best for evaluation because they tended not to confound learnability and other issues of the application itself with the issues of evaluation of the interaction technique.

From user task analysis, a subset of defensive engagement tasks were chosen, in which friendly, enemy, unknown, and neutral military objects are displayed on a map. Several scenarios were developed in which groups of friendly ships and planes remain constant from the start of the scenario, but additional enemies, unknowns, and neutrals appear during the course of the scenario. Using the testbed application, the user's goals (tasks) were to acquire and maintain an awareness of an evolving situational picture, and to appropriately allocate weapons from friendly ships to encroaching enemy planes. Specifically, a user monitored the scene, largely by panning, looking for threats. When one was observed, the user then determined the number of planes in that threat (by zooming in on textual details that were progressively disclosed as the user's head moved closer to the screen) and the number of missiles the friendly had available to fire at the enemy. The user then used a slider (not pre-screen projection) to enter the number of missiles to shoot at the enemy and fire those missiles.

As we proceeded with the design of pre-screen projection and specific tasks, we began to see its large design space of possible attributes. Each attribute could have several (and sometimes many) values, so combinatorics of the numerous design decisions quickly became very large. This meant that early design decisions must be made based on very rapid prototyping and a quick formative evaluation cycle, rather than on fuller experimental evaluation. Especially interesting design challenges were provided by two attributes, scaling and fade in/out. These were applied to three types of objects in a scene: a map, icons, and text. *Scaling* is an attribute that specifies which objects in the scene appear larger or smaller as a user moves the head toward or away from the screen. In all design iterations the map scaled, because it is the predominant spatial context for icons and text. In early designs icons did not scale (to remove any implication that icon size has meaning, which it does not in the application studied). Somewhat surprisingly, users found it difficult to zoom in on nonscaling icons, so the design was changed so that icons scaled. Also, text associated with the icons was not scaled to maximize the amount of readable text presented at any time. After several cycles of formative evaluation, users were comfortable with a final design in which the map and icons scaled, but text did not. *Fade in/out* is an attribute that specifies the gradual appearance and disappearance (progressive disclosure) of objects in a scene as a user moves the head. In all design iterations, the map and icons scaled but did not fade in/out, because of their role in maintaining spatial context. Text was fading in/out at several different levels of disclosure, based on the distance of the user's head from the physical screen, in order to display as much detailed information (in the text) as possible. In early designs, text at one level faded out completely as different text from the next level replaced it. However, users found it hard to determine which level they were in as they looked for specific information. Therefore, the design was changed to fade in only additional text, beneath (rather than on top of the location of) text displayed at prior levels. Fade-in/out of text provided some other surprises. At first, text was too densely packed, even after a user had zoomed quite far into the scene. What was thought to be a reasonable adjustment to that problem resulted in the text being too spread out. Text at first was displayed 75% opaque, but this proved to be too low contrast to read easily and also led to incomprehensible clutter when icons were close enough together so that their associated text overlapped. Again, the combinatorics of alternative designs at low levels of detail (e.g., color, font, spacing, and icon shape) was huge.

Many final detailed design decisions (e.g., distance for amplified zooming, distances for levels of disclosure) were based on trial and error, again supporting the need for a fast, highly iterative development usability engineering methodology. Thus, as pre-screen projection evolved over about a 12-month period, numerous cycles of formative evaluation were performed, some as short as 5 minutes, others lasting nearly an hour. Evolution of the design of scaling and fade in/out of text and graphics, for example, as well as virtually all other decisions about design details, came from many rounds of formative evaluation.

Once formative evaluations had identified a good design for pre-screen projection, summative evaluations began, comparing user performance and satisfaction using pre-screen projection to other interaction techniques for panning and zooming, using the defensive engagement tasks. Our study compared pre-screen projection with panning and zooming using a mouse-based interaction technique and also a Spaceball-based interaction technique. Results (Hix et al., 1999b) showed that pre-screen projection led to superior performance for more difficult tasks and, very importantly, also led users to learn a more effective strategy for panning and zooming.

Our best guesses about the design of pre-screen projection and other interaction techniques we have developed over the years were substantiated or refuted by many tight, short cycles of formative evaluation with users performing realistic tasks in the $C^2$ testbed application. The many unexpected and difficult usability issues and resulting design decisions could only have been encountered and resolved using a usability engineering methodology, such as that described, that supports this kind of fast, effective iteration.

## 5. CONCLUSION

Despite improvement in efforts to improve usability engineering of VEs, there is not nearly enough usability engineering applied during development of virtual environments. There is still a need for more cost-effective, high-impact methods. Four usability engineering techniques have been presented that have traditionally been used for GUI development and more recently modified for usability engineering of virtual environments. We have presented how these techniques may be used together, as well as given descriptions of our experiences applying these techniques to real-world and research-based development efforts.

As VEs become more mature, established interaction techniques and generic tasks for VEs will emerge, and have the potential, as they did with GUIs, to improve usability. The design space and options for VEs is enormously greater than that for GUIs, and VE applications tend to be much more complex than many GUIs. Even "standard" interaction techniques, devices, and generic tasks for VEs will help improve usability only by a small fraction. Usability engineering will continue to be a necessary process if new and exciting VEs that are, in fact, usable and useful for their users are to be created.

## 6. ACKNOWLEDGMENTS

Dr. Rudy Darken, at the Naval Postgraduate School in Monterey, California, has been a strong proponent of our work for several years; his support and encouragement are greatly appreciated. At the VR Lab of the Naval Research Lab in Washington, D.C., Dr. J. Edward Swan has led efforts on the Dragon project. The VR Lab at NRL is directed by Dr. Larry Rosenblum. Dr. Helen Gigley, of the Office of Naval Research, has funded the pre-screen projection and Dragon work described herein, as well as development of the framework of usability characteristics of VEs. Brian Amento, of AT&T, performed much of the evaluation work on pre-screen projection, which was invented by Dr. Jim Templeman, of NRL's HCI Lab. Kent Swartz and Mike McGee while Students at Virginia Tech contributed to Crumbs and Dragon, respectively. We are grateful to all these contributors, without whom this large body of work would not have been possible.

## 7. REFERENCES

Brady, R., Pixton, J., Baxter, G., Moran, P., Potter, C. S., Carragher, B., & Belmont, A. (1995). Crumbs: A virtual environment tracking tool for biological imaging. In *IEEE Symposium on Frontiers in Biomedical Visualization Proceedings* (pp. 18–25).

Del Galdo, E. M., Williges, R. C., Williges, B. H., & Wixon, D. R. (1986). An evaluation of critical incidents for software documentation design. In *Proceedings of the 30th Annual Human Factors and Ergonomics Society Conference*. Anaheim, CA: Human Factors and Ergonomics Society.

Eberts, R. E. (1999). *IE486 work design and analysis 2* [Purdue University, Online]. Available: http://palette.ecn.purdue.edu/~ie486/Class/Lecture/lect14/sld001.htm [2000, May 8].

Foley, J. D., van Dam, A., Feiner, S. K., & Hughes, J. F. (1990). *Computer graphics: Principles and practice*. Reading, MA: Addison-Wesley.

Gabbard J. L. (1997). *A taxonomy of usability characteristics for virtual environments*. Unpublished masters thesis, Virginia Tech, Department of Computer Science.

Gabbard, J. L., Hix, D., & Swan, E. J. (1999a). User-centered design and evaluation of virtual environments. *IEEE Computer Graphics and Applications, 19*(6), 51–59.

Gabbard, J. L., Swartz, K., Richie, K., & Hix, D. (1999b). Usability evaluation techniques: A novel method for assessing the usability of an immersive medical VE. In *Proceedings of Virtual Worlds and Simulation Conference* (VWSIM '99, pp. 165–170). New York: Springer-Verlag.

Gibson, J. J. (1986). *The ecological approach to visual perception*. Hillsdale, NJ: Lawrence Erlbaum Associates.

Hackos, J. T., & Redish, J. C. (1998). *User and task analysis for interface design*. New York: John Wiley & Sons.

Hix, D., Amento, B., Templeman, J. N., Schmidt-Nielsen, A., & Sibert, L. (1999b). An empirical comparison of interaction techniques for panning and zooming in desktop virtual environments. Manuscript submitted for publication.

Hix, D., & Hartson, H. R. (1993). *Developing user interfaces: Ensuring usability through product and process*. New York: John Wiley & Sons.

Hix, D., Swan, E. J., Gabbard, J. L., McGee, M., Durbin, J., & King, T. (1999a). User-centered design and evaluation of a real-time battlefield visualization virtual environment. In *Proceedings of the IEEE VR '99 Conference* (pp. 96–103). Houston, TX: IEEE Computer Society Press.

Hix, D., Templeman, J. N., & Jacob, R. J. K. (1995). Pre-screen projection: From concept to testing of a new interaction technique. In *Proceedings of CHI '95 Conference*.

Jacob, R. J. K., & Sibert, L. E. (1992). The perceptual structure of multidimensional input device selection. In *Proceedings of CHI '92 Conference*.

Nielsen, J. (1994a). Heuristic evaluation. In J. Nielsen & R. L. Mack (Eds.), *Usability inspection methods* (pp. 25–62). New York: John Wiley & Sons.

Nielsen, J. (Ed.). (1994b). *Usability engineering*. San Francisco: Morgan Kaufmann.

Nielsen, J., & Mack, R. L. (1994). Executive summary. In J. Nielsen & R. L. Mack (Eds.), *Usability inspection methods* (pp. 1–23). New York: John Wiley & Sons.

Nielsen, J., & Molich, R. (1990). Heuristic evaluation of user interfaces. In *Proceedings of CHI '90 Conference* (pp. 249–256). Seattle, WA: ACM Press.

Obeysekare, U., Williams, C., Durbin, J., Rosenblum, L., Rosenberg, R., Grinstein, F., Ramamurthi, R., Landsberg, A., & Sandberg, W. (1996). Virtual workbench: A non-immersive virtual environment for visualizing and interacting with 3D objects for scientific visualization. In *Proceedings of IEEE Visualization '96* (pp. 345–349). IEEE Computer Society Press.

Scriven, M. (1967). The methodology of evaluation. In R. E. Stake (Ed.), *Perspectives of curriculum evaluation* [Monograph of the American Educational Research Association]. Chicago: Rand McNally.

Stevens, F., Frances, L., & Sharp, L. (1997). *User-friendly handbook for project evaluation: Science, mathematics, engineering, and technology education*. NSF 93–152.

Swartz, K. (1998). *Usability issues in Crumbs: A guidelines-based evaluation*. Report to NCSA Developers of Crumbs. Final Report for Fundes Effort Unpublished. Blacksburg VA, No PaB.

Swartz, K., Thakkar, U., Hix, D., & Brady, R. (2000). Evaluating the usability of Crumbs: A case study of VE usability engineering methods. In *Proceedings of the Third International Immersive Projection Technology Conference*. Stuttgart, Germany.

Williges, R. C. (1984). Evaluating human–computer software interfaces. In *Proceedings of the International Conference on Occupational Ergonomics*. Toronto: Human Factors Conference, Inc.

# 35

# Human Performance Measurement in Virtual Environments

Donald Ralph Lampton,[1] James P. Bliss,[2] and Christina S. Morris[3]

[1] *U.S. Army Research Institute for the Behavioral and Social Sciences*
*Simulator Systems Research Unit*
*Orlando, FL 32826*

[2] *Old Dominion University*
*Associate Professor, Psychology*
*Norfolk, VA 23529*

[3] *Advanced Learning Technologies*
*Institute for Simulation and Training*
*Orlando, FL 32826*

## 1. INTRODUCTION

*"When description gives way to measurement, calculation replaces debate."*

—S. S. Stevens (1951), *Handbook of Experimental Psychology*

In this chapter, challenges to human performance measurement in general are briefly described and potential advantages and disadvantages unique to performance measurement in virtual environments (VE) are identified. Next, basic issues concerning the why, what, and how of VE performance measurement are discussed and then integrated into recommendations for when to perform various aspects of performance measurement. Emphasis is placed on psychophysiological measures because of their potential to provide a valuable complement to other measures of performance in virtual environments. Finally, properties of measurements as they relate to psychometric criteria and data integrity are presented.

The American National Standards Institute (ANSI) 1993 guide to human performance measurement lists several problems that underlie human performance measurement in the context of scientific research, and test and evaluation:

- Lack of a general theory to guide performance measurement
- The inverse relationship between operational control and realism
- The multiple dimensions of behavior
- The ambiguous relationship between objective and subjective data
- Difficulty of measuring cognitive tasks

- Lack of objective performance criteria for most tasks
- Difficulty of generalizing results to the real world

All of these problems apply to human performance measurement in VEs. In particular, the question of generalization to the real world is a critical question for measuring performance in many VE applications. Does the VE sufficiently represent the real world? Conversely, does the VE system interface allow an individual recognized as an expert in the "real world" to demonstrate that expertise in a corresponding VE?

There are several potential advantages of measuring performance in VEs. One example is the capability to precisely recreate situations or environments across different sessions at different times or places. This capability to make sure that important parameters are held constant should be of great value in training, skill-level certification, and personnel selection applications. Another advantage is the capability to record precisely, and in great detail, the user's actions in the VE system. The record can support performance analysis by expert raters or automated scoring systems.

## 2. WHY MEASURE PERFORMANCE IN VIRTUAL ENVIRONMENTS?

To facilitate effective, efficient, and safe operation of any VE system, the measurement of human performance should be a consideration during the initial design and throughout the life cycle of the system.

Performance measurement provides a basis to make meaningful comparisons involving the VE system. A systematic approach to performance measurement is necessary to allow comparisons of performance with a VE system to performance with non-VE approaches for similar applications. Along with cost-effectiveness (see chap. 28, this volume) and safety (see chaps. 29–33, 41, this volume), human performance measurement should be an integral part of an overall evaluation of a VE system.

Performance measurement is needed to support comparisons across different system configurations, or changes within a VE system. Performance measures provide a way to determine that system changes and upgrades produce desired effects. In cases where budgetary concerns mandate tradeoff decisions, performance measurement may be used to support the selection of particular hardware or software alternatives. When components of a system are upgraded, experienced users of the previous configuration may believe that their performance was better with the old system. For example, objective measurement provides a way to address claims from individuals that they preferred system performance with the old visual display system (or tracking system, rendering system, etc.) after an expensive upgrade has been implemented.

Performance measures are the foundation for objective comparisons of an individual with other individuals, norms specific to a defined group, and predetermined performance criteria. Objective measurement is a critical requirement for testing, skill-level certification, and personnel selection and assignment applications.

One of the most important reasons to measure performance is to provide meaningful feedback and knowledge of results to individual users. Holding (1965) pointed out that knowledge of results could guide users as they perform a task and could improve subsequent task performance. In addition, knowledge of results can increase motivation to perform a task. Such feedback influences motivation by allowing learners to evaluate their progress, to understand their level of competence, and to sustain their efforts to reach realistic goals.

For some VE applications in which there is not an obvious need to measure user performance, (e.g., entertainment, see chap. 55, this volume) it is possible that performance measures will assist in identifying causes of cybersickness (see chap. 30, this volume). Therefore, for almost all situations involving immersion in VEs, there would be safety (see chaps. 29–33, 41, this volume), and probably legal (see chap. 27, this volume) reasons to measure performance.

## 3. WHAT TO MEASURE

### 3.1 Overview

The National Research Council identified several areas of VE applications: design, manufacturing, and marketing; medicine and health care; teleoperation for hazardous tasks; training; education; information visualization; telecommunications and teletravel; art; and entertainment (Durlach & Mavor, 1995). Obviously, the performance measures of primary interest may differ greatly across these applications.

When devising the performance measurement system for any VE application, two levels of measures should be considered. The primary level will be determined by the specific application that the VE system is designed to address. The second level of measures will support the interpretation of the primary measures. The primary measures focus on outcome, indicating what the user accomplished in the VE system. The secondary measures help us interpret and elaborate on why performance was successful or not.

For some applications areas there will exist well-established primary measures. For these areas, the challenge will be to develop comparable measures within the VE system. For other application areas, the need to compare traditional approaches to VE-based applications may uncover inadequacies in the traditional measures. Some new VE technologies may lead to new tasks for which there are no established measures or new tasks that can be performed only in virtual environments.

For example, some aspects of architectural design performed with a VE system could be much easier to measure than real-world performance. Interim processes could be more easily observed and recreated in virtual environments. Time to complete a design or time to complete a number of alternative designs could be easily and precisely determined. Raters or judges of interim designs and the final design could walk through and interact with the designed structure. However, the ultimate performance measure of utility, style, and the multitude of other relevant measures will still be defined and judged by highly trained and experienced human observers, and applied to the real-world construction.

Although the primary performance measures of most interest will vary as a function of the specific VE application, the performance measures will share several considerations. For example, there are certain statistical or psychometric properties desirable for all measures. In addition, many of the methods of performance measurement described in this chapter can be applied to a wide range of VE applications.

Task taxonomies may provide a starting place in the identification of categories and dimensions of performance measures relevant to a specific VE application. Fleishman and Quaintance (1984) presented a comprehensive task taxonomy. Companion and Corso (1982) offered a general review of approaches to task taxonomies. Fineberg (1995) described a taxonomy specifically designed to address performance in virtual environments. An ANSI (1993) guide presented a ten-category taxonomy of performance measures. The categories were:

- Time
- Accuracy
- Amount achieved or accomplished
- Frequency of occurrence

- Behavior categorization by observers
- Consumption or quantity used
- Workload
- Probability
- Space/distance
- Errors

Many widely used global measures (such as task success rate, assembly sequence errors, or magnitude of production) have at their basis speed or accuracy. For most areas of human performance it is possible to trade off speed and accuracy. As in defining performance criteria for real-world tasks, instructions to the VE user to "perform as fast and accurate as possible" leave it to the user to decide the relative importance of those two dimensions of performance. Setting a time limit has the advantage of keeping a constant duration of immersion, a consideration in controlling or measuring the incidence and severity of cybersickness (see chap. 30, this volume).

Regardless of the VE application, interpretation of the primary measures and identification of ways to improve performance may be facilitated by a set of secondary measures. Secondary measures include performance on the subtasks and processes needed to complete the primary task, user characteristics, and conditions under which performance was measured.

Process measures include lower level processes or functions, such as sensory acuity or basic locomotion, which underlie the primary measures. Process measures are not limited to simple divisions or subtasks of the primary task. Cannon-Bowers and Salas (1997) extend the concept of process measures to include measures such as assertiveness and flexibility. The distinction between outcome measures and process measures is a function of context; that which is an outcome measure in one context can be a process measure in a larger context.

There are many candidate measures of user characteristics that may aid in interpreting performance measures. Examples are: user age, gender, handedness, previous subject matter expertise, and prior experience with computers in general and VE systems in particular. Relevant measures will vary across applications. Finally, measures of the conditions under which performance occurred might aid in the interpretation of primary measures. Examples are personal threat or environmental stress such as temperature extremes.

### 3.2 Sensation/Psychomotor Measures

Lower level performance measures, which may facilitate interpretation of higher level performance measures, include measures of sensation and perception in the appropriate sensory modalities (see chaps. 3–6, this volume), various levels of consideration of movement through the VE (locomotion, travel, and navigation, see chaps. 11 and 13, this volume), and interaction with objects (interrogation, selection, and manipulation). Gabbard and Hix (1997) presented a useful taxonomy of these VE functions. Lampton et al. (1994) developed a VE test battery that assessed several of these measures. Many of these measures should be addressed during usability testing of a VE system (see chap. 34, this volume). For some VE systems it may be advisable to maintain the ability to collect "usability" measures throughout the life cycle of the VE system.

### 3.3 Critical Incidents

The very nature of emergency situations is counterproductive to the careful documentation of what happened and why. Therefore, a structured data collection instrument should be available to document what happened and the context in which the incident occurred. For example, any incident during the use of the VE system in which the safety of individuals is placed in jeopardy should be carefully documented. For measuring performance in VE, the term critical incident should not be limited to indicating an unwanted event, but any out of the ordinary performance, whether good or bad, with the VE system.

### 3.4 Knowledge Tests

As part of the effort to differentiate the effects of the VE system per se on performance, knowledge tests can be administered to users. The tests can tap procedural and declarative knowledge. Knowledge tests can determine if the intended user has the prerequisite knowledge for worthwhile training on, or use of, the VE system.

### 3.5 Error Analyses

Error analysis should be an important part of performance measurement for most VE applications. Norman (1988) identifies two fundamental categories of errors: slips and mistakes. Slips are errors in the execution of otherwise correct plans. A mistake involves the execution of a plan as intended, but the plan was inappropriate to accomplish the overall goal. Comprehensive approaches to error analysis are presented by Reason (1990) and Senders and Moray (1991).

As VE user interfaces improve, error analysis may assume more importance relative to simpler measures of speed and accuracy. Additional measures of the judgment exhibited by the VE system user include calculated risks and intentional violations of rules or procedures. For better or worse, VE medium may encourage greater risk taking by users. Some performance measures involving judgment within VE may completely defy clear-cut scoring as right or wrong, but rather will involve evaluating idioms, which are styles of artistic expression typical of a particular medium.

## 4. PSYCHOPHYSIOLOGICAL MEASUREMENT IN VIRTUAL ENVIRONMENTS

Psychophysiological measurement has great potential to bolster measurement of human performance in VEs and can provide insight to the underlying mechanisms of performance. Psychophysiological measurement conducted while the participant is immersed does not require the participant to provide conscious feedback. Therefore, distracting self-report procedures and after-the-fact reports of questionable reliability can be avoided.

### 4.1 Psychophysiological Approaches to Measurement

Baseline measurement refers to the psychophysiological data collected before any experimental stimulus is introduced. It is the initial assessment of the physical and psychological state and traits of the individual including daily changing psychophysiological characteristics such as fatigue, concentration, and anxiety as well as long-term psychopathologies and medical problems. Both state and trait variables are interwoven in the mass of psychophysiological data and are essential to recognize for two main purposes: to separate the state and trait characteristics from the performance variables related to the virtual environment and to categorize the participant.

In the categorical approach, the purpose is to separate individuals according to a performance variable of interest. Current advances in technology and research have provided the ability for researchers to use psychophysiological baseline data to predict individuals' performance in areas such as task functional state, task performance, workload, adaptability, accuracy, and anxiety. This approach is not only used to predict future performance but also used to separate performance related variables after the experiment over and above self-report indices.

A correlation approach is often used to determine the specific psychophysiology markers characteristic of human-system variables assessed by subjective measures. The measures of interest here would be those that measure human-system variables such as sense of presence (see chap. 40, this volume) or simulator sickness (see chap. 30, this volume). When correlating these measures with psychophysiological data, one can more clearly define the human characteristics underlying the reports and use them in subsequent analyses to identify desired and relevant evoked responses to various virtual stimuli.

Measuring evoked responses is an approach used when an experimenter wishes to determine information about the system. The experimenter can separate and analyze various aspects of a condition or system that may induce immediate psychophysiological change. In VE research there is interest in determining the aspects of the experience that produce states such as arousal, attention, and distractibility, which contribute to overall performance. As researchers seek to produce an effective virtual environment, importance is placed on human responses to stimuli that are real in appearance and those virtual components that produce meaningfulness, or arousal.

Adaptation is the focus of the feedback approach (see chaps. 37–39, this volume). Adaptation is the main concept in feedback and the results of conditional change in performance are expressed through the electrophysiology of the nervous system. During real-time psychophysiological monitoring, researchers are able to identify unwanted cognitive or physiological states and then either provide indicators such as tones, or alter the environment and virtual stimuli to return the operator's psychophysiological performance to its desired state. For example, therapists are using VEs and psychophysiological monitoring with biofeedback to reduce acrophobia or agoraphobia (see chaps. 50 and 51, this volume). Also, feedback techniques have been used with pilots in order for them to maintain a necessary cognitive state of awareness (see chap. 43, this volume). Level of arousal or cognitive functioning is constantly monitored and either the task is modified, or the operators alter their own psychophysiological levels to adapt to optimal functioning.

### 4.2 Types of Measures and Performance Indicators

Most current psychophysiological techniques involve placing and securing electrodes over the areas of the body appropriate to the types of performance indicators (e.g., concentration, anxiety, and workload) to be measured.

Electroencephalography is the recording of electrocortical brainwaves using an electroencephalograph (EEG). This area of applied psychophysiology has rapidly become the most advanced psychophysiological tool for assessing human performance primarily because of its ability to specify internal cognitive functions (e.g., visualization, attention, workload) as well as most physical processing of external stimuli. Unlike other psychophysiological measures, EEG has been shown to predict and indicate a wide range of performance related variables. The brainwave recording procedure requires that electrodes be placed on the scalp to compare and measure various regions of brain activity.

Analysis of brainwave activity can provide highly sensitive measures of changes in cognitive task difficulty and working memory. When brainwave variables are analyzed in various ways, they can accurately discriminate and predict specific cognitive states. The raw EEG data are broken up into specific bands, usually ranging from 0.5 to 32 Hz, each with specialized contribution to the various performance indices.

A working-memory increase due to task load results in an increase in the frontal theta band (4–7 Hz), and a decrease in alpha band (8–12 Hz; Gevins et al., 1998). A decrease in alpha is usually indicative of fatigue and lack of attention or vigilance and has been used to reflect increased workload. Crawford, Knebel, Vendemia, and Ratcliff (1995) point out that the lower part of alpha band (7.5–9 Hz) can actually differentiate low and high sustained attention. Frontal midline (FM) theta is a useful indicator of increase in concentration and expertise (Yamada, 1998). An increase in task complexity can be determined by relative frontal and central beta (13–32 Hz) activity increase (Wilson, Swain, & Brookings, 1995). The task engagement index has been able to reflect overall task engagement best and is derived by dividing beta by alpha plus theta (beta/ (alpha + theta). The sensory–motor reflex (SMR, 12–15 Hz) is used to determine a lateralized readiness potential (LRP). This is a type of event-related potential, useful in determining the processes that prepare the motor system for action (the preparation to respond), and which initiates but inhibits the response before it takes place (Coles, 1998).

Electromyography (EMG) is the process of recording the firing of muscle units as they relate to other muscle units. The data can determine tension and even the slightest of muscle activity without any visible muscle movement. EMG can be used in combination with EEG to weed out cognitive activity from signals related to muscle movement. Tension in the forehead and eye movements can produce EMG artifact signals in EEG recording, but with the use of recorded EMG on facial locations these artifacts can be easily recognized and removed.

EMG is a measure of tension and thus can provide insight on patterns of bodily reactions, such as startle responses, tension and movement, and breathing effort. An increase in task difficulty can be identified by an increase in EMG from the neck and shoulder muscles (Hanson, Schellekens, Veldman, & Mulder, 1993).

Electrocardiogram (EKG) measures the electrical activity that spreads across the muscle of the heart. The spread of excitement through the muscle of the ventricle produces the QRS complex, a waveform in which the intervals between the R wave indicates heart rate. Heart rate deceleration is indicative of such activities as orienting to neutral or pleasant stimuli, appropriate responding, and quicker reaction-time responses. Heart rate acceleration is a characteristic response to stimuli that induce a defensive reaction or to unpleasant stimuli. Recent research suggests that cardiovascular indices can indicate individual coping strategies to external noise in order to maintain task performance (Hanson et al., 1993). Current research also indicates that respiration rate, heart rate, and electrodermal activity (EDA) correlate with degree of immersion or presence in the virtual environment (Wiederhold, Davis, & Wiederhold, 1998).

*Blood pressure* is a measure of systolic (contraction) over diastolic (relaxation) heart activity. An increase in blood pressure can indicate an arousal reaction in the form of tension, aggressive activity, or thought, occurring in performance situations that evoke threat or unpleasantly arousing stimuli. *Blood volume* is a measure of slow changes in blood volume in any limb, whereas pulse volume reflects rapid change related to heart pumping and vascular dilation and constriction. Human performance variables that include orienting responses to stimuli can be determined by a general increase or decrease in blood volume in the forehead. An increase is indicative of improved perceptual ability; a decrease suggests a defensive response to unpleasant stimuli.

*Electrodermal activity* (EDA), also known as skin conductance level, is a measure of surface skin electrical activity. The higher the sweat rises in the eccrine sweat glands (located in palm and foot soles), the lower the resistance. EDA is of interest in human performance research because it responds to both external and internal tonic activity. Eccrine glands respond to psychological stimulation instead of body temperature change and can indicate many of the same cognitive variables as the EEG. The number of skin conductance changes in a given period of time is useful in indicating the preparation of physical motor activity. A change in skin conductance and number of spontaneous skin responses are both associated with faster reaction time. An overall increase in EDA is indicative of better task performance, signal detection, and learning. A decrease in EDA is associated with increased task complexity and difficulty.

### 4.3 Data Collection Considerations

To obtain useful and reliable data, recordings must be taken in highly controlled settings considering both the individual and environmental factors that influence human physiology such as age, gender, neurological damage, noise and temperature. Moreover, recent research suggests that electromagnetic field exposure from equipment may change the psychophysiology of participants (Morris, Shirkey, & Mouloua, 2000). Andreassi (1995) discusses in depth psychophysiology in reference to human-system performance. Resources can also be found in the technical group Psychophysiology in Ergonomics (PIE) of the Human Factors and Ergonomics Society.

## 5. HOW TO MEASURE HUMAN PERFORMANCE

There are at least two performance measurement advantages inherent to VE systems. One is the ability to automatically capture aspects of the user's performance, such as operation of the input control devices. The other is the ability to provide an observer with comprehensive perspectives of the user's actions in the virtual environment.

An example of automated performance measurement is to have the system capture lower level performance, such as counting collisions with walls or objects in the virtual environment. Minimum processing should be conducted in real or near-real time to provide feedback to the user and the observer. More detailed processing should be delayed to avoid degrading system performance while the user is on the system. During use, the observer will have a checklist that addresses only those performance measures that must be scored in real time. For the most part the observer will concentrate on controlling the session and safety issues. Scoring of other performance measures should be postponed until session playback. Several raters may score playback independently. Additional replays could be conducted to resolve interrater differences.

One of the main strengths of performance measurement in VE is the ability to provide a human observer with flexible, comprehensive views of a VE system user's actions in the VE itself. In addition, for many systems it is possible to provide a simultaneous view of the user's actions in the real-world VE interface.

The U.S. Army's Simulation Network (SIMNET) system has demonstrated the powerful performance measurement advantages that VE displays can provide for training applications. SIMNET employed a Stealth Vehicle, invisible to training exercise participants, that allowed trainers and exercise controllers to view training exercises from any perspective within the virtual environment. In addition, a training mission can be replayed and viewed from any angle.

A more recent example is provided by the Fully Immersive Team Training (FITT) research system, which allows the observer/mission controller to view the unfolding training exercise from any perspective while trainees wearing head-mounted displays practice urban search and rescue missions within virtual environments. The view presented to the observer is intentionally modified from the view of the participants to aid the observer in scoring trainee performance. For example, the observer can view the unfolding training exercise from any angle, and zoom in or out. In addition, depictions of objects within the VE are intentionally modified to aid in measuring performance. For example, equipment in use by the trainees is represented to the observer at a much larger scale so it is easier for the observer to identify. The FITT system has a replay function that allows normal, slow, and fast playback. Fig. 35.1 shows the FITT replay

FIG. 35.1. The Fully Immersive Team Training (FITT) research system replay station.

station. Technical aspects of data sampling rate, filtering, capture, and storage are described in Parsons et al. (1998).

The scoring playback can be conducted with or without the user. In addition, the user may be interviewed and provided questionnaires to help interpret what occurred and why. Interviews are in general more flexible and easier for the user than questionnaires. Questionnaires are easier to administer and provide for more private responses than interviews.

## 6. WHEN TO MEASURE PERFORMANCE IN VIRTUAL ENVIRONMENTS

Some measures such as severe susceptibility to motion illness or seizures (see chaps. 29–30, this volume) should be determined before the participant is immersed in the virtual environment. These measures may be used for screening out potential users of the VE system or at least providing a cautionary alert. However, when feasible, other background measures may be taken after the participant has completed the session in the virtual environment. Scheduling paperwork for this time provides a safety period between use of the VE system and activities such as driving. Following this line of thought, even though it may be feasible to have an automated means of collecting background information that is PC-based or even VE-based, from the standpoint of mitigating cybersickness it may be preferable to have the participant complete paper and pencil questionnaires rather than prolong the amount of time spent looking at electronic displays. Because there are so many potentially important variables, when feasible a take-home package approach may be used in which the VE system user completes and returns questionnaires hours or days after using the system.

Suggested time lines for performance measurement activities before, during, and after immersion are presented in that order in Tables 35.1, 35.2, and 35.3. These tables integrate recommendations on what, how, and when to measure performance in virtual environments.

**TABLE 35.1**
Example Time Line of Preimmersion Performance Measurement Activities

Research participant or system user recruitment
- Off-site telephone screening to confirm user has at least minimally appropriate sensory and psychomotor skills and to identify safety issues
- Preimmersion health status questionnaire
- Preimmersion indices of cybersickness symptoms
- Preimmersion baseline physiological measures
- Review of participant's previous performance on the system (when appropriate)

**TABLE 35.2**
Example Time Line of Performance Measurement Activities During Immersion

- System data capture
  - Real-time or frequently updated summary statistics of performance measures
  - Displayed to participant and controller
  - Stored for replay and subsequent data analysis
- Physiological measures
- Observer/controller checklists
- Participant comments/dialog
- Critical incident intervention (when appropriate, a safety intervention, for example)

**TABLE 35.3**
Example Time Line of Performance Measurement Activities After Immersion

- Indices of cybersickness
- Postexposure physiological measures
- Replay
  - Participant's comments
  - Controller's comments
- Open interview
- Structured interview
- Questionnaires
- Take-home packet
- Follow-up interview or questionnaire
- Off-line data analyses (example: summary statistics across immersion sessions)
- Scoring of session replay and videotape by different scorers or raters
- Session to reconcile difference in scoring
- Archive performance measures in system database

## 7. PROPERTIES OF MEASUREMENTS

### 7.1 Psychometric Criteria for Performance Measures

The power to create VEs for research is both rewarding and challenging. Researchers have the ability to measure a myriad of performance variables in an automated fashion, but face unique psychometric obstacles. In traditional (non-VE) experimental contexts it can be a challenge to get enough human performance data, so researchers often resort to using surveys, questionnaires, or other measurement tools. In contrast, researchers using VEs often must make a conscious decision to limit the data generated by the computer system or to filter it after collection.

Because of the many possibilities for data capture in virtual environments, researchers are encouraged to carefully select performance measures that are psychometrically sound. Experts generally agree on two criteria by which to judge performance measures: reliability and validity (Whitley, 1996). However, sensitivity or discriminability of performance measures is also important (see Leavitt, 1991).

#### *7.1.1 Reliability*

Performance measures should tap behavior in a stable fashion. When measuring human performance in virtual environments, it is especially important to employ reliable measures because of the large and changing variety of VE hardware, software, tasks, and testing paradigms available.

Conventional explanations of reliability center around two issues: repeatability and internal consistency. Repeatability concerns whether the measure reflects performance consistently across time and testing administrations. Researchers who use VEs for training are especially interested in repeatability because performance measures that fluctuate across time or testing sessions are not comparable. Therefore, it becomes difficult to chart a trainee's progress or expertise relative to other trainees. Determining the repeatability of a performance measure is typically done by correlating measures taken during one experimental session with those taken during a subsequent session.

Internal consistency reliability indicates how well a performance measure reflects a unitary construct. For example, it is important that a measure of accuracy in a VE pick-and-place task not include an aspect of timeliness. Otherwise it would be difficult to isolate the cause of observed variability. Most researchers who study human performance calculate test–retest reliability as a substitute for internal consistency because it is more common and interpretable, and is roughly comparable to internal consistency (Whitley, 1996).

Researchers and task designers often desire to know what constitutes acceptable reliability. Although most experts have difficulty agreeing on a precise value, Whitley (1996) suggests that test–retest correlation coefficients should approximate 0.5. However, this value generally applies to paper and pencil tests. It is reasonable to assume that acceptable coefficients for other task performance data might be somewhat lower, given the relative complexity of such measures and the variability of conditions across which performance may be observed.

Researchers frequently compare task performances in VEs to those in the "real world." In some respects, data collected from immersive task performances may be more reliable than data from an actual task performance. Just as with conventional simulation, researchers using VEs have the capability to standardize aspects of the immersive experience and to hold certain task parameters (e.g., criteria for success, difficulty, and duration) constant. Loftin and Kenney's (1995) account of training astronauts to repair the Hubble Space Telescope is an example of how aspects of a VE may be standardized, augmented, or isolated to facilitate learning and retention. They constructed a VE wherein astronauts were able to rehearse major operations necessary for one of the primary repair-mission goals. The researchers were able to standardize the task parameters so that trainees could focus on crucial mission elements and exhibit meaningful, stable performance measures.

### 7.1.2 Validity

Although repeatability and consistency are necessary qualities for performance measures, researchers must also examine the validity of measures. It is crucial that performance measures be faithful indicators of the constructs they presume to reflect. As noted by Salvendy and Carayon (1997), reliability and validity must be examined together, because a measure may consistently reflect a construct of little relevance or importance.

Evidence of performance measure precision may be reflected by content validity (whether the measure adequately measures the various facets of the target construct), construct validity (how well the measured construct relates to other well-established measures of the same underlying construct) or criterion-related validity (whether measures correlate with performance on some criterion task).

All techniques for establishing validity are inferential. Whereas it is relatively easy to calculate a reliability coefficient directly from available performance data, determining the validity of measures often requires researchers to rely on evidence obtained in an indirect fashion. Frequently such evidence includes expert judgments or correlations with other marginally validated measures. Vreuls and Obermayer (1985) noted that performance measure validation was fraught with difficulties and had not been accomplished for most simulation-based training systems. Unfortunately, the difficulties observed by Vreuls and Obermayer are generally compounded in VEs; consequently, most researchers have not attempted validation of measures (but see Zeltzer & Pioch, 1996).

The optimal way to determine precision of measures is by using construct validity. It may be feasible to estimate construct validity of some VE performance measures by comparing them to measures in other contexts such as the real world (where practical) or well-established task simulations (if such simulations exist). For example, estimating the construct validity of procedural errors in a virtual mineclearing task may be possible by comparing those errors to performance in more conventional (real-world) mine-clearing training programs. However, such validation is not practical for many immersive tasks, because there is no usable standard for comparison or because the task is too dangerous, costly, or complex to be trained in a nonimmersive fashion.

As an alternative to construct validity, researchers may wish to determine criterion validity. This may be more feasible than establishing construct validity, because measures of concurrent or future performance may be readily available. For example, data concerning mineclearing expertise would presumably be available from other test scores, rankings, or safety records. Unfortunately, using future performance as a criterion requires the researcher to wait some time before obtaining comparison data. Using current data is quicker but often forces the researcher to sacrifice comparability. For example, it would be more meaningful to compare VE mine-clearing task measures with future success in mine-clearing operations. However, researchers may opt to use currently available measures (such as performance on a test of mine-clearing equipment).

To maximize convenience, researchers often judge precision by consulting with subject matter experts. Content validity procedures make intuitive sense for situations where there are no current or future data available, or where there are no other well-established measures of the construct. The disadvantage is reliance on experts who may or may not have adequate knowledge to judge performance measure acceptability.

For traditional paper-and-pencil tests, acceptable validity coefficients typically approximate 0.3 to 0.4 (Nunnally & Bernstein, 1994, p. 100). As with reliability, however, the difficulties associated with validation of VE performance measures could drive this figure lower. Yet, given comprehensive knowledge of the target task, researchers have the capability to maximize performance measure validity. There are few limits to a programmer's creative potential, given adequate computing resources. Some researchers have even been able to manipulate and measure abstract constructs like fear or trust. North, North, and Coble (1996) provide an interesting example of this, constructing a VE to treat agoraphobia (fear of open places, see chap. 51, this volume).

### 7.1.3 Sensitivity

Instructional VEs should be created so that good performers may be distinguished from poor performers. Although sensitivity of data is not always identified as a psychometric criterion by authors, it is arguably as important as reliability and validity, particularly in performance measurement situations. According to Leavitt (1991), sensitivity refers to the capability of a performance measure to adequately discriminate between experts and novices. It is important to consider sensitivity separate from reliability and validity, because performance measures may be consistent and may truly reflect the construct of interest. However, they may not reflect sufficient variability to be of use. As an example, Vreuls and Obermayer (1985) point out that low discriminability of measures severely limited the potential of some early flight simulators.

In cases where dependent measures lack sensitivity the data may reflect ceiling or floor effects, where the majority of participants score at the high or low end of the measurement range, respectively (Nunnally & Bernstein, 1994). As an example, in some of the early VE research concerning spatial navigation training, a majority of participants had problems maintaining spatial awareness and easily got lost within virtual environments. As a result, the data did not reveal much, because few people could master virtual wayfinding tasks (see chap. 24, this volume).

A common problem associated with low sensitivity is that the experimental data lack variability. Therefore, because many statistical techniques rely on variability, it may be difficult for experimenters to draw meaningful conclusions from their statistical analyses.

Unfortunately, there are no established guidelines for sensitivity measurement. Generally, it is the researcher's responsibility to examine the raw data to determine whether performance measures show acceptable levels of variability. Provided that researchers are aware of the need for sensitive measures, VEs offer the potential to optimize sensitivity. Zeltzer and Pioch (1996) provide an example of the effort necessary to ensure good sensitivity of performance measures. They describe in detail the steps they followed to construct their VE system for training submarine officers and point out the importance of consulting with subject matter experts throughout the VE development process to determine appropriate task elements to model and the degree of fidelity with which to represent them.

## 7.2 Issues Impacting Data Integrity in Virtual Environments

Weimer (1995) recently listed a number of potential threats to successful research, including the particular instrumentation used for task presentation and performance measurement. This is particularly evident when measuring human performance in virtual environments. Although immersive technology may hold promise for reliability, validity, and sensitivity of performance measures, researchers should acknowledge VE-related issues that threaten measurement rigor. The following sections are meant to expand upon Weimer's warning by identifying specific problems researchers may face when collecting data in virtual environments.

### 7.2.1 Hardware Issues

During the advent of VE technology, it was not uncommon to see processors running at a mere 66 MHz driving and updating visual displays. One consequence of slow processors was commensurate delays in screen refresh rate. As a result, participants often had to perform tasks at speeds that were unnaturally slow as they waited for the system to "catch up." Some implications of lag for task performance have been mentioned by Kenyon and Afenya (1995). In their research, participants used a high-fidelity VE setup to learn a pick-and-place task. However, excessive lag led to task performance idiosyncrasies, such as physical movements that were overly deliberate. Unfortunately, lag may also have a negative effect on performance measure sensitivity by masking performance differences between experts and novices.

In other cases, performance measures such as reaction time may not reflect true human reaction time at all, but human reaction time filtered through an overburdened processing interface. Today, the situation has improved dramatically because graphics processors are much faster. However, even the fastest processors may be bogged down by calculations required by complex environments, position tracking, and peripheral device use. Watson, Walker, Ribarsky, and Spalding (1998) discuss how task performance may be influenced by reduced system responsiveness (SR). They summarize research findings that indicate an association between low system responsiveness and degraded task accuracy and task completion speed.

In some cases, computing resources used to generate and update virtual worlds and to collect performance measures may be accessible by multiple users on a network. Unfortunately, additional users may increase the processing load on a given processor so that VE rendering occurs at variable rates. Variability in processing loads may decrease performance measure reliability within and among participants. In some cases the authors of this chapter have had to reschedule experimental sessions or arrange stand-alone machines simply to accommodate data collection.

A more problematic issue concerns situations where the functional unit is not the individual, but a team of people immersed in the same virtual environment. While technological constraints have prevented common use of such environments, there is a clear need for VEs to train teams in applied contexts like military combat, command and control, and peacekeeping (Gorman, 1991; see chap. 44, this volume). When tapping collective performance in VEs, researchers must apply the criteria of reliability, validity, and sensitivity to performance measures taken from individuals as well as from teams. Determining individual performance efficacy can be particularly challenging when task responsibilities fluctuate among team members.

When Weimer (1995) discussed instrumentation, his main concern was equipment unreliability. Performance measures taken from first-generation VE systems often reflected sporadic equipment problems such as a visual channel that malfunctioned from time to time or position trackers that drifted in and out of calibration. When equipment is unreliable, validity may also be compromised because measures may not reflect human performance on the task so much as human adaptation to equipment idiosyncrasies. Sensitivity may also be in jeopardy because of the indiscriminant nature of some equipment problems unpredictably impacting the performances of experts and novices.

Early VE equipment configurations were often uncomfortable as well as unreliable (see chap. 41, this volume). According to Durlach and Mavor (1995), the discomfort associated with early HMDs led some researchers to consider using off-head displays. Additionally, Stanney, Mourant, and Kennedy (1998) have suggested that use of some peripherals may lead to repetitive motion injuries such as carpal tunnel syndrome or tenosynovitis. Such discomfort may negatively impact performance measure reliability, because greater performance decrements are likely with prolonged peripheral use. Validity, also, may suffer because performance data may partly reflect frustration with peripheral devices. Sensitivity may be compromised because participants of all expertise levels may react differently to uncomfortable peripherals.

A related challenge faced by participants is learning to use peripheral devices to navigate and manipulate objects. As shown in early VE research by Lampton et al. (1995), such learning is not a trivial matter. Unfortunately, the reliability and stability of measures often suffer because the normal learning curve is extended to account for control device unfamiliarity. Also, participants may show reluctance to exploit the advantages of immersion (such as head rotation). Recently, researchers have begun to investigate the feasibility of using voice to interact with elements of the virtual world (Normand, Pernel, & Bacconnet, 1997). If successful, voice interaction may provide a way to overcome the performance limitations of uncomfortable, unreliable peripheral devices.

The correspondence between peripheral device movements and reactions in the virtual world is also of concern (see chaps. 11, 13, this volume). Optimally, devices used for manipulation and navigation within a virtual world should be intuitive for the user population. However, some peripheral devices may be problematic for use by participants because of issues of compatibility and ease of use. A pervasive problem is misalignment between one's physical hand and the virtual representation of it (see Groen & Verkhoven, 1998). Significant effort has been devoted to eliminating discrepancies such as these. While decrements could be expected for performance measure sensitivity and reliability, it is likely that validity may suffer worst because actions required to activate a peripheral device may not be the same as those required to accomplish a similar task in the real world.

Given the plethora of equipment manufacturers and the transient popularity of VE equipment, research participants may navigate and manipulate using Spaceballs, flight controllers, standard joysticks, gloves, or other equipment. As a result, performance measures may not be comparable from one experiment to another. Sensitivity likely suffers because experts learn to perform well using one equipment configuration but not another. Additionally, there is the potential for performance measures drawn from low-fidelity systems to have poor validity when compared with measures taken from higher fidelity VE configurations, or from real-world performance conditions. As an example, Kozak, Hancock, Arthur, and Chrysler (1993) highlighted such differences as possible reasons for lack of training transfer from a virtual to real task-performance environment. Yet Stanney, Mourant, and Kennedy (1998) attribute this lack of transfer to the choice of training task (i.e., VE training may not have significantly added to the learners' performance due to ceiling effects of an overtrained task—i.e., picking up and placing a can).

### *7.2.2 Software Issues*

Virtual environments that are used for research vary widely in terms of their complexity. Some researchers use simple models consisting of 20,000 polygons or less with elementary object dynamics, and others use models that are extremely complex (over 1,000,000 polygons) and place heavy computing demands on processing resources. Not only does model complexity differ among environments, it may differ within the same environment. Because of this variability, highly precise measures of performance may appear unreliable. For example, reaction time estimates may fluctuate widely between simple and complex environments, or at different physical locations within the same environment. Furthermore, the difference between novice and expert performance may vary with environment complexity, decreasing measurement sensitivity.

Another issue concerns fidelity, or how closely a constructed environment approximates an operational environment. According to Hays and Singer (1989), a distinction may be made between physical fidelity (the physical similarity of a constructed environment to an actual environment) and functional fidelity (the capacity for realistic interaction within a constructed environment). It is plausible to assume that a lack of either type of fidelity will compromise measurement validity. Performance measures may not reflect the constructs they were designed to reflect, because participants will not interact with the environment in a natural, spontaneous fashion. This idea is suggested by some theories of presence that suggest the realism of an environment may be judged by whether users interact realistically with it (see Sheridan, 1992; see chap. 40, this volume). There is research to suggest that learning and performance may be enhanced in well-designed VEs that are high in fidelity, because participants experience higher levels of enthusiasm and motivation than in conventional (real-world) environments (North, 1996).

For some specialized training applications, it is necessary to combine aspects of VE technology with aspects of an actual task to enhance training. However, augmented reality applications may threaten the validity and sensitivity of performance measures because the constructs indicated by the performance measures are unclear. For example, if one trainee performs worse than another, it may be difficult to isolate the reason. Performance differences may be due to aspects of the virtual environment, aspects of the real world, or aspects of the task to be trained. Similarly, validity of the measures may be in question because of the unclear link between performance measures and the constructs of interest. There is also some concern about excessive cognitive demands that augmented reality systems may place upon users (see Stedmon, Hill, Kalawsky, & Cook, 1999).

### 7.2.3 *Task Issues*

In some cases, the structure of the tasks that participants are required to perform while immersed may alter the reliability, validity, or sensitivity of the performance measures. Some immersive training situations may in effect be part-task simulation exercises because the training task is simplified in the virtual world. Such simplification may occur for a variety of reasons, including costs associated with full task programming or the desire to conserve computer-processing power. However, if researchers create VEs without considering desired behavioral outcomes (by conducting a task analysis such as that advocated by Stanney, 1995) the result may be measures with low validity. In addition, sensitivity may suffer because of the potential for ceiling effects.

Another potential threat to sensitivity concerns the fact that some research participants may be confused or overwhelmed by the immersive experience. Bliss, Tidwell, and Guest (1997) noted occasional confusion by older participants who were not familiar with computers or virtual environments. In cases such as these, the performance measures chosen may lose sensitivity due to floor effects, where most of the participants' cognitive resources are focused on adjusting to the novelty of the virtual environment, and few remain for completing the experimental task.

Because of modeling and rendering limitations, some simulations of operational tasks may lack realism. For example, in many operational environments time is a critical constraint. Other factors difficult to replicate include environmental influences such as heat, cold, vibration, noise, and the presence of other team members. If aspects of the operational task are not represented faithfully, then validity of the performance measures may suffer because they reflect skill at accomplishing a diluted task. Similarly, performance measure sensitivity may be limited if true expertise depends upon the ability to perform in a variety of environmental conditions.

In a similar vein, most real-world operational tasks include consequences for poor performance. In some cases, these consequences are quite dramatic, such as injury or death. As with most laboratory research in the behavioral sciences, it is difficult to adequately represent in VE consequences for poor performance. In many task simulations, consequences for poor performance generally resemble those of a video game. As a result, the dedication and effort of the participants may be questioned. Psychometrically, the result may be compromised performance measure validity.

### 7.2.4 User Characteristics

Researchers have found certain participants may perform better in VEs than others because of their demographic makeup or past technological experience. In addition, Howe and Sharkey (1998) recently created a method for identifying successful users of VEs based on competence and temperament. This underscores the need to select research participants with care, as noted by Whitley (1996). Particularly when using VEs to simulate task conditions or training scenarios, users must echo the target population where possible. Yet behavioral researchers often rely on a sample of convenience, usually college students. If college students are used for testing in a virtual environment, there may be the potential for a ceiling effect because of students' greater familiarity with computers. On the other hand, if a specialized sample is used there may be a danger of a floor effect because of limited experience with technology.

As demonstrated by Witmer and Singer (1994), people vary with regard to how deeply they may become psychologically immersed in a virtual environment. Some people find it quite easy to accept the VE as a distinct, separate world that they can influence. Others may find it difficult to suspend their disbelief when experiencing the same environment. Because susceptibility for presence often dictates how participants will act when immersed, researchers should assume that such differences will adversely impact performance measure reliability.

Researchers have at times noted performance differences in VEs with regard to gender. In some cases, the differences that arise are due to other related factors (i.e., males typically use computers for recreation more often and are generally less susceptible to motion sickness). Researchers should be cautious to maintain participant gender ratios that are ecologically valid. However, they should also be aware that threats to data validity and sensitivity might exist when testing gender-diverse samples.

Speed and accuracy of virtual task performances have been shown to vary as a function of both computer and video game familiarity. Therefore, it is wise to employ caution when investigating data trends. Although participant samples may echo the operational environment demographically, their experience with technology may introduce unexpected variability in performance measures, lowering measure validity and sensitivity. One related issue that is frequently observed is the influence of participant age on performance. For example, McCreary and Belz (1999) found that older adults exhibited significantly worse retention of spatial knowledge and slower task performance in VEs than younger adults.

Finally, of particular interest to researchers has been the influence of cybersickness on human performance in virtual environments (see Kolasinski, 1995; chap. 30, this volume). Accurate measurement of performance in VEs may be jeopardized by participants' varying levels of susceptibility to cybersickness and the variety of ways such sickness can manifest itself. For example, performance measures may be rendered less reliable because of the growth of cybersickness symptoms during a testing session. Validity may also suffer because participants may become too incapacitated to respond quickly or accurately. Finally, measures may lose sensitivity because experts, in contrast to novices, may be disproportionately vulnerable to cybersickness.

## 8. SUMMARY

The measurement of human performance should be a consideration throughout the design and life cycle of a VE system. An integrated approach to performance measurement should take into account the what, how, and when of performance measurement and consideration of psychometric properties of data. For most VE systems, performance measurement will involve a combination of automated data capture and observation by human raters. Other relevant chapters are: chapter 13, "Principles for the Design of Performance-oriented Interaction Techniques;" chapter 30, "Signs and Symptoms of Human Syndromes Associated With Synthetic Experiences" (i.e., cybersickness); chapter 34, "Usability Engineering of Virtual Environments;" and chapter 44, "Team Training in Virtual Environments." The cybersickness chapter is relevant in that certainly sickness can affect performance and performance can influence cybersickness. The team-training chapter presents not only an excellent discussion of team performance measurement but also describes how a proactive approach to VE design can aid performance measurement.

## 9. ACKNOWLEDGMENTS

The opinions expressed in this document are those of the authors and should not be construed as an official position of the U.S. Army Research Institute for the Behavioral and Social Sciences, the U.S. Army, or the U.S. government.

## 10. REFERENCES

American National Standards Institute. (1993). *Guide to human performance measurements*. Washington, DC: American Institute of Aeronautics.

Andreassi, J. L. (1995). *Psychophysiology: Human behavior and physiological response* (3rd ed.). New York: City University of New York, Baruch College.

Bliss, J. P., Tidwell, P. D., & Guest, M. A. (1997). The effectiveness of virtual reality for administering spatial navigation training to firefighters. *Presence, 6*(1), 73–86.

Cannon-Bowers, J. A., & Salas, E. (1997). A framework for developing team performance measures in training. In M. T. Brannick, E. Salas, & C. Prince (Eds.), *Team performance assessment and measurement: Theory, methods, and applications*. Mahwah, NJ: Lawrence Erlbaum Associates.

Coles, M. (1998). Preparation, sensory–motor interaction, response evaluation, and event related brain potentials. *Current Psychology of Cognition, 17*(4–5), 737–748.

Companion, M. A., & Corso, G. M. (1982). Task taxonomies: A general review and evaluation. *International Journal of Man–Machine Studies, 17*(8), 459–472.

Crawford, H. J., Knebel, T. F., Vendemia, L. K., & Ratcliff, B. (1995). EEG activation during tracking and decision-making tasks: Differences between low- and high-sustained attention adults. In *Proceedings of the Eighth International Symposium on Aviation Psychology* (Vol. 2, pp. 886–890). Columbus, Ohio.

Durlach, N., & Mavor, A. (1995). *Virtual reality: Scientific and technological challenges*. Washington, DC: National Academy Press.

Fineberg, M. (1995). *A comprehensive taxonomy of human behaviors for synthetic forces* IDA Paper No. P-3155. Alexandria, VA: Institute for Defense Analyses.

Fleishman, E. A., & Quaintance, M. K. (1984). *Taxonomies of human performance: The description of human tasks*. New York: Academic Press.

Gabbard, J., & Hix, D. (1997). A Taxonomy of usability characteristics in virtual environments. Masyer's thesis, Virginia Polytechnic Institute and State University, Blacksburg.

Gevins, A., Smith, M. E., Leong, H., McEvory, L., Whitfield, S., Du, R., & Rush, G. (1998). Monitoring working memory load during computer-based tasks with EEG pattern recognition. *Human Factors, 40*(1), 79–91.

Gorman, P. (1991). *Supertroop via I-Port: Distributed simulation technology for combat development and training development* (IDA Paper No. P-2374). Alexandria, VA: Institute for Defense Analysis.

Groen, J., & Verkhoven, P. J. (1998). Visuomotor adaptation to virtual hand position in interactive virtual environments. *Presence, 7*(5), 429–446.

Hanson, E. K. S., Schellekens, J. M. H., Veldman, J. B. P., & Mulder, L. J. M. (1993). Psychomotor and cardiovascular consequences of mental effort and noise. *Human Movement Science, 12*(6), 607–626.

Hays, R. T., & Singer, M. J. (1989). *Simulator fidelity in training system design*. New York: Springer-Verlag.

Holding, D. H. (1965). *Principles of training*. London: Pergamon Press.

Howe, T., & Sharkey, P. M. (1998). Identifying likely successful users of virtual reality systems. *Presence, 7*(3), 308–316.

Kenyon, R. V., & Afenya, M. B. (1995). Training in virtual and real environments. *Annals of Biomedical Engineering, 23*, 445–455.

Kolasinski, E. M. (1995). *Simulator sickness in virtual environments* (Tech. Rep. No. 1027). Alexandria, VA: U.S. Army Research Institute for the Behavioral and Social Sciences.

Kozak, J. J., Hancock, P. A., Arthur, E. J., & Chrysler, S. T. (1993). Transfer of training from virtual reality. *Ergonomics, 36*(7), 777–784.

Lampton, D. R., Knerr, B. W., Goldberg, S. L., Bliss, J. P., Moshell, J. M., & Blau, B. S. (1994). The virtual environment performance assessment battery (VEPAB): Development and evaluation. *Presence: Teleoperators and Virtual Environments, 3*(2), 145–157.

Lampton, D. R., Knerr, B. W., Goldberg, S. L., Bliss, J. P., Moshell, M. J., & Blau, B. S. (1995). *The virtual environment performance assessment battery: Development and evaluation* (Tech. Rep. No. 1029). Alexandria, VA: U.S. Army Research Institute for the Behavioral and Social Sciences.

Leavitt, F. (1991). *Research methods for behavioral scientists* p. 84. Dubuque, IA: William C. Brown.

Loftin, R. B., & Kenney, P. J. (1995). Training the Hubble Space Telescope flight team. *IEEE Computer Graphics and Applications, 15*(5), 31–37.

McCreary, F. A., & Belz, S. M. (1999). Age-related differences in navigational performance within an immersive virtual environment. In *Proceedings of the Human Factors and Ergonomics Society 43rd Annual Meeting*. Houston, TX: Human Factors and Ergonomics Society.

Morris, C., Shirkey. E. C., & Mouloua, M. (2000). QEEG characteristics of cordless vs. corded phone users (Abstract). In *Proceedings of the Human Factors and Ergonomics Society/International Ergonomics Association 44th Annual Meeting*. San Diego, CA: Human Factors and Ergonomics Society.

Norman, D. A. (1988). The psychology of everyday things. New York: Basic Books.

Normand, V., Pernel, D., & Bacconnet, B. (1997). Speech-based multimodal interaction in virtual environments: Research at the Thomson-CSF Corporate Research Laboratories. *Presence, 6*(6), 687–700.

North, M. M., North, S. M., & Coble, J. R. (1996). Effectiveness of virtual environment desensitization in the treatment of agoraphobia. *Presence, 5*(3), 346–352

North, S. M. (1996). Effectiveness of virtual reality in the motivational processes of learners. *The International Journal of Virtual Reality, 2*(1), 17–21.

Nunnally, J. C., & Bernstein, I. H. (1994). *Psychometric theory* (3rd ed., pp. 99–101). New York: McGraw-Hill.

Parsons, J., Lampton, D. R., Parsons, K. A., Knerr, B. W., Russell, D., Martin, G., Daly, J., Kline, B., & Weaver, M. (1998). Fully immersive team training: A networked testbed for ground-based training missions. In *Proceeding of the Interservice/Industry Training Systems and Education Conference*. Orlando, FL:

Reason, J. (1990). *Human error*. Cambridge, England: Cambridge University Press.

Salvendy, G., & Carayon, P. (1997). Data collection and evaluation of outcome measures. In G. Salvendy (Ed.), *Handbook of human factors and ergonomics* (pp. 1451–1470). New York: John Wiley & Sons.

Senders, J. W., & Moray, N. P. (1991). *Human error*. Hillsdale, NJ: Lawrence Erlbaum Associates.

Sheridan, T. B. (1992). Musing on teleoperations and virtual presence. *Presence, 1*(1), 120–126.

Stanney, K. M. (1995). Realizing the full potential of virtual reality: Human factors issues that could stand in the way. In *Proceedings of the 1995 VRAIS Conference* (pp. 28–34). RTP, North Carolina.

Stanney, K. M., Mourant, R. R., & Kennedy, R. S. (1998). Human factors issues in virtual environments: A review of the literature. *Presence, 7*(4), 327–351.

Stedmon, A. W., Hill, K., Kalawsky, R. S., & Cook, C. A. (1999). Old theories, new technologies: Comprehension and retention issues in augmented reality systems. In *Proceedings of the Human Factors and Ergonomics Society 43rd Annual Meeting*. Houston, TX: Human Factors and Ergonomics Society.

Stevens, S. S. (1951). Mathematics, measurement, and psychophysics. In S. S. Stevens (Ed.), *Handbook of experimental psychology* (pp. 1–49). New York: John Wiley & Sons.

Vreuls, D., & Obermayer, R. W. (1985). Human-system performance measurement in training simulators. *Human Factors, 27*(3), 241–250.

Watson, B., Walker, N., Ribarsky, W., & Spaulding, V. (1998). Effects of variation in system responsiveness on user performance in virtual environments. *Human Factors, 40*(3), 403–414.

Wiederhold, B. K., Davis, R., & Wiederhold, M. D. (1998). The effects of immersiveness on physiology. In G. Riva & B. K. Wiederhold (Eds.), *Virtual environments in clinical psychology and neuroscience: Methods and techniques in advanced patient–therapist interaction: Vol. 58. Studies in health technology and informatics* (pp. 52–60). Netherlands Antilles, Amsterdam: IOS Press.

Weimer, J. (1995). *Research techniques in human engineering* (pp. 20–45). Englewood Cliffs, NJ: Prentice-Hall.

Whitley, Jr., B. E. (1996). *Principles of research in behavioral science* (pp. 97–128). Mountain View, CA: Mayfield.

Wilson, G., Swain, C., & Brookings, J. B. (1995). The effects of simulated air traffic control workload manipulations on EEG. In *Proceedings of the Eighth International Symposium on Aviation Psychology* (Vol. 2, pp. 1025–1030). Columbus, OH.

Witmer, B. G., & Singer, M. J. (1994). *Measuring immersion in virtual environments* (Tech. Rep. No. 1014). Alexandria, VA: U.S. Army Research Institute for the Behavioral and Social Sciences.

Yamada, F. (1998). Frontal midline theta rhythm and eye blinking activity during a VDT task and a video game: Useful tools for psychophysiology in ergonomics. *Ergonomics, 41*(5) 678–688.

Zeltzer, D., & Pioch, N. J. (1996). Validation and verification of virtual environment training systems. In *Proceedings of the 1996 VRAIS Conference* (pp. 123–130). Los Alamitos, CA.

# 36

# Virtual Environment Usage Protocols


Kay M. Stanney,[1] Robert S. Kennedy,[2] and Kelly Kingdon[1]

[1]*University of Central Florida, Orlando*
*Industrial Engineering and Management Systems Dept.*
*400 Central Florida Blvd.*
*Orlando, FL 32816-2450*
*stanney@mail.ucf.edu*
[2]*RSK Assessments*
*1040 Woodcock Road, Suite 227*
*Orlando, FL 32803*
*6kennedy@bellsouth.net*


## 1. INTRODUCTION

When computer technology permeated society, few envisioned that it would disable rather than enable the workforce. In fact, the U.S. Department of Labor has concluded that musculoskeletal disorders, such as carpal tunnel syndrome, "have been the largest single job-related injury and illness problem in the United States for the last decade," disabling workers in epidemic proportions (OSHA, 2000). This has led the Occupational Safety and Health Administration to issue an Ergonomics Program standard (29 CFR 1910.900) "to address the significant risk of employee exposure to ergonomic risk factors in jobs in general industry workplaces." A proactive approach might have minimized such risks, and those venturing into virtual environment (VE) technology applications can learn from this experience.

The problems associated with VE technology are real. In fact, they could make carpal tunnel syndrome pale in comparison, because beyond the traditional problems associated with computer technology, they comprise a host of new problems (see chaps. 29–32, this volume). Virtual environment exposure can cause people to vomit (about 1%), experience nausea (about 69.9%), disorientation (about 67.5%), and oculomotor problems (about 77.7%). It can also cause sleepiness (about 41.9%) and visual flashbacks (about 11%).* Approximately 80% to 95% of those exposed to a VE report some level of symptomatology postexposure, which may be as minor as a headache or as severe as vomiting or intense vertigo (Stanney, Salvendy et al., 1998). More troubling, the problems do not stop immediately upon cessation of exposure. In fact, VE exposure is associated with aftereffects (see chaps. 31, 37–39, this volume), which can render the exposed individual ill equipped to operate in their normal environment for a period

*Statistics derived from a database of 785 participants exposed to VE for 15 to 60 minutes (Stanney, 2001).

of time after exposure. Such aftereffects bring about products liability concerns for those who use the technology without appropriate safeguards (see chap. 27, this volume). In addition, beyond the workforce, these problems may be particularly troublesome for other venues, such as entertainment-based applications (see chap. 55, this volume), which are generally less supervised than workplace applications, such as those used in training (see chaps. 43 and 44, this volume), education (see chaps. 45 and 46, this volume), and medicine (see chaps. 47–51, this volume).

The response to VE exposure varies directly with the dose (i.e., VE stimulus intensity), capacity of the individual exposed (e.g., susceptibility, experience), and exposure duration (see chap. 30, this volume). This indicates that through effective usage protocols that address the strength of the VE stimulus, screening of individuals, and usage instructions the problems associated with VE technology can be minimized.

## 2. STRENGTH OF THE VIRTUAL ENVIRONMENT STIMULUS

The "dose" of a VE stimulus is determined by technological factors that in some way disturb the internal state of the individual exposed. In particular, sickness and other adverse effects are generally thought to be due to sensory conflicts between the design of the VE stimulus and what users expect due to their previous experiences in the real world (see chaps. 30 and 31, this volume). Different VE systems have different levels of conflict, and thus system developers and administrators need to assess the strength of their VE stimulus (i.e., how likely it is to cause adverse effects during and after exposure). Once the stimulus strength is known and the capacity of those exposed has been determined, appropriate usage protocols can be established.

In establishing the strength of a VE stimulus it is beneficial to know which technological factors lead to a more intense stimulus. Measures can then be taken to reduce technological issues associated with the adverse effects of VE exposure. In general, system factors thought to influence stimulus strength include: system consistency (Uliano, Kennedy, & Lambert, 1986); lag (So & Griffin, 1995); update rate (So & Griffin, 1995; see chap. 3, this volume); mismatched interpupilary distance (IPDs) (Mon-Williams, Rushton, & Wann, 1995; Mon-Williams, Wann, & Rushton, 1993, 1995; see chap. 37, this volume); large field of view (FOV), Kennedy & Fowlkes, 1992; see chap. 3, this volume); spatial frequency content (Dichgans & Brandt, 1978, cf. Figure 6, p. 770); visual simulation of action motion (i.e., vection [Kennedy, Berbaum, Dunlap, & Hettinger, 1996; see chap. 23, this volume]); and unimodal and intersensorial distortions (both temporal and spatial). While the technological factors within any given system can produce varying adverse effects, system developers should ensure that the guidelines in Table 36.1 are heeded to minimize stimulus strength.

Focusing on the parameters in Table 36.1, system developers should identify the primary factors that are inducing adverse effects in their system. For example, in the domain of the visual system the mismatch between accommodation and vergence demands has been highlighted (Howarth & Costello, 1996; Mon-Williams et al. 1993; Wann, Rushton, & Mon-Williams, 1995). Although IPD settings may not be critical (Howarth, 1999), it can be demonstrated that a mismatch between the orientation of the optical axes of a display and the axes assumed for software generation may produce large errors (Wann, Rushton, & Mon-Williams, 1995). DiZio and Lackner (1997) identified large end-to-end visual update delays and a large FOV as significant etiologic factors in a VE that used a head-mounted display (HMD). Although FOV and display resolution may affect the usability of a display system and lead to motion sickness, they have not been shown (with current displays) to be critical factors in producing visual stress. It is pertinent to note that both common and unique stimulus factors can arise from

**TABLE 36.1**
Addressing System Factors That Influence the Strength of a VE Stimulus

- Ensure any system lags/latencies are stable; variable lags/latencies can be debilitating.
- Minimize display/phase lags (i.e., end to end tracking latency between head motion and resulting update of the display).
- Optimize frame rates.
- Provide adjustable IPD.
- When large FOVs are used, determine if it drives high levels of vection (i.e., perceived self motion) (see chap. 23, this volume).
- If high levels of vection are found and they lead to high levels of sickness, then reduce the spatial frequency content of visual scenes.
- Provide multimodal feedback that minimizes sensory conflicts (i.e., provide visual, auditory, and haptic/kinesthetic feedback appropriate for situation being simulated).

**TABLE 36.2**
Steps to Quantifying VE Stimulus Intensity

1. Get an initial estimate. Talk with target users (not developers) of the system and determine the level of adverse effects they experience.
2. Observe. Watch users during and after exposure and note comments and behaviors.
3. Try the system yourself. Particularly if you are susceptible to motion sickness, obtain a firsthand assessment of the adverse effects.
4. Measure the dropout rate. If most people can stay in for an hour without symptoms, then the system is likely benign; if most people drop out within 10 minutes, then the system is probably in need of redesign.
5. Measure. Use simple rating scales to assess sickness (see chap. 30, this volume) and visual, proprioceptive, and postural measures to assess aftereffects (see chaps. 37–39, this volume).
6. Compare. Use Table 36.3 to determine how the system under evaluation compares to other VE systems.
7. Report. Summarize the severity of the problem, specify required interventions (e.g., warnings, instructions), and set expectations for use (e.g., target exposure duration, inter-session intervals).
8. Expect drop-outs. With a high-intensity VE stimulus, dropout rates can be high.

either HMD or spatially immersive displays (SIDs or Cave Automatic Virtual Environments [CAVEs] [see Cruz-Neira, Sandin, & DeFanti, 1993]). A unique factor that is present in an HMD is lag between head movement and update of the visual display. Much like an HMD, a stereoscopic SID may also present a discord between accommodation and vergence stimuli (Wann & Mon-Williams, 1997). In addition to this, an SID may, from some viewing positions, require a small degree of aniso-accommodation (unequal focus across the two eyes). It can be demonstrated that under intensive viewing conditions, even a desktop stereoscopic display can produce aftereffects for the binocular visual system, whereas a nonstereoscopic display will not (Mon-Williams & Wann, 1998). As can be seen from the various studies reported here (see Stanney et al., 1998), the manner in which individual software and hardware components can be integrated varies greatly; thus, to establish the stimulus strength of any given system, the steps in Table 36.2 should be taken.

The comparative intensity of a VE system can be determined via reference to Table 36.3, which provides quartiles of sickness symptoms based on 29 VE studies (Cobb, Nichols, Ramsey, & Wilson, 1999; Kennedy, 2001; Kennedy, Jones, Stanney, Ritter, & Drexler, 1996; Stanney, 2001; Stanney, & Hash, 1998) that evaluated sickness via the Simulator Sickness Questionnaire (SSQ); Kennedy, Lane, Berbaum, & Lilienthal, 1993). If a given VE system is

**TABLE 36.3**
Virtual Environment Sickness Quartiles ($n = 29$)

| *Quartile* | *SSQ Score* |
|---|---|
| $25^{th}$ | 15.5 |
| $50^{th}$ | 20.1 |
| $75^{th}$ | 27.9 |
| $95^{th}$ | 33.3 |
| $99^{th}$ | 53.1 |

**TABLE 36.4**
Factors Affecting Individual Capacity to Resist Adverse Effects of VE Exposure

1. Adaptation. Set intersession exposure intervals 2 to 5 days apart to enhance individual adaptability.
2. Age. Expect little motion sickness for those under age 2; expect greatest susceptibility to motion sickness between the ages of 2 and 12; expect motion sickness to decline after 12, with those over 25 being about half as susceptible as they were at 18 years of age.
3. Gender. Expect females to be more susceptible than males (perhaps as great as three times more susceptible).
4. Anthropometrics. Consider setting VE stimulus intensity in proportion to body weight/stature.
5. Individual susceptibility. Expect individuals to differ greatly in motion sickness susceptibility and use the MHQ or another instrument to gauge the susceptibility of the target user population.
6. Drug and alcohol consumption. Limit VE exposure to those individuals who are free from drug or alcohol consumption.
7. Rest. Encourage individuals to be well rested before commencing VE exposure.
8. Ailments. Discourage those with cold, flu or other ailments (e.g., headache, diplopia, blurred vision, sore eyes, or eyestrain) from participating in VE exposure; encourage those susceptible to photic seizures and migraines, as well as individuals with preexisting binocular anomalies to avoid exposure.
9. Clinical user groups. Obtain informed sensitivity to the vulnerabilities of these user groups (e.g., unique psychological, cognitive, and functional characteristics). Encourage those displaying comorbid features of various psychotic, bipolar, paranoid, substance abuse, claustrophobic, or other disorders where reality testing and identity problems are evident to avoid exposure.

of high intensity (say the $50^{th}$ or higher percentile, with a total SSQ score of 20 or higher), significant dropouts can be expected. In VE studies, dropout rates of 20% or more are common, with about 50% of the attrition occurring within the first 20 minutes of exposure (Stanney, Lanham, Kennedy, & Breaux, 1999).

## 3. INDIVIDUAL CAPACITY TO RESIST ADVERSE EFFECTS OF VE EXPOSURE

Given the current state of knowledge, it is difficult to differentiate the interpretation of adverse effects of VE exposure according to influencing individual factors, such as age, gender, anthropometrics, drug and/or alcohol consumption, and health status. However, sufficient knowledge is available to set general guidelines regarding the capacity of those exposed (see Table 36.4).

Capacity can be defined as the capability of an individual to resist adverse effects resulting from VE exposure. The capacity of an individual to undergo VE exposure varies greatly both within an individual and between individuals. Within an individual, capacity can be reduced (e.g., via high VE stimulus intensity, extended exposure duration, consumption of drugs and/or alcohol) or enhanced (e.g., via repeated exposures and related adaptation, see chap. 31, this volume). Previous exposure to a provocative environment influences susceptibility to motion sickness. Kennedy et al. (1993) found that for repeated exposures in simulators to be effective in desensitizing individuals to sickness, the intersession interval should be short (1 week or less). This finding is consistent with other reports of motion sickness, where (1) some adaptation was retained during 7-day intervals between Slow Rotation Room exposures but not during a 30-day interval (Kennedy, Tolhurst, & Graybiel, 1965); and (2) intersession intervals in navy flight trainers of 1 day or less showed no evidence of increased tolerance, nor did those greater than 6 days apart, whereas those 2 to 5 days apart appeared to be optimum (Kennedy et al., 1993).

Between individuals, there are several individual factors related to oculomotor and disorientation effects of VE exposure, including age, gender, height, and weight.[1] (Note: Interestingly, in the data set examined nausea was not correlated to any individual factors, but solely to technical factors and exposure duration). Before the age of two, children appear to be immune to motion sickness, after which time susceptibility increases until about the age of 12, at which point it declines again. In a large study ($N > 4500$) Lawther and Griffin (1986) make this point quite clearly. Those over 25 are about half as susceptible as they were at 18 years of age (Mirabile, 1990). Whether or not this trend will hold in VE systems is not currently known, but Paige (1994) has shown increased susceptibility to vection in older adults.

Some studies have found that females generally experience greater motion sickness than males (Kennedy, Lanham, Drexler & Lilienthal, 1995). In one VE study a similar gender contrast has been reported (Kennedy, Stanney, Dunlap, & Jones, 1996). In this study the mean symptomatology for females after VE exposure was 3.4 times greater than for males. This gender difference is thought to be hormonally related (Schwab, 1954; see chap. 32, this volume). It is possible that at certain phases of the menstrual cycle a women is more susceptible to motion sickness than at others. In studies of postoperative nausea and vomiting (PONV) the severity of sickness has been found to be dependent on the stage in the menstrual cycle and if a woman is on the contraceptive pill (Beattie, Lindblad, Buckley, & Forrest, 1991; Honkavaara, Lehtinen, Hovorka, & Korttila, 1991; Ramsay, McDonald, & Faragher, 1994).

As with other poisons (see *poison theory* discussion in chap. 32, this volume), the same dosage appears to be more detrimental to those of smaller stature. Thus, the relation to height, weight, and even gender may indicate that VE stimulus intensity should be given in proportion to body weight. It must be noted, however, that there are great individual differences in susceptibility to toxins (Konz, 1997); thus, it is beneficial to use body stature in conjunction with motion sickness susceptibility to set exposure guidelines.

Susceptibility to motion sickness is thought to be influenced by several factors (Kennedy, Dunlap, & Fowlkes, 1990). Physiological predisposition has been associated with hormonal and cardiovascular factors, as well as hypersensitivity of the vestibular system. Psychological predisposition is generally attributed to personality type (e.g., propensity to report adverse effects, neuroticism, and anxiety) and cognitive style (e.g., field articulation and perceptual rigidity). Regardless of the factors that predispose one to motion sickness, an individual is generally very aware of their personal level of susceptibility; thus self-report can be used to gauge this factor.

The Motion History Questionnaire (MHQ), which was developed 30 years ago to study airsickness and disorientation due to Coriolis stimulation, is often used to assess susceptibility to motion sickness (Kennedy & Graybiel, 1965). The MHQ assesses susceptibility based on past occurrences of sickness in inertial environments. Using this multi-item paper-and-pencil questionnaire, individuals list past experiences in and preferences for provocative motion environments. Scores on the MHQ are generally predictive of an individual's susceptibility to motion sickness in physically moving environments.

Except in some clinical cases (see chaps. 49–51, this volume), VE exposure should be limited to those individuals who are free from drug or alcohol consumption. (Note: In clinical user groups, informed sensitivity to the vulnerabilities of these groups (e.g., unique psychological, cognitive, and functional characteristics) should be obtained. See (chap. 50, this volume). Both drug and/or alcohol consumption and VE exposure have been shown to degrade hand–eye coordination (see chap. 38, this volume), postural stability (see chap. 39, this volume), and visual functioning (see chap. 37, this volume). While limited research has been conducted on the combined stress of drug and/or alcohol use and VE exposure, such biodynamics (Hettinger, Kennedy, & McCauley, 1990) have been studied in the context of other environments that cause illusory self-motion (Gilson, Schroeder, Collins, & Guedry, 1972; Schroeder & Collins, 1974). If one applies the findings of these studies to VE exposure, the expectation is that adverse effects of VE exposure will be exacerbated by drug and alcohol use.

Virtual environment exposure is also known to lead to motion induced drowsiness, known as the sopite syndrome (see chap. 30, this volume). Thus, individuals should generally be well rested before commencing VE exposure. If the immune system is already compromised by ill health, this may intensify the adverse effects of VE exposure. Thus, those with cold, flu, or other ailments should be discouraged from participating.

There are some individuals who should probably not participate in VE exposure. This includes but is not limited to those susceptible to photic seizures and migraines (see chap. 29, this volume), those displaying comorbid features of various psychotic, bipolar, paranoid, substance abuse, claustrophobic, or other disorders where reality testing and identity problems are evident (see chap. 50, this volume) and those with preexisting binocular anomalies (see chap. 37, this volume). Binocular vision develops early and is relatively stable by the early school years, but the age at which children should be allowed unconstrained access to HMD-based systems is still open to debate. Similarly, adults vary in the robustness of their visual systems. It can be predicted that someone with unstable binocular vision may experience stronger postexposure effects if there are stimuli that place some stress on either the accommodation (focal) system, vergence system or the cross-links between them (Wann et al., 1995).

## 4. VE USAGE PROTOCOL

A systematic VE usage protocol can be developed that minimizes risks to users. A comprehensive VE usage protocol will consider the following factors:

1. Following the guidelines in Table 35.1, design VE stimulus to minimize adverse effects (see section 2).
2. Following the guidelines in Table 35.2, quantify VE stimulus intensity of target system and compare to quartiles in Table 35.3.
3. Following the guidelines in Table 35.4, identify individual capacity of target user population to resist adverse effects of VE exposure.
4. Provide warnings (see chap. 27, this volume) for those with severe susceptibility to motion sickness, seizures, migraines, cold, flu, or other ailments (see section 3).
5. Educate users as to the potential risks of VE exposure (see chaps. 29–33, this volume). Inform users of the insidious effects they may experience during exposure, including: nausea, malaise, disorientation, headache, dizziness, vertigo, eyestrain, drowsiness, fatigue, pallor, sweating, increased salivation, and vomiting. Depending on VE content, potential adverse psychological effects may also need to be considered (see chap. 33, this volume).

6. Educate users as to the potential adverse aftereffects of VE exposure (see chaps. 37–39, this volume). Inform users that they may experience disturbed visual functioning, visual flashbacks, as well as unstable locomotor and postural control for prolonged periods following exposure. Relating these experiences to excessive alcohol consumption may prove instructional.
7. Inform users that if they start to feel ill they should terminate their VE interaction, because extended exposure is known to exacerbate adverse effects (Kennedy, Stanney, & Dunlap, 2000).
8. Donning an HMD is a jarring experience (Pierce, Pausch, Sturgill, & Christiansen, 1999). Depending on the complexity of the virtual world, it can take 30 to 60 seconds to adjust to the new space (Brooks, 1988). Prepare users for this transition by informing them that there will be an adjustment period.
9. Adjust environmental conditions. Provide adequate air flow and comfortable thermal conditions (Konz, 1997). Sweating often precedes an emetic response; thus proper air flow can enhance user comfort. In addition, extraneous noise should be eliminated, as it can exacerbate ill effects.
10. Fatigue can exacerbate the adverse effects of VE exposure. To minimize fatigue, ensure all equipment is comfortable and properly adjusted for fit (see chap. 41, this volume). HMDs should fit snuggly and be evenly weighted about a user's head, stay in place when unsupported, and avoid uneven loading to neck and shoulder muscles. Many HMDs have adjustable head straps, IPDs, and viewing distance between the system's eyepieces and user's eyes. Ensure users optimize these adjustments to obtain proper fit. Tethers should not obstruct movements of users. DataGloves and other effectors should not induce excessive static loads via prolonged unnatural positioning of the arms or other extremities.
11. For strong VE stimuli, limit initial exposures to a short duration (e.g., 10 minutes or less) and allow an intersession recovery period of 2 to 5 days.
12. For strong VE stimuli, warn users to avoid movements requiring high rates of linear or rotational acceleration and extraordinary maneuvers (e.g., flying backward) during initial interaction (McCauley & Sharkey, 1992).
13. Throughout VE exposure, an attendant should be available at all times to monitor users' behavior and ensure their well-being. The attendant may also have to assist users if they become stuck or lost within the virtual world, as often happens (see chap. 24, this volume).
14. Indicators of impending trouble include excessive sweating, verbal frustration, lack of movement within the environment for a significant amount of time, and less overall movement (e.g., restricting head movement). Look for red flags. Users demonstrating any of these behaviors should be observed closely, as they may experience an emetic response. Extra care should be taken with these individuals postexposure. Note: It is beneficial to have a bag or garbage can located near users in the event of an abrupt emetic response.
15. Set criteria for terminating exposure. Exposure should be terminated immediately if users verbally complain of symptoms and acknowledge they are no longer able to continue. Also, to avoid an emetic response, if telltale signs are observed (i.e., sweating, increased salivation), exposure should be terminated. Some individuals may be unsteady upon postexposure. These individuals may need assistance when initially standing up after exposure.
16. After exposure, the well-being of users should be assessed. Measurements of their hand-eye coordination and postural stability should be taken. Similar to field sobriety tests, these can include measures of: balance (e.g., standing on one foot, walking an imaginary line, leaning backward with eyes closed); coordination (e.g., alternate hand

clapping and finger to nose touch while the eyes are closed); and eye nystagmus (e.g., follow a light pen with the eyes without moving the head). Do not allow individuals who fail these tests to conduct high-risk activities until they have recovered (e.g., have someone drive them home).

17. Set criteria for releasing users. Specify the amount of time after exposure that users must remain on premises before driving or participating in other such high-risk activities. In our lab a 2-to-1 ratio is used; postexposure users must remain in the laboratory twice the amount of exposure time to allow recovery.
18. Call users the next day or have them call to report any prolonged adverse effects.

## 5. CONCLUSIONS

The risks associated with VE exposure are real and include ill effects during exposure, as well as the potential for prolonged aftereffects. To minimize these risks, VE system developers should quantify and minimize VE stimulus intensity, identify the capacity of the target user population to resist the adverse effects of VE exposure, and then follow a systematic usage protocol. This protocol should focus on warning, educating, and preparing users, setting appropriate environmental and equipment conditions, limiting initial exposure duration and user movements, monitoring users and looking for red flags, and setting criteria for terminating exposure, debriefing, and release. Adopting such a protocol can minimize the risk factors associated with VE exposure, thereby enhancing the safety of users, limiting the liability (see chap. 27, this volume) of system developers and administrators.

## 6. ACKNOWLEDGMENTS

This material is based on work supported in part by the Office of Naval Research (ONR) under grant No. N000149810642, the National Science Foundation (NSF) under grants No. DMI9561266 and IRI-9624968, and the National Aeronautics and Space Administration (NASA) under grants No. NAS9-19482 and NAS9-19453. Any opinions, findings, and conclusions or recommendations expressed in this material are those of the authors and do not necessarily reflect the views or the endorsement of the ONR, NSF, or NASA.

## 7. REFERENCES

Beattie, W. S., Lindblad, T., Buckley, D. N., & Forrest, J. B. (1991). The incidence of postoperative nausea and vomiting in women undergoing laparoscopy is influenced by day of menstrual cycle. *Canadian Journal of Anesthesia, 38*(3), 298–302.

Brooks, Jr., F. P. (1988). Grasping reality through illusion: Interactive graphics serving science. *ACM SIGCHI '88 Conference Proceedings*, 1–11.

Cobb, S. V. G., Nichols, S., Ramsey, A., & Wilson, J. R. (1999). Virtual reality–induced symptoms and effects (VRISE). *Presence: Teleoperators and Virtual Environments, 8*(2), 169–186.

Cruz-Neira, C., Sandin, D. J., & DeFanti, T. A. (1993, July). Surround-screen projection-based virtual reality: The design and implementation of the CAVE. In *ACM SIGGRAPH '93 Conference Proceedings*. 135–142.

Dichgans, J., & Brandt, T. (1978). Visual-vestibular interaction: Effects on self-motion perception and postural control. In R. Held, H. W. Leibowitz, & H. L. Teuber (Eds.), *Handbook of sensory physiology: Vol. 8. Perception* (pp. 756–804). Heidelberg, Germany: Springer-Verlag.

DiZio, P., & Lackner, J. R. (1997). Circumventing side effects of immersive virtual environments. In M. Smith, G. Salvendy, & R. Koubek (Eds.), *Design of computing systems: Social and ergonomic considerations* (pp. 893–896). Amsterdam: Elsevier Science Publishers.

Gilson, R. D., Schroeder, D. J., Collins, W. E., & Guedry, F. E. (1972). Effects of different alcohol dosages and display illumination on tracking performance during vestibular stimulation. *Aerospace Medicine, 43*, 656.

Hettinger, L. J., Kennedy, R. S., & McCauley, M. E. (1990). Motion and human performance. In G. H. Crampton (Ed.), *Motion and space sickness* (pp. 411–441). Boca Raton, FL: CRC Press.

Honkavaara, P., Lehtinen, A. M., Hovorka, J., & Korttila, K. (1991). Nausea and vomiting after gynecological laparoscopy depends upon the phase of menstrual cycle. *Canadian Journal of Anesthesia, 38*(7), 876–879.

Howarth, P. A. (1999). Oculomotor changes within virtual environments. *Applied Ergonomics, 30*, 59–67.

Howarth, P. A., & Costello, P. J. (1996). Visual effects of immersion in virtual environments: Interim results from the U.K. Health and Safety Executive Study. *Digest of the Society for Information Display 27*, 885–888. CA:

Kennedy, R. S. (2001). Unpublished research data, RSL Assessments, Inc., Orlando, FL.

Kennedy, R. S., Berbaum, K. S., Dunlap, W. P., & Hettinger, L. J. (1996). Developing automated methods to quantify the visual stimulus for cybersickness. In *Proceedings of the Human Factors and Ergonomics Society 40th Annual Meeting* (pp. 1126–1130). Santa Monica, CA: Human Factors and Ergonomics Society.

Kennedy, R. S., Dunlap, W. P., & Fowlkes, J. E. (1990). Prediction of motion sickness susceptibility. In G. H. Crampton (Ed.), *Motion and space sickness* (pp. 179–215). Boca Raton, FL: CRC Press.

Kennedy, R. S., & Fowlkes, J. E. (1992). Simulator sickness is polygenic and polysymptomatic: Implications for research. *International Journal of Aviation Psychology, 2*(1), 23–38.

Kennedy, R. S., & Graybiel, A. (1965). *The Dial test: A standardized procedure for the experimental production of canal sickness symptomatology in a rotating environment* (Rep. No. 113, NSAM 930). Pensacola, FL: Naval School of Aerospace Medicine.

Kennedy, R. S., Jones, M. B., Stanney, K. M., Ritter, A. D., & Drexler, J. M. (1996). *Human factors safety testing for virtual environment mission-operation training* (Final Rep., Contract No. NAS9-19482). Houston, TX: Lyndon B. Johnson Space Center.

Kennedy, R. S., Lane, N. E., Berbaum, K. S., & Lilienthal, M. G. (1993). Simulator sickness questionnaire: An enhanced method for quantifying simulator sickness. *International Journal of Aviation Psychology, 3*(3), 203–220.

Kennedy, R. S., Lanham, D. S., Drexler, J. M., & Lilienthal, M. G. (1995). A method for certification that after effects of virtual reality exposures have dissipated: Preliminary findings. In A. C. Bittner & P. C. Champney (Eds.), *Advances in Industrial Safety VII* (pp. 263–270). London: Taylor & Francis.

Kennedy, R. S., Stanney, K. M., & Dunlap, W. P. (2000). Duration and exposure to virtual environments: Sickness curves during and across sessions. *Presence: Teleoperators and Virtual Environments, 9*(5), 466–475.

Kennedy, R. S., Stanney, K. M., Dunlap, W. P., & Jones, M. B. (1996). *Virtual environment adaptation assessment test battery* (Final Rep., Contract No. NAS9-19453). Houston, TX: Lyndon B. Johnson Space Center.

Kennedy, R. S., Tolhurst, G. C., & Graybiel, A. (1965). *The effects of visual deprivation on adaptation to a rotating environment* (Tech Rep. No. NSAM 918). Pensacola, FL: Naval School of Aviation Medicine.

Konz, S. (1997). Toxicology and thermal discomfort. In G. Salvendy (Ed.), *Handbook of Human Factors and Ergonomics* (2nd ed., pp. 891–908). New York: John Wiley & Sons.

Lawther, A., & Griffin, M. J. (1986). The motion of a sea ship and the consequent motion sickness amongst passengers. *Ergonomics, 29*(4), 535–552.

McCauley, M. E., & Sharkey, T. J. (1992). Cybersickness: Perception of self-motion in virtual environments. *Presence: Teleoperators and Virtual Environments, 1*(3), 311–318.

Mirabile, C. S. (1990). Motion sickness susceptibility and behavior. In G. H. Crampton (Ed.), *Motion and space sickness* (pp. 391–410). Boca Raton, FL: CRC Press.

Mon-Williams, M., Rushton, S., & Wann, J. P. (1995). Binocular vision in stereoscopic virtual-reality systems. *Society for Information Display International Symposium Digest of Technical Papers, 25*, 361–363.

Mon-Williams, M., & Wann, J. P. (1998). Binocular virtual reality displays: When problems occur and when they don't. *Human Factors, 40*(1), 42–49.

Mon-Williams, M., Wann, J. P., & Rushton, S. (1995). Binocular vision in stereoscopic virtual reality systems. *Digest of the Society for Information Display, 1*, 361–363.

Mon-Williams, M., Wann, J. P., & Rushton, S. (1993). Binocular vision in a virtual world: Visual deficits following the wearing of a head-mounted display. *Ophthalmic and Physiological Optics, 13*, 387–391.

Occupational Safety and Health Administration. (2000). Ergonomics Program, Federal Registration No. 65:68261-68870, CFR Title 29, Part 1910, RIN 1218-AB36 [Docket No. S-777, Online]. Available: http://www.osha.gov/ergo-temp/FED20001114.html

Paige, G. D. (1994). Senescence of human visual–vestibular interactions: Smooth pursuit, optokinetic, and vestibular control of eye movements with aging. *Experimental Brain Research, 98*, 355–372.

Pierce, J. S., Pausch, R., Sturgill, C. B., & Christiansen, K. D. (1999). Designing a successful HMD-based experience. *Presence: Teleoperators and Virtual Environments, 8*(4), 469–473.

Ramsay, T. M., McDonald, P. F., & Faragher, E. B. (1994). The menstrual cycle and nausea or vomiting after wisdom teeth extraction. *Canadian Journal of Anesthesia, 41*(9), 789–801.

Rich, C. J., & Braun, C. C. (1996). Assessing the impact of control and sensory compatibility on sickness in virtual environments. In *Proceedings of the Human Factors and Ergonomics Society 40th Annual Meeting* (pp. 1122–1125). Santa Monica, CA: Human Factors and Ergonomics Society.

Schroeder, D. J., & Collins, W. E. (1974). Effects of secobarbitol and d-amphetamine on tracking performance during angular acceleration. *Ergonomics, 17*(5), 613.

Schwab, R. S. (1954). The nonlabyrinthine causes of motion sickness. *International Record Medicine, 167*, 631–637.

So, R. H., & Griffin, M. J. (1995). Effects of lags on human operator transfer functions with head-coupled systems. *Aviation, Space, and Environmental Medicine, 66*, 550–556.

Stanney, K. M. (2001). Unpublished research data, University of Central Florida, Orlando.

Stanney, K. M., & Hash, P. (1998). Locus of user-initiated control in virtual environments: Influences on cybersickness. *Presence: Teleoperators and Virtual Environments, 7*(5), 447–459.

Stanney, K. M., Lanham, S., Kennedy, R. S., & Breaux, R. B. (1999). Virtual environment exposure drop-out thresholds. In *Proceedings of the 43rd Annual Human Factors and Ergonomics Society Meeting* (pp. 1223–1227). Santa Monica, CA: Human Factors and Ergonomics Society.

Stanney, K. M., Salvendy, G., Deisigner, J., DiZio, P., Ellis, S., Ellison, E., Fogleman, G., Gallimore, J., Hettinger, L., Kennedy, R., Lackner, J., Lawson, B., Maida, J., Mead, A., Mon-Williams, M., Newman, D., Piantanida, T., Reeves, L., Riedel, O., Singer, M., Stoffregen, T., Wann, J., Welch, R., Wilson, J., Witmer, B. (1998). Aftereffects and sense of presence in virtual environments: Formulation of a research and development agenda. *International Journal of Human–Computer Interaction, 10*(2), 135–187.

Uliano, K. C., Kennedy, R. S., & Lambert, E. Y. (1986). Asynchronous visual delays and the development of simulator sickness. In *Proceedings of the Human Factors Society 30th Annual Meeting* (pp. 422–426). Santa Monica, CA: Human Factors Society.

Wann, J. P., & Mon-Williams, M. (1997, May). Health issues with virtual reality displays: What we do know and what we don't. *Computer Graphics*, 53–57.

Wann, J. P., Rushton, S. K., & Mon-Williams M. (1995). Natural problems for stereoscopic depth perception in virtual environments. *Vision Research, 19,* 2731–2736.

# 37

# Measurement of Visual Aftereffects Following Virtual Environment Exposure

John P. Wann[1] and Mark Mon-Williams[2]

*[1]University of Reading*
*Department of Psychology*
*3, Earley Gate*
*Reading RG6 6AL, England*
*[2]University of St. Andrews*
*School of Psychology, St. Andrews*
*Fife KY16 9JU, Scotland*

## 1. INTRODUCTION

This chapter is organized in the following manner. First, a detailed consideration of the ocular-motor system is provided. This section must be reasonably detailed in order to justify the measures recommended. Second, a series of clinical tests are detailed that are useful in indicating whether the ocular-motor system has been placed under stress. These tests are used generally to explore ocular-motor status before and after using a particular virtual environment (VE) system for a given period of time. Finally, a different approach to investigating the impact of VE systems on the ocular-motor system is discussed. This final section describes experimental results that highlight some interesting features of the ocular-motor system. These results are used to illustrate the fact that VE systems can help us understand the nervous system and to argue that the most profitable aftereffect measurements result from paradigms that probe the manner in which the nervous system adapts to virtual environments.

## 2. BINOCULAR ORGANIZATION

### 2.1 Refraction

It is possible to think of the eye as an optical system where the cornea and crystalline lens have a certain refractive power with a corresponding focal length (the distance from the optical system to a principal focal point) as illustrated in Fig. 37.1. The length of the eye can be considered as being independent of the power of the optical system. If the length of the eye is such that the retina is at the focal length of the optical system, the eye is described as being *emmetropic*. If

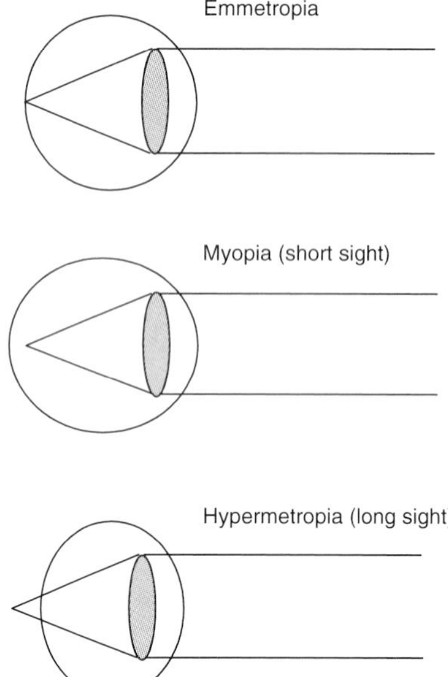


FIG. 37.1. It is possible to think of the eye as an optical system where the cornea and crystalline lens have a certain refractive power with a corresponding focal length (the distance from the optical system to a principal focal point) as illustrated. If the length of the eye is such that the retina is at the focal length of the optical system, the eye is described as being emmetropic (*upper plot*). If the retina is beyond the focal length, the eye is described as being myopic (*middle plot*), and if the retina is closer than the focal length, the eye is described as hypermetropic (*lower plot*).

the retina is closer than the focal length, the eye is described as *hypermetropic*, and if the retina is beyond the focal length, the eye is described as being *myopic*. It is possible for the optical system of the eye to cause two separate focal planes, of which both can fall behind, both can fall in front, or one can fall on either side of the retina. Such eyes are described as *astigmatic* (it is easiest to think of the astigmatic optical system being shaped like an egg instead of a sphere). Eyes that are myopic, hypermetropic or astigmatic are described as having *refractive error*. The refractive error refers to the extent to which the eye is too long or too short for the power of the optical system and is described in terms of dioptres: the reciprocal of the refractive error in meters. It is normal to express the refractive error in terms of the lens that needs to be placed in front of the eye to correct for the refractive error—myopia requires a concave (negative) lens, whereas hypermetropia requires a convex (positive) lens. Lens power is specified in dioptres, which are equivalent to the reciprocal of focal distance specified in meters.

An eye's refractive state is not constant over the life span. It has been well established that newborn babies are born with hypermetropia (approx. 2–3 dioptres) and astigmatism. Over the first 6 months of life, however, the hypermetropia and astigmatism disappear and the majority of children end up being emmetropic. This progression from hypermetropia to emmetropia is known as *emmetropisation* and is widely held to be a process under active neural control (see Saunders, 1995). Although the majority of children are emmetropic, it is reasonably common for myopia to develop between the 12 and 25 years of age (so-called late onset myopia). The etiology of late onset myopia is unknown but there is a reasonable amount of evidence that links the condition to prolonged close fixation and visual stress. Thus, it seems sensible to suggest that individuals between the ages of 12 and 25 years should beware of prolonged exposure to poor quality VE displays that require close fixation.

## 2.2 Eye Movements

Horizontal movements of the two eyes may be considered as either conjugate, when the eyes move in the same direction, or disconjugate when the eyes move in opposite directions. These two types of eye movement are also commonly referred to as version and vergence movements, but because the phonetic similarity leads to confusion, the term *conjugate* is used herein rather than version. Conjugate eye movements are responsible for maintenance of static fixation (gaze holding), rapidly changing eye fixation position (saccadic movement), or continuously maintaining fixation on a moving object (pursuit). The majority of humans are binocular, and thus conjugate eye movements require coordinated movement of both eyes. Hering (1977) originally suggested that the nervous system controls the two eyes as if they were a single organ, and this idea has subsequently received a large amount of empirical support (see Mon-Williams & Tresilian, 1998). Hering's law suggests that the same command (e.g., move 10 degrees to the left) is sent to both eyes. In order for this control arrangement to succeed, it is necessary for the peripheral ocular-motor system to adapt in response to changes between the two eyes. Thus, it appears as if the nervous system has a large degree of peripheral adaptability in order that the central command structure can remain reasonably invariant. The adaptability allows the system to respond to changes induced by neuromuscular fatigue or disease, as well as respond to changes in the viewing environment. This suggests that a crucial issue in the design of VE is the extent to which a VE system places adaptive pressures on the ocular-motor system and the manner in which the nervous system responds to those pressures. It seems reasonable to suggest that the primary adaptive pressures placed on the ocular-motor system relate to binocular vision. It is the case that a poorly designed VE headset could induce adaptation in reflexive type eye movements (e.g., the vestibular ocular reflex, see chap. 39, this volume) by introducing temporal lags between sensed changes in head movements and the resulting visual display. Nonetheless, it is difficult to see a well-designed system placing pressure on monocular viewing. Thus, the fundamental problem faced by the ocular-motor system relates to the binocular coordination between the eyes. The binocular system is, therefore, herein considered in some depth. It is important at this stage to differentiate two different types of VE systems—*biocular displays* and *stereoscopic binocular displays.* Biocular displays present identical images to the two eyes, whereas stereoscopic binocular displays present different images to the eyes (the differences, or disparities, can provide a powerful phenomenological impression of three-dimensional (3-D) depth). As demonstrated in the discussion below, the potential for visual aftereffects is far higher in stereoscopic binocular displays than in biocular displays. Indeed, a well-designed biocular display is likely to place few adaptive demands on the ocular-motor system.

## 2.3 Vergence and Accommodation

In this section, the components of the ocular-motor system, which are responsible for providing clear and single vision (accommodation and vergence eye movements, respectively), are briefly described. It is assumed initially that the eye is emmetropic. It is possible to represent the key features of accommodation and vergence in an heuristic model of the vergence and accommodation control system (see Fig. 37.2, modified from Schor & Kotulak, 1986). If an observer wishes to change fixation from a distant object to one near (or vice versa), the retinal image of the target object is initially defocused (blur describes this error of focus) and there is a fixation error between the image of the target and the fovea (disparity refers to this error of fixation). In order to bring clarity to the retinal image the eye must focus in a process known as accommodation, and to overcome disparity the eyes must change vergence angle to maintain fixation within corresponding retinal areas (if noncorresponding points of the retina are stimulated then double vision will result).

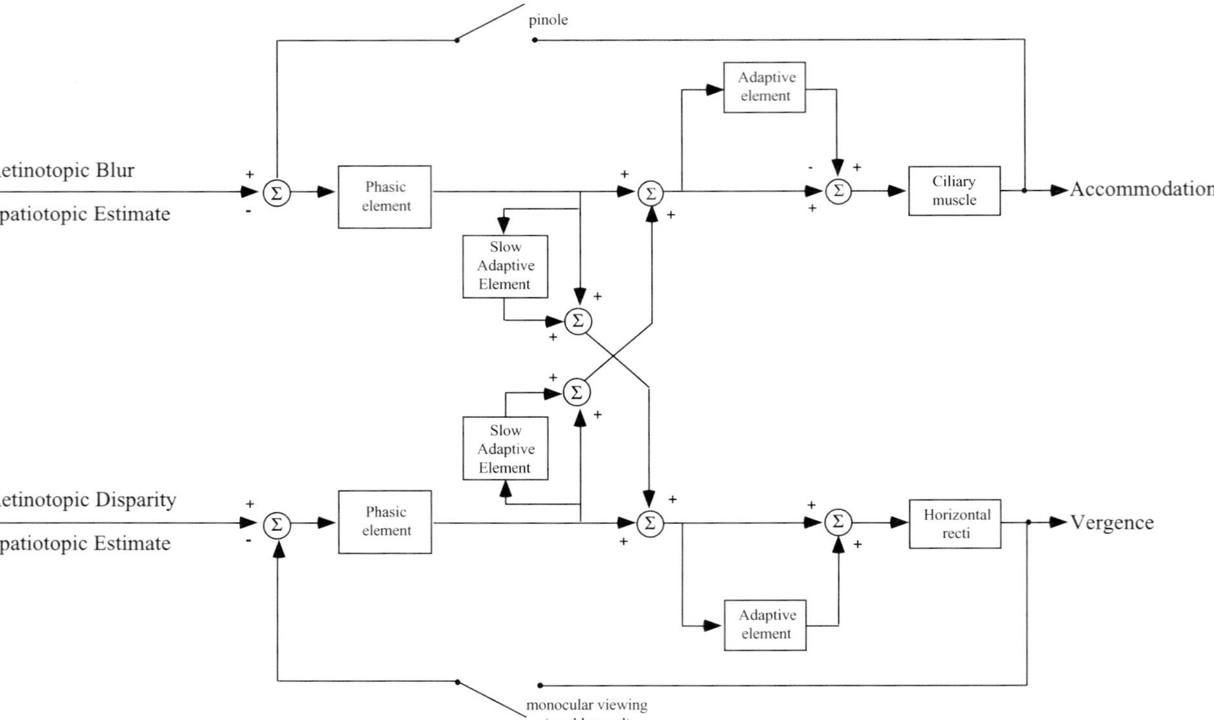


FIG. 37.2. A heuristic model of the vergence and accommodation control system (modified from Schor & Kotulak, 1986). In normal situations the systems rely on negative feedback, with a phasic element in the feed-forward pathway rapidly eliminating blur and disparity. A tonic controller is also present in both systems, and this adapts to reduce any steady-state demands placed on the phasic response component. The tonic controller ensures that the accommodation and vergence system are kept in the middle of their functional range. In order to further maximize system efficiency, the accommodation and vergence responses are neurally cross-linked so that accommodation produces vergence eye movements, whereas vergence causes accommodation. The negative feedback loop to accommodation may be opened by removing blur information (e.g., by viewing through a pinhole), and the vergence feedback loop may be opened by removing disparity (e.g., by covering one eye). The open-loop vergence bias is known as heterophoria (see text for details).

Although blur and disparity are extremely effective *retinotopic* feedback signals, they are only effective over small distances so that larger changes are driven by *spatiotopic* stimuli (Schor et al., 1992). The spatiotopically driven changes get the accommodation and vergence systems into the right "ballpark," and then retinotopically driven responses achieve precisely located fixation. Despite the importance of the spatiotopic response, it remains unclear what information provides the necessary stimuli— it is an issue of distance perception, which is still a topic of considerable debate (Mon-Williams & Tresilian, 1999).

In retinotopic conditions, the systems rely on negative feedback, with a phasic element in the feed-forward pathway rapidly eliminating blur and disparity. The feedback loop to accommodation may be negated ("opened") by removing blur information (e.g., by viewing through a pinhole), and the vergence feedback loop may be opened by removing disparity (e.g., by covering one eye). These procedures allow measurement of the bias within the system. The bias reflects the resting position of the system and can be modified through visual demand.

#### 2.3.1 Phasic Elements

Figure 37.2 illustrates that accommodation is driven (retinotopically) through blur, whereas vergence responds to disparity (see Schor, 1983, 1986, for a comprehensive overview). An initial change in vergence angle or accommodative state is initiated by a phasic element within the vergence and accommodation system, respectively. The phasic controller acts to rapidly eliminate blur and disparity in order that a clear and single image is achieved. The disparity that drives vergence is the absolute disparity of the target object (i.e., the disparity with respect to the horopter). The horizontal disparities that provide for stereopsis are derived from relative disparities. It follows that the disparity signal for vergence is not the same as that used for stereopsis.

#### 2.3.2 Tonic Elements

A tonic controller in the vergence and accommodation system can adapt to reduce any steady-state demands placed on the phasic response component (e.g., Carter, 1965; Schor, 1979). The tonic controller ensures that the accommodation and vergence system are kept in the middle of their functional range.

#### 2.3.3 Cross-links

In order to maximize system efficiency, the accommodation and vergence responses are neurally cross-linked (see Schor, 1986) so that accommodation produces vergence eye movements (accommodative vergence), whereas vergence causes accommodation (vergence accommodation). Accommodative vergence is normally expressed in terms of its ratio with accommodation (the AV/A ratio), and vergence accommodation is expressed in terms of its ratio with vergence (the VA/V ratio). The cross-coupling interactions between accommodation and vergence are stimulated by the phasic but not by the tonic control elements (Schor, 1986). Thus, the cross-links have been shown to originate after the phasic element but before the site of the tonic controller.

### 2.4 Vergence (or "Prism") Adaptation

In the initial consideration of the vergence system it was emphasized that a tonic element exists in the feed-forward pathway, and that this component acts to minimize any demands placed on the phasic controller. It follows that if the eyes maintain a convergent position for a prolonged period of time, the resting position of the eyes will shift inwards and reciprocally after prolonged divergence the resting position of the eyes will shift outward.

The tonic element means that a change in the demands placed on the vergence system will not necessarily produce visual problems: It is possible that any demands will be accommodated by the adaptable element within the vergence system. On the other hand, it has been established that symptomatic individuals have reduced tonic adaptability (Fisher, Ciuffreda, Levine, & Wolf-Kelly, 1987). A reduced ability to adapt to steady-state viewing together with large vergence demands may cause some individuals to suffer from visual fatigue. Vergence adaptation is often described as *prism adaptation*, as the easiest way of artificially inducing the adaptive behavior is to place unyoked prisms in front of the eyes. A poorly designed VE system can induce prism through a misalignment of the viewing optics in situations where the lenses are not collimated to infinity (Mon-Williams, Wann, & Rushton, 1993). In a biocular VE system, the screens can be set at any viewing distance from optical infinity to very close to the observer. The closer the viewing distance, the greater the vergence demand. It is necessary to realize that what constitutes too large a vergence demand varies from individual to individual (North & Henson, 1981) and between age groups (Winn, Gilmartin, Sculfor, & Bamford, 1994). Importantly, it has been shown that individuals with binocular vision anomalies lack the ability to adapt to induced changes in vergence bias (Henson & Dharamshi, 1982), and this has been hypothesized as being one of the causes of binocular vision problems (Schor, 1979). A well-designed biocular VE system can avoid the vergence system needing to adapt to the virtual environment.

It is worth drawing attention to another factor that can place demands on the adaptable element of the vergence system. As vertical gaze angle (the vertical orientation of the eyes with respect to the head) is changed, the effort required of the extraocular muscles becomes modified (Heuer & Owens, 1989; Heuer, Bruwer, Romer, Kroger, & Knapp, 1991). This may be readily demonstrated by fixating the tip of a pen held close (e.g., 10 cm) to the face and raising and lowering the pen. It will be noted that it is considerable more comfortable to fixate the pen when it is approximately in line with the mouth and considerably less comfortable to fixate when at eye level. The reason for these changes in comfort relate to the fact that the muscles used to lower the eyes also aid convergence, whereas the muscles used to raise the eyes aid divergence (Mon-Williams, Burgess-Limerick, Plooy, & Wann, 1999). Thus the vertical position of VE screens can alter the vergence demands placed on the ocular-motor system. It should not be assumed that the user will adjust a VE headset if the gaze angle is inappropriate. The major determinant of a head-mounted display's (HMD's) position on the head is the design of the headset together with the muscular–skeletal position of greatest comfort (see chap. 41, this volume). Once the HMD is in place it is highly unlikely that a user will equate eyestrain with the vertical location of the headset. It should also be noted that the majority of HMDs allow for initial adjustment of head position, after which time the headset is fixed in place (normally by means of a headband that is tightened).

We have previously used a software calibration routine that allows for the measurement of heterophoria within a VE headset (essentially, this routine displays a tangent screen on one screen and a vertical line on the other). Use of this routine allows a user to adjust a VE headset until the heterophoria measures reach a minimum value. Furthermore, we have described a routine that enables any demands placed on the vergence system by a VE system to be minimized (Wann, Rushton, & Mon-Williams, 1995). Such software may be extremely useful when adjusting an HMD's position on the head and ensuring that the VE system is not placing large demands on the adaptable component of vergence.

## 2.5 Accommodative Adaptation

In the same manner as the vergence system, the accommodation system has a tonic element in the feed-forward pathway in order to minimize any demands placed on the phasic controller. The easiest way of inducing accommodative adaptation is by placing refractive lenses in front of the two eyes. It follows that VE systems that use refractive lenses to present images to the human observer may induce accommodative adaptation. In general, such adaptive demands are met readily by the ocular-motor system unless the viewing distance is very close. One additional issue of accommodative adaptation that arises with VE systems is the extent to which the eyes are capable of *aniso-accommodation* (Marran & Schor, 1998). Aniso-accommodation describes the response of the ocular-motor system to unequal accommodative demands between the two eyes. Situations can arise naturally when the eyes have unequal refractive error and thus have different focusing demands. In general, large interocular refractive discrepancies lead to strabismus and amblyopia (see later), but it has been reported that the ocular system can respond to small discrepancies by adjusting the tonic accommodative resting position of the eyes independently (Marran & Schor, 1997; Marran & Schor, 1998). It is possible for a poorly designed VE system to introduce unequal focusing demands if the optical focus is not perfect for the two eyes. It is likely that such a system would place adaptive pressures on both the accommodation and vergence system. Nonetheless, such problems will only occur in poorly designed VE systems.

### 2.6 Cross-link Adaptation

Helmholtz (1924) argued that developmental change (such as increasing interocular distance and sclerosis of the crystalline lens with age) requires that the cross-links should be open to adaptation. Although some studies have provided indirect evidence for plasticity, there is no conclusive evidence that environmental pressures will directly change the normal cross-link relationship. The VA/V ratio has received relatively little attention owing to difficulties in its measurement, but it is known that the AV/A cross-link may be temporarily modified through cycloplegia, miosis, visual fatigue, and orthoptics (Christoferson & Ogle, 1956; Flom, 1960; Manas, 1955; Parks, 1958). On the other hand, it is not certain whether these transient modifications actually reflect plasticity within the cross-links or arise from an alteration in other components of the motor systems. It has been demonstrated that fatigue in the adaptable tonic component can produce predictable changes in cross-link behavior for individuals with abnormal binocular vision. Schor and Tsuetaki (1987) showed that reducing adaptability of tonic accommodation increases the AV/A and decreases the VA/V, whereas reducing adaptability of tonic vergence has the opposite effect.

What has not been determined unambiguously is whether it is possible to directly modify the cross-links in individuals with normal binocular vision. One technique for placing pressure on the cross-links is to use telestereoscopes to optically increase interocular separation, or cyclopean spectacles to negate interocular separation (see Fig. 37.3). In telestereoscopes the observer must increase vergence relative to the focal depth of a target, hence pressure is placed on the AV/A to increase and the VA/V to decrease. In contrast, cyclopean spectacles require a decrease in the AV/A and an increase in VA/V. An appropriate increase in the AV/A has been observed following the use of telestereoscopes for 30 minutes, but in conflict, a corresponding decrease in the AV/A ratio was not found after the use of cyclopean spectacles (Judge & Miles, 1985; Miles & Judge, 1982; Miles, Judge, & Optican, 1987). Cyclopean spectacles require accommodation to vary normally, whereas vergence is fixed (see Fig. 37.3), which is difficult, and hence single vision is often lost. Miles et al. (1987, p. 2585) reported that "subjects had difficulty in obtaining single, sharp images when viewing nearby objects through the cyclopean spectacles and were never able to overcome this problem entirely." To achieve adaptation, comparison is required between the ocular motor response and the perceptual goal (a clear, single target image). If the goal of single vision is consistently disrupted due to other neuromuscular factors, the cyclopean paradigm fails to provide the requisite conditions for adaptation to occur. In summary, telestereoscopes have been used to successfully increase the AV/A ratio, but attempts to reduce the AV/A ratio with cyclopean spectacles have been unsuccessful. Schor and Tsuetaki (1987)

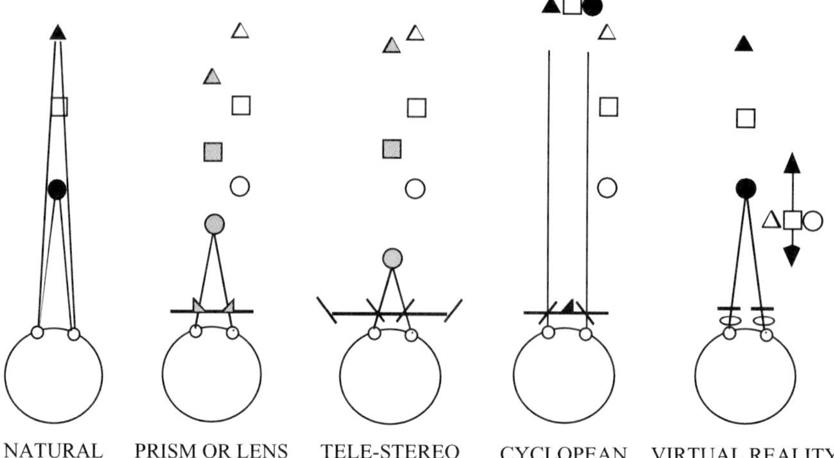


FIG. 37.3. Stimuli to accommodation and vergence in natural conditions; when a constant vergence or accommodation bias is introduced; and when viewing in optical devices. In natural conditions the stimuli to accommodation (empty symbols) and vergence (shaded symbols) are in accordance. If a lens or prism power is induced, a constant offset (bias) is introduced. It is possible for a poorly designed VE system to induce such biases (see text for details). Telestereoscopes necessitate an increase in vergence per unit change in natural unaltered accommodation. Cyclopean spectacles require normal accommodation, whereas vergence remains fixed at optical infinity. In a stereoscopic VE display, accommodation and vergence stimuli are disassociated, and the degree of disassociation depends on viewing distance (thus requiring a change in the gain of the AV/A and VA/V ratios) even when the VE system is optically perfect.

have argued convincingly that because the AV/A always rose together with a concomitant fall in the VA/V in studies using telestereoscopes, the response to the telestereoscopes might have been due to fatigue of adaptable tonic accommodation rather than direct cross-link modification.

The issue of cross-link adaptation is extremely important with regard to VE systems. Biocular VE systems will not place any adaptive demands on the cross-links. In contrast, a stereoscopic binocular system requires individuals to maintain a constant level of accommodation while changing vergence angle when changing fixation between near and far objects. The stereoscopic binocular systems thus place large adaptive demands on the ocular-motor system and raise the crucial issue of whether the cross-links themselves are open to direct adaptation.

## 2.7 Strabismus

Binocular fixation of a stationary object requires that both visual axes are aligned with an object of interest in order to eliminate double vision. Examination of the muscular and neurological systems involved in the coordination of the two eyes can be carried out using a *cover test*. The cover test consists of the covering and uncovering of each eye in turn while the fellow eye maintains fixation on a target. Any movement of the eye when its fellow is being covered indicates the presence of heterotropia (commonly referred to as strabismus or squint). Strabismus can either be convergent (esotropia) or divergent (exotropia) and can affect either eye (e.g., right esotropia). The primary sensory effect of strabismus is amblyopia in the nonfixating eye. One of the consequences of amblyopia is an increased threshold for stereopsis. Strabismus can result from adaptive pressures being placed on a compromised binocular system.

## 2.8 Heterophoria

The majority of human observers do not have strabismus (i.e., no bias is present in closed-loop situations) but do show an open-loop vergence bias. It is possible to measure the constant resting point (or bias) that exists within the vergence system by opening the normal feedback loop to vergence (e.g., by removing any disparity information). Open-loop vergence bias is known as heterophoria and may be defined as a slight deviation from perfect binocular positioning that is apparent only when the eyes are dissociated. Esophoria refers to convergent visual axes when the eyes are dissociated and exophoria to divergent axes (see Fig. 37.4). Note that the gain of the cross-links is less than one. This means that individuals are generally exophoric when fixating close targets.

It should also be noted that it is possible to observe vertical or rotational biases in the vergence system. The ocular-motor system shows a markedly lower ability to adapt to vertical or rotational deviations from perfect binocular positioning, and thus such deviations will normally produce double vision. Nevertheless, it is possible for poorly aligned VE systems to require small amounts of vertical or rotational adaptation, and thus it is necessary to test for such deviations when measuring visual aftereffects following VE exposure. In the interest of brevity, the following discussion restricts the consideration of heterophoria to the more commonly observed horizontal deviations.

## 2.9 Binocular Vision Anomalies

It is not uncommon for binocular vision to break down or fail to develop properly. Binocular vision anomalies can be classified according to the scheme in Fig. 37.5. Incomitant deviations are far rarer than comitant deviations. Incomitant deviations describe deviations that vary with the direction of gaze and arise due to direct trauma of the neuromuscular system. It is comitant deviations that are important with regard to VE systems. Heterophoria is often a precursor to strabismus (strabismus representing the situation when the system can no longer maintain binocular vision and normal binocular vision breaks down). Heterophoria can be classified in the following manner:

### *2.9.1 Esophoria*

1. Divergence weakness esophoria. Esophoria at distance > esophoria near.
2. Convergence excess esophoria. Esophoria at distance < esophoria near.
3. Basic esophoria. Esophoria at distance = esophoria near.

### *2.9.2 Exophoria*

1. Convergence weakness exophoria. Exophoria at distance < exophoria near.
2. Divergence excess exophoria. Exophoria at distance > exophoria near. This often breaks down into strabismus for distance viewing.
3. Basic exophoria. Exophoria distance = exophoria near.
4. Convergence insufficiency.

A consideration of the organization of the accommodation and vergence system provides insight into the etiology of heterophoria and strabismus. It has been well established that abnormalities of the cross-link ratios in childhood are associated with strabismus. One of the most common forms of childhood strabismus is the Donders' squint or accommodative strabismus. This strabismus arises in children who have a high hypermetropic prescription. The need to exert high levels of accommodation to produce clear vision creates a high level

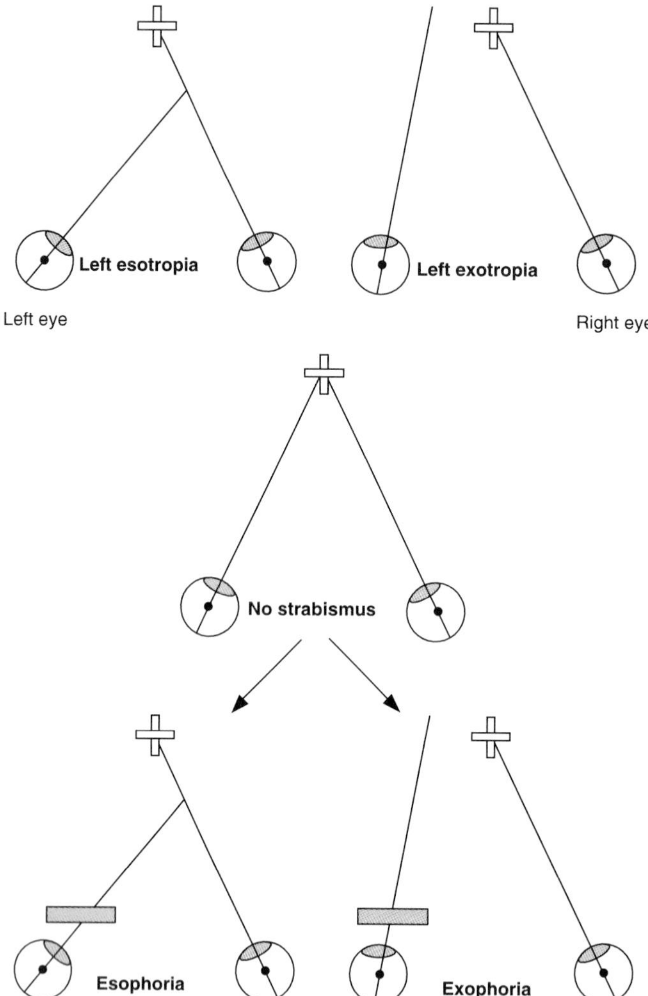


FIG. 37.4. In normal binocular viewing, the visual axes of both eyes are in alignment with a fixated target (*middle plot*). If the axes are not in alignment, the system is described as strabismic or heterotropic. The left upper plot illustrates esotropia (convergent visual axes), and right upper plot illustrates exotropia (divergent visual axes). It is possible to measure the constant resting point (or bias) that exists within the vergence system in people without strabismus by opening the normal feedback loop to vergence (e.g., by covering one eye). Open-loop vergence bias is known as heterophoria. Esophoria refers to convergent visual axes (*left lower plot*) when the eyes are dissociated and exophoria to divergent axes (*right lower plot*).

of vergence because of the accommodation-vergence cross-link. These demands ultimately result in esotropia. Note that poorly designed VE systems can induce the same high levels of accommodation as hypermetropia.

An alternative mechanism of strabismus can occur when the cross-links have too high or too low a gain. For example, a high AC/A ratio results in convergence excess and a low AC/A ratio results in convergence insufficiency. Recall that high adaptation of accommodation and low adaptation of vergence results in a low AC/A ratio and vice versa. One possible treatment of high and low AC/A ratios is thus to fatigue the highly adaptable element. In cases of convergence insufficiency, this is achieved by having patients track the tip of a pencil backward

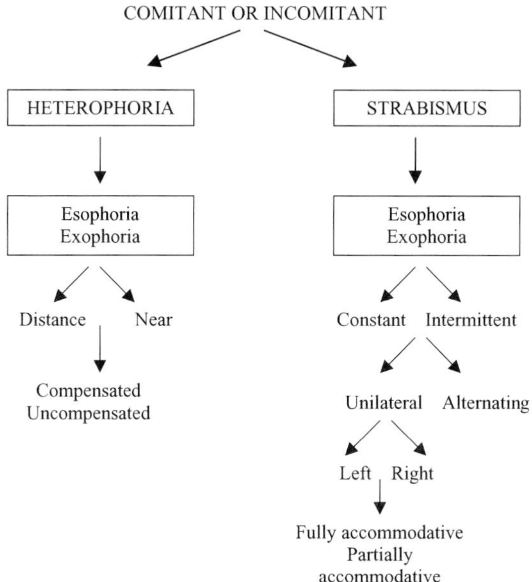


FIG. 37.5. Classification scheme for binocular visual anomalies.

and forward. Schor and Tsuetaki (1987) have shown that such treatment can restore normal cross-link function. Quick et al. (1994) have found that monkeys with esotropia show either a reduced or an abnormally high AV/A ratio. Moreover, Parks (1958) has reported a high incidence of abnormal accommodative vergence in adult humans with strabismus. The fact that the cross-links are placed under particular adaptive pressure by binocular stereoscopic VE systems has already been discussed. It is worth highlighting the fact that VE systems may also hold the key to treating binocular anomalies— the very adaptive pressures that might cause problems to a normal oculomotor system might be beneficial to a malfunctioning system.

The importance of considering the etiology of strabismus and heterophoria when evaluating VE systems relates to the pressures potentially placed by VE on the cross-links and adaptable elements of vergence and accommodation. It seems reasonable to suggest that these issues are of paramount importance when considering the use of VE systems with young children in whom binocular vision is not yet fully developed.

### *2.9.3 Position-specific Heterophoria Adaptation*

Heterophoria describes the difference in position between the two eyes when one eye is occluded. In situations where a prism is placed in front of one eye (see section 2.4), the stimulus for adaptation is reasonably constant at all direction of gaze. In situations where a magnifying lens (for example) is placed in front of one eye, the amount of adaptation required changes as the eye moves eccentrically from the optical center of the lens (Maxwell & Schor, 1994). The ocular-motor system is known to be able to adapt in a position specific manner, but the time constants of such adaptive behavior are far greater than those observed in uniform adaptation. Indeed, the initial response of the ocular-motor system to position-specific vergence pressure is to adapt uniformly in order to decrease or eliminate the demand on at least one part of the visual field (Maxwell & Schor, 1994). The initial global adaptation process is followed by a slower acting position-specific alignment. It is possible for position-specific heterophoria to be induced by a poorly designed VE system (once more through optical errors). It should also be noted that the representation of true 3-D space with two-dimensional (2-D) image planes has the potential to introduce spatial biases between the two eyes, with the biases becoming larger as gaze becomes increasingly eccentric (Wann et al., 1995). It is possible that these biases will place position-specific adaptation pressures on the ocular-motor system. It is not clear whether these small distortions have practical implications with prolonged exposure to VE systems that use flat screen projections of 3-D images.

#### 2.9.4 Summary of Some Key points

Although it is possible for a biocular VE display to promote changes in the user's visual system, adaptive pressures are only likely to be found in a system with very poor optical quality or poorly integrated tracking. In contrast, the basic design of all current stereoscopic binocular systems means that they will place adaptive pressures on the user because the binocular stimuli are not commensurate with natural viewing (Fig. 37.3). In some cases, the adaptive pressure may be minor, but sustained usage of a binocular display that presents large depth intervals is likely to produce transient aftereffects. The consequence of such effects depends on the individual's visual status and cannot be predicted without a full clinical appraisal. Where sustained usage is planned with binocular displays that present large depth intervals, prepost visual screening (see section 3) should be undertaken with candidate observers prior to unsupervised use, and any observers with pre-existing binocular anomalies should be excluded or carefully monitored. Symptoms (headache, difficulty focusing) should be used as indicators that display usage should cease, but symptoms cannot be used as the sole criteria for evaluating potential problems, hence questionnaire studies are imprecise tools for assessing visual aftereffects.

## 3. CLINICAL MEASURES OF VISUAL AFTEREFFECTS FOLLOWING VE EXPOSURE

There are two methods of measuring visual aftereffects following VE exposure. One method is to take a series of measures before and after VE use and explore whether there are any statistically reliable differences between these measures. Additionally, it is possible to compare any changes with a control condition in which the same measures are taken before and after the same period of time viewing, for example, a standard computer monitor. The alternative method is to take a series of clinical measures before and after VE use and determine whether any of these measures has shown a clinically significant change. We would argue that this technique provides a far more powerful tool and avoids the problems associated with testing the null hypothesis (a single person developing strabismus might not produce a statistically reliable change in the population means but would unarguably represent a significant aftereffect). We will outline a series of useful clinical measures that should be evaluated with regard to VE systems.

### 3.1 History and Symptoms

A general medical and ocular history should be recorded together with any symptomatic complaints. Participants should be asked to report if they are suffering from any adverse symptoms including a list of headache, diplopia, blurred vision, sore eyes, or eyestrain. Participants should be asked to report verbally any such symptoms during or following use of the VE system.

### 3.2 Vision/Visual Acuity

The term *eyesight* is herein used to describe the clarity of the retinal image. The reason for using this terminology is because the terms *vision* and *visual acuity* have specialist meanings within the ophthalmic literature: *Vision* refers to uncorrected eyesight, whereas *visual acuity* describes the level of eyesight obtained with optimum refractive correction. Participants should be assessed in their normal situation so that vision is recorded for participants who do not normally wear spectacles or contact lenses, and visual acuity for those who have (and wear) a refractive correction. It is most important to assess binocular eyesight, as disturbances of the ocular-motor system can cause a reduction in the normal advantage of binocular viewing. Eyesight is typically assessed with some form of letter chart and expressed as a fraction where the numerator refers to the distance of the chart and the denominator refers to the distance at which the smallest visible letter subtends five minutes of arc. The standard clinical chart is placed at 6 meters (or 20 feet), and a person with "good" eyesight is expected to read letters that subtend 5 minutes of arc at that distance. If someone can read the small letters near the bottom of a chart then they will typically have eyesight equal to 6/6 or better (20/20 in imperial measures), but if only the large letter on top of the chart could be read then the eyesight would be equal to 6/60 (6/12 occurs about halfway down). Traditional (Snellen) eyesight charts have major disadvantages for scientific studies. These disadvantages relate to: (1) the lines changing in unequal step sizes and (2) the lines having an unequal number of letters. These two facts make it impossible to report valid statistics for Snellen eyesight measures. For this reason we propose the use of logarithmic charts when evaluating eyesight. For example, we have used previously the Glasgow Acuity Cards designed for clinical screening (McGraw & Winn, 1993). The Glasgow Acuity Cards use a progression of letter sizes in a geometrical series with a ratio of 0.1 log units per card. Each card contains four letters of equal legibility and equivalent letter–row spacing is maintained throughout. Vision is recorded either as the log of the minimum angle of resolution or as a decimal value from 0.0 (6/60 Snellen) to 1.0 (6/6 Snellen). Use of the Glasgow Acuity Cards allows for accurate measurement and better analysis of the vision/visual acuity than the traditional Snellen chart.

It should be noted that ophthalmic charts typically measure high-contrast eyesight (i.e., a well-defined black shape or letter against a bright white background). The problem with such charts is that they do not indicate how well the visual system can process retinal information at different spatial frequencies (e.g., how well someone can see when contrast is low). A number of methods do exist, however, that allow eyesight to be assessed over a range of spatial frequencies. It is possible (in principle) for a VE system to cause a decrease in the mid-range spatial frequency sensitivity. The reason for this is that the visual system responds to an amelioration of high spatial frequencies by depressing the mid-range spatial frequencies in order to unmask the higher frequency channels (Mon-Williams, Tresilian, Strang, Kochar, & Wann, 1998). A VE system that has poor image quality effectively filters the high-spatial-frequency information from a display and thus might cause the visual system to adapt. There are a number of commercially available contrast sensitivity charts for use in a clinical setting. One such chart is the Pelli–Robson chart (Pelli, Robson, & Wilkins, 1988). This chart is wall mounted and consists of 16 letter triplets. The triplets are of constant contrast, but the contrast reduces between triplets from the top to the bottom of the chart. The Pelli–Robson chart and other available alternatives allow for easy measurement of contrast sensitivity. In all methods of eyesight measurement it is important that the recommended level of illumination is used.

### 3.3 Refractive Error

It is worthwhile measuring the refractive state of the eye before and after the use of a VE system to ensure that the refraction has not changed because of accommodative adjustment. The most common clinical method of assessing refractive status involves a *retinoscope*. The standard retinoscopy technique involves shining a light into the eye while viewing the resultant retinal reflection (the "red eye" observed in flash photography) along the visual axis through a semisilvered mirror. The practitioner then rotates the retinoscope and observes the resulting movement of the beam. If there is no resultant movement there is no refractive error; if the reflex moves in the same direction as the retinoscope (with it), then the eye is hypermetropic, and if the reflex moves in the opposite direction, the eye is myopic.

## 3.4 Ocular–motor Balance

Examination of the muscular and neurological systems involved in the coordination and integration of the two eyes can be carried out using a battery of clinical tests. The cover test is described in section 2.7, which consists of the covering and uncovering of each eye in turn while the other eye maintains fixation on the smallest target visible to both eyes. Any movement of the eye when its fellow is being covered indicates the presence of strabismus, whereas any movement of the eye when it is being uncovered shows the presence of heterophoria. Further assessment of heterophoria can be carried out using a Maddox rod at 6 meters and a Maddox wing for use at 33 centimeter in good illumination (500 lux). A "flashed" approach is the optimum method of obtaining the heterophoria reading for distance and for near. It is important to assess the horizontal, vertical, and rotational components of binocular alignment.

One useful index of binocular function is provided by *fixation disparity*: small errors of binocular fixation that do not disrupt binocular vision, as the errors do not exceed Panum's fusional areas (the corresponding retinal errors in the two eyes). It is straightforward to measure fixation disparity, but we suggest that associated heterophoria provides a much better indicator of binocular stress. *Associated heterophoria* is the term used to describe the amount of prism required to reduce fixation disparity to zero and is readily measured using a Mallett unit. (Mallett, 1974) Mallett units use a binocular lock consisting of the letters OXO and monocular markers in line with the X to demonstrate the presence of decompensated heterophoria, because viewed through Polaroid filters, the monocular markers will remain in line only in the absence of decompensated heterophoria. The existence of decompensated heterophoria provides a useful indication of binocular stress. Standard Mallett distance and near units can be used with loose trial case prisms within a trial frame in order to produce absolute alignment of the nonius lines using the minimal value of prism in 0.5 prism diopter steps for the distance and 1 prism diopter steps for near. Participants should be instructed to look for even very small movements of the nonius lines. A full review of Binocular Vision tests is presented in Pickwell (1984). An example of a Mallet unit adapted for use within a VE display is presented in Wann et al. (1995).

## 3.5 Near Point of Convergence

Clinical measurement of pursuit convergence should be taken with a vertical line target brought toward the eyes on the median plane until diplopia (double vision) is reported and/or one eye can be seen to diverge. It is also possible to assess "jump" vergence by asking participants to rapidly change fixation from a near target to a far one and vice versa.

## 3.6 Amplitude of Accommodation

The dynamic recording of accommodation is technically demanding (in contrast to the relatively straightforward techniques available for recording eye movements). It is possible, however, to use the retinoscope (see section 3.3) to determine the level of accommodation that occurs in response to a proximal target. The eye's ability to change focus from the far point to the near point of vision should be measured in dioptres using subjective report of near text becoming illegible. Amplitudes of accommodation should be checked for any abnormality. Accommodative facility can be assessed by placing ophthalmic lenses in front of the eyes and evaluating the time it takes for the eyes to relax or increase accommodation.

### 3.7 Stereopsis

There are a large number of clinical tests of stereopsis (including the TNO test, the Frisby test, the Stereo random dot E test, the Titmus random dot test, and the Lang test). Stereopsis provides a good indication of binocular status, because stereoscopic sensitivity drops rapidly when binocular vision is compromised. It is important to ensure that the tests are conducted at a standardized distance, as disparities are inversely proportional to the square of the distance. It is also important to ensure that the test is carried out under good illumination (e.g., 500 lux).

## 4. BEYOND CLINICAL MEASURES

The measures described in section 3 are proposed as the basic requirements in an applied setting for appraising any new or substantially modified display system or to screen any group of individuals who may be required to use a binocular display for sustained or repetitive periods. In addition to these measures, it is possible to undertake a more analytical approach in an experimental setting to evaluate the impact of different adaptive pressures. In this section, an example of undertaking such a study is outlined using an infrared optometer and eye tracking to appraise the response of the AV/A cross-links (section 2.3.3) to sustained and demanding usage of a conventional binocular VE system.

### 4.1 A Study of AV/A and VA/V Adaptation

The AV/A and VA/V ratio (section 2.3.3) were objectively measured before and after usage of a VE system. Two volunteer participants (AS and MB) took part in the experiments. Both were 25 years of age, emmetropic, did not take any medication, had no ophthalmic abnormalities, and were naive to the purpose of the experiment. The participants gave written consent and the experiments met the approval of the university ethics committee. Clinical measures of accommodation amplitude, stereoacuity thresholds, AV/A and VA/V ratios were all normal. LCD shutter goggles were used to present field-sequential computer generated images at 120 Hz on a high-resolution computer monitor. The participants viewed a high-contrast cross for 60 minutes in the dark oscillating from optical infinity to either 33 centimeter or 16 centimeter with accommodative demand set at the closest distance. The angular size of the target cross was kept constant as it moved in depth creating the percept of a single cross of unchanging size located at a constant egocentric distance. In the early stages of adaptation, participants had to exert voluntary vergence to eliminate double vision. At first participants had problems consistently fusing the target, but after the first 10 minutes they reported that it was easy to keep a clear single target though both complained of sore eyes and frontal headaches. Hence this presented a highly demanding task that is unlikely to occur in many occupational setting, but similar stimuli may occur if close visual monitoring is required of targets moving in depth over a sustained period (e.g., some military applications or some games). The experimental issue was what changes resulted from the sustained usage.

Dynamic accommodation was measured using a modified Canon Autoref R-1 infrared (IR) objective optometer. The autorefractor determines the position of focus in three meridians, from which the spherocylindrical refractive error can be computed to an accuracy of $\pm 0.12$D per second. Vergence was recorded with a purpose built differential IR limbal tracker, accurate to 10 arcmin under the employed experimental conditions. The IR sources were chopped at greater than 1000 Hz, with the detectors set to only recognize this frequency (thus avoiding direct current drift, reducing 1/f noise, and eliminating cross-talk between the photodiodes and the IR optometer). Cross-axis sensitivity was found to be $\pm$ 10%. The analogue output of the eye trackers was fed into a digital storage oscilloscope, which was controlled by an IBM-clone computer via an IEEE-488 interface. This system allowed inspection of the analogue signals to be made in real time so that artifacts (e.g., blinks) could be identified. Comprehensive calibration routines were carried out before preadaptation measurements and after postadaptation recordings. The ambient room illumination was constant at approximately 250 lux. Natural pupil size was monitored on an external 9-inch video monitor (Sony PVM 97) using the magnified (8.2 $\times$) image provided by the IR optometer to ensure that pupillary changes were not a confound (pupil size always exceeded 0.5 cm).

Accommodative vergence was measured by covering the left eye with an occluder. Participants fixated a high-contrast Maltese cross through a 5-dioptre Badal lens, whereas accommodative demand was altered via ophthalmic trial lenses so that the only change in the stimulus was due to accommodation. Targets were presented along the optical axis of the right eye to ensure that eye movements did not interfere with the accommodative recording. Vergence accommodation was measured using a Difference of Gaussian (DoG) stimuli. The DoG presented a blurred vertical bar of 0.4 cm, which acted as a stimulus to vergence but not to accommodation. Vergence demand was altered via ophthalmic prisms so that the only change in the stimulus was due to disparity vergence.

Accommodation and vergence responses were expressed in dioptres (D) and meter angles (MA) respectively: Both are the reciprocal of distance in meters (MA is the angle through which each eye has rotated from the primary position in order to fixate an object located 1 meter away), so 1 D or 1 MA corresponds to 100 centimeter, 2 D, or MA corresponds to 50 centimeter, etc. The normal AV/A ratio is approximately $0.60 \pm 0.2$ MA/D, and the VA/V ratio is around $0.5 \pm 0.2$ D/MA. The cross-link response was assessed using five separate measurements of accommodation and vergence for seven different stimuli conditions presented in a randomized order (to control for hysteresis). The stimuli were in unitary D or MA steps from optical infinity to six. Measurement of the AV/A and VA/V was alternated pre- and postadaptation and between participants.

## 4.2 Results

Figure 37.6 shows that cross-link interaction changed following adaptation. Regression lines are plotted and correlation coefficients are provided, although the nonlinear nature of the post-adaptation data weakens any regression analysis (the post-adaptation data for AS showed a significant, $p < 0.05$, quadratic component). Dunn's procedure (Keppel, 1982, p. 146) was carried out on the data points of interest.

The target distance did not produce exact accommodation or vergence, emphasizing the importance of measuring ocular response when computing cross-link ratios. The pre- and postadaptation vergence responses were identical for MB but showed a mean decrease of 0.28 MA for AS (maximum of 0.6), whereas the accommodation responses decreased by a mean of 0.36 D for MB and 0.54 for AS (maximum of 0.6 and 0.98, respectively). The maximum shift was always found in the stimuli where the ocular demands were greatest. These decreases suggest that some plant fatigue is occurring, but the shifts cannot explain the cross-link changes.

In participant MB it is clear that the AV/A and VA/V ratios have been reduced, thereby allowing accommodation and vergence to independently respond to blur and disparity cues. The

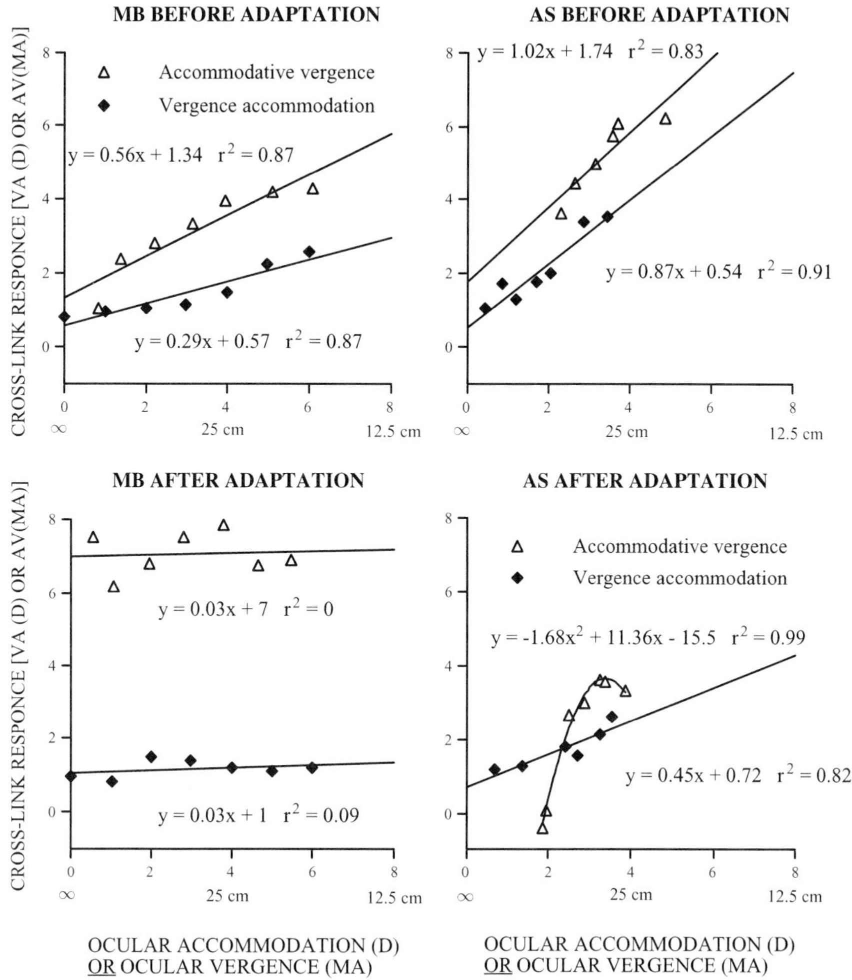


FIG. 37.6. The cross-links between accommodation and vergence before (*above*) and after (*below*) adaptation for MB (*left*) and AS (*right*). Each point is the mean of five randomly organized measurements. The standard deviation of the measurements was smaller than the plot symbol. The primary response (ocular accommodation [A in Dioptres] or vergence [V in Meter Angles]) is plotted on the abscissa with the secondary [cross-link] response (accommodative vergence [AV in MA] or vergence accommodation [VA in D], respectively) shown on the ordinate. The normal ratios may clearly be seen to change following adaptation.

variability in the AV/A response is typical of an uncoupled system, whereas the accommodation produced by the VA/V stabilizes around the bias found in the absence of accommodative stimuli.

The data from participant AS are less compelling than those of MB. On the other hand, it is clear that the cross-link responses have changed. The change in the VA/V is very similar to the clear lowering of the ratio observed in participant MB. The change in gain of the AV/A ratio is less straightforward, as the postadaptation AV/A ratio showed a significant quadratic component. It therefore appears that part of the slope has increased its gain, whereas the other part has flattened.

The data from participant MB are very compelling, suggesting that the AV/A and VA/V ratios have both decreased following exposure to the VE display. The data from participant AS are less impressive, but detailed examination shows a similar picture to the data from participant MB. The data suggest that both participants have reduced the normal action of accommodative vergence and vergence accommodation. This is the ideal response to a VE system where there is no correlation between the continually changing vergence disparity and accommodative stimulus.

## 5. CONCLUSIONS

To reiterate the summary in section 3, binocular visual displays can result in unusual visual pressures being placed on accommodation and vergence. These may be partially reduced by a change in tonic accommodation and tonic vergence, but where objects are presented with a wide range of depths, tonic shifts cannot provide a solution (Fig. 37.3). Global shifts in bias can only decrease the ocular-motor demands of uniform disparities or blur. We propose that non-uniform demands (like those produced in VE systems) require a response from the cross-links systems. The results presented in section 4 suggest that the cross-links between accommodation and vergence are directly open to adaptive pressures. These results highlight the importance of addressing fundamental issues of binocular coordination when considering the pressures placed on the ocular-motor system by VE systems. The simultaneous decrease observed in the AV/A and VA/V ratios cannot be explained (in a parsimonious manner) by an alteration in the adaptability of the respective tonic components or by plant fatigue. This may be viewed as a positive adaptation to the VE setting, but the negative consequences, highlighted by the symptoms of headache and sore eyes, should not be overlooked. Adaptation of this type is not stress free, and we would take the conservative position that it is not desirable if it can be avoided through the design of VE tasks (not just display design, but the rendered stimuli) or limiting exposure periods. Our previous suggestion that binocular VE systems may place pressure on the accommodation and vergence system was initially controversial (Mon-Williams et al., 1993; Wann et al., 1995). There is now ample evidence from a number of studies to confirm our original hypothesis. It is important therefore that such effects are monitored and managed in any setting where sustained usage of a binocular system is likely to occur.

## 6. ACKNOWLEDGMENTS

Production of this manuscript was supported by grants from the Australian Research Council; National Health and Medical Research Council, Australia; and Engineering and Physical Research Council, U.K.

## 7. REFERENCES

Carter, D. B. (1965). Fixation disparity and heterophoria following prolonged wearing of prisms. *American Journal of, 42*, 141–151.

Christoferson, K. W., & Ogle, K. N. (1956). The effect of homatropine on the accommodation-converegence association. *AMA Archives of Ophthalmology, 55*, 779–791.

Fisher, S. K., Cuiffreda, K. J., Levine, S., & Wolf-Kelly, S. (1987). Tonic adaptation in symptomatic and asymptomatic subjects *Am. J. Optom.Physiol. Opt., 64*, 333–343.

Flom, M. C. (1960). On the relationship between accommodation and accommodative convergence. I. Linearity. *American Journal of Optometry and Archives of American Academy of Optometry, 37*, 474–482.

Helmholtz, H. von (1924). *Physiological optics* (Vol. 3, pp. 191–192) New York: Dover, 1962. English translation by J. P. C. Southall for the Optical Society of America from the third German edition of *Handbuch der physiologiscen optik*, Hamburg, Voss, 1909. Original work published 1894.

Henson, D. B., & Dharamshi, B. G. (1982). Oculomotor adaptation to induced heterophoria and anisometropia. *Invest. Ophthal. Vis. Sci., 22*, 234–240.

Hering, E. (1868/1977). *The theory of binocular vision*. New York: Plenum Press.

Heuer, H., Bruwer, M., Romer, T., Kroger, H., & Knapp, H. (1991). Preferred vertical gaze direction and observation distance. *Ergonomics, 34*, 379–392.

Heuer, H., & Owens, D. A. (1989). Vertical gaze direction and the resting posture of the eyes. *Perception, 18*, 363–377.

Judge, S. J., & Miles, F. A. (1985). Changes in the coupling between accommodation and vergence eye movements induced in human subjects by altering the effective interocular separation. *Perception, 14,* 617–629.

Keppel, G. (1982). *Design and analysis: A researcher's handbook*. Englewood Cliffs, NJ: Prentice-Hall.

Mallett, R. M. J. (1974). Fixation disparity—its genesis in relation to asthenopia. *Ophthalmic Optician, 14*, 1159–1168.

Manas, L. (1955) The inconsistency of the AC/A ratio. *American Journal of Optometry and Archives of American Academy of Optometry, 32*, 304–315.

Marran, L., & Schor, C. M. (1998). Lens induced aniso-accommodation. *Vision Research, 38*, 3601–3619.

Marran, L., & Schor, C. M. (1997). Multiaccommodative stimuli in VR systems: Problems and solutions. *Human Factors, 39*, 382–388.

Maxwell, J. S., & Schor, C. M. (1994). Mechanisms of vertical phoria adaptation revealed by time-course and two-dimensional spatiotopic maps. *Vision Research, 34*, 241–251.

McGraw, P. V., & Winn, B. (1993). Glasgow acuity cards: A new test for the measurement of letter acuity in children. *Ophthal. Physiol. Opt., 13*, 400–403.

McLin, L. N., & Schor, C. M. (1988). Voluntary effort as a stimulus to accommodation and vergence. *Investigative Ophthalmology and Visual Science, 29*, 1739–1746.

Miles, F. A. (1985). Adaptive regulation in the vergence and accommodation control systems. In A. Berthoz & G. Melville-Jones (Eds), *Adaptive mechanisms in gaze control facts and theories* (pp. 81–94). Amsterdam: Elsevier Science Publishers.

Miles, F. A., & Judge, S. J. (1982). Optically induced changes in the neural coupling between vergence eye movements and accommodation in human subjects. In G. Lennerstrand, D. S. Zee, & E. L. Keller (Eds), *Functional basis of ocular motility disorders* (pp. 93–96). Oxford, England: Pergamon Press.

Miles, F. A., Judge, S. J., & Optican, L. M. (1987). Optically induced changes in the couplings between vergence and accommodation. *Journal of Neuroscience, 7*, 2576–2589.

Mon-Williams, M., Burgess-Limerick, R., Plooy, A., & Wann, J. (1999). Vertical gaze direction and postural adjustment: An extension of the Heuer model. *Journal of Experimental Psychology: Applied, 5*, 35–53.

Mon-Williams, M., & Tresilian, J. R. (1998). A framework for considering the role of afferent and efferent signals in the control and perception of ocular position. *Biological Cybernetics, 79*, 175–189.

Mon-Williams, M., & Tresilian, J. R. (1999). A review of some recent studies on the extra-retinal contribution to distance perception. *Perception, 28*, 167–181.

Mon-Williams, M., Tresilian, J. R., Strang, N., Kochar, P., & Wann, J. P. (1998). Improving vision: Neural compensation for optical defocus. *Proceedings of the Royal Society, B265*, 71–77.

Mon-Williams, M., Wann, J. P., & Rushton, S. K. (1993). Binocular vision in a virtual world: Visual deficits following the wearing of a head-mounted display. *Ophthalmic and Physiological Optics, 13*, 387–391.

North, R. V., & Henson, D. B. (1981). Adaptation to prism-induced heterophoria in subjects with abnormal binocular vision or asthenopia. *American Journal of Optometry and Physiological Optics, 58*, 746–752.

Parks, M. M. (1958). Abnormal accomodative convergence in squint. *AMA Archives of Opthalmology, 59*, 364–380.

Pelli, D., Robson, J., & Wilkins, A. (1988). The design of a new letter chart for measuring contrast sensitivity, *Clinical and Vision Science, 2*, 187–199.

Pickwell, D. (1984). *Binocular vision anomalies—Investigation and treatment*. London: Butterworth.

Quick, M. W., Newbern, J. D., & Boothe, R. G. (1994). Natural strabismus in monkeys: Accomodative errors assessed by photorefraction and their relationship to convergence errors. *Investigative Opthalmology and Visual Science, 35*, 4069–4079.

Saunders, K. J. (1995). Early refractive development in humans. *Survey of Ophthalmology, 40*, 207–216.

Schor, C. M. (1983). The Glenn A. Fry Award Lecture: Analysis of tonic and accommodative vergence disorders of binocular vision. *American Journal of Optometry and Physiological Optics, 60*, 1–14.

Schor, C. M. (1979). The influence of rapid prism adaptation upon fixation disparity. *Vision Research, 19,* 757–765.

Schor, C. M., & Kotulak, J. C. (1986). Dynamic interactions between accommodation and vergence are velocity sensitive. *Vision Research, 26*, 927–942.

Schor, C. M., & Tsuetaki, T. K. (1987). Fatigue of accommodation and vergence modifies their mutual interactions. *Investigative Ophthalmology and Visual Science, 28*, 1250–1259.

Schor, C. M., Alexander, J., Cormack, L. & Stevenson, S. (1992). Negative feedback control model of proximal convergence and accommodation. *Ophthalmological and Phsyiological Optics, 12*, 307–318.

Wann, J. P., Rushton, S., & Mon-Williams, M. (1995). Natural problems for stereoscopic depth perception in virtual environments. *Vision Research, 19*, 2731–2736.

Winn, B., Gilmartin, B., Sculfor, D. L., & Bamford, J. C. (1994). Vergence adaptation and senescence. *Optom. Vis. Sci., 71*, 1–4.

# 38

# Proprioceptive Adaptation and Aftereffects

Paul DiZio and James R. Lackner
*Ashton Graybiel Spatial Orientation Laboratory*
*Volen Center for Complex Systems*
*Brandeis University MS033*
*Waltham, MA 02454*
*dizio@brandeis.edu*
*lackner@brandeis.edu*

## 1. INTRODUCTION

Until relatively direct interfaces with brain signals become widely available, users will have to play a physically active part in controlling the virtual environment (VE). This involves moving the head, eyes, limbs, or whole body. Control and perception of movement depends heavily on proprioception, which is traditionally defined as the sensation of limb and whole body position and movement derived from somatic mechanoreceptors. This chapter presents evidence that proprioception actually is computed from somatic (muscle, joint, tendon, skin, vestibular, visceral) sensory signals, motor command signals, vision, and audition. Experimental manipulation of these signals can alter the perceived spatial position and movement of a body part, attributions about the source and magnitude of forces applied to the body, representation of body dimensions and topography, and the localization of objects and support surfaces. The effortless and unified way these qualities are perceived in normal environments depends on multiple, interdependent adaptation mechanisms that continuously update internal models of the sensory and motor signals associated with the position, motion and form of the body, the support surface, the force background, and properties of movable objects in relation to intended movements.

An understanding of these mechanisms is important to VE users because VEs will sometimes inadvertently and sometimes purposely expose active users to never before encountered combinations of sensory and motor signals and environmental constraints. In many cases, this will lead to perceptual and motor errors in the VE until internal adaptation has been updated. When a user who has adapted to a VE carries the new adaptive state back into the normal environment, he or she will experience aftereffects, usually in the form of mirror image errors to those initially made in the VE (see chap. 31, this volume). These can include deviated execution of limb and whole body movements, proprioceptive errors, misestimates of externally imposed forces, and visual and auditory mislocalizations.

This chapter will use concepts derived from laboratory studies as a basis for interpreting measurements of proprioceptive side effects and aftereffects in virtual environments. It discusses what VE conditions will alter the state of sensory and motor adaptation, what components of the movement and orientation control system will adapt, what side effects and aftereffects will result from adaptive modification of these subsystems, and what contexts will evoke aftereffects. This survey emphasizes facets of proprioception relevant to VEs involving manual performance and whole body motion tasks. A balanced presentation is attempted of both the value and the limitations of laboratory-based conceptual distinctions for making the best predictions in practice.

## 2. PROPRIOCEPTION, MOTOR CONTROL, AND SPATIAL ORIENTATION

### 2.1 Multisensory and Motor Factors

In his analysis of sensation and motor control, Sherrington (1906) coined the term *proprioceptors* for the mechanoreceptors located inside body tissues that are primarily responsive to changes within the animal itself. Examples of these receptors are spindle organs in the muscles, Golgi tendons organs, unmyelinated Ruffini endings of the joints, and various sensor types in the viscera and vasculature, which were thought to be inaccessible to external energy but well attuned to body orientation, configuration, and movement. In contrast, systems whose sensors are located on the body surface and receive external stimuli were categorized as exteroceptive. The skin mechanoreceptors are located on the boundary of the environment and the body and have overlapping proprioceptive and exteroceptive roles. They are involved in discriminating form, texture, pain, temperature, and other complex properties of the external world as well as body configuration, motion, and orientation (especially for the fingers, lips, and tongue). Vision, audition, and olfaction are exteroceptive systems, but they can also be used to monitor motion and orientation of the limbs and the whole body. Sherrington noted that the vestibular labyrinth of the inner ear (semicircular canals and otolith organs) is developmentally derived from the exteroceptive system, but it acts in concert with the proprioceptive system because it is stimulated by pressure changes and shear forces in the sensory end organs usually brought about by head and body motion and changes in orientation. His analysis acknowledges that proprioception is multisensory and overlaps with other experiential domains.

Proprioception is also directly related to motor control. Internal signals corresponding to movement commands (or the command not to move) have been posited as contributing to proprioception and spatial awareness. They have been called "the desire" (Aristotle, 1978), "the perceived effort of will" (Helmholtz, 1925), "efference copy" (Von Holst & Mittelsteadt, 1950), or "corollary discharge" (Sperry, 1950). The idea that such efference copy signals are sufficient for conscious perception of movement or position lacks strong empirical support (see McCloskey & Torda, 1975, for a review). For example, when humans are subjected to local ischemic nerve block or systemic neuromuscular block and attempt to move one of their paralyzed limbs they do not perceive any illusory movement, but instead have a sense of great heaviness (Lazlo, 1963; Melzack & Bromage, 1973). However, internal motor signals seem to interact with afferent signals in visual perception and oculomotor control. For example, fruitless attempts to move a paralyzed eye are accompanied by a great sense of effort and apparent motion and displacement of visual targets (Matin et al., 1972). In Von Holst and Mittelsteadt's (1950) terms, motor control and perception mechanisms must discriminate between "reafferent" sensory signals brought about by self-initiated movement and "exafferent" signals evoked by external events. One way for this to happen would be for an efference copy signal to cancel only the reafferent portion of the total afferent signal. This requires an internal calibration in which the afferent and efferent signals are represented in comparable units and in the appropriate proportions.

## 2.2 Sensorimotor Calibration

The correlations between afferent and efferent signals, however, are very complex. For example, there is no unique pattern of motor command and sensory feedback signals associated with even a simple single-joint arm movement. Raising the forearm 45 degrees vertically or flexing it 45 degrees in a horizontal plane involve very different efferent commands to and sensory feedback from the biceps brachii and brachialis muscles because the gravity torques are so different. The muscle command and feedback signals also differ when weights are wielded or machines are manipulated, even for the same movement amplitude and orientation of the arm. A constant efferent command will not produce the same force in fatigued and rested muscle. Self-locomotion or vehicular transport can generate accelerative loads requiring unique muscle forces for accomplishing an arm movement, relative to when the body is stationary. To complicate matters further, body acceleration activates vestibular afferent discharge, which through vestibulospinal pathways innervates skeletal muscles and modulates muscle spindle sensitivity (cf. Wilson & Melville-Jones, 1976). In order for the central nervous system to use an efference copy signal to cancel just the reafferent portion of the total afferent signal, it must first parcel the sensory signals into components due to intended movement and anticipated external loads and the efferent signals into voluntary and involuntary components.

In other words, calibration of perception and motor performance involves internalization of the relationship between combinations of sensory and motor signals in relation to body movement and orientation, external loads, environmental constraints, and internal conditions. Recalibration is required when conditions change. For example, alterations in body dimensions, strength and sensory capacity throughout the life span, loss or distortion of motor or sensory function due to illness or injury, migration to a novel environmental medium (land to water) all require adaptive accommodations. There is ample evidence that given sufficient time humans can adapt to a wide range of stable sensory–motor rearrangements produced in the laboratory or in applied technological environments such as aerospace flight, traditional simulators, teleoperation, and VEs (Held & Durlach, 1991; Kennedy, Fowlkes & Lilienthal, 1993; Lackner, 1976, 1981; Rock, 1966; Wallach, 1976; Welch, 1978; chap. 31, this volume). Adaptation is often accompanied by aftereffects that degrade performance on return to the normal environment. Such aftereffects raise questions that must be addressed by VE system designers (see Table 38.1).

## 2.3 Muscle Spindles

Muscle spindle organs provide signals correlated with the full natural range of muscle length (limb position) and velocity (Harvey & Matthews, 1961). Spindles are connected in parallel with the muscle body so that they can be unloaded when the muscle contracts on receiving input from alpha motor neurons and loaded when the muscle lengthens. Stretch receptors embedded in each spindle's visco-elastic central region project two types of sensory fibers signaling both the spindle's length and rate of change of length (primary fibers) or only its length (secondary fibers). The polar regions of each spindle are composed of contractile fibers innervated by gamma motor neurons that provide independent central regulation of their afferent sensitivity (see Fig. 38.1). Heightened gamma innervation shortens the polar regions and thereby stretches

**TABLE 38.1**
Questions Concerning the Nature of Aftereffects of VE Exposure

- What conditions will trigger adaptation?
- How does the internal calibration change?
- Can multiple states of calibration be maintained?
- In what contexts will different calibration states dominate?

These theoretical questions translate into practical ones for the VE context:

- Which VEs will elicit adaptation of proprioception, motor control, and perceived orientation?
- What kinds of aftereffects will be expressed, and under what conditions?

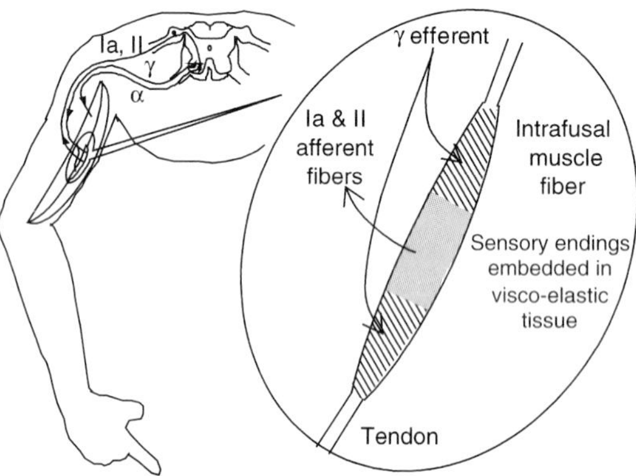


FIG. 38.1. Muscle spindles reside in parallel with extrafusal muscle fibers, which are excited by $\alpha$-motor neurons and provide the contractile force for movements. In this simplified diagram, the primary (Ia) and secondary (II) afferent fibers are shown with sensory endings on the central visco-elastic portion (gray area) of a single spindle, but in reality, type Ia and II fibers innervate different types of spindles. Primary fibers discharge in proportion to muscle length and rate of change in length, and secondary fibers discharge in relation to length. Afferent discharge can be modulated by $\gamma$ efferent activity innervating intrafusal contractile fibers (striped areas) located at the poles of each spindle, which loads or unloads the central region containing the sensory endings but does not contribute to overall muscle force.

the central region and produces a greater afferent discharge rate, for the same overall muscle length. In contrast, quiescent gamma efferent activity relaxes the spindle poles, unloads the central region, and decreases the afferent activity while muscle length remains constant.

Until relatively recently, muscle spindle organs were thought to be important in reflexive motor control but not to influence the conscious awareness of limb position. Position sense was thought to arise from receptors in the joint capsules, but this view was weakened by evidence from humans with artificial hips. Although capsular receptors are absent in artificial joints, patients postsurgery were found nevertheless to have normal position sense accuracy of their hip (reviewed by Burgess & Wei, 1982). Muscle spindle activity is still present and is implicated in their hip position sensitivity.

In relaxed mammals, the muscle spindle firing rate is very low, about 0 to 4 impulses/sec. Hagbarth and Eklund (1966) found that low amplitude mechanical vibration applied to the skin over a muscle causes reflexive contraction of that muscle by entraining spindle afferent discharge to the vibration frequency and activating the stretch reflex. This phenomenon is known as the tonic vibration reflex (TVR). The vibration technique was extended by Goodwin, McCloskey, and Matthews (1972) to demonstrate that spindle afferent signals influence conscious sense of limb position and motion. In these experiments, the biceps brachii was vibrated and the participant's unseen forearm was restrained so that reflexive contraction would not shorten the muscle and unload the spindles. All participants reported feeling an increase in the elbow joint angle. The illusion could be quantified by having participants match the perceived angle of the elbow of their vibrated arm with the other arm and measuring the discrepancy in forearm positions (see Fig. 38.2A). Similarly, triceps vibration elicited an illusory flexion of the forearm. Thus, an abnormally high discharge rate in the vibrated muscle is interpreted as lengthening of the vibrated muscle and this "lengthening" is referred to the joint(s) controlled by the muscle. If the same arm vibration experiment is done in 0 G, weightless conditions, the magnitude of the illusion is smaller than in normal 1 G conditions, and in an increased gravitoinertial force background the illusion is larger (Lackner, DiZio, & Fisk, 1992). (Note: Gravitoinertial force is the resultant of gravitational and inertial acceleration on a mass. 1 G equals the force due to acceleration due to gravity at sea level.) The novel force background affects the otolith organs, which influence muscle tone by activating gamma motor neurons in the spinal cord, heightening the spindle afferent signal in high-force backgrounds, and decreasing it in 0 G. This mechanism is part of the vestibulospinal regulation of the antigravity musculature of the body. Clearly, muscle spindle afferent and efferent signals influence proprioception and motor control.

## 2.4 The Role of Body Schema and Spatial Orientation

Vibration of the proper muscles can produce apparent displacement and motion of the head, arm, leg, and trunk (Lackner, 1988) in a dark room. The character of these vibratory myesthethic illusions demonstrates that perceptual interpretation of the afferent signal takes into account the anatomy and dimensions of the body. For example, a vibration-induced increase in biceps brachii spindle afferent discharge: (1) affects the perceived angle of the elbow joint, which is spanned by the biceps brachii; (2) alters it in a direction (extension) that agrees with the bicep's role as a flexor; and (3) evokes apparent spatial hand motion in an arc consistent with the elbow joint's hingelike motion and the length of the forearm. Some participants experience anatomically impossible hyperextension of the forearm during biceps vibration whereas others after reaching the normal limits of extension perceive a paradoxical illusion of continuous motion without further displacement (Craske, 1977). Even nonmotile appendages can undergo apparent distortion. For example, Lackner (1988) showed that if the participant holds the tip of the nose while the biceps brachii is vibrated, he or she will perceive the nose to grow in length as the forearm extends away from the face, as illustrated in Fig. 38.2B. The relatively stable, but modifiable, cortical representation of the body schema (topography and dimensions of the body) interacts with the moment-to-moment representation of body configuration based on peripheral muscle spindle and tactile signals. This conclusion is consistent with phantom limb experiences (Henderson & Smyth, 1948).

Achilles tendon vibration gives rise to a whole body movement backward relative to the support surface because of the TVR activation of the calf muscles (Eklund, 1972). When the Achilles tendons of participants restrained in a standing position are vibrated they perceive forward body pitch, sometimes in full 360-degree circles, and even though they are physically stationary, they also show nystagmoid eye movements similar to what would be evoked by vestibulo-ocular reflexes during real body tumbling. Tactile cues from environmental surfaces can modify postural vibratory myesthetic illusions. Achilles tendon vibration normally leads to apparent pivoting about the ankles, but if the participant bites a rigid dental mold then it may become the pivot point (Lackner & Levine, 1979). If a head-fixed target light is present during such illusions, it will be perceived as moving spatially relative to the current axis of rotation,

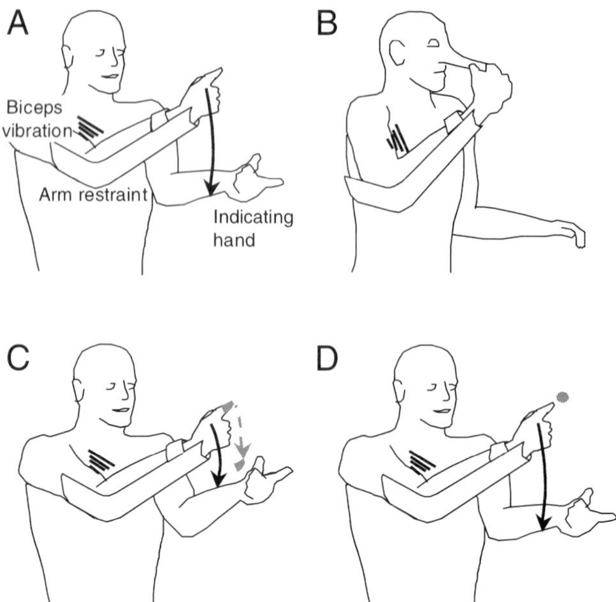


FIG. 38.2. Illusions experienced during vibration of the restrained right biceps in darkness, with different visual cues and tactile contact conditions. A: In complete darkness, biceps vibration elicits illusory extension of the restrained forearm (*solid arrow*), which can be measured if the participant points to the felt position of the hand of the vibrated arm with the other unrestrained, nonvibrated arm. B: In complete darkness, the Pinocchio illusion, elongation of the nose is experienced when it is grasped by the hand of the arm undergoing illusory elbow extension due to biceps vibration. C: When just the index finger of the vibrated right arm is made visible with phosphorescent paint (*gray*) in an otherwise dark room, the participant sees the stationary finger move down (*gray, broken arrow*) and feels the forearm displace less than in complete darkness (*compare black arrow with panel A*). D: A dim target light (*gray dot*) located 1 millimeter away from the fingertip of the restrained vibrated arm looks stationary while the concurrent apparent downward displacement of the arm is as great as in complete darkness during biceps vibration.

either the feet or the mouth. Thus, limb position, body orientation, and object localization are overlapping, interdependent representations, each derived from different combinations of muscle spindle information about a particular appendage and tactile and vestibular signals about spatial orientation. Exposure to VEs will generally not alter muscle spindle signals directly, but often will affect the user's spatial orientation, which will influence the central interpretation of proprioceptive signals, resulting in perceptual and motor errors until adaptation occurs.

## 2.5 Bidirectional Interactions of Visual and Muscle Spindle Influences

Most VEs will have prominent visual displays, so research on the registration of seen versus felt body position and orientation is very relevant. Lackner and Taublieb (1984) assessed this in experiments where they made either the whole hand or a single finger visible in a dark room by application of phosphorescent paint. When the biceps brachii was vibrated, participants felt the unseen forearm move and saw their finger or hand move as well but through a smaller distance (see Fig. 38.2C). The magnitude of felt motion of the forearm was less with the finger visible than in complete darkness, and it was least with the hand visible. In other words, "visual capture" (Hay, Pick, & Ikeda, 1965) was incomplete because the physically stationary visual finger did not prevent participants from seeing and feeling motion of their finger. Proprioceptive capture was substantial, the finger was seen to move nearly as much as the forearm was felt to displace.

If participants attempted to fixate their unseen hand as it underwent illusory downward motion their eyes moved down. However, when they fixated on a visible finger that they perceived as moving down during vibration their gaze remained spatially constant. The character and magnitude of vibration illusions changed in normal illumination conditions where participants could see their hand or finger in relation to the contours of the room, with the remainder of their arm being hidden below a screen. In this case, when the biceps brachii was vibrated, participants felt their unseen forearm move down but did not see the unoccluded finger or hand move. Felt motion of the arm was about 30% of what it had been in the dark room. These results indicate that multiple, interdependent, body-centered and spatially centered representations of hand and arm position exist, which are influenced by the visual and muscle spindle inputs. Seen and felt hand position are dissociable from each other and from motor responses. The strength of the bidirectional visual-proprioceptive influence relates to the amount of the limb that is represented. In VEs, the spatial correspondence of visual and real body positions, the level of detailed visual representation of the body and of the remainder of the virtual visual context will determine the magnitude and nature of proprioceptive and motor errors.

## 2.6 The Role of Tactile Cues in Unifying Muscle Spindle and Other Sensory Influences

When the fingertip is portrayed only by a single point of light or a punctate sound source, there are still bidirectional influences and the role of tactile contact cues is paramount. For example, if in darkness a small target light is taped to the index finger of a participant's spatially fixed arm, the stationary light will appear to displace when the biceps brachii is vibrated, but the felt motion of the arm will be less than if no target were present (Levine & Lackner, 1979). If the two forearms are restrained in a horizontal plane and vibration is applied to the right and left biceps brachii, then the apparent distance between the fingertips will increase. If target lights are attached to the opposing index fingers, participants will see them get further apart as they feel their fingers move apart (DiZio, Lackner, & Lathan, 1993). During vibration, if physical contact of a target light and finger is broken by moving the light a millimeter or more away from the finger, then the illusory visual target movement and displacement will be abolished and the illusory felt movement of the unseen limb augmented (see Fig. 38.2D). In this case, the lights seem to represent external objects instead of the tips of the fingers. During vibration eliciting apparent forearm extension or flexion, a participant will hear a sound source attached to their hand move and change spatial position in keeping with the change in apparent hand position. Breaking tactile contact with the auditory target abolishes its illusory movement during vibration. Taken together with the experiments in which various amounts of the hand were visible, it is clear that the strength of interaction among multimodal representations of self and target position depends on tactile contact cues, visual configural cues, and visual context cues.

These results have important implications for whether VEs should represent body parts with simple visual icons, high-fidelity visual representations, or haptic interfaces. In the natural world, perception and motor control are unified and accurate because there are adequate contextual cues defining how to group multisensory representations and adequate internal calibrations of how these signals should be combined. By contrast, VEs create alternate physical worlds that may require novel combinatorial rules. For example, if adequate visual form cues and tactile contact cues are present for the visual image to be perceptually grouped with the muscle spindle representation of hand position, then the visual and felt locations will influence each other. If either the visual or proprioceptive signal is inaccurate, then both perceptual representations will be biased. However, if only a visual icon is present and there isn't a contact cue with the body, then the visual object will likely be perceived as an external target rather than a representation of the hand, and the spatial perception of each may be independent.

## 3. PROPRIOCEPTIVE ADAPTATION OF THE ARM TO VISUAL DISPLACEMENT

### 3.1 Sensory Rearrangement

Visual distortion is a likely scenario in VEs involving manual tasks, and exposures will likely be prolonged and require users to be active. The retinal image may be degraded or augmented by many VE factors such as the computer model of environmental objects, the technique for tracking the user's visual perspective, the graphical rendering system, and the optics of the display system. Retinal image rearrangement may be caused by inadvertent technological limitations or deliberate augmentation of reality. A typical unplanned distortion is where the visually displayed virtual position of an object does not correspond to its haptic or auditory virtual position because one or more of the display devices is inaccurate. Improper initial alignment and random slippage of a head-mounted display's (HMD's) focal axis relative to the user's optic axis are sources of such unintentional visual inaccuracy (see chap. 3, this volume). A system latency in rendering a user's moving hand in an HMD that occludes the real hand results in a dynamic spatial dissociation of the seen and real hand positions (Ellis, Young, Adelstein, Ehrlich, 1999; Held & Durlach, 1991). A deliberate distortion is introduced, for example, when the visual image is magnified, as in the case of virtual microsurgery systems (Hunter et al., 1994). The laboratory studies reviewed below will illustrate that the type of rearrangement, activities performed, and opportunity for reafferent and exafferent sensory stimulation govern the nature, internal form and specificity of adaptation that occurs with exposure to visual rearrangements. Measuring and understanding the side effects and aftereffects of sensory–motor adaptation requires observing baseline performance before the rearrangement is introduced, the initial and subsequent performance during exposure, and the initial postexposure performance when normal conditions are restored (cf. Held, 1965).

Lateral displacement of the visual world by wedge prisms is an experimental analog of a possible visual rearrangement in a virtual environment. The vast literature on this topic (cf. Howard, 1966; Rock, 1966; Welch, 1978, for reviews) originates from Helmholtz's (1925) demonstration that small azimuthal displacements by prisms of the visual field can be adapted to in a matter of minutes. If a participant looks through base-left wedge prisms without sight of the hand, any object chosen as the target for a reaching movement will appear to the right of its true position, and a movement will be directed accordingly to the right. The prisms will make the hand as it comes into view look like it is moving more to the right than intended, away from the target, and typically participants will steer the hand nearer to the target in midcourse, but the movement endpoint will tend to be right of the target. Participants will see this gap and may also describe with surprise not seeing their arm where they feel it. It takes only 10 to 20 reaches to the same target under these conditions for participants to hit it and to move straight again, but when the prisms are removed they will reach too far to the left initially. Fig. 38.3A illustrates typical reaching paths before, during, and after exposure to wedge prisms. Improvement of performance while the world is viewed through prisms and the aftereffect when they are removed are measures of adaptation. Similar improvement and aftereffects are seen when lateral displacements are purposely introduced in a virtual environment (Groen & Werkhoven, 1998).

### 3.2 The Internal Form of Adaptation

Harris (1963) argued that when one can see only his or her arm and external targets in a featureless background, adaptation is achieved by an internally modified position sense of the arm. The participant eventually feels the arm where he or she sees it. For example, after

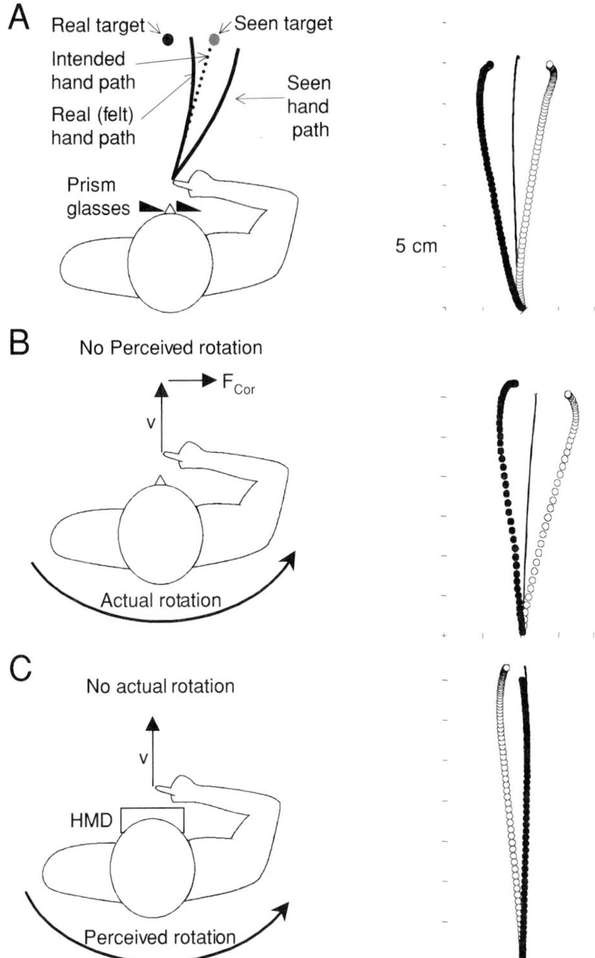


FIG. 38.3. The panels on the left are schematics of different sensory-motor rearrangements that evoke reaching errors, and with continued exposure different forms of adaptation. The panels on the right are plots (*top view*) of finger trajectory before the rearrangement (*solid line*), during the initial reach attempted in the presence of the rearrangement (*open symbols*), and during the initial reach attempted on return to normal conditions (*filled symbols*). A: Wearing prisms that shift a visual target to the right causes the initial reach to deviate to the right and to curve slightly due to corrections made when the hand comes into view. With many attempts, participants learn to make straight accurate reaches (not shown) by modifying their internal representation of felt arm position (see text). Persistence of the adaptation when the prisms are removed causes leftward end-point and curvature errors, mirror symmetric to the initial reaches made during prism exposure. B: When participants are unaware they are rotating, they do not compensate for the Coriolis forces ($F_{Cor}$) their movements (v) generate and initially make reaching errors in the direction of the Coriolis forces (open symbols). With practice they modify their motor commands as necessary to restore their baseline trajectory. When rotation ceases and they again feel stationary, they make leftward end-point and curvature errors (filled symbols). The per and postexposure reaching errors are the same in A and B, although the internal form of adaptation differs. C: Stationary participants feel like they are rotating when viewing a moving visual scene presented in a head mounted visual display (HMD). Their reaching movements deviate in the direction opposite to the Coriolis force, which would be present if they were actually rotating (open symbols). With practice they regain straight accurate movement paths by learning not to compensate for anticipated Coriolis forces. This motor adaptation is specific to the self-rotation context simulated by the HMD and does not carry over to a normal stationary context; consequently, there are no aftereffects when the HMD scene is again stationary (filled symbols).

adaptation to base-right prisms the participant will reach to the target's true position but see and feel the arm to the right of that position. The felt arm position then does not correspond to the real position, but visual and proprioceptive perception are unified and reaching movements are directed accurately to the target.

Harris recognized but rejected other possible forms of adaptation under these conditions. For example, he rejected the possibility that the adaptation is simply a conscious process of reaching to the left of where the target appears because when the prisms were removed his participants showed negative aftereffects instead of reverting to their baseline performance. Second, the adaptation is not an internal change in visual perception. Such a change would mean that a midline target optically shifted right would in the course of adaptation come to be perceived as straight ahead, and when the prisms are removed it would be registered to the left of its true position and any movement aimed at it should deviate to the left. Contrary to this, Harris found that when prisms were removed an adapted participant pointed accurately with the unadapted hand, there was no intermanual transfer. Participants positioned their adapted hand to the left of the unadapted one when asked to align the two hands in azimuth in darkness. Also contradicting a visual shift was the finding that after the prisms were removed participants made comparable errors pointing to a visual target or to an unseen sound source. An alternate explanation to Harris's is motor adaptation. Participants could, for example, change the direction in which they reach to targets without a concurrent change in the felt sense of limb position. Motor adaptation will be discussed in section 4.

The internal form of adaptation is influenced by activities performed and feedback obtained while wearing prisms and how much of one's body and the world is visible. An interesting case is walking around but keeping the arm out of sight while wearing rightward displacing prisms (Held & Bossom, 1961). At first, participants bump into things but are soon able to get around. When the prisms are removed the participants initially point incorrectly, leftward, with both arms to visual targets. They also adopt a resting head posture that is deviated in the direction of the optical displacement. These facts have been interpreted as meaning that adaptation in this context is achieved by an internal recalibration of perceived head position on the torso (Harris, 1965). In the paradigm where participants just sat and reached with one arm, only that arm had to be remapped to compensate. This illustrates that the nervous system tends to adapt in the way most specific to the conditions encountered during exposure to rearrangement. More extensive reviews on this topic are available (Dolezal, 1982; Lackner, 1981).

In the two cases presented above, the sensory rearrangement is the same but the resulting form of internal adaptive shift differs. There can be an arm–torso recalibration when the participant just sits and points to a single target in a featureless field and a head–torso recalibration when the participant walks about in the normal world. This means that detecting and understanding VE side effects and aftereffects must take into account the physical distortions introduce by the VE, the nature of the visual field, and the tasks required of the user. There is enough evidence to indicate that the form of adaptation to a sensorimotor rearrangement depends on fine details of task characteristics, but not enough to make accurate predictions for every case imaginable in practical virtual environments. Several examples not explicitly covered by the reviewed studies are VEs where the visual position of the user's hand and the whole visual scene are displaced, VEs where the virtual visual objects have real or virtual counterparts, and VEs where the hand is represented as a symbolic visual icon instead of realistically.

## 3.3 Conditions Necessary for Adaptation to Occur

Understanding what factors are necessary for adaptation to take place is important for predicting proprioceptive side effects and aftereffects from VE exposure. Held and colleagues (Held, 1965;

Held & Bossom, 1961) showed that active reaching or locomotory movements during visual rearrangement could elicit adaptation, but a laterally displaced view of one's passively moved arm or sight of the world during passive transport did not generate adaptation. This led to the idea that efference copy signals are required for adaptation. Held developed a model in which efference copy signals were stored along with their correlated reafferent sensory signals so that any active movement would call up the normally associated visual reafferent signal for comparison with the actual one. In the reaching and locomotion paradigms, discrepancies between the reactivated and current reafferent visual patterns were thought necessary for adaptation to occur. A practical implication is that greater activity within a VE should speed up adaptation.

An alternative point of view is that active movements are only superior because they enhance proprioception (Paillard & Brouchon, 1968). Consistent with the idea that active movement is not necessary, Wallach, Kravitz, and Lindauer (1963) demonstrated that partial adaptation of reaching movements to visually displaced targets occurs if an immobile observer simply views the rest of his body through displacing prisms for ten minutes. Lackner (1974) went further, showing that a discrepancy between actual visual feedback and the usual feedback associated with a voluntary movement is not sufficient for adaptation to occur, but a visual-proprioceptive discrepancy is necessary and sufficient for eliciting adaptation to laterally displacing prisms. When participants reached without sight of their arm to visually displaced dowels and contacted hidden, vertically aligned, similarly shaped extensions of these targets there was no visual feedback about the arm to compare with the visual feedback normally associated with the executed movement. Adaptation still occurred in this condition because there was a discrepancy between the visual and felt target position. Adaptation did not occur if the visual-proprioceptive discrepancy was eliminated by having participants point to the same array of prism-displaced dowels whose hidden lower halves were laterally offset to match the optical displacement. The results emphasize the importance for adaptation of sensory discordance as well as active control. They also demonstrate that adaptation is enhanced by establishing through fingertip contact an association between the unseen arm and visual targets. In VE systems, finger contact can contribute to the fidelity of the synthetic experience if vision and haptic interfaces are in register, or create side effects and aftereffects if they are discrepant.

### 3.4 Retention and Specificity of Adaptation

To assess retention of adaptation of reaching errors caused by laterally displacing prisms, Lackner and Lobovitz (1977) had participants participate in two adaptation sessions 24 hours apart. Each session had pre- and postexposure periods of reaching to virtual visual targets without sight of the arm and a prism exposure period in which participants reached to the same loci with their arm in view. A surprising finding was that in the preexposure period of the second session, reaches were deviated in the direction of the aftereffects from the previous day, although none of the participants reported any difficulty with visuomotor control between sessions outside the laboratory. This demonstrated a long-term aftereffect and raised the possibility that it was a context-specific aftereffect. To further evaluate retention and context specificity, Yachzel and Lackner (1977) gave participants six adaptation sessions over a 4-week period. The sessions were similar in that participants pointed to visual targets with their arm under a screen before and after exposure to visual displacement in which they reached with sight of their arm. Tests were conducted in two different sets of apparatus requiring different arm movements. Retention of adaptation from session to session was seen in the form of preexposure baseline shifts across the first five sessions, which were spaced two or three days apart. The size of the long-term aftereffect did not depend on what apparatus the participant was tested in. Aftereffects appeared when the participants were tested without being able to see their arm and never were noticed during daily activities. Diminished aftereffects were still evident in the sixth session, which was delayed for two weeks relative to the fifth. These results indicate that proprioceptive adaptation to displaced vision can be retained for long periods and are not necessarily context-specific. The adaptation retained does not reveal itself in daily activities unless visual cues about the arm are reduced. Long term-after effects from simulator and VE exposure are likely caused by long-term retention of adaptation, but more research is required to understand what contexts will evoke aftereffects and what sensory cues will hold them at bay.

## 4. ADAPTATION TO ALTERED FORCE BACKGROUNDS

The notion of motor adaptation introduced in section 3.2 is crucial for understanding side effects and aftereffects in real and virtual environments involving novel force backgrounds. Motion-coupled VEs rearrange the external force environment. For example, in VEs involving real body motion or visual portrayal of body motion, the background gravitoinertial forces on the body are different from the forces that would be present if the body were physically moving in the experienced fashion. This is a situation where motor adaptation occurs and individuals need to reprogram their limb movements to achieve accurate control for the current force background. This can produce aftereffects when the individual leaves the virtual environment. The situation is very different when a novel real or virtual tool or machine is introduced. In this case, individuals make errors until they learn the properties of the manipulated device. Such tasks do not lead to aftereffects outside the specific context of that device.

### 4.1 Motor Adaptation to Coriolis Force Perturbations in a Rotating Room

Traditional vehicle motion simulators often have motion bases to try to mimic features of the gravitoinertial force backgrounds of moving vehicles. For example, a rotating room generates centripetal acceleration that simulates artificial gravity, but it also generates a Coriolis force on any object moving nonparallel to the spin axis of the room. The Coriolis force is only present when an object is moving in relation to the rotating room and acts perpendicular to the direction of object motion, according to the cross-product rule: $F_{cor} = -2m(\omega \times v)$, where $m$ and $v$ are the mass and linear velocity, respectively, of the object and $\omega$ the angular velocity of the room.

When an occupant of a room rotating counterclockwise reaches forward, a rightward Coriolis force is generated on the arm deviating it rightward. Both the movement's path and end point are displaced relative to the prerotation pattern of straight movements directly to the target. Subsequent reaches during rotation return quasi-exponentially, within 10 to 20 trials, to prerotation straightness and accuracy. A new set of motor commands is being issued in order to move the hand straight to the target. When rotation stops and Coriolis forces are absent, reaching movements show curvature and endpoint deviations in the direction opposite those of the initial perturbed movements (Lackner & DiZio, 1994).

This pattern of pre-, per-, and postrotation movements resembles the pre-, per-, and post-exposure phases of a prism displacement experiment (compare Figs. 38.3A and 38.3B), but there are important differences. Adaptation to rotation involves motor remapping instead of the proprioceptive shifts that underlie some forms of adaptation to prism spectacles. Adaptation to rightward prism displacement can make movements that go straight ahead to a midline target feel like they are going rightward, and it may cause participants to make errors aligning the adapted hand with the unadapted one if it is hidden. By contrast, the true trajectory of the arm is experienced throughout the pre-, per-, and postexposure phases of adaptation to rotation.

Participants who adapt to reaching with one arm during rotation also can accurately align their left and right fingertips.

### 4.2 Motor Adaptation and Force Perception

Another special feature of motor adaptation to rotation is the alteration of force perception that accompanies it. Participants report their first movement during rotation as being deviated by a magnetic-like pull in the direction of the Coriolis force, whereas no unusual force is perceived during movements whose visual paths are perturbed by prisms. When participants adapt fully to rotation, they no longer can feel the Coriolis forces that are still present during their movements, even if their attention is called to them. They report that whatever had initially perturbed their arm is now gone and they can produce the desired movement with the same effort as before the perturbation. When rotation stops and there are no Coriolis forces during movements, participants report feeling a force on their arm that is the mirror image of the Coriolis force they had adapted to. The relationship of actual to perceived force is modified during adaptation because of an internal calibration mechanism that interrelates force feedback signals, position, and velocity signals and motor commands. Informal observations clearly show that adaptation to rotation alters the feel and control of objects and surfaces that are handled following return to a normal stationary environment. Virtual environments that alter the force environment or that involve visual simulation of body acceleration without the normal concomitant gravitoinertial forces will cause illusions in the feel of tools and affect the manual control of a vehicle or other machines.

### 4.3 Context-specific Motor Adaptation and Aftereffects in Different Force Environments

Participants who sit quietly in an enclosed rotating room turning at a constant velocity feel after about 60 seconds like they are in a completely normal, stationary environment. This is because the angular velocity sensitive semicircular canals have had time to return to their resting discharge level after acceleration to constant velocity, and the room is fully enclosed so there are no visual flow cues about rotation. Thus, participants who adapt their reaching movements in this situation learn to associate making Coriolis force compensations and receiving Coriolis force feedback with an internally registered nonrotating context. These recalibrations are carried over to the postrotation period, which also is internally registered as being a normal stationary environment, and cause aftereffects. If after adaptation to rotation, the rotating room instead of being stopped is accelerated to twice the initial speed, then after a minute the participant will again feel stationary and when reaching for the first time will make renewed endpoint and curvature errors of the same magnitude and in the same direction as the first movements at the previous speed (Lackner & DiZio, 1995). The Coriolis forces are greater at the new speed, but the motor mapping from the lower speed is carried over because the internally registered nonrotating state is the same.

Understanding whether adaptation to one rotation speed will carry over to a different speed requires the converse of the conditions provided by the rotating room where participants are rotating but feel stationary. Participants who are actually stationary can experience virtual rotation by viewing a moving visual scene in a head-mounted display. The first reaches of participants experiencing constant velocity counterclockwise rotation and displacement will be deviated leftward in path and end point (see Fig. 38.3C; Cohn, DiZio, & Lackner, 2000). The magnitude of the reaching errors is proportional to the perceived speed of self-rotation. If the participants had actually been rotating counterclockwise there would have been a rightward Coriolis force when they reached. Their leftward errors show that they had anticipated and generated muscle forces to resist the expected but absent Coriolis forces. These participants also report feeling a phantom leftward force on their arm, which has the characteristic bell-shaped profile of a Coriolis force. When repeated reaches are made during virtual rotation, the reaches become straighter and progressively more accurate, and the "force" perceived to be deviating the arm vanishes. When the visual scene is again stationary and participants feel stationary, their first postexposure reaches feel normal and go straight to a target. That is, there is no aftereffect.

These results show that the motor plan for a forthcoming reaching movement compensates for the Coriolis forces normally generated at the currently registered speed of self-rotation. In other words, we normally maintain multiple motor adaptation states that are context-specific for registered body motion. Experimental alteration of the relationship between Coriolis forces and registered body speed alters the motor compensation only for the body rotation speed at which movements are practiced and is not carried over to other speeds. Thus motor adaptation is specific to the registered context of body motion where aberrant forces were experienced. Not enough is known to allow prediction of how faithfully a moving base training simulator must reproduce an operational environment in order for training to transfer. Virtual training environments that simulate body motion with just a dynamic visual scene are unlikely to cause motor aftereffects in everyday life, but they may well produce adaptation in which participants cease to compensate or anti-compensate in terms of postural and movement control for inertial forces that will actually be present in the operational moving environment. That is, they could maladapt participants to the operational context.

## 4.4 Motor Adaptation to Environments Versus Local Contexts

The rotating room and virtual rotation provide environmental contexts in which individuals carry out all their actions. Real and virtual tools and machines create force-reflected feedback or force fields that are local contexts a user may interact with in a limited fashion. For example, humans can use a planar robotic linkage to control a cursor on a video screen in the presence of external forces generated by torque motors on the handle. Shadmehr, Mussa-Ivaldi, and Bizzi (1993) created a force field that resembles the Coriolis force field in a rotating room, such that the manipulandum pushed the hand perpendicular to its current velocity. Their participants' movements were initially perturbed, but after many hundreds of trials straight paths ending at the target characteristic of movements with a null force field were regained. This contrasts sharply with the complete adaptation participants achieve in 10 to 20 movements to Coriolis force perturbations of their reaching movements in a rotating room. Thus, learning an internal model of an electromechanical device takes much longer than recalibrating one's own unfettered movements. The nature of force perception also differs sharply during self-calibration versus internalization of a machine's force properties. As described above, perturbing Coriolis forces become perceptually transparent when motor adaptation of free reaching movements to rotation is complete. However, participants interacting with a manipulandum can still detect and describe the forces it applies even after they have learned to resist them and to move their arm in the desired path to the target. The implication of this is that learning a local force field by interacting with a real or virtual machine will not cause aftereffects in any context outside the confines of the machine. By contrast, learning to move one's arm will affect perception and performance in any local context embedded in an environment that provides the same stimuli about self-motion as the environment in which adaptation was acquired.

## 4.5 The Role of Touch Cues in Motor and Proprioceptive Adaptation

Cutaneous contact cues during a movement and at the terminus of the movement contribute to perception and adaptation of limb position and force. Continuous cutaneous contact cues throughout movements are prominent differences between Coriolis force perturbations in the rotating room and force field perturbations produced by a manipulandum. When a manipulandum perturbs a movement the muscle spindles, Golgi tendon organs, and vision can signal that the movement deviated from the intended path, and the cutaneous mechanoreceptors of the hand signal the continuous presence of a local external force. There is a systematic correlation of the cutaneous force profile and the compensatory muscular forces that need to be learned. The nervous system, following the principle of making the most parsimonious change possible, can simply learn the dynamics of the perturbing object. In adaptation to Coriolis force perturbations, there is no local cutaneous stimulation because the Coriolis force is a noncontacting inertial force applied to every moving particle of the limb. In the absence of an external agent contacting the limb, the most specific form of adaptation is an alteration in motor control of the exposed limb. This is because the central nervous system recognizes a situation requiring motor recalibration when it detects an error in movement path without an external obstruction while the movement is in progress.

The transient cutaneous stimuli occurring when the finger lands on a surface at the end of a reaching movement also influence proprioception and motor adaptation in real and virtual environments. Coriolis force induced end-point and curvature deviations of visually open-loop reaching movements in a rotating room are eliminated by adaptation within 10 to 15 movements if the reaches end on a smooth surface. However, only 50% of the initial end point error is eliminated before performance reaches a steady asymptote if the reaches end with the finger in the air. In other words, if terminal contact is denied participants will not fully adapt. This pattern shows that information obtained from finger touchdown provides information about limb position errors, which is critical for adaptation to rotation. The source of this information is the direction of shear forces generated in the first 30 ms after a reaching movement contacts a smooth surface. These shear forces are systematically mapped to the location of the finger and code where the finger is relative to the body (DiZio, Landman, & Lackner, 1999). These shear forces are about 1 N in magnitude. When participants reach in the air to objects on a virtual instrument panel their end-point variability is greater than when reaching to a similar real panel (DiZio & Lackner, forthcoming). The variability is greatest if the virtual panel is presented only in the visual mode and the reaches end in midair. It is greatly reduced by the addition of a real contourless surface in the same spatial plane as the visual virtual panel (see Fig. 38.4). Thus, minimal haptic cues, just a flat surface in the proper plane, can improve proprioception, reduce movement errors and enhance the usability of VE visual interfaces.

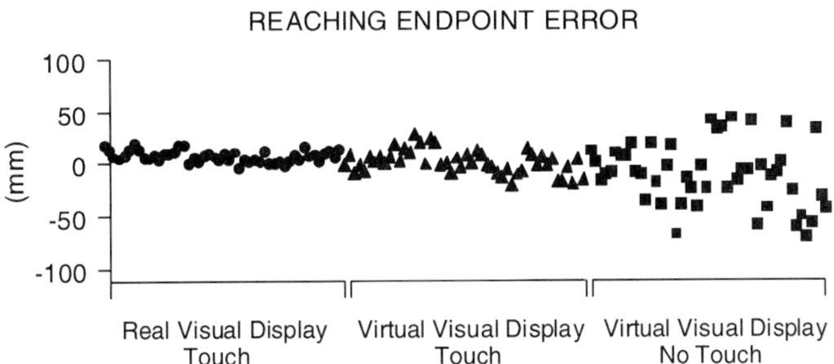


FIG. 38.4. Plot of end-point errors of repeated reaches made without sight of the arm to a single target on a real or virtual horizontal work surface. Variability is least in the first set of movements, where the target is a light-emitting diode (LED) embedded in a smooth sheet of Plexiglas with no distinguishing marks indicating the target location. Variability is greater in the second set of movements, which are aimed at a virtual target on a virtual surface programmed to coincide with the spatial location of a real Plexiglas work surface which the participant contacts at the end of each movement. The same virtual target and surface are presented without the real surface in the last set of reaches, and the variability increases markedly in the absence of physical contact. Reintroducing the real surface (not shown) restores low variability immediately.

## 5. PERCEPTION OF LIMB AND BODY MOTION DURING POSTURE AND LOCOMOTION

Control of posture and movement are highly integrative, multimodal processes, requiring sensitivity to complex environmental constraints. Touch cues play a major role in control and perception of whole body movement, the apparent stability of the environment, and attributions of causality.

### 5.1 Touch Stabilization of Posture

Fingertip contact with environmental surfaces has a stabilizing influence on standing posture. If participants standing on one foot or heel-to-toe in darkness hold their index finger on a stable surface with a force of about 0.4 N (about 41 grams), their body sway amplitude is cut in half relative to not touching (Holden, Ventura, & Lackner, 1994; Jeka & Lackner, 1994). This level of force is too low to provide mechanical stabilization, but it corresponds to the maximum sensitivity range of fingertip sensory receptors (Westling & Johansson, 1987). The horizontal and vertical forces at the fingertip fluctuate with a correlation of 0.6 to both body sway and ankle EMG activity, but lead EMG by about 150 ms and body sway by 250 to 300 ms. Light touch even attenuates body sway when the eyes are open. The stabilizing influence of touch is lost if sensory–motor control of the arm is disrupted by vibration of brachial muscles, but with the arm functioning normally touch can stabilize the body even when excessive sway is induced by ankle muscle vibration (Lackner, Rabin, & DiZio, 2000). Patients with no vestibular function who cannot stand in darkness for more than a few seconds can maintain stance for as long as desired and are as stable as control participants with eyes closed if allowed light touch of the finger (Lackner et al., 1999). These findings indicate that the finger–arm system functions as an active proprioceptive probe providing sensory information about body position and velocity. This stabilization system works with nested sensory–motor loops. A finger to brachial sensory–motor loop stabilizes the finger relative to the surface and minimizes the force changes at the fingertip. These residual fingertip force fluctuations activate the leg muscle to stabilize posture better than is possible with ankle proprioception, vision, or vestibular signals alone.

Light contact cues with a simple surface are an effective way to stabilize body sway in VEs where participants are free standing. An HMD and head-tracking system that introduces temporal distortions of visual feedback when head movements are made induces severe postural instability, and light touch suppresses this side effect (DiZio & Lackner, 1997). Postural aftereffects also occur upon return to a normal environment resulting in further postural disruption. Light touch stabilizes posture and suppresses aftereffects in such situations.

### 5.2 Touch and Locomotion

A seated individual holding his or her hands or feet in contact with a rotating floor or railing will experience self-rotation in the opposite direction (Brandt, Buchele, & Arnold, 1977; Lackner & DiZio, 1984). Actively pedaling the free wheeling floor while seated or turning the railing with a hand-over-hand motion makes the experience very powerful. Pedaling movements made without surface contact do not elicit any experience of self-motion. These demonstrations illustrate that in normal bipedal locomotion contact with the floor is very important.

The pattern of ground reaction forces during locomotion is not related to body motion through space in a univariant way. For example, walking down a hill, one is progressing forward but pushing backward. An experimental situation used by Lackner and DiZio (1988) to assess the interactions among surface contact, leg movements, and whole body motion is illustrated

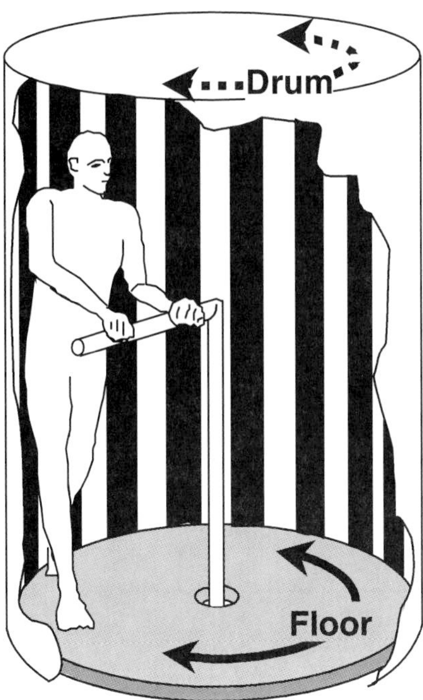


FIG. 38.5. Schematic of the apparatus used to uncouple visual motion, inertial motion, and overground motion during treadmill locomotion.

in Fig. 38.5. It is basically a VE in which the visual display and the substrate of support can be independently manipulated. A participant holding onto the world-fixed handlebar and walking in place on the backward rotating floor experiences forward motion through space. The experience is compelling and immediate if the chamber walls also are rotated backward, but it occurs in darkness (Bles & Kapteyn, 1977) and even if the stationary walls are in plain view. In the latter case, the motor and somatosensory signals from walking in place capture the visual scene, which will appear to be dragged along (Lackner & DiZio, 1988).

If the visual scene is moved backward twice as fast as the floor, a participant who is walking in place will feel the handlebar pulling him forward and sense an increase in body velocity through space. Some participants report an illusory elongation of each leg at the toe-off phase of the step cycle, and others sense that the floor has become rubbery and bounces them forward. If the visual scene motion is reversed, participants stepping forward in place will report either that they are moving backward through space and making backward stepping movements or the steps, which usually propel them forward are now propelling them backward. These effects powerfully demonstrate that afferent signals about limb movements, the body schema representation, internal calibration of body motion, and apparent environmental stability are all interrelated.

## 5.3 Proprioceptive Recalibration of Locomotion

These relationships can be recalibrated by walking in place on a circular treadmill for an extended period. After doing so for an hour in a visually normal environment, participants walk in a curved arc when trying to maintain a straight line on solid ground in darkness (Gordon, Fletcher, Melville-Jones, & Block, 1995). When passively pushed in a wheelchair, eyes closed, they can differentiate straight and curved paths as well as they can before the treadmill exposure, indicating vestibular function is not changed. These investigators concluded that the aftereffects are due to adaptive recalibration of a "podokinetic system," which through ground contact, leg proprioception, and motor copy signals provides a representation of trunk rotation relative to the stance foot. When forced not to turn by guide rails after prolonged stepping around in place, participants feel they are turning in the other direction, and they exhibit nystagmoid eye movements consistent with their perceived direction and rate of turning, not their actual body motion (Weber, Fletcher, Melville-Jones, & Block, 1997). Allowed visual feedback, the participants can walk in their desired path with no ocular nystagmus. This is an example where potentially dangerous aftereffects due to internal recalibration of proprioception in the exposure environment can be masked by the appropriate sensory information.

## 6. CONCLUSIONS

Many VEs will intentionally or inadvertently introduce users to sensorimotor rearrangements that will result in side effects, followed by adaptation, and aftereffects or maladaptive transfer on return to a normal environment. Proprioceptive adaptation is one form of adaptation associated with potential VE side effects and aftereffects. Two important facets of proprioception that can undergo adaptive modification are the sense of position and orientation and the sense of force or effort. Predicting and measuring the manifestations and underlying form of adaptation require understanding that it is not a unitary phenomenon. Proprioception involves the interplay of afferent and efferent signals about limb and body position and motion, internal representations of the body schema, and spatial orientation and representations of environmental constraints. All of these variables are labile and can undergo long-term adaptive changes in relation to the others. Reaching errors in a VE that look similar on the surface can have different causes that lead to diverse forms of adaptation and aftereffects. For example, virtual visual displacements, mismatches of the virtual gravitoinertial force environment, and novel contact force fields cause adaptation of proprioception, of motor control, and of internal models of objects, respectively. Understanding the cause and form of adaptation in a specific VE context will help predict ways of enhancing adaptation, for example, when more user activity versus sharpening the sensory discordance will help. Principles governing the specificity of adaptation are key for understanding what aftereffects will occur when a user leaves the VE and either goes about daily life or engages in the real-world operational task that was simulated. Such principles are scarce, but one important factor is feedback from contact cues during movement. Continuous and terminal cutaneous signals contribute to neural computations that partition the net force environment into functionally relevant components. For example, in adaptation of reaching movements, cutaneous signals are critical for determining whether the motor system will recalibrate or the representation of a tool's properties will be updated. Motor adaptation in a VE will carry over to any task performed with the exposed limb, but learning an internal model of a device will not deleteriously affect performance on dissimilar devices. Touch cues and visual cues can also mask potentially dangerous long-term aftereffects that appear in impoverished conditions or specific contexts.

## 7. ACKNOWLEDGMENTS

This work was supported by NAWCTSD contracts N61339-96-C-0026 and NASA grants NAG9-1037 and NAG9-1038.

## 8. REFERENCES


Aristotle. (1978). *De motu animalium* (pp. 38–42), M. C. Nussbaum, Trans. Princeton, NS: Princeton University Press.

Bles, W., & Kapteyn, T. (1977). Circular vection and human posture: 1. Does the proprioceptive system play a role? *Agressologie, 18*, 325–328.

Brandt, T., Buchele, W., & Arnold, F. (1977). Arthrokinetic nystagmus and ego-motion sensation. *Experimental Brain Research, 30*, 331–338.

Burgess, P. R., & Wei, J. Y. (1982). Signaling of kinesthetic information by peripheral sensory receptors. *Annual Review of Neuroscience, 5*, 171–187.

Cohn, J., DiZio, P., & Lackner, J. R. (2000). Reaching during virtual rotation: Context-specific compensation for expected Coriolis forces. *Journal of Neurophysiology, 83*, 3230–3240.

Craske, B. (1977). Perception of impossible limb positions induced by tendon vibration. *Science, 196*, 71–73.

DiZio, P., & Lackner, J. R. (1997). Circumventing side effects of immersive virtual environments. In M. J. Smith, G. Salvendy, & R. J. Koubek (Eds.), *Advances in human factors/ergonomics: Vol. 21. Design of computing systems* (pp. 893–397). Amsterdam: Elsevier Science Publishers.

DiZio, P., & Lackner, J. R. (forthcoming). Minimal haptic cues enable finer motor resolution during interaction with visual virtual objects.

DiZio, P., Lackner, J. R., & Lathan, C. E. (1993). The role of brachial muscle spindle signals in assignment of visual direction. *Journal of Neurophysiology, 70*(4), 1578–1584.

DiZio, P., Landman, N., & Lackner, J. R. (1999). Fingertip contact forces map reaching endpoint. *Society for Neuroscience Abstracts, 25*, 760.15.

Dolezal, H. (1982). *Living in a world transformed: Perceptual and performatory adaptation to visual distortion*. New York: Academic Press.

Eklund, G. (1972). Position sense and state of contraction: the effects of vibration. *Journal of Neurology Neurosurgical Psychiatry, 35*, 606–611.

Ellis, S. R., Young, M. J., Adelstein B. D., & Ehrlich, S. M. (1999). Discrimination of changes of latency during voluntary hand movement of virtual objects. In *Proceedings of the Human Factors and Ergonomics Society*. Houston, TX: Human Factors and Ergonomics Society.

Fisk, J., Lacker, J. R., & DiZio, P. (1993). Gravitoiner tial force level influences arm movement control. *Journal of Neurophysiology, 69*, 504–511.

Goodwin, G. M., McCloskey, D. I., & Matthews, P. B. C. (1972). The contribution of muscle afferents to kinesthesia shown by vibration induced illusions of movement and by the effects of paralysing joint afferents. *Brain, 95*, 705–748.

Gordon, C. R., Fletcher, W. A., Melvill Jones, G., & Block, E. W. (1995). Adaptive plasticity in the control of locomotor trajectory. *Experimental Brain Research, 102*, 540–545.

Groen, J., & Werkhoven, P. J. (1998). Visuomotor adaptation to virtual hand position in interactive virtual environments. *Presence, 7*, 429–446.

Hagbarth, K. E., & Eklund, G. (1966). Motor effects of vibratory muscle stimuli in man. In R. Granit (Ed.), *Muscular afferents and motor control* (pp. 177–186). Stockholm: Almqvist & Wiksell.

Harris, C. S. (1963). Adaptation to displaced vision: Visual, motor, or proprioceptive change? *Science, 140*, 812–813.

Harris, C. S. (1965). Perceptual adaptation to inverted, reversed, and displaced vision. *Psychological Review, 72*, 419–444.

Harvey, R. J., & Matthews, P. B. C. (1961). The response of de-efferented muscle spindle endings in the cat's soleus to slow extension in the muscle. *Journal of Physiology, 157*, 370–392.

Hay, J. C., Pick, H. L., & Ikeda, K. (1965). Visual capture produced by prism spectacles. *Psychonomic Science, 2*, 215–216.

Held, R. (1965). Plasticity in sensory-motor systems. *Scientific American, 213*, 84–94.

Held, R., & Bossom, J. (1961). Neonatal deprivation and adult rearrangement: Complementary techniques for analyzing plastic sensory-motor coordinations. *Journal of Comparative and Physiological Psychology, 54*, 33–37.

Held, R., & Durlach, N. (1991). Telepresence, time delay and adaptation. In S. R. Ellis, M. K. Kaiser, & A. J. Grunwald (Eds.), *Pictorial communication in virtual and real environments* (2nd ed., pp. 232–246). London: Taylor & Francis.

Helmholtz, H. (1925). *Helmholtz's treatise on physiological optics* (3rd ed., J. P. C. Southall. Ed.). Menasha, WI: Optical Society of America.

Henderson, W. R., & Smyth, G. E. (1948). Phantom limbs. *Journal or Neurology, Neurosurgery and Psychiatry, 11*, 88–112.

Holden, M., Ventura, J., & Lackner, J. R. (1994). Stabilization of posture by precision contact of the index finger. *Journal of Vestibular Research, 4*(4), 285–301.

Holst, E. von, & Mittelsteadt, H. (1950). Das reaffferenz-prinzip. *Die Naturwissenschaften, 20*, 464–467.

Howard, I. P. (1966). *Human spatial orientation*. London: John Wiley & Sons.

Hunter, I., Doukoglou, T. D., Lafontaine, S. R., Charrette, P. G., Jones, L. A. , Sagar, M. A., Mallinson, G. D., & Hunter, P. J. (1994). A teleoperated microsurgical robot and associated virtual environment for eye surgery. *Presence, 2*, 265–280.

Jeka, J. J., & Lackner, J. R. (1994). Fingertip contact influences human postural control. *Experimental Brain Research, 100*(3), 495–502.

Kennedy, R. S., Fowlkes, J. E., & Lilienthal, M. G. (1993). Postural and performance changes following exposures to flight simulators. *Aviation, Space, and Environmental Medicine, 64*, 912–920.

Lackner, J. R. (1974). Adaptation to displaced vision: Role of proprioception. *Perceptual and Motor Skills, 38*, 1251–1256.

Lackner, J. R. (1976). Influence of abnormal postural and sensory conditions on human sensory motor localization. *Environmental Biology and Medicine, 2*, 139–177.

Lackner, J. R. (1981). Some aspects of sensory-motor control and adaptation in man. In R. D. Walk & H. L. Pick (Eds.), *Intersensory perception and sensory integration* (pp. 143–173), New York: Plenum Press.

Lackner, J. R. (1988). Some proprioceptive influences on the perceptual representation of body shape and orientation. *Brain, 111*, 281–297.

Lackner, J. R., & DiZio, P. (1995). Generalization of adaptation to Coriolis force perturbations of reaching movements. *Society for Neuroscience Abstracts, 23*,

Lackner, J. R., & DiZio, P. (1994). Rapid adaptation to Coriolis force perturbations of arm trajectory. *Journal of Neurophysiology, 72*, 299–313.

Lackner, J. R., & DiZio, P. (1984). Some efferent and somatosensory influences on body orientation and oculomotor control. In L. Spillman & B. R. Wooten (Eds.), *Sensory experience, adaptation, and perception* (pp. 281–301). Clifton, NJ: Lawrence Erlbaum Associates.

Lackner, J. R., & DiZio, P. (1988). Visual stimulation affects the perception of voluntary leg movements during walking. *Perception, 17*, 71–80.

Lackner, J. R., DiZio, P., Jeka, J. J., Horak, F., Krebs, D., & Rabin, E. (1999). Precision contact of the fingertip reduces postural sway of individuals with bilateral vestibular loss. *Experimental Brain Research, 126*, 459–466.

Lackner, J. R., & Levine, M. S. (1979). Changes in apparent body orientation and sensory localization induced by vibration of postural muscles: Vibratory myesthetic illusions. *Aviation Space and Environmental Medicine, 50*, 346–354.

Lackner, J. R., & Lobovits, D. (1977). Adaptation to displaced vision: Evidence for prolonged aftereffects. *Quarterly Journal of Experimental Psychology, 29*, 65–69.

Lackner, J. R., Rabin, E., & DiZio, P. (2000). Fingertip contact suppresses the destabilizing influence of tonic vibration reflexes in postural muscles. *Journal of Neurophysiology, 84*, 2217–2224.

Lackner, J. R., & Taublieb, A. B. (1984). Influence of vision on vibration-induced illusions of limb movement. *Experimental Neurology, 85*, 97–106.

Lazlo, J. I. (1963). The performance of a simple motor task with kinaesthetic loss. *Quarterly Journal of Experimental Psychology, 18*, 1–8.

Levine, M. S., & Lackner, J. R. (1979). Some sensory and motor factors influencing the control and appreciation of eye and limb position. *Experimental Brain Research, 36*, 275–283.

Matin, L., Picoult, E., Stevens, J. K., Edwards, M. W., Young, D., & MacArthur, R. (1972). Oculoparalytic illusion: Visual-field dependent spatial mislocalizations by humans partially paralyzed with curare. *Science, 216*, 198–201.

McCloskey, D. I., & Torda T. A. G. (1975). Corollary motor discharges and kinesthesia. *Brain Research, 100*, 467–470.

Melzack, R., & Bromage, P. R. (1973). Experimental phantom limbs. *Experimental Neurology, 39*, 261–269.

Paillard, J., & Brouchon, M. (1968). Active and passive movements in the calibration of position sense. In S. J. Freedman (Ed.), *The neuropsychology of spatially oriented behavior*. Homewood, IL: Dorsey Press.

Rock, I. (1966). *The nature of perceptual adaptation*. New York: Basic Books.

Shadmehr, R., Mussa-Ivaldi, F. A., & Bizzi, E. (1993). Postural force fields of the human arm and their role in generating multi-joint movements. *Journal of Neuroscience, 13*, 45–62.

Sherrington, C. S. (1906). *The integrative action of the nervous system*. London: Charles Scribner's Sons. (Reprinted 1948, Cambridge: University Press).

Sperry, R. W. (1950). Neural basis of the spontaneous optokinetic response produced by visual neural inversion. *Journal of Comparative and Physiological Psychology, 43*, 482–489.

Wallach, H. (1976). *On perception*. New York: Quadrangle.

Wallach, H., Kravitz, J. H., & Lindauer, J. (1963). A passive condition for rapid adaptation to displaced visual direction. *American Journal of Psychology, 76*, 568–578.

Weber, K. D., Fletcher, W. A., Melvill Jones, G., & Block, E. W. (1997). Oculomotor response to rotational podokinetic

(PK) stimulation and its interaction with concurrent VOR in normal and PK adapted states. *Society for Neuroscience Abstracts, 23*(1), 470.

Welch, R. B. (1978). *Perceptual modification: Adapting to altered sensory environments*. New York: Academic Press.

Wilson, V. J., & Melvill Jones, G. (1979). *Mammalian vestibular physiology*. New York: Plenum Press.

Westling, G., & Johansson, R. S. (1987). Responses in glabrous skin mechanoreceptors during precision grip in humans. *Exprimental Brain Research, 66*, 128–140.

Yachzel, B., & Lackner, J. R. (1977). Adaptation to displaced vision: Evidence for transfer of adaptation and long-lasting aftereffects. *Perception and Psychophysics, 22*, 147–151.

# 39

# Vestibular Adaptation and Aftereffects

Thomas A. Stoffregen,[1] Mark H. Draper,[2]
Robert S. Kennedy,[3] and Daniel Compton[3]
*[1]University of Minnesota*
*Division of Kinesiology, 110 Cooke Hall*
*Minneapolis MN 55455*
*tas@umn.edu*
*[2]Air Force Research Laboratory, 2255 H Street*
*Wright-Patterson AFB, OH 45433*
*Mark.Draper@wpafb.af.mil*
*[3]RSK Assessments, Inc.*
*1040 Woodcock Road, Suite 227*
*Orlando, FL 32803*
*kennedyrs@yahoo.com*
*compton_nad@yahoo.com*

## 1. INTRODUCTION

Virtual environments (VEs) are altered environments that provide novel patterns of intersensory stimulation. The novelty may be correct (i.e., if the VE corresponds accurately to a real environment that is new to the user), or incorrect (i.e., if the VE presents perceptual stimulation that is not faithful to the simulated environment; Moroney & Moroney, 1998; Riccio, 1995; Stoffregen, Bardy, Smart, & Pagulayan, in press; cf. Stappers, Overbeeke, & Gaver, in press). In either case, adaptation is the natural response to VE systems (Stanney, Kennedy, Drexler, & Harm, 1999; see chap. 31, this volume). The majority of research interest is focused on VE-related patterns of stimulation that are not faithful to the simulated environment. Due to technological limitations, such as slow system update rates, sluggish and imprecise position-tracking sensors, and imperfect models of environmental dynamics, the user is exposed to unusual patterns of intersensory stimulation (e.g., a disparity between seen and felt limb position). The change may not be apparent to the user until critical performance errors arise while actively interacting with the virtual environment. There is substantial evidence for the existence of adaptation and aftereffects in VEs; such effects indicate altered demands on various perceptual–motor systems, such as the visual system (Stanney et al., 1999; see chap. 37, this volume), proprioceptive (see chap. 38, this volume), and the visual–vestibular systems (Draper, 1998b). With VEs, adaptation and aftereffects can also include a wide variety of subjective experiences, such as fatigue, malaise, nausea, headache, eyestrain, and drowsiness (see chap. 30, this volume). The vestibular system may be implicated in many of these experiences.

This chapter presents an introduction to the diverse area of vestibular adaptation research, focusing on its application to virtual environments. This task is not straightforward for two reasons. First, the concept of vestibular adaptation is not clearly and uniquely defined. Second, the concept seems to imply adaptation of the vestibular organ itself. It is unclear whether any meaningful form of adaptation happens in the vestibular organs; however, this issue may not be important for the design or use of virtual environments. Design implications arise from the perceptual and/or behavioral adaptations that are influenced by vestibular input. Therefore *vestibular adaptation*, as defined for this chapter, implies adaptations to vestibule-driven perceptual-motor systems that impact user behavior and task performance.

*Adaptation* has been defined as "a semipermanent change of perception or perceptual–motor coordination that serves to reduce or eliminate a registered discrepancy between or within sensory modalities or the errors in behavior induced by this discrepancy" (Welch, 1986, p. 24–33; see chap. 31, this volume). Reason and Brand (1975) described perceptual adaptation as a process in which conflicting signals from two or more perceptual systems become "normalized" over time, such that these intersensory signals are no longer perceived as being in conflict.

Exposure to a new or altered environment often causes a loss of accuracy in perception, performance, or both. With continued exposure, perception and performance can exhibit gradual improvement (i.e., adaptation), often leading to a return to baseline levels. On return to the original environment there may be a renewed error, or *aftereffect*, which itself gradually diminishes with readaptation. The sequence of events is illustrated in Fig. 39. 1. *Negative*

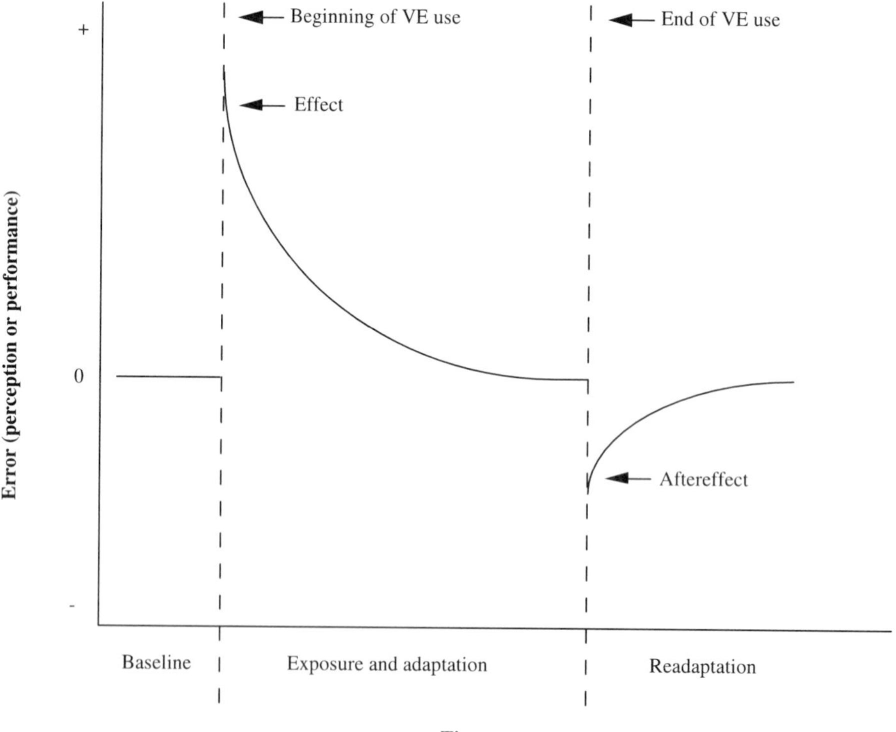


FIG. 39.1. The time course and general characteristics of adaptation and aftereffects.

*aftereffect* refers to errors observed after return to the normal environment that are in the opposite direction to errors made on initial exposure to the virtual environment (Fig. 39. 1). Research has demonstrated that more complete adaptation is generally associated with more substantial or lasting negative aftereffects (e.g., Dolezal, 1982; Welch, 1986). However, repeated exposures to the altered environment generally result in faster adaptation with reduced aftereffects (Bingham & Romack, 1999). In essence, the time constant for adaptation (and readaptation) decreases with multiple exposures.

In this chapter, the nature and measurement of vestibular adaptation as related to VEs is discussed. In section 2, an overview of the vestibular system is provided and some of the perceptual-motor phenomena that utilize vestibular input are reviewed. This review is intended for VE professionals who are unfamiliar with this area. Section 3 offers detailed descriptions of three forms of vestibular adaptation: the vestibulo-ocular reflex; the control of body posture; and past-pointing. Because the type and extent of vestibular adaptation will likely vary as a function of the specific VE involved, in section 4 a system for identifying and classifying a VE in terms of its primary visual–vestibular interactions is proposed. Section 5 considers implications of vestibular adaptation for VE applications, whereas section 6 offers a summary and conclusions. Although vestibular adaptation is believed by many to play a significant role in the occurrence of motion sickness in VE systems, this topic is more fully addressed in chapter 30 (this valume). General issues surrounding VE adaptation are discussed in chapter 31 (this valume).

## 2. THE VESTIBULAR SYSTEM

### 2.1 Overview

The vestibular apparatus is a small sensory organ in the bony labyrinth of each inner ear. Its primary function is to detect rotational and translational movements of the head and generate a corresponding response signal (see chap. 7, this volume). These signals contribute to the effective coordination of eye movements, posture and balance, and the perception of motion and orientation. Partial or complete loss of vestibular functioning makes it difficult to perform even the most basic of tasks (Howard, 1986b; Sharpe & Barber, 1993).

The vestibular apparatus consists of two principle sets of structures, the semicircular canals and the otolith organs (Fig. 39. 2), which work together to provide information on head motion and orientation. The eighth cranial nerve is the main efferent pathway for vestibular signals.

As shown in Fig. 39. 2, in each vestibular apparatus there are three semicircular canals (SCC), termed the anterior, posterior, and horizontal. These SCCs are roughly orthogonal to one another, and their collective function is to sense angular accelerations of the head (the canals may also be sensitive to higher order parameters of motion, such as jerk). These canals are each bidirectionally sensitive and, when combined across the two vestibules, form three approximately perpendicular "push–pull" pairs so as to detect angular head acceleration in any direction.

Each SCC functions as a complete and independent circuit in detecting angular acceleration (Howard, 1986b). Endolymph fluid within each SCC is prevented from free flow by the cupula, a thin elastic diaphragmlike flap. When the head rotates, the force generated by the endolymph viscosity and inertia acts against the cupula of each SCC existing within the plane of motion, causing it to deflect. This deflection bends tiny hair cells located at the base of the cupula, which generates an efferent signal. Though sensitive to rotational acceleration, the SCC efferent signal is proportional to head rotational velocity for most normal head movement frequencies (Howard, 1986b). Thus, each SCC has some characteristics of a signal integrator.

The hair cells at the base of the cupula can respond to angular accelerations as small as 0.1 deg/sec$^2$. However, with continued constant angular rotation (i.e., zero acceleration), the

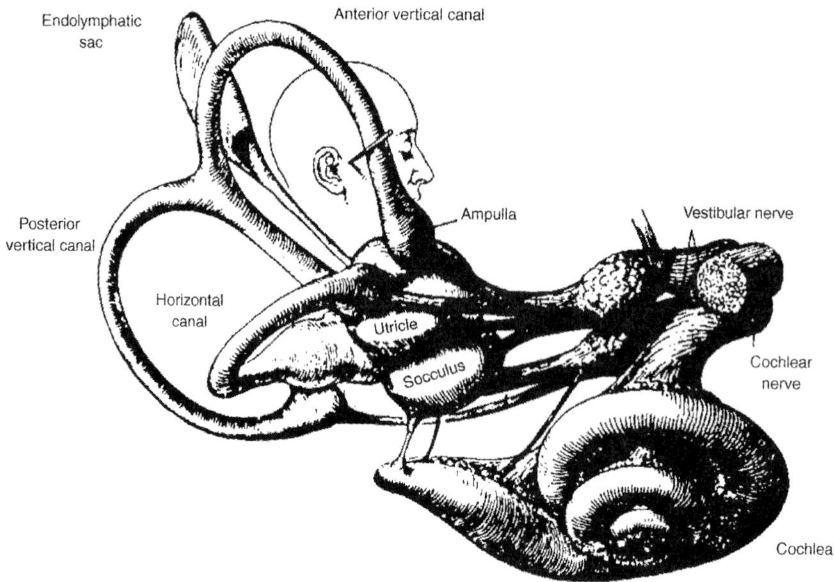


FIG. 39.2. The vestibular apparatus (From Hardy, 1934).

response diminishes. In functional terms, then, the SCCs are sensitive only to dynamic changes in head movement. The time constant for cupula deflection in humans is approximately 5 to 7 seconds (Howard, 1986b; Robinson, 1981).

In addition to the three SCCs, each vestibular apparatus also contains two otolith organs, the utricle and saccule (see Fig. 39. 2), whose activity is influenced by linear acceleration of the head. Given that the gravito-inertial force vector (i.e., the vector sum of gravity and inertial forces) constitutes an accelerative force that acts on the head, the activity of the otolith organs is influenced by static head position (i.e., head tilt) as well as dynamic linear forces (Stoffregen & Riccio, 1988). The receptor portion of these organs, termed the macula, contains many hair cells. The macula is covered with a gelatinous substance that contains tiny crystals of calcium carbonite, called *otoliths*. When there is a change in the orientation of the head relative to the gravito-inertial force vector, inertia causes the otoliths to deform the gelatinous mass; this creates a shear force that bends the receptor hair cells in the macula, generating the efferent signal. The utricle's macula is located in the horizontal plane, which means that it is strongly influenced by horizontal linear accelerations. The saccule's macula is positioned vertically and is strongly influenced by vertically directed linear accelerations (Robinson, 1981).

## 2.2 Major Influences

Stimulation of the vestibular system affects a wide variety of behaviors. A few of these behaviors are summarized below.

### *2.2.1 Gaze Stabilization*

In order to maintain successful vision, the eyes must be stabilized relative to the object of regard. The principal threat to ocular stability is head motion, as when one turns their head while attempting to maintain fixation on a given object of interest. Errors in stabilization lead to degraded visual performance, principally because the retina moves relative to the optical projection of the object of regard (this movement is widely known as *retinal image slip*).

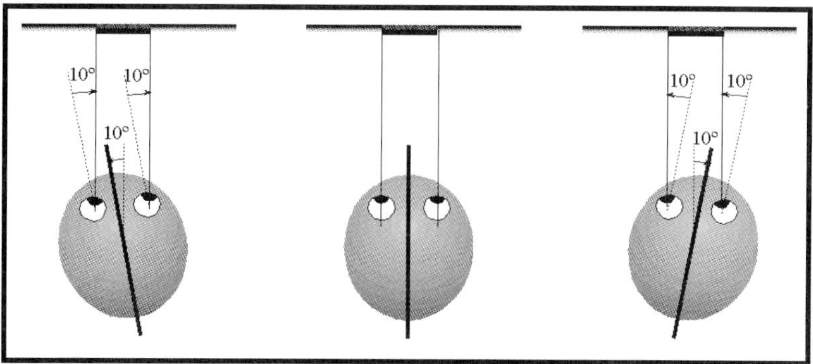


FIG. 39.3. The VOR. Left and right head rotations of 10 degrees result in a compensatory counter-rotations of the eye. Dashed lines represent original positions (for the eyes, it represents gaze direction assuming no compensation occurs for head movement); solid lines represent final positions (for the eyes, it represents gaze direction assuming a VOR gain of 1.0).

Stabilization of gaze across head movements (i.e., stabilization of the eyes relative to the illuminated environment) requires information about how the head has moved relative to the gravito-inertial force environment. This information is available to the vestibular system. Gaze stabilization across head movements is commonly studied in the context of the *vestibulo-ocular reflex* (VOR).

When the head begins to move in any direction, the vestibular apparatus senses this movement and sends velocity information directly to the oculomotor system. The oculomotor system responds by driving the eyes (conjunctively) at an approximately equal rate but opposite direction to compensate for head movement and help keep the visual image stabilized on the retina. This, in essence, is the VOR. Figure 39. 3 presents a simplistic depiction of the VOR for head rotations to the left and to the right of a center position. The VOR is a very-low-latency oculomotor response to head movements, with compensatory eye movements beginning as early as 4 ms after the onset of head rotation (Sharpe & Johnston, 1993).

Gain and phase estimates fully describe the VOR. VOR gain is defined as the ratio of slow-phase eye velocity (i.e., the response) to head velocity (i.e., the inertial stimulus). It is a relative measure of the amplitude of eye movement velocity for a given head movement. Phase (in angular degrees) describes the relative timing of eye to head motion. Phase lead indicates that the eye response precedes head movements, whereas phase lags indicate that eye response lags behind head movements.

The vestibular component alone is generally not sufficient for accurate gaze stabilization. This is because oculomotor vergence information is used to adjust the magnitude of counterrotation as a function of the distance of fixated targets (Owen & Lee, 1986). In addition, another visual component (i.e., optokinetic nystagmus) aids gaze stabilization at frequencies below about 2 Hz (Howard, 1986a; Robinson, 1981).

The vestibule is responsive to head rotation along any axis of movement. However, most studies that measure the VOR have limited rotations to a single, standard axis of motion, resulting in measurements of either the horizontal VOR (associated with yaw rotations), vertical VOR (associated with pitch rotations/tilt), and to a lesser extent the torsional VOR (associated with roll movements/tilt). Although both the SCCs and the otolith organs contribute to the rotational VOR, it is commonly believed that in many situations the otolith input is minor and transitory, whereas the SCC input dominates the response (Howard, 1986b; Robinson, 1981). Both vertical and torsional VORs can involve significant contributions from the otoliths.

#### 2.2.2 Postural/Balance System

The vestibule is known to influence the control of posture and balance. At the same time posture and balance are strongly influenced by stimulation of other perceptual systems, such as vision (i.e., optic flow), audition, and touch. Posture and balance are likely to be influenced by the overall pattern of stimulation across these systems.

Most VE systems do not include a motion base. Stimulation usually is limited to vision, audition, and/or touch. VE-based stimulation of these senses can have powerful affects on postural control and on the perception of body orientation and motion, but these systems do not directly stimulate the vestibular system. For such non-motion-base VE systems, any role of the vestibular system must come in the context of interactions between the vestibule and other perceptual systems that are directly stimulated by the virtual environment.

#### 2.2.3 Vestibular Influence on Hand–Eye Coordination: Past-pointing

In the natural world one often points at objects that are out of reach. In such situations accurate pointing is not verified by direct touch (the target is not touched), but must be verified on the basis of other types of stimulation, such as vision. The accuracy of visually guided pointing can be influenced by changes in the visual, haptic (e.g., Riley & Turvey, 2001), and vestibular systems, and by changes in relations between these three systems. Examples include pointing tasks in which the hand is visible as it moves toward a visible target (e.g., pointing finger to nose), or pointing at visible distal targets. Changes in vestibular adaptation often produce pointing errors, which are known collectively as *past-pointing*. Past-pointing was originally observed and studied as a result of the placement of refracting prisms in front of the eyes. A review of past-pointing phenomena and research can be found in Dolezal (1982). More recently, past-pointing and related effects on manual pointing have been observed among users of VE systems (see chap. 38, this volume).

#### 2.2.4 Other Influences

The vestibular system has been identified as directly or indirectly influencing autonomic function (Furman, Jacob, & Redfern, 1998; Harm, 1990). This includes various vasomotor, cardiac, gastrointestinal, and respiratory reflexes. The potential also exists for vestibular links to panic disorders, agoraphobia, othrostatic intolerance, and sleep apnea (Furman, et al., 1998). Lastly, the vestibular system is believed to play a role in the onset of motion sickness and simulator sickness (see chap. 30, this volume).

## 3. EXAMPLES OF VESTIBULAR ADAPTATION

In this section, the focus is on three manifestations of vestibular adaptation: the vestibular-ocular reflex (VOR), the control of posture, and past-pointing. Although these are not the only behaviors that are influenced by vestibular adaptation, they are relatively well studied and can be expected to be of particular relevance to VE applications.

### 3.1 Vestibulo-ocular Reflex (VOR)

#### 3.1.1 Overview/Description

As noted earlier, the VOR is a compensatory eye movement response that helps to stabilize a visual image on the retina during head motion, promoting clear vision. Adaptation of the VOR has been demonstrated in response to changes in the dynamics associated with real and virtual self-motion. Adaptation stimuli can be naturally occurring (e.g., due to the effects of disease or trauma to the vestibular and/or oculomotor systems) or generated by artificial means (e.g., wearing prescription, telescopic, or prism spectacles that change the optic flow in response to head movements). This chapter concentrates on vestibular adaptations that result from exposure to virtual environments.

As an example of VOR adaptation in VEs, consider the left side of Fig. 39.3. If the head were to rotate to the left at 10 deg/sec, the eyes would have to counterrotate 10 deg/sec to the right in order to maintain a stable retinal image of a visual target. Assuming that the VOR accounts for the all required compensation, this indicates a VOR with a gain of 1.0. If, however, the VE visual scene responded to that same leftward head movement by moving to the right at only 8 deg/sec (due to miscalibrations, image scale inaccuracies, etc.), the oculomotor response would need to be reduced through adaptation to 0.8 gain in order to maintain stable vision. Additionally, if the VE visual image were delayed in responding to a head movement (due to a system time delay), adaptation of VOR phase would likely occur to minimize retinal slip. Thus, if the motion dynamics in a VE differ from the motion dynamics in the real world, VOR adaptation will need to occur in order to minimize retinal image slip and promote stable gaze. A brief overview of this adaptation process is presented below, whereas more comprehensive descriptions can be found elsewhere (Howard, 1986a; Robinson, 1981).

The amount of VOR gain adaptation achieved depends mainly on: (1) the magnitude of retinal slip generated by the altered visual–vestibular motion stimulation (i.e., the adaptation demand); and (2) on the exposure time to the adaptation stimulus (Collewijn, Martins, & Steinman, 1983). If adaptation demand is relatively small (e.g., 30% or less), VOR gain can fully adapt within 3 to 5 minutes (Collewijn, et al., 1983). Since the majority of adaptation stimuli the VOR system will encounter in real life and within VEs will be modest in magnitude, VOR adaptation is assumed to be a fairly common phenomenon. For example, simply putting on prescription spectacles results in a 3% to 5% change per diopter of VOR gain adaptation demand (Collewijn, et al., 1983). Additionally, small visual fields, which are common in most current-generation VE head-mounted displays (HMDs), have been demonstrated to facilitate VOR adaptation nearly as well as full-field visual scenes (Shelhamer, Tiliket, Roberts, Kramer, & Zee, 1994). However, if the peripheral visual field is unobstructed such that normally occurring visual–vestibular motion relationships exist in that area (e.g., as often found in augmented reality applications), then VOR adaptation to a centrally projected adapting stimulus will be reduced (Demer, Porter, Goldberg, Jenkins, & Schmidt, 1989). Adaptation of VOR phase has also been demonstrated (Kramer, Roberts, Shelhamer, & Zee, 1998; Powell, Peterson, & Baker, 1996), but this area has received less attention from researchers.

### 3.1.2 *Measurement Techniques*

As indicated above, gaze stabilization depends on both vestibular and visual information. One way to explore gaze stabilization processes is to explore how the VOR adapts to changing motion dynamics. Due to space restrictions, the following discussion is restricted to VOR arising from on-axis rotation in the horizontal plane.

Evaluation of VOR adaptation implies measurement of VOR gain and/or phase before and after exposure to an adapting stimulus. VOR measurement generally requires that the participant's head be rotated while head and eye position data tracks are recorded (an exception to this rule, caloric testing, is often used in clinical settings). These data are then processed, filtered, and converted to velocity information so that the relative eye motion response can be standardized and quantified. The main features of VOR measurement are described below.

Certain stimuli cause the eyes to exhibit a rhythmic oscillatory pattern called *nystagmus*. There are different types of nystagmus (e.g., optokinetic, vestibular, physiological, caloric), depending on the stimulus involved. If inertial head rotation occurs in a completely dark environment, the resulting nystagmus is thought to be due solely to vestibular input and is thus termed vestibular nystagmus.

Nystagmus is recorded through use of an eye-position tracking system (e.g., electrooculography, videooculography, or search-coil contact lenses). Nystagmus appears on a strip chart as an oscillating pattern that is characteristic of a "sawtooth." For example, if the head rotates in the horizontal plane to the right, the VOR will drive the eyes to compensate by moving in the opposite direction (termed a slow phase component) at roughly the velocity of the head. This leftward movement of the eyes will continue until the eye nears the edge of its orbit, at which point the eyes will rapidly reverse direction, moving back across the center of gaze (termed a *quick-phase component*). After the quick phase, the cycle is repeated as a new slow phase occurs. Slow-phase components of vestibular nystagmus are used to determine the eye velocity component of the VOR after quick phases are removed from the data.

VOR measurement generally takes place in a completely darkened room so that there are no contributing affects of visually based image-stabilizing mechanisms. A participant is either oscillated at a fixed frequency or rotated continuously around a singular axis. Participant rotations can be passive, through the use of a rotating chair, or active through self-controlled head movements. Participants usually are given a standardized task to perform during testing (such as mental arithmetic) to reduce measurement variability. Eye position (vestibular nystagmus) and head position data are recorded continuously throughout this movement. Afterward, the slow phase nystagmus components are extracted and compared to associated head movements to determine VOR gain and phase metrics. For healthy participants performing a mental arithmetic task in the dark, average nominal VOR gain is approximately 0.6 to 0.8, and average phase angle is 0 degrees, dependent on specific head-motion frequency utilized (Howard 1986b; Robinson, 1981).

### 3.1.3 VE-related Research on the VOR

Few studies have evaluated VOR adaptation within the context of virtual interfaces. Kramer et al. (1998) demonstrated that a VE HMD could drive VOR adaptations in both gain and phase. They presented a prerecorded visual stimulus (consisting of horizontally oscillating vertical stripes) through an HMD that was attached to a rotating chair. Participants were exposed simultaneously to the oscillating stripes and passive inertial rotation in various combinations that generated specific VOR adaptation demands. Results indicated the occurrence of VOR adaptation similar to what is observed using more conventional stimuli. This research was among the first to provide evidence that videographic images, presented through a non-head-coupled HMD, can drive VOR adaptations.

Draper (1998b) focused on the potential for VOR adaptation to occur using a head-coupled HMD, a meaningful VE, active head movements, and a short exposure period (30 min). Visual image magnification factor (0.5×, 1.0×, 2.0×) was varied by manipulating geometric FOV while holding display FOV constant. The dynamics associated with the different VE image magnifications are described in Fig. 39. 4. There were significant gain adaptations when VE image magnification was 0.5× (an image minification) or 2.0× (an image magnification) but not for 1.0× (a spatially correct image, Fig. 39. 5). Thus, VE image magnification (or more appropriately, the ratio of geometric FOV to display FOV) determined VOR gain adaptation level and direction. Variations in geometric FOV produced variations in optic flow rate in response to head movements, so Draper's results also suggest that incorrectly calibrated head tracker gain settings may influence VOR adaptation to virtual environments.

In another experiment, Draper (1998a) varied system time delays (125 and 250 msec) while VE image magnification was fixed at 1.0× for a 30-min exposure period. Time delays were

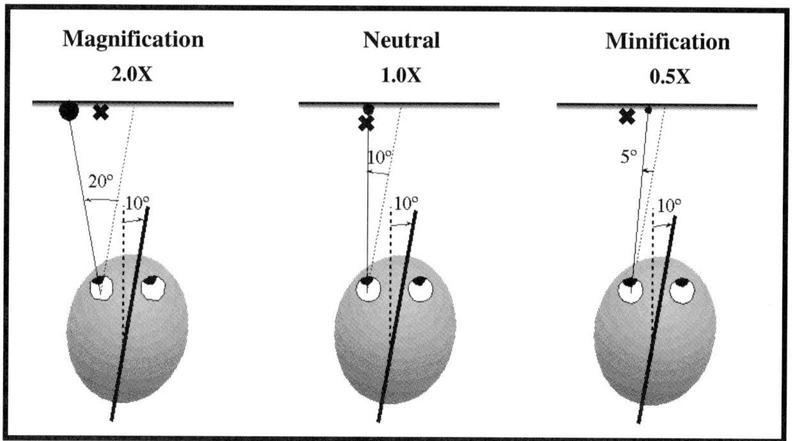


FIG. 39.4. VOR gain-change demand by VE image magnification factor, given a rightward head rotation of 10 degrees. When the VE image magnification factor is 1.0x (*center graphic*), the head-coupled visual scene response will be to counterrotate 10 degrees such that the VE will appear to remain stationary. A VE image magnification of 2.0x (*left graphic*) will result in the visual scene counterrotating 20 degrees (generating a VOR gain-increase demand), and a VE image scale factor of 0.5x (*right graphic*) will result in only a 5-degree counterrotation of the visual scene (generating a VOR gain-reduction demand). The "x" symbol in all three conditions marks eventual gaze location if the eyes were driven solely by the VOR with gain of 1.0.

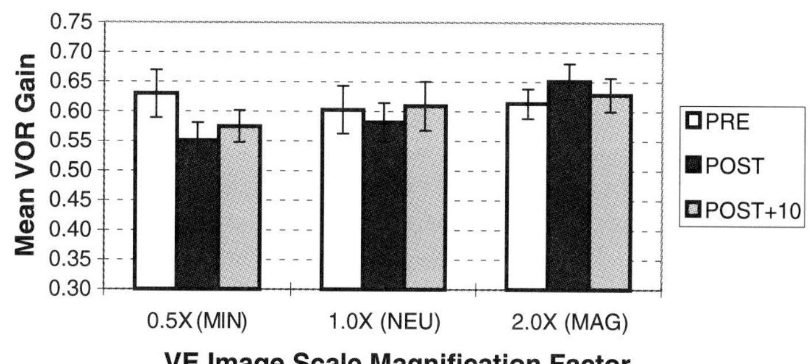


FIG. 39.5. VOR gain adaptation per VE image magnification factor (30-min exposure).

expected to influence VOR phase only. Results indicated a significant decrease in VOR gain (average 9%), in addition to an increase in phase lag of approximately 3%. Draper hypothesized that the unexpected drop in VOR gain may have resulted from the visual stimuli acting as a VOR suppression stimulus at the beginning of each head movement (due to the imposed latency).

## 3.2 Postural Effects

### *3.2.1 Overview/Description*

Exposure to virtual environments, such as flight simulators, can cause postural disequilibrium in users. If the postural disequilibrium persists beyond the time an individual is within the confines of the laboratory or system site, user safety could be compromised and product liability (see chap. 27, this volume) issues could be raised. For this reason, postural equilibrium has been proposed as a means of assessing the magnitude of aftereffects associated with VE systems.

A wide variety of techniques are used for measurement of postural motion (for a review, see Riccio & Stoffregen, 1991). Among these are kinematics of the body's center of pressure, or COP (e.g., Collins & DeLuca, 1993; Riley, Mitra, Stoffregen & Turvey, 1997), whole-body kinematics (i.e., measures of head motion; Stoffregen & Smart, 1998), joint kinematics (hip and ankle; Nashner & McCollum, 1985), and muscle activity (Nashner & McCollum, 1985). These techniques can be implemented in many ways, including qualitative (e.g., Cobb, 1999), linear quantitative (e. g., Stoffregen & Smart, 1998), and dynamical quantitative (Bardy, Marin, Stoffregen, & Bootsma, 1999). Measures of the kinesthetic position sense are also to be considered. Any of these could be used to assess adaptation to VE exposure. To date, a standardized assessment procedure has not yet been developed, and thus there is considerable opportunity for the development of new tests that may confer benefits in specific situations.

Test battery construction must focus on the accuracy and applicability of the constituent postural tasks required. Each task should be relatively unobtrusive while ensuring the comfort and safety of the participants. After such considerations, the first goal should be to ensure that each task is reliable. Unreliable tasks have no utility and should be discarded. Next, one or two tasks should be chosen to assess each domain or focus area. Tasks that consistently measure similar constructs should be refined or discarded. Finally, normative data must be established to accurately gauge any adaptive change in posture.

### 3.2.2 Measurement Techniques

Cobb (1999) reviewed a number of studies that have used qualitative measures to assess postural stability in the context of virtual environments. The majority used techniques such as the Sharpened Romberg test. In this test, the participant is instructed to stand heel-to-toe with weight equally distributed on the two feet with the chin parallel to the floor, and with the arms folded against the chest. The dependent variable is the number of seconds that this posture can be maintained. Dynamic tests are also used, such as asking participants to walk along a rail with the dependent variable being the number of steps taken. Kennedy, Berbaum, and Lilienthal (1997) evaluated postural stability following VE exposure by recording the duration that participants could stand on one leg. Although tests of this kind are simple and convenient, the data they provide are of limited precision.

More precise data can be obtained using technology to directly record body motion. One example is video recordings of stance. A variety of simple postural tests can be used with this procedure, such as a Tandem–Romberg test. This eliminates the necessity of keeping a large walking area safe for participants to traverse, and instead restricts data collection to a small, easily safeguarded area. The Tandem–Romberg stance generally restricts the majority of postural motion to lateral sway (side to side). This arrangement simplifies data reduction from videotape records. It allows one camera (placed directly behind the participant) to capture the majority of the movement. If significant movement exists outside the lateral plane, the need exists for two synchronized cameras (or another recording technique). Alternatively, a geometric shape (a reticle) of known dimensions can be used to allow calculations that characterize movement in other planes. Placing the camera at a distance equal to the participant's height allows lateral movements to be easily characterized in angular terms because the radii of the movement from the reticle to the floor and the camera are equal. Changing the camera's distance with each participant would require recalibrating the camera so that the pixel area that the reticle occupies remains constant (i.e., the viewed size of the reticle appears constant). For this reason, it is useful to keep the camera at a fixed distance of 2 meters from participants. The advantage to this method is that a camera is the only major piece of apparatus necessary during data collection. Coupled with a frame grabber for later analysis, this method is inexpensive, requires little space, and is not influenced by magnetic fields.

A second common technology used in direct measurements of postural motion is the force platform (e.g., Collins & DeLuca, 1993; Nashner & McCollum, 1985; Riley et al., 1997). These devices are becoming less expensive and easier to use. They constitute a convenient way to measure the kinetics of standing posture. Their chief limitation is their limited spatial extent. The test participant must remain on the force platform, which typically is no more than 0.5 meters square. Another method is epidermal recording of the electrical activity of muscles, or electromyography (EMG). While this method has a long history in basic research on stance (e.g., Horak & Macpherson, 1996; Nashner & McCollum, 1985), it is of limited usefulness in applied settings. Electrodes must be attached to the skin, which can be messy, and the associated wires tend to limit both the flexibility and range of movement. In addition, the interpretation of EMG data is difficult at best. This is because there is not a constant relation between the activity of muscles and the resulting motions of the body (Horak & Macpherson, 1996; Riccio & Stoffregen, 1988).

Position-tracking sensors offer another fast, easy, and precise method of collecting sway data along several dimensions. One such apparatus is a magnetic tracker. Magnetic trackers (e.g., Polhemus, Inc.; Ascension Technologies, Inc.), record the movements of a receiver/sensor relative to a central electromagnetic transmitter. Magnetic trackers can detect movements in 6 degrees of freedom (three axes of rotation and three of translation) with great accuracy. Research has demonstrated the benefits of magnetic tracking systems for recording head motion, body sway during stance (e.g., Stoffregend Smart, 1998 ), or sitting (e.g., Stoffregen, Hettinger Haas, Roe, & Smart, 2000).

Though magnetic trackers are generally very accurate, accuracy dissipates with increasing distance between the transmitter and sensor and with the inclusion of additional sensors; typical systems have an effective range of little more than one meter. Additionally, magnetic interference from other equipment may influence the data, reducing the utility of such a device in small, poorly shielded settings. The cables involved with most trackers may provide participants with feedback or compromise their stability, but a small amount of planning can alleviate this problem. In short, magnetic tracking systems offer another accurate, inexpensive method of assessing postural stability, but their utility depends upon the tester's ability to eliminate from the testing area items or materials that may interfere with the magnetic field.

### 3.2.3 VE-related Research on Posture

As noted earlier, few VE systems include a motion base. The majority of motion-base VE systems are vehicular simulators, such as flight simulators. In such systems the collection of data on postural motion is difficult. Accordingly, at present there are few studies of postural control in response to VE-generated inertial stimulation of the vestibule.

VE-related research on posture has focused on postural responses to imposed visual stimulation. Cobb (1999) measured postural stability before and after immersion in a virtual environment, consisting of an interactive video game presented via an HMD. Several different measures of postural stability were used, including tests of stance duration (i.e., Romberg test) and direct measurements of postural motion using a magnetic tracking system. Cobb reported that only the direct measurements of postural motion revealed any change in postural motion following exposure to the VE system. The study includes a helpful review of previous research relating postural stability to VE exposure.

Kennedy et al. (1997) assessed postural stability following exposure to both fixed-base and motion-base flight simulators, each of which used an external CGI display (i.e., not head-mounted). Significant correlations were found between scores on postural stability tests and measures of the subjective experience of disorientation. However, there was not a significant relation between posture and measures of motion sickness. It should be noted that Kennedy et al. used only qualitative measures of posture (i.e., stance duration) and did not include direct measurements of postural motion.

Stoffregen et al. (2000) measured postural motion of seated participants (i.e., torso sway) during exposure to visual stimulation in a dome-type fixed-base flight simulator. Magnetic tracking of sway was done during exposure to CGI imagery, and revealed substantial differences in postural motion between participants who later became motion sick and those who did not. The results of this study and of the studies by Cobb (1999) and Kennedy et al. (1997), suggest that indirect measures of postural instability, such as stance duration tests, are of limited usefulness in assessing postural effects caused by exposure to virtual environments. Direct measurement (i.e., tracking of body sway using magnetic or video-based systems) reveals the existence of VE-based effects on posture, which may be related to significant phenomena such as visually induced motion sickness.

## 3.3 Past-pointing

### *3.3.1 Overview/Description*

In many cases user interaction with altered environments (including VE systems) involves manual pointing or reaching to objects and events that are presented in a display. In such systems successful performance will depend on the user's ability to point or reach accurately (which requires proper hand–eye coordination). This is an issue because changes in visual, vestibular, and haptic adaptation are known to influence the accuracy of pointing; that is, they can induce past-pointing (which was defined in section 2.2.3). For this reason, it is important to assess the extent to which VE systems are prone to induce past-pointing.

### *3.3.2 Measurement Techniques*

Magnetic trackers may be applied to the measurement of past-pointing. The transmitter may be affixed to a participant's pointing finger via something as simple as a Velcro finger slip. This approach has the same benefits and limitations as in postural stability measurement. Magnetic trackers are small, highly accurate, and inexpensive, but may be hampered by cable placement, susceptibility to magnetic interference, and by their small range.

Another inexpensive device for measuring past-pointing changes while avoiding some of the shortcomings of magnetic trackers is a digitizing tablet with a cordless stylus. This approach involves attaching a lightweight, cordless stylus, or "pen," to a participant's finger (possibly via a Velcro finger slip). Participants would then contact a specified point on the surface on the digitizing tablet. This technique does not involve feedback from a cable (as with the magnetic tracker), but would introduce feedback from contact with the surface of the tablet. A potential solution would be to have participants close their eyes after becoming familiar with the contact point and to record the first point of contact with the tablet as their final target.

Past-pointing is easily demonstrated, yet data analysis may be fairly complex. The simplest metric is observing the magnitude of error from an established contact point. There may be utility in categorizing the error along its component axes (i.e., horizontal and vertical). Different exposures (or exposure durations) may manifest themselves strongly along one axis, while the other axis may indicate noise. Successive administrations of the test may indicate a pattern of responding that would not lend itself well to simply averaging the magnitude of error on repeated administrations. Participants may progressively move from a large deviation along one axis, to smaller deviations, and then have deviations that increase on the other side of the origin (i.e., center point). Likewise, participants may have deviations that maintain their magnitude but change their orientation/direction with regard to the central or desired contact point.

### 3.3.3 VE-related Research on Past-pointing

Research utilizing past-pointing measurement has begun to be conducted in VEs (for a review see Stanney et al., 1999). Recent studies have found an increase in pointing errors and a decrease in manual task speed following exposure to see-through HMDs. This has proved to be true for a variety of VE tasks. In experiments conducted by Stanney et al., participants interacted with targets in a VE presented via an HMD. The VE was shaped like a maze, consisting of 29 rooms and 3 long corridors. Participants performed a number of locomotion and manipulation tasks, such as grasping and moving objects located on one wall. Assessment of aftereffects was conducted immediately after VE exposure, and again 20 to 30 minutes later. Past-pointing was measured via arm movements that resulted in manual contact with a digitizing tablet. Results revealed that there were significant errors in the aiming of arm movements. In two experiments, manual reaching tended to be too high. In one of the experiments, reaching was also offset to the left. This and other recent work emphasizes the need for developing and using objective measures of post-VE exposure aftereffects so as to provide a systematic determination of the conditions under which these effects may occur.

## 4. CLASSIFICATION OF VISUAL/VESTIBULAR STIMULATION IN VIRTUAL ENVIRONMENTS

Virtual environments encompass a wide variety of interface systems including head-coupled HMDs, head-coupled and non-head-coupled interfaces to fixed displays, various degrees of motion-based and fixed-based systems, large- and small-display field of view (FOV), and active and passive control of self-movement. Each type of VE likely has characteristic patterns of motion dynamics that stimulate vestibular adaptation processes in a unique manner. Therefore, it would be helpful to categorize VEs according to their overall motion dynamic characteristics so that one might be able to better predict when specific forms of vestibular adaptation are likely to occur.

A VE's motion dynamics can be described though exploring the visual and vestibular couplings involved. One difficulty that currently exists with VE research is the lack of an accurate and consistent characterization of vestibular and visual stimulation within specific virtual environments. To properly define the motion stimuli in VEs, both visual and vestibular components must be considered, and quite possibly the most important aspect is their interaction. Described below is a classification scheme for VEs based on visual–vestibular couplings.

The idea of classifying environments according to visual and/or vestibular coupling is not new. Reason and Brand (1975) argued that all motion sickness arises from either visual–vestibular rearrangements or canal–otolith rearrangements, each with three possible permutations of the involved variables. Another example is the practice of classifying flight simulators on the basis of their tendency to induce simulator sickness (Kennedy, Drexler, & Berbaum, 1994). Kennedy, Berbaum, Dunlap, and Hettinger (1996) have also attempted to identify and quantify provocative visual stimuli in terms of optic flow velocities and accelerations. An earlier version of the classification system presented here was created by Draper (1998a) in an attempt to better generalize simulator sickness findings across virtual environments.

A critical feature of this scheme is that these classifications reflect a qualified comparison of visual–vestibular relationships in VE and real-world settings. *These categories do not imply a comparison of "visual motion inputs" to "vestibular motion inputs."* The choice of this type of classification is motivated by current debate about the status of nonidentities in stimulation of the visual and vestibular systems. Nonidentities in the simultaneous stimulation of the visual and vestibular systems may constitute a sensory "conflict" that needs to be

**TABLE 39.1**
Proposed VE Motion Dynamics Classification System

| | *Head Movement* | *Vehicular Movement* |
|---|---|---|
| Rotation | {active}<br>equivalent, altered, distinct | {active or passive}<br>equivalent, altered, distinct |
| Translation | {active}<br>equivalent, altered, distinct | {active or passive}<br>equivalent, altered, distinct |

resolved through internal processing (e.g., Howard, 1986b). Alternately, such nonidentities may constitute higher order information that can be picked up directly (i.e., without internal processing; Stoffregen & Bardy, 2001; Stoffregen & Riccio, 1988). Regardless of the outcome of this debate, it is a fact that patterns of visual–vestibular stimulation in VEs generally differ from those occurring in their real-world correlates, and these differences often stimulate forms of vestibular adaptation. The classification method herein proposed is focused on this fact.

There is often more than one type of visual–vestibular coupling in a given virtual environment. Therefore, a useful classification scheme must account for this possibility and give insight into the nature of the motion involved. Visual–vestibular relationships should be separated into those arising as a result of head movement and those arising as a result of vehicle movement, because it is through these two types of movements that most visual-vestibular interactions occur (see Table 39.1); (cf. Riccio, 1995). Each of these categories can be further segmented into *rotational* and *translational* motion.

The head movement category includes visual and vestibular stimuli that result from active, self-generated movements of the participant's head resulting from neck rotations, torso movements, or locomotion, such as walking. Head movements are actively controlled by the participant and are always assumed to have a corresponding vestibular signal because a working vestibular system is always stimulated by head motion.

The vehicle movement category applies in those situations where the participant experiences simulated motion of a vehicle (such as in driving and flight simulators), either actively controlled or passively experienced. In VEs, this vehicular motion can include 'flying' through an environment by means of a hand controller or other such device, as well as the changing of rotational views by means of an artificial control device. Within VEs, vestibular signals may not exist for certain instantiations of vehicular motion.

Now that the general categories of VE motion dynamics have been described, there must be some way to classify each category according to how well it reflects its real-world correlate. Note that the following classification method is subjective and necessarily ill defined to an extent, given the wide variety of VEs and associated motion dynamics involved. However, this system might assist the researcher in pinpointing those visual–vestibular couplings most likely to be implicated in instances of vestibular adaptation, and it will improve prediction of specific forms of vestibular adaptation across virtual environments.

Each cell in Table 39.1 is classified as having one of the following possible visual–vestibular couplings: equivalent, altered, and distinct, as compared to the real world correlate for that type of motion. *Equivalent* is used to imply that the pattern of visual-vestibular stimulation is considered to be very similar to that which occurs in the real world. *Altered* is used to denote patterns of visual–vestibular stimulation having an outward appearance of similarity with the real world correlate, but lacking equivalence due to latencies, tracking inaccuracies, optical distortions, and other causes. *Distinct* indicates visual–vestibular stimulation that differs grossly from what would occur in the real world. Usually this occurs when a VE system simply does not model real-world stimulation of one of the senses. These cases generally fall into one of two categories. Vestibular stimulation that is modeled in the absence of visual stimulation is referred to as *Vest/no Vis*, while visual stimulation that is modeled in the absence of vestibular stimulation is referred to as *Vis/no Vest*. Vis/no Vest is more common, especially for vehicular movements. The typical case is a VE system that simulates only the visual aspects of an environment.

Finally, given the potential influence of active control of movement on adaptations, there is identification of whether vehicular movement is *actively controlled* or *passively experienced* for translation as well as for rotational motion. As mentioned earlier, the head movement category always consists of active, self-generated head movements only.

This classification system organizes the major visual–vestibular couplings of all VEs according to the four quadrants of Table 39.1. Table 39.2 gives a few examples of how specific VEs would be categorized. Note the differences in visual–vestibular couplings that exist between simulators, between head-coupled VEs, and between simulators and head-coupled virtual environments.

As an example of this classification system, fixed-base domed flight simulators (Table 39. 2, row no. 1) entail visual–vestibular *equivalence* for head movements because the visual image always surrounds the pilot and therefore responds to head rotations and translations with 0 ms latency, 1.0 gain, and no phase shift (as in the real-world corelate). However, motion dynamics from vehicular movement are *distinct* from their real-world correlate, as there is no motion base to apply inertial forces associated with aircraft movements. In contrast, consider a commercially available head-coupled VE system (Table 39.2, row no. 3) that responds only to head rotations (3 DOF), with the user translating through virtual space in the direction of

**TABLE 39.2**

Classification of VEs according to specific visual-vestibular couplings involved (A: actively controlled motion. P: passively experienced motion)

| *Virtual Environment* | *Head Movement (A)* | | *Vehicular Movement* | |
|---|---|---|---|---|
| | *Rotation* | *Translation* | *Rotation* | *Translation* |
| 1) Fixed-base full visual field domed flight sim. | Equivalent | Equivalent | A:Distinct (vis/no vest) | A:Distinct (vis/no vest) |
| 2) 6 DOF motion-based full visual field domed flight sim. | Equivalent | Equivalent | A:Altered | A:Altered |
| 3) 3 DOF head-coupled (rotation only) VE (translate via hand controller in direction of gaze) | Altered | Distinct (vest/no vis) A:Distinct | N/A (same as head rotation) | A: Distinct (vis/no vest) |
| 4) 6 DOF head-coupled VE (translate via hand controller in direction of gaze) | Altered | Altered | N/A (same as head rotation) | A:Distinct (vis/no vest) |
| 5) 3 DOF (rot only) VE passive flythrough (active gaze shifts recognized) | Altered (vest/no vis) | Distinct (vis/no vest) | P:Distinct | P:Distinct (vis/no vest) |

current gaze via an artificial controller. With this system, as in the flight simulator, the user actively controls the motion, but the resulting patterns of visual–vestibular stimulation are very different. Due to latencies and minor inaccuracies associated with head-tracking systems, head rotations result in visual–vestibular stimulation that is *altered* from its real-world correlate. Head translation is *distinct* (Vest/no Vis) due to the inability of the head-tracking system to record position translations. Vehicular motion is only relevant in translation and involves visual motion cues only (i.e., *distinct*—Vis/no Vest). Thus, these two VE systems (fixed-base domed flight simulator and 3-DOF HMD interface) result in very different visual–vestibular couplings, and generalization of vestibular adaptation from one to the other may be limited at best. Correspondingly, motion sickness arising from flight simulators may differ from symptoms associated with various virtual interfaces (Stanney, Kennedy, & Drexler, 1997).

The intent of this classification scheme is to provide a basis for predicting the type and magnitude of vestibular adaptation and simulator sickness between VEs, based on the relative similarity of the visual–vestibular couplings involved. It also aids the researcher in identifying visual–vestibular couplings that most likely contributed to any occurring vestibular adaptation. Given that any complete description of a human-environment system must include elements of the task, stimuli, and individual, this framework is necessarily incomplete. However, for a particular task and individual profile, this classification scheme offers a way to specify and categorize the motion dynamics associated with VEs, while also specifying the potentially salient task variable of active versus passive control of vehicular motion.

## 5. IMPLICATIONS FOR VIRTUAL INTERFACE DESIGN AND USE

The topic of vestibular adaptation is very relevant and possibly critical to the design of effective and safe virtual interfaces. Given the current state of VE technology, the potential exists for vestibular adaptation within VEs to continue for the foreseeable future. The question becomes, What is the impact of vestibular adaptation within virtual environments? Below is a general summary of positives and negatives associated with the occurrence of vestibular adaptation in virtual environments.

Vestibular adaptation to altered environments generally benefits the human–computer system. It allows human behavior to adjust to the novel motion environment, reducing performance errors in gaze stabilization, postural control, and hand–eye control. Therefore, the faster the adaptation process, the more quickly performance will asymptote in the virtual environment. It has also been theorized that faster adaptation reduces the potential for motion sickness (Draper, 1998b; Reason & Brand, 1975; see chap. 30, this volume).

However, vestibular adaptation does not come without costs. The requirement for adaptation implies reduced performance until adaptation is complete. Additionally, substantial adaptation to a virtual interface implies a need for subsequent readaptation to the normal environment, which may entail negative aftereffects. These aftereffects (which may include incomplete gaze stabilization, postural instability, and errors in hand–eye coordination) are of primary concern to VE designers because of safety and legal issues that are involved (see chap. 27, this volume). Lastly, altered visual–vestibular couplings have been implicated in the occurrence of motion sickness in VE systems (see chap. 30, this volume). This suggests that similar stimuli may drive both vestibular adaptation and sickness.

Given that current-technology VE interfaces will likely drive vestibular adaptations, designers may wish to concentrate on speeding adaptation and readaptation processes in seeking to minimize any adverse effects. However, a far-term goal for VE designers should be to minimize all undesired adaptations that require postexposure recovery periods.

## 6. CONCLUSION

With virtual interfaces, one can no longer assume that system modification only occurs on the machine side of the human–computer boundary. Vestibular adaptation as a result of exposure to VEs has been clearly demonstrated. The benefits of vestibular adaptation include an improved ability to operate within the altered motion dynamics of the VE (including gaze stabilization, postural control, and hand–eye coordination). However, vestibular adaptation may also entail important health, safety, and legal ramifications (see chap. 27, this volume), such as the occurrence of negative aftereffects (see chap. 31, this volume) following exposure and an increased likelihood for motion sickness (see chap. 30, this volume) to occur during adaptation. Therefore, a detailed analysis of how humans are influenced by virtual interfaces must occur before these interfaces can be properly designed and administered. Stimulus rearrangements that commonly appear in virtual interfaces must be evaluated for their potential to drive vestibular adaptation processes, readaptation processes (i.e., aftereffects), and induce simulator sickness. Additionally, the consequences of these adaptations and aftereffects must be better understood. Only by fully characterizing the interface in this way can designers truly understand the health and safety issues involved.

## 7. REFERENCES


Bardy, B. G., Marin, L., Stoffregen, T. A., & Bootsma, R. J. (1999). Postural coordination modes considered as emergent phenomena. *Journal of Experimental Psychology: Human Perception and Performance, 25*, 1284–1301.

Bingham, G. P., & Romack, J. L. (1999). The rate of adaptation to displacement prisms remains constant despite acquisition of rapid calibration. *Journal of Experimental Psychology: Human Perception and Performance, 25*, 1331–1346.

Cobb, S. V. G. (1999). Measurement of postural stability before and after immersion in a virtual environment. *Applied Ergonomics, 30*, 47–57

Collewijn, H., Martins, A. J., & Steinman, R. M. (1983). Compensatory eye movements during active and passive head movements: fast adaptation to changes in visual magnification. *J Physiol, 340*, 259–286.

Collins, J. J., & De Luca, C. J. (1993). Open-loop and closed-loop control of posture: A random-walk analysis of center-of-pressure trajectories. *Experimental Brain Research, 95*, 308–318.

Demer, J. L., Porter, F. I., Goldberg, J., Jenkins, H. A., & Schmidt, K. (1989). Adaptation to telescopic spectacles: vestibulo-ocular reflex plasticity. *Invest Opththalmol Vis Sci, 30*(1), 159–170.

Dolezal, H. (1982). *Living in a world transformed.* New York: Academic Press.

Draper, M. H. (1998a). *The adaptive effects of virtual interfaces: Vestibulo-ocular reflex and simulator sickness.* Unpublished doctoral dissertation, University of Washington.

Draper, M. H. (1998b). The effects of image scale factor on vestibulo-ocular reflex adaptation and simulator sickness in head-coupled virtual environments. In *Proceeding of the Human Factors and Ergonomics Society 42nd Annual Meeting.* Chicago, IL: Human Factors and Ergonomics Society.

Furman, J. M., Jacob, R. G., & Redfern, M. S. (1998). Clinical evidence that the vestibular system participates in autonomic control. *Journal of Vestibular Research, 8*, 27–34.

Hardy, M. (1934). Observations on the innervation of the macula saculi in man. *Anatomical Record, 59*, 403–478.

Harm, D. L. (1990). Physiology of motion sickness. In G. H. Crampton (Ed.), *Motion and space sickness.* Boca Raton, FL: CRC Press.

Horak, F. B., & Macpherson, J. M. (1996). Postural orientation and equilibrium. In L. B. Rowell & J. T. Shepherd (Eds.), *Handbook of physiology* (pp. 255–292). New York: Oxford University Press.

Howard, I. P. (1986a). The perception of posture, self-motion, and the visual vertical. In K. R. Boff, L. Kaufman, & J. P. Thomas (Eds.), *Handbook of perception and human performance* (pp. 18:1–18:35). New York: John Wiley & Sons.

Howard, I. P. (1986b). The vestibular system. In K. R. Boff, L. Kaufman, & J. P. Thomas (Eds.), *Handbook of perception and human performance* (pp. 11:1–11:30). New York: John Wiley & Sons.

Kennedy, R. S., Berbaum, K. S., Dunlap, W. P., & Hettinger, L. J. (1996). Developing automated methods to quantify the visual stimulus for cybersickness. In *Proceedings of the Human Factors and Ergonomics Society 40th Annual Meeting* (pp. 1126–1130). Santa Monica, CA: Human Factors & Ergonomics Society.

Kennedy, R. S., Berbaum, K. S., & Lilienthal, M. G. (1997). Disorientation and postural ataxia following flight simulation. *Aviation, Space, and Environmental Medicine, 68*, 13–17.

Kennedy, R. S., Drexler, J. M., & Berbaum, K. S. (1994). Methodological and measurement issues for identification of engineering features contributing to virtual reality sickness, *IMAGE VII*, 245–254.

Kramer, P. D., Roberts, D. C., Shelhamer, M., & Zee, D. S. (1998). A versatile stereoscopic visual display for vestibular and oculomotor research. *Journal of Vestibular Research, 8*, 363–379.

Moroney, W. F., & Moroney, B. W. (1998). Simulation. In D. J. Garland, J. A. Wise, & V. D. Hopkin (Eds.), *Human factors in aviation systems* (pp. 358–388). Mahwah, NJ: Lawrence Erlbaum Associates.

Nashner, L. M., & McCollum, G. (1985). The organization of postural movements: A formal basis and experimental synthesis. *The Behavioral and Brain Sciences, 8*, 135–172.

Owen, B. M., & Lee, D. N. (1986). Establishing a frame of reference for action. In M. G. Wade and H. T. A. Whiting (Eds.), *Motor development in children: Aspects of coordination and control.* Dordrecht, the Netherlands: Nijhoff.

Powell, K. D., Peterson, B. W., & Baker, J. F. (1996). Phase-shifted direction adaptation of the vestibulo-ocular reflex in cat. *Journal of Vestibular Research, 6*, 277–293.

Reason, J. T., & Brand, J. J. (1975). *Motion sickness.* London: Academic Press.

Riccio, G. E. (1995). Coordination of postural control and vehicular control: Implications for multimodal perception and simulation of self-motion. In Hancock P., Flach J., Caird J., & Vicente K. (Eds.), *Local applications of the ecological approach to human–machine systems* (pp. 122–181). Hillsdale, NJ: Lawrence Erlbaum Associates.

Riccio, G. E., & Stoffregen, T. A. (1988). Affordances as constraints on the control of stance. *Human Movement Science, 7*, 265–300.

Riccio, G. E., & Stoffregen, T. A. (1991). An ecological theory of motion sickness and postural instability. *Ecological Psychology, 3*, 195–240.

Riley, M. A., Mitra, S., Stoffregen, T. A., & Turvey, M. T. (1997). Influences of body lean and vision on unperturbed postural sway. *Motor Control, 1*, 229–246.

Riley, M. A., & Turvey, M. T. (2001). Inertial constraints on limb proprioception are independent of visual calibration. *Journal of Experimental Psychology: Human Perception and Performance, 27*, 438–455.

Robinson, D. A. (1981). Control of eye movements. In V. B. Brooks (Ed.), *The handbook of physiology: Vol. 2, Part 2. The nervous system* (pp. 1275–1320). Baltimore: Williams & Wilkens.

Sharpe, J. A., & Barber, H. O. (Eds.). (1993). *The Vestibulo-ocular reflex and vertigo.* New York: Raven Press.

Sharpe, J. A., & Johnston, J. L. (1993). The vestibulo-ocular reflex: clinical, anatomic, and physiologic correlates. In J. A. Sharpe & H. O. Barber (Eds.), *The Vestibulo-ocular reflex and vertigo* (pp. 15–39). New York: Raven Press.

Shelhamer, M., Tiliket, C., Roberts, D., Kramer, P. D., & Zee, D. S. (1994). Short-term vestibulo-ocular reflex adaptation in humans 2. Error signals. *Experimental Brain Research, 100*, 328–336.

Stanney K. M., Kennedy, R. S., & Drexler J. M. (1997). Cyber sickness is not simulation sickness. In Proceedings of the Human Factors & Ergonomics Society 41st Annual Meeting (pp. 1138–1142). Santa Monica, CA: Human Factors & Ergonomics Society.

Stanney, K. M., Kennedy, R. S., Drexler, J. M., & Harm, D. L. (1999). Motion sickness and proprioceptive aftereffects following virtual environment exposure. *Applied Ergonomics, 30*, 27–38.

Stappers, P. J., Overbeeke, C. J., & Gaver, W. W. (in press). Beyond the limits of real-time realism: Uses, necessity, and theoretical foundations of non-realistic virtual reality. In M. W. Haas and L. J. Hettinger (Eds.), *Psychological issues in the design and use of virtual and adaptive environments.* Mahwah, NJ: Lawrence Erlbaum Associates.

Stoffregen, T. A., & Bardy, B. G. (2001). On specification and the senses. *Behavioral and Brain Sciences, 24*, 195–261

Stoffregen, T. A., Bardy, B. G., Smart, L. J., & Pagulayan, R. J. (in press). On the nature and evaluation of fidelity in virtual environments. In L. J. Hettinger & M. W. Haas (Eds.), *Psychological issues in the design and use of virtual and adaptive environments.* Mahwah, NJ: Lawrence Erlbaum Associates.

Stoffregen, T. A., Hettinger, L. R., Haas, M. W., Roe, M., & Smart, L. J. (2000). Postural instability and motion sickness in a fixed-base flight simulator. *Human Factors, 42*(3), p. 458–469.

Stoffregen, T. A., & Riccio, G. E. (1988). An ecological theory of orientation and the vestibular system. *Psychological Review, 95*, 3–14.

Stoffregen, T. A., & Smart, L. J. (1998). Postural instability precedes motion sickness. *Brain Research Bulletin, 47*, 437–448.

Welch, R. B. (1986). Adaptation of space perception. In K. R. Boff, L. Kaufman, & J. P. Thomas (Eds.), *Handbook of perception and human performance* (pp. 24:1–24:45). New York: John Wiley & Sons.

# 40

# Presence in Virtual Environments

Wallace Sadowski[1] and Kay Stanney[2]
*[1]IBM*
*Advisory Human Factors Engineer*
*8051 Congress Ave*
*Boca Raton, FL 33487*
*wjsadows@us.ibm.com*
*[2]University of Central Florida*
*4000 Central Florida Blvd.*
*Orlando, FL 32816-2450*
*stanney@mail.ucf.edu*

## 1. INTRODUCTION

Computer technology has been evolving for decades; thus, why should virtual environments (VE) be so fascinating? Undoubtedly, it is the conceived notion that one can step into a virtual environment, be transported to any desired place, and do things impossible in reality. Likely due to exaggerated portrayals in movies and books, to the general public the virtual experience appears limited only by one's imagination. Current VE technology falls short of this ideal; however, it does allow users to have unique experiences, such as standing inside a molecule, which were never before possible. Yet, current VEs are certainly not realistic enough to dupe someone into perceiving the VE as physical reality. Virtual environments take advantage of the imaginative ability of people to "psychologically transport" their "presence" to another place that may not exist in reality. This concept is demonstrated when one becomes engrossed in a book or movie and attends to it to the exclusion of one's surrounding environment. The primary characteristic distinguishing VEs from other means of displaying information is a focus on such immersion. Virtual environments add a dimension of physiological immersion by removing many real world sensations (e.g., obstructing the view of the real world via an head-mounted display [HMD]), while substituting sensations that would be imparted if the VE were real (e.g., the visual scenes of the virtual world).

Immersion, whether physiological or psychological in nature, is intended to instill a sense of belief that one has left the real world and is now "present" in the virtual environment. This notion of "presence" has been considered central to VE endeavors since its conception (Minsky, 1980). Presence is traditionally defined as the psychological perception of "being in" or "existing in" the VE in which one is immersed (Heeter, 1992; Sheridan, 1992; Steuer, 1992; Witmer & Singer, 1998). Barfield and Hendrix (1995) distinguish virtual presence from real-world presence as the extent to which participants believe they are somewhere different than their actual physical location while experiencing a computer generated simulation.

To date, the utility of the presence construct, either to enhance interactive design or human performance, has not been clearly established. This may be due to a lack of concise definitions and effective measures for this construct. This chapter commences by describing the concept of presence, then addresses types of presence, approaches of measuring presence, variables that influence presence, and the effects of presence on human performance. The latter may ultimately determine the utility of presence, for without benefits to performance this construct may have little value beyond ephemeral captivation.

## 2. CHARACTERIZING PRESENCE

Singer and Witmer (1997) describe presence as a perceptual flow requiring directed attention. They suggest that presence is based on the interaction of sensory stimulation, environmental factors, and internal tendencies. The psychological perception of presence in a VE is primarily thought to be the by-product of the VE's immersive and involving properties (Witmer & Singer, 1998). Kim and Biocca (1997) suggest it is fundamentally a product of two factors: (1) "arrival," or the sense of being in the VE, and (2) "departure," or the sense of not being in a virtual environment. "Arrival," or involvement in a VE, is thought to occur when one focuses energy and attention on a coherent set of stimuli or meaningfully related activities and events, presented in the virtual world. This would suggest that increasing the focus of one's attention on events portrayed in a VE enhances involvement, thereby increasing presence.

There are two differing schools of thought on what constitutes "immersion" in a virtual environment. Witmer and Singer (1998, p. 227) define immersion as "a psychological state characterized by perceiving oneself to be enveloped by, included in, and interacting with an environment that provides a continuous stream of stimuli and experiences." They also suggest that factors that affect immersion include isolation from the physical environment, perception of self-inclusion in the VE, natural modes of interaction and control, and the perception of self-movement. Witmer and Singer (1998) suggest that VEs that produce a greater sense of immersion will produce higher levels of presence.

With a different perspective, several researchers have suggested that immersion is related to the technology used to provide multimodal sensory input to users (Bystrom, Barfield, & Hendrix, 1999; Draper, Kaber, & Usher, 1998; Slater & Wilbur, 1997). For example, Slater and Wilbur (1997, p. 604) define immersion as "the extent to which computer displays are capable of delivering an inclusive, extensive, surrounding, and vivid illusion of reality to the senses of the VE participant." According to them, inclusiveness is the extent to which physical reality is shut out. Extensiveness indicates the range of sensory modalities accommodated and stimulated. Surrounding indicates the extent to which the VE is panoramic rather than limited to a narrow field of view. Vividness indicates the resolution and fidelity of stimuli for each modality and is concerned with the richness, information content, and quality of the displays. Slater and Wilbur (1997) suggest that immersion can be an objective and quantifiable description of what any particular system provides. Based on these descriptors, Slater and Wilbur characterize immersion as primarily determined by the extent to which devices and displays are capable of replicating the physiological sensations of the real world equivalent of the VE in which the person is interacting. Slater (1999) takes issue with the approach used by Witmer and Singer (1998), suggesting that their approach confounds objective, physical properties of VE technology with subjective, experiential aspects of presence. Lessiter, Freeman, Keogh, and Davidoff (2001), suggest that presence is likely related to both the subjective sense of a physical, spatial environment, as well as a personal evaluation of the appeal and naturalness or believability of the displayed environment and its content.

Regardless of how immersion is defined in regards to virtual environments, "presence" appears to be moderated by both external factors (Barfield & Hendrix, 1995; Hendrix &

Barfield, 1996a, 1996b; Prothero & Hoffman, 1995; Slater & Wilbur, 1997; Welch, Blackmon, Liu, Mellers, & Stark, 1996), as well as internal factors (Fontaine, 1992; Heeter, 1992; Witmer & Singer, 1998). Arguably, one must be afforded both technological and experiential immersion to maximize the likelihood of high levels of presence. In other words, one must be sufficiently immersed through the VE systems' hardware components and ergonomic design to enable the psychological leap of feeling "present" in the virtual environment. However, regardless of the realism of the virtual environment, if one is distracted by, or has psychological ties to, the external world, the sensation of presence may never occur and certainly will not develop to a great degree. To better understand this construct, it is important to realize that there are several different types of presence that can be elicited in a virtual environment.

## 3. TYPES OF PRESENCE

Heeter (1992) asserts that there are three different types of presence: environmental, social, and personal. *Environmental presence* is described as the extent to which the environment itself appears to acknowledge one's existence and react to their actions. Heeter contends that many VEs are unresponsive to users. The premise of *social presence* is simply that, if multiple people are immersed in the same VE, the presence of others provides further evidence that the environment "exists," and thus each participant is more prone to experience higher levels of presence. *Personal presence* is a measure of the extent to which, and the reasons why one feels like they are "in" a virtual environment. The degree to which individuals feel present in VEs varies tremendously and may never be fully understood or predictable without a comprehensive and accurate measure of presence. When presence is referred to it is generally in the context of personal presence.

## 4. VARIABLES THAT INFLUENCE PRESENCE

Just as there are several forms of presence, from environmental to social to personal, there are multiple variables that influence these forms of presence. Both individual and system variables are thought to influence the level of presence experienced in a virtual environment (Stanney et al., 1998). Some specific variables discussed include: (1) ease of interaction; (2) user-initiated control; (3) pictorial realism; (4) length of exposure; (5) social factors; and (6) system factors. Beyond these specific variables, Slater and Usoh (1993b) distinguish between internal and external factors that may contribute to a participant's sense of presence.

### 4.1 Ease of Interaction

Virtual environment users that experience difficulties navigating through or interacting with the environment, or while performing a task, will be more likely to perceive the VE as unnatural and may experience less presence in the environment. Billinghurst and Weghorst (1995) manipulated the design of VE systems, influencing the ease of interaction, and found those environmental designs that facilitated the "ease of interaction" increased the reported sense of presence in the virtual environment.

### 4.2 User-Initiated Control

Witmer and Singer (1994) suggest that the greater the level of control a user has regarding their actions in a VE, the higher the reported level of presence. Witmer and Singer explain that the immediacy of a system's response, correspondence of user-initiated actions, and naturalness of the mode of control drive this control factor. Sheridan (1992) suggests that a user's control over the relation between sensors and the environment (e.g., the ability to modify one's viewpoint) and their ability to modify the physical characteristics of a VE are two of the principle determinants of sense of presence. Similarly, Welch et al. (1996) found that interactivity (versus passive observance) was particularly influential in enhancing perceived presence.

### 4.3 Pictorial Realism

Witmer and Singer (1994) suggest that presence should increase as a function of the pictorial realism of a virtual environment. Pictorial realism, in this case, relates to the connectedness, continuity, consistency, and meaningfulness of the perceptual stimuli presented. Welch et al. (1996) found limited but positive influences of pictorial realism on presence. Wilson, Nichols, and Haldane (1997) determined that realistic visual depth cues were positively related to reported levels of presence. Snow and Williges (1998) conclude that visual display resolution and texture mapping, while significant parameters, were less influential than other factors (i.e., field of view, sound, head tracking).

### 4.4 Length of Exposure

Preliminary data indicates that presence is not enhanced by longer exposures (Stanney, 2000). This study indicates that a given VE design appears to promote a sense of presence (or not) within the first 15 minutes of exposure. It would seem reasonable, however, to expect that extending exposure duration may increase presence because longer exposures enhance other factors thought to have an effect on presence. These factors include practice with VE tasks, extent of familiarity within the VE, and level of sensory adaptation to the intersensory and sensorimotor discordances presented by the VE (Welch et al., 1996; see chap. 31, this volume). If, as has been demonstrated, prolonged VE exposure results in increased motion sickness symptoms, presence could be negatively affected (Kennedy, Stanney, & Dunlap, 2000; Stanney, Lanham, Kennedy, & Breaux, 1999), especially since presence and sickness have been found to be negatively correlated. Witmer and Singer (1998) report that combining data from four experiments resulted in a correlation between sickness and presence of $-0.426$, $p < 0.001$. Witmer and Singer (1998) suggest that symptoms accompanying sickness draw attention away from the VE and focus that attention inward, decreasing involvement in the VE and thereby reducing the sense of presence. Similarly, in a recent study using a maze shaped VE and exposures up to one hour in length, subjective presence was found to be negatively correlated ($r = -0.342$, $p < 0.001$) with sickness (Stanney, 2000). It should be noted that another study (Wilson et al., 1997) reported a positive relationship between sickness and presence using different measurement tools of sickness and presence (see discussion in Stanney et al., 1998). Stanney et al. (1998) suggested that there is the possibility that sickness and presence are both correlated (either positively or negatively) with a third factor, such as vection (see chap. 23, this volume), in that changes in one initiate changes in the other but solely via the intervening factor. Regardless of the specific relationship between them, with respect to the factors that can impede sense of presence, sickness may be one of the most deleterious. Thus, when presence is important, exposure duration should be set to minimize such adverse effects.

### 4.5 Social Factors

There is growing interest in the effects of social factors on presence (Heeter, 1992; Steuer, 1992). The premise of "social presence" is that if other people (i.e., representative avatars; see chap. 15, this volume) reside in a VE there is more evidence that the VE exists. Correspondingly, if other persons in a VE essentially acknowledge one's presence in the VE, it offers further affirmation that one actually "exists" in that environment (Heeter, 1992). Social presence may result from communicating with others verbally or by gestures, interacting with others in the environment, or through confirmation that others recognize their existence in the environment.

### 4.6 Internal Factors

Slater and Usoh (1993b) describe internal factors as individual differences in how an individual cognitively processes information provided by a VE experience. While these factors are difficult to influence, it is important to understand their affect on presence.

Slater and Usoh (1993b) used a technique known as Neurolinguistic Programming (NLP) to assess how internal factors affect virtual presence. The NLP model suggests that subjective experience is encoded in terms of three main representation systems: visual, auditory, and kinesthetic (VAK). Practitioners of NLP claim that people have a tendency to prefer one representation system to another in a given context. The visual system includes external images as well as remembered or constructed internal mental images (see chap. 3, this volume). The auditory system includes external sounds and remembered or contrived internal sounds (see chap. 4, this volume). Also included in the auditory system is internal dialogue (i.e., talking to oneself). The kinesthetic system includes tactile sensations caused by external forces acting on the body and emotional responses (specific to internal tactile and haptic sensations; see chap. 5, this volume).

The Slater and Usoh (1993b) study suggests that visually dominant people report greater levels of presence than individuals whose primary representational system is auditory or kinesthetic and that a person who tends to process information more from the first position (e.g., "I" or "my") is more likely to experience a sense of presence than those who process more from the second or third position. They also suggest that the higher the proportion of visual predicates and references used when describing a VE experience (e.g., "looking" or "seeing"), the greater the reported sense of presence, whereas the higher the proportion of auditory predicates and references the lower the reported sense of presence. For those individuals provided with a virtual body (VB), the greater the proportion of kinesthetic references and predicates the higher the reported level of presence. For those without a VB, kinesthetic references were inversely related to presence. It should be cautioned that these results were derived from a VE that stimulated primarily the visual system, with only a small amount of associated sound and no haptic interface. The critical issue is that subjective experience is encoded in terms of different representation systems and that people may prefer one representation system to another. Thus, it is important to consider the types of individuals who will use a given VE system and their preferred representational system.

### 4.7 System Factors

Slater and Usoh (1993b) describe system factors as external factors that relate to how well the system replicates the real world equivalent, how this information is presented to the user, and the how the user interacts with the virtual environment. External factors are determined wholly by the VE hardware and software that drives the display. Slater and Usoh suggest external factors that may influence reported levels of presence include high-quality, high-resolution information being presented to the participants' sensory organs in a manner that does not convey the existence of the devices or displays. The environment being presented to the participant should be consistent across all sensory information displays. The environment should be one in which the participant can interact with objects and other actors, and one that reacts to the user. They also suggest that the self-representation of the participant should include a VB that is similar in appearance to the participant's own body and responds appropriately to the participant's movements.

In a study of system factors, Hendrix and Barfield (1996a) found that the addition of stereopsis positively influenced spatial realism during interaction within the VE and resulted in reports of greater presence. The same study also found that adding head tracking and manipulating the geometric field of view (GFOV) influenced reported levels of presence, with head tracking and larger GFOVs associated with higher levels of presence. Barfield and Hendrix (1995) found that presence was enhanced with increasing update rates but was approximately constant between 15 to 20 Hz. Welch et al. (1996) also encountered a negative influence on presence when a delay in visual feedback occurred. Hendrix and Barfield (1996b) found that spatialized sounds had a significant effect on presence when compared to nonspatialized or no sounds at all. Not all of these technical factors, however, are equally influential. Snow and Williges (1998) found that field of view, sound, and head tracking were nearly three times as influential on presence as visual display resolution, texture mapping, stereopsis, and scene update rate.

It has been suggested that as more sensory modalities are stimulated presence should increase (Steuer, 1992). Sadowski (1999) reviewed olfactory stimulation in VEs and the potential of incorporating odors as a means of increasing presence and performance in virtual environments. The results of this review suggest that utilizing olfactory cues in VEs may facilitate performance on some tasks, especially those demanding recall or recognition. The human

**TABLE 40.1**
Guidelines for Supporting Presence

| | *Guideline* | *Issue* |
|---|---|---|
| Ease of interaction | Provide seamless interaction such that users can readily orient in, traverse in, and interact with the virtual environment. | Poorly designed interaction takes focus away from the experience and places it instead on motion/mechanics. |
| User-initiated control | Provide immediacy of system response, correspondence of user-initiated actions, and a natural mode of control. | Delays, discordance of users' versus effectors actions, and unnatural control devices hinder engagement in a VE. |
| Pictorial realism | Provide continuity, consistency, connectedness and meaningfulness of presented visual stimuli. | Poorly designed or displayed visual interactions may hinder engagement in a VE. |
| Length of exposure | Provide sufficient exposure time to provide VE task proficiency, familiarity with a VE, and sensory adaptation. | Avoid unnecessarily prolonged exposures that could exacerbate sickness. |
| Social factors | Provide opportunities to interact with and communicate with others verbally or by gestures. Provide confirmation that others recognize one's existence in a VE. | If one's presence in a VE is not acknowledged by others, it may hinder the perception that they "exist" in that environment. |
| Internal factors | Identify the types of individuals who will use a VE system and their preferred representational system (i.e., visual, auditory, kinesthetic). | Individual differences can render VE systems differentially effective. |
| System factors | Provide head tracking, a large field of view, sounds, stereopsis, increasing update rates, multimodal interaction, and ergonomically sound sensors/effectors to facilitate presence. | Poorly designed systems can degrade users' experience. Note: This does not suggest that "extreme realism" is required, but rather what is provided should be well designed and developed. |

olfactory system utilizes the sensation of odors for detection, recognition, discrimination, and differentiation/scaling (Barfield & Danas, 1996). Furthermore, neuroanatomically, in comparison to the visual and auditory systems, the olfactory system has many direct connections to the limbic system of the brain where emotions and affects are regulated (Kandel & Schwartz, 1985). There is strong evidence that odors can be effective in manipulating mood, increasing vigilance, decreasing stress, and improving recall and retention of learned materials (Youngblut, Johnson, Nash, Wienclaw, & Will, 1996). Knasko and Gilbert (1990) found that even the suggestion of odors affect reported levels of pleasure and mood. Thus, the powerful influences of the olfactory system can be manipulated by incorporating odors in a VE to produce a more realistic experience, thereby increasing levels of presence and enhancing performance on some tasks.

Another system factor to consider is the interplay of the ergonomic design of the VE hardware (see chap. 41, this volume). Billinghurst and Weghorst (1995) found display comfort and quality to be predictive of reported presence. If a user is distracted by, or must focus on, the VE hardware because it is uncomfortable or burdensome, the ability to simulate "being in" the VE may deteriorate (Slater & Usoh, 1993b).

From the review of factors that influence presence, general guidelines can be provided that, if effectively implemented, should enhance users' perceived presence within a virtual environment (see Table 40.1).

## 5. MEASURING PRESENCE

Although the benefits of presence have been widely discussed and touted, few researchers have attempted to systematically measure this concept and relate it to possible contributing factors. It could be that sense of presence is simply an epiphenomenon of good VE design or possibly even a distraction (Ellis, 1996; Welch, 1999). Without a valid measure of presence, however, this issue cannot be resolved. Determining how to measure presence may provide a means to establish whether presence does indeed enhance VE system interaction and a greater understanding of the factors that drive this phenomenon may result.

A valid measure of presence should be: (a) relevant—have a direct connection with presence and its components; (b) reliable—have proven test–retest repeatability; (c) sensitive—have sensitivity to variations in the variables affecting presence; (d) nonintrusive—to avoid unintentional degradation of performance and/or sense of presence; and (e) convenient—portable, low cost and easy to learn and administer (Hendrix & Barfield, 1996a; Jex, 1988).

Presence has generally been measured using subjective rating scales that require judgments comparing a VE experience to real life (Hendrix & Barfield, 1996a; 1996b; Prothero & Hoffman, 1995; Slater & Usoh, 1993a; 1993b; Slater, Usoh & Steed, 1994; Witmer & Singer, 1998), or one VE to another (e.g., Welch, Blackmon, Liu, Mellers & Stark, 1996). These rating scales generally comprise statements relating to the extent to which an individual: feels physically located in a VE; senses that a VE has become more real than the physical world; and has a sense that the VE is more than merely a mediated event and has transcended into something they have actually experienced (Lessiter, Freeman, Keogh, Davidoff, 2001). These subjective rating scales are generally relevant, sensitive, nonintrusive, and convenient. They are, however, subject to response bias and thus may be unreliable (Freeman, Avons, Pearson, & IJsselsteijn, 1999). In addition, they often assume the respondent has an understanding of the nebulous concept of "presence" (Freaman, Avons, Meddis, Pearson, & IJsselsteijn, 2000).

With the issues related to subjective rating scales, alternative forms of measurement have been investigated. Behavioral presence relates to internally measurable or externally observable responses to the VE and the presented stimuli (e.g., skin conductance, electrocardiogram

(ECG), heart rate variability). While several researchers have attempted to use behavioral or physiological measures to assess presence in virtual environments (Cohn, DiZio & Lackner, 1996; Dillon, Keogh, Freeman, & Davidoff, 2000; Wilson, et al., 1997), the results have been mixed (see section 5.2.2).

## 5.1 Subjective Measures of Presence

Subjective measures of presence include such psychological measurement instruments as: (a) rating scales (e.g., "On a scale of 1 to 7 rate how natural your interactions with the virtual environment seem?"); (b) subjective reports (e.g., "I really felt like I was in another place and forgot that I was actually in a laboratory."); (c) the method of paired comparisons (e.g., "In which of the two virtual environments did you feel most 'present'?"); and (d) cross-modality matching (e.g., "Make this music as loud as the strength of the presence you experienced in the VE with maximum music volume being equal to the highest possible presence.").

### *5.1.1 Rating Scales*

Subjective rating scales have been used in several VE research experiments to evaluate the amount of presence experienced by participants (Baños et al., 2000; Prothero & Hoffman, 1995; Slater et al., 1994; Witmer & Sadowski, 1998). Witmer and Singer (1998) developed the Presence Questionnaire (PQ), which measures presence along three subscales: (1) involvement and control; (2) naturalness; and (3) interface quality. Witmer and Singer (1998) have data from several experiments that indicate that the PQ is a reliable and valid measure of presence (see Singer & Witmer, 1999; Slater, 1999). The PQ also has shown evidence of meaningful relations with learning (see chap. 20, this volume), aftereffects (see chaps. 30–32, 37–39, this volume), and human performance (see chap. 35, this volume).

Schubert, Friedmann, and Regenbrecht (1999) used a rating scale to assess presence in a three-dimensional computer game. They found three main factors drive presence, including: (1) spatial presence; (2) involvement; and (3) realness (comparability of a VE to the real world). Similarly, Lessiter et al. (2001) identified via a rating scale (i.e., the ITC-Sense of Presence Inventory) four main factors driving presence, including: (1) physical space; (2) engagement; (3) naturalness; and (4) negative effects. They appear to be the first ones to formally incorporate the adverse effects of VE exposure into a presence measure. The three rating scales reviewed here ( Lessiter et al., 2001; Schubert et al., 1999; Witmer & Singer, 1998) all tap on similar components of presence (i.e., the physical space, involvement, naturalness, and interface quality).

Usoh, Catena, Arman, and Slater (2000) subjected the Slater et al. (1994) and Witmer and Singer (1998) measures to a "reality test" by assessing whether or not these scales provided higher scores for real experiences as compared to the diminished experiences provided by current VE technology. Usoh, Catena, Arman, and Slater (2000) found that neither scale passed the reality test (i.e., scores for real events were not higher than VE); however, the Slater et al. (1994) scale showed some marginal differences. They concluded that the utility of rating scales, while appropriate for the assessment of a single system, is limited for comparison of experiences across diverse environments.

### *5.1.2 Subjective Report*

Some VE experiments have also incorporated subjective reports as a means of obtaining information regarding presence (Hendrix & Barfield, 1996a; Slater & Usoh, 1993a). These studies used directed, but open-ended, questions to elicit participants' reactions and impressions related to presence. Since these answers were formed in the participants' own words they may serve to reduce any bias from question interpretation and Likert scale response bias. Like any open-ended questionnaire responses, however, results from such subjective reports can be difficult to generalize.

### *5.1.3 Comparison-based Prediction*

Schloerb (1995) proposed a paired-comparisons procedure in which the participants' task is to distinguish between a real-world scene and a VE simulating the same scene. Hence, if participants are unable to distinguish between the two environments, one could reasonably suggest that their sense of presence in the VE should be as strong as in the real world. Given the present state of VE technology it is unlikely that any participants would be unable to distinguish between a computer-generated VE and the real world. Therefore, this method may be more effective in comparing VEs amongst each other to determine which aspects of each environment, and which environment as a whole, has the greatest influence on facilitating or degrading presence.

Welch et al. (1996) used the method of paired comparison to determine the influence of three variables on participant's presence. Participants were presented with a pair of VEs that differed from each other in respect to: (a) whether the participant had an interactive or passive exposure; (b) whether the VE had high or low levels of pictorial realism; and (c) whether there was a short (200 msec) or long (1700 msec) delay of visual feedback (see Section 4.7). After participants viewed each pair of VEs, they reported which one produced more presence and indicated this difference on a scale of 1 to 100. Welch et al. (1996) found that participants indicated that interactivity (i.e., having active control over interaction rather than being a passive bystander) played a greater role than pictorial realism in judgments of relative presence. Similarly, Snow, and Williges (1998) used free-modulus magnitude estimation to examine the influence of a number of system parameters on perceived presence. In this method a participant is presented with a VE to which they assign a number based on their subjective impression of its intensity. The only restriction on the assigned number is that it must be positive (i.e., no experimenter defined constructs or rating scales). Then the participant assigns comparative values to successive VE systems. Using this method along with sequential experimentation and data bridging, Snow and Williges developed polynomial regression models to predict presence that could prove useful for VE system designers (e.g., prioritize field of view, sound and head tracking over other system factors).

Comparisons methods have the advantage that they do not require investigators to explain to participants what is meant by the concept of presence (Stanney et al., 1998). Unlike subjective reports and rating scales, which must often define or describe what is meant by terms, comparison-based prediction requires nonanalytical comparisons among alternatives, that is, reasoning by analogy (Klein, 1982). To standardize this method, a set of comparison cases could be established. Inferences about the level of presence in any one given system could then be generated based on how it compares to a select standardized comparison case. The benefit is that the response bias created by defining terms is avoided. There is an issue with this method in that the size of the difference that is detectable (known as the "just noticeable difference") will be dependent on the aspects of the stimuli (i.e., the real scene or comparison case versus the VE) that are compared. It is likely that individuals will be far more sensitive to differences in some dimensions (e.g., authentic sound simulation) as opposed to others (e.g., poor texture mapping) (Snow & Williges, 1998).

### *5.1.4 Cross-Modality Matching (CMM)*

The scaling of sensory experiences, or psychophysics, emerged in the late 19$^{th}$ century. Today sensory psychophysical scales are universally used and accepted in experimental psychology and tend to be reliable and stable (Stanney et al., 1998). The premise of CMM is that a person can monotonically represent the experiences of one sensory modality through another modality by producing a "subjectively equal" representation using a measure of the second sensory modality. This method is particularly effective when a concept does not readily lend itself to verbal scaling (such as when a definition of terms leads to response bias). For example, a person can characterize increasing or decreasing changes in a visual stimulus (e.g. light intensity) as they perceive them by matching these increases or decreases with an auditory stimulus (e.g. auditory intensity). Hence, it seems reasonable that the CMM technique could be utilized to provide an indicator of the overall presence experienced in a VE or utilized to determine which portion(s) of a VE provided the most influence on presence. Individual differences in representation of sensory modalities, however, can be problematic as it can lead to high response variability.

In general, the subjective measures reviewed here have the advantages of being easy to administer and interpret. However, the major drawback of such subjective measures is the problem of interrater reliability. Also, for some of these methods (i.e., rating scales and subjective reports), participants must understand the concept of what presence "is" and interpret questions uniformly. In addition, psychometric developers must ensure that the measurement tool is measuring "presence" as a whole and not only factors that contribute to it. Thus, objective measures of presence, such as physiological and behavioral measures, may be helpful in determining which individual and system factors influence presence.

## 5.2 Objective Measures of Presence

People commonly experience physiological or behavioral responses to stimuli in the real world, therefore it seems reasonable to conclude that stimuli in an immersive VE may also produce such responses. These responses may include reflexive motor acts and physiological indicators (physiometric and neurophysiological changes). Such responses could be objectively measured without participant bias and used as a primary or secondary indicator of environmental influences on perceived presence in a virtual environment.

### *5.2.1 Behavioral Measures*

First suggested by Held and Durlach (1987, 1992), measuring behaviors, such as "startle responses" (i.e., ducking or flinching), could provide indicators of presence in a virtual environment. Wilson et al. (1997) used startle responses to an unexpected event as one method of assessing presence. Other types of behaviors such as reaching for a virtual object, greeting virtual avatars, orienting responses, head positioning, or avoiding virtual hazards (i.e., visual cliffs) may suggest that participants "believe" they occupy the virtual space. Physiological and behavioral measures may be most useful when they are tailored to the experiences participants are expected to have in a virtual environment. For example, Cohn et al. (1996) demonstrated that participants responded with an automatic motor response (used as a measure of presence) in a VE designed to induce a sense of body rotation. Slater, Usoh, and Chrysanthou (1995) developed a behavioral measure by introducing contradictory information about object properties represented in both the real (via auditory cues) and virtual world (via visual cues). The extent to which users responded to the visual cues of the VE versus the contradictory auditory cues of the real world indicated their level of presence in the virtual world.

Zahoric and Jenison (1998) suggested that presence can be defined by the level to which a VE system successfully supports user action throughout the virtual world. This implies that the sensation of presence is grounded in action rather than altered psychological states. Using this action-based approach, Slater and Steed (2000) developed a virtual presence counter that is based on the number of transitions users undergo during VE exposure between two states: presence in the VE versus presence in the real world. The premise is that presence decreases as the number of such transitions increases. With this approach, a stochastic process is used to model the number of transitions, from which an estimate of the proportion of time spent within the "presence in the VE" state is derived. This approach is based on a number of simplifying assumptions (i.e., presence is a binary state; the stochastic model assumes discrete time; independence between transitions; reporting breaks in presence does not influence the incidence of such reports), which could hinder the validity of this approach.

### 5.2.2 Physiological Measures

Physiological indicators including muscular tension, cardiovascular responses, and ocular behaviors have been suggested as presence measures (Barfield & Weghorst, 1993). Bioelectric information can be obtained by electromyography (EMG), electroencephalogram (EEG), ECG, electrodermal activity (EDA), galvanic skin response (GSR), and measures of skin temperature. Measures of visual system behavior may provide a wealth of information regarding attention, alertness, and arousal. Pupilometry, eye trackers, and electro-oculograms (EOGs) have the potential to be useful tools in the isolation of presence invoking stimuli. These visual indicators may serve to identify which elements of a VE capture attention or evoke physiological changes in the person immersed. This physiological data may provide information regarding the effects of specific environmental stimuli or events experienced in a virtual environment. Strickland and Chartier (1997), although not directly investigating presence, have demonstrated the ability to utilize EEG measurements in a head-mounted display. Measuring and interpreting differences in cortical responses in real and virtual environments may lead to a better understanding of the effects of various software and hardware influences in a virtual environment.

While physiological measures are predictably related to presence, few studies have identified systematic relationships between these factors. Weiderhold, Davis, and Weiderhold (1998) found that while subjective measures indicated that flight simulations displayed via a HMD were more relaxing and higher in presence than those displayed via computer screen, there were no clear physiological relations identified. Salnass (1999) also failed to find a relationship between presence and physiological measures when assessing shared VEs with and without haptic force feedback. Conversely, Dillon et al. (2000) found greater levels of skin conductance over time using stereoscopic video presentation as compared to monoscopic presentation of a rally driving sequence; however, the stereoscopic presentation did not generate higher presence ratings. Dillon, Keogh, Freeman, and Davidoff (2001) found that heart rate is not as sensitive to manipulations in display and content design as subjective ratings of presence. These studies indicate that the drawbacks of physiological measures of presence are that collecting data can be intrusive, difficult to obtain, and the outcome is frequently unreliable. Considerable work is needed if behavioral assessment is to achieve the aforementioned criteria for a valid measure (i.e., relevance, reliability, sensitivity, unobtrusiveness, and convenience).

An objective measure of presence that can provide quantifiable differences between baseline measures, or be compared to real world situations, would be ideal. However, ensuring that physiological or behavioral responses are "directly related" to the level of presence being experienced in a VE is difficult at best. People experience varying degrees of presence while immersed in a VE, typically due to divided attention between the physical world and the virtual world (Singer & Witmer, 1997). Therefore, regardless of the technique used to measure presence, researchers must keep in mind that the point in time in which presence measures are obtained may affect the results.

Due to the inherent weaknesses in both types of measures, it may be preferable to utilize a combination of subjective and objective measures of presence to obtain a comprehensive understanding of this phenomenon. Table 40.2 provides a summary of the currently available presence measures, indicating their strengths and weaknesses.

**TABLE 40.2**
Measures of Presence

| *Presence Measure* | *Type of Measure* | *Description of Measure* | *Strengths* | *Weaknesses* |
|---|---|---|---|---|
| Rating scales | Subjective | Rate level of presence experienced in a VE. | Direct perception of user. | Interrater reliability may be weak; limited utility for comparison of experiences across diverse environments. |
| Subjective report | Subjective | Directed, open-ended questions of reactions and impressions related to presence. | Direct perception of user. | Interpreting results may be difficult due to response variability. |
| Comparison-based prediction | Subjective | Distinguish between a real-world scene and a VE simulating that scene or between alternative VE systems (standardized comparison cases may be used). | No bias due to definition of terms. | Size of just noticeable difference dependent on stimulus characteristic compared. |
| Cross-modality matching | Subjective | Represent experience of one sensory modality through another modality. | Can be used when verbal scaling is inappropriate or difficult to quantify. | Interpreting results may be difficult due to response variability. |
| Behavioral measures | Objective | Reflexive physical reactions or behaviors evoked by stimuli or events occurring in a VE. | Directly influenced by the VE, no subjective bias. | May not reflect influences of stimuli or events on total presence. |
| Physiological indicators | Objective | Physiological responses that occur when experiencing VE stimuli or events occurring within it. | Can serve to isolate influences on presence, objectively measured. | Internal or external "noise" may affect measures, intrusiveness, and reliability problems. |

## 6. RELATIONSHIP BETWEEN PRESENCE AND TASK PERFORMANCE

Once presence can be appropriately measured, its relation to other factors, such as task performance, can be assessed. Stanney et al. (1998, p. 164) suggest that although the necessity of presence to support performance in a VE is presently unclear, "the potential relation (1) has considerable face validity, (2) is suggested by perceptual and cognitive theories that indicate that positive transfer increases with the level of original learning as long as structurally similar stimuli and responses are available and required for both the training and transfer tasks

(Schmidt & Young, 1987), and (3) is supported by early but limited empirical evidence (Bailey & Witmer, 1994; Witmer & Singer, 1994)." These issues provide the impetus to explore this relationship more fully.

Performance on several VE tasks has been shown to be positively related to subjective presence, including tracking performance (Ellis, Dorighi, Menges, Adelstein, & Jacoby, 1997) and search task performance (Pausch, Proffitt, & Williams, 1997), as well as performance on a sensorimotor tasks (Maida, Aldridge & Novak, 1997). Some studies have found psychomotor task performance to be positively related to subjective presence (Witmer & Singer, 1994), although others have not (Singer, Ehrlich, Cinq-Mars, & Papin, 1995). Similarly, one study found spatial knowledge to be positively related to presence (Singer, Allen, McDonald, & Gildea, 1997), but another did not (Bailey & Witmer, 1994).

A recent study of performance on a battery of tasks, modeled after the Virtual Environment Performance Assessment Battery (VEPAB) (Lampton et al., 1994), found subjective presence to be positively correlated with both task performance and the amount of movement (i.e., yaw and roll) experienced throughout the VE interaction (Stanney, 2000). The results indicated that for tasks requiring VB movement (i.e., locomotion, choice reaction time, and object manipulation), presence was negatively correlated to performance time and movement distance, indicating that as performance time and distance decreased (and thus task efficiency increased) sense of presence increased. For stationary object manipulation tasks (i.e., where the VB did not move), there was no relation found between presence scores and performance measures.

Taken together, these studies provide early evidence that performance may indeed be positively related to presence. Greater study is needed, however, to fully characterize this relationship, with particular attention being warranted for task specific influences of presence. Although designers often aspire to create the ultimate in reality, VE systems should only be expressly designed to engender high levels of presence when the causal relationship between presence and a given VE task is empirically validated, or has clear experiential value, such as in entertainment-based VE applications (see chap. 55, this volume).

## 7. CONCLUSIONS

While there has been considerable discussion of the concept of presence, the utility of this construct has yet to be definitively determined (Stanney et al., 1998). There are even those who suggest that there is little utility to this construct and its relationship to performance. It is clear that with the current shortcoming of valid measures of presence, the fundamental research required to determine the inherent worth of this construct will be difficult to conduct properly. Stanney et al. (1998) suggest that presence may be intimately related to certain VE attributes, such as interactivity and involvement, which seem to be vital components in the creation of believable if not "realistic" virtual worlds. Thus, for certain purposes, the challenge may be to determine how best to improve the effectiveness and quality of the VE experience rather than striving for the ultimate in presence. However, for the entertainment industry (see chap. 55, this volume), increasing the sense of presence in VEs will undoubtedly continue to be a highly desirable attribute to attract consumer gamers.

## 8. ACKNOWLEDGMENTS

This material is based upon work supported in part by the Office of Naval Research (ONR) under grant No. N000149810642, the National Science Foundation (NSF) under grants No. DMI9561266 and IRI-9624968, and the National Aeronautics and Space Administration

(NASA) under grants No. NAS9-19482 and NAS9-19453. Any opinions, findings, and conclusions or recommendations expressed in this material are those of the authors and do to necessarily reflect the views or the endorsement of the ONR, NSF, or NASA.

## 9. REFERENCES

Bailey, J., & Witmer, B. (1994). Learning and transfer of spatial knowledge in a virtual environment. In *Proceedings of the Human Factors and Ergonomics Society 38$^{th}$ Annual Meeting* (pp. 1158–1162). Santa Monica, CA: Human Factors and Ergonomics Society.

Baños, R. M., Botella C., Garcia-Palacios, A., Villa, H., Perpiña, C., Alcañiz, M. (2000). Presence and reality judgment in virtual environments: A unitary construct? *CyberPsychology and Behavior, 3*(3), 327–335.

Barfield, W., & Danas, E. (1996). Comments on the use of olfactory displays for virtual environments. *Presence: Teleoperators and Virtual Environments, 5*(1), 109–121.

Barfield, W., & Hendrix, C. (1995). The effect of update rate on the sense of presence within virtual environments. *Virtual Reality: The Journal of the Virtual Reality Society, 1*(1), 3–16.

Barfield, W., & Weghorst, S. (1993). The sense of presence within virtual environments: A conceptual framework. In G. Salvendy & M. Smith (Eds.), *Human–computer interaction: Software and hardware interfaces* (pp. 699–704). Amsterdam: Elsevier Science Publishers.

Billinhurst, M. & Weghorst, S. (1995). The use of sketch maps to measure cognitive maps virtual of environments. In *Proceeding of Virtual Reality Annual International Symposium (VRAIS '95)*, pp. 40–47.

Bystrom, K.-E., Barfield, W., & Hendrix, C. (1999). A conceptual model of sense of presence in virtual environments. *Presence: Teleoperators and Virtual Environments, 8*(2), 241–244.

Cohn, V., DiZio, P., & Lackner, J. (1996). Reaching movements during illusory self-rotation show compensation for expected Coriolis forces. *Society for Neuroscience Abstracts, 22*(1), 654.

Dillon, C., Keogh, E., Freeman, J., & Davidoff, J. (2000, March). Aroused and immersed: *The psychophysiology of presence.* Paper presented at the Third International Conference on Presence, Delft Technical Museum, Delft, the Netherlands.

Dillon, C., Keogh, E., Freeman, J., & Davidoff, J. (2001, May). *Presence: Is your heart in it?* Presented at the Fourth Annual International Workshop on Presence, Tample University, Philadelphia. Available: http://homepages.gold.ac.uk.immediate/immersivetv

Draper, J. V., Kaber, D. B., & Usher, J. M. (1998). Telepresence. *Human Factors, 40*(3), 354–375.

Ellis, S. (1996). Presence of mind: A reaction to Thomas Sheridan's "Further musings on the psychophysics of presence." *Presence: Teleoperators and Virtual Environments, 5*(2), 247–259.

Ellis, S., Dorighi, N., Menges, B., Adelstein, B., & Jacoby, R. (1997). In search of equivalence classes is subjective scales of reality. In M. Smith, G. Salvendy, & R. Koubek (Eds.), *Design of computing systems: Social and ergonomic considerations* (pp. 873–876). Amsterdam: Elsevier Science Publishers.

Fontaine, G. (1992). Experience of a sense of presence in intercultural and international encounters. *Presence: Teleoperators and Virtual Environments, 1*(4), 482–490.

Freeman, J., Avons, S. E., Meddis, R., Pearson, D. E., & IJsselsteijn, W. A. (2000). Using behavioural realism to estimate presence: A study of the utility of postural responses to motion stimuli. *Presence: Teleoperators and Virtual Environments, 9*(2), 149–164.

Freeman, J., Avons, S. E., Pearson, D., & IJsselsteijn, W. A. (1999). Effects of sensory information and prior experience on direct subjective ratings of presence. *Presence: Teleoperators and Virtual Environments, 8*(1), 1–13.

Held, R. M., & Durlach, N. I. (1992). Telepresence. *Presence: Teleoperators and Virtual Environments, 1*(1), 109–112.

Held, R., & Durlach, N. (1987). *Telepresence, time delay and adaptation.* Pictorial Communication in Virtual and Real Environments (S. Ellis, M. Kaiser, A. Erunwald, Eds.), Taylor and Francis, Ltd., London.

Heeter, C. (1992). Being There: The subjective experience of presence. *Presence: Teleoperators and Virtual Environments, 1*(2), 262–271.

Hendrix, C., & Barfield, W. (1996a). Presence within virtual environments as a function of visual display parameters. *Presence: Teleoperators and Virtual Environments, 5*(3), 274–289.

Hendrix, C., & Barfield, W. (1996b). The sense of presence within auditory environments. *Presence: Teleoperators and Virtual Environments, 5*(3), 290–301.

Jex, H. (1988). Measuring mental workload: Problems, progress, and promises. In P. A. Hancock & N. Meshkati (Eds.), *Human mental workload.* Amsterdam: North-Holland.

Kandel, E., & Schwartz, J. (1985). *Principles of neural science.* New York: Elsevier Science Publishers.

Kennedy, R. S., Stanney, K. M., & Dunlap, W. P. (2000). Duration and exposure to virtual environments: Sickness curves during and across sessions. *Presence: Teleoperators and Virtual Environments, 9*(5), 466–475.

Kim, T., & Biocca, F. (1997). Telepresence via television: Two dimensions of telepresence may have different connections to memory and persuasion [Online]. *Journal of Computer Mediated Communication, 3*(2). Available: http://www.ascusc.org/jcmc/vol3/issue2/kim.html

Klein, G. A. (1982). The use of comparison cases. In *IEEE 1982 Proceedings of the International Conference on Cybernetics and Society* (pp. 88–91). Seattle, WA.

Knasko, S., & Gilbert, A. (1990). Emotional state, physical well-being, and performance in the presence of feigned ambient odor. *Journal of Applied Social Psychology, 20*(16), 1345–1357.

Lampton, D. R., Knerr, B. W., Goldberg, S. L., Bliss, J. P., Moshell, J. M., & Blau, B. S. (1994). The virtual environment performance assessment battery (VEPAB): Development and evaluation. *Presence: Teleoperators and Virtual Environments, 3*(2), 145–157.

Lessiter, J., Freeman, J., Keogh, E., & Davidoff, J. (2001). A cross-media presence questionnaire: The ITC-Sense of Presence Inventory. *Presence: Teleoperators and Virtual Environments, 10*(3), 282–297.

Maida, J., Aldridge, A., & Novak, J. (1997). Effects of lighting on human performance in training. In M. Smith, G. Salvendy, & R. Koubek (Eds.), *Design of computing systems: Social and ergonomic considerations* (pp. 877–880). Amsterdam: Elsevier Science Publishers.

Minsky, M. (1980, June). Telepresence. *Omni*, 45–51.

Pausch, R., Proffitt, D., & Williams, G. (1997). Quantifying immersion in virtual reality. *Computer Graphics Proceedings, Annual Conference Series/ACM SIGGRAPH* (pp. 13–18). Los Angeles: ACM SIGGRAPH.

Prothero, J., & Hoffman, H. (1995). *Widening the field-of-view increases the sense of presence within immersive virtual environments* (Human Interface Technology Laboratory Tech. Rep. No. R-95-4). Seattle: University of Washington.

Sadowski, W. (1999). Special report: Utilization of olfactory stimulation in virtual environments. *VR News, 8*(4), 18–21.

Salnass, E. (1999, April). *Presence in multi-modal interfaces.* Paper presented at the 2nd International Workshop on Presence, University of Essex, Colchester, England.

Schloerb, D. (1995). A quantitative measure of telepresence. *Presence: Teleoperators and Virtual Environments, 4*(1), 64–80.

Schmidt, R., & Young, D. (1987). Transfer of movement control in motor skill learning. In S. M. Cormier & J. D. Hagman (Eds.), *Transfer of learning: Contemporary research and applications* (pp. 47–79). San Diego, CA: Academic Press.

Schubert, T. W., Friedmann, F., & Regenbrecht, H. T. (1999, April). Decomposing the sense of presence: Factor analytic insights [Online]. Paper presented at the Second International Workshop on Presence, University of Essex, Colchester, England. Available: http://www.uni.jena~de/~sth/vr/insights.html

Sheridan, T. (1992). Musings on telepresence and virtual presence. *Presence: Teleoperators and Virtual Environments, 1*(1), 120–125.

Singer, M., Allen, R., McDonald, D., & Gildea, J. (1997). *Terrain appreciation in virtual environments: Spatial knowledge acquisition* (Tech. Rep. No. 1056). Alexandria, VA: U.S. Army Research Institute for the Behavioral and Social Sciences.

Singer, M., Ehrlich, J., Cinq-Mars, S., & Papin, J. (1995). *Task performance in virtual environments: Stereoscopic vs. monoscopic displays and head-coupling* (Tech. Rep. No. 1034). Alexandria, VA: U.S. Army Research Institute for the Behavioral and Social Sciences.

Singer, M., & Witmer, B. (1997). Presence: Where are we now? In M. Smith, G. Salvendy, & R. Koubek (Eds.), *Design of computing systems: Social and ergonomic considerations* (pp. 885–888). Amsterdam: Elsevier Science Publishers.

Singer, M., & Witmer, B. (1999). On selecting the right yardstick. *Presence: Teleoperators and Virtual Environments, 8*(5), 566–573.

Slater, M. (1999). Measuring Presence: A response to the Witmer and Singer Presence Questionnaire. *Presence: Teleoperators and Virtual Environments, 8*(5), 560–565.

Slater, M., & Steed, A. (2000). A virtual presence counter. *Presence: Teleoperators and Virtual Environments, 9*(5), 413–434.

Slater, M., & Usoh, M. (1993a). The influence of a virtual body on presence in immersive virtual environments. In *Virtual reality international—Proceedings of the Third Annual Conference on Virtual Reality.* London: Meckler.

Slater, M., & Usoh, M. (1993b). Representations systems, perceptual position, and presence in virtual environments. *Presence: Teleoperators and Virtual Environments, 2*(3), 221–233.

Slater, M., Usoh, M., & Chrysanthou, Y. (1995). The influence of dynamic shadows on presence in immersive virtual environments. In M. Goebel (Ed.), *Second eurographics workshop on virtual reality.* Monte Carlos: Blackwell Publishers: Eurographics Association.

Slater, M., Usoh, M., & Steed, A. (1994). Steps and ladders in virtual reality. In *ACM Proceedings of VRST '94—Virtual Reality Software and Technology.* Singapore: World Scientific Publishing. 45–54.

Slater, M., & Wilbur, S. (1997). A framework for immersive virtual environments (FIVE): Speculations on the role of presence in virtual environments. *Presence: Teleoperators and Virtual Environments, 6*(6), 603–616.

Snow, M. P., & Williges, R. C. (1998). Empirical models based on free-modulus magnitude estimation of perceived presence in virtual environments. *Human Factors, 40*(3), 386–402.

Stanney, K. M. (2000). Unpublished research data, University of Central Florida.

Stanney, K. M., Lanham, S., Kennedy, R. S., & Breaux, R. B. (1999). Virtual environment exposure drop-out thresholds. In *Proceedings of the 43rd Annual Human Factors and Ergonomics Society Meeting* (pp. 1223–1227). Santa Monica, CA: Human Factors and Ergonomics Society.

Stanney, K. M., Salvendy, G., Deisigner, J., DiZio, P., Ellis, S., Ellison, E., Fogleman, G., Gallimore, J., Hettinger, L., Kennedy, R., Lackner, J., Lawson, B., Maida, J., Mead, A., Mon-Williams, M., Newman, D., Piantanida, T., Reeves, L., Riedel, O., Singer, M., Stoffregen, T., Wann, J., Welch, R., Wilson, J., & Witmer, B. (1998). Aftereffects and sense of presence in virtual environments: Formulation of a research and development agenda. *International Journal of Human–Computer Interaction, 10*(2), 135–187.

Steuer, J. (1992). Defining virtual reality: Dimensions determining telepresence. *Journal of Communication, 42*(2), 73–93.

Strickland, D., & Chartier, D. (1997). EEG measurements in a virtual reality headset. *Presence: Teleoperators and Virtual Environments, 6*(5), 581–589.

Usoh, M., Catena, E., Arman, S., & Slater, M. (2000). Using presence questionnaires in reality. *Presence: Teleoperators and Virtual Environments, 9*(5), 497–503.

Weiderhold, B., Davis, R., & Weiderhold, M. (1998). The effects of immersiveness on physiology. In Riva, G. & Weiderhold, M. (Eds.), *Virtual Environments in Clinical Psychology and Neuroscience*. Amsterdam: Ios Press.

Welch, R. (1999). How can we determine if the sense of presence affects task performance? *Presence: Teleoperators and Virtual Environments, 8*(5), 574–577.

Welch, R., Blackmon, T., Liu, A., Mellers, B., & Stark, L. (1996). The effects of pictorial realism, delay of visual feedback, and observer interactivity on the subjective sense of presence. *Presence: Teleoperators and Virtual Environments, 5*(3), 263–273.

Wilson, J., Nichols, S., & Haldane, C. (1997). Presence and side effects: Complementary or contradictory? In M. Smith, G. Salvendy, & R. Koubek (Eds.), *Design of computing systems: Social and ergonomic considerations* (pp. 889–892). Amsterdam: Elsevier Science Publishers.

Witmer, B., & Sadowski, W. (1998). Non-visually guided locomotion to a previously viewed target in real and virtual environments. *Human Factors, 40*(3), 478–488.

Witmer, B., & Singer, M. (1994). *Measuring immersion in virtual environments* (Tech. Rep. No. 1014). Alexandria, VA: U.S. Army Research Institute for the Behavioral and Social Sciences.

Witmer, B., & Singer, M. (1998). Measuring presence in virtual environments: A presence questionnaire. *Presence: Teleoperators and Virtual Environments, 7*(3), 225–240.

Youngblut, C., Johnson, R., Nash, S., Wienclaw, R., & Will, C. (1996). *Review of virtual environment interface technology* [Institute for Defense Analysis IDA Paper No. P-3186, Online]. Available: http://www.hitl.washington.edu/scivw/IDA

Zahoric, P., & Jenison, R. L. (1998). Presence and being-in-the-world. *Presence: Teleoperators and Virtual Environments, 7*(1), 78–89.

# 41

# Ergonomics in Virtual Environments

Pamela R. McCauley Bell
*University of Central Florida*
*Industrial Engineering and Management Systems*
*4000 Central Florida Blvd.*
*Orlando, Florida 32816-2450*
*Mcbell@mail.ucf.edu*

## 1. INTRODUCTION

The growth in technical and computing capabilities have increased the ability to design complex software systems capable of producing highly immersive environments, specifically virtual environments (VEs). Although it has many definitions, a virtual environment can be described as a technologically created, immersive, and interactive space designed to invoke the sensory systems into perceiving a desired condition or situation. Advances in VE technology have created interest for applications of this technology in training, education, industrial applications and entertainment (see chaps. 42–56, this volume). The growing applications warrant ergonomic analysis, particularly as this becomes a common tool in the work environment. Just as in any work environment, it is important to understand the impact that a tool or procedure is having on users from many perspectives. Research has shown that the use of VEs is a risk in the development of ergonomic related disorders such as repetitive stress injury (Belea, 1996). Definitive techniques are available for reducing the risk of these types of injuries in the "real world," and this knowledge can be extrapolated in the virtual world before these injuries escalate.

## 2. IMPORTANCE OF ERGONOMICS IN VIRTUAL ENVIRONMENTS

Virtual environments have been applied in medical, aerospace, entertainment environments, and various other industries with the hope of even broader application. As more VE applications are developed and utilized, the likelihood of injury or risk due to preventable ergonomic factors increases. Additionally, the usability of a VE system is influenced by the ergonomic strength and usability of the physical aspects of the system (Bolas, 1994). Thus, efforts to identify and mitigate the risks associated with this technology are necessary.

## 2.1 Background

Virtual environment applications range from personalized immersion systems to public location-based systems, from devices requiring high-performance platforms to those generating desktop and laptop virtual worlds (Noor & Ellis, 1996). In a personal immersive environment the operator may wear a stereo display system, gloves, or other handheld devices accompanied by one or more position trackers. This allows the operator to experience sensory cues primarily in the visual, auditory, and tactile senses, and sometimes olfactory, gustatory, and vestibular senses as well. Location-based systems appeal to the same sensory modalities utilizing wide projection screens to create a stereo VE for individuals or groups of people. The type and degree of ergonomic concerns are clearly dependent on the type of virtual environment employed.

Virtual environment technology poses a design situation that is different from most existing human–computer interaction (HCI) design. Thus, guidelines and principles specifically tailored to the design of VEs are necessary. Limited research has been done to address such issues. One similar design situation that has been addressed to some degree is that of wearable computer systems.

Gemperle, Kasabach, Stivonic, Bauer, and Martin (1998) have codified a number of design principles for wearable computer systems. These principles focus on treating the human body and its range of motions rather than hardware or software engineering preferences. Basically, this technique of design principles recognizes that the human body is not composed of uniform blocks. Rather, it is flexible in some places, inflexible in others, and has a variety of ranges of motions based on the underlying bone, muscle, tendon, and ligament structures. The Gemperle et al. (1998) principles also reflect the general categories and subcategories of fit, feel, and safety as relevant areas of assessment for wearability. For a system to be considered wearable, the user must be able to walk, bend, crouch, reach, and climb with the equipment on. For VE applications, the user may also need to be able to lie down and run. Military training simulations

**TABLE 41.1**
Ergonomic Commonalties in Wearable Computers and Virtual Environments

| *Ergonomic Category* | *Wearable Computers* | *Virtual Environments* |
|---|---|---|
| Anthropometrics | Static and dynamic anthropometric measurements should be considered in design. | Static and dynamic anthropometric measurements should be considered in design. |
| Biomechanics | User mobility while wearing computer.<br>Impact on joint and full body loading.<br>Distribution of load. | Mobility of entire body or specific aspects of the body affected by the VE.<br>Load mobility.<br>Distribution of load. |
| Computer comfort | Design for long term wearing. | Generally, wearing time is short term. |
| Durability | Computer must be designed to be durable in a variety of harsh environments. | Generally, VEs do not have as harsh environments as do wearable military computers. |
| Cardiovascular demands | Consider weight, duration of use, task activities, and impact that these issues have on cardiovascular system. | Consider weight, duration of use, task activities, and impact that these issues have on cardiovascular system. |

may require soldiers to lie down and "crawl" forward on their stomachs or run to various parts of a building they are clearing. Nevertheless, most design factors for wearable computers can be applied to VE devices. The ergonomic principles to be considered in the design of wearable computers are closely related, but not identical to those of virtual environments. Table 41.1 shows the commonalties among the areas of ergonomic concern in designing these systems.

Although wearable computers provide some insights into ergonomic issues associated with VE systems, in general, many of the principles and guidelines developed to evaluate traditional HCI will not apply to VE systems. There are, however, those that have applicability with minimal refining (Wilson, 1997). For example, from a physical ergonomic standpoint, there exists a good body of literature that can be applied to the design of VE systems. The objective of this chapter is to offer guidance on the ergonomic techniques available for application in VE system design. This guidance is developed through review of three main areas:

- Ergonomic research needs in VE system design
- Classification of physical ergonomic issues in VE system design
- Application of ergonomics in VE system design

## 3. ERGONOMIC RESEARCH NEEDS IN VE SYSTEM DESIGN

The continual growth of VE technology suggests that the need for ergonomic research will grow. A number of research issues have been outlined as areas for future ergonomic analysis (Wilson, 1997). In this chapter, additional areas have been identified as relevant VE system design issues in need of ergonomic research. While some of these areas have been extensively addressed from a cognitive or human factors–oriented standpoint, many of them contain a strong element of physical ergonomic research need. These areas include:

- Multiple sensory channels
- Participant representation
- Head-mounted displays (HMDs)
- Navigation and orientation
- Presence and involvement (pictorial scene and steoreoscopic displays)
- Temporal constraints
- Task design within the VE system
- Biomechanics

Although each of these areas has aspects of ergonomic concern, some will be more significantly influenced through effective ergonomic application. The areas that have immediate ergonomic need include multiple sensory channels, participant representation, HMDs, task design, and biomechanics. Each of these focus areas is discussed in more detail below.

### 3.1 Multiple Sensory Channels

For comprehensive simulation of a referent environment, some researchers believe that all senses must be stimulated. Present-day VE systems are strongly dependent on visual stimulation with some form of auditory information. Additional research is being conducted to provide force and tactile feedback and to develop sensors of more than just body movement (Burdea, 1995; Hirota & Hirose, 1995; Wilson, 1997; see chaps. 5–6, this volume). Much participation from the ergonomic community is needed in the development of such advanced multimodal interaction because the risk of physical injury or trauma from such interaction is of real concern (Stanney, Mourant, & Kennedy, 1998). When stimulating multiple senses, VE equipment can become cumbersome and complex. For example, when immersive technology (i.e., and HMD) is used, obstructed vision of one's natural surroundings could lead users to fall or trip over obstacles in the real world, resulting in injury. The weight of the HMD could result in body imbalance, causing a falter or fall (Thomas & Stuart, 1992). When provided with three-dimensional audio, excessive volume could lead to noise induced hearing loss. In addition, sound cues could distract users causing them to fall during immersion. Among the challenges in producing effective virtual tactile feedback is the ability to replicate the force and/ or pressure that would be produced in an environment. For example, if a virtual keyboard is being used, the user should receive the feedback that is similar to that received when utilizing an actual keyboard. To accomplish this, types of reactive forces, textures and responses to varying levels of pressure must be understood and modeled in the VE system. Yet if such force feedback systems fail it could lead to a user being accidentally pinched, pulled, or otherwise harmed. Fortunately, precautions are in place, as most force feedback systems attenuate transmitted forces to avoid injury (Biocca, 1992).

### 3.2 Participant Representation

The representation of the user and other participants in a VE requires ergonomic consideration from the standpoint of generating realistic, occupational interaction. Generally representation of body forms and movement mechanisms in computer-aided design (CAD) modeling systems has been relatively pictorially unsophisticated. The body of ergonomic knowledge could be used to make a substantial improvement in the design of manikins and virtual body parts—or avatars within a VE (Wilson, 1997). Some of these concerns were addressed in a special issue of presence (Thalman & Thalman, 1994). More research is needed in the areas of perceptual and motor performance of participant viewing and in the ergonomics of mixed (augmented) environments. For example, if real controls are used in the VE but are required to be located via the HMD, more research is needed to improve such interactions between "worlds" so that they can be deftly performed. Additional research in this area will enhance the realism and usefulness of virtual environments.

### 3.3 Head-Mounted Displays

Head-mounted displays have improved considerably since they were first introduced. However, there still remains a good deal of research to be done on development of anthropometrically accommodating, comfortable HMDs that can be worn for extended periods. A fairly comprehensive summary of designing HMDs for users can be found in Melzer and Moffitt (1997). Their review indicated that lack of fit was the main complaint by HMD wearers. The lack of proper fit can cause irritation and discomfort when using the device. The discomfort is severe enough that the growth of VE applications could be impeded if HMD design does not improve. Participants today do not voluntarily spend long periods of time in VEs, which can be attributed to HMD discomfort. This discomfort can be attributed to a number of factors including inappropriate fit, movement obstruction from tethers, excessive HMD weight, and improper distribution of this weight, which can produce premature muscular fatigue and inhibit the quality of the experience in a virtual environment. To minimize the likelihood of these occurrences, knowledge of anthropometrics and biomechanics should be applied in the design of operator-friendly head-mounted displays. The weight of HMDs is very important, as it can impact comfort, duration of wearing, and overall quality of the VE system. Variation in weight across HMDs varies widely, weighing from 4 ounces to more than 5 pounds. In particular, the weight of the HMD should be kept to a minimum because greater loading on the head and neck can lead to an increase in neck and upper extremity discomfort. Additionally, the load should be located so that it is aligned with the spine to minimize the moment acting on the neck and lower back. In one study, as neck flexion increased from less than 55 degrees to more than 66 degrees, as could occur with a heavy HMD, reports of neck pain increased from approximately 10% to more than 50% (Hunting, Laubli, & Grandjean, 1981). In addition, limits on use should be identified for current heavier HMDs, which can weigh more than 5 pounds. With this loading, it is likely that exposure durations should be kept to a minimum to reduce head and neck strain.

### 3.4 Biomechanics

Occupational biomechanics focuses on physical exertion, muscular exertion, quantifying forces, and torques experienced by the musculoskeletal system while working, as well as quantifying, energy expenditure. For more details on the science and application of biomechanics, see Chaffin (1999). When considering equipment used in VE, a number of biomechanical concerns may surface. The use of HMDs and restrictions imposed by gloves, suits, or sensory feedback mechanisms can impact the mobility and efficiency of a VE user. Weight is one of the foremost issues in this category. Is the equipment too heavy, improperly balanced, or awkward, and, therefore, creating too much strain on the joints and muscles of the body or fatiguing the user? As noted above, unbalanced HMD weight can create strains on muscles and eventually lead to fatigue. In order to reduce the weight and strain from an HMD, some systems are now suspending HMDs from a floor or ceiling-based master arm (Pausch, Snoddy, Taylor, Watson, & Haseltine, 1996). However, there are other types of biomechanical strain, such as unnatural forces on joints or tendons that might lead to muscle pulls and strains, tendonitis, or bursitis. Chaffin (1999) provides general design considerations regarding physical activity such as work posture, work rate, and typical ranges of movement for extremities from a biomechanics perspective. The followings should specifically be addressed from a biomechanics standpoint in VE design:

- Load weight for handheld devices and HMDs
- Load balancing
- Distribution of forces
- Joint angles
- Velocity and acceleration of movement in task design
- Load task and movements place on joints and spinal system

Proper ergonomic design of these factors may lead to improved comfort and usability of physical characteristics associated with VE systems.

In evaluating the biomechanics of the VE equipment system designers should consider the likelihood and impact in long-term use. Although some VE exposures are relatively short, others may be used for extended periods of time. Similarly, the system may also be used repeatedly over short or long periods of time. In either instance, the devices must be safe for human use and not put the user at risk for joint, muscle, tendon, or fatigue problems. To avoid these risks it is important to observe the following:

- Distribute loads appropriately across the center of gravity
- Reduce force
- Improve posture
- Reduce repetitions

Principles of handgrip design are very important in VEs that utilize DataGloves as the primary means for operator input. Sanders and McCormick (1987) discuss ergonomic principles of hand tool and device design such as maintaining a straight wrist, avoiding tissue compression stress, and avoiding repetitive finger action, which can be used to enhance the design of VE DataGloves. The hand and fingers are capable of a wide range of movements and grasping actions, which depend largely on the ability to bend the wrist and rotate the forearm. Maximum grasping force can be quadrupled by changing from holding with the fingertips to clasping with the whole hand (Kroemer & Grandjean, 1998). The power of the fingers is greatest when the hand is lightly bent upward (dorsiflexion), and in contrast, grasping strength and level of skilled operation are reduced if the hand is bent downwards or turned to either side. Inclining the hand either outward (ulnar deviation) or inward (radial deviation) reduces rotational ability by approximately 50% (Tichauer, 1975). The result of research in this area suggests that from a biomechanics standpoint, the hands should be kept in line with the forearms during task performance. Specifically, for DataGlove design the wrist should be allowed to maintain neutral posture and tasks, especially those of longer duration, should require minimal grip strength. Additionally, tasks should distribute the forces applied by the hand across the fingers, rather than allocating force application to the thumb or index finger alone.

## 4. CLASSIFICATION OF PHYSICAL ERGONOMIC TECHNIQUES IN VE SYSTEM DESIGN

The objective of ergonomics is to fit the task to the human. Whether the task is considered to be an occupational effort, training, or recreational there is an opportunity for application of ergonomic principles. The classification of physical ergonomic concerns for the purposes of research and analysis is often partitioned into anthropometrics, cardiovascular, musculoskeletal, and psychomotor. Seldom is an ergonomic condition solely attributable to a single area, and, thus, analysis and treatment of ergonomic conditions generally adopt a multicategory approach.

### 4.1 Anthropometrics

Anthropometry is the science that involves the measurement of body dimensions. By measuring body lengths, girths, and breadths, one can describe, with frequency of distributions, the population's size. Anthropometric data are useful in the design of workplaces, equipment, and products relative to people's dimensions. They have been used to develop guidelines for heights, reaches, grips, and clearances. In general, the goal is to use anthropometric measurements to make workplaces, equipment, and products fit the capacities for reach, grasp, and clearance that accommodate a majority of the workforce. Thus, this science is significant in the design and operation of virtual environments.

The essence of anthropometric application in VE system design can be summarized by the "fit" of the workplaces. Do the devices anatomically conform to the human body, appendages, and sense organs in both static and dynamic motions? There are both anatomical and hardware engineering factors to consider in developing VE equipment to fit the body and deliver appropriate sense information to the user. If the equipment does not fit properly, it will not deliver the correct information to the user or feed back accurate information into the simulation. For example, if a haptic feedback system does not fit the hand well, it could produce sensations in the wrong part of the hand or fingers and, thereby, give the user incorrect information about selected objects in the virtual environment.

**TABLE 41.2**
HMD-relevant Anthropometric Dimensions for the Head (Dimensions in MM)

| | *Women* | | | | *Men* | | | |
|---|---|---|---|---|---|---|---|---|
| | *5th Percentile* | *50th Percentile* | *95th Percentile* | *Standard Deviation* | *5th Percentile* | *50th Percentile* | *95th Percentile* | *Standard Deviation* |
| Head Circumference | 523 | 546 | 571 | 15 | 543 | 568 | 594 | 15 |
| Head length | 165 | 180 | 195 | 8 | 180 | 195 | 210 | 8 |
| Head breadth | 135 | 145 | 155 | 6 | 145 | 155 | 165 | 6 |
| Interpupillary breadth | 57 | 62 | 69 | 4 | 59 | 65 | 71 | 4 |

*Source: Body Space: Anthropometry, Ergonomics and the Design of Work* ($2^{nd}$ ed.), by Stephen Pheasant

In applying anthropometric considerations to VE design, developers must understand the relationship between the equipment and the relative body component. For example, if an HMD is being designed, user population anthropometrics should be considered for relevant dimensions, such as head circumference, head length, forehead length, and distance between eyes. Obviously, no piece of equipment will accommodate every user; generally, ergonomists strive to accommodate as many users as possible. A default is to design systems to capture the anthropometric constraints of those contained within the $5^{th}$ percentile female and $95^{th}$ percentile male. Table 41.2 provides anthropometric dimensions for the head that are applicable to VE system design, specifically HMD design.

In VE system design, this degree of accommodation may not be necessary, depending on the anticipated user population and the likelihood that a variety of sizes of equipment will be made available (e.g., small, medium, and large DataGloves). The decision to have variable sizes can increase the probability of proper fit and reduce the cost associated with development. For a more comprehensive summary of HMD design guidelines, see *Head-Mounted Displays: Designing for the User* (Melzer & Moffitt, 1997).

In addition to the dimensions associated with the head, the hand is an area that requires anthropometric consideration in design of VE systems. Table 41.3 provides guidelines for anthropometric estimates of the hand for the American population.

## 4.2 Musculoskeletal Issues in VE System Design

Musculoskeletal issues are those problems that strain the muscular and skeletal systems. This is a common problem in physically intensive work places and those environments that require extensive use of a particular muscular area (e.g., hand-intensive tasks). In a VE, the likelihood for musculoskeletal problems can be increased when users are required to perform awkward movements, participate in repetitive activities, lift objects, or exert consistent muscular effort. If lifting is required in the VE, such as in picking up a rifle during a military operation, developers should be mindful of tools to measure the impact that the task is having on the operator. The National Institute for Occupational Safety and Health (NIOSH) lifting guide (Kroemer, Kroemer, & Kroemer-Elbert, 2000) is widely used in the American occupational environment to quantifying the impact of lifting and can be applied to VE system design.

**TABLE 41.3**
Anthropometric Dimensions of the Hands (All Dimensions in MM)

| | *Men* | | | | *Women* | | | |
|---|---|---|---|---|---|---|---|---|
| | *$5^{th}$ Percentile* | *$50^{th}$ Percentile* | *$95^{th}$ Percentile* | *Standard Deviation* | *$5^{th}$ Percentile* | *$50^{th}$ Percentile* | *$95^{th}$ Percentile* | *Standard Deviation* |
| Hand length | 173 | 189 | 205 | 10 | 159 | 174 | 189 | 9 |
| Palm length | 98 | 107 | 116 | 6 | 89 | 97 | 105 | 5 |
| Thumb length | 44 | 51 | 58 | 4 | 40 | 47 | 53 | 4 |
| Thumb thickness | 19 | 22 | 24 | 2 | 15 | 18 | 20 | 2 |
| Hand breadth (across thumb) | 97 | 105 | 114 | 5 | 84 | 92 | 99 | 5 |
| Hand breadth (minimum) | 71 | 81 | 91 | 6 | 63 | 71 | 79 | 5 |
| Maximum grip diameter | 45 | 52 | 59 | 4 | 43 | 48 | 53 | 3 |
| Maximum spread | 178 | 206 | 234 | 17 | 165 | 190 | 215 | 15 |
| Maximum functional reach | 122 | 142 | 162 | 12 | 109 | 127 | 145 | 11 |
| Minimum square access | 56 | 66 | 76 | 6 | 50 | 58 | 67 | 5 |

*Source: Body Space: Anthropometry, Ergonomics and the Design of Work* ($2^{nd}$ ed.), by Stephen Pheasant

The 1991 NIOSH Lifting guideline equation (NIOSH, 1992) allows, as a maximum, a "Load Constant" (LC)—permissible under the most favorable circumstances—with a value of 23 kg (51 lb). The 1991 NIOSH Lifting guideline equation is as follows:

RWL = LC*HM*VM*DM*AM*FM*CM

Where:
LC—load constant of 23 kg or 51 lb.
(Each remaining multiplier may assume a value [0,1].)
HM—the horizontal multiplier: H is the horizontal distance of the hands from the ankles (the midpoint of the ankles).
VM—the vertical multiplier: V is the vertical location (height) of the hands above the floor at the start and end points of the lift.
DM—the distance multiplier: where D is the vertical travel distance from the start to the end points of the lift.
AM—the asymmetry multiplier: where A is the angle of asymmetry, i.e., the angular displacement of the load from the medial (mid-saggital plane) that forces the operator to twist the body. It is measured at the start and end points of the lift.
FM—the frequency multiplier: where F is the frequency rate of lifting, expressed in lifts per minute.
CM—the coupling multiplier: where C indicates the quality of coupling between hand and load.

Though lifting may not be of concern in most VE systems, there are tasks such as training or other workplace augmented reality simulations where these guidelines may apply.

When considering the risks of musculoskeletal disorder (MSD) in virtual environments, analysis suggests that the most probable risk for these injuries exists for the upper extremities,

**TABLE 41.4**
Risk Factors for CTDs*

| *Occupational Risk Factors* | *Personal Characteristics* |
|---|---|
| Force | Previous CTD |
| Repetition | Habits and hobbies |
| Awkward joint posture | Diabetes |
| Hand tools | Thyroid-related problems |
| Length of work shift | Age |
| Low-frequency vibration tools | Arthritis and related joint diseases |

*Adapted from McCauley-Bell & Badiru (1996b).

neck, and back. MSDs are also referred to as cumulative trauma disorders (CTDs), overuse injuries, and a variety of other names that suggest their origin is the result of aggregate impact. Specifically, cumulative trauma disorders are injuries that primarily affect soft tissues, including muscles, tendons, tendon sheaths, joint surfaces, and nerves. Recent surges in the occurrence of upper-extremity CTDs has industry concerned and has been blamed for loss in productivity, inhibition in quality, low worker morale, and soaring workers compensation costs. The application of VE technologies should make every effort to avoid the pitfalls associated with introducing an activity that has known CTD risks.

The same considerations that are applied to occupational settings should be utilized in the design of virtual workplaces. The risks are varied, as illustrated by Crumpton and Congleton (1994), who demonstrated that the risk factors for CTDs include various occupational, personal, and organizational factors (see Table 41.4). In particular, VEs can benefit from knowledge of the occupational and personal categories. For example, in the development of systems, if extensive hand movement is required, designers can distribute the task over two hands or reduce the cycle time to minimize the amount of stress that this activity produces on the musculoskeletal system of the hand.

Proactive measures should be instituted with regard to other risk factors listed in Table 41.4; however, the most significant occupational risks to avoid are excessive repetition, awkward joint posture, and excessive force. For example, if an operator in a VE is required to perform a task wearing a DataGlove, efforts should be made to limit the amount of wrist joint deviation, force required to manipulate virtual objects, and repetitions involved with such manipulation. These are the primary factors that produce musculoskeletal injuries of the upper extremities (McCauley-Bell & Badiru, 1996b). Research has demonstrated that force, repetition and awkward posture comprise a combined weight of 0.66 of the risk level associated with task characteristics (McCauley-Bell & Badiru, 1996a).Thus, efforts should be made to minimize the presence of these conditions. Table 41.5 presents guidelines to follow when attempting to minimize the presence of CTD risk factors.

The application of these principles can help reduce the likelihood for injury in real-world workplaces, and theoretically should reduce the likelihood for injury in VE systems. If a quantifiable tool is desired for risk evaluation ratings in a proposed VE system for CTD risks, a numeric model can be utilized. For a model that can be useful in quantifying the risks of CTDs see McCauley-Bell (1999).

**TABLE 41.5**
Guidelines for Minimizing CTD Risk Factors

- Reduce repetition by increasing cycle time.
- Minimize amount of time that a person has to do a particular task.
- Provide rest breaks.
- Use ergonomic designs that minimize joint angle and force exertion.
- Minimize the amount of force required for activation of tools or controls.
- Attempt to eliminate the need for pinch strength and minimize the use of excessive grip strength. Provide palm, wrist, and forearm supports.
- Provide physical fitness training.

**Guidelines for Percent of Max Aerobic Capacity in Task Performance**

| Duration | Percent Max. Aerobic Capacity | Effort Level |
|---|---|---|
| 8 hours | 33% | Moderate |
| 1 hour | 50% | Heavy |
| 20 min | 70% | Very Heavy |
| 5 min | 85% | Extremely Heavy |

FIG. 41.1. Guidelines for maximum aerobic capacity in task performance.

### 4.3 Cardiovascular

Cardiovascular areas of concern in VE may be an issue any time a user is asked to perform an action that increases the stress level on the heart by a significant margin. This margin is a function of the heart rate over a given amount of time. Figure 41.1 provides guidelines for determining the aerobic effort level in task performance. When designing a task, efforts should be made to minimize the amount of very heavy work. In cases where this cannot be avoided, providing minimal exposure time for heavy and extremely work will reduce the likelihood of operator discomfort and injury.

### 4.4 Cognitive

Cognitive problems arise when there is either information overload or under load during information processing. This area of VE design is covered elsewhere in this volume (see chap. 20).

### 4.5 Psychomotor

Psychomotor problems include those conditions that require recognition, detection, and response to stimuli with a physical movement or a reply. Although recent technology has reduced lag time in VEs, the impact of temporal differences and perceptual lag on operator performance is an area of ergonomic consideration. Performance, user health, and acceptance of virtual technology may be at risk given the lags in current VE systems (Wilson, 1997). One of these temporal constraints includes the time lag in updating systems from a visual perspective, registering frame rate update. Richard et al. (1996) demonstrated that frame rate can have an impact on operator performance in virtual worlds. Additional temporal constraints include delay in tracking the body position or movement. Such delays could impede psychomotor performance and may be why users often experience difficulty when first trying to move about and manipulate objects in a virtual environment (Lampton et al., 1995). Thus, VE system designers cannot expect effective psychomotor performance until such technological issues are resolved.

## 5. APPLICATION OF ERGONOMICS IN VE SYSTEM DESIGN

This section provides an overview of the physical ergonomic issues to be considered in the design and use of VE technology. When considering the virtual environment, there are three broad, potentially overlapping categories of tools and interaction. These categories are sensors, effectors, and input devices:

- Sensors are those tools used to sense the behavior of the user, such as the head position tracker used to measure the user's head orientation.
- Effectors are used to stimulate user's senses, such as a miniature video display in a head-mounted display.
- Input devices, such as a mouse, DataGlove or joystick, complete the feedback and sensory loop in the virtual environment.

These three categories broadly categorize VE equipment and will be referenced as the application of ergonomic principles to VE technology is discussed.

### 5.1 Ergonomic Issues Associated With Sensors

In VEs the necessary feedback on operator status is presented through sensors. Sensors utilized in VE technology have been in the form of body suits, hand tracers, head-mounted tracers, and a host of other feedback mechanisms. These sensors detect motion and allow the system to respond with knowledge of the operator's physical state and create the illusion of presence (see chap. 40, this volume) in the virtual world. Effective application of ergonomic technology can enhance the quality of sensory information and operator comfort. The techniques for obtaining sensory feedback should not overload any sensory mode, but should instead be designed to best utilize human senses while promoting the greatest likelihood for perception and response to stimuli.

The primary means of providing sensory feedback in occupational and virtual environments are vision, audition, and tactation (i.e., touch). Some research is being done in the use of olfaction, and this can certainly be used to facilitate a more realistic environment. However, the three senses for which ergonomics consideration will be discussed include visual, auditory, and tactile.

#### *5.1.1 Ergonomics of the Visual Field*

Although other researchers have studied the aspects of design and presentation of visual information in virtual environments (see chaps. 3, 12, and 20, this volume), a few ergonomic factors are worth highlighting. When considering a user's response in a VE, the following should be noted given that the area measured in degrees can be seen by both fixated eyes:

- Horizontal: In the center, each eye is occluded by the nose. To the sides each eye can see a bit over 90 degrees.
- But color can only be seen about 65 degrees out; the farther you go to the periphery, shades of gray dominate.
- Upward: The visual field extends about 55 degrees and is limited by the eyelids.
- Color can be seen only to about 30 degrees upward.
- Downward: Vision is limited by the cheek at about 70 degrees.
- Color vision degradation begins at about 40 degrees.

For more in-depth review of the visual considerations in equipment development, see Kroemer, Kroemer, and Kroemer-Elbert (2000).

### 5.1.2 Ergonomics of the Auditory Field

To create systems that are realistic it is important to utilize, at a minimum, the visual and auditory senses. An abundance of knowledge is available and has been applied to audible issues in VE system design. The primary considerations from an ergonomic perspective would be to follow accepted occupational guidelines and use sound to add meaning to system operation. From the safety standpoint, the introduction of high levels of noise begins to be detrimental when it exceeds 90 dB(A). Guidelines for limits on exposure to noise level are shown in Table 41.6.

Thus, if the desire is to create a virtual world with noise levels comparable to an actual environment, caution should be used in the exposure of operators to db(A) levels that could produce damage.

### 5.1.3 Ergonomics of Tactation

The primary tactile sensors include the following:

- Mechanoreceptors, which sense tactation, i.e., contact touch, pressure
- Thermoreceptors, which sense warmth or cold relative to each other and the body's neutral temperature
- Noci-receptors, which sense pain including sharp pain, dull pain, etc.
- Electroreceptors, which respond to electrical stimulation of the skin

Depending on the type of training, all of these tactile sensory types could be used in VE system designs. Mechanoreceptors and thermoreceptors could be extremely useful in creating realism.

**TABLE 41.6**
Limits for Noise Exposure

| *Duration of Exposure* | *dB(A)* |
|---|---|
| 8 hours | 90 |
| 6 hours | 92 |
| 3 hours | 97 |
| 90 minutes | 102 |
| 30 minutes | 110 |

Electroreceptors could be used to stimulate the vestibular system and thus create a sense of motion (see chap. 7, this volume), whereas noci-receptors could be used in military trainers that need to train pain tolerance. Clearly, however, the latter two should be avoided unless they are an explicit element of the system design.

When using mechano-receptors and thermoreceptors, efforts should be made to minimize the amount of exposure to pressure and ensure the operator does not experience thermal discomfort. To understand the impact that mechanical or load limits may be placing on body parts, a biomechanical analysis may be performed. Such analyses calculate external and internal forces as a function of load mass, joint angle, and body movement. Section 5.2.2 provides a discussion of the necessary elements for performing a biomechanical analysis.

Understanding the thermal conditions that lead to body comfort is important in creating a realistic environment. The human body has a complex control system for maintaining deep-body-core temperature very close to 37°C (about 99°F). Designers of VE systems, and the rooms that they are situated in, should be aware that thermal conditions and body comfort are a function of four physical factors: air (or water) temperature, humidity, air (or water) movement, and the temperature of surfaces that exchange energy by radiation. Additionally, level of physical exertion can influence an operator's comfort level.

Virtual environment system designs that use thermoreceptors should maintain a comfortable environmental temperature. In order to assess the comfort level associated with an environment, two approaches can be utilized: Wet-Bulb Globe Temperature (WBGT) and subjective measures of thermal comfort. The WBGT index is generated by an instrument with three sensors, whose readings are automatically weighted and then combined (Kroemer et al., 2000). In this calculation, the combined effects of climate parameters are weighted as follows:

Outdoors: $WBGT = 0.7WB + 0.2GT + 0.1DB$

Indoors: $WBGT = 0.7WB + 0.3GT$

Where WB is the wet-bulb temperature of a sensor in a wet wick exposed to a natural air current, GT is the globe temperature at the center of a black sphere 15 cm in diameter, and DB is the dry-bulb temperature measured while the thermometer is shielded from radiation. The WBGT is often used to assess the effects of warm or high climates, but it is important to understand the metabolic rate of the operator when applying guidelines for safe WBGT temperatures. Table 41.7 shows the "safe" WBGT values given a particular metabolic rate.

In the absence of WBGT measuring equipment, subjective assessments can be performed. Two of the most common scales for assessing subjective thermal comfort include the Bedford Scale and the American Society of Heating, Refrigeration and Air conditioning Engineers (ASHRAE; Kroemer, Kroemer & Kroemer-Elbert, 1997). Each of these scales is shown in Table 41.8. To apply these scales, the operator must be queried during varying levels of task performance for thermal comfort. Prior to the analysis, the operator is to be provided with the linguistic term and corresponding numeric rating. For simplicity sake, the rating scale should be posted in view of the operator during task performance or provided at each query. At specific times during task performance, the operator is asked, "What number best describes your thermal comfort?" This information is then recorded. Efforts should be made to minimize the amount of time that operators spend at the extremes of these scales as this can inhibit successful task performance.

**TABLE 41.7**
Safe WBGT Values*

| *Metabolic Rate (M) in Watts* | *"Safe" WBGT (°C)* | |
|---|---|---|
| | *Person Acclimatized to Heat* | *Person Not Acclimatized to Heat* |
| $M \leq 117$ | 33 | 32 |
| $117 < M \leq 234$ | 30 | 29 |
| $234 < M \leq 360$ | 28 | 26 |
| $360 < M \leq 468$ | No air movement: 25 | No air movement: 22 |
| | With air movement: 26 | With air movement: 23 |
| $M > 468$ | No air movement: 23 | No air movement: 18 |
| | With air movement: 25 | With air movement: 20 |

*Adapted from Kroemer, Kroemer, and Kroemer-Elbert (2000), from ISO 7243, 1982

**TABLE 41.8**
Scales for Assessing Subjective Thermal Comfort

| *Bedford* | | *ASHRAE* | |
|---|---|---|---|
| Much too warm | 7 | Hot | +3 |
| Too warm | 6 | Warm | +2 |
| Comfortably warm | 5 | Slightly warm | +1 |
| Comfortable | 4 | Neutral | 0 |
| Comfortably cool | 3 | Slightly cool | −1 |
| Too cool | 2 | Cool | −2 |
| Much too cool | 1 | Cold | −3 |

## 5.2 Ergonomic Issues Associated With Effectors

When considering effectors such as HMDs, virtual caves, and augmented displays, a number of physical ergonomic considerations from anthropometric, biomechanical, and musculoskeletal perspectives must be considered. For example, the weight of any type of system for which the body is required to bare the load is important, as well the impact that the load produces on the musculoskeletal system and biomechanic abilities. The degree of impact will largely depend on the duration of task performance and characteristics of the load. The following areas are of particular ergonomic concern with respect to effectors:

- Anthropometric issues
- Biomechanical issues
- Musculoskeletal issues

Each of these areas is discussed in detail below.

### 5.2.1 Anthropometric Issues Associated With Effectors

When designing effectors it is important to design for the appropriate population. Such considerations include age (e.g., youth, adults, senior citizens), gender, physical abilities (e.g., handicapped issues), and the region of the body that will interface with the effector. As aforementioned, if an HMD is being designed, several linear dimensions of the head should be considered. Additional rotational anthropometric dimensions should be considered such as range of vertical and horizontal motion for the head, neck, torso, and limbs. Understanding and applying comprehensive knowledge of these anthropometric issues will result in more comfort and usability for VE systems. Anthropometric guidelines can be found in *Body Space: Anthropometry, Ergonomics and the Design of Work* (Pheasant, 1997).

### 5.2.2 Biomechanical Issues Associated With Effectors

From a biomechanical standpoint, effectors should be designed such that they are consistent with the natural movements of the body and do not cause excessive strain. Specific issues include designing the weight and distribution of loads in effectors so that they do not introduce inappropriate loads on the body. For example, an HMD should be designed so that it does not produce an inappropriate strain on the neck or head. The load should be distributed so that the head, neck, and upper body appropriately manage it. For VE systems design, a biomechanical analysis (see Chaffin, 1999) should be performed for the region of the body that utilizes the effector. This analysis will reveal the external and internal loads being placed on the joints. This analysis should be done for static conditions at a minimum and dynamic loading when the operator is required to move throughout the virtual environment. To perform a biomechanical analysis treating the body as a system of links, the following information is required:

- Mass of loads being handled
- Estimated weight of segments
- Segment lengths
- Joint angles

In the absence of specific segment weights, these values can be estimated using equations for estimating segment mass from total body weight (see Kroemer et al., 1997). After these values are determined for a given environment, the biomechanics analysis can be conducted. The basic steps for conducting a biomechanical analysis using the multisegment static model analysis includes the following:

- Treat each segment as a separate link in a kinetic chain.
- Start the analysis at the point of application of external loads (usually hands).
- Calculate forces and moments acting in opposing directions for adjacent segments.
- Proceed in sequence, solving equilibrium equations for each succeeding body segment.
- Continue until you reach the segment that supports the body (usually the feet).

This analysis will provide forces and moments acting on the joints and other specific locations (i.e., lower back). For additional details on static and dynamic biomechanical analysis, see Chaffin (1999).

### 5.2.3 Musculoskeletal Issues Associated With Effectors

Effective application of musculoskeletal concerns in effectors design should reduce the demands on the musculoskeletal system by minimizing repetition, force, and awkward posture in effector use. In the design of effectors, an analysis of the regions of the body interfacing with the system should be performed in order to understand the impact of potential strain associated with the musculoskeletal system. Musculoskeletal issues and injuries to avoid with effectors and input devices are discussed in section 5.4.

## 5.3 Ergonomic Issues Associated With Input Devices

Users have many input device options to help them navigate throughout a virtual environment. Some of these devices include but are not limited to a mouse, DataGlove, or auditory commands. Similar to computer games, the traditional mouse is often used to allow users to traverse through virtual environments. The use of the mouse is therefore a positive tool to employ because most users are familiar with this equipment and have used it before. The DataGlove is a less familiar type of input device for most users, but may be familiar to video games players. Auditory commands are less frequently used as input mechanisms due to the reliability of voice command software. The defense industry is conducting research with such devices.

### *5.3.1 Musculoskeletal Issues Associated With Input Devices*

During continual use of an input device such as a mouse or DataGlove, there will be an increased risk of upper-extremity musculoskeletal consequences. This increased risk is due to repetition and awkward joint postures required for maneuvering and manipulation. When designing a VE system using input devices, an analysis should be done to rate risk of cumulative trauma disorders (McCauley-Bell & Badiru, 1996a). In cases where extensive hand manipulation is required, systems should be designed to allow distribution of load over both limbs. Additional task issues in the design of VE systems and the injuries that can result are presented in section 5.4

## 5.4 Upper Extremity Musculoskeletal Risks for Effectors and Input Devices in VE Systems

Although there has not been an extreme prevalence of musculoskeletal disorders associated with the use of VE systems, it is important to be aware of risk factors and the types of injuries that can result. Some of the primary work related musculoskeletal disorders experienced in American industry are discussed in the following sections.

### *5.4.1 Carpal Tunnel Syndrome (CTS)*

Perhaps the most widely recognized cumulative trauma disorders (CTD) of the hand and forearm region is CTS, a condition whereby the median nerve is compressed when passing through the bony carpal tunnel. The carpal tunnel is comprised of eight carpal bones at the wrist, arranged in two transverse rows of four bones each. The tendons of the forearm muscles pass through this canal to enter the hand and are held down on the anterior side by fascia called flexor and extensor retinacula, tight bands of tissue that protect and restrain the tendons as they pass from the forearm into the hand. If these transverse bands of fascia were not present, the tendons would protrude when the hand is flexed or extended. The early stages of CTS result when there is a decrease in the effective cross section of the tunnel. Subsequently, the median nerve, which accompanies the tendons through the carpal tunnel, is compressed and the resulting condition is CTS.

Early symptoms of CTS include numbness or tingling and burning sensations in the fingers. More advanced problems involve pain, wasting of the muscles at the base of the thumb, dry or shiny palms, and clumsiness. Many symptoms first occur at night and may be confined to a specific part of the hand. If left untreated, the pain may radiate to the elbows and shoulders, causing complications such as tendonitis. Tendonitis, inflammation of tendon sheaths around a joint, is generally characterized by local tenderness at the point of inflammation and severe pain on movement of the affected joint. Tendonitis can result from trauma or excessive use of a joint and can afflict the wrist, elbow (where it is often referred to as tennis elbow), and shoulder joints.

It is important for VE system designers to understand the risk factors for CTDs, such as CTS, so that they can be minimized in their designs. In general, CTD injuries primarily affect soft tissues, including the muscles, tendons, tendon sheaths, joint surfaces, and nerves. CTDs have been referred to as repetitive motion injuries, overuse syndrome, and a host of other names that tend to imply excessive use of an area. The primary risk factors for CTDs include:

- Extensive and continuous force exertion
- Frequency or extreme repetition by a specific muscle group
- Awkward joint posture (especially awkward postures of the wrist, elbow or shoulder joints)
- Vibration (including low-frequency-vibration hand tools)
- Excessive grip or pinch strength (5 times grip)
- Extreme temperature (hot or cold extremes)
- Long work shifts without work breaks (work shifts exceeding 8 hours)

Although these risk factors are specific to the occupational environment, they have applicability in the design of VE systems, especially as they become more commonly used in training and occupational settings.

#### *5.4.2 Tenosynovitis*

Tenosynovitis is a repetition-induced tendon injury that involves the synovial sheath. The most widely recognized tenosynovitis is Dequervain's disease. This disorder affects the tendons on the side of the wrist and at the base of the thumb. Excessive joint deviation, repetition, and twisting are known risk factors. Such postures and movements should be avoided in VE system design.

#### *5.4.3 Intersection Syndrome and Dequervain's Syndrome*

Intersection syndrome and Dequervain's syndrome occur in hand intensive workplaces. These injuries are characterized by chronic inflammation of the tendons and muscles on the sides of the wrist and the base of the thumb. Symptoms of these conditions include pain, tingling, swelling, numbness, and discomfort when moving the thumb. Design of end effectors should minimize the intensity of hand-related work.

#### *5.4.4 Trigger Finger*

If the tendon sheath of a finger is aggravated, swelling may occur. If the swelling is sufficient it may result in the tendon becoming locked in the sheath. At this point, if the person attempts to move his or her finger, the result is a snapping and jerking movement. This condition is called trigger finger. Trigger finger occurs to individual fingers and results when the swelling produces a thickening on the tendon that catches as it runs in and out of the sheath. Usually, snapping and clicking in the finger occurs with this disorder. These clicks occur when one bends or straightens the fingers (or thumb). Occasionally, a digit will lock either fully bent or straightened. This condition can result from overuse of a particular digit (pointer finger used to activate a trigger on a tool), and thus, VE system design should eliminate this type of risk.

#### 5.4.5 Ischemia

Ischemia is a condition that occurs when blood supply to a tissue is lacking. Symptoms of this disorder include numbness, tingling, and fatigue depending on the degree of ischemia or blockage of peripheral blood vessels. A common cause of ischemia is compressive force in the palm of the hand. Thus, in design of end effectors, every effort should be made to reduce or eliminate the application of forces with the palm of the hand.

#### 5.4.6 Vibration Syndrome

Vibration syndrome is often referred to as white finger, dead finger, or Raynaud's phenomenon. Excessive exposure to vibrating forces and cold temperature may lead to the development of this disorder, and thus these should be avoided in VE system design. This syndrome is characterized by recurrent episodes of finger blanching due to complete closure of the digital arteries. Thermoregulation of fingers during prolonged exposure to cold is inhibited by this condition.

#### 5.4.7 Ganglion Cysts

Ganglion is a Greek word meaning "a knot of tissue." Ganglion cysts are balloonlike sacs that are filled with a jellylike material. These maladies are often seen in and around tendons, in the palm of the hand, and at the base of the finger. These cysts usually occur as a result of excessive pushing, pressing, turning, or inserting motion, or as a result of forceful hand wringing. Such activity should be avoided in VE system design.

### 5.5 Prevention of Risks

In order to reduce risk of the previously described musculoskeletal disorders, the following considerations should be applied in design and use of VE systems.

1. Reduce repetition by increasing cycle time.
2. Minimize the amount of time for performance of a task with known risks factors in a virtual environment.
3. Provide rest breaks.
4. Design and purchase input devices that meet ergonomic guidelines.
5. Minimize the amount of force required for activation of tools or controls.
6. Attempt to eliminate the need for pinch strength and minimize the use of excessive grip strength.

As more VE applications are developed, it is important to incorporate ergonomic design considerations that reduce musculoskeletal disorders. This will minimize the extensive cost and injuries associated with many occupational environments.

## 6. CONCLUSIONS

Ergonomic factors to consider in VE system design can be captured through the use of ergonomic checklists. An abbreviated list, summarizing many of the ergonomic concerns reviewed in this chapter, which can be used to determine potential problems for VE system design, is presented in Table 41.9 below.

As a general reference, the International Standards Organization (ISO 9241, see http://www.iso.ch/welcome.html) provides standards that can be used to assess many of the ergonomic

**TABLE 41.9**
Ergonomic Checklist in VE System Design

1. Do sensors inhibit operator movement?
2. Is any limb overburdened?
3. Does task require extended latter or forward reaches beyond normal reach?
4. Do seating conditions meet ergonomic considerations for back support?
5. Are dials and controls easy to view and understand?
6. Is task more than 50% repetitive?
7. Is task performance required for more than 50% of work shift?
8. Does layout lead to efficient motions?
9. Are awkward postures required?
10. Is static loading required for task performance?
11. Is excessive force required?
12. Is twisting and lifting required?
13. Is forceful exertion required at awkward postures?
14. Is noise level within ergonomic guidelines?
15. Is sound level within ergonomic guidelines?

concerns of a VE system design discussed in this chapter. This standard provides design guidelines for task design, hardware and environmental factors, and software and usability concerns that could be used to guide effective ergonomic design of VE systems.

## 7. REFERENCES


ANSI Z-365: Working Draft (1994). Available: http://mimel.marc.gatech.edu/mime/ergo/ansi/net/a2.html

Biocca, F. (1992). Virtual reality technology: A tutorial. *Journal of Communication, 42*(4), 23–72.

Bolas, M. (1994). Human factors in the design of an immersive system. *IEEE Computer Graphics and Applications*, 55–59.

Burdea, G. (1996). *Force and touch feedback for virtual reality*. New York: John Wiley & Sons.

Bureau of Labor Statistics. *Safety and health statistics—workplace injury and illness summary* [Online]. Available: http://www.bls.gov/news.release/soh.hws.htm

Chaffin, D. B. (1999). *Occupational biomechanics* (3rd Ed.). New York: Wiley-Interscience Publication.

Crumpton, L. L., & Congleton, J. J. (1994). An evaluation of the relationship between subjective symptoms and objective testing used to assess carpal tunnel syndrome. *Advances in Industrial Ergonomics and Safety, 6*, 515–519.

(1997). Cumulative trauma disorders: Identification and treatment [Online]. Available: http://www.ucfw.org/

Dellman, N. J. (1997). ISO/CEN standards on risk assessment for upper limb repetitive movements. In *Proceedings of the 13th Triennial Congress of the International Ergonomics Association* (pp. 37–39).

Dempsey, P. G. (1997). The study of work-related low-back disorders in industry: Opportunities and pitfalls. *Advances in Industrial Ergonomics and Safety, 2*, 267–270.

Dortch, H., & Trombly, C. (1990). The effects of education on hand use with industrial workers in repetitive jobs. *The American Journal of Occupational Therapy, 3*, 777–782.

Gemperle, F., Kasabach, C., Stivoric, J., Bauer, M., & Martin, R. (1998, October). Design for wearability. Paper presented at the second International Symposium on Wearable Computers, Pittsburgh.

Hirota, K., & Hirose, M. (1995) Simulation and Presentation of Curved Surface in Virtual Reality Environment Through Surface Display. *IEEE Annual Visual Reality International Symposium*, 211–216.

Hunting, W., Laubli, T., & Grandjean, E. (1981). Postural and visual loads at VDT workstations. *Ergonomics, 24*, 917–931.

ISO Standards. Available: http://www.iso.ch/welcome.html

Kroemer, K. H. E., & Grandjean, E. (1997). *Fitting the task to the human: A textbook of occupational ergonomics* (5th ed.). Bristol, PA: Taylor & Francis.

Kroemer, K. H. E., Kroemer, H. J., & Kroemer-Elbert, K. E. (1997). *Engineering physiology: Bases of human factors/ergonomics* (3rd ed.). New York: Van Nostrand Reinhold.

Kroemer, K. H. E., Kroemer, H. J., & Kroemer-Elbert, K. E. (2000). *Ergonomics: How to design for ease and efficiency* (2nd ed.). Prentice-Hall.

Lampton, D. R., Knerr, B. W., Goldberg, S. L., Bliss, J. P., Moshell, M. J., & Blau, B. S. (1995). *The virtual environment performance assessment battery: Development and evaluation* (Tech. Rep. No. 1029). Alexandria, VA: U.S. Army Research Institute for the Behavioral and Social Sciences.

McCauley-Bell, P., & Badiru, A. (1996a, May). Fuzzy modeling and analytic hierarchy processing to quantify risk levels associated with occupational injuries: Part 1. The development of fuzzy linguistic risk levels. *IEEE Transactions on Fuzzy Systems.*

McCauley-Bell, P., & Badiru, A. (1996b, May). Fuzzy modeling and analytic hierarchy processing as a means to quantify risk levels associated with occupational injuries: Part 2. The development of a fuzzy rule-based model for the prediction of injury. *IEEE Transactions on Fuzzy Systems.*

Melzer, J. E., & Moffitt, K. W. (1997). *Head-mounted displays: Designing for the user*. New York: McGraw-Hill.

Noor, A. K., & Ellis, S, R. (1996, July). Engineering in a virtual environment. *Aerospace America*, 32–37.

Pausch, R., Snoddy, R., Taylor, R., Watson, S., & Haseltine, E. (1996). Disney's Aladdin: First steps toward storytelling in virtual reality. In *SIGGRAPH '96, Proceedings of the 23rd Annual Conference on Computer Graphics* (pp. 193–203). New Orleans, LA:

Pheasant, S. (1997). *Body space: Anthropometry, ergonomics and the design of work* (2nd ed.). London: Taylor & Francis.

(1997). *Prevention of back discomfort [Online]*. Available: http://www.hermanmiller.com/us/index. bbk/US/en/WR

Richard, P., Bierebent, G., Coiffet, P., Burdea, G., Gomez, D., & Langrama, N. (1996). Effect of frame rate and force feedback on virtual object manipulation. *Presence: Teleoperators and Virtual Environments, 5*, 95–108.

Sanders, M. S., & McCormick, E. J. (1987). *Human factors in engineering and design* (6th ed.). New York: McGraw-Hill.

Stanney, K. M., Mourant, R., & Kennedy, R. S. (1998). Human factors issues in virtual environments: A review of the literature. *Presence: Teleoperators and Virtual Environments, 7*(4), 327–351.

Thalman, D., & Thalman, N. B. (1994). *Artificial life and virtual reality*. New York: John Wiley & Sons.

Thomas, J. C., & Stuart, R. (1992). Virtual reality and human factors. In *Proceedings of the Human Factors Society 36th Annual Meeting* (pp. 207–210). Santa Monica, CA: Human Factors and Ergonomics Society.

Tichauer, E. R. (1975). *The biomechanical basis of ergonomics: Anatomy applied to the design of work situations.* New York: Wiley–Interscience.

Wilson, J. (1997). Virtual Environments and ergonomics: Needs and opportunities. *Ergonomics, 40*(10), 1057–1077.

# PART VI: Selected Applications of Virtual Environments

# 42

# Applications of Virtual Environments: An Overview

Robert J. Stone
*Virtual Presence*
*Chester House*
*79 Dane Road*
*Sale M33 7BP UK*
*r.stone@vrsolns.co.uk*

## 1. INTRODUCTION

Ever since the 1970s—for many the "dawn of technological proliferation"—researchers and developers the world over have taken measures to ensure that their particular field of endeavor stays very much away from the category of "a solution looking for a problem." Very few of these measures were actually successful. The early personal computers (PCs), image processors and robots, not to mention artificial intelligence systems, speech recognition devices, and neural networks all failed to gain real-world exposure and, thus, the early application credibility they truly deserved.

For some time, the pioneers of virtual environment (VE) technologies faced similar problems. The formative U.S. attempts in the early to mid-1980s to develop immersive VE for the purposes of enhancing cockpit situational awareness (e.g., Harvey, 1987) or for astronaut training and space telepresence (e.g., Fisher et al., 1988) progressed reasonably unscathed. However, elsewhere, pioneers faced obscurity in the wake of the meteoric rise in the late 1980s and early 1990s of the immersive entertainment industry. This sector alone did more than its fair share of damage to the attempts of small, innovative companies to develop credible application demonstrators in an attempt to persuade industrialists of the power and benefits of VE and to introduce the technology into wider mainstream commercial applications. Even today, the situation is far from satisfactory, although the VE community now has an extremely exciting story to tell, as evidenced by the many applications reviewed in this chapter.

## 2. APPLICATIONS OVERVIEW

On Wednesday, May 8, 1991, at a hearing of the U.S. Subcommittee on Science, Technology and Space, a special session opened called New Developments in Computer Technology—Virtual

Reality. Part of Al Gore's (now famous—at least in VE circles) introductory message is reproduced below:

> In the case of virtual reality [VR], . . . [recognizing and exploiting new opportunities] means providing the necessary funding for technology development, not just for military applications, but for civilian applications, like education, research, and advanced manufacturing. So far, we do not seem to be doing very well. We have small, $1-million, $2-million programs scattered throughout the federal government. These programs helped create virtual reality, but they are not doing very well in applying it.
>
> —Senator Al Gore (May 1991)

Despite this gloomy early outlook, the VE community can be consoled somewhat by the fact that, uniquely, a considerable amount of first class groundwork was conducted in the early-to-mid 1990s by small innovative companies (principally in the United States and United Kingdom). Groundwork that includes projects conducted with large blue-chip or national organizations in the defense, aerospace, petrochemical, and automotive markets, producing impressive engineering demonstrations and technical/commercial results of relevance to supply chains or other related sectors. In 1993 in the United Kingdom, for example, the author's team successfully launched an industrially sponsored VE initiative called VRS (Virtual Reality and Simulation; Stone, 1996), designed to deliver applications to some 20 participating industrial sponsors, each demonstrating the potential impact of VE on their commercial and technical business practices. Groundwork included efforts on a smaller scale, such as feasibility studies with institutions dedicated to demonstrating the power of VE in communicating with and educating the general public. Just a very small sample of applications spanning these and other areas will be presented in this chapter. To set the scene:

> . . . the research results provide clear evidence that VR technology and implementation expertise have advanced sufficiently for a wide range of commercial applications to be possible, some of which seem now to be showing the prospect of substantial business benefits.
>
> —Cydata (1998, p. 1)

> There is unmistakable evidence that VR awareness, interest and activity levels in the UK have all increased significantly during the past two years. In some sectors, VR is moving into a well-accepted mainstream role; in others, it is still being driven forward by the efforts of product champions.
>
> —Cydata (2000, p. 14)

## 2.1 Engineering

The ability to convert otherwise geometrically intensive CAD data into visually acceptable, real-time interactive models capable of forming a credible input to the business process of engineering companies is a feature of VE that is now well established. Endowing those models with behavioral capabilities extracted from associated engineering databases or complex simulation code is also becoming an accepted and proven process. The applications covered by the term "engineering" are, of course, many and varied (see Fig. 42.1) and cover heavy engineering sectors of petrochemical and hydroelectric plant design or construction sites in general, to the ergonomic and aesthetic evaluation of automobile and civil aircraft interiors, from the verification of maintenance strategies for aircraft engines to the prototyping of nuclear control rooms.

One internationally active group in the petrochemical plant arena has been pushing VE technology to its limits. Fluor Corporation is one of the world's largest publicly owned consulting, engineering, and construction companies. Founded nearly 100 ago, the company now

FIG. 42.1. Example of a VE for "heavy engineering". Image courtesy Virtual Presence.

operates in more than 50 countries worldwide and employs 50,000 staff. The corporation's work includes the construction of major manufacturing facilities for companies such as Pepsi, BMW, and ICI, the design of refineries for, among others, Shell and Chevron, the development of gold mines, the building and maintenance of power stations, and major works for world leading pharmaceutical organizations.

After nearly 6 years of involvement in VE, Fluor has established a number of in-house VE prototyping systems in the United Kingdom, the Netherlands, and on the West Coast of the United States. Their decision to proceed with integrating VE into their business practice was taken after a number of quite intensive trials of the technology, assessing its potential impact in such processes as:

- **"Instant" Change,** whereby the virtual plant models could be modified efficiently during run-time by moving objects, resizing, altering valve locations, and so on.
- **Operator Training,** displaying plant walkways, access/egress routes, vessel interiors, vessel shutdown and access procedures, and potential hazard areas.
- **"Touch-and-Tell"** component tagging, whereby objects in the VE could be interrogated simply by touching the object with the virtual hand or cursor. Once touched, data such as item name, main plant item number, manufacturer, and so on would appear in a separate window.
- **Line and System Tracing,** highlighting (through simple touch-and-tell action, as above) a line or system route so that its path can be traced through the plant.
- **Automatic Virtual Pipe Route Production,** in a similar vein to line tracing, although generating the route data in the VE automatically from early isometric drawings and engineering tables.
- **Constructability,** demonstrating, by means of a virtual crane system, the power of VE as a tool for aiding constructors' and contractors' decisions in, for instance, judging the accessibility of a given area or volume during major vessel installations.

- **Internals Installation and Removal,** allowing maintenance technicians to visualize and rehearse work before entering the real vessel, for example.
- **Collaborative Working,** allowing engineers immersed within a shared virtual world to repair or maintain plant equipment cooperatively.

## 2.2 Micro- and Nano-technology

Among the earliest reported applications of VE to the field of microtechnology and molecular modeling were the scanning tunneling microscopy (STM) demonstrations staged by the United Kingdom's National Advanced Robotics Research Center (ARRC; *VR News*, November, 1992) and the GROPE IIIb Project of the University of North Carolina (UNC; e.g., Brooks, 1988; Brooks, ouh-young, Batter, & kilpatrick, 1990). The United Kingdom's ARRC VE program was the first to demonstrate the power of allowing scientists to visualize and explore the atomic surfaces of Angstrom-sized material samples through the direct linking of an STM with a powerful (then transputer-based) real-time VE graphics engine. GROPE demonstrated haptic interaction (see chaps. 5 and 6, this volume) with electrostatic molecule-substrate force simulations and nano-scale surfaces (generated from STM data). Early work at UNC utilized an Argonne Remote Manipulator (ARM) master, one of two donated from the Argonne National Laboratory, and a large field sequential display system.

Today, developments in micro- and nano-technologies are proceeding at a considerable pace, with much talk about the future of miniature robotic systems capable of performing surgery or other minimally invasive therapy (see chap. 48, this volume). Indeed, this provides for a perfect applications opportunity for future VE developers and links the field's historical pedigree—the early space telepresence work, for instance—with the design, manufacture, operation, and training requirements of today. As NASA, the United Kingdom's ARRC, and related establishments adopted a human-centered approach to the use of VE to control macro-scale robots for hazardous environments, future microrobots and even nanobots—such as the DNA "screw-driver" (Mao et al., 1999), designed for deployment within the human body—will need strict supervision by skilled operators, equipped with advanced VE equipment. Already one German company, MicroTEC GmbH, has demonstrated a micro-submarine (at EXPO 2000, Hannover), 4 mm long and 450 $\mu$m in diameter, powered by induction and using a tiny magnet integrated within the screw (www.microtec-d.com/; see Fig. 42.2). The company is developing new forms of motive power that avoid the use of the conventional screw or propeller concept. Miniature cameras are being deployed regularly for a variety of preoperative investigations, and one

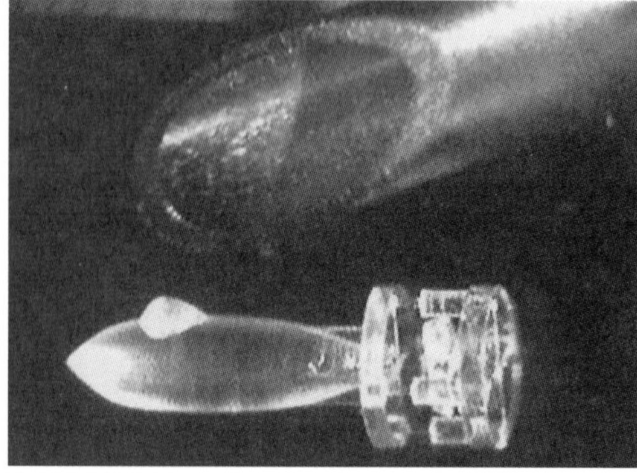

FIG. 42.2. MicroTEC's miniature submarine. Copyright micro TEC Gesellschaft für Mikrotechnologie mbH, Germany.

cannot ignore the European and U.S. developments in micro-lasers and micro-manipulators, such as Intuitive Surgical's da Vinci system (www.intusurg.com), all of which suggest a near-term revolution in the medical instrument fraternity (see chap. 47, this volume).

## 2.3 Aero and Space Engineering

Of the main engineering applications currently driving development and marketing efforts within VE companies, the aerospace sector probably ranks a close second to petrochemical and automotive and is on a par with naval and general defense developments. There are a number of "big-name" companies steering efforts in this field and, importantly, contributing to the development of VE products, many of which are still unable to service the complete organizational needs of the aerospace community, namely:

- Training (for maintenance, in particular, with future aspirations for collaborative training using networked VE)
- Prototyping (providing a VE interface for CAD design assessments and ergonomics; replacement of physical prototypes)
- Assembly planning and factory automation
- Engine and superstructure design and assessment (e.g., computational fluid dynamics data visualization, as with the NASA wind tunnel project)
- Marketing (interior cabin and cockpit designs; furnishings)

In 1992, Rolls-Royce (Aerospace, U.K.) conducted a short feasibility study to evaluate the role of VE in the future business of aero engine design, maintenance evaluation, and advanced training. At the outset, the project focused commercially on how VE technologies and design practices might be used to replace costly 1:1 scale physical preassemblies (PPAs). Working in tandem with the company's substantial investment in digital preassembly (DPA—Computervision CADDS4/5X), the first engineering application of VE was demonstrated and televised across the United Kingdom early in 1993, using the Rolls-Royce Trent 800 engine geometry. This interactive visualization was made possible by an early British parallel processing (transputer-based) computer developed by Division (now part of PTC), although one year later, real-time interaction using stereoscopic head-mounted displays (HMDs) and image projectors was demonstrated with more complex converted CADDS5X data hosted under Silicon Graphics' RealityEngine architecture.

By the spring of 1994, the virtual Trent 800 model had reached a stage whereby usability trials (see chap. 34, this volume) could be carried out (see Fig. 42.3). Potential future users from Rolls-Royce were employed as study participants and were equipped with a variety of HMDs, interactive gloves, and hand controllers. The participants were immersed in an environment that enabled them to move around and manipulate the virtual Trent geometry in a meaningful and credible fashion. Simple virtual tools were implemented, which allowed users to release bolt and bracket fittings and, with haptic (force/touch) feedback technologies in a very rudimentary state of development at that time, special visual cues were developed to assist the participants' perception of collisions between engine components.

Boeing has had an involvement with VE since 1990, the technology being a subset of the company's simulation-based design efforts. The Virtual Environments Group (Boeing Research and Technology) originally used VE for design visualization and evaluation, using a combination of CATIA with their own three-dimensional (3-D) rendering software (FlyThru). Early on, they recognized that this system (which was similar to Computervision's CADDS5/5X and PVS, or Dassault's 4D Navigation package) did not provide designers with the capability to look freely around the designs, nor did it enable the manipulation of objects. Since then,

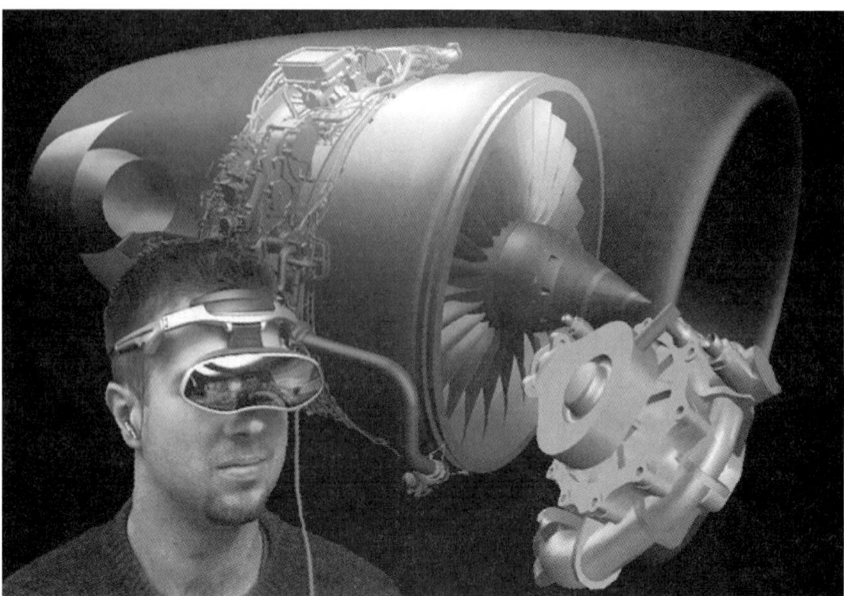

FIG. 42.3. The virtual Rolls-Royce Trent 800 Engine (1994–1995) with extracted gearbox component. Image courtesy Rolls-Royce plc and Virtual Presence.

Boeing has been evaluating and modifying proprietary VE systems. In one historic survey, the company ran a trial to assess reactions to the Boeing 777 cabin interior using VE and conventional CAD. Individuals who used the two systems manifested markedly different responses to the trials: In the CAD participants reported that they had "seen a CAD representation of the 777 interior"; in VE the report was "I've been in the 777 cabin and saw . . . " Other successful cabin interior designs have been reported by Airbus in Europe and Airline Services in the United Kingdom, both organizations having refitted aircraft virtually before committing to procuring real fabric.

More recently, Boeing researchers have been investigating the use of desktop and immersive VE as potential training delivery mechanisms for the Boeing Joint Strike Fighter (Barnett, Helbing, Hancock, Heininger, perrin, 2000). Using the replacement of an aircraft fuel valve as the target maintenance task, they found that immersive VE did not provide effective training. As well as poor depth perception (a common problem with narrow-field-of-view HMDs) and size distortion, participants reported difficulties with the interface itself (such as the manipulation of tools or small components). Barnett et al. (2000, p. 10) go on to state, "As a result of these unique features of the VR, four of the participants commented that *they focused more on interfacing with the VR than with learning the task*" (author's emphasis). Desktop VE, on the other hand, delivered comparable performance to that recorded using physical mock-ups. This problem of distraction caused by immersive VE interface technologies is not uncommon and requires strict attention when designing the virtual world content for such applications as training.

The European initiative ENHANCE (ENHanced AeroNautical Concurrent Engineering) brings together the main European civilian aeronautical companies and seeks to strengthen co-operation within the European aeronautical industry by developing common working methods that govern the European aeronautical field, defining appropriate standards and supporting concurrent engineering research. One project within ENHANCE concerns an advanced VE maintenance demonstrator that links a basic virtual mannequin with PTC's DIVISION MockUp virtual prototyping software and Sensable Technologies' PHANToM haptic feedback system.

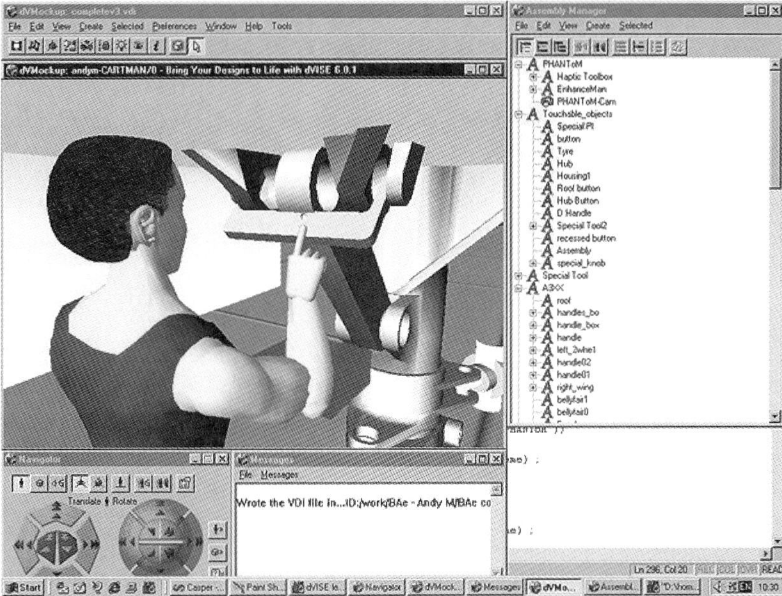


FIG. 42.4. The PTC DIVISION MockUp interface showing the PHANToM-controlled virtual mannequin interacting with an undercarriage safety lever. Image courtesy Virtual Presence.

Based on a 3-D model of a conceptual future large civil airliner, the VE demonstration involved controlling the mannequin during aircraft preparation and safety procedures (see Fig. 42.4), and in gaining access to retracted main landing gear for the purposes of wheel clearance testing.

Certain key interaction events throughout the demonstration were executed using the PHANToM. In order to define these stages clearly and to identify those procedures and events warranting the application of haptic feedback, a context-specific task analysis was carried out, as recommended in the 1999 International Standard ISO 13407 (Human-Centered Design Processes for Interactive Systems).

Turning now to space exploration, the use of VE in the U.S. space community is almost legendary, from the original NASA VIEW system, through the agency's award-winning Virtual Wind Tunnel of the early 1990s, to the use of digital panoramic and Virtual Reality Modeling Language (VRML) techniques to allow thousands to "explore" the July 1997 landing site of the *Mars Pathfinder.* It is also possible to explore the international space station on the Web, either by navigating module-by-module using digital panoramic techniques (i.e., MGI Software's Photovista; http://spaceflight.nasa.gov/gallery/vtour/index.html) or downloading files for use with contemporary games engines (e.g., http://nike.larc.nasa.gov/viss.html). The use of VE to prepare astronauts for the Hubble Space Telescope repair mission (STS-61) was itself a breakthrough, demonstrating how the technology had become a serious contender in astronaut EVA training, working alongside—and in some cases replacing—more conventional techniques based on physical mock-ups and neutral buoyancy procedures.

Indeed, NASA's history of involvement with VE technologies has served to demonstrate the successes that can be achieved when one pays full attention to the human factors issues associated with VE and associated interface technologies, including the integration of auditory cueing (see chap. 4, this volume), multisensory data representation (see chaps. 14 and 21, this volume), and speech recognition. Nowhere is this philosophy more important than in the training of air traffic controllers and NASA's most recent ATC tower simulator is exemplary in this respect (Fig. 42.5).

FIG. 42.5. NASA air traffic tower simulator. Image credit National Aeronautics and Space Administration.

Turning to Europe, in 1991, the European Space Research Technology Center, ESTEC (based in Noordwijk in the Netherlands), initiated a program of internal and contractual Research and Development to identify the suitability of VE for applications within the European space program. Toward the end of a period of research in which VE was featured in small projects dealing with telerobotics, spacecraft rendezvous, and docking (RVD) and extravehicular activity (EVA), the Simulation and Electrical Facilities Section at ESTEC set up the beginnings of a test bed that, they hoped, would link all the major technologies being developed by European contractors that boasted a simulation component. In 1992, external contracts were let to companies in France and Spain, to develop a generic VE software package and human–computer interaction (HCI), together with a high-resolution (HDTV) head-mounted display. Certainly some of the more recent demonstrations produced internally by ESTEC are quite impressive, particularly one that allows immersed users to explore the internals of the proposed Columbus Pressurized Module (Europe's "contribution" to the international space station) under visually simulated zero gravity, interacting with experimental modules using multisensory interfaces. Scientific visualization (see chap. 53, this volume) is also a key applications area for ESA's VE efforts, with visualization centers being established to deliver satellite remote sensing data to the European Community at large.

Following an early venture supported by the author's U.K. team to construct a desktop VE version of the *Mir* space station (see Fig. 42.6), Russia is also accelerating its adoption of VE in support of cosmonaut training. Despite the failure of VE (to date) to replace completely the full-size space vehicle mock-ups within the Cosmonaut Training Center (at Star City outside Moscow), networked VE workstations and PCs are being used in earnest to display real-time computer-generated orbital imagery, rather than the old practice of transmitting images of

FIG. 42.6. The VE *Mir* space station. Image courtesy ASRDC and Virtual Presence.

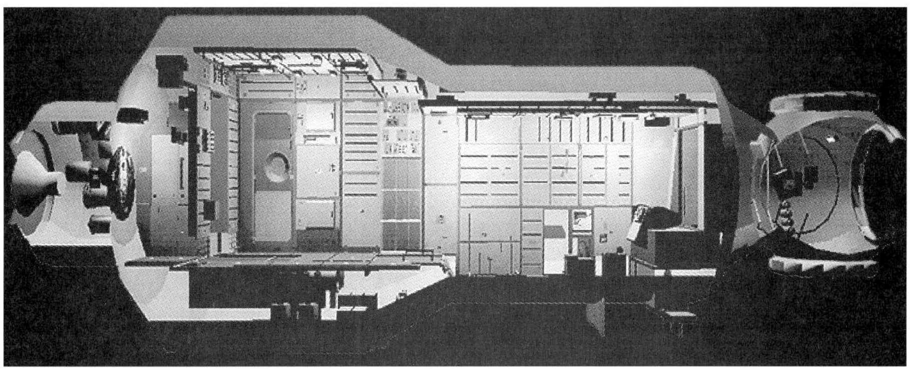

FIG. 42.7. A VE reconstruction of the Russian *Zvezda* space station module. Image courtesy Gagarin Cosmonaut Training Center (Star City).

suspended models of planets and spacecraft via TV screens located outside capsule view ports. Albatross (c. 1992) was Russia's first attempt to develop a single-channel real-time computer graphics imaging system for integration with the motion-base simulator for the *Buran* shuttle. Albatross provided a field of view of 30 degrees by 40 degrees over a simulated terrain area of 400 by 400 kilometers, with a maximum polygon count of 4,000 per scene. Although the system stood little chance of competing with Western graphics workstation developments, Albatross represented the CIS's first steps toward commercializing the many decades of mathematical modeling and simulation work for which they have become renown.

Russia's contribution to the international space station has also been the subject of a VE exercise, although the CTC still employs physical training modules, as currently exist for the *Zarya* control module and the *Zvezda* service module (both of which are in orbit at the time of writing). In cooperation with the Moscow-based Advanced Simulation Research and Development Center (ASRDC), a small group of developers within the CTC have produced a first class real-time training demonstrator of the *Zvezda* module (see Fig. 42.7), again with multiple PCs and a low-end SGI machine on a local area network. They have also been carrying out limited experiments with gestural interaction (see chap. 10, this volume) using a Polhemus Fastrak and Virtual Technologies' CyberGlove.

## 2.4 Ergonomics/Human Factors

As can be appreciated from the preceding subsections, ergonomics (or human factors) is playing an everincreasing role in engineering VE and simulation (see also Stone, 2001). There has been a marked rise in recent years in the number of commercial human mannequin packages, themselves endowed with anthropometric and biomechanics data, not to mention physiological libraries encompassing energy expenditure, clothing effects, psychomotor parameters, and so on. Not only do these packages provide the VE and human factors practitioner with a cost-effective assessment technique for engineering assessments, they also enable new industrial users of VE to do away with antiquated (and highly inaccurate) forms of "ergonomic" assessment based on toy figures, such as "GI Joe" or "Action Man". The use of ergonomics knowledge to bring a methodological credibility to many engineering projects based on VE is also growing steadily. Nowhere is this more evident than in the automotive industry where, after many years of experimenting with the technology, it is now accepted that VE in both real-time interaction and offline mannequin modes has a great deal to offer in improving the information interfaces between vehicle designers, simplified, low-cost physical mock-ups (such as seating or cockpit rigs), and end users. This close linkage of technologies with users is becoming increasingly important in a number of related applications areas (and, indeed, a requirement of such standards as ISO 13407).

For example, Fig. 42.8, although based on the Rolls-Royce Trent 800 project example discussed in section 2.3, also describes well the integrated VE–ergonomics approach adopted

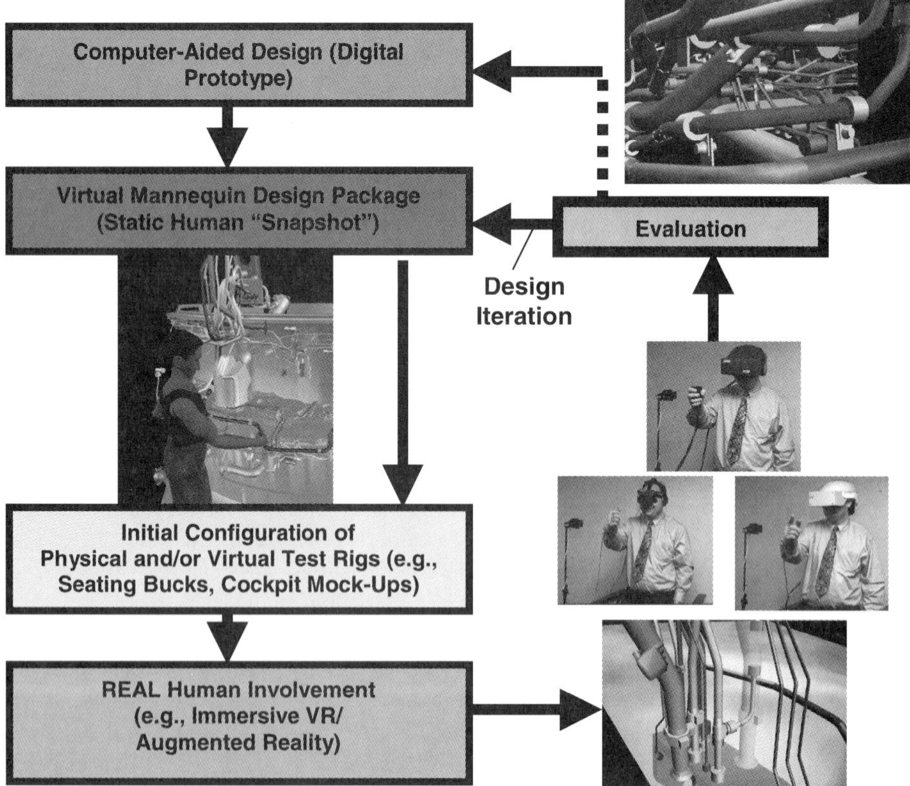


FIG. 42.8. Broad roles of mannequin simulators, physical mock-ups, and VE in the engineering design process. Image courtesy Virtual Presence.

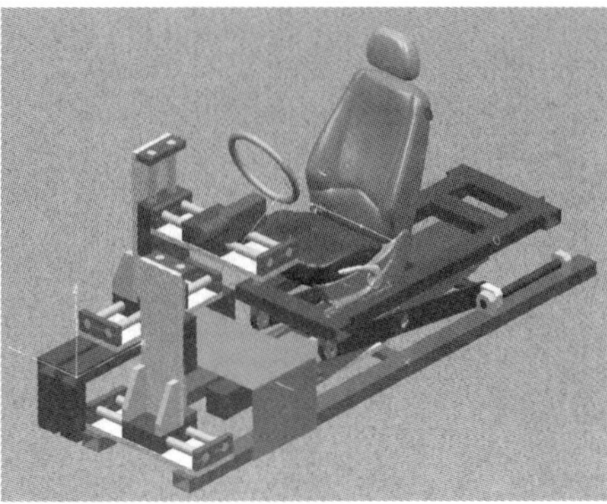

FIG. 42.9. Elasis virtual automobile interior and seat buck design concept. Images courtesy Gennaro Monacelli, project manager, Elasis Pomigliano d'Arco.

for automotive design by, for example, the Virtual Reality Center of Elasis Pomigliano d'Arco, a Fiat company in southern Italy (www.elasis.com; see also Fig. 42.9).

## 2.5 Defense

Defense establishments and military forces across the globe have long been exploiters of VE or "synthetic environments" technology, primarily in large-scale simulators designed for such activities as operations planning, war gaming, command-control-communications and intelligence ($C^3I$) and, of course tri-service pilot, navigator, and driver training (see chaps. 43–44, this volume). However, this has recently extended to part-task activities, such as seen with those military trainers that endow basic CAD or VE models of military platform subsystems with realistic behaviors, thereby enhancing the training of such procedures as maintenance, fault finding, and refit.

The favorable price–performance curve witnessed in recent years has stimulated defense procurement organizations to take a serious look at VE technology. Most have been impressed at the results, especially when compared to the limited and quite costly two-dimensional (2-D) computer-based training (CBT) packages they have been restricted to over the past decade. Virtual environments have been developed to create realistic military environments for such tasks as helicopter machine gun or ship close-range weapons training, parachuting experience, explosive ordnance disposal, naval helicopter deck landing, fire fighting, submarine and surface ship blind piloting, officer of the watch training, and many more. Just a few are detailed in the following subsections (see also chap. 43, this volume).

### *2.5.1 Airforce Avionics Training*

For example, in the United Kingdom, the Avionics Training Facility (ATF) for the RAF's *Tornado* F3 aircraft was installed late in 1999 at the Tornado Maintenance School at RAF Marham in Norfolk. The ATF is based on 10 Windows NT, triple-screen simulators that enable students to explore a virtual *Tornado* and to expose, test, and repair avionics modules ("Line Replaceable Units" or LRUs) using a variety of different items of virtual test equipment (see Fig. 42.10). Only a simple mouse-function key interface is employed.

In the short time that the ATF course has been in place, instructors at Marham have been able to reduce the time taken to complete the course—from 13 weeks using physical aircraft mock-ups to 9 weeks using VE—and the amount of downtime experienced by each student (time wasted as a result of being unable to access course equipment)—from 3 weeks to zero.

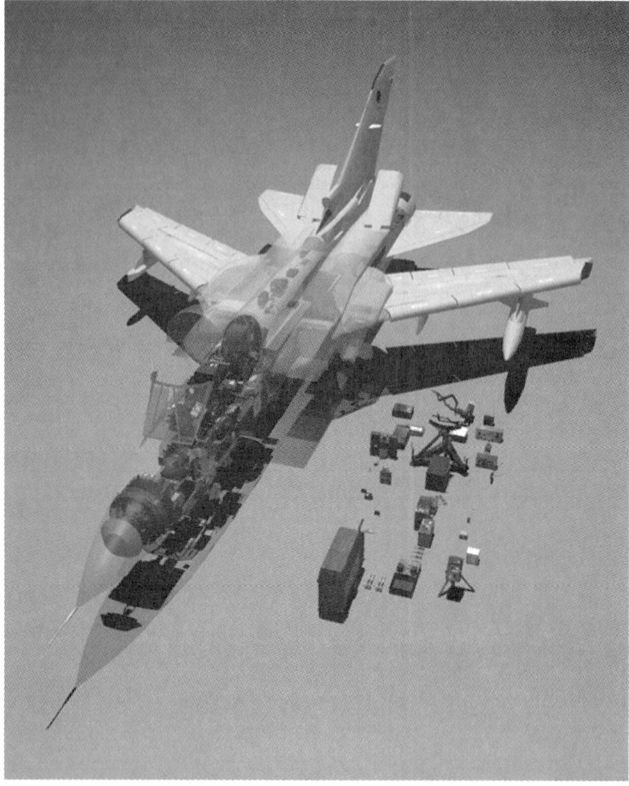

FIG. 42.10. The virtual *Tornado,* rendered transparently to show LRUs. Items of virtual test equipment are also shown. Image courtesy Virtual Presence.

Feedback from the Marham trainers suggests that, in contrast to previous courses, ATF students gain enhanced spatial and procedural knowledge relating to avionics tasks nearly 40% faster than previous mock-up or CBT-based exercises.

### *2.5.2 Close-Range Weaponry Simulation*

At the time of this writing, the U.K.'s Royal Navy is also in the process of replacing its shore-based gunnery-training outpost near Plymouth with an advanced 20 mm and 30 mm close-range weapons simulator facility near Portsmouth. The immersive VE solution equips weapon aimers, weapon directors, and course instructors with a blend of head-mounted and screen-based displays, together with virtual binoculars and tracking sensors, to create and interact with a range of target engagement scenarios—from simple sailboats and naval vessels, to drones, aircraft, and sea-skimming missiles—as if operating from a ship-based environment (see Fig. 42.11). This simulator features sea state ambient lighting and weather effects, together with own-ship pitch and roll. OpenGL has been adopted as the graphics standard, in order to deliver consistent image quality and reliable performance under the Windows NT operating system. The system will operate as a network of Windows NT workstations, each providing real-time graphics and multi-processing capabilities. The training program will be a consistent VE shared between workstations, but hosted and controlled by the main instructor consoles in a standard client–server mode.

FIG. 42.11. Early screen shot of a Royal Navy gunnery simulator environment. (Insert is view from 20 mm weapon.) Image courtesy Virtual Presence.

### 2.5.3 Royal Navy Submarine Qualification

As weapons platforms become more advanced, the inevitable reduction in real systems available for training means that computer-based lessons, many featuring VE, will become an essential tool of the military classroom. Nowhere is this problem more acute than is the case with future submarines. Here, smaller fleets will spend much more time at sea and less time in dock at the disposal of the inexperienced naval rating. The training of future submariners and dockyard support teams, then, is an excellent example where VE can deliver highly cost-effective alternatives to commandeering an otherwise fully operational boat.

A recent project conducted for the U.K. Royal Navy's flag officer submarines (FOSM) was based on the early training and vessel familiarization needs for submariners destined to serve on the United Kingdom's fleet of Trafalgar Class (SSN) boats. The results of the project are currently being given serious consideration for other submarine classes, and future platforms such as Astute and FASM (future attack submarine). Similar interests are being expressed within U.S. naval communities relating to the role of VE in NSSN and future attack sub programs. The overall Royal Navy's requirements for what is referred to as a submarine qualification (SMQ) system relate to the initial provision of a PC-based trainer that will enable students to become familiar with the layout of a target class of boat, including decks, compartments, key items of equipment, main service routes (e.g., high-pressure air), safety equipment, and so on (Stone, 2000).

A series of exercises are being developed which require students to locate and (where necessary) actuate shipboard components. Students can "walk" through the virtual submarine on predefined paths. At any point, they are free to turn, look around, and interact with "active" scene items, as appropriately identified on-screen. On entering a virtual compartment or zone of interest, the student is confronted with a virtual space—textured as necessary. The compartment possesses visually recognizable features—lockers, fire equipment, consoles, ladders, large-bore piping, and unique components (see Fig. 42.12).

For those compartments where additional detail and realism are required, panoramic VE techniques are implemented. Fixed nodes and viewpoints are linked to "higher realism

FIG. 42.12. The virtual "core" of the SMQ training system, showing compartment detail level necessary for equipment identification and system tracing. Image courtesy Virtual Presence.

databases," such that the call-up of active scene items generates static and panoramic images, which match visually with their respective locations within the geometric compartment layout. The actual paths and times taken by students in pursuit of the set tasks are recorded and made available for replay and archiving. Finally, "hot spots" within the panorama are linked to additional 3-D geometric databases (VRML), containing recognizable models of important and safety-critical equipment (e.g., high-pressure blow valves) or digitized textual/video extracts from existing training and informational sources. In the case of the 3-D models, once the hot spot has been interrogated (simple mouse click), the VRML object appears, with textured features and labels (where relevant to the identification and operational part of the SMQ task). The student is then able to initiate any animated or interactive features to demonstrate operation and, in the case where the operation of safety equipment can originate from elsewhere in the submarine (e.g., operations compartment), warning displays are also presented.

In the aftermath of the 2000 *Kursk* Russian submarine tragedy, VE is now being seriously considered in the design process and for future training régimes for the new NATO submarine rescue system (NSRS; initial study launched in 2001). Such study is designed to address a wide range of mature subsea technologies capable of producing a submersible system to replace the rescue boats of individual nations, some of which are over 20 years old. It is believed that, by using VE in the NSRS training program, the number of "wet" training dives a trainee pilot has to undertake can be cut from five to three, or possibly two. The cost of two wet dives would actually pay for a VE system. The development of such a system (see Fig. 42.13) in parallel with the general training requirements already in place for future U.K. submarines marks the beginning of the first major integrated naval application based on VE technologies.

### 2.5.4 VESUB

One of the important and widely referenced VE projects to emerge from the United States is known as the VESUB (Virtual Environment for SUBmarine Officer of the Deck ship handling training) technology demonstrator (VESUB, 1998), and was conducted at the U.S. Submarine Training Facility in Norfolk, Virginia, Naval Submarine School in Groton, and Naval Air

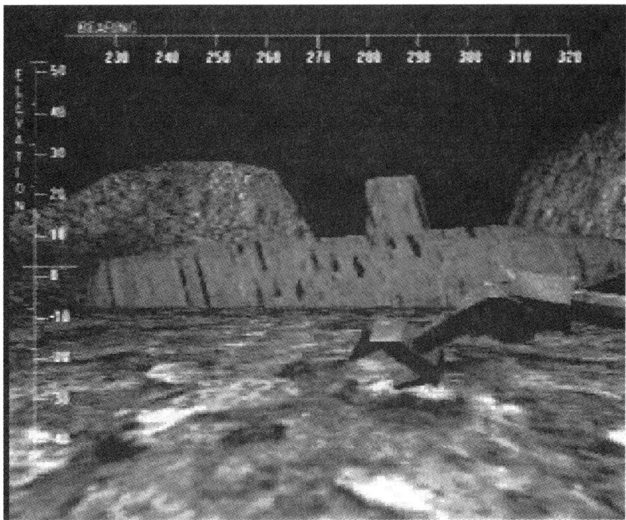

FIG. 42.13. An early immersive VE distressed submarine intervention demostrator. Image courtesy Virtual Presence.

Warfare Center Training Systems Division in Orlando, Florida (see chap. 43, this volume). Unlike many of the CBT systems in existence today, VESUB was based on immersive VE technology, using proprietary HMDs (initially the Virtual Research VR4, later n-Vision's high-resolution DataVisor), speech recognition, and a Silicon Graphics Infinite Reality Engine computing platform. Indeed, it is interesting to note that a conscious decision was taken to move away from large-screen projection technologies and more toward an HMD implementation. The immersive facilities were used to create VE views as if the user were located on the fin of a typical U.S. submarine. Some 41 VESUB users were used in study trials, all naval personnel (involved with submarines) ranging in experience from junior officers to qualified officers of the deck and commanding officers. The participants were exposed to three scenarios in the VESUB system:

- An orientation scenario to help them become familiar with the capabilities and operation of the VE system
- A training scenario comprising several ship handling tasks, such as position determination, contact location and evaluation, getting underway, maneuvering, rudder checks, man overboard, and vessel crossing ("rules of the road")
- An actual test scenario where task improvements were recorded

Results from the latter scenario demonstrated a significant improvement in learning between training and test sessions, almost regardless of experience level. For example, students demonstrated an improvement factor of 39% in checking range markers, 33% in visually checking the rudder, 57% in contact management and 44% in reaction time in a "man overboard" event (Hays, Vincenzi, seamon, & Bradley, 1998).

The need for VE classroom trainers to familiarize incoming personnel with complex military systems and subsystem layouts is incontestable. Furthermore, the success of networked simulators, courtesy of pioneering initiatives such as distributed interactive simulation (DIS), brings a major new dimension to military VE trainers. For example, networked VE allows geographically remote infantry personnel to train together (see chap. 44, this volume) in a shared virtual theater of campaign, acting as a single, coordinated battalion, complete with fighting vehicle backup, and other hardware support. It also provides the classroom educator with an important practical tool for assigning tasks and responsibilities to individual or collaborating trainees, not to mention providing a more quantifiable method of assessing the extent of trainees' learning (see chap. 35, this volume).

#### 2.5.5 Virtual Land Mine Detection Training

Turning finally to land-based activities, the use of VE to train the drivers of military vehicles or operators of weapons systems is not new. However, a unique training system based on immersive VE has been developed for the French army, to help train in the detection of land mines, using the PHANToM as the primary interaction device.

The system presents the trainee with a textured cuboid representation of the ground area to be investigated and, using a standard issue military probe attached to the PHANToM, he is required to locate potential mines by gently inserting a virtual representation of the probe into the "ground."

Once a definite contact has been made, the trainee must continue probing until a recognizable pattern of penetrations has been made (see Fig. 42.14). In addition to the visual and haptic features of this trainer, a pattern recognition system is available, which matches the trainee's penetrations with known land mine geometries. Once a pattern match has been made, a schematic of the most likely mine configuration is displayed.

### 2.6 Medicine and Surgery

During the late 1980s, many visionaries—notably at the University of North Carolina and within the Department of Defense in the United States—were developing the notion of the surgeon or consultant of the future, equipped with an HMD and rehearsing procedures in VE, from detailed inspections of an unborn fetus, through to the accurate targeting of energy in radiation therapy, even socket fit testing in total joint replacement. For many years, the United States led the field in medical VE, and some of the early conferences and exhibitions delivered many promises about how technology would revolutionize surgery in the new millennium. Many of those promises would, even today, be hard-pushed to reach reality before 2020, let alone 2001. Nevertheless, in 1995, one of the leading practical advocates of VE in the United States, Col. Richard Satava (see chap. 47, this volume), attempted to categorize achievable applications of VE in medical and surgical domains (Satava et al., 1995). He saw developments in the fields of surgical intervention and planning, medical therapy, preventative medicine, medical

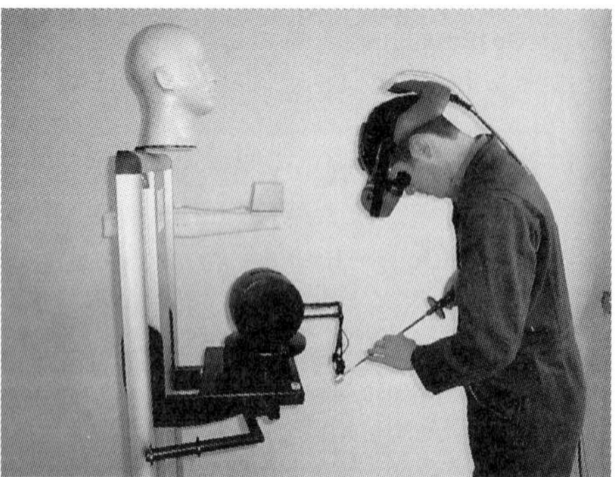

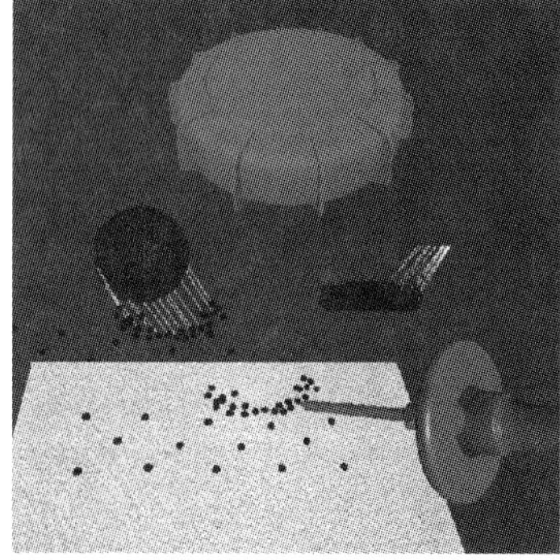

FIG. 42.14. The PHANToM system in use as part of an immersive VE trainer for land mine detection and identification. The right-hand part of the figure shows how the system's pattern recognition system can be used to present trainees with "best-fit" land mine types on the basis of ground penetration patterns. Image courtesy SimTeam.

training and skill enhancement, database visualization, and much more. Satava's original work, sponsored by the Advanced Research Projects Agency, ARPA, focused on large-scale robotic or telepresence surgery systems, using VE technologies to recreate the sense of presence (see chap. 40, this volume) for a distant surgeon when operating on, say, a battlefield casualty. However, other research efforts began to emerge across the States (and Europe) using VE in a classic simulator mode to rehearse or plan delicate operations (e.g., total joint replacement or in certain ophthalmic operations). It was then shown that one could actually use the successful virtual procedures to back up in situ performance (i.e., through the projection of 3-D graphics onto the operative site—"augmented reality").

Other programs saw the future not as either robot or surgeon, but as a combination of the two with the automated component of the operating theatre augmenting the skill of the surgeon, having been thoroughly preprogrammed in a preoperative virtual world (DiGioia, Jaramaz, O'Toole, Simon, & Kanade, 1995). Augmented reality took a step further when, as early as 1993, a magnetic resonance image (MRI) had been taken of a patient and overlaid onto a real-time video image of the head (Adam, 1994).

During the late 1980s and early 1990s, the application of VE and associated technologies to the field of medicine and surgery steadily increased, with pioneering (if, at that time, somewhat optimistic) companies such as High Techsplanations (HT Medical) and Cinémed becoming responsible for fueling the passion for "making surgical simulation real" (Meglan, 1996). However, more recent conferences and exhibitions suggest a plateau may have been reached, with some of the front-running concepts undergoing a period of consolidation through clinical validation. Tried and tested products, available at prices that are affordable to the greater majority of surgical teaching institutions, are still somewhat elusive. Nevertheless, the key uses of VE remain as Satava predicted, with new excursions into the realms of rehabilitation (see chap. 49 and 50, this volume), psychotherapy (see chap. 51, this volume), and support for disadvantaged individuals. Many projects have received academic grant support or national and continental funding (as one finds in Europe, during the Framework V Initiative, for example). Developments in technology continue to deliver more and more robust hardware and software at realistic prices. The surgical community has also gradually become more and more involved. In some countries, such as the United Kingdom, this is a painfully slow process, dominated by a small number of surgical (so-called) "high-flyer" personalities, who are producing somewhat fragmented developments of questionable value.

A chapter such as this cannot hope to cover all historical and contemporary aspects of VE and medicine and surgery under a single cover. However, the interested reader can obtain a more in-depth appreciation by selectively reading papers in the excellent publications of Westwood, Hoffman, Stredney, and Weghorst (1998; westwood et al., 1999).

Despite the coming of the PC/Windows era, bringing with it the capability of delivering highly interactive virtual medical environments (attractive to resource-limited surgical teaching bodies), the international academic research community (in the main) still shows a bias toward sophisticated anatomical and physiological simulations of the human body, hosted on graphics supercomputers. From the digital reconstruction of microtomed bodies of executed convicts (e.g., the National Library of Medicine's Visible Human Project, http://www.nlm.nih.gov/research/visible/visible_human.html) to speculative deformable models of various organs and vascular systems (e.g., Forschungszentrum Karlsruhe—http://iregt1.iai.fzk.de/KISMET, and the University of California at Berkeley's Virtual Environments for Surgical Training and Augmentation project—http://robotics.eecs.berkeley.edu/~mdownes/surgery/), the quest to deliver comprehensive "virtual humans" using dynamic visual, tactile, auditory, and even olfactory modes of interaction looks set to continue into the foreseeable future. The problem is, who in the real surgical world can afford to procure, operate and maintain such systems? More to the point, do they really offer trainee and consultant surgeons a career-enhancing advantage?

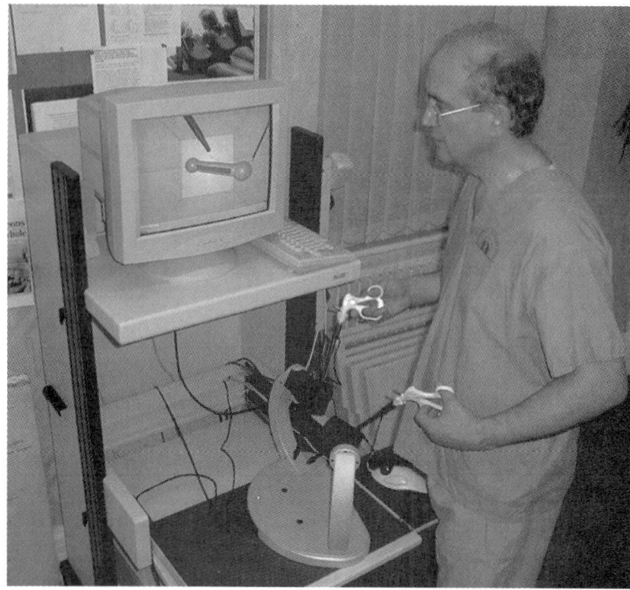

FIG. 42.15. The MIST surgical skills trainer. Image courtesy Manchester Royal Infirmary and Virtual Presence.

Virtual Presence's MIST (Minimally Invasive Surgical Trainer) system (see Fig. 42.15) has been subjected to worldwide testing by clinicians and applied psychologists (Stone, 1999a). Unlike many of its counterparts, MIST is an interactive skills trainer, utilizing symbolic task elements abstracted from the analyses of many minimally surgical interventions. The clinical tests have yielded a battery of positive objective results, one outcome being that the MIST system now forms a mandatory component of basic and advanced medical courses at the European Surgical Institute near Hamburg, covering a wide range of techniques, from cholecystectomy to thoracic surgery. The results include such features as improvements in surgical movement efficiency (actual/ideal instrument path lengths, past-pointing errors, and submovement corrections) and error reduction when MIST trainees are compared to control groups (e.g., Taffinder, McManus, Jansen, Russell, & Darzi, 1998a; Taffinder, McManus, Russell, & Darzi, 1998b; Taffinder, Sutton, Fishwick, McManus, & Darzi, 1998c). MIST task sensitivity to sleep deprivation has also been shown (Taffinder et al., 1998d), as have improvements in multiple incision performance, reduction of the fulcrum effect (perceived instrument reversal), and increased use of both hands in endoscopic tasks (Gallagher et al., 1998). In general, it can be stated that experimental results based on MIST tasks demonstrate statistically clear performance differences between novice, junior, and experienced laparoscopic cholecystectomy surgeons (Gallagher, Richie, McClure McGuigan, 1999). MIST is marketed by Mentice Medical Simulation AB (www.mentice.com).

## 2.7 Heritage

Virtual heritage may be defined as:

> . . . the use of computer-based interactive technologies to record, preserve, or recreate artefacts, sites and actors of historic, artistic, religious and cultural significance and to deliver the results openly to a global audience in such a way as to provide formative educational experiences through electronic manipulations of time and space.
>
> —Stone and Ojika (2000, p. 73)

Ever since the early 1990s, there has been a worldwide interest in the prospect of using VE to recreate historic sites and events for such purposes as education, special project commissions, and showcase features at national and World Heritage visitor centers. The power of VE, as described by Nynex researchers Stuart and Thomas in these formative years, lies with its ability to open up places and things not normally accessible to people from all walks of life, and to allow them to explore objects and experience events that could not normally be explored without "alterations of scale or time." The power of the Internet now supports interaction with remote communities and interaction with virtual (historical) actors (see chap. 16, this volume). In the context of heritage, VE goes much further, however, in that it offers a means of protecting the fragile state of some sites and can help educate visitors not so much about their history, but in how to explore, interpret, and respect them. For example, VE can display the potential damage caused by the ravages of human intervention and pollution by accelerating the simulated destructive effects to monuments, even large areas of land and sea, over short periods of time (Stone, 1999).

One can trace the impetus for such interest back even further to an impressive demonstration, staged at the Imagina Conference in Monte Carlo in February 1993. An immersed member of the French clergy was joined by a simple avatar (see chap. 15, this volume), controlled by an operator in Paris who provided him with a real-time guided tour of a virtual reconstruction of the Cluny Abbey, a building destroyed in the early 19th century. This pioneering demonstration proved conclusively that a formative virtual heritage experience could be delivered over a standard ISDN communications network. Yet it was not until 1995 that the first Virtual Heritage conference was held at the Assembly Rooms in Bath, U.K., featuring Virtual Pompeii (From Carnegie Mellon University), Virtual Lowry (from Virtual Presence) the Caves of Lascaux (from the University of Cincinnati), and the fortress at Buhen, Egypt (from ERG Engineering). At the same time, a project instigated by English Heritage was announced, to develop both "large-scale" and "Web-friendly" VE models of Stonehenge. The large-scale model, complete with a mathematically accurate model of the nighttime sky and a virtual sunrise, was presented in June of the following year at the famous London Planetarium (see Fig. 42.16).

Virtual Stonehenge was also demonstrated live at the second Virtual Heritage conference held in London in December 1996. The 1996 conference was one of computing contrasts. At one end of the spectrum were such projects as Virtual Stonehenge, the Battle of Gettysburg (from TACS, Inc.), and the tomb of Menna in Thebes (from Manchester Metropolitan University), all targeting low-to-medium-range computers. At the other end was the visually stunning (and ongoing) "supercomputer"-based works of Infobyte, notably the widely referenced

FIG. 42.16. Sunrise over virtual Stonehenge. Image courtesy Virtual Presence.

FIG. 42.17. Virtual environment of Notre Dame. Image courtesy Digitalo Design.

Coliseum project. Another advanced project was announced by Miralab (from the University of Geneva), which later resulted in a stunning virtual recreation and animation of the terracotta warriors of Xian.

Since that time, there have been a small handful of heritage events featuring VE, the most notable being staged in the closing years of the 1990s by the Japanese (Gifu)-based Virtual Systems and Multi Media Organization (VSMM: www.vsmm.org). As a result of the interest generated at the international conferences, a group of the directors of the VSMM Organization went on to establish the Virtual Heritage Network (VHN) (www.virtualheritage.net), a site rich in information relating to contacts, groups, and projects. The "mission statement" of VHN is, at the time of writing, still evolving, but attempts are being made to bring together as many examples of computer-generated heritage as possible and make those examples accessible to the world through the infrastructure of the Internet and the impressive hardware technology evolving to satisfy the domestic games industry. The trends set by, for example, Digitalo Design (Virtual Reality Notre Dame Project; www.vrndproject.com; see Fig. 42.17) are exemplary in this respect. One can now harness the power of game engines using software included with the CD one has purchased from the local PC store, backed up with tutorials, patches, and 3-D model libraries from the Web. This has given rise to a whole new breed of competent VE developers, from schoolchildren age 10 and above, to small companies using the software to service the marketing and training needs of much larger corporations. Add to this the increasing availability of low-cost, even free modeling and run-time software resources from the Web, and one can appreciate why the VHN directors claim the time for global participation in Virtual Heritage is now.

### *2.7.1 Heritage and the Ceramics Industry*

Recent developments in the British economy have prompted certain "heritage" industries to look very closely at their businesses and the prospects for improved productivity and growth in the early part of this new century. Companies such as Wedgwood and Royal Doulton, famous international, historical names in the production of quality crockery and figurines, are turning to VE in an attempt to embrace technology within their labor-intensive industries. Ceramics companies and groups, such as the Hothouse in Stoke-on-Trent, are experimenting with new haptics techniques and achieving some quite stunning results.

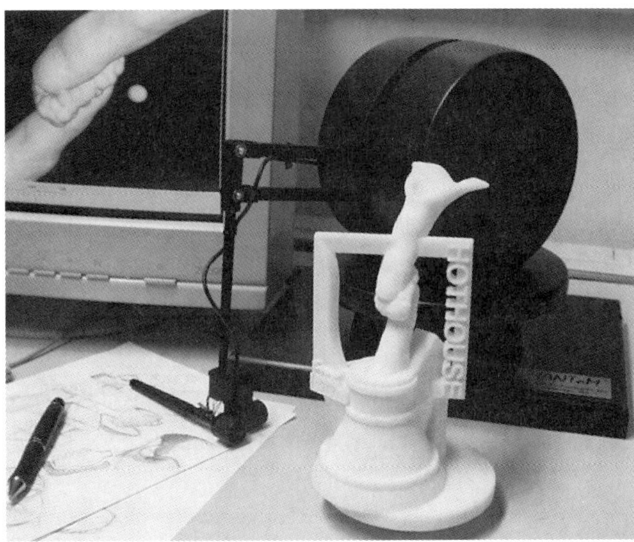


FIG. 42.18. An early example of the physical realization of a Hothouse trial sculpture using a Sensable PHANToM and free-form "digital clay." Image courtesy Hothouse.

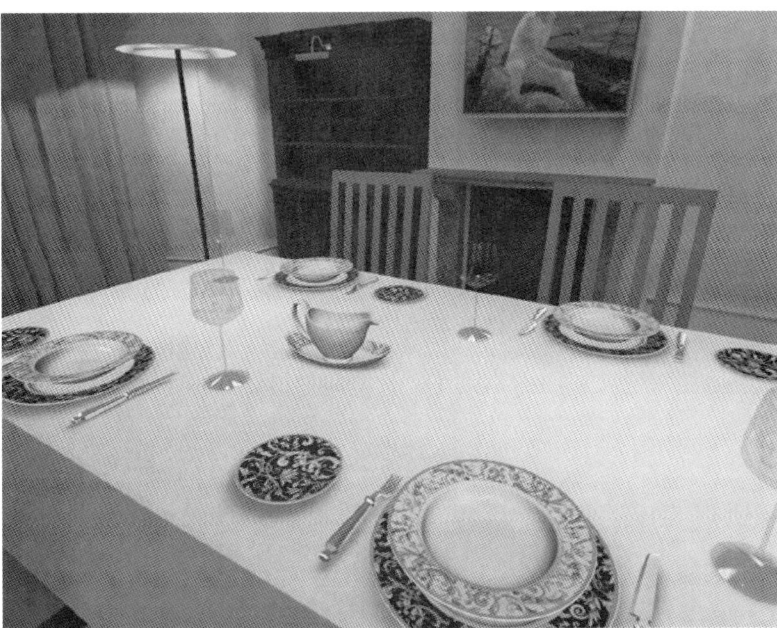

FIG. 42.19. Some of the new virtually created ceramics products in situ within a VE domestic "showroom." Image courtesy Virtual Presence.

The importance of experiments like these, however, focuses on the people who actually produce the results. Talented sculptors—people with incredible manual skills but no background in computer technology whatsoever—have, given access to Sensable Technologies, Inc.'s, PHANToM Desktop and Freeform "digital clay" products, started to produce ornate sculptures within 3 to 4 days! Then, using local industrial resources, they have used 3-D printing and stereolithography facilities to convert these virtual prototypes into physical examples (see Fig. 42.18) and high-end VE to display them in virtual showrooms and domestic settings of very high visual fidelity (see Fig. 42.19).

## 2.8 Retail

Despite its poor showing in recent VE market surveys (Cydata, 1998, 2000), the retail application domain is a natural choice for implementation of VE technology. Preoccupation with international issues such as Y2K and, in Europe, the implementation of the single currency and Euro exchange rate has (in the opinion of the author) temporarily stalled developments. There is no doubt that the success of VE applications demonstrated in the 1990s by leading supermarket chains, coupled with the low price and retail information technology (IT) compatibility of today's desktop delivery machines, make the technology an attractive and viable support to the retail business process, particularly for activities in space planning, product packaging, and consumer behavior assessment. Certainly these conclusions are borne out by activities only now getting underway in the Far East and Australia.

One of the first projects demonstrating VE as a flexible design tool for store layouts began in 1993, when the U.K.'s Cooperative Wholesale Society (CWS) sponsored the development of a virtual Late Shop model. The model enabled CWS specialists to experiment with new checkout designs, evaluate store security, and assess potential ergonomics problems faced by checkout staff and customers when handling bags full of products. It was also demonstrated for the first time that it was possible to build virtual gondola and shelf models, populated with appropriate products, by automatically converting output files from proprietary rule-based space planners, such as Spaceman or Intactix.

The well-known U.K. retailer Sainsbury's then went on to demonstrate that environmental issues surrounding a proposed supermarket site development could be assessed using VE technology. This applied not only to the visual impact of the development but also to issues such as access routes, potential congestion, car park layout planning, petrol station positioning, and local transport availability. Another challenge issued by Sainsbury's came as a result of this virtual site demonstration and, in 1994, resulted in the largest virtual store interior ever developed. Users of the virtual store could reposition and resize gondolas, change fittings and replace the entire back wall with an alternative design in a matter of seconds. Then, designers, store managers, and directors could take part in a virtual design review to determine such issues as: Were the proposed department signs visible? Where were the nearest emergency exits? Were elderly or disabled shoppers catered for?

Following on from this success, Sainsbury's, in partnership with Virtual Presence Ltd., developed a PC-based space planning tool capable of being used by existing design employees called Concept$_{VR}$. In very simple terms, Concept$_{VR}$ enables retail personnel to develop and modify new store and store fitting and style concepts on a "drag-and-drop" basis, organizing departments and products according to the organization's standards (see Fig. 42.20). Once developed, the virtual store concept can be distributed cheaply to other interested parties (such as shop fitting suppliers), using a 360-degree virtual "camera" system that takes panoramic digital pictures at predefined points in the virtual store model and, using a proprietary visualization software packages (e.g., QuickTime VR or Reality Studio), stores those nodes and pictures on CD (or dedicated Web sites) for real-time review on standard or laptop PCs. Concept$_{VR}$ has been used in a variety of projects, most recently for gas station shop design, making sure the limited space available for groceries and other nongarage items is filled to best effect, without compromising customer comfort or through-flow.

Turning now to the products themselves, for VE to contribute to the design of new items destined for supermarket sale, the retail market demands greater attention to detail and quality than is normally evident in VE demonstrations. Products are 3-D in nature. They are housed within metallic, plastic, and cardboard containers and may be covered in cellophane or some other reflective material. They have labels that need to be read and striking designs that must create on-shelf visual impact, standing out against their competitors. Products also have to be used. More recent developments, spearheaded by Virtual Presence in collaboration with many

FIG. 42.20. Concept$_{VR}$ in action, space planning for gas station retail area. Image courtesy Virtual Presence.

"household name" product developers—Lever, Nestlé, Procter & Gamble and Nestlé—mark a new era in the use of VE in the retail market. Using a methodology known as reality check, it is now possible to design, modify, and test product concepts thoroughly using VE to create realistic settings—virtual supermarkets, kitchens, or other household environments. Already poor packaging concepts have been intercepted using VE long before they have reached the, sometimes wasteful, prototyping stage. Virtual products can also be used by market research organizations for consumer testing (see Fig. 42.21), as first demonstrated by Lever and Procter & Gamble in the late 1990s. Courtesy of VE, one personal hygiene product even experienced a 2-year reduction in the time taken to reach the market from initial concept.

## 2.9 Education

Most of the inspiring and innovative developments in VE and education have happened in the United States (see chap. 45, this volume), although the United States is also home to a good number of VE educational Web sites and electronic "foundations" that do not seem to have been active since the late 1990s. Washington State University's renowned Human Interface Technology Laboratory (HIT Lab) was one of the innovators in this field, having put in place an ambitious "peripatetic VE," courtesy of sponsorship from U.S. West Foundation. The Virtual Reality Roving Vehicle (VRRV)—about the size of a transit van—was involved in taking immersive VE demonstrations developed by the HIT Lab to science classes in some 14 states in U.S. West's territory. Another respected name in the field is that of the University of East Carolina, developing VE for educational heritage teaching, nursery education, and for teaching teachers how to deal with difficult classroom situations.

In Europe, one of the most impressive feats with a loose educational theme was a Cyberspace Roadshow, staged by the German Fraunhofer Institute in the late spring of 1994, and funded by the Schweizerische Bankgesellshaft. This was not just a case of putting a handful of items of VE equipment into a small van, as happened in the States. An articulated roadshow truck was completely modified to include immersive and projection equipment, a canopy extension for some 30 seated passive viewers, and desktop VE demonstrations, including a virtual tunnel

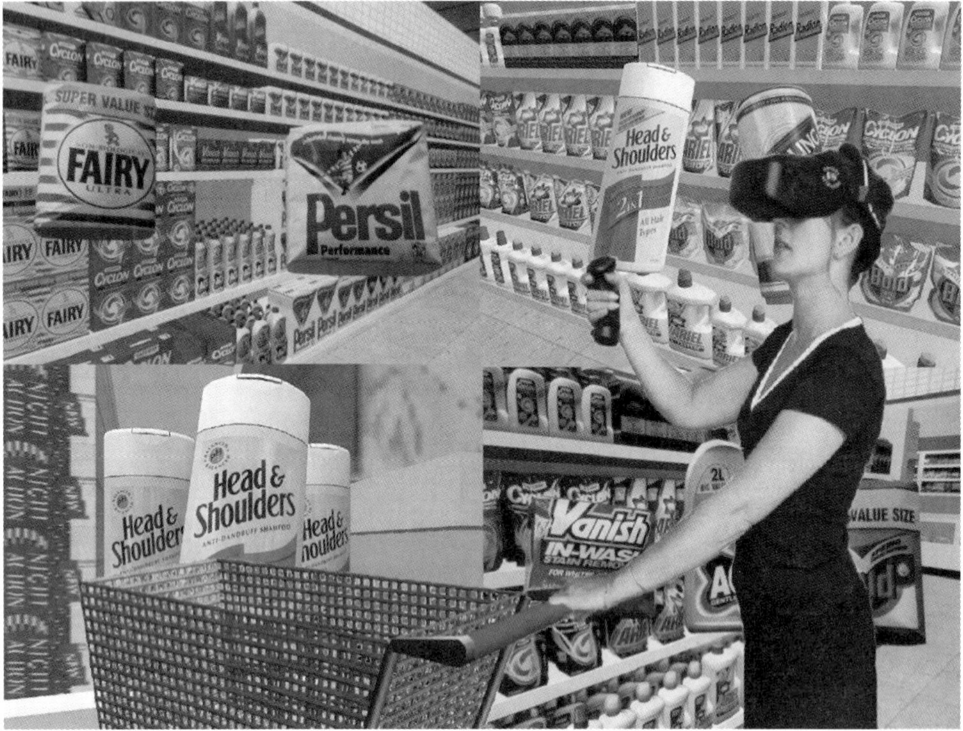


FIG. 42.21. Reality Check consumer assessment project. Image courtesy Virtual Presence.

flight, a virtual jungle, a space labyrinth, and a virtual stage with various musical and percussion instruments. One can only conclude that the budget for this, in terms of truck commissioning, equipment, development effort, power, and so on was immense. It has to be emphasized here that the Cyberspace Roadshow was carried out primarily as a special marketing event for the Swiss Bank (and the advertising agency Bosch & Butz), with the target audience being junior bankcard holders around the age of 17.

Virtual environment applications in education were in existence right at the start of popular VE activities in Europe. For example, in 1991, the late Michael Clark (then headmaster at a High School in the northeast of England), successfully canvassed local businesses to gain sponsorship for a desktop VE project he called the Intelligent City—a virtual urban scene designed to help teach schoolchildren French.

More recently, Crime Conquest was developed in the United Kingdom, using similar software to that originally procured by Michael Clark. Conquest is a desktop (PC) VE "societal trainer" that was distributed free of charge by the Greater Manchester Police Authority to some 300 schools in 10 local authorities of the greater Manchester area during September 1998. The trainer exploits VE by allowing schoolchildren to "enter the bodies" of different virtual participants. Their decisions at crime scenes involving people and property dictate what happens next. As witnesses, they have to be observant. For instance, having briefly seen the driver of a stolen vehicle, how close will their match be when invited to use the identikit computer within the VE town police station? When playing the role of a police officer, they have to make fair and just decisions. For example, are teenagers outside an elderly person's house causing a nuisance or, worse still, exchanging drugs? Initial individual user and schools-based trials of the system have been highly successful. Indeed the software is still being used at the time of writing, despite the primitive "look-and-feel" of the real-time graphics. Although the system was

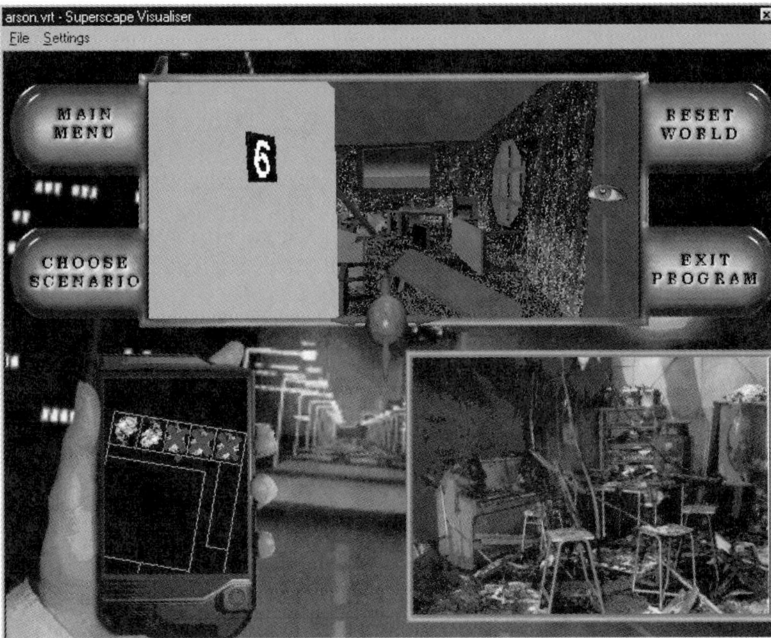


FIG. 42.22. User Interface for Crime Conquest, showing virtual schoolroom and image of real schoolroom, both subjected to arson attack. Image courtesy Virtual Presence.

originally designed for presentation to children of 11 years and older (Key Stage 3+ in the United Kingdom), children from 3 to 17 have enjoyed using the software, which has been designed in keeping with some of the aims of the IT components of the British national curriculum. Crime Conquest presents schoolchildren with 8 different scenarios—Trouble With Youths (3 variations, each escalating in severity), Shoplifting, Joyriding, Home Security, Personal Safety, and Arson (see Fig. 42.22). These can be run as individual 5- or 6-minute exercises, or as a complete "day's experience" in around 45 minutes.

So, why are educational VE pursuits so important? The importance stems not just from the Nynex perspective described earlier for Virtual Heritage (exposing VE users to communities and actors from times gone by), although this is a key benefit of using the technology. Rather the importance relates to the future users of the technologies and practices the VE community is putting in place today. For far too long now governments have been spending quite sizeable sums of money with both academic and commercial technology organizations in the vain hope that they can attract, "kick-start," and foster the interest of small- to medium-size enterprises ("SMEs" in Europe). In the United Kingdom (and to some extent in the United States) this has failed miserably. Put simply, employees, even directors of these SMEs simply do not have the time (nor the money) to be bothered with researching new technologies, especially those with a VE label. Even today, VE comes with a rather esoteric "prebranded" look and feel.

Can this situation be reversed? There are only two ways that change will come about. Either the companies concerned will be dragged, "kicking and screaming," into the new millennium by the IT demands imposed on them by the blue-chip organizations they supply. Alternatively, the change will evolve from within. In other words, their future employees will, one hopes, enter employment with a raft of skills that can be tailored reasonably painlessly to meet the company's future needs while avoiding massive organizational turmoil. This alternative depends, of course, on an appropriately educated workforce. The ability of many school children to exploit VE technologies, be it via the latest version of Epic's Unreal (as modified and used for the International Space Station model quoted in section 2.3) or Sierra's Half Life game engines, using Web-based 3-D modelers or even developing VRML worlds using free graphics and texture libraries is, in the view of the author, unquestionable. The potential for teachers to deliver many parts of their curricula effectively using VE techniques is also undeniable, although the perception of many educationalists that "I.T is hard" needs considerable work to quash. Most of today's youngsters have reached a stage where they are not only proficient in handling quite complex interactive 3-D databases, such as games and some CAD packages, they have come to expect a certain level of quality in the delivery of information using similar media, be it at school or in their first steps into full-time employment. Here then await members of a highly motivated, skilled community who will one day use the technologies our community is developing. Here, then, is a market worth developing.

### 2.10 Database and Scientific Visualization

Virtual environments have always held the potential to bring to the human eye features of living and inanimate material it would not normally see and to allow users to leave the constrained interface of the optical or digital microscope and explore such features intuitively. The future use of VE in exploring these atomic-level worlds has already been mentioned in section 2.2, in the context of micro- or nano-presence. Similarly, VE can help scientists to visualize all manner of complex artifacts and processes (see chap. 53, this volume), be they new protein chains, viral compositions, chemical drug concepts, or gaseous particles in motion (i.e., computational fluid dynamics [CFD]).

As an example of VE used for CFD visualization, NASA Ames Research Center's Numerical Aerospace Simulation Facility has developed a highly impressive facility based on VE technologies to allow users to explore numerically generated, 3-D unsteady flow fields. This award-winning facility allows users, equipped with a variety of immersive display and interaction devices (originally a BOOM [Binocular Omni Orientation Monitor] stereoscopic display and VPL's DataGlove), to view and inject smokelike virtual tracers in and around aerospace vehicles. One of NASA's sister sites (Langley) has, for international research and education purposes, now ported the concept of the Virtual Wind Tunnel on to the World Wide Web using Java (http://ad-www.larc.nasa.gov/jvwt). Another example of using low-cost (PC) computing technologies to deliver fully interactive VE representations of CFD processes was supported by the United Kingdom's Health and Safety Executive. This project assessed the design of chemistry laboratory fume cupboard handles and their effect on the propagation of potentially harmful particles back into the laboratory and toward the working scientists (see Fig. 42.23).

Figure 42.24 is a screenshot from an application developed using MUSE Technologies' $\mu$uSE (Multidimensional User-oriented Synthetic Environment) software development environment (www.musetech.com). This project successfully validated new algorithms for processing industry-standard solar-weather data. This dynamic VE synthesizes multiple data sources into a single display, thereby helping scientists to understand how fluctuations in the sun's radiance affect the earth's atmosphere. In turn, this aids in the prediction of potentially adverse situations, such as the decay of satellite orbits and, by association, problems with wireless communications and other businesses reliant on integral communications and/or remotely sensed data. Clockwise from the top left of Fig. 42.24, this stage of the application includes context-sensitive images of the sun's surface, $\mu$uSE-standard navigational instruments and a color-to-data mapping tool, context-sensitive text instruments, an image of the atmosphere's total electron content (TEC), and, finally, plots of energy content in two spectral bands coming from the sun.

Finally, VE has also proved to be an invaluable tool for displaying time-variant fused sensory data in a meaningful manner. Take for example Goodyear Tire and Rubber's application

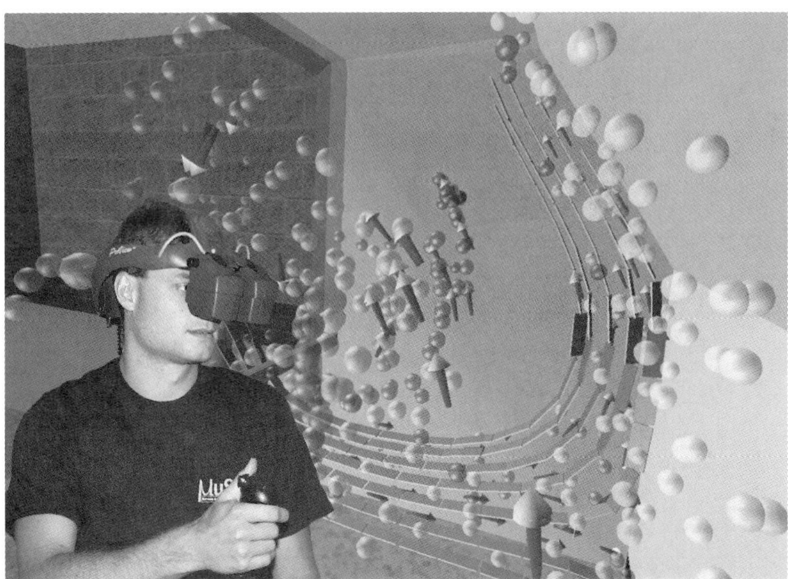

FIG. 42.23. Immersed VE user experimenting with CFD particle visualization techniques. Image courtesy Virtual Presence.

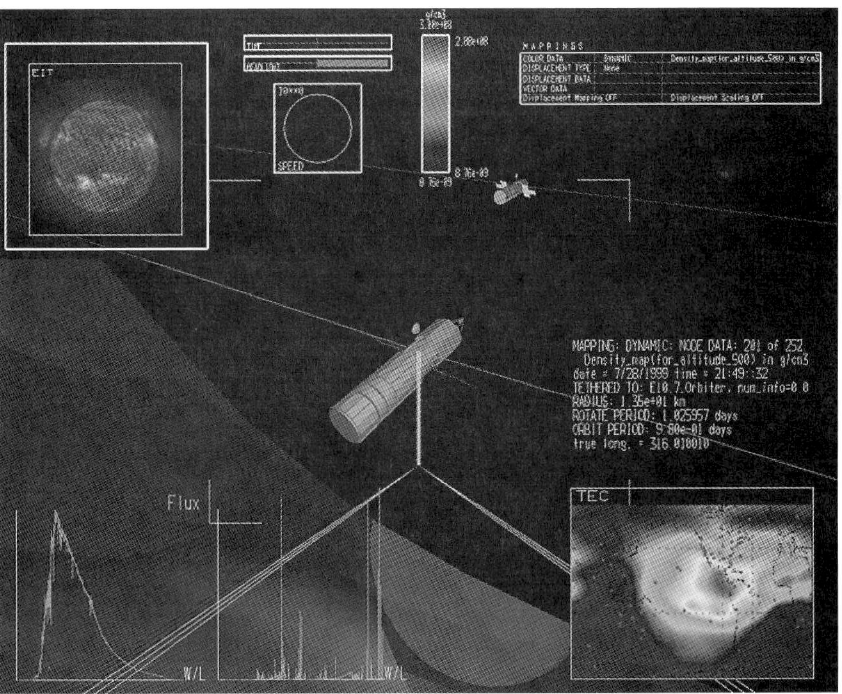


FIG. 42.24. Synthetic fused data display presenting the effects of the sun's radiance on the earth's atmosphere. Image courtesy MUSE Technologies.

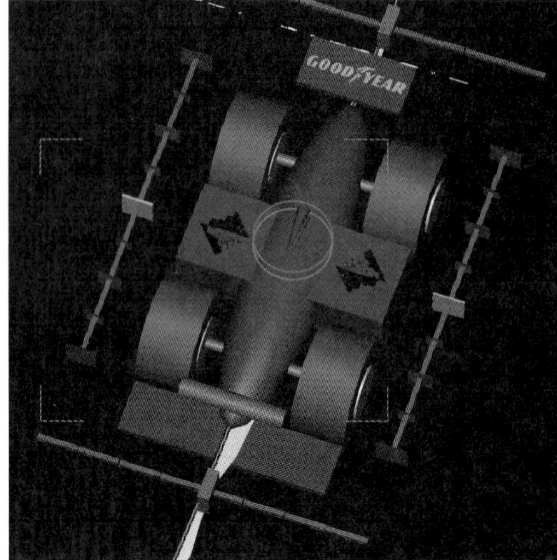


FIG. 42.25. Data fusion display depicting multiple sensor output for high-performance car and tire development. Image courtesy MUSE technologies.

for high-performance car and tire development. The real-time 3-D data display depicted in Fig. 42.25 was developed for Goodyear again using MUSE Technologies' $\mu$uSE software. The Goodyear system displays as many as 15 different types of data simultaneously, all collected at the company's test tracks.

The screenshot shown in Fig. 42.25 depicts a single racing car surrounded by instruments that help the user monitor the output of various sensors mounted on the vehicle. Users can choose any mapping of data to car component they wish. The dials at the center of the car typically show steering angle. The tires swell and shrink as the car moves along the track to indicate higher or lower pressure. The "difference bars" at the front, back, and sides of the car show load transfer across the car. The strip beneath the car is the "undertrack plot," which shows lateral acceleration.

## 3. CONCLUSIONS

To do real justice to the international VE community, a comprehensive overview section on applications would occupy the space normally demanded by a large reference book. This overview has only scratched the surface of an exciting and wide-ranging arena. Yet to many, the very fact that the arena is so wide-ranging is perceived to be a major problem in bringing stability to those product and service companies who are attempting to stay afloat in the VE community's turbulent waters.

No matter what commercial model or philosophy one subscribes to, it is a fact of life for the VE industry that there is no—and (for the foreseeable future) will be no—single "killer" market for VE, although education offers much potential for the future. More immediately, the absence of the killer application is a situation that has left many companies' boards of directors praying for salvation, having been preoccupied with trying to focus in on one or two market segments. After all, market consultants continuously bombard companies with the message that they must "focus, focus, focus" on specific products for specific markets. However, for the "killer-market-less" VE community, such focusing inevitably leads to stagnation, obscurity, and termination. It is now evident that VE companies should pay more attention to how they should structure and market the abilities of their multidisciplinary skills base to provide appropriate, timely, and relevant human-centered solutions to the problems they are asked to tackle, using VE as one tool in their multimedia armory, rather than focus on the advanced technologies per se.

It was always the case that real commercial or industrial applications of VE would help to save the field of endeavor from obscurity, just like its predecessors. The truth is that there are many more success stories and, of equal importance, stories of failure or "nonadoption" by companies (for good technical, cost, or human reasons) in existence than the VE community can, or is permitted, to report. Certainly, one cannot rely too greatly on the results of recent (and rather expensive) VE market surveys, as many have questioned the validity and reliability of the community samples from which their conclusions have been drawn. One has to rely on well-documented case studies when they emerge in order to help the technology diffusion process, as described further by the author and Professor Peter Swann in chapter 28. To this end, the remaining applications chapters (see chaps. 43–55, this volume) dealing with specific applications represent only the start of what promises to become a rewarding decade of commercially meaningful VE success stories.

What is certain is that, during the closing 2 years of the last millennium, the VE community has witnessed a technological revolution that should—stress should—ensure its presence in some shape or form for at least the next 2 decades.

## 4. REFERENCES


Adam, J. A. (1994). Medical electronics. *IEEE Spectrum, 31*(1), 70–73.

Barnett, B., Helbing, K., Hancock, G., Heininger, R., & Perrin, B. (2000). An evaluation of the training effectiveness of virtual environments. In *Proceedings of the Interservice/Industry Training, Simulation and Education Conference* (I/ITSEC 2000). Orlando, FLo:

Brooks, F. P. (1988). Grasping reality through illusion—Interactive graphics serving science. *Proceedings of CHI '88,* 1–11.

Brooks, F. P., Ouh-Young, M., Batter, J. J., & Kilpatrick, P. J. (1990). Project GROPE—Haptic displays for scientific visualisation. *Computer Graphics, 24*(4), 177–185

Cydata. (1998). U.K. business potential for virtual reality. [Executive summary of market research study conducted on behalf of the U.K. Government's Department of Trade and Industry, Online]. Available: www.ukvrforum.org.uk/resources/index.htm

Cydata. (2000). U.K. business potential for virtual reality [U.K. VR forum publication—executive summary, Online]. Available: www.ukvrforum.org.uk/resources/index.htm

DiGioia, A. M., Jaramaz, B., O'Toole, R. V., Simon, D. A., & Kanade, T. (1995). Medical robotics and computer-assisted surgery in orthopaedics: An integrated approach. In *Interactive technology and the new paradigm for healthcare* (pp. 88–90). Washington, DC: IOS Press.

Fisher, S., Wenzel, E. M., Coler, C., & McGreevy, M. (1988). Virtual interface environment workstations. In *Proceedings of the Human Factors Society 32nd Annual Meeting.*

Gallagher, A. G., McClure, N., McGuigan, J., Crothers, I., & Browning, J. (1998, January). *Virtual reality training in laparoscopic surgery: A preliminary assessment of minimally invasive surgical trainer virtual reality (MIST*VR). Paper presented at Medicine Meets Virtual Reality 6, San Diego, CA.

Gallagher, A. G., Richie, K., McClure, N., & McGuigan, J. (1999). *Psychomotor skills assessment of experienced, junior and novice laparoscopists with virtual reality.* Internal Note: Northern Ireland Centre for Endoscopic Training and Research, Queens University, Belfort.

Harvey, D. (1987). VCASS: A second look at the super cockpit. *Rotor and Wing International, 21*, 32–33 and 63.

Hays, R. T., Vincenzi, D. A., Seamon, A. G., & Bradley, S. K. (1998). *Training effectiveness evaluation of the VESUB technology demonstration system* (Tech. Rep. No. 98-0003). Orlando, FL: Naval Air Warfare Center Training Systems Division.

Mao, C., Sun, W., Shen, Z., & Seeman, N. C. (1999). A nanomechanical device based on the B-Z transition of DNA. *Nature,* 397, 144–146.

Meglan, D. (1996, November). Making surgical simulation real. *Computer Graphics,* 37–39.

Satava, R. M., Morgan, K., Sieburg, H. B., Mattheus, R., & Christensen, J. P. (1995). Medicine 2001: The king is dead. In *Interactive technology and the new paradigm for healthcare* (pp. 334–339). Washington, DC: IOS Press.

Stone, R. J. (1999a). The opportunities for virtual reality and simulation in the training and assessment of technical surgical skills. In *Proceedings of Surgical Competence: Challenges of Assessment in Training and Practice* (pp. 109–125). London: Royal College of Surgeons of England.

Stone, R. J. (1999b, October). Virtual heritage. *UNESCO World Heritage Review*, 18–27.

Stone, R. J. (2001). Virtual reality. Virtual and synthetic environments: Technologies and applications. In W. Karwowski, (Ed.), *International encyclopedia of ergonomics and human factors*. London: Taylor & Francis.

Stone, R. J. (1996). Virtual reality: A british success story. *Science in Parliament, 53*(5), 1–3.

Stone, R. J. (2000). Virtual reality training for submariners: Assessing the feasibility. *Engineering Technology*, 16–19.

Stone, R. J., & Ojika, T. (2000). Virtual heritage: What next? *IEEE Multimedia, 7*(2), 73–74.

Taffinder, N., McManus, I., Jansen, J., Russell, R., & Darzi, A. (1998a). An objective assessment of surgeons' psychomotor skills: Validation of the MIST*VR* laparoscopic simulator. *British Journal of Surgery, 85*(Suppl. 1), 75.

Taffinder, N., McManus, I., Russell, R., & Darzi, A. (1998b). An objective assessment of laparoscopic psychomotor skills: The effect of a training course on performance. *Surgical Endoscopy, 12*(5), 493.

Taffinder, N., Sutton, C., Fishwick, R. J., McManus, I. C., & Darzi, A. (1998c). Validation of virtual reality to teach and assess psychomotor skills in laparoscopic surgery: Results from randomised controlled studies using the MIST*VR* laparoscopic simulator. In Westwood, Hoffman (Eds.), *Medicine meets virtual reality* (pp. 124–130). Washington, DC: IOS Press.

Taffinder, N. J., McManus, I. C., Gul, Y., Russell, R. C. G., Darzi, A. (1998d). Effect of Sleep Deprivation on Surgeons' Dexterity on Laparoscopy Simulator. *The Lancet*. 352. 1998. 1191.

VESUB. (1998). Available: http://www.ntsc.navy.mil/Programs/Tech/Virtual/VESUB/vesubrpts.htm

Westwood, J. D., Hoffman, H. M., Robb, R. A., Stredney, D., & Weghorst, S. J. (Eds.). (1999). *Medicine meets virtual reality 7:* Vol. 62. *Studies in health technology and informatics*, Washington, DC: IOS Press.

Westwood, J. D., Hoffman, H. M., Stredney, D., & Weghorst, S. J. (Eds.). (1998). *Medicine meets virtual reality 6:* Vol. 50. *Studies in health technology and informatics*, Washington, DC: IOS Press.

# 43

# National Defense*

Bruce W. Knerr,[1] Robert Breaux,[2] Stephen L. Goldberg,[1] and Richard A. Thurman[3]

[1] *U.S. Army Research Institute, Simulator Systems Research Unit*
*12350 Research Parkway, Orlando, FL 32826-3276*
*Bruce_Knerr@stricom.army.mil*
*Stephen_Goldberg@stricom.army.mil*
[2] *Naval Air Warfare Center, Training Systems Division*
*12350 Research Parkway, Orlando, FL 32826-3275*
*BreauxRB@navair.navy.mil*
[3] *Western Illinois University*
*Instructional Technology and Telecommunications*
*37b Horrabin Hall, Macomb, Illinois 61455-1390*
*RA-Thurman@wiu.edu*

## 1. INTRODUCTION

### 1.1 Why Are Defense Organizations Interested in Virtual Environments?

During the last decade of the 20th century, the U.S. Department of Defense has been required to conduct an increasing number of missions while its monetary and personnel resources have been decreasing. Many of these missions represent significant departures from the services' primary warfighting mission. In addition to the combat of the Gulf War, missions have been conducted to enforce peace agreements (by enforcing embargoes and "no-fly" zones and to separate warring parties), and to provide humanitarian assistance and disaster relief. The increasing need to perform operations that require skills other than combat is illustrated by recent actions in Haiti, Somalia, and Bosnia. Missions such as these are often carried out as part of a multinational force. As a result, military units and personnel must be prepared to perform more tasks under a greater variety of conditions than ever before. In this context of having to do more with less, the use of virtual environment (VE) systems is seen as a means to enhance preparedness for a wide range of activities while reducing operating costs.

For purposes of this chapter, the definition of VE being used by the North Atlantic Treaty Organization (NATO) Research Study Group 28 is adopted. This group is investigating human factors issues in the use of VE for military purposes. Their definition of VE (they use the term *virtual reality*) is (Ehrlich, Goldberg, & Papin, 2000):

*Opinions expressed are those of the authors and do not represent an official position of the Department of Defense.

> Virtual reality is the experience of being in a synthetic environment and the perceiving and interacting through sensors and effectors, actively and passively, with it and the objects in it, as if they were real. Virtual Reality technology allows the user to perceive and experience sensory contact and interact dynamically with such contact in any or all modalities.

This covers immersive VE systems that are directly perceivable by participants through their senses. It does not include training on electronic systems that use sensor (e.g., radar) data as their stimuli.

Virtual environments can enhance preparedness by providing training opportunities that cannot be made available in the real world. These systems can reduce training costs by replacing the use of actual equipment and weapons, and even aging, expensive simulators, with simulators and simulations that are relatively inexpensive to operate and maintain. Modern military weapons systems, such as planes, ships, and tanks, are expensive to operate. Sophisticated or "smart" weapons are not only so expensive that they can rarely be fired for training purposes, they are in many cases designed to be employed at such long distances that there are few locations (ranges) where they can safely be fired for training. A growing population and concern for the environment make it more difficult to continue to operate existing ranges, and in most cases makes their expansion impossible. Many military tasks, such as high-speed, low-altitude flight, are inherently dangerous. Others, such as handling ships in adverse weather, can only be practiced infrequently in the real world because the appropriate conditions rarely occur.

Many existing simulators are expensive to operate. They were designed with relatively slow computer technology and complex instructor stations. These simulators were adequate for their time, but they now are becoming expensive to support and maintain. More recent simulator designs are more efficient and effective due to improvements in computer technology, new concepts in use of spare parts, and use of intelligent tutors that reduce instructor requirements.

Effectiveness in combat results not just from soldiers or pilots fighting as individuals, but from coordinated activities of different types of units and weapons systems, such as armor, infantry, and artillery (see chap. 44, this volume). Increasingly, emphasis is being placed on training for such combined arms and multiservice operations. With the many operations the U.S. armed forces are currently engaged in across the globe, training systems need to be capable of being taken to forward areas to maintain individual and collective proficiency.

### 1.2 Scope of the Chapter

Because of the breadth of national defense activity in virtual simulation, the scope of this chapter must necessarily be limited to a general overview of the VE-based training programs in the U.S. Army, Air Force, and Navy in which participants have a direct view of a virtual space. The discussion will be limited to United States programs or programs that the United States is conducting jointly with other countries.

## 2. CURRENT AND FUTURE MILITARY APPLICATION AREAS FOR VIRTUAL ENVIRONMENTS

This section describes a variety of proposed application areas for the use of VE in the military. Each area is described in terms of the specific training challenge and way in which VE is being or could be used to meet this challenge.

## 2.1 Training

### 2.1.1 *Embedded Training*

Embedded training is training provided by capabilities built into or added onto operational systems to enhance and maintain the skill proficiency of operators and maintainers. Embedded training capabilities may be fully embedded within the operational system, or they may be designed as components that can be added and removed as required. The concept of embedded training is not new. Its use in the military dates from at least the early 1970s. However, it has become feasible in more systems as onboard computers and digital information displays become more common. An advantage of embedded training is that it provides a training capability wherever the operational system is located. With these deployable systems, soldiers can train while temporarily assigned to a foreign country, or sailors can train while onboard ship at sea. This makes embedded training especially appropriate for sustainment or refresher training. Early applications of embedded training by the army and navy have focused on embedding training on computer driven systems that utilized electronic sensor data, for example, anti-submarine warfare or Aegis system training or fire control data (Army TACFIRE Computer). Today research is under way to utilize VE technology to stimulate the direct view optics and optical sensors of weapon systems, that is, to insert virtual targets into those displays, thereby turning combat vehicles into virtual simulators.

### 2.1.2 *Unit/Collective Training*

In addition to the use of simulators for training individual skills, groups of simulators are being used increasingly to train team and collective skills that military operations require (see chap. 44, this volume). Collective training systems employing VE (e.g., Army Close Combat Tactical Trainer, Air Force Distributed Mission Trainer) are being employed to meet this need. Networking operational systems that possess embedded training capabilities can provide collective training.

A variation to training intact teams and units is the use of intelligent agents (see chap. 15, this volume) to simulate missing team members. The Navy Landing Craft Air Cushion (LCAC) vehicle has multiple crew stations, but a training class may not always contain members of each station. That is, a small class of four students may contain only navigators. Thus, intelligent agents can serve the role of engineer and pilot in the VE so that the navigator student can train with a full crew.

### 2.1.3 *Soldier and Leader Training*

The use of VE for combat training has predominantly focused on training personnel who fight from within combat vehicles, such as tanks, aircraft, and ships. However, soldiers, marines, and security personnel who operate and fight on foot can also benefit from training in VEs to meet growing responsibilities and increasing challenges. These include new missions, changing tactics, and increasingly sophisticated equipment. New noncombat missions involve more diverse types of operations and more complex and dynamic rules of engagement than do combat missions. When combined with dispersed units and a high level of media attention, this creates an environment in which the actions and decisions of small-unit leaders and individual soldiers, such as firing into a hostile crowd, can have far-reaching impacts. A variety of new equipment is being developed and evaluated to improve the effectiveness and survivability of these foot soldiers. This includes wearable computers, global positioning systems, radios, digital information displays, video capture capability, and night-vision equipment. Such sophisticated equipment will permit soldiers at lower echelons to obtain more information about the tactical situation, command more firepower, and be separated further from their parent unit. Evolving tactics, made possible by the new communications and weapons systems, will shift information and decision making to lower level leaders. Soldiers will have to become more capable of either autonomous or small group action. They will also need to be more adaptive and flexible. Immersive VE can represent environments in which soldiers and leaders can practice these new tactical and decision-making skills. Virtual environments are also well suited to the incorporation of training strategies and techniques (such as intelligent tutoring systems), which can enhance decision-making skills.

## 2.2 Mission Rehearsal

While training is conducted to develop skills which will generalize to a variety of combat situations, mission rehearsals are conducted to prepare a unit to conduct a specific operation. Thus a unit might train building assault procedures with the goal of becoming more proficient in assaulting a variety of types of buildings (high-rise office building, townhouses, etc.), with the expectation that at some future time they will be assigned a similar task. The same unit might rehearse rescuing terrorist-held hostages from a specific embassy immediately before conducting that rescue. Mission rehearsal also contributes to the planning process. What is learned from the rehearsal can be used to revise the plan for the operation.

A critical element of a mission rehearsal system is representation of the physical environment in which the operation is to take place. This can be simple or elaborate. The movie *The Dirty Dozen* showed the use of scale models to rehearse an attack on an enemy-held chateau. Full-scale physical mock-ups have also been used. In Operation Desert Storm, allied forces rehearsed assaults on trench lines built from Iraqi designs. During the Vietnam War, U.S. special forces prepared for an assault on a North Vietnamese prisoner of war camp with a full-scale physical mock-up. For security reasons, the mock-up had to be disassembled whenever a Soviet reconnaissance satellite was scheduled to pass over the area. Obviously, security is another requirement for a mission rehearsal system. Desirable features include the capability to include simulated enemy and neutral forces, modify the representation as more information about the situation is obtained, and move the system to the location where in the unit is based (i.e., transportability). Virtual simulations have tremendous potential to provide a mission rehearsal capability. Aircraft simulators are already being used to rehearse air missions.

## 2.3 Evaluation of New Concepts and Equipment

The military is constantly changing to counter new threats to national security, to assume new roles (e.g., humanitarian assistance), or to take advantage of new technology. Usually, these changes are most visible in the fielding of new weapons systems and equipment, but they may also be reflected in new or revised policy, organizational structure, tactics, or training. In an effort to speed the process of acquiring new equipment, the military is placing more emphasis on the use of models and simulations to support equipment design and evaluation. Virtual simulations make it possible to perform "human in the loop" experiments and evaluations before hardware prototypes are built. Virtual simulations can be used to replace mock-ups and prototypes to design operator interfaces. Using simulated equipment in a simulated combat setting can help identify and resolve issues such as workload and task allocation. Virtual simulations can also be used to develop and evaluate procedures for the use of the equipment. Finally, virtual simulations can be used to train operators of new equipment prior to its delivery to the receiving unit.

### 2.4 Performance Measurement

One of the frequent shortcomings of the real world as a training environment is that objective measurement of the conditions under which a task is performed or of the performance of the trainee is difficult. Virtual environments, on the other hand, can provide measurement opportunities that would not be possible in the real world (see chap. 35, this volume). This is because computers are utilizing movement, firing, environmental and communication data to drive the simulation. The same data is invaluable for measurement and performance assessment. Objective performance measures can be derived in judgment tasks, such as ship handling, that traditionally have eluded objective assessment. Intelligent tutoring systems can use measurement data to assess student progress in the same way an expert would and determine appropriate further instruction. The navy is building just such a system to teach shiphandling skills (see chap. 20, this volume).

### 2.5 Knowledge Elicitation

The conning officer of a naval vessel must make complex decisions based on the perceived motion of their own ship, as well as that of possible collision hazards around their ship. Although some of the skills required to perform this task may be articulated by an expert ship handler, studies discussed in chapter 20 (this volume) indicate that experts do not easily articulate the entire task. Perhaps this is because experts have automated the task to a degree that they are now unable to describe how they do it. Nevertheless, it may be possible to elicit knowledge about performance of the task by observing the actions of expert ship handlers, and inferring relationships between the expert's actions and outcomes of the ship-handling task. As described in the preceding paragraph, VE offers an environment in which environmental states, expert behaviors, and resulting states can be carefully measured. Then, through mathematical techniques of drawing inference, the system can extract relationships and draw conclusions about what the expert attends to and what actions he takes in performance of the task. The U.S. Navy is exploring these techniques using a VE ship-handling training system.

## 3. BEHAVIORAL SCIENCE CHALLENGES TO THE USE OF VIRTUAL ENVIRONMENTS

The military services have been conducting behavioral science research in the use of VE for a number of years. This section describes the major behavioral science challenges to the use of VE for the military purposes described above. It is an overview and thus is not intended to provide comprehensive reviews of either the VE research conducted by defense organizations or the research in a particular content area (e.g., cybersickness).

### 3.1 Management of Simulator Sickness or Cybersickness

Simulator sickness, or cybersickness, can potentially degrade training effectiveness and performance, as well as affect the well being of trainees (see chap. 30, this volume). Ensuring that symptoms do not reach a disruptive level needs to be a major consideration in the design, development, and use of military VE systems. Research conducted to date suggests some recommendations for reducing symptoms, but better guidelines are needed.

In a report summarizing 13 VE experiments conducted by the Army Research Institute during the 1990s, Knerr et al. (1998) reported that simulator sickness symptoms were sufficiently severe that 5.6% of participants withdrew from an experiment prior to its completion. Rates for individual experiments varied from 0% to 25%. Other researchers have reported withdrawal rates of 10% (Garris-Reif & Franz, 1995) and 0% to 5% (Regan, 1995). Similar results are reported in chapter 30 (this volume).

Of particular relevance to the practical use of VE are findings indicating that, all other things being equal, an individual is most susceptible to simulator sickness during the first session with a simulator. For most tasks, the symptoms of inexperienced trainees decline during subsequent sessions. Lampton, Kraemer, Kolasinski, and Knerr (1995), Lockheed Martin (1997), and Regan (1995) have all reported fewer and less severe symptoms of simulator sickness with subsequent sessions (one or more days apart; see also chap. 30, this volume.) This suggests that trainees are adapting to the VE (see chap. 31, this volume), and that particular attention needs to be paid to the first session to avoid conditions that are likely to produce symptoms. Knerr et al. (1998) proposed that the first session should: be brief (inoculate against simulator sickness); be designed to minimize sickness-producing activities such as rapidly slewing the field of view, collisions, and viewing objects at very short distances; and avoid depicting self movement not under the participant's control. The widely used antimotion sickness drug hyoscine hydrobromide is effective in combating simulator sickness (Regan, 1995). Its effects on learning, however, have not been studied.

### 3.2 Presence/Immersion and Training Effectiveness and Transfer

Presence is defined as the subjective experience of being in one place or environment (the computer-generated environment), even when one is physically situated in another (the actual physical locale; see chap. 40, this volume). It has been proposed that presence can make training in simulated environments seem more realistic and increase user involvement in the tasks and situations presented in the virtual environment. This in turn should improve task performance in the VE and promote transfer of VE-acquired skills to the real world (see chap. 19, this volume). One way to improve training effectiveness and transfer, therefore, is to manipulate directly those factors that increase presence, such as interface characteristics. Although the sense of presence can be measured quantitatively (Witmer & Singer, 1998), evidence linking presence to either interface characteristics or training effectiveness is weak. In particular, there is limited evidence that a trainee who feels highly immersed in his environment learns more or faster than one who does not. However, it is necessary to make a distinction between the level of presence reported by different individuals in response to being immersed in the same environment (as represented by correlating presence scores and performance) and different levels of presence reported by the same individual in different environments. The relationship between presence and task performance in the latter case is yet to be tested.

### 3.3 Use of Autonomous Agents for Training

Autonomous agents and computer-controlled entities can enhance the VE training experience in a variety of ways (see chap. 15, this volume). In simulated combat scenarios, they can control the actions of enemy or other friendly "platforms" such as ships, aircraft, or armored vehicles. When coupled with the use of human avatars, they can substitute for live human role players in training scenarios that cover a range of activities from dismounted combat through peacekeeping to language training. There is a substantial history of development of forces of this type, also referred to as computer-generated forces (CGF) or semiautomated forces (SAF). Efforts are under way to develop semiautomated entities and fully automated agents that can represent both mounted and dismounted friendly and enemy ground forces (see, for example, Science Applications International Corporation and Lockheed Martin Information Systems Company, 1998). Research is also under way to make the behaviors of agents reflect more realistic human behavior (e.g., Gillis, 1998) by incorporating cognitive models that include variables such as sleep deprivation, amount of training and experience level. Autonomous agents can also serve as automated tutors or coaches, which can provide guidance to a trainee when needed, perform tasks when so delegated, and monitor the overall training session to ensure training objectives are being met. Breaux and Stanney (1999) have discussed the use of agents as tutors for ship-handling training and spatial knowledge acquisition.

### 3.4 Virtual Environment Training Strategies

There is an increasing body of evidence that students can acquire in VE and transfer to the real world, both route knowledge (Bliss, Tidwell, & Guest, 1997; Witmer, Bailey, & Knerr, 1995) and configuration knowledge (Darken & Sibert, 1993, 1996; Wilson, Foreman, & Tlauka, 1997). Magee (1997) and Hays, Vincenzi, Seamon, and Bradley (1998) found that their VE simulators improved ship-handling skills, which have a spatial component. One interesting aspect of these experiments is that they did not rely on sophisticated instructional strategies or features to obtain effective training, but instead used relatively simple and straightforward approaches. Hays et al. (1998) used structured training scenarios with coaching and feedback provided by a live instructor. Magee (1997) used recording and playback of two-dimensional (2-D) and three-dimensional (3-D) views, which could include graphic display of ship tracks and alphanumeric speed, heading, and related information. Instructors also had features such as stop, rewind, and fast-forward for use in reviews. Darken and Sibert (1996) found that use of certain design features (maps and gridlines) improved configuration knowledge (see chap. 24, this volume). Beyond these few experiments, however, there has been little investigation of how VEs can be used to their best advantage in training spatial knowledge. The strategies and techniques that are best suited for maximizing the effectiveness of VE for training is an open question.

### 3.5 Training Geographically Dispersed Team Members

Although military VE research to date has focused primarily on training individual skills, there is a strong need for the use of VE for training collective or team skills. Small infantry units and their leaders require increased training for noncombat missions and in the tactical use of increasingly sophisticated weapons and equipment. This will require either training multiple soldiers simultaneously in a shared VE, or training one soldier (the leader) while simulating subordinates and peers, probably with computer-generated forces (agents). Effective use of VE for either of these scenarios requires resolution of a number of additional issues, including: selection and use of appropriate training features and strategies; development of VEs to facilitate human interaction necessary for team training; determining the extent of system lags (fixed and variable) and their impact on training effectiveness; identifying the requirements for avatars that are used to represent team members in VE; and determining the impact on performance and training effectiveness of reduced nonverbal communication (posture, facial expression) among team members.

### 3.6 Computer Recognition of Human Gestures

Many military tasks require that commands be given through gestures, either because they must be performed in high-noise environments (ground combat or the deck of an aircraft carrier), or performed in silence. Unlike gestures for signed speech, military gestures are typically expansive. For example, a leader signals his unit to increase speed by raising his clenched fist to shoulder level, then rapidly thrusting it upward until it is fully extended and returning it to its original position. This motion is repeated rapidly several times. Military gestures must be recognizable when made from different positions (standing, kneeling, or prone) and orientations to the viewer. Computer recognition of these gestures could contribute to both training to perform the gestures correctly and training performance of the tasks that require their use. Computer recognition of gestures would allow live soldiers to control the behavior of computer generated forces through gesture commands.

### 3.7 Determination of Interface Requirements for Different Applications

No VE configuration replicates reality exactly. Nearly all use some "work-around" solutions, such as the use of a joystick to control locomotion, a head-mounted display (HMD) with lower resolution and smaller field of view than the human eye, or the use of enlargement or color to make distant objects visible. It is necessary to determine, for any application, the most cost-effective interface that provides the necessary cues and permits appropriate responses without using work-arounds that detract from learning and task performance by the user or trainee. Different methods of interaction, particularly of locomotion (see chap. 11, this volume), need to be evaluated in terms of the cognitive demand they place on the user and, therefore, their likely impact on training effectiveness.

For example, there are methods for self-locomotion that maintain a constant relationship between real world and VE orientation (Grant & Magee, 1997; Lockheed Martin, 1997; Singer, Ehrlich, & Allen, 1998; Templeman, 1997). The user turns in place in the VE as he or she would in the real world, and therefore receives the same kinesthetic and vestibular cues as in the real world. This aspect of the interface can be used in conjunction with a variety of techniques for moving forward, backward, or sideways. The results of locomotion experiments suggest that whereas trainees learn to move through a VE using any of a variety of well-designed interface devices, use of a "walking simulator" may produce better spatial learning than joysticks, gaze-directed, and hand-pointing controls. However, it is not clear whether this is because the walking simulator provides better cues for changes in direction, better cues for distance traveled, or because it places less cognitive demand on the trainees, allowing them to attend more carefully to other cues. There is a need for research that assesses the relative contributions of more realistic turning and movement cues to distance estimation, route learning, or configuration learning.

### 3.8 Effects of Prolonged or Repeated VE Use

There are numerous applications for VE that would require the user to be immersed in the VE for a long duration or for repeated sessions over an extended period. As indicated in chapter 30 (this volume) little is known about the distinct symptoms that may result from extended or prolonged use, or the rate and retention of adaptation (see chap. 31, this volume) that might be expected.

### 3.9 Effects of Using VE on a Moving Platform

There are applications for VE for use aboard moving platforms such as aircraft and ships. Presenting imagery derived from sensor and computer systems which is not coordinated with the perceived motions of the vessel or aircraft will eventually cause vestibular/ocular conflict and motion sickness (see chap. 30, this volume). Additionally, lag between real motions of a vehicle and the corresponding motions of virtual imagery can also induce motion sickness. Controlling a remotely operated vehicle (ROV), such as an unmanned aerial vehicle or an unmanned underwater vehicle, while aboard a moving platform such as a ship presents a complex interplay of motion. The imagery from the ROV will be moving in accordance with its aerodynamics or hydrodynamics, yet the operator will also be experiencing motions of the platform they are riding. There is no direct research into the effects of conflicts and coordination of such VEs for operators on moving platforms.

### 3.10 Accurate and Fast Position- and Orientation-Tracking Systems

Systems that can track (see chap. 8, this volume) the position and orientation of body parts, weapons, and other tools accurately and rapidly are required for two purposes: computer recognition of human gestures (as discussed above) and tracking the position of weapons. Humans can respond to stimuli with speeds that vary from a few to hundreds of milliseconds, using motions that can vary dramatically in speed, range, and complexity. Technological capabilities lag human performance capabilities in this area. Moreover, some VE equipment encumbers and hampers rapid task performance. Matching human motor capabilities in this area, particularly that of skilled performers, is and will continue to be extremely challenging.

## 4. MAJOR VE R&D PROGRAMS

This section describes several major military VE R&D programs. It will not be exhaustive. Performing organization, objectives, and major accomplishments will be described for the lay reader.

### 4.1 Simulation Networking (SIMNET)/Close Combat Tactical Trainer

The traditional method for collective training of army mechanized forces had been to conduct exercises in the field with military equipment. The Defense Advance Research Programs Agency's (DARPA) Simulation Networking (SIMNET) project provided a virtual simulation alternative to field training. It was a research project designed to demonstrate simulator networking technology within a workable trainer (Thorpe, 1987). SIMNET demonstrated the capability of man-in-the-loop simulators to create a virtual battlefield on which meaningful collective training could be accomplished (Alluisi, 1991). SIMNET was an R&D system that incorporated for the first time a number of concepts that would be carried over to later virtual training systems. It was a distributed simulation, which did not rely on a central computer. Each battlefield entity maintained its own independent worldview or record of the location movement and firing of each entity on the SIMNET battlefield. These worldviews were updated by protocol data units that were generated each time an entity within the SIMNET network moved or fired. SIMNET pioneered low-cost graphics generators and a low fidelity approach to representing vehicle controls in SIMNET simulators. SIMNET aspired to an 80% fidelity solution. SIMNET also developed the first SAF. Originally SIMNET simulators on opposing sides were to see each other as enemy vehicles. This approach was dropped in favor of development of predominantly computer controlled opposing forces that employed threat, not U.S., tactics.

Two hundred and forty-seven SIMNET simulators were eventually fielded in the United States, Germany, and Korea. As a trainer, SIMNET had a number of shortcomings (Burnside, 1991). Many collective tasks could not be completely performed. The SIMNET VE was limited to daylight, and SIMNET terrain databases had limited details. SIMNET was fielded without an After Action Review (AAR) System for measuring performance and providing feedback on a unit's performance, and without training support packages consisting of task-based training scenarios. As a result of these deficiencies, the army launched a follow-on to SIMNET, the

Close Combat Tactical Trainer (CCTT). CCTT is a product of the U.S. Army's Simulation Training and Instrumentation Command (STRICOM) and its Project Manager, Combined Arms Tactical Trainer (PM, CATT).

CCTT is not an R&D project and it has met all the requirements of the army procurement process. As of this writing it has been fielded at Fort Hood, Texas, Fort Benning, Georgia, and Fort Knox, Kentucky. Fielding will continue over the next few years at sites in the United States, Germany, and Korea. Mobile CCTT vans are also being fielded for training Army National Guard units. CCTT and SIMNET are based on the same networking concept, distributed interactive simulation (DIS). CCTT improves on many of the deficiencies the army identified in SIMNET. For example, CCTT can represent day and night and weather conditions. CCTT simulators contain all of the switches and dials present in the actual vehicle (although they may not all be functional). CCTT databases are more varied and contain a higher level of detail than SIMNET databases. CCTT uses standardized DIS protocols that SIMNET did not, and CCTT SAF are based on a thorough analysis of friendly and enemy collective tasks.

CCTT provides an excellent simulation environment. Less attention was paid to upgrading the training features of SIMNET (Goldberg, Mastaglio, & Johnson, 1995). While SIMNET had no AAR System, the AAR capability fielded with CCTT is limited at best (an improved AAR system is currently being planned). Structured training support packages containing scenarios that stressed mission execution over planning were developed in response to a need for training on which to base CCTT's operational test (Flynn, Campbell, Myers, & Burnside, 1998). A structured training approach in CCTT is being institutionalized through development of a Commander's Integrated Training Tool (CITT), which allows army commanders to build new scenarios or modify old ones to meet their training objectives.

CCTT is a large distributed simulation training system. Its effectiveness and popularity as a training tool will determine how extensively VE technology will be applied to meeting a wide range of potential army collective training applications.

### 4.2 Virtual Environment Training Technology (VETT)

The navy has launched a multifaceted research program, Virtual Environment Training Technology, directed at furthering instructional science while at the same time providing effective training solutions. The program is currently exploring a variety of approaches for training perceptual–cognitive skills, such a ship handling.

When asked how a sailor maneuvers a ship in the ocean relative to other ships or objects a ship's captain would likely say, "They have developed the 'seaman's eye.'" This concept is described in detail in chapter 20 (this volume). Training this perceptual–cognitive construct requires a package of technologies. Visualization, metaphors, knowledge extraction, behavioral modeling, collaborative learning, and details of the human–computer interface each form a portion of the infrastructure required to fulfill the potential of VE to teach mental models that form the basis of perceptual–cognitive tasks.

Visualization (see chap. 53, this volume) provides one technique for focusing a large amount of data into meaningful relationships (Gershon, Eick, & Card, 1998). Feedback on performance, suggestions for interventions, and diagnosis of a student's weak areas are each amenable to visual presentations in a training environment. Visual display theories suggest that such feedback should be organized and displayed in a manner that it is congruent with the methods in which one scans the environment. For example, visual primitives such as texture, color, length, and slope are encoded more efficiently than shape, area, orientation, and containment (Lohse, 1997). Visual displays that differentiate by color may thus be more readily perceived than those demarcated via shape. Further, 3-D representations have been found to be differentially superior to 2-D, with benefits in processing found with continuous as opposed to discrete data (Lee & MacLachian, 1986) and with lateral and altitude tracking accuracy as opposed to air speed (Haskell & Wickens, 1993).

Designing visualization displays according to such theories and findings may facilitate problem solving by streamlining information processing and reducing cognitive demands on users (Woods, 1984). Recent research in identifying display techniques for facilitating problem solving involves the utilization of different display modalities (i.e., visual, audio, tactile; Turk & Robertson, 2000). Mills and Noyes (1999) suggest that early studies support the notion that the multimodal aspect of VEs may result in enhanced human performance. For instance, Breaux, Martin, and Jones (1998) suggest that a primary advantage of VE technology is that it enables a training environment to be manipulated by both the instructor and the student to facilitate the integration of knowledge into an appropriate mental model, thereby enhancing learning and improving performance. It is also generally accepted that dynamic or interactive visual displays, such as VEs, facilitate a user's understanding of complex relations between represented parameters (Kalawsky, 1993). Moreover, it is theorized that VE technology may be an effective visualization tool for teaching and learning because it makes better use of the human central nervous system's ability to handle a rich set of sensory inputs (e.g., sight, sound, touch; Doyle & Cruz-Neira, 1999). There is a need to better understand the utility of extending visualizations into multimodal presentations. Doctoral fellowship studies are currently seeking to identify multimodal visualization principles that can be incorporated into design guidelines for training systems.

People naturally invoke metaphors when learning new principles by placing a new concept in the conceptual framework of something they already know about and understand (Carroll & Thomas, 1982). Based on this natural tendency, VE system designers should try to identify and adopt into their designs likely metaphorical frameworks to assist in training effectiveness. In general, metaphors have been broadly adopted into different fields of research and practice to aid learning. Physicists use metaphors to get a conceptual handle on the complexity of physics, educators make extensive use of metaphors in the classroom, and politicians are constantly in search of more powerful metaphors to position their proposals more effectively (Judge, 1993). Metaphors can provide a framework of credible associations to increase the probability that relationships in different domains will be conceived according to the patterns defined by the metaphors. In considering the role of metaphors in interactive system design, Dent, Klein, and Eggleston (1987) suggested metaphors are an aid to:

1. Systems designers as a source of organization and a decision guide about how to represent information, such as training material
2. Users or trainees as a means of directing their attention to important information needed for skilled action, especially under time pressure

Metaphors have been used in literature for thousands of years as a teaching mechanism. To see how they may work in VE, consider the "seaman's eye" training problem. Relative motion of ships at sea is somewhat analogous to relative motion of cars on a highway. Understanding overtaking and passing land vehicles may simplify explanation of the similar concept at sea and make learning faster and more efficient. If sailors can relate the tendency of ships to be drawn together due to the venturi effect at sea to the experience of passing an 18-wheeler and at the nose-to-nose point feeling that same venturi effect, then they may be able to avoid coming too close when they first "feel" the suction between two ships. The task is simplified when analogy is used, but again, students begin to focus on key relationships much quicker in the training process when cued by an analogy. That is, with the use of an analogy the concept of "slide" can be taught much quicker because there is a structure with which to link it. An icy road may include sliding about, and ships also have that tendency. Thus, analogies build and can be linked to teach complex concepts rather quickly.

The general goal of metaphoric design is thus to develop VE training systems that are predictable based on prior knowledge and experience. Through the application of this approach, human–computer interaction has evolved into a relatively standard set of interaction techniques, including typing, pointing, and clicking. This set of "standard" interaction techniques is evolving, with a transition from graphical user interfaces to perceptual user interfaces that seek to more naturally interact with users through multimodal and multimedia interaction (Turk & Robertson, 2000). Leveraging these evolving techniques has the potential to significantly advance the design of VE training systems.

Knowledge extraction still presents a costly challenge to development of training systems. Virtual environment systems that provide rich environmental cues are becoming cheaper and cheaper. The obvious link is to capture the behavior associated with the environment as a first approximation to the knowledge and skill needed by the expert in performing a task. For example, fuzzy set theory has been used to analyze an expert ship handler's performance within a synthetic visual environment. Relationships are derived between the commands of the human and cues available in the environment using fuzzy set theory. Those are then presented to the expert in a question and answer exploratory session to assist the expert in articulating what is important in performance of the task. This technique can save months of iterative interviews with expert performers, which is the traditional approach to knowledge extraction. Comparisons are under way between traditional cognitive task analysis and the use of fuzzy set theory for VE training systems.

Behavioral modeling can be contrasted with statistical modeling for virtual environments. That is, physics-based modeling of trajectories or impact is the standard technique used in simulation. Variation is introduced using random numbers. However, human variation is more systematic than random. For example, when one is exhausted, response time decreases. It does not fluctuate randomly but instead is generally depressed. Anxiety, fear, and stress each exert influence in a systematic way but not one describable by existing physics principles. Accurate behavioral modeling is being pursued so that as a trainee interacts with intelligent agents in a VE, it becomes difficult for the student to distinguish synthetic foe from human foe. The training session becomes more believable and more valuable. Doctoral fellowship studies are exploring insertion of anxiety and personality into VE agent behaviors.

Many complex training activities occur in groups. Expert teams are more than collections of expert individuals and exhibit certain skills such as common jargon, shared objectives, and recognition of individual differences. Some of the more subtle differences include eye contact, body posture, hand gestures, physical proximity, eye movements, and speech volume and speed. Virtual environments allow training metaphors that do just that—provide perspective from a teammate's vantage point. Thus, certain team skills such as sharing perspective and viewing common objectives from different vantage points are a simple matter for VE systems. Application of VE to training environments requiring collaborative learning is under way.

Human computer interface characteristics weave these issues into a single VE fabric. Full immersion seems to require not only visual but also haptic, auditory, thermal, olfactory, and possibly other senses. Work on the interaction of these and their relative contribution to the sense of presence awaits an adequately complex test bed. Until it is developed, the Virtual Environment Training Program is exploring interactions based on the current state of the art and its feasibility to enhance specific training applications.

## 4.3 Virtual Environment Research for Infantry Training and Simulation

SIMNET and CCTT both emphasize the training of the army's mechanized mounted force. The need for dismounted soldiers to be represented in virtual simulations and the need to train dismounted small unit leaders in difficult environments such as built up areas led to a decision in the early 1990s to initiate a program to investigate the use of VE for training and mission rehearsal for soldiers who fought on foot. Although VE was seen as having the potential to immerse infantry soldiers directly in simulations, it was recognized that effective training requires more than just VE hardware and software. It also requires identification of: the types of tasks for which VE training is most appropriate and capable; the characteristics of VE systems that are required to provide effective training; and the training strategies that are most appropriate for use with virtual environments. In order to develop this knowledge, the U.S. Army Research Institute for the Behavioral and Social Sciences (ARI), with the support of the Institute for Simulation and Training (IST), initiated a program of experimentation to investigate those issues in 1992. Following an initial analysis of the task requirements for dismounted soldier training and a review of previous VE training research, four experiments were conducted to investigate interface effects on the capabilities of participants to perform simple tasks in virtual environments. Variables investigated included the type of control device, amount of task practice, stereoscopic versus monoscopic HMDs, and type of display device (monitor, Boom, or HMD). Three experiments were performed that addressed the effectiveness of VE for teaching route and configuration knowledge of large buildings, and the transfer of this knowledge to the real world. The results of these experiments led to a program of basic research on distance estimation in virtual environments. The next phase of the research investigated the use of VE to represent exterior terrain for training both land navigation skills (identifying landmarks and learning routes) and terrain knowledge. Finally, research was conducted investigating the use of VE for training more complex tasks. An experiment examining the effects of self-representation on performance has been completed. An experiment involving the training of two-person hazardous materials teams has been completed, and an experiment involving distributed team training is under way. Overall, the program has conducted 16 experiments involving over 600 human participants. Knerr et al. (1998) provides an overview of the results of the program.

In 1999, the army initiated a cooperative program involving ARI, STRICOM, and two elements of the Army Research Laboratory (Human Research and Engineering Directorate and the Information Systems and Technology Directorate) to address a number of critical technological challenges that prevent implementation of high-fidelity dismounted-soldier simulation. Similar to the issues identified by the navy and discussed above, these challenges include: limited field of view and resolution of visual display systems; simulating locomotion; tracking weapons and body positions; realistic performance of computer-controlled dismounted friendly and enemy soldiers; simulation of night equipment and sensor images; lack of dynamic terrain and structures; development of appropriate training strategies and methods; assessing individual and unit performance; developing training materials quickly and easily; and determining transfer of training from virtual to live environments.

The product of this effort, which will be completed in 2002, will be an integrated prototype system that includes the following components and capabilities: a locomotion platform, which provides realistic perception of movement and accurate energy expenditure; a visual system, which can simulate a variety of night vision sensors and equipment accurately; "intelligent" computer-controlled forces to represent enemy, friendly, and neutral forces; dynamic terrain, including damage to structures and rubble; and features to enhance the effectiveness of training and mission rehearsal.

### 4.4 Air Force–Distributed Mission Training

The U.S. Air Force is also pursuing use of VE technologies for training. Distributed Mission Training (DMT) is an aircrew training system acquisition program that evolved out of research conducted by the Air Force Research Laboratory simulation center at the former Williams

Air Force Base in Mesa, Arizona. DMT employs state-of-the-art distributed simulation technology and advanced low-cost flight simulators. The simulators are networked electronically with other aircrews located at distant bases. DMT permits pilots to remain at their home units while "flying." Pilots are learning and practicing individual and team skills in a synthetic battlespace. DMT improves the quality and availability of training, reduces aircraft operation and maintenance costs, and increases the availability of scarce resources, such as training ranges.

Crane (1999) reports on a study that defined the state-of-the-art of DMT technologies and explored their use as a synthetic training environment. The study simultaneously employed 15 virtual (human-in-the-loop) aircrews, along with dozens of constructive (computer-generated) simulations over a wide area network in order to document the technical capabilities necessary for combat mission training and to evaluate training strategies. Training missions were conducted via a secure, wide area network, using Distributed Interactive Simulation (DIS) protocols. In these missions, virtual players distributed throughout the continental United States interacted with simulated (constructive) friendly and enemy fighter aircraft, helicopters, and ground vehicles, plus enemy surface-to-air threats. Training missions were conducted for offensive air-to-air, defensive air-to-air, and close air support. At the end of 5 days of DMT-based training, participants were asked to identify the technical successes and shortfalls of the concept and to assess the potential for DMT to improve future air force training. Crane (1999) reports that overall, participants believe that aircraft and DMT-based training complement each other, with DMT being most effective for training multiship, air-to-air engagements, particularly against multiple enemy aircraft (bandits). In addition, DMT was rated as highly effective in its ability to train leadership skills. Evidently, since multiship, air combat missions are rarely practiced in fighter aircraft it is difficult for younger pilots to gain leadership experience. Instructors reported taking advantage of DMT-based training by having less experienced pilots practice their leadership skills by acting as flight leads.

DMT is a new training medium for the air force. While early studies suggest several advantages to DMT-based training, there are still challenges to full-scale implementation of DMT. Similar again to the aforementioned navy and army challenges, technical challenges of fully realizing the potential of DMT include: (a) visual-system resolution so low that it prevents trainees from determining aspect angle of friendly and enemy aircraft at real-world distances; (b) terrain databases too small to accurately represent a theater-wide environment; (c) simulator–operator stations that often constrain the way instructors want to train or make the interface with the VE much too complex; and (d) constructive entities that do not behave and respond realistically. However, despite these technical hurdles, DMT is already showing itself as the medium of choice for air warrior training.

## 5. FUTURE DIRECTIONS

In general, we expect that in the near future the U.S. military will continue to experience the same pressures that were discussed at the beginning of this chapter: more missions, different kinds of missions, and a continued need to constrain costs while maintaining a high level of preparedness. In addition, the U.S. military will be making increased use of digital systems to obtain and transmit more information about the battlefield to more participants than ever before. Virtual environments will be used to help design these systems, and train those who will use them. The training may well be embedded in operational systems themselves (e.g., ships, planes, and tanks).

If VE can be shown to provide cost-effective training, its use for training and mission rehearsal is likely to increase. Empirical proof of effectiveness in at least some operational training settings, not just in the laboratory, is likely to be critical to acceptance and increased use.

## 6. REFERENCES


Alluisi, E. A. (1991). The development of technology for collective training: SIMNET, a case history. *Human Factors, 33*(3), 343–362.

Bliss, J. P., Tidwell, P. D., & Guest, M. A. (1997). The effectiveness of virtual reality for administering spatial navigation training to firefighters. *Presence: Teleoperators and Virtual Environments, 6*(1), 73–86.

Breaux, R. B., Martin, M., & Jones, S. (1998). Cognitive precursors to virtual reality applications. In *Proceedings of SimTecT* (pp. 21–26). Adelaide, Australia: The SimTecT 98 Organising and Technical Committee.

Breaux, R., & Stanney, K. (1999). Use of agents to train spatial knowledge in virtual environments. In *Proceedings of the IEEE Virtual Reality '99 Conference*. Houston, TX: IEEE Press.

Burnside, B. L. (1991). Assessing the capabilities of training simulations: A method and simulation networking (SIMNET) application (ARI Res. Rep. No. 1565). Alexandria, VA: U.S. Army Research Institute for the Behavioral and Social Sciences.

Carroll, J. M., & Thomas, J. C. (1982). Metaphor and the cognitive representation of computing systems. *IEEE Transactions on System, Man, and Cybernetics, 12*, 107–116.

Crane, P. (1999, May). *Designing training scenarios for distributed mission training*. Paper presented at the tenth International Symposium on Aviation Psychology, Columbus, OH.

Darken, R. P., & Sibert, J. L. (1996). Navigating large virtual spaces. *International Journal of Human–Computer Interaction, 8*(1), 49–72.

Darken, R. P., & Sibert, J. L. (1993). A toolset for navigation in virtual environments. *Proceedings of ACM User Interface Software and Technology*, 157–165.

Dent, C., Klein, G., & Eggleston, R. (1987). *Metaphor casting of information display requirements* (Tech. Rep. No. ADA196203). Ft. Belvoir, VA: Defense Technical Information Center.

Doyle, P., & Cruz-Neira, C. (1999). Virtual reality and visualization in education. *Syllabus: New Directions in Education Technology, 12*(9), 18–22.

Ehrlich, J., Goldberg, S. L., & Papin, J. P. (Eds). (2000). *The capability of virtual reality to meet military training needs* (RTO-MP-54). Neuilly-Sur-Seine Cedex, France: North Atlantic Treaty Organization Research and Technology Organization.

Flynn, M. R., Campbell, C. H., Myers, W. E., & Burnside, B. L. (1998). *Structured training for units in the Close Combat Tactical Trainer: Design, development and lessons learned* (ARI Res. Rep. No. 1727). Alexandria, VA: U.S. Army Research Institute for the Behavioral and Social Sciences.

Garris-Reif, R., & Franz, T. M. (1995). Simulator sickness and human task performance in conventional and virtual environments. In A. C. Bittner & P. C. Champney (Eds.), *Advances in industrial ergonomics and safety 7*. Seattle, WA: Taylor & Francis.

Gershon, N., Eick, S. G., & Card, S. (1998, March/April). Information visualization. *Interactions*, 9–15.

Gillis, P. D. (1998). Realism in computer-generated forces command entity behaviors. In *Proceedings of the Seventh Conference on Computer Generated Forces and Behavioral Representation*, 455–466. Orlando, FL: University of Central Florida.

Goldberg, S. L., Mastaglio, T. W., & Johnson, W. R. (1995). Training in the Close Combat Tactical Trainer. In R. J. Seidel & P. R. Chatelier (Eds.), *Learning without boundaries: Technology to support distance/distributed learning*. New York: Plenum Press.

Grant, S. C., & Magee, L. E. (1997, December). *Navigation in a virtual environment using a walking interface*. Paper presented at the NATO RSG-18 Workshop on Capability of Virtual Reality to Meet Military Requirements, Orlando, FL.

Haskell, I. D., & Wickens, C. D. (1993). Two- and three-dimensional displays for aviation: A theoretical and empirical comparison. *International Journal of Aviation Psychology, 3*(2), 87–109.

Hays, R. T., Vincenzi, D. A., Seamon, A. G., & Bradley, S. K. (1998). *Training effectiveness evaluation of the VESUB technology demonstration system* (Tech. Rep. No. 98-0003). Orlando, FL: Naval Air Warfare Center Training Systems Division.

Judge, A. (1993). *Metaphors as an unexplored catalytic language for global governance* [On line]. Available: www.uia.org/uidocs/govwfsf.html

Kalawsky, R. S. (1993). *The science of virtual reality and virtual environments*. Wokingham, England: Addison-Wesley.

Knerr, B. W., Lampton, D. R., Witmer, B. G., Singer, M. J. Parsons, K. A., & Parsons, J. (1998). *Recommendations for using virtual environments for dismounted soldier training* (ARI Tech. Rep. No. 1089). Alexandria, VA: U.S. Army Research Institute for the Behavioral and Social Sciences.

Lampton, D. R., Kraemer, R. E., Kolasinski, E., M., & Knerr, B. W. (1995) *An investigation of simulator sickness in a tank driver trainer* (Res. Rep. No. 1684). Alexandria, VA: U.S. Army Research Institute for the Behavioral and Social Sciences.

Lee, J. M., & MacLachian, J. (1986). The effects of 3D imagery on managerial data interpretation. *Management Information Systems Quarterly, 10*(3), 257–269.

Lockheed Martin Information Systems. (1997). *Dismounted warrior network front-end analysis experiments final report.* Orlando, FL: Author.

Lohse, G. L. (1997). Models of graphical perception. In M. Helander, T. K. Landauer, & P. Prabhu (Eds.), *Handbook of human–computer interaction* (pp. 107–135). Amsterdam: North-Holland.

Magee, L. M. (1997). Virtual reality simulator (VRS) for training ship handling skills. In R. J. Seidel & P. R. Chatelier (Eds.), *Virtual reality, training's future: Perspectives on virtual reality and related emerging technologies.* New York: Plenum Press.

Mills, S., & Noyes, J. (1999). Virtual reality: An overview of user-related design issues. *Interacting with Computers, 11*(4), 375–386.

Regan, C. (1995). An investigation into nausea and other side-effects of head-coupled immersive virtual reality. *Virtual Reality, 1*(1), 17–32.

Singer, M. J., Ehrlich, E. A., & Allen, R. C. (1998). *Effect of a body model on performance in a virtual environment search task.* (Tech. Rep. No. 1087). Alexandria, VA: U.S. Army Research Institute for the Behavioral and Social Sciences.

Science Applications International Corporation and Lockheed Martin Information Systems Company. (1998). *DISAF MOUT Enhancements Final Report* [CDRL AB02, Online]. Available: http://www.stricom.army.mil/PRODUCTS/DWN/warrior1.html

Templeman, J. (1997, December). *Performance-based design of a new virtual locomotion control.* Paper presented at the NATO RSG-18 Workshop on Capability of Virtual Reality to Meet Military Requirements, Orlando, FL.

Thorpe, J. A. (1987). The new technology of large scale simulator networking: Implications for mastering the art of warfighting. In *Proceedings of the Ninth Interservice Industry Training System Conference* (pp. 492–501). Washington, DC: American Defense Preparedness Association.

Turk, M., & Robertson, G. (2000). Perceptual user interfaces. *Communications of the ACM, 43* (3), 33–34.

Wilson, P. N., Foreman, N., & Tlauka, M. (1997). Transfer of spatial information from a virtual to a real environment. *Human Factors, 39*(4), 526–531.

Witmer, B. G., Bailey, J. H., & Knerr, B. W. (1995). Training dismounted soldiers in virtual environments: Route learning and transfer. (Tech. Rep. No. 1022). Alexandria, VA: U.S. Army Research Institute for the Behavioral and Social Sciences.

Witmer, B. G., & Singer, M. J. (1998). Measuring presence in virtual environments: A presence questionnaire. *Presence: Teleoperators and Virtual Environments, 7*(3), 225–240.

Woods, D. D. (1984). Visual momentum: A concept to improve the cognitive coupling of person and computer. *International Journal of Man–Machine Studies, 21*, 229–244.

# 44

# Team Training in Virtual Environments: An Event-based Approach*

Eduardo Salas,[1] Randall L. Oser,[2]
Janis A. Cannon-Bowers,[2] and Eleni Daskarolis-Kring[3]
*[1]Institute for Simulation and Training*
*University of Central Florida*
*3280 Progress Drive, Orlando, FL 32826-0544*
*esalas@pegasus.cc.ucf.edu*
*[2]Naval Air Warfare Center, Training Systems Division*
*12350 Research Parkway, Orlando, FL 32826-3224*
*OserRL@navair.navy.mil*
*Cannon-BowJA@navair.navy.mil*
*[3]University of Central Florida*
*P.O. Box 161390, Orlando, FL 32816*
*ekring@drc.com*

## 1. INTRODUCTION

As technological capabilities and the complexity of tasks and environments have broadened, likewise, the need for effective synchronicity and coordination of activities among members of crews, groups, teams, and collectives has increased. Decision making in these complex environments is often characterized by severe time stress, high stakes, uncertainty, vague goals, and many organizational constraints (Cannon-Bowers, Salas, & Pruitt, 1996; Orasanu & Connolly, 1995). At the core of performance success or failure is the human operator, making effective training a necessity. These factors also dictate that training for interactions among humans and between humans and machines should be conducted in a manner resembling real-life contexts (Oser, Gualtieri, Cannon-Bowers, & Salas, 1999). At the same time, growing restraints on resources have made it imperative that training methods be effective, efficient, and economical.

Virtual environments (VE) are one example of advancing technological possibilities that have allowed for expansions in training strategies. In fact, VEs are multifaceted and have the potential for individual and team training. Virtual environment technology immerses an

*The views of the authors do not necessarily represent the policy of the U.S. government.

individual or team in a perceptually realistic and interactive environment that is generated by a computer. Users are able to interact with autonomous agents (see chap. 15, this volume, i.e., virtual teammates) and/or human team members. Various combinations of multiple trainees and trainers, human or virtual, may be brought together for a training exercise in a simulated environment (Stiles, et al., 1996). This is significant, because most complex tasks require effective teamwork.

Although considerable efforts have been invested in improving VE technology, little is known about how to best use this technology for training team skills. Moreover, a strong theoretical framework to support learning in VE is lacking. Without such a framework to guide how the technology should be utilized for training, positive transfer of knowledge and skills (see chap. 19, this volume) to the job cannot be ensured (Salas & Cannon-Bowers, 1997).

The purpose of this chapter is to examine these and related issues concerning the use of virtual environments, specifically virtual teammates (VTMs) and their impact on training and learning environments. More specifically, a focus is put on an event-based approach to training (EBAT, Oser, Cannon-Bowers, Dwyer, & Salas, 1997a). Along with VE technologies, the use of a learning methodology may provide a possible solution to training in dynamic settings. This chapter initially defines related team training and technology concepts. An overview of the range of possibilities of VEs and VTMs in developing learning environments and EBAT formats follows. Finally, guidelines for the use of VEs for team training are described.

## 2. ASPECTS OF TEAM TRAINING

### 2.1 What Makes Up a Team?

Teams are a group of two or more people who work together interdependently to complete a task or to achieve a specified goal (Baker & Salas, 1997). Team members coordinate their work in a simultaneous or sequential style in order to complete a task, using specific skills and knowledge, and shared concepts or mental models of the desired goal (McIntyre & Salas, 1995; Salas, Dickinson, Converse, & Tannenbaum, 1992).

Teams and the individual members that make up teams are characterized by the level of competencies that they have, relative to what would be required to successfully complete a task or a goal. These team competencies fall into three categories. They include knowledge, skills, and attitudes (KSA). Knowledge includes the theories, principles, and concepts needed for effective team performance. Skills are the required behaviors and actions needed for thorough and successful completion of a task. Attitudes describe the team members' affective views, both individually and collectively, on their abilities and motivation to accomplish these goals (Cannon-Bowers, Tannenbaum, Salas, & Volpe, 1995). If a team's knowledge and skills are weak and their attitudes are unrealistic, the outcome will be different than expected. Thus, for effective team performance, it is important to (1) know the team's current level of competence (i.e., have clear and detailed analysis of the KSAs required for effective team functioning) and (2) design training to bridge the team's actual level of team competencies to the desired level.

Recently, Cannon-Bowers et al. (1995) have delineated an extensive list of the aspects of knowledge, skills, and attitudes necessary for effective team performance. Table 44.1 is a representative collection of some of those KSAs.

### 2.2 What Are Virtual Environments?

The terms *virtual reality*, *simulation*, and *modeling* are often used synonymously with virtual environments. Overall, all these terms refer to the immersion of the user into a simulated

**TABLE 44.1**
Sample of Team KSAs*

| | *KSA Issues Relevant to VE and VTM* |
|---|---|
| Knowledge | Accurate and shared task models |
| | Task sequencing |
| | Accurate problem solving |
| | Team role interaction patterns |
| | Team orientation |
| Skills | Adaptability |
| | Shared situational awareness |
| | Mission analysis |
| | Communication |
| | Decision making |
| Attitudes | Motivation |
| | Collective efficacy/potency |
| | Shared vision |

*Adapted from Cannon-Bowers et al. (1995).

environment, where the user interacts with computer-generated images (Shebilske, 1993). In fact, the phrase *virtual environment* has a very general meaning. Depending on which publication is read, it may cover any simulated environment of an actual or ideal setting in which players interact with each other and/or with computerized images. These interactions may occur through written text, audio or visual communication for computer-mediated collaboration, and knowledge sharing. For the purposes of this chapter, virtual environments are defined as simulated environments creating "synthetic sensory experiences " (Kalawsky, 1993, p. 5) of an actual or theoretical physical setting in which players interact with computerized images and possibly each other through audio, visual, haptic, and computer technology for collaboration and knowledge sharing. Cognitively, VEs are supported by their multisensory representation and computerized feedback sound, graphics, and force feedback. As defined in Kalawsky (1993), VE interactions are multifaceted with three-dimensional (3-D) and real-time interactions.

Furthermore, Washburn and Ng (1997) suggest that simulations, such as VEs, must be flexible, reconfigurable, and scalable. They explain that VEs should be adaptable to the location of use and to the environment where the task is being tested. Scenarios, number of team members, and level of difficulty should be readily modifiable. Equally important, the system should be scalable for the individual or team, have local or distributed training capabilities, and wide-area networking options. Through "telepresence," instructors should be able to observe and interact with trainees within an immersive system, allowing for training in distributed situations and within several contexts (Washburn & Ng, 1997).

## 2.3 What Are Virtual Team Members (VTM)?

Virtual team members (VTMs) are multifunctional autonomous agents (Stiles et al., 1996; see chap. 15, this volume) or simulated images of humans within a VE that function in the role for which they are programmed. They can work as leaders, instructors, mentors, colleagues, and teammates. They can be full-body images or disembodied images of hands and head. Their purpose is to assist in supplying training possibilities to large-scale and possibly geographically distributed audiences.

The use of VEs and VTM can assist in supporting the necessary skills and cognitive processes needed for accurate decision making and problem solving. Trainees can work in an environment representative of the real world and interact with other teammates, both virtual and human, in real time, allowing them to train and gain experience as if being exposed to actual events and scenarios. VTM can emulate human responses and team dynamics with more realistic feedback. This opportunity is beyond what any other technology is able to contribute to training. Furthermore, through programmed instruction, skills required to effectively make decisions and solve problems in a real-life context (Cannon-Bowers & Bell, 1997) can be trained using the capabilities of VEs and VTMs.

Given that current technology is making it possible for simulated human operators to contribute in similar ways as human team members, then perhaps for training in VE it may be necessary to rethink the definition of teams to include VTMs (Zachary, Cannon-Bowers, Burns, Bilazarian, & Krecker, 1998). Traditionally, only human contributors were included in the definition; however, VTMs are likely to play an increasing, yet often inconspicuous, role in teams in the future. In terms of training, current technology has allowed for the use of both human and virtual members to make up a team, thereby expanding the possibilities and frequency of training. As such, revising this definition to include human and virtual players working together toward a common goal in VE training settings is warranted.

## 3. LEARNING ENVIRONMENT: ENVIRONMENTS FOR EFFECTIVE TRAINING

Although VEs have the potential to facilitate training, technology alone cannot ensure successful training outcomes (Salas & Cannon-Bowers, 1997). To effectively provide training, an environment established to promote skill acquisition and learning is necessary (Salas & Cannon-Bowers, 2000). Such environments should focus on the development of core team competencies, that is, knowledge, skills, and attitudes (KSAs; Oser, Dwyer, Cannon-Bowers, & Salas, 1998). In the design of the learning environment, several issues must be taken into account. These include the targeted training audience, task requirements, training environment, and associated interrelated components. The following framework on learning environments (see Fig. 44.1) provides a synopsis of the work of Oser, Cannon-Bowers, and colleagues (Oser et al., 1997a; Oser et al., 1998; Oser, Gualtieri, Cannon-Bowers, & Salas, 1999).

### 3.1 Training Audience: Who Is the Training For?

The training audience is at the center of the learning environment; thus, its needs must be clearly identified. Each team is composed of unique characteristics. An individual team's characteristics play a significant role in the manner in which a learning environment is constructed. These characteristics include: the degree to which team members hold a shared understanding of tasks and their teammates; the team's organizational structure; and the proximity of team members to each other.

More specifically, team members should have shared concepts/mental models of the process and subtasks in reaching a goal. The higher the level of shared understanding (i.e., shared mental models) between team members, the more efficient and successful goal completion (Cannon-Bowers & Salas, 1997; Cannon-Bowers, Salas, & Converse, 1993). The more team members can understand, predict, and act upon each other's and the team's needs in carrying on their responsibilities, the more efficiently the team will operate in completing a task.

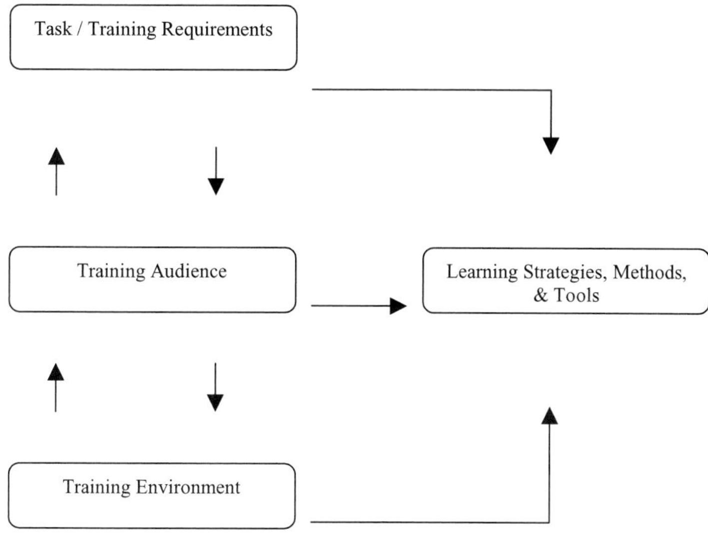


FIG. 44.1. A conceptual model of a learning environment.

Team training aiming at strengthening shared mental models has shown improvements in decision making (Blickensderfer, Cannon-Bowers, & Salas, 1998; Serfaty, Entin, & Johnston, 1998).

The team's structure is also significant in establishing the learning environment needs. The composition of team members within the team is a determining factor in how the learning environment should be designed. Different feedback mechanisms may be needed. A self-managed team may have different team dynamics than a team with a leader. In reality, organizations and institutions, and the teams within them, are hierarchical in nature. So, team-leader training may also be an issue (Smith-Jentsch, Zeisig, Acton, & McPherson, 1998). Adaptability and shared mental models are key for the members of a team to work together in an effective and efficient manner. This is especially true for teams made up of crew with diverse roles and expertise. More specifically, the methods used for information exchange and decision making are primary factors in determining the basis for the design of the learning environment. Naturally, whether team members are within the same location or have to collaborate from distributed locations influences the design of the learning environment.

### 3.2 Task Requirements: What Is Being Trained?

Another component in developing an effective learning environment is the characteristics that define a task. Teams must function under circumstances that often involve rapidly evolving situations, high stakes, time stress, uncertainty, vague goals, and information ambiguity. Within that context, although many factors are relevant to a task, three significant aspects include its information requirements, task length, and focus. First, however, it is important to take into consideration what type of training is required, that is, procedural versus strategic, and novel versus routine. Procedural and routine tasks are often short term and consist of planned and clearly defined steps. Novel and strategic tasks, on the other hand, often involve situational awareness, ambiguity, and problem-solving techniques. The sources of information requirements for monitoring and decision making can range from procedural to planning, management, and situation assessment. Team members need to be able to synthesize and coordinate information from a range of sources into a meaningful form.

In addition, the task may require an immediate or future focus. In time-critical situations, immediate actions are necessary and usually rely on automatic responses. Feedback is typically instant. In situations where tasks are related to future actions, necessary plans and strategies are developed over a longer time period and feedback is often not received until generated ideas are implemented. In either case, incorporation of feedback into strategy modifications is essential.

### 3.3 Training Environment: Under What Conditions?

The conditions of the training environment make up another variable in the learning environment design. Training environment characteristics include training frequency, length of the training cycle, level of realistic effect, and number of locations coming together in a training exercise.

How often training will take place is one component of the learning environment design. The length of the training cycle from initiation to completion is established in a learning environment to optimize the effectiveness of training. Likewise, the level of simulation should be realistic enough to allow a user to perceptually feel immersed in the setting (see chap. 40, this volume). Distracting and unrealistic characteristics should be eliminated. Yet completely realistic visuals may not be necessary, and at times counterproductive.

In considering another aspect of the training environment, distributed training (DT) has become a feasible option for convenience, economic reasons, and reliability in training. With the same training possibly being necessary in more than one location or with people in various parts of the world needing training on a specific topic, DT has certain advantages. Knowing whether the training is local or distributed plays a role in the design of the learning environment.

## 4. EVENT-BASED APPROACH TO TRAINING: STRATEGY FOR EFFECTIVE TRAINING

One avenue for structuring in a learning environment is an event-based approach to training. EBAT provides the strategies, methods, and tools that are essential for an effective learning environment (Salas & Cannon-Bowers, 1997) in a structured and measurable format for training and testing specific KSAs. In order to precisely control, measure, and benefit from user responses to events, a cyclical layout of EBAT has been established (Cannon-Bowers, Burns, Salas, & Pruitt, 1998).

EBAT allows for training and testing through simulated scenarios, where planned events are introduced to test specific skills that would be required in real-life situations. Systematic linkages of the strategies, methods, and tools of EBAT have been used to establish learning environments in a wide array of settings (Oser et al., 1998). Some of the competencies trained with EBAT have included situation assessment, decision making, planning, and resource management (Oser et al., 1997a; Oser, et al., 1998). At the same time, learning environments and EBAT methods have resulted in measures that are psychometrically strong and that have improved performance in a variety of settings (Dwyer, Oser, & Fowlkes, 1995; Fowlkes, Lane, Salas, Franz, & Oser, 1994; Johnston, Smith-Jentsch, Cannon-Bowers, 1997). EBAT success has been demonstrated in research of aviation events (Fowlkes et al., 1994), command and control (Johnston, Cannon-Bowers, & Smith-Jentsch, 1995), multiservice environments (Dwyer, Oser, Salas, & Fowlkes, 1999), and virtual environments (Hays, Vincenzi, Seamon, &

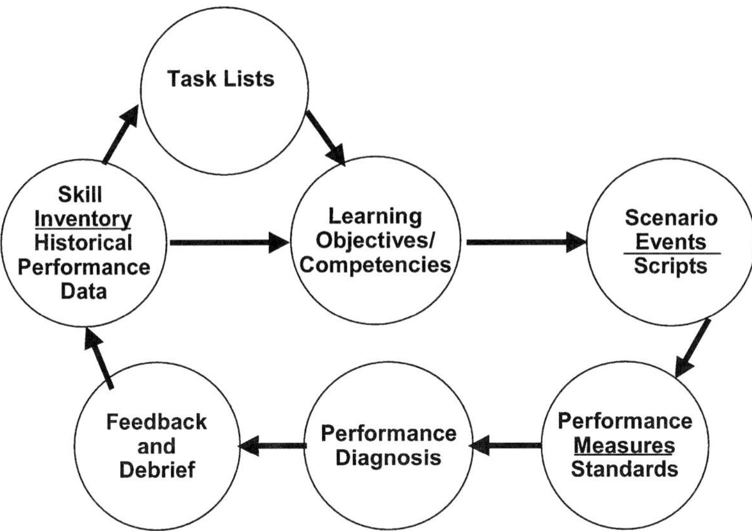


FIG. 44.2. Components of the EBAT cycle.

Bradley, 1998). Specifically, it has been shown that for effective training, performance measurements are required (Cannon-Bower & Salas, 1997) and that those measurements can provide, beyond description and evaluation, diagnostic information (Cannon-Bowers et al., 1998).

The concepts of EBAT are reviewed in this section. This review is based on research conducted by Cannon-Bowers and Salas (1997), Cannon-Bowers, Burns, Salas, and Pruitt (1998), Oser et al. (1998), Johnston et al. (1997), Fowlkes et al. (1994), Dwyer, Fowlkes, Oser, Salas, and Lane (1997), and, Dwyer et al. (1995). The seven steps of the EBAT cycle are depicted in Fig. 44.2.

### 4.1 Skill Inventory/Performance Data

Through techniques such as task analysis, cognitive-task analysis, and team-task analysis, a team's level of KSA core to performance are determined. Other techniques used can include knowledge elicitation procedures. Past behavioral records of trainees are used to establish the required skills for successful completion of a task and to identify typical deficiencies in those skills. It is important to know where trainees stand in comparison to behavioral representations of a realistic and desired course of action toward an event.

### 4.2 Learning Objectives/Competencies

Based on the skill inventory, learning objectives and required competencies are then outlined. As specific KSA deficiencies are determined, then associated learning objectives and requisite competencies become the focus of the training. These can be related to a specific task or can cover general competencies ranging over many tasks. Objectives and competencies become the root-source when developing scenario events.

## 4.3 Scenario Scripts/Trigger Events

Based on the task analysis, the learning objectives, and previous performance measurements, a scenario is developed. "Trigger events," or conditions, are designed for each learning objective to test for possible deficiencies in that specific training objective. Events can vary in number, location/timing in the exercise, and in degree of difficulty. For a realistic and continuous effect and for precise data collection, the scenario must be modifiable to user's responses in real time. The specifics of the scenario and event design are included in a master scenario event list (MSEL). For effective outcomes, the goal is to design events that will test the user's mastery level of specific skills based on the learning objective.

## 4.4 Performance Measures/Standards

Performance measures and standards must be designed to assess the user's mastery level and will assist in explaining why users performed as they did (see chap. 35, this volume). Effective measures, which are directly related to events, involve strategies and tools used and vary according to objectives being tested. Measurement methods can be through observation, semiautomated, or automated techniques. It is recommended that a multifaceted approach be used toward measurement. Examining the outcomes and processes of decision making through data collection is vital to feedback (see Fig. 44.3). Outcome measurements contribute general information on performance, whereas process measurements provide diagnostic information on the causes of performance levels. The strategies and tools used should be capable of detecting patterns and trends in performance and should provide related diagnostic feedback.

## 4.5 Performance Diagnosis

Based on the information obtained from performance measures and standards, it is possible to diagnose performance. These diagnoses provide insight into competencies. Causes of performance problems are examined by looking at possible deficiencies in skills, knowledge, and competencies. Diagnostic mechanisms must be based on linkages to the EBAT stages of Performance Measures and Standards (see section 4.4), Scenario Scripts and Trigger Events (see section 4.3), Learning Objectives and Competencies (see section 4.2), and Skill Inventory and Performance Data (see section 4.1), so that conclusions drawn can be relevant and mapped back to required competencies.

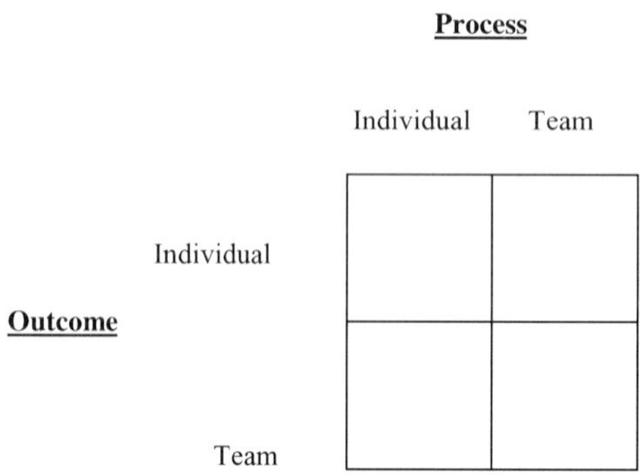


FIG. 44.3. Individual and team issues in process and outcomes.

### 4.6 Feedback and Debrief

Based on the diagnosis, constructive and specific feedback directly related to learning objectives must be provided, either immediately or at the completion of the task. This process will then enable users to focus on changes necessary for improved performance. The timing of feedback depends on the skill being tested and the performance level sought (Smith-Jentsch et al., 1998). After-action reviews (AAR) cover the debriefing and feedback information. By focusing on the learning objectives, events, and performance measures, the internal consistency of a training exercise is enhanced.

## 5. INCORPORATE INTO FUTURE TRAINING AND ARCHIVE DATA IN A MEANINGFUL MANNER

Lessons learned from the exercises can be addressed in future training exercises. Knowledge gained on deficiencies in the trainees' abilities and on their present mental model of the event can be incorporated into the training design. The focus of learning/training objectives should be on increasing user's knowledge and not merely repeating information or skills that have already been mastered. Over time, the pattern within archived data will assist in establishing a behavioral representation, or "norm," for comparisons.

The following sections discuss a vision of possible uses of VEs and VTMs in learning environments and EBAT for addressing team training issues. It is important to note that currently the technology to use VE and VTMs as fully interactive and intelligent learning tools is still at its early stages. As previously discussed, it is important for effective training to have both adequate tools and a well-structured theoretical basis. Thus, for successful merging of tools and learning theories a common vision of an intelligent, instructional VTM should be established. This chapter does not address the technical, programming, and modeling aspects of VEs and VTMs (see chaps. 12, 13, and 15, this volume). This chapter attempts to project a vision of how VEs and VTMs could be used within strong learning theories to enhance effective team training once the technology is available.

## 6. USING VE AND VTM TO ESTABLISH EFFECTIVE LEARNING ENVIRONMENTS

As with all educational technology, the theoretical framework for educational VEs should be based on learning theories, such as active learning, training through apprenticeship, and collaboration-based learning principles (Grant, McCarthy, Pontecorvo, & Stiles, 1991). In most contexts, these learning principles can be most fully implemented in situations where one-on-one attention or close monitoring of a group is available. For large-scale and distributed situations, these efforts are difficult, if not impossible, to achieve. As a result, the use of VTM can be advantageous. While some VTMs can be used for individual training, other VTMs or human instructors can work with groups or teams.

In order to have realistic apprenticeshiplike training, three VTM characteristics should be present. The behaviors of VTM should be responsive to changes in the VE caused by outside events or human team members or other VTMs. VTMs should be able to have an understanding of the relationships and rational of tasks, subtasks, and associated steps in order to provide appropriate direction to trainees. Lastly, VTMs should be designed with an understanding of the relationships between task roles to be able to encourage development of collaboration and cross training (Stiles et al., 1996). These aspects have implications for the design of VTMs.

Other areas that should be taken into account are the VE interactions of human to human, human to VTM, VTM to VTM, human to scene, and VTM to scene in terms of representation and visual and verbal communication (Stiles et al., 1996). The following sections delineate how VTMs and VEs affect the learning framework presented previously.

## 6.1 Training Audience: Who Is the Training For?

Virtual environments and VTMs can adapt to the unique characteristics and needs of a given training audience. Virtual teammates can function within and facilitate training with team issues, such as shared mental models, organizational structure, and proximity of team members in a flexible manner to meet trainees' needs. VTMs can make individual attention available for a trainee or team's unique needs. If one or few individuals or teams need further instruction in a specific area, VTMs can provide this training without slowing the overall training process for other trainees or requiring human instructors to attend to specific needs. These individuals or teams can then join the larger group when ready, which, in turn, may be led by another VTM.

As mentioned, VE and VTM technology can assist in the development of shared mental models. If the desired mental model is programmed into the VTM and the VTM requires trainees to respond accordingly, eventually the trainee should learn to make predictions and develop a similar mental model to meet expectations. Realistic features of VEs may build mental models that more accurately represent real-life scenarios and therefore better prepare the player to respond appropriately (Cannon-Bowers et al., 1993). In return, consistency offered by the VE makes it easier to pinpoint errors in players' mental models. Additionally, VTMs can play any roles assigned to them, such as functioning as team members in order to force human team members into other roles to assist cross-training and shared mental model development, or simply replacing missing human teammates.

At the same time, differences in the needs of teammates working in either close or distant proximity from each other can be addressed through VEs and VTMs. Virtual teammates and VEs can help to focus on skills required for teamwork and collaboration from distributed sites. Hence, they can help one understand, as well as train, distributed teamwork. On the other hand, VTMs can provide distributed training on a general basis while giving individuals additional instruction as needed.

## 6.2 Task Requirements: What Is Being Trained?

Task characteristics include information requirements, task length, and focus. Skills, such as coordinating and synthesizing information for monitoring, situation assessment, and decision making can be modeled by a VTM. The VTM can then test the trainee and provide feedback. More so, the VTM could easily be programmed to demonstrate or train for procedural, strategic, novel, or routine tasks. Communication and demonstration of desired concepts could be displayed through video, audio, or with computer messages. Moreover, the use of VE and VTM will allow for repeated training in time critical situations that may otherwise be inaccessible. Interactions with VTM would allow for continuity of information flow at real-time speed.

## 6.3 Training Environment: Under What Conditions?

Other than in real-life situations, no other environment can provide a more realistic training environment than simulation. With issues of locations, equipment and team member availability, safety, desire for consistency, and economical constraints, VEs and VTMs appear to be viable options. They can provide consistent, controlled, and repetitive training through their own simulated environment and equipment. The length of the training cycle, level of realistic effect, and number of locations coming together for a training exercise are easily manipulated

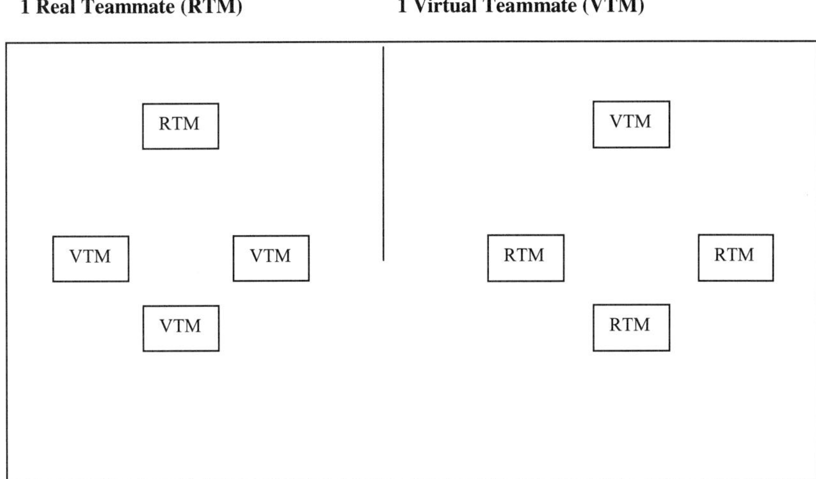


FIG. 44.4. The involvement of VTM in teams can range from one to all but one, depending on training need.

in the context of VEs and virtual teammates. For example, the STEVE (Soar Training Expert for Virtual Environments) VTM (Stiles et al., 1996), can be available in full body or with just head and hands. By allowing this option, the STEVE VTM can demonstrate a procedural task without obstructing the view of the equipment he is discussing or other relevant items (Stiles et al., 1996). In addition, advancements in VEs and VTMs can help bring people together from throughout the world to train, collaborate, and communicate. Virtual environments can provide a common location and training ground. As another advantage, VTMs can work as instructors or fill-ins for missing colleagues. The use of VTMs in a training environment can range from using only one in a team to using all VTMs and only one human trainee.

### 6.4 Interrelated Components of a Learning Environment: How Should It Be Trained?

The use of VE and virtual teammates technology allows more flexibility in training while at the same time providing a means to more accurately test and control for specific training strategies and related measurement methods. Specific cognitions (knowledge) and behaviors (skills) can be reinforced with VEs and VTMs. Use of VEs and VTMs gives the advantage of moving the VTM from instructor to colleague as needed to allow trainees to attempt different options and learn to self-correct when wrong. The systematic, deliberate approach provided by VEs more clearly points out proficiencies and deficiencies in an individual's or team's knowledge and abilities and the direct impact of these traits on the process and outcome of actions taken.

Within the learning environment, EBAT can be used to strategically test for and train specific required skills. The ways that VTMs and VEs could be used within EBAT are described below.

## 7. VE AND VTM CONTRIBUTIONS TO AN EVENT-BASED APPROACH TO TRAINING

The following sections discuss the strengths and benefits of using VTMs for training skills and building a desired knowledge base. The application of VTMs within the EBAT cycle are then described. The well-structured and consistent form and repetition of VEs and VTMs assist in training novices to identify, categorize, and encode information. Through repeated practice in controlled environments, novices can have a number of practice opportunities on a specific situation to assist them in learning to efficiently organize information and add to their knowledge. This is supported by work by Noble, Grosz, and Boehm-Davis (1987), who found that knowledge templates are directly related to naturalistic decision making. They explain that templates develop when cues from the environment are recognized and recorded as a pattern within an individual. When cues come up again, associated and previously successful behaviors are recalled and used. Therefore, it can be assumed that the structured, repeated practice available through VEs and VTMs could assist in perfecting such templates. Moreover, Klein (1989) showed that through these organized knowledge templates, experts are more likely to successfully execute an effective solution for a given problem. This would be especially important in team situations, and in developing shared mental models of required behavior and decisions.

These same characteristics of VE and VTM can assist in precise situation assessment. Situation assessment is the capability to recognize meaningful cues, causes, and patterns in the environment (Glaser, 1986). Stout, Cannon-Bowers, and Salas (1996) defined three primary considerations for accurate situation assessment: (1) assessment of cues and cue pattern within a situation; (2) team processes for cue-related information sharing and shared mental model development; and (3) type and content of these shared mental models leading to interpretation, decision making, and action. The ability to assess a situation well could lead to reduced workload and could assist in information processing, knowledge encoding, and response facilitation (Means, Salas, Crandell, & Jacobs, 1995). According to Cannon-Bowers and Bell (1997), cue and pattern recognition is required for situation assessment. The ability to effectively link cues makes an individual or team more efficient (Means et al., 1995). Moreover, Klein (1995) points out that being able to distinguish between relevant and irrelevant cues and patterns is equally important in successful problem solving. Wickens (1984) found that internal representations of the environment and identification of causal relationships are established through underlying patterns developed from experience or training. The primary method in developing these skills is practice. By containing solution options (Cellier, Eyrolle & Marine, 1997) or through repeated practice trials (Means et al., 1995), VE simulations allow opportunity for such representations to develop.

Another decision making requirement is monitoring a system for faults. For this to be done successfully, one must understand and be familiar with the system. Because most natural settings are "fluid in nature" (Cannon-Bowers & Bell, 1997), it is vital that teams are able to collect and evaluate information from the system in a dynamic environment. Metacognition of team members, where they evaluate their own performance, is vital for self-correction and can be facilitated through appropriate feedback (Cohen, Freeman, Wolf, & Militello, 1995). Glaser (1986) suggested that these skills be developed through practice and should be encouraged in novices. These skills can be advanced in a VE or through interactions with virtual teammates by providing multiple situations in which to self-examine decisions and skills used (Oser et al., 1999). Plus, virtual environments and virtual teammates are thought to add to high interobserver reliability, internal consistency, and interexercise correlation.

While VTMs have not yet been formally used in conjunction with EBAT, the following section provides a vision of how VTMs could be used to support EBAT.

### 7.1 Skill Inventory/Performance Data

The initial step in the EBAT cycle is to accurately identify and define behavior and responses that trainees should develop. Information from previous databases is used to define the domain of skills and behavioral representations that would be preferable in a given scenario.

Information gained from these representations can be incorporated into databases of behavior patterns and used to comparatively evaluate trainees and teams for deficiencies and proficiencies in those behaviors. Virtual environments and VTMs could then be designed to focus on the specific information gained through knowledge elicitation procedures and cognitive task analysis.

### 7.2 Learning Objectives/Competencies

Given the range of potential behavioral representations and deficiencies, learning objectives can be established to support their development. As deficiencies in KSAs are determined, associated learning objectives can be targeted and programmed into VEs or through VTMs. This will enable focused training to take place. A format of the desired mental model for all trainees would be the basis of the development of the learning objectives. VTMs could be programmed to teach, test, and give feedback to the user based on the objectives. More so, the ability for consistency in repetition would ensure congruency in the repeated practice of a weak skill and in the teaching methods and strategies toward appropriate mental models. Virtual teammates could hold different positions, as team leaders or subordinates. The position of virtual teammates could change to allow for training or testing of different levels of interactions and skills. By being able to alternate the position of virtual teammates, there are greater opportunities to change team members' positions, allowing for testing of the dynamics within a group under different leadership conditions, and with various levels of expertise. Greater cross-training opportunities are also afforded, which are linked to improvements in shared mental models.

### 7.3 Scenario Events/Scripts

By far, one of the areas in EBAT that would most benefit from VEs and VTMs is scenario and event execution. Manipulations of VEs and VTMs would become significant components of the scenario/event itself. Complexities and inconsistencies of developing training for real-life settings can be avoided. At the same time, these technologies would have flexibility in their design so that, beyond what would be feasible in a real-life context, numerous variations to an event could be used. Consequently, the user could learn to adapt his or her skills to the inconsistencies of the real world. As aforementioned, VTMs could give individual attention to trainees as needed, while allowing others to continue forward. Likewise, VTMs could introduce events by function (i.e., novice versus expert), allowing the user to train under various circumstances, or they could be designed to work with trainees ranging in expertise. Training from embedded expert operator training models could be precisely and realistically replicated by simulated teammates and technology.

Virtual environments and VTMs can be programmed to introduce events with high time pressure and intensity by controlling for allotted time and number of subtasks required by each team member for successful task completion. Results of decisions and alternatives can more realistically be played out and witnessed through VEs and virtual teammates (Gualtieri, Parker & Zaccaro, 1995). Research shows that structured repetition and better control over the stimulus in simulation is made possible through computers (Oser, Cannon-Bowers, Dwyer, & Salas, 1997a). Moreover, manipulation of variables and other features is easier in a computerized environment (Oser et al., 1997a).

The use of VEs and VTMs establishes consistency in testing a number of trainees in a specific event and scenario. Likewise, in situations such as distributed training, consistency in testing can be held while still enabling players to affect the environment. Using VEs and VTMs allows for theoretical concepts to be more easily tested and for hands-on training and practice.

Other modes of communicating and reinforcing information beyond the visual benefits of VEs and VTMs include audio and computer message capabilities.

## 7.4 Performance Measures/Standards

Developing effective performance measures can also be aided by VEs and VTMs. Since a scenario event is programmed into the system, likewise, the points of desired measurement could also be programmed. Thus, at specific times, the VTM or event could signal a performance measure. More so, because behavior of the events themselves and the VTMs can be held as constants, the specific responses of users are more accurately measured. In return, VEs and VTMs can be used to guide and focus the task and to collect data from performance exercises in terms of procedures and outcomes. Because these are computerized tools, the precision of measurement itself would be higher. As data accumulates, patterns of actual behavior could be established. These then could then be compared to the behavioral representations that were the basis of the learning objectives.

## 7.5 Performance Diagnosis

Reliable data and the ability to directly test specific actions and outcomes may be attained through the use of VE and VTM technology. As discussed above, the system can be programmed to monitor for patterns and significant links of action to outcomes or responses to specific situations as it is recording data. An evolving inner cycle between an event, associated performance measures, and resulting diagnosis and modification permits adaptive aiding. Specifically, as performance patterns develop from gathered data, diagnosis of users' expertise is made. The VTM is then able to adapt to an individual or team's level (as is often done with many personal computers). Based on performance observation of team members and comparisons to standards, the VTM would adjust itself appropriately. For example, if trainees performed better than expected standards, the VTM could increase the level of difficulty of an exercise. One method could involve the VTM making many mistakes, placing trainees in a situation of having to use more complex and diverse skills to remedy the situation. On the other hand, a VTM could also decrease the level of difficulty if trainees fall below norm in order to allow them to strengthen basic skills and build up to standard. These modifications could continue throughout an exercise and conclude at the end of the allotted time or once an end point in the exercise was reached.

## 7.6 Feedback and Debrief/Link to Learning Objectives

Feedback can be immediate/real time or at the completion of an exercise and can take several forms. A VTM or the VE itself could respond to a user's actions, showing the resultant consequence or interpretation of choices made or giving an overview at the end of a scenario of all actions taken. The VTM could be used during feedback/debriefing as an instructor, mentor, or colleague. A VTM could demonstrate a task while explaining it, give further explanation and answer questions as necessary, monitor and guide the user, and then follow up with feedback on performance. Clear feedback, directly related to learning objectives, would easily be given by a VTM and virtual environment. The VTM could act as a mentor or as a teammate and give guided feedback accordingly in real time. At the completion of the exercise, feedback of the performance throughout the exercise could be given. The VTM could provide this feedback as a teammate, showing how different decisions and behaviors would have led to better outcomes. The VTM could also function as an instructor, demonstrating, discussing, asking, and answering questions to facilitate discussions between teammates.

### 7.7 Incorporate Into Future Training and Archive Data in a Meaningful Manner

The patterns of computerized information provided by VEs and VTMs can more precisely and easily be recorded than that of human, possibly largely subjective, data keeping. Insights and lessons learned regarding behavioral representations and patterns and mental models could be used to establish a standard from which to judge deficiencies and proficiencies.

## 8. APPLICATION OF VE AND VTM IN TRAINING

In summary, the use of VEs and VTMs within EBAT takes the following form. Having established desired behavioral requirements to be trained, the instructional programmer would develop and input related learning objectives and significant relationships between them. Scenarios with built-in trigger events would then be designed by the program to test for trainees' deficiencies in the learning objectives. Interactive VTMs within the VE would be able to function as tutors, collaborators, and team members, as appropriate. Their purpose would be to support trainees in learning how to perform procedural tasks by demonstrating, explaining, monitoring, and assisting the trainees. In this case, one or more human team members could participate, with the VTMs being able to observe, diagnose, and introduce suggestions. As each trainee or team interacted with the VE and VTMs, this performance would be measured and diagnosed for patterns. Based on this information, the level or type of remediation that should be presented would be determined. The VE–VTM computer technology would use adaptive aiding to adjust to the trainees' level of expertise, whether individual or team, as many personal computers have the capacity of doing. Finally, feedback during the scenario exercise and afterward would be provided based on learning objectives.

As previously discussed, VTM technology must possess some essential characteristics to be successful in the role just described. First, it is important that all actions by a team member are obvious to all other team members. In addition, to allow for VTMs to do their part in recognizing human actions and interactions, VTMs need to be developed containing models of team member's possible actions, related outcomes of those actions, and relationships between them (Johnson, Rickel, Stiles, & Munro, 1998; Rickel & Johnson, 1998).

Beyond the benefits of physically seeing VTMs through the visual component of VEs, auditory and computerized messages would contribute to communication to and from teammates. Audio messages and voice recognition technology would assist in the verbal communication component. Since the military uses specific phrases for communication during tasks, it is relatively easy to program these so a VTM may verbalize and understand this communication (Zachary et al., 1998). Nonverbal communication by VTMs is also possible through programmed gestures, facial expressions, and gazes (see chap. 10, this volume). Confirmation of understanding and turn taking continuity are two areas that benefit from these options.

VTMs would also be programmed to understand and respond to questions in various forms, communicate instructions while demonstrating a task, notice students' focus of attention, and provide redirection as needed. Communication between human trainees and VTMs could be in the form of verbal communication through voice recognition devices, digital messages, e-mail, or haptic devices (see chap. 4–6, 10, this volume). VTMs' flexibility and adaptability allow students to make mistakes, recover from them, and learn from them. With the possibilities offered by VTMs, along with the greater possibilities of VEs alone, training opportunities are increased (Johnson et al., 1998; Rickel & Johnson, 1998).

Examples of VE use and VTM behavior include Training Studio, a 3-D VE for training, and STEVE, a domain-independent, interactive VTM that can function as a mentor or as a team

**TABLE 44.2**
The Effects of Using VTM in EBAT

| *Scenario/Event-Based Approach to Training* | *EBAT Steps With VTMs* |
|---|---|
| 1. Skills inventory/performance data<br>• Take a baseline reading of current KSAs on task. | • Identify and define behavioral representations from previous databases.<br>• New information gained through VE and VTM exercises can be added to the behavioral databases. |
| 2. Learning objectives/competencies (KSA)<br>• Based on current KSA level and desired KSA level, learning objectives are established. | • Based on behavioral representations and related deficiencies, learning objectives can be developed and programmed into the VE and the VTM actions. This would allow for focused and consistent training, leading to greater shared mental models and skill building. |
| 3. Scenario events/scripts<br>• Scripts for "trigger events" are designed to test specific KSA items based on the learning objectives. | • The scenario/event itself would be displayed through the VE and the actions of the VTMs. Real-life contexts and numerous variations of them would be possible while keeping consistency in exercises and measurement variables. VTMs could be used in multiple manners. Provides repeated and consistent "hands-on" training and practice involving high-intensity and time-pressure situations, multiple procedures, various expertise levels, and problem-solving and skill adaptation. |
| 4. Performance measures/standards<br>• Outcome and process measurements are taken through observation, automatic, or semiautomatic techniques. | • Data from trainee responses would more accurately be measured with advanced computer technology, such as that used with VEs and VTMs. Behavioral patterns could easily be compared to behavioral representations of desired behavior and to the learning objectives. |
| 5. Performance diagnosis<br>• Performance levels are examined for deficiencies, patterns, and causal relationships. | • Behavioral patterns would be examined and diagnosed. Modifications to the scenario/event would be made based on the user's expertise level. |
| 6. Feedback and debrief<br>• Feedback and debriefing should be directly related to the learning objectives. | • Feedback could be immediate or at the end of the exercise by the VTM. The VTM could provide the feedback in several forms, as instructor, mentor, or teammate. |
| 7. Incorporate and archive lessons learned<br>• Data stored in a meaningful manner will collect to build a performance "norm" for future comparisons. | • Computerized information can be easily stored to be incorporated into behavioral databases to establish a "norm" for comparisons and for future training scenarios. |

member to facilitate learning of tasks that are physical and procedural (Johnson et al., 1998; Rickel & Johnson, 1998; Stiles et al., 1996).

Table 44.2 summarizes some aspects of using VE and VTM effectively in an EBAT format.

## 9. GENERAL GUIDELINES FOR USING EBAT

Table 44.3 provides recommendations based on the merging of EBAT, learning environment methodology, and VE technology. The guidelines are grouped into two general and nonexclusive categories, learning environment and virtual teammates.

**TABLE 44.3**
General Guidelines for Use of EBAT with VTMs

Learning environment characteristics

1. When establishing a learning environment to be used with VTMs, a conceptual framework should be used.
2. Establish and reinforce the learning environment through EBAT learning strategies, methods, and tools and programming of VEs and VTMs.
3. Use VTMs in a way that allows learning strategies, methods, and tools to build systematic linkages between the EBAT steps of learning objectives, scenario/event development, performance measurement and diagnosis, and feedback.
4. Appropriate feedback of trainees' performances through VTMs can be provided immediately/real time or delayed until the end the exercise.
5. Using predefined events, VTMs provide opportunities for pinpointing a training audience's deficiencies and proficiencies in learning objectives and for practicing new skills or maintaining old skills.
6. Techniques used for scenario management or the use of VTMs versus another human teammate should not be apparent to the trainee and should not in any way restrict decisions or actions made.

VTM characteristics

7. VTMs should be designed to demonstrate and build critical competencies, such as information processing, situation awareness, and reasoning and monitoring skills.
8. VTMs should have the ability to hold different ranking positions within a team and should be able to represent various levels of expertise.
9. VTMs should have the ability to demonstrate and explain tasks and procedures. VEs and VTMs may also be used to facilitate instructors with monitoring, assisting, and giving feedback to trainees.
10. VTMs can be designed to have the ability to fill in for missing team members or to provide individuals with team training. Ideally, VTMs should be inconspicuous in their interactions with other teammates.
11. Behaviors of VTMs should be responsive to changes in the VE caused by outside events, human team members, or other VTMs.
12. VTMs should be able to have an understanding of the relationships and rationale of tasks, subtasks, and associated steps in order to provide appropriate direction to trainees.
13. VTMs should be designed with an understanding of the relationships between the task roles to be able to encourage the development of collaboration and cross-training.
14. Virtual environments should be adaptable to the location of use and environment where task being tested would be used.
15. Scenarios, number of team members, and level of difficulty should be readily modifiable. The system should be scalable for individual or team, have local or distributed training capabilities, and wide area networking options. Through "telepresence," instructors should be able to observe and interact with trainees within an immersive system, allowing for training in distributed situations and within several contexts.

## 10. CONCLUDING REMARKS

Changes in the world involving global markets, geographically distributed team members, and military personnel stationed abroad have placed new demands on training. There is a greater need for effective, efficient, and economical training methods. Furthermore, training that ensures that task consistency and the integrity of learning objectives are strongly maintained is imperative. Researchers and trainers, using carefully controlled VEs depicting large scale collaborative or enemy response scenarios, can respond to these demands by being able to test one or a small number of trainees in a simulated environment depicting a real-life context.

Training, especially for dynamic and time-stressed events, is complicated. Technology alone should not be thought of as the key to successful training. Vital for effective training outcomes are consideration of the learning environment and structured methods and strategies, as found in EBAT. Discussed in this chapter were the theories behind effective learning environments, the uses of EBAT, and the contributions of VEs and VTMs to these approaches.

Both VEs and VTMs have a number of potential advantages, such as providing a realistic representation of a scenario or location, data collection, and effective, efficient, and economical training for team synchronicity and appropriate responses to real-life events. Users are able to rehearse scenarios under varying conditions relating to intercrew communications, interdependent task completion, and team dynamics. Together, effective learning environments, structured training methodologies such as EBAT, and VE and VTM tools will be able to provide innovative and effective approaches to training.

## 11. REFERENCES

Baker, D. P., & Salas, E. (1997). Principles for measuring teamwork: A summary and look toward the future. In M. T. Brannick, E. Salas, & E. Prince (Eds.), *Team performance assessment and measurement: Theory, methods, and applications* (pp. 331–356). Mahwah, NJ: Lawrence Erlbaum Associates.

Blickensderfer, E., Cannon-Bowers, J. A., & Salas, E. (1998). Cross-training and team performance. In J. A. Cannon-Bowers & E. Salas (Eds.), *Making decisions under stress* (pp. 299–311). Washington, DC: American Psychological Association.

Cannon-Bowers, J. A., & Bell, H. H. (1997). Training decision makers for complex environments: Implications of the naturalistic decision-making perspective. In C. Zsambok & G. Klein (Eds.), *Naturalistic decision making* (pp. 99–110). Hillsdale, NJ: Lawrence Erlbaum Associates.

Cannon-Bowers, J. A., Burns, J. J., Salas, E., & Pruitt, J. S. (1998). Advanced technology in scenario-based training. In J. A. Cannon-Bowers & E. Salas (Eds.), *Making decisions under stress* (pp. 365–374). Washington, DC: American Psychological Association.

Cannon-Bowers, J. A., & Salas, E. (1997). A framework for developing team performance measures in training. In M. T. Brannick, E. Salas, & E. Prince (Eds.), *Team performance assessment and measurement: Theory, methods, and applications* (pp. 45–77). Mahwah, NJ: Lawrence Erlbaum Associates.

Cannon-Bowers, J. A., Salas, E., & Converse, S. A. (1993). Shared mental models in expert decision-making teams. In N. J. Castellan, Jr. (Ed.), *Current issues in individual and group decision making* (pp. 221–246). Hillsdale, NJ: Lawrence Erlbaum Associates.

Cannon-Bowers, J. A., Salas, E., & Pruitt, J. S. (1996). Establishing the boundaries of a paradigm for decision-making research. *Human Factors, 38,* 193–205.

Cannon-Bowers, J. A., Tannenbaum, S. I., Salas, E., & Volpe, C. E. (1995). Defining team competencies and establishing team training requirements. In R. Guzzo & E. Salas (Eds.), *Team effectiveness and decision making in organizations* (pp. 333–380). San Francisco: Jossey-Bass.

Cellier, J. M., Eyrolle, H., & Marine, C. (1997). Expertise in dynamic environments. *Ergonomics, 40*(1), 28–50.

Cohen, M. S., Freeman, J., Wolf, S., & Militello, L. (1995). *Training metacognitive skills in naval combat decision making*. Arlington, VA: Cognitive Technologies.

Dwyer, D. J., Fowlkes, J. E., Oser, R. L., Salas, E., & Lane, N. E. (1997). Team performance measurement in distributed environments: The TARGETs methodology. In M. T. Brannick, E. Salas, & C. Prince (Eds.), *Team performance assessment and measurement: Theory, methods, and applications* (pp. 137–153). Mahwah, NJ: Lawrence Erlbaum Associates.

Dwyer, D. J., Oser, R. L., & Fowlkes, J. E. (1995). A case study of distributed training and training performance. In *Proceeding of the Human Factors and Ergonomics Society 39th Annual Meeting* (pp. 1316–1320). Santa Monica, CA: Human Factors and Ergonomics Society.

Dwyer, D. J., Oser, R. L., Salas, E., & Fowlkes, J. E. (1999). Performance measurement in distributed environments: Initial results and implications for training. *Military Psychology, 11*(2), 189–215.

Fowlkes, J. E., Lane, N. E., Salas, E., Franz, T., & Oser, R. (1994). Improving the measurement of team performance: The TARGETs methodology. *Military Psychology, 6,* 47–61.

Glaser, R. (1986). Training expert apprentices. In I. Goldstein, R. Gagne, R. Glaser, J. Royer, T. Shuell, & D. Payne (Eds.), *Learning research laboratory: Proposed research issues* (AFHRL-TP-85-54). Brooks Air Force Base, TX: Air Force Research Laboratory, Manpower and Personnel Division.

Grant, F., McCarthy, L., Pontecorvo, M., & Stiles, T. (1991). Training in virtual environments. In *Proceedings of the Conference on Intelligent Computer-Aided Training*. Houston, TX: Lyndon B. Johnson Space Center.

Gualtieri, J. W., Parker, C., & Zaccaro, S. (1995). *Group decision-making: An examination of decision processes and performance*. Paper presented at the Annual Meeting of the Society for Industrial and Organizational Psychology, Orlando, FL.

Hays, R. T., Vincenzi, D. A., Seamon, A. G., & Bradley, S. K. (1998). *Training effectiveness evaluation of the VESUB technology demonstration system* (NAWCTSD Tech. Rep. No. 98-003). Naval Air Warfare Center Training Systems Division.

Johnston, J. H., Cannon-Bowers, J. A., & Smith-Jentsch, K. A. (1995). Event-based performance measurement system for shipboard command teams. In *Proceedings of the first International Symposium on Command and Control Research and Technology* (pp. 274–276). Washington, DC:

Johnston, J. H., Smith-Jentsch, K. A., & Cannon-Bowers, J. A. (1997). Performance measurement tools for enhancing team decision-making training. In M. T. Brannick, E. Salas, & E. Prince (Eds.), *Team performance assessment and measurement: Theory, methods, and applications* (pp. 311–327). Mahwah, NJ: Lawrence Erlbaum Associates.

Johnson, W. L., Rickel, J., Stiles, R., & Munro, A. (1998). Integrating pedagogical agents into virtual environments. *Presence*.

Kalawsky, R. (1993). *The sciences of virtual reality and virtual environments*. Reading, MA: Addison-Wesley.

Klein, G. A. (1989). Recognition-primed decisions. In W. Rouse (Ed.), *Advances in man–machine systems research* (Vol. 5, pp. 47–92). Greenwich, CT: JAI Press.

Klein, G. A. (1995). A recognition-primed decision (RPD) model of rapid decision making. In G. Klein, J. Orasanu, R. Calderwood, & C. Zsambok (Eds.), *Decision making in action: Models and methods* (pp. 138–147). Norwood, NJ: Ablex.

McIntyre, R. M., & Salas, E. (1995). Measuring and managing for team performance: Emerging principles form complex environments. In R. A. Guzzo, E. Salas, (Eds.), *Team effectiveness and decision making in organizations* (pp. 149–203). San Francisco: Jossey-Bass.

Means, B., Salas, E., Crandell, B., & Jacobs, T. O. (1995). Training decision makers for the real world. In G. Klein, J. Orasanu, R. Calderwood, & C. Zsambok (Eds.), *Decision making in action: Models and method* (pp. 306–326). Norwood, NJ: Ablex.

Noble, D. F., Grosz, C., & Boehm-Davis, D. (1987). *Rules, schema and decision making*. Vienna, VA: Engineering Research Associates.

Orasanu, J., & Connolly, T. (1995). The reinvention of decision making. In G. Klein, J. Orasanu, R. Calderwood, & C. Zsambok (Eds.), *Decision making in action: Models and methods* (pp. 3–20). Norwood, NJ: Ablex.

Oser, R. L., Cannon-Bowers, J. A., Dwyer, D. J., & Salas, E. (1997b). *An event based approach for training: Enhancing the utility of joint service simulations*. Paper presented at the 65th Military Operations Research Society Symposium, Quantico, VA.

Oser, R. L., Cannon-Bowers, J. A., Dwyer, D. J., & Salas, E. (1997a). Establishing a learning environment for JSIMS: Challenges and considerations [CD-ROM]. In *Proceedings of the 19th Annual Interservice/Industry Training, Simulation and Education Conference* (pp. 144–153). Orlando, FL:

Oser, R. L., Dwyer, D. J., Cannon-Bowers, J. A., & Salas, E. (1998, April). *Enhancing multi-crew information warfare performance: An event-based approach for training*. Paper presented at the RTO HFM Symposium on Collaborative Crew Performance in Complex Operational Systems, Edinburgh, Scotland.

Oser, R. L., Gualtieri, J. W., Cannon-Bowers, J. A., & Salas, E. (1999). Training team problem solving skills: An event-based approach. *Computers in Human Behavior, 15*, 441–462.

Rickel, J., & Johnson, W. L. (1998). Animated agents for procedural training in virtual reality: Perception, cognition, and motor control. *Applied Artificial Intelligence*.

Salas, E., & Cannon-Bowers, J. A. (2000). The anatomy of team training. To appear in S. Tobias & J. D. Fletcher (Eds.), Training & Retraining: A handbook for buisness, industry, government, and the military (pp. 312–335). New York: MacMillan.

Salas, E., & Cannon-Bowers, J. A. (1997). Methods, tools, and strategies for team training. In M. A. Quinones & A. Ehrenstein (Eds.), *Training for a rapidly changing workplace: Applications of psychological research* (pp. 249–279). Washington, DC: American Psychological Association.

Salas, E., Dickinson, T. L., Converse, S. A., & Tannenbaum, S. I. (1992). Toward an understanding of team performance and training. In R Swezey & E. Salas (Eds.), *Teams: Their training and performance* (pp. 3–29). Norwood, NJ: Ablex.

Serfaty, D., Entin, E. E., & Johnston, J. H. (1998). Team coordination training. In J. A. Cannon-Bowers & E. Salas (Eds.), *Making decisions under stress: Implications for individual and team training* (pp. 221–246). Washington, DC: American Psychological Association.

Shebilske, W. L. (1993). Visuomotor modularity, ontogeny, and training high-performance skills with spatial instruments. In S. R. Ellis, M. K. Kraiser, A. J. Grunwald (Eds.), *Pictorial communication in virtual and real environments* (pp. 304–315). Washington, DC: Taylor & Francis.

Smith-Jentsch, K. A., Zeisig, R. L., Acton, B., & McPherson, J. A. (1998). Team dimensional training: A strategy for guided team self-correction. In J. A. Cannon-Bowers & E. Salas (Eds.), *Making decisions under stress* (pp. 271–297). Washington, DC: American Psychological Association.

Stiles, R., McCarthy, L., Munro, A., Pizzini, Q., Johnson, L., & Rickel, J. (1996). Virtual environments for shipboard training. In *Proceedings of the 1996 Intelligent Ships Symposium*. Philadelphia American Society of Naval Engineers.

Stout, R. J., Cannon-Bowers, J. A., & Salas, E. (1996). The role of shared mental models in developing team situational awareness: Implications for training. *Training Research Journal, 2*, 85–116.

Washburn, K. B., & Ng, H. C. (1997, November/December). Simulation: Virtual environments for naval training. *Surface Warfare*, 26–29.

Wickens, C. D. (1984). *Engineering psychology and human performance*. Columbus, OH: Merrill.

Zachary, W., Cannon-Bowers, J. A., Burns, J., Bilazarian, P., & Krecker, D., (1998 August). *An advanced embedded training system (AETS) for tactical team training*. Paper presented at the fourth international conference on Intelligent Tutoring Systems, San Antonio, TX.

# 45

# Virtual Environments As a Tool for Academic Learning

J. Michael Moshell and Charles E. Hughes
*University of Central Florida*
*School of Electrical Engineering and Computer Science*
*4000 Central Florida Blvd.*
*Orlando, FL 32816*
*moshell@cs.ucf.edu, ceh@cs.ucf.edu*

## 1. INTRODUCTION

This chapter concerns itself with the use of virtual environments (VEs) for learning of the kind expected to occur in schools, colleges and universities, that is, the acquisition of general problem-solving skills, mastery of facts and concepts, and improvement of the learning process itself. The chapter systematically explores the potential of VE for education as well as potential obstacles to its use. As of mid-2001, there are no commercially available VE systems deployed for regular instructional use in K–12 or university education, so this review relies on research projects for insight. In analyzing the major experiments thus far carried out, one can discern various combinations of three basic educational approaches to VE: *exploration, world building*, and *world sharing*. This chapter is not a general survey of VE-based educational research. Rather, it examines in some detail six research projects, focusing on two of these projects to explore each of the approaches above. An excellent survey of the educational use of VE technology can be found in Youngblut (1998).

## 2. USE OF VE TECHNOLOGY IN EDUCATION

### 2.1 What Is VE?

Virtual environments denote a real-time graphical simulation with which the user interacts via some form of analog control, within a spatial frame of reference and with user control of the viewpoint's motion and view direction. This basic definition is often extended by provision for multimedia stimuli (sound, force feedback, etc.) by consideration of immersive displays (i.e., displays that monopolize one or more senses), and by the involvement of multiple users in a shared simulation. However, many interesting educational systems involve only the core definition.

## 2.2 Why Might VE Matter to Education?

Ever since the first computers, people have been suggesting that well-designed simulations could provide access to learning experiences that are simply unavailable via normal means. Virtual field trips consume no jet fuel; simulated laboratories do not explode; virtual dissection kills no animals. Many personal computer (PC)–based simulations have pursued this dream, with varying results (Weller, 1996). Virtual simulations add the specific requirements of a spatial metaphor and free viewpoint motion. Why is it believed that this structure would be superior to a more abstract presentation?

Consider a chemical experiment presented via a two-dimensional (2-D) schematic graphical rendering on a computer screen. An Erlenmeyer flask contains a basic liquid solution. As the student clicks on-screen button controls, a calibrated pipette drips acid into the flask until the dye changes color, indicating neutralization. Quantities are noted and calculations performed.

The same experiment in a virtual laboratory would be located on a realistic, perhaps scarred and beat-up countertop. The user is free to look and move around the lab. Perhaps the user has to find the Erlenmeyer flask in a drawer, set it up properly, use correct safety procedures with the acid and basic solutions, and run the risk of breaking equipment if it is handled too abruptly. In short, the virtual laboratory (like a real laboratory) is intended to teach the culture of the experience, whereas the 2-D schematic laboratory teaches the conceptual core of the experience. The culture consists of the collection of constraints, risks, and procedures that a real chemist must consider in doing the job. Real-life lab experience is usually complemented by classroom instruction to provide the concepts, precisely because the cultural lab experience is so information rich that the conceptual core is obscured.

The story about the chemistry experiment obviously added more than free viewpoint motion to the abstract example. It added a world worth moving through, a context in which the user has choices. These options range from the simple selection of the correct glassware, to the decision to wander away from the lab counter and see what else is in the virtual world.

In a real lab, students make some choices based on learning styles (Gardner, 1993). Some students need to experience direct consequences of manual actions to grasp causal relationships. Others need to say what they see. Still others need to hear or read explanations. The virtual lab is intended to provide a rich set of such choices and to make them accessible in natural ways.

But if all this could be accomplished, would it actually improve education?

## 2.3 Why Might VE Be a Red Herring for Education?

There are two essential obstacles to the rosy vision described above: design and pedagogy. Let us assume the existence of affordable, reliable VE software and hardware. The creation of richly detailed virtual worlds must still some how be supported, in which the glassware clinks, the wooden desktop thunks, and, most importantly, when a users does something unexpected (e.g., drops a paper clip into a beaker of liquid), something plausible happens. There are several dilemmas:

- Building richly detailed worlds is very expensive (see chaps. 12–13, this volume). Commercial video games currently cost over $1 million to produce, and they set the standard by which kids will judge virtual worlds.
- If a fascinating, complex world is built, is it actually helping or hindering the kids' learning? Is all this intriguing simulation really bribery, or does it pay its way?

There are those (e.g., Postman, 1985) who stoutly assert that television, even the best educational television, is destroying our kids' ability to deal with abstraction. The written word requires a continuous positive act of the imagination as one reads. Television, and possibly high-fidelity VE as well, do most of the work for you. Thus, Postman would probably assert that the better (and more appealing) a design gets, the further one moves from an ideal learning environment. From a slightly different perspective, Wickens (1992) argues that natural and easy-to-use interactions may reduce a student's retention because less effort is expended.

On the other hand, there is a substantial body of scholarship (described below in more detail) that asserts the value of "situated learning," of providing tasks within a realistic story, with appealing characters and situations. Humans emulate heroes and want to be admired as new heroes. Virtual environments may provide enough story structure and detail to enable learners to remain engaged with tasks, which they would have abandoned in other forms. It may empower learners whose cognitive styles will never resonate with traditional, linear book culture. Active engagement with rich simulated worlds may help students to construct their own mental models.

The following section establishes some essential principles and terminology for discussing kinds of learning, and educational theories such as constructivism and situated learning. Sections 4 through 6 explore, respectively, the following domains:

- Exploration of prebuilt worlds as a *constructivist* learning activity
- Creation or modification of worlds as a *constructionist* learning activity
- Role-playing by multiple participants as a *situated learning* activity.

Each section includes a description of two experimental projects. A final section then summarizes the state of knowledge and poses challenges for the future.

## 3. A CONCEPTUAL FRAMEWORK

A commonly cited taxonomy for learning (Gagne, Briggs, & Wager, 1988), describes five categories of learning outcomes:

- Intellectual skills and procedural knowledge
- Verbal information
- Cognitive strategies
- Motor skills
- Attitudes

The Gagne taxonomy is one of the standard tools for instructional design; its structure reflects the traditional academic focus on reading and writing. Others give more attention to nonverbal domains. Gardner (1993) describes seven kinds of intelligence (among many):

- Verbal
- Mathematical–logical
- Spatial
- Kinesthetic
- Musical
- Interpersonal (dealing with others)
- Intrapsychic (dealing with one's own self)

These two classification systems are not easily compared. For instance, Gagne's "motor skills" denotes specific behaviors such as typing or playing golf, whereas Gardner's "kinesthetic intelligence" refers to a way of understanding systems of concepts that is based on physically manipulating and interacting with objects. For instance, Papert (1980) describes a childhood epiphany in which his manipulation of the differential axle of a toy vehicle revealed to him the nature of functional relationships.

And where in Gagne's terminology does one account for knowledge of the fact that Germany lies north and east of France? Calling this "verbal information" reflects a behaviorist focus on the traditional means of reporting knowledge by writing, but students are often asked to draw maps or otherwise indicate their mastery of spatial information. It seems necessary to generalize Gagne's category of verbal information by renaming it as "factual information."

Multisensory integration (see chaps. 14, 21–22, this volume) is integral to VE's affordances for learning and motivates many of the projects discussed below. To deal with this issue, an attempt is made to use both Gagne's and Gardner's terms to describe the kind of learning under study in each project.

## 3.1 Constructivism, Constructionism, Situated Learning, and Role-playing

The concepts of constructivism, constructionism, and situated learning drive many innovations and experiments in education, ranging from individual lesson modules to entire school reform movements (Cusick, 1997; Wang, 1998). Role-playing can serve as an intense form of situated learning. In most of the experiments in VE for education, several of these learning strategies are involved. For the purposes of this chapter, two research projects have been selected to highlight each concept.

### *3.1.1 Constructivism*

A substantial body of literature (Dede, 1995; Duffy & Jonassen, 1992; Windschitl, 1998) has established the principle that students actively build their internal models of the world rather than passively accepting data. In constructivist theory, all useful knowledge is procedural knowledge. Thus, constructivism serves as a theoretical justification for discovery learning and exploratory systems. Section 4 below explores two constructivist VE projects: ScienceSpace and the Virtual Gorilla Exhibit.

### *3.1.2 Constructionism*

Constructionism is an extension of constructivism, that latter of which is usually associated with MIT's S. Papert (Harel & Papert, 1991; Kafai & Resnick, 1996), in which students must actively create artifacts—preferably interactive ones—to fully integrate their models of the world. Papert invented the well-known LOGO language for children and continues to evangelize the doctrine of constructionism. Not all authors use Papert's terminology; the creation of virtual worlds is sometimes referred to as constructivist activity. We find the distinction useful. Section 5 reviews two constructionist VE projects the VRRV project, and the NICE virtual gardening project.

### *3.1.3 Situated Learning*

Situated learning (Cognition and Technology Group at Vanderbilt, 1993) is oriented around the idea that learning takes place best in story-based, human-centered circumstances. Rather than just computing the area of a triangle, learners are asked to help characters in a story solve an extended problem such as building a house. Many individual problems naturally arise in such a story; their solution is motivated by the normal needs of the characters in the story.

Role-playing takes situated learning one step further by involving multiple learners who are identified with specific characters in the story. Simulated situations may be competitive, cooperative, or both. Introducing multiple learners raises many challenging problems of synchronization, tasking, discipline, and resource management, but offers the potential of unleashing for educational use one of the most powerful forces in the human psyche: social interaction. Section 6 profiles two role-playing experiments, the Round Earth Project and ExploreNet.

## 4. PREBUILT WORLDS, DISCOVERY LEARNING, AND CONSTRUCTIVISM

### 4.1 ScienceSpace

One of the most thorough and extensive studies of VE in education has been underway since 1994. Led by C. Dede of George Mason University (Now at Harvard University) and B. Loftin of the University of Houston (Now at Old Dominion University), the ScienceSpace project (Dede, Salzman, Loftin, & Ash, 1999a, Dede, Salzman, Loftin, & Sprague, 1999b; Salzman, Dede, Loftin, & Ash, 1998; Salzman, Dede, Loftin, & Chen, 1999) has used high performance technology to examine the educational potential of immersive VE systems. Loftin created and directs the University of Houston's Virtual Environment Technology Laboratory, and so had access to state-of-the-art VE technology used in astronaut training programs.

#### *4.1.1 Purpose*

The focal purpose of ScienceSpace was to identify the key affordances of immersive VE and evaluate their effectiveness as means of learning complex, abstract concepts such as mass, density, and momentum. Three key affordances were of principal concern:

- Immersion
- Use of multiple frames of reference
- Multisensory cues

#### *4.1.2 Setup*

Immersion was provided by the use of a state-of-the-art Silicon Graphics RealityEngine2 image generation system, a Virtual Research VR4 head-mounted display (HMD), and Crystal River audio localization equipment. Multiple frames of reference were provided by a variety of means, the principal one being that students could either move through the world coordinate system with the object whose behavior was being studied or remain fixed in the world coordinate system, and observe the phenomena. Multisensory cues were provided by the sound system and also by the use of a special vest. The vest, originally designed for multiplayer "shoot-em" games, was capable of delivering a vibrating stimulation to the user's chest. The learner held a control device in one hand and a reference device in the other hand; both were tracked by 6 degrees-of-freedom magnetic trackers.

#### *4.1.3 Experiments*

The ScienceSpace experiments consisted of three phases. The first phase (Newton World) represented basic Newtonian mechanics. The second phase (Maxwell World) concerned electrostatic forces and fields. The third phase (Pauli World) concerned quantum phenomena. During each phase, a series of experiments began with simple pilot activities and continued through formal evaluations of learning effects. Detailed documentation is available on the first two experiments, and so they are described here.

Newton World focused on the motion of spherical masses along a one-dimensional axis. The axis was visualized as a corridor passing between two rows of columns. The columns were used in a variety of ways to help the user measure and perceive the motion of masses. For instance, the columns might change color as a mass passed by, or sounds might occur. Students controlled the action in the world using three-dimensional (3-D) tracking devices or a voice control system.

When studying energy, it was necessary to devise means of concretely rendering the quantities of kinetic and potential energy. A coiled spring was used to represent potential energy; when it was compressed, vibration in the vest increased. An artificial shadow of the mass was used to represent kinetic energy; the shadow's area represented the amount of kinetic energy the object currently possessed with respect to the fixed reference frame. (The radius of the shadow would therefore be proportional to the velocity.) Learners could either travel with one of the moving masses or view the motion from various points outside or within the corridor.

A pilot experiment yielded a series of improvements in the user interface and experimental technique. It was discovered that the affordance of multiple viewpoints was crucial to learning success. A variety of methods of selecting the next action were used (e.g., picking from menus attached to the fixed reference frame, attached to the user's other hand, or via voice commands). Voice selection was found to be the most preferred means of control. Users liked the multisensory feedback modalities and found the world easier to use when multisensory feedback was presented (see chaps. 14, 21–22, this volume).

In a second experiment, 30 high school students helped to evaluate Newton World's potential for learning. The students were pretested, and it was found that most of them confused the concepts of velocity and acceleration, and that most had trouble predicting what would happen when two masses collided.

After a session of approximately an hour's use of Newton World, students' conceptual understanding was again tested. No significant improvement was measured, suggesting that a single visit to Newton World was insufficient to transform users' mental models. However, multiple improvements in experimental technique and virtual world design were carried forward to the next phase.

In the Maxwell World experiments, learners again had the opportunity to use multiple frames of reference and multisensory feedback. Learners could see electrostatic fields in the form of equipotential lines. A "force meter" represented the magnitude and direction of force that would be applied to a charge at any point in the field as the student moved the force meter around. Using the meter, a user could release a charge and watch it move through space, or "become" the charge and go for a ride.

In the first experiments, students' length of exposure to the system was increased. Fourteen students had from one to three sessions, each of 2 hours duration. As a control, a 2-D commercially developed educational software system, EMfield, was taught. The features of Maxwell World were restricted to correspond to those available in EMfield. Pre- and posttesting indicated that users of Maxwell World developed significantly better understanding of electric fields' 3-D aspects than users of EMfield. In 2-D tests (such as the drawing of sketches related to potentials) the results were mixed. However, the overall results for Maxwell World were significantly better than those for EMfield.

#### *4.1.4 Experimenters' Conclusions*

In the later experiments, significantly more learning occurred with VE than without it. Students in all the systems were enthusiastic about their use and remained motivated throughout the experiments. However, motivation did not turn out in the statistical analyses to be a predictor of learning. Some simulator sickness (see chap. 30, this volume) was experienced, with incidence and severity increasing with the longer exposures used in Maxwell World.

#### 4.1.5 Analysis

In terms of the learning taxonomies, ScienceSpace provided a new experiential domain in which students were expected to develop novel cognitive strategies in order to establish procedural knowledge about spatial and mathematical–logical relationships in physics. Despite the lack of force feedback, the investigators hoped that kinesthetic learning would contribute to students' development of physical intuition.

The project made substantial progress in determining how to build effective virtual learning environments for physics, how to enrich multisensory feedback, and how much exposure was necessary to achieve measurable effects.

### 4.2 The Virtual Reality Gorilla Exhibit

A team of researchers at Georgia Tech developed a VE simulation of a new gorilla habitat at Zoo Atlanta. A basic dimension of this project is a concern with how a virtual experience must be coupled with background information, reflection, and various forms of information accessible during the simulation itself in order to achieve useful educational outcomes.

#### 4.2.1 Purpose

The initial experiments (Allison, Willis, Bowman, Wireman, & Hodges 1997) were designed to teach middle school students about gorilla's social interactions by letting the students play the role of an adolescent gorilla. Subsequent experiments (Bowman, Wireman, Hodges, & Allison, 1999) focused on the design of the zoo's gorilla habitat, with the intention of teaching college students about principles of design.

#### 4.2.2 Setup

The experiment is hosted on Silicon Graphics equipment, using a Virtual Research HMD and Polhemus tracking equipment. A single user experiences the simulation and interacts with the exhibit via a "stylus and tablet" virtual control device. Motion is controlled either by pointing the stylus in the direction of intended travel or by dragging a dot on a map displayed on the tablet. Software was implemented using the Simple Virtual Environment (SVE) software system developed at Georgia Tech (Kessler, Kooper, & Hodges, 1998). Animal sounds are provided through external speakers.

#### 4.2.3 Experiments

In the first series of experiments, students were provided with an experience that could be dangerous in real life, taking middle school students on a visit inside a functioning goilla society. After initially viewing simulated gorillas through a simulated viewport, the learner "moves inside" the habitat and takes up the role of an adolescent gorilla. When the learner moves about, the other gorillas react by moving away or by watching the visitor. If the visitor moves too close to a male silverback (dominant individual) or stares too long at it, the silverback becomes annoyed and will ultimately carry out a "bluff charge" and beat on his chest.

The experimenters observed that most students needed help to understand the gorilla's behavior. Initially a live human guide would stand next to the student and explain what was happening; subsequently explanatory material was built into the audio portion of the experience.

The system was also modified to include explicit "mood indicator" icons to show the emotional state of each gorilla.

The second series of experiments concerned teaching college students about habitat design. The design profession is based on well-established principles and also includes a substantial aesthetic component. The experimenters' intention was to provide a series of design examples, with annotation, somewhat like a guided visit to a Frank Lloyd Wright project. A series of audio clips are embedded in the virtual gorilla exhibit. Annotations are marked by cubical signs that play the audio annotation when selected. A few fixed signs occur in places where signs might appear in a real zoo habitat.

The map that is displayed on the virtual tablet (corresponding to the real tablet the user is holding) provides additional spatial information. The map situates the exhibit among several others in the zoo and teaches the principle of bioclimatic zones, by which exhibits from similar climates and ecosystems are located close to one another. The exhibit also includes photographs of gorillas in locations they typically occupy. These help the viewer to establish the scale of the exhibit. The virtual gorillas are not active in the design experiment because their behaviors would distract from the design lesson being taught. Also, the gorilla's automated behaviors do not include the full spectrum of activities (eating, sleeping, mating, etc.) that must be accounted for in habitat design.

#### 4.2.4 Experimenter's Conclusions

In the first series of experiments with schoolchildren, the principal lessons learned concerned the user interface and the need for explanatory support. Students tended to be concerned with specifics of navigation (see chap. 24, this volume) and could be overwhelmed by the sights and sounds of the simulated animals. The experimenters found it necessary to provide a live gorilla expert to comment on what the students were experiencing.

The system was subsequently used in a Georgia Tech class on the psychology of environmental design. The system was found easy to use. However, students did not strongly relate the audio material to the environment as was intended. As expected, the groups that used the system performed as well as or better than a control group. However, the differences were not significant, probably due to the small size of the samples. A group that experienced the annotated VE without a corresponding lecture received very little benefit.

#### 4.2.5 Analysis

This project aimed to develop factual and procedural knowledge about how gorillas live and interact, as well as about how to design habitats for zoos. It used visual, verbal, and kinesthetic modes of interaction.

The experiences with schoolchildren (needing a live expert to facilitate their learning) and with the university students (needing a lecture to exhibit measurable learning) serve as a strong cautionary note to advocates of VE as a primary means of instruction. The Georgia Tech team advocates VE as a supplementary tool to be used alongside other means of teaching.

## 5. BUILDING VIRTUAL WORLDS AS A CONSTRUCTIONIST ACTIVITY

In this section two projects are described in which the learner's major activity is to construct or extend the features of a virtual world.

## 5.1 The VRRV Project

At the University of Washington's HIT Lab (Human Interface Technology Laboratory), a team led by William Winn carried out an extensive project titled VRRV (Virtual Reality Recreational Vehicle). The basic premise was to have students in schools participate in extended world-building exercises using conventional computers. Then an "RV" (originally intended to be a motor home or recreational vehicle, but actually a van) visited the school with immersive VE equipment, so the students could experience the worlds they built. The project is described in Winn et al. (1999). This report describes one component of the VRRV project.

### *5.1.1 Purpose*

The VRRV Project was intended to explore the educational utility of students' building virtual worlds within the constructionist paradigm. The investigators analyzed several previous constructionist VE projects in order to choose experimental parameters. The key questions concerned:

- How students would choose to represent objects and processes
- How students would design spaces to help their peers understand subject matter

### *5.1.2 Setup*

Teachers from 14 schools volunteered their elementary, middle, and high schools as test sites. The teachers were trained in the use of Macromodel software, and it was concluded that the software was too difficult for elementary and some middle school students. Consequently, it was decided to have the project staff build a common world and let the younger students construct auxiliary but nonessential objects. Most of the middle schools and all the high schools designed and built worlds. In these cases the students modeled the objects and specified how they should fit into the world and interact. Project programmers then assembled the worlds and programmed necessary interactions.

Portable (PC-based) VE equipment built by Division was brought to each school so that the students could visit their own worlds. The systems included Division HMDs and magnetic tracking equipment. The project involved 365 students.

### *5.1.3 Experiments*

Except for the elementary schools and one middle school, each school designed and implemented its own world. During the planning phase, students worked in groups. They selected content, specified objects and metaphors to represent invisible objects and relationships. During the modeling phase, students drew their objects on paper and then built them with the Macromodel software. Project staff assembled the worlds and programmed behaviors. The elementary children explored and manipulated a prebuilt world, to which they had contributed some auxiliary objects. Finally, students visited the worlds they had created, performed specified tasks, and filled out questionnaires about the experience.

In a few schools, the projects related to academic subjects being studied in the schools. Thus, data was gathered to see if VE students had better knowledge of the subject matter than students in other non-VE-supported classes.

### *5.1.4 The Worlds*

Tree World (shared by the elementary schools) consisted of a tree that needed sunlight, water, and nutrients. Students pushed aside a virtual cloud to let in sunlight, pushed aside a boulder to let in a river, and fed a cow so that its manure would fertilize the tree. Students provided insects, squirrels, and birds. In this world, the students' activities could be described as constructivist but not constructionist. The purpose was to make the tree look healthier by providing for its needs, and thus to learn about biological resource cycles.

Here a few typical constructionist worlds are described. Medieval Castle World represented a number of rooms of a classic castle, with a drawbridge. Rain Forest World tells a story in which bulldozers destroy the forest; cattle are introduced and become hamburgers. Finally, the earth is replaced by a big dollar sign. Endangered Species World has animals that turn into skeletons when handled. Washington World presented a maplike visitable model of the state, with many correctly located features that revealed textual annotations when selected. Tide World was intended to show how the sun and moon produce tides.

#### 5.1.5 Experimenter's Conclusions

The authors observed two kinds of metaphor in use: objects could stand for other objects or for invisible objects or processes. Hamburgers stood for the end product of clearing the forest. Skeletons stood for the (unobservable) process of extinction. Students showed the ability to construct valid metaphors but tended to undergeneralize principles. For instance, the elementary students, when asked about plant nutrients, reported that they came from cow poop.

Similarly, students showed the ability to define processes, but when asked about the underlying principles would often cite the process itself (e.g., hamburgers are not the problem, habitat destruction is). The Washington World and Tide World showed mastery of spatial relations and of scale as a dimension of design.

Student designers had to anticipate the need for interactions (e.g., feedback for successful actions) and specify them. This required that students anticipate the errors of others.

Numerical results of posttests were analyzed to investigate enjoyment, ability to navigate, and VE sickness, as well as correlations among these measures. Those who found it easy to navigate also enjoyed the experience. Separate (teacher-designed) testing of knowledge of content indicated that less able students learned a significant amount about the world's subject matter from the VE-building experience, whereas higher performing students did not.

#### 5.1.6 Analysis

VRRV's constructionist experiments spanned a wide range of subject matter and included both factual knowledge and procedural knowledge . The procedural knowledge involved phenomena within the simulation (such as flow of resources through the nutrient cycles of the tree) and processes involved in building virtual worlds. The principal cognitive strategy involved was meta-learning (thinking about the learning process itself), in the form of explicit instructional design for the benefit of other students.

In Gardner's terms, mathematical–logical (causal loops), spatial, and kinesthetic intelligences were involved in the project. Of course, verbal skills had an essential role in planning and specifying the worlds before they were built.

In common with most other VE based projects, the VRRV experiments increased the authors' understanding of the user interface and didactic issues involved, but exhibited relatively little in the way of measurable learning.

### 5.2 The NICE Project

At the University of Illinois at Chicago, the Electronic Visualization Laboratory has developed a unique immersive VE system called the CAVE (Cave Automatic Virtual Environment;

Cruz-Neira, Sandin, DeFanti, Kenyon, & Hart, 1992; see chap. 11, this volume). Rather than using an opaque HMD, the CAVE uses large projected images on three walls and the floor. A viewer wearing stereographic glasses and a tracking device has the experience of being inside a high-resolution virtual space, but without most of the motion sickness problems (see chap. 30, this volume) associated with head-mounted displays. A handheld control device is also provided. Multiple users can simultaneously see images, although only the one with the tracking device has correct perspective and parallax.

### *5.2.1 Purpose*

Using the CAVE, the NICE (Narrative Immersive Constructionist/Collaborative Environments) project explored both constructionism and collaboration by having students design, build, and "nurture" a virtual garden (Roussos et al., 1997, 1999). In this review, the focus is on the constructionist aspects of the project. The specific learning outcomes being sought were an increased understanding of the relationships between environmental conditions and plant growth. In addition, the experimenters focused attention on the suitability of the CAVE user interface for learning by children of various ages.

### *5.2.2 Setup*

Software was provided that supported the simulation of plant life in a garden. Using a handheld controller, or "wand," students dropped a seed and a plant grew on that spot. If the plants were too close together they would not grow well. Students could grab a rain cloud, "water" a plant, or provide sunlight. Weeds could be removed to a compost heap. The garden continued to evolve, with weeds growing and animals eating plants, even if no users were present. NICE is accessible not only through the CAVE, but also through desktop 3-D VE systems and a 2-D Web-accessible interface.

Two features of VE technology of particular educational interest are its ability to reveal hidden phenomena and help users interpret them. In NICE, for instance, users can go underground to see what is happening there as well as control the flow of time. In addition, a plant that is receiving too much water is shown holding an umbrella.

An unusual feature of the NICE system is that a story is automatically constructed, which describes each action of the learner. This story is expressed in simple English sentences, with some words (e.g., sun) replaced by colorful icons. The story is printed for the child to take home.

### *5.2.3 Experiments*

The basic premise of NICE was to allow self-directed exploratory activities and to carefully observe and analyze what happened. There was some guidance provided by teachers who helped students stay on task and contain their excitement. The experimenters gathered observations in five areas:

- Technical: usability of the interface, system hardware and software
- Orientation: navigation, presence
- Affective: engagement, confidence, interest level
- Cognitive: improvement of knowledge and understanding
- Pedagogical: how did student–student and student–teacher collaboration work?

Fifty-two students between 6 and 10 years old participated in the NICE experiments. The usual experimental setup was to have two groups of four students using two CAVE systems, coupled by a bidirectional audio link. In each group, a leader was designated to wear the tracking device and to appear in the NICE world as an avatar. The other team's avatar, and that of a third person (actually a teacher), appeared in the world to act as a guide.

Each group planned a garden before they entered the CAVE. The group then directed their leader as he or she carried out the design. The group could see the other group's avatar. Sessions lasted about 40 minutes; they were preceded by a set of pretest questions, and followed by interview questions. Students then made drawings and wrote essays.

### 5.2.4 Experimenter's Conclusions

With regard to usability: the stereo glasses and control wand, being adult-sized, did not work optimally for smaller children. Fatigue (see chap. 41, this volume) and simulator sickness (see chap. 30, this volume) were minimal, but some difficulty with orientation (see chap. 24, this volume) was reported. With regard to affective issues, students enjoyed the activity and particularly liked reading the stories that chronicled their adventures. The student leaders were more engaged with the learning task than the passive observers; the observers were more engaged with the medium itself. Improvement of knowledge and understanding was correlated with control; that is, the students who directly controlled the environment learned more than the passive observers.

Before the experience, 12% of the participants had a good understanding of environmental concepts; afterward, 35% met the same criterion. Most (73%) of the successful students were group leaders.

The project team plans to develop scalable models so that younger children can experience simplified biological phenomena and older ones can explore a richer set of interactions.

### 5.2.5 Analysis

In Gagne's terms, the NICE experiment helped students build their "procedural knowledge" about caring for plants and their "verbal information" about how plants live. In Gardner's terms, the project was directed at helping students develop their understanding of "mathematical–logical" and "spatial" relationships between growing plants and their environment.

The experiment's cognitive measures would seem to indicate that the educational efficacy of passive observation/participation in VE is limited. Interestingly, such passive conveyance is also attributed to higher levels of sickness (see chap. 30, this volume).

## 6. ROLE-PLAYING AND SITUATED LEARNING IN SHARED VIRTUAL WORLDS

Several of the projects already described were designed with explicit social dimensions. VRRV had students work together to design and build worlds, and NICE had groups of students sharing a CAVE system. In fact, NICE had two CAVE systems in use at once, with a network so that each user could see the other's avatar. However, the presence of specific characters inside the virtual world, with differentiated capabilities and assigned roles, was not central to the instructional design of those projects.

It can be argued that virtual worlds without specific, visible, differentiated characters (avatars, see chap. 15, this volume) and roles are of an essentially different nature than worlds built around avatars. There is a spectrum of abstraction in simulation-based learning. At the most abstract are simulations with no situational information, just the essence of the phenomenon at hand. Less abstract are simple simulations that are situated within a simulated world, but which are used, so to speak, "through a window." When an immersive system with free motion of viewpoint is used, the situation gets even more concrete, but the learner is still free to behave in a godlike fashion. The *behavioral schema* is oriented toward objects, and there is nothing but the simulated world's own rules to constrain one's behavior.

As soon as human forms appear in the world and begin to plausibly interact with the user, behavioral schema shift and become oriented toward shared experience. This may be the least content friendly of all forms of simulation; social interaction can be a powerful distractor from reflective thought. But students can and do learn from shared experiences. They explain things to one another and observe the result of others' experiments (Johnson & Johnson, 1987).

With younger children, one possible distraction is that they may begin to compete rather than cooperate. Instructional designers sometimes optimistically assume that activities originally conceived of as cooperative tasks will naturally be treated that way by students, but students may see the simulation as a me-win/you-lose game. The social schema must be carefully constructed so that learners can take advantage of known ways of interacting. Roles such as leader/follower, teacher/student, or allies/adversaries are obvious possibilities. Both the projects to be described below have carefully designed roles for participants.

## 6.1 The Round Earth Project

Like the NICE Project, the Round Earth Project was carried out by the Electronic Visualization Laboratory at the University of Illinois at Chicago. The same CAVE equipment was used. However, in the Round Earth Project, the interaction of two students within a shared virtual world is essential to the instructional design. The work is described in Johnson, Moher, Ohlsson, & Gillingham (1999).

### 6.1.1 Purpose

The experimenters wanted to select a problem that met the following criteria: Learning goals must be important and hard, and VE must offer some plausible enhancement to the learning process. The concept of the earth as a sphere was selected because it is in standard elementary science curricula and is known to be hard for many children to grasp.

The actual purpose of the experiment was to investigate how VE could teach concepts that are inconsistent with a user's current mental model. It is necessary both to establish the new phenomenon (in this case, the experience that "down" points toward the center of a sphere rather than perpendicular to an infinite ground-plane) and to link this knowledge to one's prior mental model. Two theories of deep learning, "transformationist" and "displacement" approaches, were investigated.

### 6.1.2 Setup

One CAVE and one ImmersaDesk system were used. An ImmersaDesk is a single-projector system arranged like a large drawing table, in contrast with the CAVE's four viewing surfaces arranged as three walls and a floor. Stereo glasses are used by both participants, and a bidirectional audio link is provided.

Two treatments were provided: Astronaut World and Earth World. Astronaut world was intended to provide a clean break from the current flat-earth experience, with an astronaut-avatar walking around a tiny spherical asteroid while the other learner observed and directed his actions from a command module. To enhance collaboration, the task (picking up fuel cells) was designed so that the observer had knowledge of what was essential for the astronaut's success. The students alternated roles so that each experienced both the god's-eye view and the immersive view.

Earth World involves a satellite mission in Earth orbit. Thus the learners experience the transition between the "flat earth" as seen before launch and the spherical planet with a curved horizon.

#### 6.1.3 Experiments

In three successive pilot studies, a total of 17 pairs of children were used. The children were predominantly African-American third graders from a Chicago school in which 93% of students' families are below the poverty line. An initial pilot study led to improvements in the user interface and controls. Students were introduced to the system and then spent 10 minutes on each task twice: for instance, mission controller, astronaut, mission controller, astronaut. The children were usually actively talking to one another. They focused on the task of finding and collecting fuel cells and did not pay much attention to the issues of up and down, gravity, and so forth. A group debriefing in front of the ImmersaDesk occurred after the interactive session.

Pre- and posttests were initially done using 2-D pictures and open-ended questions. When these showed little learning, a Play-Doh 3-D model was used with interview-style questions. It quickly became apparent that little conceptual learning was taking place. For the third pilot study, the approach was modified to focus more attention on the concepts. Only Asteroid World was used. The introductory activity was extended so that the students were given a guided tour around the asteroid. Students' hands-on time was decreased from 40 to 20 minutes. The group ImmersaDesk debriefing was replaced by individual guided inquiry with a physical globe and model of the asteroid.

#### 6.1.4 Experimenter's Conclusions

Some clear instances of learning occurred, but there was no large and consistent effect. The use of young children from a challenged population and away from their schools imposed severe logistical strains. The experimenters plan to refine the virtual world and experimental approach and analysis.

#### 6.1.5 Analysis

The experiment was designed to use the interaction between two children as the focus of learning. The concept of the round earth would presumably fall into Gagne's category of "factual knowledge." The observer was expected to notice and remark on the fact that the explorer was "upside-down." In the experiments, however, the observer was usually so focused on the task of helping the explorer to navigate that abstract concept formation seldom occurred.

Experimenters were surprised by how difficult it was to convey abstract information to children from deprived backgrounds, and (like the Virtual Gorilla Exhibit project) found themselves adding traditional teaching methods such as tangible 3-D models to their experiment.

### 6.2 ExploreNet

Not all virtual worlds are three-dimensional. Since 1994, Charles Hughes and Michael Moshell at the University of Central Florida have been exploring the educational utility of shared virtual worlds by using a relatively simple 2-D cartoonlike system called ExploreNet, which was inspired by the Habitat role-playing system (Morningstar & Farmer, 1991). Moshell and Hughes have conducted a series of experiments in various learning contexts, and developed an instructional model built around role-playing and improvisational drama.

#### 6.2.1 Setup

ExploreNet is a Windows application, running on ordinary PCs with Internet connectivity. A group of two to eight computers are set up with identical software. One of the computers is designated as a server, but also remains available for student use as a workstation. The users select a shared world and then choose the avatars or characters they will control during the simulation. One station can control one or several characters by taking turns.

Characters move by walking or flying to the location where a mouse-cursor is clicked. If the user clicks on a character or a "prop" (a noncharacter object) in the scene, a menu displays the actions that object can exhibit. If the user types the words appear over the head of the avatar currently controlled by that user on all workstations currently viewing the scene.

The principal expected advantage of ExploreNet's 2-D user interface over 3-D virtual worlds was its ease of world construction and use. ExploreNet's limited repertoire of simple behaviors allowed students to concentrate on the tasks at hand. However, there were a number of user interface problems that emerged as studies continued. Scenes are static and "hot spots" at the edge of scenes are used to move between scenes. When a character steps on a hot spot it disappears from the scene and finds itself in another scene. This proved quite confusing for many users. Ultimately it was found necessary to provide "mentor characters," which control student characters' ability to leave the scene of an intended action. Similar problems arose in dealing with the control of props, particularly when several props were close together.

### 6.2.2 Experiments

Three worlds were built and tested between 1994 and 1997. Zora's World was based on stories in the autobiography of Zora Neale Hurston, an African-American writer from Eatonville, Florida, and tested in Eatonville's Hungerford Elementary School. Dinosaur World was built in cooperation with students at Maitland Middle School and tested there. AutoLand was built to specifications provided by third grade teachers in the U.S. Department of Defense Dependents' Schools in Germany and tested there. These experiments are described in Moshell (1996) and Hughes (1996, 1997).

The key questions being investigated in these worlds were:

- How should the roles played by different students be differentiated?
- How many students can profitably interact within a simulation?
- How can "free play" and a story line, with educational goals, be reconciled?
- How can the learning experience be managed?

*6.2.2.1 Zora's World.* In the first experiment, two *mentor characters*, an owl and a dog, were provided. Other students controlled auxiliary characters (i.e., chickens and a cat). The students controlling the mentor and auxiliary characters were referred to as *cast members*. In addition, two pairs of *guest characters* were provided. The objective of the guests was to get eggs from the barn, build a fire in the back yard, and cook the eggs. But the chickens did not want to give up their eggs. Solving this problem required cooperative behavior between two guests. The mentor characters were trained to give hints as needed without giving away the entire story at once.

The cast members knew how the story was supposed to come out; they were responsible for the experience of the guests, much as uniformed staffs are responsible for the experience of guests in a theme park. Thus the entire experience involved eight students, four guests and four cast members. Usually, six computers were used, with two guests and two cast members sharing computers.

Like the Round Earth Project, Zora's World focused its experiments on underprivileged children. However, the experiments were conducted in the children's own school; in fact, they were situated in the classroom alongside the rest of the class. The six computers were set up with their screens facing the wall, so only the eight participating students could see what was going on. The other 16 children in the class knew that their turns would come; they were supposed to work on other projects in small groups while the current crew explored Zora's World.

Various age pairings were tried, with fourth-, fifth-, and sixth-grade children playing the roles of cast members and guests. It was found that sixth-grade students acting as cast members had the best chance of retaining control of the learning situation and achieving desired outcomes. Younger children would lose track of the fact that they were supposed to be managing others' learning experience and would play the game themselves. However, in general the younger children acting as guests successfully completed the game as often as the older ones.

All students viewed the game as a competitive one, though it had been designed as a cooperative experience. Students would hoard resources (firewood for instance) and bring the game to a standstill until a mentor character could convince them to share.

*6.2.2.2 DinoLand.* In this world, four individual guests represented different dinosaur species, two herbivores, a carnivore, and an egg-eating species. The only cast member played a managerial role in charge of invoking a day–night cycle. Every night represented a new generation of dinosaurs in which those species that successfully laid and defended their eggs in the previous round became stronger.

The students were extremely motivated by the competitive aspects of predator–prey relationships. One carnivore-player discovered that he could collect all the plants and starve the herbivores into submission. The use of four rather than eight characters proved to declutter the screen and keep interactions focused.

*6.2.2.3 AutoLand.* AutoLand was designed as a collaborative constructionist activity in which four guests had to find raw materials, assemble them, and create the components of an automobile. Unlike the previous two worlds, this lesson was created on the basis of a textbook chapter (specified by third grade teachers) and was provided to the teachers along with worksheets and paper maps. Because participating students were third graders and could not type, they were allowed to interact verbally (by talking to the students on adjacent computers). Instead of on-screen mentors, each student had a "shoulder buddy" who had previously been through the experience. A teacher served this role for the first group.

According to the teachers, Autoland was very successful as a supplementary classroom activity. In testing by 50 children in two classrooms, over 90% of the four-person teams succeeded in constructing the automobile within the allotted 40-minute period. A formal external evaluation was provided for, but due to transatlantic logistical difficulties was not completed.

### 6.2.3 Experimenter's Conclusions

The succession of experiments led the project away from the use of in-world cast members and toward a close correlation with subject matter that was of interest to teachers. The provision of external reading materials and in-class activities provided necessary a priori knowledge (e.g., of where in the United States one might find iron ore, coal, etc.) so that the virtual experience could serve to transform abstract concepts like north and south into real experiences.

### 6.2.4 Analysis

The ExploreNet experiments were intended to teach "attitudes" (cooperation) and "cognitive strategies" in Gagne's terms, or "interpersonal and intrapsychic" skills, in Gardner's terms. In AutoLand there were also goals related to factual information (where cars come from).

Lessons were learned about techniques for using one group of students (cast members) to guide the experience of other students, and about how to build materials that teachers would willingly use.

## 7. THE FUTURE OF VE TECHNOLOGY IN EDUCATION

From the six projects analyzed in this chapter, some common themes emerge. The most striking of them is that no project achieved significant, measurable learning in its first attempt at using VE technology, and most had not achieved clear results even after several experimental revisions. Among many possible reasons, the most plausible seem to be:

- System complexity and novelty consumed most of the students' and experimenters' attention.
- Evaluations that measure factual knowledge in traditional ways may not be very sensitive to the kinds of procedural knowledge and cognitive strategies that might develop in VE.
- The process of instructional design for virtual worlds is very immature. Tools for constructing 3-D geometric models require specialized technical knowledge and are generally difficult to use.

The explosive growth of video game technology means that the technical and cost barriers to high-performance 3-D computer graphics are disappearing. Interactive software will become easier to build, though it is difficult to predict how fast or far this trend will go. It seems likely that many more experiments will be conducted in the use of interactive simulation for learning. The authors hope that the lessons learned from the pioneering projects reported in this chapter can make the way easier for those who follow.

## 8. REFERENCES

Allison, D., Wills, B., Bowman, D., Wineman, J., & Hodges, L. (1997). The virtual reality gorilla exhibit. *IEEE Computer Graphics* and *Applications, 17*(6), 30–38.

Bowman, D. A., Wineman, J., Hodges, L. F., & Allison, D. (1999). The educational value of an information-rich virtual environment. *Presence: Teleoperators and Virtual Environments, 8*(3), 317–331.

Cognition and Technology Group at Vanderbilt. (1993). Designing learning environments that support thinking: The Jasper series as a case study. In T. M. Duffy, J. Lowyck, & D. H. Jonassen (Eds.), *Designing environments for constructivist learning*. New York: Springer-Verlag.

Cruz-Neira, C., Sandin, D., DeFanti, T. A., Kenyon, R. V., & Hart, J. C. (1992). The CAVE: Audio-visual experience automatic virtual environment. *Communications of the ACM, 35*(6), 64–72.

Cusick, P. A. (1997). The Coalition goes to school. *American Journal of Education, 105*(2), 211–221.

Dede, C. (1995). The evolution of constructivist learning environments: Immersion in distributed, virtual worlds. *Educational Technology, 35*(5), 46–52.

Dede, C., Salzman, M., Loftin, B., & Ash, K. (1999a). Using virtual reality technology to convey abstract scientific concepts. In M. J. Jacobson & R. B. Kozma (Eds.), *Learning the sciences of the 21st century: research, design, and implementing advanced technology learning environments*. Hillsdale, NJ: Lawrence Erlbaum Associates.

Dede, C., Salzman, M., Loftin, B., & Sprague, D. (1999b). Multisensory immersion as a modeling environment for learning complex scientific concepts. In N. Roberts, W. Feurzeig, & B. Hunter, *Computer modeling and simulation in science education*. New York: Springer-Verlag.

Duffy, T. M., & Jonassen, D. H. (Eds.). (1992). *Constructivism and the technology of instruction*. Hillsdale, NJ: Lawrence Erlbaum Associates.

Gagne, R. M., Briggs, L. J., & Wager, W. W. (1988). *Principles of instructional design* (3rd ed.). Forth Worth, TX: Holt, Rinehart, & Winston.

Gardner, H. (1993). *Multiple intelligences: The theory in practice*. New York: Basic Books.

Harel, I., & Papert, S. (1991). *Constructionism*. Norwood, NJ: Ablex.

Johnson, D., & Johnson, R. (1987). *Learning together and alone: Cooperative, competitive and individualistic learning*. Englewood Cliffs, NJ: Prentice-Hall.

Johnson, A., Moher, T., Ohlsson, S., & Gillingham, M. (1999). The Round Earth Project: Deep learning in a collaborative virtual world. In B. Loftin & L. Hodges (Eds). *Proceedings of the IEEE VR '99 Conference*. Houston, TX. IEEE Press.

Kafai, Y., & Resnick, M. (Eds). (1996). *Constructionism in practice*. Hillsdale, NJ: Lawrence Erlbaum Associates.

Kessler, D., Kooper, R., & Hodges, L. (1998). The simple virtual environment library: Version 2.0 user's guide. (Tech. Rep. No. GIT-GVU-98-13). Atlanta, GA: Georgia Tech Graphics, Visualization and Usability Center.

Morningstar, C., & Farmer, F. R. (1991). The lessons of Lucasfilm's Habitat. In M. Benedikt (Ed.), *Cyberspace: First steps*. Cambridge, MA: MIT Press.

Moshell, J. M., & Hughes, C. E. (1996). The virtual academy: A simulated environment for constructionist learning. *International Journal of Human–Computer Interaction, 8*(1), 95–110.

Papert, S. (1980). *Mindstorms: Children, computers and powerful ideas*. New York: Basic Books.

Postman, N. (1985). *Amusing ourselves to death: Public discourse in an age of show business*. New York: Viking.

Roussos, M., Johnson, A. E., Moher, T., Leigh, J., Vasilakis, C. A., & Barnes, C. R. (1999). Learning and building together in an immersive virtual world. *Presence: Teleoperators and Virtual Environments, 8*(3), 247–263.

Salzman, M., Dede, C., Loftin, B., & Ash, K. (1998). VR's frames of reference: A visualization technique for mastering abstract information spaces. In *Proceedings of the Third International Conference on Learning Sciences* (pp. 249–255). Charlottesville, VA: Association for the Advancement of Computers in Education.

Salzman, M. C., Dede, C., Loftin, B., & Chen, J. (1999). The design and evaluation of virtual reality–based learning environments. *Presence: Teleoperators and Virtual Environments, 8*(3), 293–316.

Wang, M. C. (1998). Models of reform: A comparative guide. *Educational Leadership, 55*(7), 66–71.

Weller, H. G. (1996). Assessing the impact of computer-based learning on science. *Journal of Research in Computing in Education, 28*(4), 461–485.

Wickens, C. (1992). Virtual reality and education. In *Proceedings of IEEE International Conference on Systems, Man and Cybernetics* (pp. 842–847). Chicago, IL: IEEE Press.

Windschitl, 1998

Winn, W., Hoffman, H., Hollander, A., Osberg, K., Rose, H., & Char, P. (1999). Student-built virtual environments. *Presence: Teleoperators and Virtual Environments, 8*(3), 283–291.

Youngblut, C. (1998). *Educational uses of virtual reality technology* (Tech. Rep. No. D-2128). Alexandria, VA: Institute for Defense Analysis.

# 46

# Development and Evaluation of Virtual Environments for Education

Sue Cobb, Helen Neale, Joanna Crosier, and John R. Wilson

*University of Nottingham*
*Virtual Reality Applications Research Team (VIRART)*
*England NG7 2RD*
*sue.cobb@nottingham.ac.uk*
*helen.neale@nottingham.ac.uk*
*jo.crosier@nottingham.ac.uk*
*john.Wilson@nottingham.ac.uk*

## 1. INTRODUCTION

Recent research suggests that virtual environments (VEs) may offer significant benefits to all types of education, especially in light of the types of learning supported by VE compared with other computing technologies (see chap. 45, this volume, for a review). However, evaluation studies have highlighted usability issues (see chap. 34, this volume) unique to virtual learning environments (VLEs), which traditional human–computer interaction (HCI) guidelines do not cover. Moreover, much of the HCI research to date has focused on mainstream applications and does not provide guidance on how to build interfaces for users with special educational needs (SEN), which is an important early market for VLEs. What is required is a structured approach to the design, development, and evaluation of VLEs, with close liaison with representative users and professionals at all stages of development to ensure adequate design of VEs to support learning. In this chapter the initial steps toward a user-centered model of VLE development and evaluation is described.

### 1.1 Computers in Education

The growth in the number of new technologies applied to education has produced many varied applications within the classroom. One of the most controversial of these has undoubtedly been the idea of introducing the microcomputer as a "teaching machine," first seen in schools in the 1950s (Fry, 1963). In most early examples, education via computer followed the behaviorist school of thought. Using teaching machines, responses were conditioned by providing a list of words or arithmetical products to which the student was required to respond, the correct answer being revealed after each attempt (Fry, 1963; Richmond, 1965). Skinner was an advocate of these machines and felt that many of the ideas developed in his study of animal learning could be applied to the task of improving instruction (Hilgard,

Atkinson, & Atkinson, 1979). At the same time the use of drill and practice programs was thought to be the best teaching structure to use for children with learning disabilities to acquire new skills (Mittler, 1979). Consistent computer feedback was also used for profoundly disabled children to learn cause and effect (Rostron & Sewell, 1984). This type of instruction did not significantly impact mainstream education, but the belief that computer-aided-learning (CAL) was an effective idea continued. The features of CAL seen to be of most benefit were: that it provides *active participation*, where the learner is actively interacting with curriculum materials; that *feedback* is provided immediately so that an error can be corrected and an incorrect belief or response is not reinforced; and there is *individualization of instruction*, where different learners can move ahead at their own pace (Hilgard et al., 1979).

The "LOGO" programming language for children was designed in accordance with Piaget's ideas that children should construct their knowledge in building blocks: "One might say the computer is being used to program the child. In my vision, the child programs the computer" (Papert, 1993, p. 5). He believes that in allowing the child to explore and follow his own path of inquisition, the child learns in a less formal and more natural way, paralleling the way in which basic functions such as walking, talking, and eating are learned. Papert was one of the early contributors to applying computer technology in education, and many regard him as revolutionary, although today computers are much more powerful than those used in his early work, and the number of educational computer programs available has increased dramatically.

The majority of educational computer programs created recently have been multimedia programs that combine text, graphics, sounds, and film. A major criticism of these types of programs is that they mimic resources already existing in schools, such as books. "Most multimedia programs fail because they add video and graphics to page turning programs" (Schank, 1994, p. 69). What appears to be lacking in these computer education applications is that they do not provide anything of significant value that is additional to current educational resources.

The outlook is somewhat different in special needs education, where a number of research projects have found that attributes of computers can support learning in a way in which traditional teaching methods do not. Bridge stated: "The educational potential of the micro-computer appears to be immense, and nowhere more so than in the field of Special Needs Education" (1987, p. 3). He attributed this to a change in role of the teacher from educator to advisor. A more active and exploratory learning style (Clariana, 1997) combined with students' control over their own pace of learning may motivate students.

## 1.2 Learning in Virtual Environments

Virtual environments allow the user to actively participate in experiences of infinite possibilities. A VE may allow the participant to see representations of scenarios familiar to them or objects that they may not be able to view in the real world. Any situation may be represented as a three-dimensional (3-D) environment, and the user may take on any role or perspective within the environment. The potential for learning may therefore be great, as specific skills may be practiced and observed from different viewpoints and information to be learned may be presented in meaningful and concrete ways.

While early educational computing allowed users to practice repetitive tasks (Solomon, 1986), VEs offer a different kind of arena for learning. They allow a greater degree of self-paced and determined exploration, as well as practice of naturalistic tasks in realistic settings (Brown, Cobb, & Eastgate, 1995; Brown, Standen, & Cobb, 1998; Cobb, Brown, Eastgate, & Wilson, 1993). Virtual environments provide facility for experiential learning by allowing the student to explore the VE at their own pace and interact with the VE in real time (Bricken, 1991; Dede, 1995; Winn, 1993).

For example, Bednar, Cunningham, Duffy, and Perry (1992) provide five main points, which outline the features of constructivist learning environments. These include: contextualizing learning; learner as an individual; authentic tasks; situating cognition, and nonprespecification of content.

Constructivism has been discussed as a learning theory on which to build an instructional technology for a number of years (Duffy & Jonassen, 1992; Papert, 1988; Perkins, 1996) and is the single most reported education theory supporting development of VEs for education (see review by Youngblut, 1998). However, in practice, the theory has only been applied loosely in the construction of a number of educational VEs (Dede, 1996; Roussos et al., 1999; Winn, 1993; see chap. 45, this volume).

A study by the authors (Neale, Brown, Cobb, & Wilson, 1999) describes a structured evaluation of VLEs to investigate to what extent these learning environments met constructivist principles (using principles developed from Jonassen, 1994). It was found that the applications supported moderate rather than extreme forms of constructivism. Different learning environments supported distinct learning principles in varying ways. This was mainly due to differences in the objectives of the VLEs. VLEs that aimed to promote communication skills were found to support collaboration and social negotiation, but contained more abstract visual content and did not represent the natural complexity of the real world accurately. Conversely, VEs that aimed to teach real world life skills represented the natural complexity of the real world more accurately and presented more authentic tasks.

Virtual environments have a number of characteristics which may promote certain behaviors that enhance special needs education, such as self-directed activity, motivation, and naturalistic learning (Neale et al., 1999). These features of VEs lend themselves to supporting learning according to a constructivist perspective.

In all of the VLEs herein investigated, it seemed necessary to include some form of behaviorist/objective learning. In the cases explored, teachers and students with a learning disability collaborated "shoulder to shoulder" using the virtual environment. Teachers always provided some directive or instructional role, this meant that students were more likely to take in important points of focus and understand the meaning of their actions.

### *1.2.1 Uses of VE in Mainstream Education*

Most of the VLEs that have been created for mainstream education have made use of specific characteristics that the technology provides. Virtual environments allow representation of things that would not normally be visible to the naked eye, for example, molecules in a compound (Byrne & Furness, 1994), inner workings of a machine (Wilson, D'Cruz, Cobb, & Eastgate, 1996), or electricity flow in a house (Middleton, 1992). The technology provides facility for taking on different perspectives within the virtual environment. In the case of the above examples, as well as taking a global view of each scenario, you might "be" an atom in the molecule, a part being manufactured, or an electron flowing through the wire. Virtual environments allow participants to explore dangerous situations safely. Health and safety issues have meant that some activities, such as science experiments involving dangerous chemicals, cannot be carried out in schools. Virtual environments allow these experiments to be recreated so that the "hands-on" element is put back into the lesson (Crosier, Cobb, & Wilson, 1999). Also, limited access to industrial plants could be overcome by creating virtual factories (Copeland, 1998; Goldsmith, 1999). Due to their 3-D nature of presentation, VEs also have great potential for demonstrating complex abstract concepts, for example, representing a magnetic field (Dede, Salzman, & Bowen-Loftin 1996).

### *1.2.2 Uses of VE in Special Needs Education (SEN)*

The potential for the use of VE technologies in special needs education has been talked about for a number of years (Middleton, 1992; Powers & Darrow, 1994). Virtual environments have subsequently been developed for users with physical disabilities and vision and hearing impairments.

Inman, Loge, and Leavens (1997) have used a head-mounted display (HMD) to train the use of electric wheelchairs for children with physical disabilities. A similar project is being carried out using desktop VEs for the education of severely disabled children, using a VE as a safe, enjoyable, scaffolded wheelchair training environment (Desbonnet, Cox, & Rahman, 1998). Virtual environments have been used to teach spatial skills to physically disabled children, and evaluation studies have found generalization of such skills to real physical environments (Stanton, Foreman, & Wilson, 1998).

The use of VE for blind users is being explored by a number of research groups (Jansson, 1998). For such applications, the VE interface must first be adapted to allow the user to receive and interact with 3-D information by using haptic, kinesthetic, and tactile devices (Colwell, Petrie, Kornbrot, Hardwick, & Furner, 1998; see chaps. 5 and 6, this volume). Studies have looked at the potential uses of VEs for education, training, Web browsing, and as navigational assistive devices. Lumbreras and Sanchez (1998) have developed 3-D acoustic interactive Hyperstories that are intended to develop the cognitive skills of blind children. Berka and Slavik (1998) suggest the use of such devices for navigation training in VE, or the extension of this device so that the user can navigate in real time in a real, unknown environment.

The use of VE applications for the rehabilitation of cognitive processes and functional abilities in clinical neuropsychology has included studies of developmental and learning disabled clinical populations (see chaps. 46 and 51, this volume). Rizzo, Wiederhold, and Buckwalter (1998) detail research in this area and review a numbers of studies where VEs have been used to promote learning for a disabled population. Despite growing numbers of VE applications within special needs and mainstream education, there is still little evidence of research concerning the profitable use of VEs with users with developmental and learning disabilities (Darrow, 1995; Youngblut, 1998). The few studies include the use of VEs for training autistic children safe street crossing (Strickland, 1996) and the communication and experiential VEs developed by the Virtual Reality Applications Research Team (VIRART) at the University of Nottingham (detailed in section 4).

There are many more possible applications that could be developed and used for successful learning. Preliminary studies have indicated a number of benefits from their use by clinical populations. This may be due to a combination of many factors, but one issue may be the resources required when attempting to develop virtual learning environments (VLEs) for such user groups.

## 2. DEVELOPMENT OF VIRTUAL LEARNING ENVIRONMENTS

There are three general phases of VLE development: specification of the application and its content, programming, and evaluation. Although these phases are distinct and progressive in themselves, development of the final product requires a good deal of iteration (see Fig. 46.1). A user-centered approach taken in the specification and design of VLEs will result in more useful and appropriate applications.

### 2.1 User-centered VLE Specification

In this section, involvement of user groups in early conceptualization stages of VE design is emphasized. At each stage, representatives of users and professionals should be involved to

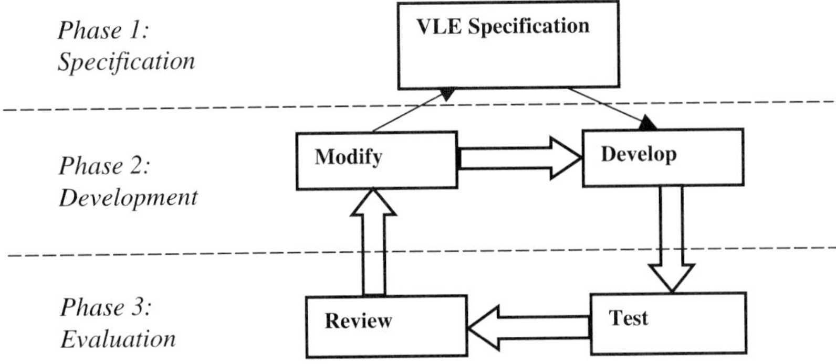


FIG. 46.1. Stages of VLE design and evaluation.

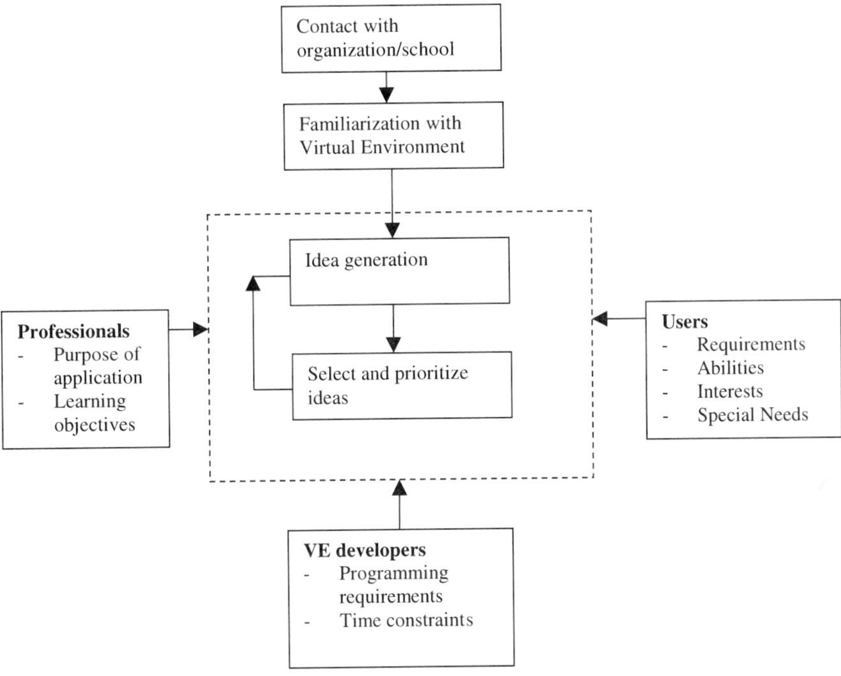


FIG. 46.2. User-centered VLE specification.

ensure that concepts are relevant and interfaces are accessible. Figure 46.2 shows how users and professionals can influence VLE development.

### 2.1.1 Initial Contact with Schools/Organizations

When designing educational VE applications, the first step is to make contact with the schools or organizations that may use the VLE. Initial contact may arise through several routes:

1. Cold contact with schools to recruit interest in a project
2. User representatives approaching developers asking them to meet a specific need
3. General interest from potential users after seeing a demonstration or publicity

It is important that VLEs are developed with a clear view of the value of VE as an aid for learning. The user-centered design approach involves teachers and professionals in the first initiation of the VE design. They will determine the purpose of the VLE, as an additional teaching aid to support existing methods or as a specific program aimed at providing learning in a unique way. In addition, they inform VE developers of the practicalities of computer-based teaching within their organization (e.g., resources available, staff interest and motivation to use the program, acceptability, etc.).

#### 2.1.2 VE Familiarization

After initial contact, the next step is to familiarize teaching professionals or care workers with VE technology and features of virtual environments. It is important to give a realistic view of VEs in accordance with the users' needs and level of available resources. Demonstration of previous applications can be used to highlight features of VEs as follows (Crosier, 1997):

1. **Representation of Objects and Environments**. VEs can be used to model all types of objects or environments. These may be representations of things that actually exist in the real world, or they may be replicas of things that either do not exist or cannot ordinarily be seen. It is possible to model an entire city, a single building, or just a single room, such as a school classroom. The example in Fig. 46.3 shows part of a virtual city, and Fig. 46.4 shows the inside of a school (Brown, Kerr, & Eynon, 1997; Kerr, 1995).
2. **Different views of the environment**. In VEs the user can take different perspectives; in many cases these are linked with movement constraints. For example, when driving a car around a virtual city, if the user moves the joystick left, the steering wheel will turn counter-clockwise and the car will turn to the left (see Fig. 46.5). In Fig. 46.6 the user takes on the characteristics of a skiier (Brown, Kerr, & Wilson, 1997; Brown, Stewart, & Mallet, 1997). In this case a constraint is applied that prevents the virtual object (which contains the user viewpoint) from moving backward, as this would be impossible on a real ski slope. It is not always the case, however, that movement is restricted in this way. It may be that the user should be able to interact with an environment in ways they would not be able to in the real world, for example, "flying" above the ground or walking through walls. This has been highlighted in special needs education where wheelchair-confined students can "walk" through walls.

FIG. 46.3. A virtual city.

FIG. 46.4. Inside a school.

FIG. 46.5. Virtual driving.

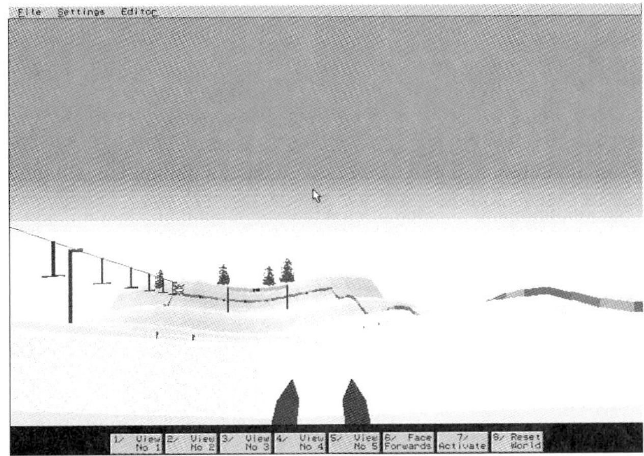


FIG. 46.6. Virtual skiing.

3. **Real-time interaction**. An important attribute of VEs is the fact that they allow real-time interaction between the user and environment. In a virtual training environment, the user can respond to a ringing telephone by clicking on the receiver with the mouse to pick up the phone (Fig. 46.7). This is also an example of how an environment can be created to elicit certain actions from the user. The kitchen environment in Fig. 46.8 includes a number of different objects with which a user can interact. The taps can be switched on, tea and coffee can be prepared, and the oven can be operated.

4. **Split-screen interfaces**. Virtual environments are completely versatile and so can be designed to suit application, task, or user specifications. An example is to split the screen or overlay menus that provide additional functions in a similar way to multimedia interfaces. Figure 46.9 shows an interface layout where a virtual hand signer and Makaton communication symbol is permanently displayed on the left side of the screen. The user can move around the VE on the right side of the screen to explore and interact with the objects represented by the Makaton symbol to gain an understanding of their meaning and function (Brown et al., 1995). Another example is where written instructions provided in text, icons, or picons (i.e., picture icons) are displayed around the outside of the screen (Cobb & Brown, 1997). Further

FIG. 46.7. Answer the telephone.

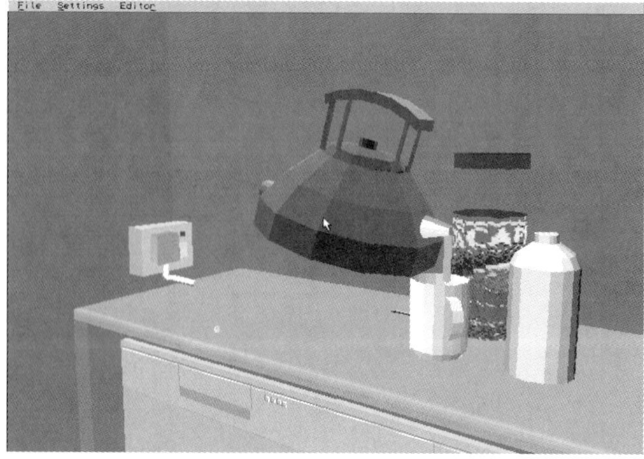


FIG. 46.8. Cooking in a virtual kitchen.

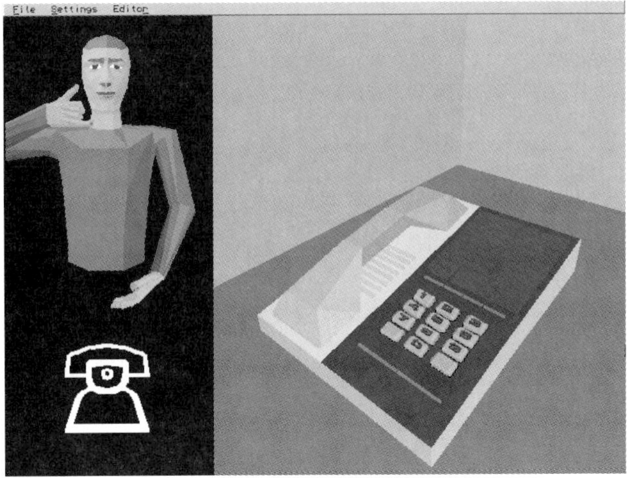


FIG. 46.9. Learning Makaton symbols and sign.

FIG. 46.10. Health and safety at work.

information can be gained by clicking on the forms on the left, which when selected are viewed in full screen and can be read (Fig. 46.10).

5. **Reality and "superreality."** One of the potential advantages of VEs is that they allow behaviors not normally possible in real life. Figure 46.11 shows how a laser has been represented (Brown, Micropoulos, & Kerr, 1996). The user can navigate with a mouse or joystick to observe the laser from different perspectives and can actually look inside the laser beam where the wavelengths of the light are represented. Figure 46.12 shows a representation of a factory that demonstrates how an entire production run can be simulated. Various aspects of the run can be altered and the consequences observed. For example, quantities of product can be altered or machines can be moved. It is also possible to see the internal workings of the machines to gain an understanding of how the process works. This VE was created as a training environment to demonstrate specific features of VE to representatives from the U.K. manufacturing industry (Cobb, D'Cruz, & Wilson, 1995). This led some companies to develop their own virtual environments (Wilson et al., 1996).

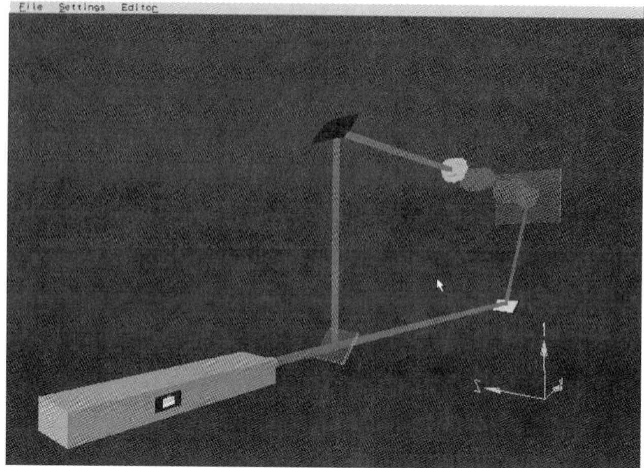


FIG. 46.11. Laser physics.

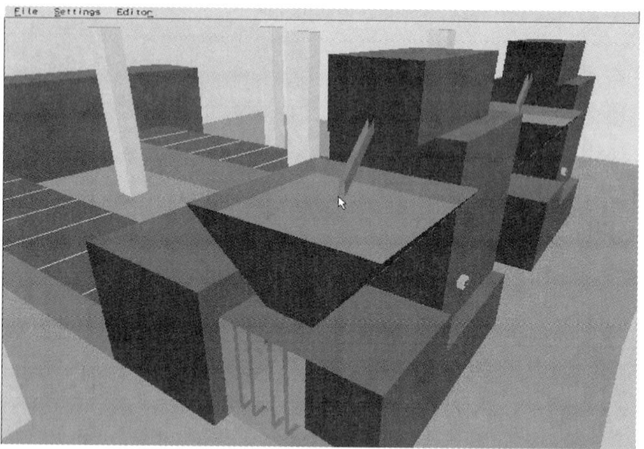


FIG. 46.12. Virtual factory.

During this familiarization phase, it is important that teachers and/or care workers are given the opportunity to explore exemplar VEs and become accustomed to participating in them. With a realistic understanding of what a VE is and the current capabilities and potentials of VLEs, both developers and potential users can work together to develop ideas for future educational VE applications. This ensures that ideas for VEs are relevant and useful to educational organizations, as well as being feasible (i.e., within the constraints of current VE technology).

### 2.1.3 Application Idea Generation

Once teachers and support workers have been made aware of VE technology and its capabilities, they are in a better position to be able to comment on how it might be useful to them. A brainstorming session can be held to generate ideas for specific applications of interest. Teachers and support workers will have responsibility for using the technology on a day-to-day basis and its integration within existing teaching methods. It can also be useful to involve users at this stage, where permissible. Users can provide insights into areas that they are interested in learning about, issues they want to more fully understand, and where current teaching methods are inadequate. In this way the VLE produced should be derived from an authentic learning need expressed by the users themselves or the professionals responsible for teaching these skills and issues (Crosier & Wilson, 1998).

### 2.1.4 How to Select and Prioritize Ideas

The early stages of idea generation frequently end up with more possible VEs than can practically be developed. Prioritization of these ideas should involve a number of professional teachers or care workers, who will have a sense of what learning or development objectives are most important for the target user group. In the case of mainstream education, decisions may be governed by curriculum restrictions.

Technical and practical information also needs to be provided by researchers and VE developers, regarding what ideas are feasible in terms of time and technology capabilities. Virtual environment scenarios should be developed which maximize VE capabilities, so consideration needs to be given to other technologies or methods which may be equally useful in teaching the subject area so as not to "reinvent the wheel."

Involvement of relevant groups of professionals at this stage of VE development is critical; it gives them insight into *why* the VE is being developed and provides them an opportunity to contribute information that could render the VE more relevant to their students or clients. Just as important, their involvement in early stages of development means that host or client organizations will be much more understanding and helpful when their time and cooperation are needed in the later stages of VE development. Since VLEs must not be one-off initiatives but should be embedded into schools and other institution's general operations, early two-way dialogue with organizations will clear the path for ongoing success.

### 2.1.5 Involvement of End Users

It is axiomatic for human factors professionals that involvement of end users results in better developments and development processes. These procedures originate from participatory design of the workplace (Noro & Imada, 1991; Wilson, 1995; Wilson & Haines, 1997). The involvement of end users in the design process gives information as to how the end product will actually be used and puts it in a real-world context. When there is a distinct user population, with characteristics far removed from the designers of the product, there may be an increased need for user involvement in the development process. The field of HCI has recently adopted these design philosophies, and in the 1990s the design of technology for children has become more prominent (Druin, Stewart, Proft, Bederson, & Hollan, 1997; Druin, 1999; Druin & Solomon, 1995; Scaife et al. 1997). The involvement of older students in the design of educational technology may not be as essential in terms of the development of an "easy to use interface" as it is for other user groups such as young children. However, the way in which the technology is used, including the environment in which it is used and how the program is used (e.g., by one student, by the student and teacher together, by a group of students) may have an impact on how it should be designed. For example, in a school setting, sessions with the computer are often held in a designated "computer lab." The students may have weekly 1-hour sessions within which they may use the computer. The software designed should take account of this and have enough content to keep the students occupied for the entire session.

The contribution of people with a disability to the research process has considerable importance as they have differing demands on the final product. Minkes, Townsley, Weston, and Williams (1995, p. 97) argued that involvement of people with learning difficulties resulted in: "research which more clearly benefits the people whom it is supposed to help." Edwards and Hancock (1995, p. 365) also promote the involvement of disabled users in HCI research: "It is vital to involve potential users in the design process so that their needs (and not the able-bodied designer's perception of their needs) can be addressed." The benefits of involving representative users in research and product development has been echoed in the Virtual Life Skills Project (Brown, Neale, Cobb, & Reynolds, 1999), also presented in this chapter as a case example (see section 4). Its production focused on users' needs and gave people with a learning disability the opportunity to develop and test a product, which they will use with professionals as part of their development. This resulted in a VLE that is both useful and usable and also improved confidence and skills of the individuals involved in the project (Meakin et al., 1998).

## 2.2 Building a Virtual Learning Environment

There is no single correct approach to building a virtual environment. In practice, the approach taken depends on what the VE is intended for, how well formalized the idea is, how specific users are in their requirements, and the experience and programming style of the individual VE programmers involved. What is most important is that the communication channel between user (or user champion) and programmer works well. In some cases this may not be a direct one-to-one communication channel but involve several links between different people such as teachers, support workers, researchers, and programming managers.

Development of VEs for education is an iterative process. Wilson and Eastgate (see chap. 17, this volume) detail a general framework for the development of virtual environments. Initial brainstorming with teachers or support workers provides the context for the learning environment, specific learning objectives, and an outline of the layout and visual representation of learning scenarios. During these discussions sketch-style storyboards are drawn up, which are then presented to VE programmers, who examine them and make recommendations back to the review or steering group about what they think is feasible within the time scale allocated. This may require the review group to prioritize features of the VE or to identify particular learning objectives that should be the main focus of the VLE. In our experience, we have found that initial user requirements are often based on what they already know about VE capabilities; having seen familiarization demonstrations, users will ask for similar scenarios.

Therefore, it is useful at this stage to have a discussion between programmers and the review group, at which the programmer suggests alternative ways to achieve required effects or ways to reduce programming effort, but does not tell users what their objectives are. It is important that user objectives for a new VLE are not derived from what they have already seen, but that they are guided into stating their actual requirements. The level of programmer involvement in initial contact with users and professionals will affect their level of understanding of user needs. There is often a different level of communication required between the programmer and user, for example, more detailed questions about the design of a layout or a task within a VLE. If VE developers are involved from the start of the process then they will have a greater understanding of the messages that the VLE should put across and how this should be done.

The design of any computer program may be directed by the use of relevant guidelines. There are a number of HCI principles and rules, which may be used to guide interface design. A number of high-level and widely applicable principles can be useful to VLE developers. General principles that should be followed by VLE designers include: know the user population; reduce cognitive load; engineer for errors; and maintain consistency and clarity (Preece, 1994). An in-depth knowledge of the user population is essential to follow all of the other principles. For example, the level of cognitive load that is appropriate will be dependent on the user (children, adult with special needs, teachers, etc.).

Many design rules developed within general HCI are not relevant to VE design because of the different features and objectives of VEs as compared to graphical user interfaces (GUIs) or more traditional computer interfaces. For example, a user of a VE can navigate in 3-D space (see chap. 24, this volume) and interact in a more "naturalistic" way with representative of their real-world counterparts (see chap. 13, this volume). These advanced navigational and interaction techniques require new directions in usability engineering (see chap. 34, this volume). Kaur, Sutcliffe, and Maiden (1998) have found that conventional interface guidelines are only partially applicable for designers of virtual environments. Designing for user populations who do not conform to general user assumptions compounds this situation.

There are a number of general guidelines and observations about VE design, a few of which concern representation of objects (Slater & Usoh, 1995), qualities of a virtual human (Thalmann & Thalmann, 1994), and communicating depth (Pimentel & Teixeira, 1993). A number of techniques for the design of user interactions have emerged such as support for guiding the user (Wilson et al., 1995) and support for object alignment (Buck, Grebner, & Katterfeldt, 1996). However, little experimental work or evaluation has been done, and there are few comparisons or established standards (Kaur et al., 1999).

There have been few studies into HCI for disabled users. Among these, Edwards (1995) reports on the work of Scadden and Vanderheiden (1988) and how they have considered the design of computers and operating systems for people with a range of disabilities. The populations covered included users with physical disabilities, visual and hearing impairments, and people sensitive to seizures. They recommended a series of steps for reducing cognitive barriers to computer use by such populations:

- Simplify language in task instructions
- Provide online help
- Make displays simple

Cress and Goltz (1989) compiled a more detailed list of design guidelines for improving computer accessibility for persons with cognitive disabilities. This list uses data from rehabilitation and educational literature and is intended to guide the design of computer hardware and software. The recommendations focus on input/interface control, presentation format, and information content and prompts.

The appropriateness of such guidelines for the development of VEs is not always clear. Perhaps, as the user is presented with a more interactive and realistic medium than the traditional computer interfaces for which these guidelines were derived, the VE interface will become more intuitive and easier to understand. Example guidelines include the recommendation for the visual and audio complexity of the output display to be minimized, as well as inclusion of online memory aids and help screens. During the development of VLEs for special needs education, developers have aimed to make VEs as naturalistic as possible, partially eliminating the need for online help. The display should allow the user to recognize the situation and to gain appropriate educational knowledge from the VE; however, this will not always result in a "simple" display.

## 3. EVALUATION OF VIRTUAL LEARNING ENVIRONMENTS

This section discusses the evaluation of VLEs, how this is being done and attempts to identify how it differs from evaluation of other learning technologies. The evaluation of VLEs may focus on two main issues. First, the design of the VLE itself may be evaluated: Does it contain relevant information and tasks for the intended user groups, and is it usable? Second, do users learn anything from using the VLE? A number of frameworks exist for the evaluation of learning technologies, such as Jones, Scanlon and Blake (1998) and Oliver (1997), and these may provide a useful basis upon which to base evaluation studies of VLEs.

In a number of cases VLEs have been compared with traditional methods of teaching and learning. The aim of technology development has frequently been to produce a program for a medium, which some hope will be better for teaching and learning than traditional methods. However, it is very difficult for evaluators to prove that something is better. Jones et al. (1998) report that the educational experience often changes when educational technologies are used, meaning that it is impossible to make direct comparisons with traditional methods as something different, not better or faster, is learned. This different experience may be even more pronounced when comparing VLEs with traditional teaching methods or computing systems. Interacting with a VLE allows the user to experience situations, visualize in 3-D, and interact in a more natural and realistic manner. Comparative studies have shown on a number of occasions that VLEs will provide a very different learning environment and features of both VLEs and traditional teaching methods should be combined in order to reap the benefits from both types of systems. For example, a comparative study carried out on CDS (Conceptual Design Space) by Georgia Institute of Technology, reported in Youngblut (1998), found that there were different strengths to the use of VEs and traditional CAD systems for an architectural design task. One of the main findings was that qualities of VE should be combined with traditional methods to exploit the strengths of both. These apparent differences between VLEs and traditional learning systems make it clear that it is important in evaluating VLEs to concentrate on the experience of using the VLE in order to understand its unique impact on learning.

Evaluations of VLEs have, to date, been limited by the size and focus of the research projects within which they have been based, and by the maturity of VE technology. The review of educational uses of VEs by Youngblut (1998) includes descriptions of the types of evaluations that were carried out on VLEs up until 1997. The majority of evaluation studies was informal and employed only one iteration of evaluation. Many early studies focused on the usability of VEs, asking basic questions such as can the VE be used by students and which input devices are preferable? Many of the studies report attitudinal data to find out if the students enjoy using VEs to learn.

A number of VLEs have been more extensively evaluated using multiple strands of evaluation, an approach that has been advocated for such an immature technology (Rose, 1995). Roussos et al. (1999) describe the NICE project, where multiple measures of performance were taken in order to examine usability, orientation, attitudes, cognitive skill, pedagogical learning, and collaborative behavior. The ScienceSpace VE has been subject to a number of successive evaluations, looking at, among other aspects, students' interactions with the VE, expert ratings, and knowledge assessments. The findings have informed the VLE development process and are reported in detail in Moshell and Hughes (see chap. 45, this volume).

Evaluations of VLEs for users with special needs have not been as widely reported as those in mainstream education, and this may be because there are fewer VE developments in this area. This is also a very difficult area to evaluate. A disabled subject population, which may find it more difficult to use the VLE, may compound the usual difficulties of evaluation. There will therefore be more usability issues, and user feedback about the VE may be more difficult to obtain. Strickland (1996) examined the use of VE for education of children with autism. This took the form of an informal observation of two young children's use of a virtual environment. Neither child could communicate well enough to describe their thoughts, and so their actions were interpreted as indicating different responses to the virtual environment. VIRART have conducted a succession of evaluation studies of VLE use with learning disabled users. Early informal studies looked at whether students with learning disabilities could access VLEs and how they could be used in support of standard teaching methods (Brown et al., 1995; Cromby, Standen & Brown, 1996). Further studies had a more formal and focused aim, such as Crosier (1996), who compared input devices for use with VEs, and Standen, Cromby, and Brown (1998), where transfer of skills from the VLE to the real world was assessed and compared with students that had no VE training. A more structured evaluation approach was adopted by Neale et al. (1999), who examined in detail the process of students interactions with VLE.

This study was based mainly on collection of observational data, as interpretation of subject's feedback was limited (i.e., participants were all young children, aged between 7 and 11, with severe learning disabilities).

Formative evaluation and iterative development of VLEs should ensure a more useable and useful product. In this section, a series of questions are presented, which researchers will face when deciding how to evaluate a VLE. The actual evaluation strategy used will be dependent on the answers to each of these questions; due to the interdependent nature of the questions, this may be a highly iterative process.

### 3.1 What to Evaluate?

The first issue evaluators of VLEs will have to ask themselves will be what do they want to evaluate. One way to determine this is to look at the rationale behind the development of the technology. The evaluator should find out what the VLE is intended to "teach," in which settings, and with which populations. These issues should have been decided in the early stages of VLE development, with the involvement of representative users and teaching/training professionals (as detailed in section 2).

The application of traditional methods for measuring knowledge acquisition has not found clear learning effects in a number of cases. Moshell and Hughes (see chap. 45, this volume) have reported that this "weak learning effect" may be due to a brief period of use of the technology, difficulty of use, and different types of learning (process and strategies rather than facts). Therefore, in deciding what to examine, the researcher may use a pilot study. By examining what happens when the VLE is used, important factors, not taken into account during the design of scenarios to support learning, may be highlighted. For example, the study may focus on usability and attitudes. If the VE is neither usable nor enjoyable it will not be used, and so the educational aims will not be achieved. There may even be a broader range of effects from the use of educational technology (Draper, Brown, Henderson, McAteer, 1996).

### 3.2 When to Evaluate?

A VE may be evaluated a number of times throughout its development cycle. Earlier in the cycle, evaluation may take the form of informal reviews, initially with research colleagues and later more formal meetings with educational professionals and representative end users. Pilot studies early on in the development cycle may be used to highlight the elements important to assess in the project and identify and develop useful methods to do so. These will be applied in more detailed summative evaluations that may take place further into the development cycle (see chap. 34, this volume).

### 3.3 Where to Evaluate?

The two basic settings for evaluation are the laboratory and the field. All the usual advantages and disadvantages of each are relevant, with the added need to consider the effects of either bringing young children or people with special needs into the laboratory or else of the researchers entering their environment. In the laboratory it is possible to control extraneous factors and to manipulate independent variables to allow stringent experimental designs to be followed. This can also be a criticism of this setting, as it is suggested that results often cannot be transferred to the real-world setting or that all the real variables of interest are those that cannot be manipulated in any valid sense. Moreover, for some individuals, the unfamiliarity of the research laboratory can represent a potentially hostile environment in itself, and this could negatively affect their attitudes and learning.

Access to classrooms or working environments for field evaluations may be limited, but can give good ecological feedback for VLE design. The types of evaluation methods used in-situ need to be carefully considered. For example, audio recording may not be successful where there are high levels of background noise. School and college classrooms often provide a challenging space in which to work, as the researcher has to fit in with class timetables as well as contending with other classroom activities and noise. However, field work gives an insight into how the application will eventually be used and may be most useful in making sure the VLE will be practical and usable.

## 3.4 Who to Evaluate?

Deciding who participates in the evaluation study will depend on when the evaluation takes place and the scale of the study, as well as upon what is to be measured. For example, teachers would identify whether educational objectives are satisfactorily represented in the VLE, pupils can provide usability data, and both might provide attitudes towards the VLE. The particular evaluation participants may vary at different stages of the evaluation cycle; different population samples may be used in parallel to look at different aspects of the same VLE.

### *3.4.1 Development Team As Evaluators*

The initial review of the VLE will be executed by the technology development team. This is commonly an informal run-through of the VLE, when bugs are checked for, the running order is verified, and consistency is examined.

### *3.4.2 Representative Users As Evaluators*

There may be difficulties in involving representative users for a number of reasons, for instance:

- Access may be limited due to busy school timetables or resistance from teachers, care workers, authorities etc.
- Insufficient communication skill among the user populations limits subjective feedback.

In spite of these difficulties, there are a great number of benefits to involving representative user populations in many stages of evaluation. Some research projects have tried to minimize the difficulties faced from working with a disabled population by testing VEs with participants not representative of the intended user population. This is often used as a first step in the evaluation process (Jansson, 1998; Jones, Scanlon, & Blake, 1998). It is often both quicker and easier to get colleagues and university students to test the virtual environment. However, the depth and validity of information that may be gathered when representative users are involved in the evaluation process will help to contribute to a more accessible, appropriate, and useful VLE.

### *3.4.3 Educational Professionals*

It may be advantageous to involve educational professionals in the evaluation phase. They have an intimate knowledge of the learning objectives that should be represented in the VLE, are likely to use the VLE themselves, and will be responsible for determining how and when the VLE will be used in-situ. Educational and learning disabilities professionals can respond to interviews and questionnaires, providing a great deal of detailed information in a short space of time.

### 3.5 How to Evaluate?

Once decisions have been made regarding the when, where, who, and what, appropriate evaluation methods must be chosen. A number of different methods may be applied to complement each other and obtain relevant information. As few guidelines exist for the development of VEs, evaluation methods tend to be based on those which have been applied extensively in HCI, education, and social science research (Dane, 1990; Oppenheim, 1992; Preece, 1994; Robson, 1993). We have begun to apply some of these methods and modify them so that they meet more specific VE evaluation needs. Deciding which methods to employ will be dependent on the educational objectives of the VLE, the intended use and user populations, and the time available for evaluation. Time is often at a premium when the evaluation stage of VE development is left until last.

Qualitative data, collected from observation studies, interviews, and questionnaires, can be illuminating, as they are used to describe and interpret what is going on. This information may be particularly useful in the examination of a recently developed VLE, as there may be little previous knowledge as to how the VLE will be used. The majority of evaluation frameworks for educational technologies and VLEs advocate a combination of methods that complement each other (Oliver, 1997). By using different methods, multidimensional aspects of VLE use may be investigated (Roussos et al., 1999).

The current immaturity of VEs in general and VLEs in particular has impacted on the types of evaluations that have been carried out. The VLE evaluations described by Moshell and Hughes (see chap. 45, this volume) reinforce this view, as the most useful results found came from informal observed phenomena, not formal evaluations. However, the field may move further toward that of educational technology evaluation, where formal comparative studies are carried out in order to show that the technology is educationally effective, which is essential if it is to be widely adopted in and educational setting.

## 4. CASE EXAMPLE: THE VIRTUAL LIFE SKILLS PROJECT

Following an early series of VLEs developed for special needs (Brown et al., 1995; Cobb et al., 1993) funding was obtained from the U.K. National Lottery Charities Board to build a virtual city. This aimed to support experiential learning environments within which adults and older children could practice a range of life skills in preparation for independent living (Brown et al., 1999). Three modules were constructed: a house (Fig. 46.13), supermarket (Fig. 46.14),

FIG. 46.13. A virtual house.

FIG. 46.14. A virtual supermarket.

FIG. 46.15. Ordering at a bar in a virtual cafe.

and cafe (Fig. 46.15), which were linked together via a bus transport system (Fig. 46.16). A variety of learning scenarios were placed into each module (26 in total), including: choosing appropriate clothing for the weather conditions; using the bathroom (highlighting issues of privacy); dealing with emergencies (e.g., fire); creating an icon-based shopping list; finding items in the supermarket; paying for goods at the checkout; choosing the right bus; crossing the road safely; putting food away in the kitchen; preparation of a simple meal (including safety issues in the kitchen and use of appliances); and ordering in the cafe.

Content and presentation of these learning scenarios were determined in consultation with a user group comprised of children and adults with learning disabilities who met regularly to brainstorm ideas and feedback on design ideas. This user group was involved with development all throughout the project, culminating in them presenting their own research paper, describing their involvement in the project, at a European academic conference (Meakin et al., 1998).

The involvement of a number of different community groups, including special needs schools and colleges, as well as adults in supported housing, meant that the project involved users from many areas ensuring the final product would have as wide an audience as possible. The user group and steering group also included members from different disciplines. This allowed people from different backgrounds, ages, and abilities to be involved in production, development, and evaluation, thereby having a real influence on the product.

## 4.1 The User Group

Fifteen representative users from the community groups previously mentioned worked together with a facilitator to provide effective advocacy to form the user group. Members of this group

FIG. 46.16. Catching a bus.

were selected to ensure a balance of age, gender, disability, and cultural heritage. The group met every month to brainstorm ideas about what components they thought should be included in a virtual city and what tasks they would like to practice in each part of the virtual environment. The group initially identified 38 modules they wanted to be included, ranging from a cinema and a pub, to a railway station and a health center. The user group decided which of these were important to them and what they wanted to be able to do in each module.

The first drafts of the VE storyboards were drawn at these meetings. The group's ideas were represented by cartoon-style storyboards, which were used to promote discussion about a topic or object using an easy to understand visual design aid.

## 4.2 The Steering Group

A project steering group, was comprised of professionals who work with learning disabled clients or students. They worked for a number of organizations, including those from which the users were drawn. Their main role was to identify specific user requirements and learning objectives appropriate to the skills to be developed. Initial meetings highlighted differences of communication, terms, and focus of interest between the disciplines. This forced the groups to bring consistency to their way of working and use effectively their diverse range of knowledge and experience.

## 4.3 User-centered Design

The ideas generated at user group meetings, represented by storyboards, were taken to the Steering Group. Their role was to identify how suitable learning objectives associated with the scenarios defined by the user group could be conveyed in a virtual environment. A final storyboard, detailing layout, structure, learning objectives, and dialogue was then drawn up for each component.

During the construction phase, members of the user group and steering group regularly reviewed developments to ensure that they were being built within the original vision of the storyboards. When completed, a program of testing was carried out to assess the suitability of these VLEs for learning of independent living skills by the user group. Evaluations were carried out by users and professionals using a variety of measures; including performance of tasks under experimental conditions, observations of usability, questionnaires, and informal interviews. The results produced recommendations for design changes to each of the VLEs.

### 4.4 Evaluation of the Virtual City

Evaluation of the project assessed usability and access of the VEs themselves, as well as skill learning and transfer to the real world. Fourteen participants completed a 7-week experimental program comprising observation of skills in the real environment (supermarket, cafe, or bus), 5 weeks use of the VE, and another visit to the real environment. Results from this and other projects show that despite support worker expectations of usability problems, users with learning disabilities are able to access and use virtual environments, and often take control over navigation and interaction decisions (Neale et al., 1999). High levels of enjoyment and motivation were expressed by users and some evidence of transfer of training was recorded, although it was not expected that high levels of skill learning would be acquired in such a short training period (Cobb, Neale, & Reynolds, 1998).

A contribution to the success of these projects has undoubtedly been the user-centered design approach. This ensures that issues of access and usability are taken into account during development and also that presentation of information is given in an interesting and appropriate manner. An added bonus to the project was that some individuals from the user group expressed personal gain from being involved in a research project (e.g., decision making, discussion, and presentation skills). Meakin et al. (1998), written by the user group, describes the user group's involvement in the Virtual Life Skills project. It is uncommon in the wider scientific community for users to be an integral part of the design process, but in this case, user group involvement ensured the relevance and accessibility of the end product. What proved important was that the users developed a sense of ownership of the project and that their needs and interests were represented in the design of the virtual city.

## 5. CONCLUSIONS

This chapter has discussed the importance of user involvement at all stages of design, development, and evaluation of virtual learning environments. As a culmination of a series of VLE development projects in which professionals from mainstream and special needs education worked with VE developers at VIRART (Brown et al., 1995; Cobb et al., 1993; Crosier et al., 1999; Neale et al., 1999), the Virtual Life Skills project, presented as a case example in section 4, provided the opportunity for a broad range of professionals and users within the community to collaborate in VLE development (Brown, Kerr, & Bayon, 1998; Brown et al., 1999; Cobb, Neale, & Reynolds, 1998; Meakin et al., 1998). An initial User-Centered Design Model, based mainly upon this particular project, is shown in Fig. 46.17 and summarized here.

It can be seen that there are a number of iterative loops within the process. At the outset of the project, the interests of end users and the professionals who will apply the VLE are gathered (via the user group and steering group). These two groups work together to define the objectives and content for the VLE. During this process storyboards are drawn up, and the user group and steering group work with VE developers adapting the storyboards until they provide a specification, which is both technically feasible and meets the requirements of the user group and Steering Group.

Once the VLE specification has been agreed, VE developers start the creation of VE modules. It is important that there is facility for regular review of VLE development to ensure that developers have interpreted storyboards appropriately. It is recommended that review meetings be held at least once every month, but the more frequently project members can review VLE development the better. It is often the case that, when members of the user group and Steering Group actually see their ideas put into effect in VLE development, they change and/or add to their original requirements. This may be because the VLE is not developing as they had

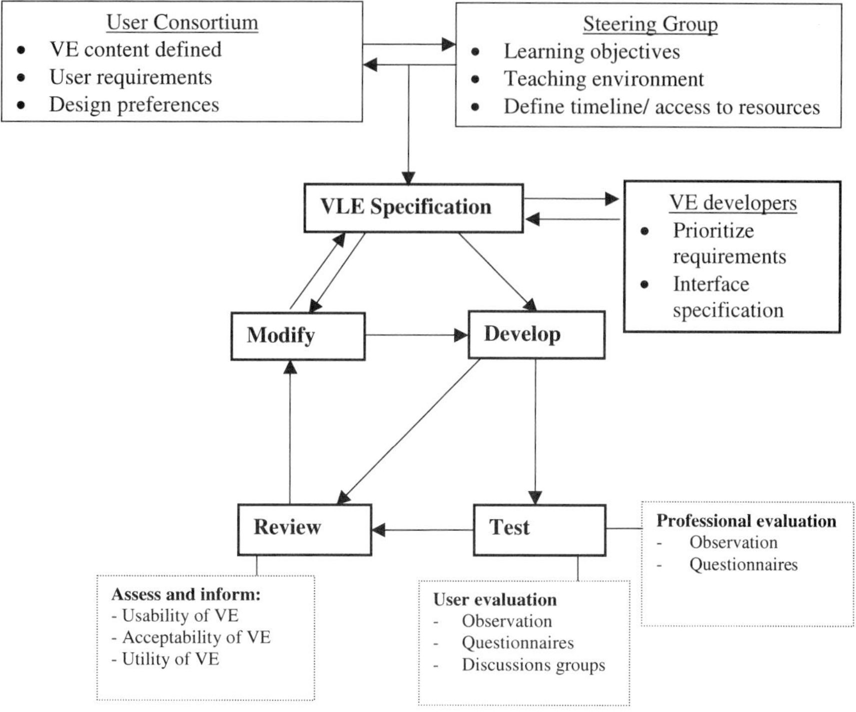


FIG. 46.17. User-centered design and evaluation model.

envisaged, or because they get new ideas as they see what can be done. They may also become more confident in expressing their views as they become more familiar with VE technology. This phase of VLE development can therefore be extremely creative, and it is worthwhile taking time to complete this phase fully.

There is an iterative cycle between VE building, review of VLE development by representatives of the user group and steering group, and modifications made to VLE modules, with reference back to VLE specification, and back into continued development of the VLE. This forum allows what may be the first opportunity for project members to discuss specific features of VLE design. Decisions made at this stage can have a huge impact on usability and acceptance of the final VLE. This chapter has indicated that existing HCI design guidelines are limited in their application for VLE design. Involving users and professionals in regular informal review and evaluation sessions can help to highlight and resolve potential barriers to usability.

When VLE development is complete, formal evaluations can take place. These evaluations feed into VLE review, which can then be used to provide recommendations for additional design modifications. Early evaluation studies provide feedback on general usability and acceptance of the VLE. If this takes place during project development, the user group and steering group will be able to comment on the results and may be able to offer suggestions for design modifications. It is not advisable to leave all of the formal evaluations until the end of the project, as there may be insufficient time and resources to make any necessary amendments.

Evaluation studies may include a number of measures in order to capture information from a number of sources to address questions such as usability, enjoyment, and educational effectiveness. It is important to take context into account, as this may greatly impact upon how the technology is used. For example, the number of users sharing a computer and length of time users have to explore in a classroom situation, as compared with home study, will affect the effectiveness of the final design.

In the case example presented for special needs education, user involvement led to many benefits over and above those provided by the VLE itself. Users expressed positive outcomes relating to increased self-esteem, confidence, and motivation as a result of participating in a research development project (Brown et al., 1999). Moreover, it was clear that users valued the fact that researchers had bothered to ask them what they wanted (Meakin et al., 1998).

The model/framework presented in this chapter has allowed us to develop useful and usable VLEs. However, it will be constantly updated as we refine and improve our ways of developing VLE systems. For example, we plan to design and evaluate new VLEs with increased numbers of user groups from different organizations, cultures, and countries. This may require additional iterations with the model to update or refine for different user groups.

One ongoing project that has adopted a similar framework of development aims to develop VEs to support social skills learning for adults and adolescents with Autistic Spectrum Disorders—the AS Interactive project (Parsons et al., 2000). In the development of the first VE prototype groups of autism training practitioners, teachers, and end users have been involved in design, through workshops, reviews and observation studies. The model has been adapted to meet the specific needs of this project, this has included the development of design criteria derived from expert literature reviews as well as refining evaluation methods so that they are appropriate for autistic users. The framework was also adopted in a three-year research project to develop a virtual environment for teaching radioactivity at secondary schol level. The resulting VE, Virtual RadLab, was developed with the co-operation of thirty teachers from sixteen schools in Nottingham, UK. Teachers were closely involved in the development process, making initial decisions about the most suitable use of virtual reality for support of secondary science teaching—radioactivity—and advising the research team on design and content of the Virtual RadLab (Crosier, Cobb & Wilson, 1999). As well as the highly iterative design process, Virtual RadLab also underwent a series of school-based evaluation studies (Crosier, Cobb & Wilson, in press; Crosier, Cobb & Wilson, 2000). As a results of the research, funding was obtained to produce and distribute 3,000 copies of Virtual RadLab to schools across England. This has allowed teachers and students across the country to experience VEs first hand, and also to recommend areas of application for VEs in the future.

Only with careful design, development, and evaluation of software can an understanding of the impact of well-designed, appropriate, and useful VLEs in the general field of education be gained. Content, its representation in the VE, which takes into account qualities of VE technology and intended users, need to be considered in the development of VLE systems. In this chapter, a framework for carrying out such a development process is introduced. Only when evaluations have shown that the content covered is both appropriate, well presented, and the VLE is usable can the true value of VLEs as a learning tool be comprehended.

## 6. ACKNOWLEDGMENTS

The authors would like to thank all of the users and professionals who have worked with us in the development of virtual learning environments. These include infant, primary and secondary schools in Nottinghamshire, Nottingham Social Services, Mencap, and further education centres supporting special needs. Particular thanks go to David Brown, who initiated the Virtual Life Skills project. Also to David Stewart of the Shepherd School in Nottingham, and Hugh Reynolds of Metropolitan Housing Trust, Nottingham, who have provided us with access and insight into the needs of users within the community. The Virtual Life Skills project was funded by the U.K. National Lottery Charities Board. The work of Joanna Crosier is funded by a Ph.D. studentship from the EPSRC.

## 7. REFERENCES


Bednar, A. K., Cunningham, D., Duffy, T. M., & Perry, J. D. (1992). Theory into practice: How do we link? In T. M. Duffy & D. H. Jonassen (Eds.), *Constructivism and the technology of instruction*. Hillsdale, NJ: Lawrence Erlbaum Associates.

Berka, R., & Slavik, P. (1998). Virtual reality for blind users. In P. Sharkey, D. Rose, & J. I. Lindstrom (Eds.), *The Second European Conference on Disability, Virtual Reality, and Associated Technologies* (pp. 89–98). Skovde, Sweden: University of Reading.

Bricken, M. (1991). Virtual reality learning environments: Potentials and challenges. *Computer Graphics, 25*(3), 178–184.

Bridge, W. (1987). *Microcomputers and special needs education*. Unpublished doctoral dissertation, University of Nottingham.

Brown, D. J., Cobb, S. V., & Eastgate, R M. (1995). Learning in virtual environments (LIVE). In R. A. Earnshaw, J. A. Vince, & H. Jones (Eds.), *Virtual reality applications* (pp. 245–252). London: Academic Press.

Brown, D. J., Kerr, S. J., & Bayon, V. (1998). The development of the virtual city. In P. Sharkey, D. Rose, & J. I. Lindstrom (Eds.), *The Second European Conference on Disability, Virtual Reality and Associated Technologies* (pp. 89–98). Skovde, Sweden: University of Reading.

Brown, D. J., Kerr, S., & Eynon, A. (1997). New advances in virtual environments for people with special needs. *Ability, 19*, 8–11.

Brown, D. J., Kerr, S., & Wilson, J. R. (1997). Virtual environments in special-needs education. *Communications of the ACM, 40*(8), 72–75.

Brown, D. J., Mikropoulos, T. A., & Kerr, S. J. (1996). A virtual laser physics laboratory. *VR in the Schools, 2*(3), 3–7.

Brown, D. J., Neale, H. R., Cobb, S. V. G., & Reynolds, H. R. (1999). Development and evaluation of the virtual city. *The International Journal of Virtual Reality* [Special issue]. Vol 4, no. 1, 28–41.

Brown, D. J., Standen, P. J., & Cobb, S. V. (1998). Virtual environments, special needs and evaluative methods. In G. Riva, B. K. Wiederhold, & E. Molinari (Eds.), *Virtual environments in clinical psychology: Scientific and technological challenges in advanced patient–therapist interaction* (pp. 91–103). IOS Press.

Brown, D. J., Stewart, D., & Mallet, A. (1997). Virtual rooms. *The SLD Experience, 16*, 15–16.

Buck, M., Grebner, K., & Katterfeldt, H. (1996). Modelling and interaction tools for virtual environments. In *Proceedings of the Virtual Reality World '96 Conference*. Stuttgart, Germany: IDG Conference & Seminars.

Byrne, C., & Furness, T. (1994). Virtual reality and education. *IFIP Transactions, Computer Science and Technology, 58*, 181–189.

Clariana, R. B. (1997). Considering learning style in computer assisted learning. *British Journal of Educational Technology, 28*(1), 66–68.

Cobb, S. V. G., & Brown, D. J. (1997). Promotion of health and safety at work using a virtual factory. *Proceedings of Virtual Reality in Education and Training, VRET'97*

Cobb, S. V., Brown, D. J., Eastgate, R. M., & Wilson, J. R. (1993). Learning in virtual environments (LIVE). In *Proceedings of the Science for Life Conference*. University of Keele, England:

Cobb, S. V. G., D'Cruz, M. D., & Wilson, J. R. (1995). Integrated manufacture: A role for virtual reality? *International Journal of Industrial Ergonomics, 16*, 411–425.

Cobb, S. V. G., Neale, H. R., & Reynolds, H. (1998). Evaluation of virtual learning environments. In P. Sharkey, D. Rose, J. I. Lindstrom (Eds.), *The Second European Conference on Disability, Virtual Reality and Associated Technologies* (pp. 17–23). Skovde, Sweden: University of Reading.

Colwell, C., Petrie, H., Kornbrot, D., Hardwick, A., & Furner, S. (1998). The use of haptic virtual environments by blind people. In P. Sharkey, D. Rose, & J. I. Lindstrom (Eds.), *The Second European Conference on Disability, Virtual Reality, and Associated Technologies* (pp. 99–104). Skovde, Sweden: University of Reading.

Copeland, P. (1998). *Development of a virtual power station (VPS) to teach GCSE students*. Unpublished undergraduate thesis, University of Nottingham, Department of Manufacturing Engineering and Operations Management.

Cress, C. J., & Goltz, C. C. (1989). Cognitive factors affecting accessibility of computers and electronic devices. In *Proceedings of the twelth Annual Resna Conference* (pp. 25–26). New Orleans.

Cromby, J. J., Standen, P. J., & Brown, D. J. (1996). The potentials of virtual environments in the education and training of people with learning disabilities. *Journal of Intellectual Disability Research, 40*(6), 489–501.

Crosier, J. (1996). *Experimental comparison of different input devices into virtual reality systems for use by children with severe learning difficulties.* Unpublished undergraduate thesis, University of Nottingham, Department of Manufacturing Engineering and Operations Management.

Crosier, J. (1997). *Information booklet for schools: Virtual reality: Attributes and applications* (VIRART Rep. No. 97/148). Nottingham, England: University of Nottingham.

Crosier, J. K., Cobb, S., & Wilson, J. R. (1999, March). Virtual reality in secondary science education—Establishing teachers' priorities. Paper presented at *CAL '99*, London.

Crosier, J., Cobb, S., & Wilson, J. R. (1999). Virtual Science Experiments—Teacher Defined VR Education. *Paper presented at the CAL '99*, The Institute of Education, London, UK.

Crosier, J. K., & Wilson, J. R. (1998, July). *Teachers' priorities for virtual learning environments in secondary science.* Paper presented at *VRET '98*, London.

Crosier, J. K., Cobb, S., & Wilson, J. R. (in press). Key lessons for the design and integration of virtual environments in secondary science. *Computers & Education.*

Crosier, J. K., Cobb, S. V. G., & Wilson, J. R. (2000). Experimental Comparison of Virtual Reality with Traditional Teaching Methods for Teaching Radioactivity. *Education and Information Technologies, 5*(4), 1–15.

Dane, F. C. (1990). *Research Methods.* CA: Brooks/Cole.

Darrow, M. (1995). Increasing research and development of virtual reality in education and special education: What about mental retardation? *VR in the Schools, 1*(3), 5–8.

Dede, C. (1995). The evolution of constructivist learning environments: Immersion in distributed, virtual worlds. *Educational Technology, 35*(5), 46–52.

Dede, C. (1996). Wired classrooms. *Issues in Science and Technology, 12*(3), 8–9.

Dede, C., Salzman, M., & Bowen-Loftin, R. (1996). ScienceSpace: Virtual realities for learning complex and abstract scientific concepts. In *Proceedings of VRAIS '96* (pp. 246–253). Silicon Valley, Santa Clava, CA, USA.

Desbonnet, M., Cox, S. L., & Rahman, A. (1998). Development and evaluation of a virtual reality–based training system for disabled children, In P. Sharkey, D. Rose, J. I. Lindstrom (Eds.), *The Second European Conference on Disability, Virtual Reality and Associated Technologies* (pp. 177–182). Skovde, Sweden: University of Reading.

Draper, S. W., Brown, M. I., Henderson, F. P., & McAteer, E. (1996). Integrative evaluation: An emerging role for classroom studies of CAL. *Computers and Education, 26*, 1–3, 33–39.

Druin, A. (1999). Cooperative inquiry: Developing new technologies for children with children. In *Human Factors in Computing Systems: CHI, 99* (pp. 223–230). ACM Press.

Druin, A., & Solomon, C. (1995). Designing educational computer environments for children. *CHI '95 Tutorial.* Denver, CO: ACM Press.

Druin, A., Stewart, J., Proft, D., Bederson, B., & Hollan, J. (1997). KidPad: A design collaboration between children, technologists, and educators. In *Proceedings of ACM CHI '97* (pp. 463–470). ACM Press. Atalanta, Georgia.

Duffy, T. M., & Jonassen, D. H. (1992). Constructivism: new implications for instructional technology. In T. M. Duffy & D. H. Jonassen (Eds.), *Constructivism and the technology of instruction*, Hillsdale, NJ: Lawrence Erlbaum Associates.

Edwards, A. D. N. (1995). Computers and people with disabilities. In A. Edwards (Ed.), *Extra-ordinary human–computer interaction* (pp. 19–43). Cambridge, England: Cambridge University Press.

Edwards, A. D. N., & Hancock, R. (1995). Resources. In A. Edwards (Ed.), *Extra-ordinary human–computer interaction* (pp. 361–373). Cambridge, England: Cambridge University Press.

Fry, E. B. (1963). *Teaching machines and programmed instruction: An introduction.* New York: McGraw-Hill.

Goldsmith, N. (1999). *Development and evaluation of an interactive virtual food processing line.* Unpublished undergraduate thesis. University of Nottingham, Department of Manufacturing Engineering and Operations Management.

Hilgard, E. R., Atkinson, R. L., & Atkinson, R. C. (1979). *Introduction to psychology* (7th ed.). New York: Harcourt Brace Jovanovich.

Inman, D. P., Loge, K., & Leavens, J. (1997). VR education and rehabilitation. *Communications of the ACM, 40*(8), 53–58.

Jansson, G. (1998, September). Can a haptic force feedback display provide visually impaired people with useful information about texture roughness and 3D form of virtual objects? In P. Sharkey, D. Rose, J. I. Lindstrom (Eds.), *The Second European Conference on Disability, Virtual Reality and Associated Technologies* (pp. 105–112). Skovde, Sweden: University of Reading.

Jonassen, D. H. (1994). Thinking technology: Toward a constructivist design model. *Educational Technology, 34*(4), 34–37.

Jones, A., Scanlon, E., & Blake, C. (1998). Reflections on a model for evaluating learning technologies, In M. Oliver (Ed.), *Innovation in the evaluation of learning technology* (pp. 25–41). University of North London.

Kaur, K., Maiden, N., & Sutcliffe, A. (1999). Interacting with virtual environments: An evaluation of a model of interaction. *Interacting with Computers, 11*, 403–426.

Kaur, K., Sutcliffe, A., & Maiden N. (1998, December). Applying interaction modelling to inform usability guidance for virtual environments. Position paper for *The First International Workshop on Usability Evaluation for Virtual Environments*, De Montfort University, Leicester, England.

Kerr, S. (1998). *Virtual reality in the assessment of disabled housing*. Unpublished master's thesis, University of Nottingham.

Lumbreras, M., & Sanchez, J. (1998). 3D aural interactive hyperstories for blind children. In P. Sharkey, D. Rose, J. I. Lindstrom (Eds.), *The Second European Conference on Disability, Virtual Reality and Associated Technologies* (pp. 119–128). Skovde, Sweden: University of Reading.

Meakin, L., Wilkins, L., Gent, C., Brown, S., Moreledge, D., Gretton, C., Carlisle, M., McClean, C., Scott, J., Constance, J., & Mallett, A. (1998). User group involvement in the development of a virtual city. In P. Shoakey, D. Rose, J. I. Lindstrom (Eds). *The Second European Conference on Disability, Virtual Reality and Associated Technologies* (pp. 1–9) Skovde, Sweden: University of Reading.

Middleton, T. (1992, January). Advanced technologies for enhancing education. *Journal of Microcomputer Applications*, 1–7.

Minkes, J., Townsley, R., Weston, C., & Williams, C. (1995). Having a voice: Involving people with learning difficulties in research. *British Journal of Learning Disabilities, 23*, 94–97.

Mittler, P. (1979). *People not patients: Problems and policies in mental handicap* Meuthen.

Neale, H. R., Brown, D. J., Cobb, S. V. G., & Wilson, J. R. (1999). Structured evaluation of virtual learning environments for special needs education. *Presence: Teleoperators and Virtual Environments, 8*(3), 264–282.

Noro, K., & Imada, A. S. (Eds.). (1991). *Participatory ergonomics*. London: Taylor & Francis.

Oliver, M. (1997). A framework for evaluating the use of educational technology. Internal Report. London: The Learning Centre, University of North London.

Oppenheim, A. N. (1992). *Questionnaire design, interviewing and attitude measurement*. London: Pinter Publishers.

Papert, S. (1988). The conservation of Piaget: The computer as grist for the constructivist mill. In G. Foreman & P. B. Pufall (Eds.), *Constructivism in the computer age*. Hillsdale, NJ: Lawrence Erlbaum Associates.

Papert, S. (1993). *Mindstorms* (2nd ed.). Harvester Wheatsheaf, New York.

Parsons, S., Beardon, L., Neale, H. R., Reynard, G., Eastgate, R., Wilson, J. R., Cobb, S. V., Benford, S., Mitchell, P., & Hopkins, E. (2000). Development of social skills amongst adults with Asperger's Syndrome: The AS Interactive project. In Sharkey, P. (ed.) *International Conference on Disability, Virtual Reality and Associated Technologies*. University of Reading, Sardinia, pp. 163–170.

Perkins, D. N. (1996). Minds in the 'hood. In B. G. Wilson, *Constructivist learning environments: Case studies in instructional design* (pp. v–viii). NJ: Educational Technology Publications.

Pimentel, K., & Teixeira, K. (1993). *Virtual reality: Through the new looking glass*. New York: McGraw-Hill.

Powers, D., & Darrow, M. (1994). Special Education and virtual reality: Challenges and possibilities. *Journal of research on computing in education, 27*(1), 111–121.

Preece, J. (1994). *Human–Computer interaction*. Addison-Wesley Longman. Harlow, England.

Richmond, W. K. (1965). *Teachers and machines: An introduction to the theory and practice of programmed learning*. Collins.

Rizzo, A. A., Wiederhold, M., & Buckwalter, J. G. (1998). Basic issues in the use of virtual environments for mental health applications. In G. Riva (Ed.), *Virtual environments in clinical psychology and neuroscience* (pp. 21–43). Amsterdom, IOS Press.

Robson, C. (1993). *Real world research*. Oxford, England: Blackwells Publishers.

Rose, H. (1995). *Assessing learning in VR: Towards developing a paradigm virtual reality roving vehicles (VRRV)*. (Tech. Rep. No. TR-95-1). Seattle: University of Washington, Human Interface Technology Laboratory.

Rostron, A., & Sewell, D. (1984). *Microtechnology and special education: Aids to teaching and learning*, Croom Helm.

Roussos, M., Johnson, A., Moher, T., Leigh, J., Vasilakis, C., & Barnes, C. (1999). Learning and building together in an immersive virtual world. *Presence: Teleoperators and Virtual Environments, 8*(3), 247–263.

Scadden, L. A. & Vanderheiden, G. C. (1988). *Considerations in the design of computers and operating systems to increase their accessibility to persons with disabilities* (Working document of the industry/government cooperative initiative on computer accessibility). Madison: University of Wisconsin, Trace Center. Reproduced in part in A. D. N. Edwards (1995), *Computers and people with disabilities*. In A. Edwards (Ed.), *Extra-ordinary human–computer interaction* (pp. 19–43). Cambridge, England: Cambridge University Press.

Scaife, M., Rogers, Y., Aldrich, F., & Davies, M. (1997). Designing for or designing with? Informant design for interactive learning environments. In *Human Factors in Computing Systems: CHI '97* (pp. 343–350). ACM Press. Atlanta, Georgia.

Schank, R. C. (1994, Spring). Active learning through multimedia. *IEEE Multimedia*, 69–78.

Slater, M., & Usoh, M. (1995). Modelling in immersive virtual environments: A case for the science of VR. In R. A. Earnshaw, J. A. Vince, & H. Jones (Eds.), *Virtual reality applications*. London: Academic Press.

Solomon, C. (1986). *Computer environments for children—A reflection on theories of learning and education*. Cambridge, MA: MIT Press.

Standen, P. J., Cromby, J. J., & Brown, D. J. (1998). Playing for real? *Mental Health Care*, 11/12, 412–415.

Stanton, D., Foreman, N., & Wilson, T. N. (1998). Uses of virtual reality in training: Developing the spatial skills of children with mobility impairments. In G. Riva, B. K. Wiederhold, & E. Molinari (Eds.), *Virtual environments in clinical psychology: Scientific and technological challenges in advanced patient–therapist interaction* (pp. 219–233). Amsterdom, IOS Press.

Strickland, D. (1996). A Virtual reality application with autistic children. *Presence: Teleoperators and Virtual Environments, 5*(3), 319–329.

Thalmann, N. M., & Thalmann, D. (1994). Introduction: Creating artificial life in virtual reality. In N. M. Thalmann & D. Thalmann (Eds.), *Artificial life and virtual reality*. Chichester, England: John Wiley & Sons.

Wilson, J. R. (1995). Ergonomics and participation. In J. R. Wilson & E. N. Corlett (Eds.), *Evaluation of human work: A practical ergonomics methodology* (2nd ed., pp. 1071–1096). London: Taylor & Francis.

Wilson, J. R., Brown, D. J., Cobb, S. V., D'Cruz, M. D., & Eastgate R. M. (1995). Manufacturing operations in virtual environments (MOVE). *Presence: Teleoperators and Virtual Environments, 4*(3), 306–317.

Wilson, J. R., D'Cruz, M. D., Cobb, S. V. G., & Eastgate, R. M. (1996). *Virtual reality for industrial application: Opportunities and limitations*. Nottingham, England: Nottingham University Press.

Wilson, J. R., & Haines, H. M. (1997). Participatory ergonomics. In G. Salvendy (Ed.), *Handbook of human factors and ergonomics* (2nd ed., pp. 490–513). Chicester, England: John Wiley & Sons.

Winn, W. (1993). *A conceptual basis for educational applications of virtual reality* (Tech. Rep. No. HITL-TR-93-9). Seattle: University of Washington, Human Interface Technology Laboratory.

Youngblut, C. (1998). *Educational uses of virtual reality technology* (Tech. Rep. No. D-2128). Alexandria, VA: Institute for Defense Analyses.

# 47

# Medical Applications of Virtual Environments*

Richard M. Satava, M.D.[1] and Shaun B. Jones, M.D.[2]

[1] *Yale University School of Medicine*
*40 Temple Street, Suite 3-A*
*New Haven, CT 06510*
*richard.satava@yale.edu*
*and*
*Advanced Biomedical Technologies*
*Defense Advanced Research Projects Agency (DARPA)*
[2] *Uniformed Services University of Health Sciences (USUHS)*
*and*
*Pathogen Countermeasures*
*Defense Advanced Research Projects Agency (DARPA)*

## 1. INTRODUCTION

The benefits of virtual environments (VEs) to health care can be summarized in a single word: revolutionary. The "Information Age" has yet to arrive for medicine, and the infrastructure put in place by other disciplines and industries can be used to leapfrog into the next generation of health care. The most pervasive aspect of VE will be the core technology of interactive three-dimensional (3-D) visualization (see chap. 53, this volume). Those aspects of VE that are relative technologies, such as whether the experience is immersive, augmented (see through), or whether the display is an head-mounted display (HMD), 3-D video monitor, or room-size Cave Automatic Virtual Environment (CAVE), or whether the experience is on a local computer or distributed over the Internet, are analogous to peripherals that can be used to customize the VE application to best suit the health care provider or patient. While there are many different ways to classify the medical applications, one that incorporates the essentials for the practice of medicine provides an excellent framework. The key components are diagnosis, therapy, education and training, and the medical record. However, it is essential that the implementation of these applications is totally integrated and that the full spectrum of health care is incorporated. One method to accomplish these two goals is through the 3-D representation of the patient as a virtual person "medical avatar" (Satava, 1998, p. 29) or Holographic Medical Electronic Representation or

*The information contained herein reflects the opinion of the authors and in no manner represents the official opinion of the Department of the Navy or Department of Defense.

holomer™. The greater the amount of data used to recreate the "information equivalent" of the person, the higher the fidelity of the representation and the more accurate and useful for patient care.

The origin of digitally representing a person derives from two sources: Negroponte's (1995) concept of "bits instead of atoms," whereby an object in the real world (atoms) can be represented in the information world by a computer image (bits), and Ackerman's (1998) program from the National Library of Medicine, the Visible Human, which is an actual person who was scanned in three modalities (computed tomography, magnetic resonance imaging, and phototomography) to provide a full 3-D-information-equivalent computer image (Spitzer & Whitlock, 1992). As more and more of the medical technologies become information-based, it will be possible to represent a patient with higher fidelity to a point that the image may become a surrogate for the patient—the holomer. The practice of medicine can take advantage of the power of the science of modeling and simulation and optimize the patient care by first performing on the surrogate until the best solution is achieved, then providing personalized quality care. The initial accomplishments will be at the macro level for the whole body, organ, and tissue systems; then further levels of glandular structure, molecular, biochemical, and genetic information can be added. With each increase in information, the validity of the model will improve and hence the quality of patient care and value of the model. In order to illustrate the power of this approach, the following scenario is offered as a reference framework of how the total spectrum of care can be completely integrated. For the purpose of identification, this framework is called the Doorway to the Future, and may require 20, 50, or even 100 years to achieve (Satava, 1998, p. 30). It is not possible to ever complete such a project, because modeling the human organism can always have more information derived for greater fidelity.

As a patient visits a physician, they will pass through a doorway in which numerous sensors and imagers are embedded, such as CT or MRI scan, ultrasound, and other technologies. Until this is accomplished, previsit scans will be accumulated. When the person sits at a desk, noninvasive vital signs and biochemical data will be acquired, just as pulse oximetery is used today to acquire heart rate, oxygen saturation, and other values. All information and images will be collected in a database, and then between the physician and patient a holographic image of the person (holomer) will be displayed, just as Dimensional Media Associates suspended holographic image (Chinnock, 1995; Fig. 47.1). This will be the medical record, a visual database with all of the information contained within the image. It is simultaneously the

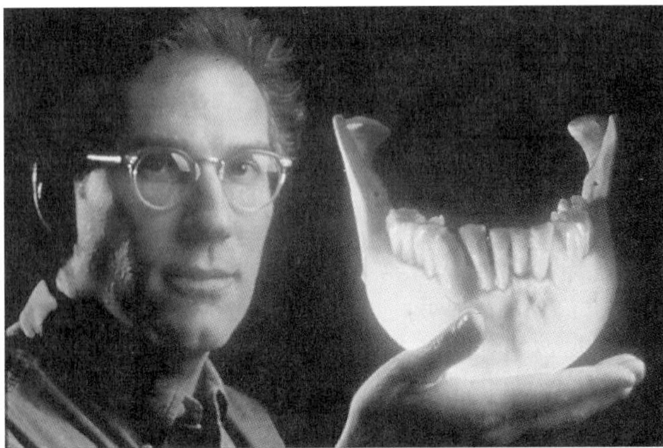

FIG. 47.1 Suspended holographic image of a mandible. Courtesy of Alan Sullivan, Dimensional Media Associates, New York, NY.

interface—an intuitive method of query. After the patient provides the history of illness and physical exam, the physician can interact with the image (similar to visual databases today) to obtain information. If the person complains of right upper quadrant pain, the image can be made transparent and the liver directly queried for relevant laboratory data such as liver enzymes, alkaline phospatase, or bilirubin, as demonstrated by Engineering Animation, Inc. (Sellberg & Kerr, 1998). If there are any abnormalities noted, the physician can use the image for diagnosis by performing a virtual endoscopic procedure to look for abnormalities. There are numerous physicians performing the clinical comparison of conventional video endoscopy to virtual endoscopy (see below, virtual endoscopy). If an abnormality is found, the physician can use the image for patient education to show and explain the patient's problem—the image is now a patient education tool. If the abnormality will require a complicated operative procedure, the image can be used for preoperative planning, as done by Taylor, Hughes, and Zarins (1998), or imported into a surgical simulator, like Levy (1996), and used to optimize the surgical procedure for that specific patient. At the time of surgery, the image can be imported into the procedure using data fusion with a video image of real-time surgery, providing assistance with intraoperative stereotactic navigation, as performed at the Brigham Women's Hospital by Jolesz and Shtern (1992). Finally, when the patient returns for follow-up weeks after the procedure, a complete scan is once again performed, the database will perform data fusion and digital subtraction like E. Grimson (personal communication, 1997), with the difference being automatic outcomes analysis. The power of this scenario is that it provides a framework for the technologies that encompasses the full spectrum of health care, and integrates the information equivalent of the patient into a single record continuous across time. It can be updated as needed and made available to the patient on a personal credit card sized record, like the military's personal information card, or perhaps contained on a secure webserver on the Internet and available for global consultation through telemedicine.

The VE technologies and applications that support this framework can be discussed within the context of the provision of medical care, mainly for diagnosis, therapy, and education and training. However, there is overlap in many of these areas, for education and training is increasingly being embedded into the actual devices that are being used for diagnosis and therapy. An example is ultrasound (Fig. 47.2), where the real-time image from a procedure is replaced by an archived image of an actual patient, and the student scans a mannequin as an ultrasound simulator with the same equipment and in precisely the same manner as an actual patient (Meller, 1997). It is imperative to understand that the utilization of VEs is in its infancy, and the following applications are just the beginning of a new direction. Many are in the laboratory investigation, beta prototype, or clinical investigation stage; few have been inserted into routine clinical practice. The furthest developed of the applications are undergoing rigorous and stringent testing and evaluation for technical and clinical efficacy, for Food and Drug Administration (FDA) approval and cost-effectiveness. It is unclear which will succeed in this arduous process for full-time clinical practice.

## 2. DIAGNOSIS

Diagnosis using virtual endoscopy is one of the areas that will achieve clinical efficacy in the earliest time frame. The National Institutes of Health and the National Cancer Institute have targeted virtual colonoscopy as the pilot study to compare standard video colonoscopy to virtual colonoscopy for colon polyps and cancer screening. Should this prove effective, virtual endoscopies of any of the other organ systems will follow. There are over two million standard video colonoscopic procedures performed each year, and over 75% are "normal," the majority due to screening for cancer. A virtual endoscopic procedure is performed by using a

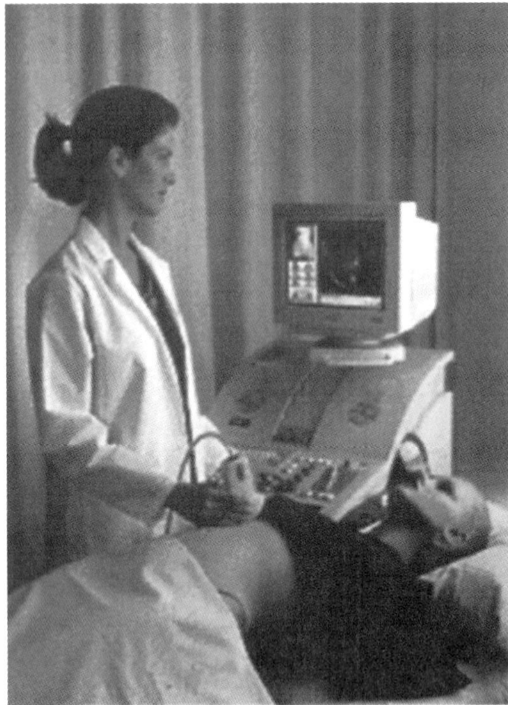

FIG. 47.2 UltraSim, the ultrasound simulator, displays patient-derived ultrasound anatomy and pathology on a monitor in response to motions of an actual transducer, precisely mimicking an ultrasound examination. Courtesy of Garrick Herrmann, MediSim, Inc., Ft. Lauderdale, FL.

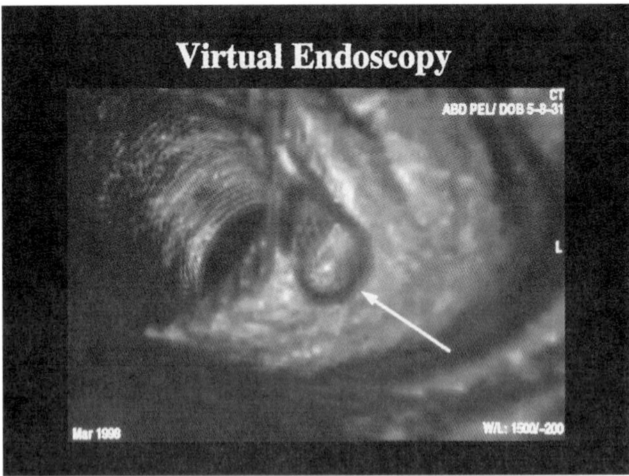


FIG. 47.3 Virtual colonoscopy. Courtesy of Dr. James Brink, M.D., Yale University School of Medicine, New Haven, CT.

standard CT scan or MRI scan, reconstructing the organ of interest into a 3-D model, and then performing a fly through of the lumen (Fig. 47.3). Typical examples of effected organs include the colon, stomach, esophagus, tracheo-bronchial tree (bronchoscopy), sinus bladder, ureter and kidneys (cystoscopy), pancreas, or biliary tree. There are numerous advantages to such a diagnostic tool as compared to inserting actual instruments (endoscopes) into the body. First, all endoscopic procedures are invasive, and thus the patient must be sedated and during the procedure they are subject to complications (small though they are) such as perforation, bleeding, and others. The cost for a typical colonoscopy is significant, for it must include a dedicated suite, personnel, specialized equipment, consumable materials, medications, and administrative staff. A virtual colonoscopy is constructed from a helical CT scan of the abdomen, is completely noninvasive, and thus without known complications. The actual cost is less than one third of video colonoscopy because it is performed in the same place and manner as all imaging modalities, utilizes the same staff, and has no consumable materials. The advantages of the video colonoscopy is that a therapeutic procedure can be performed if an abnormality is found, whereas any abnormality on a virtual colonoscopy must be referred for true colonoscopy for the therapy, such as polypectomy or biopsy of suspected cancer. Virtual colonoscopy has the advantage of being able to view the colon from any angle inside the lumen (small polyps can be hidden behind folds or twists in the colon during actual colonoscopy), or being able to fly "outside" of the lumen to see if a tumor has spread to the surrounding tissues or lymph nodes. In addition, the virtual image can be manipulated, actually opening up the tubular structure and lying it flat like a terrain map, to look for abnormalities (Meller, 1997). The wall thickness can be measured for early flat cancers, not usually detectable by standard colonoscopy (S. Napel, personal communication, 1999).

The limitations to virtual colonoscopy at this time are significant and will need substantial research to reach their full potential. These limitations include accurate color and texture mapping of the tissues (Hohne and Bernstein, 1986). A preliminary step has been taken by John Kerr of Engineering Animation, Inc. (Sellberg & Kerr, 1998), in which a color look-up table has been created using the visible human data set, comparing Hounsfield units of the CT scan to color values from the phototomographs. The preliminary images, using a liver model, are promising but are not accurate enough to be useful for clinical diagnosis. Just as the early attempts to convert black-and-white cinema films into colored films have progressed to excellent representations today, advances need to be initiated to create accurate color and texture for endoscopic images. Another limitation is removal of all the stool within the colon to prevent false diagnoses of abnormalities. Solutions under investigation include more effective purgatives or drinking a fluid that will tag the stool, allowing it to be digitally subtracted during the virtual reconstruction. Finally, the level of resolution of the image today is at 0.3 mm for the helical CT scanners. This is usually adequate for gross lesions, though full 3-D reconstruction does not give this accuracy. For lesions greater than 3 mm in size there is a 75% detection rate, and those greater than 5 mm have a 95% detection rate (Beaulieu et al., 1998). This is becoming comparable to detection rates of standard colonoscopy. As imaging modalities improve, so to will resolution and detection rate. Perhaps next generation systems will acquire multiple modality images (CT, MRI, ultrasound, etc.), and then using data fusion, provide more information than can be obtained with a single image. The extraordinary challenges of automatic segmentation, registration, accurate tissue identification, and data fusion provide nearly unlimited research opportunities, which can provide both early and long-term benefits for clinical application to patient care.

Another diagnostic application is that of preoperative planning using the virtual image of patient specific data. There are numerous areas that have had initial investigation, especially in the neurosurgical, orthopedic, maxillo-facial , plastic surgery, and vascular arenas. Stereotactic neurosurgery has been clinically implemented for over a decade, and central to its success has been accurate localization of abnormalities in the brain through 3-D reconstructions from CT and MRI scans. Using fiducial markers on the skull or by placement into a frame that provides stereotactic reference, neurosurgeons have full 3-D models of the brain to precisely plan both the exact position, as well as safe pathway to removal or ablation of deep-seated brain lesions (Fig. 47.4). The current generation systems are using bony landmarks of the skull as frameless fiducial references for the preoperative model, and then updating the image with real-time open MRI scanning as the stereotactic neurosurgical procedure is being performed (Beaulieu et al.,

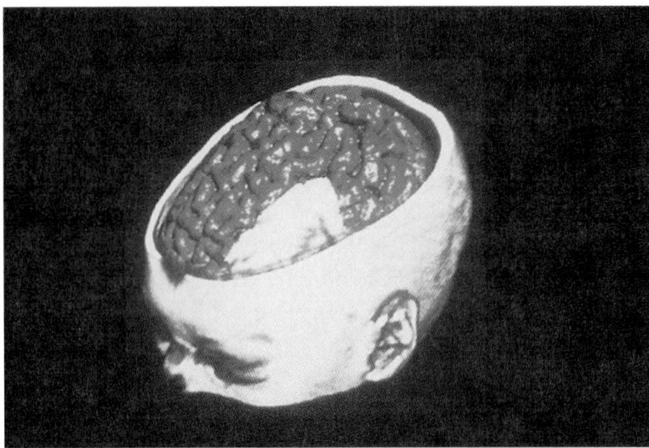

FIG. 47.4 Three dimensional representation of a patient with a brain lesion. Courtesy of Dr. James Duncan, Yale University School of Medicine, New Haven, CT.

FIG. 47.5 RoboDoc, precision surgical robot for hip replacement surgery. Courtesy of Dr. William Barger, M.D., University of California, Davis.

1998; Jolesz, 1997). In orthopedic surgery, the areas of hip and knee replacement are maturing as clinical tools. One of the earliest robotic surgical devices, RoboDoc for hip replacement (Fig. 47.5), uses a computer program called OrthoDoc to preplan the procedure (Paul, 1992). Taking orthogonal views of the patient's hip as well as manufacturer supplied models of the prosthesis, an accurate preoperative match can be obtained. The coordinates of the placement for the prosthesis are fed into a robotically controlled system to bore out the center of the bone to mount the prosthesis. Normally, a surgeon using handheld surgical instruments can create a cavity into which the prosthesis fits with approximately 75% of the surface contact to the bone. Using RoboDoc, the cavity can be precision crafted so the prosthesis has 96% surface contact, providing a better fitting and longer lasting replacement. Similarly, Anthony DiGioia of Shadyside Hospital in Pittsburgh is doing preoperative planning for replacements using HipNav (DiGioia, Jaramaz, & Colgan, 1998). For maxillo-facial reconstruction of congenital or traumatic abnormalities, Altobelli et al. (1993) of Brigham and Women's Hospital and recently Montgomery, Stephanides, Schendel, and Ross (1998) of Stanford University have been using 3-D images from CT scans. The face is a symmetrical structure, so abnormalities on one side of the face or skull have a mirror image portion of the face that is normal. By comparing the normal side of the face to the defect, an accurate reposition or replacement for the malformation can be planned. In the area of defect, the mirror-image model can be exported to a computer-assisted design and computer-assisted manufacturing (CAD/CAM) system, which can precisely manufacture an appropriate prosthesis for implantation and reconstruction. The new generation systems are able to use 3-D stereolithography to automatically construct a 3-D prosthesis directly from the CAD/CAM information (Bill et al., 1995). Drs. Pieper and Rosen of Dartmouth University (Pieper, Laub & Rosen, 1995) have developed a preoperative planning tool for plastic surgery reconstruction of skin lesions on the face. Using finite element modeling of the layers of the skin, portions of the skin can be removed, and then the results of reapproximating the skin edges using different techniques can be visualized for optimizing cosmetic appearance. Liver surgery is a very complicated and demanding procedure because of the multiple different pathways for arteries, veins, the portal system, and biliary tree. Marescaux et al. (1998) of the Telesurgical Institute of Strasbourg, France, is using patient-specific CT scans of the liver to preoperative plan these complicated procedures (Fig. 47.6). In vascular surgery, Taylor and Zarin of Stanford University (Taylor et al., 1998) have approached the difficult problem of choosing the best vascular grafting for complicated vascular lesions of the aorta and lower extremities. From digital MRI angiography, 3-D representation of the vascular tree, patient specific lesion, and fluid dynamics (flow) can be modeled. With this specific anatomy, as well as characteristics of the vascular graft prosthesis, numerous different possibilities for bypassing specific vascular obstructions can be tried, with each model predicting the probable flow to the lower extremities. The accuracy of these predictive models will soon be validated using animal and clinical trials.

The power of preoperative planning is analogous to testing and evaluation through computer simulation in many industrial sectors, from aerodynamics and ergonomics in the aerospace and automobile industries to building structure and interior appearance in the architecture and design fields. The ability to try numerous different possibilities, optimize the procedure for the

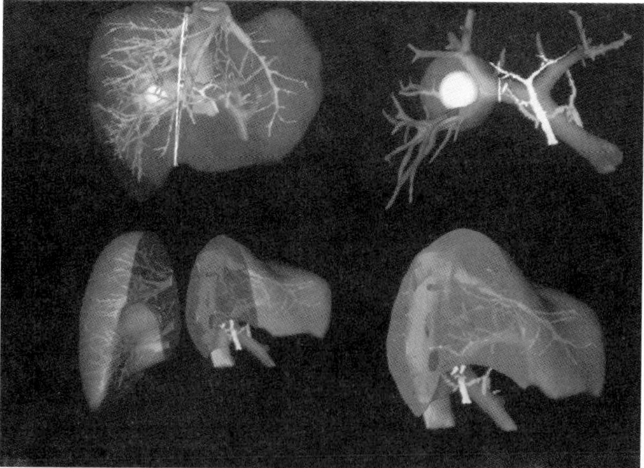

FIG. 47.6 Liver surgery simulation demonstrates any of the various hepatic anatomies, as well as patient-specific pathology. Courtesy of Dr. Jacques Marescaux, Telesurgical Institute, Strasbourg, France.

best choice, and then apply that choice to the patient will significantly increase the quality of health care. The question remains as to whether it is cost-effective and who will pay for the added expense of the preoperative planning. Without these computer assisted planning tools, the cost of "planning" an operative procedure is included as the physician's fee, the payment for a surgical procedure includes the surgeon's experience in choosing the best option based on personal experience. It is uncertain if third-party payers for health care will be willing to pay extra for very expensive tests (CT-generated 3-D representations) and planning procedures unless cost savings or long-term improved outcomes from the preplanning of the surgery can be a demonstrated.

## 3. THERAPY

In the area of therapy, virtual environments usually include computer-assisted surgery, image guided surgery, and telesurgery, because the interface through which the procedure is being performed includes an image of the patient rather than directly viewing the patient's anatomy. The display is often simply a monitor; however, head-mounted displays or overlay screens have also been used (see chap. 3, this volume). In computer-assisted surgery, the purpose is to obtain preoperative planning information that is entered into a computer program, and during the procedure the surgeon is assisted by a robotic device to perform a part of the procedure beyond human physical limitations. An example is RoboDoc mentioned above, with 96% accuracy in prosthesis placement. Telesurgery, or more accurately dexterity enhanced surgery, uses a video monitor with stereoscopic real-time video image to enhance the precision and dexterity of the surgeon through direct telemanipulation. Under these circumstances, the surgeon's full attention is on the virtual image of the patient image, and remote manipulators carry out precisely the hand motions of the surgical task. The system is in the same operating room next to the patient for the purpose of improving the performance of the surgeon (dexterity enhanced surgery); however, the procedure could just as easily be at a distant place (telesurgery, see chap. 48, this volume). The systems are comprised of a remote manipulator system with surgical instruments that are mounted (and can be interchanged) on the end of a pair of robotic arms. The instruments are placed inside the body in the proper position by the surgeon, who then sits at a surgical workstation next to the patient and uses input devices (modified handles from surgical instruments) to directly control the end effectors. The systems include force feedback for haptic response, 3-D vision on the monitor for depth perception, scaling of hand motion to increase dexterity, and filtering of tremor (at the 8–14 Hz frequency) to increase precision. Research is being conducted to include motion tracking so the system can pace the motion of the beating heart and therefore perform surgery on a beating heart without needing to place the patient on an heart–lung machine while performing the procedure. There have been no remote telesurgery procedures performed on patients; however, two systems have performed the procedures in the same room as the patient. In 1997, Himpens and Cardiere in Brussels, Belgium (Himpens, Leman, & Cardiere, 1998), used the Intuitive Surgical, Inc., system to perform the first telesurgery gall bladder operation on a patient. They also performed a number of other procedures, such as Nissen fundoplication and arterio-venous fistula construction. In 1998, Carpentier, Loulmet, Aupecle, Berrebi, and Relland (1999) of Paris, France, began performing heart operations using this system and has successfully performed over 150 of the procedures. A second system by Computer Motion, Inc., called Zeus is similar to the above system, but focuses more on performing microsurgical procedures (Fig. 47.7). In 1998, Margossian et al. (1998) performed the first procedure in the United States with a reanastomosis of a fallopian tube using the Zeus system. Both systems are currently undergoing controlled clinical trials on selected patient populations.

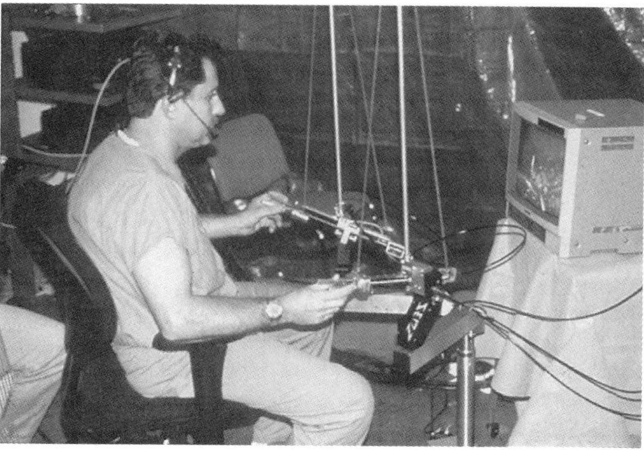

FIG. 47.7 The telesurgery system Zues, which scales hand motion and removes tremor. Courtesy of Yulun Wang, Computer Motion, Inc., Goleta, CA.

Psychiatry and rehabilitation are areas where immersive VEs have had important clinical applications (see chaps. 49–51, this volume). Lamson (1994) of Kaiser Permanente has been using VE for what he calls "virtual therapy" in patients with various forms of phobia. By recreating representations of the Golden Gate Bridge or an elevator, people with fear of heights can be deconditioned by repeatedly being exposed in small increments to the virtual threatening environment and allowed to conquer their fears without physical danger to themselves. After 1 year there is a 85% success, and some of the patients who regress are easily given further virtual therapy and they respond very well. A similar system has been developed at Georgia Technical Institute by Hodges et al. (1995), creating a variety of different environments to test acrophobia and other phobias. An interesting success has been achieved with a patient with arachnophobia (fear of spiders) by Rothbaum et al. (1995) at the University of Washington. In rehabilitation, Greenleaf and Tovar (1994) have provided a virtual world for wheelchair patients to travel through. By giving the patient ample opportunity to practice without harm to themselves, they can then begin using a wheelchair in a real environment.

An interesting project by D. Zeltzer (personal communication, 1999) is beginning called the Empathy Network. In this project, the concept is to provide the patient's family and loved ones an opportunity to experience the disability that the patient is suffering, such as memory lapses or visual disturbances secondary to minimal brain injury or stroke. A commercial project entitled In My Steps has been successful in creating the fatigue that a cancer patient suffers while under chemotherapy (Moran, 1998). The family member dons an HMD and sits on a stepping machine. The scenario is in the patient's home (Fig. 47.8), where the patient must prepare a small breakfast while waiting for delivery of important prescriptions. The person must pedal (walk) to the kitchen and put on a tea kettle and toast. By the time they have "walked" to the kitchen and put the kettle on the stove, they are fatigued (from the heavy pedaling). The door bell then rings. Even though they are pedaling hard, their walking progress is extremely slow, and by the time they open the door they see the prescription delivery truck departing without leaving their medications. At that moment the tea kettle begins whistling, so they must pedal back to the kitchen, and just as they reach the kettle the toast begins to burn. By this time the person is totally exhausted from the pedaling. There is no empiric data from this project; however, the subjective reports have had uniformly high ratings for imparting the sense of fatigue and frustration that their loved ones must be suffering.

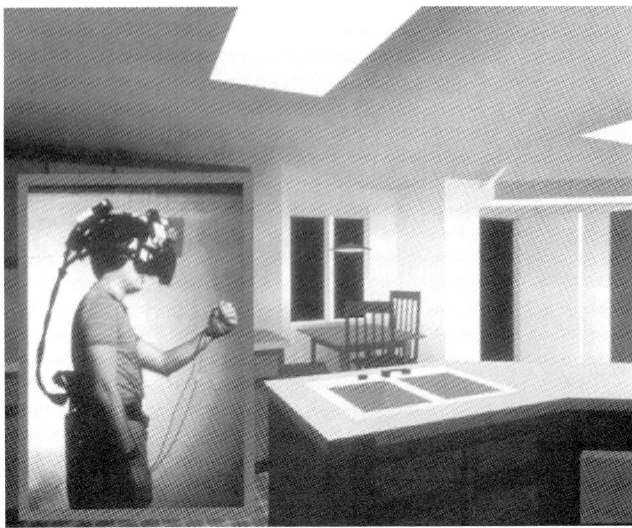

FIG. 47.8 An earlier representation of a virtual kitchen typical of the environments that could be used for various psychiatric and rehabilitative therapies. Courtesy of Jaron Lanier, VPL, Inc, Redwood City, CA.

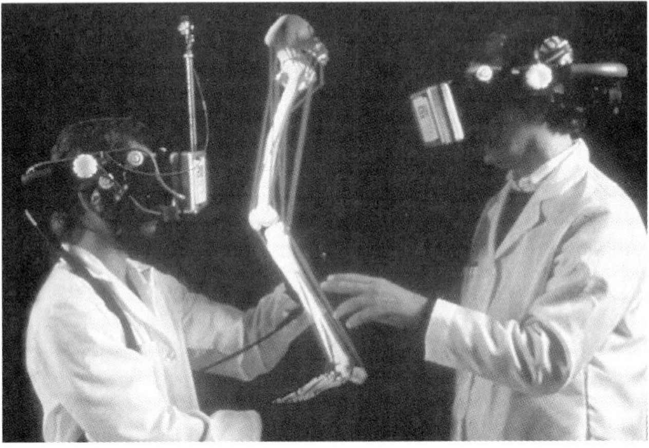

FIG. 47.9 Tendon Transplant simulator with superposition of model of leg that is actually being seen on the head-mounted displays. Courtesy of Dr. Joseph Rosen, Dartmouth Medical Center, Hannover, NH.

## 4. EDUCATION AND TRAINING

Education and training is the largest early application (see chaps. 43–46, this volume). In the beginning of the 1990s, Delp and Rosen (Delp & Zajac, 1992) created one of the first VE applications in medicine—a tendon transplant simulator (Fig. 47.9). Using an HMD and DataGlove as the interface, they created a mechanical and anatomic model of the skeleton of the lower limbs that used red lines to represent muscles and tendons. The limbs had mechanical properties and kinematics of the legs, which could be animated to simulate walking. By moving the insertion point of a muscle's tendon to a different point (to plan for a tendon transplant operation to correct a patient with a gait disorder), the model was then animated and

FIG. 47.10 Cholecystectomy simulator based on graphic drawings of upper abdominal organs. Courtesy of author.

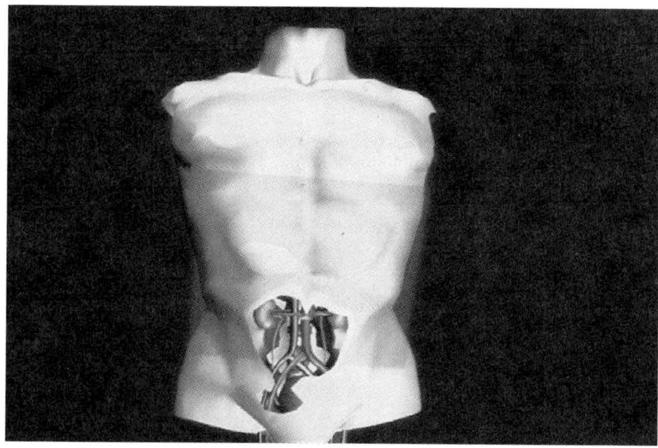

FIG. 47.11 Higher resolution graphic representation of the abdomen with morphing. Courtesy of Gregory Merrill, HT Medical, Inc., Rockville, MD.

the gait observed to predict what the result of the tendon transplant might be. In 1991, Satava and Lanier (Satava, 1993) made the first abdominal simulator (Fig. 47.10), created of simple graphic representations of the liver, stomach, colon, pancreas, gall bladder, and biliary tree. Once again, with an HMD and DataGlove, it was possible to fly around the organs and pick up and move them. A few simple surgical instruments were included, allowing for a very abstract representation of the surgical procedure of removing a gallbladder. By 1993, Merrill of High-Techsplanations had added increased graphic realism and introduced morphing to demonstrate changing of tissues in response to interaction (Merrill et al., 1995). While appearing much more realistic, the deformations were based on approximations rather than measurement of properties of actual tissues (Fig. 47.11). At about the same time Sinclair et al. (1995) developed a very practical ophthamology simulator, demonstrating that a sophisticated operative procedure such as removal of a cataract could be realistically reproduced (Fig. 47.12). The landmark was 1994, with Ackerman's (1998) program from the National Library of Medicine and the release of the Visible Human Project (performed by Drs. Spitzer and Whitlock [1992]). For the first time visual representation was based on actual anatomic data from a person rather than graphic

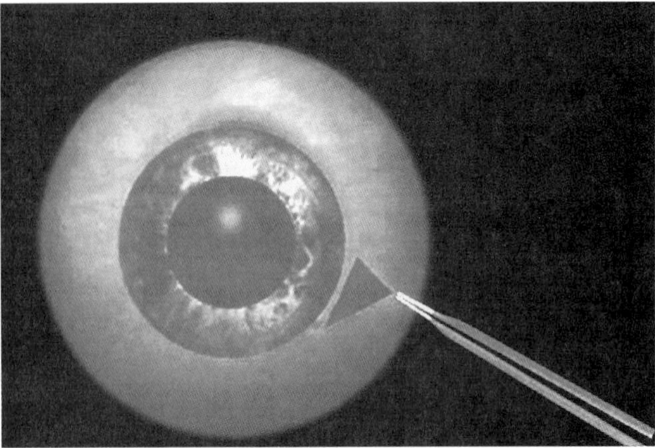

FIG. 47.12 Eye surgery simulation of removal of a cataract. Courtesy of Drs. Sinclair and Peifer, Georgia Technology Institute, Atlanta.

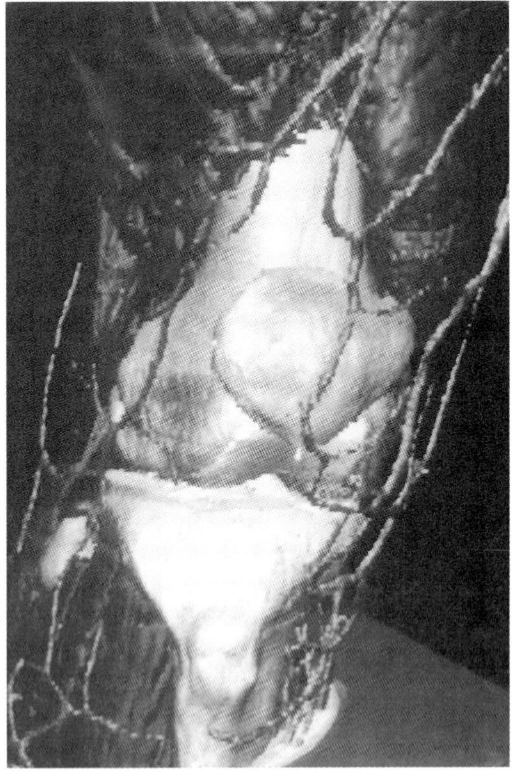

FIG. 47.13 Anatomically precise reconstructed knee from the Visible Human Project. Courtesy of Dr. Victor Spitzer, M.D., University of Colorado Medical Center, Denver.

approximations of organs and tissues (Fig. 47.13). However, the database is massive—over 150 megabytes—so at the present time it is not possible to render the entire representation in real time interactive simulation. But the Visible Human does provide a standard reference for interoperability of different simulations.

The next step was with Delp's (1998) Limb Trauma Simulator for the military Combat Casualty Care training program. Taking the upper thigh from the Visible Human and adding

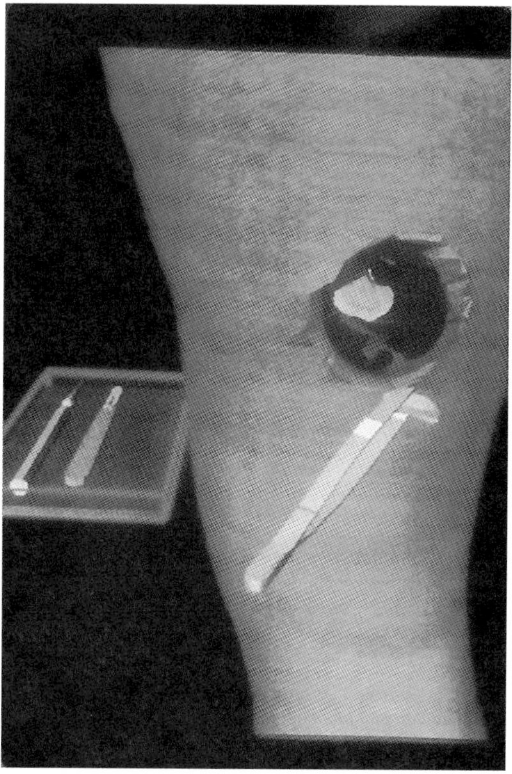

FIG. 47.14 The Limb Trauma Simulator with gunshot wound to the thigh. Courtesy of Dr. Scott Delp, Ph.D., Musculographics, Inc., Chicago, IL.

to it the properties of muscle, ballistic damage, shrapnel, bleeding, and other parameters of wounding, a simulation of a gunshot wound to the thigh was created (Fig. 47.14). The purpose was to provide an alternative to wounding animals for training in combat casualty care. The simulator provides both visual representation of the wound as well as interactive control of bleeding, debridement, and hemostasis using input devices with force feedback for haptic control. However, by the time all of the interactive properties and haptic interface (see chaps. 5 and 6, this volume) have been added, the visual fidelity of the image is reduced to a relative low-resolution representation rather than photorealistic.

As computer power increases, virtual representation will eventually have both visual fidelity, with physical-based properties derived from scientifically measured tissue and real-time interactivity at a minimum of 30 frames per second with a latency of image generation of less than 50 milliseconds. The Limb Trauma Simulator is currently under evaluation at the U.S. Army medical training laboratory of the Uniformed Services University of Health Sciences (USUHS) by Kaufmann, Rhee, and Burris (1999). In 1996, Levy collaborated with Engineering Animation, Inc., to create a hysteroscopy simulator (Levy, 1996). This system not only integrated visual and haptic input, but the image was derived from patient-specific data (Fig. 47.15). Before performing a hysteroscopic procedure on a patient, Dr. Levy imported the image from the patient's CT scan and was able to practice numerous different procedures to optimize the surgery for that specific patient. By making the task to be performed rather simple and developing the software for a personal computer platform, Merrill et al. (1995) of HT Medical created a photorealistic simulation of a central venous catheter insertion (Fig. 47.16), demonstrating that it is possible to produce a commercial system that is both realistic and cost-effective. Using training methodologies for terrain following and target acquisition

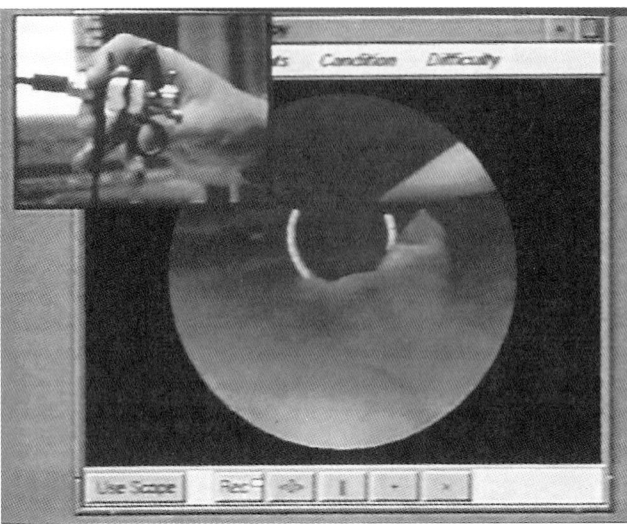


FIG. 47.15 The Hysteroscopy Simulator demonstrates patient-specific imagery for practice on an individual patient's pathologic anatomy. Courtesy of Dr. Jeffrey Levy, Engineering Animation, Inc., Ames IA.

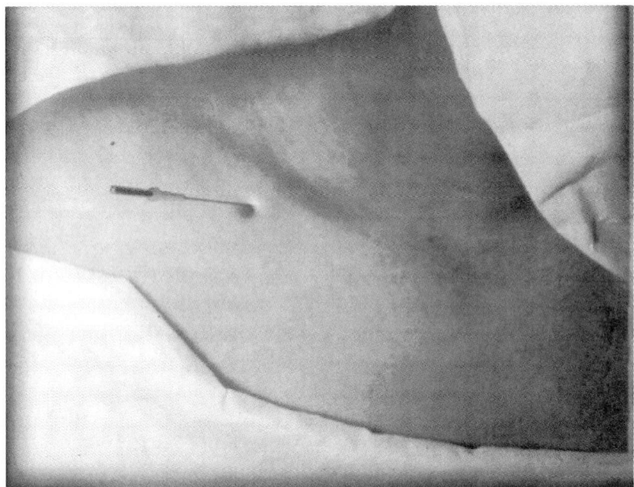

FIG. 47.16 The IV insertion simulator demonstrating high visual fidelity and haptic input on an inexpensive personal computer–based platform. Courtesy of Gregory Merrill, HT Medical, Rockville, MD.

developed by the military, Edmond, Wiet, and Bolger (1998) of the University of Washington have introduced the graphics overlay as an aid to instructing a student on the correct path for a surgical procedure. A sinus surgery simulation reconstructs the nasal cavity (Fig. 47.17) with all its irregular protrusions (turbinates), and a series of circles is used as the "cross hairs" for the student to follow in order to navigate through the nasal passages to a designated pathology (target). This is very similar to the navigation system used by fighter pilots to follow terrain and identify targets of interest.

By 1998, Raibert and Playter of Boston Dynamics, Inc., constructed the Anastomosis Simulator (Raibert, Playter, & Krummel, 1998). In addition to increasing visual fidelity and tissue

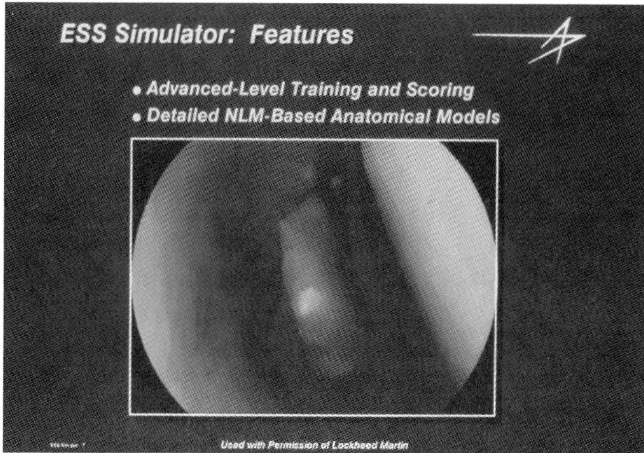


FIG. 47.17 Sinus endoscopy simulator provides internal anatomy of the nasal sinuses with overlay of circles that are path planning aids to endoscopic navigation for a surgical procedure. Courtesy of Dr. Charles Edmond, University of Washington, Seattle.

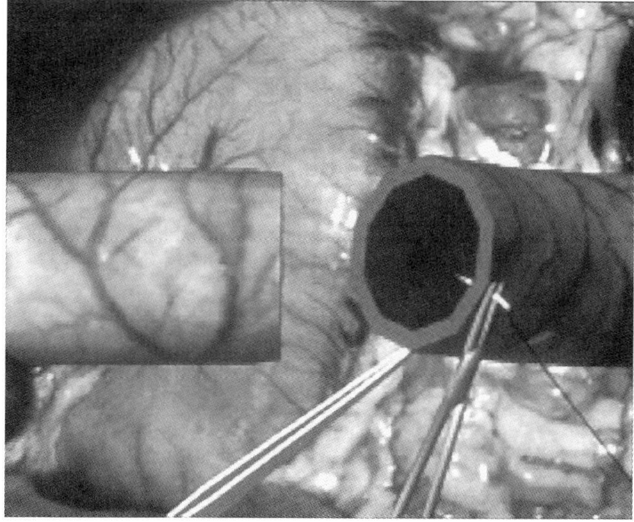

FIG. 47.18 The Anastomosis Simulator incorporates real-time tracking of hand motions by the graph in the upper left corner to give continuous feedback to the student on level of performance. Courtesy of Dr. Marc Raibert, Boston Dynamics, Inc., Boston, MA.

interaction with haptic input, this system incorporates a testing and evaluation component (Fig. 47.18). The input devices (Phantom Haptic Input Device), which track the precise motion, position, pressure, and timing in order to provide accurate force feedback, also output the information to a graphic display that tracks in real time the student's performance in regard to accuracy, position, pressure, and time to completion. At the end of the practice session, a scorecard of the student's performance is tabulated and printed out. This can be used both for real-time feedback for training, as well as evaluation for certification. The next steps are to provide a curriculum that will support the simulation technology and to develop metrics for evaluation of performance and outcomes analysis. By working with the American Board of Orthopedic Surgeons, Raibert has developed a first-generation arthroscopy simulator (Fig. 47.19),

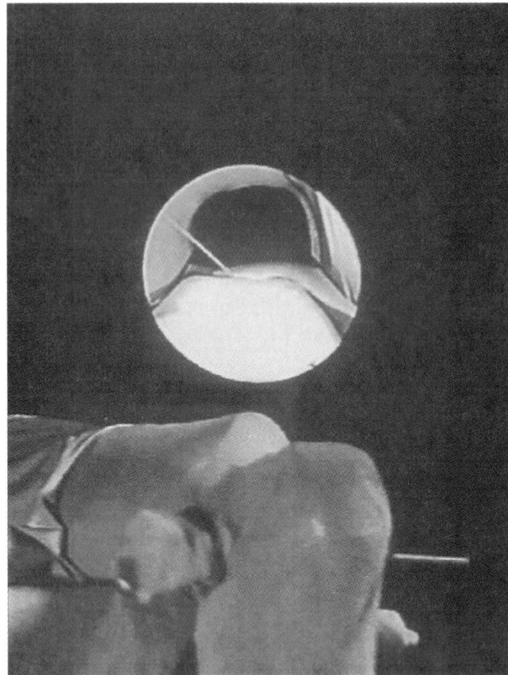

FIG. 47.19 Knee Arthroscopy simulator with mannequin leg and high-fidelity internal knee anatomy. Courtesy Dr. Marc Raibert, Boston Dynamics, Inc., Boston, MA.

thus involving those authorities who are involved in determining the efficacy of simulation for testing and certification.

There are a number of projects that focus at medical-student-level education. Imielìnska at Columbia University (Imielìnska, Laino-Pepper, & Thuman, 1998) of the Vesalius project is providing a combination of CD-ROM and Web-based VEs for pelvic and skull anatomy. Heinrichs of Stanford University (Heinrichs & Dev, 1999) is leading a team to understand pelvic and gynecologic anatomy through VE system training. Hoffman of University of California, San Diego (Hoffmann & Murray, 1999), has a curriculum based on a virtual representation of the organ systems called the Virtual Reality Multi-Media System, or VirtualizeR. The module for the biliary tree is complete, with a full 3-D hepato-biliary system (liver, gallbladder, and bile ducts) and interactive links to all relevant information. For example, while highlighting the gallbladder, the student can access normal and abnormal anatomy, typical patient history and physical exam, histology, pathology, radiologic images, surgical procedures, and a full range of other information, as well as the capability to fly through or practice procedures on the anatomic representations. The image is both the educational content and the interface to the information.

## 5. CHALLENGES

There remain a number of challenges and barriers to implementing VEs more extensively in medical applications. In all areas, there is the issue of cost and maturity of the systems. Most of the applications are still simplistic, requiring a substantial amount of "suspension of disbelief." The equipment is still rudimentary, from HMD to CAVE (see chaps. 3 and 11, this volume) to haptic input device (see chaps. 5–6, this volume). Nearly every system requires a dedicated staff or at least a computer technician to keep the system running smoothly. The majority of applications do not have a large number of patients or students for which it is applicable. The following looks at individual areas and identifies specific challenges.

In order to create image-based medical records (holomer™), there is the challenge of acquiring enough data noninvasively and automatically inserting it into a database. There is neither a consensus database, nor are the literally hundreds of databases interoperable. In terms of displaying the information as an image, there are the challenges of automatic segmentation of the CT, MRI, and ultrasound images, reconstruction of 3-D structures, accurate registration of images, and then data fusion of not only image data, but embedding the physiologic, biologic, and other data to the appropriate source. If the intent is to have an Internet-based record, there have been a number of simplistic Web-based medical records (all of which are incompatible with any other database), however, there is no method to automatically download specific data.

There are a number of overarching problems that are nontechnical but which have definite possible solutions. One of the largest challenges is coming to an agreement on a standard computerized database (computerized medical record), which can easily and transparently accept all forms of medical information (images, text, data signals, equations, etc.) for automatic archiving. It must have a powerful retrieval engine to function as an intuitive interface for access to information. The next challenge is to agree on a standard method of displaying the information, which is independent of the display device. One suggested possibility is the visual medical record in the form of a holomer™ (see chap. 15, this volume), which can be viewed on a standard monitor, as a hologram, with an HMD, in a CAVE, or with any other display device. For universal access to archived data, there appears to be consensus that a Web-based medical record can provide a unifying platform; however, as in all Internet applications, there are the issues of data accuracy, security, and privacy.

In creating 3-D images for diagnostic purposes, such as virtual endoscopy, the most important and technically difficult barrier is real-time automatic segmentation of organ systems and tissues. The issue of accurate color and texture to be derived from the acquired images has only been proposed in preliminary demonstration. There is no consensus on the best method of displaying the image. In order to aid radiologists in interpretation, automatic recognition and interpretation by decision support systems needs to be developed. To be able to have a robust diagnostic tool that rivals that of current radiologic standards, imaging technologies need to improve the resolution of the image significantly, possibly as much as an order of magnitude.

In the therapeutic implementation of telesurgical and dexterity enhancing systems, there remains the issues of latency when attempting to implement at a distance; there must be a lag time of less than 50 milliseconds for the entire system (see chap. 48, this volume). For increasing precision and dexterity, improvement is needed in the degrees of freedom of systems, accuracy of tracking of position (at least 1 mm accuracy, and for some procedures such as neurosurgery, 0.1 mm accuracy), see chap. 8, this volume), and appropriate scaling of haptics (see chaps. 5 and 6, this volume). In current computer-assisted and robotic systems, although the interface is dramatically better than current laparoscopic systems, they are far from ideal. Some unresolved system specific issues include automatic initialization and registration of the system before beginning the operative procedure, acceptable hand–eye axis, most appropriate display (HMD, monitor, binocular optical system, etc.), a tool changing system, and method of control for subsystems (voice activated, menu driven, etc).

For education and training applications, there are the same technical issues listed above for image generation in 3-D visualization in diagnosis (automatic segmentation, registration, etc.), as well as the interface issues. In addition, the biggest challenges are those of content—of developing an appropriate curriculum that can take advantage of the technology. Of all the applications, education requires that the system be intuitive and easy to use, especially to be usable by students without long periods required simply to learn the system in order to begin studying (see chap. 34, this volume). Content must be developed in order to have metrics that are able to measure technical skills and performance in such a way as to give meaningful outcomes analysis. To date, none of these metrics have been developed, and until they are derived through consensus and evidence-based testing, implementation of training systems will not be accepted by residency training programs or certification bodies.

Nontechnical challenges to the implementation of VEs are those that prevent development of the application in spite of technical feasibility. In implementation of such a diverse set of applications, which must be interoperable (such as simulators which can be used on student workstations or be integrated into telepresence surgery systems), the most important factor is standardization. To date none of the various systems are interoperable. An attempt has been made to approach such integration by requiring the 3-D visualization projects to be compatible with the Visible Human data set. This does not guarantee standardization; however, it provides a common reference from which to start. As systems become more complex, the questions of safety and liability apply, as well as certification to use the systems (see chap. 27, this volume). Not only are mechanical systems under stringent evaluation from the medical device division of the FDA, but also the computer programs that handle data or control the mechanical devices.

Once a device or software program is approved for use, a major challenge is getting third-party payers to reimburse for the cost of procedures (e.g., virtual endoscopy) or paying extra for a procedure because more advanced equipment is used, as in image-guided or computer-assisted surgery. Health care insurance companies or managed care organizations delay implementation of a procedure until it can be shown to reduce costs. The current mantra for new technology, whatever the type, is cost-benefit ratio; yet conducting the long-term clinical trials with clearly definable outcomes analysis causes a prolonged delay in payment for a service. On the other hand, a number of new technologies will not be accepted by physicians or patients because of unfamiliarity, conventional prejudice, or outright ignorance of the benefits. There is a natural skepticism against anything new and an almost inborn resistance to change. It required almost a new generation of radiologists to accept the new digital images instead of film-based radiology; it is uncertain if it will require decades to accept 3-D visualization and virtual endoscopy. Finally, there are a number of applications that raise serious bioethical and moral issues (see chap. 33, this volume). Should a psychiatrist subject a patient to a threatening VE that could possibly cause irreversible damage when the capabilities of the state of the art is unknown? Is it permissible for a surgeon to operate on a patient (using telesurgery) if the surgeon has not met the patient (in a distant city) or is not physically present in the operating room should something go wrong—are we developing a "tele-itinerant surgeon"? Thus, there are many nontechnical hurdles that can impede acceptance of what is otherwise an excellent technical solution to a difficult problem.

## 6. CONCLUSION

This overview has presented numerous specific VE applications that are emerging in the medical field; however, the majority are in the laboratory or clinical investigation stage. It will take significant effort to move these technologies into commercial success and therefore routine clinical use. As in many other fields, standardization and interoperability are some of the major issues whose solution will dramatically increase the implementation of the technology. Of all the technical challenges, auto-segmentation, registration, and data fusion are the most difficult, but these have the possibility for producing the greatest amount and most rapid acceptance of VEs as a clinically useful tool. Although there needs to be orders of magnitude increase in computer power for near realistic visual fidelity, the complexity of the human being is such that there will always be the need for further fidelity, whether on the macroscopic, microscopic, or molecular level. Yet the final determinant is the end user; therefore all VEs in use today need to become much more user friendly (see chap. 34, this volume). They must be turnkey solutions that are extremely robust and fault tolerant and that require nearly no technical support or continual maintenance.

The VE applications above provide a framework of what is possible today and give rise to speculation as to what would be possible by extending the capabilities described. The implications of a 3-D computer-generated representation of a specific person (a "digital me") or holomer$^{TM}$ that can act as a surrogate for optimizing (and possibly predicting) individual patient care is extraordinary. Virtual environments are the tools and methodology to create such an interactive information representation, a step in the direction of representing the complexity of human and biologic systems in a manner more clearly understandable and that eventually will be usable and practical for each person. The more devices developed that can acquire information about a person (whether handheld imagers or noninvasive biosensors continuously worn in our clothing or embedded in the body), the richer the holomer and the more accurate the results from modeling and simulation, preoperative planning, or intraoperative assistance. Not only is this relevant to patient care for medical conditions, but at all levels, including school age. Imagine the power of each child having a holomer that "grows up" with them, which they carry on a credit card device and use in class. By inserting their holomer into a virtual environment, they can learn health and nutrition by observing the consequences to their holomer. For example, by implementing the "smoking" module, the child's holomer could grow a cancer, get bronchitis and emphysema, and decrease its ability—as a prediction of what would happen with long-term smoking. These types of "what if" scenarios can also be used to engage people into complex intellectual dilemmas, such as bioethical and medical ethical issues of cloning, research on embryos, and alternative forms of therapy. Theoretically, if a "generic" or standardized holomer could be created, the early phases of clinical trials (on drugs or devices) or virtual "crash dummies" could replace some of the extremely expensive and high-risk testing and evaluation occurring today. Other more practical components of VE, such as customizing individual prostheses (or in the not too distant future, creating the model for the instructions to "grow" replacement organs through tissue engineering) will result from an extension of the emerging field of medical stereolithography. The applications in psychiatry or virtual therapy (see chaps. 50 and 51, this volume) are to treat disease; perhaps with insight, a VE in the home can be created to ameliorate other disorders before they erupt. By implementing VE for shared experiences, there are a number of pragmatic research goals that can soon be achieved to enrich, through high-bandwidth communications of Internet 2, a much higher sense of presence with tele-immersion for telemedicine consultations. The ease of use within these VEs needs to make the interactivity more transparent, perhaps by increasing the use of voice commands (see chap. 4, this volume) and other intuitive interfaces (see chap. 13, this volume).

It is apparent that there are numerous other directions in diagnosis, therapy, and education and training that could be speculated, and the above are listed as examples of technically feasible projects as next steps in the evolutionary process. Yet the medical discipline is unique, in that final result will impact on a living, breathing human being. The challenge will be to focus upon those applications that can make a quality-of-life difference for each and every patient.

## 7. REFERENCES

Ackerman, J. M. (1998). The visible human project. *Proceedings of IEEE, 86*, 504–511.

Altobelli, D. E., Kikinis, R., Mulliken, J. B., Cline, H., Lorensen, W., & Jolesz, F. (1993). Computer-assisted three-dimensional planning in craniofacial surgery. *Plastic and Reconstructive Surgery, 92*, 576–585.

Beaulieu, C. F., Napel, S., Daniel, B. L., Ch'en, I. Y., Rubin, G. D., Johnstone, I. M., & Jeffrey, R. B. (1998). Detection of colonic polyps in a phantom model: Implications for virtual colonoscopy data acquisition. *Journal of Computer Assisted Tomography, 22*, 656–663.

Bill, J. S., Reuther, J. F., Dittmann, W., Kubler, N., Meier, J. L., Pistner, H., & Wittenberg, G. (1995). Stereolithography in oral and maxillofacial operation planning. *International Journal of Oral Maxillofacial Surgery, 24*, 98–101.

Carpentier, A., Loulmet, D., Aupecle, B., Berrebi, A., & Relland, J. (1999). Computer-assisted cardiac surgery [Letter]. *Lancet, 353*, 379–380.

Chinnock, C. (1995). Holographic 3-D images float in free space. *Laser Focus World*, 22–24.

Delp, S. (1998). *Limb trauma simulator*. [Online]. Available: http://www.musculographics.com

Delp, S. L., & Zajac, F. R. (1992). Force and moment generating capacity of lower limb muscles before and after tendon lengthening. *Clinical Orthopedics and Related Research, 284*, 247–259.

DiGioia, A. M., Jaramaz, B., & Colgan, B. D. (1998). Computer-assisted orthopaedic surgery. Image-guided and robotic assistive technologies. *Clinical Orthopedics, 354*, 8–16.

Edmond, C. V., Wiet, G. J., & Bolger, B. (1998). Surgical simulation in otology. *Otolaryngology Clinics of North America, 31*, 369–381.

Greenleaf, W. J., & Tovar, M. A. (1994). Augmenting reality in rehabilitation medicine. *Artificial Intelligence in Medicine, 6*, 289–299.

Heinrichs, W. K., & Dev, P. (1999). Interactive pelvic anatomy [Online]. Available: http://summit.stanford.edu/welcome.html and http://www.nlm.nih.gov/research/visible/vhp_conf/heinrich/abstract.htm

Himpens, J., Leman, G., & Cardiere, G. B. (1998). Telesurgical laparoscopic cholecystectomy. *Surgical Endoscopy, 12*, 1091.

Hodges, L. F., Rothbaum, B. O., Kooper, R., Opdyke, D., Meyer, T., North, M., de Graaff, J. J., & Williford, J. (1995). Virtual environments for treating the fear of heights. *IEEE Computer, 28*, 27–34.

Hoffman, H., & Murray, M. (1999). Anatomic Visualize: Realizing the vision of a VR-based learning environment. In J. D. Westwood, H. M. Hoffman, R. A. Robb, & D. Stredney (Eds.), *Medicine meets virtual reality: The convergence of physical and informational technologies: Options for a new era in healthcare* (Vol. 62, pp. 134–141). Amsterdam: IOS Press.

Hohne, K. H., & Bernstein, R. (1986). Shading 3-D images from CT using gray-level grading shading. *IEEE Transcript of Medical Imaging*, 5–45.

Imielı̀nska, C., Laino-Pepper, L., & Thuman, R. (1998). The vesalius project [Online]. Available http://www.vesalius.com/

Jolesz, F. A. (1997). Image-guided procedures and the operating room of the future. *Radiology, 204*, 601–612.

Jolesz, F. A., & Shtern, F. (1992). The operating room of the future. *Proceedings of the National Cancer Institute Workshop, 27*, 326–328.

Kaufmann, C., Rhee, P., & Burris, D. (1999). Telepresence surgery system enhances medical student surgery training. In J. D. Westwood, H. M. Hoffman, R. A. Robb, & D. Stredney (Eds.), *Medicine meets virtual reality: The convergence of physical and informational technologies: Options for a new era in healthcare* (Vol. 62, pp. 174–179). Amsterdam: IOS Press.

Lamson, R. (1994). Virtual therapy of anxiety disorders. *CyberEdge Journal, 4*, 6–8.

Levy, J. S. (1996). Virtual reality hysteroscopy. *Journal of American Association of Gynecological Laparoscope, 3*(4, Suppl.), S25–S26.

Marescaux, J., Clement, J. M., Tassetti, V., Koehl, C., Sotin, S., Russier, Y., Mutter, D., Delingette, H., & Ayache, N. (1998). Virtual reality applied to hepatic surgery simulation: The next revolution. *Annals of Surgery, 228*, 627–634.

Margossian, H., Garcia-Ruiz, A., Falcone, T., Goldberg, J. M., Attaran, M., Miller, J., & Gagner, M. (1998). Robotically assisted laparoscopic tubal anastomosis in a porcine model: A pilot study. *Journal Laproendoscopic and Advanced Surgical Techniques, 8*, 69–73.

Meller, G. (1997). A typology of simulators for medical education. *Journal of Digital Imaging, 10*(3, Suppl. 1), 194–196.

Merrill, J. R., Merrill, G. L., Raju, R., et al. (1995). Photorealistic interactive 3-D graphics in surgical simulation. In R. M. Satava, K. Morgan, et al. (Eds.), *Interactive technology and the new medical paradigm for health care* (pp. 244–252). Washington, DC: IOS Press.

Montgomery, K., Stephanides, M., Schendel, S., & Ross, M. (1998). A case study using the virtual environment for reconstructive surgery, *IEEE Visualization '98*, Research Triangle Park, NC:

Moran, L. (1998). Doctors walk in patient's footsteps via virtual reality technology—Experience captures fatigue patients often can't describe. *Telemedicine and Virtual Reality, 3*, 107.

Negroponte, N. (1995). *Being digital*. Cambridge, MA: MIT Press.

Paul, H. A. (1992). Image-directed robotic surgery. In *Proceedings of Medicine Meets Virtual Reality*. San Diego, CA: Aligned Management Associates.

Pieper, S., Laub. D., & Rosen, J. (1995). A finite element analysis facial model for simulating plastic surgery. *Plastic Reconstructive Surgery, 96*, 1100–1105.

Raibert, M., Playter, R., & Krummel, T. M. (1998). The use of a virtual reality haptic device in surgical training. *Academic Medicine, 73*, 596–597.

Rothbaum, B. O., Hodges, L. F., Kooper, R., Opdyke, D., Williford, J., & North, M. M. (1995). Effectiveness of computer-generated (virtual reality) graded exposure in the treatment of acrophobia. *American Journal of Psychiatry, 152*(4), 626–628.

Satava, R. M. (1998). *Cybersurgery*. New York: John Wiley & Sons.

Satava, R. M. (1993). Virtual reality surgical simulator: The first steps. *Surgical Endoscopy, 7*, 203–205.

Sellberg, M., & Kerr, J. (1998). Available: http://eai.com

Sinclair, M. J., Peifer, J. W., Haleblian, R., Luxenberg, M. N., Green, K., & Hull, D. S. (1995). Computer-simulated eye surgery: A novel teaching method for residents and practitioners. *Ophthalmology, 102*, 517–521.

Spitzer, V. M., & Whitlock, D. G. (1992). Electronic imaging of the human body. Data storage and interchange format standards. In M. W. Vannier, R .E. Yates, & J. J. Whitestone (Eds.), *Proceedings of Electronic Imaging of the Human Body Working Group* (pp. 66–68).

Taylor, C. A., Hughes, T. J. R., & Zarins, C. H. (1998). Finite element modeling of blood flow in arteries. *Computer Methods—Applied Mechanical Engineering, 158*, 155–196.

# 48

# Virtual Environment–Assisted Teleoperation

Abderrahmane Kheddar,[1] Ryad Chellali,[2] and Philippe Coiffet[3]

[1] *Université d'Evry Val d'Essonne—IUP*
*Laboratoire Systèmes Complexes*
*40 rue du Pelvoux CE 1455 Courcouronnes*
*F-91020 Evry Cedex, France*
*kheddar@iup.univ-evry.fr*

[2] *Ecole des Mines de Nantes*
*4, rue Alfred Kastler*
*BP 20722, F-44307 Nantes Cedex, France*
*ryad.chellali@emn.fr*

[3] *Laboratoire de Robotique de Paris*
*10-12 avenue de l'Europe*
*F-78140 Vélizy, France*
*coiffet@robot.uvsq.fr*

## 1. INTRODUCTION

Teleoperation is the technology of robotic remote control. Teleoperation systems synergistically combine humans and machines, and this man–machine link is what differentiates these systems from their predecessors. Teleoperators are used in place of robots when the latter are unable to overcome, by themselves, either the evolutionary situation of performing a task or its dexterity demands.

The first teleoperation systems were built after the Second World War for needs in nuclear activities. These early systems used a master–slave concept with two symmetrical arms. The master arm is handled by the operator; while the slave replicates the operator's motions at the location where the task being performed. In the earliest systems, master and slave were mechanically connected. Later systems were electrically powered, affording the possibility of any distance between master and slave. In the 1980s, computers were introduced as control systems opening the way to computer-aided teleoperation (CAT). Contemporary CAT utilization has been deeply modified by the emergence of virtual environment (VE) technology.

In early systems, the absence of sophisticated electronics and computers obliged a symmetrical mechanical device to correctly transfer motions from the operator to the slave device. Nevertheless, a great aid to manipulation came from the integration of force feedback, allowing the operator to "feel" what he or she manipulated remotely. The introduction of computers led to further enhancements in the design of master and slave devices. Whereas the slave kept its mechanical structure, the master could be reduced in size, as well as transformed into a joystick with force feedback, or into a set of different mini-systems that were easy for the operator to move. However, these new master interfaces, although better adapted to the human, did not resolve the general difficulty of correctly performing remote tasks. Remote operations require visual access to the activities being conducted in the slave environment. Classic cameras, even stereoscopic ones, did not succeed in adequately informing the operator. The operator thus often became disoriented (see chap. 24, this volume), lost sight of his subject of interest, and became afflicted with motion sickness–like symptoms (see chap. 30, this volume) after several minutes of work. Virtual environments, with their ability to provide partial or total immersion (see chap. 40, this volume), promise to seriously improve this situation. The great interest in VE-enhanced teleoperation is based on expected improvement in teleoperation efficiency, from gains in both ergonomics (see chap. 41, this volume) and user friendliness (see chap. 34, this volume), as well as an information feedback point of view.

## 2. PROBLEMATIC TOPICS

Teleoperation has matured both technological and from a conceptual point of view. Yet, there are still many problems that have yet to achieve a satisfactory or convincing solution. These problems are summarized below:

- **The unreachable ideal transparency.** After stability (i.e., in terms of a control theory point of view, stability means that a system will remain close to its original motion trajectory if slightly perturbed away from it), one of the most important teleoperator characteristics is transparency, which is a kind of action/sensing fidelity measurement index (see Laurence in IEEE TRA, 1993). Although a trade-off between stability and transparency has been found, such solutions have largely been derived from a pure control theory point of view. It is an established fact that ideal transparency can never be achieved by conventional bilateral control unless it is redefined by other criteria or conceived differently. This limitation comes mainly from (1) the operator action and the feedback being conveyed through the master–slave chain before reaching the target task and the operator channels, respectively, and (2) this chain including dynamics which cannot be neglected or compensated for without compromises in stability or operator safety. Challenging bilateral coupling seems to be one way to deal with this problem.
- **Time delay in control of remote systems.** Communication time delay between master and slave constitutes one of the most crucial problems in teleoperation. Indeed, time delay affects not only transparency, because the operator actions and feedback are delayed, but also stability. In conventional control solutions (those not using VE techniques), constant and variable time delays have been solved by Anderson and Spong (1988) by means of a strategy resulting in a clever "damping" design via low-level control. This method has been more elegantly formulated by Niemeyer and Slotine (1991) and astutely adapted, recently, for the case of nonstationary time delay. Kosuge (1996) and Oboe and Fiorini (1998) also proposed solutions to deal with variable time delay. Nevertheless, it is utopia to try to keep force feedback from the remote environment when delays are of the order of seconds (IEEE TRA, 1993).
- **Taking into account human factors.** Industrial robotics is sometimes opposed to teleoperation solutions from a flexibility viewpoint. Yet teleoperation flexibility is in many ways

dependent on operator adaptation to the teleoperation system. Indeed, to perform a task the operator must be trained and specialized. How to quantify human factors is also a well-known old problem, which still has no satisfactory qualitative and quantitative evaluation methodology, yet means of assessing VE usability are under development (see chap. 34, this volume).

- **Slave autonomous behavior and man–machine shared control.** Many tentative solutions concerning man–machine cooperation, that is to say, shared control architectures and semi-autonomous teleoperators, have been proposed. The problems of autonomy sharing, modeling, and quantification are narrowly linked to the degree of robot autonomy and to the level of operator intervention, which can range from full manual to symbolic (supervisory) control (Sheridan, 1992). Although some issues do exist, this point is important and raises an unanswered question about the relation between man's intervention methods and robot autonomy level.
- **Reliability and safety.** Use of high-powered master–slave devices is unquestionably as dangerous as today's industrial robots. In addition, since the operator works very often in the vicinity of powered axes, dangers in addressing system integrity, losses of functionality, and device or sensor degradations exist (Struges, 1996). Powered manipulation arms are inherently slow devices; no quick or jerky motion should be attempted in using them. Unlike computers, the manipulator arms, if used improperly, have the physical capability of destroying themselves and their environments.

## 3. VIRTUAL ENVIRONMENT–BASED TELEOPERATION

The manifold appropriate solutions that VE offers, in order to refine solutions to the interrelated specific teleoperation problems, make it difficult to infer a standard way VE could assist teleoperation. This difficulty paradoxically constitutes the strength and cleverness of VE techniques. Therefore, the remainder of this chapter concentrates on the state of the art in VE-based teleoperation, including examples of the capabilities of some teleoperator systems and current challenges in VE-assisted teleoperation schemes. Topics covered include: coping with time delay; enhancing computer aided teleoperation; improving bilateral and shared control; human-centered architectures; enhancing sensory feedback to the human operator; and improving operator safety with human factors issues.

### 3.1 Coping With Time Delay

Predictive graphic displays offer a solution to dealing with time delay inducing instability in bilateral force reflection teleoperation. Predictive graphic display development was pioneered by the efforts and results of the Massachusetts Institute for Technology's Man–Machine Laboratory. Noyes and Sheridan (1984; see Sheridan's review in IEEE TRA, 1993) developed a predictive display where the manipulator was displayed as a stick graphic figure overlaid on the delayed video image of the manipulator. Experimental results have proven the efficiency of the proposed strategy. Bejczy et al. (1990, quoted in Kim, 1996), improved the stick figure with a high-fidelity graphic predictive display they called the phantom robot. Two graphic models (wire frame and solid) accurately depicting the actual slave robot can be overlaid on the delayed video feedback. The operator chooses, according to the context, one of the two graphic robot representations (wireframe or solid) to act as a predictor. Since the graphic representation of the actual slave robot is faithfully reproduced, accurate virtual to video superimposition on a common display window was made feasible thanks to a VE calibration method thoroughly discussed in Kim (1996).

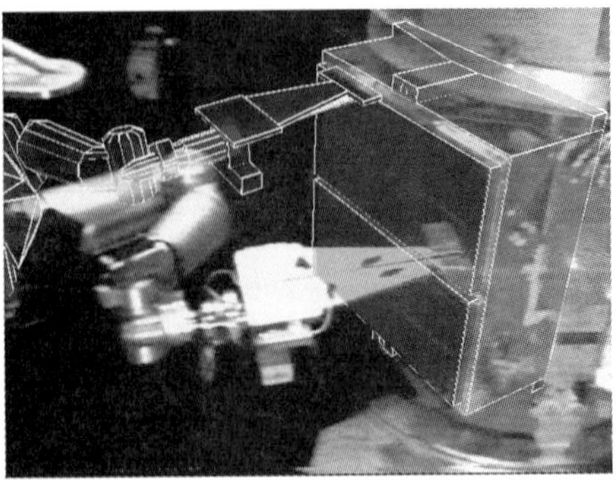
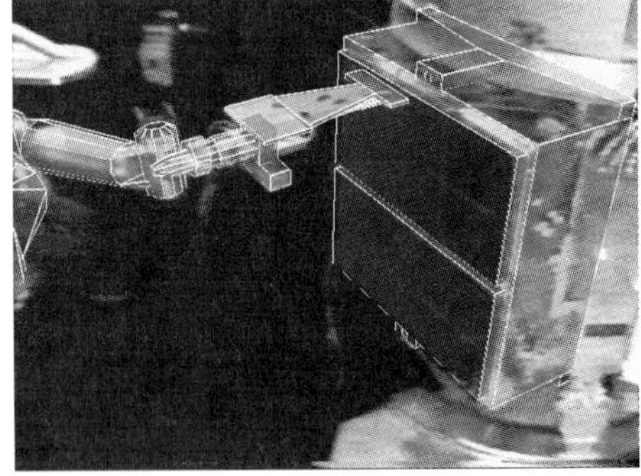

FIG. 48.1. Predictive display by means of augmented reality techniques for space teleoperation. Courtesy of Kim (1996), Jet Propulsion Laboratory, NASA.

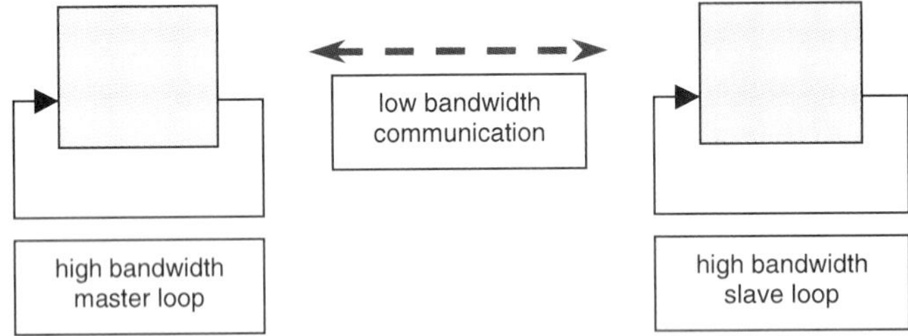


FIG. 48.2. A high-level view of teleprogramming (adapted from Funda et al., 1992).

Figure 48.1 shows a teleoperation instance in a space operation simulation case. The predictive "phantom robot" reacts in real time and smoothly to the operator commands. When subsequent predicted actions are considered satisfactory, trajectory commands are sent to the actual robot. This method, although enhancing operator performance, still uses a move-and-wait strategy in which moving concerns the virtual robot and waiting concerns actual and virtual robot matching. It must be noted that force display is obviously prohibited because only the model of the robot is available. To maintain force feedback, however, Kotoku proposed using a VE representation of the entire remote location and its components (including the slave robot) as a predictive display (Kotoku, 1992). In this case, the system consists of: the master force feedback arm; the VE, including a virtual robot actually coupled to the master arm, the actual slave robot and its environment; and communication media linking the VE to the actual robot controller. The virtual robot is operated within the VE in a similar manner as if the virtual robot were real. Indeed force feedback is artificial, generated from the VE computer simulation, and flows from estimated interaction forces between the virtual robot and the virtual environment. Since object geometric modeling is based on a polyhedral representation, the artificial force feedback algorithm is based on polyhedron interpenetration calculations, including collision detection and other well-known features used in computer graphics simulation (Anderson, 1994). In fact, this system splits the bilateral loop into two local loops, the master bilateral loop coupling the operator to an estimated remote environment (VE) and the slave loop managing actual robot interaction with its real environment provided that the interrelated (i.e., embedded local autonomy, necessary degrees of freedom, and suitable sensors instrumentation) are convenient.

In the above cited predictors, distant robot control is achieved according to the online predictive planned trajectory. An alternative clever way to perform teleoperation in the presence of considerable time delay is teleprogramming. As shown in Fig. 48.2, the principle of teleprogramming is nearly similar to VE predictive teleoperation. Indeed no bilateral coupling is achieved between the master arm and slave robot. The master site is set up with a VE coupled with the master arm. This fact allows high-bandwidth virtual robot operation with immediate visual and haptic feedback. The slave arm is also a high-bandwidth closed loop between the remote robot and its environment and concerns the classical sensory feedback needed for the controller.

Compared with the predictive display method, teleprogramming is peculiarly different. Indeed, on the one hand, robot instructions are symbolic and somehow flexible rather than low level and accurate. Subsequently this requires some robot autonomy. It also makes teleoperation nicely continuous (i.e., not move-and-wait). On the other hand, feedback from the remote location to the master site matters to the execution status. In addition to providing real-time interaction, the virtual representation software continuously monitors the slave robot and any object in its grasp for collision detection or contacts with the virtual environment. Subsequent macro-commands are generated by an integrated software, based on a priori task knowledge and a predefined command language. To prevent execution errors, due to virtual model uncertainties, guarded commands (motion, force, etc.) are generated, rather than absolute or static commands. This teleprogramming architecture, together with experimental results, are thoroughly described in Funda, Lindsay, and Paul (1992).

Another variance of predictive display and teleprogramming was developed at the German Aerospace Center (DLR). The Hirzinger team conducted the first actual space experiments on a space robot technology called ROTEX, which flew with the space-shuttle COLUMBIA from April 26 to May 6, 1993 (see Hirzinger et al. in IEEE TRA, 1993). Among the four operational modes of the ROTEX are teleoperation from ground using a predictive display and tele-sensor-programming. Both operational modes are based on a virtual simulation of the ROTEX with its remote space laboratory features (see Fig. 48.3). The predictive display contains a model of the up and down communication link delay, as well as a model of the actual states of the real robot and its environment features, more peculiarly moving objects. Measured object poses are compared with estimates as issued by an extended Kalman filter. This filter predicts, and graphically displays, the situation that will occur after the up-link delay has elapsed and allows the performed action loop to be closed by operator control, via shared control, or purely in an autonomous mode. This kind of prediction has been made possible due to a nearly perfect world model together with knowledge of the dynamics of objects under zero gravity. The tele-sensor-programming operational mode is a kind of teleprogramming but differs in that it is nearly teaching-by-showing, as applied to robot offline programming. In the proposed tele-sensor-programming, complex tasks are split up into elemental moves for which a certain constraint-frame and simulated sensor type configuration holds. This provides the actual remote robot with simulated sensory data that refer to relative positions between its

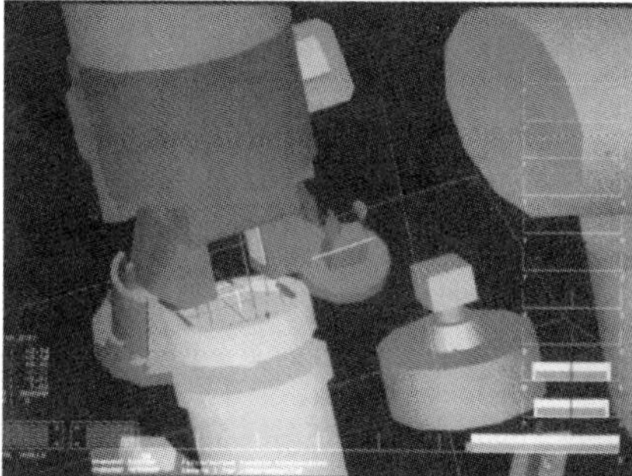

FIG. 48.3. The DLR MARCO system: Overview of the telesensor-programming interface (*left*); the ROTEX experiment VE setup; and the gripper sensors simulation (*right*). Courtesy of Hirzinger, DLR.

sensorized gripper and the environment. This compensates for any kind of inaccuracies in the absolute positions of robot and real world. These operational modes were successfully tested for multiple space laboratory tasks, including grasping of floating objects and assembly tasks.

The DLR is conducting ongoing research toward a unified concept for a flexible and highly interactive programming teleoperation station. The actual system, called MARCO, is designed as a 2-in-2 layer concept representing a hierarchical control structure ranging from the planning to executive layer. User layers are comprised of the task and operation modules, whereas the robotic expert layers module concerns elementary operations, actuators commands, and sensor phase. In the frame of a DLR–NASDA (Japanese Aerospace Agency) joint program, called German ETS-VII Experiment (for GETEX), the above modular task-directed programming scheme (MARCO) was successfully involved (April 1999) in the recent unmanned teleoperation of NASDA's ETS-VII free-floating space robot (ETS–VII, 1999).

Other ETS-VII teleoperation experiments have been conducted from the ground (ETS-VII, 1999). The purpose of the ETS-VII mission is to examine the ability of canonical tasks to support and confirm future space investigations such as building and operation of the international space station, inspection and repair of orbiting satellites, and planetary exploration. Teleoperation tasks include: an onboard satellite antenna assembling experiment, truss structure teleoperation, servicing, orbital unit replacement, add-on tools exchange, free floating target satellite capture, and visual inspection. ETS-VII ground robot control systems use operator aids based on VE techniques to assist telemanipulation. These include: predictive computer graphics, shared control capabilities, imaginary guide planes (a kind of a core-shaped virtual wall) to guide the robot arm motion to a desired position and to inhibit undesirable motions, various force controllers (for local compensation of geometric discrepancies), visualization and verification of motion commands using a motion simulator, multimodal interfaces, a visual aid system for direct teleoperation using a predictive force method, and teleoperation through virtual force reflection (including potential force fields, virtual force as physical constraints, adaptive virtual force by probing environment, etc., ETS-VII, 1999).

Figure 48.4 shows screen snapshots related to the ETS-VII teleoperation experiments. In conclusion, predictive displays and teleprogramming seem to be an attractive solution to deal with time-delayed teleoperation. Many other recent advanced teleoperation applications, such as mobile robots (an earlier application domain of teleprogramming initiated by a joint effort of the Laboratoire d'Automatique et d'Analyse des Systèmes—LAAS at Toulouse, France, and the GRASP Laboratory at the University of Pennsylvania; see also the mobile robot teleoperation section in this chapter), subsea robots, and even flying robots also use teleprogramming and graphic predictive displays. In terms of pure control theory, predictive displays and teleprogramming could likely be seen as an implementation, in the frame of teleoperation, of the well-known Smith predictor controller, proposed in 1957.

## 3.2 Enhancing Computer-aided Teleoperation

Computer-aided teleoperation (known also under the name of tele-assistance) can benefit from the high standards of excellence set for the modern human–machine interfaces associated with VE technology. Resourceful developments and solutions of VE-aided teleoperation that have potential for enhancing CAT include: virtual fixtures, active guides, and graphical programming metaphors.

One of the applications of VE techniques in CAT is directed toward fitting abstract perceptual information within the human–machine interface. The work proposed by Rosenberg (1992) explores the design and implementation of computer generated entities know as "virtual fixtures." Virtual fixtures are defined by Rosenberg (1992, p. 4) as: "abstract sensory information overlaid on top on the reflected sensory feedback from the workspace which is completely independent of all sensory information from the workspace." Although they

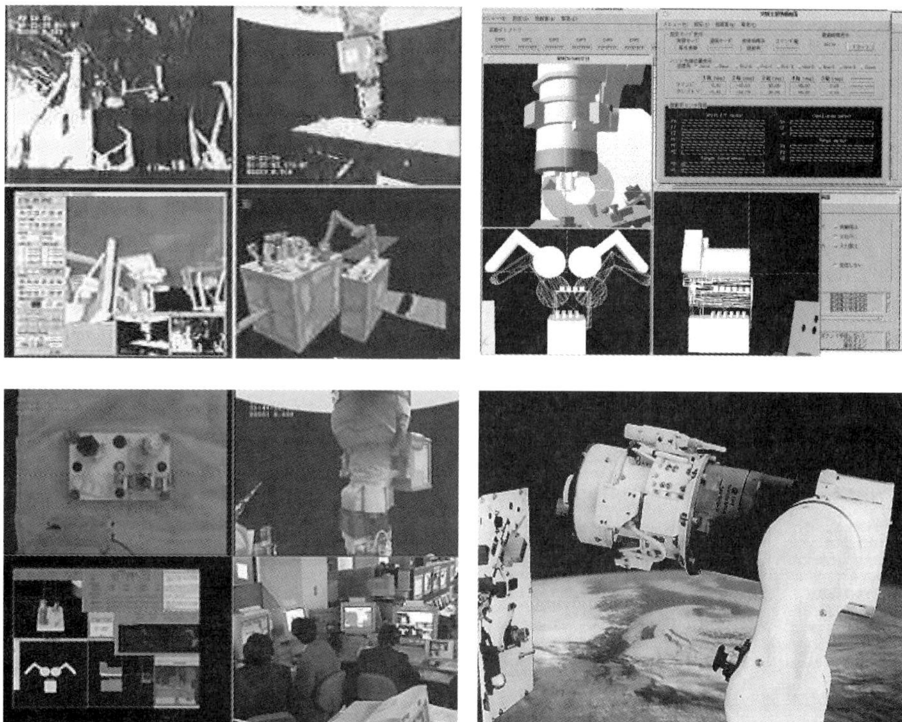

FIG. 48.4. The ETS-VII unmanned space teleoperation experiment. Clockwise: target satellite handling experiment; displayed CG image of simulation (predictive simulation system); an image from an onboard TV camera; and the whole setup. Courtesy of NASDA, http://oss1.tksc.nasda.go.jp/ets-7.

functionally embody fixtures in the real world, there are many benefits inherent to virtual fixtures, as compared to real physical ones, because they are computer generated. Indeed, virtual fixtures can be extended to include other modalities, such as visual (see chap. 3, this volume), auditory (see chap. 4, this volume), and even tactile (see chaps. 5 and 6, this volume) sensations used alone or in cross-modal combination (see chap. 21, this volume). Rosenberg highlights these advantages by an instance consisting of plotting a straight line. Obviously, the use of a ruler enhances human operator performance in plotting a straight line as compared to when no ruler is used. This may be seen as what has been proposed earlier in the frame of CAT, namely the possibility to freeze some of the robot degrees of freedom while constraining the operator to control the slave robot in the remaining ones. Virtual fixture are different, since they offer a more powerful and more flexible tool, which does not act on the slave robot directly. Indeed, more attention is given to operator assistance rather than the robot, through artificial sensory feedback overlaid on top of the sensory feedback from a remote teleoperation work site, which serves as a perceptual aid for task performance. How to make virtual percepts using perceptual rather than physical parameters is the main issue being investigated in using virtual fixtures. How does it work? Simply, as the operator interacts with virtual fixtures (via haptic or other defined modalities), appropriate reactive sensations are computed and fed back locally to the operator's perceptual channels. According to Rosenberg, abstraction can be used to operate on virtual fixtures, thereby enhancing interactivity (see section 3.5). Rosenberg has demonstrated that performance could be enhanced by 70% in telerobotic tasks that use virtual fixtures. Thus, virtual fixtures might be used to reduce operator load, facilitate supervisory control, ease the control of complex tasks, and even to "compensate" for performance degradations due to time delay.

Sayers (1999) used virtual fixtures to automatically generate low-level robot commands appropriate to the tasks to be performed. Although Sayers' team gave an implementation within a teleprogramming context, the developed strategies can be used in an augmented reality context as well. In their experiment, synthetic fixtures present the operator with task dependent and context sensitive visual and force cues. No attempt was made to provide realistic force feedback. Instead, the intention was to supply the operator with force and visual cues that can best aid task performance.

In Fig. 48.5, screen snapshots show an instance highlighting the use of virtual fixtures in a VE teleprogramming application. In the upper row of Fig. 48.5, a single-point fixture (represented as a cross) is used to help the operator bring the end point of the robot held tool to a point in space. When the telerobot tool is operated near the desired location, the virtual cross fixture is activated and pulls the robot held tool toward the cross-fixture location. In this case, the virtual fixture has not been defined to completely constrain operator (thus robot) motion. The system anticipates the operator's next desired action. In the same instance, another fixture allowing the robot to achieve circular motions is activated just after the operator decides to leave the cross fixture location. In the lower row of Fig. 48.5, a set of fixtures were defined to aid the operator in achieving flat face-to-face contacts. Each time the operator moves near the surface, the appropriate face-to-face virtual fixture is activated.

Kheddar (1997), attempted to give a unified formalism to the virtual fixtures metaphor and named it "active virtual guide" in the frame of a hidden robot teleoperator (see section 3.4). Under this metaphor, virtual fixtures are classified in three categories:

1. *Pure operator assistance:* This group includes the well known graphic metaphors used in CAT (e.g., sensory substitutions, etc.), in which virtual guides are not directly linked to robot control. Their role is mainly focused on assisting the operator to intuitively perform desired tasks.
2. *Pure remote robot control assistance:* This group includes virtual mechanism concepts, or any other virtual metaphor, induced in the low-level control necessary for strict execution of a real task by the actual robot.
3. *Operator and robot shared assistance:* This group includes those fixtures that are dedicated at the same time to operator and robot assistance in terms of robot autonomy sharing. Actual task completion results are issued from a combination of the virtual task designed by the operator and an autonomous module linked to robotic tasks. Hence, this category is split into three subclasses: autonomous function; semi-autonomous function; and collaborative function.

Virtual guides lead then to a unified structure composed of the following items:

- *Attachment:* A particular spot associated to the virtual guide. The virtual guide can either be statically attached (to any frame within the VE, object, robot controller part, etc.) or dynamically attached, appearing on a specific event (collision detection between objects or robot interaction with its environment, etc.) in a specific spot.
- *Effect zone:* A virtual guide may be associated with an effect area (volume, surface, part of the robot reach space, etc.), which may play the role of an action zone or an attractive/repulsive field within which the virtual guide acts.
- *Activation condition:* For each virtual guide an activation condition is allocated.
- *Function:* It defines the functionality, thus the reason for existence of the guide.
- *Inactivation condition:* While true, renders the guide ineffective by a set of specific actions. Any whole structured virtual guide may be completely removed, attached, or replaced offline or online (during teleoperation).

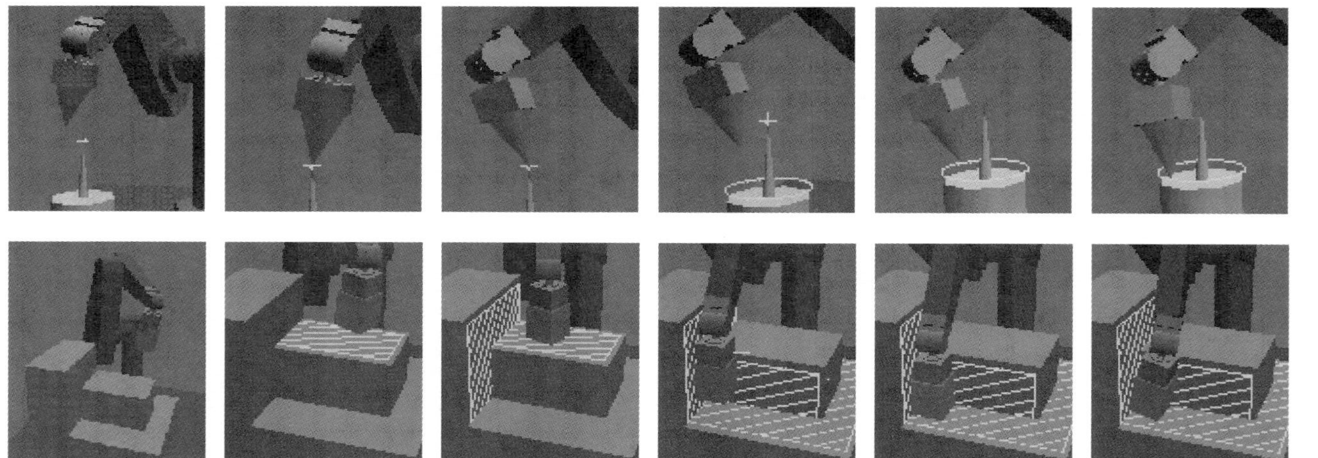

FIG. 48.5. The use of synthetic fixture in teleprogramming-based teleoperation. Courtesy of Sayers (1999), the GRASP Laboratory, University of Pennsylvania.

In Fig. 48.6, a teleoperated task consisting of grasping an object has been simulated with a cylindrical guide. The same geometry guide can serve as a collaborative (this case instance), semi-autonomous, or autonomous guide. In an autonomous style, the virtual object can be assumed grasped when one of the operator fingers touches the handle (unrealistic but easy operator grasp), which triggers an autonomous robotic grasp. In a semi-autonomous instance, the virtual object operator grasp is based upon physical realistic laws. When the virtual operator grasp is stable, the robot triggers an autonomous grasp. In a collaborative mode, operator grasping is realistic, the guide effect being used just to align the robot axis according to the object handle frame, thus freezing the guide robot gripper orientation to that of the handle and leaving the robot controlled according to the operator translational hand movement.

Graphic programming by means of VE technique is one of the most attractive paradigms used in advanced teleoperation architectures. On the one hand, graphic programming may be seen as one of the potential solutions to the well-known operator—telerobot shared control and autonomy problem. On the other hand, it might be seen as the ultimate way to fulfill both user friendly and human factors performance with high ergonomic characteristics. Some instances are presented bellow to give a better idea of how graphic programming can carry out the above cited points.

Figure 48.7 represents the task line and motion guide paradigm, which is an original form of teleoperation proposed by Backes et al. (1998). Motion guides consist of fashioning paths intuitively, in the robot virtual environment. Then, the actual telerobot is constrained to follow these paths when this is possible. Subsequent continuous commands sent to the actual telerobot are only one dimensional: forward, back, or halt along the motion guide. Motion guides are graphically represented by paths (curves and lines), and the motion direction is specified using arrow icons. Specification and modification of motion guides is achieved interactively offline or online (during teleoperation). Task line is a metaphor to design subtasks attached to motion guides, which is represented in the VE by icons. A task line is a collection of command sequences that are executed when the telerobot reaches the task line location. Among the general commands found in a task line, reactive sequences can be included to be performed upon specific events related by means of internal or external robot sensory capabilities (see Fig. 48.7).

At the Sandia National Laboratories, graphical programming tools based on three-dimensional (3-D) graphics models (called Sancho, see Fig. 48.8) are being developed to be used as an intuitive operator interface to program and control complex robotics systems, including teleoperation for nuclear waste cleanup (Small & McDonald, 1997). Sancho uses a general set of tools to implement task and operational behavior and works in four major steps:

1. The tasks are defined, by the user, by means of menus and graphical picks in the simulation interface (define/select step). In this step, a 3-D world model is used to simplify task definition; indeed, direct object selection is a primary method for defining task work location and to control task execution. At this stage, task definitions are based essentially on two paradigms: (a) operation-based, by which robot tasks are subsequently derived from a set of individual sequence executions that consist of the task to be accomplished, and (b) task-based, less flexible than the first and in which case the task results from a set of predefined useful goals (predefined subtasks).
2. The system plans and tests the adopted solutions under operator monitoring and supervision (plan/simulate step).
3. If the adopted plans are considered satisfactory by the operator, a network based robot program of the approved planning is sent to the robot controller (approve/download step).
4. Finally, the Sancho system monitors the robot and updates the simulator's environment with the data fed back from the actual robot environment (see Fig. 48.8).

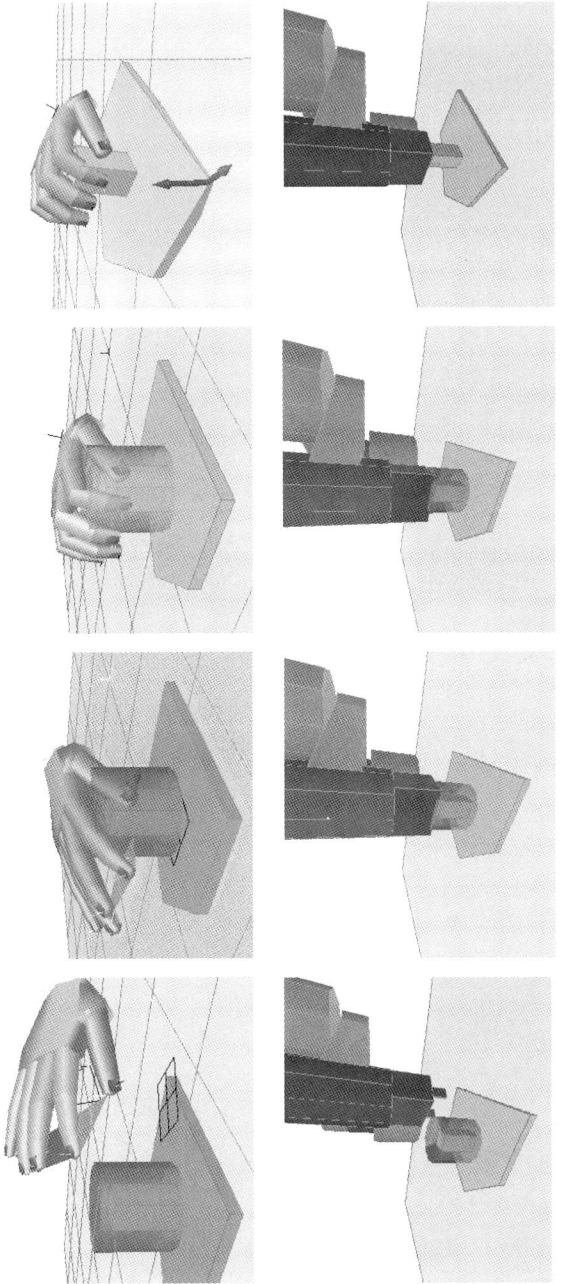

FIG. 48.6. The use of a collaborative active virtual guides in a grasping assist task for both the operator (*upper row*) and the slave robot (*lower row*) in the frame of the hidden robot concept–based teleoperation (Kheddar, 1997).

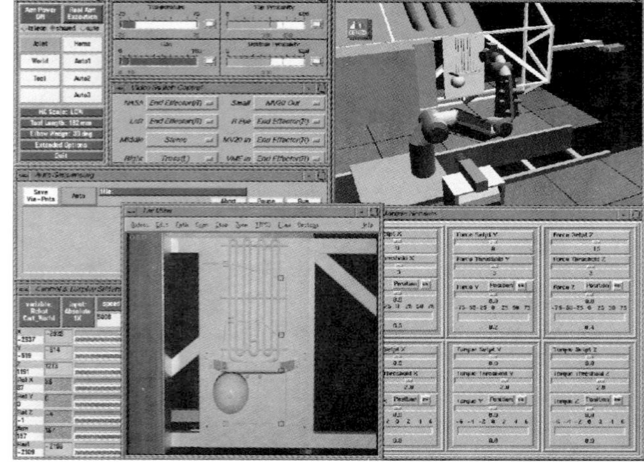

FIG. 48.7. Task line and motion guide: A graphical programming paradigm for space teleoperation. A surface inspection task (*left*). The whole teleoperation test-bed system (*right*). Courtesy of Backes, Jet Propulsion Laboratory, NASA.

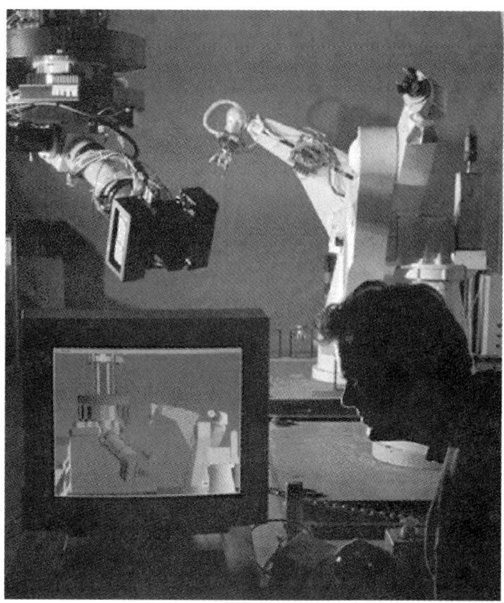

FIG. 48.8. The Sancho graphic-based robot programming experimental system. Courtesy of the Sandia National Laboratories.

Another graphic programming system case, called TAO 2000, was developed by the French Nuclear Center (see Fig. 48.9). TAO's graphic programming interface foundation is a hierarchical mission decomposition (i.e., a functional objective decomposition) into a set of processes or methods. A process finds an expression in a set of operations (abstract functions); each operation is then decomposed into a set of tasks (i.e., generic functions); and each task is described by a set of actions (i.e., physical functions), which are easily mapped into telerobot executions (physical forms). The whole process constitutes an exploitable generic skill database. For a given mission (for instance, gate inspection), the operator selects a set of processes from a predefined skill database (e.g., an unscrewing process) to achieve each stage of the mission (the unscrewing step) and maps it onto the concerned object (one of the gate screws). The mapping (map unscrewing skill to the pointed gate screw) constitutes an operation that will be executed by the telerobot agent. The operation consists of a set of field containing selected symbolic tools, such as: positioning area, an active virtual guide, recovery procedure, calibrating procedure, and agent behavior mode (autonomous, semiautonomous, or full manual teleoperation, Fournier, Gravez, Foncuberta, & Volle, 1997).

The TAO 2000 system is equipped with the PYRAMIDE 3D modeling system. As shown in Fig. 48.9, only the functional objects (i.e., those involved in the specific mission—gate inspection) are rendered to unburden operator mission preparation.

## 3.3 Improving Bilateral and Shared Control

Using a virtual intermediary world between the master and slave robot offers additional paths in control architecture design. All the same, the human operator can gain assistance from low-level control. Virtual environment techniques contribute subtly to low level control, as well as provide a potential way to efficiently carry out shared control. From the remote robot control point of view, the great advantage of reflex control is the absence of high-level reasoning before an action takes place. It is a direct link between information and action without passing

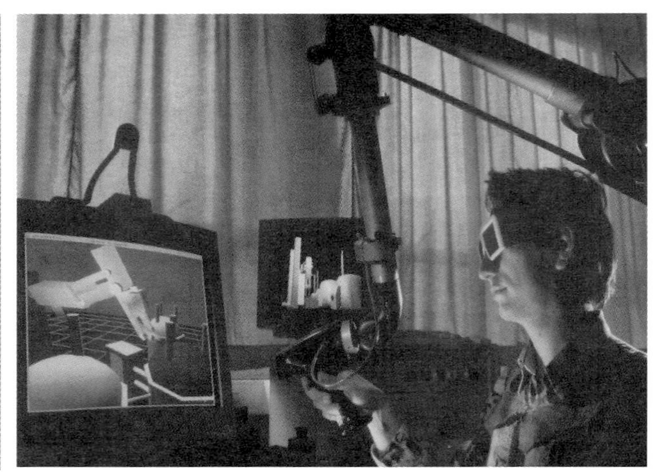


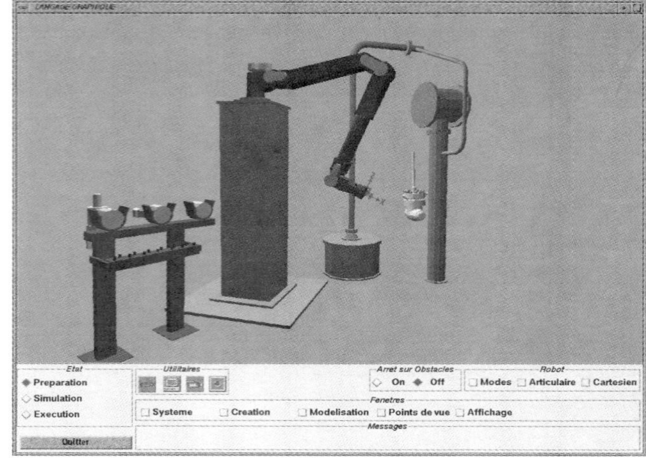

FIG. 48.9. TAO 2000: A VE-based graphical programming for nuclear maintenance (gate inspection mission case, (*left*); the whole system with the MA23 master force feedback arm (*right*). Courtesy of Gravez and Fournier, Advanced Teleoperation Service, CEA, France.

through a decision-making stage. Reflex behaviors are necessary to take advantage of robot capabilities and to make robot control easier for the operator. Moreover, when time delays prohibit bilateral coupling between the master and slave, reflex behavior becomes an intrinsic property to allow robot autonomy. Nevertheless, because decision is built into the information sensed, this inclusion of information is only possible if the environment is well defined. Therefore it is necessary to know beforehand, by some other means, the characteristics of the environment components involved in the teleoperation tasks. This unsolved problem relates to sensing and analysis of data. The use of VE may significantly improve this situation. Although the environment is unstructured it is, generally, not totally unknown. Thus, applying VE to this problem means that two categories of virtual objects may be created. The first category refers to the minimal meddling of what is known about the environment. An example is that of a robot pushing against a surface of unknown hardness. If the robot is guided under force control and the surface is actually very soft, then the system becomes unstable. Even if the surface is very hard this may also lead to instability. Traditionally, this is solved by adding a mechanical compliance at the end of the robot arm that absorbs energy. But if, instead, real environment hardness $K_e$ is replaced by a virtual hardness $K_v$, which is much larger than $K_e$ and is given a priori, then the robot, which applies a force $F_c$ on the surface, becomes stable. Hence, an unknown characteristic of the real environment can be replaced by a known virtual characteristic, which assures a good interaction with the contacted surface (Fraisse, Pierrot, & Dauchez, 1993). The same principle could probably be extended to most characteristics of unstructured environments.

The second category of virtual objects is the previously presented virtual fixtures or active virtual guides. While the concept has been previously applied in computer assisted teleoperation, it can be extended to enhance robot autonomy and to improve operator assistance through low-level control. As stated, the principle consists of introducing geometrical artifacts along which the robot is constrained to move. This has been expressed through the virtual mechanisms principle.

Kosuge et al. (1995) proposed an alternative control algorithm for telerobotic systems. For a given task, a task-oriented passive virtual tool is designed so that its dynamic behavior matches that of an ideal actual tool to be used as an interface between the remote robot and the object involved by the given task. Indeed, both robot maneuverability and stability are improved, while indirectly making bilateral control easier for the operator. As part of the TAO 2000 teleoperator project (see section 3.2 and Fig. 48.9), Joly and Andriot (1995) derived bilateral control laws from a simulated virtual ideal mechanism. Compared to Kosuge's, et al. (1995) work, the virtual mechanisms suggested here are connected to both the master and slave arms via springs and dampers. Using this, it is possible to impose any motion constraint to the teleoperator, including nonlinear constraints (complex surfaces and shapes) involving coupling between translations and rotations. This idea is being applied to most haptic interface controllers (see chaps. 5 and 6, this volume) interacting with virtual environments.

The research previously cited shows that VE techniques are not limited to operator assistance, but could also serve as an intermediary in achieving remote tasks by replacing problems related to man–robot–environment interactions with man–robot–virtual mechanism–environment interactions.

### 3.4 Human-centered Architectures

The purpose of any teleoperator is not the perfection of a master design, or an adequate executing machine, or the control architecture, or even the present remote environment state. The purpose is rather a future environment state expressed through its transformation. Indeed, only this transformation is of interest. On the other hand, almost any complex task to be remotely performed can be broken down into subtasks that can be expressed by a set of elementary moves for which a certain sensory-based state holds (i.e., a set of relations between the motion space and sensors space). An important remark can also be formulated to justify the design of human-centered telerobotic architectures; the way human operators perform tasks is in most cases not the most suited way for robots, and vice versa (Kheddar, 1997). This basic idea constitutes the foundation of various so-called human-centered architectures.

In the "hidden robot" telerobotic architecture proposed by Kheddar (1997), the goal of an ideal teleoperation system is defined by the possibility to build an intermediate world keeping only a functional copy of the real remote environment adapted to the desired task transformation. The part of the system devoted to the execution of the task must involve additional transformations implicating the intermediate world as a real one. Thus, the proposed teleoperation scheme leads to the design of two separate subsystems and the development of the link layer necessary for their connection. Therefore, to the operator, a representation of the real environment (with the possibility of changing objects location, shape, etc.) is made to suit his ergonomic requirements. Moreover, this representation handles what is necessary to be adapted to operator skill and dexterity, free from any constraint or transparency compromise inherent to bilateral (real or virtual) robot control. Following the above cited consideration, it is easy to see that in any intermediate functional representation the first object to be eliminated will be the "picture" of the remote system moving the real operational tool. There are at least three advantages of such an approach compared to the use of a classical VE based representation. First, the master station components (VE plus haptic device and sensory feedback) are adapted to direct task performance by the operator without any interference due to robot control. Second, the slave station is adapted to the task achievement by the robot without any direct interference due to the operator. Finally, bilateral data exchange is performed according to clever cross-modality-based task transformations.

At the Institute of Robotic Research (IRF), Germany, a clever VE system was developed for teleoperation. It allows for intuitive control of a multirobot system and several other automation devices with a standardized, easy-to-understand user interface. The general aim of the development was to create a universal frame for projective virtual reality, in which it is possible to "project" tasks, carried out by the user in the virtual world, into the real world with the help of robots or other automation components. In a projective virtual world the user can act as in the real world. Therefore, this kind of man–machine interface reaches the previously not achieved standard of intuitive operability, even for complex automation devices. The less the user has to know about the real automation device, the better the design of the man–machine interface.

In the projective virtual reality approach, with the help of robots changes made in the virtual world are "projected" in the real world. Thus, this approach builds a bridge between VE and robotic automation technology by providing techniques for the connection of these two fields (Freund & Rossmann, 1999), see Fig. 48.10).

One of the current human-centered architecture technologies is telepresence or telesymbiosis (Vertut & Coiffet, 1986). When coupled to other ways of reproducing natural human actions at remote places, telepresence provides users with the possibility of creating, with different degrees of realism, the notion of presence (see chap. 40, this volume). Telepresence has been defined by Sheridan (1992, p. 6), as a human–machine system in which the human operator receives "sufficient information about the teleoperator and the task environment, displayed in a sufficiently natural way, that the operator feels physically present at the remote site." Very similar to immersion-based virtual environments (see chap. 40, this volume), telepresence strives to achieve an actual feeling of presence at a remote real location. The end goals of both telepresence and immersion-based VE are fundamentally the same—a human interface that

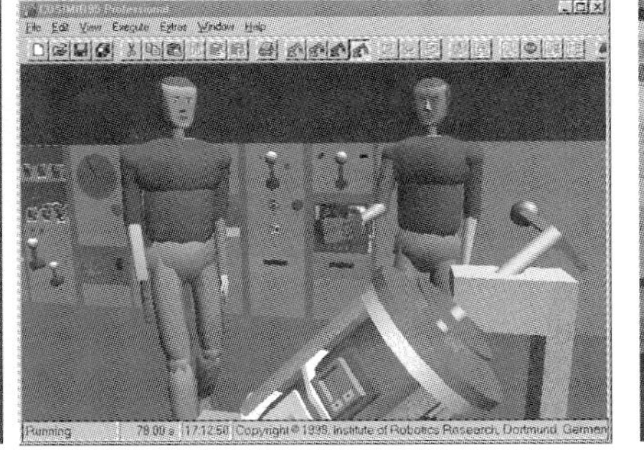

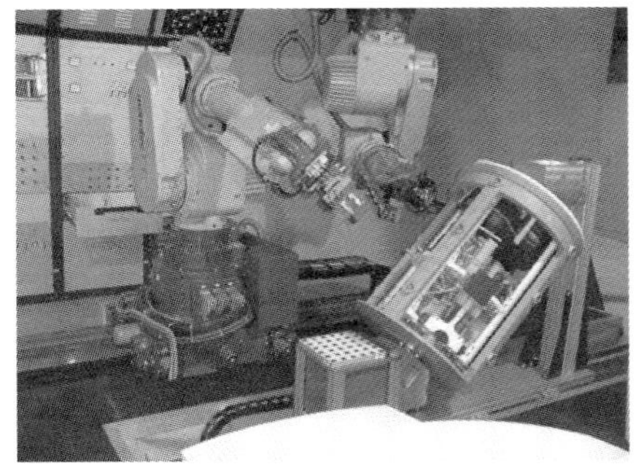

FIG. 48.10. Projective Virtual Reality concept for space and complex task telecontrol. From left to right: the operator interface (an immersion-based interfacing); the operator virtual projection into virtual avatars; and the actual robots mapping the virtual avatars actions into actual ones within the remote location. Courtesy of Freund and Rossmann, the Institute of Robotics Research, Dortmund, Germany.

allows a user to take advantage of natural human abilities when interacting with an environment other than one's direct surroundings. In the case of teleoperation in real environments, this is achieved by projecting the operator's skill and dexterity while reflecting sensory feedback so realistically that the operator feels present at the remote site. Indeed, telepresence borrows a lot from teleoperation technology because a remote physical device is needed to act on the remote environment while feeding back sensory information to the operator. The experience of being fully present at a real-world location remote from one's own physical location is obviously afforded by a high degree of transparency and realism. Someone experiencing transparent telepresence would therefore be able to behave, and receive stimuli, as though at the remote site. As for teleoperation, for any telepresence system there are three essential components: (1) the operator site technology; (2) the communication link; and (3) the remote site technology. Therefore, video (see chap. 3, this volume), audio (see chap. 4, this volume), and haptic (see chaps. 5 and 6, this volume) display systems, such as head-mounted displays (HMDs), auto-stereoscopic display screens, stereo headphones, and gloves or other devices equipped with touch feedback, may all be used by the operator. Control also needs to be exercised by the human operator and devices such as head and body tracking devices (see chap. 8, this volume), joysticks, master hands and arms in the form of gloves and exoskeleton structures, and other application specific controllers (see chaps. 11 and 13, this volume), are used. Contrary to teleoperation systems, telepresence VE techniques could be seen as being more concerned with interfacing technology and not interfacing strategy. If realistic and fully transparent feedback is the top consideration, illusion-based artifacts may be useless (see chap. 22, this volume).

In the RCAST laboratory at the University of Tokyo, Tachi's team is conducting studies on Tele-Existence, (Tachi, 1998). Tele-Existence is an advanced type of telepresence system that enables an operator to perform remote manipulation tasks dexterously with the feeling that he or she exists in the remote environment where the robot is working. He or she can "tele-exist" in the VE, which a computer generates, and be able to operate the real environment through the virtual space. One of the distinguishing characteristics of the system is the use of VE to allow operator self-projection. Two self-projection characteristics can be defined. The first one, defined by the authors, concerns the drawing of an operator's legs, body, arms, hands, and so on from which the operator feels as if he or she is self-projected onto the virtual human in the virtual environment. The operator "mapped" to the virtual robot can freely move around in the building, though the real robot can move only in restricted areas (see Fig. 48.11).

The second self-projection function that might be understood from using a VE tele-existence is the possibility to allow the operator, in an actual environment telepresence frame, to switch from a realistic representation to a virtual one so that he or she can plan strategy to exercise telepresence tasks thanks to a better understanding of the real environment from one's virtual representation. In this case the operator is split into two operators: the one who can explore ("what–if" strategies) the remote location without an actual, but rather a virtual, physical action ported on it (obviously a virtual representation is needed); and the one who can actually act on the real environment (a virtual representation is not necessarily needed in this case).

### 3.5 Enhancing Sensory Feedback to the Human Operator

Improving sensory feedback deals with using VE as a means to compensate for lack of sensory feedback from the remote site, that is to say, superimposing or combining artificial sensory feedback to the real one. With regards to the master site, in the presence of time delay, virtual environments were either used as an overlay to actual feedback (predictors case) or used as a

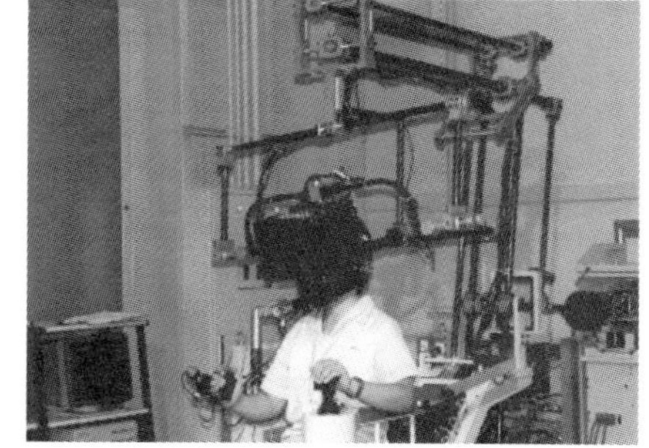

FIG. 48.11. TeleSAR (TELE-existence Slave Anthropomorphic Robot) system. From left to right; the master station, virtual environment tele-existence, and real environment tele-existence. Courtesy of Professor Tachi, RCAST, the University of Tokyo, Japan.

whole virtual intermediary to allow local feedback with respect to operator sensory bandwidths (for instance teleprogramming). Using VE as an overlay to real feedback is also known as augmented reality. Even in the absence of time delays, augmented reality may be helpful to improve corrupted sensory feedback information from the remote location. For instance, when vision feedback is defective (undersea boisterousness or cloudiness, smoky working area, or any poor-visibility environment) an augmented or VE display will not prevent the operator from driving the remote robot. The effectiveness of this kind of sensory feedback support has been proven in experiments by Oyama, Tsunemoto, Tachi, and Inoue (1993) and Mallem, Chavand, and Colle (1992). The virtual fixtures introduced by Rosenberg are also a powerful tool to improve sensory feedback. As stated previously, a virtual fixture is essentially a sensory overlay to actual feedback from the remote environment to increase operator performance. Moreover, fixtures can be invisible if the operator gains no benefits from visual cues, they can be viewed as a synthetic solid virtual object if rich visual cues are useful for the task, and can even be turned into a transparent glassy solid if visual cues are important and the operator does not want an occluded workspace. Virtual fixtures can be conceived as selective visual filters to block particular distraction, enhance contrast, provide depth cues, and even magnify a part of the workspace. Indeed, one could imagine the huge number of possible combinations that could be achieved by overlapping real sensory feedback in order to improve actual sensory feedback.

Sensory feedback enhancement can also stand out through different sensory substitutions (see chap. 21, this volume). For instance, the combination of visual fixtures, artificial sound, and tactile stimuli could substitute for a lack of actual sensory feedback from the remote robot location. In addition, using a whole virtual representation in the case of teleprogramming could be seen as a kind of sensory improvement because it maintains force feedback allowing more intuitive task achievement (even if this is within a VE). As far as a virtual representation is adopted, the number of sensory feedback modalities could be increased and combined in different ways thanks to various I/O interface technologies. For instance, the ROTEX experiment lacks force feedback, though a VE was used to perform space robot teleoperation. Instead, a SpaceBall was used for the virtual/real robot tele-sensor-programming and virtual interaction forces were monitored and substituted into visual cues to be displayed to the operator on a screen. Hence, force feedback is not always necessary if adequate substitution over the operator interface is conceived to tackle this shortcoming.

### 3.6 Improving Safety and Human Factors

In most teleoperator schemes using VE as an additional intermediary, operator safety could be addressed in two ways:

- The first is preventive and seems to imply the possibility to affix safety functionality integrated within the VE by means of computer programming.
- The second way deals with the possibility to fit the VE within human sensory capabilities so that teleoperation tasks could be achieved thanks to less risky, user-friendly, I/O devices derived from VE technology (3-D mouse, space balls, sensing gloves, desktop force feedback devices, etc.).

Obviously, both ways could be combined and additional features may be added to the remote site, namely by exploiting robot autonomy and enhancing its perceptual issues.

There might be various strategies to act on the VE by adding functions to improve operator safety (see chap. 41, this volume). One of them is straightforward: The operator would have

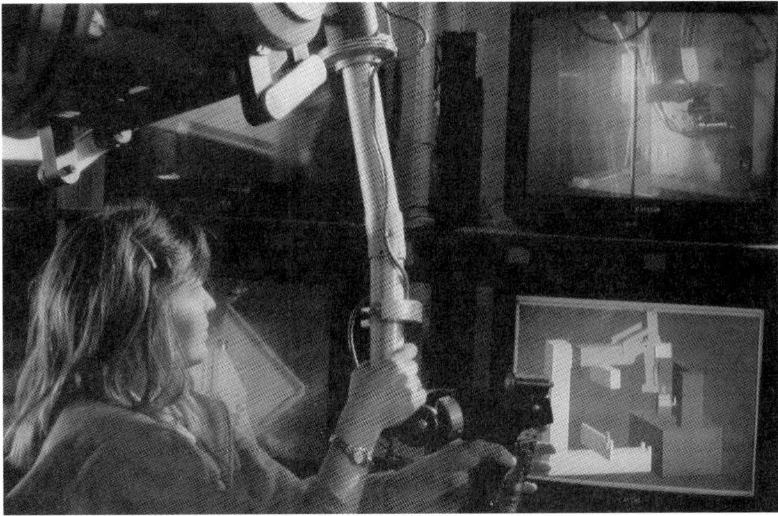

FIG. 48.12. Improving both visual sensory feedback and safety for the operator: The VE representation enhances visual cues and acts as a filter using an active collision avoidance algorithm developed for teleoperation purposes. Courtesy of Fournier, TAO 2000 Project, the French Nuclear Center.

the opportunity to simulate a teleoperation task before actually performing it. This is a kind of a "what–if" strategy. Another more transparent strategy is to prevent eventual collisions by means of clever active collision avoidance in the VE, as has been proposed for the before cited TAO 2000 system (see section 3.2 and Fig. 48.9). Extending this last principle to more general tasks, the VE can be fashioned to include functionally a kind of "intelligent" filter. The latter may guide or prevent situations that may compromise the operator's safety (see Fig. 48.12).

In general, for direct bilateral master–slave coupling, the safety risk is the aggregate likelihood of master, slave, and communication media damage (or functionality loss). Since an intermediate VE is used: (1) the antagonistic well-known transparency-stability problem is subsequently shifted to a local man–VE transparency problem without compromising any of the slave stability; thus, (2) one can state that safety risk is reduced of master dysfunction or a VE crash. The first could be diminished by using more user-friendly interfaces (see chap. 34, this volume, as is the case for many teleoperator instances cited herein). Subsequently, additional mapping functions are added to derive telerobotic commands (as is the case of the hidden robot architecture). In this case, the VE must offer rich and astute sensory feedback modalities, however, potentially restricting experiential telepresence, because this could be counterproductive. Indeed, in a telerobotic system a certain detachment may be desirable to keep the operator from becoming totally immersed in a manual control phase such that he might resist returning to higher level monitoring (this prohibits the use of HMDs in many actual teleoperator systems). As far as a VE crash is concerned, this problem connects software engineering (i.e., one must foresee recovery procedures to restart teleoperation from the crash state) with debugging facilities (e.g., frequent VE state saving). While safety and sensory feedback improvement receives special attention (e.g., http://www.ornl.gov/rpsd/humfac/baseframe.html), no generic performance taxonomy exists for classical teleoperator evaluation (though a VE usability taxonomy does exist, see chap. 34, this Volume). Yet one can state that in essence VE-assisted teleoperation is a move toward user-friendly, more refined control, rich multimodal feedback, but it is still difficult to give a qualitative and a quantitative measurement to these improvements. Concepts for teleoperation can be derived from human factor evaluation obtained from more general application results (Stanney, Mourant, & Kennedy, 1998).

## 4. VIRTUAL ENVIRONMENTS AS A POWERFUL TOOL FOR SPECIAL PURPOSE TELEOPERATORS

There are many special teleoperation applications among which VE plays a considerable role, either because without VE techniques, these applications would not be feasible, or because VE brings a considerable improvement and/or contribution.

### 4.1 Telesurgery

Improvement in the precision and reliability of robotic systems has won the trust of physicians and medical personnel. Indeed, applications of robotics in the medical field relate to surgery (see chap. 47, this volume), rehabilitation (see chaps. 49 and 50, this volume), general services including laboratory and prosthetics orthotics. Likewise, VE technology is a part of many medical applications. Because, on the one hand, an entire chapter is devoted to medical application of VE (see chap. 47, this volume), and, on the other hand, general medical robotics goes beyond the scope of this book, this section focuses on VE-based teleoperation aspects as applied to medicine and more specifically to telesurgery. Since the essence of surgery are precision and motion coordination of surgeons, it is not surprising that robotics—the technology of controlled motion—is investigated and widely used in operating wings. Because of the special safety needed, surgical robots are more "assistive" than "active." This means that in terms of robot control, the surgeon may always be in the control loop.

In minimally invasive surgery (MIS; e.g., laparoscopy, thoracoscopy, arthroscopy), robots serve as telemanipulators used to guide micro-instruments. An operation is performed with instruments and viewing equipment inserted into the body through small incisions made by the surgeon, in contrast to open surgery, which uses large incisions. Like arthroscopes used in orthopedics, MIS requires endoscopes. Endoscopic surgery has many common points with teleoperation, because the surgical environment could be seen as "remote," with sensing and manipulation projected via the endoscope and other long instruments. Nevertheless, endoscopes resemble mechanically coupled master–slave systems, i.e., action/feedback is directly mapped from the surgeon to the patient's organs. The well-known limitations of master-slave systems are hampering surgeon's abilities (Tendick, Jennings, Tharp, & Stark, 1993). In this mechanical "bilateral control" scheme, VE techniques enhance tactile sensory capabilities or augment visual feedback. Eventual use of robots as an intermediary for handling surgical tools is called telesurgery. Therefore, telesurgery is based on existing solutions from computerized teleoperation, and VE applications likely contribute in many ways to fulfill reduced dexterity, work space, and sensory input and feedback. Instances of actual telesurgery systems are cited in chap. 47 (this volume). Also, because tools are manipulated through medical robotic systems, the patient could actually be at a remote location from the surgeon. The other advantage gained from telesurgery is the use of a robot in a semi-autonomous mode. Indeed robots can move with a very low speed and can find less constrained paths thanks to their own haptic sensory or haptic sensory information (see Chaps. 5 and 6, this volume) fed back to the surgeon. This implies less healthy tissue damage for patients, resulting in shorter recovery time and reducing surgeon stress.

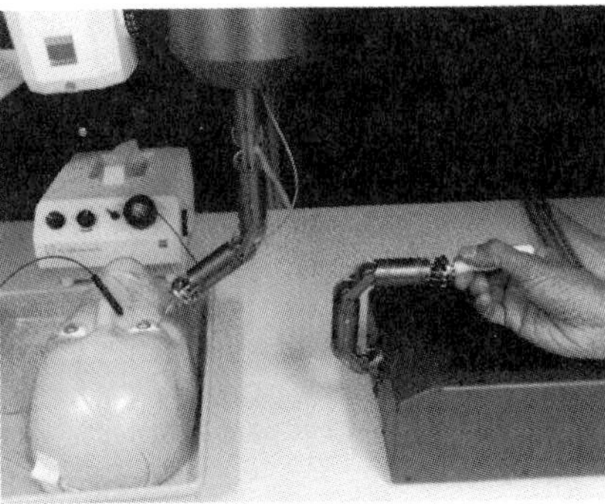

FIG. 48.13. Master–slave micro-telesurgery system. Courtesy of the RAMS Project at the Jet Propulsion Laboratory.

Teleoperation technology has also been investigated in the development of emerging microsurgery systems. An instance is the Robot Assistance Micro Surgery (RAMS) system, developed at JPL/NASA (see Fig. 48.13, Charles et al., 1997). The RAMS system is a 6 degrees-of-freedom master–slave telemanipulator with:

(1) different control schemes including direct telemanipulation, which includes task-frame referenced haptic feedback and shared automated control of robot trajectories;
(2) facilities such as physical scale of state-of-the-art microsurgical procedures; and
(3) features to enhance manual positioning (e.g., procedures such as manual positioning and tracking in the face of myoclonic jerk and tremor that limit most surgeons' fine-motion skills).

Another example demonstrating the VE contribution in tele-microsurgery is the prototype system developed by Hunter et al. (1993), for eye surgery applications. This system includes two force reflecting interfaces (a shaft shapedlike microsurgical scalpel) to control the left and right limbs of the slave microsurgical robot. A stereo camera system is used to relay visual feedback, on a worn helmet, to the surgeon. The camera position is controlled through the surgeon's head movements, which lead to interactive changes of the camera point of view (see Fig. 48.14).

There are many applications where surgical assistance robots can be considered as VE/computer guided robots through picture processing and augmented reality. The latter technique contributes to surgeon's operation strategy improvement. Enhancement of surgical environments with image overlay has been proposed for online use in neurosurgery (stereotactic brain surgery), orthopedics (like hip replacement), microsurgery, obstetrics, plastic surgery, and other specialties. In these applications, the physician can view medical images or computer generated graphics overlaid on and registered with the patient. For example, in neurosurgery a rendering of a brain tumor can be displayed inside the patient's head during surgery, providing localization and guidance to a surgeon. Three-dimensional image overlay capabilities can be used in place of, or to complement, telerobotic systems.

There are also many investigations to use telepresence and/or teleoperation techniques allowing a surgeon to operate at a location remote from his real physical location. But because of ethical and safety reasons, even if telerobots were used clinically, human assistants at the remote site would certainly be needed.

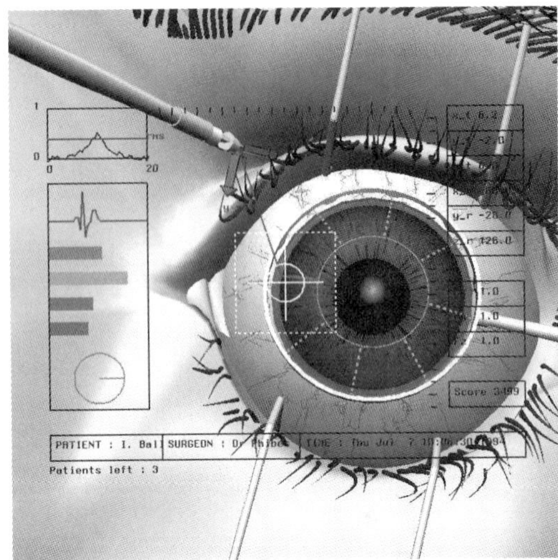


FIG. 48.14. Screen snapshot of a VE eye representation overlaid with additional information being used for online telesurgery as well as for training and simulation. Courtesy of Auckland University, New Zealand.

## 4.2 Teleoperation at Micro and Nano Scale

There is an important demand to reveal the nature of much smaller worlds. Micro and nano (abridged $\mu n$) system technology, including $\mu n$-robotics, is becoming a challenging area of research because of its potential applications. Interrelated applications concern: industry ($\mu n$: assembly, sensors, actuators, and mechanics, miniaturization); information technology (disk storage with high density, memory, semiconductor, and integrated circuits); biotechnology, biomedical, and genetic sciences (genes, biological particles and DNA manipulation; repairing or understanding mechanisms, cell handling, noninvasive eye and plastic surgery); and chemistry and materials (fabrication of $\mu n$structures and man-made materials, study of related quantum effect devices).

Consideration of $\mu n$-specific problems, in addition to task application, tools, and interconnection technologies specific requirements, lead obviously to many flexible $\mu n$manipulation concepts: purely manual teleoperation; automated; and robotics by means of flexible and cooperating $\mu n$-robots. Up-to-date, scanning tunneling microscopy (STM), scanning electron microscopy (SEM), and atomic force microscopy (AFM) seem to be the common tools for scanning and manipulating at the $\mu n$-scale. Each of the above microscopes has a range of applications linked to the remote environment nature and the kind of desired $\mu n$-tasks. Each has specific limitations while being used in an actual $\mu n$-manipulation. For instance, biological samples, such as cells, cannot be visualized using a SEM because their electrical properties can change during envisaged $\mu n$-manipulations.

Ideal performance requirements are such that a human operator manipulates in the normal-size world $\mu n$-parts and performs tasks (such as cutting, grasping, transportation, assembly, scratching, digging, and stretching), which have a direct similar mapping at the $\mu n$-world. Indeed, construction of the $\mu n$-manufacturing world is dependent on solutions adhering to the following constraints:

- The working environment must be perceivable by the operator and information in the processing scene must be transmitted accurately to the operator. As far as $\mu n$-tasks are concerned;

tools must be arranged in the observing area (colocality); bilateral magnification must be stable and fully transparent; and direct and natural perception is required with 3-D movements, dynamic images, sound, and haptics.

- The grasping, release, or assembly of $\mu n$-items in the $\mu n$-world need perspicacious procedures, which are completely different from those commonly used in the macro-world. Gripping with forceps may not be adequate in many cases. Indeed, vacuum-assisted gripping by electric power or fluid and release under vibration, the use of $\mu n$-forces and adhesion, the use of plunger mechanism, and other astute techniques are mostly used. Moreover, the success and performance of $\mu n$-grasping, release, or assembly functionality together with $\mu n$-manipulation are, in many cases, dependent on operator skill.
- The operational remote environment is actually a hostile environment to humans because $\mu n$-world components behavior is very complex to understand and to manipulate. Moreover, since humans operate based on macro world model physics, tasks cannot be easily executed by the operator. In the $\mu n$-world, mass and inertial forces are negligible and $\mu n$-interaction forces are non linear and resultant forces from van der Waals, capillary, electrostatic, pull-off, rubbing phenomena, and even radiation forces of light exceed the gravity.
- Task taxonomies might include operational sequences (e.g., positioning, assembling, grip, release, adjust, fix-in-place, push, pull, etc.) and processing steps (e.g., cutting, soldering, gluing, drilling, twisting, bending, etc.).

To link the macro world to the $\mu n$, an interface that can match the two physical worlds and compensate for human operator inaccuracy is indispensable.

From the vision feedback point of view, VE can bring a considerable contribution. Indeed, the field of view or scanning is in most cases restricted to a small area and the distance between $\mu n$-objects, and the lens or the probe is very small. Moreover, on the one hand, scanning (which is on the order of seconds to minutes in some cases) does not allow online imaging. On the other hand, since the same single probe is used for both functions, scanning and $\mu n$-manipulating, these cannot physically be achieved in parallel whatever the scanning speed. Finally $\mu n$-operation are executed, in general, within the field of view. For these reasons, a 3-D VE topology can be built and displayed to the user. A 3-D virtual $\mu n$-world can be a global view of the real $\mu n$-world or restricted to the actual working area, according to the operator needs. Static or intuitively manipulated multiple views are then allowed in real time. As well, remote features may be augmented with multiple contrasting colors to present data in a comprehensive and easily interpreted manner.

The lack of direct 3-D visual feedback from the $\mu n$-world on the one hand, and the fragility of the manipulated $\mu n$-objects on the other hand, make force feedback an essential — even unavoidable —component of the macro/$\mu n$-world interface. Indeed it is primordial to understand well the condition of the probe during operation. An excessive force applied on a $\mu n$-object may lead to a non-negligible degree of probe or object deformation, destroy the $\mu n$-object, or make the $\mu n$-object flip away. As well-known, $\mu n$-interaction is not reproducible enough to automate $\mu n$-procedures. Hence teleoperation control mode seems to be attractive; however, the traditional bilateral control through master–slave coupling has to face many severe problems: understanding of $\mu n$-dynamics together with reliable modeling; the effect of various nonlinear forces is completely different from the macro world (attractive, repulsive, and adhesion forces, etc.); bilateral stable, transparent, robust scale mapping, and others. Thus, it is not always possible to adapt easily conventional bilateral methods, moreover, monitoring small forces of 1 $\mu$N to 1 $n$N range needs very accurate sensors. Thus, solutions using a VE-based intuitive interface, which hides the details of performing complex tasks using SPM in combination with 3-D topography, seem to be an attractive solution. As for assisted teleoperation at the macro world, the interface would include virtual tools or virtual effective probes (Finch et al., 1995)

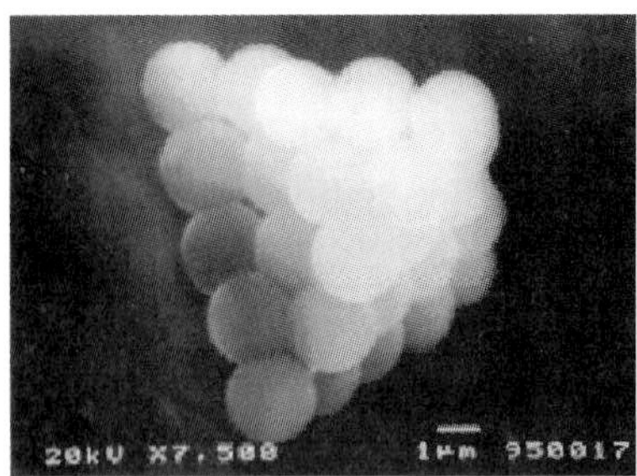


FIG. 48.15. Micro-Pyramid constructed with a $\mu$-handling robot.

together with a 3-D representation used as a functional intermediary to map operator actions to the $\mu n$-world (rough to fine methodology) and vice versa. Then, the degree of abstraction of the commands would be determined by the capabilities of the control system. Hereafter, some developed prototype systems are mentioned.

At the University of Tokyo, Sato's team developed one of the most advanced $\mu n$-teleoperators (Sato, 1996). As a haptic interface, the system utilizes a touch sensor screen and specially designed, pencil-shaped master manipulator to enable sensitive bilateral $\mu n$-teleoperation thanks to its lightweight and small inertia during movements. A $\mu$-handling result (see Fig. 48.15 at right, courtesy of Professor Sato, the University of Tokyo) is a Micro-Pyramid constructed with a $\mu$-handling robot inside an SEM. This system is currently being enhanced by a more user-friendly interface based on bilateral behavior media for $\mu n$-teleoperation.

Other instances are from the Nanomanipulator research group at the University of North Carolina (Finch et al., 1995). Their system is using advanced VE techniques as an interface enhanced with virtual tools (grabbing, scaling, flying, etc.), virtual measuring fixtures (marked mesh, etc.), standard VCR functions (replay, save, etc.), and virtual modification tools. The main applications are concerned with biological studies and virus manipulations (see Fig. 48.16), the PHANToM is a force feedback desktop mechanism for VE applications, http://www.sensable.com).

At the same university, studies are being conducted in the frame of molecular docking. Indeed, today's computer-aided molecular design software (Sibyll™, MolMol™, etc.) lacks user-friendly interfaces and algorithmic search of the configuration space is extremely costly. In return for advanced VE techniques and an "intelligence augmentation" man–machine system (see Fig. 48.17), chemists and drug designers would gain in efficiency and time (Ouh-young, Pique, Hughes, Srinivasan, & Brooks, 1988).

Another tele-nano-manipulation system is being developed in the Hashimoto laboratory at the University of Tokyo. This system is using an AFM for both scanning and manipulating at the nano scale. As a user-friendly interface, the tele-nano-manipulator is doted with a 1-degree-of-freedom haptic interface and a 3-D topology builder for nano manipulations (Sitti & Hashimoto, 1999). The controller is actually a bilateral mode based on impedance shaping to allow the operator to feel the forces from the nano world. Different control strategies utilizing the haptic interface have been tested, and results are thoroughly described in Sitti and Hashimoto (1999; see Fig. 48.18.)

## 4.3 Mobile Robots Teleoperation

Since long ago, mobile robotics had the connotation of autonomy and artificial intelligence, and was not concerned with teleoperation. Early mobile robots were designed with a "teleoperation" mode so a human operator could unilaterally control them during transportation,

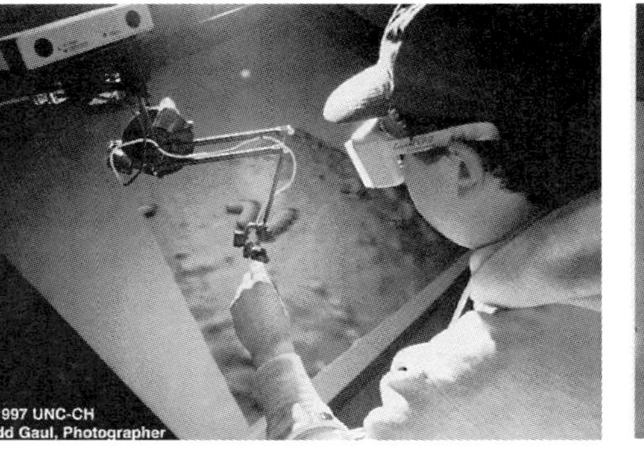

FIG. 48.16. Different *n*-manipulators configurations (*from left to right*): PHANToMT-based desktop configuration; a WorkBench together with the PHANToMT force feedback stylus; and the Argone robotic master arm (*right*). Courtesy of the nanoManipulator research group, University of North Carolina.

FIG. 48.17. Molecular docking using advanced VE interfaces. Courtesy of the IDock research group, University of North Carolina.

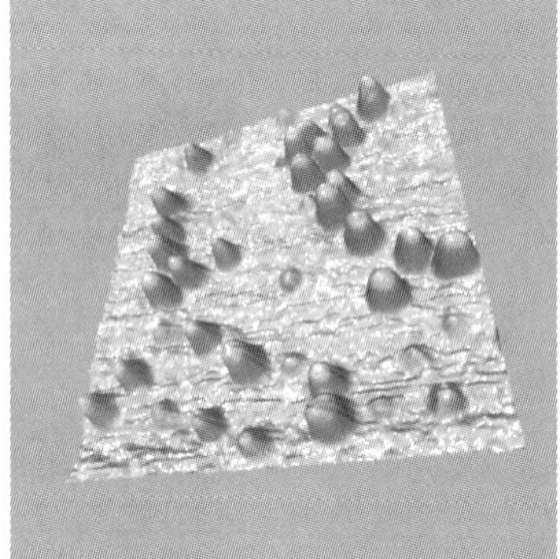

FIG. 48.18. Tele-nano-manipulation using 1 degree-of-freedom haptic interface (*left*), scanned synthetic converted AFM picture (*right*). Courtesy of Sitti and Hashimoto, the University of Tokyo.

and setup phases. Difficulties in achieving completely autonomous behavior (see chap. 15 this volume) in terms of auto-decision and of a mobile robot unilateral teleoperation mode turns out not to be simple when the robots are conceived as a complex structure (subsea and flying robots), with specific locomotion (such as many legs) or propelling mechanisms, or when refined feedback using the operator haptic channel is attempted. Pioneering work in the latter was done by Clement, Fournier, Gravez, and Morillon (1988). Teleoperation mode can be considered as an intrinsic characteristic of advanced autonomous mobile robots. We emphasize this proposition with a simple example; the conceiver of the well-known Honda Humanoid robot reported: "... this robot is completely autonomous" and with insistence declared: "But, it can be teleoperated as well." We stress that this specificity tends to draw closer a classical teleoperation field researchers utterance: "... this robot is teleoperated, but, it can be autonomous as well." As far as mobile robot teleoperation is concerned, VE techniques can bring similar advantages as for classical robots. From the related literature, many VE-based mobile robot architectures have been proposed. Keywords that seem to be a standard in the way VE techniques are used include: off-line mission planning; highly interactive environment for abstract programming at various levels; simulation and prediction of complex missions with a virtual teleoperated vehicle; robot computer-assisted design; and training, safety, and user friendly control through augmented reality (superposition of virtual and real environments may reveal discrepancies or sensors defects, etc.).

In the work of Komoriya and Tani (1990), an actual robot together with a VE set around it constitute a simulation system considered to lie between the full computer simulation and the full actual system. This hybrid system provides an efficient testbed for developed control laws and planning algorithms, without any damage risks for the robot.

Another VE technique for teleoperation has been proposed by Kheddar (1997), which allows prediction and bilateral robot teleoperation control with multimodal feedback, including haptics. In this proposed scheme, operator commands will involve desired robot reorientation, speed, and acceleration, embedded tool(s) control, global inertia redistribution, and other factors. It has been suggested that pertinent parameters that the operator may feel include: stability margins, inertial forces, applied torque, joint limit margins, contact points and forces, related obstacles distance, and robot attitude. All these parameters are computed locally (i.e., based on the virtual model of the robot and its environment) and fed back to the operator locally thanks to a suitable sensory substitution. The appropriate sensory substitution is chosen to stimulate the operator sensory channel so that subsequent telerobotic actions are reflex generated. In this case, the VE constitutes a filter to intuitively control safely the robot since actual robot controls are sent only if some selected parameters fit within allowed margins. Nevertheless bilateral control is permitted only if the communication time delay between the master and robot is small enough (less than 1 sec). In the case of larger time delays, the VE is functionally used similarly to arm teleoperation (i.e., as a predictive or a teleprogramming interface). This is thoroughly explained in the following cases.

The exceptional instance highlighting the use of VE techniques in mobile robot teleoperation is indubitably the Mars Path Mission (a NASA Discovery exploring Mars planet). Indeed, a VE-technology-based supervision and control workstation was designed to remotely command the *Sojourner* rover, which landed on Mars by July 4th, 1997. A graphical user interface supplied numerous available rover commands (i.e., macro-operations with respective various parameters, such as Calibrate Heading with Sun, Capture Images, Drive, GoTo WayPoint, etc.). Offline, a 3-D terrain model is processed from both previous mission accumulated stereo images obtained from an embedded IMR camera (a stereoscopic imager) and the partial VE already constructed from previous missions. Hence, using a simple SpaceBall as an I/O device together with a stereo rendering of the virtual working environment and a virtual model of the Sojourner rover, the operator designs plans and actions to be achieved by means of a set of graphic programming metaphors. Subsequently, when the simulation results are considered to be satisfactory, a control code is generated and transmitted, by means of the Deep Space Network (up to 6 $\sim$ 11 sec time delay, thus prohibiting manual teleoperation), to the actual Sojourner rover on Mars (see Fig. 48.19).

More enhanced interfaces and telerobotic technology are being developed for future unmanned exploration of the "Red Planet" to provide scientists with a telepresence interface for real-time interaction with, and interpretation of, the returned geophysical data (Backes, Kam, & Tharp, 1999).

In the frame of subsea robotics, Paul's team at the GRASP Laboratory applied the teleprogramming concept they developed in the frame of a subsea ROV equipped with a robotic arm (Sayers, 1999). An experiment has been conducted with the Deep Submergence Laboratory of the Woods Hole Oceanographic Institution (WHOI Laboratory). Bandwidth of some Kbits/sec and round-trip communication delays of 7 seconds are typical of the subsea acoustic transmission mean. The operator disposes of a VE remote site wherein control or supervision tasks can be achieved. Remote slave actions are governed by those resulting from the master station. Eventual discrepancies are resolved from a continuous comparison between the virtual (simulated) and the real internal and external embedded sensors. Subsequent mismatches and diagnosis are deduced by the operator that is responsible for the suitable recovery procedures to be taken (see Fig. 48.20).

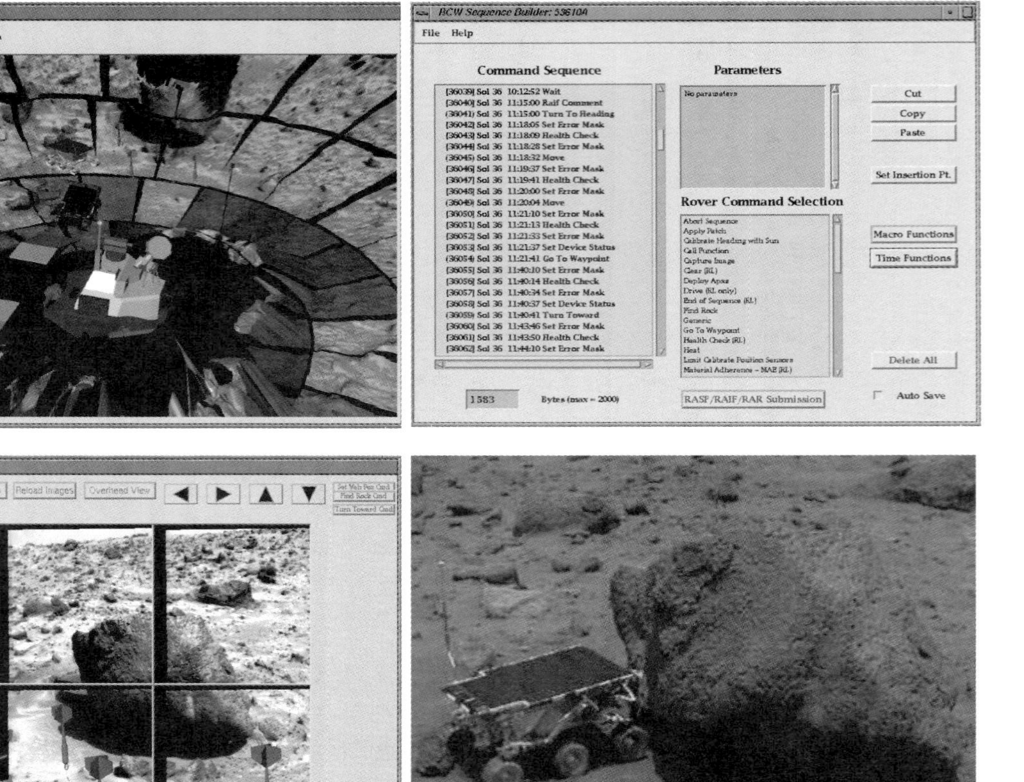


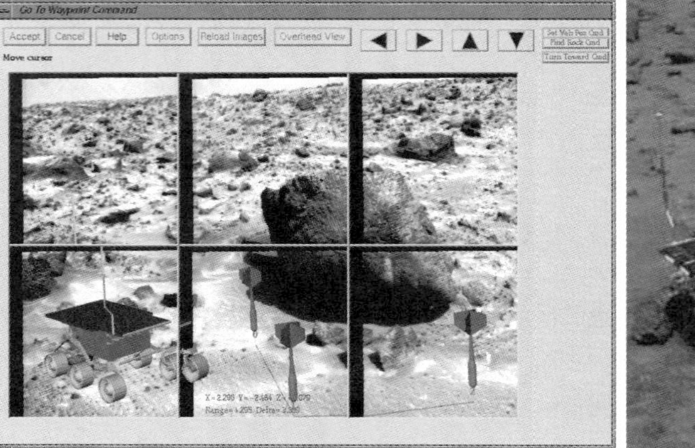

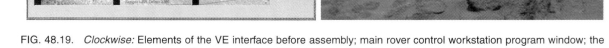

FIG. 48.19. *Clockwise:* Elements of the VE interface before assembly; main rover control workstation program window; the actual Sojourner rover at work on Mars; and driver's VE interface with Waypoints shown. Courtesy of the Jet Propulsion Laboratory, NASA, http://robotics.jpl.nasa.gov/tasks/scirover/homepage.html.

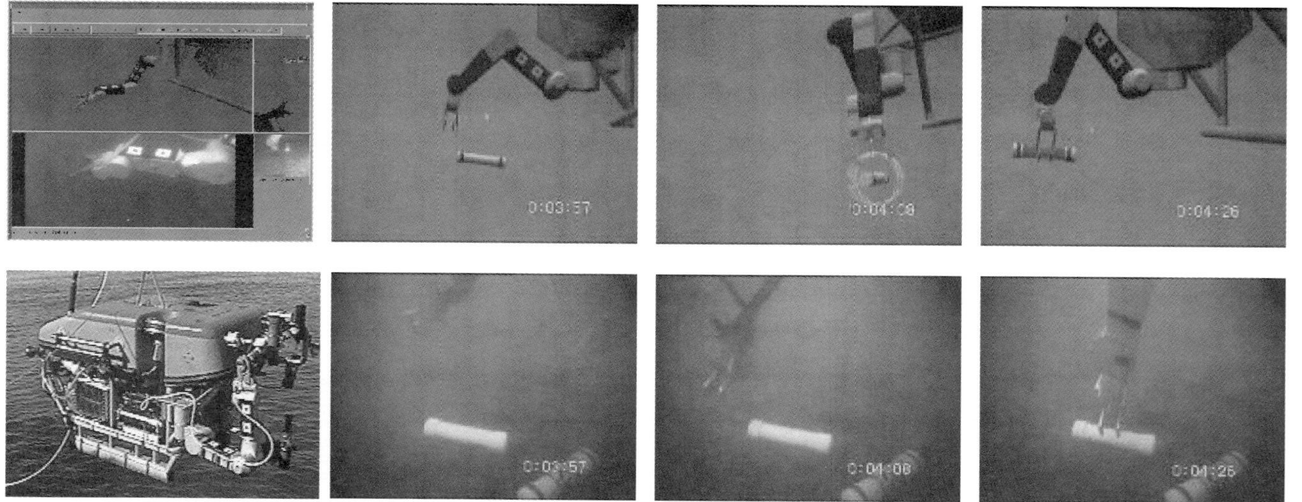


FIG. 48.20. Teleprogramming as a mean for subsea ROV teleoperation. Upper row pictures design of the master station, lower row pictures show snapshots of the video feedback from the slave (JASON ROV) environment. Courtesy of Sayers, the GRASP laboratory, University of Pennsylvania and the WHOI Laboratory.

At the French Institute of Research and Exploitation of the Sea (IFREMER), the control of robots for intervention, reliability of these systems, and dynamic stabilization of the machines at low speed are the aspects under investigation. A VE technique named VESUVE (Virtual Environment for SUbsea VEhicles) has been developed for 3-D subsea scenes simulation and visualization.

VESUVE offers a general framework for creation, visualization, real-time animation, and interactive manipulation of any kind of virtual subsea scene involving deep sea vehicles. The main objective of VESUVE is to study, implement, and evaluate VE technologies for subsea vehicles. Its potential areas of application are: engineering of subsea vehicles (visual simulation for integration, test, and validation of the system, real-time visualization of CAD data, visual simulation for pilot training or scientist familiarization); operation of subsea vehicles (3-D piloting aid for ROV or towed vehicles, 3-D navigation aid, telepresence, visual simulation for mission rehearsal or debriefing, teleprogramming for autonomous or semi-autonomous vehicles); and subsea robotics R&D (virtual lab for experimental subsea robots in research fields like control, mission programming, navigation, fault detection diagnosis and recovery, and real-time presentation of simulation results, see Fig. 48.21).

Obviously there are still many other remote mobile robotic teleoperation systems using telepresence and VE interfaces, such as the VEVI (a distributed Virtual Environment Vehicle Interface) developed at the NASA Ames Research Center's Intelligent Mechanisms Group (Hine et al., 1995), adopted to the Dante mission at Alaska, Antarctica, and other locations, and the advanced computer project to assess Chernobyl damage. We do hope that these few instance being used on actual hazardous environment missions have shown the benefits that VE brings to advanced teleoperator schemes in the special mobile robot teleoperation context.

### 4.4 Web-based Teleoperation

Some of the most challenging purposes for today's multimedia systems are to include haptic data exchange (see chaps. 5 and 6, this volume) and to allow physical teleworking, through the network (Internet World Wide Web, for instance, see chap. 16, this volume). Currently various remote robots can be controlled through the Internet network by means of any Internet browser. A number of Web addresses are available at: http://ranier.oact.hq.nasa.gov/telerobotics_page/realrobots.html.

Control interfaces are using Java Applets with different available command buttons allowing any user connected to the Internet to experience actual remote control of robots (see Fig. 48.22)—arms, mobile cameras, telescopes, and even toy trains!. Tasks are generally concerned with object assembly, exploration (of museums), among others. The user may see the result of his actions via continuous video feedback (in general, the refresh rate is too slow due to network traffic).

Many laboratory experiments involving teleprogramming through the Internet and diverse other network protocols have been conducted. Many of these, such as the ROTEX experiment and subsea teleoperation, have already been mentioned. An advanced ATM Network has been used by the Fukuda team in Japan for a multimedia telesurgery context. The satellite network has also been used in a telesurgery experiment between the Bejczy team at the Jet Propulsion Laboratory and Rovetta team at the Politecnico di Milano, Italy (references are quoted in Kheddar, Tzafestas, Coiffet, Kotoku, & Tanie, 1998).

Other experiments have been conducted between the Laboratoire de Robotique de Paris and the Mechanical Engineering Laboratory involving the control of parallel multirobots through a single operator and VE interface based on the hidden robot concept (Kheddar et al., 1998).

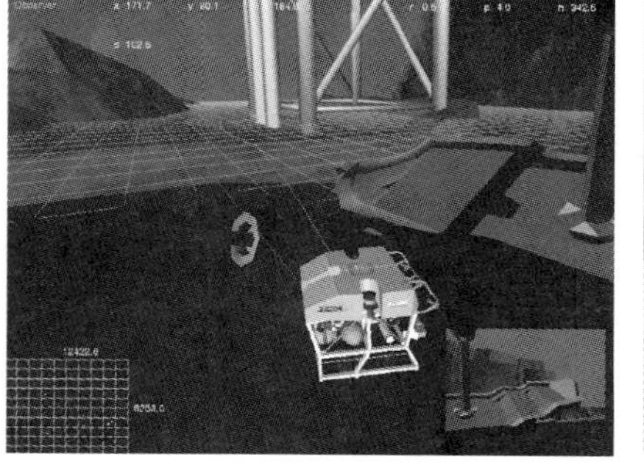

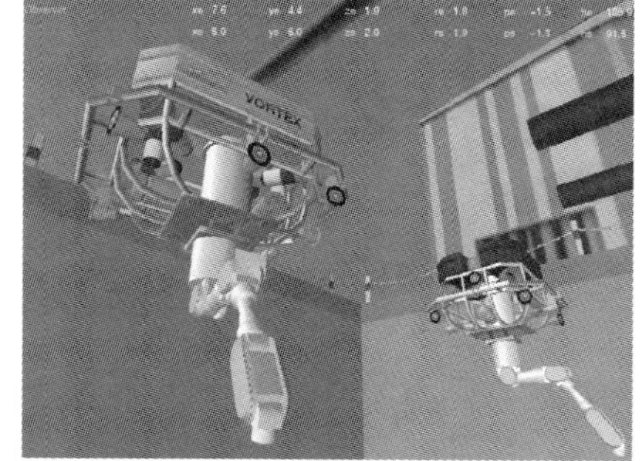


FIG. 48.21. VISUVE case studies: Cooperative ROVs and control simulation; offshore planning; and the *Titanic* mission simulation, an aid for the actual remote control of the VICTOR 6000 ROV. Courtesy of the IFREMER, France, http://www.ifrem.fr/ditidsiw3_uk/engins/vesuve.html.

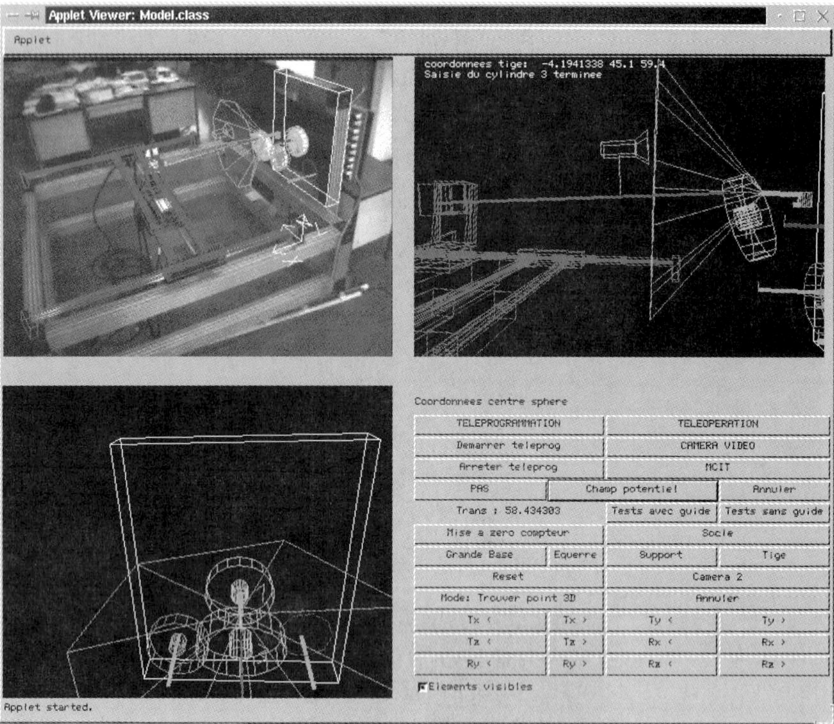


FIG. 48.22. An Internet Web-based teleoperation interface using augmented reality and online assistance by means of virtual fixtures. Courtesy of Otmane, the Complex Systems Laboratory, Evry University, France.

With different robots controlled in parallel, a common intermediary functional representation is imperative and enhances the interest of the proposed concept.

As shown in Fig. 48.23, the teleworking experiment consisted of a four-piece puzzle assembly within a fence on a table. The real remote assembly operation was to be performed by slave robots (one situated in Japan and three in France, for the first experiment, one in France for the experiment described herein). The operator performs the virtual puzzle assembly using his own hand, skill, and naturalness. Visual and haptic feedback is local and concerns only the graphic representation of the remote task features without any remote robot. Operator–VE interaction parameters are sent to another workstation in order to derive robot actions (graphically represented for software check and results visualization) and do not involve direct operator action/perception. Video feedback was kept for safety and error recovery purposes.

Web-based teleoperation has a direct impact on industry; possible applications include telemaintenance, telemonitoring, and telesupervision. As an instance, an European project for maintenance system based on telepresence for remote operators was launched in September 1996 (Bisson & Conan, 1998). The resulting technology enabled remote users to train themselves to deal with maintenance tasks by connecting to a "virtual showroom" where they could learn maintenance procedures through computer augmented video based telepresence and augmented reality techniques.

The teleoperation environment consisted of (1) an extended prototype of the equipment to be maintained, that is, the target equipment, customized to allow remote control and extensive diagnostic, and (2) auxiliary devices such as manipulators and vision systems, used

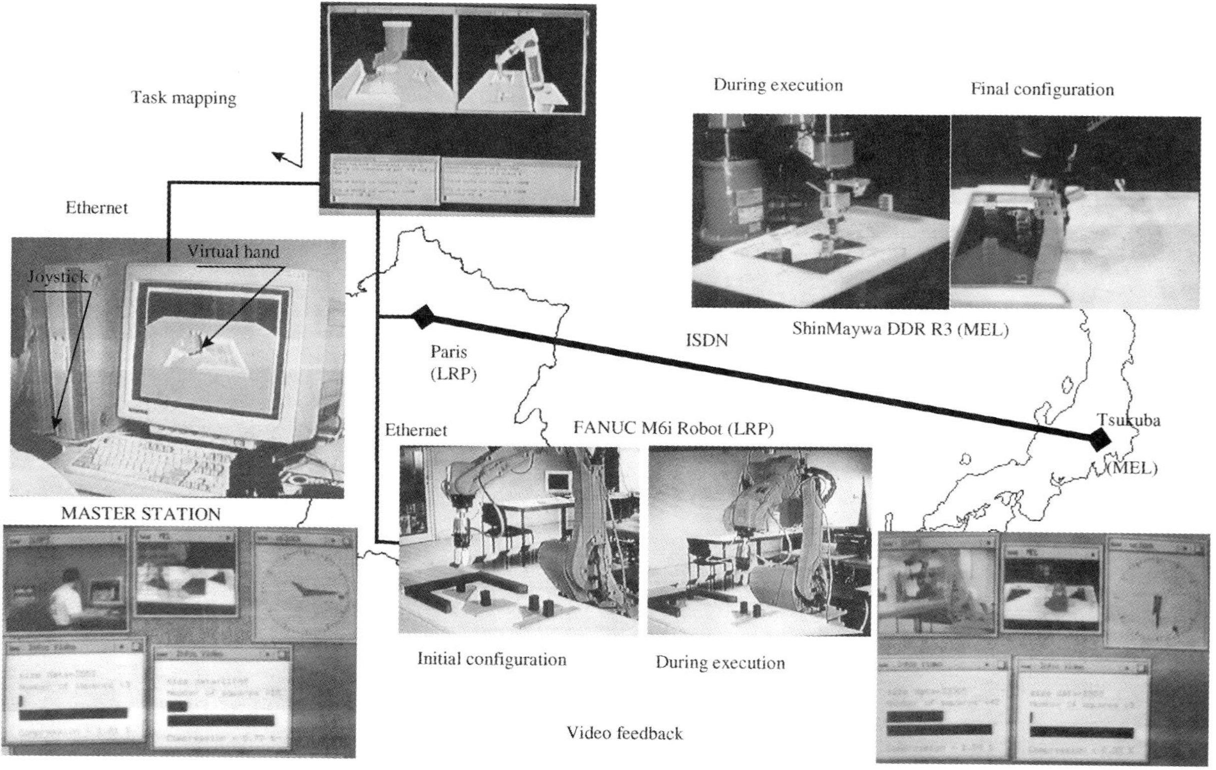


FIG. 48.23. Parallel long-distance multirobots teleoperation. Courtesy of the Labaratoire de Robotique de Paris and the robotics department of the Mechanical Engineering Laboratory.

both for equipment servicing and for enhancing user interoperability. Marking techniques allow registration of a virtual model of the video designed equipment provided by the remote camera. The virtual model is annotated with links to a multimedia database containing specific information (e.g., operating and maintenance instructions, functional data describing regular, or faulty conditions). Speech recognition is used as well to allow spoken commands. This system is dedicated to the training of complex maintenance and installation scenarios for industrial companies that cannot afford on-site complex training equipment (see Fig. 48.24).

NASA's Jet Propulsion Laboratory and Ames Research Center are developing a next-generation ground system called RoCS for (Rover Control Station) for use on planetary rover missions. The RoCS architecture includes a module called WITS for Web Interface for TeleScience (Backes et al., 1999). The latter serves two purposes: (1) an interface for scientists to use from their home institutions to view and download data and generate commands for the remote rover; and (2) to make the interface available to the public domain: http://robotics.jpl.nasa.gov/tasks/scirover or http://mars.graham.com/wits. Thus, any person connected to the server could downlink images, generate rover waypoints, science targets, and commands, which, however, cannot be sent to the actual or true rover.

## 5. PROGNOSIS AND CONCLUSION

Undoubtedly there is still much to say concerning the prosperous manner in which robotics can benefit from VE techniques (Burdea, 1999; see Burdea & Coiffet in Nof, 1999). It is impressive to note that, a cleverly conceived yet "simple" VE intermediary representation contributes to solving the time-delay problem, offers ingenious metaphors for both operator assistance and robot autonomy sharing problems, enhances operator sensory feedback through multiple sensory modalities admixtures, enhances operator safety, offers a huge possible combination of strategies for remote control and data feedback, shifts the well-known antagonistic transparency/stability problem into an operator/VE transparency one without compromising the slave stability, offers the possibility to enhance—in terms of pure control theory—remote robot controllers, allows new human-centered teleoperation schemes, permits the production of advanced user-friendly teleoperation interfaces, makes possible the remote control of actual complex systems, such as mobile robots, nano and micro robots, surgery robots, and others. Subsequently, VE techniques seem to provide the ultimate magic solution solving all the hard teleoperation problems at a glance. Nevertheless, this considerable opening gives rise to additional new problems.

One problem deals with VE modeling techniques (see chap. 12, this volume). There is a grand demand for robust software, which can reproduce the VE with high fidelity and sufficiently rich realism to allow both real-time user-friendly operator interaction and easy implementation of necessary metaphors, such as virtual guides, that allow operator assistance and enhance local sensory feedback and robot autonomy sharing in an intuitive way (i.e., without setting the operator at a determined control layer). Nevertheless, in some cases, namely when remotely controlled systems are complex, there is a need for a virtual representation that cleverly keeps only functional aspects of the complexity of the remote environment, thereby unburdening the operator and facilitating control or supervision. This aspect is only possible if a clever bilateral mapping exists between real environments and their virtual functional representations.

The other hard problem with using VE technology is the error detection and recovery inherent to VE—real-environment discrepancies. This problem has received little attention from the teleoperation researchers community. It could be tackled from two ways: a high-fidelity error detection strategy (eventually involving the operator) that matches what is expected to be

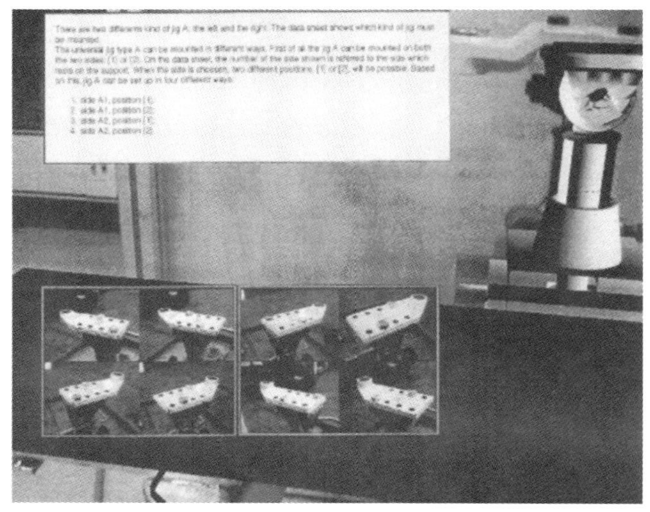

FIG. 48.24. Teleoperation environment. An XYZ table plus yaw pitch camera to explore the maintenance equipment (*left*). Output (AR picture) from the MAESTRO, answer to query: "What are the different jig types?" Courtesy of Bisson, Thomson CSF, France.

done in the VE and what is actually done in the real environment; and a strategy for recovery based on robot autonomy and local sensory robot perception. If this problem can be solved, a merely partial virtual representation of the real environment could be acceptable.

In conclusion, teleoperation efficiency was almost stagnant during several decades since the discovery of the master–slave structure by Goertz in the 1950s (Vertut & Coiffet, 1985), and that was in spite of technical improvements and the use of computers. It seems that VE brings total renewal through the development of telepresence or tele-existence concepts. Now, the desired efficiency could be a reality, allowing not only a clear improvement in traditional applications such as nuclear or space activities, but opening numerous new fields of interest. Thanks to VE technology, teleoperation/telepresence could soon represent a major percentage of robotics applications.

## 8. REFERENCES


Anderson, R. J. (1994). Teleoperation with virtual force feedback. Lecture notes in *Control and information sciences 200: Experimental robotics III*. (pp. 366–375). New York: Springer-Verlag.

Anderson, R. J., & Spong, M. (1988). Bilateral control of teleoperators with time delay. In *Proceedings of the IEEE Conference on Decision and Control*. (pp. 167–173). Austin, USA.

Backes, P. G., Kam, S. T., & Tharp, G. K. (1999). The Web interface for telescience. *Presence: Teleoperators and Virtual Environments, 8* (5), 531–539.

Backes, P. G., Peters, S. F., Phan, L., & Tso, K. S. (1998). Task lines and motion guides. *Presence: Teleoperators and Virtual Environments, 7* (5), 494–502. Available: http://robotics.jpl.nasa.gov/people/backes/homepage.html

Bejczy, A. K., Kim, W. S., & Venema, S. (1990). The Phantom robot: predictive displays for teleoperation with time delay. In *Proceedings of the IEEE International Conference on Robotics and Automation*. (pp. 546–551).

Bisson, P, & Conan, V. (1998). The MAESTRO Project: an augmented reality environment for telemaintenance. In *Proceedings of the 6 éme Journées de travail du GT réalité virtuelle*. Issy-les-mounlinaux, France.

Burdea, G. C. (1999). Invited review: The synergy between virtual reality and robotics. *IEEE Transactions on Robotics and Automation, 15* (3), 411–422. Available: http://www.caip.rutgers.edu

Burdea, G. C., & Coiffet, Ph. (1999). Virtual reality and robotics. In S. Y. Noy (Ed.), *Handbook of robotics*. New York: John Wiley & Sons.

Charles, S., Das, H., Ohm, T., Boswell, C., Rodriguez, G., Steele, R., & Istrate, D. (1997). Dexterity-enhanced telerobotic microsurgery. In *Proceedings of the International Conference on Advanced Robotics* (pp. 5–10). Monterey, CA. Available: http://robotics.jpl.nasa.gov/tasks/rams/homepage.html

Clement, G., Fournier, R., Gravez, Ph., & Morillon, J. (1988). Computer-aided teleoperation: From arm to vehicle control. In *Proceedings of the IEEE International Conference on Robotics and Automation* (pp. 590–592). Philadelphia, Pennsylvania:

ETS-VII. (1999). Space Robot sessions on ETS-VII experiments and results. In *Proceedings of the International Conference on Advanced Robotics* (TA1-A, pp. 243–273; TP1-A, pp. 329–360; TP2-A, pp. 409–436). Tokyo, Japan:

Finch, M., Chi, V. L., Taylor, II, R. M., Falvo, M., Washburn, S., & Superfine, R. (1995). Surface modification tools in a virtual environment interface to a scanning probe microscope. In *Proceedings of ACM Symposium on Interactive 3D Graphics* (pp. 13–19). New York. Available: http://www.cs.unc.edu/nano/etc/www/nanopage.html).

Fournier, R., Gravez, Ph., Foncuberta P., & Volle C. (1997). MAESTRO hydraulic manipulator and its TAO-2000 control system, In *Proceedings of the seventh ANS Topical Meeting on Robotics and Remote Systems*. Augusta, GA:

Fraisse, P., Pierrot, F., & Dauchez, P. (1993). Virtual environment for robot force control. In *Proceedings of the IEEE International Conference on Robotics and Automation* (pp. 219–224). Atlanta, GA: IEEE Press.

Freund, E., & Rossmann J. (1999). Projective virtual reality: Bridging the gap between virtual reality and robotics. *IEEE Transactions on Robotics and Automation, 15* (3), 411–422. Available: http://www.irf.de/welc_eng.htm

Funda, J., Lindsay, T. S., & Paul, R. P. (1992). Teleprogramming: Towards delay-invariant remote manipulation. *Presence: Teleoperators and Virtual Environments, 1* (1), 29–44.

Hint, B., Hontalas, P., Fong, T., Piguet, L., Nygren, E., & Kilne, A. (1995). VEVI: A virtual environment teleoperations interface for planetary exploration. In *Proceedings of the SAE 25th International Conference on Environmental Systems* (preprints). San Diego, CA: Available: http://img.arc.nasa.gov/VEVI

Hunter, I. W., Doukoglou, T. D., Lafontaine, S. R., Charrette, P. G., Jones, L. A., Sagar, M. A., Mallison, G. D.,

& Hunter, P. J. (1993). A teleoperated microsurgical robot and associated virtual environment for eye surgery. *Presence: Teleoperators and Virtual Environments*, *2* (4), 265–280. Available: http://biorobotics.mit.edu and http://www.esc.auckland.ac.nz/Groups/Bioengineering

IEEE TRA. (1993). *IEEE Transactions on Robotics and Automation* [Special issue], *9* (5).

Joly, L. D., & Andriot, C. (1995). Imposing motion constraints to a force reflecting telerobot through real-time simulation of a virtual mechanism. In *Proceedings of the IEEE International conference on robotics and automation* (pp. 357–362). Nagoya, Japan:

Kheddar, A. (1997). *Teleoperation based on the hidden robot concept*. Unpublished doctoral dissertation, University of Paris, 6, France.

Kheddar, A., Tzafestas, K. S., Coiffet, Ph., Kotoku, T., & Tanie, K. (1998). Multi-robot teleoperation using direct human hand actions. *Advanced robotics*, *11* (8), 799–825.

Kim, W. S. (1996). Virtual reality calibration and preview/predictive display for telerobotics. *Presence: Teleoperators and Virtual Environments*, *5* (2), 173–190. Available: http://robotics.jpl.nasa.gov/people/kim/csv/homepage.html

Komoriya, K., & Tani, K. (1990). Utilization of virtual environment system for autonomous control of mobile robots. In *Proceedings of IEEE International Workshop on Intelligent Motion Control* (pp. 439–444). Istanbul, Turkey: IEEE Press.

Kosuge, K., Itoh, T., Fukuda, T., & Otsuka, M. (1995). Tele-manipulation system based on task-oriented virtual tool. In *Proceedings of the IEEE International Conference on Robotics and Automation* (pp. 351–356). Nagoya, Japan: IEEE Press.

Kosuge, K., Murayama, H., & Takeo, T. (1995). Bilateral feedback control manipulation system with transmission time delay. In *Proceedings of the International Conference on Intelligent Robots and Systems*. Vol. 3 (pp. 1380–1385). Osaka, Japan.

Kotoku, T. (1992). A predictive display with force feedback and its application to remote manipulation system with transmission time delay. In *Proceedings of the IEEE/RSJ International Conference on Intelligent Robots and Systems* (pp. 239–246). Raleigh, NC: IEEE Press: Available: http://www.mel.go.jp

Mallem, M., Chavand, F., & Colle, E. (1992). Computer-assisted visual perception in teleoperated robotics. *Robotica, 10*, 93–103. Available: http://www.univ-evry.fr/labos/cemif/index.html

Niemeyer, G., & Slotine, J-J. (1991). Stable adaptive teleoperation. *The IEEE Journal of Oceanic Engineering, 16*(1), 152–162.

Noyes, M. V. & Sheridan, T. B. (1984). A novel predictor for telemanipulation through time delay. In *Proceedings of the Annual Conference on Manual Control*. NASA Ames Research Center, Moffelt Field, CA.

Ouh-young, M., Pique, M., Hughes, J., Srinivasan, N., & Brooks Jr., F. P. (1988). Using a manipulator for force display in molecular docking. In *Proceedings of the IEEE International Conference on Robotics and Automation* (pp. 1824–1829). Philadelphia: IEEE Press.

Oyama, E., Tsunemoto, N., Tachi, S., & Inoue, T. (1993). Experimental study on remote manipulation using virtual reality. *Presence: Teleoperators and Virtual Environments*, *2* (2), 112–124.

Rosenberg, L. B. (1992). *The use of virtual fixtures as perceptual overlays to enhance operator performance in remote environments*. (Tech. Rep. No. AL-TR-1992-XXX). Wright Patterson Air Force Base, OH: U.S.A.F. Armstrong Laboratory.

Sato, M. (1996). Micro/nano manipulation world. In *Proceedings of the IEEE/RSJ International Conference on Intelligent Robotics and Systems* (pp. 834–841). Available: http://www.lssl.rcast.u-tokyo.ac.jp:80/~tomo

Sayers, C. (1999). *Remote control robotics* [online]. Springer Verlag Ed. Available: http://www.cis.upenn.edu/~sayers

Sheridan, T. (1992). *Telerobotics: Automation and human supervisory control*. Cambridge, MA: MIT Press.

Sitti, M., & Hashimoto, H. (1999). Teleoperated nano-scale object manipulation. In *Recent advances on mechatronics*. Springer-Verlag.

Small, D. E., & McDonald, M. J. (1997). Graphical programming of telerobotic tasks. In *Proceedings of the Seventh Topical Meeting on Robotics and Remote System* (pp. 3–7). Augusta, GA: Available: http://www.sandia.gov/LabNews/LN10-25-96/robot.html

Stanney, K. M., Mourant, R. R., & Kennedy, R. S. (1998). Human factors issues in virtual environments: A review of the literature. *Presence: Teleoperators and virtual environments*, *7* (4), 327–351.

Struges, R. H. (1996). A review of teleoperator safety. *International Journal of Robotics and Automation*, *9* (4), 175–187.

Tachi, S. (1998). Real-time remote robotics—Towards networked telexistance. *IEEE Computer Graphics and Applications*, *18* (6), 6–9. Available: http://www.star.t.u-tokyo.ac.jp

Tendick, F., Jennings, R. W., Tharp, G, & Stark, L. (1993). Sensing and manipulation problems in endoscopic surgery: Experiment, analysis, and observation. *Presence: Teleoperators and Virtual Environments*, *2* (1), 66–81. Available: http://robotics.eecs.berkeley.edu/ mcenk/medical

Vertut, J., & Coiffet, Ph. (1985). *Teleoperation and robotics: Applications and technology*. Englewood Cliffs, NJ: Prentice-Hall.

# 49

# Use of Virtual Environments in Motor Learning and Rehabilitation

Maureen K. Holden[1] and Emanuel Todorov[2]
[1]*Massachusetts Institute of Technology*
*Department of Brain and Cognitive Sciences*
*E 25-526b, Cambridge, MA 02139*
*holden@ai.mit.edu*
[2]*Gatsby Computational Neuroscience Unit*
*University College London 17 Queen Square*
*London WC1N 3AR England*
*emo@gatsby.ucl.ac.uk*

## 1. INTRODUCTION

The purpose of this chapter is to describe how virtual environments (VEs) can be utilized to facilitate motor learning in normal subjects and to enhance the rehabilitation of disabled subjects.

The chapter begins with a review of relevant literature on the use of VEs for training sensorimotor tasks in normal populations in two main areas, spatial–motor and navigation learning studies. Next, clinical literature on the potential and actual applications of VEs to the field of rehabilitation is reviewed. Advantages and disadvantages for the use of VEs, discovered through these studies, are identified and discussed.

Next, a system designed by the authors is described. The system provides a new method for motor learning, "by imitation," within a virtual environment. A number of key design issues that were encountered during development of the system, along with the solutions chosen to address them, are described. Finally, summaries of results to date are provided, based on using the system to train normal subjects and patients with stroke on selected motor tasks. The chapter concludes with a short description of ongoing work and future directions.

## 2. LITERATURE REVIEW: MOTOR TRAINING USING VEs IN NORMAL SUBJECTS

In recent years, there has been great interest in using VEs for learning various types of motor and spatial tasks (Darken, Allard, & Achille, 1998; Kozak, Hancock, Arthur, & Chrysler, 1993; Regian, Shebilske, & Monk, 1992). This interest has been driven by a number of perceived advantages of virtual over real-world training. These advantages are both practical and theoretical in nature. Practical advantages are discussed first; theoretical advantages are discussed later in the context of system design in section 4.1.

## 2.1 Practical Advantages of VE Training

Practical advantages of using VEs for training include safety, time, space and equipment, cost efficiency, and documentation considerations:

1. Safety—in practicing a virtual task that is potentially dangerous (such as flying a plane for a normal subject or lifting a pan of hot water for a disabled subject) there are no negative consequences to failure
2. Time—the training task can (in theory) be quickly altered to become easier or more difficult
3. Space and Equipment—requirements may be minimal compared to the real environment (for example, training military personnel on a "virtual" vs. a real ship)
4. Cost Efficiency—the possibility for fewer training personnel (or therapists in the case of patients) while still maintaining a similar outcome level following training or treatment. This may be achieved by semiautomating training on the computer, allowing subjects to work independently some of the time. In medical applications (see chap. 48, this volume), telemedicine has the potential to reduce the cost of providing care to remote areas while improving access for remote areas to higher quality of care
5. Documentation—automatic scoring and report systems can be developed to allow easy monitoring of progress of subjects using the VE training system.

Obviously, the potential to apply VEs for training to all sorts of sensorimotor tasks in fields as diverse as defense, medicine, industry, and sports is huge. In looking at the literature on VE applications to motor learning—broadly defined—in normal subjects, one finds that much work is geared toward very complex motor skills that have a large cognitive component (such as military training for flying planes or navigating ships (Kennedy, Lanham, Massey, Drexler, & Lilienthal, 1995; Lawson, Rupert, Guedry, Grissett, & Mead, 1997; Theasby, 1992), or a very high level of eye–motor coordination (such as VE training aimed at virtual surgical training for physicians (Docimo, Moore, & Kavoussi, 1997; see chap. 48, this volume). The issues important to researchers studying these types of problems are often quite different from those faced by the researcher whose goal is to develop a system that would be useful in helping disabled subjects relearn simple movement skills like lifting a cup or turning a key in a lock, or to train skills that are more predominantly perceptual–motor in nature, such as a tennis swing.

In fact, relatively few studies in the literature were found on using VEs to train such tasks. The most relevant studies fell into two classes: (1) studies of spatial–motor learning such as the "pick and place" task (Kozak et al., 1993) or procedural console-operations tasks (Regian et al., 1992); and (2) studies of learning to navigate in small- (Waller, Hunt & Knapp, 1998) or larger-scale environments (Ruddle, Payne, & Jones, 1999; Wilson, Foreman, & Tlauka, 1997; see chap. 24, this volume).

## 2.2 Spatial–Motor Learning in VE

Kozak et al. (1993) trained participants to move a series of cans, positioned in a line from left to right directly in front of them, to matching discs placed 6 inches forward of the start location. The sequence was then repeated in reverse order so that the cans ended up at the original starting location. Three groups were used; one group trained in a virtual environment, one trained on the real world task, and one group received no training but only testing. Participants who trained in the real world were significantly better than either the VE or no-training groups.

Participants who trained in the VE were no better at performing the real-world task following training than were participants who received no training.

However, participants who trained in VE *did* improve their performance *during* VE, at a rate comparable to participants who trained in the real world. That is, they learned the *virtual* task, but learning the virtual task did not *transfer* to improved performance on the comparable real-world task. Proposed reasons for the lack of transfer included subtle differences in the grasp required in the two situations (VE vs. real world), system time lag, differences in the field of view in which the hand could be seen, lack of tactile and acoustic feedback, and difficulty aligning some participants with different body sizes so that all the virtual cans could be reached. If participants were using the method of loci as a learning strategy (Bower, 1972; in this method, participants associate motor responses with imaged spatial locations), then slight differences in displayed locations relative to the participant's arm in the two conditions could have interfered with learning and transfer of skill to the real-world test (see chap. 19, this volume).

A later study by Richard et al. (1996) examined the influence of several of the factors identified by Kozak et al. (1993) as possible factors that interfered with transfer of learning. Richard et al. (1996) examined the effect of different graphics frame rates and viewing modes on grasp acquisition time, and the effect of force feedback and added acoustic signals on force regulation ability during grasp in a virtual environment. They found no difference in object capture times (time to grasp a virtual object) between stereo and mono displays for frame rates of 28 or 14 frames/sec (fps). However, when frame rate fell below 7 fps, stereo viewing increased performance by 50%. (Note that currently available graphics hardware allows higher frame rates at lower cost.) Direct haptic feedback (see chaps. 5 and 6, this volume) was found to be superior to pseudo-feedback (reduced error rates and improved performance of a force regulation task). The lowest error scores and task completion times occurred when direct haptic feedback was augmented by an acoustic signal that indicated the initial contact of the virtual hand with the virtual ball.

Regian et al. (1992) trained participants in a procedural learning task (a complex 17-step console operation task) using a virtual environment, and found that all 31 participants in the experiment were able to learn the task using VE for training. The purpose of the experiment was to compare the effects of two types of instructional strategies on learning, therefore, transfer to real-world performance was not tested. However, this study did serve to establish that it was possible to learn motor skills in a virtual environment.

### 2.3 Motor Learning and Transfer

The most essential issue in considering the usefulness of VE training for motor control in normal subjects, as well as for rehabilitation of patients, is the ability to transfer the performance that is learned in the virtual world to performance of that same task in the real world. It was surprising to see that many studies examined learning in a virtual environment but did not measure performance on similar real-world tasks following training (Gillner & Mallot, 1998; Ruddle, Payne, & Jones, 1999). Perhaps this occurs because researchers are aware of the many technical factors that account for discrepancies between the virtual and real worlds, and the goal of many studies is to understand these factors better so that the fidelity of VEs can be improved. However, without testing transfer to real-world performance, much effort could be consumed in improving fidelity in VE systems beyond the level that is really necessary to accomplish effective training.

### 2.4 Navigational Learning

Two studies using VEs to teach spatial-navigation knowledge found that under certain conditions training transferred well to the real-world environment (Waller et al., 1998;

Wilson et al., 1997). In the first of these studies, Waller et al. (1998) examined the effect of six different training environments on participants' ability to perform blindfolded in a real world 14 × 18-feet maze environment. The conditions compared were: (1) no training; (2) real-world training, (3) map; (4) VE desktop; (5) VE short immersive (1 min); and (6) VE long immersive (2 min). Participants ($n = 120$) received six consecutive bouts of training on one of these conditions. Each training bout was followed by a test trial on the real-world maze. Results showed that for early learning (defined as performance on test trials 1 and 2), real-world training was significantly better than all other methods, and although all other conditions were equal statistically there was a trend to worse performance for participants trained in VEs ($x = 270$ sec. for VE participants; 163 sec. for real-world trained participants). However, by the sixth trial, mean times for the real and long-immersed VE conditions were statistically equal and significantly shorter than for the other two VE conditions (desktop and short immersive). Although this implies superiority for immersive VE over desktop, it is not clear that the superior performance for the immersive condition would have held if participants had trained in the other conditions for equivalent amounts of time as in the long-immersive VE condition. In examining the issue of transfer from VE training to real-world performance this study is of interest because it shows that both desktop and immersive VE methods were capable of improving performance on a similar real-world task. In addition, when controlled for time spent in training, there was no difference in training effectiveness of the desktop and immersive VE methods.

Wilson et al. (1997) examined navigational learning in a large environment via exploration of a real multistory building and a to-scale computer simulation of the same building. The computer simulation was presented to participants using a desktop VE system. The computer keyboard was used to navigate. A control group received no training. The results showed that participants who received the simulation training were able to transfer the three-dimensional (3-D) spatial knowledge learned in the VE to a variety of spatial measures in the real world. For many measures, the VE trained participants performed as well as participants who trained in the real environment, and for most measures the VE participants performed better than the control group. This study is of interest because it shows that learning and transfer can occur with less sophisticated desktop systems.

### 2.5 Cybersickness and Negative Aftereffects

Another issue that stands out in many studies of VE training performed on normal subjects is the relatively common incidence of negative side effects (see chaps. 29–32, this volume). These side effects include cybersickness, altered visual-motor and sensorimotor coordination, and postural disequilibrium (Cobb, 1999; Nichols, 1999; Stanney, Kennedy, Drexler, & Harm, 1999; Stanney et al., 1998). These side effects have the potential to negatively affect motor coordination in the real-world posttraining (see chaps. 37–39, this volume).

Such problems are of even greater concern for patient populations, especially if central nervous system function is impaired due to disease or injury. Minor alterations in postural equilibrium or eye–hand coordination that could safely be handled by a subject with a normal nervous system, might, in a patient with impaired nervous system function, result in a fall due to loss of balance or inability to catch oneself with an arm that reaches to the incorrect location in space for support. Because the topic of negative aftereffects is covered in detail elsewhere in this volume, an extensive review is not provided here. However, in the sections on design (see sections 4, 5.1, and 6.1) a discussion is provided on how findings from research in this area have influenced design decisions in the system we are presently using. As more is learned about the exact causes of these side effects through studies on normal subjects, design changes in future systems and improved training routines will be better able to prevent the occurrence of unwanted side effects in normal, and, hopefully, disabled subjects as well.

## 3. CLINICAL LITERATURE: VIRTUAL ENVIRONMENT USE IN REHABILITATION

In the field of rehabilitation, VE application is still in its infancy. Much of the clinical literature published to date on VEs in rehabilitation has been focused on describing the potential applications of the method to different areas of rehabilitation (Greenleaf & Tovar, 1994; Kuhlen & Dohle, 1995; Rose, Attree, & Johnson, 1996; Wilson, Foreman, & Stanton, 1997). However, relatively few reports of actual use and effectiveness of VE with patient populations have been published. While many investigators are working in this field, much of the work is still under development or just beginning to move from the development to the clinical testing phase. In the first section below, different areas of application that are under development are briefly described to give the reader a broad sense of the field. In the second section, several reports relevant to motor learning of upper extremity tasks in disabled subjects are reviewed.

### 3.1 Overview of VE Applications in Rehabilitation

Proposed applications of VE to the field of rehabilitation fall into four major categories: (1) use in measurement/diagnostic tests; (2) assistive technology; (3) social and entertainment applications; and (4) training of impaired functions. Training of impaired functions may be divided into training of tasks which are primarily perceptual–motor or cognitive in nature.

#### *3.1.1 Applications for Measurement and Diagnostic Tests*

Kuhlen and Dohle (1995) describe a system designed to aid physicians in the analysis and diagnosis of movement disorders, such as different types of pareses and apraxias. With this system, movements of patients performing different motor tasks are recorded, then trajectories of these movements can be displayed in 3-D, viewed from different angles, and precisely quantified. Virtual environments can also be used to interactively explore medical imaging data such as MRI and PET scans of the brain, allowing physicians to gain a clearer understanding of the location and size of lesions. When paired with clinical evaluations from patients, such information should enhance the ability to predict structure–function relationships in the brain. Rose et al. (1996) propose using VE to make neuropsychological assessments of cognitive function in patients with brain injury more ecologically valid. This could be done in VEs by measuring cognitive function in the context of realistic everyday functions, such as cooking in a virtual kitchen. Another advantage of VE testing in brain injured subjects would be the ability to distinguish the contribution of cognitive vs. motor impairments in task performance. Because patients with brain injury often have both cognitive and motor impairments, these two factors are mixed together in real-world testing. Virtual environment testing could allow users to focus on only cognitive factors by reducing the motor parts of the task in VE to one simple movement.

#### *3.1.2 Applications for Assistive Technology*

Assistive technology in general is designed to help disabled persons use different types of technology to substitute for functional abilities they have lost. Virtual environments would be one more tool to extend the possibilities for such individuals. For example, telerobotics could be applied to allow users with very little motor control to manipulate objects in their environment by interfacing with a robot device. Warner, Anderson, and Johanson (1994) are working on a system to allow quadriplegics such control. In another example, Greenleaf and Tovar (1994) describe a Virtual Receptionist system that allows people with speech or motor impairments to perform the job of receptionist: answering and making phone calls, and other secretarial tasks. Two groups have worked on systems that use a VE DataGlove to record sign language gestures, then translate these gestures into written or spoken words by using a speech synthesizer (Kramer & Leifer, 1989; Vamplew & Adams, 1992). Such a system would allow deaf individuals to converse freely with hearing persons who are not knowledgeable in sign language.

#### 3.1.3 Social and Entertainment Applications

Individuals with severe disabilities often experience social isolation due to difficulties with transportation, energy levels, and access. Virtual environments have been proposed as a way of allowing handicapped individuals at remote sites to join together and interact in a virtual world (Smythe, 1993).

#### 3.1.4 Training of Impaired Functions

In the field of locomotor training, a prototype device which can superimpose objects into the field of view of patients with Parkinson's disease, while still allowing them to see the real world, has shown promise as a method to counteract the slow shuffling gait which is characteristic of patients with this disorder (Emmet, 1994; Reiss & Weghorst, 1995). A method to teach children with orthopedic impairments to operate motorized wheelchairs has been developed by Inman, Peaks, Loge, and Chen (1994). Another VE system designed to train standing posture and dynamic balance control in the elderly is currently being tested (Cunningham, 1999). The system is based on video capture in a nonimmersive environment.

A method for treating strabismus (a deviation of the eye position in the socket, often due to weakness in one or more eye muscles) by exercising eye muscles in VE has been proposed (Lusted & Knapp, 1992). By attempting to keep objects in a VE aligned, the eye muscles could be exercised; the difficulty could be calibrated based on the degree of misalignment of the resting eye axis of the patient.

Finally, methods are being developed to assess and train patients with acquired brain injuries on cognitive function tasks (Christiansen, et al. 1998; Pugnetti et al., 1995). The method being developed by Christiansen et al. (1998) uses a virtual kitchen to assess performance on daily tasks, such as preparing a can of soup. The focus of the task is cognitive, as the subject uses a mouse click or joystick to navigate in the environment and accomplish each step of the task in sequence. That is, subjects are not required to move a virtual (or real) arm to actually do the task as they would in the real world. Such a test can help therapists determine the degree to which cognitive vs. motor deficits contribute to a patient's impaired function on a variety of activities of daily living (see chap. 50, this volume).

### 3.2 Reports on VE Training Results in Rehabilitation

#### 3.2.1 Spatial Learning

Children with severe motor disabilities often have poorly developed spatial awareness (Foreman, Orencas, Nicholos, Morton, & Gell, 1989). One significant contributing factor to this deficit is thought to be reduced ability to explore the environment in an active way secondary to the mobility impairment. Virtual environment technology has been proposed as a way to offer such children a chance for active, independent spatial exploration.

Wilson, Foreman and Tlauka (1996) decided to test whether such training in a VE would result in improved spatial knowledge in a comparable real world setting. The authors used a desktop VE system to train large-scale spatial knowledge of a two-story building. Subjects ($n = 10$) were children with a variety of disabling conditions, including spina bifida, cerebral palsy, and muscular dystrophy. All relied on a wheelchair as their prime method of mobility, though three subjects could walk short distances with a walker. They ranged in age from 7 to 11 years. A control group of eight healthy adults with a mean age of 23 years received no VE training, but underwent the same testing as the children. Children received one training session. They explored a virtual model of the actual building they were in for the experiment, with the goal of finding fire extinguishers and a fire exit door. Immediately following VE training they were tested on knowledge of fire extinguishers and fire door locations in the real building. This was done in two ways. First, while still in the training room with the computer, subjects used a pointer to aim at the expected location. This was followed by a wheelchair tour of the real building during which subjects were asked to guide the experimenter to two specific locations. Results showed that all the subjects were able to point to and find desired locations in the real world and to do so significantly better than control subjects who received no training.

The results of this experiment are significant for our work because the system used was a relatively simple VE desktop display. Subjects had no vestibular or kinesthetic feedback during their exploration of the virtual world; the feedback was purely visual in nature. Despite these limiting factors, the exposure of only one session resulted in significant learning.

#### *3.2.2 Motor Learning*

Researchers at Rutgers (Burdea, Deshpande, Liu, Langrona, & Gomez, 1997) have developed an interesting line of work using a VE system designed for hand rehabilitation and diagnosis. The system uses a PC workstation, Rutgers Master II force feedback glove, and a Multipurpose Haptic Control Interface, which allow measurement of grasping forces applied to 16 regions of the hand. The system was initially designed as a diagnostic tool for hand rehabilitation, then developed further to include rehabilitation capabilities. WorldToolKit was used to create exercise routines modeled after standard hand rehabilitation exercises. These include squeezing a ball, compressing "silly putty," isolated finger flexion/extension, and functional tasks of pegboard insertion and ball throwing. The interaction with virtual objects is accomplished with different hand gestures, e.g., grasp, lateral pinch, pointing, release. A desktop display is used. The forces generated by the patient in the virtual movement are fed back through the glove with a mechanical feedback bandwidth of 10 to 20 Hz. The system is currently undergoing clinical trials at Stanford University. Telerehabilitation capability has recently been developed for the system (Popescu, Burdea, Bouzit, & Hentz, 2000), and plans are under way to extend the rehabilitation capability of the system to the elbow and knee.

Several additional studies on motor learning in normal subjects (Todorov, Shadmehr, & Bizzi, 1997) and subjects with stroke (Holden, Todorov, Callahan, & Bizzi, 1999; Piron, Dam, Trivello, Iaia, & Tonin, 1999; Piron et al., 2000) utilize the VE system we have developed. The results of these studies are therefore presented following a description of the "learning by imitation" system.

## 4. DESIGN OF A SYSTEM FOR "LEARNING BY IMITATION" IN VIRTUAL ENVIRONMENTS

### 4.1 Theoretical Considerations

Observations of human movement suggest that the nervous system uses some sort of abstract motor plan, and fills in details needed to instantiate that plan in specific situations (Schmidt, 1975). These observations fall into two categories:

- Certain geometric features of movement trajectories (individual writing style, shape of speed profiles) are preserved when movement is executed with different actuators (left or right arm, foot), requiring very different muscle activation patterns and dynamic interactions with the environment (Atkeson & Hollerbach, 1985; Lacquaniti, Terzuolo, & Viviani, 1983; Morasso, 1981).

• If external forces repeatedly distort the movement trajectory or the subject perceives the trajectory as being distorted (due to distorted visual feedback), the motor system often adapts so that the (perceived) movement trajectory returns to normal (Lackner & DiZio, 1994; Shadmehr & Mussa-Ivaldi, 1994; Wolpert, Ghahramani, & Jordan, 1995).

Such observations have led a number of investigators to believe that the geometric (kinematic) aspects of a movement form a major component of the abstract motor plan. If this is the case, the acquisition and rehabilitation of motor skills should be enhanced by training methods that somehow create an internal representation of the desired kinematic pattern (Schmidt & Young, 1991). An obvious way to do this is to learn by imitation: show people a detailed movement, ask them to reproduce that movement, and provide feedback on the mismatch between desired and actual movement.

While this can be accomplished with traditional methods, such as demonstration or video playback, imitation learning systems utilizing VEs offer a number of advantages. Some of these practical advantages were discussed earlier and are common to VE systems in general, for example, space and equipment requirements are less, and mistakes made during practice have no negative consequences. The more important advantage of using VE technology for imitation learning, however, is its potential to enhance motor learning. This is due to several factors. The most important factor is the unique capability for real time feedback in the very intuitive and interpretable form that VE provides. Other advantages are that learners can see their own movement attempts in the same spatial frame of reference as that of the teacher (unlike practice with a real coach or therapist) and that the task can be simplified in early stages of learning, allowing the learner to focus on key elements of the task. Training environments can also be customized for different therapeutic purposes and the system designed to help the learner detect and correct errors more rapidly. Finally, practice, an essential element in motor learning, can be facilitated by making the task fun.

### 4.2 Design Considerations for Normal Subjects

The basic idea for implementing learning by imitation is illustrated in Fig. 49.1. A computer monitor is mounted at eye level with the screen facing down. A see-through mirror is mounted 20 centimeters below the screen. Thus, images displayed on the screen appear in a workspace below the mirror (in 3-D when stereo glasses are used). The user of the system can see both his hand and computer-generated images, in the same physical space (the picture is taken from the user's point of view while the user is looking down through the mirror). The user can "insert" his hand in the virtual hand and attempt to match the posture exactly. The virtual hand can display an animation of a desired movement that the user attempts to follow repeatedly. Both the animation and real movement occur simultaneously (overlaid or next to each other), making any deviations between the two immediately obvious.

While this would be the ideal form of imitation learning, it is hard to implement such training systems because existing technology does not always permit superimposing virtual images on the real world. In cases where the workspace is large and monitors/mirrors cannot be easily mounted, a head-mounted display (HMD) would have to be used to present the virtual image. That image will move with the head of the user unless head-tracking is used for correction. Our experience with head-tracking (even with highly accurate sensors, such as Optotrak 3020) suggests that it is extremely difficult to fool the human visual system (see chap. 3, this volume), virtual objects always look unstable due to tracking/calibration

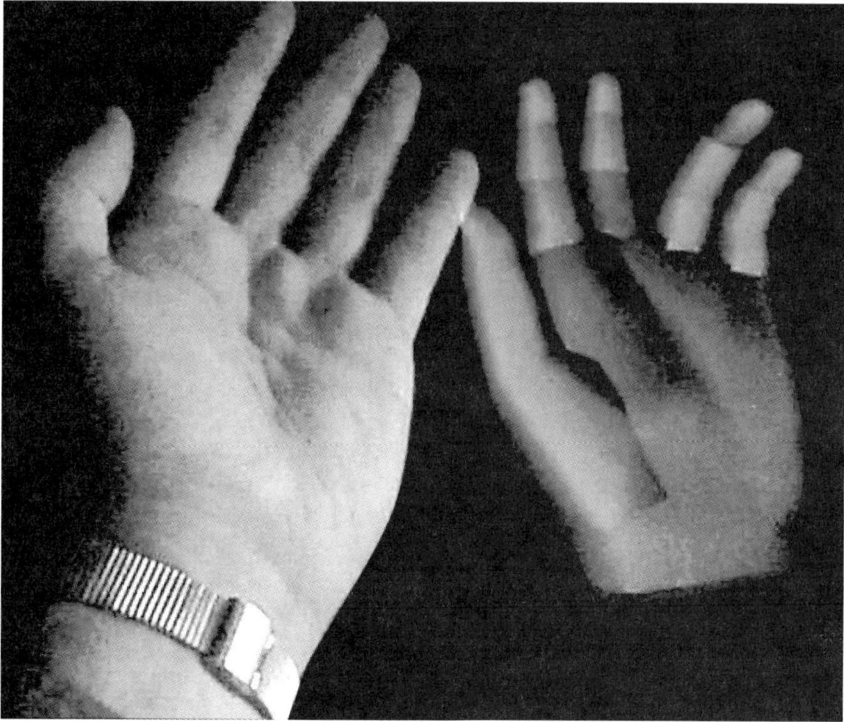

FIG. 49.1. A photograph taken from the user's point of view, while the user is looking down through a semitransparent horizontal mirror. A monitor mounted above the mirror (facing down) projects virtual images that appear aligned with the physical hand in the same workspace. Note that appropriate illumination has to be used in order to see both hands.

errors and delays. This may not be a major problem when virtual objects are used to provide abstract information; however, in our case we are dealing with a real-time visuo-motor loop and a motor learning system that is accustomed to a perfectly "calibrated" physical world. So the ideal implementation was left for the future, and we focused on more feasible alternatives where both the desired and actual movements are presented in a virtual environment.

## 4.3 Design Considerations for Rehabilitation Subjects

In considering the kinds of system features that would make a VE system useful for motor training in patients with neurological impairments, one can begin by asking the question, What is different about training patients as compared to normal subjects?

In contrast to healthy subjects, where the focus of VE training is often complex perceptual-motor skills, such as flying a plane, in rehabilitation the goal is usually to relearn very simple motor tasks—ones we all take for granted—and to do so in the face of a variety of sensory, motor, and cognitive deficits. The exact type of deficit can vary tremendously from patient to patient, even with the same diagnosis. To accommodate for this variety, the first characteristic one needs to think about is flexibility so that the training can easily be adjusted to the needs of different patients. Another major focus is to facilitate transfer of motor learning that occurs in the virtual world to performance in the real-world. Flexibility in system design helps ensure that the virtual tasks created resemble real tasks as much as possible, particularly in terms of the movement required to perform the task.

Deficits that patients have can make learning even simple movements daunting. Thus, VE must be utilized in a way that will allow enhanced feedback to be provided to patients. The idea of augmented feedback is not a new one. But applying it in the context of VE in combination with learning by imitation of a virtual teacher, in real time, does differ from techniques used in the past. Thus, a number of ways for patients to detect the exact kinds of errors they are making during attempted movements have been incorporated into our system. Simply trying to imitate a desired movement may not be enough, particularly for patients with a sensory deficit who cannot "feel" exactly what their arm is doing. (Note, however, many of the features we have designed with patients in mind could well enhance motor learning in normal subjects as well.)

What about graphics displays? Although very complex, realistic, or fanciful graphics and fast-paced game formats are appealing to normal subjects for VE training, they may be overwhelming for patients. When the brain itself has been affected, as in stroke or acquired brain injury (in contrast to disabilities where the brain is intact, like spinal cord injuries), processing of all this information may be difficult. In contrast, a very simple display, with few movements, may help a patient to focus on the task at hand and enhance learning.

Next, the cardinal rule in medicine must be considered: "First, do no harm." The issue of possible negative side effects, such as cybersickness (see chap. 30, this volume), altered eye–motor coordination (see chap. 37, this volume), and postural disequilibrium (see chap. 39, this volume), must be carefully considered. To date, there seem to be fewer reports of negative side effects using desktop display systems for virtual environments. Most of the problems seem related to the immersive environment created by HMD-based systems (Cobb, 1999; Nichols, 1999; Stanney et al., 1999; Stanney et al., 1998). It seems obvious that feelings of disorientation, postural disequilibrium, or aftereffects in arm control that produce past-pointing could be much more dangerous for disabled subjects, in whom a fall and injury could more easily be elicited than in normal subjects. Another problem is that in an HMD, the patient cannot see the therapist or her instructions. Even with see-through HMDs, problems with postexposure alterations in eye–hand coordination and felt limb position have been noted (Biocca & Rolland, 1998; Rolland, Biocca, Barlow, & Kancherla, 1995). Even without aftereffects, HMDs present some problems. The extra weight imposed on the head will be much more difficult for patients to handle, because trunk and postural control muscles are often adversely affected by neurological impairments. Also, if a HMD is used without a head-position tracker (and the visual display is thus fixed relative to the head), the patient may become disoriented due to inaccurate cues for real-world horizontal and vertical, and begin tilting to one side while sitting or standing.

Finally, practical issues are important. Keeping costs down, making systems easy to use (see chap. 34, this volume) and minimizing the amount of equipment attached to the patient will offer the greatest potential for success and acceptance of VEs as a standard treatment procedure in rehabilitation clinics.

Because of many of the factors listed above, we have designed our system to be used with a desktop display. The display can also be projected on a large screen instead of a monitor to give a greater feeling of immersion (see chap. 40, this volume). A significant disadvantage of the desktop display is poor depth visualization. This can be compensated for in part by training features that allow playback of the recorded performance of a patient, then rotating the display in 3-D in any desired direction to show the patient where his or her errors are and how to correct them.

## 4.4 System Components

To date we have built and tested two virtual environments—one for training Ping-Pong shots and the other for rehabilitation of stroke patients (Bizzi, Mussa-Ivaldi, & Shadmehr, 1996; Holden et al., 1999; Todorov et al., 1997). The features common to both systems are described first, followed by a more detailed discussion of each system and experimental results.

Components described in sections 4.4.1 to 4.4.3 were used in our first two experiments. Based on results in these experiments, components listed in sections 4.4.4 to 4.4.6 were developed and are currently being tested.

In the two systems described below, both desired and actual movements were displayed in real time in a VE using a desktop computer. The main concern with using purely virtual training was that transfer from virtual to real tasks has sometimes been difficult to achieve in prior studies. To avoid such failure, every effort was made to "integrate" the VE into the physical context where the task normally takes place.

### 4.4.1 Scenes and Teacher

We created virtual worlds closely matching the desired task, that is, containing relevant objects with appropriate scale and distances among them. Instead of a fully immersive environment, a computer monitor was used that was properly aligned with the physical setting (as shown in Fig. 49.2). In the Ping-Pong experiment described below, training in the VE was interleaved with practice of the real task.

A human expert/therapist executed a desired movement that was recorded with a motion sensor (Polhemus FASTRAK) and imported in the simulation. An animated object (the Teacher object) displayed the desired movement. Another animated object (the User object) displayed, with minimum delay, the position and orientation of a physical object held by the user. The user was instructed to repeatedly produce movements such that the User and Teacher objects remained perfectly aligned at all times. Additional examples of scenes are shown in Figs. 49.3, 49.6, and 49.8. To aid the acquisition of the desired movement, a number of additional features were implemented, as described below. Movement parser, score calculation, and semiautomated training features were developed based on our experience with the system in experiments 1 and 2. They are being evaluated in our current experiments.

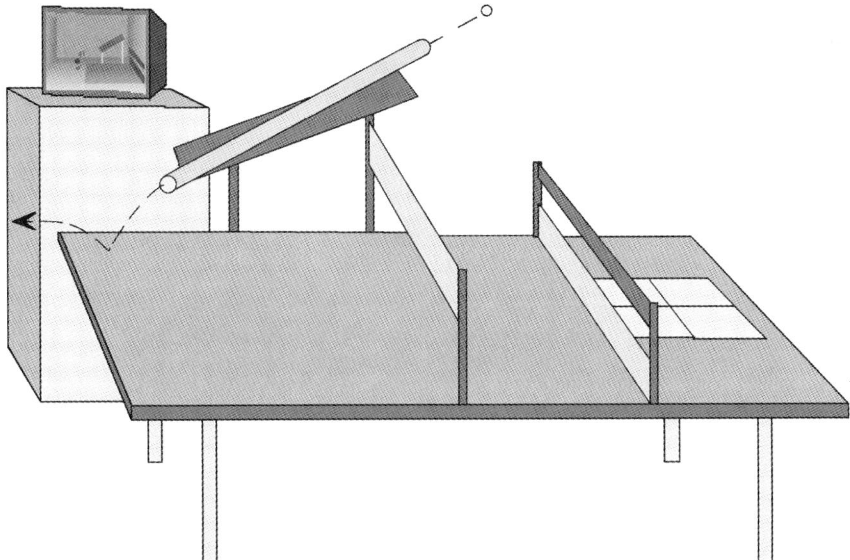

FIG. 49.2. A schematic diagram of the Ping-Pong experimental setup. Participants were standing in front of the table, hitting Ping-Pong balls with a paddle held in the left hand and trying to send them to the target zone. The balls had to pass through the opening and under the obstacle in the middle of the table. During training participants were standing in the same place required for the real task and looking at the monitor. Courtesy of Journal of motor Behavior.

#### 4.4.2 Practice Modes

In the main practice mode, the Teacher animation was displayed repeatedly, while the user attempted to synchronize with it and track the desired movement accurately. It was possible to speed up or slow down the prerecorded animation, trigger the animation when the user started moving (see Movement parser), or display a static 3-D trace of the entire Teacher movement along with the animation (see Fig. 49.6).

It was also possible to work in passive mode, in which the Teacher "followed" the User while still remaining on the prerecorded trajectory. This was accomplished by rewinding the prerecorded animation (at each point in time) to the video frame in which the Teacher object was as close as possible to the current position and orientation of the User's object. This provided a way for the user to learn the spatial path of the movement separately from the velocity profile. An active component was also included in this mode, where the Teacher moved a small distance along its trajectory if the User made no movements for a certain period of time. This provided guidance especially helpful for patients.

#### 4.4.3 Augmented Feedback

In both practice modes, the mismatch between User and Teacher movements could be emphasized by connecting the two objects with lines or plotting circular arrows in the direction of the orientation error. The movements could be recorded and replayed later, and sound cues could be added to help timing. The visual display of either the teacher or patient trajectories could be altered so that the trajectory appeared as a wire frame or solid object. The size and frame frequency of teacher and patient "objects" could also be varied.

#### 4.4.4 Movement Parser

The repeated Teacher animation described above required the user to synchronize with the system, which was difficult for some patients. To alleviate this problem we developed an automated movement parser, which monitored the FASTRAK sensor data and determined online when the user started and stopped moving. This information was used to trigger the Teacher animation and also to provide sound cues.

We have found it rather difficult to implement a movement parser that works reliably. After experimenting with various designs, we opted for a semiautomatic solution: The system provides a number of settings specifying velocity thresholds, distance thresholds, and other features and ways of combining them in Boolean expressions. These settings are adjusted by an expert user for each desired trajectory being trained with the system.

#### 4.4.5 Score Calculation

An online scoring system has been developed to provide objective feedback as to how well the user is executing the desired movement. Since different movements require different scoring techniques, a general-purpose algorithm, with a number of parameters that an expert user could set, was implemented. The algorithm takes into account characteristics of the movement (average speed, duration, smoothness, etc.) as well as "closeness" to the desired trajectory. The latter is specified by entering weights for different types of error: position, orientation and velocity deviations, and different temporal misalignments.

#### 4.4.6 Semiautomated Training

It was clear from the beginning that a fully automated training schedule would be unrealistic. In both training and rehabilitation, the system was originally operated by an expert instructor/

therapist who controlled transitions between different scenes, training modes, and feedback features. The expert was guided to a large extent by observing subject/patient performance. In our most recent version, however, a semiautomated "script" feature was implemented. The script feature allows specification, prior to training, of a series of scenes and teacher and feedback settings that progress automatically through a preset number of repetitions, simulating a treatment or training session. The scores computed online can be linked to this function so that the script will "jump ahead" if the task is too easy for the subject. The level of difficulty can be adjusted by changing the score value set as the goal. In general, our intention was to utilize the valuable expertise of a human instructor/therapist as much as possible while automating routine tasks and providing new sources of real-time feedback and motivation.

## 5. EXPERIMENT 1: TRAINING A COMPLEX MOTOR SKILL IN NORMAL SUBJECTS

### 5.1 System Design

Our first system was designed for training normal subjects to hit Ping-Pong balls and send them into a target zone (Todorov et al., 1997). The setup is shown in Fig. 49.2.

Participants were standing in front of the table, holding a Ping-Pong paddle in their left hand (to make the task more difficult). A ball was dropped through the tube shown in the figure, at an angle and speed corresponding to an intermediate to difficult shot. The task was to hit the ball after it bounced from the table and send it in the target area through the opening above the net. In the second experiment an additional obstacle was added in the middle of the table, in which case the ball had to pass under it.

During training the participants looked at the monitor, which was properly aligned with the physical setup. The desired movement was recorded from an expert hitting a real ball on the real task. When the recorded expert movement was transferred to the virtual environment the spatial alignment was preserved. In other words, in order to track the desired movement in the virtual environment, the participant had to make a movement in the real-world that would have accomplished the real task.

A snapshot of the simulation is shown in Fig. 49.3. Shadows, illumination cues, and occlusions were used to create a realistic simulation. The solid paddle shows the User position and orientation; the transparent paddle shows the Teacher recording.

The ball shown in the figure only interacted with the Teacher paddle; its dynamic simulation was adjusted so that the impact looked realistic. Participants were not allowed to interact with the simulated ball during training, since our simulation of the ball dynamics was not completely accurate. One might wonder then, why include the ball in the first place if it does not seem to provide any additional information? It turned out, however, that including the ball was crucial—without it learning in the simulator did not transfer to the real task (see later discussion).

During the development and fine-tuning of the system we experienced two main problems. Participants occasionally confused the Teacher and User objects. This resulted in an intriguing but very undesirable phenomenon: when the User was falling behind the Teacher and instead thought he was ahead, the User movement slowed down further, causing an even bigger error. To avoid such confusion the two paddles were made as distinct as possible; the transparent Teacher paddle shown in Fig. 49.3 solved the problem.

The other problem was synchronizing the two animations. We attempted to use a version of the movement parser described above, but that did not produce the desired effect. In order to make the parser robust (i.e., avoid signaling movement onset due to sensor noise), reasonably high-velocity thresholds had to be used. As a result the system detected movement onset with

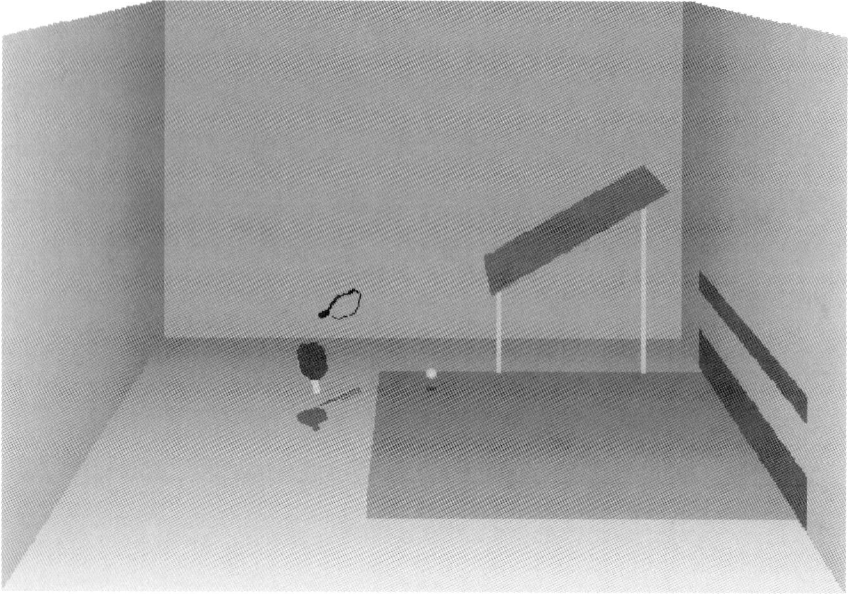

FIG. 49.3. A snapshot of the simulation. The transparent paddle displayed a prerecorded (teacher) animation, which the user attempted to follow in real time with the solid paddle. The simulated ball only interacted with the teacher paddle. The display was refreshed at about 40 Hz. Courtesy of Journal of Motor Behavior.

a delay, so the Teacher animation started after the User movement. Such timing misalignment was very detrimental, the participants spent most of their time trying to compensate for it rather than learning the teacher trajectory. This problem was solved by simply animating the Teacher at fixed intervals and asking participants to synchronize with it.

## 5.2 Experimental Results

In Experiment 1, volunteer college students were randomly assigned to one of three groups: a pilot group with 13 participants, a control group with 20 participants, and a training group with 19 participants. Each participant was introduced to the apparatus and the task, and given 10 practice balls, the score for which was not recorded. Then a baseline was recorded over 50 trials, which lasted approximately 10 minutes. Participants in all groups were able to see where the ball landed and therefore knew their score, but were not given any other feedback. After the first block the control group was given standard coaching by the experimenter, who was an experienced player. Coaching included discussion of what the participant was doing wrong, demonstration, and extra practice balls.

Participants in the pilot and training groups were trained in the simulator for the same amount of time as the coaching given to the control group. Training started with a slow version of the teacher trajectory, followed by 1 to 2 minutes of passive mode training. After that only the main mode (repeating the desired movement with the teacher) was used. The only difference between the pilot and training groups was that the pilot participants were not shown a ball in the virtual environment, whereas the participants in the training group could see the Teacher paddle hitting a simulated ball.

Results are shown in Fig. 49.4. Analysis of variance showed that all groups started with similar performance on the first 50 trials (no significant differences were found). At the end of the experiment the pilot group did not have better performance compared to the control group;

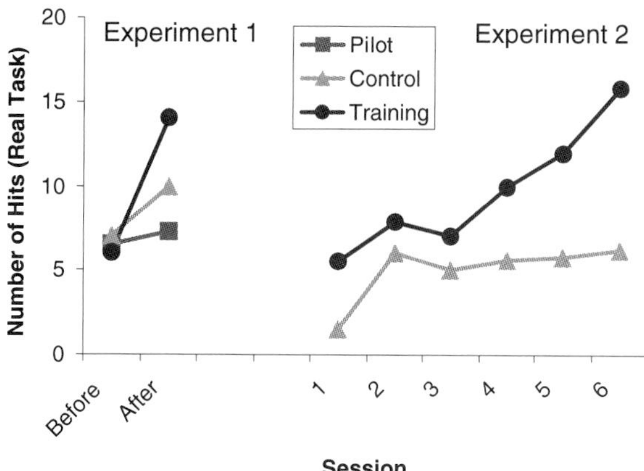


FIG. 49.4. Summary of experimental results, in numbers of successful trials (in Experiment 1 the maximum was 50; in Experiment 2 the maximum was 60).

pilot participants were actually worse, and their improvement was smaller (the differences were close to being significant). The training group, however, had significantly higher scores on the second block of trials ($p = 0.02$) and more improvement ($p = 0.01$) compared to the control group.

In Experiment 2, the task was made more difficult by increasing the speed of the bouncing ball and introducing the obstacle in the middle of the table (Fig. 49.2). This time six sessions (two sessions per day in three consecutive days) were included. In each session, control participants played 30 practice balls, followed by 60 test balls (scores in Fig. 49.4). Training participants used the simulator for the same amount of time that it took the controls to play the 30 practice balls. Then the training participants were also tested on 60 balls in the real task. Again, the results indicated significantly higher ($p = 0.02$) performance in the training compared to the control group on the last session. The improvement of the training group was also significantly higher ($p = 0.05$).

In summary, in two experiments involving more than 70 participants we demonstrated that training in the VE resulted in better performance than a comparable amount of coaching or extra practice on the real task.

It should be stressed, however, that while the observed performance gains are rather encouraging, our analysis revealed a complex relationship between the teacher trajectory and success on the task. On one hand, the increased performance of the training group suggests that learning a teacher trajectory is useful. On the other hand, if accurate reproduction of the teacher trajectory was related to task performance, one would expect to see within the training group a correlation between each participant's performance level and how well that participant followed the teacher trajectory. No such correlation was found. The trajectories used on the real task deviated significantly from the teacher trajectory, and for most participants these deviations increased over the three days (Fig. 49.5) while performance was improving. Brisson and Alain (1996) found a related result in a simpler interception task. These authors showed that training to reproduce a teacher trajectory improved performance, but the details of the teacher trajectory did not make a difference. In other words, imitating a teacher trajectory in the context of the task helps, but not because the exact trajectory being imitated was learned. The exact reason for this improvement remains unclear.

Furthermore, only when the teacher trajectory was coupled with a simulation of a ball did the training participants improve in the real task. This unexpected result suggests that seemingly minor changes in the simulator can dramatically reduce its efficacy, whereas details of the teacher trajectory remain identical. One possible explanation for this effect is that the ball

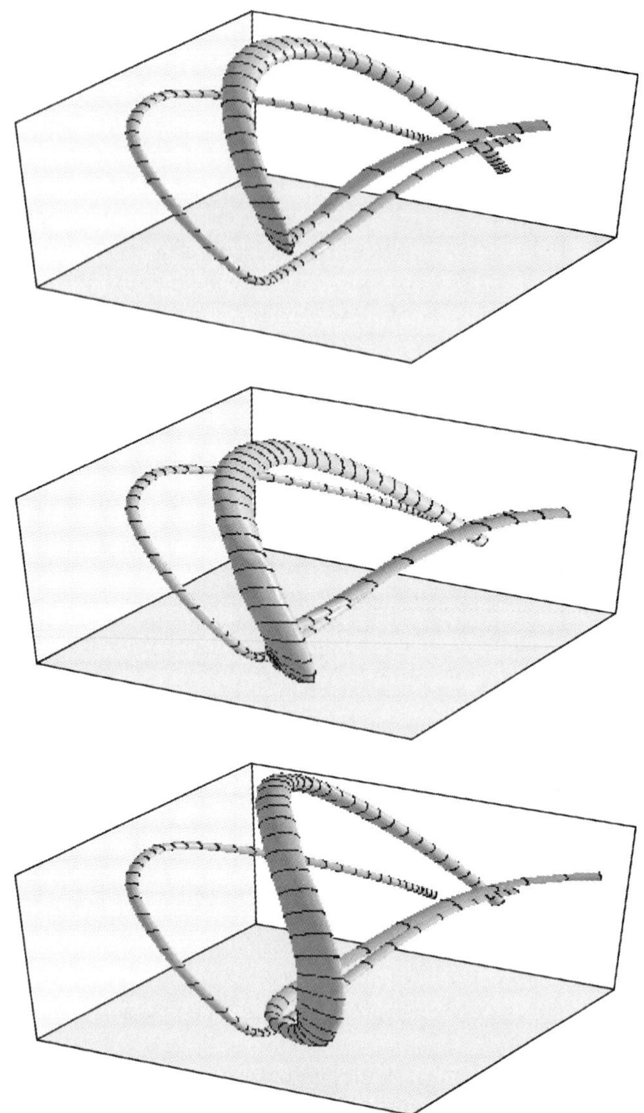

FIG. 49.5. Trajectories of one training subject on the real task, days 1 to 3 from top to bottom. The thin curve is the teacher trajectory. The thick curve is the average subject's trajectory on that day, where the width corresponds to one standard error of the mean. Black rings mark equally spaced points in time; that is, wider ring spacing corresponds to higher velocity. Note that the deviation between teacher and subject movements (on the real task) actually increases with practice, except for the region of impact with the ball (lower part). Courtesy of Journal of Motor Behavior.

simplified the synchronization task, because its flight could have been used to predict when the Teacher paddle started moving.

## 5.3 Ongoing and Future Work

So far we have focused on training end-effector trajectories, that is, only the paddle was displayed rather than the movement of the entire arm or body. While this is a good place to start, it certainly does not exhaust the possibilities. As Fig. 49.1 suggests, this form of training can be used for more complex trajectories involving multiple body parts. For example, consider the game of golf. In this case there is a well-defined end effector (the golf club); however, training the movement of the golf club alone is probably insufficient. Instead, golf coaches insist that movement of the entire body is important. Another good example of a task where whole body movement matters is dance; in this case there is no end effector to begin with.

If our training method is to be extended to movements of complex objects with multiple joints, the problem of confusing the Teacher and User objects becomes much more severe. Preliminary attempts to resolve this issue in the system shown in Fig. 49.1 indicate that more advanced visualization techniques are needed before the motion of the entire hand can be tracked successfully. Once User and Teacher hands are overlaid, there are so many small discrepancies that it is difficult to focus on any one of them and correct for it. One method we are considering is an automated procedure that selects body parts with the largest errors, and highlights them by making all other parts semitransparent, for example.

In the case of hand posture, we recently studied a matching task (Todorov & Ghahramani, 2000) where participants were shown computer generated images of random hand postures and asked to reproduce them as accurately as possible. Hand joint angles were recorded using a Cyberglove. The surprising finding was that subjects were very inaccurate in matching individual joint angles; the average correlation coefficients ($R^2$) of desired and actual values ranged between 0.05 and 0.2. It turned out, however, that certain changes in overall hand shape (i.e., linear combinations of joints identified through Canonical Correlation Analysis) were matched better than any single joint; the first five such combinations had correlation coefficients of about 0.8! This result suggests that in the case of multiarticulate bodies, the human visual system may extract overall shape rather than individual joint angles. Information about such visually salient shape changes can be useful in the future for designing better feedback systems.

## 6. EXPERIMENT 2: REHABILITATION IN PATIENTS WITH STROKE

### 6.1 System Design

The second system was designed for the purpose of arm rehabilitation. The training method was very similar to the Ping-Pong study, however the rationale behind it was somewhat different. Consider a stroke patient who is currently incapable of lifting the impaired arm above shoulder level. What "desired movement" should be chosen for such a patient? There is an infinite variety of tasks (involving objects above shoulder level), which the patient is incapable of executing. This is where the skill of an experienced therapist is required. Based on therapy evaluation, the therapist will determine the family of movements that are likely to be useful for a particular patient and the family of tasks where such movements could arise naturally. In other words, we are now starting with a desired movement and constructing a task around it. This is the opposite to training a specific motor skill, where the task is primary and the desired movement is the one used by experts on that task.

Fig. 49.6 illustrates the main features of the system. The trace corresponds to a desired movement, which the User (cube located near the "hand") is attempting to follow. The simplest way to create a task compatible with a given movement is to insert objects along the movement trajectory; in this case the donut was placed around the desired path, a second cube marked the beginning and end of the movement, and the task was defined as starting from the second cube and passing through the donut.

One major technical problem experienced with the rehabilitation system was alignment of the patient with the virtual environment. A scene contains a number of virtual objects and

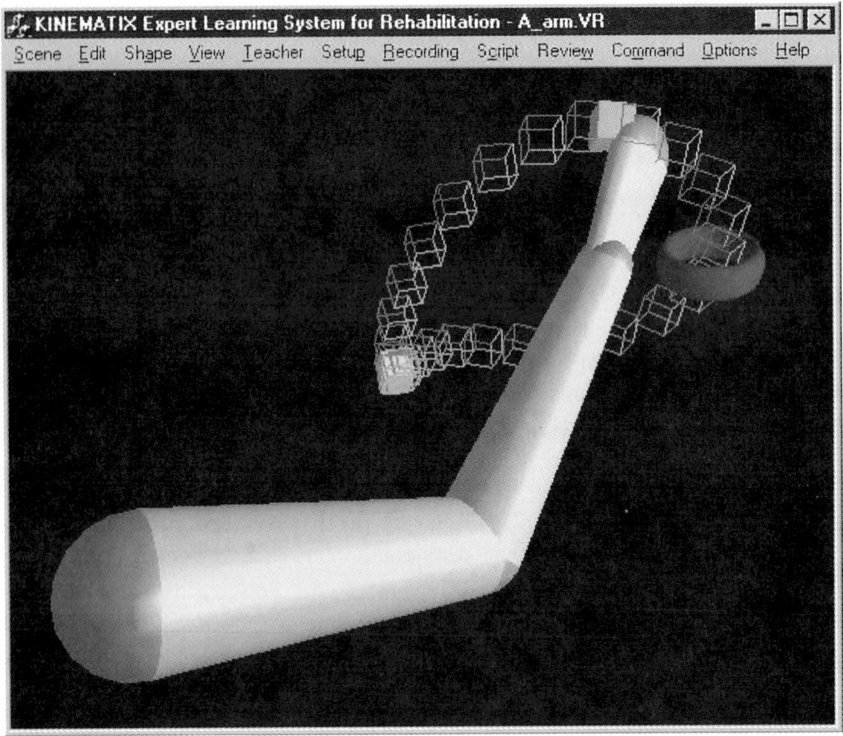


FIG. 49.6. An example of a virtual scene in the rehabilitation system (not used for training). The arm of the patient was tracked in real time, joint angles were estimated, and the virtual arm was rendered at about 40 Hz. The trace shows the teacher trajectory, for a task defined as starting from the solid cube (near elbow) and passing through the donut.

a teacher recording whose positions and orientations are determined in a coordinate system centered at the transmitter of the FASTRAK (and oriented along its axes). The sensor readings are also defined in that coordinate system. Thus, if a virtual object corresponding to some sensor location is inserted, the scene saved, and the sensor positioned at the same location in physical space on the next day, it will appear on top of the virtual object, assuming the transmitter did not move. Similarly, if the scene is adjusted so that the starting position and target correspond to the desired arm configurations of the patient, that correspondence will hold as long as the body of the patient is in exactly the same position and orientation relative to the transmitter.

In general, such alignment was not always easy to achieve. Although we attempted to position the patient in the same place relative to the computer in each training session, small differences inevitably occurred. Even small differences in patient position affected the alignment of the patient and teacher trajectories. While normal subjects could easily move to realign themselves, patients were slower and had much more difficulty with this task. Therefore, these alignment problems were resolved in software.

### *6.1.1 Scene Alignment*

One of the objects in the virtual scene (corresponding to the desired starting position and orientation of the movement) was defined as the "starting position object." To align the scene, the physical sensor was placed at the new starting position (and orientation). The system applied a rigid body transformation to the entire scene (i.e., a translation and a rotation) such that the old starting position object moved to the current sensor location. All other objects (and the teacher) moved accordingly so that relative distances and orientations in the scene were preserved. Note that such a transformation is uniquely defined.

The problem with this procedure was that typically the whole scene tilted. In such cases, a different alignment command was used that aligns only the positions of the physical sensor and the starting position object, but leaves the orientation of the scene intact.

### 6.1.2 Teacher Alignment

Sometimes the scene was aligned properly, but the patient was incapable of assuming the starting arm configuration of the teacher because of weakness, decreased range of joint motion, or muscle contractures. In such cases the system adapted the teacher movement so that it started at a new starting position (one achievable by the patient) but ended in the old ending position, or passed though a previously specified target.

The algorithm used to solve this problem defined a transformation (which included both translation and rotation) that sent the beginning teacher frame into the new starting position. Also, the transformation could be scaled to allow us to apply the complete transformation, half of it, and so forth (for translations, scaling was straightforward; for rotations the quaternion representation was computed and the rotation angle was scaled; the rotation axis was unchanged). For each frame of the teacher animation, we determined how much of the transformation should be applied. For the beginning frame, the number was 1 (i.e., the beginning frame is transformed completely to match the new starting position). For the target frame the number was 0 (i.e., it was left unchanged). For intermediate frames we interpolated between 0 and 1. To preserve the shape of the speed profile of the teacher movement, the interpolation was based on distance from the target rather than time. Figure 49.7 illustrates the procedure. The transformed teacher movements produced by this algorithm were surprisingly similar to the original.

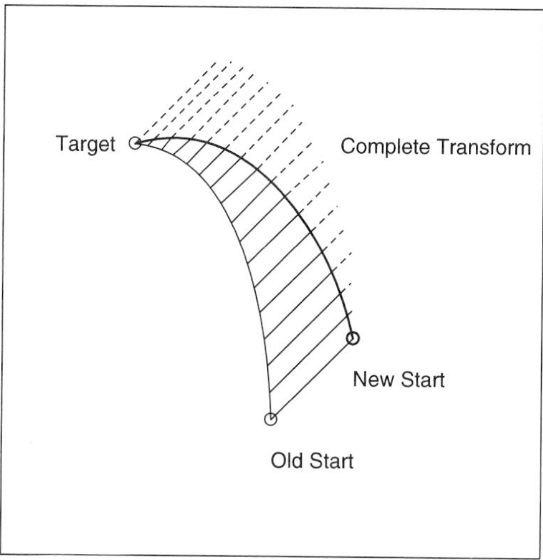


FIG. 49.7. Illustration of the teacher alignment algorithm. The straight lines correspond to applying a transformation that takes the old starting position to the new starting position. To each frame of the teacher animation, the same transformation is applied, scaled by the normalized distance (1–0) away from the starting position. This procedure results in a new teacher movement that looks very similar to the original but starts at the new desired position.

## 6.2 Experimental Results

In a pilot study (Holden et al., 1999), two patients were treated with hemiplegia using a computer-generated VE to train upper extremity reach in the impaired limb. The goal of the experiment was to answer three questions:

1. Can hemiplegic subjects improve in a virtual task following virtual practice?
2. Does learning which occurs in a VE transfer to a similar real task?
3. Does learning in a VE transfer to related but untrained real tasks, or to functional activities not specifically trained?

Subject 1 (S1) was a 76-year-old male, 3.5 years post left (L) cerebrovascular accident (CVA) due to thrombotic occlusion of the left internal carotid artery (ICA), confirmed by magnetic resonance imaging (MRI). He had resultant right hemiparesis, significant expressive aphasia, but excellent receptive abilities. This stroke was his first, with no evidence of bilateral or brain-stem stroke. He displayed no evidence of ongoing motor recovery in the right arm.

Subject 2 (S2) was a 76-year-old female, 1.5 years post right CVA due to thrombotic occlusion of the right internal carotid artery, confirmed by computerized tomography (CT) scan. She had resultant left hemiparesis. This stroke was her first, with no evidence of bilateral or brain-stem stroke. She displayed no evidence of ongoing motor recovery in her left arm.

We selected only one movement for training in the virtual environment, with the criterion that it be a functional, goal-oriented movement that highlighted typical motor control problems seen in patients with stroke. To train this movement a reaching task was devised, in which the subject held a styrofoam "envelope" (using a lateral grasp), then extended the arm to place the "envelope" in a "mailbox" slot.

Next, we created a series of scenes in the virtual environment. The scenes had a one-to-one spatial correspondence with the real world and were displayed on a desktop computer. They were simple, containing only a virtual mailbox and two virtual envelopes. One envelope was a "teacher" who performed the correct movement over and over again (see Fig. 49.8). The teacher animation was a recording of a well-practiced normal subject performing the virtual task. The second envelope was a virtual representation of the real envelope that the patient held and moved during practice. Thus, the patient could match the endpoint trajectory of his or her movement with that of the teacher during training, in real time. (Movements of the patient were monitored using a Polhemus FASTRAK, then displayed on the computer in the context of the virtual scene.) The scenes progressed from easy to more difficult in order to train the movement in a sequential fashion. The sequence used was near and far reach, first with forearm pronation, then forearm neutral, and lastly with supination. This resulted in three different hand orientations: palm down; palm facing right (when left arm was trained) or left (when right arm was trained); and palm up. The endpoint of the near reach for each different hand orientation was set at the distance midpoint of the normal trajectory for the far reach for that hand orientation.

The "teacher" movement could be altered in a number of ways designed to facilitate learning. The animation could be adjusted in speed to any level slower or faster than original; it could be paused at any point, displayed as a trace (solid line vs. moving object), or hidden from view entirely. The teacher could also be made to "follow" the trajectory of the subject, that is, temporal information could be eliminated, and the learner could focus exclusively on learning the spatial elements of the movement. Additional features (audio and visual) were designed to assist the patient with error detection, timing, and positioning. A model of the entire arm, as well as the held object, could also be displayed on the screen in real time, if desired. This was used intermittently in training to correct the patient's tendency to try to achieve the task with compensatory movement patterns such as excessive shoulder abduction with elbow flexion, which were ineffective.

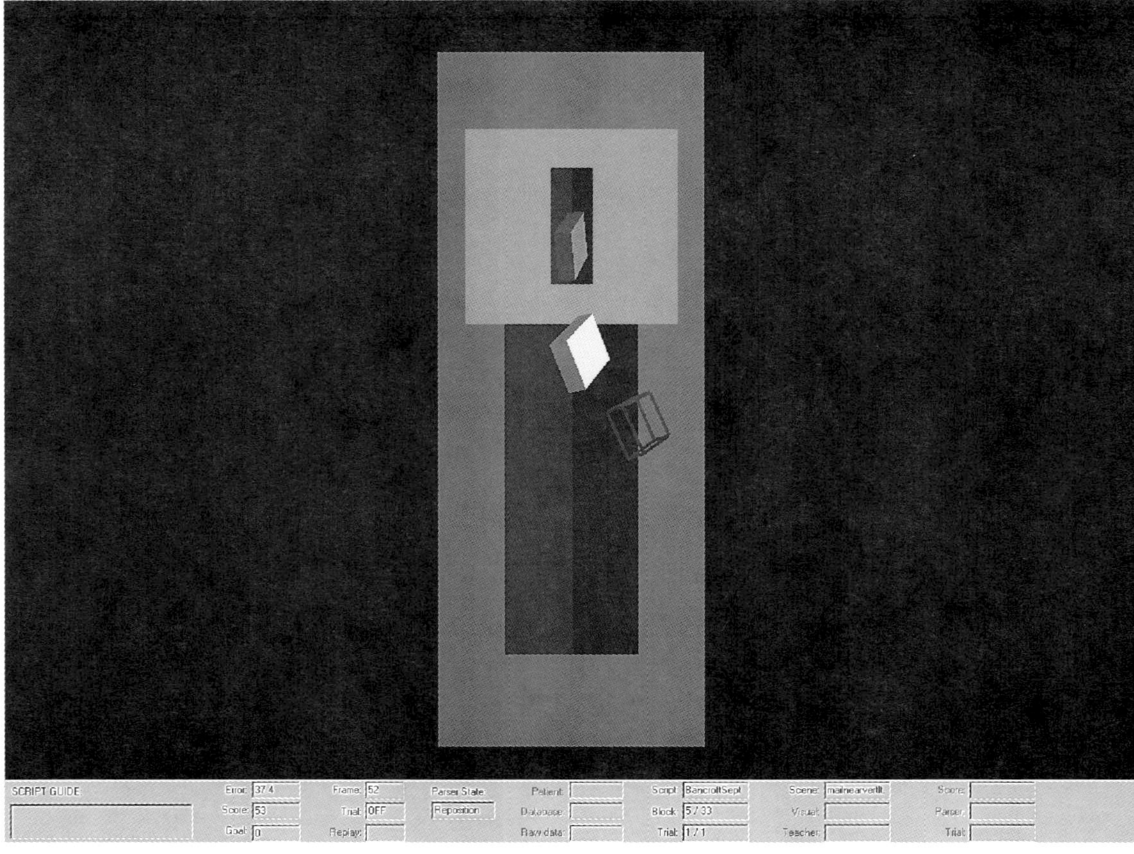


FIG. 49.8. Example of a “mailbox” scene used for VE training. The “teacher” envelope is shown entering the target slot. (In VE, the entire trajectory is animated, and objects appear in color against a black background.) Below the teacher, the white “patient” envelope is shown; this envelope moves when the patient moves his or her hand. Below the patient envelope, a wire-frame rectangle identifies the start position, used to align the teacher and patient trajectories at movement onset. Courtesy of Neurology Report.

Each subject received a series of 16 treatment sessions of 1 to 2 hours duration, conducted by a physical therapist. The planned treatment frequency was two times per week, although subjects actually came less frequently than this due to problems with transportation and minor illnesses. Virtual performance improved over the 16 sessions, as measured by assessing how many scenes a subject could progress through in a session. (Once subjects could perform the movement in a scene correctly for three consecutive trials, they progressed to the next scene.) Real-world performance in the "mailbox" task showed improved reaching in both the trained and untrained parts of the workspace for both subjects. The distance and orientation error for the "envelope" relative to the slot was calculated (see Fig. 49.9 for real-world test setup.)

S1 showed an average pre-/post-decrease in reach excursion error of 18 cm (i.e., 64% reduction in error); S2 showed a decrease of 9 cm, representing a 50% reduction in error (Fig 49. 10). The 9 and 18 cm gains represented roughly a 25% improvement in reach excursion. In both subjects, some of the largest improvements occurred in untrained parts of the workspace. S1 showed better transfer to upper regions of the workspace; S2 showed better transfer to lower regions (Figs. 11 and 12). Similar, but smaller magnitude results were found for hand orientation errors.

Several findings in this study have important clinical implications. The most important finding was that subjects not only improved on the virtual task, but also showed transfer of that improvement to similar real-world tasks—both trained and untrained.

## 6.3 Ongoing and Future Work

The VE training system we have developed is currently being utilized in Italy in a clinical study of stroke patients who have upper extremity impairments (Piron et al., 1999; Piron et al., 2000). This study is an ongoing randomized, controlled clinical trial designed to evaluate the effectiveness of VE training as a complementary therapy for motor rehabilitation following stroke. Experimental group subjects receive 1 hour of VE therapy plus 1 hour of conventional physical therapy, daily for 4 weeks (20 two-hour treatments total); control group subjects are treated for 2 hours of conventional therapy daily for 4 weeks (20 two-hour treatments total). Both acute (less than 6 months following stroke) and chronic (6–72 months following stroke) duration patients are being studied. Therapy in both virtual and conventional sessions is focused on the arm. All patients are evaluated before and after training, using the Fugl-Meyer scale for motor recovery (Fugl-Meyer, Jaasko, Leyman, Olsson, & Steglind, 1975) and the Functional Independence Measure (Hamilton, Granger, Sherwin, Zielezny, & Tashman, 1987), a scale that measures the amount of assistance needed to perform activities of daily living (such as eating and dressing). Forty-three subjects (28 experimental and 15 control) have been tested to date. Preliminary analysis suggests that the subjects who received combined virtual and conventional therapy improved more than subjects who received conventional therapy alone, on both the test of motor recovery (Fugl-Meyer arm subscore) and the functional scale. However, the study is still ongoing, and detailed statistical testing has not been performed. The response to the VE training on the part of both therapists and patients who have used the system has been quite positive. No negative side effects have been reported, and both patients and therapist feel it provides enhanced motivation for the rehabilitation program.

We have recently begun a second study utilizing the VE system to study aspects of motor learning and motor generalization in patients with stroke. In addition to knowledge about generalization, we hope to gain further practical experience with the newly developed features of our system (scoring, movement parser, and semiautomated functions.)

An extension of this work, currently being developed, is a networked virtual environment that will allow the therapist to work with patients remotely. The current software is being modified so that new virtual scenes can be loaded remotely, alignment commands can be

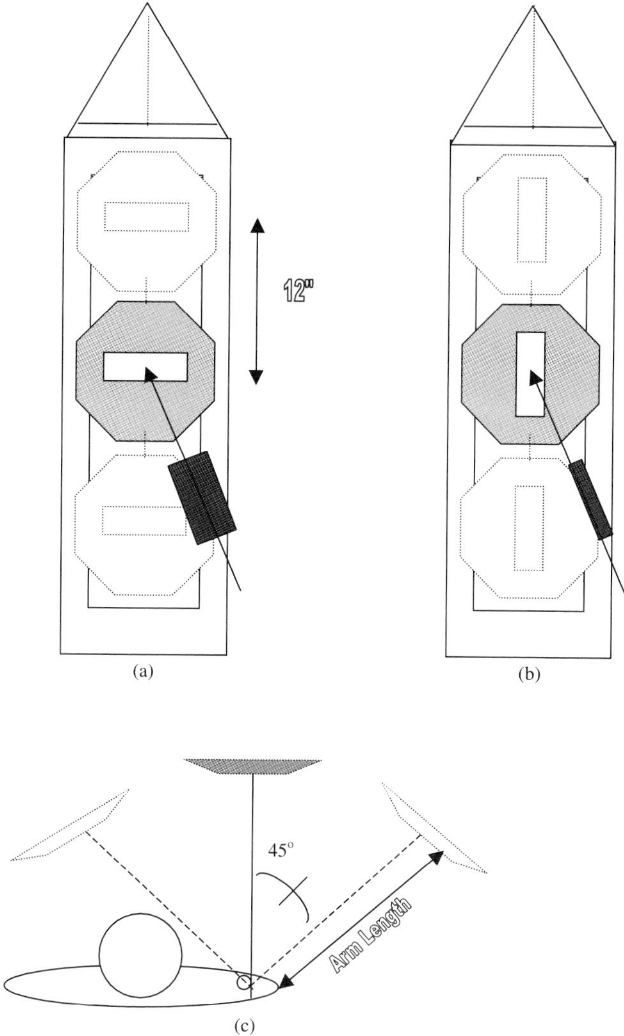


FIG. 49.9. Schematic of real-world reaching setup used to test subjects before and after training in a similar virtual task: (a) a 3-D "mailbox" with movable target slot (1″ × 5″) horizontal orientation was used for supinated and pronated reach attempts. The "envelope" (4″ × 8′ × 1/4″) is shown moving toward the target. The center target was positioned at a location corresponding to the hand position when the upper extremity was in 90-degree shoulder flexion/neutral abduction, and elbow fully extended (i.e., the distance equal to arm length and height equal to the height of the acromion in sitting position). For testing upper and lower workspace reaching, the slot was positioned 12 inches above and below the center position. Target positions later trained in VE are shown in gray; untrained locations are shown in white, (b) vertical slot orientation was used for reaching with forearm in neutral position; (c) testing in transverse plane (view from above); slot was positioned 45 degrees medial (adducted reach) and lateral (abducted reach) to the center position (shoulder-centered frame of reference). Only the center forward position (gray) was trained in VE. Courtesy of Neurology Report.

applied, and key feedback features can be turned on and off. We are also incorporating an Internet videophone to provide an audio-visual link between patient and therapist.

A third ongoing project is a study that uses the VE system to train subjects with acquired brain injury (ABI) to learn a task-oriented movement with the upper extremity (Holden, Dettwiler, Dyar, Niemann & Bizzi, 2001). The task we are presently working with is pouring from a cup

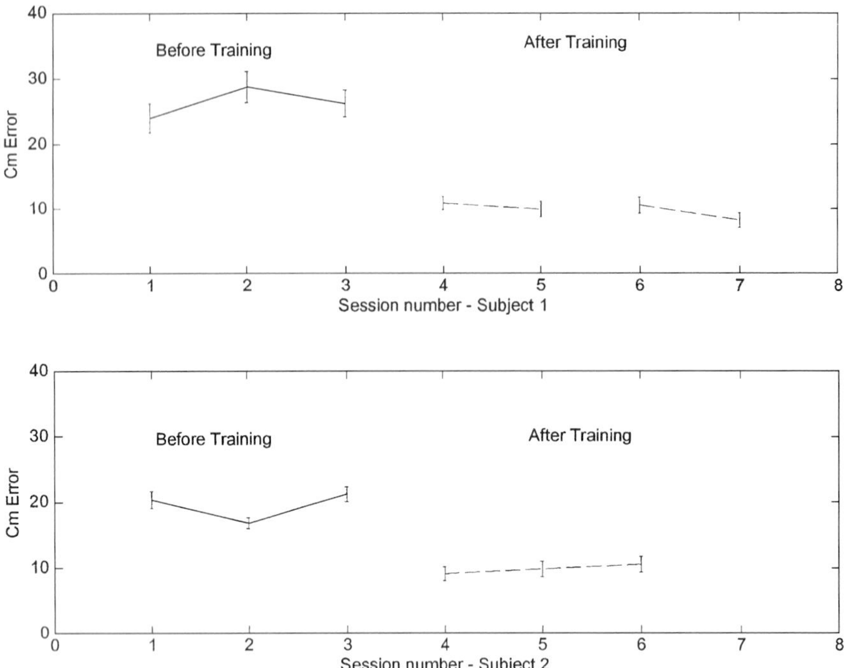


FIG. 49.10. Performance on real-world reaching task pre- and post–virtual training. *Top panel:* The mean error scores +/− one standard deviation for distance (cm) from target (averaged across all nine target positions) plotted by session number for subject 1, involved arm. Sessions 1 to 3 (solid line) equal to pretraining scores; sessions 4 and 5 (dotted line) equal to scores after eight training sessions; session 6 and 7 (dotted line) equal to scores after 16 training sessions. The magnitude of the pre- and postdifferences would be considered statistically significant, based on nonoverlap of the standard deviation values. Values for the noninvolved arm are not shown, but were ~ 1 cm with little variability for all sessions, with the exception of session 1, when error value was ~ 5 cm with +/−2 cm standard deviation. *Bottom panel:* Mean error scores for distance (cm) from target (averaged across all nine target positions) plotted by session number for subject 2, involved the arm. Sessions 1 to 3 (solid line) equal to pretraining scores; sessions 4 to 6 (dotted line) equal to scores after 16 training sessions. Values for noninvolved arm are not shown, but were ~ 1 cm with little variability for all sessions. (The values for the noninvolved arm do not represent true "errors," i.e., subjects could readily perform the task; rather, they result from the sensor position on the envelope.) Courtesy of Neurology Report.

held in the impaired hand, at different locations in the workspace. During training, subjects hold a real cup in their hand while practicing the pouring movement and viewing their virtual performance (and that of the teacher) on the computer screen. One purpose of this study is to gain more knowledge about how different clinical symptoms and impairments affect the ability of subjects with ABI to benefit from training with the system, and to identify equipment and software adaptations that might enhance the ability of subjects with acquired brain injuries to improve from motor training in virtual environments.

In the future, we would like to make the VE used in our system more interactive and also incorporate a haptic interface (see chaps. 5 and 6, this volume). This would allow training of a greater variety of movements in more interesting ways. Currently static objects define the task, and the only moving parts are the teacher and patient animations. One can imagine instead a gamelike scenario, where the objects respond to patient movements in various ways. For many patients, the response would need to be very simple, but for others we could adapt state-of-the-art computer games with advanced graphics, sound effects, realistic physics engines, and

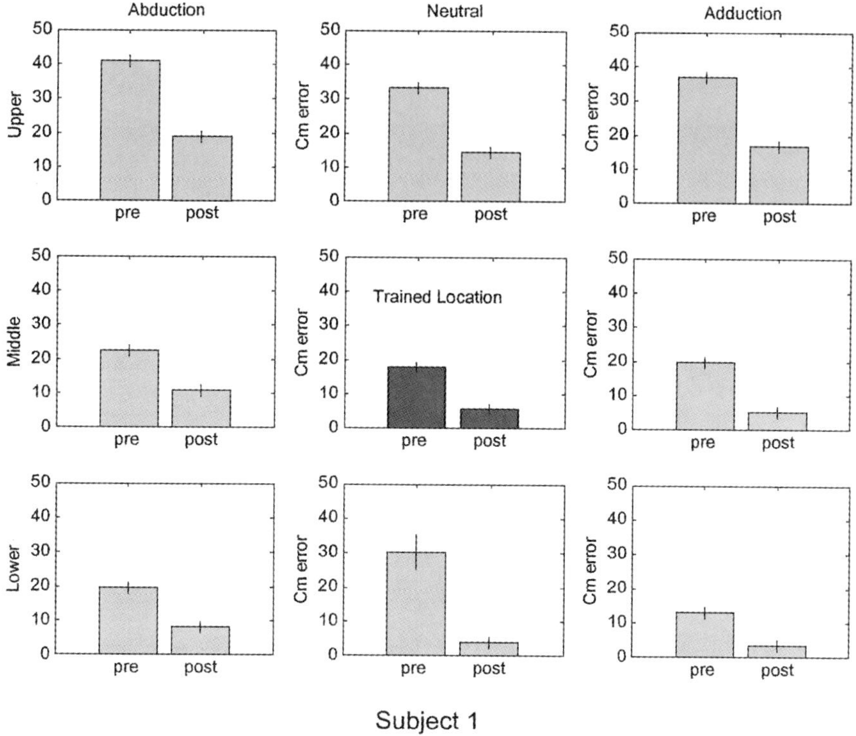


FIG. 49.11. Distance errors (cm) +/− one standard deviation for real-world reaching task, plotted by target location in the workspace, for subject 1, involved arm. Values are means across three sessions pre- and four sessions posttraining on a similar virtual task. Error scores for the trained movement are in the center of the middle row. Note that all locations show a reduction in error despite no specific training in those locations. Note better transfer to upper parts of the workspace for this subject. Values for noninvolved arm (not shown) were close to 1 cm and were nearly identical pre- and posttraining with the exception of upper adduction, which had a 10 cm mean error pretraining, 1 cm post. (The 1 cm error score for noninvolved arm was a function of the sensor position on the "mail" piece, and not due to inability of the subject to perform the task.) Courtesy of Neurology Report.

entertaining story lines to appropriate therapeutic levels. The problem is that current computer games are designed with traditional input devices in mind, such as a keyboard, joystick, or mouse. For the purpose of motor rehabilitation, games are needed that accept input from 3-D motion capture hardware and use it in some meaningful way. If such games become available in the future they could be used not only for entertainment, but also as therapeutic tools.

## 7. SUMMARY

In this chapter, first literature relevant to spatial-motor learning in VE in normal subjects was reviewed, as well as the literature on clinical applications of VE in rehabilitation. Next, a system that we developed to enhance motor learning in healthy and disabled individuals was described. The system is based on learning by imitation and uses a VE to implement training routines. The chapter concluded with a description of our experimental results to date and a discussion of design issues that arose during use of the system, along with a brief description of ongoing work and future research directions.

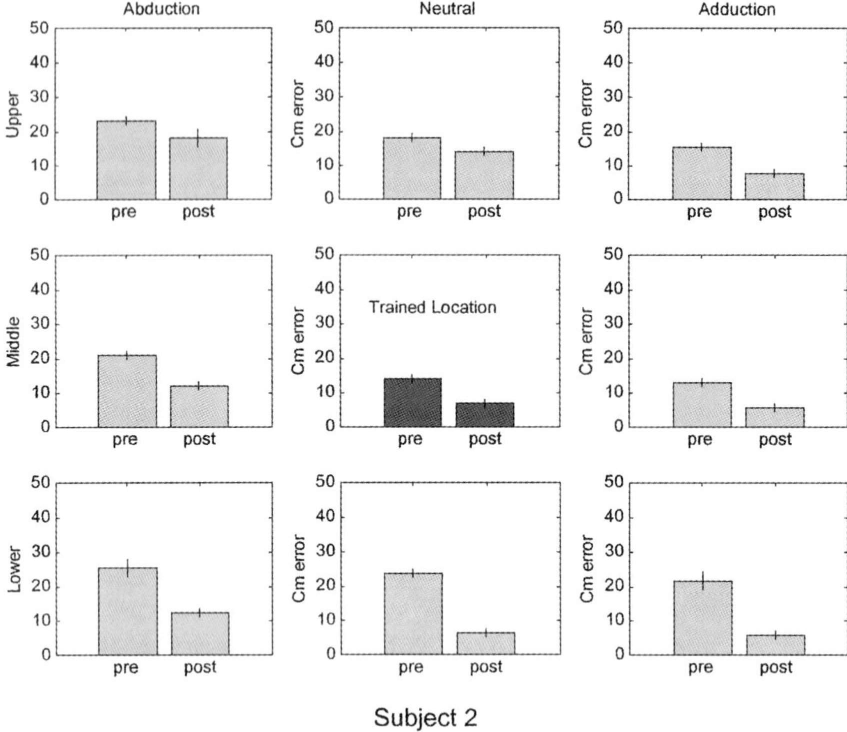


FIG. 49.12. Distance errors (cm), +/− one standard deviation, for real-world reaching task, plotted by target location in the workspace, for subject 2, involved arm. Values are means across three sessions pre- and posttraining on a similar virtual task. The trained movement is in the center of the middle row. Note that most locations show a reduction in error despite no specific training in those locations. Note better transfer to lower parts of the workspace for this subject. Noninvolved arm values are not shown, but were ~ 1 cm for all locations and essentially identical for pre- and posttests. Courtesy of Neurology Report.

## REFERENCES

Atkeson, C., & Hollerbach, J. (1985). Kinematic features of unrestrained vertical arm movements. *Journal of Neuroscience 5*, 2318–2330.

Biocca, F. A., & Rolland, J. P. (1998). Virtual eyes can rearrange your body: Adaptation to virtual eye location in see-through head-mounted displays. *Presence*, *7* (3), 262–277.

Bizzi, E., Mussa-Ivaldi, F. A., & Shadmehr, R. (1996). *System for Human Trajectory Learning in Virtual Environments* (U.S. Patent No. 5,554,033, Massachusetts Institute of Technology, September 10, 1996).

Bower, G. H. (1972). A selective review of organizational factors in memory. In E. Tulving & W. Dondson (Eds.), *Organization of memory* (pp.93–137). New York: Academic Press.

Brisson T., & Alain C. (1996). Should common optimal movement patterns be identified as the criterion to be achieved? *Journal of Motor Behavior*, *28*(3), 211.

Burdea, G., Deshpande, G., Liu, N., Langrana, N., & Gomez, D. (1997). A virtual reality–based system for hand diagnosis and rehabilitation. *Presence*, *6*(2), 229–240.

Christiansen, C., Abreu, B., Ottenbacher, K., Huffman, K., Masel, B., & Culpepper, R. (1998). Task performance in virtual environments used for cognitive rehabilitation after traumatic brain injury. *Arch. Phys. Med. Rehabil.*, *79*, 888–892.

Cobb, S. V. G. (1999). Measurement of postural stability before and after immersion in a virtual environment. *Applied Ergonomics*, *30*, 47–57.

Cunningham, D. (1999, Dec/Jan). Therapy for new millennium. *Rehab Management*, 84–85.

Darken, R. R., Allard, T., & Achille, L. B. (1998). Spatial orientation and wayfinding in large-scale virtual spaces: An introduction. *Presence*, *7*(2), 101–107.

Docimo, S. G., Moore, R. G., & Kavoussi, L. R. (1997). Telerobotic surgery is clinical reality: Current experience with telementoring in adults and children. *Presence*, *6*(2), 173–178.

Emmett, A. (1994). Virtual reality helps steady the gait of Parkinson's patients. *Computer Graphics World*, *17*, 17–18.

Foreman, N., Orencas, C. Nicholas, E., Morton, P., & Gell, M. (1989). Spatial awareness in seven- to eleven-year-old physically handicapped children in mainstream schools. *European Journal of Special Needs Education*, *4*, 171–178.

Fugl-Meyer, A. R., Jaasko, L., Leyman, I., Olsson, S., & Steglind, S. (1975). The post-stroke hemiplegic patient: A method for evaluation of physical performance. *Scandinavian Journal of Rehabilitation Medicine*, *7*, 13–31.

Gillner, S., & Mallot, H. A. (1998). Navigation and acquisition of spatial knowledge in a virtual maze. *Journal of Cognitive Neuroscience*, *10*(4), 445–463.

Greenleaf, W., & Tovar, M. A. (1994). Augmenting reality in rehabilitation medicine. *Artificial Intelligence in Medicine*, *6*, 289–299.

Hamilton, B. B., Granger, C. V. Sherwin, F. S., Zielezny, M., & Tashman, J. S. (1987). A uniform national data system for medical rehabilitation. In M. J. Fuher (Ed.), *Rehabilitation outcomes: Analysis and measurement*. (pp. 137–147) Baltimore: Brookes.

Holden, M., Dettwiler, A., Dyar, T., Niemann, G. & Bizzi, E. (2001). Retraining of movement control in patients with acquired brain injury using a virtual environment. In J. D Westwood et al. (Eds.) Medicine Meets Virtual Reality 2001(pp. 192–198). Amsterdum: IOS Press.

Holden, M., Todorov, E., Callahan, J., & Bizzi, E. (1999). Virtual environment training improves motor performance in two patients with stroke. *Neurology Report*, *23*(2), 57–67.

Inman, D., Peaks, J., Loge, K., & Chen, V. (1994). Teaching orthopedically impaired children to drive motorized wheelchairs in virtual reality. In *Proceedings of the second Annual International Conference on Virtual Reality and Persons with Disabilities*. San Francisco, CA: Calif State Univ. Press.

Kennedy, R. S., Lanham, D. S., Massey, C. J., Drexler, J. M., & Lilienthal, M. G. (1995). Gender differences in simulator sickness incidence: Implications for military virtual reality systems. *SAFE J.*, *25*, 69–76.

Kozak, J. J., Hancock, P. A., Arthur, E. J., & Chrysler, S. T. (1993). Transfer of training from virtual reality. *Ergonomics*, *36*, 777–784.

Kramer, J., & Leifer, L. (1989). The talking glove: A speaking aid for nonvocal deaf and deaf–blind individuals. Paper presented at *RESNA, twelth Annual Conference*, New Orleans, LA.

Kuhlen, T., & Dohle, C. (1995). Virtual reality for physically disabled people. *Computers in Biology & Medicine*, *25*(2), 205–211.

Lackner, J. R., & DiZio, P. (1994). Rapid adaptation to Coriolis force perturbations of arm trajectory. *Journal of Neurophysiology*, *72*, 299–313.

Lacquaniti, F., Terzuolo, C., & Viviani, P. (1983). The law relating the kinematic and figural aspects of drawing movement. *Acta Psychologica*, *54*, 115–130.

Lawson, B. D., Rupert, A. H., Guedry, F. E., Grissett, J. D., & Mead, A. M. (1997). The human–machine interface challenges of using virtual environment (VE) displays aboard centrifuge devices. In *Proceedings of the Seventh International Conference on Human–Computer Interaction*, (Vol. 2, pp. 945–948). San Francisco, CA; Amsterdam: Elsevier Science Publishers.

Lusted, H. S., & Knapp, B. R. (1992). Biocontrollers for the physically disabled: A direct link from the nervous system to computer. In *Proceedings of Virtual Reality and Persons with Disabilities*. Los Angeles: Calif State University Press.

Morasso, P. (1981). Spatial control of arm movements. *Experimental Brain Research*, *42*, 223–227.

Nichols, S. (1999). Physical ergonomics of virtual environment use. *Applied Ergonomics*, *30*, 79–90.

Piron, L., Dam, M., Trivello, E., Iaia, E., & Tonin, P. (1999). Virtual environment training ameliorates motor deficits in post-stroke patients. Paper presented at *American Academy of Neurology Annual Conference*, Toronto, Canada.

Piron, L., Trivello, E., Cenni, F., Iaia, E., Dam, M., & Tonin, P. (2000, January). Motor training in virtual environment for post-stroke and traumatic brain injured patients. *Eighth Annual Medicine Meets Virtual Reality Conference*, Newport Beach, CA.

Popescu, V. G., Burdea, G. C., Bouzit, M., & Hentz, V. R. (2000). A virtual-reality-based telerehabilitation system with force feedback. *IEEE Transactions on Information Technology in Biomedicine*, *4*(1), 45–51.

Pugnetti, L., Mendozzi, L, Motta, A., Cattaneo, A., Barbieri, E., & Brancotti, A. (1995). Evaluation and retraining of adults for cognitive impairments: Which role for virtual reality technology? *Computers in Biology & Medicine*, *25*(2), 213–227.

Regian, J. W., Shebilske, W. L., & Monk, J. M. (1992). Virtual reality: An instructional medium for visual–spatial tasks. *Journal of Communication*, *42*, 136–149.

Richard, P., Birebent, G., Coiffet, P., Burdea, G., Gomez, D., & Langrana, N. (1996). Effect of frame rate and force feedback on virtual object manipulation. *Presence*, *5*(1), 95–108.

Reiss, T., & Weghorst, S. (1995). Augmented reality in the treatment of Parkinson's disease. In K. Morgan, R. M. Satava,

H. B. Sieburg, R. Mattheus, J. P. Christensen (Eds.), *Interactive Technology and the New Paradigm for Healthcare* (pp. 415–422). Amsterdam: IOS Press.

Rolland, J. P., Biocca, F. A., Barlow, T., & Kancherla, A. (1995). Quantification of adaptation to virtual-eye location in see-through head-mounted displays. In *Virtual Reality Annual International Symposium '95* (pp. 55–66). Los Alamitos, CA: IEEE Computer Society Press.

Rose, F. D., Attree, E. A., & Johnson, D. A. (1996). Virtual reality: An assistive technology in neurological rehabilitation. *Current Opinion in Neurology, 9*, 461–467.

Ruddle, R. A., Payne, S. J., Jones, D. M., (1999). Navigating large-scale virtual environments. *Presence, 8*, 57–68.

Schmidt, R. (1975). A schema theory of discrete motor skill learning. *Psychological Review, 82*, 225–260.

Schmidt, R., & Young, D. (1991). Methodology for motor learning: A paradigm for kinematic feedback. *Journal of Motor Behavior, 23*, 13–24.

Shadmehr, R., & Mussa-Ivaldi, S. (1994). Adaptive representation of dynamics during learning of a motor task. *Journal of Neuroscience, 14*, 3208–3224.

Smythe, P. (1993). Use of networked VR to assist people with special needs. In H. J Murphy (Ed.), *Proceeding of the First Annual International Conference or Virtual Reality and Persons with Disabilities* (pp. 113–117). Northridge: California State University Press.

Stanney, K. M., Kennedy, R. S., Drexler, J. M., & Harm, D. L. (1999). Motion sickness and proprioceptive after effects following virtual environment exposure. *Applied Ergonomics, 30*, 27–38.

Stanney, K., Salvendy, G. Deisigner, J., DiZio, P., Ellis, S., Ellison, E., Fogleman, G., Gallimore, J., Hettinger, L., Kennedy, R., Lackner, J., Lawson, B., Maida, J., Mead, A., Mon-Williams, M., Newman, D., Piantanida, T., Reeves, L., Riedel, O., Singer, M., Stoffregen, T., Wann, J., Welch, R., Wilson, J., Witmer, B. (1998). Aftereffects and sense of presence in virtual environments: Formulation of a research and development agenda. *International Journal of Human–Computer Interaction, 10*(2), 135–187.

Theasby, P. J. (1992). The virtues of virtual reality. *GEC Review, 7*(3), 131–145.

Todorov, E., & Ghahramani, Z. (2000). Degrees of freedom and hand synergies in manipulation tasks. Abstract P-28, *18th Annual mtg of Neural Control of Movement*, Key West, FL: Neural Control of Movement Society.

Todorov, E., Shadmehr, R., & Bizzi, E. (1997). Augmented feedback presented in a virtual environment accelerates learning of a difficult motor task. *Journal of Motor Behavior, 29*(2), 147–158.

Vamplew, P., & Adams, P. (1992). The SLARTI system: Applying artificial neural networks to sign language recognition. In *Proceeding of Virtual Reality and Persons with Disabilities*. Los Angeles: California State University Press.

Waller, D., Hunt, E., & Knapp, D.(1998). The transfer of spatial knowledge in virtual environment training. *Presence, 7*(2), 129–143.

Warner, D., Anderson, T., & Johanson, J. (1994). Bio-cybernetics—A biologically responsive interactive interface. In *Proceedings of Medicine Meets Virtual Reality*. San Diego, CA:

Wilson, P. N., Foreman, N., & Stanton, D. (1997). Virtual reality, disability and rehabilitation. *Disability and Rehabilitation, 19*(6), 213–220.

Wilson, P. N., Foreman, N., & Tlauka, M. (1996). Transfer of spatial information from a virtual to a real environment in physically disabled children. *Disability and Rehabilitation, 18*, 663–637.

Wilson, P. N., Foreman, N. & Tlauka, M. (1997). Transfer of spatial information from a virtual to a real environment. *Human Factors, 39*(4), 526–531.

Wolpert, D., Ghahramani, Z., & Jordan, M. (1995). Are arm trajectories planned in kinematic or dynamic coordinates? An adaptation study. *Experimental Brain Research, 103*, 460–470.

# 50

# Virtual Environment Applications in Clinical Neuropsychology

Albert A. Rizzo,[1] J. Galen Buckwalter,[2] and Cheryl van der Zaag[1]

*[1] University of Southern California*
*Integrated Media Systems Center and School of Gerontology*
*3715 McClintock Ave. MC-0191*
*Los Angeles, CA. 90089*
*arizzo@usc.edu*
*vanderz@usc.edu*
*[2] Southern California Permanente Medical Group*
*100 South Los Robles Ave.*
*Pasadena, CA. 91101*
*Galen.x.Buckwalter@kp.*

## 1. INTRODUCTION

Virtual environments (VEs) have undergone a transition in the past few years that has taken from the realm of expensive toy into that of functional technology. This emerging computer-driven simulation technology appears to be well matched to the assessment and rehabilitation needs of persons with cognitive impairments and functional disabilities due to various forms of central nervous system (CNS) dysfunction. The capacity of VE technology to create dynamic three-dimensional (3-D) stimulus environments, within which all behavioral responding can be recorded and measured, offers clinical assessment and rehabilitation options that are not available using traditional neuropsychological methods. This work has the potential to advance the scientific study of normal cognitive and behavioral processes and to improve our capacity to understand, measure, and treat the impairments typically found in clinical populations with CNS dysfunction. Clinical populations that could benefit from VE approaches include persons with cognitive and functional impairments due to traumatic brain injury (TBI), neurological disorders, learning/developmental disabilities, as well as the elderly. This chapter begins with a brief description of the relevant clinical populations and an introduction to neuropsychological assessment (NA) and cognitive rehabilitation, (CR), with rationales presented for the use of VE for these purposes. The literature on research done thus far in these areas is then reviewed. The final section of this chapter details the basic clinical, human factors, and theoretical issues for the rational and effective targeting, development, and implementation of these applications.

## 2. BACKGROUND

Virtual environment technology is increasingly being recognized as a useful tool for the study, assessment, and rehabilitation of cognitive processes and functional abilities. Much like an aircraft simulator serves to test and train piloting ability, virtual environments can be developed to present simulations that target human cognition and behavior in normal and impaired populations. In this regard, a growing number of laboratories are developing research programs investigating the use of VEs for these purposes, and initial exploratory studies reporting encouraging results are now beginning to emerge. Virtual environment applications are now being developed and tested that focus on component cognitive processes including: attention, executive functions, memory, and spatial abilities. Functional VE training scenarios have also been designed to test and teach instrumental activities of daily living such as street crossing, automobile driving, meal preparation, supermarket shopping, use of public transportation, and wheelchair navigation. These initiatives have formed a foundation of work that provides support for the feasibility and potential value of further development of neuropsychological VE applications. If the associated technology continues to advance in the areas of computing power, graphics and image capture, display technology (see chap. 3, this volume), immersive audio (see chap. 4, this volume), haptics (see chaps. 5 and 6, this volume), wireless tracking (see chaps. 8 and 9, this volume), gesture recognition (see chap. 10, this volume), voice recognition, intelligent agents (see chap. 15, this volume), and VE authoring software (see chap. 12, this volume), then more powerful and naturalistic VE scenarios will be possible. These advances could result in more readily available desktop-powered VE systems (see chap. 13, this volume) with greater sophistication and responsiveness. Such increases in access would allow for more widespread application of VE technology for clinical purposes and promote the independent replication of research findings needed for scientific progress in this field.

Indeed, mainstream researchers and clinicians in neuropsychology are "wanting" for these advances in VE technology—whether they realize it or not. For example, in a recent National Institutes of Health (NIH) Consensus paper entitled "Rehabilitation of Persons with Traumatic Brain Injury (TBI)," two recommendations were made that suggest research directions that VE technology appears well poised to address. The report recommended that "Innovative study methodologies that enhance the ability to assess the effectiveness of complex interventions for persons with TBI should be developed and evaluated." (p. 27) and that "Innovative rehabilitation interventions for TBI should be developed and studied..." (p. 26, National Institute of Health [NIH], 1998a). Further, indirect support for VE technology's potential contribution to neuropsychology can be implied in an *American Psychologist* article by the well-respected intelligence theorist, Sternberg (1997). In this paper, he contends that with the exception of "cosmetic" changes, the field of cognitive ability testing has progressed very little in the last century in contrast to the advancements seen in other technologies over this same period. He posits that "dynamic" interactive testing provides a new option that could supplement traditional "static" tests. The "dynamic" assessment approach requires the provision of guided performance feedback as a component in tests that measure learning. This method appears well suited to the assets available with VE technology. In fact, VEs might be the most efficient vehicle for conducting dynamic testing in an ecologically valid or "real-world" manner while still maintaining an acceptable level of experimental control. A more specific recommendation for future research into the possibilities of VE technology appeared earlier in a 1995 NIH report of the National Advisory Mental Health Council (NIH, 1995). In this report, the impact of VE on cognition was specifically cited with the recommendation that "Research is needed to understand both the positive and negative effects of such participation on children's and adults' perceptual and cognitive skills ... (p. 51, NIH, 1995). These observations suggest that the disciplines of NA and CR are fertile ground for developing the innovative applications that are possible with VE technology.

The current status of VE technology applied to clinical populations, while provocative, is still limited by the small (but growing) number of controlled studies in this area. This is to be expected, considering the technology's relatively recent development, its high initial development costs, and the lack of familiarity with VE technology by established researchers employing the traditional tools and tactics of their fields. In spite of this, a nascent body of work has emerged that can provide knowledge for guiding future research efforts. Although much of the work does not involve the use of fully immersive head-mounted displays (HMDs), studies reporting 3-D projection screen and PC-based flat-screen approaches are providing useful information necessary for the reasoned development and implementation of VE technology with clinical CNS populations. This chapter will provide a rationale for the application of VE technology in the areas of neuropsychological assessment and cognitive rehabilitation, review the VE literature targeting cognitive–functional processes in clinical CNS populations, and discuss the issues that are relevant for the rational development of VEs designed for neuropsychological applications. Although considerable information is available on the use of VEs with unimpaired populations (as evidenced in the numerous chapter in this volume), our focus will be primarily on VE applications targeting clinical groups having CNS-based cognitive–functional impairments.

## 2.1 Clinical Populations

CNS dysfunction, resulting in cognitive and functional impairments, can occur through a variety of circumstances. The most frequent causes include TBI due to accidents, neurological disorders, developmental and learning disorders, as well as complications from medical conditions and procedures. The resulting impairments commonly involve processes of attention, memory, language, spatial abilities, higher reasoning, and functional abilities. Significant emotional, social, vocational, and self-awareness components that typically co-occur can also further complicate these areas. Because of the pervasive nature of CNS dysfunction, the cost to individuals and society is significant.

TBI is the most common cause of CNS dysfunction and is broadly defined as brain injury resulting from externally inflicted trauma. Such injury is often the result of automobile accidents, falls, sports accidents, and bullet wounds. In the United States, estimates range from 500,000 to two million new cases per year (National Center for Injury Prevention and Control, 1999). The peak age of incidence is in the 15- to 24-year range (closely followed by the birth to 5 years group). In addition to the cost of human suffering, one estimate places the economic costs in terms of medical care, rehabilitation, and lost work potential at $48.3 billion annually (NIH, 1998b).

Neurological disorders that cause CNS dysfunction include Alzheimer's disease (AD), vascular dementia, Parkinson's disease, Huntington's disease, cerebral palsy, epilepsy, and multiple sclerosis. In addition, other relatively common causes of CNS dysfunction include strokes, drug reactions, thyroid disease, nutritional deficiencies, tumors, alcoholism, and infections. Alzheimer's disease has been estimated to afflict nearly four million Americans, or between 2% to 4% of the population over the age of 65. It is estimated that the prevalence of AD doubles every five years beyond 65 to the point that nearly half of all people 85 and older display symptomatology. Alzheimer's disease is the third most expensive disease in the United States (following heart disease and cancer), with associated costs close to $100 billion per year. With the increase in life expectancy, it is estimated that the number of Americans aged 85 and over will double by the year 2030 (NIA *Progress Report on Alzheimer's Disease*, 1998), an estimate that has alarming social, economic, and public health implications.

Approximately three million Americans also suffer with some degree of disability from stroke (Gresham, Duncan, Stason et al., 1995). Although a stroke can occur at any age, for every decade after the age of 55 the risk doubles. Of the nearly 500,000 individuals annually who have a first-time stroke, 55% experience varying degrees of disability, including a range of deficits in language, cognition, and motor functions. The total cost of stroke to the United States is estimated to range from $30 to $43 billion per year (National Stroke Association, 1997).

Many others, particularly the young, experience cognitive–functional impairments due to various developmental and learning disabilities. Estimates place the number of children receiving special education services at between 2.5 to 4 million (Barkley, Fisher, Edelbrock, Smalish, 1990). Rates of other childhood learning disorders, such as attention deficit hyperactivity disorder (ADHD) and reading disorders, push estimates even higher. Methodological complications preclude precise estimates of the cost of ADHD to society, but according to 1995 statistics, additional national public school costs for these students may have exceeded $3 billion. Taken together, the above outlined statistics suggest a significant clinical population of persons with CNS dysfunction that may be better served by the types of advanced assessment and rehabilitation tools that are possible via the emerging application of VE technology.

## 2.2 Brief Introduction to Neuropsychological Assessment

In the broadest sense, neuropsychology is an applied science that evaluates how specific activities in the brain are expressed in observable behaviors (Lezak, 1995). Effective NA is a prerequisite for both the treatment and scientific analysis of any CNS-based cognitive–functional impairment. The NA of persons with CNS disorders serves a number of functions. These include the determination of a diagnosis, provision of normative data on the status of impaired cognitive and functional abilities, production of information for the design of rehabilitative strategies, and measurement of treatment efficacy. Neuropsychological assessment also serves to create data for the scientific understanding of brain functioning through the examination of measurable sequelae that occur following brain damage or dysfunction. Our understanding of brain morphology and activity has undergone a revolution in the past three decades that is akin to the revolution seen in microtechnology. However, the increase in our knowledge of the genetics, chemistry, molecular biology, and "physics" of the brain is mitigated by our understanding of the behavior that is related to specific brain activity. The fact that postmortem studies of AD have identified the entorhinal cortex as the area where the pathological changes of AD are first noted (Braak, Braak, & Bohl, 1993) is of little clinical value unless the cognitive and behavioral processes that are serviced by this region can be identified. Once such processes are identified, it becomes possible to diagnose more effectively and intervene at an earlier stage of this neurodegenerative process. Virtual environment technology offers the potential to develop human performance testing environments that could supplement traditional NA procedures and conceivably produce new methodologies that support earlier diagnosis by improving on standards for psychometric reliability and validity.

The assessment of cognitive abilities has a long and, at times, explosive history. In 1921, at a contentious symposium on what comprises intelligence, the historian and experimental psychologist Edwin Boring asserted that "intelligence is what the tests test" (Boring, 1923, p. 35). This notion of making a test and then accepting the results as indicative of an aspect of brain performance may be the classic case of putting the cart before the horse, but to some extent it is a pragmatic reality of our ability, or inability, to measure behaviors. Neuropsychology has proceeded by reducing complex behaviors to component cognitive domains. Domains of cognitive functioning identified by neuropsychologists are numerous and many subcomponents have been identified and studied. Some examples of more global cognitive processes include attention, executive functions, memory, language processing, spatial abilities, problem solving, and higher level abstract reasoning. Neuropsychological assessment also attempts to define and measure functional processes, which are assumed to be comprised of the integration of component cognitive processes that are manifested in everyday human behavior. Functional processes might include such instrumental activities of daily living (IADLs) as meal preparation, using transportation, shopping, housework, vocational activities, and handling finances. These cognitive–functional processes are commonly assessed using such NA methods as psychometric testing, behavioral observation, and an examination of past historical information.

The measurement of cognitive–functional processes is based on two criteria: reliability and validity (see chap. 35, this volume). Reliability is the capacity of an instrument to consistently obtain the same results. Validity is concerned with how well an instrument actually measures what it purports to measure. Traditional assessment methodology, primarily based on the use of pencil and paper tests, presents the neuropsychologist with both reliability and validity problems. The reliability of these instruments is adversely affected by the variability of administration procedures due to differences in examiners, testing environment (i.e., lighting, room size, background noise), quality of the stimuli presented to the subject, and by inevitable scoring errors. The validity of traditional methods is attenuated by the fact that some tests require multiple cognitive domains for successful completion, and thus it remains unclear what specific cognitive domain is being evaluated. It is likely that both reliability and validity can be improved on with the use of VE technology. Reliability can be enhanced by better control of the perceptual environment, more consistent stimulus presentation, and by more precise and accurate scoring. Virtual environments may also improve on the validity of measurement via the quantification of more discrete behavioral responses, allowing for the identification of more specific cognitive domains.

Traditional NA testing has also been criticized as limited in the area of ecological validity, which is concerned with the degree of relevance or similarity that a test has relative to the "real" world (Neisser, 1978). While existing neuropsychological tests obviously measure some behavior mediated by the brain, it is difficult to say with any certainty how performance on a contrived assessment task relates to complex performance in an "everyday" functional environment. Cognitive–functional performances could be tested in simulated "real-world" VE scenarios, thereby improving on the ecological validity of measurement. In this way, the complexity of stimuli found in naturalistic settings could be presented while still maintaining the experimental control required for rigorous scientific analysis. Thus, results would have greater clinical relevance and could have direct implications for the development of functional cognitive rehabilitation approaches. However, it is also important to recognize that in all cases it may not be desirable for VEs to fully "mimic" reality. Another strength of VEs for assessment purposes may include the capacity to present scenarios that include features not available in the "real world." This would be the case when "cueing" stimuli are presented to examine at what level "augmentive" information can be used by patients to assist in compensatory strategies aimed at improving day-to-day functional behavior.

It is possible that the use of VE technology could revolutionize our approach to neuropsychological assessment. While formidable problems remain, the potential is impressive. The possibility of linking VE assessment with advanced brain imaging and psychophysiological techniques (Aguirre & D'Esposito, 1997; Decety et al., 1994; Pugnetti, Mendozzi, Barbieri, Rose, & Attree, 1996) may allow neuropsychology to reach its stated purpose, that of determining unequivocal brain-behavior relationships. While pragmatic concerns need to be addressed in order for this technology to come close to meeting this lofty goal, the benefits that could be accrued appear to justify the effort.

## 2.3 Brief Introduction to Cognitive Rehabilitation

Many definitions of what CR involves have been put forth, with Parente and Herrmann (1996) describing it in general terms as "the art and science of restoring . . . mental processes after injury to the brain" (p. 1). Sohlberg and Mateer (1989) define CR as "the therapeutic process of increasing or improving an individual's capacity to process and use incoming information so as to allow increased functioning in everyday life" (p. 3). As with NA, both component cognitive processes and functional activities of daily living are typically targeted with cognitive rehabilitation. Cognitive processes are often further broken down into subcomponents and addressed. For example, attention can be partitioned into focused, sustained, selective, alternating, and divided components that are targeted with different tactics (Sohlberg & Mateer, 1989). Likewise, memory is often broken down into traditional cognitive psychology categories such as iconic, working, procedural, declarative, prospective, and episodic memory domains. IADLs are also targeted with functional CR strategies. Between the complexity of the subject matter and the nascent status of work in this area, considerable controversy exists as to the relative effectiveness of various CR approaches (Wilson, 1997). Rather than debating the merits of any specific CR approach, this discussion will focus primarily on how VE technology may be a useful tool for administering a wide range of CR strategies.

Cognitive rehabilitation approaches can differ based on a variety of conceptual criteria (Kirsch, Levine, Lajiness-O'Neill, & Schnyder, 1992). For the purposes of describing the application of VE technology to CR, these conceptual dimensions can be "collapsed" into two general domains, restorative approaches that focus on the systematic retraining of component cognitive processes (i.e., attention, memory, etc.) and functional approaches that emphasize the stepwise training of observable behaviors, skills, and IADLs. These domains can be viewed as opposite ends of a continuum of methods, with many specific CR approaches falling somewhere between these poles. The restorative and functional approaches to CR have different methods and goals. The primary objective of the restorative approach is to retrain individuals how to *think*, whereas the emphasis of the functional approach is to teach individuals how to *do*. For example, the treatment direction for a 20-year-old with mild head injury may primarily have a restorative focus and target component thinking processes with a goal of improving cognitive flexibility. By contrast, an elderly patient with dementia may be better suited to a functional approach targeting compensatory, environment-centered goals needed to maintain independent living. An example of the differential emphasis contained within these approaches can be seen in the area of memory rehabilitation. The restorative approach emphasizes a "drill and practice" method in which the person is challenged to attend to, remember, and/or manipulate increasingly more difficult pieces of information contingent on success (Lynch, 1992) and hence, cognitive ability is expected to improve, much like a muscle gets stronger with increased exercise. Stimuli that are easily quantifiable and can be gradually increased in difficulty level are often used with this approach (i.e., lists of letters, numbers, words, sentences, directions, geometric designs, etc.). Functional approaches generally focus on training IADLs, such as practicing a sequence of events to prepare for work in the morning or a set of structured steps for completing day-to-day functional activities. These strategies attempt to rehabilitate the person by training the actual IADL skills using well-practiced routines within the target environment, sometimes incorporating compensatory mental "prosthetic" devices, such as alarms and reminder notes, located in strategic positions around the environment.

Specific weaknesses have been identified in both of these approaches. One often cited criticism of restorative methods is the reliance on test materials or tasks that are essentially artificial and have little relevance to real-world functional cognitive challenges. This criticism holds that "memorizing" increasingly difficult lists of words or activities within a therapy or school environment does not support the transfer or generalization of memory ability to the person's real-world situation (Chase & Ericsson, 1981; O'Connor & Cermack, 1987). The fundamental criticism of functional methods is that the learning of standard stereotyped behaviors to accomplish IADLs assumes that the person lives in a static world where life demands do not change and that the person's underlying cognitive processes are not addressed. This is believed to limit the flexible and creative problem solving required to adjust to and think through changing circumstances in the real world (Kirsch et al., 1992).

The application of VE technology for the rehabilitation of cognitive–functional deficits could serve to limit the major weaknesses of both the restorative and functional approaches, and actually produce a systematic treatment method that would integrate the best features from both methods. In essence, it may be possible for a VE application to provide systematic restorative training within the context of functionally relevant, ecologically valid simulated environments that optimize the degree of transfer of training (see chap. 19, this volume) or generalization of learning to the person's real-world environment. Virtual environments could also serve to provide a more controlled and systematic means for separately administering restorative or functional techniques when this direction is deemed appropriate. An analysis of the suitability of VE technology in meeting the minimum criteria for both restorative and functional approaches can be found in a previous paper (Rizzo, 1994).

It should also be noted that underlying the goals of both of these conveniently termed treatment directions (thinking vs. doing) is the concept of *neural plasticity*. Neural plasticity refers to the capacity of the brain to reorganize or repair itself following injury, through various mechanisms (i.e., axonal sprouting, glial cell activation, denervation supersensitivity, and metabolic changes) in response to environmental stimulation. Recognition of neural plasticity in response to both environmental enrichment and impoverishment has its roots in the animal literature (Renner & Rosenzweig, 1987), and detailed reviews of this increasingly favored view of the brain can be found elsewhere (Rose & Johnson, 1992). Consequently, it can be appreciated that the stimulation or "enrichment" provided by both restorative and functional approaches may each have some effect on physical brain structure, and hence training with both methods would be assumed to affect brain plasticity. If this view is accepted, stimulating virtual training environments would seem well suited to support this process and new approaches to CR would be warranted.

## 3. REVIEW OF VE APPLICATION AREAS TARGETING CLINICAL POPULATIONS

Virtual environment technology may offer specific advantages for the assessment and rehabilitation of cognitive–functional processes (see Table 50.1). While encouraging on a theoretical level, the utility of this technology still needs to be substantiated via systematic empirical research with normal and clinical populations. It is also necessary to determine whether these advantages add value to existing technologies targeting these aims. Nonimmersive computerized testing and training tools have been available for some time and a case can be made that they offer many of the same advantages found with virtual environments. It is therefore imperative that research be conducted to determine the incremental value of VE-specific assets (immersive naturalistic or supranormal interaction) in this area. Over the last 6 years, a growing number of researchers have begun the initial work of exploring the use of VE technology for applications designed to target cognitive–functional performance with populations having CNS dysfunction. While a good deal of this initial work employs non-HMD flat-screen environments, these less immersive systems have produced useful results, which could inform future applications that attempt to leverage the assets available with the more immersive approaches. In addition, with certain clinical populations, less immersive systems may be sufficient for the designated

**TABLE 50.1**
VE Advantages for Neuropsychological Assessment and Cognitive Rehabilitation Applications

1. The presentation of ecologically valid testing and training scenarios and/or cognitive challenges, which are difficult to present using other means (i.e., dynamic, interactive 3-D stimuli).
2. Total control and consistency of stimulus delivery.
3. The presentation of hierarchical and repetitive stimulus challenges, which can be varied from simple to complex, contingent on success.
4. The provision of cueing stimuli or visualization tactics (i.e., selective emphasis) designed to help guide successful performance within a "dynamic" testing or errorless learning paradigm.
5. The delivery of immediate performance feedback in a variety of forms.
6. The capacity to pause assessment and training for discussion or other means of instruction.
7. The option for self-guided exploration and independent testing and training when deemed appropriate.
8. The modification of sensory presentation and response requirements based on the user's impairments (i.e., movement, hearing, and visual disorders).
9. The capacity for complete performance recording.
10. The availability of a more naturalistic, or intuitive, performance record for review and analysis by the user.
11. The design of safe learning environments, which minimize risks due to errors.
12. The introduction of "gaming" factors into the learning situation to enhance motivation.
13. The ability to create low-cost functional training environments.
14. The creation and integration of virtual human representations (avatars/agents) to foster interaction and to enhance sense of presence.

purposes. Although the breadth of the clinical VE literature pales by comparison to work done in the testing and training area (see chaps. 43–46, this volume) with normal populations, the initial efforts with applications designed for impaired clinical groups are encouraging and will be reviewed in this chapter. It should also be noted in advance that some applications reviewed under one category could conceivably be described in another. For example, much of the work subsumed under the "memory" category involves research on spatial memory in VEs modeled from real environments and could as easily be discussed in the "spatial ability" or "functional" sections. An effort is made in these cases is to categorize applications based on the cognitive area or clinical population issue of primary focus to the researcher. In this way, an application that in one view may address training of visuospatial wayfinding (see chap. 24, this volume), may in fact be covered in the "memory" section because the researcher's emphasis appears to be on the rehabilitation of memory of an amnesic patient (although it is acknowledged that part of the memory targeted is spatial, and the application addresses a functional target—learning a rehabilitation unit layout).

### 3.1 Attention Processes

Little VE work has been done with this "basic" cognitive process thus far, which is surprising in view of the widespread occurrence and relative significance of attention impairments seen in a variety of clinical conditions across the human life span. Most notably, attention difficulties are seen in persons with ADHD, TBI, and as a feature of various forms of age-related dementia (i.e., AD, vascular dementia, etc.). For example, the prevalence of ADHD is estimated at 3% to 5% in school-age children in the United States and attention deficits are frequently cited as the chief disability following the occurrence of TBI (Sohlberg & Mateer, 1987). Attention difficulties also commonly occur in the elderly due to dementing conditions and may also result from a lack of oxygen or nutrients to the brain secondary to strokes and myocardial infarctions. More effective assessment and rehabilitation tools are needed to address attention abilities for a variety of reasons. In children, attention skills are the necessary foundation for future educational activities. Better attention process assessment is also vital for decision making regarding special education placement decisions, determination of the use and effectiveness of pharmacological treatments, and for treatment efficacy and outcome measurement. Persons with TBI also require focus on attention abilities as a precursor to rehabilitative work on higher cognitive processes such as memory, spatial abilities, executive functions, and problem solving. Even if higher processes are less amenable to remediation, as in cases of severe TBI, some level of attention ability is essential for vocational endeavors, functional independence, and quality of life pursuits. A more fine-grained assessment of basic attention deficits may also provide an early indicator of dementia-related symptoms and could suggest functional areas where an older person might be at risk (i.e., automobile driving, operating machinery, etc.) or where compensatory strategies may be needed in order to maximize or maintain functional independence.

Virtual environment technology appears to provide specific assets for addressing these impairments that are not available using existing methods. In this regard, our lab is currently developing an HMD-delivered VE system for the assessment and possible rehabilitation of attention processes. This work was motivated by the view that HMDs are well matched for these types of applications, as they provide a controlled stimulus environment where cognitive challenges can be presented along with the precise delivery and control of "distracting" auditory and visual stimuli. This level of experimental control could potentially allow for the development of attention assessment and rehabilitation tasks that are more similar to what is found in the real world, and hence the ecological validity of measurement and treatment methods for this process could be improved. Our first effort in this area has involved the development of a virtual classroom specifically aimed at the assessment of ADHD (Rizzo et al., 1999). A recent consensus report by the National Institutes of Health on ADHD suggests a variety of areas where better assessment tools would be of value and specifically cited the need for better definition of the nature of this disorder and an emphasis on measuring the effectiveness of intervention strategies (NIH, 1998a). These recommendations supported our interest in addressing this clinical group in our initial VE–attention application. The scenario consists of a standard rectangular classroom environment containing three rows of desks, a teacher's desk at the front, a male or female teacher, a blackboard across the front wall, a side wall with a large window looking out onto a street with background buildings, vehicles, and people, and on each end of the wall opposite the window—a pair of doorways through which activity occurs. Within this scenario, children's attention performance will be assessed while a series of typical classroom distracters (i.e., ambient classroom noise, movement of other pupils, activity occurring outside the window, etc.) are systematically controlled and manipulated within the virtual environment. The child sits at a virtual desk within the virtual classroom, and the environment can be programmed to vary with regards to such factors as seating position, number of students, gender of the teacher, and other factors. On-task attention can be measured in terms of performance (reaction time) on a variety of attention challenges that can be adjusted based on the child's age or expected grade level of performance. For example, on the simpler end of the continuum, the child can be required to press a "colored" section of the virtual desk on the direct instruction of the teacher or whenever the child hears the name of the color mentioned by the teacher (*focused* or *selective* attention task). *Sustained attention* can be assessed by manipulating the time demands of the testing. More complex demands requiring *alternating* or *divided attention* can be developed whereby the student needs to respond by pressing the "colored" section only when the teacher states the color in relation to an animal (i.e., the brown dog, as opposed to the statement "I like the color brown") and only when the word *dog* is written on the blackboard. In addition to attention-driven reaction-time performance, behavioral measures that are correlated with distractibility and/or hyperactivity components (i.e., head turning, gross motor movement) and impulsive nontask behaviors (time playing with "distracter" items on the desk) can be measured. Other scenarios (i.e., work situations, home environments, etc.) using the same logic and approach are being conceptualized to address these issues with other clinical groups. This work is currently in progress and results from our first pilot study should be available by early 2000. Figure 50.1 contains images of the Virtual Classroom.

Another form of attention disorder that has been addressed with VE technology concerns the area of visual neglect, or inattention to a specific visual field, which is sometimes seen following a stroke or TBI. Visual neglect is defined as inattention to objects or events positioned in the visual space opposite to a brain lesion. It is not a vision problem, but a disorder of the integrated functioning of vision and attention. Classic signs of neglect are combing hair on only one side of the head, reading words only from the unaffected half of a printed page, or eating food that is on one side of the plate and believing it to be empty. Standard assessments for visual neglect commonly use static two-dimensional (2-D) materials, such as requiring the patient to determine the center of lines on a piece of paper (line bisection task) or the crossing out of specific targets on a page containing an array of stimuli. Virtual environment technology offers the potential to expand traditional assessment methods by providing a controlled 3-D assessment environment, which allows for assessment of depth and motion factors and these capabilities could serve to support new NA and CR strategies. For example, Rushton, Coles, and Wann (1996) propose using VEs to remove vestibular and visual frames of reference by rotating the visual world and directing to attention, objects that are in the neglected field. In their current application, a flat-screen VE scenario consisting of a series of mazes is presented, which requires varying levels of attention and decision making for navigation (Wann, Rushton, Smyth, & Jones, 1997). This type of application can provide extended practice, systematic stimulus delivery of cueing stimuli contingent on level of impairment, and constant feedback of the type that is not available in conventional settings. Results from clinical trials with these patients are anticipated. Another group is targeting the assessment of neglect using an HMD with eye-tracking capability in order to record eye movements across an array of 3-D virtual objects (Kodgi, Gupta, Conroy, & Knott, 1999). Initial results from a pilot study of four neglect patients confirmed that the subjects made eye movements only to the objects they reported seeing. The integration of eye-tracking technology (see chap. 9, this volume) with this VE scenario is well matched to the aims of this assessment challenge and future work with moving stimuli could present new NA and CR options for this impairment.

## 3.2 Executive Functioning

The reasoning that supports the use of VE technology to address impairments with more complex attention processes can also be applied to the more integrative cognitive domain referred to as executive functions. This cognitive process generally refers to "a set of behavioural competencies which include planning, sequencing, the ability to sustain attention, resistance to interference, utilization of feedback, the ability to coordinate simultaneous activity, cognitive flexibility (i.e., the ability to change set), and, more generally, the ability to deal with novelty" (Crawford, 1998, p. 209). The complexity of this cognitive process makes assessment and rehabilitation a questionable process using traditional psychometric methodologies (Elkind, 1998; Pugnetti et al., 1995a). However, as with the attention examples cited above, the attributes of a VE offer a unique capacity to address this cognitive process. More complex VEs targeting executive functioning would be of particular value as CNS patients often perform well on traditional neuropsychological tests of specific cognitive ability (i.e., attention, memory, visuospatial, language, etc.), yet may display impairments in the ubiquitous day-to-day activities that require planning, initiating, multitasking, and self-assessment competencies that are believed to be part of this integrative function. From this perspective, VEs may offer a way

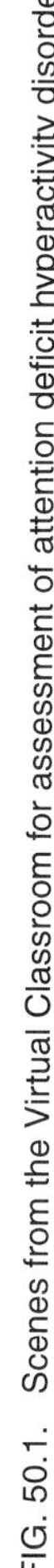

FIG. 50.1. Scenes from the Virtual Classroom for assessment of attention deficit hyperactivity disorder.

to systematically assess and rehabilitate these competencies within the context of the demands found in everyday settings modeled in a virtual environment.

One group of researchers has begun the work of developing an HMD-based VE system specifically designed for the assessment and rehabilitation of this process in persons with TBI, multiple sclerosis, and stroke (Pugnetti et al., 1995a, 1995b, 1996, 1998a, 1998b). Using a standard tool of neuropsychological assessment as a model (the Wisconsin Card Sorting Test—WCST), these researchers have created a virtual building that requires the person to utilize environmental clues in the selection of appropriate choices (doorways) in order to navigate from room to room in the structure. The doorway choices can vary according to the categories of shape, color, and number of portholes and the person is required to refer back to the previous doorway for clues as to the appropriate next choice. When the choice criterion is changed, the person is then required to shift cognitive set, analyze clues, and devise a new choice strategy. The parameters of this system are fully adjustable so that training applications can follow initial standardized assessments. In light of the large body of literature that exists on the WCST (Kolb & Wishaw, 1990), its choice as a model for this initial VE application is potentially quite useful, and results from this application have been encouraging. Pugnetti et al. (1998a) reported a study comparing a mixed group of neurological patients (MS, stroke, and TBI) with normals on performance on both the WCST and on their HMD executive function system. The results confirmed previous anecdotal observations with patient populations. While the psychometric properties of the VE task were comparable to the WCST in terms of gross differentiation of patients and controls, weak correlations between the two methods suggest that they are measuring different functions. In this regard, the VE task was seen to specify impairments earlier in the test sequence compared to the analogue pencil and paper test. The authors suggest that "this finding depends on the more complex (and complete) cognitive demands of the VE setting at the beginning of the test when perceptuomotor, visuospatial (orientation), memory, and conceptual aspects of the task need to be fully integrated into an efficient routine" (p. 160). The detection of these "integrative" difficulties for this complex cognitive function is of vital importance when predicting real-world capabilities from test results. Further support for this more ecologically valid VE assessment method can be seen in a detailed single subject case study of a stroke patient using this system. In this report (Mendozzi, Motta, Barbieri, Alpini, & Pugnetti, 1998), results indicated that the VE was more accurate in specifying executive function deficits in a highly educated patient two years following a stroke. The VE was successful in detecting deficits that had been reported to be limiting the patient's everyday performance, yet were missed using traditional neuropsychological tests. Additionally, this group has reported findings pertaining to psychophysiological variables (EEG, heart rate) that were recorded during participation in their VE system (Pugnetti et al., 1995b), and this area is continuing to be explored (L. Pugnetti, personal communication, February 8, 1999).

### 3.3 Memory

The assessment and rehabilitation of memory disorders have received considerable attention in the neuropsychological literature. This has resulted in the better specification of memory processes, elucidation of underlying CNS mechanisms, and a greater effort to determine the problems that occur in everyday life due to clinical memory disorders. Targeting the memory deficits seen with TBI, learning disabilities, and neurodegenerative disorders is also vital for immediate and long-term functional rehabilitation purposes. The diagnosis of age-associated dementia also focuses on memory processes and requires the presence of explicit verbal memory deficits as a criteria for the diagnosis of dementia established by the *DSM-IV* (American Psychiatric Association, 1994). The increased emphasis in diagnosing dementia at its very earliest stage in order to deliver treatments that could slow the progression of the disorder and maximize functional independence, increase longevity, and improve quality of life has spurred on research to discover more sensitive tools for addressing memory functioning. This is an area where VE technology could be of value for improving the specificity of memory assessment for more effective diagnosis.

When speculating on possible VE applications for CR purposes, it is also helpful to consider findings pertaining to preserved memory, or learning abilities, following brain trauma and occurring with some forms of dementia. A number of studies have shown that in persons with neurologically based memory impairment, certain memory/learning subprocesses often remain relatively intact. *Procedural*, or skill memory, is one such cognitive operation (Charness, Milberg, & Alexander, 1988; Cohen & Squire, 1980). This type of memory ability concerns the capacity to learn rule-based or automatic procedures including motor skills, certain kinds of rule-based puzzles, and sequences for running or operating things (Sohlberg & Mateer, 1989). Procedural memory can be viewed in contrast to *declarative*, or fact-based memory, which is usually more impaired following CNS dysfunction and less amenable to rehabilitative improvement. Additionally, these patients often demonstrate an ability to perform procedural tasks without *any* recollection of the actual training. This is commonly referred to as *implicit* memory (Graf & Schacter, 1985), and its presence is indicative of a preserved ability to process and retain new material without the person's conscious awareness of when or where the learning occurred. These observations provide encouragement for the idea that VEs, by way of their interactive and immersive features, could provide training environments that foster cognitive improvement by exploiting the person's preserved procedural abilities. Hence, cognitive processes could be restored via procedures practiced repetitively within an environment that contains functional real-world demands. Whether the person could recall the actual training episodes is irrelevant, as long as the learned process or skill is shown to generalize to functional situations. An additional challenge would then be to somehow translate difficult declarative (and semantic) tasks into procedural learning activities, with the goal being the restoration of more complex higher reasoning abilities.

Following the reasoning from the cognitive process scenarios described above, it is also possible to design VE approaches that could target memory abilities systematically within simulated functional environments. Efforts in this area would be particularly useful as results using traditional methods for memory rehabilitation have been inconsistent at best (Schacter & Glisky, 1986; Wilson, 1997). One possible reason for the poor results in this area may be due to the inability to maintain a patient's motivation when confronting them with a repetitive series of memory training challenges, whether using word-list exercises or real-life functional activities. Virtual environment training might address this problem by providing environments that initially utilize gaming incentives followed by the gradual fading in of functional environments with the aim of developing domain-specific memory (Glisky, 1992).

Research examining memory for objects and the spatial layout contained in a VE has been under way since the mid-1990's at the University of East London. The targeting of memory processes in environments designed to test and train clinical groups is closely related to the VE work in the area of "spatial orientation and wayfinding," and a more specific review of applications targeting these processes with unimpaired groups can be found in the chap. 24 (this volume). The UEL group has focused on specifying the types of memory that may be enhanced during a four-room house navigation task (Attree et al., 1996; Rose, 1996) with an aim toward targeting memory deficits in clinical populations. This approach uses a non-HMD flat-screen system with a joystick interface that allows one subject to navigate the house (active condition), whereas a yoked subject is simply exposed to the same journey but has no navigational control (passive condition). Both subjects are directed to seek out an object (toy car) during the exploration and differential memory performance between the two groups on spatial versus object memory of the environment is tested. In initial tests with normal populations, it was observed that the active groups showed better spatial memory for the route, whereas the passive group displayed superior object recall and recognition memory for the items viewed along the route (Attree et al., 1996). These results have been replicated by collaborators with this group (Pugnetti et al., 1998a), whereas mixed results have been reported by other non-HMD VE studies in this area, one showing active participation enhancing spatial memory (Peruch, Vercher, & Gauthier, 1995) and no differences reported in the others (Wilson, 1993; Wilson, Foreman, Gillet, & Stanton, 1997). However, the spatial orientation tasks used in these latter studies differed in the degree to which subjects were allowed to "retrace their route," and this may have mitigated the procedural training component's contribution to the observed outcomes (no difference between "active" vs. "passive" exposure; Rose et al., 1999).

Research using this scenario with clinical populations (multiple sclerosis and stroke) has produced results that suggest the value of this sort of VE application to inform NA and CR with these groups. In a recent study using this VE application, as expected, stroke patients performed more poorly than normals. What is more interesting though, is that whereas the typical spatial–content memory dissociation was found with unimpaired groups (i.e., active equal to better spatial memory; passive equal to better object memory), and the active stroke group displayed better spatial memory compared to the passive group, the stroke patients displayed no advantage on object memory while in the passive condition (Rose et al., 1999). Similar findings using the UEL scenario were also reported by Pugnetti et al. (1998b) in a study comparing multiple sclerosis (MS) patients with a normal group. These findings also could not be accounted for via relationships with standard pencil and paper memory assessment tools. This type of impairment in explicit incidental memory observed in MS patients suggests that a VE approach may be of particular value in detecting subtle deficits in these patients (Pugnetti et al., 1998b). This is in line with previous work by Andrews, Rose, Leadbetter, Attree, & Painter (1995), who found that subjects recalled half as much in a VE that contained the sort of distracters present in everyday life when compared to "static" presentations, as are typically used in standard NA methods. From this, the targeting of memory processes within a VE may provide more unique and specific information that could enhance our understanding of impaired cognitive processes, as well as underscore the potential value of immersive approaches that promote "procedural" involvement in the design of better rehabilitative memory approaches.

In this regard, Brooks et al. (1999) reported the case of a female stroke patient with severe amnesia who showed significant improvements in her ability to find her way around a rehabilitation unit following training within a VE modeled after the unit. Prior to training, the patient had lived in the unit for 2 months and was still unable to find her way around, even to places that she visited regularly. In the first part of the training, improvements on two routes were seen after a 3-week period of VE route practice lasting only 15 minutes per weekday, and retention of this learning was maintained throughout the patient's stay on the unit. In the second part of the study, the patient was trained on two more routes, one utilizing the VE and the other actually practicing on the "real" unit. Within 2 weeks the patient learned the route practiced in the VE, but not the route trained on the real unit, and this learning was maintained throughout the course of the study (Brooks et al., 1999). The authors account for this success as being due in part to the opportunity in the VE for quicker traversing of the environment than in the real world, which allowed for more efficient use of training time. Another factor in this success may be that the VE training did not contain the typical distractions normally present when real-world training is conducted, which might have impeded route learning. Additionally, VEs allow for repeated independent explorations that may be limited in the real world for persons with motoric impairments, and this asset may offer additive value for these types of spatial memory rehabilitation approaches. This asset was also alluded to by way of similar findings reported in a series of studies (Foreman, Wilson, & Stanton, 1997; Stanton, Foreman, & Wilson,

1998), where children with severe physical disabilities were able to transfer learning from a virtual school environment to the actual school environment. Further research examining these issues and other memory subprocesses with clinical populations needs to be conducted taking into account findings with normal populations concerning different levels of immersion (Lackner & DiZio, 1998; Waller, Hunt, & Knapp, 1998) and with controlled variations in the multisensory characteristics of the VE (Dinh, Walker, Hodges, Song, & Kobayashi, 1999).

Lastly, a unique VE learning and memory application has been developed by Grealy, Johnson, and Rushton (1999), who created three flat-screen VE scenarios that allow a patient to navigate environments via a recumbent bicycle and exercise equipment interface. The task requires the person to sequentially visit various objects, or locations, in the virtual environment. The authors present the interesting concept that the VE training will improve fitness level, which is hypothesized to enhance brain activation, as well as other variables relevant to rehabilitative concerns. Related support for integrating exercise in a VE to enhance the effect of CR can be found in recent reports indicating that cell growth in the hippocampus (a key brain center for memory) was promoted in mice following both mental and physical exercise (Gould, Beylin, Tanapat, Reeves, & Shors, 1999; van Praag, Kempermann, & Gage, 1999). In an initial study by Grealy et al. (1999), TBI subjects trained with the VE–exercise system performed significantly better than controls on digit coding, visual, and verbal learning tasks. These provocative findings suggest areas of application where the assets available with VE technology combined with exercise might be creatively applied to the rehabilitation of learning and memory with clinical populations.

### 3.4 Spatial Ability

Tests of spatial ability are commonly used for the study of brain behavior relationships and for measuring the consequences of various forms of CNS dysfunction. Spatial ability is commonly broken down and defined in terms of the tests used to measure its various subcomponents (i.e., spatial perception, orientation, visualization, mental rotation, etc.). However, Carpenter and Just (1986) present a general definition of this process as the ability to generate "a mental representation of a two or three-dimensional structure and then assessing its properties or performing a transformation of the representation" (p. 221). The analysis of human spatial processing is of value for diagnosing brain dysfunction in a variety of domains and for addressing the "topographical disorientation" (De Renzi, 1985) that is sometimes seen in these conditions. Also, the rehabilitation of spatial abilities may be vital to supporting the reacquisition of many functional abilities including driving, navigating environments, and other IADLs commonly found to be impaired due to CNS dysfunction. Virtual environment technology may provide unique assets for targeting spatial abilities with its capacity for creating, presenting, and manipulating dynamic 3-D objects and environments in a consistent manner and for the precise measurement of human interactive performance with these stimuli. These types of VE spatial ability assessment and rehabilitation systems may provide ways to target these cognitive processes beyond what exists with methods relying on 2-D pencil and paper representations of 3-D objects (or methods using actual real objects) that are typically found with traditional tools in this area. These traditional methods are often limited by poor depth, motion, and 3-D cues needed for proper stimulus delivery and have limited capacity for the precise measurement of responses. Virtual environments also offer the potential to address these variables in an ecologically valid manner (functional simulations) without the loss of experimental control common with naturalistic studies in this area relying on observational methods.

A number of flat-screen VE systems have been developed to target spatial ability variables. For example, results from studies of human "place learning" ability have been reported (Astur, Ortiz, & Sutherland, 1998; Jacobs, Laurance, & Thomas, 1997; Jacobs, Thomas, Laurance, &

Nadel, 1998; Sandstrom, Kaufman, & Huettel, 1998; Skelton, Bukach, Laurance, Thomas, & Jacobs, 1999), with researchers having designed VEs modeled after the Morris water task (Morris, 1981). The Morris water task was developed as a systematic method for the study of spatial abilities in rodents and requires subjects placed in a pool of water to swim from various starting points and find a fixed hidden platform located below the water surface. Since rodents are unable to see or smell the platform, they must rely solely on the spatial arrangement of cues on the walls outside of the pool in order to locate it. Reports of animal performances on this task have produced useful findings relating to spatial ability differences due to gender, age, drug effects, and brain lesions (Astur, Ortiz, & Sutherland, 1998). However, although the Morris water task is the "gold standard" for measuring place learning in animal studies, it is not a practical test for use with human subjects. Flat-screen, joystick-controlled VEs modeled after the Morris task, which allow a person to navigate an environment in search of a hidden platform, have produced a useful method for studying this process in humans. Studies using this VE have reported significant effects in the areas of aging, CNS dysfunction, and gender differences (Astur et al., 1998; Jacobs et al., 1997; 1998; Sandstrom, Kaufman, & Huettel, 1998; Skelton et al., 1999; Thomas, Laurance, Luczak, & Jacobs, 1999). For example, a recent study showed TBI patients to have significantly impaired VE place learning abilities, which correlated with self-reported frequency of wayfinding problems in everyday life (Skelton et al., 1999). Most of these studies also report a large and reliable male advantage on these tasks, and these findings are of value for understanding differences in brain function between the genders that are believed to be influenced by hormonal status (see chap. 32, this volume). Similar gender differences have been reported in a recent study comparing different levels of immersion on the transfer of spatial knowledge from a variety of training environments (no training, real world, map, VE desktop, VE immersive, and VE long immersive), where "robust gender differences in training effectiveness" in favor of males were also reported (p. 129; Waller, Hunt, & Knapp, 1998). Continuing research with clinical populations using the VE Morris Water Test is currently under way and these results are highly anticipated (K. Thomas, personal communication, June 25, 1999).

The use of flatscreen VE scenarios has also shown very promising results for training spatial orientation and navigation skills with physically disabled children (Stanton, Foreman, & Wilson, 1998). Applications in this area could be of particular value for children with physical disabilities who are limited in their ability to independently explore environments and who may consequently display impairments in cognitive mapping abilities. Foreman, Orencas, Nicholas, Morton, and Gell (1989) reported these types of impairments with a group of 10 physically disabled students having deficits in spatial awareness and cognitive mapping skills possibly due to the adoption of a "passive" navigational strategy. Other researchers have also reported similar effects due to passive navigation in unimpaired children (Hazen, 1982; McComas, Dulberg, & Latter, 1997). In a series of four studies reviewed in Stanton et al. (1998), children with physical disabilities were allowed to independently explore VEs modeled after school environments. Transfer of spatial learning was shown to generalize to the actual school that was modeled, and this learning improved with practice. Also, flexibility in the cognitive representation, or "mapping" of the environment, was inferred, as the children were able to accurately indicate the direction of objects that were not in their line of sight. Similar findings have been reported by another group that has developed a visuospatial VE scenario designed to be ultimately applied with children having motor impairments (McComas, Pivak, & Laflamme, 1998). Successful learning and transfer for object locations were found using a flat-screen "classroom" VE with unimpaired children. In this study, spatial performance equaled "real-world" training, and successful transfer to the real environment was reported. Similar improvement using a flat-screen VE training system targeting functional spatial knowledge has also been reported for teenagers with developmental disabilities in a supermarket search and navigation scenario (Cromby, Standen, Newman, & Tasker, 1996). Further details on this and related work on spatial navigation will be discussed in the "Functional Skills" section.

Taking a more component-based approach to addressing spatial ability, we have developed a suite of ImmersaDesk-delivered (Couling, 1998) applications targeting mental rotation, depth perception, manual movement stability, 3-D field dependence (3-D rod and frame test), 3-D manual tracking, and visual field specific reaction time. These scenarios were designed to leverage the 3-D interactive assets available with this type of projection-based system in the development of a series of tasks that could serve to assess and possibly rehabilitate these more molecular components of visuospatial functioning. Our initial study targeted mental rotation (MR), a well-researched visuospatial variable, which can be described as a dynamic imagery process that involves "turning something over in one's mind" (Shepard & Metzler, 1971).

In the VE system, MR was assessed via a manual spatial rotation task that required subjects to manipulate virtual block configurations. Subjects were presented with a target block configuration and the speed and efficiency of their movements to superimpose a replica design on the target was measured and recorded. All manner of angular disparity and axis combinations can be programmed into the system allowing for hierarchical presentation of cognitive challenges required for NA and CR purposes. Details on the relevance of this application and prior methodology applied to the study of mental rotation can be found in Rizzo et al. (1998b). Our initial feasibility study targeting MR tested 18- to 40-year-olds on self-reported side effect occurrence, performance change (learning) on the VE MR task, transfer of training from the VE to performance on a pencil and paper MR test (Vandenberg & Kuse, 1978), the relationships between VE performance and other "standard" NA tests of cognitive performance, and gender differences on all of these questions. A number of encouraging findings emerged, including minimal side effect occurrence (see chap. 30, this volume), good psychometric properties of the VE test (see chap. 35, this volume), provocative relationships with standard NA tests, a lack of gender differences compared to the pencil and paper measures, training improvement, and significant transfer of training (see chap. 19, this volume) with low initial pencil and paper performers (Larson et al., 1999; Rizzo et al., 1999).

We are now testing out the full VE system (all visuospatial scenarios cited above) with normal elderly subjects and comparisons of this data with CNS impaired populations (Alzheimer's, TBI) will follow. Our plan is to develop a desktop system, using shutter glasses, in order to run a comparative test to determine if these scenarios can be successfully delivered on a less expensive and more accessible platform. The aim of this work is to provide a suite of VE-delivered 3-D testing and training tools that could target the NA and CR of visuospatial processes for use by researchers and clinicians.

HMD-delivered VEs have been a staple for studying spatial processes in normal subjects (see chap. 24, this volume) as seen in the work of Wraga, Creem, and Proffitt (1997), who have also investigated mental rotation ability. Other researchers are using HMD-delivered VEs to address spatial ability deficits with clinical populations. One group at the National Rehabilitation Hospital in Washington, D.C. (Trepagnier, 1999), is examining facial scanning in autistic children using an HMD with a built-in eye-tracking system. This application may allow for a more controlled analysis of the hypothesized deficits in facial eye-scanning behavior believed to occur in this population via the systematic presentation of functional 3-D stimuli. Alpini et al. (1998) have begun using an HMD-based VE to assess spatial exploratory behavior with clinical populations having vestibular dysfunction, and a lab in Denmark (Dobrzeniecki, Larsen, Andersen, & Sperling, 1999) has commenced feasibility studies for a similar application with stroke patients. These HMD systems will offer new insights into the complex factors that comprise impairments in spatial abilities. However, considerable challenges in this area remain. Reports of early work with stroke patients have shown that they have "highly disparate and constantly changing cognitive deficits, which require virtual worlds and tasks that are adaptive to a wide range of symptoms and personality traits ..." (A. Dobrzeniecki, personal communication, June 22, 1999). Further discussion of these types of clinical user issues appears in section 4.3.

Another promising area that has recently emerged is in the application of immersive audio VEs that can deliver 3-D spatialized sound (see chap. 4, this volume) to provide cues that are relevant to the simulated physical structure of an environment (Kyriakakis, 1999; Pair, 1999). A number of applications have been developed that provide auditory inputs to supplement environmental information for persons with visual impairments (Berka & Slavík, 1998; Cooper & Taylor, 1998). One group in particular has demonstrated the feasibility of promoting spatial–cognitive mapping using immersive audio (Lumbreras & Sánchez, 1998). Their initial work has involved the development of a computer game (AudioDoom) designed to allow blind children to navigate and interact with a spatial environment solely through the use of 3-D sound stimuli. The dramatic effect of this application can be seen in a demonstration of this work where children were asked to reconstruct the characteristics of the explored audio gaming environment using Lego blocks. The resulting Lego constructions were strikingly similar to the structure of this "sound-delivered" VE, and this may illustrate the value of auditory cues in developing a spatial cognitive map in children who have never had any visual experience with the physical world. This system, as well as other similar immersive audio applications, may have additional usefulness for studying audiovisual integration and spatial image formation for sighted populations with other forms of CNS dysfunction, and further work in this area is anticipated. One possible application might include the use of functional magnetic resonance imaging (fMRI) brain mapping technology to directly study the underlying neural circuitry of audiospatial mapping. For example, Aguirre and D'Esposito (1997) used a flat-screen VE model of a town derived from a commercially available video game (Marathon2) to train subjects on landmark and survey spatial information of an environment. The subjects' knowledge was subsequently tested while undergoing fMRI brain imaging, and the findings indicated specific differences in brain circuitry activation for landmark and survey tasks. Practical constraints of fMRI testing on the delivery of HMD immersive visual stimuli limited the use of the actual VE during imaging, and instead, still images of the environment were presented via a mirror device. In view of this limitation for the delivery of active VE visual environments, the presentation of immersive audio environments via earphones during brain imaging studies may be of some interim value. In this way, auditory cues supporting spatial mapping could be presented, and concurrent brain imaging of the underlying brain circuitry might be explored. Until it becomes more feasible in the near future to present HMD-delivered visual stimuli within a brain scanning device such as this, auditory immersion studies could be conducted to shed insight on these types of spatial learning events in the brain.

### 3.5 Functional Skills/Instrumental Activities of Daily Living

Some of the VE systems targeting component cognitive processes reported on above could conceivably be reviewed here (i.e., facial scanning, virtual classroom, etc.). However, the applications reviewed in this section focus primarily on the performance of more molar behaviors required for functional activities in VEs targeting IADLs. These functional approaches emphasize the design of ecologically valid VEs to test and train more integrated behavioral repertoires in populations which are limited in their capacity for real-world independent learning. Advances in this area could be of considerable value in view of the costly and labor-intensive efforts that are required for in vivo, one-on-one approaches often applied with certain clinical populations. In this way, VEs can be used to assess functional behavior and deliver consistent, hierarchical, and safe training that could free the therapist's time to be spent in other more necessary and intensive one-on-one client contact, as needed.

Virtual environments that target specific functional skills in clinical populations have been developed in the areas of street crossing, wheelchair navigation, meal preparation, use of public transportation, driving, and for the analysis of obstacle avoidance with elderly persons at risk for falling. Strickland (1996) reported initial results on the feasibility of using an HMD with autistic children with the long-term goal of targeting VE training for safe street crossing. The two autistic children initially reported on adapted to the headset and were able to track moving automobile stimuli and select objects. Continued efforts with this population (Strickland, 1997) and with children having motor disabilities (Inman, Loge, & Leavens, 1997) are under way to target the functional behaviors required for safe street crossing in a virtual environment. Inman et al. (1997) have also reported success in training children with motor impairments to use motorized wheelchairs in a variety of simulated real-world conditions using an HMD-based VE system (see chap. 49, this volume). Promising initial usability (see chap. 24, this volume) has also been reported for a flat-screen VE scenario targeting this same purpose by another group of researchers in Ireland (Desbonnet, Cox, & Rahman, 1998). Related work is proposed for VE assessment of operating skills and route navigation abilities in adult rehabilitation patients (D. Rose, personal communication, March 4, 1999). This project will include real-world baseline and post-VE assessment to determine transfer of training (see chap. 19, this volume) to the targeted environment. Such work is important to determine patients' ability to use powered wheelchairs safely and effectively.

Food preparation skills within a virtual kitchen scenario have also been investigated. Christiansen et al. (1998) have developed an HMD-based VE of a kitchen in which persons with TBI have been assessed in terms of their ability to perform 30 discrete steps required to prepare a can of soup. Contingent upon success at each step, various auditory and visual cues can be presented to help prompt successful performance and learning. In the first reported pilot study with this system, 30 patients with closed-head injuries were able to successfully function using the HMD with minimal side effects. In addition, acceptable reliability coefficients were reported using a test–retest assessment. These researchers also report ongoing enhancements to the system regarding the delivery of more complex challenges and increased flexibility in the presentation of cueing stimuli. A similar approach was presented by Davies et al. (1998), targeting the steps required to brew a pot of coffee. Although the initial study tested unimpaired subjects (occupational therapists), the authors reported usability data that they conclude supports the potential for use of these types of VEs with clinical populations.

Virtual environments that target driving ability have also been tested with TBI and elderly populations. Liu, Miyazaki, and Watson (1999) reported that an HMD-based driving scenario successfully discriminated between a TBI and an unimpaired group and that age effects were also detected with this system. Virtual environments that target driving ability under a range of conditions that are impossible to measure in standard behind-the-wheel assessments could become invaluable for safely assessing this functional ability in persons with CNS impairments, as well as with aged populations (M. Schultheis, personal communication, February 12, 1999). For example, normal elderly were compared with early Alzheimer's patients using the Iowa Driving Simulator, and, as would be expected, significantly more "accidents" occurred in the Alzheimer's group (Rizzo et al., 1997). If comparable assessment were possible using a less expensive VE scenario, more widespread use of these types of realistic functional assessment tools would be possible (Mourant & Chiu, 1997). The rapidly increasing size of the aging population, coupled with age-related changes in cognitive–functional performance makes investigations into the feasibility of VE applications with elderly populations of considerable importance. For example, one area that might be of value for aging populations involves simple walking and obstacle avoidance. A VE system is currently being developed to train elderly subjects to step over obstacles (Jaffe, 1998). The relevance of this can be seen in the fact that preventable falls, and the hip damage that often accompanies them, are leading causes of loss of functional independence with this population. This HMD-based VE application will allow at-risk elderly persons to practice stepping over moving obstacles on a treadmill while wearing an overhead safety harness. As such, an effective VE approach to address this basic skill would be of considerable value.

One particularly noteworthy and ambitious effort to address functional life skills is the work of the University of Nottingham group in the development of the Virtual City (Brown, Kerr, & Bayon, 1998; see chap. 46, this volume). This flat-screen scenario addresses independent living skills in persons with learning and developmental disabilities, and the researchers have implemented good user-centered design principles by incorporating user group input in the selection of environments targeted in the application. These settings were designed to address the use of public transport, road safety, home safety, use of public facilities within a cafe, and shopping skills in a large supermarket. An initial study tested 20 subjects with mental retardation and addressed usability, enjoyment, skill learning, and transfer of skills to the real environment using subject's self-report and behavioral observation methods (Cobb, Neale, & Reynolds, 1998). Subjects reported high enjoyment of the task and more "ease of use" with the system than anticipated in the "expected" usability ratings determined by support workers. Transfer from the VE to the real environment for some skills was noted in this initial feasibility study. This was determined by examining changes in ratings of the level of assistance provided by support workers in the real world, both before and after a 1-hour VE session. While this report was somewhat vague on presenting specific transfer data, the authors present anecdotal evidence that transfer of learning occurred in some instances and acknowledge that a longer period of training would be needed to see significant effects in this area. Work with this evolving scenario continues and results from more systematic measurement of learning and transfer are anticipated. Such results are important to support previous observations of transfer seen with these populations. For example, Cromby et al. (1996) reported evidence of positive learning transfer for another group of developmentally disabled students using a supermarket VE modeled after an actual market. Subjects were reported to be better able to navigate within, and select specific items in the real supermarket following training in the VE and these subjects outperformed those who practiced in nonspecific virtual environments. Positive results were also reported for a public transit VE training program called Train to Travel (Mowafy & Pollack, 1995). This HMD-based VE system was developed in the mid-1990s to assess and teach persons with developmental disabilities on how to use key routes on the Miami Valley Regional Transit System. However, since the VE component was part of a comprehensive training package, including video and multimedia tutorials, the specific effect of the VE on successful independent transit use was not determinable. Future studies that methodologically target the specific effects of the VE "ingredient" are necessary to determine whether these types of programs are effective for efficiently and economically targeting functional abilities with these populations. The above-cited investigations represent essential "first-steps" in determining whether VE training can foster transfer of learning to activities of daily living. For persons whose learning abilities are challenged due to CNS dysfunction, this line of research is especially important.

## 4. BASIC ISSUES FOR DECISION MAKING ON THE DESIGN AND USE OF VES WITH CNS CLINICAL POPULATIONS

In order for VE applications targeting NA and CR to be efficiently developed, a number of basic theoretical and pragmatic issues need to be addressed. The added issues of working with clinical user groups with CNS dysfunction also require consideration with these applications. On top of standard human factors and usability concerns (see chap. 34, this volume), unique aspects seen with various clinical populations add another level of complexity to an already difficult set of challenges in the design and implementation of effective NA and CR virtual environments. The major questions for these types of applications are addressed in this section. A brief discussion of the basic issues as they relate to unimpaired user groups (while possibly redundant to more detailed reviews found elsewhere in this volume) is included in order to provide the necessary context for addressing the additional concerns that are specific to clinical populations.

### 4.1 Can the Same Objective Be Accomplished Using a Simpler Approach?

While VE technology appears to offer many advantages for neuropsychological applications, the first step for any such program is for the developer to perform a realistic cost–benefit analysis (see chap. 28, this volume). Implementation of an honest cost–benefit analysis can serve to prevent costly and misguided (though well-intentioned) system development from diverting resources from the areas where a VE approach can make a unique and useful contribution. Virtual reality has been characterized as a technology in search of an application (Wann, 1996). From this, the excitement surrounding this new technology has the potential to affect judgment of the real needs of the situation. The first question to be asked is, Does the objective that is being pursued require the expense and complexity of a VE approach, or can it be done as effectively and efficiently using simpler, less costly, and currently available methods? This "elegant simplicity" criterion requires the investigator to fully consider the objectives of the application and decide whether the development of a VE genuinely adds value beyond what already exists, or is it simply a case of technological overkill.

Virtual environments, in fact, offer immersive and interactive features (see chap. 40, this volume) that, if deemed important to a NA–CR approach, could provide valuable assets. However, if these features are not of vital importance, perhaps the use of flat-screen multimedia, "simple" video, or even already established in vivo tools, would be sufficient. For example, if the objective is to assess elderly persons for global cognitive functioning as an initial screening procedure for dementia (Alzheimer's, vascular, Parkinson's, etc.), then a very basic, standardized 30-question mental status interview/exam that measures orientation, attention, short- and long-term memory, verbal fluency, and judgment may be sufficient for this purpose. A test like this already exists, the Mini Mental State Exam (Folstein, Folstein, & McHugh, 1975), and has been shown to be of acceptable reliability and validity, inexpensive, easy to administer quickly in most settings, and provides data that can be usefully compared with norms generated over many years as a standard tool in this area. For this specific purpose, a VE approach would be redundant and inefficient. However, the same researcher may believe that probable AD could be recognized at its very earliest stage via a more systematic analysis of specific attention–memory components (perhaps iconic memory) while a person is immersed within the demands of a stimulus-rich functional virtual environment. For this purpose, a VE application may be the most efficient (and perhaps the only) method for systematically controlling the stimulus environment while precisely measuring millisecond by millisecond responses. Although the initial cost of developing such a system might be high, the earlier detection of dementing symptomatology could allow for possible treatments at an earlier stage in the disease process, leading to more informed long-term care and increased functional longevity for the client.

Following from this last point, a commonly cited advantage of VEs is the capacity to record and measure naturalistic behavior within a simulated functional scenario (Riva, 1998; Rizzo et al., 1998a; Rizzo, Wiederhold, & Buckwalter, 1998c; Rose, 1996). This asset offers the potential to collect reliable data that might have been otherwise lost using methods employing behavioral ratings from trained observers of behavior in "real-world" settings. However, behavioral observation methods that make use of video recordings for later rating may be sufficient for certain purposes. For example, this may be the case with the assessment of "rough and tumble" play that is commonly focused on in comparisons of gender differences in children, as well as with the study of aggressive behavior (Eaton & Enns, 1986). The assessment of these behaviors within a simulated environment would require a strong tactile and physical interaction component that is not currently possible in a virtual environment. While it may be possible to design VEs with the capacity to stimulate these activities, the absence of the haptic feedback loop (see chaps. 5 and 6, this volume) from actual physical contact, as well as the practical issues of using an HMD with this level of gross motor activity, makes a real-world application a relatively simpler and more effective environment for this purpose. As well, certain types of critical thinking training approaches may be more efficiently administered to a person with a mild head injury by utilizing preserved reading comprehension skills to hierarchically teach reasoning by analogy. This is a relatively straightforward approach that the client can practice at home as well as in a treatment setting. This sort of language–logic process is viewed as less amenable to a VE approach at the current state of the technology, and this is discussed further in the next section. By contrast, if the client has impairments in areas relating to active problem solving and executive functions, then VE-delivered scenarios that allow for exploration and interaction within a dynamic simulated functional environment might offer hierarchical challenges that could not be presented efficiently and consistently using traditional methods.

In summary, the examples cited above were chosen to illustrate the "elegant simplicity" criterion applied to a range of possible applications. This first step in the VE decision-making process requires one to justify the selection of this technology in contrast to simpler, presently available tools. If the application meets this criterion and is expected to add value to already existing approaches via the assets that are available with VEs, then it is time to examine the specifics of the application in light of the following issues.

### 4.2 How Well Do Current VE Assets Fit the Needs of the Cognitive/Functional Approach or Target?

As can be seen in the previous section, it is recommended that value added VE applications be developed that avoid the redundant development of something that has already been proven effective in a simpler form. The selection of appropriate cognitive–functional targets is inextricably related to this, but requires an additional understanding of the match, or "good fit criterion," between these targets and the current capabilities of VE technology. At the present time, VEs can be said to offer certain specific "attributes" (or ingredients) that would seem to be well matched for neuropsychological approaches. In general terms these include such fundamental attributes as the capacity to expose subjects to precisely administered, dynamic, 3-D, visual and auditory stimuli, as well as the capacity to involve a person within an interactive/immersive procedural activity with fuller response measurement capability. These basic attributes appear to provide the necessary ingredients for VE NA and CR applications in a number of areas and may provide "good fit" for scenarios designed to mimic real-world challenges without the loss of experimental control that often occurs in naturalistic settings. For example, as described in the attention process section, an HMD-based VE for the assessment of these processes could improve on existing methods. The attributes of an HMD-based VE appear to fit well with the need for precise stimulus control and response measurement required for attention-targeted applications. An HMD-based VE allows for the systematic simultaneous/sequential presentation of stimulus challenges and distractions within a contained setting in a manner that is not possible using traditional methods. This reasoning also holds for VEs designed to address the integrative attention control required for executive functions. The complexity of this cognitive process makes assessment and rehabilitation a questionable process using traditional psychometric methodologies (Pugnetti et al., 1995a), and VEs offer a unique capacity to address these functions. This was seen in a case study where a VE was successful in detecting deficits that had been reported to be limiting the patient's everyday performance, yet were missed using standard NA tests (Mendozzi et al., 1998).

The NA and CR of spatial ability is another area that is particularly well matched with the unique stimulus and response assets of a virtual environment. The capacity to design 3-D objects that can be presented in a specific and consistent manner and manipulated by the user contingent on a variety of task demands could be a powerful tool for these applications. The attributes of a VE appear to be well matched to these challenges, and this is reflected in the number of spatial ability applications that are currently aimed at clinical populations, as reviewed in sections 3.3, 3.4, and 3.5. On the other hand, complex reasoning abilities that have heavy requirements for language-based representation, and declarative memory may be more difficult to address at the current state of VE development, although this may well be possible in the near future. The areas addressed above are not presented as an exhaustive list of the areas where a VE might be of value. Rather, they are meant as examples of the type of analysis that we have applied to judge the fit between VE applications and NA and CR targets in this nascent field. Also, we hesitate to address target areas that do not suggest a good match with the attributes of a VE or where the technology is not in a state of readiness to add to what currently exists with traditional methodologies. Our vision of VEs may reflect gaps that others do not possess.

### 4.3 How Well Does a VE Approach Match the Characteristics of the Target Clinical User Group?

Awareness of the interaction between assessment and treatment strategies and patient characteristics specific to various clinical conditions has guided the development of mental health approaches for many years (Paul, 1969). The usefulness of VE applications across different clinical populations requires the same measured consideration of these issues that is found in user-centered design studies with normal groups. The unique psychological, cognitive, and functional characteristics that are commonly seen in different types of clinical conditions must be considered, along with an informed sensitivity to the vulnerabilities of these groups. Clinical user groups may differ in areas such as apprehensiveness to use an HMD, reality testing, capacity to learn to operate in a VE, susceptibility to side effects, and verbal reporting ability. Awareness and preparedness for these issues are necessary for ethical reasons (see chap. 33, this volume) as well as treatment efficacy concerns. Expected benefits of using a VE approach need to be tempered by a clinical vigilance for possible unanticipated consequences that could limit the applicability of a VE approach for certain clinical conditions. Thus far, the clinical application of VEs has appeared to be fairly thoughtful and rational, possibly owing to both the technology's limited availability and the professionalism of the innovators of these early applications. However, as the technology becomes more accessible, individuals who become enamored with virtual methods but lack sensitivity to the relevant clinical and ethical concerns may be in a position to develop VE applications. The resulting systems could be of questionable utility at best, and have the potential to do harm at the worst. Although such incidents have yet to appear in the clinical lore and literature, the possibility exists that these types of events may be selectively overlooked and/or go unreported.

When working with patient groups, an ethically based screening procedure is necessary to minimize the possibility of inducing harmful consequences on the patient via a VE approach. Persons displaying comorbid features of various psychotic, bipolar, paranoid, substance abuse, claustrophobic, or other disorders where reality testing and identity problems are evident, may be ill advised to participate in a virtual environment. It is also essential to anticipate possible difficulties that could arise due to specific characteristics that exist for persons with CNS dysfunction and resulting neuropsychological impairments. For example, orientation and equilibrium difficulties are often sequelae of CNS dysfunction. Proper screening and ongoing monitoring guidelines are of vital importance with such individuals. Although it is the cognitive/functional impairments that are most often the focus with these groups, significant emotional, social, and identity difficulties commonly exist within the total pattern of CNS dysfunction (see chap. 33, this volume). Although a full discussion of the complex issues involved with these clinical groups is beyond the scope of this chapter, some brief discussion of age-related user issues is possible.

Issues regarding the age of the participants should be considered in decision making on the applicability of VE technology. There now appears to be a developing literature on VE applications with young populations (Cromby et al., 1996; Escamilla, Falkinstein, Chang, Schneider, & Zeltzer, 1997; Inman et al., 1997; McComas, Pivik, & Loflamme, 1998; Stanton et al., 1998; Strickland, 1996) that can allow examination of this issue. Youthful populations, having grown up during the age of computer and video games, may be more at home with VE setups. The reports to date with youthful populations mainly involve the training of functional skills that involve learning of the spatial characteristics of a target environment, sometimes requiring navigation ability. These studies have tested children and teenagers in VEs targeting wheelchair operation, spatial navigation, and supermarket shopping. Results suggesting successful assessment, training, and transfer to the real world have been published and these studies indicate that VE cognitive–functional applications can be usefully applied to these youthful clinical groups with few reports of negative complications. The occurrence of cybersickness (see chap. 30, this volume) and aftereffects (see chaps. 31 and 37–39, this volume) has not been reported to be problematic with these younger populations, although systematic assessment has not often occurred (though it should become a mandatory component of all VE investigations). The initial concerns that children might have difficulties adjusting to an HMD (see chaps. 37 and 41, this volume) did not appear to be borne out in some of these applications. For example, positive results were reported for two autistic children adapting to an HMD (Strickland, 1996). However, mixed results were observed by Inman et al. (1997) in a wheelchair navigation training VE tested with young children. In this work, it is reported that many of the users had a preference for using a large TV monitor instead of the HMD, although this was not reported to decrease performance or interest. Generally, these applications have shown the potential for positive impact on these children's chances for functional independence in the future. This is an important factor as a good number of CNS conditions that affect cognitive–functional behavior occur with children. Also, since the peak incidence for TBI is with the 15 to 24 age group, it would appear that many "youthful" persons could potentially benefit from VE cognitive–functional applications, and this age group appears to be safe to target for these purposes. However, care should be taken when there is a question about the person's capacity to verbalize their experience of discomfort due to VE exposure. In fact, this is a concern with any age group where verbal expression is suspected to be compromised.

On the other end of the age continuum, only a few studies have reported on VE applications for older populations (over 65). For example, Liu, Miyazaki, & Watson (1999) reported on the use by the elderly of an HMD driving scenario with good results, and Rose et al. (1999) tested a group of stroke patients (average age of 65) on a flat-screen system assessing spatial versus content memory following navigation within a virtual environment. Relevant issues for this age group include limiting factors such as visual difficulties, problems with mobility and balance, and susceptibility to cybersickness (see chap. 30, this volume) and/or aftereffects (see chaps. 31 and 37–39, this volume). There is also a similar dearth of research with older populations within the general simulation technology literature that could provide some direction for VE development.

Research into this age group's susceptibility for cybersickness and aftereffects (see section 4.6) would be a useful starting point for this area. If older persons are found to satisfactorily tolerate a VE application, these tools could be of particular utility to address the progressive cognitive–functional impairments seen with the increased incidence of neurodegenerative disorders observed in older populations (see chap. 18, this volume). One area of concern pertains to the observation that traditional rehabilitation approaches for progressive CNS disorders, such as AD, have met with limited success. Virtual environments designed to deliver functional training for elderly populations may fare better, and this underscores the need to determine VE feasibility for older persons. Functional training VEs, which exploit preserved procedural learning abilities in a relatively safe environment, could serve to help maintain adequate performance of IADLs needed to maximize safe living and functional independence. By contrast, procedural VE rehabilitation approaches may be of limited value for elderly persons with certain CNS disorders. Individuals with Huntington's disease often manifest difficulties with procedural learning (Butters, Salmon, Heindel, & Granholm, 1988), and these patients may not be able to benefit from a highly procedural virtual learning environment. In this case, VE applications for this group might be more usefully aimed at *assessing* the decline of this type of "hands-on" learning as a tool for measuring disease progression or for determining the efficacy of different treatment approaches.

Thus far, the few VE applications that have targeted clinical user groups appear to have been relatively benign. This may reflect either appropriate caution in the development of these applications or selective reporting. It is important that negative reactions to VE exposure be documented and shared with the VE community. In this manner, the causes of these problems can be addressed and understood, and future methods to prevent the reoccurrence of such problems can be developed. As the field continues to develop and more objective results become available, we will be in a better position to formalize the advisements and proscriptions for VE use with different clinical populations. Until that time, rational caution mixed with thoughtful sensitivity to these issues will be required in order to ethically advance these applications.

### 4.4 What Is the Optimal Level of Presence Needed for Clinical CNS Applications?

More detailed discussions of the human factors issues pertaining to the concept of "presence" can be found elsewhere in this volume (see chap. 40, this volume). However, issues regarding presence will be briefly addressed here, as they are seen to be relevant for the conceptualization and design of VEs for NA and CR applications. The concept of presence is often cited as one of the important features of a virtual environment (Durlach & Mavor, 1995; Slater & Usoh, 1993; Stanney, 1998). Presence can be simply described as the experience a person has when in a VE of "being there." Stated another way, *presence* is "the subjective experience of being in one place or environment, even when you are physically located in another" (p. 225, Witmer & Singer, 1998). This experience is not restricted to VEs; reading a book, watching TV, and talking on the phone can all engender some level of presence. For the purposes of the present chapter, the concept of presence and how it may relate to the design and potential effectiveness of various NA and CR VEs will be examined. An understanding of the factors involved in this relationship may be important for predicting the relative value of presence for different applications. Awareness of the nature of presence "factors" that are under a developer's control in the design of VE systems could serve to guide and inform future system development targeting clinical populations. This knowledge does not formally exist at this early stage in the study of NA and CR VE applications, but it is safe to say that different cognitive–functional targets will likely have different optimal presence requirements. Such a determination will require an analysis of presence–effectiveness relationships for a range of different target applications. Intuitive thinking regarding VEs might suggest that a large "amount" of presence would be necessary for effective NA and CR scenarios and that this may require a high degree of realism. However, the capacity to develop veridical VE scenarios that match the characteristics of the "real-world" absolutely, and which might be expected to enhance the experience of presence, is not technically feasible at the current time and may never be. Fortunately, the experience of presence is not totally dependent on the level of realism (Welch, Blackmon, Liu, Mellers, & Stark, 1996). An extreme level of VE similarity with the real world may not be necessary for the experience of presence to occur or for VEs to be effective.

Witmer and Singer (1998) suggest that the psychological states of involvement and immersion are both necessary for the experience of presence. Involvement results when a person selectively focuses attention and energy on coherent stimuli, activities, and events, and can depend on the degree of meaning and significance that the person attaches to the activity as well as the "compelling" nature of the activity. Immersion is seen as related to one's perception of being enveloped by, included in, and interacting with environments that provide a continuous stream of stimuli and experiences (Witmer & Singer, 1998). Involvement seems to be more a function of the user's internal characteristics (interest, motivation, etc.) and how that relates to the "nature" of the scenario. Immersion appears more related to system attributes (for example, non-HMD vs. HMD; see chap. 40, this volume). However, these characteristics are seen to operate in an interdependent fashion, and both are believed to be necessary ingredients for the experience of presence to occur. In the process of conjecturing the value of presence for clinical VE effectiveness, it may be useful to consider the immersive and involvement components of presence and how different weightings of these factors may impact the effectiveness of various applications. The "suspension of disbelief" that occurs when one is using a VE appears, in part, due to system factors (i.e., HMD vs. non-HMD, graphic quality, computer speed, etc.) and to the individual differences of the user (i.e., degree of claustrophobic concerns, capacity for a good visual imagination, past experience with VEs, etc.; see chap. 18, this volume). The design of effective clinical VEs requires an appreciation for the complex relationship between available equipment, cognitive–functional targets, and the characteristics of the clinical population. This combination may reflect Maida's construct of "sense of engagement," as cited in Stanney et al. (1998), whereby unique levels of presence are optimal to support certain applications contingent on the interaction of user, system, and gaming-related motivational factors.

For example, mental health VE applications for fear reduction with phobic clients have initially been shown to be effective, even though the scenarios are often somewhat "cartoonish" and would never be mistaken for the real thing (see chap. 51, this volume). The effectiveness of these scenarios may be found to rely more on the HMD-based immersion component of presence than on an involvement component. This implies that for persons who have a long history of avoiding feared stimuli, a highly immersive HMD system may be necessary to keep their awareness directed toward the modeled stimuli (immersion promoting involvement) in order to encourage therapeutic exposure. These clients would not be expected to naturally be motivated to experience the stimuli (low involvement), and therefore a less immersive non-HMD flat-screen system for this application may fall short of the level of presence needed for effective therapeutic exposure to occur. However, an HMD may not necessarily be mandatory to achieve an optimal presence level or to be effective for targeting cognitive–functional processes. Navigation studies with clinical populations have shown successful functional skills learning using flat-screen systems for persons with severe learning or motor disorders (Cromby et al., 1996; Stanton et al., 1998). These results suggest that for certain populations and training objectives, an HMD may not always be necessary. However, HMDs may still be the most efficient tool to support the immersion-based component of presence and may be necessary for VE effectiveness with certain clinical targets. More immersion may be especially useful for systems that target the assessment of cognitive processes, where HMD-fostered immersion would be needed to eliminate external "distractions" that would intrude on the controlled environment (Rizzo et al., 1999). Also, applications that require HMD immersion to better replicate a dynamic environment may enhance ecological validity and promote better transfer. These areas are not meant to be exhaustive, but rather illustrate examples where the immersion component of presence is emphasized.

Certain VE applications exist where the immersive component is less important, and involvement can be promoted to support a level of presence that may be necessary for effectiveness. This is obviously seen with flat-screen computer games that create a sense of presence due to their involving features (gaming, interesting graphics, sound effects, etc.). Likewise, less immersive VEs may effectively address certain NA and CR targets by leveraging the involvement component of presence. However, users in targeted clinical groups may need to possess more motivation to participate as well as have intact attentional abilities in order to stay focused and benefit from less-immersive approaches. Virtual environment applications that target visuospatial processes and more molecular motor abilities may be usefully addressed in this manner. In this regard, our lab has developed a series of visuospatial skills assessment and training tasks that utilize the ImmersaDesk system. This system is a drafting table projection-based VE device that employs stereo glasses and magnetic head and hand tracking that offers a type of VE that is semi-immersive. The size and position of the screen gives a wide-angle view of the scenario, yet the user is also able to look around and see the real physical environment. Using this system we have developed a series of visuospatial manual tasks. Unimpaired subjects that have been assessed report the tasks to be engaging (involvement), and examiners have rarely observed distraction away from the screen. Positive assessment and learning transfer effects seen thus far may suggest that for this purpose this type of presence (lower immersion and higher involvement) is acceptable for the goals of the application. However, trials with impaired populations are just beginning and the effectiveness of this system for those groups remains to be seen.

### 4.5 Will Clinical CNS Target Users Be Able to Navigate and Interact With the VE in an Effective Manner?

The method of navigation (see chap. 24, this volume) and requirements for interaction (see chap. 13, this volume) in a VE are important factors to consider in the design of systems for use with clinical populations. Although human–computer interaction for general computer use by persons with physical disabilities has been addressed for a number of years, VE usage by persons with cognitive–functional impairments requires additional focus. Virtual environments have been characterized as an "intuitive interface" that allows a person to interact with a computer (and data) in a naturalistic fashion (Aukstakalnis & Blatner, 1992). In this regard, Wann and Mon-Williams (1996) suggest, "The goal is to build [virtual] environments that minimize the learning required to operate within them, but maximize the information yield" (p. 845). This "interactional intuitiveness" issue becomes more relevant when designing VEs applied to persons with cognitive–functional impairments. In order for these individuals to be in a position to benefit from a VE, they often must be capable of learning how to navigate and interact within the environment.

Many modes of VE navigation (DataGloves, joysticks, space balls, etc.; see chaps. 11 and 13, this volume), while easily mastered by unimpaired participants, could present problems for those with cognitive–functional difficulties. Even if patients are capable of interacting in a VE system at a basic level, the extra nonautomatic cognitive effort required to navigate may serve as a distraction and limit targeted assessment and rehabilitation processes. In this regard, Psotka (1995) hypothesizes that facilitation of a "single egocenter" found in highly immersive interfaces serves to reduce "cognitive overhead," thereby enhancing potential information access and learning. Thus far, early reports on VEs with neurological patient populations using both joystick HMD-based and non-HMD systems have produced encouraging results. For example, in studies assessing executive functioning in various clinical populations (Mendozzi et al., 1998; Pugnetti et al., 1998a), teaching supermarket navigation with developmentally disabled students (Cromby et al., 1996) and teaching spatial skills to children with cognitive and motor impairments (Stanton et al., 1998), difficulties in learning to navigate in the VEs were not reported. However, navigational interface factors were not the empirical focus of these studies, and it might be expected that as VEs are designed to address more complex cognitive–functional targets, these factors may become more problematic. More naturalistic interfaces for navigation and interaction may be required to optimize performance and improve access for patients having severe cognitive or motor impairments. For example, the use of voice recognition technology may be a useful approach for some types of navigation or interaction (Middleton & Boman, 1994). This technology may also provide a more naturalistic and ecologically valid interface for some types of tasks, in addition to improving VE access for persons with motor impairments.

Another factor of critical importance is whether the means of navigation actually affects what aspects of the VE are focused on and consequently what is measured and learned in scenarios addressing cognitive processes. This was seen to be the case in a study that looked at what types of memory were enhanced during a four-room house navigation task (Attree et al., 1996; Pugnetti et al., 1998a). As aforementioned, these studies, using a joystick interface, allowed one subject to navigate the house (active condition), whereas a yoked subject was simply exposed to the same journey but had no control (passive condition). With unimpaired subjects, differential memory performance between active versus passive groups was observed. The active group showed better spatial memory for the route, whereas the passive group displayed superior object recall and recognition memory for the items viewed along the route. Perhaps a more intuitive method of navigation may have allowed the active group to perform as well on object memory via a more equal allocation of cognitive resources. Also, this navigation method may have taxed the subjects' divided attention capacity and thereby influenced the memory results found using this paradigm. These issues may be particularly noteworthy for the development of VEs designed to address the NA and CR of attention processing.

Increased generalization of learning to "real-world" performances might also be expected as the method of navigation and object interaction more closely resembles the requirements of the "natural" or target environment. Conversely, reduced motivation may result when a person's first VE experience is characterized as "more work than it is worth" when confronted with an unnatural or awkward navigational interface. One obvious example of an effective match between the mode of VE navigation and the real-world objective is with the training of motorized wheelchair navigation skill in children with cerebral palsy (Inman et al., 1997). In this application, the controls for the motorized chair in the virtual application are essentially identical to the controls used in the child's real-world environment. This would be an "ideal" match between the VE navigation mode and the demands of the actual targeted behavior. However, this sort of one-to-one correspondence is relatively rare for VEs having more complex navigational and interactional demands. It might be interesting to note any difference in effectiveness that might occur if the learning in this scenario was directed toward the operation of a standard manually operated chair. More effective transfer of this skill would be expected with the original VE application and functional target. However, if equal criterion performance was found, it might support the idea of the robustness of training with nonidentical navigational devices and would be quite interesting indeed. In essence, development of more naturalistic interfaces could be of vital importance for precise assessment and rehabilitative targeting, and could also have implications for the generalization issues discussed in section 4.7.

### 4.6 What Is the Potential for Side Effects Due to the Characteristics of Different CNS Clinical Groups?

In order for VEs to become safe and useful tools for NA and CR applications, the potential for adverse side effects needs to be considered and addressed. This is a significant concern as the occurrence of side effects could limit the applicability of VEs for certain clinical populations. Two general categories of VE-related side effects have been reported: cybersickness (see chap. 30, this volume) and aftereffects (see chaps. 31 and 37–39, this volume). Detailed discussion of these issues and analysis regarding normal populations is covered extensively in other chapters in this volume, so the focus here will be on reports with clinical groups.

A recent review of this area (Stanney et al., 1998) targets four primary issues in the study of VE-related side effects, which may be of particular value for guiding feasibility assessments with different clinical populations. These include: "(1) How can prolonged exposure to VE systems be obtained? (2) How can aftereffects be characterized? (3) How should they be measured and managed? (4) What is their relationship to task performance?" (p. 6). These questions are particularly relevant to developers of NA and CR VEs, as these systems are primarily designed to be used by persons with some sort of defined diagnosis or impairment. It is possible that these users may have increased vulnerability and a higher susceptibility to VE-related side effects, and ethical clinical vigilance to these issues is essential. Particular concern may be necessary for neurologically impaired populations, some of which display residual equilibrium, balance, perceptual, and orientation difficulties. It has also been suggested that subjects with unstable binocular vision (which sometimes can occur following strokes, TBI, and other CNS conditions) may be more susceptible to postexposure visual aftereffects (Wann et al., 1996; see chap. 37, this volume). Unfortunately, statistics on the occurrence of side effects with clinical populations have been inconsistently reported in the published literature to date. This is an aspect of data reporting on VEs that should be changed. Some type of assessment and reporting of VE-related side effects, whether using in-house designed ratings scales or standardized subjective and objective measures (Kennedy, Lane, Berbaum, & Lilienthal, 1993), should be standard procedure for presenting results on systems in this area.

Thus far, anecdotal reports of flat-screen scenarios used with clinical populations have not indicated problems with these less-immersive systems. However, it does not appear that much systematic assessment has occurred, or in some cases the verbal report of the patients may have been compromised. Also, most clinical applications appear to use short periods of exposure (10 to 20 minutes), and this may have served to mitigate the occurrence of side effects based on the scant reporting in this literature. In one of the first studies to present systematic data for an HMD-based system used with CNS populations, 11 neurological patients were compared with 41 non-neurological subjects regarding self-reported occurrence of side effects (Pugnetti et al., 1995b). Subjects were tested in a VE specifically designed to target executive functioning with CNS populations. The results suggest a reduced occurrence of VE related side effects relative to other studies using the same assessment questionnaire, the Simulator Sickness Questionnaire (SSQ; Kennedy et al., 1993), with an overall rate of 17% for the total sample. The authors concluded that the neurological subjects appeared to be at no greater risk for developing cybersickness than the non-neurological group. In a more recent study, Pugnetti et al. (1998c) reported side effect results comparing 36 patients having mixed neurological diagnoses, with 32 normal controls for a 30-minute VE exposure using the system described above. Using a variety of self-report questionnaires, assessments were conducted prior, during, and following VE usage, and no differences were found between the groups on any of the side effect measures. It is important to note, though, that the patient group was recruited from those with stable neurological conditions (good bilateral visual acuity, no epilepsy, preserved dominant handedness, and no psychiatric, vestibular, or severe cognitive disorders), and this screening procedure may have contributed to the low level of side effect occurrence. The screening of patient groups, as was prudently done by these authors, may be the safest course of action until more specific and "cautiously" acquired data become available from more impaired populations, particularly regarding the objective assessment of perceptual aftereffects (see chaps. 37–39, this volume).

While these initial findings are encouraging, further work is necessary to specifically assess how the occurrence of side effects is influenced by factors, such as the type and severity of neurological trauma, specific cognitive impairments, prior VE exposure, length of time within the VE, and characteristics of the specific VE program. This is an essential step in determining the conditions where VEs would be of practical value with CNS clinical groups. A useful tool for monitoring VE-related side effects is the Simulator Sickness Questionnaire (Kennedy et al., 1993). While more involved "objective" measures may exist (see chaps. 37–39, this volume), particularly for the measurement of aftereffects, SSQ data is relatively simple to collect and may serve as a low-cost method to begin to specify and document the basic occurrence of VE side effects in clinical populations (see chap. 30, this volume). Until better data on these issues are obtained, extra caution may be needed with some applications. For example, since we cannot be confident regarding the absence of potential perceptual aftereffects occurring in our current study with an elderly group (over 65 years old) testing visuospatial abilities, we have money built into our grant to provide transportation to and from the test site. The worst thing for the user, the field, and us would be for an elderly person to have a car accident driving home after participating in an "experimental VE setting" where subjects were exposed to visuospatial manipulations. These concerns must be addressed in order to assure a positive course for developing VE applications for persons with CNS disorders.

### 4.7 Will VE Assessment Results and Treatment Effects Generalize or Transfer to the "Real World"?

A fundamental issue having important implications for the ultimate utility of VEs for clinical applications is the concept of generalization of measurement and treatment. In a classic review from the applied behavioral psychology literature, Stokes and Baer (1977) place strong emphasis on the need to plan and program for generalization when designing assessment and treatment interventions. This is no less important for the design and implementation of VEs for NA and CR applications. In the area of neuropsychology, improved generalization of NA and CR results is viewed as one of the primary expected benefits of using VEs (Rizzo et al., 1997). Support for this expectation can first be seen in the predecessor field of aviation simulator research. In an article on theoretical issues concerning generalization and transfer of training from aircraft simulators, Johnston (1995) cites a transfer effectiveness ratio in the aviation simulation research of .48. This ratio indicates that for every hour spent in aviation simulator training, one half hour is saved in actual aircraft training. However, while it is intuitively seductive to assume that VEs are "just another form of simulation" and that generalization will be promoted, research specific to VEs must be examined (see chap. 19, this volume). As VEs are developed to assess and treat various cognitive–functional targets, it will become increasingly important to demonstrate that the results have some relevance or functional impact on users' real-world ability. For example, initial psychological studies with VE scenarios designed to treat phobias have shown that fear reductions accrued in a VE generalize to the person's real-world functional behavior (Wiederhold & Wiederhold, 1998; see chap. 51, this volume). In these cases, clients were able to walk across real bridges, go up glass elevators, ride in aircraft, and participate in areas of everyday life in which they were limited prior to VE treatment. Such evidence of generalization of treatment from the VE to functional activities in everyday living has encouraged researchers and clinicians working in other areas with the hope that NA and CR VEs can actually produce tangible benefits for clinical populations.

Three types of generalization or transfer have been described that have relevance to the design of systems and studies investigating VE-produced transfer (Gordon, 1987). These are: (1) transfer of gains on the same materials on separate occasions; (2) improvement on similar but not identical training tasks; (3) transfer from the training environment to day-to-day functioning. From this framework, a VE for training of some hypothetical visuospatial ability would show good generalization if improvements were seen across multiple VE sessions, on pencil-and-paper measures of the skill, and observed in a real-life task such as assembling a piece of furniture or finding one's way home. These types of generalization are of similar concern for performance measurement used for assessment purposes. The emphasis on any one of these forms of generalization will depend, of course, on the goals of the application. At this stage of VE development the primary emphasis has been on generalization from a VE environment to the actual "real-world" environment. How well suited a VE is for this purpose may be related to a variety of factors and further study with clinical populations in this area is needed. For example, one investigation with unimpaired participants reported that a short period of VE training for spatial navigation of a maze was no more effective than map training, although with longer exposure time, VE training eventually surpassed real-world training (Waller et al., 1998). This finding illustrates the importance of temporal parameters that may influence learning and generalizability in virtual environments. When designing VEs for rehabilitation purposes, this factor may be particularly relevant, especially for persons with impaired learning abilities.

While support has grown for the idea that VEs can promote generalization and transfer in unimpaired groups (see chap. 19, this volume), the research with CNS groups is limited. Evidence for generalization with clinical groups has mainly appeared in the form of positive learning transfer from non-HMD virtual training settings to the real world on various navigation training tasks (Brooks et al., 1999; Cromby et al., 1996; Stanton et al., 1998). Positive generalization of results for an HMD-based VE targeting the assessment of executive functioning has also been reported in a case study (Mendozzi et al., 1998). In this case, the patient's everyday performance was reported by family members to be impaired, yet traditional neuropsychological assessment tools suggested normal functioning. By contrast, the VE was better able to detect deficits that had been reported to be limiting this patient's everyday functioning. These findings support the idea that VEs may provide generalizable results for the testing of cognitive–functional behaviors in situations that are ecologically valid. However, these investigations represent first steps in determining whether VE-based NA and CR can foster generalization and transfer to other settings, particularly the real world. Generalization concerns have been fundamental in the evaluation of the effectiveness and value of NA and CR approaches, making it essential that intuitive expectations of positive VE results in this area be, in fact, supported with quality research. This will be vital in order for a VE approach to be taken seriously for NA and CR purposes.

## 5. CONCLUSIONS

The information presented in this chap. is an attempt to describe the status of the current literature and the reasoning behind the application of VE technology for NA and CR purposes with CNS clinical populations. However, the potential benefits that could be derived from VE technology are equaled by the amount of what we do not know at this time. Although a fair amount of the VE literature targeting clinical populations consists of research designs employing good methodology and statistical analysis, much of the initial work in this area is limited by a number of factors. These include inadequate control groups, inconsistent reporting of statistics, and what appears to be a lack of a comprehensive long-term plan to address important issues. This is not meant to slight the innovative work of scientists and clinicians that have begun the difficult work of exploring the practical applications and feasibility of this emerging technology. Often with any new area of study, initial efforts consist mainly of exploratory pilot projects designed to determine potential value and serve as a springboard for future funding opportunities that will provide the support needed for more serious in-depth investigation. The current status of research on NA and CR VEs reflects this early stage of development. However, the field is at a turning point where the initial findings that hint at the potential of this technology must be further explored and developed. Awareness of the important issues and knowledge that has been uncovered thus far should support the building of more sophisticated investigations. The value of pursuing this goal is reflected in a recent National Science Foundation proposal announcement (National Science Foundation, 1998), where it is stated that, "Computer simulation has now joined theory and experimentation as a third path to scientific knowledge. Simulation plays an increasingly critical role in all areas of science and engineering . . ." (p. 1). In order translate this hopeful view into useful application, systematic research is needed to provide answers to the basic issues and questions that were discussed throughout this chapter.

It is recommended that future clinical VE studies include some recording and reporting of side effects. This can, at a minimum, be attempted by using variously available questionnaires (see chap. 30, this volume). The same recommendation can be made for the measurement of presence factors via the use of available rating scales (see chap. 40, this volume). Information regarding both side effects and presence acquired with these basic tools might provide the field with useful information that could aid in understanding the strengths and limitations of a VE approach for certain populations and for specific targets. A primary area that warrants in-depth exploration involves the issue of generalization or transfer of training (see chap. 19, this volume). This issue is at the crux of the presumed value of VE applications and requires considerable forethought when developing an experimental design for any type of assessment and treatment study, and is no less important in the study of VE approaches. There is a good number of excellent transfer studies that are coming out of the human factors literature on VEs, and these need to be carefully examined for the insight and tools that they can provide in the construction of better designs for specific NA and CR applications with clinical populations. Consideration of the specific characteristics commonly observed with different clinical populations and how they could be addressed in the design of navigation and interaction methods within a VE is also required (see chaps. 13 and 24, this volume). Finally, an assessment of the match between the assets found in a VE and the requirements of the specific cognitive–functional targets needs to be continually addressed.

The brief sampling of methodological, conceptual, and pragmatic issues presented in this chapter suggests some very basic considerations for the development and application of NA/CR VEs for use with clinical CNS populations. Virtual environment technology is at the very beginning stages of development in terms of our understanding on how to scientifically maximize its potential usefulness and value for clinical applications. As a larger body of VE-related literature develops, we will be in a better position to consider the standards that should be applied to research-based and clinically oriented NA and CR applications. Future integration of physiological monitoring, brain imaging, and eye-tracking technologies within VEs guarantee new application domains waiting to be explored once more basic parameters are addressed. The research conducted thus far applying VE technology to clinical CNS populations, while still in its infancy, has produced initial results that are encouraging. As the technical capabilities for producing these systems continue to improve, it will be necessary that our conceptual and methodological understanding of VE technology keep pace. If this concurrent progress takes place, the remaining challenges will exist only in the limits of our ethics, clinical sensitivity, and powers of imagination.

## 6. REFERENCES


Aguirre, G. K., & D'Esposito, M. (1997). Environmental knowledge is subserved by separable dorsal/ventral neural areas. *Journal of Neuroscience, 17*, 2512–2518.

Alpini, D., Pugnetti, L., Mendozzi, L., Barbieri, E., Monti, B., & Cesarani, A. (1998). Virtual reality in vestibular diagnosis and rehabilitation. In P. Sharkey, D. Rose, & J. Lindstrom (Eds.), *Proceedings of the Second European Conference on Disability, Virtual Reality and Associated Techniques* (pp. 221–228). Sköve, Sweden: University of Reading.

American Psychiatric Association. (1994). *Diagnostic and statistical manual of mental disorders* (4th ed.). Washington, DC: Author.

Andrews, T. K., Rose, F. D., Leadbetter, A. G., Attree, E. A., & Painter, J. (1995). The use of virtual reality in the assessment of cognitive ability. In I. Placencia Porrero & R. Puig de la Bellacasa (Eds.), *The European context for assistive technology: Proceedings of the second TIDE congress* (pp. 117–121). Amsterdam: IOS Press.

Astur, R. S., Ortiz, M. L., & Sutherland, R. J. (1998). A characterization of performance by men and women in a virtual Morris water task: A large and reliable sex difference. *Behavioural Brain Research, 93*, 185–90.

Attree, E. A., Brooks, B. M., Rose, F. D., Andrews, T. K., Leadbetter, A. G., & Clifford, B. R. (1996). Memory processes and virtual environments: I can't remember what was there, but I can remember how I got there. Implications for people with disabilities. In P. Sharkey (Ed.), *Proceedings of the First European Conference on Disability, Virtual Reality and Associated Technology* (pp. 117–121). Maidenhead, England: University of Reading.

Aukstakalnis, S., & Blatner, D. (1992). *Silicon mirage: The art and science of virtual reality*. Berkeley, CA: Peachpit Press.

Barkley, R. A., Fisher, M., Edelbrock, C. S., & Smalish, L. (1990). The adolescent outcome of hyperactive children diagnosed by research criteria: Vol. I. An eight-year prospective follow-up study. *Journal of the American Academy of Child and Adolescent Psychiatry, 29*, 546–557.

Berka, R., & Slavík, P. (1998). Virtual reality for blind users. In P. Sharkey, D. Rose, & J. Lindstrom (Eds.), *Proceedings of the Second European Conference on Disability, Virtual Reality and Associated Techniques* (pp. 89–98). Sköve, Sweden: University of Reading.

Boring, E. R. (1923, June). Intelligence as the tests test it. *The New Republic*, 35–37.

Braak, H., Braak, E., & Bohl, J. (1993). Staging of Alzheimer-related cortical destruction. *European Neurology, 33*, 403–408.

Brooks, B. M., McNeil, J. E., Rose, F. D., Greenwood, R. J., Attree, E. A., & Leadbetter, A. G. (1999). Route learning in a case of amnesia: A Preliminary investigation into the efficacy of training in a virtual environment. *Neuropsychological Rehabilitation, 9*(1), 63–76.

Brown, D. J., Kerr, S. J., & Bayon, V. (1998). The development of the Virtual City: A user centred approach. In P. Sharkey, D. Rose, & J. Lindstrom (Eds.), *Proceedings of the Second European Conference on Disability, Virtual Reality and Associated Techniques* (pp. 11–16). Sköve, Sweden: University of Reading.

Butters, N., Salmon, D. P., Heindel, W., & Granholm, E. (1988). Episodic, semantic, and procedural memory: Some comparisons of Alzheimer's and Huntington's disease patients. In R. D. Terry (Ed.), Aging and the brain. New York: Raven Press.

Carpenter, P. A., & Just, M. A. (1986). Spatial ability: An information processing approach to psychometrics. In R. J. Sternberg (Ed.), *Advances in the psychology of human intelligence* (Vol. 3, pp. 221–253). Hillsdale, NJ: Lawrence Erlbaum Associates.

Charness, N., Milberg, W., & Alexander, M. P. (1988). Teaching an amnesic a complex cognitive skill. *Brain Cognition, 8*(9), 253–272.

Chase, W. G., & Ericsson, K. A. (1981). Skilled memory. In J. R. Anderson (Ed.), *Cognitive skills and their acquisition*. Hillsdale, NJ: Lawrence Erlbaum Associates.

Christiansen, C., Abreu, B., Ottenbacher, K., Huffman, K., Massel, B., & Culpepper, R. (1998). Task performance in virtual environments used for cognitive rehabilitation after traumatic brain injury. *Archives of Physical Medicine and Rehabilitation, 79*, 888–892.

Cobb, S. V G., Neale, H. R., & Reynolds, H. (1998). Evaluation of virtual learning environments. In P. Sharkey, D. Rose, & J. Lindstrom (Eds.), *Proceedings of the Second European Conference on Disability, Virtual Reality and Associated Techniques* (pp. 17–23). Sköve, Sweden: University of Reading.

Cohen, N. J., & Squire, L. R. (1980). Preserved learning and retention of pattern-analyzing skill in amnesia: Dissociation of "knowing how" and "knowing that." *Science, 210*, 207–209.

Cooper, M., & Taylor, M. E. (1998). Ambisonic sound in virtual environments and applications for blind people. In P. Sharkey, D. Rose, & J. Lindstrom (Eds.), *Proceedings of the Second European Conference on Disability, Virtual Reality and Associated Techniques* (pp. 113–118). Sköve, Sweden: University of Reading.

Couling, M. (1998). The ImmersaDesk semi-immersive virtual reality workstation. *CyberPsychology and Behavior, 1*, 4.

Crawford, J. R. (1998). Introduction to the assessment of attention and executive functioning. *Neuropsychological Rehabilitation, 8*(3), 209–211.

Cromby, J. J., Standen, P. J., Newman, J., & Tasker, H. (1996). Successful transfer to the real world of skills practised in a virtual environment by students with severe learning difficulties. In P. Sharkey (Ed.), *Proceedings of the First European Conference on Disability, Virtual Reality and Associated Technology* (pp. 103–107). Maidenhead, England: University of Reading.

Davies, R. C., Johansson, G., Boschian, K., Lindén, A., Minör, U., & Sonesson, B. (1998). A practical example using virtual reality in the assessment of brain injury. In P. Sharkey, D. Rose, & J. Lindstrom (Eds.), *Proceedings of the Second European Conference on Disability, Virtual Reality and Associated Techniques* (pp. 61–68). Sköve, Sweden: University of Reading.

Decety, J., Perani, D., Jeannerod, M., Betttinardi, V., Tadary, B., Woods, R., Mazziotta, J. C., & Fazio, F. (1994). Mapping motor representations with positron emission tomography *Nature, 371*, 600–602.

De Renzi, E. (1985). Disorders of spatial orientation. In J. Frederiks (Ed.), *Handbook of clinical neurology* (Vol. 1, pp. 405–422). Amsterdam: Elsevier Science Publishers.

Desbonnet, M., Cox, S. L., & Rahman, A. (1998). Development and evaluation of a virtual reality–based training system for disabled children. *Proceedings of the Second European Conference on Disability, Virtual Reality and Associated Technologies* (pp. 177–182). Skövde, Sweden: University of Reading.

Dinh, H. G., Walker, N., Hodges, L. F., Song, C., & Kobayashi, A. (1999). Evaluating the importance of multisensory input on memory and the sense of presence in virtual environment. In L. Rosenblum, P. Astheimer, & D. Teichmann (Eds.), *Proceedings of the IEEE Virtual Reality '99 Conference* (pp. 222–228). Los Alamitos, CA: IEEE Computer Society Press.

Dobrzeniecki, A. B., Larsen, P., Andersen, J. R., & Sperling, B. (1999). *Virtual environment for cognitive assessment and rehabilitation of stroke*. Paper presented at the Seventh Annual Medicine Meets Virtual Reality Conference, San Francisco, CA.

Durlach, B. N. I., & Mavor, A. S. (1995). *Virtual reality: Scientific and technological challenges*. Washington, DC: National Academy Press.

Eaton, W. O., & Enns, L. R. (1986). Sex differences in human motor activity level. *Psychological Bulletin, 100*, 19–28.

Elkind, J. S. (1998). Uses of virtual reality to diagnose and habilitate people with neurological dysfunctions. *CyberPsychology and Behavior, 1*(3), 263–274.

Escamilla, G. M., Falkinstein, Y., Chang, C. Y., Schneider, D. I., & Zeltzer, L. K. (1997). Reducing acute pain through computer generated distraction in children. *Journal of Investigative Medicine, 45*(1), 86A.

Folstein, M. F., Folstein, S. E., & McHugh, P. R. (1975). "Mini Mental State": A practical method for grading the cognitive state of outpatients for the clinician. *Journal of Psychiatric Research, 12*, 189–198.

Foreman, N. P., Orencas, C., Nicholas, E., Morton, P., & Gell, M. (1989). Spatial awareness in seven to eleven-year-old physically handicapped children in mainstream schools. *European Journal of Special Needs Education, 4*, 171–179.

Foreman, N. P., Wilson, P., & Stanton, D. (1997). VR and spatial awareness in disabled children. *Communications of the ACM, 40*(8), 76–77.

Glisky, E. L. (1992). Computer-assisted instruction for patients with traumatic brain injury: Teaching of domain-specific knowledge. *Journal of Head Trauma Rehabilitation, 7*(3), 1–12.

Gordon, W. (1987). Methodological considerations in cognitive remediation. In M. Meier, A. Benton, & L. Diller (Eds.), *Neuropsychological rehabilitation*. New York: Guilford Press:

Gould, E., Beylin, A., Tanapat, P., Reeves, A., & Shors, T. J. (1999). Learning enhances adult neurogenesis of the hippocampal formation. *Nature Neuroscience, 2*(3), 260–265.

Graf, P., & Schacter, D. L. (1985). Implicit and explicit memory for new associations in normal and amnesic patients. *Journal of Experimental Psychology: Learning, Memory, and Cognition, 11*, 501–518.

Grealy, M. A., Johnson, D. A., & Rushton, S. K. (1999). Improving cognitive function after brain injury: The use of exercise and virtual reality. *Archives of Physical Medicine and Rehabilitation, 80*, 661–667.

Gresham, G. E., Duncan, P. W., Stason, W. B., et al. (1995). *Post-stroke rehabilitation. clinical practice guideline no. 16* (Public Health Service, Agency for Health Care Policy and Research, AHCPR Publication No. 95-0662). Rockville, MD: U.S. Department of Health and Human Services.

Hazen, N. L. (1982). Spatial exploration and spatial knowledge: Individual and developmental differences in very young children. *Child Development, 53*, 826–833.

Inman, D. P., Loge, K., & Leavens, J. (1997). VR education and rehabilitation. *Communications of the ACM, 40*(8), 53–58.

Jacobs, W. J., Laurance, H. E., & Thomas, G. F. (1997). Place learning in virtual space 1: Acquisition, overshadowing, and transfer. *Learning and Motivation, 28*, 521–541.

Jacobs, W. J., Thomas, K. G. F., Laurance, H. E., & Nadel, L. (1998). Place learning in virtual space 2: Topographical relations as one dimension of stimulus control. *Learning and Motivation, 29*, 288–308.

Jaffe, D. L. (1998, March). *Use of virtual reality techniques to train elderly people to step over obstacles*. Paper presented at the Technology and Persons with Disabilities Conference, Los Angeles, CA.

Johnston, R. (1995). Is it live or is it memorized? *Virtual Reality Special Report, 2*(3), 53–56.

Kennedy, R. S., Lane, N. E., Berbaum, K. S., & Lilienthal, M. G. (1993). Simulator sickness questionnaire: An enhanced method for quantifying simulator sickness. *International Journal of Aviation Psychology, 3*(3), 203–220.

Kirsch, N. L., Levine, S. P. Lajiness-O'Neill, R., & Schnyder, M. (1992). Computer-assisted interactive task guidance: Facilitating the performance of a simulated vocational task. *Journal of Head Trauma Rehabilitation, 7*(3), 13–25.

Kodgi, S. M., Gupta, V., Conroy, B., & Knott, B. A. (1999, November). *Feasibility of using virtual reality for quantitative assessment of hemineglect: A pilot study*. Paper present at American Academy of Physical Medicine and Rehabilitation 61st Annual Assembly, Washington, DC.

Kolb, B., & Wishaw, Q. (1990). *Fundamentals of neuropsychology* (3rd ed.). New York: W. H. Freeman.

Kyriakakis, C. (1999). Fundamental and technological limitations of immersive audio systems. In *Proceedings of the IEEE Virtual Reality '99 Conference* (pp. 941–951). Los Alamitos, CA: IEEE Computer Society Press.

Lackner, J. R., & Dizio, P. (1998). Spatial orientation as a component of presence: Insights gained from nonterrestrial environments. *Presence: Teleoperators and Virtual Environments, 7*(2), 108–115.

Larson, P. A., Rizzo, A. A., Buckwalter, J. G., van Rooyen, A., Kratz, K., Neumann, U., Kesselman, C., Thiebaux, M., & van der Zaag, C. (1999). Gender issues in the application of a virtual environment spatial rotation project. *Cyberpsychology and Behavior, 2*(2), pp. 113–124.

Lezak, M. D. (1995). *Neuropsychological assessment*. New York: Oxford University Press.

Liu, L., Miyazaki, M., & Watson, B. (1999). Norms and validity of the DriVR: A virtual reality driving assessment for persons with head injuries. *CyberPsychology and Behavior, 2*(1), 53–68.

Lumbreras, M., & Sánchez, J. (1998). 3-D aural interactive hyperstories for blind children. In P. Sharkey, D. Rose, & J. Lindstrom (Eds.), *Proceedings of the Second European Conference on Disability, Virtual Reality and Associated Techniques* (pp. 119–128). Sköve, Sweden: University of Reading.

Lynch, W. J. (1983). Cognitive retaining using microcomputer games and commercially available software. *Cognitive Rehabilitation, 1*, 19–22.

McComas, J., Dulberg, C., & Latter, J. (1997). Children's memory for locations visited: Importance of movement and choice. *Journal of Motor Behavior, 29*, 223–229.

McComas, J., Pivik, J., & Laflamme, M. (1998). Children's transfer of spatial learning from virtual reality to real environments. *CyberPsychology and Behavior, 1*(2), 121–128.

Mendozzi, L., Motta, A., Barbieri, E., Alpini, D., & Pugnetti, L. (1998). The application of virtual reality to document coping deficits after a stroke: Report of a case. *CyberPsychology and Behavior, 1*(1), 79–91.

Middleton, T., & Boman, D. (1994). "Simon Says": Using speech to perform tasks in virtual environments. In Murphy (Ed.), *Proceedings of the Second Annual Conference on Virtual Reality and Persons with Disabilities* (pp. 76–79). Northridge: California State University.

Morris, R. G. M. (1981). Spatial localization does not require the presence of local cues. *Learning and Motivation, 12*, 239–260.

Mourant, R. R., & Chiu, S. A. (1997, April). *A virtual–environments based driving simulator*. Paper presented at the First Annual Virtual Reality Universe Conference, Santa Clara, CA.

Mowafy, L., & Pollack, J. (1995). Train to travel. *Ability, 15*, 18–20.

National Center for Injury Prevention and Control. (1999). Acute care, rehabilitation and disabilities [Online]. Available: http://www.cdc.gov.ncipc/dacrrdp/dacrrdp.htm

National Institute on Aging. (1998). *Progress report on Alzheimer's disease* (NIH Publication No. 99-3616). Washington, DC: U.S. Government Printing Office. Available: http://www.alzheimers.org/pr98.html

National Institutes of Health. (in press). Diagnosis and treatment of attention deficit hyperactivity disorder. *NIH Consensus Statement, 16*(2). Available: http://odp.od.nih.gov/consensus/cons/110/110_statement.htm

National Institutes of Health. (1998). *Rehabilitation of persons with traumatic brain injury* (NIH Consensus Statement, October 26–28) 16(1). Available: http://odp.od.nih.gov.consensus/cons/110/110statement.htm

National Institutes of Health. (1995). *Basic behavioral science research for mental health: A national investment* (A report of the U.S. National Advisory Mental Health Council, NIH Publication No. 95-3682). Washington, DC: Author.

National Institutes of Health. (1998a). *Diagnosis and treatment of Attention Deficit Hyperactivity Disorder. Consensus Statement* [Online, November 16–18], 16(2). Available: http://consensus.nih.gov

National Institutes of Health. (1998b). *Rehabilitation of persons with traumatic brain injury. Consensus Statement* [Online, October 26–28], 16(1). Available: http://consensus nih.gov

National Science Foundation. (1998). *Advanced computational research* [Proposal announcement: NSF 98-168, Online, October, 1998]. Available: http://www.usalert.com/htdoc/usoa/nsf/any/any/proc/any/nsf98168.htm

National Stroke Association. (1997). *The stroke/brain attack reporter's handbook*. Englewood, CO:

Neisser, U. (1978). Memory: What are the important questions? In M. M. Gruneberg, P. E. Morris, & R. N. Sykes (Eds.), *Practical aspects of memory* (pp. 3–24). London: Academic Press.

O'Connor, M., & Cermack, L. S. (1987). Rehabilitation of organic memory disorders. In M. J. Meier, A. L. Benton, & L. Diller (Eds.), *Neuropsychological rehabilitation*. New York: Guilford Press.

Pair, J. (1999) Virtual Environment Sound Design Primer. [Online]. Available: http://www.cc.gatech.edu/gvu/people /jarrell.pair/primer.html [July 5, 1999]

Parente, R., & Herrmann, D. (1996). *Retraining cognition: Techniques and applications*. Gaithersburg, MD: Aspen Publishing.

Paul, G. L. (1969). Outcome of systematic desensitization 2: Controlled investigations of individual treatment, technique variations and current status. In C. M. Franks (Ed.), *Behavior therapy: Appraisal and status*. New York: McGraw-Hill.

Peruch, P., Vercher, J. L., Gauthier, G. M. (1995). Acquisition of spatial knowledge through visual exploration of simulated environments. *Ecological Psychology, 7*(1), 1–20.

Psotka, J. (1995). Immersive training systems: Virtual reality and education and training. *Instructional Science, 23*, 405–431.

Pugnetti, L., Mendozzi, L., Attree, E. A., Barbieri, E., Brooks, B. M., Cazzullo, C. L., Motta, A., & Rose, F. D. (1998a). Probing memory and executive functions with virtual reality: Past and present studies. *CyberPsychology and Behavior, 1*(2), 151–162.

Pugnetti, L., Mendozzi, L., Barbieri, E., Motta, A., Alpini, D., Attree, E. A., Brooks, B. M., & Rose, F. D. (1998). Developments of a collaborative research on VR applications for mental health. In P. Sharkey, D. Rose, & J. Lindstrom (Eds.), *Proceedings of the Second European Conference on Disability, Virtual Reality and Associated Techniques* (pp. 77–84). Sköve Sweden: University of Reading.

Pugnetti, L., Mendozzi, L., Barbieri, E., Rose, F. D., & Attree, E. A. (1996). Nervous system correlates of virtual reality experience. In P. Sharkey (Ed.), *Proceedings of the First European Conference on Disability, Virtual Reality and Associated Technology* (pp. 239–246). Maidenhead, England: University of Reading.

Pugnetti, L., Mendozzi, L., Brooks, B. M., Attree, E. A., Barbieri, E., Alpini, D., Motta, A., & Rose, F. D. (1998). Active versus passive exploration of virtual environments modulates spatial memory in MS patients: A yoked control study. *Italian Journal of Neurological Sciences 19*, S424–S430.

Pugnetti, L., Mendozzi, L., Motta, A., Cattaneo, A., Barbieri, E., & Brancotti, S. (1995a). Evaluation and retraining of adults' cognitive impairments: Which role for virtual reality technology? *Computers in Biology and Medicine, 25*(2), 213–227.

Pugnetti, L., Mendozzi, L., Motta, A., Cattaneo, A., Barbieri, E., Brancotti, A., & Cazzullo, C. L. (1995b) *Immersive VR for the retraining of acquired cognitive defects*. Paper presented at the Medicine Meets Virtual Reality 3 Conference. San Diego, CA.

Renner, M. J., & Rosenzweig, M. R. (1987). *Enriched and impoverished environments: Effects on brain and behaviour*. New York: Springer-Verlag.

Riva, G. (1998). Virtual reality in neuroscience: A survey. In G. Riva, B. K. Wiederhold, & E. Molinari (Eds.), *Virtual environments in clinical psychology and neuroscience: Methods and techniques in advanced patient–therapist interaction* (pp. 191–199). Amsterdam: IOS Press.

Rizzo, A. A. (1994). Virtual reality applications for the cognitive rehabilitation of persons with traumatic head injuries. In Murphy, H. J. (Ed.), *Proceedings of the Second International Conference on Virtual Reality and Persons with Disabilities* (pp. 135–140). Northridge: California State University.

Rizzo, A. A., & Buckwalter, J. G. (1997). Virtual reality and cognitive assessment and rehabilitation: The state of the art. In G. Riva (Ed.), *Psycho-neuro-physiological assessment and rehabilitation in virtual environments: cognitive, clinical, and human factors in advanced human–computer interactions* (pp. 123–146). Amsterdam: IOS Press.

Rizzo, A. A., Buckwalter, J. G., Neumann, U., Chua, C., Van Rooyen, A., Larson, P., Kratz, K., Kesselman, C., Thiebaux, M., & Humphrey, L. (1999). Virtual environments for targeting cognitive processes: An overview of projects at the University of Southern California. *CyberPsychology and Behavior, 2*(2), pp. 89–100.

Rizzo, A. A., Buckwalter, J. G., Neumann, U., Kesselman, C., & Thiebaux, M. (1998a). Basic issues in the application of virtual reality for the assessment and rehabilitation of cognitive impairments and functional disabilities. *CyberPsychology and Behavior, 1*(1), 59–78.

Rizzo, A. A., Buckwalter, J. G., Neumann, U., Kesselman, C., Thiebaux, M., Larson, P., & van Rooyan, A. (1998b). The virtual reality mental rotation/spatial skills project: Preliminary findings. *CyberPsychology and Behavior, 1*(2), 107–113.

Rizzo, A. A., Wiederhold, M. D., & Buckwalter, J. G. (1998c). Basic issues in the use of virtual environments for mental health applications. In G. Riva, B. K. Wiederhold, & E. Molinari (Eds.), *Virtual environments in clinical psychology and neuroscience: Methods and techniques in advanced patient–therapist interaction* (pp. 21–42). Amsterdam: IOS Press.

Rizzo, M., Reinach, S., McGehee, D., & Dawson, J. (1997). Simulated car crashes and crash predictors in drivers with Alzheimer's disease. *Archives of Neurology, 54*, 545–551.

Rose, F. D., Brooks, B. M., Attree, L. A., Parslow, D. M., Leadbetter, A. G., McNeil, J. E., Jayawardena, S., Greenwood, R. J., & Potter, J. (1999). A preliminary investigation into the use of virtual environments in memory retraining of patients with vascular brain injury: Indications for future strategy? *Disability and Rehabilitation, 21*(12) 548–554.

Rose, F. D., & Johnson, D. A. (1992). *Brain injury and after*. Chichester, England: John Wiley & Sons.

Rose, F. D. (1996). Virtual reality in rehabilitation following traumatic brain injury. In P. Sharkey (Ed.), *Proceedings of the First European Conference on Disability, Virtual Reality and Associated Technology* (pp. 5–12). Maidenhead, England: University of Reading.

Rushton, S. K., Coles, K. L, & Wann, J. P. (1996). Virtual reality technology in the assessment and rehabilitation of unilateral neglect. In P. Sharkey (Ed.), *Proceedings of the First European Conference on Disability, Virtual Reality and Associated Technology* (pp. 227–231). Maidenhead, England: University of Reading.

Sandstrom, N. J., Kaufman, J., & Huettel, S. A. (1998). Males and females use different distal cues in a virtual environment navigation task. *Cognitive Brain Research, 6*, 351–360.

Schacter, D. L., & Glisky, E. L. (1986). Memory remediation: Restoration, alleviation, and the acquisition of domain-specific knowledge. In B. Uzzell & Y. Gross (Eds.), *Clinical neuropsychology of intervention*. Boston: Martinus Nijhoff.

Shepard, R. N., & Metzler, J. (1971). Mental rotation of three-dimensional objects. *Science, 171*, 701–703.

Skelton, R. W., Bukach, C. M., Laurance, H. E., Thomas, K. G. F., & Jacobs, W. J. (1999). *Humans with traumatic brain injuries show place-learning deficits in computer-generated virtual space*. Manuscript in preparation.

Slater, M., & Usoh, M. (1993). The influence of a virtual body on presence in immersive virtual environments. In *Virtual Reality '93: Proceedings of the Third Annual Conference on Virtual Reality* (pp. 34–42). Westport, CT: Meckler.

Sohlberg, M. M., & Mateer, C. A. (1987). Effectiveness of an attention training program. *Journal of Clinical and Experimental Neuropsychology 9*, 117–130.

Sohlberg, M. M., & Mateer, C. A. (1989). *Introduction to cognitive rehabilitation: Theory and practice*. New York: Guilford Press.

Stanney, K. M., Salvendy, G., Deisigner, J., DiZio, P., Ellis, S., Ellison, E., Fogleman, G., Gallimore, J., Hettinger, L., Kennedy, R., Lackner, J., Lawson, B., Maida, J., Mead, A., Mon-Williams, M., Newman, D., Piantanida, T., Reeves, L., Riedel, O., Singer, M., Stoffregen, T., Wann, J., Welch, R., Wilson, J., & Witmer, B. (1998). Aftereffects and sense of presence in virtual environments: Formulation of a research and development agenda. *International Journal of Human–Computer Interaction, 10*(2), 135–187.

Stanton, D., Foreman, N., & Wilson, P. N. (1998). Uses of virtual reality in clinical training: Developing the spatial skills of children with mobility impairments. In G. Riva, B. Wiederhold, & E. Molinari (Eds.), *Virtual reality in clinical psychology and neuroscience* (pp. 219–232). Amsterdam: IOS Press.

Sternberg, R. J. (1997). Intelligence and lifelong learning: What's new and how can we use it? *American Psychologist, 52*(10), 1134–1139.

Stokes, T. F., & Baer, D. M. (1977). An implicit technology of generalization. *Journal of Applied Behavior Analysis, 10*, 349–367.

Strickland, D. (1996). A virtual reality application with autistic children. *Presence: Teleoperators and Virtual Environments, 5*(3), 319–329.

Strickland, D. (1997). VR and health care. *Communications of the ACM, 40*(8), 78.

Thomas, K. G. F., Laurance, H. E., Luczak, S. E., & Jacobs, W. J. (1999). Age-related changes in a human cognitive mapping system: Data from a computer-generated environment. *CyberPsychology and Behavior, 2*(5), 545–566.

Trepagnier, C. (1999). Virtual environments for the investigation and rehabilitation of cognitive and perceptual impairments. *NeuroRehabilitation, 12*, 63–72.

Vandenberg, S. G., & Kuse, A. R. (1978). Mental rotations: A group test of three-dimensional spatial visualization. *Perceptual and Motor Skills, 47*, 599–604.

Van Praag, H., Kempermann, G., & Gage, F. H. (1999). Running increases cell proliferation and neurogenesis in the adult mouse dentate gyrus. *Nature Neuroscience, 2*(3), 266–270.

Waller, D., Hunt, E., & Knapp, D. (1998). The transfer of spatial knowledge in virtual environment training. *Presence: Teleoperators and Virtual Environments, 7*(2), 129–143.

Wann, J. P. (1996) Virtual reality environments for rehabilitation of perceptual-motor disorders following stroke. In P. Sharkey (Ed.), *Proceedings of the First European Conference on Disability, Virtual Reality and Associated Technology* (pp. 233–238). Maidenhead, England: The University of Reading.

Wann, J. P., & Mon-Williams, M. (1996). What does virtual reality NEED? Human factors issues in the design of three-dimensional computer environments. *International Journal of Human–Computer Studies, 44*, 829–847.

Wann, J. P., Rushton, S. K., & Mon-Williams, M. (1995). Natural problems for stereoscopic depth perception in virtual environments. *Vision Research, 19*, 2731–2736.

Wann, J. P., Rushton, S. K., Smyth, M., & Jones, D. (1997). Virtual environments for the rehabilitation of disorders of attention and movement. In G. Riva (Ed.), *Virtual Reality in neuro-psycho-physiology: Cognitive, clinical, and methodological issues in assessment and rehabilitation* (pp. 157–164). Amsterdam: IOS Press.

Welch, R. B., Blackmon, T. T., Liu, A., Mellers, B. A., & Stark, L. W. (1996). The effects of pictorial realism, delay of visual feedback, and observer interactivity on the subjective sense of presence. *Presence: Teleoperators and Virtual Environments, 5*(3), 263–273.

Wiederhold, B. K., & Wiederhold, M. D. (1998). A review of virtual reality as a psychotherapeutic tool. *CyberPsychology and Behavior, 1*(1), 45–52.

Wilson, B. A. (1997). Cognitive rehabilitation: How it is and how it might be. *Journal of the International Neuropsychological Society, 3*(5), 487–496.

Wilson, P. N. (1993). Nearly there. *Special Child, 68*, 28–30.

Wilson, P. N., Foreman, N., Gillet, R., & Stanton, D. (1997). Active versus passive processing of spatial information in a computer simulated environment. *Ecological Psychology, 9*, 207–222.

Witmer, B. G., & Singer, M. J. (1998). Measuring presence in virtual environments: A Presence Questionnaire. *Presence: Teleoperators and Virtual Environments, 7*(3), 225–240.

Wraga, M., Creem, S. H., & Proffitt, D. R. (1997). Imagined viewer vs. object rotations: Is mental rotation a general ability? [Online]. Available: http://www.Virginia.EDU/~perlab/presenta.html [July 5, 1999].

# 51

# Virtual Reality Therapy: An Effective Treatment for Psychological Disorders

Max M. North,[1] Sarah M. North,[2] and Joseph R. Coble[2]
[1]*Kennesaw State University*
*1000 Chastain Road, Kennesaw, Georgia 30144*
*max@acm.org*
[2]*Clark Atlanta University*
*223 James P. Brawly Drive, Atlanta, Georgia 30314*
*sarah@acm.org*
*jrcoble@prodigy.com*

## 1. INTRODUCTION

Virtual environment (VE) technology is currently being used to treat phobias and other psychological disorders. This chapter reports a brief collection of pioneering and current research activities that introduce an innovative paradigm in psychotherapy that is termed virtual reality therapy (VRT).

Behavioral therapy techniques for treating phobias and other psychological disorders include exposing the patient to anxiety-producing stimuli on a graded scale, known as systematic desensitization. These stimuli can be generated either through the patient's imagination or in actual real-life situations. In using systematic desensitization techniques, many patients appear to have difficulty imagining the prescribed anxiety-evoking scenes. They also seek to avoid experiencing real-life situations that trigger their fears. As a learned behavior, this avoidance lowers their anxiety, reducing public embarrassment but it does nothing to relieve them of their fears.

VRT can overcome some of the difficulties inherent in the traditional treatment of phobias. VRT, like current imaginal and in vivo modalities, can generate stimuli that are useful in desensitization therapy. VRT can provide stimuli for patients who have difficulty imagining fear-producing triggers and/or are too phobic to undergo real-life experiences in public. Unlike in vivo systematic desensitization, VRT can be performed within the privacy of a room, thus avoiding public embarrassment and violation of patient confidentiality. VRT can generate stimuli of much greater magnitude than standard in vivo techniques. Since VRT is under the control of the patients, it may appear safer to them and, at the same time, more realistic than imaginal desensitization. VRT also has the advantage of greater efficiency and economy in delivering the equivalent of in vivo systematic desensitization in the therapist's office.

The idea of using VE technology to combat psychological disorders was conceived within the Human–Computer Interaction Group at Clark Atlanta University in November 1992. Since then we have successfully conducted pilot experiments in the use of VE technologies for the treatment of specific phobias: fear of flying (North & North, 1994; North, North, & Coble, 1996b, 1996a, 1997); fear of heights (North & North, 1996; North et al., Coble, 1996b; Rothbaum et al., 1995); fear of being in certain situations (such as a dark barn, an enclosed bridge over a river, and in the presence of an animal [e.g., a black cat] in a dark room, North et al., 1995b, 1995a, 1995a, 1996c); fear of public speaking (North et al., 1996b, 1997); and obsessive–compulsive disorder.

# 2. APPLICATIONS OF VRT

## 2.1 Fear of Flying

Two case studies show the effectiveness of VRT in treating the fear of flying. The first experiment, conducted in late November 1992, had to do with a 32-year-old married woman, a human–computer interaction group researcher. She was diagnosed and treated for fear of flying using an existing virtual scene. The virtual scene was a simulated city running on a Silicon Graphics computer. This scene originally was created for research on an innovative navigational technique for virtual environments. The subject participated in eight sessions, each lasting about 30 minutes. She reported a high level of anxiety at the beginning of each session, gradually reporting lower anxiety levels after remaining in the virtual situation for a few minutes, and eventually reported an anxiety level of zero. To investigate the transfer effect of VRT to the real world she was flown, in the company of the therapist, in a helicopter for approximately 10 minutes at low altitude over a beach on the Gulf of Mexico. As with the VRT sessions, she reported some anxiety at the beginning but rapidly grew more comfortable. Now the subject comfortably flies for long distances and experiences much less anxiety (North & North, 1994; North et al., 1996b, 1996a, 1997).

In September 1995, a 42-year-old married man who conducts research at Clark Atlanta University sought treatment for his fear of flying. The subject's anxiety and avoidance behaviors were interfering with normal vocational and personal activities. For example, he was unable to travel to professional conferences, visit relatives, or take a vacation by air. The subject, accompanied by a virtual therapist, was placed in the cockpit of a VE helicopter and flown over a simulated city for five sessions. The modified 11-point (0 for complete calm and 10 for complete panic) Subjective Units of Discomfort (SUD) scale was used to measure the degree to which the subject was affected by VRT. In VRT treatment, the subject's anxiety usually increased as he was exposed to more challenging situations and decreased as the time in those new situations was increased. The subject experienced a number of physical and emotional anxiety-related symptoms during the VRT sessions, including sweaty palms, loss of balance, and weakness in the knees. The VRT resulted in both a significant reduction in anxiety symptoms and the ability to face the phobic situation in the real world. The subject at this time is able to fly with reasonable comfort (North et al., 1996a, 1996b, 1997).

Virtual environment software developed for treatment of fear of flying was used in two additional case studies. Hodges et al. (1999) and Wiederhold, Gevirtz, and Wiederhold (1998) reported the successful treatment of two female patients suffering from this phobia. During 20 minutes of VRT, the patients wore a head-mounted display (HMD) and viewed three-dimensional (3-D) computer-generated images of several flying scenes, including sitting in a plane with the engines off, sitting in a plane with the engines on, pushing back to taxi, taxiing down the runway, taking off, flying in good weather, flying in turbulent weather, and landing. Self-report scores on questionnaires relating to fear of flying and attitudes toward flying were positively changed after the VRT session.

## 2.2 Agoraphobia

We conducted a controlled comprehensive study on the effectiveness of VE technology in the treatment of psychological disorders, beginning in1993. Specifically, we assessed the effectiveness of VEs in the treatment of agoraphobia, the fear of being in places or situations from which escape may be difficult or embarrassing. Sixty participants were selected for this study. Thirty were placed in the experimental group and 30 were placed in the control group.

The VE system for this study consisted of a stereoscopic HMD (CyberEye, General Reality Company), an electromagnetic head-tracker (Flock of Birds, Ascension Technology), and a DataGlove (Power Glove) worn by the participant to allow interaction with objects in the virtual environment. Interactive imagery was generated by software developed at the Human–Computer Interaction Group and Virtual Reality Technology Laboratory at Clark Atlanta University, executed on a Silicon Graphics Workstation and Pentium Personal Computer (VREAM Virtual Reality Development Software). Several VE scenes were created for use in the therapy sessions, including these:

- A series of Balcony Scenes, consisting of four balconies attached to a tall building. The balconies were at ground level, second-floor level (6 meters), 10-floor level (30 meters), and 20-floor level (60 meters).
- Empty Room Scene, consisting of four walls, ceiling, and floor. The room was 4 by 6 meters in size, with only one door (entrance and exit). There were no windows or furniture in this room.
- Dark Barn Scene, consisting of a barn in an open field. The interior of the barn was black (simulating darkness). The barn had a wide door so participants could exit as quickly as they wished. There were several dark-colored objects inside the barn.
- Dark Barn with a Black Cat Scene. A black cat was simulated within the dark barn. The black cat was placed on top of an object but was not visible from outside the barn. In order to see the black cat, the participant had to enter the barn and look to the right.
- Covered Bridge Scene, containing a bridge (10 meters high) that ran across a river, had walls on each side and a ceiling. There were two windows on each of the walls. Dark colors were used to simulate a closed environment.
- Elevator Scene, consisting of an open elevator (no walls or ceiling) located in the atrium of a 49-story hotel.
- Canyon with Bridges Scene, containing three bridges of different heights spanning the canyon from one side to the other. A river ran through the bottom of the canyon. The bridges varied not only in height but also in apparent steadiness. The lowest two bridges (7 and 50 meters) appeared safe and solid. The highest bridge (80 meters) was a rope bridge with widely spaced wooden slats as the flooring.
- Balloons Scene, consisting of three hot-air balloons at different heights. The first balloon was 20 meters high, the second balloon 30 meters high, and the third balloon 40 meters high. A one-floor building and a four-floor building were in the scene.

Only participants in the experimental group were exposed to the VRT treatment. In the initial laboratory session, which lasted approximately 20 minutes per participant, the VRT participants were familiarized with the VE equipment. During this session, VRT participants also eliminated any of the eight scenes described above which did not cause anxiety and ranked the remaining scenes from least to most threatening. For the VRT participants' subsequent eight sessions, which were 15 minutes each for 8 weeks (one session per week), individual VE desensitization was conducted in a standard format. The first session began with the least threatening level of the participants' least threatening scene. The SUD was administered every 5 minutes. Participants progressed systematically through each level of a scene, and then moved to their next most threatening scene. This progress was totally under the control of the participant, with the exception that if a participant's SUD score was two or less, the experimenter urged the participant to move up to the next level or next scene. Each session after the first began at the scene and level at which the participant's previous session ended. After eight weeks, a posttest of all participants in the experimental and control groups was conducted to obtain Attitude Towards Agoraphobia Questionnaire (ATAQ) and SUD scores.

VRT appeared to be effective in the treatment of individuals with agoraphobia. Negative attitudes toward agoraphobic situations decreased significantly for the VRT group but not for the control group. The average SUD scores decreased steadily across sessions, indicating habituation.

For a complete and detailed report of this study, readers may refer to the publications in the *Virtual Reality Therapy Book* (North & North, 1996), *CyberEdge Journal* (North, North, & Cobb, 1995b), *International Journal of Virtual Reality* (North, North, & Cobb, 1995a), or *Presence: Teleoperators and Virtual Environments* (North, North, & Cobb, 1996c).

## 2.3 Acrophobia (Fear of Heights)

Twenty college students took part in the first acrophobia study, conducted in collaboration with others (Rothbaum, Hodges, Kooper, et al., 1995a; Rothbaum, Hodges, Opdyke, et al., 1995b). The goal of this study was to investigate the efficacy of VE graded exposure in the treatment of acrophobia. Twenty college students who had reported fear of heights were randomly assigned to VE-graded exposure treatment or to a waiting-list comparison group. Sessions were conducted individually over 8 weeks. Outcome was assessed by using measures of anxiety, avoidance, attitudes, and distress associated with exposure to heights before and after treatment. Significant differences between the participants who completed the VE treatment and those in the waiting list were found on all measures.

Another case study also demonstrated the effectiveness of VRT in the treatment of acrophobia. After obtaining informed consent, the subject was asked to rank order a list of acrophobic situations according to the degree of anxiety arousal. During the subject's first session he was familiarized with VE technology through several demonstrations. For the subject's subsequent eight sessions, which were between 15 and 28 minutes each, individual VRT treatment was conducted in a standard format. The first session began with the least threatening level, which was at ground level near a bridge crossing a river in the middle of a simulated town. The SUD was administered periodically every 2 to 5 minutes. The progress was totally under the control of the subject, with the exception that if the subject's SUD score was zero the experimenter urged the subject to move up to the next level of the scene. At one month after treatment, the subject was asked to complete a 10-point scale (including degrees for worsening symptoms) rating the degree to which his acrophobia symptoms had changed since a pretreatment test (SUD). The results indicated significant habituation of the subject with respect to both anxiety symptoms and avoidance of anxiety-producing situations. Thus, we concluded that VRT treatment was successful in reducing fear of heights (North & North, 1996; North, 1996b).

## 2.4 Fear of Public Speaking

The fear of public speaking, sometimes referred to as social phobia, is a common disorder that affects many people in the world. Research into this widespread phobia was conducted through the collaboration of Clark Atlanta University (CAU), the U.S. Army Research Laboratory, and Boeing Computer Services, with special technical assistance from the Speech Improvement Company, Inc. The research may have far reaching benefits. With the assistance of the Speech Improvement Company, it may have a positive impact on clinical sessions of individuals suffering from the fear of public speaking. Additionally, the general population may benefit from this research because new technology could provide greater access to a safe, confidential, and economical approach to the treatment of psychological disorders.

Participants in the research were recruited from CAU Introductory Psychology classes. After an extensive two-stage screening process, 16 participants were selected from the pool. They were assigned to two treatment components: virtual reality therapy and a comparison group. Selected participants reported experiencing anxiety and avoidance behavior that was interfering with their normal activities. They were unable to participate in social gatherings, classes, or professional conferences.

Apparatus for this study consisted of a Pentium-based computer and an HMD with head-tracker (Virtual–I/O). Modeling was done by VREAM Company under the direction of the researchers. A model of an auditorium located in the CAU Research Science Building was created. The virtual auditorium is 48 feet wide, 100 feet long, and 55 feet high. The seating area has three sections of chairs and can accommodate over 100 people. Specialized features created for the facility include a virtual wooden podium with a speaker's stand. An amplifier with direct connection to the VE software and hardware were used in the therapy sessions. This enabled participants to hear the echo of their voices. Simulation of the real echo in the auditorium was created by a headphone attached to the HMD. The treatment schedule consisted of five weekly sessions. The sessions lasted 10 to 15 minutes.

Two assessment measures were used in this study. The first measure was the Attitude Towards Public Speaking (ATPS) questionnaire, which contains six items for assessment that range from 0 to 10 on a semantic differential scale. The second measure used was the SUD scale.

Symptoms experienced by participants during VRT included an increase in heart rate, feeling a lump in the throat, dry mouth, sweaty palms, loss of balance, weakness in the knees, and other symptoms. These symptoms also appeared in the studies that dealt with the treatment given for acrophobia, agoraphobia, and fear of flying.

Virtual reality therapy appears to be very effective in reducing self-reported anxiety. The VRT treatment resulted in both a significant reduction of anxiety symptoms (SUD and ATPS measurements) and the ability to face phobic situations in the real world. At this time several of the participants can comfortably speak in front of a crowd with improved confidence.

## 2.5 Posttraumatic Stress Disorder (PTSD)

Posttraumatic stress disorder (PTSD) is one of the most disabling psychopathological conditions affecting military veterans. An estimated 830,000 U.S. veterans currently have symptoms of chronic combat-related PTSD. Hodges et al. (1996) report an experiment in which virtual reality exposure (VRE) is used as an alternative to typical imaginal exposure treatment for Vietnam combat veterans with PTSD. A patient was exposed to two virtual environments, a virtual Huey helicopter flying over a virtual Vietnam countryside and a clearing surrounded by jungle. The patient was encouraged to approach the landed aircraft and fly over jungle, river, and rice paddies. The fly-through included sounds of the rotary blades, combat noises, jungle noises, explosions, and men's voices. The pretreatment and posttreatment assessment technique was used for this case study. The patient experienced a 34% decrease on clinician-rated PTSD and a 45% decrease on self-rated PTSD after VRE treatment.

## 2.6 Claustrophobia

Claustrophobia is a cousin to agoraphobia but has its own peculiar characteristics. Botella et al. (1998) report VE to be effective the treatment of patients suffering from claustrophobia. The selection of claustrophobia for study was based on its unique characteristics. Clinically, claustrophobia is the functional equivalent of agoraphobia; however, there are fewer avoidance situations (Barlow, 1988; Booth & Rachman, 1992). Moreover, by itself claustrophobia may place many limits on its victims and severely impair the quality of their life. This is particularly evident in that closed places are a part of daily life. People commonly encounter elevators, planes, tunnels, cabins, tests for medical diagnosis and therapy such as computed tomography scan and magnetic resonance imaging and the like.

To increase the levels of difficulty in the "claustrophobic" environment, two different settings were created. The first was a house and the second an elevator. In each setting, several different environments existed that permitted researchers to construct performance hierarchies, with degrees of increasing difficulty.

These were the settings for VE scenes in this experiment:

1. A 4 × 5 meter room that has a door that exits to an external terrace and 2 × 4 meter garden. There is also a large window with blinds in the room. When and window are open, a blue sky can be seen and the sound of birds can be heard. The door, window, and blinds can be opened and closed in three stages. Another door in the room opens to a second room or second claustrophobic setting.
2. A 3 × 3 meter room in which there is no furniture or windows. The ceiling and floor are darker with a wood texture that gives the sensation of greater closure. In the center of the room there is as lectern with several buttons that permit the patient to interact with the virtual environment. The patient can also close the door of the room in three stages. Once it is closed, the patient can block it by pushing one of the buttons on the lectern. Additionally, a wall in the room can be displaced at the will of the patient by making a loud noise. The wall can also be moved by manipulating buttons on the lectern and by keeping the patient enclosed in a 1-meter-wide space.
3. A wide entry with a large window in which the large window cannot be opened; however, the window permits one to see plants growing outside. From this entry, the patient has access to the elevator by pressing a button.
4. The elevator is designed to offer different claustrophobic possibilities and threats. These threats take into account various parameters (i.e., size, placing, and the possibility of blocking the elevator). All of the possibilities created by the big elevator may be realized through the use of a small elevator.

At first, the hardware used in the experiment was a Silicon Graphics Indigo High Impact computer graphics workstation. To extend the commercial use of this technology, the VE system used in clinical psychology was adapted and used with a less-expensive personal computer. Currently, work is being conducted on a PC Pentium Pro with 128 Mg of RAM and a graphic card AccelEclipse of 32 Mg, which is being used as a hardware platform. The software used are Windows NT and World Up (Sense8).

This research has demonstrated the effectiveness of the software. It can be stated, with confidence, that the claustrophobic context can activate a high degree of anxiety in patients and through virtual exposure victims of claustrophobia can be threatened and helped to overcome this debilitating phobia.

These results were achieved by using virtual exposure as a unique technique. This therapy was not used in conjunction with any other psychological treatment techniques. Therefore, VE appears to be very useful as a therapeutic treatment strategy.

### 2.7 Obsessive–Compulsive Disorders

Approximately five million people suffer from obsessive–compulsive disorder (OCD) in the United States alone. One in 50 Americans suffers from OCD, and because of the disorder they are forced to alter the normal activities in their lives. Men, women, and children of all races and socioeconomic backgrounds are affected by OCD.

Fear of contamination, fear of making mistakes, and fear of harm to another person are the most common obsessions. Common compulsions include cleaning and washing, arranging and organizing, collecting, counting, and repeating. The most common treatments for OCD are medication and behavior therapy.

Most victims of the disorder are ashamed of their uncontrollable repetitive behavior and are very concerned about what other people, who have seen their obsessive behavior, think of them. VRT has the potential to reduce a patient's embarrassment by providing privacy. VRT also offers the patient a chance to concentrate on the problem. This process may help the patient reduce information overload. It also assists the patient in releasing cognitive resources to seek an alternative to the obsessive–compulsive behavior.

North and North (2000) combined their VRT technical advancements with a program developed by Marc Degroot of Immersive Company in California to create a treatment for individuals suffering from OCD. This treatment was evaluated in a case study. Following an extensive two-stage screening process, a subject was recruited from a Clark Atlanta University Introductory Information Technology class for the study. The subject was placed in a virtual room that contained several books, pencils, and other school equipment. The treatment was administered over eight weekly sessions. The sessions lasted 10 to 15 minutes apiece. The subject was asked to pick items from the virtual room that she needed for a day at school and place them in a virtual book bag. She was asked to repeat this process until she no longer had doubts about what she had collected for her classes. She was asked to leave the virtual room and go outside the virtual house. She was then asked if she was sure about the collected items in her book bag. If she had a doubt and her anxiety level was high, she was given a chance to repeat the process until her anxiety level reached zero (no anxiety). The symptoms experienced by the subject during VRT mirrored those that most OCD patients experience during a real situation. They included obsessive and compulsive thoughts and behavior, an increase in heart rate, sweaty palms, loss of balance, and weak knees.

This OCD study indicated that VRT is very effective in reducing self-reported anxiety. The VRT treatment resulted in both a significant reduction of anxiety symptoms (SUD and ATOCD measurements) and the ability to face phobic situations in the real world. The treated subject now reports the ability to comfortably go about her daily schooling without excessive thoughts about the items she needs to have for each class.

### 2.8 Attention Deficit Disorders

Attention deficit disorder (ADD) is common among children. Current research indicates that between 5% and 10% of children suffer from this disorder. Many of these children are highly intelligent and creative despite wandering attention and impulsiveness. Because they have problems with short-term memory and short attention spans, they have a hard time completing tasks. They have trouble concentrating, being easily distracted, mostly by sights and sounds. They especially exhibit problems in a group situation (Hunsucker, 1988). In general, ADD patients' problems are physical, academic, behavioral, emotional, and social.

Traditional treatment for ADD includes medications such as Methylphenidate (Ritalin), D-Amphetamine (Dexedrine), and Pemoline (Cylert). Behavioral management also includes punishment and reward techniques.

Several research studies reported in this chapter have shown that VRT provides specific stimuli that can be used in removing distractions and providing environments that get the subjects' attention and increases their concentration. Virtual environments technology can hold a patient's attention for a longer period of time than other methods can. Virtual scenes may enhance the patient's short-term memory and increase attention span. Researchers at CAU's Virtual Reality Technology Laboratory are developing several classroom situation scenes to extensively test these hypotheses. An informal pilot study conducted in this lab showed that a patient's mild ADD symptoms decreased while under VRT. Although we have used only visual stimuli, the preliminary observations are very encouraging. There is good evidence from our previous research that this kind of experience may be easily transferred to everyday activities. The ongoing ADD research will be extended to include other stimuli such as auditory and tactile. A scenario for social interaction is also under development to test the effects of VRT in teaching patients interactions needed to improve their social skills.

## 2.9 Autism

Virtual reality therapy has the potential of helping children with autism. This is a disorder that involves an internal distortion of the external world. Unfortunately, there are few if any effective treatment methods for children with the kind of serious learning problems that autism produces. Experiments with VRT indicate it may offer some progress toward an effective treatment. Virtual reality therapy allows controlled distortion of the environment in a way that almost matches the perception of an individual. Two intensive case studies were conducted to determine if children with autism could respond to VRT and to investigate if they might benefit from environments that such therapy can provide. Both cases showed that VRT can provide a sense of presence (see chap. 40, this volume), which in turn allows a safe and controlled learning environment for individuals afflicted by autism (Strickland, 1996).

## 2.10 Body Image Experience

How one views his or her own body can have an effect on one's health. Eating disorders, one of the most common pathological concerns of Western society, have long been associated with alterations in perceptual–cognitive representations of the body. Many studies have shown that the perception of one's body and the experiences associated with it represent one of the key problems of anorexic, bulimic, and obese people. These perceptions also have a strong influence on therapy. Unfortunately, the treatment of body experience problems is not well defined. Both cognitive and behavioral approaches have an influence on the level of body awareness. Virtual Environment for Body Image Modification (VEBIM) treatment consists of a set of tasks aimed at altering a patient's body image. It attempts to integrate both cognitive and behavioral therapeutic approaches within an immersive virtual environment. A study on a preliminary sample of sixty normal subjects was conducted to test the efficacy of this approach. The study showed that even a short-term application of VEBIM can partially reduce the level of body dissatisfaction without any major side effects. Nevertheless, an alteration of one's body image toward a more realistic perception might be decisive for the long-term outcome (Riva, 1997a, 1997b).

### 2.11 More Research on Virtual Reality Graded Exposure Therapy

Wiederhold and her associates have established the Center for Advanced Multimedia Psychotherapy (CAMP) in San Diego, California. The Center assists patients who suffer from phobias by using state-of-the-art VRT systems to combat a variety of phobias (Wiederhold & Wiederhold, 1998).

At CAMP they have investigated the use of virtual reality graded exposure therapy (VRGET) for the treatment of fear of flying (Wiederhold, Gevirtz, & Wiederhold, 1998). In one of the first studies, three groups were compared. One group received traditional imaginal exposure therapy (IET). The participants in this group, with the therapist's assistance, constructed an individualized fear hierarchy and progressed through the hierarchy based on SUDSs levels. One group, who received VRGET, also progressed through the virtual world based on SUDs levels. The other group, which received virtual reality exposure therapy plus physiological feedback (VRETpm), was progressed through the virtual world based on skin resistance levels. Heart rate, respiration rate, skin resistance, peripheral skin temperature, and brain wave activity were monitored and measured for all three groups. However, the pm group received information on physiology (biofeedback). Prior to beginning exposure therapy, all three groups were given two sessions of relaxation and breathing retraining to use as a coping mechanism while performing the exposure therapies and to use later while actually taking a real airplane flight. All three groups completed questionnaires relating to fear of flying, attitudes toward flying, and state and trait anxiety prior to beginning relaxation training, after two sessions of relaxation training, and after six sessions of exposure therapy. A 3-month posttreatment follow-up was done with all participants to see if flying behavior had changed following treatment. The study was completed in February 1999. The group receiving VRETpm was 100% successful (with success measured as ability to fly with no medication and no alcohol when contacted at 3 month posttreatment follow-up). The VRGET group receiving no physiological feedback was 80% successful, and the IET group was 20% successful.

A study that began in October 1998 is looking at the efficacy of using VRGET to treat the fear of driving. The same protocol used in the fear of flying study is being utilized in the fear of driving study. Two sessions of relaxation training are followed by six sessions of exposure therapy. This study does not, however, include an imaginal group.

A third study that began in February 1999 is investigating the use of VRGET in patients with neurovestibular disorders (see chap. 7, this volume). It is hoped that VRGET will help desensitize these patients to complex visual stimuli, which often provoke dizziness and vertigo.

## 3. VRT ISSUES

### 3.1 Assertions Concerning VRT

We can now make these assertions concerning virtual reality therapy based on the data collected and subjects' verbal reports concerning the VRT experiments.

- A person's experience of a situation in a virtual environment may evoke the same reactions and emotions as the experience of a similar real-world situation.

People who are agoraphobic in the real world are also agoraphobic in the virtual world. This assertion is based on results of our research studies on psychological treatment categories. When subjected to virtual phobic-invoking situations, participants exhibited the same types of responses as would be exhibited in real-world situations. These responses included anxiety, avoidance, and physical symptoms.

Subjects were repeatedly asked to rate their current level of anxiety on a SUD scale. The relatively high SUD scores at the beginning of each treatment session indicated that the subjects' fear structures were invoked, and SUD scores (and thus fear levels) gradually decreased as subjects remained in the virtual scene.

Actual behavior and verbalization were a second measure of anxiety. Examples of common subject behavior included subjects tightly gripping support rails and displaying reluctance to let go of the rails. Verbal expressions included such comments as these: "The higher I get, the more worried I get," "I am really there!" "It feels like being in a real helicopter," "I am afraid to fall down!" "I do not like this at all!" "I am scared!" and "I feel like I am actually on the fiftieth floor!"

Physical symptoms reported by subjects included shakiness in the knees, heart palpitations, tenseness, sweaty palms, and dizziness.

- A person may experience a sense of virtual presence similar to the real world even when the VE does not accurately or completely represent the real-world situation.

Remarkably, subject reactions consistent with phobic stimuli were experienced in spite of the fact that their virtual experience did not correspond to the real-world experience in several ways. All environments were visually extremely less detailed than a real scene, and some environments included much simpler auditory and tactile cues, such as engine sound and vibration of the Apache AH64 helicopter in the fear of flying study.

As stated previously, the subjects reported a number of physical and emotional anxiety-related symptoms, such as dizziness, sweaty palms, heart palpitations, etc. Subjects would likely not have reported these feelings if they did not perceive that they were experiencing a realistic situation, even though the VEs were not exact copies of real world scenes.

- Each person brings his or her own background into a VE experience.

Perception is in many ways just as much a product of previous experiences as of current stimulation. Each subject is a unique, specific individual, with independent experiences of reality, which are unique and different from the objective world, or so-called world of reality. The implication for VE is that the sense of virtual presence (see chap. 40, this volume) is dependent not only on the physical qualities (resolution, realism, interactivity, lag time) of the experience provided by the virtual environment, but also on what the participant psychologically brings to the environment. The very nature of the act of perception causes each person to react differently to the same real or virtual experience.

This was evidenced by SUD, ATAQ, and ATPS scores and verbal comments of subjects. Just as different individuals may react differently to a real world experience, subjects exhibited different reactions to the same virtual world experience. This point was clearly demonstrated by the variety of responses among subjects to the same phobic stimuli of the virtual scene. Several subjects went through several levels of the phobic situations without reporting any significant anxiety. On the other hand, many subjects reported differing amounts of anxiety in different levels of the virtual scene. There was major variation in the amount of time spent in each level of the virtual scenes by different subjects.

- Experience with a VE increases a participant's sense of virtual presence.

The idea that a sense of presence may increase with experience has been suggested by several researchers (Held & Durlach, 1992; Loomis, 1992; 1993; Naiman, 1992; see chap. 40, this volume). Our experiments verified this hypothesis in that the longer subjects stayed in the virtual scene the deeper they were pulled into the virtual world and the greater their sense of virtual presence.

Based on subjects' SUD, ATAQ, and ATPS scores and verbal comments during the experiments, subjects initially felt some level of virtual presence in the phobic situation, and their sense of virtual presence increased over time or at worst was maintained during all the sessions.

- The sense of presence in virtual and physical environments is constant and subjects have to give up the sense of presence in one environment (e.g., physical environment) to achieve a stronger sense of presence in the other one (e.g., virtual world).

This assertion is based on the data (SPSVP—Sense of Presence Scale in the Virtual and the Physical environments questionnaire—and SUD) collected from subjects. Specifically, the SPSVP was designed to assess one's sense of presence of the VE and physical environment, sense of interactivity with the VE system, and one's perception of the real world in comparison to the virtual environment. The subjective measures of sense of presence in the VE increased gradually during each session. The subjective measures of sense of presence of the physical environment while attending the VE decreased gradually within and between sessions. These results led to the conclusion that the longer subjects remained in the VE, the higher was the subject's sense of presence in the VE (even when using very minimal stimuli), while the sense of presence of the physical environment decreased. This supports a theory that the total sense of presence is constant, and subjects have to divide their overall sense of presence between the virtual and real worlds.

- Subject concentration increases significantly in the virtual world as compared to the physical world, when the subject has enough interaction to develop a strong sense of virtual presence.

Each subject's interest level in the learning study was determined by a 10-point scale instrument administered at the end of each experiment. The scores ranged from very weak to very strong. The interest level and sense of control level in the virtual world were always higher than the scores in the physical world.

It was obvious that each subject was excited, enthusiastic, and eager to be in the virtual environment, rather than the physical environment. The main conclusion of this research was that memory span increased significantly in the VE as compared to the span in the physical environment, and that learner's motivation and interest levels may be maintained longer in the virtual environment. We hypothesize that at least a part of this effect may be due to the simplicity of the virtual environment, which provides less distraction to the learner.

- A person's perceptions of real-world situations and behavior may be modified based on his experiences within a virtual world.

The use of virtual environments is intended to augment human intelligence by either increasing or modifying a person's intellectual understanding of the structure or nature of objects or tasks (Bajura, Fuchs, & Ohbuchi, 1992). A VE can also modify users' perceptions of real-world situations and thus behavior in those situations.

This could readily be seen in the reports of subjects who exposed themselves to real-world phobic situations after receiving VRT treatment. In these cases, what was learned and experienced in the VE was transferred to real-world perception and behavior.

## 3.2 Safety

There are risks associated with virtual reality technology, as pointed out by Stanney (1995; see chap. 29 and 36, this volume), and definite steps must be taken in treatment to minimize these risks. According to Stanney, subjects at risk for psychological harm are primarily those who suffer from panic attacks, those with serious medical problems such as heart disease or epilepsy, and those who are (or have recently been) taking drugs with major physiological or psychological effects. Questions regarding these situations must be asked as a part of the screening process, and persons with these characteristics must be excluded from VRT experiences. Also, some people experience symptoms ranging from headache to epileptic seizure when exposed to visual stimuli that flickers at 5 to 30 Hz (see chap. 29, this volume). In VRT, no frame update rates in this range must be used. Furthermore, patients have to be closely observed by therapists at all times, and if there is evidence of any significant physical or psychological distress, both the patient and the therapist must have the ability to quickly terminate the virtual reality session.

Therapists are urged to ask patients to sit on a chair rather than stand up; use a modified HMD so patients can see their physical body partially; choose HMDs with a narrower field of view; and most importantly keep sessions brief (between 15 and 20 minutes long). This configuration reduces the degree of immersion (see chap. 40, this volume) but increases the physical and psychological safety of the patients (see chaps. 29–32 and 36–39, this volume). There is still a great need for research in this area, and we strongly recommend that researchers take appropriate steps in minimizing patient risks.

It must also be noted that symptoms of anxiety while under VRT are often distinctly different but may overlap with those of cybersickness (see chap. 30, this volume). The anxiety symptoms evoked under VRT are the same as the real-world experience of the patient and include shortness of breath, heart palpitations (irregular or rapid heartbeat), trembling or shaking, choking, numbness, sweating, dizziness or loss of balance, feeling of detachment, being out of touch with self, hot flashes or chills, loss of control, abdominal distress, and nausea. It is thus important that therapists distinguish between the two syndromes.

## 3.3 Complexity of VRT

The differences between VRT and simple desensitization and exposure therapy are important. VRT appears to be oriented more toward neurophysiological information processing theory and the accelerated integrative information processing paradigm presented by Frances Shapiro. Thus far, research indicates that VRT works very well with subjects who suffer from various kinds of phobias. Intuitive observation has also led to the belief that more than desensitization is at work during VRT treatment. Patients who were immersed in the virtual world typically would not communicate with therapists residing in the physical world. They appeared to be reliving their previous disturbing or anxiety-provoking experiences even though the virtual world did not accurately match their existential world. They would usually look repeatedly at the same simple object or objects within the virtual world. In general, VRT appears to provide a link between the reality of the client and the objective world. However, at this time there is no concrete or empirically based evidence to explain why and how VRT works. Thus, there is a great need for researchers to investigate the psychological mechanics of VRT.

# 4. DISCUSSION AND CONCLUSIONS

Results of the experiments reported herein indicate that VRT might be effective in reducing self-reported anxiety of subjects with a variety of psychological disorders. It should also be noted that these studies do not speak to the relative effectiveness of VRT as compared to conventional treatments (e.g., imaginal systematic desensitization) as the control group used in most of these studies was simply a no-treatment group.

Although results of the research projects covered in this chapter are impressive, additional research is needed to more thoroughly explore the effectiveness of VRT and extend it to other psychological disorders. These studies must allow for both objective and subjective measurement of anxiety to ensure the validity of research outcomes.

The groundbreaking efforts reported herein only scratch the surface of a vast uncharted area. It will take many more innovations, and many more trials and errors to arrive at the point of fully utilizing VE in the treatment of a majority of psychological disorders. Perhaps it will take even longer to develop VRT systems that can be used by patients as self-healing machines.

## 5. ACKNOWLEDGMENTS

The research projects described in this chapter were sponsored by several grants from Boeing Computer Services (virtual systems department), U.S. Army Research Laboratory under contract DAAL03-92-6-0377, and supported by the Emory University and Georgia Institute of Technology Biomedical Technology Research Center (for the collaborative research of the fear of heights). The views contained in this document are those of the authors and should not be interpreted as representing the official policies of the U.S. government, either expressed or implied.

## 6. REFERENCES

Bajura, M., Fuchs, H., & Ohbuchi, R. (1992). Merging virtual objects with the real world: Seeing ultrasound imagery within the patient. *Computer Graphics, 26*(2), 203–210.

Barlow, D. H. (1988). *Anxiety and its disorders: The nature and treatment of anxiety and panic*. New York: Guilford Press.

Booth, R., & Rachman, S. (1992). The reduction of claustrophobia—1. *Behaviour Research and Therapy, 30*, 207–221.

Botella, C., Baños, R. M., Perpiñá, C., Villa, H., Alcañiz, M., & Rey, A. (1998). Virtual reality treatment of claustrophobia: A case report. *Behaviour Research and Therapy*.

Held, R. M., & Durlach, N. I. (1992). Presence. *Presence: Teleoperators and Virtual Environments, 1*(1), 109–112.

Hodges, L. A., Rothbaum, B. O., Watson, B. A., Kessler, G. D., & Opdyke, D. (1996). Virtually conquering fear of flying. *IEEE Computer Graphics and Applications, 16*(6), 42–49.

Hodges, L. F., Rothbaum, B. O., Alarcon, R., Ready, D., Shahar, F., Graap, K., Pair, J., Herbert, P., Gotz, D., Wills, D., & Baltzell, D. (1999). Virtual Vietnam: A virtual environment for the treatment of Vietnam War veterans with post-traumatic stress disorder. *CyberPsychology and Behavior, 2*(1).

Hunsucker, G. (1988). *Attention deficit disorder*. Fort Worth, TX: Forrest Publishing.

Loomis, J. M. (1992). Distal attribution and presence. *Presence: Teleoperators and Virtual Environments, 1*(1), 113–119.

Loomis, J. M. (1993). Understanding synthetic experience must begin with the analysis of ordinary perceptual experience. In *IEEE Symposium on Research Frontiers in Virtual Reality* (pp. 54–57). San Jose, CA: IEEE Press.

Naiman, A. (1992). Presence, and other gifts. *Presence: Teleoperators and Virtual Environments, 1* (1), 145–148.

North, M. M., & North, S. M. (1994). Virtual environments and psychological disorders. *Electronic Journal of Virtual Culture, 2*(4), 37–42.

North, M. M., & North, S. M. (1996). Virtual psychotherapy. *Journal of Medicine and Virtual Reality, 1*(2), 28–32.

North, M. M., North, S. M., & Coble, J. R. (1995a). An effective treatment for psychological disorders: Treating agoraphobia with virtual environment desensitization. *CyberEdge Journal, 5*(3), 12–13.

North, M. M., North, S. M., & Coble, J. R. (1995b). Effectiveness of virtual environment desensitization in the treatment of agoraphobia. *International Journal of Virtual Reality, 1*(2), 25–34.

North, M. M., North, S. M., & Coble, J. R. (1996a). Application: Psychotherapy, flight fear flees. *CyberEdge Journal, 6*(1), 8–10.

North, M. M., North, S. M., & Coble, J. R. (1996b). Effectiveness of virtual environment desensitization in the treatment of agoraphobia. *Presence: Teleoperators and Virtual Environments. 5*(3) 346–352.

North, M. M., North, S. M., & Coble, J. R. (1996c). *Virtual reality therapy: An innovative paradigm.* CO: IPI Press.

North, M. M., North, S. M., & Coble, J. R. (1997). Virtual environment psychotherapy: A case study of fear of flying disorder. *Presence: Teleoperators and Virtual Environments. 6*(1). pp. 127–132.

North, M. M., North, S. M., & Coble, J. R. (1997). *Virtual reality therapy combating fear of public speaking.* Manuscript submitted for publication.

North, M. M., & North, S. M. (2000). Virtual reality combats obsessive-compulsive disorders (OCD). *Medicine Meets Virtual Reality 2000 Conference.* San Diego, Calfornia.

Riva, G. (1997a). *Virtual reality in neuro-psycho-physiology.* IOS Press. Amsterdam, Netherlands.

Riva, G. (1997b). The virtual environment for body image modification (VEBIM): Development and preliminary evaluation. *Presence: Teleoperators and Virtual Environments. 6*(1), 106–117.

Rothbaum, B., Hodges, L., Kooper, R., Opdyke, D., Williford, J., & North, M. (1995a). Effectiveness of computer-generated (virtual reality) graded exposure in the treatment of acrophobia. *American Journal of Psychiatry, 152*(4), 626–628.

Rothbaum, B. O., Hodges, L. F., Opdyke, D., Kooper, R., Williford, J. S., & North, M. M. (1995b). Virtual reality graded exposure in the treatment of acrophobia: A case study. *Journal of Behavior Therapy, 26*(3), 547–554.

Stanney, K. (1995). Realizing the full potential of virtual reality: Human factors issues that could stand in the way. In *IEEE Proceedings of Virtual Reality Annual International Symposium '95.* NC: IEEE Press.

Strickland, D. (1996). A virtual reality application with autistic children. *Presence: Teleoperators and Virtual Environments, 5*(3). pp. 319–329.

Wiederhold, B. K., & Wiederhold, M. D. (1998). A review of virtual reality as a psychotherapeutic tool. *CyberPsychology and Behavior 1*(1), 45–52.

Wiederhold, B. K., Gevirtz, R., & Wiederhold, M. D. (1998). Fear of flying: A case report using virtual reality therapy with physiological monitoring, *CyberPsychology and Behavior 1*(2), 97–103.

# 52

# Applications of Systems Design Using Virtual Environments

Roy C. Davies

*Division of Ergonomics and Aerosol Technology*
*Department of Design Sciences*
*Lund Institute of Technology*
*University of Lund*
*Box 118, SE 22100 Lund, Sweden*
*roy.c.davies@ieee.org*

## 1. INTRODUCTION

How can virtual environment (VE) technology be used for design? Many designers are beginning to realize the potential of VE for design, some have been using the tool for many years. However, there are still many people and companies who could benefit from this technology but do not know where to begin. In many cases, people are even unsure as to which questions they must find answers to. This chapter addresses these issues and provides a number of key questions to consider in the procurement of a VE based design solution. A classification scheme of VE solutions for design is put forward and is illustrated by examples from current design situations. These examples serve also to provide the reader with a starting point for their own solution.

*Design* can be described as the process of taking a number of often conflicting criteria for the creation of an object such as form, function, cost, and manufacturability and marrying these together to form a whole (can one say that any man-made object is not designed?). A system is a complex entity involving many parts that have to function together. *Systems design*, therefore, involves putting together a complex entity with conflicting requirements to function harmoniously. This can naturally be a time consuming process, involving many people, physical prototypes, and much expense.

Computerized design and visualization based on VE technology is being increasingly used to simplify this process and at the same time reduce costs, wastage, and product development time. This chapter is aimed at people interested in learning where VE technology can be applied to systems design and includes examples from real design situations.

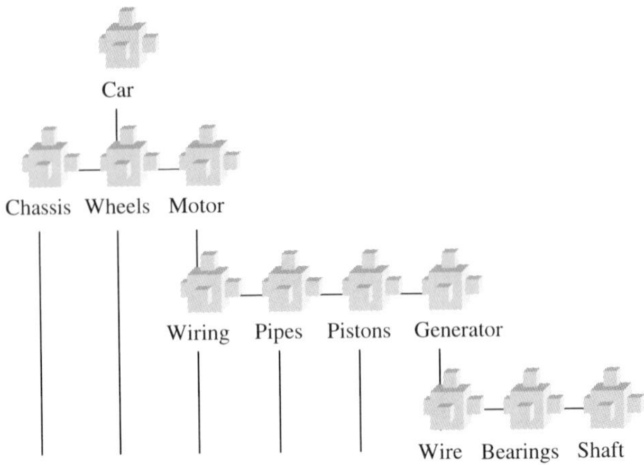


FIG. 52.1. Part of the system hierarchy of a car.

## 1.1 What Is a System?

*System* is an overused word. In essence, any collection of parts put together into a larger functional whole can be considered a system, such as an artwork, a car, computer, factory, or refinery. A system may be composed of a hierarchy of subsystems, which in turn may themselves be composed of subsystems and individual parts. So, for example, a car is composed of wheels, motor, chassis etc, with the motor being comprised of pistons, generator, wiring, pipes etc, with the generator containing wire, bearings, a shaft etc (see Fig. 52.1).

When designing such a system, a subsystem may be used ready made or designed anew. So, a new model of car may, in fact, be based on the same chassis as a previous model with a motor from a completely different vehicle. Reusing subsystems like this saves both development time and money.

## 1.2 Design

According to Rheinfrank, Evenson, and Wulff (1994, p. 78), design can be used for "facilitating meaningful innovation" and plays an important role in providing "useful tools and meaningful cultural artifacts" from a combination of "understanding what people do and think" and "innovative technology." In this sense, design is considered part of the entire production process, from ideas to final product, and is not just an aesthetic afterthought. A system evolves, ideally, through an iterative process of prototypes, involving actual users, designers, and other experts until a satisfactory result has been achieved (Norman, 1988). In many cases, constraints are imposed that contradict optimal user design (such as cost and manufacturability), which must become part of the total design.

Furthermore, experts from diverse fields may be responsible for parts of the system, and they need to be able to communicate their ideas to the other people in the design team. An architect needs to be able to communicate their ideas to the builders and a common language is sought—in this case, plans and diagrams. The design process can, in itself, become a common artifact, around which communication takes place.

One major distinguishing factor between design processes is the level of end users participation. The end users are those people who will be using the final product, such as driving the car, living in the house, or wearing the shoes. In a few cases, they may also be the people using the VE tool, for example, to plan a building or to experience a concept design or artwork.

In some situations, such as the design of a new concept car, experts answer for the needs of end users using experience from previous designs and knowledge about the intended market. In other situations, such as the design of a new rehabilitation aid (see chaps. 49 and 50, this volume), end-user involvement is critical if the final result is to match the user's needs and be accepted. In this case, the design process must help to bridge the gap between the design experts and end users in such a way as to ensure that the end users are able to express their desires and can communicate these to the experts.

At what stage can design enter into the system development process? One can argue that design is present at all stages, so perhaps it is more useful to define the stages of a system conceptualization and manufacture in terms of design terminology.

A concept design is often formed before the physical system. Such designs push the barriers of conservatism in order to try new ideas. This process begins in the minds of the designers and expresses itself in rough sketches, drawings, and other informal material, later becoming more concrete, and in some cases resulting in concept sculptures or even working models. If practical considerations as to cost and manufacturability have been taken into account, these are only loosely bound, because the point of a concept design is to give the imagination free reign.

Product design is the purposeful design of an object for a particular market and must not only take into consideration form and function, but also cost, manufacturability, intended market, and so forth. Some design situations may concentrate on a particular criterion of the design, such as usability (see chap. 34, this volume), ergonomics (see chap. 41, this volume), or accessibility for disabled people (see chap. 49, this volume).

The object being designed also affects the design process. Small simple objects require few resources and minimal time to design. Combining many components into an interconnected system, on the other hand, often requires repeated prototypes, even partial manufacturing and testing for compatibility, consuming greater time and resources.

### 1.3 Virtual Environments and Computer-aided Design

Computer-aided design (CAD) can be considered the forerunner and heavyweight cousin of today's active design VE systems. Typical CAD programs are aimed at the single expert user, whereas VE design tools allow novices (end users, for example) to design in a more intuitive and natural way and interdisciplinary design teams to communicate ideas using the tool as the medium. Large libraries of objects can be made available without the need for expensive storage spaces. Objects can be reused and new designs can be created from the components of a previous design.

Alternatively, the term *computer-aided design* can also imply a tool that aids in the design of a system through automated procedures. For example, a tool may allow one to design a factory by positioning prefabricated and functioning elements such as conveyor belts and machinery. Or the form of a car may be interactively sculptured by hand.

Virtual environments can be used for everything from visualization (see chap. 53, this volume) and presentation of ideas to interactive design and brainstorming, from concept design to final product specification. Objects can possess all the attributes associated with a real object such as form, color, material, movement, weight, and sound. Furthermore, a virtual object can be "superreal," with extra attributes such as textual annotations, manufacturing instructions, multiplicity (alternative designs for different situations), history (playback of the design process), antigravity, and physically impossible form changes. Changes can be made interactively and more easily than if the object was physical and more prototypes can be tested than financially possible otherwise. Design alternatives and their consequences can be immediately visualized with the computer providing feedback about structural problems such as stress distribution (Vance, 1998) and object fit. Ergonomic testing can occur, with either people models (virtual humans) being placed in the environment or end users interacting with the virtual model. Once a product design is complete further processing can take place, with the tool finally passing specifications on to the manufacturer, such as which and how much material to use, as well as design alternatives given available resources. If the object being designed is going to remain immaterial, as in the case of futuristic concept designs or computerized art, then connections to reality can be completely severed, allowing designs to break the laws of physics and achieve greater freedom of expression.

Interaction with the model can take advantage of the user-friendly input devices usually associated with VEs such as DataGloves, head-mounted displays (HMDs; see chap. 3, this volume), stereo audio (see chap. 4, this volume), and haptic feedback (see chaps. 5 and 6, this volume). Designers can interact directly with the model, perhaps moving large objects with a single movement or making shape changes by waving a hand. It must be noted, however, that current VE systems promise a higher degree of usability than can be delivered—this is still an area of active research (see chap. 34, this volume). Moreover, a VE system can be programmed to take an active part in the design process, such as:

- Performing background analyses and presenting the results within a virtual environment (for instance, finite element analysis, structural strength calculation, loading on a human body, and field of view)
- Making suggestions on issues such as form, manufacturability, color, and lighting
- Providing tools for the testing of ergonomic considerations
- Connecting parts in such a way as to minimize material usage (such as pipe and cable drawing)
- Automatically generating designs based on specifications

Networking VEs allows designers in distant locations to collaborate, coming together in the same virtual environment. Alternatively, if the design is a sequential process from one person to the next, the computer can aid by providing a track through the design sequence and annotations from one designer to subsequent ones.

However, despite the potential of VEs to support interactive design and to have a truly three-dimensional (3-D) interface, the standard tried-and-true process is still very common. One typically begins by modeling in a CAD package, importing the models into a VE package, adding movements, sound, and other features not possible in the CAD program, then allowing end users to use or view the resulting virtual environment.

## 1.4 Classification of Virtual Environments Design Tools

Ten characteristics of VE design tools are described below for the classification of the form and usage of VEs in a given design situation. These characteristics have been chosen to be as independent from each other as possible, though other classification schemes are equally possible. The choice of a VE design tool depends on the desired design situation, budget, and project goals. Through the example situations described later, the reader can work backward, classifying his or her own design situation to allow the definition of the system most appropriate to their needs. Each characteristic is headed by a number of questions to aid in the decision process.

### *1.4.1 Task*

What do you want to do? What are your objectives?

It is important that a detailed task analysis is performed, with particular regard to the expected role of the VE system. Many questions must be answered to provide the basis for further analysis, such as: What is the actual design task? Is the purpose to design a product or concept? Or is the task to document a design process in the real world? Is the intention to compare and contrast a number of prototypes? Or to test aspects such as ergonomics and manufacturability? Is the user to be allowed into the design process? In what way? Is the intention to sell an idea or product that needs to be viewed in the right context?

### *1.4.2 Concreteness*

Do you want to design a concept or something concrete?

Is the design of an object a product that is to be finally manufactured, or is the design for conceptualization around a theme? The former requires a greater attendance to practical details such as manufacturability, structural strength, and space overlap, whereas the latter allows such considerations to be ignored. The form of the tool also varies; product design requires a certain amount of "intelligent" programming and support tools, whereas the concept design tool must allow free-form design and not hamper the designer.

### *1.4.3 End-User Participation*

Will you involve end users in the design team?

Level of end-user participation defines, to some extent, the ease of use that a VE design tool will have. Expert designers have time to learn a complex tool, whereas end users may have to use the tool without any former training. On the one end of the scale is participatory design, in which the end users themselves have a key role in the design team with experts providing support and backup. At the other end of the scale, knowledge of the needs of end users is invested in the expert designers themselves, and end user input is restricted to perhaps a predesign survey or a postdesign evaluation. Both methods are equally valid in appropriate situations.

Participatory design is advantageous when the tacit knowledge invested in the people who work day-to-day in a particular situation must be captured. So, for example, in the redesign of a factory in a process industry, involving the factory-floor workers ensures that knowledge about how one can sense a smoothly running machine (a combination of sound, vibration, smell, and visual cues) is incorporated into the VE design of the centralized control room. Furthermore, when large changes are to take place in a work situation, involving the people who are to work in the new environment in the design process increases the acceptance of these changes. User involvement can take a variety of forms, from appraisal of an expert modeled and animated 3-D virtual model with ensuing discussion to active design using a VE design tool. Moreover, designs can be worked on over a long period of time and discussed among a larger group than is possible in traditional design situations. In all cases, the situation must always be controlled by design experts to help reach the desired goal, though the designer's role becomes that of a facilitator, with their own ideas taking no more weight than any others in the design team. Using VEs for participatory design is often justified by considering it a medium that supports discussion across diverse backgrounds and work competencies using a common, visual language.

In situations where designers regard the end user as a collection of typical characteristics such as a size, arm length, wrist torque, and so forth, expert-driven design works well. In this case, typical end users may be called on for evaluation, but the bulk of the work progresses in a team of experts. Even so, VEs aid in the discussion across backgrounds and work competencies, allowing engineers, designers, and administrators to communicate using a common visual language.

A variant on the participatory design situation is where the tool itself is expert driven, but the design group consists primarily of end users who instruct the expert on what to do or build. This situation can alleviate the problems with having a large group simultaneously trying to use a VE design tool or when the end users lack competence to use the tool directly, and thus must spend too much time simply learning the tool rather than concentrating on the design task.

### 1.4.4 Level of Expertise

What is the level of expertise of the members of your design team? In which areas?

Related to the above scale is that of intended user group of the design tool. This defines how complex the tool can be, the format of the user interface, and the tasks that it can support (such as level of detailed design, coloring, and interactivity). Some tools may even be equipped with two modes, novice and expert—the former supporting the standard, easy-to-understand functions (such as moving, rotating and resizing objects), the latter allowing an expert greater modeling freedom. Generally, if the tool is to be used for participatory design, the user group must be considered novice, whereas expert designers will require greater functionality. This, in turn, defines the types of design that can be achieved, with end-user designs often being less realistic, but with a high level of description, and expert designs being precise to the millimeter and realistic.

### 1.4.5 Simultaneity

How long is the design process? Are the members of your design team in the same geographical location?

Cooperative design is supported using VEs to allow separation of the participants in both time and space:

- *Same time and place*: The designers are in the same place at the same time, each with their own computer or sharing a display and control device. Communication is direct through normal speech with the virtual design acting as a point of commonality.
- *Same time, different place*: The designers are separated by distance but able to participate at the same time. In this case, cooperation may be between individuals or groups working together with the VE acting as the communication medium transmitting speech and body movements.
- *Different time, same place*: Design is often a long process, with several people taking part at different times using the same equipment. The VE can function as the communication medium, recording changes and annotations from the designers and keeping track of who has made which changes.
- *Different time and space*: Similarly, if the virtual design is made available on a network, a combination of the above two can occur with people meeting in the VE from distant locations and also able to see other's changes made previously with annotations.

### 1.4.6 Degree of Immersion

Do you require a fully immersive VE system, or is a desktop system adequate? What is your budget? What resources do you currently have available?

Virtual environment hardware can range from purely desktop (standard personal computer screen, keyboard, and mouse) to fully immersive systems with sight, sound, touch, and even smell information supplied by the computer (see chaps. 11–13, this volume). In these environments one can either be alone or interact with others. Desktop systems and projector systems intended for an audience are best suited for design groups working together, particularly if coupled to human-friendly input devices. Single-person immersive systems tend to exclude the potential for cooperative design unless each person has their own hardware. Ease of use varies, with many tools requiring the design process to be on a desktop system and final visualization in a more immersive environment.

### 1.4.7 Interactivity

Do you intend use the VE as part of a presentation? Do you want the audience to sit passively or take an active role? Will the VE tool be a stand-alone application to be shipped to customers? What should they be able to do?

A VE can be viewed passively with automated fly or walk-throughs, perhaps even presented as a video; can allow simple interaction such as viewpoint maneuvering and object manipulation; can permit structural changes, say the repositioning of furniture; or be a fully interactive design tool with the ability to create new objects and alter existing ones (see chap. 13, this volume). Most current VE tools require prior modeling of objects in a traditional CAD package but then allow all positional changes to be made interactively (see chap. 12, this volume).

### 1.4.8 Realism

What are the important aspects of your design? Which of these need to be "real"? In what way?

Many designers strive for picture-perfect realism in their models, both physical and virtual (see chap. 40, this volume). This is appropriate for the visualization of, say, a concept car or a new building (seen chap. 42, this volume). In other situations, realistic behaviors of objects are more appropriate, such as in the design of a functioning factory or the testing of a driver's controls. Realistic objects can help the user overcome some of the hardware limitations causing problems in the perception of scale and space in desktop VE systems. However, too much realism too early in the design phase can be a distraction, causing designers to focus on irrelevant detail and detracting from the high-level design task. Similarly, end users may not have the competence or time to design the smallest details and thus need either vast libraries of detailed objects or a small library of representative simple objects (Davies, 2000).

### 1.4.9 System Control

Do you want the computer to take an active role in the design process? If so, how?

The computer managing the VE can also be programmed to perform other tasks such as structural analysis, fluid flow, object collision and overlap, calculating distances between points, or reach envelopes for virtual humans. These can be represented graphically for the user, thereby supporting the design process. Such tools do not take control away from the designer, but rather enhance his or her abilities. Powerful tools also exist for terraforming, realistic cloud control, and others to enhance the realism of the environment and to decrease nonproductive work of designers, allowing them to concentrate on the design task.

Alternatively, designers may have produced a VE to be tested by end users in which some alterations can be made to allow contrasting of a variety of designs. These alterations are usually limited within the boundaries set out by the designers. Similarly, the computer may control the entire design, allowing the user only limited direction choices, such as is the case of some art design programs.

### 1.4.10 Object Library

Will your VE be constructed from scratch, or can objects be taken from a library? What do want your end users, audience, or customers to be able to do?

One of the powers of a VE-based design tool is the ability to access vast libraries of objects. These can either be static objects classified according to some scheme such as function (office furniture, factory fittings) or form (tables, chairs) or be more dynamic in nature with generic objects being modifiable. For example, one can choose to place a table, specifying the number of legs, size and shape of the top, and the finish.

Alternatively, the system can require that some objects be created from scratch. This is often the case with unusual shapes or where a new form is necessary. The normal procedure is then to design the object in a CAD package and import it into the design environment, thus being suitable only for expert designers.

However, if the purpose of the design is for concept formation, simple objects may suffice as substitutes for more complex ones. Indeed, complex objects may detract from the design task as the users tend to focus on irrelevant detail (Davies, 2000).

## 2. APPLICATIONS

The potential for VEs to support the design process is great; however, there has been a certain reluctance among companies to invest in this technology due to cost and doubtful returns on their investment (Wilson, Cobb, D'Cruz, & Eastgate, 1996; see chap. 28, this volume). This initial sluggishness is, however, beginning to subside due to the successes of pioneer companies and decreasing costs of VE systems (see chap. 42, this volume). Instead, companies are now faced with the challenge of selecting the correct VE tool and obtaining the competence to use it effectively.

Many VE-based design tools begin with university research, gradually expanding into the marketplace as the usability level becomes acceptable and a need is recognized. The majority of standard VE programs have begun this way, with resulting continuous improvement. All support the expert designer in building realistic and functional environments; some even allow interactive design in immersive environments by nonexpert users, though this is less usual.

This section describes a cross section of design applications using virtual environments. Reference to specific hardware and software platforms has been avoided to ensure generality. Furthermore, each application area is analyzed and important features are highlighted with reference to the above classifications.

### 2.1 Design of Land Vehicles and Aircraft

#### *2.1.1 Background*

Virtual environments are well suited for the design of large and complex systems that constitute a vehicle. In such designs, many people with diverse areas of expertise are involved over a long period of time (Östman, 1998). Companies using VEs for vehicle design are Volvo, John Deere, Ford, Chrysler, General Motors, British Aerospace, Aerospatiale, and Boeing, to name a few.

Design in this area includes concept formulation, perhaps looking at the outer form of a car (Is it pleasing to the eye? What is the wind drag coefficient?, etc.), planning the actual construction sequence of the final product, and testing issues such as order of assembly and part fit. Furthermore, people may also be considered within the design (see Fig. 52.2), perhaps as drivers or machinery controllers, determining whether there is enough standing room or if controls can be reached and activated. In some cases, reaching a control, perhaps a lever, can result in a body position that does not allow enough torque to be applied to actually perform required actions. In other situations, the factor of interest is more obscure, such as sitting comfort over a long period of time or reachability of driver controls for people of varying body dimensions.

Another problem is assembling parts into small areas, such as in the engine bay. Wires and pipes have to be drawn in a way that is physically possible, keeps the amount of material used to a minimum, allows mounting in the first place and fixing at later stages, and, in many cases, the mounting must not be in the way, or indeed, completely hidden.

FIG. 52.2. Vehicle design using VE technology. Image Courtesy of Opticore.

Sometimes, the usage of a vehicle within a certain environment is of interest, in which case not only is the vehicle modeled but also a suitable place for it to be driven in. This allows ergonomic testing in an ecologically valid situation. Combining with physical artifacts can further heighten the sense of realism (see chap. 40, this volume), particularly if the physical controls are coupled to the computer system.

### 2.1.2 The User Group

Most companies do not have the resources or expertise to develop their own VE systems, though some in the vehicle industry are large enough to affect the development of VE software with their requirements. The majority of design is by experts, usually experienced CAD and 3-D modelers. Desired behaviors may be programmed into the VE model (see chap. 12, this volume), for example, to give controls some functionality or to allow the dynamic alteration of seat colors. Virtual humans are frequently used to test reachability, field of view, and other characteristics. People other than designers may then interact with the model in a limited way, activating predefined behaviors or visualizing the environment and providing comments to help in the prototype evaluation. Often the virtual model is used to sell an idea to customers or management, or to try out a variety of alternatives.

### 2.1.3 Rationale for Using Virtual Environments

The main reason for using VEs instead of physical mock-ups is cost saving, reduced development time, and improved quality (Östman, 1998). A greater number of more realistic prototypes can be assessed in quicker time using less material (thus reducing time to market). Some parts of an environment may still need to be built, such as the driver's seat, to test ergonomic considerations not possible in a VE due to lack of tangibility of the virtual models. These physical models can be combined with a VE to provide a realistic setting. Concept vehicles can easily be created and outfitted with standard fittings from a library. Designs can be contrasted side by side to compare the relative merits and demerits of each. The design process can be retraced and a person's input reassessed to determine whether the design could progress in a different direction.

Designs can also be easily worked on by multidisciplinary teams of people located in remote parts of the company when kept virtual. As in any distance work, this reduces the need for travel and allows expertise to be spread more effectively.

### 2.1.4 Classification

Generally, due to the broad nature of the different companies using VEs for vehicle design, all possible variations exist, depending on need and the design task:

*Task*: The task is usually to design a manufacturable and working vehicle or aircraft that fulfils design criteria such as safety, functionality, and appearance, and sometimes involves testing with end users. Alternatively, the task may be to develop a new idea, a concept vehicle to show to the public or to provide inspiration within the company.

*Concreteness*: Vehicle design can range from concept to production. Concepts can be produced that may not necessarily be possible in reality. Aircraft can now be totally designed and tested in a VE, resulting in plans (and 3-D models) from which the real aircraft is then built.

*End-User Participation*: End-user participation is uncommon in the actual design of vehicles, though they may be called in to test a selection of half-ready alternative designs. In some situations, such as the testing of driver controls, end users may be asked to test functionality. The value of bringing end users early into the design process is slowly being realized.

*Level of Expertise*: Design teams are typically composed of experts with the tool being used to aid communication, speed up the design process, and reduce prototyping costs. For these reasons, the tools tend to be a combination of intuitive interfacing using advanced VE hardware (typically a large multiprojector system with stereo and DataGlove) and complex design tools.

*Simultaneity*: The design of vehicles can cover the entire scale of simultaneity. Typically, several designers work on the same model over a period of time. The most common method is to work on a shared design, with an individual's changes and comments being tracked by the system. Some work has also been performed in allowing several designers to participate in the same environment at the same time, though this is less common.

*Degree of Immersion*: This can range from desktop systems, where designers view a VE in a window on the screen and ordinary graphical user interfaces (GUIs), to fully immersive 3-D systems (see chaps. 11–13, this volume). The former is typically used for the actual design and testing, whereas the latter is more common for visualization and presentation of the resulting design.

*Interactivity*: Depending on the level of immersion and the users of the system, interactivity can range from full control over the design for experts in a typically nonimmersive environment to control of only the programmed behaviors for testers in a fully immersive environment (see chap. 13, this volume). Full control is provided by standard VE software packages and allows all object attributes to be modified.

*Realism*: The standard means of managing realism and detail level is to include only those details relevant to the design. For example, in testing of different fabrics on the seats in an airplane, none of the objects are given realistic movements, nor are objects outside the cabin modeled. However, the seats are able to show the fabric patterns in a realistic and convincing manner. In the design of a new car shape, only the outer form is modeled, but this must be in a fluid and natural way, without any jagged lines or flat patches usually associated with 3-D graphics.

*System Control*: Ordinary VE packages do not provide many automated tools, though features such as collision detection and point-to-point measurement are usually standard (see chap. 12, this volume). Tracking of changes and annotations are also common; however, extra features such as automatic connection of, say, engine components needs to be programmed. Virtual humans are frequently used to test field of view and reachability of controls, thus taking the place of end-user testing.

*Object Library*: For designers of vehicles and aircraft, a specific object library consisting of parts from previous designs is beneficial. Standard parts can be included in the library, allowing a new vehicle to use as many of the standard fittings as possible, thus reducing the need for new production lines or orders. New objects are usually modeled in a CAD package and imported into the design.

#### 2.1.5 Summary

Current VE systems for vehicle design have shown great benefits in increasing the competitive edge. Advantages are a reduction in production time, quicker time to market, less costly prototypes, and a greater distribution of expertise. High-end graphics computers are used, the cost of which is offset by the savings over previous techniques. However, designers are let down by the intangibility of virtual models and the complexity of software programs. These will improve, however, as work on tactile and haptic feedback (see chaps. 5 and 6, this volume) progresses and interaction becomes more intuitive and moves from two-dimensional (2-D) to 3-D GUI (see chap. 13, this volume). A problem may be that as more companies use virtual design, the competitive edge may be lost.

### 2.2 Architectural and Interior Design

#### 2.2.1 Background

Interior and exterior design—the design of person-size spaces, such as workplaces and the buildings they are housed in, is one of the most common applications for VE technology. At its simplest form, interiors can be modeled by experts—architects, designers and VE technicians—and shown to potential customers or users. The intention may be to entice people into buying a particular product by showing it in a realistic and immersive way (see Fig. 52.3) or to perform an evaluation for further refinement of the design. In this case, visualization is the key task and the system being designed is usually a personal space, say the interior of an aircraft or a house, or the exterior view of a building as it will look in current settings. A VE allows one

FIG. 52.3. Britannia Airways "360" cabin design concept prototyped using a VE image. Courtesy of Virtual Presence (U.K.).

FIG. 52.4. Visualization of the view from a reception area. Courtesy of Design@Work, University of Lund, Sweden.

to see how the environment will look for people untrained in reading plans (see Fig. 52.4) and allows comparison of alternative designs and modification of details, such as color and wallpaper design.

At the other end of the scale, VEs are used to aid end users in the design of their own environments (Davies, 2000; Ehn et al., 1996; Wilson, 1999). Three-dimensional visualization has been used for many years, often with models presented as static pictures or animation sequences (Akselsson et al., 1994) that users are asked to evaluate. Similarly, 3-D visualization is used to aid in the discussion between parties involved in the design of environmental adaptations for disabled people (Eriksson, 1998; Eriksson & Johansson, 1996). Both of these have allowed end users to participate in the design process and produced results that are functional and well received. Virtual environments for active design by end users themselves, however, is a less well-solved problem and still the area of much research (Davies, 2000; Wilson, 1999). In all these cases, the goals are to give end users, designers, architects, and others a common language in the communication about the design of an environment and to bring out the user's tacit knowledge of their environment (see Fig. 52.5). One of the main problems is finding a tool that can be used by nonexperts without hours of training. Previously, noncomputerized design tools have been employed, such as full-scale modeling, advanced meeting techniques, and role-play to activate end users' tacit knowledge and as a communication medium (Ehn et al., 1996). These allow direct participation in the design process and are hands-on. However, the full-scale laboratory imposes the following limitations: difficulty in changing color and lighting; ease of alteration (considerable physical effort is required to dismantle and rebuild rooms and to model an object); difficulty in documenting and reproducing visualizations; building very large environments; portability; and showing results after the laboratory has been emptied again. Additionally, all participants must come together in the same time and place—not always possible for a company or process industry (Davies, 2000). This is where a VE solution can be of help, with its user-friendly interface devices and time and space independence. The important considerations are how people will have to use the environment and the actions they will have to perform. Furthermore, using a VE allows a projection of oneself into the

FIG. 52.5. Pausing for a break during the participatory design of a workplace. Courtesy of Design@Work, University of Lund, Sweden.

intended environment and the enacting of scenarios, either through immersive technology or by placement of life-size virtual humans.

### 2.2.2 The User Group

The primary users are VE designers themselves, using the standard CAD and VE design tools. Visualization may take place in an immersive setting, involving end users, either as an audience, being shown the highlights of a particular design, or actively in control of the viewpoint, being allowed to explore the space and form their own impressions.

Active design with users in control is more problematic since a careful balance must be struck between the capabilities of the tool and its corresponding complexity. If the tool is too complex, users feel disinclined to use it and tend to opt for less cognitively loading discussion aids such as pen and paper (Davies, 2000).

### 2.2.3 Rationale for Using Virtual Environments

Virtual environment technology has proven useful as an aid in visualization—many companies make money by architectural design using this, often with adjustable parameters such as decoration, furniture layouts, exterior cladding, and building size. The main advantage over traditional plan drawings and mock-ups is the heightened feeling of presence (see chap. 40, this volume)—of being able to see the environment from all angles and being able to unambiguously communicate with end users or customers.

The people who will have to live with an environment are more likely to understand and accept the architect's limitations if these can be shown graphically and if they have a chance of affecting the design before construction. Furthermore, getting the design right the first time reduces the need for expensive rebuilding, as in the case of adaptations for disabled people (Eriksson & Johansson, 1996).

Architecture is very often about feeling. Plan drawings just do not create a feeling for the size and scale of a room—it is not possible, particularly for an ordinary person, to get a feel for the space. However, when immersive VEs are used, observers are able to get inside the environment, feel the place in relation to their own body space, and even try out an environment themselves, perhaps looking to see what the view from a window is or trying to squeeze into a kitchen. When a virtual building is placed in a picture of the real environment it becomes possible to assess the impact on the surroundings.

As with all design situations, one can also try many more designs using a VE than with mock-ups, build very large environments, and even try out scenarios in the environments to see whether, for example, the room layout is optimal or what would happen in an evacuation situation.

### *2.2.4 Classification*

*Task*: The task in interior design is usually to build and visualize a planned real environment. In some cases, the place may already exist and is being redesigned, but more often the intention is to visualize a new idea or design in a more realistic way than plan drawings and to aid in communication among the interested parties.

*Concreteness*: The design is usually of an actual environment and is thus not merely a concept. On the other hand, an architectural office may use VE technology to sell their design or win a contract. Thus, the VE begins as a concept, but may become a truly "concrete" design. In the case of participatory design, end users manipulate objects in the VE that symbolize real objects. In many cases these objects represent more complex forms, such as a cube for a machine, and the design progresses at a high conceptual level.

*End-User Participation*: End users—people who are going to live or work in an environment—are often included at the final design phase, choosing among a selection of prototypes. In some cases their comments are fed back into a new round of designs, resulting in new prototypes to be tested, or their suggestions are taken into consideration during the actual drawing up of architectural plans. Fully participatory design, where the end user is essentially given *carte blanche*, is becoming more common, particularly in Scandinavia (Davies, 2000; Ehn et al., 1996) and England (Wilson, 1999; see chap. 17, this volume). Halfway houses are common, where the end user is able to manipulate some features of the design, perhaps altering the layout of a shop or the fabric on a chair.

*Level of Expertise*: In most cases the worlds are designed by experts, which are then shown to novice end users. Very few situations allow end users to design the desired environment themselves completely from scratch.

*Simultaneity*: Architectural design is often a group activity, with people sharing the same view of the VE and using it as a discussion starting point. In some cases, a design may be placed on the Internet for others to see and comment on.

*Degree of Immersion*: Immersive environments that allow several people to participate at once are best for group discussions. Either projection walls or surround projection are often used, but HMDs are unlikely due to the closed-in nature and discouragement of group discussion.

*Interactivity*: For designers, standard VE tools are used with standard 2-D interfaces, giving total control over the environment, but not necessarily from within. If end users are involved, the level of interactivity regularly allows the ability to walk around and interact with the environment, such as opening doors, and operating taps. In some cases, the repositioning of objects or changing of appearance from within is allowed.

*Realism*: Level of realism has to be carefully controlled if end users are participating in the actual design. If there is too much detail, people will focus on the wrong aspects. If there are misleading details, then their perception of the environment may be incorrect. A careful balance must be struck between speed of rendering and which details to include. Objects of prime interest may need to be highly detailed, whereas others can simply be boxes, perhaps with a digital photo from the real environment pasted on its side.

*System Control*: The computer is usually not required to do much more above the visualization of an environment, though some care may need to be taken with large environments to remove all parts that are not currently visible from calculation (such as above and below floors) to speed up computer processing. Realistic movement of objects is often desirable, not just that objects look right; they need to work as in reality too. For example, it should be possible to open doors and objects should be affected by gravity and have momentum. The viewpoint should move in a natural way, perhaps with a selectable eye-height, being fixed to the ground, and able to climb stairs.

*Object Library*: If the environment is being created by expert designers, a library of favorite objects is beneficial, tuned to the type of environment being built (such as a home or office). If end users are to design the environment, one can either incorporate a large comprehensive library that includes all possible objects (Eriksson & Johansson, 1996) or use representative objects and require participants to fill in the details from imagination and through a shared history (Davies, 2000).

#### *2.2.5 Summary*

Many visualizations of residential buildings or offices are successfully performed around the world every year by architectural firms and housing agencies. However, there are still many interface problems to be solved before such organizations can buy off-the-shelf VE packages that allow end users an active role in the design process. There is also a certain reluctance to give away the design to end users before architectural plans are finished—a fear, perhaps of losing control over the design and involving people who are not experts in architectural design. This is one of the greatest misunderstandings—VE technology is a tool that can be used by all people involved in the design process—architect, builders, and end users to aid communication. However, the process needs to become mode democratic rather than autocratic.

Another problem is that VEs do not allow the feeling of recreating an environment completely. Missing are tactile (see chaps. 5 and 6, this volume) and audible (see chap. 4, this volume) feedback. Everything appears ghostly and can be passed through, and realistic sound effects such as echo of footsteps and background noises that help create a feeling of space are generally not modeled. Furthermore, participatory design using VEs will be improved when several people can act at once—current systems require that people must take turns or see each other as avatars (see chap. 15, this volume) in the VE and cannot easily converse and cooperate. These will change as the technology advances. Interaction with the VE tool can also be problematic, and improvements are continuously being sought to make immersive design easier for end users.

### 2.3 Planning and Monitoring of Construction Projects

#### *2.3.1 Background*

Working paper drawings are typically used to communicate a building design to a contractor. The contractor's planners then have the task of taking these drawings and creating schedules and plans for the construction process—a process requiring much intuition, imagination, and judgment (Adjei-Kumi & Retik, 1997). The planners use software that aids project management by supporting formulation and evaluation of project schedules in a symbolic and abstract manner. The output is hard to interpret and does not realistically support the visualization of schedules, thus requiring great experience and knowledge of perceived characteristics

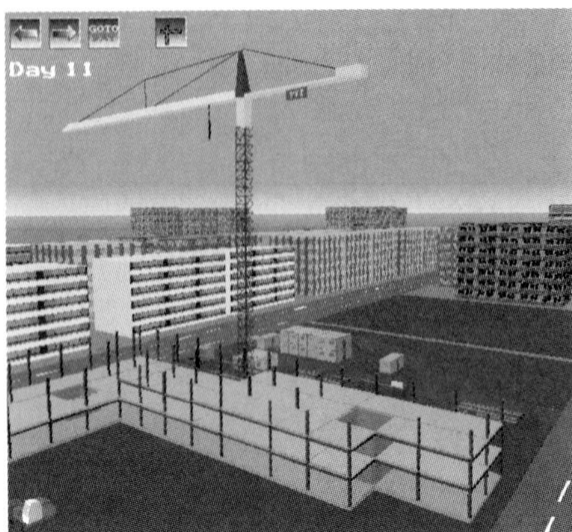


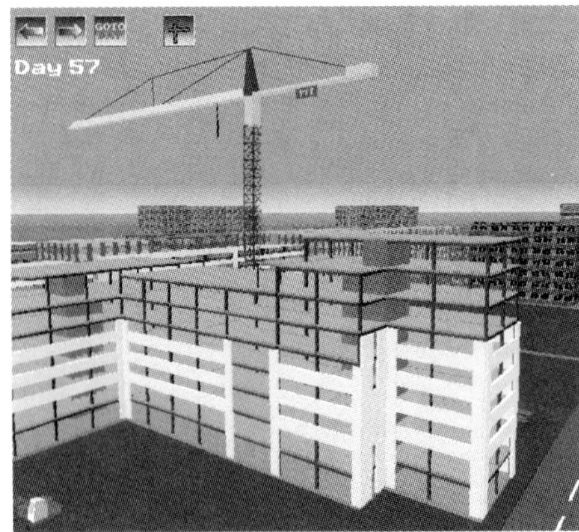


FIG. 52.6. Visualization of stages in a construction process. Courtesy of the Department of Civil Engineering, University of Strathclyde, Glasgow.

and spatial relationships between various components. Furthermore, communication between planners and other parties involved becomes prone to misunderstandings, leading to potentially dangerous errors in the construction process.

Current visualization techniques for building construction concentrate on the final product (see section 2.2), not the process, whereas construction planning tasks deal with the entire project from beginning to end. Current computer tools cannot capture or display the construction process, hence there is a need for a tool that supports both visualization of schedules for planners and communication between planners and others during the construction process.

Recent research in England (Adjei-Kumi & Retik, 1997; Retik, 1997) has concentrated on producing computerized tools for visual planning of the construction process using VE techniques to allow realistic visual presentation of construction project information (see Fig. 52.6). These tools allow planners to both assemble a building in stages from a library of ready-made objects and to replay the construction process to facilitate visualization, show others, and allow checking against the actual progress of the real building. Within the framework of the drawings given to the contractor, planners are more or less free to design the construction process as they see fit—in this case, the design takes the form of construction schedules rather than the building form itself.

### 2.3.2 The User Group

The primary users are planners, with experience using planning and construction tools and an expected high level of computer competence.

### 2.3.3 Rationale for Using Virtual Environments

Due to the programming capabilities of VEs, a tool can be made that fulfills the above criteria and that can be used on the construction site or at the contractor's offices. Simulation of time steps can be programmed, and the building can be viewed from all angles, including inside. As a communication medium, 3-D visualization is less ambiguous than plans, drawings, and schematics, though it must currently remain a complementary tool.

Reduced misunderstanding is expected to lead to fewer errors in the construction process, reducing the need for redesign or partial deconstruction in extreme cases, thus saving time and materials and shortening the entire construction time.

### 2.3.4 Classification

*Task*: The task is to construct a building virtually, record the schedule of the construction process over time, and be able to review this process for both discussion of the construction process and monitoring of subsequent construction.

*Concreteness*: The building itself is a concrete object; however, planners do not determine its form. Rather, planners focus on the process of construction, which, in this case, is the object of design. This is a more abstract entity than usually built using a VE, as it is essentially the passage of time that is to be modeled. Nevertheless, there is a definite process, resulting in a concrete object, not a concept design.

*End-User Participation*: End users, those who are to populate the building, are not considered in this stage, though in the initial design of the building, architects may have brought them in. However, the people using the tool itself are different from the tool designers, thus are end users of a different nature.

*Level of Expertise*: The users of the VE tool are experienced in planning construction and are hopefully motivated to learn a new tool, especially if it is expected to facilitate their job and reduce the chance of mistakes.

*Simultaneity*: The tool may be used by several people at one time as a discussion focus for the construction of a building, but it is not intended to be used by planners in different locations, nor necessarily by different people at different times in the design process—though the tool does not exclude these possibilities.

*Degree of Immersion*: For greater portability and reduced cost, such a tool is not intended to be immersive. Nor does an immersive system facilitate group work and discussion.

*Interactivity*: In the design of the construction process, the tool allows full control of the placement of objects representing the parts of a building and the manipulation of time. For visualization, navigation around the building site and the ability to step through time are provided.

*Realism*: The detail required for this tool must concentrate on the parts of the building. To be useful to planners, it must be able to accurately depict shapes of building parts (such as girders, fasteners, wall segments, etc.) so that it can accurately portray problems that may occur due to assembly in the wrong order. On the other hand, realism of the graphics is not so important. Lighting and shadowing are irrelevant, as are sound effects, realistic detail, and picture perfect textures on objects. Thus, a simple PC-based VE system is adequate.

*System Control*: The system is required to understand concepts such as linkages of components, gravitation (or rather that objects must sit on other objects), and a breakdown of working areas to trace the intended progress of the work. It is also required to link to other modules, such as commercially available construction scheduling software to display schedules, resources, costs, and other features in the traditional way (bar charts, networks, etc).

*Object Library*: The crux of this tool is the existence of a large object library from which planners can select the parts they require and the ability to import other elements that may already exist in CAD format. The library, in this case consists of two types of elements—process-based and product-based. Product-based elements include cranes, mixers, scaffolding, fences, huts, among others, and can be directly placed into the environment as 3-D models. Process-based elements describe activities. These are automatically connected by the system and require the integration of technical knowledge and experience. These can also be placed interactively during the planning process.

### 2.3.5 Summary

The work to date has highlighted a major weakness, that VE design tools are only as good as their virtual library and that this must be large in order to be useful. Similar problems have been found in other design tools using VEs, such as in the adaptation of environments for disabled people (Eriksson & Johansson, 1996; see chap. 49, this volume). Nevertheless, such a planning and monitoring tool has been successfully designed and tested, and further work is in progress to incorporate greater automation and design aids, such as an expert system to advise on rectification of deviations from the planned schedule.

A further point to note is the success in using PC-based desktop VE technology; immersion is not necessary, making such a tool affordable to construction companies. Careful concentration on only essential features to support the design process has allowed the peeling away of unnecessary complexity and spurious detail.

## 2.4 Industrial Plant Design

### 2.4.1 *Background*

Design of industrial plants requires not only knowledge of chemical and mechanical processes, but also how the various parts can be linked together. Such a plant is a highly complex and sophisticated entity, with a multitude of parts interconnected by pipes and wires and requiring access by operators and service personnel.

Previously, physical models of industrial plants were built and used as discussion pieces for design teams, with the design process being primarily performed using 2-D CAD drawings. All object placements and connections were done by hand, a complex task prone to errors and nonoptimal solutions. Furthermore, going between the schematics and plans and the physical model was a difficult process of identification and location. Model adjustment was, of course, time-consuming and expensive.

A tool for the design of industrial plants based on VE technology has been successfully used in a variety of industries, including offshore oil and gas, chemical and pharmaceutical production, power generation, and paper milling (see Fig. 52.7). This is a highly sophisticated and complex tool designed for experts to use and provides many automated features.

### 2.4.2 *The User Group*

The primary users are plant designers with training and experience in chemical and mechanical processes pertinent to the type of plant being designed. Secondary users may be others that the design is shown to, potential investors or politicians, for example. An interactive design

FIG. 52.7. Designing an industrial plant. Courtesy of Cadcentre. (a) Cadcentre pioneered the use of wide-screen immersive visualization in the process and power industries, seen here in action at the company's Visual Engineering Centre in Cambridge, England. Plant model courtesy of Conoco/Brown & Root. (b) Using live design data, Cadcentre's REVIEW Reality photorealistic plant visualization software gives engineers and nonengineering staff alike a unique impression of what the finished plant will look like, in this case, a pharmaceutical plant in North America.

can also be used for later identification of faults and for coupling the plan drawings with the locations of elements in the real model.

### 2.4.3 Rationale for Using Virtual Environments

As an interactive medium, a VE-based tools offers a significant improvement over prior techniques of plant design. Entire factories can be assembled by selecting components from a library of objects and placing them within a designated space. Cable and pipe drawing are automatic, thus reducing the chance of error due to misconnection or unnecessarily crooked pipes to fit around cables and pipes already placed. Adjustments can be performed in real time with the results immediately visible in the factory design and on the automatically generated schematics. Furthermore, components can be coupled to a database and thus interactively interrogated for information. Likewise, individual components can be isolated, viewed, and shown within the context of the entire system.

### 2.4.4 Classification

*Task*: This tool is used primarily to design working industrial plants and secondarily to provide information throughout the lifetime of an installation for maintenance and modification purposes. Components can be placed and connected into a (theoretically) working system. These can then be interactively interrogated to provide information for designers, builders, and maintenance personnel.

*Concreteness*: The design task is very concrete and usually results in a functioning factory or plant. In some cases a plant may be designed to show others, with the VE tool acting as a high-technology viewer of the proposed plant, perhaps to convince investors of the benefits and safety of the proposed system.

*End-User Participation*: There is almost no end-user (e.g., plant operator) participation. The design is driven mainly by experts with knowledge of how and where people must be able to fit into the design (e.g., gangways, room for turning valves, etc).

*Level of Expertise*: A high level of expertise is required of users. Not only does one need to know about the plant and process being designed, but the tool also requires a certain degree of modeling experience. On the other hand, many tasks are automated, which significantly increases the ease-of-use and reduces the complexity of the tool. Additionally, VE technology has been used to extensively improve the interactivity of the system.

*Simultaneity*: Many people can participate, both across time and space.

*Degree of Immersion*: The hardware used ranges from ordinary desktop PCs, or even portable systems, to fully immersive multiprojector studios and is usually determined by the budget and complexity of the system to be designed. As with most design tools, however, it is unusual for HMDs to be used due to the closed-in nature and loss of contact with others on the design team.

*Interactivity*: Plant design is highly interactive. The key is a tight coupling between direct manipulation in 3-D of the entire plant, individual objects, and other information such as part specifications and schematics. This can be automatically linked from the 3-D interface to 2-D windows on the computer screen to Web pages showing 3-D images.

*Realism*: Models are static, with no actions associated with them. The aim is to design, not simulate a process.

*System Control*: As previously described, there is a high degree of automation. Indeed, this is necessary if such a tool is to be of any advantage over existing techniques.

*Object Library*: There is a large object library, which can be complemented by CAD models. Building a plant becomes a simple process of dragging and placing system components on a "concrete" base.

### 2.4.5 Summary

Using VEs for plant design is a good example of a commercial application based on standard, albeit advanced technology, which has shown consistent advantages for its users. Nevertheless, improvements are continuously being made as new hardware and software become available.

## 2.5 Others

This section contains descriptions of some other applications of VE technology to design. These are kept short, because they are similar to previously described applications, are on the fringe of systems design, or there has been very little work performed in the area.

### 2.5.1 Artistic Design

Within the artistic community, VEs are commonly used to create virtual art galleries to display ordinary static art. However, in some cases it is used as the actual medium for the artwork itself (Shaw, 1995), for designing physical artworks, or for exploring other facets of artistic interconnectedness, such as between art, design, and architecture (Kiechle, 1996).

Artwork usually intends to produce a pleasing, controversial, or perhaps thought-provoking experience for the observer. There is not necessarily a product to be developed, nor limitations of space and time to be adhered to. In fact, many interactive artworks deliberately break these barriers (e.g., Sommerer & Mignonnneau, 2000). Virtual environment technology is used both for construction and interaction with virtual artworks, perhaps with many participants, and as a tool in the design and documentation of real, physical art (Kiechle, 1996). Artworks can be considered a system in that they are composed of many interacting parts built using a hierarchical structure.

Virtual environment tools, CAD packages, and other 3-D design programs are used by artists to assemble the system or for the construction of physical artworks. The primary users are artists—expert users—with the audience being able to perhaps navigate through the constructions or make predefined alterations within the boundaries of the experience. Ordinary VE technology is used for the construction phase; however, the potential for immersive systems to create experiences far beyond those of, say, an ordinary sculpture are well recognized by art designers (e.g., Shaw, 2000) but are seldom realized due to the cost of such systems. It is expected that more interactive art will be constructed in the future, perhaps for perusal over the Internet or available in art galleries, with maybe even the ability to share the experience with people at other locations around the world.

### 2.5.2 Small Things

Many small devices such as telephones and video players can be designed using VE tools. Prototype interfaces can be constructed and tested with end users for comprehension and ease of use; parallel prototypes can be contrasted and the optimal design chosen. Typically, ordinary CAD and VE packages are used to construct the devices, with output being via an ordinary screen, or even a stereoscopic display. When this is combined with tactile feedback (see chaps. 5 and 6, this volume), highly realistic interaction can occur, giving the testing a high degree of ecological validity.

### 2.5.3 Ergonomic Design

Improvement of VE design tools for architectural and interior design is occurring due to the gradual unification of virtual human modeling programs and VE technology. This is increasing the ability to put people into the environment being designed, try different body sizes according to race, gender and disability, and run routines to automatically test factors such as turning space, angle of lift and the corresponding stress on the body, glare, and so forth. The next step will be to allow real people to interact with virtual humans in a realistic manner, using them as role-players in a scenario being played out in the VE (see chap. 15, this volume).

## 3 CONCLUSION

This chapter has shown the potential of VE for systems design and allowed the reader to see where it can fit within his or her own design process. A number of common features and useful rules of thumb have been identified; however, a few further comments are necessary.

It is essential to have a clear idea of how VE technology can help in the design of your system. What is the task to be performed? And can one buy a ready-made solution or reuse a solution from another field? Many general-purpose VE systems can be purchased, but these require significant work to create a useful tool for a specific design situation. This cost must be added to that of hardware and software. Alternatively, one can often buy an entire package from a VE consultancy company, which will deliver a ready-made solution of hardware and (programmed) software as well as support and service. However, in either situation, be aware that there are many technical limitations and compatibility issues to be resolved, small conversion programs can be surprisingly expensive, and current practices may have to be severely modified (or even discarded) to incorporate VE design.

Next, one must consider who is to use the tool. If end users are to be involved in some way: Will they be required to interact directly with the tool, or will they be chaperoned by a design expert? This significantly affects the degree of ease of use and selection of display methods and input devices. Even if end users are not to be directly involved, what is the CAD and 3-D modeling expertise of those who will use the system? End-user participation seems to be beneficial where the clients are those who will be expected to live or work in the environment being designed. Increased involvement appears to improve acceptance of the results, providing a feeling of ownership for the design, and allows for the capture of tacit knowledge not necessarily available to the designer.

The level of immersion, interactivity, and realism are closely bound to the design task. Tasks where attention to detail is essential, developing a feeling for the form of an object (say a car) or being able to get a feel for a space, require often-expensive immersive technology, a high degree of interactivity, and realism. If, however, such attention to pictorial detail nor immersion is required, excellent results are obtained by small systems, often based on ordinary PC technology. This is indeed borne out by work in other areas such as VEs for rehabilitation (see chap. 49, this volume), where expensive immersive technology is beyond the budget of most medical institutions and positive results have been obtained with smaller systems. Furthermore, the so-called ordinary PC technology of today is equivalent to the high-end graphics workstations of a couple of years ago—and this trend shows no sign of flattening off, implying that only extreme cases require a large investment.

Finally, a good recommendation is to see how one's competitors are using VE technology. If they are not—why not? Do they lack vision, or are the predicted financial gains too uncertain (Wilson et al., 1996)? If they merely lack vision, then maybe it is time to be a pioneer—VEs are here to stay, with some developers now reaching over a decade of experience and many companies reporting substantial savings in time and resources, as well as a significant competitive advantage.

## 4. REFERENCES

Adjei-Kumi, T., & Retik, A. (1997). A library-based 4D visualisation of construction processes. *Proceedings of IEEE*, 315–321.

Akselsson, K. R., Bengtsson, P., Eriksson, J., Johansson, C., R., Johansson, G., I., & Klercker, J. (1994). Computer-aided planning of production and working environment. In G. E. Bradley & H. W. Hendrick (Eds.), *Human factor in organizational design and management* (vol. 4, pp. 499–504). Amsterdam: Elsevier Science Publishers.

Cadcentre. (2000). *Products and Service—Plant design and viewing solutions* [Online]. Available: http://www.cadcentre.com/product/design.htm [2000, June 25]

Davies, R. C. (2000). *Developing a virtual reality tool for participatory design of real environments—A multiple-case study*. Manuscript submitted for publication.

Ehn, P., Brattgård, B., Dalholm, E., Davies, R. C., Hägerfors, A., Mitchell, B., & Nilsson, J. (1996). The envisionment workshop—From visions to practice. In *Proceedings of the Participatory Design Conference* (pp. 141–152). Cambridge, MA: MIT Press.

Eriksson, J. (1998). *Planning of environments for people with physical disabilities using computer-aided design*. Unpublished doctoral dissertation, Lund Institute of Technology.

Eriksson, J., & Johansson, G. I. (1996). Adaptation of workplaces and homes for disabled people using computer-aided design. *International Journal of Industrial Ergonomics, 17*(2), 153–162.

Kiechle, H. (1996). *Amorphous constructions* [Online]. Available: http://www.vislab.usyd.edu.au/staff/horst [2000, June 25]

Norman, D. (1988). *The design of everyday things*. New York: Doubleday.

Östman, J. (1998). *Virtual reality and virtual prototyping—Applications in the American car industry* (Overseas Rep. No. 9806). Swedish Office of Science and Technology.

Retik, A. (1997). Planning and monitoring of construction projects using virtual reality, *Project Management, 3*, 28–31.

Rheinfrank, J., Evenson, S., & Wulff, W. (1994). Design as common ground. In D. E. Mahling, & F. Arefi (Eds), *Cognitive aspects of visual languages and visual interfaces* (pp. 77–102). Amsterdam: Elsevier Science Publishers.

Shaw, J. (1995). Exhibit review: Jeffrey Shaw's golden calf—Art meets virtual reality and religion [*Leonardo*, Online]. Available: http://www-mitpress.mit.edu/e-journals/Leonardo/reviews/shankencalf2.html [2000, June 25]

Shaw, J. (2000). *Place—A user's manual* [Online]. Available: http://www.kah-bonn.de/1/17/0e.htm [2000, June 25]

Sommerer, C., & Mignonnneau, L. (2000). *Christa and laurent—Works* [Online]. Available: http://www.mic.atr.co.jp/~christa/WORKS, [2000, June 25]

Vance, J. M. (1998). A virtual environment for engineering design optimization. In *Proceedings of 1998 NSF Design and Manufacturing Grantees Conference* (pp. 27–28). Monterrey, Mexico:

Wilson, J. R., Cobb, S. V. G., D'Cruz, M. D., & Eastgate, M. D. (1996). *Virtual reality for industrial applications: Opportunities and limitations*, Nottingham, England: Nottingham University Press.

Wilson, J. R. (1999). Virtual environments applications and applied ergonomics. *Applied Ergonomics, 30*, 3–9.

# 53

# Information Visualization in Virtual Environments

Steve Bryson
*NASA Ames Research Center*
*Numerical Aerospace Simulation Division*
*MS T27B-1*
*Bldg. T27B Room 126*
*Moffett Field, CA 94035*
*bryson@nas.nasa.gov*

## 1. INTRODUCTION

Virtual environments provide a natural setting for a wide range of information visualization applications, particularly when the information to be visualized is defined on a three-dimensional (3-D) domain (Bryson, 1996). This chapter provides an overview of the issues that arise when designing and implementing an information visualization application in a virtual environment. Many design issues that arise (e.g., issues of display or user tracking) are common to any application of virtual environments. In this chapter, the focus is on those issues that are special to information visualization applications, as application issues of wider concern are addressed elsewhere (see chaps. 42–55, this volume).

### 1.1 What Is the Problem?

In this chapter, information visualization is considered to be the visual investigation of numeric data. This data may represent real-world observations or the result of simulation. The data may express tangible, real-world quantities, such as water temperature, or may represent abstract quantities, such as stock market prices. Data is herein restricted to "information" rather than "objects" in order to avoid having to discuss representation of conventional real-world objects. Using this restriction allows the current discussion to concentrate on those aspects of information visualization that are unique to nonrealistic, abstract representations of information. Of course there is considerable overlap between "information" and "object" (e.g., representation of topographic data). If that representation is abstract in order to bring out some feature of the topography, the discussion in this chapter is relevant. If the topography is to be represented realistically, then the discussion here may not apply.

Virtual environments (VEs) provide unique opportunities for information visualization (Bryson, 1996; Cruz-Neira et al., 1993; Lin, Loftin, & Nelson, 2000). The inherent three-dimensional (3-D) nature of VEs makes them a natural choice for investigation of multidimensional data domains. The natural, anthropomorphic interface suggested in VEs inspires natural, intuitive interfaces for the investigation of data. This intuitive interface may be particularly suitable for visualizing highly complex data sets, where phenomena in the data cannot be effectively displayed all at once. In such situation an intuitive 3-D interface may allow for rapid investigation of data employing a What's happening here? paradigm.

This chapter explores how VEs can be effectively used for information visualization. After characterizing the information visualization process, an examination of how information visualization environments differ from other, more "real-world"-oriented applications is provided. Using these observations, implementation strategies tailored for information visualization in VEs are provided, including issues of run-time software architectures, distribution, and control of time flow. These implementation strategies will be driven by consideration of the human factors of interaction. The chapter closes with a discussion of the challenging aspects of information visualization in virtual environments.

## 1.2 The Data Analysis Pipeline

In order to describe special issues that arise in the implementation of VE information visualization systems, a conceptual model of an information visualization system is required. There are many ways to conceptualize information visualization, and we make no claim to the most complete or optimal conceptualization. The following, however, has been found very informative when considering implementation issues.

The information visualization process is considered as a pipeline, which in its most generic form starts with the data to be visualized. From this data visualization primitives are extracted. These primitives may consist of vertices in a polygonal representation, text for a numerical display, or a bitmap resulting from, for example, a direct volume representation. Primitive extraction typically involves many queries for data values. The extracted primitives are then rendered to a display. This pipeline allows user control of all functions, from data selection through primitive extraction to rendering. This pipeline is shown in Fig. 53.1.

As an example of this pipeline in operation, consider streamlines of a vector field (see, e.g., Bryson, 1998). Given a starting point of a streamline, data (the vectors at that point) are accessed by the streamline algorithm. The vector value is then added (sometimes with a complex high-accuracy algorithm) to the starting point, creating a line primitive. This process is iterated to build up a (typically curved) line with many vertices. These vertices are the streamline's extracted representation. These vertices are then rendered in the visualization scene. The extraction of these primitives may involve significant computation even though the data may exist as a precomputed file. Computations like those in this example will turn out to be a significant issue in the implementation of information visualization in virtual environments.

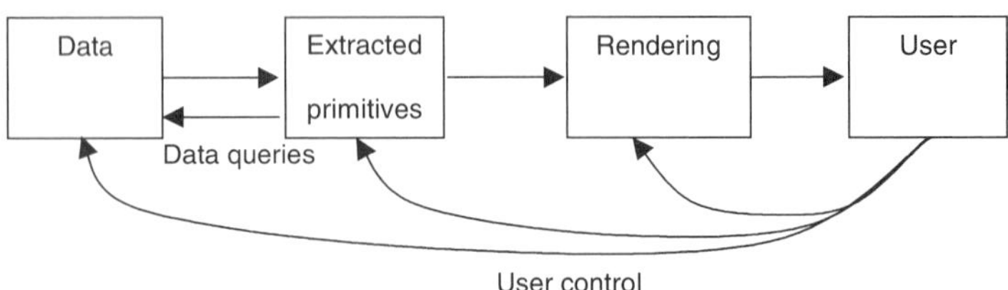


FIG. 53.1. The data analysis pipeline.

### 1.3 Why Perform Information Visualization in a Virtual Environment?

Virtual environments offer several advantages for many classes of information visualization (Bryson, 1996; Cruz-Neira et al., 1993; Lin et al., 2000; Song & Norman, 1993). The inherent 3-D interface makes VEs a natural setting for information that exists on a 3-D domain. Examples of such data include 3-D scientific simulations, such as fluid flow, electromagnetic fields, or statistical data indexed by location in 3-D space. Three-dimensional interaction techniques common in VEs (see chap. 13, this volume) provide natural ways to control such features as visualization selection and data manipulation. In addition, experience has shown that one of the greatest advantages of information visualization in VEs is the inherent "near-real-time" responsiveness required by head-tracked direct-manipulation virtual environments (see chap. 8, this volume). This responsiveness allows rapid queries of data in regions of interest. Maintaining this responsiveness in an information visualization environment is the most challenging aspect of such an application and will be one of the primary foci of this chapter. When designed well, the combination of 3-D display, 3-D interaction, and rapid response creates an intuitive environment for exploration and demonstration.

Figures 53.2 and 53.3 show examples of VEs for the visualization of information arising from scientific computation. Figure 53.2 shows a system developed at the University of Houston for the exploration of geophysical flows (Lin et al., 2000). Figure 53.3 shows the Virtual Windtunnel, developed at NASA Ames Research Center, which is used to investigate results of simulations in computational fluid dynamics (Bryson & Levit, 1991; Bryson, Johan, & Schlect, 1997). Both examples exhibit the use of multiple visualization extracts in a single environment. All of these extracts are interactive and can be changed under user control.

Multimodal interfaces, using sound (see chap. 4, this volume), touch, and force (see chaps. 5 and 6, this volume) displays for VEs have been explored for information visualization in virtual

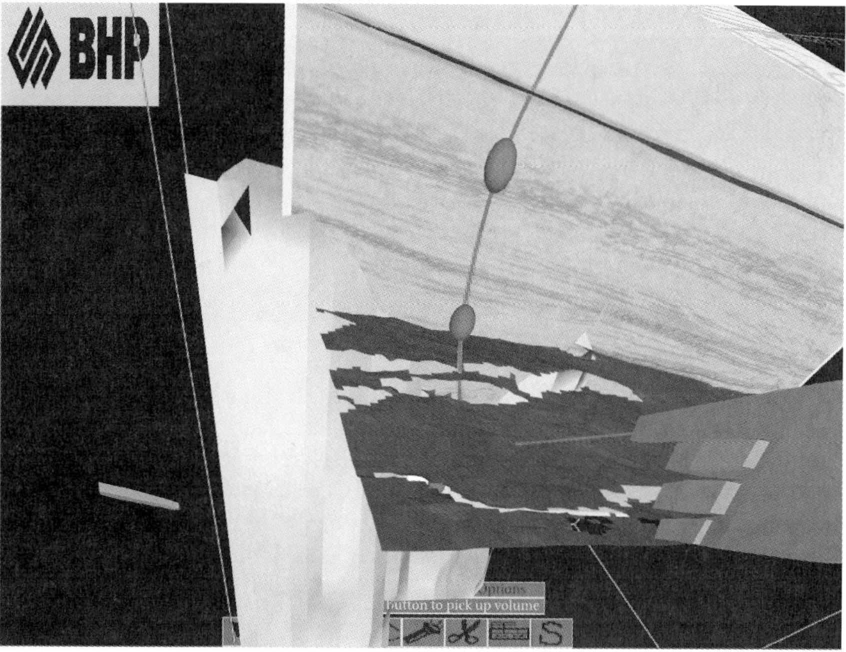


FIG. 53.2. An information visualization VE for exploration of geological data sets. Courtesy of the University of Houston/Bowen Loftin.

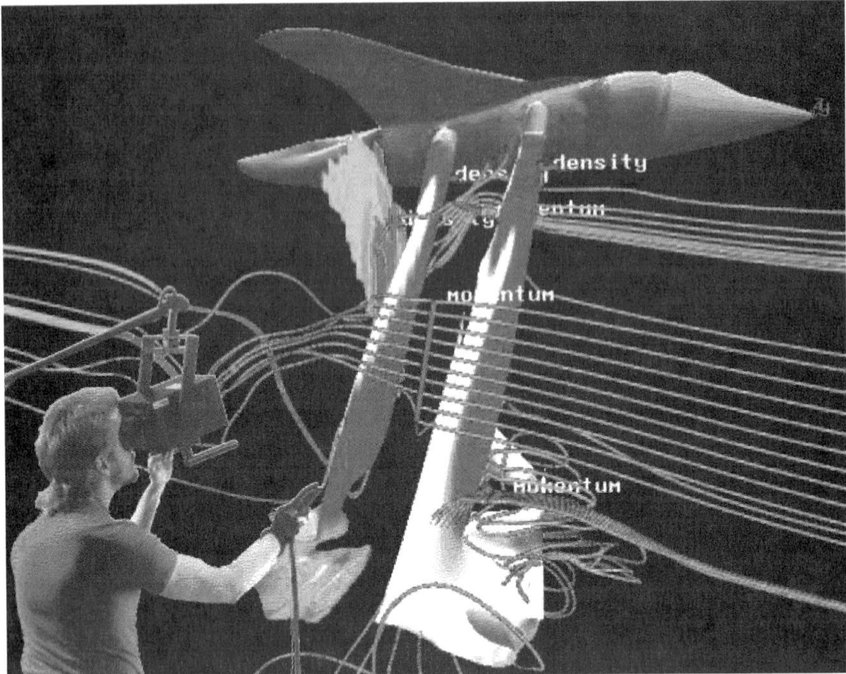


FIG. 53.3. The Virtual Windtunnel, a VE for visualization of Computational Fluid Dynamics simulation results.

environments (e.g., Taylor, Robinett, Chi, Brooks, & Wright, 1993). With the possible exception of force displays, however, the efficacy of such interfaces has not been demonstrated.

## 1.4 How Information Visualization Differs from Other VE Applications

The design and development of an information visualization application within a VE is different from most VE application development (see chaps. 42–55, this volume). Information visualization environments are often abstract and involve large amounts of data access and computation in response to a query. For time-varying data, different senses of time arise, in which the data may evolve more slowly or even backward relative to user time. Such differences between information visualization environments, and more conventional VE applications can be generalized into the following areas:

- **Greater flexibility in graphical representation**: The inherently abstract nature of information implies opportunities for simpler, faster graphics, such as representing a streamline as a simple polyline. Conventional applications, such as training (see chaps. 43 and 44, this volume) or entertainment (see chap. 55, this volume), are typically more concerned with realistic graphical environments and so may have less flexibility in the selection of graphical representation. Of course this is not a universally applicable rule, as some representations of information, such as direct volume rendering, can be very graphically intensive.
- **A large amount of numerical computation may be required**: Although information visualization typically addresses pre-existing data, visualization extracts may themselves require considerable computation. Streamlines or isosurfaces in the visualization of continuous vector and scalar fields are well-known examples that require large amounts of computation. Statistical or data mining techniques for abstract data may also require computation. As more

sophisticated data analysis techniques are used in VEs more computational demands can be expected.

• **A large amount of data access may be required**: Visualization extracts require access to data, and some extracts require more data access than others. Complete isosurfaces, for example, require traversing an entire data set (at a particular time step for time-varying data) for the computation of the data set. Time-varying data sets can be extremely large, requiring access to hundreds of gigabytes of data in a single analysis session, albeit a single time step at a time.

• **There may be various senses of time**: Several senses of time can arise in an information visualization system, particularly when addressing time-varying data sets. While some of these senses of time correspond to conventional time in other VE applications, completely new ways of thinking of time arise from the fact that a user may wish to manipulate time flow in a time varying environment. This issue and its impact are discussed in much more detail in section 4.

These differences between an information visualization VE and other applications have major impacts on the design of the VE system.

## 2. SYSTEM ARCHITECTURE ISSUES SPECIFIC TO INFORMATION VISUALIZATION IN VIRTUAL ENVIRONMENTS

There are several issues that arise in the design and implementation of VEs for information visualization. In this section, some of these issues are examined in detail. First, however, a classification of the types of interaction that may occur, which in turn depends on the time flow of the data, is provided.

### 2.1 Classification of Interaction and Time Flow

There are two design questions that must be answered before considering an appropriate implementation of an information visualization application in virtual environments:

1. **Is the user allowed to interactively query data at run time, generating new visualization extracts**? If so, the system will likely have user-driven data accesses and computation to support extraction of new visualization geometry.
2. **Is the data time-varying**? If so, there will be at least two senses of time in the virtual environment: user time and data time. The flow of data time may be put under user control so that it may be slowed, stopped, reversed, or at times randomly accessed.

These questions are independent, and both must be answered in order to determine which kind of implementation strategy will be appropriate. There are four combinations that arise from the answers to these two questions:

• **Noninteractive, non-time-varying data**: This is the simplest information visualization environment, where the visualization geometry can be extracted ahead of time and displayed as static geometry in a head-tracked virtual environment. No data access or computation issues occur in this case. The user may be able to rotate or move the static geometry. The design issues that arise in this case are common to all VE applications and so will not be considered further in this chapter.

• **Noninteractive, time-varying data**: In this case the visualization extract geometry can be precomputed as a time series, which may be displayed as a 3-D animation in the virtual environment. The user may be given control over the flow of the animation, to slow it down, stop it, or reverse direction. Such user-controlled time flow implies a special interface, which may include rate controls to determine the speed and direction of the data time or a time step control for random access of the extracted geometry at a particular data time. When the extracts are defined for discrete time steps, interpolation in time may be desirable to allow for a smooth flow between time steps. What kind of interpolation is appropriate will be highly domain and application dependent. Except for these considerations, however, the issues that arise in this case are common to most VE applications and so will not be considered further in this chapter.

- **Interactive, non-time-varying data**: In this case the data does not change in time, but the user specifies the visualization extracts at runtime. In a VE, such extracts may be specified via a direct-manipulation interface where the user either specifies a point or manipulates an object in 3-D space. The visualization extract may require significant computation, which will have an impact on the responsiveness of the system. This impact is discussed at length in section 2.2. When visualization extracts do not change they are typically not recomputed.
- **Interactive, time-varying data**: For this type of environment, the data itself is changing with time, so any existing visualization extracts must be recomputed whenever the data time changes. This can result in a large amount of computation for each data time step, the implications of which are discussed below in section 2.3. In addition, the user may be given control over the flow of data time, allowing data time to run more quickly or slowly, or in reverse. The user may wish to stop time and explore a particular time step. When time is stopped the system should act like an interactive, non-time-varying data environment, allowing visualization extracts to be computed in response to user commands.

## 2.2 Interaction Versus Data Time Scales

Responsive interaction plays two important roles in virtual environments:

- **Fast graphics** is required to permit the head-tracked display required for a strong sense of object presence and immersion (see chap. 40, this volume).
- **Fast response** to user commands are required for a direct manipulation interface paradigm, where the user is allowed to directly "pick up and move" objects in the virtual environment.

In information visualization environments, there is the added requirement that the user be able to move a data probe displaying a visualization extract and see that extract update in near real time as appropriate to its new position. If the data probe's extract updates sufficiently quickly, the user will have a sense of exploring the data in real time.

How fast the graphics and interaction response must be turns out to be both application and domain dependent. A VE for training manual tasks, such as a flight simulator, requires response times of less than $\frac{1}{30}$ of a second. Longer response times have been observed to train users to expect delays that may not exist in the real world and lead to incorrect operation (Sheridan & Ferrill, 1974). Information visualization, however, does not typically require fidelity to any real-world time scales. In fact it is often desirable that the information being visualized be presented with very different time scales from the real world, with data time being either slowed down or sped up depending on the user's needs.

While fidelity to real-world time scales is not a guide to the required responsiveness of an information visualization environment, human factors considerations will set performance requirements. Three human factors issues turn out to be important for information visualization in virtual environments:

- **Graphics update rate**: How fast must the graphical display update rate (graphical animation rate) be to preserve a sense of object presence and/or immersion? By graphical update rate we mean the rate at which frames are drawn to the frame buffer(s), not the display device's refresh rate.
- **Interaction responsiveness**: How quickly must objects in the environment respond to user actions in order to maintain a sense of presence and direct manipulation? This requirement applies to the overall head-tracked display as well as ability to interact with objects.
- **Data display responsiveness**: How fast must interactive data display devices, such as a data probe, update to give the user a sense of exploring the data?

The relationships and differences between these time scales are subtle. Graphics update rate will limit interaction and data display responsiveness because the interactive displays cannot be presented faster than the graphics update rate. Update rate and responsiveness, however, are very different kinds of measures: Update rate is measured in frames/second, while responsiveness is the *latency*, measured in seconds—the time interval between a user action and when the system's response is displayed. This latency is determined by all processes triggered by the user's actions, from reading the user tracking devices through processing the user's commands through possible subsequent computation to the display of the result. Experience has shown that the limits on these time scales are as described below (see Bryson, 1996; Sheridan & Ferrill, 1974):

- **The graphics update rate must be greater than 10 frames/second**. While faster update rates are desirable, 10 frames/second is sufficient to maintain a sense of object presence even though the discrete frames of the display are easily perceptible. Slower update rates result in a failure of the sense of object presence, eliminating the enhanced 3-D perception advantages of a virtual environment.
- **Interaction responsiveness must be less than 0.1 seconds**. While lower latencies and faster responsiveness is desirable, a latency of 0.1 seconds is fast enough to give the user a good sense of control of objects in the virtual environment. Longer latencies typically cause the user to experience unacceptable difficulty in selecting and manipulating objects in 3-D space.
- **Data display responsiveness must be less than about one third of a second**. While faster responsiveness is desirable, a data display latency of one third of a second maintains a sense of "exploring" the environment, though the user may use slow movements to adapt to this relatively long latency. Longer latencies in data display require such slow movements on the part of the user as to lose usability (see chap. 34, this volume).

The graphics update rate and interaction responsiveness requirements are similar: One graphics frame of latency is allowed for to maintain good responsiveness for user interaction. The data display responsiveness requirement, however, is less restrictive. The difference in latency requirement between interactivity and data displays is due to the fact that user interaction (selecting, acquiring, and moving objects) is a manual task driven by the human factors of manual control, whereas observing data display during movement is an intellectual task in which the user observes what happens as a data probe is moved through space.

Interactive time-varying environments potentially present a difficult challenge in meeting the above requirements: All nonprecomputed visualization extracts must be computed whenever the time step changes. Furthermore, when the time step changes, all computations must take place before any of the extracts can be displayed so that extracts from different data time steps are not displayed at the same time. Experience has shown that it is acceptable for the data time step to change quite slowly, so long as the 10 frames/second graphical update rate and the one third of a second data responsiveness requirements are met.

These observations suggest that graphical display and user interaction be treated independently of the visualization. For example, interaction and visualization display may be treated independently by having simple graphical tools in the VE that respond to the user without waiting for data to be displayed (Bryson et al., 1997; Herndon & Meyer, 1994). These tools may trigger associated data displays, but these data displays should occur asynchronously from the display of the tool, allowing the tool to respond to the user's input as quickly as possible.

A simple crosshair in 3-D space, with an associated streamline of a vector field, is an example of such a tool. In this example, the user can "pick up and move" the crosshair, which will be very responsive (within the limits of the graphical update rate) due to its simple graphical nature. When this crosshair is moved it will trigger the computation of a streamline at its current location. This streamline may take much longer to compute and display, after which time the tool will trigger a new streamline from its new location. If the process computing the streamline runs asynchronously from the process handling user interaction and graphical display, interaction responsiveness will be only slightly impacted by the computation of the streamline (assuming a preemptive multitasking operating system). This example shows that the difference in time scales between interaction responsiveness and data display responsiveness has strong implications for the run-time architecture of the information visualization system. These implications for the overall system architecture are discussed in section 2.3. The method of triggering computations outlined in this simplified scenario will be discussed in more depth in section 4.

### 2.3 System Architecture

The observations in the previous section imply that any implementation of an interactive information visualization VE in which computation of visualization extracts takes place should contain at least two asynchronous processes: a graphics and interaction process and a visualization extract computation process. More generally, the graphics and interaction task may be performed by a group of processes we shall call the interaction (process) group, one or more processes for graphics display and possibly separate processes for reading and processing user tracking data. Similarly, the visualization extraction task may be performed by several processes, called the computation (process) group, possibly operating on multiple processors in parallel. These groupings are chosen because processes in the interaction group all have the same 10 frames/second and 0.1 second latency requirements, whereas the computation group has the one third of a second latency requirement. This process structure decouples display and computation so that a slow computation does not slow down the display process and the speed requirements of the display process do not limit the computation.

The interaction process group passes user commands to the computation process group, which triggers the computation of visualization extracts. These resulting extracts are passed back to the interaction process group for display. This basic architecture is outlined in Fig. 53.4.

## 3. DISTRIBUTED IMPLEMENTATION

Distributed data analysis can be highly desirable, particularly in one or more of the following circumstances:

- Data exists on a remote system, and it is inconvenient to move the data to the local visualization system.

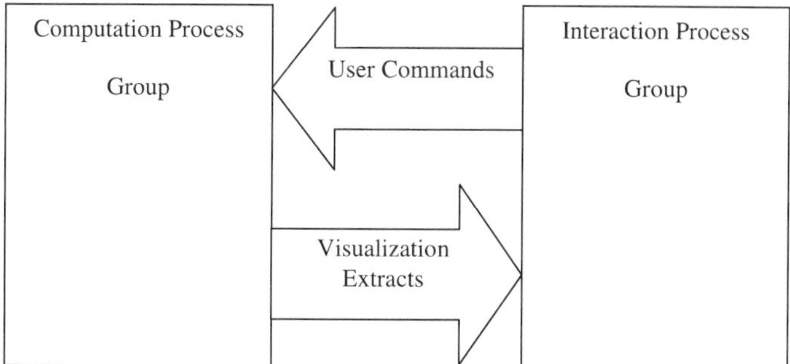


FIG. 53.4. Run-time process architecture of an information visualization system for interactive and/or time-varying data.

- Desired visualization extracts require more computation than the local visualization system can easily deliver.
- Data is the result of a concurrently running computation on a separate high-performance computational system.
- Remote collaboration is desired, where several clients have (effective) access to the same data.

The use of separate, asynchronous computation and interaction process groups communicating via buffers as described in section 2.3 facilitates a distributed implementation, where the process groups exist on separate, possibly remote machines communicating over a network.

## 3.1 Distribution Strategies

There are several strategies for the distribution of data analysis. These strategies are determined by selection of where to place which operations in the data analysis pipeline of Fig. 53.1. These strategies and their advantages and disadvantages are as follows:

- **Remote data, local extraction, and rendering**: In this option data exists on a remote system, and individual data values are obtained over a network as required by the local visualization extraction algorithm. This strategy has the advantage of architectural simplicity, with all visualization activities taking place on the local system, as if the data were local. This strategy has the disadvantage that it requires a network access each time data is required, which can be time-consuming. There are techniques to overcome this disadvantage, such as clever prefetching, where data is delivered in groupings that anticipate expected future queries. For some information types the remote data distribution strategy may be too slow for a VE interface, for example, when analyzing continuous fields that arise in computational fluid dynamics, where large visualization extracts may require many data accesses for their computation. For other applications, however, where data accesses are relatively few, remote data distribution may be a viable strategy.
- **Remote data and extraction, local rendering**: With this strategy visualization extraction occurs on a remote system, typically the same system that contains the data. In an interactive system the extraction computations are in response to user commands passed from the user's local system. The results of the extraction, typically geometrical descriptions of 3-D visualization objects, are transmitted to the user's local system for rendering. Architecturally, this strategy maps closely to the run time architecture illustrated in Fig. 53.4, with the computation process group on (one or more) remote machines and the display and

interaction process on the local system. This strategy has the advantage that the extractions algorithms are "close to the data," so data access is not a bottleneck. It also has the advantage of local rendering, which allows head-tracking display for each participant as required in virtual environments (see chap. 8, this volume). The disadvantages of this strategy include the fact that response to user commands requires a round trip over the network, and the requirement that the user's local system be capable of rendering the extract's geometry.

- **Remote data and extraction, distributed rendering**: This strategy is a variation of the above remote data and extraction, local rendering strategy. In this case the rendering commands occur on the remote system and are passed as distributed rendering calls to the user's local system for actual rendering. A local client program is still required to read the user's trackers and process and send user commands. This strategy has advantages and disadvantages similar to the remote data and extraction, local rendering strategy, except that the network round-trip time is now part of the graphics display loop. This may introduce unacceptable delays into head-tracking responsiveness.
- **Remote data, extraction and rendering**: This strategy places all the operations of the data analysis pipeline on the remote system(s), with the final rendered frames returned to the user's local system over the network. This strategy has the advantage that very powerful remote systems can be used when the user has a very low-power local system. The disadvantage is that the rendered frames can be large, for example, for a $1{,}024 \times 1{,}024$ 24-bit RGB-alpha display requires a four-megabyte frame buffer, and two such frame buffers are required for stereo display. This implies an 80-megabyte transfer every second in this example to maintain the frame rate of 10 frames per second. This bandwidth requirement is beyond most available large-area networks. There are also serious issues of latency control in this strategy because the network time is part of the display responsiveness loop. There are, however, situations where the local system is incapable of the kinds of rendering desired and this strategy may be the only viable option. Direct volume rendering is an example when this strategy may provide the optimal choice.

The above strategies are not exclusive: one may have remote visualization extraction taking place on one remote system while the data resides on a third system.

### 3.2 Remote Collaboration Strategies

Once a system is distributed, the opportunity arises for remote collaboration, where two or more noncolocated users examine the same data together. Strategies for remote collaboration are related to, but different from, distribution strategies. Next, common remote collaboration strategies are briefly summarized:

- **Distributed data**: This collaboration strategy places copies of the data to be examined on all participants' client systems. Collaboration is implemented by passing either user commands or computed visualization extract results between each of the participants' systems. The primary advantage of this strategy is that the software used is similar to stand-alone versions of the same systems. The main disadvantages include the difficulty of ensuring synchronization among the participants, and the requirement that each participant's system be capable of storing data and computing (at least locally generated) visualization extracts. Many distributed collaborative VE systems, such as military training systems (see chap. 43, this volume), utilize the distributed data collaboration strategy.
- **Data server**: This collaboration strategy builds on the remote data distribution strategy. The data typically resides on a single remote system and is accessed by the participant's local system as needed. The visualization extracts are computed locally and are communicated in the same manner as in the distributed data collaboration strategy above.

- **Visualization extract server**: This collaboration strategy builds on the remote extraction distribution strategy, in which visualization extracts are computed on a remote system, which is typically where the data being examined is stored. The extracts are sent to each participant's system for local rendering. The advantages of this strategy include:
  - As there is only one set of extracts associated with each set of data, synchronization is greatly simplified.
  - Local rendering allows each participant to render the visualization extracts from a local point of view, as required for head-tracked displays.
  - The extract server system can arbitrate conflicting user commands.

The visualization extract server collaboration strategy has the disadvantage of poor scaling to large numbers of users, though this problem will be alleviated when reliable multicast technologies become available. The other disadvantages of this strategy are the same as those for the remote extraction distribution strategy.

- **Scene replication**: This collaboration strategy has a privileged user whose view is presented to the other participants. This collaboration strategy is similar to the remote rendering distribution strategy. This strategy has the same advantages and disadvantages as the remote data, extraction and rendering distribution strategy, with the added disadvantage that all participants will see the same view thereby precluding the use of head tracking for all participants.

## 4. TIME FLOW AND CONTROL

When visualizing time-varying data, a user may desire to control the flow of time, for example, make time speed up, slow down, or reverse. Furthermore, to view a particular phenomenon in more detail at a specific point in time, the user may stop the time flow. These non-real-world manipulations of the flow of time require several new ways of describing time in a virtual environment. The discussion in this section closely follows Bryson and Johan (1996), where more details can be found.

### 4.1 Requirements of Time-Varying Interactive Visualization

There are several design principles that must be considered before developing time-varying interactive visualization applications:

- **Correct simultaneity**: All visualization extracts displayed at the same time should have been computed for the same data time, unless explicitly directed otherwise by the user.
- **Completeness**: All visualization extracts requested for a particular data time should be displayed unless the user directs otherwise.
- **Time flow control**: The user should be able to change the rate and direction of the flow of data time steps.
- **Interactive exploration**: If the user slows or stops the flow of time, interactive exploration should still be allowed. This implies that new visualization extracts may be computed even if the data time step does not change.
- **Minimize delay**: The system should display the most recently computed visualization extracts unless otherwise explicitly directed by the user.

### 4.2 Types of Time

In time-varying information visualization environments, time assumes an unusual and subtle meaning. There are several senses of time that arise in such environments.

- **User time**: The time that the user experiences, the "wall clock" time.
- **Data display time**: The displayed time step, defined as the time step of the data from which currently visible visualization objects were extracted.
- **Data time**: The time step from which visualization extracts have been computed. Data time may or may not be equal to data display time.
- **Frame time**: The number of times a set of visualization extracts have been computed for a given time step. Frame time is measured by counting the number of times the computation process has been called for a given data time.

One should distinguish between data display time and data time so that when data time is moving forward visualization extracts can be computed from the next data time while the current data display time is presented to the user. The next data time may not be related to the current data display time; for example, the user may choose a particular data time for the next display frame unrelated to the current data display time.

The purpose of the frame time counter is to support the scenario in which the advance of data time is stopped and the user wishes to interactively explore the current time step's data. By stamping the visualization extracts with their frame time in addition to their data time the most recently computed extract may be displayed, maximizing interaction responsiveness. The use of frame time is discussed in the next section on buffering strategies.

## 4.3 Buffering Strategies

In the run-time architecture described in section 2.3, the computation process group produces visualization extracts that are displayed to the user by the interaction process group. The method of communication between these groups is driven by the principles outlined in section 4.1 and is implemented using the time parameters defined in section 4.2.

The most difficult case is when the information visualization system supports several independent visualization objects, such as streamlines and isosurfaces, which can be independently specified and displayed. This shall therefore be assumed the case in the current discussion. It is further assumed that the graphics process keeps a list of visualization extracts to be displayed. It is desirable that visualization extracts that do not change be displayed without recomputation. The buffers that pass the visualization extracts from the computation process to the graphics process should be capable of passing only those extracts that are newly computed without destroying those older extracts that should continue to be displayed. This requirement is most easily met if buffering takes place at the visualization object level, where each visualization object manages its own buffer.

There are two approaches to buffering, *local time* and *time-cached buffering*. Local-time buffering does not retain visualization extracts over time, while time-cached buffering saves visualization extracts for possible reuse when appropriate.

### 4.3.1 Local-Time Buffering

In local-time buffering, only the most recently computed visualization extracts are displayed and there is no attempt to cache visualization objects from previous time steps. This approach to buffering is appropriate to non-time-varying data because new visualization techniques are generated only in response to user commands. Local-time buffering should be implemented so that: (1) the most recently computed extracts are available, and (2) the computation process does not have to wait while a visualization extract is displayed. Conventional double-buffering techniques, where one has a *write buffer* into which computed visualization extracts are stored and a *read buffer* out of which previously computed extracts are displayed, only partially meet these requirements. Consider the case where the computation process is faster than the graphics process. This will often happen in non-time-varying environments where perhaps one visualization extract is being computed in response to a user command, whereas all current visualization extracts need to be displayed. When using a double-buffer approach and the computation process is faster than the graphics process, the write buffer will be filled before the graphics process's read buffer has been used and released for writing. This will cause the computation process to wait until the read buffer has been released. Such waiting will result in a delay if the user is "sweeping" a visualization object in space, producing a continuous stream of visualization extracts to be computed.

A better approach is to use a triple-buffer strategy, where three buffers are available. During typical operation, the computation process will fill one or more of these buffers, so (after the buffers have been filled) there may be two or more buffers marked "read." Each of these buffers will be marked with the time frame number defined in section 4.2, as well as the data time number if the data is time varying. In order to minimize delay the graphics process will choose the buffer with the highest time frame number, which will be the visualization extract most recently computed. In the meantime, if the computation process wants to proceed it can compute the response to the most recent user commands, overwriting the older of the two available read buffers, thus minimizing latency. In this way a triple-buffer strategy allows local-time buffering to meet the requirement to minimize delay.

In a time-varying data environment, the buffers will also be stamped with the time step in which the extract was computed. The graphics process then selects buffers from the time step being currently displayed in addition to selection of the highest data frame time. This assures that the requirement of "correct simultaneity" of section 4.1 is met.

#### *4.3.2 Time-Cached Buffering*

In a time-varying data environment one may wish to cache previously computed visualization extracts. When this is done, if the time step from which those visualization extracts were computed is encountered again, those extracts can be displayed without recomputation.

Time-cached buffering requires local-time buffering within each time step. When the time step advances the most recent visualization extracts for that time step can be stored. Then, when that time step is reselected, either by user command or because the time flow is periodic, those extracts can be displayed without recomputation if their specifying data (e.g., seed positions for streamlines or values for isosurfaces) are still valid.

## 5. TIME-CRITICAL TECHNIQUES

One of the prime requirements of VEs is responsiveness. In section 2.2, the performance requirements for various aspects of an information visualization application within a VE were discussed. These requirements can be difficult to meet in light of the possibly complex graphics and extensive computation required for computation of visualization extractions. One is often faced with a conflict between the requirements of a complete or accurate visualization and the requirements for responsiveness and fast graphical display rates. Although accuracy is often critical in an information visualization environment, users often prefer fast response with a known degradation in accuracy for purposes of exploration. When a phenomenon of interest is found in the more responsive but less accurate mode, the user can request that this phenomenon be recomputed and displayed more slowly with higher accuracy. The automatic resolution of the conflict between accuracy and responsiveness, finding the appropriate balance, is known as *time-critical design*, the topic of this section.

## 5.1 The Time-Critical Philosophy

Time-critical design attempts to automate the process of finding a balance between required accuracy and responsiveness. This approach is very different from real-time programming, which guarantees a particular result in a specified time. Real-time programming typically operates in a fixed, highly constrained environment, whereas time-critical programs are typically highly variable. This variability is particularly evident in an information visualization environment, where the data and extracts computed and displayed may vary widely within a single user session. Time-critical design does not guarantee a particular result, instead it delivers the "best" result possible within a given time constraint. A successfully designed time-critical system will provide a graceful degradation of quality or accuracy as the time constraint becomes more difficult to meet.

Time-critical design for a particular aspect of a program begins with defining a cost and benefit metric for the task to be completed. The task is then parameterized in a way that controls both costs and benefits. When the cost and benefit of a task are known as a function of the task's parameters before that task is performed, the appropriate choice of parameters is selected to maximize the cost–benefit ratio. There are often many tasks to be performed in an environment, and the benefit of a particular task can be a function of the state of that environment. The solution of this problem is often approached as a high-dimensional constrained optimization problem, maximizing the total cost–benefit ratio for the sum of the tasks to be performed given the constraint of the total time allowed for all tasks.

As discussed below, however, it is often very difficult to know the benefit and cost of a task before that task is performed. In such situations, hints provided by the user or simple principles, such as assuming equal benefit within a set of tasks, are often used.

## 5.2 Time-Critical Graphics

Time-critical techniques were pioneered in computer graphics (Funkhouser & Sequin, 1993), where objects were drawn with higher or lower quality depending on such benefit metrics as position in the field of view and distance from the user. Such implementations often used multiple representations of an object at varying levels of detail. In information visualization, however, many visualization extracts are already in a minimal form, such as streamlines defined as a set of points. There are opportunities for graphical simplification in information visualization, however. For example, it may be the case that many more primitive elements are used to define a surface than is necessary for its display. An isosurface may have regions that are close to flat but the algorithm used to derive the isosurface may create many surface elements in that flat region. Display of that surface would be faster if the flat region were represented by fewer surface elements (see, e.g., Chen et al., 1999, and subsequent papers in that volume). A surface may also be represented in simplified form until it became the focus of attention, where small variations from flatness may be important. Unfortunately, algorithms that identify such opportunities for surface simplification are computationally intensive and may therefore be unsuited to recomputation in every frame.

From similar considerations, it may be concluded that unlike general computer graphics based on precomputed polygonal models, the use of time-critical graphics in information visualization will be highly dependent on the domain-dependent specifics of the visualization extracts. It may be very difficult, for example, to assign a benefit to a particular extract, especially when that extract may extend to many regions of the user's view. Although simple benefit metrics, such as the location of the extract on the screen, may be helpful, one should keep in mind that the user's head may be pointed in one direction while the user's eyes may be scanning the entire view. Such scanning is to be expected in an information visualization environment where the scene may contain many related, extended objects.

From these considerations there are few generalizations that can be drawn about the use of time-critical graphics techniques in information visualization. Opportunities do arise, however. Simple examples of time-critical graphics techniques that may be useful in information visualization include:

- Simplified representations, such as wireframe rather than polygonal rendering
- Surfaces represented as 2-D arrays, where simplified versions of the surface may be obtained by rendering every *n* point in the array
- Techniques that have been developed for time-critical direct volume rendering (e.g., Wan, Kaufman, & Bryson, 1999).

A more general approach to time-critical graphics is to use multiresolution representations of data or resulting visualization extracts. This is a new area of research, however, with relatively few results at the time of this writing.

### 5.3 Time-Critical Computation

Computation of visualization extracts can provide several opportunities for time-critical design (Bryson & Johan, 1996) because such computation is often the most time-consuming aspect of a visualization system. As in the case of time-critical graphics, the specifics of time-critical computational design will be highly dependent on the nature of the extract computed. Several general observations, however, can be made.

- Both the cost and benefit of a visualization extract can be very difficult to estimate a priori based on its specification and control parameters, especially since the extent of an extract is difficult to predict based on its specification alone.
- The cost of an extract can be roughly defined as the time required for its computation. Experience has shown that this cost does not vary widely between successive computations.
- Parameterizing the extent or resolution of the computation may most easily control the cost of a visualization extract. Techniques that begin their computation at a particular location in space, such as a streamline emanating from a point in space, lend themselves well to controlling their extent in space, which controls the time required by their computation. The cost of visualization techniques that rely on abstract or indirect specification is more effectively controlled by varying resolution.
- Other ways to control the cost of the visualization extract include choice of computational algorithm and error metrics for adaptive algorithms. These control parameters have a more discrete nature and may be set by the user or set automatically via specific trigger criteria.

Given that the benefit of a visualization extract is hard to predict, one may treat all extracts as having equal benefit unless specified by the user. In combination with the observation that costs do not change dramatically in successive computations, this allows the following simple solution to the problem of choosing the control parameters. For simplicity, the situation in which all visualization extract's costs are controlled by limiting their extent is considered. Here, each extract computation is assigned a time budget, and each extract's computation proceeds until its time budget is used up. Then the time taken to compute all extracts is compared to the overall time constraint. Each extract's time budget is divided by a scale factor determined by the total actual time divided by the total time constraint. This scale factor may have to take into account any parallel execution of the extract computations. If the time required to compute all extracts is greater than the time constraint, this will result in smaller visualization extracts that will take less time to compute. If the extracts become too small, a faster but less accurate computational algorithm may be chosen. If the time required to compute all extracts is smaller than the time constraint, the time budget of each extract is increased, resulting in larger visualization extracts. A similar approach may be used to choose the resolution with which visualization extracts may be computed.

It may be evident from this discussion that the types of control parameters one may use in time critical design will be highly dependent on the nature of the extract and how it is computed. Clever approaches can significantly enhance the time-critical aspects of a visualization technique. As an example, consider isosurfaces of a 3-D scalar field. Such isosurfaces are traditionally specified by selecting a value, resulting in a surface showing where that value is attained in the scalar field. Controlling the cost of a traditional isosurface by limiting the time of the computation will have unpredictable results. In the conventional marching cubes algorithm for computing isosurfaces the entire data set is traversed. If the marching cubes algorithm is stopped before completion and before regions of the field where the isovalue is attained are traversed, no isosurface will appear at all. Controlling the cost of the marching cubes algorithm by controlling the resolution with which the data set is traversed is possible, but this strategy does not provide fine control and may result in a significant degradation in quality. A different approach to isosurfaces, *local isosurfaces* (Meyer & Globus, 1993), directly addresses this problem. Rather than choosing an isovalue, for a local isosurface the user chooses a point in the dataset, from which the isovalue is determined as the value of the scalar field at that point. The isosurface is then computed (via a variation on the marching cubes algorithm) so that it emanates from that point in space and is spatially local to the user's selection. The cost of a local isosurface in controlled by computing the isosurface until that isosurface's time budget has been used up. Two examples of local isosurfaces can be seen in Fig. 53.3.

## 6. CASE STUDY: THE VIRTUAL WINDTUNNEL

The Virtual Windtunnel (VWT, Bryson & Levit 1991; Bryson et al. 1997) is a VE for the analysis of time-varying fluid-flow data sets that are the results of computational fluid dynamics computations. VWT supports distributed collaborative operation as well as several types of data analysis (see Fig. 53.3). In this section the implementation strategy of VWT is outlined.

VWT is designed to address the case of interactive, time-varying data from section 2.1 and therefore uses the system architecture described in section 2.3. There are several processes in the display and interaction process group, including a graphics process and multiple processes to read user tracking. The computation process group uses multiple processes to perform computations in parallel when multiple processors are available. VWT represents the visualizations as object classes and maintains lists of object instantiations for both computation and display. For distributed operation, the remote data and extraction, local rendering strategy of section 3.1 is used in combination with the visualization extract server collaboration strategy of section 3.2.

Time-flow control fully implements the discussion in section 4, using time-local buffering with the triple-buffering technique (Bryson & Johan, 1996). Each visualization object contains the triple buffer structure in the object definition, and each object manages its own buffer. Time-critical computation is implemented for several of the visualization techniques using the simple scaling strategy described in section 5.3. When the data is not time varying, the computation of streamlines and local isosurface extracts is reentrant, so in subsequent time frames these extracts may be enlarged if their defining data have not changed. Time-critical computation of the streamline extracts implements a choice of several computational algorithms, listed here in order of increasing accuracy and decreasing speed: one-step Euler integration; second-order Runge Kutta; and adaptive fourth-order Runge-Kutta. Which algorithm is used is automatically chosen based on the size of the streamline relative to a user-controlled *preferred size*. When the streamline falls significantly below the preferred size, a faster but less accurate streamline computation algorithm is chosen. Details can be found in Bryson (1998).

VWT has been tested in distributed collaborative mode with as many as four participants, two in Washington, D.C., and two in California (a distance of approximately 2,400 miles), with the data and visualization extract server in California. Communications in this test were over a conventional network. When examining a non-time-varying data set, the performance requirements of 0.1 second latency were maintained. Time-varying data sets also maintained the latency performance requirement for a small number of visualization extracts, however, as more and larger extracts were used (e.g., large isosurfaces) network bandwidth constraints became significant.

## 7. CHALLENGES

While several information visualization systems have been implemented in a VE setting, there are several challenges that have been encountered, which at the time of this writing do not have satisfactory solutions.

- **Large data sets**: Many information visualization applications involve very large data sets. Examples from computational fluid dynamics include time-varying data sets hundreds of gigabytes in size. Examples from the NASA Earth Observing System are expected to generate a terabyte of data a day. Accessing this data within the performance constraints of VE is beyond the reach of current technology as of 2000. Higher bandwidth and lower latency access to mass storage devices will help, but the tendency in the data-generating community is to use higher computational power to generate higher resolution and therefore lager data sets. Techniques to preextract interesting features from such data sets hold the greatest promise as a solution to this problem.
- **Representation of abstract data**: Most information visualization systems that have been successfully implemented in VEs are in domains that map closely to the real world, such as fluid flows or geophysics. Other domains involve the use of more abstract data. Appropriate mappings of abstract data into a 3-D VE is an area of ongoing research.
- **High-dimensional data**: Most information visualization systems that have been successfully implemented in VEs have data that is defined in 3-D space. Many information domains have data defined on a higher dimensional domain. Visualizing this data in two dimensions has been an ongoing area of research. We expect that visualizing this data in the inherently 3-D environment of VEs would be of benefit. Mapping these higher dimensions into the three dimensions of VEs is a problem that has yet to be thoroughly explored.
- **Time-critical distributed design for distributed environments**: Time-critical design for distributed VEs, where network traffic and latency times must be considered, is an unexplored area (see chap. 16, this volume). This challenge becomes more complex in the context of distributed computation, where different extracts are computed on separate systems.

## 8. CONCLUSIONS

Information visualization is an active and fruitful area of VE research and development. The advantages of information visualization in VEs include:

- 3-D display
- 3-D user interaction
- An intuitive interface that facilitates the exploration of complex data

In order to implement an effective information visualization application in a virtual environment, however, issues of responsiveness and fast updates must be addressed. These issues may be resolved via use of appropriate system architectures, design based on human factors issues, appropriate time control for time-varying data, implementation of time-critical techniques whenever possible, and appropriate choices for distributed implementations. The details of how these solutions are implemented will be highly dependent on the target domain and the specifics of the visualization techniques used.

## 9. REFERENCES


Bryson, S. (1996). Virtual environments in scientific visualization. *Communications of the ACM, 39*(5), 62–71.

Bryson, S. (1998). Time-critical computational algorithms for particle advection in flow visualization. In *Late-breaking hot topics proceedings, IEEE Visualization '98* (pp. 21–24). Research Triangle Park, NC: IEEE Press.

Bryson, S., & Levit, C. (1991). The virtual wind tunnel: An environment for the exploration of three-dimensional unsteady flows. In *Proceedings of IEEE Visualization '91* (pp. 17–24). San Diego, CA: IEEE Press.

Bryson, S., & Johan, S. (1996). Time management, simultaneity and time-critical computation in interactive unsteady visualization environments. In *Proceedings of IEEE Visualization '96* (pp. 255–261). San Francisco, CA: IEEE Press.

Bryson, S., Johan, S., & Schlecht, L. (1997). An extensible interactive framework for the Virtual Windtunnel. In *Proceedings of the Virtual Reality Annual International Symposium '97* (pp. 106–113). Albuquerque, NM: IEEE Press.

Chen, B., Swan, J. E., III, Kuo, E., & Kaufman, A. (1999). LOD-sprite techniques for accelerated terrain rendering. In *Proceedings of IEEE Visualization '99* (pp. 291–298). San Francisco, CA: ACM Press.

Cruz-Neira, C., Leigh, J., Barnes, C., Cohen, S., Das, S., Englemann, R., Hudson, R., Papka, M., Siegel, L., Vasilakis, C., Sandin, D. J., & DeFanti, T. A. (1993). Scientists in Wonderland: A report on visualization applications in the CAVE virtual reality environment. In *Proceedings of IEEE Symposium on Research Frontiers in Virtual Reality* (pp. 59–66). San Jose, CA: IEEE Press.

Funkhouser, T. A., & Sequin, C. H. (1993). Adaptive display algorithm for interactive frame rates during visualization of complex virtual environments. In *Computer Graphics: Proceedings of SIGGRAPH '93* (pp. 247–254), Anaheim, CA: ACM Press.

Herndon, K. P., & Meyer, T. (1994). 3-D widgets for exploratory scientific visualization. In *Proceedings of User Interface Software Technology '94* (pp. 69–70), Marina del Rey, CA: ACM Press.

Lin, C.-R., Loftin, R. B., & Nelson, Jr., H. R. (2000). Interaction with Geoscience data in an immersive environment. In *Proceedings of IEEE Virtual Reality 2000*. New Brunswick, NJ: IEEE Press.

Meyer, T., & Globus, A. (1993). *Direct manipulation of isosurfaces and cutting planes in virtual environments* (RNR Tech. Rep. No. RNR-93-019). Moffett Field, CA: NASA Ames Research Center.

Sheridan, T. B., & Ferrill, W. R. (1974). *Man–machine systems*. Cambridge, MA: MIT Press.

Song, D., & Norman, M. L. (1993). Cosmic Explorer: A virtual reality environment for exploring cosmic data. In *Proceedings of IEEE Symposium on Research Frontiers in Virtual Reality*. San Jose, CA: IEEE Press.

Taylor, R. M., Robinett, W., Chi, V. L., Brooks, Jr., F. P., & Wright, W. (1993). The Nanomanipulator: A virtual reality interface for a scanning tunnelling microscope. In *Computer Graphics: Proceedings of SIGGRAPH '93* (pp. 127–134). Anaheim, CA: ACM Press.

Wan, M., Kaufman, A., & Bryson, S. (1999). High-performance presence-accelerated ray-casting. In *Proceedings of IEEE Visualization '99* (pp. 379–386). San Francisco, CA: ACM Press.

# 54

# Virtual Environments in Manufacturing

John P. Shewchuk, Kyung H. Chung, and Robert C. Williges
*Virginia Polytechnic Institute and State University*
*Department of Industrial and Systems Engineering*
*Blacksburg, VA 24061*
*shewchuk@vt.edu; kychung@vt.edu; williges@vt.edu*

## 1. INTRODUCTION

By considering both the sensory and cognitive capabilities and limitations of humans, significant advances can be made in applying virtual environment (VE) technology to manufacturing systems. An integrated approach, which includes feasibility analysis, iterative design, and systematic evaluation, is needed to realize this potential. In order to provide appropriate information support, activity analysis needs to be conducted. A three-step manufacturing activity analysis procedure is presented to evaluate feasibility, determine potential benefits, and to establish the costs and benefits of using VE technology. A limited number of demonstrations and applications of VE technology using visualization (see chap. 53, this volume), simulation, information provision, and telerobotics (see chap. 48, this volume) have been made in manufacturing. These applications relate to tasks including the design of manufacturing activities, design of manufacturing facilities, training, execution of control and monitoring activities, execution of planning activities, and execution of physical processing activities. Only approximately 10% of these applications, however, have been evaluated empirically. The most promising areas for future application and research include manufacturing facilities and product design (see chap. 52, this volume), training (see chaps. 43 and 44, this volume), and execution of physical processing activities.

Applications of VE technology to manufacturing are just beginning to emerge. These applications are directed toward using VE technology as a means of improving performance for a wide variety of manufacturing tasks involving humans. The goal of this chapter is to review recent VE applications in this domain in order to isolate areas where additional applications and research have the potential for improving modern manufacturing.

## 1.1 Manufacturing Systems

Manufacturing is the process by which raw materials and purchased items are transformed into finished goods of greater value. Every type of tangible good available to society, from consumer durable goods to electronics, computers, and automobiles, results from manufacturing. The importance of manufacturing to the economic well-being of a nation cannot be underestimated. Higher levels of manufacturing activity lead to higher standards of living, as evidenced by such nations as the United States, Canada, Great Britain, France, Germany, Switzerland, and Japan. Manufacturing accounts for about 20% of the gross national product (GNP) in the United States, whereas primary industries (agriculture, forestry, mining, etc.) account for less than 5% of the GNP (Groover, 1996).

In order to manufacture items, manufacturing systems are required. A *manufacturing system* is an organized collection of machines, equipment, and humans that function in a predetermined manner to convert inputs into finished goods in response to actual or anticipated demand. Various types of manufacturing systems are found in different industries, depending on the nature of the products to be made and the demand for them. Fixed-position systems are those where the facilities required for manufacturing (machines, tools, etc.) are brought to the item for each operation. This approach is used for very large items, such as aircraft and ships. Product-oriented systems (flow shops) are those where the processing equipment is arranged in a line (or similar configuration) and items flow down the line one after another with finished items coming off the end of the line. These are the easiest systems to operate and control, and are used when product variety is low and demand is high (mass production). Process-oriented systems (job shops) are those where similar machines are grouped together into different departments (machining department, inspection department, etc.) and items travel from department to department as required for processing. Such systems are able to accommodate the largest possible variety of different products, but items take the longest to produce. Job shops are used when product variety is high and demand is low. Group technology systems are those where similar products are grouped together into part families, equipment is formed into one or more groups (or cells), and each part family is assigned to a specific machine group for production. This approach is a compromise between product-oriented and process-oriented systems and is used when product variety and demand are both medium. Finally, flexible manufacturing systems consist of multiprocess, numerically controlled machines that are linked together via an automated material handling system and are under hierarchical computer control. These systems have a large range of processing capabilities and can change from one product to another quickly and easily. Flexible manufacturing systems are best suited for mid-volume, mid-variety production where different products may be required in random order and in small order quantities.

## 1.2 Framework for Investigating Virtual Environments in Manufacturing

In order to discuss applications of VEs in manufacturing, some type of framework is needed. All manufacturing systems are comprised of three basic components, each of which can be considered a system. The *physical processing system* is responsible for the actual physical conversion of inputs into finished goods. The *planning system* is responsible for the generation of plans establishing how to employ the physical processing system. These plans consist of what products to make as well as when, how, and in what quantities. Finally, the *control and monitoring system* is responsible for ensuring that the plans generated by the planning system are properly executed in the physical processing system. This often requires real-time decision making to be performed and feedback to be sent back to the planning system.

Williams (1992) stated that a manufacturing system could be viewed in two fundamentally different ways. The first is the *functional view*, where one is concerned with what activities have to be performed and under what conditions they are performed. Though there are many different activities in manufacturing, one finds that every activity involves either a transformation of information or a transformation of matter and/or energy. Activities of the former type are known as informational activities, whereas the latter are known as physical activities. One can differentiate between three different types of informational activities: decisional, information acquisition, and information support. Decisional activities are those that result in the creation of new information (e.g., designs, plans) from existing information. Information acquisition activities are those in which new information is acquired. Such information can come from outside the system or from within the system via data acquisition (e.g., temperature sensors, position sensors) or data input (e.g., keypad, touch screen). Information support activities are those that provide information as required for execution of other activities. Information communications, storage, and presentation are all necessary for providing information support.

The second way to view manufacturing systems is the *implementation view*, where one is concerned with which resources (machines, equipment, computers, people, etc.) are used to implement the various manufacturing activities specified in the functional view. There are two main categories of resources in manufacturing systems: humans and mechanisms. When a human is involved in activity execution, the activity is done manually. If no human is involved, the activity is automated. Note that humans usually require one or more mechanisms to assist in activity execution. For example, a planner may need a calculator to execute a decisional activity, whereas a shop-floor worker may need a screwdriver to assemble two items together or a cart to move heavy items.

The functional and implementation views, as well as the basic components of manufacturing systems, can be combined into a classification of components, resources, and activities for manufacturing systems, as shown in Table 54.1. For simplicity, only the most fundamental planning activities are considered. The planning activities and control and monitoring activities are all decisional in nature; any information needed for executing these activities must be obtained via information support and information acquisition activities. The physical processing activities are all physical in nature, and in a similar manner rely on information support and information acquisition activities to provide any information needed.

Williams (1992) also discusses the concept of a manufacturing system life cycle. The three stages of his manufacturing system life cycle that are most relevant to VE applications are system design, system manifestation, and system operation. During the *design stage*, each component of the manufacturing system must be designed, both in terms of activities (functional view) and resources (implementation view). In the *manifestation stage*, the manufacturing system design becomes reality: resources are purchased, installed, and tested, and human operators are trained to perform the various manufacturing activities for which they are responsible. Following the manifestation stage is the *operation stage*, where the various manufacturing activities are executed, with the given resources, in order to obtain finished items of various types in response to demand. All of the activities shown in Table 54.1 occur during the operation stage.

Putting these three elements—functional and implementation views, basic components, and life-cycle stages—together results in the specification of the different tasks involved in manufacturing over the life cycle of a system, as shown in Table 54.2. For the purpose of investigating VEs in manufacturing, six major tasks involved in manufacturing can be considered over the life cycle of a system. Note that planning, control and monitoring, and physical processing under the operation stage of manufacturing are not combined into one task in Table 54.2 because applications of VE technology can vary greatly across these three tasks. We use the term *task* rather than activity to distinguish between those activities that are executed during the operational stage and those that are executed at different stages of the life cycle of manufacturing systems, as summarized in Table 54.2.

**TABLE 54.1**
Resources Employed and Activities Performed in Manufacturing Systems

| *Manufacturing System Component* | *Manufacturing Resources (Implementation View)* | *Manufacturing Activities (Functional View)* |
|---|---|---|
| *Planning* | **Planning Facility** | **Planning Activities** |
| | Buildings, rooms, areas | Product design: products to be made. |
| | Computers | Process planning: operations required to make the products. |
| | Data communications/ storage devices | Production planning: when to make the various products, in what quantities, and with which resources. |
| | Data acquisition/input devices | Shop-floor scheduling: how the various items are scheduled for production in the physical processing facility. |
| *Control and Monitoring* | **Control and Monitoring Facility** | **Control and Monitoring Activities** |
| | Buildings, rooms, areas | Manufacturing activity control: control of all physical activities in the manufacturing facility to ensure that plans are realized. |
| | Computers | Quality control and inspection: activities performed to ensure items are produced to specification. |
| | Data communications/ storage devices | Testing: activities performed to ensure that items function as intended. |
| | Data acquisition/input devices | Inventory control: control of inventories (raw materials, work-in-process, and finished items). |
| *Physical Processing* | **Physical Processing Facility** | **Physical Processing Activities** |
| | Buildings, rooms areas | Fabrication: activities that change the dimensions, geometry, and/or one or more properties of an item directly (e.g., machining). |
| | Processing machines, equipment, tools, fixtures | Assembly/disassembly: activities that change the dimensions, geometry, and/or one or more properties of an item indirectly, i.e., by combining the item with other items (e.g., welding). |
| | Material handling equipment | Material handling: activities that transform physical items in space (e.g., moving items from one machine to another, shipping). |
| | Material storage equipment | Material storage: activities that transform physical items in time (e.g., storage of items between operations). |
| | Data communications/ storage devices | Maintenance: activities that keep the manufacturing process functioning. |
| | Data acquisition/input devices | |

**TABLE 54.2**
Manufacturing Tasks Over the Life Cycle of a Manufacturing System

| *Life-Cycle Stage* | *Manufacturing System View* | *Manufacturing System Component* | *Manufacturing Task* |
|---|---|---|---|
| Design | Functional | Planning, control and monitoring, physical processing | Design of manufacturing activities |
| | Implementation | Planning, control and monitoring, physical processing | Design of manufacturing facilities |
| Manifestation | Functional | Planning, control and monitoring, physical processing | Training for manufacturing activity execution |
| Operation | Functional | Planning | Execution of planning activities |
| | | Control and monitoring | Execution of control and monitoring activities |
| | | Physical processing | Execution of physical processing activities |

## 1.3 Role of Humans in Manufacturing

Humans are always present in manufacturing systems. The manner and extent to which humans are involved in executing the various manufacturing tasks may vary greatly from one manufacturing system to another. The role of humans in manufacturing is dependent on several parameters. Both the type of manufacturing system and nature of the products being made must be considered. For example, flow shop operation requires relatively few decisional activities but many physical activities per unit time, whereas job shop operation is the opposite. Certain decisional and/or physical activities are present in some manufacturing systems but not others.

In addition, Williams (1992) defines the *automatability, humanizability*, and *extent of automation* limits as additional parameters for determining the role of humans in manufacturing. The automatability limit defines the absolute extent to which tasks can be executed by automated means. This limit results from the fact that certain tasks may require capabilities beyond those found in currently available technology. For example, humans must handle decisional tasks requiring thought, creativity, and other cognitive activities, as well as physical tasks that require human capabilities (e.g., intricate assembly operations). The humanizability limit defines the absolute extent to which humans can execute tasks. This limit results from the fact that certain tasks may require computational capabilities, speed of response, physical strength, endurance, and so forth, beyond human capabilities. The difference between the automatability and humanizability limits defines the extent to which choice exists in using humans to execute manufacturing tasks. If the difference is large, much choice exists, and execution could range from highly automated (closer to automatability limit) to highly manual (closer to humanizability limit). If the difference is small, there is little choice: whether or not tasks are executed by humans or by automated means is largely predetermined. The actual allocation of tasks to humans and mechanisms, where choice exists, determines the extent of automation for the task set.

Once the manufacturing tasks to be done by humans are established, one must establish how. Two distinct problems must be tackled: task assignment and task design. Task assignment establishes what tasks are to be done by which workers. Multitasking is common in manufacturing where human operators are routinely called on to perform a variety of time-sharing tasks, such as job scheduling, inventory management, machine setup, and inspection. Pesante, Woldstad, and Williges (1997) investigated characteristics of an inspection task that could facilitate multitasking. The ability to employ multitasking depends on the tasks involved. Task design, on the other hand, establishes exactly how the human operator is to perform the required task(s). This, in turn, will depend on the task requirements and human capabilities and limitations.

## 1.4 Virtual Environments

Virtual environments, in the broadest sense, are tools used to assist humans in task execution by providing information and feedback. They can be viewed as a means of executing information support activities. Milgram and Drasic (1997) describe a reality–virtuality continuum for characterizing VE representations. The four major components of this continuum include: reality, augmented reality, augmented virtuality, and virtuality. This classification might provide some intuition to determine potential applications of VE by linking a given application domain to a VE class along this continuum.

The "reality" class defines the environment as consisting solely of real objects observed via a video display. Telepresence using video cameras (see chap. 48, this volume) are included in this level of virtuality to improve an operator's perception and to provide a natural interface.

Augmented reality (AR) is an enhanced, or augmented, environment where computer-generated images are added to the real environment. Most AR systems are coupled with see-through head-mounted displays (HMDs) and head trackers. Task-related information or graphics are provided from the computer to support the operators who are executing tasks in the real environment. The success of AR applications depends mainly on the way information is transmitted to the operators rather than the immersive feeling or fidelity within the virtual environment (see chap. 40, this volume). This class of applications can support tasks requiring physical manipulation of real objects.

Augmented virtuality is almost the same as AR, except the primary VE is enhanced through some additional real-world image. Augmented virtuality provides a partially immersive environment where real physical objects interact in a virtual world. An example of augmented virtuality is where a user reaches forward in the virtual world and grasps a virtual object on a VE workbench that provides a large projection screen, or in a Cave Automatic Virtual Environment (CAVE) that provides a VE room.

Virtuality is a completely immersed environment and is commonly referred to as "virtual reality." Many applications are implemented using an HMD to provide complete immersion. The main uses of virtuality are training and applications requiring simulation that do not allow the operator to interact with real objects. The success of these applications depends on the level of immersive feeling and the degree of fidelity of the VE system. For the purpose of this chapter, a distinction is made between four different methods by which VEs can provide information support:

- *Visualization*: A VE visualization occurs when the user interacts with three-dimensional (3-D) objects in a VE and the interaction is in terms of navigation only (see chap. 24, this volume). In other words, the user can move through the VE world as desired, scrutinizing virtual objects from different positions, angles, and orientations. The virtual scene is continuously updated according to the user's viewpoint; therefore, the user can feel and experience the VE to a greater level than possible for a passive observer. This capability results in an ideal workspace for various engineering applications. Although visualization requires 3-D representation of objects in the VE, the difference between visualization and traditional 3-D computer-aided design (CAD) is that visualization concentrates on the high degree of perception rather than on the pure quality of the graphics (Wilson, Cobb, D'Cruz, & Eastgate, 1996). In VE visualizations, the world may be either static or dynamic (see chap. 53, this volume). In the latter case, system behavior in the VE is presented according to some predefined script, as the user has no control over the objects. This is essentially equivalent to simulation animation via a VE, where the term *simulation* is used in the operations-research sense of the word (i.e., objects interacting with one another, as part of a system or process, in a scripted manner established by specified input parameters).

- *Simulation*: In a VE simulation, the user interacts with virtual objects in a virtual or augmented world, and the virtual objects act like the real ones in response to unscripted user actions. Users can be immersed in the VE, where they interact directly with the elements of the simulated VE world. This allows users to experience simulated situations directly, they can touch virtual objects and change their state in the same way as they do in the real world. This results in an ideal environment for evaluating the design of equipment, systems, and tasks that either do not exist in the real world or for which actual usage would be impractical or cost-prohibitive (see chap. 52, this volume). The critical component is the interaction between users and the VE (i.e., feedback from the VE to the user's physical world). Simulation via VEs requires more than realistic visualization; it also requires realistic physical behavior on the part of the virtual objects. Therefore, most VE simulations employ visual (see chap. 3, this volume), auditory (see chap. 4, this volume), and/or haptic (see chaps. 5 and 6, this volume) displays to support realistic interaction between users and the VE system.
- *Information Provision*: Virtual environment technology can be used to guide and support workers to complete their work more efficiently by means of information provision. Task-related information, such as instructions and data, can be conveyed to workers both when and where needed via VE technology. Augmented-reality displays are a promising approach for providing information for execution of real-world tasks because they allow users to see both virtual and real objects together. In this approach, virtual objects are used to present required task-related information to the user directly.
- *Telerobotics*: Telerobotics involves remote control of robots (see chap. 48, this volume) and is used when the user interacts with virtual or real objects in an unscripted manner. This interaction causes certain activities to occur in the real world via a robot or some other electromechanical device. These activities, in turn, affect virtual objects via feedback and provide a sense of telepresence. The fundamental purpose of telepresence is to extend an operator's sensory-motor facilities and problem solving abilities to a remote environment (Rosenberg, 1993). Telerobotics is typically used to remove an operator from a hazardous environment. However, the current generation of telerobotics does not convey the necessary level of perception and awareness to an operator for accurate control and decision making. To overcome this limitation, a VE display can reconstruct the remote environment so that the operator can control robots in the real environment via a virtual interface.

The ability to use a given VE method for a particular information support activity depends on the informational requirements of the activity. Thus, in order to employ VEs to aid in activity execution, it is necessary to examine both the informational requirements and ability of the VE method to satisfy these requirements. Two important attributes, from the VE perspective, are the information modality and the volume of information per transaction. The information modality is the principle factor in establishing the VE technology and hardware required, but a certain VE method may be cost-prohibitive or even infeasible for certain information modalities when applied to certain tasks. Volume of information per transaction is one of the principle factors for establishing the level of "reality" associated with any virtual, or augmented, world. When information requirements are high, reduced presentation resolution (e.g., rendering of graphical objects), and hence reduced reality, is necessary.

## 2. VIRTUAL ENVIRONMENTS FOR MANUFACTURING

The value of VE technology for a given manufacturing task must be considered in terms of applicability, technological and human considerations, and potential benefits and cost considerations. Each of these three items is discussed separately before reviewing the current applications of VE technology to manufacturing.

### 2.1 Applicability of Virtual Environments to Manufacturing

Each of the six fundamental types of manufacturing tasks requires information support for execution, and VEs have been defined as tools for providing informational support to humans. Thus, VEs have the potential for aiding humans in executing tasks of each type in manufacturing. The choice among different VE methods to be used for different tasks depends upon the informational requirements of the task. As each of the six types of manufacturing tasks has different information requirements, the VE methods that are suitable for each will also differ. Table 54.3 shows which VE methods have been applied to each of the six task types.

Visualization can be employed for almost all types of manufacturing tasks but is particularly well suited to the design of manufacturing facilities and the execution of planning activities. One of the most common planning activities for which visualization is employed is product design. The process of designing and validating products via a real-time graphics system that allows the user to be immersed in and interact with the product is known as *virtual prototyping* (Flaig & Thrainsson, 1996). Although VE simulation can be used for various design and planning activities (see chap. 52, this volume), it holds particular promise for training operators to execute manufacturing activities. This potential is magnified by the fact that training is an essential component of all manufacturing activities. Information provision is particularly well suited for aiding in manufacturing activity execution and can lead to enhanced performance. This performance enhancement can in turn reduce operator workload and facilitate multitasking. Chung, Williges, and Shewchuk (1999), for example, show how AR can be used to provide task-related information for thickness inspection, where performance is measured by task completion time and magnitude of measurement error. As shown in Table 54.3, telerobotics is employed only in the execution of physical processing activities in manufacturing.

**TABLE 54.3**
Applicable VE Methods for Each Type of Manufacturing Task

| *Manufacturing Tasks* | *VE Methods* | | | |
|---|---|---|---|---|
| | *Visualization* | *Simulation* | *Information Provision* | *Telerobotics* |
| Design of manufacturing activities | * | | | |
| Design of manufacturing facilities | * | | | |
| Training for manufacturing activity execution | * | * | | |
| Execution of planning activities | * | * | | |
| Execution of control and monitoring activities | * | * | * | |
| Execution of physical processing activities | | | * | * |

## 2.2 Considerations in Applying Virtual Environments to Manufacturing

Several considerations need to be addressed when applying VE technology to manufacturing. Dai (1998), for example, discussed the intersection of VE graphics, CAD, and telecommunications technologies as a means of supporting concurrent engineering in integrated product development. He describes the activities of a German working group using VE technology for improving primarily interactive visual prototyping and simulation in a variety of industrial applications. Likewise, Wilson et al. (1996) summarized work on the Manufacturing Operations in Virtual Environments (MOVE) program, which surveyed and provided demonstrations of VE applications to British industry. They viewed the primary role of VEs as that of providing interaction (see chap. 13, this volume) and a feeling of presence (see chap. 40, this volume) to visualization techniques used in industry. The most feasible industrial applications were related to planning and design, operations, training and communication (see chaps. 43 and 44, this volume) and integrated product development (see chap. 52, this volume).

Two aspects of human performance that must be considered in VE system development efforts are the sensory and information processing capabilities and limitations of workers. Most applications of VE technology in industry only deal with the visual representation of the virtual environment. Other sensory modalities such as auditory, vestibular, olfactory, kinesthetic, proprioceptive, locomotion, cutaneous, and gesture representations in VEs still need to be explored in manufacturing applications (see chaps. 4–11, this volume). Information processing capabilities and limitations deal with cognitive functions (see chap. 20, this volume) of the human such as spatial perception, attention sharing, memory, error analysis, decision making, workload analysis, and process control. All of these functions can affect human performance in manufacturing environments that involve VE systems.

## 2.3 Task Analysis

Based on our six manufacturing tasks and four VE methods for providing information (see Table 54.3), a methodology for establishing potential benefits and cost using VEs for a given manufacturing task can be developed. As shown in the Fig. 54.1 decision flow diagram, the methodology consists of three steps: evaluate overall feasibility (Step 1); determine potential benefits (Step 2); and conduct a cost–benefit analysis (Step 3).

The applicability of VEs to a given manufacturing task is evaluated in Step 1. A VE is applicable if (1) one or more information support activities are needed for task execution, and (2) at least one of these activities can be performed via one or more VE methods. Thus, the information requirements must be established and compared to the capabilities of each VE method. The result of this step will determine whether or not VEs are applicable for a given manufacturing task, and if so, the VE methods that can be used.

During Step 2, the potential benefits of VEs are determined. The objective is to use a VE to provide the necessary information for a given task in order to improve task execution (reduced time, improved quality, etc.; see chap. 35, this volume). As shown in Fig. 54.1, the potential for such improvements can be established in two ways. The first of these, analytical evaluation, consists of estimating potential benefit based on the impact of the information on overall task performance. For example, if the presentation of a high-quality, dynamic graphic image is of paramount importance to task execution, the potential benefit of using a VE for this task is high. On the other hand, the potential benefit is low if either the presentation of such an image will likely have only a minor impact on performance or a VE is not well suited to presenting such images. The second of these, experimental evaluation, consists of obtaining prototype/production VE hardware, establishing a procedure, and performing experiments to

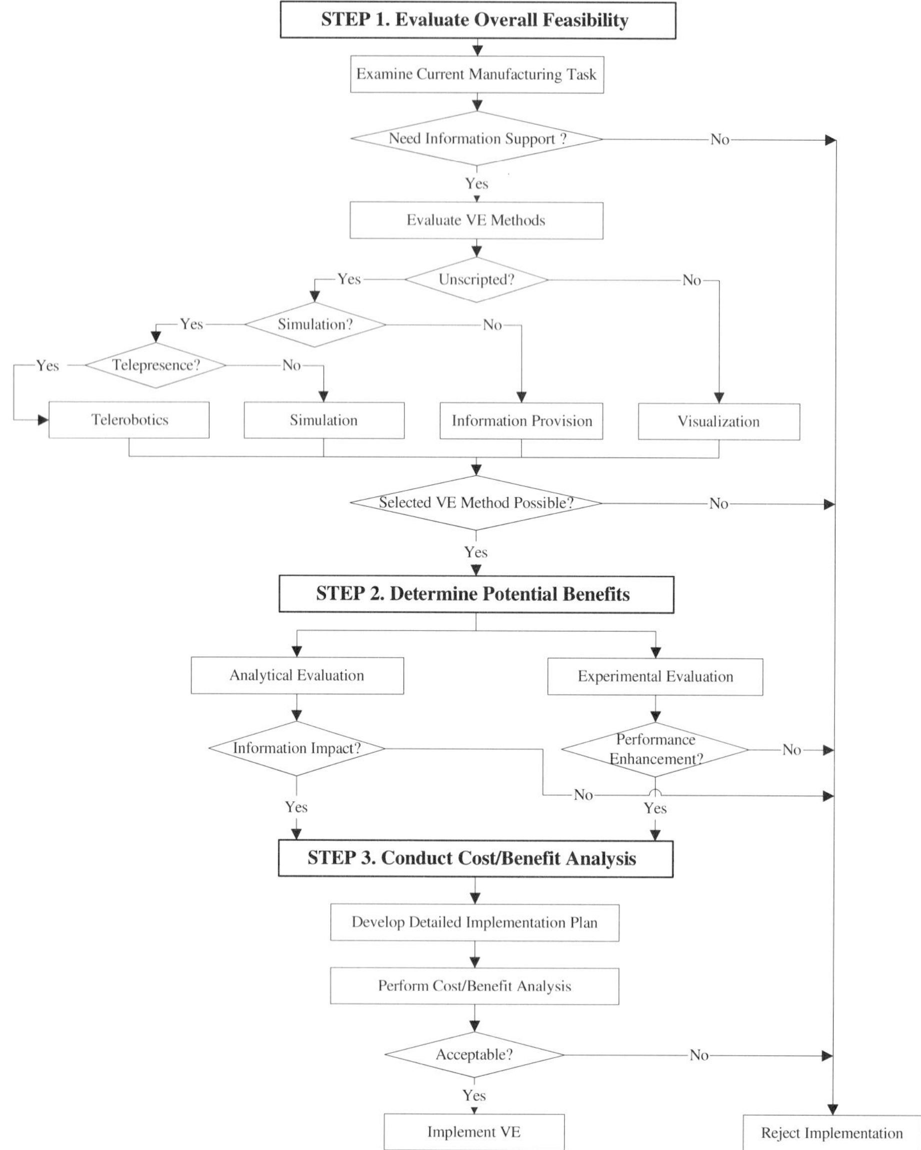


FIG. 54.1. Decision Process for Analyzing Manufacturing Tasks for Application of Virtual Environments.

establish quantitatively the performance benefit under experimental conditions. This gives an estimate of the potential benefit if the VE hardware and procedure were employed in actual production conditions. The results of Step 2 will provide quantitative and/or qualitative measures establishing the potential benefit of employing each of the applicable VE methods for the given task.

Virtual environments should actually be used, however, only if the benefits outweigh the costs (i.e., VE hardware, procedure development, training, and lost production time during implementation). Thus, a cost–benefit analysis must be performed in Step 3 (see chap. 28, this volume). This can only be done after a detailed plan of the VE implementation has been developed so that costs can be estimated. Once this has been done, the cost–benefit analysis will be either subjective or objective depending on how the potential benefits were quantified in Step 2. Given that cost–benefit analysis provides a sufficiently positive result from an economic point of view, VE technology is a good candidate for implementation.

The main goal of the above three-step procedure is to ensure that VEs are used only if they provide concrete benefits and these benefits outweigh the costs involved. If these conditions are not met, it is better to continue using traditional methods (paper, traditional computer presentations, etc.) for supplying the required information.

## 3. CURRENT APPLICATIONS OF VIRTUAL ENVIRONMENTS IN MANUFACTURING

The application of VE technology to manufacturing is an emerging area, and about 50 applications appear in the scientific literature. As shown in Table 54.4, most applications exist in product design. Since the current focus of a VE is mainly on visual information, applications employing haptic feedback (see chaps. 5 and 6, this volume) to increase fidelity and presence (see chap. 40, this volume) are not well represented.

### 3.1 Design of Manufacturing Activities

Visualization and simulation are useful VE methods for designing and evaluating manufacturing activities. These methods are applicable for finding possible problems in a manufacturing process and for the design and redesign of the manufacturing process and activity. Evaluating current manufacturing procedures and activities via a virtual environment can determine optimal work methods and workplace design. For example, Arzi (1997) used VE technology to find the best design solution for a workstation. The results showed that VE technology allowed the designer to assess the operator's view, test the ease and the comfort of doing a task, and evaluate various critical functions for the given task.

Several applications of VEs for the design of vehicle assembly were found in the literature. Davies, Medbo, Engström, and Akselsson (1999) suggested that VE technology was applicable for an ergonomic study of the car body assembly task (see chap. 52, this volume). Volkswagen developed assembly and disassembly models that can be used for designing the assembly process (Purschke et al., 1998). They built a simplified virtual factory where robots assembled and welded the parts of a vehicle. The model showed that collisions could occur between robots and vehicle components. This application was useful to evaluate manufacturing feasibility and assembly sequence. The Virtual Reality Laboratory at the University of Michigan (1999) was used to study both the virtual modeling of ship production processes and automotive assembly robots. The results indicated that an immersive walk-through of the full-scale representation of a complex ship structure not only revealed severe design flaws but was also useful in determining the assembly sequences, welding accessibility, and movement of equipment inside the ship. The application of a VE in automotive robot assembly showed similar results to the virtual ship modeling. Modeling with animated robots permits checking of layout efficiency, correctness of robot programming, robot reach, clearances, and collision events.

### 3.2 Design of Manufacturing Facilities

A VE provides an ideal environment to visualize and simulate the design of a manufacturing facility. A key attribute in a VE application to facility design is the ability to maneuver around a 3-D environment in the same manner as one does in the real world. This attribute can provide

**TABLE 54.4**
Summary of Current VE Applications to Manufacturing Tasks and Activities

| *Life-Cycle Stage* | *Manufacturing Task* | *Activity* | *VE Method* | *References* |
|---|---|---|---|---|
| *Design* | Design of manufacturing activities | | Visualization | Arzi, 1997<br>Davies et al., 1999<br>Purschke et al., 1998<br>University of Michigan, 1999 |
| | Design of manufacturing facilities | | Visualization | Krishnamurthy et al., 1999<br>Wilson et al., 1996 |
| *Manifestation* | Training for manufacturing activity execution | | Simulation | Adams & Lang, 1995<br>Göbel, 1996<br>Gomes & Baacke, 1998<br>Pere et al., 1996 |
| | | | Visualization | Wilson et al., 1995<br>Wilson et al., 1996 |
| *Operation* | Execution of planning activities | Product design | Visualization | Beier, 1994<br>Bryson, 1997<br>Buck, 1998<br>Chu et al., 1999<br>Dai, 1998<br>Dai & Göbel, 1994<br>Finger et al., 1997<br>Flaig & Thrainsson, 1996<br>Flanagan & Earnshaw, 1997<br>Harrison et al., 1993<br>Purschke et al., 1998<br>University of Michigan, 1999<br>Wesche et al., 1997<br>Wilson et al., 1995<br>Zachmann, 1998 |
| | | | Simulation | Kalawsky, 1993<br>Lehner & DeFanti, 1997<br>Maxfield et al., 1998<br>Vance, 1997, 1998 |
| | | Process planning | Visualization | Bickel, 1998<br>Cobb et al., 1995<br>Vance, 1999 |
| | | Production planning | Visualization | No references found |
| | | Shop-floor scheduling | Visualization | Frölich et al., 1997 |
| | Execution of control and monitoring activities | Manufacturing activity control | Information provision | No references found |
| | | Quality control and inspection | Information provision | Chung et al., 1999 |
| | | | Visualization | Purschke et al., 1998 |
| | | Testing | Information provision | Kesavadas, 1996 |

*(continued)*

**TABLE 54.4**
*continued*

| *Life-Cycle Stage* | *Manufacturing Task* | *Activity* | *VE Method* | *References* |
|---|---|---|---|---|
| | | | Visualization | Purschke et al., 1998<br>Wesche et al., 1997 |
| | | Inventory control | Simulation | No references found |
| | Execution of physical processing activities | Fabrication | Telerobotics | Cannon & Thomas, 1997<br>Kheddar et al., 1997<br>Milgram et al., 1993<br>Natonek et al., 1994<br>Rosenberg, 1993<br>Tezuka et al., 1994 |
| | | Assembly/ Disassembly | Information provision | Caudell, 1994<br>Caudell & Mizell, 1992<br>Gupta et al., 1997<br>Nash, 1999<br>Sharma & Molineros, 1997<br>Sims, 1994 |
| | | Material handling | Telerobotics | Cannon & Thomas, 1997<br>Tezuka et al., 1994 |
| | | Material storage | Information provision | No references found |
| | | Maintenance | Information provision | ECRC, 1994<br>Feiner et al., 1993<br>Pimentel & Teixeira, 1995<br>Tezuka et al., 1994 |

a realistic perception of the visual scene as well as the behavior of objects. Krishnamurthy, Shewchuk, and McLean (1999) describe how Virtual Reality Modeling Language (VMRL) can be used to model and study manufacturing facilities in this manner.

A virtual factory is often used in the early phases of facilities layout. Users can identify problems by watching the manufacturing process during a walk-through in the virtual factory. Wilson et al. (1996) described the VE project of the Rover Powertrain Group to explore the visualization of plant layouts in Great Britain. Generic shop-floor layouts, conveyor and lift systems, cranes, storage, and distribution depots were included in a virtual factory (see chap. 46, this volume). They found that the use of VEs was a valuable aid to industrial and manufacturing engineers in facility planning and design. The virtual model showed proposed changes to building and equipment placement. Their model proved useful for communication, potential problem identification, and logistic considerations.

## 3.3 Training for Manufacturing Activity Execution

By working with virtual objects in a VE training simulation, users can improve their skills and knowledge just as if they were working with objects in the real world. The richer perceptual cues and feedback from a VE can facilitate positive transfer of VE-based skills to real-world tasks (see chap. 19, this volume). The VE training simulation results in more affordable, interactive, efficient, and safer training, and provides a more comprehensive experience for the user than training under real conditions.

Motorola developed a VE for training pager assembly workers (Adams & Lang, 1995). To avoid the expense of shutting down actual production facilities, a VE was developed to replicate equipment and provide effective training. The VE included a conveyor system, three robotic work cells for pager assembly, and inspection and laser marking systems. An evaluation of training effectiveness was conducted on three groups (non-VE, desktop VE, and HMD). The HMD group resulted in the best performance for most tasks. During posttraining evaluation, the HMD trainees claimed not only an immersed feeling but also freedom from reluctance to learning in the VE system.

The Fraunhofer Institute developed a virtual generator assembly scenario that includes visual collision cues to guide workers. Their virtual assembly allows users to grab and assemble objects in a VE using a DataGlove. This system served as a training tool for maintenance and design support at BMW and Mercedes-Benz (Göbel, 1996). Recently, BMW implemented two additional application scenarios for assembly. These include the installation of door locks and window regulator components into doors (Gomes & Baacke, 1998). Their application models used sound to compensate for lack of force feedback (see *sensorial transposition*, chap. 21, this volume). These studies showed that visible collision detection (e.g., overlay of virtual objects) was very important for assembly tasks (see chap. 12, this volume). Pere, Gomez, Langrana, and Burdea (1996) studied the use of haptic information for the assembly of an engine case. A desktop VE system was integrated with a force feedback device. They suggested that this type of feedback allows the user to feel virtual tools and makes training tasks more realistic. Rolls Royce and Associates used a VE for safety training (see chap. 42, this volume). They developed a training program to enable efficient maintenance of complex machinery in hazardous operational environments. The VE system allowed maintenance engineers to familiarize themselves with location, layout, and physical restriction of equipment (Wilson, Brown, Cobb, D'Cruz, & Eastgate, 1995; Wilson et al., 1996).

The costs and benefits of flexible VE training are evident for various applications (see chap. 28, this volume). However, most of the current implementations are still somewhat restricted. The number of states and objects in VE simulation are limited, behaviors of objects are primitive (e.g., no gravity for falling objects), and only predetermined scenarios are supported. To be an effective training tool, Johnson, Rickel, Stiles, and Munro (1998) recommend that VE simulators should support delivery of information to facilitate effective learning and not just concentrate on the fidelity of the renderings and the accuracy of simulated behaviors (see also chap. 20, this volume). Integration of intelligence and behavior-oriented design into the VE simulation process is needed.

## 3.4 Execution of Planning Activities

Planning activities include product design, process planning, production planning, and shop-floor scheduling (see Table 54.1). Virtual environment technology can be used to visualize or simulate product design and manufacturing planning and can result in significant savings in cost and time by early detection of errors in the product design and manufacturing process. Most applications use HMDs, VE workbenches, and/or CAVE equipment to provide a perception of immersion for the user. However, some applications employ desktop computers for user comfort and convenience.

### 3.4.1 Product Design

Virtual prototyping is a widely used term to describe product design that is based on VE technology. Virtual prototyping provides an alternative concept for the design-test-evaluation cycle by eliminating the fabrication of physical prototypes. It can also support product design and presentation through simulations and analysis so that designers can evaluate the product functions by manipulating and modifying the model directly (Dai & Göbel, 1994). A VE is an effective analysis tool for problems where complex spatial arrangements are dominant and humans have to interact with these arrangements. Because immersion facilitates realistic responses from users, VEs can help designers identify possible design problems and improve the usability of final products (see chap. 34, this volume). Expected benefits include improved product design, savings in cost and time, and better market response through reduced design cycles.

The automotive industry has already recognized the potential of VEs as a tool to improve the design process (see chap. 42, this volume). Virtual environment technology might provide an efficient way of prototyping automotive interiors and can be used for design analysis and human factors studies (Beier, 1994; Buck, 1998; Flanagan & Earnshaw, 1997). The Virtual Reality Laboratory at the University of Michigan (1999) is developing a virtual prototype of a concept car. A VE provides a tool for evaluating such designs in full scale without building physical prototypes. Volkswagen is also studying the virtual product clinic that is conceptually similar to the concept car approach. A vehicle concept, which includes an ergonomic mockup, is tested in the virtual product clinic (Purschke et al., 1998). In Great Britain, Kalawsky (1993) developed an interactive virtual car interior. Designers could sit inside the virtual car on a real car seat and reach out and grab objects such as the steering wheel.

Another class of applications in the automotive industry is 3-D visualization, which demands a fine granularity of visualization and interaction with the virtual scene rather than navigating through a virtual space. A VE workbench offers an efficient way to develop this type of application. The Fraunhofer Institute has developed a VE for visualization of air flow around a car (Zachmann, 1998). Wesche, Wind, and Göbel (1997) summarized a VE project at Daimler-Benz that simulates fluid dynamics. Engineers gained a much deeper insight into the structure of the velocity field inside a cylinder during the injection process. NASA has also been involved in a similar application to depict a virtual wind tunnel for visualization of air flow around an aircraft (Bryson, 1997; see chap. 53, this volume). This model allowed two users, each with independent viewpoints, to reach into the simulation and manipulate the visualization tools interactively. Some applications that do not require fully immersive VEs, such as part design (e.g., motor vehicle engine thermostat housing), can use a desktop VE system. Desktop VE has an advantage in terms of graphic quality, flexibility, user comfort and convenience, and longer work times with lower cost (Wilson et al., 1995).

Virtual prototyping has also been used to facilitate communications in distributed environments. Maxfield, Fernando, and Dew (1998) studied a distributed collaborative engineering environment system that allows members of a geographically dispersed multidisciplinary team to collaborate during product design. Lehner and DeFanti (1997) also discussed a distributed VE communication project between the National Center for Supercomputing Applications and the German National Research Center for Information Technology. They developed a collaborative virtual prototyping system for Caterpillar so that engineers in Belgium and the United States could work together on vehicle designs in a distributed virtual environment. Finger et al. (1997) studied the integration of virtual and physical prototyping technologies to support rapid design and rapid manufacturing. Their experimental system could be accessed through the Internet.

One of the major problems of virtual prototyping is the link between VE and CAD systems (Dai, 1998; Harrison, Jagues, & Strickland, 1993). The generation of a 3-D model rendered by computer graphics is an extremely elaborate and time-consuming task. Impressive VE demonstrations are frequently based on geometric models that have been developed over a long time period (Beier, 1994). Although these designs are 3-D in nature, user inputs are often restricted to two-dimensional (2-D) devices (i.e., keyboard and mouse). Several researchers are examining new interfaces that enable interaction in an intuitive and easy way for visualizing, creating, and editing 3-D product models based on VE technology (Chu, Dani, & Gadh, 1999; Flaig & Thrainsson, 1996; Harrison, Jagues, & Strickland, 1993). In a related study, Vance (1997, 1998) studied a VE where a designer can change the shape of design by using a DataGlove. She optimized engineering design by reviewing stress distribution as the shape was changed.

### *3.4.2 Process Planning*

Process planning can be supported by an interactive VE that merges a fully functional factory-floor process simulator with a 3-D factory model. Engineers might have a better understanding of functionality in the manufacturing process by using real-time simulation in a VE and evaluating their plans according to the simulation results. For this reason, Bickel (1998) emphasized the need to integrate VE tools with simulation software and demonstrated several simulation-based applications. One application simulated the automobile manufacturing process. A virtual factory was built to replicate real manufacturing sequences, and the improved material flow saved Audi millions of Deutsche marks.

Vance (1999) is developing a virtual factory that employs a discrete-event simulation program. The users are able to examine part flow through the factory by varying different simulation parameters. Cobb, D'Cruz, and Wilson (1995) developed a virtual factory on a desktop VE representing the manufacture of a consumer product that could be used to support process planning as well as facility layout. Various stages of the manufacturing process were featured, including design, manufacturing, and testing. The virtual factory allowed demonstrations such as factory walk-throughs, visualization of the manufacturing process, and design modification. Users could observe finished parts on conveyor lines and bottlenecks in the manufacturing process. A national survey of this application showed that respondents from 170 companies considered the walk-through to be useful. But respondents were mixed in their opinions of the usefulness of two other applications, citing that the current user interface was difficult to use and the system was not suitable for complex modeling.

### *3.4.3 Production Planning*

No applications of VE technology to production planning were found in the literature. Production planning determines when to make the various products, in what quantity, and with which resources. It is mainly concerned with quantitative data rather than visual information. Therefore, VE applications that rely primarily on visual displays may be somewhat limited.

### *3.4.4 Shop-Floor Scheduling*

When scheduling the assembly of a ship or airplane, a scheduler must cognitively step through the assembly sequence and visualize the changes over time. A scheduler primarily uses Gantt charts or network diagrams to represent the construction schedule and static 3-D models to represent the ship or airplane. A VE, however, allows a new way to experience complex assembly through an interactive and dynamic modeling environment. Such an application allows a scheduler to get immediate and consistent feedback by integrating a 3-D model with scheduled activities. Frölich, Fischer, Agrawala, Beers, and Hanrahan (1997) developed an application that combined 3-D modeling and scheduling for building construction on a VE workbench. The application facilitated successful interaction and collaboration between two users who were sharing the virtual environment.

## 3.5 Execution of Control and Monitoring Activities

Control and monitoring activities include manufacturing activity control, quality control and inspection, testing, and inventory control (see Table 54.1). Applications of VEs for visualization, simulation, and information provision are appropriate. Since users have to interact with real objects directly or indirectly, see-through HMDs or video-based AR displays may be employed for information provision while performing such activities. In addition, several simulation applications using a VE workbench for control and monitoring exist.

### 3.5.1 *Manufacturing Activity Control*

No applications of VE technology to manufacturing activity control were found in the literature. Manufacturing activity control requires constant evaluation of activities and conditions in a manufacturing facility with respect to pre-established plans. Employing VEs to provide the necessary real-time status information and relevant plans could dramatically increase the ability of humans to perform this task effectively.

### 3.5.2 *Quality Control and Inspection*

One of the few projects that has shown real benefits for the use of a VE was the application of AR for thickness inspection (Chung et al., 1999). Performance differences among manual, computer-aided, and AR-aided inspection were measured. Results indicated that while all three methods were similar in terms of accuracy, AR-aided inspection was two to three times faster that the other two methods. Additionally, although inspection time increased with part shape complexity for the manual and computer-aided inspections, it was unaffected by part shape complexity for AR-aided inspection. Purschke et al. (1998) described a surface inspection application at Volkswagen. Virtual models offered not only high-quality shape and surface rendering, but also some physical features such as shininess, contour changes, and functional features (e.g., opening and closing of doors). Their application has already been incorporated into the development process of an automobile.

### 3.5.3 *Testing*

Virtual prototyping or visualization is often used for testing during the iterative design process. These test data are used to evaluate the design of a product (see chap. 34, this volume). Kesavadas (1996) developed a virtual template that is a graphic replica of a real template to carry out various design checks. Workers can check design factors (i.e., drilled holes, bores, and slots of various diameters) by merging the prototype of a component with a set of virtual templates in an AR environment. Purschke et al. (1998) and Wesche et al. (1997) described how Daimler-Benz and Volkswagen are using a VE system for crash and airbag testing where engineers can move around the vehicle and observe the effects of crash-state computations (e.g., structure behavior for a specific time frame).

### 3.5.4 *Inventory Control*

No applications of VE technology to inventory control were found in the literature. However, VE simulation could easily be used for warehouse control.

## 3.6 Execution of Physical Processing Activities

Physical processing activities include fabrication, assembly and disassembly, material handling, material storage, and maintenance (see Table 54.1). Telerobotics and information provision are appropriate VE methods for the execution of these activities. Virtual environment technology has been used to improve the perception of teleoperators or to provide information required for a given task (see chap. 48, this volume). Since users have to interact with real objects directly or indirectly, a see-through HMD or video-based AR display can be used. Some applications use multimodal displays (i.e., visual, auditory, and haptic displays) and focus on natural interfaces.

### 3.6.1 Fabrication

Robots have been used in fabrication for many years. Virtual environment technology is expected to provide enhanced perception and a natural interface for telerobotics and robotics. Since the current generation of teleoperated robots does not convey a sense of real presence, operators cannot obtain the necessary level of awareness about remote sites. In order to enhance the operator's perception and understanding within remote sites, improvements in depth perception of teleoperated robots through AR visual feedback have been investigated. Milgram, Zhai, Drascic, and Grodski (1993) studied an overlaid virtual stereographic display to enhance absolute depth judgment and perception when using augmented reality. Another limitation of telerobotics is that feedback to operators has not been designed to map human kinematics accurately. Operators must adapt their reflexes, dexterity, and skill to the robot. Kheddar, Tzafestas, and Coiffet (1997) and Tezuka, Goto, Kashiwa, Yoshikawa, and Kawano (1994) studied a natural control interface and direct manipulation using a DataGlove. Rosenberg (1993) used haptic and auditory feedback to provide additional information in the form of a virtual fixture to reduce the demands on visual cues when an operator controls a robot. He conducted an experiment to evaluate peg insertion performance and demonstrated up to a 70% improvement in operator performance using haptic and auditory virtual fixtures as compared to no virtual fixtures.

Natonek, Flückiger, Zimmerman, and Baur (1994) developed a VE system in which users can determine trajectories of robots in an intuitive way. The results showed that VEs are a powerful interface for robotic system control; users without programming experience could successfully define robot trajectories using the VE system. Cannon and Thomas (1997) also introduced virtual tools with collaborative robotic control concepts that enable users with no expertise in robotics to point and direct robots and machines to complete specific tasks.

### 3.6.2 Assembly/Disassembly

Augmented reality can provide workers with direct information access when performing manufacturing or assembly tasks. Such information includes assembly guides, templates, drawings, parts lists, and locating marks. Workers can see real parts and augmented information together via a see-through HMD. Fabrication and assembly in the aircraft industry are usually done by hand because they are too complex for automation (Caudell & Mizell, 1992). For instance, a commercial airplane has about 1,000 wire bundles ranging from 2 feet to 120 feet in length, and each bundle can consist of hundreds of wires. A significant source of expense and delay in aircraft manufacturing is that workers must constantly refer to several complex drawings during assembly. Boeing used a see-through HMD for four such assembly applications, including wiring formboard, wiring connection assembly, composite cloth layout, and wire frame display (Caudell & Mizell, 1992; Caudell, 1994; Sims, 1994). A test in 1995 using a prototype resulted in a 20% to 50% improvement in wiring assembly performance even though the application was limited by lack of resolution accuracy and long-range head-position tracking (Nash, 1999).

Sharma and Molineros (1997) investigated an efficient information presentation scheme for augmented stimuli in assembly. Assembly of compact disks was developed on a video-based desktop VE system using computer-vision algorithms to track the position and orientation of each part. An assembly graph showed different assembly sequences to guide a person through various stages of the assembly task. Gupta, Sheridan, and Whitney (1997) studied a multimodal VE system for assembly analysis. They developed a system that tracked collisions and contacts of parts (i.e., overlay of virtual objects) using auditory and haptic feedback. Peg-in-hole assembly (most common in metal products) was used for evaluating task performance. The results showed, however, that multimodal VE assembly required twice the time of the real assembly task. This additional time might be due to the inappropriate task application and/or the immaturity of the haptic feedback technology used in the multimodal display.

#### *3.6.3 Material Handling*

No applications of VE technology to material handling were found in the literature. Material handling is a physical activity that does not include high-level decisions; therefore, applications where realistic force feedback is not present are somewhat limited. Telerobotics might be an appropriate method for improving material handling, and some research studies related to fabrication (Cannon & Thomas, 1997; Natonek et al., 1994) are also applicable for material handling.

#### *3.6.4 Material Storage*

No applications of VE technology to material storage were found in the literature. Augmented reality could be employed, however, to help workers locate items in storage, ascertain stock rotation patterns, and perform other tasks.

#### *3.6.5 Maintenance*

One of the first applications of AR in manufacturing was maintenance assistance. Feiner, Macintyre, and Seligmann (1993) demonstrated a maintenance and repair application for a laser printer. See-through HMDs hold promise for providing data without requiring maintenance engineers to take their hands off their work. The European Computer-Industry Research Center, or ECRC (1994), used AR for automobile maintenance and repair. A mechanic was assisted by an AR system while repairing a vehicle engine. Part names, functions, and a maintenance record could be annotated by the system. Pimentel and Teixeira (1995) described a similar scenario for aircraft engine maintenance. An aircraft mechanic could use an AR system to view virtual information panels that overlaid the real image of a jet engine. By tracking hand position as compared to where parts were located inside the engine, a mechanic could "see through" the machinery to the workplace without disassembly.

Another potential application area is maintenance using telerobots in hazardous environments. Tezuka et al. (1994) developed an experimental VE system to assemble and disassemble machines for maintenance using a robot arm manipulated by a DataGlove, as well as the use of hand gesture recognition (see chap. 10, this volume). They found that the time lag between a DataGlove and robot arm was an important factor in determining the performance of this system.

Although VEs are usually considered for isolated task applications, often various VE methods can be used in several application areas simultaneously. Perhaps the greatest benefit can be achieved by integrating VE technology into manufacturing as a whole, rather than by improving one isolated manufacturing task with a specific application. For example, product design is facilitated and tested by virtual prototyping and visualization, process planning and scheduling are aided by VE visualization, and production workers are supported by training and information-provision applications of AR to enhance performance. Virtual environment technology should become just another tool that can be used to pursue continuous improvement in design and manufacturing by facilitating the communication of ideas and concepts across the entire manufacturing system and product life cycle.

## 4. FUTURE APPLICATIONS OF VIRTUAL ENVIRONMENTS IN MANUFACTURING

Although relatively few real applications of VEs (and even fewer empirical studies that evaluate these VEs) exist, the potential for using VEs in manufacturing is great based on current demonstrations. Most of the current technical literature deals primarily with expert opinion, industrial surveys, and isolated demonstration examples that have not received systematic evaluation. Surprisingly, only about 10% of the nearly 50 references listed in Table 54.4 provided empirical evaluations of their VE applications. As Kies, Williges, and Rosson (1998) point out, most human–computer interface applications are developed in an iterative design fashion that involves both formative and summative evaluation (see chap. 34, this volume). Formative evaluation occurs during design iteration to form the basis of redesign and uses primarily subjective methods based on user opinions and expert observations based on ethnographic methods. Summative evaluation is a formal procedure for evaluating final application efficacy and is usually based on controlled testing using experimental design methods to collect performance metrics and subjective opinions of users.

What is needed is an integrated and focused approach combining research, demonstration, and rigorous formative, as well as summative evaluation of VE applications to manufacturing. Based on our overview, the areas of facilities and product design, training, and execution of physical processing activities hold promise for significant research and applications leading to improved performance in manufacturing.

### 4.1 Facilities/Product Design

Interactive simulation coupled with interactive virtual prototyping can improve the design of manufacturing facilities and products. This includes all aspects of the manufacturing facility from computer, communications, and data-storage systems to processing equipment, material handling equipment, and the layout of these items in the facility. The use of VE simulation procedures with interactive visualization capabilities is a logical extension of current industrial use of 3-D CAD technology. However, Wilson et al. (1996) note an incompatibility between current CAD packages that provide high-resolution 3-D graphics and current VE software that provides interactive and immersive visualization. They also note usability difficulties of VE software, few working demonstrations of VE applications, and little evidence of added value for the use of VE in industry. Once software support is available to facilitate the importing of CAD files directly into VE software, increased value in the use of simulation should be realized in all aspects of facilities and product design.

### 4.2 Training

Perhaps the greatest potential for VE applications to manufacturing is in the area of training, because training is a component of all of manufacturing activities. The potential for using a VE as a training simulator has already been demonstrated (see chaps. 43 and 44, this volume). What has not been demonstrated is the effectiveness of this training technique as compared to other training procedures. Cost-effectiveness of training needs to be considered. For example, training workers to perform multitasking can be cost-prohibitive if real production systems are used. Virtual environment simulation of production systems might be quite cost-effective for training purposes, however. Likewise, the appropriate level of simulation fidelity that results in the maximum positive transfer of training to on the job performance must be established (see chap. 19, this volume). Both psychological and physical fidelity of VE simulation training must be considered and evaluated (see chap. 35, this volume).

### 4.3 Execution of Physical Processing Activities

The major application area of VE technology is the improvement of performance in execution of physical processing activities. Maintaining the appropriate level of operator workload is critical in advanced manufacturing systems. Insufficient workload could result in reduced attention span and reaction time, whereas excessive workload may lead to reduced performance or failure. Virtual environment technology has the potential for playing a major role in human-operator workload management applications in manufacturing.

Worker performance in executing various manufacturing activities can be improved through the judicious use of VE technology. For example, assembly workers need to be given information as to what the item looks like before and after assembly, a detailed description of how and in what order assembly tasks are to be done, a list of tools required, and disposition instructions. This information is textual and graphic in nature and requires a low-to-medium volume of transactions. Augmented reality could be employed to provide this information. The worker could wear a see-through HMD when picking up a part to examine. The next item to be assembled could appear as a phantom on the part, and the instructions could float in space beside this part. The needed tool and type of fastener could then be listed. When the worker looks at the bin of different fasteners, an arrow could appear in space pointing to the correct bin. By reducing information acquisition time, VE technology could improve both throughput and product quality. The more customized products become, the more these advantages will be realized.

## 5. CONCLUSIONS

Based on our review, the appropriate application of VE for information support holds great promise for manufacturing. An integrated approach is needed to realize this potential. The proposed three-step activity analysis would evaluate feasibility, determine potential benefits, and conduct a cost–benefit analysis of using VE technology. Once a VE application is considered feasible, the implementation should follow an iterative design process that incorporates formative and summative evaluation procedures. Both the appropriate level of virtuality and the allocation of functions between humans and the manufacturing process must be considered. Systematic research is needed to determine the advantages and disadvantages of differing levels of virtuality with respect to the manufacturing framework described in this chapter. This research can lead to improved function allocation between humans and mechanisms in manufacturing systems. Applications of VE technology to manufacturing activity control, inventory control, and material handling (including shipping) are particularly needed. Research is needed to develop guidelines and models of the appropriate use of VE technology to improve manufacturing performance in an integrated manner.

## 6. REFERENCES

Adams, N., & Lang, L. (1995). VR improves Motorola training program. *AI Expert, 10*(5), 13–14.

Arzi, Y. (1997). Methods engineering: Using rapid prototype and virtual reality techniques. *Human Factors and Ergonomics in Manufacturing, 7*(2), 79–95.

Beier, K. P. (1994). Virtual reality in automotive design and manufacturing. In *Proceedings of Convergence '94 International Congress on Transportation Electronics* (pp. 141–147). Warrendale, PA: Society of Automotive Engineers.

Bickel, D. (1998). Virtual reality (VR)—New methods for improving and accelerating vehicle development. In F. Dai (Ed.), *Virtual reality for industrial application* (pp. 105–122). New York: Springer.

Bryson, S. (1997). The virtual windtunnel on the virtual workbench. *IEEE Computer Graphics and Applications, 17*(4), 15.

Buck, M. (1998). Immersive user interaction within industrial virtual environments. In F. Dai (Ed.), *Virtual reality for industrial application* (pp. 39–59). New York: Springer.

Cannon, D. J., & Thomas, G. (1997). Virtual tools for supervisory and collaborative control of robots. *Presence: Teleoperators and Virtual Environments, 6*(1), 1–28.

Caudell, T. P. (1994). Introduction to augmented and virtual reality. In *Proceedings of Telemanipulator and Telepresence Technologies* (pp. 272–281). Bellingham, WA: SPIE—International Society for Optical Engineering.

Caudell, T. P., & Mizell, D. W. (1992). Augmented reality: An application of heads-up display technology to manual manufacturing processes. In *Proceedings of the Hawaii International Conference on System Science* (pp. 659–669). Los Alamitos, CA: IEEE Press.

Chu, C. C., Dani, T., & Gadh, R. (1999). *Shape generation and manipulation in a virtual reality–based CAD system* [Online]. Available: http://www.usc.edu/dept/ise/NSF/proceedings/edu.html [1999, May 16]

Chung, K. H., Shewchuk, J. P., & Williges, R. C. (1999). An application of augmented reality to thickness inspection. *Human Factors and Ergonomics in Manufacturing, 9*(4), 331–342.

Cobb, S., D'Cruz, M. D., & Wilson, J. R. (1995). Integrated manufacture: A role for virtual reality? *International Journal of Industrial Ergonomics, 16*, 411–425.

Dai, F. (1998). Introduction-Beyond Walkthroughs. In F. Dai (Ed.), *Virtual reality for industrial application* (pp. 1–9). New York: Springer.

Dai, F., & Göbel, M. (1994). Virtual prototyping—An approach using VR-techniques. In *Proceedings of Computers in Engineering* (pp. 311–316). New York: American Society of Mechanical Engineers.

Davies, R. C., Medbo, L., Engström, T., & Akselsson, R. (1999). *Work and work place design using empirical shop floor information and virtual reality techniques* [Online]. Available: http://www.ie.lth.se/IE/Roy/papers/IEA97/WorkplaceDesign.html [1999, February 14]

European Computer-Industry Research Center (1994). *Augmented reality for mechanical maintenance and repair* [Online]. Available: http://www.ecrc.de/research/uiandv/research.html [1999, May 22]

Feiner, S., Macintyre, B., & Seligmann, D. (1993). Knowledge-based augmented reality. *Communications of the ACM, 36*(7), 53–62.

Finger, S., Baraff, D., Siewiorek, D. P., Weiss, L., Witkin, A., Sundararajan, V., Wright, P., Cutkosky, M., & Prinz, M. C. (1997). Rapid design: Integrating virtual and physical prototypes. In *Proceedings of 1997 NSF Design and Manufacturing Grantees Conference* (pp. 41–42). Arlington, VA: National Science Foundation.

Flaig, T., & Thrainsson, M. T. (1996). Virtual prototyping—New information techniques in product design. In *Proceedings of 1996 ASME Design Engineering Technical Conferences and Computers in Engineering Conference* (pp. 96–DETC/DFM-1417). New York: American Society of Mechanical Engineers.

Flanagan, W. P., & Earnshaw, R. (1997). Meeting the future at the University of Michigan media union. *IEEE Computer Graphics and Applications, 17*(3), 15–19.

Frölich, B., Fischer, M., Agrawala, M., Beers, A., & Hanrahan, P. (1997). Collaborative production modeling and planning. *IEEE Computer Graphics and Applications, 17*(4), 13–15.

Göbel, M. (1996). Industrial application of VEs. *IEEE Computer Graphics and Applications, 16*(1), 10–13.

Gomes, A., & Baacke, P. (1998). Experience with virtual reality technologies in the prototyping process at BMW. In F. Dai (Ed.), *Virtual reality for industrial application* (pp. 151–158). New York: Springer.

Groover, M. P. (1996). *Fundamentals of modern manufacturing: Materials, processes, and systems*. Upper Saddle River, NJ: Prentice-Hall.

Gupta, R., Sheridan, T., & Whitney, D. (1997). Experiments using multimodal virtual environments in design for assembly analysis. *Presence: Teleoperators and Virtual Environments, 6*(3), 318–338.

Harrison, D., Jagues, M., & Strickland, P. (1993). Design by manufacture simulation using a glove input. In K. Warwick, J. Gray, & D. Roberts (Eds.), *Virtual reality in engineering* (pp. 91–103). London: Institution of Electronic Engineers.

Johnson, W. L., Rickel, J., Stiles, R., & Munro, A. (1998). Integrating pedagogical agents into virtual environments. *Presence: Teleoperators and Virtual Environments, 7*(6), 523–546.

Kalawsky, R. S. (1993). *The science of virtual reality and virtual environment*. New York: Addison-Wesley.

Kesavadas, T. (1996). Augmented reality based interaction: Application to design and manufacturing. In *Proceedings of 1996 ASME Design Engineering Technical Conferences and Computers in Engineering Conference* (pp. 96–DETC/DFM-1268). New York: American Society of Mechanical Engineers.

Kheddar, A., Tzafestas, C., & Coiffet, P. (1997). The hidden robot concept—High-level abstraction teleoperation. In *Proceedings of the 1997 IEEE/RSJ International Conference on Intelligent Robot and Systems* (Vol. 3, pp. 1818–1824). Los Alamitos, CA: IEEE Press.

Kies, J. K., Williges, R. C., & Rosson, M. B. (1998). Coordinating computer-supported cooperative work: A review of research issues and strategies. *Journal of the American Society for Information Science, 49*(9), 776–791.

Krishnamurthy, K. R., Shewchuk, J. P., & McLean, C. R. (1999). A hybrid manufacturing system modeling environment using VRML. In J. J. Mills & F. Kimura (Eds.), *Information infrastructure systems for manufacturing* (pp. 163–174). Boston, MA: Kluwer Academic Publishers.

Lehner, V. D., & DeFanti, T. A. (1997). Distributed virtual reality: Supporting remote collaboration in vehicle design. *IEEE Computer Graphics and Applications, 17*(2), 13–17.

Maxfield, J., Fernando, T., & Dew, P. (1998). A distributed virtual environment for collaborative engineering. *Presence: Teleoperators and Virtual Environments, 7*(3), 241–161.

Milgram, P., & Drasic, D. (1997). Perceptual effects in alignment virtual and real objects in augmented reality displays. In *Proceedings of Human Factors and Ergonomics Society Annual Meeting* (pp. 1239–1243). Santa Monica, CA: Human Factors and Ergonomics Society.

Milgram, P., Zhai, S., Drascic, D., & Grodski, J. (1993). Applications of augmented reality for human–robot communication. In *Proceedings of 1993 IEEE/RSJ International Conference on Intelligent Robots and Systems, IORS '93* (Vol. 3, pp. 1467–1472). Los Alamitos, CA: IEEE Press.

Nash, J. (1999). *Wiring the jet set* [Online]. Available: http://www.wired.com/wired/5.10/wiring.html [1999, May 20]

Natonek, E., Flückiger, L., Zimmerman, T., & Baur, C. (1994). Virtual reality: An intuitive approach to robotics. In *Proceedings of Telemanipulator and Telepresence Technologies* (pp. 260–270). Bellingham, WA: SPIE—International Society for Optical Engineering.

Pere, E., Gomez, D., Langrana, N., & Burdea, G. (1996). Virtual mechanical assembly on a PC-based system. In *Proceedings of 1996 ASME Design Engineering Technical Conferences and Computers in Engineering Conference* (pp. 96–DETC/DFM-1306). New York: American Society of Mechanical Engineers.

Pesante, J. A., Williges, R. C., & Woldstad, J. C. (1997). The effects of multitasking on quality inspection in advanced manufacturing. In *Proceedings of the Sixth Annual Industrial Engineering Research Conference* (pp. 113–118). Atlanta, GA: Institute of Industrial Engineers.

Pimentel, K., & Teixeira, K. (1995). *Virtual reality through the new looking glass*. New York: Intel/McGraw-Hill.

Purschke, F., Rabatje, R., Schulze, M., Starke, A., Symietz, M., & Zimmermann, P. (1998). Virtual reality (VR)—New methods for improving and accelerating vehicle development. In F. Dai (Ed.), *Virtual reality for industrial application* (pp. 105–122). New York: Springer.

Rosenberg, L. B. (1993). Virtual fixtures: Perceptual tools for telerobotic manipulation. In *Proceedings of IEEE Virtual Reality Annual International Symposium* (pp. 76–82). Piscataway, NJ: IEEE Press.

Sharma, R., & Molineros, J. (1997). Computer vision–based augmented reality for guiding manual assembly. *Presence: Teleoperators and Virtual Environments, 6*(3), 292–317.

Sims, D. (1994). New realities in aircraft design and manufacture. *IEEE Computer Graphics and Applications, 14*(2), 91.

Tezuka, T., Goto, A., Kashiwa, K., Yoshikawa, H., & Kawano, R. (1994). A study on space interface for teleoperation system. In *Proceedings of the Third IEEE International Workshop on Robot and Human Communication RO-MAN '94* (pp. 62–67). Piscataway, NJ: IEEE Press.

University of Michigan. (1999). *Selected projects* [Online]. Available: http://www-vr1.umich.edu/projects,html [1999, May 16]

Vance, J. M. (1997). Virtual environment for engineering design optimization. In *Proceedings of 1997 NSF Design and Manufacturing Grantees Conference* (pp. 23–24). Arlington, VA: National Science Foundation.

Vance, J. M. (1998). Virtual environment for engineering design optimization. In *Proceedings of 1998 NSF Design and Manufacturing Grantees Conference* (pp. 27–28). Arlington, VA: National Science Foundation.

Vance, J. M. (1999). *CAREER: Virtual reality techniques to improve product design* [Online]. Available: http://www.usc.edu/dept/ise/NSF/proceedings/ed.html [1999, May 16]

Wesche, G., Wind, J., & Göbel, M. (1997). Visualization on the Responsive Workbench. *IEEE Computer Graphics and Applications, 17*(4), 10–12.

Williams, T. J. (1992). *The Purdue enterprise reference architecture*. Research Triangle Park, NC: Instrument Society of America.

Wilson, J. R., Brown, D. J., Cobb, S. V., D'Cruz, M. M., & Eastgate, R. M. (1995). Manufacturing operations in virtual environments (MOVE). *Presence: Teleoperators and Virtual Environments, 4*(3), 306–317.

Wilson, J. R., Cobb, S. V., D'Cruz, M. M., & Eastgate, R. M. (1996). *Virtual reality for industrial application: Opportunities and limitations*. Nottingham, England: Nottingham University Press.

Zachmann, G. (1998). VR—Techniques for industrial applications. In F. Dai (Ed.), *Virtual reality for industrial application* (pp. 13–27). New York: Springer.

# 55

# Entertainment Applications of Virtual Environments

Eric Badiqué,[1] Marc Cavazza,[2] Gudrun Klinker,[3] Gordon Mair,[4] Tony Sweeney,[5] Daniel Thalmann,[6] and Nadia M. Thalmann[7]

*[1]European Commission, DG INFSO—IST Programme*
*Avenue des Nerviens 105, Office 6/014, B-1040 Brussels*
*Eric.Badique@cec.eu.int*
*[2]School of Computing and Mathematics*
*University of Teesside, Borough Road*
*Middlesbrough, UK, TS1 3BA*
*M.O.Cavazza@tees.ac.uk*
*[3]Technische Universität München*
*Fachbereich Informatik (H1)*
*Arcisstr. 21, Munich, Germany, D-80333*
*klinker@in.tum.de*
*[4]University of Strathclyde—DMEM, James Weir Building*
*75 Montrose Street, Glasgow, UK, G1 1XJ*
*g.m.mair@strath.ac.uk*
*[5]National Museum of Photography, Film & Television*
*Bradford, United Kingdom, BD1 1NQ*
*a.sweeney@nmsi.ac.uk*
*[6]Computer Graphics Lab*
*Swiss Federal Institute of Technology (EPFL)*
*DI/LIG, Lausanne, Switzerland, CH-1015*
*Daniel.Thalmann@epfl.ch*
*[7]University of Geneva, 24, rue General Dufour*
*Geneve 4, Switzerland, CH-1211*
*Nadia.Thalmann@cui.unige.ch*

*"If you're looking for what's next in online technology and commerce, just follow the gamers."*
—(Berry, 1997)

---

Disclaimer: The opinions expressed in this chapter are those of the authors and do not necessarily reflect the views of their respective organizations.

## 1. INTRODUCTION

Virtual environments (VEs) as we know them today originated in the 1970s with simple computer-based flight simulators, which continued to be developed through the 1980s and 1990s. This form of VE was overtaken by the emergence of VE video games in the entertainment market, both in amusement arcades and at home on video games consoles.

Entertainment-based VEs, however, are nothing new. At the Paris Fair in 1900, three high-tech virtual experiences were presented to the public. The Trans-Siberian Railway Panorama enlivened existing dioramas and panoramas by introducing motion and treated visitors to a 45-minute experience reenacting a 14-day rail journey between Moscow and Peking. A number of horizontal belts and painted screens driven at high speed gave the illusion of traveling while being comfortably seated in a train coach. The Mareorama added a motion platform to simulate a cruise from Nice to Constantinople, and the Cineorama Air-Balloon Panorama presented for the first time a full-surround movie shot from a balloon rising over Paris. Fifty years later, the Scopitone, Experience Theater, Sensorama Simulator, and Telesphere Mask offered remarkably visionary ways to transport people through alternative realities, either in large groups or individually (Dodsworth, 1998).

Today, virtual environments are at the core of most successful games and other entertainment platforms. This chapter does not seek to establish an exhaustive list of technologies, systems, and applications available. This area is moving much too fast for this to be realistic. Besides, most of the technologies relevant to games and entertainment applications, in general, have been covered in previous chapters. Instead, this chapter will survey the major entertainment areas in which VEs are an essential element and attempt to provide some guidance on what the future may bring.

Current VE entertainment applications include games, whether they are stand-alone or networked, and online communities created around games. They increasingly offer various services reaching much beyond pure entertainment, such as interpersonal communication forms and online electronic transactions. They also include game characters and avatars (see chap. 15, this volume), pointing the way to artificial life and populated persistent worlds. High-end systems, such as location-based entertainment (LBE) centers and digital theme parks, are also addressed, offering some of the most exciting virtual experiences to date.

The chapter also examines the strong historical links between military simulation and entertainment and important issues, such as violence and pornography in entertainment, are briefly discussed. A final section attempts to guide the reader through future prospects for entertainment applications based on virtual environments. What is the future of content? What are the trends in interactivity? How will theme parks and LBEs evolve? What about distributed games? What can research in augmented reality (AR) and telepresence offer to entertainment? When will we be able to fully immerse ourselves in highly compelling shared VEs to play, experience an impossible ride, or simply communicate with others in a completely new way?

Because they are developed in a very demanding and competitive environment, games have always been at the forefront of the technology wave. Games and entertainment systems are a major driving force for the development of VE-based systems and their applications in other sectors. This chapter aims at explaining why.

## 2. GAMES, ONLINE GAMES, AND VIRTUAL COMMUNITIES

Computer games are as financially successful as film, video, and pop music. They may run on arcade machines, personal computers (PCs), gaming consoles and palmtop devices. They are interactive, providing an experience of story line, time, and pacing that is entirely different from the linear content of film. Computer gaming is a hugely influential popular culture. Many games aim just for sheer entertainment. Others may also be educational, intellectually challenging, or emotionally engaging. We are arguably still in the early days of computer gaming. Some see the move from two-dimensional (2-D) to three-dimensional (3-D) computer graphics paralleling how movies evolved from silents to talkies in the 1920s.

Early games were 2-D, and were viewed from the top or side. For years, characters jumped endlessly from one flat platform or level to another to score points—some still do! In 1980, Battlezone was the first true 3-D gaming world, and ever since then virtual environments have been the setting of most successful games. Games developers are working on innovative forms of computer-generated "beings," artificial life, and intelligence (see chap. 15, this volume). Along with interface technologies such as speech recognition, these will enable more sophisticated interaction and independent behavior in gaming characters. Better imaging quality and video headsets will provide more visually realistic, immersive worlds.

The success of computer games has been due to the availability of low-cost ubiquitous machines. The Nintendo 64 platform provided the first truly interactive 3-D video game machine and used a Silicon Graphics state-of-the-art semiconductor to achieve a low-cost, high-performance, high-volume product. Because it is important to get the most out of technology, games are mostly based on proprietary technology, guaranteeing appropriate reaction speed for fighting games. The difficulty for game producers is to have one platform for several games to ensure sufficient revenues while keeping enough room for flexibility for constant innovation in order to keep games attractive. Progress is constant. As an example, recent advances are taking the visual reality of 3-D scenes to unprecedented heights with real-time special effects. Games are set to become more elaborate, engaging, cooperative, intense, and cinematic—while not forgetting their sense of fun!

## 2.1 Gaming Genres

Many computer games have been born, flourished for a moment and then been lost to history. Those that survive generally belong to a few popular types or genres. The genres discussed in this section demonstrate some of the visual styles, game-playing formulas, and intentions typical of contemporary games.

### *2.1.1 Shoot-'em-ups and Beat-'em-ups*

A staple games diet is relentless, fast-paced combat action, backed-up by awesome arsenals of weapons and powers. Run around killing everything that moves before it kills you! Game play consists of fast eye–brain–hand reflexes focused on control-button combinations. These offer some of the most intense gaming experiences—but leave your intellect behind.

### *2.1.2 Flying, Racing, and Sports*

These games range from cartoon capers offering slapstick thrills and spills, to serious simulators of military jets or F1 racers that can hardly be called games at all. Such simulations represent the pinnacle of realism in computer games—not just in image quality but also in the authenticity with which they mimic the dynamics of the vehicles and gadgets involved.

### *2.1.3 Strategy and Puzzle Games*

These simulations require long-term game plans for managing resources and negotiating problems. Players build and run complex, dynamic structures, such as cities or ecosystems, with populations and maybe armies for conquest. Some require only mechanical learning of the rules, but the best present challenging scenarios akin to the real world. Puzzlers are often more restful, involving ingenuity and problem solving.

### 2.1.4 Adventures and Role-playing Games (RPGs)

Somewhere between shoot-'em-ups and strategies, adventures let you control characters engaged in action-based quests. They involve exploration, puzzles, obstacles and enemies to overcome. In more complex RPGs, the main character is your alter ego—a personality that evolves as you journey through a maze of plots and subplots.

### 2.1.5 Online Games

Online games are the latest trend and offer completely new and exciting perspectives to gaming. Popular consoles are now equipped with modems for online multiplayer gaming, and the Internet takes the game experience to new levels. There are more than a dozen online gaming services available today—Interactive Magic's shock force, Segasoft's 10six, Sony's Everquest, CyberPark, Heat.net, Engage, Battle.net, MPG Net, Dwango, TEN, Kali, Mpath, Ultima online, and others—and new entrants are around the corner (Berry, 1997).

So far, shoot-'em-ups have dominated networked gaming. Internet connections allow remote players to compete in shared multiplayer gaming worlds through avatars. Cyber wars are being organized starting at a given time, complete with battle moderators, prebattle briefing, and daily war reports.

Teams of players often interact and collaborate with each other through additional audio and text chat channels. The best games have sophisticated 3-D rendering and the ability to custom design your own worlds. Since the Internet is not yet able to provide the kind of low latency necessary to fast shooting games, online games tend to emphasize strategy and collaboration in what can be seen as a positive trend. Like for many other Internet businesses, the perfect business models for such games is still being sought and few game companies make money. Most online games offer licenses for a day or for unlimited use. Some rely on commercial sponsors.

Online, graphic RPGs are increasingly popular. Thousands of people can play simultaneously in massive, mysterious worlds full of intrigue and conflict. These environments go beyond games and offer a constantly changing world where players chart their own destiny. They choose their appearance, personality, and place in the world. They become involved in intrigues, fulfill quests, kill scores of monsters, or even other players.

Some games offer an entire world with diverse species, economic systems, alliances, and politics. One can choose from a variety of races and classes, and begin a journey in any number of cities or villages throughout virtual continents. One can learn skills, earn experience, acquire treasure and equipment, meet friends, and encounter enemies. The multiplayer online gaming community explores and extends the digital frontier, pushing new technology to its performance limit and in effect creating new opportunities for building immersive virtual communities.

In the search for killer applications appealing to a broader population base, gambling and virtual sex are hot contenders. Gambling is a very large business and cyberspace is gambling's next frontier. Transposed to the Net, gambling could be everywhere and nowhere at once. Cybercasinos offer the perfect shelter from taxes, and if ever legalized, gambling from home is likely to drive masses of people to virtual worlds. For now, online casinos simulate the excitement of gambling—without real betting. They aim to recreate the "Las Vegas experience", with the hope of being well positioned for the day when online wagering is legalized (Virtual Vegas).

### 2.1.6 Interactive Virtual Worlds

New cyber settlements allow users to own 3-D virtual real estate with personal chat rooms, message boards, and clubs. Friends can be invited to come over and socialize. These "cyber towns" can have their own economy and social structure. One can become a proper citizen, take jobs in the community, and earn community "money" (community credits), which can be used in the virtual community mall. Some cyber towns are fully functioning communities with an elected mayor and city council, city guides, and security officers. There is even a virtual reality time (VRT), suggested to be the common time zone for all virtual environments, wherever you are around the world.

These virtual worlds offer places to meet, venues for online events, message boards, and shopping malls. They overcome distances and open limitless possibilities for the development of new markets and shopping experiences opened 24 hours a day. Conducting business in an entertainment-like setting is part of the attraction. Online communities are bound to have a huge impact on e-commerce, education, health, and leisure. Ultimately, they may also change democracy, as more direct contact with government and other citizens may lessen the role of elected representatives.

Online communities are particularly interesting because they offer a radical alternative to real-life communities. They involve large numbers of geographically dispersed people. They have access to and strive on an infinite source of information. They involve people who are mostly complete strangers but yet are friendly with each other. Online communities have their roots in text-based multiuser games and communities (i.e., multiuser domains [MUDs], MUDs object-oriented [MOOs], and Web-based MOOs [WOOs]). These environments have been successful in building functioning communities. Inhabited worlds have added the rendering power developed for games such as Doom and Quake. There are many inhabited virtual worlds on the Internet, although these are still early days for some. Some of them are work in progress and still have problems with communication between avatars. Some are also very sparsely populated, and one can feel pretty lonely while exploring them (Damer, Gold, Marcelo, & Revi, 1997). But others are already fully functioning business and entertainment portals with large user communities (Blaxxun Interactive, Cybertown). They integrate with popular World Wide Web browsers and require a set of Java plug-ins, running on any platform supporting Java.

The technology involved is not necessarily very sophisticated. Some interactive worlds do not always make use of available VE features. Instead, they concentrate on a large range of consumer platforms at modem speed. Some companies specialize in the use of proprietary or industry standard "photographic" VE software to enable virtual travel around the world.

Avatars can be personalized and are able to exchange cards with any user met online. Messages can be sent, private chats opened, and voice conversation supported. An open architecture allows users to build their own world. Since the potential of user creativity in virtual worlds is huge, there has been a major change in the way content is created (see chap. 25, this volume).

But this is just the beginning. Navigation in virtual worlds can still be difficult because it is mostly based on a set of keyboard commands (see chap. 24, this volume). Lack of collision detection (see chap. 12, this volume) is often a source of confusion in Virtual Reality Modeling Language (VRML) worlds. Some worlds are still too artificial. As more worlds are built, artistic quality will improve and progress in interface technologies will ensure that navigation in online virtual worlds is as easy as in popular games such as Doom.

These new interfaces naturally appeal to the younger generation. Although most people have only recently got used to 2-D interfaces, the generation brought up on Quake, Doom, and Nintendo 64 is used to navigation through complex 3-D environments and sophisticated behaviors. They demand metaphors based on real place, and the Web is giving them what they want.

The home-based playing, chatting, socializing, and playing experience is increasingly being integrated within multipurpose infotainment appliances offering radio, TV, computing, and VE-based services. Virtual communities are also approaching convergence with traditional media programming. Some include network cameras and live 3-D broadcasts for special events. They also feature new modes of interaction, such as 3-D chat worlds, text to speech, and audio conferences with artists and personalities. The result is a new media for which the frontier between "storytelling," as in current broadcasting, and "communication" will progressively blur, leaving us with a continuum of image-rich and flexible "infocommunication" services somewhere between plain old television and multimodal telepresence.

## 3. POPULATING VIRTUAL ENVIRONMENTS FOR ENTERTAINMENT: AUTONOMOUS ELEMENTS, ARTIFICIAL LIFE, AND VIRTUAL HUMANS

Just as in real life, the characters one meets in virtual environments are often essential to the quality of the experience. Empty virtual environments would quickly loose their attractiveness. Characters are naturally central to games, but they are also moving to the forefront in other applications of VEs, whether they support new forms of human-to-human or human-to-machine communication or they are meant to increase the realism of complex environments. This section surveys a few of the most popular game stars and takes us through some of the latest research on the way to virtual humans.

### 3.1 Game Characters

Computer games have their own virtual superstars—leading characters that have helped sell games by the millions. Good characters can be vital to the success of a game. They can help draw players into the story line and identify with the fun and tension that game play offers. Successful characters also build brand loyalty, making sequels easier to market and supporting spin-off merchandising.

ATARI's 1980 Pac-Man was perhaps the first true computer-game character. The middle-aged Italian plumber Mario is one of the most successful game characters of all. Mario first appeared in 1981 in the arcade game Donkey Kong and has since become as important to Nintendo as Mickey Mouse is to Disney.

Lara Croft, star of Tomb Raider, is widely credited with introducing a whole new audience to computer gaming since she first appeared in the mid-1990s. Lara's British developers, Core Design, "manage" her like a film star. They even keep a manual on her that records her personality, dislikes, and favorite colors, music, and food.

Character modeling and movement in 2-D and 3-D are constantly improving, but it is not just looks that count. Some games go to great lengths to create believable identities and convincing personalities, even for fantasy creatures. Characters have aims, motives, desires, sadness, humor, problems, backgrounds, and relationships with other characters. Players can explore these as the story line and characters themselves evolve.

The interactive structure of games can undermine the use of some traditional storytelling techniques that help establish characters. Whether games will ever need—or could achieve—the rich characterization that traditional linear media can achieve is open to question. Perhaps this would only be needed in games in which the whole aim was to explore the nature of the characters themselves. In the end, the primary reason for any game character's existence and success will continue to be that the game itself is fun to play.

### 3.2 Virtual Humans

In the games market, constant innovation is required in order to prevent sales of games from falling off. Convincing simulated humans in games have been identified by the industry as a way of giving a fresh appearance to existing games and enabling new kinds of game to be produced. The ability to embed the viewer, as an avatar (Capin, Pandzic, Magnenat, Thalmann, & Thalmann, 1999), in a dramatic situation created by the behavior of other, simulated, digital actors will add a new dimension to existing simulation-based products for education and entertainment.

In most popular games, the level of artificial intelligence of digital creatures is still limited. Recently, however, a few games have appeared based on artificial life principles. The most well known is Creatures, from Cyberlife. Artificial life in VEs is a new promising approach, and several researchers have proposed innovative algorithms and systems (Magnenat-Thalmann & Thalmann, 1994). Sims (1994) describes a system for the evolution and coevolution of virtual creatures that compete in physically simulated 3-D worlds. Pairs of individuals enter one-on-one contests in which they contend to gain control of a common resource. The winners receive higher relative fitness scores allowing them to survive and reproduce. Realistic dynamics simulation, including gravity, collisions, and friction, restricts the actions to physically plausible behaviors. The morphology of these creatures and the neural systems for controlling their muscle forces are both genetically determined, and the morphology and behavior can adapt to each other as they evolve simultaneously. In the ALIVE system (Maes, Darrell, Blumberg, & Pentland, 1995), the virtual world is inhabited by inanimate objects as well as agents. Agents are modeled as autonomous behaving entities, which have their own sensors and goals and which can interpret the actions of participants and react to them in "interactive time." Several digital creatures have been created: autonomous fish living in a physically modeled virtual marine world (Tu & Terzopoulos, 1994); virtual actors, with seemingly emotionally responses driven by random noise functions; and autonomous dogs (Blumberg & Galyean, 1995). The system IMPROV (Goldberg, 1997) allows production of 3-D VEs in which human-directed avatars and agents interact with each other in real-time, through a combination of procedural animation and scripts.

One of the most challenging VE applications is real-time simulation and interaction with autonomous actors, especially in the area of games and cooperative work. Systems such as the Virtual Life NETwork (VLNET; Capin, Pandzic, Magnenat-Thalmann, Capin, & Thalmann, 1997; Pandzic, Magnenat-Thalmann, Capin, & Thalmann, 1997) support a networked shared VE that allows multiple users to interact with each other and their surrounding in real time. Users are represented by 3-D virtual human actors who serve as agents to interact with the environment and other agents. The agents have a similar appearance and behavior to real humans, to support the sense of presence (see chap. 40, this volume) of the users in the environment. In addition to user-guided agents, the environment can also be extended to include fully autonomous human agents used as a friendly user interface to different services, such as navigation. Virtual humans can also be used in order to represent currently unavailable partners, allowing asynchronous cooperation between distant partners.

Users should animate their virtual human representation in real-time. However, virtual human control is not straightforward; complexity of virtual human representation needs a large number of degrees of freedom to be tracked. In addition, interaction with the environment increases this difficulty even more. Therefore, virtual human control should use higher level mechanisms to animate the representation with maximal facility and minimal input. Virtual human control can be divided into three classes:

1. *Avatars*: Basic motion data of the virtual human are modified directly by providing new degrees of freedom values directly (e.g., by sensors attached to the body).
2. *User-guided virtual humans*: A user guides the virtual human by defining tasks to perform, and the virtual human uses its motor skills to perform actions by coordinated joint movements (e.g., walk, sit).
3. *Autonomous virtual humans*: The virtual human is assumed to have an internal state, which is built by its goals and sensor information from the environment, and the participant modifies this state by defining high-level motivations and state changes (e.g., turning on vision behavior).

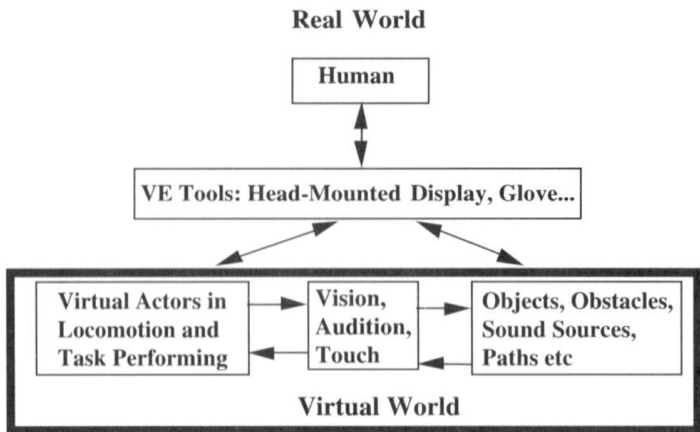


FIG. 55.1. Real and virtual sensors.

Participants are naturally aware of the actions of the virtual humans through VE tools, such as a head-mounted display (HMD), but one major problem to solve is to make virtual actors conscious of the behavior of participants. Virtual actors should sense the participants through their virtual sensors (see Fig. 55.1). For the interaction between virtual humans and real ones, gesture and facial expression recognition (see chap. 10, this volume) is a key issue for new high-end games. For example, Emering, Boulic, and Thalmann (1998) describe a project in which a real person represented by an avatar can fight against an autonomous actor. The motion of the real person is captured using a Flock of Birds. Gestures are recognized by the system, and the information is transmitted to the virtual actor, who is able to react to the gestures.

Artificial intelligence (AI) is increasingly making its way into games and other forms of entertainment. It primarily affects characters' behavior and offers radical new forms of game play. New AI-based games feature characters that can cope with complex situations and evolve, for example, moving up the social ladder throughout a game session. Others can learn gestures and movements during the course of the game. Artificial life is here and some of the first commercial successes were due to games.

## 4. LOCATION-BASED ENTERTAINMENT AND VIRTUAL ENVIRONMENTS IN THEME PARKS

Before making their way to the home on low-cost platforms such as consoles and personal computers, virtual environments have flourished in entertainment centers and theme parks. The following section takes a look at some of the most successful 'high-end' VEs so far.

### 4.1 Location-based Entertainment Centers

Location-based entertainment (LBE) blends the amusement park ride and video game. Pods or cockpits are installed in a thematic setting, and passengers discover fantasy worlds via computer-based graphics, sound, input, and various display technologies. LBE centers started in the 1990's as an outgrowth of video arcade and military simulators.

The most popular LBE games are flight and driving simulators. Two-seat flight simulators allow players to experience and perform maneuvers that replicate those of an actual fighter aircraft. A flight crew is composed of two people, a pilot and a weapons system officer. Cockpits are networked to allow teams to wage war in a simulated battle environment. The experience is complete. Cockpit controls include foot pedals, throttle, military-style joystick, and monitors. Team members cooperate to accomplish a mission and are in radio contact. Surround Sound

FIG. 55.2. Indy racing car. Courtesy of Illusion, Inc.

and high-resolution graphics guarantee high realism with the sound of explosions, crashes, and gunfire.

LBE provide powerful dramatic immersive experiences. An important part of their fun comes from communicating with other players and through collaboration. LBE centers are fast becoming some of the most visible applications of interactive digital graphics and visualization. In some places, LBE centers have given birth to a subculture. They provide a specific attraction, often accompanied by a cafe, bar, or restaurant. When players come out of a cockpit they have shared an experience and talk about it with newly made friends. But despite very sophisticated technology, current scenarios are usually very basic and juvenile. Players are mostly young males, and these centers capitalize on the fact that most young males enjoy seek-and-destroy games. The emphasis is on quick action and high tension. The content is mostly competition-based and most LBEs as they exist today can be seen as "boy toys." This is changing as content evolves to address a broader user group.

## 4.2 Virtual Environments in Theme Parks

Theme parks offer the best of VE entertainment in the form of simulation films, interactive rides, and immersion theaters, such as the popular IMAX. Immersion theaters have been around for three decades. They are designed to immerse large audiences within a high-resolution movie, projected on a very large screen so that the image field of view completely dominates peripheral vision. Immersion theaters were long used for short documentary films or for education purposes. They are now increasingly moving to entertainment content with more commercial potential. They also often provide environmental scenography as a background for live actors, motion-based simulations, and special effects. But production costs tend to limit their use to high-end theme parks. Digital technology may provide solutions to this cost issue in the future (Dodsworth, 1998).

Simulation films usually place the audience inside a vehicle (e.g., train, plane, or car) and give them a few minutes of a sensation-rich ride. Hydraulic systems animate the entire floor of the theater, synchronized with the video image. Motion-enhanced rides are increasingly interactive. Some rides have seamless transition between video-disk-based real imagery walk-throughs and fully interactive rides. Since such amusement park rides require waiting time, this time is generally used to create an atmosphere, define the context, and explain what visitors have to do during interactive sequences. Giving a pre-immersion background story, giving clues and concrete goals to fulfill are important elements.

There are many examples of popular games and rides experienced in LBE centers, theme restaurants, bars, family entertainment centers, or theme parks. Their number is growing by the day, only limited by people's imagination and availability of venture capital to make them happen.

Thanks to advances in interface technologies and to a positive evolution in the content of attractions, a new generation of rides is now emerging. There has been a move away from competition to an emphasis on experience or story line, as opposed to a competitive objective. Here is a snapshot of some of the best second-generation rides available today.

#### *4.2.1 Sport*

Simulators are targeting many sports, including soccer, basketball, racquetball, table tennis, ice hockey, tennis, and darts. Video-based, unencumbered interfaces allow players to measure themselves against a team of opponents by moving their body and interacting with images. These highly interactive games challenge physical coordination, tactical decision making, and cooperation between players.

Distributed sport simulation is the subject of interesting experiments. A virtual tennis game over a network (Molet et al., 1999) was shown in the opening and closing ceremonies of Telecom Interactive '97 in Geneva, Switzerland. The tennis game involved two participants and one umpire. The participants, one in Geneva, the other in Lausanne, were represented by directly controlled virtual humans wearing Flock of Birds magnetic trackers attached to their bodies. They were using immersive HMDs to visualize the scene. The autonomous umpire used her synthetic vision to watch the game and decided the state of the game and score. She updated the state of the match when she "heard" a ballcollision event (ball–ground, ball–net, ball–racket) according to what she saw and her knowledge of the tennis game. She communicated her decisions and the state of the game by "spoken words" (sound events, see Fig. 55.3).

#### *4.2.2 Thrill Rides*

Being able to fly is a universal dream. A virtual hang gliding simulator (Dreamality Technologies, Inc.) allows users to experience the thrill of hang gliding without leaving the ground. It brings a level of realism that permits virtual pilots to experience this sport without the risk or expense. The rider scores points for finding thermals and gaining altitude, the length of flying time, and for making successful landings. The scoring system also deducts points for crashes and collisions with weather phenomena and objects such as planes, jets, birds, balloons, ultralites, or buildings. When using an HMD, head tracking can be used to enable the rider to look in any direction and view that portion of the visual scene. Different high-quality sounds and various noises are also experienced while flying. The sensation of wind in the face is also provided. These simulators can be networked to other simulators, allowing each rider to see other riders and interact with them.

Directly derived from military training systems, parachuting in the safety of an indoor simulator is also possible. In one such system, JumpZone, the floor drops away, and the guest drops several inches as the ride starts. As the ride is activated, the noise of the wind and the receding airplane is heard, and wind effects start. As soon as the ripcord is pulled, the guest hears the sound of the chute unfurling and feels the harness jerk (see Fig 55.4).

FIG. 55.3. Anyone for tennis? Courtesy of EPFL/UniGe.

FIG. 55.4. JumpZone. Courtesy of Illusion, Inc.

The ultimate thrill may be virtual skyboarding, where users interact by moving their body in real space thanks to camera-based unencumbered virtual environments. A tracking system allows a player to interact with on-screen background and foreground graphics. Images and game-play are viewed via three television monitors, and players direct a flying sky board through a futuristic landscape to earn points and progress to higher levels. Moving in front of a video camera, players see themselves on-screen, where they control their movement flying through and interacting with a computer generated landscape. The experience tests a user's agility and speed. In the future it may lead to very extreme physical experiences through the use of full 360-degree rotational interfaces such as VE gyroscopes.

#### 4.2.3 Disney Quest

Some of the best examples of immersive entertainment are Disney Quest's indoor interactive theme parks (Disney Quest). These multiple-story centers combine Disney content with leading-edge immersive technology to create what is claimed to be a new kind of entertainment. Such centers offer various entertainment zones to explore and discover rich VEs or play games based on popular Disney characters. Guests can embark on adventures that let them actually enter the story and become a part of it.

A creation zone even allows the design of one's own roller-coaster ride before experiencing it on Cyberspace Mountain, a 360-degree pitch-and-roll simulator. Rides such as the Virtual Jungle Cruise or Aladdin's Magic Carpet are very popular. Virtual environments are used as a new medium to tell stories. Content and interactivity are the key. The shows contain a large number of synthetic characters and narrative storylines.

In Aladdin's Magic Carpet, visitors wear HMDs and pilot a flying carpet sitting on a motorcycle-style seat. The seat is mounted on a movable base, allowing it to pitch up and down following the driver's instructions. Stereo ambient sound, binaural recorded sound, and eight channels of localized sound ensure total audio immersion.

#### 4.2.4 Experiential Rides

Building on game technology, new entertainment options are becoming available. Instead of fast reactions and strategy, they emphasize discovery and novel experiences. Some VE installations are far away from the thrill-seeking VE games found in malls and theme parks.

From water beds to gyroscopes and hydraulic units, a variety of platforms provide a new kind of travel into cyberspace and into virtual worlds, where one can experience intense sensory stimulation or where two or more players can interact with each other in challenging play/reality situations. Capitalizing on the success of "riding" games (previously discussed), virtual gliders enable you to fly into canyons with no other purpose than the observation of virtual insects or fellow gliders.

Perfectly merging auditory and visual VE, other experiences use cues extracted from recorded music to create geometry and control object behaviors within a virtual world. Virtual objects respond in sync with music, creating a rich, multisensory experience. Such tools allow directors and designers to develop visually active, immersive environments that merge 3-D models with sound, such as a music video, in which users are completely immersed and free to explore a world generated by music. In essence, these new virtual environments explore new art forms.

## 5. VIRTUAL ENVIRONMENTS, INTERACTIVE MOVIES, AND THE FUTURE OF STORYTELLING

Nakatsu and Tosa (1997) proposed a model in which virtual environments, movies, and video games will soon converge into a single kind of interactive media. Such a convergence of traditional media and VE, however, raises several issues in terms of interactivity and user experience (Nakatsu & Altman, 1997). The next sections discuss required additions to current VE technologies that could assist in realizing the next generation of interactivity. After analyzing user behavior, some concepts are introduced that can support analysis of next-generation interactivity. This framework is related to the process of interactive content creation and authoring. The discussion concludes with a review of technical challenges. The present situation in interactive entertainment and storytelling will be mainly illustrated by examples from recent computer games, while state-of-the-art VE applications can be found in other chapters (see chaps. 42–55, this volume).

In future interactive media, the traditional distinction between actor, director, and spectator will tend to disappear (Benford, Greenhalgh, & Craven, 2001). Bolter and Grusin (1999) noted that in current interactive games, players are both directors and spectators. This is, for instance, announced by the way some games offer a third-person perspective on the player characters, such as Lara Croft (Tomb Raider) or Solid Snake (Metal Gear Solid). And it can be observed that the games with personified avatars have also introduced a variety of camera movements during the action itself. Another relevant paradigm consists in alternating active gaming phases and replay phases where the player is a spectator of his previous game. This has already been introduced in games in which the strain was too important for the user to fully enjoy the global action as a visual scene. The intense playing phases of some video games (e.g., Gran Turismo or Ace Combat) generate corresponding replay sequences of the same length, in which the user can still control the viewpoint and camera effects. These games offer an interesting model for interactive media in light of the current discussion on the trade-off between playing and authoring. This feature can be part of the solution to a well-known problem—that interactivity could involve the spectator far beyond his own wishes and make interactive media an exhausting rather than relaxing experience.

There are two major concepts that should be analyzed in connection with interactivity. The first one is user experience/user involvement. How can the user both interact with the scene and with the plot? How can he interact and be a spectator at the same time? The second one is adaptive storytelling, which should support interactive experience from the storytelling perspective by offering models and formalisms for content description, thereby enabling user-controlled nonlinear storytelling.

There are currently no complete theories of interactivity that would be suitable for the analysis of digital media. For the brief analysis outlined here, we propose to ground the analysis of interactivity on the concepts of causality and determinism. The notion of causality is central to many epistemological discussions and has been extensively investigated in the field of artificial intelligence. It should thus not come as a surprise that it surfaces here, as AI should be an important enabling technology for interactive media, not only through artificial actors but also through nonlinear storytelling management, natural language interfaces, user-dependent customization, and other means. What causality essentially corresponds to is that user actions should determine changes, not so much in the virtual world in which the action now takes place but in the plot itself, which can involve more complex and abstract representations. This trend would be supported by the emerging field of intelligent VEs, which brings back AI into VE systems and applications.

It can be observed that the situation in current video games only reflects a single movie category—action movies. This is largely because of the kind of actions that are made possible with current input devices, which are physical actions. Being able to interact on the plot at a more abstract level is less common, for instance by engaging in conversation with synthetic actors whose intentions and beliefs would relate actions to the narrative structure of the plot. The current situation in interactive entertainment is largely that of full determinism. The same actions produce the same effects, as initial conditions are repeated through various stages of the game. The representation behind the plot is still a treelike representation, with branching nodes being associated with key events. Interactivity thus takes place only at these key events.

In order to implement more sophisticated forms of causality, an appropriate level of modeling has to be found to make it possible to dynamically propagate the consequences of user actions in terms of the plot itself. When considering the simplest computer games, such as driving simulators or shoot-'em-ups, there is no autonomous storytelling. The interaction loop is a simple one in which the player is constantly solicited to interact and in which there are direct consequences of those actions. It thus appears that there is a trade-off between storytelling and full real-time plot generation. Let us examine the benefits and drawbacks of both extremes in terms of game-play and user acceptance. A full dynamic computation would allow an infinite number of stories and endings to be generated (provided this would be technically feasible on the basis of physical simulation, which is unlikely) but would on the other hand demand permanent user involvement. The main advantage of operating within an interaction framework is that users can decide when to be involved. It is indeed a property of storytelling itself that some key events largely decide the further unfolding of the plot, and this is reflected by the specific meaning that some actions take for the spectator. In other words, some properties of the content itself could facilitate its formalization in an interactive computer system.

Therefore some causality in storytelling should exist to support user involvement, just like (a simplified) physical causality supports game-play in first generation computer games. At this stage, one realizes the delicate balance between determinism and creativity. This important concept deserves a more detailed discussion. Several arguments can minimize the drawbacks of determinism. The first one is simply to say that entertainment is just emerging from noninteractive movies and that any way of choosing alternate plots, even if deterministic, would represent some progress in terms of spectator's experience (for a receptive audience). The second argument is that predictability, not determinism, is the actual problem. In the real world, system complexity and the inaccessibility of some initial conditions can make deterministic systems nonpredictable. In virtual worlds, this arises from the fact that users have no access to the underlying principles that control event generation. The final paradox is that some form of determinism is required for the story to be understandable and that narratology is, to some extent, based on the determinism of narrative functions. The novelty of interactive storytelling should thus take place in the framework of such a deterministic formalism. That is to say, interactive storytelling will be a matter of substituting some narrative functions to others, under the supervision of the spectator.

Let us consider, as an example, the unfolding of a James Bond movie. This has been the object of narrative analysis since Barthes (1966). Barthes, analyzing some phases of the James Bond movie *Goldfinger*, described a subset of narrative functions and gave an example of how alternative actions might have resulted in a different course of action, still compatible with the storytelling formalism. Following traditional concepts in narratology established by Propp (1968), Barthes introduced the notion of *sequence* as the basic unit of storytelling. A sequence is an elementary ordered set of actions. For instance, offering a cigarette involves the following elementary actions: offering, accepting, lighting, and smoking. Schank and Abelson (1977) developed similar concepts in the field of AI for story understanding, which were termed *scripts*. However, while Schank and Abelson put emphasis on the script as a normal course of action that enables recognition of a classic situation from the occurrence of its basic elements, Barthes considers that elementary actions within a sequence act as *dispatchers*, serving as a basis for story variability. This can be reinterpreted in terms of interactive storytelling. The main problem is to be able to authorize interactions at crucial stages of the plot and to confer an implicit meaning to user intervention. This implicit meaning is often accessible to the user through the common knowledge of narrative structures that popular culture has assembled. Note that, through this knowledge, the user will have some awareness of the nature of his intervention, yet this will not be a deterministic process, as he might miss many elements of the plot. So the spectator can look like he is acting perfectly as he tries to "help the hero," although in reality his actions will be highly detrimental.

The forms of interaction and user involvement that can be conceived in the perspective of VE and storytelling convergence can now be summarized:

- *Physical interaction with the world.* As VE systems are strongly based on physical worlds, one might think that their support of interactivity should be through actual physical simulation. But, apart from being largely out of reach in terms of generalized physical modeling, this would simply miss the main point of storytelling, which is to allow characters to be the main source of determinism. Designers should thus not restrict themselves to physical interaction, but also consider interactivity as involving information exchange with artificial actors (see chap. 15, this volume), which in turn influences their beliefs and plans, hence the story as a whole. In this case, it is more AI techniques than rendering and simulation that are going to support interactivity.
- *Conversational interaction with one of the characters.* This would appear to be the most natural and acceptable way of interacting with characters. The user would give instructions or advice to the main characters (e.g., "do not open that door"), thus changing the course of the plot. Apart from the well-known difficulties of speech recognition and natural language understanding, the main technical problem would be to control the phases at which such interaction would take place.
- *High-level plot alteration.* This is actually the degree zero of interaction. For example, early interactive movies allowed the choice of different paths on a tree-structured story, based on a given action being chosen by the audience from a list. That such a primitive approach should be used for next generation systems can only be justified by the desire of some spectators to actually control future events or impose constraints on the scenario (such as "Character X should not die").

The convergence of VE and storytelling also brings an important consequence that it is possible to navigate in the virtual world outside of the main scene. This is a dramatic extension from just being able to change viewpoints and perspective on the scene itself, as it disrupts the traditional conception of storytelling being based on a visible set. It thus allows altering the future course of action by interacting with world objects outside the main focus of the story at the expense of the user's spectator role. Such deferred interaction, however, would consist in the user altering the course of action, not through interfering with the action itself, but by changing the structure of the physical world in which the action will take place.

The convergence of VE and storytelling could represent a major transformation of both media and entertainment technologies. However, many technical challenges that go far beyond the development of new VE or interface technologies can be identified. It appears that new theories of storytelling should be developed in order to describe computer formalisms that can support interactivity while still accommodating traditional narrative content.

## 6. MILITARY SIMULATION AND GAMES

Many computer games find their inspiration in military simulation, whether these are flight simulators or war games. In both cases, they replicate, in a reduced and more accessible form, actual military systems, not only flight simulators but also control and command systems. The latter also belong to VE technology, as they eventually implement user-centered interaction with 3-D world models (Cavazza, Bonne, Pernel, Pouteau, & Prunet, 1995).

While it has been reported that defense organizations have shown interest in using gaming software for training purposes (one famous example being the use of the Doom engine, adapted for simple infantry training), the main rationale for the use of gaming technology has been its performance–cost ratio. However, in most cases, the technology transfer is still more likely to take place from military systems toward entertainment applications, though recently technological convergence between military and entertainment systems has been addressed (Zyda & Sheehan, 1997).

Military simulation is the most relevant technology for entertainment systems. One of the major endeavors of modern military simulation is the achievement of realistic large-scale distributed battlefield simulations (see chap. 43, this volume). Many R&D programs are part of this long-term effort, such as the Synthetic Theatre of War (STOW) and Command FORces (CFOR) initiatives.

These include several technologies that are relevant to distributed gaming as well, including efficient distributed simulation using minimal bandwidth and population by both human actors and computer-generated forces, which demand the development of credible synthetic opponents. The simulation of intelligent behavior has certainly been one of the main achievements of military simulation programs. Furthermore, this research has addressed both individual and intelligent collective behavior, such as the coordination of synthetic forces towards a common goal (see chap. 44, this volume). One such example of intelligent behavior modeling for synthetic forces has been the use of the SOAR system. Recently, a simplified version of SOAR has been adapted to the popular Quake video game to control behavior of nonplayer characters by a research group at the University of Michigan (van Lent & Laird, 1999).

Another recent example of the convergence between games and distributed simulation has been the development of the DIS-lite protocol, a simple version of the Distributed Interactive Simulation (DIS) protocol. This has been adopted by a commercial video game offering distributed realistic tank combat simulation over the Internet (Spearhead from Zombie/Interactive Magic).

The main characteristic of military simulation is its total realism, whereas games are governed by a trade-off between game-play and realism. Some simplifications occurring in current games derive from technical choices, and others are required to make the game accessible. A fully accurate simulator would require the same skills, hence the same amount of training as a professional, which might put off a significant fraction of players. Simplification in games finds its origin in the need to get the whole process accessible to a single player on a reasonable time scale. It thus tends to reduce complexity and time scales simultaneously. This in turn enables a single player, not only to control many different entities, but also to perform situation assessment on his own, whereas in the real world this is achieved by specialized teams whose efforts are supported by complex information systems. Computer games usually ignore some essential components of modern warfare, such as electronic warfare, communications, intelligence, and detection, which would appear less spectacular than actual phases of air or ground combat.

Technically speaking, realism is supported by a set of computing techniques, which can be briefly summarized as real-time distribution protocols and architectures, realistic physical modeling (also based on real-world data), terrain modeling (with large-scale geographical data), and, most important, intelligent behavior modeling for synthetic forces. One example familiar to gamers is path planning on a simulated terrain. While simple algorithms like A* can compute a path on the sole basis of geometrical constraints, important additional constraints exist on the battlefield, such as line of sight and direction from which a tank platoon can approach an enemy position (Hepplewhite & Baxter, 1998). These techniques can be straightforwardly ported to computer games to increase realism, credibility of synthetic opponents, and game play. The whole set of behavioral techniques that have been developed for tactical simulation and synthetic forces could be transferred to computer games to improve their realism and make them more challenging.

However, the advent of large-scale distributed gaming can potentially make realistic simulations a viable option, as players would tend to specialize, very much like in the real world. The fact that the frontier between actual military simulation and large-scale gaming may eventually disappear could probably have significant implications for military doctrines, but at this stage this still remains a science-fiction scenario. Distribution (see chap. 16, this volume) can thus have a major impact on game play, depending on the availability of large-scale distribution techniques. The same applies to intelligent behavior of synthetic opponents (see chap. 15, this volume). Synthetic actors were originally designed to populate distributed VEs that could not accommodate a large number of users, whether because of limited distribution capabilities or 24 hours availability of such human players. With the advent of large-scale distributed simulation, the population of human users will increase dramatically, thereby creating a need to redefine the role of synthetic actors towards more specialized assistants.

Apart from having inspired a significant number of computer games, military applications have paved the way for important technical developments, like distributed systems, synthetic opponents, and a large set of AI techniques, which will play a major role in next generation entertainment systems. The possible convergence of gaming and military simulations is attracting growing attention—the actual data on which these systems will be used would eventually constitute the main difference between the two applications. The U.S. Army is already working on a "holodeck," a super simulator that will leverage virtual cinematography and video game technology to create realistic 3-D scenes of actual inhabited locations worldwide (Verton & Caterinicchia, 2000).

## 7. POLITICS OF GAMES

Film, television, and video have long lived with the politics of regulation and censorship. Contentious matters such as violence, sex, and stereotyping are becoming bigger issues for computer gaming too, particularly as it moves toward more intense, graphic realism.

Like early rock'n'roll, gaming is often seen as a potentially subversive popular culture. Some think that because gaming involves active participation, rather than passive viewing as in film, it can have more damaging and addictive psychological effects (see chap. 33, this volume). Others feel that game play is mainly physical activity, and less dangerous than cinema's ability to arouse powerful emotions.

The computer games industry is largely self-regulating (see ESRB), and games companies such as Sega, Sony, and Nintendo voluntarily publish guidelines on what they deem acceptable. Such policies have caused games to be cut or dropped entirely. For example, Sensible Software's game Sex and Drugs and Rock'n'Roll, based on the theme of a rising rock star that falls for every imaginable vice on the road to stardom, was shelved because it could not find a publisher.

Games publishers frequently tailor content to differing sensibilities among global markets. The Japanese games Samurai Spirits 3 and Resident Evil, for example, reduced their bloodletting for release into the U.S. and U.K. markets. Computer gaming can also be affected by legislation. In the U.K. game titles can come under the scope of the Video Recordings Act, which deals with gross violence or sexual activity. The British Board of Film Classification can ban games it finds unacceptable.

Yet such policing of computer games can be tricky. In the mid-1990s, the game Carmageddon invited players to drive cars recklessly and score points by mowing down innocent bystanders. The game was banned, and could only be released after zombies replaced human pedestrians. But the resulting publicity helped elevate Carmageddon to cult status. Shortly after its release, software patches that restored it to its original status could be found on the Internet.

Pornography has not tended to be a major issue for games, as it has for the Internet. However, games are notorious for gender stereotyping. A late-1990s U.K. television advertisement for the game Zelda symbolized gaming's virtual world of muscle-bound male leads and story lines by asking Wilt thou get the girl, or play like one? Traditionally, the vast majority of computer games players have been adolescent males. Precisely why is the subject of much debate, maybe it is simply to do with different male and female attitudes to popular culture, technology, and computers. In any case, games developers have long drawn upon male stereotypes and fantasies to appeal to this audience.

One of the key games icons of the 1990s, Lara Croft, star of Tomb Raider, bears many of the hallmarks of gender stereotyping. Her curvaceous female form is pure fantasy. Some see the way players watch and control her from above or behind as "voyeuristic." Yet she has also been hailed as one of the first strong female games characters, a model of 1990s "girl-power"—intelligent, tough, independent, and in control.

As games companies seek to expand their markets, games are likely to be carefully designed to appeal to specific target groups. To attract a wider female audience, characters may be used that females are more likely to identify with. The success of Barbie games has already shown the potential of that market. Some games are designed to appeal to girls and boys. The cute and quirky Pac-Man was a massive hit with both sexes. The game play in Sega's Sonic the Hedgehog involves rescuing and protecting creatures rather than terminating their virtual lives.

There is even a two-player version of Sonic, where players wait for each other to catch up and collaborate in solving problems. This social and cooperative "female" gaming model may well appeal to girls more than the competitive and combative "male" gaming model that seems to appeal to boys. Perhaps the ideal gender-neutral game would combine the two ideas, pitting the "female" and the "male" models against each other in the same arena—one called real life!

## 8. VIRTUAL ENVIRONMENTS "FUTURES"

What is the future of entertainment? How pervasive will VEs, be? In the following section, some of the trends in new forms of content and interactivity are reviewed. Prospects for future LBEs and high-end rides in theme parks are discussed. Research in AR and telepresence are examined, which point to some exciting possibilities for future entertainment. The section concludes with a peek inside the living (and playing) room of the future.

### 8.1 Content

Whether in arcade games, LBEs, or high-end theme park rides, the prime element of interactive experiences is content and not technology. An attractive background story is a prerequisite. Good games and rides are able to bring users to the point where the interface becomes transparent and where they focus only on task performance. Sound is a key element for complete immersion but the need for high-fidelity multimodal cues can be reduced by engaging the user in a complex series of real-time tasks.

The future may belong to goal-oriented entertainment, which emphasizes the journey rather than the objective. Rides will become more experiential, allowing us to explore or unwrap a story hidden beneath a rich environment. The main question for content creators is what is in guests' heads during the experience; finding the answer requires part science, part art, and part sociology.

As the new language of interactive entertainment is being developed, edutainment content will become more important. Museums, science centers, zoos, and aquariums have a need for novel information delivery systems and will want to capitalize on new digital entertainment techniques. The potential for growth is enormous.

Thanks to online virtual communities, we are about to see an explosion of decentralized competition for content creation. This is already noticeable in the musical area and when extended to 3-D and video content will lead to a complete revolution in the way content is produced and distributed (see chap. 25, this volume). Content will also evolve to appeal to a wider population. Interactive entertainment or edutainment, if aimed at culture, health, or well-being could appeal to everyone, including senior citizens.

### 8.2 Interactivity

Beyond the navigation level, interactivity is still essentially an open field. Attempts at interactive plots have not always been successful (Benford, Greenhalgh, & Craven, 2001). There is clearly still a lot of room for passive media, as most people still want to be told stories. This may change as the new "Nintendo generation" grows up. Today's children are used to navigating complex 3-D spaces and interacting with content. They expect more than static displays and television monitors. They want immersion and interaction. They want to explore and discover, and simulation is the dynamic medium that will satisfy them.

Group interactivity in immersive theaters and simulation films is even more of a challenge (e.g., Avatar Farm, an inhabited television event, collaboratively performed over the Web). But it is attracting a lot of interest since people want shared, social experiences when they go for out-of-home entertainment. People usually come in a group to enjoy themselves, to have fun together, to share an experience that draws them closer to each other. Beyond the ride itself, the experience should help people enjoy the company of each other. Games should either be true group games or single interaction activities that others are encouraged to share. People should laugh and talk and bump and touch and never lose track of the other people they came with, no matter how absorbing the activity. This type of multiuser interactivity is now being rolled out in large reality-center simulators for up to 100 participants. Each player controls the action through individual controls and decisions are taken on a majority basis (DePinxi Interactive Experiences).

### 8.3 LBE Centers and Theme Parks

Future LBE centers will introduce new kinds of immersive technologies, higher quality texture mapping, and improved motion platforms. New interfaces, such as voice (see chap. 4, this volume), touch (see chaps. 5 and 6, this volume), position (see chap. 8, this volume), gaze (see chap. 9, this volume), and gesture (see chap. 10, this volume) will be introduced. They will also be networked (see chap. 16, this volume), so that players from around the world can challenge each other in the same virtual world. LBEs are starting to use large-scale networked systems based on DIS. LBE experiences will also be personalized. Transponder badges will give player preferences to the system and help track performance improvements.

But LBE games are also likely to evolve from the fighting and driving thematic to exploration and discovery. LBEs offer a chance to develop new kinds of collaboration-focused activities supported by advanced technology (Dodsworth, 1998). New centers may create a legitimate entertainment format, interesting to adults, with a story to escape into; perhaps similar to Internet-based RPGs but improved with high-end audiovisual and haptic interfaces still out of reach of home users for some time.

As they spread in more accessible locations, take advantage of advanced networking infrastructure, and leverage better content targeted to a broader audience, LBEs will create opportunities for new frameworks, scenarios, and situations. Social multiplayer network experiences may recreate the entertainment format of our ancestors; collectively told, locally created interactive storytelling with Technicolor fidelity (Dodsworth, 1998).

The future may belong to combinations of real, brick-and-mortar theme parks, with online, broadband virtual worlds. Online players will build a character in the virtual world and when they come to the park, they can literally log into it. As they roam the park, transponders pick them up and they can interact with virtual characters through large, high-resolution screens. Such a real-virtual amusement park is currently being developed around the *Wizard of Oz* story. The real park will be built in Kansas and in addition to thousands of real virtual it hopes to help attract hundreds of millions of people in the virtual world as soon as broadband Internet allows for it (see Oz). This demostrates how real and virtual worlds are likely to be intertwined in the future.

## 8.4 Players and Spectators

An important aspect of group entertainment is the spectator's role. Today's audience wants more direct participation even if not everyone wants to be a direct participant. People also enjoy being active spectators, which offers a less threatening forum to express themselves than direct participation. The best examples are sports such as football, in which only a few participants play although tens of thousands share an exciting experience. There is an important potential market to be developed for providing a substantial and rewarding spectator "experience" in the digital entertainment area (Zyda & Sheehan, 1997).

## 8.5 Augmented Reality for Entertainment

Augmented reality is a relatively new field of research by which a user's view of the real world is augmented with additional information from a computer model (Azuma, 1997; Ohta & Tamura, 1999; see chap. 48, this volume). Users can work with and examine real 3-D objects while receiving additional information about those objects or the task at hand. By bridging the gap between VE and the real world, it occupies a central position in the reality–virtuality continuum (Milgram & Colquhoun, 1999), exploiting technologies and methods developed in the VE and sensing domains.

Augmented reality, although currently still in its infancy, is maturing rapidly. Since its inception as a new, distinct, research area (Caudell & Mizell, 1992), it has seen an impressive rate of growth. Depending on many information technologies, such as mobile wearable computing, wireless networking, computer graphics/VE, and sensing that are all showing rapid maturations as well, AR is likely to quickly become a commodity in the next millennium. Demonstrations and systems presented in this handbook are still prototypical. But they should be seen as early indicators of a new user interface paradigm yet to fully emerge. Entertainment applications of AR are no exception. Most of them are yet to be invented.

Augmented reality can provide a set of rather sophisticated approaches focusing on "digital workbenches" (Krueger et al., 1995; MacKay, Velay, Carter, Ma, & Pagani, 1993; Ullmer & Ishiii, 1997; Wellner, 1993). Such approaches require special setups, typically consisting of a back-projected near-horizontal screen serving as a table plus several tracking devices, such as magnetic trackers and cameras. Digital workbenches, especially in combination with physical objects facilitating new user interaction schemes, can be very suitable for educational purposes in museums and schools. One demonstration concerned a map of the MIT campus, identifying and marking various landmarks by physical replicas and computing distances between them.

Similar concepts have been explored using a less specialized computing environment (Klinker, Stricker, & Reiners, 1999a). On a regular table, users can combine real and virtual objects to create elaborate "mixed mock-up" scenes. Real and virtual objects are each identified by a special marker (like a bar code) and can be moved about by users. A computer with a video camera tracks object motions and augments the scene with virtual objects.

Applications in museums and schools can thereby enact "what-if" scenarios, using dollhouses, small physical replicas of historical battle scenes, or construction kits. Such systems will certainly be used for entertainment once content creators discover them and let free their imagination.

Other scenarios involve outdoor tourist applications. A prototypical system has demonstrated how users can walk about in an area, such as a campus, and gain background information about buildings they are looking at (Feiner, MacIntyre, & Höllerer, 1999). In this system, a PC in a backpack tracks the user's current position and viewing direction with a differential GPS antenna and a magnetometer/inclinometer. It accesses information pertaining to buildings in view wirelessly from the Intranet of Columbia University and presents information both on a head-worn display and, in more detail, on a small handheld monitor via which the user can also request more specialized information. Such prototypical applications can be extended toward scenarios in which tourists roam historical sites of ancient cities, requesting information about buildings, seeing them "virtually" reconstructed or repainted, as well as finding out what the opening hours for special shows are and when the next bus will be at a nearby bus stop to take them to a different site.

Relatively inexpensive game applications of AR technology have been demonstrated. In their simple setup involving a small workstation equipped with a video camera and a monitor or (optionally) a $500 HMD, they allow users to play board games like Tangram and tic-tac-toe with a computer. In contrast to conventional computer games, these AR-based versions allow users to play games on a real game board, laid out on a table in front of them. They place real stones on the board. Via the video camera, the computer keeps track of what the user is doing. On the monitor or inserted into the HMD, the video picture is augmented with additional information and comments, thereby indicating to the user its own next move in a tic-tac-toe game or hints where to place the next Tangram piece (Klinker et al., 1999a,b).

An AR AiR Hockey ($AR^2$ Hockey) system has been developed as a case study of real-time collaborative AR for human communication. In demonstrations at several occasions, such as Siggraph '98, this system was tested by thousands of visitors. Air hockey is a game in which two players hit a puck with mallets on a table. In $AR^2$ Hockey, the puck is virtual, whereas the two mallets are real. Users use the mallets to shoot a virtual puck into the opponent's goal. The game is played on a real table. Each player wears an optical see-through HMD to see the puck superimposed on the table. Users' heads are tracked both by magnetic sensors and by cameras on the HMDs (Tamura, Yamamoto, & Katayama, 1999).

The Kids Room at the MIT Media Lab (Bobick et al., 1999) was set up as a one-time experiment for a 6-week period in 1996. It was a fully automated, interactive narrative playspace for children. Using images, lighting, sound, and computer vision action recognition technology, a child's bedroom was transformed into an unusual world for fantasy play. Objects in the room became characters in an adventure, and the room itself actively participated in the story, guiding and reacting to the children's choices and actions. Through voice, sound, and image the Kids Room entertained and stimulated the minds of the children.

### 8.6 Telepresence for Entertainment

There is no doubt that a highly immersive system would be extremely attractive to theme parks and other entertainment centers. Being able to go into a booth and vicariously visit exotic locations throughout the world using sensor platforms (surrogate heads) located at remote sites would be immensely attractive. Sites should be chosen that provide the maximum stimulus to visual and aural senses (i.e., activities need to be happening, such as people moving about and talking). Examples of fitting locations might be the Grand Canyon, Great Wall of China, Great Barrier Reef, Times Square, Niagara Falls, or even just busy city centers or beaches.

One could use an intelligent avatar that could be available to every individual user. The avatar would provide all necessary information to the user about the remote site and could be interrogated for further information. Other computer generated text and graphics could be integrated with live video (e.g., an external view of an Egyptian pyramid could be augmented with computer generated internal views).

Sports and entertainment events, such as the Olympics and music concerts, could all be visited using telepresence. The user would be able to experience the live atmosphere and have control over the viewing by zooming, panning, and tilting cameras. Although this might seem akin to digital interactive television, the "immersive" element would make the experience unique.

Combining telepresence experiences with educational themes would also be possible. For example, a famous battle site could be visited by telepresence as it is today. However, the user would then be able to overlay the actual battle and its development on the live image, considerable content creation would obviously be necessary (e.g., a prerecorded film using actors or a VE animation).

A commercial use, bordering on edutainment, would be the use of telepresence by travel agents to allow potential customers to view holiday resorts live and overlay, where necessary, temperature charts, flight times, and enhancing QuicktimeVR or other 360-degree still and video images.

For reasonable immersion, stereoscopic full color vision with a wide field of view and full binaural or environmental sound effects may be necessary (see chap. 40, this volume). For telepresence utilizing telecommunications systems this obviously creates bandwidth problems, especially where interactivity is necessary. If a motorized camera platform is being used at the remote site this also means that every simultaneous visitor to the site requires exclusive use of their own platform for the duration of the visit. Good quality stereo viewing systems are still expensive, but binaural or surround systems are reasonable. Another factor that will influence cost is whether the home site will be designed for single or multiple users.

Telepresence for entertainment is certainly possible today but probably still too expensive to be commercially attractive. Improvements in bandwidth availability, cost, home site visual and aural display technology, and remote site sensor platform design are all necessary before the adoption of telepresence for entertainment becomes commercially attractive.

## 8.7 A Holodeck at Home?

The ultimate VE would be a computer interface that allows users to perform tasks as easily as they do in the real world. This would require building interfaces that effectively add natural manipulation (see chap. 13, this volume) and locomotion (see chap. 11, this volume) to visual immersion. Science fiction has given us attractive visions of this superinterface, of which the Star Trek holodeck is probably the most popular. The holodeck is envisioned as a next-generation VE simulation system that would allow people to experience multisensorial virtual worlds including sight, hearing, touch, and even smell. The holodeck would allow virtual travel anywhere, anytime. In the movie Star Trek, crew members used the holodeck to play, exercise, travel, and even discover historical sites that disappeared centuries ago. One of the essential features of the holodeck is the presence of characters with whom the user can interact in a natural way.

We are still far from being able to realize this vision. Building a holodeck will require breakthroughs in sensors, robotics, and mechanical interfaces, not to mention significant progress in graphical interfaces and computational simulations. The key will be to attain latency levels far more stringent than anything available today. It will also mean adapting movie making and storytelling skills to create realistic simulations.

Not only would it be the ultimate interface to experience telepresence, learn, or communicate, a holodeck would also be the perfect entertainment platform and one could argue that some games are the closest venue to a holodeck available today. But when the dream becomes reality, players will be able to fully immerse themselves in virtual worlds to experience rich, multisensory adventures. These very compelling experiences will make current motion platform rides feel like black-and-white silent cinema does today. We will finally be able to fly.

## 9. ACKNOWLEDGMENTS

We thank the following individuals for their contributions and support: Fiona Allan (European Commission), Lia Brosseau (Illusion, Inc.), Philippe Chiwy (DePinxi), Karen Drasler (Disney Quest), Richard Gallery (Philips/S3), Anne Hohenberger (SGI), Steve Hunter (Dreamality Technologies, Inc.), Thierry Nabeth (INSEAD), Randy Pausch (Carnegie Mellon University), Frank van Reeth (Limburg University Center, Belgium), Peter Schickel (Blaxxun), Kay Stanney (University of Central Florida) and the four anonymous reviewers whose constructive comments were much appreciated.

## 10. REFERENCES

Avatar Farm, September 2000. [Online]. Available: http://www.crg.cs.nott.ac.uk/events/avatarfarm

Azuma, R. (1997). A survey of augmented reality. *Presence: Teleoperators and Virtual Environments, 6*(4), 355–385.

Barthes, R. (1966). Introduction à l'analyse structurale des récits. *Communications*, vol. 8, pp. 1–27.

Berry, C. (1997, October). [Online]. "The bleeding edge" *Wired*. Available: http://www.wired.com/wired/archive/5.10/es_gaming.html

Benford, S., Greenhalgh, C., & Craven, M. (2001). Producing television shows in collaborative virtual environments. *Interactions, 3*(1), 13–14.

Blaxxun Interactive, June 2001. [Online]. Available: http://www.blaxxun.com/

Blumberg, B., & Galyean, T. A. (1995). Multi-level direction of autonomous creatures for real-time virtual environment. In *SIGGRAPH '95 proceedings* (pp. 47–54). New York: ACM Press.

Bolter, J. D., & Grusin, R. (1999). *Remediation. Understanding new media*. Cambridge, MA: MIT Press

Bobick, A., Intille, S., Davis, J., Baird, F., Pinhanez, C., Campbell, L., Ivanov, Y., Schütte, A., & Wilson, A. (1999). The KidsRoom: A perceptually-based interactive and immersive story environment [Online]. *Presence: Teleoperators and Virtual Environments, 8*(4), pp. 367–391. Available: http://vismod.www.media.mit.edu/vismod/demos/kidsroom/info.html

Capin T., Pandzic I., Magnenat-Thalmann N., & Thalmann D. (1997). Virtual human representation and communication in the VLNET networked virtual environments. *IEEE Computer Graphics and Applications, 17*(2), 42–53.

Capin T., Pandzic I., Magnenat-Thalmann N., & Thalmann D. (1999). *Avatars in networked virtual environments* . New York: John Wiley & Sons.

Caudell T., & Mizell, D. (1992). Augmented reality: An application of heads-up display technology to manual manufacturing processes. In *HICCS '92 Proceedings*, University of Hawaii.

Cavazza, M., Bonne, J.-B., Pernel, D., Pouteau, X., & Prunet, C. (1995). Virtual reality for command and control environments. In *FIVE '95 Conference Proceedings*. London: Queen Mary and Westfield College.

Cybertown. Civilization for the Virtual Age, 2000. [Online]. Available: http://www.cybertown.com/

Damer, B., Gold, S., Marcelo, K., & Revi, F. (1997). *Inhabited virtual worlds in cyberspace* [Online]. Available: www.digitalspace.com/papers/vwpaper/vw98chap.html

DePinxi Interactive Experiences, June 2001. [Online]. Available: http://www.depinxi.be

Disney Quest, June 2001. [Online]. Available: http://disney.go.com/DisneyQuest/Orlando/home.html

Dodsworth, C. (1998). *Digital illusion: Entertaining the future with high technology*. Wokingham, England: Addison-Wesley.

Dreamality Technologies, June 2001. [Online]. Available: http://www.dreamalitytechnologies.com/

Emering, L., Boulic, R., & Thalmann, D. (1998). Interacting with virtual humans through body actions. *IEEE Computer Graphics and Applications, 18*(1), 8–11.

Entertainment Software Rating Board, June 2001. [Online]. Available: http://www.esrb.org

Feiner, S., MacIntyre, B., & Höllerer, T. (1999). Wearing it out: First steps toward mobile augmented reality systems. In Y. Ohta & H. Tamura (Eds.), *Mixed reality—Merging real and virtual worlds* (pp. 363–375). Berlin: Springer-Verlag.

Goldberg, A. (1997). IMPROV: A system for real-time animation of behaviour-based interactive synthetic actors. In R. Trappl & P. Petta (Eds.), *Creating personalities for synthetic actors* (pp. 58–73). Berlin: Springer-Verlag.

Hepplewhite, R. T., & Baxter, J. W. (1998). Planning and search techniques to produce terrain-dependent behaviours. In *Proceedings of the ECAI98 Workshop on Intelligent Virtual Environments*. Brighton, England: University of Brighton.

Illusion, Inc., June 2001. [Online]. Available: http://www.illusioninc.com/

INSEAD. Virtual Environments [Online]. Available: http://www.insead.fr/Encyclopedia/ComputerSciences/VR/vr.htm

Klinker G., Stricker, D., & Reiners, D. (1999a). Augmented reality technology for exterior construction. In W. Barfield & T. Caudell (Eds.), *Augmented reality and wearable computers*. Mahwah, NJ: Lawrence Erlbaum Associates.

Klinker, G., Stricker, D., & Reiners, D. (1999b). Optically based direct manipulation for augmented reality. *Computers & Graphics, 23*(6).

Krueger W., Bohn, C.-A., Froehlich, B., Schueth, H., Strauss, W., & Wesche G. (1995). The responsive Workbench: A virtual work environment. *IEEE Computer, 28*(7), 42–48.

MacKay, W., Velay, G., Carter, K., Ma, C., & Pagani D. (1993). Augmenting reality: Adding computational dimensions to paper. *Communications of the ACM, 36*(7), 96–97.

Maes, P., Darrell, T., Blumberg, B., & Pentland, A. (1995). The ALIVE system: Full-body interaction with autonomous agents. In *Computer Animation '95 Proceedings* (pp. 11–18). Los Alamitos, CA: IEEE Press.

Magnenat-Thalmann, N., & Thalmann, D. (1994). *Artificial life in virtual reality*. New York: John Wiley & Sons.

Metal Gear Solid, June 2001. [Online]. Available: http://www.metalgear.com/index2_hi.html

Milgram, P., & Colquhoun, Jr., H. (1999). A taxonomy of real and virtual world display integration. In Y. Ohta & H. Tamura (Eds.), *Mixed reality—Merging real and virtual worlds* (pp. 5–30). Berlin: Springer-Verlag.

Molet, T., Aubel, A., Capin, T., Carion, S., Lee, E., Magnenat-Thalmann, N., Noser, H., Pandzic, I., Sannier, G., & Thalmann, D. (1999). Anyone for tennis? *Presence: Teleoperators and Virtual Environments, 8*(3), 140–156.

Nakatsu, R., & Altman, E. (1997). *Interactive movies: Techniques, technologies, and contents* (SIGGRAPH '97 Tutorial Course Notes). Los Angeles, USA, ACM Press.

Nakatsu, R., & Tosa, N. (1997). Toward the realization of interactive movies—Inter Communication Theater: Concept and system. In *Proceedings of the International Conference on Multimedia Computing and Systems* (pp. 71–77). Ottawa, Canada. IEEE Computer Society Press.

Ohta, Y., & Tamura, H. (1999). Mixed reality—Merging real and virtual worlds. Berlin: Springer-Verlag.

Oz. (1997). On the road to Oz, An interview with Dan Mapes, May 2000 [Online]. Available: http://www.e3dnews.com/e3d/Issues/200005-May/lead.html

Pandzic I., Magnenat-Thalmann, N., Capin, T., & Thalmann, D. (1997). Virtual life network: A body-centered networked virtual environment. *Presence: Teleoperators and Virtual Environments, 6*(6), 676–686.

Propp, V. (1968). *Morphology of the folktale* (2nd Rev. ed., L. Scott, Trans). Austin: University of Texas.

Schank, R., & Abelson, H. (1977). *Scripts, plans, goals and understanding*. New York: Addison-Wesley.

Sims, K. (1994). Evolving virtual creatures. In *SIGGRAPH '94 Proceedings* (pp.15–22). New York: ACM Press.

Tamura, H., Yamamoto, H., & Katayama, A. (1999). Steps towards seamless mixed reality. In Y. Ohta & H. Tamura (Eds.), *Mixed reality—Merging real and virtual worlds* (pp. 59–80). Berlin: Springer-Verlag.

Tu, X., & Terzopoulos, D. (1994). Artificial fishes: Physics, locomotion, perception, behavior. In *SIGGRAPH '94 Proceedings, Computer Graphics* (pp. 42–48). New York: ACM Press.

Ullmer, B., & Ishii, H. (1997). The metaDESK: Models and prototypes for tangible user interfaces. In *ACM Symposium on User Interface and Technology* (UIST '97), (pp. 223–232). New York: ACM Press.

Van Lent, M., & Laird, J. (1999). *Learning from observation in computer games* [Online]. Available: http://ai.eecs.umich.edu/people/vanlent/summary.html

Verton, D., & Caterinicchia, D. (2000). Army wants to harness power of the Matrix. *Federal Computer Week* [Online]. Available: http://www.cnn.com/2000/TECH/computing/05/02/army.matrix.idg/index.html

Wellner, P. (1993). Interacting with paper on the digital desk. *Communications of the ACM, 36*(7), 87–96.

Zyda, M., & Sheehan, J. (Eds.). (1997). *Modeling and simulation: Linking entertainment and defense* [Online]. Available: http://bob.nap.edu/readingroom/books/modeling/

# PART VII: Conclusion

# 56

# Virtual Environments: History and Profession

Richard A. Blade[1] and Mary Lou Padgett[2]
[1] *University of Colorado at Colorado Springs*
*13631 E. Marina Dr., No. 302, Aurora, CO 80014*
*rblade@mail.uccs.edu*
[2] *Padgett Computer Innovations, Inc.*
*1165 Owens Road, Auburn, AL 36830*
*m.padgett@ieee.org*

## 1. INTRODUCTION

Although this handbook largely focuses on the current status and future vision of virtual environments (VE) technology, it would be unjust to develop such a work without providing acknowledgement to the pioneers whose vision and innovativeness laid the groundwork for the science, technology, and applications described here. This chapter provides a very brief historical overview of some of the key people and major milestones that led to the current state of the art. In addition, readers of this handbook may be left wondering where they can gain more information about the VE profession. This chapter provides a number of such references, including major application Web links, associated periodicals, major conferences and trade shows, major organizational players, and general and historical references.

## 2. A BRIEF HISTORY OF VIRTUAL ENVIRONMENTS

Not surprisingly, the science fiction literature envisioned the generation of artificial or illusory environments before there was any idea of the technology to accomplish such a feat. For example, a famous 1950 science fiction story by Ray Bradbury (Bradbury, 1976) called *The Veldt* depicted a playroom where children experienced an African landscape. The animals ultimately ate the parents. The first attempts at developing such a technology used electronics, but not computers, which at that time were much too primitive. In 1956, Morton Heilig, a filmmaker, developed *Sensorama* (U.S. Patent No. 3,050,870), a mechanical virtual display device (see Fig. 56.1). Perhaps the first head-mounted display (HMD) was developed by Philco in 1961, which permitted remote viewing via a video camera (Kalawsky, 1993).

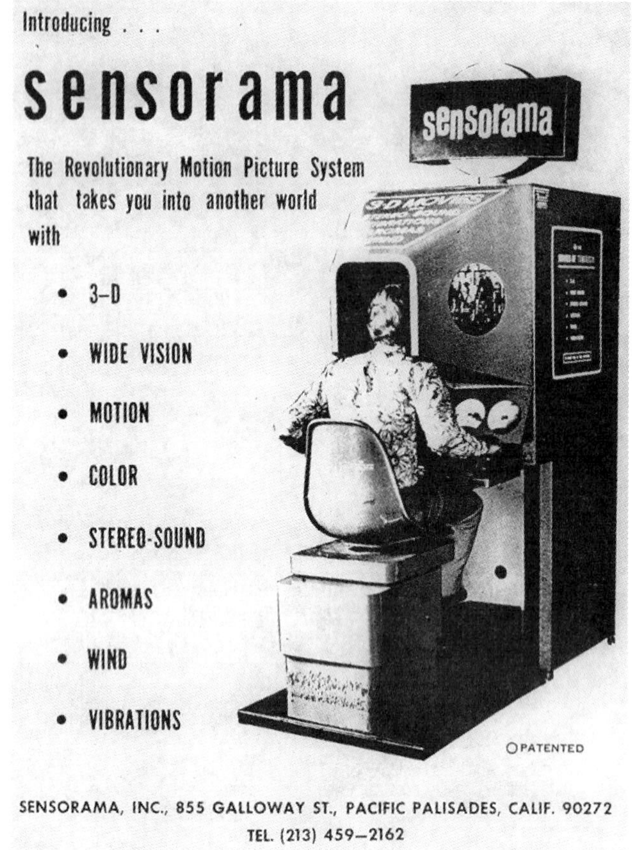


FIG. 56.1. Poster advertising Helig's Sensorama.

In the early 1960s computer graphic technology was beginning to be developed. The technology for computer-generated VEs probably began in 1963, when Dr. Ivan Sutherland developed the first interactive computer graphics system, called *Sketchpad*, while a graduate student at the Massachusetts Institute of Technology (MIT; Sutherland, 1965). Sketchpad created highly precise engineering drawings that could be manipulated, duplicated, and stored. In 1968, Sutherland joined with David Evans to build an HMD at the University of Utah (Sutherland, 1970). Shortly thereafter, Thomas Furness, working at Wright-Patterson Air Force Base, developed an HMD that he called a "visually coupled system" (Furness, 1969). Furness went on to develop a virtual cockpit flight simulator in 1981 (Furness, 1988). Since the early developments of Sutherland and Furness, there has been continual evolution in HMD design by various government organizations and private companies, including Honeywell and Hughes.

While Sutherland and Furness were working on the technical aspects, Myron Krueger was studying the artistic and psychological aspects of VEs at the University of Wisconsin. Krueger first concocted *Glowflow* in 1969, a kind of artistic light and sound show that was controlled by a computer but involved no computer graphics (Krueger, 1977). In *Glowflow*, audiences actively participated in the show by moving around on pressure-sensitive plates embedded in the floor, though most were unaware of the controlling mechanism. The following year Krueger developed the much more elaborate *Metaplay* with a grant from the National Science Foundation and the university's computer science department. This system involved 800 pressure-sensitive switches and an $8 \times 10$-foot rear projection screen on which people viewed video images of themselves from a video camera, superimposed on computer graphics generated by a minicomputer. During the next several years Krueger developed ever more elaborate artistic works that provided interactive experiences to audiences, such as *Videoplace*, which immerses an individual into a computer-generated world inhabited by other human and virtual participants, in which the laws of cause and effect can be composed from moment to moment (Krueger, 1985). Krueger is credited with coining the term *artificial reality* in 1973 (Krueger, 1985).

In 1974 personal computers were introduced, and in 1977 the first glove device for controlling a computer was developed. One of the first commercial gloves, was VPL Research's *DataGlove*, which was invented by Thomas Zimmerman (U.S. Patent No. 4,542,291) and operated by a programming language developed by Jaron Lanier (Jones, Rheingold, 1991). Also in 1977, Kit Galloway and Sherri Rabinowitz created virtual space with the Satellite Arts Project, implementing the vision of Arthur C. Clark and Marshall McLuhan's global village, interconnecting the people of the world via electronic communication. In what Galloway and Rabinowitz called *Hole in Space*, they set up a large video screen and camera in public spaces in New York City and Los Angeles to effectively videoconference in a group setting. A decade later they went on to found the *Electronic Café* (http://www.ecafe.com/) in Santa Monica, California, allowing videoconferencing on an individual basis.

By 1979 researchers at MIT began developing spatial data management systems, and in the early 1980s educators began interactive control of video playback with Apple II computers for use in computer-aided instruction (CAI). One of the group, Scott Fisher, developed the technology of stereographic displays using an HMD. In 1981 those researchers, including Fisher, produced the *Aspen Movie Map*, in which a person could navigate, using a touch-sensitive display screen, along 20 miles of streets in Aspen, Colorado (Fisher, 1982). This did not use an HMD; rather, it was an elaborate 'CAVE'-type (Cave Automatic Virtual Environment; see chap. 11, this volume) system that projected scenes on the surrounding walls of a room.

Also in the early 1980s the Defense Advanced Research Projects Agency (DARPA) funded a global war game simulator called "*simulation networking*," or *SIMNET*, that broke new ground in large-scale networks. Using SIMNET and its later upgrade, Distributed Interactive Simulation (DIS), hundreds of soldiers all over the world then—and now—can sit in tank, helicopter, and fighter bomber simulators, testing their combat skills against one another in real time, much like a video game, but very realistic. The cost savings and safety of such simulators will undoubtedly continue to drive the development of military battle simulators ever more in the future (see chap. 43, this volume).

In 1984 the movie *The Last Starfighter*, written by Jonathan R. Betuel and produced by Edward Denault and Gary Edelson, merged live action with computer graphics to save millions of dollars in production costs. Also in 1984, William Gibson wrote a science fiction novel, *Neuromancer*, in which he introduced the now-ubiquitous term *cyberspace*. In 1986 Lucasfilm, a firm founded from the revenues of the 1976 *Star Wars* movie series, began developing computer movies. Since that time the movie industry has steadily increased its use of computer graphics to reduce costs and create astounding visual effects.

In 1985 Jaron Lanier and Jean-Jacques Grimaud founded VPL Research, Inc., to produce state-of-the-art human interface devices (Kalawsky, 1993). Lanier, who coined the term *virtual reality*, was an interesting, creative, and eccentric character that the public came to identify so closely with the technology that many incorrectly viewed him as the founder of virtual environments. In the middle and late 1980s VPL Research became famous for making what was intended to be the first consumer grade VE hardware: the DataGlove and the *EyePhone*. Unfortunately, in a few years VPL Research came on hard financial times and Lanier, the company, and the products essentially disappeared.

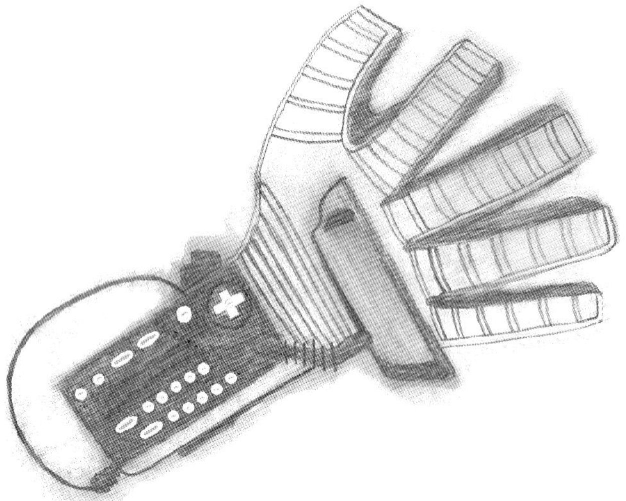

FIG. 56.2. Mattel's PowerGlove. Drawing Courtesy of Kay Stanney.

Also in 1985 a team under Dr. Frederick P. Brooks, Jr., began experimenting with three-dimensional (3-D) perception of molecules at the University of North Carolina. In the next decade that project developed into the *Nanomanipulator*, where a person wearing an HMD can manipulate single atoms or small groups of atoms by a force-feedback arm electronically connected to a scanning tunneling electron microscope (Taylor et al., 1993; see chap. 42, this volume).

In 1988 Chris Gentile, of Abrams/Gentile Entertainment, developed the *PowerGlove* for Nintendo Home Entertainment System (see Fig. 56.2; Gardner, 1989). Licensed and marketed by Mattel, the PowerGlove became a best-selling toy in 1989 and 1990, but was discontinued at the end of 1991 because it failed to produce expected revenues. That was probably the first truly consumer-grade VE hardware. Since that time, the PowerGlove has remained a prized item for VE professionals and amateur enthusiasts alike.

In 1989 and the early 1990s the general public became aware of the concept of VEs and their potential, and enough people began working in the field that the first professional VE conferences were held. Fakespace developed the *BOOM*[1] (Binocular Omni Orientation Monitor), an HMD mounted on a boom for tracking position (see Fig. 56.3). Sense8 Corporation and the Human Interface Technology Laboratory (HIT Lab) were formed. AutoDesk demonstrated their PC-based VE CAD system at SIGGRAPH. W. Industries in the United Kingdom began to sell *Virtuality*, a complete VE system for interactive, interpersonal gaming, and Division Industries began to sell VE systems.

In 1993 SGI announced their *Reality Engine*, a computer capable of running significant VE applications, beginning the current era of rapidly multiplying applications as significant computer power has developed on relatively inexpensive personal computers. In 1996 VE was ported to the Internet (see chap. 16, this volume), with the creation of Virtual Reality Modeling Language (VRML), and today there exists a plethora of hardware and software that permits easy development of VE applications. Finally, in a well-publicized application of VE, astronauts were trained to do repairs on the Hubble space telescope using a full scale VE simulator constructed at Goddard Space Flight Center (see chap. 42, this volume).

[1]BOOM is a registered trademark of Fakespace Labs, Inc.

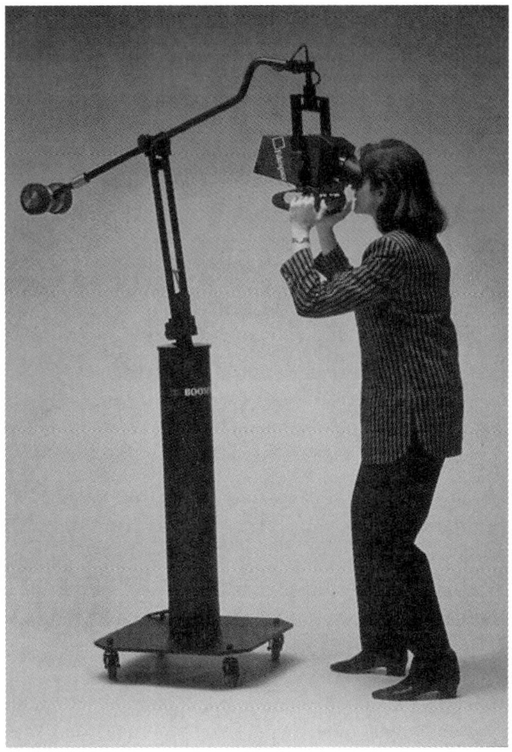

FIG. 56.3. Fakespace's Boom HMD. Photo Courtesy of Fakespace Labs, Inc., Mountain View, California.

Today, applications of VE technology include architectural design with walk-throughs, VE fantasy games, VE flight simulations, and educational materials, including human anatomy walk-throughs and map fly-arounds (see chap. 42, this volume). NASA plans virtual exploration of the solar system. The entertainment industry not only creates many movie effects with computer graphics, but provide VEs in the form of passive and interactive movie rides that include motion simulation (see chap. 55, this volume). The military continues to develop simulators for all aspects of training and is working on telerobotics for battlefield medical treatment (see chap. 43, this volume). The medical industry trains surgeons using virtual patients and "telesurgery" is expected to be prevalent some time in the future (see chap. 47, this volume). Psychologists use VEs to desensitize against phobias (see chap. 51, this volume). Disabled persons use VEs to train wheelchair operation as well as a host of other rehabilitative applications (see chaps. 49 and 50, this volume). Financial professionals use VEs to visualize stock market and other financial trends.

There are major challenges, however. Most researchers in the field express disappointment in the current state of specialized hardware necessary to support VE applications. True consumer grade HMDs still do not exist, and even the expensive ones used by researchers are sadly lacking. Tracking methods and force feedback devices are still primitive, and more sophisticated means of getting information from the body to the electronics need to be developed, such as signal processing of electrical signals emanating directly from the nervous system. Clearly, though, no one questions the potential and ultimate development of VE, including applications not yet conceived.

## 3. MAJOR APPLICATION AREAS

Many VE applications have developed over the past decade (see chaps. 42–55, this volume). Provided below is a short list of some of the main applications and associated reference sites. Links to Web sites for most can be found at the Virtual Reality Resources Web site: http://ils.unc.edu/houseman/applications.html/:

**Scientific Visualization**

- The CAVE, Electronic Visualization Lab, University of Illinois at Chicago, http://evlweb.eecs.uic.edu/
- "Chemical Flooding in a Virtual Environment"—a paper by Wes Bethel, Lawrence Livermore Lab, http://www-vis.lbl.gov/publications/avs94/avs94.html/.
- Lateiner Dataspace—research at Lateiner Dataspace Labs on high-speed volume visualization, discrete physical simulation techniques, and Dataspace distributed datastructures, http://www.dataspace.com/.
- The Nanomanipulator, Department of Computer Science, University of North Carolina at Chapel Hill, http://www.cs.unc.edu/Research/nano/index.html/.
- Virtual Reality Room (VROOM) Exhibits, Electronic Visualization Lab, University of Illinois at Chicago, http://www.evl.uic.edu/EVL/VROOM/HTML/OTHER/HomePage.html/.

**Architecture and Design**

- Conceptual Design Space Project, Graphics Visualization Center, Georgia Tech, http://www.cc.gatech.edu/gvu/virtual/CDS/.
- The Design Virtual Environment Project, Graphics Visualization Center, Georgia Tech, http://www.cc.gatech.edu/gvu/virtual/DVE/DVE.html/.
- Human Virtual Environment System, NASA Lyndon B. Johnson Space Center, http://jsc-web-pub.jsc.nasa.gov/fpd/SHFB/GRAF/graf7.html/.
- Kitchen Design Showroom, Division and Matsushita Electric Works, http:// www.mew.co.jp/e-index.html/
- Virtual Design Studio '95, Human Interface Technology Lab, University of Washington, http://www.hitl.washington.edu/.
- "Virtual Reality: Architecture and the Broader Community," a paper by Kate McMillan of the School of Architecture, University of New South Wales, Australia, is a comprehensive study of the issues surrounding the application of VE to architecture, http://www.fbe.unsw.edu.au/research/student/VRArch/.

**Education and Training**

- Combat Training, NPSNET Research Group, Naval Postgraduate School, http:// www.cs.nps.navy.mil/research/rescat/rescat.html/.
- Hubble Space Telescope Repair Training System, STB VR Lab, NASA Lyndon B. Johnson Space Center, http://setas-www.larc.nasa.gov/HUBBLE/hst.html/.
- The Learning Center, Human Interface Technology Lab, University of Washington, http://www.hitl.washington.edu/projects/learning_center/text.html/.
- The Networked Virtual Art Museum, Studio for Creative Inquiry, Carnegie Mellon University, http://www.cmu.edu/studio/.
- Project ScienceSpace, STB VR Lab, Johnson Space Center, http://cgi.media.hku.hk/virtual_curr/virtual_science.html/.

- Virtual Environment Testbed, Institute for Simulation and Training, University of Central Florida, http://www.vsl.ist.ucf.edu/.

### Entertainment

- Atlantis Cyberspace—Virtual Reality Entertainment Centers, http://vr-atlantis.com/.
- DisneyQuest—Walt Disney World's interactive theme park, http://disney.go.com/disneyquest/index.html.
- Virtual World Wide Web—Virtual World Entertainment's Web site links to all of its games, such as BattleTech and Red Planet, http://www.virtualworld.com/.
- W Industries' Virtuality—one of the first location-based VE entertainment systems, http://www.virtualworld.com/.
- Activeworlds.com, Inc.—Developers of 3-D virtual worlds to be distributed over the Internet, http://www.activeworlds.com/.

### Manufacturing

- A Virtual Environment for Automotive Design, General Motors Research and Development Center, http://www.generalmotors.com/company/gm_exp_live/events/concept_vehicles/virback.htm/.
- The Virtual Backhoe/Wheel Loader, NCSA Virtual Reality Lab, http://www.ncsa.uiuc.edu/Vis/.
- "Virtual Reality in Manufacturing—Case Studies," by Sandy Ressler, NIST, http://ovrt.nist.gov/projects/mfg/mfgVRcases.html/.

### Medicine

- Medical Telepresence, Department of Computer Science, University of North Carolina at Chapel Hill, http://www.cs.unc.edu/~mcmillan/telep.html.
- Medical Telepresence Interfaces, Human Interface Technology Lab, University of Washington, http://www.hitl.washington.edu/projects/medicine/telemedicine.html/.
- Minimally Invasive Surgery, Fraunhofer Institute (IAO), Germany, http://www.iao.fhg.de/.
- Treatment of Acrophobia, Graphics Visualization Center, Georgia Tech, http://www.cc.gatech.edu/gvu/virtual/Phobia/.
- Treatment of Parkinson's disease, Human Interface Technology Lab, University of Washington, http://www.hitl.washington.edu/research/parkinsons/.
- Ultrasound Examination, Department of Computer Science, University of North Carolina at Chapel Hill, http://www.cs.unc.edu/Research/ProjectSummaries/ultrasou.pdf.

## 4. PERIODICALS FOR PROFESSIONALS

There are a handful of journals and other publications that focus on VE science, technology, and applications. Provided below is a list of the main publications and associated reference sites:

- *Human–Computer Interaction*, published quarterly by Lawrence Erlbaum Associates, Inc. This academic journal publishes theoretical, empirical, and methodological articles on user psychology and computer system design as it affects the user; http://www.erlbaum.com/Journals/Journals/HCI/hci.htm.

- *The International Journal of Human–Computer Interaction*, published quarterly by Lawrence Erlbaum Associates, Inc. This academic journal addresses the cognitive, social, health, and ergonomics aspects of work with computers and emphasizes both the human and computer science aspects of the effective design and use of computer interactive systems; http://ijhci.engr.ucf.edu.
- *The International Journal of Virtual Reality*, a multimedia (CDROM with each issue) journal of research and applications, published irregularly but nominally four times per year. Though published in English, the journal emphasizes international involvement and cooperation; http://www.ijvr.com.
- *Presence*: *Teleoperators and Virtual Environments*, published by MIT Press, an academic journal published bimonthly, provides a scientific forum for current research and advanced ideas on teleoperators and virtual environments; http://mimsy.mit.edu/presence/.
- *RealTime Graphics*, published by CGSD Corporation in Mountain View, California, a newsletter published ten times a year, providing in-depth information on the technology of real time imaging and providing coverage of industry news; http:www.realtimegraphics.com.
- *VR News*, a monthly newsletter published digitally in the United Kingdom by Cydata Ltd., emphasizing developments in the virtual reality industry; http://www.vrnews.com.

## 5. MAJOR CONFERENCES AND TRADE SHOWS

There are a number of conferences and trade shows that focus on VE science, technology, and applications. Provided below is a list of many of these forums:

- A/E/C SYSTEMS: computer and management show for design construction, utilities, and manufacturing industries; http://www.aecsystems.com/.
- ACM Conference on Human Factors in Computing Systems (CHI); http://www.acm.org/sigchi/.
- ACM Multimedia Conference; http://www.acm.org/sigmm/.
- Advances in Automotive and Transportation Technologies and Practice for the 21st Century (ISATA); http://www.isata.com/.
- Amusement Trades Exhibition International (ATEI): a trade show for the entertainment industry; http://www.atei.co.uk/.
- Annual IASTED International Conference on Computer Graphics and Imaging (CGIM); http://www.iasted.com.
- ASME CEI: organized by the ASME Virtual Environments and Systems Technical Committee; http://www.asme.org/. http://www.asme.org/divisions/cie/
- Autonomous Agents Conference (AGENTS); http://www.cs.washington.edu/research/agents99/.
- Games Developers Conference (CGDC); http://www.gdconf.com/ or http://www.gamasutra.com/.
- Conference on Computer Animation; http://www.miralab.unige.ch/.
- Conference on Computer Generated Forces and Behavioral Representation (CGF&BR); http://www.sisostds.org/cgf-br/index.htm.
- Defense Modeling and Simulation Office Industry Days (DMSO); http://www.dmso.mil/.
- Electronic Entertainment Expo (E3 EXPO): a trade show for new multimedia, entertainment, and educational products; http://www.e3expo.com/.
- Eurographics Workshop on Virtual Environments (EGVE): focusing on augmented reality; http://www.cg.tuwien.ac.at/conferences/egve99/.

- Human–Computer Interaction (HCI) International; http://hcii2001.engr.wisc.edu/.
- ICAT: the oldest international conference on virtual environments, sponsored by the Virtual Reality Society of Japan and the University of Tokyo; http://www2.vsl.gifu-u.ac.jp/vrsje/.
- *IEEE Real Time Technologies and Applications Symposium (RTAS)*; http://cs-www.bu.edu/pub/ieee-rts/Home.html/.
- IEEE Virtual Reality Conference (IEEE VR); http://www.ieee-vr.org.
- Industrial Virtual Reality Symposium; http://www.mae.buffalo.edu/symposiums/ASME/imece99.html.
- International Conference in Central Europe on Computer Graphics and Visualization (WSCG); http://wscg.zcu.cz/.
- International Conference on Disability, Virtual Reality and Associated Technologies (ICDVRAT); http://www.cyber.reading.ac.uk/P.Sharkey/WWW/ecdvrat.
- International Conference on Virtual Reality Modeling Language and Web 3D Technologies (VRML); http://www.c-lab.de/web3d2001/.
- International Conference on Visual Computing Interaction, Modeling, Rendering, Animation, and 3-D Environments (ICVC); http://www.ncst.ernet.in./icvc99/.
- International Immersive Projection Technology Workshop (IPTW '99), Stuttgart, Germany; http://vr.iao.fhg.de/index.en.html/.
- International Workshop on Presence; http://www.presence-research.org/presence2000.html/.
- Interservice/Industry Training, Simulation, and Education Conference (I/ITSEC), Orlando, Florida; http://www.iitsec.org.
- ITEC: a major conference on equipment and simulation for education and training; http://www.itec.co.uk/.
- IVR: a comprehensive VR exhibition in Japan; http://www.reedexpo.co.jp/ivr/.
- Medicine Meets Virtual Reality (MMVR): the foremost conference on virtual reality and health care; http://www.amainc.com/index.html/.
- Modeling and Simulation (MS); http://www.iasted.com/.
- Modeling and Simulation Crosstalk Conference, sponsored by the International Test and Evaluation Association; www.itea.org/.
- SIGGRAPH: a very large exhibition and conference on all aspects of computer graphics, http://www.siggraph.org/conferences/.
- Society for Information Display Symposium (SID); http://www.sid.org/.
- SPIE Annual Meeting and Exhibition, http://www.spie.org/.
- Symposium on Augmented and Virtual Reality (AVR); http://www.graphicslink.demon.co.uk/IV2001/AVR.htm.
- Symposium on Virtual Reality Software and Technology (VRST); http://www.cs.ucl.ac.uk/staff/m.slater/VRST99/. See also http://www.siggraph.org/calendar/calendar.html.
- Symposium on Visualization (VisSym), sponsored jointly by EUROGRAPHICS and the IEEEE; http://www.cg.tuwien.ac.at/conferences/VisSym99/cfp.html.
- Technology and Persons With Disabilities; http://www.csun.edu/cod/.
- Trade Show of the International Association of Amusement Parks and Attractions (IAAPA): the latest VE and simulation rides, http://www.iaapa.org. International Association of Amusement Parks and Attractions, 1448 Duke Street, Alexandria, VA 22314.
- Trends in Leisure and Entertainment (TiLE 1999); http://www.andrich.com/tile/.
- United Kingdom Virtual Reality Special Interest Group Conference (UKVRSIG): includes preconference on intelligent agents; http://www.crg.cs.nott.ac.uk/groups/ukvrsig/.
- Virtual Systems and Multimedia (VSMM); http://www.vsmm.org/. International Society on Virtual Systems and MultiMedia Virtual System Laboratory, Gifu University, 1-1 Yanagido, Gifu 501-1193 JAPAN.

## 6. MAJOR ORGANIZATIONS

There are a number of professional organizations that focus on VE science, technology, and applications. Provided below is a list of some of the primary endeavors:

- Advanced Interfaces Group, University of Manchester Department of Computer Science, England; http://aig.cs.man.ac.uk/.
- Army Research Institute; http://www-ari.army.mil/.
- Computer Graphics and User Interfaces Lab, Columbia University; http://www.cs.columbia.edu/graphics/.
- Computer Graphics Lab, Ecole Polytechnique Federale de Lausanne (Swiss Federal Institute of Technology), Switzerland; http://ligwww.epfl.ch/.
- Department of Computer Science, University of North Carolina at Chapel Hill; http://www.cs.unc.edu/.
- Electronic Systems Technology Office of the Advanced Research Projects Agency (ARPA). Although not a research lab, this agency provides information on research it funds at several institutions in the United States; http://www.arpa.mil/.
- Electronic Visualization Lab, University of Illinois at Chicago; http://evlweb.eecs.uic.edu/.
- Graphics Visualization Center, Georgia Institute of Technology; http://www.gvu.gatech.edu/.
- Human Interface Technology Lab, University of Washington; http://www.hitl.washington.edu/.
- I-CARVE Lab (Laboratory for Integrated Computer Aided Research in Concurrent and Virtual Design), University of Wisconsin Department of Mechanical Engineering; http://www.engr.wisc.edu/me/.
- Information Systems Department, Technical University of Delft, the Netherlands; http://www.its.tudelft.nl/.
- Institute for Simulation and Training Visual Systems Lab, University of Central Florida, http://www.vsl.ist.ucf.edu/.
- Iwata Lab, University of Tsukuba, Japan; http://intron.kz.tsukuba.ac.jp/.
- Korean Advance Institute for Science and Technology; http://dangun.kaist.ac.kr/.
- Navigating and Acting in Virtual Environments (NAVE) Research Group, Department of Computer Science and Institute of Cognitive Science, University of Colorado at Boulder; http://psych-www.colorado.edu/ics/home.html/.
- NPSNET Research Group, Naval Postgraduate School; http://www.nps.navy.mil/.
- Mobile Aeronautics Education Lab Virtual Reality Station, NASA; http://www.grc.nasa.gov/WWW/MAELVRSTATION/.
- Graphics Research and Analysis Facility (GRAF), NASA Lyndon B, Johnson Space Center; http://jsc-web-pub.jsc.nasa.gov/fpd/SHFB/GRAF/GRAF_Home.html.
- Studio for Creative Inquiry, Carnegie Mellon University; http://www.cmu.edu/studio/index.html/.
- User Interface Group, University of Virginia Computer Science Department; http://www.cs.virginia.edu/or http://www.cs.cmu.edu/~stage3/.
- Virtual Reality Lab, National Center for Supercomputing Applications; http://www.ncsa.uiuc.edu/.
- Vision and Autonomous System Center (VASC), Carnegie Mellon University, http://www.vasc.ri.cmu.edu/.

## 7. REFERENCES


Bradbury, R. (1976). The veldt. In *The Illustrated Man*. New York: Bantam Doubleday Dell.

Fisher, S. (1982). Viewpoint dependent imaging: An interactive stereoscopic display. In S. Benton (Ed.), *Processing and display of three-dimensional data (Proceedings of SPIE*, p. 367).

Furness, III, T. A. (1969). The application of head-mounted displays to airborne reconnaissance and weapon delivery. In *Proceedings of Symposium for Image Display and Recording* (Tech. Rep. No. TR-69-241). Wright-Patterson Air Force Bose, OH: U.S. Air Force Avionics Laboratory.

Furness, III, T. A. (1988). Harnessing virtual space. *Society for Information Display Digest*, 4–7.

Gardner, D. L. (1989). The PowerGlove. *Design News, 45*(23), 63–68.

Gibson, W. (1984). *Neuromancer*. New York: Simon & Schuster.

Jones Telecommunications and Multimedia Encyclopedia. [Online]. Available: http://www.digitalcentury.com/encyclo/update/lanier.html

Kalawsky, R. S. (1993). *The science of virtual reality and virtual environments*. Wokingham, England: Addison-Wesley.

Krueger, M. (1977). Responsive environments. In *Proceedings of the National Computer Conference* (pp. 423–433).

Krueger, M. (1983). *Artificial reality*. Reading, MA: Addison-Wesley.

Krueger, M. (1985). VIDEOPLACE: A report from the Artificial Reality Laboratory. *Leonardo, 18*(2),

Lanier, J. Personal web site: http://www.advanced.org/jaron/

Rheingold, H. (1991). *Virtual Reality*. London: Secker & Warburg.

Sutherland, I. E. (1965). The ultimate display. In *Proceedings of the IFIP Congress* (Vol. 2, pp. 506–508). New York:

Sutherland, I. E. (1970). Computer displays. *Scientific American, 222*(6), 56–81.

Taylor II, Russell M., Robinett, W., Chi, V. L., Brooks, Jr., F. P., Wright, W. V., Williams, R. S., & Snyder, E. J. (1993, August). The Nanomanipulator: A virtual-reality interface for a scanning tunneling microscope. *Computer Graphics* (*Proceedings of SIGGRAPH '93*), 127–134.

# Author Index

Note: (fig) denotes a figure; (tab) denotes a table; italic page numbers denote a multiauthor work; n denotes a note

## B

## C

### D

## E

## I

## J



## K

## M

### N

## O



## P

## S

T

# Subject Index

Note: (fig) denotes a figure; (tab) denotes a table

## A

### B

## C

## D

## E

## I

## J

## P

## T

## W



## Y